Periodic Table of the Elements

Legend: 4 — Atomic number; Be — Symbol for element

Period	1	2	3	4	5	6	7	8	9	10	11	12	13	14	15	16	17	18
1	1 H																	2 He
2	3 Li	4 Be											5 B	6 C	7 N	8 O	9 F	10 Ne
3	11 Na	12 Mg											13 Al	14 Si	15 P	16 S	17 Cl	18 Ar
4	19 K	20 Ca	21 Sc	22 Ti	23 V	24 Cr	25 Mn	26 Fe	27 Co	28 Ni	29 Cu	30 Zn	31 Ga	32 Ge	33 As	34 Se	35 Br	36 Kr
5	37 Rb	38 Sr	39 Y	40 Zr	41 Nb	42 Mo	43 Tc	44 Ru	45 Rh	46 Pd	47 Ag	48 Cd	49 In	50 Sn	51 Sb	52 Te	53 I	54 Xe
6	55 Cs	56 Ba	57 La	72 Hf	73 Ta	74 W	75 Re	76 Os	77 Ir	78 Pt	79 Au	80 Hg	81 Tl	82 Pb	83 Bi	84 Po	85 At	86 Rn
7	87 Fr	88 Ra	89 Ac	104 Rf	105 Db	106 Sg	107 Bh	108 Hs	109 Mt	110 Ds	111 Rg	112 Cn	113 Nh	114 Fl	115 Mc	116 Lv	117 Ts	118 Og

6 Lanthanides	58 Ce	59 Pr	60 Nd	61 Pm	62 Sm	63 Eu	64 Gd	65 Tb	66 Dy	67 Ho	68 Er	69 Tm	70 Yb	71 Lu
7 Actinides	90 Th	91 Pa	92 U	93 Np	94 Pu	95 Am	96 Cm	97 Bk	98 Cf	99 Es	100 Fm	101 Md	102 No	103 Lr

Genetic Code

First letter	Second letter: U	C	A	G	Third letter
U	UUU Phe	UCU Ser	UAU Tyr	UGU Cys	U
U	UUC Phe	UCC Ser	UAC Tyr	UGC Cys	C
U	UUA Leu	UCA Ser	UAA Stop	UGA Stop	A
U	UUG Leu	UCG Ser	UAG Stop	UGG Trp	G
C	CUU Leu	CCU Pro	CAU His	CGU Arg	U
C	CUC Leu	CCC Pro	CAC His	CGC Arg	C
C	CUA Leu	CCA Pro	CAA Gln	CGA Arg	A
C	CUG Leu	CCG Pro	CAG Gln	CGG Arg	G
A	AUU Ile	ACU Thr	AAU Asn	AGU Ser	U
A	AUC Ile	ACC Thr	AAC Asn	AGC Ser	C
A	AUA Ile	ACA Thr	AAA Lys	AGA Arg	A
A	AUG Met	ACG Thr	AAG Lys	AGG Arg	G
G	GUU Val	GCU Ala	GAU Asp	GGU Gly	U
G	GUC Val	GCC Ala	GAC Asp	GGC Gly	C
G	GUA Val	GCA Ala	GAA Glu	GGA Gly	A
G	GUG Val	GCG Ala	GAG Glu	GGG Gly	G

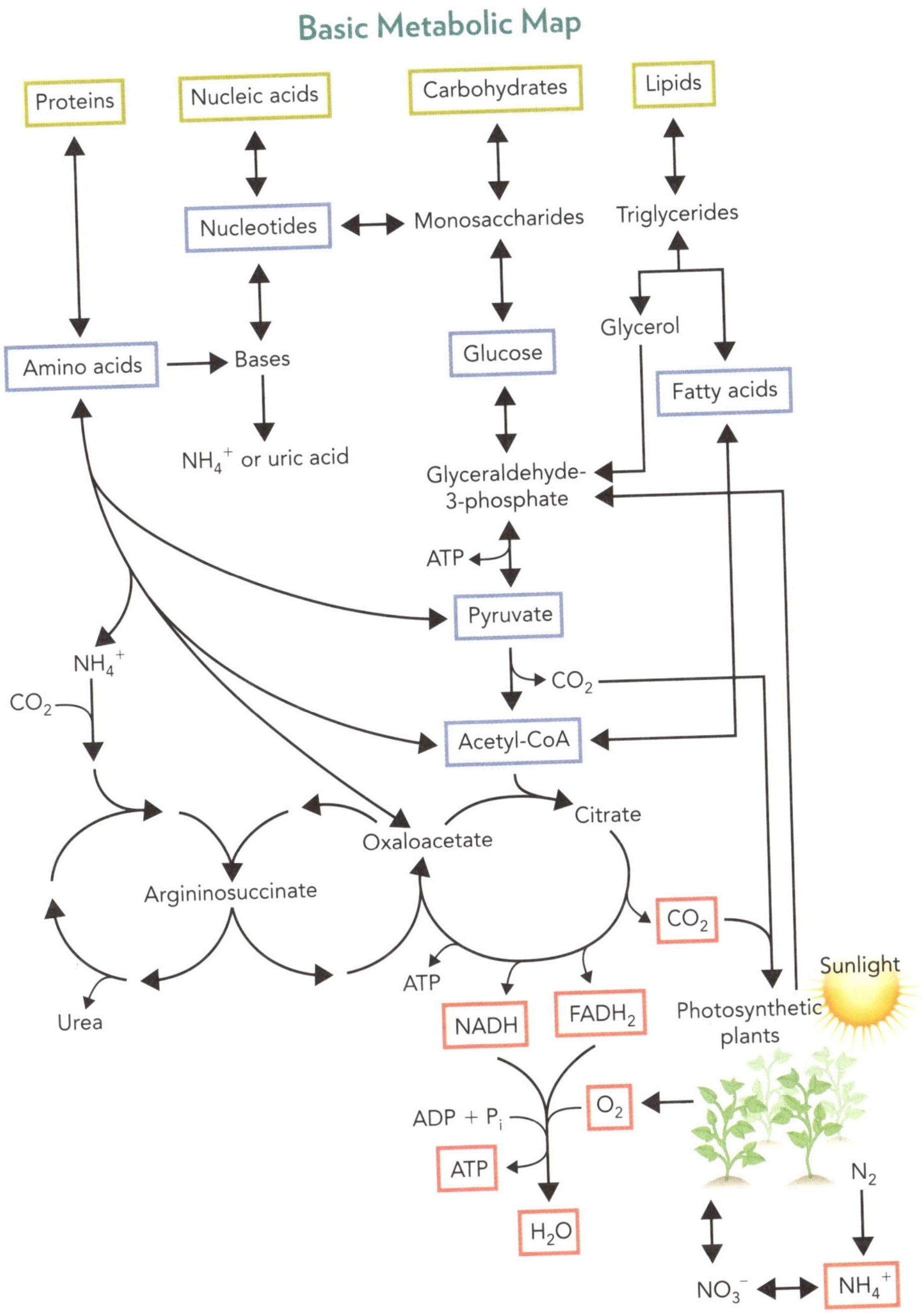
Basic Metabolic Map
Proteins
Nucleic acids
Carbohydrates
Lipids
Nucleotides
Monosaccharides
Triglycerides
Glycerol
Amino acids
Bases
Glucose
Fatty acids
NH_4^+ or uric acid
Glyceraldehyde-3-phosphate
ATP
Pyruvate
NH_4^+
CO_2
CO_2
Acetyl-CoA
Citrate
Oxaloacetate
Argininosuccinate
CO_2
ATP
Urea
NADH
$FADH_2$
Sunlight
Photosynthetic plants
$ADP + P_i$
O_2
ATP
H_2O
N_2
NO_3^-
NH_4^+

Common Equations Used in Biochemistry

$$\Delta G = \Delta H - T\Delta S \qquad (2.11)$$

$$\Delta G^\circ = -RT \ln K_{eq} \qquad (2.12)$$

$$\Delta G = \Delta G^\circ + RT \ln Q \qquad (2.14)$$

$$\Delta G = RT \ln(C_2/C_1) + ZF\Delta\psi \qquad (11.1)$$

$$\Delta G^{\circ\prime} = -nF\Delta E^{\circ\prime} \qquad (10.2)$$

$$\text{EC} = \frac{[\text{ATP}] + 0.5[\text{ADP}]}{[\text{ATP}] + [\text{ADP}] + [\text{AMP}]} \qquad (2.15)$$

$$K_w = 1.0 \times 10^{-14}\,\text{M}^2 = [\text{H}^+][\text{OH}^-] \qquad (2.16)$$

$$\text{pH} = \log \frac{1}{[\text{H}^+]} = -\log[\text{H}^+] \qquad (2.17)$$

$$\text{p}K_a = \log \frac{1}{K_a} = -\log K_a \qquad (2.19)$$

$$\Delta E^{\circ\prime} = (E^{\circ\prime}_{e^-\ \text{acceptor}}) - (E^{\circ\prime}_{e^-\ \text{donor}}) \qquad (10.1)$$

$$E = E^{\circ\prime} + \frac{RT}{nF} \ln \frac{[e^-\ \text{acceptor}]}{[e^-\ \text{donor}]} \qquad (10.3)$$

$$v = k[\text{S}] \qquad (7.1)$$

$$k = \frac{v}{[\text{S}]} = \frac{k_B T}{h} e^{-\Delta G^\ddagger/RT} \qquad (7.2)$$

$$\text{E} + \text{S} \underset{k_{-1}}{\overset{k_1}{\rightleftharpoons}} \text{ES} \xrightarrow{k_2} \text{E} + \text{P} \qquad (7.3)$$

$$K_m = \frac{k_{-1} + k_2}{k_1} \qquad (7.4)$$

$$v_0 = \frac{v_{max}[\text{S}]}{K_m + [\text{S}]} \qquad (7.5)$$

$$v_0 = \frac{k_{cat}[\text{E}_t][\text{S}]}{K_m} \qquad (7.7)$$

BIOCHEMISTRY

Roger L. Miesfeld
University of Arizona

Megan M. McEvoy
University of California, Los Angeles

W. W. Norton & Company
New York London

W. W. Norton & Company has been independent since its founding in 1923, when William Warder Norton and Mary D. Herter Norton first published lectures delivered at the People's Institute, the adult education division of New York City's Cooper Union. The firm soon expanded its program beyond the Institute, publishing books by celebrated academics from America and abroad. By midcentury, the two major pillars of Norton's publishing program—trade books and college texts—were firmly established. In the 1950s, the Norton family transferred control of the company to its employees, and today—with a staff of four hundred and a comparable number of trade, college, and professional titles published each year—W. W. Norton & Company stands as the largest and oldest publishing house owned wholly by its employees.

Printed in Canada
First Edition

Editor: Betsy Twitchell
Associate Managing Editor, College: Carla L. Talmadge
Editorial Assistant: Taylere Peterson
Associate Director of Production, College: Benjamin Reynolds
Managing Editor, College: Marian Johnson
Managing Editor, College Digital Media: Kim Yi
Media Editor: Kate Brayton
Media Project Editor: Jesse Newkirk
Associate Media Editor: Cailin Barrett-Bressack
Media Editorial Assistant: Victoria Reuter
Marketing Manager, Biology: Lauren Winkler
Design Director: Rubina Yeh
Photo Research and Permissions Manager: Ted Szczepanski
Permissions Manager: Megan Schindel
Permissions Clearer: Elizabeth Trammell
Composition: codeMantra
Illustrations: Imagineering—Toronto, ON
Manufacturing: Transcontinental

Permission to use copyrighted material is included alongside the appropriate images.

Library of Congress Cataloging-in-Publication Data

Names: Miesfeld, Roger L., author. | McEvoy, Megan M., author.
Title: Biochemistry / Roger L. Miesfeld, Megan M. McEvoy.
Description: First edition. | New York : W.W. Norton & Company, [2017] |
Includes bibliographical references and index.
Identifiers: LCCN 2016029046 | **ISBN 9780393977264 (hardcover)**
Subjects: | MESH: Biochemical Phenomena
Classification: LCC QP514.2 | NLM QU 34 | DDC 612/.015—dc23 LC record available at https://lccn.loc.gov/2016029046

W. W. Norton & Company, Inc., 500 Fifth Avenue, New York, NY 10110-0017
wwnorton.com
W. W. Norton & Company Ltd., 15 Carlisle Street, London W1D 3BS

1 2 3 4 5 6 7 8 9 0

To my academic mentors who taught me the importance of communicating science using clear and concise sentences—David C. Shepard, Norman Arnheim, Keith R. Yamamoto, and Michael A. Wells—and to my family for their patience and support.

—Roger L. Miesfeld

To the many people who have fostered my development as a scientist and educator, particularly my mentors Harry Noller, Kathy Triman, Jim Remington, and Rick Dahlquist, and to my family and friends who make every day a joy.

—Megan M. McEvoy

Brief Contents

Contents

PART 3 Energy Conversion Pathways

Preface

This book was conceived more than 15 years ago when W. W. Norton editor Jack Repcheck popped his head into Roger Miesfeld's office one sunny afternoon in Tucson, Arizona. Jack had just seen Roger's new textbook on molecular genetics in the bookstore and had been impressed with the illustrations. He said, "Dr. Miesfeld, how would you like to author a full-color textbook that takes the same visual approach to biochemistry as you did for the topic of molecular genetics?" And with those fateful words began a conversation, and then the creation of a textbook that focuses on how biochemistry relates to the world around us without relying on rote memorization of facts by students. In 2011, Roger's colleague at the University of Arizona and next-door-office neighbor, Megan McEvoy, who is also an instructor of a large biochemistry service course, mentioned that she would be eager to work on a textbook that would improve pedagogy in the field. Thus, this project, which began years ago with a simple question, has resulted in the publication of the first truly new biochemistry textbook in decades.

Meanwhile, we (Roger and Megan) have been teaching biochemistry to undergraduate, graduate, and medical school students for nearly 40 years combined and have loved every minute of it—seriously. During this time, we noticed that many biochemistry textbooks seemed to sidestep a very basic question in the minds of most students: "Why do I need to learn biochemistry?" To answer this question in the classroom, we developed a number of story lines that revolve around a simple premise: how it works and why it matters. We used the assigned textbook to fill in the details for our students but used the in-class lectures to provide the context the students needed to see the big picture. During this same time, the Internet became much more accessible so that it was almost trivial to find the name of an enzyme in a metabolic reaction or the equation required for calculating changes in free energy.

But despite the ease with which "info-bytes" could be obtained, and often simply memorized, what still required thought was integration of these pieces of information to fully understand concepts such as allosteric regulation of an enzyme, rates of metabolic flux, or the importance of weak noncovalent interactions in assembling gene transcription complexes. We challenged the students in our classes to approach each biochemical process—especially those that are conceptually the most difficult—to answer the questions how does it work and why does it matter to me. The "it" could be a cancer drug that inhibits an enzyme, an external stimulus that activates a signaling pathway and controls blood sugar, or a biochemical assay that measures gene expression levels. We told them that to answer the how it works part, they would have to explain the biochemical process in clear and concise language, while the why it matters part required them to make it relevant to their own life experience.

As we collected more and more of these "how and why" examples over the years, it became clear to us that our biochemistry textbook should focus on presenting core concepts in a relatable way centered around three themes: (1) the interdependence of energy conversion processes, (2) the role of signal transduction in metabolic regulation, and (3) biochemical processes affecting human health and disease. The pedagogical foundation for each of these themes is that molecular structure determines chemical function. In developing the outline for the book, we ignored the urge to write it like an automobile owner's manual in which all of the parts are listed first (proteins, lipids, carbohydrates, nucleic acids), and then the function of the car (metabolic pathways) is described by assembling the parts in a systematic way (easy to memorize).

Instead, we chose to organize the book using five core blocks (collections of chapters, or parts) that consist of modules (individual chapters) made up of concept-based submodules (numbered chapter sections) with limited, focused, unnumbered subsections. The five core blocks we chose are "Part 1: Principles of Biochemistry" (Chapters 1–3), "Part 2: Protein Biochemistry" (Chapters 4–8), "Part 3: Energy Conversion Pathways" (Chapters 9–12), "Part 4: Metabolic Regulation" (Chapters 13–19), and "Part 5: Genomic Regulation" (Chapters 20–23). This organization provides the student with an opportunity to work through related concepts before moving on to new ones. For example, what is needed to understand protein structure and function is presented in Part 2, including how proteins function as enzymes or as relay partners in a signal transduction pathway. In Part 4, carbohydrate structure and function (Chapter 13) and carbohydrate metabolism (Chapter 14) are paired together, as are lipid structure and function (Chapter 15) and lipid metabolism (Chapter 16), while the structure of nitrogen-based biomolecules and their metabolism are presented together in Chapters 17 (amino acids) and 18 (nucleotides).

The figures in our book have been paramount since the very beginning; indeed, it was a commitment by W. W. Norton to a modern art program that hooked Roger in the first place. So we created each chapter starting with a collection of 30–40 hand-drawn illustrations or Web images that were complemented with molecular renderings based on Protein Data Bank (PDB) files and with photographs of people, places, or things. At the beginning of each chapter section, the topic is presented broadly, and then the reader is led into the themed concepts. With regularity, examples of everyday biochemistry are woven into the story line to provide an opportunity to step back for a moment and see the relevance of the topic to life around us. In our classes, we tell the students to use the everyday biochemistry examples as a way to make it personal, rather than as more info-bytes to memorize. The point of these examples is to generate excitement about biochemistry so that the student can get through the more difficult concepts knowing there is a good reason to push ahead—it is likely to be relevant.

Instructors may engage students more fully in the beauty of the world's biological diversity using this book's chemical framework, which frequently rises into the cellular level. One could follow our sequence through Parts 1–5 as we do in our classes or mix and match using a sequence that works best for the instructor. Students can likewise use our book as a biochemistry reference and read sections individually without having to read the book cover to cover. There are plenty of online materials and ancillary tools that have been developed for instructors and students, and we urge you to take full advantage of them.

Finally, we encourage you to look for new examples of everyday biochemistry and send the details to us so that we can add them to the collection for future editions.

Roger L. Miesfeld
Megan M. McEvoy

Authors' Tour of the Book Features

The Only Textbook That Makes Visuals the Foundation of Every Chapter

Every figure in this textbook originated in our biochemistry lectures, and our preparation of each chapter involved creating the figures we wanted to include *first* and then writing the text of the chapter to fit those figures. The result is a book in which the figures and the text are inseparable from one another; they are one learning tool that strengthens students' understanding of how biochemical processes and structures work. Specifically:

- We've made sure that key chapter figures help students see how biochemistry functions in context. For example, Figure 9.3 in Chapter 9 provides a basic metabolic map that emphasizes the major biomolecules in cells and the interdependence of pathways. On the basis of this detailed figure, Figure 9.4 and similar figures in subsequent chapters of Parts 3 and 4 present simplified, iconic metabolic maps that clearly divide pathways into two discrete groups: those linked to energy conversion (red) and those linked to metabolite synthesis and degradation pathways (blue).

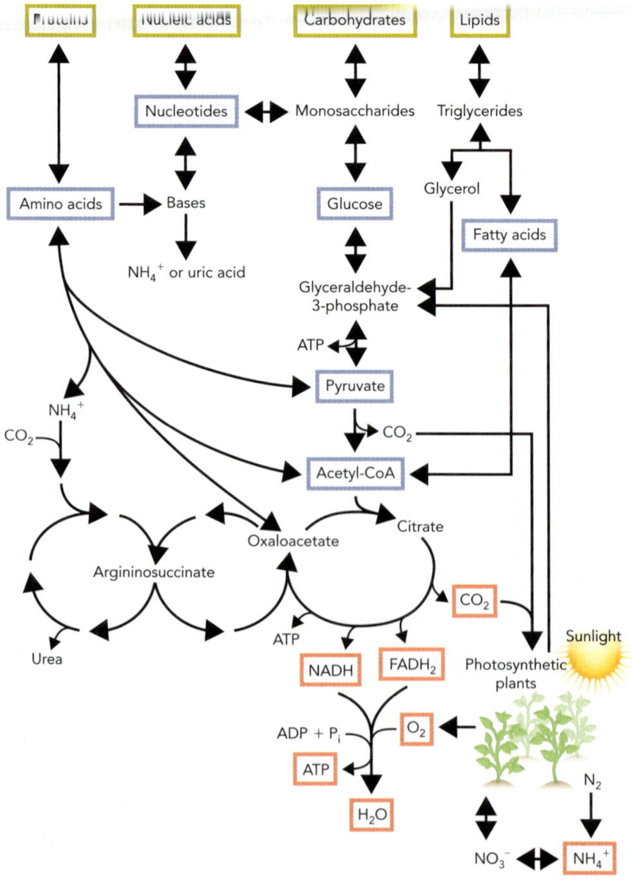

- We've included hundreds of vibrant, precise, and information-rich molecular representations. These figures in the text are paired with state-of-the-art 3D interactive versions in the online homework.

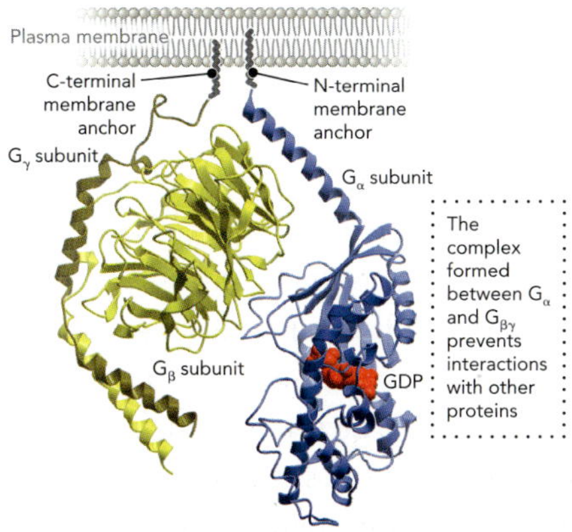

- In the digital resources available to instructors, we are making available cutting-edge process animations—many reflecting state-of-the-art 3D technology—that will strengthen students' understanding of challenging biochemical processes.

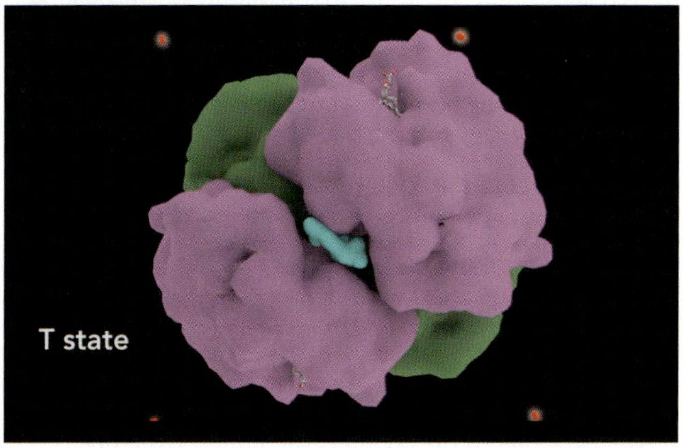

- We've added abundant in-figure text boxes, numbered steps, and icons to help students navigate the most complex biochemical processes. Figure 7.35 provides a good example of our art program's pedagogical value: It clearly illustrates a complex four-step reaction through numbered steps, descriptive captions, and a thorough complementary explanation in the text.

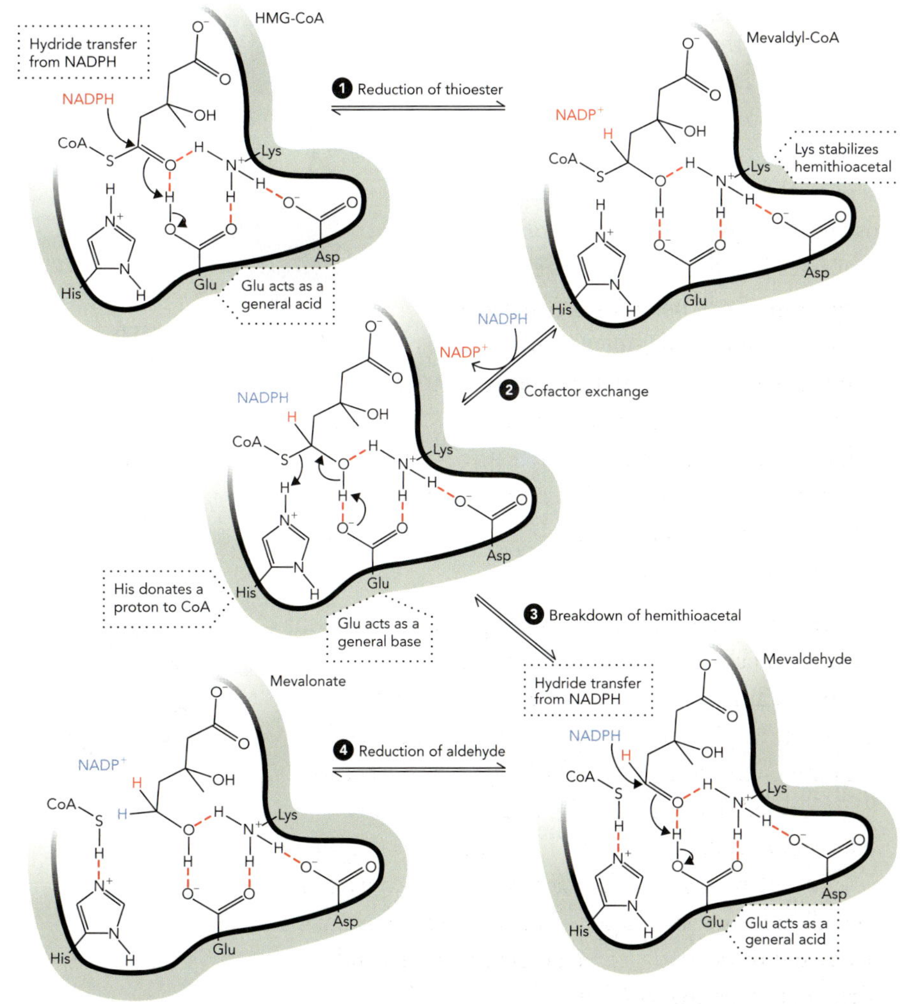

Clear Explanations and a Distinctive Chapter Sequence Help Students Make Connections between Concepts

Our distinctive chapter sequence highlights connections between key biochemical processes, encouraging students to move beyond mere memorization to consider *how* biochemistry works.

- In Part 1, we introduce essential, unifying concepts that are interwoven throughout the chapters that follow: hierarchical organization of biochemical complexity; energy conversion in biological systems; the chemical role of water in life processes; the function of cell membranes as hydrophobic barriers; and the central dogma of molecular biology from a biochemical perspective.
- As a capstone to the chapters on protein structure and function (Part 2), we present signal transduction (Chapter 8) as the prototypical example of how proteins work to mediate cellular processes.
- The topical sequence in Parts 3 and 4 underscores the importance of energy conversion as the foundation for all other metabolic pathways, introducing enzyme regulation of metabolic flux as a central theme. In Part 3, we present the pathways involved in energy conversion processes before presenting degradative and biosynthetic pathways in Part 4. This helps students see complex processes and connections between concepts more clearly.
- We present the biomolecular structure and function of carbohydrates, lipids, amino acids, and nucleotides in Part 4 in the context of their metabolic pathways. This integrated approach encourages students to associate biochemical structure with cellular function in a way that promotes deeper understanding.
- Rather than an encyclopedic list of individual reactions that can obscure students' understanding of the important concepts, in Parts 3 and 4 we emphasize the regulation of 10 major (and broadly representative) metabolic pathways, with a special emphasis on the human diseases associated with these pathways.

Unmatched Emphasis on Applications and Biomedical Examples Motivates Learning by Helping Students Connect the Material to both Their Majors and Their Everyday Experience

We know from our teaching that students can be equally engaged by biomedical examples and examples of biochemistry in the world around them. So throughout this book we've reinforced key biochemical concepts with applied examples that show why biochemistry matters.

- Each chapter-opening vignette provides an introduction to a biochemical application connected to the chapter's central topic. Later, we ask students to reexamine the application in light of their newly acquired knowledge of the biochemistry behind it. For example, the opening vignette for Chapter 22 examines how an ingenious laboratory method enabled study of soil bacteria that were previously impossible to culture in the lab, which led to discovery of a new antibiotic. Another example is the opening vignette for Chapter 13, which visually presents the biochemistry behind the commercial product Beano.

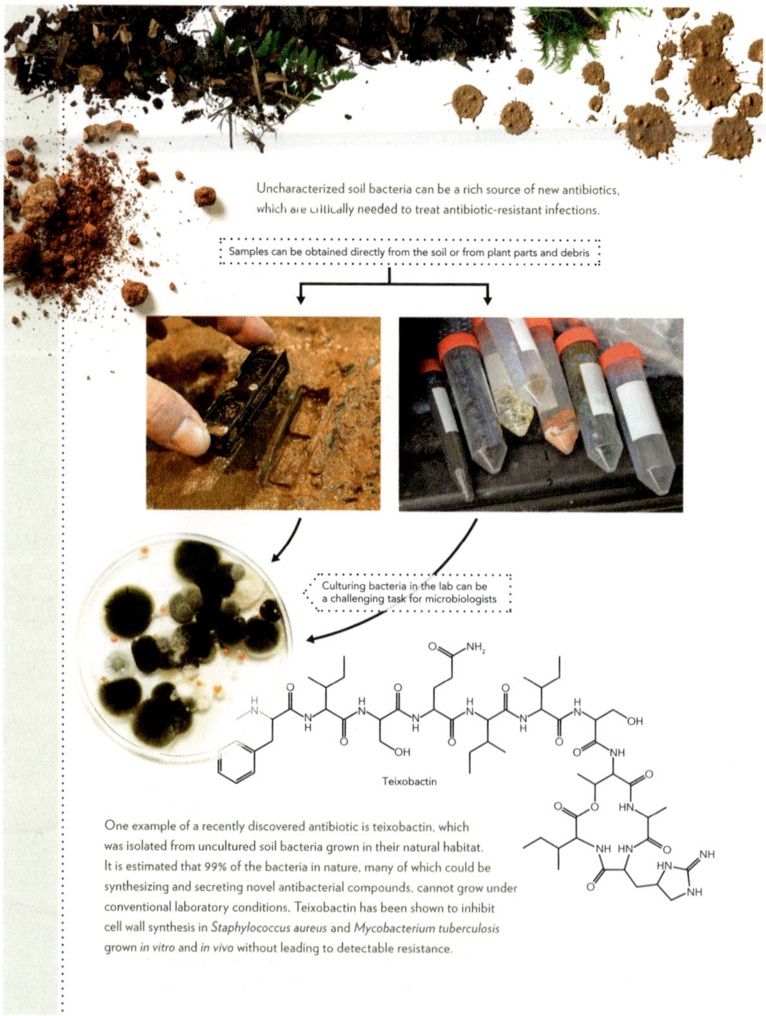
Uncharacterized soil bacteria can be a rich source of new antibiotics, which are critically needed to treat antibiotic-resistant infections.
Samples can be obtained directly from the soil or from plant parts and debris
Culturing bacteria in the lab can be a challenging task for microbiologists
Teixobactin

One example of a recently discovered antibiotic is teixobactin, which was isolated from uncultured soil bacteria grown in their natural habitat. It is estimated that 99% of the bacteria in nature, many of which could be synthesizing and secreting novel antibacterial compounds, cannot grow under conventional laboratory conditions. Teixobactin has been shown to inhibit cell wall synthesis in *Staphylococcus aureus* and *Mycobacterium tuberculosis* grown *in vitro* and *in vivo* without leading to detectable resistance.

- Real-life examples from nature help students understand how structure (of a protein, lipid, carbohydrate, or nucleic acid) affects function, an important takeaway insight we stress in our biochemistry courses. A great example is the discussion in Chapter 2 concerning antifreeze proteins in fish and insects that live in extreme cold. Threonine amino acids in these proteins line up perfectly with ice crystals and thus prevents them from growing within the animals.

- We distributed human health examples, particularly discussions of human disease, throughout the text. These are especially relevant for the many students planning to pursue careers in medicine or other health-related professions. A prominent example occurs in Chapter 21—the description of a degenerative disease of the retina called retinitis pigmentosa, which is caused by defects in the RNA splicing machinery. This is a surprise to students, who expect that most human disease is the result of enzyme defects.

Thoughtful Pedagogy and Assessment Promotes Mastery of Biochemical Concepts

We feel strongly that myriad boxes and sidebars in textbooks distract from the content of the chapters and are rarely read by students. As a result, this book has a design that is clean and uncluttered.

- A Concept Integration question and its answer occurs at the end of each numbered chapter section. This feature prompts students to think critically about what they're reading and to synthesize concepts in a meaningful way.

concept integration 5.1

A frog species was found to contain a cytosolic liver protein that bound a pharmaceutical drug present at high levels in effluent from a wastewater facility. Describe how this protein could be purified.

The first step in purifying an uncharacterized protein is to develop a method to detect it specifically, such as an enzyme activity assay or binding assay. In this case, the protein is known to bind to a small molecule (pharmaceutical drug), and this binding activity can be used to develop a protein detection assay. The assay could be based on protein binding to the drug that has been radioactively labeled or it might be possible to develop a fluorescently labeled version of the drug that has an altered absorption or emission spectrum as a function of specific protein binding. The next step would be to use cell fractionation, centrifugation, and a combination of gel filtration and ion-exchange column chromatography to enrich for drug binding activity relative to total protein in the frog liver extract. A final step would be to develop an affinity column that contains the drug covalently linked to a solid matrix and use this column to bind specifically, and then elute, the high-affinity binding protein. The purity of the protein would be assessed by SDS-PAGE at several steps within the purification protocol.

concept integration 14.3

Why does it make physiologic sense for muscle glycogen phosphorylase activity to be regulated by both metabolite allosteric control and hormone-dependent phosphorylation?

Muscle glycogen phosphorylase is allosterically activated by AMP, which signals low energy charge in the cell. High AMP levels also indicate a need for glycogen degradation and release of glucose substrate for ATP generation to support muscle contraction. Both ATP and glucose-6-P are allosteric inhibitors of muscle glycogen phosphorylase activity and signal a ready supply of chemical energy without the need for glycogen degradation. Both types of allosteric regulation occur rapidly on a timescale of seconds in response to sudden changes in AMP, ATP, and glucose-6-P levels. Allosteric control by metabolites provides a highly efficient means to control rates of glycogen degradation in response to the immediate energy needs of muscle cells. In contrast, hormonal regulation of muscle glycogen phosphorylase activity by glucagon and epinephrine is a delayed response (occurring on a timescale of hours), resulting in phosphorylation and activation of the enzyme after neuronal and physiologic inputs at the organismal level. Similarly, insulin signaling, which inhibits muscle glycogen phosphorylase activity through dephosphorylation, is also a delayed response at the organismal level and depends on multiple physiologic inputs. Taken together, allosteric regulation of muscle glycogen phosphorylase activity provides a rapid-response control mechanism to modulate muscle glucose levels, whereas hormonal signaling requires input from multiple stimuli at the organismal level and provides a longer-term effect on enzyme activity through covalent modifications.

- We know the quality and quantity of end-of-chapter problems is an important litmus test for many instructors when reviewing textbooks. Our end-of-chapter material includes a plentiful, balanced mix of basic Chapter Review questions and thought-provoking Challenge Problems.

- Online homework is becoming a more and more powerful learning tool for biochemistry courses. Norton's Smartwork5 online homework platform offers book-specific assessment through a wide array of exercises: art-based interactive questions, critical-thinking questions, application questions, process animation questions, and chemistry drawing questions, as well as all of the book's end-of-chapter questions. We are particularly excited to be the first to offer interactive 3D molecular visualization questions within the homework platform. Everything the student needs to interrogate a molecular structure is embedded in Smartwork5 using Molsoft's ICM Browser application.

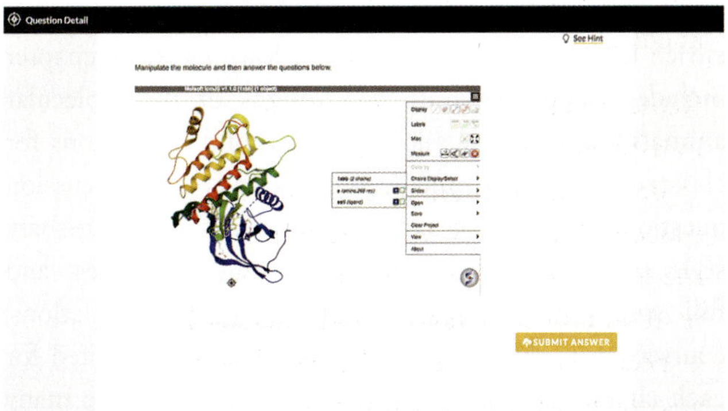

Resources for Instructors and Students

Smartwork5

This dynamic and powerful online assessment resource uses answer-specific feedback, a variety of engaging question types, the integration of the stunning book art, 3D molecular animations, and process animations to help students visualize and master the important course concepts. Smartwork5 also integrates easily with your campus learning management system and features a simple, intuitive interface, making it an easy-to-use online homework system for both instructors and students.

3D Molecular Animations

Eleven photorealistic 3D molecular animations based on PDB files were created by renowned molecular animator Dr. Janet Iwasa from the Department of Biochemistry at the University of Utah College of Medicine. Janet brings some of the most difficult concepts in biochemistry to life in stunning detail. These animations are available to students in coursepack assessments and through the ebook and are available with associated assessments for instructors to assign in the Smartwork5 homework system. Links to the animations are available to instructors at wwnorton.com/instructors.

Process Animations

Twenty process animations showcase the complex topics that students find most challenging. The animations are available to students in mobile-compatible format in the coursepack and the ebook, as well as online. Assessments written specifically for the animations are included in Smartwork5. Links to the animations are available to instructors at wwnorton.com/instructors.

Ultimate Guide to Teaching with Biochemistry

This enhanced instructor's manual will help any professor enrich his or her course with active learning. Each chapter includes sample lectures, descriptions of the molecular animations with discussion questions and suggestions for classroom use, multimedia suggestions with discussion questions, an active learning activity, a think–pair–share style of activity, book-specific learning objectives, and full solutions. A list of other resources (animations, coursepack resources, and so forth) will also be listed for each chapter to ensure instructors are aware of the many instructor-provided materials available to them. Activity handouts will be available for download at wwnorton.com/instructors for easy printing and distribution.

Coursepacks

Available at no cost to professors or students, Norton Coursepacks for online or hybrid courses are available in a variety of formats, including Blackboard, Desire2Learn (D2L), and Canvas. With just a simple download from the instructor's website, instructors can bring high-quality Norton digital media into a new or existing online course. Content is fully customizable and includes chapter-based assignments with high-quality visual assessments, perfect for distance learning classes or assignments between classes. The coursepack for *Biochemistry* also features the full suite of animations, vocabulary flashcards, and assignments based on 3D animations as well as art from the book—everything students need for a great out-of-the-classroom experience.

PowerPoint Presentations and Figures

PowerPoint slide options meet the needs of every instructor and include lecture PowerPoint slides providing an overview of each chapter, five clicker questions per chapter, and links to animations. There is also a separate set of art PowerPoint slides featuring every photograph and drawn figure from the text. In addition, the PDB files used as the basis for many of the molecular structures in the book are available for download.

Test Bank

The Test Bank for *Biochemistry* is designed to help instructors prepare exams quickly and effectively. Questions are tagged according to Bloom's taxonomy, and each chapter includes approximately 75 multiple-choice and 25 essay questions. Five to ten questions per chapter use art taken directly from the book. In addition to tagging with Bloom's, each question is tagged with metadata that places it in the context of the chapter and assigns it a difficulty level, enabling instructors to easily construct tests that are meaningful and diagnostic.

Ebook

Available for students to purchase online at any time, the *Biochemistry* ebook offers students a great low price, exceptional functionality, and access to the full suite of accompanying resources.

Acknowledgments

This book was a very long time in the making, and it would not have been possible without the hard work, dedication, and care of dozens of people. To begin with, we would like to thank our editors at Norton, the late Jack Repcheck, Vanessa Drake-Johnson, Michael Wright, and last but certainly not least, Betsy Twitchell. Your combination of vision, patience, and persistence kept us going even when the going was rough. Our deepest gratitude to project editor Carla Talmadge, the "master of the schedule," for keeping the innumerable moving parts of our book organized and in forward motion. Our developmental editor, David Chelton, is, simply put, a rock star, and we were so lucky to work with him through the many years that it took to find the perfect balance of chemistry, biology, and everyday biochemistry examples that make this book so remarkable. It can't be easy to copyedit a book this big, but Christopher Curioli brought a level of skill and expertise that was truly remarkable. We owe a huge debt of gratitude to Elyse Rieder, who miraculously tracked down every photograph our hearts desired, and to Ted Szczepanski for being with her every step of the way. We were very fortunate to work with incredibly talented designer Anne DeMarinis on the book design, chapter openers, and cover. It is through Anne's vision that our thousands of pages of manuscript became the beautiful book you're holding in your hands. We must thank the unsung heroes of this project, editorial assistants Taylere Peterson, Katie Callahan, Courtney Shaw, Cait Callahan, Callinda Tayler, and the many who came before them for their hours spent posting files, making copies, mailing proofs, and countless other essential tasks. Production manager Ben Reynolds adeptly managed the process of translating our raw material into the polished final product; for that he has our deepest thanks. The amazing folks at Imagineeringart.com Inc. deserve medals for living up to our high standards for every figure and every page in our book regardless of how many times we sent the artwork back for just one more tweak until we considered it perfect. Thank you to Wynne Au Yeung, Alicia Elliott, and the rest of the Imagineering team.

We have an absolutely tireless team at Norton creating the print and digital supplementary resources for our book. Media editor Kate Brayton, associate editor Cailin Barrett-Bressack, and media assistant Victoria Reuter worked on every element of the package as a team, and the content meets our very high standards as a result. Thank you also to Kim Yi's media project editorial group for the invaluable work they do shepherding content through many stages of development. We thank everyone involved in Norton's sales and marketing team for their unflagging support of our book. Roby Harrington deserves a special shout-out: Roby made a number of trips to Tucson (usually in the winter) to meet with Roger at a local coffee shop on University Boulevard and ask him one more time, "Why is it taking so long?" We thank Roby and the other Norton editors for responding positively to Roger's enthusiasm and extending the deadline again and again. It paid off. Finally, we thank Drake McFeely, Julia Reidhead, Stephen King, Steve Dunn, and Marian Johnson for believing in us all these years.

The original figures we developed for this book, and the end of chapter review questions and challenge problems, have been used in our classes at the University of Arizona for well over a decade, which means we have had the benefit of constructive feedback from literally thousands of students. We truly appreciate each and every one of these comments as they helped guide the book's development.

We thank our three contributing authors for helping us draft the final chapters in our book—Kelly Johanson, Scott Lefler, and John W. Little. Your effort was the x-factor that got us over the finish line, and for that you have our eternal gratitude. We also want to acknowledge the late Professor Michael A. Wells of the University of Arizona who provided W. W. Norton with the first set of PDB files for homework questions that were similar in many ways to the current set of Smartwork5/Molsoft questions we have today. In addition, we thank Dr. Andrew Orry at Molsoft, LLC (La Jolla, California), who provided personal guidance on how best to use Molsoft's ICM Browser Pro rendering program to create the stunning molecular images we have included in the book and the online materials.

Finally, we thank each and every one of the biochemists who reviewed chapters in our text throughout the years. Your feedback—sometimes positive, sometimes not—has been absolutely invaluable to the development of this book. We are deeply grateful for your willingness to give us your time so that we can benefit from your experience.

Paul D. Adams, University of Arkansas, Fayetteville
Mark Alper, University of California, Berkeley
Richard Amasino, University of Wisconsin–Madison

Christophe Ampe, Ghent University
Rhona Anderson, Brunel University London
Ross S. Anderson, The Master's College
Eric Arnoys, Calvin College
Kenneth Balazovich, University of Michigan
Daniel Alan Barr, Utica College
Dana A. Baum, Saint Louis University
Robert Bellin, College of the Holy Cross
Matthew A. Berezuk, Azusa Pacific University
Steven M. Berry, University of Minnesota, Duluth
John M. Brewer, University of Georgia
David W. Brown, Florida Gulf Coast University
Nicholas Burgis, Eastern Washington University
Bruce S. Burnham, Rider University
Robert S. Byrne, California State University, Fullerton
Yongli Chen, Hawaii Pacific University
Jo-Anne Chuck, University of Western Sydney
Karina Ckless, SUNY Plattsburgh
Lindsay R. Comstock-Ferguson, Wake Forest University
Maurizio Costabile, University of South Australia
Sulekha Coticone, Florida Gulf Coast University
Rajalingam Dakshinamurthy, Western Kentucky University
S. Colette Daubner, St. Mary's University
Dan J. Davis, University of Arkansas
John de Banzie, Northeastern State University
Frank H. Deis, Rutgers University
Paul DeLaLuz, Lee University
Rebecca Dickstein, University of North Texas
Karl-Erik Eilertsen, University of Tromsø
Timea Gerczei Fernandez, Ball State University
Matthew Fisher, Saint Vincent College
Robert Ford, The University of Manchester
Christopher Francklyn, University of Vermont
Laura Frost, Florida Gulf Coast University
Matthew Gage, Northern Arizona University
Donna L. Gosnell, Valdosta State University
Nora S. Green, Randolph-Macon College
Neena Grover, Colorado College
Peter-Leon Hagedoorn, Delft University of Technology
Donovan C. Haines, Sam Houston State University
Christopher S. Hamilton, Hillsdale College
Gaute Martin Hansen, University of Tromsø
Lisa Hedrick, University of St. Francis
Newton P. Hilliard, Jr., Texas Wesleyan University
Jason A. Holland, University of Central Missouri
Charles G. Hoogstraten, Michigan State University
Holly A. Huffman, Arizona State University
Tom Huxford, San Diego State University
Constance Jeffery, University of Illinois at Chicago
Bjarne Jochimsen, Aarhus University
Jerry E. Johnson, University of Houston
Joseph Johnson, University of Minnesota, Duluth
Michael Kalafatis, Cleveland State University
Margaret I. Kanipes-Spinks, North Carolina A&T State University
Rachel E. Klevit, University of Washington
James A. Knopp, North Carolina State University
Andy Koppisch, Northern Arizona University
Peter Kuhlman, Denison University
Harry D. Kurtz, Jr., Clemson University
Thomas Leeper, University of Akron
Linda A. Luck, SUNY Plattsburgh
Lauren E. Marbella, University of Pittsburgh
Darla McCarthy, Calvin College
Eddie Merino, University of Cincinnati
David J. Merkler, University of South Florida
Leander Meuris, Ghent University
Rita Mihailescu, Duquesne University
Frederick C. Miller, Oklahoma Christian University
David Moffet, Loyola Marymount University
Debra M. Moriarity, The University of Alabama in Huntsville
Andrew Mundt, Wisconsin Lutheran College
Fares Z. Najar, The University of Oklahoma
Odutayo O. Odunuga, Stephen F. Austin State University
Edith Osborne, Angelo State University
Darrell L. Peterson, Virginia Commonwealth University
William T. Potter, The University of Tulsa
Joseph Provost, University of San Diego
Tanea T. Reed, Eastern Kentucky University
James Roesser, Virginia Commonwealth University
Gordon S. Rule, Carnegie Mellon University
Wilma Saffran, Queens College
Michael Sehorn, Clemson University
Robert M. Seiser, Roosevelt University
David Sheehan, University College Cork
Kim T. Simons, Emporia State University
Kerry Smith, Clemson University
Charles Sokolik, Denison University
Amy Springer, University of Massachusetts, Amherst
Jon Stewart, University of Florida
Paul D. Straight, Texas A&M University
Manickam Sugumaran, University of Massachusetts, Boston
Janice Taylor, Glasgow Caledonian University
Peter E. Thorsness, University of Wyoming
Marianna Torok, University of Massachusetts, Boston
David Tu, Pennsylvania State University
Marcellus Ubbink, Leiden University
Peter van der Geer, San Diego State University
Kevin M. Williams, Western Kentucky University
Nathan Winter, St. Cloud State University
Ming Jie Wu, University of Western Sydney
Shiyong Wu, Ohio University
Wu Xu, University of Louisiana at Lafayette
Laura S. Zapanta, University of Pittsburgh
Yunde Zhao, University of California, San Diego
Brent Znosko, Saint Louis University
Lisa Zuraw, The Citadel

About the Authors

Roger L. Miesfeld is a professor and department head in the Department of Chemistry and Biochemistry at the University of Arizona in Tucson. Dr. Miesfeld's research focus for the past 30 years has been on regulatory mechanisms governing signal transduction in eukaryotic cells. For much of this time, his lab investigated steroid hormone signaling in human disease models, primarily cancer (leukemia and prostate cancer) and asthma. More recently, his research group has been studying metabolic regulation of blood meal metabolism in vector mosquitoes that transmit the dengue and Zika viruses (*Aedes aegypti*). Their current efforts are aimed at identifying mosquito-selective and bio-safe small-molecule inhibitors of processes regulating mosquito eggshell synthesis. Dr. Miesfeld has taught a variety of undergraduate, graduate, and medical school biochemistry courses over the years and now teaches the largest undergraduate biochemistry courses at the University of Arizona. He has authored two other textbooks, *Applied Molecular Genetics* and *Biochemistry: A Short Course*, and was the recipient of the University of Arizona Honors College Faculty Excellence Award.

Dr. Miesfeld received his BS and MS degrees in cell biology from San Diego State University, and his PhD in biochemistry from Stony Brook University. He was a Jane Coffin Childs Postdoctoral Fellow in the Department of Biochemistry and Biophysics at the University of California, in San Francisco, before becoming a faculty member at the University of Arizona in 1987.

Megan M. McEvoy is broadly trained as a protein biochemist and structural biologist, and her research work is primarily concerned with how metal ions are handled in microbial systems. She is interested in the general area of how metal ions are acquired when needed or eliminated when in excess. Her work focuses on studies of protein–protein interactions and conformational changes and how metal ions are specifically recognized by proteins. Dr. McEvoy has taught numerous undergraduate biochemistry courses, including courses for majors, nonmajors, and honors students. Along with Dr. Miesfeld, she taught the nonmajors biochemistry courses at the University of Arizona for many years.

Dr. McEvoy received her BS degree in biochemistry and molecular biology from the University of California, Santa Cruz, and her PhD in chemistry from the University of Oregon. She started her career at the University of Arizona as an assistant professor in the Department of Biochemistry and Molecular Biophysics, then became an associate professor in the Department of Chemistry and Biochemistry. She is now a professor in the Department of Microbiology, Immunology, and Molecular Genetics at the University of California, Los Angeles.

About the Authors

BIOCHEMISTRY

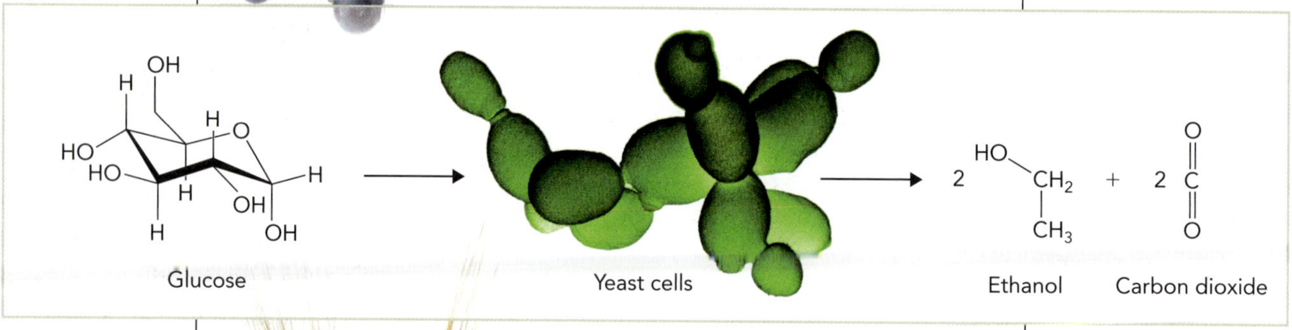

Grapes and barley are the sources of sugar and natural flavors that are metabolized by live yeast cells to produce alcoholic wine and beer, respectively.

1

Principles of Biochemistry

◀ In the late 1800s, chemists in Europe sought to uncover the chemical basis for alcoholic fermentation in hopes of improving the quantity and quality of beer and wine production. In 1897, the German chemist Eduard Buchner discovered that an extract of yeast cells could be used *in vitro* (outside a living cell) to convert glucose to carbon dioxide and ethanol under anaerobic conditions. The discovery that some yeast proteins could function as chemical catalysts in the fermentation reaction ushered in the modern era of biochemistry.

CHAPTER OUTLINE

CREDITS: GRAPES: ARTJAZZ/SHUTTERSTOCK; WINE: SOMCHAI SOM/SHUTTERSTOCK; YEAST CELLS: DAVID M. PHILLIPS/SCIENCE SOURCE, COLORIZATION BY JESSICA WILSON; BARLEY: ANMBPH/SHUTTERSTOCK; BEER: MTSARIDE/SHUTTERSTOCK.

The birth of modern biochemistry can be traced to the end of the 19th century, when chemists discovered that cell extracts of brewer's yeast contained everything necessary for alcoholic fermentation. That is, processes associated with living organisms could actually be understood in terms of fundamental chemistry. The reductionist approach of breaking open cells and isolating their components for use in *in vitro* chemical reactions continued for most of the 20th century. During this time, scientists made numerous discoveries in cellular biochemistry that transformed our understanding of the chemical basis of life. These advances included describing the chemical structure and function of the major classes of biomolecules: nucleic acids, proteins, carbohydrates, and lipids. Moreover, thousands of metabolic reactions that direct molecular synthesis and degradation in cells were characterized in bacteria, yeast, plants, and animals. Knowledge gained from these biochemical studies has been used to develop pharmaceutical drugs, medical diagnostic tests, microbial-based industrial processes, and herbicide-resistant plant crops, among other things.

The field of biochemistry enjoyed tremendous growth in the 1970s, when techniques were developed to manipulate deoxyribonucleic acid (DNA) based on an experimental approach that became known as recombinant DNA technology. This achievement led to the creation of the first biotechnology company in 1977, which later went on to use recombinant DNA technology to produce human insulin in bacteria. The following 20 years were an explosive time for biochemical research. In addition to the development of more sophisticated biochemical tools, scientists achieved vast improvements in protein purification and structure determination as a result of new instrumentation and computational power.

Modern biochemistry encompasses both organic chemistry and physical chemistry, as well as areas of microbiology, genetics, molecular biology, cell biology, physiology, and computational biology. In this introductory chapter, we first present an overview of modern biochemistry. We then describe three biochemical principles that together provide a framework for understanding life at the molecular level:

1. The hierarchical organization of biochemical processes within cells, organisms, and ecosystems underlies the chemical basis for life on Earth.
2. DNA is the chemical basis for heredity and encodes the structural information for RNA and protein molecules, which mediate biochemical processes in cells.
3. The function of a biomolecule is determined by its molecular structure, which is fine-tuned by evolution through random DNA mutations and natural selection.

In Chapter 2, we describe three additional biochemical principles:

4. Biological processes follow the same universal laws and thermodynamic principles that govern physical processes.
5. Life depends on water because of its distinctive chemical properties and its central role in biochemical reactions.
6. Biological membranes are selective hydrophobic barriers that define aqueous compartments in which biochemical reactions take place.

1.1 What Is Biochemistry?

Biochemistry aims to explain biological processes at the molecular and cellular levels. As its name implies, biochemistry is at the interface of biology and chemistry. It is a hands-on experimental science that relies heavily on quantitative analysis of data. Biochemists are interested in understanding the structure and function of biological molecules. Biochemical research often involves mechanistic studies that focus on hypothesis-driven experiments designed to answer specific biological questions. Examples include determining how a group of proteins catalyze the synthesis of a complex biomolecule or why biological membranes have different physical properties depending on their chemical composition.

One of the first biochemical processes to be investigated was **fermentation**: the conversion of rotting fruit or grain into solutions of alcohol through the action of yeast. The Egyptians knew as early as 2000 B.C. that crushed dates produce both an intoxicating substance (ethanol) and a caustic acid (acetic acid). The Greeks used "zyme" (yeast) to produce gas (carbon dioxide) in bread and turn grapes into wine. Through the 17th and 18th centuries, great scientific debates centered around the question whether fermentation was the result of an ethereal "vital life force" present in living cells or instead was based only on the fundamental laws of chemistry and physics that govern the physical world. Some scientists reasoned that if fermentation could be shown to occur outside of a living cell, it would provide evidence that a vital life force was not required for this chemical process.

Numerous attempts by Louis Pasteur and others to prepare cell-free extracts from yeast cells failed, which some interpreted to mean that a vital life force was indeed required for fermentation. The turning point came in 1897, when the German chemist Eduard Buchner (**Figure 1.1**) demonstrated that carbon dioxide and ethyl alcohol could in fact be produced from sugar using brewer's yeast extracts in an *in vitro* reaction. Buchner published his observations and proposed that fermentation required the "ferments of zyme," now known as **enzymes**, which function as catalysts to drive the *in vitro* reactions. Buchner's work set a foundation for the field of biochemistry, where *in vitro* studies are the cornerstone for numerous advances in medical science.

Figure 1.1 Biochemical reactions are often studied or used in *in vitro* systems. Eduard Buchner (1860–1917) was the first to demonstrate that cell-free yeast extracts could accomplish *in vitro* fermentation of sugar into alcohol and carbon dioxide, a discovery that led to the birth of modern biochemistry. Buchner was awarded the 1907 Nobel Prize in Chemistry for his groundbreaking research on *in vitro* fermentation.
HULTON ARCHIVE/GETTY IMAGES.

As is often the case in an experimental science such as biochemistry, several arbitrary decisions led to the success of Buchner's extracts. First, where Pasteur had used glass to grind up yeast and release the fermentation "juices," Buchner chose to use quartz mixed with diatomaceous earth (*kieselguhr*) to prepare the extract. This choice was a good one because it avoided making the extract alkaline and inactive, which occurs when yeast proteins come in contact with glass. Second, after trying a variety of preservatives to prevent coagulation, Buchner decided to use a 40% sucrose solution, not realizing at the time that this would provide the necessary glucose for alcoholic fermentation. Lastly, Buchner used a strain of yeast called *Saccharomyces cerevisiae*, provided by the local brewery in Munich, to prepare an undiluted cell-free extract. This strain of yeast turned out to work much better than yeast strains available in Paris, where Pasteur had done his experiments years earlier. Although it might appear from this that Buchner's accomplishment of *in vitro* alcoholic fermentation was the result of luck, his optimized protocol was developed only after many failed attempts. Indeed, Buchner's systematic approach to solving the problem of inactive cell-free extracts is a classic example of experimental biochemistry.

As we shall see shortly, all living cells contain enzymes. These biomolecules, either protein or ribonucleic acid (RNA), function as reaction catalysts to increase the rates

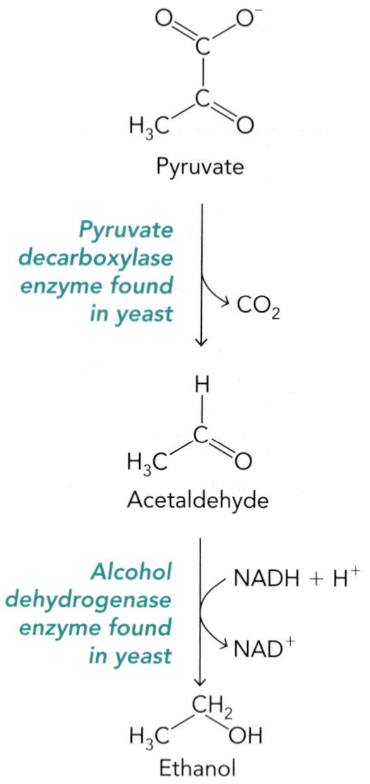

Figure 1.2 The yeast enzymes pyruvate decarboxylase and alcohol dehydrogenase are responsible for converting pyruvate, a product of glucose metabolism, into alcohol and carbon dioxide.

of biochemical reactions dramatically. Enzymes are responsible for aerobic respiration, fermentation, nitrogen metabolism, energy conversion, and even programmed cell death. Two key enzymes are required for the fermentation of glucose by yeast. The first is pyruvate decarboxylase, which converts pyruvate, a breakdown product of glucose, into acetaldehyde and carbon dioxide (CO_2). The second is alcohol dehydrogenase, an enzyme that reduces acetaldehyde to form ethanol (**Figure 1.2**).

Following the lead of Buchner and others, biochemists throughout much of the 20th century focused on systematically dismantling each of the chemical reactions required for cellular life. Almost half of this book describes the biochemical reactions and metabolic pathways (functionally related chemical reactions in cells) elucidated by early biochemists (Chapters 9–19). The rest of the book is devoted to biochemical discoveries made primarily since the 1970s, focusing on the structure and function of proteins (Chapters 4–8) and the biochemistry of genetic inheritance (Chapters 20–23). Both of these modern advances in biochemistry can be traced to the *Eureka!* moment in 1953 when James Watson and Francis Crick solved the molecular structure of DNA.

Biochemistry, like genetics and cell biology, is a core discipline in the life sciences. Biochemistry provides the underlying chemical principles guiding discoveries in medicine, agriculture, and pharmaceuticals. A molecular understanding of chemical reactions in living cells and of how cells communicate to one another in a multicellular organism has led to a dramatic increase in expected human life spans through improved health care, food production, and environmental science. Biochemistry is also a powerful applied science that uses advanced experimental methods to develop *in vitro* conditions for exploiting cellular processes and enzymatic reactions. Examples include the development of new pharmaceutical drugs based on the knowledge of biochemical processes under pathologic conditions, as well as diagnostic tests that detect these abnormalities (**Figure 1.3**). Improved detergents based on enzymatic reactions and the faster ripening of fruits and vegetables using ethylene gas are other examples of applied biochemistry. Moreover, environmental science has benefited from advances in biochemistry through the development of quantitative field tests that can provide vital information about changes in fragile ecosystems due to industrial or biological contamination.

It is an exciting time to be learning biochemistry! Indeed, in this current "Age of Biology," no field is more centrally positioned to exploit this new era. Technological advances in microanalytical chemical methods such as mass spectrometry and enhanced techniques to render high-resolution images of biomolecular structures provide immense opportunity for new discoveries in biochemistry. Chemists, life scientists, and health-field professionals with a firm understanding of the role that biochemistry plays in the chemical nature of life are certain to have a distinct advantage in applying biological discoveries made during the next 50 years.

concept integration 1.1

How did *in vitro* alcoholic fermentation provide evidence for the "chemistry of life"?

Eduard Buchner's *in vitro* experiment in 1897 used a yeast cell-free extract to convert glucose into ethanol and CO_2, thereby providing the first compelling evidence that a "vital force" was not required for alcoholic fermentation. Moreover, this landmark biochemical experiment suggested that conventional chemical reactions were likely to be the molecular basis for life itself and stimulated 50 years of research to prove it.

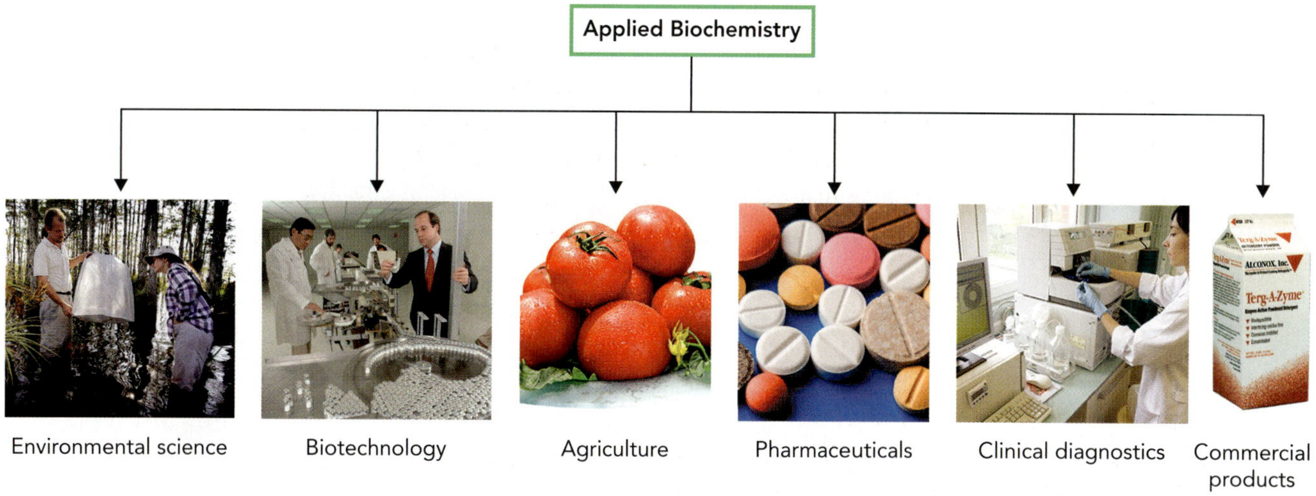

Figure 1.3 Applied biochemistry uses a basic understanding of biochemical principles to guide advances in agriculture, medicine, and industry. ENVIRONMENTAL SCIENCE: EMILY MICHOT/MIAMI HERALD/MCT VIA GETTY IMAGES; BIOTECHNOLOGY: ROGER RESSMEYER/CORBIS; AGRICULTURE: TOHRU MINOWA/A.COLLECTIONRF/GETTY IMAGES; PHARMACEUTICALS: DIMA SOBKO/SHUTTERSTOCK; CLINICAL DIAGNOSTICS: JAVIER LARREA/AGEFOTOSTOCK; COMMERCIAL PRODUCTS: ©ALCONOX, INC.

1.2 The Chemical Basis of Life: A Hierarchical Perspective

We have seen that biochemistry is an interdisciplinary science that brings together many concepts from chemistry, cell biology, and physiology. This integrated approach to molecular life science makes biochemistry very important, but it also means that the student needs to master many terms and definitions. In this section, we review seven levels of biochemical hierarchy—or levels of organizational complexity—that encompass the chemistry of life and use terminology that you will encounter throughout the book.

The foundation of this hierarchy is chemical elements and functional groups (**Figure 1.4**). Next, chemical groups are organized into **biomolecules**, of which there are four major types in nature: amino acids, nucleotides, simple sugars, and fatty acids. Then, higher-order structures of biomolecules form **macromolecules**, which can be chemical polymers such as proteins (polymers of amino acids), nucleic acids (polymers of nucleotides), or polysaccharides such as cellulose, amylose, and glycogen (polymers of the carbohydrate glucose).

Organization of macromolecules and enzymes into **metabolic pathways** is the next hierarchical level. These pathways enable cells to coordinate and control complex biochemical processes in response to available energy. Examples of metabolic pathways include glucose metabolism (glycolysis and gluconeogenesis), energy conversion (citrate cycle), and fatty acid metabolism (fatty acid oxidation and biosynthesis). Metabolic pathways function within membrane-bound cells. The membranes create aqueous microenvironments within the cells for biochemical reactions involving metabolites and macromolecules.

Cell specialization, the next level of organizational complexity, allows multicellular organisms to exploit their environment through **signal transduction** mechanisms that facilitate communication between cells. Organisms represent the subsequent level, as they consist of large numbers of specialized cells, allowing multicellular organisms to respond to environmental changes. One way multicellular organisms

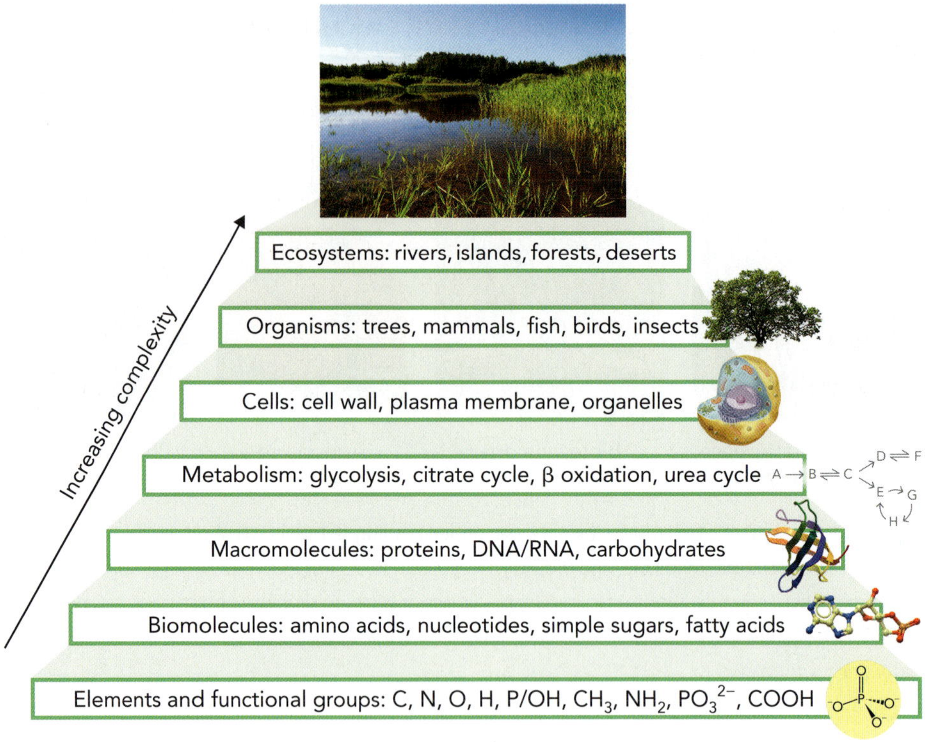

Figure 1.4 A summary of the hierarchical organization and chemical complexity of living systems, including the seven hierarchical levels, along with examples of organizational complexities within these levels.
ECOSYSTEM: JACOBH/ISTOCK/360/GETTY IMAGES; TREE: VISUALL2/SHUTTERSTOCK.

are able to adapt to change is through signal transduction mechanisms that facilitate cell–cell communication. Finally, cohabitation of different organisms in the same environmental niche creates a balanced **ecosystem**, characterized by shared use of resources and waste management. As you will see, the field of biochemistry incorporates the study of chemical life at all levels of this hierarchy.

Elements and Chemical Groups Commonly Found in Nature

Almost 100 chemical elements are found in nature, and chemists have organized them into the periodic table according to their atomic properties. The distribution of these elements in living systems is very different from that in the physical world. In particular, more than 97% of the weight of most organisms consists of just six elements: hydrogen, oxygen, carbon, nitrogen, phosphorus, and sulfur (**Table 1.1**). The vast majority of this mass comes from hydrogen and oxygen, most of which is present as H_2O (the human body is 70% water). In addition to the six most abundant elements, trace elements such as zinc, iron, manganese, copper, and cobalt are required for life, primarily as cofactors in proteins. Essential ions include calcium, chloride, magnesium, potassium, and sodium, many of which play key roles in cell signaling and neurophysiology. The amount of carbon in living organisms is disproportionately high, being 100 times more abundant in the human body than in Earth's crust.

Although the abundance of elements in biological systems is quite different from the abundance of elements in Earth, biochemical reactions are no different from other chemical reactions with regard to bond properties and reaction mechanisms. As you learned in introductory chemistry, covalent bonds form when two atoms share unpaired electrons in their outer shells. The strength of a covalent bond depends on the relative affinities of the two atoms for electrons, the distance between the bonding electrons and the nucleus of each atom, and the nuclear charge of each atom. For example, water, ammonia, carbon dioxide, and carbonic acid are formed by covalent bonds between

Table 1.1 ELEMENTAL COMPOSITION OF THE HUMAN BODY AS A PERCENTAGE OF DRY WEIGHT

			Additional trace elements (<0.1%)	
Element	**Symbol**	**Percent dry weight (%)**	**Element**	**Symbol**
Carbon	C	62	Manganese	Mn
Nitrogen	N	11	Iron	Fe
Oxygen	O	9	Cobalt	Co
Hydrogen	H	6	Copper	Cu
Calcium	Ca	5	Zinc	Zn
Phosphorus	P	3	Selenium	Se
Potassium	K	1	Molybdenum	Mo
Sulfur	S	1	Iodine	I
Chlorine	Cl	<1	Fluorine	F
Sodium	Na	<1	Chromium	Cr
Magnesium	Mg	<1	Tin	Sn

Note: These values exclude the contribution of oxygen and hydrogen to the large amount of water in the human body (70% by weight).

H, O, N, and C (**Figure 1.5**). Hydrogen requires two electrons to complete its outer shell, whereas O, N, and C each require eight electrons. Ions such as hydronium ion, H_3O^+, ammonium ion, NH_4^+, and bicarbonate ion, HCO_3^- are formed by the gain of a proton and loss of an electron (or vice versa), so as to maintain a complete outer shell. Double bonds are stronger than single bonds, as more energy is required to break a double bond (**Table 1.2**).

The chemical nature of life on Earth is based on the element carbon (Figure 1.5). Molecules containing carbon are called organic molecules, and organic chemistry is the study of carbon-based compounds. Indeed, early biochemists were often organic chemists who became interested in "biological" chemistry. Carbon has a unique ability

Figure 1.5 Covalent bonds result from sharing of an electron pair between two atoms. **a.** H, O, N, and C all have unpaired electrons in their outer shell that can participate in bond formation. Unpaired electrons are shown as red dots and paired electrons as black dots. **b.** The arrangement of electron sharing for some common biomolecules. Covalent bonds occur when unpaired electrons in each of two atoms interact, forming an electron pair that is shared between the atoms.

a.

Atom	Number of unpaired electrons
H·	1
:Ö·	2
:N̈·	3
·Ċ·	4

b.

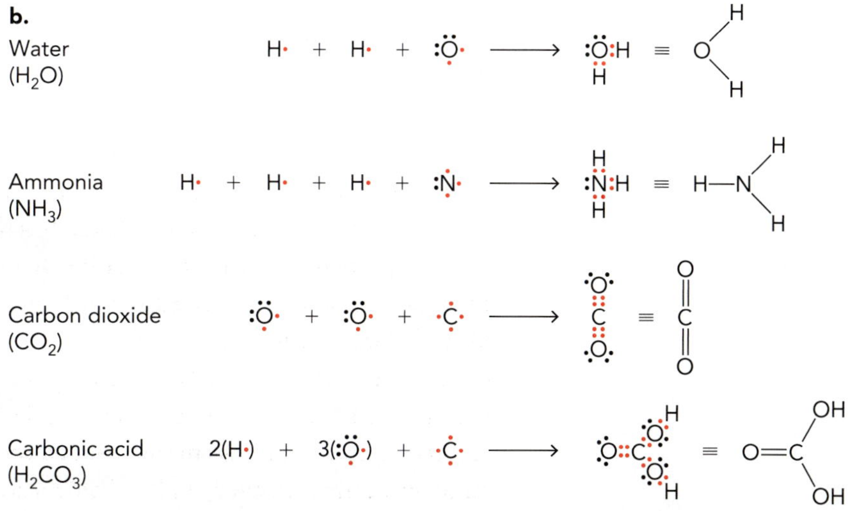

Table 1.2 BOND ENERGIES AND BOND LENGTHS OF COMMON COVALENT BONDS IN NATURE

Type of bond	Bond energy (kJ/mol)	Bond length (Å)	Type of bond	Bond energy (kJ/mol)	Bond length (Å)
C—C	346	1.54	P—O	335	1.63
C═C	602	1.34	P═O	544	1.50
C—N	305	1.47	N—N	167	1.45
C═N	615	1.29	N═N	418	1.25
C—O	358	1.43	O—H	459	0.96
C═O	799	1.20	N—H	386	1.01
C—H	411	1.09	P—H	322	1.44

Note: 1 angstrom (Å) = 10^{-10} meter.

to form up to four stable covalent bonds because of its four unpaired electrons, which means that a chain of carbon atoms can serve as a backbone for the assembly of a variety of organic molecules.

The most common carbon bonds in biomolecules are C—C, C═C, C—H, C═O, C—N, C—S, and C—O bonds. Four single bonds to a carbon atom are arranged in a tetrahedron, as in methane, CH_4 (**Figure 1.6**). This tetrahedral arrangement has an angle of 109.5° between the bonds and an average bond length of 1.5 **angstroms (Å)** (10^{-10} meter). In the simplest molecule that contains a carbon–carbon single bond, ethane (C_2H_6), the bond angles are very near the tetrahedral value and rotation can occur around each single bond including the carbon–carbon bond. In molecules with double-bonded carbon atoms (C═C), such as ethylene (C_2H_4), all the atoms are in the same plane and the bond angles are approximately 120°. Rotation does not readily occur around the carbon–carbon double bond, and therefore the atoms are largely fixed in position relative to each other.

Hydrogen, oxygen, carbon, nitrogen, phosphorus, and sulfur combine into functional groups, which are responsible for many of the chemical properties of biomolecules. The most abundant functional groups in biomolecules are amino (NH_2), hydroxyl (OH), sulfhydryl (SH), phosphoryl (PO_3^{2-}), carboxyl (COOH), and methyl (CH_3) groups (**Figure 1.7**).

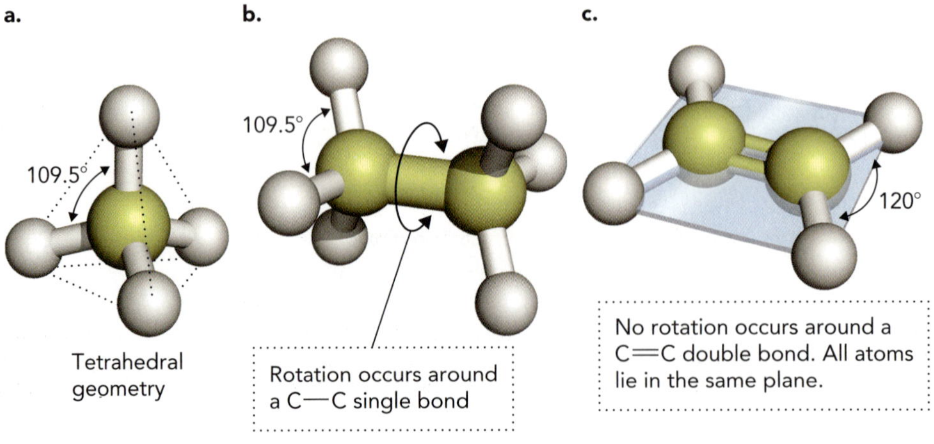

Figure 1.6 Covalent bonds containing carbon can vary in their characteristics. **a.** Carbon has four unpaired electrons in its outer shell and can form four covalent bonds in a tetrahedral arrangement at angles of 109.5°. **b.** Carbon–carbon single bonds (C—C) can rotate freely relative to each carbon atom. **c.** Rotation around a carbon–carbon double bond (C═C)is restricted, and therefore the atoms are held in place with respect to each other. The bond angles are approximately 120°.

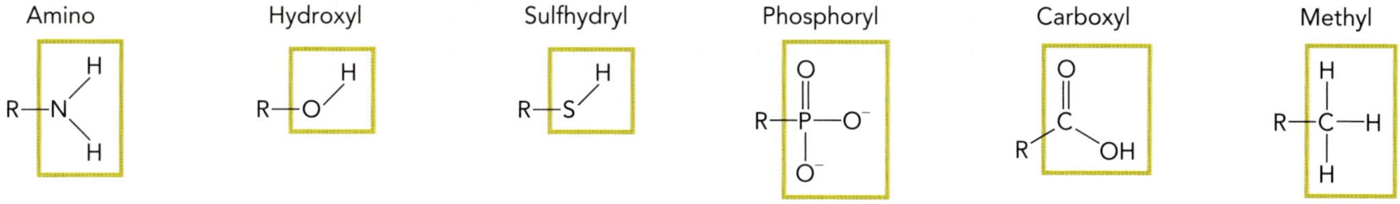

Figure 1.7 Six chemical groups are very commonly found in biomolecules. The methyl group has a single protonation state. However, the amino, hydroxyl, sulfhydryl, phosphoryl, and carboxyl groups may have different protonation states from what is shown, depending on the nature of other atoms in the vicinity. R represents the rest of the molecule to which the functional group is attached.

Four Major Classes of Small Biomolecules Are Present in Living Cells

The essential elements and functional groups required for life are contained within four major classes of small biomolecules in cells. These are (1) amino acids, (2) nucleotides, (3) simple sugars, and (4) fatty acids (**Figure 1.8**). All of these biomolecules are described in more detail later in this book, but we introduce them briefly here to provide an overview of their structures and functions in living cells.

Amino acids are nitrogen-containing molecules that function primarily as the building blocks for **proteins.** In the process of protein synthesis, amino acids are covalently linked into a linear chain to form **polypeptides**. Proteins are mixed polymers of the different amino acids, and the function of each protein is determined by the sequential arrangement of amino acids along the polypeptide chain. The amino acids differ from one another in the side chains attached to the central carbon. Glycine is the smallest amino acid and contains a hydrogen atom as the side chain (see Figure 1.8). Besides contributing to the structure and function of proteins, glycine is also necessary for the synthesis of heme, an iron-containing molecule required for hemoglobin function in red blood cells. The amino acid glutamate and derivatives of the amino acid tyrosine are important signaling molecules in the brain and function as neurotransmitters. The amino acids glutamine and alanine are required for metabolic

Figure 1.8 Four major classes of small biomolecules are contained in all living cells.

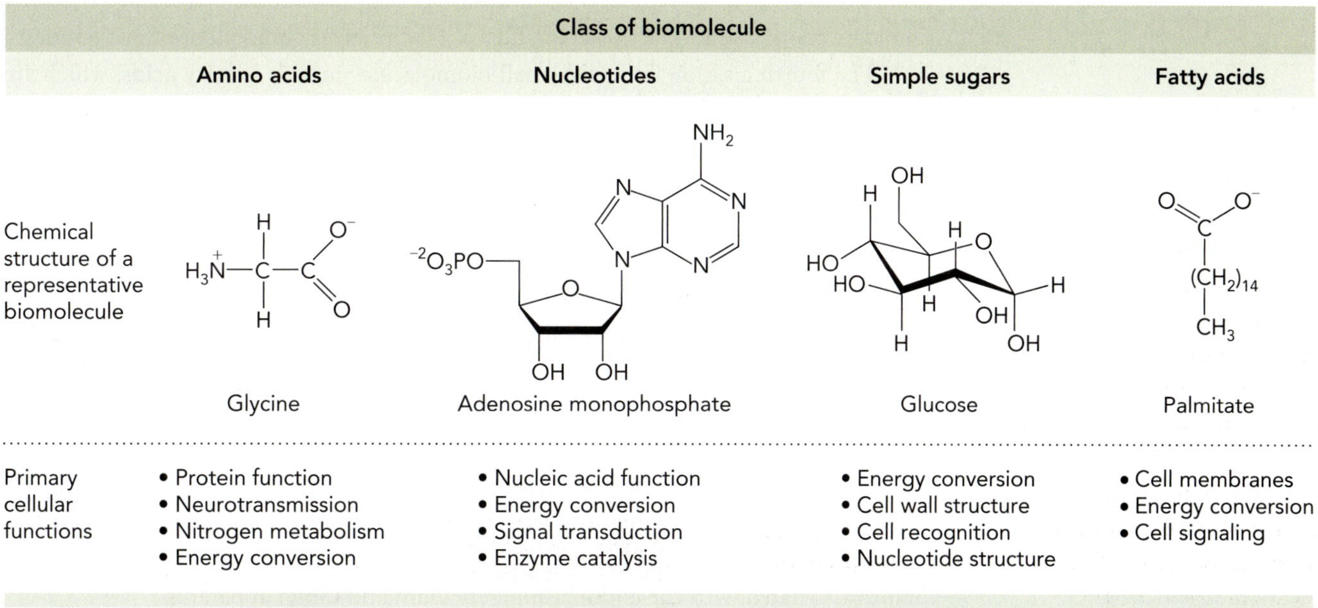

pathways involved in nitrogen metabolism. Amino acids derived from the degradation of skeletal muscle proteins can also be a source of energy for the rest of the body under conditions of fasting or starvation.

Nucleotides consist of a nitrogenous base, a five-membered sugar (ribose or deoxyribose), and one to three phosphate groups (see Figure 1.8). The nucleic acids **deoxyribonucleic acid (DNA)** and **ribonucleic acid (RNA)** are formed from assembly of nucleotides into linear chains. The order of the five nucleotide bases—adenine, guanine, cytosine, thymine, and uracil—in nucleic acids is responsible for imparting biological specificity to nucleic acids. The nucleotide adenosine triphosphate (ATP) functions as the "energy currency" of the cell through phosphoryl group transfer to other molecules, thus providing a driving force for reactions to occur. Other important nucleotides in cells are cyclic adenosine monophosphate (cAMP) and cyclic guanosine monophosphate (cGMP), both of which are signaling molecules that control metabolic and physiologic processes in living organisms. The coenzymes acetyl-coenzyme A (acetyl-CoA), nicotinamide adenine dinucleotide (NAD^+; oxidized form), and flavin adenine dinucleotide (FAD; oxidized form) are nucleotides that work in combination with proteins to help carry out chemical reactions. In Chapter 8, we will examine the signaling functions of cAMP and cGMP, and in Chapter 10, we will look at the involvement of acetyl-CoA, NAD^+, and FAD in citrate cycle reactions.

The third important class of biomolecules in living cells is **simple sugars.** These compounds are formed only of carbon, oxygen, and hydrogen, with a 2:1 ratio of hydrogen to oxygen atoms, as in water. Historically, for this reason these compounds are also known as **carbohydrates**, a term that refers to both simple sugars and polymers of sugars. The simple sugars are also called monosaccharides or disaccharides ("saccharide" is derived from the Latin word for sugar, *saccharum*). Glucose ($C_6H_{12}O_6$) is a monosaccharide involved in energy conversion reactions, cell signaling, and cell structure (see Figure 1.8). Oxidation of glucose by enzymatic reactions in cells releases energy that can be captured in the form of ATP and used to drive other chemical reactions. Glucose is also the building block for cellulose, which is the structural component of plant cell walls; glycogen, which is an energy storage form of carbohydrate in animals; and amylose (starch), which is the primary form of stored energy in plants. Additionally, we will see that glucose derivatives are important in cell recognition when they are covalently attached to proteins (glycoproteins) or lipids (glycolipids) on the cell surface. Another abundant monosaccharide, ribose ($C_5H_{10}O_5$), is the sugar component of nucleotides.

The fourth class of abundant small biomolecules in cells is **fatty acids**, which are **amphipathic** molecules (polar and nonpolar chemical properties contained within the same molecule). Fatty acids consist of a carboxyl group (polar) attached to an extended hydrocarbon chain (nonpolar). Saturated fatty acids such as palmitic acid contain no C═C double bonds in the hydrocarbon chain [$CH_3(CH_2)_{14}CO_2H$], whereas the polyunsaturated fatty acid eicosapentaenoic acid contains five C═C double bonds [$CH_3(CH_2CH{=}CH)_5(CH_2)_3CO_2H$]. Fatty acids in living cells primarily act as components of plasma membrane lipids, which form a hydrophobic barrier separating the aqueous phases of the inside and outside of cells. The most abundant lipids in cell membranes are **phospholipids**, which generally contain a simple organic molecule attached to a negatively charged phosphoryl group and two fatty acids. Besides the plasma membrane, eukaryotic cells (plant and animal cells) contain a variety of intracellular membranes consisting of fatty acid–derived lipids. These include the nuclear membrane, the inner and outer mitochondrial and chloroplast membranes, and membranes associated with the endoplasmic reticulum and Golgi apparatus.

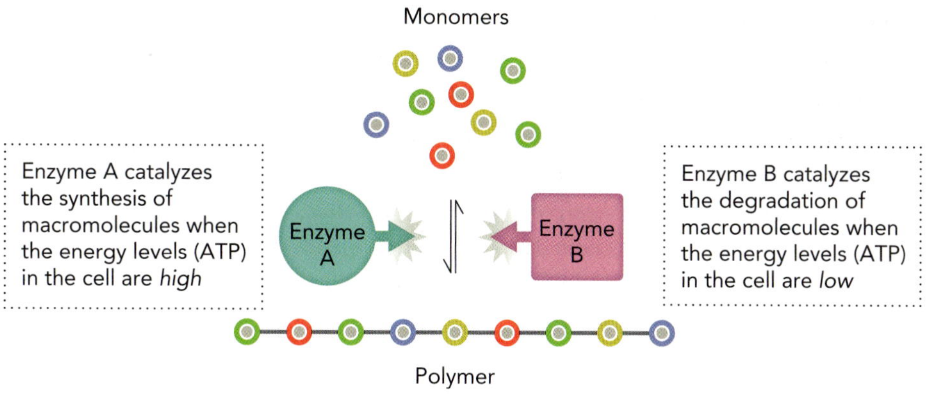

Monomers
Enzyme A catalyzes the synthesis of macromolecules when the energy levels (ATP) in the cell are *high*
Enzyme A
Enzyme B
Enzyme B catalyzes the degradation of macromolecules when the energy levels (ATP) in the cell are *low*
Polymer

Figure 1.9 The processes of assembly of macromolecular polymers from monomers and disassembly of macromolecular polymers into monomers are often controlled by similar but distinct enzymes that are directly or indirectly regulated by energy levels in the cell. ATP levels in the cell are a measure of available energy because a large number of biochemical reactions depend on phosphoryl transfer energy made available from ATP hydrolysis.

Another important function of fatty acids in eukaryotes is as a storage form of energy, which is made possible by their highly reduced state. Fatty acids yield chemical energy upon oxidation in mitochondria. Used for energy storage in this way, fatty acids are converted to **triacylglycerols** and sequestered in the adipose tissue of animal cells, whereas plants store triacylglycerols in seeds. Triacylglycerols are neutral (uncharged) lipids that contain three fatty acid esters covalently linked to glycerol. Lastly, fatty acids and fatty acid–derived molecules have recently been shown to be important signaling molecules that bind to nuclear receptor proteins. In this way, fatty acids regulate lipid and carbohydrate metabolism, inflammatory responses, and cell development.

Macromolecules Can Be Polymeric Structures

The most common structural arrangement of small biomolecules is in the form of polymers, which create large macromolecules. The two most abundant polymers in cells are nucleic acids, which consist of covalently linked nucleotides, and proteins, which are made up of covalently linked amino acids in the form of polypeptides. Simple sugars can also be linked into polymeric structures, forming a type of carbohydrate called **polysaccharides**. The most common polysaccharides in nature are cellulose, chitin, starch, and glycogen.

The enzymatic reactions that assemble and disassemble polymers must be regulated to control these processes in response to cellular conditions. One of the most important determinants of this regulatory process is the availability of chemical energy in the form of ATP, which is required for assembly of many macromolecules, not just nucleic acids. In general, when ATP levels in the cell are high, energy is available for the synthesis of polymeric macromolecules; however, when ATP levels in the cell are low, then degradation of polymeric macromolecules is favored (**Figure 1.9**).

It is important to recognize that the unique chemical properties of independent macromolecular polymers are a function of the chemical complexity of the monomeric units. For example, DNA polymers contain combinations of four different deoxyribonucleotides linked together through **phosphodiester bonds** (**Figure 1.10**). Because DNA and RNA have polarity in which the 5′-phosphoryl and 3′-hydroxyl groups on the ribose sugars are distinct, the sequential arrangement of monomers along the nucleic acid chain has functional significance in terms of information content. Indeed, a DNA octamer (eight linked nucleotides) can have any one of 65,536 (4^8) different sequence

Figure 1.10 DNA is a polymeric macromolecule consisting of nucleotides that are covalently linked through a phosphodiester bond (shown in red), which connects the 3′-carbon of one nucleotide to the 5′-carbon of a second nucleotide. DNA polymers have chemical polarity, meaning that the 5′-phosphoryl and the 3′-hydroxyl termini are distinct.

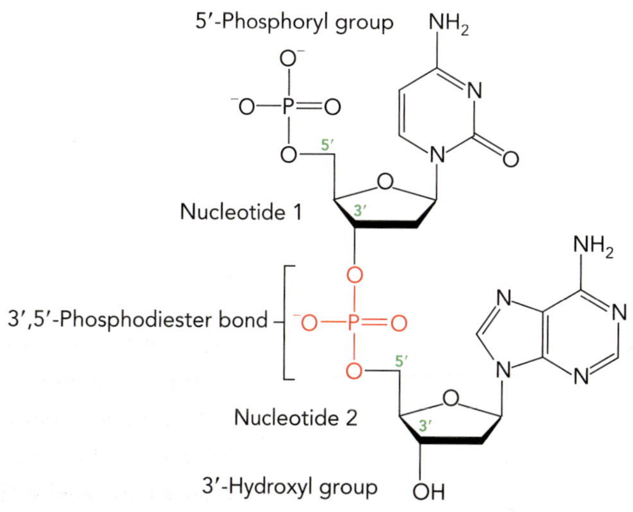

5′-Phosphoryl group
Nucleotide 1
3′,5′-Phosphodiester bond
Nucleotide 2
3′-Hydroxyl group

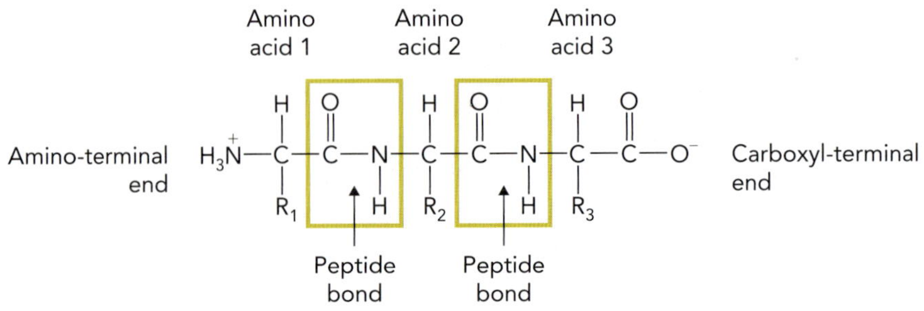

Figure 1.11 Proteins are polymeric macromolecules consisting of amino acids that are covalently linked through peptide bonds (highlighted by boxes). The chemical properties of proteins are determined by the different side chains in the amino acids (shown here as R_1, R_2, and R_3).

combinations, as each of the eight positions can have any one of the four deoxyribonucleotides (5′-phosphate esters of deoxyadenosine, deoxyguanosine, deoxycytosine, or deoxythymidine). In the case of proteins, the complexity of the monomeric units is even greater: Any of 20 different amino acids can be linked together by a covalent bond called a **peptide bond** (**Figure 1.11**). Polypeptide chains also have polarity, as they contain an amino group on one end and a carboxyl group on the other. Therefore, the number of octamer polypeptides that can theoretically be assembled with any one of the 20 amino acids at each position is a staggering 20^8, or 2.58×10^{10} different protein sequences. However, the actual number of different polypeptide sequences encountered biologically is much smaller than the theoretical number because not all combinations of amino acids have useful structural and functional properties due to differences in the size and chemistry of their side chains.

Polysaccharides can consist of mixtures of different simple sugars or just repeating units of the monosaccharide glucose. However, the type of covalent bond linking adjacent sugar molecules in polysaccharides can make a big difference in terms of chemical properties. For example, the covalent linkage between glucose units in **amylose** is an $\alpha(1\rightarrow4)$ glycosidic bond, whereas in cellulose, it is a $\beta(1\rightarrow4)$ glycosidic bond (**Figure 1.12**). Moreover, the glucose units in cellulose are rotated 180° relative to one another, unlike the glucose units in amylose. The presence of $\beta(1\rightarrow4)$ or $\alpha(1\rightarrow4)$ glycosidic bonds in cellulose and amylose, respectively, is important to humans in terms of obtaining metabolic fuel from foods containing these two types of carbohydrates. The reason is that we have a salivary enzyme named α-amylase that degrades amylose (found in such food staples as rice and potatoes) into glucose monomers, which are then oxidized to generate ATP. However, humans lack the enzyme cellulase, which is needed to hydrolyze the $\beta(1\rightarrow4)$ glycosidic bond in cellulose. Therefore, humans cannot directly obtain glucose from most plant parts.

Chitin, another abundant carbohydrate polymer found in nature, is a major component in the exoskeletons of many invertebrates (insects and crustaceans) and in the cell walls of some types of fungi. Chitin consists of repeating *N*-acetylglucosamine units, a derivative of glucose, linked together by the same type of $\beta(1\rightarrow4)$ glycosidic bonds found in cellulose. Many types of bacteria contain the enzyme chitinase, which is able to cleave the $\beta(1\rightarrow4)$ glycosidic bond in chitin and thereby facilitate decomposition processes.

Nucleic acids, proteins, and polysaccharides adopt three-dimensional conformations that are essential to their functions. These structures will be discussed in detail in later chapters of the book (nucleic acids in Chapter 3; proteins in Chapter 4; and polysaccharides in Chapter 13). For now, it is important to recognize that the sequence of the building blocks (e.g., the amino acids) imparts critical information to specify the overall structure of the molecule. Proteins, with a repertoire of 20 amino acids that can be sequentially arranged, adopt vastly different structures depending on the sequence of the amino acids; examples of this are given later in the chapter. It is also important

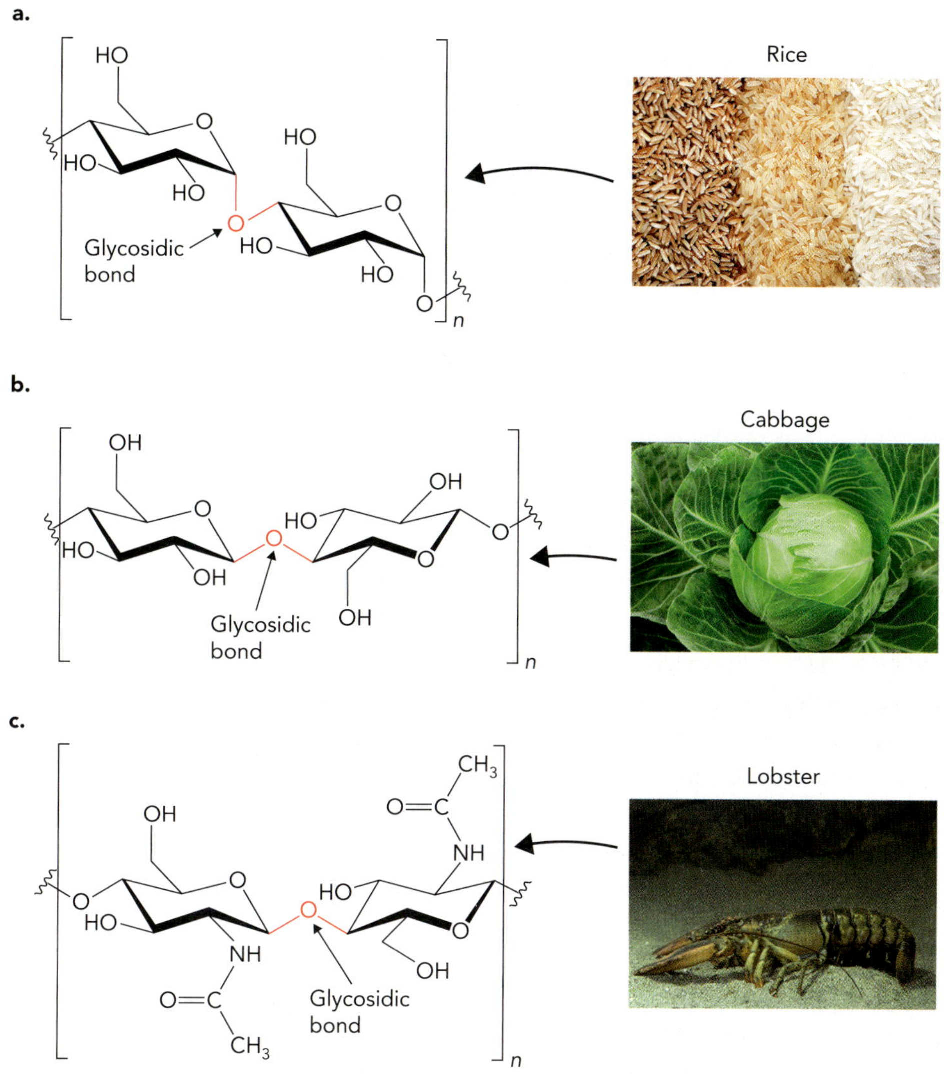

Figure 1.12 The polysaccharides amylose and cellulose contain the same repeating unit of glucose, but differ in the structure of the glycosidic bond linking adjacent units (shown in red). **a.** Amylose (starch) is a polymer of glucose containing α(1→4) glycosidic bonds between glucose units. PHOTO: INNA FELKER/SHUTTERSTOCK. **b.** Cellulose, contained in plant cell walls, is identical in composition to amylose; however, the glycosidic bonds between glucose units are in the β(1→4) configuration, and adjacent glucose residues are rotated 180°. PHOTO: USYNINA/SHUTTERSTOCK. **c.** Chitin is the primary carbohydrate component in the exoskeletons of insects and crustaceans. Chitin contains a β(1→4) glycosidic bond linking adjacent *N*-acetylglucosamine units. PHOTO: BRECK P. KENT/ANIMALS ANIMALS.

to recognize that biological macromolecules are structurally dynamic, and their overall shapes or conformations can change in response to changes in the environment or their interactions with other molecules.

Metabolic Pathways Consist of Linked Biochemical Reactions

Small biomolecules serve as both reactants and products in biochemical reactions within cells—reactions necessary for life-sustaining processes. In this context, we refer to these molecules as **metabolites**. The catalysts that drive these biochemical reactions are enzymes, which consist of protein or RNA macromolecules.

In enzyme-mediated reactions, the products of one reaction are often the reactants for other functionally related reactions. Indeed, the thousands of reactions required for sustaining life in a cell are interdependent and highly regulated to maximize efficient use of limited metabolic resources.

The emerging discipline of **systems biology** attempts to describe complex chemical reaction networks in cells. This approach uses mathematical models that reflect **metabolic flux** (the rate at which reactants and products are interconverted in a metabolic pathway) in response to environmental or physiologic conditions. However, formulation of accurate models is extremely difficult because of current limitations in measurement of metabolite concentrations inside living cells. In addition, unknown factors contribute to the size of metabolite pools under different conditions. Therefore, although systems biology is still being developed to provide an overall picture of cellular metabolism, it is more instructive to focus on a limited number of biochemical reactions that have been highly characterized and which provide a basis for understanding the chemistry of life. Using this approach, sets of biochemical reactions have been grouped together into metabolic pathways that have pedagogical value for historical and conceptual reasons. Examples of these "classic" metabolic pathways are glycolysis (Chapter 9), the citrate cycle (Chapter 10), the pentose phosphate pathway and gluconeogenesis (Chapter 14), fatty acid oxidation and synthesis (Chapter 16), and the urea cycle (Chapter 17).

To understand how a metabolic pathway is organized, let's first look at a mini-pathway consisting of just two reactions (**Figure 1.13**). The reactants citrulline

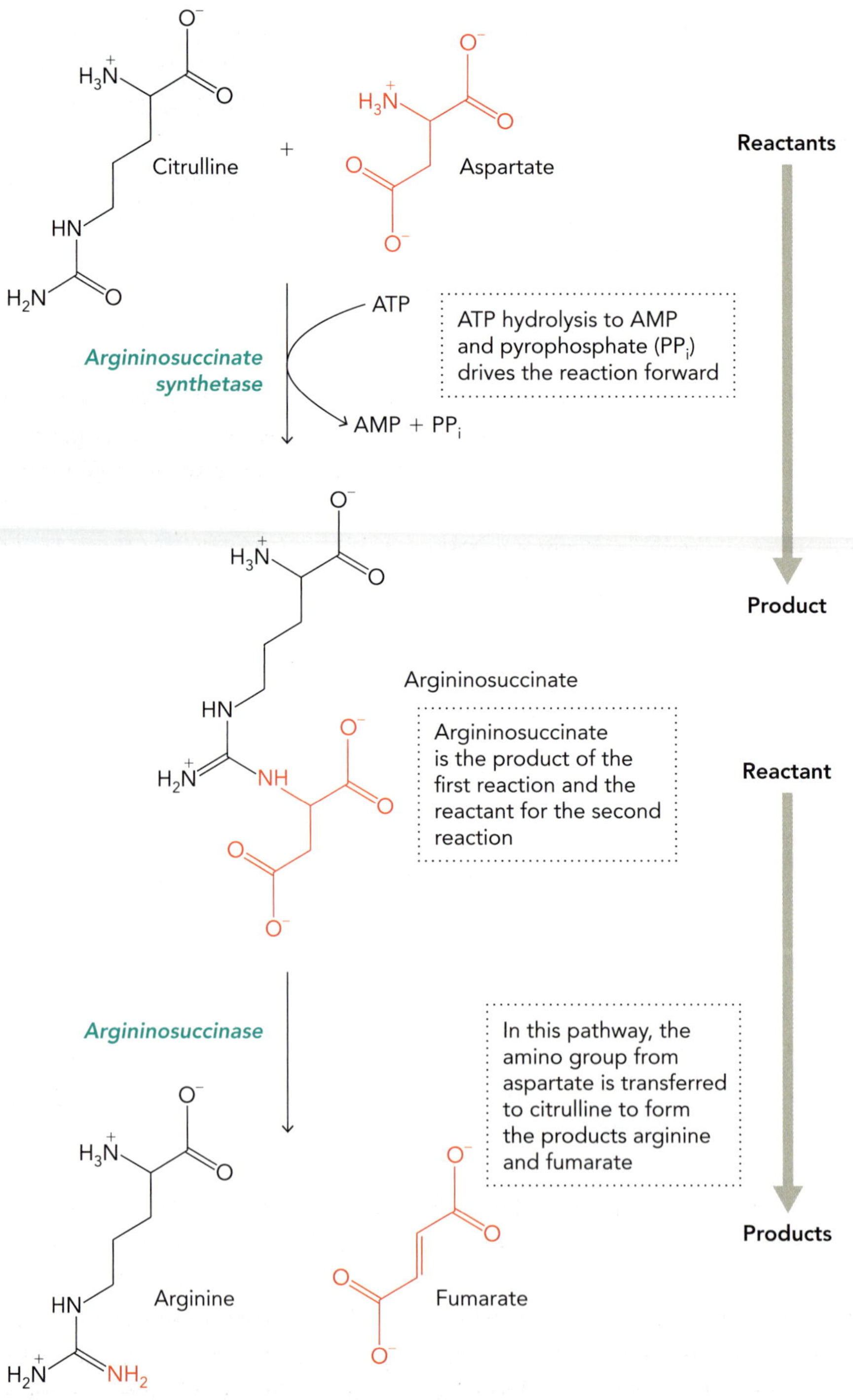

Figure 1.13 Metabolic pathways consist of linked biochemical reactions in which the product of one reaction is the reactant for another reaction. In this example, the two urea cycle enzymes, argininosuccinate synthetase and argininosuccinase, function in a mini-pathway that converts citrulline and aspartate into arginine and fumarate through formation of argininosuccinate.

and aspartate are metabolites in the urea cycle, which is a metabolic pathway in liver cells responsible for nitrogen excretion in animals. The protein enzyme argininosuccinate synthetase catalyzes a condensation reaction that forms the product argininosuccinate, using energy made available by the hydrolysis of ATP. In the second reaction, argininosuccinate is cleaved by the enzyme argininosuccinase to form the products arginine and fumarate. The chemical difference between citrulline and arginine is the addition of a single amino group obtained from aspartate; however, in order for this addition to occur, argininosuccinate has to function both as a product and a reactant. (The biochemical roles of argininosuccinate synthetase and argininosuccinase in catalyzing these two urea cycle reactions is described in Chapter 17.) Note that the fermentation reaction that converts pyruvate to ethanol and CO_2 by the enzymes pyruvate carboxylase and alcohol dehydrogenase (see Figure 1.2) is another example of a two-reaction mini-pathway. In this case, acetaldehyde is the key pathway intermediate serving as both a product and a reactant.

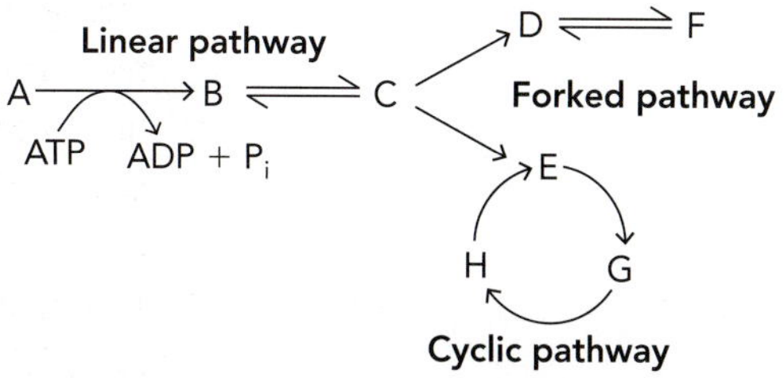

Figure 1.14 Metabolic pathways are linked together by shared products and reactants. At least three types of pathways are common: linear pathways, forked pathways, and cyclic pathways. Some reactions, such as from B to C, are reversible (double arrows). Other reactions, such as from A to B, are essentially irreversible, as a result of reaction coupling to ATP hydrolysis or chemical instability of the reactant. P_i = inorganic phosphate.

Figure 1.14 illustrates a more complex metabolic pathway, in which enzymes interconvert metabolites using reversible and irreversible reactions. In some reactions, as in the conversion of A to B, hydrolysis of ATP is used to drive the reaction toward product formation. This figure shows three types of linked reactions commonly found in metabolism. The most common types are **linear metabolic pathways,** in which each reaction generates only a single product, which is a reactant for the next reaction in the pathway. In contrast, **forked pathways** usually generate two products, each of which undergoes a different metabolic fate. Lastly, **cyclic pathways** contain several metabolites that regenerate during each turn of the cycle, serving as both reactants and products in every reaction. Two examples of cyclic pathways described later in the book are the citrate cycle and the urea cycle, in which the reactant for the first reaction is the product of the last reaction. Note that the flux of metabolites through metabolic pathways depends on both the activity of each enzyme in the pathway and the intracellular concentrations of reactants and products, which affects the directionality of reactions as a function of mass action (Le Châtelier's principle).

Structure and Function of a Living Cell

Many biochemical processes, such as metabolic pathways, are common to a variety of organisms. Bacteria and yeast are single-cell organisms that contain many of the same metabolic pathways found in our own cells. Bacteria (and archaebacteria) are prokaryotic organisms (**prokaryotes**), which possess all the essential genetic information needed for autonomous life in a nutrient-rich environment. Yeast are eukaryotic organisms (**eukaryotes**), which—like the cells of other fungi and all plants and animals—are more highly evolved and contain specialized cellular structures that create microenvironments for biochemical reactions. Far from being "bags of enzymes and DNA," living cells are highly ordered structures of internal components surrounded by a lipid membrane. Living cells obtain energy from the Sun or from oxidation–reduction reactions to support metabolic processes.

Bacteria are often about 1 μm in diameter and are surrounded by a cell wall that encases a lipid-rich plasma membrane (**Figure 1.15a**). The bacterial cell wall itself is coated with either a **capsule** or a slime layer that aids in enabling the bacterium to attach to other cells or solid surfaces. The **cytoplasm** of the bacterial cell contains all of the enzymes required for cell metabolism, as well as the **chromosome**, which consists of DNA compacted with nucleic acid binding proteins to minimize its size. The bacterial chromosome is circular and localized to a region in the cell called the **nucleoid**.

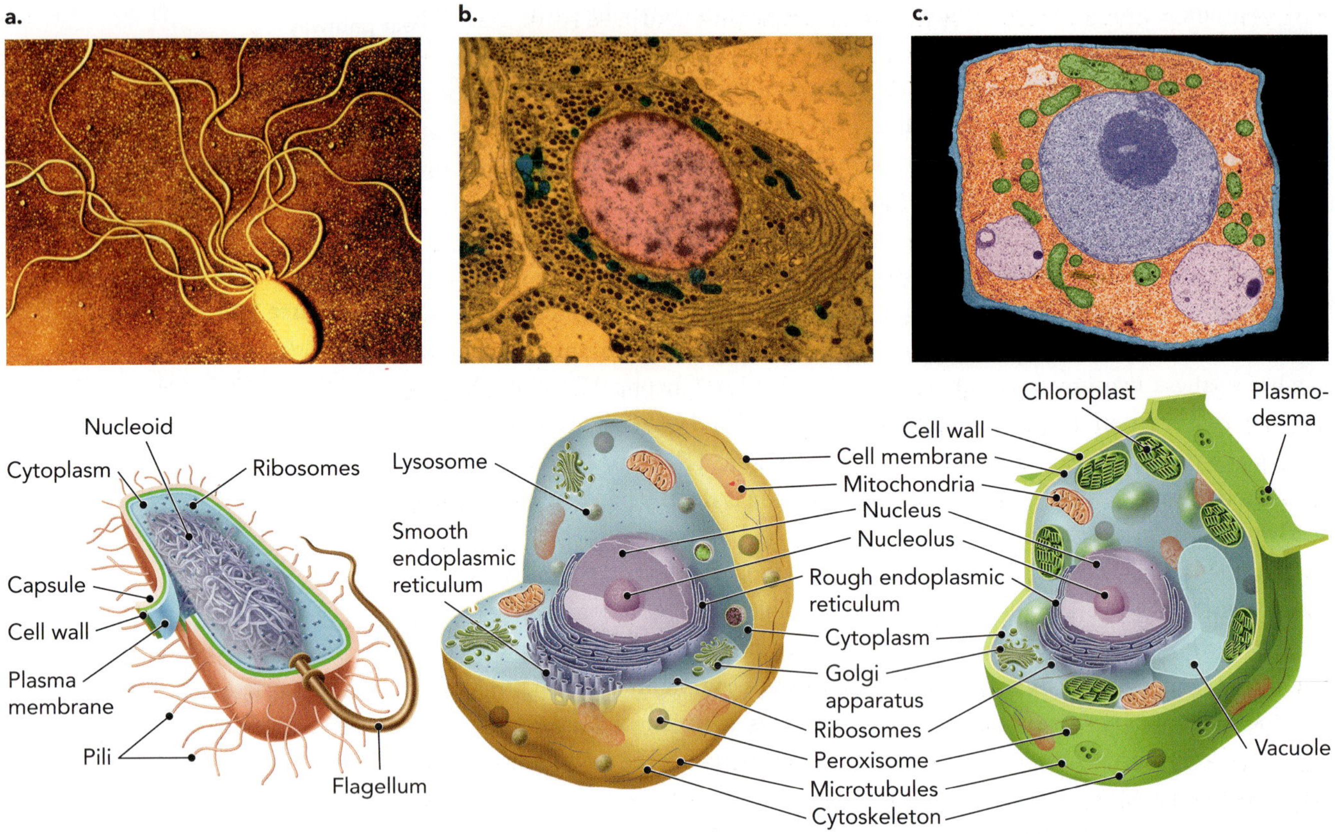

Figure 1.15 The cellular structures present in prokaryotic and eukaryotic cells are shown here. Note that the diameter of eukaryotic cells is as large as ~10–100 μm, whereas the diameter of prokaryotic cells is only ~1 μm. **a.** Bacteria are prokaryotic cells with a cytosolic compartment surrounded by a plasma membrane that forms a barrier separating the cell from the environment. Most bacteria contain a single circular chromosome and move using flagellar structures or pili located on the outside of the cell. PHOTO: FRANCIS LEROY, BIOCOSMOS/SCIENCE SOURCE. **b.** Animal cells are a type of eukaryotic cell and contain numerous intracellular membrane-bound compartments, which create microenvironments for biochemical reactions. Membrane-bound organelles in all types of eukaryotic cells include mitochondria, lysosomes, and peroxisomes, which are subcellular sites for specific metabolic reactions. PHOTO: BIOPHOTO ASSOCIATES/SCIENCE SOURCE. **c.** Plant cells are eukaryotic cells that contain chloroplasts, which convert light energy into chemical energy by the process of photosynthesis. Plant cells also contain large vacuoles, which are responsible for maintaining metabolite pools. The plasma membrane of plant cells is surrounded by a cell wall consisting of cellulose, which provides structural integrity to the plant. PHOTO: BIOPHOTO ASSOCIATES/SCIENCE SOURCE.

The cytoplasm also contains proteins involved in cell division and the assembly of extracellular structures, such as **flagella** and **pili**, which are used for cell movement. Many types of bacteria contain one or more circular DNA molecules called **plasmids,** which may encode genes involved in cell mating or antibiotic resistance. The plasmid replicates independently of the bacterial chromosome. Recombinant DNA methods make use of bacterial plasmids for gene cloning.

Eukaryotic cells are about 10–100 times larger than most bacteria (**Figure 1.15b**). The **genome**, which includes all of the encoded genes and other DNA elements specifying the genetic composition of prokaryotic and eukaryotic cells, is also much larger in eukaryotic cells than in prokaryotic cells owing to the increased complexity of eukaryotic organisms. Eukaryotic DNA is packaged with proteins to form a structure called **chromatin** that is contained within a membrane-bound region of the cell called the **nucleus**. The genomic DNA of eukaryotes is divided into linear chromosomes, which in most eukaryotes can number from 10 to 50 unique chromosomes per cell. The nucleus also contains a region called the **nucleolus,** which is where ribosomes are assembled from ribosomal RNA and protein. **Ribosomes** are large RNA–protein complexes that mediate protein synthesis in prokaryotic and eukaryotic cells. The structure of eukaryotic cells is maintained by the **cytoskeleton**—a network of intracellular filaments consisting of assemblies of proteins.

Several types of membrane-bound organelles are present in the cytosol of eukaryotic cells, and each has specialized functions related to the proteins and enzymes contained within them. Three organelles abundant in most types of eukaryotic cells are **mitochondria**, which are responsible for many of the metabolic reactions involved in energy conversion and production of ATP, and **peroxisomes** and **lysosomes**, which are involved in degradation and detoxification of macromolecules brought into the cell from the outside. Plant cells also contain **chloroplasts**, which are organelles that convert light energy into chemical energy, and very large membrane-bound **vacuoles**, which store metabolites and are functionally similar to lysosomes (**Figure 1.15c**). Both mitochondria and chloroplasts contain their own genomic DNA, which encodes proteins required for organelle function. Other organelles include the rough and smooth **endoplasmic reticulum**, which are highly invaginated membrane structures that sequester ribosomes for protein synthesis. Also, the **Golgi apparatus** is a membranous structure involved in protein translocation within the cell and in facilitating protein secretion at the plasma membrane. Later in the book, we will discuss the biochemical functions and processes associated with each of these organelles.

The fact that animal cells lack a rigid cell wall allows them to alter the shape of the plasma membrane, using components of the cytoskeleton called **microtubules**. These cellular cables enable the cell to move by extending the plasma membrane in one direction while retracting it at the opposite end of the cell (**Figure 1.16**). (Note that plant cells also have microtubules, but the rigid cell wall prevents this kind of movement.) It is thought that mobile ancestral eukaryotic cells were predators that sought out and engulfed other cells as a source of nutrients. This behavior can still be seen in mammalian macrophage cells, which are protective cells in the immune system that seek out and destroy invading microorganisms or abnormal cells based on identification of foreign cell surface proteins. The **endosymbiotic theory** proposes that eukaryotic cells evolved about 1.5 billion years ago as a result of large predatory cells engulfing aerobic bacteria, which eventually gave rise to mitochondria. The symbiotic bacteria were able to use O_2 in the atmosphere as an electron acceptor in oxidation–reduction reactions, which provided a form of chemical energy for the synthesis of ATP. The predatory cells benefited from the extra ATP produced by the symbiotic bacteria, which in turn were rewarded with a nutrient-rich environment. Supportive evidence for the endosymbiotic theory comes from DNA sequence similarities between mitochondrial and bacterial genomes that have been analyzed using computational methods.

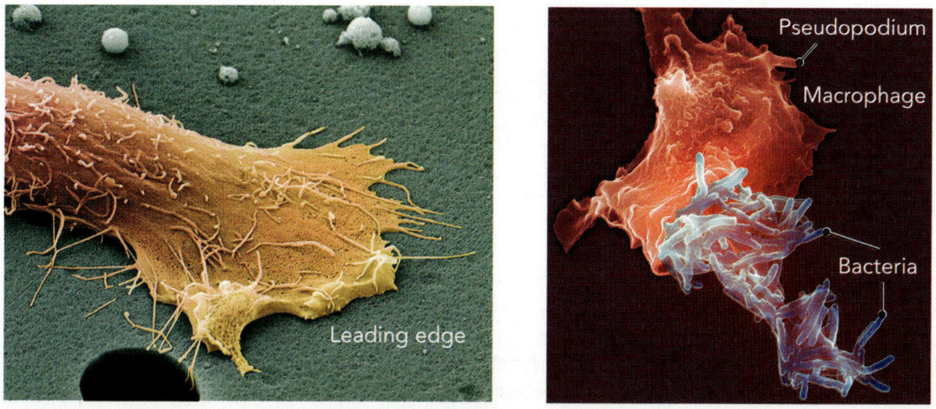

Figure 1.16 Animal cells can change shape by altering the plasma membrane using intracellular protein filaments (microtubules) that function as structural cables. **a.** Cell migration involves the extension of cytoskeletal structures at the forward (leading) edge of the cell with simultaneous retraction at the opposite end. The red staining indicates the location of an antibody bound to the protein myosin, which is needed for the movement of the cytoskeleton. STEVE GSCHMEISSNER/SCIENCE SOURCE. **b.** Macrophages are immune cells in animals that engulf microorganisms such as bacteria. Microphages use pseudopodia to recognize foreign or abnormal cells based on cell surface proteins. SPL/SCIENCE SOURCE.

Multicellular Organisms Use Signal Transduction for Cell–Cell Communication

Biochemical processes can reach far beyond the single cell. In a multicellular organism, individual cells need to communicate with one another in response to environmental changes using signal transduction. As described in Chapter 8, signal transduction involves the binding of small molecules, functionally defined as **ligands**, to proteins called receptors. The receptors can change their shape in response to ligand binding to affect intracellular activity. In multicellular organisms, biological ligands are most often metabolites or hormones that may travel through the body in association with carrier proteins; however, ligands can also be cell surface molecules that stimulate cell–cell interactions. Through the process of signal transduction, events outside the cell, such as changes in concentrations of metabolites, can lead to differences in intracellular processes to adapt to these changing conditions.

The majority of receptor proteins are integrated into plasma membranes in such a way that the binding domain of the protein is oriented toward the extracellular space, with the signal-transducing domain located inside the cytosol (**Figure 1.17**). Because the plasma membrane forms a hydrophobic barrier that physically separates the inside and outside of the cell, transmembrane receptor proteins must be able to transduce extracellular signals across the membrane upon ligand binding. Receptor proteins, like other cellular proteins, are not rigid molecules, but can undergo changes in shape (conformational changes) when bound by a ligand or another protein. Receptor-mediated signal transduction mechanisms involve ligand-induced conformational changes in the receptor protein. This results in increased enzyme activity in the cytoplasmic tail of the receptor, or formation of an intracellular protein complex. The "downstream" biochemical processes affected by receptor activation can include altered ion flux across membranes, phosphorylation of other cell signaling proteins, and altered levels of protein synthesis. When extracellular ligand concentrations decrease, the ligand can dissociate from the receptor, returning the receptor to an inactive, nonsignaling state. This ensures that intracellular signaling reflects extracellular ligand concentrations (see Figure 1.17).

The circulatory system of animals, and its counterpart in plants and invertebrates, is the physiologic mechanism whereby biomolecules travel throughout the organism. In addition to transporting signaling molecules, the circulatory system has the

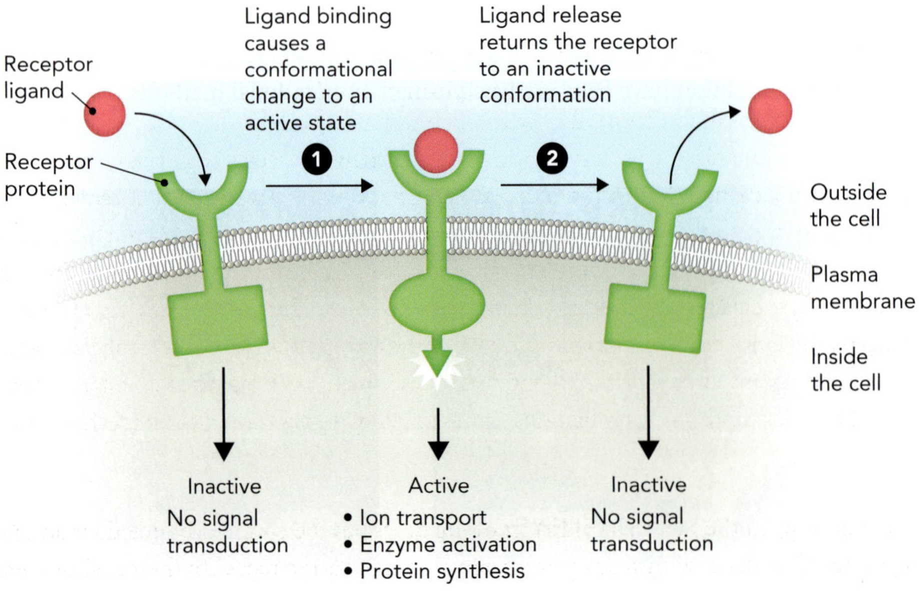

Figure 1.17 Cell functions in multicellular organisms are coordinated by signal transduction mechanisms involving ligand activation of cellular receptor proteins. In this example, an extracellular ligand is shown binding to a membrane-bound receptor protein, resulting in conformational changes that activate the receptor. The activated receptor can influence processes inside the cell in response to the ligand, examples of which include ion transport, enzyme activation, and protein synthesis. When ligand concentrations decrease, the ligand is released from the receptor protein, allowing it to return to the inactive (nonsignaling) conformation.

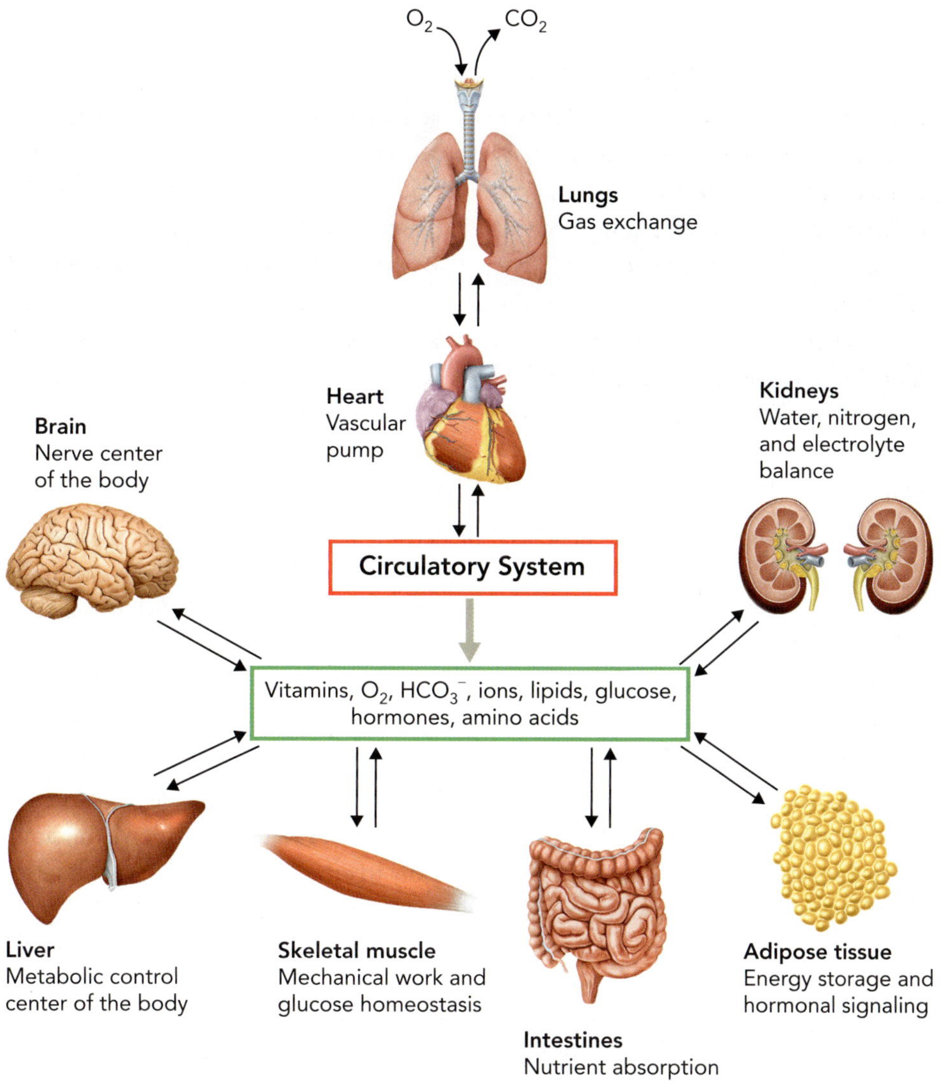

Figure 1.18 Human physiology provides an example of how multicellular organisms depend on a circulatory system to distribute metabolites between specialized tissues and organs. Gas exchange in the lungs provides the oxygen required for aerobic metabolism and at the same time discharges the waste gas carbon dioxide. Signal transduction pathways coordinate cellular responses within tissues.

important role of distributing metabolic fuel obtained from the diet or produced in photosynthetic tissues. In the case of animals, the two most abundant types of metabolic fuels in the circulatory system are carbohydrates, in the form of glucose, and lipids, primarily as free fatty acids or triacylglycerols. Plants use light energy to synthesize the disaccharide sucrose (glucose + fructose) in the leaves and then distribute this carbohydrate to other plant parts through the phloem sap.

Figure 1.18 illustrates how the circulatory system in humans connects specialized tissues and organs in the body, all of which perform a defined function that contributes to the life of the organism. The primary tissues and organs involved in controlling metabolic functions in humans are the brain (nerve center), the liver (metabolic control center), skeletal muscle (mechanical work and glucose homeostasis), intestines (nutrient absorption), adipose tissue (energy storage and hormonal signaling), and the kidneys (water, nitrogen, and electrolyte balance). In addition, the heart pumps blood throughout the circulatory system, and the lungs exchange oxygen and carbon dioxide gases with the atmosphere to keep tissues and organs alive. The coordination of these tissues and organs is essential to their overall function, and this coordination is maintained by signal transduction mechanisms.

The Biochemistry of Ecosystems

The top rung on the hierarchical ladder of life describes the complex interactions among organisms that occur within ecosystems. How organisms interact with their environment

a.

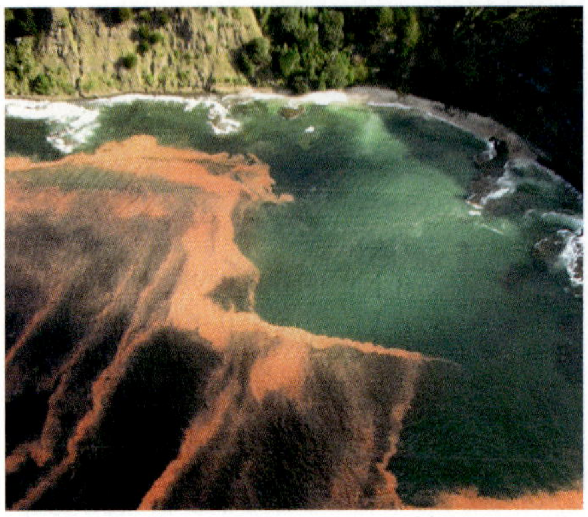

b.

Figure 1.19 Deadly algal blooms in aquatic environments are an example of the importance of ecosystem interactions at the biochemical level. **a.** Algal blooms, also called red tides, such as this one that occurred off the coast of Maine, lead to reduced levels of oxygen in the water. NIWA/PHOTO BY MIRIAM GODFREY. **b.** Fish kills can occur in algal bloom regions as a result of oxygen depletion, following bacterial decomposition of the organic algal material. J. EMILIO FLORES/CORBIS.

and with each other can have beneficial or detrimental effects on life at the level of local or global ecosystems. Aquatic environments, for example, are extremely sensitive to changes in ecosystem dynamics, as seen by the effects of algal blooms on fish populations in localized regions (**Figure 1.19**). Algae are microscopic photosynthetic organisms that mostly live freely in freshwater or saltwater. If a rapid increase in algal growth occurs in a confined area such as a bay or a lake, it can lead to a biochemical imbalance in the ecosystem, resulting from elevated rates of bacterial decomposition of the organic algal material. In some cases, bacterial decomposition can cause acute oxygen depletion resulting in massive fish kills, especially if the water is warm and water circulation is minimal. Some forms of algae produce toxins that are deadly to other aquatic organisms and even to humans when the toxins accumulate in filter-feeding shellfish that people consume.

Harmful algal blooms occur when nutrient levels increase in an aquatic ecosystem at a time when water temperatures and sunlight exposure are optimal for growth. The sudden changes that stimulate an algal bloom can occur naturally as a result of seasonal variations in environmental conditions or can occur from industrial discharges that directly increase nitrogen or phosphate levels in the water. Understanding environmental factors that contribute to algal blooms and finding safe ways to control them in sensitive aquatic environments requires an understanding of key biochemical processes at multiple levels within the ecosystem.

concept integration 1.2

How does the molecule glucose fit into the seven hierarchical levels that define the chemical basis of life on Earth (see Figure 1.12)?

Glucose is a *biomolecule* that contains three of the most abundant *elements* in living systems: carbon, oxygen, and hydrogen. Amylose and cellulose are *polymers* of glucose, and glucose biosynthesis is maintained by highly regulated *metabolic pathways*. Glucose is the primary component of plant cell walls and can be stored for energy needs in plant and animal *cells* in the form of starch and glycogen, respectively. The circulatory systems of *multicellular organisms*, such as mammals, transport glucose between tissues, which primarily use this metabolite for energy conversion processes (glycolytic pathway to generate ATP). Flowers contain nectar, a rich source of glucose (and fructose), which is used by honeybees as a nutrient source; when honeybees retrieve nectar, they cross-pollinate plants to help build a healthy *ecosystem*.

1.3 Storage and Processing of Genetic Information

Prior to the 1940s, most biochemists thought that proteins, rather than DNA, were the key biomolecules that encoded the genetic information of life, passing it from one generation to the next. The reasoning behind this idea was that proteins were known to be more chemically complex than nucleic acids. After all, proteins consist of as many as 20 different amino acids compared to just four different nucleotide bases in DNA. However, in the 1940s and 1950s, experiments using bacteria and bacterial viruses showed that DNA alone was sufficient to pass on genetic information.

At about this same time, an X-ray diffraction technique was developed that allowed biochemists to collect diffraction data from biomolecules—data that could be used to build molecular models reflecting the locations of atoms. Because of the potential to gain insight into the chemical basis of life, several labs began to collect X-ray diffraction data using purified DNA. Some of the best DNA X-ray data were collected by Rosalind Franklin, a chemist at King's College in London, England, who used calf thymus DNA to prepare very thin DNA fibers that could be used to collect a set of X-ray diffraction data (**Figure 1.20a**).

Shortly after it was reported in 1952 that DNA from a bacterial virus was sufficient to promote viral replication in infected cells, a fierce competition to determine the structure of DNA began between the famed American biochemist Linus Pauling and biochemists at Cambridge University in England. Realizing that the quality of Franklin's DNA X-ray data might help solve the structure, Maurice Wilkins, an associate of Franklin's, shared Franklin's data with James Watson and Francis Crick at Cambridge. Watson and Crick used Franklin's data, along with other information about the chemical properties of DNA (water content of DNA fibers and ratios of nucleotide bases), to build a wire-frame model of a DNA double helix (**Figure 1.20b**). Their *Eureka!* moment came on February 28, 1953, when they realized that the positioning of nucleotide bases along the DNA double helix (Pauling had proposed a DNA triple helix) could explain how genetic information was copied by the cell machinery and passed on to daughter cells. In their excitement, Watson and Crick entered The Eagle pub that day in Cambridge and proclaimed with confidence that they had just discovered the "secret of life." Their pivotal discovery led to the 1962 Nobel Prize in Physiology or Medicine for Wilkins, Watson, and Crick. (The Nobel Prize is not awarded posthumously, so Franklin was sadly excluded: She had died 4 years earlier from cancer at the age of 37.)

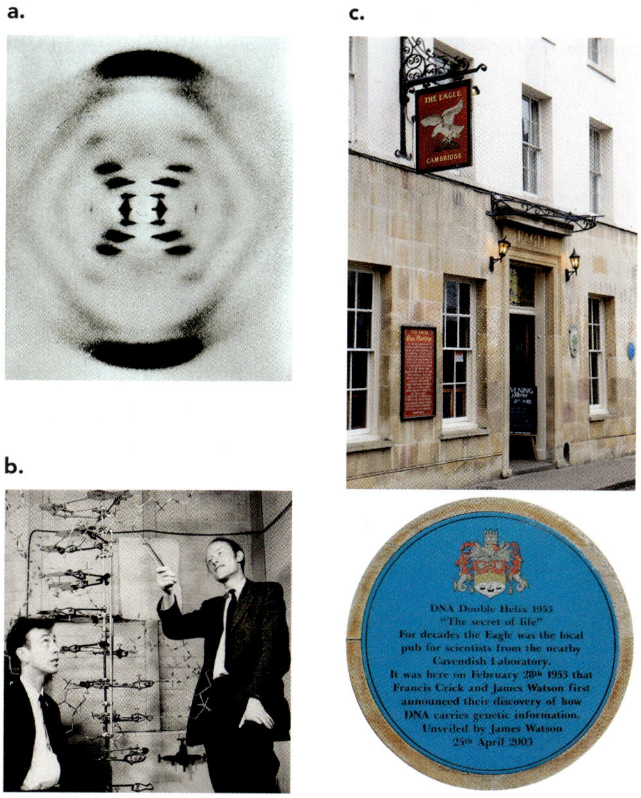

Figure 1.20 James Watson and Francis Crick used X-ray diffraction data from Rosalind Franklin to help solve the structure of DNA. **a.** Franklin's original X-ray diffraction data were collected from very thin fibers of purified DNA obtained from calf thymus tissue. The regular pattern and spacing of the diffraction spots, plus information about the chemical composition of DNA, gave Watson and Crick the information they needed to construct a model of DNA as a double helix with the nucleotide bases directed toward the interior of the helix. SCIENCE SOURCE. **b.** Watson and Crick admiring the wire-frame model of DNA they built in the Cavendish Laboratory at Cambridge University. A. BARRINGTON BROWN/SCIENCE SOURCE. **c.** The Eagle pub in Cambridge, England, where Watson and Crick first proclaimed that they had uncovered the "secret of life" by solving the double helix structure of DNA. TOP: ALISTAIR LAMING/ALAMY; BOTTOM: AWE INSPIRING IMAGES/SHUTTERSTOCK.

Genetic Information Is Stored in DNA as Nucleotide Base Pairs

The structure and function of nucleic acids is covered in Chapter 3, and the biochemistry of genetic information transfer from DNA to RNA to proteins is described in Part 5 of this book. However, we need to review the basic concepts of genetic information transfer here to lay the foundation for related topics in other chapters that rely on a basic understanding of these processes.

The building blocks of nucleic acids are nucleotides, which consist of (**Figure 1.21a**)

- a nucleotide base (adenine, guanine, cytosine, thymine, or uracil);
- a five-carbon ribose or deoxyribose sugar; and
- one or more phosphate groups.

Note that when referring to specific atoms of the nucleotide, each carbon of the sugar is numbered with a prime symbol to distinguish them from the carbons of the base.

Deoxyribonucleotides are the monomeric units of DNA and lack a hydroxyl group on the carbon at the 2′ position (C-2′) of the ribose sugar, whereas ribonucleotides in RNA contain a hydroxyl group in this same position. As originally proposed by Watson and Crick and later confirmed by high-resolution X-ray crystallography, the DNA double helix contains two single polynucleotide strands that interact noncovalently to form a duplex. These strands are noncovalently associated through hydrogen bonds between the nucleotide bases, forming a duplex structure. Hydrogen bonds, discussed in more detail in Chapter 2, are a type of weak noncovalent interaction in which a hydrogen atom is shared between polar groups. The deoxyribonucleotide **base pairs** in DNA consist of guanine hydrogen-bonded to cytosine (G-C or C-G base pair) and adenine hydrogen-bonded to thymine (A-T or T-A base pair) (**Figure 1.21b and c**). RNA lacks the nucleotide base thymine and instead contains the nucleotide base uracil, which hydrogen bonds with adenine (A-U or U-A base pair) (**Figure 1.21d**). Because the chemical structures of nucleic acids are very similar, hybrid molecules of DNA and RNA can form, provided they contain complementary nucleotide bases to form G-C,

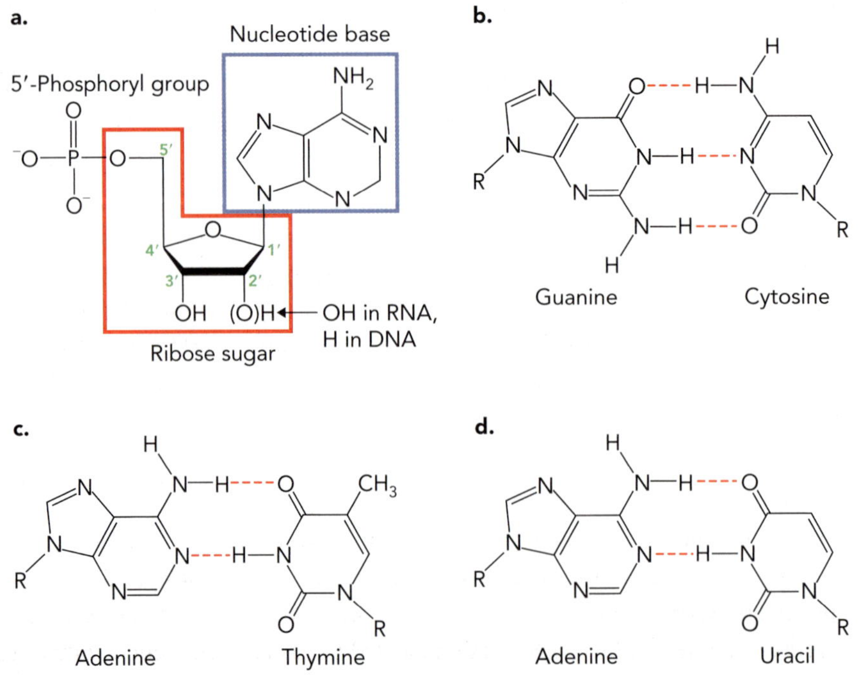

Figure 1.21 Nucleotides are the building blocks of nucleic acids. **a.** The common nucleotides in DNA and RNA contain a nucleotide base attached to the 1′-carbon of the ribose sugar and a phosphoryl group attached to the 5′-carbon. The 2′-carbon of ribonucleotides contains a hydroxyl group, whereas deoxyribonucleotides have a hydrogen atom in this position. **b.** G-C base pairs contain three hydrogen bonds. **c.** A-T base pairs in DNA–DNA hybrids contain two hydrogen bonds. **d.** A-U base pairs in DNA–RNA duplexes or RNA–RNA duplexes contain two hydrogen bonds. Hydrogen bonds are shown as dashed red lines between the base pairs.

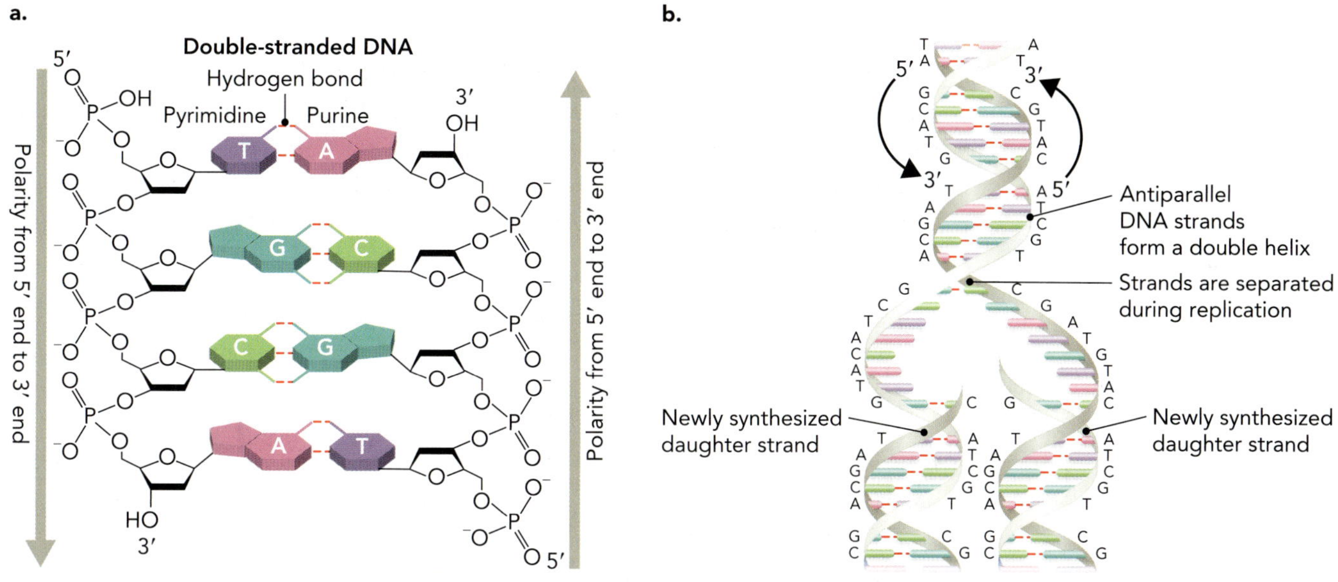

Figure 1.22 The nucleotide base pairs in the DNA double helix explain how genetic information can be passed down through genetic inheritance. **a.** The two antiparallel DNA strands are held in register by hydrogen bonds between the G-C and A-T base pairs. **b.** The DNA double helix is a right-handed helix that can be copied through complementary base-pairing to produce two exact DNA replicas.

A-T, or A-U base pairs. As described in Chapter 3, the structure of the DNA double helix is stabilized by both hydrogen bonds between nucleotide bases on opposite DNA strands and **base stacking** interactions between the aromatic rings of the nucleotide bases within the interior of the DNA helix.

Strands of DNA are formed from covalent linkages of deoxyribonucleotides through a phosphodiester bond between 3′-carbon of one deoxyribonucleotide and 5′-carbon of another deoxyribonucleotide (**Figure 1.22**). The polarity of each DNA strand is determined by the bonds between the ribose and phosphate groups, which form the sugar–phosphate backbone. Two antiparallel DNA strands are held in register relative to each other by the hydrogen bonds between the nucleotide base pairs. The spacing between these pairs is accommodated in the three-dimensional structure by the formation of a right-handed helix. The breakthrough that Watson and Crick made in 1953 was to realize that the double-stranded DNA molecule must contain chemical information stored in the sequence of nucleotide base pairs that can be "copied" into new DNA strands by following the G-C and A-T base pair rule. Although they did not know at the time how the chemical code in each DNA strand was "translated" into a protein (Crick helped figure this out later), they were quick to realize that the elegant simplicity of the G-C and A-T base pairs in the DNA double helix was, in fact, the chemical basis of genetic inheritance.

Information Transfer between DNA, RNA, and Protein

The **central dogma** of molecular biology describes how genetic information stored in DNA is used to direct the biological processes in cells (**Figure 1.23**). The basic principle of the central dogma is that each organism contains a set of DNA molecules that together encode all of the information needed to sustain life. Genetic inheritance

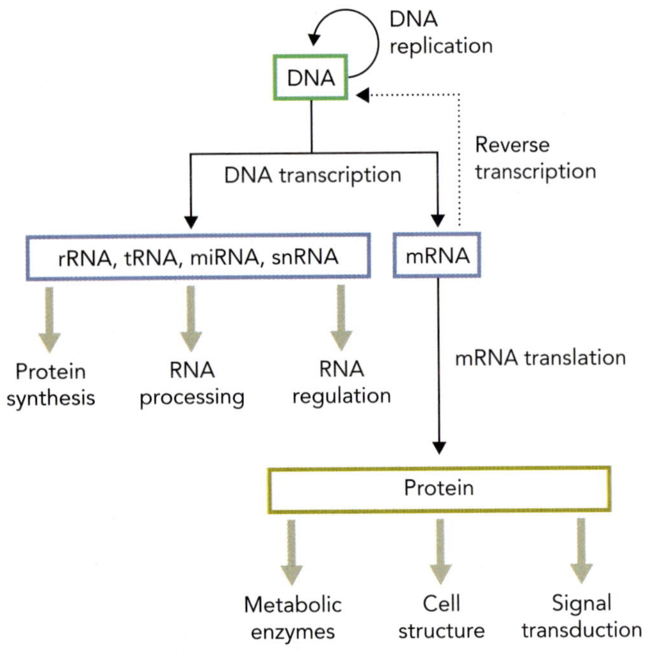

Figure 1.23 The central dogma of molecular biology describes the transfer of information between nucleic acids and proteins. The genome is copied during cell division by the process of DNA replication. A variety of RNA products are generated by DNA transcription, one of which is mRNA, which is used to synthesize proteins by mRNA translation. The majority of RNA produced by DNA transcription is required for protein synthesis (rRNA and tRNA), RNA processing (snRNA), and regulation of gene expression or protein synthesis (miRNA). Under some conditions, RNA molecules can also be converted back into DNA by reverse transcription.

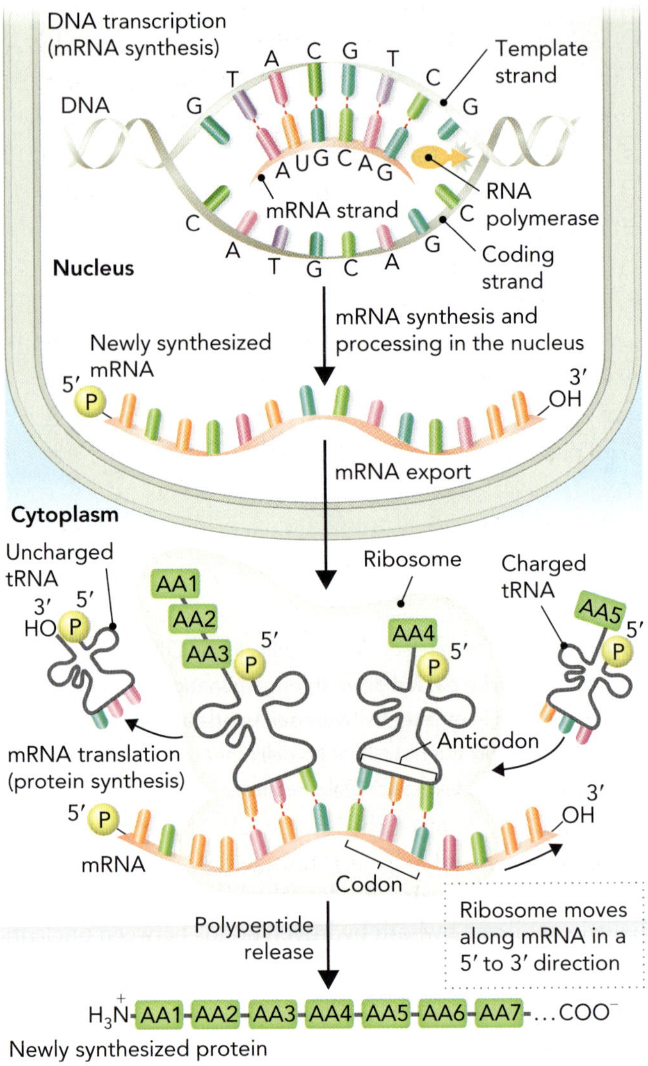

Figure 1.24 DNA sequences are transcribed into mRNA by the enzyme RNA polymerase in the nucleus of eukaryotic cells. RNA polymerase synthesizes mRNA using ribonucleotides that form complementary base pairs with deoxyribonucleotides in the template strand of DNA. The DNA sequence of the coding strand is the same as the RNA sequence in the mRNA, with the exception of uridine replacing thymidine. The mRNA is then exported to the cytoplasm, where it is translated into protein. Translation of mRNA requires a protein-synthesizing complex, which consists of ribosomes, tRNAs, and the mRNA. Charged tRNAs bring amino acids into the ribosome complex, where they form complementary base pairs with the mRNA through codon–anticodon hydrogen bonds. Once the amino acid has been covalently linked to the growing polypeptide chain, the uncharged tRNA exits the ribosomal complex, making room for the next charged tRNA. AA = amino acid.

requires that DNA be faithfully copied by a process called **DNA replication** and that segments of DNA, known as **genes,** be converted into RNA by **DNA transcription.** Genes are functional units of DNA that are defined by the RNA products they produce. Genes account for the biochemical properties of living cells.

A subset of the gene-encoded RNA molecules are called **messenger RNA (mRNA)** molecules. These are used as templates for protein synthesis in a process referred to as **mRNA translation.** As we noted earlier, proteins are considered the workhorses of living cells: they carry out metabolic chemical reactions, serve as components of cell structures, and act as signaling molecules, among other functions. In addition to mRNA, cells also contain **small nuclear RNA (snRNA)** molecules, which are involved in RNA processing, and **micro RNA (miRNA)** molecules, which regulate gene expression and mRNA translation. The most abundant RNA molecules in cells are **ribosomal RNA (rRNA)** molecules and **transfer RNA (tRNA)** molecules, both of which are required for protein synthesis in addition to mRNA. The function of each of these RNA molecules is described in Chapter 22. A fourth biochemical process defined by the central dogma of molecular biology is that of **reverse transcription**, which converts RNA into DNA under special conditions, most often related to virus replication.

As described earlier, the DNA content of an organism is called a genome. The suffix "-ome" refers to set or collection, so a genome is a set of genes. The collection of DNA transcripts (RNA products) generated by DNA transcription is called a **transcriptome.** The transcriptome can refer to all possible RNA products present in an organism or cell type or to just those generated under defined conditions, such as the transcriptome of embryonic liver cells or that of yeast cells cultured in galactose-supplemented media. Similarly, the **proteome** is the collection of proteins produced by mRNA translation, either in the entire organism or under special conditions.

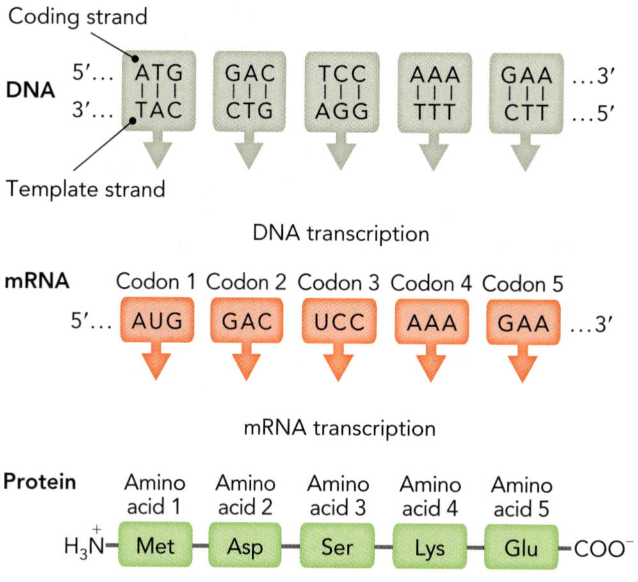

Figure 1.25 The relationship between the DNA sequence of a gene and the amino acid sequence of the protein product is shown. Using the genetic code, the primary amino acid sequence of a protein can be determined from the DNA sequence of the coding strand, which has the same 5′ to 3′ polarity of the corresponding mRNA transcript.

In eukaryotic cells, the process of DNA transcription takes place in the cell nucleus and leads to the production of mRNA transcripts. These transcripts are exported to the cytoplasm, where they are translated into polypeptide chains (**Figure 1.24**). In prokaryotic cells, DNA transcription and mRNA translation take place in the cytosol, with mRNA molecules being translated as soon as they are synthesized. Note that the DNA sequence in the **template strand** forms complementary base pairs with the ribonucleotide sequence in the mRNA strand. (For example, the DNA sequence 3′-TAC-5′ corresponds to 5′-AUG-3′ in RNA.) In contrast, the DNA sequence in the nontemplate, or **coding strand**, is identical to the mRNA sequence, with the exception of uracil replacing thymine (**Figure 1.25**). After export to the cytoplasm, the mRNA is translated by ribosomes, which consist of proteins and rRNA. The ribosomes serve as binding sites for charged tRNA molecules, which deliver amino acids to the protein-synthesizing complex. We describe the fascinating biochemistry of ribosome-mediated protein synthesis in Chapter 22.

concept integration 1.3

How did solving the molecular structure of DNA in 1953 provide a molecular explanation for the central dogma of molecular biology?

When James Watson and Francis Crick built their model of the DNA double helix, they discovered that the hydrogen-bonding between G-C and A-T nucleotide base pairs would allow an exact replica of a DNA strand to be created. Thus, the "secret of life" was encoded by a genetic language that could be passed on to daughter cells. This led to the elucidation of the genetic code and what became known as the central dogma of molecular biology, which states that DNA encodes RNA, which is translated into proteins. This knowledge has led to great strides in our understanding of basic life processes and gave rise to the so-called Age of Biology, characterized by an understanding of the basic mechanisms underlying developmental biology, neurobiology, and metabolic regulation.

1.4 Determinants of Biomolecular Structure and Function

An inherent principle in both biology and chemistry is that structure determines function. Moreover, biological structures are governed by evolutionary processes that impact function. This general principle, which holds true for macromolecules, cells, and organisms, can be seen in both the simplicity of the DNA double helix and the complexity of proteins. The structure and function of the DNA double helix provides an ideal solution to the problem of storing genetic information in a way that can be quickly copied (DNA replication) and accessed (DNA transcription) with high fidelity. The use of only four different nucleotide bases (A, G, C, T) to encode the required information for life and the arrangement of the sugar–phosphate backbone relative to the hydrogen-bonded base pairs is a perfect mix of form and function.

Proteins, in contrast, need to carry out a variety of biochemical functions that require a vast array of molecular structures. The evolutionary driving force creating these diverse protein structures is nucleotide changes in the coding sequences of genes, resulting from random mutation and **natural selection**. These nucleotide changes can include single nucleotide base substitutions—such as changing a guanine-containing nucleotide to an adenine-containing nucleotide—as well as larger DNA rearrangements resulting from deletions or chromosomal rearrangements. If the DNA change leads to a beneficial change in the encoded protein through its altered structure and function, then the gene sequence is maintained through selective pressure (that is, natural selection). However, as we will see later in our discussions of signal transduction and metabolic regulation, nucleotide changes in gene coding sequences can also lead to detrimental changes in protein structure, resulting in pathologic conditions.

For example, suppose a single nucleotide substitution (mutation) occurs in a **wild-type** (fully functional) protein-coding sequence, changing the codon for the negatively charged amino acid glutamate (GAA) into that of the positively charged amino acid lysine (AAA). This change can result in a functional defect that blocks the catalytic activity of the protein (**Figure 1.26**). If this protein is required for transmission of an extracellular signal through a receptor protein in the plasma membrane or controls a critical metabolic pathway, then the defect could result in alterations at the cellular and even organismal level.

Many human diseases are caused by mutations in DNA. If the mutation is passed from the parents to their offspring, then the mutation is contained within the DNA of a **germ-line cell** (egg and sperm cells in eukaryotes) and is referred to as an inherited genetic disease. However, if the DNA mutation occurs during the lifetime of the organism in a **somatic cell** (all cells that are not germ cells), then the consequences of this mutation—the disease phenotypes—are limited to the individual organism and are not inherited by its offspring. Tay–Sachs disease (a disorder of the nervous system) is an example of an inherited genetic disease, whereas most forms of cancer, such as the most common types of lung and colon cancer, are the result of DNA mutations that occur in damaged somatic cells. Inherited genetic diseases are the common causes of many of the metabolic disorders examined in Part 4 of this book, whereas DNA-damaging mutations occurring in somatic cells are discussed in Part 5.

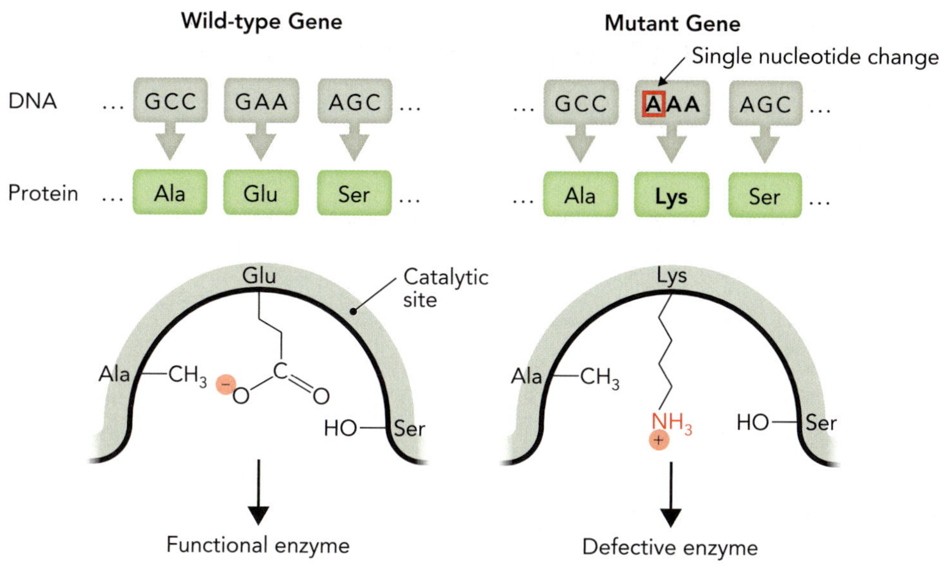

Figure 1.26 Single nucleotide changes in DNA can have profound effects on the structure and function of encoded proteins. The guanine-to-adenine mutation shown here changes the codon that specifies the amino acid glutamate (Glu), found normally in the wild-type protein, to the codon for the amino acid lysine (Lys) in the mutant form of the same protein. This results in the introduction of a positive charge in the catalytic site of the protein, inhibiting its activity.

Evolutionary Processes Govern Biomolecular Structure and Function

Several models have been proposed to explain how life on Earth settled on a scheme that follows the central dogma of DNA → RNA → protein. One possible scenario is that RNA was actually the precursor to DNA because it could both function catalytically itself (ribozyme) and also direct the synthesis of proteins. In this RNA world model, the advent of deoxyribonucleotides led to DNA becoming the molecular database of genetic information because polydeoxyribonucleotides are chemically more stable than polyribonucleotides. The cause of this instability of ribonucleotides is the hydroxyl group on the 2′-carbon of the ribose sugar, which can readily cyclize with the 3′-carbon and thereby cleave the sugar–phosphate backbone.

We can compare the DNA sequences of different organisms alive today to get a glimpse into our evolutionary past. This type of evolutionary analysis has been made possible by advances in high-throughput DNA sequencing, which over the past 20 years has given rise to a new field of computer-based biological research called **bioinformatics**. Phylogenetic trees can now be based on DNA sequence similarities, rather than on outward morphological appearances. The so-called Tree of Life consists of three domains referred to as Archaea, Bacteria, and Eukaryota (**Figure 1.27a**). Clusters of organisms in a phylogenetic tree indicate evolutionary relatedness, whereas branch points imply common ancestors. The Web-based Tree of Life project provides detailed descriptions of these proposed relationships (see http://tolweb.org).

Although DNA-based phylogenetic trees are generally similar to morphologically based phylogenetic trees, bioinformatic analyses have found some surprises within the plant and animal kingdoms. For example, based on morphologies alone, it was thought that chimpanzees and gorillas were more closely related to each other than they were to

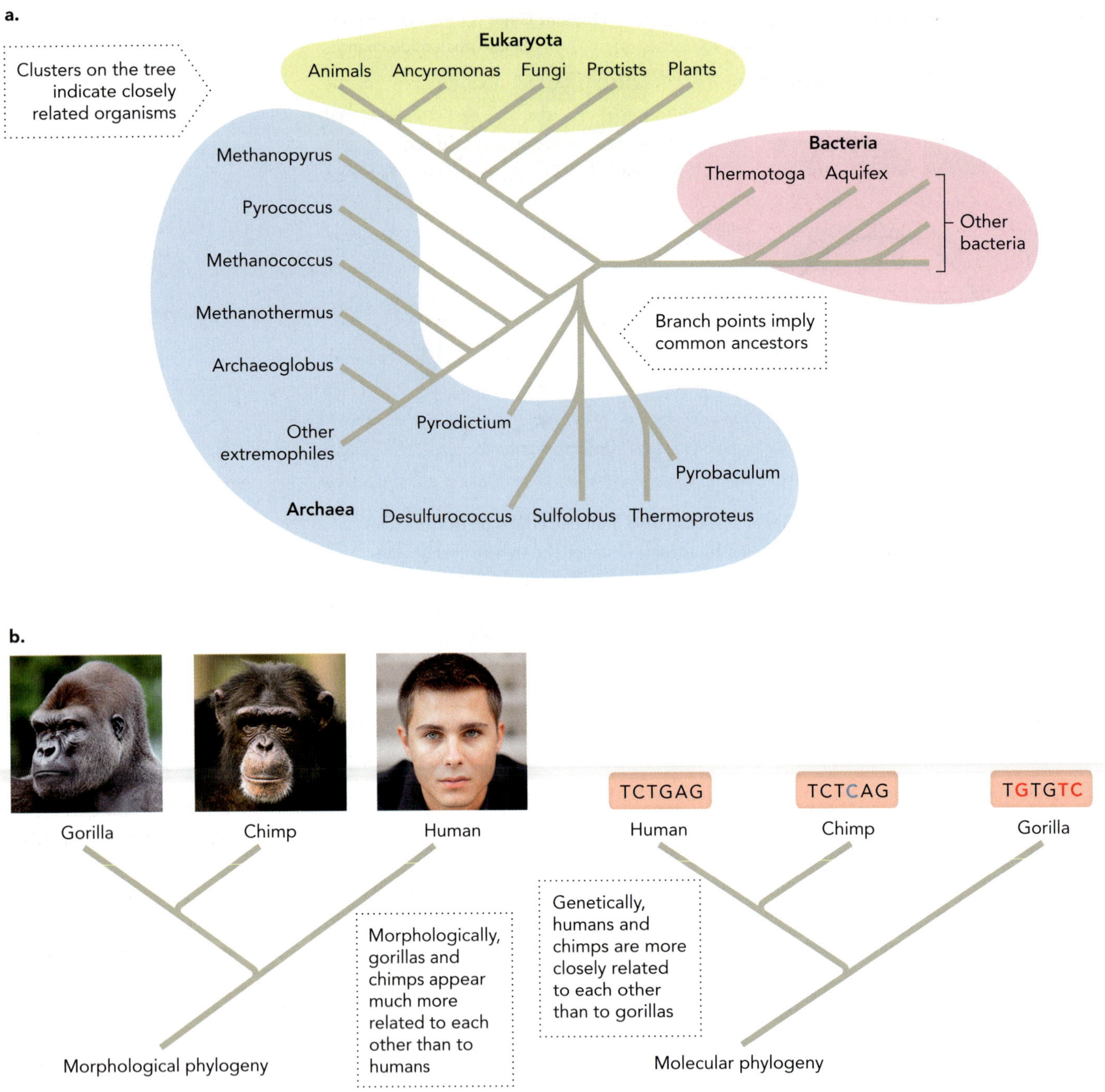

Figure 1.27 Phylogenetic trees represent evolutionary relatedness among species. **a.** In this phylogenetic tree, constructed using comparisons of nucleic acid sequences, the length of each branch denotes the length of evolutionary time separating species from a common ancestor, represented by a fork in the tree. The two prokaryotic groups of Bacteria and Archaea are much more diverse than the multicellular Eukaryota such as animals and plants. **b.** Evolutionary relationships can also be represented with a cladogram, which is a type of phylogenetic tree that shows branching relationships, though the length of the branch does not represent evolutionary time. On the basis of appearance (morphology) alone, scientists thought that chimpanzees and gorillas were more closely related to each other than either one was to humans (left). However, using DNA sequence analysis, it was found that gorillas are actually a distant relative to humans and chimpanzees (right). GORILLA: STEVE BLOOM IMAGES/ALAMY; CHIMP: GERRY ELLIS/GETTY IMAGES; HUMAN: FELIX MIZIOZNIKOV/SHUTTERSTOCK.

humans. However, molecular comparisons of human, chimpanzee, and gorilla proteins and DNA over the past 40 years clearly indicate that chimpanzees are much more similar to humans than they are to gorillas (**Figure 1.27b**). Indeed, bioinformatic analysis of the chimpanzee and human genomes shows that up to 98% of our DNA is identical to that of chimpanzees and ~30% of our proteins have the exact same amino acid sequences as those of chimpanzees. In light of such a high degree of DNA sequence similarity between the two genomes, it is interesting that one of the fundamental differences between humans and chimpanzees is altered levels of specific gene transcripts in the brain. This finding suggests that gene expression, rather than gene function, may account for the increased neuronal development and enhanced cognitive skills of humans compared to chimpanzees.

Evolution is the ultimate biochemical experiment, in that random changes in the DNA sequences of genes are tested over time to see which ones are most beneficial. Indeed, one way to gain insight into the function of a protein is to compare its amino acid sequence to those of other proteins to see if conserved regions appear that might suggest an important function. This can be done using the genetic code to convert the DNA sequence in the coding strand of a gene into the inferred amino acid sequence of the encoded protein (see Figure 1.25). Protein-coding sequences evolve much more slowly than noncoding sequences because protein function is tightly linked to protein structure, which in turn depends on the amino acid sequence (**Figure 1.28a**). Moreover, DNA sequences encoding amino acids that are critical to protein function—for example, amino acids in the active site of an enzyme—evolve more slowly than coding sequences specifying amino acids required only for overall protein structure.

Highly conserved gene sequences that encode proteins with the same function in different organisms are called **orthologous genes** and are thought to have arisen from a common ancestral gene. An example is the glucose-6-phosphate dehydrogenase gene,

a.

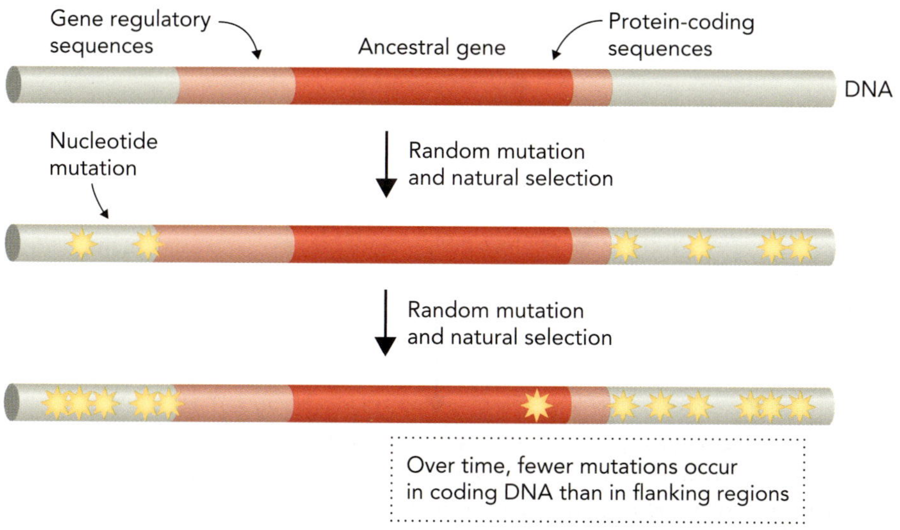

b.

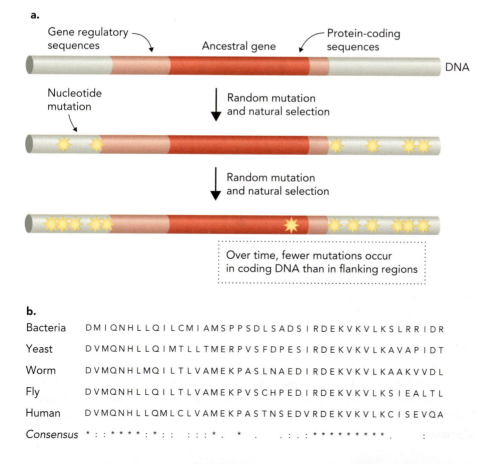

```
Bacteria   D M I Q N H L L Q I L C M I A M S P P S D L S A D S I R D E K V K V L K S L R R I D R
Yeast      D V M Q N H L L Q I M T L L T M E R P V S F D P E S I R D E K V K V L K A V A P I D T
Worm       D V M Q N H L M Q I L T L V A M E K P A S L N A E D I R D E K V K V L K A A K V V D L
Fly        D V M Q N H L L Q I L T L V A M E K P V S C H P E D I R D E K V K V L K S I E A L T L
Human      D V M Q N H L L Q M L C L V A M E K P A S T N S E D V R D E K V K V L K C I S E V Q A
Consensus  * : : * * * * : * : :   : : : * .   *   .     . : . : * * * * * * * * * .         :
```

Figure 1.28 Gene sequences evolve much more slowly than DNA sequences located between genes because natural selection provides a mechanism to preserve useful gene functions. **a.** Random nucleotide mutations that have no major effect on gene expression and function accumulate in flanking DNA at much higher rates than deleterious mutations accumulate within genes, the latter being eliminated by natural selection. **b.** The glucose-6-phosphate dehydrogenase gene is an orthologous gene in bacteria, yeast, worms, flies, and humans, with very high amino-acid-sequence conservation in regions of the protein required for enzymatic function. The consensus sequence identifies amino acid residues within the aligned sequences that are identical (*), chemically conserved substitutions (:), or chemically related substitutions (.).

which encodes a highly similar amino acid sequence in bacteria, yeast, worms, flies, and humans (**Figure 1.28b**). The conservation of structure and function in this protein has been maintained over ~3 billion years of evolution, dating back to the appearance of prokaryotic cells. The protein glucose-6-phosphate dehydrogenase is important in providing cells with a biomolecule called nicotinamide adenine dinucleotide phosphate (NADPH; reduced form), which is required for oxidation–reduction reactions in cells. Glucose-6-phosphate dehydrogenase also has a critical role in the metabolism of ribose sugars, which are required for the backbone of nucleic acids. The fact that the amino acid sequence for this protein has gone unchanged for so long is evidence of its importance and of the long-term success of its efficient functioning.

A second way in which evolution affects protein structure and function is through **gene duplication**, which leads to the appearance of new genes in the genome (**Figure 1.29a**). The duplication of an ancestral gene leads to three distinct outcomes during the process of speciation. If having a second copy of the gene provides an evolutionary advantage by doubling the amount of protein produced (two genes are better than one), then the second gene is maintained with essentially the same amino acid sequence. However, if one gene copy is sufficient, then the second gene copy can diverge through random mutations, either acquiring a new function that is beneficial or simply becoming lost as a result of deleterious mutations.

Bioinformatic analyses suggest that in many cases, the second gene is, in fact, selectively retained and evolves to encode a protein with a related but distinct function. Related genes within a species are called **paralogous genes** and are considered members of a gene family. Many examples of paralogous genes occur in the human genome. One is the nuclear receptor gene family, which encodes steroid receptor proteins that function as transcription factors (**Figure 1.29b**). Another example of paralogous genes is the globin gene family, which encodes proteins involved in oxygen transport. Note

a.

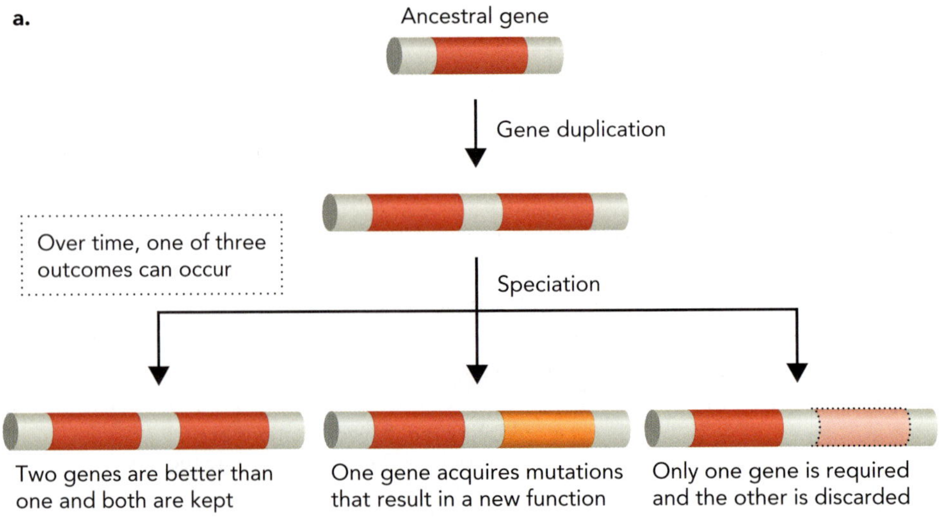

Ancestral gene
Gene duplication
Over time, one of three outcomes can occur
Speciation
Two genes are better than one and both are kept
One gene acquires mutations that result in a new function
Only one gene is required and the other is discarded

b.

Human GR	C L V C S D E A S G C H Y G V L T C G S C K V F F K R A V E G Q H N Y L C	Paralogs (Human GR, Human PR, Human FXR)
Human PR	C L I C G D E A S G C H Y G V L T C G S C K V F F K R A M E G Q H N Y L C	
Human FXR	C V V C G D R A S G Y H Y N A L T C E G C K G F F R R S I T K N A V Y K C	Orthologs (Human FXR, Mosquito ECR)
Mosquito ECR	C L V C G D R E S G Y H Y N A L T C E G C K G F F R R S V T K N A V Y C C	
Consensus	* : * . * * * * * . . : * . * * : * * : * : : : * *	

Figure 1.29 Gene duplications can give rise to paralogous genes with similar functions. **a.** DNA rearrangements can result in gene duplications that lead to one of three outcomes, depending on the additive effect of the two genes. In many cases, the second gene acquires mutations that result in the generation of a related gene called a paralog, which has similar functions. **b.** Three paralogous genes in the human nuclear receptor gene family are the glucocorticoid receptor (GR), progesterone receptor (PR), and farnesoid X receptor (FXR). The human FXR gene is the ortholog to the mosquito ecdysone receptor (ECR), which is also a nuclear receptor. Portions of the amino acid sequences of these four related nuclear receptor proteins are shown here.

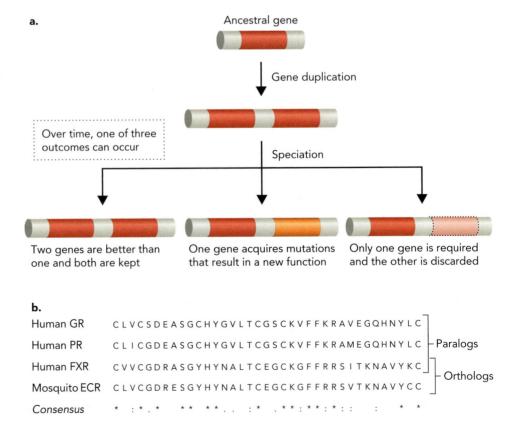

that because the ancestral gene evolved before it was duplicated, paralogous genes also have orthologous genes in other species. For example, the ecdysone receptor in mosquitoes is an invertebrate ortholog to the human farnesoid X receptor, a member of the nuclear receptor gene family that regulates bile acid metabolism (Figure 1.29).

Protein Structure–Function Relationships Can Reveal Molecular Mechanisms

The amino acid sequence of a protein determines its structure, owing to the chemical properties of amino acid side chains and the peptide bond. Therefore, two proteins with high sequence conservation at the amino acid level are likely to be similar in both three-dimensional structure and biochemical function. This is true for the enzyme glucose-6-phosphate dehydrogenase and the nuclear receptor proteins, shown in Figures 1.28 and 1.29. However, proteins with very different amino acid sequences can still have similar overall three-dimensional structures, which may or may not correspond to similar biochemical functions. For example, the ribbon structures shown in **Figure 1.30** were rendered by a computer modeling program that traces the polypeptide backbone of the protein, displaying the overall shape of the protein as a combination of helical and strand structural elements with connecting loops. We describe the biochemistry behind these protein structural components in Chapter 4, but for now, just look at the ribbon models as accurate representations of the shapes of the polypeptide chains within proteins.

You can see in Figure 1.30a that the protein structure of the enzyme ribonucleotide reductase has remained remarkably similar over the several billion years of evolutionary time separating bacteria and mammals. Both of these protein

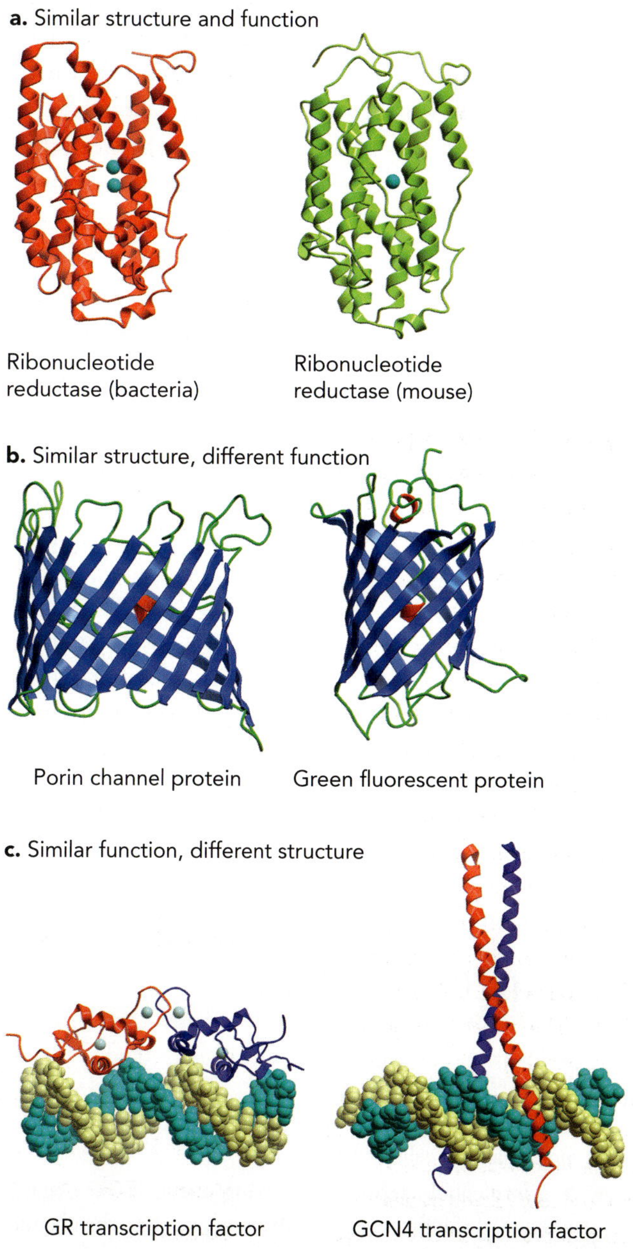

Figure 1.30 Comparative relationships between molecular structure and function are shown. The structures in this figure and similar structures appearing throughout this book were generated using the MolSoft ICM Browser rendering application based on experimental data deposited in the Protein Data Bank (PDB). **a.** The ribonucleotide reductase proteins from bacteria and mice share significant amino acid sequences in the catalytic site and retain an α-helical structure most likely found in the ribonucleotide reductase ancestral protein. BASED ON PDB FILES FOR BACTERIAL (1R2F) AND MOUSE (1XSM) RIBONUCLEOTIDE REDUCTASE PROTEINS. **b.** The bacterial porin channel protein and jellyfish green fluorescent protein have no amino acid sequence similarity, yet both have a β-barrel structure as a key functional element. BASED ON PDF FILES 1PRN (PORIN CHANNEL PROTEIN) AND 1GFL (GREEN FLUORESCENT PROTEIN). **c.** The rat glucocorticoid receptor (GR) and yeast GCN4 protein are unrelated in amino acid sequence and have very different three-dimensional structures. However, they both function as transcription factors that bind specific DNA sequences with high affinity using α-helical regions. BASED ON PDB FILES 1YSA (GCN4 LEUCINE ZIPPER PROTEIN) AND 1GLU (GLUCOCORTICOID RECEPTOR).

structures consist of bundled α helices with metal ions in the catalytic site of the enzyme. Ribonucleotide reductase is responsible for the biochemical reaction that converts ribonucleotides to deoxyribonucleotides and is considered an ancient protein on the basis of its highly conserved structure and central role in nucleic acid biochemistry.

In contrast, the bacterial porin channel protein has a basket-like structure, called a β barrel, which is also found as the major structural element in the jellyfish green fluorescent protein (Figure 1.30b). However, the function of the β-barrel structure in these two proteins is quite different. In the case of the porin channel protein, it forms a water-filled pore in the outer plasma membrane of bacteria, which allows solutes to exchange with the environment. In the green fluorescent protein, the β barrel acts as a scaffold that supports the tripeptide chromophore required for autofluorescence.

A third example of protein structure–function relationships is the rat glucocorticoid receptor (Figure 1.30c), which is a dimeric transcription factor that uses two short α helices to form high-affinity contacts with nucleotide bases in the DNA double helix. The DNA-binding α helices are held in place by zinc atoms, which form contacts with four amino acids in the protein. The yeast GCN4 transcription factor protein, which shares no amino acid sequence similarity with the glucocorticoid receptor, also binds DNA with high specificity and affinity using two α helices. However, this functionally related protein uses two very long α helices with multiple noncovalent contacts between them to accomplish the same task.

The take-home message from the three examples of protein structure in Figure 1.30 is that although protein structure and function are intimately related, you need a working knowledge of both structure and function to have a clear understanding of molecular reaction mechanisms. Moreover, even though it is often possible to predict the structure and function of proteins encoded by closely related orthologous or paralogous genes, it is not so easy to predict protein function from structure alone, and vice versa.

concept integration 1.4

How do evolutionary processes affect protein structure and function?

Random mutations and natural selection lead to changes in the DNA sequences of genes and therefore to the corresponding amino acid sequences of the encoded proteins. Because the primary amino acid sequence of a protein determines its three-dimensional structure and function, these evolutionary processes give rise to regions of high similarity of amino acid sequences between related proteins from different species. Gene duplication and natural selection further enhance the diversity of protein structure and function by permitting nucleotide changes in one of the two gene copies without loss of the original function.

chapter summary

1.1 What Is Biochemistry?

- Biochemistry aims to explain biological processes at the molecular and cellular levels.
- *In vitro* alcoholic fermentation using yeast cell-free extracts was one of the first experiments demonstrating the chemical basis of life.
- Biochemical applications have led to the development of new pharmaceutical drugs, advances in medical diagnostics, the rise of the biotechnology industry, and improvements in agricultural and environmental sciences.

1.2 The Chemical Basis of Life: A Hierarchical Perspective

- The six elements that predominate in nature are H, O, C, N, P, and S, which together form the common chemical groups NH_2, OH, SH, PO_3^{2-}, COOH, and CH_3.
- The four major classes of small biomolecules are amino acids, nucleotides, simple sugars, and fatty acids.
- The most abundant macromolecules in nature are polymers of nucleotides (DNA, RNA), amino acids (proteins), and the simple sugar glucose (cellulose, amylose, glycogen).
- Metabolic pathways consist of linked biochemical reactions in which the product of one reaction is the reactant for another.
- Living cells are highly ordered structures surrounded by a lipid membrane; they obtain energy from the Sun or from oxidation–reduction reactions to support metabolic processes.
- Organisms consist of many types of specialized cells that respond to changes in the environment by communicating with each other using a biochemical process called signal transduction, which involves the binding of molecules to receptor proteins, thus affecting the signaling activity of the receptors.
- Within ecosystems, organisms undergo complex interactions with one another, which can only be understood by studying key biochemical processes.

1.3 Storage and Processing of Genetic Information

- Deoxyribonucleotide base pairs in DNA consist of guanine hydrogen-bonded to cytosine (G-C and C-G) and adenine hydrogen-bonded to thymine (A-T and T-A). RNA lacks the nucleotide base thymine and instead contains the nucleotide base uracil, which forms hydrogen bonds with adenine (A-U and U-A).
- The right-handed DNA double helix contains two antiparallel strands stabilized by the formation of hydrogen bonds between G-C and A-T base pairs and by base stacking in the interior of the DNA helix.
- DNA replication makes faithful copies of DNA using G-C and A-T base pairing. DNA transcription makes complementary RNA copies of protein-coding sequences called mRNA molecules, which are translated into proteins by tRNA and ribosomes.

1.4 Determinants of Biomolecular Structure and Function

- Biological structure and function are governed by evolutionary processes that affect function. This general principle holds true for macromolecules, cells, and organisms and can be seen in both the simplicity of the DNA double helix and the complexity of proteins.
- The evolutionary driving force for creating diverse protein structures is nucleotide changes in the coding sequences of genes, resulting from random mutation and natural selection.
- Orthologous genes are functionally related genes that have been evolutionarily conserved between species. Paralogous genes are functionally related genes present in the same species that have arisen from gene duplication.
- Proteins with high sequence conservation at the amino acid level usually have similar three-dimensional structures and biochemical functions. However, proteins with very different amino acid sequences can also have similar overall structures, which may or may not correspond to similar biochemical functions.
- Proteins in solution are in constant motion as a result of the formation and disruption of noncovalent interactions; therefore, molecular models of protein structures reveal very little about dynamic changes in protein structure that are likely to be involved in regulating protein function.

biochemical terms

(in order of appearance in text)

fermentation (p. 5)
enzyme (p. 5)
biomolecule (p. 7)
macromolecule (p. 7)
metabolic pathway (p. 7)
signal transduction (p. 7)
ecosystem (p. 8)
angstrom (Å) (p. 10)
amino acid (p. 11)
protein (p. 11)
polypeptide (p. 11)
nucleotide (p. 12)
deoxyribonucleic acid (DNA) (p. 12)
ribonucleic acid (RNA) (p. 12)
simple sugar (p. 12)
carbohydrate (p. 12)
fatty acid (p. 12)
amphipathic (p. 12)
phospholipid (p. 12)
triacylglycerol (p. 13)
polysaccharide (p. 13)
phosphodiester bond (p. 13)
peptide bond (p. 14)
amylose (p. 14)
chitin (p. 14)
metabolite (p. 15)
systems biology (p. 15)
metabolic flux (p. 15)
linear metabolic pathway (p. 17)
forked pathway (p. 17)
cyclic pathway (p. 17)
prokaryote (p. 17)
eukaryote (p. 17)
capsule (p. 17)
cytoplasm (p. 17)
chromosome (p. 17)
nucleoid (p. 17)
flagella (p. 18)
pilus (p. 18)
plasmid (p. 18)
genome (p. 18)
chromatin (p. 18)
nucleus (p. 18)
nucleolus (p. 18)
ribosome (p. 18)
cytoskeleton (p. 19)
mitochondria (p. 19)
peroxisome (p. 19)
lysosome (p. 19)
chloroplast (p. 19)
vacuole (p. 19)
endoplasmic reticulum (p. 19)
Golgi apparatus (p. 19)
microtubule (p. 19)
endosymbiotic theory (p. 19)
ligand (p. 20)
base pair (p. 24)
base stacking (p. 25)

central dogma (p. 25)
DNA replication (p. 26)
gene (p. 26)
DNA transcription (p. 26)
messenger RNA (mRNA) (p. 26)
mRNA translation (p. 26)
small nuclear RNA (snRNA) (p. 26)
micro RNA (miRNA) (p. 26)
ribosomal RNA (rRNA) (p. 26)
transfer RNA (tRNA) (p. 26)
reverse transcription (p. 27)
transcriptome (p. 27)
proteome (p. 27)
template strand (p. 27)
coding strand (p. 27)
natural selection (p. 28)
wild-type (p. 28)
germ-line cell (p. 28)
somatic cell (p. 28)
bioinformatics (p. 29)
orthologous gene (p. 31)
gene duplication (p. 32)
paralogous gene (p. 32)

review questions

1. What experiment did Eduard Buchner perform in the late 19th century concerning yeast? What has this done for modern biochemistry?
2. What types of molecules are enzymes? Name three physiologic processes that involve enzymes.
3. Briefly describe the field of biochemistry and its goals.
4. What are the four major types of biomolecules?
5. What are the seven levels of biochemical hierarchy in increasing order of complexity?
6. What element must a molecule contain to be considered organic? Why is this element so critical to the formation of organisms?
7. Name four important functions of nucleotides.
8. What are the three basic parts of nucleotides?
9. What is base stacking and how does it contribute to the stability of the DNA helix? What other factor helps to stabilize this helix?
10. Identify three types of RNA and describe their functions.
11. Explain how DNA mutations can have deleterious effects.
12. Explain why gene duplication is evolutionarily important.

challenge problems

1. Write out the chemical reactions in fermentation that are responsible for (a) carbonation in beer and (b) the ethanol in vodka; include the names of the enzymes required. (c) Is acetaldehyde a reactant or product in these reactions?
2. Eduard Buchner is credited with discovering reaction conditions required for *in vitro* fermentation using protein extracts prepared from yeast. Louis Pasteur, at the urging of the local wineries in France, had also tried to reproduce fermentation *in vitro* using yeast cell extracts, but to no avail. Describe three things that Buchner did differently from Pasteur that are thought to have contributed to his success.
3. Calculate the total number of possible dodecanucleotides that can be synthesized using the four nucleotides found in RNA. What is the maximum number of peptide sequences that can be encoded by this collection of dodecanucleotides using all 20 amino acids? (Assume that none of the trinucleotides are termination codons.)
4. Explain why amylose and cellulose, both polymers of glucose, are not of equal nutritional value to humans. Explain why horses, which have similar digestion enzymes encoded in their genomes to those of humans, can live on cellulose-based foods for long periods of time and humans cannot. Use an Internet search to answer this question.
5. Peptide hormones, such as insulin and glucagon, are extracellular molecules that circulate in the blood and transmit signals to the inside of target cells by binding to transmembrane receptor proteins. Considering that both insulin and glucagon are secreted from the pancreas into the blood and have equal access to all cells in the body, explain why insulin activates signaling pathways in both liver cells and skeletal muscle cells, whereas glucagon activates signaling pathways in liver cells but not in skeletal muscle cells. Use an Internet search to answer this question.
6. Use the genetic code to answer these questions. On the basis of the double-stranded DNA molecule shown below, what is the sequence of the corresponding mRNA transcript (label the 5′ and 3′ termini)? What is the amino acid sequence of the encoded pentapeptide (label the NH_3^+ and COO^- termini)?

Coding strand 5′ – AAAAAATTTAAATTT – 3′
Template strand 3′ – TTTTTTAAATTTAAA – 5′

7. Random mutations occur in DNA at a high frequency, although most are repaired quickly and do not alter the DNA sequence in germ cells (sperm and egg cells). However, occasionally a DNA mutation will occur in the germ line and be passed on to the next generation through mating. Explain why most variations in DNA sequence between individuals of the same species are found outside of the protein-coding regions of genes.
8. Explain how proteins encoded by paralogous genes can be less similar to each other in structure and function than corresponding proteins encoded by orthologous genes.
9. The amino acid sequence of a protein determines its three-dimensional structure, which in turn determines its function. The amino acid sequences of the bacterial and mouse ribonucleotide reductase proteins shown in Figure 1.30a are ~20% identical, and both proteins have the same structure and function. Moreover, the amino acid sequence of the bacterial porin channel protein in Figure 1.30b is ~10% identical to the amino acid sequence of the jellyfish green fluorescent protein, and although both have the same β-barrel three-dimensional structure, the functions of these two proteins are very different. Lastly, the two DNA-binding proteins shown in Figure 1.30c have amino acid sequence identities of ~10% and have similar functions but very different three-dimensional structures. Based on these relationships, what could you predict about the structure and function of two proteins that were <20% identical in their amino acid sequences? What could you predict about the structure and function of two proteins that were >80% identical in their amino acid sequences?

suggested reading

Books and Reviews

Fruton, J. S. (1999). *Proteins, enzymes, genes: the interplay of chemistry and biology*. New Haven, CT: Yale University Press.

Lagerkvist, U. (2006). *The enigma of ferment: from the philosopher's stone to the first biochemical Nobel Prize.* Singapore: World Scientific.

Patthy, L. (1999). *Protein evolution*. Oxford: Blackwell.

Petsko, G. A., and Ringe, D. (2004). *Protein structure and function*. London: Blackwell.

Rabinow, P. (1996). *Making PCR: a story of biotechnology*. Chicago, IL: University of Chicago Press.

Varmus, H. E., and Weinberg, R. A. (1993). *Genes and the biology of cancer*. New York, NY: Scientific American Press.

Watson, J. D. (1980). *Norton Critical Editions: the double helix*. (G. Stent, Ed.). New York, NY: Norton.

Watson, J. D. (2000). *A passion for DNA: genes, genomes, and society*. Cold Spring Harbor, NY: Cold Spring Harbor Press.

Primary Literature

Landsberg, J. H. (2002). The effects of harmful algal blooms on aquatic organisms. *Reviews in Fisheries Science, 10*, 113–390.

Maher, B. (2009). Evolution: biology's next top model? *Nature, 458*, 695–698.

The formation of ice crystals inside living organisms can damage membrane integrity and lead to cell death. The yellow mealworm (*Tenebrio molitor*) larva contains high levels of an antifreeze protein during winter months. Pairs of threonine residues within a repeating sequence of 12 amino acids in the antifreeze protein provide the hydrogen bonding interactions with water molecules that prevent ice crystal growth. Each of the threonine amino acids in the yellow mealworm antifreeze protein has a hydroxyl group that can form hydrogen bonds with H_2O at the leading edge of the ice crystal.

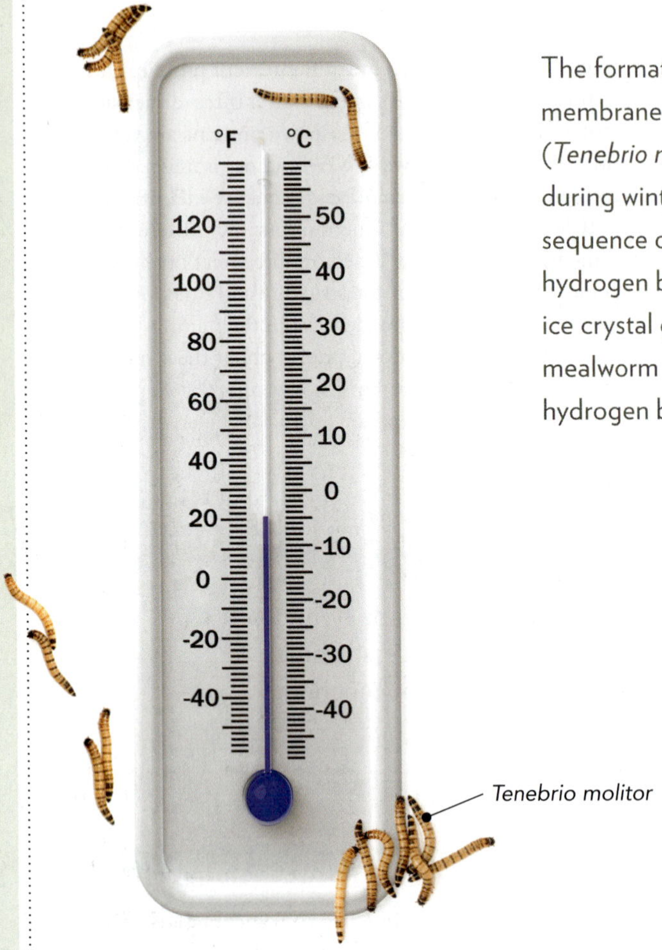

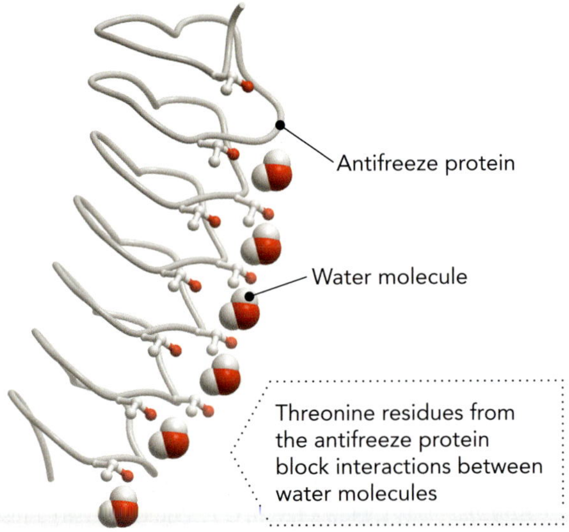

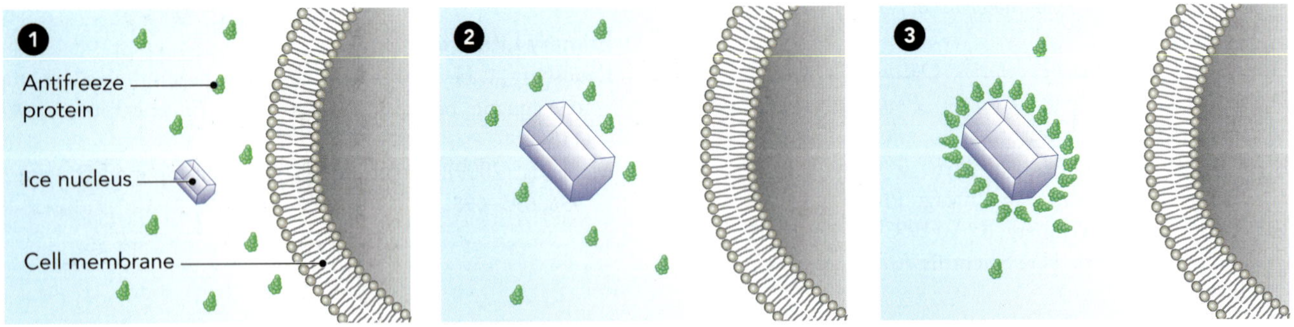

Temperatures drop below freezing and antifreeze protein is expressed

Ice crystals begin to grow and the antifreeze protein binds to the surface of the largest crystals

Antifreeze proteins coat the ice crystal and block its growth to prevent cell death

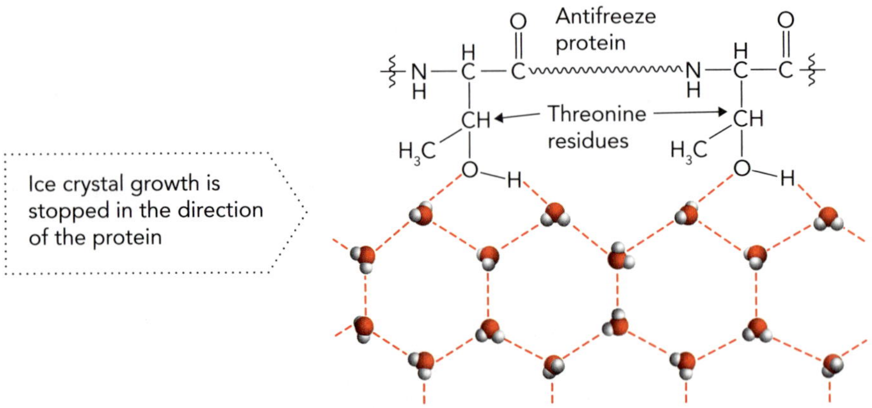

2

Physical Biochemistry

Energy Conversion, Water, and Membranes

Water is critical for life and is considered the universal solvent because of its unique properties. However, when temperatures drop below the freezing point of water, ice crystals form and can be lethal to cells. Cold-blooded organisms that live in cold climates have evolved mechanisms to cope with freezing temperatures, one of which is the synthesis of antifreeze proteins that block ice crystal growth. The larval antifreeze protein contains multiple stretches of 12 amino acids, repeating the sequence threonine-cysteine-threonine. The placement of the threonines in the repeating amino acid sequence positions hydroxyl groups at precisely the spacing needed to maximize hydrogen bond formation with the water at the leading edge of the ice crystal, and thereby blocks further growth of the crystal lattice.

CHAPTER OUTLINE

2.1 Energy Conversion in Biological Systems

- Sunlight is the source of energy on Earth
- The laws of thermodynamics apply to biological processes
- Exergonic and endergonic reactions are coupled in metabolism
- The adenylate system manages short-term energy needs

2.2 Water Is Critical for Life Processes

- Hydrogen bonding is responsible for the unique properties of water
- Weak noncovalent interactions in biomolecules are required for life
- Effects of osmolarity on cellular structure and function
- The ionization of water

2.3 Cell Membranes Function as Selective Hydrophobic Barriers

- Chemical and physical properties of cell membranes
- Organization of prokaryotic and eukaryotic cell membranes

CREDITS: THERMOMETER: PINCAREL/ISTOCK/GETTY IMAGES PLUS; MEALWORMS: BADAHOS/ISTOCK/GETTY IMAGES PLUS; ANTIFREEZE PROTEIN: BASED ON PDB FILE 1EZG.

In this chapter, we describe energy conversion in biological systems, the chemical properties of water, and the structure and function of biological membranes. We group these topics together because they are all major topics within the subdiscipline of physical biochemistry. This discussion will complete the presentation of the six biochemical principles referred to at the start of Chapter 1.

Energy conversion in biological systems, referred to as **bioenergetics**, is observed in processes that transform solar energy into chemical energy and in the interconversion of chemical energy through the oxidation and reduction of molecules. Chemical energy is used by organisms to perform work, which is necessary for cells to survive.

To understand the relationship between energy conversion and work in living systems, we will describe the physical principles that underlie this process. After a discussion of the first and second laws of thermodynamics, we describe a useful term for quantifying energy conversion reactions known as Gibbs free energy. This is followed by an examination of the unusual properties of water, which are important for the weak noncovalent interactions that are common in biological systems. In this context, we review pH, pK_a, and buffers. Finally, we describe biological membranes, which function as selective hydrophobic barriers that partition aqueous environments into specialized compartments. Biological membranes are a complex arrangement of proteins, lipids, and carbohydrates, which together determine the structure and function of the membrane. Living cells use energy to maintain differential solute concentrations across biological membranes.

2.1 Energy Conversion in Biological Systems

Photosynthetic organisms affect almost all biological processes on Earth, either directly or indirectly. Oxidation–reduction reactions in these organisms convert solar energy into chemical energy. They use this chemical energy to sustain life during daylight hours and to produce carbohydrates from CO_2 that can be stored as metabolic fuel for use at night. All other organisms obtain chemical energy from their environment, which in many cases means consuming the organic materials produced by photosynthetic organisms and using them as metabolic fuel for aerobic respiration. Energy conversion in living systems is required for three types of work: (1) *osmotic work*, in the form of maintaining differential solute concentrations across biological membranes; (2) *chemical work*, in the form of biosynthesis and degradation of organic molecules; and (3) *mechanical work*, in the form of muscle contraction in animals (**Figure 2.1**).

Living organisms need a constant input of energy to put off death as long as possible. This is because life depends on maintaining a highly ordered steady state called **homeostasis**, which requires energy. An organism in **equilibrium** with its environment is no longer alive; indeed, in order to survive, organisms must maintain a steady state with respect to temperature, concentrations of biomolecules, and other parameters that are far from equilibrium. When an organism can no longer maintain homeostasis using energy conversion processes to perform work, the intracellular concentrations of water, essential ions, and macromolecules begin to equilibrate with the surroundings, and the organism dies. Indeed, the fundamental reason all living organisms need an input of energy is to delay reaching equilibrium with the environment.

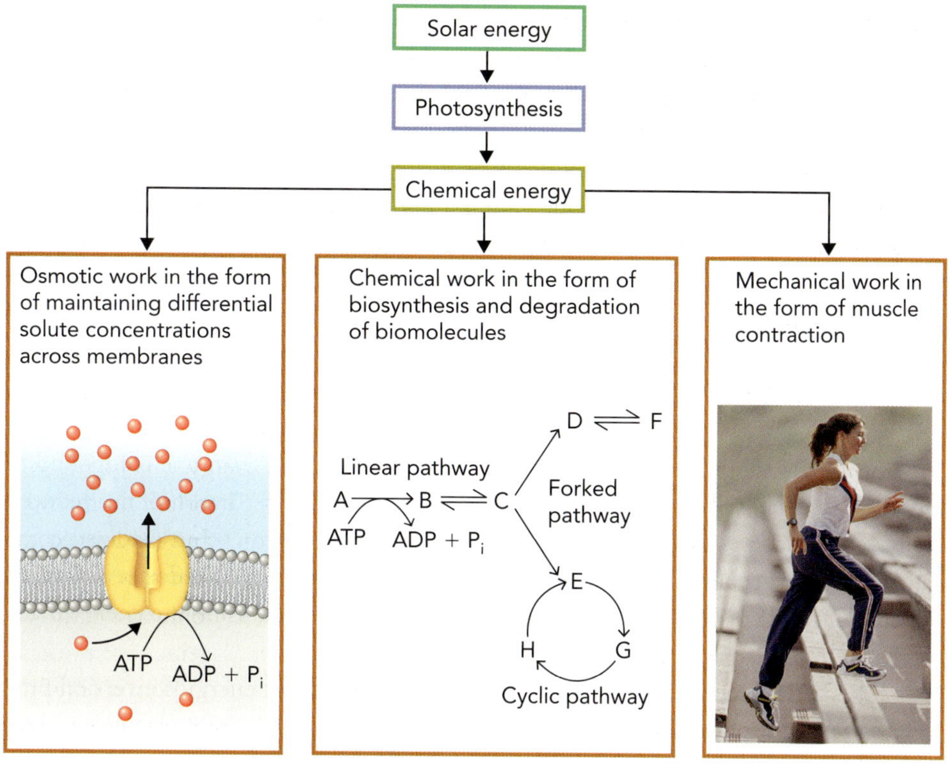

Figure 2.1 Photosynthesis converts solar energy into chemical energy, which can then be converted into the three types of work cells need to perform to survive: osmotic work, chemical work, and mechanical work. PHOTO: DAVID MADISON/GETTY IMAGES.

Sunlight Is the Source of Energy on Earth

The ultimate power source for life on Earth is **solar energy**, which is produced by the conversion of hydrogen to helium through thermonuclear reactions in the Sun. The subsequent release of this energy is in the form of electromagnetic radiation, including visible light (**Figure 2.2**). Solar energy provides all the energy required for photosynthetic autotrophs and heterotrophs to inhabit Earth. **Photosynthetic autotrophs** use solar energy to oxidize H_2O and produce O_2, generating chemical energy in the form of glucose ($C_6H_{12}O_6$). The plant uses this glucose at night for metabolic fuel to sustain **aerobic respiration**. **Heterotrophs** cannot convert solar energy into chemical energy directly and therefore depend on photosynthetic autotrophs to generate the O_2 and glucose needed for aerobic respiration and to provide those essential nutrients required for life that heterotrophs cannot synthesize themselves. As described in Chapter 12, the process of oxidizing H_2O to capture chemical energy and generate O_2 is called **photosynthesis**, whereas the conversion

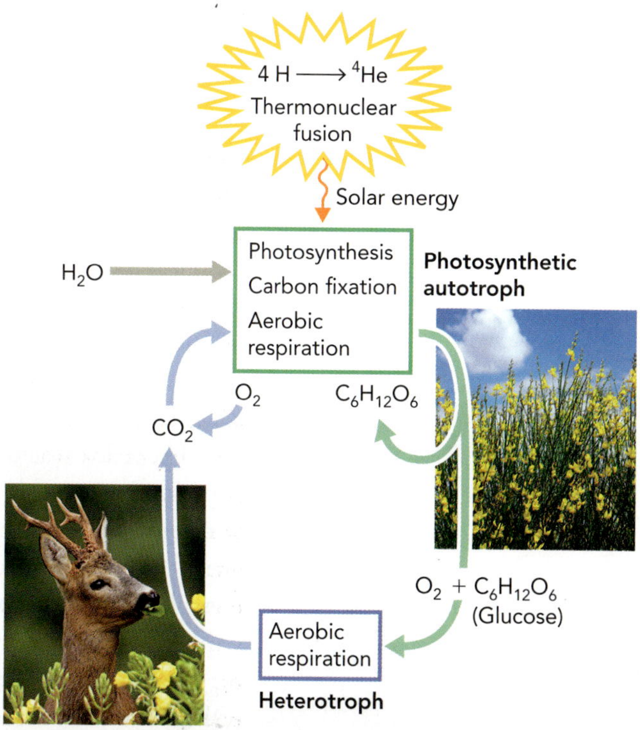

Figure 2.2 Photosynthetic autotrophs use solar energy produced by thermonuclear fusion reactions in the Sun for the process of photosynthesis, which provides the necessary energy for life during the daylight hours and for carbon fixation. Both heterotrophs and photosynthetic autotrophs use O_2 and glucose ($C_6H_{12}O_6$) for the process of aerobic respiration, which is a form of chemical energy conversion. Oxidation of H_2O by photosynthetic organisms is the primary source of O_2 in our atmosphere. DEER: JUNIORS/JUNIORS/SUPERSTOCK; FLOWERS: INACIO PIRES/SHUTTERSTOCK.

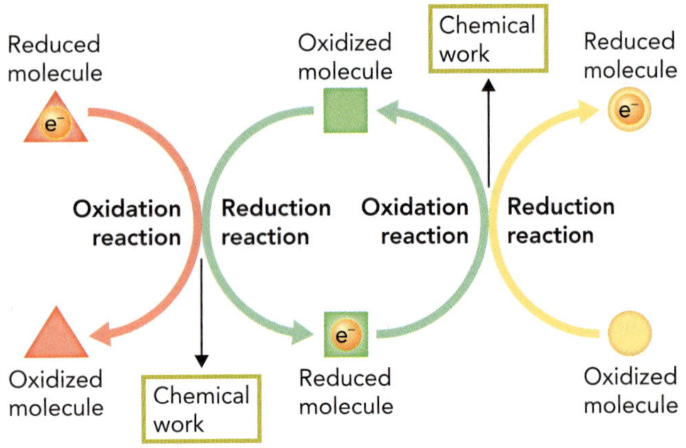

Figure 2.3 Coupled oxidation–reduction (redox) reactions in biological systems involve the transfer of electrons through a redox circuit consisting of oxidized and reduced compounds. The result is provision of energy for chemical work.

of CO_2 to organic compounds is referred to as **carbon fixation**. The most abundant photosynthetic autotrophs in the **biosphere** (the Earth environment containing living organisms) are vascular plants, single-cell algae, and photosynthetic bacteria.

Both photosynthesis and aerobic respiration interconvert energy using a series of linked oxidation–reduction reactions. In these reactions, electrons are transferred from a molecule of higher electrochemical potential to one of lower electrochemical potential. (Recall that oxidation is the loss of electrons and reduction is the gain of electrons.) A series of linked oxidation–reduction reactions, often called **redox reactions**, transfers electrons from one compound to another in sequential fashion. Because electrons do not exist free in solution, a reduced compound becomes oxidized only when it transfers an electron to an oxidized compound, which becomes reduced (**Figure 2.3**). The importance of redox reactions in biochemical processes is that chemical work can be performed using the energy made available by electron transfer.

The initial biochemical event required for all subsequent energy conversion processes in our biosphere is the absorption of light energy by pigment molecules, such as chlorophyll, present in photosynthetic organisms. Light absorption causes **photooxidation** of chlorophyll, in which an electron is transferred from the chlorophyll molecule to an acceptor molecule. The acceptor molecule then passes the electron to another acceptor molecule of lower electrochemical potential. The electrons lost from chlorophyll photooxidation are replaced by the oxidation of 2 H_2O, which generates 4 e^- + 4 H^+ + O_2. The absorption of four photons of light results in the transfer of four e^- through a series of redox reactions, causing the reduction of two molecules of an electron carrier molecule called plastoquinone (PQ) to generate plastoquinol (PQH_2) (**Figure 2.4a**; molecular structures are shown in Chapter 12, Figure 12.26).

Absorption of another four photons of light causes another photooxidation event, ultimately leading to the reduction of two molecules of the coenzyme nicotinamide adenine dinucleotide phosphate ($NADP^+$; oxidized form) to produce 2 NADPH + 2 H^+. (The structures of NADPH and some other redox compounds are shown in Chapter 7.) Additionally, the light-induced flow of four electrons through this photosynthetic electron transfer system results in the formation of ~3 ATP molecules through the process of **photophosphorylation**, which traps the redox energy in a useful chemical form. The chemical bond energy in NADPH and ATP is used to convert CO_2 to organic molecules using the process of carbon fixation, which provides heterotrophs with an indirect means to harness solar energy. Amazingly, the simple absorption of light by chlorophyll molecules in cyanobacteria and plants sets off a cascade of biochemical events that supports *essentially all life* on our planet!

The redox reactions of aerobic respiration are fundamentally similar to those of photosynthesis, except in this case, oxidation of glucose by enzyme-mediated reactions transfers 24 e^- to the coenzymes nicotinamide adenine dinucleotide (NAD^+; oxidized form) and flavin adenine dinucleotide (FAD; oxidized form). Through **oxidative phosphorylation**, the oxidation of NADH and $FADH_2$ provides redox energy for the synthesis of ~32 ATP for each glucose molecule oxidized (**Figure 2.4b**). The electrons are passed through the electron carrier ubiquinone (Q), which is reduced to ubiquinol (QH_2). The electrons in QH_2 are donated to O_2, the final electron acceptor, to produce

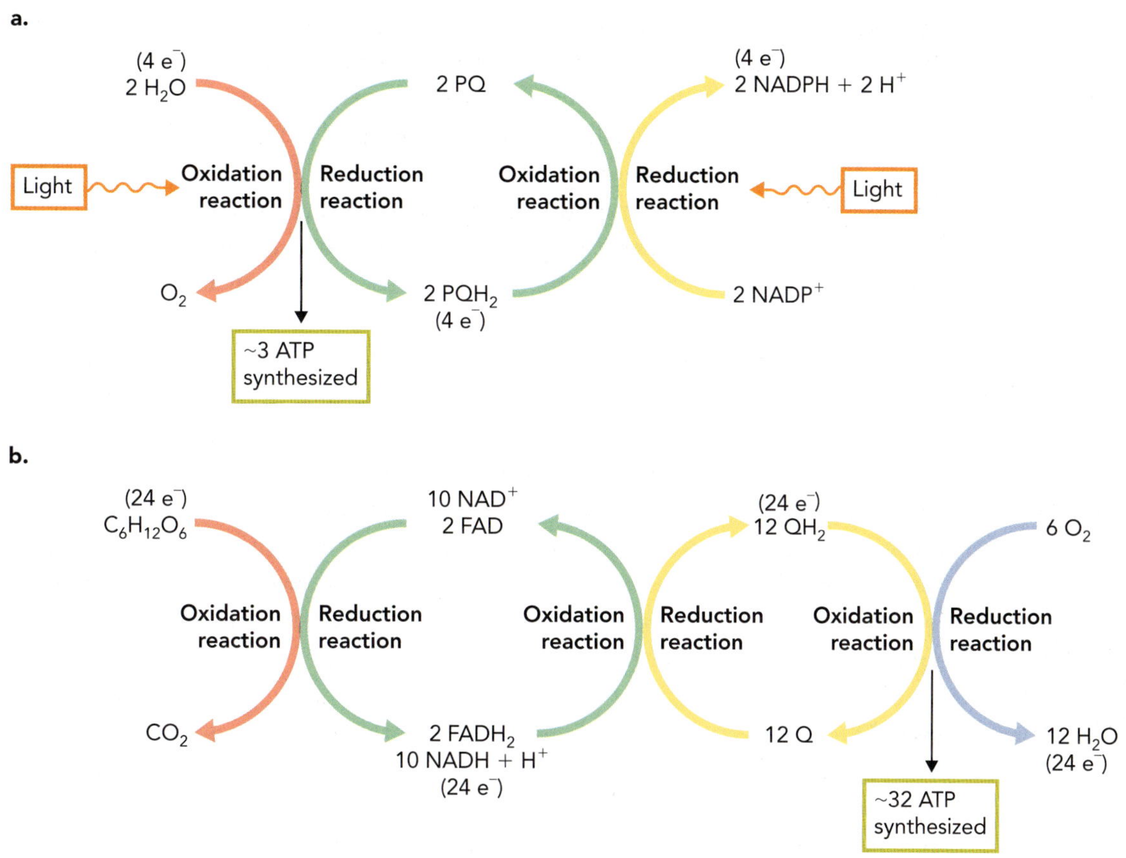

Figure 2.4 Photosynthesis and aerobic respiration are the two major energy conversion pathways in living organisms. Both processes involve a series of coupled redox reactions that result in the synthesis of ATP, which can be used for chemical work. **a.** Photosynthesis converts light energy into chemical energy. The redox energy captured in this process is used for the synthesis of ATP through a process called photophosphorylation. **b.** Aerobic respiration oxidizes glucose to CO_2 and H_2O to capture 24 e^- in the form of the reduced coenzymes NADH and $FADH_2$. Oxidation of NADH and $FADH_2$ provides redox energy for the synthesis of ATP through a process called oxidative phosphorylation.

12 H_2O. The biochemistry of aerobic respiration (glycolysis, citrate cycle, electron transport system, and ATP synthesis) is described in Chapters 9, 10, and 11.

The Laws of Thermodynamics Apply to Biological Processes

Biological processes follow the same universal laws and thermodynamic principles that govern physical processes. To understand better how energy is used in biological systems, we need to review basic thermodynamic principles in the context of a system and its surroundings.

The term *system* refers to a collection of matter in a defined space, whereas the term *surroundings* refers to everything else. The system and surroundings together constitute the *universe*. A system can be anything we choose to study: a test tube with reaction components, a living cell, or even a whole organism or ecological niche. Systems may be one of three types: (1) an *open system*, in which matter and energy are freely exchanged with the surroundings; (2) a *closed system*, in which energy is exchanged with the surroundings but matter is not; and (3) an *isolated system*, in which neither matter nor energy are exchanged with the surroundings. Biological systems are open systems, as both matter (nutrients and waste products) and energy (primarily heat) are exchanged with the surroundings.

Biochemists use two main laws to describe biological processes in thermodynamic terms. The **first law of thermodynamics** states that the total amount of energy in the universe remains the same, even though the form of energy may change. Put another way, energy can neither be created nor destroyed, only transformed (converted). The **second law of thermodynamics** states that in the absence of an energy input, all spontaneous processes in the universe tend toward dispersal of energy (disorder), and moreover, that the measure of this disorder, called **entropy**, is always increasing in the universe. It is important to note that all biological energy conversion processes (as well as all mechanical energy conversion processes) are less than 100% efficient; some of the converted energy is lost as heat rather than used for work.

Applying thermodynamic principles to biological systems provides a way to quantify life processes in terms of energy and, moreover, to determine if a process is spontaneous (favorable). Here we define the first and second laws of thermodynamics in quantitative terms and then relate them to biochemical processes using the concepts of Gibbs free energy (G) and the equilibrium constant (K_{eq}).

First Law of Thermodynamics The first law of thermodynamics states that energy cannot be created or destroyed, only converted from one form to another. In mathematical terms, we first define the energy change in a system (ΔE) as the difference between the final (E_{final}) and initial ($E_{initial}$) energy states. The first law of thermodynamics says that this energy change is equal to the difference between the heat (q) transferred into the system from its surroundings, and the work (w) done by the system on the surroundings (using the convention that w is positive for work done by the system):

$$\Delta E = E_{final} - E_{initial} = q - w \quad (2.1)$$

Most energy conversions in the physical world involve changes in pressure (P) or volume (V); however, a living cell must convert energy under conditions of constant pressure and volume. To measure the energy change in a system, we need to introduce a term called **enthalpy (H)**, which refers to the heat content of a system. Enthalpy is defined by the equation

$$H = E + PV \quad (2.2)$$

We can express the change in enthalpy (ΔH) in terms of the change in energy (ΔE) with respect to pressure and volume by

$$\Delta H = \Delta E + \Delta PV \quad (2.3)$$

Therefore, under biological conditions, where pressure and volume do not change and no work is done, we can combine Equations 2.1 and 2.3 and show that ΔH is a function of ΔE, which is a measure of heat (q):

$$\begin{aligned} \Delta E &= \Delta H - \Delta PV = q - w \\ &= \Delta H - 0 = q - 0 \\ &= \Delta H = q \end{aligned} \quad (2.4)$$

In a closed system such as a **bomb calorimeter**, in which a compound is combusted by a spark at constant volume in the presence of pure oxygen (completely oxidized), the amount of heat exchanged between the reaction chamber and a surrounding

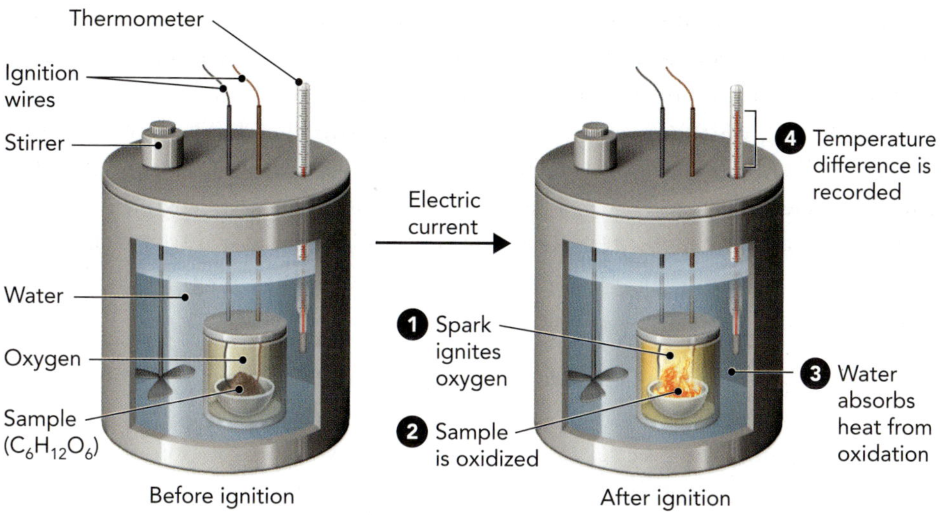

Figure 2.5 A bomb calorimeter is a device that permits an indirect measurement of heat content (enthalpy) in a compound by recording changes in the temperature of water surrounding a reaction chamber held at constant volume. Complete oxidation occurs by igniting a spark in the presence of pure O_2. In this example, glucose ($C_6H_{12}O_6$) is converted to CO_2 and H_2O in an exothermic reaction that raises the temperature of the surrounding water.

water jacket is a measure of ΔH (**Figure 2.5**). A reaction that gives off heat is called **exothermic** and has a negative ΔH value, whereas a reaction that absorbs heat is called **endothermic** and has a positive ΔH value.

The standard enthalpy value for a chemical compound is determined by the number and type of chemical bonds. Therefore, depending on how a chemical reaction changes the number and type of bonds from the reactants to the products, the reaction is either exothermic or endothermic. For example, the combustion of glucose is an exothermic reaction and produces heat, CO_2, and H_2O. The temperature increase of the surrounding water is a measure of the potential energy of the glucose, as reflected in the number and type of chemical bonds. A **calorie (cal)** is a unit of energy that was originally defined by the amount of heat energy required to raise 1 g of water by 1 °C using a calorimeter. Energy is also expressed in the International System of Units (SI system) as the unit **joule (J)**, where 1 calorie = 4.184 J. Note that in nutritional sciences, calorie with a capital "C" actually refers to a kilocalorie (kcal), so 1 Calorie = 1 kcal = 4.184 kJ. Unfortunately, strict adherence to the capitalization convention does not occur, so when "calories" are used, the intent must be inferred from context. Though in the biochemical literature units of calories are still in use, in this text, we use the SI unit joule.

We can write an equation that describes the complete oxidation of glucose in which the heat produced is at constant pressure (q_P):

$$C_6H_{12}O_6 + 6\ O_2 \rightarrow 6\ CO_2 + 6\ H_2O + \text{Heat}$$

$$\text{Heat} = \Delta E = \Delta H = q_P \qquad (2.5)$$

The change in enthalpy ΔH between two states (glucose + 6 O_2 and 6 CO_2 + 6 H_2O) is independent of the path taken. Therefore, the experimental value of 15.7 kJ/g glucose, which was determined by bomb calorimetry, is equal to the amount of energy released from glucose when it is oxidized by a cell through the process of aerobic respiration. Knowing the amount of potential energy in a molecule of glucose, however, does not tell us if glucose oxidation is a favorable reaction. For that, we need to determine the overall change in Gibbs free energy, which also includes the contribution of entropic effects, as described in the second law of thermodynamics.

Second Law of Thermodynamics The second law of thermodynamics states that all spontaneous processes in the universe tend toward dispersal of energy in the absence of energy input. This concept is defined by entropy (S), or, more accurately, the change in entropy (ΔS), which is a measure of the spreading of energy. The more dispersal of energy that occurs in a system, the higher the value of entropy. The entropy of the universe is always increasing and is equal to the entropy of the system plus the entropy of its surroundings, as defined by the equation

$$\Delta S_{\text{universe}} = \Delta S_{\text{system}} + \Delta S_{\text{surroundings}} > 0 \tag{2.6}$$

Entropy is sometimes described as a measure of disorder; however, ambiguity in what exactly is meant by "disorder" has led to some confusion. Therefore, it is conceptually easier (and more precise) to think about entropy as a dispersion of energy, or how energy is spread out. In chemistry, it is perhaps easiest to visualize entropy as a consideration of energy due to motion, such as translational, rotational, or vibrational energy.

A familiar example that can help you understand the concept of entropy is to consider the transition between ice and liquid water, a process that happens spontaneously at room temperature (**Figure 2.6**). But why does ice melt at room temperature? To answer this question, we need to consider both the enthalpic and entropic

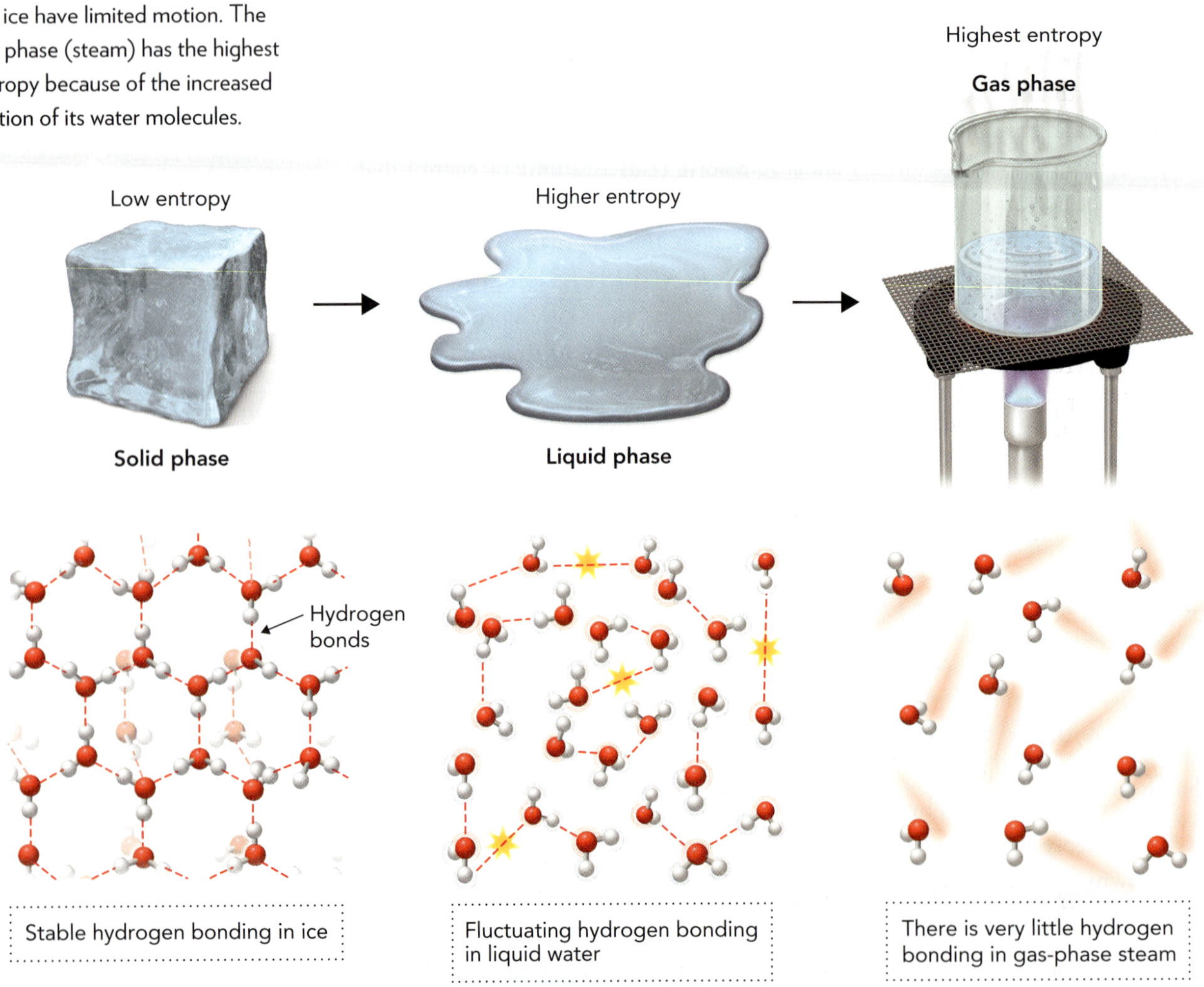

Figure 2.6 Ice melting at room temperature is a process dominated by the second law of thermodynamics. Solid water (ice) has lower entropy than liquid water because the water molecules in the ice have limited motion. The gas phase (steam) has the highest entropy because of the increased motion of its water molecules.

contributions in this process (as described in the next subsection). However, for this example, the entropic term dominates, so our discussion here centers around changes in the entropy of this process. The entropic contribution comes from the fact that the H_2O molecules in ice crystals interact extensively through hydrogen bonding and have relatively little motional energy (a low-entropy state). However, the H_2O molecules in liquid water have fluctuating hydrogen bonds that are made and broken continually, resulting in much more motional energy (a high-entropy state). In other words, the dispersal of energy is greater in liquid water than in ice. Thus, melting of ice is favorable from an entropic standpoint because the entropy of the system increases. To reverse this process to form ice requires an input of energy—specifically, the electricity used by a freezer to lower the temperature of an ice-cube tray. Put another way, the input of electricity in the freezer restrains the ice crystals from melting by limiting the dispersal of energy (motional energy is restricted).

Similarly, the metabolic energy required for sustaining life restrains the natural tendency of the molecules within the organism to disperse their energy, as dictated by the second law of thermodynamics. Two examples of increased entropy in living systems are (1) the degradation of large biomolecules into a larger number of smaller molecules through the process of decay, and (2) the flow of ions through a channel in a cell membrane from a region of high ionic concentration to one of low concentration during an action potential (for example, a nerve impulse).

Entropy varies with temperature: the higher the temperature, the more dispersion of energy in the system. The boiling of water produces steam, in which the energy is more spread out than in liquid water because the molecules in steam are constantly moving around and changing their positions (see Figure 2.6). With this in mind, we can now quantify entropy, as defined by the equation

$$S = k_{\mathrm{B}} \ln W \tag{2.7}$$

in which k_{B} is the Boltzmann constant (1.3806×10^{-23} J/K), and W is the number of energetically equivalent ways of arranging components in a system. Thus entropy, S, has units of J/K. (Recall that K, kelvin, is a unit of temperature where 273 K = 0 °C.) Although it is not possible experimentally to determine W in a biological system, we can describe the change in entropy (ΔS) as a function of the temperature (T) of the system under constant pressure using the expression

$$\Delta S \geq \frac{q_{\mathrm{P}}}{T} \tag{2.8}$$

We can rewrite this equation by combining Equations 2.5 and 2.8 to show that

$$\Delta S \geq \frac{\Delta H}{T} \tag{2.9}$$

In other words, the change in entropy between two states is a function of the change in enthalpy (number and types of bonds broken or formed) and the temperature of the system.

Gibbs Free Energy and the Equilibrium Constant In 1878, the American theoretical physicist J. Willard Gibbs described a way to determine if a chemical reaction is favorable or unfavorable under a given pressure and temperature using a function he called free energy. The **Gibbs free energy (G)** is defined as the difference between the enthalpy (H) and the entropy (S) of a system at a given temperature (T):

$$G = H - TS \tag{2.10}$$

Absolute values for G, H, and S cannot be determined, but we can measure the change between two states. Therefore, the most useful form of the Gibbs free energy equation is

$$\Delta G = \Delta H - T\Delta S \tag{2.11}$$

If the Gibbs free energy change (ΔG) value for a reaction is less than zero ($\Delta G < 0$), then the reaction is favorable in the forward direction and **exergonic**. However, if the ΔG value is greater than zero ($\Delta G > 0$), then the reaction is unfavorable in the forward direction and **endergonic**. A reaction in which ΔG is equal to zero ($\Delta G = 0$) is at equilibrium, meaning that the rate of formation of products is equal to the rate of formation of reactants, so no net change occurs in the concentrations of products and reactants. Note that in the Gibbs free energy equation, we need to consider the contributions from ΔH, ΔS, and temperature to determine whether the reaction is spontaneous or favorable in a given direction ($\Delta G < 0$). For example, a reaction that is favorable from an enthalpic standpoint ($\Delta H < 0$) is not necessarily spontaneous, as this depends on the entropic contribution and temperature as well, as shown in **Table 2.1**.

The **standard free energy change (ΔG°)** is useful as a reference point for comparing chemical reactions under a defined set of conditions. We define it as the free energy change, at constant pressure (1 atm, or 101.3 kPa) and room temperature (298 K), when going from the condition where all reactants and products are present at 1 M concentration to the condition of equilibrium concentrations. Thus, the standard free energy change is a constant for a given reaction. The value of $\Delta G°$ is related to the enthalpy and entropy changes under standard conditions, as shown by

$$\Delta G° = \Delta H° - T\Delta S°$$

Biochemists have found it useful to define **biochemical standard conditions**, where in addition to the above conditions, the pH is constant at pH 7 ($[H^+] = 10^{-7}$ M) and the concentration of H_2O is constant at 55.5 M. (Also, when Mg^{2+} is involved in a reaction, its concentration is assumed to be constant at 1 mM.) Thus, the biochemical standard free energy change, denoted with a prime symbol ($\Delta G°'$), is the free energy change under biochemical standard conditions and is also a constant for a given reaction.

Table 2.1 SPONTANEITY OF A REACTION

If the sign of ΔH is:	If the sign of ΔS is:	Then the sign of ΔG will be:	Spontaneous?
Negative (favorable)	Positive (favorable)	Negative	Yes $\Delta G < 0$ at all temperatures
Positive (unfavorable)	Negative (unfavorable)	Positive	No $\Delta G > 0$ at all temperatures
Negative (favorable)	Negative (unfavorable)	Negative when T is low ($T < \Delta H/\Delta S$) Positive when T is high ($T > \Delta H/\Delta S$)	Depends on T $\Delta G < 0$ only when $T < \Delta H/\Delta S$
Positive (unfavorable)	Positive (favorable)	Negative when T is high ($T > \Delta H/\Delta S$) Positive when T is low ($T < \Delta H/\Delta S$)	Depends on T $\Delta G < 0$ only when $T > \Delta H/\Delta S$

Note: The spontaneity of a reaction is determined from the combination of the changes in the enthalpy and entropy as described by the Gibbs free energy equation, $\Delta G = \Delta H - T\Delta S$.

The magnitude of the standard Gibbs free energy is a measure of how far the standard state is from equilibrium. The standard Gibbs free energy is directly linked to a reaction's equilibrium constant (K_{eq}) by the expression

$$\Delta G^\circ = -RT\ln K_{eq} \tag{2.12}$$

where R is the gas constant (8.314 J/mol K), and T is the absolute temperature. Thus, the **equilibrium constant (K_{eq})** is a measure of the directionality of a reaction under standard conditions, where all products and reactants start at 1 M and proceed to their equilibrium concentrations. For example, for the reaction

$$\text{A} + \text{B} \rightleftharpoons \text{C} + \text{D}$$

K_{eq} is defined by the concentrations of A, B, C, and D when the reaction has reached equilibrium, using the relationship

$$K_{eq} = \frac{[\text{C}]_{eq}[\text{D}]_{eq}}{[\text{A}]_{eq}[\text{B}]_{eq}} \tag{2.13}$$

(the brackets [ ] denote concentration). If $K_{eq} > 1$, the reaction proceeds spontaneously to form C and D (left to right as written), as this results in a negative value of ΔG°. If $K_{eq} < 1$, the reaction favors the formation of A and B (right to left as written), as the value of ΔG° is positive in this case. For any reaction, we can determine the ΔG° value experimentally by setting up the reaction under standard conditions (starting with 1 M of each solute), and then allowing it to proceed to equilibrium. Once the reaction has reached equilibrium, we measure the concentrations of all reactants and products ($[\text{A}]_{eq}$, $[\text{B}]_{eq}$, and so forth) and use this value of K_{eq} to determine ΔG° using Equation 2.12.

For the same reaction $\text{A} + \text{B} \rightleftharpoons \text{C} + \text{D}$ under conditions where reactants and products are not at 1 M initial concentrations (as would be true under cellular conditions), Gibbs defined the following relationship to determine the actual free energy change:

$$\Delta G = \Delta G^\circ + RT\ln Q \tag{2.14}$$

where the mass-action ratio, Q, is the ratio of initial concentrations of products over reactants:

$$Q = \frac{[\text{C}]_i[\text{D}]_i}{[\text{A}]_i[\text{B}]_i}$$

and the subscript i denotes initial concentrations. Equation 2.14 shows us that the actual free energy change for a given reaction (ΔG; Gibbs free energy change) is the sum of the standard free energy change ($\Delta G^{\circ\prime}$; our reference value, a constant) and a term reflecting any initial conditions.

It is important to note that Q reflects the actual initial concentrations of reactants and products and does not equal K_{eq} except when the reaction is at equilibrium. When the reaction is at equilibrium, $\Delta G = 0$, and Equation 2.14 becomes

$$0 = \Delta G^\circ + RT\ln\frac{[\text{C}]_{eq}[\text{D}]_{eq}}{[\text{A}]_{eq}[\text{B}]_{eq}}$$

$$\Delta G^\circ = -RT\ln K_{eq} \tag{2.12}$$

The importance of free energy values in biochemistry is that we can use them to predict if a reaction is favorable or unfavorable, given the characteristic ΔG° value for the reaction and the initial concentrations of each reactant and product. If $\Delta G < 0$, the reaction proceeds spontaneously from left to right as written. If $\Delta G > 0$, the reaction proceeds spontaneously from right to left as written. However, you should be aware of two important points regarding the spontaneity of a reaction. First, on the basis of the Gibbs equation (2.14), the actual free energy change of a reaction inside a living cell can be favorable ($\Delta G < 0$) even if the characteristic ΔG° for the reaction is unfavorable ($\Delta G^\circ > 0$) under standard conditions. This is because the actual concentrations of reactants and products are important in determining reaction spontaneity. When product concentrations are lower than reactant concentrations, the second term of Equation 2.14 will be negative: $RT \ln([\text{products}]/[\text{reactants}]) < 0$. If this term is negative and has a larger absolute value than ΔG°, then the actual free energy change ΔG will be negative, and the reaction will be spontaneous under these conditions. The second point is that the ΔG value reveals the directionality of a reaction, but not the rate at which the reaction occurs. Though a reaction may be spontaneous, it does not necessarily occur on a reasonable timescale. To determine rates of reactions, we need to consider the kinetics of a reaction, as we will describe in Chapter 7 for enzyme-catalyzed reactions.

Exergonic and Endergonic Reactions Are Coupled in Metabolism

This brings us to the question of how a reaction that is unfavorable (endergonic) can occur in living systems. The answer is that an endergonic reaction can be coupled to an exergonic reaction through a common intermediate such that the overall change in free energy is favorable (exergonic).

The thermodynamic basis for coupling endergonic and exergonic reactions is that for two reactions that share a common intermediate (the product of the first reaction is a reactant in the second reaction), the ΔG° value is equal to the sum of the ΔG° values for the two separate reactions:

$$\text{Reaction 1: A + B} \xrightleftharpoons{\text{Enzyme 1}} \textbf{C} \qquad \Delta G^\circ_1 = +3 \text{ kJ/mol (endergonic)}$$

$$\text{Reaction 2: } \textbf{C} + \text{D} \xrightleftharpoons{\text{Enzyme 2}} \text{E} \qquad \Delta G^\circ_2 = -10 \text{ kJ/mol (exergonic)}$$

$$\text{Net reaction: A + B + D} \rightleftharpoons \text{E} \qquad \Delta G^\circ_1 + \Delta G^\circ_2 = \Delta G^\circ_3 = -7 \text{ kJ/mol (exergonic)}$$

Although the first reaction is unfavorable ($\Delta G^\circ_1 > 0$), the formation of product E occurs because the combined standard free energies of reactions 1 and 2 are favorable ($\Delta G^\circ_3 < 0$). One way to think about how reaction 2 affects reaction 1 in this example is to realize that because C is being converted into E (in the presence of D), the concentration of C is lower than the equilibrium concentration for the isolated reaction 1. To replace the C that is continually being consumed by reaction 2, the equilibrium of reaction 1 shifts toward more product formation. In living cells, metabolic pathways are formed from multiple coupled reactions. By coupling reactions in a metabolic pathway, a reaction that would be unfavorable in isolation can be driven in the forward direction.

One of the most common types of coupled reactions is one in which ATP is used to make an overall reaction favorable. ATP contains two phosphoanhydride bonds, sometimes referred to as "high-energy phosphate bonds" (~P), which can be hydrolyzed to yield adenosine diphosphate and inorganic phosphate (ADP + P_i) or adenosine monophosphate and pyrophosphate (AMP + PP_i; **Figure 2.7**). Note, however, that the term "high-energy phosphate bond" can be misleading—phosphoanhydride bonds are no different from any other chemical bond and follow the laws of thermodynamics.

The change in biochemical standard free energy ($\Delta G^{\circ\prime}$) for cleavage of the phosphoanhydride bond between the β and γ phosphates of ATP is −30.5 kJ/mol and that of the phosphoanhydride bond between the α and β phosphates is −32.3 kJ/mol. It is important to recognize, however, that although hydrolysis reactions were used to calculate these $\Delta G^{\circ\prime}$ values, it is not the cleavage of these phosphoanhydride bonds that provides energy for coupled metabolic reactions (bond cleavage actually requires energy). Rather, the transfer of a phosphoryl or adenylyl (AMP) group to a reactant generates a highly reactive intermediate. ATP-coupled reactions take place within the active site of an enzyme, which accelerates the rate of a reaction by providing an ideal chemical environment for product formation (see Chapter 7). In these enzyme-mediated coupled reactions, the phosphorylated or adenylated chemical intermediates are the compounds that function as the shared intermediate in the two reactions.

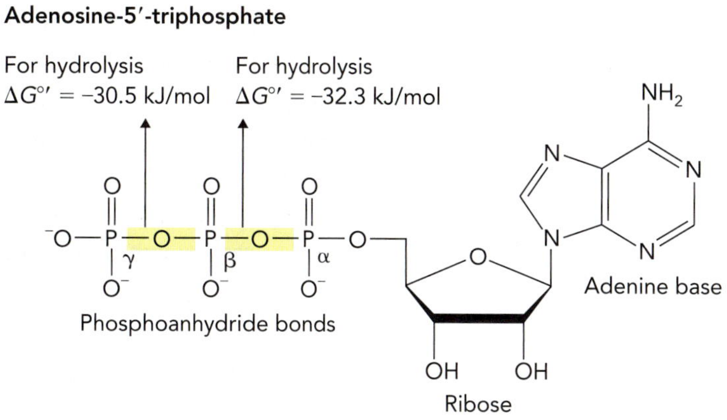

Figure 2.7 ATP is a carrier of chemical energy in living systems by virtue of its two phosphoanhydride bonds, which each contain ~30 kJ/mol of potential energy ($\Delta G^{\circ\prime}$ = −30.5 kJ/mol and $\Delta G^{\circ\prime}$ = −32.3 kJ/mol).

A good example of an ATP-coupled reaction, in which the γ-phosphoryl group of ATP is used in a phosphoryl transfer reaction to generate a highly reactive intermediate, is the conversion of the amino acid glutamate to glutamine by the enzyme glutamine synthetase (**Figure 2.8**). In the first step, the γ-phosphoryl group of ATP is transferred to a carboxyl group on glutamate, forming an intermediate in which the phosphoryl group is covalently linked to the glutamyl group. This reaction intermediate is then converted to glutamine in the second step when an amine group from the ammonium ion (NH_4^+) replaces the phosphoryl group, releasing P_i. The bioenergetics of this enzyme-mediated, ATP-coupled reaction can be broken down into two reactions, in which an unfavorable endergonic reaction ($\Delta G^{\circ\prime}$ = +14.2 kJ/mol) is coupled to a favorable exergonic reaction ($\Delta G^{\circ\prime}$ = −30.5 kJ/mol) to give an overall favorable $\Delta G^{\circ\prime}$ of −16.3 kJ/mol:

$$\text{Glutamate} + NH_4^+ \rightarrow \text{Glutamine} \qquad \Delta G^{\circ\prime} = +14.2\ \text{kJ/mol}$$

$$\text{ATP} \rightarrow \text{ADP} + P_i \qquad \Delta G^{\circ\prime} = -30.5\ \text{kJ/mol}$$

$$\text{Glutamate} + NH_4^+ + \text{ATP} \rightarrow \text{Glutamine} + \text{ADP} + P_i \qquad \Delta G^{\circ\prime} = -16.3\ \text{kJ/mol}$$

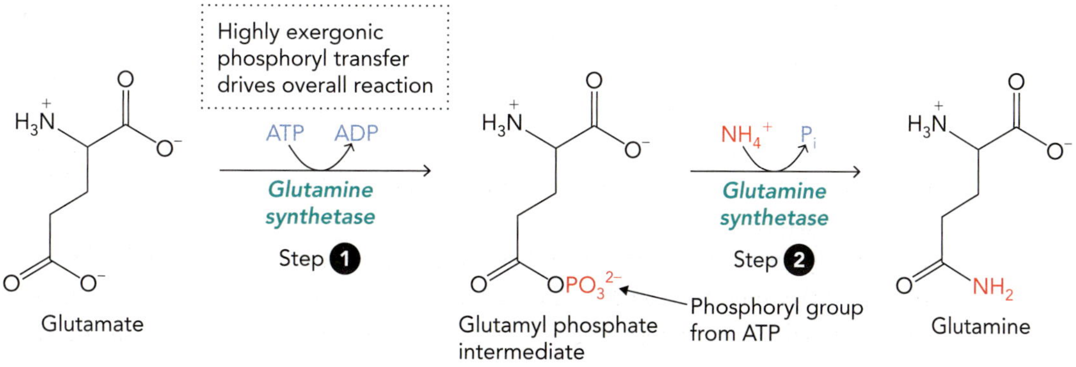

Figure 2.8 The glutamine synthetase reaction is a two-step reaction. In the first step, the γ-phosphoryl group of ATP is transferred to glutamate, forming a glutamyl phosphate intermediate, and ADP is released. In the second step, NH_4^+ reacts with glutamyl phosphate to generate glutamine and release of P_i.

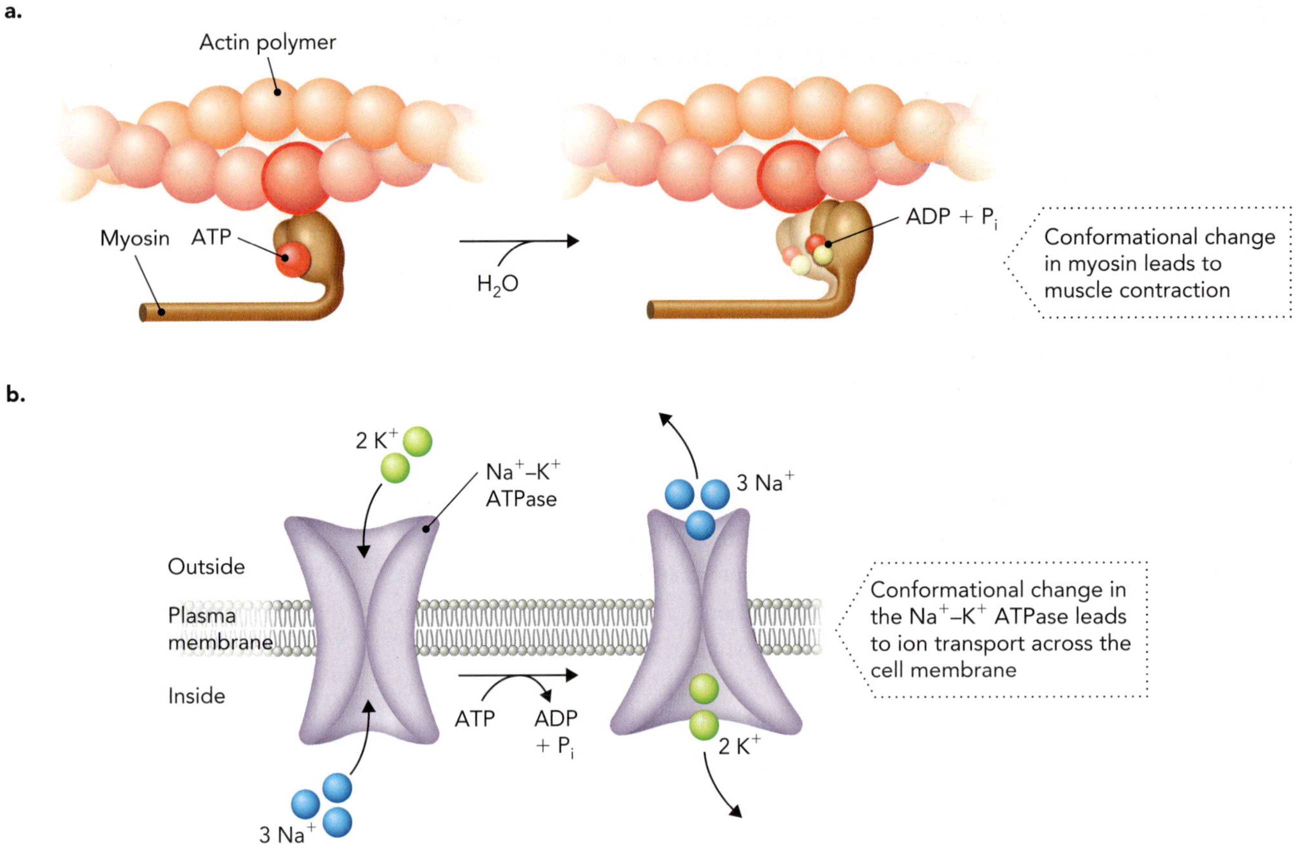

Figure 2.9 ATP hydrolysis can provide energy for protein conformational changes by converting chemical energy to mechanical energy. **a.** ATP binding to myosin protein in skeletal muscle leads to a large conformational change upon ATP hydrolysis, resulting in muscle contraction. **b.** ATP binding to the Na^+–K^+ ATPase transporter protein induces a conformational change that leads to the exchange of Na^+ and K^+ ions across the plasma membrane.

Although transfer reactions such as these account for the majority of ATP-driven reactions in cells, phosphoanhydride bond energy in ATP is also used for cellular work through the binding of ATP to proteins, which undergo a conformational change upon ATP hydrolysis. Two examples of these energy-converting ATP hydrolysis reactions are muscle contraction and ion transport across cell membranes (**Figure 2.9**). As described in some detail in Chapter 6, hydrolysis of ATP bound to the myosin protein causes a large conformational change in the head region of the protein, resulting in muscle contraction by pulling one actin muscle fiber past another (see Figure 6.67). We will also see in Chapter 6 that ATP hydrolysis induces a conformational change in the Na^+–K^+ ATPase transporter protein, facilitating the energy-driven exchange of 2 K^+ ions on the outside of the cell for 3 Na^+ ions on the inside of the cell (see Figure 6.46).

The following chemical properties of ATP account for the large standard free energy change that occurs when a phosphoanhydride bond is cleaved:

1. Electrostatic repulsion between the charged phosphoryl groups destabilizes ATP. Repulsion is reduced on hydrolysis, and therefore the products of ATP hydrolysis are more stable than ATP itself, which favors the hydrolysis reaction.

2. The released phosphate ion has more possible resonance forms than when it is covalently attached to adenylate. Entropically, this favors the free phosphate ions compared to ATP or ADP.
3. The phosphate ion and ADP have a greater degree of solvation than ATP. This means that the phosphate ion and ADP form hydration layers and are more stable than ATP.

Thermodynamically, the hydrolysis of ATP is favorable under standard conditions, but it is important to realize that ATP is a kinetically stable compound and does not rapidly undergo hydrolysis in the absence of a catalyst. This is due to a relatively large energy barrier that must be overcome in order for hydrolysis to occur, as we will describe in Chapter 7. Indeed, it is primarily in the environment of an enzyme active site where this energy barrier is decreased to the point where cleavage of the phosphoanhydride bond is favored because of the chemistry of the enzyme functional groups. This makes sense because in order for ATP to be the energy currency of the cell, it not only needs to be highly reactive in the context of an enzyme-mediated reaction, but also it must be chemically stable to serve as a form of energy storage.

The Adenylate System Manages Short-Term Energy Needs

Because ATP plays such an important role in the cell as a source of free energy for coupled reactions and mechanical work, the cell needs to maintain ATP levels within a fairly narrow range to avoid a metabolic catastrophe. This is done by interconverting ATP, ADP, and AMP, using several key phosphoryl transfer reactions that together constitute the **adenylate system**. To see why the adenylate system is important, consider that a person weighing 70 kg requires ~100 mol of ATP every day, based on the energy content of food. The molecular mass of ATP is 507 g/mol, which means we hydrolyze as much as 50 kg of ATP every day. Rather than synthesizing our own weight in ATP on a daily basis, it is much more efficient to recycle adenylate forms by re-forming ATP from ADP, AMP, and P_i.

The most common reaction for re-forming ATP is for ADP + P_i to be converted to ATP by the enzyme ATP synthase, which is a component of the oxidative phosphorylation and photophosphorylation reaction pathways (see Chapter 11). However, because some reactions lead to production of AMP rather than ADP, AMP must first be phosphorylated to generate the ADP needed for ATP synthesis. This reaction is catalyzed by the enzyme adenylate kinase, which transfers the γ-phosphoryl group from ATP to AMP, generating 2 ADP. The 2 ADP generated by adenylate kinase can then be phosphorylated to 2 ATP by ATP synthase, giving the net reaction of AMP + 2 $P_i \rightarrow$ ATP:

$$\begin{array}{rcl} \text{AMP} + \text{ATP} & \xrightleftharpoons{\text{Adenylate kinase}} & 2\ \text{ADP} \\ 2\ \text{ADP} + 2\ P_i & \xrightleftharpoons{\text{ATP synthase}} & \text{ATP} + \text{ATP} \\ \hline \text{Net reaction: AMP} + 2\ P_i & \rightleftharpoons & \text{ATP} \end{array}$$

Structure–function analysis of the adenylate kinase enzyme has shown that the protein undergoes a large conformational change upon AMP and ATP binding. This conformational change, analogous to the closing of the lid on a trash can, results in both the trapping of the substrates and the exclusion of water, thereby preventing a

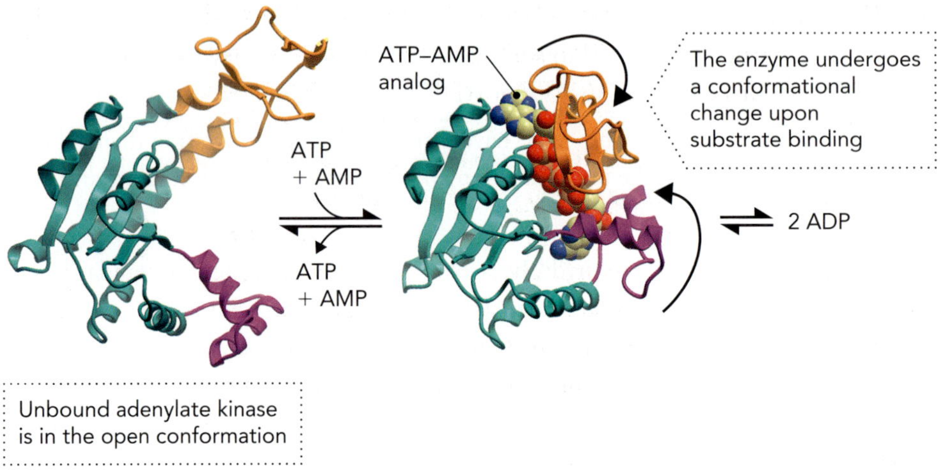

Figure 2.10 Adenylate kinase is a highly conserved enzyme that plays a central role in maintaining ATP levels in the cell. Molecular structure of the *Escherichia coli* adenylate kinase enzyme is shown in both the unbound (open) and substrate-bound (closed) forms. The substrate is an ATP–AMP analog called Ap_5A [P1,P5-di(adenosine-5′)-pentaphosphate]. BASED ON PDB FILES 4AKE (OPEN) AND 1AKE (CLOSED).

wasteful ATP hydrolysis reaction (**Figure 2.10**). The importance of adenylate kinase in energy conversion can be seen in the fact that multiple **isozymes** (functionally related enzymes encoded by different genes) and **isoforms** (functionally distinct proteins transcribed from the same gene) of adenylate kinase have been identified in a variety of organisms.

Because ATP is the high-energy form of the adenylate system, we can use the ratio of the concentration of ATP to the concentration of ADP and AMP in the cell at any given time as a measure of the energy state of the cell. This relationship is expressed in terms of the **energy charge (EC)**, which takes into account the number of phosphoanhydride bonds available for work:

$$\text{EC} = \frac{[\text{ATP}] + 0.5[\text{ADP}]}{[\text{ATP}] + [\text{ADP}] + [\text{AMP}]} \tag{2.15}$$

If the adenylate system components are present in the cell at the same concentration, so that [ATP] = [ADP] = [AMP], then EC = 0.5. However, most cells are found to have an EC in the range 0.7–0.9, which means that the concentration of ATP is higher than that of ADP or AMP (**Figure 2.11**). For example, under steady-state conditions, the EC value in rat hepatocytes is 0.8 based on the adenine nucleotide concentrations [ATP] = 3.4 mM, [ADP] = 1.3 mM, and [AMP] = 0.3 mM:

$$\text{EC} = \frac{[\text{ATP}] + 0.5[\text{ADP}]}{[\text{ATP}] + [\text{ADP}] + [\text{AMP}]}$$

$$= \frac{3.4 \text{ mM} + 0.5(1.3 \text{ mM})}{3.4 \text{ mM} + 1.3 \text{ mM} + 0.3 \text{ mM}} = 0.8$$

Figure 2.11 ATP, ADP, and AMP concentrations vary as a function of energy charge. Under physiologic conditions, the EC ranges from approximately 0.7 to 0.9. When EC is near 0.7, ATP levels are relatively low, and ADP levels are near maximum. In contrast, when EC is at 0.9, ATP levels are near maximum, and AMP levels are very low. (Note: The curves were constructed assuming the adenylate kinase reaction is at equilibrium and that K_{eq} is 1.6.)

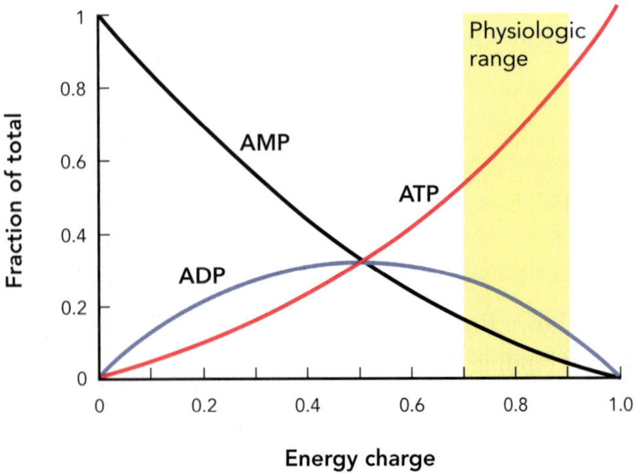

A cell maintains an EC between 0.7 and 0.9 by regulating metabolic flux (the flow of metabolites through metabolic pathways in living cells) via pathways that generate and consume ATP. Photosynthetic autotrophs use sunlight as their source of energy for ATP production. Heterotrophs use nutrients present in their diet as a source of metabolic fuel—in the form of carbohydrates, proteins, and lipids—to synthesize ATP. Most organisms use stored metabolic fuels as a source of energy when other forms of energy are not readily available. Extracting energy from metabolic fuels is the function of **catabolic pathways**, which convert energy-rich compounds into energy-depleted compounds. In the process, catabolic pathways generate ATP and reduced coenzymes (NADH, NADPH, and $FADH_2$; **Figure 2.12**). These high-potential-energy compounds are used for the biosynthesis of biomolecules through **anabolic pathways.**

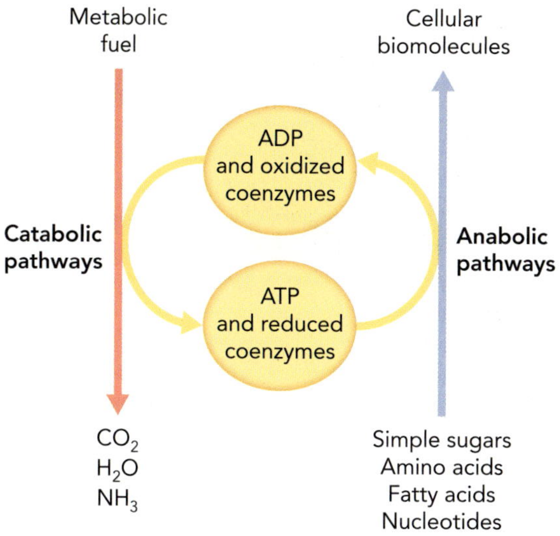

Figure 2.12 A balanced flux through catabolic and anabolic pathways provides chemical energy in the form of ATP and reduced coenzymes to maintain the steady state (homeostasis).

It is important to note that regulatory processes control the activity of key enzymes in catabolic and anabolic pathways in order to stabilize the EC and maintain homeostasis. For example, when EC levels decrease because of sustained flux through anabolic pathways, then enzymes responsible for ATP synthesis become activated, and the flux through catabolic pathways increases (**Figure 2.13**). In most organisms, this means degrading metabolic fuel stored in the form of carbohydrates or lipids. Alternatively, when EC levels are elevated because of photosynthesis or high levels of nutrients after a meal, then enzymes that control flux through anabolic pathways are activated to take advantage of the available ATP and replenish stored metabolic fuel. As described in later chapters, the two primary mechanisms of enzyme regulation in the context of metabolic control are (1) bioavailability (compartmentation within the cell and altered rates of protein synthesis and degradation) and (2) control of catalytic efficiency through protein modification (covalent modifications and noncovalent binding of regulatory molecules).

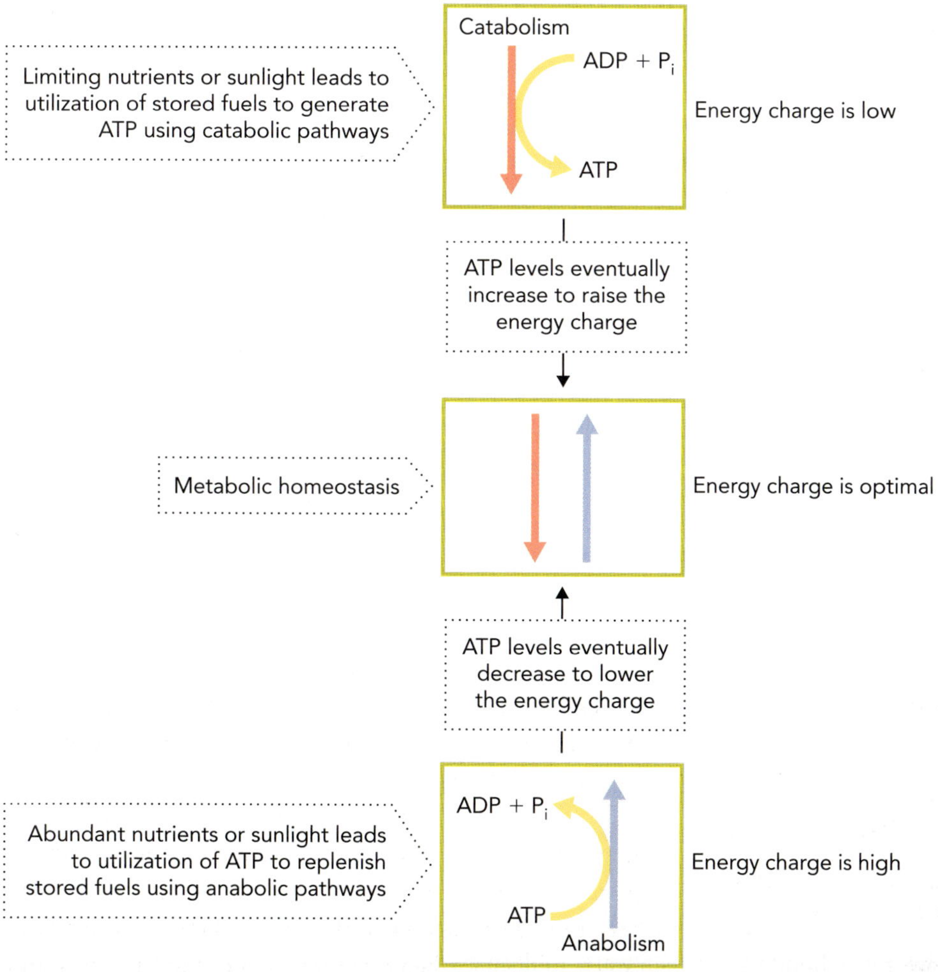

Figure 2.13 Energy charge in the cell is maintained at homeostatic levels by regulating enzymes that control flux through catabolic and anabolic pathways. Under conditions of low nutrients or low light (photosynthetic organisms), enzymes in catabolic pathways are activated to degrade stored metabolic fuels and generate ATP. This increases the EC as ATP accumulates. In contrast, when nutrients or light are abundant, then ATP is used to replenish supplies of stored metabolic fuel, and the EC decreases.

concept integration 2.1

Explain in thermodynamic terms how solar energy sustains life on Earth.

The first law of thermodynamics states that energy is neither created nor destroyed, only converted from one form to another. This principle applies to thermonuclear fusion reactions that take place in the Sun and release electromagnetic energy that travels to Earth, where it is converted to chemical energy by the processes of photosynthesis and carbon fixation. Additional energy conversion processes in living cells transform this chemical energy into useful work in the forms of osmotic work, chemical work, and mechanical work, all of which are needed to prevent cells from reaching equilibrium with the environment. This is necessary because the second law of thermodynamics states that entropy in the universe (dispersal of energy) is always increasing, and therefore without the input of energy to restrain entropy, the highly ordered structures of organisms at the molecular, cellular, and genetic levels will fail and the organism will die. Photosynthetic autotrophs obtain the energy they need to restrain entropy and avoid environmental equilibrium directly from the Sun, whereas heterotrophs depend on photosynthetic autotrophs (and other heterotrophs) to obtain the chemical energy they need for survival.

2.2 Water Is Critical for Life Processes

We have seen that solar energy is the ultimate source of energy for life on Earth. However, life on Earth would not be possible without water. Life as we know it depends on water because of its distinctive chemical properties and its central role in biochemical reactions. Indeed, more than 70% of the mass of most cells is water, and microenvironments in which water is plentiful have the highest abundance and diversity of plant and animal life. Where there is water, there is life.

Water is a simple molecule consisting of one oxygen atom and two hydrogen atoms (H_2O), but it has some distinctive properties that make it uniquely essential for life as we know it. Three unusual physical and chemical properties make water especially important in sustaining life processes:

1. Water is less dense as a solid than as a liquid, which is why ice floats. If ice were to sink when it froze, the sinking of ice in the oceans would result in an upwelling of cold water from the ocean floor, which would also sink when frozen, continuing the cycle until the oceans had completely frozen over.
2. Water is liquid over a wide range of temperatures, particularly the temperatures that are largely found on the surface of Earth. This property of water is critical to aquatic life and ultimately the oxygen content of our atmosphere, which depends on photosynthetic algae in the oceans.
3. Water is an excellent solvent because of its hydrogen-bonding capabilities and polar nature.

To appreciate the role of water in biological systems, we take a closer look at hydrogen bonding between water molecules, and then examine osmolarity as it relates

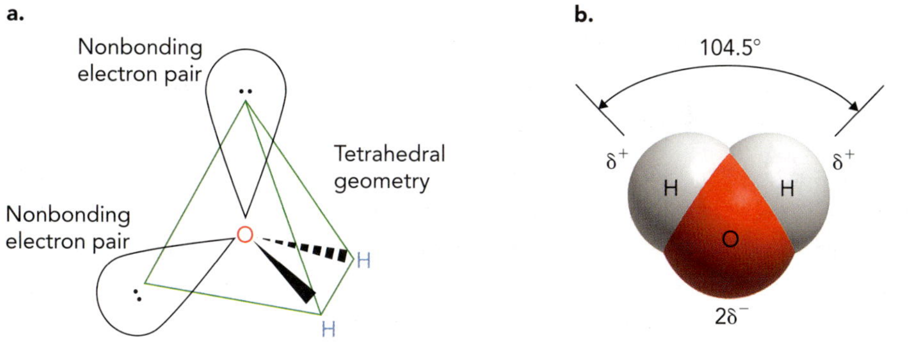

Figure 2.14 Water contains an oxygen atom and two hydrogen atoms. **a.** The four sp^3 orbitals of the oxygen atom contain two electrons that form covalent bonds with hydrogen atoms and two pairs of unshared electrons. **b.** A space-filling model of water, showing the van der Waals radii of the oxygen and hydrogen atoms, the partial charges on each atom, and the angular separation between the two hydrogen atoms (104.5°).

to cell structure and function. We end this section with a discussion of the ionization properties of water, including a brief review of the pH scale, acid–base chemistry, and buffers.

Hydrogen Bonding Is Responsible for the Unique Properties of Water

The molecular structure of H_2O explains many of its unusual properties. Each hydrogen atom in H_2O shares an electron pair with oxygen, with a bond angle of 104.5° between the two hydrogens (**Figure 2.14**). This bond angle indicates that the four sp^3 hybrid orbitals of oxygen are positioned nearly at the angles of a tetrahedron. Because oxygen is more electronegative than hydrogen, this results in a partial negative charge on the oxygen ($2\delta^-$) and a partial positive charge on each hydrogen ($1\delta^+$ and $1\delta^+$). The charge distribution of H_2O makes it a polar molecule and also permits the formation of four hydrogen bonds.

Water can participate in noncovalent interactions with another water molecule through a type of bond called a **hydrogen bond** (**Figure 2.15a**). A hydrogen bond can form when a hydrogen that is covalently attached to an electronegative atom, such as oxygen, is in proximity to another electronegative atom, such that the hydrogen is "shared" between the two electronegative atoms. Hydrogen bonds, which are described in more detail later in this chapter, are a relatively weak interaction compared to covalent bonds but are important because of their ubiquity in biological systems, in particular

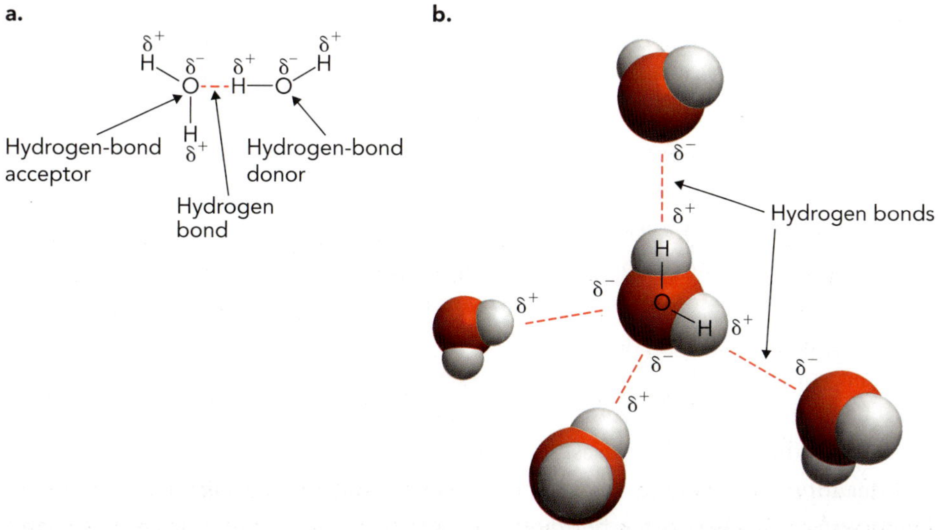

Figure 2.15 The electronic polarity of water enables it to function as both a hydrogen-bond donor and a hydrogen-bond acceptor. **a.** The water molecule on the right is serving as the donor for the hydrogen bond, and the molecule on the left is serving as the acceptor. **b.** The four surrounding H_2O molecules are positioned at the vertices of a tetrahedron with a center-to-center distance of 2.84 Å.

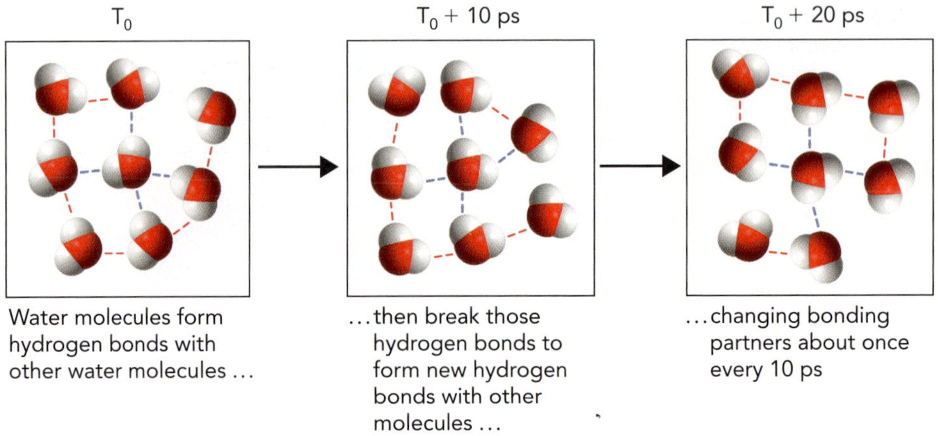

T_0
$T_0 + 10$ ps
$T_0 + 20$ ps
Water molecules form hydrogen bonds with other water molecules ...
...then break those hydrogen bonds to form new hydrogen bonds with other molecules ...
...changing bonding partners about once every 10 ps

Figure 2.16 Hydrogen bonding between H_2O molecules in water is transient. In liquid water, H_2O molecules form and break hydrogen bonds approximately every 10 ps. On average, one H_2O molecule is hydrogen bonded to ~3.4 other H_2O molecules at a time.

in water. Each water molecule can donate two hydrogens for hydrogen bonds and can accept two hydrogens for hydrogen bonds through the lone pairs of electrons on oxygen, thus forming up to four total simultaneous hydrogen bonds (**Figure 2.15b**). Four hydrogen-bonded H_2O molecules form a tetrahedron, with a center-to-center distance between molecules of 2.84 Å (0.284 nm). Although the bond strength of a single hydrogen bond is relatively weak (~20 kJ/mol), the four hydrogen bonds together lead to the high viscosity, boiling point, and melting point of water compared to those of other molecules of similar molecular mass. For example, the boiling point of H_2O is 100 °C, whereas the boiling point of ammonia (NH_3) is −33.5 °C.

Although most H_2O molecules in liquid water are connected through hydrogen bonds at any given time, the lifetime of these hydrogen bonds is extremely short because hydrogen bonds are relatively weak. Indeed, about once every 1–10 picoseconds (ps), an H_2O molecule forms a hydrogen bond with another H_2O molecule, breaks that bond, rotates, and forms a new hydrogen bond with a different H_2O molecule (**Figure 2.16**). This constant formation and breakage of hydrogen bonds between water molecules has been called "flickering clusters."

A related phenomenon resulting from the hydrogen bonding of water molecules in an electric field is that of **proton hopping** (**Figure 2.17**). In this case, hydrogen bonding holds H_2O molecules in close proximity to one another, such that in an electric field, the loss of a proton (H^+) from a hydronium ion (H_3O^+) initiates a series of hydrogen bond "trades" between adjacent H_2O molecules. As a result, the H^+ ion seems to move along a "water wire" to form a hydronium ion at the other end. (As described later in this chapter, H_3O^+ ions form in solution between a free proton and water, yielding $H^+ + H_2O \rightleftharpoons H_3O^+$.) Proton hopping is a very fast process, as it relies on the formation and breakage of hydrogen bonds, not the actual movement of an ion over long distances. This phenomenon was originally proposed to explain the observed high rate of movement of H^+ ions toward the anode (negatively charged electrode) in an electric field compared to the diffusion rate of Na^+ ions.

The unusual geometry of hydrogen bonds between H_2O molecules in an ice crystal is the reason why ice floats. The density of ice (0.92 g/mL) is less than the density of liquid water (1.0 g/mL), primarily because H_2O molecules in ice crystals have the maximum number of four hydrogen bonds between molecules, thus creating a regular tetrahedral open-lattice structure. The flotation of ice is critical to aquatic life because ice melts as the temperature rises, resulting in only a seasonal variation in the quality and quantity of the aquatic environment (**Figure 2.18**). If ice were more dense than liquid water and sank to the bottom of a lake or ocean, it would be insulated from

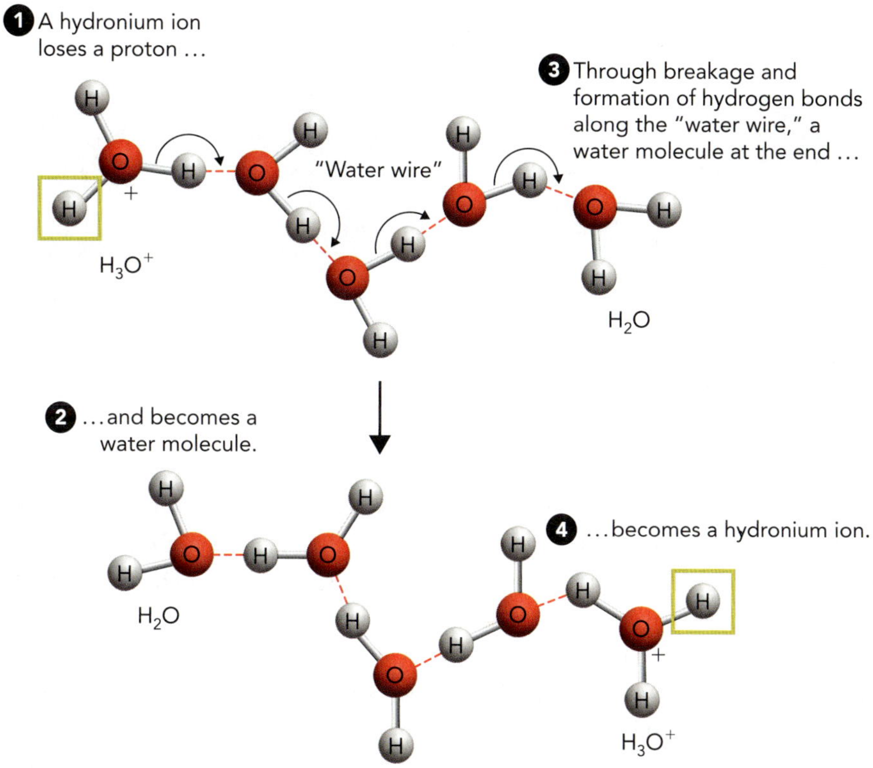

1 A hydronium ion loses a proton ...
H_3O^+
"Water wire"
3 Through breakage and formation of hydrogen bonds along the "water wire," a water molecule at the end ...
H_2O
2 ...and becomes a water molecule.
H_2O
4 ...becomes a hydronium ion.
H_3O^+

Figure 2.17 A "water wire" forms when hydrogen-bonded H_2O molecules in an electric field pass H^+ ions from one H_2O molecule to the next through a process called proton hopping. A hydronium ion (H_3O^+) readily forms when H_2O accepts an H^+.

temperature changes on the surface and not undergo a seasonal thaw. If this were to happen, most of the aquatic environment would eventually be encased in solid ice.

Organisms that cannot maintain a constant body temperature above 0 °C risk damage due to the formation of ice crystals inside cells, where the rigid structure of ice would destroy cellular integrity. Many organisms that need to survive in subfreezing conditions contain special antifreeze proteins that prevent the formation of ice crystals. These antifreeze proteins are found in high abundance during cold weather in certain types of fish, such as winter flounder, and in the hemolymph of grain beetle larvae (yellow mealworm). Antifreeze proteins interact with H_2O at the edge of large ice crystals, thus preventing the ice crystals from growing larger. These proteins often use threonine residues on the surface of the protein to make hydrogen bonds with H_2O and thereby counteract the addition of more H_2O molecules to the crystal. As described in the chapter opener, the antifreeze protein of the yellow mealworm contains regularly spaced threonine residues that surround the water molecules on all sides. Interestingly, the structure of antifreeze proteins from a variety of organisms indicates that this ability to survive freezing temperatures evolved independently through a mechanism called convergent evolution (different solutions to the same problem). For example, as shown in **Figure 2.19**, although the mealworm antifreeze protein shows a compact arrangement of threonine residues on the surface, the flounder (a fish) antifreeze protein is a single α helix with regularly spaced threonine residues along one face of the helix. These two protein structures are different from those of the antifreeze proteins from the spruce budworm and the ocean pout, which also contain threonine residues on the surface. Studies have shown that large amounts of antifreeze proteins in the cell can reduce the freezing point of water up to 6 °C in some cases.

Figure 2.18 Ice floats because the crystal structure of water is an open lattice, with each H_2O molecule hydrogen-bonded to four other molecules. Krill are a vital food source for marine animals and survive the Arctic and Antarctic winters because ice floats, but importantly, this ice melts each summer and does not accumulate. ARDEA/FERRERO, JOHN PAUL/ANIMALS ANIMALS.

Krill under a layer of ice

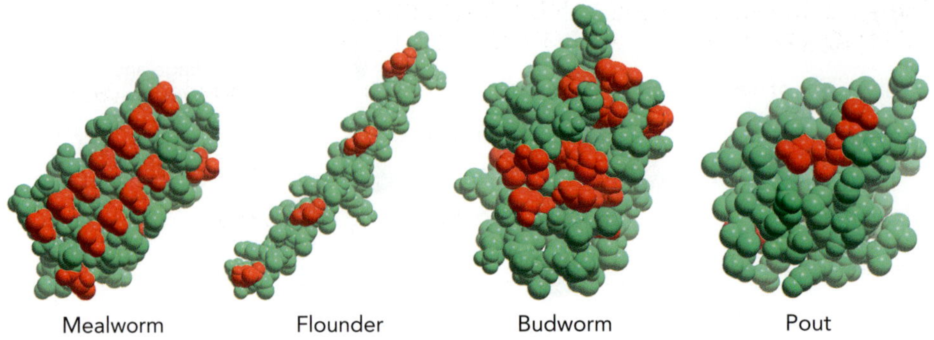

Figure 2.19 Insect antifreeze proteins prevent the formation of ice crystals in the hemolymph by disrupting hydrogen bonding between H_2O molecules. The structures of antifreeze proteins from four different organisms show that although they all contain numerous threonine residues on the protein surface (shown in red), the overall structure of the proteins is quite different. This represents a classic example of convergent evolution. BASED ON PDB FILES 1EZG (MEALWORM), 1WFB (FLOUNDER), 1EWW (BUDWORM), AND 1KDF (POUT).

Because of the aqueous environment of cells, water plays an important role in the solubility of biomolecules. Solubility is defined as the ability of a solute to dissolve to homogeneity in a solvent such as water. For example, NaCl crystals dissolve in water because the Na^+ and Cl^- ions form weak ionic interactions with H_2O molecules, which effectively compete with ionic interactions within the crystalline salt (**Figure 2.20**). The orientation of the charged ions with the polar water molecules results in the formation of a hydration layer around each ion, thus preventing the ions from rejoining the crystal. Solubility of NaCl in water is energetically favorable because of the combined large increase in entropy (Na^+ and Cl^- are less ordered in solution) and small change in enthalpy (strong NaCl ionic bonds are replaced by many weak ionic interactions with H_2O), resulting in an overall decrease in the free energy of the system ($\Delta G < 0$).

Weak Noncovalent Interactions in Biomolecules Are Required for Life

Covalent bonds are required to hold together the chemical structures of biomolecules. However, biochemical reactions—and indeed, life processes—depend on weak interactions characterized by noncovalent bonds. These weak interactions are responsible for large-scale intramolecular and intermolecular structures. For example, the three-dimensional structures of proteins and nucleic acids depend on multiple intramolecular weak interactions between amino acids and nucleotides, respectively, in polymeric molecules. Intermolecular weak interactions are crucial for many enzyme–substrate interactions; hormones binding to hormone receptors; and stability of the DNA double helix.

The importance of noncovalent bonds in nature is that they permit unstable structures to exist for short periods of time, during which biochemical reactions can take place. The four basic types of weak noncovalent interactions encountered in biochemistry are (1) *hydrogen bonds*, (2) *ionic interactions*, (3) *van der Waals interactions*, and (4) *hydrophobic effects*.

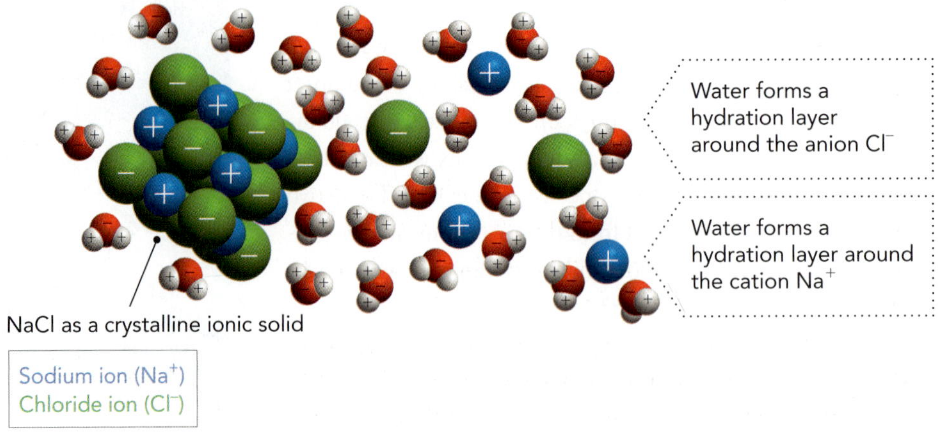

Figure 2.20 NaCl crystals dissolve in water because H_2O molecules effectively compete with ionic bonds between Na^+ and Cl^- to form new ionic interactions. Ionic interactions occur between the partial charges on hydrogen (δ^+) and oxygen (δ^-) in water and the charges on the Cl^- and Na^+ ions, respectively.

Table 2.2 BOND ENERGIES AND BOND LENGTHS IN COMMON NONCOVALENT BONDS IN NATURE

Type of bond	Bond energy (kJ/mol)	Bond length (Å)
Hydrogen bonds	10–30	~2.5–3.0 (between nonhydrogen atoms)
Ionic interactions	20–80	~2.0–2.5
van der Waals interactions	1–10	~1.2–1.6

Hydrogen Bonds Water is not the only molecule that forms hydrogen bonds. In general, hydrogen bonds can form between a hydrogen atom on an electronegative donor group and another electronegative atom that serves as a hydrogen-bond acceptor. Essentially all hydrogen-bond donors are hydrogens bonded to O or N because these are the only situations that result in a partially positive charge on the hydrogen strong enough to interact with the negative charge of an acceptor atom (usually O, N, or S). Note that because a C—H bond is not polar under most circumstances, hydrogen bonds do not generally form with C—H groups. The strength of a hydrogen bond depends on the angle and the distance between the atoms. The length of a hydrogen bond (between the two nonhydrogen atoms) is about twice the length of a covalent bond, which is reflected in the reduced strength of this noncovalent bond (**Table 2.2**). **Figure 2.21** shows some of the most common hydrogen-bond donors and acceptors in biomolecules.

Ionic Interactions Weak interactions between oppositely charged atoms or groups are called **ionic interactions**, which are a type of electrostatic interaction. For example, the attraction between a positively charged amino group NH_3^+ and a negatively charged carboxylate group COO^- in a protein is an ionic interaction. Moreover, many macromolecules in the cell contain a net charge (either positive or negative), as do small ions such as Na^+, K^+, Cl^-, and HPO_4^{2-}.

The strength of an ionic interaction depends on the environment of the ions and on the distance between them. Unlike hydrogen bonds, the angle of the ionic interaction does not affect the strength of the interaction. Electrostatic interactions are strongest within hydrophobic environments, such as a hydrophobic pocket on a protein, where water molecules cannot shield the charges and thus weaken the interaction. Sometimes ionic interactions in proteins are called salt bridges, although the strengths of these interactions are typically much less than those found in NaCl crystals, for example. Repulsive forces between two like-charged particles are another type of "interaction" (though in this case it is actually repulsion) that also contributes to the overall structure of biomolecules.

van der Waals Interactions A third type of weak interaction occurs between the dipoles of nearby electrically neutral molecules, which are named **van der Waals interactions** after the Dutch physicist who first described this phenomenon. Polar molecules have dipole moments, resulting from unequal sharing of electrons between atoms. However, even nonpolar molecules or atoms have temporary dipole

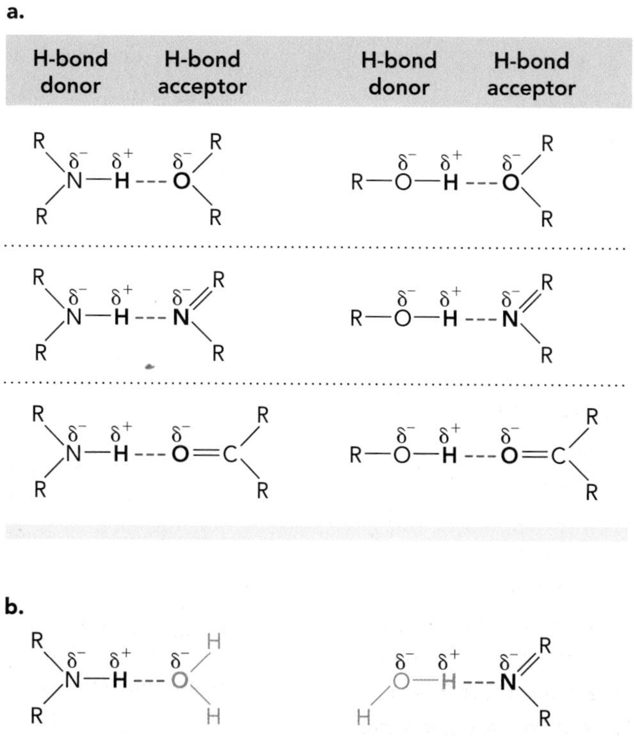

Figure 2.21 In biomolecules, hydrogen bonds most often form between oxygen- or nitrogen-containing molecules. The bond involves a partial positive charge (δ^+) in the donor hydrogen and a partial negative charge (δ^-) in the acceptor atom. **a.** Representative hydrogen bonds involving O and N atoms in biomolecules. **b.** Water molecules (blue) can serve as either donors or acceptors for hydrogen bonds with other molecules, as shown here with an amide.

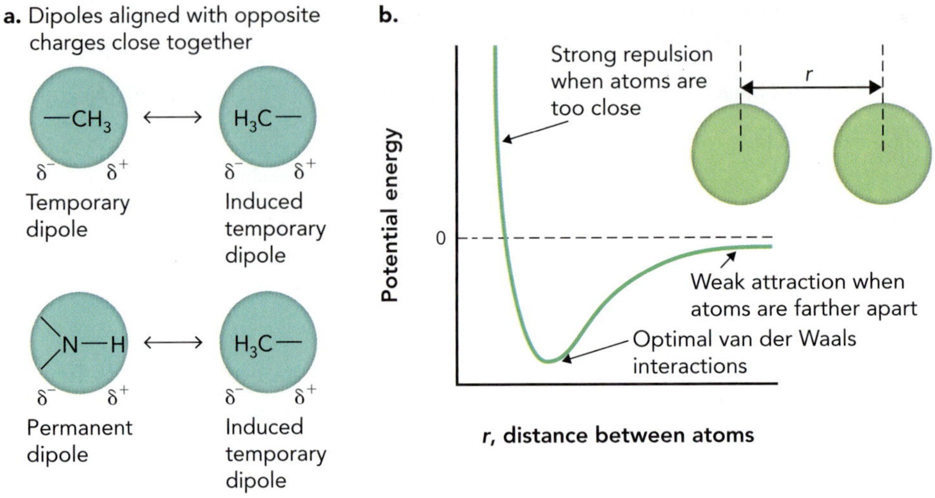

Figure 2.22 van der Waals interactions occur between electrically neutral molecules. **a.** Motions of electrons in nonpolar molecules can result in temporary dipoles, as shown for a methyl group (upper left). Polar molecules, such as amino groups (lower left), have permanent dipoles. Either of these dipoles can induce a dipole in a neighboring molecule, as shown for the methyl groups on the right, or interact with a permanent dipole from a polar molecule (not shown). Nearby dipoles that are aligned with opposite charges close together will attract. **b.** The potential energy of a van der Waals interaction depends strongly on distance. If two atoms are too close, they experience repulsion. If they are too far away, they interact only weakly. The optimal distance for van der Waals interactions, resulting in the lowest potential energy, occurs at a distance slightly greater than the length of the covalent bond that would form between the two atoms.

moments because of fluctuations in electron clouds. If dipoles are momentarily aligned with opposite signs at an appropriate close distance, they can result in a van der Waals interaction (**Figure 2.22a**). Though van der Waals forces between two atoms are much weaker (~5 kJ/mol) than a hydrogen bond (~20 kJ/mol; Table 2.2), they are significant in biology because many van der Waals interactions can occur simultaneously with a variety of atoms and thus can have great cumulative strength.

van der Waals interactions depend strongly on the distance between the two atoms (**Figure 2.22b**). Repulsion occurs between atoms that are brought too close to each other because their electron orbitals cannot occupy the same space. The van der Waals interactions are most favorable when the atoms are at a distance slightly greater than when they are covalently bonded. When the distance between the atoms increases further, the van der Waals interaction energy decreases.

Each atom has a characteristic **van der Waals radius,** which can be used to calculate the approximate volume occupied by an atom and to approximate when atoms are within van der Waals contact distance. The experimentally determined van der Waals radii of the six common elements found in biomolecules range from 1.2 to 1.8 Å (**Figure 2.23a**). Computer algorithms use these calculated van der Waals radii to graphically illustrate the space-filling packing of atoms in a three-dimensional structure (**Figure 2.23b**). Note that the van der Waals radii of covalent bonds are depicted as overlapping spheres.

Hydrophobic Effects A fourth type of weak "interaction" is due to the tendency of hydrophobic molecules to pack close together away from water. Fully reduced hydrocarbon molecules are nonionic and nonpolar, so they cannot form hydrogen bonds with

a.

Element	van der Waals radius (Å)
Hydrogen	1.20
Oxygen	1.52
Nitrogen	1.55
Carbon	1.70
Sulfur	1.80
Phosphorus	1.80

b.

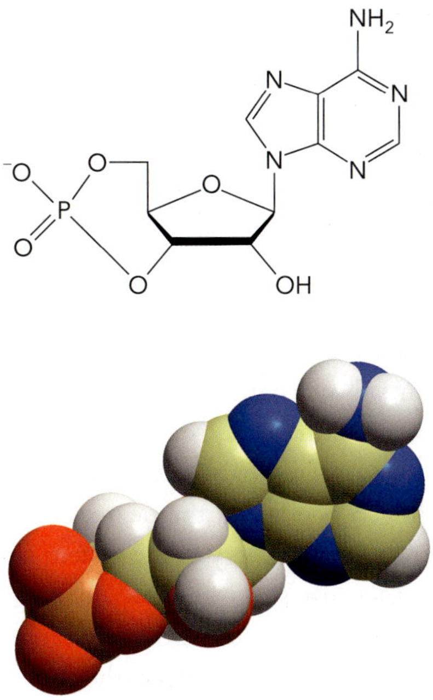

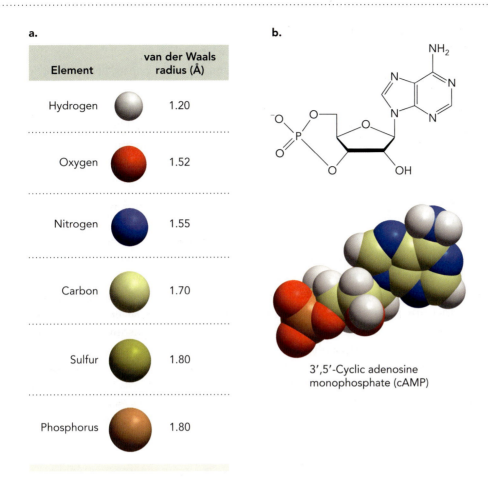

Figure 2.23 van der Waals radii are used to calculate the approximate volumes occupied by each atom. **a.** Calculated van der Waals radii (in angstroms) of common elements found in biomolecules. The colors shown for each element are a modified Corey–Pauling–Koltun (CPK) scheme used throughout the book. **b.** Structural model of cyclic AMP (3′,5′-cyclic adenosine monophosphate) based on van der Waals radii for each atom. The chemical structure of cyclic AMP is shown at top and its space-filling model is shown at bottom. The element colors are the same as in panel **a**.

water. Such molecules are referred to as **hydrophobic**, or "water fearing" (in Greek, *hydros* means "water" and *phobos* means "fear"). In contrast, the "water loving" property of polar substances is described by the term **hydrophilic** (*philia* means "friendship" in Greek). Because hydrophobic molecules cannot form hydrogen bonds with water, the water that surrounds a hydrophobic region becomes more ordered to satisfy its hydrogen-bonding potential through interactions with other water molecules, forming cage-like structures (**Figure 2.24**). This is energetically unfavorable, largely because of the decrease in entropy of the water molecules through the restriction of their motion. When hydrophobic regions cluster together, they have less exposed surface area, and as

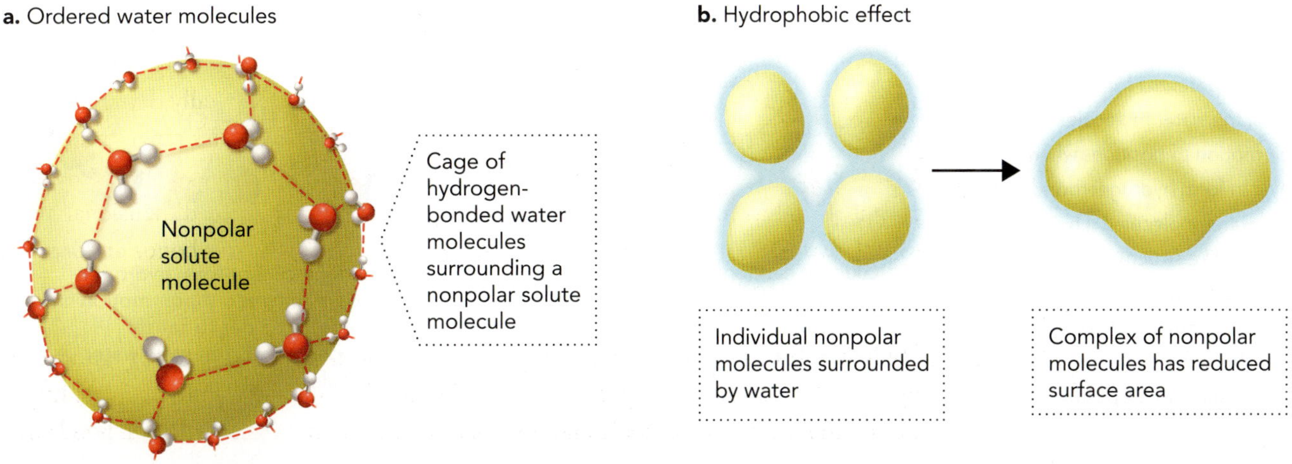

Figure 2.24 The formation of cage-like structures of H_2O molecules around uncharged nonpolar complexes in solution results in decreased entropy, which can be partially offset by the hydrophobic effect. **a.** Reduced degrees of freedom for H_2O molecules near the surface of a nonpolar molecule result in a decrease in entropy. **b.** It is energetically favorable to reduce the surface area of nonpolar molecules exposed to H_2O molecules, thereby restricting the motion of fewer water molecules. This hydrophobic effect results in the formation of complexes of nonpolar molecules in aqueous solutions.

a result, fewer water molecules need to be ordered. This is a more energetically favorable state, so hydrophobic regions do tend to pack together.

Weak hydrophobic effects are not the same as other noncovalent interactions because they are the result of avoiding water rather than a true molecular attraction. Nevertheless, hydrophobic effects play an important role in biomolecular structure and function, especially with regard to protein-folding reactions (see Chapter 4).

Biomolecules with polar chemical groups readily dissolve in water because of ionic interactions and hydrogen bonding with H_2O molecules. Consider what happens to hydrogen bonds when an uncharged polar substance such as glucose is put into water (**Figure 2.25**). Glucose can form hydrogen bonds with multiple H_2O molecules, and essentially no change in the motional energy of the water occurs when glucose is added. In terms of thermodynamics, introduction of polar substances into water has very little effect on the enthalpy (ΔH) or entropy (ΔS) of the system because the number of hydrogen bonds and the degrees of freedom are essentially the same. Therefore, the change in free energy of the system is negligible.

In contrast to glucose, uncharged nonpolar molecules such as limonene (an organic odorant molecule present in orange and lemon peels) disrupt hydrogen bonds between H_2O molecules without forming new hydrogen bonds (Figure 2.25). The water adjacent to the nonpolar molecule can only form hydrogen bonds with other

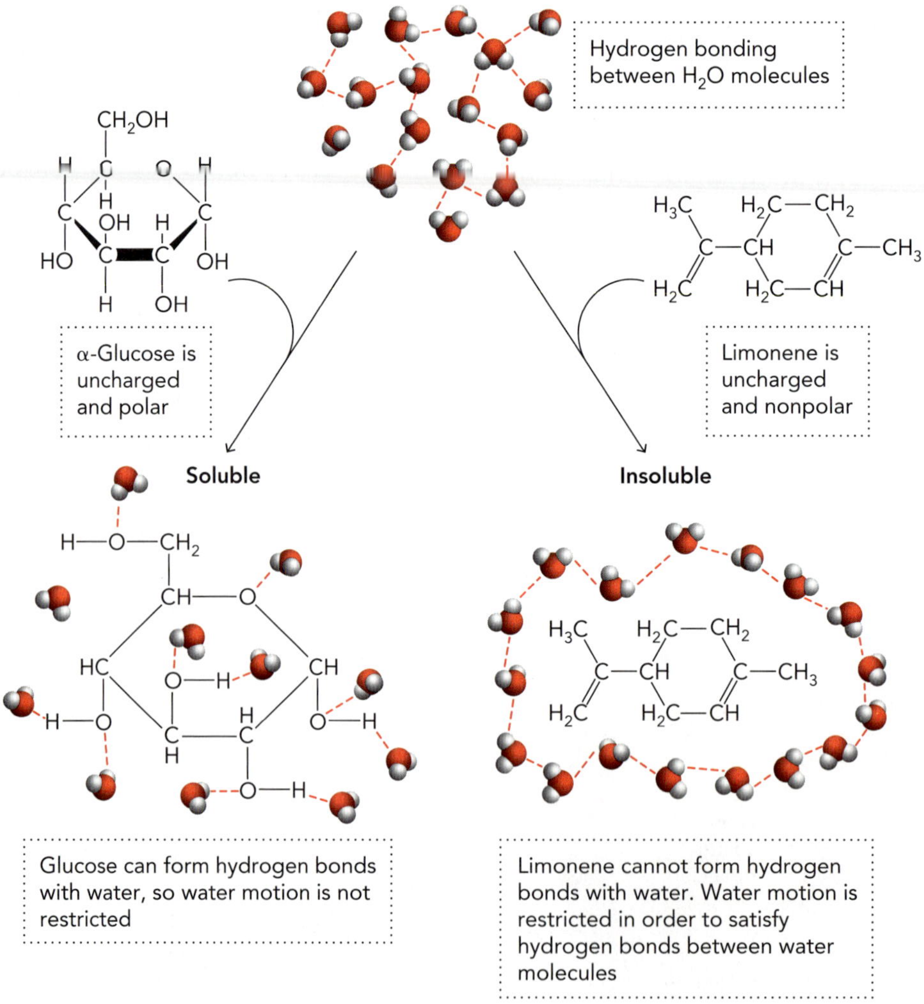

Figure 2.25 Introduction of uncharged polar and nonpolar substances into water alters the hydrogen-bonding networks among H_2O molecules. The monosaccharide α-glucose is a polar molecule that can form multiple hydrogen bonds with H_2O molecules; it therefore has little effect on ΔG because ΔH and ΔS are largely unchanged. In contrast, the nonpolar molecule limonene cannot form hydrogen bonds with water. For water to satisfy its potential for hydrogen bonds, the water molecules restrict their motion to positions where they can form hydrogen bonds with other water molecules. This results in a cage-like structure of water molecules around the limonene molecules, resulting in a decrease in ΔS (from the restriction of water motion) and a more positive ΔG.

water molecules; thus, H_2O molecules in close proximity to hydrophobic compounds form a water shell in which their motion is restricted, resulting in decreased entropy. This leads to an overall change in free energy that is unfavorable ($\Delta G > 0$). By reducing the size of the hydration layer through clustering of the nonpolar molecules, this ordering is minimized, and thus is a less unfavorable situation. You can see this happening when you vigorously shake a bottle of salad dressing containing oil and vinegar and then let it stand for a few minutes. The small oil droplets, which form as a result of mechanical mixing, continually merge into larger nonpolar hydrophobic complexes until the aqueous and oil phases are completely separated.

In biological systems, the hydrophobic portions of biomolecules are often clustered together away from water. For example, leucine and isoleucine are hydrophobic amino acids often found in the interior of folded proteins, thus reducing the surface area exposed to water (**Figure 2.26**). Hydrophobic effects between nonpolar amino acids in proteins play a major role in the proper folding of newly synthesized proteins. **Figure 2.27** shows how nonpolar amino acids collapse into the core of the folded protein, leaving polar amino acids on the surface to increase protein solubility. As nonpolar amino acids pack closer together due to hydrophobic effects, additional weak interactions between polar and nonpolar amino acids (hydrogen bonds, ionic interactions, van der Waals interactions) serve to stabilize the overall three-dimensional structure of the protein.

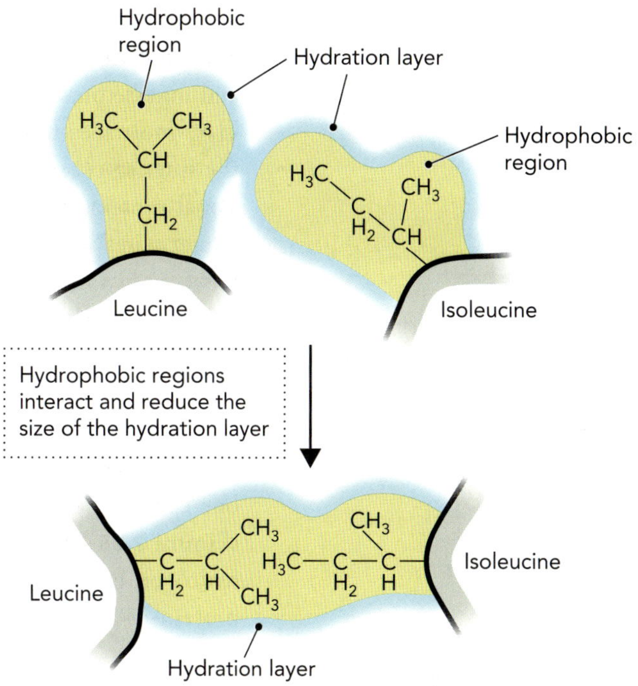

Figure 2.26 Hydrophobic effects are an important class of weak interactions that affect the structure and function of biomolecules. Leucine and isoleucine are two hydrophobic amino acids that pack together in the interior of proteins to minimize exposure to water, thus minimizing the ordered hydration layer.

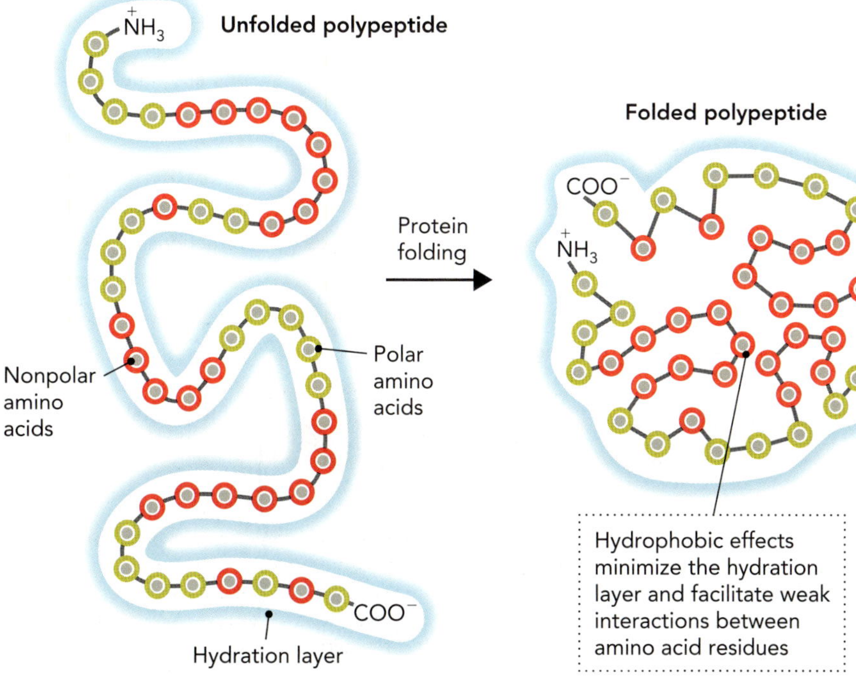

Figure 2.27 Hydrophobic effects play an important role in the folding of globular proteins. The hydration layer around the polypeptide is minimized by the hydrophobic nonpolar amino acids gathering toward the interior of the folded protein—a result of the hydrophobic effect.

Although the interiors of most soluble proteins are largely hydrophobic because of the hydrophobic effect, not all H_2O molecules are excluded from the interior of proteins and may be present to serve a particular function for the protein. For example, some H_2O molecules that are hydrogen bonded to amino acid side chains located within the interior core of proteins play an important role in stabilizing the three-dimensional protein structure. Moreover, hydrogen-bonded H_2O molecules can also form "water wires" that traverse a protein complex, as seen in Figure 2.17. Such arrangements of H_2O molecules have been found in the crystal structures of proteins involved in a process called proton pumping, which moves protons across the chloroplast and mitochondrial membranes in response to redox-driven energy conversion reactions (Chapters 11 and 12).

Collectively, the noncovalent weak interactions that take place in aqueous solution play enormous roles in the structure and function of biomolecules. As one example, multiple weak interactions occur when two proteins form a complex (**Figure 2.28**). A combination of hydrophobic effects, van der Waals interactions, hydrogen bonds, and electrostatic interactions permit the formation of a specific protein–protein interaction. Most important, because these are noncovalent interactions, the complex can quickly dissociate as a result of chemical changes in the environment or from modifications to one or both of the molecules. The formation of protein complexes through multiple weak interactions is commonly found in multi-subunit enzymes that catalyze biochemical reactions and in protein oligomers that assemble and disassemble as a function of monomer concentration or

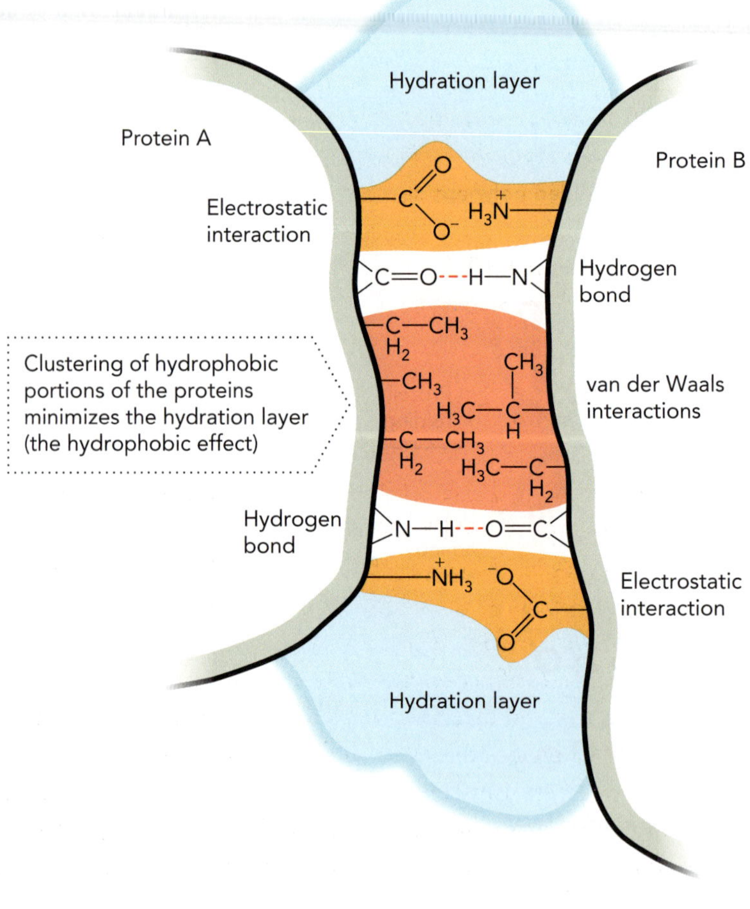

Figure 2.28 Multiple noncovalent weak interactions are involved in the formation of biomolecular complexes. van der Waals interactions occur between nearby regions, such as in the hydrophobic regions that have clustered together to minimize the ordering of water through the hydrophobic effect. Hydrogen bonds are common both within and between proteins. Electrostatic interactions are often closer to the surface.

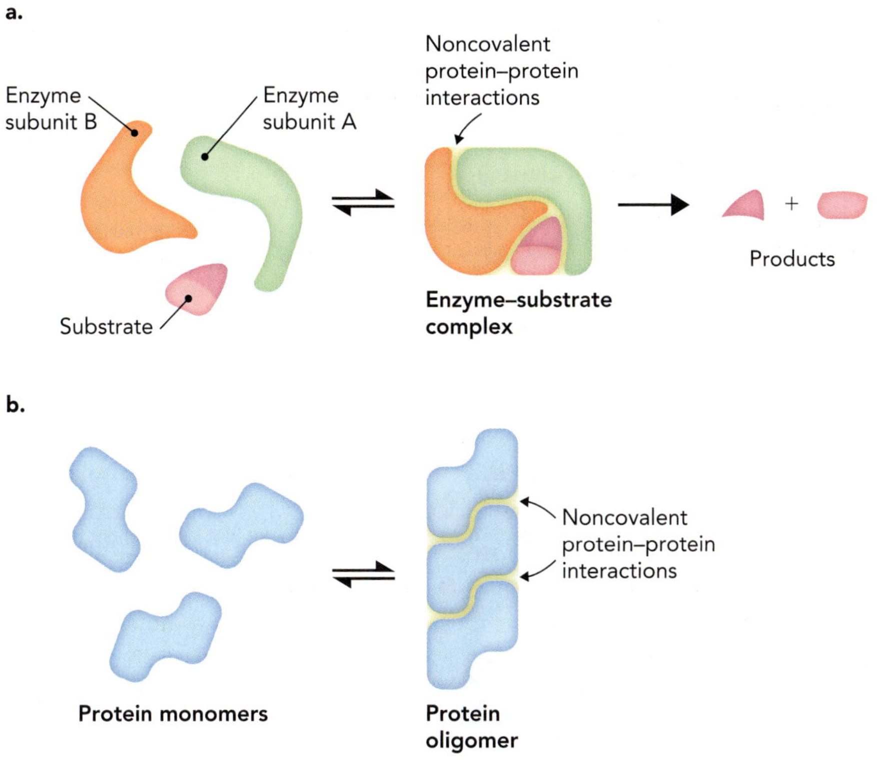

Figure 2.29 The formation of many types of protein complexes is facilitated by reversible, weak noncovalent interactions. **a.** The formation of a protein dimer (two subunits) by noncovalent interactions at the interface of the two subunits can create an enzyme that catalyzes the cleavage of a reactant into two products. **b.** The generation of protein oligomers is often the result of noncovalent interactions between monomer subunits.

modification (**Figure 2.29**). We will see numerous examples of complex formation through multiple weak interactions when we examine the structures and functions of nucleic acids and proteins later in the book.

Effects of Osmolarity on Cellular Structure and Function

Because most biochemical reactions take place in aqueous (water) solutions, it is important to study the characteristics of these solutions. The concentration of a solute in solvent has an effect on the **colligative properties** of the solution. Colligative properties refer to a collection of related properties—such as freezing point depression, boiling point elevation, vapor pressure lowering, and osmotic pressure—that depend on the number of solute particles. When comparing a 1 molal solution (1 mol of solute in 1,000 g of solvent) to pure solvent at 1 atm pressure, it can be shown that if the solvent is water, then the freezing point of the solution is lowered by 1.86 °C, the boiling point is increased by 0.54 °C, and the vapor pressure is lowered in a way that depends on temperature. The reason for these changes in physical properties of the solution is that the presence of solute molecules lowers the effective concentration of solvent molecules. This makes it more difficult to form solvent crystals and to convert solvent molecules from the liquid phase to the gas phase.

The effects of solutes on the colligative properties of a solution depend only on the number of solute particles in the solution, not the chemical properties or molecular masses of the solute particles. Therefore, a 1 molal solution of NaCl in water has almost a twofold greater effect on the colligative properties of water than does a 1 molal solution of glucose, because most of the NaCl ionizes into Na^+ and Cl^- ions, whereas glucose does not dissociate into smaller components when dissolved in water.

Because only a few organisms are able to exist under conditions where cell temperatures are near the freezing point or boiling point of water, solute effects on these two colligative properties of water are of minimal importance for biology. However,

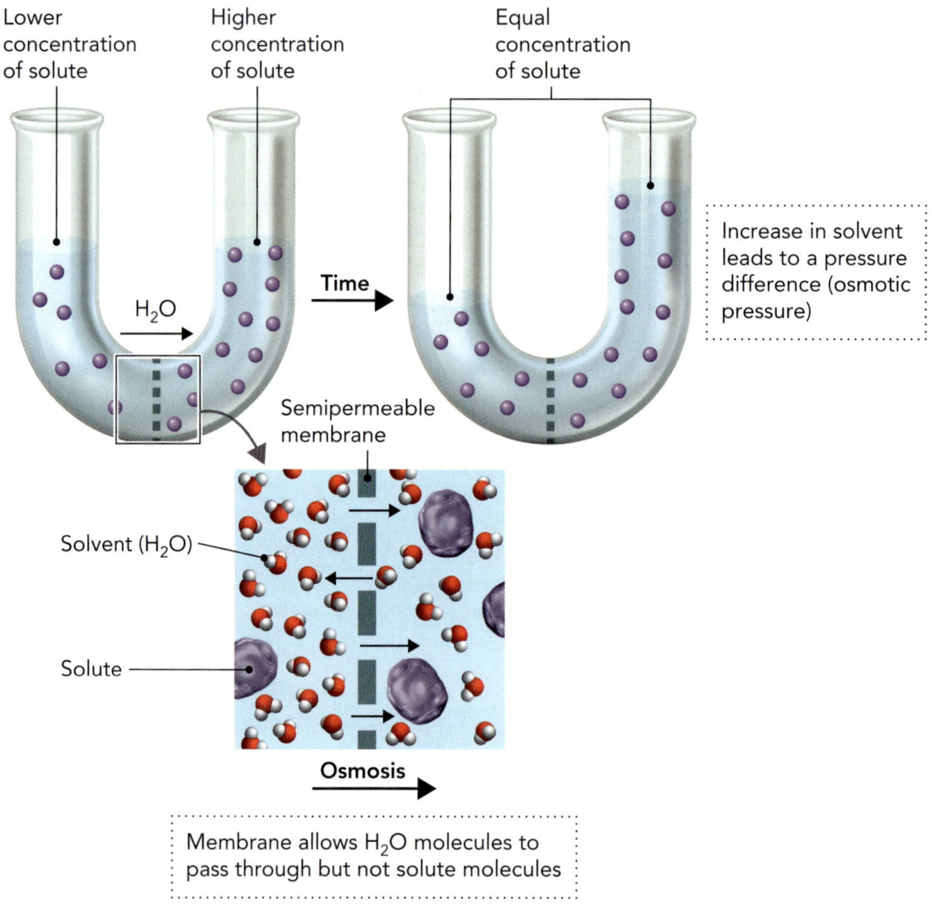

Figure 2.30 Osmosis is the diffusion of H_2O molecules from a solution of high H_2O concentration to one of low H_2O concentration. The separation of two solutions of different solute concentrations by a semipermeable membrane results in the net movement of H_2O molecules as a function of time, with H_2O molecules passing into the solution containing the higher solute concentration. Once the system has reached equilibrium, the concentration of solute (and H_2O) on both sides of the membrane is equal. In this experiment, the semipermeable membrane permits H_2O molecules to pass through, but not solute molecules.

the solute effect on **osmotic pressure** is very important for life because the concentration of biomolecules inside cells is often different from that of the surrounding environment. Osmotic pressure is a consequence of **osmosis**, which refers to the diffusion of solvent molecules from a region of lower solute concentration to one of higher solute concentration. If two solutions of different solute concentrations are separated by a semipermeable membrane—which allows H_2O molecules to pass through but not solute molecules—then H_2O molecules will diffuse by osmosis from the solution of low solute concentration (higher H_2O concentration) into the solution of high solute concentration (lower H_2O concentration; **Figure 2.30**). The net effect is that solute concentrations become equal in the two solutions (equal molal concentration of solute) as a consequence of water diffusion across the membrane.

We can determine the osmotic pressure of a solution experimentally by measuring the amount of pressure required to counteract osmosis across a semipermeable membrane (**Figure 2.31**). For example, 5.6 atm pressure is needed to counteract osmosis from a solution of pure water into a 0.25 molal solution at 25 °C. However, 22.4 atm pressure is required to counteract osmosis against a 1.0 molal solution under the same conditions, demonstrating that osmotic pressure is proportional to solute concentration. As with other colligative properties, osmotic pressure is a function of only the number of solute molecules, not their chemical or physical properties. This is important in living systems where ions, metabolites, biomolecules, and macromolecules all contribute to the solute concentration of aqueous solutions.

In living systems, the semipermeable membrane that controls osmotic pressure is the plasma membrane that surrounds cells and separates the aqueous environments of the cytosol and the extracellular milieu. If isolated **erythrocytes** (red blood cells) are placed in a **hypotonic** solution, which contains a lower solute concentration than that inside the cell (higher H_2O concentration), then osmosis causes H_2O molecules to diffuse into the cell, causing it to swell to the point of bursting (**Figure 2.32**). In contrast, if an erythrocyte is placed in a **hypertonic** solution, which contains a higher solute concentration than that inside the cell (lower H_2O concentration), then the cell shrinks because H_2O molecules diffuse out of the cell. Under normal conditions, erythrocytes in blood maintain their cell size because the solute concentration in blood is similar to that of the cytosol of the cell, making blood an **isotonic** solution relative to the cytosol (Figure 2.32). In this case, the amount of H_2O diffusing into the cell is equal to the amount of H_2O diffusing out of the cell.

Important solutes in human blood are glucose and the protein serum albumin, which functions as a carrier molecule for fatty acids. Normal glucose levels in blood are in the range of ~5 mM, and albumin, which accounts for 60% of the free protein in serum, is present in blood at about 0.5 mM. Note that animal cells within tissues, such as liver cells, are surrounded by interstitial fluid, which is isotonic relative to the cell cytosol. Interstitial fluid contains many of the same solutes as blood and bathes the surrounding tissues in nutrients, oxygen, signaling molecules, and ions.

Not all cells are surrounded by an isotonic medium and therefore must cope with the effects of osmotic pressure on cell structure and cell function. For example, protists, such as paramecia, live in freshwater that is hypotonic compared to the cytosol. Paramecia are unicellular eukaryotic organisms that contain one or more intracellular **contractile vacuoles**, which actively remove water from the cytosol to offset the inward

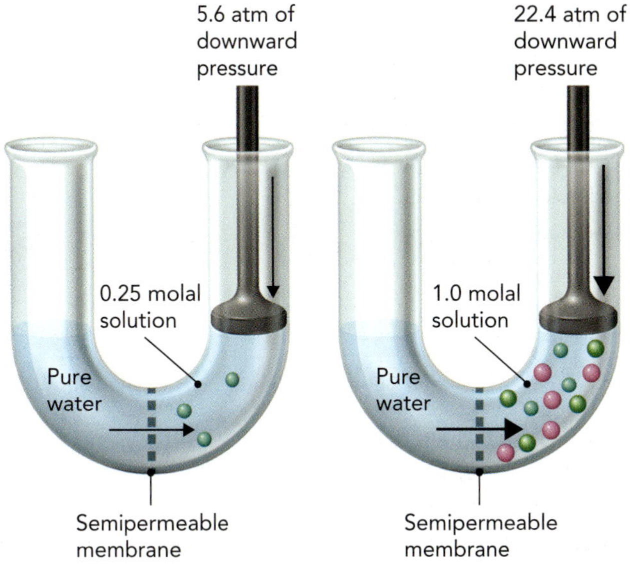

Figure 2.31 Osmotic pressure refers to the amount of pressure required to counteract the movement of H_2O across a semipermeable membrane. The only factor that affects osmotic pressure is the concentration of solute in the solution.

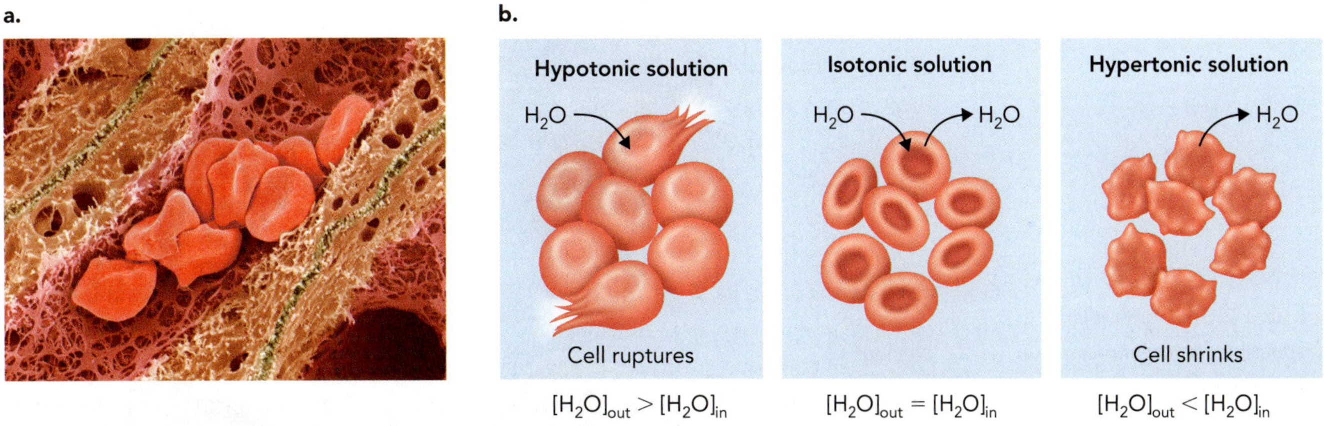

Figure 2.32 The osmolarity of a solution affects the size and shape of isolated erythrocytes. **a.** Photomicrograph of human erythrocytes moving through a capillary. SUSUMU NISHINAGA/SCIENCE SOURCE. **b.** Schematic illustration showing the movement of H_2O molecules across the erythrocyte plasma membrane as a function of osmolarity. A hypotonic solution has a lower solute concentration than that of the cytosol, and therefore a higher H_2O concentration, which leads to H_2O diffusing into the cell and causing it to lyse (rupture). In contrast, a hypertonic solution has a higher solute concentration than that of the cytosol, and H_2O diffuses out of the cell. The solute concentration in an isotonic solution is the same as that of the cytosol, and H_2O diffusion into and out of the cell is balanced.

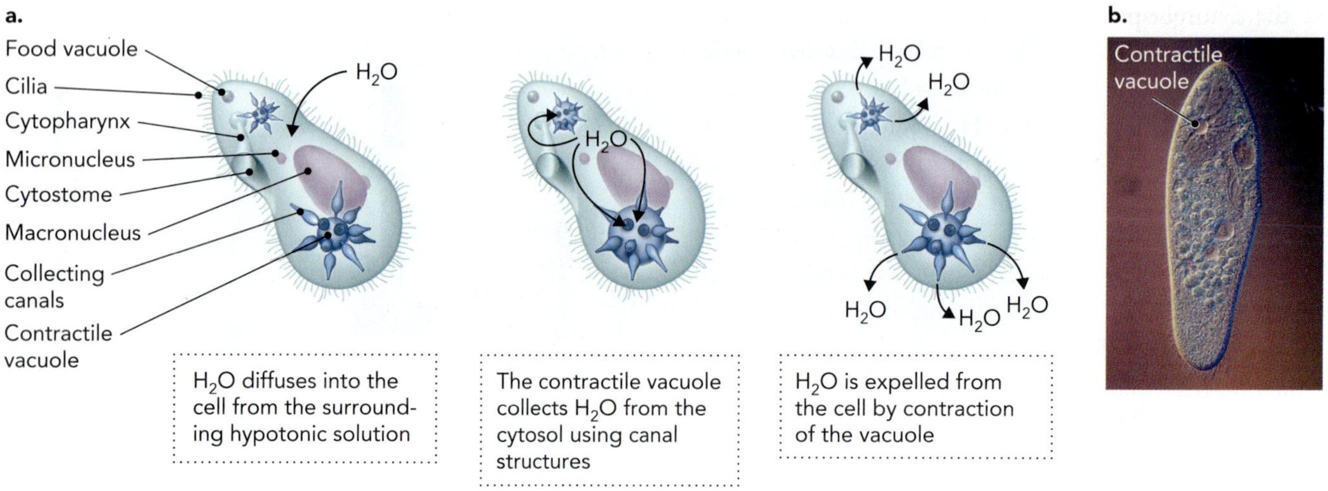

Figure 2.33 The contractile vacuoles of freshwater protists such as the paramecium collect and expel water that accumulates in the cytosol. **a.** The contractile vacuole collects H_2O molecules from the cytosol using canal structures that deliver H_2O to the vacuole body. Periodic expulsion of the H_2O from the vacuole maintains safe osmotic pressure in the cell and prevents cell lysis. **b.** Photomicrograph of a paramecium showing a contractile vacuole in the cytosol. MICHAEL ABBEY/VISUALS UNLIMITED.

diffusion of H_2O by osmosis (**Figure 2.33**). Although the molecular mechanism by which contractile vacuoles collect and expel water from the cell is not completely understood, studies have shown that inhibition of contractile vacuole function leads to cell swelling and lysis.

Plant cells, fungi, and bacteria are also surrounded by a hypotonic environment; however, they avoid the damaging effects of osmotic pressure by encasing the fragile plasma membrane within a rigid cell wall. Plant and bacterial cell walls consist of a network of covalently linked carbohydrates (and other components) that provide structural support to cells and protection against harsh environmental conditions. The effect of osmotic pressure on cell volume is critical in many types of plants because it provides the energy needed to overcome the effects of gravity (**Figure 2.34**). Water is stored in a cytosolic vacuole (central vacuole) that shrinks or expands in response to available water, thus creating the necessary osmotic pressure that provides turgor (stiffness) to

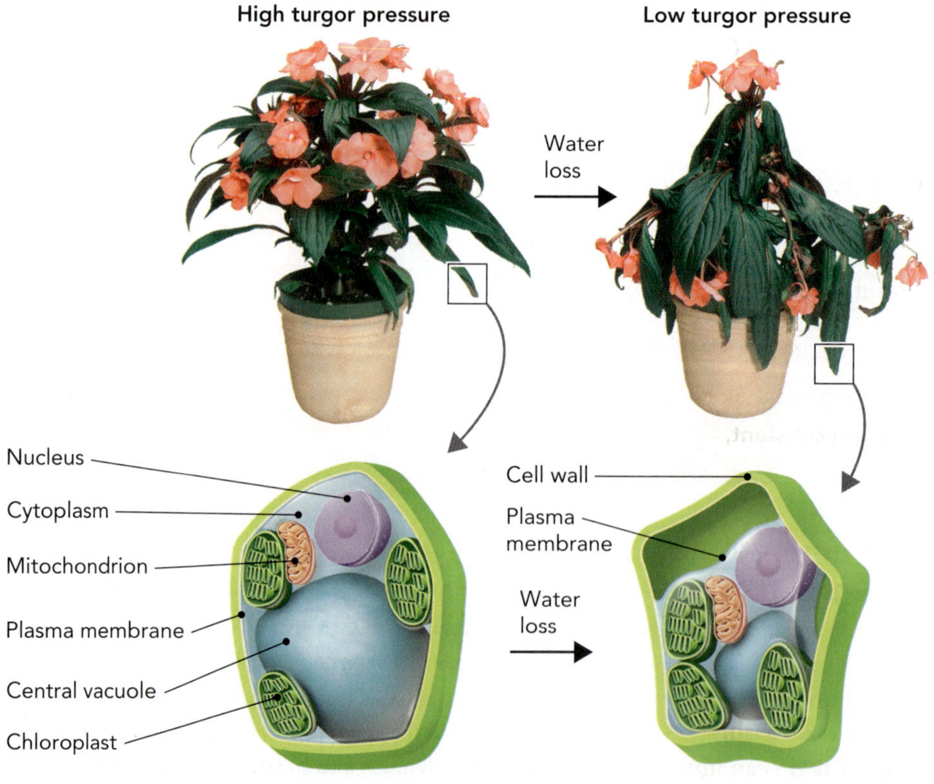

Figure 2.34 Osmosis is responsible for turgor pressure in plant cells, which are protected from cell lysis by a rigid cell wall that surrounds the plasma membrane. Plant cell shape is determined by the amount of water stored in the central vacuole. Plants wilt when turgor pressure decreases because of lack of water. PLANTS: NIGEL CATTLIN/ALAMY.

the plant. Turgor pressure is important for extending the leaves to maximize photon absorption and photosynthesis and to keep flowering parts upright to attract pollinating insects.

The Ionization of Water

So far in this section, we have talked about water molecules. However, a small amount of H_2O ionizes in solution to form a **hydrogen ion, H^+** (proton), and a **hydroxyl ion, OH^-**, both of which are important components of biochemical reactions. Hydrogen ions do not actually exist free in solution, but rather are hydrated to form **hydronium ions (H_3O^+)**.

Ionization of H_2O is a reversible reaction that takes place when two hydrogen-bonded water molecules undergo a bond rearrangement to form H_3O^+ and OH^-:

$$2\,H_2O \rightleftharpoons H_3O^+ + OH^-$$

This reaction is often written more simply as the ionization of H_2O to form H^+ and OH^-:

$$H_2O \rightleftharpoons H^+ + OH^-$$

We will follow this convention in the book, but remember that the true species is H_3O^+ and not simply H^+. Like all equilibrium reactions, the ionization of H_2O can be described by an equilibrium constant, K_{eq}:

$$H_2O \rightleftharpoons H^+ + OH^- \qquad K_{eq} = \frac{[H^+][OH^-]}{[H_2O]}$$

When pure water is subjected to an electric field, the H^+ ions migrate toward the cathode, and OH^- ions migrate toward the anode. By measuring the strength of this ionic current, we can calculate a K_{eq} value for the ionization of H_2O, which at 25 °C is 1.8×10^{-16} M. Using this K_{eq} value, along with the concentration of H_2O, which is approximately 55.5 M (1,000 g/L divided by 18.0 g/mol), we can calculate the concentration of H^+ and OH^- ions at equilibrium as follows:

$$K_{eq} = \frac{[H^+][OH^-]}{[H_2O]}$$

Rearrange the equation,

$$K_{eq}[H_2O] = [H^+][OH^-]$$

then substitute the numerical values of K_{eq} and H_2O concentration:

$$(1.8 \times 10^{-16}\,\text{M})(55.5\,\text{M}) = [H^+][OH^-] = 1.0 \times 10^{-14}\,\text{M}^2$$

Finally, replace the constant value of the multiplicands (K_{eq})(55.5 M) with the **water ionization constant, K_w**:

$$K_w = 1.0 \times 10^{-14}\,\text{M}^2 = [H^+][OH^-] \qquad (2.16)$$

Because the ionization of H_2O generates equimolar amounts of H^+ and OH^-, we can solve for the concentration of H^+ and the concentration of OH^- at equilibrium by taking the square root of K_w, showing that $[H^+] = [OH^-] = 1 \times 10^{-7}$ M:

$$\sqrt{K_w} = \sqrt{1.0 \times 10^{-14}\,\text{M}^2} = 1 \times 10^{-7}\,\text{M} = [H^+] = [OH^-]$$

The relationship between the concentration of H^+ and the concentration of OH^- is such that we can use the value of K_w to calculate H^+ and OH^- ion concentrations when additional concentrations of H^+ or OH^- are added to water. Recall from general chemistry that NaOH completely ionizes in water to yield Na^+ and OH^-, and similarly, HCl completely ionizes in water to yield H^+ and Cl^-. Therefore, by knowing the concentration of either NaOH or HCl in a water solution, we can determine the concentrations of H^+ and OH^- ions in the solution. Here are two examples to illustrate this point.

Example 1: What are the concentrations of H^+ and OH^- in a solution of 0.02 M NaOH?
Solution:

$$0.02 \text{ M NaOH} = 2 \times 10^{-2} \text{ M} = [\text{OH}^-]$$

$$K_w = [\text{H}^+][\text{OH}^-] = 1.0 \times 10^{-14} \text{ M}^2$$

$$[\text{H}^+] = \frac{K_w}{[\text{OH}^-]} = \frac{1.0 \times 10^{-14} \text{ M}^2}{2 \times 10^{-2} \text{ M}} = \mathbf{5 \times 10^{-13} \text{ M}}$$

Example 2: What are the concentrations of H^+ and OH^- in a solution of 150 mM HCl?
Solution:

$$150 \text{ mM HCl} = 0.15 \text{ M HCl} = 1.5 \times 10^{-1} \text{ M} = [\text{H}^+]$$

$$K_w = [\text{H}^+][\text{OH}^-] = 1.0 \times 10^{-14} \text{ M}^2$$

$$[\text{OH}^-] = \frac{K_w}{[\text{H}^+]} = \frac{1.0 \times 10^{-14} \text{ M}^2}{1.5 \times 10^{-1} \text{ M}} = \mathbf{6.7 \times 10^{-14} \text{ M}}$$

The important concept here is that because the water ionization constant K_w is always equal to $1 \times 10^{-14} \text{ M}^2$, the concentrations of H^+ and OH^- are reciprocally related. When the concentration of H^+ is high, the concentration of OH^- is low, and vice versa. In most biological systems, the concentrations of H^+ and OH^- ions are each around 1×10^{-7} M, which is considered to be a neutral solution.

The pH Scale Expressing very low concentrations of H^+ and OH^- ions is cumbersome. In the early 1900s, the Danish chemist Søren Peter Lauritz Sørensen developed the term **pH** to provide a more convenient value for H^+ concentrations in solution. pH is the abbreviation for the Latin term *pondus hydrogenii*, meaning "potential hydrogen," and is the log of the inverse of $[H^+]$, which can be converted to $-\log[H^+]$:

$$\text{pH} = \log \frac{1}{[\text{H}^+]} = -\log[\text{H}^+] \tag{2.17}$$

Using this definition, an H^+ concentration of 1×10^{-7} M is equal to a pH of 7:

$$\text{pH} = -\log(1 \times 10^{-7} \text{ M}) = -(-7) = 7$$

A change in one pH unit is equivalent to a 10-fold change in H^+ concentration because of the $\log_{10}$ conversion. For example,

$$[\text{H}^+] \text{ of } 1 \times 10^{-8} \text{ M} = -\log(1 \times 10^{-8} \text{ M}) = -(-8) = \text{pH } 8$$

$$[\text{H}^+] \text{ of } 1 \times 10^{-9} \text{ M} = -\log(1 \times 10^{-9} \text{ M}) = -(-9) = \text{pH } 9$$

On the basis of Equation 2.17 and the product of the H^+ and OH^- concentrations, you can see that most measured pH values of a solution will range from pH = 0 (1.0 M $[H^+]$) to pH = 14 (1×10^{-14} M $[H^+]$). This range is called the pH scale (**Figure 2.35**). Solutions with pH values less than 6.5 are acidic (high H^+ concentration), whereas solutions with pH values greater than 7.5 are basic (low H^+ concentration). Solutions with a pH between 6.5 and 7.5 are considered to be neutral ($[H^+] \approx [OH^-]$). The H^+ concentration of a solution is most commonly measured using a pH meter or with colorimetric indicators such as litmus paper.

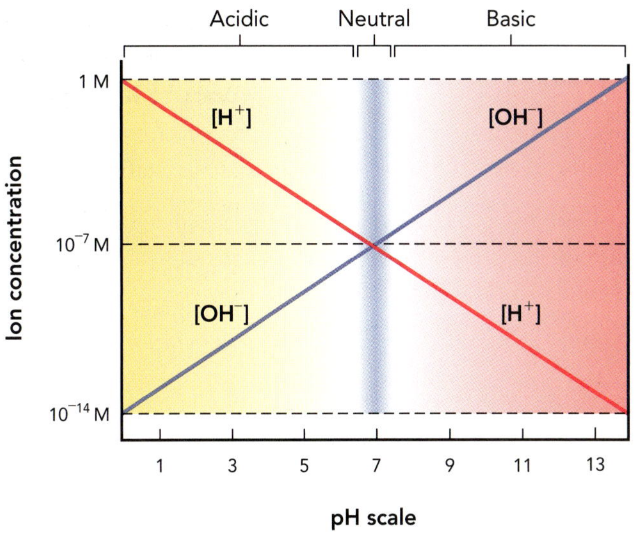

Figure 2.35 pH is related to concentrations of H^+ and OH^-. Note that H^+ and OH^- concentration values are reciprocal across the pH scale. Acidic solutions have a pH below 6.5, with $[H^+] > [OH^-]$, whereas basic solutions have a pH above 7.5, and $[H^+] < [OH^-]$.

Review of Acids, Bases, and pK_a The pH of a solution is determined by the H^+ concentration, which in turn depends on the concentration of solutes that function as acids and bases. Recall that acids are proton donors, whereas bases are proton acceptors. In biological systems that must maintain a neutral pH (pH ~7), the concentration of acids and bases in solution is very important. Unlike strong acids and bases, which completely ionize in aqueous solutions, biological systems contain numerous compounds that function as **weak acids** and their **conjugate bases**, which are only partially ionized. The strength of the tendency to lose or gain a proton in water is determined by the chemical properties of a compound and can be represented by the following chemical equation, where HA is the weak acid and A^- is the conjugate base:

$$HA + H_2O \rightleftharpoons H_3O^+ + A^-$$

To simplify this equation, we can use H^+ in place of H_3O^+ and rewrite the reaction as

$$HA \rightleftharpoons H^+ + A^-$$

By definition, a weak acid dissociates H^+ less readily than does a strong acid. We can calculate a value that describes the affinity of an acid for the dissociable H^+, using an equilibrium equation in which K_{eq} for the reaction is called the **acid dissociation constant, K_a**:

$$K_{eq} = \frac{[H^+][A^-]}{[HA]} \tag{2.18}$$

Weak acids have a low K_a value because the HA concentration is high, whereas strong acids have a high K_a value because most of the acid is in the dissociated form (A^-). However, because the K_a value is usually very small for biologically relevant acid–base conjugates, biochemists use the same convention as pH, in which K_a is expressed as **pK_a**:

$$pK_a = \log \frac{1}{K_a} = -\log K_a \tag{2.19}$$

Therefore, weak acids have a high pK_a value (low K_a), and strong acids have a low pK_a value (high K_a). For example, acetic acid has a pK_a of 4.7, whereas ammonia has a pK_a of 9.25. This means that acetic acid loses its H^+ much more easily than ammonia when a base (proton acceptor) is nearby.

We can calculate the pH of a solution from the concentration of an acid–base conjugate pair by making the following connections. First, rearrange Equation. 2.19 to describe K_a in terms of H^+ concentration:

$$K_a = \frac{[H^+][A^-]}{[HA]}$$

$$[H^+] = \frac{K_a[HA]}{[A^-]}$$

Then take the negative log of both sides,

$$pH = -\log[H^+] = -\log K_a - \log\frac{[HA]}{[A^-]}$$

Now invert $-\log\,[HA]/[A^-]$ to change the sign,

$$pH = -\log[H^+] = -\log K_a + \log\frac{[A^-]}{[HA]}$$

Substituting for pK_a, we arrive at the **Henderson–Hasselbalch equation:**

$$pH = pK_a + \log\frac{[A^-]}{[HA]} \tag{2.20}$$

The Henderson–Hasselbalch equation is very useful in biochemistry because we can use it to calculate the pK_a of a weak acid at a given pH if the concentrations of HA and A^- are experimentally determined, as shown in the following example.

Example 3: What is the pK_a of a weak acid if the ratio of the unprotonated form to the protonated form is 50 in an aqueous solution at pH 6.5?
Solution: Substitute the given data directly into a slightly rearranged Henderson–Hasselbalch equation:

$$pH = pK_a + \log\frac{[A^-]}{[HA]}$$

$$pK_a = pH - \log\frac{[A^-]}{[HA]}$$

$$pK_a = pH - \log\frac{50}{1} = 6.5 - \log 50 = 6.5 - 1.7 = 4.8$$

Conversely, as shown in the next example, we can use the Henderson–Hasselbalch equation to determine the amount of conjugate acid (protonated) and conjugate base (deprotonated) at a given pH if we know the pK_a value of the acid–base conjugate.

Example 4: How much lactic acid [proton donor, $CH_3CH(OH)COOH$] is present in an aqueous solution at pH 5.1 if the concentration of lactate [proton acceptor, $CH_3CH(OH)COO^-$] is 0.2 M and the pK_a of lactic acid is 3.9?
Solution: This calculation also requires rearranging of the equation:

$$pH = pK_a + \log\frac{[\text{Proton acceptor}]}{[\text{Proton donor}]}$$

$$\log\frac{[\text{Lactate}]}{[\text{Lactic acid}]} = pH - pK_a$$

$$\log\frac{[\text{Lactate}]}{[\text{Lactic acid}]} = pH - pK_a = 5.1 - 3.9 = 1.2$$

$$\frac{[\text{Lactic acid}]}{[\text{Lactate}]} = 10^{1.2} = 15.8$$

$$\frac{0.2\text{ M}}{[\text{Lactic acid}]} = 15.8$$

$$[\text{Lactic acid}] = \frac{0.2\text{ M}}{15.8} = 0.013\text{ M}$$

The Henderson–Hasselbalch equation illustrates two basic concepts in acid–base chemistry. First, the pK_a of an acid can be determined by finding the pH of a solution in which 50% of the acid molecules are deprotonated and 50% are protonated (ratio of 1). This is because the log of 1 is equal to 0, so pH = pK_a when 50% of the molecules are protonated and 50% deprotonated. That is, when $[A^-] = [HA]$, then

$$\text{pH} = \text{p}K_a + \log\frac{[A^-]}{[HA]} = \text{p}K_a + \log 1 = \text{p}K_a + 0$$

$$\text{pH} = \text{p}K_a \text{ when } [A^-] = [HA]$$

Second, by knowing the pK_a of an acid and the pH of the solution, we can predict whether the acid–base conjugate is mostly protonated or mostly deprotonated, based on the Henderson–Hasselbalch equation:

When pH > pK_a then the $[A^-]/[HA]$ ratio is greater than 1 and $[A^-] > [HA]$.

When pH < pK_a then the $[A^-]/[HA]$ ratio is less than 1 and $[HA] > [A^-]$.

This makes sense because at high pH, the OH^- concentration in the aqueous solution is high, which provides a base that pulls the H^+ off of the conjugate acid, resulting in the formation of H_2O ($H^+ + OH^-$). Conversely, at low pH, the H^+ concentration is high, which helps keep the acid in the protonated form, as $[H^+] >> [OH^-]$.

Titration Curves and Biological Buffers We can use the principles of acid–base chemistry in the lab to determine the amount of acid in a solution by performing a titration experiment. Graphs of the results, called titration curves, can also be used to find the pK_a of a weak acid.

A **titration curve** is a plot of experimental data showing the pH of a solution as a function of the amount of base added, usually in the form of OH^- ions from the ionization of NaOH. A titration curve of the weak acid acetic acid in solution is relatively flat at pH values near the pK_a of 4.76 because the added OH^- is acting as a proton acceptor to remove H^+ from acetic acid to form H_2O (**Figure 2.36**). On the basis of the Henderson–Hasselbalch equation (Equation 2.21), pK_a = pH at the midpoint of the titration curve because this is where the concentration of A^- is equal to the concentration of HA. At pH values significantly below the pK_a, the dissociable H^+ on acetic acid is not easily removed, so the pH rises sharply as OH^- is added and removes H^+ from solution to form H_2O. Similarly, at pH values significantly above the pK_a, the pH rises sharply again because the dissociable

Figure 2.36 The titration curve of acetic acid is shown here. The pK_a can be determined from the midpoint of the titration curve and corresponds to the pH at which the concentrations of acid and conjugate base are equal. The term *equivalents* describes the amount of base (OH^-) that is required to deprotonate the acid. Also shown is the buffer range of acetic acid, which corresponds to one pH unit above and below the pK_a.

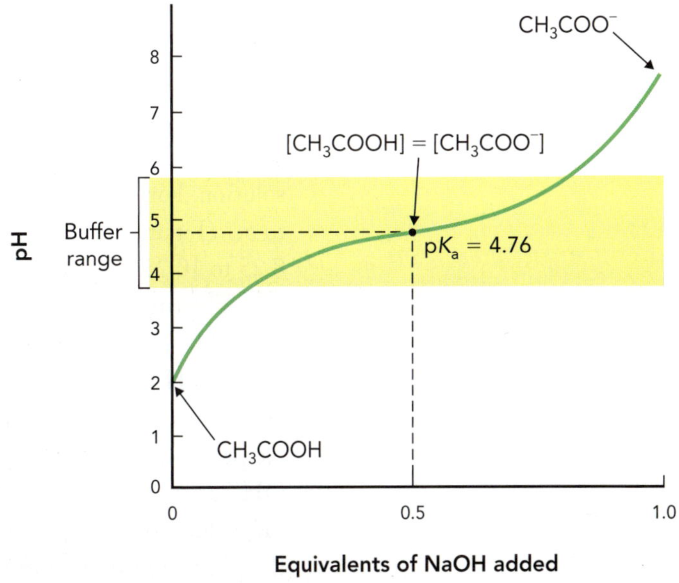

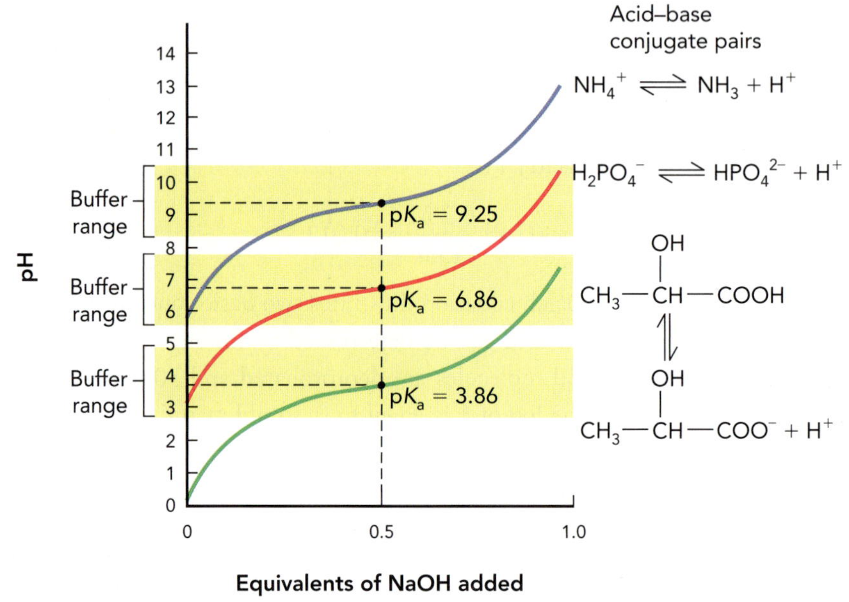

Figure 2.37 The titration curves of weak acids have similar shapes. The titration curves, pK_a values, and buffering ranges are shown for ammonium ion (NH_4^+), dihydrogen phosphate ion ($H_2PO_4^-$), and lactic acid [$CH_3CH(OH)COOH$].

H^+ from acetic acid has been removed, and the added OH^- combines with H^+ in solution to generate H_2O.

Because there is a one-to-one relationship between the titratable H^+ from the weak acid and the amount of OH^- added, titration curves reveal how much acid is in a solution. For example, if 0.1 M NaOH is required to reach the midpoint of the titration curve (pK_a), then you know that the solution contains 0.2 M of the weak acid because 50% of the acid has been dissociated. The amount of added OH^- required to reach the pK_a and dissociate 50% of the H^+ from the acid is denoted as 0.5 equivalents. The amount of OH^- required completely to dissociate H^+ from the acid is 1.0 equivalent and is equal to the amount of acid present in the solution at the start of the titration. Note that the pK_a for different acids determines the pH profile of the titration curve, but the shape of the curves for acids with a single dissociable H^+ are identical (**Figure 2.37**).

Titration curves can be used to identify pH ranges in which acid–base conjugate pairs in solution are able to function as buffers. **Buffers** are aqueous solutions that resist changes in pH because of the protonation or deprotonation of an acid–base conjugate pair, which is present at a high enough concentration to absorb small changes in H^+ or OH^- concentration. The ability of an acid or base to resist changes in pH is referred to as the buffering capacity. The pH range of a buffering system is generally taken to be ~1 pH unit above and below the pK_a because this is where sufficient amounts of the weak acid and conjugate base are present to buffer against increased H^+ or OH^- in solution. For example, the buffering capacity of acetic acid is in the pH range of 3.7 to 5.7 (Figure 2.36), whereas the buffering capacity of ammonia is in the pH range of 8.25 to 10.25 (Figure 2.37).

The titration curve of a **polyprotic acid**—a weak acid with more than one dissociable H^+—is much more complex because the acid has more than one pK_a, representing the pH values at which each H^+ is 50% dissociated. One of the most important weak acids in biological systems is phosphoric acid, H_3PO_4, which has three dissociable protons, and therefore three pK_a values (**Figure 2.38**). Indeed, three equivalents of OH^- are required to convert H_3PO_4 to phosphate ion, PO_4^{3-}, the conjugate base. The pK_a values for each of the three dissociable forms of phosphoric acid are identified by the pH corresponding to the addition of 0.5, 1.5, and 2.5 equivalents of base.

a.

$$H_3PO_4 \rightleftharpoons H_2PO_4^- + H^+ \quad pK_{a1} = 2.1$$

$$H_2PO_4^- \rightleftharpoons HPO_4^{2-} + H^+ \quad pK_{a2} = 7.2$$

$$HPO_4^{2-} \rightleftharpoons PO_4^{3-} + H^+ \quad pK_{a3} = 12.4$$

b.

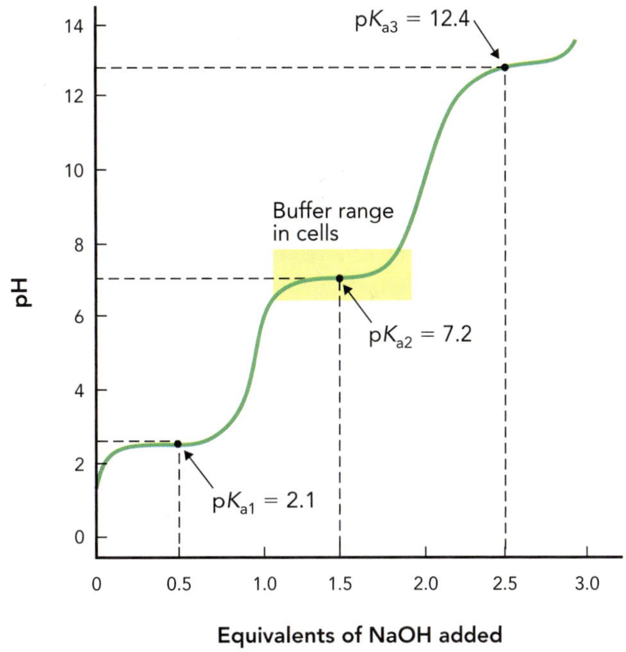

Figure 2.38 Phosphoric acid is a polyprotic acid that contains three disscociable protons. **a.** Equilibrium reactions for each acid–base conjugate pair of phosphoric acid, with the corresponding pK_a values. **b.** Titration curve of phosphoric acid at 25 °C. Note that $H_2PO_4^-$ and HPO_4^{2-} serve as the weak acid and conjugate base, respectively, in biological buffering systems that function near a neutral pH.

The acid–base conjugate pair of $H_2PO_4^-$ and HPO_4^{2-} has a pK_a value of 6.8 at 37 °C (note that pK_a values depend on the temperature) and plays an important role in biological systems because it functions as a buffering system at neutral pH values. Buffered biological systems are critical to cellular life because even small changes in pH can have profound effects on the activity of enzymes, which depend on an optimal pH to maintain optimal catalytic activity (see Chapter 7). Inorganic phosphate and phosphoryl groups on biomolecules such as nucleic acids both contribute to the buffering capacity of phosphoryl acid in biological systems.

A second example of a biological buffering system is that of the carbonic acid–bicarbonate buffer system in the blood plasma of animals, which functions to maintain blood pH levels at 7.40. In this buffering system, the weak acid is carbonic acid (H_2CO_3) and the conjugate base is bicarbonate HCO_3^-:

$$H_2CO_3 \rightleftharpoons H^+ + HCO_3^-$$

Although the pK_a for this reaction at 37 °C is only 6.1, the carbonic acid–bicarbonate conjugate pair is able to function as the primary buffering system in blood because it is in equilibrium with three other processes that together adjust H_2CO_3 and HCO_3^- levels in response to changes in serum H^+ levels (**Figure 2.39**). One of these processes is

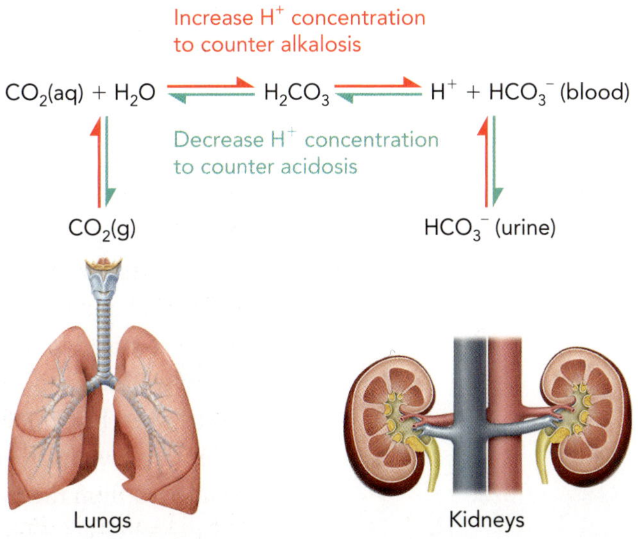

Figure 2.39 The bicarbonate buffering system in animals plays a key role in maintaining blood pH levels through exchange of CO_2(g) and regulated excretion of HCO_3^-. When pH levels drop (acidosis), the blood H^+ concentration is reduced by increasing the breathing rate (increased exhalation of CO_2) and by decreasing HCO_3^- excretion. In contrast, when pH levels rise (alkalosis), the blood H^+ concentration is increased by slowing the breathing rate (decreased exhalation of CO_2) and by increasing HCO_3^- excretion.

a reversible reaction catalyzed by the enzyme carbonic anhydrase, which interconverts H_2CO_3 with dissolved $CO_2(aq)$ and H_2O:

$$H_2CO_3 \underset{}{\overset{\text{Carbonic anhydrase}}{\rightleftharpoons}} CO_2(aq) + H_2O$$

The other two processes are exchange of $CO_2(aq)$ in the blood with atmospheric $CO_2(g)$ in the air spaces of the lungs, and the excretion of HCO_3^- (urine) or retention of HCO_3^- (blood) through the kidneys:

$$CO_2(aq) \overset{\text{Lungs}}{\rightleftharpoons} CO_2(g)$$

$$HCO_3^-\ (\text{blood}) \overset{\text{Kidneys}}{\rightleftharpoons} HCO_3^-\ (\text{urine})$$

The bicarbonate buffering system is governed by **Le Châtelier's principle** of mass action, which states that the equilibrium of a reaction shifts in the direction that reduces a change imposed on it by external sources. Therefore, if serum pH falls below 7.4, a condition called **acidosis**, then the equilibrium of the carbonic acid–bicarbonate reaction needs to be shifted toward H_2CO_3 formation to decrease H^+ concentration and thereby increase the pH. This is done by lowering $CO_2(aq)$ through hyperventilation [exhaling $CO_2(g)$] and by decreasing excretion of HCO_3^- by the kidneys. Alternatively, if serum pH rises above 7.4, a condition called **alkalosis**, then the equilibrium of this reaction is shifted toward H^+ and HCO_3^- formation to increase H^+ concentration and lower the pH. This is accomplished by reducing the breathing rate to increase levels of $CO_2(aq)$ and by increasing excretion of HCO_3^- in the urine.

concept integration 2.2

Why is hydrogen bonding between water molecules so important to life on Earth?

Water has the property of being able to act as both a hydrogen-bond donor and hydrogen-bond acceptor. This property makes water a "universal" solvent, which is critical to biochemical reactions that require soluble reactants, products, and enzyme catalysts to proceed on a biological timescale. Moreover, hydrogen bonding between H_2O molecules explains hydrophobic effects, which dictate the structure and function of biomolecules containing nonpolar groups, such as lipids and proteins. Because nonpolar groups are insoluble in water, the mixing of polar and nonpolar compounds in water is energetically unfavorable due to the formation (through hydrogen bonds) of cage-like H_2O structures around nonpolar groups, which decrease entropy in the system. Hydrophobic effects offset this unfavorable energy condition by minimizing the interaction of nonpolar groups with water to reduce the size of hydration layers. In the case of lipids, this leads to the formation of lipid bilayers, whereas with proteins, it promotes the folding of polypeptide chains into stable organized structures. Finally, hydrogen bonding between H_2O molecules explains why ice is less dense than water, which is critical to the survival of aquatic life in subzero environments. The lower density of solid ice compared to that of liquid water is due to the open lattice structure that forms when H_2O molecules hydrogen bond with the maximum number of donors and acceptors (two hydrogen bonds originating from the oxygen atom and one hydrogen bond each from the hydrogen atoms).

2.3 Cell Membranes Function as Selective Hydrophobic Barriers

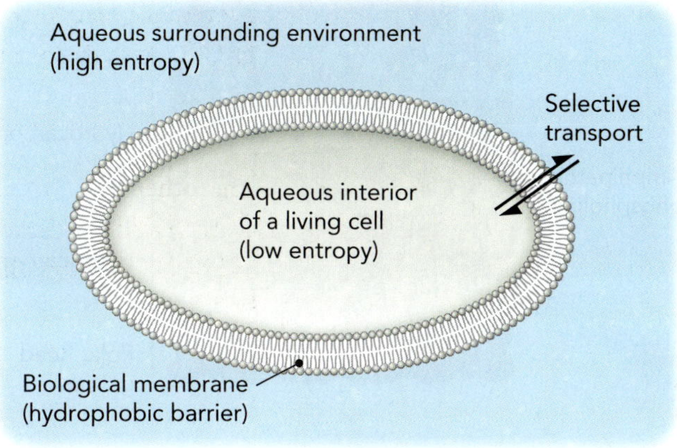

Figure 2.40 Biological membranes partition aqueous environments to permit life processes to maintain regions of low entropy within the high-entropy state of the surrounding environment. Because the solvent of life on Earth is water, biological membranes function as hydrophobic barriers.

Life processes require a way to protect fragile biological systems from the harsh physical and chemical properties of their environment. Cells achieve this by creating boundaries where processes can take place without interference from the surroundings. This partitioning separates highly organized compartments of low entropy (the interior of cells) from disorganized regions of high entropy (the environment), while still permitting the exchange of nutrients and waste (**Figure 2.40**). Because the solvent of life on Earth is water, such selective barriers need to partition aqueous compartments, which is best done using nonpolar components with hydrophobic properties. These physical barriers in living systems are **biological membranes.**

Biological membranes contain, among other things, amphipathic lipid molecules that self-assemble into a bilayer structure. An amphipathic molecule is one that possesses both hydrophobic and hydrophilic chemical properties (in Greek, *amphi* means "both" and *pathos* means "suffering"). The most abundant class of amphipathic lipids in biological membranes is phospholipids, which have a polar charged head group (hydrophilic) and long nonpolar hydrocarbon tails (hydrophobic) (**Figure 2.41**). In an aqueous environment, phospholipids organize into **phospholipid bilayers**, such that

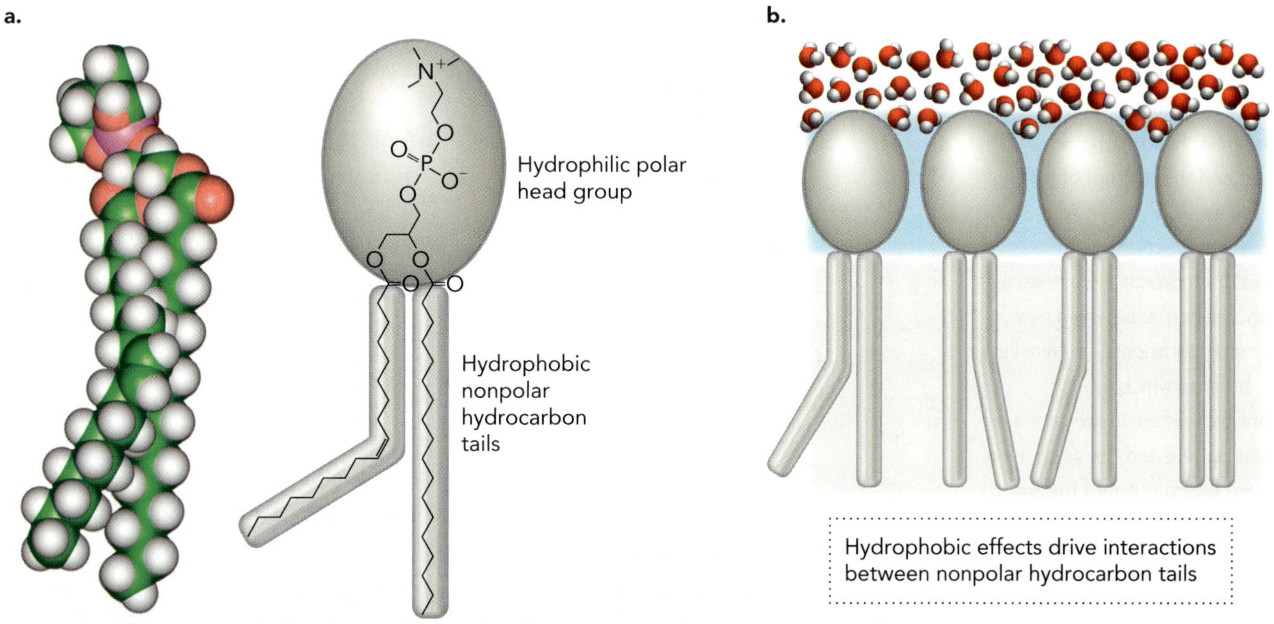

Figure 2.41 Phospholipids are amphipathic biomolecules containing a polar head group and two nonpolar hydrophobic tails. **a.** Phosphatidylcholine is an amphipathic glycerophospholipid found in many types of biological membranes. The hydrophilic head group contains numerous oxygen molecules that can function as hydrogen-bond acceptors with H_2O, along with two charged groups (phosphate and nitrogen) that can form ionic interactions with H_2O. **b.** Hydrophobic effects induce phospholipids to form complexes in which the nonpolar hydrophobic tails associate with one another to minimize exposure to H_2O, while the polar head groups form hydrogen bonds with H_2O molecules.

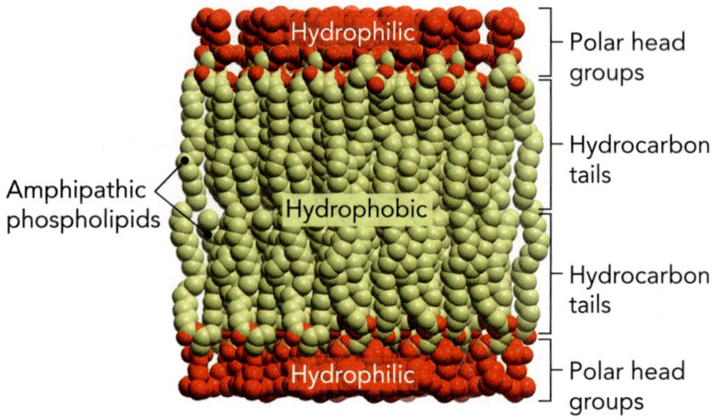

Figure 2.42 A side view of a phospholipid bilayer is shown here. The hydrophobic tails of phospholipids, an abundant constituent of biological membranes, interact with one another to form a hydrophobic barrier between two aqueous compartments.

hydrophilic polar head groups orient toward the aqueous environment, and hydrophobic nonpolar hydrocarbon tails form a water-impermeable barrier in the interior of the membrane (**Figure 2.42**).

Chemical and Physical Properties of Cell Membranes

The organization of phospholipids within cell membranes is a function of both their amphipathic chemical properties and the aqueous environment of the cell. Artificial phospholipid membranes can be synthesized *in vitro* by mixing them with water. Depending on the process by which the phospholipids and water are mixed and on the concentration of phospholipids, the phospholipids can form lipid monolayers and bilayers or assemble into **micelles** and liposomes (**Figure 2.43**). Lipid monolayers are formed by slowly adding phospholipids to the surface of water, resulting in the formation of a thin film that has the polar head groups oriented toward the water and the hydrophobic tails at the air interface. If phospholipids are added to water and the mixture is agitated, the phospholipids can form lipid bilayers or micelles in which the polar head groups face outward toward the aqueous environment and the hydrophobic tails face inward to avoid contacting H_2O molecules.

It is also possible to form bilayer structures, called **liposomes**, by vigorous agitation of phospholipid mixtures in water. Liposomes form when pieces of the lipid

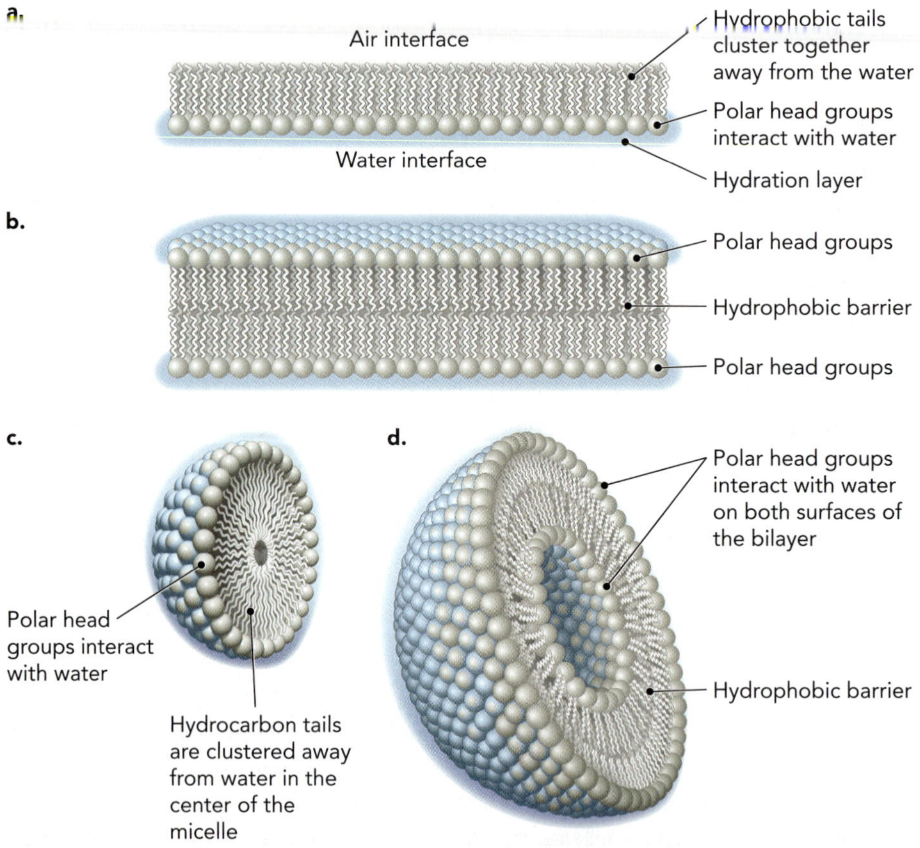

Figure 2.43 Hydrophobic effects cause amphipathic phospholipids to associate into four different types of complexes when mixed with water. **a.** Phospholipid monolayers have the polar head groups in the water and the hydrophobic tails in the air. **b.** Phospholipid bilayers are characteristic of biological membranes and create a hydrophobic barrier between two aqueous compartments. **c.** Micelles are structures in which the hydrophobic tails are in the center of a globular sphere and the polar head groups are facing outward toward the water, as shown in this cutaway view. Note that phospholipid micelles are actually not perfect spheres because the hydrophobic tails cannot pack symmetrically into the center region of the micelle. **d.** Liposomes are spherical structures bounded by a lipid bilayer and contain an aqueous center, as shown in this cutaway view.

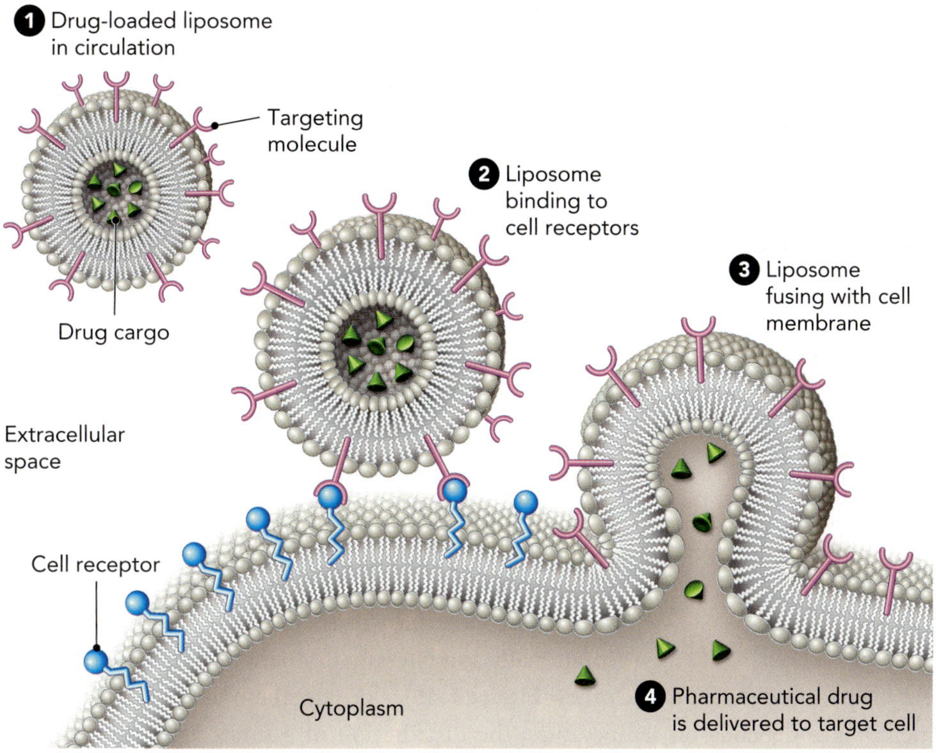

Figure 2.44 Liposomes can be synthesized that function as drug-delivery systems, fusing with the membrane of target cells through binding to extracellular receptor proteins.

bilayer break off into smaller patches and capture an aqueous droplet inside the vesicle. If the aqueous droplet contains pharmaceutical drugs or fluorescent molecules, they will be trapped inside the liposome, which can then serve as a carrier that delivers "cargo" to cells by fusing with the plasma membrane (**Figure 2.44**). By embedding hydrophobic targeting molecules in the lipid bilayer, liposomes can be made that bind with high selectivity to receptor molecules located on the surface of specific cell types.

The fluidity of the membrane depends on the chemical properties of the phospholipid components. The chemical properties of the head groups can be quite variable with respect to size and charge. For example, some phospholipid head groups are glycosylated (have carbohydrates attached), which makes them much more bulky. In the fatty acid tails, a carbon–carbon double bond introduces a "kink," which decreases the interactions between the fatty acid tails. Therefore, phospholipid membranes containing fatty acid tails with a high degree of saturation (ratio of C—C bonds to C=C bonds) are less fluid than membranes composed of phospholipids with fatty acid tails having C=C double bonds.

Other components in addition to the phospholipids affect membrane fluidity. As an example, cholesterol, a small amphipathic molecule with a rigid nonpolar ring system connected to a hydroxyl group, can be a component of biological membranes. At low temperatures, pure phospholipid membranes form a gel-like substance with crystalline characteristics, which is converted to a liquid state at higher temperatures. When small amounts of cholesterol are added to a phospholipid bilayer, it prevents the saturated hydrocarbon chains from closely packing. This has the effect of decreasing the temperature at which the gel-like state occurs and thereby helps maintain the liquid state at physiologic temperatures. However, cholesterol can also have the opposite effect of decreasing membrane fluidity when the ratio of cholesterol to phospholipid

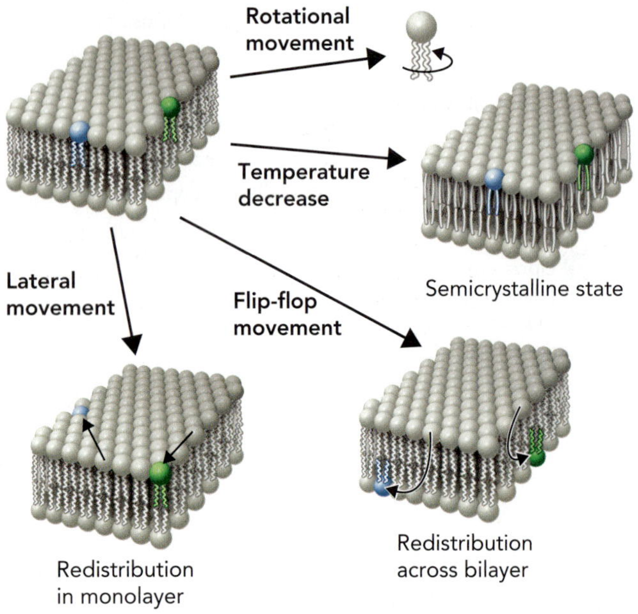

Figure 2.45 At physiologic temperatures, the membrane is fluid, and individual phospholipids can move rotationally, laterally, and transversely (flip-flop). At low temperatures, the lipid bilayer becomes semicrystalline and forms a gel-like solid.

is increased (up to a factor of ~2.0) due to the rigid ring structure of cholesterol.

The ratio of cholesterol to phospholipid also influences the thickness of the membrane, depending on the types of phospholipids in that region of the membrane. Increased ratios of cholesterol to phospholipid result in membrane thickening because the polar head group on cholesterol is quite small (just a hydroxyl group), and the rigid nonpolar sterol ring must fit in between the fatty acid tails, which causes distortion of the membrane.

Lipid molecules in membranes are able to move rotationally and laterally throughout the membrane and even flip from one side to the other (**Figure 2.45**). The lateral mobility of lipids within membranes depends on temperature, the distribution of saturated and unsaturated lipids and cholesterol, and the density and types of proteins embedded in the membrane. The transverse flipping of phospholipids from one side of the lipid bilayer to the other is normally very slow; however, it has been found that a class of ATP-dependent membrane proteins, called **flippases**, uses energy available from ATP hydrolysis to catalyze phospholipid flipping.

Biological membranes made up only of phospholipids would be nothing more than chemical barriers between two aqueous compartments, which would limit the types of interactions that could occur between the cell and its surroundings. Because most of the biomolecules required for life are polar in composition—glucose, for example—they cannot simply diffuse across a hydrophobic barrier. Therefore, ways to allow passage of these molecules across membranes are needed. Biological membranes are actually a complex mixture of phospholipids and proteins, which together impart structural and functional characteristics to membranes.

Many membrane proteins function as transmembrane pores or energy-driven gates that open and close to permit solute exchange. Just as lipids move around in membranes, so do these membrane-transport proteins. In 1970, Michael Edidin showed that membrane proteins of mouse and human cells can intermix after two cells are fused together, as shown in **Figure 2.46**. This experiment was done using a eukaryotic virus called Sendai virus, which promotes membrane fusion and the formation of large cells containing all of the contents of the fusion partners, including two cell nuclei. Fused cells such as these are called heterokaryon cells. By using fluorescent molecules (antibodies that uniquely recognize mouse and human cell surface proteins), it was

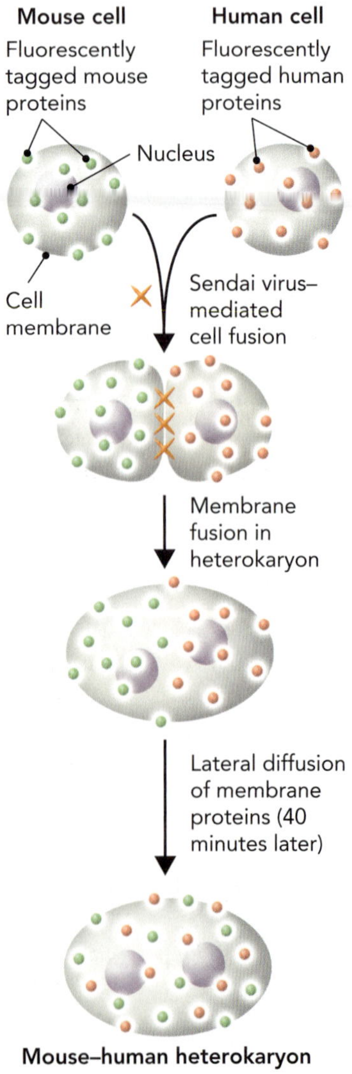

Figure 2.46 Cell fusion experiments demonstrate that proteins are able to diffuse laterally within the plasma membrane. In this experiment, fluorescently tagged antibodies were used specifically to monitor the position of mouse and human proteins on the surface of cells before and after virus-mediated fusion. By 40 minutes after cell fusion, the majority of labeled mouse and human proteins had redistributed within the membrane of these heterokaryon cells, providing evidence for lateral diffusion.

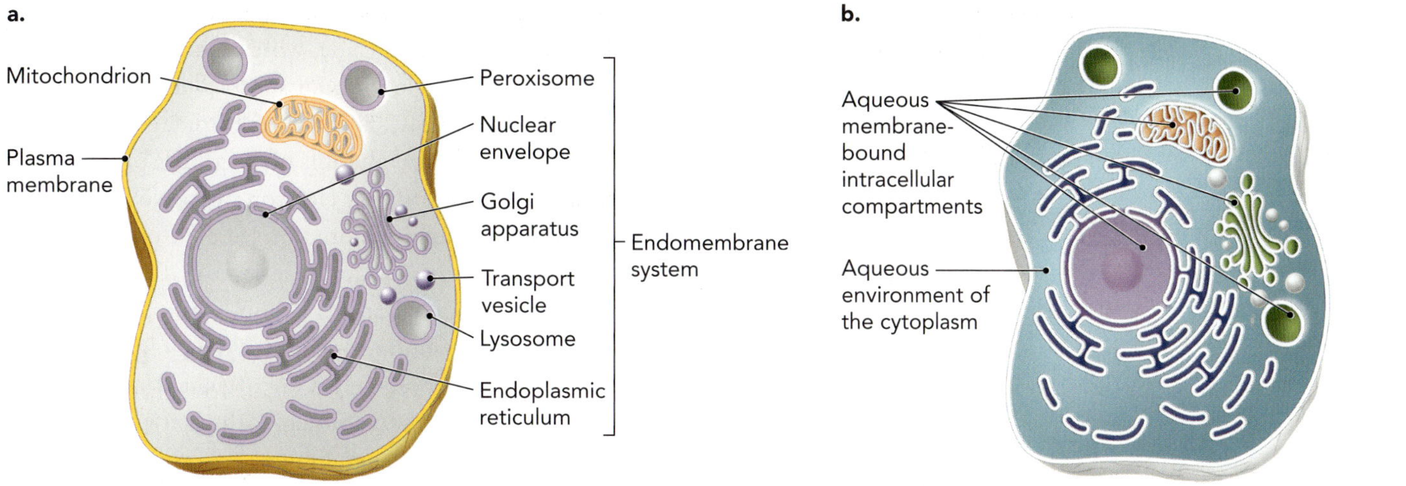

Figure 2.47 Eukaryotic cells contain three major classes of biological membranes that separate the cell from the surroundings or form compartments within a cell. **a.** Schematic representation of an animal cell identifying (1) the plasma membrane, which separates the contents of the cell from the surrounding environment; (2) the endomembrane system, which is a network of intracellular membranes connected to the nuclear envelope that transport cellular components throughout the cytoplasm; and (3) the mitochondrial membrane, as well as the chloroplast membrane in plant cells, which are the energy-conversion factories of eukaryotic cells. **b.** Intracellular membranes in eukaryotic cells serve as selective hydrophobic barriers that separate aqueous compartments within the cell.

observed that membrane proteins are free to move laterally throughout the membrane shortly after the membranes fuse.

Organization of Prokaryotic and Eukaryotic Cell Membranes

Chemical differences between membrane lipids and proteins embedded within the phospholipid bilayer influence the biological functions of cell membranes. The overall organization of biological membranes also affects function. For example, some membranes are embedded within each other to create a double membrane structure (bacterial and cell organelle membranes), whereas others are layered as a result of membrane invaginations to create a stacking arrangement (endoplasmic reticulum and the Golgi apparatus).

Eukaryotic cells contain three major classes of biological membranes: (1) the plasma membrane surrounding every cell (plant cells are also protected by a carbohydrate cell wall); (2) endomembranes, which consist of structurally related intracellular membrane networks and vesicles; and (3) mitochondrial and chloroplast membranes, which contain the energy-converting enzymes required for life. These membrane systems separate distinct aqueous compartments within the cytoplasm of the cell (**Figure 2.47**).

The **endomembrane system** is an intracellular network of lipid bilayers that is used to exchange material through vesicle transport. Through this system, the nuclear envelope is linked to the plasma membrane through a series of stacked membranes and diffusible membrane-enclosed vesicles (**Figure 2.48**). The nuclear envelope compartmentalizes genomic DNA and consists of two membranes that are fused together (except in regions where nuclear pores permit the direct exchange

Figure 2.48 The endomembrane system is a network of lipid bilayers that are either continuous with each other or exchange material through vesicle fusion. See the text for details of the endomembrane system.

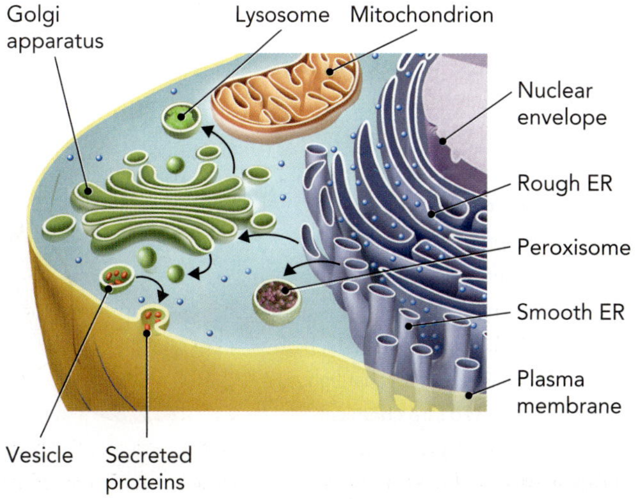

of biomolecules between the cytoplasm and the nucleus). The nuclear envelope is connected directly to the endoplasmic reticulum (ER), which is a large collection of invaginated membranes that enclose a single internal space. The ER, of which there are two types (rough ER and smooth ER), is the site of lipid, carbohydrate, and protein biosynthesis in the cell. Biochemical analysis of rough ER preparations show that the "knobs" observed on the membrane by electron microscopy are ribosomes involved in the synthesis of membrane-anchored or secreted proteins. Peroxisomes, which contain enzymes involved in redox reactions and biosynthesis, are a type of ER-derived vesicle. The Golgi apparatus is another set of interconnected membrane sacs within the endomembrane system. The Golgi apparatus is the location in eukaryotic cells where carbohydrates are attached to lipids and proteins by enzymes. Lysosomes, which contain numerous hydrolytic enzymes, are one of the vesicle types derived from Golgi membranes. Lysosomes fuse with endocytic vesicles to degrade macromolecules imported into the cell by the process of exocytosis.

Mitochondria and chloroplasts are the energy-converting organelles of eukaryotic cells, having evolved from endosymbiotic bacteria about 2 billion years ago. The lipid bilayers contained within these organelles are the location of membrane-embedded proteins that convert redox energy or light energy into chemical energy in the form of ATP.

As shown in **Figure 2.49**, the inner mitochondrial membrane is highly convoluted, which increases the membrane surface area, and thereby the energy-converting capacity of this organelle. The outer mitochondrial membrane surrounds the inner membrane to create the intermembrane space, which is critical to harnessing the energy-converting power of the proton motive force, as we will see in Chapter 11. The mitochondrial matrix contains enzymes required for aerobic respiration.

The membrane organization in chloroplasts is quite different, but the principles of energy conversion using proton motive force are the same as in mitochondria (see Chapter 12). In the case of chloroplasts, light-harvesting proteins in the thylakoid membrane convert light energy into proton motive force, which is used for ATP synthesis. Thylakoid disks are stacked into structures called grana, which fill the stromal space, a chloroplast compartment defined by the inner chloroplast membrane (Figure 2.49). The biochemistry of these chloroplast membranes and the light-harvesting reactions of photosynthesis are discussed in Chapter 12.

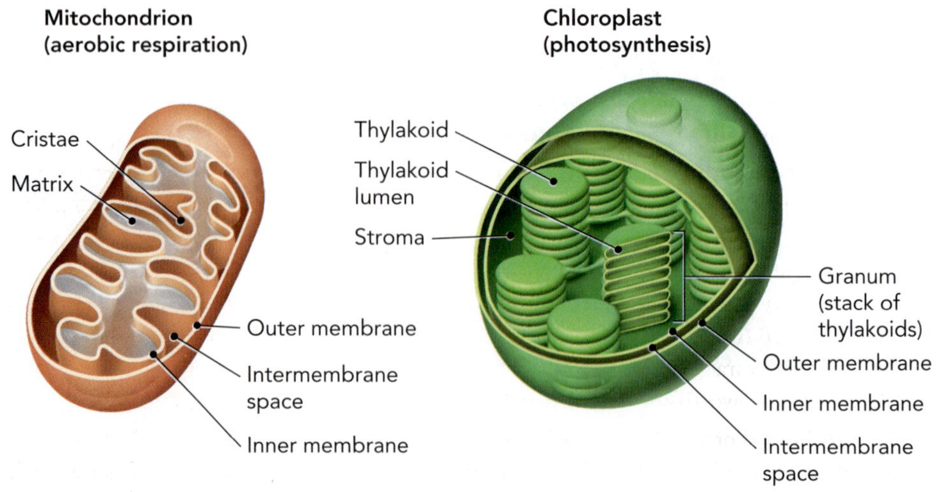

Figure 2.49 Mitochondria and chloroplasts are organelles that convert redox energy or light energy into chemical energy. The inner mitochondrial membrane is a proton-impermeable barrier that functions in oxidative phosphorylation to convert proton motive force into chemical energy. The thylakoid membrane of the chloroplast has a similar role in energy conversion, except that light energy is converted into chemical energy through the process of photosynthesis.

concept integration 2.3

How would a hydrophilic drug be loaded into a hydrophobic artificial liposome, and how would lipid composition affect liposome properties?

Liposomes are formed by mixing lipids and water under conditions that promote partitioning of amphipathic lipid molecules into lipid bilayers with polar head groups oriented toward water and hydrophobic tails of the lipids oriented away from water. The large lipid bilayers can be broken into smaller pieces using mechanical forces or sonication, which leads to formation of spherical liposomes. If the aqueous solution contains hydrophilic drug molecules, then they will be trapped in the interior of the liposome during the manufacturing process. The biochemical properties of liposomes are affected by the lipid composition; for example, the fluidity of the liposome membrane is determined by the concentration of cholesterol in the lipid mixture. The ratio of cholesterol to phospholipids in the lipid mixture may also affect the thickness of the liposome membrane, which could alter liposome properties with regard to the efficiency of drug delivery.

chapter summary

2.1 Energy Conversion in Biological Systems

- Organisms use energy obtained from the environment to maintain homeostatic conditions that are far from equilibrium; reaching equilibrium with the environment is equivalent to death.
- Solar energy is converted to chemical energy through the processes of photosynthesis and carbon fixation; chemical energy is converted by cells into useful work (osmotic work, chemical work, and mechanical work).
- The primary energy conversion processes in cells are oxidation–reduction (redox) reactions that involve the transfer of electrons between molecules.
- All biological processes follow the laws of thermodynamics. The first law of thermodynamics states that energy cannot be created or destroyed, only converted from one form to another. The second law of thermodynamics states that entropy (S), or the dispersion of energy, in the universe is always increasing.
- Enthalpy (H) is defined as the heat content of a molecule and is reflected in the number and type of chemical bonds. Exothermic reactions give off heat, and the change in enthalpy (ΔH) is negative; endothermic reactions absorb heat, and ΔH is positive.
- Changes in Gibbs free energy (ΔG) describe the spontaneity of a reaction in terms of absolute temperature (T) and changes in enthalpy (ΔH) and entropy (ΔS), using the relationship $\Delta G = \Delta H - T\Delta S$. Exergonic reactions ($\Delta G < 0$) are favorable, whereas endergonic reactions ($\Delta G > 0$) are unfavorable for the forward reaction.
- The equilibrium constant (K_{eq}) can be used to determine the standard free energy (ΔG°) of a reaction using the equation $\Delta G^\circ = -RT \ln K_{eq}$, in which K_{eq} is the ratio of the equilibrium concentrations of products over reactants, R is the gas constant, and T is the temperature in kelvins.
- The ΔG° value for a given reaction is a constant and is determined experimentally from the K_{eq} using standard conditions (298 K, 1 atm pressure) and an initial 1 M concentration of reactants and products.
- The ΔG° value of coupled reactions is equal to the sum of the ΔG° values for each individual reaction. Cleavage of a phosphoanhydride bond in ATP is often coupled to an otherwise unfavorable reaction to drive the reaction forward (overall $\Delta G^\circ < 0$).
- The energy charge (EC) of the cell reflects the relative concentrations of ATP, ADP, and AMP. When the EC is high, anabolic pathways (biosynthesis) are favored because ATP is plentiful; when the EC is low, catabolic pathways (degradation) are favored.

2.2 Water Is Critical for Life Processes

- Water has three properties that make it essential for life on Earth: (1) the solid form of water is less dense than the liquid form, which is why ice floats; (2) water is liquid over a wide range of temperatures; and (3) water is an excellent solvent because of its hydrogen-bonding abilities and polar properties.
- The molecular structure of H_2O gives rise to a separation of charge, with two partial negative charges on the oxygen atom ($2\delta^-$) and one partial positive charge on each of the hydrogens (δ^+ and δ^+).
- Extensive hydrogen bonding between H_2O molecules gives water its unusually high viscosity, boiling point, and melting

point compared to those of other molecules of a similar molecular mass.

- The transfer of H^+ between H_2O molecules is called proton hopping and gives rise to an electric current through a "water wire."
- Biochemical processes depend on weak noncovalent interactions, which permit structures to exist for short periods of time. The four basic types of weak noncovalent interactions in nature are hydrogen bonds, ionic interactions, van der Waals interactions, and hydrophobic effects.
- The addition of nonpolar compounds to water breaks hydrogen bonds between H_2O molecules without replacing them and leads to the formation of ordered cage-like H_2O structures, which is energetically unfavorable.
- Hydrophobic effects result from nonpolar compounds associating with each other to minimize the amount of H_2O that must be ordered at the interface between the hydrophobic region and H_2O.
- The colligative properties of aqueous solutions (freezing point depression, boiling point elevation, vapor pressure lowering, and osmotic pressure) are affected by the concentration of solute molecules, not their chemical properties. Osmotic pressure is especially important in biological systems where solute concentrations modulate cell size as a result of water diffusion across a semipermeable plasma membrane.
- The ionization of water gives rise to hydrogen ions, H^+, which are protons, and hydroxyl ions, OH^-. Protons do not exist free in solution, but instead combine with H_2O to generate hydronium ions, H_3O^+.
- The water ionization constant $K_w = [H^+][OH^-] = 1.0 \times 10^{-14}\ M^2$. The values of H^+ concentration and OH^- concentration are reciprocally related, and $[H^+] = [OH^-] = 1.0 \times 10^{-7}$ M in pure water.
- The pH scale runs from 0 to 14 and is a convenient method to describe H^+ concentration in aqueous solutions, using the expression $pH = -\log[H^+]$. Solutions with $pH < 6.5$ are considered acidic ($[H^+] > [OH^-]$), solutions with $pH > 7.5$ are basic ($[H^+] < [OH^-]$), and neutral solutions have a pH in the range of 6.5 to 7.5 ($[H^+] \approx [OH^-]$).
- Acids are proton donors, and bases are proton acceptors. The ionization reaction of an acid–base conjugate pair can be written as $HA \rightleftharpoons H^+ + A^-$.
- The acid dissociation constant, K_a, is derived from the ionization reaction of an acid and can be defined as $pK_a = -\log K_a$. Acid–base conjugate pairs with low pK_a values are able to dissociate protons at low pH, whereas acid–base conjugate pairs with high pK_a values dissociate protons at high pH.
- The Henderson–Hasselbalch equation relates pH and pK_a and can be used to determine the ratio of HA (proton donor) and A^- (proton acceptor) at a given pH if the pK_a is known:

$$pH = pK_a + \log \frac{[A^-]}{[HA]}$$

- In aqueous solutions with pH values below the pK_a, the acid–base conjugate pair is mostly in the protonated form ($[HA] > [A^-]$), whereas in aqueous solutions with pH values above the pK_a, the acid–base conjugate pair is mostly in the unprotonated form ($[A^-] > [HA]$).
- A titration curve is a plot of experimental data showing the pH of a solution as a function of the amount of base added. Titration curves can be used to determine the amount of acid in a solution and to identify the pK_a of a weak acid.
- Buffers are aqueous solutions that resist changes in pH because of the protonation or deprotonation of an acid–base conjugate pair present at a high enough concentration to absorb small changes in H^+ or OH^- concentration.
- The acid–base conjugate pairs of phosphoric acid and carbonic acid are two of the biological buffers that function to keep pH values in the neutral range.

2.3 Cell Membranes Function as Selective Hydrophobic Barriers

- Separation of aqueous compartments by hydrophobic lipid bilayers permits regulation and specialization of biochemical processes. Selective exchange of nutrients and toxic waste products across cell membranes requires transmembrane proteins.
- The major components of cellular membranes are phospholipids, which are amphipathic molecules that contain both hydrophobic (water-fearing) and hydrophilic (water-loving) chemical groups. Phospholipids form lipid monolayers at the air–water interface or, upon vigorous mixing, generate lipid bilayers, micelles, and vesicles.
- Lipids can move laterally within cell membranes by simple diffusion; however, the fluidity of membranes can differ depending on temperature and lipid composition.
- Eukaryotic cells contain three major membrane types: (1) a plasma membrane that surrounds the entire cell; (2) an endomembrane system of cytoplasmic membrane structures; and (3) organelle membranes in mitochondria and chloroplasts that function in energy conversion processes.
- The endomembrane system is a network of lipid bilayers that includes the nuclear envelope, smooth and rough endoplasmic reticulum, the Golgi apparatus, and vesicles carrying catabolic enzymes (lysosomes and peroxisomes).
- Mitochondria and chloroplasts are subcellular organelles in eukaryotic cells that contain membrane-embedded proteins, which carry out energy-converting reactions leading to the production of ATP.

biochemical terms

(in order of appearance in text)
bioenergetics (p. 40)
homeostasis (p. 40)
equilibrium (p. 40)
solar energy (p. 41)
photosynthetic autotroph (p. 41)
aerobic respiration (p. 41)
heterotroph (p. 41)
photosynthesis (p. 41)
carbon fixation (p. 42)
biosphere (p. 42)
redox reaction (p. 42)
photooxidation (p. 42)
photophosphorylation (p. 42)
oxidative phosphorylation (p. 42)
first law of thermodynamics (p. 44)
second law of thermodynamics (p. 44)
entropy (p. 44)
enthalpy (H) (p. 44)
bomb calorimeter (p. 44)
exothermic (p. 45)
endothermic (p. 45)
calorie (cal) (p. 45)
joule (J) (p. 45)
Gibbs free energy (G) (p. 48)
exergonic (p. 48)
endergonic (p. 48)
standard free energy change ($\Delta G°$) (p. 48)
biochemical standard conditions (p. 48)
equilibrium constant (K_{eq}) (p. 49)
adenylate system (p. 53)
isozyme (p. 54)
isoform (p. 54)
energy charge (EC) (p. 54)
catabolic pathway (p. 55)
anabolic pathway (p. 55)
hydrogen bond (p. 57)
proton hopping (p. 58)
ionic interaction (p. 61)
van der Waals interaction (p. 61)
van der Waals radius (p. 62)
hydrophobic (p. 63)
hydrophilic (p. 63)
colligative properties (p. 67)
osmotic pressure (p. 68)
osmosis (p. 68)
erythrocyte (p. 69)
hypotonic (p. 69)
hypertonic (p. 69)
isotonic (p. 69)
contractile vacuole (p. 69)
hydrogen ion (H^+) (p. 71)
hydroxyl ion (OH^-) (p. 71)
hydronium ion (H_3O^+) (p. 71)
water ionization constant (K_w) (p. 71)
pH (p. 72)
weak acid (p. 73)
conjugate base (p. 73)
acid dissociation constant (K_a) (p. 73)
pK_a (p. 73)
Henderson–Hasselbalch equation (p. 74)
titration curve (p. 75)
equivalent (p. 76)
buffer (p. 76)
polyprotic acid (p. 76)
Le Châtelier's principle (p. 78)
acidosis (p. 78)
alkalosis (p. 78)
biological membrane (p. 79)
phospholipid bilayer (p. 79)
micelle (p. 80)
liposome (p. 80)
flippase (p. 82)
endomembrane system (p. 83)

review questions

1. Give examples of three types of work cells perform that require energy conversion.
2. Why are redox reactions so critical in terms of the survival of organisms?
3. What is the process by which ATP is formed by a phosphorylation reaction using energy released from the redox reactions in photosynthesis, and why is this important?
4. Write out the equation for the change in Gibbs free energy. Briefly describe the meaning of this equation.
5. How can reactions that are unfavorable (endergonic) under standard conditions occur in living cells?
6. What three properties of water make it so vital for sustaining life? Explain them.
7. Name and briefly describe the four basic types of weak interactions encountered in biochemistry.
8. What are colligative properties?
9. What kind of scale is the pH scale? What values are considered acidic, neutral, and basic?
10. What does it mean for a molecule to be amphipathic? Why are amphipathic lipids important for life?
11. Name and briefly describe four complexes of phospholipids formed in aqueous environments. What role does the hydrophobic effect play in the formation of these complexes?
12. Name and briefly describe the three major classes of eukaryotic biological membranes.

challenge problems

1. Explain why plants at night can be thought of as animals in terms of the similarities in their bioenergetic needs.
2. How does the second law of thermodynamics apply to energy conversion systems in a living organism?
3. The change in free energy between reactants and products can be used to determine if a reaction is spontaneous.
    a. What is meant by the free energy terms $\Delta G°'$ and ΔG?

b. Write an equation that describes $\Delta G^{\circ\prime}$ when a reaction is at equilibrium.

c. What effects are there on ΔG, $\Delta G^{\circ\prime}$, or K_{eq} if an enzyme is present in the reaction? Explain.

4. The equilibrium constant K_{eq} for the reaction $A \rightleftharpoons B$ is 1×10^5 at 25 °C.

a. If you started with a solution containing 1.000 M of A and 1 mM of B and let the reaction proceed to equilibrium, what would be the equilibrium concentrations of A and B?

b. Calculate $\Delta G^{\circ\prime}$ for this reaction.

c. In cells at steady state, the concentration of A is 0.05 mM and the concentration of B is 15 mM. Calculate ΔG for the reaction $A \rightleftharpoons B$.

5. Here are five hypothetical reactions:

1. $A \rightleftharpoons B$	$\Delta G^{\circ\prime} = -5$ kJ/mol
2. $B \rightleftharpoons C + D$	$\Delta G^{\circ\prime} = +8$ kJ/mol
3. $B \rightleftharpoons E$	$\Delta G^{\circ\prime} = -10$ kJ/mol
4. $E \rightleftharpoons F$	$\Delta G^{\circ\prime} = +5$ kJ/mol
5. $D \rightleftharpoons F$	$\Delta G^{\circ\prime} = +1$ kJ/mol

a. From these five reactions, it is possible to write two metabolic pathways that could generate the metabolite F starting with metabolite A. What are these two pathways?

b. Which pathway is most likely to proceed, based on the overall $\Delta G^{\circ\prime}$ value?

6. The energy charge (EC) of a cell reflects the amount of phosphoryl transfer energy available from ATP. The levels of ATP, ADP, and AMP in a cell are maintained in part by the adenylate kinase reaction, which interconverts $ATP + AMP \rightleftharpoons ADP + ADP$ ($\Delta G^{\circ\prime} \approx 0$).

a. Calculate the energy charge in a cultured cancer cell line found to have the following adenylate concentrations: 1.25 mM ATP, 0.35 mM ADP, and 0.12 mM AMP.

b. What is the K_{eq} of the adenylate kinase reaction in this cancer cell line, assuming the reaction is at equilibrium under the growth conditions used?

7. Explain how energy charge affects metabolic flux through anabolic and catabolic pathways.

8. Hydrolysis of ATP is favorable ($\Delta G^{\circ\prime} = -30.5$ kJ/mol), but the addition of inorganic phosphate (P_i) to glucose to yield glucose-6-phosphate (the first step in glycolysis) is unfavorable ($\Delta G^{\circ\prime} = +13.8$ kJ/mol). Show how these two reactions can be coupled to favor formation of glucose-6-phosphate and allow glycolysis to proceed. Include the overall $\Delta G^{\circ\prime}$ value for the reaction.

9. Why is ΔS negative when a nonpolar molecule such as limonene is dissolved in water?

10. In solid NaCl, both the sodium and chloride ions are held in a rigid crystalline lattice and are immobile in an electric field. Frozen water (ice) also forms a lattice structure; however, in the presence of an electric field, significant proton mobility occurs in ice. Explain.

11. Name the four noncovalent interactions that occur within and between biomolecules that facilitate life processes at the molecular level. Which of these noncovalent interactions directly or indirectly involve H_2O molecules?

12. The total carbonate pool in blood plasma (blood without red blood cells) is 0.025 M and consists of both HCO_3^- HCO_3^- and CO_2(aq). The pK_a for the dissociation of H_2CO_3 to produce $HCO_3^- + H^+$ at 37 °C is 6.1. Because H_2CO_3 is readily produced in blood from dissolved $CO_2(aq) + H_2O$, in a reaction catalyzed by the enzyme carbonic anhydrase (see Figure 2.39), CO_2(aq) can be considered the conjugate acid and HCO_3^- the conjugate base in the bicarbonate buffering system.

a. What is the ratio of HCO_3^- and CO_2(aq) in blood plasma at pH 7.4?

b. What are the individual concentrations of CO_2(aq) and HCO_3^- under these same conditions?

13. Membrane proteins in a cell can be covalently labeled with a fluorescent compound. Photobleaching of the cells with a laser initially creates a nonfluorescing spot in the plasma membrane that can be observed by fluorescence microscopy. Explain why this nonfluorescing spot disappears over time.

14. The pK_a for a typical long-chain fatty acid is ~5. Explain why long-chain fatty acids can form micelles in solutions with pH > 7 but are insoluble in solutions with pH < 5.

smartwork5

If your instructor assigns homework with Smartwork5, access it here: digital.wwnorton.com/biochem.

suggested reading

Books and Reviews

Angermayr, S. A., Hellingwerf, K. J., Lindblad, P., and de Mattos, M. J. (2009). Energy biotechnology with cyanobacteria. *Current Opinion in Biotechnology*, *20*, 257–263.

Di Paolo, D., Pastorino, F., Brignole, C., Marimpietri, D., Loi, M., Ponzoni, M., and Pagnan, G. (2008). Drug delivery systems: application of liposomal anti-tumor agents to neuroectodermal cancer treatment. *Tumori*, *94*, 246–253.

Frey, T. G., and Mannella, C. A. (2000). The internal structure of mitochondria. *Trends in Biochemical Sciences*, *25*, 319–324.

Lynden-Bell, R. M. (2010). *Water and life: the unique properties of H_2O*. Boca Raton, FL: CRC Press.

Mayer, A. (2002). Membrane fusion in eukaryotic cells. *Annual Review of Cell and Developmental Biology*, *18*, 289–314.

Nicholls, D. G., and Ferguson, S. J. (2002). *Bioenergetics 3*. New York, NY: Academic Press.

Segal, I. H. (1976). *Biochemical calculations* (2nd ed.). New York, NY: Wiley.

Primary Literature

Arora, K., and Brooks, C. L., III. (2007). Large-scale allosteric conformational transitions of adenylate kinase appear to involve a population-shift mechanism. *Proceedings of the National Academy of Sciences USA*, *104*, 18496–18501.

Liu, K., Jia, Z., Chen, G., Tung, C., and Liu, R. (2005). Systematic size study of an insect antifreeze protein and its interaction with ice. *Biophysical Journal*, *88*, 953–958.

Oakhill, J. S., Scott, J. W., and Kemp, B. E. (2012). AMPK functions as an adenylate charge-regulated protein kinase. *Trends in Endocrinology and Metabolism*, *23*, 125–132.

Oaknin, A., Barretina, P., Perez, X., Jimenez, L., Velasco, M., Alsina, M., Brunet, J., Germa, J. R., and Beltran, M. (2010). CA-125 response patterns in patients with recurrent ovarian cancer treated with pegylated liposomal doxorubicin (PLD). *International Journal of Gynecological Cancer*, *20*, 87–91.

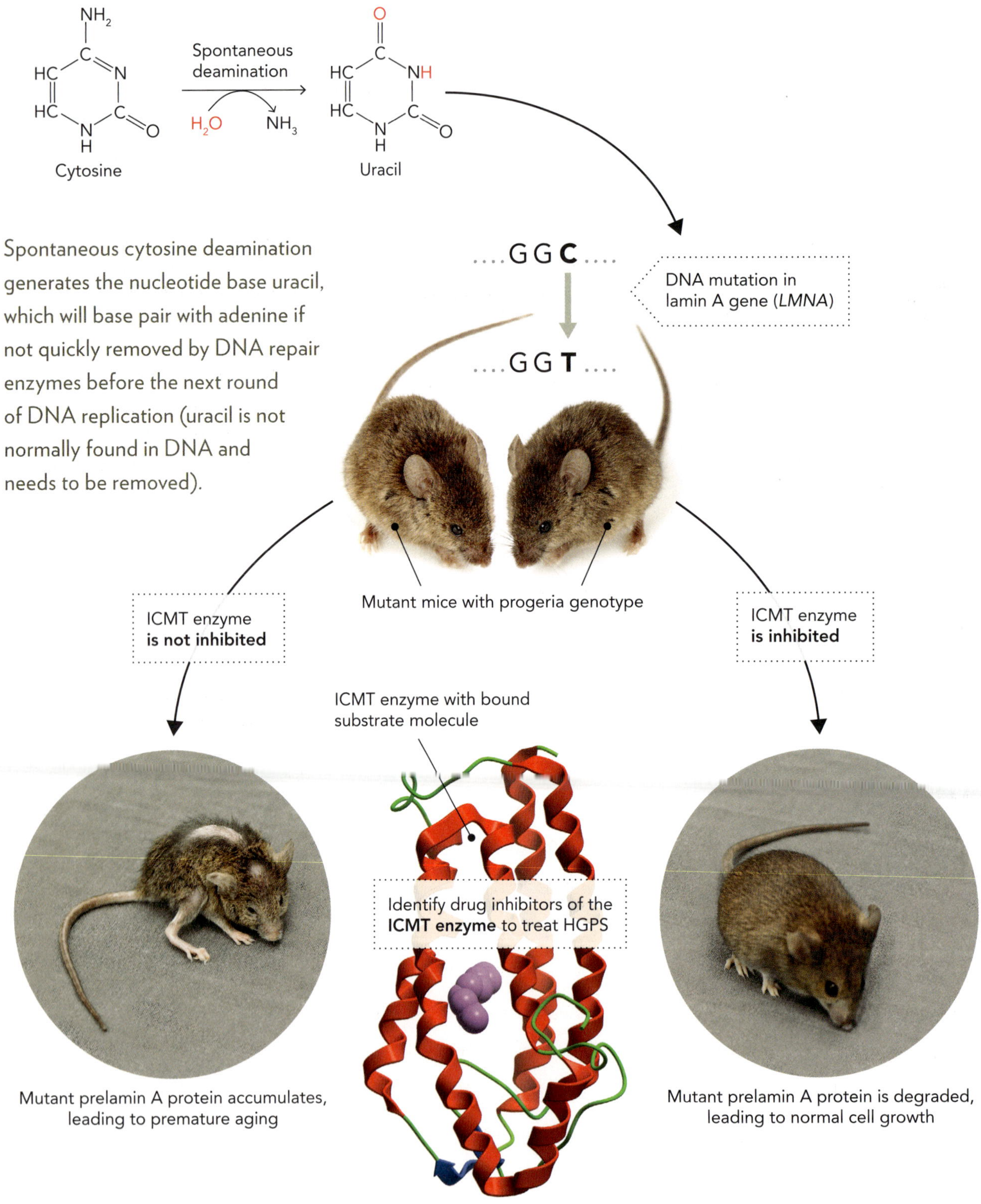

NH_2
HC
N
HC
C
N
H
O
Cytosine
Spontaneous deamination
H_2O
NH_3
O
C
HC
NH
HC
C
N
H
O
Uracil
Spontaneous cytosine deamination generates the nucleotide base uracil, which will base pair with adenine if not quickly removed by DNA repair enzymes before the next round of DNA replication (uracil is not normally found in DNA and needs to be removed).
....GGC....
....GGT....
DNA mutation in lamin A gene (*LMNA*)
Mutant mice with progeria genotype
ICMT enzyme **is not inhibited**
ICMT enzyme **is inhibited**
ICMT enzyme with bound substrate molecule
Identify drug inhibitors of the **ICMT enzyme** to treat HGPS
Mutant prelamin A protein accumulates, leading to premature aging
Mutant prelamin A protein is degraded, leading to normal cell growth

This newly formed U-A base pair then becomes a T-A base pair in the subsequent round of DNA replication; thus, the conversion C→U→T effectively occurs on the same strand of DNA (see Figure 3.22). Using a mouse model of HGPS, it was discovered that inhibition of the enzyme ICMT prevents progeria symptoms. The molecular structure of a related ICMT enzyme from the organism *Methanosarcina acetivorans* is shown with a substrate analog in the enzyme active site.

3

Nucleic Acid Structure and Function

◀ DNA sequencing has allowed scientists to identify genetic mutations that cause human disease, such as mutations causing Hutchinson–Gilford progeria syndrome (HGPS).

In HGPS, a single nucleotide substitution of C→T at position 1824 in the coding sequence of the lamin A gene (*LMNA*) causes aberrant RNA splicing and production of a truncated prelamin A protein. When the defective prelamin A protein is modified by the enzyme isoprenylcysteine carboxyl methyltransferase (ICMT), it is not processed to become a functional lamin A protein and accumulates at the nuclear membrane and inhibits cell division. Researchers found that by inhibiting the activity of the ICMT enzyme in a mouse model of HGPS, the defective prelamin A protein does not accumulate, and many of the progeria-associated symptoms are reduced. If pharmaceutical drugs can be developed that inhibit the activity of the ICMT enzyme in humans with the same *LMNA* mutation as that studied in the mouse model, then disease treatment may be possible.

CHAPTER OUTLINE

3.1 Structure of DNA and RNA

- Double-helical structure of DNA
- DNA denaturation and renaturation
- DNA supercoiling and topoisomerase enzymes
- Structural differences between DNA and RNA
- Nucleic acid binding proteins

3.2 Genomics: The Study of Genomes

- Genome organization in prokaryotes and eukaryotes
- Genes are units of genetic information
- Computational methods in genomics

3.3 Methods in Nucleic Acid Biochemistry

- Plasmid-based gene cloning
- High-throughput DNA sequencing
- Polymerase chain reaction
- Transcriptome analysis

CREDITS: MICE, TOP: BRACKISH_NZ/SHUTTERSTOCK; MICE, BOTTOM: M. X. IBRAHIM ET AL. (2013). TARGETING ISOPRENYLCYSTEINE METHYLATION AMELIORATES DISEASE IN A MOUSE MODEL OF PROGERIA. *SCIENCE, 340(6138)*, 1330–1333. DOI: 10.1126/SCIENCE.1238880. EPUB 2013 MAY 16. © 2013 AMERICAN ASSOCIATION FOR THE ADVANCEMENT OF SCIENCE; ICMT ENZYME: BASED ON PDB FILE 4A2N.

A new era of biochemistry and biotechnology began in 1953 when James Watson and Francis Crick, using the X-ray diffraction data of Rosalind Franklin, built a model of the three-dimensional structure of DNA. In Chapter 1, we reviewed the nucleotide building blocks of DNA and RNA and described how they fit in with the central dogma of molecular biology. Here in Chapter 3, we examine the nucleotide components of DNA and RNA in more detail as we begin to understand the interdependence of biomolecular structure and function. We will see that the properties of the nucleic acids DNA and RNA depend on the biochemistry of the nucleotides of which they are composed.

We first look at the molecular structures of DNA and RNA at the chemical level and explore how the higher-order structures of these molecules influence their properties and interactions with other biomolecules. Next, we examine the organization of genomes and some of the bioinformatic tools that have been developed to analyze them. Finally, we end the chapter with a biochemical explanation of four key methods in nucleic acid biochemistry that have been developed to analyze and manipulate genomes.

3.1 Structure of DNA and RNA

Before exploring the structures of DNA and RNA, it is important to review the building blocks of these molecules. DNA and RNA are formed from nucleotides that are linked together through a phosphodiester backbone into a linear chain (see Figure 1.10). Nucleotides are composed of a nitrogenous base, a five-carbon ribose or deoxyribose sugar, and one or more phosphate groups (**Figure 3.1a**). You can see that a **nucleoside** consists of a base and a sugar, whereas a nucleotide is a phosphorylated nucleoside. The bases in DNA and RNA are either purines, which contain nine atoms in the heterocyclic rings, or pyrimidines, which contain six atoms in the ring. **Figure 3.1b** shows the common nitrogenous bases found in DNA and RNA and the numbering system for the atoms within the bases.

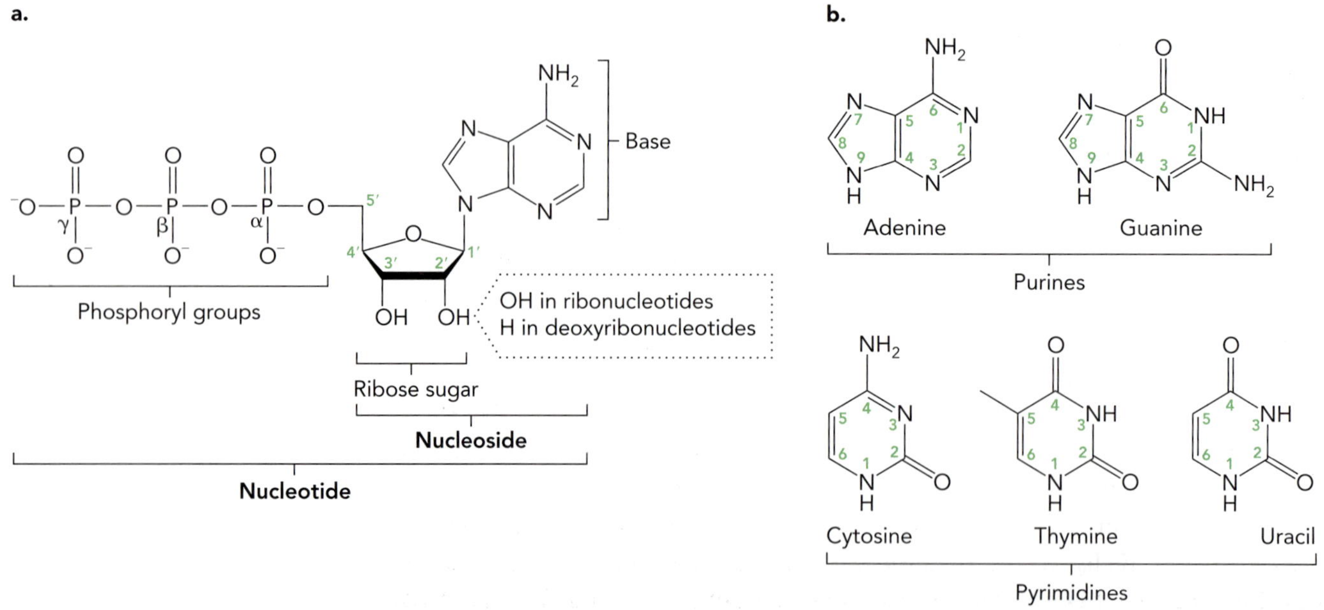

Figure 3.1 Nucleotides are the building blocks of DNA and RNA. **a.** Nucleotides consist of a nitrogenous base, a ribose or deoxyribose sugar, and one or more phosphoryl groups linked to the sugar. ATP is shown here as an example of a nucleotide. Numbered carbons in the ribose ring are labeled as prime (′) to distinguish them from numbered atoms in the nucleotide base. Ribonucleotides contain a hydroxyl group on C-2′ of ribose, which is lacking in deoxyribonucleotides. The phosphoryl groups are denoted as α, β, and γ, starting with the phosphoryl group closest to the sugar ring. **b.** The common bases in DNA and RNA are the purines adenine and guanine and the pyrimidines cytosine, thymine (DNA only), and uracil (RNA only). The numbering for the atoms within the bases is shown.

Table 3.1 NOMENCLATURE FOR THE NUCLEOSIDES AND NUCLEOTIDES CONTAINING THE FIVE MOST COMMON NITROGENOUS BASES

Nucleic acid	Base	Nucleoside	Nucleotide (nucleoside-5′-monophosphate)
RNA	Adenine (A)	Adenosine	Adenosine-5′-monophosphate, or adenylate (AMP)
	Guanine (G)	Guanosine	Guanosine-5′-monophosphate, or guanylate (GMP)
	Cytosine (C)	Cytidine	Cytidine-5′-monophosphate, or cytidylate (CMP)
	Uracil (U)	Uridine	Uridine-5′-monophosphate, or uridylate (UMP)
DNA	Adenine (A)	Deoxyadenosine	Deoxyadenosine-5′-monophosphate, or deoxyadenylate (dAMP)
	Guanine (G)	Deoxyguanosine	Deoxyguanosine-5′-monophosphate, or deoxyguanylate (dGMP)
	Cytosine (C)	Deoxycytidine	Deoxycytidine-5′-monophosphate, or deoxycytidylate (dCMP)
	Thymine (T)	Deoxythymidine	Deoxythymidine-5′-monophosphate, or deoxythymidylate (dTMP)

Using these bases, we can form the structures of many nucleoside and nucleotide derivatives, the most common of which are listed in **Table 3.1**. As shown in this table, ATP is named adenosine-5′-triphosphate, and AMP is adenosine-5′-monophosphate. All of the nucleoside monophosphates also have generic names that end in the suffix "-ylate." For example, adenosine-5′-monophosphate is called adenylate, and cytosine-5′-monophosphate is called cytidylate. The deoxyribonucleotides are abbreviated with a "d" at the beginning of the abbreviation to distinguish them from the ribonucleotides. For example, deoxyadenosine-5′-monophosphate is abbreviated dAMP. Note that deoxythymidine and deoxythymidylate are sometimes referred to as **thymidine** and thymidylate, respectively, as the thymine base is only found in deoxyribonucleotides.

Double-Helical Structure of DNA

DNA and RNA, like other biological macromolecules, have several levels of structure that influence their functions. The **primary structure** of the molecule refers to the unique arrangement of deoxyribonucleotides or ribonucleotides arranged in a single chain. At the level of **secondary structure**, two complementary strands of DNA bind together (or anneal) through base pairing in an antiparallel fashion. (RNA can also have secondary structure, as we will discuss later in this section.)

The DNA double helix is the secondary structure that most people recognize. Although we sometimes think of DNA existing only as this helical structure, in fact DNA must form more intricate structures in order to fit within the cell. These more complex structures also play a role in controlling DNA replication and the transcription of genes.

Figure 3.2 illustrates how the genomic DNA in both prokaryotes and eukaryotes is condensed to fit within the nucleoid or nucleus, respectively. Although all organisms must condense their DNA, the mechanism of condensation and the final form of the condensed structure differs. In prokaryotes, the circular chromosome is compacted in a manner reminiscent of what happens when a rubber band is twisted. Compaction of the bacterial chromosome is explained later in the chapter when we describe supercoiled DNA. Bacteria also contain small regions of additional DNA,

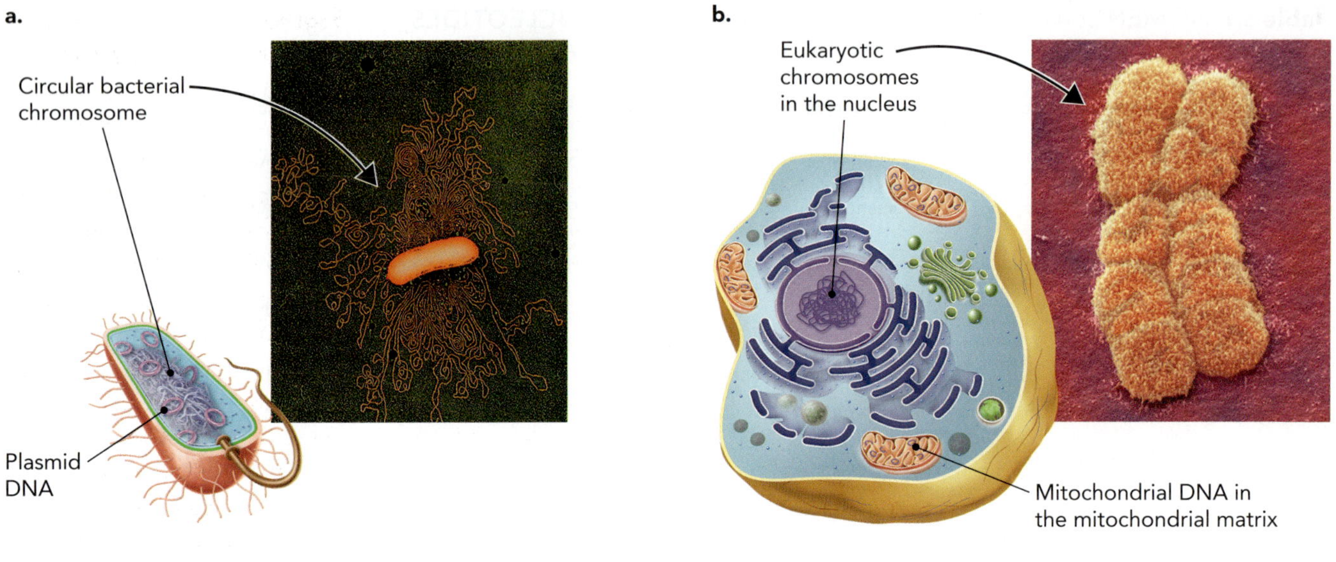

Figure 3.2 Genomic DNA is organized in both prokaryotes and eukaryotes. **a.** Bacterial genomic DNA forms a circular chromosome packaged within the nucleoid of the cell. An electron micrograph (right) shows the DNA released from a bacterium (orange cylinder). This DNA must be significantly condensed to fit within the intact organism. PHOTO: G. MURTI/SCIENCE SOURCE. **b.** Eukaryotic DNA is located in the nucleus of the cell. DNA strands are tightly wrapped around a protein component to form the well-recognized shape of the chromosome (right) observed in cells undergoing mitotic cell division. PHOTO: ANDREW SYRED/SCIENCE SOURCE.

known as plasmids (see Section 3.3), which can be compacted as well. In eukaryotes, the storage system for genomic DNA is more complex and involves a large protein component for support.

Compact storage of DNA is necessary because of the large amount of genetic information contained in every cell. Prokaryotic chromosomes can range in size from 0.6 million base pairs (megabase pairs, or Mbp) to more than 10 Mbp. Every base pair is separated from its neighboring base pair by approximately 0.34 nm (3.4 Å); therefore, prokaryotic DNA would be up to 3.4×10^6 nm (3.4 mm) in length if it were a linear double helix. A typical bacterial cell is 1 μm in diameter and 2–8 μm in length, which means its length is 1,000 times smaller than the extended length of its genomic DNA. Eukaryotic DNA is even larger. The human genome contains more than 6 billion base pairs per cell, thus a human cell would need to be more than 2 meters in diameter to contain that amount of DNA if it were not compacted. To understand how the complex DNA structures shown in Figure 3.2 are formed, we need to know more about the chemistry behind the formation of the DNA double helix.

Figure 3.3 shows the basic structure of a typical right-handed double helix formed by two complementary, antiparallel strands of DNA. The strands are often referred to as the coding (or sense) strand and the template strand. These strands are distinguished by considering that if the DNA were transcribed into RNA, the coding strand would have the same base sequence as the RNA transcript, while the template strand would have a complementary sequence to the transcribed RNA. As the two strands wind in a helix, grooves are formed on the outer portion of the helix. On one side, the phosphodiester backbones of each strand are close together, yielding a smaller or **minor groove**. On the opposite side, the backbones are farther apart, resulting in a larger or **major groove**. Because the major groove is much wider than the minor groove, it is often the site where proteins specifically bind to DNA (see Chapter 23).

As you can also see in Figure 3.3, A-T and G-C base pairs line the center of the molecule. The distance between sugar residues on opposing strands is the same for an A-T or G-C base pair, so any arrangements of nucleotides give the same helical structure within a small region. Over larger regions, the sequence of base pairs can have an effect on secondary structure, as we will see later. The hydrophobic nature of the bases leads to their sequestration in the interior of the molecule, whereas the polar sugar–phosphate backbone is exposed to water.

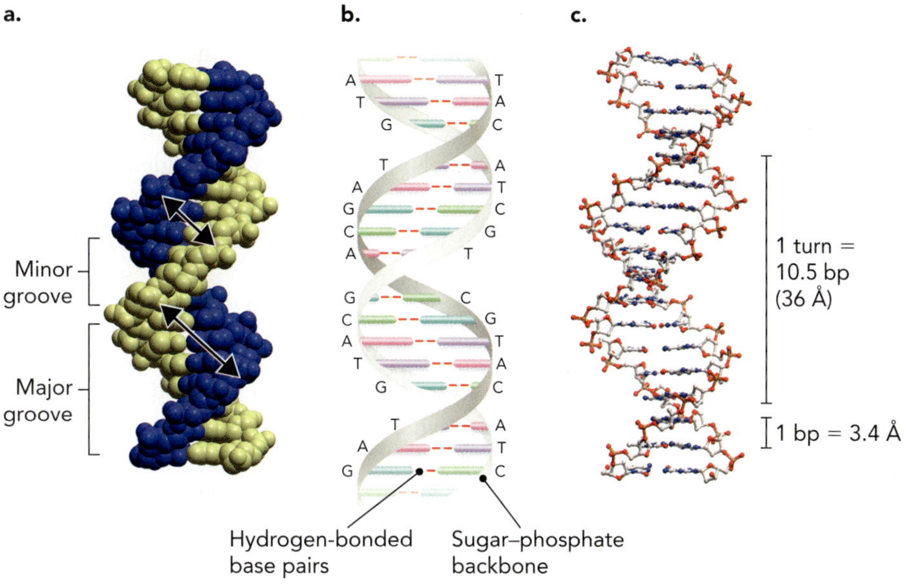

Figure 3.3 Two complementary antiparallel strands of DNA often from a double-helical structure, as shown here. **a.** A space-filling model of the double helix is shown based on the crystal structure of the common physiologic form of DNA (B-form DNA, or B-DNA). One strand is colored gold, and the other is blue. The minor and major grooves are indicated where the backbones are closer together or farther apart, respectively. BASED ON PDB FILE 1BNA. **b.** The hydrogen-bonded base pairs (A-T) and (C-G) are at the center of the helix, while the backbones of each strand, shown as ribbons, are at the exterior. **c.** The structural properties of B-form DNA can be observed in the ball-and-stick model of the same DNA molecule shown in panel **a**. The distance between adjacent base pairs is 3.4 Å. A complete turn of the helix in B-form DNA contains 10.5 base pairs (bp), or 36 Å.

Each turn of the standard double helix contains an average of 10.5 base pairs, resulting in a 36-Å helical turn (Figure 3.3). Analysis of crystallized DNA prepared from longer oligonucleotides has revealed that the turn distance can range from 28 to 42 Å, indicating that many local variations can occur in the DNA structure. These variations may be influenced by the base composition of a particular region, possibly enhancing the recognition of certain sequences in the major groove by DNA binding proteins.

In Chapter 1, we saw that Watson and Crick made a breakthrough in forming their model of DNA structure when they realized the importance of nucleotide base pairs. This discovery was due in part to the research of Erwin Chargaff, who determined in 1950 that although the base composition differed among organisms, there were approximately equal numbers of adenine and thymine residues and approximately equal numbers of guanine and cytosine residues. In other words, the ratios of A to T and G to C are each close to 1. The fact that A = T and C = G is known as **Chargaff's rule.** Watson and Crick realized that this equality in bases meant that pairs must exist, with each A pairing with T and each C pairing with G. The challenge was to understand how these base pairs formed and what forces held them together.

Watson and Crick manipulated their wire-frame model to pair these bases while keeping the overall structure in agreement with Rosalind Franklin's X-ray diffraction data. Eventually, they realized that in an antiparallel arrangement, the base pairs lined up to allow the formation of three hydrogen bonds in a G-C pair and two hydrogen bonds in an A-T pair. As shown in **Figure 3.4**, a G-C base pair forms its three hydrogen bonds between the NH_2 at C-2 on guanine and the C-2 carbonyl

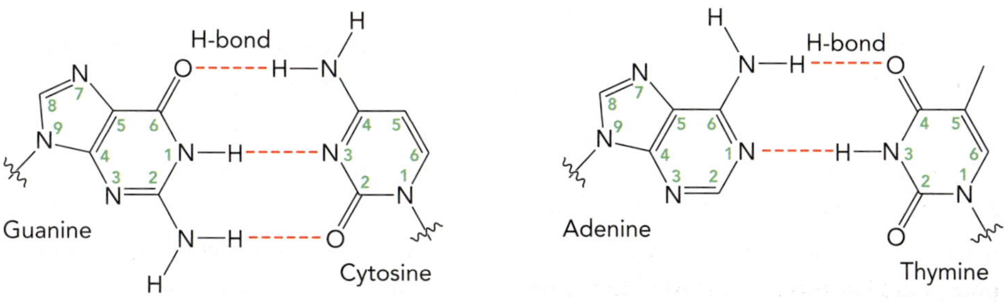

Figure 3.4 Hydrogen bonding interactions form G-C and A-T base pairs. In DNA, three hydrogen bonds form between guanine and cytosine, and two hydrogen bonds form between adenine and thymine. The numbering of the atoms within the bases is indicated. Hydrogen bonds are shown as dashed red lines between the bases.

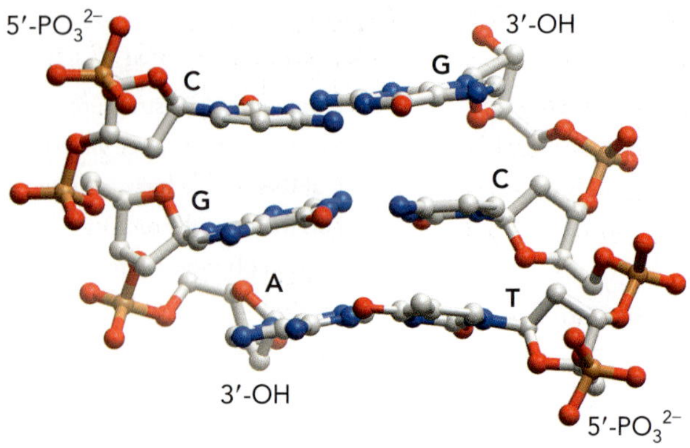

Figure 3.5 Two antiparallel strands can form base pairs. In this model of a trinucleotide segment of DNA, the DNA sequence on the left strand reads 5′-CGA-3′. The complementary nucleotides 5′-TCG-3′ are on the opposite strand. (Note that by convention, DNA sequences are listed in the 5′ to 3′ direction unless otherwise indicated.) The phosphate backbones are on the outside edges of the molecule, placing the base pairs in a stacked arrangement in the center. BASED ON PDB FILE 1BNA.

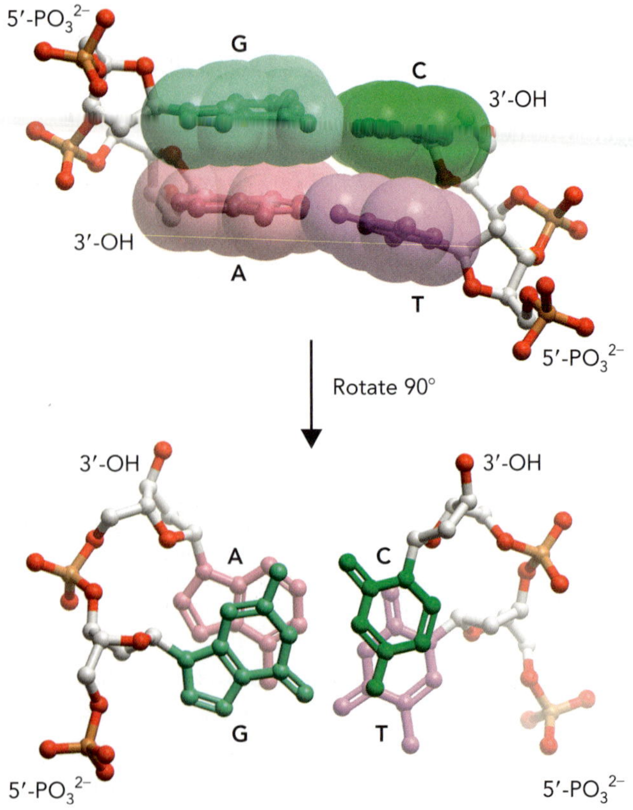

Figure 3.6 Nucleotide base stacking stabilizes the DNA double helix. Ball-and-stick dinucleotide segments of DNA are shown with the electron cloud interactions (π–π) indicated by the space-fill shadow around each base. A rotation of 90° (bottom) shows the overlap of the nucleotide bases. BASED ON PDB FILE 1BNA.

on cytosine; the N-1 H of guanine and N-3 of cytosine; and the C-6 carbonyl of guanine and the NH_2 on C-4 of cytosine. The two hydrogen bonds between an A-T base pair form between the N-1 of adenine and the N-3 H of thymine and between the NH_2 at C-6 of adenine and the C-4 carbonyl of thymine. The standard A-T and G-C base pairs with these hydrogen bonds are also referred to as Watson–Crick base pairs.

Hydrogen bonding is optimized in Watson–Crick base pairing, but this does not mean that other base pair configurations do not exist or that all possible hydrogen-bonding pairs are satisfied when DNA is in this conformation. The unpaired hydrogen-bond donors and acceptors are available to interact with other molecules or with DNA binding proteins. As we will see later in this section, other base pairs can form in a DNA or RNA helix and within an RNA molecule, both with these standard nucleotides and with other, modified forms of the bases.

Orientation of a base to allow for hydrogen bonding with another base requires that they be arranged in a planar fashion, parallel to the adjacent bases on the same strand and located in the interior of the helix (**Figure 3.5**). In this position, the base pairs are stacked upon each other and within van der Waals contact distance, which is referred to as base stacking (**Figure 3.6**). Stacking of the DNA bases in the interior of the helix provides stability to the molecule through the hydrophobic effect and van der Waals interactions (see Chapter 2).

When adjacent bases are at their van der Waals distance, an attraction arises between them and provides enthalpic stability. Though each interaction is relatively weak on its own, these van der Waals forces are additive and therefore contribute a significant amount of stability to the molecule. Base stacking interactions are further enhanced by the hydrophobic effect. In a double helix, the bases are sequestered such that they interact primarily with each other but not water. Therefore, the water does not need to be organized around these hydrophobic molecules, which is entropically favorable. Hydrogen bonding of the bases also provides some stability to a helix; however, because the hydrogen bonds can form in both the helical state (with another base) and the unfolded state (with water molecules), hydrogen bonding contributes little to overall stability of the helix. Most of the stability of a DNA double helix is provided by base stacking. Under physiologic conditions, the overall free energy change favors the helical state for most DNA molecules.

Table 3.2 lists the base stacking energies of possible dinucleotide base pairs. You can see that the stacking energy is more negative (more stable) for complementary dinucleotides containing at least one G-C base pair because of more favorable interactions for these base pairs. You can also see that differences in stacking energy depend on the orientation and order of the base pairs. For example, the G-C/C-G interaction is more stable (–9.37 kJ/mol) than the C-G/G-C interaction (–9.07 kJ/mol) because the right-handed twist of the helix allows more favorable interactions for the first set of stacked base pairs.

The orientations of base pairs and the ribose–phosphate backbone are not fixed. Some rotational constraints affect movement, but each base has the ability to twist and rotate to some extent. The ribose–phosphate backbone also has some rotational freedom. This freedom can result in alternative conformations of the double helix, depending on the sequence of the region and the conditions of the aqueous environment.

The double-helical structure proposed by Watson and Crick was based on the X-ray diffraction pattern of hydrated DNA under conditions similar to those found in the cell. Determination of the crystal structure of DNA molecules confirmed this structure, now known as B-form DNA (B-DNA), and it is thought to be the most stable conformation for the helix under physiologic conditions. However, crystallization and X-ray diffraction of different DNA sequences under different conditions has revealed alternative forms of the molecule (**Figure 3.7**). These three forms have

Table 3.2 BASE STACKING ENERGIES OF ADJACENT BASE PAIRS

Stacked dimer[a]	Stacking energy[b] (kJ/mol)
G-C C-G	−9.37
C-G G-C	−9.07
G-C G-C	−7.70
C-G A-T	−6.07
G-C T-A	−6.03
G-C A-T	−5.44
C-G T-A	−5.36
A-T A-T	−4.18
A-T T-A	−3.68
T-A A-T	−2.43

[a]The hyphens indicate base pairs. The top base pair is stacked on the bottom base pair, such that the left-hand dinucleotide reads 5′ to 3′ down the left side, and the right-hand dinucleotide reads 3′ to 5′ down the right side; for example, the fourth entry is a 5′ CA 3′ dinucleotide base paired with a 3′ GT 5′ dinucleotide.

[b]The stacking energy is given as the nearest-neighbor standard free energy change in 1 M NaCl.

BASED ON J. SANTALUCIA, JR., AND D. HICKS (2004). THE THERMODYNAMICS OF DNA STRUCTURAL MOTIFS. *ANNUAL REVIEW OF BIOPHYSICS AND BIOMOLECULAR STRUCTURE*, 33, 415–440.

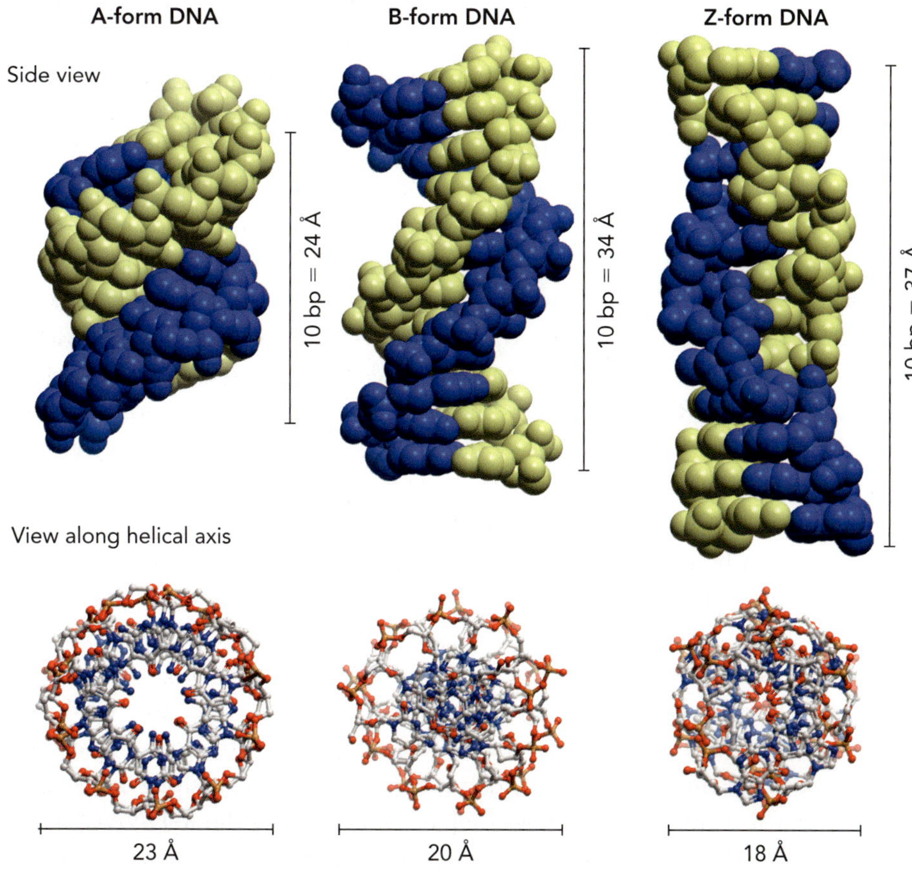

Figure 3.7 The A, B, and Z conformations are three arrangements of the DNA double helix. Space-fill models are shown with one strand colored blue and the other colored yellow. Each structure shown here was generated from the crystallization of a 12-base-pair sequence.
THE A-DNA STRUCTURE IS BASED ON PDB FILE 1ANA; B-DNA ON PDB FILE 1BNA; AND Z-DNA ON PDB FILE 2DCG.

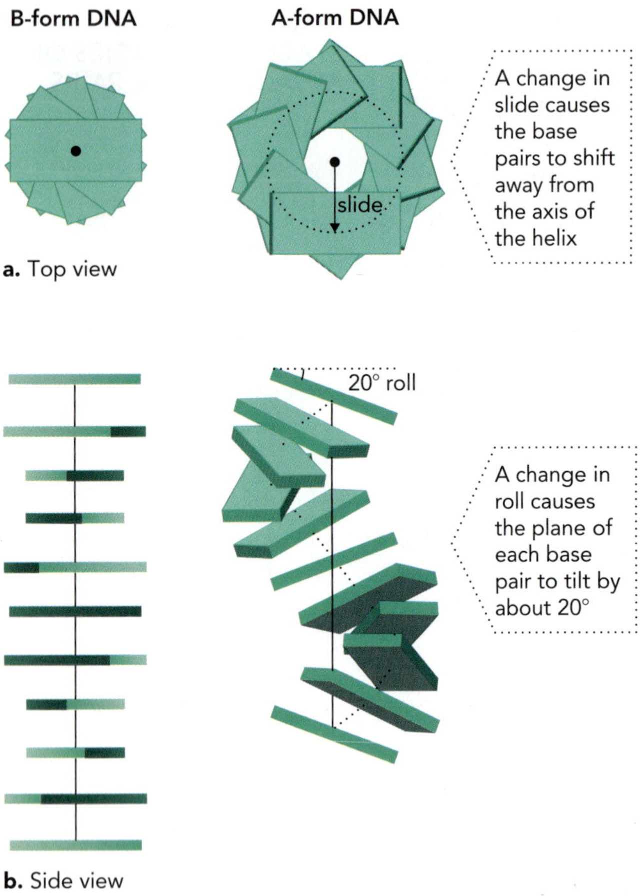
B-form DNA
A-form DNA
A change in slide causes the base pairs to shift away from the axis of the helix
slide
a. Top view
20° roll
A change in roll causes the plane of each base pair to tilt by about 20°
b. Side view

Figure 3.8 B-form DNA can be converted into A-form DNA by altering the orientation of the bases. **a.** View down the axis of B-form (left) or A-form (right) DNA. Each base pair is represented by a rectangle. The dot represents the axis of the helix. **b.** The view from the side of the B-form (left) or A-form (right) DNA helix. The line represents the axis of the helix. Changes in slide and roll of the base pairs affect DNA conformation.

identical covalent linkages between the nucleotides, but several structural differences exist from changes in conformation and noncovalent interactions. The most obvious difference is overall shape: A-form DNA (A-DNA) has the shortest and widest (23 Å in diameter) helix, and Z-form DNA (Z-DNA) is the most narrow (18 Å in diameter). The shape differences are reflected in the number of bases per helical turn. The wide helix of A-DNA allows 11 base pairs per turn, slightly more than B-DNA. The stretched nature of Z-DNA allows for an additional base pair, giving it 12 base pairs per helical turn.

Both the A and B conformations are right-handed helical structures, but the Z form is a very different conformation (see Figure 3.7). Although the two strands in Z-DNA do run antiparallel and exhibit the same Watson–Crick base pairing, the helix is left-handed rather than right-handed. The DNA bases are arranged in a zigzag fashion, giving the helix its name. This conformation can be formed with sequences containing an alternating purine/pyrimidine sequence of deoxynucleotides; those with alternating dC and dG are most likely to form the Z structure. The biological significance of Z-DNA is not fully understood, but this conformation has been detected *in vivo*, as have examples of proteins that preferentially bind Z-form DNA.

We can think of the A-, B-, and Z-DNA conformations in Figure 3.7 as idealized structures. The structure of genomic DNA *in vivo* most closely resembles B-form DNA. The A conformation, although not found as frequently in DNA, is known to occur in X-ray structures of RNA duplexes and in DNA–RNA hybrid molecules. It is important to understand that A-, B-, and Z-DNA are not completely unrelated structures and that a large stretch of genomic DNA does not have a perfectly regular or identical structure. Regions within a stretch of genomic DNA may form structures more closely resembling A- or Z-DNA, depending on the sequence and presence of protein factors. It is also possible for B-DNA to transition to A- or Z-DNA without extreme changes in the environmental conditions.

B-DNA can be converted into A-DNA by altering both the roll and slide of the adjacent bases (**Figure 3.8**). The change in roll (angle of the base pairs relative to the helical axis) results in a tilt of ~20° in the plane of the base pairs relative to B-DNA. Changing slide (displacement of the base pairs from the helical axis) between two adjacent bases results in a change in the opposite ribose residue to a C-3′ endo conformation in which C-3′ is above the plane formed by the other ribose atoms. (In B-DNA, the ribose ring is in the C-2′ endo conformation.) The resulting double helix has a deeper major groove and shallower minor groove, giving the A-DNA a shorter and wider appearance compared to B-DNA. A-DNA cannot bind water molecules as easily as does B-DNA, which explains why the A-DNA structure was first identified in a dehydrated DNA strand. Although A- and B-form DNA are traditionally depicted as two different structures, intermediates between the two forms have been shown to exist under very mild conditions, indicating that the transition from one form to another may occur as part of natural structural variation in the genome.

DNA Denaturation and Renaturation

In Chapter 1, we explained that DNA is the carrier of genetic information. In order for this information to be accessed and used, the two strands of a double helix must be separated within a specific region. Strand separation allows DNA replication or transcription complexes to bind and copy or transcribe one or both strands. In the laboratory, heating or addition of acid or alkali agents to alter the pH can separate DNA strands in solution. In the cell, strand separation is carried out by enzymes known as helicases (see Chapter 20).

In both cases, strand separation requires that the conditions be altered in a region of DNA to unwind the double helix and disrupt the hydrogen bonds between the base pairs. The ease with which this process occurs is governed by the base composition of the molecule and the length of the molecule, and it can also be influenced by the solution conditions, including ionic strength.

To monitor the strand separation of a DNA molecule, we can take advantage of the physical and chemical properties of the bases. All nucleotide bases consist of aromatic rings that absorb light in the ultraviolet (UV) range; therefore, nucleotides or nucleic acids containing these bases also absorb UV light. Although the four nucleotides have slightly different absorbance maxima (**Figure 3.9**), in practice, for mixtures of nucleotides or DNA molecules, absorbance is commonly measured at 260 nm, where all bases have a strong absorbance.

The absorbance of DNA at 260 nm can be used to monitor the **denaturation** (also called melting) of the double helix into two individual strands because of a property known as the **hyperchromic effect**. Double-stranded DNA absorbs less light at 260 nm than that absorbed by single-stranded molecules (due to the stacking interactions in double-stranded DNA). Therefore, as the DNA unwinds and denatures, its absorbance increases. We can monitor this increase in absorbance and use it to determine

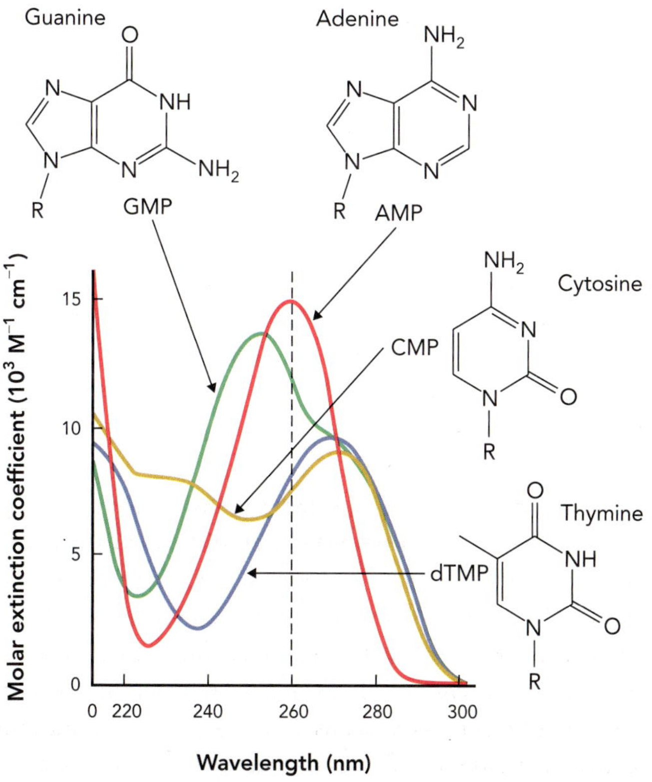

Figure 3.9 Nucleotides absorb light in the ultraviolet range. In these absorbance spectra of four common nucleotides (GMP, CMP, AMP and dTMP), the molar extinction coefficient is measured on the y axis, and the wavelength is on the x axis. The absorbance of a DNA molecule or a mixture of nucleotides is commonly measured at 260 nm, which is a wavelength where all the bases have a strong absorbance.

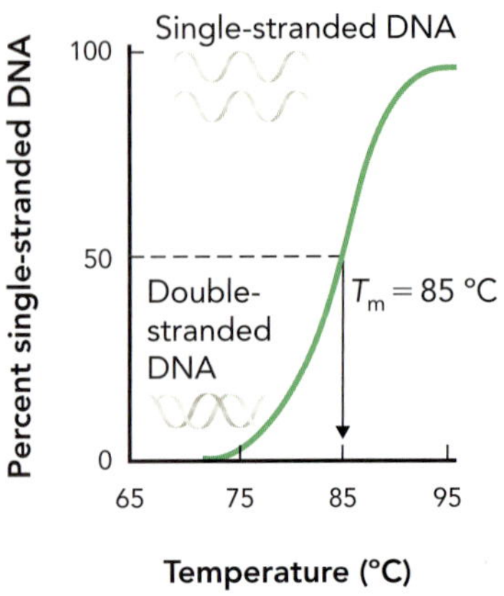

Figure 3.10 DNA absorbance at 260 nm can be used to measure the melting temperature (T_m). The melting curve plots percent single-stranded DNA (determined from relative absorbance at 260 nm) on the y axis and temperature on the x axis. As temperature increases, a DNA molecule unwinds, and the two strands separate. The absorbance of DNA increases as it becomes single stranded. The T_m is found at the midpoint of this curve and is the temperature at which half of the DNA molecules are denatured.

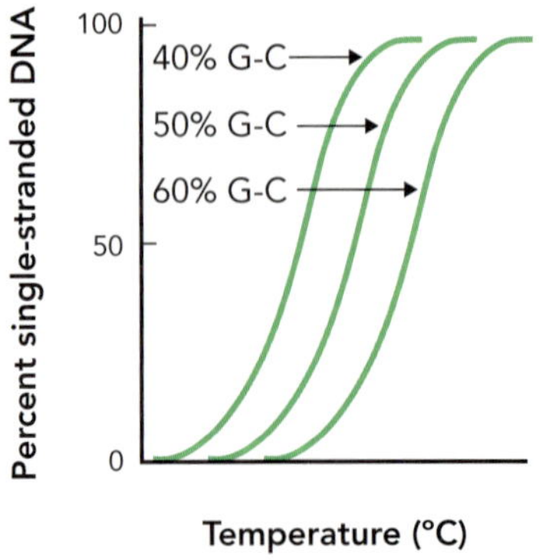

Figure 3.11 Melting temperature is influenced by G-C content. A melting curve of percent single-stranded DNA plotted against temperature is shown here for three different fragments of DNA of the same length in solutions of the same ionic strength. As the G-C content increases from 40% to 60%, there is a corresponding increase in the T_m.

the amount of DNA that has been denatured under a particular set of conditions. **Renaturation** (also called annealing) of two DNA strands to form a helix can also be monitored using this same method.

It can be important to determine the **melting temperature** (T_m) for a region of DNA when using certain methods in nucleic acid biochemistry. We define T_m as the temperature at which half of the DNA molecules are denatured ("melted") to the single-stranded state, and half of the molecules are double stranded. We can identify this point by monitoring absorbance changes at 260 nm that reflect the transition from double-stranded to single-stranded DNA as the temperature increases. The temperature at the midpoint in the resulting curve is the T_m (**Figure 3.10**).

The sigmoidal shape of the melting curve is due to the fact that for relatively short segments of DNA, the DNA molecules are either all double stranded or all single stranded, and there are essentially no partially folded molecules. Once the DNA molecule starts to unfold, the transition to the single-stranded state occurs cooperatively. Such cooperative transitions, which can be thought of as "all-or-nothing," are observed in numerous biological systems. (Cooperative changes in protein structure will be discussed in detail in Chapter 5.) The steepness of the curve indicates that the transition between the two states occurs over a relatively narrow temperature range. For DNA, the cooperative transition between states occurs because the initial formation of a small number of base stacking interactions is energetically unfavorable. However, once this nucleus of structured DNA forms, the addition of more base stacking interactions gives increased stabilization. This cooperative folding and unfolding for DNA molecules only occurs for DNA molecules of less than ~100 bp. In biological systems where DNA segments are usually much longer, the DNA can have small regions of single-stranded DNA, while the remainder remains in the double-stranded, or duplex, state. This is important because small regions of DNA often need to be separated—for duplication of a chromosome, for example—but the entire length of the DNA does not need to be separated at one time.

Each unique double-stranded DNA sequence has a unique T_m. We know from the previous section that sequences rich in G-C have more stability than A-T base pairs primarily due to more favorable base stacking interactions (not because of differences in the number of hydrogen bonds). Therefore, sequences with numerous A-T base pairs are more easily disrupted, and less heat energy is required to dissociate the base stacking interactions. The effect of G-C content on T_m is significant: As the G-C content increases, so does the T_m (**Figure 3.11**).

The fact that DNA regions rich in A-T are more easily denatured than regions with a higher G-C content is biologically significant. In genomic DNA, regions rich in A-T are often the initiation sites for DNA replication or transcription because the ease of unwinding and separating DNA strands in such regions allows replication or transcription to begin with a lower input of energy.

The length of double-stranded DNA and the presence of positively charged ions in solution also affect the T_m (**Figure 3.12**). As we discussed earlier, several forces act to stabilize the double helix. The longer the strand of DNA, the greater the stabilizing forces that are present, and therefore more heat energy is required to dissociate the strands. *In vivo*, it is important that only a region undergoing active replication or transcription be exposed and denatured. The energy required to unwind long stretches of DNA prohibits the entire chromosome from unwinding when a small region is exposed.

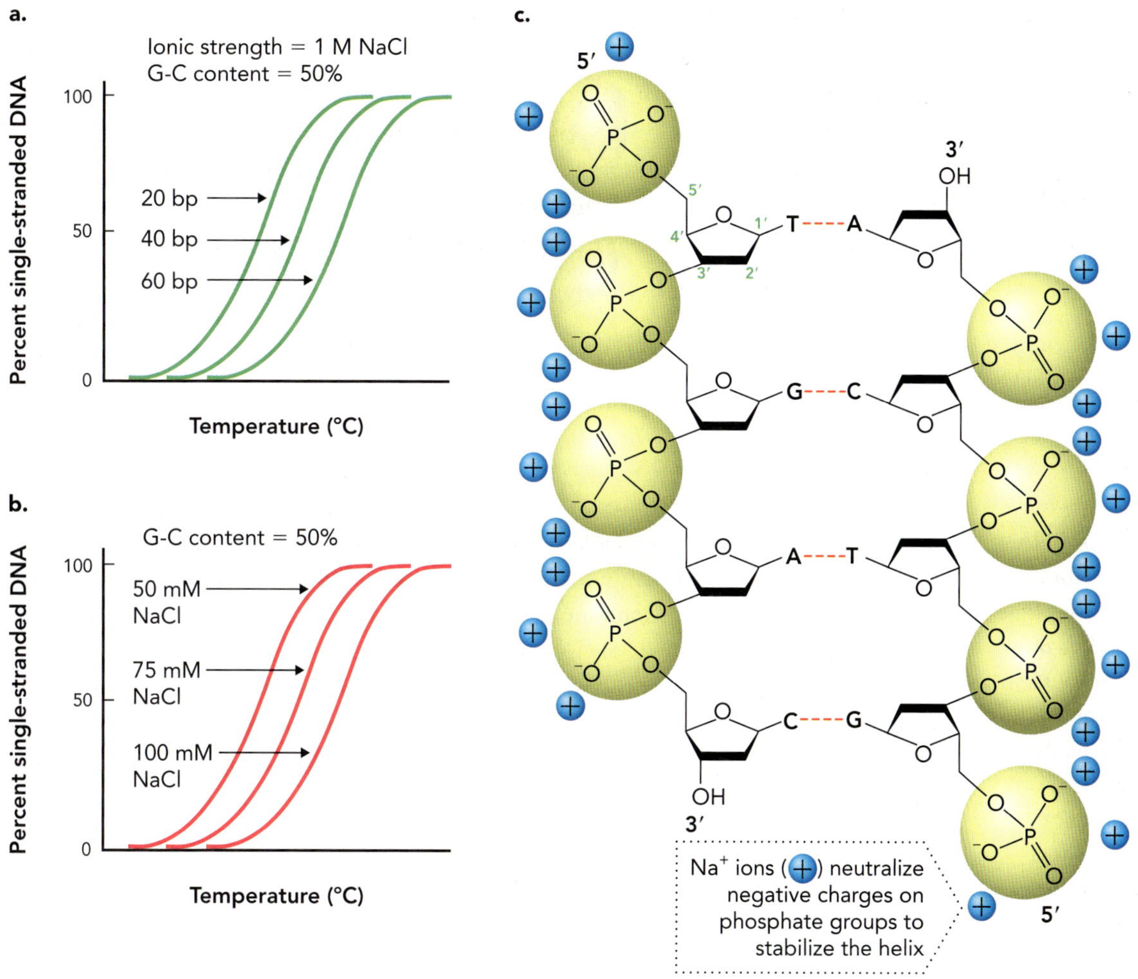

Figure 3.12 Melting temperature is influenced by strand length and ionic strength. **a.** Melting curves of percent single-stranded DNA plotted against temperature for three different-length fragments of DNA with identical G-C content and in solutions of the same ionic strength. As the length of the DNA strand increases from 20 bp to 60 bp, the T_m increases. **b.** Melting curves of percent single-stranded DNA plotted against temperature for DNA in solutions of three different ionic strengths but with identical G-C content and length. As the Na^+ content increases from 50 mM to 100 mM, the T_m increases. **c.** Na^+ ions can bind to the PO_3^{2-} groups in the DNA backbone. This can help stabilize the helix and raise T_m.

Finally, the ionic strength of a DNA solution can alter T_m. Positively charged ions can bind to the sugar–phosphate backbone and neutralize the negative charges on the phosphate groups. At very low ionic strength, these negative charges can repel one another and cause unstacking of the bases. Neutralizing these charges leads to an increase in stability of the helix, and thus more heat energy is required for denaturation, as seen by an increase in T_m.

DNA Supercoiling and Topoisomerase Enzymes

Now that we have studied the secondary structure of DNA, we can examine the topological properties of DNA and see how it can be further organized and compressed. This explains how long molecules of DNA can fit within the typical dimensions of a cell nucleus.

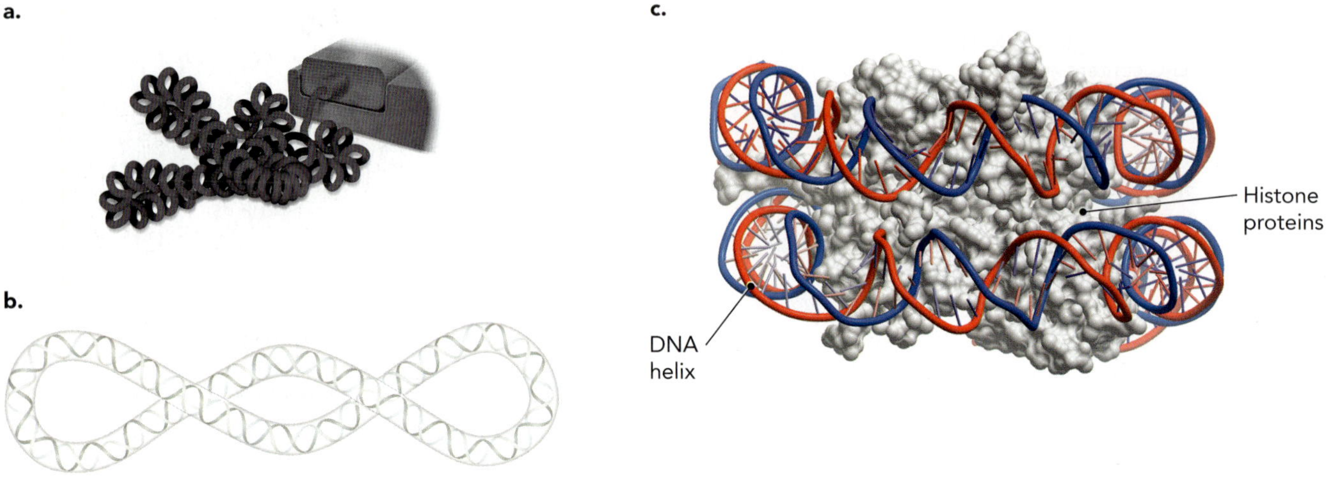

Figure 3.13 Supercoiled DNA is generated by twisting of the double-stranded helix. **a.** A twisted telephone cord is an example of a supercoiled structure. Because the cord is fixed on either end, twisting causes the cord to wrap around itself. **b.** Circular DNA, as found in prokaryotic plasmids, often twists upon itself creating supercoils. **c.** Wrapping double-helical DNA around histone proteins also creates supercoils. BASED ON PDB FILE 3AFA.

DNA Supercoiling The majority of DNA molecules inside a cell are folded up on themselves to create a structure known as a **supercoil**. **Figure 3.13a** shows an electrical cord found on a vintage telephone, which can easily become supercoiled, much like a DNA helix.

Supercoiling is found in both prokaryotic and eukaryotic DNA. As mentioned earlier, circular DNA can coil on itself (**Figure 3.13b**), similar to what happens when you twist a rubber band. The areas where the double helix crosses itself are the supercoils. In eukaryotes, wrapping of the DNA around histone proteins (discussed later) also results in supercoiling (**Figure 3.13c**).

Supercoiling can be right- or left-handed, referring to the direction of the twist. Right-handed twist produces a negative supercoil (**Figure 3.14a**), while a left-handed twist results in positive supercoiling. Electron micrographs (**Figure 3.14b**) illustrate the difference between a relaxed, circular DNA molecule and that same molecule after supercoiling. Note that use of the term *circular* to describe a DNA molecule indicates that it is a closed loop; it does not necessarily relate to the shape of the molecule *in vivo* or *in vitro*.

To understand the structural changes caused by supercoiling, we must first examine the topological properties that govern these structures. Each circular DNA molecule

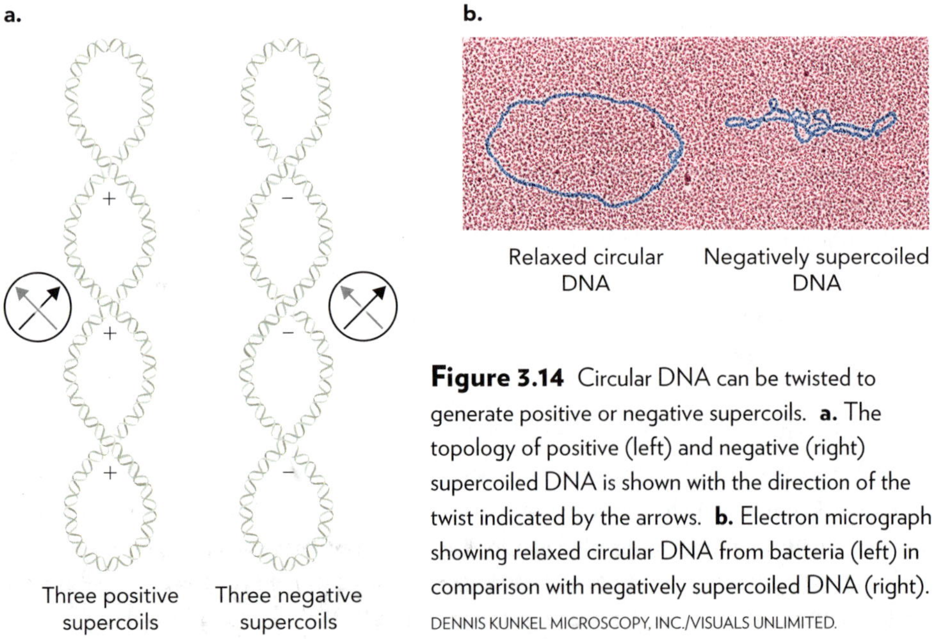

Figure 3.14 Circular DNA can be twisted to generate positive or negative supercoils. **a.** The topology of positive (left) and negative (right) supercoiled DNA is shown with the direction of the twist indicated by the arrows. **b.** Electron micrograph showing relaxed circular DNA from bacteria (left) in comparison with negatively supercoiled DNA (right). DENNIS KUNKEL MICROSCOPY, INC./VISUALS UNLIMITED.

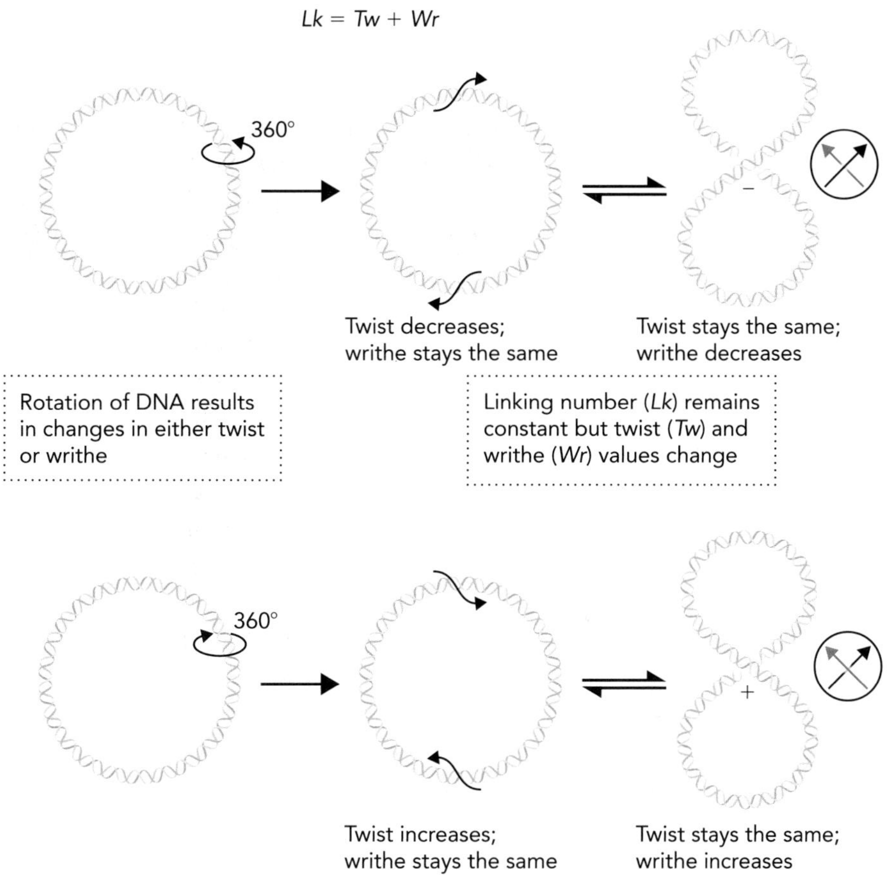

Lk = Tw + Wr
360°
−
Twist decreases; writhe stays the same
Twist stays the same; writhe decreases
Rotation of DNA results in changes in either twist or writhe
Linking number (Lk) remains constant but twist (Tw) and writhe (Wr) values change
360°
+
Twist increases; writhe stays the same
Twist stays the same; writhe increases

Figure 3.15 DNA topology is defined by the linking number (*Lk*), twist (*Tw*), and writhe (*Wr*). For a circular piece of DNA, if it is rotated the twist and writhe can each change, but the total linking number remains constant.

has a **linking number** (*Lk*), defined as the number of times a strand of DNA winds in the right-handed direction around the helix axis when the axis lies in an imaginary plane. The linking number has two components: twist (*Tw*) and writhe (*Wr*):

$$Lk = Tw + Wr$$

As long as the DNA backbone is not disrupted, total *Lk* remains constant. A change in *Tw*, which measures the winding of DNA strands around each other, must be accompanied by a corresponding change in *Wr*, which measures the crossing of the DNA strands (**Figure 3.15**). Both *Tw* and *Wr* can be changed by stretching or bending the DNA, but these manipulations do not change the number of helical turns, which means that total *Lk* remains constant.

To change the topology of DNA, one strand must first be cut and then resealed after adding or removing turns. For example, **Figure 3.16** shows relaxed DNA with $Lk = 50$. We can calculate *Lk* by knowing the average number of base pairs per helical turn (10.5) and the total number of base pairs (in this case, 525). Therefore, $Lk = 525/10.5 = 50$. In relaxed DNA, the axis of the double helix is not coiled, so $Wr = 0$ and $Tw = 50$. If one strand is cut, either positive or negative supercoiling can be induced by adding or removing four turns. Removing four turns results in a decrease in *Lk* and *Tw* to 46, whereas adding four turns increases both values to 54. When the cleaved strand is resealed, the helix twists to reach the conformational equilibrium disrupted by the addition or removal of turns. After reaching equilibrium, DNA that had four turns removed forms a negative supercoil of four turns ($Wr = -4$). The DNA with added turns forms a positive supercoil of the same number ($Wr = +4$). The change in *Wr* is reflected in a corresponding change in *Tw*. The negatively and positively supercoiled molecules now have $Tw = 50$. The two final structures are known as **topoisomers** because they differ in *Lk* only.

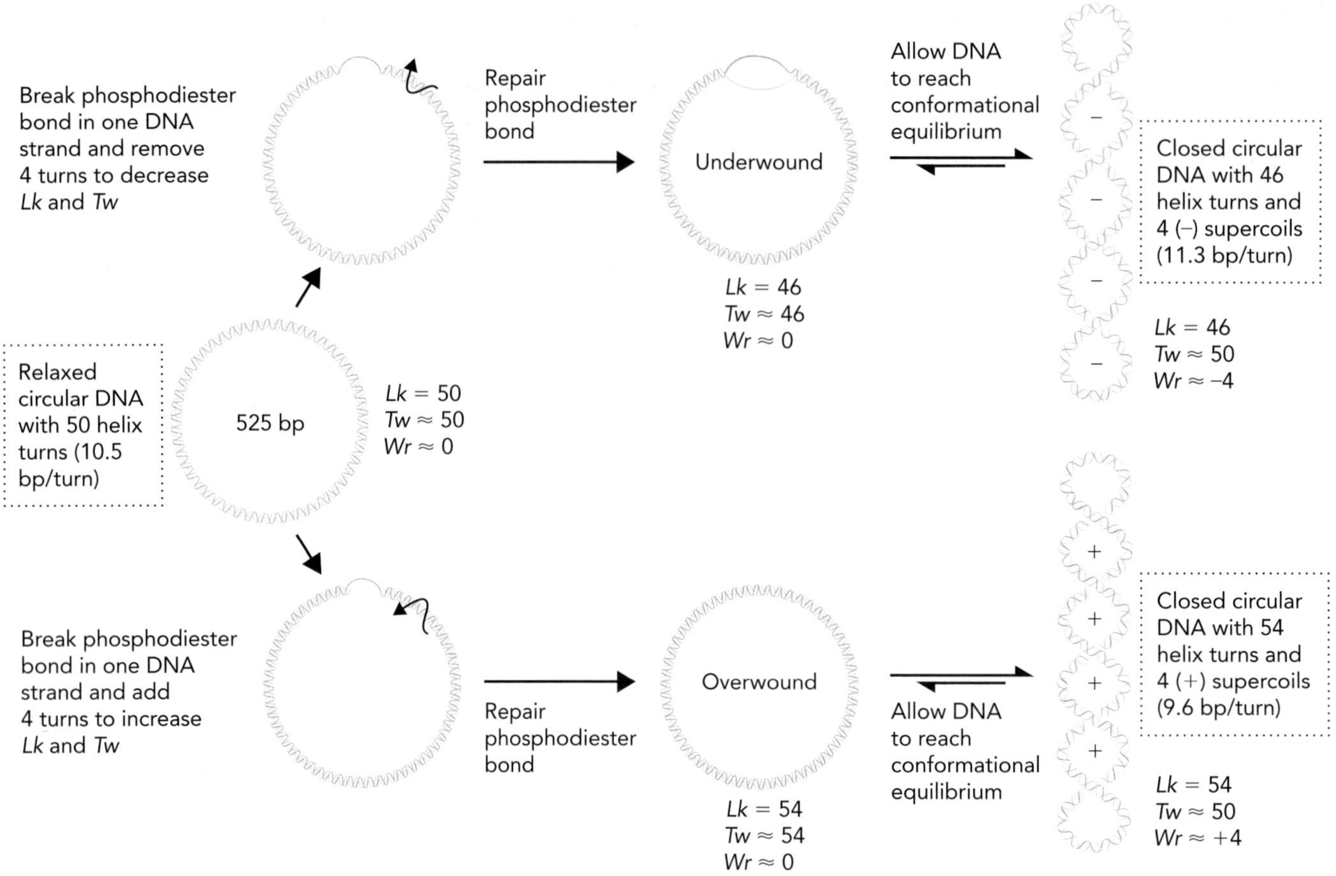

Figure 3.16 Supercoils are induced after cleavage and resealing of a DNA strand. In this example, positive or negative supercoils are generated in a 525-bp piece of circular DNA. First, one strand of DNA is cleaved by breaking the phosphodiester bond between two adjacent bases. Top branch: If we remove four turns by unwinding the cut strand before resealing, it gains four negative supercoils after reaching conformational equilibrium and is underwound. Bottom branch: If instead we add four turns before resealing the cut strand, the DNA gains four positive supercoils and is overwound. Removing or adding turns changes the linking number from 50 to 46 or 54, respectively.

Topoisomerases Supercoiling can also occur as the result of circular DNA wrapping around **histone proteins** to generate a DNA–protein structure called a **nucleosome** (**Figure 3.17**). Histones are small proteins that provide structural support for the DNA and are also involved in some DNA functions, as we will see later. (In Section 3.2, we discuss more about the assembly of nucleosome complexes into a larger structure known as chromatin.) Although eukaryotic DNA is linear rather than circular (with the exception of mitochondrial DNA), we can look at the interaction between circular DNA and histones as an example of how negative supercoils can be induced.

In order for the helix to wrap tightly around the histone, one turn must be removed. As we have just seen, removal of a turn from the helix results in a negative supercoil. The DNA is not cleaved when wrapped around the protein, so *Lk* must remain constant. To maintain a constant *Lk*, the negative supercoil induced by histone wrapping must be balanced with a positive supercoil in the free end of the loop. Enzymes known as **topoisomerases** can relieve positive supercoiling through cleavage and resealing of

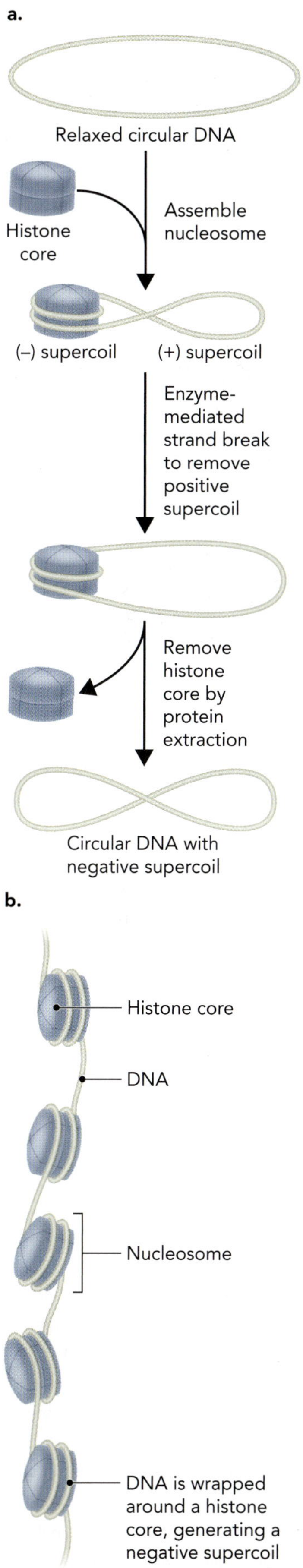

Figure 3.17 DNA is supercoiled through assembly into nucleosomes. **a.** A relaxed, circular piece of DNA is wrapped around a histone core, which induces both positive and negative supercoiling. The positive supercoil is removed by an enzyme-catalyzed double-stranded break. When the DNA is resealed and released from the histone core, the negative supercoil remains intact. **b.** A linear strand of DNA wrapped around a histone core generates a negative supercoil.

the DNA. In the example shown in Figure 3.17, relaxation of the positive supercoil leaves the negative supercoil intact and leads to a reduction in *Lk* in the DNA molecule after the histone protein is removed.

The presence of negative supercoils in chromosomal DNA is advantageous to both DNA replication and transcription. In order for these processes to occur, the DNA strands must be unwound and separated. This process is made easier by DNA that contains fewer turns; that is, negatively supercoiled DNA (underwound). In fact, most genomic DNA exists in a negatively supercoiled state rather than a relaxed state.

Although the presence of negative supercoils facilitates DNA replication and transcription, these processes themselves promote the formation of positive supercoils. **Figure 3.18a** illustrates the basic mechanism by which supercoils are induced in DNA during a process such as transcription. Genomic DNA is fixed on both ends; if one end is unwound while the other end remains fixed, the region in between the two ends continues to twist. This process adds turns, resulting in a positive supercoil.

During transcription, the double helix is unwound in a small region to allow RNA polymerase to access the DNA template strand, forming a transcription bubble (**Figure 3.18b**). Because both ends of the DNA are fixed, supercoiling occurs both upstream and downstream of the transcription bubble. Unwinding at the transcription bubble removes turns, resulting in negative supercoiling upstream. At the same time, the downstream region gains turns, causing positive supercoils to keep *Lk* constant.

As RNA polymerase continues down the DNA strands, positive supercoiling can increase to the point where it inhibits unwinding of the region for replication or transcription. To remove the added turns (change *Lk*), it is necessary to cleave and reseal one or both strands of DNA. Topoisomerases catalyze the cleavage of one or both strands and relax positive-supercoiled regions, allowing DNA to return to its relaxed state.

DNA topoisomerases can be divided into two classes. Type I topoisomerases break only one strand of DNA and reduce the supercoiled region by one turn ($Lk = -1$). Type II topoisomerases break both strands and reduce the supercoiled region by two turns ($Lk = -2$). The example shown in **Figure 3.19** of type I topoisomerase activity results in a region of negative-supercoiled DNA, whereas type II activity would yield a section of relaxed DNA containing no supercoils. Type I enzymes are generally monomeric, meaning a protein consisting of a single polypeptide chain binds to the DNA strand. Type II enzymes are genetically dimeric and cleave both DNA strands using two identical polypeptide chains, which each bind one strand of DNA. Prokaryotes and eukaryotes have different forms of these enzymes, although the overall reaction is very similar.

Type II topoisomerase (Topoisomerase II) enzymes are important to the process of replication and transcription because they relieve the positive supercoiling generated

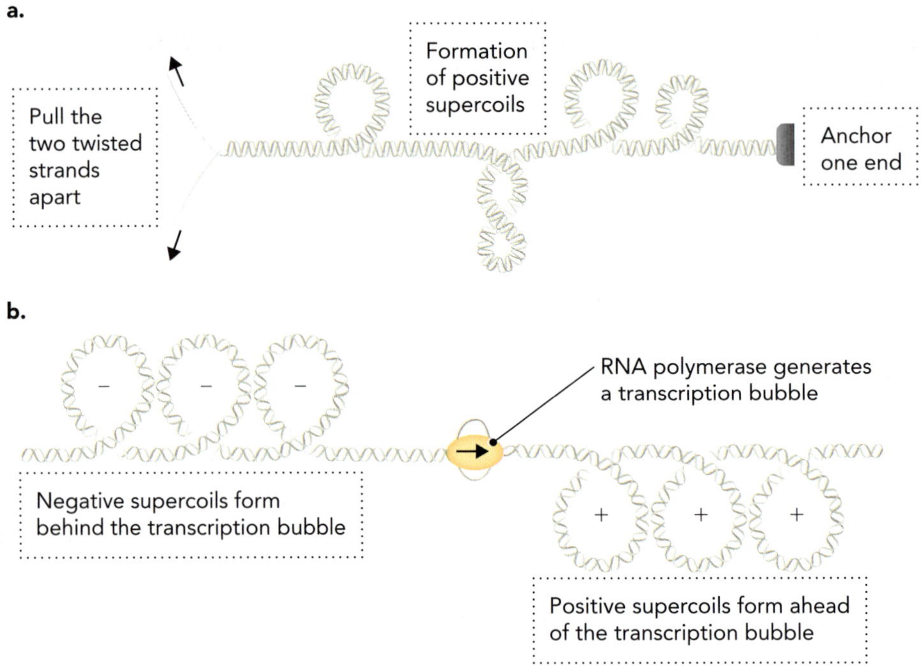

Figure 3.18 DNA supercoiling is induced during transcription by RNA polymerase. **a.** The mechanism of supercoil formation is shown here. During transcription, the helix is unwound at one end but remains fixed at the other. The region between the unwound and fixed ends becomes supercoiled. **b.** A region of genomic DNA undergoing transcription. The helix is unwound in the region surrounding the polymerase (which is called the transcription bubble). This generates negative supercoils behind the actively transcribing region while the region ahead develops positive supercoils.

downstream of the replication or transcription bubble. Inhibition of these enzymes at particular steps in the reaction pathway can lead to unrepaired double-strand breaks, which trigger cell death. **Figure 3.20** shows the overall reaction catalyzed by type II topoisomerases, along with inhibitors that have been found to block parts of this pathway. Inhibitors can prevent the formation of the initial protein–DNA complex (aclarubicin); prevent DNA cleavage (merbarone and ICRF-187); or stabilize the cleaved complex, preventing resealing of DNA (etoposide).

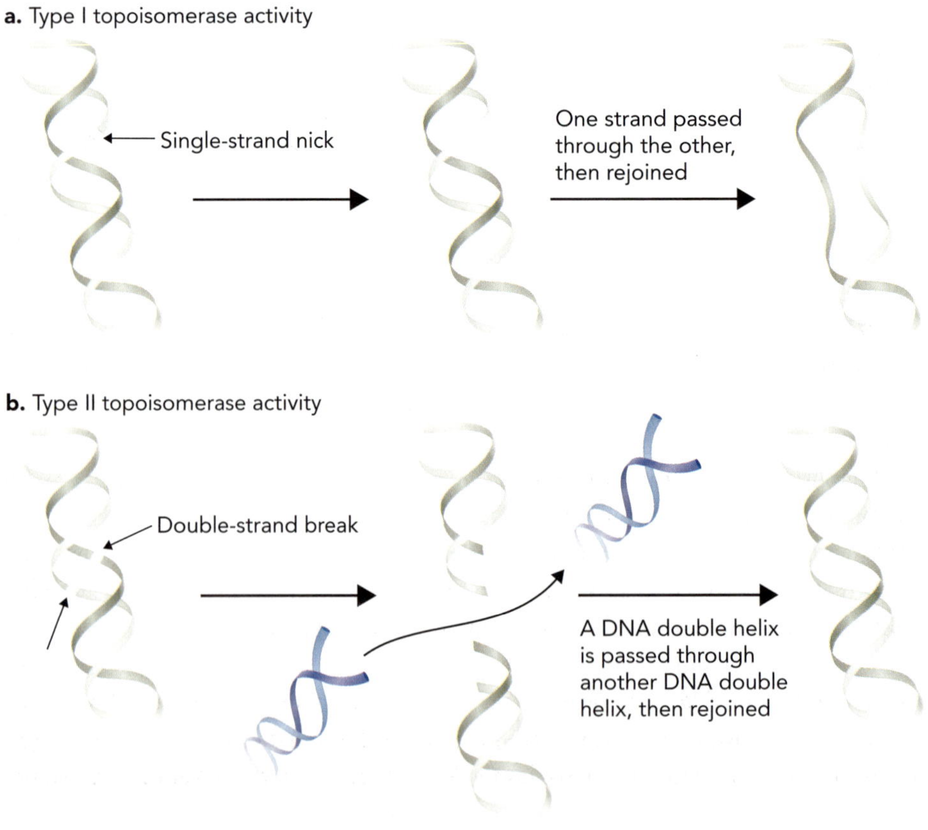

Figure 3.19 Topological strain caused by supercoiling is relieved by DNA type I and II topoisomerases. **a.** Type I topoisomerase activity, showing the nick generated in one strand. After one strand passes through the other, the nick is resealed to leave an unwound section of DNA. **b.** Type II topoisomerases generate a double-strand break as shown here, allowing another region of helix to pass through the first. After resealing, the original double helix remains intact.

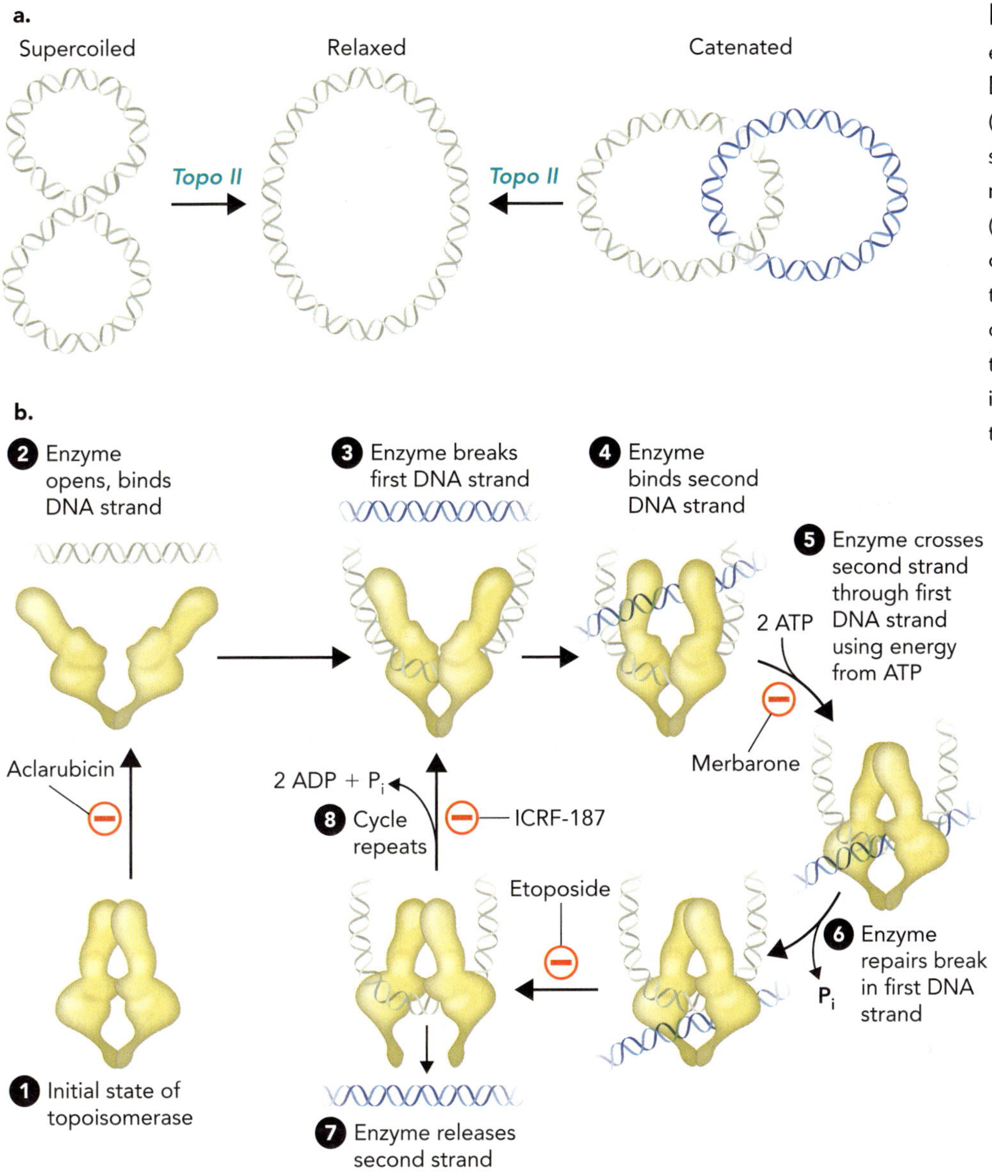

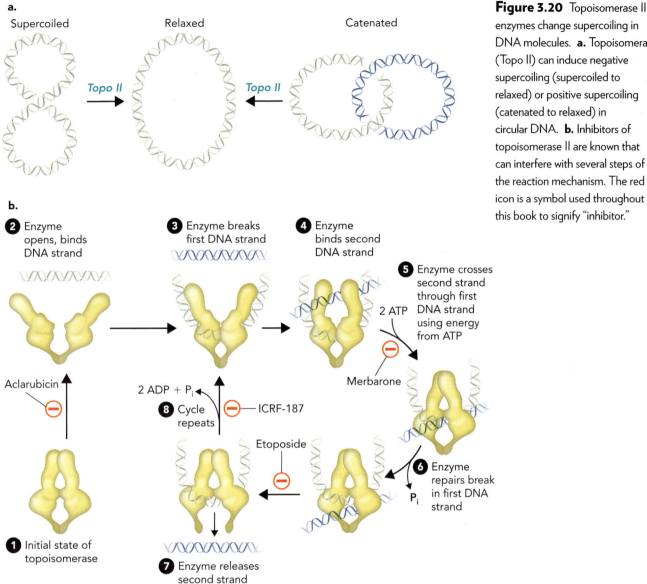

Figure 3.20 Topoisomerase II enzymes change supercoiling in DNA molecules. **a.** Topoisomerase II (Topo II) can induce negative supercoiling (supercoiled to relaxed) or positive supercoiling (catenated to relaxed) in circular DNA. **b.** Inhibitors of topoisomerase II are known that can interfere with several steps of the reaction mechanism. The red icon is a symbol used throughout this book to signify "inhibitor."

Structural Differences between DNA and RNA

It has been proposed that life began with RNA, an idea called the "RNA world" hypothesis. (This was first proposed in 1967 by Carl Woese.) Why then has DNA evolved to be the carrier of genetic information? To answer this question, we can compare the two molecules and examine the differences in their building blocks and structures.

The primary structures of DNA and RNA differ in three ways: (1) presence of a hydroxyl group on the 2′-carbon of the ribose ring in RNA, whereas DNA lacks the hydroxyl group at this position; (2) use of the base uracil in RNA instead of thymine as in DNA; and (3) complex intrastrand structures in RNA can form catalytic molecules called ribozymes, but the DNA double helix does not have this property.

The 2′-Hydroxyl Group in RNA Facilitates Spontaneous Degradation The presence of a 2′-hydroxyl group in RNA leads to the possibility of an autocleavage reaction that can disrupt the phosphodiester backbone. Under alkaline conditions,

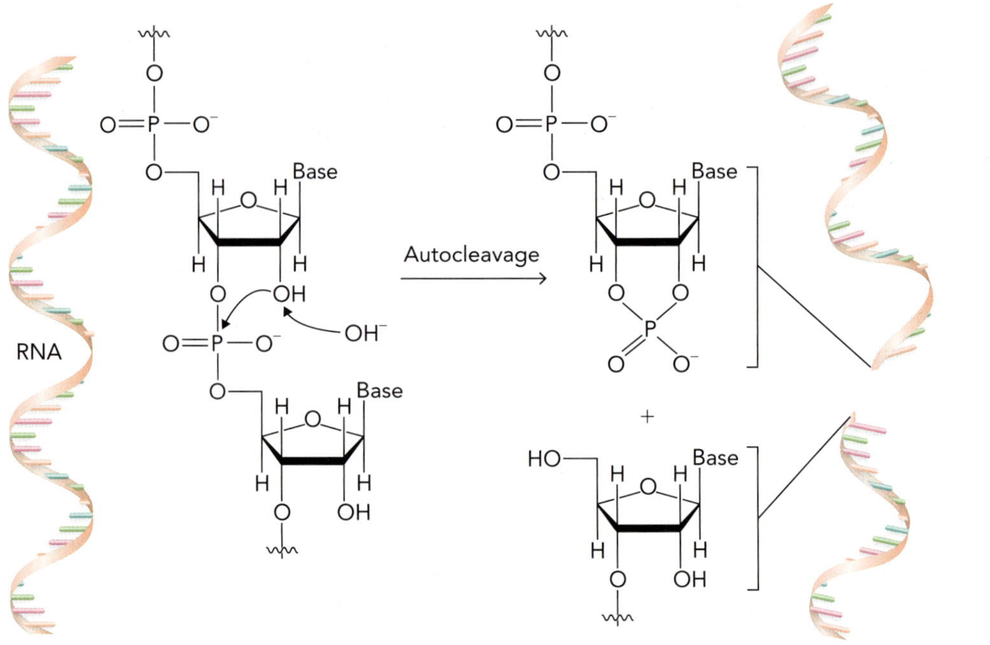

Figure 3.21 Alkaline-mediated autocleavage makes RNA chemically unstable. The RNA backbone is shown here with a close-up of two adjacent bases. Alkaline conditions facilitate a base-catalyzed nucleophilic attack on the phosphate, breaking the phosphodiester bond.

free hydroxyls in the environment induce deprotonation of the 2′-hydroxyl on RNA, resulting in an intramolecular nucleophilic attack by the 2′-hydroxyl on the adjacent phosphate in the backbone. This reaction results in the cleavage of the phosphodiester backbone (**Figure 3.21**). In contrast, the backbone in DNA is protected from autocleavage because DNA does not have a 2′-hydroxyl group.

Although the instability of RNA makes it unsuitable as a storage molecule for genomic information, this same instability is an advantage in the process of transcription and translation. To regulate gene expression tightly, messenger RNA (mRNA) must be quickly translated into protein or it will be degraded. If mRNA were stable, protein expression would continue to occur well after a gene is transcribed, even if the cellular conditions no longer required that particular protein. The short half-life of mRNA allows a tight coupling between gene expression and protein synthesis in prokaryotes. In eukaryotes, mRNA is modified to increase its stability as it travels from the nucleus to the cytoplasm (see Section 3.2). Other forms of RNA, such as transfer RNA (tRNA) and ribosomal RNA (rRNA), must be stable for longer periods of time inside the cell. These RNAs fold into distinct structural elements that inhibit autocleavage.

RNA Contains Uracil and DNA Contains Thymine The presence of thymine in place of uracil in DNA helps to maintain genetic information. The reason is that spontaneous deamination of cytosine to uracil is a common reaction in cells. For example, the human genome with its ~3 billion base pairs (~6 billion bases, of which ~1.5 billion are cytosine) undergoes ~40 spontaneous cytosine deaminations per day based on the measured rate constant of spontaneous cytosine deamination in *Escherichia coli* of 3×10^{-13}/s. Adenine and guanine can also undergo spontaneous deamination, but with a much lower frequency. In DNA, any cytosine deamination producing uracil that is not repaired before the next round of DNA replication will change a C-G base pair to a U-A base pair, which in the subsequent round of DNA replication will become a T-A base pair.

As shown in **Figure 3.22**, after the first round of DNA replication following spontaneous cytosine deamination, an adenine will be inserted on the opposite DNA strand to base pair with the uracil. In the second round of replication, the uracil will be replaced with a thymine, thereby converting a C-G base pair to a T-A base pair. If this mutation occurs in somatic cells during mitosis, then the mutation only affects that cell and its progeny. However, if the mutation occurs during meiosis in gamete cells, then the mutation affects the genetic lineage of the organism.

Obviously, failure to remove uracil from DNA after spontaneous cytosine deamination significantly increases the chance of accumulating deleterious mutations. Organisms have enzymes designed to identify and remove uracil residues in DNA to prevent such mutations. Therefore, the limited use of uracil as a pyrimidine base only in RNA provides a biochemical safety mechanism to ensure that these uracil DNA repair enzymes repair spontaneous cytosine deamination in DNA, as uracil should never be present in DNA.

RNA Can Form Catalytic Molecules Called Ribozymes

Like DNA, RNA forms double-stranded helices, but often it also forms more complicated structures with interactions occurring between distant parts of the molecule. Some of these structures act to stabilize RNA by preventing autocleavage of the backbone. However, unlike DNA, some RNA molecules have catalytic activity: These RNA molecules are called **ribozymes**. One example of a ribozyme is ribonuclease P (RNaseP), which cleaves nucleic acids. RNaseP consists of a 417-nucleotide strand of RNA that is folded into an intricate three-dimensional structure. As you can see in **Figure 3.23**, RNaseP contains several examples of the more complicated secondary structures that RNA can form. A linear stretch of RNA base pairs is known as a duplex and has an A-form helical structure. One side of the duplex can contain one or more unpaired bases that extrude from the structure as a bulge. An unpaired region in the middle of a duplex forms a loop, while a duplex that ends in a loop is referred to as a hairpin. These structures are essential to the function of RNaseP (see Chapter 21).

RNA Contains Modified Nucleotide Bases The RNA secondary structures that have just been described can interact to form three-dimensional structures such as found in tRNA, which functions in protein synthesis. This relatively small RNA molecule has different secondary structures, long-range interactions, and modified

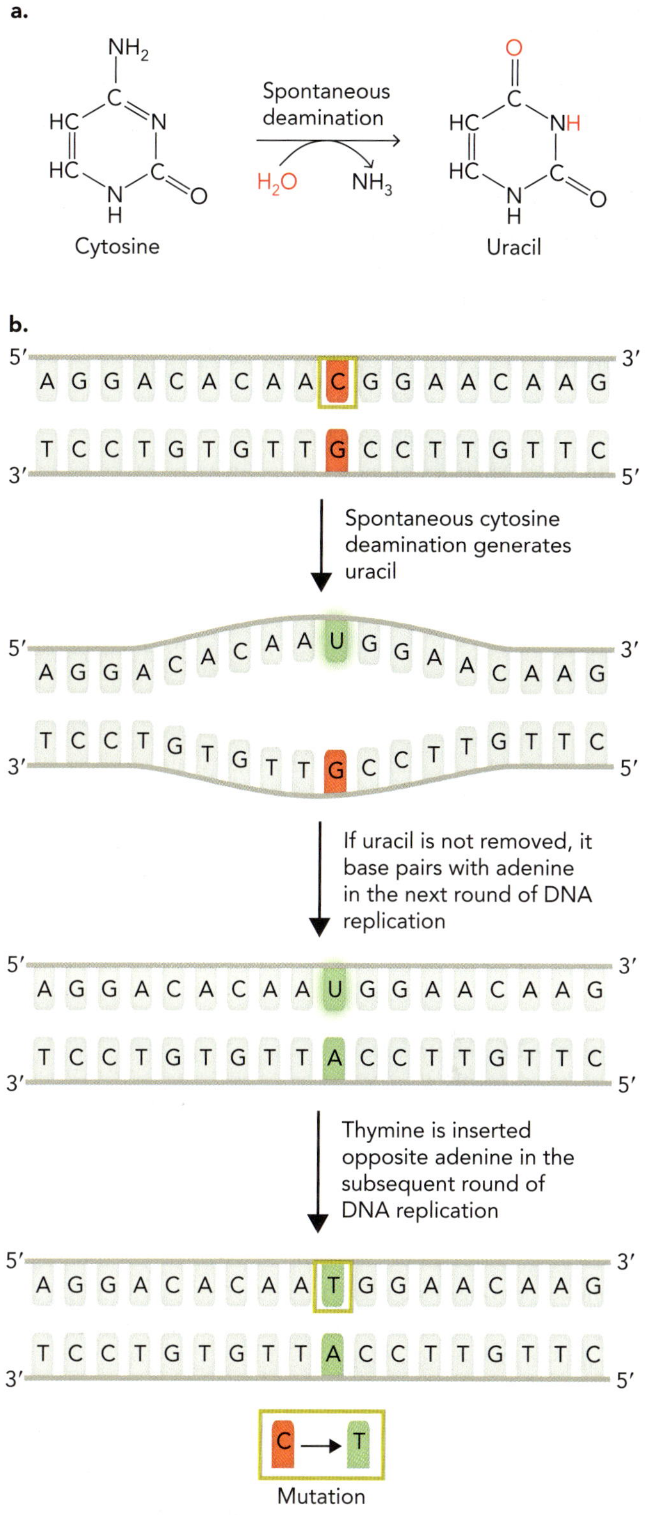

Figure 3.22 Deamination of cytosine can be mutagenic if not repaired. **a.** Spontaneous deamination of cytosine generates uracil. **b.** Deamination of cytosine to generate uracil produces a mismatch in the DNA strands. During the first round of replication following the deamination, an adenine is inserted in the DNA strand opposite the uracil to provide the correct hydrogen-bonded pair. In the next round of replication, a thymine is inserted to pair with the adenine, thereby converting a C-G base pair to a T-A base pair.

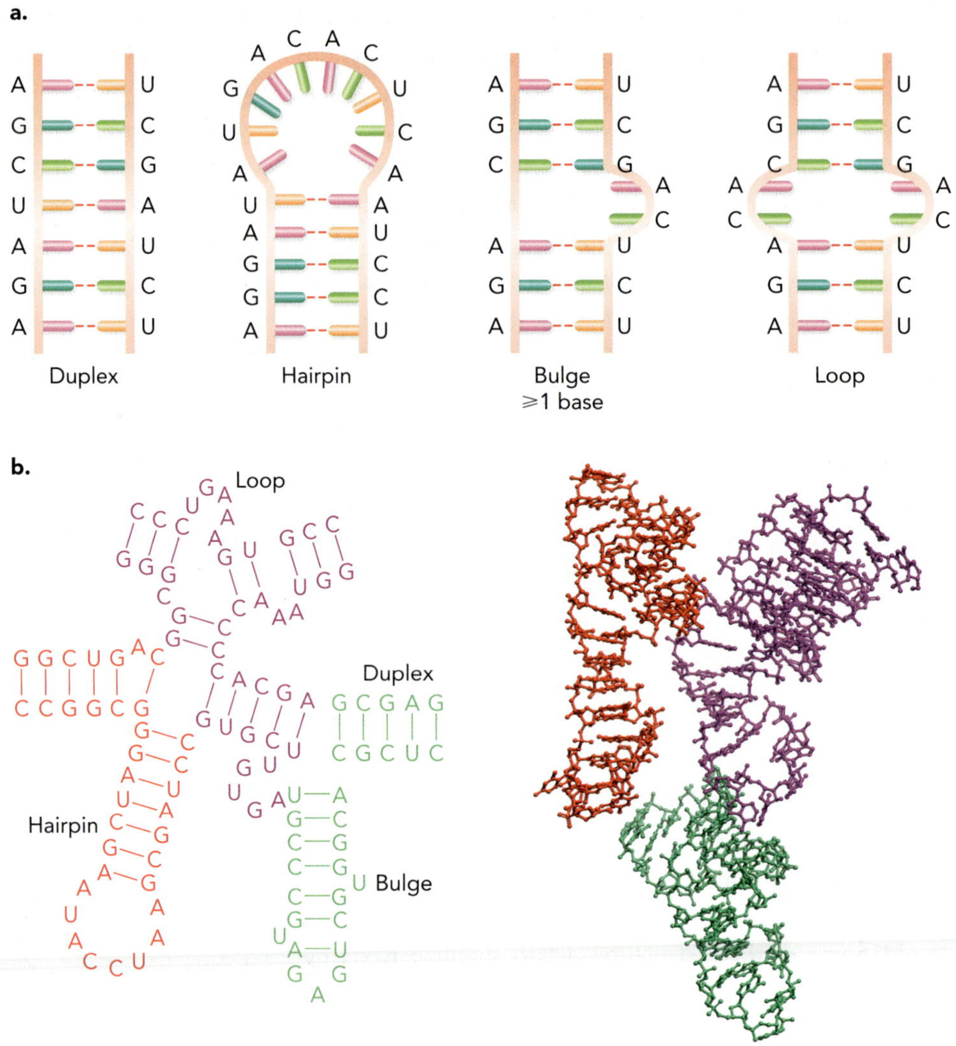

Figure 3.23 The ribozyme RNaseP contains several examples of RNA secondary structure in RNA. **a.** Schematic representations of duplex, hairpin, bulge, and loop structures. **b.** The base sequence of a portion of the RNaseP ribozyme is shown in a two-dimensional representation on the left. Examples of the secondary structures are labeled. The three-dimensional structure of a portion of RNaseP is shown in a ball-and-stick model on the right. Different structural domains are shown in distinct colors. BASED ON PDB FILE 2A64.

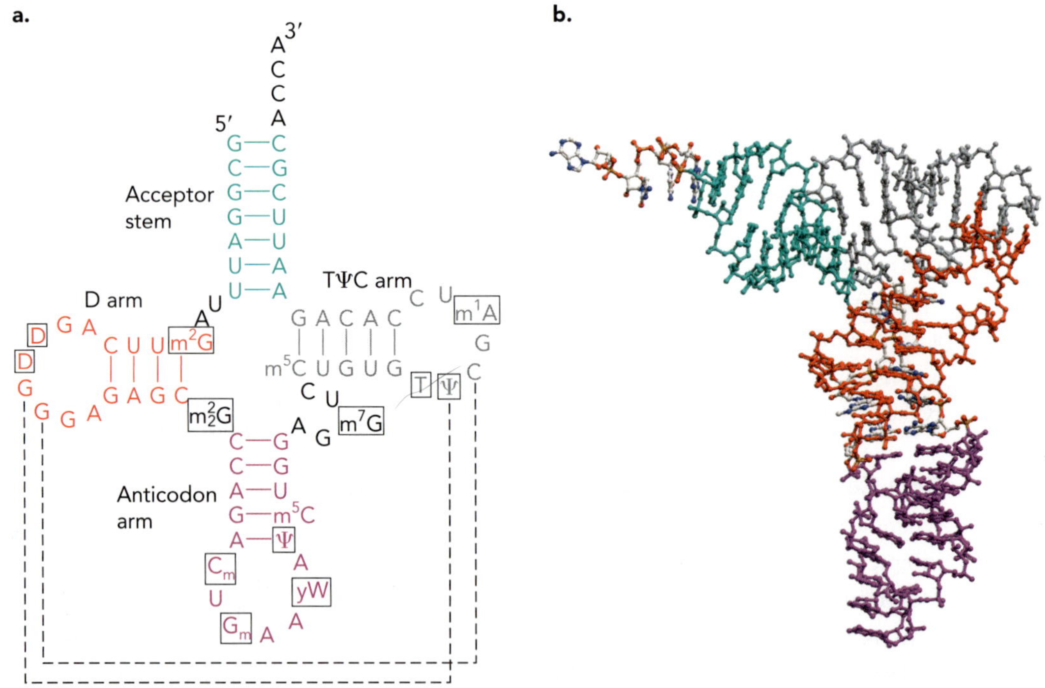

Figure 3.24 The structure of tRNA is important for its function. **a.** Secondary structure of tRNA. The acceptor stem, D arm, anticodon arm, and TΨC arm are labeled and colored separately. The boxed residues are nonstandard ribonucleotides (see Figure 3.25). Several long-range interactions are found in tRNA molecules. Two of these, which form a G-C base pair and a G-Ψ base pair, are shown by dotted lines. **b.** Tertiary structure of tRNA forms an L-shaped molecule. The helices of the D arm (red) and anticodon arm (purple) are stacked to form a continuous helix. The helices of the acceptor stem (cyan) and TΨC arm (gray) are also coaxially stacked. The boxed area shows the region of the interactions indicated with dotted lines in panel **a.** BASED ON PDB FILE 1EHZ.

bases that are vital to the structure and function of the molecule. tRNA needs to maintain a particular tertiary fold in order to interact correctly with the ribosome and with the enzymes that add the amino acids to the ends of the tRNA. (Protein synthesis and details of tRNA structure are discussed in Chapter 22.) Two-dimensional and three-dimensional representations of a tRNA structure are shown in **Figure 3.24**. Stacking of the helices and interactions between the loops help to stabilize this L-shaped structure. The modified RNA bases shown in **Figure 3.25** contribute to stabilizing the three-dimensional structure of tRNA and to mediating tRNA functions.

Several modified nucleosides—such as N^2,N^2-dimethylguanosine and 1-methyladenosine—contain a methyl group(s) in place of hydrogen on a former hydrogen-bonding N donor. Removing these potential hydrogen bonds can prevent certain structural combinations and allow others. The addition of hydrophobic methyl groups may also contribute to base stacking and can help stabilize the RNA structure. Modification of uridine to pseudouridine adds an additional hydrogen-bond donor to the ring. Alteration of the functional groups allows more variety in the types of structures that form and can help to constrain a particular tertiary structure.

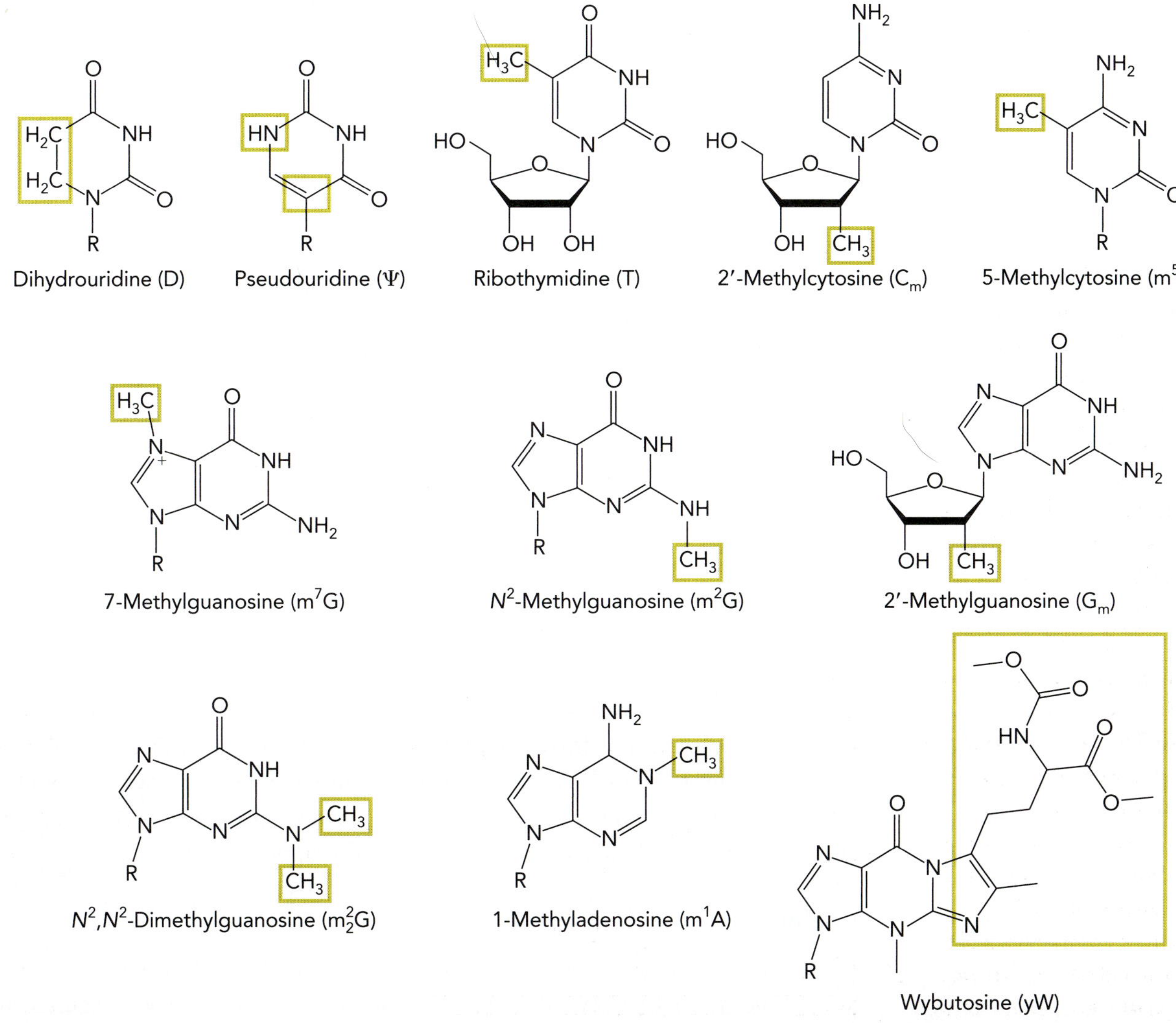

Figure 3.25 Structures of the modified nucleosides found in the tRNA structure in Figure 3.24 are shown here, along with their standard abbreviations. The R-group indicates the ribose sugar. The modification sites of the bases are highlighted.

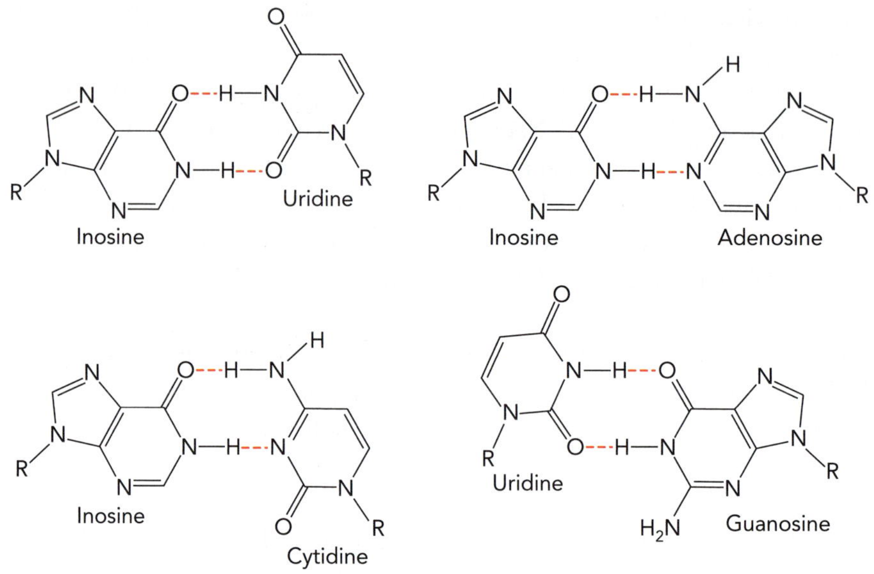

Figure 3.26 In RNA, the nucleosides uridine, adenosine, and cytidine can form base pairs with inosine, and guanosine can base pair with uridine. The R-group indicates the ribose sugar.

RNA also contains nucleotides that are modified to serve a functional role more than a structural one. In the process of translation, a tRNA anticodon must match up with each codon in mRNA to add the correct amino acid to a growing protein chain. Many tRNA anticodons contain the nucleoside inosine (I), which is able to base pair with the nucleosides cytidine, uridine, or adenosine (**Figure 3.26**). With this flexibility, a single tRNA can encode the same amino acid using three different codons (see Chapter 22). For example, a tRNA with the anticodon 3′-GAI-5′ could pair with the codons 5′-CUA-3′, 5′-CUU-3′, or 5′-CUC-3′, all of which code for the amino acid leucine.

Unconventional Base Pairing in RNA and DNA Both RNA and DNA can participate in base pairing outside of the normal Watson–Crick model by forming triplet and quadruplet interactions. Triplet interactions can occur between a single-stranded region of DNA or RNA and another RNA, DNA, or RNA–DNA duplex molecule. The single strand binds to a complementary region of the duplex exposed in the major groove. A single triplet interaction can form or, if particular sequences are present, a triple helix may form (called a triplex). Triplex-forming regions are generally found in a polypurine/polypyrimidine region, and the single strand must contain a U(T)-C, G-A, or G-U(T) sequence. The single strand can bind either parallel or antiparallel to the duplex. **Figure 3.27a** shows two examples of triplet interactions. Triplet interactions composed only of RNA assist with the formation of secondary and/or tertiary structures. Triplexes formed by a combination of DNA and RNA may have a variety of structural and functional roles in the cell, including transcriptional regulation, RNA processing, formation of nucleosomes, and DNA repair. Quadruplex structures can form among guanine bases found in particular G-rich DNA sequences (**Figure 3.27b**). Four guanine bases can hydrogen bond to form a planar structure called a guanine tetrad. The guanine tetrads can stack in several ways, either within a single strand of DNA or between multiple strands, to form structures called G-quadraplexes (**Figure 3.27c**). These structures are thought to be involved in transcriptional regulation.

Nucleic Acid Binding Proteins

So far, we have discussed how the structures of DNA and RNA influence their functions. Now we turn to another key feature in both molecules: the ability to be recognized and

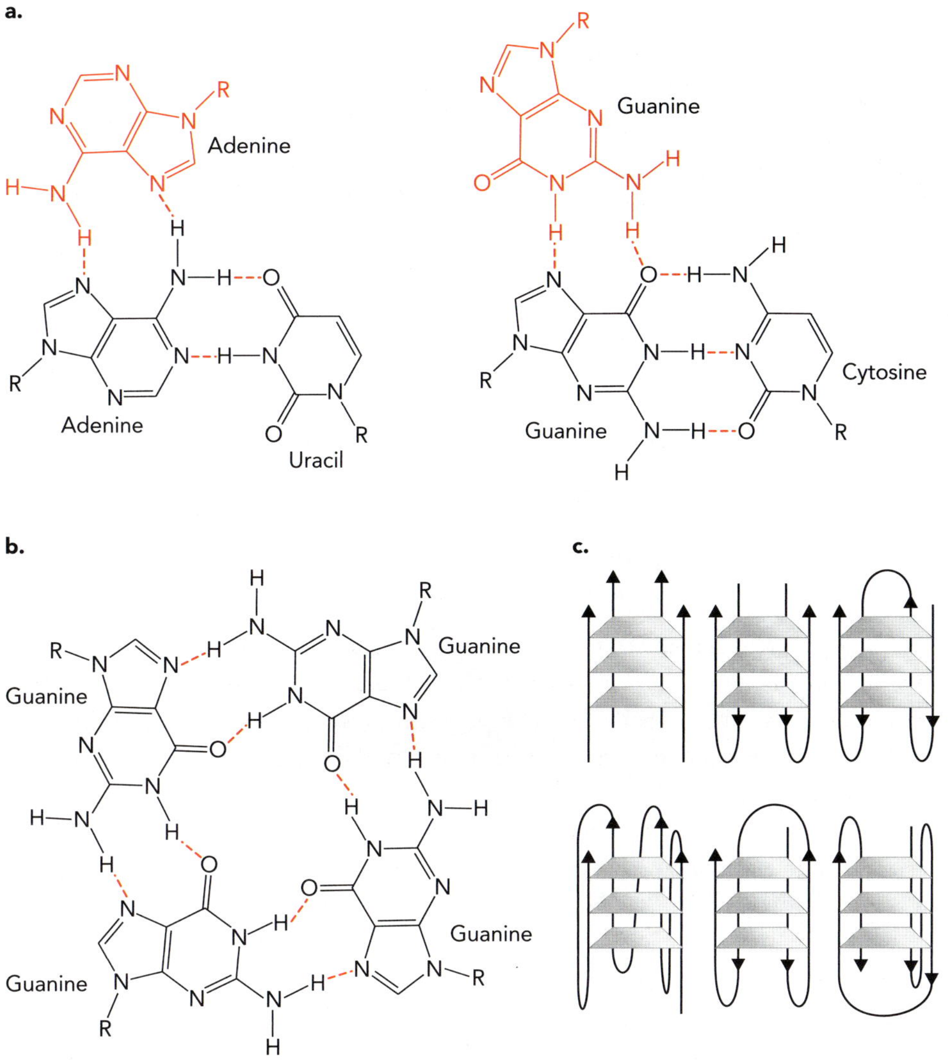

Figure 3.27 Unconventional base pairing can occur in RNA and DNA. **a.** Two examples of purine–purine–pyrimidine triplet interactions. In each case, the standard Watson–Crick base pair is shown in black, and the base that is forming the triplet interaction is shown in red. **b.** G-quadruplex interactions are formed by interactions of four guanine bases among particular DNA sequences. **c.** Guanine tetrads (shown as squares) can stack to form G-quadruplexes. In the examples shown here, the strands can be either parallel or antiparallel, and one or more strands can be involved.

bound by proteins. We explored this briefly in the discussion of topoisomerases. Here, we examine several examples of DNA binding proteins to identify some similarities and differences in the way proteins interact with DNA and RNA. In Chapter 23, we describe noncovalent interactions between sequence-specific DNA binding proteins and nucleotide bases in much more detail.

We have seen that eukaryotic DNA must interact with histones in order to be packaged within the cell. DNA–histone complexes provide an example of one type of DNA–protein interaction. This interaction is not sequence specific; that is, histone binding does not require a specific combination of nucleotide bases. Several histones bind to one DNA molecule, forming a repeating unit called a nucleosome (**Figure 3.28**). Each nucleosome consists of a region of double-stranded helical DNA bound to eight histone proteins: two copies each of H2A, H2B, H3, and H4. H1, or the linker histone, connects to the outside of this structure and helps keep the ends of the DNA in place.

Large, positively charged sections in the protein mediate histone binding to DNA. These positive charges interact with the negatively charged DNA backbone, bringing the two structures together. Once DNA and histones have bound together through the attraction of opposite charges, hydrogen bonds form between nucleotide bases in DNA and amino acid side chains in the protein, which further stabilize the interaction.

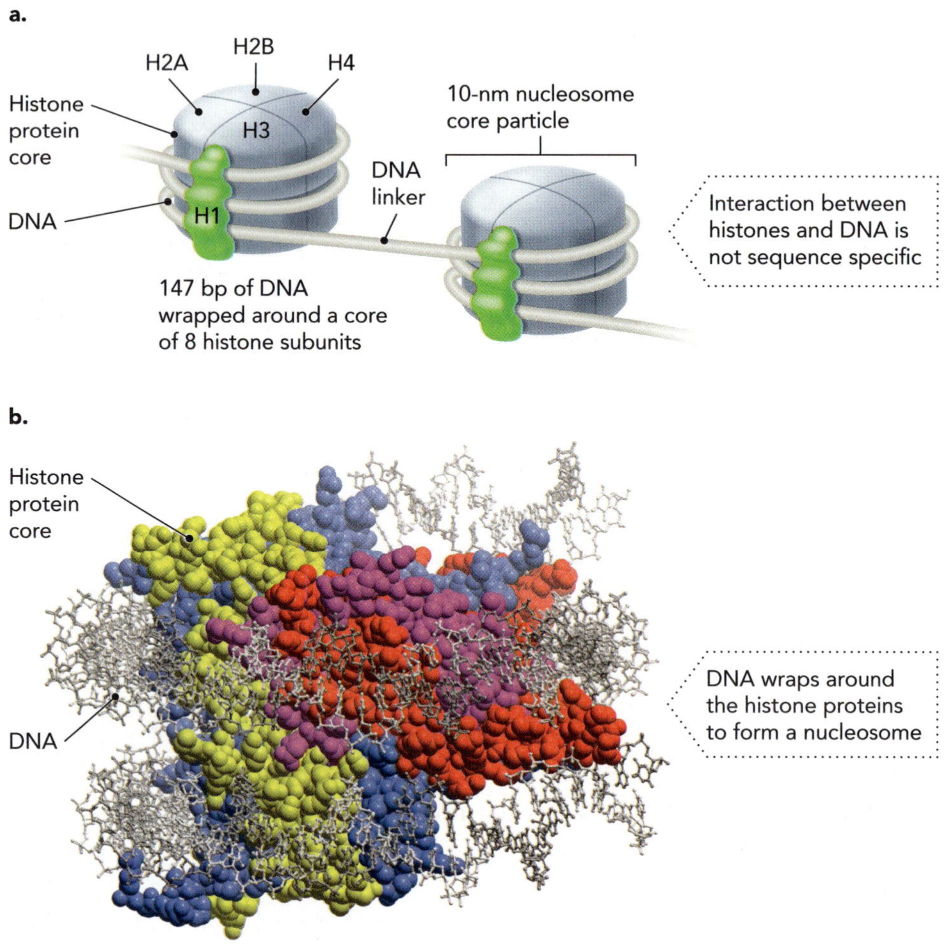

Figure 3.28 Histone proteins bind to DNA in a sequence-independent manner. **a.** Two nucleosome core particles connected by a linker DNA region. Each nucleosome particle consists of 147 bp of double-stranded DNA wrapped around two each of the histone proteins H2A, H2B, H3, and H4, with H1 (green) on the outside of the structure. **b.** Molecular structure of a single nucleosome without H1 protein. DNA is shown in ball-and-stick style, and the histones are shown in space-filling representation with each of the four histone proteins colored differently. BASED ON PDB FILE 3AFA.

Another example of a sequence-independent interaction is the binding of single-stranded DNA binding protein to DNA (**Figure 3.29**). As the name implies, this protein preferentially binds single-stranded DNA. Cellular processes such as DNA replication, repair, recombination, and transcription must first denature a local region of DNA. It is important that these single-stranded regions be protected so that they do not prematurely anneal or form structures such as hairpins or triplexes that would inhibit the desired reaction. Single-stranded DNA is also sensitive to cleavage by nucleases. The role of single-stranded DNA binding proteins in DNA replication will be discussed in more detail in Chapter 20.

All organisms contain some form of single-stranded DNA binding protein that binds and protects single-stranded regions of DNA. The single-stranded DNA binding protein consists of four identical subunits, and the single-stranded DNA wraps around all four. Like histones, single-stranded DNA binding proteins contain a positively charged region that attracts the negatively charged DNA backbone, and this interaction is not sequence specific. These proteins are able to discriminate between single-stranded DNA and RNA because the binding pocket is so tightly arranged that the 2′-hydroxyl of RNA excludes it from binding.

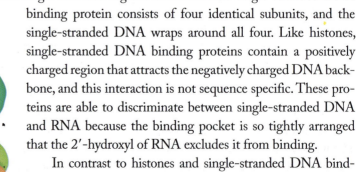

In contrast to histones and single-stranded DNA binding proteins, which serve more general functions in the cell, many DNA binding proteins interact only with specific DNA sequences. One example is the lac repressor protein, which will be discussed in detail in Chapter 23. This protein is a type of

Figure 3.29 The structure of single-stranded DNA binding protein is shown bound to DNA, with the single DNA strand shown in ball-and-stick style and the protein subunits shown as space-filling structures. BASED ON PDB FILE 3ULP. The four subunits of single-stranded DNA binding protein are also shown in cartoon style where the gray line represents single-stranded DNA.

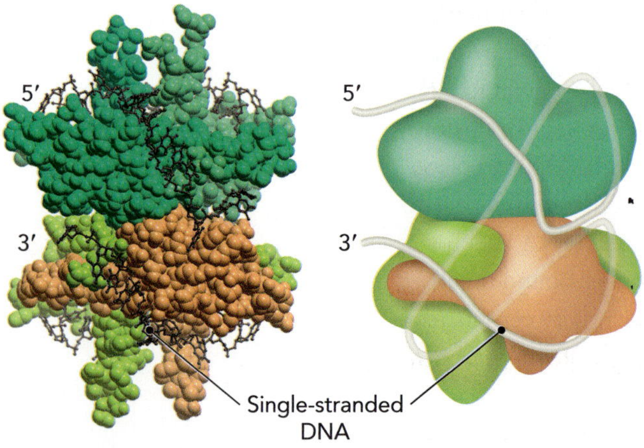

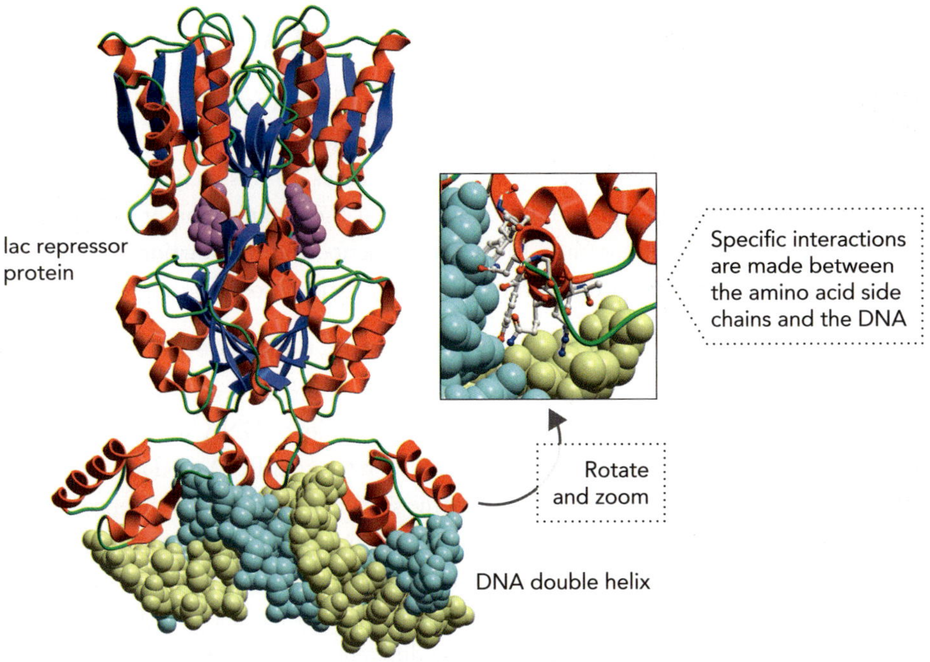

Figure 3.30 The lac repressor protein dimer binds to DNA sequence specifically. The DNA is shown as a blue-and-yellow space-fill structure and the protein as a ribbon structure. The close-up of the DNA–protein interaction (inset) shows the amino acid side chains in a ball-and-stick form. The side chains form hydrogen bonds and other interactions with the DNA bases and phosphate backbone to stabilize the interaction. BASED ON PDB FILE 1EFA.

negative regulator that binds to a specific region in the bacterial genome to prevent transcription of the *lac* operon in *E. coli*. The lac repressor protein has several discrete functional regions, one of which binds DNA. *In vivo*, it binds DNA as a homotetramer, meaning that four identical copies of the protein associate, and each binds an identical DNA sequence.

Figure 3.30 shows the molecular structure for a lac repressor–DNA complex consisting of only half of the normal complex (one DNA strand and a lac repressor homodimer). Each monomer, or single copy of the protein, has a region that sits within the major groove of DNA. Notice that the region of the protein that fits within the groove looks like a helix. This is a type of protein secondary structure commonly found in regions of proteins that contact DNA because the amino acid side chains on the outside of the helix often have numerous hydrogen-bonding groups. If we look more closely at this interaction, we can observe some of these groups and their association with the DNA strand (see Figure 3.30, inset). The amino acid arrangement in this area allows for the optimum formation of hydrogen bonds with only the correct DNA sequence. The recognition sequence in the DNA must be repeated for the second half of the protein to bind as well, which also increases the specificity of the interaction. We describe lac repressor binding to DNA in more detail in Chapter 23.

concept integration 3.1

Why are type II topoisomerase inhibitors more potent anticancer drugs than type I topoisomerase inhibitors? What might explain the side effects of nausea and hair loss in patients treated with anticancer drugs?

Type II topoisomerase inhibitors bind to the enzyme and prevent rejoining of double-stranded DNA after cleavage, which is required in the process of DNA replication. Because broken DNA fragments induce cell death, these damaged cancer cells are rapidly removed by the process of apoptosis. In contrast, type I topoisomerase inhibitors often block cancer cells from further cell division rather than killing them. The side effects of nausea and hair loss encountered with most chemotherapeutic drugs targeting DNA replication are the result of killing rapidly dividing cells elsewhere in the body, such as intestinal cells and hair follicle cells.

3.2 Genomics: The Study of Genomes

We have now examined the basic structures of DNA and RNA and discussed fundamental principles of how proteins can interact with these structures. We also began to explore what accounts for DNA condensation into chromosomes. In this section, we study the structure of chromosomes in more detail and look at the organization of the genomic regions within these chromosomes. Finally, we describe the information that can be acquired through the study of these chromosomes using bioinformatic approaches.

Genome Organization in Prokaryotes and Eukaryotes

As we have seen, the genome—the complete set of genes in an organism—can be very large. Recall from Chapter 1 that genes are functional units of DNA and are defined by the RNA products they produce, which in most cases refers to mRNA products coding for proteins. It would stand to reason that the genome size of an organism is influenced by the number of genes, but genome studies have shown that this relationship is not directly proportional. For example, *Saccharomyces cerevisiae* (budding yeast) has approximately 6,600 genes, while *Drosophila melanogaster* (the common fruit fly) has about 17,000 genes—a 2.6-fold increase over *S. cerevisiae*. However, the difference between the two genome sizes is much larger. The *S. cerevisiae* genome is 1.2×10^6 base pairs, whereas the *D. melanogaster* genome is 1.2×10^8 base pairs, or a factor of 10^2 larger than the yeast genome (**Figure 3.31**). In addition, genome size varies within groups of organisms; for example, the genome sizes in insects, fish, amphibians, and flowering plants differ by 100- to 1,000-fold.

In addition to genes, all genomes contain some amount of DNA that is not transcribed into a known functional RNA. Prokaryotic genomes tend to have fewer of these regions and are therefore more compact and gene rich. Eukaryotic genomes, in contrast, appear to have a larger proportion of DNA sequences that have no known function, with only a small fraction of the human genome devoted to protein-coding sequences. Despite this small percentage of protein-coding gene sequences in the human genome, it was recently discovered that a large amount of the human genome is actually transcribed into noncoding RNA molecules (see Chapter 21). These transcribed, noncoding RNA sequences are found in the genomic regions that separate genes and within the genes themselves. Eukaryotic genomes also contain repetitive DNA sequences, both short (3–100 nucleotides) and long (1,000–10,000 nucleotides), which are sometimes found in large tandem arrays spread throughout the genome.

Regardless of the genome size, the cellular DNA must undergo some type of condensation to fit within the cell. As we saw earlier, the prokaryotic genome is often contained within a single circular chromosome that is supercoiled to fit within the cell. Eukaryotic genomes are much larger and therefore are split into a series of linear chromosomes, which are condensed with the help of protein components. Just as the number of genes does not directly correlate with the size of the genome, the number of chromosomes in an organism does not necessarily reflect the size of the genome or complexity of the organism. You might expect that the yeast strain *S. cerevisiae* would have fewer chromosomes than the fruit fly *D. melanogaster*, as both the gene number and genome size are smaller. In fact, yeast has four times as many chromosomes as the fruit fly.

The first stage of eukaryotic DNA condensation is mediated by histone proteins, which results in the

Figure 3.31 Schematic representations of the size of the genomic DNA for a variety of organisms are shown, measured in number of base pairs. The bars represent the range of genome sizes found within groups of organisms.

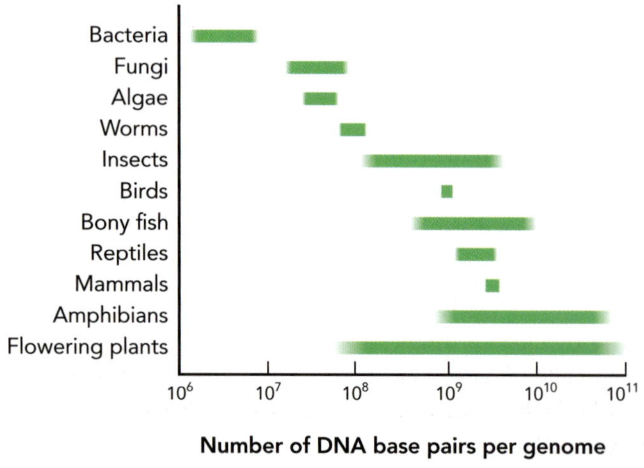

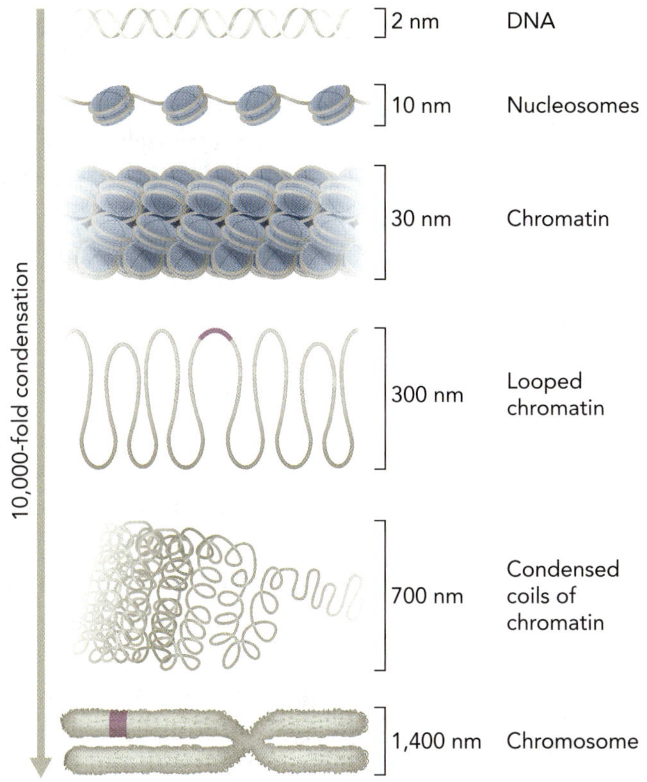

Figure 3.32 Condensation of eukaryotic DNA into chromosomes reduces the size of the DNA. The progression from double-stranded DNA to a chromosome is shown with the size of each structure indicated at the right. The assembly of nucleosomes facilitates the tight compaction of mitotic chromosomes. The overall effect of DNA assembly into the mitotic chromosome is the compaction of DNA by 10,000-fold.

formation of chromatin, a term used to describe genomic DNA that has condensed using protein–DNA complexes. As shown in **Figure 3.32**, a double-stranded region of DNA is wrapped around these histones, forming a nucleosome particle. Nucleosome particles are then packed together to form a 30-nm chromatin fiber, representing a 12- to 14-fold compaction of DNA. Although this represents a significant compaction of the DNA, a segment of DNA containing ~40 million bp would still be close to 100 μm in length, or four times larger than the average eukaryotic cell.

The next stage of condensation depends on the region of DNA and the phase of the cell cycle. For example, during interphase, or the resting phase between cell division, the chromosomes are less condensed than in mitosis, in which the cell is dividing. Mitotic chromosomes are more easily visualized in a cell because they are tightly compacted, and as a result, their structures are more often shown as a representative chromosome structure (Figure 3.32). The tight compaction of mitotic chromosomes is necessary for the proper segregation of DNA during cell division.

It is important to keep in mind that although genomic DNA must be condensed to fit within a cell, it must also be arranged in such a way that specific regions can be exposed to allow enzymes and other proteins to bind and initiate DNA replication and transcription. We also know that genomic DNA is composed of both coding and noncoding sequences. Therefore, a gene-rich region of chromatin would need to be less condensed to allow access by DNA binding proteins that regulate transcription. Here we see another example of the relationship between structure and function in biomolecules. Less condensed, gene-rich chromatin, or **euchromatin**, offers more sites for protein binding than **heterochromatin**, which consists of more condensed regions composed of mostly noncoding DNA (**Figure 3.33**). Covalent modifications of the

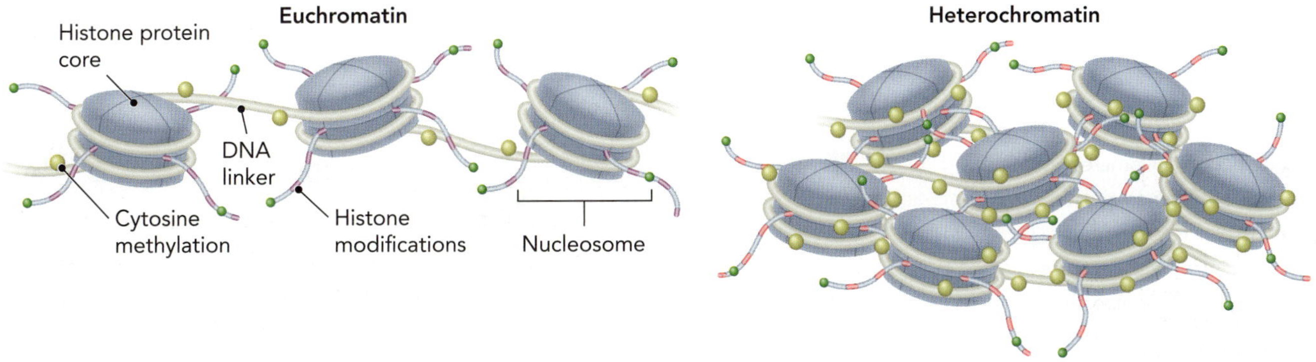

Figure 3.33 Condensation of DNA can result in euchromatin or heterochromatin. The extensions from the histones represent protein modifications (methylation, acetylation, phosphorylation), and the circles on the DNA strands represent sites where cytosine bases are methylated. These histone modifications are described in more detail in Chapter 20.

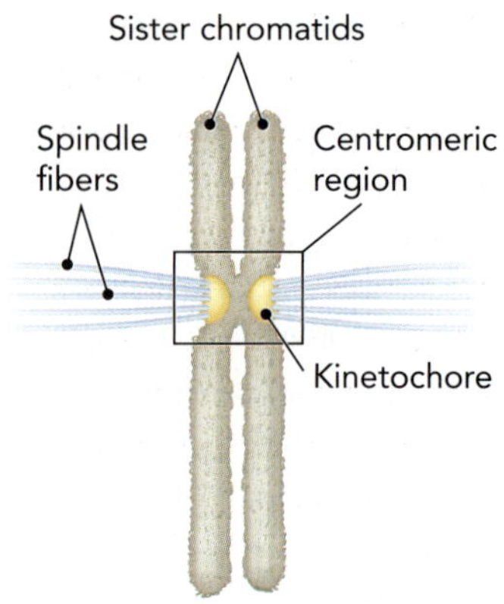

Figure 3.34 A schematic drawing of a mitotic chromosome is shown here. The process of chromosome separation is facilitated by the centromere, which is composed of heterochromatin. Microtubules from the mitotic spindle attach to the kinetochore. During cell division, the sister chromatids are separated when drawn in opposite directions through forces exerted on the mitotic spindle.

histones and DNA by cellular enzymes influence whether the DNA is in the more condensed form or not (see Chapter 20).

Although the majority of heterochromatin does not contain genes, some regions of heterochromatin play an important role in chromosomal division and maintenance. In **Figure 3.34**, you can see the structure of a mitotic chromosome. The region of connection between sister chromatids is composed of heterochromatin and is known as the **centromere. Sister chromatids**, which are two identical copies of replicated DNA contained in a mitotic chromosome, remain attached at the centromere until the process of cell division. The centromere is also the place of assembly for the **kinetochore**, a protein complex necessary for proper separation of the chromosomes during cell division. The exact composition of proteins attached at the kinetochore can vary with the cell cycle and can include more than 80 different proteins. The microtubules of the mitotic spindle attach to the kinetochore, and are used to pull the sister chromatids in opposite directions during cell division.

Another important region of heterochromatin is found at the ends of chromosomes, in regions known as **telomeres.** More than 70 years ago it was observed that the ends of chromosomes have a unique structure, and it was proposed at that time that these structures prevented chromosomes from fusing together at the ends. It is now known, however, that telomeres function to maintain the length of chromosomes after replication. Telomeres are often compared to the plastic pieces on the ends of shoelaces; without these plastic protectors, the shoelace begins to unravel. Telomeres are composed of short, repetitive DNA sequences with a high G content (in vertebrates, the DNA sequence is 5′-TTAGGG-3′). **Figure 3.35** shows how these G-rich structures are able to form loops that help to protect the end of the chromosome from degradation. Several protein complexes associated with the telomere stabilize these loop structures. Additionally, these G-rich regions can form G-quadruplex structures (see Figure 3.27). Telomere replication and maintenance is described in Chapter 20.

Genes Are Units of Genetic Information

Now that we have discussed the structure and organization of chromosomes, we can take a closer look at the organization of the coding regions within these chromosomes. Although some features are common to all genes, some differences appear between the genes of prokaryotes and eukaryotes. **Figure 3.36** shows the basic organization of a gene

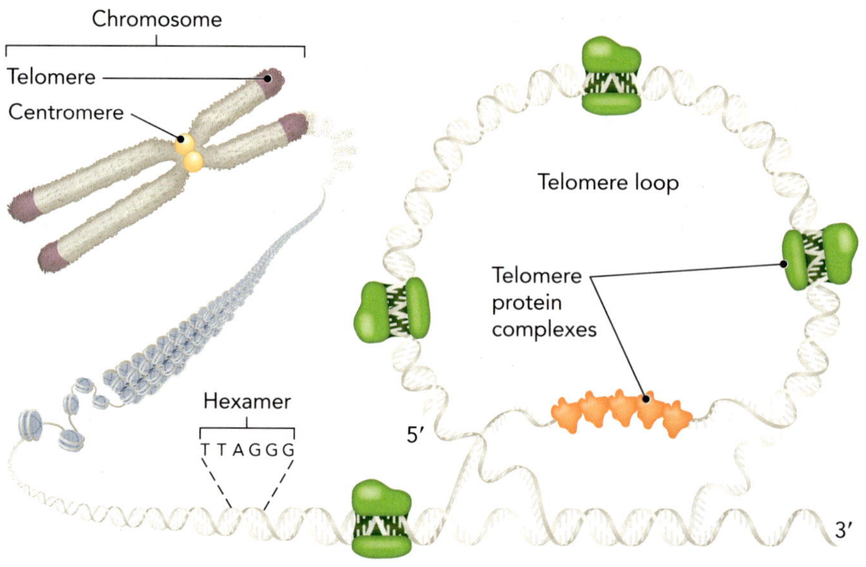

Figure 3.35 Telomeres protect the ends of chromosomes from degradation. The vertebrate hexameric telomere sequence 5′-TTAGGG-3′ is repeated thousands of times within the telomeric region. Protein complexes bind to telomeres to stabilize the loops.

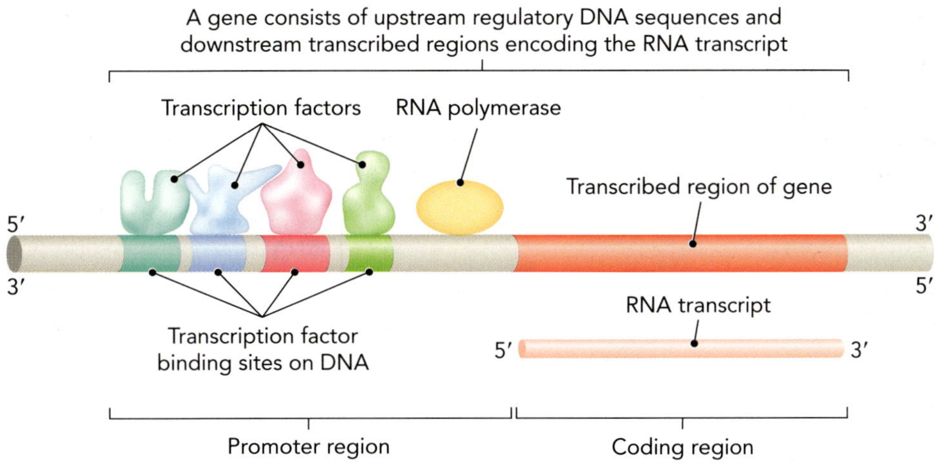

Figure 3.36 Genes are regions of DNA that contain a coding sequence for functional biomolecules. Here, the basic structure of a gene is shown on a region of double-stranded DNA. The promoter region contains binding sites for transcription factors and RNA polymerase and is most often on the 5′ side of the coding region. The RNA product generated by DNA transcription of the gene is shown below the DNA strands.

divided into two components: the **promoter** and the coding sequence. The promoter is a specific DNA sequence that occurs upstream, or 5′, of the coding sequence. The promoter is necessary for interactions with **transcription factors,** which are proteins that recognize specific DNA sequences and facilitate transcriptional initiation at gene promoters by recruiting RNA polymerase.

Figure 3.37 shows an example of prokaryotic gene structures. Prokaryotic genes can be either of two types: monocistronic or polycistronic. A **monocistronic** gene contains a promoter followed by a single protein-coding sequence. In a **polycistronic** gene, a promoter is followed by multiple coding regions. The polycistronic gene is transcribed into a single mRNA, but translation of this mRNA results in the production of three different proteins, one from each coding region. The short regions of DNA between the coding regions of polycistronic genes contain instructions to halt translation between each, which allows the proteins to be produced as separate units.

Polycistronic genes are not organized at random; their gene products frequently have a related function or are part of a regulatory unit in the cell. Polycistronic genes that contain coding sequences for proteins involved in a single biochemical process or pathway are known as **operons,** such as the *lac* operon discussed in Chapter 23.

Many genes in eukaryotes contain **exons,** or coding regions, that may be separated by noncoding sequences known as **introns. Figure 3.38** shows an example of this organization. From the left, the first region indicated encodes upstream regulatory sequences where transcriptional regulatory proteins bind. These upstream regulatory sequences

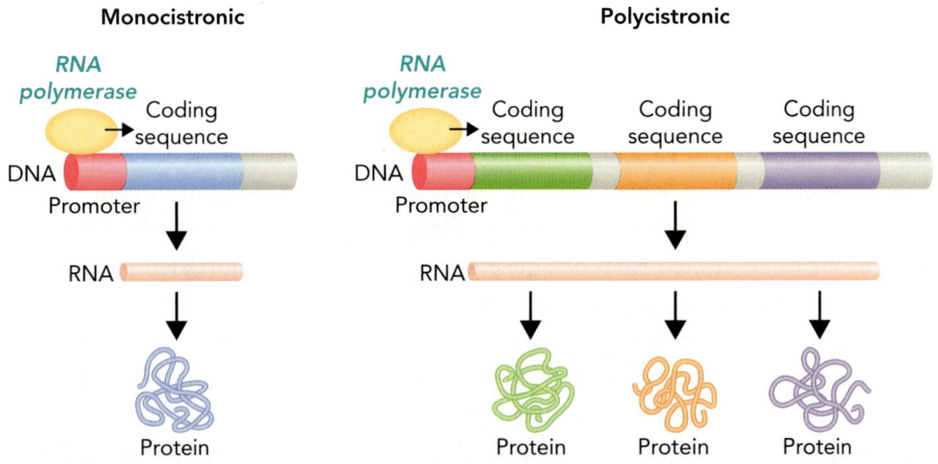

Figure 3.37 Two types of sequence organization are shown for prokaryotic genes. Monocistronic prokaryotic genes encode one protein product in a single RNA transcript, whereas polycistronic prokaryotic genes encode multiple protein products in a single RNA transcript.

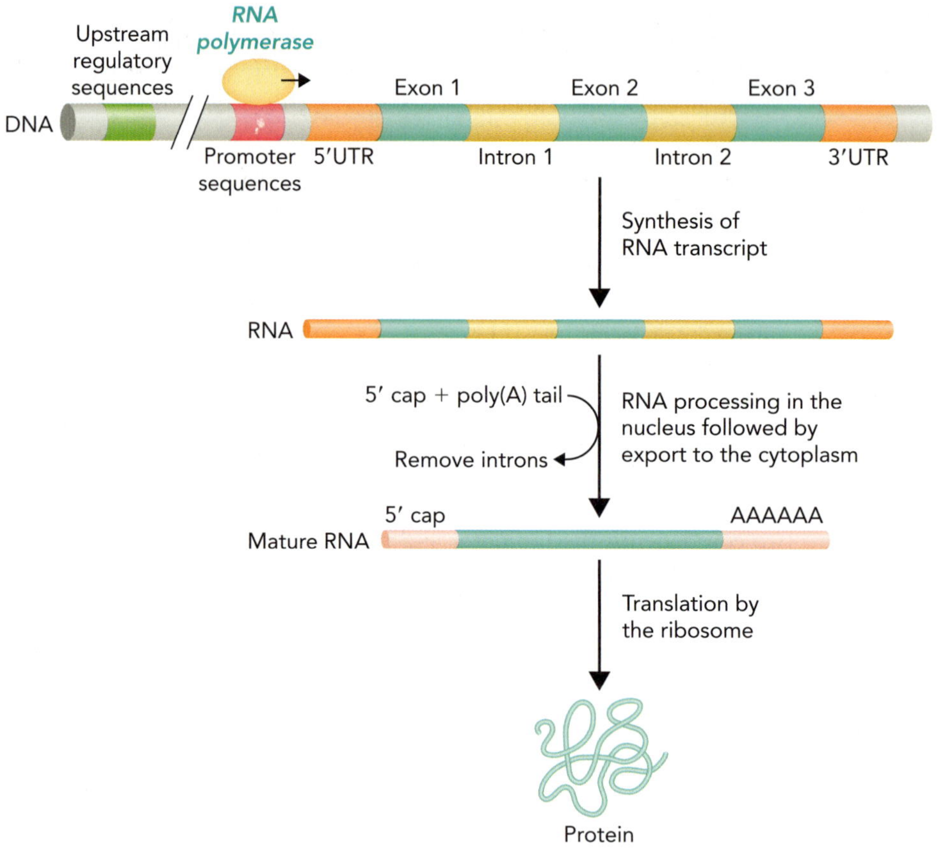

Figure 3.38 The organization of a typical eukaryotic gene is shown. The structure of a eukaryotic gene contains the upstream regulatory sequences and the promoter sequences, as well as the 5′UTR upstream of the coding sequence. In this example, the coding sequence consists of three exons and two introns. The precursor RNA transcript contains the intron and exon sequences, as well as the 5′UTR and 3′UTR. Concurrent with synthesis of the precursor mRNA by RNA polymerase, the transcript undergoes RNA processing to remove intron sequences, add a 5′ cap, and add a 3′ poly(A) tail before being exported from the nucleus to the cytoplasm. See Chapter 23 for more detail.

can be located hundreds or many thousands of base pairs upstream or downstream of the activated promoter; the only requirement is that they be located on the same chromosome as the promoter. The promoter region is separated from the coding sequence by a region known as the 5′ untranslated region, or **5′UTR**. As the name implies, this region is not translated but rather contains sequences that when transcribed into RNA will facilitate interaction with the protein translational machinery. At the other end of the coding sequence is another untranslated region, the 3′ untranslated region, or **3′UTR**. This region contains sequences necessary for the termination of transcription by RNA polymerase.

After the primary RNA transcript has been synthesized, the initial processing of RNA occurs. This process includes the addition of a covalent nucleotide modification at the 5′ terminus of mRNA, which is termed a **5′ cap**, and polyadenylation (the addition of adenine-containing nucleotides) at the 3′ end, which is termed a **3′ poly(A) tail.** Both of these features contribute to the stability and translational efficiency of mRNA and are described in Chapter 21.

The exons of a single gene often code for different functional domains of a protein. This separation of structural units has allowed for the evolution of new proteins through **exon shuffling**, which refers to the mixing and matching of protein-coding sequences into novel genes. As shown in **Figure 3.39a**, genetic recombination can cause exons to be deleted from one gene and inserted into an unrelated gene. Transcription and translation of these recombined genes results in the production of a novel protein. Recombination of exons is not always beneficial to an organism, as the new protein product may have a negative effect on the growth or regulation of a cell.

It is also possible for a single coding region to produce more than one protein product through alternative splicing of the exons. This process is illustrated in **Figure 3.39b**,

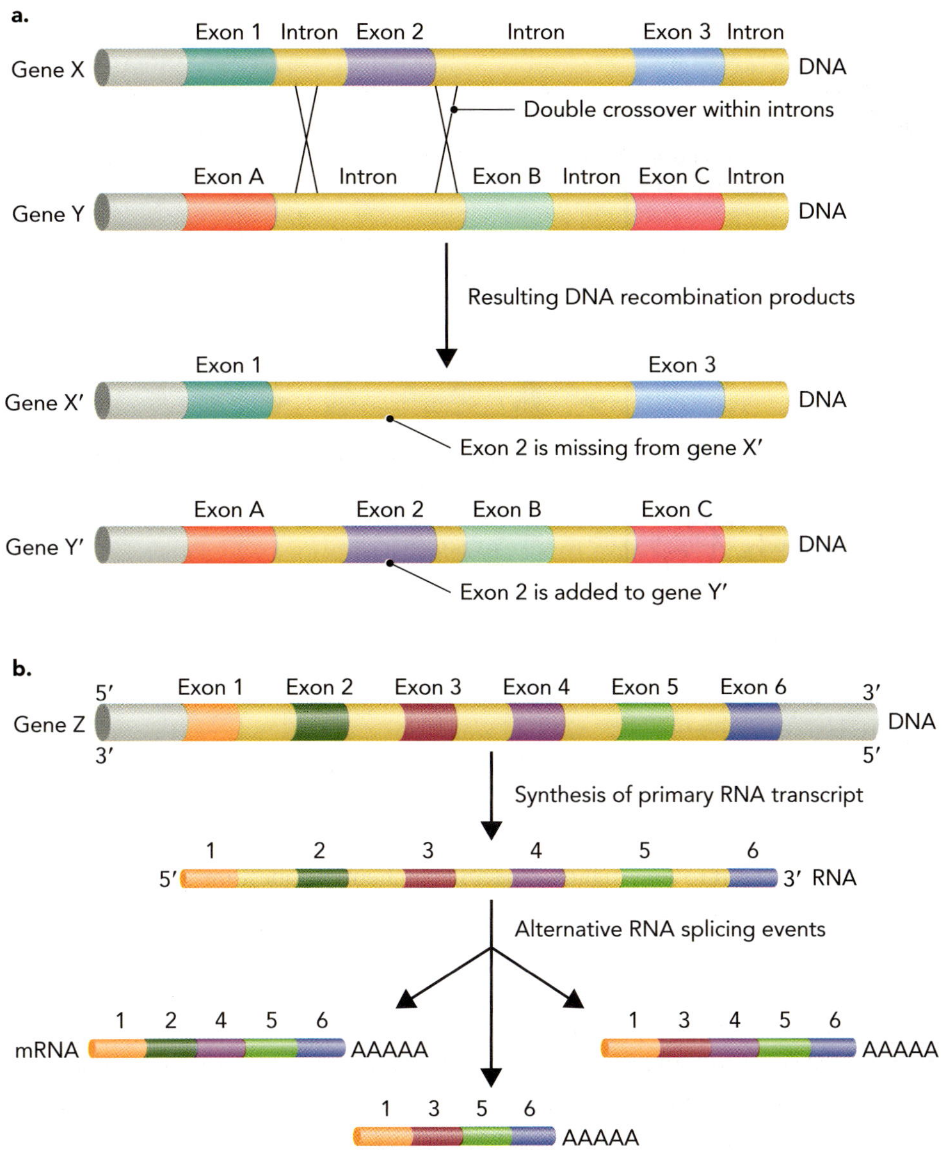

Figure 3.39 Exon shuffling and alternative splicing promote genetic diversity. **a.** In this example of exon shuffling, alternative recombination within the introns of two genes results in loss of one intron in gene X′ and gain of this same intron in gene Y′. **b.** Alternative splicing of gene Z can lead to three different mRNA products, which will encode three unique proteins.

where it can be seen that a gene with six different exons undergoes alternative splicing, resulting in the production of three different mRNAs. Each exon codes for a specific region of the protein, so combining these regions in a different order can produce proteins with alternative structures and/or functions. Alternative splicing is caused by the presence of different splice sites within the intron–exon junction (see Chapter 21).

Computational Methods in Genomics

Even the simple genome of a bacterium or yeast contains millions of base pairs, so it would be impossible to examine these nucleotide sequences by hand. However, computational tools have allowed scientists to assemble genome sequences for a significant number of organisms, and these sequences are freely available to the public for use by other research laboratories. In Section 3.3, we discuss some of the methods used to generate these sequences. Here, we focus on the tools that can be used to analyze them. With these tools, we can find differences caused by mutation or normal genetic variation or determine how the nucleotide sequence of a gene or genomic region has changed with evolution.

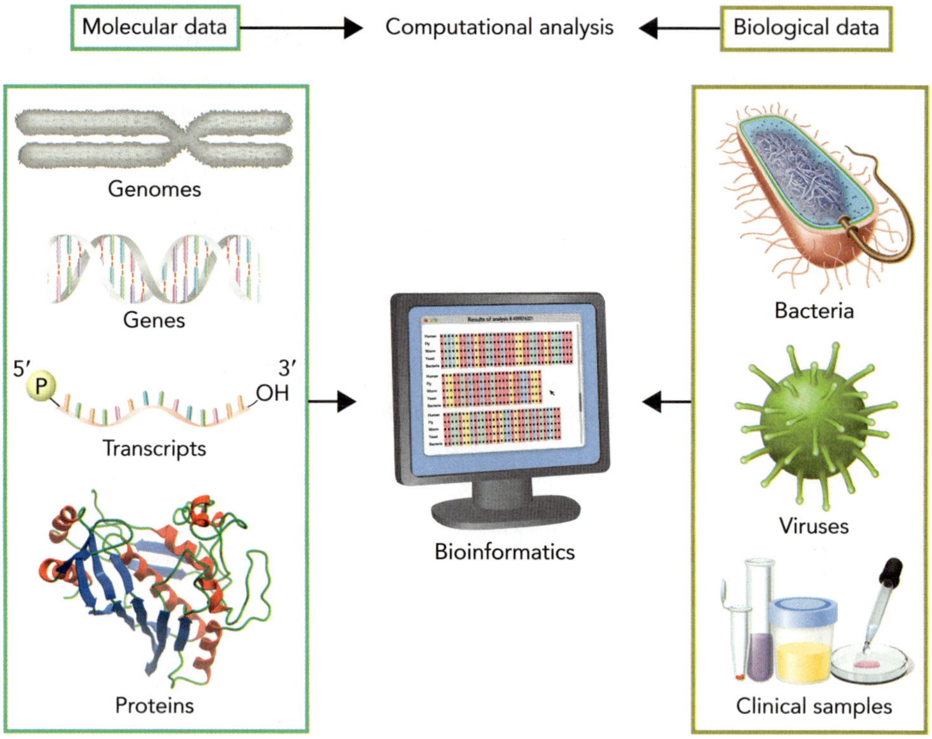

Figure 3.40 The field of bioinformatics uses computational tools to analyze data generated from molecular and biological samples.

Figure 3.40 shows an example of how the information available in public databases might be used to investigate a specific biochemical question. If scientists were interested in finding a way to block the activity of a cellular protein, it would be helpful to know its amino acid sequence, which can easily be determined from the sequence of the DNA that codes for it. Once the amino acid sequence is known, it can be compared to other protein sequences in a variety of databases. In some cases, it may be possible to find a protein with a similar sequence whose structure has been determined. The known structure and the identity between the two sequences can be used to develop a model of the protein under study and predict interactions between this protein and other molecules.

The National Center for Biotechnology Information (NCBI) hosts the most frequently used online resources for genomic and proteomic analysis. NCBI archives the genomic sequences of more than 10,000 organisms and provides tools to analyze these sequences and their transcripts and protein products. As shown in **Figure 3.41**, one method of analysis is to compare the sequence under study to other related sequences. This type of analysis can easily be completed through use of the NCBI Basic Local Alignment Search Tool (BLAST).

The tools for genome analysis have allowed scientists to make significant progress in the diagnosis and treatment of disease. For example, **Figure 3.42** shows pictures of two children with **Hutchinson–Gilford progeria syndrome (HGPS)**, the most common in a group of fatal disorders that cause rapid aging in children. This syndrome was first characterized by Dr. Jonathan Hutchinson in 1886 and Dr. Hastings Gilford in 1904. HGPS is a rare disorder, occurring in 1 in 4 million to 8 million newborns. However, it has captured public notice because of its drastic effects on the children's appearance and the information it may provide scientists about the normal aging process. In 2003, a joint effort by the National Human

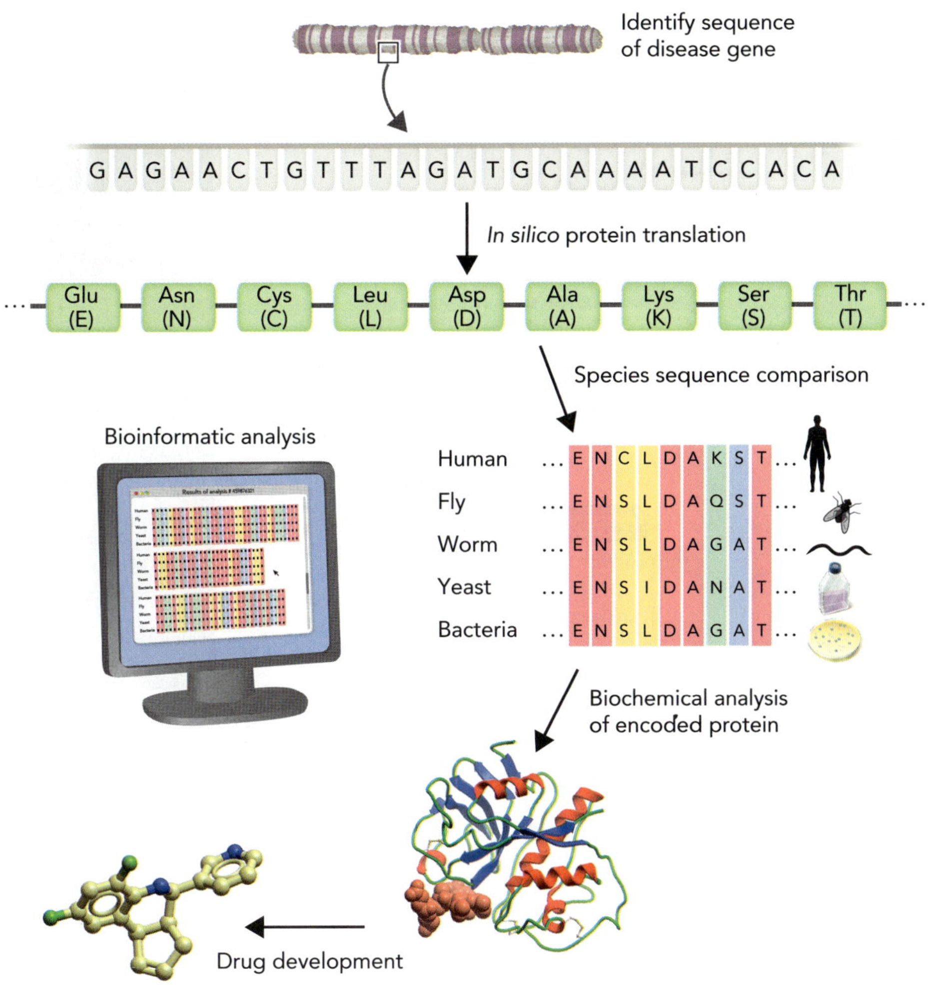

Figure 3.41 Bioinformatics is used in biochemistry to discover the function of an unknown gene, such as a human disease gene that can be targeted with pharmaceutical drugs. In this example, a region of a human chromosome was sequenced to determine whether a disease-causing mutation was present. DNA sequencing and subsequent bioinformatic analysis, comparing this human gene sequence to similar genes in other organisms, provided insight into the function of the encoded protein based on evolutionary conservation. Biochemical characterization of the protein and drug development could ultimately provide a treatment for the disease.

Genome Research Institute (NHGRI), the Progeria Research Foundation, and several research laboratories identified a single gene that was the site of mutation in those affected with HGPS.

The most common mutation leading to HGPS in humans is the nucleotide change C→T at nucleotide position 1824 in the coding sequence of exon 11 of the lamin A gene (*LMNA*). Most C→T mutations in DNA arise from cytosine deamination events that are not repaired prior to DNA replication (see Figure 3.22). As shown in **Figure 3.43**, a single mutation at this position leads to aberrant splicing of exon 11 to exon 12. When the exons are combined in the mutated RNA, 150 nucleotides are missing. Translation of this mRNA into protein results in a shortened prelamin A precursor protein. The key to understanding the biochemical basis for HGPS is that

a.

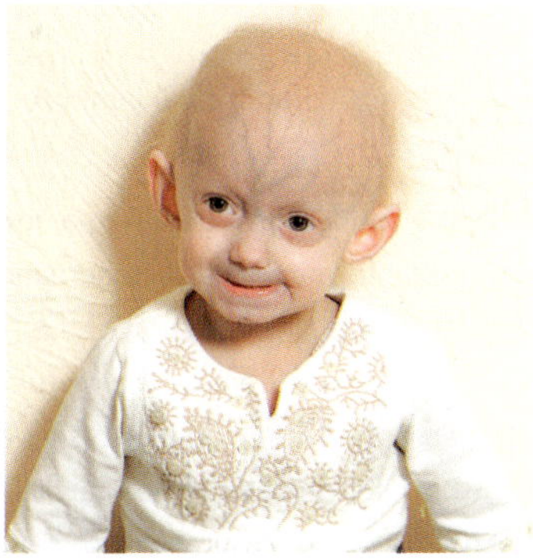

b.

Figure 3.42 Hutchinson–Gilford progeria syndrome is a fatal disorder that causes accelerated aging in children. **a.** A child with classic HGPS features, such as baldness, aging skin, a face and jaw size that is small compared to the rest of the body, and a pinched nose. JEREMY DURKIN/REX/NEWSCOM. **b.** Dr. Francis Collins, one of the scientists who contributed to the discovery of the HGPS mutation, is shown here in 2003 at a press conference announcing the discovery. With him is John Tacket, a 15-year-old with HGPS. AP PHOTO/GERALD HERBERT.

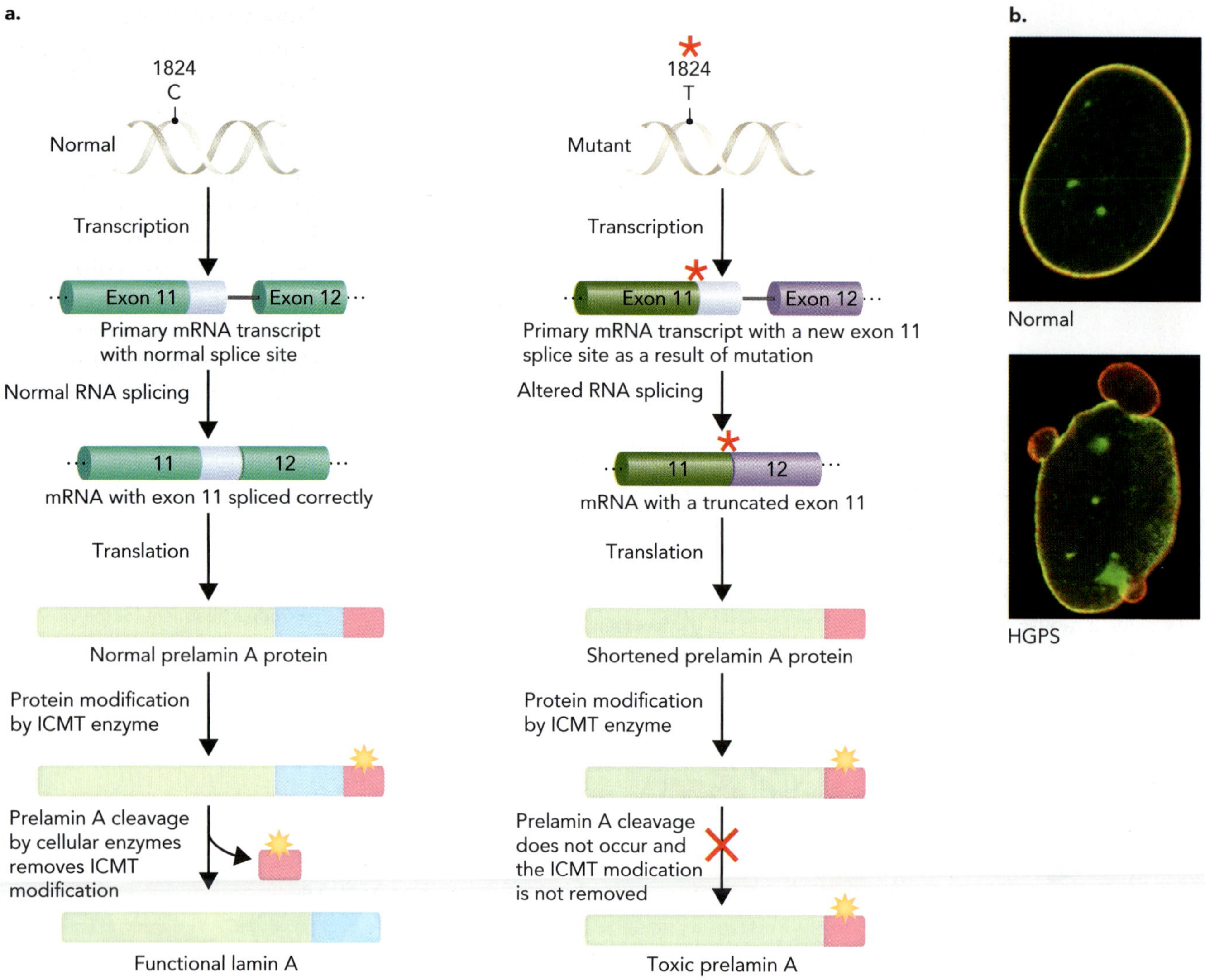

Figure 3.43 Introduction of a new splice site by a C→T mutation at nucleotide 1824 of the lamin A gene *(LMNA)* was mapped using bioinformatics. **a.** A schematic of the effect of the HGPS mutation on RNA splicing is shown next to the same process in an unaffected cell. Introduction of the new splice site leads to a deletion in the prelamin A precursor protein and also causes the protein to retain a toxic posttranslational modification that is normally removed in the processing of full-length lamin A. The red asterisk signifies mutation. ICMT = isoprenylcysteine carboxyl methyltransferase. **b.** Fluorescent staining of lamin proteins in a normal (top) and an HGPS (bottom) cell nucleus. In normal cells, the lamin proteins are located around the edge of the nucleus, shown here by the yellow staining. In HGPS cells, the mutant, unprocessed prelamin A protein (red staining) does not colocalize with the other lamin proteins, causing cell abnormalities associated with premature aging. T. SHIMI, V. BUTIN-ISRAELI, AND R. D. GOLDMAN (2012). THE FUNCTIONS OF THE NUCLEAR ENVELOPE IN MEDIATING THE MOLECULAR CROSSTALK BETWEEN THE NUCLEUS AND THE CYTOPLASM. *CURRENT OPINION IN CELL BIOLOGY, 24*, 71–78. © 2011 ELSEVIER LTD. ALL RIGHTS RESERVED.

the prelamin A precursor protein is modified by the enzyme isoprenylcysteine carboxyl methyltransferase (ICMT), which covalently attaches a methyl group to the C-terminal carboxylate of prelamin A (Figure 3.43). This C-terminal protein modification is normally removed from prelamin A by protease cleavage, thus generating the functional lamin A protein; however, the mutant prelamin A precursor protein is not cleaved and retains the modification, which leads to accumulation of the mutant protein in the nuclear membrane. The disease symptoms of HGPS are caused by the buildup of this altered protein, which damages the structure and function of the nuclei, leading to premature cell death.

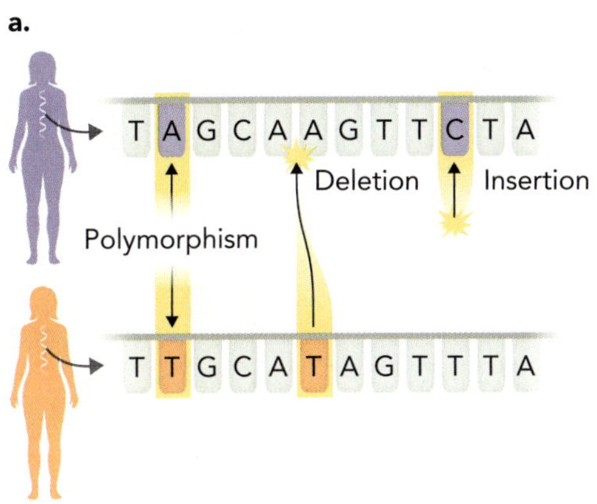

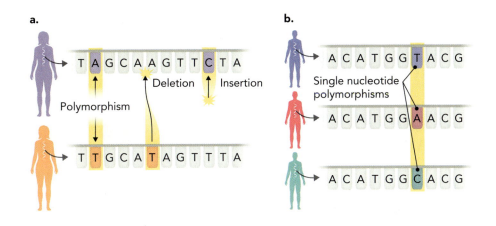

Figure 3.44 DNA sequence polymorphisms are small changes in the DNA sequence and can be used to generate a molecular fingerprint of an individual. **a.** Differences found when comparing the DNA of two individuals can be caused by polymorphisms, insertions, or deletions in the DNA sequence. **b.** DNA polymorphisms often occur as a single nucleotide change, as shown here with the variable nucleotide highlighted.

The discovery of the mutation responsible for HGPS using bioinformatics has allowed for an earlier diagnosis of the disease and the development of new potential treatments. Patients suspected of having HGPS based on possible disease symptoms can be screened for presence of the mutation using DNA analysis methods, and the diagnosis can be confirmed while the child is only a few months old. Scientists are now using a mouse model of the disease to understand better how accumulation of the toxic prelamin A protein leads to advanced aging and to test potential treatments. One promising treatment is through the inhibition of the ICMT enzyme, as without the modification by this enzyme, the nuclear buildup of the mutant prelamin A protein is reduced (see the chapter-opening illustration).

The hunt for the HGPS mutation was complicated by the fact that it is very rare and most often occurs spontaneously, which means it is not inherited from either parent (untreated HGPS children have an expected life span of 10–15 years). Searching for a new mutation in a gene or genome is complicated because many genetic variations can be found when comparing the gene sequences of individuals. Single nucleotide changes in the genome can occur when an error is made during DNA replication and is not repaired. Additional rounds of DNA replication incorporate the change into the genome, and if this change occurred in germ-line cells, then this DNA alteration will be passed on to the next generation. These **single nucleotide polymorphisms (SNPs)** can occur at random (**Figure 3.44**), and more than 1 million SNPs have been identified in the human genome. If an SNP occurs in a noncoding region, it is often silent; that is, it has no effect on the organism. SNPs that occur in the coding region of a gene or in splice sites, as with HGPS, can result in mild or significant phenotypic changes.

DNA sequencing can determine the genetic makeup of an individual and can be used to diagnose disease, as we have just seen. Interest is also increasing in use of this "DNA fingerprint" to tailor the treatment of disease to fit the genetic material of individuals through an approach called **precision medicine**. **Figure 3.45** illustrates how this knowledge might be applied in treating a group of patients with the same disease diagnosis. Using genetic differences to modify disease treatments could help increase the effectiveness of these treatments while decreasing unwanted side effects.

Deletions and insertions in the genome can cause a type of polymorphism known as a **short tandem repeat (STR)**. In the beginning of this section, we learned that the noncoding regions of DNA often contain repetitive sequences. At many chromosomal loci, these regions are composed of repeats of a core sequence, called **variable number**

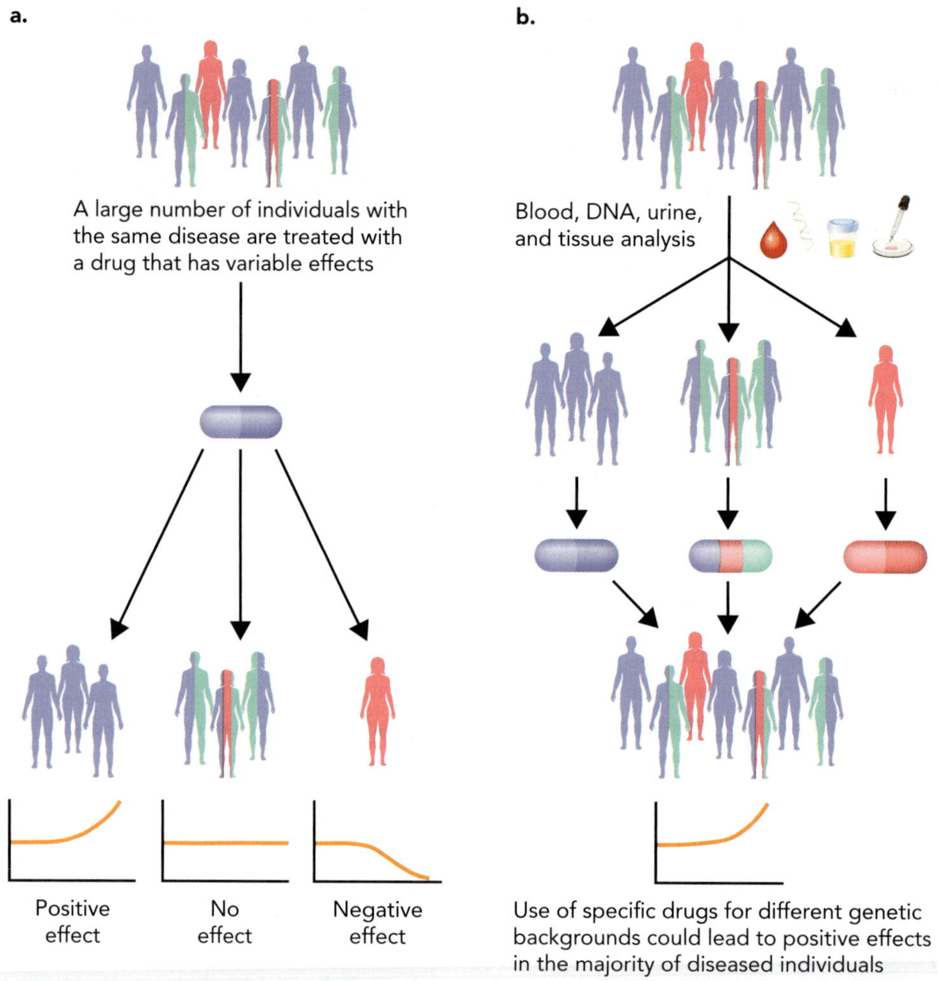
a.
A large number of individuals with the same disease are treated with a drug that has variable effects
Positive effect
No effect
Negative effect
b.
Blood, DNA, urine, and tissue analysis
Use of specific drugs for different genetic backgrounds could lead to positive effects in the majority of diseased individuals

Figure 3.45 Bioinformatic analysis is ushering in the age of precision medicine. This diagram illustrates how a DNA fingerprint could be used to identify the best treatment for diseased individuals with different genetic backgrounds. Data from large clinical studies would then be used to predict the best treatment regimen for a patient having a known DNA fingerprint to optimize a positive drug response and minimize toxic side effects.

tandem repeats (VNTRs), and the number of repeats is inherited from either parent. If two individuals have a significant number of VNTRs in common, then they are likely related. As shown in **Figure 3.46**, DNA from a crime scene can be matched to a specific individual if the two samples contain the same or a significant number of the same VNTRs.

The polymorphic differences in DNA can also be used to track the evolution of pathogens over time and their spread within a population. For example, if a virus with similar symptoms is identified in two individuals, then the DNA sequence can be used to determine if the two viruses are the same. This type of analysis is important for determining common strains of the influenza virus present in the population each year so that an effective vaccine can be formulated. DNA sequence analysis can also be useful when a new viral strain emerges from a mutated animal virus, such as a pig influenza virus. In this case, it may be possible to track the spread of a virus across time and space if its DNA genome is mutating rapidly, as is often the case.

As illustrated in **Figure 3.47**, DNA sequencing of viral DNA present in the blood of patients with disease symptoms in different areas of the world can be used to determine how two independent mutations gave rise to related strains. Moreover, this bioinformatic information can trace the infection trajectory by correlating the viral DNA sequence with the known travel activities of individuals with similar viral infections. Similar approaches were used to track the swine H1N1/09 viral pandemic in 2009, which spread rapidly owing to the ease of international travel.

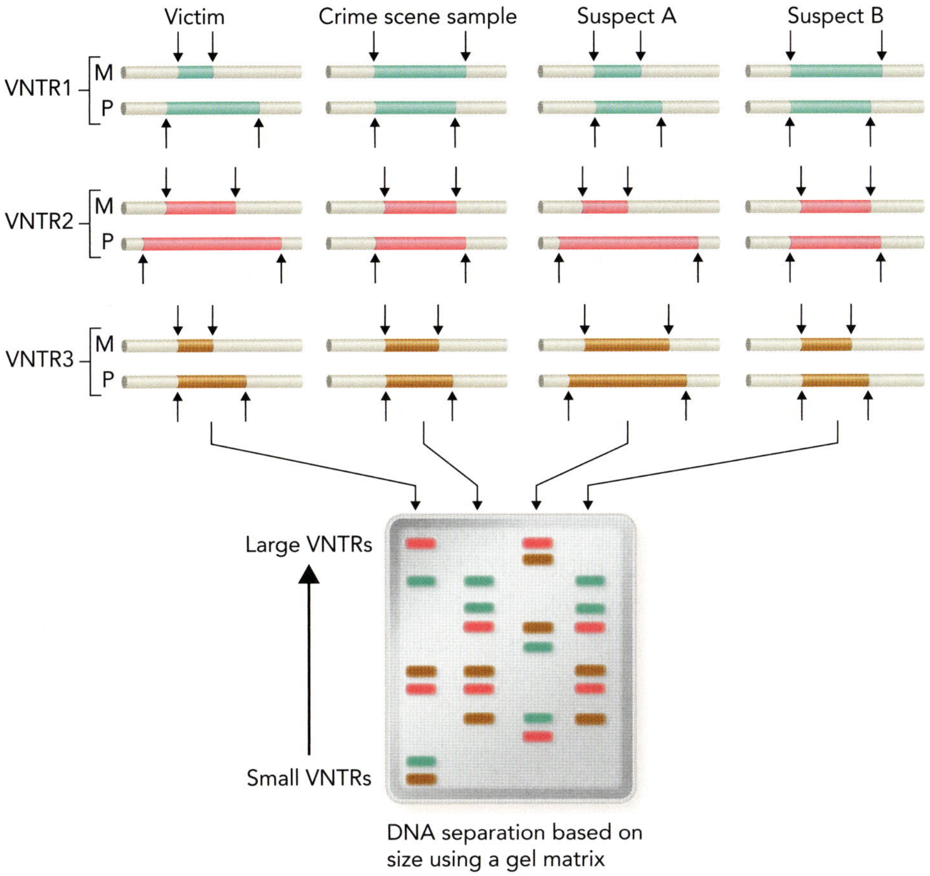

Figure 3.46 Polymorphic differences in DNA can be used to identify individuals linked to a crime scene. This illustration shows how three distinct VNTRs can be used to match a DNA sample. The colored bars in each chromosome represent the length of each type of VNTR. In all samples, the top chromosome represents maternal (M) DNA, and the bottom represents paternal (P) DNA. For each sample, the segments of DNA, which are colored according to the VNTRs in this example, are separated by size in a gel matrix. Samples that have the same distribution of DNA sizes are likely from the same individual.

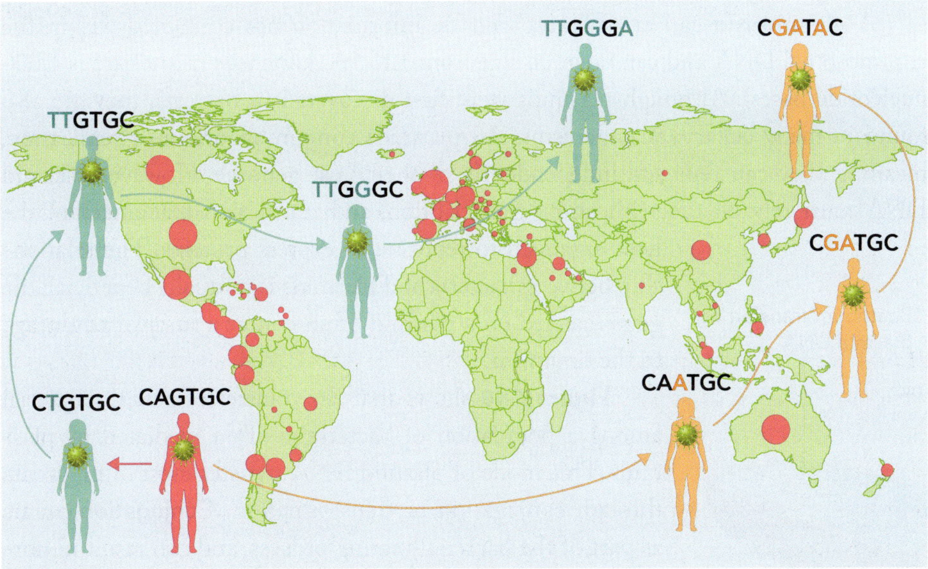

Figure 3.47 The spread of infectious disease can be mapped using DNA sequencing and bioinformatics. A rapidly evolving (mutating) animal virus that infects a human in a single event can be tracked by following changes in viral sequences over time and space. In the case illustrated here, independent mutations gave rise to two related viral strains that continued to evolve as more and more individuals became infected. Disease tracking uses maps that illustrate the sites of viral outbreaks and indicate the magnitude by the size of the circles.

concept integration 3.2

Explain why organisms accumulate DNA nucleotide changes throughout their lives, but these types of DNA changes are not inherited by their offspring. How does this affect DNA forensic analysis?

Changes in the DNA sequence of somatic cells, such as cells in the skin or liver tissue cells, are not passed down to the organism's offspring because only DNA in gamete cells (sperm and egg) are inherited. Therefore, DNA alterations must occur in cells of the reproductive organs, such as testes and ovaries in humans, to be inherited by future generations. Somatic mutations are unlikely to affect DNA forensic analysis that is based on length of tandem repeats in STR and VNTR sequences because the primary determinant is number of repeats in the tandem array. In the case of forensic DNA analysis that is using SNPs or direct DNA sequencing, the analysis is only reliable for inherited DNA alterations.

3.3 Methods in Nucleic Acid Biochemistry

The ability to manipulate, modify, and determine the sequence of DNA is critical to biotechnology. Techniques in nucleic acid biochemistry are fundamentally based on our knowledge of enzymology, molecular biology, genetics, and microbiology. In this section, we describe the basic principles of some of these techniques and explain how they are used to investigate a variety of biochemical processes that are described elsewhere in the book.

Plasmid-Based Gene Cloning

One of the most useful tools in protein biochemistry is the ability to isolate genes and express the protein product in an exogenous system. At the heart of this technology are the self-replicating, circular pieces of DNA known as plasmids.

A plasmid is a carrier of "extra" genetic information not contained within the chromosomal DNA and can range in size from 1 to 100 **kilobases** (a kilobase is 1,000 nucleotide bases). Although plasmids were first discovered in bacteria, they are also found in many eukaryotic organisms. All plasmids contain an origin of replication, meaning they can replicate independently of their host genome. However, plasmid DNA must rely on the replication, transcription, and translation machinery of the host organism to replicate or express proteins. The relationship between bacteria and plasmids is mutually beneficial; the genes carried on a plasmid often confer a survival advantage to the organism.

Figure 3.48 DNA can be transferred between organisms in three major ways: conjugation, transformation, and transduction. A bacterial cell is shown with both the chromosomal and plasmid DNA indicated.

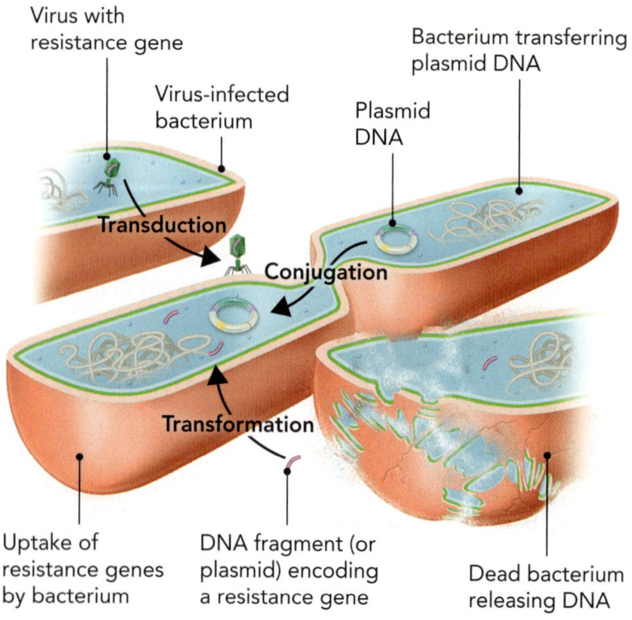

Figure 3.48 shows that the transfer of genetic material among a population of bacteria is often mediated by plasmids. This mode of plasmid DNA transfer most often occurs through conjugation or transformation. **Conjugation** occurs as part of the bacterial mating process, and can result in horizontal gene transfer. The donor bacterium connects with the recipient and transfers a copy of the plasmid. The plasmid is replicated prior to transfer so that at the end of the mating process, both bacteria contain an identical copy.

Gene fragments or plasmids can also be transferred between bacteria through **transformation**. In this process, DNA released into the environment by dead bacteria is obtained by other bacteria in the immediate vicinity through a poorly understood mechanism involving the bacterial cell wall. Random DNA fragments from the environment can combine with existing plasmid DNA inside the living recipient bacterial cell to generate a new plasmid.

Bacterial cells can gain genetic material through yet another process called **transduction.** In this case, a virus mediates the transfer of genetic material between the bacterial cells. Viruses that infect bacteria through transduction are known as **bacteriophages** and are useful tools in nucleic acid biochemistry and molecular biology.

Once scientists had determined the structure and properties of plasmids, they quickly realized that plasmids could be useful tools in the laboratory. For example, plasmids were modified for use as cloning vectors (we will discuss the cloning process at the end of this section). The first cloning vector was pBR322, a 4,361-bp plasmid named after the two postdoctoral researchers, Bolivar and Rodriguez, who generated the construct in Herbert Boyer's lab at the University of California, San Francisco. As shown in **Figure 3.49a**, pBR322 contains an origin of replication and genes that confer resistance to the antibiotics ampicillin and tetracycline. **Figure 3.49b** shows a typical plasmid cloning vector in use today. It contains a replication origin and an antibiotic-resistance gene, as well as a segment of DNA called a **multiple cloning site (MCS).** The multiple cloning site segment of DNA can be cleaved by sequence-specific endonucleases called restriction enzymes to facilitate gene cloning.

The multiple cloning sites can be used to insert foreign DNA fragments into the plasmid. This process produces **recombinant DNA**, which refers to DNA molecules from multiple sources that have been combined by laboratory methods. When foreign

Figure 3.49 Plasmid DNA cloning vectors contain functional elements that facilitate propagation of recombinant molecules in antibiotic-resistant bacteria. **a.** pBR322 was the first plasmid DNA cloning vector used for recombinant DNA methods. Insertion of foreign DNA into the antibiotic-resistance genes for tetracycline and/or ampicillin disrupts these genes. Therefore, the loss of antibiotic resistance provides a functional genetic screen to determine whether the DNA insertion has been successful. **b.** A common plasmid DNA cloning vector containing antibiotic-resistance genes, an origin of DNA replication conferring high plasmid copy number per cell, and a multiple cloning site (MCS). **c.** Insertion of foreign DNA into the multiple cloning site permits blue–white selection screening owing to disruption of *lacZ* and loss of β-galactosidase enzyme activity.

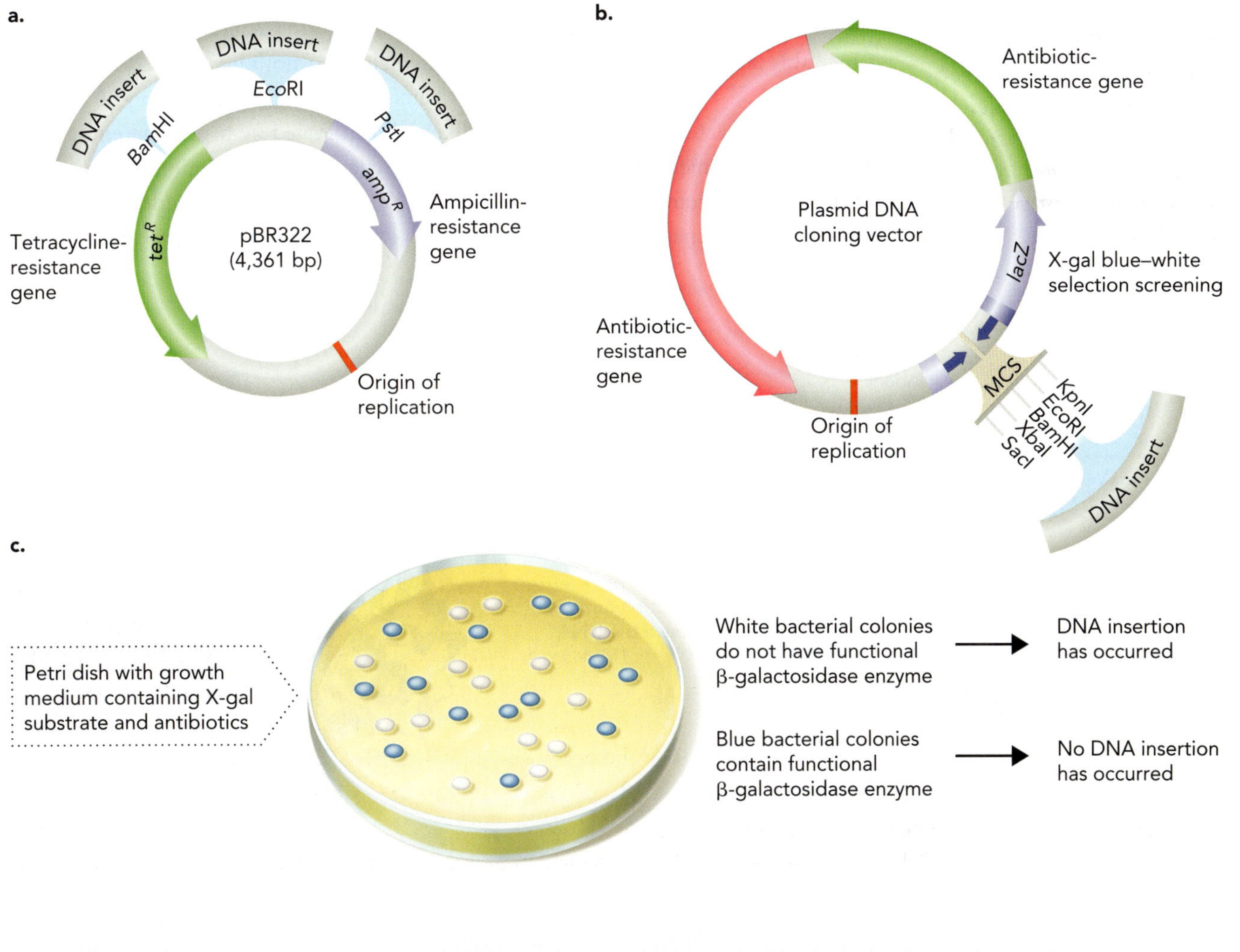

DNA is inserted into one of these restriction sites, it disrupts that antibiotic-resistance gene and makes it nonfunctional. Thus, the loss of antibiotic resistance can be used as a screen to determine whether the DNA insertion has been successful.

Many plasmid cloning vectors also include a sequence encoding the bacterial β-galactosidase gene (*lacZ*), which is useful for another type of genetic screening. The β-galactosidase enzyme cleaves its substrate, 5-bromo-4-chloro-3-indoyl-β-D-galactopyranoside (X-gal), into a blue product that is visible by eye. Therefore, bacterial colonies that contain functional β-galactosidase appear blue when X-gal is present (**Figure 3.49c**). If foreign DNA is inserted into *lacZ*, enzyme function is lost, and the bacterial colonies that contain the DNA insertion are white. This process to determine whether the cloning has been successful is called blue–white screening. Bacteria maintain and replicate the plasmid DNA cloning vector because of the survival advantage gained by antibiotic resistance.

Along with antibiotic resistance conferred by plasmids, bacteria also have evolved mechanisms to protect against bacteriophage infection, which was first characterized in the 1950s. This process is a natural defense system in bacteria that requires two types of enzymes: (1) **DNA methylases** that methylate DNA at specific sequences, and (2) **restriction endonucleases** that cleave DNA at specific sequences. DNA methylases transfer a methyl group to C-5 of a cytosine or to N-6 of an adenine within a DNA sequence recognized by the enzyme. As shown in **Figure 3.50**, restriction endonucleases cleave the phosphodiester backbone of DNA at specific sequences. Importantly, methylated DNA is not cleaved by the corresponding restriction endonuclease that recognizes and cleaves the same DNA sequence that is methylated by a DNA methylase enzyme. For example, the *Eco*RI methylase enzyme recognizes the sequence 5′-GAATTC-3′ and transfers a methyl group to the first adenine base (Figure 3.50). Once this occurs, the *Eco*RI restriction endonuclease is no longer able to cleave this methylated 5′-GAATTC-3′ sequence. Therefore, when a bacteriophage injects its DNA into a bacterial cell, the DNA is likely to be cleaved by restriction endonucleases before it is methylated by DNA methylase enzymes that function to protect the bacterial DNA from restriction endonuclease cleavage. Restriction endonucleases are named after the species from which they were isolated. Roman numerals indicate the

Figure 3.50 Bacterial cells can be protected against bacteriophage infection by DNA methylases and restriction endonucleases. **a.** *Eco*RI restriction enzyme and *Eco*RI methylase each recognize a 6-bp region of DNA with the sequence 5′-GAATTC-3′. Methylation of this sequence on the first adenine protects the DNA from cleavage by *Eco*RI endonuclease. **b.** Molecular structure of a protein dimer of the restriction enzyme *Hind*III, bound to its cognate double-strand DNA sequence. BASED ON PDB FILE 3A4K.

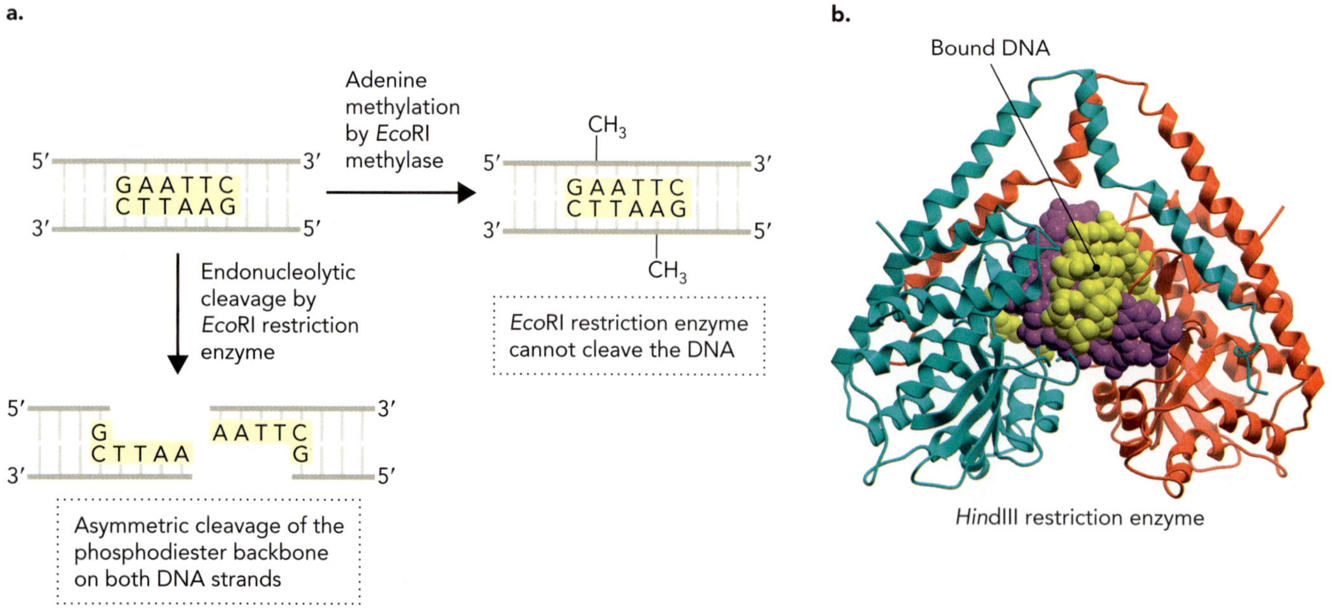

number of enzymes isolated from the same strain. For example, *Hind*III was the third enzyme isolated from *Haemophilus influenzae* strain d.

Three classes of bacterial restriction endonucleases have been characterized: type I, type II, and type III. Type II restriction endonucleases have become a useful tool in gene cloning applications because they cleave DNA within their respective recognition sequences. By contrast, type I and III restriction endonucleases cleave DNA outside their respective recognition sequences. The type I and III enzymes require ATP for hydrolysis, whereas the type II enzymes do not. **Figure 3.51** shows examples of several type II restriction endonucleases and the DNA sequences they recognize. You can see that each sequence on one strand is identical to the sequence on the complementary strand when read 5′ to 3′, which means they are palindromes. A **palindrome** is a string of letters or numbers that reads the same in both directions; for example, the word *radar* is a palindrome. Palindromic recognition sequences for restriction endonucleases are usually four to eight base pairs in length, with six base pairs being the most common length. Figure 3.51 shows examples of restriction endonucleases with four-base-pair recognition sequences (*Hpa*II, *Bst*KTI, *Hae*III), as well as restriction endonucleases with six-base-pair recognition sequences. The shorter the recognition sequence, the more frequently it is found in the genome. Consider that a four-base-pair recognition

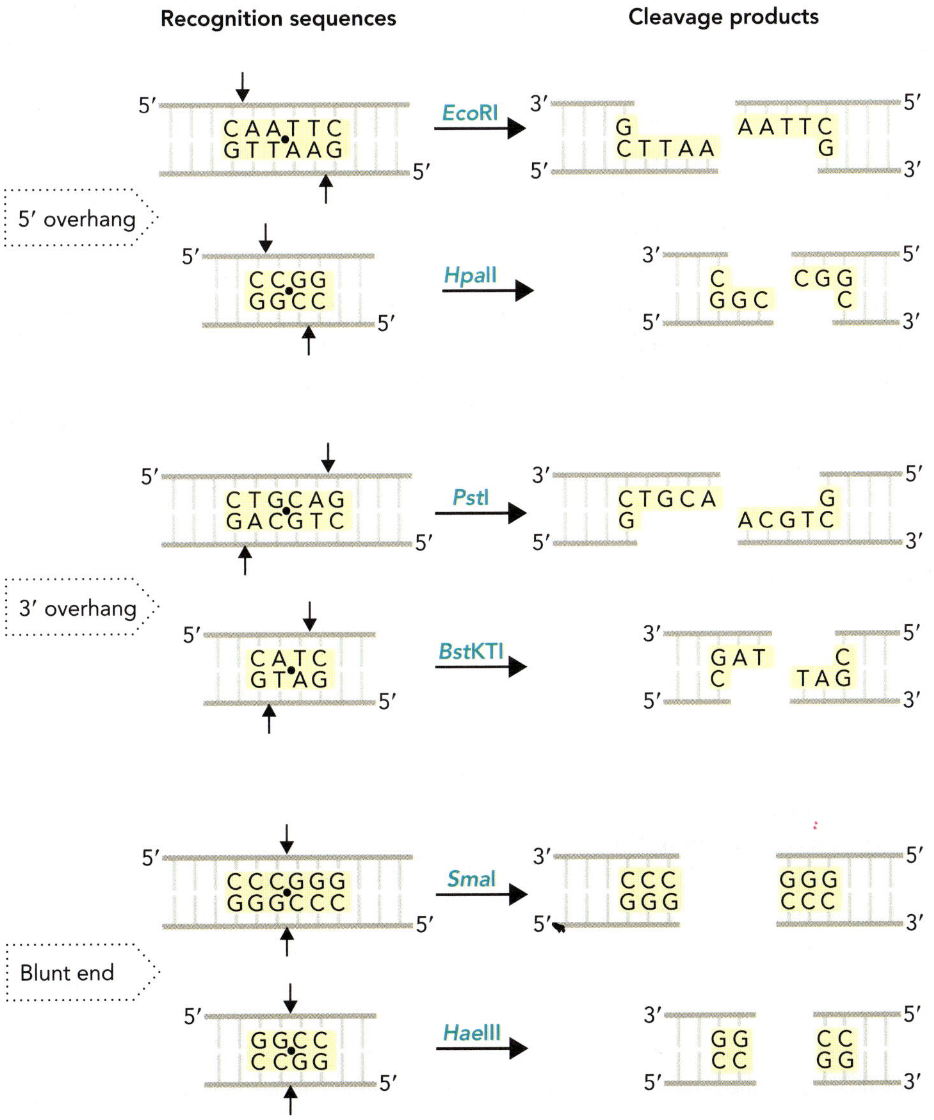

Figure 3.51 Restriction endonucleases recognize palindromic sequences in DNA and cleave double-stranded DNA to generate 5′ overhangs, 3′ overhangs, or blunt ends. Six different restriction endonuclease recognition sequences are shown, with the axis of symmetry indicated by the black dot in the middle of each sequence. The arrows on the top and bottom of the recognition sequences indicate the cleavage sites.

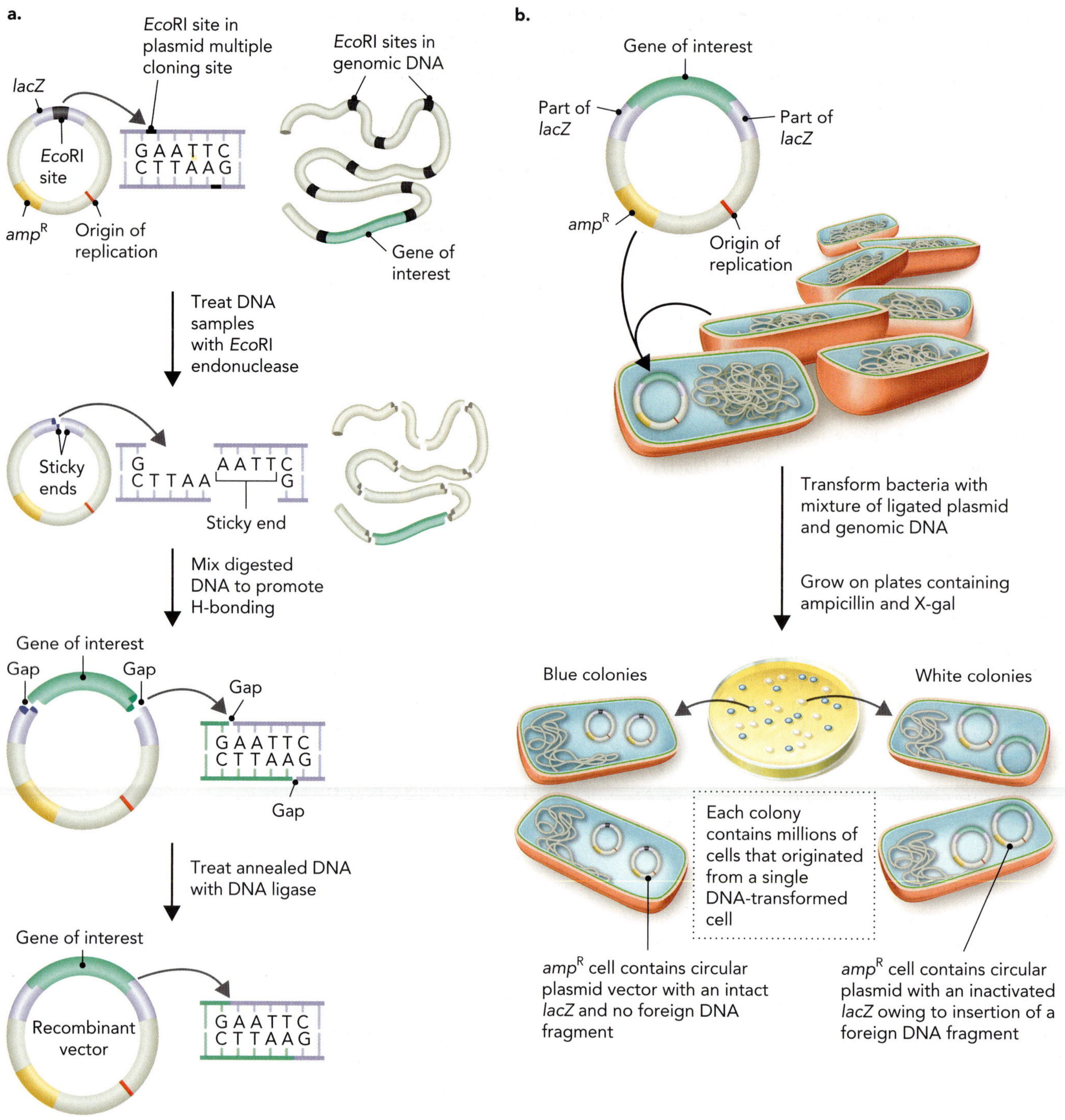

Figure 3.52 Genomic DNA cloning is based on methods collectively called recombinant DNA technology. **a.** This example shows cloning achieved with compatible ends generated from *Eco*RI restriction digests of DNA. Both the plasmid and genomic DNA are digested with *Eco*RI, annealed to promote hydrogen bonding between complementary DNA sequences from the overhanging ends, and then covalently joined with DNA ligase. Note that the plasmid multiple cloning site is located in the middle of *lacZ*. Because the process is not 100% efficient, a mixture of DNA results, which includes unligated DNA, self-ligated genomic DNA, self-ligated plasmid DNA (these are not illustrated), and the correct product, ligated plasmid–genomic recombinant DNA. **b.** The mixture of DNA is added to a bacterial culture to initiate transformation. The transformed bacteria are grown on culture plates containing both ampicillin to select for the presence of replicating plasmid DNA and the β-galactosidase substrate X-gal to permit for blue–white colony screening.

sequence should occur on average once in every 256 base pairs in DNA (4^4), whereas a six-base-pair recognition sequence should occur on average once every 4,096 base pairs (4^6). More than 600 restriction endonucleases are available commercially, with a wide range of sequence recognition sites.

Restriction endonuclease cleavage of DNA can occur at the ends or middle of a sequence. A sequence cleaved in the middle results in a blunt end (*Sma*I, *Hae*III). If the DNA backbone is cleaved after the first or before the last nucleotide, the DNA has a 5′ (*Eco*RI, *Hpa*II) or 3′ (*Pst*I, *Bst*KTI) single-stranded overhang. DNA with 5′ or 3′ overhangs are referred to as having "sticky" ends because if the free ends interact with a homologous single-stranded DNA sequence, they form hydrogen bonds between the bases and "stick" together. Blunt ends do not have a strong tendency to interact with other blunt ends, as they have no unpaired bases at the two junctions.

How are restriction endonucleases and purified plasmid DNA used to clone foreign DNA? As we saw earlier, commercially available plasmids have been designed to contain a stretch of unique restriction sites in the multiple cloning site. If the gene of interest and plasmid are cleaved with the same restriction enzyme, then the matching cohesive ends can be covalently linked together by the enzyme **DNA ligase**, which is an enzyme that catalyzes the formation of a phosphodiester bond. As shown in **Figure 3.52**, the recombinant DNA plasmid can be transferred into bacterial cells by the process of plasmid transformation using a buffer containing divalent cations and heat shock. The mixture of DNA and transformed bacterial cells are then grown on the appropriate antibiotic-containing culture plates. Only cells that contain the plasmid are able to grow owing to the antibiotic-resistance gene expressed on the plasmid. These cells can then be selected and grown in liquid media to amplify or make more copies of the cloned DNA. Blue–white screening can be used to determine which cells contain a plasmid with a DNA insertion.

It is also possible to clone gene sequences using mRNA as a starting material. This technique is most often used to generate a library of all actively transcribed genes in a cell. **Figure 3.53** illustrates the process of generating such a library of gene sequences using mRNA transcripts. First, mRNA is isolated from the cell and converted back into a double-stranded sequence using the enzyme **reverse transcriptase** to generate **complementary DNA (cDNA)**. Reverse transcriptase is an RNA-dependent DNA polymerase that was first discovered in retroviruses. The production of cDNA begins with the first strand of cDNA being synthesized by reverse transcriptase using a poly(dT) oligonucleotide primer that anneals to the 3′ poly(A) tail of eukaryotic mRNA. This oligonucleotide primer contains 12–20 thymine-containing deoxyribonucleotides (deoxythymidine; dT). Reverse transcriptase completes the single-strand cDNA when it reaches the 5′ end of the mRNA transcript, leaving behind an RNA–DNA hybrid duplex. Treatment of the RNA–DNA hybrid with the enzyme ribonuclease H cleaves only the RNA strand and produces short RNA fragments that are still annealed to the cDNA strand (Figure 3.53). These RNA fragments serve as primers for DNA synthesis of the second strand of cDNA by the enzyme DNA polymerase. In the final step of double-stranded cDNA synthesis, the enzyme DNA ligase is used to seal the single-strand gaps left behind in the second strand of cDNA. The next step in cDNA cloning is to treat the double-stranded cDNA—and the plasmid vector—with the same restriction endonuclease to generate compatible ends for annealing and ligation. The ligated DNA is then transformed into bacterial cells, and antibiotic-resistant colonies are isolated and characterized as described earlier (see Figure 3.52).

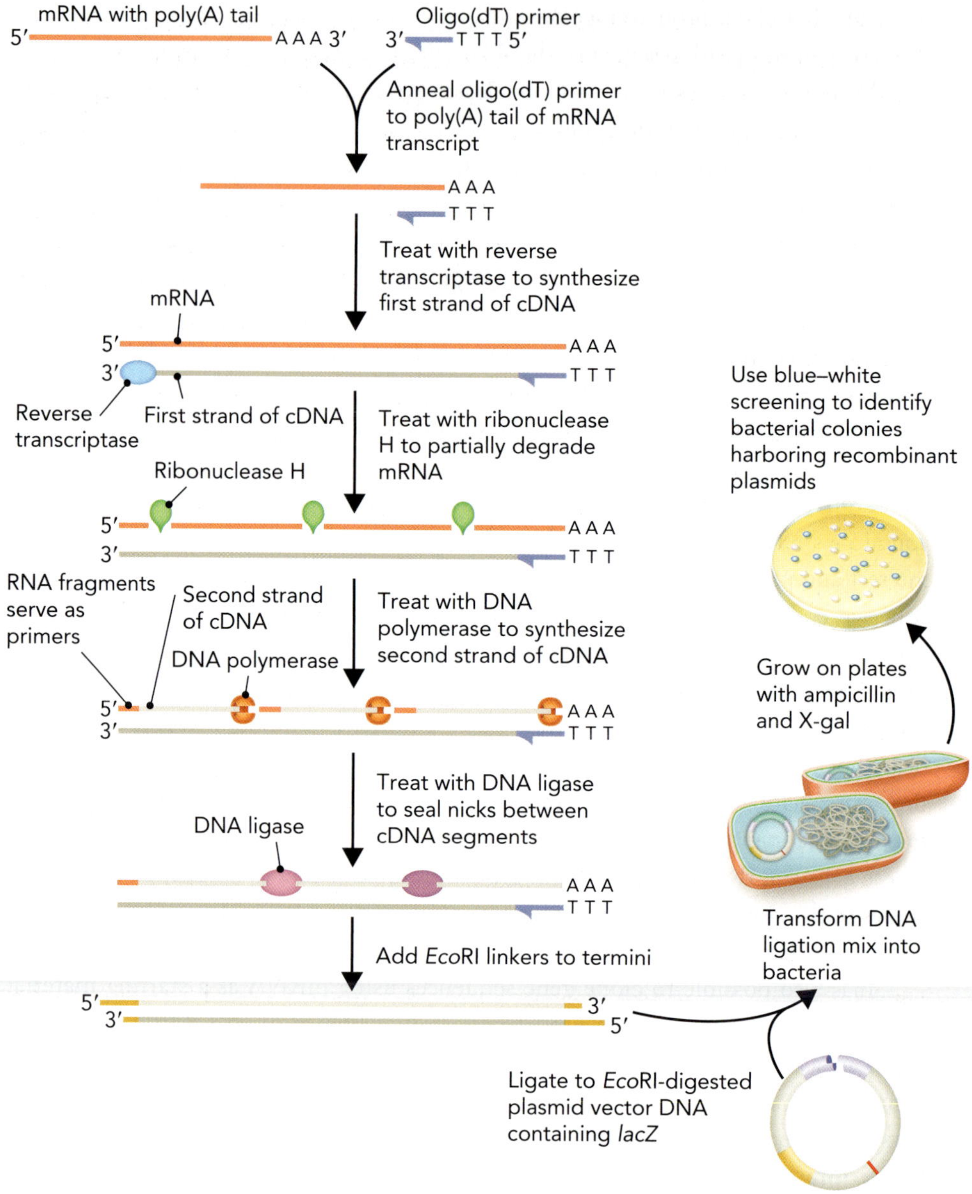

Figure 3.53 Protein-coding sequences can be cloned using mRNA that is converted to double-stranded complementary DNA (cDNA) using the enzymes reverse transcriptase, ribonuclease H, DNA polymerase I, and DNA ligase. Once the cDNA has been purified, it can be cleaved with a restriction enzyme such as *Eco*RI to facilitate plasmid cloning as described in Figure 3.52.

High-Throughput DNA Sequencing

We have seen that examining the DNA sequence of an individual can be useful for disease diagnosis and in some cases can guide disease treatment. In 1980, the Nobel Prize in Chemistry was awarded to scientists from two different universities who independently developed different methods of DNA sequencing. In 1977, Allan Maxam and Walter Gilbert at Harvard University developed a chemical degradation method that could be applied to DNA sequencing. At the same time, Frederick Sanger at the University of Cambridge developed a DNA sequencing method called chain termination, which is based on *in vitro* DNA synthesis. Although the Maxam and Gilbert method of chemical DNA sequencing is no longer in use because of its low efficiency, and there are now techniques that are an improvement over the Sanger method, the Sanger DNA sequencing method is presented here as an elegant example of applied biochemistry.

The **Sanger DNA sequencing** method is based on the use of limiting amounts of **dideoxynucleoside triphosphates (ddNTPs)** in a primer-dependent DNA synthesis reaction. (The abbreviation NTP is used to refer to any nucleoside triphosphate.) The four ddNTPs used in the *in vitro* DNA sequencing reaction differ from the deoxynucleoside triphosphates (dNTPs) by the presence of a hydrogen at the ribose 3′-position instead of a hydroxyl group. (Recall from Section 3.1 that the ribose 3′-hydroxyl group is necessary for the formation of the phosphodiester backbone.) DNA polymerase, which catalyzes the formation of this backbone *in vivo* and *in vitro*, does not distinguish between dNTPs and ddNTPs, so it can incorporate a ddNTP into the growing DNA chain. The key to this original Sanger DNA sequencing method was that limited incorporation of ddNTPs into newly synthesized DNA molecules terminated the reaction randomly and thereby generated a pool of chain-terminated DNA molecules that each differed by one nucleotide in length. The original Sanger method used radioactively labeled nucleotides, but a more modern adaptation uses fluorescent labels to distinguish the nucleotides, as described in the following paragraphs.

As shown in **Figure 3.54**, a sequencing reaction begins with a region of single-stranded DNA, usually obtained by denaturation of the double-stranded DNA, which acts as a template for the sequencing reaction. DNA polymerase requires a double-stranded region to initiate the replication process, so a short single-stranded oligonucleotide is added to the reaction as a primer, which anneals to its complementary sequence in the single-stranded DNA template. In addition to the DNA template and DNA primer, the reaction mixture also contains the enzyme DNA polymerase and a 300:1 mixture of dNTPs to ddNTPs to ensure sufficient chain elongation to generate DNA sequencing information out to ~500 nucleotides. The four ddNTPs are each fluorescently labeled so that the color indicates which nucleotide has been added to the end of the fragment.

When the DNA synthesis reaction has run to completion, a mixture of chain-terminated DNA fragments is obtained, and these fragments are separated by size using a technique called capillary gel electrophoresis. In this technique, DNA fragments of different sizes have different mobilities through a gel-filled capillary tube, with smaller DNA fragments moving at a higher rate through the gel matrix. A laser coupled to a detector system is used to identify each of the four distinct fluorescent tags on the DNA molecules as they pass in front of the detector. The fluorescence data obtained from the pool of chain-terminated DNA molecules is then analyzed to infer the combined nucleotide sequence of the original DNA template.

Polymerase Chain Reaction

DNA cloning and sequencing methods often require DNA amplification using a technique called the **polymerase chain reaction (PCR)**. This method is based on *in vitro* DNA replication to generate multiple copies of a specific target DNA segment. The replication reaction is repeated through multiple cycles to exponentially amplify the DNA template. As shown in **Figure 3.55** (p. 137), each cycle of PCR amplification duplicates the original DNA template, and therefore the number of DNA molecules increases in each cycle by 2^n, where n is the cycle number. Amplification for 20 cycles results in a 2^{20} increase, which generates a theoretical maximum of 1,048,576 identical DNA molecules starting from a single template. The time required for each cycle is generally only a few minutes, meaning that this amplification can be accomplished in a matter of hours.

As shown in **Figure 3.56a** (p. 138), each PCR cycle has three temperature phases. First, the DNA is heated to denature and separate the strands. Then, the temperature is lowered to facilitate annealing of the primers to each strand, usually in the temperature range 55–65 °C, depending on the length and G-C content of each

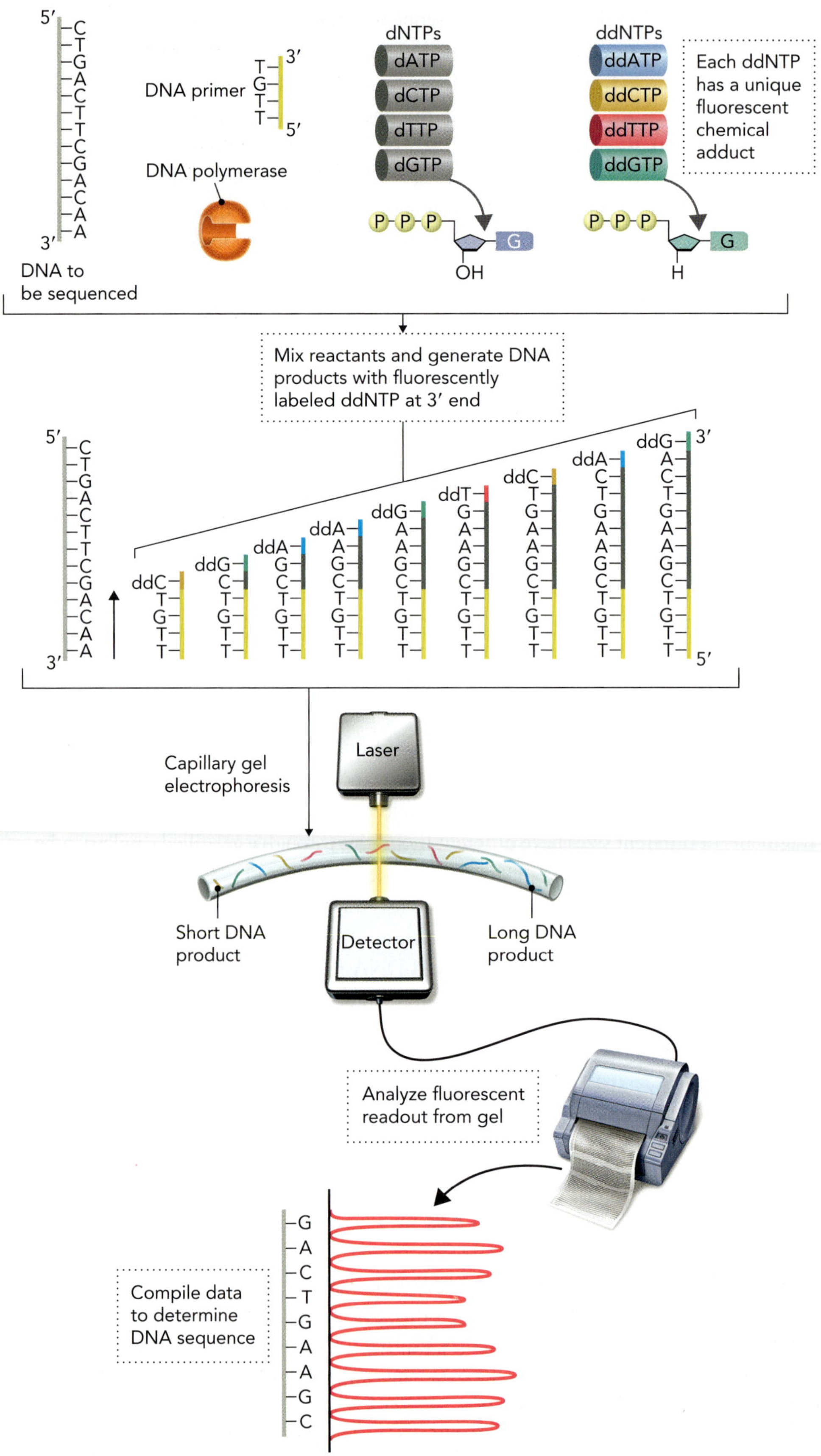

Figure 3.54 The chain-termination method can be used to determine the sequence of a region of DNA. Fluorescently labeled ddNTPs and unlabeled dNTPs are mixed with the DNA to be sequenced, along with DNA primer and DNA polymerase. During DNA synthesis, random incorporation of a ddNTP in the DNA product terminates synthesis and incorporates a different fluorescent label for each ddNTP. The synthesis products are separated according to size by capillary gel electrophoresis, and laser excitation identifies the fluorescent color associated with each DNA product. The DNA sequence is determined by comparing the fluorescence color with the size of the fragment.

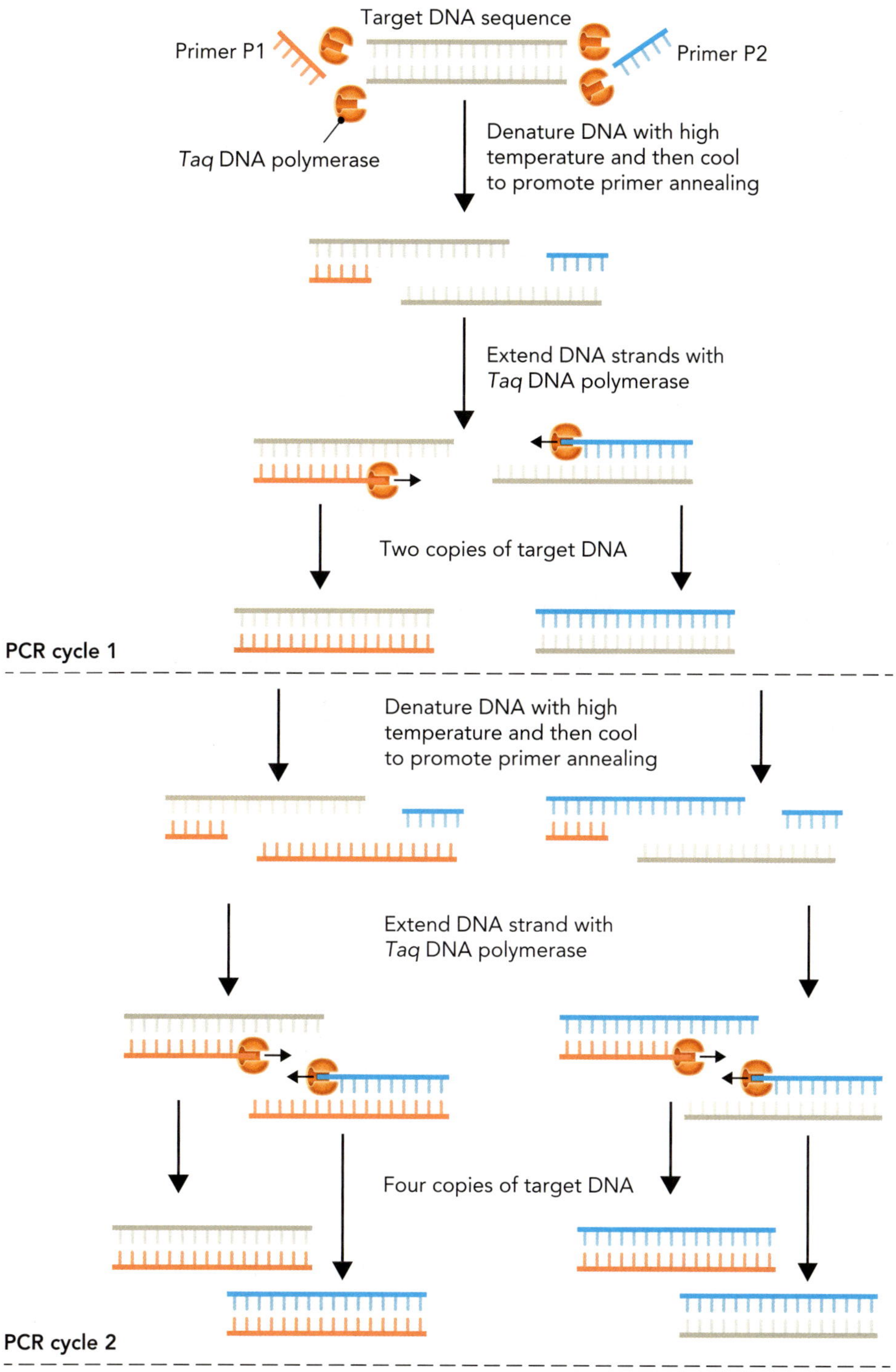

Figure 3.55 The polymerase chain reaction is an *in vitro* method of DNA amplification. Two cycles of polymerase chain reaction are shown. The target DNA is denatured by high temperature, then the temperature of the reaction is lowered to allow primers complementary to each strand to anneal. DNA polymerase catalyzes the extension of the primer to produce a sequence complementary to each template strand. At the end of the first cycle, the original DNA molecule has been doubled in number, with each molecule containing one new DNA strand and one strand from the target DNA. At the end of the next PCR cycle—which doubles the number of DNA molecules fed into it—two of the four DNA molecules contain only newly synthesized DNA, whereas two contain one new DNA strand and one target DNA strand.

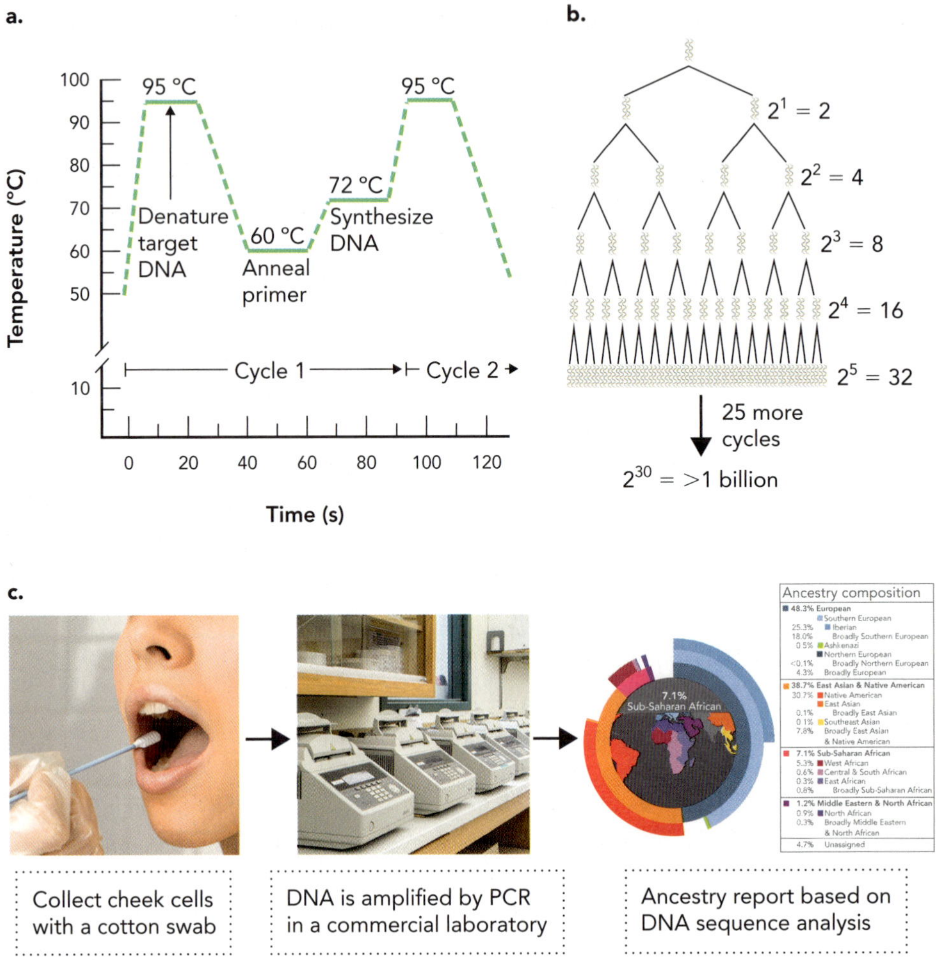

Figure 3.56 The PCR amplification cycle consists of three temperature phases and can lead to a billion-fold amplification of target DNA molecules. **a.** Each PCR cycle requires a brief temperature denaturation step (phase 1), followed by primer annealing (phase 2), and then DNA synthesis (phase 3). **b.** Exponential increases in DNA synthesis yield 2^n products, where n = cycle number, assuming 100% efficiency of each cycle. **c.** The commercial applications of PCR are numerous, including human DNA genotyping using cheek cells as a source of target DNA to deduce ancestry. CELL COLLECTION: PETER DAZELEY/GETTY IMAGES; LABORATORY: © 2014 ALASKA DEPARTMENT OF FISH AND GAME. USED WITH PERMISSION; ANCESTRY REPORT: BASED ON 23ANDME, INC.

primer. The annealing temperature is based on the primer melting temperature (T_m) and can be increased or decreased within a small range to affect the stringency of the annealing process. Finally, temperature is raised to 72 °C to allow extension of the primer and DNA synthesis under temperature conditions that ensure sequence specificity in primer annealing and high-fidelity amplification of the target DNA. Because PCR instruments can rapidly shift reaction temperatures between each phase, and because of use of specialized DNA polymerase enzymes that are highly efficient, each complete PCR cycle can be completed in under 2 minutes. As shown in **Figure 3.56b**, as many as 1 billion (10^9) DNA molecules can be produced after 30 PCR cycles, with efficiencies of 98% or higher. The many applications of PCR include DNA ancestry analysis, which requires only a simple swab of cheek cells as the source of target DNA, followed by target-specific DNA sequencing (**Figure 3.56c**).

Under normal circumstances, the high temperature required to denature DNA would also lead to denaturation of DNA polymerase. To perform many cycles of PCR, scientists have taken advantage of the existence of polymerases isolated from thermophilic organisms. In 1966, Thomas Brock isolated the bacterium *Thermus aquaticus* from a hot spring in Yellowstone National Park. Because of the high temperature in the native environment of *T. aquaticus*, the enzymes produced by this organism were stable at elevated temperatures. Scientists developing PCR turned to the DNA polymerase from *T. aquaticus*, which is now commonly called ***Taq* DNA polymerase**, and found it was successful in catalyzing the replication reaction through repeated temperature cycles.

Transcriptome Analysis

It is often important to determine the level of gene expression in a cell. Data from gene expression assays can be used to determine how diseases such as cancer affect metabolic pathways or cellular processes. Gene expression assays can also provide information about the response of a cell to a drug or environmental stimuli.

The two primary methods used to analyze the complexity (variety) and abundance of RNA transcripts in a cell under various conditions are (1) gene expression microarrays and (2) next-generation transcriptome assembly using RNA sequencing (RNA-seq). **Gene expression microarrays** provide a readout of transcript abundance using a predetermined collection of complementary DNA sequences attached to a solid surface. **RNA-seq** takes an unbiased approach by using PCR amplification to generate cDNA fragments that are then sequenced by high-throughput DNA sequencing. Moreover, RNA-seq provides a readout of all transcripts from the same gene, which permits the identification of alternatively spliced RNA products.

The most commonly used commercial microarray platform is called GeneChip, which is an array chip made by the company Affymetrix. As shown in **Figure 3.57a**, an array chip consists of more than 6 million unique, known DNA sequences, which fit on a chip that is only 1.2 cm on each side. Each of the locations on the array chip contains millions of identical single-stranded DNA molecules that are 25 nucleotides in length. Together, these small DNA sequences cover up to ~2,500 nucleotides of each gene, with about 100 unique gene sequences for each mRNA transcript. Fluorescently labeled complementary RNA (cRNA) fragments (probes) prepared from mRNA isolated from a cell sample are incubated with the array chip under hybridizing conditions. After washing away the unbound cRNA molecules, the annealed fluorescent probes are excited with a laser. The fluorescence intensity is measured for each location on the chip, which corresponds to a known gene sequence. Because the number of identical DNA molecules in each location is most often greater than the number of cRNA molecules corresponding to that sequence in the sample, the fluorescence intensity can be used as a relative measure of transcript abundance.

As shown in **Figure 3.57b**, the cRNA probes are prepared from the RNA sample using a cDNA template that is first synthesized with reverse transcriptase and ribonuclease H as described earlier (see Figure 3.53). The subsequent cRNA synthesis is then performed using the enzyme **T7 RNA polymerase**, which is a highly efficient RNA polymerase derived from the bacteriophage T7. Fragmentation of the labeled cRNA probes facilitates rapid hybridization to the 25-nucleotide-long DNA sequences on the array chip.

In contrast to gene expression microarrays, RNA-seq uses a combination of cDNA synthesis, PCR amplification, and high-throughput DNA sequencing to determine the identity and abundance of all RNA transcripts in a cell sample. Because RNA-seq is not based on a predetermined set of DNA sequences, it provides a complementary approach to array chip analysis, as well as the ability to identify low-abundance, alternatively spliced transcripts. As shown in **Figure 3.58** (p. 141), mRNA is isolated from cell samples and fragmented to produce small RNA molecules that are converted to cDNA using standard procedures (see Figure 3.53). After the addition of terminal

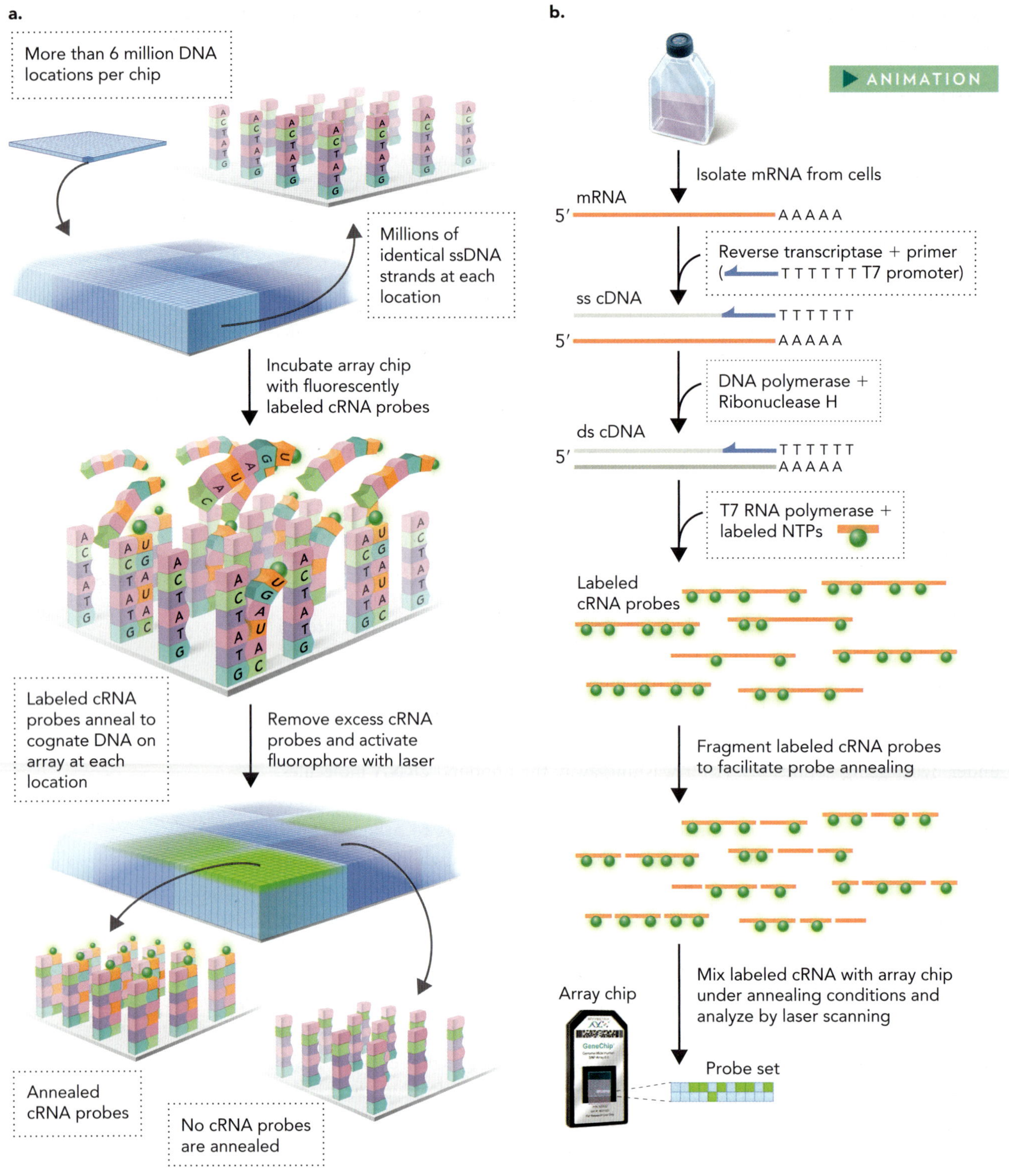

Figure 3.57 Gene expression analysis can be performed using a gene array containing millions of short DNA strands chemically attached to a solid support. **a.** The process by which gene arrays are used to identify the abundance of transcripts within a sample is illustrated here. After hybridization with labeled cRNA probes, the intensity of the fluorescence signal generated by laser scanning at each location on the gene array is used to determine the abundance of the transcripts. **b.** Labeled cRNA probes are synthesized *in vitro* using cDNA templates containing the T7 gene promoter sequence linked to each cDNA molecule. cRNA synthesis is initiated by the addition of labeled nucleoside triphosphates (NTPs) and the enzyme T7 RNA polymerase to the *in vitro* reaction. To facilitate rapid cRNA annealing to the single-stranded DNA sequences on the gene array, the cRNA probes are fragmented. ss = single-stranded; ds = double-stranded.

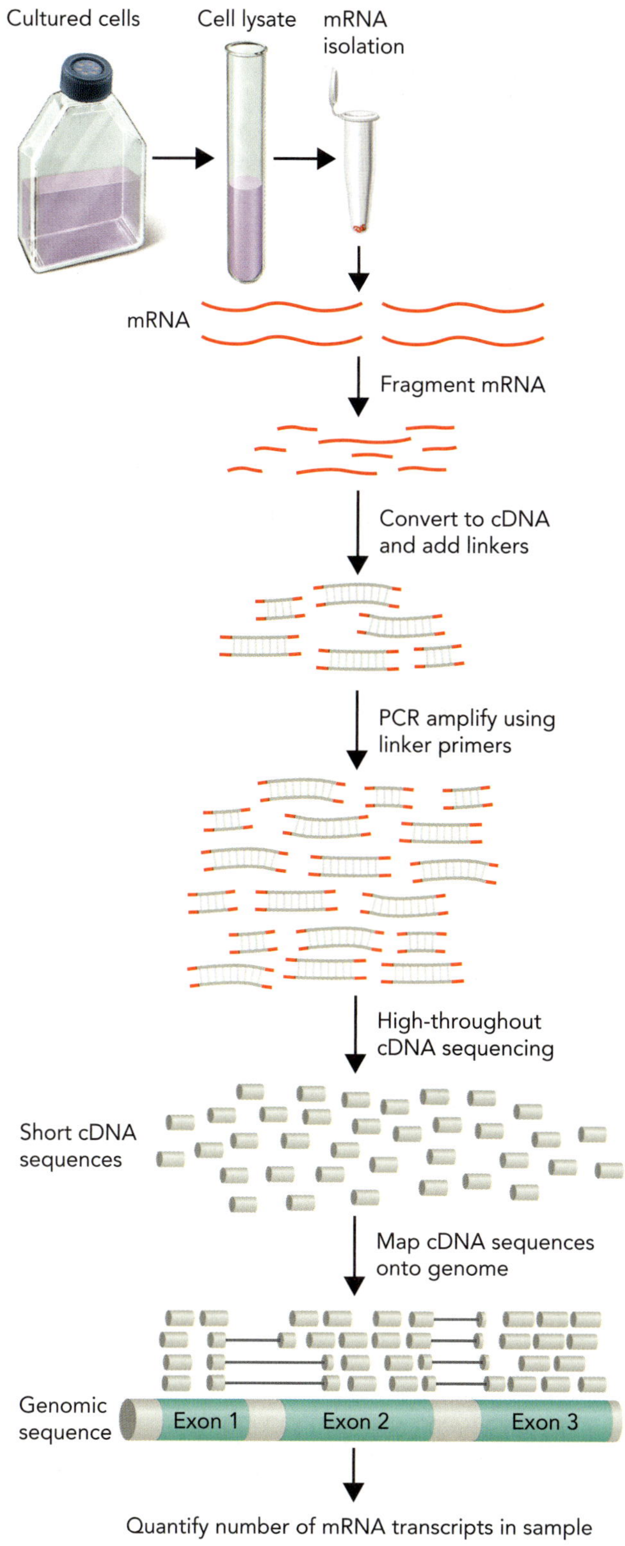

Figure 3.58 RNA-seq is an unbiased transcriptome analysis method based on high-throughput cDNA sequencing and gene mapping. The number of cDNA sequences detected corresponds to the abundance of that RNA transcript in the original sample.

DNA linkers to the cDNA, the pooled molecules are amplified by PCR and then subjected to high-throughput cDNA sequencing. The resulting data set is analyzed using bioinformatics, and a transcript map is assembled using the genome sequence of the organism. The identity and abundance of all transcripts in the RNA sample are computed from the number of corresponding cDNA sequences in the data set relative to internal-control RNA sequences.

Transcriptome analysis using gene arrays has been shown to be complementary to RNA-seq analysis when both methods are used to analyze the same RNA samples. In fact, gene arrays and RNA-seq are powerful techniques that can be used together to identify differences between two RNA populations; for example, determining how physiologic or pharmaceutical stimuli affect gene expression in a given cell type. As shown in **Figure 3.59**, analysis by gene expression microarrays and RNA-seq of two RNA samples, isolated from cells that are either treated with a steroid hormone or left untreated, provides two independent measurements of transcript levels. In cases where these two methods have been compared in differential gene expression studies, it has been shown that quantitatively similar results are found.

concept integration 3.3

How can the four biochemical techniques discussed in this section be used together to explore the function of a newly discovered gene?

To understand the role of a gene completely, it is necessary to examine the gene using a variety of techniques. If even a short region of the gene sequence is known, PCR can be used to amplify the gene from genomic DNA. After determining the DNA sequence, it can be compared to the sequences of genes from different organisms to help identify introns and other regulatory regions. The possible protein products from this gene can be predicted and also compared to similar proteins in other organisms. The coding sequence can then be amplified from mRNA and cloned into a plasmid designed for protein expression. Gene expression microarrays can be used to determine the expression level of this gene under different growth conditions and these results compared to the predicted functions.

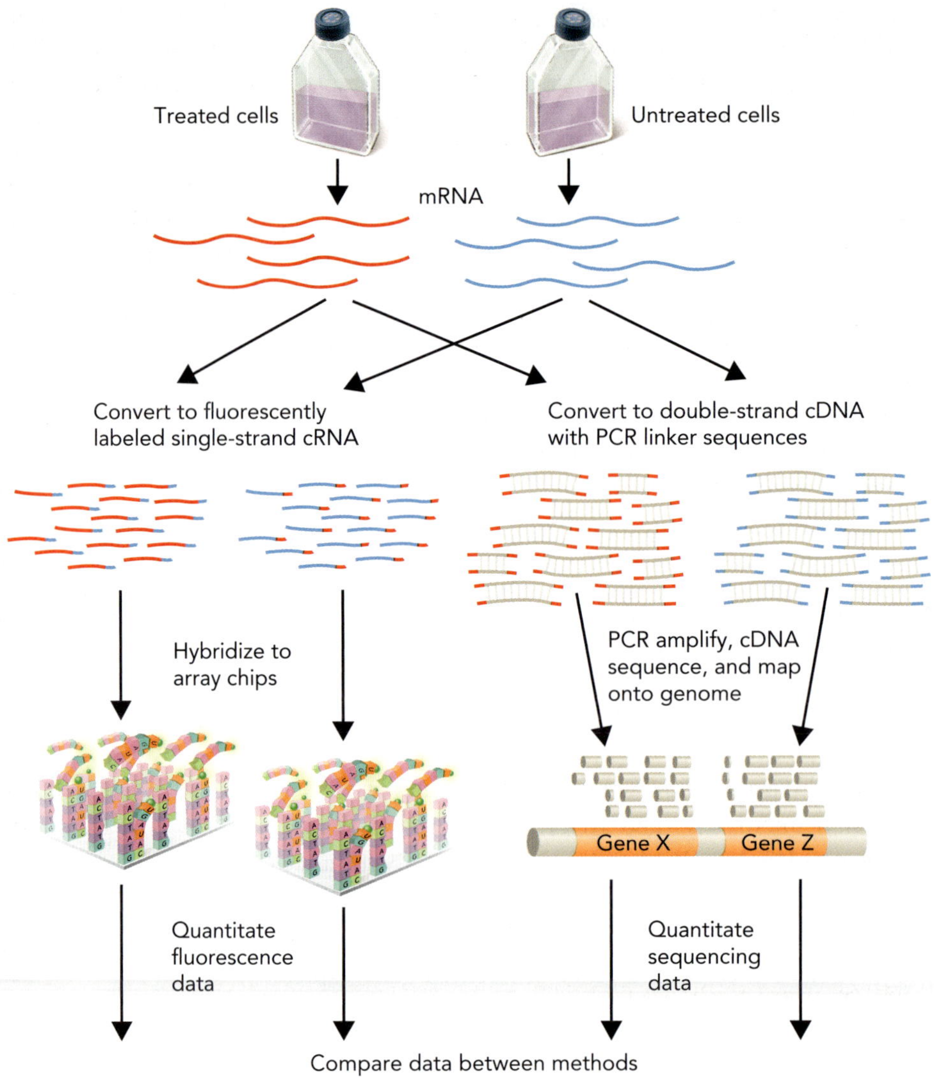

Gene ID	Transcripts (treated) Array	RNA-seq	Transcripts (untreated) Array	RNA-seq
Gene X	1,849	1,770	825	913
Gene Z	433	421	415	398

Figure 3.59 A hypothetical experiment comparing the results of a differential gene expression study using both gene microarrays and RNA-seq. In this example, gene X is found to be expressed at a higher level in the treated cells (~1,800 transcripts per cell) than in the untreated cells (~875 transcripts per cell), regardless of the analysis method used. In contrast, gene Z is expressed at about the same level in both treated and untreated cells (~400 transcripts per cell), suggesting that gene Z transcription is unaffected by the treatment.

chapter summary

3.1 Structure of DNA and RNA

- Nucleotides are the building blocks of DNA and RNA. Nucleotides consist of a nitrogenous base, a ribose or deoxyribose sugar, and one or more phosphate groups.
- DNA is formed from nucleotides containing the bases adenine, thymine, guanine, and cytosine. RNA is formed from nucleotides containing the bases adenine, uracil, guanine, and cytosine. RNA contains ribose sugars, and DNA contain 2′-deoxyribose sugars.
- The DNA double helix is formed through pairing of nucleotides. Adenine bases hydrogen bond with thymine bases. Guanine bases hydrogen bond with cytosine bases. The two strands of DNA are oriented in an antiparallel fashion.
- Stacking of base pairs on adjacent base pairs in a helix leads to stabilization of the helix through van der Waals interactions and the hydrophobic effect. Base stacking is the primary source of stabilization for the double helix.

- DNA can adopt three double-helical conformations: A-DNA, B-DNA, and Z-DNA. B-DNA is the primary conformation under most physiologic conditions.
- Denaturation and renaturation refer to the process of separating and reannealing, respectively, two strands of double-helical DNA.
- The melting temperature of DNA is influenced by its base composition, the ionic strength of the solution, and the length of the DNA molecule.
- The absorbance of nucleotides at 260 nm can be used to monitor the transition between single-stranded and double-stranded states.
- The coiling of a coiled (helical) structure such as DNA results in a supercoil. Supercoils can be described by their twist (*Tw*) and writhe (*Wr*) values.
- DNA supercoiling is induced by DNA replication and transcription; it is relieved by topoisomerases. Type I topoisomerases only break one strand of DNA at a time. Type II topoisomerases cleave both strands of DNA.
- The 2′-hydroxyl group on RNA can lead to an autocleavage reaction. This property makes RNA unsuitable as a storage form for genetic material. DNA, which lacks the 2′-hydroxyl group, is a chemically more stable molecule.
- Catalytic RNA molecules are called ribozymes. RNaseP is an example of a ribozyme.
- Common secondary structures in RNA include duplexes, bulges, loops, and hairpins.
- RNA molecules, excluding mRNA, often contain modified nucleotides that aid in adoption of the correct three-dimensional structure or perform other functional roles.
- Both DNA and RNA can form triplex base pairs. G-quadruplexes can form in G-rich regions.
- DNA and RNA binding proteins can bind in either a sequence-independent or sequence-dependent manner. Proteins largely contact the DNA or RNA backbone through sequence-independent interactions. In the case of sequence-dependent interactions, specific contacts are made between the protein and the DNA or RNA nucleotide bases.

3.2 Genomics: The Study of Genomes

- Condensation of genomic DNA differs in prokaryotes and eukaryotes but is necessary in both to compact the DNA within a cell.
- Condensation of eukaryotic DNA is facilitated by histones. Nucleosomes are DNA–histone complexes, which can be further condensed to form chromatin.
- Chromatin can be found as either euchromatin (less condensed) or heterochromatin (more condensed). The switch between these states is influenced by covalent modifications of the histones or DNA.
- DNA in chromosomes is condensed more than 10,000-fold compared to the extended double-helical structure. A mitotic chromosome consists of a central region of heterochromatin called the centromere.
- The ends of chromosomes contain telomeres, which are repetitive DNA sequences. Telomeres function to maintain the length of chromosomes after replication.
- Genes contain regulatory elements and transcribed regions that encode an RNA transcript.
- Prokaryotic genes may have a single promoter but multiple coding regions. These polycistronic genes often encode gene products with related functions.
- Each coding sequence in eukaryotes is under the control of a separate promoter. Regulatory sequences in eukaryotes are generally more extensive than those in prokaryotes.
- Eukaryotic RNA transcripts are usually modified by 5′ capping, 3′ polyadenylation, and splicing.
- Eukaryotic genes often contain both noncoding regions (introns) and coding regions (exons). Alternative splicing of these exons generates multiple mRNA and protein products from a single gene.
- Bioinformatic tools can be used to compare the genomes of different organisms and predict the structure of a gene or function of a protein.
- Single nucleotide polymorphisms (SNPs) in DNA can be used to diagnose disease, explore genetic diversity, and identify individuals in forensic DNA analysis.
- Short tandem repeats (STRs) and variable number tandem repeats (VNTRs) are repetitive regions that can be used as markers to compare DNA from different individuals.

3.3 Methods in Nucleic Acid Biochemistry

- Plasmids are circular DNA molecules that are self-replicating in bacteria and normally encode antibiotic-resistance genes as a self-defense mechanism. Engineered plasmids are used for gene cloning methods.
- Bacteria contain enzymes that protect against bacteriophage infection: DNA sequence–specific DNA methylases and restriction endonucleases. DNA methylation blocks restriction endonuclease cleavage at recognition sites; thus, bacteriophage DNA is cleaved, but the bacterial DNA is protected.
- Restriction endonucleases can leave three different classes of DNA termini after cleavage: 5′ overhangs, 3′ overhangs, or blunt ends.
- Genes can be inserted into plasmids by cleaving the DNA with restriction endonucleases and resealing with DNA ligase. Bacterial colonies containing the recombinant plasmid are identified by genetic screening.
- The Sanger DNA sequencing method is based on chain termination through incorporation of dideoxynucleotides, which lack a hydroxyl group at the ribose 3′-position and block phosphodiester bond formation.
- PCR is an automated method of amplifying DNA sequences. Temperature cycling is used to denature and

reanneal the DNA, and extension of DNA primers is accomplished by thermostable DNA polymerases.

- The complexity and variety of RNA transcripts in a cell sample can be analyzed by gene expression microarrays and by RNA-seq. Gene expression microarrays measure the abundance of predetermined sequences, whereas RNA-seq provides unbiased data on all transcripts.

biochemical terms

(in order of appearance in text)

nucleoside (p. 92)
thymidine (p. 93)
primary structure (p. 93)
secondary structure (p. 93)
minor groove (p. 94)
major groove (p. 94)
Chargaff's rule (p. 95)
denaturation (p. 99)
hyperchromic effect (p. 99)
renaturation (p. 100)
melting temperature (p. 100)
supercoil (p. 102)
linking number (p. 103)
topoisomer (p. 103)
histone protein (p. 104)
nucleosome (p. 104)
topoisomerase (p. 104)
ribozyme (p. 109)
euchromatin (p. 117)
heterochromatin (p. 117)
centromere (p. 118)
sister chromatids (p. 118)
kinetochore (p. 118)
telomere (p. 118)
promoter (p. 119)
transcription factor (p. 119)
monocistronic (p. 119)
polycistronic (p. 119)
operon (p. 119)
exon (p. 119)
intron (p. 119)
5′UTR (p. 120)
3′UTR (p. 120)
5′ cap (p. 120)
3′ poly(A) tail (p. 120)
exon shuffling (p. 120)
Hutchinson–Gilford progeria syndrome (HGPS) (p. 122)
single nucleotide polymorphism (SNP) (p. 125)
precision medicine (p. 125)
short tandem repeat (STR) (p. 125)
variable number tandem repeat (VNTR) (p. 125)
kilobase (p. 128)
conjugation (plasmid) (p. 128)
transformation (plasmid) (p. 128)
transduction (viral) (p. 129)
bacteriophage (p. 129)
multiple cloning site (MCS) (p. 129)
recombinant DNA (p. 129)
DNA methylase (p. 130)
restriction endonuclease (p. 130)
palindrome (p. 131)
DNA ligase (p. 133)
reverse transcriptase (p. 133)
complementary DNA (cDNA) (p. 133)
Sanger DNA sequencing (p. 134)
dideoxynucleoside triphosphate (ddNTP) (p. 134)
polymerase chain reaction (PCR) (p. 135)
Taq DNA polymerase (p. 138)
gene expression microarray (p. 139)
RNA-seq (p. 139)
T7 RNA polymerase (p. 139)

review questions

1. What is the difference between a nucleoside and a nucleotide?
2. In what way was Chargaff's rule an essential part of Watson and Crick's discovery of the double helix structure?
3. What are the three forms of double-helical DNA? Discuss their similarities and the major differences.
4. What are the topological properties of DNA, and how are they related to each other? Which of these can be changed without cleaving the DNA backbone?
5. Describe the primary structure and secondary structure of DNA.
6. What is the difference between type I and type II topoisomerases with respect to DNA cleavage?
7. How are topoisomerases involved in the process of chromatin formation and DNA replication?
8. Why is DNA more stable than RNA in the cell?
9. What is the hyperchromic effect, and how can it be monitored *in vitro*?
10. Give an example of a modified base found in RNA. Why are these bases biologically useful?
11. What is the first stage of eukaryotic DNA condensation?
12. What is the difference between euchromatin and heterochromatin?
13. What are the functions of centromeres and telomeres in cell division?
14. What are the two types of gene organization found in prokaryotes?
15. Describe the difference between intron and exon sequences in eukaryotic genes.
16. What is plasmid DNA, and what is its function in bacteria?
17. Describe the essential biochemical steps in a PCR cycle.
18. What three types of ends can be generated through DNA cleavage by restriction endonucleases?
19. Why are dideoxynucleoside triphosphates required for Sanger DNA sequencing?

challenge problems

1. Why does hydrogen bonding between base pairs contribute little to overall helix stability?
2. Etoposide is a chemotherapeutic agent successfully used to treat a variety of cancers (see Figure 3.20). However, treatment with etoposide can result in chromosomal rearrangements or deletions that result in secondary leukemia. How does this occur?
3. What is the biological rationale for having both sequence-dependent and sequence-independent DNA and RNA binding proteins in the cell?
4. The first cloned mammal was created by transferring the nucleus of a somatic cell into an unfertilized egg. This accomplishment represented a significant advance in cloning technology, but there was concern that cloning animals using this method would result in animals that aged prematurely. What was the basis for this concern?
5. Why is the promoter region of a gene often A-T rich?
6. How does a lack of introns in prokaryotic genes contribute to their faster growth?
7. A scientist can use a variety of bioinformatic tools to compare the DNA of individuals suffering from a genetic disease to the DNA of unaffected individuals. Why, then, do scientists try to identify the gene involved in the disease before searching for a specific mutation?
8. DNA remaining at crime scenes is often used as evidence that a particular individual is responsible for the crime. In some cases, investigators have used DNA from relatives of a suspect to test against DNA samples from a crime scene. Why is it possible to use a relative's DNA to link a suspect with a crime? Would the case be more convincing with DNA from a sibling, parent, or child or is the DNA from any relative acceptable?
9. Why is it important that plasmids contain antibiotic-resistance genes when they are used for cloning in a laboratory?
10. How would the G-C content of a primer affect the annealing temperature used for PCR?
11. The T_m of a DNA strand can be calculated by hand using the formula

    $$2\ ^\circ\text{C (number of A + T)} + 4\ ^\circ\text{C (number of G + C)} = T_m\ ^\circ\text{C}$$

    Using this formula, calculate the T_m for the following DNA sequence:

    CTTTCACAGCCACTATCCAGCGGTAC

    This formula has several limitations and is not useful for sequences longer than 14 bp. Use an Internet search to find an online T_m calculator. Use this calculator to find the T_m of the above sequence. Using information from your search, identify three factors that can affect the T_m.
12. A student is tasked with cloning a specific gene-coding sequence from a human cancer cell for his undergraduate research project. The objective is to produce the corresponding protein in bacterial cells so that sufficient protein can be obtained for *in vitro* biochemical studies. The student decides to use primers specific to the 5′ and 3′ ends of the human gene based on a genomic DNA sequence he obtained from a BLAST bioinformatic analysis. After PCR amplification, plasmid cloning, and bacterial transformation, the student discovers that only short polypeptides are produced, rather than the full-length protein. Another undergraduate student in the same lab mentions that he should have cloned the gene sequence using mRNA instead of genomic DNA. Why should he follow her suggestion, and what exactly does he need to do?

suggested reading

Books and Reviews

Brock, T. D. (1997). The value of basic research: discovery of *Thermus aquaticus* and other extreme thermophiles. *Genetics*, *146*, 1207–1210.

Champoux, J. J. (2001). DNA topoisomerases: structure, function, and mechanism. *Annual Review of Biochemistry*, *70*, 369–413.

Miesfeld, R. L. (1999). *Applied molecular genetics*. New York, NY: Wiley-Liss.

Rabinow, P. (1996). *Making PCR: a story of biotechnology*. Chicago, IL: University of Chicago Press.

Primary Literature

Ibrahim, M. X., Sayin, V. I., Akula, M. K., Liu, M., Fong, L. G., Young, S. G., and Bergo, M. O. (2013). Targeting isoprenylcysteine methylation ameliorates disease in a mouse model of progeria. *Science*, *340*, 1330–1333.

Shen, J-C., Rideout, W. M., and Jones, P. A. (1994). The rate of hydrolytic deamination of 5-methylcytosine in double stranded DNA. *Nucleic Acids Research*, *22*, 972–976.

Introducing the GFP gene into cells enables researchers to track the GFP-marked cells in real-time without harming the organism. For example, identifying which cells are expressing the GFP protein in a whole mouse used for physiological studies, or identifying GFP-positive cancer cells for the purpose of drug development.

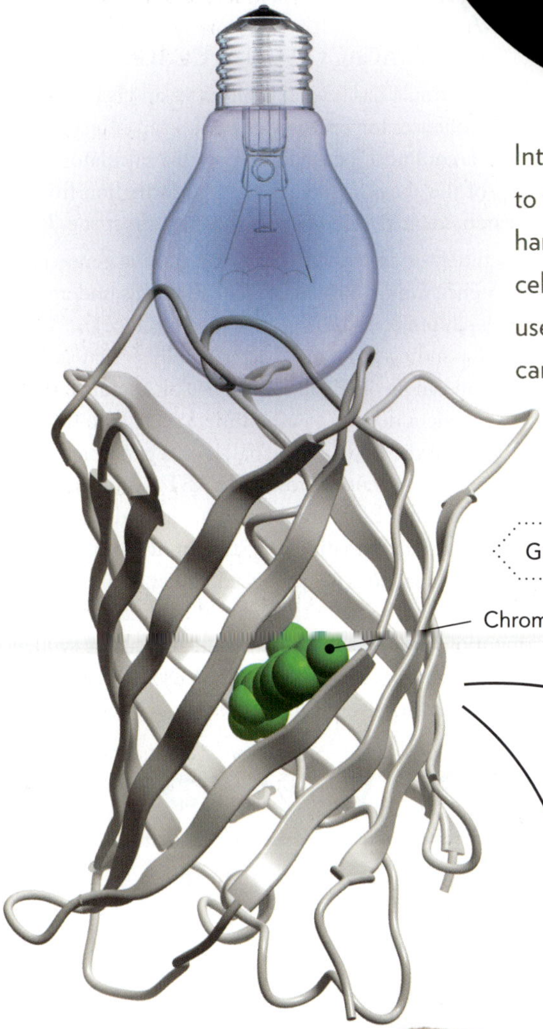

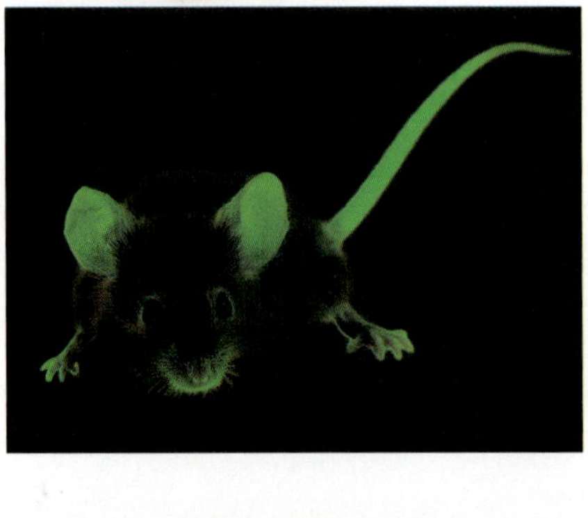

In blue light, mouse glows green in places where modified genes produce GFP

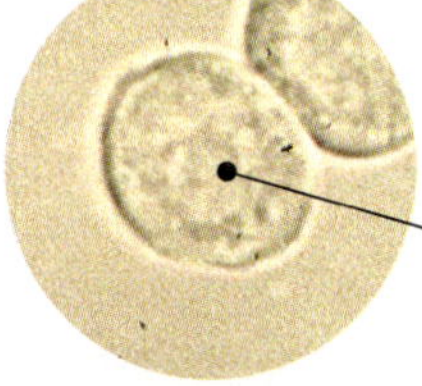

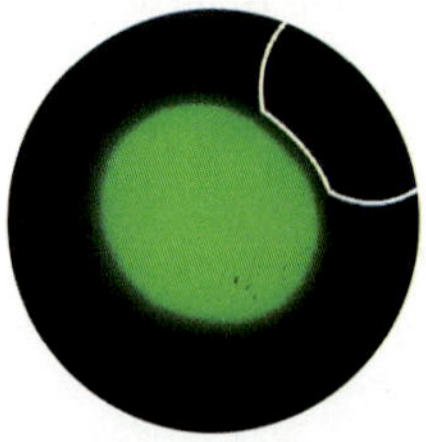

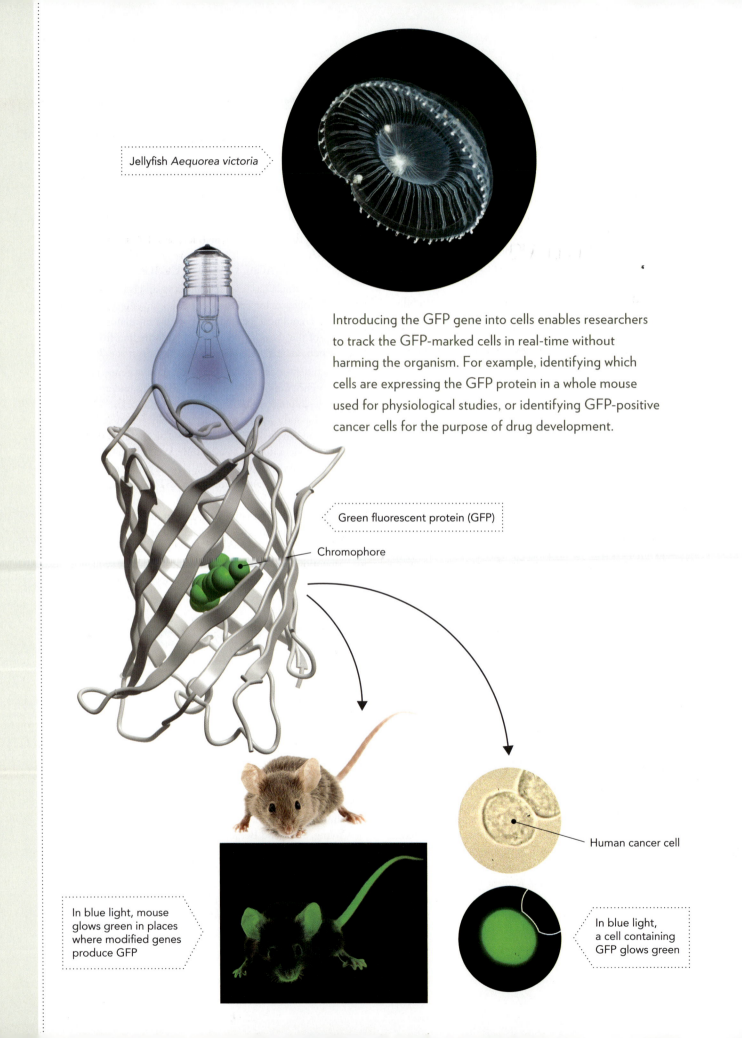

4

Protein Structure

The core biochemical principle that protein structure determines protein function is illustrated by green fluorescent protein (GFP), which is an autofluorescent protein isolated from the Pacific Northwest jellyfish *Aequorea victoria.* GFP is a β-barrel protein containing in its interior its own chromophore (part of a molecule responsible for producing color). The chromophore forms spontaneously through a chemical reaction involving three amino acid residues. Expression of the cloned GFP gene in most cells results in the emission of green light (~510 nm) after excitation with blue light (~470 nm). In other words, as long as the protein maintains its structure, it performs the same function, regardless of the organism from which it originates.

The use of GFP as a fluorescent marker in live cells has allowed cell biologists and physiologists to study cellular processes in real time. Use of this technique has become an important part of experimental biochemistry and cell biology, leading to the awarding of the 2008 Nobel Prize in Chemistry to the three pioneers of this method: Martin Chalfie, Osamu Shimomura, and Roger Y. Tsien.

CHAPTER OUTLINE

4.1 Proteins Are Polymers of Amino Acids

- Chemical properties of amino acids
- Peptide bonds link amino acids together to form a polypeptide chain
- Predicting the amino acid sequence of a protein using the genetic code

4.2 Hierarchical Organization of Protein Structure

- Proteins contain three major types of secondary structure
- Tertiary structure describes the positions of all atoms in a protein
- Quaternary structure of multi-subunit protein complexes

4.3 Protein Folding

- Protein-folding mechanisms can be studied *in vitro*
- Chaperone proteins aid in protein folding *in vivo*
- Protein misfolding can lead to disease

CREDITS: JELLYFISH: SSBLAKELY/WIKIMEDIA COMMONS; LIGHTBULB: EVGENY TERENTEV/GETTY IMAGES; GFP: BASED ON PDB FILE 1GFL; MICE: S-F/SHUTTERSTOCK.

Nearly every biochemical process that takes place in living cells requires a protein to catalyze a chemical reaction, regulate a biological response, or provide structural support to the cell. We introduced proteins in Chapter 1 as polymers of 20 different amino acids, linked together in a specific sequence as dictated by genetic information encoded in genes. The precise amino acid sequence of a protein is responsible for its three-dimensional structure, which in turn determines the biochemical function of the protein.

The term *protein* comes from the Greek word *proteios*, which means "primary," and was first used in 1838 by the Swedish chemist Jöns Jakob Berzelius in his communications with Gerrit Jan Mulder. Mulder, a Dutch physiologic chemist, is credited with recognizing the primary role of proteins in plant and animal metabolism. He was the first to use the term *protein* in a publication, specifically his 1838 manuscript entitled "On the Composition of Some Animal Substances."

To help you appreciate just how critical proteins are to life, Part 2 of this book is devoted to the topic of protein biochemistry. We begin in Chapter 4 by presenting the biochemical properties of amino acids and how they affect the three-dimensional structures of proteins. Chapter 5 describes common methods used in the laboratory to study protein structure and function. (It is from these studies that scientists learned the structural concepts we discuss in Chapter 4.) In Chapter 6, we examine ligand binding and conformational changes in proteins using well-known systems that serve as paradigms for understanding these fundamental functions common to many proteins. Chapter 6 demonstrates the idea that molecular structure determines biochemical function. We then go on in Chapter 7 to describe the function of enzymes, which are largely proteins and responsible for catalyzing the majority of chemical reactions in living cells. Finally, in Chapter 8 we integrate what we have learned about protein structure and function to examine the process of signal transduction, which depends on proteins that function as signal transducers. The process of biochemical signal transduction allows cells to communicate with one another and to respond to environmental cues using receptor proteins, enzymes, and gene regulatory proteins (transcription factors).

We begin our examination of protein structure with a quote from Max Perutz, a pioneer in the field of protein biochemistry (**Figure 4.1**). Perutz describes how Linus Pauling's general chemistry textbook inspired him as a young graduate student to solve the molecular structure of hemoglobin, the oxygen transport protein in blood.

> By such examples, Pauling's book fortified my belief, already inspired by J. D. Bernal, that knowledge of three-dimensional structure is all important and that the functions of living cells will never be understood without knowing the structures of the large molecules composing them. (M. F. Perutz. 2003. *I Wish I'd Made You Angry Earlier*, p. 166. Cold Spring, NY: Cold Spring Harbor Laboratory Press)

This conviction that molecular structure provides insight into molecular function is a primary driving force in modern biochemistry. Linus Pauling himself was one of the most influential chemists of the 20th century, with major contributions to our understanding of chemical bonds and to protein secondary structures. Both Perutz and Pauling were awarded the Nobel Prize in Chemistry during their careers, Pauling in 1954 and Perutz in 1962.

Figure 4.1 Linus Pauling and Max Perutz were two of the most influential protein biochemists of the 20th century: They discovered core principles underlying protein structure and also provided the intellectual link between protein structure and function. **a.** Linus Carl Pauling (1901–1994) is shown lecturing on the structure of an α helix, a fundamental component of proteins first described by him. THE PRINT COLLECTOR/ALAMY. **b.** Max Ferdinand Perutz (1914–2002) was a key contributor to the experimental determination of the first three-dimensional protein structures and is shown here with an early wire-frame model of a protein. AP PHOTO.

a.

b.

4.1 Proteins Are Polymers of Amino Acids

Twenty common amino acids are found in nature, each with a unique chemical composition. Amino acids can be covalently linked together by a type of amide bond called a peptide bond (see Figure 1.11). Longer amino acid polymers are called **polypeptide chains**, or simply proteins, and short polypeptides (usually less than 40 amino acids) are called peptides, or **oligopeptides**.

The precise arrangement of chemically distinct amino acids in the polypeptide chain, along with their unique chemical and physical properties, dictates both the structure and function of a protein. For example, some amino acids have chemical side groups that are hydrophobic, whereas others are more hydrophilic. Seven of the 20 common amino acids have ionizable chemical groups in their side chains that can participate in acid–base reactions that take place in the catalytic cleft of enzymes. Amino acid side chains can be large and structurally bulky, with long aliphatic groups or aromatic rings, or they can be quite small and easily pack into tight spaces.

As illustrated in **Figure 4.2**, when a polypeptide chain is exposed to the aqueous environment of a cell, the hydrophobic amino acids in the chain tend to cluster in the interior of the protein, minimizing their interactions with water (see Figures 2.26 and 2.27). In contrast, hydrophilic amino acids, which readily form hydrogen bonds with water, are often oriented toward the aqueous interface on the protein surface. These hydrophobic effects act as a driving force for the protein to assume its three-dimensional shape. The macromolecular protein structure is stabilized by the formation of large numbers of weak noncovalent interactions between atoms within the protein. This energetically favorable folding of the linear polypeptide chain into a stable three-dimensional structure ultimately determines protein function, as explained in more detail in Section 4.3. Much of Chapters 4 and 6 are devoted to explaining this relationship between amino acid sequence, protein structure, and protein function because it is so critical to understanding biochemical processes.

To appreciate the diversity of proteins in nature, consider the possible arrangements of amino acids in an oligopeptide containing just 10 **residues** (amino acids within a polypeptide chain are called residues). Each position in the oligopeptide can have any one of the 20 amino acids, so the total number of possible protein sequences is 20^{10}, or about 10 trillion different sequence combinations for these 10 residues.

Figure 4.2 The chemical properties of amino acid side chains play a crucial role in determining the three-dimensional structure of proteins. Because of hydrophobic effects, hydrophilic amino acid residues (for example, serine, cysteine, and aspartate) are found predominantly on the surface of the protein, whereas hydrophobic amino acid residues (such as leucine) are found in the interior of the protein. The figure highlights the chemical structures and locations within the protein structure of four amino acid residues in the larger polypeptide chain.

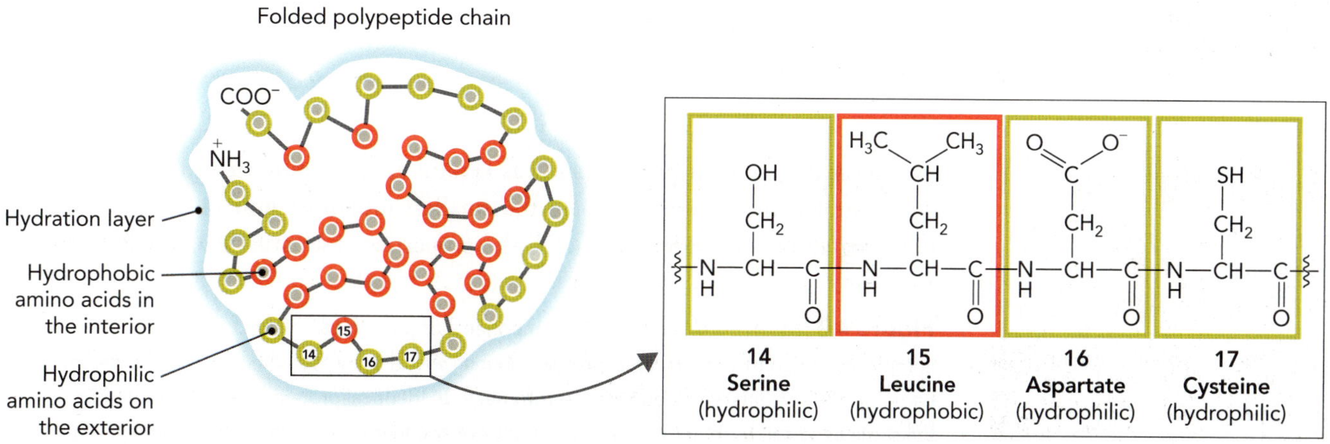

Table 4.1 REPRESENTATIVE PROTEINS IN NATURE

Protein name	Cell function	Number of amino acids	Molecular mass (D)	Number of subunits in complex
Human thioredoxin	Redox regulation	105	11,737	1
Chicken egg white lysozyme	Carbohydrate metabolism	129	14,313	1
Horse myoglobin	Oxygen transport	153	16,951	1
Rabbit γ-interferon	Signal transduction	288	33,842	2
Bacterial triose phosphate isomerase	Carbohydrate metabolism	510	53,944	2
Human hemoglobin	Oxygen transport	574	61,986	4
Thale cress (plant) Hsp70 protein	Protein folding	650	71,101	1
Bacteriophage RNA polymerase	RNA synthesis	883	98,885	1
Minnow glucocorticoid receptor	Signal transduction	1,490	163,750	2
Yeast pyruvate carboxylase	Carbohydrate metabolism	2,252	245,456	4
Bacterial glutamine synthetase	Nitrogen metabolism	5,616	621,264	12
Human titin protein	Muscle contraction	26,926	2,993,428	1

Most proteins contain 100–1,000 amino acids, and therefore it is no wonder that proteins are capable of such a vast array of biochemical structures and associated functions.

Table 4.1 lists representative proteins that together demonstrate the range of functions and size among proteins found in nature. The average molecular mass of an amino acid residue within a protein is ~110 daltons (D), which means that a polypeptide chain containing 500 amino acids has a molecular mass of about 55 kilodaltons (kDa). Table 4.1 shows that some functional protein complexes consist of more than one polypeptide chain. Each polypeptide chain is called a protein **subunit**. These subunits form complexes that are most often associated with each other through multiple noncovalent weak interactions, although some protein complexes are held together by covalent linkages. Protein subunits can be encoded by different genes and contribute unique chemical or structural properties to the protein complex or they can consist of identical subunits encoded by the same gene (**Figure 4.3**). Recent studies of protein complexes within living cells suggest that most proteins function as components of multi-subunit complexes. Throughout this book, we will see many examples of proteins functioning within large complexes.

Chemical Properties of Amino Acids

Amino acids were identified by biochemists in the early 1900s as the products of acid hydrolysis of abundant proteins such as keratin and albumin. In 1907, Emil Fischer demonstrated that amino acids are the smallest monomer unit in proteins by synthesizing an oligopeptide that had all the properties of a naturally occurring protein.

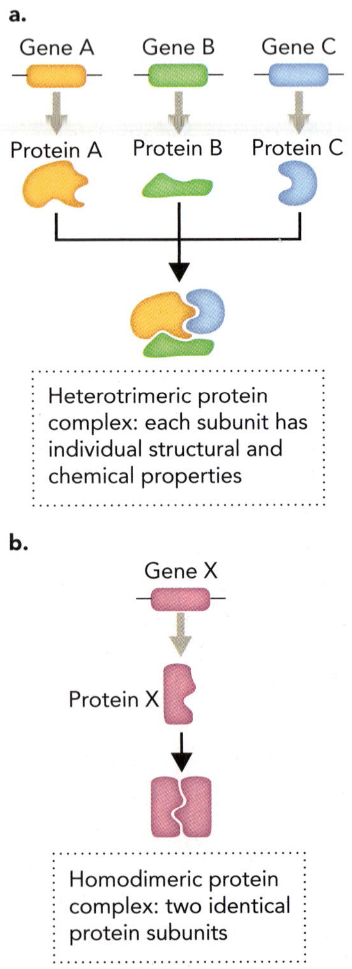

Figure 4.3 Most proteins in cells are components of protein complexes. **a.** Protein complexes can consist of individual protein subunits encoded by separate genes, in this case a complex of three different subunits (heterotrimer). **b.** Identical protein subunits encoded by the same gene can form, for example, a complex of two identical subunits (homodimer).

Figure 4.4 shows the tetrahedral structure of a generic amino acid, in which R refers to one of the 20 different side chains commonly found at this position. These functional groups, also called **amino acid side chains**, differ in chemical structure, and can differ in polarity and charge. The central carbon atom (C_α; referred to as the α carbon) of every amino acid except glycine is an asymmetric chiral center having four different substituents arranged in a tetrahedron. (Two of the substituents in glycine are hydrogen atoms, so the glycine C_α is not a chiral center.) Amino acids are often called **α amino acids** because they have a primary amino group (often called the α amino group) and a carboxyl group attached to the C_α. One exception to this arrangement is the amino acid proline, in which the amino group is incorporated into a pyrrolidine ring attached to the C_α.

The pK_a of the α carboxyl group of amino acids is ~2.3; therefore, the α carboxyl group of free amino acids is deprotonated at pH 7, resulting in a negative charge. In contrast, the pK_a of the α amino group is ~9.7, which means that at pH 7, the nitrogen is protonated and carries a positive charge. For alanine, which contains a methyl group as the side chain bonded to the C_α, at pH 0, both the α carboxyl and α amino groups are fully protonated, giving alanine a net +1 charge (cation), whereas at pH 14, both groups are fully deprotonated, and alanine has a net charge of −1 (anion). The **isoelectric point** (pI) is the pH at which the amino acid carries no net charge. In the case of alanine, the pI is 6.0. At this pH, alanine is also a **zwitterion**, which means that it is an electroneutral molecule that contains both positive and negative charges. Note that alanine is dipolar at its pI of 6.0; however, it carries no net charge and therefore does not migrate in an electric field.

Approximate pI values can be calculated as the arithmetic mean of two pK_a values. For alanine, the pI is calculated as $[pK_1 + pK_2]/2 = pI$, or $[2.3 + 9.7]/2 = 6.0$. For molecules with more than two titratable groups, the pI can be approximated by calculating the average of the two pK_a values that govern the formation of the neutral species. The titration curve for the amino acid aspartate is shown in **Figure 4.5**.

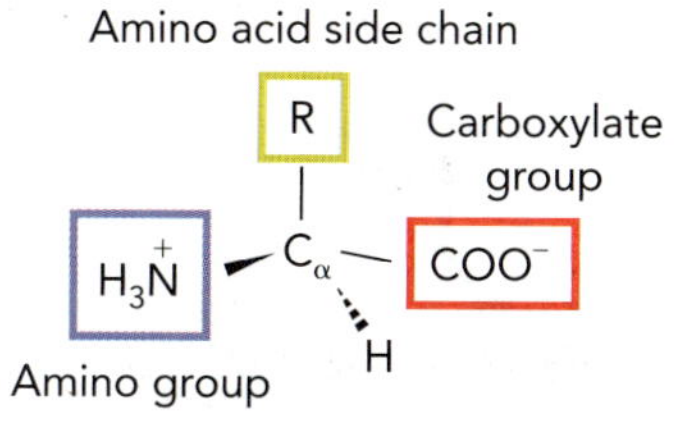

Figure 4.4 Amino acids have a tetrahedral bond structure. All amino acids (with the exception of glycine) contain a chiral α carbon (C_α) with four functional groups attached: a hydrogen, an amino group, a carboxylate group, and an amino acid side chain.

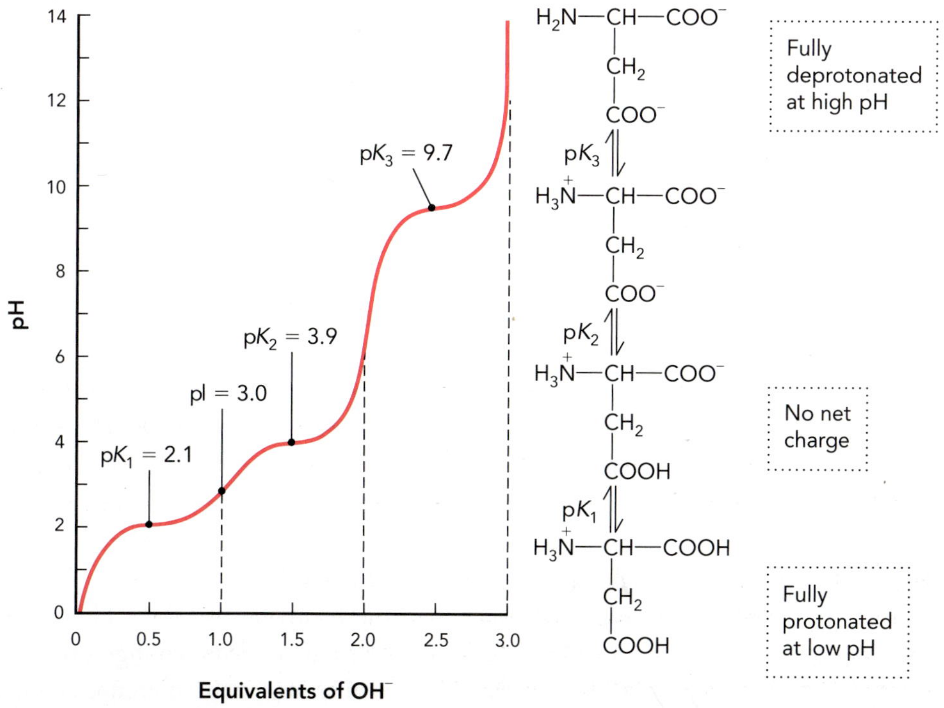

Figure 4.5 The amino acid aspartate has three titratable ionic groups: an α carboxyl group with a pK_1 of ~2, a side-chain carboxyl group with a pK_2 of ~3.9, and an α amino group with a pK_3 of ~9.7. The isoelectric point (pI) is the pH at which the amino acid carries no net charge. The zwitterion is a charged form of the free amino acid that has no net charge and is present at its highest concentration when pH = pI. The approximate pI for aspartate can be calculated from the arithmetic mean of pK_1 and pK_2 because these are the two pK_as for the ionizable groups that govern the formation of the neutral species.

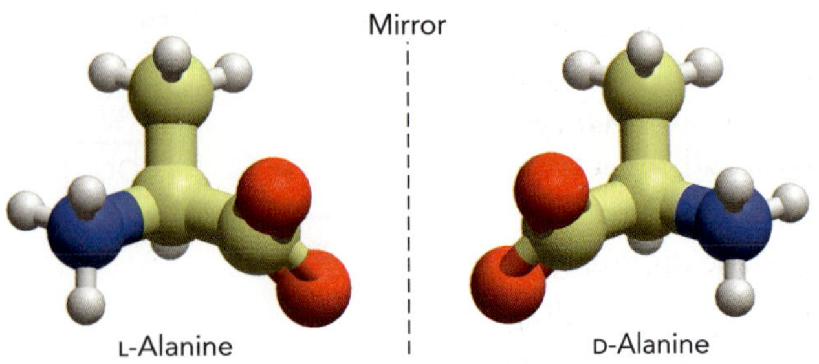

Figure 4.6 Amino acids are chiral compounds. On the basis of the nomenclature developed in organic chemistry to describe the stereoisomers of glyceraldehyde, the common amino acids found in nature are L-amino acids. In the L-amino acid configuration, if the α carbon (C_α) is at the center, with the side chain projecting upward and the H pointed toward the back, then the amino and carboxyl groups are on the left and right, respectively, as shown here for L-alanine.

Aspartate, an amino acid that has three titratable groups, has pK_a values of ~2.1 for the α carboxyl group, ~3.9 for the side-chain carboxyl group, and ~9.7 for the α amino group. The charge on this amino acid is either +1, 0, −1, or −2, depending on the pH. The two pK_a values that govern the formation of the neutral species are pK_1 and pK_2, therefore the pI of aspartate can be approximated by $[2.1 + 3.9]/2 = 3.0$. Note that this is an approximation because the pK_a values are approximations themselves, and additionally the other titratable groups will have an effect on the pI, though this is minimized if the pK_a values are well separated. Accurate determination of pI values must be done experimentally through a technique such as isoelectric focusing (see Chapter 5).

The amino acid alanine, as well as the other 18 amino acids with a chiral center, can exist in two distinct **stereoisomer** conformations (**Figure 4.6**). The L and D stereoisomers are mirror images of each other, and their structures cannot be superimposed. Stereoisomers with this property are called **enantiomers.** All amino acids found in proteins are of the L configuration; however, a few short bacterial peptides have been found that contain D-amino acids. The L and D stereoisomer nomenclature used for amino acids is based on the chemical definitions developed in organic chemistry for the stereoisomers of L-glyceraldehyde and D-glyceraldehyde. It stands to reason that synthesizing proteins with mixtures of L and D amino acids would be an energy burden to the cell because two sets of enzymes would be required for protein synthesis (enzymes bind substrates stereospecifically). It is not clear why amino acids in proteins are all L-amino acids rather than all D-amino acids. One possible explanation is that a selection for L-amino acids over D-amino acids occurred by chance early on in evolution, and that this outcome became fixed to minimize energy requirements needed for the maintenance of multiple stereospecific enzymes.

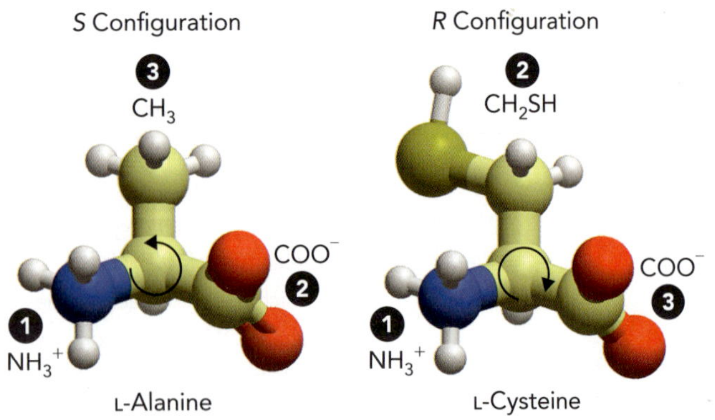

Figure 4.7 The *RS* system can also be used to describe the stereoisomer configuration of amino acids. Based on the atomic numbers of atoms attached to the C_α or to the carbon atom attached to the C_α, the amino acid is either in the *S* or *R* configuration. All L-amino acids except L-cysteine are in the *S* configuration.

Another method of describing the absolute spatial configuration of a molecule, called the ***RS* system**, can also be applied to amino acids. Using this nomenclature, molecules have an absolute *S* configuration if the hierarchical arrangement of chemical groups around the chiral C_α follows a counterclockwise direction (*S* is Latin for *sinister*, which means "left"). This hierarchy is based primarily on the atomic numbers of atoms attached to the chiral C_α. If more than one carbon is linked to the chiral center, then the numbering system reflects the atoms bonded to each of these carbons. In **Figure 4.7**, if you look down toward the group with the lowest atomic number (H in the case of amino acids), L-alanine has an absolute configuration of *S* (counterclockwise) because the oxygen atoms on the COO^- group have a higher atomic number than the hydrogens on the CH_3 group. In contrast, L-cysteine has an absolute configuration of *R* because the hierarchical arrangement of groups around the C_α follows a clockwise orientation (*R* is Latin for *rectus*, which means "right"). The reason is that the sulfur atom attached to the side-chain carbon has a higher atomic number than the oxygen atoms in the carboxyl group. It turns out that cysteine is the only L-amino acid in the *R* configuration—all other L-amino acids are in the *S* configuration.

A series of single-letter abbreviations for the 20 amino acids was developed by Margaret Dayhoff in the 1960s to aid in computational analysis of protein-coding genes. Dayhoff, a biochemist at Georgetown University and pioneer in the field of bioinformatics, came up with the single-letter amino acid code using the mnemonics listed in **Table 4.2**. As many of the amino acids begin with the same letter—for example, the letter "a" in alanine, aspartate, asparagine, and arginine—she needed to find a way to assign single letters that made sense to her. The single-letter amino acid code is a convenient way to compare the amino acid sequence of related proteins (see Figure 1.29) and to denote protein mutations at specific residues. For example, the mutation in human hemoglobin that causes sickle cell anemia occurs at position 6

Table 4.2 THE DAYHOFF SINGLE-LETTER AMINO ACID CODE

	Amino acid abbreviations		
Amino acid (or residue in protein)	Three-letter abbreviation	Single-letter abbreviation	Mnemonic for single-letter abbreviation
Glycine	Gly	G	Glycine
Alanine	Ala	A	Alanine
Valine	Val	V	Valine
Leucine	Leu	L	Leucine
Isoleucine	Ile	I	Isoleucine
Proline	Pro	P	Proline
Methionine	Met	M	Methionine
Phenylalanine	Phe	F	Fenylalanine
Tryptophan	Trp	W	tWyptophan
Tyrosine	Tyr	Y	tYrosine
Serine	Ser	S	Serine
Threonine	Thr	T	Threonine
Cysteine	Cys	C	Cysteine
Aspartic acid	Asp	D	asparDic acid
Glutamic acid	Glu	E	gluEtamic acid
Asparagine	Asn	N	asparagiNe
Glutamine	Gln	Q	Q-tamine
Histidine	His	H	Histidine
Lysine	Lys	K	(before L)
Arginine	Arg	R	aRginine

of the polypeptide chain and converts a glutamate (E) to valine (V). This mutation therefore may be written as E6V.

One way to learn the individual chemical and physical properties of the 20 different amino acids is to group them into subfamilies based on similar chemical characteristics. Although several different schemes could be used for this purpose, we will divide the 20 amino acids into four general subfamilies that emphasize the chemical interaction of amino acids with the aqueous environment of the cell:

1. *charged amino acids*: aspartate, glutamate, lysine, arginine, histidine
2. *hydrophilic amino acids*: serine, threonine, cysteine, asparagine, glutamine
3. *hydrophobic amino acids*: glycine, alanine, proline, valine, leucine, isoleucine, methionine
4. *aromatic amino acids*: phenylalanine, tyrosine, tryptophan

Note that some amino acids could fit into more than one group. For example, the charged amino acids are also polar; similarly, the aromatic amino acids have hydrophobic properties. Glycine, in contrast, does not fit very well into any of these four groups because its hydrogen side chain is chemically inert. Nevertheless, we include it in the hydrophobic subfamily because its side chain is more hydrophobic than polar, and it certainly is not a charged or aromatic amino acid.

Charged Amino Acids Four amino acids have ionizable groups in their side chains, with pK_a values significantly higher or lower than neutral pH. This results in a charge at pH 7 (**Figure 4.8**).

The side-chain carboxyl groups in **aspartate** and **glutamate** have a pK_a of ~4.0 and therefore carry a negative charge at pH 7. The conjugate acid (protonated) forms of aspartate and glutamate are referred to as aspartic acid and glutamic acid, respectively. **Lysine** has a long aliphatic side chain with a terminal amino group that has a pK_a of ~10.5 and is usually positively charged at pH 7. Similarly, **arginine** carries a guanidino group with a pK_a of ~12.5 and is also often positively charged at pH 7. It is important to note, however, that the side-chain pK_a of an amino acid residue within a protein can be very different from its value for a free amino acid in water. The reason is that the chemical properties of nearby functional groups within the folded protein may alter the pK_a to maintain energetically favorable conditions. For example, although the pK_a of the side-chain carboxyl group in an aspartate residue on the surface of a protein is close to ~4, this same carboxyl group has a higher pK_a if the aspartate residue is buried within a hydrophobic pocket. This happens because the generation of a negative charge (loss of the proton) is unfavorable in the absence of a compensating positive charge or polar interaction. This leads to the carboxyl group having a higher affinity for protons under these conditions, resulting in an elevated pK_a (higher pH is required to remove the proton).

A fifth amino acid, **histidine**, has an ionizable imidazole group with a pK_a for the free amino acid near neutral pH (pK_a of ~6.0). Therefore, histidine residues can function as both a hydrogen donor and a hydrogen acceptor at neutral pH. Indeed, histidine is often an active participant in enzyme reactions driven by acid–base interactions. In proteins, the specific environment around the histidine residue determines

Name	Molecular structure	Chemical structure	Molecular mass (D)	Percent frequency in proteins (%)	Comments
Negatively charged					
Aspartate Asp D	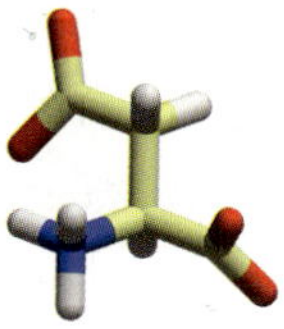	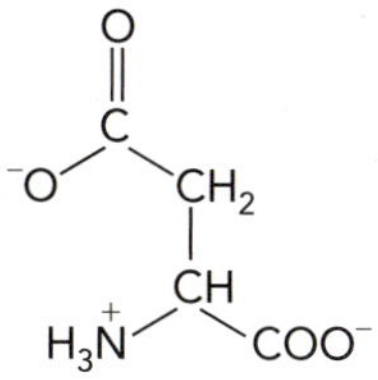	133	5.3	Ionizable side chain has a pK_a of ~3.9. In proteins, Asp often has a net negative charge at pH 7.
Glutamate Glu E	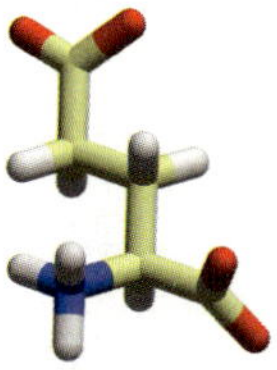	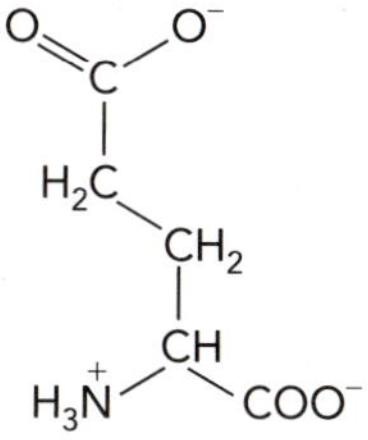	147	6.3	Ionizable side chain has a pK_a of ~4.1. In proteins, Glu often has a net negative charge at pH 7.
Positively charged					
Lysine Lys K	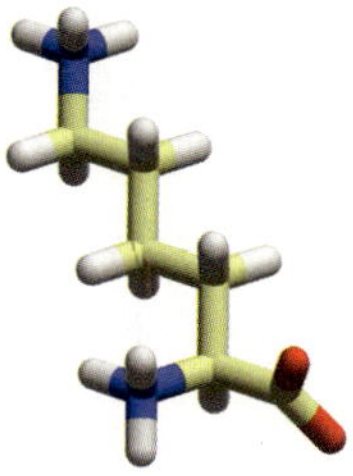	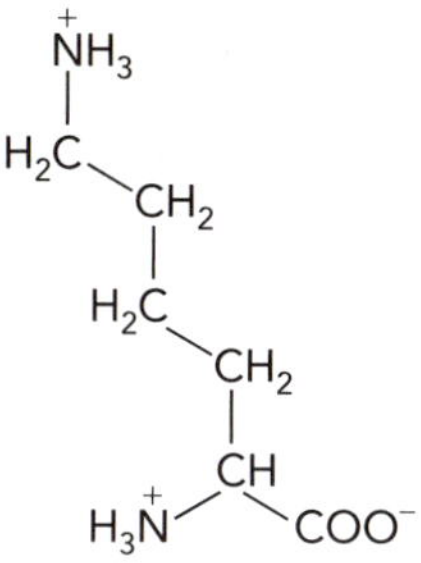	146	5.9	Ionizable side chain has a pK_a of ~10.5. In proteins, Lys often has a net positive charge at pH 7.
Arginine Arg R	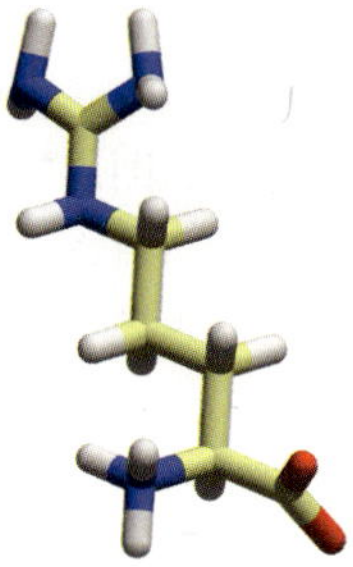	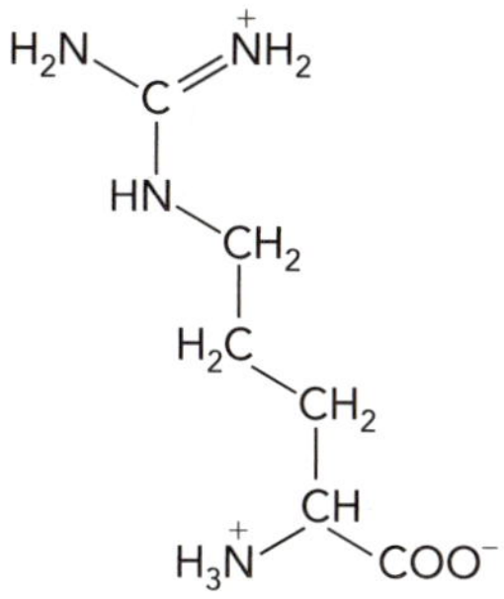	174	5.1	Ionizable side chain has a pK_a of ~12.5. In proteins, Arg often has a net positive charge at pH 7.
Histidine His H	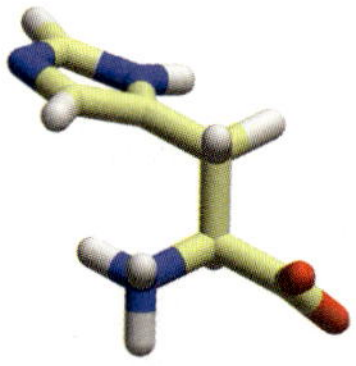	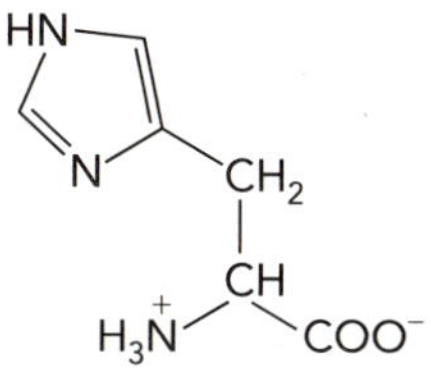	155	2.3	Ionizable side chain has a pK_a of ~6.0. In proteins, the imidazole ring in His can be either positively charged or neutral depending on the environment.

Figure 4.8 There are five amino acids with commonly charged side chains. Aspartate and glutamate are often negatively charged, whereas lysine and arginine are often positively charged. Histidine may be either positively charged or neutral depending on its surrounding environment.

Table 4.3 IONIZATION STATES AND APPROXIMATE pK_a VALUES OF TITRATABLE GROUPS WITHIN PROTEINS

Amino acid	Group[a]	Acid	⇌	Base		Typical pK_a[b]
Aspartic acid Glutamic acid	Side-chain carboxyl group	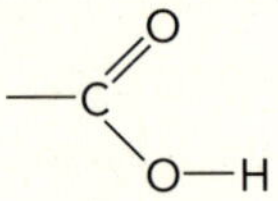	⇌	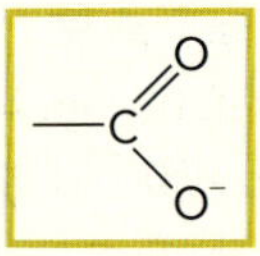	+ H^+	4.0
Histidine	Imidazole group	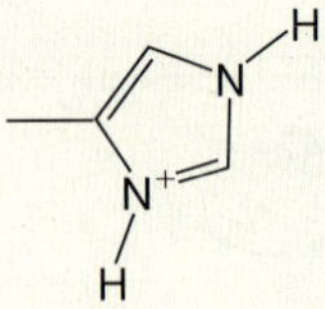	⇌	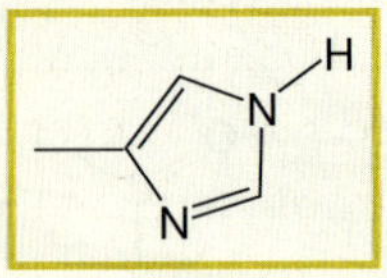	+ H^+	6.0
Cysteine	Thiol group	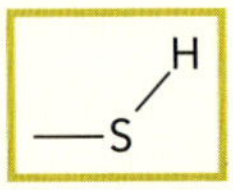	⇌	—S^-	+ H^+	8.3
Tyrosine	Aromatic hydroxyl group	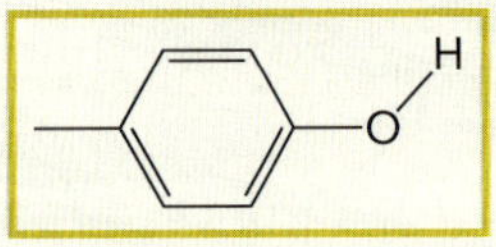	⇌	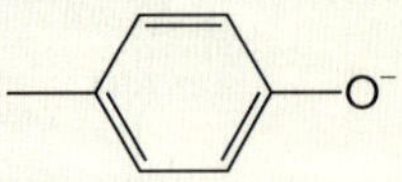	+ H^+	10.1
Lysine	ε amino group	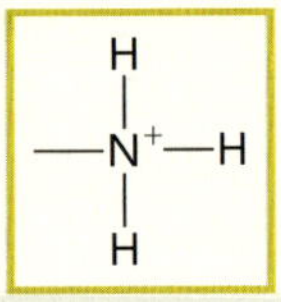	⇌	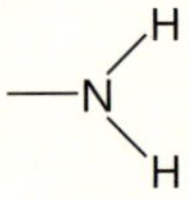	+ H^+	10.5
Arginine	Guanidino group	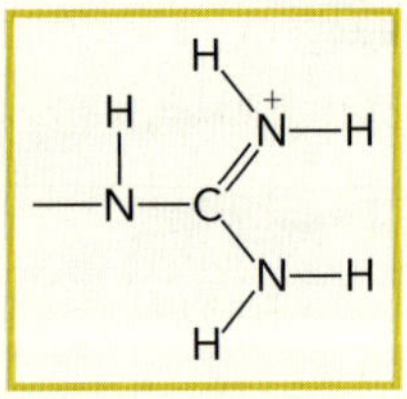	⇌	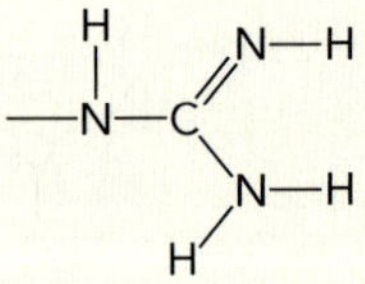	+ H^+	12.5
Terminal residues	α carboxyl group	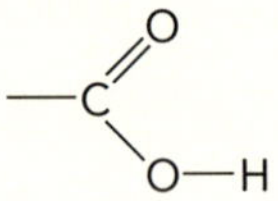	⇌	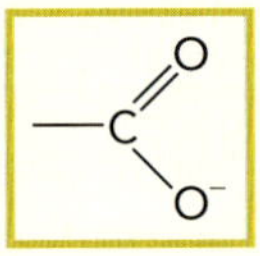	+ H^+	3.1
	α amino group	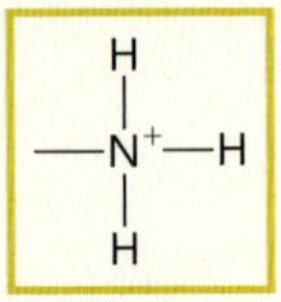	⇌	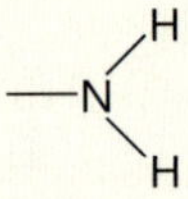	+ H^+	8.0

[a]Outline boxes identify the more abundant chemical species at pH 7.

[b]These pK_a values are affected by the local chemical environment of the amino acid side chains, so the actual pK_a may be higher or lower, depending on the protein. The pK_a values are dependent on the microenvironment of the ionizable group within the protein and the physiologic conditions in the cell (ionic strength and temperature).

whether it is positively charged or neutral. **Table 4.3** lists the approximate pK_a values for the titratable groups of amino acids within proteins.

Hydrophilic Amino Acids The five charged amino acids just described are hydrophilic; however, another five amino acids are hydrophilic but generally uncharged at physiologic pH. The amino acids in this subfamily—serine, threonine, cysteine, asparagine, and glutamine—all contain functional groups that participate in hydrogen bonding with H_2O and have relatively short aliphatic side chains (**Figure 4.9**). **Serine** and **threonine** both contain a hydroxyl group (–OH). This not only gives these two amino acids their hydrophilic property, but also within proteins these residues can be phosphorylated by enzymes called **serine/threonine kinases**, which use ATP as a phosphate donor (**Figure 4.10**). Phosphoryl groups on biomolecules are removed by another class of enzymes, called **phosphatases**. Phosphorylation of amino acid residues occurs after the protein is synthesized (posttranslational modification) and is a key regulatory switch in

Figure 4.9 There are five polar amino acids that can readily form hydrogen bonds, but are not usually charged at physiologic pH. These are classified as hydrophilic amino acids and consist of serine, threonine, cysteine, asparagine, and glutamine.

Name	Molecular structure	Chemical structure	Molecular mass (D)	Percent frequency in proteins (%)	Comments
Serine Ser S	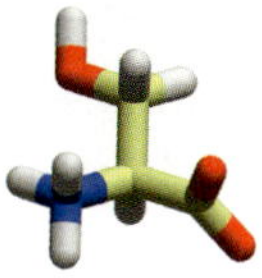	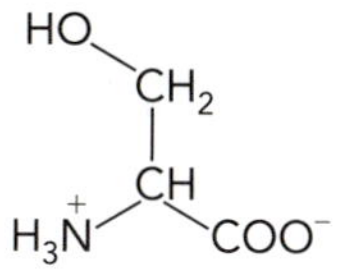	105	6.8	Hydroxyl group in Ser can form hydrogen bonds and may be a substrate in proteins for kinase-mediated phosphorylation
Threonine Thr T	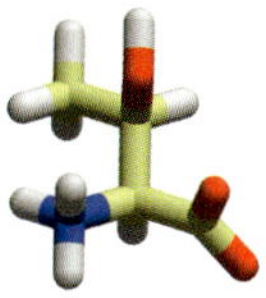	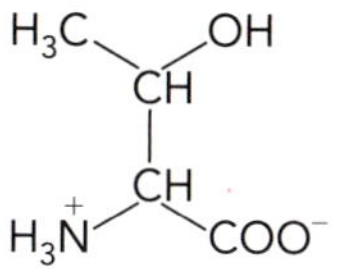	119	5.9	Chemically similar to Ser in that the hydroxyl group can form hydrogen bonds and may be phosphorylated by kinases
Cysteine Cys C	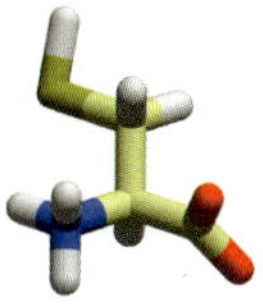	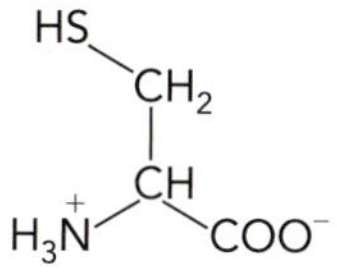	121	1.9	The ionizable side chain has a pK_a of ~8.3 in proteins. The sulfhydryl can form a disulfide bond with other cysteines and can form weak hydrogen bonds. It is the only L-amino acid with an absolute configuration of *R*.
Asparagine Asn N	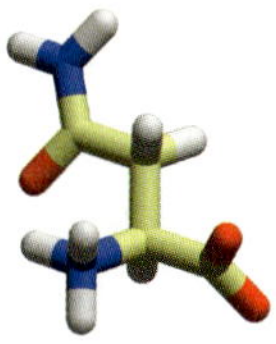	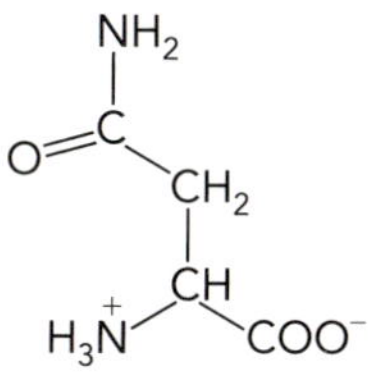	132	4.3	Often found on the surface of globular proteins, the side chain of Asn can form numerous hydrogen bonds. It is structurally similar to the charged amino acid aspartate.
Glutamine Gln Q	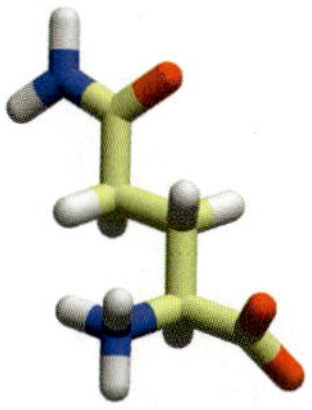	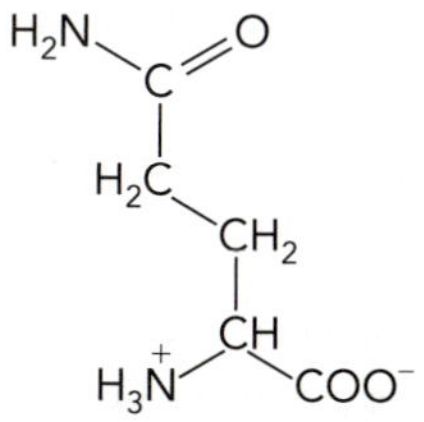	146	4.2	Like asparagine, the side chain of Gln can form numerous hydrogen bonds. It is structurally similar to the charged amino acid glutamate.

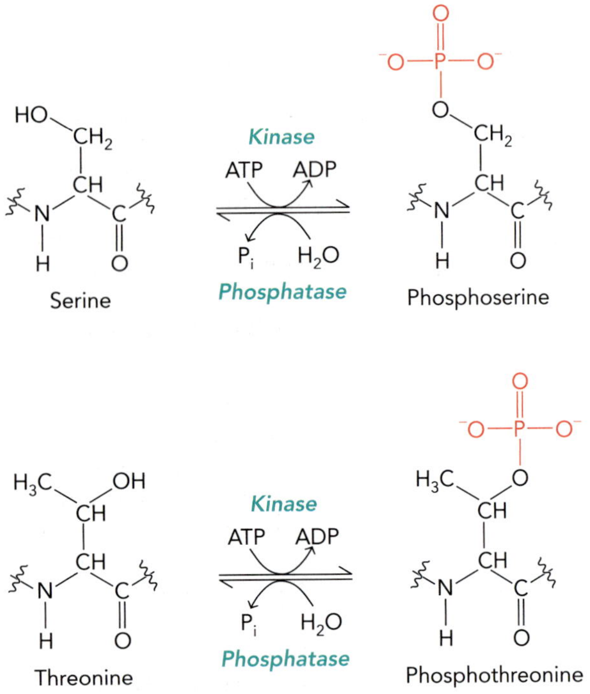

Figure 4.10 The hydroxyl group on serine and threonine residues can be phosphorylated by enzymes called serine/threonine kinases, which use ATP as a phosphate donor. Phosphatases are enzymes that remove phosphoryl groups from biomolecules by catalyzing a hydrolysis reaction.

many types of signal transduction pathways. As serine and threonine residues are commonly on the surface of proteins because of their hydrophilic property, they are often readily accessible for these phosphorylation and dephosphorylation reactions.

The amino acids **glutamine** and **asparagine** are also hydrophilic and the polar groups in the side chain can both donate and accept hydrogens to form hydrogen bonds. An NH group in the side chain can function as a hydrogen-bond donor, and the oxygen atom can be a hydrogen-bond acceptor. The carbon skeleton of glutamine is the same as glutamate; however, an oxygen in the terminal carboxyl group of glutamate is replaced with NH_2. The same is true about the related structures of asparagine and aspartate. Glutamine is important in nitrogen metabolism because it buffers toxic ammonia by transporting nitrogen to the liver, where it can be used for urea production.

The final amino acid in this group of uncharged hydrophilic amino acids is **cysteine**, which is one of two amino acids containing sulfur (the other is methionine). Cysteine is a weak acid, and the sulfhydryl group is ionizable with a pK_a of ~8.3 in proteins (see Table 4.3). Importantly, cysteine has the most highly reactive amino acid side chain in proteins. As shown in **Figure 4.11**, when two cysteine residues are located near each other in the three-dimensional structure of a protein, they can be oxidized to form a **disulfide bond**, also called a **disulfide bridge**. In an oxidizing environment, disulfide bonds can be important in stabilizing the three-dimensional structures of proteins and can also be used to form covalent linkages between two polypeptide subunits in a protein complex.

Hydrophobic Amino Acids The seven amino acids that lack polar groups are the most hydrophobic, and as such are usually found near the interior of proteins in an aqueous environment. Nonpolar amino acids play an important role in protein folding by facilitating the collapse of hydrophobic groups toward the center of a folded protein due to the hydrophobic effect in water, as described in Chapter 2. The side chains of the amino acids **isoleucine**, **leucine**, and **valine** are all hydrocarbon chains, as seen in **Figure 4.12**. The hydrophobic properties and small size of valine make it ideal for dense packing in the hydrophobic core of proteins. Leucine is the most abundant amino acid found in proteins, about two times more abundant than would be expected (observed level of 9.1% compared to an expected level of 5%), and is a major contributor to the hydrophobic core of proteins. **Methionine** is also very hydrophobic, and although it contains a sulfur atom like cysteine, the sulfur in methionine is unreactive.

The three other amino acids in this group are not nearly as hydrophobic as the rest, but we include them here because they lack polar groups and do not contain aromatic rings. The first of these is

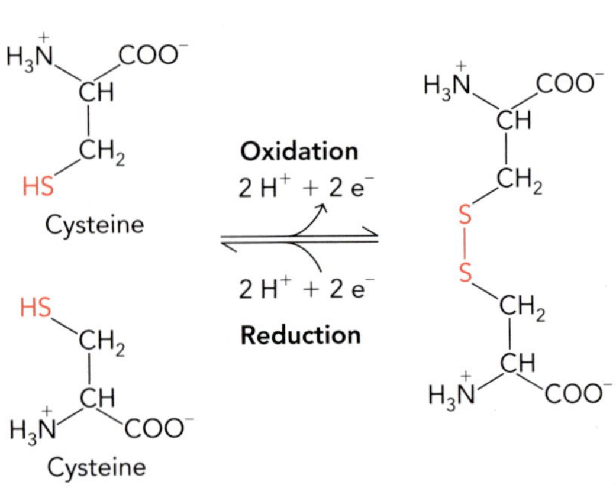

Figure 4.11 The oxidation of two cysteine residues creates a covalent linkage called a disulfide bond, also called a disulfide bridge. A reduction reaction can readily restore the two sulfhydryl groups.

Name	Molecular structure	Chemical structure	Molecular mass (D)	Percent frequency in proteins (%)	Comments
Glycine Gly G	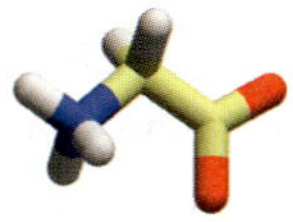	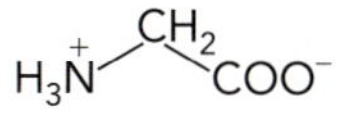	75	7.2	The smallest amino acid; the hydrogen side chain makes this the most chemically neutral of all the amino acids
Alanine Ala A	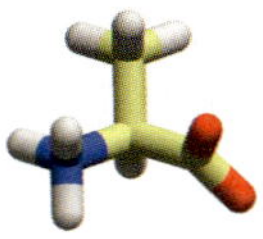	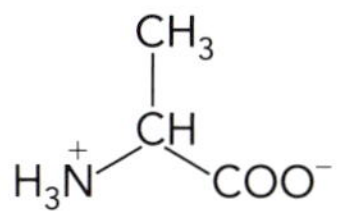	89	7.8	Minimally hydrophobic, Ala is one of the more abundant amino acids in proteins
Proline Pro P	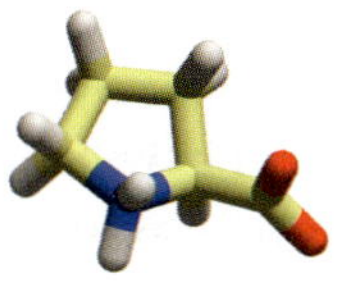	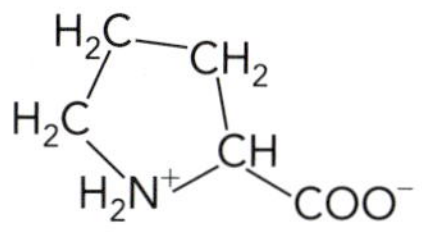	115	5.2	Cyclic ring limits the conformations that proline can adopt. In a polypeptide, the backbone nitrogen lacks a hydrogen, which limits hydrogen bonding.
Valine Val V	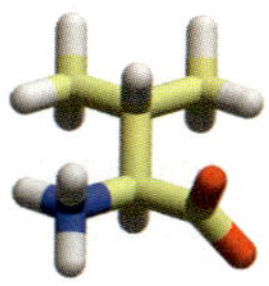	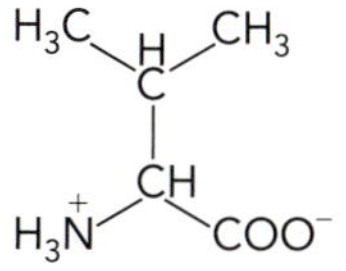	117	6.6	Small hydrophobic amino acid often found in the hydrophobic core of globular proteins
Leucine Leu L	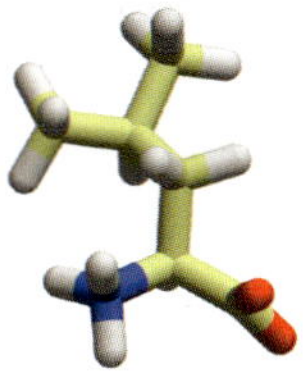	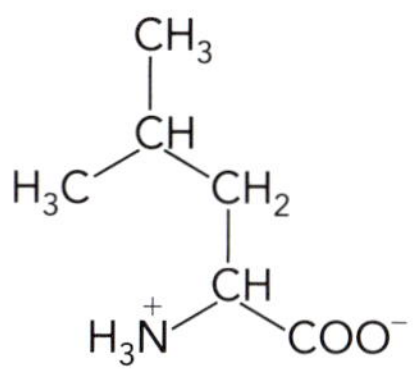	131	9.1	Leu is the most abundant amino acid in proteins and plays a major role in promoting hydrophobic interactions in the core of globular proteins
Isoleucine Ile I	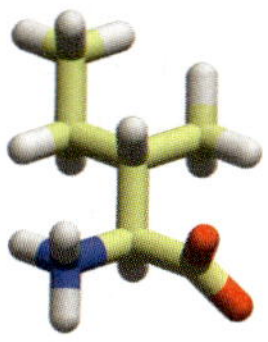	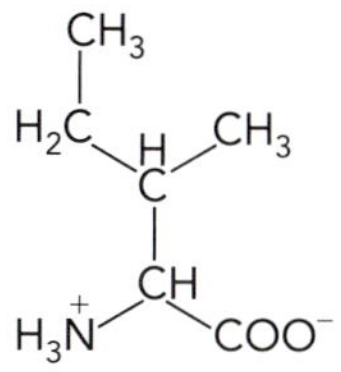	131	5.3	Often found in close proximity to leucine and other hydrophobic amino acids in proteins
Methionine Met M	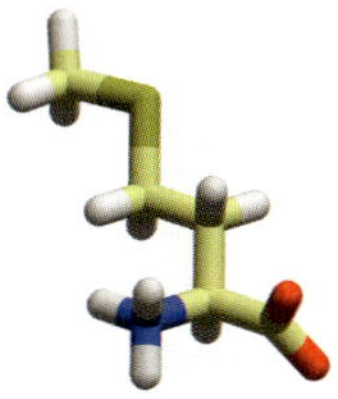	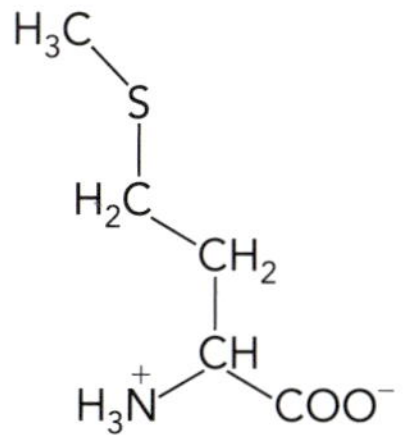	149	2.3	Unlike the sulfur in Cys, the sulfur in Met is unreactive. Met is often the amino-terminal amino acid in nascent polypeptides.

Figure 4.12 There are six amino acids that have hydrophobic properties as compared to other amino acids. These are alanine, proline, valine, leucine, isoleucine, and methionine. Glycine, which is not considered hydrophobic in the same sense as the other amino acids in this group, is included here because it is the most chemically neutral of all 20 amino acids.

glycine, which is the smallest of the 20 amino acids and the least chemically active. Because of its small side chain, glycine can adopt more conformations than other amino acid residues, and therefore it is often found in regions of the polypeptide chain that form sharp turns in the three-dimensional structure. **Alanine** is also relatively small, although the CH_3 group does make it more hydrophobic than glycine. Even though glycine and alanine contribute little to the chemical function of proteins, their small side chains make them ideal for high-density packing of the protein structure. This physical property explains why glycine and alanine together constitute 15% of all amino acids in proteins. The last amino acid in this group is **proline**, which contains a pyrrolidine ring that incorporates the amino nitrogen. When incorporated in a polypeptide chain, the nitrogen of proline lacks a hydrogen, and therefore, unlike the other amino acid residues, cannot act as a hydrogen-bond donor. The ring structure also restricts the conformations that proline residues can adopt in a polypeptide chain. For these reasons, proline residues are usually found in regions of the protein that separate two primary components of protein structure; namely, α helices and β sheets. As will be described in Section 4.2, α helices and β sheets are stabilized by intrastrand and interstrand hydrogen bonding, respectively, which is incompatible with proline residues.

Aromatic Amino Acids The fourth subfamily of amino acids are those with bulky aromatic side chains. These are phenylalanine, tyrosine, and tryptophan (**Figure 4.13**). A common feature of these aromatic amino acids is their absorbance of ultraviolet light in the range 250–280 nm. This property is exploited to quantitate protein

Figure 4.13 There are three aromatic amino acids: phenylalanine, tyrosine, and tryptophan. These amino acids absorb ultraviolet light and provide a useful way to measure protein concentration in solutions.

Name	Molecular structure	Chemical structure	Molecular mass (D)	Percent frequency in proteins (%)	Comments
Phenylalanine Phe F	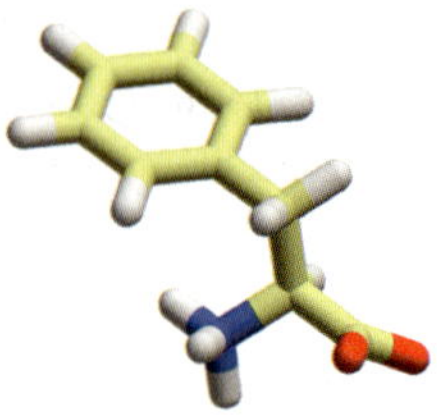	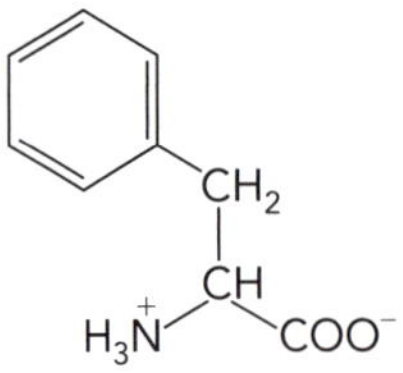	165	3.9	Phenyl ring is hydrophobic and chemically inert. Unlike Tyr and Trp, Phe only weakly absorbs ultraviolet light.
Tyrosine Tyr Y	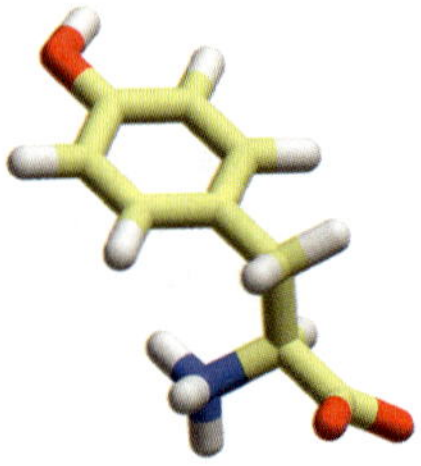	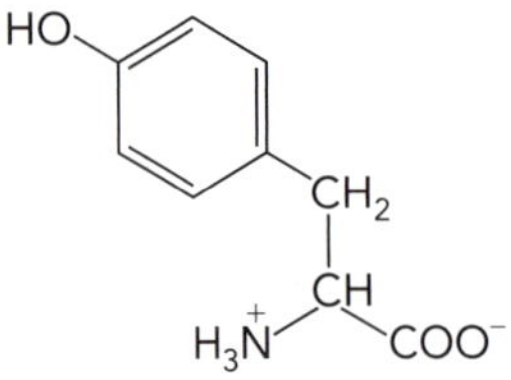	181	3.2	The ionizable hydroxyl group (pK_a ~10.1 in proteins) forms hydrogen bonds making Tyr an amphipathic amino acid having both hydrophobic and hydrophilic properties. Tyr absorbs light at 280 nm.
Tryptophan Trp W	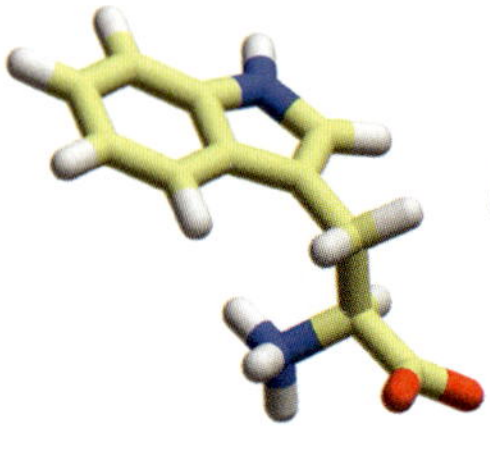	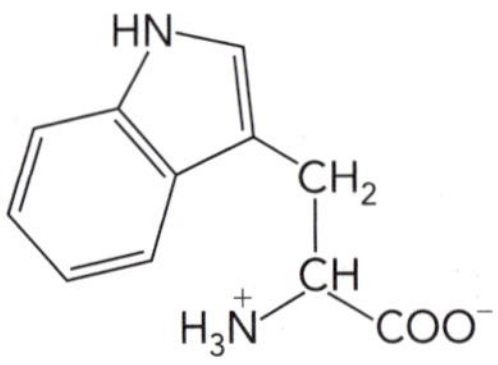	204	1.4	Trp is the largest amino acid and strongly absorbs light at 280 nm. The ring nitrogen forms weak hydrogen bonds, which makes the Trp indole amphipathic.

levels in solution by using spectroscopy to measure light absorption at 280 nm. As shown in **Figure 4.14**, **tryptophan** has the highest molar absorbance among the three amino acids and is most responsible for absorbance at 280 nm. Tryptophan is the largest of the 20 amino acids, with a molecular mass of 204 D. The nitrogen in the aromatic ring of tryptophan can form weak hydrogen bonds. **Tyrosine** also absorbs light at 280 nm, although to a lesser extent. The hydroxyl group on the phenyl ring of tyrosine makes this otherwise hydrophobic amino acid more polar through its hydrogen-bonding capability. In fact, both tyrosine and tryptophan are considered amphipathic amino acids in that they contain both hydrophobic and hydrophilic chemical groups. Amphipathic refers to a dual chemical nature—the molecule is part hydrophobic, or nonpolar, and part hydrophilic, or polar (*amphi* is Greek for "both"). Because of this property, tyrosine and tryptophan are often found in regions of the protein that transition between the aqueous interface at the exterior of the protein and the hydrophobic core.

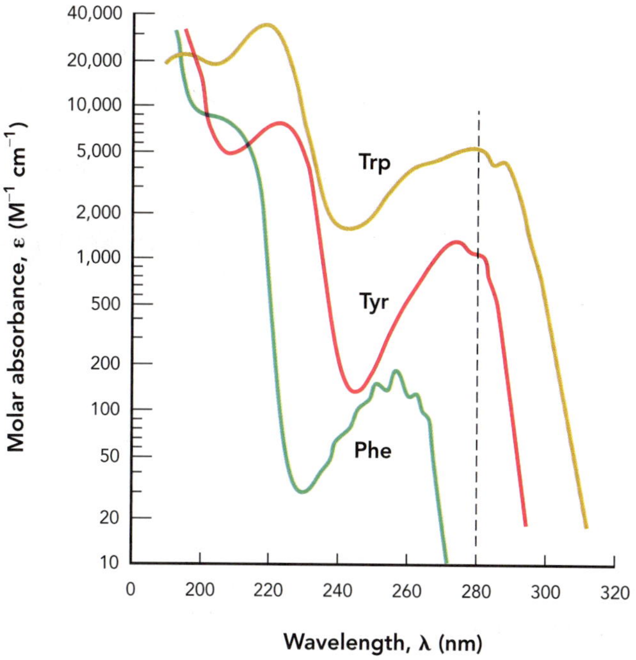

Figure 4.14 This graph illustrates absorbance spectra of the aromatic amino acids tryptophan, tyrosine, and phenylalanine. Note that molar absorbance (ε) is shown on a logarithmic scale; thus tryptophan absorbs about five times more light than tyrosine at 280 nm (dotted line), which is the wavelength often used to gauge protein levels in *in vitro* solutions.

As in serine and threonine, the hydroxyl group in tyrosine can be phosphorylated by kinase enzymes and dephosphorylated by phosphatase enzymes. Indeed, the insulin receptor is a tyrosine kinase that phosphorylates tyrosine residues in target proteins, which is a key step in the insulin signaling cascade in liver and skeletal muscle cells (see Chapter 8). The tyrosine hydroxyl group has a pK_a of ~10.1 and is usually protonated at neutral pH.

Phenylalanine is structurally related to tyrosine, lacking only the hydroxyl group on the aromatic ring. Phenylalanine is very hydrophobic and has a lower level of ultraviolet absorbance than either tryptophan or tyrosine (Figure 4.14). The absorbance spectrum of phenylalanine is also very different in that its absorbance maximum is at 257 nm with no measurable absorbance at 280 nm, which is the wavelength routinely used to measure protein concentrations. Although phenylalanine residues in proteins are considered chemically inert, the free amino acid plays important roles in animal metabolism. Phenylalanine, for example, is the chemical precursor to tyrosine, which is metabolized into numerous neurotransmitters and pigments (see Chapter 17).

The optical properties of aromatic amino acids have been exploited by nature to generate **autofluorescent** proteins such as the **green fluorescent protein** (GFP) of the Pacific Northwest jellyfish *Aequorea victoria* (**Figure 4.15**). This 238-amino-acid protein is synthesized in the photoorgans of the jellyfish and emits a green light after absorption of blue light. GFP has two peak excitation wavelengths, one at 395 nm and another one at 470 nm. The peak emission wavelength is 509 nm. Inside the jellyfish light organ, the blue light (with a wavelength of 460 nm) that stimulates GFP is emitted by a fluorescent protein called aequorin, which contains a tightly associated chromophoric ligand named coelenterazine (Figure 4.15). Aequorin fluorescence is stimulated by calcium binding in response to neuronal signaling in the jellyfish. The resulting blue light it emits, in turn, stimulates GFP fluorescence. GFP fluorescence can also be stimulated directly in the laboratory by illumination with light in the range 450–490 nm.

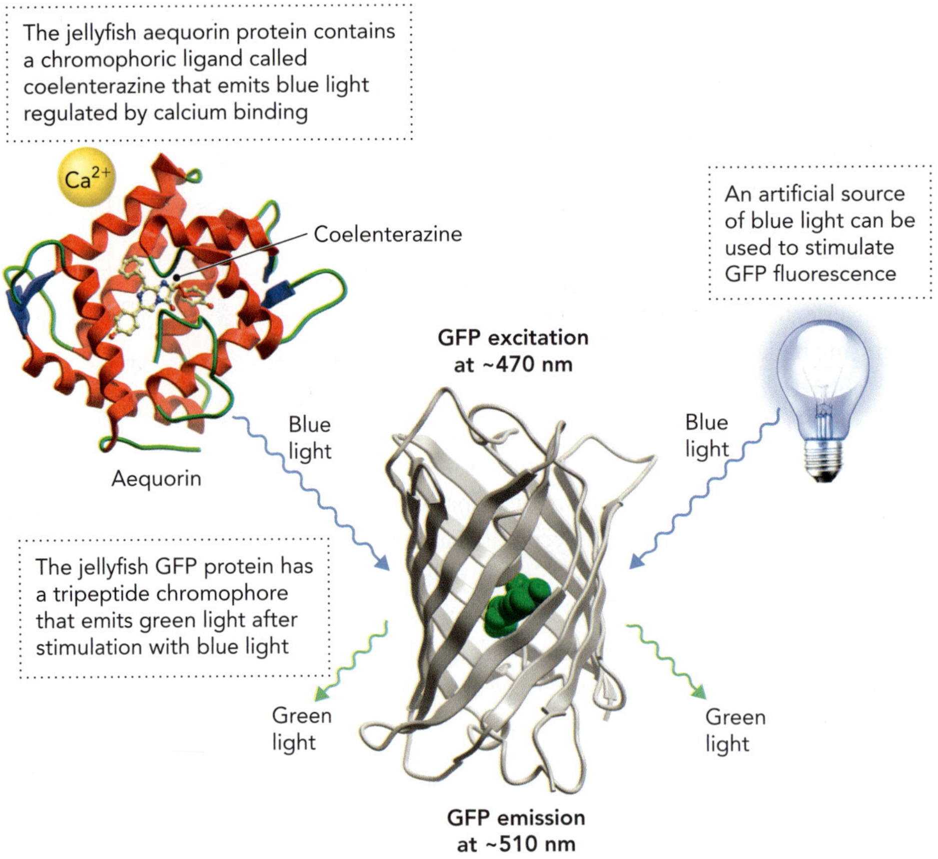
The jellyfish aequorin protein contains a chromophoric ligand called coelenterazine that emits blue light regulated by calcium binding
Ca^{2+}
Coelenterazine
Aequorin
An artificial source of blue light can be used to stimulate GFP fluorescence
GFP excitation at ~470 nm
Blue light
Blue light
The jellyfish GFP protein has a tripeptide chromophore that emits green light after stimulation with blue light
Green light
Green light
GFP emission at ~510 nm

Figure 4.15 GFP fluorescence is stimulated by absorption of blue light by the tripeptide chromophore of the protein. The source of blue light in the jellyfish light organ is another bioluminescent protein called aequorin, which has an associated chromophoric ligand (coelenterazine) that emits light regulated by calcium binding. GFP fluorescence can also be stimulated by illumination with external light of the proper wavelength. BASED ON PDB FILES 1EJ3 (AEQUORIN) AND 1GFL (GFP).

Surprisingly, unlike aequorin, which has a chromophoric ligand, the GFP chromophore is generated directly within the protein through modification of three amino acid residues, one of which is tyrosine. As shown in **Figure 4.16**, a spontaneous cyclization and oxidation reaction—involving oxygen and adjacent serine, tyrosine, and glycine residues—creates the chromophore. The realization that the chromophore is auto-generated by amino acids within the protein itself and that GFP fluorescence can be stimulated by an exogenous light source led to the use of GFP as a molecular marker in live cells. In practice, cells expressing the recombinant GFP protein are illuminated with light at a wavelength of 470 nm, resulting in peak emission at 510 nm that is detected as a green light.

Peptide Bonds Link Amino Acids Together to Form a Polypeptide Chain

Peptide bonds are formed by a condensation reaction between the carboxylic acid and amino groups of two amino acids, which leads to the release of H_2O (**Figure 4.17**). Peptide bond formation is catalyzed inside cells by the RNA component of ribosomes. Because the condensation reaction is unfavorable on its own ($\Delta G > 0$), energy input is required during protein synthesis to catalyze peptide bond formation. As described in Chapter 22, this energy is provided by phosphoanhydride bond energy present in ATP and GTP. Even though peptide bond hydrolysis (the reverse reaction) is energetically more favorable than condensation, proteins are remarkably stable in water. The average half-life of a peptide bond in water is about 10 years, but it can be much longer, depending on the chemistry of the two linked amino acids and their chemical environment in the protein. Because the peptide bond is so stable, protein degradation in biological systems depends on enzymes called **proteases**, which catalyze peptide bond hydrolysis.

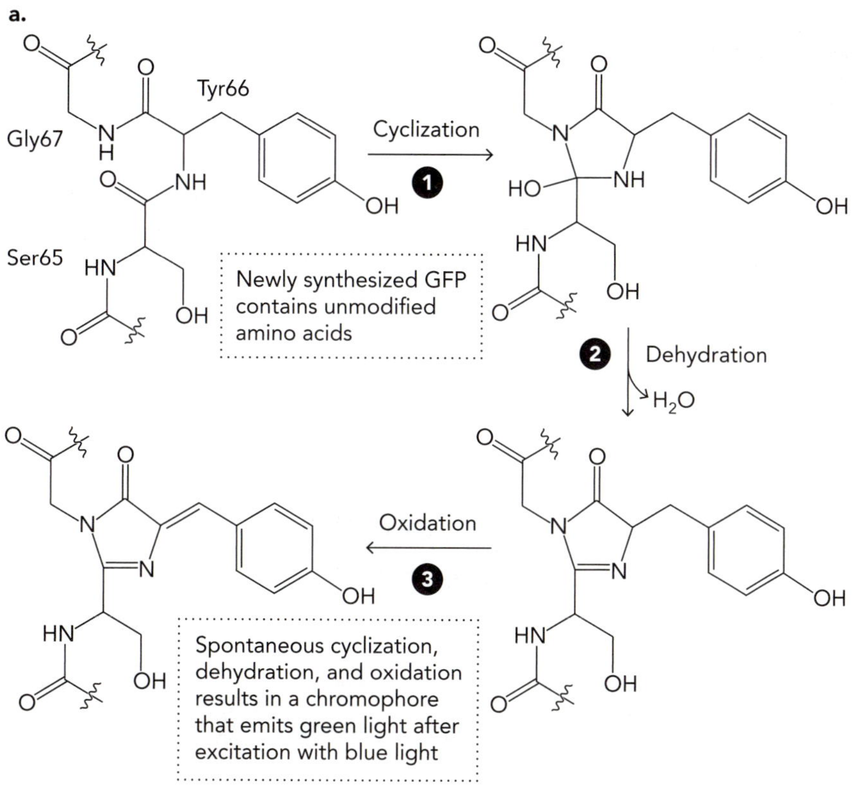

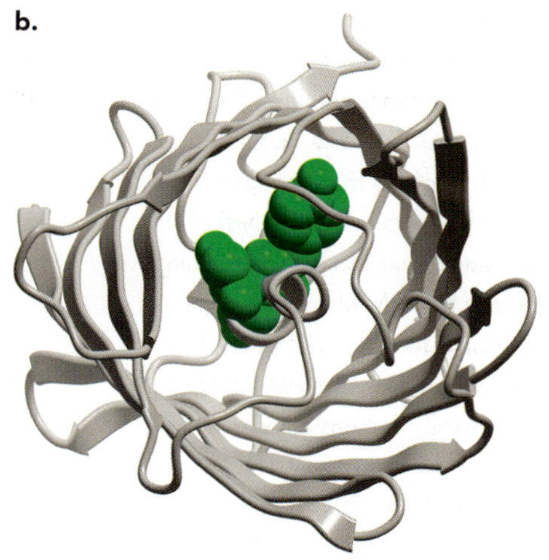

Figure 4.16 The GFP chromophore forms spontaneously in the folded protein by cyclization, dehydration, and reduction reactions involving three amino acids. **a.** The three amino acid residues that form the GFP chromophore are serine (Ser65), tyrosine (Tyr66), and glycine (Gly67). **b.** Molecular structure of the GFP protein, showing the tripeptide chromophore in the center of the protein in space-filling style. BASED ON PDB FILE 1GFL.

By convention, amino acid sequences are specified from the amino-terminal to the carboxyl-terminal direction. **Figure 4.18** shows the chemical and molecular structures of a peptide segment having the sequence Asn-Tyr-Arg-His. These four amino acids are linked together by three peptide bonds and correspond to residues 167 through 170 of the 453-amino-acid rat phenylalanine hydroxylase protein. In the full protein, the first amino acid in a polypeptide chain contains the free amino terminus (NH_3^+), whereas the last amino acid corresponds to the carboxyl end (COO^-). In this short peptide, Asn167 is on the amino-terminal end of the sequence, and His170 is on the carboxyl-terminal end. The molecular structure of this segment of the phenylalanine hydroxylase protein (Figure 4.18b) illustrates how the side chains of each amino acid residue are oriented with respect to the polypeptide **backbone** (—N—C_α—C—N—C_α—C—). Moreover, you can see that each amino acid contributes to the chemical properties of the protein, in that Tyr168, Arg169, and His170 all contain ionizable groups (see Table 4.3), and Asn167 is a hydrophilic amino acid that can form hydrogen bonds with its side chain.

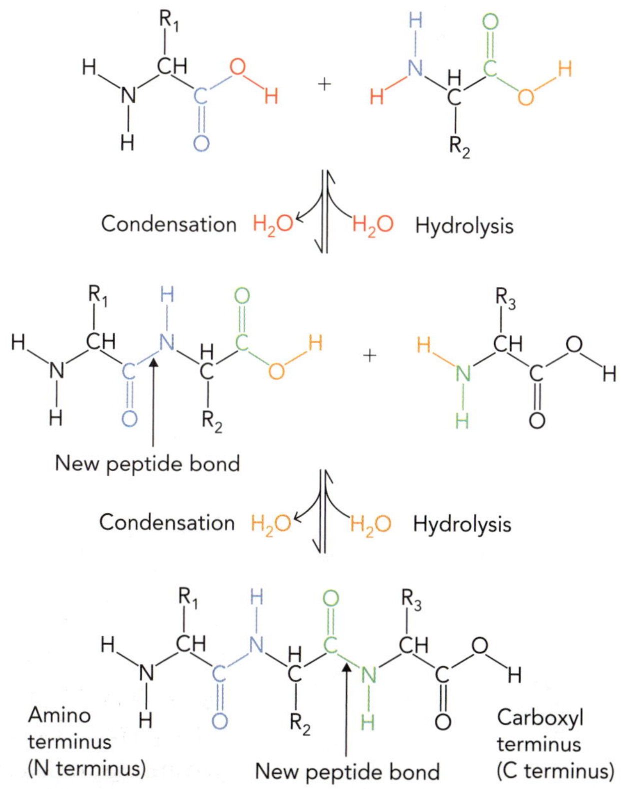

Figure 4.17 Amino acids are joined together by peptide bonds. Peptide bond formation is a condensation reaction. In cells, a similar reaction is catalyzed by ribosomes. Peptide bond cleavage is a hydrolysis reaction. In cells, peptide bond cleavage is catalyzed by enzymes called proteases. The growing peptide has a single amino terminus (N terminus) and a single carboxyl terminus (C terminus).

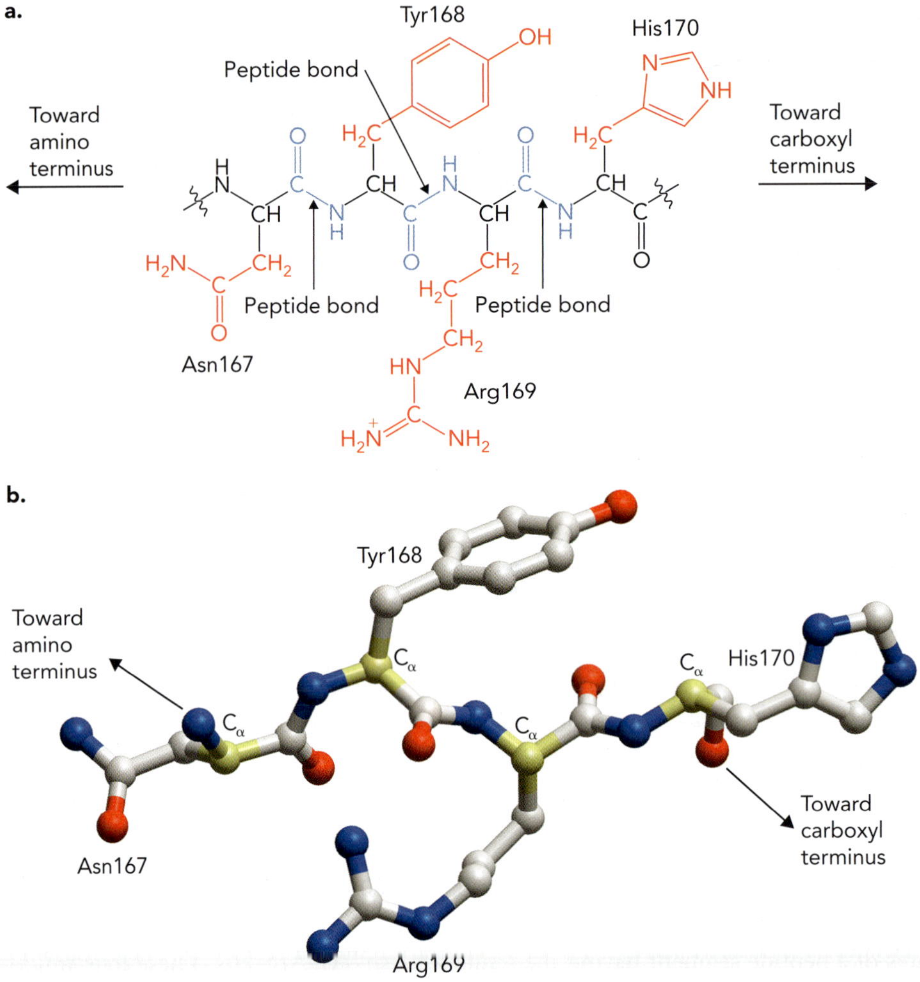

Figure 4.18 A four-amino-acid segment of a polypeptide, which is linked together by three peptide bonds, is shown here. **a.** Chemical structure of the tetrapeptide Asn167-Tyr168-Arg169-His170 showing the amino acid side chains in red. **b.** Molecular structure of this same tetrapeptide within the phenylalanine hydroxylase protein with the C_α carbons colored in gold. BASED ON PDB FILE 1PHZ.

The peptide bond itself is rigid and has partial double bond character because of resonance in the C—N and C═O bonds. Evidence for resonance comes from the intermediate bond length of 1.32 Å for the peptide bond compared to 1.45 Å for a typical C—N bond and 1.25 Å for a C═N bond. As shown in **Figure 4.19**, six atoms lie within a plane defined by the peptide bond. These six atoms include the four atoms in the peptide linkage (CONH) and also the two C_α atoms in the adjacent amino acid residues. The partial double bond character of the peptide bond restricts rotation around the bond between the carbonyl carbon and amide nitrogen and constrains the peptide bond to either the cis or trans configuration. The trans configuration, in which the flanking C_α atoms are on opposites sides of the peptide bond, as shown in Figure 4.19, predominates in proteins. However, rotation can occur around the bonds to the C_α atoms. The torsional angle between the amide nitrogen and the C_α (N—C_α)

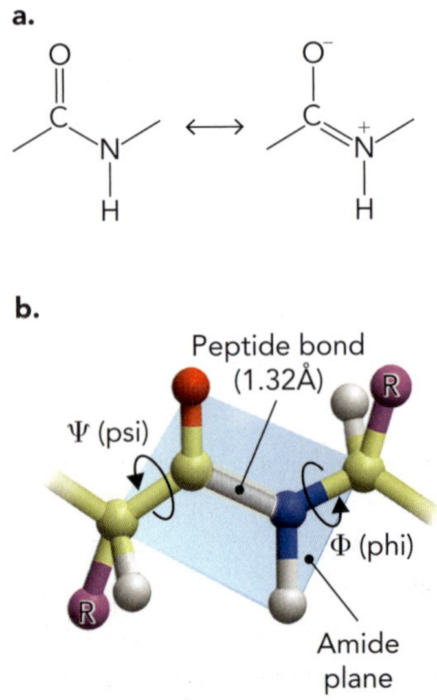

Figure 4.19 The peptide bond has partial double bond character. **a.** Resonance results in the partial double bond character of the peptide bond. **b.** The partial double bond character of the peptide bond restricts rotation around this bond and constrains six atoms to lie within the same plane, as indicated by the rectangle. However, rotation can occur around the two bonds that flank C_α. The single bond between the amide nitrogen and the C_α (N—C_α) allows rotation, as does the single bond between the adjacent C_α and the carbonyl carbon (C_α—C) of the peptide bond. These two torsional angles on either side of the peptide bond are called the ϕ (phi) and ψ (psi) angles, respectively.

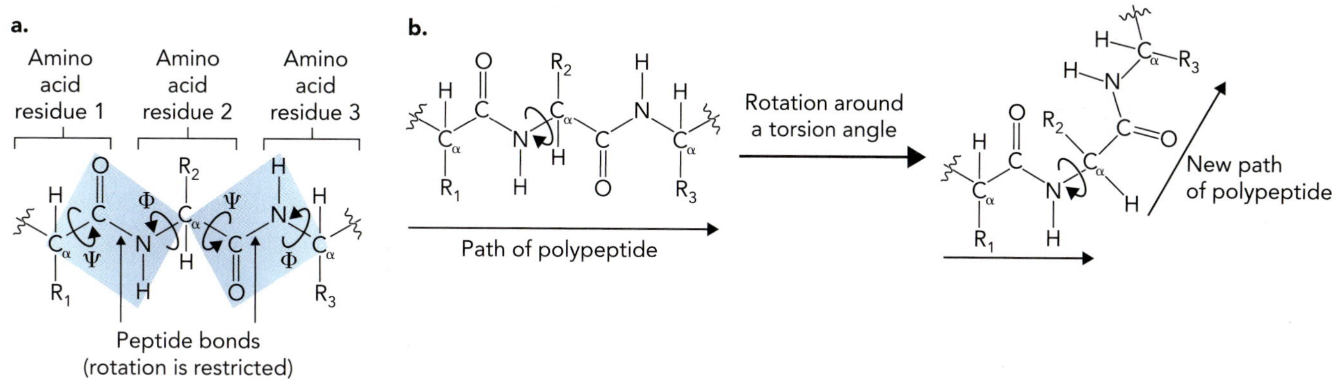

Figure 4.20 Adjacent amide planes rotate through torsional angles, which together determine the overall structure of a protein. **a.** The ϕ and ψ torsional angles on either side of the C_α orient adjacent amide planes relative to each other. **b.** Rotation around a torsional angle changes the path of the polypeptide chain, as shown here for a rotation around a single ϕ angle.

is called **ϕ** (**phi**), and the torsional angle between the C_α and the carbonyl carbon (C_α—C) is known as **Ψ** (**psi**). The ϕ and ψ angles determine the structure of the peptide backbone as a function of bond rotation around the C_α (**Figure 4.20**).

The allowable ϕ and ψ angles for any two amino acids in a dipeptide can be calculated within the limits of steric interference using the van der Waals radii of atoms in the respective residues. This determination was first done in 1963 by Gopalasamudram Ramachandran, who identified the energetically favored combinations of ϕ and ψ angles in a large number of dipeptide combinations. **Figure 4.21** shows the results of these calculations, which can be plotted as ϕ versus ψ angles and is called a **Ramachandran plot** or diagram. Ideal combinations of ϕ and ψ angles that do not result in steric hindrance are seen as darkly colored areas in the Ramachandran plot. Lighter-colored regions depict ϕ and ψ combinations that are slightly less ideal. Uncolored regions correspond to ϕ and ψ angle combinations that are not favored because of steric hindrance. Using structural information from actual protein structures, it is found that most amino acid residues have combinations of ϕ and ψ angles that fall in the colored areas of the Ramachandran plot (Figure 4.21b). However, because of the small size of the side chain in glycine (hydrogen), glycine residues can adopt more combinations of ϕ and ψ angles than can the other amino acids and are found dispersed throughout the Ramachandran plot (Figure 4.21b).

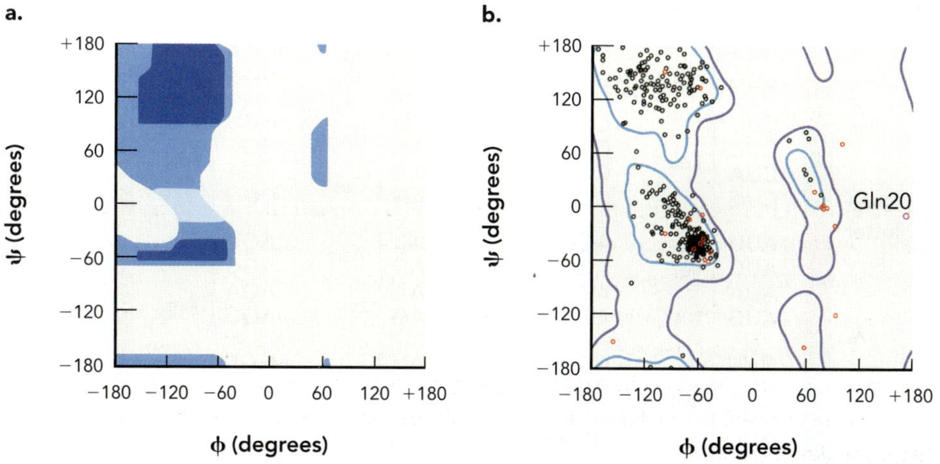

Figure 4.21 A Ramachandran plot shows the allowable combinations of ϕ and ψ angles for amino acid residues on the basis of steric hindrance. **a.** A schematic Ramachandran plot showing ϕ and ψ angle combinations that are theoretically possible in proteins. The dark blue areas show combinations of angles that are most compatible, light blue areas cover angles that are less favorable, and uncolored areas are energetically unfavorable. **b.** The combinations of ϕ and ψ angles measured from an actual protein generally fall in the most favored regions of the Ramachandran plot. In this example, the combinations of ϕ and ψ angles for each residue of the 453-residue rat phenylalanine hydroxylase protein are shown as dots. Combinations for glycine residues, which can favorably adopt more conformations than other residues, are shown as red dots. Proline residues and residues preceding proline are not shown. Occasionally, non-glycine residues fall outside the most energetically favorable regions, such as Gln20 in this example. BASED ON PDB FILE 1PHZ.

Predicting the Amino Acid Sequence of a Protein Using the Genetic Code

Before we look in detail at protein structure, we first need to consider how modern protein biochemistry uses **DNA bioinformatics** as a platform to investigate protein structure and function. Isolating and characterizing large amounts of individual proteins from bacterial, animal, or plant tissue was the mainstay of protein biochemistry 50 years ago, but today protein structure and function studies most often start with a DNA sequence. The reason is that protein sequences are encoded by DNA, and the complete DNA sequences of thousands of organisms are now available through public domain databases. This list of genomes includes not only humans, but essentially every model organism used in biological research (flatworm, yeast, fruit fly, zebra fish, frog, mouse, canine, mustard plant, and so forth), along with a large number of diverse microorganisms.

The central dogma of biology states that DNA sequences are transcribed into complementary RNA sequences, which are then translated into proteins based on triplet codons specified by the genetic code (see Figure 1.25). The genetic code consists of 64 triplet codons, of which 61 codons specify amino acids, and 3 codons correspond to termination codons (**Figure 4.22**). As uracil replaces thymine in RNA, the RNA genetic code can be converted to the DNA genetic code by replacing uracil (U) with thymine (T).

Figure 4.23 illustrates how gene sequence information in DNA is used to deduce the corresponding amino acid sequence of an encoded protein based on bioinformatic approaches. (This method is often termed *in silico* to reflect the use of silicon-chip computers for performing the calculations.) Assuming that we know the orientation of the gene, then we can use the genetic code to predict three different protein sequences for a segment. The three possible reading frames generated from a DNA sequence correspond to triplet codons set in register with the first, second, or third nucleotide in the sequence. If one of three termination codons (TAA, TAG, TGA) is found within the DNA sequence of a particular reading frame (Figure 4.23), then it suggests that this register is either incorrect or the sequence represents the carboxyl-terminal end of the protein. This straightforward approach makes it easy to identify the reading frame with the longest protein-coding sequence. Once this is done, the correct reading frame needs to be verified by biochemical analysis of the encoded protein product.

	Second letter				
First letter	**U**	**C**	**A**	**G**	**Third letter**
U	UUU Phe UUC Phe UUA Leu UUG Leu	UCU Ser UCC Ser UCA Ser UCG Ser	UAU Tyr UAC Tyr UAA Stop UAG Stop	UGU Cys UGC Cys UGA Stop UGG Trp	U C A G
C	CUU Leu CUC Leu CUA Leu CUG Leu	CCU Pro CCC Pro CCA Pro CCG Pro	CAU His CAC His CAA Gln CAG Gln	CGU Arg CGC Arg CGA Arg CGG Arg	U C A G
A	AUU Ile AUC Ile AUA Ile AUG Met	ACU Thr ACC Thr ACA Thr ACG Thr	AAU Asn AAC Asn AAA Lys AAG Lys	AGU Ser AGC Ser AGA Arg AGG Arg	U C A G
G	GUU Val GUC Val GUA Val GUG Val	GCU Ala GCC Ala GCA Ala GCG Ala	GAU Asp GAC Asp GAA Glu GAG Glu	GGU Gly GGC Gly GGA Gly GGG Gly	U C A G

Figure 4.22 The RNA genetic code identifies amino acids specified by 61 of the 64 possible triplet codons. Redundancy exists in the RNA genetic code in that multiple triplet codons can specify the same amino acid. The triplet codons denoted as "stop" function as termination signals in protein synthesis.

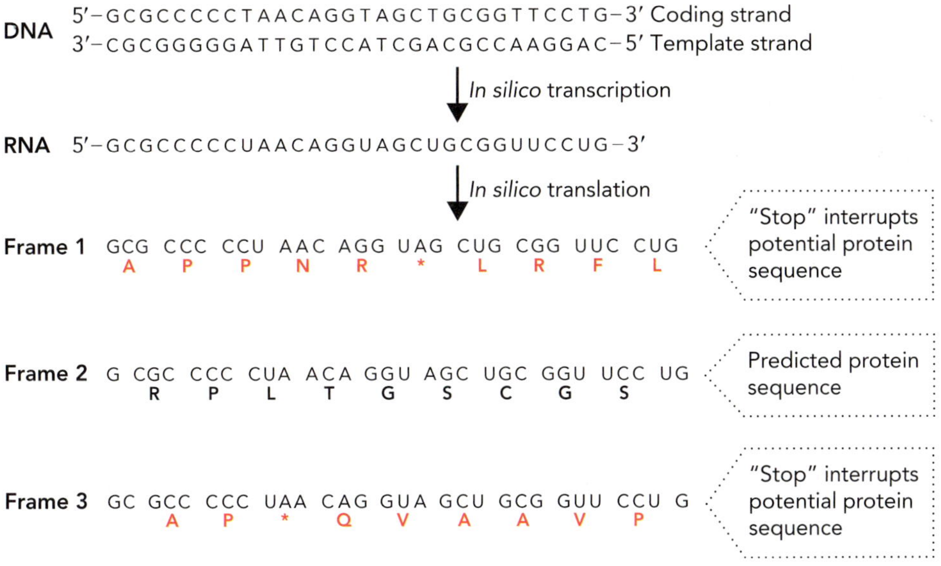

Figure 4.23 DNA bioinformatics can be used to predict the amino acid sequence of a protein by examining the three possible reading frames derived from the coding strand. The absence of termination sequences, indicated by asterisks, in reading frame 2 in this example suggests that this may be the correct amino acid sequence of the encoded protein.

Using bioinformatics to predict the amino acid sequence of a protein has the potential to reveal nucleotide mutations that affect protein structure and function. As shown in **Figure 4.24**, three types of single nucleotide changes can result in altered amino acid sequences in proteins. These are **missense mutations, nonsense mutations,** and **frameshift mutations.** A fourth nucleotide change, called a **silent mutation,** has no effect on the amino acid sequence because of redundancy in the genetic code, and thus the new codon specifies the same amino acid. Missense mutations lead to a different amino acid in the protein sequence, and nonsense mutations introduce a termination codon. Frameshift mutations are due to nucleotide insertions or deletions that change the triplet codon register, and thereby alter the amino acid sequence in the new reading frame until a termination codon is reached.

The informational relationship between the DNA sequence of a gene and the amino acid sequence of the encoded protein is the key to understanding molecular evolution because it links random mutations to protein function. Moreover, biochemists and molecular geneticists have been able to exploit this relationship to uncover critical determinants of protein structure and function. This has been done using protein

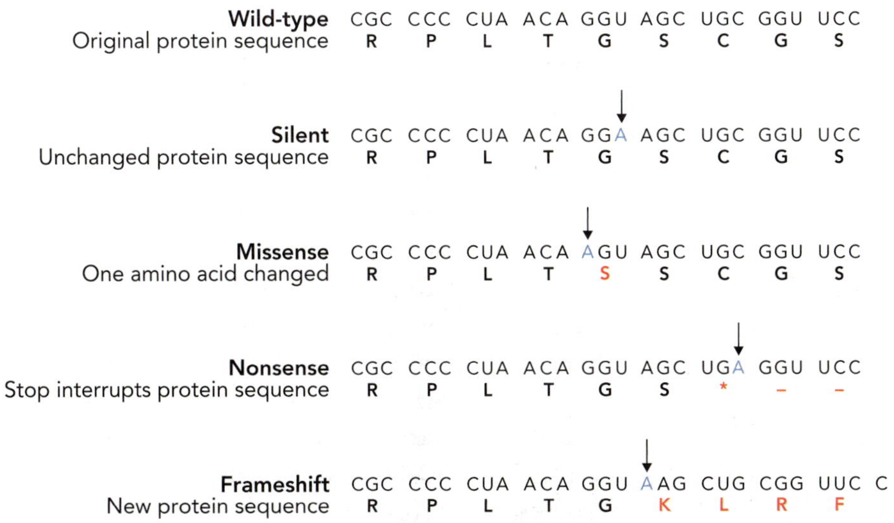

Figure 4.24 DNA bioinformatics can be used to predict amino acid mutations that could alter protein structure and function. Single nucleotide mutations in the DNA can cause missense, nonsense, or frameshift mutations that change the amino acid sequence of the protein. Silent mutations change the DNA sequence without altering the amino acid sequence, owing to redundancy in the genetic code.

biochemistry to analyze the effects of naturally occurring mutations on protein function and by using recombinant DNA technology to create specific mutations.

concept integration 4.1

How does the amino acid sequence of a protein determine its chemical properties?

The primary structure of a protein, which describes the amino acid sequence, determines its function through the various chemical properties of the amino acid side chains, which give rise to the overall three-dimensional structure of the protein. Amino acids are covalently linked together by peptide bonds formed between the carboxylic acid group of one amino acid and the amine group of another. The 20 common amino acids found in proteins can be divided into four chemical groups: charged amino acids (Asp, Glu, Lys, Arg, His); hydrophilic amino acids (Ser, Thr, Cys, Gln, Asn); hydrophobic amino acids (Gly, Ala, Pro, Val, Leu, Ile, Met); and aromatic amino acids (Phe, Tyr, Trp). The various chemical properties of the amino acid side chains determine the chemical properties and the three-dimensional structures of proteins.

4.2 Hierarchical Organization of Protein Structure

A core principle of protein biochemistry is that protein structure and function are determined by the primary amino acid sequence. The chemical properties of the amino acid side chains contribute to the protein's fold and function. A second key concept is that a large number of relatively weak noncovalent interactions between atoms within a polypeptide chain account for the stabilized structure of folded proteins in their native environment (aqueous cytosol or lipid membrane). As described in Chapter 2, weak noncovalent interactions in biological systems consist of hydrogen bonding, van der Waals forces, and ionic interactions, as well as hydrophobic effects. Stronger forces can also stabilize protein structures—for example, disulfide bridges can form between cysteine residues or electrostatic interactions can occur between amino acids and metal ions (Mg^{2+}, Zn^{2+}, Fe^{2+})—however, these are the exceptions rather than the rule.

One of the advantages of weak interactions in biological systems is that they permit flexibility. In terms of protein structure, this means that proteins are able to change their shape by breaking and re-forming weak interactions. It is a misconception to think that proteins are rigid, crystalline-like structures, even though molecular models are often derived from X-ray crystallography data. Quite the contrary, proteins are flexible and undergo conformational changes on both small (<1 Å) and large (>10 Å) scales. **Figure 4.25a** shows the results of a molecular dynamics simulation that predicts flexibility within a cell signaling protein called Abl, which is mutated in certain types of cancer. This simulation shows that some parts of the protein have very similar structures among the collection of conformations, whereas other regions show large variabilities. In addition to these small fluctuations, some proteins, such as the iron binding protein lactoferrin, undergo large conformational changes as a result of substrate binding (**Figure 4.25b**). This large claw-like movement results from the binding of two Fe^{3+}

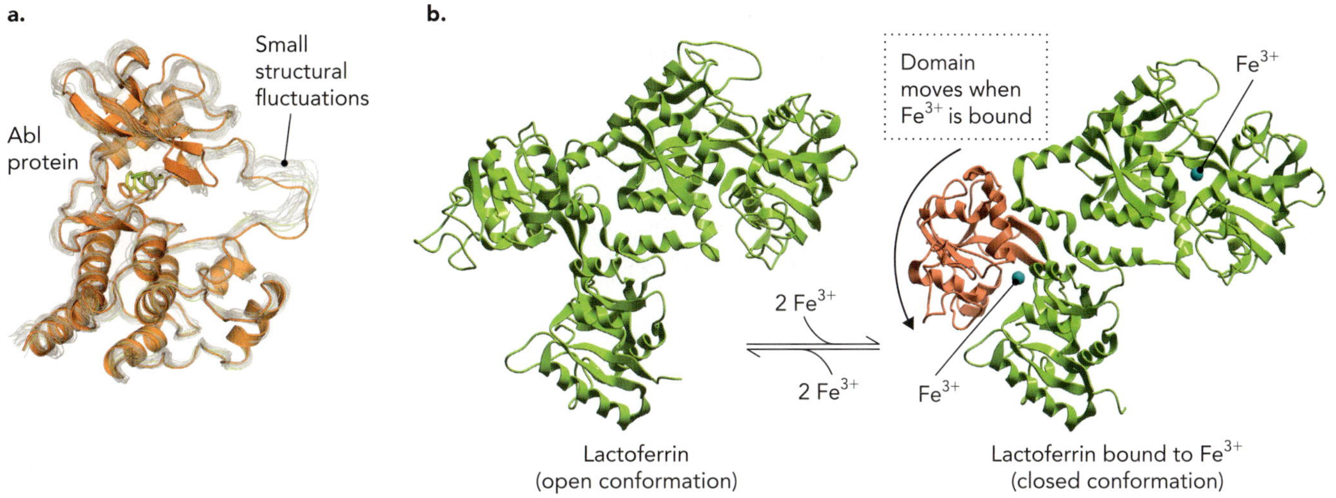

Figure 4.25 Proteins are flexible and can undergo both small and large conformational changes as a result of dynamic alterations in noncovalent interactions. **a.** Molecular simulation of conformational flexibility in the kinase domain of the human signaling protein Abl, showing that some regions are predicted to be more flexible than others. BASED ON M. HOFF (2006). CONFORMATIONAL VARIATIONS OF A KEY ENZYME OFFER CLUES TO CANCER-DRUG RESISTANCE. *PLOS BIOLOGY*, *4*(5), E 166. **b.** Some proteins, such as the iron binding protein lactoferrin, undergo large conformational changes in response to ligand binding (protein domain colored in red). BASED ON PDB FILES 1LFG (OPEN) AND 1LFH (CLOSED).

ions to different regions of the protein. Throughout this book, we will see many examples of this type of "induced fit" conformational change in protein structure.

We should emphasize here that the amino acid sequence of a protein not only determines its three-dimensional structure and flexibility but also directly affects its physical and chemical properties. For example, the distribution of charged amino acids on the surface of a protein facilitates its interactions with other biomolecules. As shown in **Figure 4.26a**, the surface charge of a restriction enzyme that cleaves DNA has patches of positive charge that strengthen interactions with the negatively charged phosphate backbone of the DNA substrate. Depending on the function of the protein, specialized chemical environments within the interior of the protein can occur, formed by the arrangement of amino acid side chains. This is especially true of enzymes that catalyze chemical reactions within internal pockets or clefts, which are lined with reactive groups

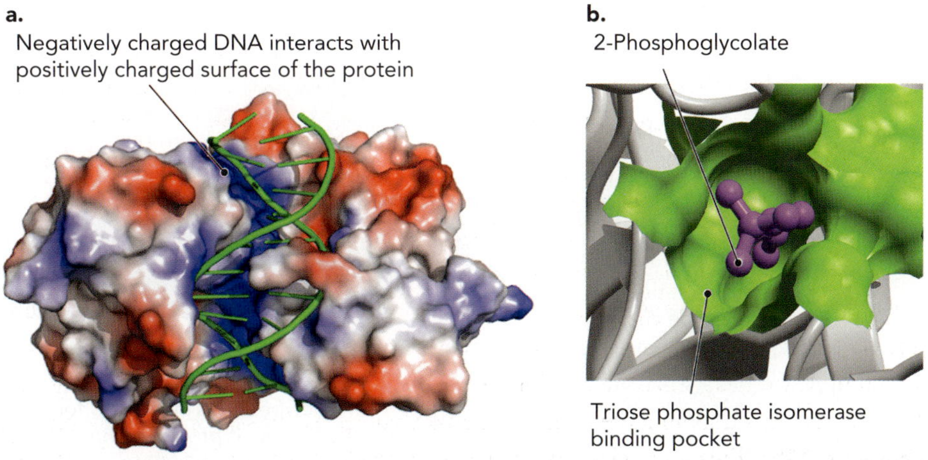

Figure 4.26 The arrangement of amino acid side chains along the polypeptide chain directly affects the surface chemistry and ligand binding properties of proteins. **a.** A molecular model of the bacterial restriction enzyme *Bam*HI, showing the electrostatic potential (positive is blue and negative is red) on the surface of the protein. The positive charges create a binding pocket for the negatively charged phosphate backbone of DNA (green). BASED ON PDB FILE 1BHM. **b.** Enzymes such as triose phosphate isomerase often contain internal binding pockets, which create optimal chemical environments for catalysis. The van der Waals radii of nearby atoms form a substrate binding pocket for the inhibitor 2-phosphoglycolate. BASED ON PDB FILE 1AMK.

from nearby amino acid residues. The molecular structure in **Figure 4.26b** shows the substrate-binding pocket for 2-phosphoglycolate in the active site of the metabolic enzyme triose phosphate isomerase. The green mask outlines the van der Waals radii of nearby atoms in amino acids surrounding the substrate binding pocket.

Now that we have a basic chemical understanding of the amino acid building blocks of proteins and have seen how amino acid residues collectively determine the structure and function of proteins, we are ready to examine the hierarchical organization of protein structure, which consists of four levels (**Figure 4.27**).

▶ ANIMATION

1. The **primary structure** (1°) of a protein refers to the amino acid sequence, which determines how the polypeptide backbone folds into an energetically stable three-dimensional structure (Figure 4.27a). We have been describing elements of protein primary structure throughout Section 4.1.
2. The **secondary structure** (2°) of a protein refers to the regular repetitive arrangement of local regions of the polypeptide backbone. The polypeptide backbone is commonly found in a limited number of secondary structure arrangements because of physical constraints that limit ϕ and ψ angles (the Ramachandran plot) and because of limits imposed by the energetics of protein folding, in which hydrogen bonds of the backbone groups must be satisfied. Figure 4.27b shows the three major secondary structures in proteins: β strands, α helices, and β turns. The graphical representation of the polypeptide backbone associated with each of these secondary structures is called a **ribbon diagram**. Ribbon diagrams depict β strands as arrows, α helices as coils, and β turns—as well as the irregular loops that connect these elements—as ropes. As described in Chapter 5, ribbon diagrams are rendered by computer graphics algorithms, which interpret the coordinates of the

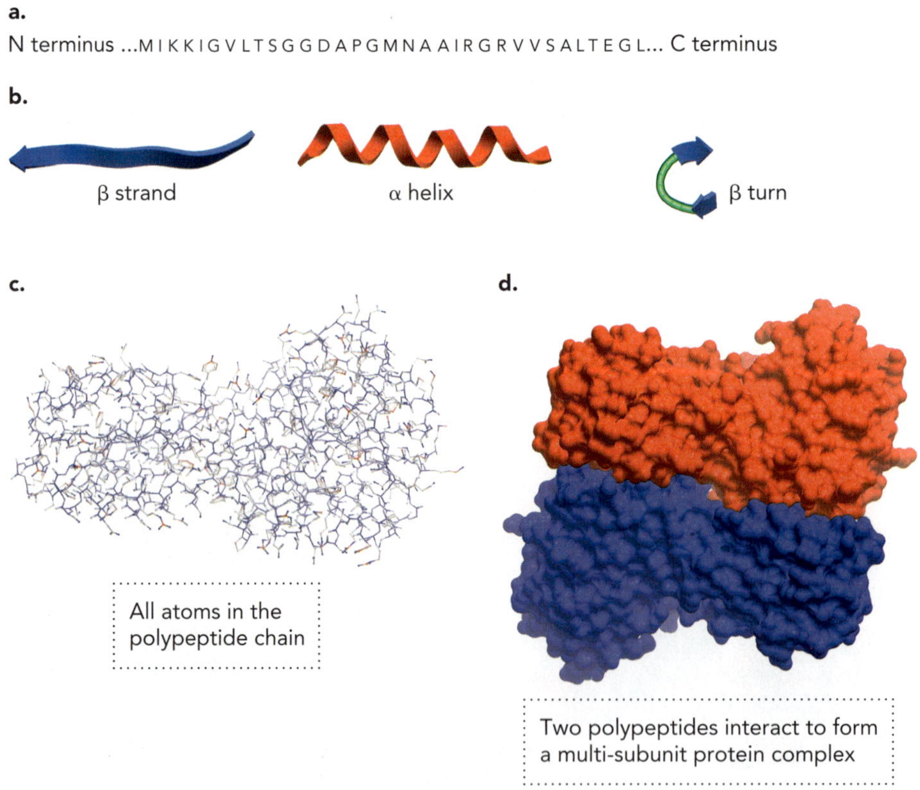

Figure 4.27 There are four parts to the hierarchical organization of protein structure. **a.** The primary structure of a protein is the amino acid sequence of the polypeptide chain. It is conventionally written left to right, beginning with the N-terminal amino acid residue. **b.** The secondary structure describes the local arrangement of the polypeptide backbone into regular, repeating conformations. The three most abundant secondary structures in proteins are β strands, α helices, and β turns. **c.** The tertiary structure is the complete arrangement of all the atoms in the polypeptide chain. **d.** Quaternary structure refers to the number and organization of multiple protein subunits within a protein complex. Shown is a homodimer of the enzyme phosphofructokinase.

ALL STRUCTURES IN THIS FIGURE ARE BASED ON PDB FILE 6PFK.

atomic structures generated from X-ray crystallography or nuclear magnetic resonance data to generate three-dimensional molecular models.

Collections of secondary structures are arranged into **protein folds**, which describe the path of the polypeptide chain. Hundreds of distinct protein folds have been observed in protein structures, which can contain various combinations of α helices and/or β strands. As described later, four of the more common protein folds are the four-helix bundle fold, Greek key fold, Rossmann fold, and TIM barrel fold.

3. The **tertiary structure** (3°) of a protein includes the spatial location of all the atoms in the polypeptide chain (Figure 4.27c). With this information, the path of the polypeptide through space and the positions of the side chains are known.
4. Protein complexes consisting of more than one polypeptide chain are organized into **quaternary structures** (4°), as found in multi-subunit protein complexes. The quaternary structure of multi-subunit protein complexes can involve multiple copies of the same polypeptide chain or single copies of different polypeptide chains (Figure 4.27d).

Proteins Contain Three Major Types of Secondary Structure

Almost all proteins contain combinations of two or more secondary structure elements, which are largely independent of the primary amino acid sequence. These secondary structure elements fall into the following groups: (1) α helices, (2) β strands and β sheets, and (3) β turns. The reason these specific structural groups occur in so many different proteins is their exceptional stability, which is due to a combination of hydrogen bonding and geometry.

The α Helix One of the most common elements of secondary structure in proteins is the right-handed **α helix**. The reason that α helices are so common is that a given polypeptide chain, with the right mix of amino acids, can fold into an α-helical structure spontaneously, stabilized by numerous intrastrand hydrogen bonds. As you will see shortly, this is in contrast to **β strands**, the second major type of secondary structure, which are stabilized by hydrogen bonding between β strands (interstrand) to generate **β sheets**. As shown in **Figure 4.28**, a trace of the peptide backbone in an α helix

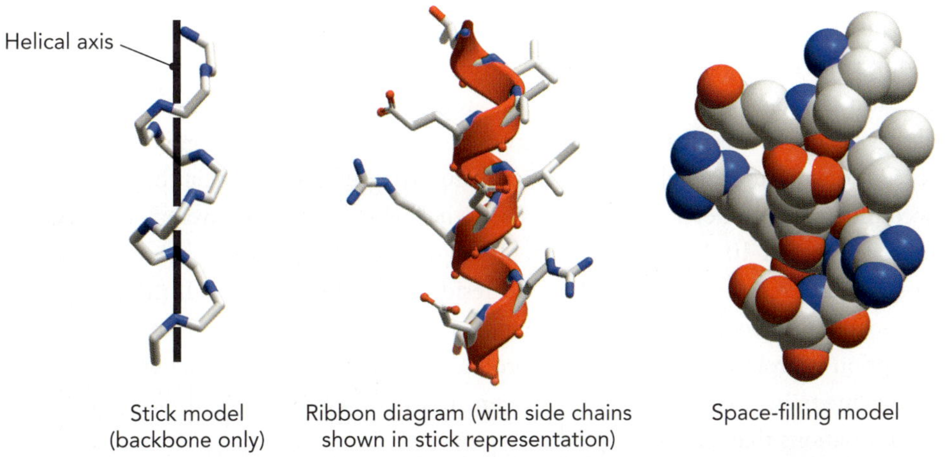

Figure 4.28 The α helix secondary structure in proteins is a right-handed helix that places the amino acid residue side chains on the outside of the helix relative to the polypeptide backbone. Each of these models shows the same α-helical region using different computer-rendering styles: stick (without the amino acid side chains included); ribbon, including stick representation of the side chains; and space-filling styles. Note that the helix does not have any empty space along its axis. BASED ON PDB FILE 1PFK.

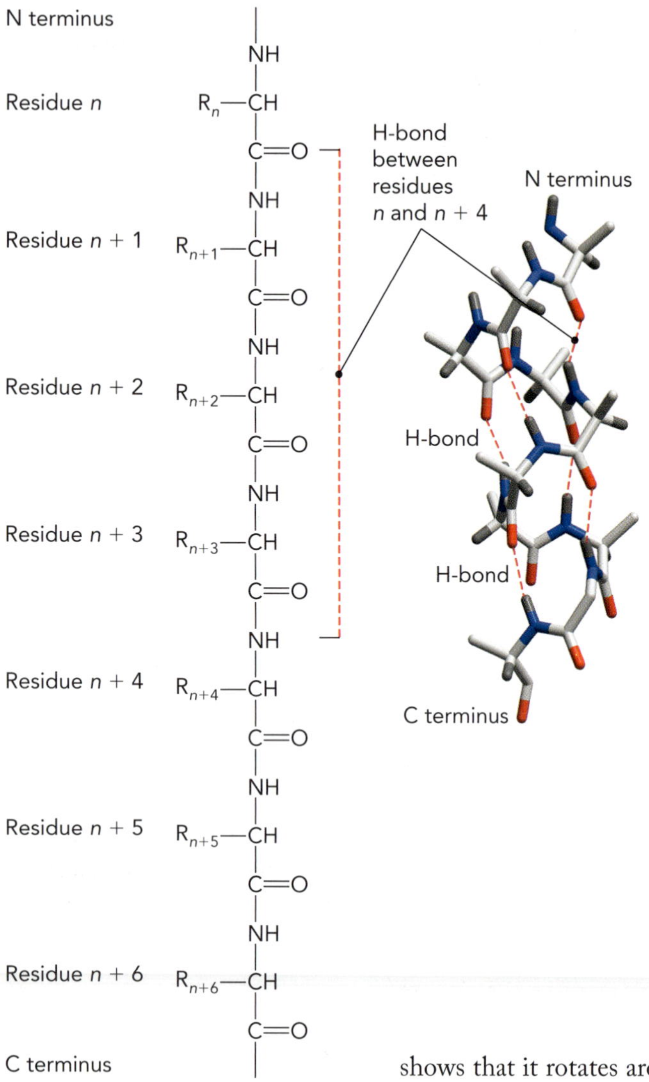

Figure 4.29 The α helix structure is stabilized by intrastrand hydrogen bonding between the carbonyl oxygen in one amino acid residue and the hydrogen atom on the amide nitrogen in an amino acid four residues away. BASED ON PDB FILE 1PFK.

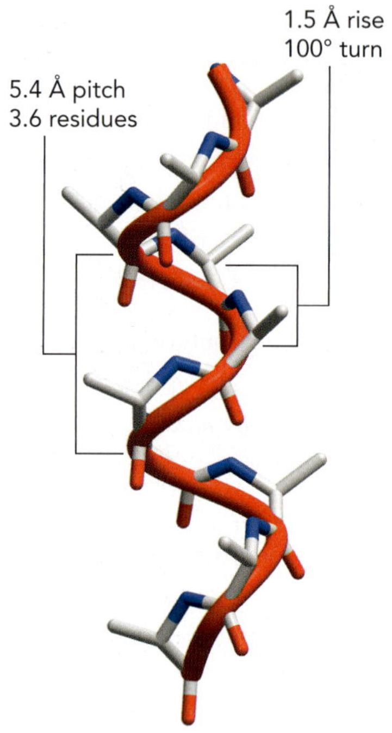

Figure 4.30 The α helix has 3.6 residues per turn of the helix, and therefore each residue corresponds to a 100° turn of the helix. The vertical distance for one consecutive turn of the helix is 5.4 Å. The rise from one amino acid residue to the next along the helix is 1.5 Å. BASED ON PDB FILE 1PFK.

shows that it rotates around an imaginary central rod in a right-hand direction. Only right-handed α helices are abundant in proteins. Although it is also possible to model a left-handed helix, it is much less stable and therefore found only infrequently in short stretches. The ribbon diagram in Figure 4.28 displays the peptide backbone as a coiled red ribbon and shows how the amino acid side chains, drawn as sticks, radiate outward. The space-filling model of this same α-helical structure reveals the close packing of atoms in each of the amino acid residues.

Figure 4.29 shows the hydrogen bonds that form in an α helix between the carbonyl oxygen (residue *n*) and the hydrogen atom attached to the nitrogen in the peptide bond located four amino acids away (*n* + 4). The length of these hydrogen bonds is optimal with an N—O distance of 2.8 Å. This arrangement results in 3.6 residues per turn of the helix (one residue per 100° rotation) and a pitch of 5.4 Å (the distance along the axis between each turn; **Figure 4.30**). All hydrogen bonding backbone atoms (except the most terminal on either end) form hydrogen bonds within the helix, which accounts for the inherent stability of the α-helical structure. An average α helix in a protein is about 10–14 residues long, although naturally occurring helices are found as short as 5 residues and as long as 50 residues. Note that although the side chains of most amino acids can be accommodated within the structure of an α helix, one exception is proline. The nitrogen in proline lacks a hydrogen and therefore does not contribute to hydrogen bonding (see Figure 4.12). Moreover, the rigid ring restricts the conformations that can be adopted.

Figure 4.31 This photograph was taken in 1962 and shows Max Perutz on the left holding his balsa wood model of hemoglobin and John Kendrew on the right holding his wire frame model of myoglobin. Perutz and Kendrew shared the 1962 Nobel Prize in Chemistry for their studies of globular proteins. HULTON-DEUTSCH COLLECTION/CORBIS.

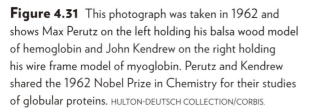

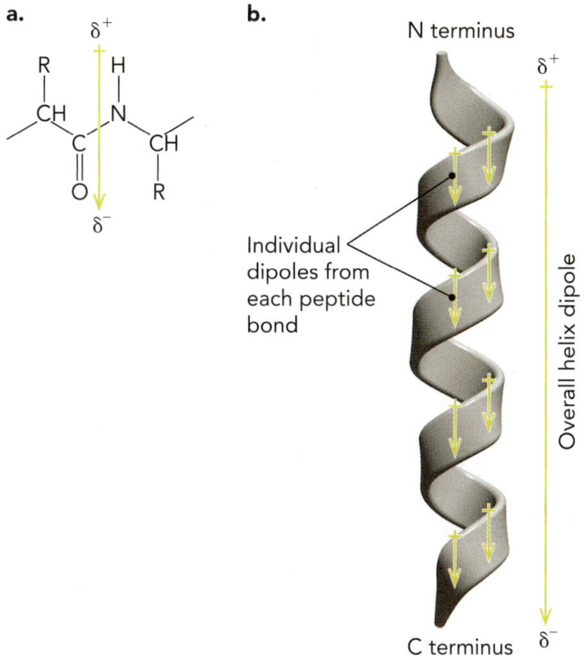

Figure 4.32 Amino acids and α helices have dipole moments. **a.** A single amino acid has a dipole moment across the peptide bond. **b.** The amino acid dipole moments add together to create an overall dipole moment in α helices, with a partial positive charge on the N-terminal end and a partial negative charge on the C-terminal end. BASED ON PDB FILE 1PFK.

The structure of the α helix was first predicted by Linus Pauling and Robert Corey in 1951 using models they built based on the X-ray diffraction pattern of keratin. Their model for the α helix was validated 7 years later when the structure of whale myoglobin was determined by John Kendrew, a biochemist at Cambridge University in England. The myoglobin protein structure revealed eight α helices that had structural properties consistent with the original model. John Kendrew and Max Perutz, a molecular biologist at Cambridge University, shared the 1962 Nobel Prize in Chemistry for solving the molecular structures of myoglobin and hemoglobin, respectively (**Figure 4.31**).

The resonance structure of the peptide bond causes a dipole moment because of the slight separation of charge between the electronegative carbonyl oxygen and the electropositive amide nitrogen (**Figure 4.32a**). As the amino acids in an α helix are held in alignment by hydrogen bonding, the combined effect of these dipole moments in the peptide bond is a dipole moment for the entire α helix (**Figure 4.32b**). This results in a slight positive charge on the amino-terminal end of the α helix and a corresponding negative charge on the carboxyl-terminal end. One of the consequences of this charge separation is that α helices are often further stabilized by compensating charged amino acids at either end of the helix. For example, negatively charged aspartate or glutamate residues are often found at the amino-terminal end, whereas positively charged lysine or arginine residues are more often located on the carboxyl-terminal end. Another consequence of the dipole moment in α helices is that the termini tend to be located on protein surfaces, where the charge can be neutralized by interactions with H_2O or charged ions such as phosphate or Mg^{2+}.

Many α helices are amphipathic, meaning that one side of the helix is hydrophobic and the other side is hydrophilic. This chemical property is important because α helices may need to be able to interact with both the hydrophobic core of the protein and the aqueous environment at the surface of the protein. An amphipathic α helix is generated when amino acids with hydrophilic or hydrophobic properties are positioned every three to four residues along the polypeptide backbone. This is consistent with the physical parameters of an α helix that contains 3.6 residues per 360° rotation (see Figure 4.30). **Figure 4.33** shows an amphipathic α helix found in the enzyme phosphofructokinase-1, which contains hydrophilic amino acids at positions 1 (Thr145), 4 (Glu148), 7 (Asp151), 8 (Arg153), 10 (Arg155), 11 (Asp156), and 12 (Thr157). It also has hydrophobic amino acids located at positions 2 (Val146), 3 (Val147), 5 (Ala149), 6 (Ile150), and 9 (Leu154). Because of the spacing needed to position hydrophilic or hydrophobic chemical groups on one face of the α helix, we can predict

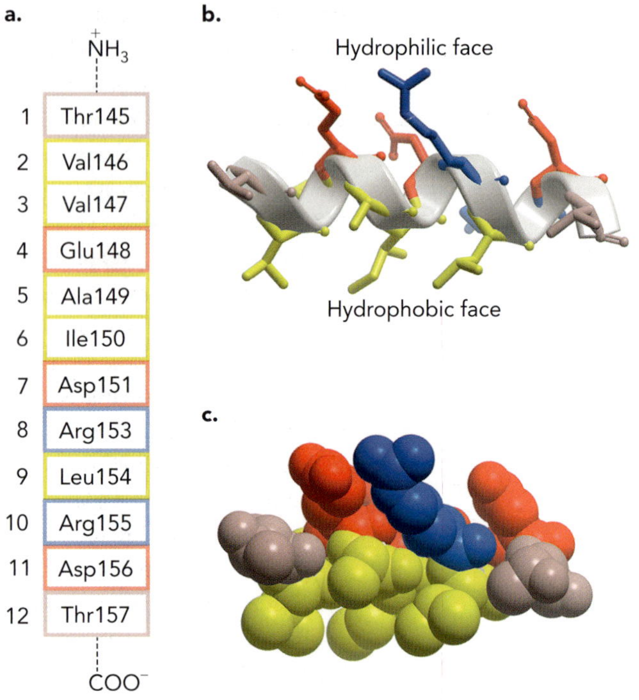

Figure 4.33 An amphipathic α helix has hydrophobic and hydrophilic residues on opposite faces of the helix. **a.** The primary amino acid sequence of a segment in the phosphofructokinase-1 protein (amino acid residues 145–157), showing the placement of hydrophilic (red, blue, and mauve) and hydrophobic (yellow) residues. **b.** Ribbon structure of this same α-helical segment in the protein. **c.** Space-filling model of the amphipathic α helix in phosphofructokinase-1.

BASED ON PDB FILE 1PFK.

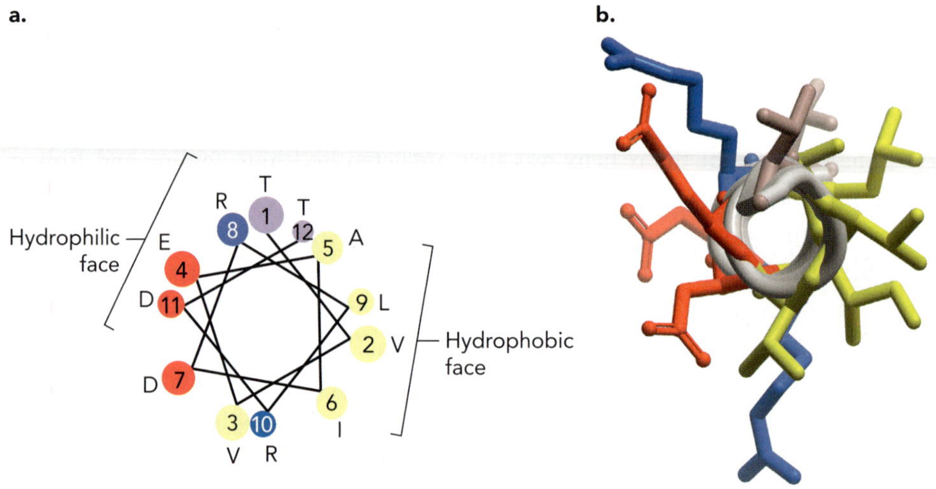

Figure 4.34 Given the sequence of amino acid residues, we can use a helical wheel to predict the presence of an amphipathic α helix in a protein based on the pattern of hydrophilic and hydrophobic amino acids. **a.** In a helical wheel representation, we are looking down the helical axis on a helix that has been flattened into two dimensions. In this example, the polypeptide has sequence TVVEAIDRLRDT. Starting with amino acid 1 (T) at the top of the wheel, each residue is placed sequentially on the helical wheel 100° around the circle from the previous residue. Thus, residue 1 (T) is placed at number 1 on the wheel, residue 2 (V) is placed at position 2, and so forth. The decreasing diameters of the amino acid labels indicate residues that are farther away from the initial residues in the helix. The completed helical wheel indicates that the hydrophobic residues lie primarily on one face of the helix and the hydrophilic residues are primarily on the other face, demonstrating that this sequence forms an amphipathic helix. **b.** Looking down the helical axis, the stick representation of the helix with the sequence shown in panel **a** demonstrates the hydrophobic side chains (yellow) primarily on one face of the helix and the hydrophilic side chains (red, blue, and mauve) on the other face.

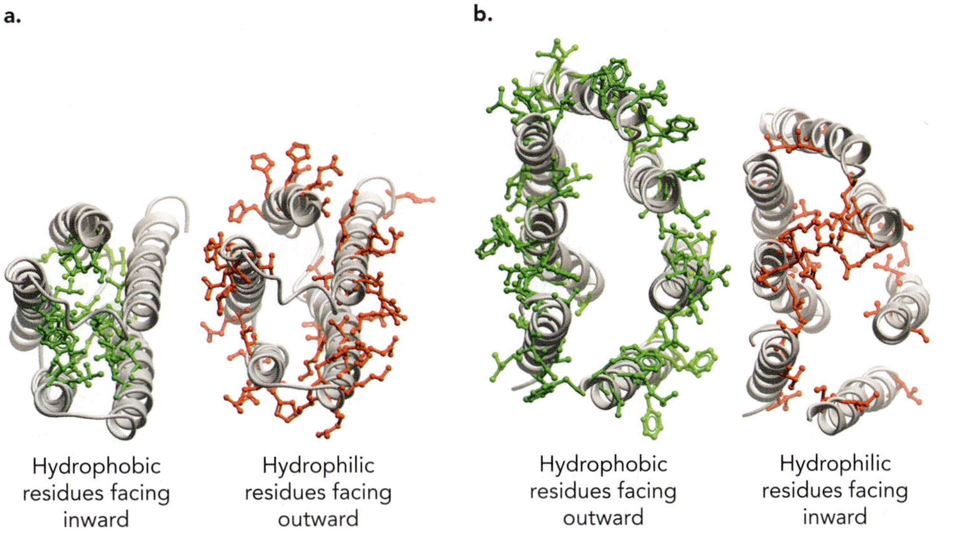

Figure 4.35 Hydrophobic residues of amphipathic α helices can be oriented inward or outward, depending on the environment of the protein. **a.** This apolipoprotein is a soluble protein containing five amphipathic α helices in which the hydrophobic residues (green) are pointing inward, with hydrophilic residues (red) pointing outward toward the solvent. BASED ON PDB FILE 1AEP. **b.** Rhodopsin is a transmembrane protein with seven amphipathic α helices that have hydrophobic residues (green) pointing outward toward the hydrophobic membrane and hydrophilic residues (red) pointing inward. BASED ON PDB FILE 1BRD. The highlighted hydrophilic amino acids are Asp, Glu, His, Lys, Arg, Thr, and Ser, and the highlighted hydrophobic amino acids are Ile, Leu, Phe, Trp, and Val.

the existence of an amphipathic α helix from the primary amino acid sequence by using a representation called a helical wheel (**Figure 4.34**).

Amphipathic α helices can pack closely together with each other. In this way, the hydrophobic and hydrophilic surfaces can face inward or outward, depending on the cellular environment. Two examples of densely packed amphipathic α helices are found in the apolipoproteins and the rhodopsin protein (**Figure 4.35**). Apolipoproteins are components of lipid transport particles that carry cholesterol and triacylglycerols (fats) from the liver to the tissues. The amphipathic α helices found in apolipoproteins have hydrophobic amino acid residues oriented inward with the hydrophilic amino acid residues facing outward. This makes sense because lipid transport particles carry hydrophobic fats through the aqueous environment of the blood. In contrast, the amphipathic α helices located in the membrane protein rhodopsin—a signaling protein that absorbs light—have just the opposite arrangement. As rhodopsin is embedded in lipid membranes, the hydrophobic amino acid residues face outward toward the fatty acid tails of the lipid bilayer, and many hydrophilic amino acid residues face inward.

The β Strand and β Sheet Another protein structure that Pauling and Corey predicted from their models was the β strand, so-called because it is an extended polypeptide chain, and it was the second protein structure they proposed (α helix being the first). As shown in **Figure 4.36**, the amino acid side chains within a single

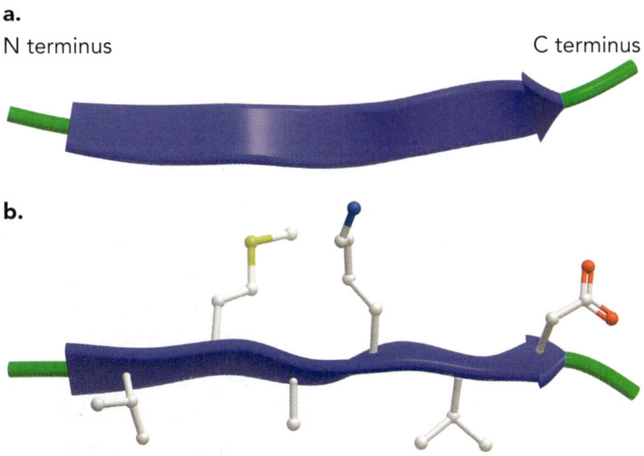

Figure 4.36 A β strand consists of an extended polypeptide chain. **a.** The ribbon style of protein secondary structures depicts β strands as arrows in which the arrowhead is on the C-terminal end of the polypeptide chain. Note that β strands are often twisted rather than being completely planar. **b.** The same β strand as in panel **a** is rotated approximately 90° to show that adjacent amino acid side chains are on alternate sides of the β strand. This is due to the extension of the polypeptide backbone and the tetrahedral nature of amino acids. BASED ON PDB FILE 1W4V.

β strand are positioned above and below the polypeptide backbone because the chain is in a fully extended conformation. In the ribbon style of protein modeling, β strands are depicted as arrows with the arrowhead pointing toward the carboxyl terminus. The distance between adjacent amino acids in β strands is 3.4 Å, which is much farther apart than in the compact α helix, where amino acids are spaced 1.5 Å apart (see Figure 4.30).

The Pauling and Corey model of β strands predicted that they would form stable secondary structures called β sheets, which are held together by hydrogen bonding between backbone NH and CO groups on separate strands. These interstrand hydrogen bonds between β strands are distinct from the intrastrand hydrogen bonds that hold α helices together. Single β strands are not often energetically stable in folded proteins and are almost always found as components of either **antiparallel β sheets,** in which the β strands lie in opposite directions, or **parallel β sheets,** in which the direction of the polypeptide chain is in the same amino to carboxyl orientation in all β strands (**Figure 4.37**). The hydrogen-bond arrangement is more stable for antiparallel β sheets than for parallel β sheets, which explains why antiparallel β sheets are more common in proteins. Because of the way the polypeptide backbones align within a β sheet, they are sometimes called **β pleated sheets** because they resemble

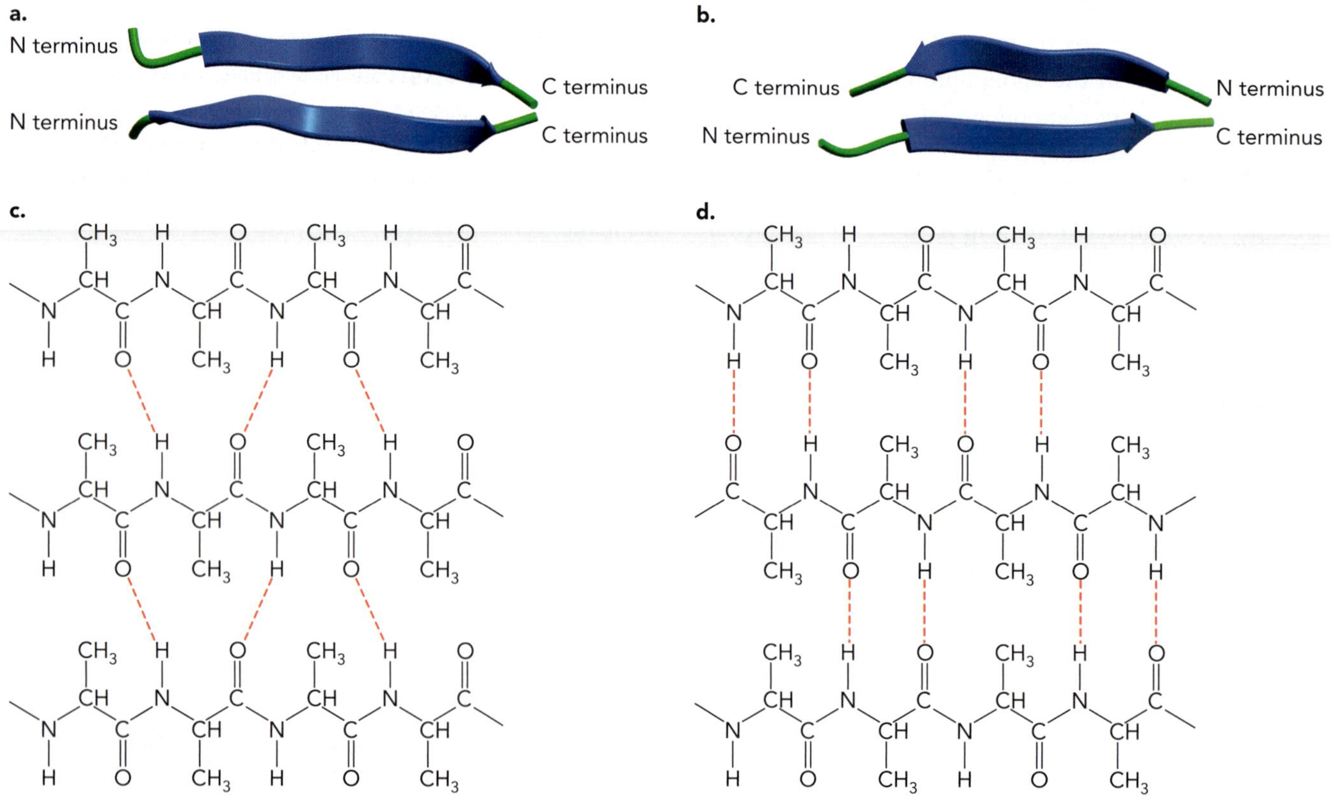

Figure 4.37 Multiple β strands form β sheets in proteins through interstrand hydrogen bonds. **a.** Ribbon representation of a parallel β sheet. BASED ON PDB FILE 1PFK. **b.** Ribbon representation of an antiparallel β sheet. BASED ON PDB FILE 1RGS. **c.** Hydrogen bonds in parallel β sheets. Hydrogen bonds are formed between backbone NH and CO groups on separate strands and serve to stabilize β-sheet structures. **d.** Hydrogen bonds in antiparallel β sheets. The hydrogen bonds in antiparallel β sheets are slightly more optimal with respect to length and geometry than in parallel β sheets, resulting in more stability for antiparallel β sheets.

pleated fabric (**Figure 4.38**). The side chains are oriented outward at each of the alternating peaks and troughs of the β pleated sheet.

β sheets are rarely flat, uniform structures, but rather are twisted to various extents, usually in a right-ward direction, to reduce steric hindrance between the amino acid side chains. (To visualize the effect, imagine holding a piece of paper by the ends and rotating one end relative to the other.) Because of the flexibility in the peptide backbone, the twisted β strands remain linked to each other through hydrogen bonding. Fatty acid binding protein, which transports fatty acids through the cytosol, is predominantly one large antiparallel β sheet that is twisted around to form a hydrophobic binding pocket for the fatty acid (**Figure 4.39**). Although the ribbon model of fatty acid binding protein suggests that the extended β sheet forms an open, basket-like structure, the space-filling model reveals a well-packed protein core. The fatty acid substrate is actually buried deep within the protein, which shields it from the aqueous environment of the cell.

The β Turn and Loops Most proteins are globular in shape, as a result of frequent turns or loops in the polypeptide chain that connect β strands and α helices. The most common types of turns are called **β turns**, which connect two β strands in an antiparallel β sheet (**Figure 4.40**). Two common structurally distinct types of β turns have been

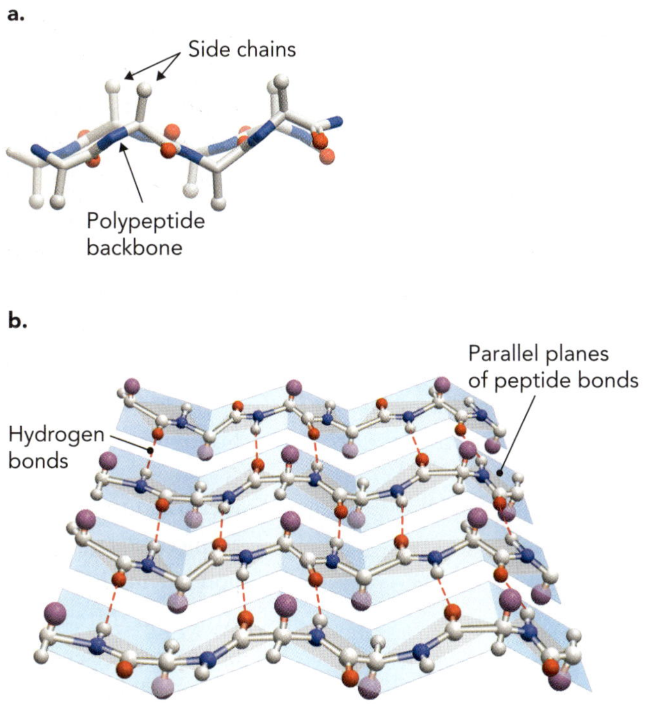

Figure 4.38 The term "β pleated sheet" in reference to β sheets comes from the observation that they resemble pleated fabric. **a.** Each β strand in a β sheet is aligned in a similar way, with the amino acid side chains protruding above and below the polypeptide backbone. BASED ON PDB FILE 1HDI. **b.** Illustration of a β sheet to show its resemblance to pleated fabric.

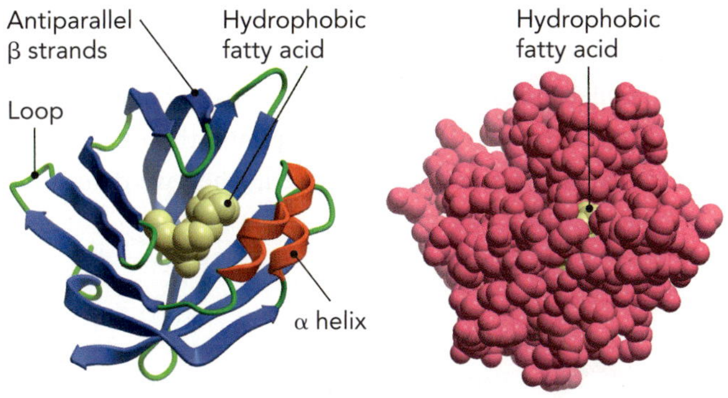

Figure 4.39 β sheets are usually twisted rather than flat, as seen here in a fatty acid binding protein. The ribbon model of fatty acid binding protein shows the fatty acid cargo within a large hydrophobic pocket formed by 10 antiparallel β strands. The space-filling model of the protein in the same orientation shows that the fatty acid molecule is protected from the aqueous environment of the cell. BASED ON PDB FILE 1HMT.

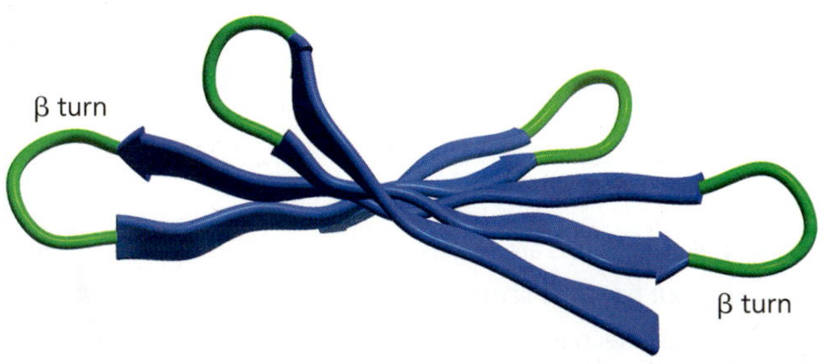

Figure 4.40 The fatty acid binding protein contains numerous β turns connecting the β strands surrounding the hydrophobic binding pocket. BASED ON PDB FILE 1HMT.

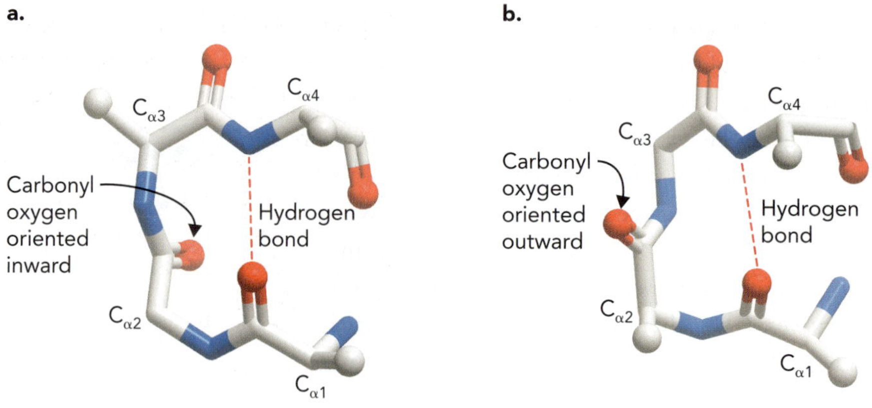

Figure 4.41 Type I and type II β turns differ in the orientation of the peptide plane between the second and third amino acid residues in the turn. **a.** Structure of a type I β turn showing that the carbonyl oxygen is oriented inward. **b.** Structure of a type II β turn showing that the carbonyl oxygen is oriented outward. The α carbons are numbered relative to the amino-terminal residue ($C_{\alpha 1}$). BASED ON PDB FILE 1HMT.

identified in proteins: type I and type II β turns. Both of these turns consist of four amino acids in which the carbonyl oxygen of the first amino acid residue is hydrogen bonded to the nitrogen atom of the fourth amino acid residue (**Figure 4.41**). The structural difference between type I and type II β turns is that the peptide plane between the second and third residues has the carbonyl oxygen flipped in toward the middle of the turn in type I, whereas the carbonyl oxygen is facing outward in a type II turn. Glycine is often the second or third residue in β turns because the hydrogen-atom side chain allows glycine to adopt the unusual dihedral angles necessary to make tight turns.

Loops are irregular segments that connect elements of secondary structure or are found at protein termini and usually range from 6 to 20 residues in length (**Figure 4.42**). Although loops lack a regular recognizable structure, they can actually be quite rigid because of multiple weak interactions between closely packed atoms. Loops are usually found on the surfaces of proteins and often contribute to the specificity of protein–protein interactions through the chemistry of the amino acid side chains.

Geometric Limitations of Secondary Structure We can now revisit the Ramachandran plot to see how the limitations of ϕ and ψ angles for the 20 different amino acids contribute to the secondary structure of proteins. As shown in **Figure 4.43**, α helices, β sheets (parallel and antiparallel), and β turns all fall within the allowable ranges of ϕ and ψ angles (compare to Figure 4.22) because these secondary structures minimize steric hindrance of the side chains. Indeed, when we determine

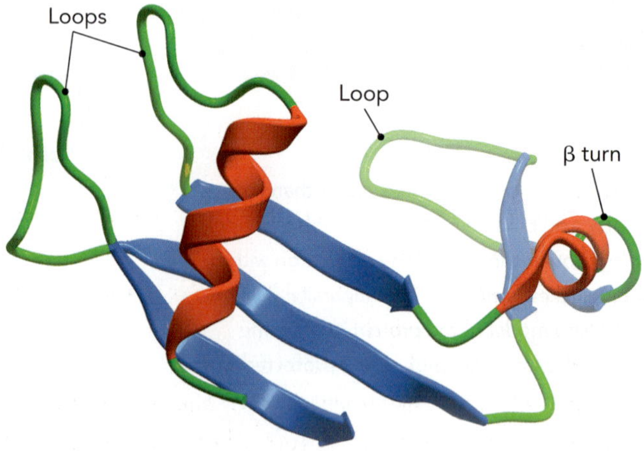

Figure 4.42 Loops are larger than turns but have the same function of connecting α helices and β strands. Loops vary in length and structure and are usually located on the surfaces of proteins, where amino acid side chains can interact with solvent molecules and other proteins. BASED ON PDB FILE 1FS3.

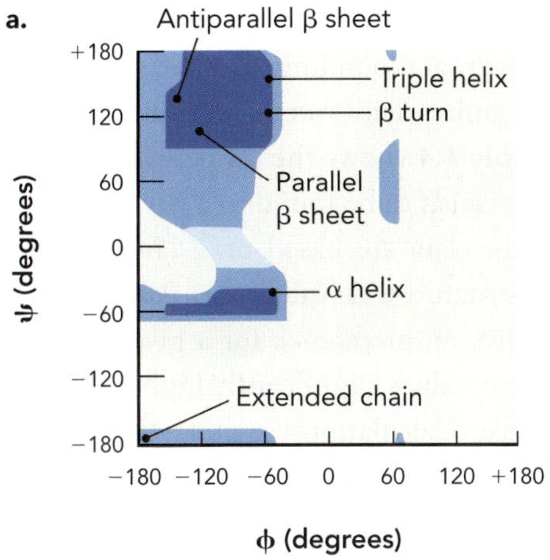

Figure 4.43 The common secondary structures in proteins contain amino acid residues with ϕ and ψ angles that minimize steric hindrance and are thus found in the allowed (low energy) regions of the Ramachandran plot. **a.** Location of various secondary structures within an idealized Ramachandran plot. Note that α helices, β sheets (parallel and antiparallel), and β turns (type I and type II) are found in the most favorable regions of the Ramachandran plot, as seen by the dark-shaded areas. **b.** The estrogen receptor ligand binding domain is mostly composed of α helices with connecting loops. The Ramachandran plot for this protein shows that most amino acid residues have ϕ and ψ combinations within the region of the plot corresponding to the α-helical conformation. BASED ON PDB FILE 1A52. **c.** The green fluorescent protein mostly consists of β strands with connecting loops. For this protein, most ϕ and ψ combinations are found within the region of the Ramachandran plot corresponding to β strands. BASED ON PDB FILE 1EMA.

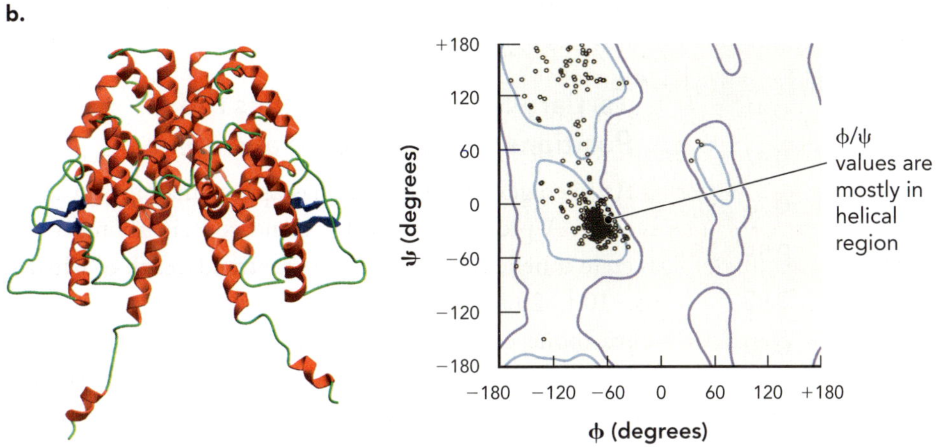

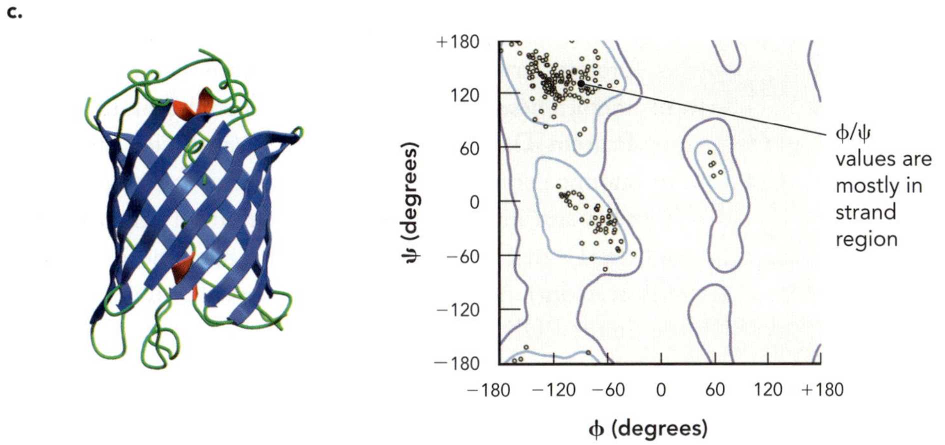

the ϕ and ψ angles of proteins, based on the solved molecular structure, most of the residues have ϕ and ψ angles within the most favorable regions of the Ramachandran plot (Figure 4.43). Major elements of secondary structures are common in proteins both because the ϕ and ψ angles for these structures minimize steric hindrance and because the hydrogen bonds between the CO and NH groups of the polypeptide backbone are maximized in these elements.

Not surprisingly, the size and chemistry of the different amino acid side chains influences the probability that they occur in any one of the three major types of secondary structures (α helix, β sheet, β turn). For example, proline is not often

Table 4.4 PROPENSITY OF AMINO ACIDS TO OCCUR IN THE THREE MAJOR TYPES OF SECONDARY STRUCTURE

Amino acid	α helix	β strand	β turn
Glu	**1.59**	0.52	**1.01**
Ala	**1.41**	0.72	0.78
Leu	**1.34**	1.22	0.57
Met	**1.30**	1.14	0.52
Gln	**1.27**	0.98	0.84
Lys	**1.23**	0.69	**1.07**
Arg	**1.21**	0.84	0.90
His	**1.05**	0.80	0.81
Val	0.90	**1.87**	0.41
Ile	1.09	**1.67**	0.47
Tyr	0.74	**1.45**	0.76
Cys	0.66	**1.40**	0.54
Trp	1.02	**1.35**	0.65
Phe	1.16	**1.33**	0.59
Thr	0.76	**1.17**	0.90
Gly	0.43	0.58	**1.77**
Asn	0.76	0.48	**1.34**
Pro	0.34	0.31	**1.32**
Ser	0.57	0.96	**1.22**
Asp	0.99	0.39	**1.24**

Note: Scores above 1.0 indicate a preference for that class of secondary structure. The three groupings of amino acids show those most often found in α helices, β strands, and β turns as indicated by boldface values.

BASED ON R. H. WILLIAMS ET AL. (1987). SECONDARY STRUCTURE PREDICTIONS AND MEDIUM RANGE INTERACTIONS. *BIOCHIMICA ET BIOPHYSICA ACTA, 26,* 200–204.

found in α helices or β sheets because the nitrogen cannot contribute to hydrogen bonding. Similarly, glycine is the most common amino acid found in β turns because of its small size. **Table 4.4** shows the propensities for any one of the 20 amino acids to be found in a particular secondary structure; the data are based on a large number of solved protein structures. A value of 1.0 means that the amino acid shows no preference for a given secondary structure, whereas values significantly higher than 1.0 indicate a preference. Calculations such as these, along with sequence similarities to known proteins for which structural information is available, are used in bioinformatics to predict the possible secondary structure of proteins with unknown structures.

Tertiary Structure Describes the Positions of All Atoms in a Protein

An average-length protein has approximately 300 amino acids. A protein of this size could contain about 10 discrete α helices and 10 discrete β strands, each of which is ~10 to 20 amino acids long. The rest of the polypeptide backbone consists of turns and loops. A description of the protein's tertiary structure includes the arrangement of secondary structural elements within a protein and the positions of the side chains.

To understand how protein structures determine function, biochemists have analyzed large numbers of protein structures, looking for common themes among them. These structures are available in the public-domain **Protein Data Bank (PDB)**, a Web-based archive of atomic coordinates of protein structures that are assembled and maintained by a consortium of scientists from the United States, Europe, and Japan (see http://www.rcsb.org). Analyses of large numbers of protein structures in the PDB have identified four general classes of protein structures: (1) predominantly α helix; (2) predominantly β sheet; (3) α/β combined, which are often intermixed α helices and β strands; and (4) α + β, consisting of α-helical regions adjacent to β sheets. Examples of these four classes of structures are shown in **Figure 4.44**.

The tertiary structure of large proteins can include more than one structural **domain**, which is defined as an independent folding module within the polypeptide chain. Protein structural domains can be identified by analyzing the three-dimensional structure of a protein or by biochemical methods that use protease enzymes to cleave proteins at exposed linkages, releasing individual domain fragments. Pyruvate kinase, a metabolic enzyme in the glycolytic pathway (Chapter 9), is an example of a multidomain protein in which four discrete structural domains can be identified (**Figure 4.45**).

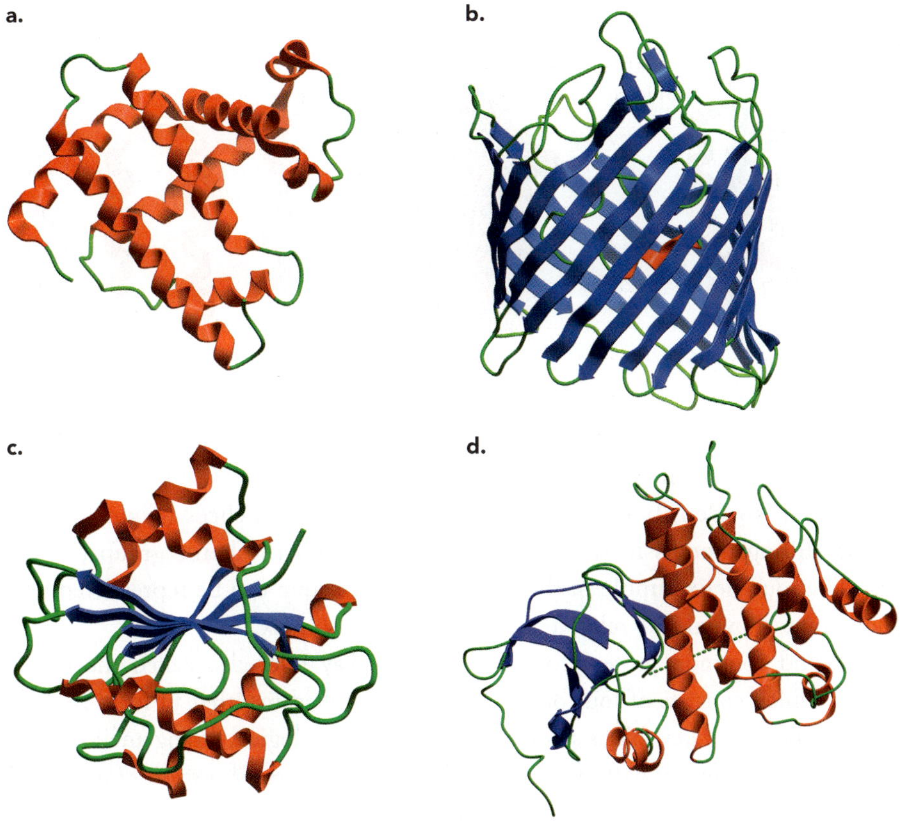

Figure 4.44 Four major classes of structures have been identified in proteins. **a.** John Kendrew's whale myoglobin protein was found to contain all α helices. BASED ON PDB FILE 1MBN. **b.** The OmpF bacterial porin protein consists of predominantly β-sheet structure. BASED ON PDB FILE 3NSG. **c.** Lipase contains an α/β tertiary structure. BASED ON PDB FILE 4TGL. **d.** This domain of a tyrosine kinase has a region consisting of β strands with another region of α helices, and is called a mixed α + β structure. BASED ON PDB FILE 3D7U.

Tertiary Structures Can Consist of Recognizable Protein Folds We can describe structural domains of proteins in terms of the topological arrangement of secondary structures. An analysis of thousands of protein structures contained within the PDB database reveals that most tertiary structures consist of recognizable protein folds. Within these folds often lie smaller defined structural units referred to as **motifs.** It is currently estimated that there are about a thousand recognizable protein folds within the ~20,000+ protein-coding genes in the human genome. To categorize these protein folds, a formal system of hierarchical organization of protein structures, called the **Structural Organization of Proteins (SCOP)**, was developed in 1995 by Cyrus Chothia and colleagues at the Medical Research Council labs in Cambridge, England. In the SCOP system, protein folds are organized into superfamilies, families, and domains. The SCOP system is a useful way to annotate newly solved protein structures in the PDB. In summary, tertiary protein structures can consist of discrete structural domains, which can be built from recognizable protein folds and motifs. Defining tertiary protein structures by their ensemble of protein folds provides a framework for describing structure and function within the context of protein evolution.

One of the most common protein folds contains four α helices linked together to form a **four-helix bundle**. The heme binding domain of cytochrome b—a redox protein in the mitochondrial electron transport system—is an example of a four-helix bundle that provides a hydrophobic pocket for the heme group. A comparison of the

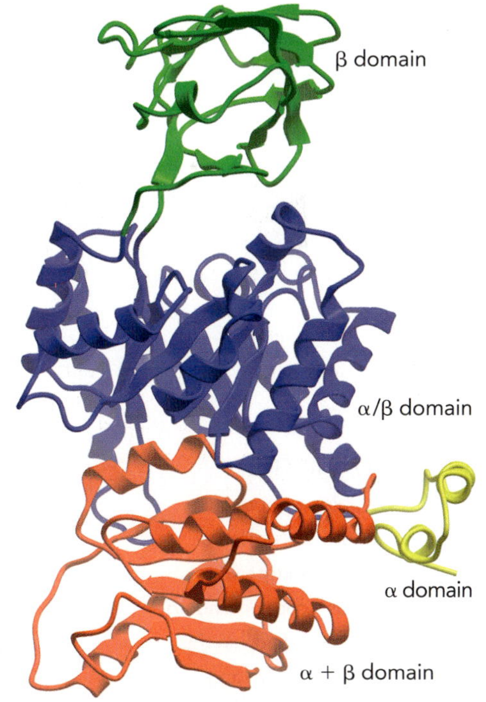

Figure 4.45 The metabolic enzyme pyruvate kinase is a large protein that consists of a single polypeptide chain with four discrete domains. These include a nine-stranded β domain (green), an α/β domain (blue), a short α-helical domain (yellow), and an α + β (red). BASED ON PDB FILE 1PKN.

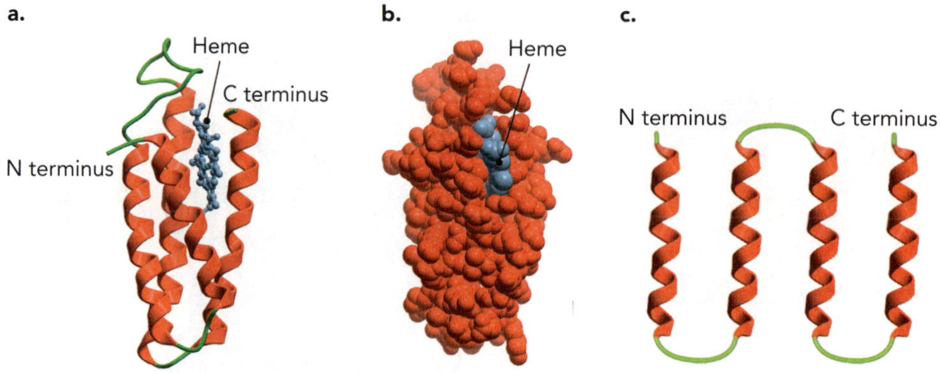

Figure 4.46 One of the most common protein folds is the four-helix bundle, seen here in the heme binding portion of a cytochrome b protein. **a.** Ribbon structure of the four-helix bundle in a cytochrome b protein showing the position of a prosthetic heme group. BASED ON PDB FILE 1LM3. **b.** Space-filling model of the same protein structure shown in panel **a**. **c.** Topology map of the four-helix bundle contained in the cytochrome b protein shown in panel **a**.

two molecular models of cytochrome b in **Figure 4.46** illustrates the value of using ribbon models to identify secondary structures, such as α helices (Figure 4.46a), and space-filling models to show atom packing around the heme group (Figure 4.46b). The four-helix bundle in the cytochrome b protein can be represented by a topology map showing the relative positions of each α helix.

A second common protein fold is the **Greek key fold**, also called the Greek key motif, which consists of four or more β strands linked together to form β-sheet structures. The Greek key fold is so named because its topological arrangement resembles the Greek key design pattern found in ancient pottery. As shown in **Figure 4.47**, the topology map of the Greek key fold in a bacterial copper-binding protein consists of four antiparallel β strands.

A more complex protein fold, called the **Rossmann fold**, is found in a number of proteins that bind nucleotides. Many types of proteins that function as dehydrogenase enzymes have a Rossmann fold that binds the coenzyme nicotinamide adenine dinucleotide (**Figure 4.48**). The Rossmann fold is named after Michael Rossmann, a protein biochemist at Purdue University, who first noticed this common topological arrangement in the nucleotide binding domains of numerous proteins. The topology of the Rossmann fold consists of two regions of alternating α helices and β strands that fold together into a compact structural domain.

Some protein folds are very large and serve as structural scaffolds for enzyme active sites or as binding surfaces to facilitate bimolecular interactions. The **TIM barrel fold** is an alternating α-helix/β-strand fold, also called an **α/β barrel**, which was first identified in the glycolytic enzyme triose phosphate isomerase (TIM for triose phosphate isomerase). This compact domain of ~200 amino acid residues contains eight or more radial β strands that form the interior of the protein. This twisted parallel β sheet

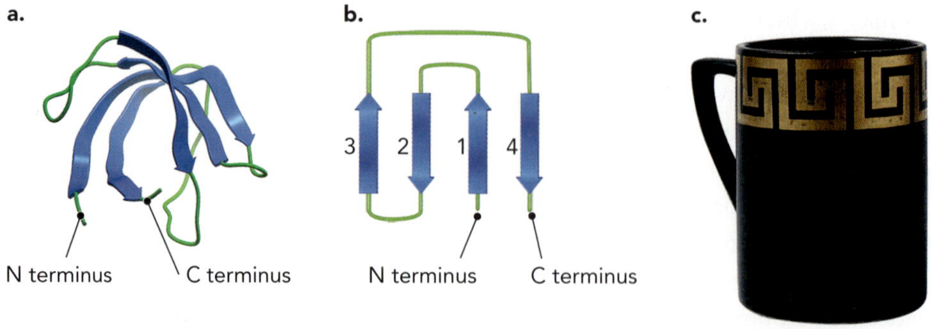

Figure 4.47 The Greek key fold resembles a popular Greek art design. **a.** Molecular structure of a Greek key fold in the bacterial CusF copper binding protein. Note that strand 1 in this particular protein is quite long, and it begins in the front of the fold and terminates in the back. BASED ON PDB FILE 1ZEQ. **b.** Topology map of the Greek key fold contained in the CusF protein in panel **a**. **c.** Example of the Greek key design used to decorate pottery. DE AGOSTINI/GETTY IMAGES.

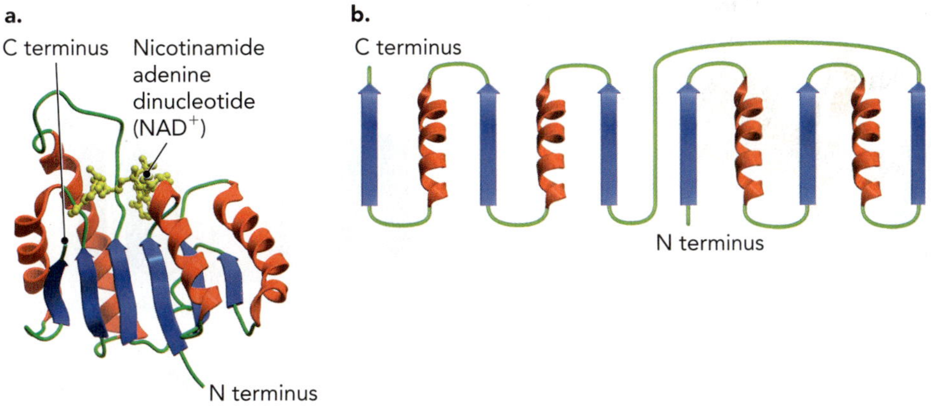

Figure 4.48 The Rossmann fold contains an arrangement of α helices and β strands that is found in many nucleotide binding proteins. **a.** Molecular structure of a protein containing a Rossmann fold, bacterial lactate dehydrogenase, with bound nicotinamide adenine dinucleotide. BASED ON PDB FILE 2FNZ. **b.** Topology of the bacterial lactate dehydrogenase enzyme, which forms a Rossmann fold.

is surrounded by eight α helices that face outward toward the aqueous environment (**Figure 4.49**). In the case of the enzyme triose phosphate isomerase, the substrate for the reaction binds to a catalytic active site at the top of the TIM barrel. The TIM barrel structure is a common protein fold that has been found in a large number of metabolic enzymes. More than 400 proteins in the PDB have been classified by the SCOP system as having a TIM barrel fold.

The **FERM domain fold** is another example of a large protein fold. The FERM domain fold was named for the four different cytoskeletal proteins that were shown to contain this large, ~300-amino-acid fold. The four proteins are Band 4.1, ezrin, radixin, and moesin, which together spell out the acronym FERM (the "F" comes from the number four in 4.1). The FERM domain fold consists of three distinct structural components (F_A, F_B, F_C), which could also be thought of as subdomains (**Figure 4.50**). The F_A and F_C structural components are mostly β sheet, whereas the F_B structural component contains only α helices. The FERM domain fold binds to membrane-associated proteins or glycolipids and serves to anchor proteins to the cytoplasmic face of the plasma membrane. As evidence that different amino acid sequences can specify the same secondary structures (see Table 4.4), the amino acid sequences of the ezrin and Band 4.1 proteins are less than 30% identical, yet both proteins have very similar

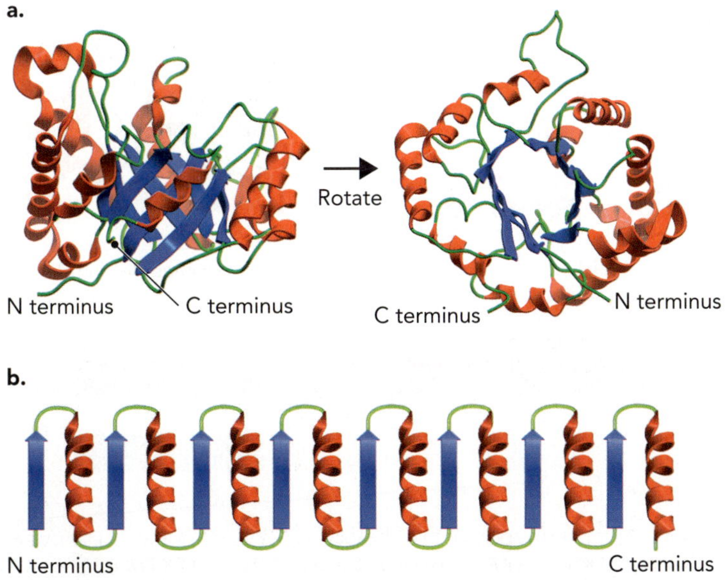

Figure 4.49 The TIM barrel fold is an alternating α-helix/β-strand structural motif that was first identified in the metabolic enzyme triose phosphate isomerase. **a.** Molecular structures of the TIM barrel in the enzyme triose phosphate isomerase. BASED ON PDB FILE 1WY1. **b.** Topology map of the TIM barrel fold contained in the triose phosphate isomerase protein in panel **a**.

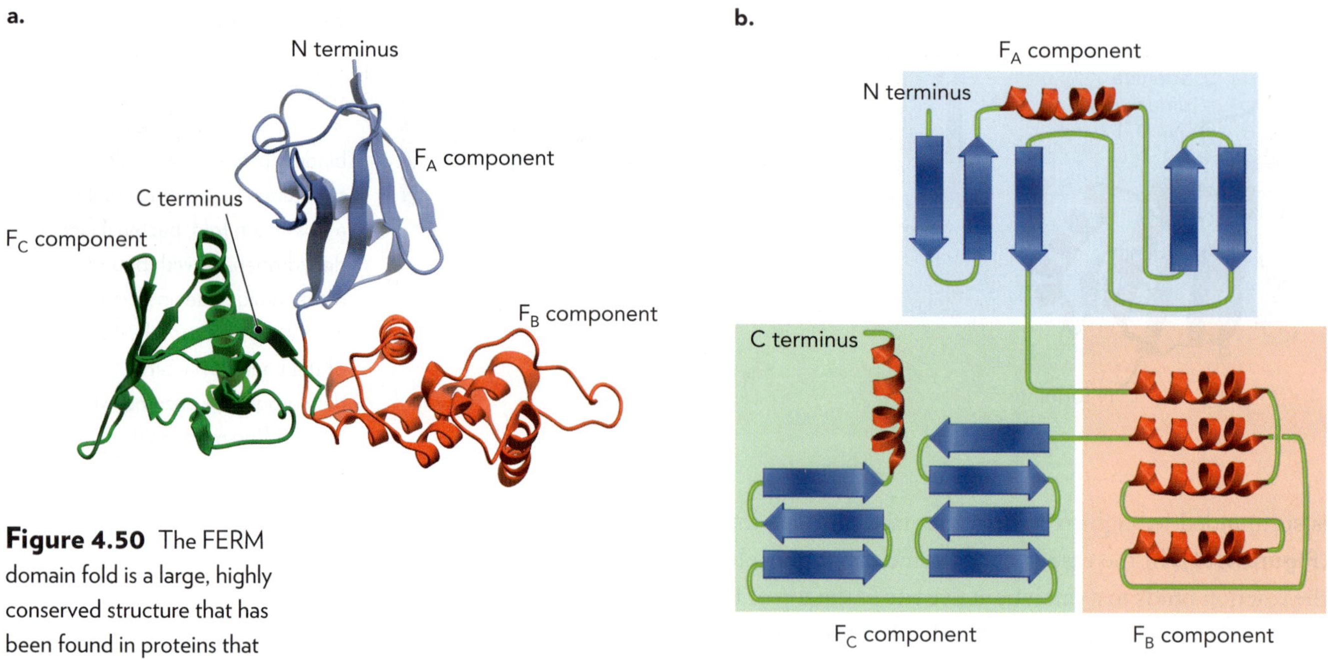

Figure 4.50 The FERM domain fold is a large, highly conserved structure that has been found in proteins that interact with components of the plasma membrane. **a.** Molecular structure of a FERM domain fold in the cytoskeletal protein radixin showing three distinct structural components: F_A, F_B, and F_C. BASED ON PDB FILE 1GC7. **b.** Topology map of the FERM domain fold in panel **a**.

FERM domain folds (**Figure 4.51**). The conserved structure of these two FERM domain folds demonstrate how missense nucleotide mutations can accumulate in the gene-coding sequences without altering the structure and function of the protein see Figure 4.24).

Proteins with a FERM domain fold also contain one or more additional structural components that reflect the evolutionary history of the corresponding gene. Radixin is a FERM domain fold protein that functions as a molecular bridge linking membrane-associated proteins such as the intercellular adhesion molecule 1 receptor or the membrane glycolipid phosphatidylinositol-4,5-bisphosphate to the actin cytoskeleton of cells (**Figure 4.52a**). The FERM domain fold in radixin is located

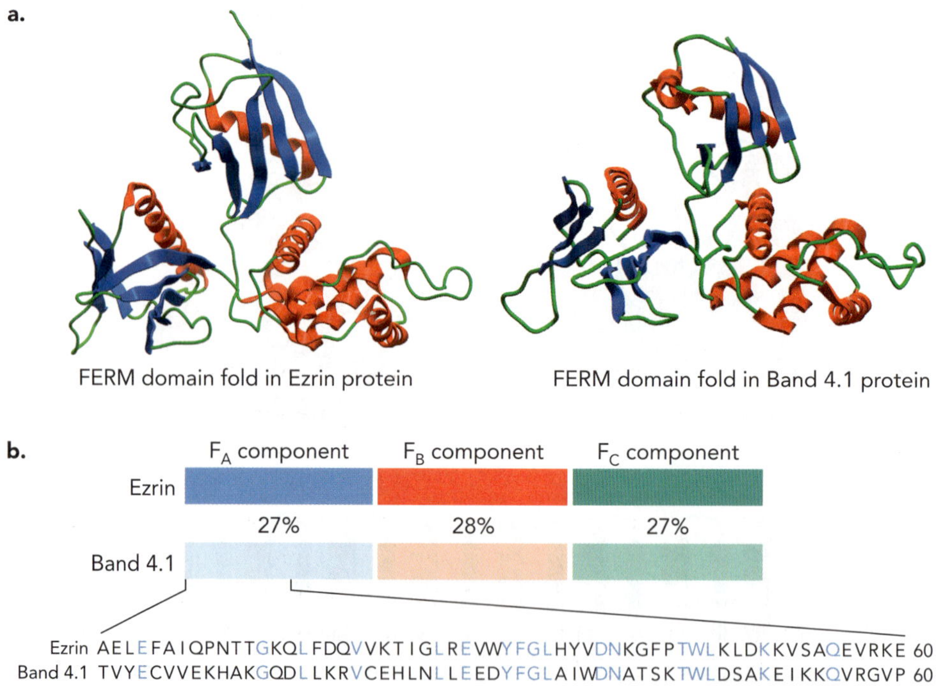

Figure 4.51 Dissimilar amino acid sequences can encode similar FERM domain folds. **a.** FERM domain folds in the ezrin and Band 4.1 proteins. BASED ON PDB FILES 1NI2 (EZRIN) AND 1GG3 (BAND 4.1). **b.** Sequence identity comparisons between the ezrin and Band 4.1 proteins shown in panel **a** reveal less than 30% amino acid sequence identity between the two proteins in this region. A portion of the amino acid sequence in the F_A structural component of the human ezrin and Band 4.1 proteins is shown. Identical amino acids between the two proteins are highlighted in blue (17 of 60 amino acids are identical in this region).

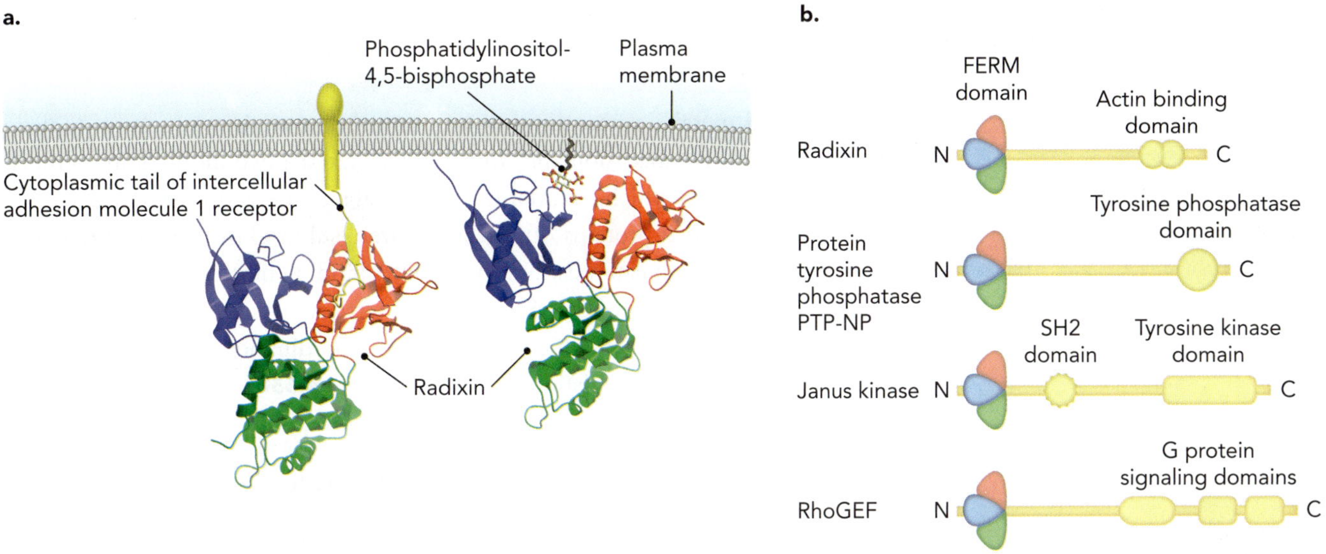

Figure 4.52 The evolutionarily conserved FERM domain fold functions as a binding module for multifunctional proteins associated with the cytoplasmic face of the plasma membrane. **a.** The FERM domain fold of the radixin protein can bind to either the cytoplasmic tail of the intercellular adhesion molecule 1 receptor or a carbohydrate group on membrane lipids called phosphatidylinositol-4,5-bisphosphate. Note that different regions of the radixin F_A, F_B, and F_C structural components are involved in these distinct binding functions. BASED ON PDB FILES 1J19 (ICAM RECEPTOR) AND 1GC6 (PHOSPHATIDYLINOSITOL-4,5-B ISPHOSPHATE). **b.** Most FERM domain fold proteins contain a second region in the C-terminal portion of the protein that functions in cytoskeletal organization or cell signaling. N = N terminus; C = C terminus.

at the N terminus, whereas the C terminus of the radixin protein is an actin binding domain. Other FERM domain fold proteins also bind to membrane-associated proteins or glycolipids through the FERM fold; however, the C-terminal regions of many of these proteins encode distinct functions (**Figure 4.52b**). For example, the protein tyrosine phosphatase PTP-NP contains a FERM domain fold as well as a C-terminal tyrosine phosphatase domain (an enzyme that removes phosphates from tyrosine residues in proteins). In contrast, the FERM domain fold protein Janus kinase has both a tyrosine kinase domain (adds phosphates to tyrosine residues in proteins) and an SH2 domain, which has a high affinity for phosphotyrosine residues. Similarly, RhoGEF has an N-terminal FERM domain fold but a series of C-terminal segments that function together to activate G protein signaling in cells, as described in Chapter 8.

Tertiary Structures Can Be Stabilized by Disulfide Bonds and Metal Ions The majority of tertiary structures found in proteins are stabilized by numerous weak interactions, but some tertiary structures are stabilized by covalent disulfide bonds or by coordinated metal ions. Disulfide bonds are formed when two nearby cysteine residues are oxidized (see Figure 4.11), resulting in the formation of a disulfide bridge. The Brazilian scorpion TS1 neurotoxin is a small polypeptide of 61 amino acids that contains four disulfide bridges (**Figure 4.53a**). Two of these disulfide linkages connect the single α helix to one of the β strands, and the other two linkages hold the loop portions of the protein in place. Disulfide bridges are common in small, secreted

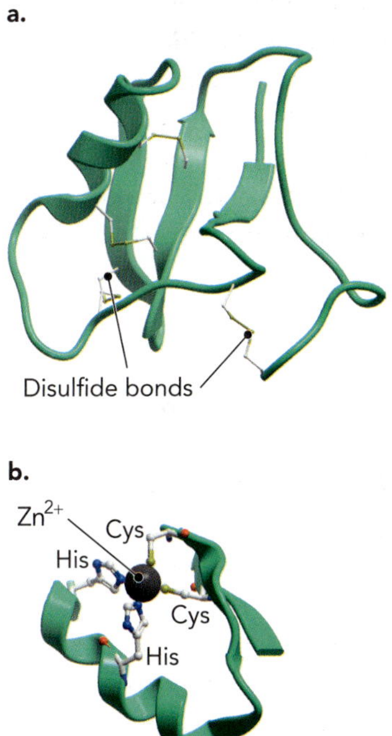

Figure 4.53 Tertiary structures can be stabilized by disulfide bonds between cysteine residues or by metal ions. **a.** The scorpion toxin TS1 contains four disulfide bonds that stabilize this secreted protein. BASED ON PDB FILE 1B7D. **b.** Molecular structure of a zinc finger domain, showing metal ion interactions with two cysteine and two histidine residues. BASED ON PDB FILE 1ZNF.

proteins found in oxidizing environments, such as toxins and growth factors, which require stabilized tertiary structures to retain functional activity in dilute aqueous environments.

Coordinated metal ions are another way to stabilize tertiary structures that do not readily fold into compact domains. Iron and zinc are the two most common metal ions in proteins, both of which commonly form metal ion interactions with cysteine or histidine residues. Redox proteins often contain iron–sulfur clusters that function to exchange electrons during catalysis, and some DNA binding proteins contain zinc as a coordinating metal ion. One of the zinc-mediated tertiary structures in DNA binding proteins is called a **zinc finger**, which contains four coordination sites to the protein through two cysteine and two histidine residues (**Figure 4.53b**).

Quaternary Structure of Multi-subunit Protein Complexes

Many proteins exist in cells as part of larger protein complexes, which can consist of multiple identical polypeptides or a collection of distinct polypeptides encoded by different genes (see Figure 4.4). The number and overall arrangement of the polypeptide chains within these complexes is the protein's quaternary structure. The simplest type of quaternary structure is an oligomer called a **homodimer**, which contains two identical protein subunits. If the two subunits are derived from distinct polypeptides (different gene products), then the complex is called a **heterodimer**. Larger homoprotein and heteroprotein complexes containing three (trimer), four (tetramer), or six (hexamer) subunits are also common.

Quaternary structures are stabilized by multiple protein–protein interactions occurring at the interface of protein subunits. These interaction surfaces are physically and chemically complementary to one another, with numerous noncovalent bonds stabilizing the complex. In some cases, quaternary structures are held together by disulfide bridges formed between cysteine residues located on different protein subunits.

Quaternary structures provide increased functionality to proteins in at least three ways.

1. Multiple subunits can provide structural properties not present in individual subunits. The best example of this is fibrous proteins, such as keratin, which consists of homodimers of long α-helical subunits that function as a single structural element to form hair, nails, hooves, and horns.
2. Multiple protein subunits provide a mechanism for regulation of protein function through conformational changes that alter the protein subunit interfaces. This is a common occurrence in signaling proteins, which respond to external stimuli such as ligand binding to receptor proteins or to phosphorylation of serine, threonine, or tyrosine residues on one of the subunits.
3. Protein complexes can significantly increase the efficiency of biochemical processes by bringing linked functional components into close proximity. For example, multi-subunit protein complexes make it possible to synthesize long hydrocarbon chains in an assembly-line fashion (fatty acid synthase complex), convert the potential energy of a membrane proton gradient into chemical energy (ATP synthase complex), and degrade proteins that are marked for disposal by covalent modification (proteasome complex).

Multi-subunit Complexes: Fibrous Proteins Fibrous proteins are multi-subunit complexes that constitute some of the most abundant proteins in nature, including keratin, silk fiber, and collagen. As shown in **Figure 4.54**, keratin is a homodimer of two helical polypeptides that are wrapped around each other to form what is called a **coiled coil**. The keratin helix itself is right-handed; however, the coiled coil dimer is formed from the helices wrapping around each other in a left-handed direction. The ~300-amino-acid core polypeptide of each keratin subunit consists of a 7-amino-acid repeating unit, in which the first, and usually the fourth, amino acid is hydrophobic. This repeat provides a hydrophobic strip that rotates anticlockwise around the surface of the helix. Two helices have to twist around each other in order to line up those two hydrophobic strips. Thus the helices pack together with an extensive hydrophobic interface. The keratin helix contains ~3.5 amino acids per turn and is slightly more extended than an α helix, which has 3.6 amino acids per turn (see Figure 4.29).

The strength of keratin comes from covalent cross-linking between coiled coil dimers as a result of disulfide bridge formation between cysteine residues (Figure 4.54b). Many different types of keratins are found in nature, which differ primarily in the number of cysteine residues. Hair, for example, is flexible and has fewer disulfide bridges between coiled coil dimers than fingernails, which are rigid and more durable because of extensive disulfide cross-linking between fibers. One of the strongest keratin structures known is rhinoceros horn, which is made up of keratin subunits containing 18% cysteine.

The disulfide bonds in keratin can be broken with reducing agents, such as ammonium thioglycolate. This compound is used in combination with an oxidizing agent in hair products to introduce artificial curls through the breaking and re-forming of

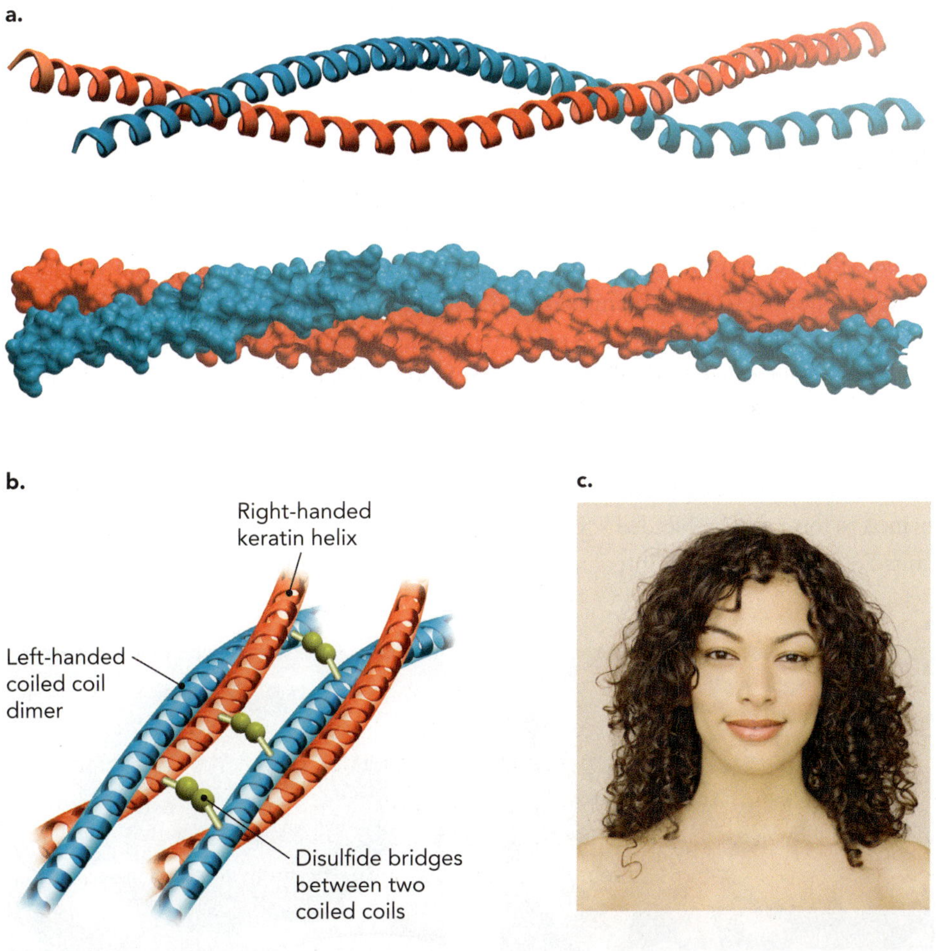

Figure 4.54 Keratin is a fibrous coiled coil protein consisting of two helical polypeptides covalently linked by disulfide bridges. **a.** Molecular structure of a coiled coil region in the myosin protein. The two helices pack together tightly, as can be seen in the space-filling view. BASED ON PDB FILE 2FXM. **b.** Two coiled coil dimers of keratin are covalently linked by disulfide bridges. **c.** Products used to make hair curly contain reducing agents that break the disulfide bonds between keratin helices and then, under appropriate conditions, induce new disulfide bonds to hold the curls in place. ERIC BEAN/GETTY IMAGES.

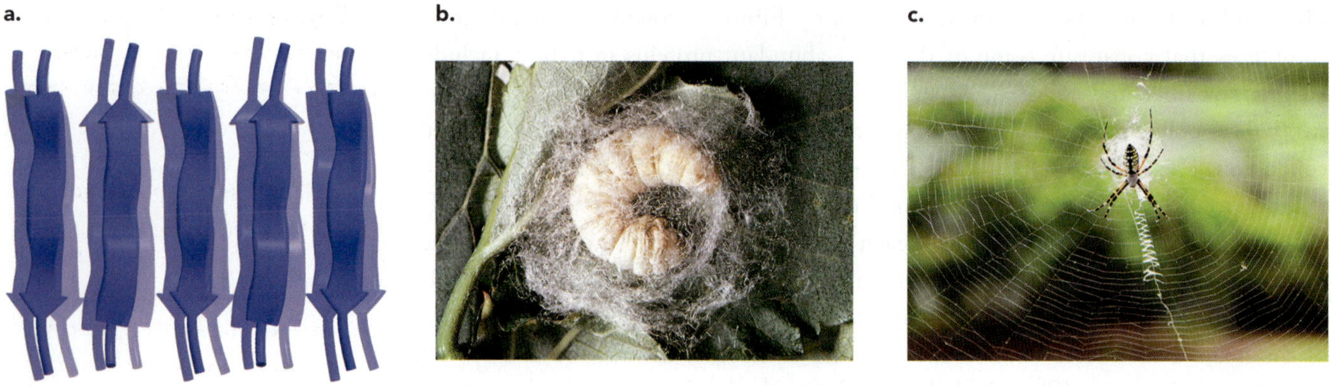

Figure 4.55 Silk is a fibrous protein consisting of three subunits, one of which consists almost entirely of β sheets, which collectively give silk its strength. **a.** Molecular model of the overlapping β sheets in the heavy-chain subunit of silkworm silk. BASED ON PDB FILE 1SLK. **b.** The silk used in fabric comes from the cocoon of the Japanese silkworm *Bombyx mori*. © STEPHEN DALTON/MINDEN PICTURES/CORBIS. **c.** Spider silk glands produce a silk thread that is stronger than a steel wire of the same diameter. TETRA IMAGES/GETTY IMAGES.

disulfide bonds (Figure 4.54c). The chemically induced curls are eventually diluted out by new hair that grows from the scalp and takes on the shape of keratin fibers specified by the amino acid sequence of the corresponding keratin genes.

Silk fibroin is another abundant fibrous protein that consists of multiple protein subunits with a repeating secondary structure, in this case β sheets. Silkworm fibroin from the insect *Bombyx mori* is a heterotrimer of three proteins: one is a 350-kDa **fibroin heavy chain** containing large numbers of β sheets, and the other two are smaller proteins of ~30 kDa each. One of these protein subunits is the **fibroin light chain**, which is covalently linked to the heavy chain by disulfide bonds. The other silk fibroin protein subunit is the P25 glycoprotein, which is associated with the heavy chain–light chain heterodimer through noncovalent interactions. The fibroin heavy-chain polypeptide contains a large number of alanine and glycine residues. The small sizes of these side chains allow the β sheets to layer very closely with each other and thus allow tight packing of the protein subunits.

The amino acid sequence of the fibroin heavy chain reveals stretches of polyalanine residues (Ala-Ala-Ala), followed by hundreds of repeats of the tripeptide Gly-Gly-X, in which X is often Tyr, Gln, or Ser. This repeating polypeptide sequence facilitates the formation of β strands, which fold into β sheets that can organize into large arrays consisting of many stacked β sheets (**Figure 4.55**). Silk is one of the strongest biopolymers known and is extraordinary in its combined properties of durability and flexibility. Spider silk, for example, is stronger than metal wires of equal diameter, and yet it can bend (but not stretch) to accommodate the spinning of an intricate spiderweb.

Collagen is yet another type of protein complex that can form a biological fiber. Collagen is primarily found in soft tissue that holds bones together and is the major constituent of tendons and cartilage (**Figure 4.56**). Collagen is the protein component in the cornea of the eye, which is the clear tissue on the outer front surface of the eyeball. Similar to keratin, collagen is a protein complex consisting of long helical subunits that are intertwined to form a right-handed triple-helix fiber. In the case of collagen, three helical subunits are held together by interstrand cross-links involving hydroxylated lysine residues.

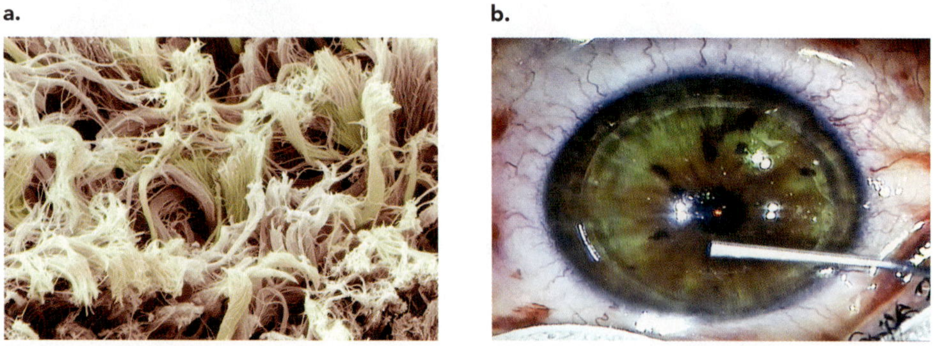

Figure 4.56 Collagen is the major protein in connective tissue and in the cornea of the eye. **a.** Tendons in soft connective tissue consist of large numbers of cross-linked collagen strands, which attach skeletal muscle to bone. STEVE GSCHMEISSNER/GETTY IMAGES. **b.** The clear cornea of the eye is made up almost entirely of collagen fibers. Cornea transplant surgery replaces the damaged cornea of a patient with a grafted cornea from a donor. UIG VIA GETTY IMAGES.

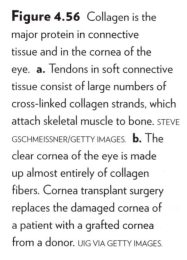

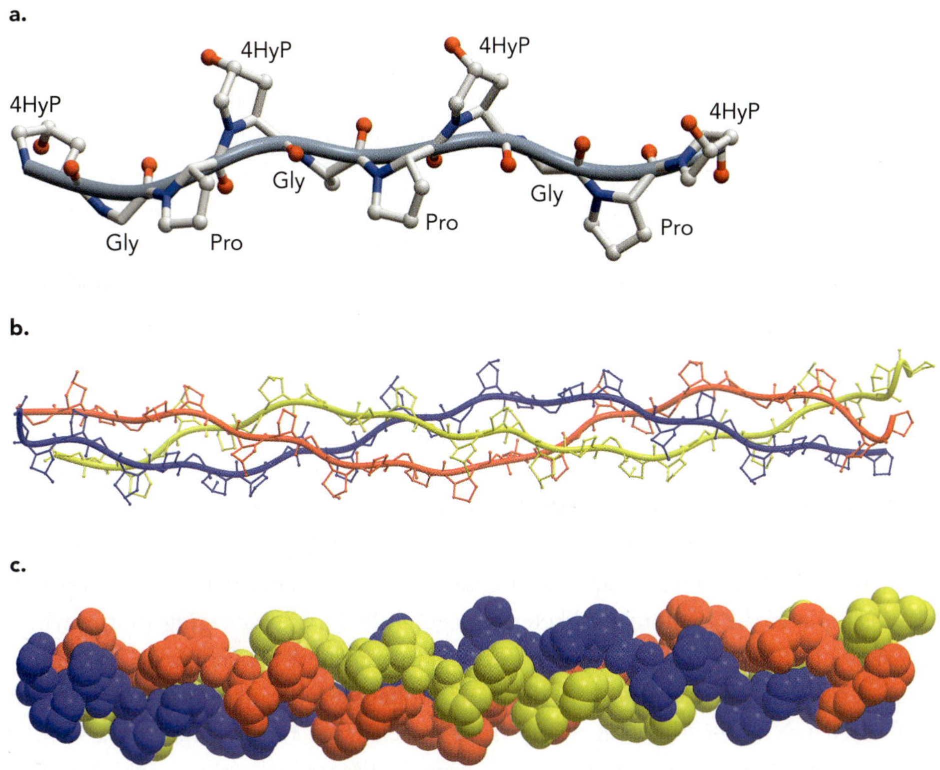

Figure 4.57 Collagen consists of three intertwined left-handed helices. **a.** The molecular structure of a single collagen-like polypeptide chain shows a repeating 4HyP-Gly-Pro tripeptide, giving rise to a left-handed helix with three amino acids per turn. **b.** Molecular structure of the triple helix shown in a wire and ribbon model. **c.** In the assembled protein, the left-handed helical structure of each protein subunit is stabilized by interstrand hydrophobic interactions, resulting in a tightly packed right-handed triple helix. BASED ON PDB FILE 1CAG.

Importantly, the collagen helix is not an α-helix structure, but instead each polypeptide is a stretched-out left-handed helix with three residues per turn (**Figure 4.57**). One of the most common repeating units in collagen is a tripeptide with the sequence Gly-X-Y, in which X is usually Pro and Y is often 4-hydroxyproline (4HyP). In the triple helix , hydrogen bonds form between the glycine NH and proline (or 4-hydroxyproline) CO, however the Pro and 4HyP residues stabilize the triple helix structure through hydrophobic interactions. This is a distinct method from that of an α helix, in which the core of the helix is stabilized by hydrogen bonding between atoms in the peptide bond (see Figure 4.29).

The 4HyP residues in collagen are derived from hydroxylation of proline by the enzyme prolyl hydroxylase, which uses ascorbate (vitamin C) as a cofactor (**Figure 4.58**). This is an important biochemical reaction because 4HyP is required

a.

b.

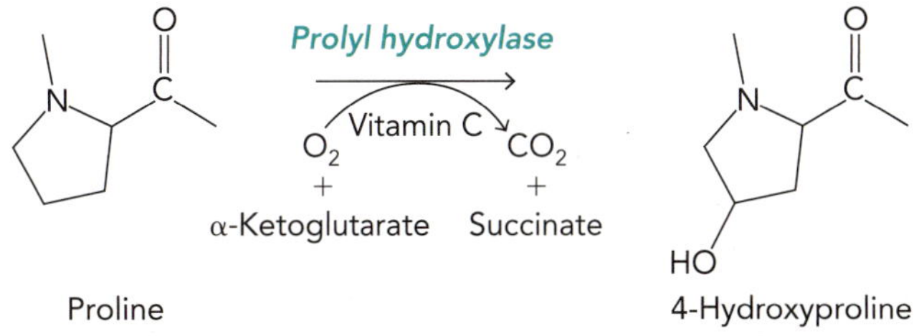

Figure 4.58 Ascorbate, also known as vitamin C, is a required cofactor for the enzymatic reaction converting proline to hydroxyproline. **a.** Citrus fruits, such as lemons and limes, and potatoes are all rich in vitamin C. LIMES: ZAZA STUDIO/SHUTTERSTOCK; POTATOES: ALEKSANDR BRYLIAEV/SHUTTERSTOCK. **b.** Prolyl hydroxylase is the enzyme that adds a hydroxyl group to proline to form hydroxyproline in a reaction using vitamin C. Hydroxyproline modifications provide stability to the collagen helix, which is required to prevent the symptoms of scurvy.

to form strong noncovalent interactions within the collagen helix. Diets deficient in vitamin C can lead to a condition called **scurvy**, in which the connective tissue in skin and blood vessels breaks down due to structural defects in collagen. British sailors in the 17th and 18th centuries were especially susceptible to scurvy because of the long voyages they took under conditions in which fresh fruit and vegetables were unavailable. In one of the first examples of a controlled clinical trial, Dr. James Lind, a surgeon in the British Royal Navy, found in 1747 that sailors with citrus fruits in their diet recovered from scurvy faster than those on any other diet he tested. This treatment for scurvy was soon adopted, and British sailors became known as *limeys* because limes were frequently used as a source of vitamin C. It was later found that potatoes are also a good source of vitamin C, and as potatoes can easily be stored for long periods of time, they became a staple in the British diet.

Multi-subunit Complexes: Globular Proteins In addition to these fibrous proteins with extended structures, a variety of globular proteins with more compact, somewhat spherical shapes also consist of multi-subunit protein complexes. An example is the homomeric bacteriophage ϕ 29 DNA packaging motor, which consists of 12 identical α subunits that together form a concentric α_{12} homododecamer complex (**Figure 4.59**). This bacteriophage packaging motor guides newly replicated DNA into the viral capsid prior to cell lysis. The asymmetric α subunit contains two complementary quaternary protein interaction surfaces, which permit adjacent subunits to form interlocking junctions. Because of the shape of these two interaction surfaces, the homododecamer closes up on itself to form a donut-shaped macromolecule with a hole in the middle for the DNA molecule to transit.

Two examples of heteromeric quaternary structures in globular protein complexes are the heterotrimeric G protein complex involved in cell signaling and the heterotetrameric hemoglobin protein complex required for O_2 transport in animals (**Figure 4.60**). The G protein complex contains one subunit each of three polypeptides, designated α, β, and γ, which form a heterotrimeric ($\alpha_1\beta_1\gamma_1$) signaling complex that can interact with membrane receptor proteins. As described in Chapter 8, conformational changes in the G protein–coupled receptor cause the G protein complex to dissociate into two signaling components: the G_α subunit and the $G_{\beta\gamma}$ heterodimer. The weak noncovalent interactions at the G_α and $G_{\beta\gamma}$ interface within the complex are critical to the on/off signaling function of G proteins.

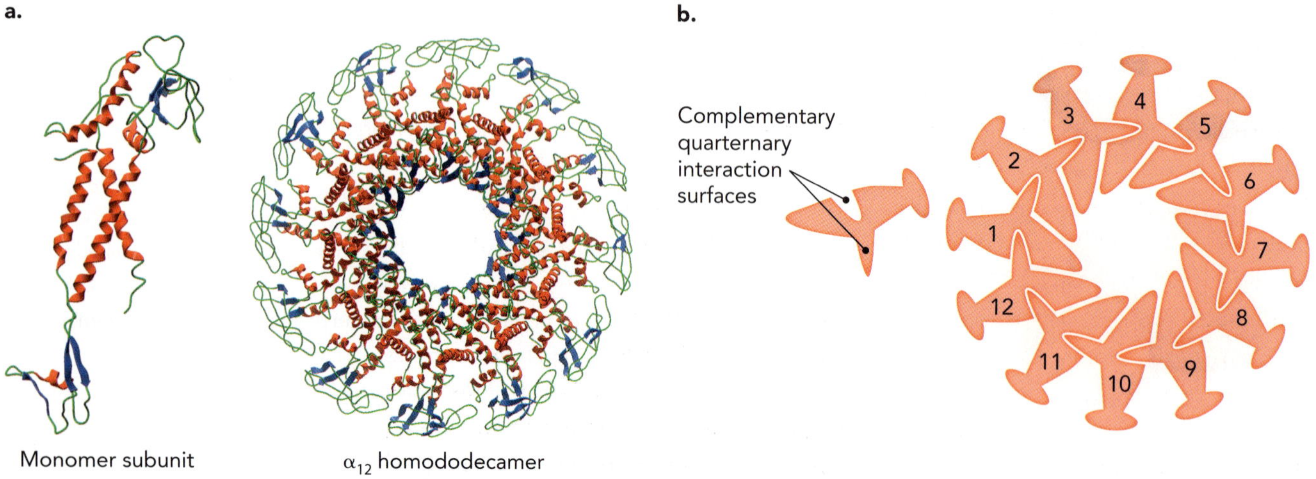

Figure 4.59 The bacteriophage ϕ 29 DNA packaging motor is a protein complex containing 12 identical α subunits, each containing two protein interaction sites. **a.** Molecular structures of a single protein subunit in the bacteriophage ϕ 29 DNA packaging motor and the dodecamer protein complex. BASED ON PDB FILE 1FOU. **b.** Schematic drawing of the packaging motor protein subunits, highlighting the two complementary quaternary protein interaction sites that facilitate formation of the dodecamer protein complex.

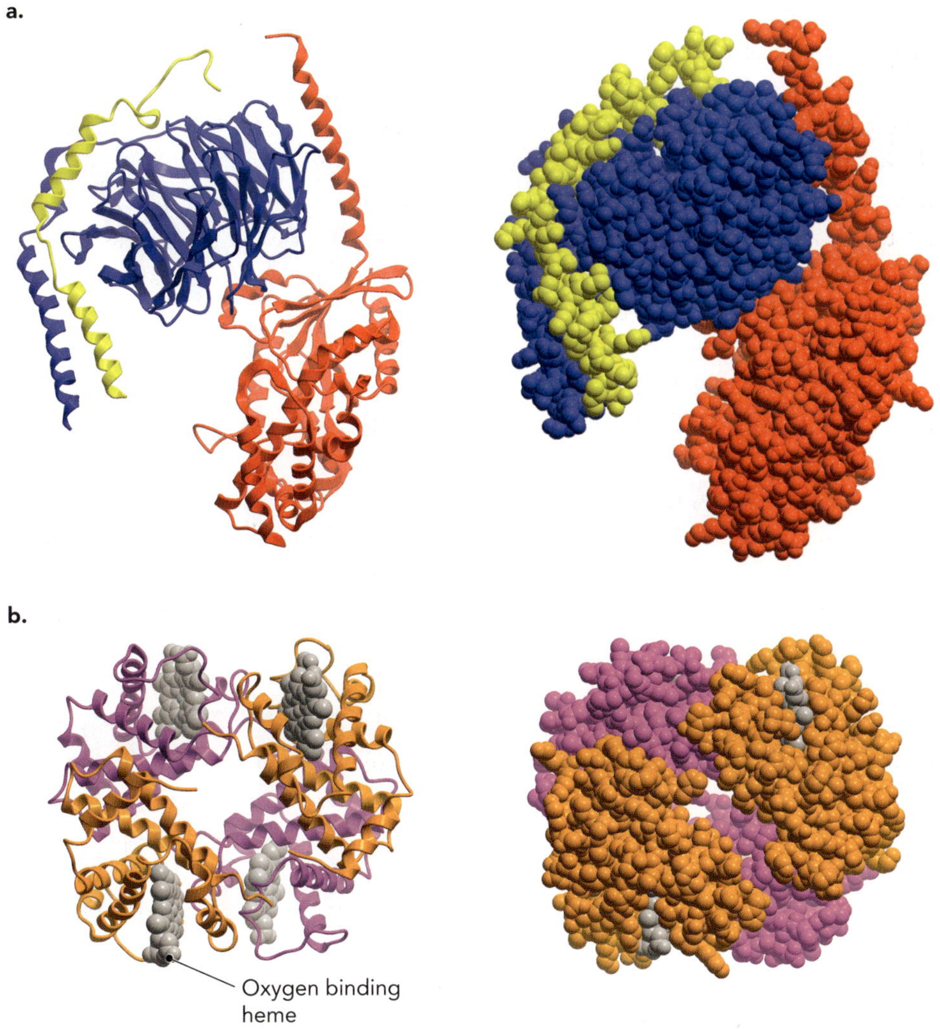

Figure 4.60 Heteromeric quaternary protein complexes contain more than one type of protein subunit. **a.** The G protein signaling complex is an example of a heterotrimeric quaternary complex containing one subunit each of three distinct polypeptides ($\alpha_1\beta_1\gamma_1$). BASED ON PDB FILE 1GG2. **b.** Hemoglobin is a heterotetrameric quaternary complex containing two copies each of an α and a β protein subunit, giving rise to an $\alpha_2\beta_2$ heterotetramer. Each subunit contains a heme group (gray) where oxygen binds. BASED ON PDB FILE 1A3N.

Unlike the G protein complex, which contains three different protein subunits, the hemoglobin complex is formed from two identical heterodimers, each consisting of one α subunit and one β subunit, to form an $\alpha_2\beta_2$ heterotetramer. The α and β subunits are very similar in structure, and each has an iron-containing heme group that binds oxygen. As described in Chapter 6, all four hemoglobin subunits undergo conformational changes as a result of O_2 binding to the iron center of the heme group.

Immunoglobulin proteins are another example of a quaternary protein complex that consists of two copies each of related protein subunits. As shown in **Figure 4.61**, immunoglobulin proteins, functionally known as **antibodies**, contain two heavy-chain subunits (H) and two light-chain subunits (L) to form an H_2L_2 heterotetramer. The biological function of antibodies is to bind foreign molecules (non-self), called **antigens**, which are then destroyed by other immune cells that recognize antibody–antigen complexes and engulf them. Typically, antigens are bacteria, viruses, or pollen molecules that antibodies recognize as foreign. The entire ~150-kDa H_2L_2 protein complex is a Y-shaped macromolecule that contains two identical antigen binding sites located near the N termini of the heavy- and light-chain subunits. Because of the many disulfide bonds stabilizing the tertiary and quaternary structures of immunoglobulin protein complexes, mild

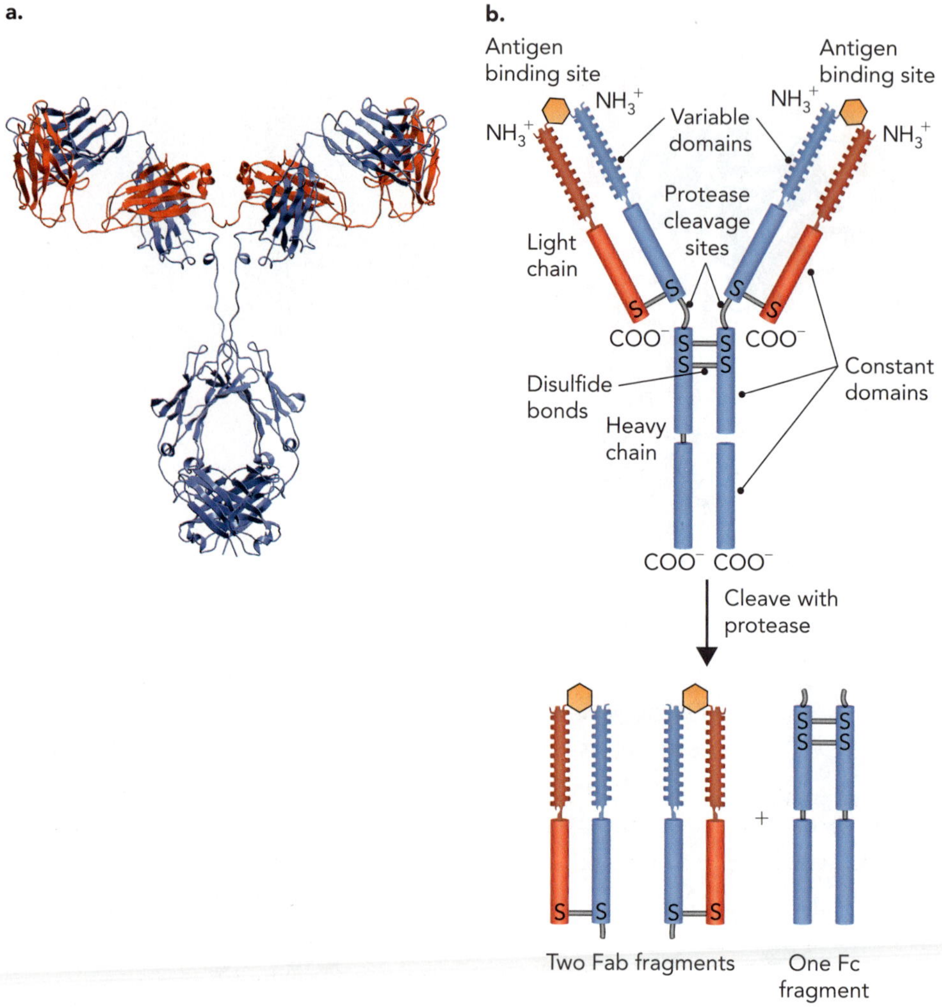

Figure 4.61 Immunoglobulin protein complexes contain two heavy (H) chains and two light (L) chains to form a H_2L_2 complex. **a.** Molecular structure of a human immunoglobulin G tetrameric protein complex. BASED ON PDB FILE 2IG2. **b.** Schematic representation of an immunoglobulin G protein complex containing two identical heavy-chain protein subunits and two identical light-chain protein subunits. The N-terminal regions of the heavy- and light-chain polypeptides are variable in sequence and form an antigen binding site. Mild protease treatment of immunoglobulin molecules produces two Fab fragments, containing the antigen binding sites, and one Fc fragment, containing the constant domains of the heavy-chain protein subunits.

treatment of antibody molecules with a protease, such as papain, releases two types of immunoglobulin protein subfragments, which retain their structure and function (Figure 4.61). One type is called the antigen binding fragment (Fab), which contains the entire light chain plus the N-terminal half of the heavy chain. The second type is called the crystallization fragment (Fc), which contains the C-terminal half of the two heavy chains. The functional descriptions of Fab and Fc fragments are derived from early biochemical studies in which it was found that antigen specificity resides in the Fab fragments and that the Fc fragment readily formed crystals that could be studied by X-ray crystallography.

The two light-chain polypeptides of immunoglobulin complexes are ~25 kDa in mass and contain two distinct protein domains. One is called the **variable domain,** which binds antigen molecules, and the other is the **constant domain.** The variable and constant domains of immunoglobulin molecules all have the same basic protein fold, called an **immunoglobulin fold** (Ig fold). The immunoglobulin fold is an all-β structure consisting of two β sheets forming a two-layered sandwich with a Greek key fold (**Figure 4.62**). In addition to the intermolecular disulfide bonds that covalently link the heavy and light chains, most immunoglobulin folds also contain intramolecular disulfide bonds between the two β sheets.

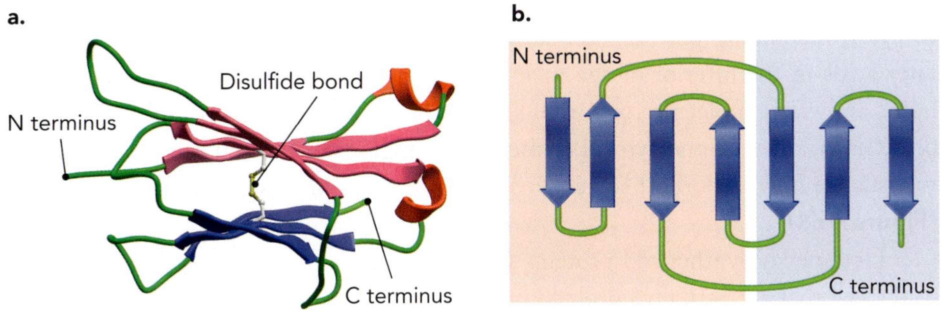

Figure 4.62 The immunoglobulin fold contains a two-layered sandwich with a Greek key fold. **a.** Molecular structure of an immunoglobulin fold in the light-chain protein subunit of an immunoglobulin molecule. In this case, a disulfide bond connects two β sheets. BASED ON PDB FILE 1IGT. **b.** Topology map of the immunoglobulin fold protein structure shown in panel **a**.

concept integration 4.2

What are protein folds and how do they relate to protein function?

The three-dimensional structure of proteins can be described in hierarchical terms beginning with the primary structure, which is the amino acid sequence. The secondary structure of a protein describes the regular, repetitive structures that form local arrangements of the backbone, such as α helices, β strands, and β turns. Tertiary structures describe the location of every atom in the protein. Finally, quaternary structures are formed by subunit interactions between separate polypeptide chains. Folds are a description of the path the polypeptide follows in three-dimensional space. Protein folds can be evolutionarily conserved structures that function as molecular modules within proteins. For example, the immunoglobulin folds found in antibody proteins stabilize the overall structure of these large protein complexes and also form the antigen binding site. Similarly, the Rossmann fold, which is characterized by alternating α helices and β strands, forms a nucleotide binding pocket that has been adopted by numerous proteins throughout evolution and used as a discrete functional module.

4.3 Protein Folding

We have seen that the primary amino acid sequence of a protein determines both its three-dimensional structure and its biochemical function. With this in mind, how do newly synthesized polypeptide chains fold into a stable three-dimensional structure? To begin to answer this question, it is important to remember that proteins are not rigid structures, but instead are flexible molecules that are constantly moving and fluctuating, as described by classic Newtonian laws. Indeed, with every atom in the protein moving as much as 0.1 Å in any direction, the three-dimensional structure of a protein is constantly adjusting and wriggling (see Figure 4.25). This flexibility is not only important for protein function but also provides a way for proteins to test different three-dimensional structures during the folding process.

Another thing to keep in mind is that cellular proteins are in contact with solvent, which has a major influence on the way a protein folds into its final three-dimensional structure. For cytosolic proteins, the solvent is water, which forms numerous hydrogen bonds with N and O atoms near the surface of the protein and, most important, drives hydrophobic residues toward the interior of the protein. In contrast, membrane proteins are surrounded both by hydrophobic side chains of fatty acids in the interior of the membrane and by polar groups of the membrane lipids and water molecules on either side of the membrane. Molecular modeling of proteins localized to the cytosol or membrane regions of the cell can provide a way to visualize how important solvent interactions are to protein structure and folding (**Figure 4.63**).

How newly synthesized proteins fold, and moreover, what happens to proteins in the cell that are partially unfolded or misfolded are important questions in protein biochemistry. Although the answers are not yet clear, some general principles about the formation of stable tertiary structures have begun to emerge.

1. First, the primary amino acid sequence of a protein leads to a limited number of possible folding pathways that result in a stable protein structure with a low energy state. The idea that protein folding must follow some energy minimization "path" was first pointed out by Cyrus Levinthal in 1968, when he estimated that it would take billions of years for a protein of only 100 amino acids to sample even three possible ϕ and ψ angles for every amino acid if the folding process were random. Levinthal reasoned that because proteins of this size can fold in a matter of milliseconds, preferential folding paths must be available that dominate the folding process and thereby avoid the time-consuming sampling of unstable protein structures.
2. A second key principle in protein folding is that the free energy difference (ΔG) between the folded (native) and unfolded (denatured) state of the protein must be favorable ($\Delta G < 0$) in order to drive the folding process. Protein denaturation (unfolding of tertiary structure) and renaturation (refolding of tertiary structure) studies have revealed that protein folding is an equilibrium between the folded and unfolded states, with the equilibrium shifted toward the folded state for most proteins under the favorable conditions of the cellular environment. In particular, at constant temperature and pressure, the two contributing factors to ΔG are changes in enthalpy (ΔH) and changes in entropy (ΔS). Two important sources of entropy changes contribute to the folding process. When the polypeptide folds, it adopts a more limited number of conformations than it does in the unfolded state, which is entropically

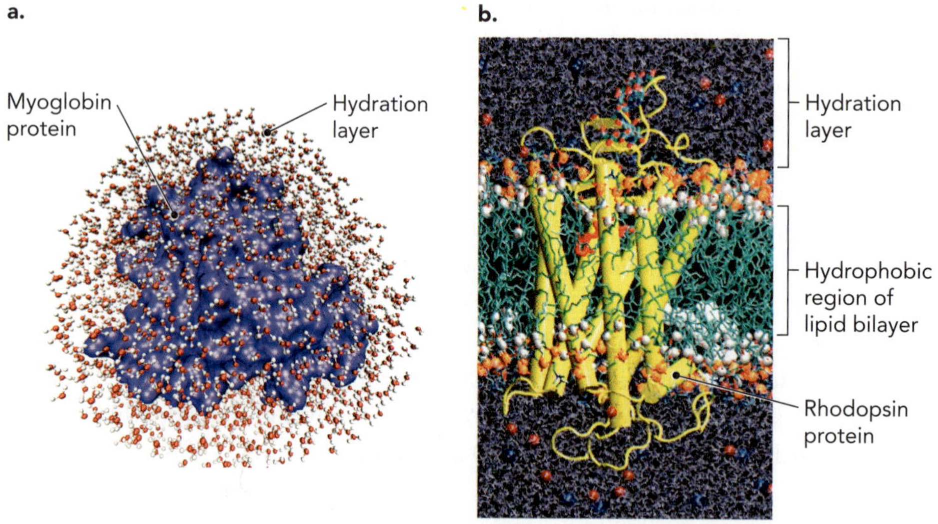

Figure 4.63 Protein structure is influenced by its surrounding chemical environment. **a.** Model of a protein in an aqueous solvent. This computer model shows the globular structure of myoglobin protein surrounded by ~2,000 water molecules, which form a hydration layer around the protein. H. FRAUENFELDER ET AL. (2009). A UNIFIED MODEL OF PROTEIN DYNAMICS. *PROCEEDINGS OF THE NATIONAL ACADEMY OF SCIENCES, 106*(13), 5129–5134. DOI: 10.1073/PNAS.0900336106. **b.** Membrane proteins are surrounded by the hydrophobic environment of the plasma membrane. They are also exposed to the aqueous environment where they protrude from the membrane, as shown here in a computer model of rhodopsin protein. T. HUBER AND T. P. SAKMAR (2008). RHODOPSIN'S ACTIVE STATE IS FROZEN LIKE A DEER IN THE HEADLIGHTS. *PROCEEDINGS OF THE NATIONAL ACADEMY OF SCIENCES, 105*(21), 7343–7344. DOI: 10.1073/PNAS.0804122105. © 2008 NATIONAL ACADEMY OF SCIENCES, USA.

unfavorable. However, a favorable entropic change results from the increase in the disorder of surrounding water molecules when the protein adopts its folded conformation. The clustering of hydrophobic side chains away from water, known as the hydrophobic effect, is driven by the increase in disorder of the water molecules (see Figure 2.25). For most proteins, the entropic contribution from the hydrophobic effect is the major driving force for protein folding.

3. A third important principle in protein folding is that the *in vivo* folding process inside the cell can be very different from the *in vitro* folding process that takes place under laboratory conditions. The primary reason for this is that *in vitro* protein-folding experiments are often performed under dilute conditions (less than 10 μM), so as to minimize the formation of protein aggregates that can form when hydrophobic regions of protein molecules interact. In contrast, the protein concentration in most cells is up to 30 times higher than this (300 μM) without inducing deleterious protein aggregation. Another difference is that *in vivo* protein folding can take place as the polypeptide chain is being synthesized, whereas *in vitro* experiments use the full polypeptide chain.

   The molecular basis for high-fidelity protein folding and refolding *in vivo* is that special helper proteins called **chaperones** provide favorable interaction surfaces that facilitate the formation of stable three-dimensional structures. Chaperone proteins perform three main functions in cells: (1) they help newly synthesized proteins fold properly (they "chaperone" them during the folding process); (2) they rescue misfolded proteins; and (3) they disrupt protein aggregates.

Before we look more closely at what is known about protein folding *in vitro* and *in vivo*, let's consider just how important the primary amino acid sequence is to determining the three-dimensional structure of a protein. In one of the more dramatic examples of this principle, Robert Sauer and his colleagues at the Massachusetts Institute of Technology used the bacteriophage Arc repressor protein as a model system to show that small changes in the polypeptide sequence can result in large structural alterations. As shown in **Figure 4.64**, they found that by simply swapping a hydrophobic core residue at position 12 (leucine) of the protein, with an adjacent polar residue at position 11 (asparagine), they were able to convert a β strand to a short helix. In both the normal

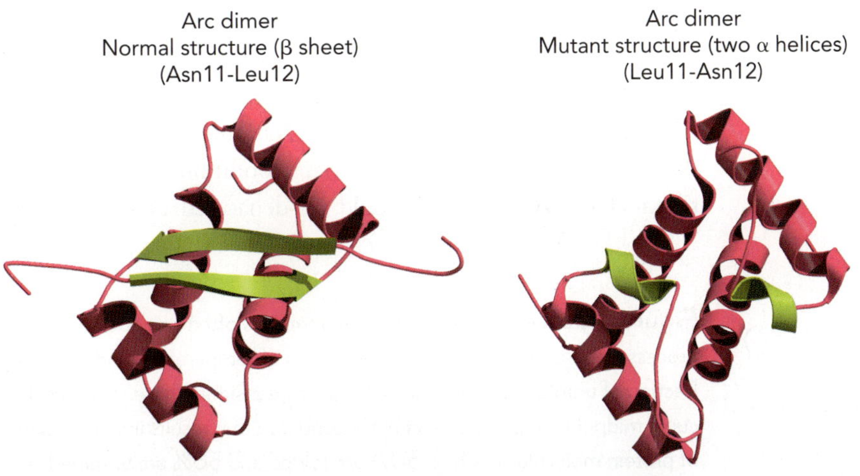

Figure 4.64 A protein fold may be altered through minor amino acid sequence changes. The normal Arc repressor protein is a dimer in solution, with a β-sheet structure formed from residues 9–14 (yellow). Reversing the positions of Asn11 and Leu12 creates a mutant in which the β-sheet is converted to two short helices without altering the structure of any other region (pink). A driving force for this structural change is hydrophobic packing of the Leu residues toward the interior of the protein dimer. BASED ON PDB FILES 1ARR (WILD-TYPE) AND 1QTG (MUTANT).

(Asn11-Leu12) and mutant (Leu11-Asn12) proteins, the hydrophobic leucine residues in the dimer pack toward the interior of the protein, but differences in their position alter the local secondary structure. This result clearly demonstrates the role of primary amino acid sequence in determining the three-dimensional structure of a protein, and moreover, that new protein folds can evolve rapidly through minor sequence changes. Studies such as these aid in our understanding of protein evolution and provide insights into the relationship between amino acid sequence and protein structure that could be useful in deriving predictive algorithms of protein folding or design of new proteins.

Protein-Folding Mechanisms Can Be Studied *In Vitro*

Because protein folding is an equilibrium between the native and denatured states, and because it is such a rapid process (milliseconds to seconds), most of what we know about protein folding comes from studying how proteins unfold under denaturing conditions. These studies involve unfolding a globular protein by disrupting noncovalent interactions within the protein using heat, pH (pH < 5 or pH > 10), or chemicals such as salts or urea. Reducing agents, such as β-mercaptoethanol (BME) or dithiothreitol, which convert disulfide bonds into two sulfhydryl groups, can also destabilize the tertiary structure.

One way to study protein unfolding and folding *in vitro* is to use fluorescence spectroscopy to monitor changes in the fluorescence of aromatic amino acids, primarily tryptophan, in the folded versus the unfolded state. A more informative method is to use **circular dichroism (CD)**, which measures differences in the absorption of right-handed versus left-handed circularly polarized light. Using wavelengths between 190 and 250 nm, it is possible to use CD spectroscopy to measure the relative amounts of α helix, β sheet, and unstructured regions in a protein sample.

Figure 4.65 shows an example of a protein-unfolding curve, in which the relative amount of unfolded protein in the sample increases as a function of increasing denaturant concentration or temperature. An important feature of this unfolding curve is that a sharp transition appears between the folded and unfolded state of the protein over a narrow range of conditions. This physical property indicates that unfolding this protein is a cooperative process, meaning that once the protein begins to unfold, it continues to convert to the unfolded state, so that there are essentially no partially folded proteins. At the **transition curve midpoint (T_m)**, 50% of the proteins are fully folded and 50% are fully unfolded. A variety of biochemical techniques and computer simulations have been used to analyze *in vitro* protein folding, and it is now well established that both protein folding and unfolding are cooperative processes for most domains in proteins.

Unfolding a protein is easily done *in vitro*, but is it possible to refold a denatured protein and recover its function? Critical *in vitro* protein-folding studies were performed in the 1950s when Christian Anfinsen, a biochemist at the National Institutes of Health, demonstrated that under appropriate conditions, the small enzyme **ribonuclease A** (RNaseA) could be denatured and then refolded.

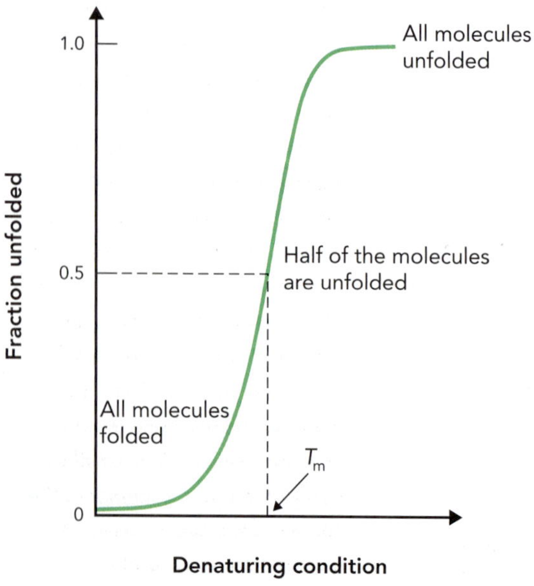

Figure 4.65 Unfolding of a protein domain is often a cooperative process. Increased denaturant concentration or temperature leads to a higher fraction of unfolded protein through a sharp transition phase. The transition curve midpoint, T_m, is the denaturing condition that results in a population of protein molecules in which 50% are folded and 50% are unfolded.

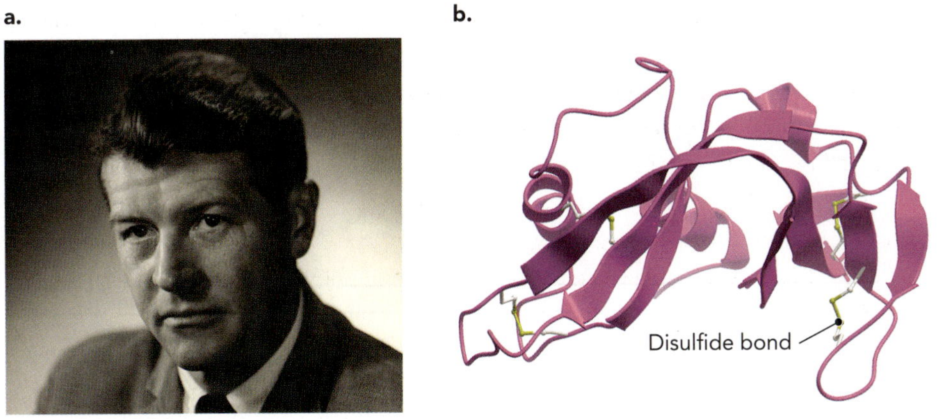

Figure 4.66 All of the biochemical information required for protein folding resides in the primary amino acid sequence. **a.** Christian Anfinsen was the first to demonstrate experimentally that the primary amino acid sequence of a protein encodes all of the structural information required for protein function. COURTESY OF THE NIH RECORD AND THE NATIONAL INSTITUTES OF HEALTH. **b.** Ribonuclease A is a small globular protein with four disulfide bonds. Anfinsen showed that it could be denatured and then refolded *in vitro*. BASED ON PDB FILE 1FS3.

These experiments convincingly showed that all of the biochemical information required for protein folding resides in the primary amino acid sequence. These studies led to Anfinsen being awarded a share of the 1972 Nobel Prize in Chemistry (**Figure 4.66**).

As illustrated in **Figure 4.67**, Anfinsen denatured purified RNaseA in the presence of 8 M urea and reduced the four disulfide bonds using β-mercaptoethanol (BME) to generate the unfolded, enzymatically inactive polypeptide. Urea, $CO(NH_2)_2$, unfolds proteins by disrupting the polar interactions within the protein but does not alter the covalent structure. When Anfinsen removed the urea and BME by dialysis in the presence of oxygen, he was able to show that the protein refolded correctly, based on the recovery of full enzymatic activity. Importantly, if he first removed only the BME, and then under oxidizing conditions removed the urea, he found that the resulting protein was only 1% active. The reason for this was that the eight cysteine residues formed random disulfide bonds during the refolding process, and the resulting "scrambled" proteins were trapped in misfolded states 99% of the time. The observed 1:100 ratio of active to inactive RNaseA proteins was entirely consistent with the probability that all eight

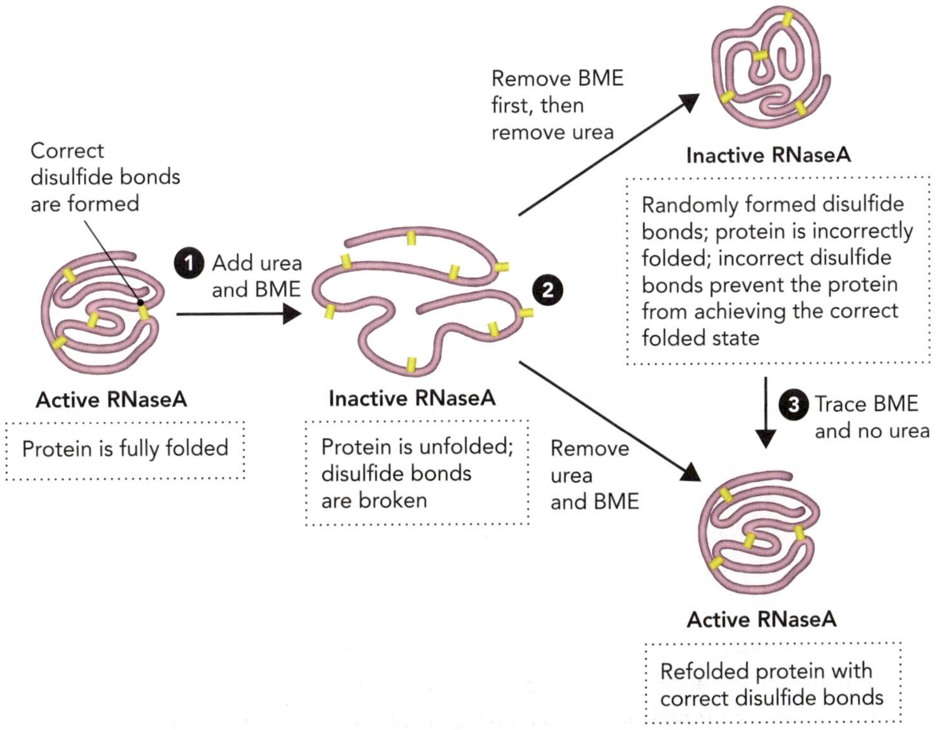

Figure 4.67 This diagram shows the steps Anfinsen used to demonstrate that proper refolding of the ribonuclease protein led to full recovery of enzyme activity. Ribonuclease protein was denatured in the presence of urea and BME, which disrupted the noncovalent interactions and the four disulfide bonds, resulting in loss of enzyme activity. If the urea and BME were removed at the same time by dialysis, the correct disulfide bonds were formed, and the enzyme regained full activity. However, if the BME was removed first, and then the urea, the protein incorrectly refolded and remained inactive due to incorrect disulfide bond formation. The inactive misfolded protein could be reactivated by adding back trace amounts of BME in the absence of urea, which allowed the disulfide bonds to break and re-form as the correct noncovalent interactions occurred within the protein.

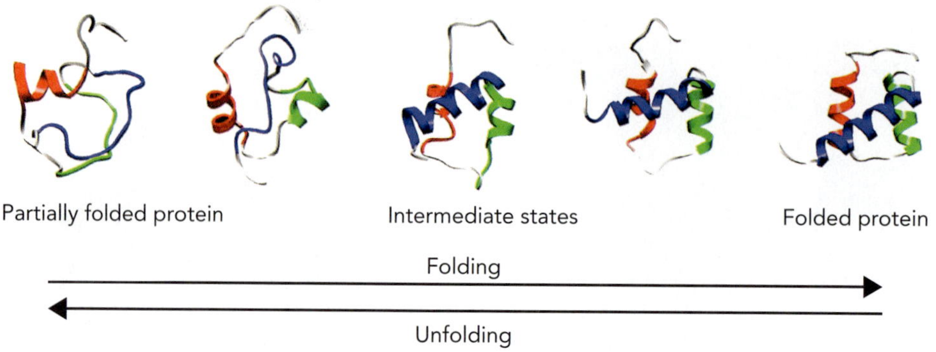

Figure 4.68 Molecular dynamic simulations can be used to model protein unfolding, as shown here to depict a possible folding and unfolding pathway for the DNA binding domain of the *Drosophila* Pit1 protein. This computer simulation model predicts that α-helical structures form early as part of intermediate structures called molten globules, which then rearrange to form the correct tertiary interactions that are characteristic of the native protein. W. BANACHEWICZ ET AL. (2011). MALLEABILITY OF FOLDING INTERMEDIATES IN THE HOMEODOMAIN SUPERFAMILY. *PROCEEDINGS OF THE NATIONAL ACADEMY OF SCIENCES, 108*(14), 5596–5601. DOI: 10.1073/PNAS.1101752108. © 2011.

cysteines would form the correct disulfide bonds. As final proof that the primary amino acid sequence of RNaseA determines its three-dimensional structure, Anfinsen added trace amounts of BME to the scrambled protein preparations to disrupt disulfide bonds. Over time, the protein was able to regain enzymatic activity as more and more of the molecules refolded into the active form with the proper disulfide bonds (Figure 4.67).

Experimental data based on *in vitro* unfolding and refolding experiments along with molecular simulation data using computational approaches have led to several proposed folding pathways for small globular proteins. A pioneer in these studies is Alan Fersht, a protein biochemist at the Medical Research Council labs in Cambridge, England, who used a variety of techniques to predict how small globular proteins are likely to fold. His group investigated how small protein domains involved in DNA binding unfold under optimal *in vitro* conditions and then used these data to propose a protein-folding pathway. These computer simulations suggested that secondary structures initially form, in this case, as three independent α helices; however, tertiary structures are lacking. Intermediate states are then converted into the stable three-dimensional structure through the formation of long-range noncovalent contacts between the three α helices (**Figure 4.68**).

Three models have been proposed to describe how globular proteins fold in the aqueous environment of the cell (**Figure 4.69**). One favored model is the hydrophobic collapse model, which proposes that hydrophobic residues first form the interior of the protein through loosely defined tertiary structures, characterized as a **molten globule**. The close proximity of residues in the molten globule facilitates the formation of well-ordered secondary and tertiary structures through van der Waals interactions and hydrogen bonding. In contrast, the framework model proposes the opposite sequence of events, in which local secondary structures form independently in the first phase, leading to the formation of tertiary structures in the second phase. A third protein-folding model, called the nucleation model, suggests that random interactions give rise to a localized region of correct three-dimensional structure, which serves to seed the formation of all the surrounding secondary and tertiary structures. It is likely that for any given protein, one or more of these folding mechanisms may be involved in promoting the final folded state of the protein.

Chaperone Proteins Aid in Protein Folding *In Vivo*

Although many cellular proteins are thought to fold *in vivo* following the same general folding pathways required for *in vitro* folding, some proteins appear to require the assistance of chaperones to complete the folding process. Chaperone proteins are highly conserved throughout evolution and have been characterized in bacteria, yeast, plants, and animals. Several different types of chaperone proteins occur. The two most common are the clamp-type and chamber-type chaperones, which bind to misfolded proteins and use ATP hydrolysis as an energy source to facilitate correct folding.

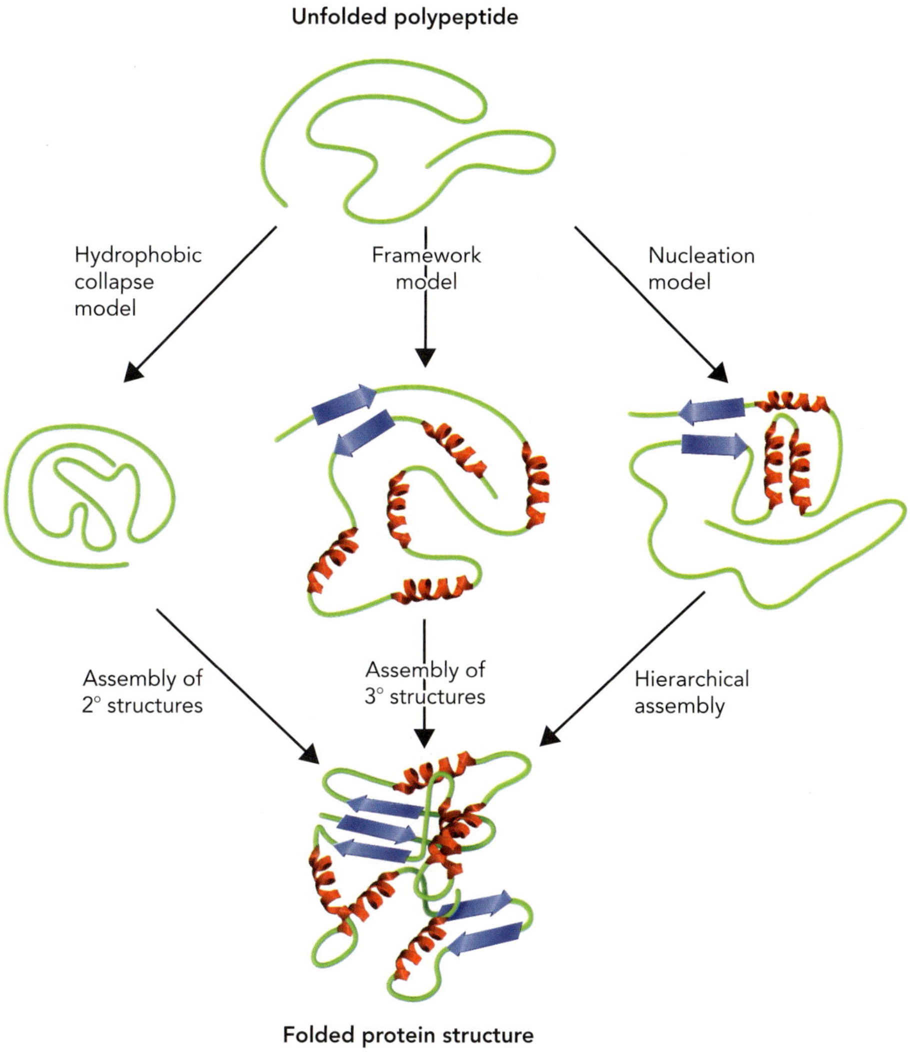

Figure 4.69 Three proposed models of protein folding pathways. The hydrophobic collapse model predicts that the clustering of hydrophobic side chains of amino acid residues drives the formation of a hydrophobic core that facilitates secondary and tertiary structure formation. The framework and nucleation models propose that local secondary structures form before tertiary structures, either as independent structures (framework model) or driven by localized interactions between secondary structures (nucleation model).

Clamp-type chaperone proteins are represented by the **heat shock family** of proteins, which were initially discovered as highly expressed proteins in heat-treated cells. Because increased temperature can denature most proteins, heat shock chaperone proteins provide a way for the cell to recover by helping proteins refold when the temperature returns to normal. Importantly, chaperone proteins are expressed in cells under all growing conditions and are required for protein folding even when cells are not heat treated.

The best-characterized clamp-type chaperone protein is **heat shock protein 70 (Hsp70)**, so named because its molecular mass is 70 kDa. ATP binding to the nucleotide binding site of Hsp70 induces a conformational change that allows unfolded or misfolded proteins to bind to the Hsp70 substrate binding site through contacts with the hydrophobic groups. When the bound ATP is hydrolyzed to ADP + P_i, a second conformational change in Hsp70 releases the partially refolded protein substrate. The protein can then rebind to Hsp70 and continue the folding process as ADP is released and replaced by ATP. Hsp70 proteins are often components of larger chaperone complexes that contain additional heat shock protein family members. **Figure 4.70** shows the molecular

Figure 4.70 The molecular structure of the peptide binding domain of the bacterial Hsp70 chaperone protein HscA is shown here with an oligopeptide bound to the unfolded protein recognition site. BASED ON PDB FILE 1U00.

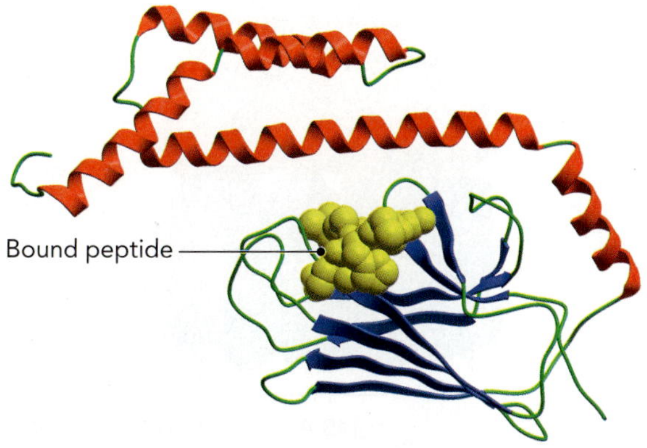

structure of the substrate binding domain of a bacterial Hsp70 protein family member called HscA, which has a small peptide bound within the clamp region.

▶ ANIMATION

The GroEL–GroES protein complex from bacteria is an example of a chamber-type chaperone. This large protein machine—sometimes called a chaperonin to distinguish it from Hsp70-type chaperone proteins—was initially discovered through genetic experiments that identified proteins required for bacterial survival after heat treatment. The GroEL component of the complex consists of 14 identical polypeptide subunits arranged in two rings of seven subunits each. Each ring houses a protein-folding chamber (**Figure 4.71**). A single GroEL subunit is 547 amino acids in length and consists of three protein domains: the substrate binding domain, an intermediate domain, and the ATP binding domain. The substrate binding and intermediate domains form the walls of the protein-folding chamber. The ATP binding domain separates the two chambers and controls the opening and closing of the chamber. GroES subunits are much smaller and function as the "cap" for the GroEL–GroES complex. There are seven identical GroES subunits in the cap, each of which consists of 97 amino acids. The GroES cap binds to only one end of the chaperone complex at a time and is required to complete the protein-folding cycle. The entire GroEL–GroES complex is very large, with a length of 180 Å, an outer diameter of 140 Å, and an inner chamber of ~40 Å in diameter, depending on the conformation of the GroEL ring. Altogether, the molecular mass of the 21-subunit GroEL–GroES complex is an impressive 850 kDa.

As shown in **Figure 4.72**, the four-step GroEL–GroES cycle aids in the refolding of two polypeptides, one in each of the two chambers, and requires the hydrolysis of 14 ATP molecules. In the first step, GroES and 7 ATP bind to the upper GroEL ring and trap an unfolded polypeptide inside the chamber. This is followed by a conformational change in the lower chamber that releases a partially folded protein, GroES, and 7 ADP. After a few seconds, the 7 ATP in the upper chamber are hydrolyzed, causing a conformational change in the complex that facilitates both protein folding in the upper chamber and reactivation of the lower chamber. In the final step of the cycle, 7 ADP, GroES, and the partially folded protein are released from the upper chamber. Taken together, the upper and lower GroEL chambers undergo the same reaction steps; however, the synchrony of the reaction cycle is offset by two steps.

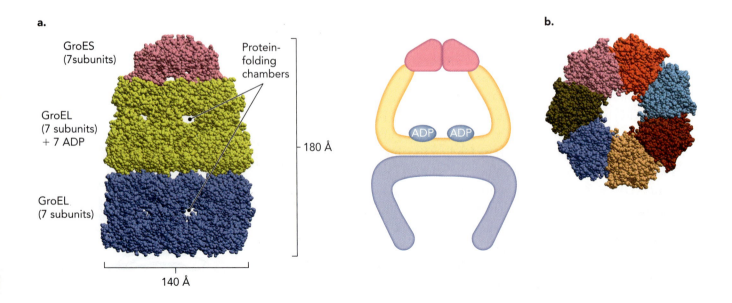

Figure 4.71 Molecular structure of the bacterial GroEL-GroES chaperonin protein complex. The entire protein complex consists of 14 identical GroEL protein subunits arranged in two concentric rings of seven subunits each and a smaller GroES protein complex that contains seven subunits and serves as a cap. **a.** Side view of GroEL-GroES complex, also shown in cartoon representation. **b.** End view of GroEL subunits.

BASED ON PDB FILE 1AON.

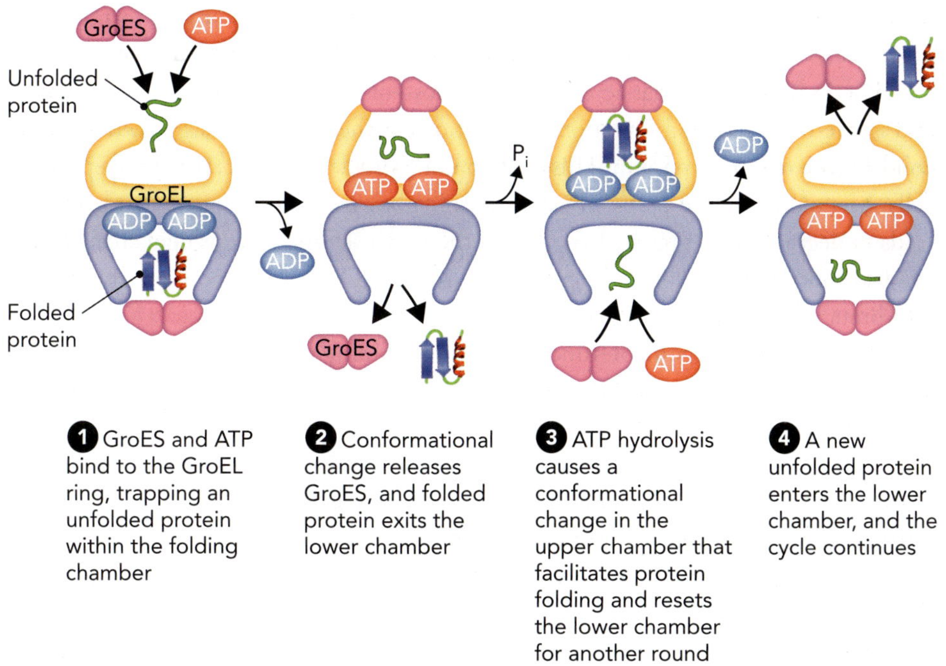
GroES
ATP
Unfolded protein
GroEL
ADP
ADP
Folded protein
ADP
ATP
ATP
GroES
Pᵢ
ADP
ADP
ATP
ADP
ATP
ATP
1 GroES and ATP bind to the GroEL ring, trapping an unfolded protein within the folding chamber
2 Conformational change releases GroES, and folded protein exits the lower chamber
3 ATP hydrolysis causes a conformational change in the upper chamber that facilitates protein folding and resets the lower chamber for another round
4 A new unfolded protein enters the lower chamber, and the cycle continues

Figure 4.72 The GroEL–GroES protein-folding cycle. Binding of GroES and 7 ATP to the GroEL ring traps an unfolded protein in the chamber, causing a conformational change in the complex that leads to the release of GroES, 7 ADP, and the partially folded protein from the other chamber. Similarly, ATP hydrolysis in one chamber causes a conformational change that facilitates protein folding in that chamber and reactivation of the other chamber. Note that unfolded protein can only bind to the GroEL subunits in the absence of bound ADP and ATP.

Protein Misfolding Can Lead to Disease

It has been estimated that up to 30% of proteins that are synthesized in cells need to be refolded or degraded to prevent the accumulation of nonfunctional proteins. If these misfolded proteins are not refolded or removed, they can interfere with the function of normal proteins through the formation of protein aggregates. Numerous human diseases have now been associated with defects in protein folding (**Table 4.5**). Many of these protein-folding diseases are due to mutations in the protein-coding sequence of the gene, whereas others have been shown to be the result of accumulated misfolded proteins that contain no amino acid changes. Misfolded proteins are either degraded, which

Table 4.5 REPRESENTATIVE PROTEIN-FOLDING DISEASES IN HUMANS

Disease	Protein	Amino acid sequence	Protein structure
Cystic fibrosis	Cystic fibrosis transmembrane conductance regulator	Mutant	Degraded
Hypercholesterolemia	Low-density lipoprotein	Mutant	Degraded
Phenylketonuria	Phenylalanine hydroxylase	Mutant	Degraded
Tay–Sachs	Hexosaminidase	Mutant	Degraded
Sickle cell anemia	Hemoglobin	Mutant	Aggregated
Huntington's	Huntingtin	Mutant	Aggregated
Kennedy's	Androgen receptor	Mutant	Aggregated
Alzheimer's	Amyloid β	Normal	Aggregated
Creutzfeldt–Jakob and kuru	Prion protein (PrP)	Normal	Aggregated
Gerstmann–Sträussler–Scheinker	Prion protein (PrP)	Mutant	Aggregated

reduces the level of functional protein in the cell (loss of function), or they form protein aggregates, which interfere with cellular processes (gain of function; **Figure 4.73**).

Cystic fibrosis is a classic example of a loss-of-function protein-folding disease and is due to amino acid mutations or deletions in the cystic fibrosis transmembrane conductance regulator protein. This protein is a chloride ion channel in lung epithelial cells. Misfolding of the cystic fibrosis transmembrane conductance regulator protein leads to degradation of more than 90% of the newly synthesized proteins before they can be properly inserted into the plasma membrane. The result is pulmonary dysfunction brought on by fluid imbalances in the lungs.

An example of a gain-of-function protein-folding disease is Huntington's disease, which is due to the expansion of triplet nucleotides (CAG) within the gene encoding the huntingtin protein. As the CAG triplet codon specifies the amino acid glutamine, the mutant huntingtin protein acquires an increase in the number of glutamine residues, called a **polyglutamine track expansion**. The expanded polyglutamine track in the mutant huntingtin protein is associated with protein aggregation, which leads to neuronal cell death and the onset of a neurodegenerative disease.

In contrast to Huntington's disease, which is due to changes in the protein coding sequence, gain-of-function protein-folding disease may also be caused by the accumulation of misfolded proteins that have no amino acid mutations. For example, in Alzheimer's disease, cleavage of a normal protein product produces a hydrophobic protein fragment that can aggregate. This misfolded protein can accumulate in brain cells and form large protein fibers that ultimately cause neuronal cell death. Patients with Alzheimer's disease have regions in the brain containing **amyloid plaques**, which are hallmarks of the disease. These plaques contain amyloid β protein aggregates that can be identified by differential staining techniques or by amyloid β protein antibodies (**Figure 4.74**).

Transmissible spongiform encephalopathies (TSEs) are also due to the accumulation of misfolded proteins that form large aggregates in brain tissue. It was originally thought that TSEs were caused by slow-growing infectious viruses contained in brain tissue. It is now generally accepted, however, that these diseases are almost certainly due to misfolded proteins that accumulate in neuronal tissues.

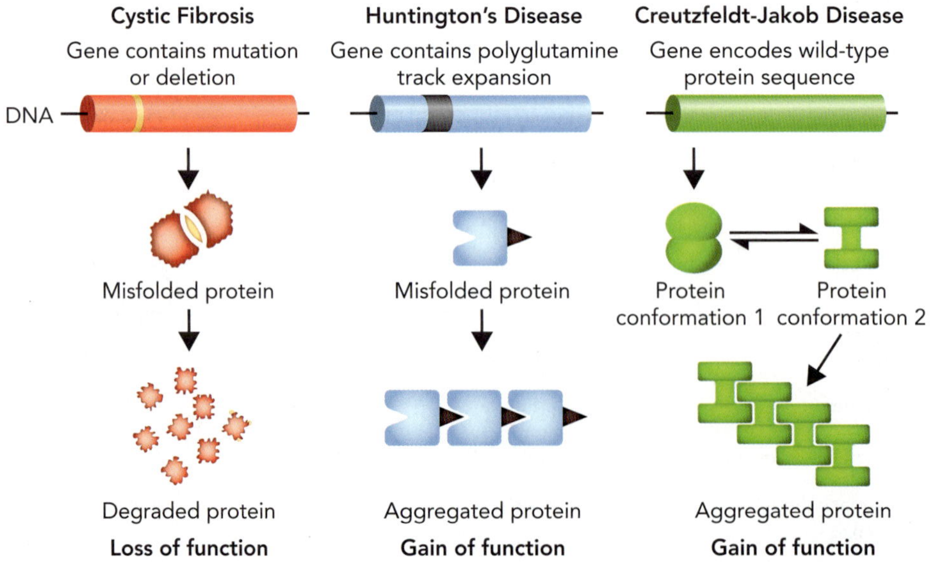

Figure 4.73 Protein-folding diseases can be the result of genetic mutations or increased levels of misfolded proteins encoded by wild-type genes. Degradation of a misfolded protein, as seen in cystic fibrosis, is considered a loss-of-function mutation because the activity of the protein is missing from the cell. In contrast, protein aggregation is considered a gain-of-function mutation because the protein adds a process to the cell. Gain-of-function protein-folding diseases can be the result of amino acid mutations (Huntington's disease) or accumulation of misfolded wild-type proteins (Creutzfeldt-Jakob disease).

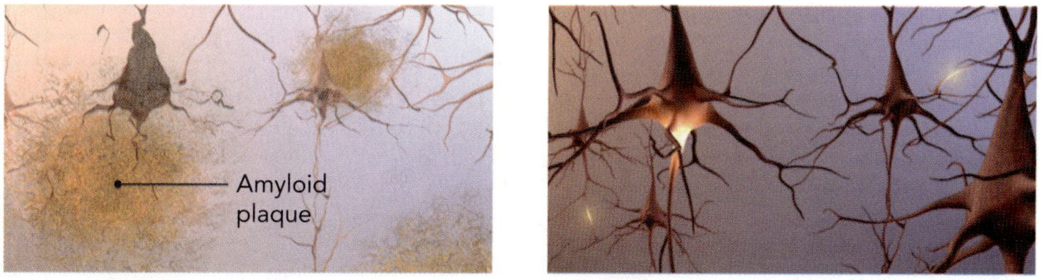

Alzheimer cells

Healthy cells

Figure 4.74 An amyloid plaque can be clearly seen in the brain tissue of a rabbit model of Alzheimer's disease. Neuronal cell death in areas surrounding amyloid plaques is responsible for the dementia and other neuronal abnormalities observed in patients with Alzheimer's disease.

The best studied of the TSE proteins is **prion protein** (PrP), which is responsible for scrapie disease in sheep. Kuru and mad cow disease are two other transmissible prion-related diseases that have been described. Kuru was first discovered in Papua New Guinea and was found to be transmitted by cannibalism among tribesmen. Mad cow disease is transmitted through livestock and can infect humans that consume infected meat. A rare and spontaneous form of kuru in humans is called Creutzfeldt–Jakob disease, which is a progressive neurologic disease characterized by the lethal accumulation of prion protein aggregates in the brain.

Normal cellular PrP protein, denoted as PrP^{c}, is found in the brains of most mammals and presumably functions as a cell surface glycoprotein. However, according to the **prion hypothesis** proposed by Stanley Prusiner at the University of California Medical School in San Francisco, PrP^{c} protein can be converted to an infectious, misfolded version of the same protein called PrP^{Sc} ("Sc" for scrapie), which forms large protein aggregates in the cell. The structural conversion of a correctly folded PrP^{c} protein into a misfolded PrP^{Sc} protein that readily aggregates is poorly understood.

Figure 4.75a shows the molecular structure of a portion of the normal human PrP^{c} protein as determined in solution using NMR spectroscopy. It can be seen that the C-terminal region of the polypeptide forms a stable tertiary structure, whereas the N-terminal portion is unstructured. One way to investigate possible structures of prion aggregrates that may form in human cells is to study prion-like proteins from other organisms. As shown in **Figure 4.75b**, a prion-like protein called HET-s found in the fungus *Podospora anserina* has been shown by NMR spectroscopy to form a β-solenoid structure consisting of extended β sheets. The β-solenoid structure found in aggregated HET-s proteins may represent amyloid fibrils that form in human neurodegenerative diseases.

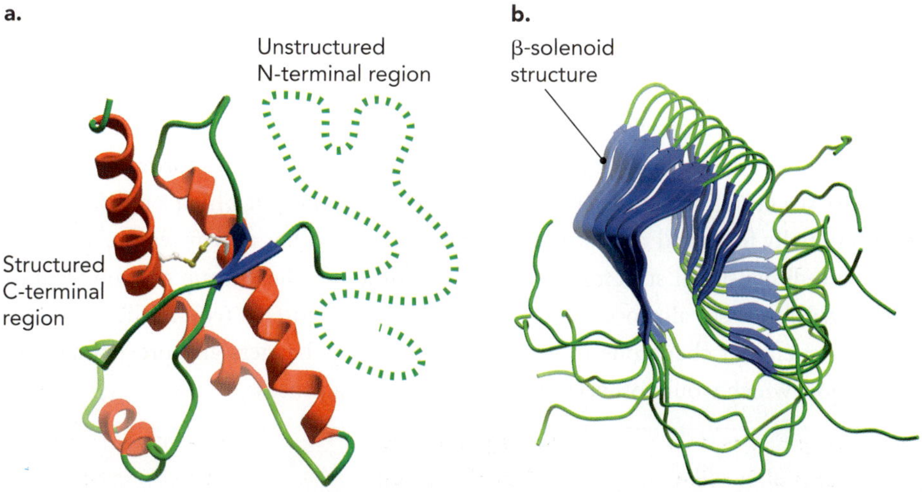

Figure 4.75 NMR structures of prion proteins are shown here. **a.** Structure of normal human PrP^{c} protein in monomeric form. The C-terminal portion of human PrP^{c} is structured, while the N-terminal region is disordered (represented by a dotted line). It is thought that the flexible N terminus is involved in PrP^{Sc} infection ability. BASED ON PDB FILE 1QM1. **b.** Structure of an amyloid fibril formed by the prion-like domain of the fungal protein HET-s. The extended β sheets in HET-s protein aggregate to form a β-solenoid structure that may be representative of aggregated prion proteins associated with neurodegenerative disease. BASED ON PDB FILE 2RNM.

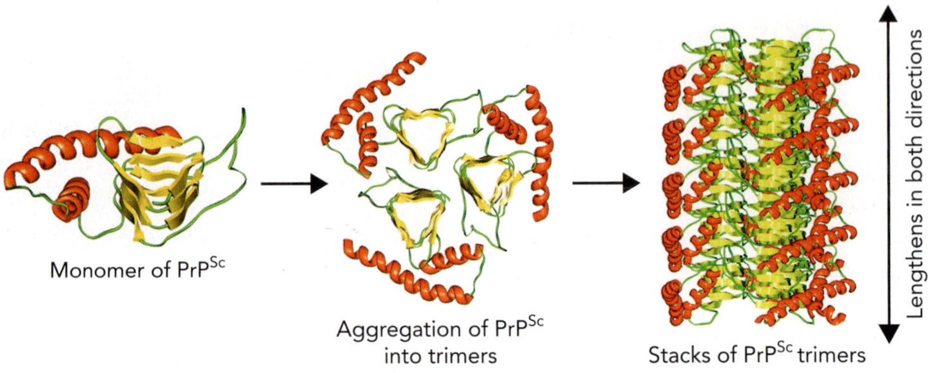

Figure 4.76 Fred Cohen and Stanley Prusiner proposed that misfolded PrP^{Sc} prion protein monomer contains stacked β sheets that promote trimer formation and subsequent aggregation of these PrP^{Sc} trimers. The prion model suggests that stacks of PrP^{Sc} trimers form long macromolecular fibrils, which lead to cell death and the accumulation of plaques in the brain. C. GOVAERTS ET AL. (2004). EVIDENCE FOR ASSEMBLY OF PRIONS WITH LEFT-HANDED β-HELICES INTO TRIMERS. *PROCEEDINGS OF THE NATIONAL ACADEMY OF SCIENCES, 101*, 8342–8347. © 2004 NATIONAL ACADEMY OF SCIENCES, USA.

It is thought that the flexible N-terminal region of the PrP^c protein is responsible for the structural switch from PrP^c to PrP^{Sc} (Figure 4.75a). Fred Cohen and Stanley Prusiner have proposed that this region changes from a mostly α-helical structure in PrP^c to a stacked β sheet in the PrP^{Sc} form, which may be similar to the β-solenoid structure formed by the fungal HET-s protein (Figure 4.75b). Once these β sheets form, the PrP^{Sc} protein is incorporated into large protein aggregates that eventually lead to neuronal cell death (**Figure 4.76**). Factors that shift the protein-folding equilibrium from PrP^c to PrP^{Sc} are not yet known; however, the infectious nature of PrP^{Sc} proteins indicates that physical interactions between PrP^c and PrP^{Sc} may play a role.

concept integration 4.3

What are the molecular determinants of protein folding *in vitro* and *in vivo*?

Protein folding *in vitro* appears to be a cooperative process based on the rapid transition between the folded and unfolded states over a small range of conditions in protein denaturation studies. Three models have been proposed to explain cooperative protein folding *in vitro*: (1) the hydrophobic collapse model, in which clustering of the hydrophobic residues drives the formation of an initial loosely defined structure; (2) the framework model, which proposes that secondary structures form first, leading to tertiary structure formation; and (3) the nucleation model, in which a portion of the protein initially folds and then stimulates the folding of other nearby regions until the final three-dimensional structure is stabilized. Although these same thermodynamically driven processes also occur *in vivo*, many cellular proteins require the assistance of ATP-dependent chaperone proteins to lower energy barriers and prevent subunit aggregation, which would otherwise inhibit the folding process.

chapter summary

4.1 Proteins Are Polymers of Amino Acids

- Proteins are polypeptides consisting of amino acids covalently linked together by peptide bonds. Peptide bonds are formed by a condensation reaction between the carboxylic acid group of one amino acid and the amine group of a second amino acid.
- Twenty different amino acids are commonly found in proteins. Their stereochemistry is that of L-amino acids with a C_α chiral center, except for glycine, the smallest of the 20 amino acids, which has a C_α that is not chiral.
- The chemical properties of the amino acid side chains determine the structure and function of proteins. The 20 amino acids can be divided into four chemical groups based on shared properties: charged, hydrophilic, hydrophobic, and aromatic.
- The peptide bond is rigid with partial double bond characteristics and defines a flat plane containing six atoms. Rotation around the N—C_α and the C_α—C bonds is defined by the ϕ (phi) and ψ (psi) torsional angles, respectively. Rotation is limited by steric hindrance of amino acid functional groups, as shown by a Ramachandran plot.
- Modern protein biochemistry often uses DNA sequence information as a starting point for studies of protein structure and function. The genetic code is used to predict the amino acid sequence of a protein.
- Genetic mutations in DNA can alter the amino acid sequence of a protein and lead to defects in protein structure and function, resulting in disease.

4.2 Hierarchical Organization of Protein Structure

- The three-dimensional structure of a protein is defined by four hierarchical levels: primary (amino acid sequence), secondary (α helix, β sheet, β turn), tertiary (positions of all atoms with the protein), and quaternary (subunit interactions).
- α helices are stabilized by intrastrand hydrogen bonding between N—H and C═O groups along the polypeptide backbone. β sheets are stabilized by interstrand hydrogen bonding between N—H and C═O groups along the polypeptide backbone.
- Tertiary structures describe the positions of all the atoms within the polypeptide. The arrangements of α helices, β sheets, β turns, and polypeptide loops compose the protein fold. Examples of folds include four-helix bundles, the Greek key fold, and the Rossmann fold. It is thought that all protein structures in nature can be described by combinations of only a few thousand protein folds.
- Tertiary structures are stabilized by weak noncovalent interactions. However, some tertiary structures are also stabilized by disulfide bridges between cysteine residues or are stabilized by metal ions such as zinc and iron that coordinate with residues in the protein.
- Quaternary structures consist of two or more protein subunits that can be identical or different. Quaternary structures provide increased structural integrity, as demonstrated in fibrous proteins, regulatory functions commonly found in large protein complexes, and increased enzyme efficiency provided by nearby catalytic sites.

4.3 Protein Folding

- High-fidelity protein folding is critical to protein structure and function and is governed by three principles: (1) protein folding must follow a preferred path of energy minimization; (2) the change in free energy between the folded and unfolded states must be favorable ($\Delta G < 0$) for folding to occur; and (3) mechanisms of *in vitro* and *in vivo* folding may be different, as chaperone proteins are often required for *in vivo* protein folding to occur.
- Protein folding *in vitro* is a cooperative process based on protein-unfolding curves, which show a sharp transition between the number of molecules in the folded and unfolded states as a function of increased temperature or denaturant concentration. The transition curve midpoint, T_m, corresponds to the condition when 50% of the molecules are folded and 50% are unfolded.
- Anfinsen's classic *in vitro* folding experiment using purified ribonuclease A demonstrated that all of the information required for functional protein folding is contained within the primary amino acid sequence.
- Chaperone proteins function *in vivo* to assist in *de novo* protein folding, rescue unfolded proteins, and disrupt nonfunctional protein aggregates. The two major types of ATP-dependent chaperone proteins are the clamp type, as exemplified by Hsp70, and the chamber type, of which the bacterial GroEL–GroES complex is a prime example.
- Defects in protein folding have been associated with increased rates of protein degradation (loss of function) and generation of large protein aggregates (gain of function). Numerous examples of human diseases caused by protein misfolding have been characterized.
- The prion hypothesis proposed by Stanley Prusiner states that scrapie-related protein-folding diseases, such as Alzheimer's, kuru, Creutzfeldt–Jakob, and mad cow disease, are the result of infectious protein particles that induce misfolding in protein molecules that contain no mutations in the amino acid sequence. Molecular models suggest that abnormal β-sheet structures in misfolded prion protein molecules lead to the formation of large protein aggregates consisting of trimeric complexes.

biochemical terms

(in order of appearance in text)
polypeptide chain (p. 149)
oligopeptide (p. 149)
residue (p. 149)
subunit (p. 150)
amino acid side chain (p. 151)
α amino acid (p. 151)
isoelectric point (p. 151)
zwitterion (p. 151)
stereoisomer (p. 152)
enantiomer (p. 152)
RS system (p. 152)
aspartate (p. 154)
glutamate (p. 154)
lysine (p. 154)
arginine (p. 154)
histidine (p. 154)
serine (p. 157)
threonine (p. 157)
serine/threonine kinase (p. 157)
phosphatase (p. 157)
glutamine (p. 158)
asparagine (p. 158)
cysteine (p. 158)
disulfide bond (p. 158)
disulfide bridge (p. 158)
isoleucine (p. 158)
leucine (p. 158)
valine (p. 158)
methionine (p. 158)
glycine (p. 160)
alanine (p. 160)
proline (p. 160)
tryptophan (p. 161)
tyrosine (p. 161)
phenylalanine (p. 161)
autofluorescent (p. 161)
green fluorescent protein (p. 161)
protease (p. 162)
backbone (p. 163)
ϕ (phi) and ψ (psi) torsional angles (p. 165)
Ramachandran plot (p. 165)
DNA bioinformatics (p. 166)
missense mutation (p. 167)
nonsense mutation (p. 167)
frameshift mutation (p. 167)
silent mutation (p. 167)
primary structure (p. 170)
secondary structure (p. 170)
ribbon diagram (p. 170)
protein fold (p. 171)
tertiary structure (p. 171)
quaternary structure (p. 171)
α helix (p. 171)
β strand (p. 171)
β sheet (p. 171)
antiparallel β sheet (p. 176)
parallel β sheet (p. 176)
β pleated sheet (p. 176)
β turn (p. 176)
Protein Data Bank (PDB) (p. 180)
domain (p. 180)
motif (p. 181)
Structural Organization of Proteins (SCOP) (p. 181)
four-helix bundle (p. 181)
Greek key fold (p. 182)
Rossmann fold (p. 182)
TIM barrel fold (p. 182)
α/β barrel (p. 182)
FERM domain fold (p. 183)
zinc finger (p. 186)
homodimer (p. 186)
heterodimer (p. 186)
coiled coil (p. 187)
fibroin heavy chain (p. 188)
fibroin light chain (p. 188)
scurvy (p. 190)
immunoglobulin (p. 191)
antibody (p. 191)
antigen (p. 191)
variable domain (p. 192)
constant domain (p. 192)
immunoglobulin fold (p. 192)
chaperone (p. 195)
circular dichroism (CD) (p. 196)
transition curve midpoint (T_m) (p. 196)
ribonuclease A (p. 196)
molten globule (p. 198)
heat shock family (p. 199)
heat shock protein 199 (Hsp70) (p. 199)
polyglutamine track expansion (p. 202)
amyloid plaques (p. 202)
prion protein (p. 203)
prion hypothesis (p. 203)

review questions

1. What is the isoelectric point of a protein? What is the specific term that describes amino acids at this point?
2. Explain how the pK_a of an amino acid can differ within a folded protein compared to that of the free amino acid in water.
3. Classify all of the amino acids into the categories charged, hydrophobic, hydrophilic, and aromatic.
4. What is a peptide bond and how is it formed?
5. Explain the physical relevance of ϕ (phi) and ψ (psi) angles.
6. Name and briefly describe the four levels of protein structure.
7. Explain how an α helix can be amphipathic.
8. Explain why secondary structures are so prevalent in proteins.
9. What are the three most common amino acids found in α helices, β strands, and β turns, respectively?
10. What are three ways in which quaternary structures can provide increased functionality for a protein?
11. What is thought to be the dominant mechanism of protein folding? Why can it not be random?
12. What are the enthalpic and entropic factors that lead to the stabilization of a protein upon protein folding?
13. Name and briefly describe the three proposed mechanisms of how globular proteins fold in aqueous environments.
14. What are the two phenotypical consequences produced by protein-folding diseases?

challenge problems

1. Consider a pentapeptide with the sequence Lys-Tyr-Glu-Asn-His (K-Y-E-N-H). Using the pK_a values in Table 4.3, make a table to show the approximate charges (if any) on the terminal groups and the side chain of each residue in the peptide at pH 7 and at pH 11. What would be the approximate net charge on this pentapeptide at pH 7 and at pH 11?
2. The polarity of the solvent and other environmental factors can affect the pK_a of a weak acid. Suppose the α amino group of a protein has a pK_a of about 8.0 when it is exposed to H_2O on the outside of a protein.
   a. Would you expect the pK_a to be higher or lower than 8 if the group were buried in the hydrophobic interior of the protein? Explain.
   b. This same α amino group in the hydrophobic interior of the protein has the opportunity to form an ionic bond in that hydrophobic environment with a carboxylate group in the side chain of a charged Asp residue. Under these conditions, how would the pK_a of this α amino group compare with the pK_a of the α amino group in the hydrophobic interior of the protein without a nearby Asp residue to form this ionic bond?
3. In the closely packed interior of the tertiary structure of an enzyme, an alanine residue was changed by mutation to a valine, leading to a loss of enzyme activity, although that residue was not directly involved in the catalytic function of the enzyme. However, activity was partially regained when an additional mutation at a different position in the primary structure changed an isoleucine residue to a glycine. Based on the structure of the amino acid side chains of alanine, valine, isoleucine, and glycine, explain how the first mutation Ala→Val likely caused a loss of activity, and the second mutation in another region of the protein, Ile→Gly, resulted in a partial recovery of enzyme activity.
4. Briefly explain how proteins manage to "neutralize" the polarity of main-chain carbonyl O and amide NH groups that have to be buried in the hydrophobic interior of the protein when the protein folds.
5. Using three-letter abbreviations, name the amino acid residues in the partial protein sequence shown in the illustration that follows. If this region of the protein were part of an α helix, identify one atom or group that would participate as a hydrogen-bond donor to stabilize the structure of the α helix, and label it "Donor." Identify the one atom or group that would participate as a hydrogen-bond acceptor and label it "Acceptor."

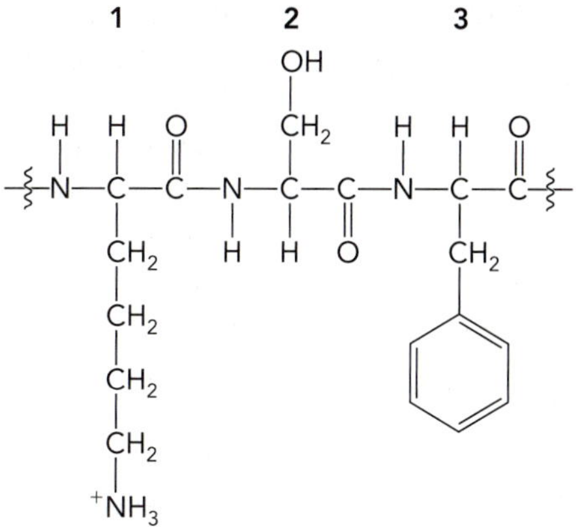

6. Name the four types of noncovalent interactions that stabilize tertiary and quaternary structures of proteins.
7. Answer the following on the basis of the tetrapeptide structure shown here.
   a. Name the residues in the peptide.
   b. Draw boxes to identify the peptide bonds between residues 1 and 2 and between residues 3 and 4.
   c. Draw a rectangle to identify the six atoms that are coplanar with the peptide bond between residues 2 and 3.
   d. Predict a pH range that would result in the ionization state of this tetrapeptide.

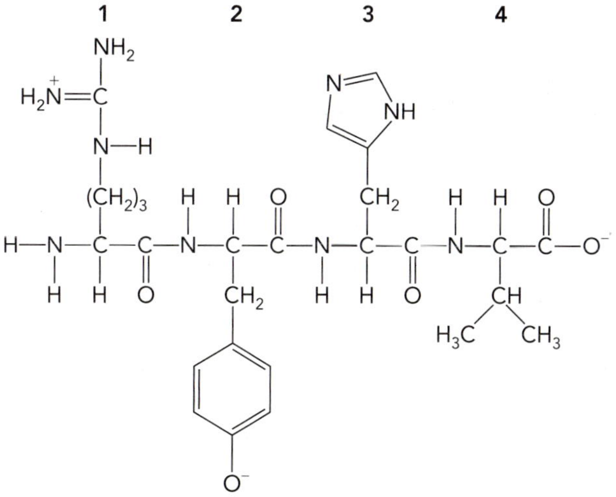

8. Briefly explain in terms of the thermodynamics of protein folding why the folded structures of water-soluble globular proteins have extensive secondary structure.
9. Consider the two polypeptide strand backbones derived from a large protein that are shown in the illustration that

follows. Label the N-terminal to C-terminal (N→C) direction of each polypeptide strand and determine if these are parallel or antiparallel β strands. Draw dashed lines to identify the hydrogen bonds that stabilize the β sheet.

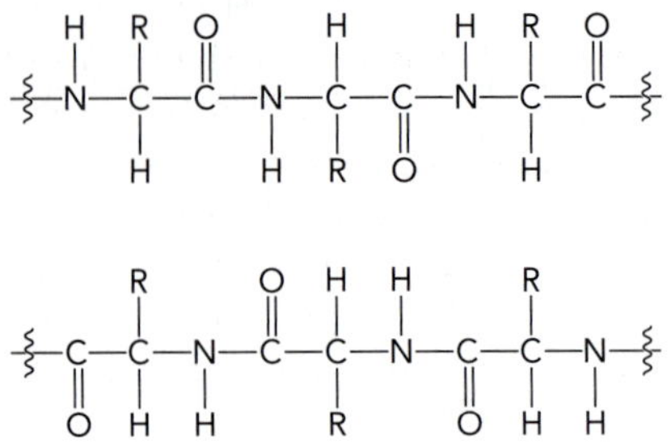

10. The ionization of the amino acid Glu is characterized by three weak acid groups with $pK_1 = 2.1$, $pK_2 = 4.1$, and $pK_3 = 9.9$. The following illustration shows four possible ionic forms of Glu (**a–d**) that might occur in solution and a titration curve for Glu. Which of these ionic forms represents the zwitterion of Glu? Identify the structure (by letter) that corresponds to the ionic form of Glu that predominates at each pH value indicated by the arrows. If two ionic forms are present equally, select two choices.

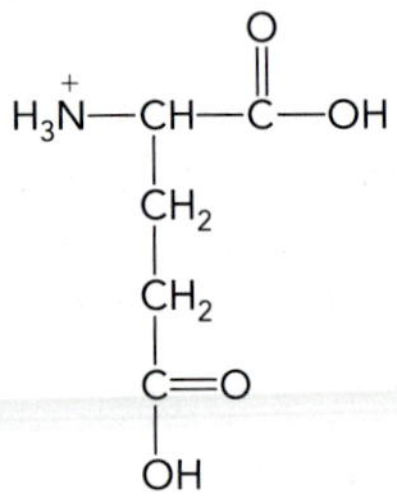

a.

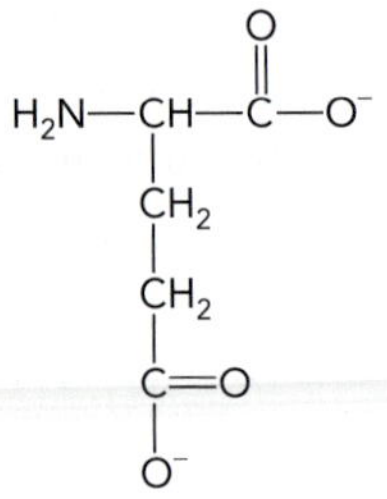

b.

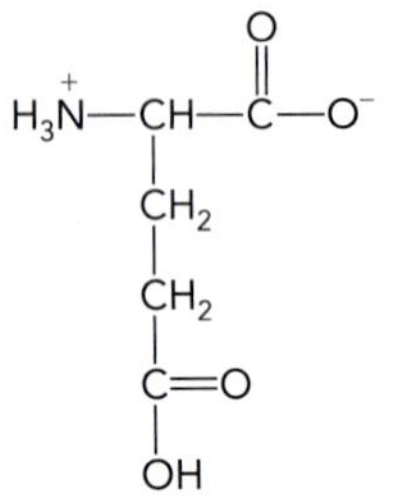

c.

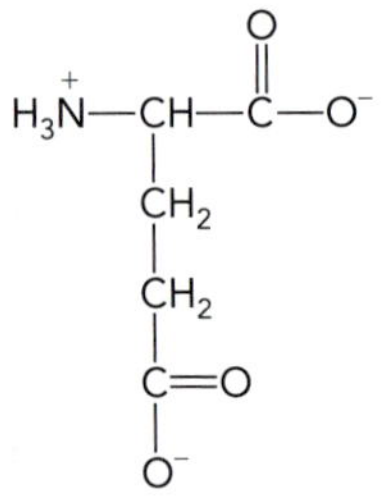

d.

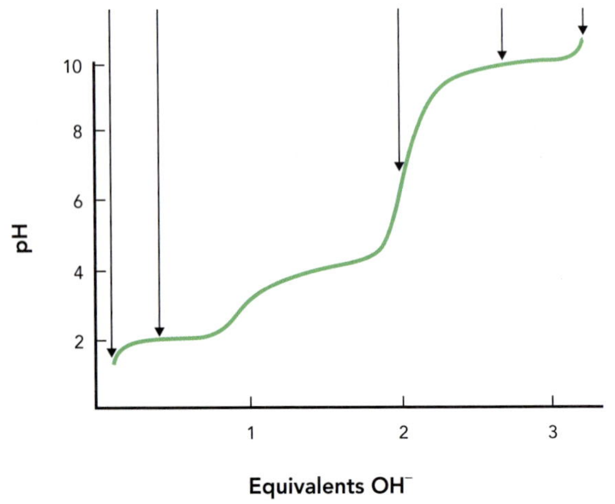

11. In the small peptide Cys1-Asn-Cys2-Lys-Ala-Pro-Cys3-Ala-Arg-Cys4-Gln-His, there are three possible arrangements of disulfide bonds between two pairs of cysteine residues that will result in two disulfide bonds. One of these combinations corresponds to disulfide bonds between Cys1 and Cys3 and between Cys2 and Cys4. What are the other two combinations of cysteine pairs that would result in two disulfide bonds? To identify the correct arrangement of disulfide bonds, the peptide was treated with the protease trypsin, which generated two oligopeptides under nonreducing conditions that preserve the disulfide bonds. Considering that trypsin cleaves peptide bonds on the C-terminal side of Arg and Lys residues, what is the correct arrangement of disulfide bonds that must exist in the small polypeptide to result in two oligopeptide products?

12. Explain the roles of urea and β-mercaptoethanol in Anfinsen's experiments on protein folding using the protein ribonuclease. What was the most important conclusion resulting from this experiment that earned Anfinsen the Nobel Prize?

13. A homopolymer of lysine residues (polylysine) can adopt an α-helical conformation or is unfolded, depending on the pH of the solution. Predict whether the conformation of polylysine would be α-helical or unfolded at pH values of 1, 7, and 11. Explain your reasoning.

14. Answer the following questions based on these four primary protein sequences:
    1. Asp-Gln-Leu-Glu-Lys-Glu-Leu-Gln-Ala-Leu-Glu-Lys-Glu-Leu-Ala
    2. Phe-Gln-Ile-Asp-Met-Glu-Leu-Lys-Val-Asn-Leu-Asp-Phe-Arg-Ala
    3. Ala-Gln-Tyr-Gly-Pro-Asn-Leu-Phe-Ala-Val-Ile-Lys-Asn-Cys-Ala
    4. Phe-Asn-Ser-Val-Leu-Gln-Asp-Ile-Glu-Gln-Phe-Met-Ser-Cys-Ala

    a. Which sequence looks as if it could include a β turn (reverse turn)? Explain your reasoning.

    b. Which sequence looks as if it could form a β strand, with one surface facing the interior of the protein and the other surface exposed to water? Explain your reasoning.

    c. Which sequence looks as if it could form an α helix that would participate in a coiled coil structure within a protein? Explain your reasoning.

    d. Using the helical wheel drawn in Figure 4.34, which sequence looks as if it could form an amphipathic α helix? Which amino acids form the hydrophobic and hydrophilic faces of this amphipathic α helix?

suggested reading

Books and Reviews

Boehr, D. D., and Wright, P. E. (2008). Biochemistry. How do proteins interact? *Science*, *320*, 1429–1430.

Branden, C., and Tooze, J. (1999). *Introduction to protein structure* (2nd ed.). New York, NY: Garland Publishing.

Chiti, F., and Dobson, C. M. (2006). Protein misfolding, functional amyloid, and human disease. *Annual Review of Biochemistry, 75*, 333–366.

Dinner, A. R., Sali, A., Smith, L. J., Dobson, C. M., and Karplus, M. (2000). Understanding protein folding via free-energy surfaces from theory and experiment. *Trends in Biochemical Sciences*, *25*, 331–339.

Eisenberg, D. (2003). The discovery of the alpha-helix and beta-sheet, the principal structural features of proteins. *Proceedings of the National Academy of Sciences USA*, *100*, 11207–11210.

Fersht, A. R. (2008). From the first protein structures to our current knowledge of protein folding: delights and skepticisms. *Nature Reviews Molecular Cell Biology*, *9*, 650–654.

Gianni, S., Guydosh, N. R., Khan, F., Caldas, T. D., Mayor, U., White, G. W., DeMarco, M. L., Daggett, V., and Fersht, A. R. (2003). Unifying features in protein-folding mechanisms. *Proceedings of the National Academy of Sciences USA*, *100*, 13286–13291.

Gregersen, N. (2006). Protein misfolding disorders: pathogenesis and intervention. *Journal of Inherited Metabolic Disease*, *29*, 456–470.

Klug, A. (2002). Structural biology and biochemistry. Retrospective: Max Perutz (1914–2002). *Science*, *295*, 2382–2383.

Patthy, L. (2007). *Protein evolution* (2nd ed.). Oxford: Wiley-Blackwell.

Petsko, G. A., and Ringe, D. (2004). *Protein structure and function*. London, England: Blackwell.

Prusiner, S. B. (2012). A unifying role for prions in neurodegenerative diseases. *Science*, *336*, 1511–1513.

Remington, S. J. (2006). Fluorescent proteins: maturation, photochemistry and photophysics. *Current Opinion in Structural Biology*, *16*, 714–721.

Richardson, J. S. (2000). Early ribbon drawings of proteins. *Nature Structural Biology*, *7*, 624–625.

Sadowski, M. I., and Jones, D. T. (2009). The sequence-structure relationship and protein function prediction. *Current Opinion in Structural Biology*, *19*, 357–362.

Service, R. F. (2008). Problem solved* (*sort of). *Science*, *321*, 784–786.

Whitford, D. (2005). *Proteins: structure and function* (1st ed.). Hoboken, NJ: Wiley.

Primary Literature

Andreeva, A., Howorth, D., Chandonia, J. M., Brenner, S. E., Hubbard, T. J., Chothia, C., and Murzin, A. G. (2008). Data growth and its impact on the SCOP database: new developments. *Nucleic Acids Research*, *36*, D419–425.

Diakowski, W., Grzybek, M., and Sikorski, A. F. (2006). Protein 4.1, a component of the erythrocyte membrane skeleton and its related homologue proteins forming the protein 4.1/FERM superfamily. *Folia Histochemica et Cytobiologica*, *44*, 231–248.

Govaerts, C., Wille, H., Prusiner, S. B., and Cohen, F. E. (2004). Evidence for assembly of prions with left-handed β-helices into trimers. *Proceedings of the National Academy of Sciences USA*, *101*, 8342–8347.

Hamada, K., Shimizu, T., Yonemura, S., Tsukita, S., and Hakoshima, T. (2003). Structural basis of adhesion-molecule recognition by ERM proteins revealed by the crystal structure of the radixin-ICAM-2 complex. *EMBO Journal*, *22*, 502–514.

Lang, D., Thoma, R., Henn-Sax, M., Sterner, R., and Wilmanns, M. (2000). Structural evidence for evolution of the beta/alpha barrel scaffold by gene duplication and fusion. *Science*, *289*, 1546–1550.

Mayor, U., Guydosh, N. R., Johnson, C. M., Grossmann, J. G., Sato, S., Jas, G. S., Freund, S. M., Alonso, D. O., Daggett, V., and Fersht, A. R. (2003). The complete folding pathway of a protein from nanoseconds to microseconds. *Nature*, *421*, 863–867.

Newlove, T., Atkinson, K. R., Van Dorn, L. O., and Cordes, M. H. (2006). A trade between similar but nonequivalent intrasubunit and intersubunit contacts in Cro dimer evolution. *Biochemistry*, *45*, 6379–6391.

Stohr, J., Weinmann, N., Wille, H., Kaimann, T., Nagel-Steger, L., Birkmann, E., Panza, G., Prusiner, S. B., Eigen, M., and Riesner, D. (2008). Mechanisms of prion protein assembly into amyloid. *Proceedings of the National Academy of Sciences USA*, *105*, 2409–2414.

Wasner, C., Lange, A., Van Melckebeke, H., Siemer, A. B., Riek, R., and Meier, B. H. (2008). Amyloid fibrils of the HET-s prion form a β solenoid with a triangular hydrophobic core. *Science*, *319*, 1523–1526.

Zahn, R., Liu, A., Luhrs, T., Riek, R., von Schroetter, C., Lopez Garcia, F., Billeter, M., Calzolai, L., Wider, G., and Wuthrich, K. (2000). NMR solution structure of the human prion protein. *Proceedings of the National Academy of Sciences USA*, *97*, 145–150.

Biochemists often use bacterial cells that have been engineered through recombinant DNA techniques to produce large quantities of the protein they want to study

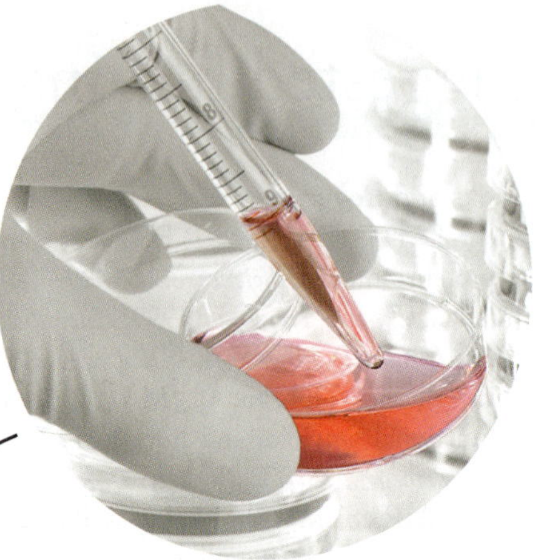

Cell extracts prepared from tissue culture cells can also be used for protein characterization

Recombinant protein expressed in *E. coli* culture

Proteins in cell extracts are visualized by SDS-PAGE

Mammalian cells propagated in tissue culture

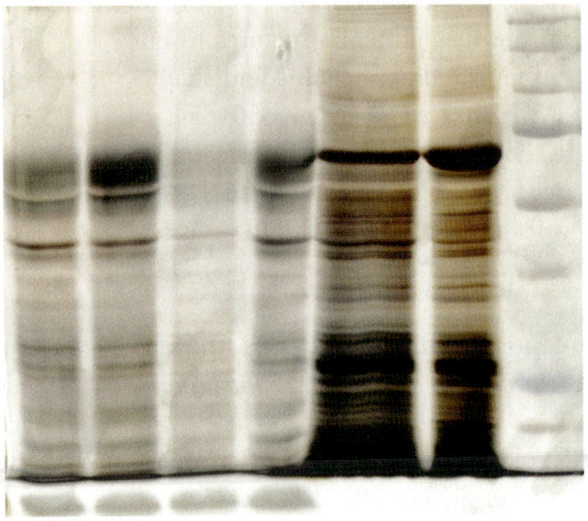

Fundamental information on the protein's sequence, structure, and location in the cell can be determined once the protein has been purified.

Biochemical methods lead to the isolation of purified proteins

The identity of unknown proteins can be determined by peptide characterization on a mass spectrometer

X-ray crystallography determines the molecular structure of the protein

Antibodies generated from an immune response to purified proteins can be linked to fluorescent markers, which are used to detect the locations of proteins within cells

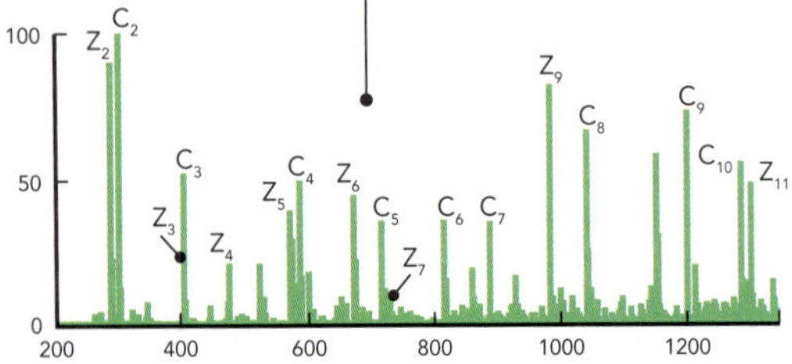

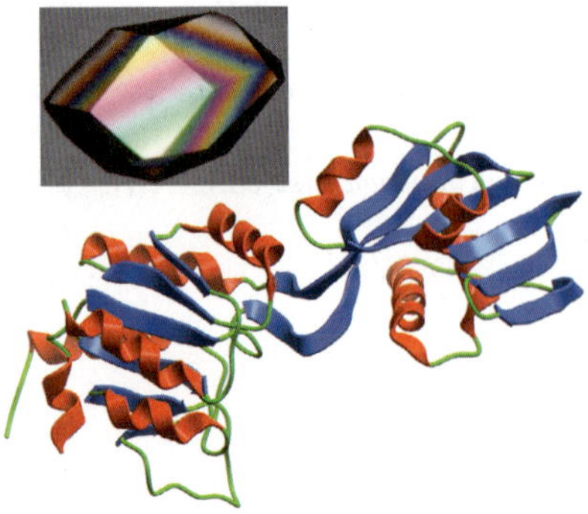

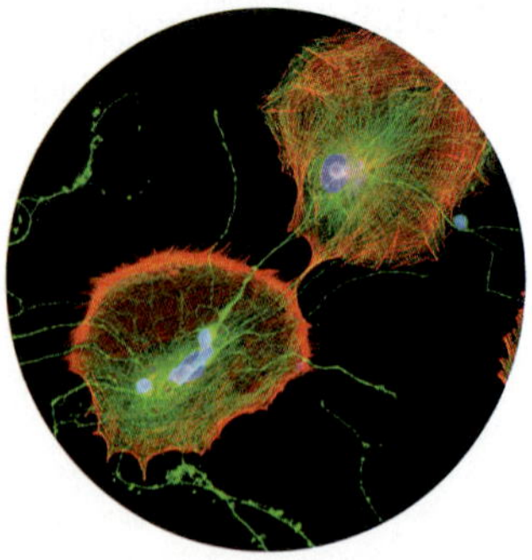

5

Methods in Protein Biochemistry

◀ Protein purification is a necessary first step in determining the molecular structure and function of a protein. The source of proteins to be characterized can be tissue homogenates, mammalian tissue culture cells, or bacterial cultures that express specific recombinant proteins.

Sodium dodecyl sulfate–polyacrylamide gel electrophoresis (SDS-PAGE), followed by staining with a protein-specific dye, provides a way to visualize the molecular mass distribution of proteins in the cell extract or to track a particular protein during purification. The physical and chemical properties of any given protein will dictate the optimized purification protocol, which often requires trial and error experimentation. Often, multiple protein purification steps, such as column chromatography, are combined to yield a homogeneous sample of purified protein.

Characterization of purified proteins can include determining the three-dimensional structure of the protein using X-ray crystallography or nuclear magnetic resonance spectroscopy. The identity of unknown proteins can be determined using the technique of mass spectrometry. Lastly, purified protein can be used as an antigen to produce polyclonal or monoclonal antibodies in animals that are then used for a variety of biochemical and cell biological analyses.

CHAPTER OUTLINE

Much of biochemistry in the 20th century focused on the development of methods to isolate biomolecules in a pure form to perform molecular analysis. Because of the central role of proteins in biochemical processes, many of these techniques focused on protein purification and analysis. Purified proteins have formed the basis for a large body of biochemical work, including the determination of structures, as we discussed in Chapter 4, and functional analyses, as described in Chapter 6 and later chapters. In this chapter, we give a brief overview of the main biochemical methods used to purify and characterize proteins and provide a working vocabulary of this important branch of applied biochemistry.

We begin by presenting the basic protein purification techniques used in a modern biochemistry laboratory and then examine chemical methods used to sequence and synthesize oligopeptides. Then we describe the two primary approaches biochemists use to decipher the three-dimensional structure of a protein: X-ray crystallography and nuclear magnetic resonance (NMR) spectroscopy. Lastly, we look at the utility of protein-specific antibodies as biochemical tools to characterize protein structure and function. We should emphasize that a typical protein purification scheme might not use all the steps we discuss, but a biochemist can pick and choose those methods that might be useful in any particular situation.

5.1 The Art and Science of Protein Purification

Proteins are involved in the vast majority of biochemical reactions, whether as reactant or catalyst (enzyme). A biological sample might contain as many as 10,000–100,000 different proteins. The challenge of protein purification is to exploit the chemical and physical properties that distinguish a particular protein from all the other proteins present in the same sample.

Humans have only about 20,000 protein-coding genes, but differences in RNA processing and protein modifications result in the production of many more proteins with distinct biochemical structures and functions. The entire constellation of proteins in an organism is called the proteome. Recent developments in high-throughput protein biochemical methods, collectively called proteomics, are now providing fresh insights into how changes in an organism's physiology alters the proteome. These physiologic changes can be in response to environmental signals such as nutrient availability or in response to cell damage and disease. We limit our discussion here to the basic biochemical methods that are commonly used to analyze individual proteins, but the same principles of protein biochemistry apply to proteomic methods on a much larger scale.

Protein purification is often called an "art and science" because it relies on exploiting the inherent qualitative and quantitative differences in the biochemical properties of individual proteins. Conditions that work for the purification of one protein may not work for any others. Therefore, biochemists need to revise protocols continually, using their own experience and intuition, to develop a successful purification strategy.

The essential component of all protein purification strategies is a specific biochemical assay that will uniquely identify a protein of interest among all other proteins

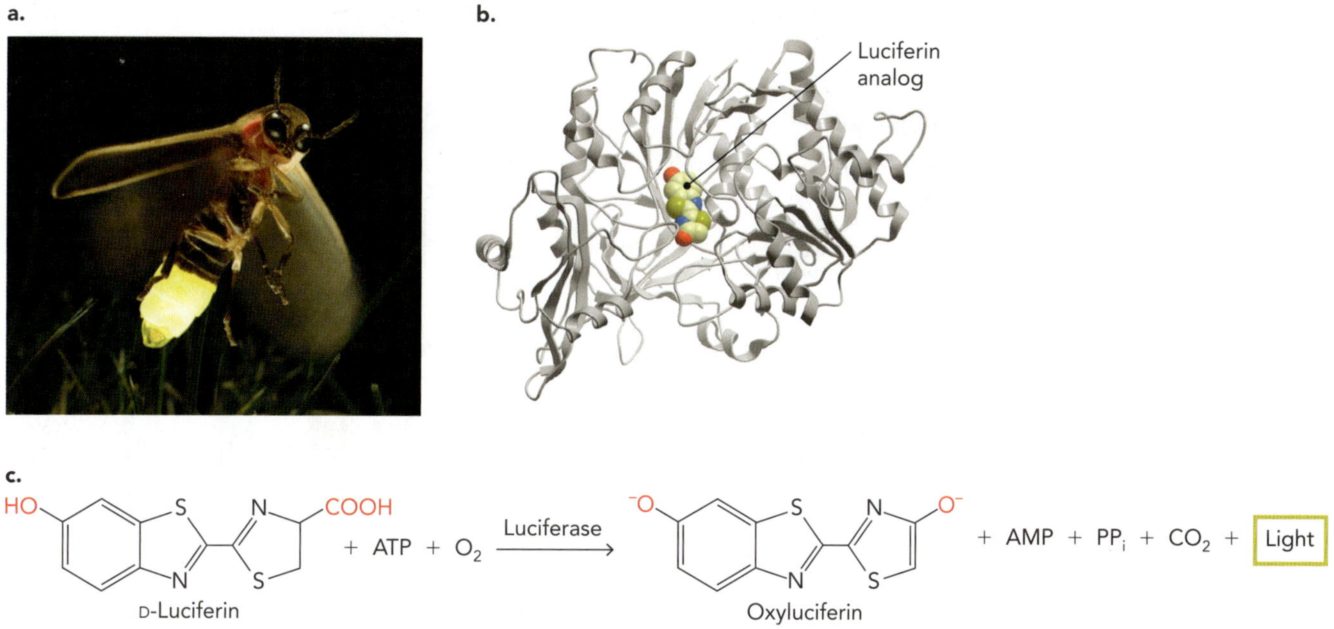

Figure 5.1 The firefly enzyme luciferase converts D-luciferin to oxyluciferin in an ATP-dependent reaction that produces light. **a.** Fireflies are a type of beetle belonging to the family Lampyridae. The light organ located under the abdomen produces light in a blinking pattern by regulating the luciferase reaction. © PHIL DEGGINGER/ANIMALS ANIMALS. **b.** Molecular structure of the luciferase enzyme from the Japanese firefly *Luciola cruciata*, showing the location of a luciferin analog in the enzyme active site. BASED ON PDB FILE 2D1R. **c.** The enzymatic reaction catalyzed by luciferase produces light, which is used as the biochemical assay to detect luciferase protein. PP_i = pyrophosphate.

contained in the sample to be analyzed. These assays often detect the product of a chemical reaction performed by the protein of interest. One of the earliest biochemical assays was designed to detect the protein **luciferase** (an enzyme found in bioluminescent organisms) when it is mixed with its substrate **luciferin**. In the presence of ATP, luciferase catalyzes a reaction that oxidizes D-luciferin to produce oxyluciferin and, in the process, generates visible light (**Figure 5.1**). Luciferase and luciferin were first discovered in 1887 by the French physiologist Raphael Dubois, who mixed a protein extract prepared from the bioluminescent mollusk *Pholas dactylus* with a heat-inactivated extract prepared from the same organism (**Figure 5.2**). Because the enzyme luciferase is inactivated by heat denaturation but the organic substrate luciferin is not, the heat-treated extract supplies only the substrate for the reaction. From monitoring the production of light after the addition of the substrate to different protein samples, he was able to detect the presence of luciferase in crude protein extracts.

Developing a highly specific and sensitive biochemical assay is the starting point for protein purification schemes. These schemes often rely on differences in solubility, size, net charge, and substrate binding affinity between the protein of interest and other proteins present in the starting material, as described in the following commonly used protein separation techniques.

Cell Fractionation

The source of proteins for most purification schemes is freshly isolated cells (such as bacterial cells grown in a liquid culture) or tissues. To prepare soluble protein that is suitable for a biochemical assay, the cells must first be disrupted using methods that do

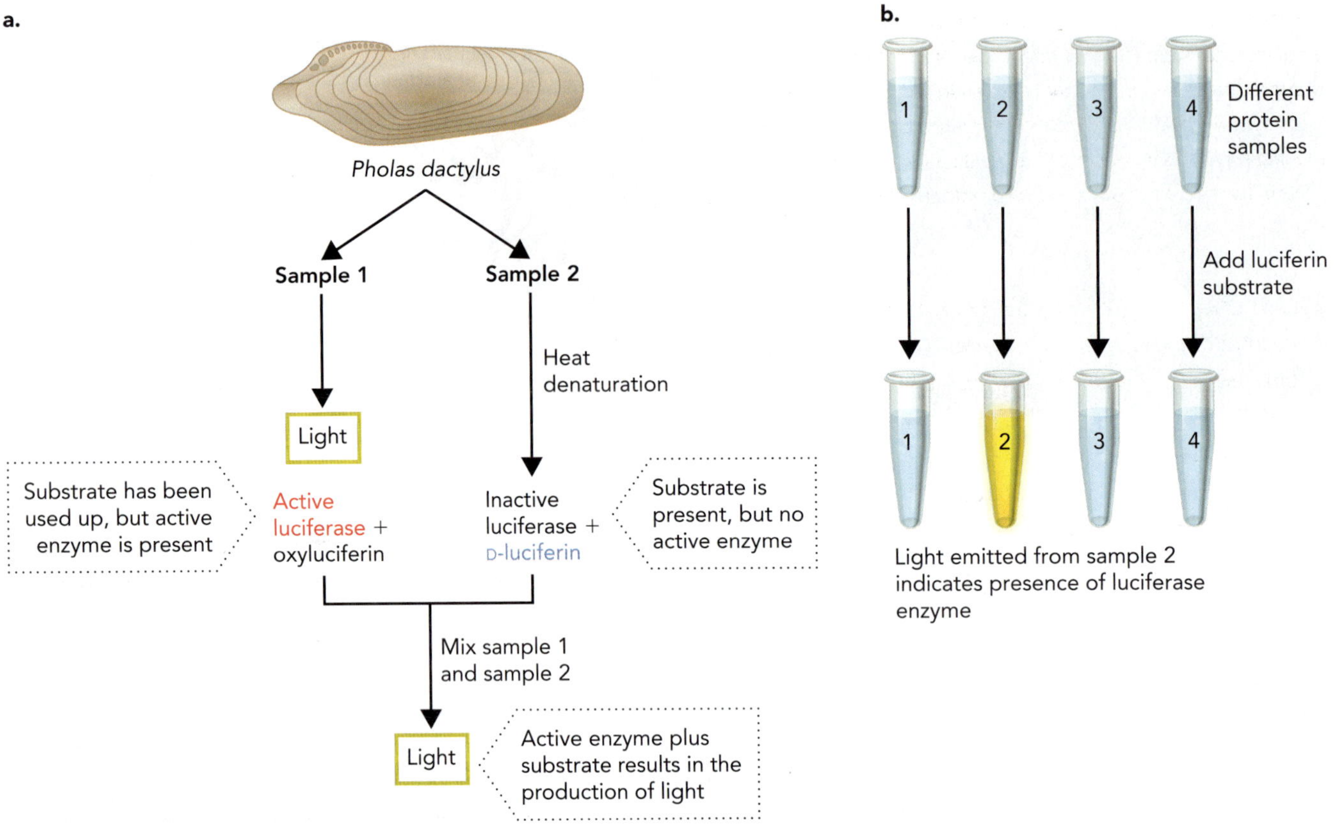

Figure 5.2 Light production in the presence of the substrate luciferin indicates the presence of the luciferase enzyme. **a.** Raphael Dubois was the first to develop a method for detecting functional proteins in a crude extract using an enzymatic assay. In his assay, a protein extract from the light organ of the marine clam *Pholas dactylus* was prepared and divided into two samples. Sample 1 was incubated under optimal conditions to convert all of the D-luciferin into oxyluciferin, whereas sample 2 was treated with heat to inactivate the luciferase enzyme, which left the substrate D-luciferin intact. At this point, individually neither sample can produce light; however, by mixing the pretreated samples together, the active enzyme in sample 1 was able to convert the D-luciferin in sample 2 into oxyluciferin and produce light. **b.** The addition of luciferin substrate to different protein extracts can be used to detect the presence of the luciferase enzyme through the production of light. In this example, four different protein samples are being tested for the presence of luciferase. Only sample 2 emits light after the addition of the luciferin substrate, indicating luciferase enzyme is present only in this sample.

not inactivate the protein. As illustrated in **Figure 5.3**, this generally involves breaking open cells that are suspended in an isotonic solution (osmotically neutral) to produce a **cell extract**, also called a cell homogenate, which retains protein activity based on the chosen biochemical assay. Cell suspensions are prepared by mincing tissue mechanically or by using enzymes that break down connective tissue to release separated cells. If the cells contain a rigid cell wall (for example, plant cells, yeast cells, or some types of bacteria), the cell wall is removed by enzymatic treatment to generate a membrane-bound cell that can be homogenized. The most commonly used homogenization techniques are (1) **sonication**, which disrupts cell membranes through the vibrational effects of ultrasonic waves; (2) shearing, which involves use of either a tight-fitting Teflon plunger in a glass vessel, a syringe, or a mechanical device called a **French press** to force the cells through a small opening; and (3) incubation of the cell sample with mild detergents, which disrupt cell membranes.

Once samples are prepared that contain the target protein (protein of interest), the next step is to increase the concentration and purity of this protein in the samples.

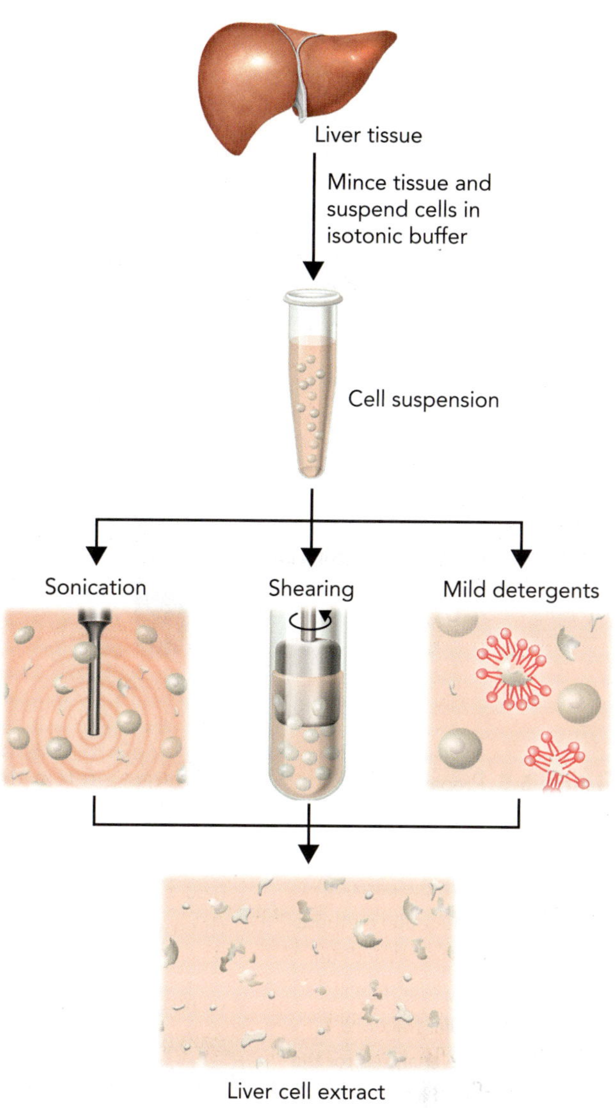

Figure 5.3 The first step in protein purification is to prepare a cell extract. Starting with a cell suspension, the cells can be disrupted to release their contents by sonication, shearing, or treatment with mild detergents. The method used to prepare the initial cell extract needs to be gentle enough to preserve protein function as determined by a biochemical assay.

This is often done by preparative **centrifugation**, which is a means of separating large macromolecules into fractions called the pellet and the supernatant. Centrifugal forces are applied by spinning solution samples in a rotor at high speeds (**Figure 5.4**) and particles of different sizes or densities will sediment at different rates. The larger or more dense particles will sediment first. Sedimentation is also affected by the density of the solvent, as well as by the size and shape of the solute molecules. Depending on the rotational speed and radius of the rotor, preparative centrifuges can produce centrifugal forces that are $\sim 10^3$ to $\sim 10^5$ times the force of gravity (g) on Earth.

In a standard preparative centrifugation protocol, it is possible to obtain four fractions from a eukaryotic cell extract based on the time and centrifugal force applied at each step (**Figure 5.5**). These four fractions contain nuclei, mitochondria, components of the plasma membrane, and the cytosol. It is important to note that because each fraction contains less total protein than the beginning cell extract, enrichment of the target protein occurs relative to the starting material. This enrichment can be expressed as a **specific activity**, which is the total amount or activity of

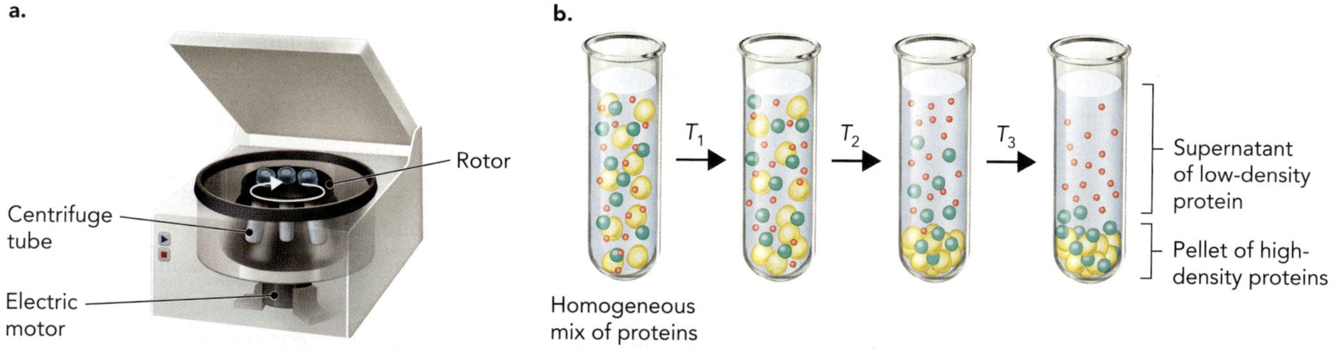

Figure 5.4 Centrifugation separates macromolecules on the basis of density, centrifugal force (multiple of g), and the centrifugation time. **a.** A centrifuge consists of a rotational motor, a refrigerated chamber to reduce heat generation during the spin, and a rotor connected to the motor that holds the balanced centrifuge tubes. The radius of the rotor and the speed of the motor determine the centrifugal force applied to the centrifuge tube. **b.** At a given centrifugal force, high-density macromolecules collect at the bottom of the centrifuge tubes before low-density macromolecules, forming a high-density pellet. In this example, T_1, T_2, and T_3 refer to increased amounts of centrifugation time.

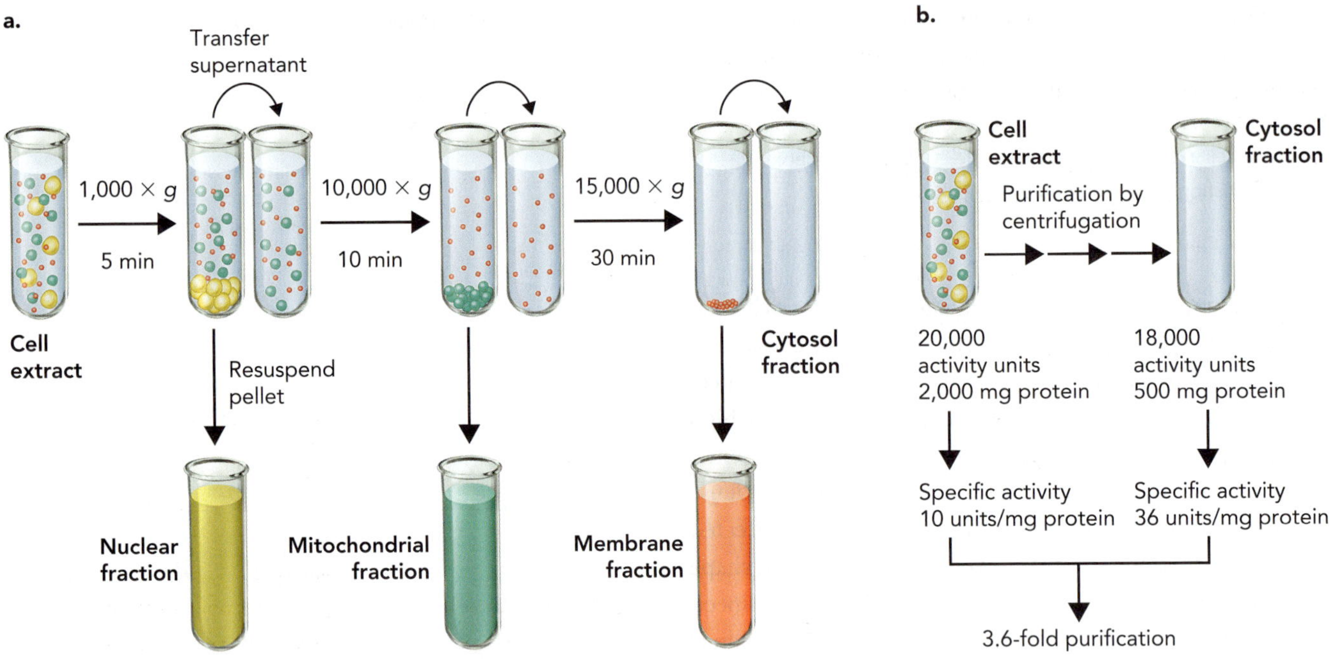

Figure 5.5 Preparative centrifugation can be used as an initial step in protein purification. **a.** On the basis of differential centrifugation methods using increasing *g* forces and run times, it is possible to obtain at least four subcellular fractions, which can each be analyzed for the target protein using a biochemical assay. **b.** Specific activity refers to the purity of the protein and is defined by the ratio of activity units to total protein. For example, after differential centrifugation, a 10% decrease in activity, combined with a 75% decrease in total protein, can result in a 3.6-fold purification of the target protein.

the target protein divided by the total amount of protein in the fraction. The total amount of protein can be estimated by measurement of the sample absorbance at 280 nm (A_{280}), which is the wavelength where tryptophan residues absorb most strongly (see Figure 4.14). Note that while A_{280} measurements are commonly used to estimate the total amount of protein, the A_{280} value is strongly dependent on the composition of the protein, in particular the number of tryptophan residues. In most cases, the amount of target protein is expressed as an activity unit. For example, if the target protein is the enzyme luciferase, then the protein activity unit would be the amount of light produced per time as measured by a luminometer. By keeping track of both the activity units and amount of total protein at each fractionation step, it is possible to calculate the increase in specific activity and the extent of purification (Figure 5.5).

The next protein separation method usually tried after centrifugation is to exploit differences in the solubility of the target protein relative to most other proteins in the fraction. The most common technique is called **salting out**, which involves adding increasing amounts of a saturated salt solution to the protein solution. Ammonium sulfate solutions are commonly used in this approach. As the NH_4^+ and SO_4^{2-} ions compete with the protein to form ionic interactions with H_2O molecules, newly exposed nonpolar surfaces of the protein interact, causing the formation of insoluble protein aggregates. Proteins have different solubility characteristics in ammonium sulfate solutions, so an ideal concentration of ammonium sulfate can be determined that will maximally separate the target protein from the rest of the proteins in either the aggregated or soluble portion of the solution. Protein aggregates formed under these conditions are often fully functional when resolubilized. Therefore, salting out can be an effective coarse separation technique leading to a high yield of functional protein.

After ammonium sulfate precipitation, the ammonium sulfate is removed using a diffusion-based technique called **dialysis**, which leaves the protein in a buffer of the proper ionic strength and pH. This procedure uses dialysis tubing, which is a semipermeable membrane made up of reconstituted cellulose that has a pore diameter smaller than the size of the target protein (**Figure 5.6**). The small pore size allows the free exchange of buffers and salts across the membrane but prevents the passage of the protein. In this technique, the protein solution is added to the inside of a dialysis bag made by clamping off both ends of a section of dialysis tubing. The dialysis bag is suspended in a large volume of an appropriate buffer, where the protein remains inside the dialysis bag, but the salts and small molecules, including acids and bases, on the inside and outside of the dialysis bag equilibrate. When dialysis is performed against a large volume of buffer or several changes of buffer, this results in a significant dilution of the ammonium sulfate that was originally inside the dialysis bag and the proteins are resolubilized.

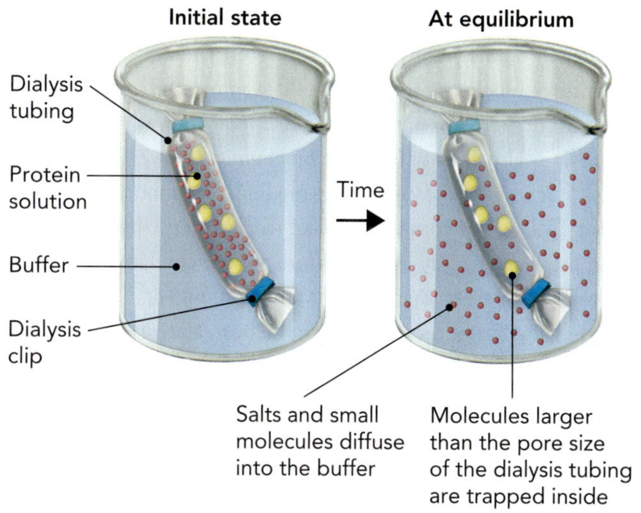

Figure 5.6 Dialysis uses a semipermeable membrane to allow equilibration of small molecules across the membrane. By equilibrating against fresh dialysis buffer in the beaker several times, the ammonium sulfate salt is effectively diluted from inside the dialysis bag and replaced with an appropriate buffer. The trapped proteins can then be removed from the dialysis bag and further purified.

Column Chromatography

A commonly used protein purification method that is both efficient and reliable is called **column chromatography**. In this method, the column (a glass or plastic cylinder) is packed with a solid matrix and it has an aqueous buffer reservoir at the top and a fraction-collecting device at the bottom. A buffered solution containing a protein extract is added to the top of the column by gravity or peristaltic pumping. Once the solution has entered the matrix, the same buffer without protein, sometimes called running buffer, is continually added to the column. Based on differential physical or chemical interactions between the proteins and the column matrix, subsets of proteins in the mixture exit the column at different times: these fractions are collected separately in smaller containers (**Figure 5.7**).

ANIMATION

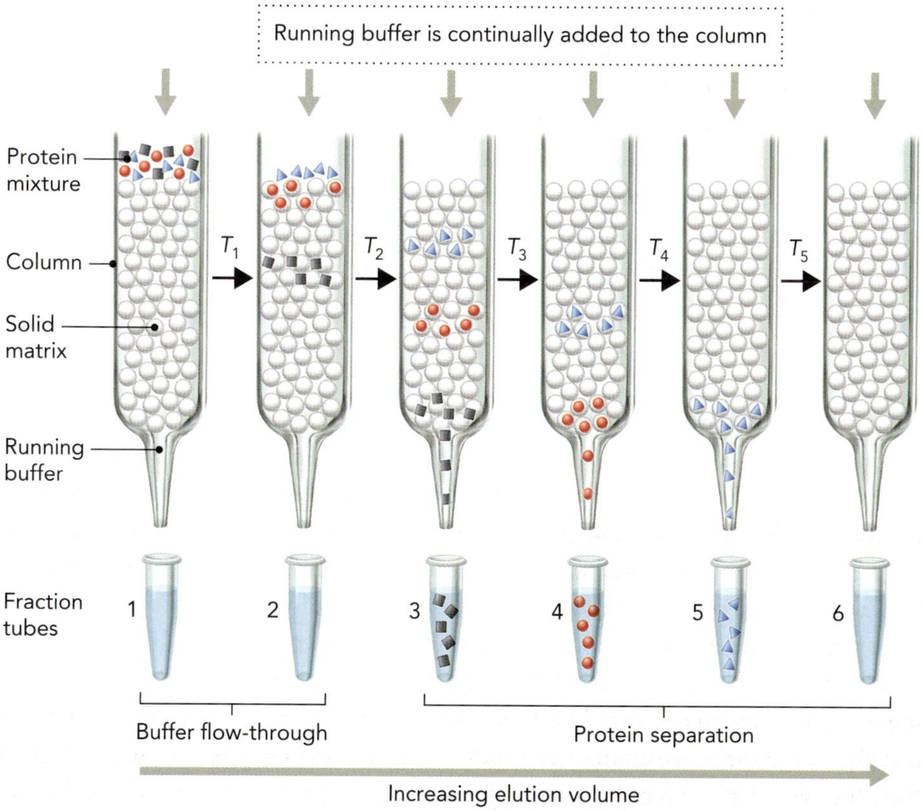

Figure 5.7 Column chromatography separates proteins on the basis of differential physical or chemical interactions with a solid gel matrix. As the proteins flow through the column, the proteins that interact poorly with the matrix are eluted first from the column and can be separated away from the other proteins, which elute more slowly. In this example, fractions 1, 2, and 6 contain buffer that has flowed through the column. Fraction 3 contains the first protein to elute from the column (black). Fraction 4 contains the next protein (red) to elute, followed by the last protein (blue) in fraction 5.

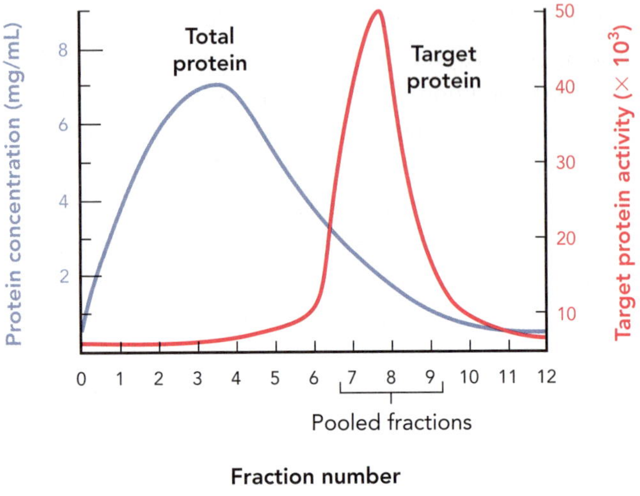

Total protein
Target protein
Protein concentration (mg/mL)
Target protein activity (× 10³)
Pooled fractions
Fraction number

Figure 5.8 The presence of a given target protein in column chromatography fractions is determined by a sensitive biochemical assay. In this example, the amount of total protein in the fractions is indicated by the blue line. The activity from the biochemical assay for each fraction is indicated by the red line. Though fractions 7–9 do not contain the most total protein, they contain the most activity, and therefore are enriched in the protein of interest. By comparing the amount (or activity) of the target protein in each fraction relative to the total amount of protein, it is possible to calculate the degree of purification. Fractions with the highest yield of target proteins are combined for further use.

For most chromatographic columns, the column matrix constitutes about 65% of the column volume. The running buffer fills the remaining 35% of the column volume, which is referred to as the **void volume.** Commonly, the volume of each fraction collected is about 10% of the total column volume, so approximately three fractions are collected per void volume. Each of the column fractions is analyzed for the presence of the target protein using a sensitive biochemical assay, such as an enzyme assay. The fractions that have significant amounts of the target protein are pooled together (**Figure 5.8**). Depending on the separation of the target protein relative to that of other proteins in the sample and on the type of column chromatography techniques that are used, this step can lead to significant increases in the specific activity of the protein sample.

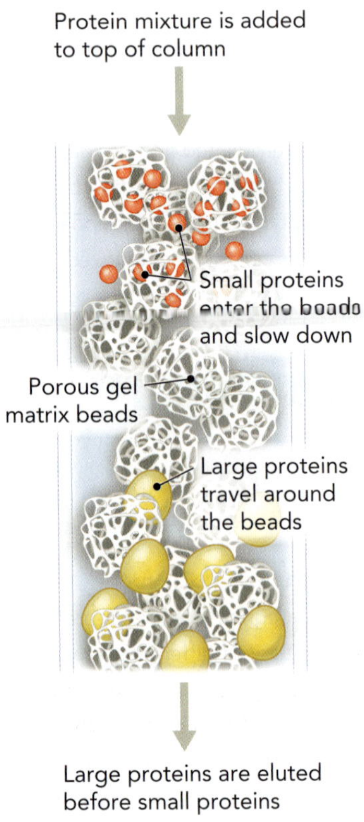

Protein mixture is added to top of column
Small proteins enter the beads and slow down
Porous gel matrix beads
Large proteins travel around the beads
Large proteins are eluted before small proteins

Figure 5.9 Gel filtration chromatography separates proteins on the basis of size. Large proteins cannot enter the pores of the gel matrix beads and elute from the column first. The smaller proteins elute later because they are slowed down by entering the beads.

Gel Filtration Chromatography There are three major types of column chromatography, each of which exploits a different physical or chemical property of proteins. The first of these is **gel filtration chromatography**, also called **size-exclusion chromatography**. This method uses porous carbohydrate beads made of dextran or agarose that separate proteins on the basis of size. The carbohydrate beads are about 0.1 mm in diameter and contain pores that allow small proteins to enter the beads, whereas large proteins flow around them (**Figure 5.9**). The net result is that large proteins flow through the column faster than small proteins, with medium-sized proteins coming off the column in intermediate fractions. Gel filtration chromatography is essentially a race of proteins through an obstacle course, with the largest proteins coming out ahead of the smaller proteins because they go around the obstacles not through them.

We can estimate the molecular mass of a protein (or intact protein complex) by gel filtration chromatography using columns containing gel matrix beads with different size-exclusion properties. If the target protein is globular in shape, then its elution profile (rate of elution) is proportional to its molecular mass. For example, large globular proteins loaded onto a gel filtration column elute early in the process if the size of the protein is larger than the pore size of the gel matrix; that is, the large proteins are excluded from the bead interior. In contrast, small globular proteins elute later in the process because small proteins enter the bead interior and are impeded. It is possible to estimate the molecular mass of an uncharacterized protein using gel filtration—assuming the protein is globular in shape—by comparing its

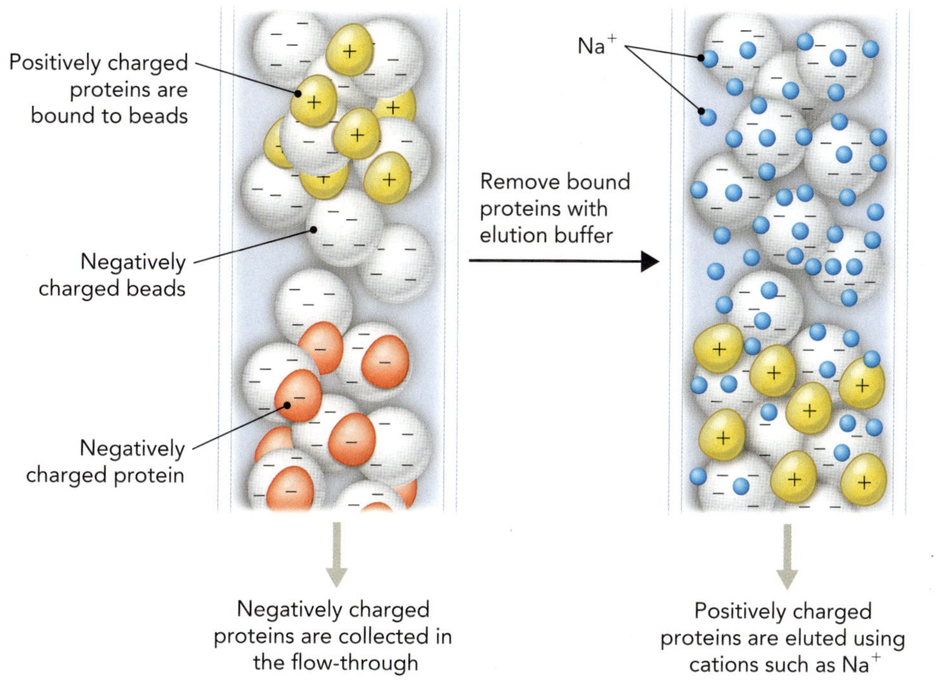
Positively charged proteins are bound to beads
Negatively charged beads
Negatively charged protein
Na+
Remove bound proteins with elution buffer
Negatively charged proteins are collected in the flow-through
Positively charged proteins are eluted using cations such as Na+

Figure 5.10 Ion-exchange chromatography separates proteins on the basis of charge. Depending on the type of resin used in the column, charged proteins are differentially retained or flow through the column. Addition of an ionic elution buffer in the final wash step releases the bound proteins from the column.

elution profile to that of proteins with known molecular masses, which serve as standards.

Both protein purification and molecular mass determination can also be performed using **high-performance liquid chromatography (HPLC)**, which is a high-resolution version of gravity-based gel filtration chromatography. The HPLC column matrix consists of smaller particle beads and leads to greater separation between proteins of similar size. Because of the tighter packing of the small particle beads, high pressure is required to force buffer and protein through the column.

Ion-exchange Chromatography The second type of column chromatography, **ion-exchange chromatography**, is a purification method that exploits charge differences between proteins. Net positively charged or net negatively charged proteins are bound differentially, depending on the charge on the column matrix and the pH of the buffer (**Figure 5.10**). Two commonly used ion-exchange matrices are a positively charged anion-exchange matrix called diethylaminoethyl (DEAE) cellulose and a negatively charged cation-exchange matrix called carboxymethylcellulose (CMC; **Figure 5.11**). An anion-exchange resin binds negatively charged proteins (anions), whereas a cation-exchange resin binds positively charged proteins (cations). Once the protein sample is loaded onto the column, the oppositely charged protein and resin interact, and the matrix is extensively washed with loading buffer to remove unbound protein. An **elution buffer** is then added to the column: This buffer contains a high concentration of an appropriate competing ion such as Na^+ or Cl^- that displaces the bound protein. The eluted fractions are assayed for the target protein, then pooled and dialyzed for further use.

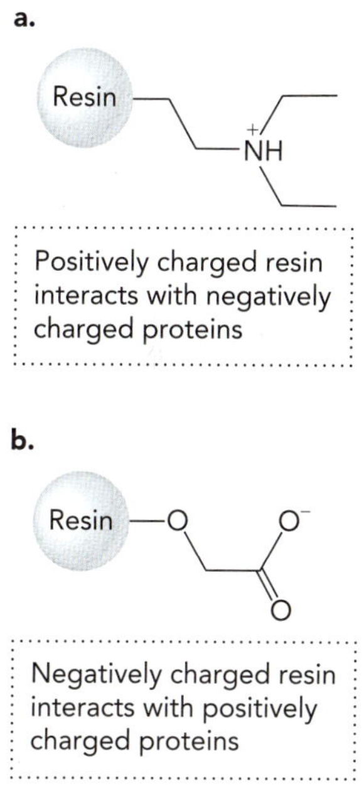
a.
Resin
NH
Positively charged resin interacts with negatively charged proteins
b.
Resin
O
O−
O
Negatively charged resin interacts with positively charged proteins

Figure 5.11 Chemical structure of two commonly used ion-exchange resins. **a.** The anion-exchange resin DEAE cellulose is positively charged and interacts with negatively charged proteins. **b.** The cation-exchange resin CMC is negatively charged and interacts with positively charged proteins.

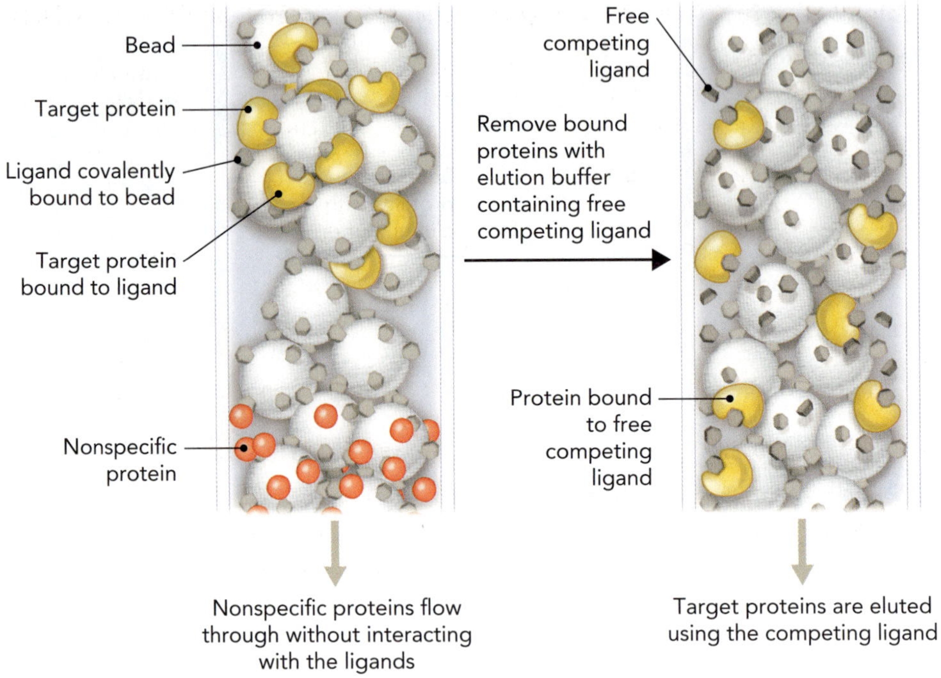

Figure 5.12 Affinity chromatography separates proteins on the basis of specific ligand interactions. The target protein specifically interacts with the ligand that is covalently bound to the resin, whereas nonspecific proteins flow through the column without interacting. Eluting the target protein from the affinity column involves either adding large amounts of a competing ligand to the elution buffer or disrupting the binding interaction with changes in salt or pH.

Affinity Chromatography The third type of column chromatography used to purify proteins is **affinity chromatography.** This method exploits specific binding properties of the target protein to separate it from other cellular proteins that lack this binding function (**Figure 5.12**). Typically, a high-affinity ligand for the target protein is covalently linked to a matrix bead, and then the protein sample is passed through the column under optimal binding conditions of pH and ionic strength. The target protein specifically associates with the column matrix under these conditions, whereas proteins without affinity for the ligand, called nonspecific proteins, flow through the column. To release the target protein from the column, an elution buffer is applied that contains an excess of a competing ligand.

Affinity chromatography can only be used if the target protein has a binding function that can be exploited by covalently linking an appropriate functional group to the matrix beads. For example, a receptor protein can be purified on an affinity column if the ligand for the receptor can be obtained in large quantities and used to prepare a column. A different type of column used in affinity chromatography is the antibody column, which isolates antigenic proteins present in the protein mixture. Another example of a commonly used affinity column is the Ni^{2+} chelating column, which is used to purify recombinant proteins that have been bioengineered to contain a series of histidine residues, which strongly coordinate divalent metals. The advantage of this approach is that the designed protein bypasses many of the trial-and-error steps in development of a purification protocol for a new protein.

Table 5.1 illustrates how a combination of protein purification steps can collectively lead to overall enrichment of a target protein by several thousand-fold relative to other proteins in the mixture. By keeping track of the total amount of protein and the units of target protein activity based on the biochemical assay, it can be seen that the specific activity of the protein preparation increases ~3- to 30-fold with each step. The affinity chromatography step is most effective in this example, owing to its high specificity for the target protein. Although high purity is an important consideration in designing a protein purification protocol, the overall yield of purified protein is also

Table 5.1 PURIFICATION OF A TARGET PROTEIN USING A COMBINATION OF BIOCHEMICAL METHODS

Protein purification method	Total protein (mg)	Total units of activity	Specific activity (units/mg)[a]	Multifold increase in purity[b]	Total yield (%)[c]	Total multifold purification[d]
Preparation of a crude cell extract	2,000	20,000	10	—	100	1.0
Subcellular fractionation using differential centrifugation	500	18,000	36	3.6	90	3.6
Ammonium sulfate precipitation	180	16,000	89	2.5	80	8.9
Gel-filtration chromatography	45	13,000	289	3.2	65	28.9
Ion-exchange chromatography	15	11,500	767	2.7	57	76.7
Affinity chromatography	0.4	9,900	24,750	32.3	49	2,475.0

[a]Specific activity is calculated by dividing the total units of activity by the total protein.

[b]Multifold increase in purity is calculated by dividing the specific activity at a given step by the specific activity from the previous step.

[c]Total yield is calculated by dividing the total units of activity at a given step by the initial units of activity, then multiplying by 100 to obtain a percentage.

[d]Total multifold purification is calculated by dividing the specific activity at a given step by the initial specific activity.

critical because this determines how much material can be obtained for biochemical experimentation. In most procedures, a low yield of a highly purified protein can be compensated for by increasing the amount of starting material.

Gel Electrophoresis

Throughout the purification process, the specific activity of the protein solution can provide a good estimation of the relative purity of the target protein. However, in order to approximate the molecular mass of a protein and to find out if the purified protein includes more than one polypeptide chain, a separation technique called **gel electrophoresis** is very useful.

As illustrated in **Figure 5.13**, **polyacrylamide gel electrophoresis (PAGE)** separates proteins on the basis of charge and size. The method uses a frame-supported molecular sieve made of polyacrylamide gel that is placed in a buffer tank where an electric field can be applied. The polyacrylamide gel matrix and electrophoresis buffer often contain the amphipathic molecule **sodium dodecyl sulfate (SDS)**, which gives the proteins a net negative charge so they migrate toward the positive electrode (anode) in the bottom chamber of the buffer tank. This technique is referred to as **SDS-PAGE**. The association of the hydrophobic tail of SDS with the nonpolar regions of the protein results in denaturation of the protein, so that migration through the matrix is not affected by the original shape of the protein. The denatured protein is coated by negative charges derived from the sulfate group on SDS. It has been determined that about one molecule of SDS associates with every two amino acid residues (1.4 g of SDS for every gram of protein). This significant negative charge contributed by the SDS molecules effectively masks the inherent charges on the amino acid residues, and therefore the charge-to-mass ratio is essentially constant between different proteins. As a result, the migration of a protein through an SDS-PAGE gel is approximately a function of molecular mass.

Unlike size-exclusion column chromatography, in which small proteins are trapped by the carbohydrate beads and travel more slowly through the matrix than large molecules, small proteins migrate faster in an SDS-PAGE gel because they can more easily maneuver through the matrix. Improved physical separation of proteins with similar

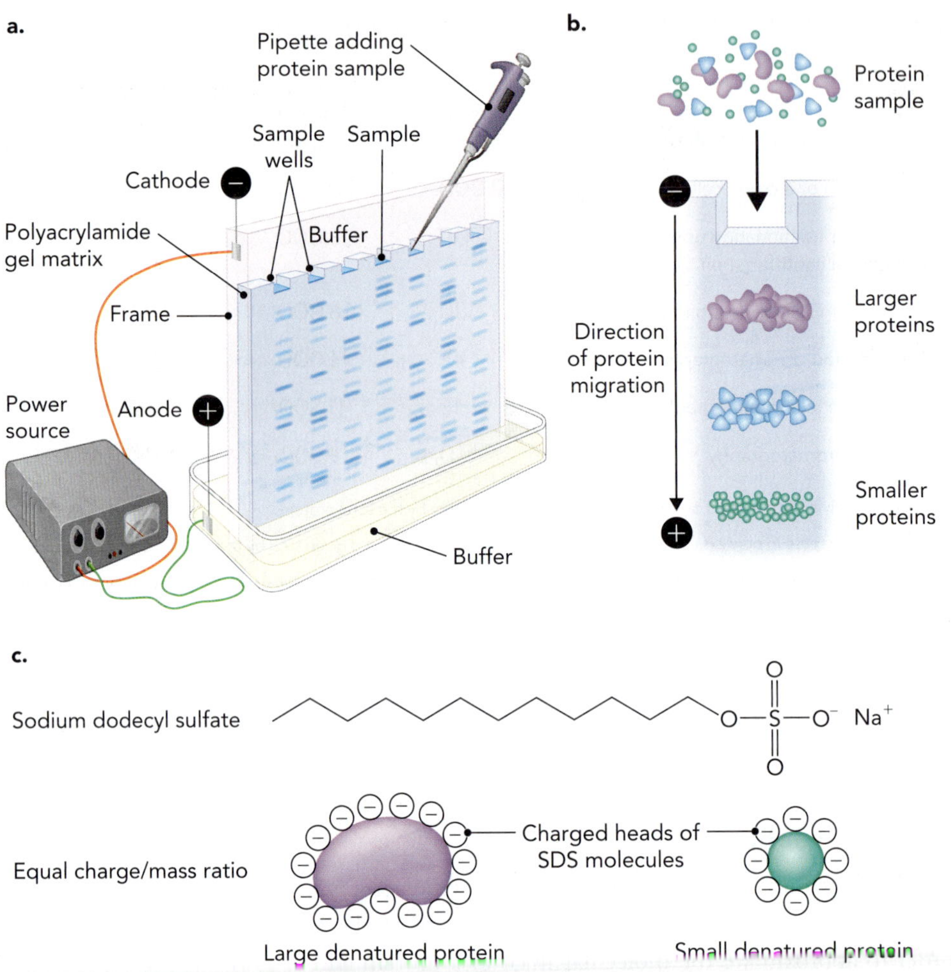

Figure 5.13 PAGE is used to separate proteins on the basis of their mass. The proteins are given a uniform negative charge relative to mass by noncovalent association with SDS. **a.** SDS-PAGE is performed using a vertical apparatus that contains buffer in the top and bottom chambers. The two chambers are connected by the gel matrix, which is sandwiched between two plates of glass or other rigid material constituting the frame. Samples are added to the wells with a pipette. An external power supply is attached and current flows between the two chambers by passing through the gel matrix, resulting in migration of the negatively charged proteins toward the anode side of the gel chamber. **b.** Small proteins migrate faster through the buffer-saturated polyacrylamide gel matrix than large proteins. **c.** SDS is an amphipathic molecule that associates with hydrophobic regions of proteins, thereby giving them an overall negative charge and equivalent charge-to-mass ratio.

▶ ANIMATION

molecular masses in an SDS-PAGE gel can be achieved by altering the density of the polyacrylamide matrix within the gel. As illustrated in **Figure 5.14a**, large proteins resolve well (are effectively separated) in gels with a low percentage of polyacrylamide, whereas small proteins resolve best in high-percentage gels. The apparent molecular mass of an unknown protein can be determined by comparing its migration distance in the gel to that of known proteins that serve as molecular-mass markers (**Figure 5.14b**).

We saw in Chapter 4 that many proteins are components of large protein complexes stabilized by numerous noncovalent interactions or consist of multiple polypeptide chains covalently linked by disulfide bridges. To separate the complexes into individual polypeptide chains, proteins are usually subjected to denaturing SDS-PAGE gel. The SDS itself disrupts noncovalent interactions, but complete denaturation of the proteins may require that they be solubilized in sample loading buffer that also contains β-mercaptoethanol (a reducing agent to break disulfide bonds) and then heated for 5 minutes in a boiling water bath. The sample loading buffer also contains 15% glycerol, which adds density to the protein sample so it sinks to the bottom of the sample well, and a negatively charged dye called bromophenol blue, which is used to monitor the rate of electrophoresis.

As an alternative to denaturing SDS-PAGE, intact protein complexes can also be characterized using nondenaturing conditions in the absence of SDS, a method referred to as **native PAGE**. In this separation procedure, proteins migrate in the electric field on the basis of their intrinsic charge at the pH of the electrophoresis buffer and on the basis of the molecular mass and overall shape of the protein. The sample loading buffer does not contain SDS or β-mercaptoethanol, and the protein is not

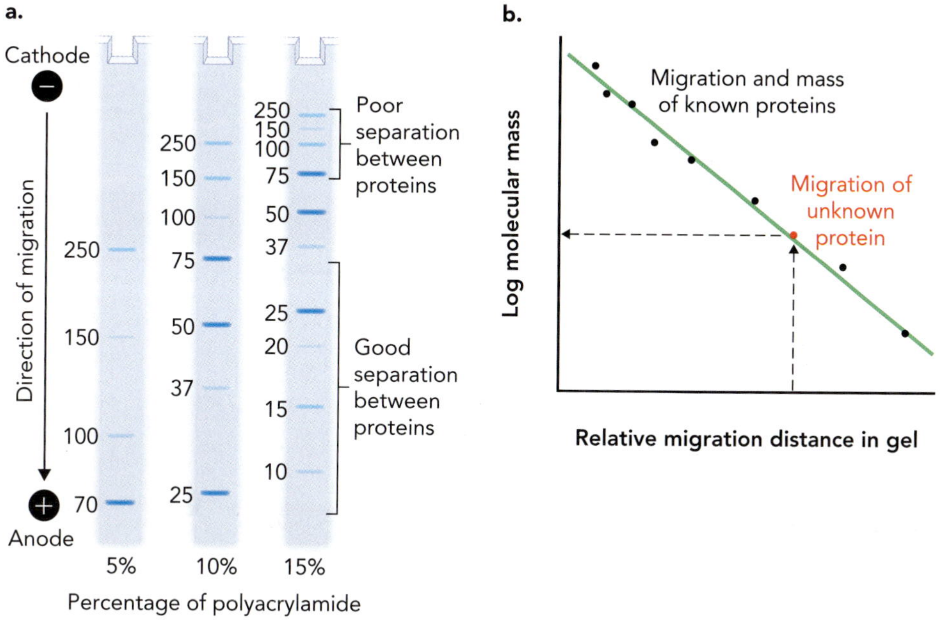

Figure 5.14 Negatively charged SDS–protein complexes migrate through the gel matrix at a rate that is proportional to their molecular mass. **a.** The migration distance of proteins as a function of time is affected by the percentage of polyacrylamide in the gel matrix. Large proteins are best separated by SDS-PAGE using low-percentage polyacrylamide gels, whereas small proteins resolve better in high-percentage polyacrylamide gels. Molecular masses are shown here in kilodaltons. **b.** By plotting the log molecular mass of known proteins versus migration distance in an SDS-PAGE system, it is possible to estimate the mass of an unknown protein based on its migration distance in the same gel.

heat denatured prior to loading. This approach can give information on the charge or conformation of a protein under a specific set of conditions.

Proteins that have been separated by native PAGE or SDS-PAGE can be visualized by staining with a dye such as Coomassie Brilliant Blue G-250, which forms ionic interactions with basic groups in proteins (**Figure 5.15**). The original Coomassie Blue dye was developed by the textile industry to dye wool; however, in the 1960s biochemists discovered it was an ideal stain for visualizing proteins separated by SDS-PAGE. (The textile chemists who developed the dye coined the name Coomassie after a West African city called Kumasi, which is located in modern-day Ghana.)

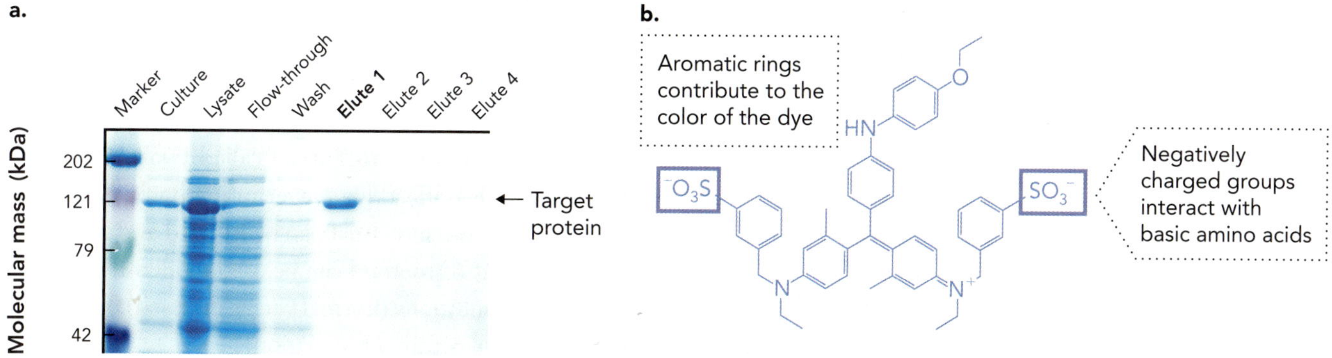

Figure 5.15 Proteins separated by SDS-PAGE are visualized by staining with a high-contrast protein dye. **a.** SDS-PAGE analysis of a protein purification scheme using affinity column chromatography. The gel was stained with Coomassie Brilliant Blue G-250, which shows the location of proteins as bands within the gel. In this example, the gel includes (from left to right) a set of standard proteins as molecular-mass markers (Marker); samples of the initial cell culture (Culture); the lysate after the cells were disrupted and centrifuged (Lysate); the proteins that flowed through the affinity column without interacting (Flow-through); the proteins that washed off the affinity column (Wash); and four elution fractions (Elute 1–4). The ~110-kDa target protein is primarily concentrated in the first elution fraction. The purity of the target protein has significantly increased through the use of the affinity column: The lysate has many other proteins present in addition to the target protein, whereas in the first elution fraction the target protein is almost completely pure. COOMASSIE-STAINED SDS-PAGE GEL PROVIDED BY ALBERTO RASCON AT THE UNIVERSITY OF ARIZONA. **b.** Chemical structure of the protein dye Coomassie Brilliant Blue G-250. The negatively charged sulfate groups interact with the basic amino acid residues in proteins.

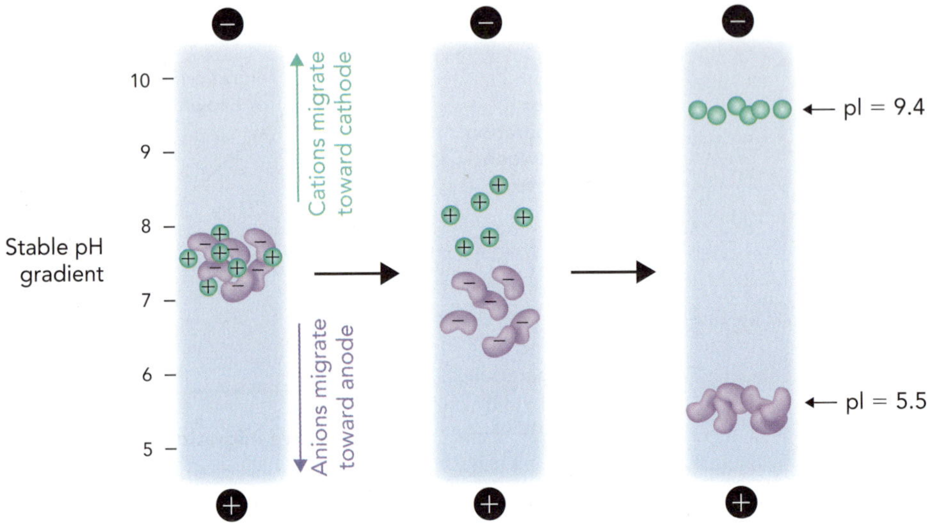

Figure 5.16 Isoelectric focusing takes advantage of the isoelectric point (pI) of a protein to separate it physically from other proteins within a gel matrix. A protein mixture is applied to a strip gel that contains a stable pH gradient. In the presence of an electric field, proteins migrate toward the oppositely charged electrode within the gel until they reach a point in the pH gradient where they have no net charge; that is, where pH = pI.

A similar polyacrylamide gel separation technique called **isoelectric focusing** separates proteins on the basis of charge as a function of pH (**Figure 5.16**). Rather than use SDS to equalize the charge-to-mass ratio, the inherent charge on a protein—at a given pH—determines the protein's physical migration in an electric field applied to a stable pH gradient. The pH gradient is established through a mixture of organic acids and bases (ampholytes) within the strip gel. The sample is applied to the gel, and when the electric field is turned on, the proteins migrate through the gel on the basis of their charge. At a pH below its isoelectric point (pI), the protein has a net positive charge, and at a pH above its isoelectric point, it has a net negative charge. The overall charge on the protein is affected by the pH of the environment, so the protein will stop migrating through the gel when it reaches a pH value where its net charge is zero (its isoelectric point) due to gaining or losing protons on its functional groups. Isoelectric focusing gels can be used to determine the isoelectric point of a protein based on the position of the protein band in the pH gradient.

Isoelectric focusing can be paired with SDS-PAGE to separate proteins on the basis of both pI and molecular mass using a technique called **two-dimensional polyacrylamide gel electrophoresis (2-D PAGE)**. This powerful method can be used to separate thousands of proteins into discrete spots that can be isolated and biochemically analyzed (**Figure 5.17**). The 2-D PAGE method can be used to identify changes in the proteome of cells under different conditions (normal versus diseased) or to compare the proteomes of different individuals within a population.

Comparing multiple 2-D PAGE gels to identify subtle differences in the abundance or migration of individual proteins can be difficult and tedious. Therefore, high-throughput 2-D PAGE methods have been developed using a single gel with differentially labeled protein samples. One such method is called **two-dimensional differential in-gel electrophoresis (2-D DIGE)**, which uses the fluorescent dyes Cy3 and Cy5 to distinguish two protein samples run on the same 2-D PAGE gel (**Figure 5.18**). In one version of this method, two closely matched protein samples are covalently labeled with Cy3 (protein sample 1) or Cy5 (protein sample 2) and then mixed together and separated by 2-D PAGE. The two closely matched protein samples can be from the same cells that were treated in different ways; for example, cultured in the absence or presence of hormone

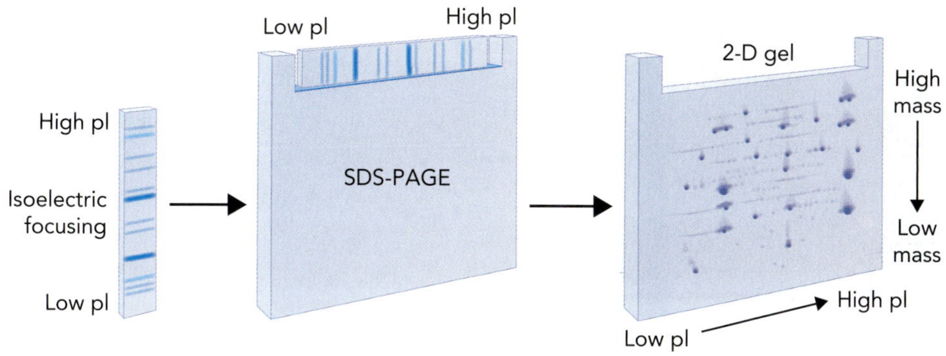

Figure 5.17 Two-dimensional polyacrylamide gel electrophoresis (2-D PAGE) is used to separate proteins on the basis of both pI and molecular mass. In this method, a protein mixture is separated in the first dimension by isoelectric focusing. Then the proteins are further separated in the second dimension by standard SDS-PAGE. The protein spots on the 2-D gel can be excised from the gel and identified by Edman degradation or mass spectrometry (see Section 5.2).

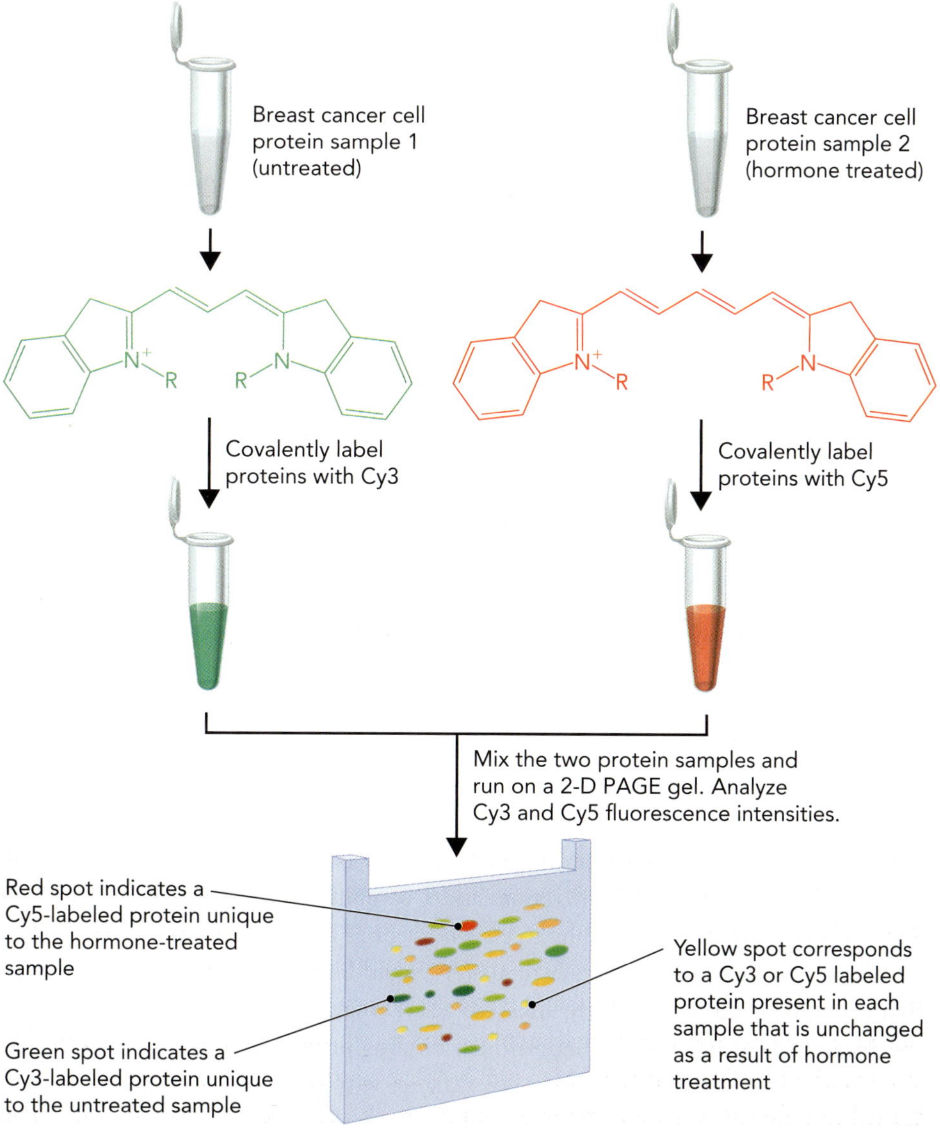

Figure 5.18 A proteomic application of 2-D PAGE is called two-dimensional differential in-gel electrophoresis (2-D DIGE), which allows for the detection of proteins that differ in abundance, charge, or molecular mass. Closely matched protein samples (for example, from untreated and hormone-treated samples) are individually labeled with Cy3 or Cy5 fluorescent dyes, mixed together, and then analyzed by a single 2-D PAGE gel. A protein unique to one sample shows a single fluorescent color, whereas proteins common to both samples appear as shades of yellow that are a mixture of green and red fluorescence.

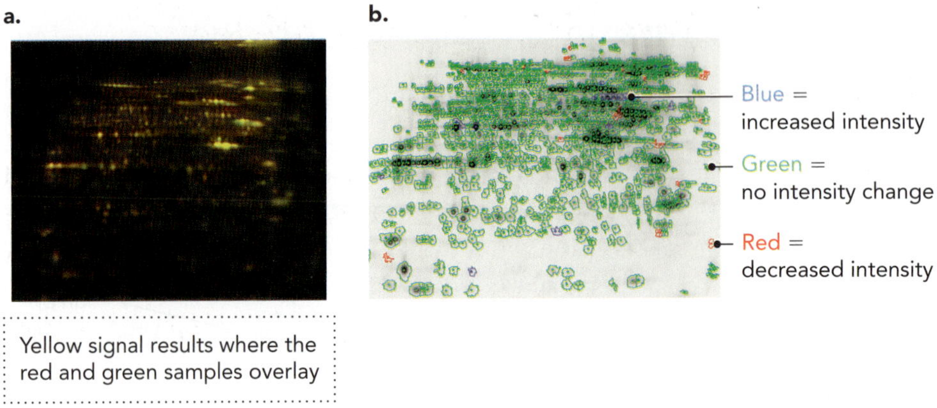

Figure 5.19 Example of protein analysis using 2-D DIGE to identify 48 proteins that differ in abundance out of ~1,300 protein spots that were detected using this method. The two protein samples are from primary human eosinophils that were cultured in the absence or presence of a glucocorticoid hormone prior to covalently labeling the untreated proteins with Cy3 or the hormone-treated proteins with Cy5. **a.** The 2-D DIGE experiment shows an abundance of yellow spots, which result from the overlay of red and green fluorescent spots. **b.** The computer analysis uses false colors to identify spots with relative intensity differences; in this case, blue is increased intensity, green is no intensity difference, and red is decreased intensity. Of the approximately 1,300 spots, 1,250 are colored green, 27 are red, and 21 are blue. 2-D DIGE ANALYSIS DATA PROVIDED BY SUSAN KUNZ AT THE UNIVERSITY OF ARIZONA.

or samples taken at different times after a single treatment. By scanning the 2-D PAGE gel for fluorescence emission at 570 nm (Cy3) or 670 nm (Cy5), the location and relative abundance of the proteins in each sample can be determined. When the two data sets are merged, relative changes in protein abundance at identical positions in the gel can be quantified (**Figure 5.19**). The intensity of the fluorescence emission is an indication of protein abundance, with more abundant proteins having a more intense fluorescent signal.

concept integration 5.1

A frog species was found to contain a cytosolic liver protein that bound a pharmaceutical drug present at high levels in effluent from a wastewater facility. Describe how this protein could be purified.

The first step in purifying an uncharacterized protein is to develop a method to detect it specifically, such as an enzyme activity assay or binding assay. In this case, the protein is known to bind to a small molecule (pharmaceutical drug), and this binding activity can be used to develop a protein detection assay. The assay could be based on protein binding to the drug that has been radioactively labeled or it might be possible to develop a fluorescently labeled version of the drug that has an altered absorption or emission spectrum as a function of specific protein binding. The next step would be to use cell fractionation, centrifugation, and a combination of gel filtration and ion-exchange column chromatography to enrich for drug binding activity relative to total protein in the frog liver extract. A final step would be to develop an affinity column that contains the drug covalently linked to a solid matrix and use this column to bind specifically, and then elute, the high-affinity binding protein. The purity of the protein would be assessed by SDS-PAGE at several steps within the purification protocol.

5.2 Working with Oligopeptides: Sequencing and Synthesis

Today, genomics plays an ever-increasing role in characterizing the structure and function of individual proteins through expression of cloned gene sequences. However, in many cases it is still necessary to identify an unknown purified protein by determining its amino acid sequence biochemically. This is especially true in many proteomic methods where proteins can be separated by 2-D PAGE and are then subjected to amino acid sequence analysis using very small amounts of sample. Moreover, the ability to synthesize short peptides *in vitro* for use in biochemical assays or for antibody production (see Section 5.4) is another important application of protein biochemistry.

In this section, we briefly review the most commonly used methods for biochemical analysis of oligopeptides. We begin with protein identification by Edman degradation, describe mass spectrometry analysis, and conclude with a description of solid-phase peptide synthesis.

Edman Degradation

A chemical method to deduce the amino acid sequence of short polypeptides was first developed in the early 1950s by Frederick Sanger, a biochemist at Cambridge University, using the reagent 1-fluoro-2,4-dinitrobenzene. The protein sequencing method developed by Sanger was based on chemical modification of the amino-terminal residue by 1-fluoro-2,4-dinitrobenzene, followed by acid hydrolysis of the protein and identification of the dinitrophenyl-amino acid derivative using chromatography. This method was labor intensive, requiring cleavage of proteins into small polypeptides and addition of fresh protein after each reaction because the acid hydrolysis destroyed the sample. Indeed, determining the amino acid sequence of insulin, a feat that earned Sanger the 1958 Nobel Prize in Chemistry, required more than 100 g of purified insulin protein though the protein only contains 51 amino acid residues.

Sanger switched to studying nucleic acids in the 1960s, and in 1975 he developed a method to sequence DNA using DNA polymerase and 2′,3′-dideoxynucleotides (described in Section 3.3). This second breakthrough in biopolymer sequencing earned him the 1980 Nobel Prize in Chemistry, which he shared with Walter Gilbert and Paul Berg (**Figure 5.20**). Sanger is the only person to date to have been awarded two Nobel Prizes in Chemistry.

Pehr Edman, a Swedish biochemist, improved Sanger's protein sequencing method by modifying the N-terminal amino acid tagging reaction using phenylisothiocyanate (PITC) rather

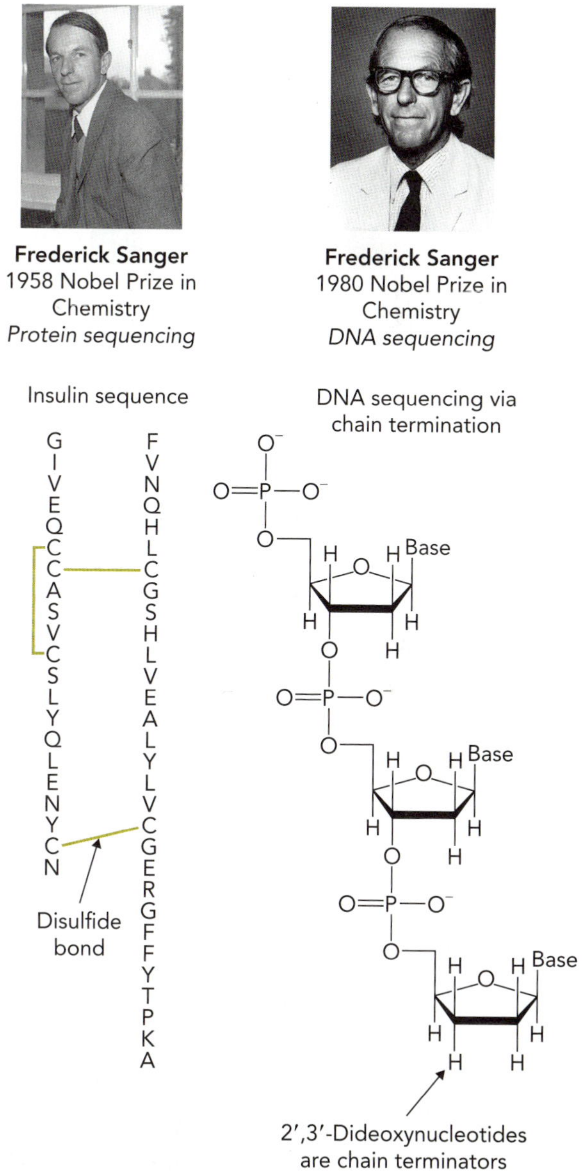

Figure 5.20 Frederick Sanger (1918–2013) was awarded the 1958 Nobel Prize in Chemistry for development of a chemical cleavage method to sequence oligopeptides. He also shared the 1980 Nobel Prize in Chemistry for his enzyme-based method for 2′,3′-dideoxynucleotide DNA sequencing. 1958: KEYSTONE/GETTY IMAGES; 1980: SCIENCE SOURCE.

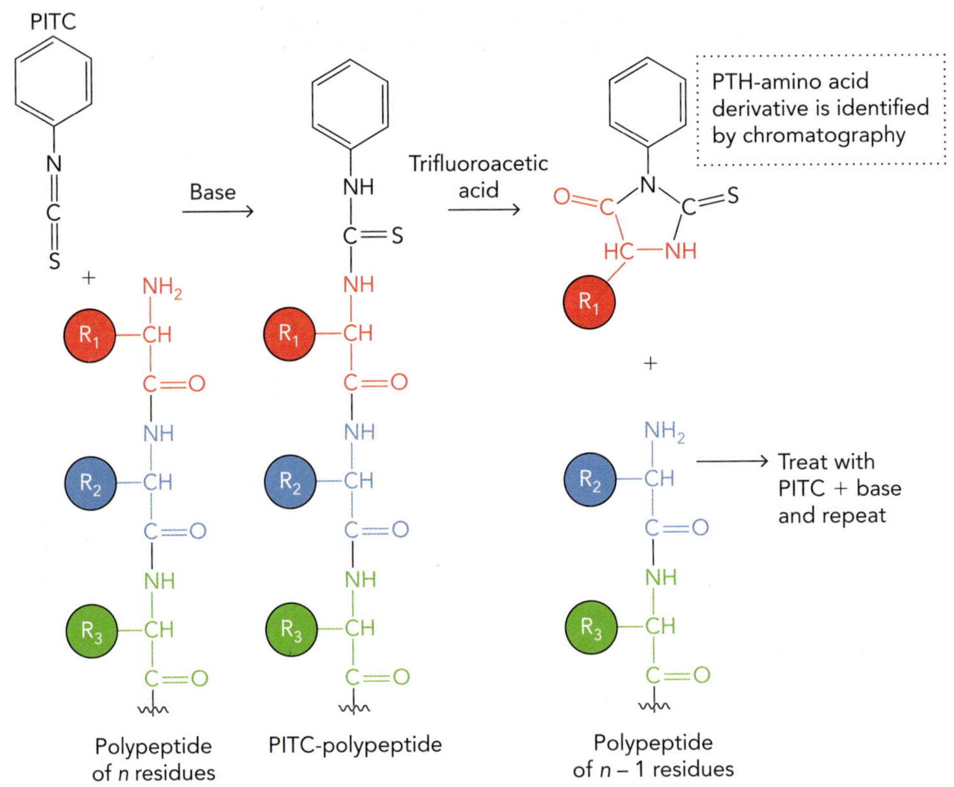

Figure 5.21 The Edman degradation method provides protein sequence information based on chemical determinations of N-terminal amino acids. The polypeptide is chemically tagged at the N terminus with PITC in the presence of base. Acid hydrolysis yields a PTH-amino acid derivative, which can be identified by HPLC or mass spectrometry using PTH-amino acid standards.

than 1-fluoro-2,4-dinitrobenzene (**Figure 5.21**). In the **Edman degradation** procedure, the PITC-modified peptide is treated with trifluoroacetic acid. This cleaves the polypeptide bond between the first and second amino acid, releasing a thiazolinone derivative and the original polypeptide—less one amino acid. The thiazolinone derivative is isolated by an organic extraction and converted to a phenylthiohydantoin (PTH) derivative, which is identified by chromatography against known PTH-amino acid standards. The next amino acid in the polypeptide chain is identified by repeating the N-terminal tagging reaction—using the same protein sample—followed by cleavage with trifluoroacetic acid and identification by chromatography. The advantage of the Edman degradation procedure over Sanger's original protein sequencing method is that it does not require the input of additional protein after each round of cleavage. Automated instrumentation based on the Edman degradation procedure can sequence an oligopeptide of up to 50 amino acid residues using less than 1 μg of starting material (10–100 pmol of oligopeptide).

Because of limitations in the efficiency of each reaction in the Edman degradation procedure, it is not possible to obtain the sequence of a polypeptide much longer than 50 amino acids in a single set of reactions. Therefore, most proteins must be cleaved into smaller polypeptides that are individually purified and then subjected to Edman degradation. This cleavage of the protein into smaller fragments can be done with protease enzymes such as **trypsin**, which cleaves polypeptide chains on the carboxyl side of lysine or arginine residues, or **chymotrypsin**, a protease that cleaves on the carboxyl side of tyrosine, tryptophan, and phenylalanine residues.

A popular strategy for using Edman degradation is to obtain overlapping fragments that can be reassembled based on shared amino acid sequences. Initially, a portion of the purified protein is cleaved with trypsin to obtain one set of fragments (referred to as tryptic fragments). Then, another portion of the protein preparation is cleaved with

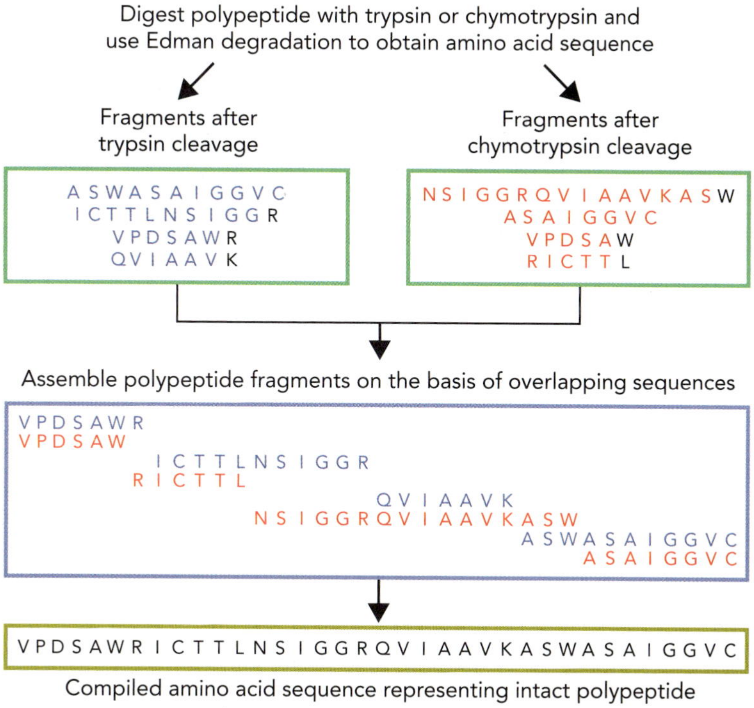

Figure 5.22 Differential protease cleavage is used to generate small overlapping polypeptides that can be sequenced by Edman degradation. Trypsin cleaves on the carboxyl side of lysine (K) and arginine (R) residues, whereas chymotrypsin cleaves on the carboxyl side of tyrosine (Y), tryptophan (W), phenylalanine (F), leucine (L), and methionine (M) residues.

chymotrypsin (**Figure 5.22**). Not all proteins have convenient cleavage sites for trypsin and chymotrypsin that provide key overlap fragments for reassembly; thus, sometimes it is necessary to use other protease enzymes or chemical reagents. For example, a secreted protease from the bacterium *Staphylococcus aureus*, called V-8 protease, cleaves proteins on the carboxyl side of aspartate and glutamate residues. Another common reagent is cyanogen bromide, which cleaves on the carboxyl side of methionine residues.

Mass Spectrometry

The second method of protein sequencing involves the application of mass spectrometry to measure the mass of small peptide fragments. **Mass spectrometry** measures the mass-to-charge ratio (m/z) of molecules, which can then be used to deduce the molecular mass. The basic operating principle is that in a mass spectrometer, an applied magnetic field exerts a force on the molecule, which directly affects its acceleration. This relationship is described by Newton's second law, $F = ma$, in which the applied force (F) equals the product of the molecule's mass (m) and acceleration (a). By measuring the molecule's acceleration along a curved path and knowing the applied force, it is possible to calculate the molecule's mass.

The application of mass spectrometry to protein sequencing was made possible by the development of methods to ionize peptides without disintegrating them. One technique is called **electrospray ionization (ESI)**, which releases polypeptides—usually polypeptide fragments of a trypsin-cleaved protein (tryptic fragments)—out of a small metallic capillary at high voltage, under conditions that cause the solvent containing the peptides to rapidly evaporate. This process generates a highly charged molecule in the gas phase. A second method for generating peptide ions is called **matrix-assisted laser desorption/ionization (MALDI)**, in which tryptic fragments are embedded in a light-absorbing matrix. The fragments are released as charged molecules after exposure to the flash of a laser. A detector in the mass spectrometer measures the acceleration of the ions and thereby determines their mass.

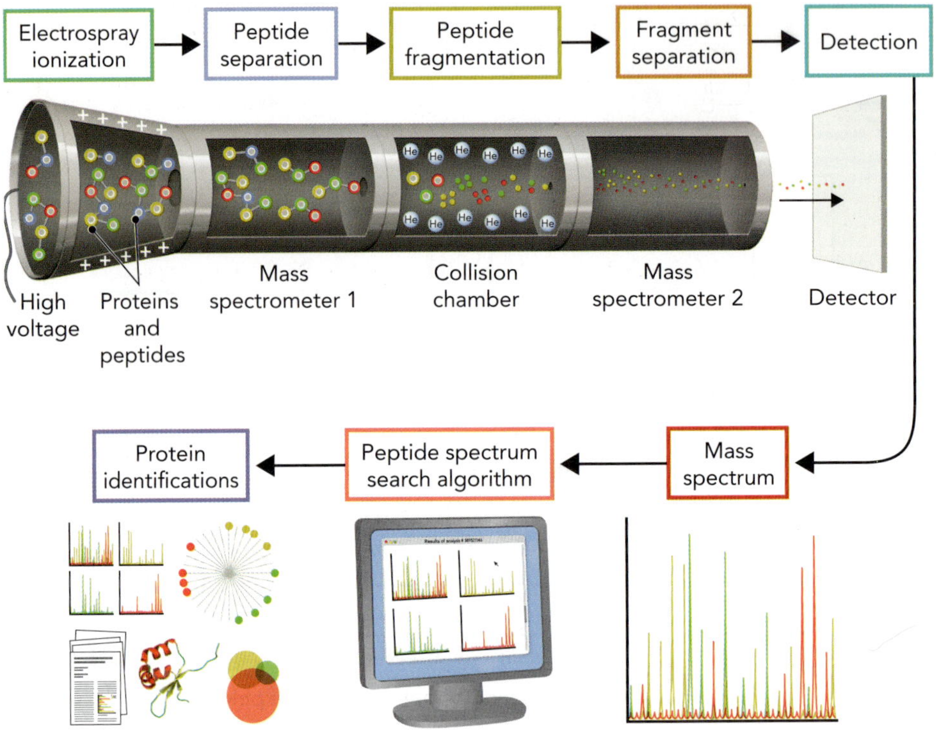

Figure 5.23 Tandem mass spectrometry is used to predict the amino acid composition of tryptic fragments using bioinformatic approaches. Tryptic fragments enter the first mass spectrometer after electrospray ionization. A narrow range of masses is selected and allowed to enter the collision chamber, which uses helium to fragment the peptides into smaller molecules. Size-selected peptide ions enter the second mass spectrometer, which is connected to a detector and the data are recorded as a spectrum. In the final step, predicted masses of *in silico* tryptic fragments generated from a genomic DNA sequence database are matched up to the actual peptide masses obtained from mass spectrometry. High-probability matches are tested directly using protein biochemistry methods.

The identity of a protein can be determined by a method called tandem mass spectrometry, which uses two mass spectrometers. As illustrated in **Figure 5.23**, the first mass spectrometer selects out individual peptide fragments produced by electrospray ionization. The second mass spectrometer determines the masses of subfragments generated by a collision chamber placed between the two mass spectrometers. The resulting mass spectrum is a plot of intensity versus mass-to-charge ratios. The mass spectrum can be used to determine the amino acid composition of each fragment by comparing the precisely determined masses of the peptides to the predicted masses of peptides generated *in silico* (by computer) using protein sequences contained in genome databases. Because trypsin cleaves on the carboxyl side of lysine and arginine residues, it is possible to match experimentally determined molecular masses of tryptic fragments to those of predicted tryptic fragments contained in the database (Figure 5.23).

Solid-Phase Peptide Synthesis

The technique of **solid-phase peptide synthesis** is used routinely to synthesize peptide antigens for antibody production and to manufacture peptide-based therapeutic drugs to treat a variety of diseases. The basic strategy for solid-phase peptide synthesis was worked out in 1962 by Bruce Merrifield, a biochemist at Rockefeller University. It involves adding one amino acid at a time to the peptide through a covalent linkage on the amino terminus. Note that this is the opposite direction of *in vivo* polypeptide synthesis, in which amino acids are added one at a time to the carboxyl terminus of the growing polypeptide chain.

As illustrated in **Figure 5.24**, solid-phase peptide synthesis can be broken down into five steps. In step 1, the C-terminal amino acid in the peptide (AA1) is attached to a resin molecule that serves as the column matrix, and the N-terminal blocking group, abbreviated as Fmoc (9-fluorenylmethoxycarbonyl), is removed by treatment with a base. In step 2, the incoming amino acid (AA2), which is blocked on the N-terminal

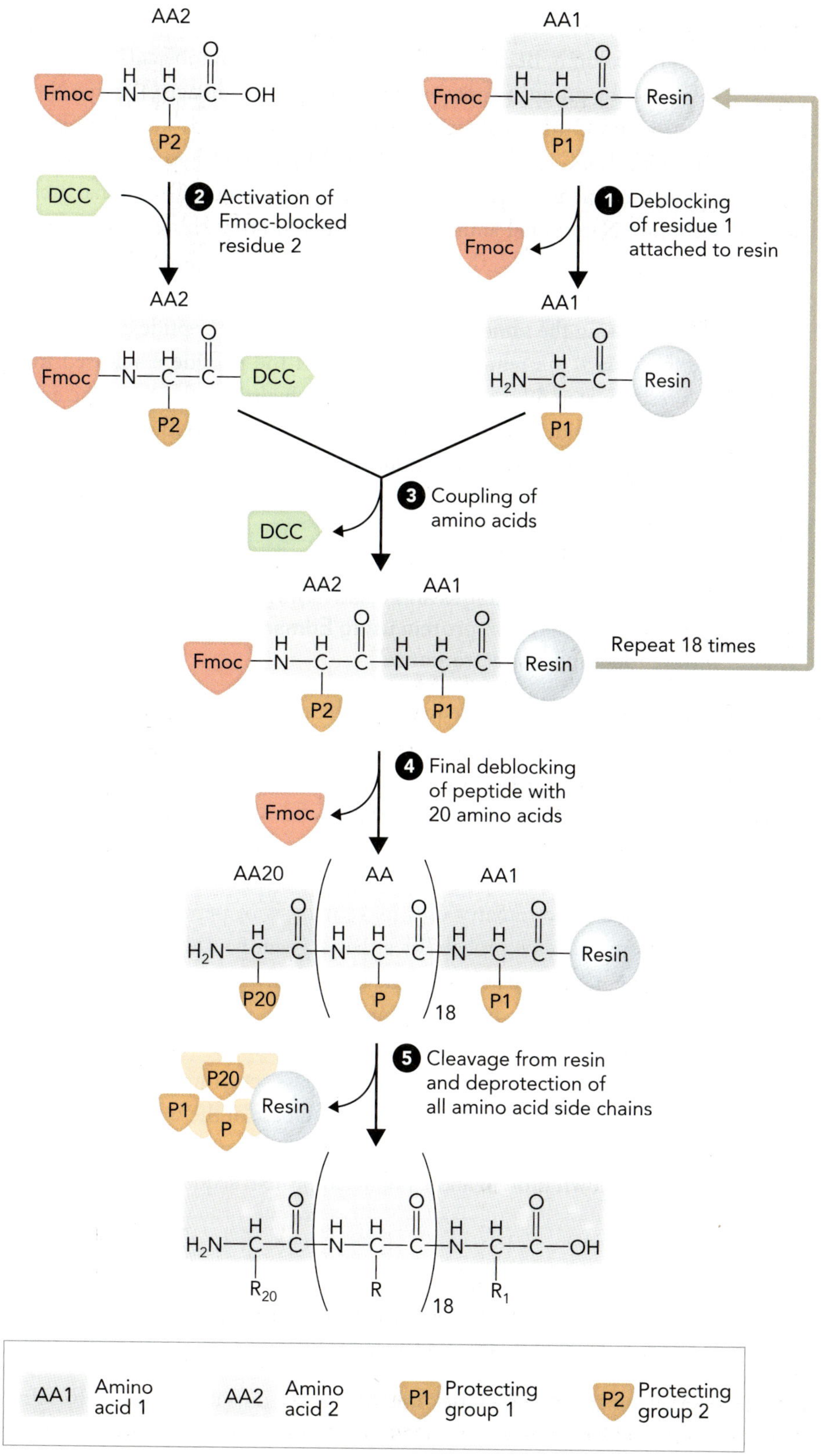

Figure 5.24 Solid-phase peptide synthesis can be automated to generate polypeptides of up to 25 amino acid residues efficiently in a few hours. The five-step procedure begins with the C-terminal amino acid being covalently attached to a solid matrix (resin), and Fmoc is removed from the amino group. This permits covalent linkage to an activated carboxyl group (DCC) on the incoming Fmoc-blocked amino acid. Note that the side chains of each amino acid must also be appropriately blocked by protecting groups to prevent aberrant side reactions (shown here as P1, P2, P, P20). After this initial coupling, an additional 18 coupling reactions are performed to generate a 20-amino-acid-residue polypeptide. In the final step, the protecting groups are removed, and the polypeptide is released from the matrix.

side with Fmoc, is activated at the carboxyl group by dicyclohexylcarbodiimide (DCC) and then added to the column. Because the side chains of the amino acids are chemically blocked by protecting groups to prevent aberrant covalent linkages, the free N terminus of the resin-bound C-terminal amino acid residue is the only available reactive group for the incoming DCC-activated amino acid. As shown in step 3, this chemical strategy leads to very efficient coupling between the resin-bound C-terminal amino acid and the incoming N-terminal amino acid. After column washing to remove any unreacted Fmoc–amino acid–DCC, the Fmoc group on the dipeptide is removed by base treatment. Then, the next Fmoc–amino acid–DCC is added to the column, generating a tripeptide by using the same sequential steps. Once the peptide is completed, all of the protecting groups are removed in step 4, and the peptide is released from the solid support resin in step 5 by ester cleavage using hydrogen fluoride. The entire process of solid-phase peptide synthesis is automated and can routinely be used to generate peptides of 15 to 25 amino acid residues in length in a matter of hours.

concept integration 5.2

What are some advantages and disadvantages of determining the identity of an unknown protein using Edman degradation compared with tandem mass spectrometry?

Edman degradation can determine the sequence of the amino terminal end of a protein. The technique is based on labeling the N-terminal amino acid with phenylisothiocyanate, cleaving it off with trifluoroacetic acid, and then identifying the phenylthiohydantoin moiety using chromatography and known amino acid standards. The disadvantage of using Edman degradation is that it only sequences proteins from the amino terminal end and will ultimately only cover a small region of the total protein sequence. Protein identification using mass spectrometry uses mass analysis of peptide fragments and computer algorithms to calculate probabilities that a given peptide corresponds to a predicted protein using genome databases. The biggest advantage of mass spectrometry over Edman degradation for protein identification is that complex mixtures of protein subfragments can be analyzed without prior purification, which cuts down on sample preparation time. However, a disadvantage of mass spectrometry is that protein identification requires access to a genomic sequence database of the species being studied. Also, protein identification is indirect because it is based only on probabilities of a match between the unknown protein and a predicted protein sequence.

5.3 Protein Structure Determination

In 1953, James Watson and Francis Crick used the X-ray diffraction data of Rosalind Franklin to model the structure of DNA, as described in Chapter 3. However, the first protein molecular structure was not solved by X-ray crystallography until 1957, when John Kendrew reported a low-resolution model of the monomeric oxygen transport molecule myoglobin (**Figure 5.25**). Two years later, Kendrew determined an atomic-level structure of myoglobin, which was the same year that Max Perutz reported a low-resolution structure of the tetrameric oxygen transport protein hemoglobin. Kendrew and Perutz, both at the Medical Research Council in Cambridge,

a.

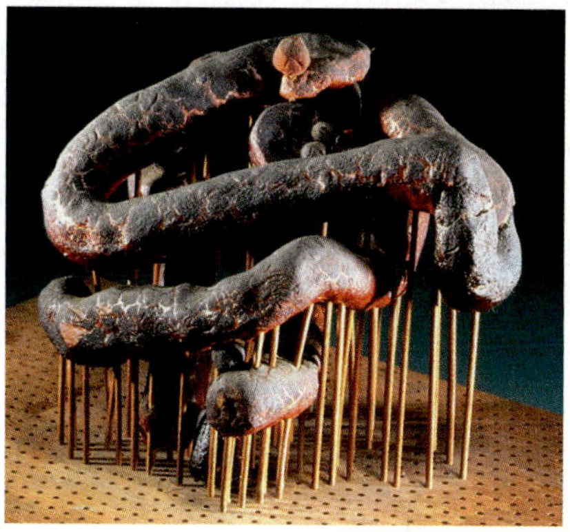

b.

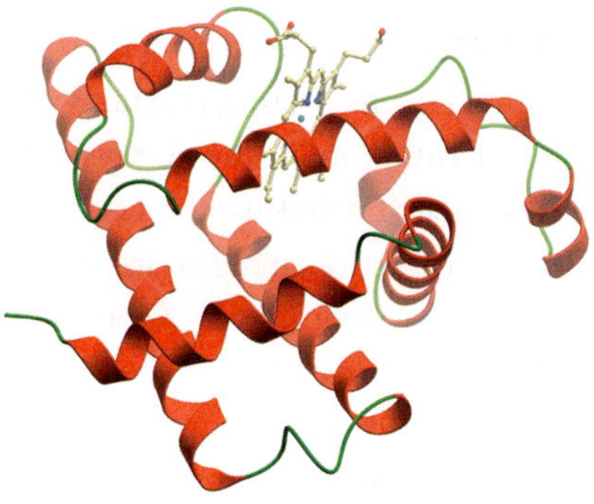

Figure 5.25 John Kendrew was the first to use X-ray diffraction data of a protein crystal to build a low-resolution, three-dimensional model of a protein structure. SSPL VIA GETTY IMAGES. **a.** This Plasticine model shows Kendrew's early structure of the sperm whale myoglobin protein. **b.** Molecular structure of sperm whale myoglobin shown in a modern ribbon representation. BASED ON PDB FILE 1MBO.

England, shared the 1962 Nobel Prize in Chemistry for their accomplishments. This was the same year that Watson and Crick were awarded the Nobel Prize in Physiology or Medicine for discovery of the structure of DNA.

Today, biochemists have two primary methods available for determining the molecular structure of proteins: X-ray crystallography and nuclear magnetic resonance (NMR) spectroscopy. We begin this section with a brief comparison of these two methods, then we discuss each one in turn, starting with X-ray crystallography.

X-ray crystallography is based on the diffraction of X-rays by protein crystals. The intensities and directions of the diffracted X-rays can be used to determine the three-dimensional distribution of electron density arising from the protein's atoms. In contrast, **NMR spectroscopy** detects nuclear spin properties of certain atoms (^{1}H, ^{13}C, ^{15}N) to deduce their relative locations. In simple terms, X-ray crystallography provides a snapshot in time of protein structure because it gives atomic positions in a static protein crystal lattice. NMR spectroscopy, in contrast, uses a high-strength magnetic field to gather information on the relative positions of certain atoms in a protein solution. In cases where it has been possible to use both X-ray crystallography and NMR spectroscopy to solve the structure of the same protein, it has been found that the two methods are often complementary and provide similar structures (**Figure 5.26**). X-ray crystallography is currently the more common of the two methods because full structure determination by NMR spectroscopy is generally applicable to only relatively small proteins (less than ~40 kDa), though in favorable cases, this size limit may approach ~100 kDa.

a.

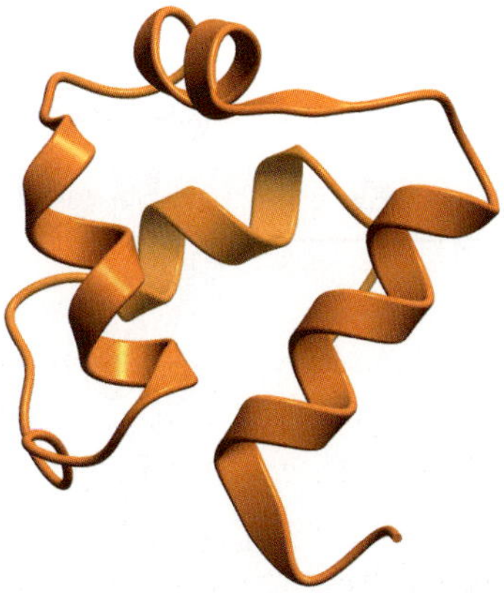

b.

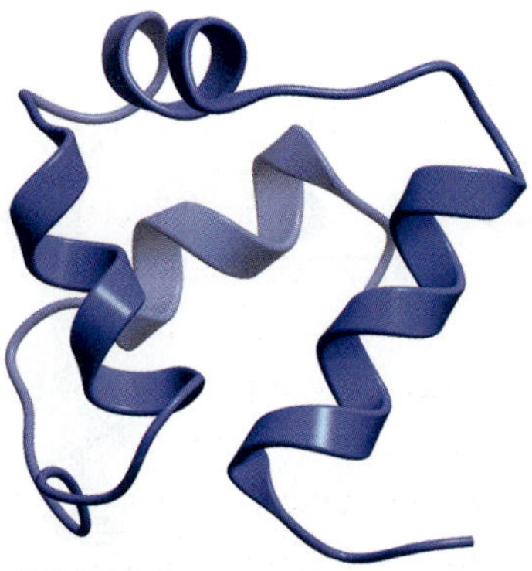

Figure 5.26 The three-dimensional structures of the bacteriophage Cro repressor protein derived from X-ray crystallography and NMR spectroscopy are similar. **a.** The Cro repressor structure determined by X-ray crystallography is shown. **b.** The NMR structure of the Cro repressor is represented as an average of 20 individual structures that agree with the experimental NMR data. BASED ON PDB FILES 2CRO (X-RAY) AND 1ZUG (NMR).

X-ray Crystallography

Figure 5.27 illustrates the principle of X-ray crystallography. An X-ray beam is aimed at a homogeneous protein crystal. When the beam interacts with the electron-dense regions of the protein, the X-rays are diffracted and scatter in different directions. The X-rays diffracted by the crystal are captured using electronic X-ray detectors. Multiple diffraction images are collected from successive rotations of the crystal, measuring the intensity of X-rays scattering in all directions. Much as a lens can recombine scattered light to form an image, these measurements of the scattered X-rays can be combined using computational methods to form an image. In the case of crystallography, the image is a three-dimensional map of the electron-dense regions of the crystal (called an electron density map). A model of the protein is then built to match the regions of electron density, which is rendered by a computer graphics program. As an example, **Figure 5.28** shows several steps in the X-ray structure determination for the protein nitrophorin 2. This protein is contained in the saliva of the blood-sucking insect *Rhodnius prolixus*, which transmits the pathogen responsible for human Chagas disease. Analysis of the X-ray diffraction pattern led to an electron density map of the nitrophorin 2 protein, revealing the location of an iron heme group that binds nitric oxide and plays a role in host vasodilation during blood feeding by the insect.

The two most difficult steps in X-ray crystallography are (1) growing diffraction-quality crystals and (2) determining the phases of the diffracted X-rays. Crystals are needed in order to provide sufficient mass for scattering X-rays, as an individual protein is too small. These crystals also need to be highly ordered so that

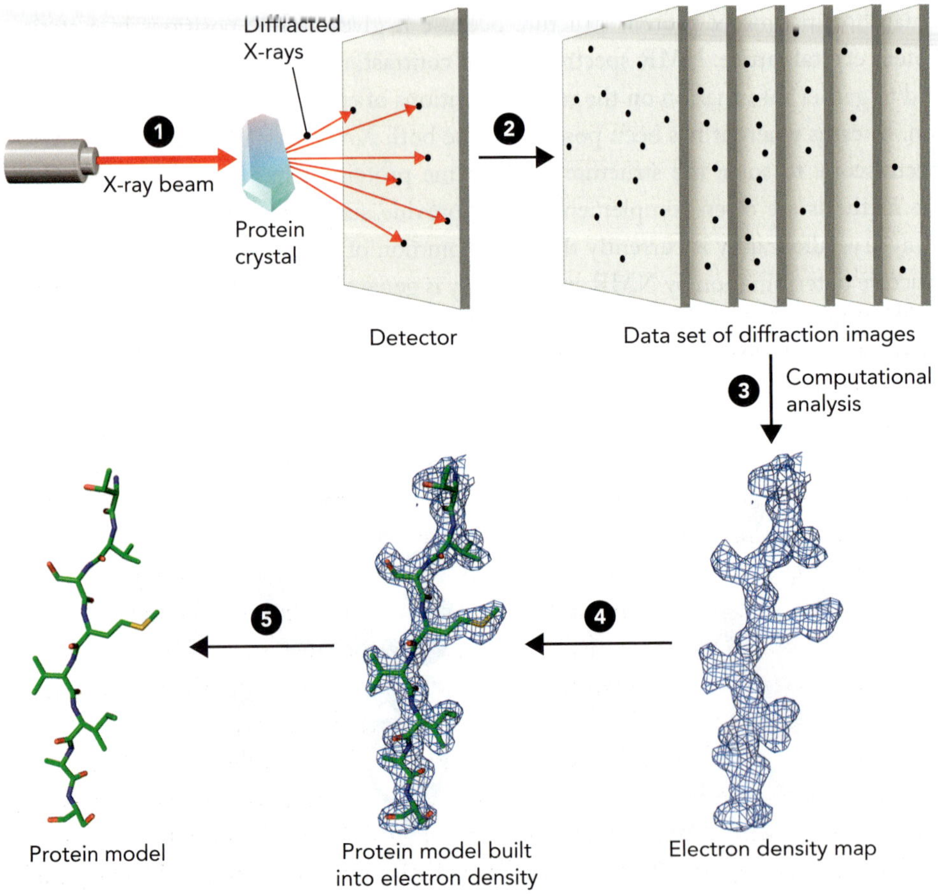

Figure 5.27 The determination of a protein's structure by X-ray crystallography involves several steps. A beam of X-rays is directed at a protein crystal. The X-rays are diffracted by the crystal, leading to a pattern of spots on the diffraction image. The positions and intensities of these spots provide information on the electron-dense regions of the crystal. Multiple diffraction images are collected and then computationally analyzed to form a map of the electron density. A model of the protein is built by fitting the polypeptide chain into the regions of electron density. In this example, only a small portion of the protein is shown. BASED ON PDB FILE 1T68.

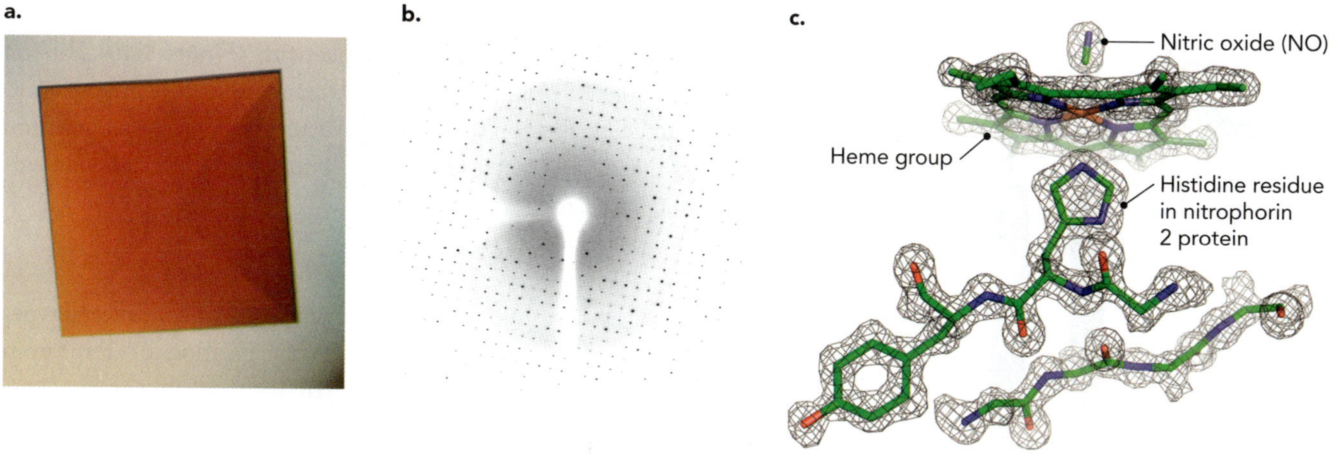

Figure 5.28 The structure of nitrophorin 2 was determined by X-ray crystallography. **a.** A nitrophorin 2 protein crystal in buffer. The crystal dimensions are approximately 1 mm × 1 mm × 1 mm. **b.** An X-ray diffraction image collected from nitrophorin 2 protein crystals. **c.** Electron density map of the heme region in nitrophorin 2, which reveals the binding of nitric oxide (NO) to the iron atom of the heme. The electron density of only a portion of the nitrophorin 2 protein is represented here as "chicken wire" (a three-dimensional contour map), and the protein model is shown in stick representation within the electron density. NP2 X-RAY DIFFRACTION DATA AND ELECTRON DENSITY MAP PROVIDED BY ANDRZEJ WEICHSEL AT THE UNIVERSITY OF ARIZONA. BASED ON PDB FILE 1T68.

the diffraction pattern can be interpreted. Protein crystallization is a trial-and-error process with no guarantee of success. Crystallization of proteins that may have multiple conformational states is particularly difficult. Success in obtaining crystals of these types of proteins is usually through "trapping" the protein in a particular state by the addition of a ligand or interacting molecule in the crystallization solution.

Once diffraction data have been obtained from the crystals, the phases of the diffracted X-rays need to be determined before structure determination can proceed. Because X-rays have wave properties, they can add together or cancel each other out—processes that affect the intensities of the spots on the diffraction images. Phase determination often requires comparison to X-ray diffraction by a second crystal containing an electron-dense atom such as mercury or selenium (referred to as a heavy atom derivative) using a procedure called **isomorphous replacement**. Alternatively, if the structure of a similar protein has previously been determined, this can be used as a starting point for solving the phase problem using an approach called **molecular replacement**.

Well-ordered crystals give rise to high-resolution diffraction patterns. In turn, these diffraction patterns determine the level of detail in the electron density maps and, therefore, the accuracy of the resulting protein models. The resolution of a structure is usually given in angstroms, where the smaller the resolution number, the greater the amount of detail that can be seen in the structure. Some structures, such as Kendrew's first model of whale myoglobin (see Figure 5.25), are of relatively low resolution. In such a case, the path of the polypeptide chain may be apparent, but the details of the side chains may be missing. Structures of 3 Å resolution (or smaller) generally have electron density maps of sufficient quality to show the details of the side chains.

Even though we may know the full structures of proteins, we often represent these structures with the ribbon style of protein structure modeling, which was first developed in 1979 by Jane Richardson at Duke University. Richardson's ribbon style of

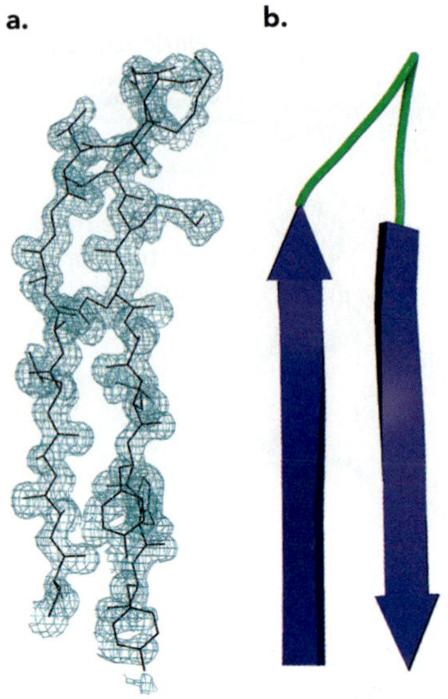

Figure 5.29 High-resolution electron density maps show the positions of both backbone and side chain atoms in the polypeptide. **a.** An electron density map of an antiparallel β-strand region in the nitrophorin 2 protein. The resolution of the map is 1.45 Å. At this resolution, the details of the side chains are apparent in the electron density map. The protein model is shown in stick representation. **b.** Computer rendering of the ribbon structure from the same region shown in panel **a.** BASED ON PDB FILE 1T68.

protein models is informative for describing protein structure–function relationships because it reveals the location of secondary structure elements (**Figure 5.29**). Ribbon models are essentially cartoons of protein molecular structures, but it is important to remember that they are actually based on the locations of atomic nuclei, usually within a very small error of less than 10^{-10} meter (1 Å), and therefore have tremendous predictive value in biochemical research.

NMR Spectroscopy

NMR spectroscopy takes advantage of the intrinsic magnetic properties of several types of atoms—most often ^1H, ^{15}N, and ^{13}C—to determine relative locations of atoms from highly concentrated solutions of purified proteins. NMR instruments contain large magnetic fields that align the nuclear spins from NMR-active nuclei, including the stable (non-radioactive) isotopes ^{15}N and ^{13}C. Short radio-frequency pulses are used to perturb these nuclear spins. Computers collect information about the perturbations and the return to the ground state and use these data to interpret the electronic environment of each nucleus.

The size of proteins that can be studied by NMR is limited because large molecules reorient slowly in solution, which means that their signals can average out and be lost during the course of the NMR experiment. Although modern techniques can be used to study some properties of very large proteins (~500 kDa), full protein structure determination by NMR is currently limited to proteins that are less than ~100 kDa. Additionally, because NMR spectroscopy is a relatively insensitive technique, high protein concentrations (~0.1–0.5 mM) are needed, which may not be possible for all proteins.

Various combinations of pulses can be used in an NMR instrument to collect different kinds of information, such as short-range (less than ~5 Å) distances between atoms or subsets of atoms (for example, only ^1H atoms that are covalently bonded to ^{15}N, not ^{13}C). The different pulse sequences generate NMR spectra that may have one axis (one-dimensional), two axes (two-dimensional), or more (three- or four-dimensional), often based on the nuclei observed. Data from numerous NMR experiments are combined to calculate models of the proteins that satisfy all the distance and bonding constraints that are determined from the NMR data.

Because measurements of distances between atoms are only approximate, several structures can be calculated that agree equally well with the NMR data. Therefore, NMR data result in a family of structures, as shown in **Figure 5.30**, though for simplicity's sake often a single average structure is shown. Because the proteins are in solution, not immobilized in a crystal, NMR spectroscopy is especially useful for studying protein dynamics, such as conformational changes as a result of ligand binding, or protein dynamics over time, such as in protein-unfolding experiments.

concept integration 5.3

What is the primary difference between the types of protein samples used in X-ray crystallography and in NMR spectroscopy, and how does this difference introduce distinct limitations to eachmethod?

Protein samples used for X-ray crystallography are ordered arrays of protein molecules in a crystal, whereas NMR samples are concentrated protein solutions. Each of these techniques provides useful data about protein structure; however, caveats exist. In the case of X-ray crystallography, protein crystals may be difficult to obtain, because with a new protein, it is not known what conditions may facilitate crystallization.

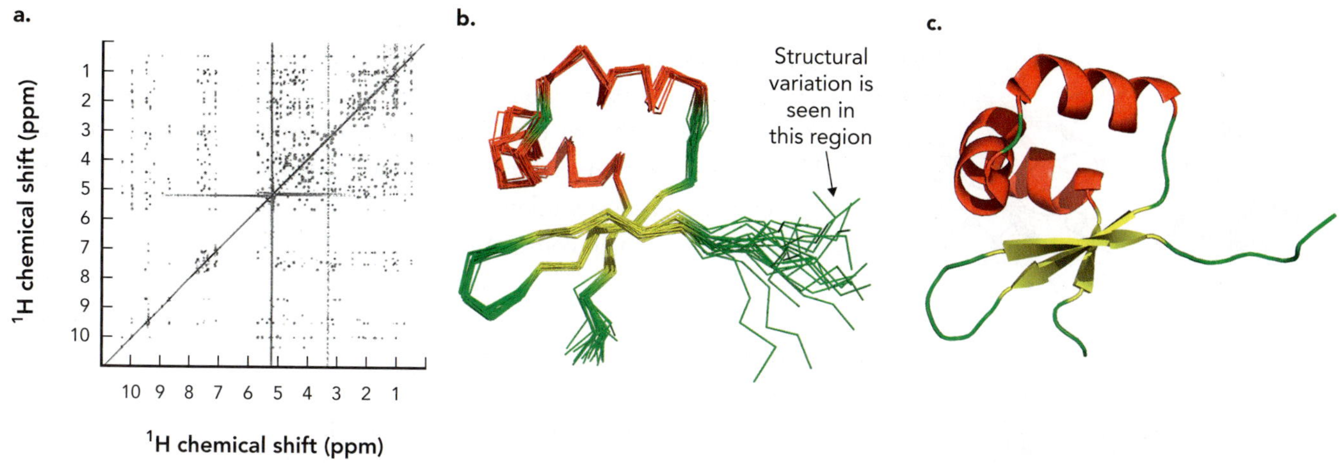

Figure 5.30 NMR spectroscopy provides information on the relative locations of atoms within a protein based on their nuclear spin properties within a magnetic field. **a.** An example of NMR data collected from the lambda Cro protein. In this two-dimensional NMR experiment, each axis shows the ^{1}H chemical shift in parts per million (ppm) relative to a reference. The peaks that are off the diagonal indicate the distances between two ^{1}H nuclei. From analysis of these peaks and from related data of other experiments, constraints of many distances between atoms within the protein can be determined. From this information, the three-dimensional structure can be calculated. **b.** The polypeptide backbone is shown for the 20 calculated structures that are consistent with the NMR data. These structures are very similar to each other, with the exception of the N-terminal region of the protein, which has a great deal of structural variability. **c.** The average of the 20 structures of lambda Cro protein determined from NMR spectroscopy is shown as a ribbon model. NMR SPECTROSCOPY DATA AND MOLECULAR STRUCTURES PROVIDED BY MATT CORDES AT THE UNIVERSITY OF ARIZONA. BASED ON PDB FILE 2A63.

Furthermore, proteins that have disordered regions or exist in multiple conformations are especially difficult to pack in the ordered, regular pattern that results in a crystal. An advantage is that there is no theoretical size limitation of the proteins studied by X-ray crystallography. NMR spectroscopy uses samples of proteins in solution; however, the high protein concentration (~0.1–0.5 mM) needed cannot be obtained for all proteins. Because large proteins reorient more slowly in solution, signal loss from larger proteins means that full structure determinations from this technique are restricted to relatively small proteins and protein complexes. The advantage of NMR is that conformational changes or dynamic fluctuations can readily be observed.

5.4 Protein-Specific Antibodies Are Versatile Biochemical Reagents

As we described in Chapter 4, antibodies are large multi-subunit proteins that bind antigen molecules, such as other proteins, with high specificity and affinity (see Figure 4.61). This property of antibodies makes them very useful reagents in biochemistry because they can identify specific proteins within complex protein mixtures in cell extracts and tissue samples. Before describing how antibodies are used as biochemical tools to study proteins, we will examine the relationship between antibody structure and function.

Antibody proteins are produced by cells in the immune system called B cells. It is important to note that each B cell makes only a single type of antibody. Normally, B cells that produce antibodies recognizing our own proteins (self) are eliminated during development to prevent autoimmune disease. Human B cells are capable of producing two classes of immunoglobulin (Ig) light chains, referred to as λ and κ

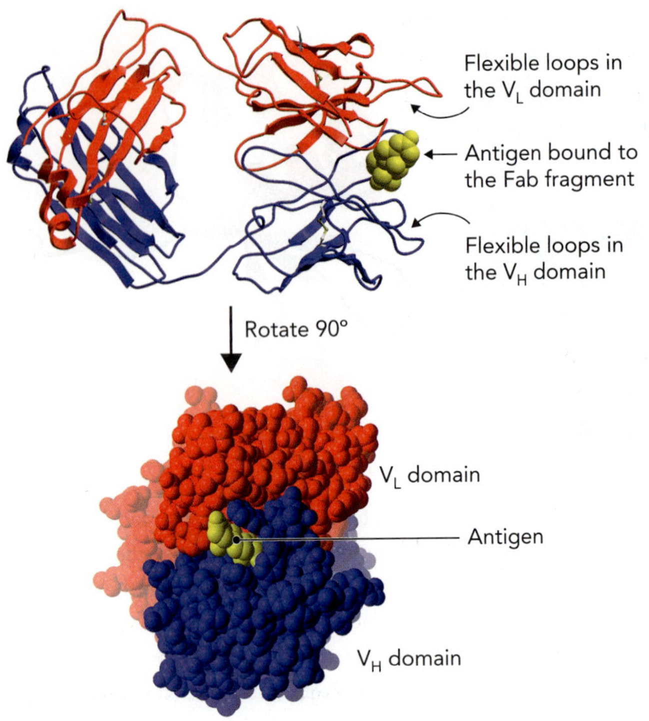

Figure 5.31 Molecular structure of an immunoglobulin Fab fragment, showing the binding interactions between the heavy-chain and light-chain variable domains and their cognate antigen. BASED ON PDB FILE 1DBB.

light chains, and five classes of Ig heavy chains, denoted μ (IgM), α (IgA), δ (IgD), ε (IgE), and γ (IgG). The variable domains of the immunoglobulin polypeptides are referred to as V_L and V_H to specify the light-chain and heavy-chain polypeptides, respectively. The light-chain constant domain is referred to as C_L, and the three heavy-chain constant domains are C_{H1}, C_{H2}, and C_{H3}. The Fab fragment contains the high-affinity antigen binding site, which consists of amino acid side chains located in flexible loop regions of both the V_L and V_H domains. The side chains from these loop regions make noncovalent contacts with the antigen (**Figure 5.31**). The discrete molecular interactions between an antibody and an antigen involve variable-domain amino acid residues and a specific site on the antigen called the epitope. An antigen can have multiple epitopes and therefore can interact with multiple different antibody molecules (**Figure 5.32**). Epitopes on protein antigens are often localized to polar surface groups contained within loop regions.

Note that theoretically, an individual B cell could express any combination of light-chain and heavy-chain polypeptides to produce one of 10 different antibody complexes (either a λ or κ light chain paired with one of the five different heavy chains). However, this would not generate the antibody diversity required to protect us from the tens of millions of foreign antigens we encounter throughout our lives. Instead, the major mechanism underlying antibody diversity is a series of DNA recombination events. These actions join different variable-domain coding sequences to

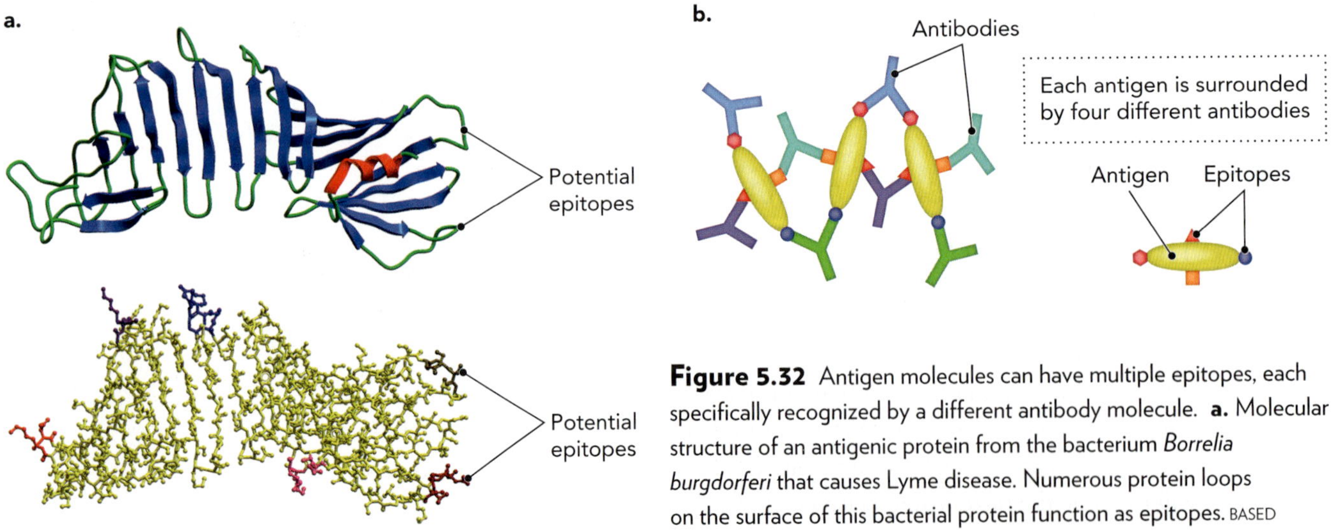

Figure 5.32 Antigen molecules can have multiple epitopes, each specifically recognized by a different antibody molecule. **a.** Molecular structure of an antigenic protein from the bacterium *Borrelia burgdorferi* that causes Lyme disease. Numerous protein loops on the surface of this bacterial protein function as epitopes. BASED ON PDB FILE 1FJ1. **b.** Large networks of antibody–antigen molecular interactions can form as part of the immune response. In this example, the bacterial antigen contains four distinct regions that function as epitopes, which are recognized by four different B-cell antibodies.

small DNA segments on the same chromosome. As described in Chapter 20, these random DNA recombination events between different immunoglobulin gene DNA segments, in concert with inexact DNA joining processes that introduce additional coding sequence differences, can generate a multitude of unique antibody protein complexes in humans.

Generation of Polyclonal and Monoclonal Antibodies

Protein-specific antibodies are generated by immunizing animals with purified peptides under conditions that initiate an immune response. The protein antigen is usually a polypeptide produced *in vivo* using a cloned gene in a bacterial expression system or a small peptide generated by solid-phase peptide synthesis. When an animal is injected with the protein antigen, its immune system recognizes the injected polypeptide as a foreign molecule, resulting in the amplification of B-cell populations that express antibodies with a high affinity for the antigen. This is the same biological response elicited by vaccinations, in which an antigenic protein from a pathogen, such as a bacterium or virus, is injected into the animal to stimulate the production of neutralizing antibodies.

Two types of antibodies are used in biochemical research: polyclonal and monoclonal. A **polyclonal antibody** is a heterogeneous mixture of immunoglobulin proteins that recognize one or more epitopes on an antigenic protein. A **monoclonal antibody** is a homogeneous immunoglobulin species that recognizes only a single epitope on an antigenic protein. The procedures used to generate useful polyclonal and monoclonal antibody reagents are quite different, as we now describe.

Antigen-specific polyclonal antibodies are obtained by first immunizing laboratory animals (usually a rabbit, chicken, or goat) with a purified protein or oligopeptide antigen, and then isolating blood samples after giving multiple immunizations (booster shots) over a 6- to 8-week period. Once the red blood cells are removed from the blood sample, the remaining animal serum contains large amounts of both antigen-specific and nonspecific antibodies. Antigen-specific polyclonal antibodies can be purified using affinity chromatography, in which the protein or oligopeptide antigen is covalently linked to Sepharose beads (**Figure 5.33**). Using appropriate binding and wash conditions, the majority of serum proteins and nonspecific antibodies can be removed

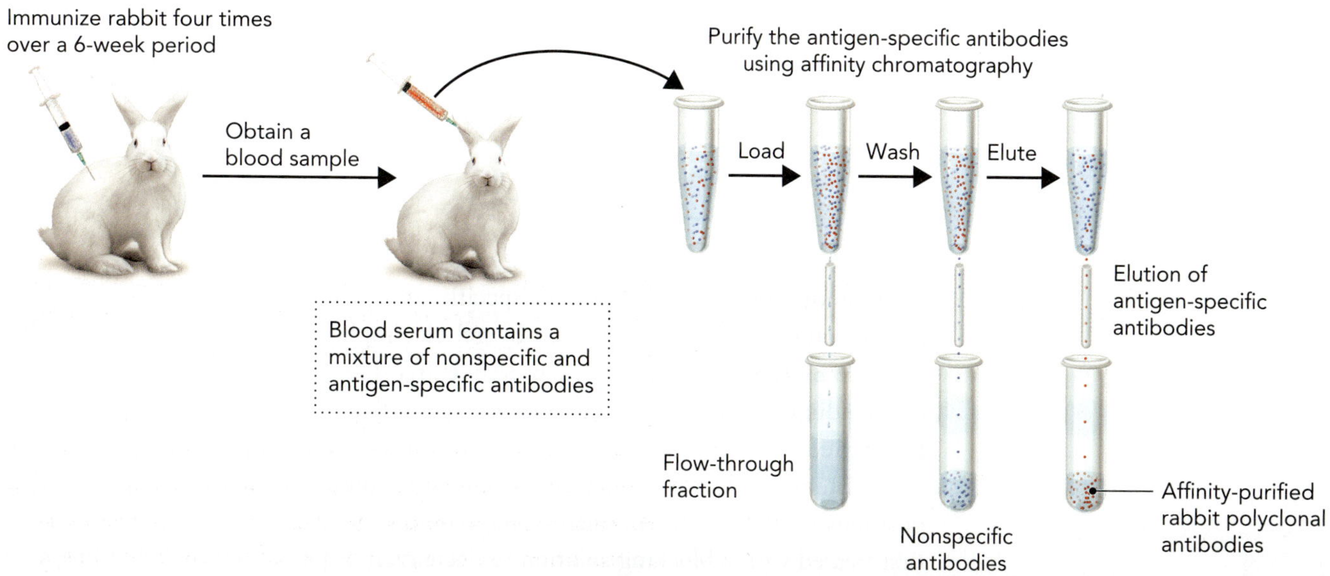

Immunize rabbit four times over a 6-week period
Obtain a blood sample
Blood serum contains a mixture of nonspecific and antigen-specific antibodies
Purify the antigen-specific antibodies using affinity chromatography
Load
Wash
Elute
Flow-through fraction
Nonspecific antibodies
Elution of antigen-specific antibodies
Affinity-purified rabbit polyclonal antibodies

Figure 5.33 Generating antigen-specific polyclonal antibodies using immunized rabbits is a process that takes several weeks. Multiple injections of a rabbit with a purified protein or oligopeptide antigen over a 6- to 8-week period induces an immune response that leads to the expansion of B-cell populations secreting antigen-specific antibodies. Because the serum fraction is a mixture of antibodies, the antigen-specific antibodies are often purified using affinity chromatography.

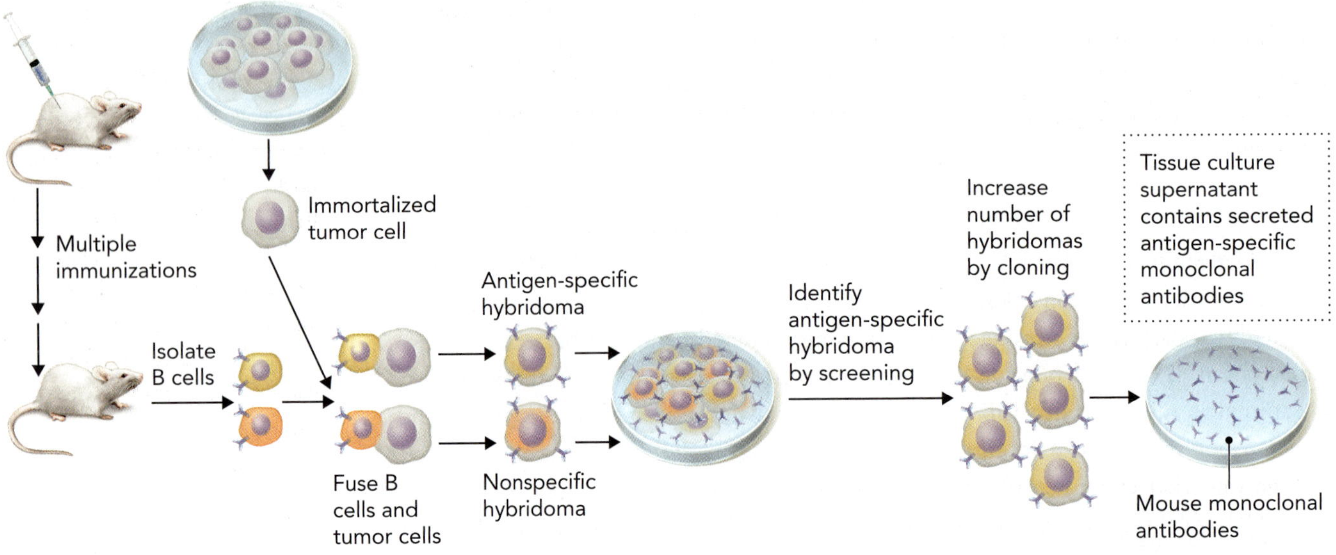

Figure 5.34 Immunized mice are used to generate monoclonal antibodies. The antibody-producing B cells are isolated from the mouse spleen and are used to create an immortalized cell line. Once a hybridoma clone is identified that secretes an antigen-specific antibody, it can be expanded in culture and used to make an unlimited supply of antibody.

prior to elution of the antigen-specific antibodies with a low-pH buffer. Depending on the purity and binding characteristics of polyclonal antibodies, they can be used for a variety of immunologic methods.

Monoclonal antibodies are a second type of antibody and are made by immunizing mice with a polypeptide antigen, similar to the method used to immunize rabbits (booster shots are given every few weeks). However, in this case, antibody-expressing B cells are isolated from spleen tissue and fused with long-lived cells (called immortalized tumor cells) to produce **hybridoma cells**, which can be cultured *in vitro* and screened for antibody production (**Figure 5.34**). Each hybridoma cell secretes a single antibody species, which recognizes only a single epitope on the protein or oligopeptide antigen. Therefore, all of the clonal hybridoma cell lines established in tissue culture need to be analyzed individually for antigen-specific binding activity. Although this screening process is time consuming, and there are no guarantees that an appropriate hybridoma cell line will be identified, a successful process provides an unlimited supply of highly specific antibody for immunologic assays.

In general, polyclonal antibodies are less labor intensive to generate and are usually sufficient for biochemical research; however, the polyclonal antibody supply is limited because the immunized animal eventually dies. In contrast, monoclonal antibodies are well suited for commercial or clinical applications because the process provides unlimited supplies of these antibodies from immortalized hybridoma cell lines. Moreover, monoclonal antibodies provide the ability to monitor quality continually using standardized diagnostic measures.

Western Blotting

A common application of protein-specific antibodies in biochemical research is the detection of proteins separated by gel electrophoresis using a technique called **Western blotting**. In this method, protein extracts are separated by one-dimensional or two-dimensional polyacrylamide gel electrophoresis, and then the proteins are transferred out of the gel matrix onto a nitrocellulose filter membrane (**Figure 5.35**). The proteins bind with high affinity to the nitrocellulose membrane and retain their positions on the membrane relative to those on the original gel. The membrane is first treated with a blocking solution to decrease nonspecific antibody binding, then

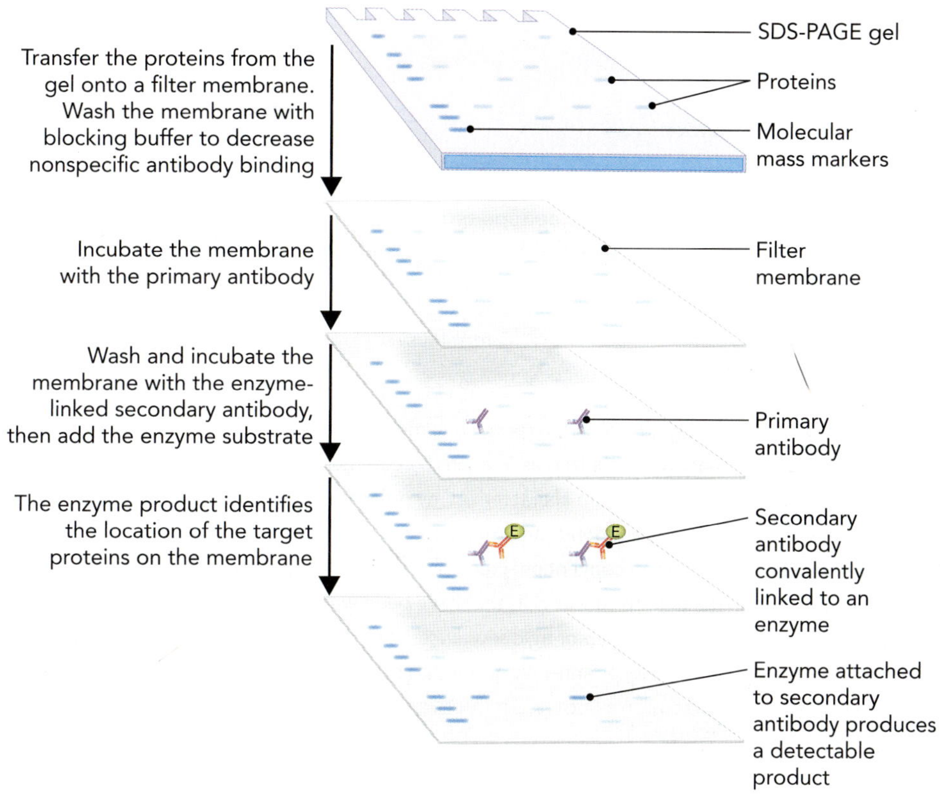
Transfer the proteins from the gel onto a filter membrane. Wash the membrane with blocking buffer to decrease nonspecific antibody binding
Incubate the membrane with the primary antibody
Wash and incubate the membrane with the enzyme-linked secondary antibody, then add the enzyme substrate
The enzyme product identifies the location of the target proteins on the membrane
SDS-PAGE gel
Proteins
Molecular mass markers
Filter membrane
Primary antibody
Secondary antibody convalently linked to an enzyme
E
Enzyme attached to secondary antibody produces a detectable product

Figure 5.35 Western blotting is used to identify proteins that have been separated by SDS-PAGE (one-dimensional or two-dimensional) and are recognized by an antigen-specific antibody. Faithful transfer of proteins from the polyacrylamide gel to a nitrocellulose filter membrane permits the information obtained by SDS-PAGE to be combined with immunologic analysis using antigen-specific antibodies.

it is incubated with the protein-specific antibody, called the **primary antibody**, under the conditions that facilitate antigen–antibody interactions. Unbound primary antibody is removed by washing, and then the membrane is incubated with a **secondary antibody** (detection antibody) that recognizes the primary antibody as an antigen. For example, if the primary antibody is a mouse monoclonal antibody, then the secondary antibody is a rabbit polyclonal antibody, called a rabbit anti-mouse secondary antibody, which binds to mouse immunoglobulin. An important feature of the secondary antibody is that it is covalently linked to an enzyme, which catalyzes a chemical reaction that can be detected by a colorimetric or fluorometric assay when substrate is added. Most secondary antibodies are covalently linked to the enzyme **horseradish peroxidase**. By comparing the position of the secondary antibody on the membrane (as visualized by the product of the horseradish peroxidase enzymatic reaction) to those of the proteins on the original polyacrylamide gel, it is possible to determine which protein samples contain the antigenic protein recognized by the primary antibody.

Let's look at an application of Western blotting in a typical protein biochemical method. Recombinant DNA technology provides a convenient way to add protein-coding sequences of highly antigenic peptides, called **epitope tags**, to the protein-coding sequences of cloned genes. Commercially available high-affinity antibodies that are specific for epitope tags make it possible to use immunologic methods for analysis of gene-encoded proteins, and therefore eliminates the need to make a unique antibody for the protein under analysis. For example, the aspartate-rich amino acid sequence DYKDDDDK (FLAG epitope) is a high-affinity epitope tag that can be added to the N-terminal or C-terminal end of a protein-coding sequence using the corresponding DNA sequence. This FLAG epitope is recognized by a commercially available anti-FLAG antibody, which can be used as the primary antibody in Western blots

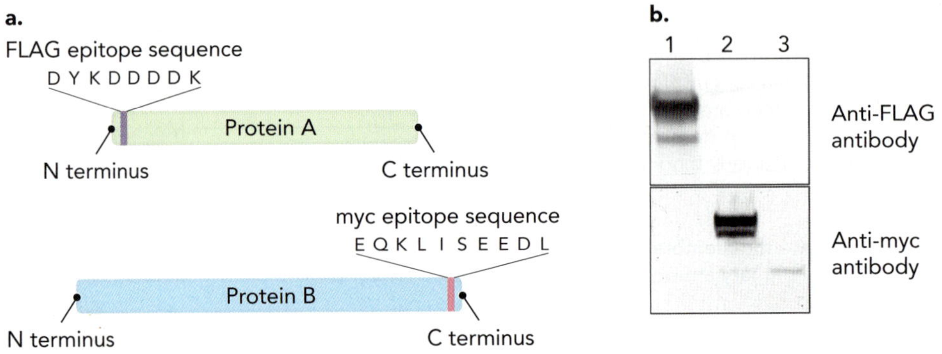

Figure 5.36 Epitope tagging is a molecular genetic method used to modify proteins with specific amino acid sequences that are recognized as antigens by commercially available antibodies. **a.** The FLAG and myc epitope sequences are characterized by polar amino acid residues (glutamate and aspartate). They are highly antigenic, and their coding sequences are easily added to protein-coding sequences of genes using conventional recombinant DNA techniques. **b.** Western blot results using protein extracts from cancer cells that have been engineered to express epitope-tagged proteins containing the FLAG or myc sequences. The same filter membrane was incubated at different times with anti-FLAG or anti-myc antibodies to detect antigenic proteins in three different cancer cell extracts. WESTERN BLOT DATA PROVIDED BY SUSAN KUNZ AT THE UNIVERSITY OF ARIZONA.

containing protein extracts from cells expressing the recombinant protein (**Figure 5.36**). A similar antibody tag is the myc epitope amino acid sequence, EQKLISEEDL, which contains several glutamate residues and is recognized by a commercial anti-myc antibody. Most epitope tags contain a high density of polar residues, ensuring that they will stay on the protein surface and be accessible as antibody targets in folded proteins.

Immunofluorescence

Protein-specific antibodies can also be used to detect proteins within cells using a technique called **immunofluorescence**. In this application, primary antibodies are incubated with cells that have been cross-linked to a microscope slide in such a way as to preserve cell architecture. (The process is much like the incubation step in a Western blot.) Following a wash step, the cells can either be visualized directly using fluorescence microscopy, if the primary antibody is fluorescently labeled, or further incubated with a secondary antibody that contains a fluorescent chemical group.

Figure 5.37 shows an example of immunofluorescence in which a primary antibody against the cytoskeletal protein cytokeratin 8 was incubated with human cells fixed to a microscope slide with the cross-linking agent formaldehyde. The anti-cytokeratin 8 antibody is recognized by a secondary antibody that has a covalently linked fluorescent group, which emits red light after laser excitation. A second fluorescent probe was used in this experiment to identify actin protein in the cytoskeleton by emission of green light after laser excitation. This green fluorescent probe is not an antibody, but rather a protein called phalloidin, which is one of the toxic substances in *Amanita phalloides*—a poisonous mushroom—that has been covalently linked to a fluorescent group.

Enzyme-Linked Immunosorbent Assay

The protein-specific binding properties of antibodies make them ideal biochemical reagents for identifying antigenic proteins that may be present at low levels in biological samples, such as trace amounts of protein in blood or urine samples. The

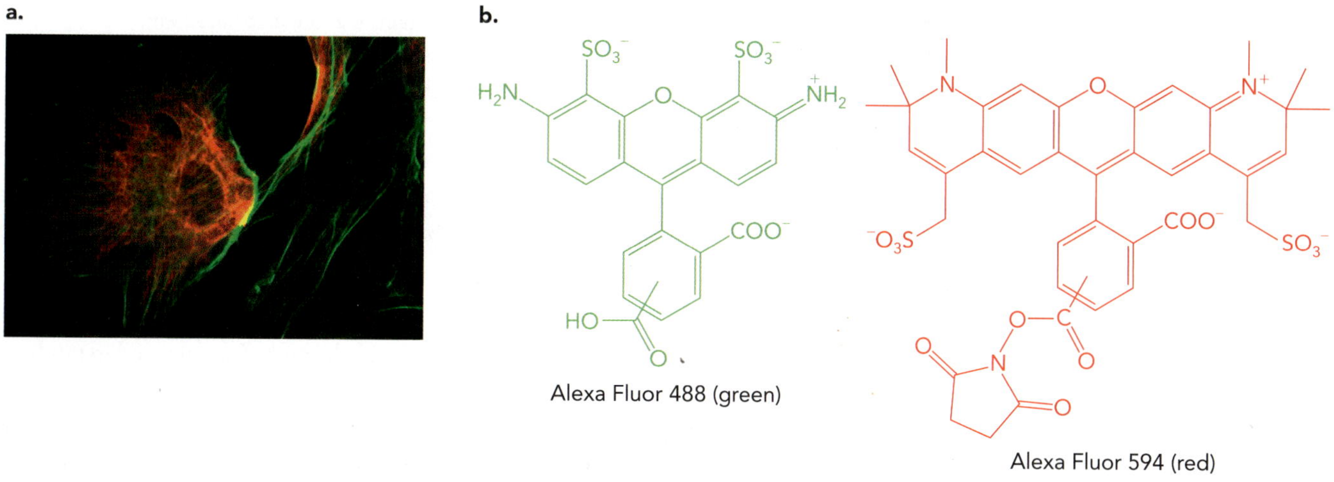

Figure 5.37 Immunofluorescence is an antibody-based technique used to identify proteins in cells that have been chemically treated in a way that preserves cell architecture. **a.** Immunofluorescence of a formaldehyde-fixed cancer cell that has been incubated with a mouse monoclonal antibody, which recognizes the cytoskeletal protein cytokeratin 8. The anti-mouse IgG secondary antibody in this case contained a covalently linked fluorochrome called Alexa Fluor 594, which emits red light upon laser excitation. The green fluorescence in this slide preparation is a result of phalloidin protein, which is fluorescently labeled with the fluorochrome Alexa Fluor 488, binding to the actin protein, which forms long filaments inside the cell. **b.** The chemical structures of the fluorochromes Alexa Fluor 488 and Alexa Fluor 594. IMMUNOFLUORESCENCE DATA PROVIDED BY ANNE CRESS AT THE UNIVERSITY OF ARIZONA.

enzyme-linked immunosorbent assay (ELISA) illustrated in **Figure 5.38** is an example of one such high-throughput biochemical assay.

ELISA uses two different protein-specific monoclonal antibodies to detect antigenic proteins in up to 96 different samples at a time. In this "sandwich" ELISA method, a primary monoclonal antibody, which recognizes a single epitope on the antigenic protein, is covalently attached to the bottom of a 96-well microtiter plate. This is referred to as the "capture" antibody. Dilutions of an aqueous sample are then added to each well and incubated for several hours to promote antigen–antibody binding. After washing away unbound proteins from the sample, a second primary monoclonal antibody, called the "detection" antibody, is added to the well. It is important that the detection antibody recognizes a distinct epitope on the antigenic protein in order for the molecular sandwich to form. Finally, an antibody that recognizes the detection antibody as an antigen, and which is linked to an enzyme such as horseradish peroxidase, is added to the wells, along with a chromogenic or fluorogenic substrate for the horseradish peroxidase enzyme assay. Samples that contain the antigenic protein will be positive for the horseradish peroxidase reaction product, which can be detected using a spectrophotometric 96-well plate reader.

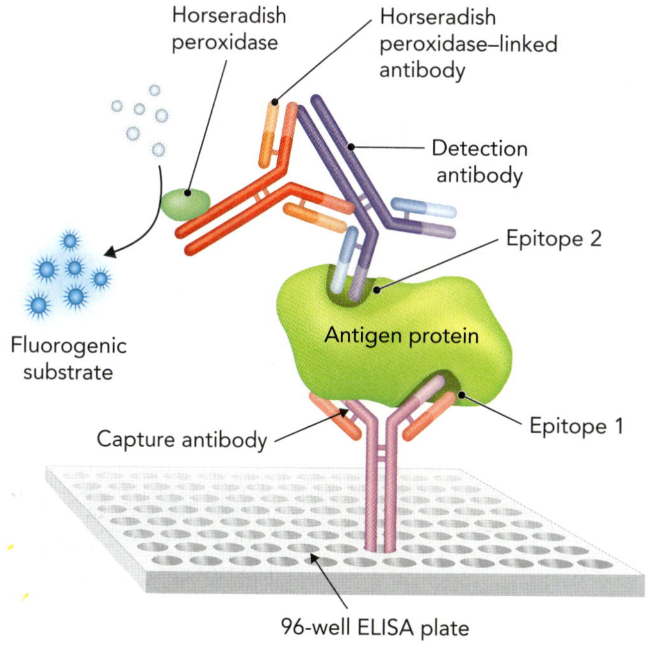

Figure 5.38 The enzyme-linked immunosorbent assay (ELISA) makes use of highly specific monoclonal antibodies that can be used in a high-throughput format to detect small amounts of antigen in aqueous samples.

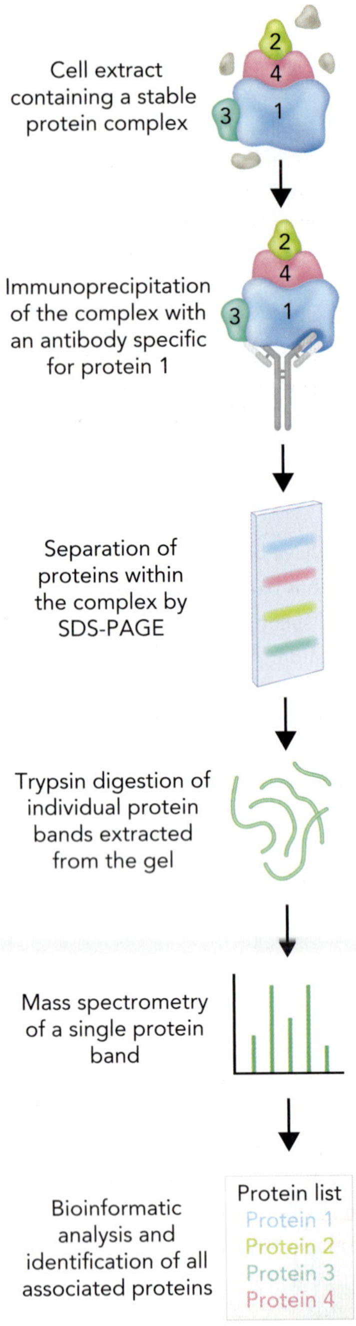

Figure 5.39 Antibodies can be used to identify proteins that are associated with protein antigens in large cellular complexes by combining immunoprecipitation with mass spectrometry. Immunoprecipitation uses antibodies that are covalently linked to carbohydrate beads and then applied in a co-immunoprecipitation protocol that enriches for a particular protein and any other proteins with which it is associated. Conventional SDS-PAGE and mass spectrometry are then used to identify all of the proteins in the complex. In this example, each protein is excised separately and analyzed by mass spectrometry as shown.

Immunoprecipitation

Biochemistry is an experimental science, and breakthroughs in our understanding of biological processes often come from technological advances that increase the specificity and sensitivity of a biochemical assay. In some cases, a combination of two or more such biochemical assays leads to new opportunities for discovery. One example of such a combinatorial approach is the use of high-affinity antibodies to isolate protein antigens that are components of macromolecular protein complexes. By combining the technique of **immunoprecipitation** with mass spectrometry, it is possible to identify proteins within large cellular complexes that associate with target proteins.

Immunoprecipitation is a variation of affinity purification in which a monoclonal antibody is covalently linked to a carbohydrate bead. The antibody–bead combination is used to separate protein antigens from other proteins physically using low-speed centrifugation. By optimizing the binding and wash conditions, it is possible to enrich the protein solution for intact macromolecular protein complexes containing the target protein and its associated proteins (a method called co-immunoprecipitation). In the most common version of this protein purification strategy, co-immunoprecipitated proteins are first separated by SDS-PAGE and then extracted from the gel as protein bands, digested with trypsin, and analyzed by mass spectrometry (**Figure 5.39**). A further modification of this approach is to use epitope-tagged recombinant proteins, which are expressed in tissue culture cell lines and used as "bait" to "fish out" interacting cellular proteins, which are then identified by mass spectrometry.

concept integration 5.4

Describe how epitope tagging can be used to identify unknown cellular proteins that specifically associate with a target protein *in vivo*.

Epitope tagging is a method in which antigenic peptide fragments are added to any protein for which the encoding DNA is available, thereby permitting immunologic analysis using a "universal" antibody that recognizes this specific antigenic sequence. The epitope tag sequences are usually added to the N-terminal or C-terminal region of a polypeptide to avoid disruption of protein function. In the given situation, the antigenic sequence DYKDDDDK could be added to the cloned target protein using recombinant DNA technology, and then the engineered target protein could be expressed in cells using a molecular genetic approach. Using the FLAG commercial antibody to immunoprecipitate recombinant DYKDDDDK-tagged proteins from cell extracts, it should be possible, using mass spectrometry, to identify cellular proteins that associate with the target protein.

chapter summary

5.1 The Art and Science of Protein Purification

- The "art and science" of protein purification refers to the need to develop customized strategies that exploit unique chemical and physical properties of a target protein in order to separate it from other cellular proteins with minimal loss of biochemical activity.
- Protein purification strategies usually require the use of a highly sensitive and specific biochemical assay that identifies the target protein against a background of all other cellular proteins. *Specific activity* refers to the ratio of protein activity units to the total amount of protein in the sample.
- Differential centrifugation is used to fractionate proteins contained in a cell extract based on the applied centrifugal force and length of centrifugation time. The resulting separation of pellet and supernatant can be used to isolate fractions enriched for proteins localized to nuclei, mitochondria, membranes, and cytosol.
- Column chromatography is a macromolecular separation technique that is essential to most protein purification strategies. Three common types of column chromatography are gel filtration chromatography, ion-exchange chromatography, and affinity chromatography.
- SDS-PAGE is a protein separation technique that uses sodium dodecyl sulfate (SDS) to give proteins a uniform charge-to-mass ratio, which allows their physical separation in polyacrylamide gel electrophoresis (PAGE) on the basis of molecular mass.
- Isoelectric focusing is a protein separation technique that is based on the differential isoelectric points of proteins. In the presence of a pH gradient and electric current, charged proteins migrate in the gel until they reach a point at which pI = pH and they carry no net charge.
- Two-dimensional polyacrylamide gel electrophoresis (2-D PAGE) separates proteins by combining two techniques: isoelectric focusing in one dimension and SDS-PAGE in the second dimension. One variation of 2-D PAGE, called two-dimensional differential in-gel electrophoresis (2-D DIGE), uses a mixed sample of fluorescently labeled proteins to identify alterations in protein abundance as a function of cell treatment.

5.2 Working with Oligopeptides: Sequencing and Synthesis

- Edman degradation is a protein-sequencing method that uses chemical labeling and cleavage, in conjunction with amino acid standards, to identify N-terminal amino acids sequentially. By fragmenting a polypeptide into small fragments using differential protease digestion, it is possible to deduce the amino acid sequence of the entire protein.
- Protein identification by mass spectrometry uses polypeptide fragmentation and high-resolution mass analysis to predict amino acid sequence based on comparison of the measured mass to a predicted mass using computer algorithms and whole-genome databases.
- Solid-phase peptide synthesis is a method to generate oligopeptides of up to 25 amino acids using successive rounds of covalent linkage, washing, and deblocking to add amino acids one at a time to a resin-attached C-terminal amino acid. Once the oligopeptide is synthesized, it is released from the resin.

5.3 Protein Structure Determination

- X-ray crystallography uses a focused X-ray beam directed at a protein crystal to obtain a diffraction pattern. From analysis of the position and intensities of the diffraction spots, an electron density map of the protein crystal can be calculated. Protein models are then built to match the regions of high electron density.
- NMR spectroscopy is used to determine the relative locations of atoms in a purified protein solution. From NMR spectra, a family of three-dimensional structures can be calculated that are in agreement with the data.

5.4 Protein-Specific Antibodies Are Versatile Biochemical Reagents

- Protein-specific antibodies are useful reagents in protein biochemistry because they identify target proteins with high affinity and specificity.
- Polyclonal antibodies are generated by injecting host animals with an antigenic protein or oligopeptide to induce an immune reaction and stimulate antibody production. Polyclonal antibodies are isolated from blood serum and represent a collection of antigen-selective antibodies that recognize multiple epitopes and are often affinity purified to remove nonspecific antibodies.
- Monoclonal antibodies are produced by injecting host animals with an antigenic protein or oligopeptide to induce an immune reaction and stimulate antibody production. Spleen cells are isolated from the immunized animal and fused with an immortalized tumor cell line to generate immortalized mouse hybridoma cells, which produce a single antibody that recognizes a particular epitope.
- Common applications of protein-specific antibodies include Western blotting, immunofluorescent staining of cells and tissues, enzyme-linked immunosorbent assays (ELISAs), and protein immunoprecipitation.

biochemical terms

(in order of appearance in text)
luciferase (p. 213)
luciferin (p. 213)
cell extract (p. 214)
sonication (p. 214)
French press (p. 214)
centrifugation (p. 215)
specific activity (p. 215)
salting out (p. 216)
dialysis (p. 216)
column chromatography (p. 217)
void volume (p. 218)
gel filtration chromatography (p. 218)
size-exclusion chromatography (p. 218)
high-performance liquid chromatography (HPLC) (p. 219)
ion-exchange chromatography (p. 219)
elution buffer (p. 219)
affinity chromatography (p. 220)
gel electrophoresis (p. 221)
polyacrylamide gel electrophoresis (PAGE) (p. 221)
sodium dodecyl sulfate (SDS) (p. 221)
SDS-PAGE (p. 221)
native PAGE (p. 222)
isoelectric focusing (p. 224)
two-dimensional polyacrylamide gel electrophoresis (2-D PAGE) (p. 224)
two-dimensional differential in-gel electrophoresis (2-D DIGE) (p. 224)
Edman degradation (p. 228)
trypsin (p. 228)
chymotrypsin (p. 228)
mass spectrometry (p. 229)
electrospray ionization (ESI) (p. 229)
matrix-assisted laser desorption/ionization (MALDI) (p. 229)
solid-phase peptide synthesis (p. 230)
NMR spectroscopy (p. 233)
isomorphous replacement (p. 235)
molecular replacement (p. 235)
polyclonal antibody (p. 239)
monoclonal antibody (p. 239)
hybridoma cell (p. 240)
Western blotting (p. 240)
primary antibody (p. 241)
secondary antibody (p. 241)
horseradish peroxidase (p. 241)
epitope tag (p. 241)
immunofluorescence (p. 242)
enzyme-linked immunosorbent assay (ELISA) (p. 243)
immunoprecipitation (p. 244)

review questions

1. How many protein-coding genes are contained in the human genome, and why are more than this number of proteins actually found in humans?
2. What are the two general properties of proteins that directly influence the development of a purification strategy?
3. Name and briefly describe the three most commonly used homogenization techniques.
4. What is specific activity?
5. Name and briefly describe the three major types of column chromatography.
6. Describe the protein separation technique sodium dodecyl sulfate–polyacrylamide gel electrophoresis (SDS-PAGE).
7. What was the drawback to Sanger's method of protein sequencing?
8. Explain the process of Edman degradation.
9. Briefly describe the principles of mass spectrometry.
10. Describe the process of solid-phase peptide synthesis.
11. What is an advantage of X-ray crystallography over nuclear magnetic resonance spectroscopy for protein structure determination?
12. Briefly describe the technique of X-ray crystallography as it applies to protein structure determination.
13. What are often the two most difficult steps in X-ray crystallography?
14. Briefly describe the technique of NMR spectroscopy as it applies to protein structure determination.
15. What are antibodies?
16. Briefly name and describe the two types of antibodies used in biochemical research.
17. What are epitope tags, and what are they used for?
18. Describe the enzyme-linked immunosorbent assay (ELISA).

challenge problems

1. A target protein was purified through several steps as shown in the table here. Answer the following questions about this purification scheme as they relate to the "?" values.

Step	Method	Total protein yield (mg)	Total units of activity	Target protein specific activity (units/mg protein)
1	Crude extract (lysed cells)	5×10^5	1×10^{10}	?
2	Gel filtration chromatography	?	8×10^9	8×10^4
3	Anion-exchange chromatography	?	5×10^9	5×10^6
4	Affinity chromatography	20	4×10^9	?

a. What is the specific activity of the target protein after the first and fourth purification steps?

b. Calculate the total protein yield (in milligrams) after the second and third purification steps.

c. Calculate the overall percentage yield of the target protein (final activity/initial activity).

d. Using gel filtration chromatography, the target protein activity was associated with a protein of 100,000 kDa. To get a more accurate molecular-mass estimation of the target protein, a small sample of the protein present in the most active fraction after step 4 was analyzed by SDS-PAGE, and as shown in the graph, the relative mobility of the most abundant protein (red circle) was plotted, along with that of known molecular-mass standards (black circles). What is the molecular mass of this purified protein based on SDS-PAGE, and how do you reconcile this with the molecular-mass estimate from gel filtration chromatography?

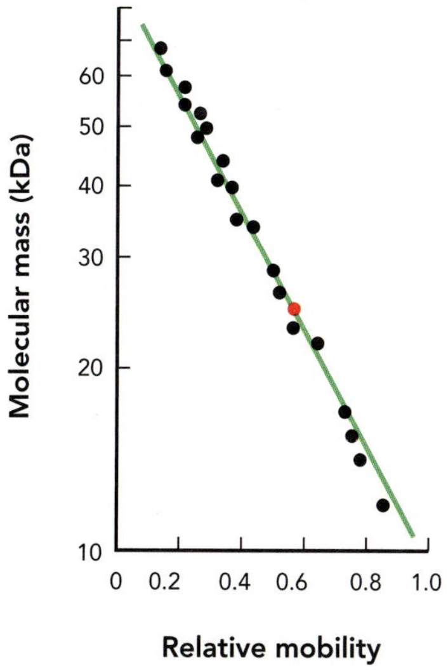

2. Consider five proteins with the properties shown in the following table. Answer the questions below about these proteins and justify your answers.

Protein	Subunit mass (Da)	Native mass (Da)	Isoelectric point (pI)
A	10,000	40,000	8.1
B	15,000	120,000	5.3
C	20,000	20,000	7.2
D	25,000	75,000	4.5
E	30,000	60,000	6.2

a. Which protein would elute last from a gel filtration chromatography column under nondenaturing conditions?

b. Which protein would migrate the slowest in an SDS-PAGE gel?

c. Which protein would elute last from an anion-exchange column (DEAE gel matrix) using running buffer at pH 6.5?

d. What is the likely quaternary structure for protein D on the basis of the data in the table?

3. Precipitation of proteins can be accomplished by either heating a protein solution, followed by centrifugation, or by adding ammonium sulfate to the solution and then centrifuging the precipitate. With respect to an enzyme that was being purified by an activity assay, how would precipitation of this enzyme by these two methods affect the next purification step after resolubilization of the pelleted protein?

4. Compare the order of migration of a small (10 kDa) and large (80 kDa) protein through a size-exclusion column and an 8% SDS–polyacrylamide gel.

5. Cytochrome c protein (pI = 12) and cytochrome c peroxidase protein (pI = 5) can be isolated from yeast mitochondria. Both proteins are apparently "bound" to a DEAE ion-exchange column in a low-ionic-strength (10 mM) buffer at pH 7. When a buffer with increasing amounts of salt (a salt gradient) is applied to the column, cytochrome c elutes before cytochrome c peroxidase. From this information, can you determine whether cytochrome c is bound directly to the DEAE column or is more likely indirectly bound through association with cytochrome c peroxidase? Justify your answer using the pI values, the pH, the net charge on the proteins, and the charge on the ion-exchange column.

6. Your research project requires that you purify protein A away from proteins B and C using a combination of gel filtration chromatography and ion-exchange

chromatography. Describe a purification scheme using just these two columns that would accomplish your goal with the least amount of protein dilution. Include in your answer the types of running buffers you would use (ionic strength and pH) and the expected order of elution of the proteins from the two columns, given the physical properties of the proteins in the following table:

Protein	Isoelectric point (pI)	Molecular mass (kDa)
A	4.2	40
B	4.4	70
C	8.0	44

7. A functional assay was designed to isolate a putative wound-healing growth factor present in plant extracts made from an exotic flower found high in the forest canopy of the Amazon rain forest. Initial field studies of the purified factor identified a tetradecapeptide (14 amino acid residues) that consisted of 2 mol of glycine and 1 mol each of 12 other amino acids—none of which were alanine, histidine, leucine, serine, threonine, tryptophan, or valine on the basis of available amino acid standards. Although the field laboratory was set up to use Edman degradation to deduce partial amino acid sequences, the efficiency of the reaction limited its application to tetrapeptides and smaller. Therefore, the Edman degradation data were augmented by a combination of proteolytic cleavage assays with trypsin, chymotrypsin, and V-8 protease, as well as chemical cleavage with cyanogen bromide. Using the following information collected by the field biochemist, determine the most likely sequence of this tetradecapeptide, written left to right from the N-terminal residue using the single-letter amino acid code. Explain your reasoning.

1. Cleavage with trypsin yielded a hexapeptide, a septapeptide, and a single amino acid that was identified with amino acid standards as glutamine.
2. Cleavage with chymotrypsin yielded a hexapeptide, a pentapeptide, and a tripeptide with the amino acid sequence G-I-F.
3. Cleavage with V-8 protease yielded a heptapeptide, a tripeptide with the sequence P-R-Q, and a tetrapeptide with the sequence G-Y-N-D.
4. Cyanogen bromide chemical cleavage yielded a decapeptide and a tetrapeptide with the sequence G-I-F-M.

8. Most X-ray crystallography analyses are now performed at centralized government labs that provide high-powered X-ray beams, which are able to resolve electron densities better than what can be done with smaller institutional instruments, and without the high costs of maintaining an in-house facility. However, even with this improved data collection methodology, the rate-limiting step in X-ray crystallography is often the ability to grow diffractable crystals.

a. If a protein of interest is not amenable to crystallization, even after testing a large number of buffer conditions and temperatures, what is the most likely explanation?

b. What are some of the advantages and disadvantages of obtaining diffractable protein crystals using an orthologous protein from another species?

c. If no suitable orthologous proteins were available, how might you modify the protein of interest to increase your chances of obtaining diffractable crystals?

d. If none of these alternative approaches result in the isolation of diffractable crystals, what other method could possibly be used to obtain the molecular structure of your protein? What is the primary advantage and disadvantage of this other method relative to X-ray crystallography?

9. What is the biochemical difference between polyclonal and monoclonal antibodies? Why do you think that monoclonal antibodies, but not polyclonal antibodies, are suitable for human diagnostic and clinical applications?

10. Antibodies can be useful reagents to analyze specific target proteins both *in vitro* and *in vivo*; however, not all antibodies work equally well for both types of studies. Explain why an antibody might be able to detect its target protein in rat liver cell extracts by Western blotting of an SDS-PAGE gel but fail to detect the same protein using immunofluorescent staining of rat liver cell tissue sections.

smartwork5

If your instructor assigns homework with Smartwork5, access it here: digital.wwnorton.com/biochem.

suggested reading

Books and Reviews

Aebersold, R., and Mann, M. (2003). Mass spectrometry-based proteomics. *Nature*, *422*, 198–207.

Celis, J. E., and Geri, K. (2009). *Cell biology assays: essential methods*. New York, NY: Academic Press.

Fruton, J. S. (1999). *Proteins, enzymes, genes: the interplay of chemistry and biology*. New Haven, CT: Yale University Press.

Lane, D., and Harlow, E. (1998). *Using antibodies: a laboratory manual* (2nd ed.). Cold Spring Harbor, NY: Cold Spring Harbor Press.

Rupp, B. (2009). *Biomolecular crystallography: principles, practice, and application to structural biology* (1st ed.). New York, NY: Garland Publishing.

Russell, D. W., and Sambrook, J. (2006). *The condensed protocols from molecular cloning: a laboratory manual.* Cold Spring Harbor, NY: Cold Spring Harbor Press.

Sanger, F. (1988). Sequences, sequences, and sequences. *Annual Review of Biochemistry*, *57*, 1–28.

Scopes, R. K. (1994). *Protein purification: principles and practice* (3rd ed.). New York, NY: Springer.

Smejkal, G. B. (2004). The Coomassie chronicles: past, present and future perspectives in polyacrylamide gel staining. *Expert Review of Proteomics*, *1*, 381–387.

Primary Literature

Andersen, J. F., and Montfort, W. R. (2000). The crystal structure of nitrophorin 2. A trifunctional antihemostatic protein from the saliva of *Rhodnius prolixus*. *Journal of Biological Chemistry*, *275*, 30496–30503.

O'Farrell, P. H. (1975). High resolution two-dimensional electrophoresis of proteins. *Journal of Biological Chemistry*, *250*, 4007–4021.

Van Dorn, L. O., Newlove, T., Chang, S., Ingram, W. M., and Cordes, M. H. J. (2006). Relationship between sequence determinants of stability for two natural homologous proteins with different folds. *Biochemistry*, *45*, 10542–10553.

Flower from the African climbing oleander

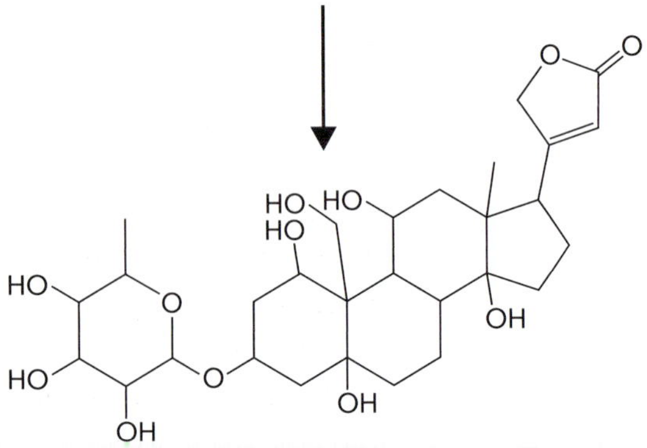

Ouabain

African tribesmen used plant extracts containing ouabain to make poison arrows for hunting

Ouabain is a poison found in the seeds of the climbing oleander plant (*Strophanthus preussii*). It inhibits the Na^+–K^+ ATPase membrane transport protein, which is required for neuromuscular signaling to control contraction in vital organs such as the heart and lungs.

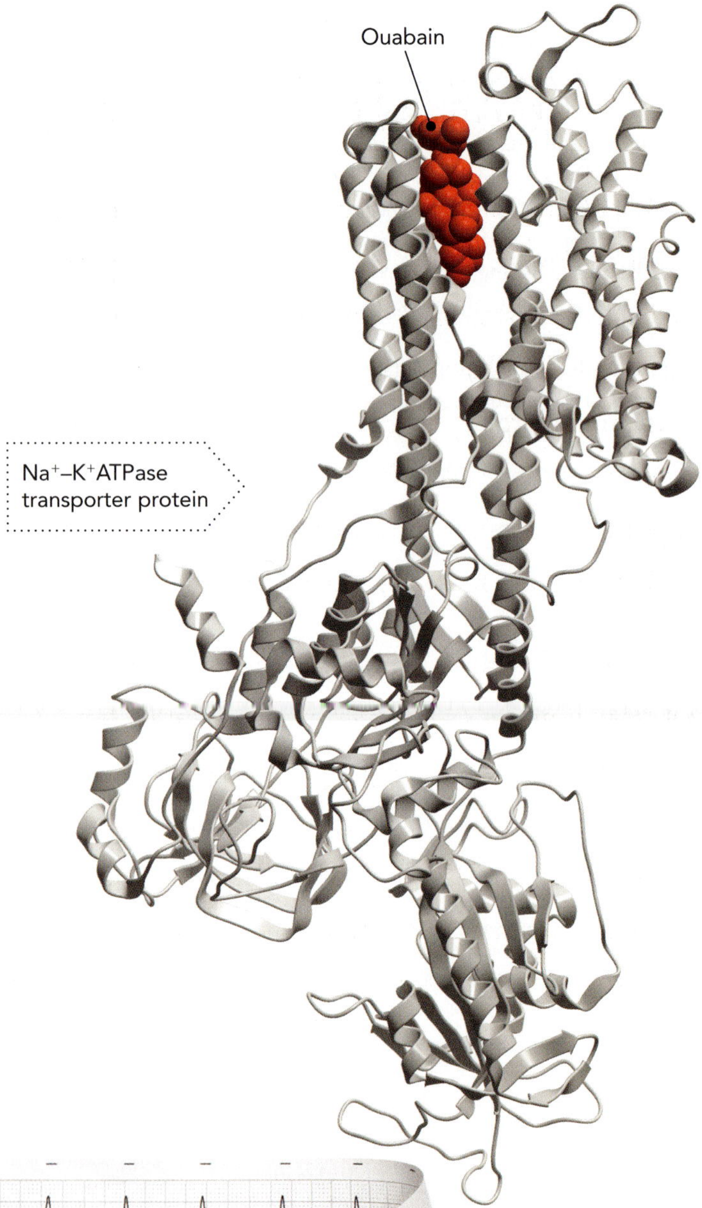

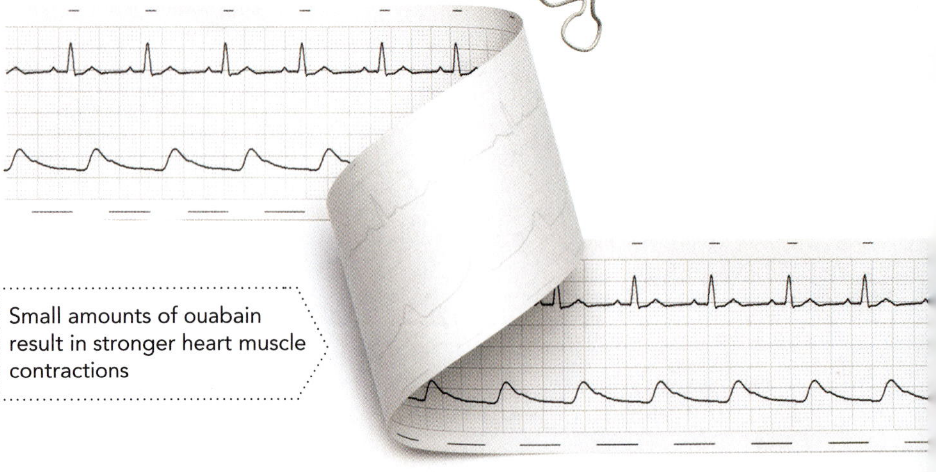

Small amounts of ouabain result in stronger heart muscle contractions

6

Protein Function

◀ Tribesmen in Africa used an extract from the climbing oleander, *Strophanthus preussii*, to poison the tips of their arrows. The active substance is ouabain, which is a specific inhibitor of the Na^+–K^+ ATPase transporter protein. This protein is required to maintain proper sodium and potassium ion concentrations across neuronal cell membranes. Structure–function relationships that govern ion transport by membrane proteins have been elucidated at the molecular level using X-ray crystallography.

The effect of ouabain is that the heart muscle contracts but cannot relax in order to produce another contraction. As a poison, this effect is strong enough to kill a hippopotamus. However, in small, controlled doses, ouabain can act as a heart stimulant when the muscle has trouble contracting. The study of how ouabain works has increased our understanding of the function and importance of the Na^+–K^+ ATPase transporter protein.

CHAPTER OUTLINE

CREDITS: FLOWERS: PINKANNJOH/BLICKWINKEL/ALAMY STOCK PHOTO; TRIBESMAN: IMAGNO/GETTY IMAGES; ELECTROCARDIOGRAM: PICTURE PARTNERS/ALAMY STOCK PHOTO.

Proteins, the workhorses of living cells, are responsible for many cellular functions. These functions include catalyzing biochemical reactions, organizing cell structures, transporting biomolecules, transducing cellular signals, and managing genetic information. As discussed in Chapter 4, the function of a protein is determined by its three-dimensional structure. We now explore this relationship in more detail by first describing the five major functional classes of proteins and then looking at how protein structure mediates protein function.

We will describe three examples of protein structure and function: (1) regulated oxygen transport by myoglobin and hemoglobin proteins; (2) passive and active membrane transport by aquaporins and ATP-dependent ion-channel proteins; and (3) skeletal muscle contraction in response to calcium-induced conformational changes in actin and myosin proteins. Two main concepts are conveyed in this chapter: protein functions arise from precise structural interactions within and between proteins; and the binding of small molecules (ligands) to a protein can induce changes in conformation that enable the protein to perform its function.

6.1 The Five Major Functional Classes of Proteins

Figure 6.1 illustrates the five major functional classes of proteins in cells:

1. *metabolic enzymes,* which are reaction catalysts that control metabolic flux;
2. *structural proteins,* which maintain the integrity of cell structures and promote changes in cell shape;
3. *transport proteins,* which facilitate movement of molecules both within and between cells;
4. *cell signaling proteins,* which transmit extracellular and intracellular signals by functioning as molecular switches;
5. *genomic caretaker proteins,* which maintain the integrity and accessibility of genomic information.

These five protein classes are not totally inclusive—some proteins do not fit in any of these categories, and other proteins fulfill more than one function. But these five classes do embody the most abundant proteins in cells. Many of these protein types were first introduced in Chapter 4 and are well represented throughout the book.

Metabolic Enzymes

Most enzymes are proteins that function as chemical catalysts, which provide an optimal environment for the rapid conversion of reactants to products. RNA molecules can also function as enzymes, although they represent a smaller class of enzymes in the cell. **Metabolic enzymes** catalyze biochemical reactions involved in energy conversion pathways and are responsible for the synthesis and degradation of macromolecules. As with all chemical catalysts, enzymes are not consumed by the chemical reaction, and they increase the reaction rate without altering the equilibrium concentrations of products and reactants.

The primary mechanism by which enzymes function as chemical catalysts is to lower the activation energy of a reaction and thereby increase the rate of product

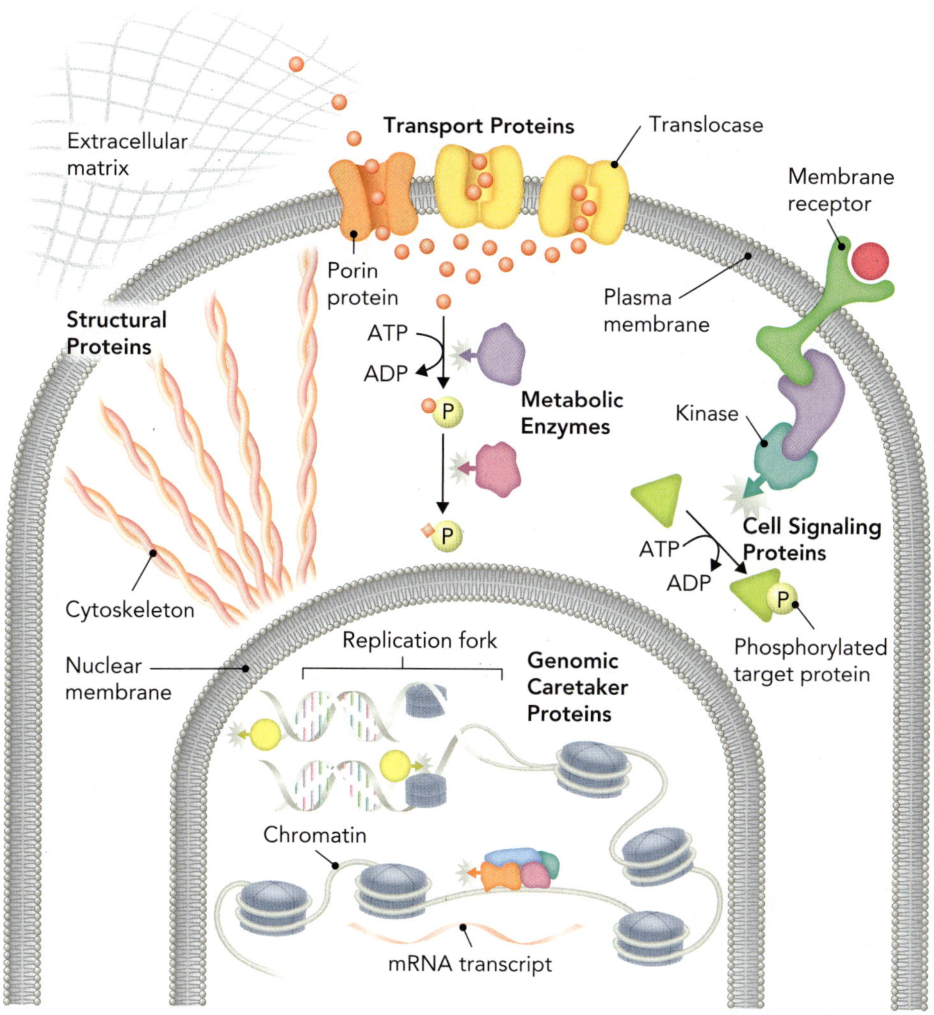

Figure 6.1 Proteins can be categorized into at least five functional classes on the basis of their primary roles in the cell. These five classes are (1) metabolic enzymes, which catalyze chemical reactions involved in energy conversion in cells; (2) structural proteins, which provide an intracellular and extracellular network of protein cables and scaffolds; (3) transport proteins, which passively or actively move small molecules across otherwise impermeable cellular membranes; (4) cell signaling proteins, which transmit extracellular signals into the cell and between cells; and (5) genomic caretaker proteins, which maintain and regulate genetic information in cells.

formation, as we discuss in detail in Chapter 7. The amino acid side chains of enzymes specify the shape and chemical environment of the **enzyme active site** and play a direct role in lowering the activation energy of a reaction. Metabolic enzymes have been shown to increase the rate of product formation up to 100,000-fold compared to that of the uncatalyzed reaction. Most enzymes are components of multi-subunit protein complexes, containing from 2 to 60 homogeneous or heterogeneous protein subunits.

The biochemical name of a metabolic enzyme is often based on the name of one of the reaction components and a description of the reaction mechanism. For example, the enzyme malate dehydrogenase catalyzes a reaction that oxidizes (dehydrogenates) the citrate cycle intermediate malate. The reaction forms oxaloacetate using the coenzyme nicotinamide adenine dinucleotide (NAD^+; oxidized form) as shown in **Figure 6.2**. Other representative metabolic enzymes that we will describe later include phosphofructokinase-1, pyruvate dehydrogenase, acetyl-CoA carboxylase, and thymidylate synthase.

Structural Proteins

Structural proteins are the most abundant proteins in living organisms and function as the architectural framework for individual cells and for tissues and organs. **Cytoskeletal proteins** are structural proteins that are responsible for cell shape, cell migration, and

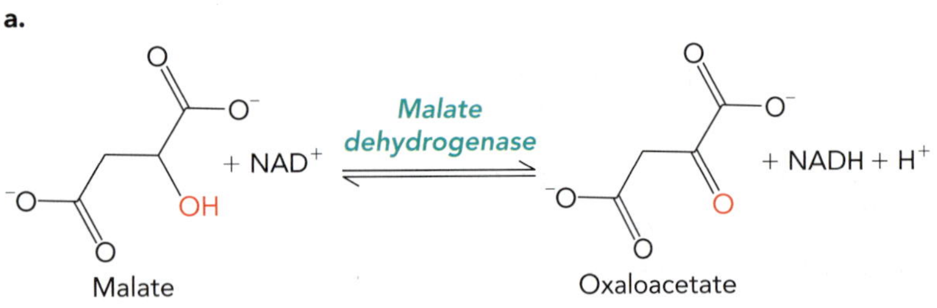
a.
Malate
dehydrogenase
+ NAD+
+ NADH + H+
OH
Malate
Oxaloacetate

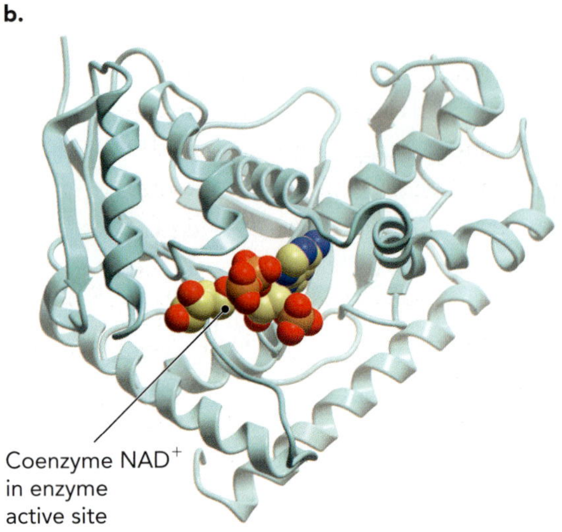
b.
Coenzyme NAD+
in enzyme
active site

Figure 6.2 Malate dehydrogenase is a metabolic enzyme that oxidizes malate to form oxaloacetate. **a.** The conversion of malate to oxaloacetate is catalyzed by the enzyme malate dehydrogenase and requires the coenzyme NAD^+. **b.** To increase the rate of the reaction, the enzyme uses the coenzyme NAD^+ (shown in Corey–Pauling–Koltun [CPK] style), which is noncovalently bound to the active site. BASED ON PDB FILE 2CVQ.

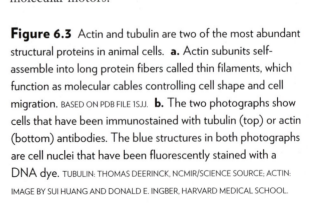

cell signaling. Actin and tubulin are two of the most abundant cytoskeletal proteins in animal cells, both of which form long polymers consisting of hundreds of self-assembled monomeric subunits (**Figure 6.3**). Actin polymers are called **thin filaments,** and tubulin polymers are called microtubules. Microtubules not only contribute to cell shape but also act as "roads" inside cells for the movement of organelles and chromosomes during the process of cell division. **Intermediate filaments** are another type of cytoskeletal protein complex that is critical to cell structure and function. Intermediate filament proteins include vimentin, which specifies cell shape; laminin, which provides structural support to the nuclear envelope; and keratin, the protein component of hair, skin, and nails.

Collagen is a major structural protein in animals and is the primary component of connective tissue (see Section 4.2 in Chapter 4). Collagen is responsible for the strength of tendons, cartilage, bone, and teeth. Muscle tissue consists almost entirely of structural proteins functioning as molecular cables, which mediate muscle movement through calcium-stimulated muscle contraction. Two of these muscle structural proteins are **actin** and **myosin.** Later in this chapter, we describe how actin–myosin macromolecular protein complexes function as molecular motors.

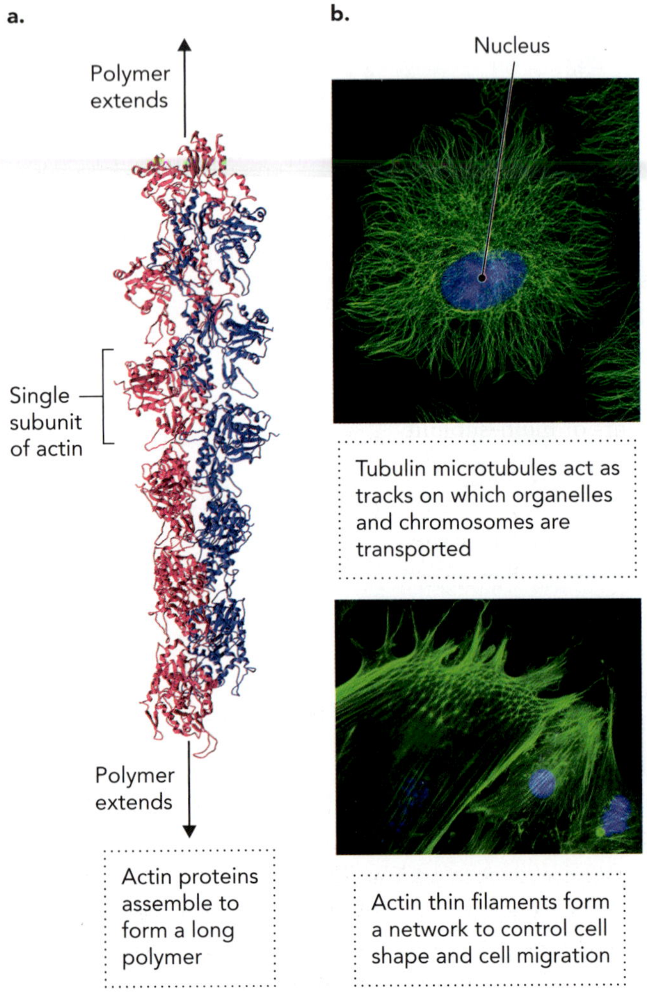
a.
Polymer
extends
Single
subunit
of actin
Polymer
extends
Actin proteins
assemble to
form a long
polymer
b.
Nucleus
Tubulin microtubules act as
tracks on which organelles
and chromosomes are
transported
Actin thin filaments form
a network to control cell
shape and cell migration

Figure 6.3 Actin and tubulin are two of the most abundant structural proteins in animal cells. **a.** Actin subunits self-assemble into long protein fibers called thin filaments, which function as molecular cables controlling cell shape and cell migration. BASED ON PDB FILE 1SJJ. **b.** The two photographs show cells that have been immunostained with tubulin (top) or actin (bottom) antibodies. The blue structures in both photographs are cell nuclei that have been fluorescently stained with a DNA dye. TUBULIN: THOMAS DEERINCK, NCMIR/SCIENCE SOURCE; ACTIN: IMAGE BY SUI HUANG AND DONALD E. INGBER, HARVARD MEDICAL SCHOOL.

Transport Proteins

The plasma membrane of a cell forms an important physical and chemical barrier between the cytosol and the extracellular environment. This lipid bilayer is ~40 Å thick and contains numerous membrane-spanning **transport proteins**, which permit polar and charged molecules to enter or exit the cell. There are two basic classes of membrane transport proteins: (1) Energy-independent **passive transporter proteins** allow molecules to move across a membrane in response to chemical gradients; and (2) energy-dependent **active transporter proteins** require energy, usually from ATP hydrolysis or an ionic gradient, to induce protein conformational changes that open or close a gated channel.

Porin proteins are passive membrane transport proteins that selectively permit small molecules such as H_2O or glycerol to pass through the membrane. **Ion channels** are another type of passive transporter. These transport proteins are responsible for the diffusion of charged ions such as H^+, K^+, Na^+, Cl^-, and Ca^{2+} across the membrane.

In contrast to passive transporters, active transporter proteins use energy, often from ATP hydrolysis, to induce protein conformational changes that pump small molecules or ions against a concentration gradient. For example, the Ca^{2+}-ATPase transporter protein pumps Ca^{2+} across membranes (**Figure 6.4**). It is able to maintain Ca^{2+} concentration up to 10,000 times higher on one side of the membrane relative to that on the other side. This is important in cardiac muscle cells, for example, where each neuronal heartbeat signal leads to a rapid increase in cytosolic Ca^{2+} concentrations of more than 100-fold to stimulate muscle contraction. After each contraction, the Ca^{2+}-ATPase transporter protein quickly pumps the cytosolic Ca^{2+} back into an intracellular compartment, called the sarcoplasmic reticulum, so that the heart muscle is able to contract again in less than a second.

Although much more specialized, and not as universally abundant as membrane-embedded transport proteins, several soluble transport proteins also occur in nature. These include the oxygen-transporting proteins myoglobin and hemoglobin in animals and leghemoglobin in plants. (We discuss myoglobin and hemoglobin in more detail in Section 6.2.) Another important class of soluble transport proteins is the lipid-binding transport proteins, which carry hydrophobic metabolites through aqueous intracellular and extracellular environments. A very abundant protein of this type in mammals is serum albumin, an ~65-kDa protein that transports fatty acids through the circulatory system.

The examples of membrane-embedded and soluble transport proteins that we examine in this chapter share an important feature: When small molecules (ligands) bind to the proteins, they induce changes in the conformations of the proteins. These conformational changes are key to the protein's function. You will see that this

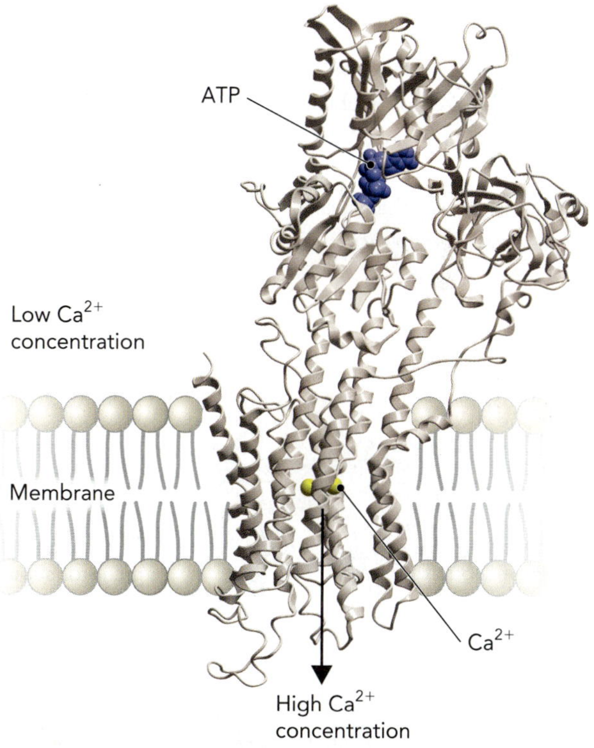

Figure 6.4 The Ca^{2+}-ATPase transporter protein uses ATP hydrolysis to pump Ca^{2+} ions across cell membranes against a steep concentration gradient. The molecular structure of the sarcoplasmic reticulum Ca^{2+}-ATPase transporter protein is shown here with ATP bound to the nucleotide domain and two Ca^{2+} ions inside the membrane channel. BASED ON PDB FILE 1VFP.

same connection between ligand binding and conformational changes exists with many types of proteins, including enzymes and cell signaling proteins.

Cell Signaling Proteins

The ways in which cells communicate with one another and how cells interpret environmental signals that affect growth and viability are only recently beginning to be understood at the molecular level. Although many of the details still need to be worked out, it is clear that proteins play the central role in these signal transduction processes. As described in Chapter 8, the three major types of cell signaling proteins are **membrane receptors**, **nuclear receptors**, and **intracellular signaling proteins**, which include protein kinases and phosphatases, intermolecular adaptor proteins, and site-specific proteases.

There are two large classes of membrane receptors: the G protein–coupled receptors and receptor tyrosine kinases. Membrane receptors are the molecular targets for more than half of all pharmaceutical drugs on the market. For example, adrenergic receptors are G protein–coupled receptors that bind epinephrine-related ligands, whereas the insulin receptor is a receptor tyrosine kinase. Another class of membrane receptors is the growth hormone receptors, such as the erythropoietin receptor. These receptors, as well as many other types of receptors, form dimers in the membrane upon binding of a growth hormone polypeptide (**Figure 6.5**). Receptor dimerization on the cell surface stimulates the formation of an intracellular protein kinase complex. This complex binds to the cytoplasmic tails of the receptor and transduces the growth hormone signal.

Nuclear receptor proteins are also major targets for pharmaceutical drugs. These soluble receptor proteins are transcription factors that regulate gene expression in response to ligand binding (see Chapter 8). Important nuclear receptor proteins include steroid receptors, two of which are the estrogen and progesterone receptors. These are the molecular targets for the steroids that are contained in birth control pills.

The majority of intracellular signaling proteins function as molecular switches, which undergo conformational changes in response to incoming signals, such as receptor activation. For example, heterotrimeric G proteins are membrane-associated signaling proteins that directly activate protein targets, such as the enzyme adenylate cyclase, in response to activation of G protein–coupled receptors.

One of the most widely used cytosolic signal transduction mechanisms in nature is reversible protein phosphorylation. Protein kinases—such as mitogen-activated protein kinase (MAP kinase) and protein kinase A (PKA)—phosphorylate serine and threonine residues on downstream target proteins in response to upstream receptor activation signals. Similarly, the Src kinase is a signaling protein that phosphorylates tyrosine residues on target proteins. Another important kinase in cell signaling is phosphoinositide-3 kinase, which is activated by insulin

Figure 6.5 The erythropoietin receptor is a typical membrane-bound growth hormone receptor with two subunits, which associate into a homodimer to form a single hormone-binding site. The erythropoietin hormone, shown here in space-filling style, binds to the extracellular domains of two erythropoietin receptor subunits, which are shown in ribbon style. The transmembrane and intracellular regions are drawn schematically in blue. BASED ON PDB FILE 1CN4.

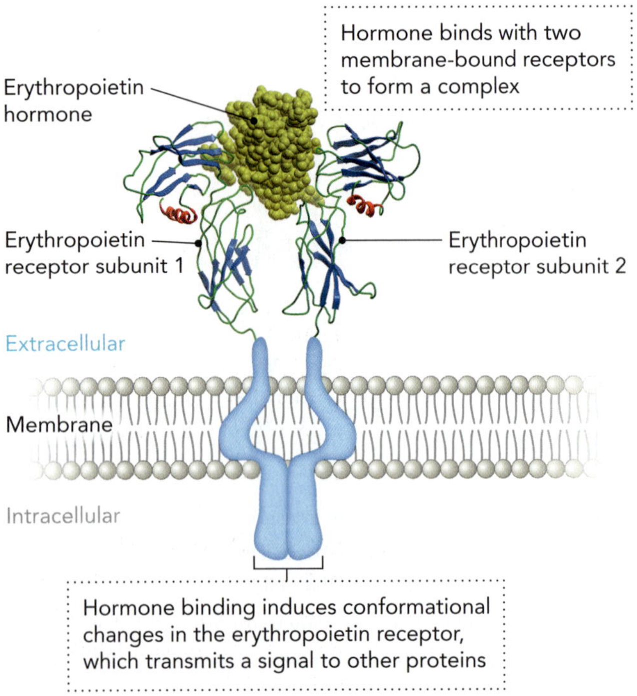
Hormone binds with two membrane-bound receptors to form a complex
Erythropoietin hormone
Erythropoietin receptor subunit 1
Erythropoietin receptor subunit 2
Extracellular
Membrane
Intracellular
Hormone binding induces conformational changes in the erythropoietin receptor, which transmits a signal to other proteins

receptor signaling. It phosphorylates the carbohydrate functional group of phosphatidylinositol, which is a phospholipid found in some types of membranes.

Genomic Caretaker Proteins

The function of **genomic caretaker proteins** is to ensure that the integrity of genomic DNA is maintained throughout the life of the cell and that the expression of specific genes is tightly controlled and reflects the biochemical needs of the organism. The action of some of these proteins is important in repairing mutations in the DNA of reproductive cells, which will otherwise be inherited by the offspring.

One major class of genomic caretaker proteins is the enzymes required for DNA replication, repair, and recombination. At least five different types of DNA polymerase enzymes occur in prokaryotic and eukaryotic cells. Most of these function within a large protein complex and use a DNA template to guide the synthesis of nascent DNA strands based on G-C and A-T base pairing. Other proteins involved in DNA synthesis are single-stranded binding proteins, DNA ligase, topoisomerase, and DNA primase, which function to assist DNA polymerase by preparing the DNA template for replication (see Chapter 20).

Besides undergoing replication in dividing cells, DNA is in constant need of repair because of environmental insults that cause DNA breakage and chemical modification of nucleotide bases or even base excision. DNA repair proteins include photolyase enzymes, which repair thymine dimers that are formed as a result of ultraviolet radiation, and a variety of alkyltransferase enzymes, which specifically replace inappropriately methylated nucleotide bases.

DNA recombination occurs when DNA strands from nearby or distant sites recombine as a result of sequence-specific DNA cleavage or because of random DNA breaks or misalignment of DNA strands. The recombination proteins RecA and RecBCD are examples of two well-characterized genome caretaker proteins in *Escherichia coli*. As shown in **Figure 6.6**, the RecBCD heterotrimer protein complex

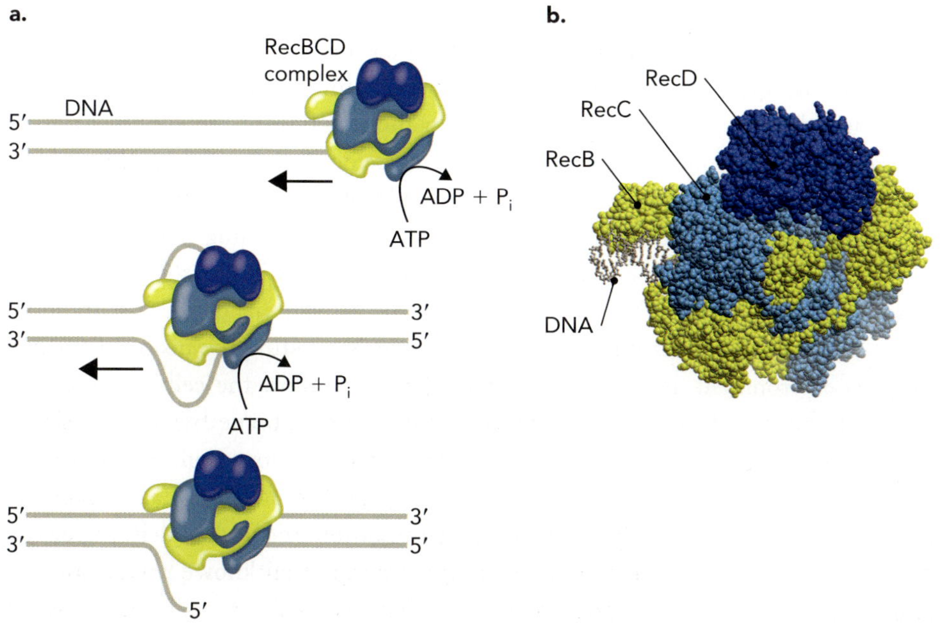

Figure 6.6 The bacterial RecBCD protein complex is a genomic caretaker protein. **a.** RecBCD binds to DNA, unwinds the DNA helix, and cleaves one of the two strands to facilitate DNA repair or recombination. **b.** The molecular structure of the RecBCD complex is shown here bound to DNA. BASED ON PDB FILE 1W36.

binds tightly to double-stranded DNA and unwinds the DNA double strand using an ATP-dependent function called a helicase. Ultimately, this reaction generates a single-strand binding site for the RecA protein, which mediates DNA repair and recombination.

A second major class of genomic caretaker proteins is involved in gene expression; that is, the activation and inhibition of gene-specific RNA synthesis, also called gene transcription. Similar to DNA polymerases, RNA polymerases copy the information stored in DNA into nascent nucleic acid strands using the rules of base pairing. There are three primary RNA polymerase enzymes in eukaryotes, and each of these makes an RNA copy of a specific type of DNA sequence. For example, RNA polymerase I transcribes ribosomal RNA genes; RNA polymerase II transcribes protein-coding genes; and RNA polymerase III is responsible for synthesizing a variety of small RNA molecules encoded in DNA. However, in order for RNA polymerases to transcribe specific DNA sequences into RNA, a whole cadre of proteins, called transcription factors, is required. These proteins either bind DNA directly or form protein complexes with other transcription factors and RNA polymerases.

Some transcription factors activate gene expression by promoting the activity of RNA polymerases at specific sites on genomic DNA. Others inhibit gene expression by interfering with DNA–protein interactions. An important subclass of gene regulatory proteins is involved in chromatin remodeling, which is known to play a role in the accessibility of transcription factors to specific DNA sequences. For example, histone acetylase and deacetylase enzymes covalently modify histone proteins that are associated with DNA, resulting in gene activation or gene repression, respectively (see Chapter 23).

concept integration 6.1

Explain why proteins are called the workhorses of living cells.

Proteins mediate or regulate nearly all aspects of cell structure and function. For example, of the four major classes of macromolecules in a living cell (proteins, nucleic acids, lipids, carbohydrates), only proteins are capable of synthesizing and degrading the other three, primarily through the function of metabolic enzymes. Proteins are also responsible for the energy conversion processes of oxidative phosphorylation and photosynthesis that generate ATP, the energy currency of the cell. Proteins provide the structure of cells, either directly by forming filamentous networks that define cell shape and facilitate cell migration or indirectly by catalyzing reactions that build carbohydrate-based cell walls in plants and bacteria. In addition, although lipids constitute the physical barrier separating the inside of the cell from its environment, membrane-embedded proteins allow the cell to communicate with the outside world. This can occur through membrane transport proteins that function as pores or pumps for the translocation of ions and metabolites or by membrane receptors that function as cell signaling molecules. Lastly, proteins are required to maintain the integrity and propagation of DNA and RNA, which together carry the information for the characteristics of all known organisms and viruses on Earth.

6.2 Globular Transport Proteins: Transporting Oxygen

The oxygen binding proteins **myoglobin** and **hemoglobin** were the first globular proteins to be characterized at the molecular level using X-ray crystallography. Moreover, they were the first proteins for which structural data were used to understand biochemical function. Because so much is known about myoglobin and hemoglobin as oxygen binding proteins, they provide classic examples of how protein structure mediates protein function.

Structure of Myoglobin and Hemoglobin

When Max Perutz and John Kendrew began their pioneering studies in the 1940s on hemoglobin and myoglobin, respectively, they needed large quantities of purified protein for growing crystals and for conducting biochemical analyses. Perutz obtained hemoglobin protein from horses and human placentas, and Kendrew isolated myoglobin from whale muscle. Hemoglobin is the major protein in red blood cells (erythrocytes) and accounts for 35% of the dry weight of these cells. Whale muscle is a rich source of myoglobin because of the need to provide O_2 to the tissues during long ocean dives to collect food.

In both myoglobin and hemoglobin, O_2 reversibly binds to an iron atom contained in a porphyrin ring that is tightly bound to the protein. The Fe^{2+} porphyrin complex is called **heme** and functions as a prosthetic group (**Figure 6.7a**). The heme is needed because no amino acid side chains are ideally suited to the reversible binding of oxygen. Oxygen will only bind to this prosthetic group when the iron is in the +2 oxidation state. By binding heme in a pocket of the protein, myoglobin and hemoglobin help prevent oxidation of iron, which would reduce oxygen binding ability. When iron is in the +2 oxidation state (ferrous) with O_2 bound, it has a bright red color, as seen in fresh blood or meat. However, in dead cells the color of meat or blood changes from red to brown over time as the iron is oxidized to the ferric (+3) state (**Figure 6.7b**). Both the heme and protein components must function together to bind and release oxygen suitably.

Myoglobin and hemoglobin are both heme-utilizing oxygen binding proteins. However, hemoglobin transports heme-bound O_2 from the lungs to the tissues through the circulatory system, whereas myoglobin is concentrated in muscle tissue and functions

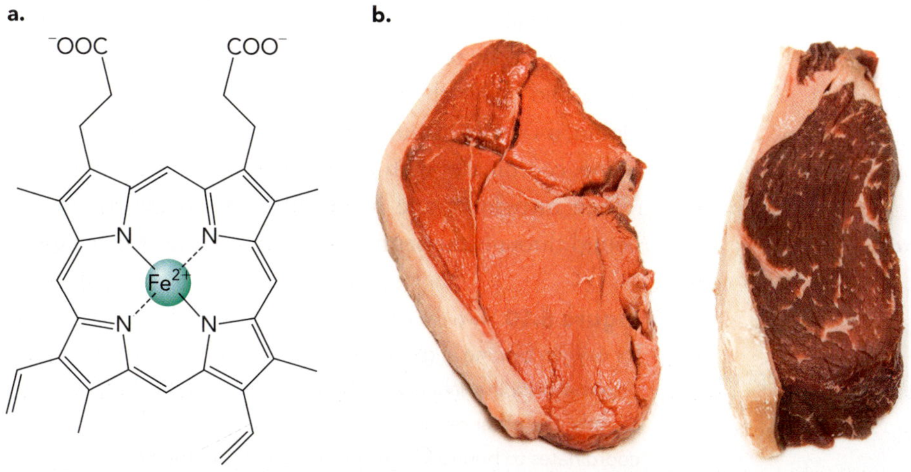

Figure 6.7 The heme of globin proteins contains Fe^{2+} in a porphyrin ring. **a.** Chemical structure of Fe^{2+}-protoporphyrin IX (heme). **b.** Fresh meat has a red color from the Fe^{2+} in hemoglobin. In dead cells, the iron oxidizes to Fe^{3+}, resulting in the brown color seen in old meat.
EDWARD WESTMACOTT/SHUTTERSTOCK.

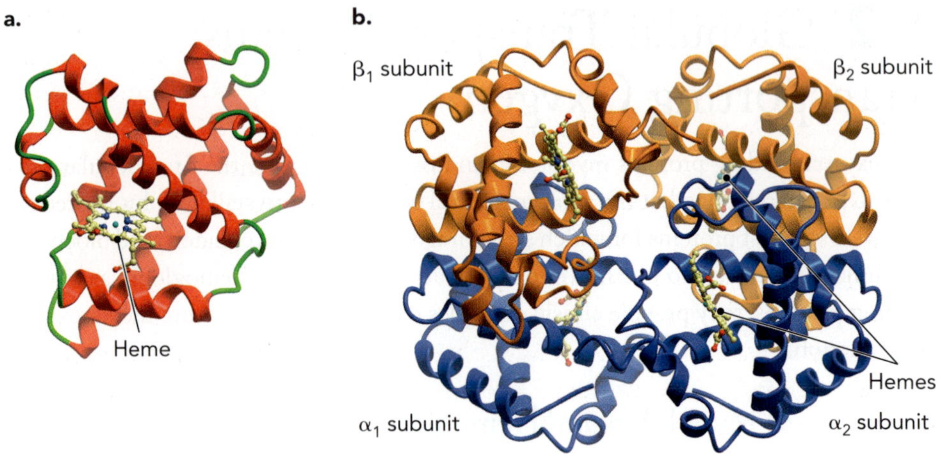

Figure 6.8 The myoglobin polypeptide and the hemoglobin polypeptide subunits each consist of eight α helices with a single molecule of heme bound by each polypeptide. **a.** Molecular structure of myoglobin showing the location of the iron porphyrin ring. BASED ON PDB FILE 1MBO. **b.** Molecular structure of the tetrameric hemoglobin protein complex, which contains a total of four heme groups. BASED ON PDB FILE 1A3N.

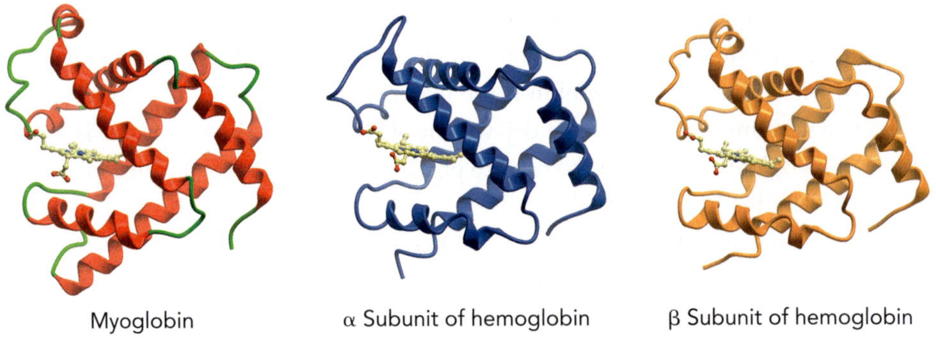

Figure 6.9 The globin fold of myoglobin and that of each of the two types of hemoglobin subunits are structurally very similar, even though the amino acid sequence similarity between myoglobin and either the α or β subunit of hemoglobin is low. BASED ON PDB FILES 1MBO (MYOGLOBIN) AND 1A3N (HEMOGLOBIN).

▶ ANIMATION

as a storage depot for O_2. **Figure 6.8** shows that myoglobin consists of a single polypeptide chain with one heme group, whereas hemoglobin contains four polypeptides: two α subunits (α_1, α_2) and two β subunits (β_1, β_2), each with its own heme group. The hemoglobin tetrameric structure is considered a dimer of heterodimers ($\alpha_1\beta_1$ and $\alpha_2\beta_2$), which has functional significance in terms of subunit interactions. Although less than 20% of the amino acids are identical between myoglobin and either the α or β hemoglobin subunits, all three polypeptides share the same protein fold, called a **globin fold**, which contains eight α helices (**Figure 6.9**).

The Fe^{2+} in the heme group of globin proteins has six coordination bonds: four of these are with nitrogens in the plane of the porphyrin ring, and the remaining two are above and below the ring. In both myoglobin and hemoglobin, a critical histidine residue, called the **proximal histidine**, coordinates with the Fe^{2+} at one of these positions. When myoglobin and hemoglobin are oxygenated, O_2 is bound through the sixth coordination bond (**Figure 6.10**).

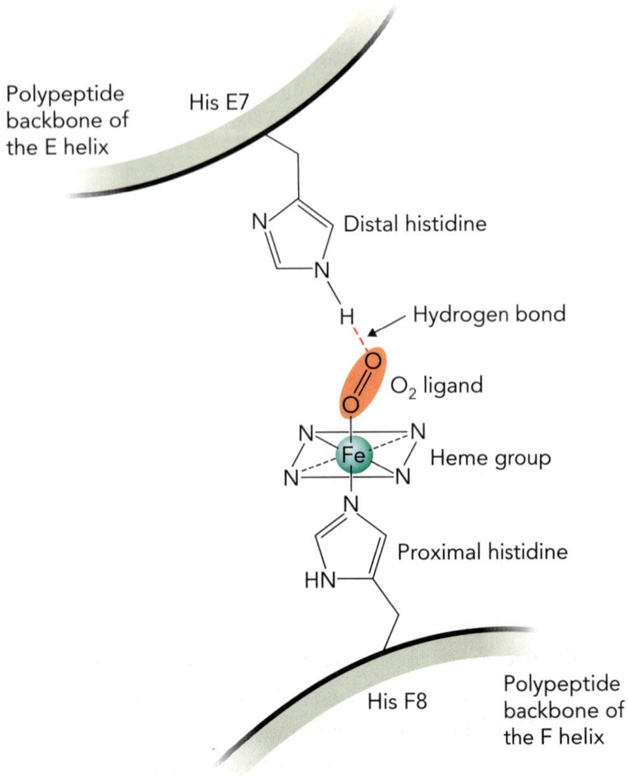

Figure 6.10 Two critical histidine residues are important for oxygen binding in globin proteins. One residue (the proximal histidine, labeled His F8) coordinates to the Fe^{2+} directly; the other residue (the distal histidine, labeled His E7) coordinates to bound O_2, which coordinates to the Fe^{2+}.

A **distal histidine** in globin proteins forms a hydrogen bond with the O_2 and stabilizes its interaction with the heme group. The eight α helices of globin proteins are named A through H, with the proximal His residue referred to as His F8 because it is the eighth residue on the F helix. Similarly, the distal His residue is denoted as His E7 because it is the seventh residue on the E helix. This labeling scheme enables comparison of different globin proteins that have similar structures, even though they may have little sequence similarity.

Figure 6.11 illustrates that in the absence of O_2, the Fe^{2+} is not in the plane of the porphyrin because the ion is too large and the heme is puckered. However, upon O_2 binding, the shared electrons result in a slightly smaller atomic radius for the Fe^{2+}, allowing it to move into the plane of the porphyrin ring. Once this happens, the position of His F8 is translocated closer to the heme, pulling the F-helix polypeptide backbone with it. Analysis of the molecular structures of oxyhemoglobin and deoxyhemoglobin in the vicinity of the heme revealed that a displacement of 0.6 Å in the position of His F8 results in a 1 Å tilt in the F helix (**Figure 6.12**). The movement of the F helix is crucial to the overall structural changes that take place upon oxygen binding in hemoglobin, as discussed in the next subsection.

Another diatomic molecule, carbon monoxide (CO), can also bind to the heme group in myoglobin and hemoglobin. Although CO contains a rigid triple bond that creates steric hindrance between the carbon and the distal histidine, the affinity of CO for the Fe^{2+} in hemoglobin is 200 times higher than that of O_2. This extremely high affinity of CO for the heme Fe^{2+} means that CO displaces O_2 and thus cuts off oxygen delivery to the tissues. This is the molecular basis of carbon monoxide poisoning.

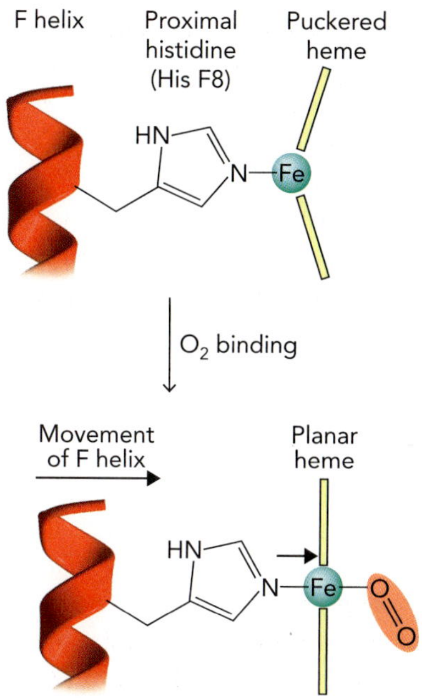

Figure 6.11 In the absence of O_2, the heme group is puckered because of the larger atomic radius of the Fe atom. (The hemes are shown schematically here in an edge-on view.) Oxygen binding to the heme group reduces the radius of the Fe^{2+}, forming a planar heme. Because of the bond between Fe^{2+} and the histidine, formation of a planar heme pulls His F8 toward the heme group and results in movement of the F helix.

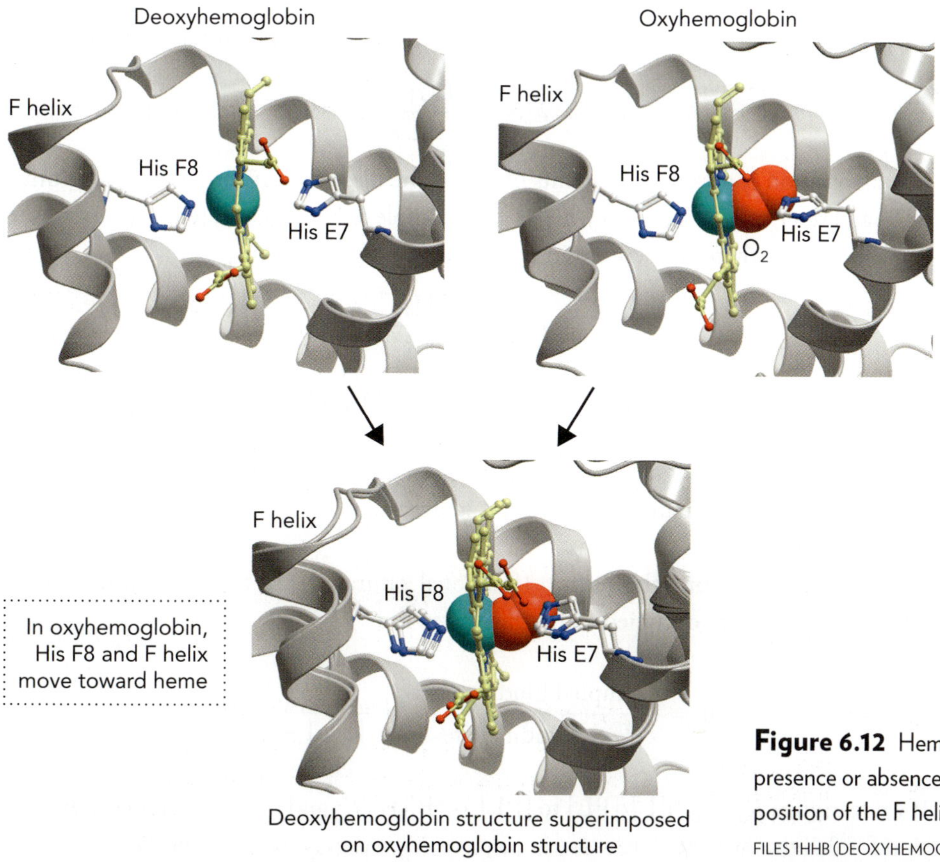

Figure 6.12 Hemoglobin structure changes in the presence or absence of oxygen. Note the different position of the F helix in the presence of O_2. BASED ON PDB FILES 1HHB (DEOXYHEMOGLOBIN) AND 1HHO (OXYHEMOGLOBIN).

Function and Mechanism of Oxygen Binding to Heme Proteins

Reversible binding of O_2 to the heme Fe^{2+} of hemoglobin and myoglobin is the key to their functions. The binding is reversible because the protein's structure changes under different conditions and results in altered affinities for the oxygen ligand. A ligand is a molecule that binds to another (usually larger) molecule. Reversible binding of ligands is a common feature of many biomolecules, and therefore understanding how affinities are measured and can change are important to understand function. We will first discuss general features of protein–ligand interactions and then describe the particular case of O_2 binding to myoglobin and hemoglobin.

For any protein and ligand, we can write an equation for their binding as

$$\mathrm{P} + \mathrm{L} \rightleftharpoons \mathrm{PL}$$

Here, P is the protein, L is the ligand, and PL is the protein–ligand complex. Referring to the equation above, we can write an expression for the equilibrium constant, K_{eq}, for this reaction as

$$K_{eq} = \frac{[\mathrm{PL}]}{[\mathrm{P}][\mathrm{L}]}$$

in which the products of the reaction are in the numerator and the reactants are in the denominator. Note that to write an expression for K_{eq}, the direction of the reaction has to be defined. For reversible binding reactions, it is often advantageous to refer to more specific equilibrium constants—the association constant, K_a, or the dissociation constant, K_d—because for these constants, the direction of the reaction is implied in their names. (Note that though the same notation is used, the association constant K_a is not to be confused with the *acid dissociation* constant K_a that was discussed in Chapter 2.) The **association constant, K_a**, refers to the reaction for the binding (or association) of the protein and ligand, which means that the PL complex is the product, and therefore

$$K_a = \frac{[\mathrm{PL}]}{[\mathrm{P}][\mathrm{L}]}$$

The association constant K_a has units of M^{-1}. The **dissociation constant, K_d**, has units of M and refers to a reaction in which the uncomplexed (or dissociated) species are the product and are therefore written in the numerator as

$$K_d = \frac{[\mathrm{P}][\mathrm{L}]}{[\mathrm{PL}]}$$

Biochemists often compare dissociation constants for different reactions to assess their relative affinities. When two K_d values are compared, the larger value of K_d indicates that more of the dissociated species are present (that is, the numerator is increased relative to the denominator). Therefore, a larger value of K_d indicates a lower affinity between two molecules.

We can also define the term θ (fractional saturation) as the fraction of protein binding sites that are occupied using the expression

$$\theta = \frac{\text{Occupied binding sites}}{\text{Total binding sites}} = \frac{[\mathrm{PL}]}{[\mathrm{PL}] + [\mathrm{P}]}$$

If we rearrange $K_d = [\mathrm{P}][\mathrm{L}]/[\mathrm{PL}]$ to $[\mathrm{PL}] = [\mathrm{P}][\mathrm{L}]/K_d$ and substitute into the expression for θ above, we get

$$\theta = \frac{\left(\frac{[P][L]}{K_d}\right)}{\left(\frac{[P][L]}{K_d}\right) + [P]}$$

We then rearrange to get

$$\theta = \frac{[L]}{[L] + K_d}$$

As you can see from this equation, when the ligand concentration equals K_d, the value of θ is 0.5. A plot of the fractional saturation versus the concentration of free ligand (θ versus [L]) produces a **hyperbolic curve** as shown in **Figure 6.13**. The K_d value can be determined from these types of binding curves by determining the concentration of ligand at which θ equals 0.5.

The binding curves for two different proteins that bind to the same ligand (L) are shown in **Figure 6.14**. You can see that the binding curves are hyperbolic for both proteins yet have different shapes reflecting their different affinities for the ligand. One protein (protein A) has a higher affinity for the ligand, which is indicated by a smaller value for K_d. In other words, a smaller value for K_d means that it takes less ligand to achieve half-saturation of the protein. Correspondingly, a larger value of K_d means that it takes more ligand to achieve half-occupancy of the protein's binding sites. By determining K_d values for a protein and different ligands under a variety of conditions, we can learn what factors influence a protein's interaction with a ligand to

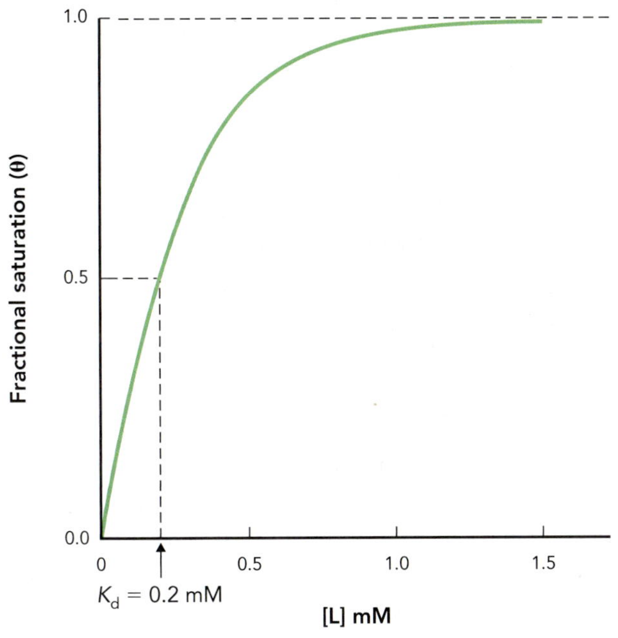

Figure 6.13 A plot of fractional saturation, θ, versus the concentration of free ligand, [L], results in a hyperbolic curve for a simple protein–ligand interaction. The dissociation constant, K_d, can be determined from this plot by determining the concentration of ligand at which θ = 0.5. Note that the curve levels off at higher ligand concentrations as fractional saturation approaches 1 (a horizontal asymptote).

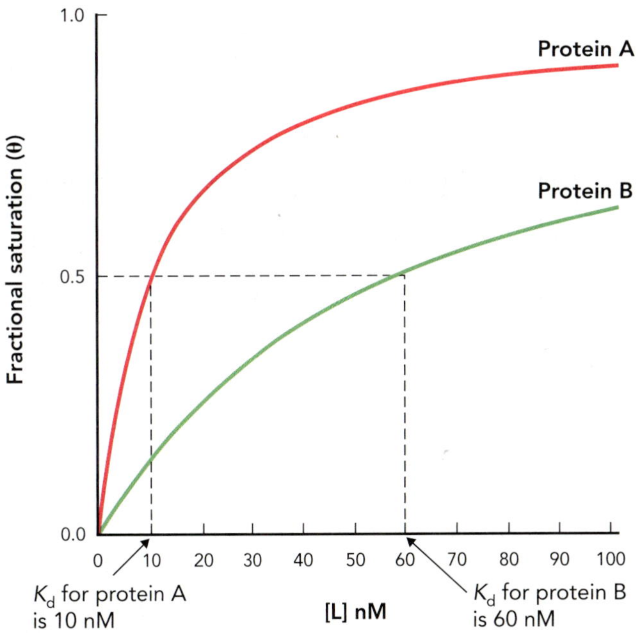

Figure 6.14 Ligand binding curves are shown for two proteins (protein A and protein B) that bind to the same ligand (L). Both binding curves are hyperbolic curves and will approach θ = 1 at higher ligand concentrations than are shown in this plot. Protein A has a higher affinity for ligand than that of protein B as indicated by the lower dissociation constant for protein A (10 nM) than for protein B (60 nM).

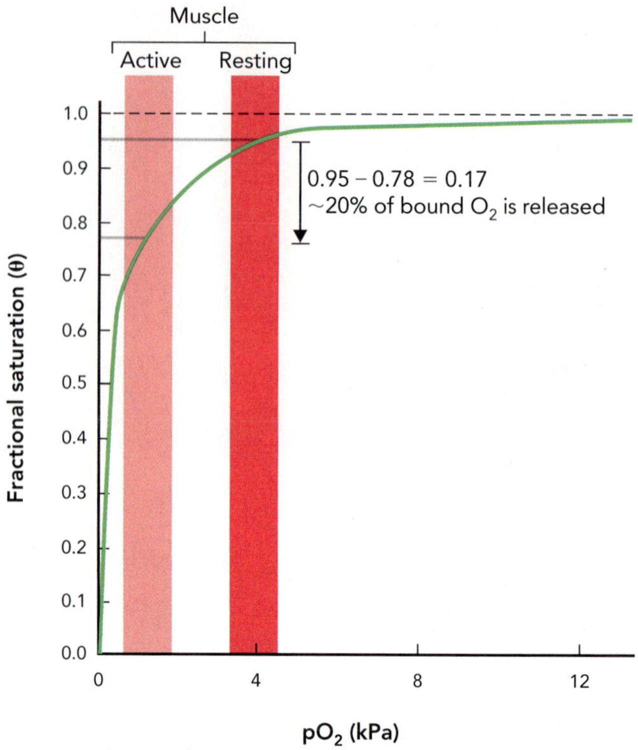

Figure 6.15 The oxygen binding curve for myoglobin is hyperbolic in shape and shows that the binding site is greater than 95% saturated at pO_2 values above 4 kPa. The ranges of pO_2 in active muscle and resting muscle are shown.

▶ ANIMATION

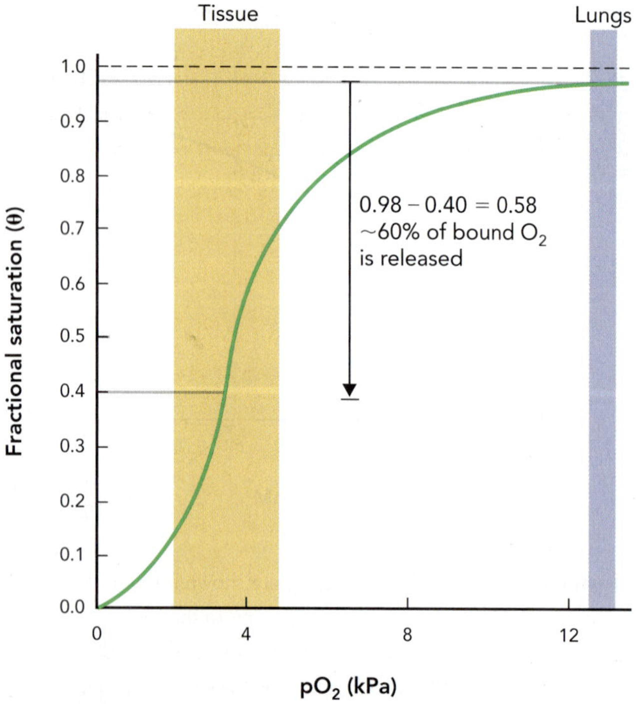

Figure 6.16 Oxygen binding to hemoglobin produces a sigmoidal curve due to cooperative binding.

generate stronger or weaker binding. This knowledge can help elucidate which ligands are physiologically important, for example, or the conditions that affect protein function.

For oxygen binding to myoglobin, the reaction can be written as

$$\text{Mb} + \text{O}_2 \rightleftharpoons \text{MbO}_2$$

Here, MbO_2 is the oxymyoglobin, and Mb is the deoxymyoglobin.

In this case, θ (fractional saturation) is the fraction of oxygen binding sites on myoglobin that are occupied, which can be written as

$$\theta = \frac{\text{Occupied binding sites}}{\text{Total binding sites}} = \frac{[\text{MbO}_2]}{[\text{MbO}_2] + [\text{Mb}]}$$

By plotting the fraction of sites bound (θ) versus the partial pressure of oxygen, pO_2 (a measurement of the concentration of O_2 in solution, using the unit pascal; Pa), we can see that the O_2 binding profile of myoglobin produces a hyperbolic curve as expected (**Figure 6.15**). (Note that when using partial pressure as in pO_2, the half-saturation value is referred to as P_{50} instead of K_d. P_{50} thus refers to the partial pressure of oxygen that results in a fractional saturation value of 0.5.) The pO_2 in resting muscle is about 4 kPa, whereas it is about 1.3 kPa in muscle that is actively exercising. This curve demonstrates that when oxygen concentrations are high, as in resting muscle, myoglobin O_2 binding sites are almost saturated. When oxygen concentrations decrease, as in active muscle, fewer myoglobin binding sites are occupied. The difference in fractional saturation of myoglobin under these two physiologic conditions is ~20% (θ goes from 0.95 to 0.78), indicating the physiologic function of myoglobin as an oxygen storage protein. Myoglobin binds oxygen when the oxygen concentrations are high and releases oxygen when the oxygen concentrations decrease as O_2 is used by active muscle. Having readily available O_2 is critical for muscle activity because O_2 is used by cells to form ATP through the biochemical process of oxidative phosphorylation (see Chapter 11).

In contrast to myoglobin, the shape of the O_2 binding curve of hemoglobin is **sigmoidal**, with a sharp decrease in O_2 saturation at low values of pO_2 (**Figure 6.16**). The primary function of hemoglobin is to deliver O_2 obtained from the lungs to the tissues, which is indicated by this oxygen saturation curve. You can see this from the fact that O_2 saturation of hemoglobin drops from 0.98 to 0.40 between the pO_2 of the lungs and the tissues, resulting in an ~60%

decrease in θ. Therefore, at the high oxygen concentrations in the lungs, hemoglobin is almost fully bound to O_2, but under the conditions of low pO_2 found in the tissues, much less O_2 is bound. You can think of this as hemoglobin "collecting" O_2 in the lungs and "delivering" it to the tissues. The sigmoidal shape of the oxygen binding curve for hemoglobin is indicative of positive **cooperative binding**, meaning that the binding of the first ligand (O_2 in this case) to the protein complex facilitates the binding of additional ligands on the same protein.

The binding of O_2 to globin proteins is a good example of ligand-induced protein conformational changes. In the context of hemoglobin structure and function, the two most important ligands are O_2 and 2,3-bisphosphoglycerate. There are three key features of ligand–protein interactions:

1. Ligand binding is a *reversible process* involving noncovalent interactions.
2. Ligand binding induces or stabilizes *structural conformations* in target proteins.
3. The equilibrium between ligand-bound protein and ligand-free protein can be altered by the binding of *effector molecules*, which induce conformational changes in the protein that increase or decrease ligand affinity.

As described earlier, O_2 binding to hemoglobin stabilizes a protein conformation in which the Fe^{2+} is planar with the heme group, thus fixing the position of His F8 and ultimately the orientation of the entire F helix (see Figure 6.12). These conformational changes within a subunit also affect the quaternary structure of the entire hemoglobin protein. Perutz referred to the O_2 unbound and bound forms of hemoglobin as the *T state* (tense) conformation for deoxyhemoglobin and the *R state* (relaxed) conformation for oxyhemoglobin, which have significant conformational differences (**Figure 6.17a**). A dramatic demonstration of how large these oxygen-induced conformational changes are in the hemoglobin protein structure may be seen when protein crystals of deoxyhemoglobin are exposed to O_2—the crystals shatter!

How do structural changes in the F helix affect the quaternary structure of the entire $\alpha_2\beta_2$ hemoglobin complex, and how do these changes promote cooperative binding of O_2? Perutz proposed that small movements in the F helix, resulting from changes in the position of the Fe^{2+} atom in the heme due to O_2 binding, are translated to the entire protein complex through changes in noncovalent contacts at the

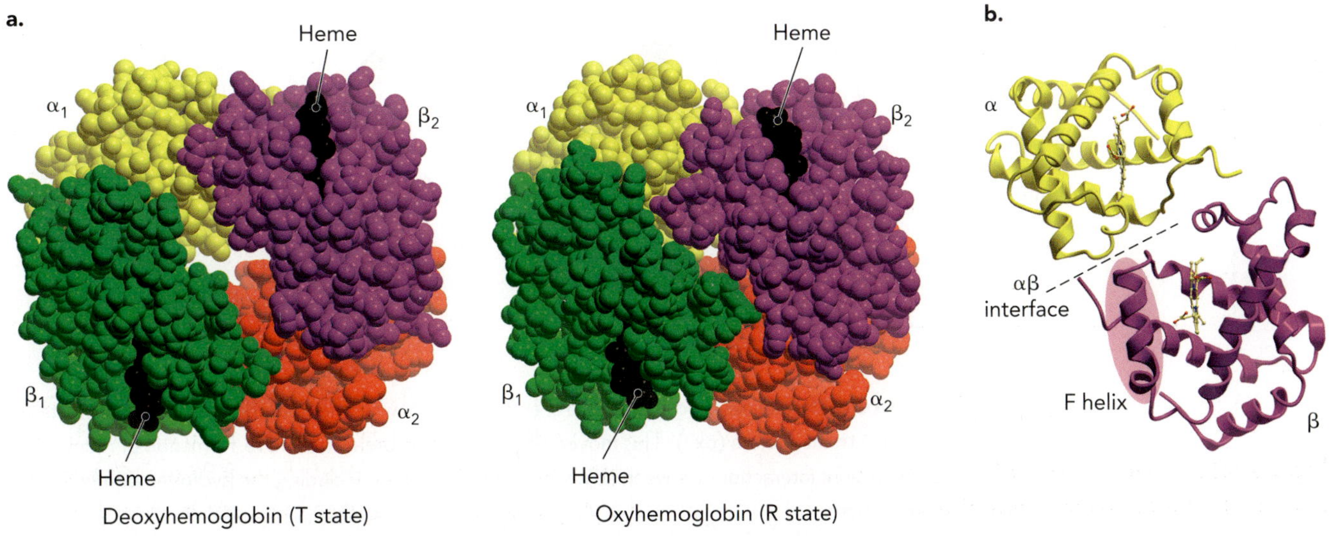

Figure 6.17 Hemoglobin structure changes when oxygen is bound. **a.** The space-filling representations of the tetramer in the T and R states show that significant structural changes occur upon O_2 binding. **b.** Small changes in the position of the F helix alter contacts at the interface between the α and β subunits. Here, only two of the four subunits of deoxyhemoglobin are shown for clarity. BASED ON PDB FILES 1HHB (DEOXYHEMOGLOBIN) AND 1HHO (OXYHEMOGLOBIN).

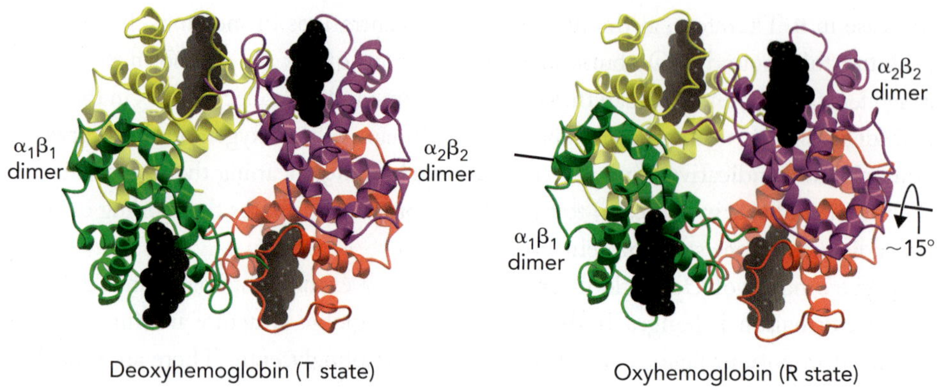

Figure 6.18 Oxygen binding in hemoglobin induces a rotational movement of approximately 15° of one αβ heterodimer relative to the other. BASED ON PDB FILES 1HHB (DEOXYHEMOGLOBIN) AND 1HHO (OXYHEMOGLOBIN).

interface of the α and β subunits (**Figure 6.17b**). These small structural changes in one region of the protein lead to large structural changes in the entire hemoglobin molecule. These conformational changes alter O_2 affinity in all four globin subunits and form the molecular basis for the observed cooperativity in O_2 binding in hemoglobin.

One way to think about how the hemoglobin structure changes between the deoxyhemoglobin (T state) and oxyhemoglobin (R state) conformations is to examine how the α helices of one αβ dimer are repositioned relative to the second αβ dimer. By drawing an axis through the protein, you can see the repositioning of one αβ heterodimer with respect to the other (**Figure 6.18**). This rotation between the T and R states alters as many as 50 noncovalent interactions—consisting of hydrogen bonds, van der Waals interactions, and ion pairs—that lie at the interface between the $\alpha_1\beta_1$ and $\alpha_2\beta_2$ dimers. An example of how distinct sets of noncovalent interactions stabilize the T and R conformations is shown in **Figure 6.19**. In human hemoglobin in

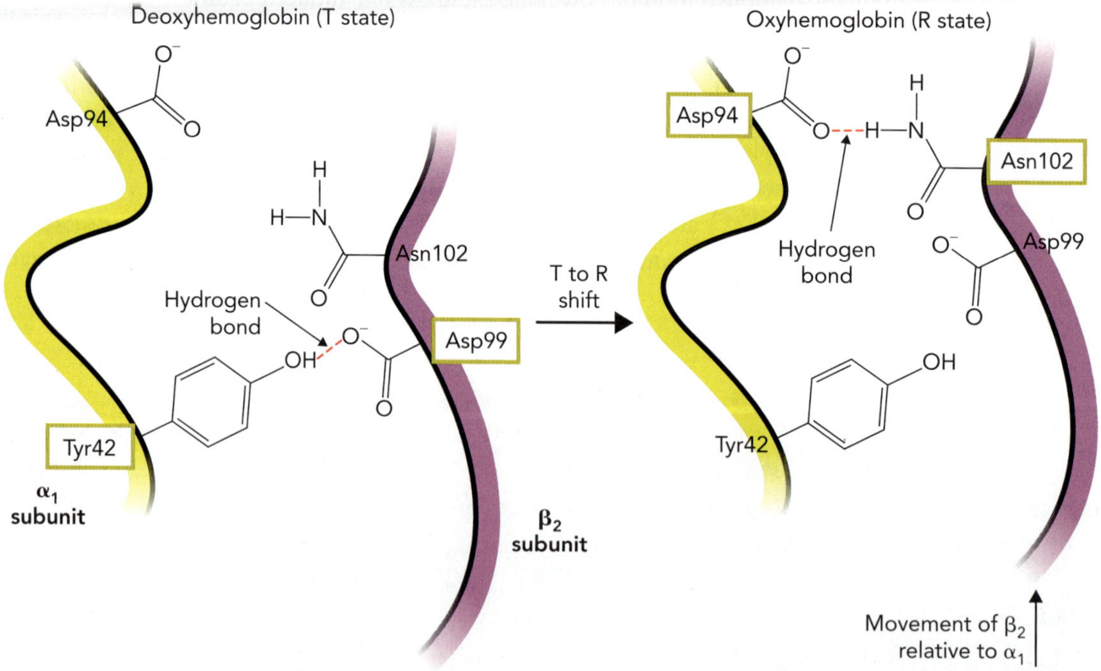

Figure 6.19 Helices reposition in the α and β subunits when hemoglobin goes from the T state (deoxy) to the R state (oxy). This movement involves the breakage and re-formation of many noncovalent interactions between adjacent α and β subunits, including the hydrogen bonds shown here.

the T state, a hydrogen bond forms between Tyr42 in the α_1 subunit and Asp99 in the β_2 subunit. However, upon oxygen binding, the two helices slide past each other, and a new hydrogen bond forms between Asp94 of the α_1 subunit and Asn102 of the β_2 subunit in the R state.

The Perutz model suggests that communication occurs between the O_2 binding sites, which is achieved through propagated conformational changes. Thus, the T-state conformation has a different affinity for O_2 than that of the R-state conformation. Indeed, affinity measurements of O_2 binding have shown that the fourth O_2 binds with an affinity that is 100 times higher than that of the first O_2. But how does the protein transition between the low-affinity and high-affinity states? Two models have been proposed to explain this cooperative behavior of proteins: the concerted model and the sequential model. These models are likely not mutually exclusive because experimental data derived from protein structure–function studies suggest that a mixture of both the concerted and sequential mechanisms is a more likely representation of actual events.

The Concerted Model In the **concerted model** of **allostery**, proposed by Jacques Monod, Jeffries Wyman, and Jean-Pierre Changeux, the protein complex exists in either the T state or the R state, as illustrated in **Figure 6.20a** for a tetrameric protein. The word *allostery* is derived from Greek and means "other shapes." The population of molecules in each of these two states is affected by ligand binding. You can see in the concerted model that ligand binding to a single subunit in a tetrameric protein complex helps to stabilize the R-state conformation, and thus more of the protein complexes are found in the R state than for the unliganded protein complex. Note that the ligands can bind to either the T-state or the R-state conformation; however, high-affinity binding is only associated with protein subunits in the R-state conformation. Moreover, all complexes are symmetrical, such that they do not contain mixtures of subunits in the T and R states. The concerted model, also known as the **symmetry model**, predicts that protein subunits in the R-state conformation can exist in the absence of bound ligands, though the T state is expected to dominate the population.

The Sequential Model The **sequential model** of allostery, proposed by Daniel Koshland and colleagues, states that binding of ligand to one subunit does not convert the entire complex to the R state, but rather alters the affinity of adjacent subunits. This still favors conversion from the T- state conformation to the R state when ligands bind (**Figure 6.20b**). In the most permissive sequential model,

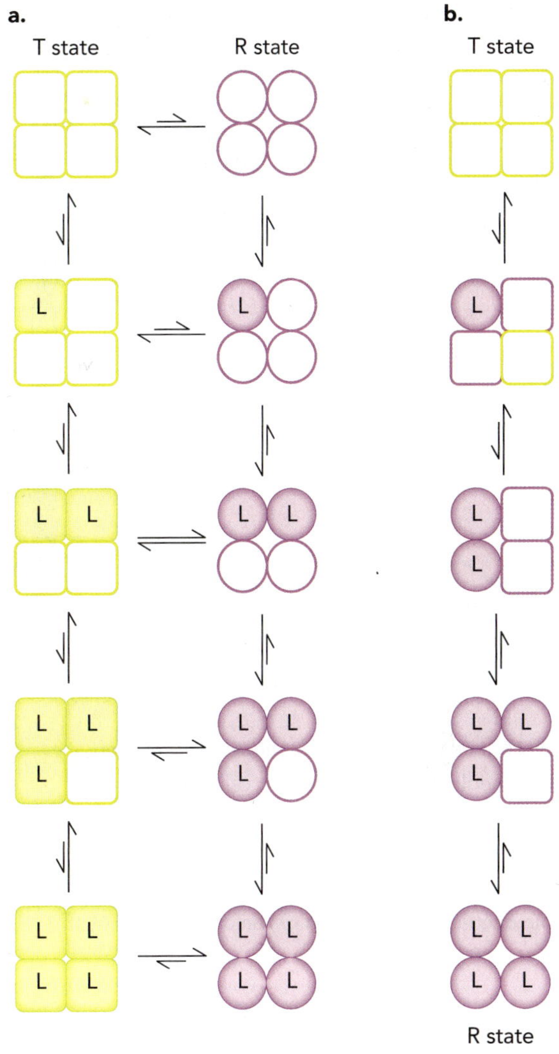

Figure 6.20 The concerted and sequential models, schematically represented here for a tetrameric protein, have been proposed to explain the cooperative binding behavior of proteins. **a.** The concerted model of Monod, Wyman, and Changeux predicts that the T and R states are in equilibrium. The T state is favored when little or no ligand is bound (shown as "L" for ligand). Binding of ligand increases the population of the R state. **b.** The sequential model of Koshland and colleagues proposes that ligand binding to one subunit of a tetrameric protein induces a conformational change such that nearby subunits are more likely to bind ligand and shift to the R state. Note that as the T state can exist with bound ligand and the R state can exist without ligand, there are actually 20 more possible configurations of this tetrameric complex than are shown here (though these will be relatively small percentages of the population of molecules).

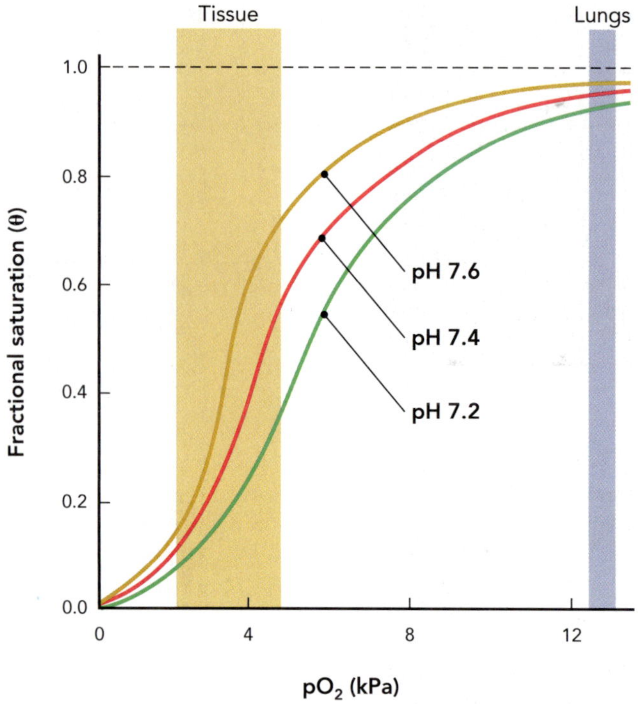

Figure 6.21 The observed pH dependence of oxygen binding to hemoglobin is part of what is called the Bohr effect. The ranges of pO_2 in the tissues and in the lungs are shown. Note that at the lower pH of 7.2 normally found in the tissues, the oxygen saturation (θ) of hemoglobin decreases, resulting in O_2 unloading.

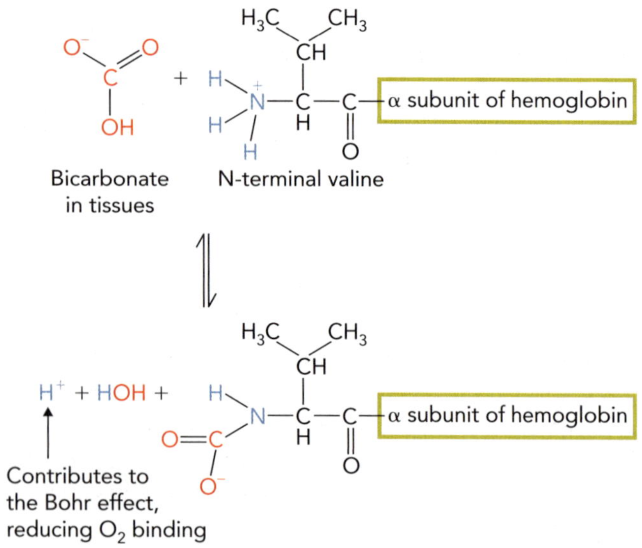

Figure 6.22 Carbon dioxide, in the form of bicarbonate, generates a carbamate group on the amino-terminal residue (Val) of all four hemoglobin subunits. This reversible reaction also releases a proton that contributes to the Bohr effect.

the protein complex can contain subunits in both the T-state and R-state conformations, and moreover, they can exist in either the ligand-bound or ligand-free mode (although the T state has low affinity for the ligand). The sequential model predicts that as the number of R-state conformation subunits in a complex increases, the more likely it is that the remaining T-state subunits will be in the R-state conformation and bind ligand.

Allosteric Control of Oxygen Transport by Hemoglobin

Oxygen binding to hemoglobin is affected by the biochemical environment of lung cells and the peripheral tissues. Indeed, the biochemical basis for O_2 transport by hemoglobin from the lungs to the tissues is differential O_2 affinity for hemoglobin in lung cells (where O_2 concentration is high) compared to peripheral tissues (where O_2 concentration is low). This differential O_2 affinity is the result of effector molecules, which bind to hemoglobin and induce conformational changes that shift the equilibrium toward the R state or the T state. Because these molecules alter structural conformations to favor one function or another, they are called allosteric effectors.

Oxygen is a positive allosteric effector of hemoglobin, meaning that the binding of a single O_2 molecule to one globin subunit shifts the conformational equilibrium from the T state toward the R state. This increases O_2 affinity in other subunits (see Figure 6.20). Oxygen binding to hemoglobin represents *homotropic* allostery because O_2 binding affects the binding of other O_2 molecules (these ligands are the same). Three other allosteric effectors that modulate O_2 affinity of hemoglobin are CO_2, H^+, and 2,3-bisphosphoglycerate. However, these three molecules function as negative allosteric regulators because they shift the conformational equilibrium from the R state toward the T state, reducing hemoglobin's affinity for oxygen. These molecules all mediate *heterotropic* allostery, which refers to the mechanism whereby allosteric effectors bind to secondary sites in the protein. This in turn affects ligand binding at the primary site (O_2 binding site).

The Bohr Effect One of the by-products of aerobic respiration is CO_2, which needs to be removed from the tissues. Some of the CO_2 dissolves in the blood as gas, but most of it is hydrated by the enzyme carbonic anhydrase to form highly soluble HCO_3^-:

$$CO_2 + H_2O \xrightleftharpoons{\text{Carbonic anhydrase}} HCO_3^- + H^+$$

You can see from this reaction that, in addition to HCO_3^-, carbonic anhydrase also produces H^+, which causes a decrease in blood pH from 7.6 in the lungs to 7.2 in the tissues. This is very important because the α and β subunits of hemoglobin can be protonated at several key residues, and therefore pH changes will affect the stabilization of either the T state or the R state. We can see the effect of protonation on O_2 binding by examining the oxygen saturation of hemoglobin as a function of pH. As shown in **Figure 6.21**, oxygen saturation of hemoglobin at the pO_2 of tissues is significantly decreased at pH 7.2 relative to pH 7.6.

The heterotropic allosteric effector CO_2 also reversibly binds to hemoglobin and leads to the formation of additional ion pairs that stabilize the T state. The binding of CO_2 to hemoglobin occurs through the formation of a carbamate group at the amino-terminal residue of each globin subunit (**Figure 6.22**). This reversible reaction in red blood cells involves bicarbonate (HCO_3^-) and produces an H^+ that further decreases O_2 affinity.

The pH and CO_2 dependence of O_2 binding is called the **Bohr effect** after Christian Bohr, a Danish physiologist who first reported it in 1904. This discovery is said to have been an inspiration to Christian's son, Niels Bohr, who won the 1922 Nobel Prize in Physics for his elucidation of atomic structure and quantum mechanics. Niels's son, Aage Bohr, was born that same year (1922) and continued in his father's and grandfather's footsteps by becoming a famous scientist himself. Just 13 years after the death of Niels Bohr, Aage Bohr was awarded a share of the 1975 Nobel Prize in Physics for his work on atomic nuclei vibrations (**Figure 6.23**).

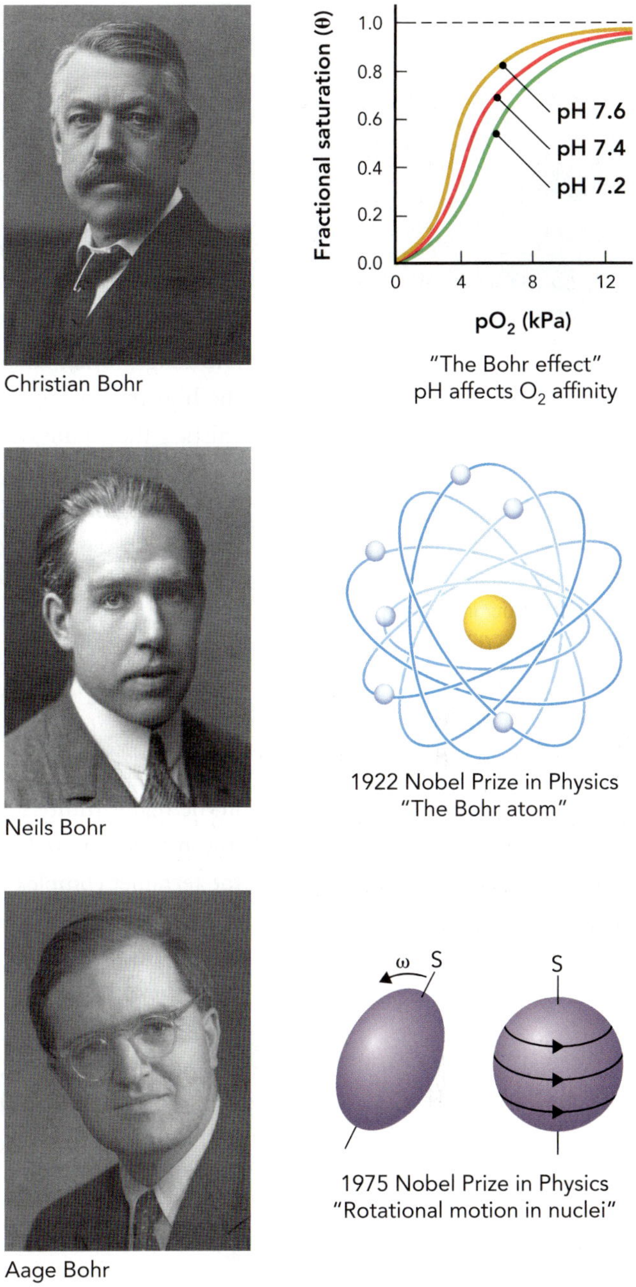

Figure 6.23 The Bohr family had three generations of prominent scientists. Christian Bohr discovered the effect of pH and CO_2 on O_2 binding by hemoglobin; his son Niels Bohr was awarded the 1922 Nobel Prize in Physics; and Niels's son Aage Bohr (Christian's grandson) was awarded a portion of the 1975 Nobel Prize in Physics. C. BOHR: NIELS BOHR ARCHIVE, COPENHAGEN; N. BOHR: UNIVERSAL HISTORY ARCHIVE/GETTY IMAGES; A. BOHR: THE ROYAL DANISH ACADEMY OF SCIENCES AND LETTERS.

Function of 2,3-Bisphosphoglycerate Erythrocytes (red blood cells) contain the small, three-carbon compound 2,3-bisphosphoglycerate (2,3-BPG). 2,3-BPG binding traps hemoglobin (Hb) in the T state by interacting with numerous residues in the center of the hemoglobin tetramer. **Figure 6.24** shows where 2,3-BPG binds between the two β subunits. The R state of hemoglobin has a smaller central cavity (see Figure 6.17), so 2,3-BPG binds preferentially to the T state. 2,3-BPG is a classic heterotropic negative allosteric regulator because binding of 2,3-BPG to a secondary site inhibits binding of O_2 to the primary sites. Note that only one molecule of 2,3-BPG binds to hemoglobin in the T state; however, this has the effect of inhibiting O_2 binding to all four subunits.

Figure 6.25 shows the chemical structure of 2,3-BPG, which has a net negative charge due to the carboxyl and

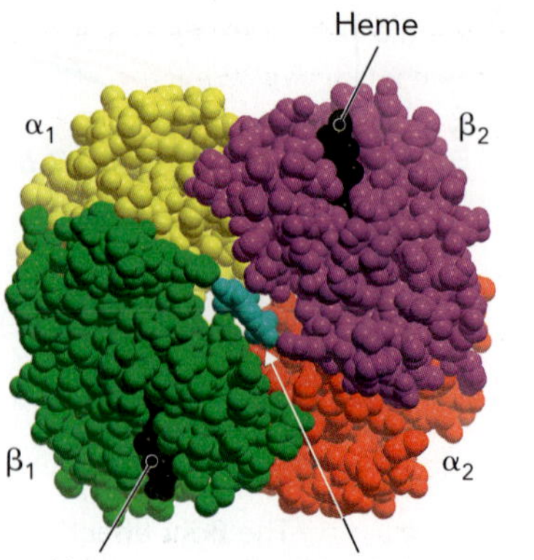

Figure 6.24 2,3-BPG binds to the central cavity of T-state hemoglobin, as shown in this molecular structure. The 2,3-BPG molecule (blue) makes contact with both the β_1 and β_2 subunits. BASED ON PDB FILE 1B86.

phosphoryl groups. The binding of 2,3-BPG to the β subunits is stabilized by ionic interactions between 2,3-BPG and three positively charged residues on each β subunit (His2, Lys82, and His143).

We are now ready to put together all this information and see how oxygen transport from the lungs to the tissues is controlled by allosteric mechanisms. **Figure 6.26** illustrates this principle by showing that the high concentration of O_2 in the lungs favors the R state, whereas a lower O_2 concentration in the tissues favors the T state. Because 2,3-BPG concentration is constant in red blood cells, the equilibrium shift between the R and T states is facilitated by the difference in pO_2 between the tissues and lungs, rather than by a change in 2,3-BPG concentration. As mentioned earlier, the higher levels of CO_2 and H^+ in the tissues also contribute in a significant way to shifting the equilibrium in favor of the T state (the Bohr effect).

Notably, although 2,3-BPG is present in both adult and embryonic red blood cells, fetal hemoglobin consists of a tetramer complex in which the normal adult β subunit is replaced by a special fetal hemoglobin subunit called γ. An important difference between the β and γ subunits is the replacement of His143 with Ser143, which eliminates two of the six positive charges in the 2,3-BPG binding site (see Figure 6.25). This single amino acid difference reduces the affinity of 2,3-BPG for the $\alpha_2\gamma_2$ complex compared to that for the normal adult $\alpha_2\beta_2$ complex. As illustrated in **Figure 6.27**, this decrease in 2,3-BPG affinity facilitates transfer of O_2 from the mother's hemoglobin to the fetal hemoglobin because fetal hemoglobin can obtain more O_2 if more hemoglobin molecules are in the high-affinity R state. Soon after birth, red blood cells in the newborn baby begin to synthesize more β than γ subunits, such that by 6 months of age, more than 90% of the tetramer complex in the red blood cells of babies contains the β subunit.

2,3-BPG is also involved in physiologic adaptation to high altitudes, where the pO_2 is lower than at sea level. Within a few hours of switching from low altitudes to altitudes above 4,000 meters, the body begins to synthesize more 2,3-BPG. This results in a larger percentage of oxygen unloading in the tissues due to a shift in the equilibrium to the T state. As shown in **Figure 6.28**, even though increased levels of 2,3-BPG in red blood cells means that less oxygen saturation occurs at high altitude (4,500 meters), the more efficient delivery of oxygen to the tissues offsets this effect. 2,3-BPG levels return to normal within a day after returning to lower altitudes.

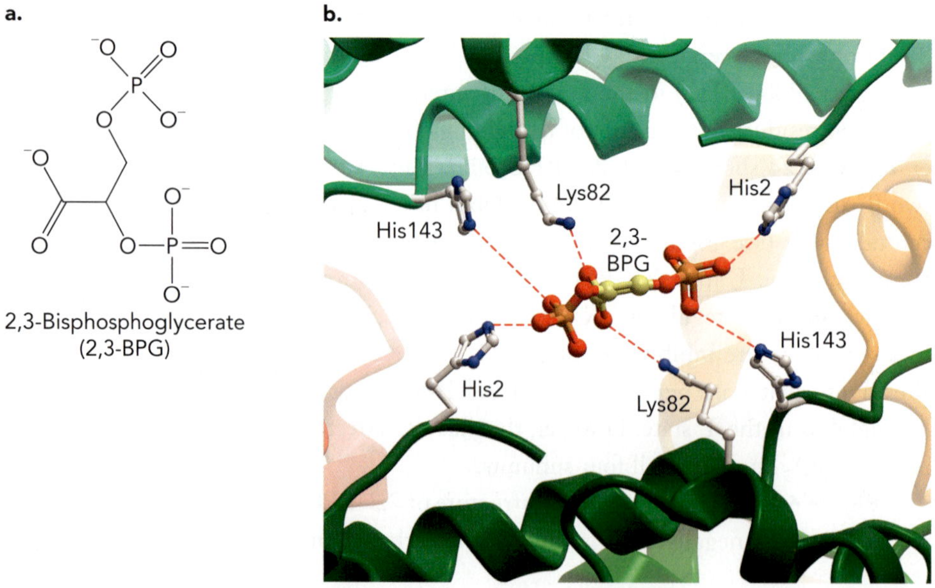

Figure 6.25 2,3-BPG is a negatively charged molecule that interacts with positively charged residues in hemoglobin. **a.** Chemical structure of 2,3-BPG. Note the negative charges on the carboxyl and phosphate groups. **b.** The molecular structure of the hemoglobin–2,3-BPG complex shows the ionic interactions between 2,3-BPG and the Lys and His residues on the β_1 and β_2 subunits. BASED ON PDB FILE 1B86.

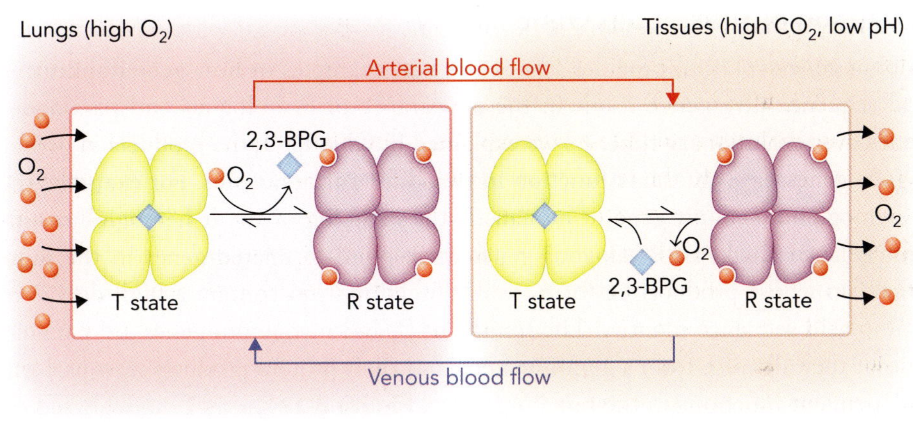

Figure 6.26 Allosteric regulation of hemoglobin functions in delivering O_2 from the lungs to the tissues. High O_2 concentrations in the lungs favor the R-state conformation of hemoglobin, and under these conditions, 2,3-BPG molecules are mostly unbound. However, in the tissues, the combination of reduced O_2 levels and the Bohr effect favor the T-state conformation, which is stabilized by 2,3-BPG binding.

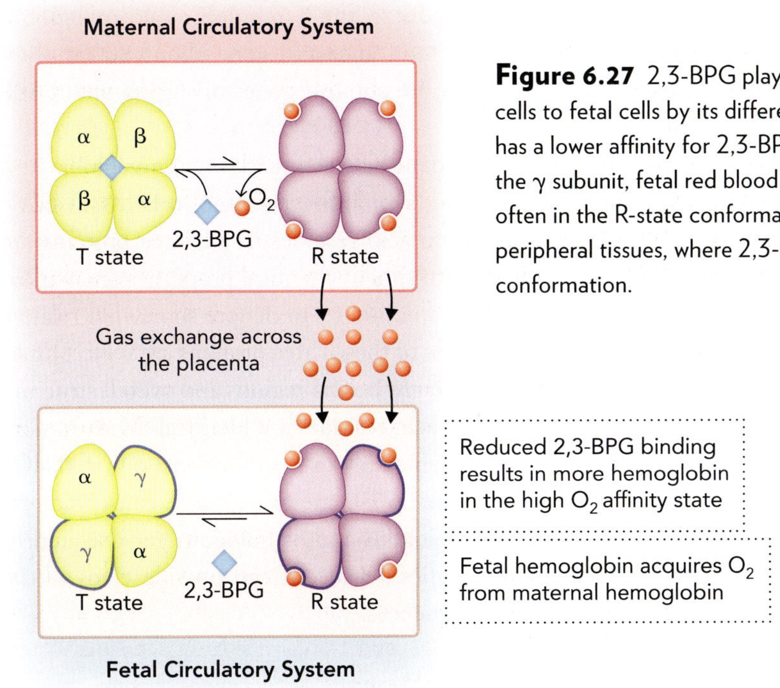

Figure 6.27 2,3-BPG plays a role in facilitating O_2 transport from maternal cells to fetal cells by its differential binding to $\alpha_2\beta_2$ versus $\alpha_2\gamma_2$. Because $\alpha_2\gamma_2$ has a lower affinity for 2,3-BPG due to the His143→Ser143 difference in the γ subunit, fetal red blood cells have hemoglobin molecules that are more often in the R-state conformation compared to maternal red blood cells in the peripheral tissues, where 2,3-BPG binding to hemoglobin stabilizes the T-state conformation.

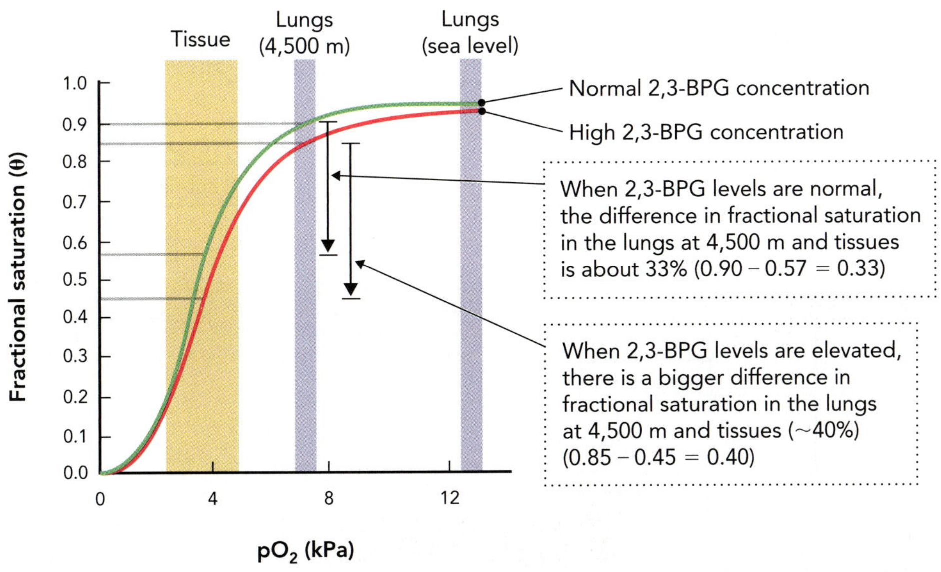

Figure 6.28 Changes in 2,3-BPG affect oxygen binding to hemoglobin. The oxygen saturation (θ) of hemoglobin shifts in response to 2,3-BPG concentration. At high altitudes, humans make more 2,3-BPG, which results in increased O_2 unloading in the tissues, as evidenced by the greater difference in fractional saturation at increased 2,3-BPG levels.

Evolution of the Globin Gene Family

Globin gene evolution provides one of the best examples of how gene duplication and sequence divergence result in the appearance of orthologous and paralogous genes over evolutionary time. As we explained in Chapter 1, the products of orthologous genes have the same function in two different organisms. For example, the α hemoglobin subunit gene in humans is the ortholog of the α hemoglobin subunit gene in chickens. Paralogous genes correspond to related genes in the same organism whose products perform similar functions and contain a high degree of amino acid sequence identity. The human α, β, and myoglobin genes are paralogs in that they all arose from a common gene, but their protein products now perform independent functions in the same organism. **Figure 6.29** shows a model based on sequence similarity and protein structure–function relationships that depicts how gene duplication in an early ancestor could have given rise to the human globin gene family. On the basis of sequence relatedness, the hemoglobin and myoglobin genes appear to have diverged from a common ancestral gene about 800 million years ago, with the human α and β hemoglobin subunit genes diverging about 300 million years later.

Figure 6.30 shows an amino acid alignment between the human myoglobin and hemoglobin proteins that was prepared using a bioinformatic tool called ClustalW. In this bioinformatic analysis, identical amino acid residues in all three proteins are denoted by an asterisk (*), and amino acids with similar chemical properties are marked by a colon (:) to denote closely related or by a period (.) to denote somewhat related. Note that even though the molecular structures of these three proteins are very comparable, as evidenced by the conservation of the eight helical regions and overall structure (see Figure 6.9), less than 20% of the amino acid residues are identical. Most important, the critical His E7 and His F8 residues, which play a key functional role in the O_2 binding properties of globin proteins (see Figure 6.10), are two of these conserved residues. Computational analysis to predict paralogous and orthologous gene sequences is a powerful tool in the field of bioinformatics because it can provide clues to the functions of uncharacterized gene-coding sequences.

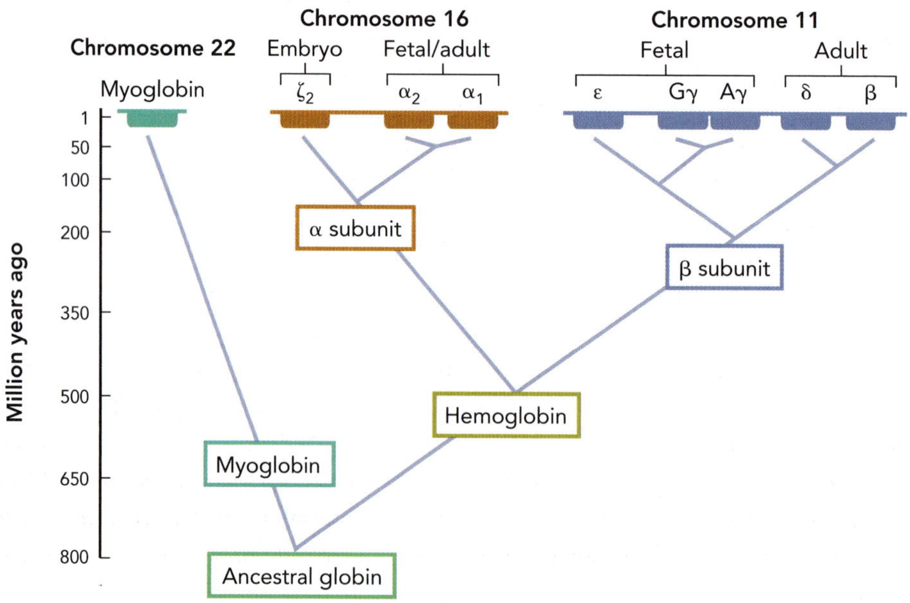

Figure 6.29 Gene duplication is the most likely explanation for globin gene evolution. Forks in the lineage correspond to approximate time of divergence (in millions of years). Protein evolution models are based on analyzing amino acid sequence similarities. Note that the α and β hemoglobin genes are on different chromosomes (11 and 16). The notations δ, ε, ζ_2, Aγ, and Gγ represent other forms of hemoglobin that are not discussed in this text.

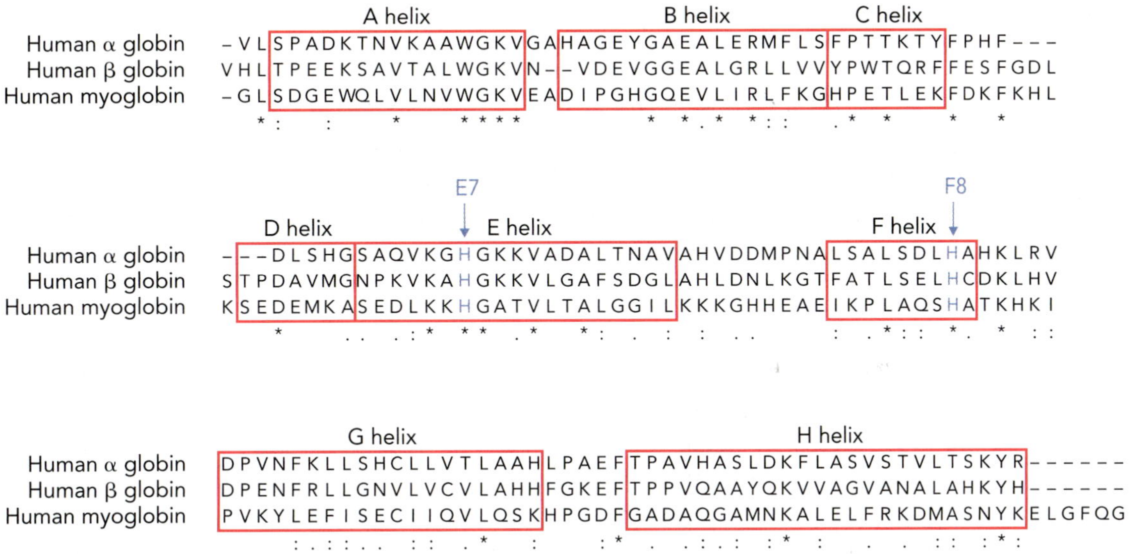

Figure 6.30 Amino acid sequence alignments of human myoglobin and hemoglobin proteins were generated by the bioinformatic analysis program ClustalW. The dashes in the sequences indicate where theoretical gaps have been introduced to allow better alignments. The asterisk (*) indicates identical amino acids in all sequences, the colon (:) indicates a conservative substitution, and the period (.) indicates a semiconservative substitution. The eight α helices are outlined with red boxes, and the key His E7 and His F8 residues required for O_2 binding are shown in blue.

Globin Gene Mutations Can Cause Human Anemia Another way to examine protein evolution is to compare DNA sequences between individuals within a population to identify amino acid alterations arising from specific genes. Rather than sequence an entire genome, we can simply examine the primary sequence of selected genes or DNA segments to make comparisons between individuals. When this was done with human globin genes, it was discovered that hundreds of amino acid variations occur in the α and β hemoglobin subunits in the population. Indeed, it is estimated that 5% of the human population contains point mutations in one copy of either the α or β hemoglobin subunit gene. Some of these mutations cause amino acid alterations in hemoglobin proteins that can lead to serious clinical symptoms, such as anemia. **Anemia** is the clinical term for reduced O_2 transport efficiency from the lungs to the tissues, arising from altered hemoglobin function or reduced numbers of red blood cells per unit of blood. Anemia can cause fatigue, shortness of breath, and cognitive decline. **Table 6.1** lists representative amino acid changes in human α and β globin proteins that cause anemia because of the resulting defects in hemoglobin structure and function.

One of the best-characterized human globin mutations is an amino acid substitution of valine for glutamate at position 6 of the β-globin polypeptide (β-E6V), which causes the recessive genetic disease **sickle cell anemia**. The mutated β-globin gene encodes the sickle cell β polypeptide, which is called β_S. The tetramer of hemoglobin containing $\alpha_2\beta_{S2}$ subunits is called hemoglobin S, or HbS. The presence of HbS in red blood cells is pathologic under low O_2 concentrations, primarily in the tissues such as the kidney and the spleen, where erythrocytes circulate more slowly and are often destroyed.

Table 6.1 REPRESENTATIVE LIST OF AMINO ACID SEQUENCE ALTERATIONS THAT HAVE BEEN IDENTIFIED IN β AND α HEMOGLOBIN SUBUNITS IN HUMANS

Globin disease	Genotype	Phenotype
β subunit		
Hemoglobin S (sickle cell)	Glu6 (A3)→Val6	Globin protein aggregation
Hemoglobin Hammersmith	Phe42 (CD1)→Ser42	Loss of heme binding
Hemoglobin Savannah	Gly24 (B6)→Val24	Protein misfolding
Hemoglobin Milwaukee	Val67 (E11)→Glu67	Loss of O_2 transport function
Hemoglobin Kansas	Asn102 (G4)→Thr102	Destabilizes R state
Hemoglobin Yakima	Asp99 (G1)→His99	Destabilizes T state
α subunit		
Hemoglobin Bibba	Leu136 (H19)→Pro136	Destabilizes tetramer
Hemoglobin St. Lukes	Pro95 (G2)→Arg95	Destabilizes tetramer
Hemoglobin Philadelphia	Tyr35 (C1)→Phe35	Destabilizes tetramer

Note: The name of each hemoglobin deficiency reflects its discovery. The position of a mutated amino acid within the eight α helices is shown in parentheses.

The alteration of residue 6 from glutamate to valine results in a hydrophobic amino acid residue on the surface of the protein. **Figure 6.31** shows the molecular structure formed between two HbS tetramers of deoxyhemoglobin as a result of interactions between β_S-Val6 of one HbS tetramer and β_S-Phe85 and β_S-Leu88 of another HbS tetramer via the hydrophobic effect. Deoxyhemoglobin HbS (but not oxyhemoglobin HbS) is responsible for the generation of long chains of complex protein polymers, which cause erythrocytes to be distended into a sickle-like structure. These rigid distorted cells clog microcapillaries and cut off oxygen supply, causing tissue damage and pain (**Figure 6.32**). The reason for HbS polymer formation only under low O_2 conditions, when hemoglobin is in the T state, is because β_S-Phe85 and β_S-Leu88 are not in the correct positions to interact with the β_S-Val6 residue when hemoglobin is in the oxygenated R-state conformation.

Polymerization of deoxyhemoglobin HbS tetramers only occurs when both β subunits are the β_S variant (mixed

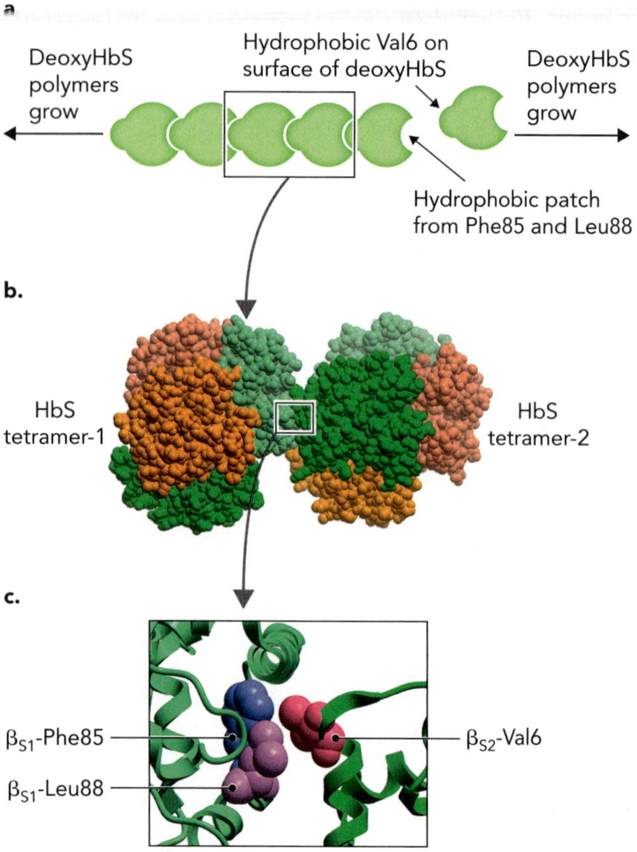

Figure 6.31 Deoxyhemoglobin with the Val6 mutation (deoxyHbS) can form extended polymers through association of hydrophobic residues. **a.** Representation of the polymer formed by deoxyHbS through the interaction of hydrophobic residues. **b.** Space-filling model of a deoxyHbS complex, consisting of two tetramers ($\alpha_2\beta_{S2}$) that form strong interactions between β_S subunits in each tetramer. BASED ON PDB FILE 2HBS. **c.** Structure near the hydrophobic pocket of the deoxyHbS complex shown in panel **a.** Note that the β_S-Val6 residue in one HbS tetramer can interact with the β_S-Phe85 and β_S-Leu88 residues in the other HbS tetramer.

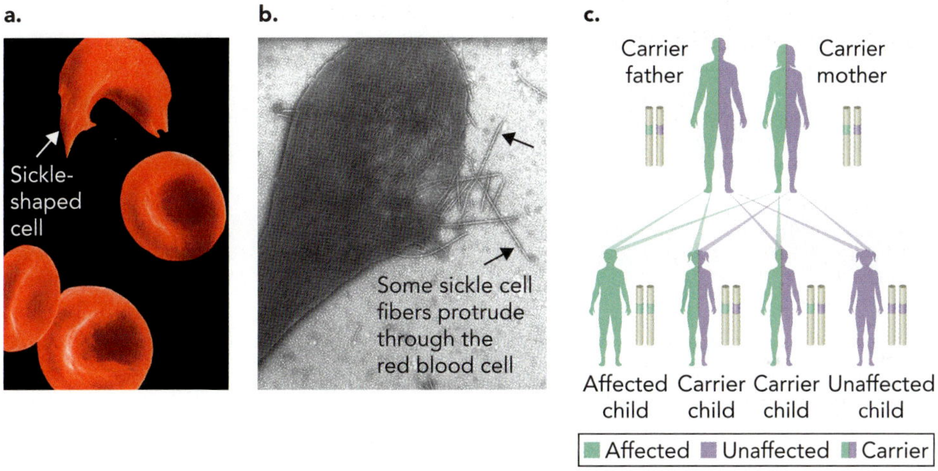

Figure 6.32 Sickle cell anemia is characterized by deformed red blood cells, which have a sickle appearance due to the accumulation of intracellular deoxyHbS fibers. **a.** Photomicrograph of normal and sickle-shaped red blood cells. DR. STANLEY FIEGLER/GETTY IMAGES. **b.** Electron micrograph showing a red blood cell that is completely filled with deoxyHbS fibers. T. WELLEMS AND R. JOSEPHS (1979). CRYSTALLIZATION OF DEOXYHEMOGLOBIN S BY FIBER ALIGNMENT AND FUSION. *JOURNAL OF MOLECULAR BIOLOGY, 135*(3), 651–674. © 1979 PUBLISHED BY ELSEVIER. **c.** Sickle cell anemia is an autosomal recessive genetic disease, which means that an affected individual needs to inherit two copies of the defective gene from heterozygous parents (carriers).

$\alpha_2\beta_S\beta$ tetramers do not polymerize). Sickle cell anemia is an autosomal recessive genetic disease, and therefore the symptoms are most severe in homozygous individuals, in whom both copies of the globin gene are mutated. Heterozygous individuals contain one normal globin gene and one β_S gene; thus, the amount of normal hemoglobin protein in erythrocytes is high enough to disrupt the formation of long HbS protein polymers and prevent clinical sickle cell anemia. Notably, sickle cell disease is not observed in homozygous individuals until several months after birth, when the γ subunit of fetal hemoglobin is replaced by the β_S globin subunit. On the basis of this observation, clinical studies have shown that sickle cell disease can be partially treated in adults by the chemical hydroxyurea, which elevates expression of the fetal gene for the γ subunit through an unknown mechanism. The presence of even low amounts of the fetal globin subunit in the erythrocytes of sickle cell patients is apparently sufficient to disrupt some of the HbS polymer formation.

The Hemoglobin β_S Mutation Is Associated with Malaria Resistance It was discovered more than 50 years ago that sickle cell anemia occurs at much higher rates than expected in areas with tropical climates, especially in Africa. The reason for this was not understood until it was found that these same geographic areas overlap with regions of high **malaria** incidence, suggesting that the two observations may be linked (**Figure 6.33**). African malaria is caused by a protozoan pathogen called *Plasmodium falciparum*, which is transmitted from female *Anopheles* mosquitoes to humans during the blood-feeding process. The malarial parasite invades human liver cells, where it replicates and is released into the blood, resulting in invasion of red blood cells and completion of its life cycle.

It has been hypothesized that individuals who are heterozygous for the β_S mutation, which does not cause sickle cell anemia, are more resistant to malaria than are people with normal hemoglobin. A rationale for this proposal is that *Plasmodium* infection of human red blood cells leads to a reduction in pH, which induces aggregation of HbS tetramers and cell sickling. Because these mis-shapen cells are removed by the spleen, the *Plasmodium*-infected cells are preferentially depleted from circulation, thereby protecting heterozygous β_S individuals from severe malaria. If this hypothesis is correct, then the advantage of maintaining the β_S mutation in the population to protect a large number of heterozygotes from malaria must be greater than the cost associated with small population declines due to homozygous individuals who die from sickle cell anemia.

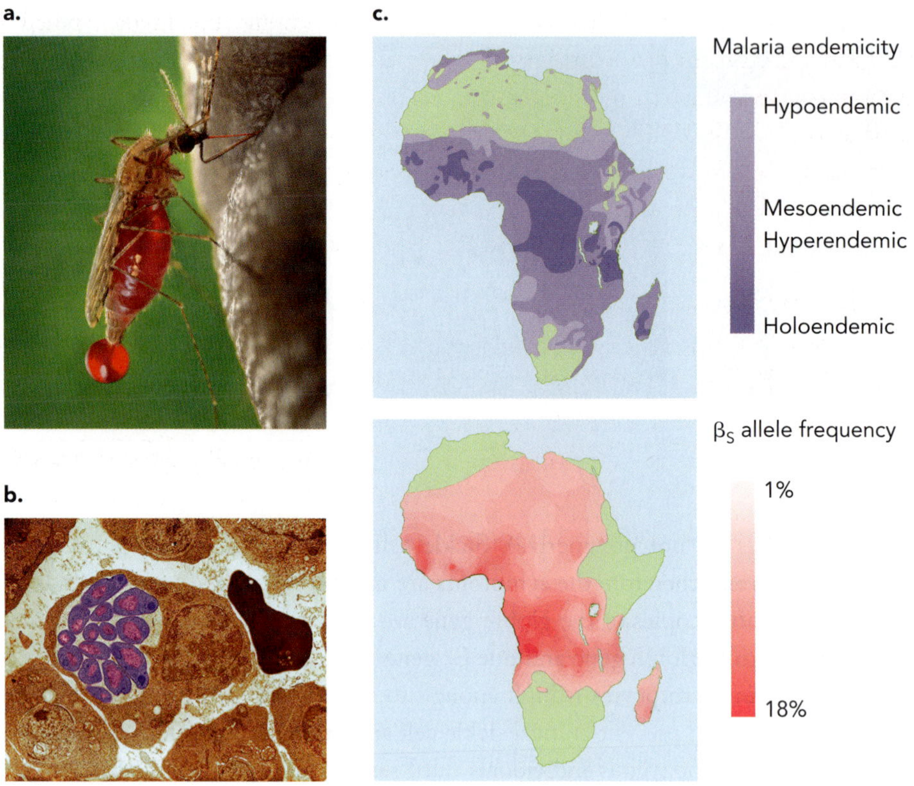

Figure 6.33 The presence of the β_S gene in human populations is associated with resistance to malaria. **a.** The primary mosquito species responsible for transmission of malaria in Africa is *Anopheles gambiae*. CDC/JAMES GATHANY. **b.** The pathophysiologic symptoms of African malaria result from infection of human cells with the *Plasmodium falciparum* parasite, which is contained in mosquito saliva. This color-enhanced transmission electron micrograph shows malarial parasites infecting a human red blood cell. MARY MARTIN/SCIENCE SOURCE. **c.** Malaria incidence in Africa overlaps with the distribution of the β_S gene in the population.

concept integration 6.2

How does allosteric control of hemoglobin protein structure regulate oxygen transport?

Allosteric control of hemoglobin protein structure is mediated by O_2, which is both a ligand and a positive homotropic effector molecule, and by the negative heterotropic effectors CO_2, H^+, and 2,3-bisphosphoglycerate (2,3-BPG). The reversible binding of O_2 to the heme group in one globin subunit induces favorable conformational changes in the protein, which promote O_2 binding in the other three globin subunits through altered noncovalent interactions at the subunit interfaces. In contrast, binding of CO_2, H^+, and 2,3-BPG to sites outside of the heme pocket results in unfavorable conformational changes that inhibit O_2 binding. A shift in the equilibrium between the R-state conformation (high O_2 binding) and the T-state conformation (low O_2 binding) of hemoglobin is controlled by the local concentrations of O_2, CO_2, and H^+ in the lungs and tissues to maximize O_2 transport efficiency.

6.3 Membrane Transport Proteins: Controlling Cellular Homeostasis

Up to this point, we have focused primarily on globular proteins, which function as soluble proteins inside or outside the cell. Now we turn our attention to a very important class of proteins that function as integral components of biological membranes; their function, however, is not structural support for membranes, but to act as regulators of molecules passing through a membrane.

Recall from Chapter 2 that membranes consist of amphipathic lipids, typified by phospholipids. Proteins inserted into membranes must be able to associate with the hydrophobic environment of the membrane. They do this by having nonpolar amino acids on the outside of the protein, facing the hydrophobic portion of the bilayer, while at the same time keeping polar residues pointing inward, where they interact within the protein (see Figure 4.35).

There are three major types of membrane proteins in cells:

1. *membrane receptor proteins,* which are involved in transducing extracellular signals across the plasma membrane through ligand-induced conformational changes (see Chapter 8);
2. *membrane-bound metabolic enzymes*, in particular, membrane proteins embedded in the inner mitochondrial membrane and chloroplast thylakoid membrane, which are involved in oxidation–reduction reactions and ATP synthesis (see Chapters 11 and 12);
3. *membrane transport proteins,* which facilitate the movement of polar molecules across the hydrophobic membrane, either in response to a concentration gradient (passive transporters) or through an energy-dependent transport mechanism (active transporters).

Membrane transport proteins are the subject of this section because they provide an excellent example of the interdependence of protein structure and function.

Membrane Transport Mechanisms

Biomolecules cross cell membranes by one of two routes, depending on their chemical characteristics. Hydrophobic molecules *diffuse* across lipid bilayers in response to a concentration gradient, moving from an environment of high concentration to one of low concentration. For example, steroid hormones are cholesterol-derived biomolecules that have been shown to cross the plasma membrane by diffusing through the lipid bilayer in response to a concentration gradient. In contrast, polar molecules must be *transported* across cell membranes by membrane proteins, which provide a mechanism to shield the transported molecule from the interior of the hydrophobic membrane.

There are two classes of membrane transport proteins:

1. *Passive membrane transport proteins*: These proteins facilitate biomolecule movement, often very selectively, across a membrane in the same direction as the concentration gradient and do not require an external energy source.
2. *Active membrane transport proteins*: These proteins translocate biomolecules across cell membranes, often against a concentration gradient, using energy conversion processes such as ATP hydrolysis (**Figure 6.34a**).

Membrane proteins can also be described as either carriers or channels. Active transporter proteins function as carriers that are so-named because they can be thought of as "carrying" one or a few biomolecules at a time from one side of a membrane to the other. Carriers, which are often proteins, move molecules across a membrane with (usually) high specificity and at rates that are much slower than that of diffusion. Because carriers transport molecules relatively slowly, they can be saturated at high substrate concentrations, much like an enzyme (**Figure 6.34b**). Passive transporters can either be carriers or channels. Channels serve as semiselective pores through which

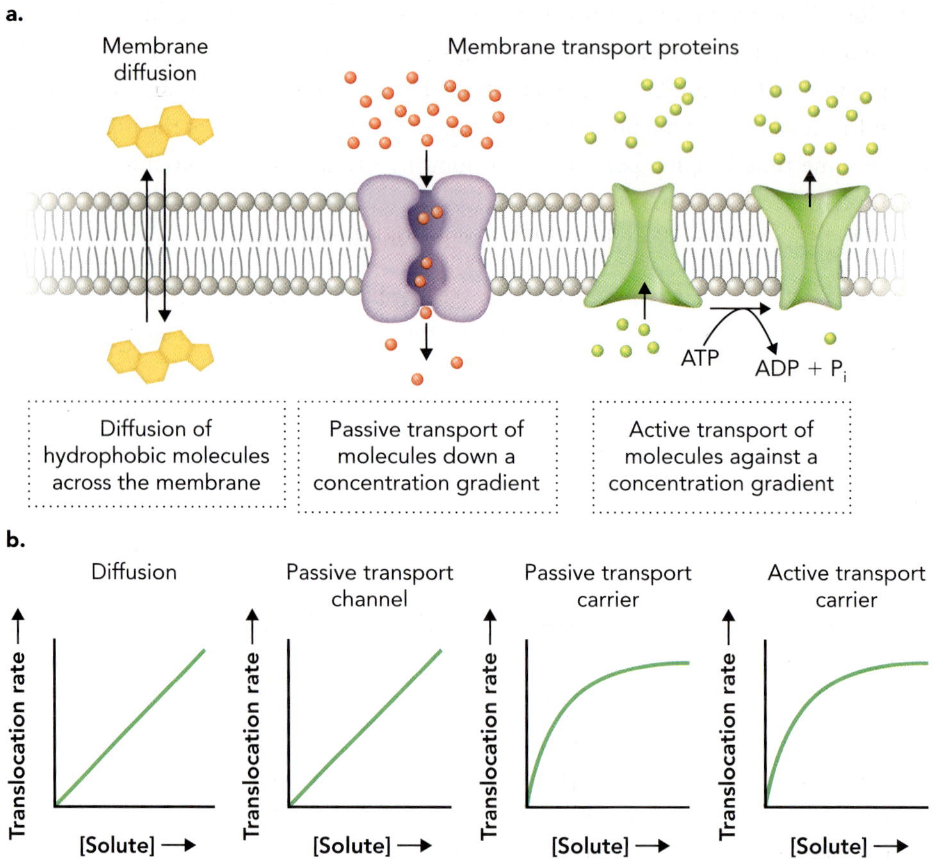

Figure 6.34 Hydrophobic biomolecules can diffuse across cell membranes in response to a concentration gradient, whereas polar molecules must be transported across the hydrophobic membrane by passive or active membrane transport proteins. **a.** Schematic diagram of diffusion and transport mechanisms in cells. **b.** The relationship between the solute concentration and the translocation rate differs for different transport types. For simple diffusion of hydrophobic molecules and for passive channels, the rate of transport increases as solute concentrations increase. For both active and passive carriers, the rate of transport reaches a maximum at high substrate concentrations when transporter function becomes rate-limiting.

the transported molecules flow at rates that can approach the diffusion-controlled limit. Channels cannot usually be saturated with substrates, and therefore the rate of transport will increase with increasing substrate concentrations (Figure 6.34b).

What determines if a molecule can cross the cell membrane using a passive membrane transport mechanism or if it must be actively transported by an energy-dependent mechanism? The answer is that it depends on the change in free energy required to traverse the membrane. If the free energy change is negative, $-\Delta G$, for transport in a given direction, then the molecule can use a passive transport mechanism; if the free energy change is positive, $+\Delta G$, then the molecule requires an active transport mechanism. An active transport mechanism provides an input of free energy to make the overall reaction spontaneous ($-\Delta G$) in a given direction. The ΔG for membrane transport can be calculated using the following equation, in which R is the gas constant, T is temperature in Kelvin, C_1 and C_2 are solute concentrations on the two sides of the membrane (C_1 is the concentration at the starting point and C_2 is the concentration at the destination), Z is the electrical charge of the solute molecule, F is the Faraday constant, and ΔV is the membrane potential across the membrane, in volts:

$$\Delta G = RT \ln\left(\frac{C_2}{C_1}\right) + ZF\Delta V$$

Free Energy Changes in Membrane Transport: Passive Transport We can use this free energy equation to look at an energy-independent sugar transporter (passive transporter protein), which facilitates sugar import into cells under conditions where the extracellular concentration of the sugar molecule (C_1) is 100 times higher than the intracellular concentration (C_2) under optimal growth conditions. This is considered a large downhill gradient that runs from the outside of the cell toward the inside of the cell. Because the sugar molecule is uncharged, the electrical potential drops out of the equation ($Z = 0$, so $ZF\Delta V = 0$). Thus, ΔG is determined only by the first term, which includes the magnitude of the concentration gradient across the membrane, the gas constant R (8.314×10^{-3} kJ/mol K), and the temperature of 37 °C (310 K):

$$\begin{aligned}\Delta G &= RT\ln\left(\frac{C_2}{C_1}\right) + ZF\Delta V \\ &= (8.314 \times 10^{-3}\ \text{kJ/mol K})(310\ \text{K})\ \ln\left(\frac{1}{100}\right) + 0 \\ &= (2.6\ \text{kJ/mol})(-4.6) \\ &= -12.0\ \text{kJ/mol}\end{aligned}$$

On the basis of the calculated $-\Delta G$ value, sugar transport from the outside to the inside of the cell through this passive membrane transport protein is favorable under these conditions.

Free Energy Changes in Membrane Transport: Active Transport We will now consider the energy needed to move a Ca^{2+} ion from the cytosol of a muscle cell to the inside of a membrane-bound organelle called the sarcoplasmic reticulum, a process that is critical to muscle contraction. The normal Ca^{2+} concentration in skeletal muscle cells is 0.1 μM in the cytosol (C_1) and 1.5 mM inside the sarcoplasmic reticulum (C_2). Transport of Ca^{2+} from the cytosol into the sarcoplasmic reticulum represents a steep uphill concentration gradient of 1 to 15,000, which needs to be overcome for transport to occur. As Ca^{2+} is charged, an electrical potential will exist because of the differential charge distribution across the sarcoplasmic reticulum membrane. Therefore, to calculate the total change in free energy associated with Ca^{2+} transport across this membrane, we need to consider both the concentration gradient and the electrical potential. This is done by using values of +2 for the charge on the Ca^{2+} ion, 96.5 kJ/V mol for the Faraday constant (F), and 50 mV (0.050 V) for the observed membrane potential:

$$\begin{aligned}\Delta G &= RT\ln\left(\frac{C_2}{C_1}\right) + ZF\Delta V \\ &= (8.314 \times 10^{-3}\ \text{kJ/mol K})(310\ \text{K})\ \ln\left(\frac{1.5 \times 10^{-3}\ \text{M}}{1.0 \times 10^{-7}\ \text{M}}\right) \\ &\quad + (+2)(96.5\ \text{kJ/V mol})(0.050\ \text{V}) \\ &= (2.6\ \text{kJ/mol})(9.6) + 9.7\ \text{kJ/mol} \\ &= 34.7\ \text{kJ/mol}\end{aligned}$$

From this calculation, we can see that both the concentration gradient and electrical potential disfavor transport in this direction (both terms have positive ΔG values), and the overall change in free energy is positive (unfavorable). To overcome this unfavorable free energy change, cells achieve the transport of Ca^{2+} from the cytosol to the sarcoplasmic reticulum by using an active membrane transport protein. This transporter uses the energy equivalent of ATP hydrolysis (~50 kJ/mol under cellular conditions) to make the overall reaction favorable (negative ΔG), thereby removing calcium from the cytosol so that the muscle cell is able to undergo multiple rounds of contraction. Later in this chapter, we will examine the mechanism by which the sarcoplasmic reticulum Ca^{2+} membrane transport protein harnesses the energy made available by ATP hydrolysis to pump cytosolic Ca^{2+} across the sarcoplasmic reticulum membrane.

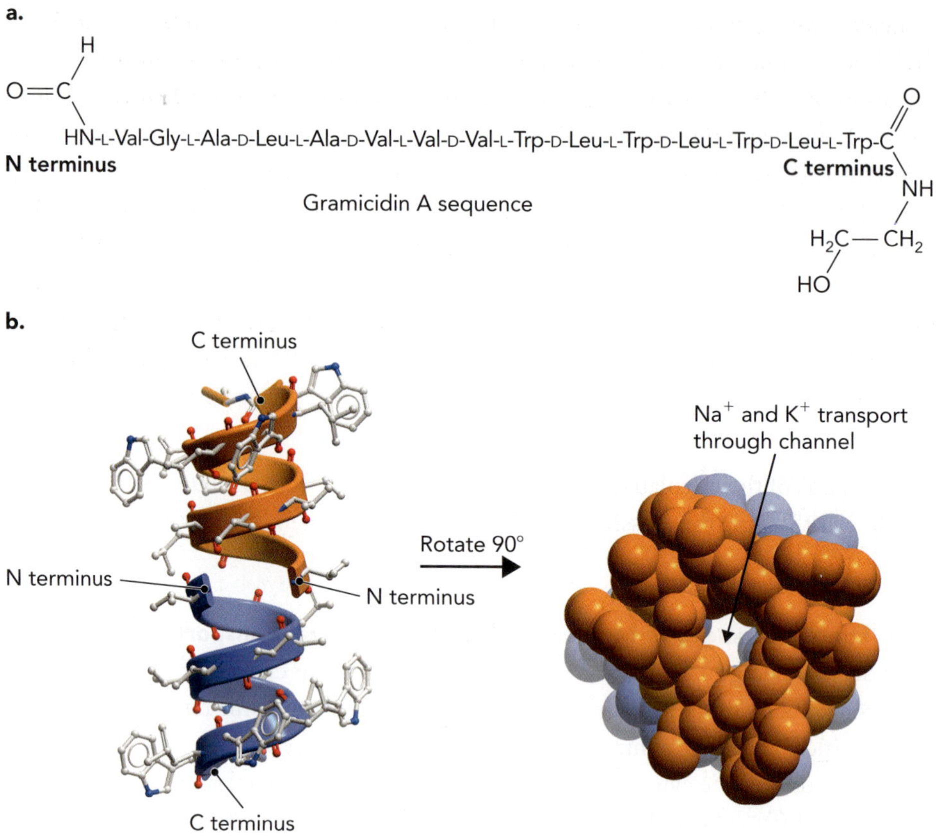

Figure 6.35 Gramicidin A is an antibiotic that functions as a membrane channel, permitting the passive transport of Na^+ and K^+ ions. **a.** Gramicidin A contains alternating L and D amino acids linked by peptide bonds. **b.** The molecular structure of the gramicidin A channel shows that it consists of two molecules stacked as head-to-head helices, with hydrophobic residues facing outward toward the membrane. This unusual helix has a channel down its center. BASED ON PDB FILE 1JNO.

Structure and Function of Passive Membrane Transport Proteins

The smallest passive membrane transporters are not gene-encoded proteins, but rather small polypeptides produced by bacteria from D and L amino acids. One of the best-characterized biomolecules in this class is the antibiotic **gramicidin A**, which functions as an ion channel that permits Na^+ and K^+ ions to leak out from bacterial cells. As shown in **Figure 6.35**, gramicidin A is a linear polypeptide consisting of 15 amino acid residues and forms an unusual right-handed helix. Two molecules of gramicidin A stack in a head-to-head arrangement, forming a single membrane-spanning channel with hydrophobic chemical groups facing outward toward the membrane. Unlike the common α helix in protein structures, which has a tightly packed center, the gramicidin A helix contains a channel 4 Å in diameter that permits the passive transport of Na^+ and K^+ ions.

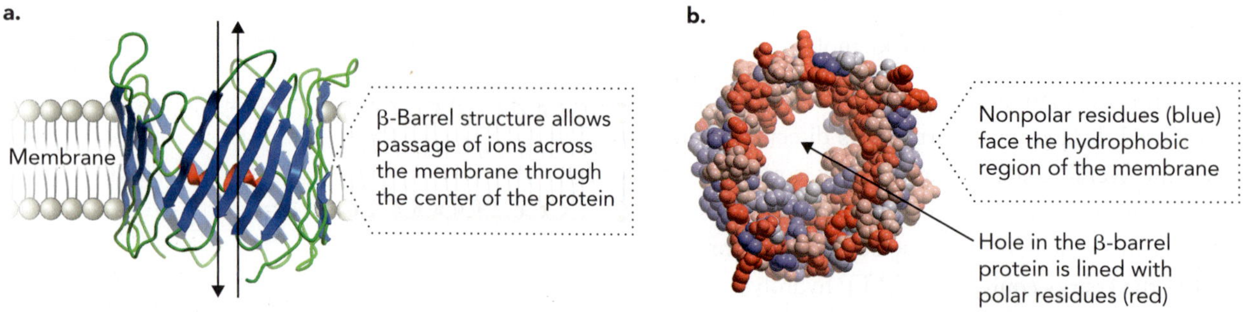

Figure 6.36 The human mitochondrial voltage-dependent anion channel (VDAC) protein is a passive membrane transporter in the porin protein family. **a.** This transport protein contains a transmembrane β-barrel structure with 19 β strands surrounding a large interior channel ~3 nm in diameter. **b.** Top view of a space-filling model of the same porin protein, with hydrophobic nonpolar residues oriented mostly outward (valine, leucine, isoleucine) and hydrophilic and charged groups oriented mostly inward (serine, threonine, arginine, lysine, aspartate). BASED ON PDB FILE 2JK4.

Bacterial Porin Proteins The outer membrane of some types of bacteria, mitochondria, and chloroplasts contains large numbers of passive membrane transport channel proteins, called **porins**, which share a common structural motif known as the β barrel (**Figure 6.36a**). Amino acids that are adjacent in sequence in a β strand have side chains on opposite sides of the strand, so a sequence with alternating polar and nonpolar residues within the primary sequence results in hydrophobic residues facing outward toward the cell membrane and hydrophilic and charged residues facing the interior of the porin channel. The hydrophilic residues can interact with the polar molecules that are being transported (**Figure 6.36b**). Porin proteins are organized as homotrimers in the membrane, as in the *E. coli* maltoporin (LamB) protein. This protein transports sugar molecules across the bacterial outer membrane in response to a concentration gradient (**Figure 6.37**).

Porin proteins can be either relatively nonselective or highly selective, depending on the number of β strands, which determines the inner diameter of the porin channel, and the chemical properties of the amino acid side chains that line the channel. Some of the more selective porin channels contain binding sites for substrate carrier proteins that pick up ions and small molecules in the periplasmic space and deliver them to primary active transporter proteins located in the inner bacterial membrane (**Figure 6.38**). (The **periplasmic space** is an aqueous compartment located between the outer and inner membranes of Gram-negative bacteria.) A good example of a selective porin protein is the Omp32 protein of the bacterium *Delftia acidovorans*, which contains numerous positively charged arginine residues within the channel. These residues increase the protein's specificity for anions over cations by a ratio of 20:1 (**Figure 6.39**). The Omp32 porin protein transports not only anions such as Cl^- across the outer membrane but also malate, an organic acid with two negatively charged carboxyl groups at pH 7.

K^+ Channel Protein The β-barrel structure of porin proteins would seem to be the ideal shape for a membrane transport protein because of the cylindrical interior cavity. However, it turns out that most membrane transport proteins consist of several α helices, which fit together to form a substrate-selective transmembrane channel.

One of the best examples of this structure–function relationship is the K^+ channel protein, which passively transports K^+ ions across bacterial and eukaryotic cell membranes. In 1998, Rod MacKinnon and his colleagues at Rockefeller University in New York were the first to elucidate the molecular structure of a K^+ channel protein. Using X-ray crystallography, they characterized the K^+ channel protein of *Streptomyces lividans* bacteria and made an important discovery. When they analyzed the structure of the tetrameric protein complex, they found that the observed 10,000 to 1 selectivity of

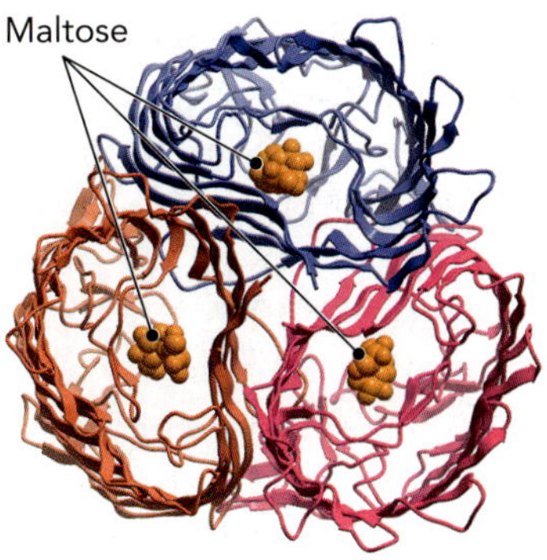

Figure 6.37 The *E. coli* maltoporin protein is a typical homotrimeric porin complex. The space-filling molecule in the interior channel of each maltoporin subunit is maltose. BASED ON PDB FILE 1MPQ.

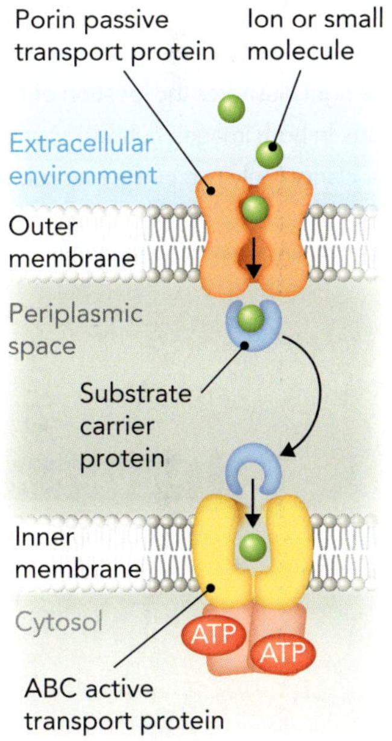

Figure 6.38 Some selective porin proteins in the outer membrane of Gram-negative bacteria contain binding sites for substrate carrier proteins. These proteins ferry transported molecules through the periplasmic space to ATP binding cassette (ABC) active transport proteins in the inner membrane and can be limiting for the rate of passive transport (see Figure 6.34).

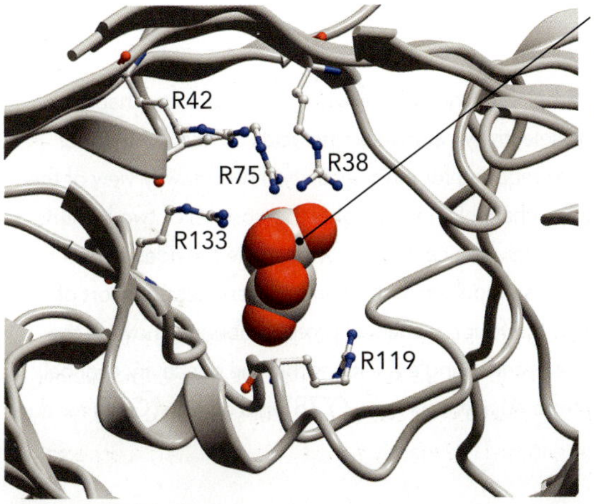

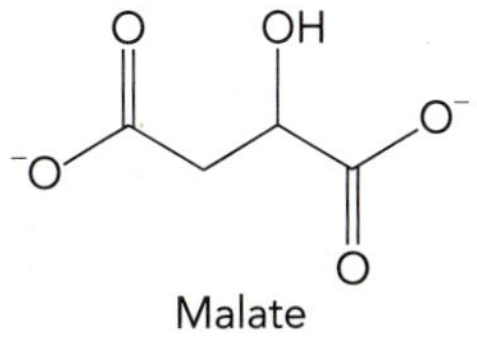

Figure 6.39 The Omp32 protein of *Delftia acidovorans* is a bacterial porin protein that is highly selective for anions such as Cl^- and molecules such as malate. Numerous arginine residues line the channel interior to facilitate anion entry and transport across the membrane. BASED ON PDB FILE 2FGQ.

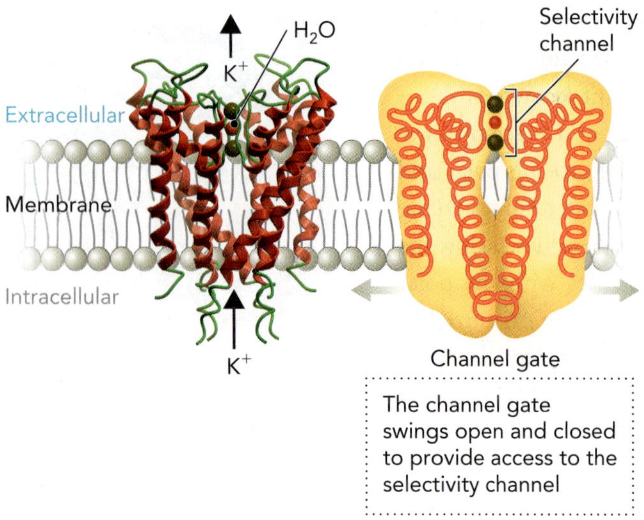

Figure 6.40 The *Streptomyces lividans* K^+ channel protein is a homotetrameric complex that contains a selectivity channel in the interior of the protein. The protein structure shown here is in the closed conformation and contains two K^+ ions and one H_2O molecule within the selectivity channel. BASED ON PDB FILE 1BL8. The schematic drawing on the right illustrates the location of the selectivity channel and channel gate. In both images, cutaway views of the protein complex are used for clarity (only two of the four subunits are illustrated).

K^+ ions over Na^+ ions was due to a narrow opening within the interior of the K^+ channel protein complex, called the **selectivity channel** (**Figure 6.40**). They proposed that when a channel gate swings open in response to environmental stimuli, both hydrated K^+ and Na^+ ions can enter a large chamber on the cytoplasmic side of the protein complex. Surprisingly, primarily K^+ ions are able to pass through the selectivity channel and exit on the extracellular side of the inner membrane (periplasmic space), even though K^+ and Na^+ ions differ in atomic radius by only 0.4 Å. Thus, other factors need to be considered in addition to the atomic radii in order to understand the selectivity of this channel.

How do the amino acids within the selectivity channel of the K^+ channel protein function as a chemical filter that distinguishes between K^+ and Na^+ ions? The answer is that the main-chain carbonyl oxygen atoms are positioned in such a way that they provide a favorable desolvation energy for K^+ ions. The K^+ ions shed their hydration layers (eight H_2O molecules associate with each K^+ ion in solution) in order to interact with eight carbonyl oxygens from the protein (**Figure 6.41**). The reason why Na^+ ions do not pass through the selectivity channel is that they have a slightly smaller atomic radius, so that the exchange of H_2O

▶ ANIMATION

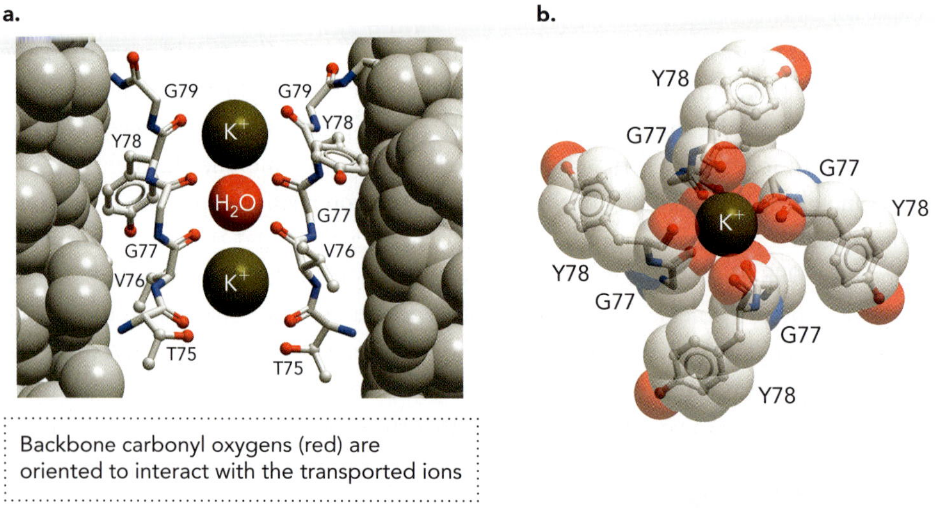

Figure 6.41 The placement of backbone carbonyl oxygen atoms within the selectivity channel of the K^+ channel protein is responsible for ion selectivity. **a.** Five amino acids on each of the four protein subunits contribute the carbonyl oxygens required for ion selectivity. This cutaway view of the molecular structure shows two subunits of the K^+ channel and highlights the position of two K^+ ions and one H_2O molecule within the selectivity channel. The central H_2O molecule provides a spacer between two nearby K^+ ions, which would otherwise repel each other and prevent net transport of K^+ ions through the channel. **b.** An end view of the four K^+ channel protein subunits, showing the interactions between eight carbonyl oxygen atoms and a single K^+ ion within the selectivity channel. Here, the eight carbonyl oxygens are derived from adjacent tyrosine (Y78) and glycine (G77) residues in the four subunits. BASED ON PDB FILE 1BL8.

molecular interactions for carbonyl oxygen interactions is unfavorable. In other words, it is too costly in desolvation energy for Na^+ to give up its six ion-associated H_2O molecules in exchange for weaker interactions with the carbonyl oxygen atoms that are inappropriately spaced for Na^+. Moreover, because the solvated Na^+ ion is larger than the 3 Å diameter of the selectivity channel, it cannot pass through the channel while still associated with its H_2O molecules. Once again we see that slight subtleties in protein structure lead to important characteristics of the protein's function.

Aquaporins The elegant structure–function analysis of a bacterial K^+ channel protein earned Rod MacKinnon a share of the 2003 Nobel Prize in Chemistry. The other recipient of the Nobel Prize in Chemistry that year was Peter Agre, a biochemist at Johns Hopkins University, who discovered **aquaporin** proteins, a major class of passive membrane transport proteins. Aquaporins are a highly conserved protein family responsible for transporting H_2O molecules across hydrophobic cell membranes. Aquaporins are found in many different membrane systems, including human red blood cells, sweat glands, and kidney cells. Humans have 11 different aquaporin genes, all of which encode a protein with six transmembrane α helices; these six helices form the water channel through the membrane (**Figure 6.42**). Most aquaporins exist as tetrameric protein complexes, with each subunit transporting up to 3 billion H_2O molecules per second in response to an osmotic gradient. Although many aquaporins are selective for H_2O, some aquaporins are able to transport small molecules, such as urea and glycerol, preferentially.

The selectivity function in aquaporin proteins is determined by two short α helices that protrude into the channel, creating a constriction point that narrows the channel, much like the center of an hourglass (**Figure 6.43a**). The diameter at the constriction

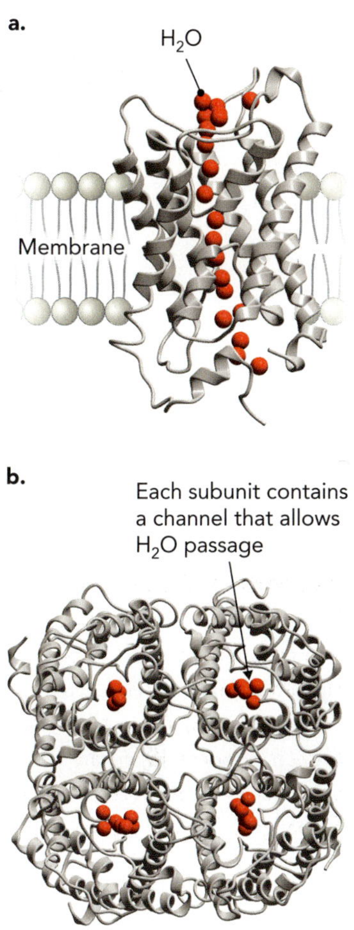

Figure 6.42 Human aquaporin 5 is expressed in salivary and sweat glands and in the cornea of the eye. **a.** The water channel of one monomer of aquaporin 5 is at the center of six transmembrane α helices. The channel is densely packed with H_2O molecules that pass through the channel in response to an osmotic gradient. **b.** Homotetrameric aquaporin 5 contains a water channel in each subunit. In this view, the structure is rotated by 90 degrees relative to the structure shown in panel **a.** BASED ON PDB FILE 3D9S.

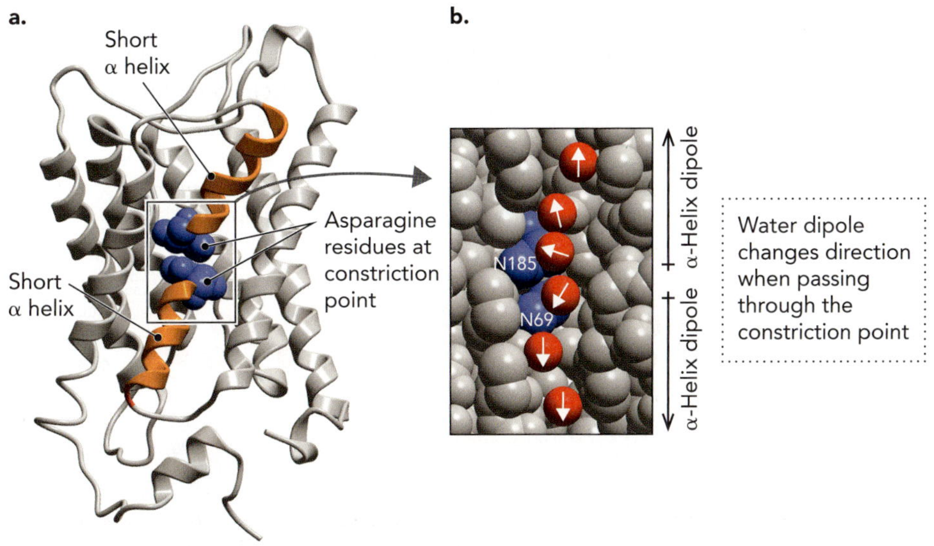

Figure 6.43 The constriction point within the water channel of aquaporin proteins is formed by two short α helices, each of which contains a conserved asparagine residue at the N terminus. **a.** Molecular structure of human aquaporin 5, highlighting the two α helices that together form the constriction point within the channel. The two Asn residues, N69 and N185, are shown in space-filling representation. **b.** Reorientation of H_2O molecules as they pass through the constriction point disrupts the proton wire by breaking hydrogen bonds between adjacent H_2O molecules. Both hydrogen bonding to the amino group of the Asn residues and charge repulsion aided by α-helix dipoles contribute to reorientation of the H_2O molecules. BASED ON PDB FILE 3D9S.

point where these two α helices meet is as small as 2.8 Å, which is just wide enough for a single H_2O molecule to pass through. The N-terminal sequence in each of these α helices is a conserved Asn-Pro-Ala (NPA) tripeptide. This tripeptide contributes to the structure of the constriction point, and its Asn residue forms hydrogen bonds with H_2O to aid in selectivity. Remarkably, although H_2O is permitted to pass through the aquaporin channel, protons are excluded, both as hydronium ions (H_3O^+) and by H^+ jumping (see Figure 2.17 in Chapter 2). Disruption of this "proton wire" within the channel is an important feature of aquaporins because proton movement across the cell membrane would alter the membrane potential by redistributing positive charges.

Two structural features of the water channel seem to explain why the proton wire is short-circuited at the constriction point: One feature is the conserved Asn residue within the NPA sequence, which forms hydrogen bonds with the H_2O molecules and causes them to reorient as they pass through the narrow opening. This reorientation within the channel breaks hydrogen bonds that link H_2O molecules together and thereby short-circuits the proton wire. The other feature in H_2O selectivity is the inverted α-helix dipoles, which converge at the constriction point and contribute to the reorientation of H_2O molecules by electrostatic repulsion (**Figure 6.43b**).

Active Membrane Transport Proteins Require Energy Input

We have seen that passive membrane transport proteins, such as porins, move molecules across membranes in the same downhill direction as the electrochemical gradient. However, active membrane transport proteins are carriers that move molecules uphill against an electrochemical gradient. To do this, they need an input of energy.

Active membrane transport proteins fall into two basic classes. The first is *primary* active transporters, which use an energy conversion process, such as ATP hydrolysis, to "pump" molecules across the membrane. The two most abundant types of primary active transporters in cells are the **P-type transporters** (phosphorylated transporters) and the **ATP binding cassette (ABC) transporters**. Both of these proteins use ATP hydrolysis to drive large conformational changes in the protein complex. In contrast,

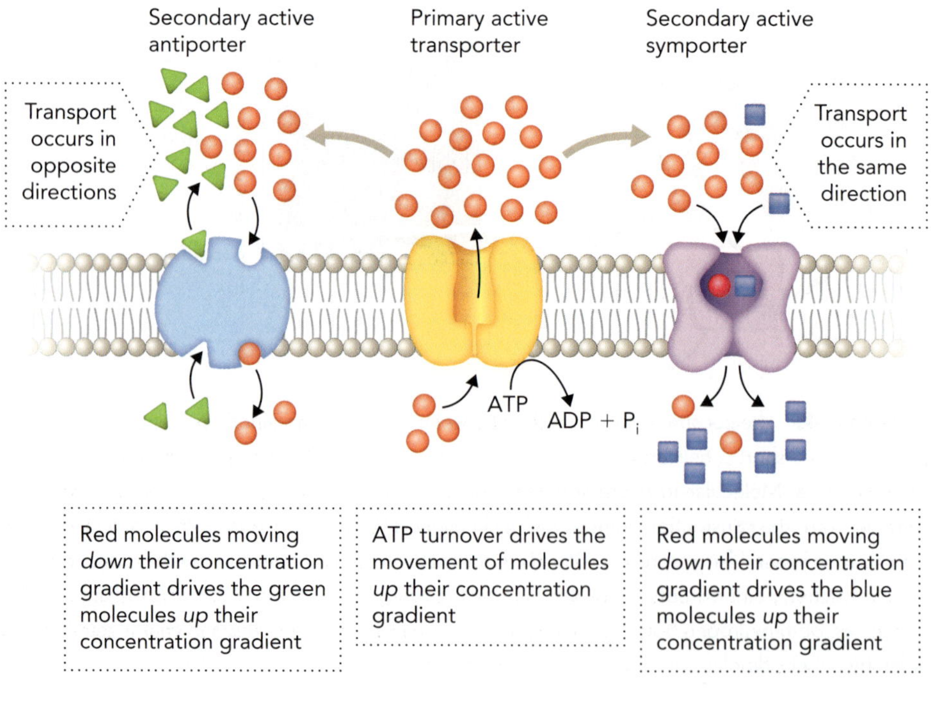

Figure 6.44 Proteins that mediate active membrane transport are either primary active transporters or secondary active transporters. In many cases, the energy used by secondary active transporters to pump molecules against an electrochemical gradient is actually derived from ATP-driven primary active transporters. An antiporter moves molecules across a membrane in opposite directions, whereas a symporter co-transports molecules in the same direction.

secondary active transporters use the energy available from a downhill electrochemical gradient for one molecule to co-transport a second molecule against an uphill electrochemical gradient. This co-transport mechanism is usually coupled to an ATP-dependent primary active transport mechanism, which establishes a concentration gradient. An **antiporter** moves molecules across a membrane in opposite directions, whereas a **symporter** co-transports molecules in the same direction (**Figure 6.44**).

The ability to influence the function of active membrane transport proteins explains the effect of several mood-altering drugs and has become an important area of biomedical research. For example, active transporters are targets of the widely used pharmaceutical drugs Prilosec and Zoloft and of the natural product cocaine, which is found in coca leaves (**Figure 6.45**).

- Prilosec (omeprazole) is an inhibitor of the gastric proton pump, which is responsible for transporting H^+ into the stomach and lowering the pH of gastric juices to aid in digestion. The gastric proton pump is a P-type primary active transporter named **H^+–K^+ ATPase**. This protein exchanges K^+ for H^+ across the parietal cell membrane, using energy from ATP hydrolysis. An overly active H^+–K^+ ATPase pump can lead to abnormally high H^+ concentrations and damage the stomach lining, causing peptic ulcers. Omeprazole is a prescribed treatment for several conditions involving too much stomach acid, including ulcers, heartburn, and acid reflux.

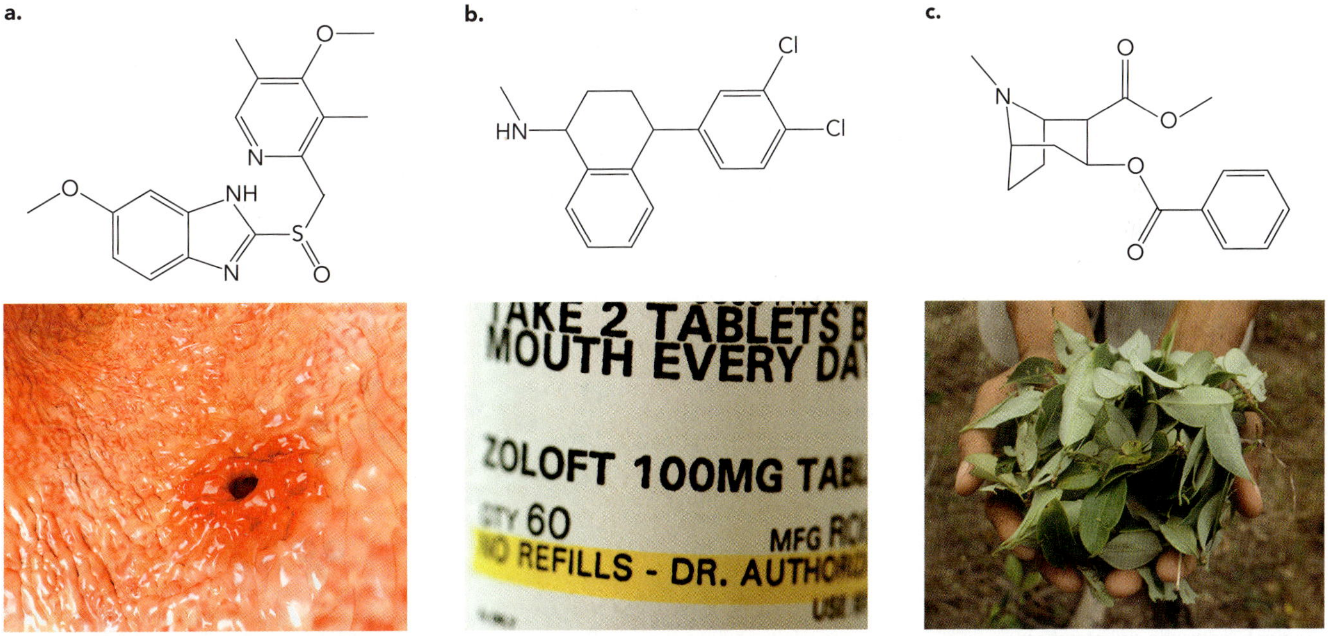

Figure 6.45 A number of pharmaceutical drugs and natural products inhibit primary and secondary active membrane transport proteins. **a.** Prilosec (omeprazole) inhibits the gastric H^+–K^+ ATPase protein, a primary active transporter that pumps protons into the stomach to lower the pH. An overactive H^+–K^+ ATPase can lead to acid-induced damage in the stomach lining. PHOTO: JUAN GARTNER/SCIENCE PHOTO LIBRARY/ALAMY. **b.** Zoloft (sertraline) inhibits the serotonin transporter protein, a secondary active transporter responsible for removing serotonin from neuronal synapses. Sertraline stimulates serotonin signaling in the brain and is associated with feelings of well-being and happiness. PHOTO: CHRIS GALLAGHER/GETTY IMAGES. **c.** Cocaine, derived from coca leaves, inhibits the dopamine transporter protein and like sertraline causes psychological changes as a result of increased neurotransmitter signaling. PHOTO: CARLOS CARRION/SYGMA/CORBIS.

- Zoloft (sertraline) is one of many antidepressant drugs known as **serotonin-selective reuptake inhibitors**, which target the serotonin transporter proteins. This secondary active symporter couples Na^+ ion transport with the reuptake of serotonin from the synaptic cleft into presynaptic neurons. The primary active transporter responsible for the Na^+ ion gradient is called **Na^+–K^+ ATPase**, another P-type active transporter. Sertraline inhibition of serotonin transporter proteins results in elevated serotonin levels in neuronal synapses and subsequent activation of the serotonin receptor. Sertraline treatment of depression is often associated with prolonged feelings of well-being and happiness.
- Similar to sertraline, cocaine also inhibits a secondary symporter protein in neuronal cells that depends on the Na^+–K^+ ATPase primary active transporter. The protein that cocaine binds to and inhibits is the dopamine active transporter. Its inhibition leads to increased levels of synaptic dopamine and overstimulation of dopamine receptors. The euphoric feeling that elevated dopamine levels produce can result in addictive behavior and cocaine abuse.

Primary Active Transporters: P-Type Transporters One of the first primary active transporters to be discovered was the Na^+–K^+ ATPase membrane transport protein, which is a carrier responsible for maintaining an electrochemical gradient across most animal cell membranes. This integral membrane protein uses phosphoryl transfer energy from ATP hydrolysis to drive protein conformational changes. The result is the export of three Na^+ out of the cell for every two K^+ ions that are imported into the cell. As shown in **Figure 6.46**, this unequal ion transport leads to a high intracellular K^+ concentration and high extracellular Na^+ ion concentration. It also creates a charge difference, leading to a membrane potential of about −70 mV. This membrane potential is critical to neuronal cell function and axon firing. In addition, the differential Na^+ and K^+ ion concentration is used in a variety of cell types to drive secondary active transport systems via symporter or antiporter mechanisms (see Figure 6.44). The importance of the Na^+–K^+ ATPase transporter proteins in cell function can be seen by the fact that up to 25% of the ATP hydrolyzed every day by animals at rest is used to maintain the Na^+–K^+ electrochemical gradient.

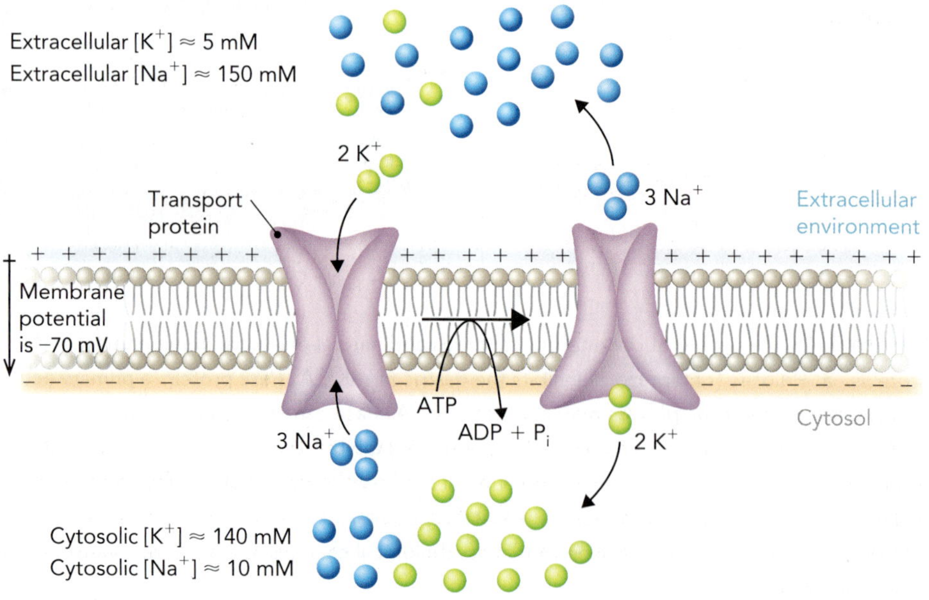

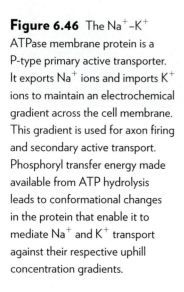

Figure 6.46 The Na^+–K^+ ATPase membrane protein is a P-type primary active transporter. It exports Na^+ ions and imports K^+ ions to maintain an electrochemical gradient across the cell membrane. This gradient is used for axon firing and secondary active transport. Phosphoryl transfer energy made available from ATP hydrolysis leads to conformational changes in the protein that enable it to mediate Na^+ and K^+ transport against their respective uphill concentration gradients.

The Na^+–K^+ ATPase is an integral membrane protein containing an α-helical transmembrane domain (M) and three distinct functional domains: an ATP binding domain (N), a regulatory domain (A), and a phosphoryl domain (P). **Figure 6.47** shows the molecular structure of the porcine Na^+–K^+ ATPase. This structure was determined using purified protein that was crystallized in the presence of Rb^+ ions to mimic K^+ and tetrafluoromagnesate as a substitute for phosphate groups. You can see that in this protein conformation, the two Rb^+ ions are located within the ion channel of the M domain, and the phosphoryl group analog is bound to the P domain.

The rabbit skeletal muscle sarco/endoplasmic reticulum **Ca^{2-}-ATPase** (SERCA) is another P-type primary active transporter protein that has been studied at the molecular level. SERCA is a very abundant protein in animal muscle cells, where it is responsible for transporting Ca^{2+} ions from the cytoplasm into the lumen of the **sarcoplasmic reticulum (SR)** to promote muscle relaxation. As illustrated in **Figure 6.48**, the Ca^{2+} transporting activity of SERCA is controlled by a second membrane protein called phospholamban. When phospholamban is in the unphosphorylated state, it inhibits Ca^{2+} transport. Muscle relaxation is initiated when phospholamban is phosphorylated by the enzymes protein kinase A or Ca^{2+}/calmodulin kinase II and dissociates from SERCA. This leads to a conformational change in SERCA that permits Ca^{2+} uptake from the cytosol. After an ATP hydrolysis step and dissociation of ADP, a channel opens up on the

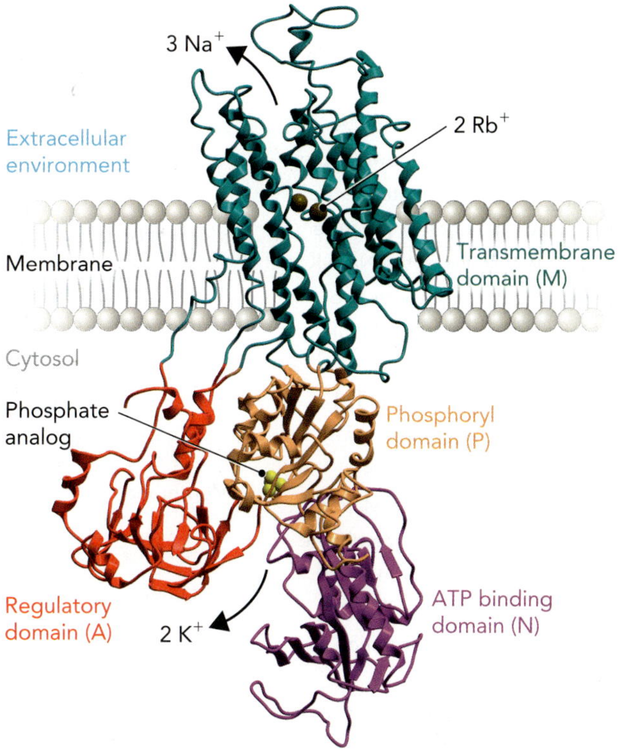

Figure 6.47 The porcine Na^+–K^+ ATPase is a P-type primary active transport protein. Binding of ATP to the N domain leads to phosphorylation of the P domain and subsequent conformational changes in the A domain. These changes facilitate the export of three Na^+ ions and import of two K^+ ions. Rubidium (Rb^+) and tetrafluoromagnesate (phosphoryl group analog) were used in this experiment to mimic the presence of K^+ in the M domain and P-domain phosphorylation, respectively. BASED ON PDB FILE 3B8E.

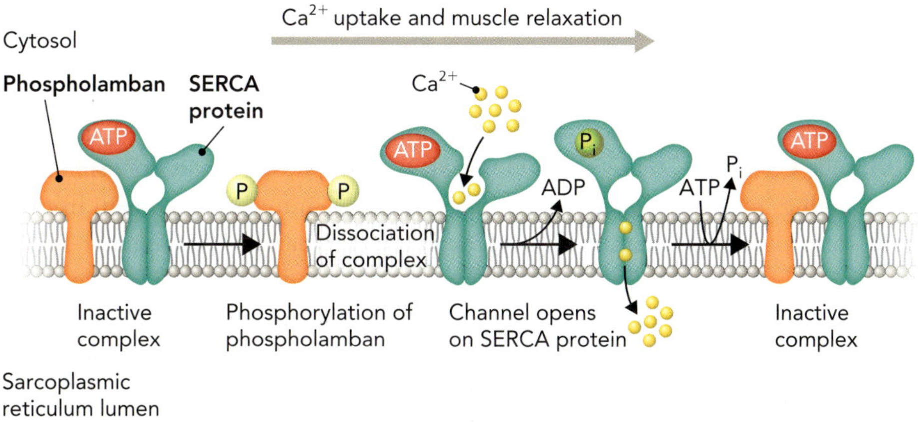

Figure 6.48 Transport of Ca^{2+} from the cytosol to the sarcoplasmic reticulum lumen by SERCA is required for muscle relaxation. Phospholamban is an inhibitor of SERCA activity in the resting state; however, this inhibitory activity is blocked when phospholamban is phosphorylated by protein kinase A or Ca^{2+}/calmodulin kinase II. Calcium transport from the cytosol to the sarcoplasmic reticulum lumen requires the binding and hydrolysis of ATP in a multistep process (see Figure 6.49).

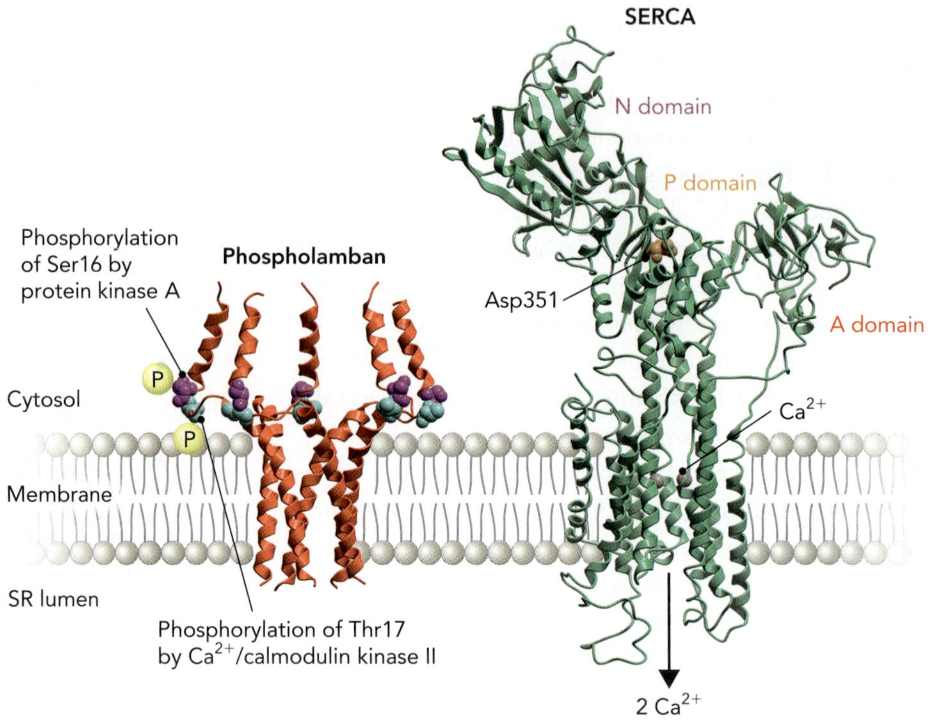

Figure 6.49 The molecular structures of rabbit skeletal muscle SERCA transporter protein and phospholamban, a SERCA regulatory protein are shown as they are expected to be oriented in the membrane. The domain structure of SERCA is similar to the Na^+–K^+ ATPase. SERCA transports two Ca^{2+} from the cytosol to the SR lumen after muscle contraction. Phosphorylation of the Asp351 residue in the P domain by ATP is required for Ca^{2+} pumping. Phospholamban phosphorylation on each of the five transmembrane α helices by protein kinase A and Ca^{2+}/calmodulin kinase II leads to phospholamban dissociation and SERCA activation (see Figure 6.48). BASED ON PDB FILES 1SU4 (SERCA) AND 1ZLL (PHOSPHOLAMBAN).

luminal side of the protein, and Ca^{2+} is released. ATP binding and release of inorganic phosphate (P_i) returns SERCA to its resting-state conformation in a complex with phospholamban.

The SERCA transporter protein consists of the same general domain structure as the Na^+–K^+ ATPase transporter protein, with the N, P, and A domains facing the cytosol and the M domain containing 10 transmembrane α helices (**Figure 6.49**). Biochemical and X-ray structure analyses have shown that SERCA transports two Ca^{2+} for every ATP that is hydrolyzed and that the site of phosphorylation in the P domain is an aspartate residue (Asp351). Phospholamban is a small protein of 52 amino acids that forms a pentameric complex when inserted into membranes. The SERCA inhibitory activity of phospholamban is reversed by phosphorylation of Ser16 and Thr17 by the cell signaling enzymes protein kinase A and Ca^{2+}/calmodulin kinase II, respectively. Inactivation of phospholamban by Ca^{2+}/calmodulin kinase II makes sense because Ca^{2+}/calmodulin kinase II is activated by high levels of cytosolic Ca^{2+}. Ca^{2+}/calmodulin kinase II, in turn, phosphorylates phospholamban, resulting in SERCA-mediated transport of Ca^{2+} into the SR lumen. In contrast, protein kinase A is activated by **epinephrine** (adrenaline), the "fight-or-flight" hormone. Protein kinase A also phosphorylates phospholamban, leading to Ca^{2+} import and muscle relaxation. The net effect is that epinephrine signaling improves recovery time between each muscle contraction, thereby increasing the rate of muscle contraction, which is needed in times of acute stress.

How does phosphorylation of Asp351 in the P domain of SERCA contribute to Ca^{2+} transport from the cytosol to the sarcoplasmic reticulum lumen? The answer to this question comes from Poul Nissen and his colleagues at Aarhus University in Denmark, who solved the structure of several different SERCA protein conformations in the presence of ATP, ADP, and phosphate metabolite mimics. As shown in **Figure 6.50**, SERCA exists in two distinct protein conformations called E1 and E2, which are distinguished by the orientation of the A domain relative to the N and

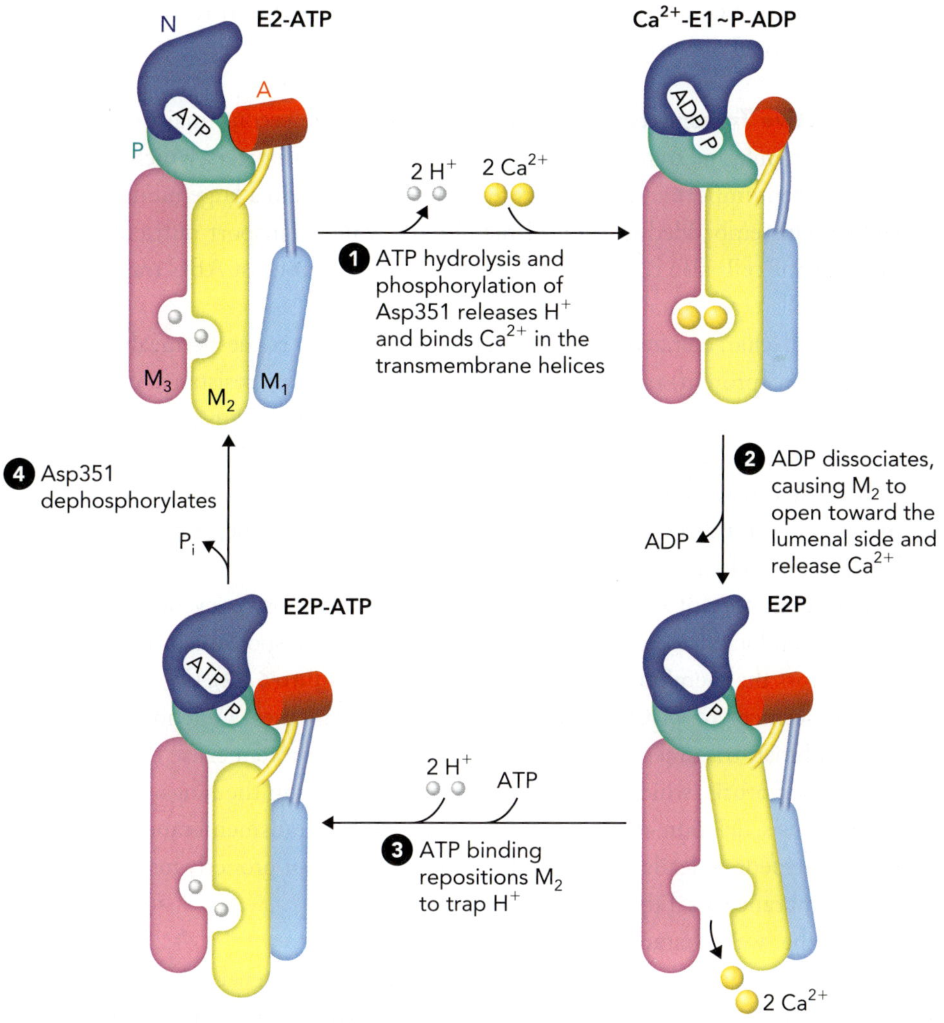

Figure 6.50 The four-step Ca^{2+} transport model shown here is based on distinct SERCA protein structures, determined under different binding conditions. BASED ON C. OLESEN ET AL. (2007). THE STRUCTURAL BASIS OF CALCIUM TRANSPORT BY THE CALCIUM PUMP. *NATURE, 450*, 1036–1042 AND PDB FILES 2BY4 (E2-ATP), 3BA6 (CA^{2+}-E1~P-ADP), 3B9B (E2P), AND 3B9R (E2P-ATP).

P domains. Based on four related E1 and E2 structures, Nissen proposed a four-step model of Ca^{2+} transport by SERCA. In step 1, two H^+ (or possibly three H^+) are exchanged for two Ca^{2+}. The two Ca^{2+} bind to a region in the M domain flanked by two α-helical subdomains, referred to here as M_2 (transmembrane helices 3 and 4) and M_3 (transmembrane helices 5–10). This exchange takes place when the protein is in the E2-ATP conformation and the M_1 subdomain (transmembrane helices 1 and 2) moves away from M_2 to trap the Ca^{2+} ions in an M_2–M_3 binding pocket.

Phosphorylation of Asp351 generates an intermediate structure referred to as Ca^{2+}-E1~P-ADP. This structure is converted to E2P in step 2 when ADP dissociates. In this new conformation, M_2 moves away from M_3 to create an opening on the lumenal side of the sarcoplasmic reticulum membrane. Once this happens, Ca^{2+} ions exit the binding pocket, and several H^+ ions enter. In step 3, ATP binds to the N domain, inducing the E2P-ATP conformation. Now domain M_2 moves back into place and seals off the opening, thereby trapping the H^+ ions in the binding pocket. Finally, in step 4, dephosphorylation of Asp351 returns the protein to the E2-ATP conformation, which has an opening on the cytosolic side of the sarcoplasmic reticulum membrane to permit another round of Ca^{2+} and H^+ exchange.

This four-step structure–function model of Ca^{2+} transport by SERCA is reminiscent of a Japanese Yosegi puzzle box, which requires multiple interdependent rearrangements of the locking wood components to open the sealed chamber

Figure 6.51 The sequential subdomain movements required in the SERCA Ca^{2+} transport cycle are analogous to the interdependent steps needed to open the inner chamber of a Japanese puzzle box. HAKONE MARUYAMA INC.

(**Figure 6.51**). Notice how these complex protein structures lead to very simple and efficient functional operations.

Primary Active Transporters: ABC Transporters The second major type of primary active membrane transport proteins are the ABC transporters, which function as ATP-dependent import and export proteins. Bacteria contain a large number of ABC *import* proteins embedded in the cell membrane, which transport nutrients into the cell. Eukaryotic cells also contain ABC import proteins, but most ABC transporters in eukaryotic cells *export* small molecules out of the cell.

One well-characterized ABC transporter in humans is the permeability glycoprotein (P-glycoprotein) transporter, also called the multidrug resistance protein. This export protein removes toxic compounds from the cell. Although this is normally a beneficial function, multidrug resistance proteins present a problem in cancer treatment because cancer cells can become resistant to chemotherapy drugs by increasing the number of multidrug resistance proteins in the cell membrane, removing the drugs before they get a chance to work.

Another important ABC transporter protein in humans is the cystic fibrosis transmembrane conductance regulator protein. This protein is responsible for transporting Cl^- ions across the membrane of lung epithelial cells (**Figure 6.52**). Cl^- ions help control the movement of H_2O molecules across cell membranes. Functional defects in the cystic fibrosis transmembrane conductance regulator protein cause the pulmonary disease **cystic fibrosis**, which results from thickened mucus in the lungs. In most cases of cystic fibrosis, deletion of Phe508 in the cystic fibrosis transmembrane conductance regulator polypeptide leads to a misfolded protein that is not properly inserted into the plasma membrane (see Table 4.5).

Two other well-characterized ABC transporter proteins are the *Archaeoglobus fulgidus* molybdate transporter and the *E. coli* maltose transporter. Both of these transporters are homodimeric import proteins that contain binding sites for periplasmic substrate carrier proteins, a transmembrane domain, and a nucleotide binding domain (**Figure 6.53**). As with the P-type primary active transporters, the conformational changes required for membrane transport are driven by ATP hydrolysis. However, rather than forming a phosphorylated intermediate as part of a complex transport cycle, as seen with the SERCA protein (see Figure 6.50), ATP hydrolysis by ABC transporters induces a large conformational change that converts the protein from an outward-facing transporter to an inward-facing one. Another difference between P-type and ABC transporters is that the formation of the ATP catalytic sites within the two nucleotide binding domains requires an initial conformational change, which brings two ATP binding half-sites together to generate the catalytic site. Once this happens, ATP is hydrolyzed and a second conformational change discharges the transported molecule into the cytosol.

Figure 6.52 Immunofluorescence staining of human lung epithelial cells with an anti-cystic fibrosis transmembrane conductance regulator antibody localized to epithelial cells. Mutant cystic fibrosis transmembrane conductance regulator proteins associated with the pulmonary disease cystic fibrosis fail to localize to the epithelial cell membrane. X. WANG, C. LYTLE, AND P. M. QUINTON (2005). PREDOMINANT CONSTITUTIVE CFTR CONDUCTANCE IN SMALL AIRWAYS. *RESPIRATORY RESEARCH, 6*(7).

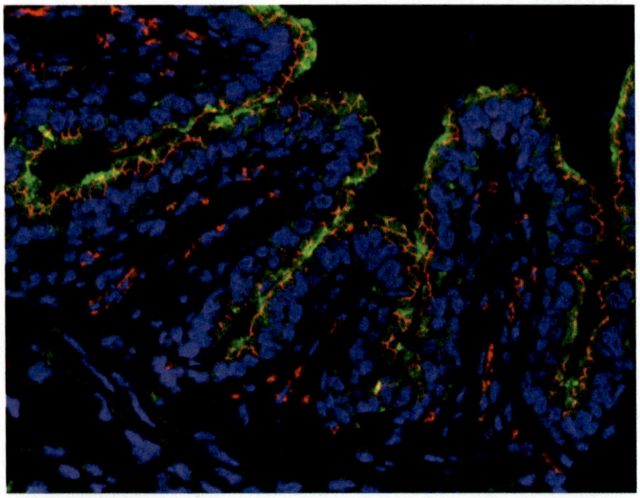

From X-ray structures of several different ABC transporters that were determined in the presence or absence of bound substrate and ATP, a three-step import model was proposed to explain metabolite import into bacterial cells (**Figure 6.54**). In step 1, substrate carrier protein binds to the transmembrane domains on the periplasmic side of the membrane. This induces a conformational change that

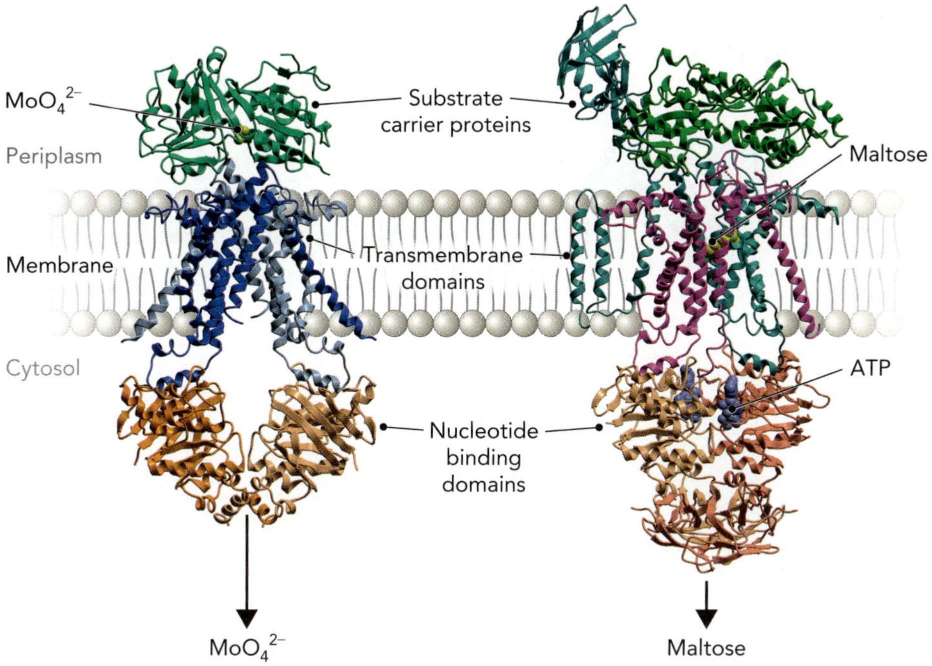

Figure 6.53 Most bacterial ABC transporters are import proteins consisting of two identical subunits. Left: The ABC transporter protein from *Archaeoglobus fulgidus* imports both molybdate and tungstate into the cell. BASED ON PDB FILE 2R6G. Right: The *E. coli* maltose transporter imports the disaccharide maltose into cells. BASED ON PDB FILE 2ONK.

exposes an internal substrate binding pocket. In addition to altering the conformation of the transmembrane domain, this structural rearrangement also brings the nucleotide binding domains together to form two ATP binding sites. Step 2 is initiated by hydrolysis of ATP, which causes a second conformational change in the protein. This change opens up the substrate binding site on the cytoplasmic side of the membrane to eject the substrate. In step 3, the ADP and P_i in the nucleotide binding domains are replaced by an incoming molecule of ATP, returning the transporter protein to the resting state.

An important feature of ABC transporters is that substrate binding and ATP hydrolysis are required to induce conformational changes that both close off the substrate binding pocket to the periplasmic space and open up an exit route to the cytoplasm. However, because an uphill substrate concentration gradient exists across the membrane, the substrate binding site can only be accessible to one side or the other at any given time. In addition, a favorable energy process must occur that ensures directional movement of the substrate out of the chamber and into the cytosol after

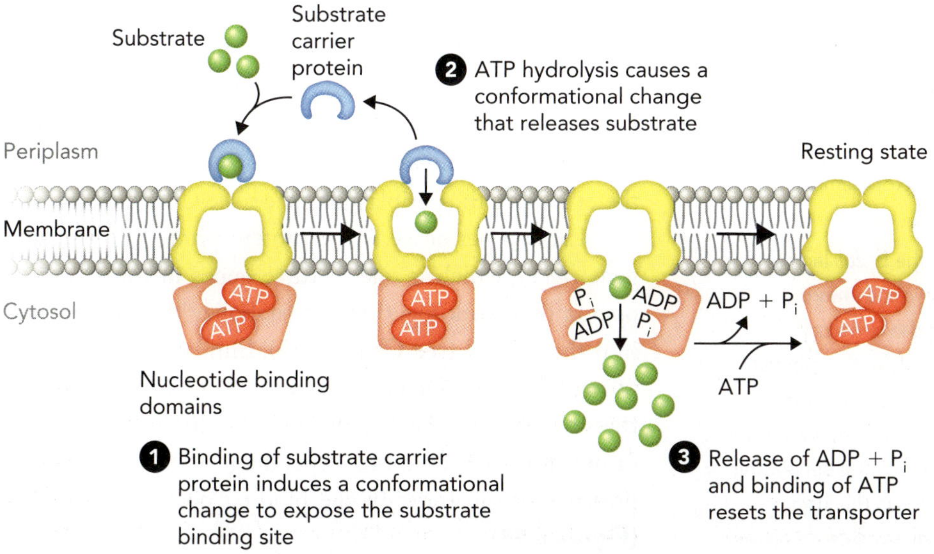

Figure 6.54 Bacterial ABC transporter proteins use a three-step process to import substrate molecules into the cytosol.

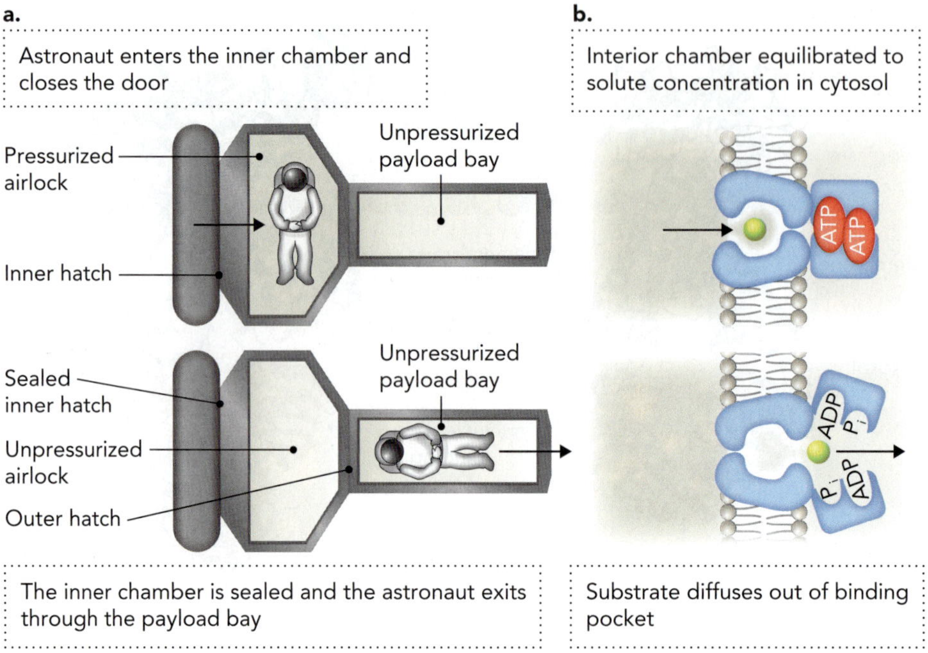

Figure 6.55 ABC membrane transport proteins are analogous to an airlock mechanism in which only one door is open at a time to prevent equilibration across an impermeable barrier. **a.** An airlock permits an astronaut to go from a pressurized cabin inside the spacecraft to the outside environment without equilibration. **b.** An ABC transporter prevents solute equilibration across the membrane by only allowing solute access to one side of the membrane at a time.

ATP hydrolysis. An intermediate step may be required in which the interior chamber is first equilibrated to the solute concentration on the cytosolic side of the membrane to facilitate substrate diffusion out of the binding pocket. This is analogous to a two-door airlock mechanism used in spacecraft (**Figure 6.55**).

Secondary Active Transporters Secondary active membrane transport proteins do not use the energy available from ATP hydrolysis directly to transport molecules against a concentration gradient. Rather, they depend on the stored potential energy from a concentration gradient that was generated by ATP hydrolysis or redox energy.

The *E. coli* **lactose permease** protein is an example of a secondary active transporter. It functions as a symporter, using the potential energy available from a steep proton gradient across the bacterial inner membrane to transport lactose from the periplasmic space to the cytosol. (The proton gradient is established by the electron transport system, which uses metabolic redox energy to translocate H^+ from the cytosol to the periplasmic space; see Chapter 11.)

Lactose permease is the protein product of the lacY gene (*lacY*), which is one of three genes in the *E. coli lac* operon that work together to metabolize lactose imported into the cytosol (see Chapter 23). Similar to ABC transporter proteins, lactose permease consists of multiple transmembrane α helices, which are either in the outward or inward conformation, depending on the absence or presence of substrates in the binding pocket, respectively (**Figure 6.56**).

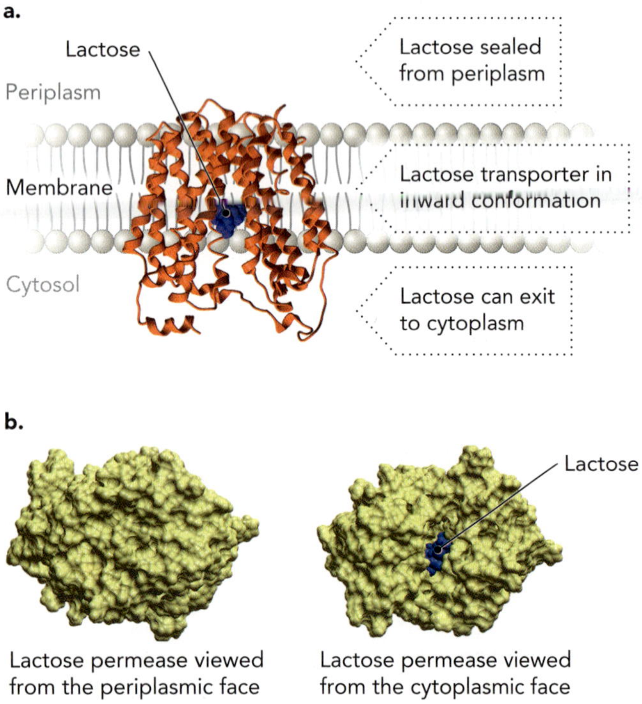

Figure 6.56 The molecular structure of the *E. coli* lactose permease transporter protein is shown here in the inward conformation. **a.** The ribbon model shows the α-helical transmembrane region, with a lactose molecule in the substrate binding site. In this orientation, the lactose is able to exit the binding site and enter the cytosol. **b.** The two molecular surface representations illustrate that substrate access to the periplasmic space is blocked when the lactose permease is in the inward conformation with an opening to the cytosol. BASED ON PDB FILE 1PV7.

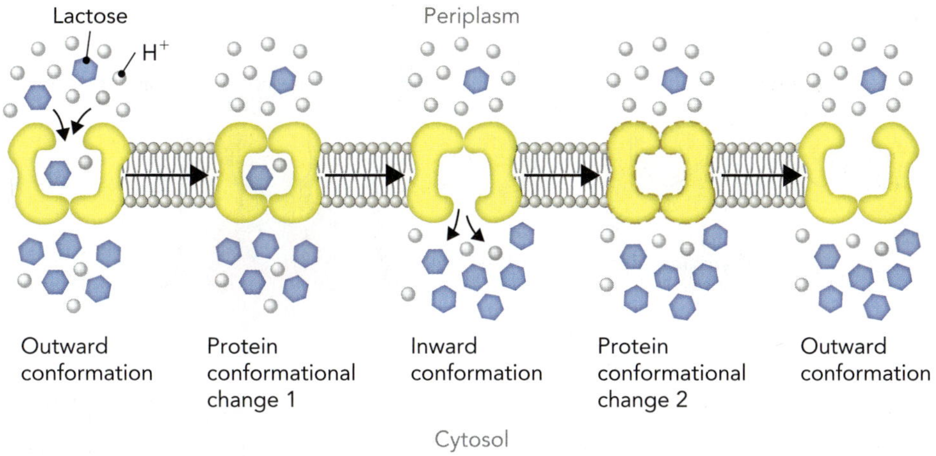

Figure 6.57 The lactose permease transporter imports lactose into bacterial cells using a symporter mechanism. The process harnesses the potential energy available from a steep proton gradient to move lactose across the inner bacterial membrane. The outward and inward protein conformations are regulated by the presence or absence of substrate bound to the transporter.

A simple model of lactose transport by lactose permease is shown in **Figure 6.57**. It can be seen that at least two protein conformations are required to facilitate lactose and H^+ transport from the periplasm to the cytosol. The initial conformational change is triggered by the binding of lactose and H^+ molecules to the substrate binding sites. This binding converts the outward conformation to the inward conformation. The second conformational change takes place when the substrate binding site is empty, resulting in a switch from the inward conformation back to the outward conformation. The precise mechanism controlling these two conformational changes is not well understood, but it presumably involves protonation and deprotonation of amino acid side chains in the substrate binding site.

An important secondary active transporter in humans is the **Na^+–I^- symporter** in thyroid gland cells, which imports one iodide ion (I^-) for every two sodium ions that are transported (**Figure 6.58**). The Na^+ gradient across the thyroid follicular cells is established by the Na^+–K^+ ATPase primary active transporter, which provides the potential energy for the Na^+–I^- symporter. Iodine is an essential nutrient that is required for the biosynthesis of two major thyroid hormones—thyroxine (T_4) and triiodothyronine (T_3)—that play important roles in metabolic regulation and tissue

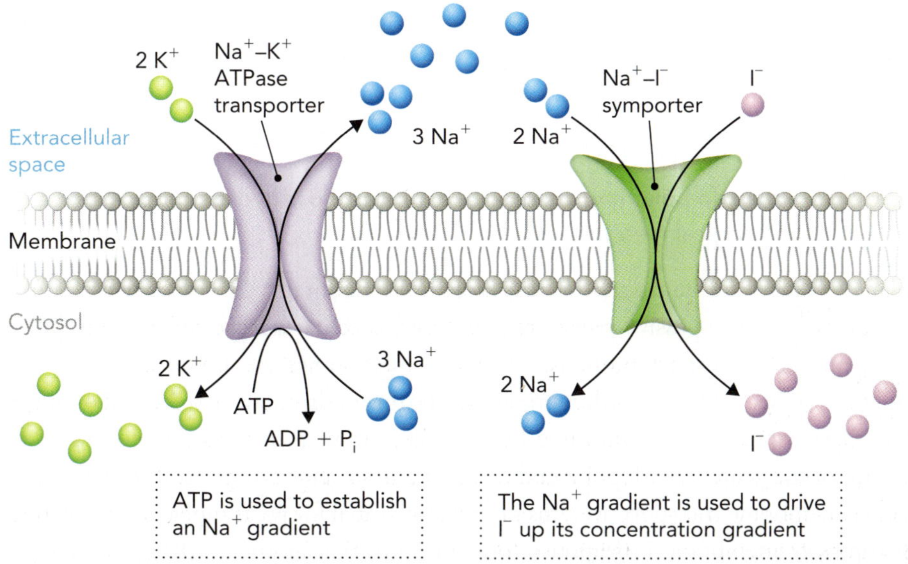

Figure 6.58 The Na^+–I^- symporter is a secondary active transporter. It uses the potential energy available from an Na^+ gradient across the thyroid follicular cell membrane to import iodine ion (I^-) into thyroid cells so that it can be used for thyroid hormone synthesis. The Na^+ gradient is maintained by the ATP-dependent Na^+–K^+ ATPase primary active transporter.

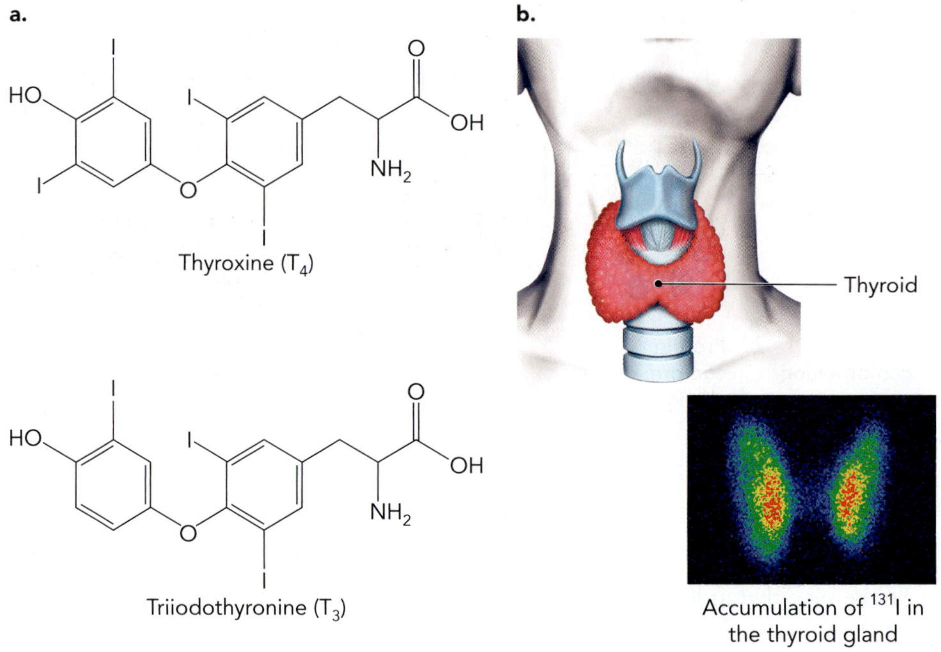

Figure 6.59 Uptake of radioactive iodine by thyroid cells is used to diagnose and treat thyroid gland abnormalities. **a.** Chemical structures of the two most abundant thyroid hormones: thyroxine (T_4) and triiodothyronine (T_3). **b.** Thyroid gland structure and function can be monitored by imaging radioactive iodine (^{131}I) in the neck area. THYROID: BSIP/UIG VIA GETTY IMAGES; ^{131}I IMAGE: DR. A. LEGER/ISM/PHOTOTAKE.

growth (**Figure 6.59a**). In order to provide sufficient dietary iodine for T_4 and T_3 synthesis, the Na^+–I^- symporter concentrates I^- up to 40-fold inside the thyroid follicular cells relative to levels in the blood. In humans, insufficient levels of dietary iodine can lead to the development of goiter—a massively enlarged thyroid gland protruding from the neck. To prevent iodine disorders, most countries add iodine to certain food staples, such as table salt, to provide safe levels of dietary iodine. Although rare, a few cases of genetic mutations in the Na^+–I^- symporter have been reported; these resulted in similar low thyroid uptake of I^- and the development of goiter.

Iodide uptake by the Na^+–I^- symporter in thyroid cells is used clinically both as a diagnostic tool and as a treatment. The thyroid gland has two lobes, which surround the windpipe in the front of the neck. They can be visualized after ingestion of radioactive iodine (^{131}I), which is monitored by an instrument that detects radioactive decay (**Figure 6.59b**). In cases of hypothyroidism (low iodine uptake), iodine supplements are given to the patient, whereas in hyperthyroidism (enlarged thyroid gland due to Graves disease or thyroid cancer), thyroid cells are eliminated using therapeutic doses of radioactive iodine. The radioactive iodine accumulates in the thyroid gland and causes DNA damage.

concept integration 6.3

How do polar and charged substances cross hydrophobic biological membranes?

Polar substances are transported across biological membranes by one of two classes of membrane proteins: passive transporter proteins or active transporter proteins. Without membrane transport proteins, biological membranes are impermeable to substances as small as H^+ ions. Membrane transport proteins may be α-helical or consist of a β barrel structure. In both cases, hydrophobic amino acids are oriented toward the exterior of the protein, with polar and charged amino acids lining the channel interior. Passive membrane transport proteins provide a channel for polar molecules to cross the hydrophobic membrane in response to a concentration gradient. Often, these channels are selective and contain constriction points that limit which molecules can pass through. In contrast, active membrane transport proteins require energy input, such as ATP hydrolysis, to drive protein conformational changes, which in effect selectively pump polar molecules across the membrane. Some active transporters pump more than one type of molecule across the membrane, which can either be in the same direction (symporter) or in the opposite direction (antiporter). For example, the Na^+–K^+ ATPase active membrane transport protein pumps Na^+ out of cells and K^+ ions into cells each time an ATP molecule is hydrolyzed.

6.4 Structural Proteins: The Actin–Myosin Motor

People have studied muscle contraction for hundreds of years, and over that time, numerous models have been proposed to explain how muscular force can be exerted at the cellular and molecular level. The breakthrough idea came in 1954 from two independent papers that were published in the same issue of *Nature* but used different microscopy techniques to show how muscle fibers slide past one another during contraction. One paper was by Andrew Huxley and Rolf Niedergerke at the University of Cambridge in England, and the other was by Hugh Huxley (unrelated) and Jean Hanson at the Massachusetts Institute of Technology in Cambridge, Massachusetts (**Figure 6.60**).

The data presented in these two papers gave rise to the **sliding filament model**, which stated that muscle filaments, consisting of actin and myosin proteins, slide over one another during the muscle contraction phase using chemical energy provided by

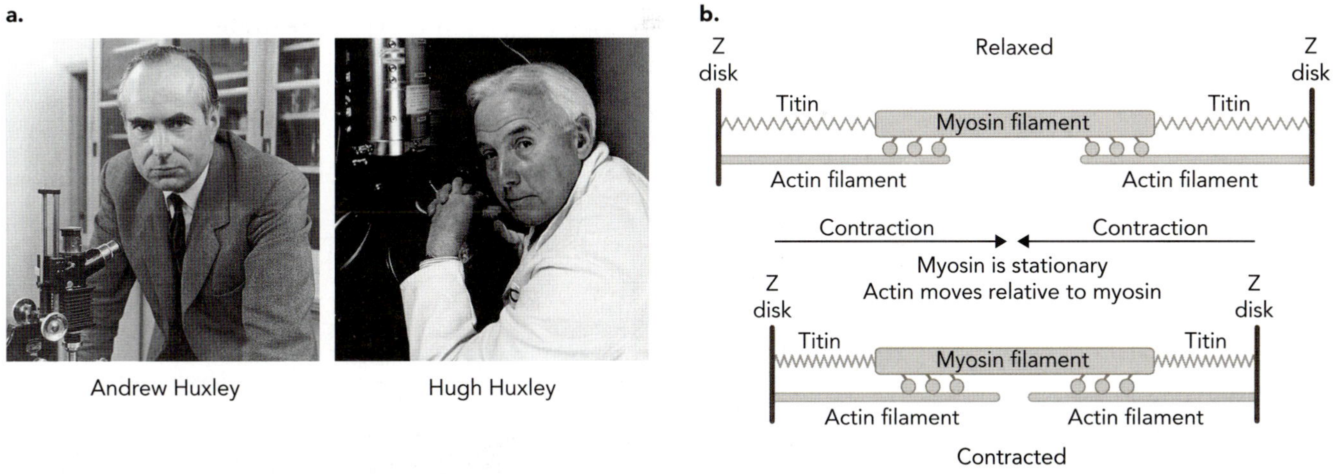

Figure 6.60 **a.** Andrew Huxley and Hugh Huxley independently proposed that actin and myosin filaments must slide past one another during muscle contraction. Hugh Huxley went on to provide much of the early biochemical evidence that supported what became known as the sliding filament model. A. HUXLEY: KEYSTONE/GETTY IMAGES; H. HUXLEY: REX USA/PHOTOFUSION/REX. **b.** A schematic diagram of the sliding filament model, which proposes that myosin filaments remain stationary while actin filaments move relative to the myosin filaments.

ATP hydrolysis. During the next 30 years, Hugh Huxley, along with other physiologists and biochemists, struggled to understand how the actin and myosin proteins could accomplish this feat. Before we look at the molecular details of the sliding filament model, we first need to acquaint ourselves with the arrangement of thick and thin filaments within muscle fibers and the protein components required for skeletal muscle contraction.

Structure of Muscle Cells

Muscle cells are not individual cells, as is common in other tissues, but rather are large fused cells called **myoblasts**. Myoblasts contain many nuclei and share a common plasma membrane known as a sarcolemma (**Figure 6.61**). Invaginations in the sarcolemma create a network of T tubules, which provide the necessary exchange of extracellular O_2 and nutrients required for muscle contraction. Myoblasts contain numerous bundles of smaller fibers, called **myofibrils**, which are surrounded by the sarcoplasmic reticulum. The primary active transport protein SERCA is localized to the sarcoplasmic reticulum membrane and is critical to modulating intracellular Ca^{2+} levels (see Figure 6.49).

Myosin and actin are the two most abundant proteins in myofibrils and are organized into substructures called thick and thin filaments, respectively. The overlapping arrangement of thick and thin filaments forms a repeating structure in skeletal muscle, called a **sarcomere**, which decreases in length during muscle contraction. The sarcomere is bordered by two protein structures known as Z disks, which contain the proteins vimentin, desmin, and α-actinin. Each Z disk functions as an anchor for the thick and thin filaments to permit muscle contraction as the filaments slide past one another.

Individual thick filaments contain hundreds of myosin molecules, which are arranged in such a way that the fibrous "tails" are associated in the middle of the

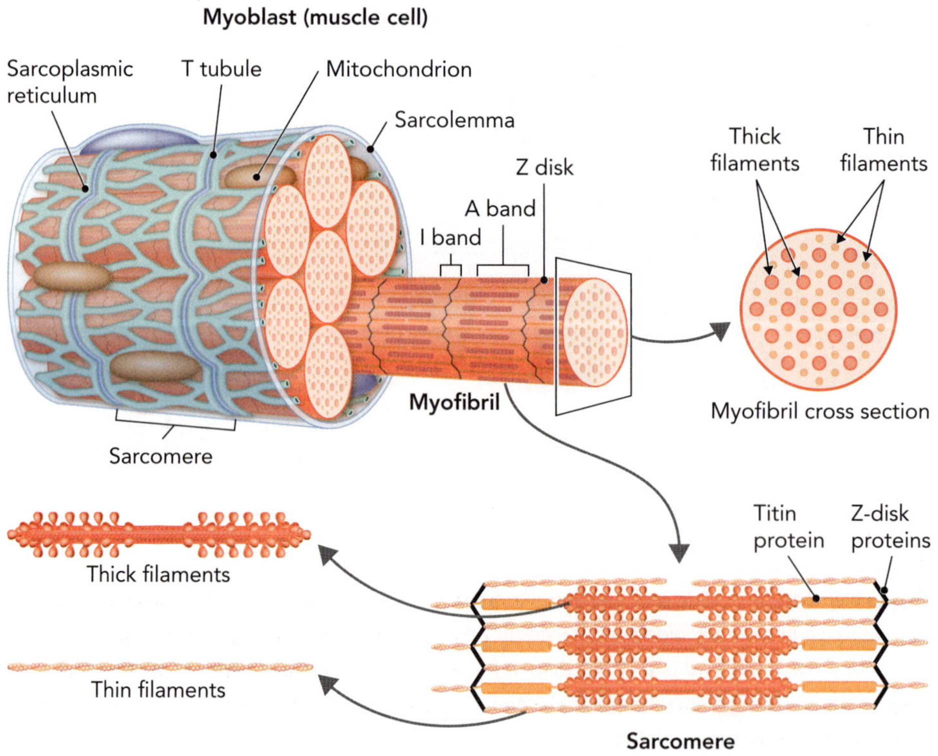

Figure 6.61 Muscle cells contain numerous myofibrils, consisting of interlocking thick and thin filaments arranged into repeating units called sarcomeres. Thick filaments are made up of myosin protein subunits, whereas thin filaments contain actin, tropomyosin, and troponin proteins. Z-disk proteins serve as anchors for the thick and thin filaments.

thick fiber, with the globular "heads" at either end of the fiber. Thin filaments consist primarily of actin, a globular protein of ~375 amino acid residues that self-assembles into long polymers to form the filament, which also has directionality ("plus" and "minus" ends). Actin is tightly associated with a second thin filament protein called **tropomyosin**, which is a coiled coil α-helical protein (see Chapter 4) arranged head-to-tail along the entire length of the actin polymers. Both actin and tropomyosin bind to a third protein, called **troponin**, a heterotrimeric protein complex that mediates Ca^{2+} regulation of muscle contraction. A cross-sectional view of a myofibril shows that each thick filament associates with up to six thin filaments. Moreover, a single thin filament contacts as many as three thick filaments (see Figure 6.61).

The Sliding Filament Model

The sliding filament model is illustrated in **Figure 6.62**. It can be seen that the arrangement of thick and thin filaments within a single sarcomere appears as dark bands in electron micrographs of muscle tissue: The large A band in the middle of the sarcomere consists of thick filaments as well as overlapping thick and thin filaments, whereas the two I bands identify the locations of thin filaments. Another important protein in the I band is **titin**, which functions as a flexible spring connecting myosin proteins to Z-disk proteins.

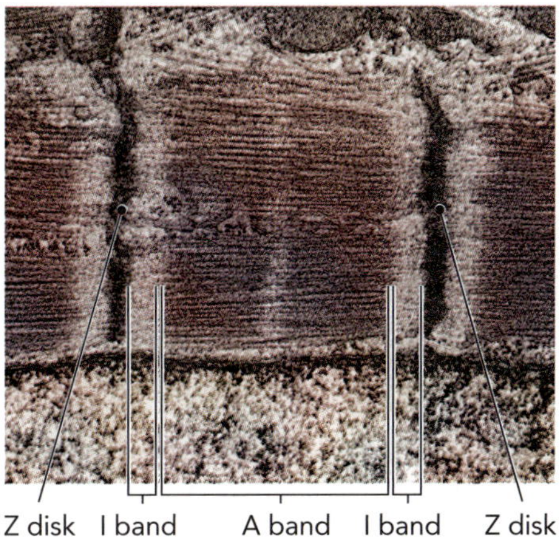

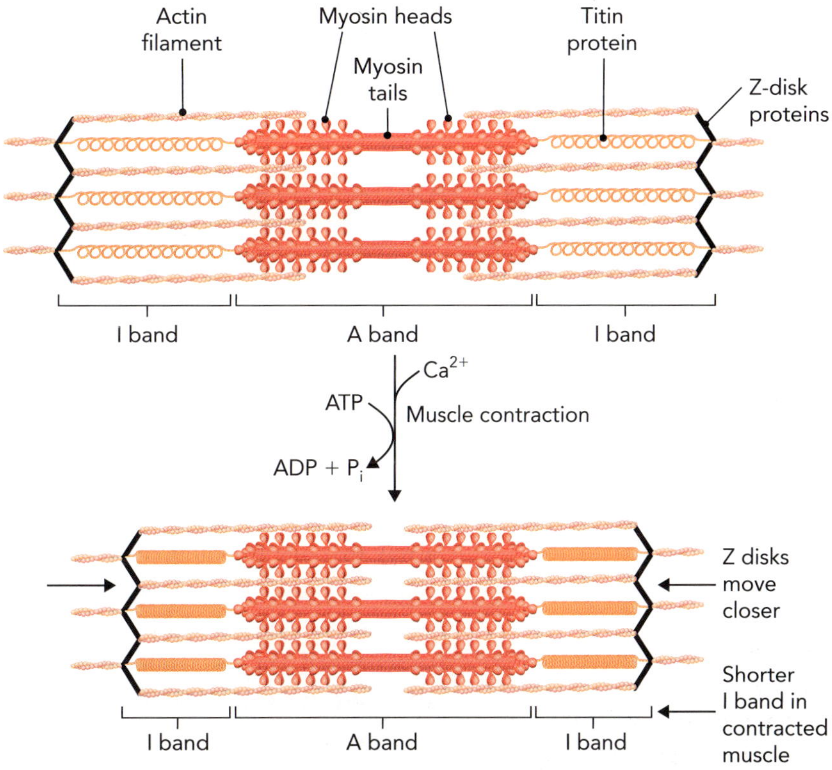

Figure 6.62 The sliding filament model illustrates the molecular basis for muscle contraction. **a.** Electron micrograph of a single sarcomere in skeletal muscle. It consists of two Z disks and two I bands, with a dark region in the middle called the A band. DR. DAVID PHELPS/GETTY IMAGES. **b.** During muscle contraction, thick and thin filaments slide past one another to shorten the distance between adjacent Z disks. This mechanism explains why the width of the A band remains constant during muscle contraction, whereas the widths of the I bands are reduced.

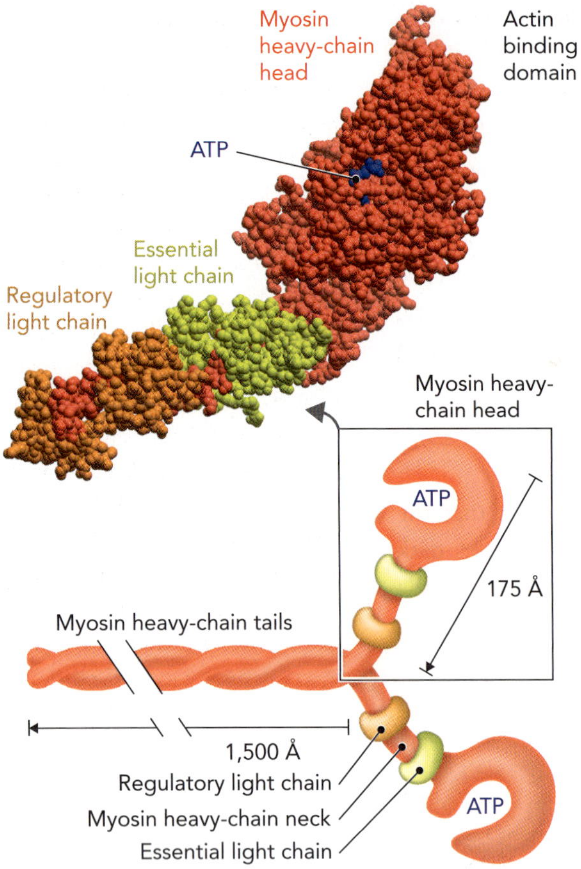

Figure 6.63 The myosin complex in thick filaments contains six polypeptide chains. The complex has three structural regions termed the head, neck, and tail regions. The tails intertwine to form a long coiled coil. The two head groups bind and hydrolyze ATP, which regulates their binding to actin subunits in the thin filaments. BASED ON PDB FILE 1S5G.

ANIMATION

Muscle contraction is initiated by neuronal stimulation of muscle cells at neuromuscular junctions, leading to the release of Ca^{2+} from the sarcoplasmic reticulum. Ca^{2+} binding and ATP hydrolysis drive conformational changes in the thin and thick filaments, causing their relative movement. Because of the polarity of both myosin thick filaments and actin thin filaments, when the thick and thin filaments slide past one another, the Z-disk proteins on either end of the sarcomere are brought closer together, and the titin protein "spring" is compressed. This motion can be observed by electron microscopy, which shows that the decreased I band width observed in contracted muscle compared to relaxed muscle is associated with little or no change in the width of the A band (Figure 6.62).

How do Ca^{2+} binding and ATP hydrolysis result in the physical movement of thick and thin filaments past one another, causing muscle contraction? We can find the answer by examining the molecular structure and function of skeletal muscle proteins. As shown in **Figure 6.63**, the myosin complex consists of two copies each of three polypeptide chains: myosin, regulatory light chain, and essential light chain. Structurally, the myosin complex can be divided into three regions: the tails, which have an extended helical structure; the heads, which have a globular structure; and the necks, which connect the heads and tails. The tail region of one myosin polypeptide interacts with the tail of another myosin polypeptide to form a coiled coil. Further interactions with the tail regions of other myosin molecules form bundles that result in thick filaments. The myosin head is a globular protein structure that extends into a helical neck region that joins the tail. The essential light chain and the regulatory light chain proteins associate with the neck region. The head domains bind ATP and undergo large conformational changes in response to ATP hydrolysis and P_i release. These conformational changes affect the affinity of the head group for actin in the thin filaments, as well as the relative positions of the head groups with respect to the myosin tails, as discussed later.

The structures of thin filament proteins are shown in **Figure 6.64**. The actin subunits assemble to form a right-handed polymer. The minus ends of these actin filaments are capped by tropomodulin, a protein that regulates the filament length by preventing the association of additional actin subunits. The troponin complex is associated with the coiled coil structure of a single tropomyosin α-helical complex in such a way that one troponin complex is spaced every seven actin subunits. The troponin complex, considered the regulatory component of the myofibril, contains three functional subunits: TnT, which binds to tropomyosin; TnI, which inhibits myosin binding to actin; and TnC, which binds to Ca^{2+} and controls muscle contraction. Figure 6.64 also shows the structures of two adjacent immunoglobulin folds within the titin spring protein. This globular protein fold represents one of several types of functional units encoded by this gigantic protein, which changes in length by as much as 10-fold during muscle contraction. Titin is the longest known polypeptide in nature, with 34,350 amino acid residues and a molecular weight of more than 3 million Da.

Now that we have an idea of how the skeletal muscle proteins are assembled within the thick and thin filaments, we can examine how these assemblies work together in muscle contraction. To move the thin and thick filaments with respect to one another,

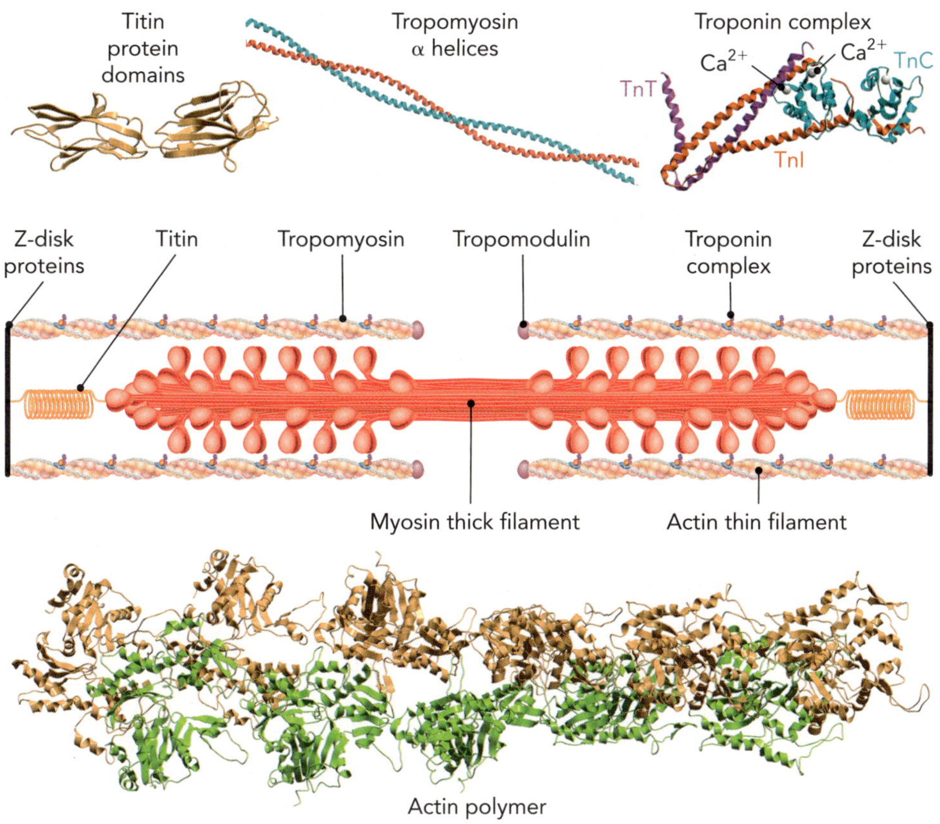

Figure 6.64 The molecular structures of the thin filament proteins actin, tropomyosin, and troponin are shown along with a portion of titin. Actin monomers assemble to form a right-handed duplex that associates with tropomyosin, which forms a left-handed coiled coil. Troponin is a heterotrimer that binds Ca^{2+} and regulates muscle contraction. Titin protein functions as a spring, which connects myosin filaments to Z-disk proteins and contains large numbers of adjacent immunoglobulin domains. Also shown are the locations of tropomodulin, an actin filament capping protein, and the myosin thick filament. PROTEIN MOLECULAR STRUCTURES BASED ON PDB FILES 3B5U (ACTIN), 1J1D (TROPONIN), 2EFS (TROPOMYOSIN), AND 2J8O (TITIN).

the head groups of myosin need to bind and release the actin subunits, coupled with the conformational changes that "ratchet" one protein along with respect to the other. Both Ca^{2+} and nucleotides affect protein conformations, as well as the interactions between proteins.

We first look at the effects of Ca^{2+}. **Figure 6.65** shows how tropomyosin and troponin are bound to the actin polymers in relaxed muscle in such a way that

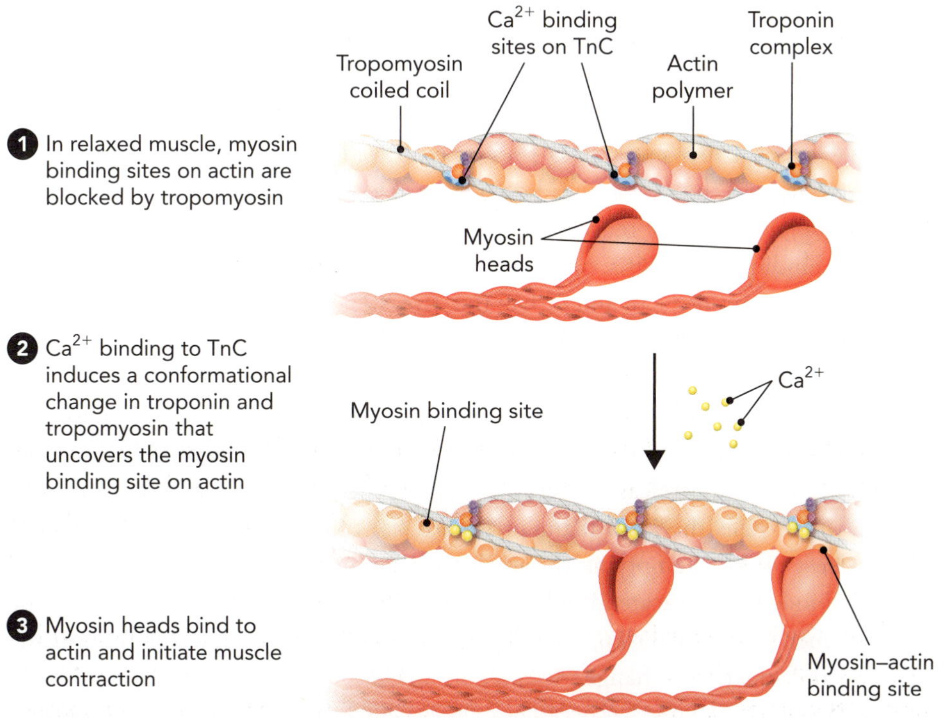

Figure 6.65 Calcium controls muscle contraction by binding to troponin, inducing a conformational change in the thin filament structure that exposes myosin binding sites on actin subunits.

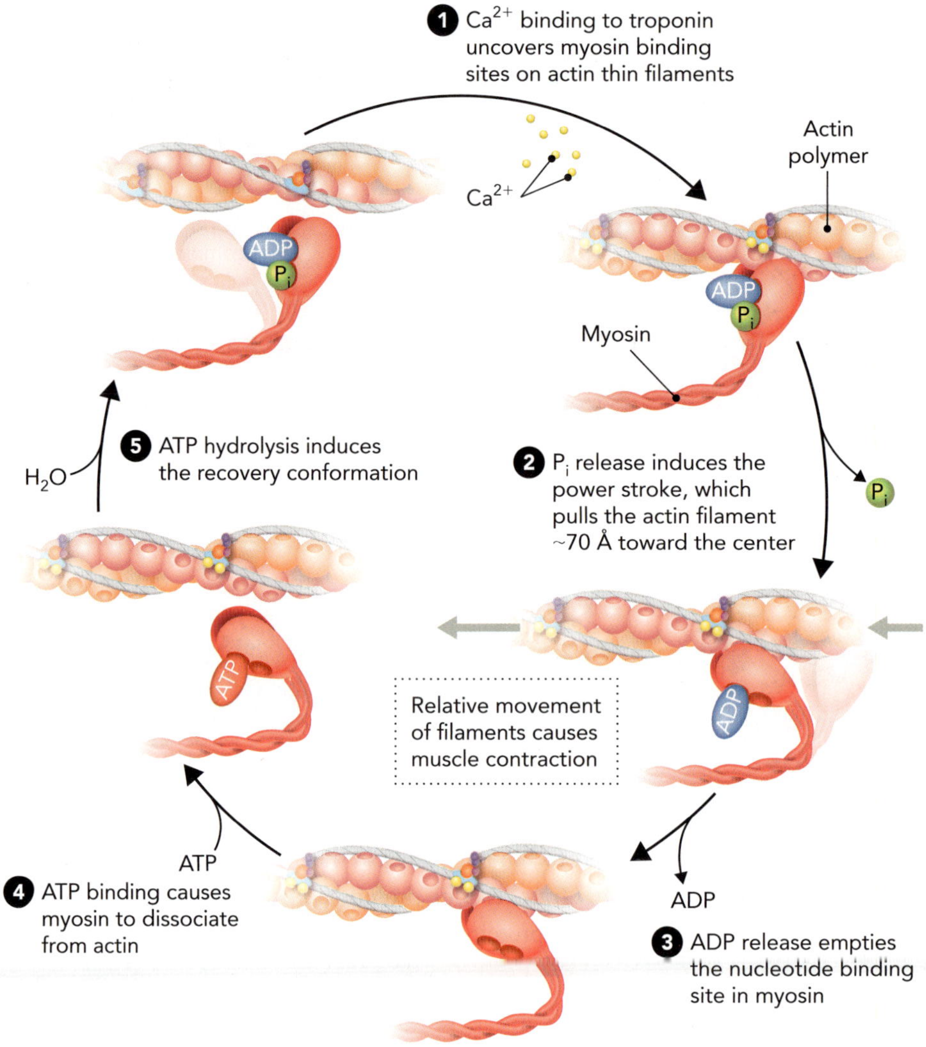

1 Ca^{2+} binding to troponin uncovers myosin binding sites on actin thin filaments
Ca^{2+}
Actin polymer
ADP
P_i
Myosin
2 P_i release induces the power stroke, which pulls the actin filament ~70 Å toward the center
P_i
ADP
Relative movement of filaments causes muscle contraction
ADP
3 ADP release empties the nucleotide binding site in myosin
ATP
4 ATP binding causes myosin to dissociate from actin
ATP
H_2O
5 ATP hydrolysis induces the recovery conformation

Figure 6.66 The five-step actin–myosin reaction cycle explains the molecular basis of the sliding filament model of muscle contraction.

tropomyosin blocks the myosin binding site on individual actin subunits. However, in the presence of Ca^{2+}, the troponin complex undergoes a conformational change that moves the TnI and TnC subunits and a large portion of the tropomyosin molecule away from the myosin binding sites on either side of the actin polymer duplex. Once this happens, the myosin head domain is able to bind tightly to the actin thin filament in preparation for muscle contraction.

ATP binding, hydrolysis, and release are key to the conformational changes that take place in myosin and drive the five-step actin–myosin reaction cycle (**Figures 6.66 and 6.67**). This cycle ultimately pulls the actin thin filament ~70 Å toward the center of the sarcomere every time the myosin head undergoes an ATP-dependent power stroke. As shown in Figure 6.66, the first step involves binding of myosin head domains, containing ADP + P_i in the nucleotide binding site, to actin subunits. In step 2, P_i release induces the power-stroke conformational change in myosin, which pulls the actin filament toward the center of the sarcomere. Step 3 is the release of ADP, emptying the nucleotide binding site. In step 4, a new ATP molecule binds to the vacated binding site, causing myosin to be released from the actin polymer. In the final step, this ATP molecule is hydrolyzed, which induces the recovery conformational change.

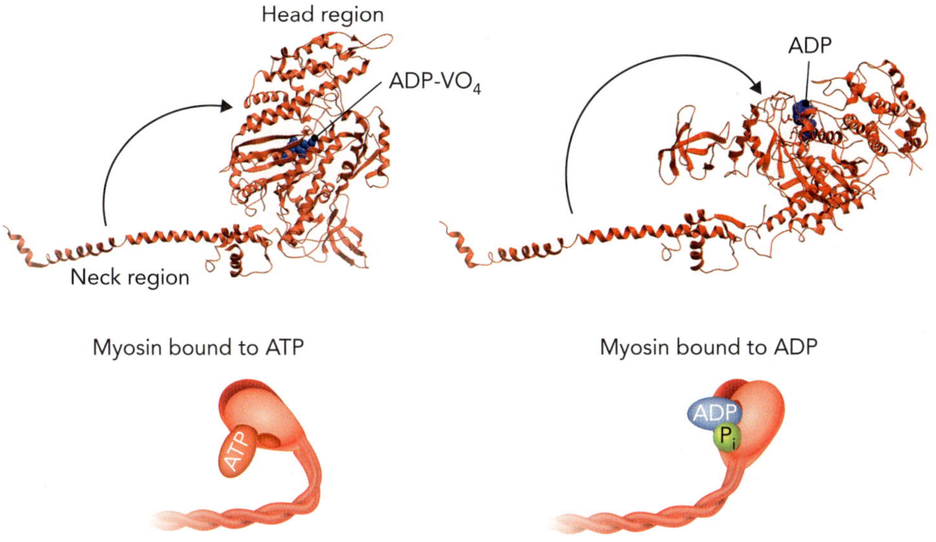

Figure 6.67 The conformation of the myosin head region with respect to the neck region changes dramatically, depending on the nucleotide state. The structure on the left shows the conformation of the head and neck regions of myosin when it is bound to ADP-VO_4, which is a molecular mimic of the transition state between the ATP-bound and (ADP + P_i)-bound conformations. The structure on the right shows the head and neck regions of myosin when it is bound to ADP. These structures represent only two of the many structures that the myosin protein adopts during the cycle of muscle contraction.
BASED ON PDB FILES 1DFL AND 1B7T.

The five steps of the actin–myosin reaction cycle can be summarized as follows:

1. After Ca^{2+}-induced changes in the structure of the thin filament, myosin heads containing ADP + P_i in the nucleotide binding site will bind with high affinity to actin subunits.
2. Release of P_i from the myosin head causes a large conformational change in myosin (power stroke), which pulls the actin fiber along the myosin fiber.
3. ADP is released from the myosin head to free up the nucleotide binding site.
4. A new molecule of ATP binds to the myosin head, causing myosin to disengage from the actin filament.
5. ATP hydrolysis induces the recovery conformation in the myosin head, and now the ADP + P_i form of the protein is ready for another round of the reaction cycle.

Muscle relaxation occurs when neuromuscular signals are terminated and Ca^{2+} levels in the cytosol decrease as a result of SERCA-mediated active transport. When Ca^{2+} levels decrease, the myosin binding sites on actin are once again blocked. Once this happens, the thick and thin filaments slip past one another in the opposite direction as the titin spring protein "uncoils" and the sarcomere returns to its elongated state.

concept integration 6.4

What is the molecular role of ATP binding and hydrolysis in muscle cell contraction?

ATP binding, hydrolysis, and dissociation of ADP and P_i from the nucleotide binding site of the myosin head domain are molecular events that each have distinct functions in the actin–myosin reaction cycle. ATP binding to a vacated nucleotide binding site causes the myosin head to dissociate from the actin thin filament. This step is followed by ATP hydrolysis, which induces a conformational change in the myosin head domain that puts it in position to be ready for the next round of muscle contraction. When neuromuscular signaling elevates Ca^{2+} levels in the cell, myosin binding sites on actin are uncovered, and the ADP + P_i-bound form of myosin is able to bind tightly to the actin thin filament. In the most important step of the reaction cycle, P_i is released from the nucleotide binding site, inducing the power-stroke conformation in the myosin head that pulls the actin filament 70 Å along the myosin filament, thereby executing muscle contraction. Finally, with the myosin head domain in this post-power-stroke conformation, ADP is able to dissociate from the nucleotide binding site and make way for a new molecule of ATP to bind.

chapter summary

6.1 The Five Major Functional Classes of Proteins

- Metabolic enzymes are chemical catalysts that lower the activation energy of a biochemical reaction to increase the rate of product formation without altering the equilibrium constant.
- Structural proteins are often assembled into long filaments that form cytoskeletal structures involved in cell migration, chromosomal segregation, and muscle cell contraction.
- Transport proteins span membranes and function as selective pores. Passive transporter proteins allow diffusion of molecules down a concentration gradient; active transporter proteins act as energy-dependent pumps that transport molecules against a concentration gradient.
- Cell signaling proteins respond to changes in the extracellular environment by undergoing conformational or chemical changes that function to regulate cellular processes. A common mechanism to control signal transduction in cells is the phosphorylation and dephosphorylation of signaling proteins.
- Genomic caretaker proteins maintain the integrity of genetic information encoded in DNA and control gene expression. DNA metabolizing enzymes are required for DNA replication, repair, and recombination. RNA metabolizing enzymes are required for RNA transcription, processing, and stability.

6.2 Globular Transport Proteins: Transporting Oxygen

- Myoglobin has a single polypeptide chain and functions as an O_2 storage protein in muscle tissues. Hemoglobin is an O_2 transport protein that contains two α-globin subunits and two β-globin subunits; together they form a heterotetramer ($\alpha_2\beta_2$) capable of binding and transporting four O_2 molecules at a time from the lungs to the tissues.
- Each globin polypeptide contains a single heme group that reversibly binds O_2 as a ligand. Two histidine residues in the protein—His F8 (proximal histidine) and His E7 (distal histidine)—play a critical role in O_2 binding to the heme.
- A plot of the fraction of occupied O_2 binding sites in myoglobin as a function of the O_2 concentration (partial pressure of O_2) generates a hyperbolic binding curve. An O_2 binding curve for hemoglobin generates a sigmoidal curve because of cooperative O_2 binding to the four globin subunits.
- Hemoglobin without bound O_2 molecules (deoxyhemoglobin) is in the T-state conformation (tense), whereas hemoglobin with bound O_2 (oxyhemoglobin) is in the R-state conformation (relaxed). The T-state conformation has a low affinity for O_2, and the R-state conformation has a high affinity for O_2.
- The T→R conformational shift is triggered by small structural changes in the F helix. These changes occur when O_2 binds to the Fe^{2+} atom and reduces its effective atomic radius so that it can move into the plane of the heme. This translocation of the Fe^{2+} atom displaces the proximal histidine (His F8) and tilts the entire F helix, resulting in numerous conformational changes throughout the tetrameric complex.
- Allosteric mechanisms control the relative amounts of hemoglobin in the T and R conformational states by shifting the equilibrium. The O_2 molecule is a positive homotropic allosteric regulator that facilitates the binding of additional O_2 molecules to other globin subunits by shifting the equilibrium toward the R state (oxyhemoglobin). Conversely, CO_2, H^+, and 2,3-bisphosphoglycerate (2,3-BPG) are all negative heterotropic allosteric regulators that shift the equilibrium toward the T state (deoxyhemoglobin).
- The Bohr effect describes the pH and CO_2 dependence of O_2 binding to hemoglobin, in which decreased pH and increased CO_2 lead to decreased O_2 binding. The molecular basis of the Bohr effect is the protonation of key residues at the subunit interfaces, resulting in stabilization of the T-state conformation at low pH and CO_2 binding to the N terminus of the β subunits. Elevated pH causes

deprotonation of these same residues and favors the R-state conformation.

- 2,3-BPG is a negatively charged metabolite that binds to positively charged Lys and His residues at the interface of the two β subunits in deoxyhemoglobin, thereby shifting the equilibrium toward the T-state conformation.
- Sickle cell anemia is a genetically inherited blood disorder caused by a single amino acid change (E6V) in the β subunit of hemoglobin. Interactions due to the hydrophobic effect between β_S subunits on different hemoglobin tetramers leads to the formation of large polymeric fibers that result in defective O_2 transport by erythrocytes.

6.3 Membrane Transport Proteins: Controlling Cellular Homeostasis

- Membrane proteins must be able to associate with the hydrophobic environment of the lipid membrane by orienting nonpolar amino acid residues toward the outside of the protein. The three major types of membrane proteins are membrane receptor proteins, membrane-bound metabolic enzymes, and membrane transport proteins.
- Hydrophobic biomolecules are able to diffuse across cell membranes, whereas polar and charged molecules must be transported across the membrane by proteins. There are two classes of membrane transport proteins: energy-independent passive membrane transport proteins, and energy-dependent active membrane transport proteins.
- The value of ΔG for membrane transport can be calculated using the following equation, in which R is the gas constant, T is temperature (in Kelvin), C_1 is the starting-point solute concentration, C_2 is the destination solute concentration, Z is the electrical charge of the solute, F is the Faraday constant, and ΔV is the membrane potential (in volts):

$$\Delta G = RT\ln\left(\frac{C_2}{C_1}\right) + ZF\Delta V$$

- Porins are passive membrane transport proteins consisting of β-barrel structures that provide a channel for ions and small molecules to pass through the outer membrane of Gram-negative bacteria. The Omp32 porin protein is an example of a selective bacterial transport protein that uses positively charged Arg residues within the channel to permit anions, but not cations, to enter.
- The K^+ channel protein is an α-helical passive membrane transport protein found in both prokaryotic and eukaryotic cells that displays a 10,000-fold selectivity for K^+ ions over Na^+ ions. The molecular basis for this exquisite selectivity is the specific placement of carbonyl oxygen atoms within the channel, which interact precisely with desolvated K^+ ions but not Na^+ ions and allow passage through the opening.
- Aquaporins are tetrameric passive membrane transport proteins that provide a very efficient means for H_2O molecules to pass through biological membranes in response to an osmotic gradient. Selectivity within the channel is achieved by a physical restriction imposed by two short α helices containing Asn residues.
- Primary active membrane transport proteins require energy input, such as ATP hydrolysis, to drive protein conformational changes required for their "pumping" function. In contrast, secondary active membrane transport proteins use the potential energy available in a downhill concentration gradient to transport other molecules across the membrane.
- Two primary active membrane transport proteins, Na^+–K^+ ATPase and skeletal muscle SERCA, are both P-type active transporters that use the energy available in ATP hydrolysis to translocate ions across membranes.
- ABC transporters are another type of *primary* active transport protein that use ATP hydrolysis to drive the protein conformational changes required for pumping ions and small molecules across membranes. ABC transporters are homodimeric complexes containing two ATP binding half-sites that are activated by ligand-induced conformational changes; ATP hydrolysis restores the resting-state conformation.
- The *E. coli* lactose permease protein is a bacterial secondary active membrane transport protein that uses potential energy in a proton membrane gradient to import lactose into the cell via a symporter mechanism. Similarly, the human Na^+–I^- symporter protein is also a secondary active membrane symporter; however, it uses the Na^+ gradient maintained by the Na^+–K^+ ATPase primary active transporter to pump I^- ions into thyroid gland cells for use in thyroid hormone synthesis.

6.4 Structural Proteins: The Actin–Myosin Motor

- The sliding filament model of muscle contraction proposes that Ca^{2+}- and ATP-mediated conformational changes in proteins that make up the thick and thin filaments in muscle cells lead to muscle contraction as a result of filaments physically sliding past one another.
- Thick filaments contain hundreds of myosin protein molecules arranged tail to tail within the fiber, in such a way that their globular head domains are oriented toward the two ends. Titin, the largest protein found in nature, connects the two ends of the thick filaments to anchor proteins located in regions of the sarcomere called the Z disks.
- Thin filaments consist primarily of polymerized actin proteins that serve as myosin binding sites during muscle contraction. Two other proteins in the thin filaments are tropomyosin, a dimeric α-helical protein, and troponin, a heterotrimeric protein complex that binds Ca^{2+} and regulates muscle contraction.
- Muscle contraction is initiated by neuromuscular signals that lead to Ca^{2+} release from the sarcoplasmic reticulum, an intracellular membrane-bound compartment that contains high levels of the primary active membrane transport protein SERCA. The constant pumping of Ca^{2+} back into the sarcoplasmic reticulum by SERCA ensures that muscle relaxation occurs once the neuromuscular signals are turned off.

biochemical terms

(in order of appearance in text)

metabolic enzyme (p. 252)
enzyme active site (p. 253)
structural protein (p. 253)
cytoskeletal protein (p. 253)
thin filament (p. 254)
intermediate filament (p. 254)
actin (p. 254)
myosin (p. 254)
transport protein (p. 255)
passive transporter protein (p. 255)
active transporter protein (p. 255)
ion channel (p. 255)
membrane receptor (p. 256)
nuclear receptor (p. 256)
intracellular signaling protein (p. 256)
genomic caretaker protein (p. 257)
myoglobin (p. 259)
hemoglobin (p. 259)
heme (p. 259)
globin fold (p. 260)
proximal histidine (p. 260)
distal histidine (p. 261)
association constant, K_a (p. 262)
dissociation constant, K_d (p. 262)
hyperbolic curve (p. 263)
sigmoidal (p. 264)
cooperative binding (p. 265)
concerted model (p. 267)
allostery (p. 267)
symmetry model (p. 267)
sequential model (p. 267)
Bohr effect (p. 269)
anemia (p. 273)
sickle cell anemia (p. 273)
malaria (p. 275)
gramicidin A (p. 280)
porin (p. 281)
periplasmic space (p. 281)
selectivity channel (p. 282)
aquaporin (p. 283)
P-type transporter (p. 284)
ATP binding cassette (ABC) transporter (p. 284)
antiporter (p. 285)
symporter (p. 285)
H^+–K^+ ATPase (p. 285)
serotonin-selective reuptake inhibitor (p. 286)
Na^+–K^+ ATPase (p. 286)
Ca^{2+}-ATPase (p. 287)
sarcoplasmic reticulum (SR) (p. 287)
epinephrine (p. 288)
cystic fibrosis (p. 290)
lactose permease (p. 292)
Na^+–I^- symporter (p. 293)
sliding filament model (p. 295)
myoblast (p. 296)
myofibril (p. 296)
sarcomere (p. 296)
tropomyosin (p. 297)
troponin (p. 297)
titin (p. 297)

review questions

1. List the five major functional classes of proteins.
2. Describe the functional role of enzymes in metabolic reactions.
3. What are the two most abundant cytoskeletal proteins in animal cells?
4. Describe the two classes of membrane transport proteins.
5. Describe the functional role of genomic caretaker proteins.
6. What are the two oxygen-binding globin proteins in animals and what are their functions?
7. Why is carbon monoxide toxic to animals?
8. What do the oxygen-binding curves of myoglobin and hemoglobin reveal about the oxygen-binding properties of these two proteins?
9. What are three key features of ligand–protein interactions?
10. What is the difference between fetal and adult hemoglobin? Why is this important?
11. What is the cause of sickle cell anemia, and why are heterozygous carriers of sickle cell anemia more resistant to malaria than individuals who do not carry the mutation?
12. Describe three types of membrane proteins.
13. Describe the two types of active transporter proteins.
14. What is the biochemical basis for Zoloft treatment of depression?
15. Describe the effects of Ca^{2+} binding and ATP hydrolysis on muscle contraction.
16. Describe the five steps of muscle contraction.

challenge problems

1. Name three heterotropic effectors that alter the O_2 binding affinity of hemoglobin. For each, state whether the effector is a positive or negative regulator of O_2 binding affinity. Why is O_2 considered a positive homotropic effector of O_2 binding affinity for hemoglobin?
2. Blood that has been stored for some time becomes depleted of 2,3-bisphosphoglycerate (2,3-BPG). What problems, if any, would this cause upon transfusion into a patient? Why?
3. Bird hemoglobins are tetrameric and very similar in structure and function to mammalian hemoglobins. However, in some bird species, O_2 binding affinity to hemoglobin is not regulated by 2,3-BPG, but rather by a different compound that functions as a 2,3-BPG analog. Answer the following questions on the basis of the

mechanism by which 2,3-BPG regulates the O_2 binding affinity of human hemoglobin.

a. Considering the chemical and physical properties of the following compounds, which is the most likely candidate for the 2,3-BPG analog in bird red blood cells?

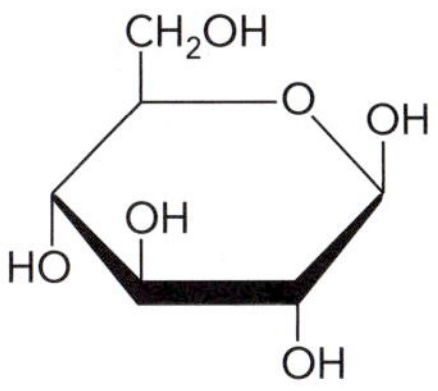

Glucose

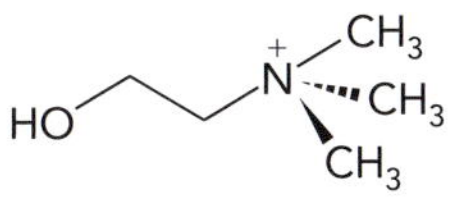

Choline

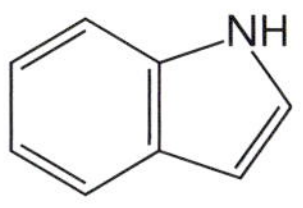

Indole

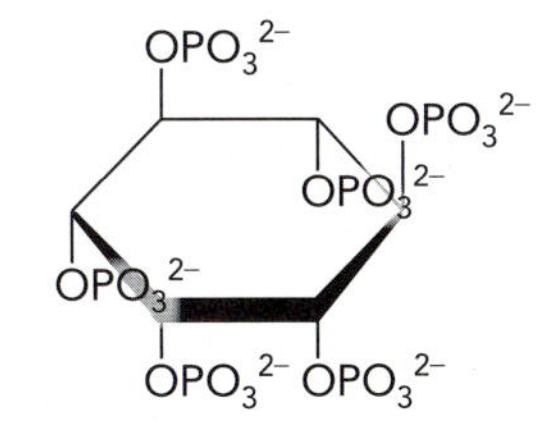

Inositol hexaphosphate

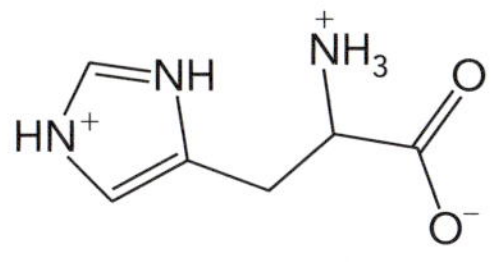

Histidine

b. The bird 2,3-BPG analog binds to hemoglobin in the same way 2,3-BPG binds to mammalian hemoglobin. Briefly describe where in the structure of the tetrameric bird hemoglobin you would expect the compound to bind and by what type of bonds and/or interactions.

c. Would you expect the compound to increase or decrease the O_2 binding affinity of bird hemoglobin?

4. The four graphs here show the same reference curve for O_2 binding by normal human adult hemoglobin at pH 7.4 in the presence of a physiologic concentration of 2,3-BPG. On each graph, sketch an additional O_2 binding curve for the Hb as described in **a–d**, and briefly explain how this Hb differs from the Hb reference in terms of changes in the R-state–T-state equilibrium and P_{50} (pO_2 at 50% saturation of O_2 binding sites).

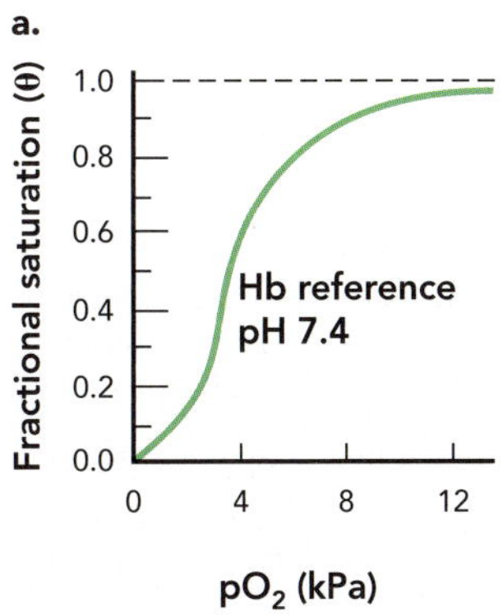

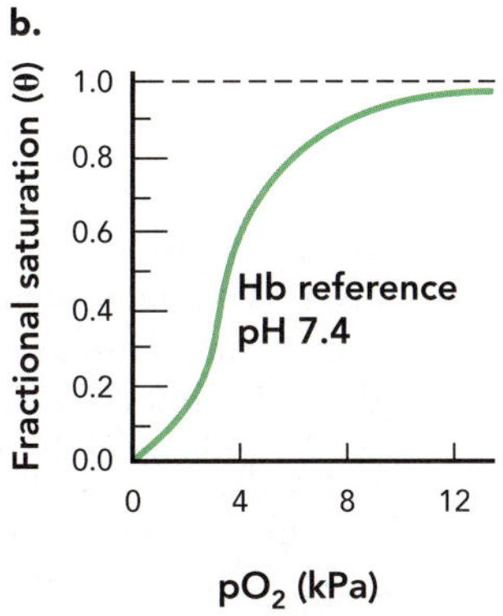

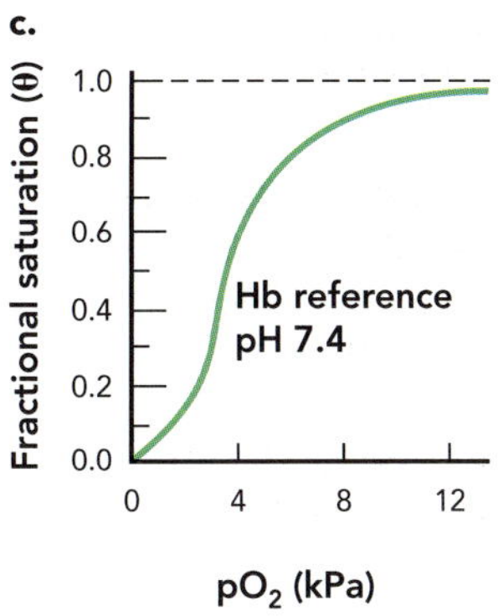

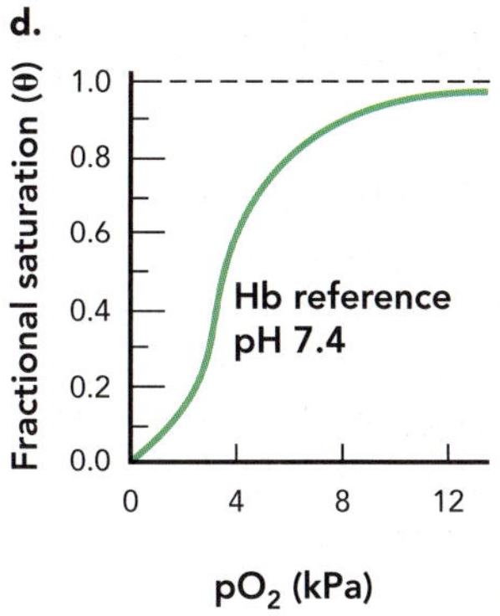

a. For graph **a**: Hemoglobin with all of the 2,3-BPG removed from the O_2 binding assay.

b. For graph **b**: Hemoglobin at pH 7.2 rather than pH 7.4.

c. For graph **c**: Hemoglobin mutant in which the predominant form of the protein is monomeric, with a very low binding affinity for O_2.

d. For graph **d**: Fetal hemoglobin in which His143 in the β subunit is replaced by Ser143 in the γ subunit.

5. Would you expect the α and β subunits of hemoglobin to have more or fewer hydrophobic amino acids than myoglobin? Why?

6. Amino acid substitutions at the α–β interface of hemoglobin can change the relative stabilities of the R and T forms of hemoglobin. For example, in one mutant form of hemoglobin, the hydrogen bond that is primarily responsible for stabilizing the R state is lost. Would you expect this hemoglobin to have a higher or lower than normal O_2 binding affinity? Explain.

7. Describe the major structural differences between fetal hemoglobin (HbF) and maternal adult hemoglobin (HbA). Explain how these structural differences affect the ligand binding properties of the two hemoglobins and what impact this has on O_2 delivery from the maternal red blood cells to the fetal red blood cells in the placenta.

8. Which of the following sequences would most likely be found in a transmembrane β strand in the structure of a bacterial porin protein? Explain your choice.

a. -Arg-Asn-Ser-Ile-Phe-Met-Lys-Glu-Gly-

b. -His-Leu-Phe-Ala-Val-Asp-His-Lys-Asp-

c. -Lys-Gln-Asp-His-Arg-Ser-Asn-Gly-Glu-

d. -Asn-Val-Lys-Met-Glu-Ile-Arg-Leu-Gln-

e. -Phe-Met-Leu-Val-Ala-Ile-Phe-Val-Leu-

9. Each human red blood cell has about 2×10^5 aquaporin monomers. If water flows through the membrane at 5×10^8 molecules per aquaporin tetramer per second, and the volume of a red blood cell (RBC) is 5×10^{-11} mL, how rapidly would an RBC lose half its volume if put in a high-salt environment? The density of water is 55.5 mol/L.

10. Glucose is transported across the cell membrane by Glut proteins, which function as primary energy-independent passive transporters. Under normal physiologic conditions when the concentration of serum glucose outside the cell, $[\text{glucose}]_{\text{outside}}$, is 5 mM, what is the free energy change for transport of serum glucose into the cell at 37 °C, considering that the glucose concentration inside the cell, $[\text{glucose}]_{\text{inside}}$, is 0.1 mM?

11. Answer the following questions about the bioenergetics of ion transport across the cell membrane by the Na^+–K^+ ATPase transporter protein at 37 °C under conditions in which the membrane potential is 70 mV (inside of the cell is negative relative to the outside) and the ion concentrations are as follows:

$[K^+]_{\text{outside}}$ = 5 mM
$[K^+]_{\text{inside}}$ = 140 mM
$[Na^+]_{\text{outside}}$ = 150 mM
$[Na^+]_{\text{inside}}$ = 10 mM

a. What is the energy requirement to transport 2 mol of K^+ across the membrane?

b. What is the energy requirement to transport 3 mol of Na^+ across the membrane?

c. On the basis of the proposed mechanism of ion transport by the Na^+–K^+ ATPase transporter protein, explain how the hydrolysis of a single ATP is sufficient to transport 2 mol of K^+ into the cell *and* 3 mol of Na^+ out of the cell, considering that the amount of energy available from ATP hydrolysis is only ~50 kJ/mol under cellular conditions, and as much as ~72 kJ could be required for both transport processes.

12. The lactose permease transporter of *E. coli* is a secondary active transporter protein that can accumulate lactose inside the cell against a concentration gradient. This symporter mechanism is driven by proton motive force that serves as the energy source for transport. If the ΔG for the proton gradient across the membrane is −10 kJ/mol, what is the maximum concentration ratio of lactose that can be achieved at 37 °C?

13. The globular head domain of myosin has binding sites for actin and ATP that are far from each other in the tertiary structure of the protein, yet the protein cannot bind both ATP and actin at the same time. In terms of two distinct myosin protein conformational states that preferentially bind actin *or* ATP with high affinity, explain how ATP binding reduces the binding affinity of myosin for actin and vice versa.

14. Briefly describe the chain of events in a skeletal muscle cell that occurs between a nerve impulse stimulating influx of Ca^{2+} from the sarcoplasmic reticulum into the cytosol and contraction of the muscle.

smartwork5

If your instructor assigns homework with Smartwork5, access it here: digital.wwnorton.com/biochem.

suggested reading

Books and Reviews

Branden, C., and Tooze, J. (1999). *Introduction to protein structure* (2nd ed.). New York, NY: Garland Publishing.

Dohan, O., De la Vieja, A., Paroder, V., Riedel, C., Artani, M., Reed, M., Ginter, C. S., and Carrasco, N. (2003). The sodium/iodide symporter (NIS): characterization, regulation, and medical significance. *Endocrine Reviews, 24*, 48–77.

Gadsby, D. C. (2007). Structural biology: ion pumps made crystal clear. *Nature, 450*, 957–959.

Geeves, M. A., and Holmes, K. C. (2005). The molecular mechanism of muscle contraction. *Advances in Protein Chemistry, 71*, 161–193.

Hollenstein, K., Dawson, R. J., and Locher, K. P. (2007). Structure and mechanism of ABC transporter proteins. *Current Opinion in Structural Biology, 17*, 412–418.

Huxley, H. E. (2004). Fifty years of muscle and the sliding filament hypothesis. *European Journal of Biochemistry, 271*, 1403–1415.

Jones, P. M., O'Mara, M. L., and George, A. M. (2009). ABC transporters: a riddle wrapped in a mystery inside an enigma. *Trends in Biochemical Sciences, 34*, 520–531.

Kaback, H. R., Sahin-Toth, M., and Weinglass, A. B. (2001). The kamikaze approach to membrane transport. *Nature Reviews, 2*, 610–620.

Oldham, M. L., Davidson, A. L., and Chen, J. (2008). Structural insights into ABC transporter mechanism. *Current Opinion in Structural Biology, 18*, 726–733.

Petsko, G. A., and Ringe, D. (2004). *Protein structure and function*, London, England: Blackwell Publishing.

Sansom, M. S., Shrivastava, I. H., Ranatunga, K. M., and Smith, G. R. (2000). Simulations of ion channels—watching ions and water move. *Trends in Biochemical Sciences, 25*, 368–374.

Strandberg, B. (2009). Chapter 1: building the ground for the first two protein structures: myoglobin and haemoglobin. *Journal of Molecular Biology, 392*, 2–10.

Whitford, D. (2005). *Proteins: structure and function.* Hoboken, NJ: Wiley.

Zeuthen, T. (2001). How water molecules pass through aquaporins. *Trends in Biochemical Sciences*, *26*, 77–79.

Primary Literature

Berenbrink, M. (2006). Evolution of vertebrate haemoglobins: histidine side chains, specific buffer value and Bohr effect. *Respiratory Physiology & Neurobiology*, *154*, 165–184.

Bers, D. M., and Despa, S. (2009). Na/K-ATPase—an integral player in the adrenergic fight-or-flight response. *Trends in Cardiovascular Medicine*, *19*, 111–118.

de Groot, B. L., and Grubmüller, H. (2001). Water permeation across biological membranes: mechanism and dynamics of aquaporin-1 and GlpF. *Science*, *294*, 2353–2357.

Hollenstein, K., Frei, D. C., and Locher, K. P. (2007). Structure of an ABC transporter in complex with its binding protein. *Nature*, *446*, 213–216.

Jiang, Y., Lee, A., Chen, J., Cadene, M., Chait, B. T., and MacKinnon, R. (2002). Crystal structure and mechanism of a calcium-gated potassium channel. *Nature*, *417*, 515–522.

Morth, J. P., Pedersen, B. P., Toustrup-Jensen, M. S., Sørensen, T. L., Petersen, J., Andersen, J. P., Vilsen, B., and Nissen, P. (2007). Crystal structure of the sodium-potassium pump. *Nature*, *450*, 1043–1049.

Olesen, C., Picard, M., Winther, A. M., Gyrup, C., Morth, J. P., Oxvig, C., Møller, J. V., and Nissen, P. (2007). The structural basis of calcium transport by the calcium pump. *Nature*, *450*, 1036–1042.

Penman, B. S., Pybus, O. G., Weatherall, D. J., and Gupta, S. (2009). Epistatic interactions between genetic disorders of hemoglobin can explain why the sickle-cell gene is uncommon in the Mediterranean. *Proceedings of the National Academy of Sciences, USA*, *106*, 21242–21246.

Smirnova, I. N., Kasho, V., and Kaback, H. R. (2008). Protonation and sugar binding to LacY. *Proceedings of the National Academy of Sciences USA*, *105*, 8896–8901.

Tajkhorshid, E., Nollert, P., Jensen, M. O., Miercke, L. J., O'Connell, J., Stroud, R. M., and Schulten, K. (2002). Control of the selectivity of the aquaporin water channel family by global orientational tuning. *Science*, *296*, 525–530.

Vandecaetsbeek, I., Trekels, M., De Maeyer, M., Ceulemans, H., Lescrinier, E., Raeymaekers, L., Wuytack, F., and Vangheluwe, P. (2009). Structural basis for the high Ca^{2+} affinity of the ubiquitous SERCA2b Ca^{2+} pump. *Proceedings of the National Academy of Sciences USA*, *106*, 18533–18538.

Zachariae, U., Kluhspies, T., De, S., Engelhardt, H., and Zeth, K. (2006). High resolution crystal structures and molecular dynamics studies reveal substrate binding in the porin Omp32. *Journal of Biological Chemistry*, *281*, 7413–7420.

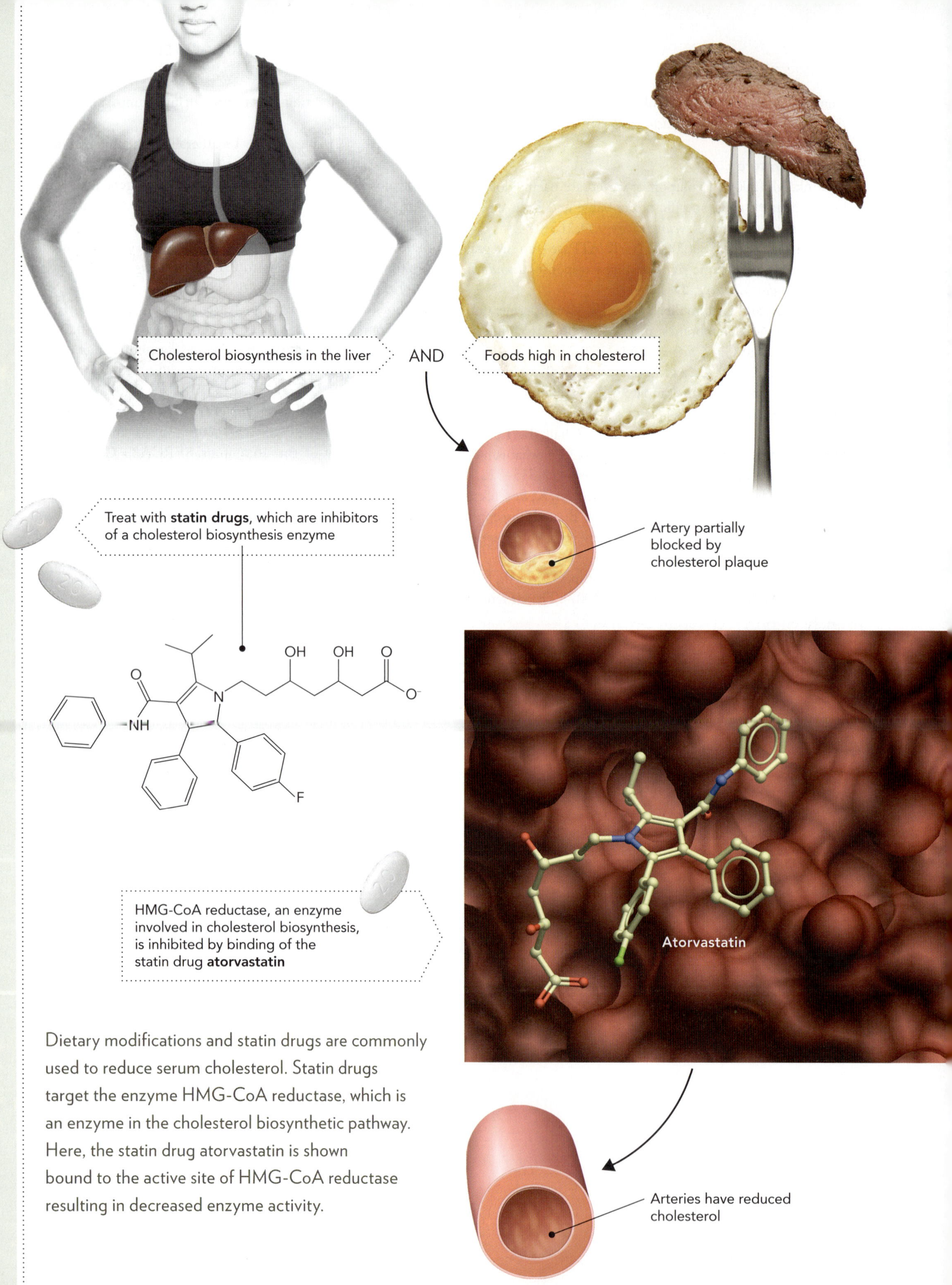

Cholesterol biosynthesis in the liver
AND
Foods high in cholesterol
Treat with statin drugs, which are inhibitors of a cholesterol biosynthesis enzyme
Artery partially blocked by cholesterol plaque
OH
OH
O
O
N
NH
O⁻
F
HMG-CoA reductase, an enzyme involved in cholesterol biosynthesis, is inhibited by binding of the statin drug atorvastatin
Atorvastatin
Arteries have reduced cholesterol

Dietary modifications and statin drugs are commonly used to reduce serum cholesterol. Statin drugs target the enzyme HMG-CoA reductase, which is an enzyme in the cholesterol biosynthetic pathway. Here, the statin drug atorvastatin is shown bound to the active site of HMG-CoA reductase resulting in decreased enzyme activity.

7

Enzyme Mechanisms

◀ High serum cholesterol can result from diets that are rich in cholesterol-containing foods or from biosynthesis of cholesterol in the liver. Studies have shown that excess serum cholesterol is associated with blockage of arteries and cardiovascular disease. High serum cholesterol is often treated with statin drugs. These drugs function by inhibiting the enzyme 3-hydroxy-3-methyl-glutaryl-coenzyme A (HMG-CoA) reductase, which catalyzes an early step in the cholesterol biosynthetic pathway.

Inhibition of HMG-CoA reductase lowers intracellular cholesterol levels by reducing biosynthesis. Importantly, when intracellular cholesterol is lowered, this triggers increased uptake of cholesterol from the serum and thereby reduces the circulating cholesterol that contributes to arterial plaques and heart disease.

CHAPTER OUTLINE

Life depends on an enormous number of highly integrated and regulated chemical reactions. However, on their own, these reactions do not always happen quickly. Enzymes are the necessary biological catalysts that accelerate the rates of these chemical reactions under physiologic conditions. Indeed, without enzymes, metabolic reactions would be so slow—some requiring millions of years—that life would not be possible.

Enzymes increase the rates of chemical reactions, but they are not consumed by the reactions, nor do they alter the equilibrium concentration ratio of substrates and products. Most enzymes are proteins, although some enzymes, called ribozymes, are composed of RNA (see Chapter 21). Enzyme kinetics, discussed in this chapter, is a quantitative analysis of enzyme function that can be used to probe reaction mechanisms and to compare the efficiency of closely related enzymes under a variety of conditions.

We begin this chapter with an overview of enzymes and their structure and function, then proceed to discuss key concepts in enzyme reaction mechanisms and enzyme kinetics. The chapter concludes with a description of the most common enzyme regulatory mechanisms. Enzymes provide the catalytic power for living cells to perform chemical reactions on a timescale of microseconds to minutes, but enzyme function must be highly regulated to avoid biochemical chaos.

In keeping with the protein biochemistry focus of Part 2 of this book, we present enzyme mechanisms in the context of protein structure and function. Throughout the chapter, we emphasize the chemical properties of functional groups that directly or indirectly contribute to the catalytic efficiency of enzymes.

7.1 Overview of Enzymes

The discovery that proteins can function as enzymes—once called the *ferments of zymes* (from yeast)—was not made until the 1930s, nearly 100 years after proteins were first characterized. The reluctance to accept the notion that proteins can function as biological catalysts was rooted in the way biochemists approached the study of biomolecules. Early biochemists thought of proteins as structural components in cells and focused their research on the most abundant proteins, such as keratin and albumin, which could be readily purified and analyzed by acid hydrolysis. At the same time, other biochemists were interested in measuring rates of chemical reactions in cell-free extracts and isolating organic reaction intermediates formed by mysterious catalytic "enzymes." These two areas of research did not overlap or intersect for quite a long time.

The understanding that most enzymes were proteins did not come about until a controversial report by James Sumner in 1926. Based on his experiments, in which he was able to purify and crystallize the enzyme urease from jack beans, he proposed that urea hydrolysis was mediated by a protein enzyme (**Figure 7.1**). Sumner's work

a.

b.

c.

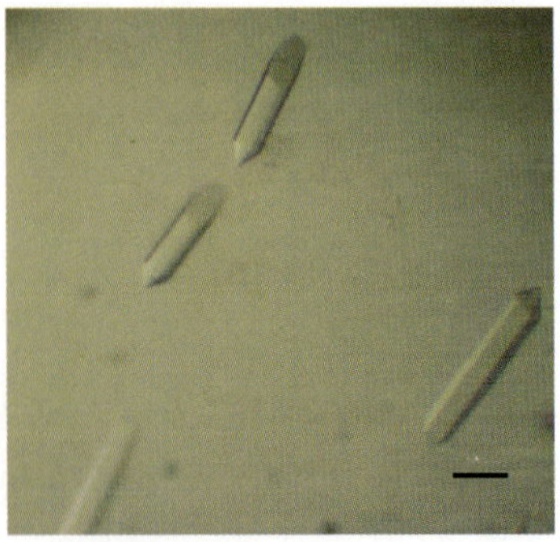

Figure 7.1 James Sumner provided the first evidence that a protein can function as an enzyme when he purified and crystallized the enzyme urease from jack bean protein extracts. Crystallization of the protein was an important step to demonstrate its purity. **a.** James Sumner (1887–1955) was a biochemist at Cornell University when he first purified urease in 1926. SPL/SCIENCE SOURCE. **b.** Jack beans (*Canavalia ensiformis*) contain high levels of the enzyme urease. SEAN SPRAGUE/ALAMY. **c.** Protein crystals of purified urease isolated from jack beans. A. BALASUBRAMANIAN AND K. PONNURAJ (2009). PURIFICATION, CRYSTALLIZATION, AND PRELIMINARY X-RAY ANALYSIS OF UREASE FROM JACK BEAN *CANAVALIA ENSIFORMIS. ACTA CRYSTALLOGRAPHICA*, 65, 949–951. HTTP://JOURNALS.IUCR.ORG. © 2009 INTERNATIONAL UNION OF CRYSTALLOGRAPHY.

was followed 3 years later by that of John Northrop, who was able to prepare the digestive enzyme pepsin in a highly purified form. With pure preparations of proteins, Sumner's earlier discovery that proteins can function as enzymes could be confirmed. Sumner and Northrop, along with the virologist Wendell Stanley, shared the 1946 Nobel Prize in Chemistry for their combined work on protein purification and characterization.

Our current understanding of how enzymes catalyze biochemical reactions follows from these early studies and relies heavily on analyzing three-dimensional protein structures in the context of enzyme kinetics. However, as described in Chapter 6, the three-dimensional structure of a protein does not by itself reveal the dynamic processes required for it to function. Nothing exemplifies this more than the structure of enzymes. Up until 1958, biochemists thought that the observed high specificity of enzyme-mediated catalysis was best explained by rigid physical and chemical complementarity between the reactant, usually referred to as the substrate, and the enzyme. This older view, called the **lock and key model** of enzyme specificity, was first described by the German chemist Emil Fischer in 1894. However, the lock and key model could not explain how enzymes are regulated or how substrates can bind to sites buried deep within the interior of a protein.

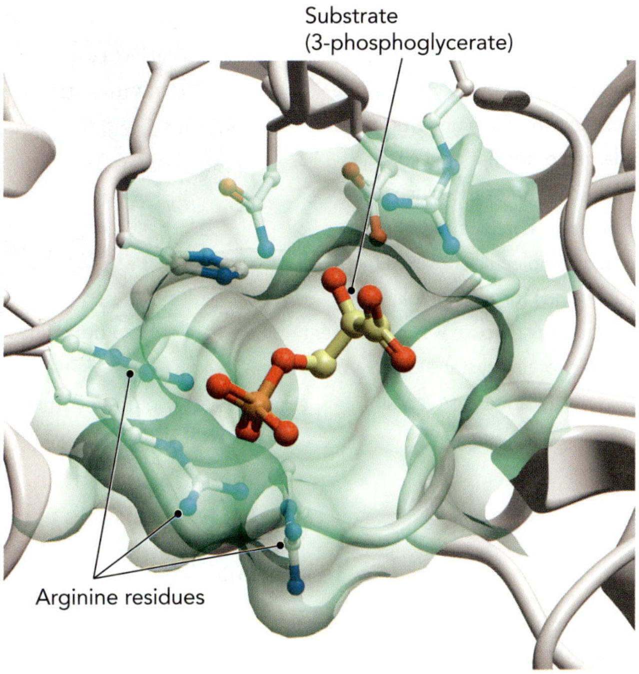

Figure 7.2 Enzyme active sites provide chemical environments that facilitate catalytic reactions by excluding excess solvent and bringing the reactive functional groups of the enzyme into close proximity to the substrate. The molecular structure of the enzyme phosphoglycerate kinase, shown here, highlights the substrate binding site of the enzyme (shaded in green), in which several key amino acid residues help orient the substrate through electrostatic interactions with the phosphate group. BASED ON PDB FILE 3C39.

In 1958, Daniel Koshland proposed the **induced-fit model** of enzyme catalysis. In this model, the enzyme is analogous to a glove that has a three-dimensional shape, but is flexible and able to accommodate an equally flexible hand, which represents the substrate. A major advantage of the induced-fit model is that it permits a much larger number of weak interactions to occur between the substrate and enzyme, as a result of structural adjustments in the enzyme–substrate complex that occur upon binding. These numerous weak interactions between an enzyme and its cognate substrate provide binding energy that contributes to the catalytic activity of enzymes.

Before we discuss enzyme mechanisms in more detail, we want to emphasize three critical aspects of enzyme structure and function.

1. *Enzymes usually bind substrates with high affinity and specificity.* Enzyme active sites are physical pockets or clefts in an enzyme where the substrates bind and catalytic reactions take place. As shown in the structure of the enzyme phosphoglycerate kinase (**Figure 7.2**), the active site provides a protective microenvironment away from bulk solvent. Here, the substrate can bind to the enzyme's functional groups that participate in the catalytic reaction.

2. *Substrate binding to the active site induces structural changes in the enzyme.* These changes result in a large number of weak interactions between the substrate and the enzyme (hydrogen bonds, ionic interactions, and van der Waals interactions) and facilitate the structural changes needed to form the product. A classic example of the induced-fit model of substrate binding occurs in hexokinase, a glycolytic enzyme that phosphorylates glucose using ATP (**Figure 7.3**). Binding of the glucose substrate to the hexokinase active

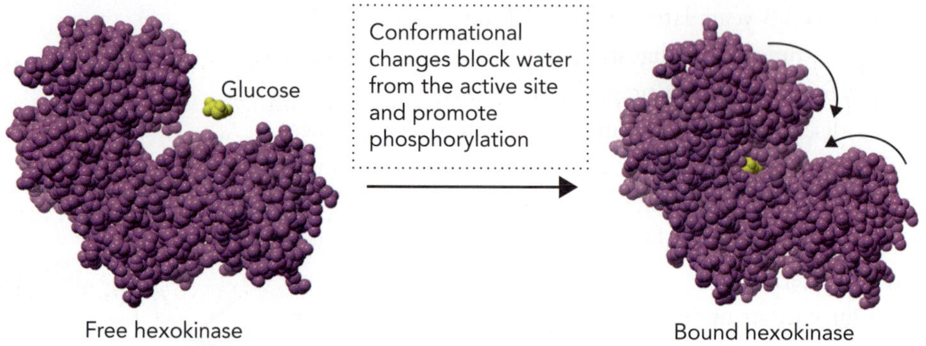
Glucose
Conformational changes block water from the active site and promote phosphorylation
Free hexokinase
Bound hexokinase

Figure 7.3 An example of the induced-fit model of enzyme catalysis is the glycolytic enzyme hexokinase, which undergoes a large conformational change upon binding of the glucose substrate. The location of the glucose molecule shown in the free (unbound) form of hexokinase is arbitrary, as glucose was not present in the protein crystal. BASED ON PDB FILES 1IG8 (FREE) AND 3B8A (BOUND).

site results in a large conformational change in the enzyme. This change excludes H_2O from the active site and facilitates glucose phosphorylation by ATP. In most enzymes, however, substrate binding leads to more subtle changes in protein structure, with the most significant changes occurring in the vicinity of the enzyme active site.

3. *Enzyme activity is highly regulated in cells.* Enzyme regulation is necessary to maximize energy balance between catabolic and anabolic pathways (see Chapter 2) and to alter cell behavior in response to environmental stimuli. The two primary modes of enzyme regulation are bioavailability and catalytic efficiency. **Bioavailability** refers to the amount of enzyme present in the cell as a result of regulated gene expression and protein turnover. **Catalytic efficiency** is a quantitative measure of enzyme activity. The catalytic efficiency of enzymes can be controlled by the binding of regulatory molecules or by covalent modification—most often phosphorylation of Ser, Thr, or Tyr residues. **Figure 7.4** shows the metabolic enzyme glycogen phosphorylase, whose catalytic efficiency is stimulated both by covalent modification (phosphorylation of a serine residue) and by the noncovalent binding of adenosine monophosphate (AMP), a regulatory molecule. Glycogen phosphorylase is a highly regulated enzyme in liver and muscle cells that controls the amount of glucose released from stored glycogen in response to the energy needs of the cell or the organism.

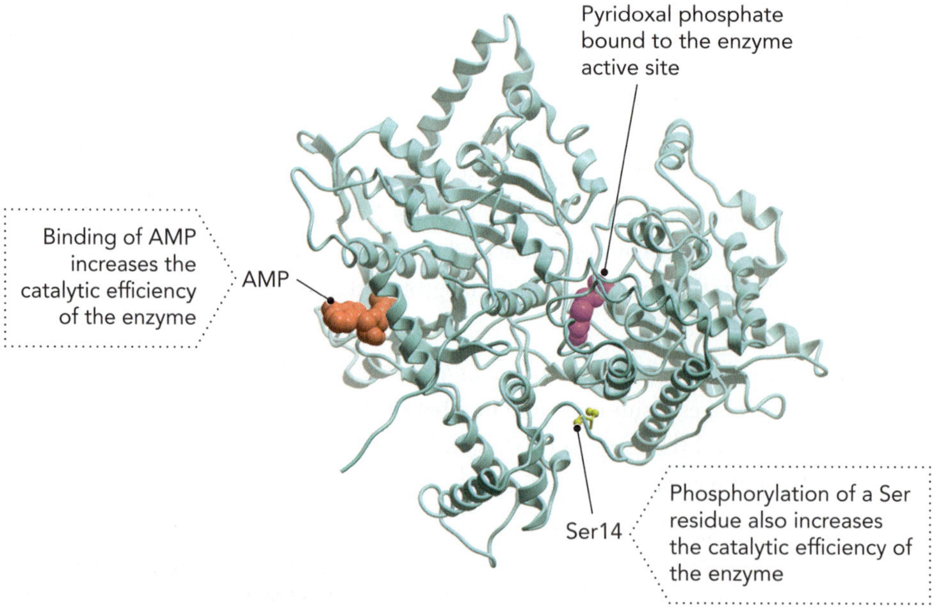
Pyridoxal phosphate bound to the enzyme active site
Binding of AMP increases the catalytic efficiency of the enzyme
AMP
Ser14
Phosphorylation of a Ser residue also increases the catalytic efficiency of the enzyme

Figure 7.4 The catalytic efficiency of glycogen phosphorylase is increased by noncovalent binding of allosteric regulators, such as AMP, and by covalent attachment of a phosphoryl group to Ser14. The bound coenzyme pyridoxal phosphate is shown in the active site. Note that in glycogen phosphorylase, the regulatory sites are not directly at the active site. However, binding of AMP or phosphate at the regulatory sites causes conformational changes that affect the catalytic efficiency in the active site. BASED ON PDB FILE 8GPB.

Enzymes Are Chemical Catalysts

Like all chemical catalysts, enzymes alter the rates of reactions ($A \rightleftharpoons B$) without changing the ratio of substrates and products at equilibrium. Instead, catalysts simply decrease the time it takes to reach equilibrium. Catalysts increase the rates of reactions in both directions by the same amount ($A \rightarrow B$ and $B \rightarrow A$). Thus, the rates of reactions are affected without changing the final equilibrium or the overall change in free energy (ΔG) of the reaction.

Consider, for example, the thermodynamically favorable conversion of hydrogen peroxide (H_2O_2) to water and oxygen (gas):

$$2\,H_2O_2 \rightleftharpoons 2\,H_2O + O_2$$

The decomposition of H_2O_2 occurs very slowly at room temperature—so slowly that it would take about 3 years for 1 mol of H_2O_2 to decompose to 1 mol of H_2O and 0.5 mol of O_2. This amount of time represents the **half-life, $t_{1/2}$,** of the reaction, meaning the time it takes for half of the reactant to decompose. The $t_{1/2}$ of this reaction can be decreased considerably by adding a small amount of free iron (Fe^{2+}/Fe^{3+}) as a chemical catalyst. Under these conditions, the $t_{1/2}$ is only 11.6 minutes, representing an increase in the rate of decomposition by five orders of magnitude (the rate is 10^5 times faster in the presence of ferric ion).

Now, H_2O_2 is made in small amounts inside cells, where it is a highly reactive and toxic compound that must be rapidly eliminated to avoid damage to cellular components. The enzyme catalase has evolved to play the role of a cellular H_2O_2 detoxifying agent. It catalyzes a very efficient decomposition reaction involving an Fe^{3+} porphyrin ring located at the end of a narrow channel that connects the outside of the enzyme to the internal active site (**Figure 7.5**). The rate of H_2O_2 decomposition in the presence of catalase is amazingly fast, so that millions of molecules of H_2O_2 are decomposed per second per molecule of catalase. The 10^{15}-fold enhancement of the enzyme-catalyzed reaction over the uncatalyzed reaction protects cells from the toxic effects of H_2O_2 through rapid conversion to H_2O and O_2.

How does adding an enzyme catalyst increase reaction rates? To answer this question, we need to introduce the concept of chemical transition states. The **transition state theory** developed by Henry Eyring in the 1930s states that the conversion of substrate to product involves a high-energy transition state in which a molecule can either become a product or remain a substrate. This high-energy state is very

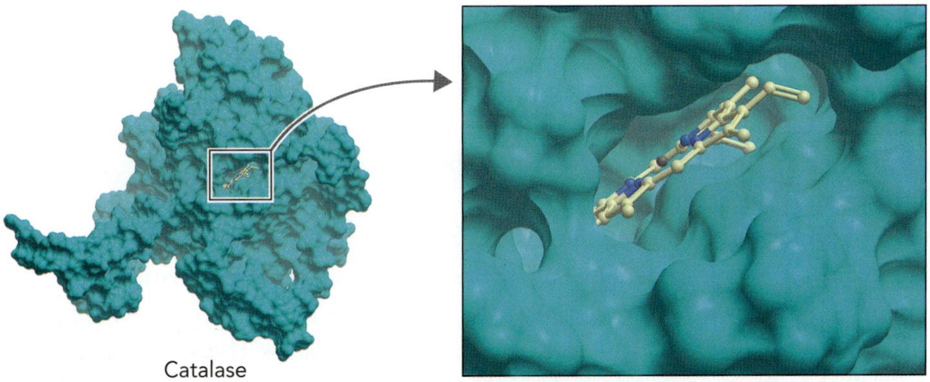

Iron-containing heme is buried in a pocket in the enzyme active site

Figure 7.5 The enzyme human erythrocyte catalase, shown in space-filling representation, has a heme group buried within the enzyme active site. The Fe atom in the porphyrin ring is required for an electron transfer step that takes place during the reaction. Amino acid residues in the active site also contribute to the reaction by assisting in proton movement and the oxidation of Fe^{3+} to Fe^{4+} and by interactions with the reaction intermediates. The result is that the catalase reaction is far more efficient than the reaction catalyzed by free iron. BASED ON PDB FILE 1QQW.

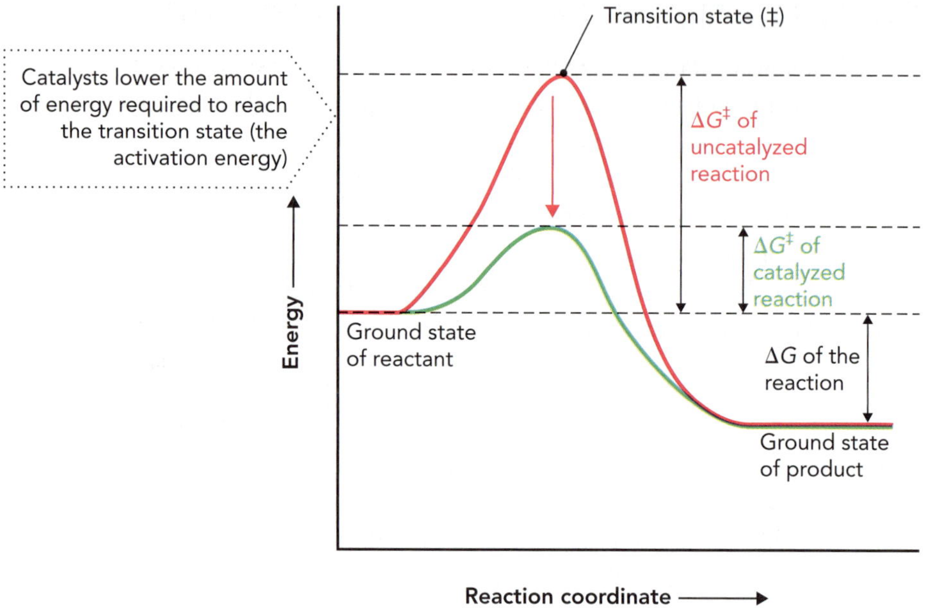

Figure 7.6 A reaction coordinate diagram for a catalyzed chemical reaction and an uncatalyzed chemical reaction is shown here. The difference in energy between the ground state of the reactant and the ground state of the product is ΔG. However, for the reaction to occur, the reactant must first reach the transition state (‡), and the energy required for the reactant to do this is called the activation energy, denoted $\Delta G^{\ddagger}$. Adding a catalyst to the reaction lowers the activation energy by providing an alternative path to product formation.

unstable, and only a few molecules in a reaction mixture are able to achieve this state at any one time. When they do, molecules are only in the transition state for 10^{-15} second—basically the time of a single atomic vibration. Molecules in the transition state are not reaction intermediates that can be physically isolated, but rather molecules that have attained an energy level that must be reached in order to convert from substrate to product.

Figure 7.6 describes the free energy of a chemical reaction, in which we plot $A \rightarrow B$ using a **reaction coordinate diagram**. As described in Chapter 2, the ΔG of a reaction is the change in free energy between the ground states of the reactant and the product, an energy value that is not changed by the presence of catalyst. Transition state theory dictates that a reactant must first reach an energy level required for the transition state, symbolized by a "double dagger" (‡), before it can form product. For example, a reaction might need enough energy to break a molecular bond so that product can form. As seen in Figure 7.6, the energy required to reach the transition state, also called the **activation energy**, denoted as $\boldsymbol{\Delta G^{\ddagger}}$, is the difference in energy between the ground state of the reactant and the transition state.

The function of a catalyst is to lower $\Delta G^{\ddagger}$ by providing more favorable reaction conditions. This has the effect of lowering the transition energy barrier, thereby increasing the rate of the reaction without altering the overall change in free energy, ΔG. Enzymes like catalase are excellent catalysts because they provide a highly reactive Fe^{3+} porphyrin ring within a protected environment. The enzyme readily promotes product formation as a function of the reduced transition state energy barrier. Catalase reduces $\Delta G^{\ddagger}$ from +71 kJ/mol in the uncatalyzed reaction to just +8 kJ/mol and, in so doing, increases the reaction rate 10^{15}-fold. It is important to note that the decomposition of H_2O_2 is highly favorable regardless of whether or not a catalyst is included in the reaction, as the free energy change (ΔG) of the reaction is approximately −95 kJ/mol. However, the rate of the reaction is dramatically affected by the addition of the enzyme catalase because it provides a chemical environment that makes it easier for H_2O_2 to reach the transition state. (Later in this chapter, we will describe the attributes of this favorable chemical environment within the enzyme active site.)

Table 7.1 REPRESENTATIVE METAL-ION COFACTORS IN ENZYMES

Cofactor	Representative enzymes	Role in catalysis
Fe^{2+}	Cytochrome oxidase	Oxidation–reduction
Mg^{2+}	Hexokinase	Helps bind ATP
Mn^{2+}	Ribonucleotide reductase	Oxidation–reduction
Cu^{2+}	Nitrite reductase	Oxidation–reduction
Zn^{2+}	Alcohol dehydrogenase	Helps bind the substrate
Ni^{2+}	Urease	Required in the catalytic site
K^{+}	Pyruvate kinase	Increases enzyme activity
Se	Glutathione peroxidase	Oxidation–reduction
Mo	Xanthine oxidase	Oxidation–reduction

Cofactors and Coenzymes

Proteins are the primary enzymes in living systems, but they often require **cofactors** to aid in the catalytic reaction mechanism within the enzyme active site. Cofactors provide additional chemical groups to supplement the chemistry of the enzyme's amino acid functional groups when they are not sufficient to mediate a particular catalytic mechanism. An enzyme with a bound cofactor is called a **holoenzyme**, whereas removal of the cofactor produces an inactive **apoenzyme** (enzyme without cofactor).

Enzyme cofactors include a variety of inorganic ions such as Fe^{2+}, Mg^{2+}, Mn^{2+}, Cu^{2+}, and Zn^{2+}, which are often required in enzymes that catalyze redox reactions. **Table 7.1** lists the most common metal-ion cofactors found in enzymes and their roles in these reactions. Many nucleic acid metabolizing enzymes, such as RNA and DNA polymerases, require Zn^{2+} in the active site, whereas Mg^{2+} is often associated with ATP hydrolyzing enzymes. In fact, one of the reasons mercury (Hg^{2+}) is so toxic is that it can replace Zn^{2+} in a nonproductive way in a variety of enzymes. Copper (Cu^{2+}) is another important metal-ion cofactor in enzymes. Coordination of a Cu^{2+} ion by His residues in the active site of the *Alcaligenes faecalis* nitrite reductase enzyme positions the metal cofactor so that it can bind nitrite in the correct orientation and thereby mediate its reduction to form nitric oxide (**Figure 7.7**).

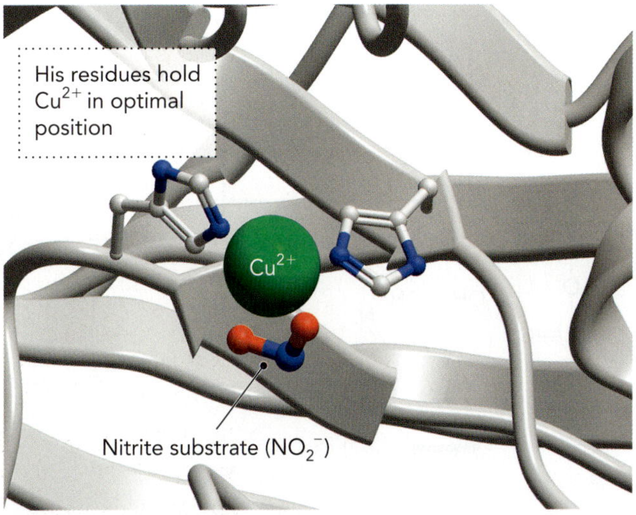

His residues hold Cu^{2+} in optimal position
Cu^{2+}
Nitrite substrate (NO_2^-)

Figure 7.7 Nitrite reductase is an enzyme present in several types of soil bacteria that recycle nitrogen in the environment. As seen here in the molecular structure of the enzyme's active site, the Cu^{2+} ion is coordinated to two histidine residues that function to hold the metal cofactor in an optimal position for catalyzing the reduction of the substrate nitrite (NO_2^-) to form the product, nitric oxide. BASED ON PDB FILE 1SJM.

Enzyme cofactors with organic components are called **coenzymes**. Coenzyme molecules can be loosely associated with the enzymes or very tightly bound, even covalently attached to the enzyme. Coenzymes provide additional chemical flexibility for facilitating the catalytic reaction, which the protein component of the enzyme may be unequipped to do. Coenzymes that are permanently associated with enzymes, such as the heme group of catalase, are called **prosthetic groups**.

Coenzymes are usually derived from vitamins and were first discovered as biomolecules required for health. One of the coenzymes we will encounter frequently in metabolism

Table 7.2 REPRESENTATIVE COENZYMES AND THEIR VITAMIN SOURCES

Coenzyme	Vitamin source	Representative enzyme	Role in catalysis
Nicotinamide adenine dinucleotide (NAD^+/NADH)	Vitamin B_3 (niacin)	Lactate dehydrogenase	Oxidation–reduction
Flavin adenine dinucleotide (FAD/$FADH_2$)	Vitamin B_2	Succinate dehydrogenase	Oxidation–reduction
Thiamine pyrophosphate (TPP)	Vitamin B_1	Pyruvate dehydrogenase	Aldehyde group transfer
Biotin	Vitamin B_7	Pyruvate carboxylase	Carboxylation
Coenzyme A (CoA)	Pantothenic acid	Acetyl-CoA carboxylase	Acyl group transfer
Tetrahydrofolate (THF)	Folate	Thymidylate synthase	Single-carbon transfer
Pyridoxal phosphate (PLP)	Vitamin B_6	Aspartate aminotransferase	Amine group transfer
Lipoamide	Lipoic acid	Pyruvate dehydrogenase	Two-carbon transfer
Cobalamin	Vitamin B_{12}	Methylmalonyl-CoA mutase	Alkyl group transfer

is **nicotinamide adenine dinucleotide (NAD^+/NADH)**, which is derived from the vitamin niacin and is a required component in many redox reactions involving dehydrogenase enzymes. Another important coenzyme is **thiamine pyrophosphate**, which is derived from vitamin B_1. A deficiency of thiamine pyrophosphate in the diet results in the human disease beriberi, characterized by neurologic disorders. Thiamine pyrophosphate is a required coenzyme in decarboxylation reactions, such as that catalyzed by the pyruvate dehydrogenase complex (see Chapter 10). Some common coenzymes and their vitamin sources are listed in **Table 7.2**.

Lactate dehydrogenase is an important enzyme in cells. It converts pyruvate to lactate during anaerobic respiration under low O_2 conditions or in cells lacking mitochondria, such as erythrocytes. The reduction of pyruvate to form lactate is a typical metabolic redox reaction in which the coenzyme NADH is concomitantly oxidized to generate NAD^+. In this case, NADH binds to the enzyme near the active site and functions as a transient electron carrier that donates a pair of electrons to the redox reaction, resulting in reduction of the enzyme substrate (**Figure 7.8**). The NAD^+ product formed by the lactate dehydrogenase reaction is used by other enzymatic reactions

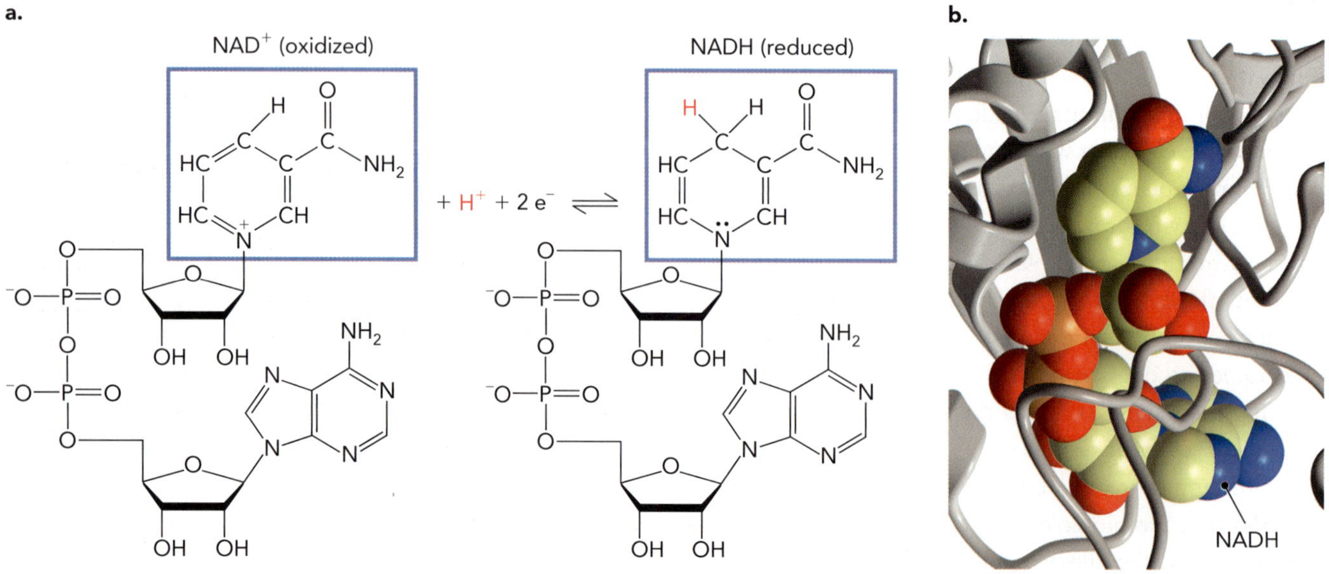

Figure 7.8 Nicotinamide adenine dinucleotide is a coenzyme in the lactate dehydrogenase reaction. **a.** Chemical structure of nicotinamide adenine dinucleotide in the oxidized (NAD^+) and reduced (NADH) forms. **b.** Molecular structure of part of the lactate dehydrogenase enzyme from the organism *Plasmodium falciparum,* with an NADH molecule bound to the active site. BASED ON PDB FILE 1TC2.

in the cell as a coenzyme that accepts a pair of electrons in redox reactions that oxidize enzyme substrates. NAD^+ and NADH are sometimes called co-substrates because they are altered in the course of the reaction through the donation or acceptance of electrons. For another round of catalysis to take place, they need to be regenerated to their original state via another reaction.

Some coenzymes are covalently linked to amino acid functional groups in enzymes and serve as integral components in the catalytic reaction. One example of this is the attachment of lipoic acid to a specific lysine residue in enzymes that catalyze redox and acyl transfer reactions. The oxidized form of this coenzyme is called **lipoamide**, and the reduced form is dihydrolipoamide (**Figure 7.9**). In the pyruvate dehydrogenase complex, which requires three different enzyme subunits in a multistep reaction to decarboxylate pyruvate to form acetyl-coenzyme A (acetyl-CoA), the dihydrolipoyl transacetylase enzyme performs the acyl transfer portion of this reaction with the lipoyl group. (In Chapter 10, we will look at this reaction in some detail.) Remarkably, this highly favorable reaction (essentially irreversible) also requires the participation of three additional coenzymes (flavin adenine dinucleotide, coenzyme A, and thiamine pyrophosphate), all of which are transiently associated with the pyruvate dehydrogenase enzyme complex.

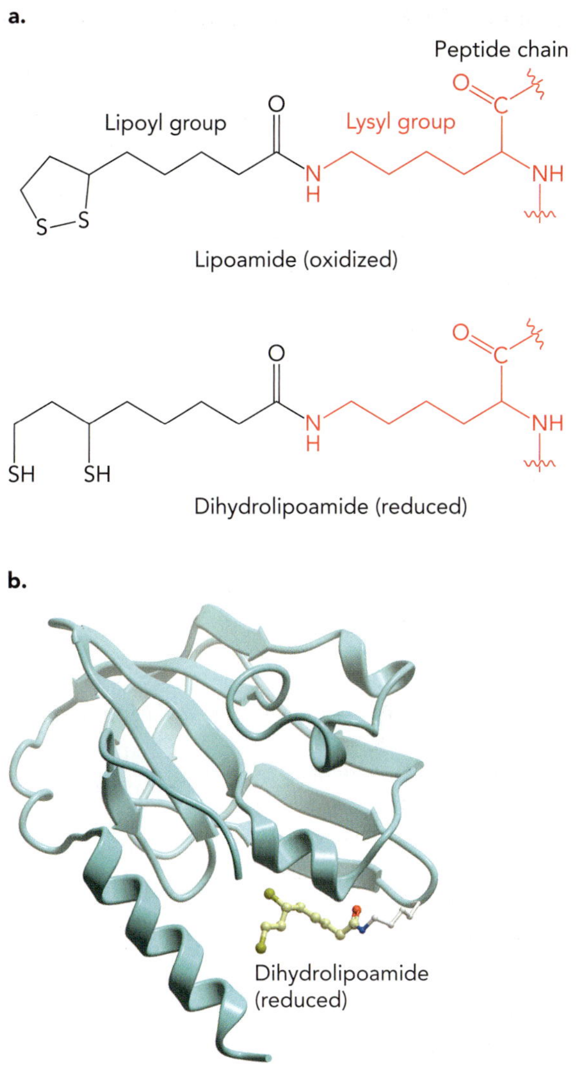

Figure 7.9 Lipoamide is a covalently attached coenzyme that plays a key role in several decarboxylase reactions. **a.** A lipoyl group is attached to the amino group of lysine residues in enzymes, shown here in both the oxidized (lipoamide) and reduced (dihydrolipoamide) forms. **b.** The pea glycine carboxylase enzyme contains a lipoamide coenzyme that protrudes into the enzyme active site. This protein structure contains the reduced form of lipoamide, called dihydrolipoamide. BASED ON PDB FILE 1DXM.

Enzyme Nomenclature

The metabolism of glucose to pyruvate by enzymes in the glycolytic pathway—an important energy conversion pathway in cells—requires 10 enzymatic reactions. Because each reaction step involves a different enzyme, in this one pathway you need to become familiar with 10 different enzyme names. Fortunately, biochemists have developed a systematic nomenclature to name enzymes on the basis of their characterized function.

Most proteins that function as enzymes end with the suffix "-ase" to denote that the protein is an enzyme. In addition, the substrate, or a description of the biochemical function of the enzyme, is usually included in the name. For example, the common name for the protein that converts urea to ammonia and carbon dioxide is *urease*. The term *hexokinase* refers to an enzyme that phosphorylates hexose sugars (a *kinase* is a phosphoryl-transferring enzyme). Not all common names are that useful; for example, the enzyme name *catalase* doesn't reveal anything about hydrogen peroxide decomposition except to signify that this protein is a *catalyst*—like all enzymes.

To improve the usefulness of enzyme nomenclature, a functional classification system has been adopted by the International Union of Biochemistry and Molecular Biology (IUBMB). This classification system is based on six classes of enzymatic reactions, as listed in **Table 7.3**. Each of these six enzyme classes has been further subdivided into subclasses and sub-subclasses. Strictly speaking, the IUBMB system does not distinguish between proteins, but rather between enzymatic reactions. This is apparent from the fact that evolutionarily related proteins from different organisms, which catalyze the same chemical reaction in their respective species, have the same IUBMB number. Other databases have been developed specifically to name proteins, one of which is the Universal Protein Resource (UniProt).

Table 7.3 THE INTERNATIONAL UNION OF BIOCHEMISTRY AND MOLECULAR BIOLOGY ENZYME CLASSIFICATION SYSTEM

Number	Enzyme class	Type of reaction	Generic enzymes
1	Oxidoreductase	Oxidation–reduction, transfer of H or O atoms	Oxidases, dehydrogenases
2	Transferase	Transfer of functional groups; e.g., methyl, acyl, amino, phosphoryl	Kinases, transaminases
3	Hydrolase	Formation of two products by hydrolyzing a substrate	Peptidases, lipases
4	Lyase	Cleavage of C—C, C—O, C—N, and other bonds by means other than hydrolysis or oxidation	Decarboxylases, carboxylases
5	Isomerase	Intramolecular rearrangements, transfer of groups within molecules	Mutases, isomerases
6	Ligase	Formation of C—C, C—O, C—S, or C—N bonds using ATP cleavage	Synthetases

The IUBMB system provides both an official name (systematic name) and a classification number beginning with the letters EC (Enzyme Commission), as well as a method for keeping track of common names (alternative names). To see how the IUBMB system distinguishes between closely related enzyme activities, let's examine the IUBMB number of the enzyme hexokinase. This enzyme catalyzes the phosphorylation of glucose to form glucose-6-phosphate during the first step in glycolysis (see Chapter 9). The IUBMB classification number for hexokinase is EC 2.7.1.1. This EC number is based on classification "2" as a transferase, subclass "7" because it transfers a phosphoryl group, and sub-subclass "1" because the phosphoryl transfer involves an alcohol acceptor group on glucose. The last digit in EC 2.7.1.**1** denotes that it is the first enzyme activity named in this category.

The importance of the IUBMB classification number is illustrated by the example of an enzyme related to hexokinase, called glucokinase, which catalyzes a similar phosphotransferase reaction. However, in the case of glucokinase, its affinity for glucose is much lower than that of hexokinase, and it has a more limited substrate specificity (glucokinase cannot phosphorylate fructose). On the basis of the distinct biochemical properties of glucokinase compared to those of hexokinase, the IUBMB number for glucokinase is EC 2.7.1.**2**. Hexokinase and glucokinase have different amino acid sequences and are encoded by two distinct genes on different chromosomes in the human genome.

Although the IUBMB system is useful in clearing up confusion associated with using historical or "generic" enzyme names, it can become cumbersome when trying to distinguish between enzymes encoded by the same gene, but that differ in amino acid sequence due to alternative mRNA splicing or differential translational start sites. These distinctions are best sorted out by genomic and proteomic databases, as described in Part 5 of this book.

concept integration 7.1

How do enzymes function as biological catalysts?

Enzymes, like all catalysts, lower the activation energy ($\Delta G^{\ddagger}$) of a reaction without affecting the overall change in free energy (ΔG). Enzymes do this by providing an optimal environment for chemical catalysis, called the enzyme active site. Enzymes increase the rates of reactions so that the reactions will happen on a biologically appropriate timescale.

Enzyme functional groups help bind and orient the substrate(s), and in some cases chemical cofactors are used to provide functionality that may be lacking from amino acid side chains. Enzyme activity is regulated by cellular conditions, and enzymes usually have specific cognate substrates. These properties ensure that the appropriate reaction takes place at the appropriate time in the cell.

7.2 Enzyme Structure and Function

The formal definition of a catalyst is that it increases the rate of a chemical reaction without changing the chemical equilibrium. Moreover, a catalyst is not consumed by the reaction. Enzymes increase the rate of a reaction inside cells in three major ways:

1. *They stabilize the transition state*, and thus lower the activation barrier.
2. *They provide an alternative path for product formation*, which could involve formation of stable reaction intermediates that are covalently attached to the enzyme.
3. *They orient the substrates appropriately for the reaction to occur*, thus reducing the entropy change of the reaction.

Enzymes use all of these strategies to some extent—sometimes in combination for the same reaction—with the net result being an increased rate of reaction on a biological timescale.

Raising effective substrate concentrations or elevating the reaction temperature or pressure can also lead to increased reaction rates. In fact, industrial chemical reactions are often made more economical by simply manipulating the reaction conditions within the vessel. However, in nature, where time, space, pressure, and temperature are all rigidly constrained within a cell, enzymes must rely on natural selection to optimize molecular structure and function to achieve increased rates of reactions under physiologic conditions.

To understand how enzymes function as catalysts, it is first necessary to describe the general properties of enzyme active sites. Then we will use this information to examine some of the most common enzyme reaction mechanisms found in cells.

Physical and Chemical Properties of Enzyme Active Sites

Enzymes function as catalysts by providing a physical and chemical environment that promotes product formation. They do this by increasing the local concentration of substrates through selective binding—which also orients substrates in an optimal configuration for functional group interactions—and by providing reactive groups that can participate in the chemistry of the reaction itself.

Figure 7.10 illustrates a reaction in which two substrates form one product, A + B → C. The structure of the enzyme active site provides a geometric and chemical complementarity that favors the binding of substrates in a way that is both selective and productive (substrate reactive

Figure 7.10 The enzyme active site provides an optimal physical and chemical environment that promotes product formation. Random collisions between two substrates are often unproductive due to misalignment of reactive groups required for product formation. The enzyme active site contains binding sites to select substrates and align the reactive groups correctly. This is accomplished by multiple weak interactions through polar and nonpolar regions in the substrates and enzyme. Many substrates bind to enzymes using an induced-fit mechanism, in which the enzyme structure changes to accommodate substrate binding, though not too tightly or products won't be released. Chemical groups present in the enzyme or cofactors are in close proximity to the substrates. Product release frees up the enzyme to bind new substrate molecules.

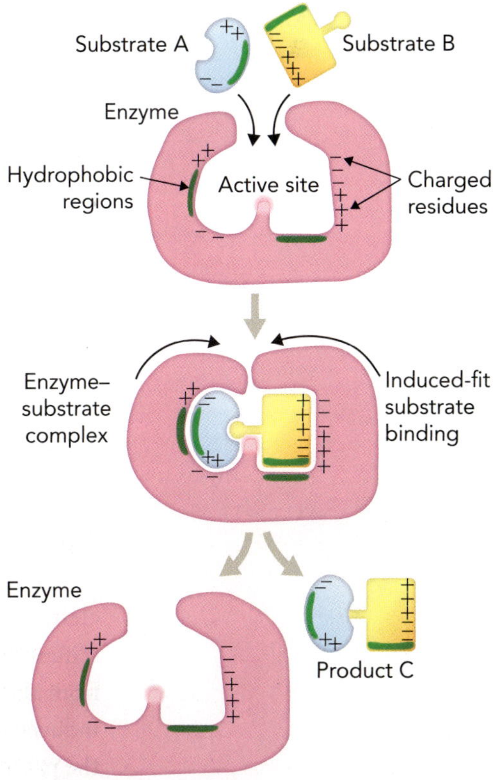

groups are in close proximity). Moreover, through the cumulative effect of multiple non-covalent interactions between the substrates and enzyme, substrates can have a relatively high binding affinity for the enzyme active site. It is important, however, that this binding not be too tight, especially between the enzyme and the product; otherwise, it will have a negative effect on reaction rates by interfering with product release and binding of new substrate molecules.

Let's now examine three specific physical and chemical properties of enzyme active sites that contribute to their catalytic properties: (1) the sequestered micro-environment of the active site; (2) binding interactions between the substrate and the enzyme that facilitate formation of the transition state; and (3) the presence of catalytic functional groups.

The Active Site Microenvironment The features of an enzyme active site that contribute to a decrease in the $\Delta G^{\ddagger}$ of the reaction are, for the most part, independent of the specific catalytic mechanism:

1. Enzyme active sites provide an optimal orientation of the substrate(s) relative to reactive chemical groups.
2. Enzyme active sites exclude excess solvent (H_2O) that can interfere with the reaction.

An example of the first feature, how enzymes provide an optimal spatial orientation of substrates within the active site, is the metabolic enzyme aldolase. The aldolase reaction in the gluconeogenic pathway converts two phosphorylated three-carbon compounds into one bisphosphorylated six-carbon compound. This reaction is highly favorable under standard conditions with a $\Delta G^{\circ\prime}$ of −23.8 kJ/mol:

$$\text{Glyceraldehyde-3-phosphate} + \text{Dihydroxyacetone phosphate} \rightleftharpoons \text{Fructose-1,6-bisphosphate}$$

Amino acid functional groups within the aldolase active site position the substrates in a way that favors aldol condensation. The reaction mechanism involves formation of an enzyme-linked covalent intermediate between a lysine residue in the active site and the substrate dihydroxyacetone phosphate (**Figure 7.11**). Once this intermediate forms, glyceraldehyde-3-phosphate binds to the active site. The condensation reaction is favored because of the proximity of glyceraldehyde-3-phosphate to both the reactive C-3 carbon in the intermediate and to amino acid residues in the enzyme active site, which participate in an acid–base chemical reaction mechanism (described in Section 7.3). After the reaction is complete, the fructose-1,6-bisphosphate product diffuses out of the active site. Note that this same aldolase reaction is favored in the opposite direction (aldol cleavage of fructose-1,6-bisphosphate to form glyceraldehyde-3-phosphate and dihydroxyacetone phosphate) when flux through the glycolytic pathway is high due to elevated levels of glucose in the cell, coupled with a need for ATP production (see Chapter 9). Thus, this same enzyme can catalyze the reaction in either direction, depending on conditions in the cell.

The second general feature of most enzyme active sites is their location in clefts on the protein surface or near the interior of the protein, which functions to sequester the substrates away from excess water. Most H_2O molecules are excluded from the enzyme active site in one of two ways. The first mechanism is through an induced-fit mechanism of substrate binding, in which conformational changes in the protein create significant shape and chemical complementarity between the substrate and enzyme. These conformational changes eject excess H_2O molecules from

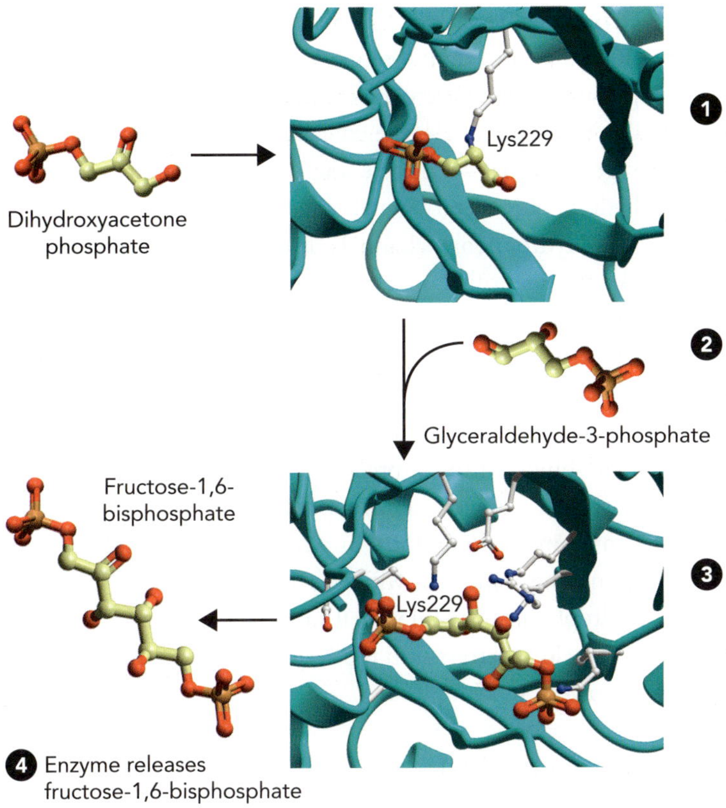
Dihydroxyacetone phosphate
Lys229
1 Dihydroxyacetone phosphate forms an intermediate with Lys229 in the enzyme active site
2 Glyceraldehyde-3-phosphate binds to the active site
Glyceraldehyde-3-phosphate
Fructose-1,6-bisphosphate
Lys229
3 A condensation reaction takes place in the active site, forming fructose-1,6-bisphosphate
4 Enzyme releases fructose-1,6-bisphosphate

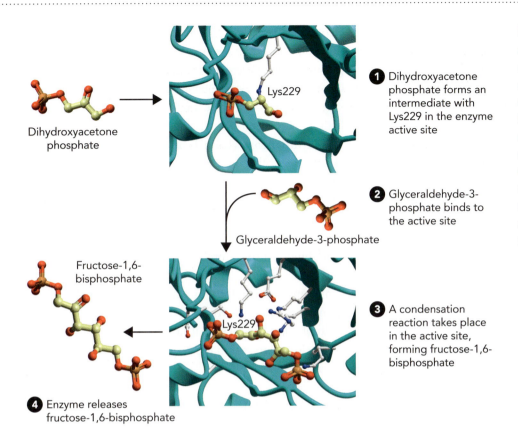

Figure 7.11 The aldolase reaction illustrates the importance of favorable spatial arrangements of substrates within the enzyme active site to promote a reaction. The formation of a covalent substrate–enzyme intermediate in the first step of this reaction contributes to the decreased $\Delta G^{\ddagger}$. BASED ON PDB FILES 1J4E (ALDOLASE WITH COVALENTLY BOUND DIHYDROXYACETONE PHOSPHATE) AND 4ALD (ALDOLASE WITH FRUCTOSE-1,6-BISPHOSPHATE IN THE ACTIVE SITE).

the active site. The binding of glucose by the enzyme hexokinase is a good example of this induced-fit mechanism (see Figure 7.3). A closer look at the hexokinase active site reveals that numerous H_2O molecules occupy the substrate binding site in the absence of glucose because this region of the enzyme is exposed to solvent (**Figure 7.12**). However, upon glucose binding, several polar residues within the active site are brought closer together and form hydrogen bonds with the glucose molecule instead of forming hydrogen bonds with H_2O. Thus, water is displaced from the active site. The importance of this substrate-induced conformational change in the active site is that it prevents phosphoryl transfer from ATP to H_2O, which would be a waste of metabolic energy. Instead, the terminal phosphate of ATP is used to convert glucose to glucose-6-phosphate.

The second mechanism used by enzymes to prevent excess H_2O from entering the active site is to sequester the active site away from the surface, with substrate accessibility controlled by a gated hydrophobic channel. The hydrophobic channel effectively limits H_2O entry to the active site, while still allowing access by hydrophobic substrates. One way to determine how hydrophobic substrates gain access to a buried

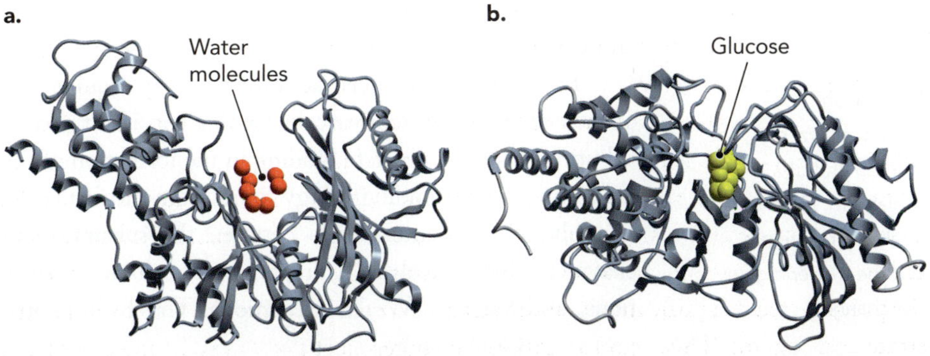
a.
Water molecules
b.
Glucose

Figure 7.12 The enzyme active site of hexokinase excludes H_2O molecules upon glucose binding, thereby preventing nonproductive phosphoryl transfer from ATP to H_2O. **a.** In the absence of glucose, H_2O molecules (shown as red spheres) occupy the solvent-exposed active site. BASED ON PDB FILE 1IG8. **b.** A conformational change is induced upon glucose binding at the active site. Glucose replaces water in the active site by occupying a similar volume and using similar weak interactions for binding. Therefore, the H_2O molecules are expelled from the active site. BASED ON PDB FILE 3B8A.

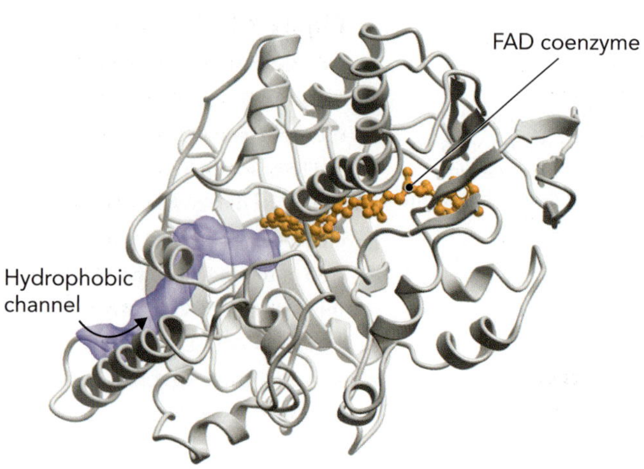

Figure 7.13 Hydrophobic substrate channels prevent H_2O molecules from entering a buried enzyme active site. The structure of a bacterial fatty acid isomerase enzyme is shown with a flavin adenine dinucleotide (FAD) coenzyme bound in the internal active site. The hydrophobic substrate channel is highlighted by the surface outline of a C_{24} PEG molecule. The arrow shows the likely entry pathway of a fatty acid substrate. BASED ON PDB FILE 2B9W.

enzyme active site is to run computer modeling simulations that use energy minimization calculations to predict possible substrate entry and exit channels. This has been done for several P450 enzymes, which are heme monooxygenases that synthesize and degrade a variety of mostly hydrophobic dietary metabolites.

Hydrophobic channels leading to active sites can also be identified in a structure by mixing the enzyme with linear hydrocarbon chains of various lengths, usually **polyethylene glycol (PEG)** derivatives. The PEG binding site in the structure is resolved using X-ray crystallography. As shown in **Figure 7.13**, a portion of a C_{24} PEG molecule binds within a predicted hydrophobic channel of a fatty acid isomerase enzyme that connects the active site, containing a bound flavin adenine dinucleotide (FAD) coenzyme molecule, with the protein surface. This long hydrophobic channel is well suited for entry of the normal fatty acid substrate into the active site but is not hospitable to solvent H_2O molecules.

Stabilization of the Transition State Now that we have seen how the enzyme active site functions, we can study how stabilization of the transition state within the active site leads to increased reaction rates of enzyme-catalyzed reactions. We first consider the equilibrium reactions that define substrate (S) binding to the enzyme (E) and releasing product (P). In this three-step process, substrate binding to the enzyme (E + S) leads to the formation of an enzyme–substrate complex (ES). This is followed by conversion of the enzyme-bound substrate to an enzyme-bound product (EP). Finally, the product is released from the enzyme (E + P):

$$\mathrm{E + S} \rightleftharpoons \mathrm{ES} \rightleftharpoons \mathrm{EP} \rightleftharpoons \mathrm{E + P}$$

The reaction coordinate diagram in **Figure 7.14** illustrates the effect of the enzyme on reducing the activation energy ($\Delta G^{\ddagger}$) for the enzyme-catalyzed reaction compared to that for the uncatalyzed reaction. Again, note that the overall free energy of the reaction (ΔG) does not change in the presence of the enzyme, and that the path taken by the enzyme-catalyzed reaction involves the formation of ES and EP complexes.

To explain the reduced activation energy requirement of the enzyme-catalyzed reaction, we first consider the contribution made by substrate binding in the formation of the ES complex. This binding energy comes from the formation of multiple weak interactions between the substrate and enzyme, which involves hydrogen bonds, ionic interactions, and van der Waals interactions. The hydrophobic

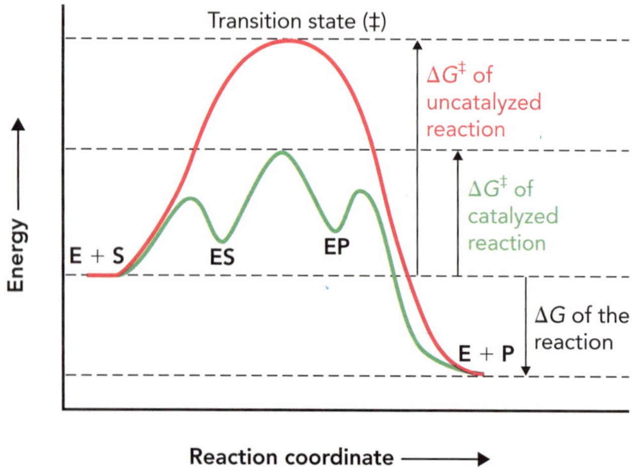

Figure 7.14 In this reaction coordinate diagram, an enzyme-catalyzed reaction is compared to the uncatalyzed reaction. In the enzyme-catalyzed reaction, the ES and EP intermediates are local minima in the energy diagram. Note that the activation energy for the enzyme-catalyzed reaction is less than that for the uncatalyzed reaction, but the free energy change (ΔG) for the reaction $S \rightarrow P$ does not change.

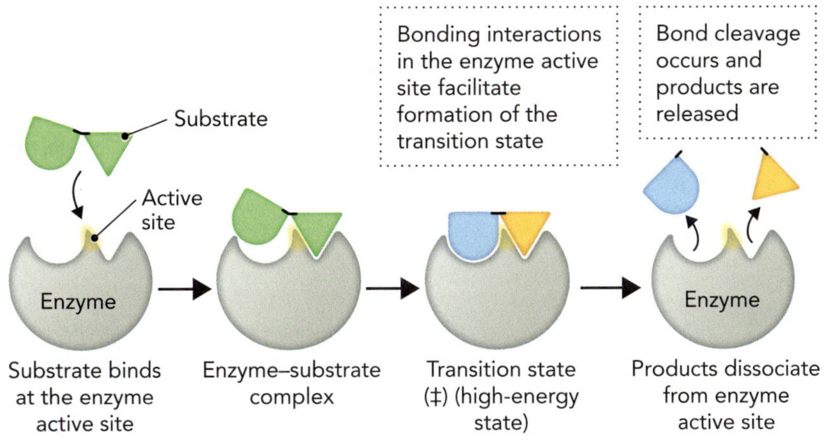
Bonding interactions in the enzyme active site facilitate formation of the transition state
Bond cleavage occurs and products are released
Substrate
Active site
Enzyme
Enzyme
Substrate binds at the enzyme active site
Enzyme–substrate complex
Transition state (‡) (high-energy state)
Products dissociate from enzyme active site

Figure 7.15 The enzyme active site facilitates formation of the transition state. The substrate binds at the active site of the enzyme. Conformational changes in the substrate aided by bonding interactions in the active site result in the formation of the transition state. Products are released after bond cleavage.

effect also plays an important role because of an increase in the entropy of H_2O molecules that were released from either the enzyme or substrate upon substrate binding. However, substrate binding is only part of the explanation for a reduced $\Delta G^{\ddagger}$ value.

The next step, the conversion of ES $\rightleftharpoons$ EP, requires the breaking and forming of bonds that distinguish the substrate from the product. If the enzyme active site shares more complementarity with the substrate than with the product (as first proposed in the lock and key model of enzyme action), then it would favor the substrate within the active site and limit the ES $\rightleftharpoons$ EP reaction. Alternatively, if amino acid residues within the enzyme active site make the most contacts (best geometric and chemical complementarity) during the transition state of the reaction, then breaking of bonds during the reaction requires no direct input of energy by the protein. This proposal is called the transition state stabilization model.

One way to think about how binding energies and transition state stabilization work together to lower the activation energy of a reaction is to imagine that formation of the ES complex initiates a series of bond formations that accumulate as the reaction proceeds toward the transition state (the midpoint of the ES $\rightleftharpoons$ EP reaction). The gain in free energy resulting from bond formation between the substrate and the enzyme is used to lower the activation energy required to reach the transition state (**Figure 7.15**).

The transition state stabilization model also explains why product release occurs at the end of the reaction (EP $\rightleftharpoons$ E + P); namely, because the enzyme active site has a higher affinity for the transition state than for either the product or the substrate.

Support for the transition state stabilization model comes from analyzing the binding of stable molecules that mimic the proposed transition state, called **transition state analogs**, to enzyme active sites. **Figure 7.16** shows an example of this idea. Adenosine deaminase is a critical enzyme involved in purine degradation (see Chapter 18). This enzyme deaminates adenosine to form the purine inosine, a process that is proposed to occur through direct addition of water to the purine ring resulting in a tetrahedral transition state. The transition state analog, 6-hydroxy-1,6-dihydropurine ribonucleoside, has a similar tetrahedral arrangement at position 6 of the purine ring, and thus binds very tightly to the enzyme. In fact, the affinity of the enzyme for this transition state analog is estimated to be 10^{-13} M, which exceeds the affinity of the enzyme for substrate or product by approximately a factor of 10^8. As discussed later in this chapter, nonhydrolyzable transition state analogs are often very effective enzyme inhibitors.

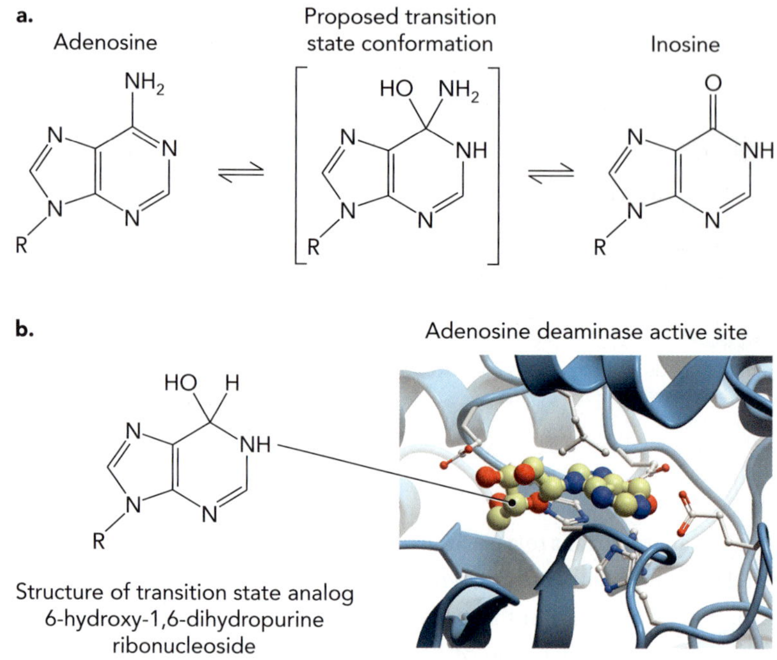

Figure 7.16 Tight binding of transition state analogs to enzyme active sites supports the transition state stabilization model. **a.** Proposed adenosine deaminase reaction, showing the predicted transition state conformation. **b.** Molecular structure of the mouse adenosine deaminase enzyme, with a transition state analog bound to the active site. Note the similar structures of the transition state analog 6-hydroxy-1,6-dihydropurine ribonucleoside and the proposed transition state shown in panel **a.** BASED ON PDB FILE 1KRM.

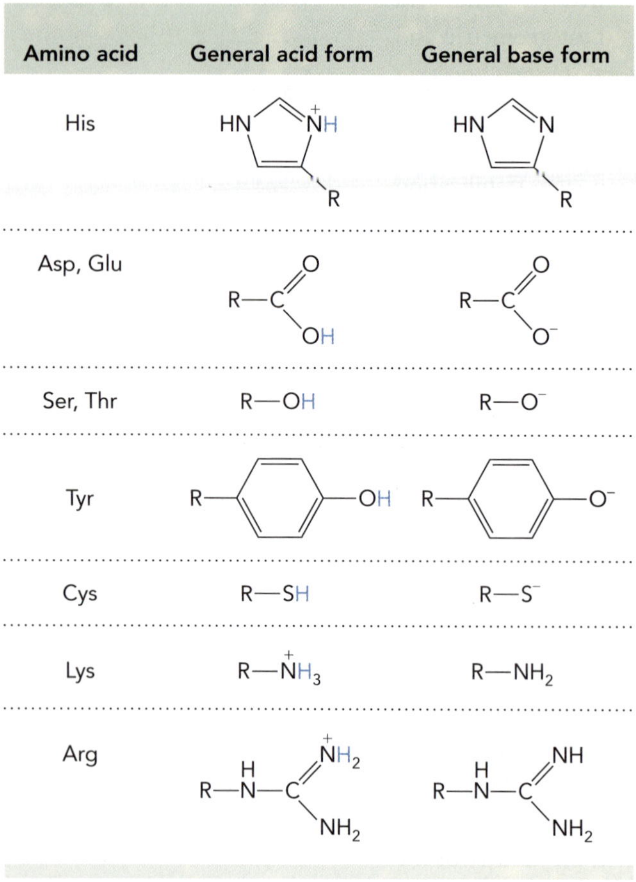

Amino acid
General acid form
General base form
His
Asp, Glu
Ser, Thr
Tyr
Cys
Lys
Arg

Figure 7.17 Amino acid groups commonly involved in enzyme-mediated acid–base catalysis. The acid form is the proton donor (shown in blue), and the base form is the proton acceptor.

Presence of Catalytic Functional Groups Once substrates are bound to the protected enzyme active site away from solvent, often through a mechanism that involves stabilization of the transition state, it is the job of nearby reactive chemical groups on amino acids or coenzymes to catalyze the reaction. The three most common catalytic reaction mechanisms in the enzyme active site are (1) acid–base catalysis; (2) covalent catalysis; and (3) metal-ion catalysis.

Acid–base catalysis. Many enzyme reactions involve proton transfer through acid–base mechanisms that involve addition or removal of a proton. Two types of acid–base catalysis can occur in enzyme reactions: (1) specific acid–base catalysis, which involves water, or (2) general acid–base catalysis, in which the proton transfer involves a functional group. General acid catalysis refers to the donation of a proton by the enzyme, whereas general base catalysis refers to the removal of a proton by the enzyme. **Figure 7.17** shows amino acid residues that commonly function in general acid–base catalysis.

The histidine side chain is often involved in enzyme-mediated acid–base catalysis because the imidazole ring has a pK_a near ~7 in most proteins and can therefore be found in both the protonated and deprotonated states. This means that depending on the local charge distribution in the enzyme active site, histidine can function as either a proton donor or a proton acceptor at physiologic pH. **Figure 7.18** illustrates the role of two histidine residues (His12 and His119) within the active site of the enzyme pancreatic ribonuclease. The residues are involved in promoting the cleavage of an RNA substrate molecule through a general acid–base catalysis mechanism.

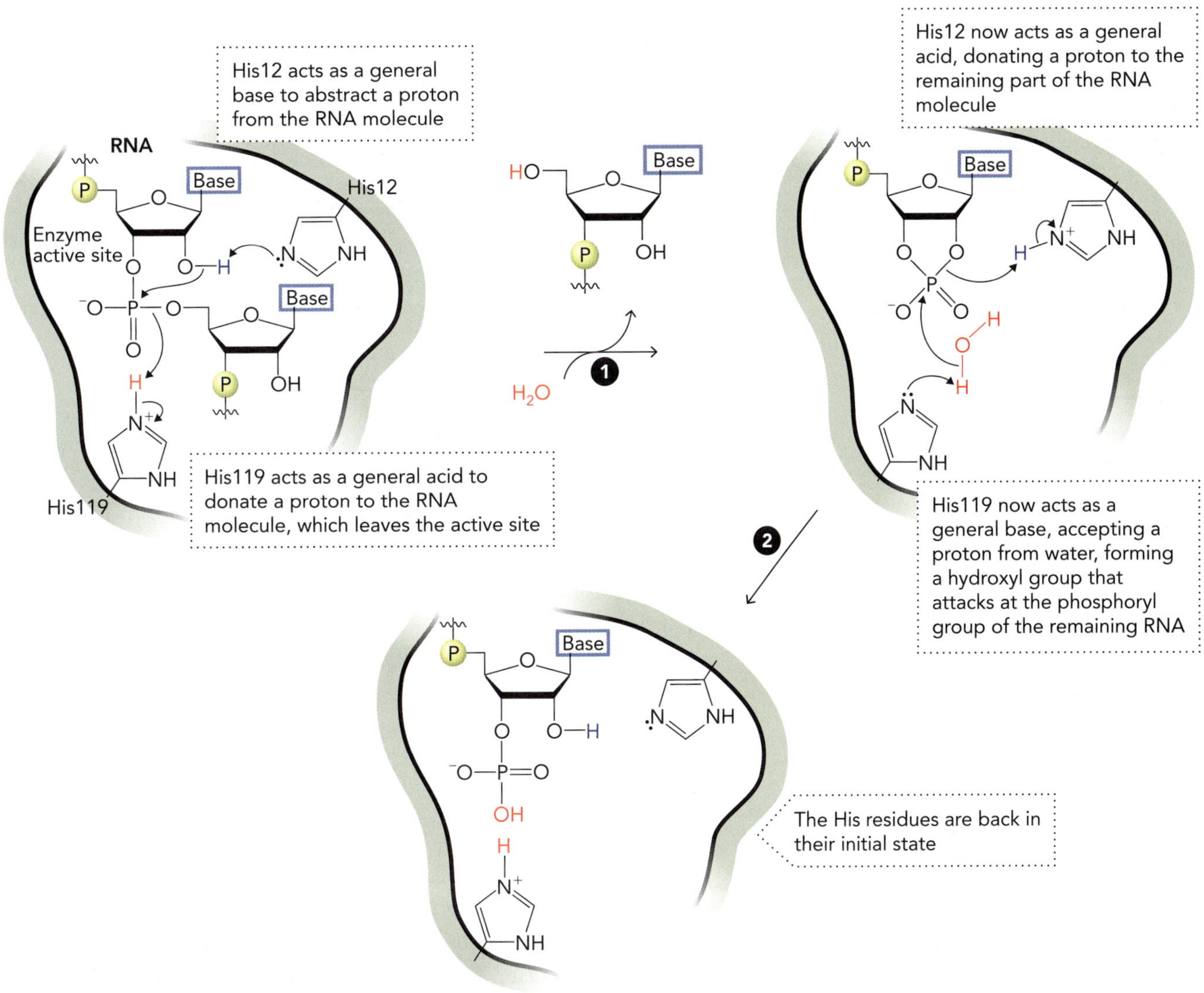

His12 acts as a general base to abstract a proton from the RNA molecule
RNA
Enzyme active site
Base
His12
P
OH
His119
His119 acts as a general acid to donate a proton to the RNA molecule, which leaves the active site
HO
H_2O
1
His12 now acts as a general acid, donating a proton to the remaining part of the RNA molecule
His119 now acts as a general base, accepting a proton from water, forming a hydroxyl group that attacks at the phosphoryl group of the remaining RNA
2
The His residues are back in their initial state

Figure 7.18 The mechanism of RNA cleavage by pancreatic ribonuclease is shown here. Two histidine residues function in a general acid–base catalysis mechanism that requires the addition of H_2O.

Covalent catalysis. Formation of a transient covalent bond between the substrate and the enzyme can be used to create an unstable intermediate that promotes the catalytic reaction. Usually a nucleophilic (electron-rich) group on the enzyme attacks an electrophilic (electron-deficient) center on the substrate to form a covalent enzyme–substrate intermediate. Consider the following two-step reaction:

Step 1	A–X + Enzyme → Enzyme–X + A
Step 2	Enzyme–X + Y → X–Y + Enzyme
Net reaction	Enzyme + A–X + Y → Enzyme + X–Y + A

In this case, Y can be a second substrate (or H_2O) that reacts more strongly with the unstable enzyme–X intermediate than it would with A–X. In other words, the enzyme uses two faster reactions instead of one slower reaction to achieve an overall increased rate. Note that the active enzyme is regenerated after the formation of the X–Y product.

One example of covalent catalysis is the formation of 1,3-bisphosphoglycerate from glyceraldehyde-3-phosphate by the glycolytic enzyme glyceraldehyde-3-phosphate dehydrogenase. This important reaction in carbohydrate metabolism illustrates the central role of the coenzyme NAD^+ in mediating product formation. Substrate binding initiates a nucleophilic attack by a sulfhydryl group in the enzyme on

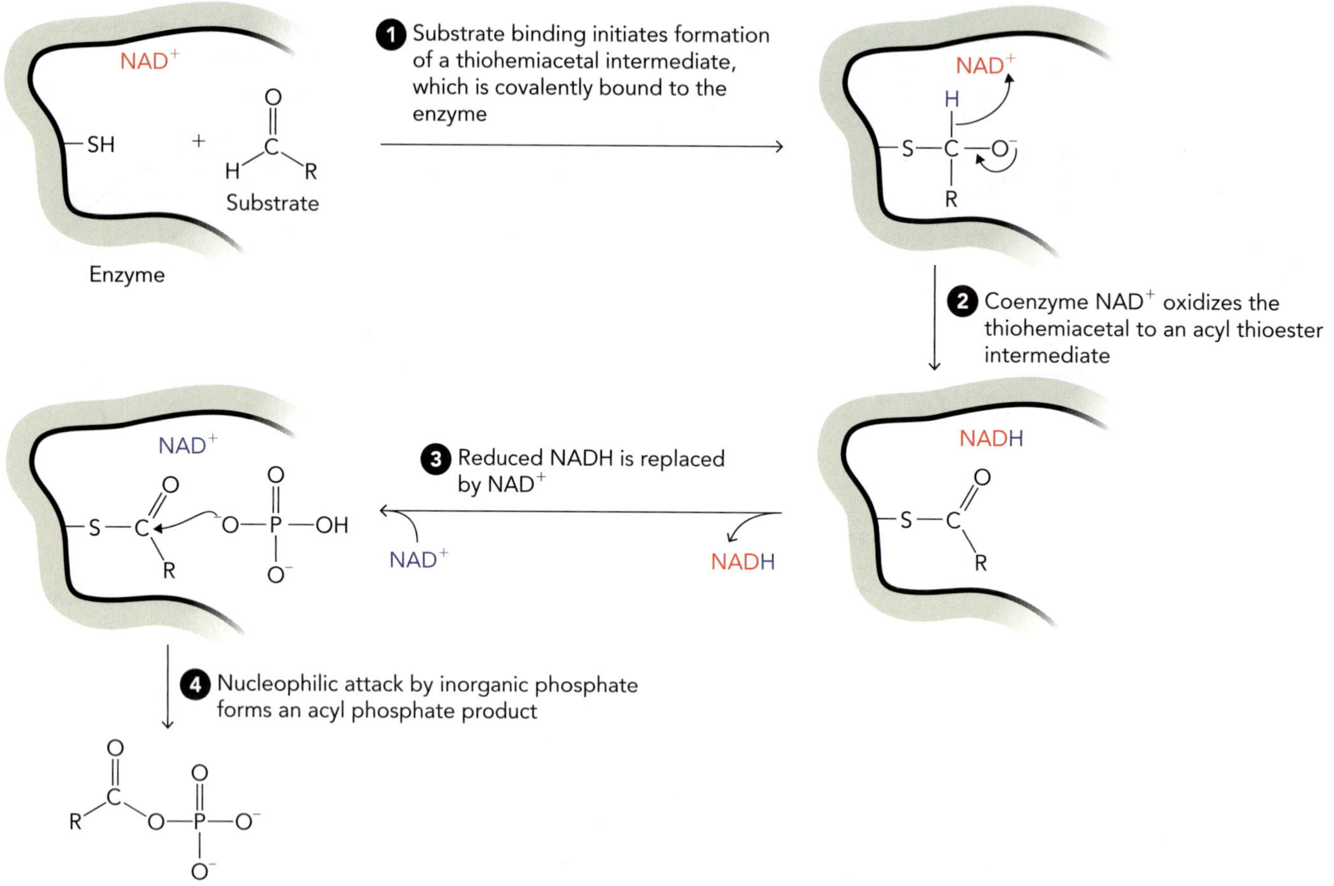

Figure 7.19 The formation of 1,3-bisphosphoglycerate from glyceraldehyde-3-phosphate by the glycolytic enzyme glyceraldehyde-3-phosphate dehydrogenase is an example of covalent catalysis involving the coenzyme NAD^+. The sulfhydryl group from a cysteine residue of the enzyme forms a covalent bond with glyceraldehyde-3-phosphate to form the thiohemiacetal intermediate.

the substrate, forming a thiohemiacetal enzyme intermediate. As shown in **Figure 7.19**, the coenzyme NAD^+ functions as a co-substrate and oxidizes the thiohemiacetal to form an acyl thioester intermediate. The reduced coenzyme NADH is then replaced by NAD^+ in the active site, and inorganic phosphate (HPO_4^{2-}) attacks the thioester intermediate, resulting in the formation of the acyl phosphate product. We will look at this reaction in more detail in Chapter 9 when we discuss glycolytic enzymes.

Metal-ion catalysis. A large number of enzymes require metal ions to function as enzyme cofactors in the catalytic reaction (see Table 7.1). Positively charged metal ions function to promote proper orientation of bound substrates and to shield or stabilize negative charges through electrostatic interactions. Some metal ions mediate redox reactions through reversible changes in oxidation state. Enzymes with loosely bound metal ions are called metal-activated enzymes, whereas enzymes with tightly bound metal ions are called **metalloenzymes**. Metal-activated enzymes bind alkali and alkaline earth metals present in solution, most often Mg^{2+}, Ca^{2+}, Na^+, and K^+. Metalloenzymes usually contain Zn^{2+}, Fe^{2+}, Fe^{3+}, Mn^{2+}, Cu^{2+}, or Co^{2+}.

Carbonic anhydrase, an important enzyme in gas exchange and blood pH balance, is a metalloenzyme that uses a Zn^{2+} ion to catalyze the reversible reaction of bicarbonate formation from carbon dioxide, as shown in **Figure 7.20**. Through metal ion coordination bonds, the Zn^{2+} lowers the pK_a of water and facilitates formation of a nucleophilic hydroxyl group (OH^-) that attacks the substrate (CO_2). The active enzyme is regenerated by the binding of another H_2O molecule. The molecular structure of carbonic anhydrase reveals the positions of three conserved histidine residues in the substrate binding pocket that help coordinate the Zn^{2+} ion (**Figure 7.21**).

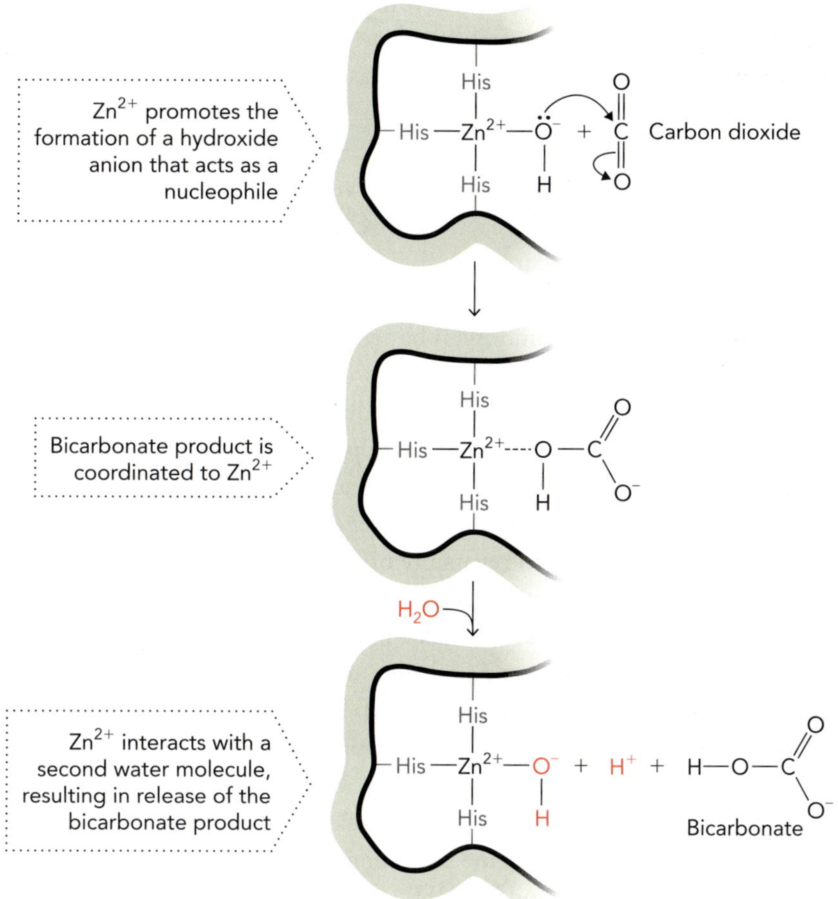

Zn2+ promotes the formation of a hydroxide anion that acts as a nucleophile
His
His—Zn2+—O⁻ + C
His H
Carbon dioxide
Bicarbonate product is coordinated to Zn2+
H2O
Zn2+ interacts with a second water molecule, resulting in release of the bicarbonate product
H+
H—O—C
Bicarbonate

Figure 7.20 The carbonic anhydrase reaction uses a coordinated Zn^{2+} ion as a catalytic group that supports the nucleophilic attack on the CO_2 substrate, yielding the bicarbonate product.

Enzymes Perform Work in the Cell

Enzymes are essential to life because they enable chemical reactions in cells to occur on a useful timescale and under physiologic conditions characterized by neutral pH (pH 7–8), low pressure (1 atm), and relatively low temperature (25–40 °C). The thousands of enzyme-mediated reactions that take place inside cells can be grouped into three general categories on the basis of the work they accomplish:

- *Coenzyme-dependent redox reactions* are responsible for much of the energy conversion that takes place in cells. These essential redox reactions are central components of three major metabolic pathways described later in the book: (1) the citrate cycle, (2) the electron transport system, and (3) the photosynthetic light reactions.
- *Metabolite transformation reactions* are essential steps in all biosynthetic (anabolic) and degradative (catabolic) pathways. These reactions include a variety of isomerization and condensation reactions, as well as hydrolysis and dehydration reactions.
- *Reversible covalent modification reactions* attach or remove molecular tags to biomolecules to alter their recognition properties and regulate cellular processes. Two of the most common

Figure 7.21 Human carbonic anhydrase uses three histidine residues to coordinate the Zn^{2+} ion (blue sphere) in the active site. BASED ON PDB FILE 5CAC.

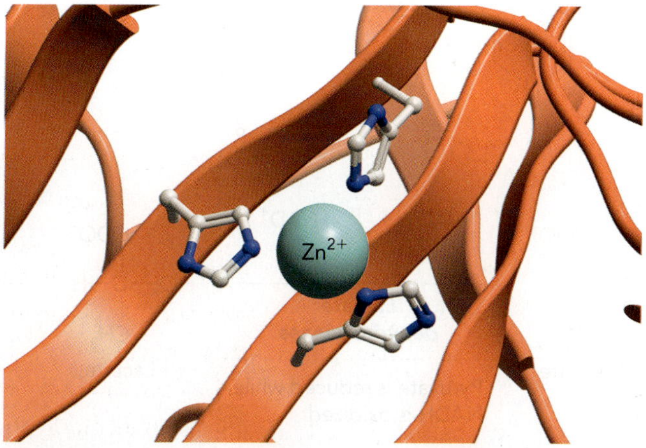

Zn2+

of these reversible reactions in eukaryotic cells are the phosphorylation and dephosphorylation of signaling molecules and the methylation and demethylation of DNA at cytosine bases in the regulatory regions of genes.

Coenzyme-Dependent Redox Reactions We first introduced biochemical oxidation and reduction (redox) reactions in Chapter 2, when we discussed principles of bioenergetics (see Figure 2.3). Simply put, oxidation is a loss of electrons and reduction is a gain of electrons. Moreover, when one compound is oxidized by an enzyme, another compound must be reduced, meaning that redox reactions are coupled. The reason is that free electrons do not exist stably in solution, but instead pass from compounds that have a tendency to be electron donors to those that function as electron acceptors. Most important, the physical and chemical properties of amino acid functional groups located within the enzyme active site can alter the redox potential of bound substrates. However, common amino acid side chains, which are unable to reversibly bind or accept electrons, are not sufficient to catalyze the redox reaction, and coenzymes are required.

In general, redox reactions that take place at C—O bonds use the coenzymes NAD^+/NADH (nicotinamide adenine dinucleotide) and **$NADP^+$/NADPH (nicotinamide adenine dinucleotide phosphate)**. Redox reactions at C—C bonds are usually mediated by the coenzymes **flavin adenine dinucleotide (FAD/$FADH_2$)** and **flavin mononucleotide (FMN/$FMNH_2$)**. Redox reactions requiring these coenzymes involve the transfer of 2 e^-, either as a pair of electrons in the case of NADH and NADPH or 1 e^- at a time through the formation of semiquinone intermediates, as in the free-radical forms $FADH^•$ and $FMNH^•$.

Figure 7.22 illustrates two enzymatic reactions that use coenzymes as electron carriers during redox reactions that are involved directly or indirectly in energy conversion. The enzyme lactate dehydrogenase catalyzes a redox reaction that interconverts pyruvate, the product of the glycolytic pathway, and lactate, a metabolite that accumulates during anaerobic respiration. The reaction requires the conjugated redox pair NAD^+/NADH. In the direction shown, pyruvate is reduced to form lactate, while at the same time, NADH is oxidized to generate NAD^+. Similarly, the citrate cycle

Figure 7.22 Enzyme-mediated redox reactions involved in energy conversion often require coenzymes, which function as electron carriers. **a.** Lactate dehydrogenase uses the coenzyme NADH to form lactate. NADH is oxidized while pyruvate is reduced. **b.** The enzyme succinate dehydrogenase forms fumarate in a redox reaction requiring the coenzyme FAD. FAD is reduced while succinate is oxidized.

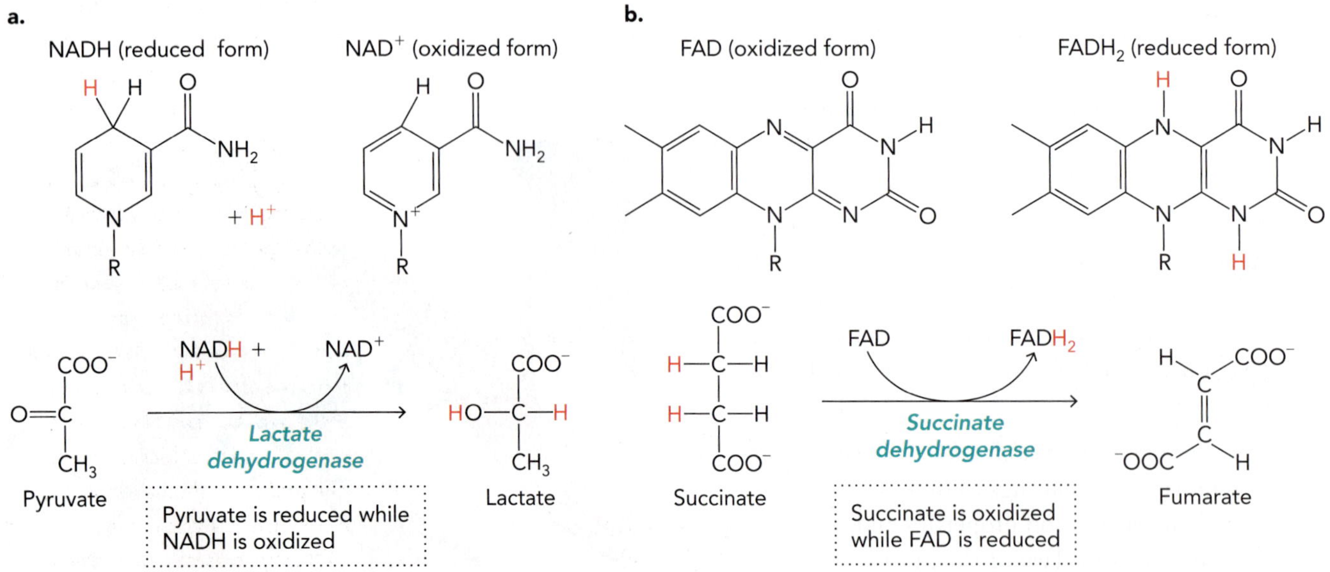

enzyme succinate dehydrogenase uses the conjugated redox pair FAD/$FADH_2$ to interconvert succinate and fumarate. In the citrate cycle, the oxidation of succinate forms fumarate, while the reduction of FAD generates $FADH_2$. Note that each reaction involves the transfer of a pair of electrons.

Metabolite Transformation Reactions An important biochemical reaction in cells is the chemical transformation of metabolites to generate reactive intermediates, which are necessary components of anabolic and catabolic pathways. The variety of enzyme-mediated reactions in this category reflects the chemical diversity of organic compounds, but three types of these reactions are the most common in metabolic pathways:

1. **Isomerization** reactions do not change the molecular formula of the product compared to that of the substrate.
2. **Condensation** reactions combine two substrates to form a larger molecule, with the loss of a smaller molecule, usually water.
3. **Hydrolysis** or **dehydration** reactions cleave a substrate to two products by the addition or removal of water.

For example, **Figure 7.23a** illustrates the isomerization of dihydroxyacetone phosphate to form glyceraldehyde-3-phosphate in a reaction catalyzed by the glycolytic enzyme triose phosphate isomerase. The molecular formula of these two metabolites, $C_3H_7O_6P$, is identical, and under physiologic conditions in the cell, this isomerization reaction is readily reversible.

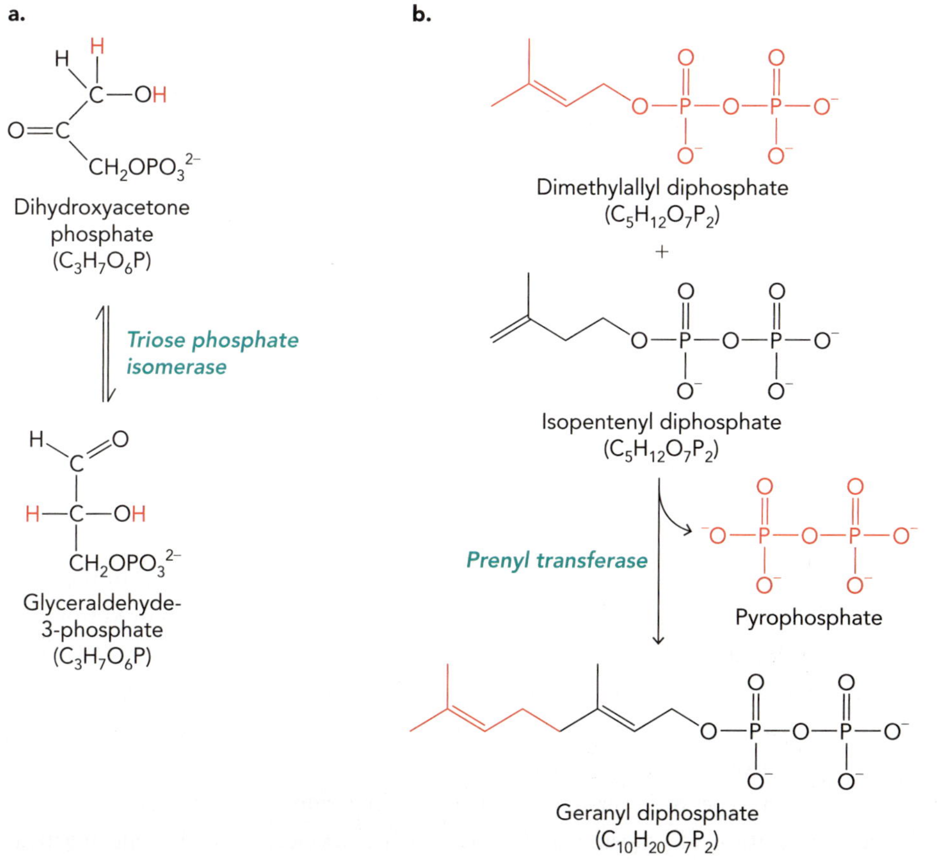

Figure 7.23 Metabolite transformation reactions are essential components of anabolic and catabolic pathways. **a.** Triose phosphate isomerase catalyzes an isomerization reaction that interconverts the phosphorylated C_3 metabolites dihydroxyacetone phosphate and glyceraldehyde-3-phosphate. Isomerization reactions generate a product that has the same molecular formula as the substrate. **b.** Condensation reactions in metabolic pathways are often characterized by the combination of two substrates with the same molecular formula, which function as building blocks for the synthesis of larger biomolecules. The prenyl transferase reaction shown here is a step in the cholesterol biosynthetic pathway.

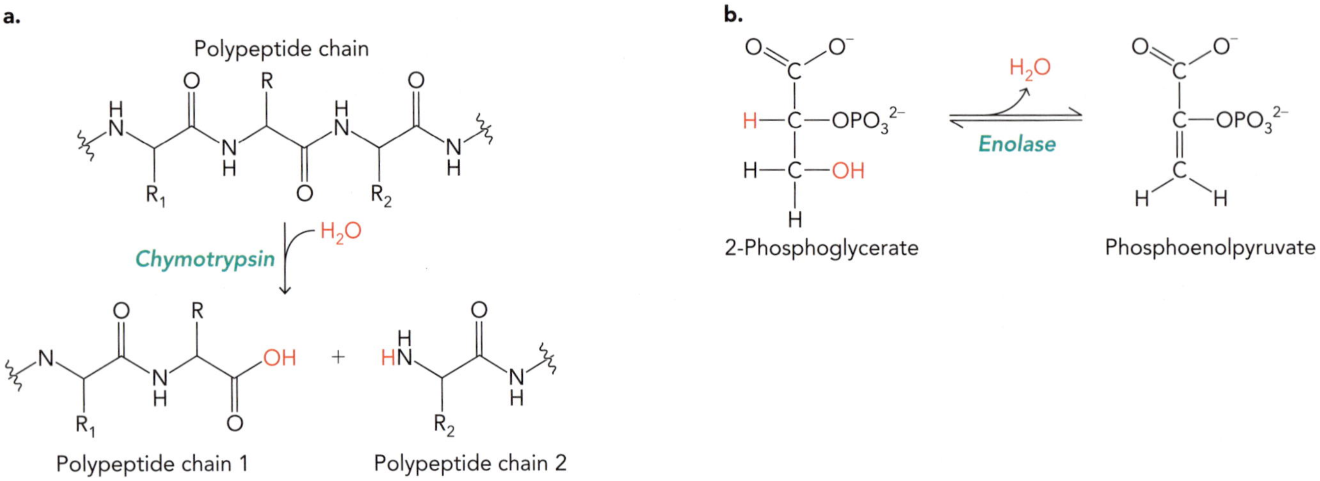

Figure 7.24 Numerous reactions in metabolic pathways involve H_2O directly as a substrate or product. **a.** Hydrolysis reactions often result in chemical cleavage, as seen here in the reaction catalyzed by the protease enzyme chymotrypsin. **b.** The removal of H_2O from 2-phosphoglycerate to form phosphoenolpyruvate, an enzymatic reaction in the glycolytic pathway, is a dehydration reaction catalyzed by the enzyme enolase.

In the condensation reaction, two compounds function as building blocks for larger biomolecules. One example is the condensation of dimethylallyl diphosphate ($C_5H_{12}O_7P_2$) and isopentenyl diphosphate ($C_5H_{12}O_7P_2$) to form geranyl diphosphate ($C_{10}H_{20}O_7P_2$), a reaction catalyzed by the enzyme prenyl transferase (**Figure 7.23b**). Multiple condensation reactions containing C_5-derivative compounds are a hallmark of the cholesterol biosynthetic pathway. Notably, the dihydroxyacetone phosphate and glyceraldehyde-3-phosphate isomers in the triose phosphate isomerase reaction can also be combined in a condensation reaction mediated by the enzyme aldolase (see Figure 7.11). This reaction leads to the formation of fructose-1,6-bisphosphate, an intermediate in the gluconeogenic pathway.

Water is present in cells at very high concentrations and behaves as if it were a pure substance. One of the reasons for the central role of H_2O in life processes is that it is a critical component of so many enzymatic reactions. Not only does it participate in acid–base catalysis mechanisms, but also it can be directly involved in enzyme reactions as a substrate or product—the third type of metabolite transformation reactions listed earlier. As shown in **Figure 7.24a**, the enzyme chymotrypsin, which belongs to a class of enzymes called **endoproteases**, uses H_2O as a substrate in a hydrolytic cleavage reaction that breaks a peptide bond in proteins. In contrast, dehydration reactions remove H_2O from substrates, as is the case when the glycolytic intermediate 2-phosphoglycerate is converted to phosphoenolpyruvate by the enzyme enolase (**Figure 7.24b**).

Reversible Covalent Modification Reactions A large number of enzymatic reactions in cells are involved in turning molecular switches on and off as a mechanism to control cell signaling and gene expression. In the first type of molecular switching reactions, kinase enzymes add a phosphoryl group to signaling molecules, which alters their activity or recognition by other signaling molecules. In many cases, the phosphorylated signaling molecules are themselves proteins, and the kinase enzyme uses a phosphoryl transfer reaction to add the γ-phosphoryl group from ATP to the side-chain hydroxyl of Ser, Thr, or Tyr residues. The phosphoryl group is removed from the signaling molecule by phosphatase enzymes in the cell, which use a hydrolysis reaction to release inorganic phosphate, P_i. As shown in **Figure 7.25a**, kinase-mediated phosphorylation can also be used to activate phospholipids that function as cell signaling molecules embedded in the plasma membrane.

A second type of reversible covalent modification in most eukaryotic cells is the methylation of cytosine in DNA, which is associated with gene inactivation (**Figure 7.25b**). DNA methyltransferase enzymes transfer a methyl group

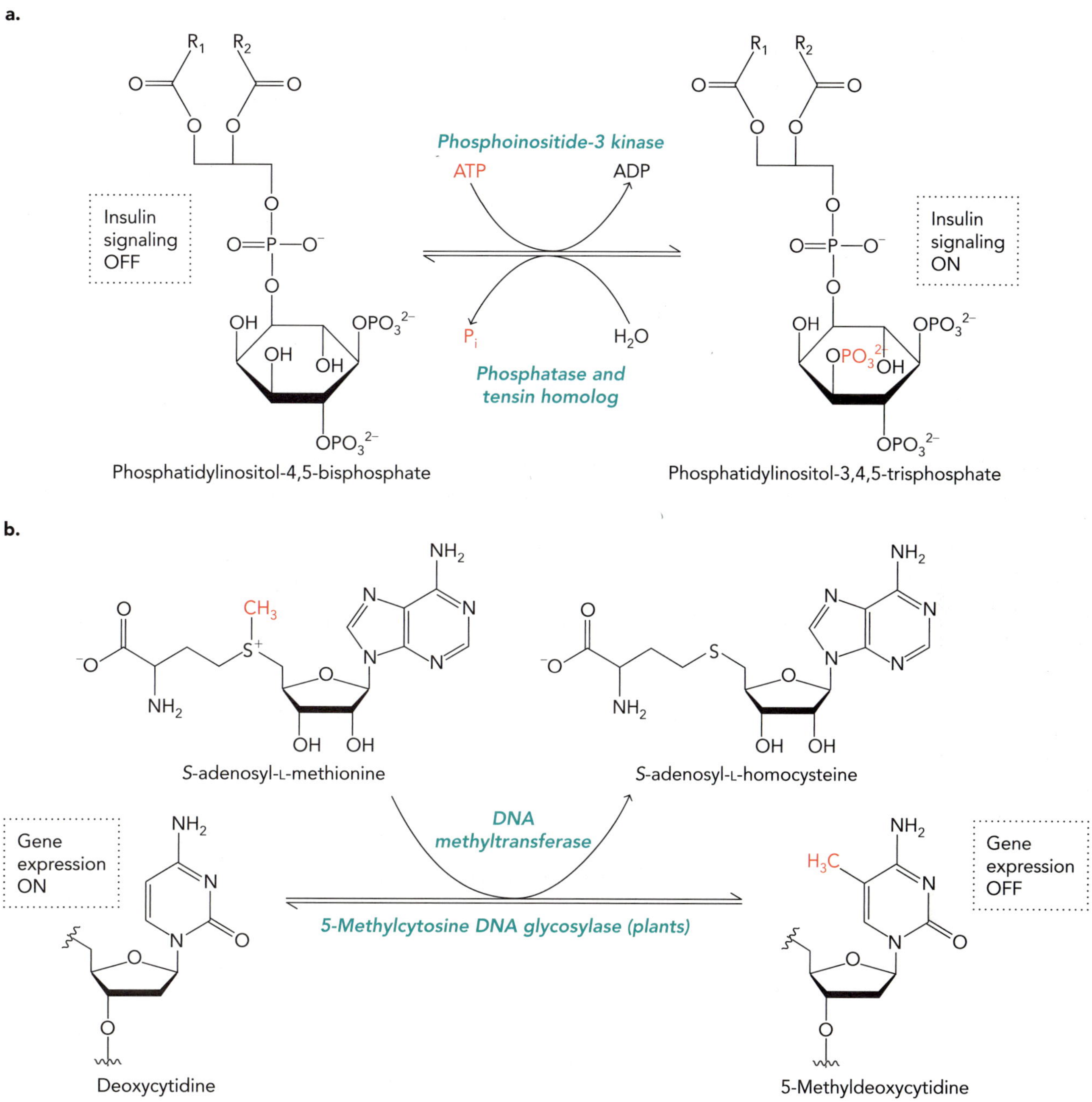

Figure 7.25 Reversible covalent modifications of biomolecules by enzymes in the cell function as molecular switching signals that control a variety of cellular processes. **a.** Insulin signaling leads to stimulation of phosphoinositide-3 kinase activity. This enzyme then phosphorylates phosphatidylinositol-4,5-bisphosphate on C-3 of inositol to generate phosphatidylinositol-3,4,5-trisphosphate. The insulin signaling pathway is turned off when the enzyme phosphatase and tensin homolog (PTEN) removes this phosphate. **b.** DNA methyltransferases use the metabolite *S*-adenosyl-L-methionine as a methyl donor to add a CH_3 group to deoxycytidine in DNA on the C-5 position of the cytosine base, which generates 5-methylcytosine. The demethylation reaction is associated with gene activation. In plants, the base containing the CH_3 group is removed through a DNA repair mechanism that involves a specific glycosylase enzyme.

from ***S*-adenosyl-L-methionine** to the C-5 position of the base, generating **5-methylcytosine.** The presence of 5-methylcytosine in the regulatory region of genes is associated with decreased rates of transcriptional initiation. This is most likely a result of 5-methylcytosine binding proteins that recognize this form of DNA and recruit transcriptional silencing proteins to the gene promoter (see Chapter 23). Removal of the methyl group from 5-methylcytosine is associated with gene activation, although the precise mechanism of this demethylation reaction is not understood in most organisms.

concept integration 7.2

How do the physical and chemical properties of an enzyme active site lower the activation energy ($\Delta G^{\ddagger}$) of a metabolite transformation reaction?

The sequestered environment of the enzyme active site provides an opportunity for (1) selective binding of substrates; (2) optimal configuration of multiple substrates to position substrate reactive groups correctly near each other; and (3) close proximity of the substrates to reactive groups within amino acids or coenzymes. Together, these properties of the enzyme active site lower the $\Delta G^{\ddagger}$ of a reaction by using a combination of catalytic mechanisms; for example, stabilizing the transition state or providing an alternative pathway to product formation.

The condensation reaction catalyzed by the enzyme aldolase is a good example of how the enzyme active site lowers the $\Delta G^{\ddagger}$ of a metabolite transformation reaction. In this reaction, a lysine residue in the aldolase active site forms an enzyme-linked covalent intermediate with the substrate dihydroxyacetone phosphate, which allows the second substrate, glyceraldehyde-3-phosphate, to bind to the active site in an optimal configuration. The result promotes condensation through an acid–base reaction mechanism involving aldolase amino acid side chains.

7.3 Enzyme Reaction Mechanisms

We have observed that enzymes function by creating ideal conditions for chemical reactions, providing an optimized microenvironment for catalysis to occur. We have also observed that enzymes stabilize transition states or provide an alternative pathway to product formation, thus lowering the activation energy. In this section, we examine in more detail three proposed enzyme reaction mechanisms. These models are based on molecular analysis of protein structures and on results of *in vitro* kinetic studies using a variety of high-affinity inhibitors and nonhydrolyzable substrates. The use of X-ray crystallography to determine the structures of enzymes in the presence and absence of bound pseudosubstrates has permitted biochemists to view enzymes "caught in the act" of catalysis.

The enzyme reaction mechanisms used by chymotrypsin, enolase, and HMG-CoA reductase provide classic examples of multistep reactions. Together, these examples reinforce two core concepts in enzymology: (1) substrates bind to enzyme active sites through weak noncovalent interactions, which orient amino acid functional groups within close proximity to substrate reactive centers; and (2) enzymes use conventional catalytic reaction mechanisms that follow basic principles of organic chemistry.

a.

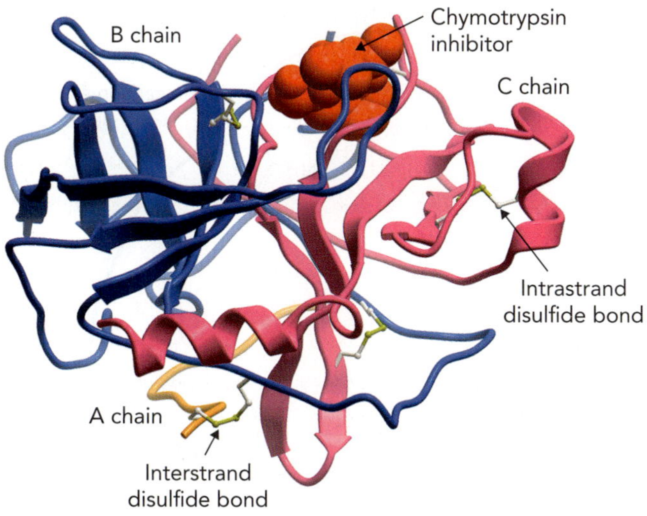

b.

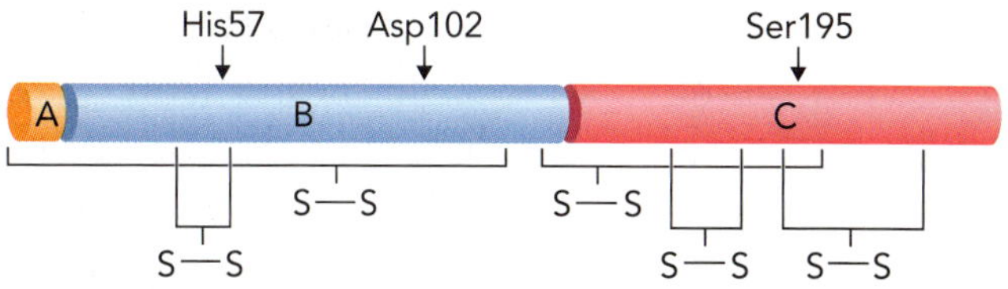

Figure 7.26 Bovine chymotrypsin consists of three polypeptide chains linked together by disulfide bonds. **a.** Ribbon structure of chymotrypsin, highlighting the two interstrand and three intrastrand disulfide bonds. It also shows the location of the enzyme active site, indicated by the space-filling model of a small chymotrypsin inhibitor. One of the disulfide bonds is behind the inhibitor and is not visible in this orientation. BASED ON PDB FILE 1GGD. **b.** Map of the A, B, and C chains of chymotrypsin, illustrating the disulfide bonds and the catalytic triad residues in the active site.

Chymotrypsin Uses Both Acid–Base Catalysis and Covalent Catalysis

Chymotrypsin is a member of a family of endoproteases called **serine proteases,** which function in digestion to cleave the peptide backbone of dietary proteins (see Figure 7.24a). The enzyme consists of three individual polypeptide chains, which started out as one nascent polypeptide chain that was cleaved into three pieces to form the functional enzyme. The three chains (A, B, and C) are covalently linked by disulfide bonds, and the enzyme active site sits on the surface of the protein, where it can readily access the large polymer substrate (**Figure 7.26**).

The chymotrypsin enzyme reaction mechanism involves both covalent catalysis and acid catalysis. One of the key features of the chymotrypsin reaction, and indeed of all serine proteases, is a set of three amino acids called the **catalytic triad**, which forms a hydrogen-bonded network required for catalysis. In chymotrypsin, these three amino acids are Ser195, located on the C chain, and His57 and Asp102 on the B chain (Figure 7.26). Ser195 is a catalytic residue that forms an enzyme–substrate covalent intermediate, whereas His57 and Asp102 function together to convert Ser195 into a highly reactive nucleophile. **Figure 7.27** shows the location of the catalytic triad within

Figure 7.27 The catalytic triad in the chymotrypsin enzyme active site consists of the amino acid residues Asp102, His57, and Ser195. **a.** Molecular structure of the bovine chymotrypsin active site, with the three amino acids in the catalytic triad shown in stick representation. BASED ON PDB FILE 1GGD. **b.** A hydrogen-bond network is established among the Asp102, His57, and Ser195 residues of the chymotrypsin catalytic triad. Through an interaction with Asp102, the proton from Ser195 is transferred to the His57 imidazole ring as the nucleophilic reaction occurs.

a.

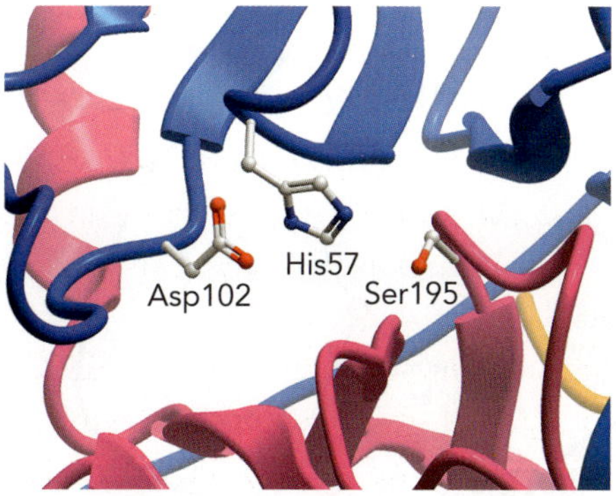

b.

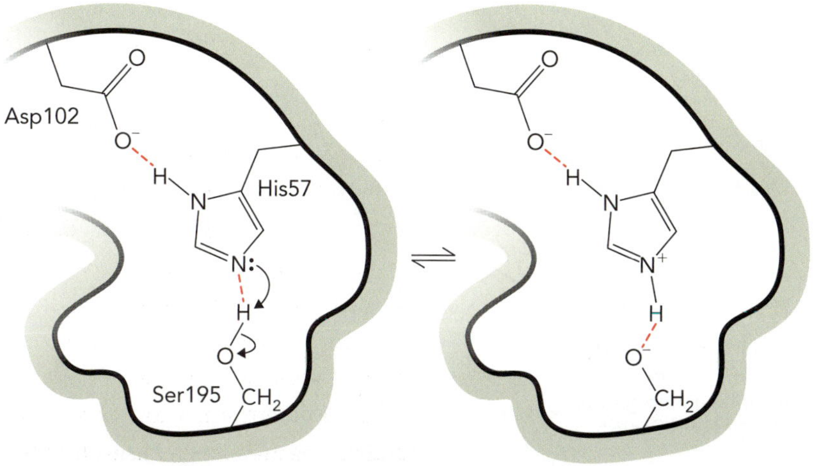

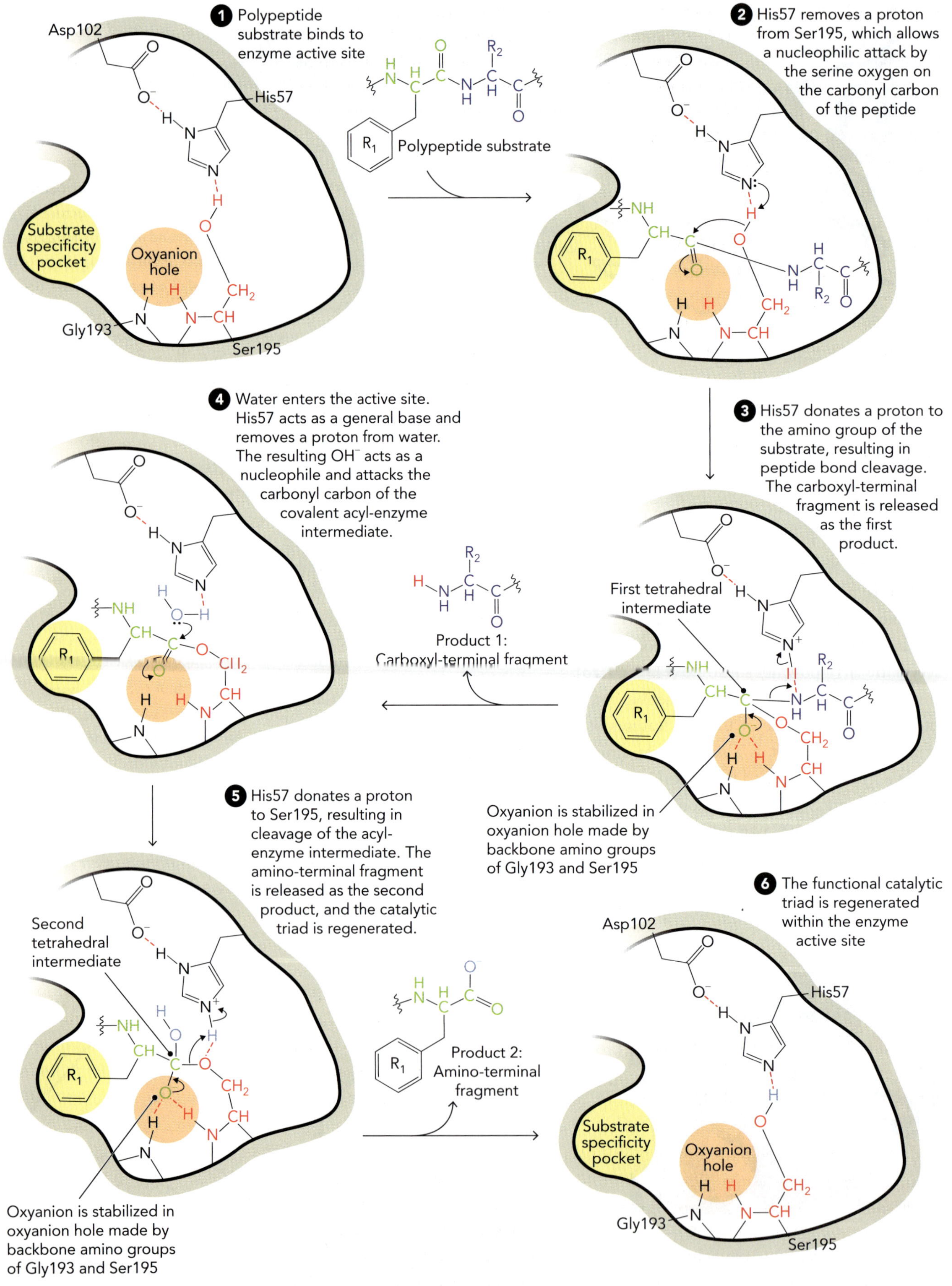

1 Polypeptide substrate binds to enzyme active site
Asp102
His57
Substrate specificity pocket
Oxyanion hole
Gly193
Ser195
Polypeptide substrate
2 His57 removes a proton from Ser195, which allows a nucleophilic attack by the serine oxygen on the carbonyl carbon of the peptide
3 His57 donates a proton to the amino group of the substrate, resulting in peptide bond cleavage. The carboxyl-terminal fragment is released as the first product.
First tetrahedral intermediate
Oxyanion is stabilized in oxyanion hole made by backbone amino groups of Gly193 and Ser195
Product 1: Carboxyl-terminal fragment
4 Water enters the active site. His57 acts as a general base and removes a proton from water. The resulting OH⁻ acts as a nucleophile and attacks the carbonyl carbon of the covalent acyl-enzyme intermediate.
5 His57 donates a proton to Ser195, resulting in cleavage of the acyl-enzyme intermediate. The amino-terminal fragment is released as the second product, and the catalytic triad is regenerated.
Second tetrahedral intermediate
Oxyanion is stabilized in oxyanion hole made by backbone amino groups of Gly193 and Ser195
Product 2: Amino-terminal fragment
6 The functional catalytic triad is regenerated within the enzyme active site

Figure 7.28 The proposed catalytic mechanism of chymotrypsin can be broken down into two discrete phases. Phase 1 includes the first three steps of the reaction, which includes formation of an acyl-enzyme intermediate and cleavage of the scissile peptide bond, releasing the carboxyl-terminal polypeptide fragment. In phase 2, steps 4 and 5 lead to cleavage of the covalent bond between the substrate and Ser195 to release the amino-terminal polypeptide fragment and regenerate the enzyme (step 6).

the enzyme active site and the organization of the hydrogen-bond network formed between Asp102, His57, and Ser195.

The proposed enzyme reaction mechanism of chymotrypsin consists of six steps, which can be divided into two phases. In the first phase, a covalent acyl-enzyme intermediate is formed between Ser195 and the polypeptide substrate, which promotes cleavage of the **scissile peptide bond** (site of cleavage). Then the carboxyl-terminal polypeptide fragment is released. In the second phase, the enzyme is regenerated after a series of steps that results in deacylation and release of the amino-terminal polypeptide fragment. Let's now look at each of these six steps individually, as illustrated in **Figure 7.28**.

Step 1 of the reaction begins with binding of the substrate to the enzyme active site after a productive interaction between the aromatic side chain (R_1 in Figure 7.28) of the polypeptide substrate and the nearby specificity pocket, which functions to properly align the substrate with the catalytic triad. In *step 2*, a proton is transferred from Ser195 to His57, which allows a nucleophilic attack by the oxygen of Ser195 to occur on the carbonyl carbon of the polypeptide backbone. The result is the formation of a covalent, transient **tetrahedral intermediate**, which is thought to resemble the transition state conformation. This tetrahedral intermediate has a C—O bond that is longer than a double bond, which allows the negatively charged oxygen (called the oxyanion) to fit into a region of the active site called the **oxyanion hole**. Here, the oxyanion forms hydrogen bonds with the backbone NH groups of Ser195 and Gly193. In *step 3* of the reaction, the imidazole ring of His57 functions as an acid catalyst by donating a proton to the nitrogen of the peptide bond, facilitating cleavage. The carboxyl-terminal fragment of the polypeptide (product 1) leaves the active site, and the amino-terminal fragment of the polypeptide remains bound to the enzyme as a covalent acyl-enzyme intermediate.

Once the carboxyl-terminal fragment of the polypeptide substrate exits the active site, H_2O is able to enter and donate a proton to His57. This results in the generation of a free OH^- in *step 4* that attacks the carbonyl carbon on the acyl-enzyme and leads to the formation of a second tetrahedral intermediate stablized by the oxyanion hole. *Step 5* occurs when the now protonated imidazole ring of His57 donates a proton that cleaves the covalent bond of the acyl-enzyme intermediate. Then, the amino-terminal fragment of the polypeptide (product 2) is released, and the functional catalytic triad is regenerated within the enzyme active site (*step 6*).

One of the important features of serine proteases is that the size of the enzyme active site of three closely related proteins—chymotrypsin, trypsin, and elastase—imparts a degree of substrate specificity. Chymotrypsin, for example, contains a hydrophobic region in the substrate binding pocket that can accommodate proteins with aromatic amino acids adjacent to the scissile peptide bond. In contrast, a region of the substrate binding pocket of trypsin is much deeper and contains a negatively charged Asp

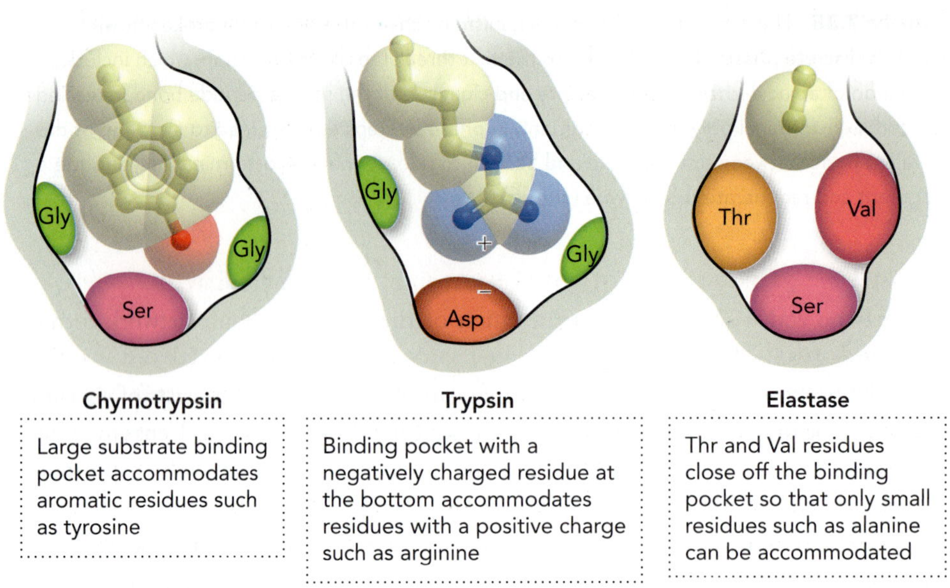

Gly
Gly
Ser
Chymotrypsin
Large substrate binding pocket accommodates aromatic residues such as tyrosine
Gly
Gly
+
−
Asp
Trypsin
Binding pocket with a negatively charged residue at the bottom accommodates residues with a positive charge such as arginine
Thr
Val
Ser
Elastase
Thr and Val residues close off the binding pocket so that only small residues such as alanine can be accommodated

Figure 7.29 Regions of the substrate binding pockets in serine proteases reflect specificity for amino acids adjacent to the scissile bond. Note that the Ser and Asp residues shown in this figure are different amino acids from the ones in the catalytic triad.

residue at the bottom (**Figure 7.29**). This substrate specificity pocket explains why trypsin cleaves proteins at peptide bonds bordering positively charged lysine and arginine residues. Elastase does not contain a functional substrate specificity pocket because instead of the glycine residues located at the opening of the chymotrypsin and trypsin substrate specificity pockets, elastase has amino acids, such as threonine and valine, that effectively close off this region. This structure is consistent with the observation that the natural substrate of elastase, the fibrous protein elastin, is rich in glycine and alanine residues. These relatively small amino acids do not require a separate substrate binding pocket for proper alignment of the scissile bond with the serine of the catalytic triad.

Enolase Uses Metal Ions in the Catalytic Mechanism

Enolase is an enzyme that catalyzes the dehydration of 2-phosphoglycerate (2-PG) to form phosphoenolpyruvate in the glycolytic pathway (see Figure 7.24b). It also catalyzes the reverse reaction in gluconeogenesis. The formation of phosphoenolpyruvate is the penultimate step of the glycolytic pathway, as phosphoenolpyruvate—whose phosphoryl group has high transfer energy—serves as the phosphoryl group donor to produce ATP from ADP.

Enolase functions as a dimer with each monomer containing one active site. The active site is located within a cleft of the alpha/beta domain of the monomer, as shown in **Figure 7.30**. Enolase is a metalloenzyme: Each active site contains two divalent metal ions, which are absolutely required for the reaction.

Enolase, like chymotrypsin and many other enzymes, uses a combination of catalytic strategies in order to carry

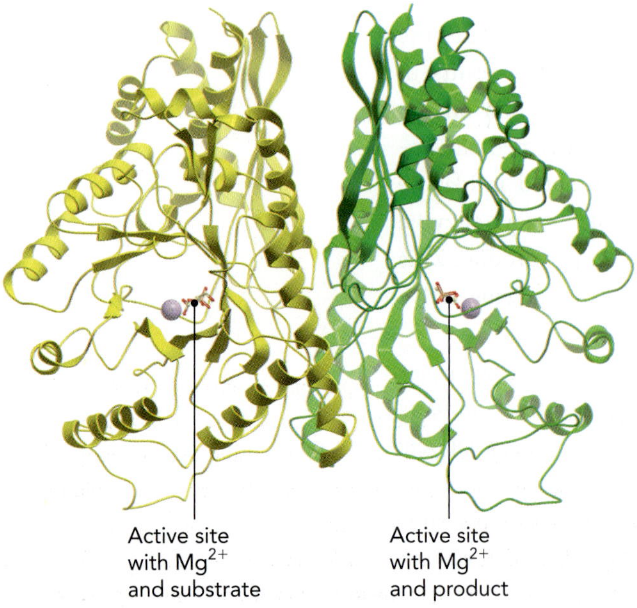

Figure 7.30 Enolase is a dimeric enzyme with one active site per monomer. In the monomer on the left (yellow), the substrate 2-phosphoglycerate is shown in stick representation, and one Mg^{2+} ion is shown as a lavender sphere. In the monomer on the right (green), the product phosphoenolpyruvate is shown in stick representation, and one Mg^{2+} ion is shown as a lavender sphere. The second Mg^{2+} ion in each active site is not shown. BASED ON PDB FILE 2ONE.

out its reaction. Mechanistic studies have shown that enolase uses general acid–base catalysis as well as metal-ion catalysis in its reaction mechanism.

The dehydration reaction takes place in two steps, as illustrated in **Figure 7.31**. In the first step, Lys345 functions as a general base, which removes the proton at the C-2 position of the substrate, generating an intermediate with a carbanion at C-2. The negative charge of the carbanion is delocalized to the carboxylate group, which is stabilized by ionic interactions with the Mg^{2+} ions. In the second step, Glu211 functions as a general acid, donating a proton to the OH leaving group, resulting in the elimination of H_2O and formation of phosphoenolpyruvate.

The divalent Mg^{2+} ions play several important roles in this reaction. First, they serve to bind and orient the substrate in the active site. Furthermore, while normally the proton at the C-2 position is not acidic and would be difficult to remove in the first step of the reaction, the strong ionic interactions of the substrate with the Mg^{2+} ions serve to make the proton at the C-2 position more acidic, facilitating its removal by Lys345. After abstraction of this proton, the Mg^{2+} ions also stabilize the increased negative charge on the intermediate.

In addition to the essential Mg^{2+} ions, other residues at the active site also contribute to the overall mechanism (**Figure 7.32**). Many of these residues are important for the formation of weak interactions. Lys396 and Gln167, for example, act in concert

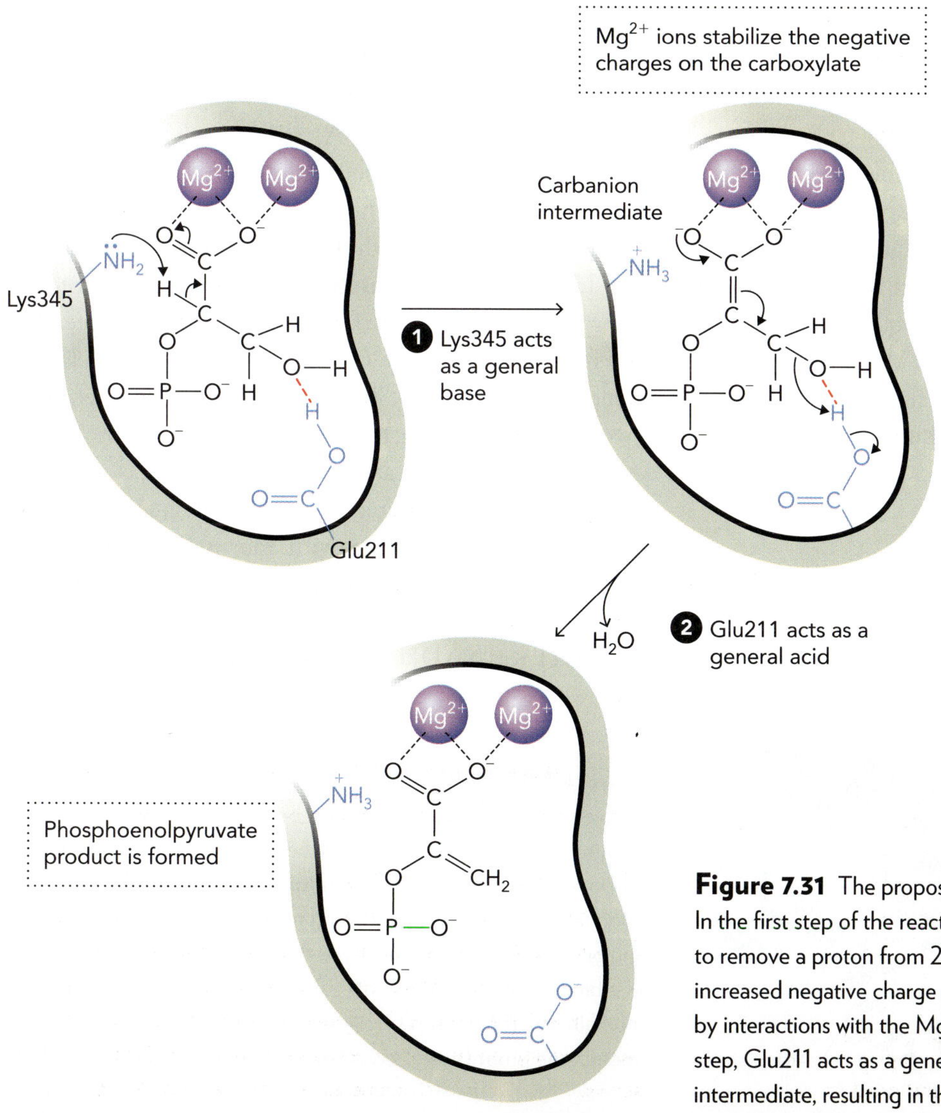
Mg2+ ions stabilize the negative charges on the carboxylate
Lys345
Glu211
Carbanion intermediate
1 Lys345 acts as a general base
2 Glu211 acts as a general acid
H2O
Phosphoenolpyruvate product is formed

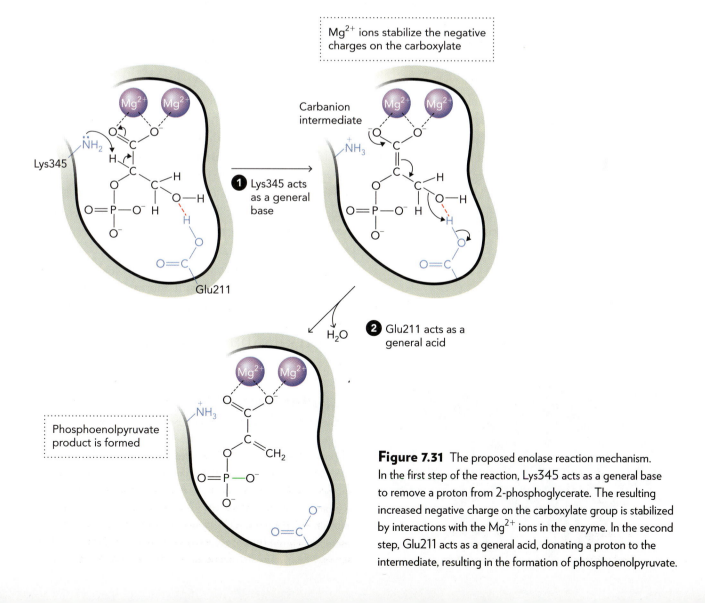

Figure 7.31 The proposed enolase reaction mechanism. In the first step of the reaction, Lys345 acts as a general base to remove a proton from 2-phosphoglycerate. The resulting increased negative charge on the carboxylate group is stabilized by interactions with the Mg^{2+} ions in the enzyme. In the second step, Glu211 acts as a general acid, donating a proton to the intermediate, resulting in the formation of phosphoenolpyruvate.

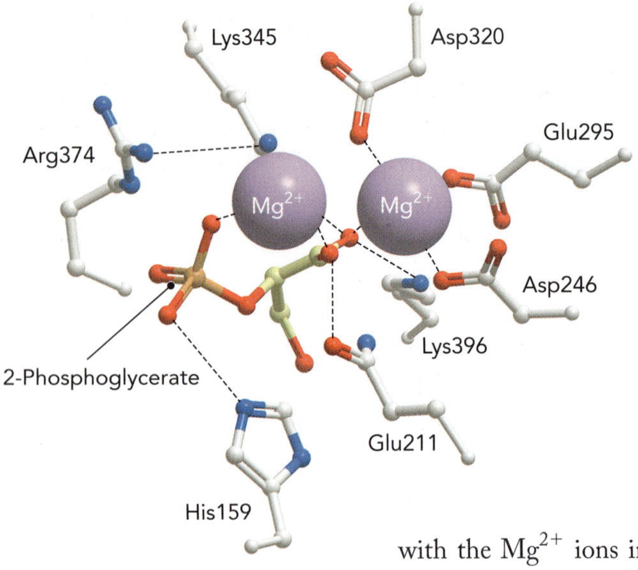

Figure 7.32 Residues in the active site of enolase are precisely arranged for recognition of the substrate and to promote catalysis. Lys396 and Gln167 form ionic interactions with the carboxylate group of the substrate 2-phosphoglycerate. His159 makes a hydrogen bond with the phosphoryl group of the substrate. Arg374 interacts with Lys345 to help produce Lys345 as a general base. Negatively charged aspartate and glutamate residues help bind the positively charged metal ions, which are shown as lavender spheres. Because this illustration is a two-dimensional representation of a three-dimensional structure, the lengths of the dashed lines indicating the bonds vary greatly. BASED ON PDB FILE 2ONE.

with the Mg^{2+} ions in making ionic interactions with the carboxylate group of the substrate. Other residues participate in the activation of acid–base groups. The formation of a hydrogen bond between His159 and the phosphate group of the substrate contributes to the acidification of the C-2 proton. As another example, interactions between Arg374 and the amino group of Lys345 serves to set up Lys345 as a general base for the start of the reaction.

The Mechanism of HMG-CoA Reductase Involves NADPH Cofactors

HMG-CoA (3-hydroxy-3-methylglutaryl-coenzyme A) reductase is an enzyme in the biosynthetic pathway for cholesterol and other isoprenoids (see Chapter 16). This enzyme is an important pharmaceutical target for the treatment of hypercholesterolemia because reduced enzyme activity results in both direct and indirect mechanisms that reduce serum cholesterol. HMG-CoA reductase inhibition directly reduces the amount of cholesterol that is produced in liver cells (see Figure 16.37 in Chapter 16). When intracellular cholesterol is reduced, the indirect benefit is that more cell surface cholesterol receptors are produced (see Figure 16.50 in Chapter 16). These receptors bind cholesterol-containing particles in the serum, thus decreasing the total cholesterol in the serum and thereby decreasing the risk of heart disease.

The HMG-CoA reductase enzyme is a membrane-bound tetrameric enzyme with four active sites located between monomer interfaces as shown in **Figure 7.33**. Each active site binds the substrate HMG-CoA and the cofactor NADPH (or NADH for some organisms). The enzyme

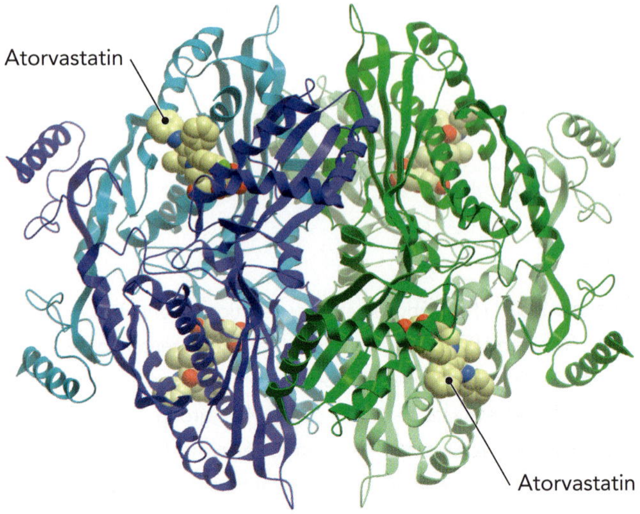

Figure 7.33 HMG-CoA is a tetrameric enzyme with four active sites located at the interface between monomeric subunits. The catalytic portion of the enzyme is shown here in ribbon representation with an enzyme inhibitor, the statin drug atorvastatin, shown in space-filling representation. The statin drug inhibits the enzyme by binding to the active site and blocking the binding of the HMG-CoA substrate. The transmembrane domain that localizes the enzyme to the endoplasmic reticulum membrane is not shown. BASED ON PDB FILE 1HWK.

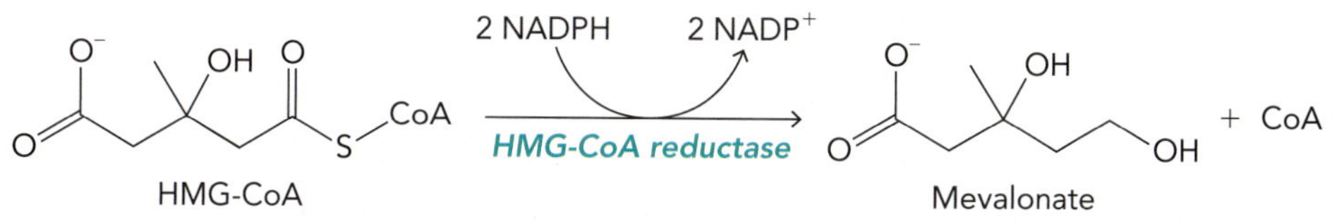

Figure 7.34 HMG-CoA reductase catalyzes the four-electron reduction of HMG-CoA to mevalonate and CoA using two NADPH molecules as cofactors.

catalyzes the four-electron reduction of HMG-CoA into CoA and mevalonate (**Figure 7.34**). Mevalonate is a precursor in the production of terpenes and steroids, such as cholesterol.

The HMG-CoA reductase mechanism has been elucidated from numerous kinetic studies and crystallographic structures, aided by computational studies. The reduction of the HMG-CoA thioester is achieved by two hydride transfer steps involving two NADPH cofactors. The active site of the protein is important for both binding and orienting HMG-CoA and NADPH so hydride transfer can occur, and it also provides functional groups to stabilize the transition state.

The overall reaction mechanism of HMG-CoA reductase can be described in four steps as shown in **Figure 7.35**:

1. The hydride ion from NADPH attacks the carbonyl carbon of HMG-CoA, developing an oxyanion transition state, which is stabilized by a lysine side chain in the active site. A glutamate residue acts as a general acid to protonate the oxyanion, resulting in the formation of a hydroxyl group in the intermediate, mevaldyl-CoA. At the end of the first step, the thioester of HMG-CoA has been reduced to a mevaldyl-CoA hemithioacetal.
2. Protein conformational changes trigger the exchange of $NADP^+$ for NADPH. The second NADPH is now in position for another hydride transfer reaction.
3. The glutamate side chain in the active site now acts as a general base to deprotonate the hydroxyl group producing an aldehyde, mevaldehyde. As the thiolate bond is broken, the active site histidine donates a proton to CoA forming the reduced CoA molecule (CoA-SH).
4. The hydride from the second NADPH molecule attacks the carbonyl center of the aldehyde, and the glutamate residue donates a proton to the oxygen of the aldehyde. This step results in the formation of the reduced mevalonate species and $NADP^+$.

The HMG-CoA reductase mechanism presented here is largely consistent with the available data, but there are still outstanding questions that will require additional studies. For example, there has been some debate over whether the NADPH cofactor exchange step happens before or after the hemithioacetal is broken down. If the NADPH exchange happens after mevaldehyde is formed, then it would be expected that conformational changes that promote cofactor exchange may allow the aldehyde species to be released from the active site to some extent. Moreover, the details of the enzyme conformational changes that take place during cofactor exchange are also an active area of research. It is anticipated that answers to these mechanistic details of HMG-CoA reductase structure and function will aid in the development of more specific inhibitors of this biomedically relevant enzyme.

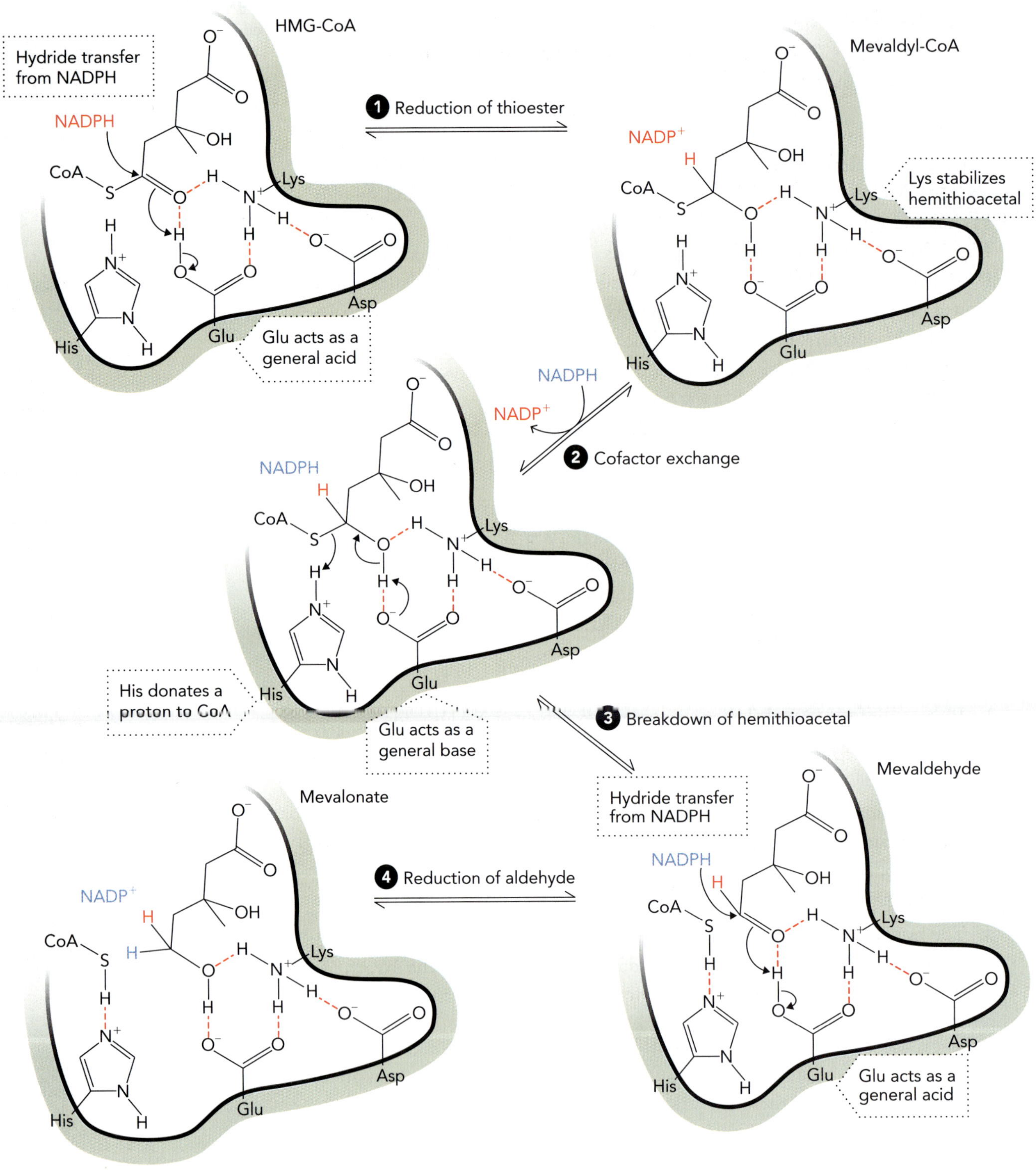

Figure 7.35 HMG-CoA reductase converts HMG-CoA to mevalonate in a four-step reaction using two NADPH cofactors. The residues shown are Glu83, Lys267, Asp283, and His381 of the soluble HMG-CoA reductase enzyme from *Pseudomonas mevalonii*. The details of each step are described in the text.

concept integration 7.3

What mechanisms of catalysis are shared by the chymotrypsin and enolase reactions?

The reaction mechanisms for both enzymes involve general acid–base catalysis. The chymotrypsin mechanism involves the formation of a covalent intermediate, which does not occur in the enolase reaction. Enolase requires divalent metal ions for catalysis, whereas the chymotrypsin mechanism relies solely on functional groups from the enzyme active site. Both reactions may also use transition state stabilization, although this is clearer in the case of the chymotrypsin reaction than it is for enolase.

7.4 Enzyme Kinetics

Enzyme kinetics is the quantitative study of the rates of chemical reactions performed by enzymes. An understanding of enzyme kinetics is important for determining the mechanism by which a reaction takes place, the effects of regulatory molecules (including metabolites or toxins), and the effects of enzyme variants under a defined set of conditions.

Enzyme kinetics is best characterized as the quantitative analysis of reaction rate data obtained with purified enzymes under defined laboratory conditions. For example, to determine how efficient a mutant enzyme is under various conditions of temperature, pH, and substrate concentration or to understand mechanisms of enzyme inhibition, we need to measure enzyme reaction rates quantitatively. Enzyme kinetics involves relating reaction rates to free energy and equilibrium. In Chapter 2, we introduced the concept of the change in free energy (ΔG) and showed how it relates to the concentrations of reactants and products of a reaction. We now describe enzyme kinetics in terms of reaction rates (v, velocity) and substrate concentration ([S]). With a quantitative understanding of how enzyme activity relates to protein function, we can define the catalytic efficiency of different enzymes to gain insight into their role as biological catalysts.

Enzymes, like all catalysts, function by increasing the rate at which a reaction reaches equilibrium. As already mentioned, enzymes catalyze reactions by lowering the activation energy ($\Delta G^{\ddagger}$) without altering the change in free energy of the reaction (ΔG). It is important to note that $\Delta G^{\ddagger}$ is not directly related to ΔG because catalyzed and uncatalyzed reactions have different $\Delta G^{\ddagger}$ values without affecting ΔG (see Figure 7.6).

Relationship between $\Delta G^{\ddagger}$ and the Rate Constant k

There are many parameters to consider when studying enzyme kinetics and numerous variations in types of reactions. As a starting point, we will first consider how the initial substrate concentration, [S], relates to the activation energy, $\Delta G^{\ddagger}$, for a **first-order reaction** (a reaction in which the rate varies as the first power of the reactant concentration) in which a single substrate is converted to a single product, S → P. To do this, we define the **velocity of the reaction, *v***, as the product of the **rate constant of a reaction, *k***, and [S]:

$$v = k[\text{S}] \tag{7.1}$$

The rate constant k reflects how quickly a substrate molecule is converted to product ($S \rightarrow P$) as a function of time under a defined set of conditions. For a first-order reaction, k has units of second^{-1} (s^{-1}). At a given substrate concentration, the velocity is directly related to the rate constant and refers to the units of product formed per unit time; for example, molarity per second ($M\ s^{-1}$). This will be important in the next section when we describe Michaelis–Menten enzyme kinetics.

In the description of enzyme mechanisms earlier in the chapter, we stated that the activation energy of a reaction, $\Delta G^{\ddagger}$, is lowered by the presence of a catalyst, resulting in an increased reaction rate. This relationship can be expressed mathematically by relating velocity v to activation energy:

$$v = \frac{k_B T}{h}[S]e^{-\Delta G^{\ddagger}/RT}$$

In this equation, k_B is the Boltzmann constant (1.38×10^{-23} J K^{-1}), and h is Planck's constant (6.63×10^{-34} J s). The value of $k_B T/h$ at 25 °C is equal to $6.2 \times 10^{12}\ s^{-1}$. We can rewrite this equation in terms of the rate constant, $k = v/[S]$:

$$k = \frac{v}{[S]} = \frac{k_B T}{h} e^{-\Delta G^{\ddagger}/RT} \qquad (7.2)$$

On the basis of this equation, we can see why an enzyme-catalyzed reaction has a higher reaction rate, k, than that of an uncatalyzed reaction. Specifically, a lower $\Delta G^{\ddagger}$ value produces a higher (inverse and exponential) reaction velocity v, as defined by k and [S].

Many reactions in metabolism are bimolecular and involve two substrates or two molecules of the same substrate. These reactions are **second-order reactions** in which the reaction rate is proportional to the product of the substrate concentrations. The units of a second-order rate constant are molarity^{-1} second^{-1} ($M^{-1}\ s^{-1}$). A bimolecular reaction consisting of two substrates, S and Y, can be written as

$$S + Y \rightarrow P$$

The rate equation for this reaction is

$$v = k[S][Y]$$

By experimentally measuring the disappearance of the substrates S and Y independently, we can see that $S \rightarrow P$ and $Y \rightarrow P$ are each first-order reactions, but that the overall conversion of $S + Y \rightarrow P$ is a second-order reaction.

Michaelis–Menten Kinetics

To characterize the catalytic properties of an enzyme under experimental conditions, it is necessary to know the substrate concentration during the course of the reaction. However, because substrate is constantly being consumed to generate product, substrate concentration is not constant, which affects the rate of the reaction over time. A common simplifying approach is to measure the reaction rate at the beginning of the reaction (**initial velocity, v_0**) before the substrate concentration has changed significantly. If working under conditions where [S] >> [E], at the beginning of the reaction (approximately within the first 60 seconds in typical reactions) the substrate concentration changes only minimally and can thus be considered constant.

We can measure initial velocities from plots of product formation versus time by taking the slope of the tangent at the beginning of each reaction (**Figure 7.36a**). The initial velocities vary with the substrate concentration when substrate concentrations

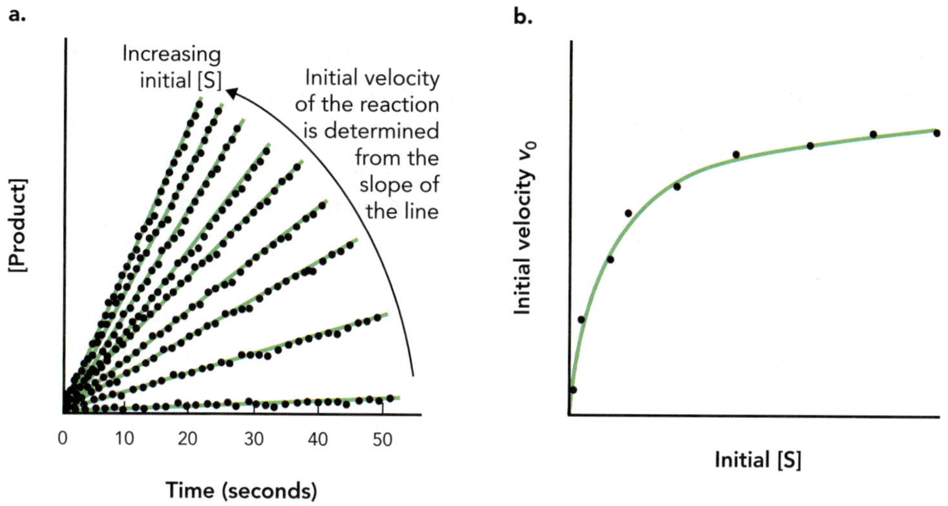

Figure 7.36 Michaelis–Menten kinetic parameters are obtained from experimental data under steady-state conditions. **a.** From a plot of product [P] formation versus time for different substrate concentrations [S], the initial velocity v_0 for each reaction is determined from the slope of the tangent at the early part of the reaction. **b.** Plotting v_0 versus [S] produces a hyperbolic curve if the enzyme reaction is first order and follows Michaelis–Menten kinetics.

▶ ANIMATION

are relatively low. When substrate concentrations become sufficiently high, the initial velocities no longer show significant changes. At this point, the reaction is said to have reached the **maximum velocity, v_{max},** where the reaction velocity cannot increase any further, even with the addition of more substrate. A plot of initial velocity versus substrate concentration shows the curve leveling off as v_{max} is approached (**Figure 7.36b**).

For an enzyme to convert a substrate (S) to a product (P), an enzyme–substrate (ES) complex must first be formed. Then the chemistry occurs to convert substrate to product, and finally product is released. This process can be described as

$$E + S \underset{k_{-1}}{\overset{k_1}{\rightleftharpoons}} ES \underset{k_{-2}}{\overset{k_2}{\rightleftharpoons}} EP \underset{k_{-3}}{\overset{k_3}{\rightleftharpoons}} E + P$$

We can make several simplifying assumptions to facilitate the study of enzyme-catalyzed reactions of this type, which are described in the kinetic model developed by Leonor Michaelis and Maud Menten in 1913. In the **Michaelis–Menten kinetic model**, first we consider the reaction at an early time when no appreciable product has been generated. Therefore the back reaction, where ES forms from enzyme-bound product (EP) with a rate constant k_{-2}, is negligible. Second, product released is assumed to be a rapid step in the process, so the conversion of EP to E + P does not need to be explicitly considered. Thus the reaction scheme can be reduced to

$$E + S \underset{k_{-1}}{\overset{k_1}{\rightleftharpoons}} ES \xrightarrow{k_2} E + P \qquad (7.3)$$

Another assumption in Michaelis–Menten kinetics, as defined in 1925 by G. E. Briggs and James Haldane, is that the **steady-state condition** is reached quickly, such that the concentration of ES is relatively constant after an initial reaction time. By working under conditions where [S] >> [E], the concentration of ES remains approximately constant. **Figure 7.37** plots the time course of an

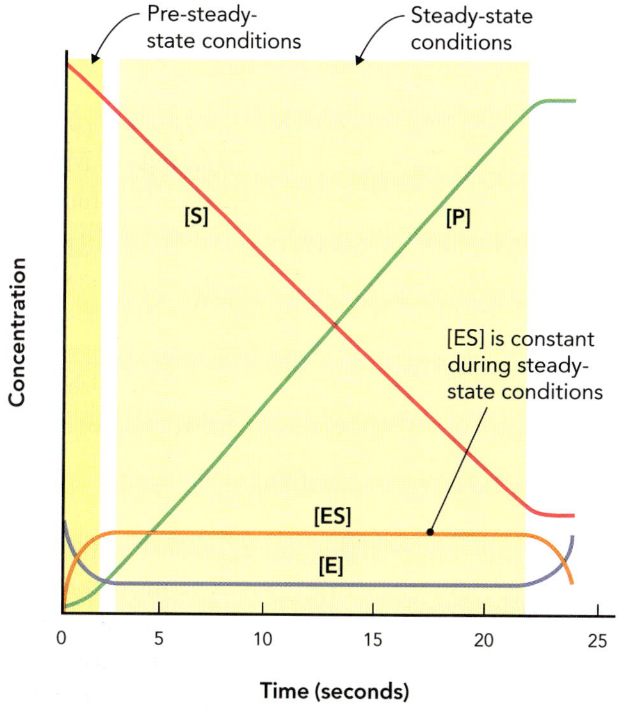

Figure 7.37 The relative concentrations of reaction components change over the course of an enzyme reaction. After the initial burst of ES complex formation during the pre-steady-state period, which often occurs on a timescale of milliseconds, the reaction reaches the steady-state phase, characterized by constant levels of [ES] and free [E], with a linear increase in [P]. The accumulation of [P] during the steady-state phase provides rate data as a function of initial [S].

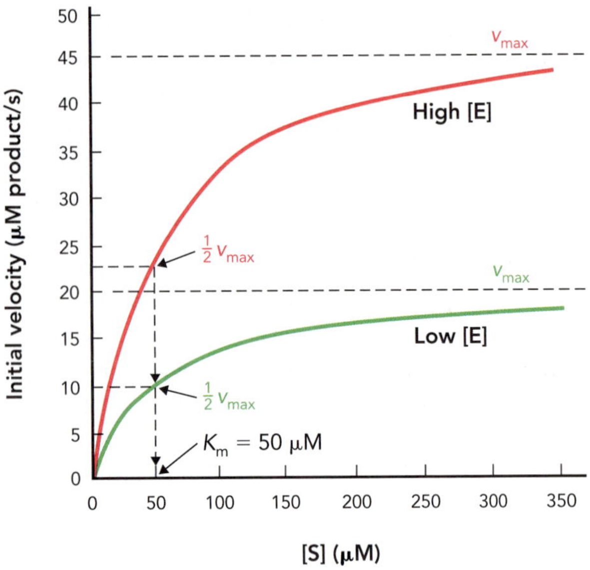

Figure 7.38 We can determine the Michaelis–Menten parameters K_m, v_{max}, and $\frac{1}{2}v_{max}$ by plotting kinetic data derived from enzyme activity assays. The K_m for any given enzyme is independent of enzyme concentration when enzyme concentration [E] is limiting for the reaction.

enzyme reaction as a function of concentration for each component in the reaction. The steady-state condition is reached when the concentration of ES is constant, even though the concentration of substrate is decreasing and that of product is increasing. Note that the k_1 and k_{-1} rate constants are determined during the **pre-steady-state condition,** at a time when ES complex formation is linear over time, and changes in the concentration of substrate are negligible. Analyzing the pre-steady-state condition experimentally requires a very rapid enzyme assay method called **stop-flow kinetics** that permits measurements in the millisecond range. It is important to note that Michaelis–Menten kinetics are valid only for enzyme reactions under steady-state conditions that contain only one substrate and one enzyme (first-order reaction). Although most enzymes have more than one substrate, these enzymes can still be studied using the Michaelis–Menten approach, as long as only one substrate is varied at a time.

To understand how we can use Michaelis–Menten kinetics to analyze enzyme function, we first need to define the **Michaelis constant, K_m,** which relates the rate constants k_1, k_{-1}, and k_2 (describing the rates of breakdown and formation of the ES complex) for a given enzyme reaction:

$$K_m = \frac{k_{-1} + k_2}{k_1} \tag{7.4}$$

We can determine the K_m value for an enzyme experimentally from a plot of initial velocity versus substrate concentration under conditions where [S] >> [E], and the reaction is at steady state (**Figure 7.38**). If the enzyme reaction is first order and follows Michaelis–Menten kinetics, then plotting v_0 data (concentration/time) versus [S] produces a hyperbolic curve. The experimentally determined value of K_m is defined as the substrate concentration at which the reaction rate is half of its maximum value, $\frac{1}{2}v_{max}$. A low value of K_m means that an enzyme has high catalytic activity at low substrate concentration. As shown in Figure 7.38, the K_m, v_{max} and $\frac{1}{2}v_{max}$ values can all be extracted from a plot of initial velocity versus substrate concentration.

The **Michaelis–Menten equation** (see derivation in **Box 7.1**) describes the hyperbolic curve in Figure 7.38 using [S], v_{max}, and K_m to define v_0 as follows:

$$v_0 = \frac{v_{max}[S]}{K_m + [S]} \tag{7.5}$$

The importance of the Michaelis–Menten equation is that it relates initial velocity v_0 to the maximum velocity v_{max} as a function of the substrate concentration and the Michaelis constant K_m. These values are very useful in describing the properties of an enzyme under a defined set of conditions.

For enzyme reactions that follow Michaelis–Menten kinetics, v_0 does not increase appreciably at high substrate concentration, and moreover, v_0 is proportional to enzyme concentration, as shown in Figure 7.38. This makes sense because the reaction rate is directly related to the amount of enzyme (when enzyme concentration is limiting), whereas the K_m of an enzyme is an inherent property of the geometric and chemical complementarity between amino acids in the enzyme active site and the chemical structure of the substrate.

Box 7.1 DERIVATION OF THE MICHAELIS-MENTEN EQUATION

For an uncatalyzed reaction

$$S \underset{k_{-1}}{\overset{k_1}{\rightleftharpoons}} P$$

the rate of product formation is proportional to the substrate concentration, so that the rate of production of [P] is k_1[S]. Thus for an uncatalyzed reaction, increases in the substrate concentration always result in increases in the rate, and no maximum velocity is reached. In contrast, enzyme-catalyzed reactions reach a maximum velocity at high substrate concentration, giving rise to the characteristic curves as shown in Figure 7.38.

Michaelis and Menten developed an equation to describe the kinetic characteristics of enzyme-catalyzed reactions. This equation describes the relationship of substrate concentrations to initial velocity, as the simple model for an uncatalyzed reaction is inadequate to explain these data.

Consider the following reaction, in which substrate (S) is converted to product (P) via an enzyme (E)-catalyzed reaction:

$$E + S \underset{k_{-1}}{\overset{k_1}{\rightleftharpoons}} ES \underset{k_{-2}}{\overset{k_2}{\rightleftharpoons}} EP \underset{k_{-3}}{\overset{k_3}{\rightleftharpoons}} E + P$$

Michaelis and Menten made several simplifying assumptions that facilitate the treatment of enzyme-catalyzed reactions. The first of these assumptions is that when working at the early stages of the reaction, no appreciable enzyme–product complex (EP) has yet accumulated. Thus, the back reaction of EP to enzyme–substrate complex (ES) can be neglected. Furthermore, Michaelis and Menten assumed that once substrate is converted to product, the release of product from enzyme is fast. So this backward step to form ES from E + P can be neglected as well. Thus, the reaction above simplifies to

$$E + S \underset{k_{-1}}{\overset{k_1}{\rightleftharpoons}} ES \xrightarrow{k_2} E + P$$

The three ways in which the concentration of ES can change are through the breakdown of ES to E and P (which occurs with a rate k_2[ES]), the dissociation of ES to E and S (rate = k_{-1}[ES]), or through the formation of ES from E and S (rate = k_1[E][S]). Michaelis and Menten used a further assumption, originally developed by Briggs and Haldane, that the reaction is being measured under steady-state conditions, where the concentration of ES is not changing. Therefore, the rates of formation and breakdown of ES must be equal.

$$\text{Rate of formation of ES} = \text{Rate of breakdown of ES}$$

$$k_1[E][S] = k_{-1}[ES] + k_2[ES]$$

$$k_1[E][S] = (k_{-1} + k_2)[ES]$$

Under these conditions, where the concentration of ES is not changing, they defined the Michaelis constant K_m, which balances the rates of formation and breakdown of ES:

$$K_m = \frac{k_{-1} + k_2}{k_1}$$

Another assumption in Michaelis–Menten kinetics is that the total amount of enzyme (E_t) stays constant throughout a reaction. Though the total amount of enzyme in a reaction does not change, once the reaction starts some free enzyme

[E] binds to substrate to become [ES], so the total amount of enzyme is $[E_t] = [E] + [ES]$. Rearranging this equation in terms of [E] gives

$$[E] = [E_t] - [ES]$$

Substitute this expression for [E] into

$$k_1[E][S] = (k_{-1} + k_2)[ES]$$

to get

$$k_1([E_t] - [ES])[S] = (k_{-1} + k_2)[ES]$$

Distribute the term:

$$k_1[E_t][S] - k_1[ES][S] = (k_{-1} + k_2)[ES]$$

Add $k_1[ES][S]$ to both sides:

$$k_1[E_t][S] - k_1[ES][S] + k_1[ES][S] = (k_{-1} + k_2)[ES] + k_1[ES][S]$$

$$k_1[E_t][S] = k_{-1}[ES] + k_2[ES] + k_1[ES][S]$$

$$k_1[E_t][S] = (k_{-1} + k_2 + k_1[S])[ES]$$

Divide by $(k_{-1} + k_2 + k_1[S])$ to yield

$$\frac{k_1[E_t][S]}{(k_{-1} + k_2 + k_1[S]} = [ES]$$

Divide both numerator and denominator by k_1:

$$\frac{(k_1/k_1)[E_t][S]}{\{(k_{-1}+k_2)/k_1\} + (k_1[S]/k_1)} = \frac{[E_t][S]}{(k_{-1}+k_2)/k_1 + [S]} = [ES]$$

$$[ES] = \frac{[E_t][S]}{(k_{-1} + k_2)/k_1 + [S]}$$

Recall that

$$K_m = \frac{(k_{-1} + k_2)}{k_1}$$

Therefore, K_m can be substituted into the equation above to yield

$$[ES] = \frac{[E_t][S]}{K_m + [S]}$$

The only way product can be formed is through [ES], and thus the initial velocity of the reaction is

$$v_0 = k_2[ES]$$

Substituting the expression obtained for [ES] into $v_0 = k_2[ES]$ gives

$$v_0 = \frac{k_2[E_t][S]}{K_m + [S]}$$

However, the reaction is limited by the amount of enzyme. The maximum velocity v_{max} occurs when all the enzyme is in the form [ES] such that

$$v_{max} = k_2[E_t]$$

By substitution of this equation for v_{max} into the initial velocity expression, we obtain the *Michaelis–Menten equation*:

$$v_0 = \frac{v_{max}[S]}{K_m + [S]}$$

This expression relates the initial velocity of a reaction to the substrate concentration, maximum velocity, and the Michaelis constant.

For the simple reaction shown earlier, where

$$E + S \underset{k_{-1}}{\overset{k_1}{\rightleftharpoons}} ES \xrightarrow{k_2} E + P$$

under the special condition where product formation is slow, such that $k_{-1} >> k_2$, we can reduce K_m to

$$K_m = \frac{k_{-1}}{k_1} = K_d$$

This is defined as the dissociation constant, K_d, of the ES complex (see Chapter 6). A high K_d value, measured in units of molar concentration (M; typically micromolar to millimolar), means that the enzyme has a low affinity for the substrate because a higher substrate concentration is required to reach K_m. However, it is important to remember that experimentally determined values for v_{max} and K_m do not provide information about the number of discrete steps involved in the reaction or the rates of individual steps. In fact, for most enzymes, the reaction steps are not as simple as defined here, and K_m is not a measure of the enzyme affinity for substrate. For example, when $k_2 >> k_{-1}$, making $K_m = k_2/k_1$, then K_m is not equal to K_d. However, even though K_m values may not represent K_d for a particular enzyme, K_m is a value that can be experimentally determined and is therefore a useful comparative measure for enzymes under various conditions.

Although Michaelis–Menten parameters are commonly derived computationally by analyzing experimental data directly, it is sometimes useful to use an algebraic transformation of the Michaelis–Menten equation, called the **Lineweaver–Burk equation**, to draw a double reciprocal plot of the enzyme data. The Lineweaver–Burk equation is derived from the Michaelis–Menten equation by taking the reciprocal of both sides of the equation and rearranging:

$$\frac{1}{v_0} = \frac{K_m}{v_{max}[S]} + \frac{1}{v_{max}} \tag{7.6}$$

This rearrangement is useful because we can fit the data points with a straight line. As shown in **Figure 7.39**, we can obtain Michaelis–Menten parameters from a **Lineweaver–Burk plot** by determining the slope of the line to obtain the ratio of K_m/v_{max}; the y-axis intercept to find $1/v_{max}$; and the x-axis intercept to determine the value of $-1/K_m$. Note that using the Lineweaver–Burk plot allows a determination of v_{max} (through extrapolation) even under conditions of nonsaturating concentrations of substrate. You will see the usefulness of Lineweaver–Burk plots for enzyme kinetic analysis later in this chapter when we describe mechanisms of enzyme inhibition.

Figure 7.39 We can plot enzyme kinetic data as a straight line by using the Lineweaver–Burk equation. Note that the x-axis intercept ($-1/K_m$) is the same for both sets of enzyme reactions, confirming that K_m is not affected by enzyme concentration under these conditions.

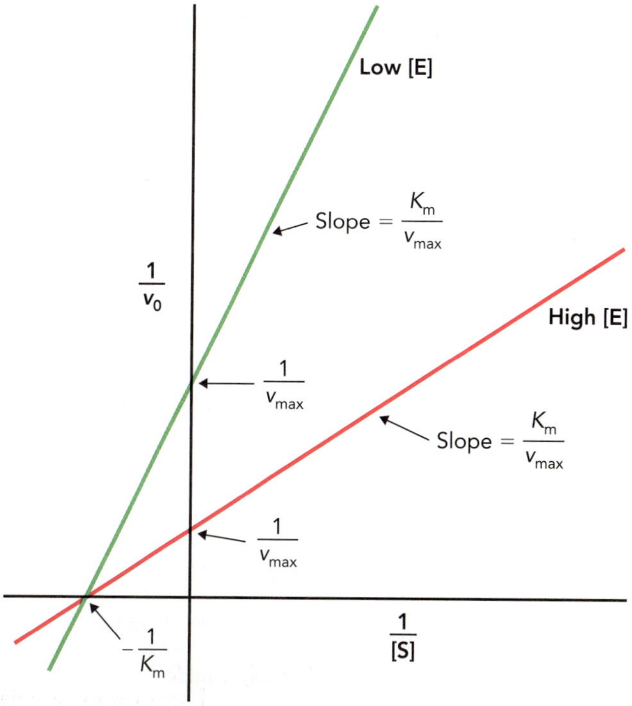

Enzymes Have Different Kinetic Properties

Michaelis–Menten kinetics helps us determine how an enzyme functions and what its reaction rate is. However, other comparisons of enzyme properties can be helpful in putting together the picture of how efficiently enzymes function under different conditions of pH and temperature.

Catalytic Efficiency of Enzymes Another way to compare enzymes is to determine the maximum catalytic activity under saturating levels of substrate. This value, called the **turnover number**, is defined as $\mathbf{k_{cat}}$. The k_{cat} for a reaction is the value of the rate-limiting step. In the Michaelis–Menten equation, $k_{cat} = v_{max}/[E_t]$, where $[E_t]$ equals the total amount of enzyme in the reaction ([E] + [ES]). Substituting k_{cat} into the Michaelis–Menten equation gives

$$v_0 = \frac{k_{cat}[E_t][S]}{K_m + [S]}$$

The turnover number k_{cat} provides information about the maximum catalytic rate, but it does not reveal how well the enzyme works. Two enzymes may have the same k_{cat} value, but the efficiency of the reaction can be different. For example, consider two enzymes E1 and E2 that have the same k_{cat} value. If E1 increases the rate of reaction 10^{12}-fold over the uncatalyzed reaction, whereas E2 only increases the reaction rate 10^8-fold, you can conclude that E1 has a higher catalytic efficiency than E2.

The catalytic efficiency of two enzyme reactions or of the same enzyme with two different substrates can be represented by the **specificity constant**, defined as $\mathbf{k_{cat}/K_m}$. Under conditions of low substrate concentrations ($[S] \ll K_m$), we can rewrite the rate equation as

$$v_0 = \frac{k_{cat}[E_t][S]}{K_m} \tag{7.7}$$

This expression makes k_{cat}/K_m an apparent second-order rate constant because v_0 depends on the concentrations of two reactants: E_t and S. It is apparent here that there is an upper limit to the rate of the reaction, which is the rate of encounter of E and S. In aqueous solution, this limit is approximately 10^8 to 10^9 M^{-1} s^{-1}, which is the diffusion-controlled limit. **Table 7.4** lists K_m, k_{cat}, and specificity constants for several enzymes.

As shown in Table 7.4, catalase has an extremely high turnover rate (4×10^7 s^{-1}); however, it also has a high K_m (1.1 M) and is therefore not as catalytically efficient as some other enzymes with a much lower turnover rate. The data in Table 7.4 also illustrate the difference in kinetic properties when the same enzyme reacts with different

Table 7.4 MICHAELIS–MENTEN PARAMETERS FOR REPRESENTATIVE ENZYMES AND SUBSTRATES

Enzyme	Substrate	k_{cat} (s^{-1})	K_m (M)	k_{cat}/K_m (M^{-1} s^{-1})
Triose phosphate isomerase	Glyceraldehyde-3-phosphate	4.3×10^3	4.7×10^{-4}	2.4×10^8
Acetylcholinesterase	Acetylcholine	1.4×10^4	9.0×10^{-5}	1.5×10^8
Fumarase	Fumarate	8.0×10^2	5.0×10^{-6}	1.6×10^8
Fumarase	Malate	9.0×10^2	2.5×10^{-5}	3.6×10^7
Carbonic anhydrase	CO_2	1.0×10^6	1.2×10^{-2}	8.3×10^7
Carbonic anhydrase	HCO_3^-	4.0×10^5	2.6×10^{-2}	1.5×10^7
Catalase	H_2O_2	4.0×10^7	1.1	4.0×10^7
Urease	Urea	1.0×10^4	2.5×10^{-2}	4.0×10^5
Ribonuclease	Cytidine cyclic phosphate	8.0×10^2	8.0×10^{-3}	1.0×10^5
Pepsin	Phe-Gly peptide	5.0×10^{-1}	3.0×10^{-4}	1.7×10^3

substrates. In the case of the enzyme fumarase, the K_m for fumarate is fivefold lower than it is for malate, yet the k_{cat} values for these two substrates are nearly the same. This results in a specificity constant for fumarate that is about four times higher for the enzyme than when malate is the substrate. These kinetic properties of fumarase are consistent with the fact that fumarate is the preferred substrate of the enzyme in the citrate cycle reactions (malate is the product of this reaction), as described in Chapter 10.

Effect of pH and Temperature on Enzyme Activity Rates of enzymatic reactions are affected by both pH and temperature, reflecting changes in the structure and chemistry of the active site under differing conditions. With regard to pH, seven amino acids have ionizable side chains that gain or lose a proton depending on the local environment and effective pK_a. As a result, changes in pH can alter the chemistry of the active site and disrupt critical tertiary structures within the protein, potentially resulting in denaturation. **Figure 7.40** shows that the optimal pH for an enzyme reflects the normal physiologic conditions in which it operates. Pepsin, the pancreatic protease found in the acidic environment of the stomach, has an optimal pH around 1.6, whereas the liver enzyme glucose-6-phosphatase has an optimal pH of 7.8. Arginase, an important liver enzyme involved in the urea cycle, has an optimal pH of 9.7.

Figure 7.41 shows the activity curves for two enzymes as a function of reaction temperature. Temperature affects both the catalytic properties of the enzyme and the stability of the protein structure. Increased temperatures lower the activation energy by increasing the free energy of the ground state; however, the protein structure may be destabilized at high temperature. Thus, the temperature optimum for a reaction balances catalytic efficiency and structural stability. Similar to the natural difference in pH optimums of pepsin and arginase, the DNA polymerase from a thermophilic bacterium, which normally lives at temperatures as high as 80 °C, has a much higher temperature optimum than the DNA polymerase of *Escherichia coli*, which lives in the human intestine at a comfortable 37 °C.

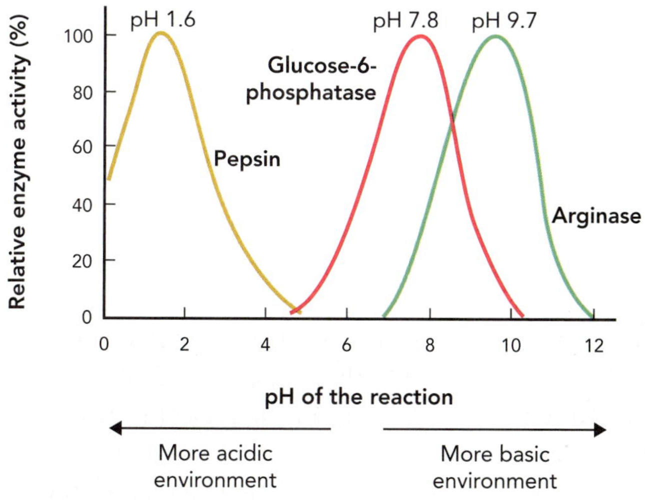

Figure 7.40 The optimal pH of an enzyme reaction reflects the chemical environment of the active site and is consistent with the physiologic role of the enzyme.

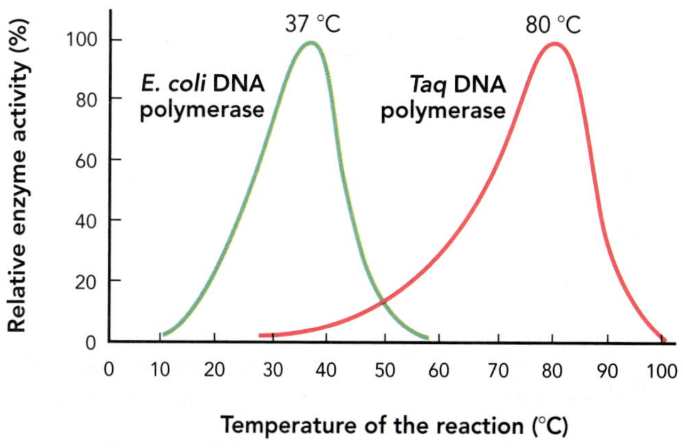

Figure 7.41 *E. coli* DNA polymerase and *Taq* DNA polymerase from the thermophilic bacterium *Thermus aquaticus* have different optimal temperatures for activity, reflecting their environmental niches.

concept integration 7.4

What can enzyme kinetics reveal about enzyme mechanisms?

Enzyme kinetics is an experiment-based approach used to study enzyme function and provides critical insight into the mechanism of an enzyme-catalyzed reaction. The primary goal in enzyme kinetics is to use quantitative analysis directly to compare the rates of reactions under various conditions of substrate concentration, temperature, and pH to understand catalytic mechanisms. Enzyme kinetics can also be used to discover the molecular basis for diseases in which a mutant enzyme is expressed in cells, but it is not known what accounts for decreased product formation. Michaelis–Menten kinetics is a mathematical description of enzyme reactions that take place at steady state. Measurements of maximum reaction velocity (v_{max}), substrate concentration at half-maximal velocity (K_m), and turnover number (k_{cat}) are parameters that can be used to compare related enzymes and to determine the potency of enzyme inhibitors. In the case of a disease caused by a defective enzyme, quantitative differences in v_{max}, K_m, and k_{cat} between the wild-type and mutant enzymes can provide clues that could lead to the development of new treatment strategies.

7.5 Regulation of Enzyme Activity

One of the reasons enzymes make such great biological catalysts is that they are proteins, which means they are subject to multiple levels of biochemical regulation. This regulation could be stimulatory, resulting in an overall increase in enzyme activity, or it could be inhibitory, resulting in a decrease in enzyme activity.

Enzyme regulation is mediated by two primary mechanisms: (1) bioavailability with regard to the amount of enzyme in different tissues and cellular compartments, and (2) control of catalytic efficiency through protein modification (covalent and noncovalent chemical bonds). For example, a positive regulatory mechanism could involve an increase in enzyme activity as a result of increased enzyme synthesis or could be due to enhanced catalytic efficiency as a result of conformational changes in the enzyme active site. Negative regulation would be due to opposite mechanisms; that is, decreased enzyme bioavailability or disturbance of the protein structure.

As shown in **Figure 7.42**, the biochemical processes affecting enzyme bioavailability are RNA synthesis (gene transcription); RNA processing (alternative joining of coding regions—exons—of the mRNA, which will result in different protein products); protein synthesis (amount of enzyme produced); protein degradation (protein turnover); and protein targeting (organelle sublocalization or membrane insertion). These regulatory control points are illustrated in Figure 7.42 and described in more detail in Part 5 of this book. Because Part 2 of the book focuses on protein structure and function, in this section we present enzyme regulatory mechanisms that affect catalytic efficiency. The three primary mechanisms that affect catalytic efficiency are (1) binding of regulatory molecules, (2) covalent modification, and (3) proteolytic processing (Figure 7.42). We will discuss control of activity by the binding of regulatory molecules from two different perspectives: First, we discuss molecules that function as enzyme inhibitors; then, we discuss how an enzyme can be activated or inhibited by allosteric regulators and use the bacterial

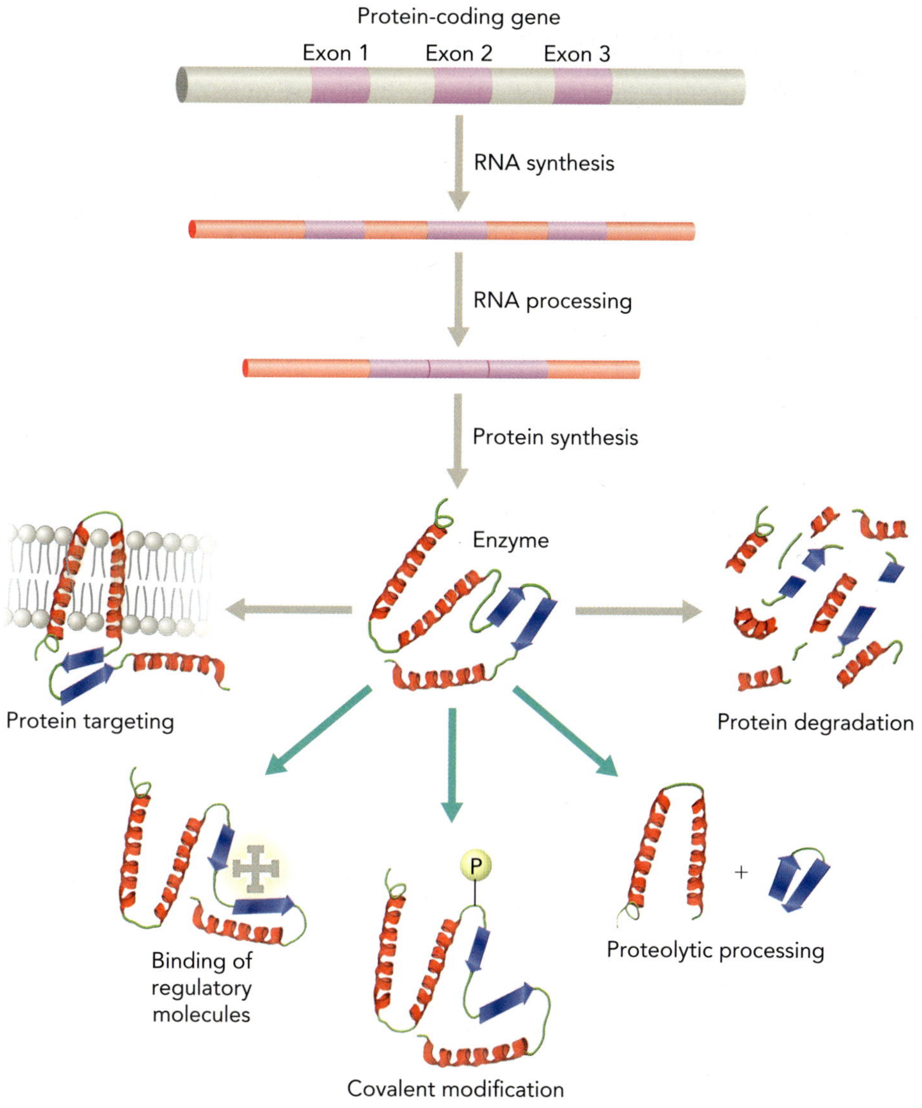

Figure 7.42 Enzyme bioavailability and catalytic efficiency are the two main enzyme regulatory mechanisms. Enzyme bioavailability is controlled by biochemical processes involved in protein synthesis and localization (gray arrows), whereas catalytic efficiency is determined by protein modifications (teal arrows).

enzyme aspartate transcarbamoylase as an example. Most enzymes are regulated by more than one mechanism, which function together to provide a fine-tuned level of control.

Mechanisms of Enzyme Inhibition

Enzyme inhibition is used in cells as a regulatory mechanism to control enzyme activity. Enzyme inhibitors are also used in *in vitro* studies to learn about the catalytic mechanisms of enzymes. Enzymes are subject to both **reversible inhibition**, due to the noncovalent binding of small biomolecules or proteins to the enzyme subunit, and **irreversible inhibition**, in which the inhibitory molecule forms a covalent bond (or, less commonly, very strong noncovalent interactions) with catalytic groups in the enzyme active site. These two types of inhibition can be distinguished by analyzing the effect of enzyme dilution on the level of inhibition. Specifically, the effect of reversible inhibitors can be decreased by diluting the enzyme reaction, whereas irreversible inhibitors are not affected by dilution because the covalent bond remains intact independent of enzyme and inhibitor concentrations.

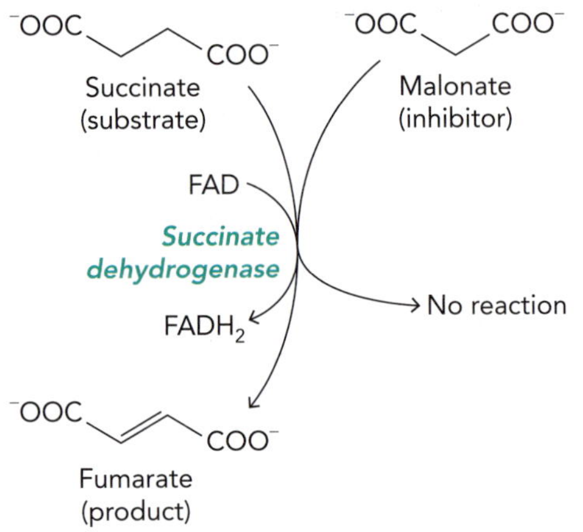

Figure 7.43 Malonate is structurally related to succinate and functions as a reversible inhibitor by competing with succinate for binding to the succinate dehydrogenase active site. Because malonate has only one methylene group, the oxidation reaction cannot proceed, and the enzyme is inhibited.

Malonate is an example of a reversible inhibitor of the enzyme succinate dehydrogenase. It functions by competing with succinate, the normal substrate of the enzyme, for active site binding (**Figure 7.43**). Under conditions where malonate levels are higher in the cell than levels of succinate, the succinate dehydrogenase reaction is inhibited by excess malonate binding to the active site. However, when malonate levels decrease, succinate binds to the enzyme more often, and the rate of fumarate production increases.

In contrast, irreversible inhibitors effectively "kill" the enzyme by forming a tight complex with the enzyme, sometimes through formation of a covalent nonhydrolyzable enzyme-inhibitor complex. The α_1-antitrypsin inhibitor is an example of a regulatory protein that irreversibly inactivates serine proteases, such as elastase, through formation of a covalent complex. **Figure 7.44** illustrates how the irreversible enzyme inhibitor **diisopropylfluorophosphate**—a compound that forms a covalent link with specific reactive serine residues—blocks protease and phospholipase enzymes. Diisopropylfluorophosphate is not found in cells but is often used in *in vitro* studies to identify reactive active site serine residues.

Reversible inhibition is the more common mechanism used in cells. It can also be exploited as a strategy to develop structure-based pharmaceutical drugs because the effects can be reversed by drug withdrawal. Three classes of reversible inhibitors have been characterized: **competitive inhibitors, uncompetitive inhibitors**, and **mixed inhibitors. Figure 7.45** is a schematic depiction of the molecular mechanisms and kinetic parameters of reversible enzyme inhibitors.

Competitive Inhibitors A competitive inhibitor is defined as a molecule that inhibits substrate binding at the active site (Figure 7.45a). A classic competitive inhibitor binds directly to the active site. However, some competitive inhibitors only partially obstruct the substrate from binding, either through steric clash or through partial occupancy of the active site by similar binding groups. Other competitive inhibitors bind at sites on the enzyme other than the active site, but result in the inability of the active site to bind substrate. For example, an inhibitor might cause a conformational change that closes the active site.

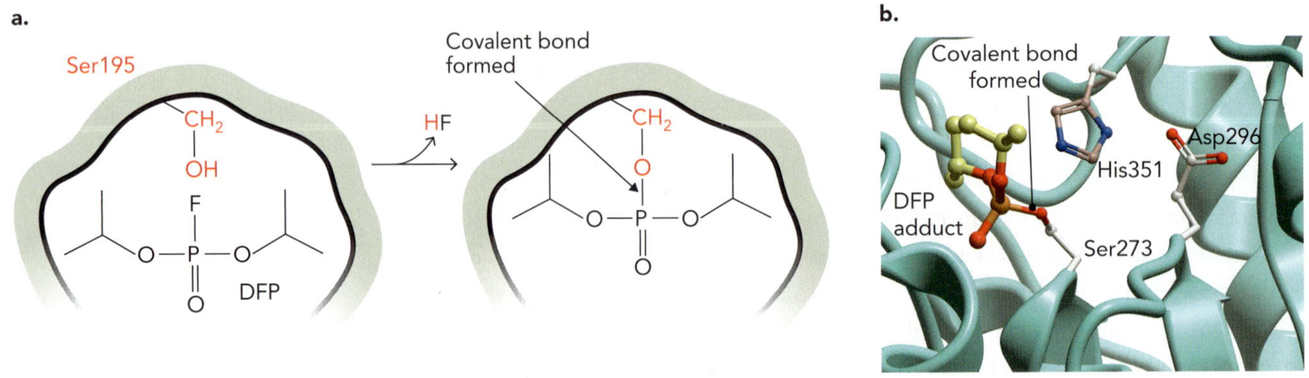

Figure 7.44 Diisopropylfluorophosphate (DFP) is an irreversible inhibitor of enzymes that contain catalytic serine residues in the active site. **a.** Schematic representation of a covalent link that forms between diisopropylfluorophosphate and the catalytic Ser195 residue located in the active site of the serine protease chymotrypsin. **b.** Molecular structure of the human phospholipase A_2 enzyme, showing the location of a diisopropylfluorophosphate adduct formed with Ser273 in the active site. The other two residues contributing to the catalytic triad in this enzyme active site are His351 and Asp296. BASED ON PDB FILE 3F9C.

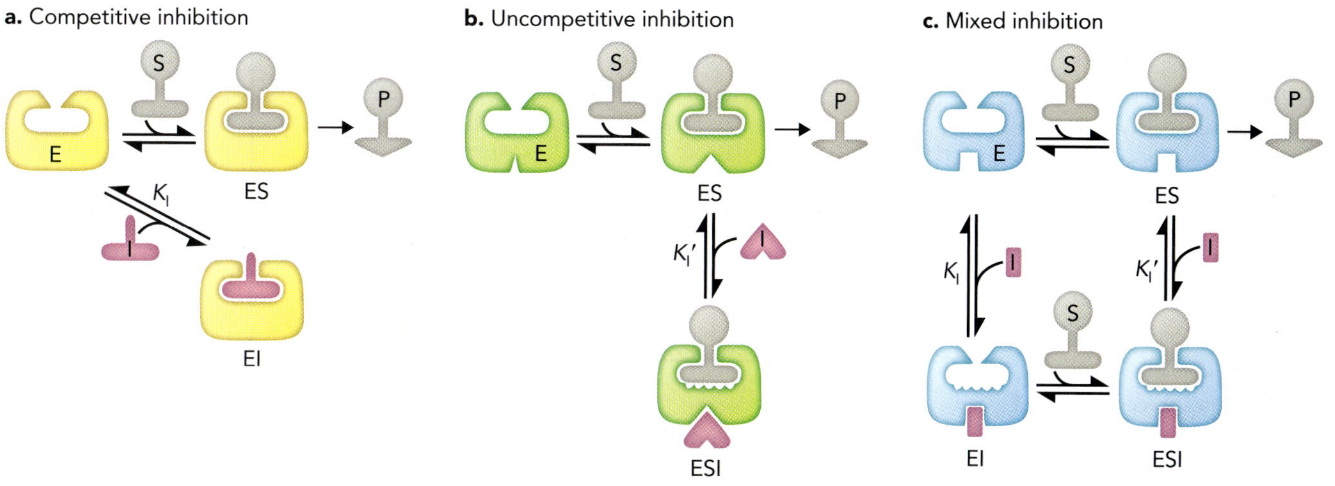

Figure 7.45 Three distinct classes of reversible enzyme inhibitors have been characterized. **a.** Competitive inhibition is characterized by inhibitors that bind to the free enzyme and inhibit substrate binding at the active site. Competitive inhibitors often bind to the active site or to a site that overlaps the active site on the enzyme. The equilibrium dissociation constant for inhibitor binding is defined by K_I (see text). **b.** Uncompetitive inhibition is characterized by inhibitors that bind only to the enzyme–substrate complex. **c.** Mixed inhibitors bind to both the enzyme and the enzyme–substrate complex. In all cases, inhibitor-bound complexes do not catalyze product formation. E = enzyme; I = inhibitor; P = product; S = substrate; EI = enzyme–inhibitor complex; ES = enzyme–substrate complex; ESI = enzyme–substrate–inhibitor complex.

The affinity of an enzyme for inhibitor can be described by the equilibrium dissociation constant K_I as follows:

$$\mathrm{E} + \mathrm{I} \rightleftharpoons \mathrm{EI} \qquad K_I = \frac{[\mathrm{E}][\mathrm{I}]}{[\mathrm{EI}]}$$

Note that this is similar to the equilibrium dissociation constant K_d discussed for general protein–ligand complexes in Chapter 6; however, K_I specifically refers to the equilibrium dissociation constant for an enzyme–inhibitor complex. In competitive inhibition, the substrate concentration required to reach $\frac{1}{2}v_{max}$ increases (an increase in apparent K_m; $K_{m\text{-app}}$), because more substrate is required to form the same amount of ES complex. This can be stated as

$$K_{m\text{-app}} = K_m\left(1 + \frac{[\mathrm{I}]}{K_I}\right)$$

This equation takes into account both the concentration of inhibitor, [I], and the dissociation constant of the enzyme–inhibitor complex, K_I. In the presence of a competitive inhibitor, the Michaelis–Menten equation becomes

$$v_0 = \frac{v_{max}[\mathrm{S}]}{K_m\left(1 + \frac{[\mathrm{I}]}{K_I}\right) + [\mathrm{S}]}$$

Here you can see that the effectiveness of a competitive inhibitor depends on the concentrations of inhibitor and substrate and the relative affinities of the enzyme for inhibitor and substrate.

A plot of reaction velocity versus substrate concentration shows the effect of adding a competitive inhibitor (**Figure 7.46a**). Note that because the inhibitor competes for binding to the active site, the apparent K_m value for the enzyme is shifted to the right as a function of increasing inhibitor concentration. With addition of sufficiently high concentrations of substrate, the effects of the inhibitor can be overcome, and therefore competitive inhibitors do not affect the maximum velocity, v_{max}, of the reactions. The effect of a competitive inhibitor on enzyme kinetics can be seen more easily by plotting the data as a Lineweaver–Burk plot (**Figure 7.46b**). In this double reciprocal plot, it is clear that v_{max} is not affected by inhibitor concentration, whereas the apparent K_m of the enzyme increases.

Structure-based drug design is an active area of research in which knowledge of an enzyme's structure is used to devise an inhibitor with shape and chemical

Figure 7.46 A competitive inhibitor affects the apparent K_m of an enzyme but not the v_{max}. **a.** Plot of initial velocity (v_0) versus substrate concentration ([S]) in the absence and presence of inhibitor. With increasing concentration of inhibitor ([I]), the apparent K_m of the enzyme increases, reflecting the requirement of higher substrate concentration to reach $\frac{1}{2}v_{max}$. **b.** Lineweaver–Burk plot of the data shown in panel **a.** Note that v_{max} is unaffected by an increase in inhibitor concentration (*y*-axis intercept remains constant), whereas the apparent K_m increases, as shown by the decrease in the $-1/K_{m\text{-app}}$ value at the *x*-axis intercept.

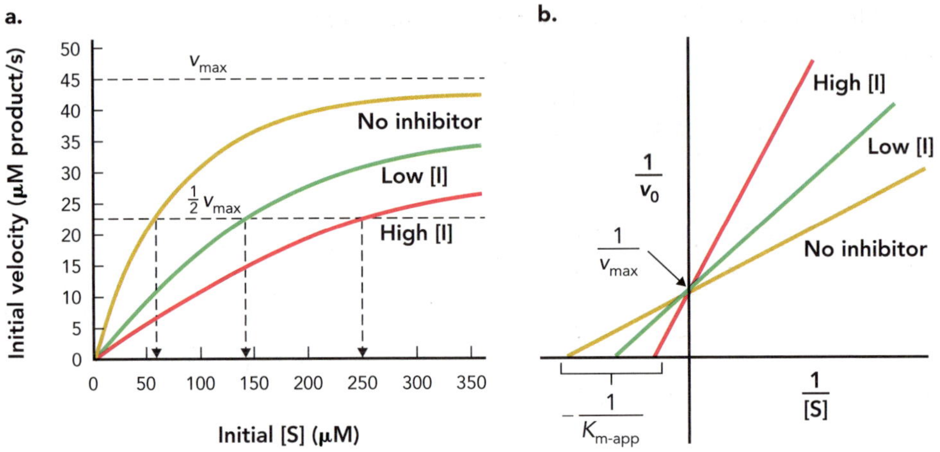

complementarities to the enzyme's active site. An example of structure-based drug design using these ideas is the development of competitive inhibitors for the aspartate protease enzyme in human immunodeficiency virus (HIV). This viral enzyme is required for proper maturation of infectious particles. The enzyme cleaves viral proteins at specific sites between Phe-Pro or Tyr-Pro residues, using a pair of catalytic Asp residues in the active site. As shown in **Figure 7.47**, the HIV protease inhibitors indinavir and saquinavir contain a component in their structures that mimics the natural Phe-Pro dipeptide substrate. Both of these competitive inhibitors bind tightly and reversibly to the enzyme active site, which is located at the dimer interface (**Figure 7.48**).

Uncompetitive Inhibitors Uncompetitive inhibitors bind to enzyme–substrate complexes and alter the active site conformation, thus rendering the enzyme less catalytically active (see Figure 7.45b). An uncompetitive inhibitor does not bind to the free enzyme.

Uncompetitive inhibitors for enzymes with a single substrate are rare. More often, uncompetitive inhibitors act upon enzymes with multiple substrates that are added sequentially. Even in situations with high substrate concentrations, some of the enzyme remains in a nonproductive ESI form. Furthermore, because some ES complex is used to make the enzyme–substrate–inhibitor (ESI) complex, the apparent K_m decreases as well. Thus, the net effect of an uncompetitive inhibitor is to decrease both K_m and v_{max}. However, because K_m and v_{max} decrease by the same factor, the slope of the line K_m/v_{max} in a Lineweaver–Burk plot is unaffected.

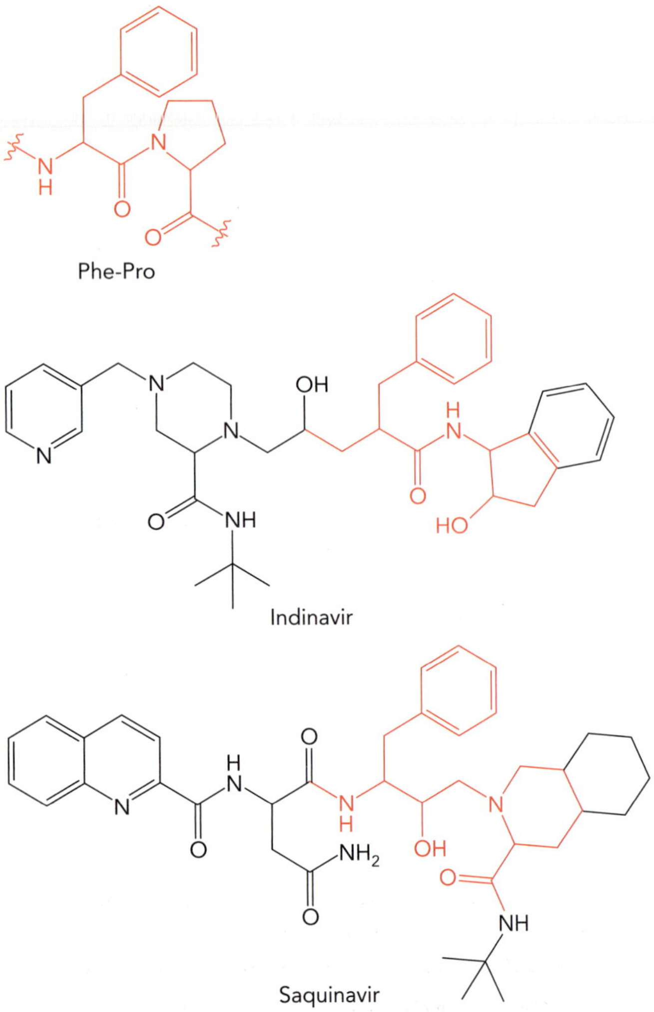

Figure 7.47 Indinavir and saquinavir are competitive inhibitors of the HIV aspartate protease enzyme required for maturation of viral proteins. The regions of the chemical structures of these inhibitors that mimic the HIV protease substrate Phe-Pro dipeptide are highlighted.

The Michaelis–Menten equation for an uncompetitive inhibitor is

$$v_0 = \frac{v_{max}[S]}{K_m + \left(1 + \frac{[I]}{K_I'}\right)[S]}$$

Here, the dissociation constant K_I' reflects the affinity of inhibitor to the enzyme–substrate complex:

$$ES + I \rightleftharpoons ESI \qquad K_I' = \frac{[ES][I]}{[ESI]}$$

The effect of uncompetitive inhibition on enzyme activity can be seen in a Lineweaver–Burk plot over a range of inhibitor and substrate concentrations (**Figure 7.49**).

Mixed Inhibitors Mixed inhibitors are similar to uncompetitive inhibitors in that they bind to sites distinct from the active site. The main difference is that mixed inhibitors can bind to both the enzyme and the enzyme–substrate complex (see Figure 7.45c), which can be seen as a mixture of both competitive and uncompetitive inhibition and thus is termed *mixed*. An unproductive EI or ESI complex is formed in which the enzyme's catalytic activity is decreased, potentially through structural alterations of the catalytic residues. As with uncompetitive inhibition, because some enzyme remains in an unproductive ESI complex, high concentrations of substrate cannot overcome the presence of a mixed inhibitor, resulting in a decreased v_{max}. Because some enzyme is bound to inhibitor, the overall effect is that less enzyme appears to be present. The Michaelis–Menten equation for a mixed inhibitor is

$$v_0 = \frac{v_{max}[S]}{K_m\left(1 + \frac{[I]}{K_I}\right) + [S]\left(1 + \frac{[I]}{K_I'}\right)}$$

A mixed inhibitor decreases v_{max} and may increase or decrease K_m, depending on the relative K_I and K_I' values (relative inhibitor affinity for free enzyme and ES complex, respectively). **Figure 7.50a** shows the Lineweaver–Burk plots for two mixed inhibitors.

A special rare case of mixed inhibition is called **noncompetitive inhibition**, where the inhibitor has equal affinity for both the free enzyme and the ES complex ($K_I = K_I'$). In the case of noncompetitive inhibition, the Michaelis–Menten equation becomes

$$v_0 = \frac{v_{max}[S]}{(K_m + [S])\left(1 + \frac{[I]}{K_I}\right)}$$

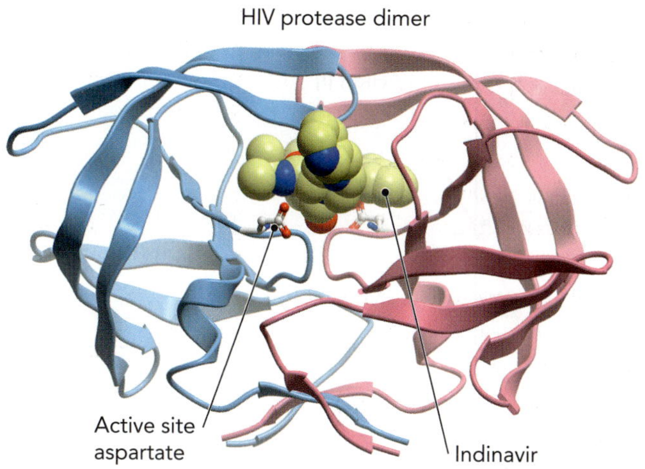

Figure 7.48 Structure of an HIV protease dimer with the competitive inhibitor indinavir bound to the active site. The catalytic aspartate residues located in each monomer are observed in the enzyme active site. BASED ON PDB FILE 1K6C.

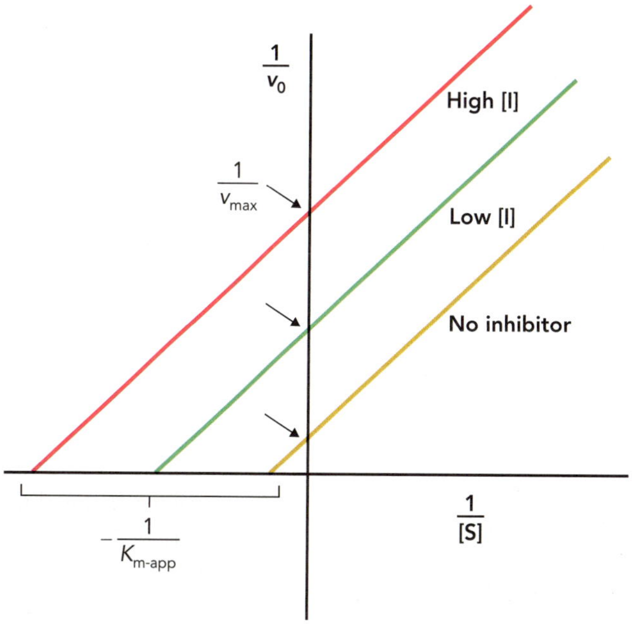

Figure 7.49 Uncompetitive inhibitors decrease both the v_{max} and apparent K_m kinetic parameters by the same factor. Note that unlike competitive inhibition, uncompetitive inhibition is not overcome by increasing substrate concentration.

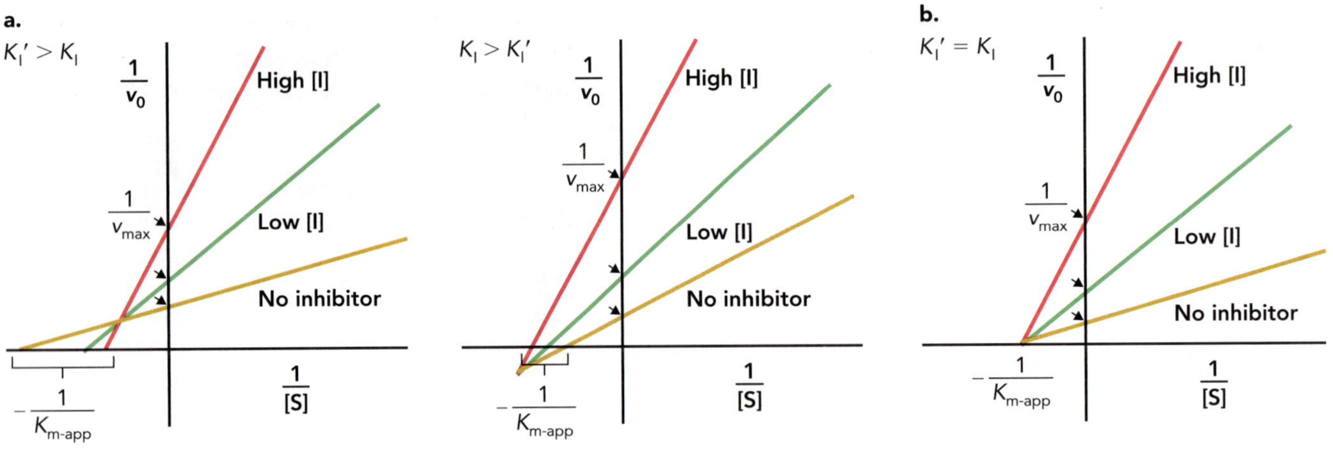

Figure 7.50 **a.** Mixed inhibitors decrease v_{max} and increase or decrease K_m, depending on the relative values of K_I and K_I'. In the plot on the left, where K_I' is greater than K_I, the K_m is increased, indicating decreased affinity of enzyme for substrate in the presence of inhibitor. **b.** Noncompetitive inhibitors decrease v_{max}, though the K_m value is unaffected.

Figure 7.50b shows the Lineweaver–Burk plot for a noncompetitive inhibitor. The K_m value is unaffected, as both the free enzyme and the enzyme–inhibitor complex can bind to substrate.

Both uncompetitive and mixed inhibitors affect multi-substrate enzymes and often can be used to distinguish different types of enzyme reaction mechanisms. For example, an inhibitor that binds to a site occupied by one substrate (S1) may influence binding of another substrate (S2) to its binding site. This would be an example of uncompetitive or mixed inhibition with regard to S1 binding. By altering the concentration of substrates and inhibitors, enzyme kinetics can be used to determine if substrate binding is ordered (S1 binds before S2) or random (either S1 or S2 can bind first).

Allosteric Regulation of Catalytic Activity

The first step in a metabolic pathway is often controlled by a regulated enzyme to maximize the efficient use of metabolic intermediates. For example, if the end product of a metabolic pathway accumulates in the cell, it is prudent to decrease production of the metabolite by inhibiting the first step in the pathway. In the simplest case of **feedback inhibition**, the end product of a pathway functions as an inhibitor of the first enzyme in the pathway. Similarly, if metabolic products of a particular pathway are in short supply, molecules that function as metabolic activators can bind to and stimulate the activity of key regulated enzymes.

The bacterial enzyme aspartate transcarbamoylase (ATCase) is one of the best examples of regulation of an enzyme by metabolic products of its associated pathway or related metabolites. It is also a classic example of allosteric regulation of catalytic activity. ATCase is one of the key regulated enzymes in the pyrimidine biosynthetic pathway that leads to the production of cytidine triphosphate (CTP; see Chapter 18). ATCase catalyzes the formation of *N*-carbamoyl-L-aspartate from carbamoyl phosphate and aspartate, which is an early step in the CTP synthesis pathway (**Figure 7.51**). ATCase is feedback inhibited by the product of the pathway, CTP, and is activated by ATP, as described shortly. Both CTP and ATP are allosteric regulators of enzyme activity. (Recall that allosteric regulation was first described in Chapter 6 in the context of oxygen binding to hemoglobin and the role of 2,3-bisphosphoglycerate [2,3-BPG]

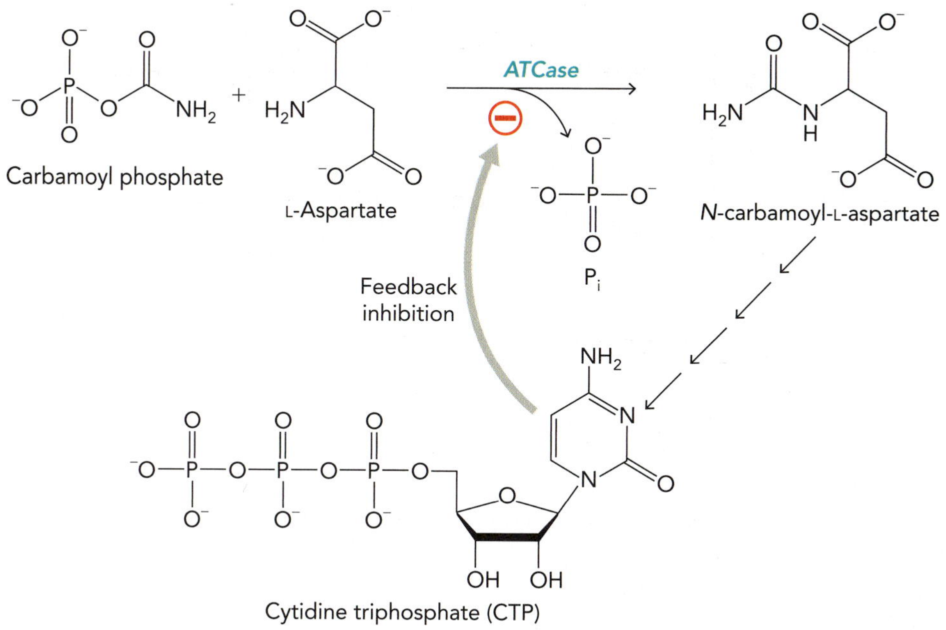

Figure 7.51 ATCase activity is feedback inhibited by CTP, the end product of the pyrimidine pathway.

in shifting the equilibrium between the T state [low oxygen affinity] and R state [high oxygen affinity].) Allosteric regulation allows for a higher degree of control of enzyme activity.

Accumulation of CTP in the cell is an indication that nucleic acid biosynthesis is saturated, which results in feedback inhibition of ATCase enzyme activity by CTP. In contrast, ATP is a purine-containing nucleotide, and the cell must maintain equivalent amounts of purine-containing and pyrimidine-containing nucleotides for DNA synthesis. Moreover, high levels of ATP increase the energy charge of the cell and favor DNA synthesis (purine-containing and pyrimidine-containing nucleotides are needed for DNA synthesis; see Chapter 18). Therefore, ATP-mediated stimulation of ATCase activity ensures that sufficient amounts of CTP will be available.

ATCase is a multi-subunit protein containing both catalytic (C) and regulatory (R) subunits with an overall organization of C_6R_6. ATCase is an enzyme that shows cooperativity, in that the binding of the reaction substrates carbamoyl phosphate and aspartate to an active site in one catalytic subunit of the enzyme increases the affinity of the other active sites for their substrates. (This situation is analogous to the cooperative effects that O_2 binding has on oxygen saturation of the hemoglobin tetramer; see Chapter 6.) The substrates carbamoyl phosphate and aspartate can also be called homotropic allosteric activators, as they affect the binding of the same molecules at other active sites. Cooperative enzymes such as ATCase do not follow standard Michaelis–Menten kinetics because substrate binding by one subunit changes the substrate affinity of other subunits. This is apparent in the sigmoidal shape of the activity curve, as shown in **Figure 7.52**. This sigmoidal shape shows that enzyme activity can increase or decrease significantly over a relatively small range of substrate concentrations.

CTP and ATP, which are not substrates for the ATCase-catalyzed reaction, bind to the regulatory subunits

Figure 7.52 ATCase is a cooperative enzyme and is allosterically regulated by ATP and CTP. The initial velocity (v_0) versus substrate concentration curve of ATCase activity shifts in the presence and absence of allosteric effectors. In the absence of both regulators, the behavior of the enzyme is shown by the curve marked "control." Note that the shape of the activity curves are sigmoidal (reflecting cooperatively in the enzyme) and therefore not amenable to analysis by Michaelis–Menten kinetics.

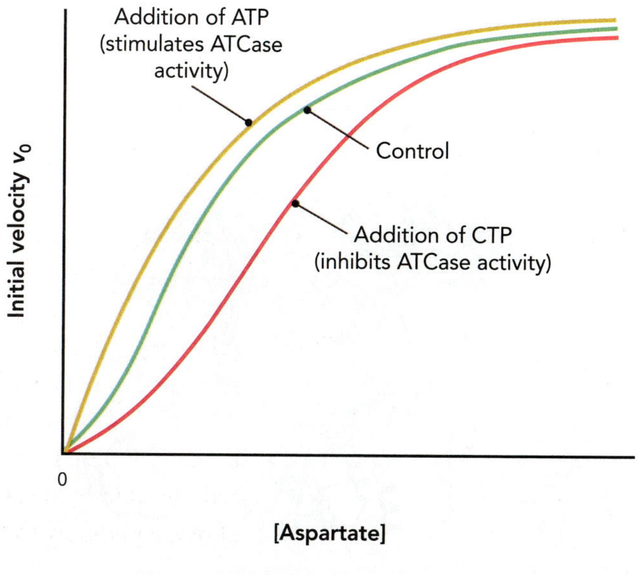

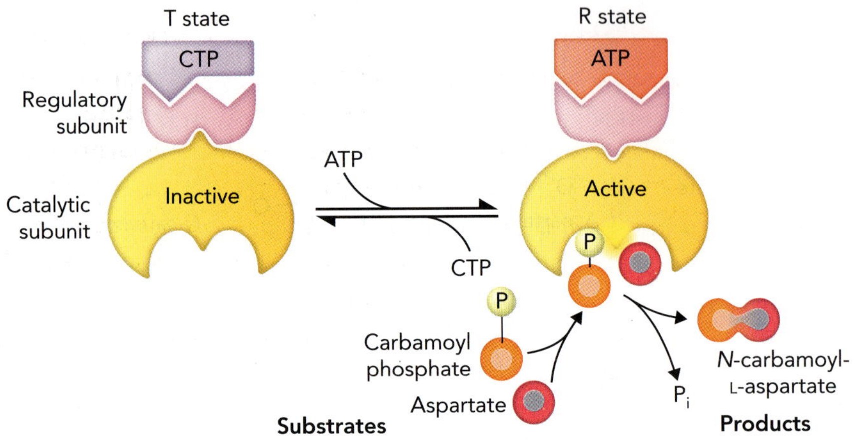

Figure 7.53 Allosteric regulation of ATCase activity by CTP and ATP. Binding of ATP to the regulatory subunit shifts ATCase to the activated R state, which catalyzes the formation of *N*-carbamoyl-L-aspartate from carbamoyl phosphate and aspartate. CTP binding to the regulatory subunit stabilizes the inactive T-state conformation.

▶ ANIMATION

Figure 7.54 In this molecular structure of an ATCase trimer (C_3R_3), a bisubstrate analog (*N*-phosphonacetyl-L-aspartate) is bound in each of the three catalytic subunits. The regulatory subunits each contain a coordinated zinc ion, which has been shown to be required for stability of the tertiary structure. In ATCase, the regulatory molecules bind at a site distant from the active site. Ligand binding at the regulatory sites induces conformational changes that affect activity at the catalytic sites. BASED ON PDB FILE 1Q95.

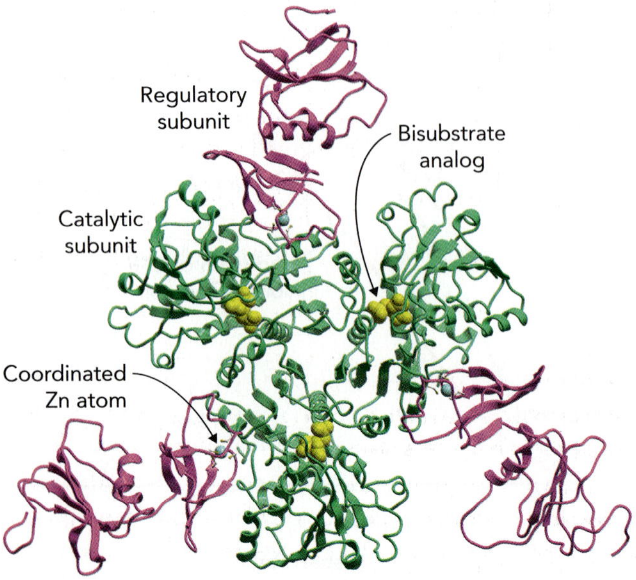

and affect enzyme activity. Binding of CTP or ATP to the regulatory subunits induces a conformational change in the protein that affects catalytic activity at the active site. **Figure 7.53** illustrates how CTP and ATP shift the equilibrium of ATCase between the inactive state (T state) and the active state (R state). CTP and ATP are considered heterotropic allosteric regulators (similar to 2,3-BPG for hemoglobin) because their binding affects the binding of different molecules (the substrates) to other sites. The addition of ATP or CTP to the reaction mixture will shift the enzyme activity curve to the left or right, respectively, reflecting changes in enzyme activity (see Figure 7.52).

Figure 7.54 shows the structure and organization of three ATCase dimers, each containing one catalytic subunit (C) and one regulatory subunit (R), forming a C_3R_3 trimeric protein complex. This structure illustrates the relationship between the regulatory nucleotide binding site and the active site. Though the regulatory and catalytic sites reside in separate subunits and are separated from each other in the assembled protein, regulation is achieved through binding-induced conformational changes that are propagated through the subunits. Two C_3R_3 complexes associate to form the functional C_6R_6 ATCase complex, which is shown in **Figure 7.55** in both the T-state and R-state conformations. The most notable difference in these two conformations is the increased separation of catalytic subunits in the R conformation. This rearrangement eliminates a region of steric interference between two catalytic subunits in the T conformation. Careful inspection reveals that the catalytic subunits have also rotated in opposite directions relative to each other. Although not evident from the T-state and R-state conformations shown in Figure 7.55, the geometry of the active site in the R-state conformation stimulates catalysis by bringing the carbamoyl phosphate and aspartate substrates into close proximity.

As you have now seen, regulatory molecules can either activate or inhibit enzyme activity. For example, in the case of ATCase, CTP is a negative heterotropic allosteric regulator, which could also be called an enzyme inhibitor. But CTP is not a competitive inhibitor, as it does not bind at the active site. Positive and negative regulators can also be called activators or inhibitors, respectively.

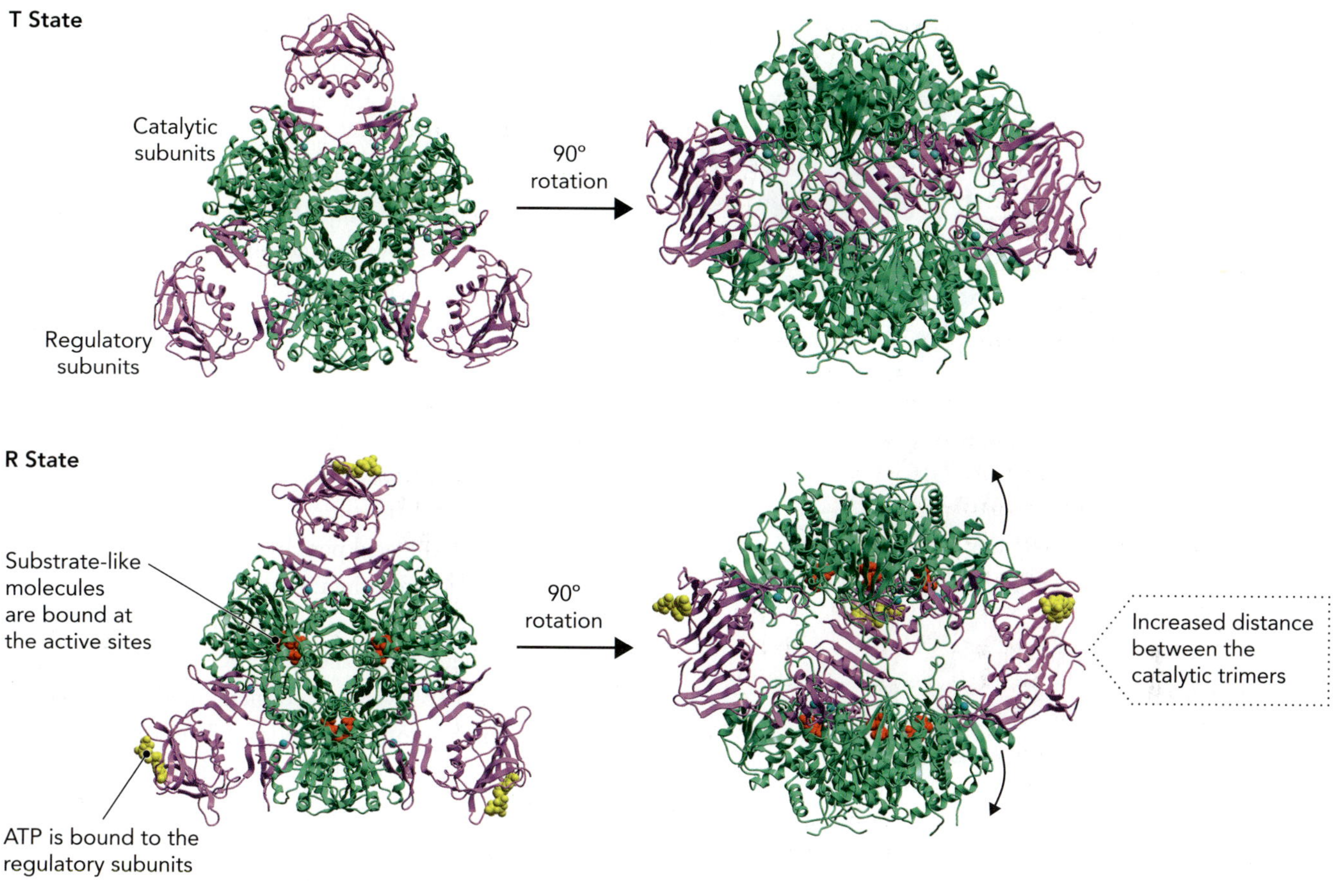

Figure 7.55 The molecular structures of ATCase in the T-state and R-state conformations are shown here in two orthogonal views. The six regulatory subunits are shown in magenta, and the six catalytic subunits are shown in green. In the R-state conformation (bottom), the allosteric activator ATP (yellow) is shown bound to three of the six regulatory sites. The R-state molecule has substrate analogs (phosphonoacetamide and malonate) bound in each of the six active sites. The transition between the T and R states involves significant movements between the subunits, resulting in conformational changes at the active site that affect reactivity. BASED ON PDB FILES 6AT1 (T-STATE CONFORMATION) AND 7AT1 (R-STATE CONFORMATION).

Covalent Modification of Enzymes

As we described earlier in the context of reversible covalent modifications (see Section 7.2), a large number of enzymes catalyze reactions involving the covalent linkage of functional groups to target biomolecules in the cell. In some cases, enzymes are themselves the target of covalent modification, and the addition or removal of functional groups on these enzymes can regulate their catalytic efficiency. The functional groups involved in this form of enzyme regulation can be phosphoryl groups, methyl or acetyl groups, and adenylyl or uridylyl groups.

The most common of these covalent modifications in enzymes is the phosphorylation of Ser, Thr, and Tyr residues by **kinase** enzymes, which add a large, negatively charged group to the enzyme through the addition of a phosphoryl group (PO_3^{2-}). This can alter the chemistry or structure of the enzyme active site and thereby regulate catalytic activity. Phosphoryl groups are removed from proteins by phosphatase enzymes, which hydrolyze the ester linkage between the phosphate and the amino acid side chain. Some enzymes are activated by phosphorylation, whereas other enzymes are inactivated by phosphorylation. This process of covalent modification by phosphorylation–dephosphorylation results in a functional on/off switch for enzyme activity in much the same way it does for signal transduction (see Figure 7.25).

An example of how phosphorylation controls enzyme activity is the regulation of glycogen phosphorylase. This enzyme degrades glycogen—a storage form of glucose in liver and muscle cells—to generate glucose for energy conversion reactions associated with the glycolytic pathway. In the unphosphorylated T-state conformation, glycogen phosphorylase is inactive, whereas it is active in the phosphorylated R-state conformation. Similar to ATCase, the regulatory site in glycogen phosphorylase is not directly at the active site

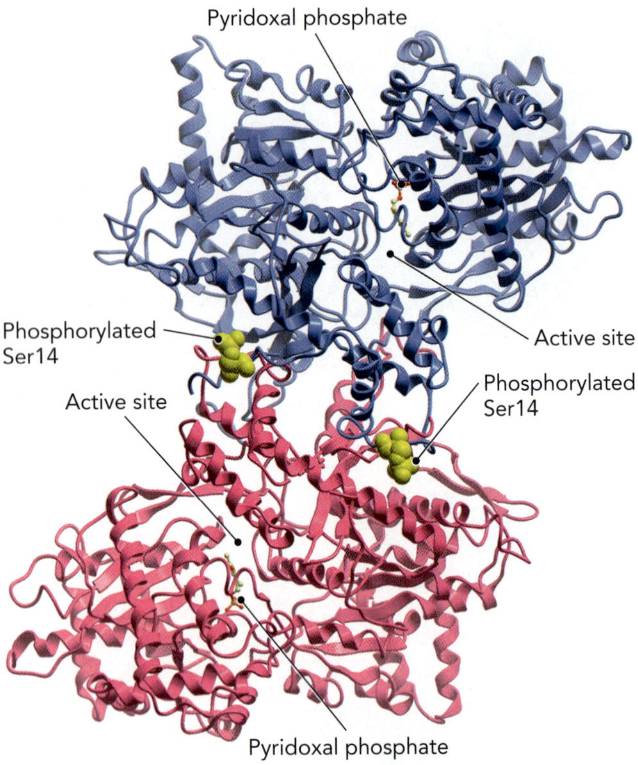

Figure 7.56 The molecular structure of the glycogen phosphorylase homodimer in the phosphorylated R-state conformation is shown here. The coenzyme pyridoxal phosphate is bound in the enzyme active site. The phosphorylated Ser residues in each monomer, shown in yellow, are at the subunit interface and are distinct from either active site. See also Figure 7.4. BASED ON PDB FILE 1GPA.

(**Figure 7.56**), but phosphorylation causes structural changes that are propagated to the active site to affect activity. Phosphorylation of Ser14 by the enzyme phosphorylase kinase shifts the equilibrium toward the R state; dephosphorylation of Ser14 by the enzyme protein phosphatase 1 shifts the equilibrium toward the T state (**Figure 7.57**).

Another type of covalent modification of enzymes that regulates catalytic efficiency is the addition and removal of nucleoside monophosphate (NMP) groups on tyrosine residues. A striking example of this regulatory mechanism is the inactivation of the *E. coli* glutamine synthetase enzyme by **adenylylation** of Tyr397 (**Figure 7.58**). Glutamine synthetase plays an integral role in bacterial nitrogen metabolism by catalyzing an ATP-dependent reaction incorporating free NH_4^+ into the amino acid pool by converting glutamate to glutamine. Adenylylated glutamine synthetase is in the inactive T-state conformation, and the deadenylylated form of glutamine synthetase is in the active R-state conformation. Unlike many other reversible covalent modification reactions, the adenylylation and deadenylylation of glutamine synthetase is catalyzed by the same multi-subunit enzyme in *E. coli*, called glutamine synthetase adenylyltransferase.

This raises a question: If the same enzyme catalyzes both reactions, how are the two opposing activities regulated to prevent futile cycling? The answer is through **uridylylation** of glutamine synthetase adenylyltransferase on Tyr51 by another bifunctional enzyme called uridylyltransferase. Under conditions of high ATP and elevated levels of the metabolite α-ketoglutarate, the *uridylylation activity* of uridylyltransferase is stimulated, resulting in uridylylation of glutamine synthetase adenylyltransferase and activation of its *deadenylylation activity* (**Figure 7.59**). However, when P_i and glutamine levels are elevated in the cell, the *deuridylylation activity* of uridylyltransferase is stimulated, leading to the deuridylylation of glutamine

Figure 7.57 Covalent modification can act as an on/off switch for enzyme activity. Phosphorylase kinase catalyzes the phosphorylation reaction at Ser14 that stimulates the catalytic efficiency of glycogen phosphorylase (active R-state conformation). Protein phosphatase 1 catalyzes the dephosphorylation reaction that shifts the equilibrium toward the inactive T-state conformation. The hormones glucagon and epinephrine result in stimulation of the activity of phosphorylase kinase, whereas the hormone insulin results in stimulation of the activity of protein phosphatase 1 (see Chapter 14).

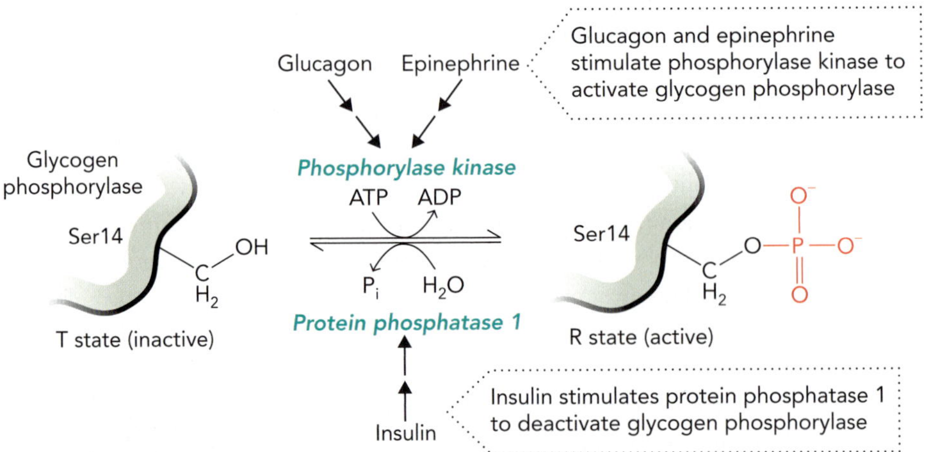

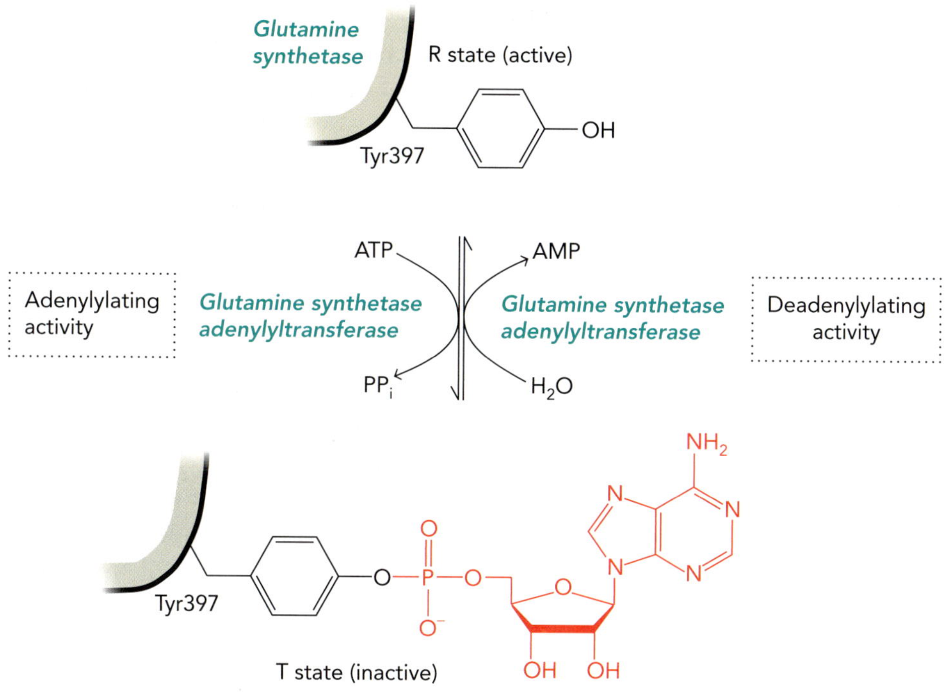

Figure 7.58 The glutamine synthetase adenylyltransferase protein complex catalyzes a reaction that covalently attaches an AMP moiety to a tyrosine residue on glutamine synthetase, which inhibits enzyme activity. A different subunit in the glutamine synthetase adenylyltransferase protein complex catalyzes the reverse reaction, which deadenylylates glutamine synthetase to activate the enzyme.

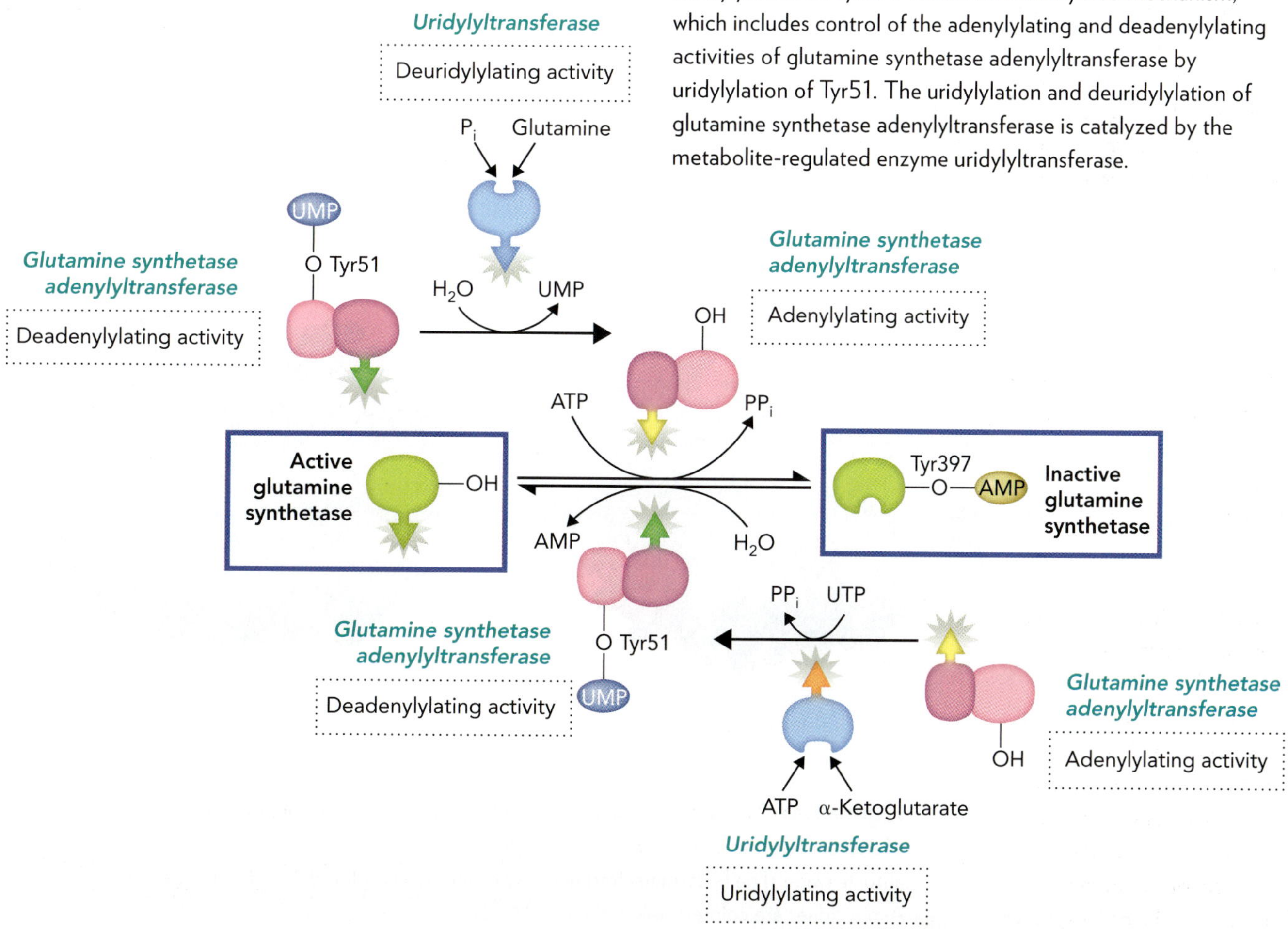

Figure 7.59 Regulation of glutamine synthetase activity by adenylylation on Tyr397 involves a multilayered mechanism, which includes control of the adenylylating and deadenylylating activities of glutamine synthetase adenylyltransferase by uridylylation of Tyr51. The uridylylation and deuridylylation of glutamine synthetase adenylyltransferase is catalyzed by the metabolite-regulated enzyme uridylyltransferase.

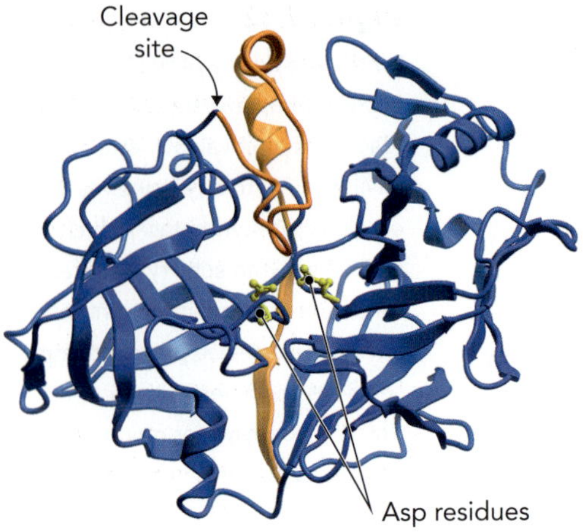

Figure 7.60 Pepsinogen is a zymogen that catalyzes an autocleavage reaction to create the active form of the enzyme pepsin. The N-terminal 44 amino acids (orange) that block access to the active site must be removed from pepsinogen by autocleavage to generate the active enzyme. The two catalytic Asp residues (yellow) in the active site are shown. BASED ON PDB FILE 2PSG.

synthetase adenylyltransferase and activation of its *adenylation activity*. This multilayered regulatory scheme is described in more detail in Chapter 17, when we describe regulation of nitrogen metabolism.

Enzymes Can Be Activated by Proteolysis

Most proteins fold into an active conformation as a concomitant step in protein synthesis. However, some enzymes—in particular, proteases involved in digestion and blood clotting—require subsequent processing by proteolytic cleavage to become fully active. If proteases were synthesized in their active form, they would digest cellular proteins indiscriminately and cause damage to the cell. Therefore, inactive precursor proteins called **zymogens**, or proenzymes, are synthesized with an active site that is inaccessible to protein substrates. The zymogen is converted to the active form of the protease by either an autocleavage reaction or by a trans cleavage reaction mediated by another protease. In many cases, the proteolytic processing reaction removes an N-terminal protein fragment present in the zymogen that prevents entry of the substrate to the enzyme active site.

For example, the removal of a 44-amino-acid N-terminal segment of pepsinogen, the zymogen, generates the proteolytically active form of the enzyme pepsin (**Figure 7.60**). This autocleavage reaction of pepsinogen, which occurs as both an intramolecular and intermolecular reaction, is stimulated by the acidic environment of the stomach (pH 1.5) and leads to the formation of an accessible substrate binding cleft (**Figure 7.61**). Unlike the reversible regulatory mechanisms discussed earlier, regulation of enzyme activity by proteolytic cleavage is an irreversible process.

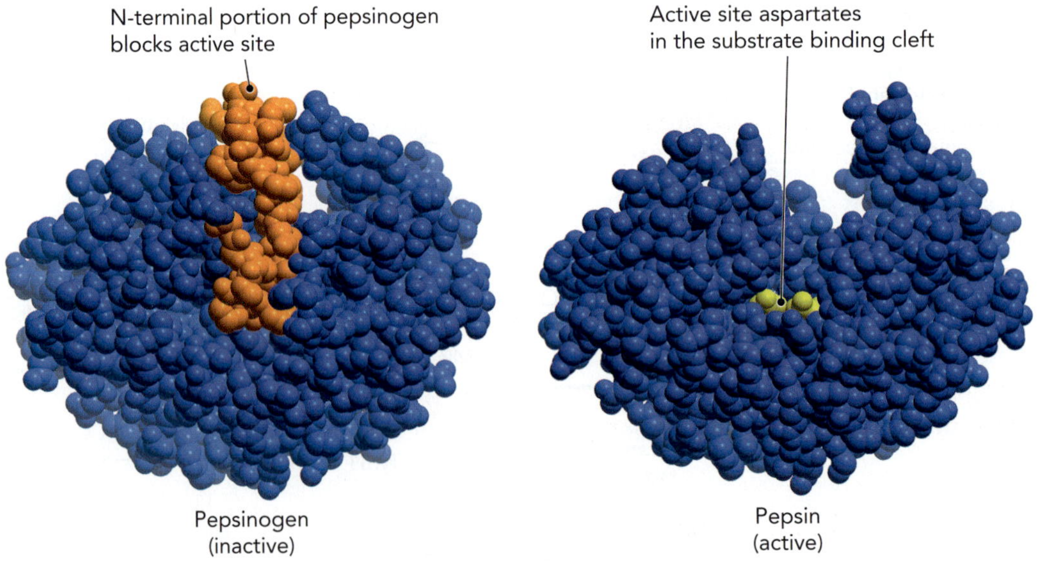

Figure 7.61 The catalytic Asp residues in the pepsin active site are located at the bottom of a large cleft that is accessible after the N-terminal fragment of pepsinogen is removed by autocleavage. The N-terminal fragment in pepsinogen is colored orange, and the catalytic Asp residues are colored yellow. BASED ON PDB FILES 2PSG (PEPSINOGEN) AND 5PEP (PEPSIN).

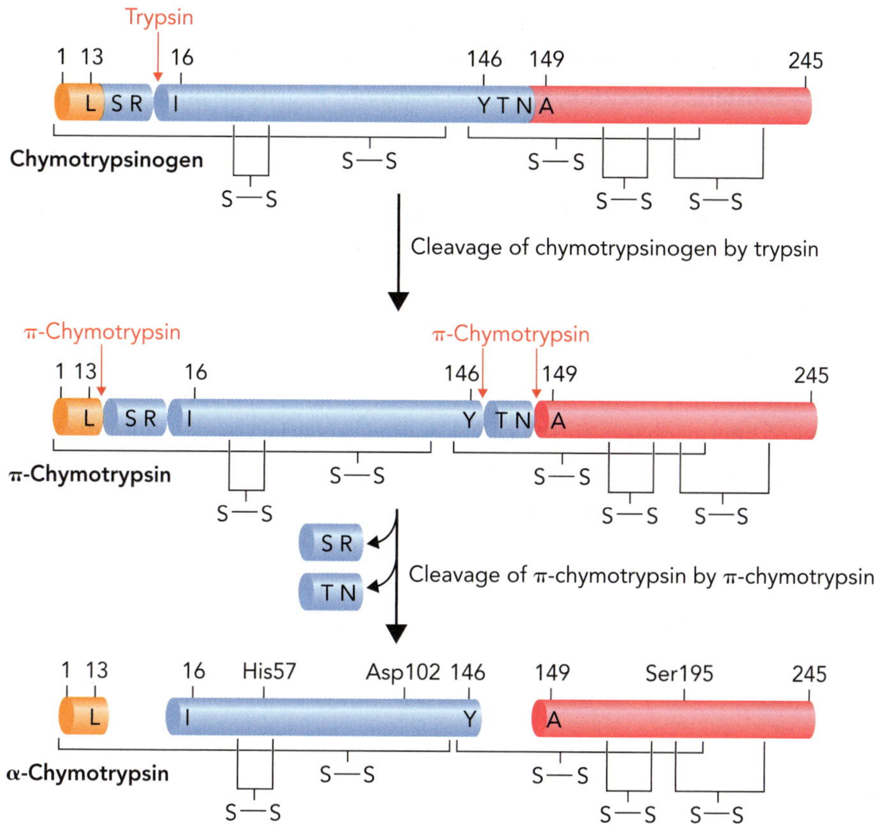

Figure 7.62 Chymotrypsin is generated by proteolytic processing of chymotrypsinogen into three polypeptide chains as shown by the colored bars. Trypsin cleaves chymotrypsinogen (inactive zymogen) to create π-chymotrypsin, which cleaves other π-chymotrypsin molecules to generate the fully active α-chymotrypsin enzyme. The relative positions of the five disulfide bonds and the three amino acids in the catalytic triad are shown (see also Figure 7.26).

Another example of enzyme regulation by protein processing is the conversion of chymotrypsinogen to its active form chymotrypsin, after four proteolytic cleavage events. These cleavages generate a serine protease consisting of three separate polypeptides held together by disulfide bonds (see Figure 7.26). Chymotrypsin is a digestive enzyme produced in the pancreas and secreted into the small intestine, where it digests small polypeptides that exit the stomach.

As shown in **Figure 7.62**, the first chymotrypsinogen cleavage reaction is catalyzed by the enzyme trypsin, which breaks the peptide bond on the C-terminal side of Arg15. The product of this trypsin cleavage reaction is a moderately active serine protease called π-chymotrypsin, which then functions in trans to cleave other π-chymotrypsin proteins on the C-terminal side of Leu13, Tyr146, and Asn148. The final product of these proteolytic processing events is the fully active form of the enzyme, called α-chymotrypsin. The key to this regulatory mechanism is the site-specific cleavage of chymotrypsinogen by trypsin, which is itself activated by a serine protease called enterokinase. Enterokinase is released into the duodenum after proteolytic cleavage of the membrane-bound enterokinase zymogen.

concept integration 7.5

How is the catalytic activity of an enzyme reversibly regulated to facilitate on/off control?

Enzyme activity in the cell is regulated by both bioavailability and catalytic efficiency. Regulating enzyme bioavailability can be a rather slow process, but in contrast, reversibly regulating the catalytic efficiency of an enzyme can be rapid and is an efficient means to control signal transduction and metabolic processes tightly. The three primary mechanisms of reversible regulation in the cell are competitive inhibition, allosteric regulation, and most forms of covalent modification. An example of reversible competitive inhibition is feedback control, where a downstream metabolite competes for substrate binding, thereby leading to decreased product formation. As product levels drop because of competitive feedback inhibition, the level of downstream metabolite decreases and the competitive inhibition is reversed. Another mechanism of feedback control is negative allosteric regulation, in which the metabolite is an allosteric effector molecule rather than a competitive inhibitor for substrate binding. In this case, the negative allosteric regulator need not be a specific metabolite in the pathway, but could be ATP or AMP, both of which are indicators of energy charge in the cell. Finally, reversible covalent modification of serine, threonine, and tyrosine residues by phosphorylation and dephosphorylation is another common mechanism of enzyme regulation. The presence or absence of a phosphoryl group can control catalytic efficiency by altering the structure and function of the enzyme. Kinases are phosphorylating enzymes, and phosphatases are dephosphorylating enzymes.

chapter summary

7.1 Overview of Enzymes

- Enzymes are biological catalysts that alter reaction rates without changing the overall ΔG or K_{eq} and are not consumed by the reaction.
- Substrates often bind with high affinity and specificity to an enzyme's active site, which is a cleft or pocket in the protein structure where the catalyzed reaction takes place.
- Induced fit is a term that describes how substrate binding to enzymes is often associated with structural changes in the enzyme. These changes maximize the number of weak noncovalent interactions between the substrate and amino acid residues in the enzyme active site.
- Enzyme activity is highly regulated in cells to maximize energy balance between anabolic and catabolic pathways and to alter cell behavior in response to environmental stimuli.
- According to transition state theory, a reactant must first reach an energy level required for chemical transformation before the product can be formed.
- The activation energy ($\Delta G^{\ddagger}$) is the difference between the ground state energy of the reactant and the transition state energy. Enzymes lower $\Delta G^{\ddagger}$ by providing a favorable physical and chemical environment in the active site to promote catalysis.
- Cofactors and coenzymes provide additional reactive groups to the enzyme active site that complement the limited chemistry of amino acid side chains. Cofactors are inorganic ions, whereas coenzymes are small organic molecules originally discovered as vitamins.
- The IUBMB enzyme classification system provides a standard nomenclature for enzymes. It is based on a hierarchical numbering system beginning with one of six classes of reactions (redox reactions, transferase reactions, hydrolase reactions, lyase reactions, isomerase reactions, and ligase reactions).

7.2 Enzyme Structure and Function

- Enzymes lower the activation energy ($\Delta G^{\ddagger}$) of a reaction in three different ways: (1) stabilizing the transition state, which lowers the activation barrier; (2) providing an alternative path for product formation through reaction intermediates; and (3) orienting the substrates appropriately for the reaction to occur.
- Stabilizing the transition state is one of the key mechanisms of enzyme catalysis and is the molecular basis for tight binding of transition state analogs, which often function as enzyme inhibitors.
- Functional groups in the active site mediate three main types of catalytic reaction mechanisms: (1) acid–base catalysis, (2) covalent catalysis, and (3) metal-ion catalysis.
- Metal ions are important enzyme cofactors and promote catalysis by aiding in substrate orientation, mediating redox reactions, and shielding or stabilizing negative charges through electrostatic interactions.

- Enzymes perform three main types of work in the cell: (1) coenzyme-dependent redox reactions associated with energy conversion; (2) metabolite transformation reactions to interconvert metabolites in anabolic and catabolic pathways; and (3) reversible covalent modification reactions to control cell signaling processes and enzyme activity.
- Enzyme-catalyzed redox reactions in the cell often require coenzymes such as NAD^+/NADH, $NADP^+$/NADPH, FAD/$FADH_2$, or FMN/$FMNH_2$. These redox reactions involve the transfer of a pair of electrons or a single electron through a radical intermediate.
- Metabolite transformations in metabolic pathways most often involve isomerization reactions, condensation reactions, or hydrolysis or dehydration reactions.
- One of the most common types of reversible covalent modification reactions in cells is the addition and removal of a phosphoryl group in biomolecules. Enzymes that attach phosphoryl groups are called kinases, and enzymes that remove phosphoryl groups are called phosphatases.

7.3 Enzyme Reaction Mechanisms

- Chymotrypsin is a serine protease that cleaves peptide bonds using a combination of acid–base and covalent catalysis. In addition, a tetrahedral intermediate is formed that resembles the transition state conformation.
- A key feature of serine proteases is the presence in the enzyme active site of three amino acids called the catalytic triad, which consists of a catalytic serine residue plus histidine and aspartate residues that function to convert the serine residue into a highly reactive nucleophile.
- Enolase catalyzes the dehydration of 2-phosphoglycerate to form phosphoenolpyruvate in a two-step mechanism that involves both acid–base and metal-ion catalysis. The metal ions in this reaction are necessary for ionic interactions with the substrate and intermediate.
- HMG-CoA reductase, an enzyme that catalyzes an early step in cholesterol biosynthesis, uses two NADPH coenzymes to achieve catalysis.

7.4 Enzyme Kinetics

- Enzyme kinetics is the quantitative analysis of reaction rate data obtained with purified enzymes and defined laboratory conditions. We can use enzyme kinetic parameters to compare the catalytic efficiency of related enzymes under a variety of conditions.
- The velocity v of an enzyme reaction is the product of the rate constant k and substrate concentration [S], where k refers to the rate at which $S \rightarrow P$ under standard conditions.
- Michaelis–Menten enzyme kinetics provides a way to analyze a first-order reaction under steady-state conditions in order to relate the initial velocity v_0 to the maximum velocity v_{max}, substrate concentration [S], and Michaelis constant K_m. K_m is experimentally determined as the concentration of substrate required to attain $\frac{1}{2}v_{max}$.
- The values of v_{max} and K_m for an enzyme reaction are obtained from experiments in which data are collected under steady-state conditions when the concentration of the enzyme–substrate complex [ES] is minimally changing (substrate binding to enzyme is rate limiting). Product formation is measured over time for several different initial substrate concentrations.
- Plotting experimental rate data as initial velocity v_0 (which is the slope of the line [P]/time) versus initial [S] produces a Michaelis–Menten plot that is hyperbolic if the enzyme reaction follows simple Michaelis–Menten kinetics.
- A Lineweaver–Burk plot is a double reciprocal plot of enzyme kinetic data that transforms the Michaelis–Menten plot into a linear plot that can be used to estimate values for v_{max} and K_m.
- The calculated efficiency of an enzyme is called the turnover number, k_{cat}, which is a measure of how well an enzyme functions in the reaction. Turnover number is defined as $k_{cat} = v_{max}/[E_t]$.
- Enzyme reaction rates are affected by pH and temperature, which reflect physical and chemical changes in the active site under suboptimal conditions.

7.5 Regulation of Enzyme Activity

- Enzyme regulation is mediated by both enzyme bioavailability (amount of enzyme in the cell and where it is located) and catalytic efficiency (how well an enzyme works).
- Catalytic efficiency of an enzyme is regulated by reversible and irreversible inhibition, allosteric control, covalent modification, and proteolytic processing.
- The three types of reversible and irreversible inhibition are (1) competitive inhibition, (2) uncompetitive, and (3) mixed inhibition, which can be distinguished from each other using enzyme kinetic data.
- The three most common ways that enzymes are regulated by covalent modification are the addition and removal of (1) phosphoryl groups, (2) methyl or acetyl groups, and (3) NMP groups, primarily adenylyl and uridylyl groups.
- Zymogens are inactive proenzymes that are irreversibly processed by proteolysis to generate the active form of the enzyme.

biochemical terms

review questions

1. What three structure–function relationships make enzymes efficient catalysts?
2. Describe two ways in which enzymes are regulated.
3. How do enzymes increase rates of biochemical reactions?
4. Describe the functional roles of enzyme cofactors.
5. In what three ways do enzymes lower the activation energy ($\Delta G^{\ddagger}$) of a reaction?
6. Describe the structure and function of the catalytic triad in chymotrypsin.
7. Briefly describe the two phases of the proposed mechanism for the chymotrypsin reaction.
8. Explain how changes in pH can affect enzyme activity.
9. What are the three classes of reversible inhibitors? Briefly describe them and their effects on apparent K_m and v_{max}.
10. What is the mechanism of feedback inhibition?
11. What is the most common type of covalent modification to enzymes?

challenge problems

1. In most enzymes, the active site consists of only a few residues. Why is the rest of the protein necessary?
2. If you wanted to improve the catalytic efficiency of an enzyme, would you mutate amino acid residues to increase binding affinity for the substrate or increase the binding of the transition state? Explain.
3. Why can't an enzyme use an induced-fit mechanism to achieve catalytic perfection?
4. It was found that for the reaction $A \underset{k_R}{\overset{k_F}{\rightleftharpoons}} B$, the forward rate constant (k_F) in the absence of enzyme was $1 \times 10^{-2}\ s^{-1}$, whereas the k_F in the presence of enzyme was $5 \times 10^{6}\ s^{-1}$.
   a. Calculate the rate enhancement provided by the enzyme for this chemical reaction.
   b. If the equilibrium constant (K_{eq}) for the $A \rightleftharpoons B$ reaction is 1×10^3 in the absence of enzyme, what is the K_{eq} in the presence of enzyme? Explain.

5. Explain the following observation: Enzyme A has a very broad pH optimum and exhibits the same catalytic activity at pH 6.5 as at pH 8.5. However, a competitive inhibitor, X, is effective at pH 6.5, but not at pH 8.5.

6. How does the addition of an enzyme to a chemical reaction affect each of the following parameters (no effect, increase, or decrease)?

   a. Standard free energy of the reaction

   b. Activation energy of the reaction

   c. Initial velocity of the reaction

   d. Equilibrium constant of the reaction

   e. Time to reach equilibrium

7. A mutation in the active site of an enzyme results in a large increase in stabilization of the ES complex, while there is no change in the stabilization of the transition state complex. What effect will this have on the rate of product formation?

8. Consider the simple reaction below, in which $k_{F(uncat)}$ with no catalyst is 10^{-5} s^{-1}, $k_{F(cat)}$ with an enzyme catalyst is 10^{7} s^{-1}, and $k_{R(uncat)}$ with no catalyst is 10^{-2} s^{-1}:

$$S \underset{k_R}{\overset{k_F}{\rightleftharpoons}} P$$

   a. What is K_{eq} for the reaction?

   b. What is $k_{R(cat)}$ with enzyme catalyst?

   c. The enzyme that catalyzes this reaction has one active site per molecule. What is the k_{cat} value for this enzyme if the v_{max} value is 5×10^{3} M s^{-1} at an enzyme concentration of 2×10^{-6} M?

9. The activity of chymotrypsin changes as the pH changes in the range of pH 5–9, as shown on the following graph. From your understanding of the chemical mechanism of the chymotrypsin reaction, explain the pH effect on chymotrypsin activity by answering parts **a**, **b**, and **c**.

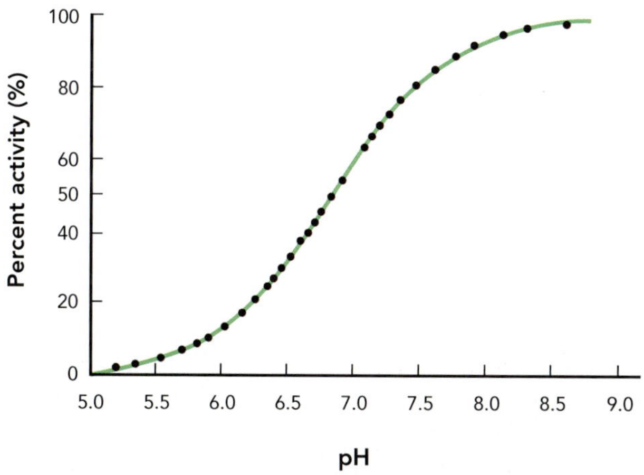

   a. What amino acid functional group on which residue is most likely responsible for the effect of pH?

   b. What is the apparent pK_a of that functional group? Indicate on the graph how you deduced the pK_a.

   c. Briefly explain in terms of the chymotrypsin mechanism why the activity increases as the pH increases in this pH range.

10. Identify the two chemical structures that represent the first and second tetrahedral intermediates in the chymotrypsin-catalyzed cleavage of a peptide.

a.

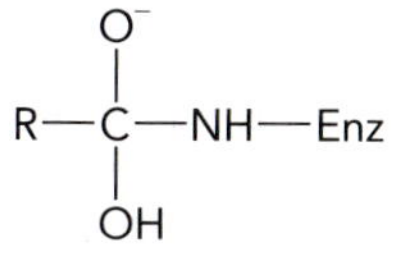

d.

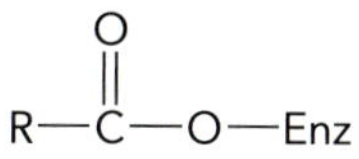

b.

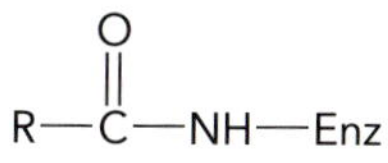

e.

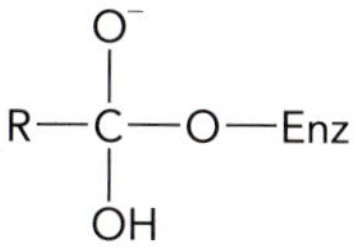

c.

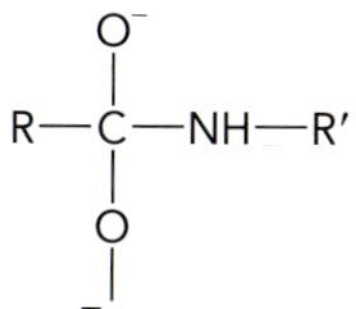

11. The protease papain, which can be isolated from papaya fruit, is a monomer with one active site per protein molecule. The activity of purified papain enzyme was measured at various substrate concentrations and plotted as initial velocity on the following graph. Considering that the enzyme concentration was held constant in these experiments at 1.2×10^{-7} M, answer the following questions about this enzyme.

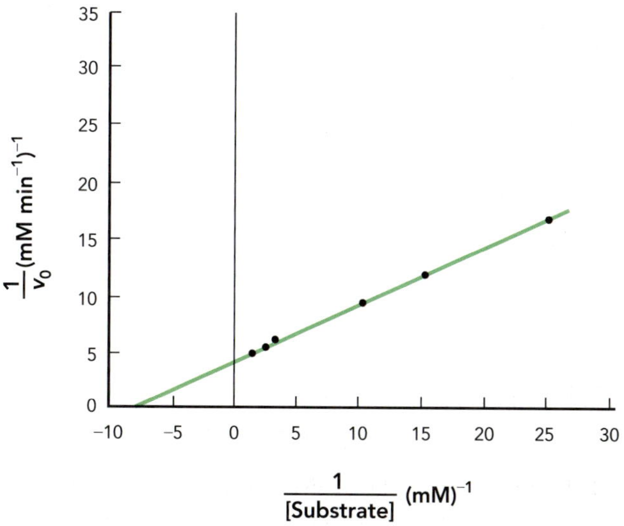

   a. Calculate K_m for this substrate.

   b. Calculate v_{max} for this substrate.

   c. Calculate k_{cat} for papain with this substrate.

   d. What fraction of total enzyme has substrate bound (θ) when the substrate concentration is 2.5×10^{-4} M?

12. An enzyme has a single active site at which it can bind and hydrolyze either X or Y; however, the enzyme cannot bind X and Y at the same time. Answer the following questions regarding the K_m and v_{max} of this enzyme.

a. Will the K_m for X be affected if Y is present in the reaction mixture? Explain.

b. Will v_{max} for X be affected if Y is present in the reaction mixture? Explain.

c. Is it possible for v_{max} and v_{max}/K_m to show a different dependence on pH? Explain.

13. Determine the following values using the accompanying Lineweaver–Burk plots (double reciprocal plots) for an enzyme in the absence (− I) and in the presence (+ I) of an inhibitor.

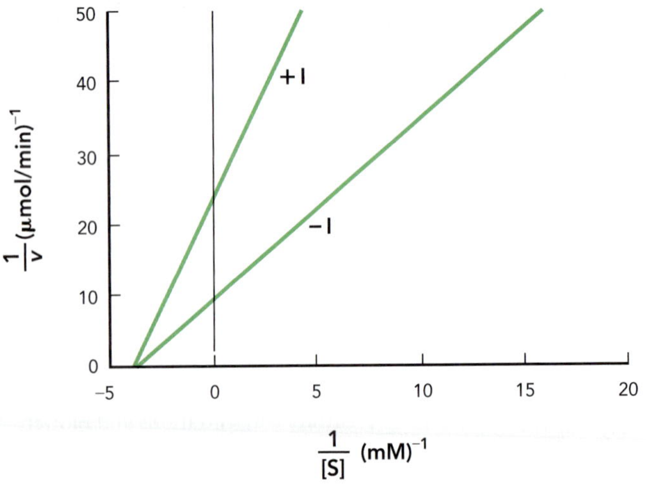

a. v_{max} in the absence of inhibitor

b. v_{max} in the presence of inhibitor

c. K_m in the absence of inhibitor

d. K_m in the presence of inhibitor

e. Type of inhibition (explain)

14. The following graph shows the activity of an enzyme that functions as a homotetramer in which substrate concentration is plotted versus v_0/v_{max} (ratio of the initial velocity relative to the maximum velocity). Use this graph to draw three new curves that reflect enzyme activity under the following conditions. (Label the curves E + Act, E + Inh, and E-mut.)

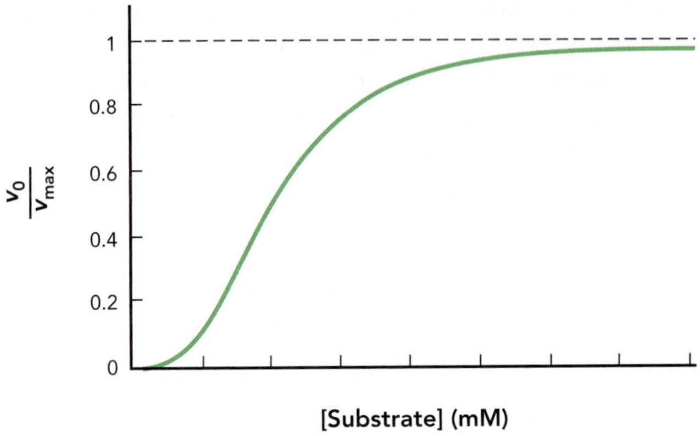

a. The enzyme in the presence of a heterotropic allosteric activator (E + Act).

b. The enzyme in the presence of a heterotropic allosteric inhibitor (E + Inh).

c. A mutant enzyme (E-mut) in which the predominant structure in solution consists of monomers that have a higher affinity for substrate than that of the tetrameric enzyme.

d. Show on the plot how to determine the K_m of E-mut.

15. Methanol and ethanol are both oxidized by alcohol dehydrogenase (ADH). Methanol is poisonous because it is oxidized by ADH to form the highly toxic compound formaldehyde. The adult body has 40 L of water, throughout which these alcohols are rapidly and uniformly distributed. The densities of both alcohols are 0.79 g/mL. The K_m of ADH for ethanol is 1.0×10^{-3} M, and for methanol the K_m is 1.0×10^{-2} M. The molecular mass of methanol is 32 g/mol and that of ethanol is 46 g/mol.

If an individual consumed 100 mL of methanol (a lethal amount), how much 100-proof whiskey (50% ethanol by volume) must this person consume in order to reduce the activity of ADH toward methanol to 5%? Assume the K_I for ethanol = K_m.

suggested reading

Books and Reviews

Fersht, A. (1999). *Structure and mechanism in protein science.* New York, NY: W. H. Freeman.

Kirk, O., Borchert, T. V., and Fuglsang, C. C. (2002). Industrial enzyme applications. *Current Opinion in Biotechnology, 13,* 345–351.

Kuser, P., Cupri, F., Bleicher, L., and Polikarpov, I. (2008). Crystal structure of yeast hexokinase PI in complex with glucose: a classical "induced fit" example revised. *Proteins, 72,* 731–740.

Nagel, Z. D., and Klinman, J. P. (2009). A 21st century revisionist's view at a turning point in enzymology. *Nature Chemical Biology, 5,* 543–550.

Norvell, A., and McMahon, S. B. (2010). Cell biology. Rise of the rival. *Science*, *327*, 964–965.

Peracchi, A. (2001). Enzyme catalysis: removing chemically "essential" residues by site-directed mutagenesis. *Trends in Biochemical Sciences*, *26*, 497–503.

Petsko, G. A., and Ringe, D. (2004). *Protein structure and function*. London, England: Blackwell.

Purich, D. L. (2001). Enzyme catalysis: a new definition accounting for noncovalent substrate- and product-like states. *Trends in Biochemical Sciences*, *26*, 417–421.

Ringe, D., and Petsko, G. A. (2008). How enzymes work. *Science*, *320*, 1428–1429.

Schwartz, S. D., and Schramm, V. L. (2009). Enzymatic transition states and dynamic motion in barrier crossing. *Nature Chemical Biology*, *5*, 551–558.

van Beilen, J. B., and Li, Z. (2002). Enzyme technology: an overview. *Current Opinion in Biotechnology*, *13*, 338–344.

Whitford, D. (2005). *Proteins: structure and function*. Hoboken, NJ: Wiley.

Zalatan, J. G., and Herschlag, D. (2009). The far reaches of enzymology. *Nature Chemical Biology*, *5*, 516–520.

Primary Literature

Balasubramanian, A., and Ponnuraj, K. (2009). Purification, crystallization and preliminary X-ray analysis of urease from jack bean (*Canavalia ensiformis*). *Acta Crystallographica*, *65*, 949–951.

Chook, Y. M., Ke, H., and Lipscomb, W. N. (1993). Crystal structures of the monofunctional chorismate mutase from *Bacillus subtilis* and its complex with a transition state analog. *Proceedings of the National Academy of Sciences USA*, *90*, 8600–8603.

Haines, B. E., Wiest, O., and Stauffacher, C. V. (2013). The increasingly complex mechanism of HMG-CoA reductase. *Accounts of Chemical Research*, *46*, 2416–2426.

Liavonchanka, A., Hornung, E., Feussner, I., and Rudolph, M. G. (2006). Structure and mechanism of the *Propionibacterium acnes* polyunsaturated fatty acid isomerase. *Proceedings of the National Academy of Sciences USA*, *103*, 2576–2581.

Zhang, E., Brewer, J. M., Minor, W., Carreira, L. A., & Lebioda, L. (1997). Mechanism of enolase: the crystal structure of asymmetric dimer enolase-2-phospho-D-glycerate/enolase-phosphoenolpyruvate at 2.0 Å resolution. *Biochemistry*, *36*, 12526–12534.

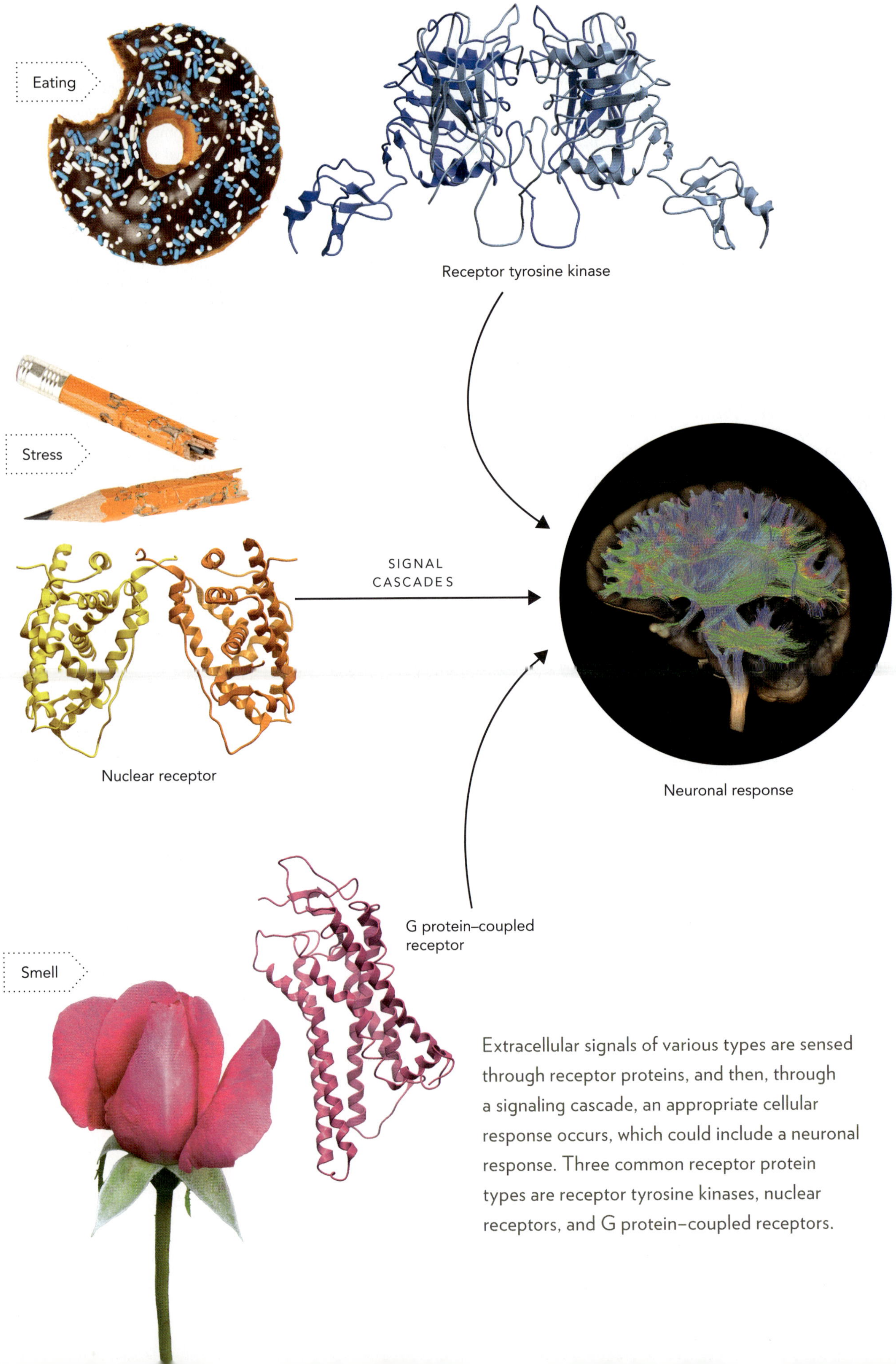

Extracellular signals of various types are sensed through receptor proteins, and then, through a signaling cascade, an appropriate cellular response occurs, which could include a neuronal response. Three common receptor protein types are receptor tyrosine kinases, nuclear receptors, and G protein–coupled receptors.

8

Cell Signaling Systems

◀ Cell signaling systems link the extracellular environment to changes in cellular behavior. Several classes of receptor proteins sense different types of extracellular inputs and, in response, stimulate a variety of intracellular biochemical processes. The common feature of these signaling systems is that they amplify signals through enzyme-mediated reactions. This signaling occurs through ligand binding to receptor proteins, which causes conformational changes in the receptor proteins that affect the biochemical activities of intracellular proteins. This process is similar to that of ligand binding–induced conformational changes in the proteins we studied in Chapters 6 and 7.

CHAPTER OUTLINE

CREDITS: DONUT: BRYAN SOLOMON/SHUTTERSTOCK; RECEPTOR TYROSINE KINASE: BASED ON PDB FILE 2DTG; BRAIN: SCOTT CAMAZINE/SCIENCE SOURCE; G PROTEIN-COUPLED RECEPTOR: BASED ON PDB FILE 3DQB; ROSE: ALESIKKA/ISTOCKPHOTO/GETTY IMAGES; NUCLEAR RECEPTOR: BASED ON PDB FILE 3CLD; BROKEN PENCIL: DEVONYU/ISTOCKPHOTO/GETTY IMAGES.

In this chapter, we conclude our study of the central role of protein structure and function in biochemical processes. In particular, we examine how proteins work together to create highly sensitive and specific biochemical pathways that respond to extracellular signals. Understanding how signaling proteins control cell functions is a major focus area in biochemical research—and provides a foundation for our discussion of metabolic regulation later in the book.

We begin the chapter with an overview of basic concepts in cell signaling systems and then examine the biochemistry of receptor proteins. These proteins serve as the primary link between extracellular signals and intracellular biochemical responses.

We could choose from among many different cellular processes to illustrate the biochemical principles of cell signaling. We have chosen to focus on signaling pathways stimulated by the physiologic mediators glucagon, epinephrine (adrenaline), epidermal growth factor, insulin, tumor necrosis factor, and glucocorticoids (cortisol). There are two reasons for these choices. First, signaling pathways stimulated by these biomolecules are some of the best-characterized signaling systems in humans, which helps in explaining how biochemistry relates to everyday life. Second, we will encounter many of these signaling pathways again in Parts 3 and 4 of the book, when we describe 10 major metabolic pathways found in most eukaryotic organisms. Thus, we present the upstream signaling mechanisms here and focus on the downstream targets later.

With this strategy in mind, we first present an overview of the most abundant classes of receptor proteins in eukaryotes. We then describe in some detail the structure and function of G protein–coupled receptors, receptor tyrosine kinases, tumor necrosis factor (TNF) receptors, and nuclear receptors. Each of these receptor classes provides a representative example of how protein structure and function play key roles in transmitting extracellular signals to intracellular targets that control homeostasis.

8.1 Components of Signaling Pathways

Biological diversity is based on complexity, and with this complexity comes the need for high-level organization. We are familiar with the critical role of communication in organizing individuals to function as a group, such as a sports team or a college class, to accomplish a common goal. Similarly, cells must also be able to communicate with one another about subtle changes in the environment, or the intracellular milieu, that are occurring over time.

For single-cell organisms, such as bacteria or yeast, communication between cells often signals a change in reproductive behavior in response to the availability of nutrients in the surrounding environment. Along with the evolution of multicellular organisms, cell signaling processes became more complex, allowing cells in different tissues the bodies to communicate physiologic changes that affect the entire organism.

The molecular basis of this communication process is a biochemical pathway. This usually consists of protein structural changes initiated by the binding of extracellular messenger molecules (ligands) to **receptor proteins**, which triggers a cellular response. The name given to the receptor protein often reflects the ligand that activates it. For example, glucagon binds to the glucagon receptor, and insulin binds to the insulin receptor. Another type of cell–cell communication in multicellular eukaryotic organisms is mediated by lipophilic ligands derived from cholesterol, which pass through the plasma membrane and bind to nuclear receptor proteins. For example, estrogen, cortisol, and vitamin D bind to the estrogen receptor, the glucocorticoid receptor, and the vitamin D receptor, respectively.

The term signal transduction refers to the biochemical mechanism responsible for transmitting extracellular signals across the plasma membrane and throughout the cell. A transducer is a device that changes the nature of a signal, such as changing a pressure or temperature reading into an electrical signal. In biochemistry, signal transduction changes information into a chemical signal. Signal transduction often culminates in the covalent or noncovalent modification of intracellular **target proteins**, which are different from receptor proteins and control a variety of cellular responses. For example, flux through metabolic pathways, ion flow across the plasma membrane, cell motility, and gene expression are all modulated by signal transduction.

As you will see in this chapter, the effect of activating a receptor protein involves one or more of these biochemical responses:

1. *covalent protein modifications,* such as phosphorylation and dephosphorylation reactions;
2. *protein conformational changes,* resulting from high-affinity noncovalent binding interactions between adaptor proteins; and
3. *altered rates of protein expression,* which can occur at the transcriptional level (RNA synthesis and turnover) or translational level (protein synthesis and turnover).

Another important concept in this chapter is the **cell signaling pathway**. This term describes the linked set of biochemical reactions that are initiated by ligand-induced activation of a receptor protein and terminated by a measurable cellular response. For example, the insulin signaling pathway is activated by elevated levels of insulin in the blood. The insulin molecules bind to and activate insulin receptors on the surface of target cells. This, in turn, triggers a series of biochemical reactions involving the activation of intracellular protein kinases, which regulate glucose homeostasis in the cell.

Early steps in a signaling pathway are called upstream events, whereas later steps, such as target protein modification, are referred to as downstream events. As shown in **Figure 8.1**, in addition to extracellular ligands that bind to receptor proteins and function as **first messengers**, signal transduction also involves **second messengers** and a variety of intracellular **signaling proteins**, which function together to transmit, amplify, and terminate the signal.

To appreciate just how important signal transduction is to multicellular organisms, it is useful to compare the number of signal transduction genes encoded in the human, the

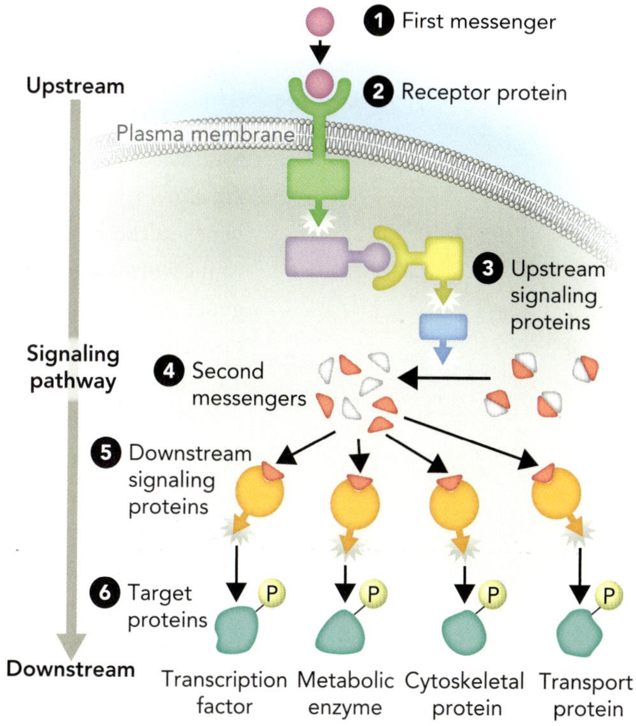

Figure 8.1 This illustration of a simplified signaling pathway shows the relationship among receptors, first and second messengers, and signaling proteins in transmitting the biochemical signal to target proteins. The generation of second messengers by an upstream signaling protein results in signal amplification through the activation of one or more downstream target proteins. In this example, the downstream signaling protein is a kinase enzyme that phosphorylates a variety of cellular proteins.

Table 8.1 ESTIMATED NUMBERS OF GENES IN DIFFERENT CLASSES OF SIGNAL TRANSDUCTION GENES IN THE GENOMES OF HUMANS, ROUNDWORMS (*C. ELEGANS*), AND YEAST (*S. CEREVISIAE*)

Gene class (by product)	Number of genes in human genome	Number of genes in roundworm genome	Number of genes in yeast genome
G protein–coupled receptors	616	284	3
Nuclear receptors	59	183	1
Peptide hormone receptors	101	6	0
Protein kinases	501	415	119
Peptide hormones	207	16	0
G protein superfamily	186	88	29
Ligand-gated ion channels	183	239	13
Protein phosphatases	100	156	22
Cyclic nucleotide phosphodiesterases	25	6	2
Total number of signal transduction genes	1,978	1,393	189
Total number of protein-coding genes in genome	21,787[a]	22,227[b]	6,606[c]
Percentage of signal transduction genes	9.0%	6.3%	2.8%

[a]Based on the National Center for Biotechnology Information (NCBI) Human Genome Project annotation.

[b]Based on the WormBase annotation.

[c]Based on the *Saccharomyces* Genome Database (SGD) annotation.

roundworm (*Caenorhabditis elegans*), and the yeast (*Saccharomyces cerevisiae*) genomes. As shown in **Table 8.1**, bioinformatic analyses reveal that 9% of the protein-coding genes in the human genome are involved in signal transduction compared to ~6% in roundworms and ~3% in yeast cells. Not included in this table are the hundreds of genes coding for transcription factors (gene regulatory proteins) that function as the downstream targets in many signaling pathways.

The majority of the genes of the classes listed in Table 8.1 encode either receptor proteins (G protein–coupled receptors, peptide hormone receptors, nuclear receptors) or protein kinases, which are the two types of signaling proteins we describe in some detail in this chapter. Humans also have a large number of genes encoding peptide hormones—examples of which include insulin, glucagon, epidermal growth factor, and tumor necrosis factor—as well as signaling proteins called G proteins. Two of the most extensively studied G proteins are the heterotrimeric G proteins (α, β, γ) and the Ras proteins, which function in signaling pathways linked to a large number of human disease states. Table 8.1 also shows that both humans and roundworms contain a large number of genes encoding ligand-gated ion channels, which is consistent with the role of these transmembrane proteins in neuronal signaling. In addition, protein phosphatases and cyclic nucleotide phosphodiesterases (enzymes involved in modulating signal transduction pathways) are well represented in these genomes. Finally, you can observe from Table 8.1 that the number of signal transduction genes encoded in the yeast genome is significantly lower than that in humans or in roundworms, with most of the 189 yeast signaling genes encoding protein kinases. This difference in signaling gene complexity reflects the reduced need for cell–cell communication in populations of yeast compared to that needed within the bodies of roundworms and humans (**Figure 8.2**).

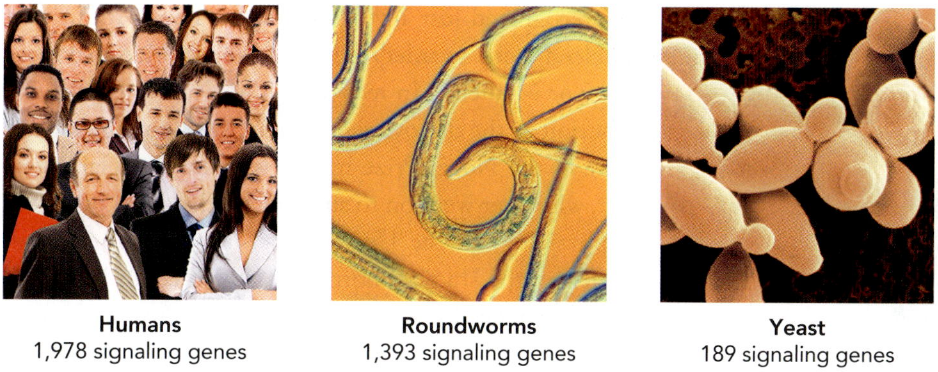

Figure 8.2 Populations of humans, roundworms, and yeast have different communication needs between individual cells and between organisms, which is reflected in the number of signaling genes encoded by their genomes. HUMANS: EDHAR/SHUTTERSTOCK; ROUNDWORMS: SINCLAIR STAMMERS/SCIENCE PHOTO LIBRARY/CORBIS; YEAST: SCIMAT/GETTY IMAGES.

Small Biomolecules Function as Diffusible Signals

Research into cell signaling pathways began through a series of isolated experiments that were performed before researchers knew exactly what they were studying. Near the end of the 19th century, physiologists began to experiment with animal tissue extracts, describing a variety of diverse biological effects that resulted from injecting these tissue extracts into live animals. Analysis of these rather crude studies gave rise to the idea that small biomolecules contained in tissues act on specific organs through some unknown signaling mechanism.

One example of these early experiments in endocrinology (endocrinology is the study of glands and hormones in the body) was described in 1889 by the French physiologist Charles Brown-Séquard. Brown-Séquard widely publicized the invigorating results he obtained from daily self-injections of an extract prepared from canine testicles. Although Brown-Séquard's data were generated by uncontrolled experiments and likely were influenced by the placebo effect (perceived benefit from a treatment rather than a direct result of the treatment), his experiments laid the groundwork for a much more rigorous endocrinology study done ~30 years later in Toronto, Canada. In 1922, Frederick Banting, a Canadian surgeon at the University of Toronto, and Charles Best, his research assistant, first showed that a small biomolecule present in dog pancreatic cell extracts—a compound they called **insulin**—could control blood glucose levels and be used to treat human diabetes (**Figure 8.3**).

Banting and Best performed their groundbreaking work in the lab of Prof. John Macleod, where they developed a technique to purify insulin. Insulin is a small protein, easily degraded by proteolytic enzymes present in the pancreatic tissue extract. When they injected their animal-derived insulin into a 14-year-old diabetic boy, they found that his blood glucose levels quickly decreased, and this decrease seemed to alleviate some of his disease symptoms. The importance of this discovery

Figure 8.3 Dr. Frederick Banting and Charles Best discovered the hormone insulin through their research work at the University of Toronto. **a.** Best (left) and Banting (right) initially used pancreatic protein extracts from laboratory dogs to identify insulin but soon afterward found that they could isolate high quantities of active insulin from bovine pancreas. FOTOSEARCH/GETTY IMAGES. **b.** Banting and Best sold the patent to insulin to the University of Toronto for one dollar. In turn, the university gave pharmaceutical companies the right to mass-produce insulin free of royalties to expedite its use in the clinic. SANOFI PASTEUR CANADA (CONNAUGHT CAMPUS) ARCHIVES, TORONTO, ON. **c.** Photograph of a modern-day insulin pump, which delivers synthetic human insulin directly to the bloodstream to treat the symptoms of diabetes. CLICK AND PHOTO/SHUTTERSTOCK.

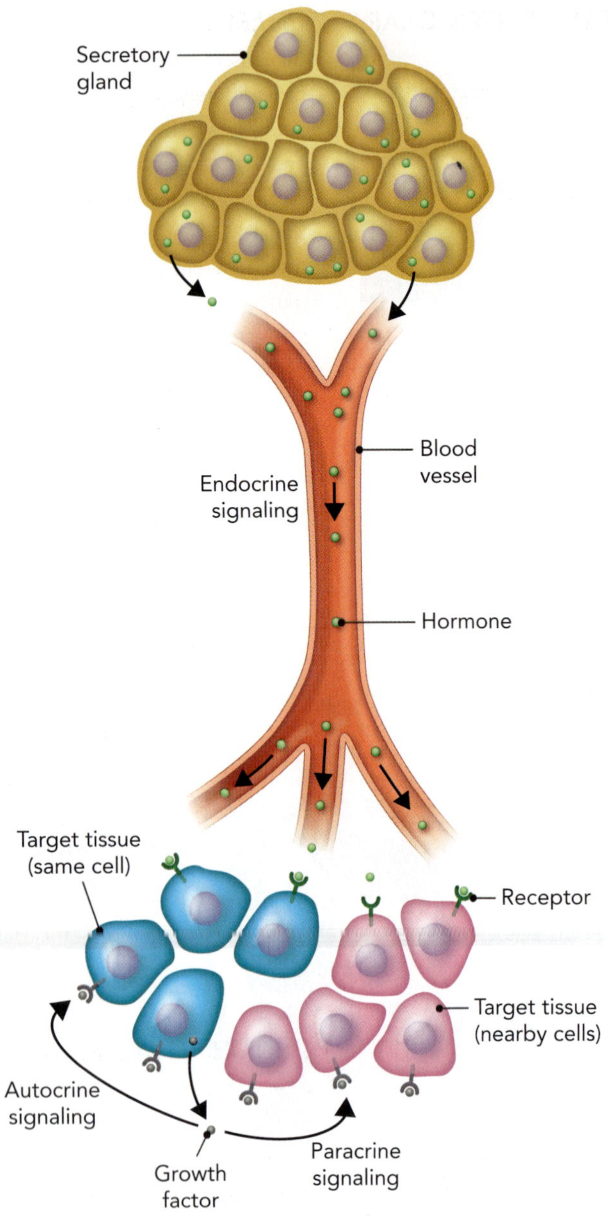

Figure 8.4 Hormones function as first messengers through endocrine, paracrine, and autocrine mechanisms. Endocrine hormones are produced by secretory glands that export hormones into the circulatory system. Eventually, these hormones come in contact with specific receptor proteins in target tissues. Autocrine and paracrine hormones are small peptides that function over short distances to activate receptors on nearby cells (paracrine hormones) or to activate receptors on the same cell (autocrine hormones).

was recognized just 1 year later when Drs. Banting and Macleod were awarded the 1923 Nobel Prize in Physiology or Medicine. In recognition of Charles Best's contribution to the discovery of insulin, Dr. Banting split his share of the Nobel Prize money with his loyal lab assistant, who later went on to succeed Macleod as a professor of physiology at the University of Toronto. Their work pioneered the study of hormones, which control the first step in many cell signaling pathways.

Hormones and Other First Messengers **Hormones** are defined as biologically active compounds that are released into the circulatory system, where they are able to come in contact with hormone receptors contained in target cells. Hormones are considered first messengers in cell signaling because they initiate the receptor-activating signal that gives rise to a physiologic response (see Figure 8.1).

Hormones can act at a distance through **endocrine** mechanisms or function locally as **paracrine** or **autocrine** signals (**Figure 8.4**). Endocrine hormones, such as insulin and the reproductive steroid hormones estrogen and testosterone, are produced by glands that secrete hormones into the circulatory system. These endocrine hormones come in contact with receptor proteins located in the cells of target tissues. In contrast, autocrine and paracrine hormones are small, secreted peptides that function locally to activate receptor proteins on nearby cells (paracrine) or receptor proteins on the same cell (autocrine).

First messengers are not always peptide hormones; they can also be soluble gases such as nitric oxide or neurotransmitters such as acetylcholine. **Table 8.2** lists several representative first messengers, along with their origins, targets, and biochemical effects. **Figure 8.5** shows the molecular structures of the first messengers nitric oxide, epinephrine (adrenaline), β-estradiol, acetylcholine, insulin, and epidermal growth factor.

One of the most interesting first messengers in human physiology is **nitric oxide (NO)**, which is generated in cells by the enzyme nitric oxide synthase through the oxidative deamination of the amino acid arginine. Rapid diffusion of NO into smooth muscle cells results in muscle relaxation, vasodilation, and increased blood flow. NO signaling is mediated by activation of the soluble enzyme guanylate cyclase, which generates the second messenger cyclic GMP. As an example of evolutionary adaptation, NO signaling is used by blood-sucking insects to increase blood flow during feeding. As shown in **Figure 8.6** (p. 378), the kissing bug, *Rhodnius prolixus*, delivers NO to its victims by injecting heme-containing proteins called nitrophorins, which carry

Table 8.2 REPRESENTATIVE FIRST MESSENGERS THAT HAVE BEEN CHARACTERIZED WITH RESPECT TO SPECIFIC SIGNALING PATHWAYS

First messenger	Origin	Target	Biological response
Acetylcholine	Nerve cells	Muscle cells	Muscle contraction
Cortisol	Adrenal gland	Immune cells, liver cells	Anti-inflammatory; glycogen degradation
Epidermal growth factor	Many cells	Many cells	Cell proliferation
Epinephrine (adrenaline)	Adrenal medulla	Heart cells, liver cells	Increased pulse rate; glycogen degradation
Estradiol	Ovarian tissues (females only); adrenal glands	Female reproductive tissues, brain cells	Female development; behavior
Glucagon	Pancreatic α cells	Liver cells	Glycogen degradation
Insulin	Pancreatic β cells	Muscle cells, liver cells, adipose cells	Glucose uptake
Ions and small molecules (e.g., Ca^{2+}, NO, CO_2)	Many sources, both intracellular and extracellular	Most prokaryotic and eukaryotic cells have specific receptor proteins	Membrane depolarization; intracellular signaling; gene expression
Metabolites (amino acids, nucleotides)	Nutrients in the diet or released from other cells in the organism	Many cells, including liver, muscle, adipose	Energy conversion reactions; neurotransmission; gene expression
Prostaglandins	Many cells	Many cells	Inflammation
Testosterone	Testes (males only); adrenal glands	Male reproductive tissues, brain cells	Male development; behavior

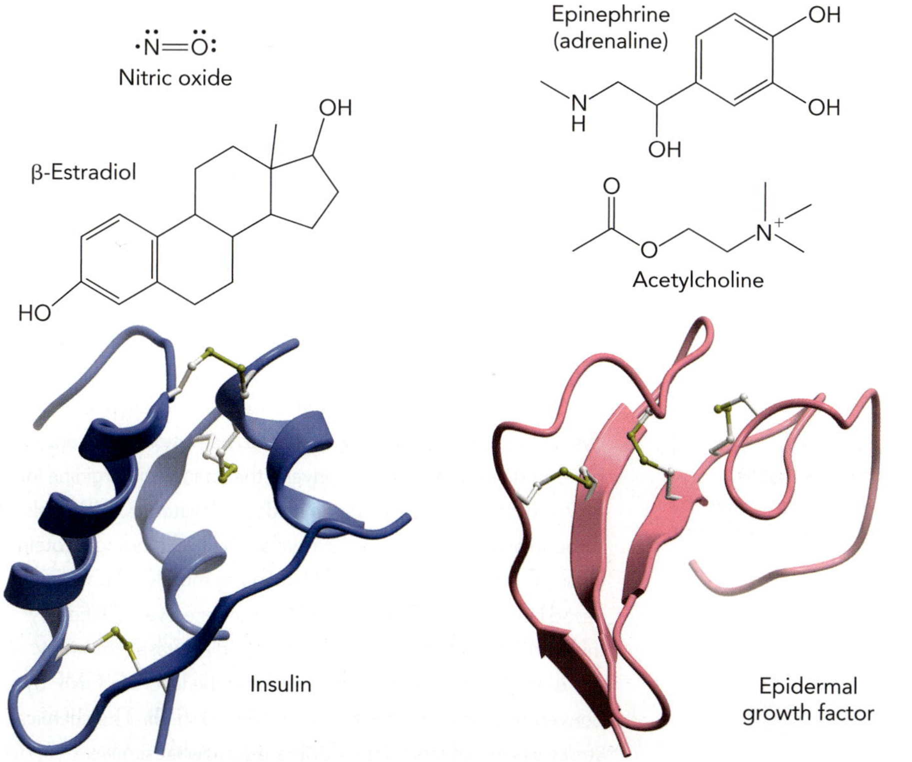

Figure 8.5 The structures of some representative first messengers important in biochemistry are shown here. Nitric oxide is a soluble gas (derived from arginine) that activates signaling pathways by diffusing across cell membranes and directly activating signaling proteins. Epinephrine, also known as adrenaline, is a metabolite of the amino acid tyrosine. β-Estradiol is a steroid hormone and the most abundant estrogen in women. Acetylcholine is a neurotransmitter in both the peripheral and central nervous systems. Insulin and epidermal growth factor are small proteins that are stabilized by disulfide bridges, which are shown in stick representation.

BASED ON PDB FILES 1TRZ (INSULIN) AND 1HGU (EPIDERMAL GROWTH FACTOR).

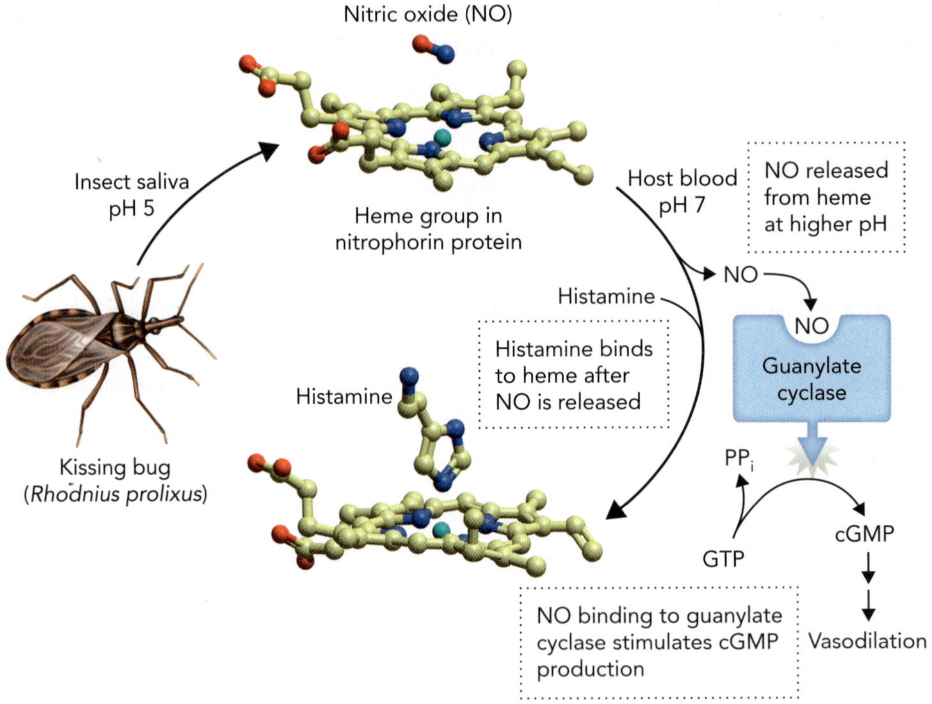

Figure 8.6 Nitrophorin is a heme-containing protein found in the saliva of the kissing bug, *Rhodnius prolixus*. At the low pH of 5 present in insect saliva, NO is bound to the heme of the nitrophorin protein. However, in the host blood at pH 7, NO is released, causing vasodilation through activation of guanylate cyclase and production of the second messenger cyclic GMP. The host's histamine—an anti-inflammatory molecule that is released upon tissue damage—binds to the nitrophorin heme group at pH 7, resulting in decreased anti-inflammatory responses and an extended blood-feeding period for the insect. PP_i = pyrophosphate. BASED ON PDB FILE 1ERX (NITROPHORIN).

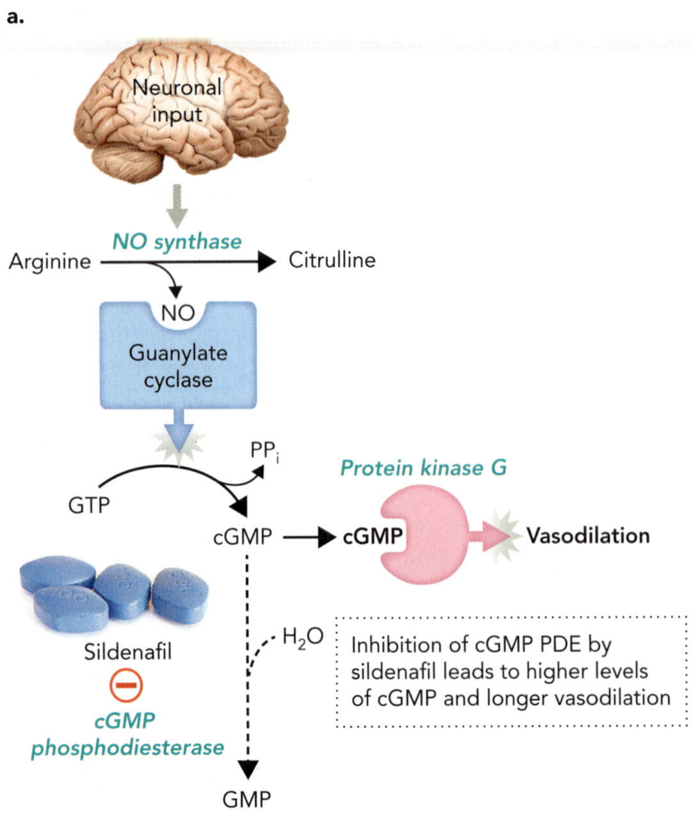

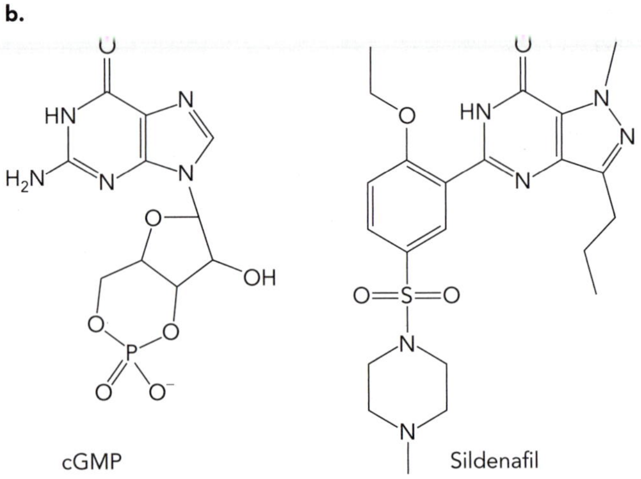

Figure 8.7 Sildenafil, marketed as Viagra, is an inhibitor of cGMP phosphodiesterase and is used to treat erectile dysfunction. **a.** Neuronal input leads to activation of the enzyme nitric oxide synthase, which converts the amino acid arginine into NO and citrulline. This reaction leads to elevated cGMP levels through activation of guanylate cyclase by NO binding. Protein kinase G is activated by cGMP, leading to muscle relaxation and vasodilation. Sildenafil prolongs NO-mediated vasodilation by inhibiting the activity of cGMP phosphodiesterase (shown by the dotted line), an enzyme that reduces the level of cGMP by converting it to GMP. PHOTO: UROS POTEKO/ALAMY. **b.** The chemical structures of sildenafil and cGMP are somewhat similar.

NO into the wound along with the insect's saliva. Nitric oxide readily dissociates from the nitrophorin heme in the microcapillaries of the host as a result of elevated pH. Notably, histamine, an inflammatory molecule in the host blood, competes with NO for heme binding in the nitrophorin protein. Therefore, while the insect delivers NO to stimulate vasodilation, histamine is taken up by nitrophorin proteins to dampen the victim's immune response. This combination of effects allows the insect to maximize its feeding time.

Second Messengers Second messengers are small, nonprotein intracellular molecules whose functional role is to amplify receptor-generated signals. One of these second messengers is 3′,5′-cyclic guanosine monophosphate, or **cyclic GMP (cGMP)**, which is produced from guanosine triphosphate (GTP) by the enzyme guanylate cyclase. Second messenger signaling through cGMP is the molecular basis for synaptic transmission in light-stimulated vision and for the control of arterial blood flow through vasodilation. Indeed, one of the blockbuster drugs developed in the 1990s was a cGMP analog called sildenafil, which is marketed as Viagra.

Sildenafil was originally developed to treat heart disease by inhibiting **cGMP phosphodiesterase**, an enzyme that hydrolyzes cGMP. Although it was not a very effective cardiovascular drug, it was found to have an interesting side effect that was exploited as a treatment for erectile dysfunction—sildenafil causes prolonged penile erection under the appropriate conditions. As illustrated in **Figure 8.7**, after neuronal stimulation by the brain and activation of the enzyme nitric oxide synthase, nitric oxide levels increase. This leads to stimulation of guanylate cyclase activity and production of cGMP. Elevated cGMP levels in penile tissue results in stimulation of protein kinase G (PKG) activity, resulting in vasodilation. By inhibiting the activity of cGMP phosphodiesterase, sildenafil treatment causes an increase in the steady-state levels of cGMP and, as a consequence, prolonged vasodilation. Two other vasodilating drugs that treat erectile dysfunction by a similar mechanism are tadalafil (Cialis) and vardenafil (Levitra). It is important to note that in order for vasodilating drugs in this class to work effectively in the target tissue, neuronal stimulation and activation of guanylate cyclase must occur first to produce sufficient levels of cGMP.

Another well-characterized second messenger is 3′,5′-cyclic adenosine monophosphate, or **cyclic AMP (cAMP)**. **Figure 8.8** shows that cAMP is produced from ATP by the enzyme **adenylate cyclase** and is hydrolyzed by the enzyme **cAMP phosphodiesterase**. As noted earlier, a hallmark of signal transduction through second messengers is signal amplification, which is illustrated in **Figure 8.9** for the case of cAMP. Receptor activation of adenylate cyclase generates cAMP, which in turn binds to and activates a downstream signaling protein called **protein kinase A (PKA)**. The signal is further amplified by PKA through phosphorylation of numerous target proteins, one of which

Figure 8.8 Steady-state levels in cells of the second messenger cAMP are controlled by the combined activities of the receptor-stimulated enzymes adenylate cyclase and cAMP phosphodiesterase.

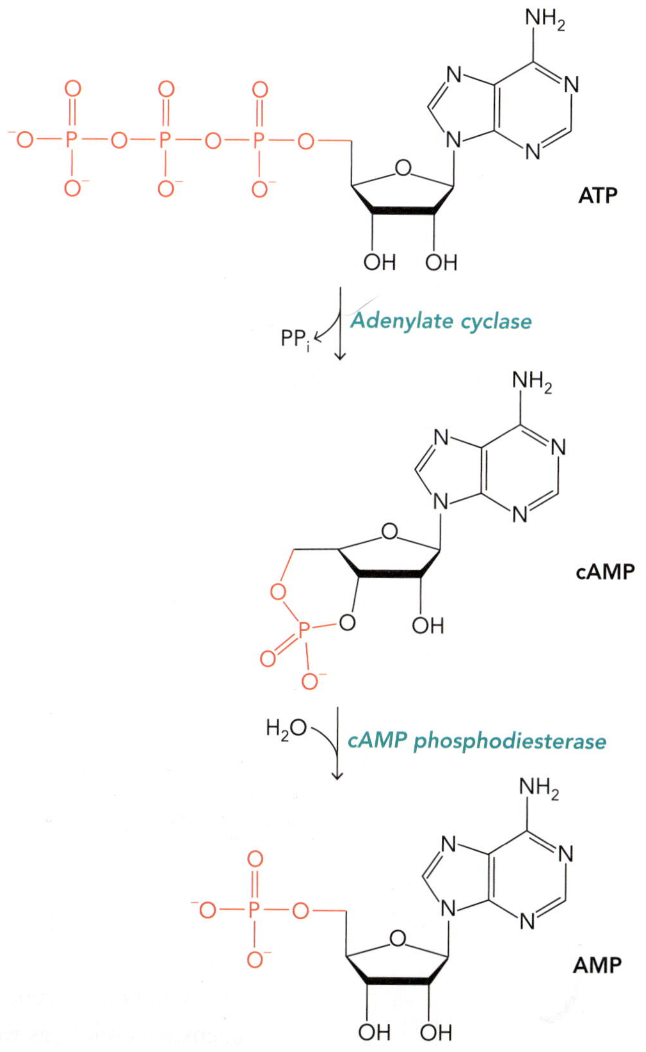

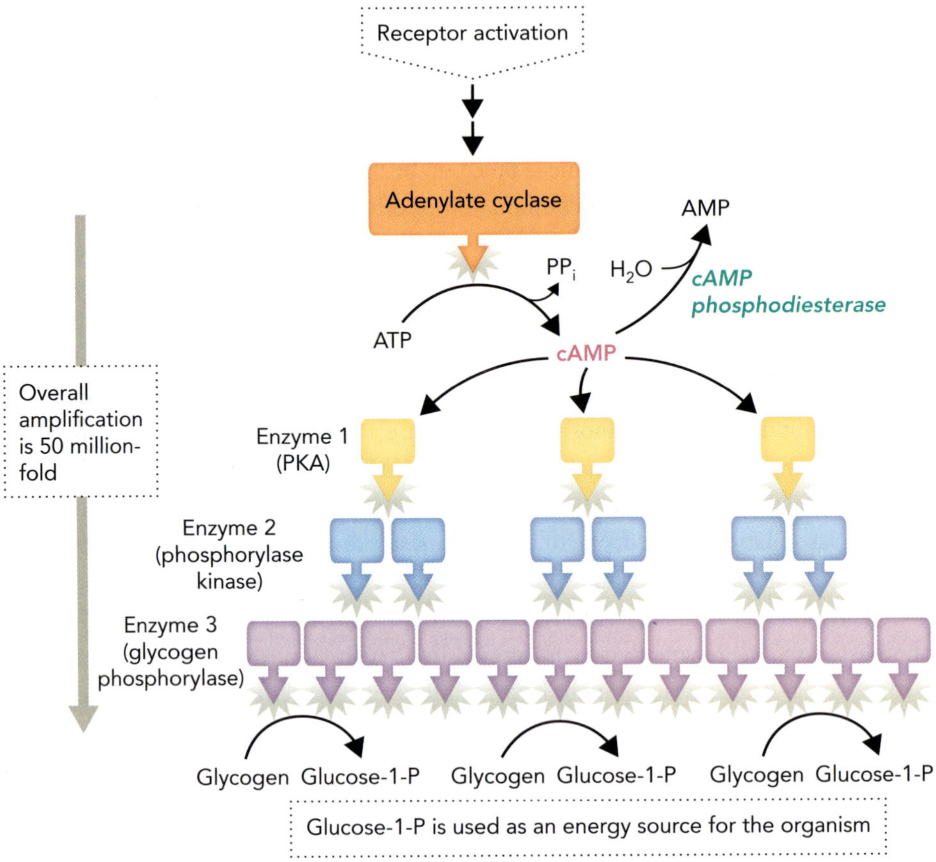

Figure 8.9 Signal amplification by second messengers often involves enzymes that catalytically activate downstream signaling proteins, many of which are also enzymes. In the example shown here, receptor activation stimulates adenylate cyclase activity to generate cAMP, which in turn, activates protein kinase A (PKA). Enzyme 2 is phosphorylase kinase and enzyme 3 is glycogen phosphorylase. As described in Chapter 14, glycogen phosphorylase catalyzes a cleavage reaction that releases glucose-1-phosphate (glucose-1-P) from the free ends of glycogen polymers. The overall result is the quick release of energy for the organism.

is another kinase called phosphorylase kinase. As described in some detail in Chapter 14, phosphorylase kinase activates a third enzyme in the cAMP signaling pathway called glycogen phosphorylase, the enzyme responsible for the phosphorolysis reaction releasing glucose from glycogen.

If each step of this amplification scheme involved 100 molecules, for example, then activation of one adenylate cyclase enzyme would generate 100 cAMP, which would activate 50 PKA molecules (activation of one PKA molecule requires two cAMP molecules; see Figure 8.26). This reaction would lead to the phosphorylation and activation of 5,000 phosphorylase kinase molecules (each PKA molecule activates 100 phosphorylase kinase molecules). If each phosphorylase kinase enzyme then phosphorylated 100 glycogen phosphorylase molecules, and they each released 100 glucose-1-phosphate molecules from glycogen, the maximum amplification resulting from receptor activation of one adenylate cyclase enzyme would be 10^2 (cAMP) $\times$ 0.5 (PKA) $\times$ 10^2 (phosphorylase kinase) $\times$ 10^2 (glycogen phosphorylase) $\times$ 10^2 (glucose) = 50,000,000. Activation of PKA by the second messenger cAMP also leads to phosphorylation of numerous transcription factors that regulate gene expression—another form of signal amplification.

Three other second messengers in signal transduction pathways are **diacylglycerol (DAG)**, **inositol-1,4,5-trisphosphate (IP_3)**, and calcium ion (Ca^{2+}). As illustrated in **Figure 8.10**, the intracellular levels of DAG and IP_3 are controlled by the activity of a membrane-associated enzyme called **phospholipase C**. Phospholipase C hydrolyzes the membrane phospholipid **phosphatidylinositol-4,5-bisphosphate (PIP_2)** to form both DAG and IP_3. The enzymatic activity of phospholipase C is regulated by activation of a membrane receptor using a mechanism similar to receptor-mediated activation of adenylate cyclase, as we will describe shortly.

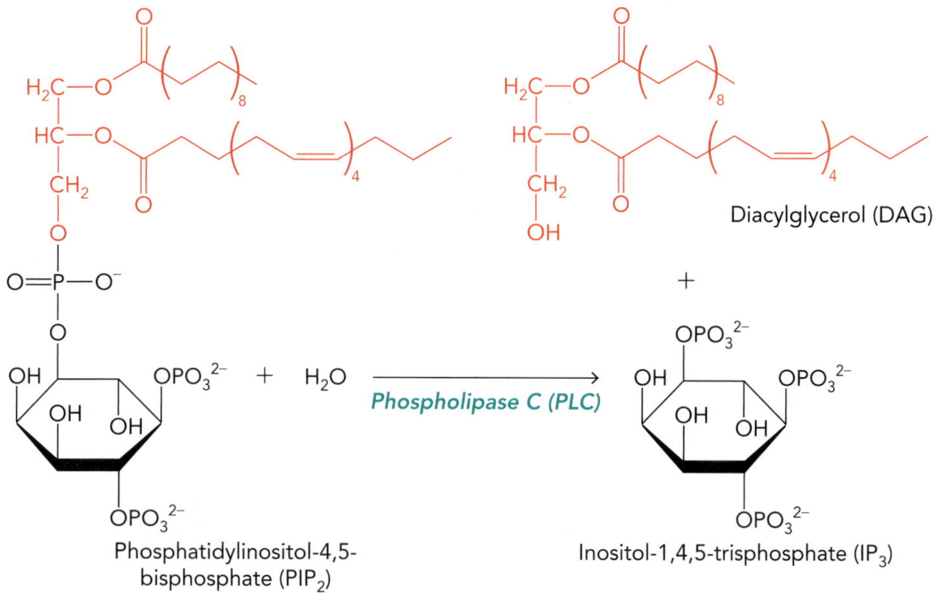

Figure 8.10 Phospholipase C hydrolyzes phosphatidylinositol-4,5-bisphosphate to generate the second messengers diacylglycerol and inositol-1,4,5-trisphosphate.

Once PIP_2 is hydrolyzed by phospholipase C, the newly generated second messenger DAG binds to and activates protein kinase C (PKC), which phosphorylates downstream targets (**Figure 8.11**). In addition, the second messenger IP_3 activates calcium channels located on the endoplasmic reticulum, leading to the release of Ca^{2+} from the endoplasmic reticulum and a rapid increase in cytoplasmic Ca^{2+} levels. Intracellular signaling by cytosolic Ca^{2+} involves activation of numerous Ca^{2+} binding proteins. One of these is **calmodulin**, a signaling protein that binds to and activates a wide variety of target proteins. Most PKC proteins bind two or more Ca^{2+} ions, whereas calmodulin binds four Ca^{2+} ions and undergoes a large conformational change when it binds to target proteins (**Figure 8.12**).

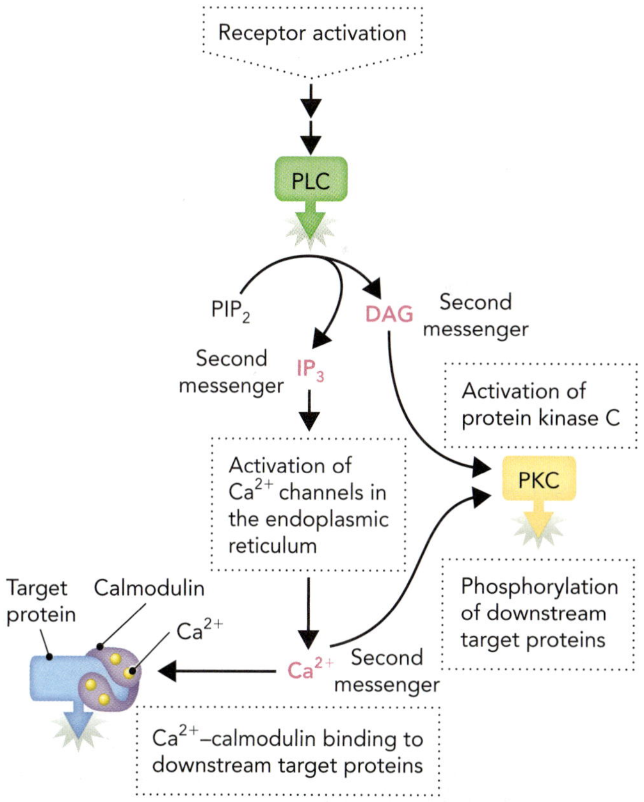

Figure 8.11 Receptor-mediated activation of phospholipase C leads to an increase in IP_3 and DAG levels and stimulation of downstream signaling. DAG binds to and stimulates the activity of PKC, which phosphorylates target proteins. IP_3 activates Ca^{2+} channels on the endoplasmic reticulum and causes an increase in cytoplasmic Ca^{2+} levels. The activities of PKC and calmodulin are both regulated by Ca^{2+} binding.

Receptor Proteins Are the Information Gatekeepers of the Cell

Higher eukaryotes contain five abundant classes of receptor proteins, four of which we focus on in this chapter: (1) G protein–coupled receptors, (2) receptor tyrosine kinases, (3) tumor necrosis factor receptors, and (4) nuclear receptors (**Figure 8.13**). All but the nuclear receptors are transmembrane proteins that bind extracellular ligands. We briefly describe the fifth class of receptor proteins shortly.

The signaling pathways initiated by each of these four receptor classes use distinct mechanisms to transduce the downstream signal. For example, activation of **G protein–coupled receptors (GPCRs)** results in the dissociation of the heterotrimeric G protein complex. This leads to activation of enzymes such as adenylate

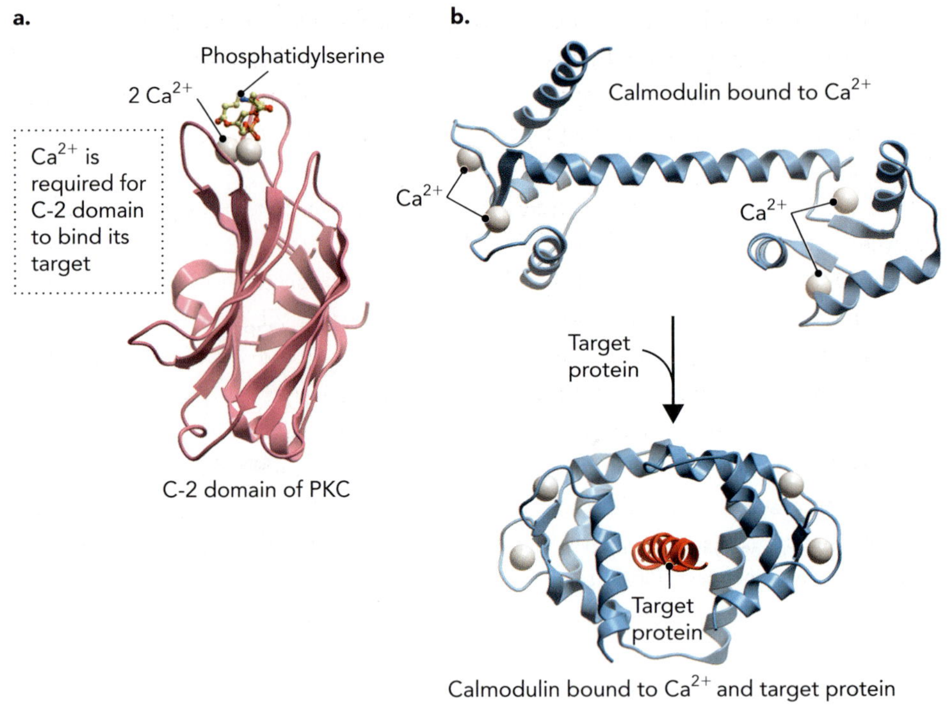

Figure 8.12 Calcium binding stimulates the activity of PKC and calmodulin. **a.** Two Ca^{2+} ions bind to PKC in the C-2 domain and facilitate association of PKC with the plasma membrane. Association with the membrane occurs through direct interactions of Ca^{2+} with phosphatidylserine, which is a phospholipid component of cell membranes. BASED ON PDB FILE 1DSY. **b.** Molecular structure of calmodulin in the absence and presence of a peptide analog of a target protein. Four Ca^{2+} ions bind to each molecule of calmodulin. Calmodulin undergoes a large conformational change when it binds a target protein. BASED ON PDB FILE 1CLL (UNBOUND) AND 1CDL (BOUND).

cyclase and phospholipase C that generate second messenger molecules. In contrast, **receptor tyrosine kinases (RTKs)** are enzymes containing an extracellular domain that binds ligands and an intracellular domain that phosphorylates tyrosine residues in target proteins, including tyrosines within the receptor itself. As you will see shortly, these phosphotyrosine residues create docking sites for intracellular signaling proteins, which function as molecular adaptors. The **tumor necrosis factor (TNF) receptors** transmit extracellular signals by forming receptor trimers, which direct the

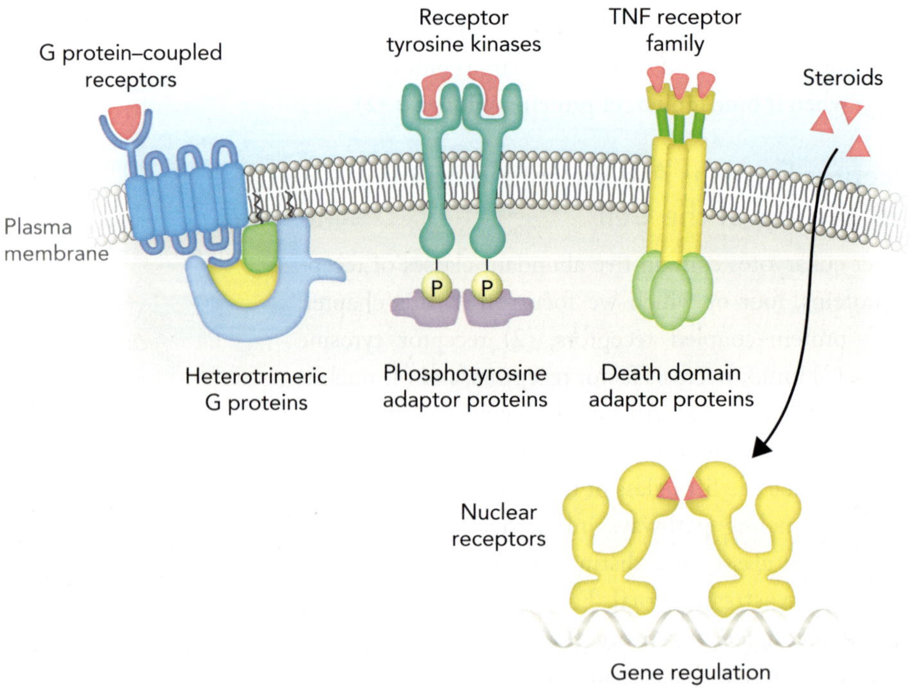

Figure 8.13 Four abundant classes of eukaryotic receptor proteins exist in a variety of cell types. Three of these receptor classes are transmembrane proteins: (1) G protein–coupled receptors, which signal through heterotrimeric G proteins; (2) receptor tyrosine kinases, which signal through adaptor proteins that bind phosphotyrosine residues on the receptors; and (3) the tumor necrosis factor (TNF) receptor family, which signals through adaptor proteins that contain a protein binding module called a death domain. The fourth class of eukaryotic receptor proteins, nuclear receptor proteins, are transcription factors that reside in the cytoplasm or nucleus and regulate gene expression in a ligand-dependent manner.

association of cytosolic adaptor protein complexes. This class of membrane receptors regulates signaling pathways that control inflammation and apoptosis (programmed cell death). Finally, nuclear receptors are ligand-regulated transcription factors that modulate gene expression through protein–DNA and protein–protein interacting functions.

The fifth class of abundant receptors in higher eukaryotes is **ligand-gated ion channels**, which control the flow of K^+, Na^+, and Ca^{2+} ions across cell membranes in response to ligand binding. Ligand-gated ion channels are functionally similar to the passive membrane transport proteins described in Chapter 6 that transport ions (see Figure 6.44). One of the best-characterized ligand-gated ion channels is the **nicotinic acetylcholine receptor**, which mediates ion transport in response to the neurotransmitter acetylcholine. A second type of acetylcholine receptor is the muscarinic acetylcholine receptor, which is a GPCR protein (see Figure 17.64). The function of the nicotinic acetylcholine receptor is to transmit physiologic signals across neuromuscular junctions in response to neuronal stimuli (**Figure 8.14**). Release of acetylcholine from vesicles in the presynaptic neuron occurs when a nerve impulse, which is transmitted by membrane depolarization, travels down the nerve cell axon. Acetylcholine crosses the neuromuscular synapse and binds to α subunits in nicotinic acetylcholine receptors, which are located in the plasma membrane of muscle cells. Ligand binding triggers a conformational change that opens the ion channel and allows Na^+ and K^+ ions to flow into the muscle cell, resulting in membrane depolarization. As described in Section 6.4, membrane depolarization in

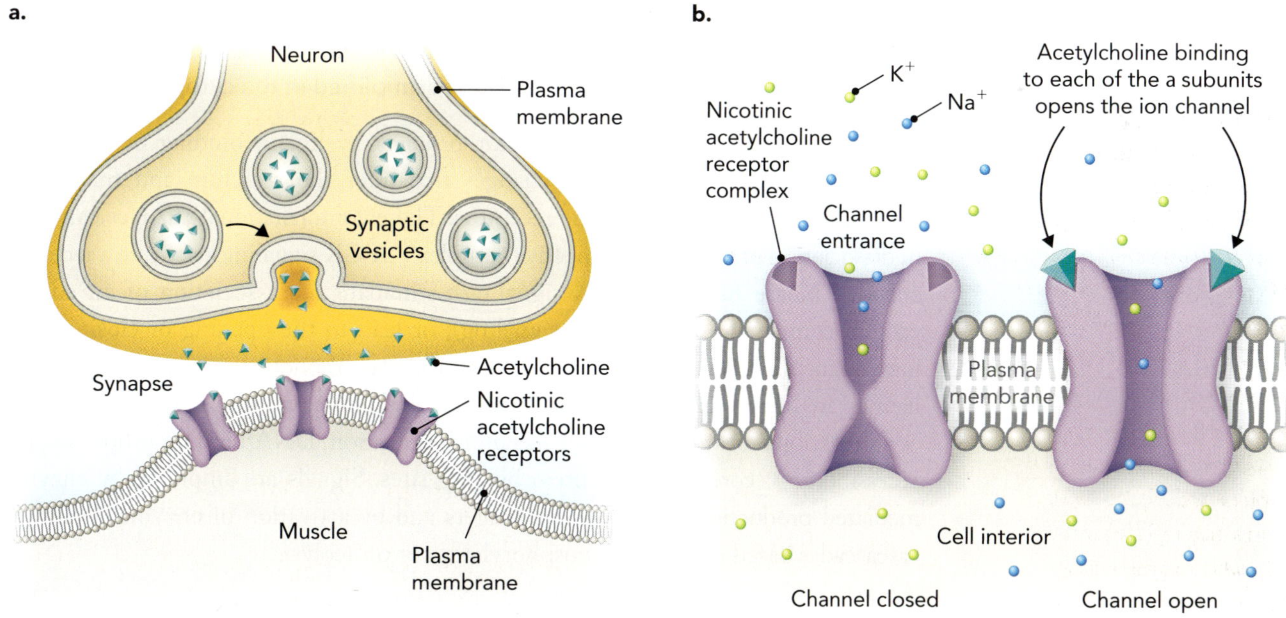

Figure 8.14 Nicotinic acetylcholine receptors are ligand-gated ion channels that transmit neuronal signals from nerve cells to muscle cells. **a.** Membrane depolarization in nerve axons stimulates release of the neurotransmitter acetylcholine from vesicles that fuse with the plasma membrane near the synaptic cleft. Acetylcholine diffuses across the synapse and binds to acetylcholine receptors on muscle cells, triggering membrane depolarization and muscle contraction. **b.** Acetylcholine binding to the α subunits of the nicotinic acetylcholine receptor causes a conformational change in the protein complex. This change opens the ion channel and allows Na^+ and K^+ ions to flow across the membrane and depolarize the cell.

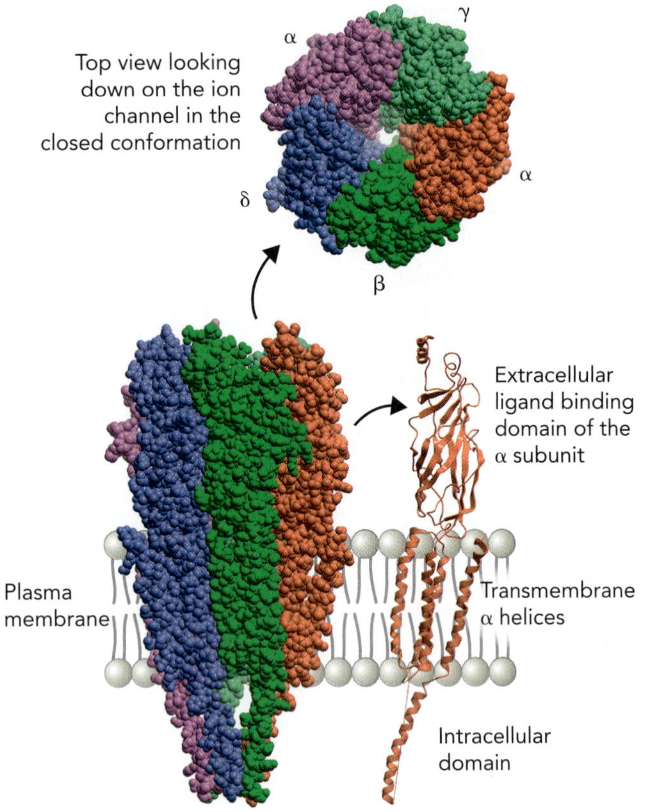

Figure 8.15 The molecular structure of the nicotinic acetylcholine receptor is shown here in the closed conformation. The five subunits of the $\alpha_2\beta\gamma\delta$ pentameric complex each contain three domains: the extracellular domain, the transmembrane domain, and the intracellular domain. The two identical α subunits contain acetylcholine binding sites located near the subunit interfaces. BASED ON PDB FILE 2BG9.

muscle cells induces Ca^{2+} release from the sarcoplasmic reticulum, and the binding of the released calcium causes a conformational change in muscle proteins that initiates muscle contraction.

Figure 8.15 shows the molecular structure of the nicotinic acetylcholine receptor isolated from the electric ray *Torpedo marmorata*. This pentameric transmembrane protein has a subunit composition of $\alpha_2\beta\gamma\delta$, in which acetylcholine binding sites are located at the domain interfaces in both α subunits. The central ion channel is formed by the convergence of transmembrane α helices present in each of the five subunits. Acetylcholine binding to the receptor opens the ion channel by inducing a conformational change that rotates the transmembrane α helices relative to one another.

concept integration 8.1

What is the mechanism of signal transduction across the plasma membrane, and how are signals amplified in the cytoplasm?

Stimulation of transmembrane receptor proteins by ligand binding or absorption of light induces conformational changes that alter the structure and function of cytoplasmic-facing regions of the receptors. These conformational changes can result in dissociation of a preformed inactive complex (GPCR signaling); activate a receptor-encoded kinase function (RTK signaling); stimulate the association of an adaptor protein complex (TNF receptor signaling); or lead to the opening or closing of an ion channel (ligand-gated ion channels). In the case of nuclear receptor signaling, the ligands are lipophilic and cross the membrane by diffusion, leading to ligand-induced conformational changes that alter receptor localization, DNA binding properties, and accessibility of coregulatory protein binding sites. Signals are amplified by enzyme-mediated production of second messengers and by activation of enzyme-dependent cascascades based on protein phosphorylation or proteolysis.

8.2 G Protein–Coupled Receptor Signaling

The human genome encodes a large number of G protein–coupled receptors, most of which are involved in sensory perceptions such as vision, taste, and smell. G protein–coupled receptors share an evolutionarily conserved structure consisting of a central core that encodes seven transmembrane α helices. The bundled arrangement of the

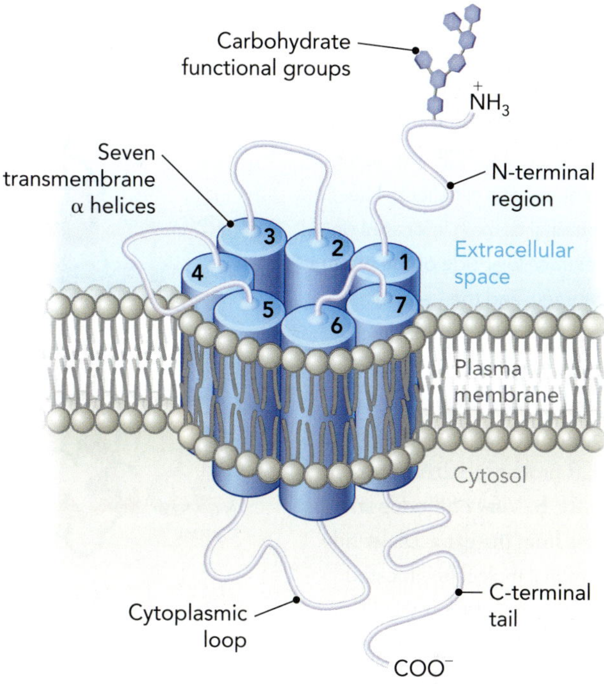

Figure 8.16 G protein–coupled receptors contain seven transmembrane α helices and are oriented with the N terminus on the outside of the cell and the C terminus on the inside. The N-terminal domain of most G protein–coupled receptors contains one or more carbohydrate functional groups. The α-helical regions are shown here as cylinders and are connected by polypeptide segments.

α helices orients the amino terminus of the receptor protein toward the extracellular space, with the carboxyl terminus exposed to the cytosol (**Figure 8.16**). Because the seven transmembrane helices traverse in and out of the plasma membrane, GPCRs have been called serpentine receptors (*serpentine* is Latin for "snake"), but they are more commonly referred to simply as GPCRs. (Refer to **Table 8.3** for a list of acronyms used in this section.)

Many GPCRs are glycoproteins that contain carbohydrate functional groups covalently attached to the extracellular domain. These carbohydrate moieties on membrane proteins contribute to their molecular recognition properties. The first GPCR to be analyzed at the molecular level by X-ray crystallography was **rhodopsin**, which contains a bound retinal molecule that absorbs light. The molecular structure of squid rhodopsin protein

Table 8.3 ABBREVIATIONS, PROTEIN NAMES, AND FUNCTIONS IN G PROTEIN–COUPLED RECEPTOR SIGNAL TRANSDUCTION SYSTEMS

Abbreviation	Protein name	Function
AC	Adenylate cyclase	Enzyme that produces cAMP
βAR	β-Adrenergic receptor	Membrane-spanning receptor that binds epinephrine
βARK	β-Adrenergic receptor kinase	Specific G protein–coupled receptor kinase enzyme that acts on β-adrenergic receptors
CAMK	Ca^{2+}/calmodulin kinase	Enzyme that phosphorylates and inhibits glycogen synthase
G_α	G_α protein	G protein subunit with GTPase activity
$G_{\alpha\beta\gamma}$	Heterotrimeric G protein	Complex of G_α, G_β, and G_γ subunits associated with GPCRs
GAP	GTPase activating protein	Protein that stimulates GTPase activity of G proteins to inhibit signaling
GEF	Guanine nucleotide exchange factor	Protein that promotes GDP–GTP exchange to activate signaling
GPCR	G protein–coupled receptor	Membrane-spanning receptor associated with intracellular G proteins
GRK	G protein–coupled receptor kinase	Enzyme that phosphorylates GPCRs
PDE	Phosphodiesterase	Enzyme that breaks down cAMP or cGMP
PH	Pleckstrin homology domain	Adaptor domain that binds membrane lipids
PKA	Protein kinase A	cAMP-dependent enzyme with kinase activity
PKC	Protein kinase C	Lipid-activated enzyme with kinase activity
PLC	Phospholipase C	Enzyme that produces diacylglycerol and inositol triphosphate
RGS	Regulator of G protein signaling	GTPase activating protein that is associated with GPCRs

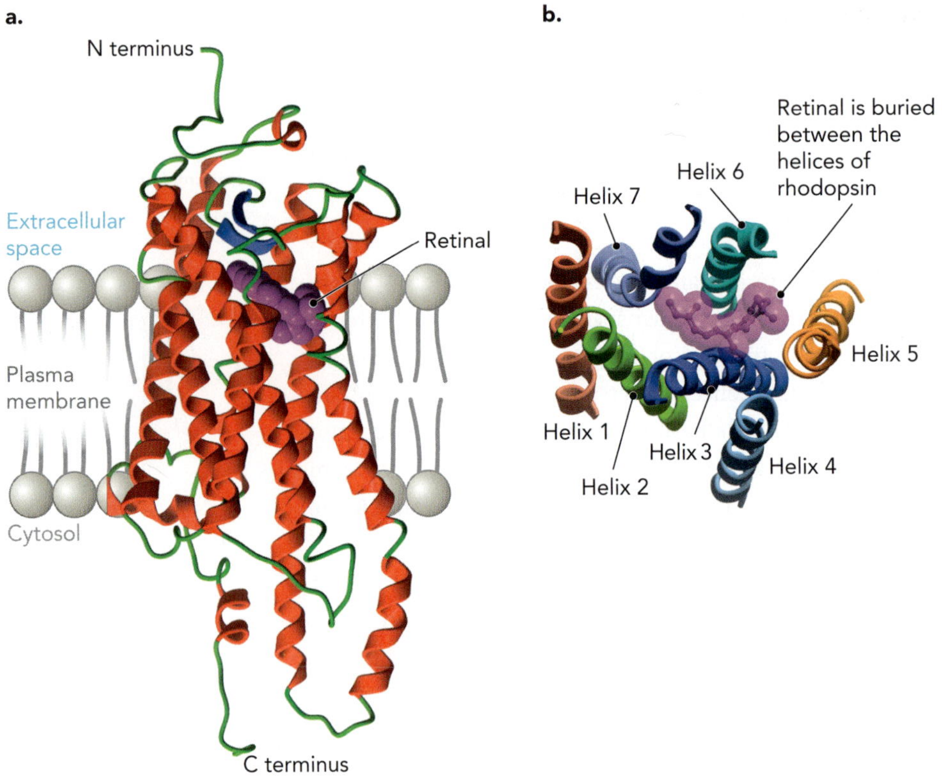

Figure 8.17 Squid rhodopsin protein is a GPCR that detects light through a tightly bound retinal molecule. **a.** Rhodopsin is a prototypical GPCR with seven transmembrane α helices. The N-terminal domain is located on the extracellular side of the plasma membrane, and the C-terminal domain is on the cytoplasmic side. Absorption of light energy by the retinal molecule induces a conformational change in the GPCR protein. BASED ON PDB FILE 2Z73. **b.** View of rhodopsin helices from the extracellular side. The retinal molecule is located within the receptor protein in a hydrophobic pocket formed by the seven transmembrane α helices. BASED ON PDB FILE 2Z73.

is shown in **Figure 8.17**. Observe that the retinal light sensor is localized within the interior of the receptor, formed by the seven transmembrane α helices. Rhodopsin transduces this light signal to cytosolic signaling proteins through receptor-mediated conformational changes. In other GPCRs, ligand binding on the extracellular side of the receptor leads to conformational changes on the cytosolic side of the receptor, as shown in **Figure 8.18** for the β2-adrenergic receptor (βAR). These cytosolic conformational changes affect the interactions of the receptor with intracellular signaling proteins, leading to changes in their activity.

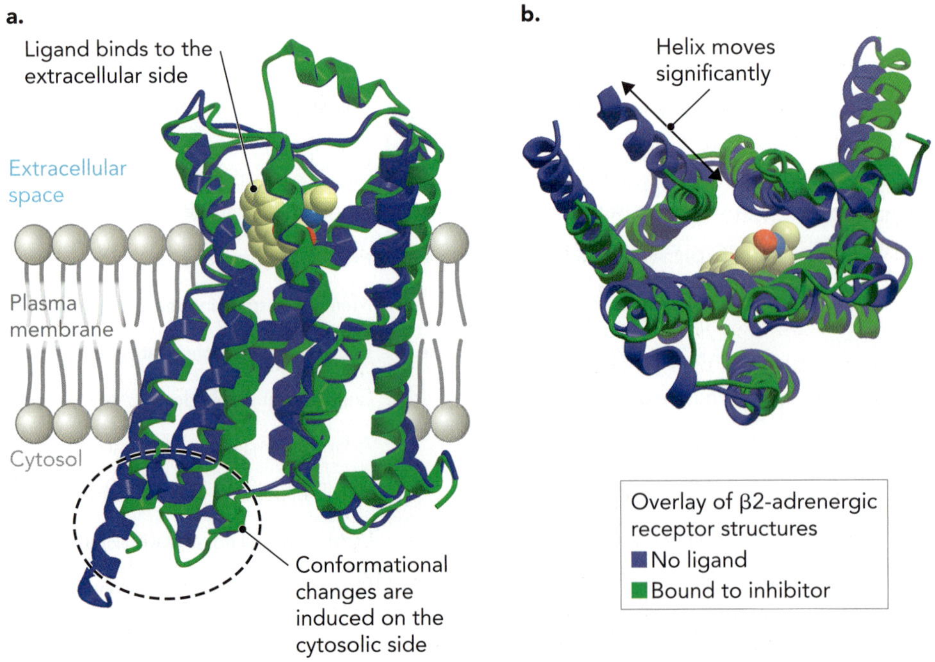

Figure 8.18 The β2-adrenergic receptor is a GPCR that binds epinephrine. **a.** Binding of a ligand (here shown as the inhibitor carazolol) to the extracellular side of the receptor causes conformational changes on the cytosolic side. One region on the cytosolic side that undergoes significant movement is highlighted. BASED ON PDB FILES 3SN6 (UNBOUND) AND 2RH1 (BOUND TO CARAZOLOL). **b.** Looking from the cytosolic side, it is clear that one of the transmembrane helices has shifted in position in response to ligand binding. BASED ON PDB FILES 3SN6 (UNBOUND) AND 2RH1 (BOUND TO CARAZOLOL).

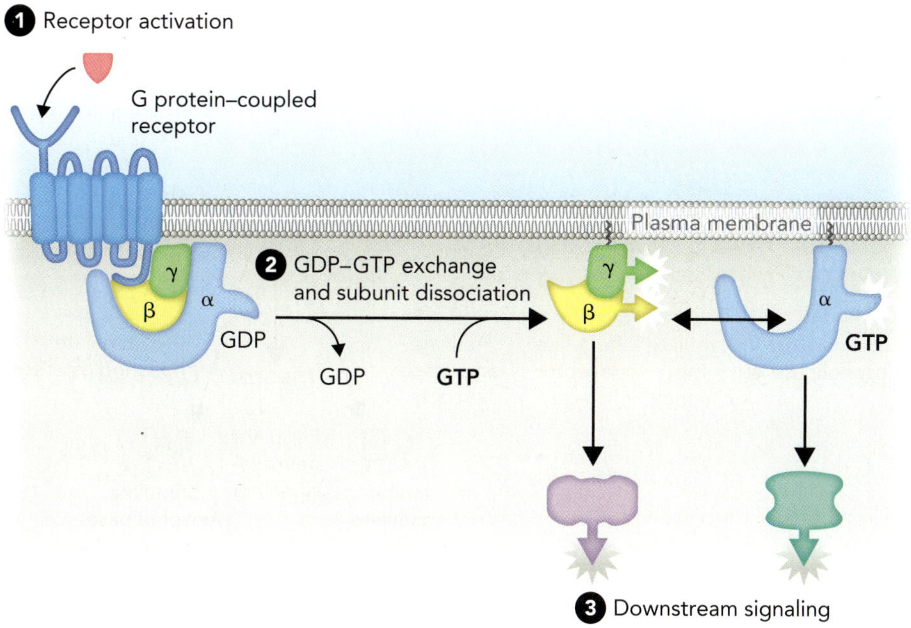

Figure 8.19 Receptor activation of heterotrimeric G proteins leads to subunit dissociation and regulation of downstream signaling proteins. The three steps in G protein–coupled receptor signaling are (1) ligand-induced conformational changes in the GPCR; (2) receptor-mediated stimulation of guanine nucleotide exchange (GTP replaces GDP) in the G_α subunit, leading to dissociation into G_α–GTP and $G_{\beta\gamma}$ complexes; and (3) regulation of downstream effector processes by the G_α–GTP and $G_{\beta\gamma}$ complexes.

GPCRs Activate Heterotrimeric G Proteins

GPCRs transmit extracellular signals to the cytoplasm through direct interaction with a membrane-bound protein complex called a **heterotrimeric G protein**, which consists of one each of G_α, G_β, and G_γ subunits ($G_{\alpha\beta\gamma}$). The G_α subunit is a member of the **G protein family** of signaling proteins, which are enzymes called **GTPases**. These enzymes cleave GTP to form guanosine diphosphate plus inorganic phosphate (GDP + P_i). G proteins are in the active conformation when GTP is bound, but when GTP is hydrolyzed, the GDP-bound G proteins are in the inactive conformation.

The inactive $G_{\alpha\beta\gamma}$ complex contains GDP bound to the G_α subunit and associates with the unliganded GPCR. Ligand binding to the GPCR (or light absorption by retinal, in the case of rhodopsin) induces a conformational change in the receptor. This conformational change stimulates exchange of GDP for GTP in the G_α subunit. As illustrated in **Figure 8.19**, the G_α subunit bound to GTP is now activated and dissociates from the heterotrimeric complex. After dissociation, both the G_α subunit and the $G_{\beta\gamma}$ complex stimulate multiple downstream signaling pathways. Note that the G_α and $G_{\beta\gamma}$ proteins are anchored to the cytosolic side of the plasma membrane by covalently attached lipid moieties. **Figure 8.20** shows the molecular structure of an inactive, GDP-bound $G_{\alpha\beta\gamma}$ complex called transducin, which is associated with rhodopsin GPCRs in the absence of light.

▶ ANIMATION

The human genome contains 17 G_α, 5 G_β, and 12 G_γ genes, which could theoretically result in the formation of almost 1,000 different $G_{\alpha\beta\gamma}$ complexes ($17 \times 5 \times 12 = 970$).

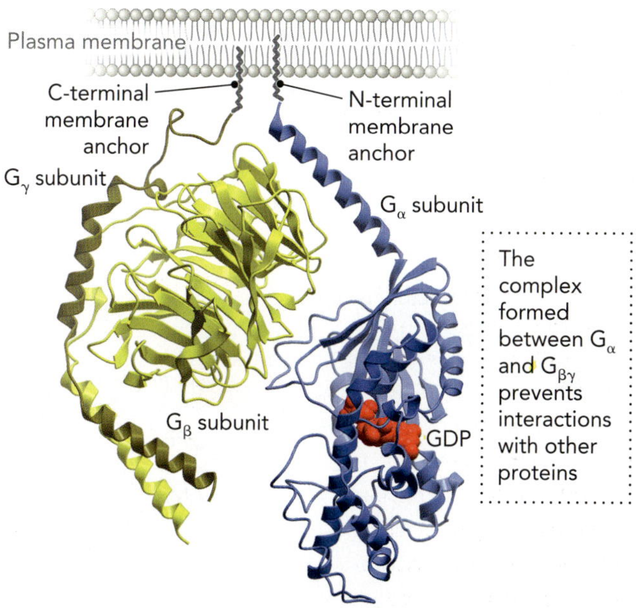

Figure 8.20 The transducin heterotrimeric G complex ($G_{\alpha\beta\gamma}$) is in the inactive state when bound to GDP as shown here. The G_α and G_γ subunits have lipid moieties covalently attached to the N-terminal and C-terminal amino acid residues, respectively, which are schematically shown here. BASED ON PDB FILE 1GOT.

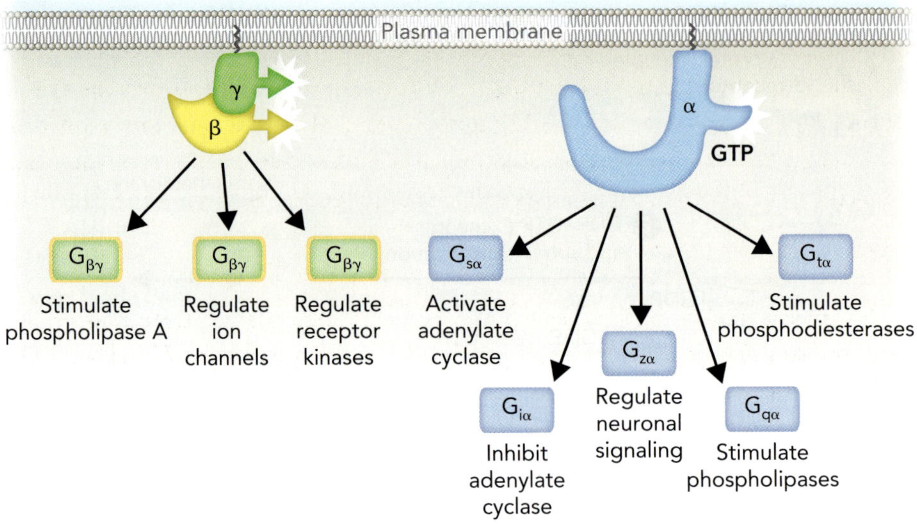

Figure 8.21 Distinct $G_{\beta\gamma}$ and G_{α} proteins can have different downstream signaling functions.

Studies have shown that each of these different signaling molecules is capable of activating distinct downstream pathways (**Figure 8.21**). For example, different $G_{\beta\gamma}$ complexes have been shown to regulate ion channels and receptor kinases and to stimulate a regulated enzyme called phospholipase A, which is an enzyme that hydrolyzes fats. Moreover, specific G_{α} subunits have been found that regulate a variety of downstream target proteins, two of which are introduced in Figure 8.21: $G_{s\alpha}$, which is a stimulatory G_{α} protein that activates the enzyme adenylate cyclase to generate the second messenger cAMP (see Figure 8.9); and $G_{q\alpha}$, which activates the enzyme phospholipase C that generates the second messengers DAG and IP_3 (see Figure 8.11).

Examples of other G_{α} proteins include one that inhibits the activity of adenylate cyclase ($G_{i\alpha}$ protein); the G_{α} protein in transducin that functions in the rhodopsin signaling pathway and activates cGMP phosphodiesterase ($G_{t\alpha}$ protein); and $G_{z\alpha}$, a G_{α} protein that regulates neuronal signaling pathways. Indeed, three of our five sensory

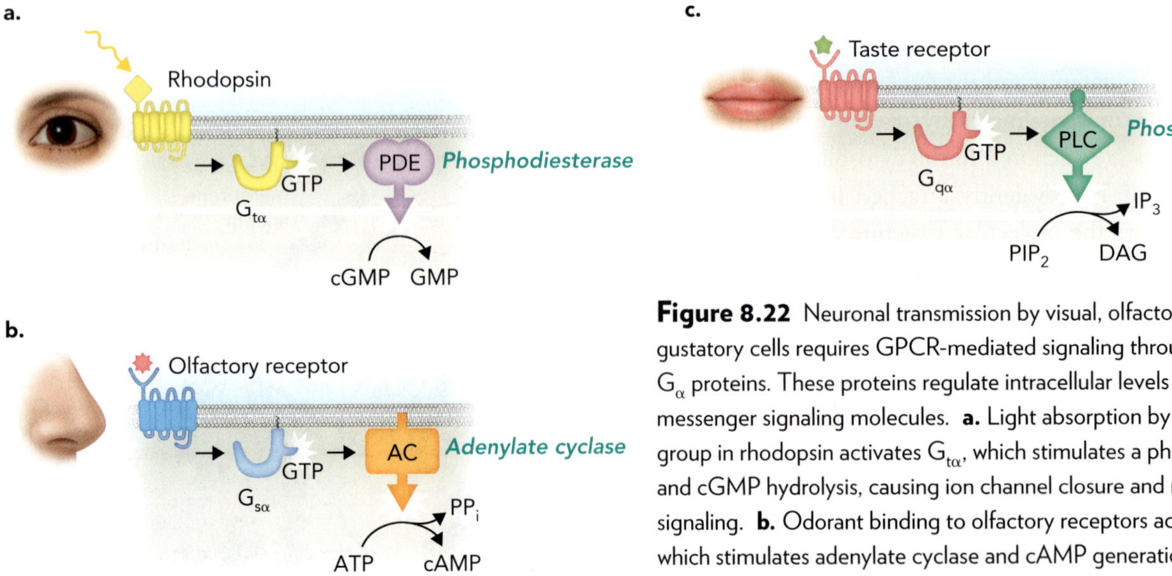

Figure 8.22 Neuronal transmission by visual, olfactory, and gustatory cells requires GPCR-mediated signaling through specific G_{α} proteins. These proteins regulate intracellular levels of second messenger signaling molecules. **a.** Light absorption by the retinal group in rhodopsin activates $G_{t\alpha}$, which stimulates a phosphodiesterase and cGMP hydrolysis, causing ion channel closure and neuronal signaling. **b.** Odorant binding to olfactory receptors activates $G_{s\alpha}$, which stimulates adenylate cyclase and cAMP generation, causing ion channel opening and neuronal signaling. **c.** Binding of sweeteners to taste receptors on tongue cells activates $G_{q\alpha}$, which stimulates phospholipase C and the generation of DAG and IP_3, causing ion channel closure and neuronal signaling.

inputs depend on signal transduction through these same G_α proteins. **Figure 8.22** shows that the perceptions of vision, smell, and taste are all mediated in part by GPCR signaling pathways, which regulate intracellular levels of second messengers. These regulated levels include reduced levels of cGMP in response to light absorption by rhodopsin (sight); elevated levels of cAMP upon binding of specific odorants to olfactory receptors (smell); and the production of IP_3 and DAG when certain sweeteners bind to specific taste receptors in gustatory cells of the tongue (taste).

GPCR-Mediated Signaling in Metabolism

As described above for sensory perception, ligand activation of GPCRs can stimulate signal transduction in different cell types through distinct pathways. However, different GPCR-mediated signals can also be integrated within a single cell type in response to multiple extracellular stimuli.

One of the best examples of this complexity is the upstream regulation of glycogen degradation in mammalian liver cells by the hormones **glucagon** and epinephrine (adrenaline). **Figure 8.23** shows how these two hormones stimulate both shared and parallel downstream signaling pathways that converge on the same phenotypic response. In this case, the response is degradation of stored glycogen in liver cells to produce glucose for export to the tissues. Both glucagon and epinephrine signal to the human liver that blood glucose levels need to rise through increased rates of glycogen degradation and glucose export (see Chapter 14).

Glucagon binds specifically to glucagon receptors and activates $G_{s\alpha}$ signaling, which stimulates adenylate cyclase activity to produce the second messenger cAMP. In contrast, epinephrine binds to both the β2-adrenergic and the α1-adrenergic receptors (receptors that bind adrenaline, that is, epinephrine), which are expressed on liver cells. Epinephrine binding to β2-adrenergic receptors increases intracellular levels of cAMP, using the same $G_{s\alpha}$ signaling pathway as does glucagon—this is an example of a shared pathway in the same cell. However, epinephrine binding to α1-adrenergic receptors on liver cells activates $G_{q\alpha}$ signaling, which stimulates phospholipase C activity to produce the second messengers DAG and IP_3. Because epinephrine can stimulate liver glycogen degradation by binding to both the β2-adrenergic and the α1-adrenergic receptors, each of which activates different G_α proteins ($G_{s\alpha}$ or $G_{q\alpha}$), epinephrine signaling in liver cells is an example of parallel signaling pathways.

Glucagon is a peptide hormone released by the pancreas and has been called the "I am hungry" hormone because it signals low blood glucose levels. Consistent with this physiologic role, glucagon receptors are primarily expressed in liver and adipocyte cells, where energy stores in the form of glycogen and triacylglycerols are located (see Chapter 19). The hormone epinephrine belongs to a class of first messengers called

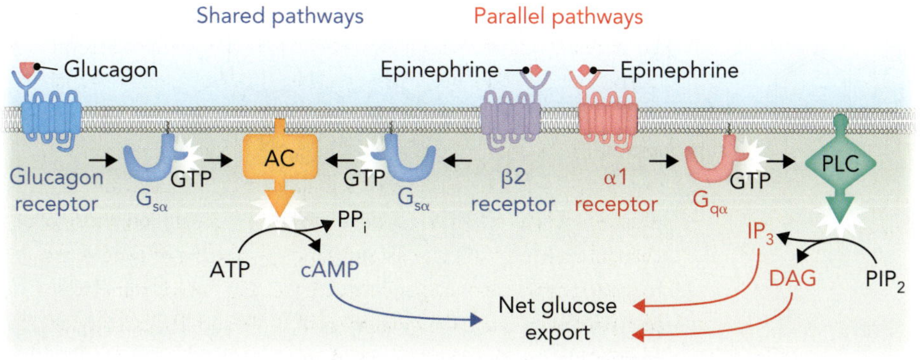

Figure 8.23 GPCR-mediated signaling pathways can be shared or parallel, as shown here in liver cells. Glucagon receptors and β2-adrenergic receptors activate a shared cAMP-mediated signaling pathway through $G_{s\alpha}$, which stimulates adenylate cyclase (AC) activity and the production of cAMP. In contrast, epinephrine binding to β2-adrenergic and α1-adrenergic receptors activates parallel pathways using $G_{s\alpha}$ and $G_{q\alpha}$, respectively. The $G_{q\alpha}$ pathway stimulates phospholipase C (PLC) activity, leading to the production of the second messengers DAG and IP_3. All three of these signaling pathways converge on target proteins in liver cells that degrade glycogen, leading to increased rates of glucose export.

catecholamines, which are derived from the amino acid tyrosine (see Chapter 17). Epinephrine is sometimes called the "fight or flight" hormone because it is released from the adrenal medulla under times of acute stress. Epinephrine signaling has many physiologic effects, including increased heart rate and mobilization of energy stores through glycogen breakdown and lipolysis.

Epinephrine and two other catecholamines, norepinephrine and dopamine, bind to two types of adrenergic receptors called the α-adrenergic and β-adrenergic receptors. The tissue distribution and physiologic responses governed by α-adrenergic and β-adrenergic receptors are diverse, controlling everything from metabolism in liver, skeletal muscle, and adipose cells to relaxation and contraction of smooth muscle. The α-adrenergic and β-adrenergic receptors have different affinities for both natural and synthetic ligands, a finding that has been exploited to develop numerous pharmaceutical drugs to treat human diseases. Adrenergic **receptor agonists** activate receptor signaling by mimicking the natural ligands epinephrine, norepinephrine, or dopamine. In contrast, adrenergic **receptor antagonists** bind to receptors with high affinity and block the binding of physiologic agonists without inducing the structural changes required for signal transduction.

Figure 8.24 shows the chemical structures of the best-characterized physiologic and pharmaceutical adrenergic receptor ligands. Isoproterenol (similar to albuterol) is a potent β2 agonist used by asthmatics to relax smooth muscle in the lungs and open airway passages. Clonidine is a selective α2 agonist used to treat high blood pressure and insomnia. Examples of adrenergic receptor antagonists include prazosin, which primarily blocks α1-adrenergic receptor signaling, and metoprolol, a β1-adrenergic receptor antagonist. Prazosin and metoprolol inhibit adrenergic signaling pathways that control vasoconstriction.

GPCR-mediated stimulation of GDP–GTP exchange in $G_{s\alpha}$, brought about by glucagon and epinephrine signaling, induces a conformational change in the $G_{s\alpha}$ protein that facilitates binding to adenylate cyclase. **Figure 8.25** shows the structural differences in G_{α} when bound to GDP or GTP. A short helical region in the $G_{s\alpha}$ subunit, called switch II, undergoes a conformational change in the presence of nucleotide.

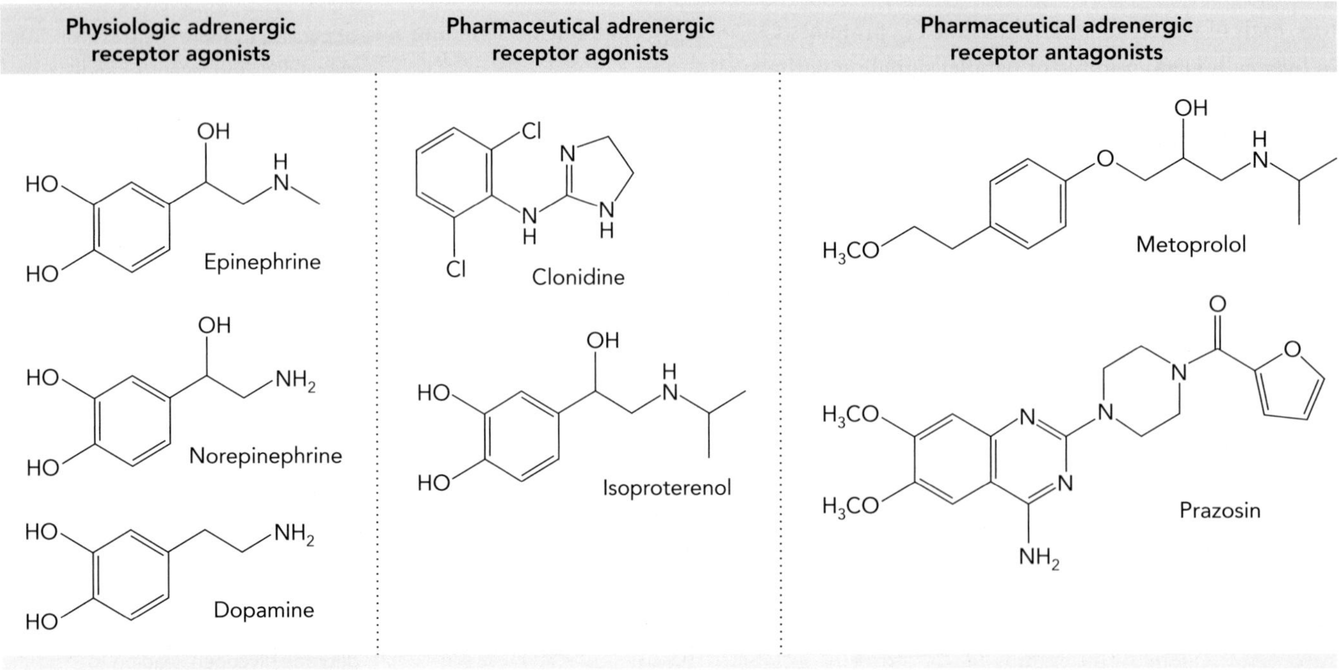

Figure 8.24 The chemical structures of representative physiologic and pharmaceutical adrenergic receptor agonists and antagonists are shown here. Note that the pharmaceutical agonists clonidine and isoproterenol are more similar in overall structure to the physiologic ligands (epinephrine, norepinephrine, and dopamine) than are the pharmaceutical antagonists metoprolol and prazosin, which are much larger molecules. This chemical difference between agonist and antagonist is a general feature of many pharmaceutical agents.

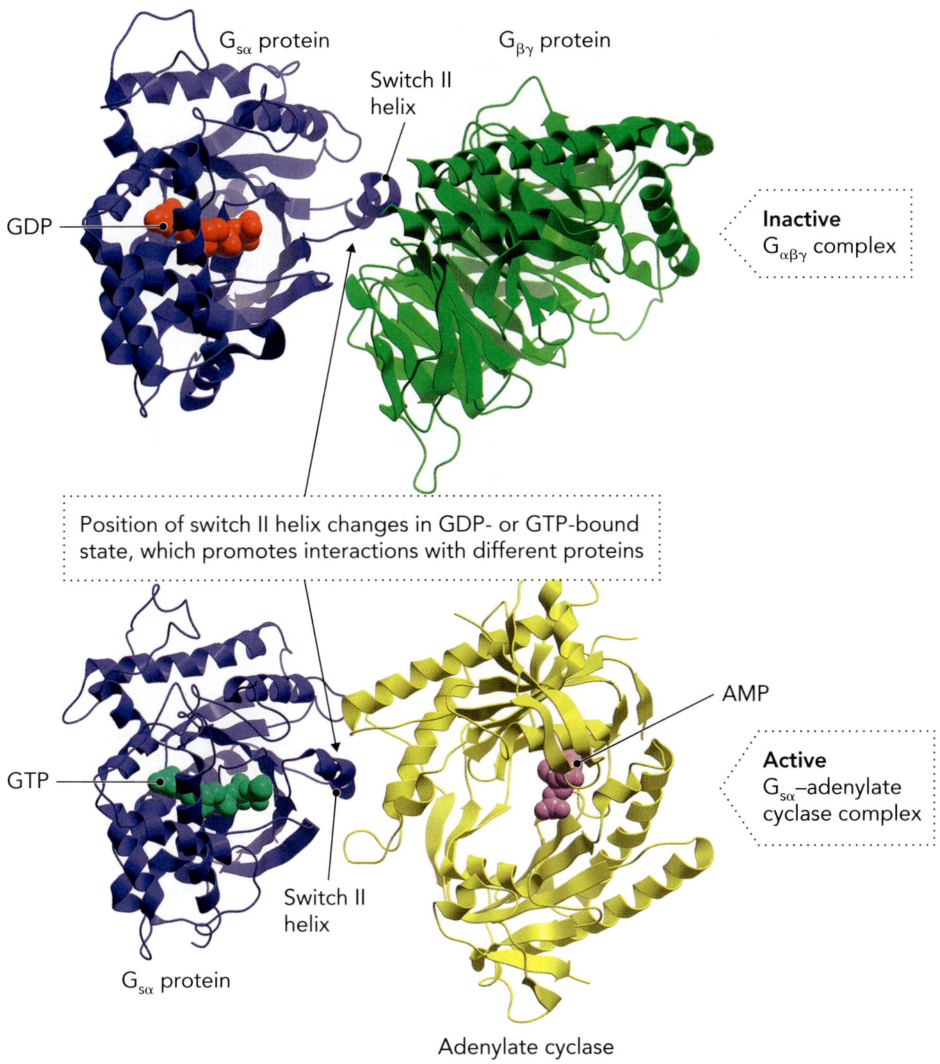

Figure 8.25 In the GDP-bound state, $G_{s\alpha}$ interacts with $G_{\beta\gamma}$ through the switch II helix region. GTP binding to $G_{s\alpha}$ induces a conformational change in the switch II helix region so that it is in a position to bind with adenylate cyclase. BASED ON PDB FILES 1GG2 ($G_{\alpha\beta\gamma}$) AND 1CS4 ($G_{s\alpha}$–ADENYLATE CYCLASE).

When $G_{s\alpha}$ is bound to GDP, the switch II helix region interacts with $G_{\beta\gamma}$. GTP binding changes the conformation of the switch II region so that now $G_{s\alpha}$ interacts with adenylate cyclase, which is critical for stimulation of adenylate cyclase activity and subsequent production of cAMP. In liver cells, the primary response to elevated cAMP levels is activation of the downstream signaling protein protein kinase A (PKA). As illustrated in **Figure 8.26**, when cAMP levels in the cell are low, PKA is inactive and exists as an R_2C_2 tetramer, consisting of two regulatory subunits (R) and two catalytic subunits (C). When two cAMP molecules bind to each regulatory subunit, the protein undergoes a conformational change, leading to disruption of the R_2C_2 tetramer and release of the catalytic subunits (active monomers). The liberated PKA monomers bind to and phosphorylate target proteins on serine or threonine residues.

The regulatory subunit of PKA contains the sequence Arg-Arg-Gly-*Ala*-Ile, which is called a pseudosubstrate sequence because it is similar to the substrate sequence recognized by the catalytic domain, Arg-Arg-Gly-*Ser*-Ile (where the Ser residue is the site of phosphorylation). Because of its similarity to the substrate sequence, this region of the regulatory subunit binds

Figure 8.26 Activation of protein kinase A by cAMP requires the dissociation of regulatory subunits and catalytic subunits. Four cAMP molecules bind to the inactive R_2C_2 complex and induce a conformational change in the regulatory (R) subunits that results in the release of two catalytically active monomers.

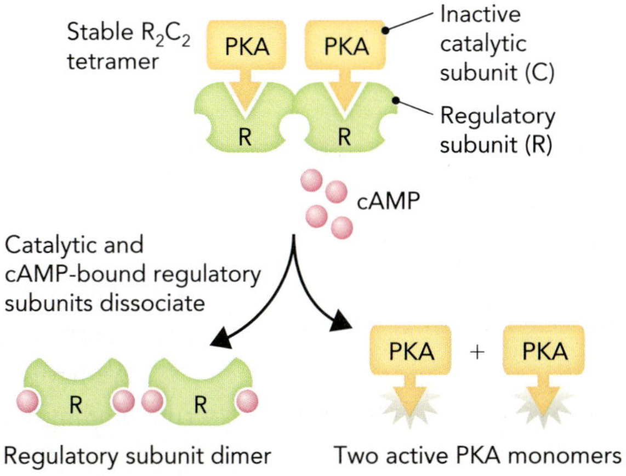

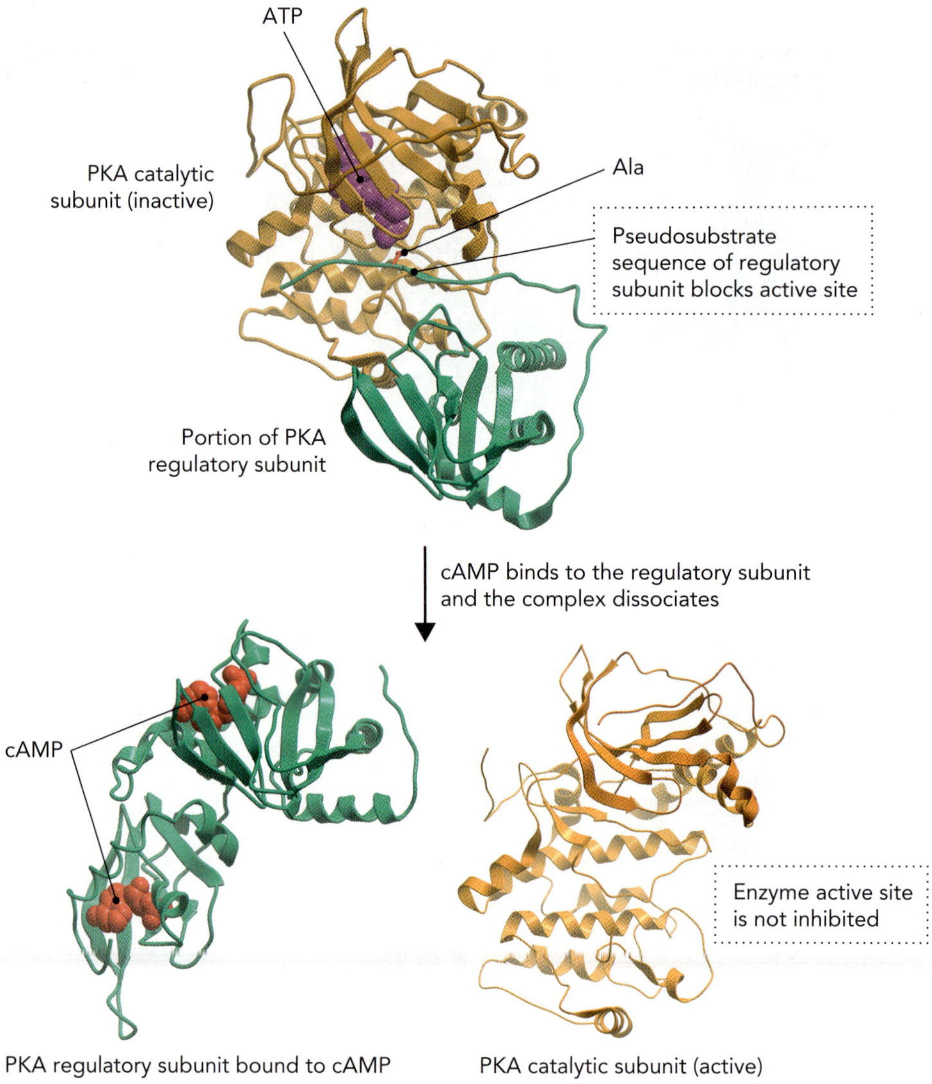

Figure 8.27 Protein kinase A (PKA) complex contains both regulatory and catalytic subunits. In the PKA complex, the pseudosubstrate sequence of the regulatory subunit binds to the active site of the catalytic complex and inhibits substrate binding. The alanine residue of the pseudosubstrate sequence is shown in red stick representation. ATP, shown in purple space-filling representation, is also in the enzyme active site. When cAMP binds to the regulatory subunit, the complex dissociates. The catalytic subunit active site is no longer blocked by the regulatory subunit. BASED ON PDB FILES 3FHI (COMPLEX WITH ATP), 1RGS (REGULATORY SUBUNIT), AND 4DFY (CATALYTIC SUBUNIT).

to the active site of the catalytic domain and inhibits it by blocking substrate binding (**Figure 8.27**). Because the pseudosubstrate sequence contains Ala instead of Ser, the regulatory subunit cannot be phosphorylated and thus remains in the active site. The binding of cAMP to the regulatory subunits provides the conformational switch that releases the pseudosubstrate sequence from the catalytic subunit. The active site of the monomeric catalytic subunit is now available to bind and phosphorylate substrates.

Figure 8.28 shows how ligand binding to glucagon receptors or β2-adrenergic receptors in liver cells stimulates $G_{s\alpha}$ signaling, inducing cAMP production by the enzyme adenylate cyclase. This series of upstream signaling events results in the activation of downstream PKA signaling, culminating in at least three distinct metabolic responses: (1) phosphorylation and inhibition of an enzyme required for glycogen synthesis (glycogen synthase); (2) phosphorylation and activation of enzymes involved in glycogen degradation and production of glucose; and (3) phosphorylation and activation of enzymes involved in glucose synthesis (gluconeogenic pathway). The combined effect of PKA signaling in liver cells is a net glucose export for use as chemical energy by other tissues.

We have just seen the upstream PKA signaling pathway initiated by epinephrine binding to liver β2-adrenergic receptors. Now let's compare it to the pathway

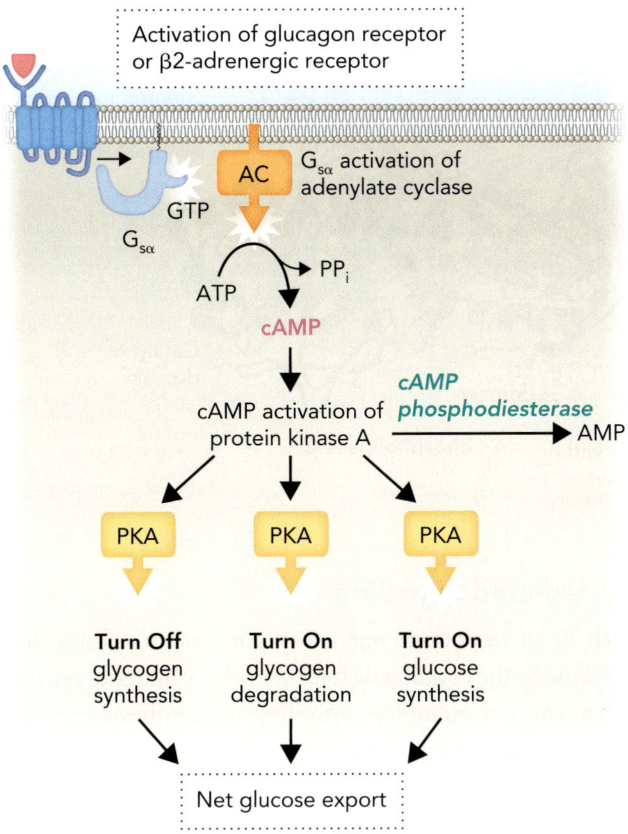

Figure 8.28 Epinephrine and glucagon signaling in liver cells stimulates $G_{s\alpha}$ and activates PKA, leading to an increase in glucose export. PKA-mediated phosphorylation of target proteins activates glucose synthesis (gluconeogenesis) and glycogen degradation, while at the same time inhibiting glycogen synthesis. The result is a net glucose export. The activity of cAMP phosphodiesterase modulates steady-state levels of cAMP by hydrolyzing it to AMP.

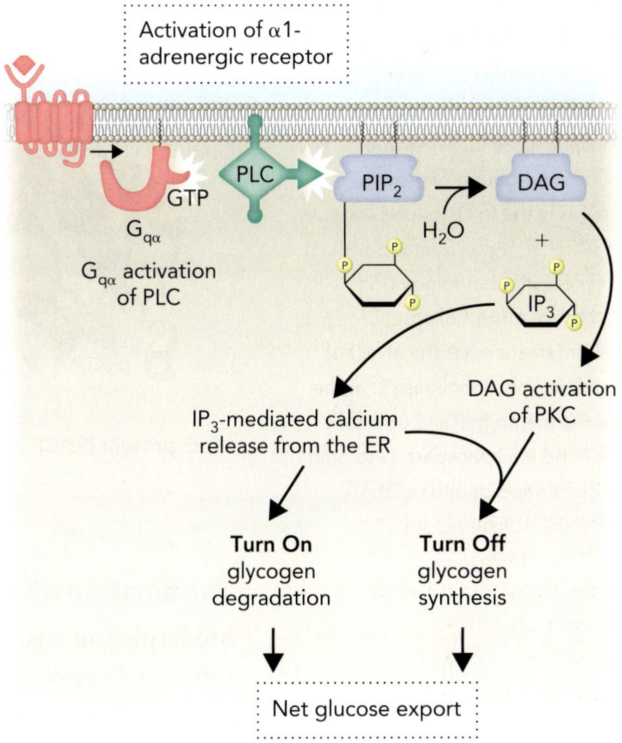

Figure 8.29 Epinephrine activation of α1-adrenergic receptor signaling in liver cells stimulates $G_{q\alpha}$ signaling. This activates phospholipase C (PLC), which hydrolyzes PIP_2 to generate downstream signaling through the second messengers DAG and IP_3. DAG activates the signaling enzyme PKC, and IP_3 stimulates Ca^{2+} release from the endoplasmic reticulum (ER). Activation of phospholipase C signaling by epinephrine leads to a net increase in glucose export as a result of increased glycogen degradation and decreased glycogen synthesis.

initiated by epinephrine binding to liver α1-adrenergic receptors. As shown in **Figure 8.29**, α1-adrenergic receptors signal through $G_{q\alpha}$, which activates the enzyme phospholipase C. PLC is a large membrane-associated protein of 110 kDa and consists of multiple domains that facilitate functional interactions with several signaling molecules. The molecular structure of a protein complex containing phospholipase C and a GTP-bound G protein called RAC—which, like $G_{q\alpha}$, is also an activator of phospholipase C—is shown in **Figure 8.30**. The catalytic domain binds to its substrate, the membrane lipid PIP_2, which is hydrolyzed by phospholipase C to form the second messengers DAG and IP_3 (see Figure 8.10). A region of the catalytic domain called the hydrophobic ridge facilitates phospholipase C binding to PIP_2 by stabilizing hydrophobic interactions with the plasma membrane. Another important domain in phospholipase C is the pleckstrin homology (PH) domain, which mediates an adaptor function that binds the G protein RAC and the membrane lipid PIP_3. We describe the structure and function of other PH domains in signaling proteins later in this chapter.

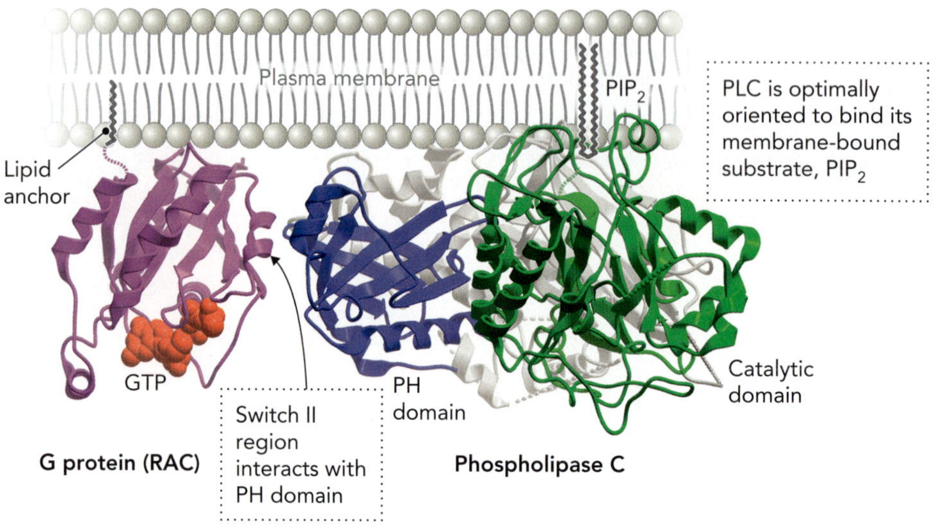

Figure 8.30 Shown is the structure of a phospholipase C protein complex containing RAC, a small G protein that activates phospholipase C. When the membrane-bound G protein RAC is in the GTP-bound state, its switch II region interacts with the pleckstrin homology (PH) domain of phospholipase C. This interaction has the effect of stabilizing phospholipase C at the plasma membrane and optimally orienting phospholipase C to bind its membrane-bound substrate, PIP$_2$. BASED ON PDB FILE 2FJU.

Termination of GPCR-Mediated Signaling

Modulation of signal strength is an important regulatory component of signaling pathways. In most cases, this is done by the combined effect of both a stimulatory mechanism and an inhibitory mechanism. For example, controlling the synthesis and degradation of cAMP and cGMP by cyclase and phosphodiesterase enzymes, respectively, is an important mechanism in sensory perception (see Figure 8.22). GPCR and G_α proteins are also subject to both stimulatory and inhibitory control.

In the case of G proteins, their signaling activity is controlled by two types of proteins. **Guanine nucleotide exchange factor (GEF)** proteins promote GDP–GTP exchange and activate signaling. **GTPase activating proteins (GAPs)** stimulate the intrinsic GTP hydrolyzing activity of G proteins to inhibit signal transduction. A GAP that functions with G proteins associated with GPCRs is specifically called a **regulator of G protein signaling (RGS)**.

The sequential stimulation of G protein signaling by GEF activity, with subsequent activation of its intrinsic GTPase activity by GAPs, is known as the **G protein cycle** (**Figure 8.31**). (Later in this chapter, we will describe a similar G protein cycle that controls the signaling activity of a G protein called Ras, which is activated by receptor tyrosine kinase signaling; see Figure 8.38 later.) **Figure 8.32** shows the human GAP RGS2 (which is specific for activating the GTPase activity of $G_{q\alpha}$ proteins) in complex with $G_{q\alpha}$–GDP. The GTPase active site in $G_{q\alpha}$ is identified by the bound GDP molecule.

Another important mechanism for terminating GPCR-mediated signals is receptor desensitization after dissociation of the $G_{\alpha\beta\gamma}$ complex. Regulatory proteins, called **G protein–coupled receptor kinases (GRKs)**, phosphorylate the GPCR cytoplasmic domain on serine and threonine residues, which marks it for recycling. In the case of β2-adrenergic receptors, the G protein–coupled receptor kinase enzyme is called **β-adrenergic receptor kinase (βARK)**, which is recruited to the membrane by the dimeric $G_{\beta\gamma}$ complex (**Figure 8.33**, p. 396). Phosphorylation of GPCRs mediated by G protein–coupled receptor kinase provides a docking site on the receptor for a second protein called **β-arrestin**, which is a transport protein that binds to the receptor and prevents it from reassociating with the $G_{\alpha\beta\gamma}$ complex.

Note that GPCRs can also be phosphorylated by downstream kinases in their own signaling pathway, which provides a mechanism for feedback inhibition. For

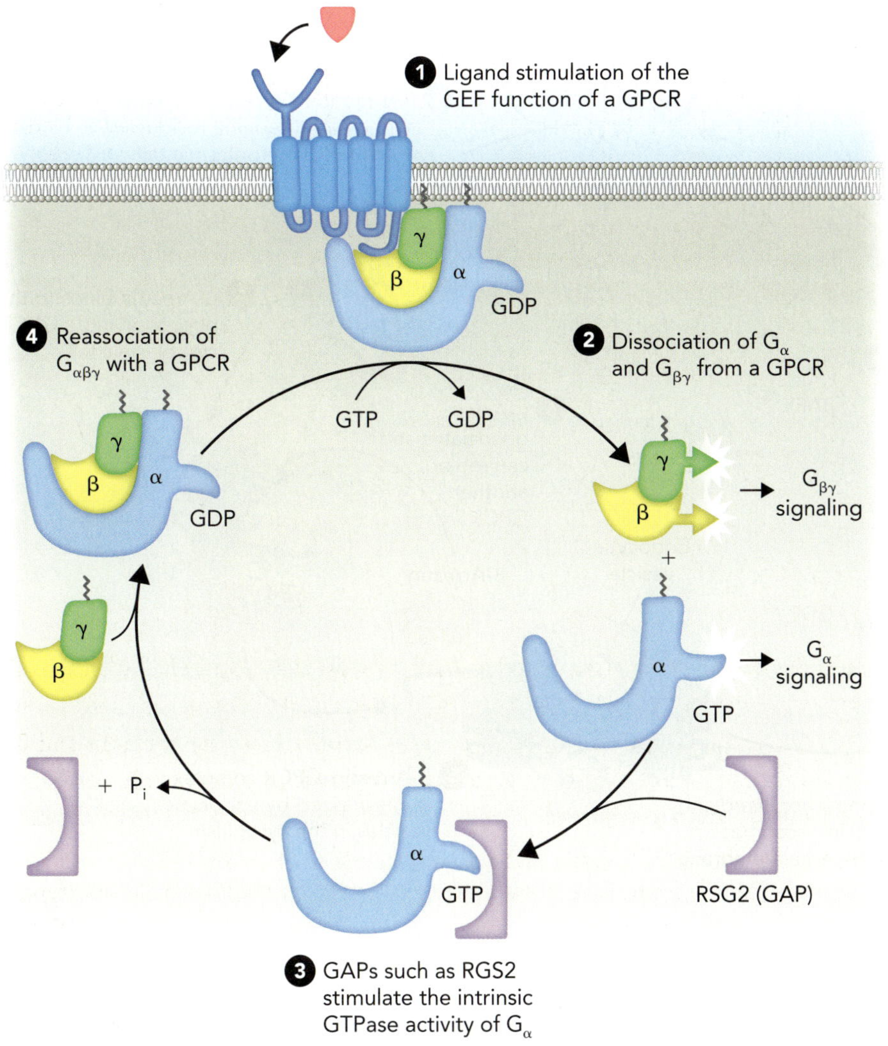
1 Ligand stimulation of the GEF function of a GPCR
γ
β
α
GDP
2 Dissociation of G_α and $G_{\beta\gamma}$ from a GPCR
4 Reassociation of $G_{\alpha\beta\gamma}$ with a GPCR
GTP
GDP
γ
β
α
GDP
γ
β
$G_{\beta\gamma}$ signaling
+
α
G_α signaling
GTP
γ
β
+ P_i
α
GTP
RSG2 (GAP)
3 GAPs such as RGS2 stimulate the intrinsic GTPase activity of G_α

Figure 8.31 The G protein cycle describes sequential stimulations by GEFs and GAPs and the associated changes in protein complexes and activity. The signaling activity of G_α proteins (and of all G proteins) is controlled by regulatory proteins called GEFs, which stimulate GDP–GTP exchange and activate signaling. In the case of GPCR signaling pathways, the ligand-bound receptor functions as a GEF for G_α proteins. Hydrolysis of GTP by the intrinsic GTPase activity in G proteins terminates G protein signaling. The GAP RGS2 functions to stimulate GTPase hydrolysis in GPCR signaling pathways.

example, PKA phosphorylates β2-adrenergic receptors on serine and threonine residues, which recruits β-arrestin to the cytoplasmic tail. As shown in Figure 8.33, the inactive β-arrestin–GPCR complexes are translocated to endocytic vesicles in the cytoplasm. In the vesicle, the receptor is dephosphorylated and either degraded by lysosomes or returned to the plasma membrane. The β-arrestin protein is left free to bind with other phosphorylated GPCR molecules. The removal of GPCRs from the plasma membrane through this recycling process provides a mechanism to control the duration and strength of the extracellular signal.

The molecular structure of a β-arrestin protein is shown in **Figure 8.34**. Observe that it contains two domains, each of which has a protein binding pocket that recognizes phosphoserine and phosphothreonine residues on target proteins. In addition to facilitating the transport of GPCRs from the plasma membrane to endosomal compartments, β-arrestin has also been found to serve as a molecular scaffold that assembles phosphorylated proteins into functional complexes.

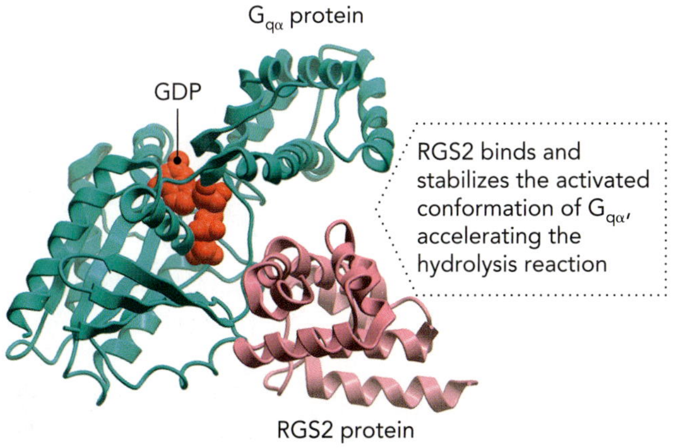
$G_{q\alpha}$ protein
GDP
RGS2 binds and stabilizes the activated conformation of $G_{q\alpha}$, accelerating the hydrolysis reaction
RGS2 protein

Figure 8.32 The ribbon diagram of the $G_{q\alpha}$–RGS2 protein complex shows a GDP molecule bound in the GTPase active site of $G_{q\alpha}$. RGS2 increases the rate of hydrolysis of GTP by binding to $G_{q\alpha}$ and stabilizing its activated conformation. RGS2 itself does not contribute any catalytic residues for the hydrolysis reaction. BASED ON PDB FILE 2V4Z.

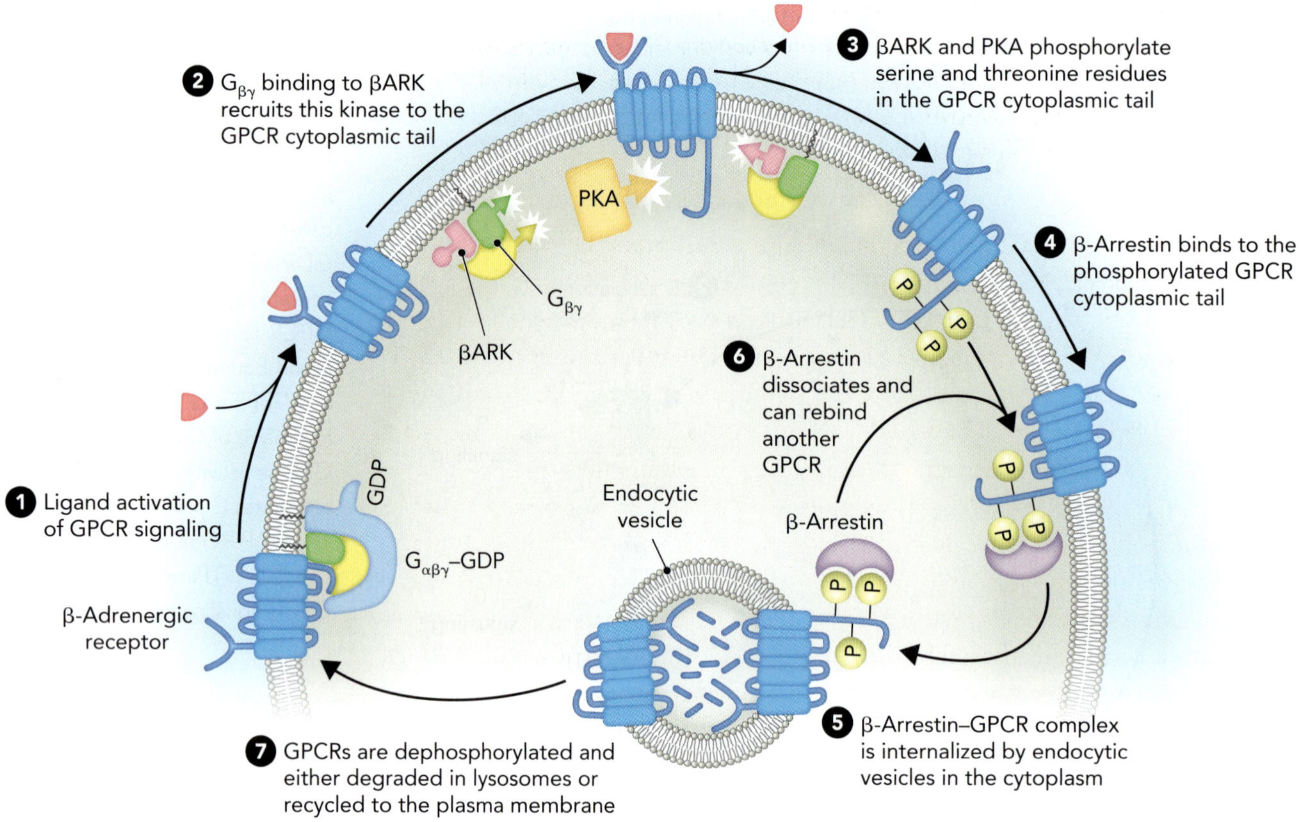

Figure 8.33 GPCR signaling is terminated by recycling the receptors through endocytic vesicles. After ligand binding and dissociation of the GPCR–heterotrimeric G protein complex, G protein–coupled receptor kinase proteins (such as βARK) phosphorylate serine and threonine residues in the GPCR cytoplasmic tail. This process generates binding sites for the endosomal transport protein β-arrestin. After GPCR dephosphorylation in endocytic vesicles, the receptor is either degraded or returned to the plasma membrane for another round of signaling.

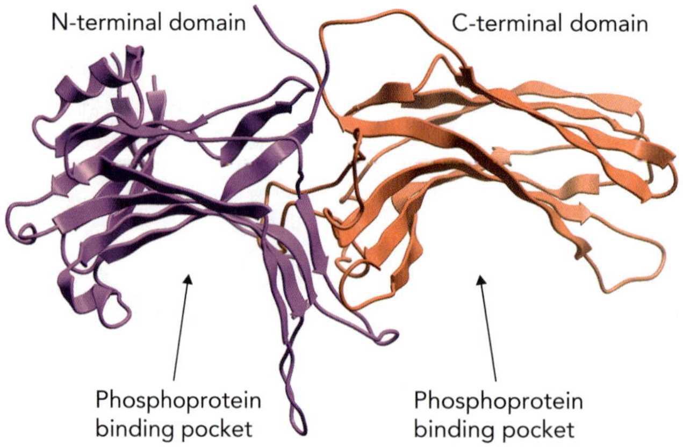

Figure 8.34 This molecular structure of a single bovine β-arrestin subunit shows the locations of phosphoprotein binding sites contained within the N-terminal and C-terminal domains. BASED ON PDB FILE 1CF1.

concept integration 8.2

What class of physiologic responses do you think accounts for the majority of the 600-plus human GPCR genes? Explain why relatively few heterotrimeric G protein genes are required to mediate GPCR signaling.

The sensitivity and specificity of sensory perception (vision, smell, taste) depend on high-affinity GPCRs that distinguish between related stimuli. For example, color vision and night vision require light-absorbing prosthetic groups covalently linked to different GPCRs, whereas olfactory receptors must contain ligand-binding-site amino acid residues that discriminate closely related airborne odorants. Also, some but not all of the gustatory receptors expressed in tongue cells are GPCRs and, similar to olfactory receptors, must be able to distinguish a variety of compounds present in food. When taking into account the large number of stimuli presented to the GPCRs responsible for perceiving vision, smell, and taste, it is easy to see why these represent the largest class of the 600-plus GPCRs in the human genome. In contrast, the human genome only encodes 17 G_α, 5 G_β, and 12 G_γ genes, which makes sense for two reasons. First, the selectivity of signal transduction depends on the receptor protein, not the intracellular signaling mechanism, as is seen in the shared signaling pathway activated by glucagon and epinephrine signaling. Second, the three heterotrimeric G proteins can theoretically combine in 900-plus different ways to form a distinct $G_{\alpha\beta\gamma}$ complex, and therefore fewer genes of each type are needed.

8.3 Receptor Tyrosine Kinase Signaling

The second major class of signaling proteins in eukaryotes is receptor tyrosine kinases (RTKs). These enzymes transmit extracellular signals by ligand activation of an intrinsic tyrosine kinase function found in the cytoplasmic tail of the receptor. (Refer to **Table 8.4** for a list of acronyms used in this section.)

Because activated RTKs are dimers, the intrinsic kinase phosphorylates tyrosine residues on the paired cytoplasmic domain through an intermolecular reaction. RTK also phosphorylates downstream signaling proteins that bind to the RTK phosphotyrosines. Some of these targets for RTK phosphorylation serve as adaptor proteins, which function as molecular bridges to bring together other proteins (**Figure 8.35**). Many of these RTK target proteins are also kinases, so their phosphorylation and activation establishes a relay signal between the receptor and a downstream phosphorylation cascade.

Two of the best-characterized RTK signaling pathways are those controlled by the epidermal growth factor receptor (EGFR) and the insulin receptor. Defects in these RTK signaling pathways have been associated with two of the most prevalent human diseases: cancer in the case of EGFR signaling and type 2 diabetes as a result of defects in insulin receptor signaling.

Epidermal Growth Factor Receptor Signaling

RTK signaling pathways were discovered by biochemists studying the *in vitro* effects of serum proteins on the growth of cancer cells. They found that most RTK ligands contained in serum were small proteins that function as growth factors, which signal

Table 8.4 ABBREVIATIONS, PROTEIN NAMES, AND FUNCTIONS IN RECEPTOR TYROSINE KINASE SIGNAL TRANSDUCTION SYSTEMS

Abbreviation	Protein name	Function
Akt	Akt protein	Enzyme with kinase activity (also known as protein kinase B; PKB)
EGFR	Epidermal growth factor receptor	RTK that binds the epidermal growth factor
ERK	Extracellular signal–regulated kinase	Enzyme with kinase activity
GRB2	Growth factor receptor–bound 2	Adaptor protein that binds epidermal growth factor receptors
IRS	Insulin receptor substrate	Adaptor protein that binds insulin receptors
MAP kinase	Mitogen-activated protein kinase	Enzyme with kinase activity
MEK	MAP/ERK kinase	Enzyme with kinase activity
PDK1	Phosphoinositide-dependent kinase	Enzyme with kinase activity
PI-3K	Phosphoinositide-3 kinase	Enzyme with kinase activity
PKB	Protein kinase B	Enzyme with kinase activity (also known as Akt)
PTB	Phosphotyrosine binding domain	Domain that binds phosphotyrosine residues
PTEN	Phosphatase and tensin homolog	Enzyme with phosphatase activity
Raf	Rapid accelerated fibrosarcoma protein	Enzyme with kinase activity
Ras	Rat sarcoma protein	G protein with GTPase activity
RasGAP	Ras GTPase-activating protein	Protein that stimulates GTPase activity of Ras
SH2	Src kinase homology-2	Adaptor domain that binds phosphotyrosine residues
SH3	Src kinase homology-3	Adaptor domain that binds proline-rich sequences
Shc	Src homology collagen	Adaptor protein that binds receptors
SOS	Son of sevenless	Guanine nucleotide exchange factor protein

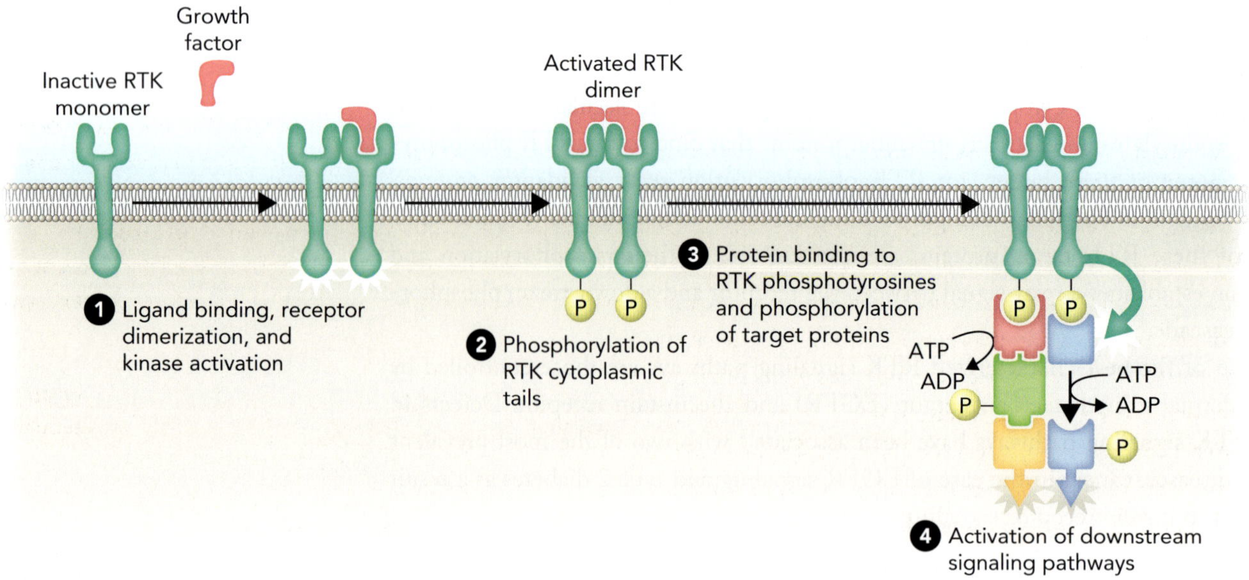

Figure 8.35 Growth factors bind to RTKs encoding an intrinsic tyrosine kinase activity, which is activated by receptor dimerization. The intrinsic tyrosine kinase is required for both autophosphorylation of the RTK cytoplasmic tail and for phosphorylation of receptor-associated target proteins, which bind to RTK phosphotyrosine residues.

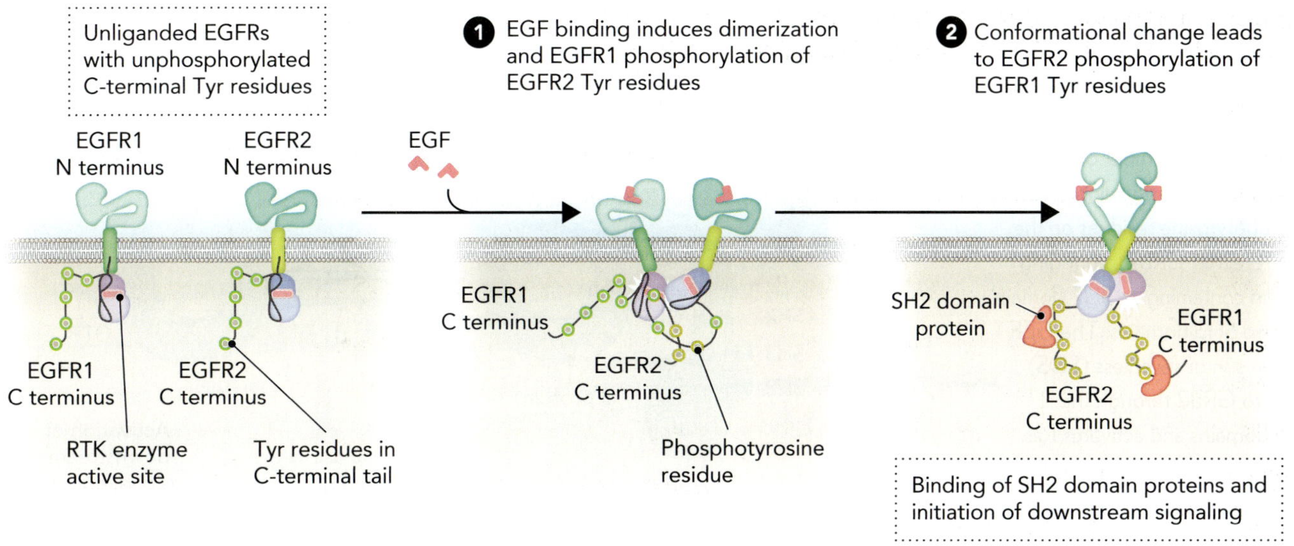

Figure 8.36 A two-step model is shown proposing how EGFR dimerization allosterically regulates tyrosine phosphorylation in the cytoplasmic tails of each monomer. In step 1, EGF binding to each monomer induces receptor dimerization that establishes a symmetric conformation in the extracellular domains (head to head) but an asymmetric conformation in the cytoplasmic domains (head to tail). This arrangement stimulates kinase activity in one receptor (EGFR1), which phosphorylates tyrosine residues in the paired receptor (EGFR2). In step 2, a large conformational change in the cytoplasmic domains leads to activation of the EGFR2 kinase and phosphorylation of tyrosine residues in EGFR1.

through paracrine or autocrine mechanisms. The downstream response in this signaling pathway is increased cell division.

One of the first serum growth factors identified was the **epidermal growth factor (EGF)**, which binds to the EGFR and stimulates receptor dimerization on the cell surface. EGF is a 53-amino-acid protein that, like insulin, is proteolytically processed from a larger precursor protein to yield an endocrine hormone with three disulfide bridges (see Figure 8.5). Structural studies using individual EGFR domains have led to a proposed dimeric structure of the ligand-bound receptors. In this structure, the extracellular domains are symmetric, but the kinase domains are arranged asymmetrically.

In a two-step model of EGF function, binding of an EGF molecule to each of two EGFRs in step 1 induces receptor dimer formation. Dimerization of the EGFRs leads to activation of kinase activity in one of the two receptor molecules (EGFR1) and subsequent phosphorylation of tyrosine residues in the cytoplasmic tail of the other receptor molecule (EGFR2) as shown in **Figure 8.36**. Phosphorylation of the five tyrosine residues in EGFR2 induces a conformational change in the receptor dimer, resulting in activation of the EGFR2 kinase domain and tyrosine phosphorylation of EGFR1 residues (step 2). **Figure 8.37** shows the structure of the extracellular domains of an EGFR dimer bound to two molecules of EGF.

Phosphotyrosine (pY) residues in RTK cytoplasmic tails form binding sites for intracellular adaptor proteins containing a **Src kinase homology-2 (SH2)** domain (or PTB domain, as discussed later). The enzyme Src kinase was first found in the Rous sarcoma virus and later shown

Figure 8.37 This view of dimeric EGFR from the extracellular side shows the location of EGF binding sites in each receptor and the extensive subunit interactions at the dimer interface. BASED ON PDB FILE 1IVO.

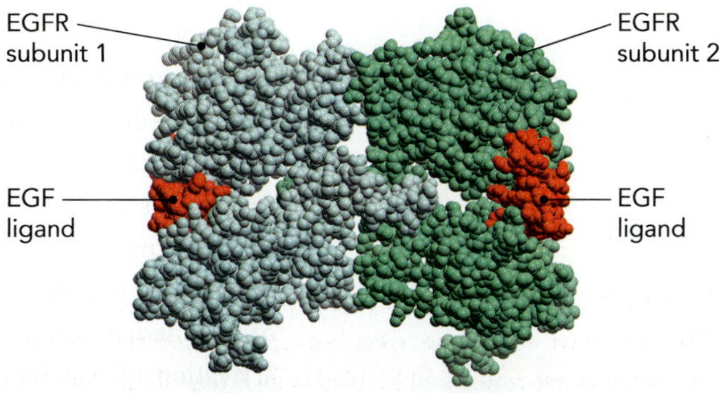

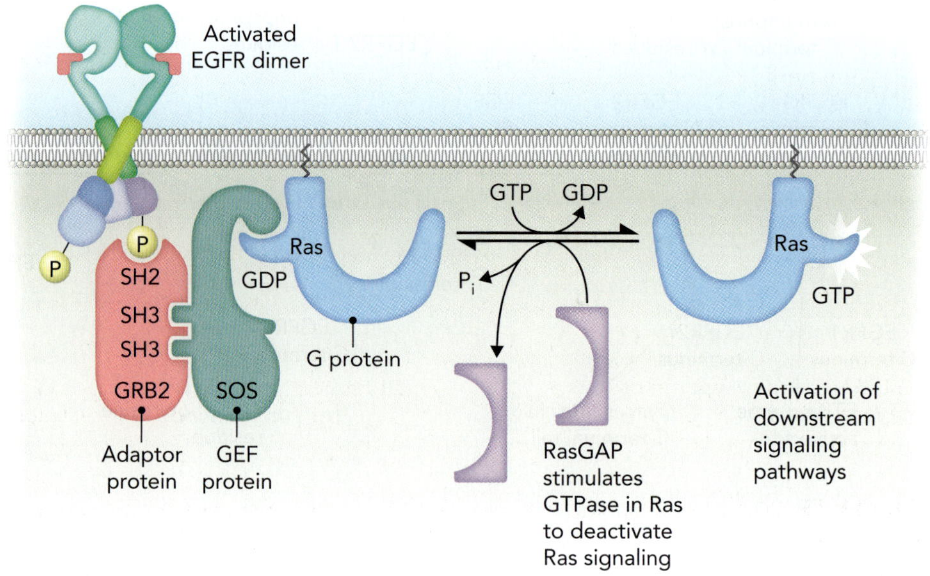

Figure 8.38 Upstream EGFR signaling involves two adaptor proteins: GRB2 and SOS. These proteins link the ligand-bound RTK to Ras, a member of the G protein family of signaling proteins. Phosphotyrosine residues on the EGFR recruit GRB2, a bridging protein containing an SH2 domain and two SH3 domains. The GEF protein son of sevenless (SOS) binds to GRB2 through the SH3 domains and activates Ras by stimulating the GDP–GTP exchange reaction. The activated Ras–GTP protein stimulates downstream signaling pathways. Ras signaling is inactivated by GAPs, such as RasGAP, which stimulates the intrinsic GTPase activity in Ras to generate the inactive-conformation Ras–GDP.

to be mutated in some types of human cancers. SH2 domains are protein segments of ~100 amino acids. They contain both a pY binding site and a specificity pocket, which recognizes amino acids a few residues away on the carboxyl-terminal side of pY. **Growth factor receptor–bound 2 (GRB2)** is an EGFR SH2 adaptor protein that binds pY residues. It also contains a specificity pocket recognizing asparagine residues at pY + 2. Many types of intracellular signaling proteins contain SH2 domains that are specific for different pY sequences. For example, the enzyme **phosphoinositide 3 kinase (PI-3K)**, which plays a key role in insulin signaling, as we will see later in this section, is an SH2-containing lipid kinase that binds pY and recognizes methionine residues at pY + 3. Note that the addend "+2" or "+3" refers to the amino acid at a position two or three amino acid residues away from the pY residue, respectively.

As illustrated in **Figure 8.38**, GRB2 binds to pY residues in the EGFR and recruits a GEF signaling protein called son of sevenless (SOS), which binds to and activates a G protein called **Ras**. Ras is a member of the same family of G protein signaling molecules as the G_α subunit of heterotrimeric G proteins (see Figure 8.20). Similar to G_α proteins, Ras signaling function is based on three key features: (1) Ras is anchored to the cytoplasmic side of plasma membranes by a covalently attached lipid moiety; (2) Ras is activated by GEF proteins; and (3) Ras contains an intrinsic GTPase activity that controls its signaling function and is regulated by GAPs.

The interaction between GRB2 and SOS requires another type of binding module called an **Src kinase homology-3 (SH3)** domain. This is an ~70-amino-acid segment that binds to specific proline-rich sequences. As shown in **Figure 8.39**, in addition to the SH2 domain that binds to pY residues in the EGFR, GRB2 contains two SH3 domains, which are called the N-SH3 domain and the C-SH3 domain and are located in the amino-terminal region and carboxyl-terminal regions, respectively. This combination of SH2 and SH3 domains in GRB2 allows it to function as a bridging protein that links activated EGFRs to GEF signaling proteins, thereby stimulating guanine nucleotide exchange in Ras proteins. Just as we saw with control of G_α-mediated signaling by the G protein cycle (see Figure 8.31), GEF-mediated activation of Ras by SOS is short-lived, because GAP proteins, such as **RasGAP**,

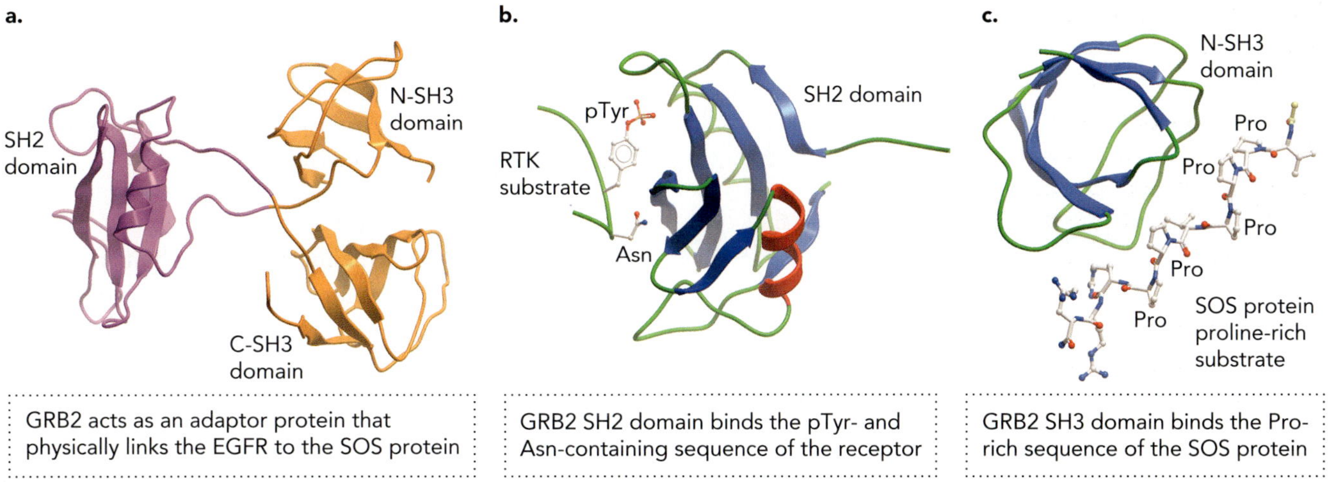
a.
N-SH3
domain
SH2
domain
C-SH3
domain
GRB2 acts as an adaptor protein that physically links the EGFR to the SOS protein
b.
SH2 domain
pTyr
RTK
substrate
Asn
GRB2 SH2 domain binds the pTyr- and Asn-containing sequence of the receptor
c.
N-SH3
domain
Pro
Pro
Pro
Pro
Pro
SOS protein
proline-rich
substrate
GRB2 SH3 domain binds the Pro-rich sequence of the SOS protein

Figure 8.39 GRB2 is an RTK adaptor protein. **a.** GRB2 consists of three separate domains: an N-terminal SH3 domain (N-SH3), a central SH2 domain, and a C-terminal SH3 domain (C-SH3). BASED ON PDB FILE 1GRI. **b.** The protein complex formed between the SH2 domain of GRB2 and a peptide substrate with the amino acid sequence Lys-Pro-Phe-*pTyr*-Val-*Asn*-Val. The GRB2 SH2 domain makes high-affinity contacts with the amino acids pTyr (phosphotyrosine) and Asn (specificity residue). BASED ON PDB FILE 1TZE. **c.** The protein complex formed between the GRB2 N-SH3 domain and a proline-rich segment from the SOS signaling protein. BASED ON PDB FILE 1GBQ.

bind to Ras and stimulate GTP hydrolysis to inactivate its signaling functions (see Figure 8.38).

How does the G protein Ras transmit the EGFR signal downstream to specific target proteins? This is done through the stable activation of a phosphorylation cascade. The cascade is mediated by a trio of related kinases, collectively called the **mitogen-activated protein kinase (MAP kinase)** pathway. (A mitogen is a compound that activates mitosis, resulting in cell division, which is one of the cellular responses to MAP kinase signaling.)

As shown in **Figure 8.40**, the first kinase in the MAP kinase pathway is **Raf**, a 74-kDa enzyme that phosphorylates target proteins on serine and threonine residues. Ras–GTP activates the serine/threonine kinase activity of Raf by recruiting it to the plasma membrane, where it can be phosphorylated by Src kinase. Once activated by phosphorylation, Raf in turn phosphorylates serine residues on **MAP/ERK kinase (MEK)**, leading to MEK phosphorylation of **extracellular signal–regulated kinase (ERK)**. Phosphorylated ERK forms a homodimer that translocates to the nucleus, where it phosphorylates several target proteins. These proteins function as transcription factors and regulate gene expression. Signaling through the EGFR pathway is terminated by both RasGAP inactivation of Ras function and by phosphatase enzymes that remove the activating phosphates from MAP kinase proteins and transcription factors.

Defects in Growth Factor Receptor Signaling Are Linked to Cancer

Human diseases can be caused by inherited genetic defects carried in the germ line or as a result of somatic mutations that occur during an individual's lifetime. Many types of human cancers are due to somatic mutations in cell signaling genes, which become damaged over decades of life as a result of insufficient DNA repair.

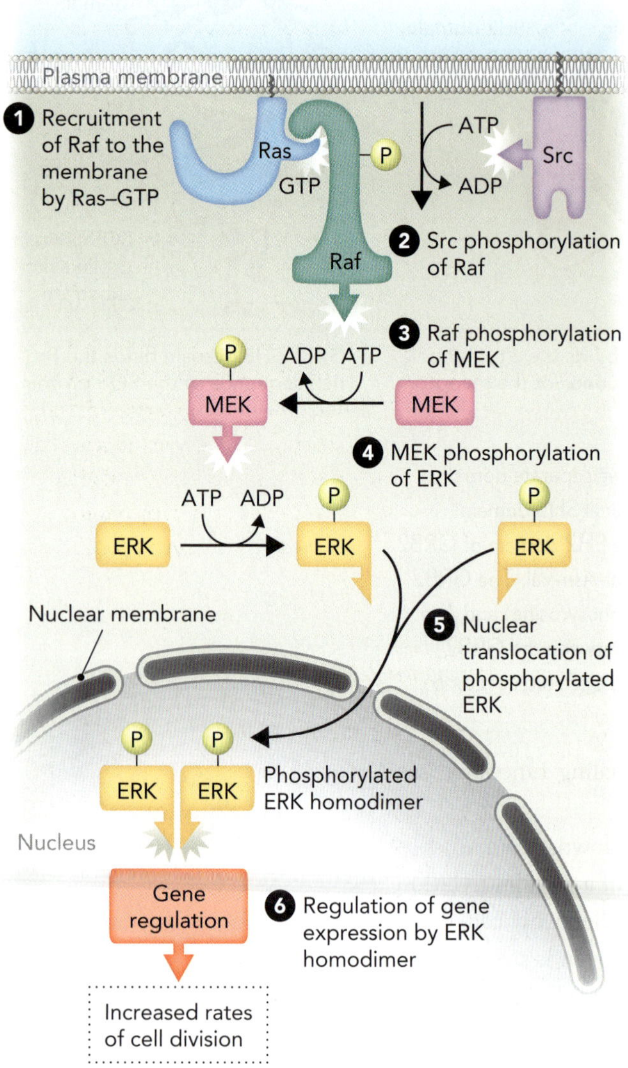

Figure 8.40 Activated Ras protein stimulates the MAP kinase signaling pathway, which initiates a phosphorylation cascade consisting of three MAP kinase proteins. The cellular response to activation of the MAP kinase signaling pathway is regulation of gene expression and increased rates of cell division.

The discovery that the most common human cancers are caused by somatic mutations in cell signaling genes was made by two virologists. In the 1970s, J. Michael Bishop and Harold E. Varmus, both physician scientists at the University of California, San Francisco, were studying animal viruses that cause tumors in the infected hosts. They found evidence for the idea that during the evolutionary history of several tumor viruses, the viruses had acquired "hitchhiker genes" from the animals they infected. Bishop and Varmus proposed that these animal-derived hitchhiker genes had accumulated mutations during many generations of viral replication and were now genetically responsible for the observed animal tumors that formed in virus-infected tissues. Bishop and Varmus called these cancer-causing genes **oncogenes** (oncology is the study of cancer) and were able to demonstrate that these viral oncogenes were in fact mutated versions of normal animal genes. This discovery led to Bishop and Varmus being awarded the 1989 Nobel Prize in Physiology or Medicine (**Figure 8.41**).

The two major types of somatic mutations found in humans are gain-of-function oncogene mutations, also called **dominant mutations**, and loss-of-function oncogene mutations, which are the result of **recessive mutations**. A dominant mutation leads to the disease phenotype anytime the mutation is present in a cell, whereas cancer caused by a recessive mutation requires that both copies of the mutated oncogene be present in the same cell (**Figure 8.42**). A dominant gain-of-function mutation is the most common type of cancer mutation and results in new activity in the cell as a result of the mutated protein. For example, a mutated intracellular signaling protein might no longer be controlled by negative-feedback regulation or a mutated growth factor receptor might be fully active even in the absence of ligand binding. Gain-of-function mutations need only affect one of the two gene copies in the cell (paternal and maternal chromosomes) to cause a deleterious phenotype such as uncontrolled cell growth observed in cancer cells. In contrast, a loss-of-function mutation must occur in both gene copies to have an effect on cell function and is therefore much more rare as a cause of cancer. One example of a loss-of-function mutation is found in retinoblastoma cancer, which is due to inactivating mutations in the retinoblastoma gene involved in cell cycle control. The retinoblastoma gene is

Figure 8.41 J. Michael Bishop, MD (left), and Harold E. Varmus, MD (right), were awarded the 1989 Nobel Prize in Physiology or Medicine for their discovery of human oncogenes. MIKKEL AALAND.

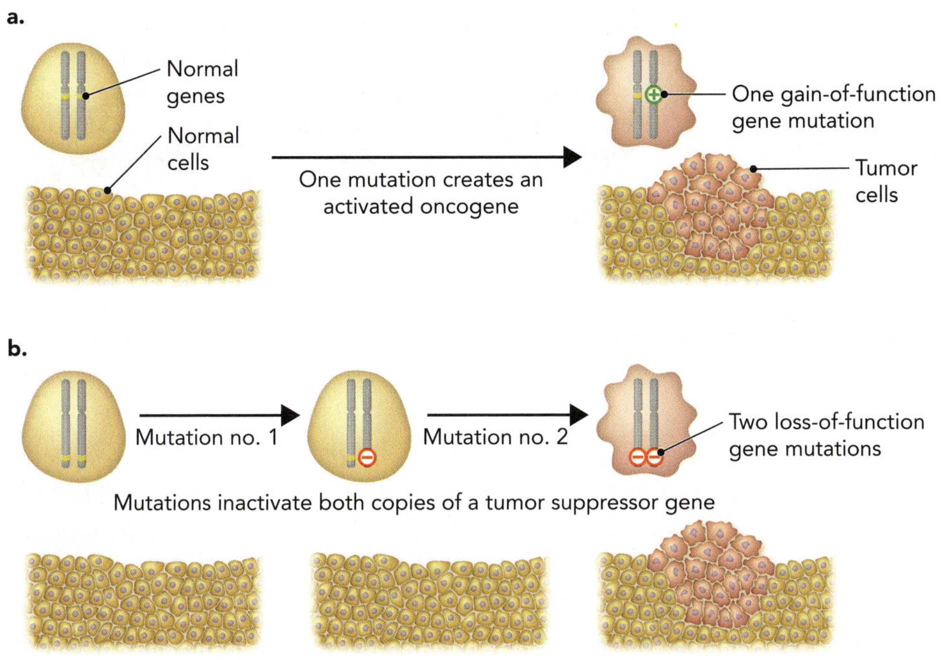

Figure 8.42 Most human cancers are the result of dominant gain-of-function mutations in cell signaling genes that occur during the lifetime of an individual. Some cancers are caused by rare recessive loss-of-function mutations in the same cell. **a.** Cancer caused by a dominant oncogene mutation requires that only one copy of the two genes be defective in a cell. **b.** Cancer caused by a recessive oncogene mutation requires that both copies of the gene be defective in the same cell. Cells with a recessive mutation in only one gene copy often have a normal phenotype.

considered a **tumor suppressor** gene because under normal conditions, it functions to inhibit uncontrolled cell proliferation.

The first isolated human oncogenes were found by identifying DNA fragments obtained from tumor samples that increased cell division rates when introduced into cultured cells. Many of the dominant mutations identified by these gain-of-function screens were found to block feedback inhibition of growth factor signaling pathways. One of the first human genes to be classified as an oncogene was mutant *Ras*, with the most common *Ras* mutations being those that decreased the intrinsic GTPase activity. Oncogenic Ras proteins are insensitive to GAP regulation—much like a stuck gas pedal on a car that has no brakes.

As illustrated in **Figure 8.43**, oncogenic Ras protein chronically stimulates the MAP kinase pathway because GTP hydrolysis and subsequent Ras inactivation does

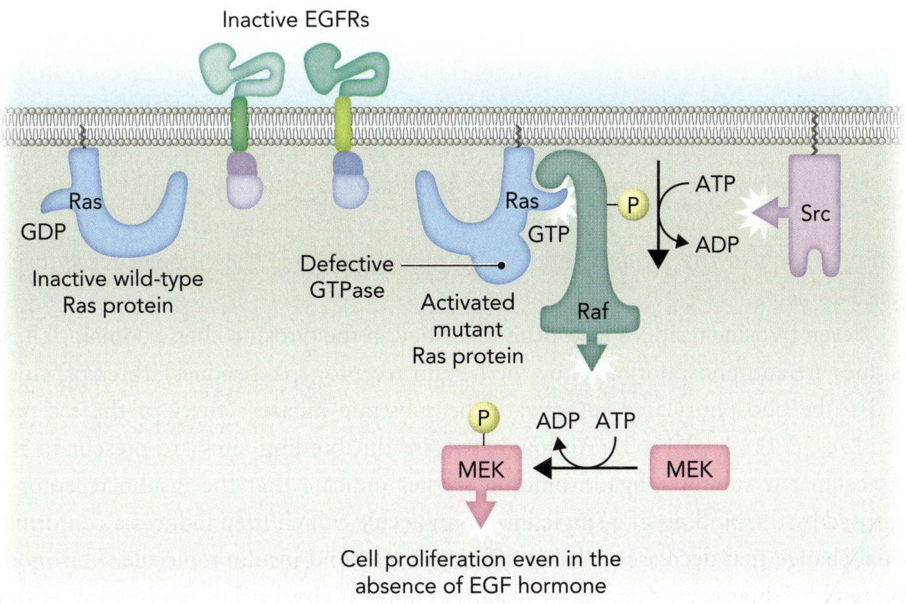

Figure 8.43 The most common oncogenic *Ras* mutations lead to defects in the intrinsic GTPase activation and thereby block Ras protein inactivation. Dominant Ras mutations in the GTPase domain lead to chronic stimulation of the MAP kinase signaling pathway, even in the absence of growth factors.

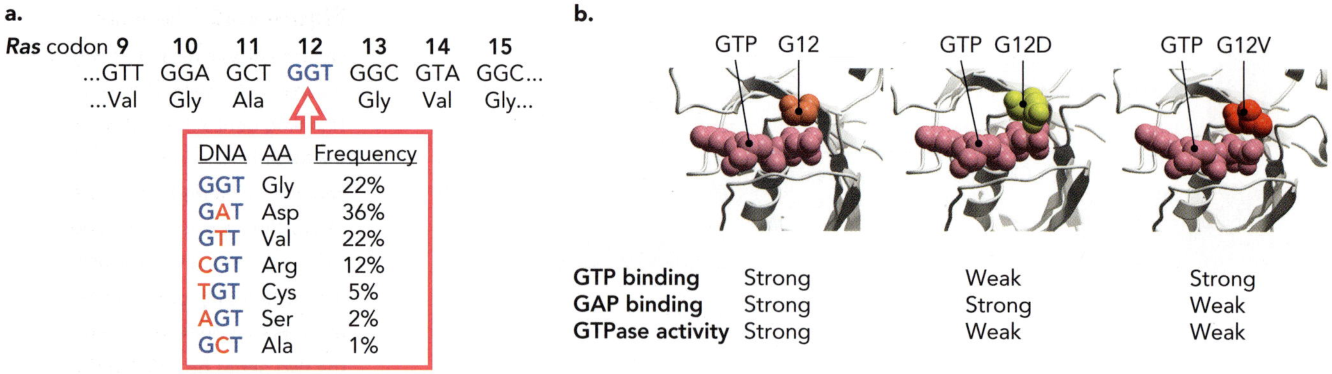

Figure 8.44 Studies have shown that ~80% of pancreatic cancers have missense mutations that decrease GTPase activity of the Ras protein. **a.** The most common *Ras* mutations found in human pancreatic cancers alter the Gly residue at codon 12 (GGT) by single nucleotide changes. **b.** Biochemical studies of the GTP binding, GAP binding, and GTPase activities of G12D and G12V Ras protein mutants have shown that structural changes in this region of the protein can alter either GTP or GAP binding, both of which decrease GTPase activity. AA = amino acid. BASED ON PDB FILES 2RGE (G12), 1AGP (G12D), AND 2VH5 (G12V).

not occur. In fact, missense mutations in codon 12 of *Ras* are found in a large majority of pancreatic cancers (~80%) and in up to 25% of all solid tumor cancers. The most common single nucleotide mutations at codon 12 in pancreatic cancers are missense mutations that convert Gly to Asp (G12D), Val (G12V), or Arg (G12R), all of which decrease the intrinsic GTPase activity of the oncogenic Ras protein (**Figure 8.44**). These types of *Ras* mutations are gain-of-function mutations and therefore need only occur in one of the two genomic copies of the Ras gene to disrupt cell cycle control. Therefore, even though these cancer cells still express an equal amount of wild-type Ras, the phenotype reflects the overstimulation of the MAP kinase pathway from the unregulated mutant Ras protein.

Insulin Receptor Signaling Controls Two Major Downstream Pathways

The second major example of receptor tyrosine kinase signaling is the pathway controlled by insulin. Diabetes is a prevalent human disease caused by defects in insulin signaling that stem from either lack of insulin production by the pancreas (type 1 diabetes) or decreased efficiency of insulin receptor signaling in target cells (type 2 diabetes). We have already seen how glucagon, the "I am hungry" hormone, increases blood glucose levels by signaling through a GPCR pathway. Now we examine how insulin, the "I just ate" hormone, decreases blood glucose levels by signaling through an RTK pathway.

The molecular architecture of the insulin receptor is shown in **Figure 8.45**. You can see that it consists of an $\alpha_2\beta_2$ tetrameric complex linked together by disulfide bonds. The α and β subunits are proteolytically processed from a single polypeptide chain and are covalently attached by an intermolecular disulfide bridge. In addition, the α subunits of the $\alpha\beta$ monomers are also linked together by disulfide bridges to form the transmembrane $\alpha_2\beta_2$ complex. The extracellular α subunits form the insulin binding region of the receptor, and the β subunits encode the cytoplasmic tyrosine kinase domains.

▶ ANIMATION

Seven tyrosine residues have been identified in the tyrosine kinase domain. These residues are autophosphorylated by the insulin receptor (pY residues), three of which need to be phosphorylated to activate the substrate kinase activity of the receptor (pY1158, pY1162, pY1163). Although two insulin binding sites are present in the extracellular α subunits, ligand binding studies indicate that the insulin receptor is activated by the binding of a single insulin molecule, which then induces a conformational change that decreases binding affinity of a second insulin molecule. This mode of ligand binding is an example of negative cooperativity, in that the binding of one

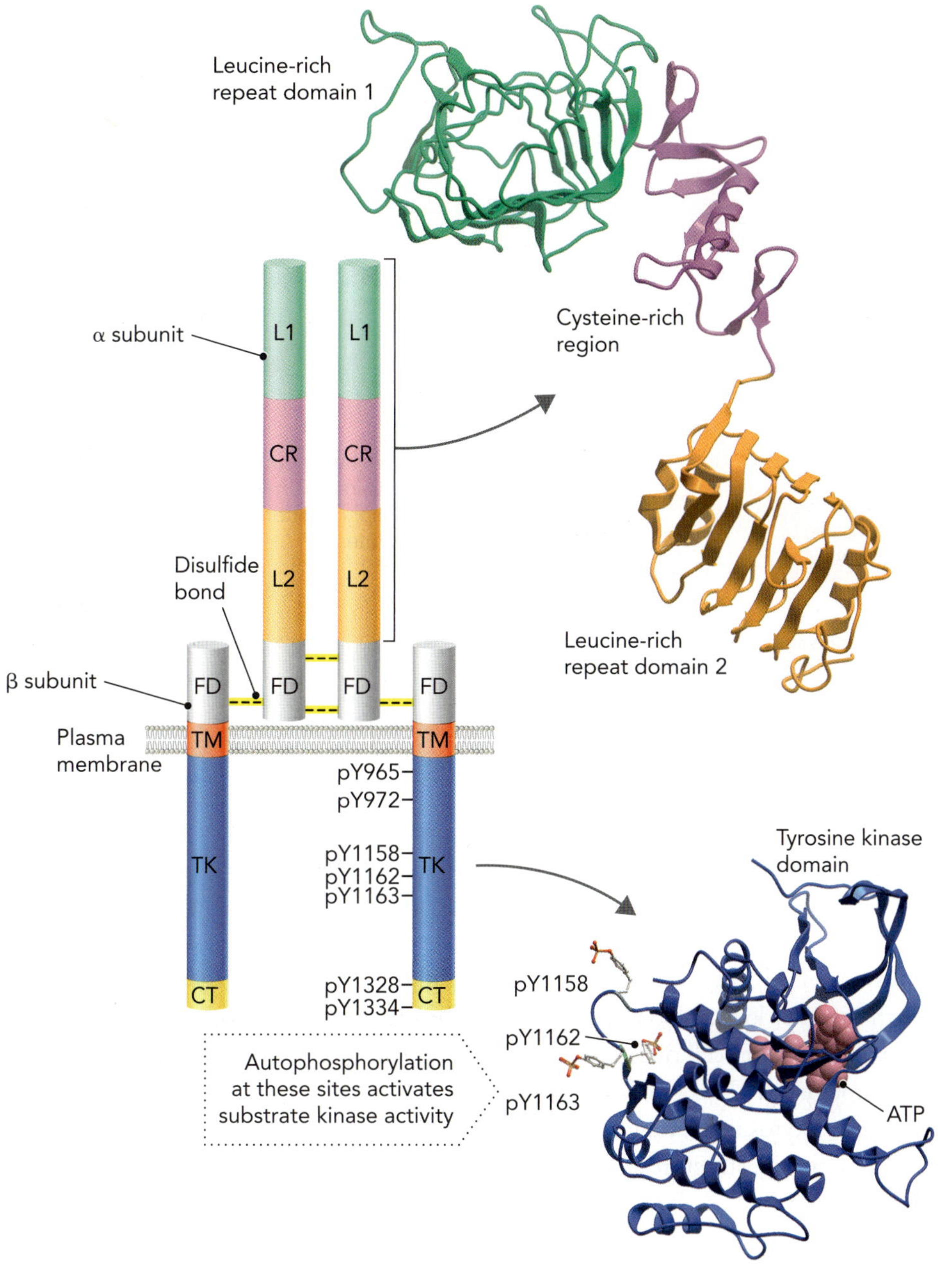

Figure 8.45 The insulin receptor is an RTK consisting of a cross-linked tetrameric $\alpha_2\beta_2$ complex. The α subunit is entirely extracellular and contains the insulin binding sites (only one molecule of insulin is required to stimulate receptor signaling); two domains that are abundant in leucine residues (termed leucine-rich repeat domains, L1 and L2); a cysteine-rich region (CR); and a fibronectin domain region (FD). The β subunit anchors the α subunit to the plasma membrane through a transmembrane region (TM) and contains the intracellular tyrosine kinase domain (TK), as well as a short C-terminal region (CT). Three tyrosine residues in the TK domain (pY1158, pY1162, pY1163) must be autophosphorylated by the ligand-bound receptor before the kinase can phosphorylate other substrates. BASED ON PDB FILES 2HR7 (L1, CR, L2) AND 3BU5 (TK).

insulin molecule inhibits the binding of a second insulin molecule. This is the opposite of what happens when one molecule of O_2 binds to the hemoglobin tetramer and increases the affinity for additional O_2 molecules (see Figure 6.18), a classic example of positive cooperativity.

Insulin binding to the insulin receptor induces a conformational change that stimulates tyrosine autophosphorylation of the β subunits through both intramolecular and intermolecular reactions. This leads to additional conformational changes in the β subunits, which stimulate insulin receptor–mediated tyrosine phosphorylation of substrate proteins. These substrate proteins bind to phosphotyrosine residues in the insulin receptor cytoplasmic domain through a **phosphotyrosine binding (PTB) domain**, as illustrated in **Figure 8.46**. One class of signaling proteins that bind to phosphorylated insulin receptors through PTB domains are called **insulin receptor substrate (IRS)** proteins. The major insulin receptor substrate protein in muscle cells is IRS-1, whereas the predominant IRS protein in liver cells is IRS-2.

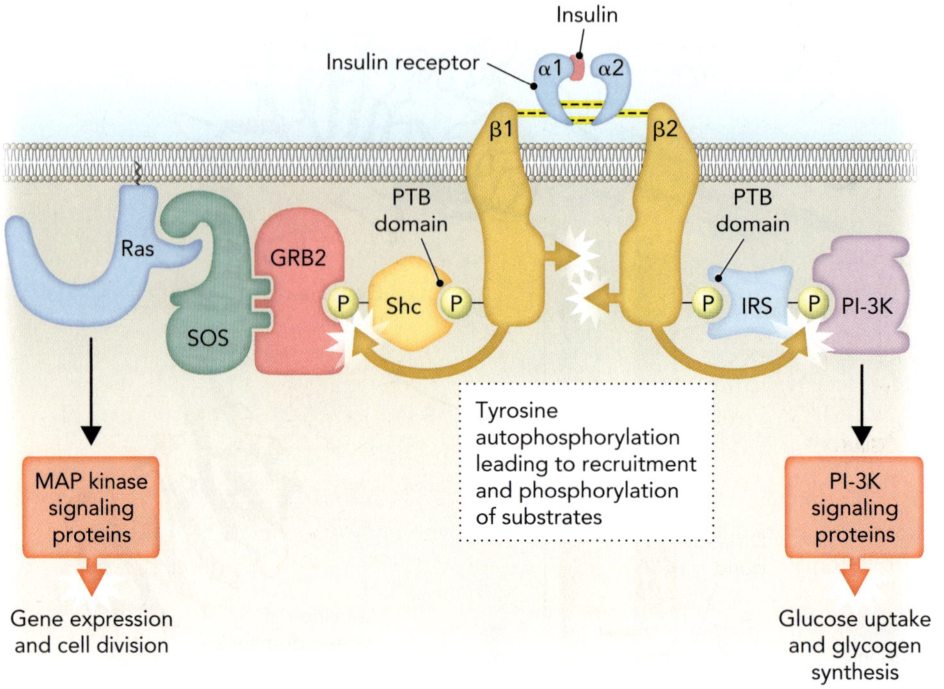

Figure 8.46 Insulin receptor tyrosine phosphorylation of IRS and Shc proteins, which bind to the receptor cytoplasmic domain through PTB domains, leads to stimulation of two downstream signaling pathways. In one branch of the insulin signaling pathway, phosphorylated IRS binds to SH2 domains on PI-3K and activates a downstream signaling pathway that leads to increased glucose uptake and stimulation of glycogen synthesis. The other branch involves binding of phosphorylated Shc protein to SH2 domains on GRB2, which activates the MAP kinase pathway, leading to altered gene expression and cell division.

Tyrosine phosphorylation of insulin receptor–bound insulin receptor substrate proteins, as well as insulin receptor phosphorylation of another signaling protein, called Src homology collagen (Shc), leads to the stimulation of two major downstream pathways. One branch of the insulin signaling pathway involves IRS-mediated activation of PI-3K. This leads to increased glucose uptake and glycogen synthesis, which together lower blood glucose levels. The other major insulin receptor signaling pathway stimulates cell division by Shc-mediated activation of the MAP kinase pathway through the GRB2–SOS–Ras signaling module. The structure and specificity of the phosphotyrosine binding sites (PTB domains) present in IRS proteins and Shc are distinct from the structure and specificity of SH2 domains (compare Figure 8.39 and **Figure 8.47**). These differences ensure that signaling modules in the PI-3K and MAP kinase pathways assemble correctly.

Tyrosine-phosphorylated insulin receptor substrate proteins bind to and activate PI-3K, which is a lipid kinase that phosphorylates PIP_2 to produce

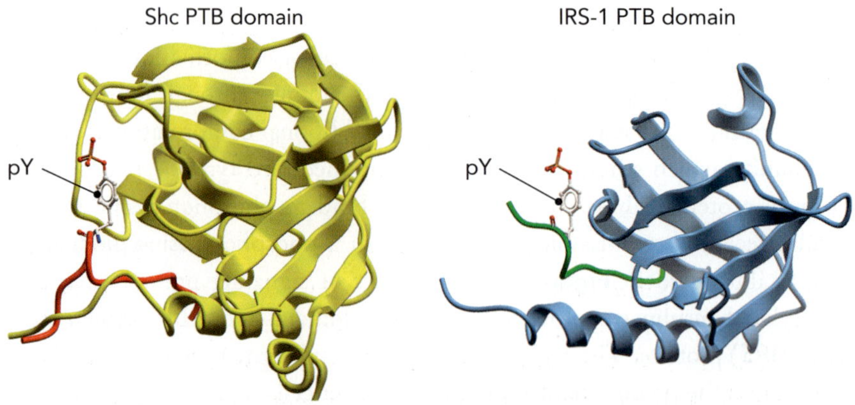

Figure 8.47 The insulin receptor adaptor proteins Shc and IRS-1 both contain phosphotyrosine binding (PTB) domains. The amino acid sequence of the phosphotyrosine binding site in the polypeptide bound to the Shc PTB domain is N-P-E-pY, whereas the amino acid sequence of the corresponding region in the polypeptide bound to IRS-1 is N-P-A-pY. PTB domains bind pY sequences but are structurally different from SH2 domains. BASED ON PDB FILES 1SHC (SHC) AND 1IRS (IRS-1).

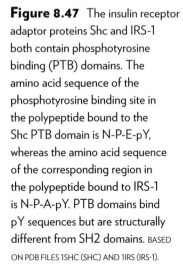

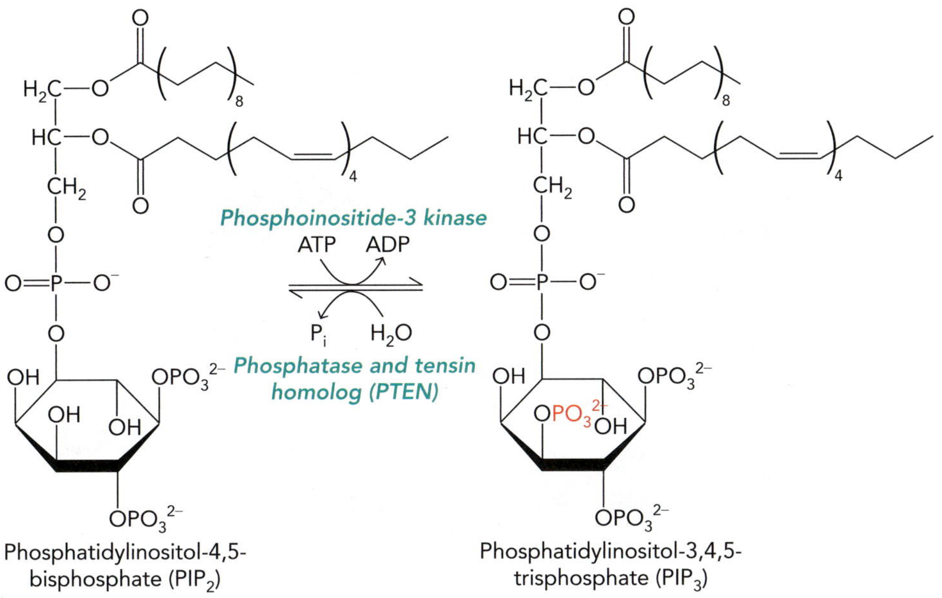

Figure 8.48 Phosphoinositide-3 kinase catalyzes the phosphorylation of PIP_2 to form PIP_3. Proteins with PH domains, such as phosphoinositide-dependent kinase and Akt, bind to PIP_3 and are thereby recruited to the plasma membrane.

phosphatidylinositol-3,4,5-trisphosphate (PIP_3) (**Figure 8.48**). PI-3K is a heterodimer consisting of a 110-kDa catalytic subunit (p110) and an 85-kDa regulatory subunit (p85). The regulatory subunit contains two SH2 domains that bind phosphotyrosine residues on insulin receptor substrate proteins. Unlike PKA, in which cAMP binding to the regulatory subunit leads to dissociation of the catalytic subunit (see Figure 8.26), binding of the PI-3K p85 regulatory subunit to phosphotyrosines on IRS-1 induces a conformational change that activates the p110 catalytic subunit without dissociating the heterodimer. You may remember that the PI-3K substrate PIP_2 is also the substrate for phospholipase C, which catalyzes a hydrolysis reaction to produce DAG and IP_3 (see Figure 8.10).

PI-3K signaling is terminated by the phosphatase enzyme **phosphatase and tensin homolog (PTEN)**, which removes the phosphate from PIP_3 to regenerate PIP_2, and thereby disrupts downstream insulin signaling. In addition to insulin receptor signaling, PI-3K can also be activated by growth factor receptor signaling in some cell types. This function regulates a downstream pathway that promotes cell survival. Indeed, mutations in both PI-3K and phosphatase and tensin homolog have been found in human tumors, enabling the cancer cells to activate this cell survival pathway in the absence of growth factor signaling. These gain-of-function mutations in cancer cells promote cell growth in the absence of regulatory signals that control cell growth. Similar gain-of-function mutations activate the MAP kinase signaling pathway through Ras mutations (see Figure 8.43).

Once PIP_3 is generated by PI-3K, the glycolipid remains in the plasma membrane and serves as a docking site for signaling proteins containing a phosphatidylinositol binding domain called a **pleckstrin homology (PH) domain.** Two proteins in the insulin signaling pathway that contain PH domains and bind to PIP_3 are phosphoinositide-dependent kinase and Akt, a serine/threonine kinase originally identified as an oncogene (cancer gene) in the Akt8 murine retrovirus (**Figure 8.49**). (Akt is also known as protein kinase B [PKB] because it shares sequence similarity with PKA and PKC.) Both phosphoinositide-dependent kinase and Akt bind to PIP_3, which facilitates phosphoinositide-dependent kinase phosphorylation of

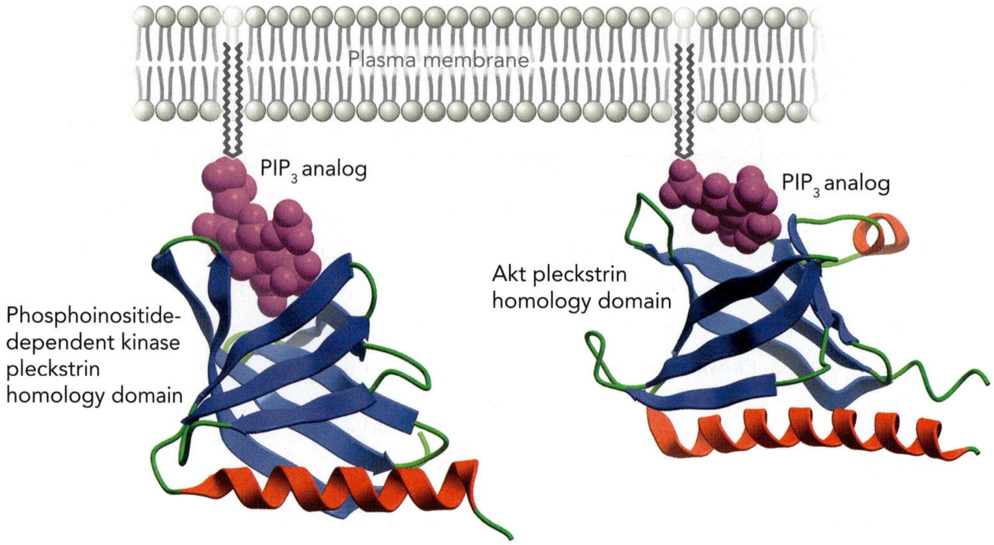

Figure 8.49 The pleckstrin homology domains in phosphoinositide-dependent kinase and Akt bind to membrane-associated lipids, which localizes them to the membrane. BASED ON PDB FILES 1W1G (PDK1) AND 1UNQ (AKT).

serine and threonine residues on Akt. Phosphorylated Akt then dissociates from PIP_3 and diffuses through the cytosol, where it phosphorylates numerous downstream target proteins.

As illustrated in **Figure 8.50** and described in more detail in Chapter 14, insulin receptor activation in liver cells stimulates glucose uptake and glycogen synthesis. These processes lower blood glucose levels by removing glucose from circulation after a carbohydrate-rich meal. Note that insulin signaling in liver cells has the exact opposite effect of glucagon signaling in that glucagon signaling increases glucose export by stimulating glycogen degradation and glucose synthesis (see Figure 8.28).

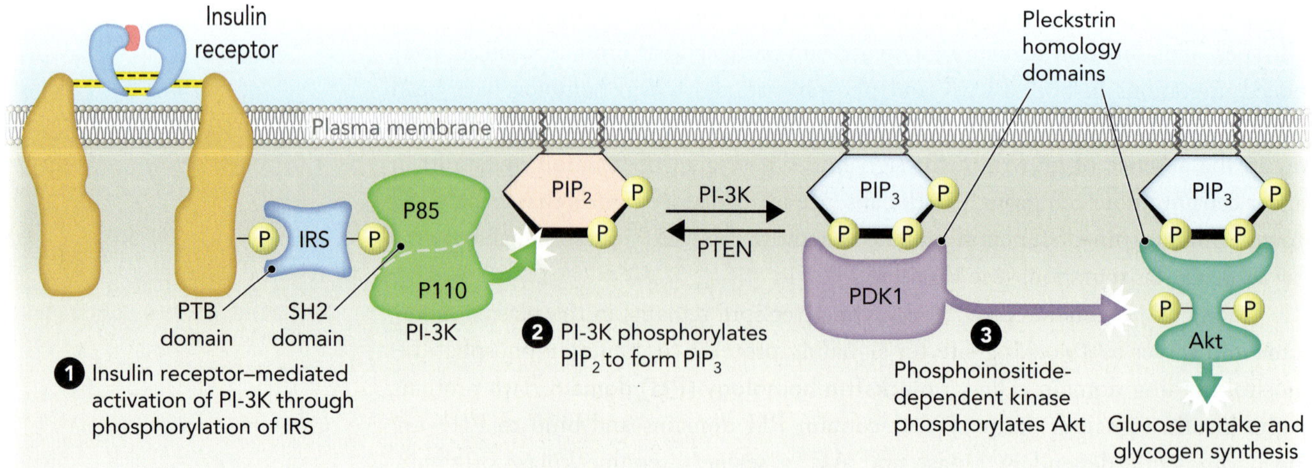

Figure 8.50 Insulin receptor signaling activates PI-3K through insulin receptor substrate (IRS) adaptor proteins, leading to the production of PIP_3. Phosphorylation and activation of Akt by phosphoinositide-dependent kinase initiates a downstream signaling pathway in liver cells that stimulates glucose uptake and glycogen synthesis.

concept integration 8.3

Describe the four types of binding domains found in signaling proteins required for EGF and insulin signaling. How do these binding domains function with regard to the localization of signaling enzymes?

The four types of binding domains required for EGF and insulin signaling are (1) SH2 domains, which bind to phosphotyrosine residues and a second amino acid located two to three residues away on the carboxyl-terminal side; (2) SH3 domains, which bind to proline-rich regions of target proteins; (3) PTB domains, which bind to phosphotyrosine residues; and (4) pleckstrin homology domains, which bind to the head groups of phosphoinositides. All of these domains function to localize cytoplasmic signaling proteins to the plasma membrane, where they are activated by other membrane-associated proteins.

8.4 Tumor Necrosis Factor Receptor Signaling

In the preceding two sections, we examined the control of glycogen degradation in liver cells by the hormones epinephrine and glucagon. This is an example of signal integration involving multiple receptors that activate shared and parallel pathways (see Figure 8.23). For our third type of major cell signaling system, we look at a different type of signal integration. In this system, a single receptor stimulates intracellular pathways with opposing cellular responses. (Refer to **Table 8.5** for a list of acronyms used in this section.)

Table 8.5 ABBREVIATIONS, PROTEIN NAMES, AND FUNCTIONS IN TUMOR NECROSIS FACTOR SIGNAL TRANSDUCTION SYSTEMS

Abbreviation	Protein name	Function
CASP3	Cysteine–aspartate protease 3	Enzyme with protease activity
CASP8	Cysteine–aspartate protease 8	Enzyme with protease activity
DD	Death domain	Protein–protein interaction domain
DED	Death effector domain	Protein–protein interaction domain
FADD	Fas-associated death domain	Adaptor protein that binds TNF receptor
FasR	Fas receptor	Trimeric membrane-spanning receptor that binds Fas ligand (FasL)
IκBα	Inhibitor of NFκB	Protein that retains NFκB in the cytosol
IKK	IκBα kinase	Enzyme with kinase activity
NFκB	Nuclear factor κ-light-chain-enhancer of activated B cells	Transcription factor consisting of p50 and p65 subunits
NIK	NFκB-inducing kinase	Enzyme with kinase activity
RIP	Receptor interacting protein	Adaptor protein
SODD	Silence of death domain	Protein that inhibits TNF receptor
TNF-α	Tumor necrosis factor-α	Extracellular protein ligand that binds TNF receptor
TNFR	Tumor necrosis factor receptor	Trimeric membrane-spanning receptor that binds tumor necrosis factor-α
TRADD	TNF receptor–associated death domain	Adaptor protein that binds TNF receptor
TRAF2	TNF receptor–associated factor 2	Adaptor protein

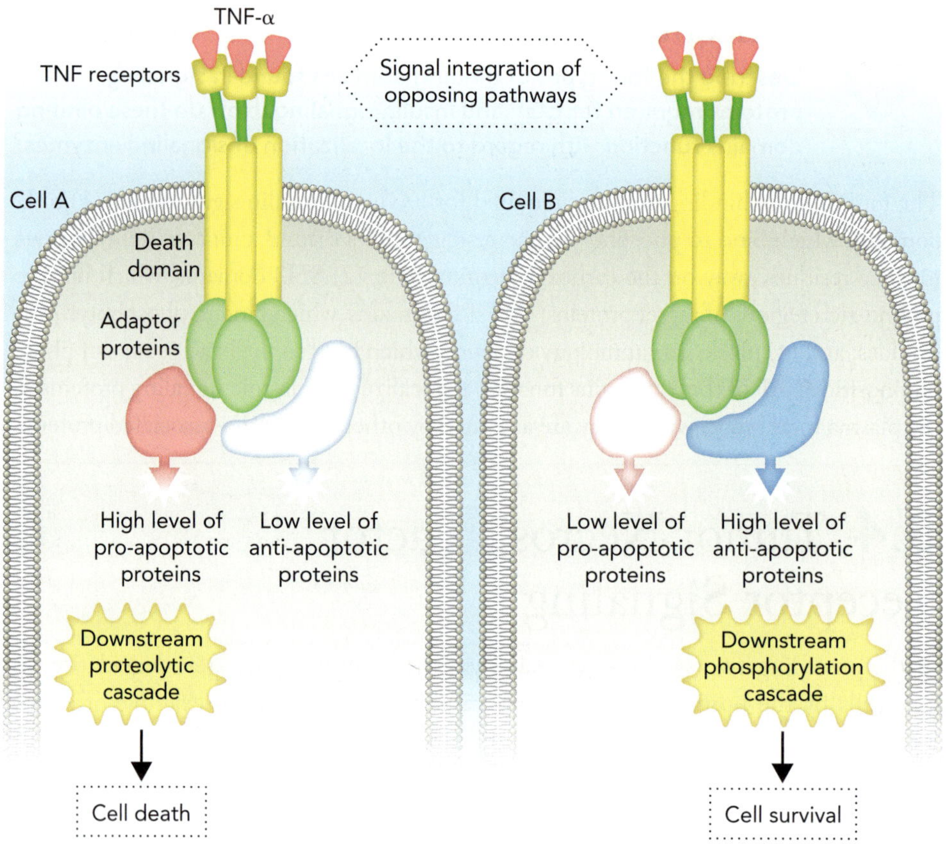

Figure 8.51 TNF receptor signaling regulates opposing cell death and cell survival pathways, with the outcome being determined by the relative abundance of downstream signaling proteins. The cell death pathway is mediated by a proteolytic cascade, whereas the cell survival pathway is regulated by a phosphorylation cascade. Death domains are protein binding modules in TNF receptors that initiate the assembly of adaptor signaling complexes.

As illustrated in **Figure 8.51**, binding of the small protein tumor necrosis factor-α (TNF-α) to the trimeric TNF receptor initiates a form of programmed cell death called **apoptosis** and at the same time stimulates a cell survival pathway. The downstream cellular response is determined by the relative levels of downstream pro-apoptotic and anti-apoptotic proteins. To understand how these opposing pathways are regulated by TNF receptor signaling, we need to examine the structure and function of the TNF receptor adaptor proteins and their downstream target proteins.

In both pathways, adaptor proteins bind to a region in the cytoplasmic tail of TNF receptors called the **death domain (DD)**, which functions as a protein–protein interaction module. As described shortly, the downstream target proteins in the cell death pathway are a class of enzymes called **cysteine–aspartate proteases (caspases)**. Once activated, these enzymes initiate a proteolytic cascade that degrades cellular proteins and kills the cell. Alternatively, the cell survival pathway is regulated by a kinase-mediated phosphorylation cascade. This process activates transcription factors, which induce the expression of genes encoding caspase inhibitory proteins.

TNF Receptors Signal through Cytosolic Adaptor Complexes

Human TNF-α is a 212-amino-acid protein expressed on the surface of immune cells. Proteolytic processing of TNF-α releases a 185-amino-acid signaling molecule that binds to TNF receptors as a homotrimeric complex. As illustrated in **Figure 8.52**, activation of the homotrimeric TNF receptor complex by TNF-α induces a conformational change in TNF receptor cytoplasmic tails that restructures the DD region, causing dissociation of an inhibitory protein called silence of death domain (SODD). In the absence of SODD proteins, the DD regions of the TNF receptor

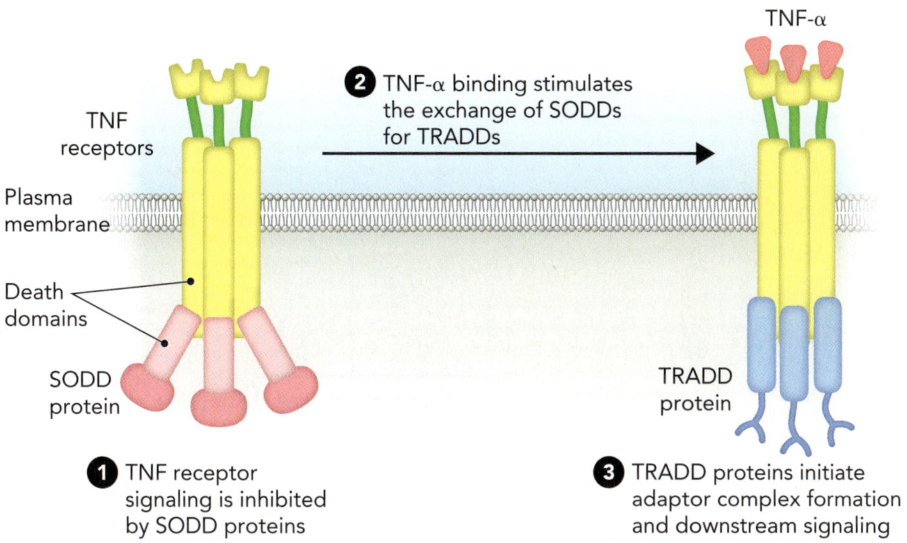

Figure 8.52 Activation of trimeric TNF receptor complexes by TNF-α binding induces a conformational change that promotes exchange of an inhibitory protein (SODD) for a downstream signaling protein (TRADD). Adaptor complex assembly on the cytoplasmic tails of TNF receptors is mediated by death domains, which function as protein–protein interaction modules.

function as high-affinity binding sites for DD regions present on the downstream signaling protein **TNF receptor–associated death domain (TRADD)**. The binding of TRADD proteins to TNF receptors creates an activated adaptor complex that binds additional downstream signaling proteins.

The TNF receptor is a member of a large family of trimeric receptors, also called death domain–containing receptors, that regulate the apoptotic pathway through the assembly of adaptor complexes utilizing the DD module. One of the best-characterized members of this receptor family is the Fas receptor, which binds to Fas ligand (FasL), a proteolytically processed signaling molecule similar to TNF-α. Activation of the Fas receptor by FasL induces apoptosis in T-cell and B-cell lymphocytes during normal development of the immune system (**Figure 8.53**).

TNF Receptor Signaling Regulates Programmed Cell Death

As illustrated in **Figure 8.54**, the TNF signaling pathway bifurcates at the level of TRADD binding to the receptor. In the apoptosis branch of the pathway, the DD region of TRADD forms a complex with the DD region of the adaptor protein **Fas-associated death domain (FADD)**. This complex is similar to the DD protein–protein interaction that has been characterized between FADD and Fas (**Figure 8.55**, p. 413). The TRADD–FADD complex then recruits procaspase 8 to the TNF receptor complex using a second protein–protein interaction module called the **death effector domain (DED)**. Formation of the TNF receptor–TRADD–FADD–procaspase 8 adaptor complex, also called the death-inducing signaling complex (DISC), stimulates an intermolecular reaction in procaspase 8 that generates the proteolytically active caspase 8 (CASP8) enzyme. Once this happens, caspase 8 cleaves procaspase 3 to generate caspase 3 (CASP3), the "executioner" caspase, which then degrades key regulatory molecules to kill the cell quickly and efficiently.

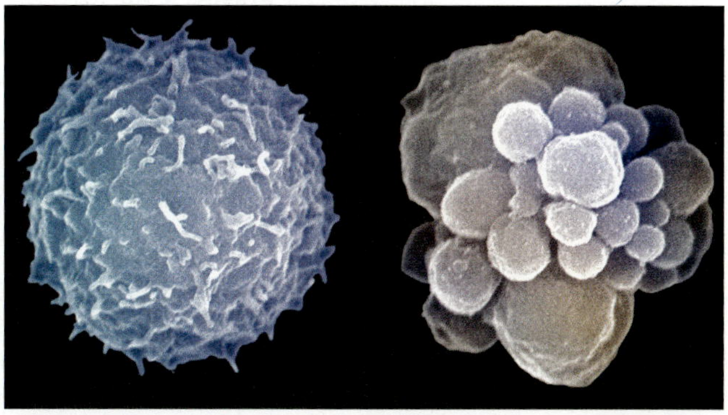

Healthy immune cell with intact plasma membrane

Dying immune cell showing plasma membrane blebbing

Figure 8.53 Fas receptor is another member of the TNF receptor family of trimeric receptors that regulates apoptosis. Fas-induced apoptosis in B and T lymphocytes is an important process during development of the human immune system. Note that *blebbing* refers to irregular bulges in the cell membrane as it separates from the cell's cytoskeleton during apoptosis. DR. GOPAL MURTI/VISUALS UNLIMITED/CORBIS.

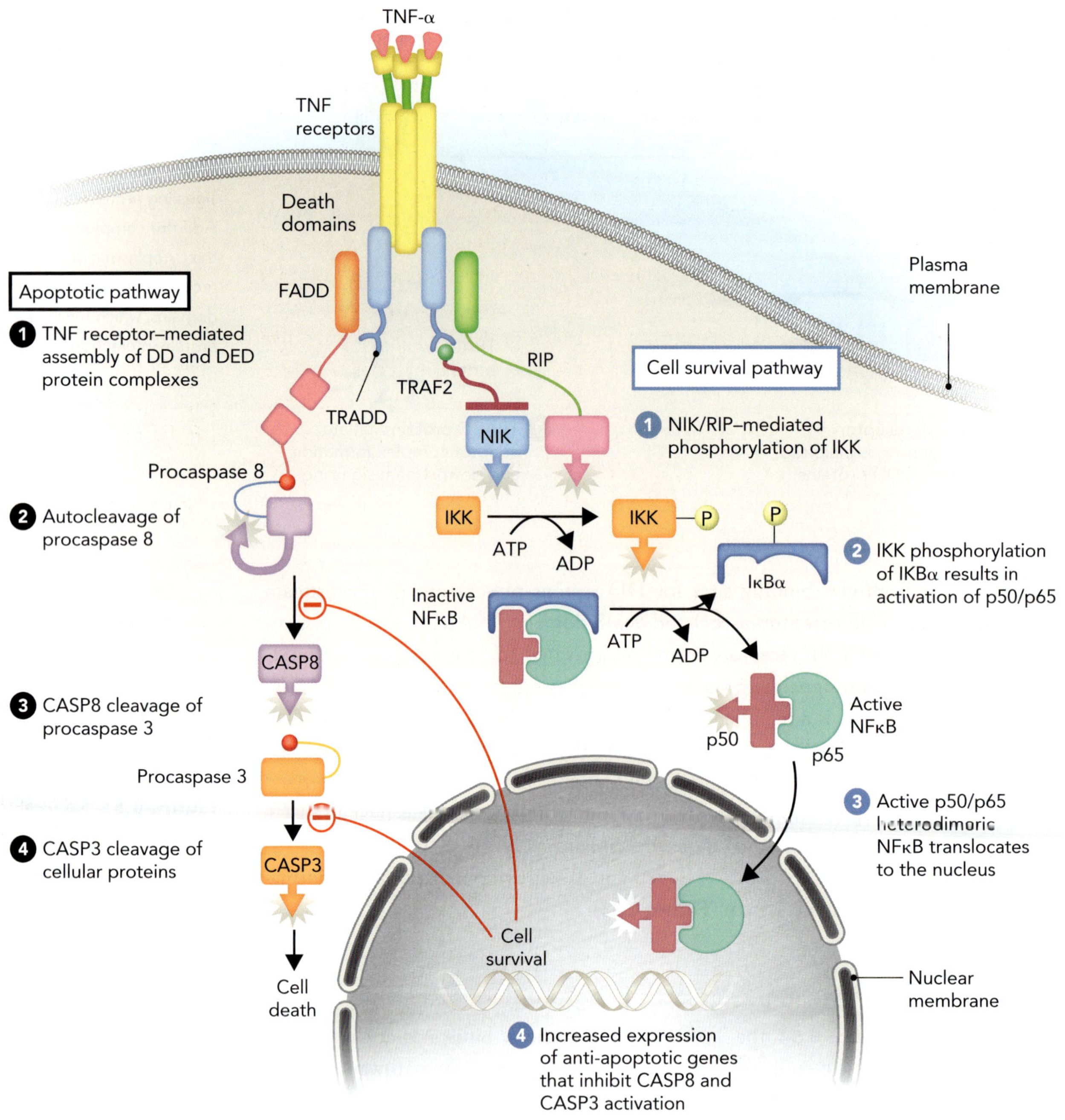

Figure 8.54 TNF-α activation of the TNF receptor stimulates both a protease cascade, leading to cell death, and a phosphorylation cascade, resulting in cell survival. The two signaling pathways diverge at the level of the TRADD protein, which contains both a DD and an N-terminal domain that binds TNF receptor–associated factor 2 (TRAF2). In the apoptosis pathway, TNF receptor activation leads to assembly of a TRADD–FADD complex that results in autocleavage of procaspase 8 through its interaction with the DED of FADD. Once caspase 8 (CASP8) is activated, a proteolytic cascade is initiated by the cleavage and activation of caspase 3 (CASP3), the "executioner" protease. TNF receptor signaling can also induce a cell survival pathway through an adaptor complex consisting of TRADD, TRAF2, and receptor interacting protein (RIP), which leads to the phosphorylation and activation of IκBα kinase (IKK). Phosphorylation of IκBα by IKK targets IκBα for degradation, which leads to activation of the transcription factor NFκB. Once the heterodimeric NFκB (p50/p65) transcription factor is activated, it enters the nucleus and induces the expression of anti-apoptotic genes that inhibit caspase 8 and caspase 3 activation.

How is caspase 3, the executioner caspase, specifically activated by caspase 8, given the exquisite substrate specificity of cysteine–aspartate proteases? The answer is through a combination of cleavage reactions involving both caspase 8 and caspase 3. Caspase 8 and caspase 3, which mediate proteolytic cleavage using a nucleophilic cysteine residue in the enzyme active site, have coevolved in such a way that the substrate recognition site for caspase 8 cleavage is contained within the coding sequence of procaspase 3—at a position that separates the two caspase 3 functional domains. As illustrated in **Figure 8.56**, the caspase 8 recognition sequence Ile-Glu-Thr-Asp (caspase enzymes cleave on the carboxyl side of Asp residues) is located between amino acid residues 172 to 175 of procaspase 3. After the initial cleavage by caspase 8, a partially active caspase 3 complex is formed between a 20-kDa precursor protein containing the active site cysteine (p17) and the mature 12-kDa subunit (p12). This intermediate form of caspase 3 then recognizes and cleaves other procaspase 3 molecules at Asp9 and Asp28 to yield the mature 17-kDa subunit (p17). The fully active caspase 3 executioner enzyme is a $(p17p12)_2$ heterotetramer containing two enzyme active sites that recognize and cleave proteins at the carboxyl Asp residue contained in the sequence Asp-Glu-Val-*Asp* (**Figure 8.57**).

In the cell survival branch of the TNF receptor signaling pathway (see Figure 8.54), TRADD recruits two adaptor proteins to the receptor complex, which function together to initiate a downstream phosphorylation cascade. One of these adaptor proteins is called **TNF receptor–associated factor 2 (TRAF2)**, which recruits NFκB-inducing kinase (NIK) to the receptor complex. The protein–protein interaction stabilizing

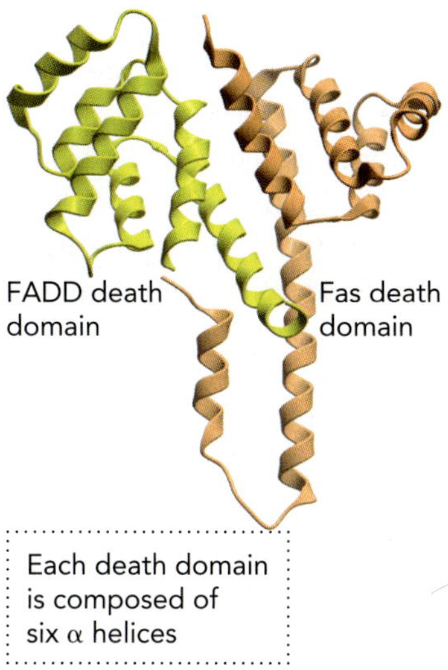

Figure 8.55 The two death domains of the TNF receptor family member Fas and the signaling protein FADD can form a molecular complex, as shown here. Death domains each contain six α helices. A similar DD interaction has been shown to exist between FADD and TRADD, two adaptors in the cell death branch of the TNF receptor signaling pathway. BASED ON PDB FILE 3EZQ.

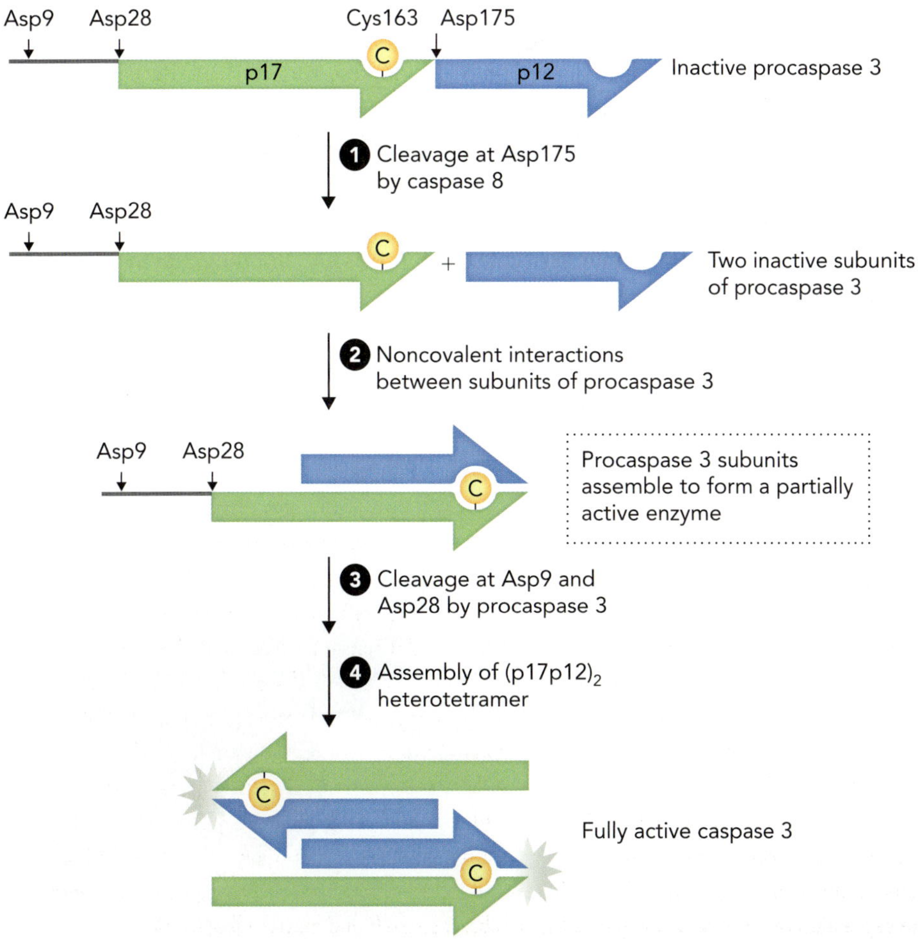

Figure 8.56 Procaspase 3 is converted to its active form through proteolytic processing. The catalytic cysteine (Cys163) is highlighted yellow and resides in the N-terminal fragment that is released by caspase 8 cleavage and processed by a partially active form of procaspase 3 to generate the p17 subunit. The fully active caspase 3 executioner protease is a $(p17p12)_2$ complex.

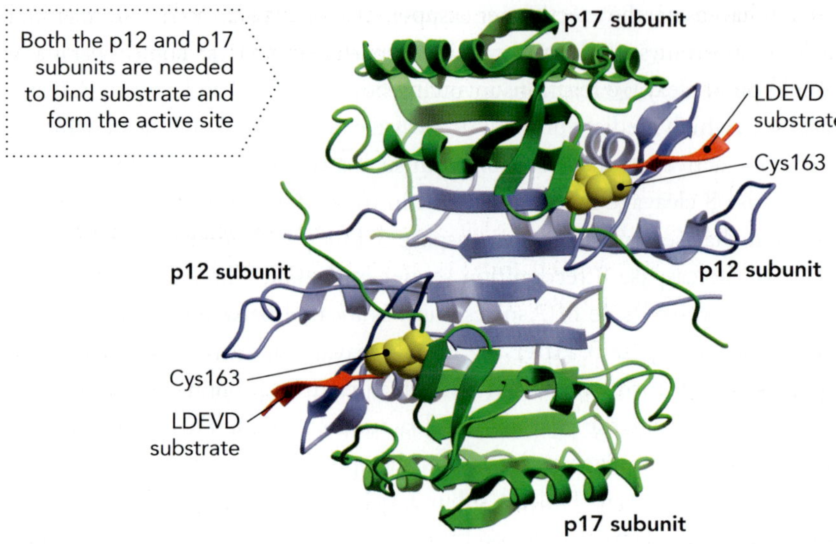

Figure 8.57 The molecular structure of an active caspase 3 heterotetramer with the pentapeptide substrate Leu-Asp-Glu-Val-Asp (LDEVD) bound in each of the two active sites is shown here. Residues from both the p12 and p17 subunits form the substrate binding site and active site. The two catalytic cysteine residues are highlighted. BASED ON PDB FILE 3EDQ.

this TRADD–TRAF2 complex involves the N-terminal domain of TRADD and the TRAF domain motif in TRAF2 (**Figure 8.58**). The other adaptor protein recruited to the TNF receptor complex is a kinase known as **receptor interacting protein (RIP)**, which binds to TRADD through a DD interaction that is functionally similar to formation of the FADD–TRADD complex. As shown earlier in Figure 8.54, NIK and RIP phosphorylate the downstream signaling protein IKK. This process, in turn, activates the transcription factor NFκB by targeting the NFκB inhibitory protein IκBα for proteasomal degradation. The net result is NFκB-mediated transcriptional activation of anti-apoptotic genes encoding proteins that inhibit caspase 8 and caspase 3 activation (see Figure 8.54). If the levels of TRAF2 and RIP in the cell (survival pathway) are higher than those of FADD or procaspase 8 (apoptosis pathway), then the cell stimulated by TNF-α will live.

Understanding the control of apoptotic signaling is an active area of biochemical research because it may provide new approaches to disease prevention and treatment. For example, most cancer cells are resistant to apoptosis because they express high levels of anti-apoptotic proteins or have lost the ability to respond to apoptotic signals, such as DNA damage. Developing methods specifically to block the function of anti-apoptotic proteins in cancer cells or to increase the expression of pro-apoptotic proteins could lead to improved drug therapies. In contrast, many types of neurodegenerative

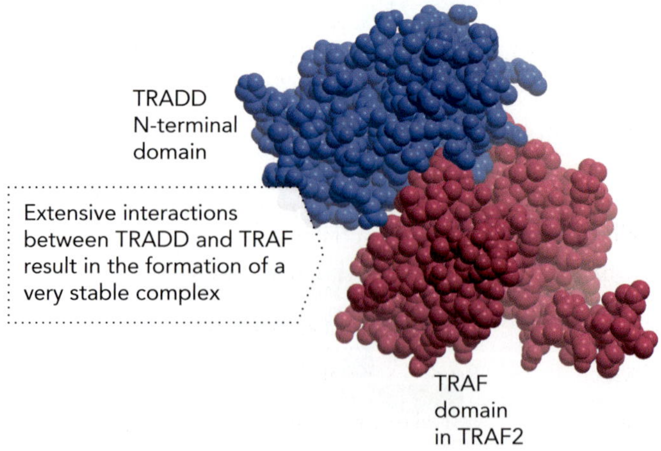

Figure 8.58 The N-terminal domain of TRADD and the TNF receptor–associated factor (TRAF) domain motif of TRAF2 associate tightly. This high-affinity binding interaction establishes the adaptor complex required for the cell survival branch of the TNF receptor signaling pathway. BASED ON PDB FILE 1F3V.

diseases are due to increased activation of apoptotic pathways, leading to neuronal cell death. Finding ways to activate cell survival signaling in neurons subject to premature death could potentially prevent disease symptoms.

concept integration 8.4

What do you think explains the existence of opposing cell death and cell survival pathways in TNF-α target cells? Isn't this a waste of valuable cell resources?

One explanation for having opposing cell death and cell survival pathways in the same cell is that it provides a "toggle switch" to kill or rescue the cell quickly, depending on the quantity and quality of stimuli. For example, if components of the cell survival pathway are expressed at sufficiently high levels when nutrients are readily available or in the presence of extracellular stimuli that activate NFκB signaling, then the cell will survive in the presence of a weak cell death signal. However, if stimuli that promote cell survival are weakened or if stimuli that promote cell death are strengthened—over a relatively short time period—the cell can be sacrificed by initiating the apoptotic pathway. This must not be a waste of resources, as many cells contain both pathways. Perhaps the reason for having both pathways triggered by the same receptor is that apoptosis is advantageous to the organism as a whole because it prevents a damaged or inappropriate cell from surviving.

8.5 Nuclear Receptor Signaling

The final class of receptor signaling proteins we will describe are the nuclear receptors, also known as intracellular receptors. Unlike the other receptor signaling proteins we have described, these receptors are not membrane bound. Nuclear receptors function as transcription factors that regulate gene expression in response to ligand binding. (Refer to **Table 8.6** for a list of acronyms used in this section.) The wide range of cell-specific physiologic responses controlled by nuclear receptors is governed by three parameters:

1. cell-specific expression of nuclear receptors and/or coregulatory proteins;
2. localized bioavailability of ligands; and
3. differential accessibility of target gene DNA sequences in chromatin to nuclear receptor binding.

Table 8.6 ABBREVIATIONS, PROTEIN NAMES, AND FUNCTIONS IN NUCLEAR RECEPTOR SIGNAL TRANSDUCTION SYSTEMS

Abbreviation	Protein name	Function
GR	Glucocorticoid receptor	Nuclear receptor that binds steroid hormones
GRE	Glucocorticoid response element	DNA sequence where glucocorticoid receptors bind
Hsp90	Heat shock protein 90	Chaperonin protein that assists protein folding
PPAR	Peroxisome proliferator–activated receptor	Nuclear receptor that forms a complex with RXR
RXR	Retinoid X receptor	Binding partner of most metabolite nuclear receptors
SRC	Steroid receptor coactivator	Coregulator of nuclear receptor activity

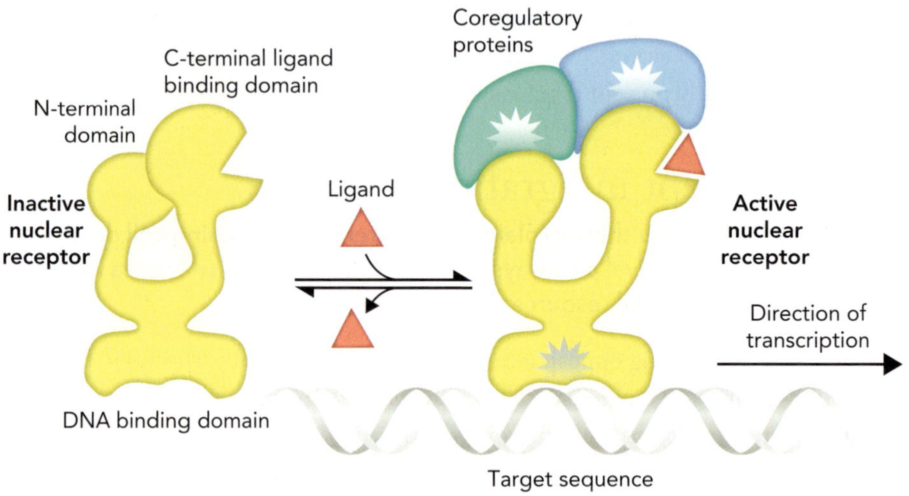

Figure 8.59 Nuclear receptors are ligand-regulated transcription factors, which bind to specific DNA sequences in target genes and recruit coregulatory proteins that modulate transcriptional initiation rates. The three functional domains of nuclear receptors are (1) the N-terminal domain, which contributes to coregulatory protein binding; (2) the DNA binding domain, which binds specific DNA sequences in target genes; and (3) the C-terminal ligand binding domain, which also encodes coregulatory protein binding sites.

The general mechanism of cell signaling through nuclear receptors involves the binding of lipophilic first messengers to a receptor's ligand binding domain. This results in protein conformational changes that remove the inhibitory intramolecular domain interactions in the unliganded receptor and activate the transcriptional regulatory functions of the receptor (**Figure 8.59**). Some nuclear receptors only bind DNA in the presence of ligand, whereas others can bind DNA in the presence or absence of ligand, but the transcriptional regulatory functions of all nuclear receptors are ligand regulated. Ligand-activated nuclear receptors recruit **coregulatory proteins**, also called coactivator or corepressor proteins, to the nuclear receptor–DNA complex through specific protein–protein interactions. These coregulatory proteins modulate rates of transcription by modifying chromatin-associated proteins through acetylation and deacetylation reactions. These reactions serve as "gene on" and "gene off" switches, respectively (see Chapter 23).

Nuclear Receptors Bind as Dimers to Repeat DNA Sequences in Target Genes

As illustrated in **Figure 8.60**, we can divide nuclear receptors into two basic types on the basis of the conserved amino acid sequences in their DNA binding domains, which affects their mode of DNA binding. In addition, these same two nuclear receptor types also tend to be activated by two broadly defined groups of chemical ligands.

The first type of nuclear receptor is represented by the **steroid receptors**. These receptors are head-to-head homodimers, enabling them to bind to inverted repeat DNA sequences similar to the consensus sequence 5′-AGAACA-3′. (A consensus sequence is the most frequent nucleotide at each position within a defined DNA sequence.) The ligands for steroid receptors are physiologic hormones derived from cholesterol. The estrogen receptor, androgen receptor, progesterone receptor, glucocorticoid receptor, and aldosterone receptor are all steroid receptors.

The second type of nuclear receptor, referred to here as **metabolite receptors**, binds to direct repeat DNA sequences similar to the consensus sequence 5′-AGGTCA-3′. The ligands for metabolite receptors are often derived from dietary nutrients, including vitamins, unsaturated fatty acids, and compounds derived from essential amino acids. Examples of metabolite receptors are the vitamin D receptor, retinoic acid receptors, **peroxisome proliferator–activated receptors (PPARs)**, and the thyroid hormone receptor. Many metabolite receptors form head-to-tail

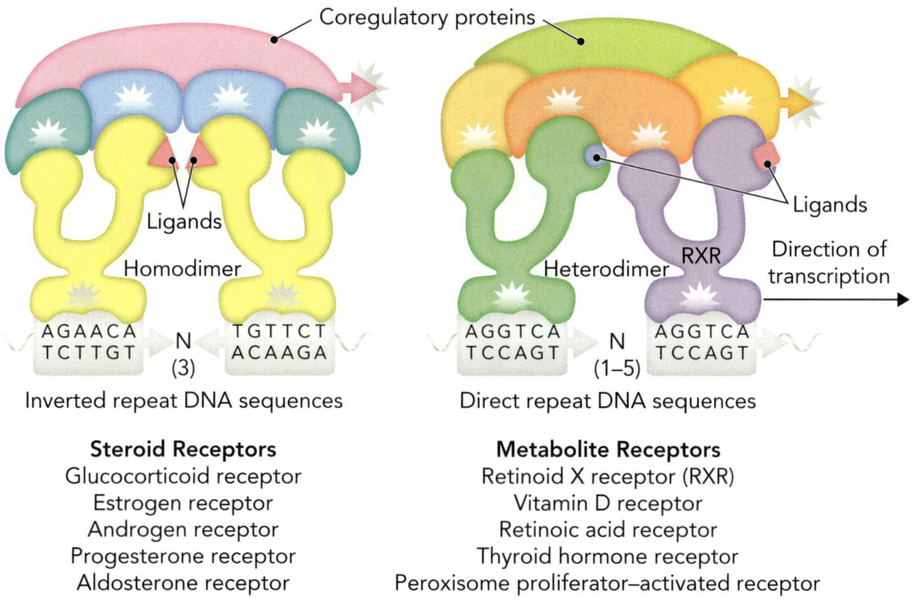

Figure 8.60 The two major types of nuclear receptors are the steroid receptors, which bind to inverted repeat DNA sequences as homodimers; and the metabolite receptors, which primarily bind to direct repeat DNA sequences as heterodimers. The consensus DNA binding sites for these two types of nuclear receptors are shown. Note that the nucleotide spacing (number of nucleotides) between inverted repeat sequences for the homodimeric steroid receptors is 3, whereas the heterodimeric complexes formed between RXR and various subtypes of metabolite receptors bind to direct repeat sequences separated by 1–5 nucleotides, depending on the receptor subtype ("N" is any nucleotide). The RXR metabolite receptor is shown here bound to the 3′ side of the repeat sequence, but in some target genes, it binds to the 5′ side.

heterodimers. The heterodimeric binding partner of most metabolite receptors is the **retinoid X receptor (RXR)**, which has the same DNA binding specificity and facilitates head-to-tail binding to direct repeat sequences. Note that some metabolite receptors bind to DNA as monomers and are functional ligand-activated transcription factors in the absence of RXR.

Figure 8.61 shows the molecular structure of an intact ligand-activated nuclear receptor heterodimer bound to a high-affinity DNA binding site. This large protein–DNA complex consists of the following: a direct repeat DNA sequence; the peroxisome proliferator–activated receptor γ (PPARγ) bound with the ligand rosiglitazone; the retinoid X receptor RXRα bound with the ligand 9-*cis*-retinoic acid; and two bound peptides, each of which contains a leucine-rich binding motif (LXXLL) found in the

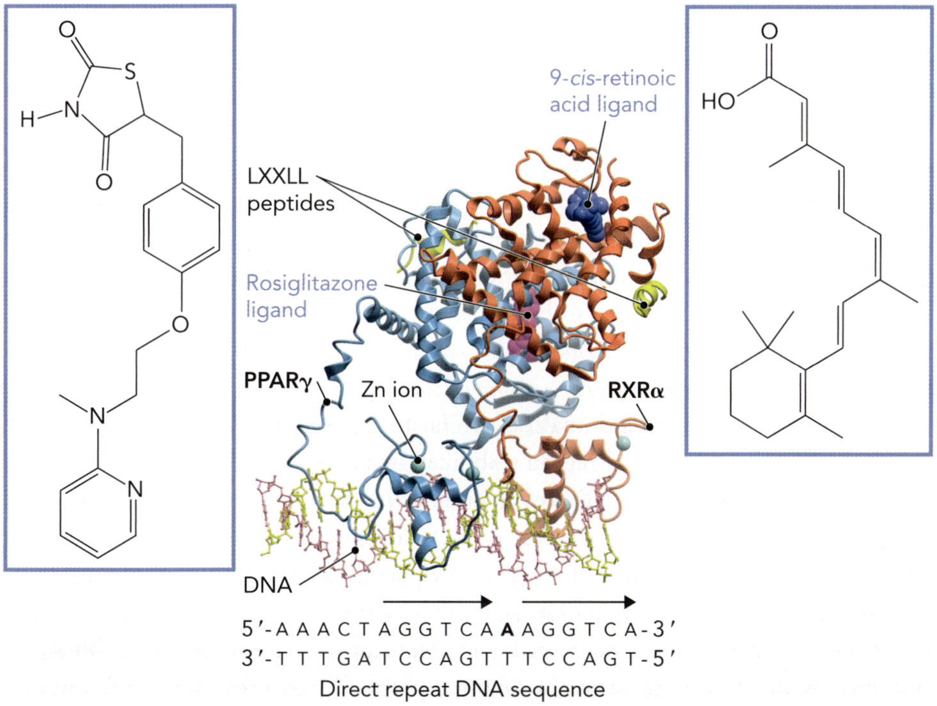

Figure 8.61 The nuclear receptor heterodimer formed between ligand-activated human PPARγ and human RXRα is shown here bound to a direct repeat DNA sequence separated by one nucleotide (highlighted A residue). The chemical structures of a PPARγ agonist, rosiglitazone, and the RXRα ligand, 9-*cis*-retinoic acid, are shown. The nuclear receptor protein–DNA complex was formed in the presence of a polypeptide derived from steroid receptor coactivator-1 (SRC-1), which is a nuclear receptor coregulatory protein that contains leucine-rich LXXLL binding motifs.

BASED ON PDB FILE 3DZY.

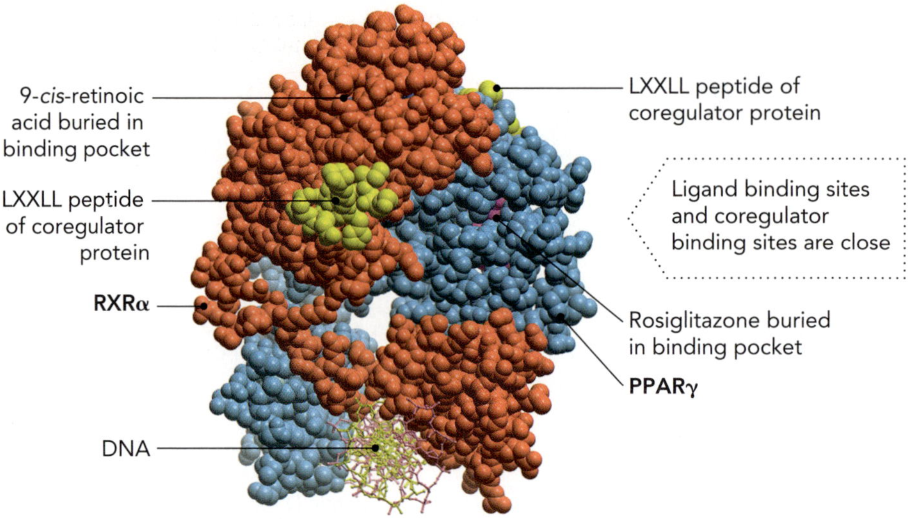

Figure 8.62 The same PPARγ–RXRα heterodimeric complex shown in Figure 8.61 is shown here in a space-filling model, viewed from the end closest to the start of gene transcription. You can see in this representation that the ligand binding pocket and the binding site for LXXLL motif–containing coactivator proteins localize to the ligand binding domain, which facilitates ligand activation of coregulatory protein recruitment. BASED ON PDB FILE 3DZY.

steroid receptor coactivator (SRC) family of nuclear receptor coregulatory proteins. You can see from the tertiary structure of PPARγ and RXRα that nuclear receptors bind to target DNA sequences using a zinc finger DNA binding motif (see Chapter 4), and that the ligand binding domain is attached to the DNA binding domain by a long linker region.

Figure 8.62 shows a space-filling representation of the PPARγ–RXRα heterodimeric complex. Here it is observed that the hydrophobic ligands are buried deep within the ligand binding domain, and that LXXLL coregulatory peptides bind to the surface of this same domain. The nearness of the ligand and coregulatory protein binding sites within the ligand binding domain makes sense because it facilitates the recruitment of coregulatory proteins to the target gene as a function of ligand-induced protein conformational changes.

Glucocorticoid Receptor Signaling Induces an Anti-inflammatory Response

The human glucocorticoid receptor (GR) is a 90-kDa protein expressed in a variety of cell types. It is required for lung development, carbohydrate metabolism in the liver, modulation of the inflammatory response, and neuronal signaling in the brain. Like all steroid hormones, glucocorticoids are derivatives of cholesterol and are transported through the blood by carrier proteins. Glucocorticoids are synthesized in the adrenal glands and are secreted in response to low blood glucose levels, usually with a peak serum concentration in the morning, prior to the first meal of the day. Glucocorticoid levels also rise in response to psychological stress, and it is thought that their anti-inflammatory action may contribute to a decreased resistance to viral infections in people who are overworked.

The GR DNA-binding domain is similar to that of PPARγ and RXRα in that it consists of a short α-helical region stabilized by two zinc ions coordinated to cysteine residues (**Figure 8.63**). We saw a similar zinc finger protein–folding motif in Chapter 4, when we looked at the role of metal ions in stabilizing tertiary structures (see Figure 4.53). As with almost all sequence-specific DNA binding proteins, binding affinity is determined by amino acid side chains in an α-helical region, which make contacts with nucleotide bases in the major groove of the DNA double helix. An obvious difference between the heterodimeric PPARγ–RXRα DNA-binding

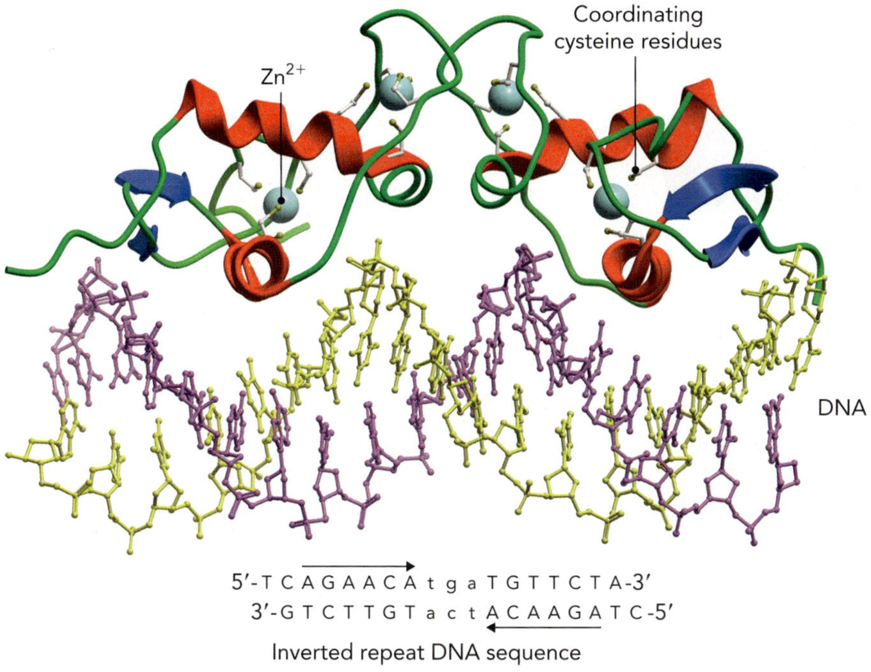

Figure 8.63 The molecular structure of a homodimer of the GR DNA-binding domain bound to an inverted repeat DNA sequence is shown here. The four cysteine residues that coordinate each zinc ion in the DNA binding domains are highlighted. BASED ON PDB FILE 1R4R.

domains and the homodimeric GR DNA-binding domains is the orientation of the zinc finger α helices relative to the repeat DNA sequences. In the case of the GR DNA-binding domains, the inverted repeat sequence must be at least partially palindromic to be recognized by a homodimer that binds in a head-to-head configuration. (A palindrome has the same sequence of characters when read in either direction; the words *madam* and *racecar* are examples of alphabetic palindromes.) In double-stranded DNA, palindromes are of complementary base pairs. As seen in Figure 8.63, the GR DNA-binding domain interacts with a DNA region containing the sequence 5′-AGAACAtgaTGTTCT-3′, where the palindromic portion of the sequence is shown in uppercase letters and the three-nucleotide spacer is in lowercase letters. The pair of zinc atoms at the top of the GR dimer stabilize the protein dimer interface, whereas the zinc atoms in the middle of the DNA binding domains stabilize the α-helical region to optimize amino acid contacts with the DNA major groove.

Figure 8.64 shows the molecular structure of a homodimer of the human GR ligand-binding domain in complex with a derivative of the asthma drug fluticasone and an LXXLL motif–containing peptide from the human coregulatory protein TIF2. Similar to the PPARγ and RXRα ligand-binding domains, the coregulatory protein binding site is on the surface of the GR ligand-binding domain and, as clearly seen in this ribbon model, is in direct contact with two α helices that extend into the hydrophobic pockets.

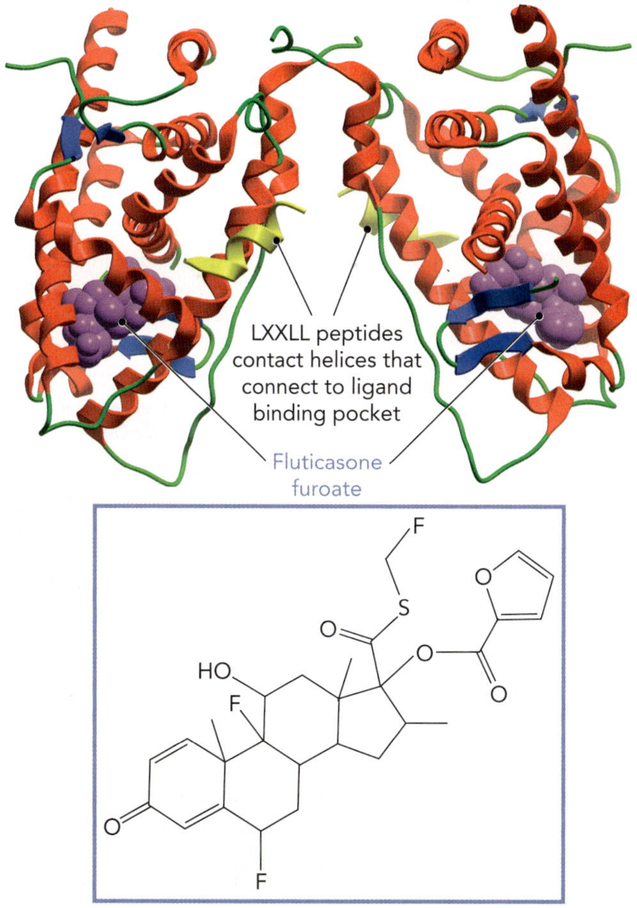

Figure 8.64 A homodimer of the GR ligand-binding domain is shown here in a complex with the synthetic anti-inflammatory corticosteroid fluticasone furoate (space-filling model) and an LXXLL motif–containing peptide from the coregulatory protein TIF2. BASED ON PDB FILE 3CLD.

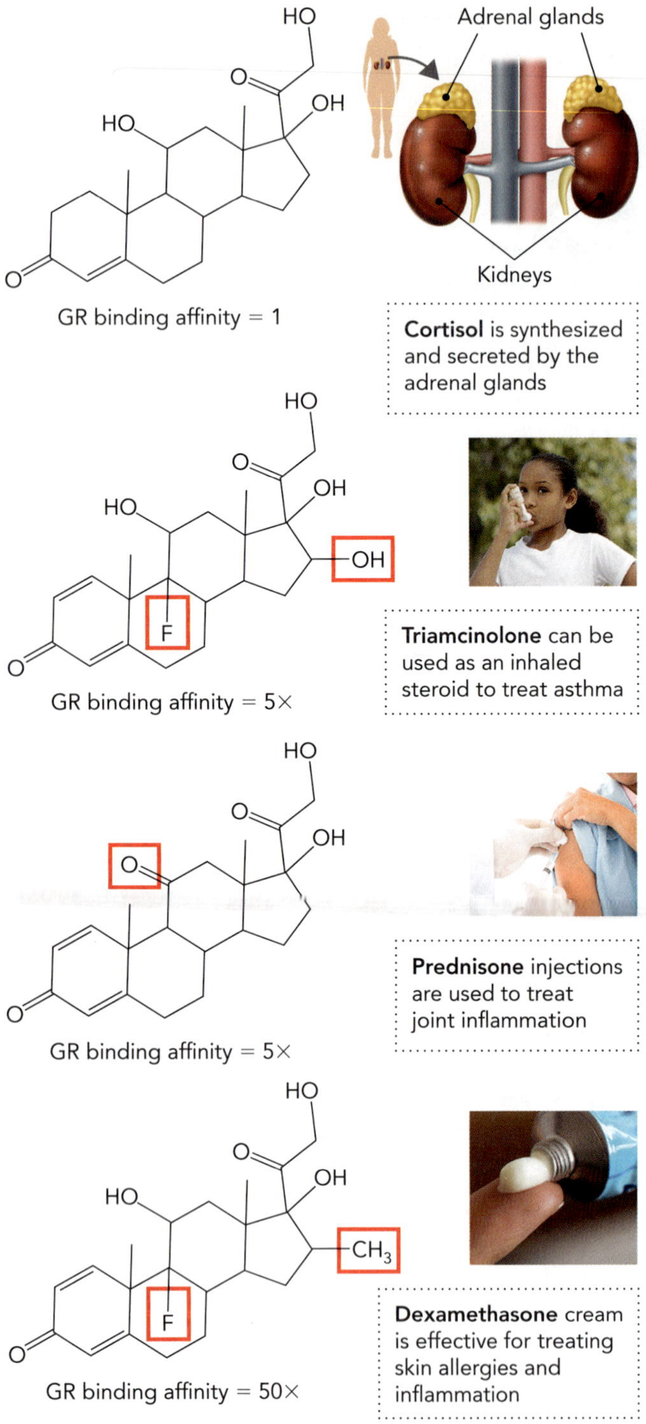

Figure 8.65 Pharmaceutical glucocorticoids are potent anti-inflammatory drugs that have a higher binding affinity for the GR than that of the physiologic steroid cortisol. Chemical modifications that distinguish these drugs from cortisol are highlighted in red boxes, and the GR binding affinities of these drugs relative to cortisol are listed. INHALER: BIKERIDERLONDON/SHUTTERSTOCK; INJECTION: DRAGON IMAGES/SHUTTERSTOCK; CREAM: SIMARIK/GETTY IMAGES.

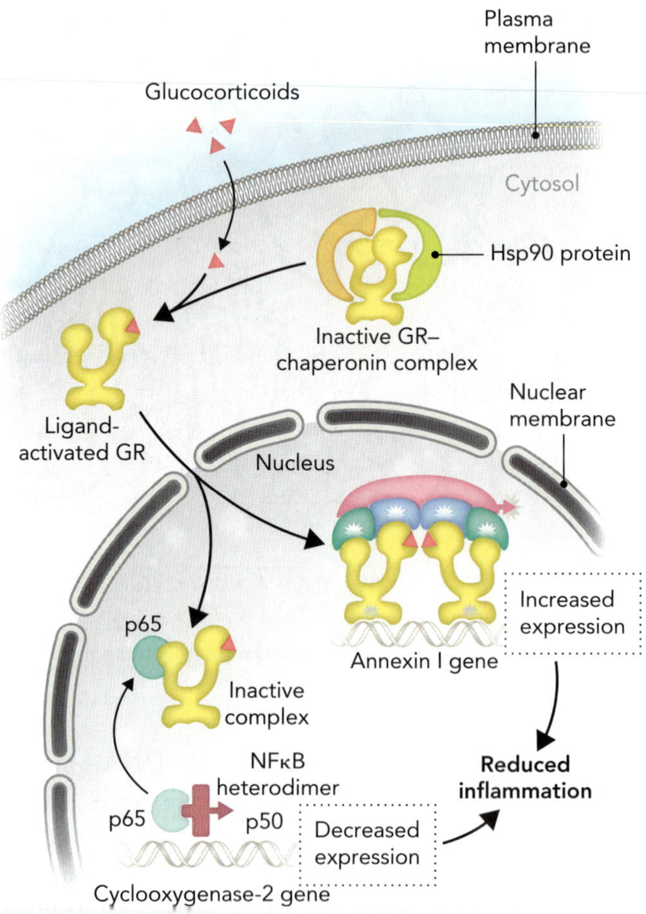

Figure 8.66 Downstream GR signaling regulates the anti-inflammatory pathway in target cells. Glucocorticoids cross the plasma membrane and bind to GR, releasing it from an inactive chaperonin complex containing the heat shock protein Hsp90. After GR nuclear translocation, the receptor both activates expression of the annexin I gene, which inhibits the inflammatory response, and blocks NFκB-mediated expression of cyclooxygenase-2, a pro-inflammatory gene, by binding to the p65 subunit of NFκB and sequestering it away from the cyclooxygenase-2 gene promoter. The net result is reduced inflammation.

Fluticasone is a potent anti-inflammatory drug that has a higher binding affinity for GR than that of the physiologic glucocorticoid cortisol, a property that dramatically increases its clinical effectiveness. As shown in **Figure 8.65**, several pharmaceutical glucocorticoids have been developed to treat inflammation, all of which have higher GR binding affinities than cortisol. In each case, only a few small changes in one or more chemical groups are all that is needed to improve the ligand binding affinity and to provide the basis for patenting—an important consideration in drug development.

How do pharmaceutical glucocorticoids function as anti-inflammatory drugs? The answer is by directly or indirectly regulating the expression of genes encoding proteins that modulate the inflammatory response. Specifically, ligand-activated GR homodimers directly induce the expression of an anti-inflammatory protein called annexin I (also called lipocortin-1), which functions to inhibit prostaglandin synthesis (prostaglandins are pro-inflammatory molecules). In addition, ligand-activated GR also inhibits the transcriptional regulatory functions of NFκB through protein–protein interactions, leading to reduced expression of cyclooxygenase-2, the protein target of aspirin and ibuprofen. As described in Section 8.4 (see Figure 8.54), the transcriptionally active form of NFκB is a heterodimer containing a 50-kDa subunit (p50), which binds DNA sequences in NFκB target genes, and a 65-kDa subunit (p65), which encodes the transcriptional regulatory functions of NFκB. As shown in **Figure 8.66**, the unliganded GR protein resides in the cytoplasm as part of a large complex containing **chaperonin proteins**, which assist in protein folding. One of the most abundant chaperonin proteins in cells is called **heat shock protein 90 (Hsp90)**, which was first discovered as a differentially expressed gene in heat-treated cells. Ligand binding to GR results in disassembly of the GR–chaperonin complex and translocation of GR into the nucleus. In the nucleus, homodimers bind to DNA sequences called **glucocorticoid response elements (GREs)** in the regulatory region of the annexin I gene. Studies have shown that monomers of ligand-activated GR, rather than homodimers, bind to the p65 subunit of NFκB, which inhibits its transcriptional regulatory activity and reduces the expression of cyclooxygenase-2.

concept integration 8.5

Why are several days required before the beneficial effects of pharmaceutical steroids are observed? What explains the relatively long-lasting effects of this class of drugs even after treatment has stopped?

Steroids bind to and activate nuclear receptor proteins, which function by regulating gene expression. Because gene transcription, RNA processing, and accumulation of the encoded protein take time, the sustained benefit of steroid treatment is not immediate. The long-lasting effects of steroids are likely due to two factors. First, decreased levels of steroids readily reduce *de novo* gene expression as a consequence of deactivated nuclear receptors; however, steady-state levels of transcript and protein may take some time to return to basal levels. Second, steroids are derived from cholesterol and are not readily degraded. Moreover, steroids are hydrophobic molecules that accumulate in nonpolar compartments in cells, such as membranes and lipid droplets. Therefore, long-term steroid treatment leads to a buildup of steroid molecules that may take some time to metabolize completely, unlike polar compounds that are eliminated as waste or more easily degraded.

chapter summary

8.1 Components of Signaling Pathways

- Receptor protein activation by ligand binding results in one or more of three biochemical responses: (1) covalent modification of target proteins; (2) protein conformational changes; and (3) altered rates of protein expression.
- A signaling pathway is a linked set of biochemical reactions that consists of upstream events occurring at or near the plasma membrane and downstream events that alter the activity or level of target proteins.
- The number of signaling genes in an organism's genome reflects the need for complex networks of communication within and between individuals.
- Hormones are biologically active compounds that bind to receptor proteins and function as first messengers through endocrine, paracrine, or autocrine mechanisms.
- The functional role of second messengers, such as cAMP, cGMP, and Ca^{2+}, is to amplify receptor-generated signals.
- cAMP production by the enzyme adenylate cyclase initiates a second messenger response characterized by signal amplification through enzyme activation.
- Receptor activation of the enzyme phospholipase C leads to the hydrolysis of phosphatidylinositol-4,5-bisphosphate (PIP_2), which generates the second messengers diacylglycerol (DAG) and inositol-1,4,5-trisphosphate (IP_3). In turn, activity of these molecules leads to increased intracellular levels of calcium ions (Ca^{2+}), another second messenger.
- Higher eukaryotes contain five abundant classes of receptor proteins: (1) G protein–coupled receptors (GPCRs); (2) receptor tyrosine kinases (RTKs); (3) tumor necrosis factor (TNF) family receptors; (4) nuclear receptors; and (5) ligand-gated ion channels.
- The nicotinic acetylcholine receptor is an Na^+–K^+ ligand-gated ion channel that transmits physiologic signals across neuromuscular junctions in response to the release of its ligand acetylcholine, a neurotransmitter.

8.2 G Protein–Coupled Receptor Signaling

- G protein–coupled receptors (GPCRs) contain seven transmembrane α helices that are oriented with the N terminus in the extracellular space and the C terminus in the cytoplasm.
- GPCRs transmit extracellular signals to the cytoplasm through direct interaction with a membrane-bound protein complex called a heterotrimeric G protein, which consists of one each of G_α, G_β, and G_γ subunits ($G_{\alpha\beta\gamma}$).
- G_α is a member of the G protein family of signaling proteins, which contain an intrinsic GTP hydrolyzing activity (GTPase) that converts the active GTP-bound protein into the inactive GDP-bound protein.
- GPCR-mediated activation of the associated heterotrimeric complex occurs when the GDP bound to the G_α subunit is replaced by GTP, resulting in dissociation of the G_α subunit from the heterotrimeric complex.
- The human genome contains multiple G_α, G_β, and G_γ genes, encoding proteins that combine to form nearly 1,000 unique $G_{\alpha\beta\gamma}$ complexes, many of which are involved in sensory perception (sight, smell, taste).
- After GPCR activation by ligand binding, the dissociated G_α subunit activates several target proteins, depending on the specific G_α subtype. In addition, distinct $G_{\beta\gamma}$ complexes activate a variety of target proteins.
- Glucagon and epinephrine signaling in liver cells is mediated by GPCRs, which stimulate downstream signaling through shared and parallel pathways.
- Glucagon binds to glucagon receptors in liver cells to initiate a $G_{s\alpha}$-mediated response that stimulates adenylate cyclase, resulting in cAMP activation of protein kinase A (PKA) and net glucose export.
- Epinephrine (adrenaline) binding to β2-adrenergic receptors activates the same $G_{s\alpha}$-mediated pathway as glucagon (shared pathway), whereas epinephrine binding to α1-adrenergic receptors activates a $G_{q\alpha}$-mediated pathway that stimulates the activity of phospholipase C and net glucose export (parallel pathway).
- Guanine nucleotide exchange factors (GEFs) promote GDP–GTP exchange and activation of G protein signaling, whereas GTPase activating proteins (GAPs) stimulate the intrinsic GTPase of G proteins and inhibit G protein signaling.
- The GEF proteins in GPCR-mediated signaling are the ligand-activated receptors themselves, and the GAPs are members of the regulator of G protein signaling (RGS) family of signaling proteins.
- GPCRs are removed from the plasma membrane and recycled to terminate signal transduction. This mechanism involves phosphorylation of the GPCR cytoplasmic tail by G protein–coupled receptor kinases (GRKs) and binding of β-arrestin protein to these phosphorylated residues.

8.3 Receptor Tyrosine Kinase Signaling

- Receptor tyrosine kinases function as homodimers that bind extracellular ligands. Ligand binding activates a cytoplasmic tyrosine kinase function.
- RTK-mediated autophosphorylation of tyrosine residues in the cytoplasmic tail generates phosphotyrosine binding sites for adaptor proteins, which link downstream signaling pathways to activated receptors.
- Epidermal growth factor receptors (EGFRs) are RTKs that form homodimers upon binding of two epidermal growth factor (EGF) molecules, thereby activating the intrinsic tyrosine kinase activity in each

receptor subunit and autophosphorylation of multiple tyrosine residues.

- SH2 domains are adaptor modules in signaling proteins that bind selectively to phosphotyrosine residues in target proteins.
- GRB2 is an SH2-containing adaptor protein that links the Ras signaling pathway to EGFR activation through binding of another adaptor protein, called SOS, to the SH3 domains in the GRB2 protein. SH3 domains bind to proline-rich sequences in target proteins.
- Ras is a G protein that is similar to the G_α subunit of heterotrimeric G proteins and is characterized by three features: (1) it is attached to the cytoplasmic face of the plasma membrane by a lipid anchor; (2) it is activated by GEFs such as SOS; and (3) it is deactivated by stimulation of its intrinsic GTPase activity by GAPs.
- Ras activation initiates a downstream signaling pathway, which consists of a phosphorylation cascade mediated by kinases in the mitogen-activated protein (MAP) family of signaling proteins. This results in gene regulation and increased rates of cell division.
- Oncogenes are cancer-causing, mutated copies of normal genes. Although they were initially discovered as "hitchhiker genes" in animal tumor viruses, they were later shown to be present in a variety of human cancers.
- Many oncogenes encode signaling proteins that contain dominant mutations (gain of function). For example, Ras proteins that are defective in GTPase activation, due to missense mutations in codon 12, stimulate cell division in the absence of growth factor signaling.
- Insulin binds to the insulin receptor, a disulfide-linked RTK. The insulin receptor consists of a transmembrane $\alpha_2\beta_2$ complex that is activated by a single insulin molecule binding to one of the α subunits; this mode of ligand binding is an example of negative cooperativity.
- Insulin receptor signaling initiates two downstream signaling pathways: one signals through Ras to activate the MAP kinase phosphorylation cascade, and the other activates the PI-3K pathway, leading to glucose uptake.
- Insulin receptor substrate (IRS) proteins bind to phosphotyrosine residues on the insulin receptor cytoplasmic tail through a phosphotyrosine binding (PTB) domain, resulting in the recruitment of SH2-containing signaling proteins.
- PI-3K phosphorylates PIP_2 to produce PIP_3, which functions as a membrane–docking site for proteins containing pleckstrin homology (PH) domains. Phosphatase and tensin homolog (PTEN) converts PIP_3 back into PIP_2 to terminate PI-3K signaling.

8.4 Tumor Necrosis Factor Receptor Signaling

- Binding of tumor necrosis factor-α (TNF-α) to the trimeric TNF receptor initiates two opposing pathways: one that leads to programmed cell death (apoptosis) and the other that promotes cell survival. The net cellular response to TNF-α signaling is determined by the relative abundance of downstream signaling proteins.
- TNF receptor signaling is mediated by the assembly of an adaptor protein complex, which consists of signaling proteins that share protein binding modules called death domains (DDs) and death effector domains (DEDs).
- TNF receptor–induced cell death involves recruitment of the adaptor proteins TRADD and FADD, leading to the activation of caspase 8 and initiation of a downstream proteolytic cascade that activates caspase 3—the executioner protease.
- Caspase enzymes function as heterotetramers that catalyze proteolytic cleavage of peptide bonds on the carboxyl side of aspartate residues using a nucleophilic cysteine residue located in the enzyme active site.
- The TNF receptor–mediated cell survival pathway involves recruitment of the adaptor kinases TRAF2–NIK and RIP to the receptor complex, which stimulates a phosphorylation cascade leading to increased expression of anti-apoptotic genes.

8.5 Nuclear Receptor Signaling

- Nuclear receptors are ligand-activated transcription factors that control a wide range of physiologic responses, as governed by (1) ligand bioavailability, (2) cell-specific expression of nuclear receptors and coregulatory proteins, and (3) accessibility of target gene DNA sequences in chromatin to nuclear receptor binding.
- Nuclear receptors bind to specific DNA sequences in target genes using two zinc finger protein folds. The receptors recruit coregulatory proteins (coactivators or corepressors) that modulate transcriptional initiation rates by modifying chromatin-associated proteins.
- There are two major types of nuclear receptor proteins, which are categorized on the basis of their mode of DNA binding: (1) steroid receptors bind as head-to-head homodimers to inverted repeat DNA sequences; and (2) metabolite receptors bind as head-to-tail heterodimers to direct repeat DNA sequences.
- Peroxisome proliferator–activated receptors are metabolite receptors that selectively bind dietary unsaturated fatty acids and form heterodimers with retinoid X receptors, which are metabolite receptors that bind 9-*cis*-retinoic acid.
- Glucocorticoids are steroid hormones that are synthesized in the adrenal glands and bind to the glucocorticoid receptor and regulate a variety of cellular responses, including inflammation, lung cell development, and carbohydrate metabolism.
- The pharmaceutical drugs prednisone, triamcinolone, and dexamethasone are glucocorticoid agonists that function as anti-inflammatory agents by activating glucocorticoid receptor signaling in target cells, leading to increased expression of the annexin I gene and decreased expression of the cyclooxygenase-2 gene.

biochemical terms

(in order of appearance in text)

receptor protein (p. 372)
target protein (p. 373)
cell signaling pathway (p. 373)
first messenger (p. 373)
second messenger (p. 373)
signaling protein (p. 373)
insulin (p. 375)
hormone (p. 376)
endocrine (p. 376)
paracrine (p. 376)
autocrine (p. 376)
nitric oxide (NO) (p. 376)
cyclic GMP (cGMP) (p. 379)
cGMP phosphodiesterase (p. 379)
cyclic AMP (cAMP) (p. 379)
adenylate cyclase (p. 379)
cAMP phosphodiesterase (p. 379)
protein kinase A (PKA) (p. 379)
diacylglycerol (DAG) (p. 380)
inositol 1,4,5-trisphosphate (IP_3) (p. 380)
phospholipase C (p. 380)
phosphatidylinositol-4,5-bisphosphate (PIP_2) (p. 380)
calmodulin (p. 381)
G protein–coupled receptor (GPCR) (p. 381)
receptor tyrosine kinase (RTK) (p. 382)
tumor necrosis factor (TNF) receptor (p. 382)
ligand-gated ion channel (p. 383)
nicotinic acetylcholine receptor (p. 383)
rhodopsin (p. 385)
heterotrimeric G protein (p. 387)
G protein family (p. 387)
GTPase (p. 387)
glucagon (p. 389)
catecholamine (p. 390)
receptor agonist (p. 390)
receptor antagonist (p. 390)
guanine nucleotide exchange factor (GEF) (p. 394)
GTPase activating protein (GAP) (p. 394)
regulator of G protein signaling (RGS) (p. 394)
G protein cycle (p. 394)
G protein–coupled receptor kinase (GRK) (p. 394)
β-adrenergic receptor kinase (βARK) (p. 394)
β-arrestin (p. 394)
epidermal growth factor (EGF) (p. 399)
Src kinase homology-2 (SH2) (p. 399)
growth factor receptor–bound 2 (GRB2) (p. 400)
phosphoinositide-3 kinase (PI-3K) (p. 400)
Ras (p. 400)
Src kinase homology-3 (SH3) (p. 400)
RasGAP (p. 400)
mitogen-activated protein kinase (MAP kinase) (p. 401)
Raf (p. 401)
MAP/ERK kinase (MEK) (p. 401)
extracellular signal–regulated kinase (ERK) (p. 401)
oncogene (p. 402)
dominant mutation (p. 402)
recessive mutation (p. 402)
tumor suppressor (p. 403)
phosphotyrosine binding (PTB) domain (p. 405)
insulin receptor substrate (IRS) (p. 405)
phosphatidylinositol-3,4,5-trisphosphate (PIP_3) (p. 407)
phosphatase and tensin homolog (PTEN) (p. 407)
pleckstrin homology (PH) domain (p. 407)
apoptosis (p. 410)
death domain (DD) (p. 410)
cysteine–aspartate protease (caspase) (p. 410)
TNF receptor–associated death domain (TRADD) (p. 411)
Fas-associated death domain (FADD) (p. 411)
death effector domain (DED) (p. 411)
TNF receptor–associated factor 2 (TRAF2) (p. 413)
receptor interacting protein (RIP) (p. 414)
coregulatory protein (p. 416)
steroid receptor (p. 416)
metabolite receptor (p. 416)
peroxisome proliferator–activated receptor (PPAR) (p. 416)
retinoid X receptor (RXR) (p. 417)
steroid receptor coactivator (SRC) (p. 418)
chaperonin protein (p. 421)
heat shock protein 90 (Hsp90) (p. 421)
glucocorticoid response element (GRE) (p. 421)

review questions

1. Describe the six functional components of a signaling pathway that is initiated outside of the cell and culminates with activation of a metabolic enzyme. Provide specific examples of each component.
2. What are the differences between endocrine, paracrine, and autocrine signaling pathways?
3. How can an extracellular signal be amplified a million-fold inside a cell?
4. Describe the structure and function of the five major types of receptor proteins.
5. Describe the structure and function of the α, β, and γ subunits of a heterotrimeric G protein.
6. What explains the diversity of GPCR-mediated signaling pathways, considering there are only three protein subunits in the heterotrimeric G protein complex?
7. What contributes to the selectivity of GPCR-mediated signaling pathways that transmit the sensory perceptions of sight, smell, and taste?
8. Describe three regulatory mechanisms that determine the efficiency of GPCR-mediated signaling in cells containing the appropriate GPCR and heterotrimeric G protein complex, required for transduction of a specific first-messenger signal.
9. How does the structure of the GRB2 upstream signaling protein function to link epidermal growth factor receptor (EGFR) activation to a downstream phosphorylation cascade?

10. Compare and contrast the EGFR and insulin receptor pathways with regard to activation of upstream signaling proteins on the cytosolic side of the plasma membrane.
11. Explain how the TNF receptor signaling pathway can stimulate both an apoptosis pathway and a cell survival pathway in the same cell. What molecular mechanism determines if a TNF-stimulated cell will ultimately live or die?
12. What are the two major classes of nuclear receptors, and what are the three molecular determinants of nuclear receptor signaling?

challenge problems

1. Second messengers serve to amplify upstream extracellular signals by dramatically increasing the number of downstream signaling events. What characterizes the biochemical processes that generate second messengers, and how does this contribute to intracellular signal transduction? Cite two specific examples of how second messengers are generated.
2. What is the net level of amplification in a signaling pathway in which there are three signal transmission processes, A, B, and C, each of which amplifies the signal 250-fold? By what percentage would this net level of amplification be reduced if the proteins required for process A were defective and only amplified the signal 100-fold instead of 250-fold?
3. What is the biochemical mechanism of sildenafil (Viagra), one of a class of drugs used to treat erectile dysfunction? What upstream signal is required for sildenafil to be effective?
4. Glucagon, the "I am hungry" hormone, is released from the pancreas when blood glucose levels are low, and it signals to the liver to export glucose through a combination of glycogen degradation (a form of stored glucose) and glucose synthesis from noncarbohydrate precursors (a process called gluconeogenesis). Epinephrine, the "fight or flight" hormone, is released from the adrenal medulla when neuronal inputs signal that metabolic energy is required quickly in the form of glucose export from the liver. Describe the two intracellular signaling pathways in liver cells that are stimulated by glucagon and epinephrine binding to their respective receptors.
5. On the basis of the definition of shared pathways and parallel pathways as illustrated in Figure 8.23 and the physiologic functions of glucagon and epinephrine, describe which second messengers would most likely be produced when (1) you are hungry for cake and (2) you almost step on a snake. Explain.
6. Explain why mutations in Ras that block its intrinsic GTPase activity result in cancer cell division even in the absence of epidermal growth factor (EGF) signaling.
7. The G protein cycle controls both G_α protein signaling and Ras signaling. Draw a G protein cycle that controls Ras signaling activity and label all of the protein components. Identify which forms of Ras are active and inactive with respect to signaling.
8. What two mechanisms ensure that GPCR-mediated signaling is turned off when extracellular signals decrease?
9. Using the epidermal growth factor receptor (EGFR) as an example, explain how protein conformational changes brought on by ligand binding, receptor dimerization, and activation of intrinsic tyrosine kinase activity can lead to autophosphorylation of tyrosine residues in the cytoplasmic tails of both receptor subunits in a receptor tyrosine kinase (RTK) homodimer.
10. The phosphoinositide-3 kinase (PI-3K) pathway is activated by insulin signaling in liver cells. Number the following statements 1–10 to order the sequence of events that lead to glucose uptake and glycogen synthesis in response to insulin signaling. The abbreviations for each signaling component of the pathway are defined in Section 8.3.

    ___ PDK1 and Akt bind to PIP_3 in the plasma membrane via PH domains.

    ___ Insulin receptor autophosphorylates tyrosine residues in the cytoplasmic tail.

    ___ Increased rates of glucose uptake and glycogen synthesis lower blood glucose.

    ___ Insulin receptor phosphorylates IRS proteins on tyrosine residues.

    ___ PI-3K phosphorylates PIP_2 to generate PIP_3.

    ___ Insulin binds to the insulin receptor and activates its intrinsic kinase activity.

    ___ IRS proteins bind to phosphotyrosines in the insulin receptor via PTB domains.

    ___ Akt is phosphorylated and activated by the serine/threonine kinase activity of PDK1.

    ___ PI-3K binds to phosphotyrosines on IRS proteins via SH2 domains.

    ___ Akt dissociates from PIP_3 and phosphorylates downstream target proteins.
11. TNF receptor signaling involves the assembly of protein complexes through shared protein–protein interaction modules, one of which is the death domain (DD) module. Identify the five DD-containing proteins

required for TNF-α signaling and briefly describe the function of each.

12. List three reasons why it makes sense that proteases, such as cysteine–aspartate proteases (caspases), function as "executioners" in the cell death pathway, rather than some other class of proteins or biomolecules.
13. The human genome encodes ~60 nuclear receptors, the vast majority of which belong to the category of metabolite receptors. What are the primary differences between the steroid receptors and metabolite receptors with regard to DNA binding and subunit organization of the active dimers? What might be some advantages of the metabolite receptors compared with the steroid receptors in terms of the specificity of ligand-dependent target gene expression?
14. Glucocorticoids regulate a variety of cell-specific physiologic responses, ranging from anti-inflammatory action in immune cells to stimulation of carbohydrate synthesis in liver cells. Given the following three observations, what explains the cell specificity of glucocorticoid signaling? (1) Glucocorticoid receptors are expressed in essentially all human cell types. (2) Glucocorticoid hormones are endocrine signaling molecules that are circulated throughout the entire body. (3) The DNA sequence in every cell is 99.99% identical.

smartwork5

If your instructor assigns homework with Smartwork5, access it here: digital.wwnorton.com/biochem.

suggested reading

Books and Reviews

Gomberts, B. D., Kramer, I. M., and Tatham, P. R. (2009). *Signal transduction* (2nd ed.). San Diego, CA: Academic Press.

Khorasanizadeh, S., and Rastinejad, F. (2001). Nuclear-receptor interactions on DNA-response elements. *Trends in Biochemical Sciences, 26*, 384–390.

Pierce, K. L., Premont, R. T., and Lefkowitz, R. J. (2002). Seven-transmembrane receptors. *Nature Reviews. Molecular Cell Biology, 3*, 639–650.

Rhen, T., and Cidlowski, J. A. (2005). Antiinflammatory action of glucocorticoids—new mechanisms for old drugs. *New England Journal of Medicine, 353*, 1711–1723.

Spehr, M., and Munger, S. D. (2009). Olfactory receptors: G protein-coupled receptors and beyond. *Journal of Neurochemistry, 109*, 1570–1583.

Weinberg, R. A. (2007). *The biology of cancer* (1st ed.). New York, NY: Garland Science.

Youngren, J. F. (2007). Regulation of insulin receptor function. *Cellular and Molecular Life Sciences, 64*, 873–891.

Primary Literature

Aggerbeck, M., Ferry, N., Zafrani, E. S., Billon, M. C., Barouki, R., and Hanoune, J. (1983). Adrenergic regulation of glycogenolysis in rat liver after cholestasis. Modulation of the balance between alpha 1 and beta 2 receptors. *Journal of Clinical Investigation, 71*, 476–486.

Al-Mulla, F., Milner-White, E. J., Going, J. J., and Birnie, G. D. (1999). Structural differences between valine-12 and aspartate-12 Ras proteins may modify carcinoma aggression. *Journal of Pathology, 187*, 433–438.

Chandra, V., Huang, P., Hamuro, Y., Raghuram, S., Wang, Y., Burris, T. P., and Rastinejad, F. (2008). Structure of the intact PPAR-γ–RXR-α nuclear receptor complex on DNA. *Nature, 456*, 350–356.

Cherezov, V., Rosenbaum, D. M., Hanson, M. A., Rasmussen, S. G., Thian, F. S., Kobilka, T. S., Choi, H. J., Kuhn, P., Weis, W. I., Kobilka, B. K., and Stevens, R. C. (2007). High-resolution crystal structure of an engineered human β2-adrenergic G protein-coupled receptor. *Science, 318*, 1258–1265.

Hicks, S. N., Jezyk, M. R., Gershburg, S., Seifert, J. P., Harden, T. K., and Sondek, J. (2008). General and versatile autoinhibition of PLC isozymes. *Molecular Cell, 31*, 383–394.

Hirsch, J. A., Schubert, C., Gurevich, V. V., and Sigler, P. B. (1999). The 2.8 Å crystal structure of visual arrestin: a model for arrestin's regulation. *Cell, 97*, 257–269.

Kim, C., Xuong, N. H., and Taylor, S. S. (2005). Crystal structure of a complex between the catalytic and regulatory (RIα) subunits of PKA. *Science, 307*, 690–696.

Kimple, A. J., Soundararajan, M., Hutsell, S. Q., Roos, A. K., Urban, D. J., Setola, V., Temple, B. R., Roth, B. L., Knapp, S., Willard, F. S., and Siderovski, D. P. (2009). Structural determinants of G-protein α subunit selectivity by regulator of G-protein signaling 2 (RGS2). *Journal of Biological Chemistry, 284*, 19402–19411.

Lambright, D. G., Sondek, J., Bohm, A., Skiba, N. P., Hamm, H. E., and Sigler, P. B. (1996). The 2.0 Å crystal structure of a heterotrimeric G protein. *Nature, 379*, 311–319.

Maignan, S., Guilloteau, J. P., Fromage, N., Arnoux, B., Becquart, J., and Ducruix, A. (1995). Crystal structure of the mammalian Grb2 adaptor. *Science, 268*, 291–293.

Murakami, M., and Kouyama, T. (2008). Crystal structure of squid rhodopsin. *Nature, 453*, 363–367.

Ogiso, H., Ishitani, R., Nureki, O., Fukai, S., Yamanaka, M., Kim, J. H., Saito, K., Sakamoto, A., Inoue, M., Shirouzu, M., and Yokoyama, S. (2002). Crystal structure of the complex of human epidermal growth factor and receptor extracellular domains. *Cell, 110,* 775–787.

Park, Y. C., Burkitt, V., Villa, A. R., Tong, L., and Wu, H. (1999). Structural basis for self-association and receptor recognition of human TRAF2. *Nature, 398,* 533–538.

Park, Y. C., Ye, H., Hsia, C., Segal, D., Rich, R. L., Liou, H. C., Myszka, D. G., and Wu, H. (2000). A novel mechanism of TRAF signaling revealed by structural and functional analyses of the TRADD-TRAF2 interaction. *Cell, 101,* 777–787.

Scheerer, P., Park, J. H., Hildebrand, P. W., Kim, Y. J., Krauss, N., Choe, H. W., Hofmann, K. P., and Ernst, O. P. (2008). Crystal structure of opsin in its G-protein-interacting conformation. *Nature, 455,* 497–502.

Scott, F. L., Stec, B., Pop, C., Dobaczewska, M. K., Lee, J. J., Monosov, E., Robinson, H., Salvesen, G. S., Schwarzenbacher, R., and Riedl, S. J. (2009). The Fas-FADD death domain complex structure unravels signaling by receptor clustering. *Nature, 457,* 1019–1022.

Smith, B. J., Huang, K., Kong, G., Chan, S. J., Nakagawa, S., Menting, J. G., Hu, S. Q., Whittaker, J., Steiner, D. F., Katsoyannis, P. G., et al. (2010). Structural resolution of a tandem hormone-binding element in the insulin receptor and its implications for design of peptide agonists. *Proceedings of the National Academy of Sciences USA, 107,* 6771–6776.

Tesmer, J. J., Dessauer, C. W., Sunahara, R. K., Murray, L. D., Johnson, R. A., Gilman, A. G., and Sprang, S. R. (2000). Molecular basis for P-site inhibition of adenylyl cyclase. *Biochemistry, 39,* 14464–14471.

Unwin, N. (2005). Refined structure of the nicotinic acetylcholine receptor at 4 Å resolution. *Journal of Molecular Biology, 346,* 967–989.

Watson, L. C., Kuchenbecker, K. M., Schiller, B. J., Gross, J. D., Pufall, M. A., and Yamamoto, K. R. (2013). The glucocorticoid receptor dimer interface allosterically transmits sequence-specific DNA signals. *Nature Structural & Molecular Biology, 7,* 876–883.

Zhang, X., Gureasko, J., Shen, K., Cole, P. A., and Kuriyan, J. (2006). An allosteric mechanism for activation of the kinase domain of epidermal growth factor receptor. *Cell, 125,* 1137–1149.

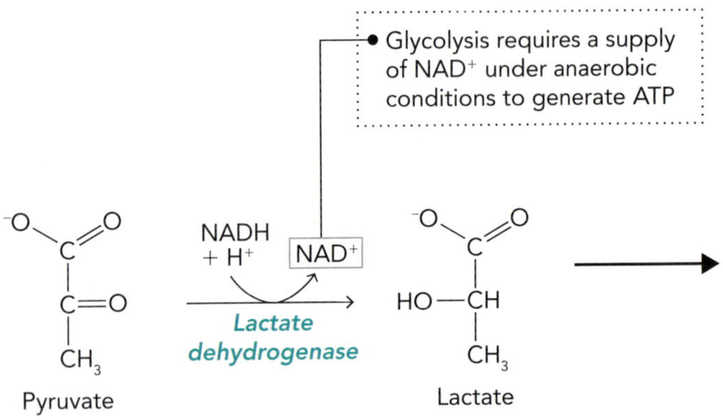

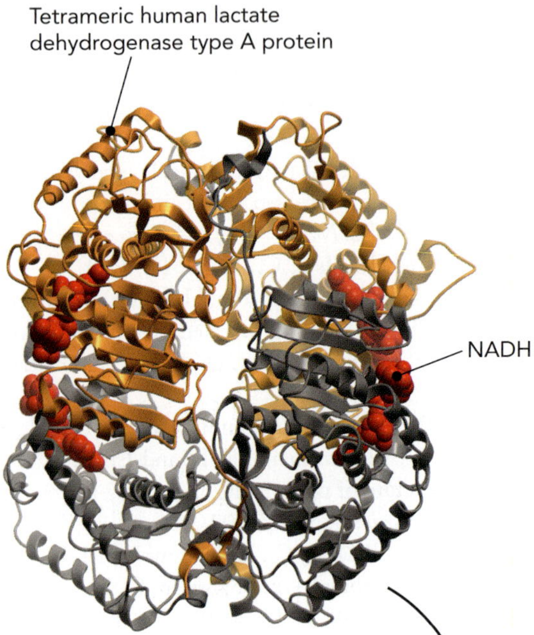

Human skeletal muscle tissue section after reaction with lactate dehydrogenase type A antibody

The enzyme lactate dehydrogenase catalyzes a reaction that reduces pyruvate to lactate and, in the process, oxidizes NADH to generate NAD^+ for the glycolytic pathway. The lactate dehydrogenase type A enzyme is expressed as a tetramer at high levels in human skeletal muscle cells. NADH (shown in space-filling representation) is bound to the active site of each of the four lactate dehydrogenase type A subunits. The activity of lactate dehydrogenase type A enzyme is required in skeletal muscle cells to maintain ATP production under anaerobic conditions, such as when running.

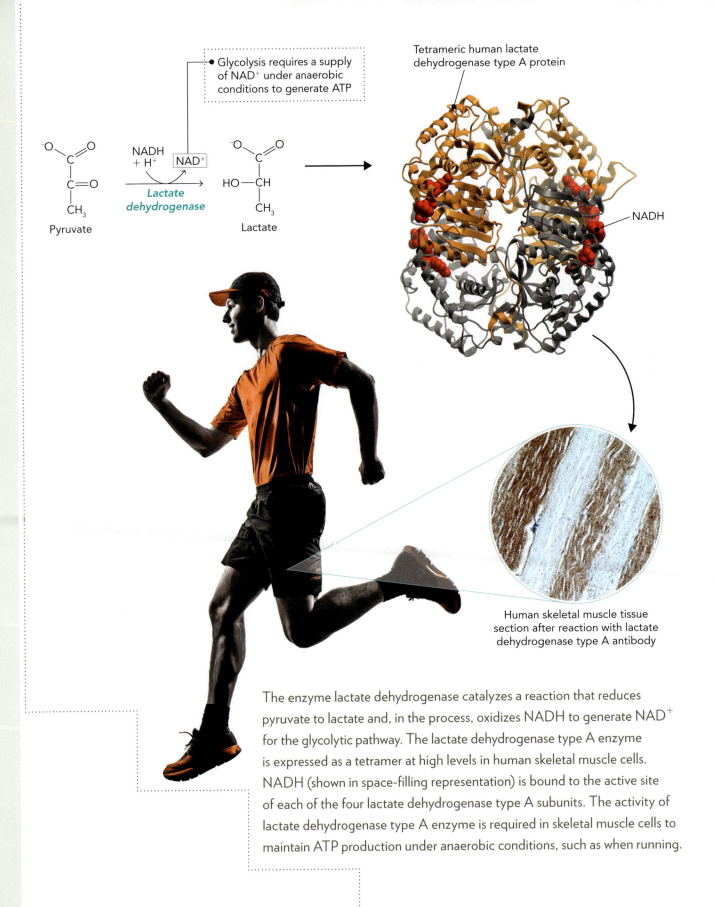

9

Glycolysis

A Paradigm of Metabolic Regulation

◀ In cells under aerobic conditions (O_2 is abundant), the glycolytic enzyme glyceraldehyde-3-phosphate dehydrogenase obtains its required cofactor NAD^+ from reactions occurring at the mitochondrial inner membrane that are involved in metabolite transport. However, under anaerobic conditions (O_2 is depleted), the NAD^+ required for glycolysis is supplied by a reaction catalyzed by the cytosolic enzyme lactate dehydrogenase, which reduces pyruvate to form lactate. If NAD^+ becomes limiting under anaerobic conditions, then ATP production by the glycolytic pathway is reduced, and cell function is severely impaired. The human lactate dehydrogenase enzyme in skeletal muscle cells is a tetramer of four identical subunits, each of which contains a catalytic active site. During intense physical activity, such as climbing stairs, skeletal muscle cells become depleted of O_2, but the lactate dehydrogenase reaction regenerates NAD^+ for the glycolytic pathway to maintain ATP production for muscle cell contraction.

CHAPTER OUTLINE

CREDITS: PROTEIN STRUCTURE: BASED ON PDB FILE 1I10; HUMAN SKELETAL MUSCLE REACTED WITH LDHA ANTIBODY: PROTEINTECH GROUP, INC.; MAN RUNNING: OSTILL/ISTOCK/GETTY IMAGES PLUS.

Metabolism refers to the collection of biochemical reactions in a free-living organism that convert chemical energy into work. On Earth, **metabolism** depends primarily on the initial conversion of solar energy to chemical energy by the process of photosynthesis.

Back in Chapter 1, we defined metabolic pathways as linked biochemical reactions. To study how complex metabolic pathways function together in a cell to maintain life far from equilibrium, we divide our presentation of 10 major metabolic pathways into two parts. In Part 3 of this book, we examine four essential metabolic pathways involved directly in energy conversion—primarily the synthesis of ATP from either simple sugars (Chapters 9–11) or solar energy (Chapter 12). In Part 4, we describe six additional key metabolic pathways that use this ATP, as well as redox reactions, to synthesize and degrade metabolites.

We begin this journey in Section 9.1 by presenting an overview of metabolism, which includes looking more closely at how redox mechanisms account for many energy-converting reactions. This is followed by a detailed look at the first of these 10 metabolic pathways: **glycolysis**, which is considered to be one of the oldest metabolic pathways in terms of evolutionary time. Our presentation of glycolysis begins in Section 9.2 with a look at the primary metabolic fuel for this pathway; namely, simple sugars. In Section 9.3, we examine each of the biochemical reactions required to yield two net ATP for every molecule of glucose that enters the glycolytic pathway. Then in Section 9.4, we discuss three ways in which metabolic regulation in glycolysis is controlled by enzyme activity levels or substrate availability. We end the chapter in Section 9.5 by describing the metabolic fates of pyruvate—the final product of glycolysis—under aerobic and anaerobic conditions.

9.1 Overview of Metabolism

Catabolic pathways are the collection of enzymatic reactions in the cell that lead to the degradation of macromolecules and nutrients for the purpose of energy capture, usually in the form of ATP and reducing power (NADH and $FADH_2$). In contrast, anabolic pathways use energy available from the hydrolysis of ATP and the oxidation of reducing equivalents (primarily NADPH) to synthesize biomolecules for the cell (see Figure 2.12). It is important to realize that catabolic and anabolic pathways are active in the cell at the same time, and many metabolites serve as both substrates and products for different enzymes. Sometimes the catabolic pathways are more active than the opposing anabolic pathways, leading to depletion of stored energy and biomolecules. At other times, when energy stores are plentiful and the necessary building blocks are available, anabolic pathways predominate.

The flux of metabolites through catabolic and anabolic pathways is determined by two primary factors:

1. *level of enzyme activity,* which is controlled by enzyme levels (gene transcription, protein synthesis, and protein turnover), catalytic activity (binding of regulatory molecules and covalent modification), cellular location (membranes or organelles); and
2. *level of substrates,* which are obtained from the diet and released from energy stores in the body.

As described in Chapter 7, enzymes function by providing a suitable reaction environment—the enzyme active site—that lowers the energy of activation for a given reaction. Enzymes cannot change the equilibrium of a reaction but instead function as catalysts that increase reaction rates in both directions. As you will see, some reactions in metabolism normally function near equilibrium and are therefore controlled by substrate concentration, whereas other reactions are catalyzed by highly regulated enzymes that function far from equilibrium.

One of the best ways to understand how flux through various catabolic and anabolic pathways changes in response to substrate concentration and enzyme activity levels is to look at glucose metabolism in the liver before and after breakfast (**Figure 9.1**). Early in the morning, before your first meal, blood glucose levels begin to decline after a night of "fasting," which triggers glucagon release from the pancreas. Glucagon signaling in liver cells activates both a catabolic pathway (glycogen degradation) and an anabolic pathway (gluconeogenesis) and at the same time inhibits the catabolism of glucose by the glycolytic pathway. This activity causes a release of glucose from the liver into the bloodstream, where it travels to cells throughout your body as an energy source. However, within an hour of eating a bowl of cereal and drinking a cup of fruit juice, your insulin levels increase due to elevated blood glucose. Insulin signaling in the liver leads to stimulation of glucose uptake and glycogen synthesis, as well as an increase in glucose catabolism by the glycolytic pathway (Figure 9.1). These metabolic changes remove glucose from the bloodstream and stores it in the liver for later use. Therefore, when you "break your fast" by eating in the morning, you initiate a transient shift in flux through these various metabolic pathways until blood glucose levels stabilize at ~5 mM. The four pathways shown in Figure 9.1 are active in liver cells all the time, with the only change being the relative metabolite flux through each pathway in response to glucose concentrations and hormone activation of key enzymes.

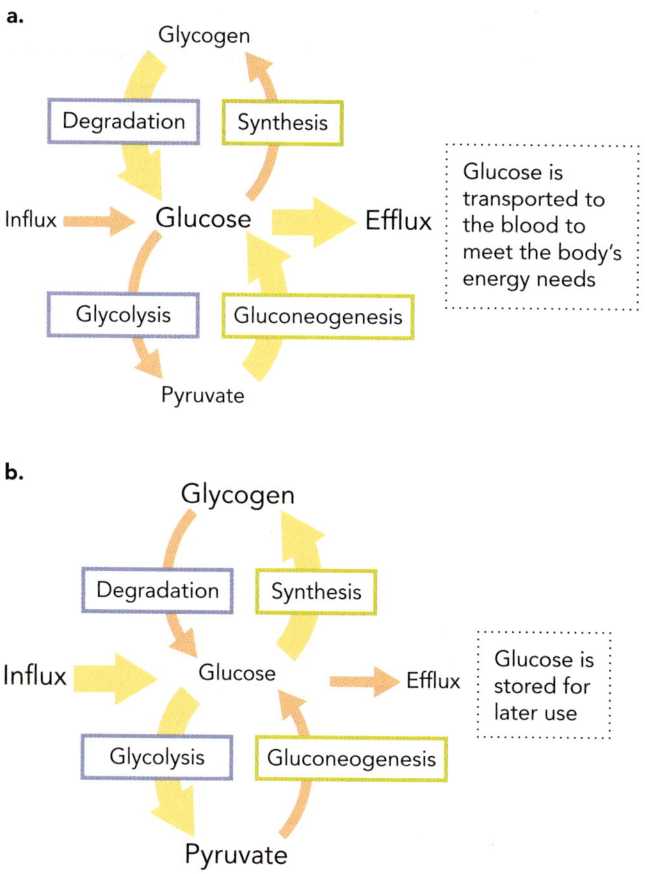

Figure 9.1 Substrate concentration and enzyme activity levels control metabolic flux through opposing pathways. **a.** Before breakfast, when blood glucose levels are low from a night of fasting, glucagon levels are high and insulin levels are low. This leads to an increased rate of glucose efflux from the liver. **b.** After breakfast, when blood glucose levels have risen, glucagon levels decrease and insulin levels increase. This causes stimulation of glucose influx and altered rates of flux through glucose metabolizing pathways.

The 10 Major Catabolic and Anabolic Pathways in Plants and Animals

The breakfast scenario in Figure 9.1 gives the take-home message for Parts 3 and 4 of this book: *Metabolic pathways are highly interdependent and are exquisitely controlled by enzyme activity levels and substrate concentration.*

As you will see, even though we examine one pathway at a time for pedagogical purposes, the key to understanding metabolic integration in terms of nutrition, exercise, and disease is learning how cells control metabolic flux *between* pathways. Put another way, when studying metabolism it is important to keep sight of the "forest through the trees" by keeping in mind the collection of metabolic pathways in an organism (the entire forest), even though we discuss only one pathway at a time (a grove of trees).

Figure 9.2 shows two different types of metabolic maps that have been developed. The stylized map in Figure 9.2a was designed by Richard Wheeler, who wanted to illustrate that metabolic maps are a bit like subway maps with metabolites

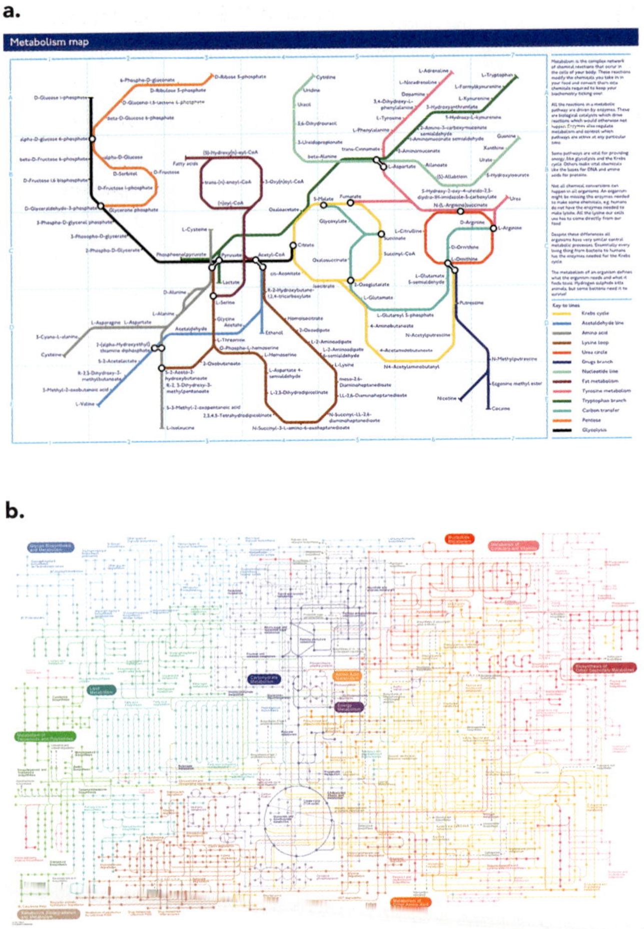

Figure 9.2 Metabolic pathway charts can be drawn in a number of ways, and many are accessible online. **a.** A stylized map of metabolism illustrating the similarities between people moving through subways and metabolites moving through metabolic pathways. RICHARD WHEELER. **b.** The Kyoto Encyclopedia of Genes and Genomes (KEGG) metabolic map developed at Kyoto University. KANEHISA LABORATORIES.

being representative of people—both are in a hurry to get somewhere. Metabolic pathways can best be studied using Web resources that can be queried and linked to a variety of databases. An example of such an interactive Web-based map is shown in Figure 9.2b. This metabolic map was developed by a group of biochemists at Kyoto University in Japan, who created a Web resource called **KEGG (Kyoto Encyclopedia of Genes and Genomes)**. This database links enzymes to substrates and products using an interactive grid. Each dot represents a primary metabolite, and the connecting lines correspond to series of enzymatic reactions.

Although both of the metabolic maps in Figure 9.2 provide a means to investigate specific reactions within all known metabolic pathways, they contain too much detail for a systematic approach to the study of metabolism. Therefore, we will use the simplified metabolic map shown in **Figure 9.3**, which is designed to highlight the interdependence of major anabolic and catabolic pathways. It also illustrates the hierarchical nature of metabolism, consisting of four classes of macromolecules (proteins, nucleic acids, carbohydrates, and lipids); six primary metabolites (amino acids, nucleotides, fatty acids, glucose, pyruvate, and acetyl-CoA); and seven small biomolecules (NH_4^+, CO_2, NADH, $FADH_2$, O_2, ATP, and H_2O), all of which we will encounter frequently in our study.

Figure 9.4 (p. 434) shows that we can divide the basic metabolic map into two distinct pathway groups, each of which is divided into subgroups. Chapters 9–12 cover the pathways grouped in red, which focus on energy conversion reactions. We begin with three pathways that oxidize simple sugars through a series of redox reactions, culminating in ATP synthesis (**Table 9.1**, p. 435). We then present a fourth pathway represented by photosynthesis and carbon fixation, which plants use to convert light energy to carbohydrates in the form of starch. Chapters 13–18 cover the pathways grouped in blue, which together control the synthesis and degradation of carbohydrates, lipids, amino acids, and nucleotides. We describe six sets of pathways in this group, which are found collectively in plants and animals. Finally, in Chapter 19 we describe how these various pathways are integrated at the organismal level, and we use diet, exercise, and metabolic disease in humans as a framework for discussion.

We start off the discussion of each of the 10 major sets of metabolic pathways shown in Table 9.1 by asking and answering these four key questions:

1. What does the pathway accomplish for the cell?
2. What is the overall net reaction of the pathway?
3. What are the key enzymes in the pathway?
4. What is an example of this pathway in everyday biochemistry?

The purpose of these four questions is to provide a link back to the big picture (question 1) and at the same time to elucidate the essential elements of a pathway in terms of

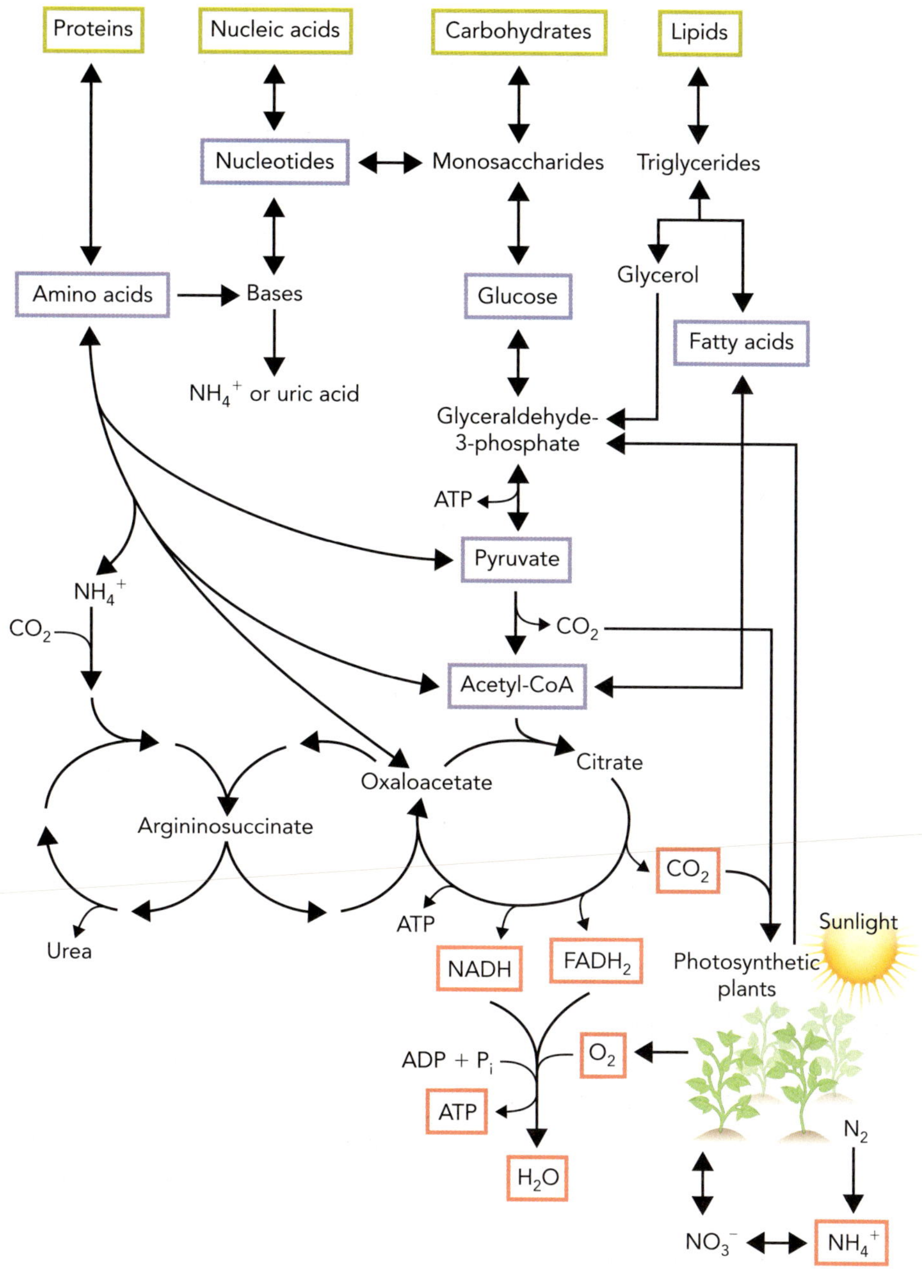

Figure 9.3 A basic metabolic map emphasizing the major biomolecules in cells and the interdependence of pathways. The four classes of macromolecules are shown in yellow boxes and the key metabolic intermediates in most cells are denoted by blue boxes. Frequently encountered small molecules appear in red boxes.

the net reaction (question 2) and the key enzymes of the pathway (question 3). Finally, we illustrate how the particular metabolic pathway relates to our everyday lives (question 4) and do so in a way that hopefully makes it memorable.

Metabolite Concentrations Directly Affect Metabolic Flux

Before examining the glycolytic pathway in detail, we first need to revisit the concept of bioenergetics introduced in Chapter 2. However, this time, we need to consider more carefully the effect of substrate and product concentrations in coupled reactions. In particular, we want to examine how these concentrations directly affect the overall Gibbs free energy change calculation for a defined metabolic pathway.

We begin by defining the term *flux*, which, in the context of a metabolic pathway, is the rate at which substrates and products (metabolites) are interconverted.

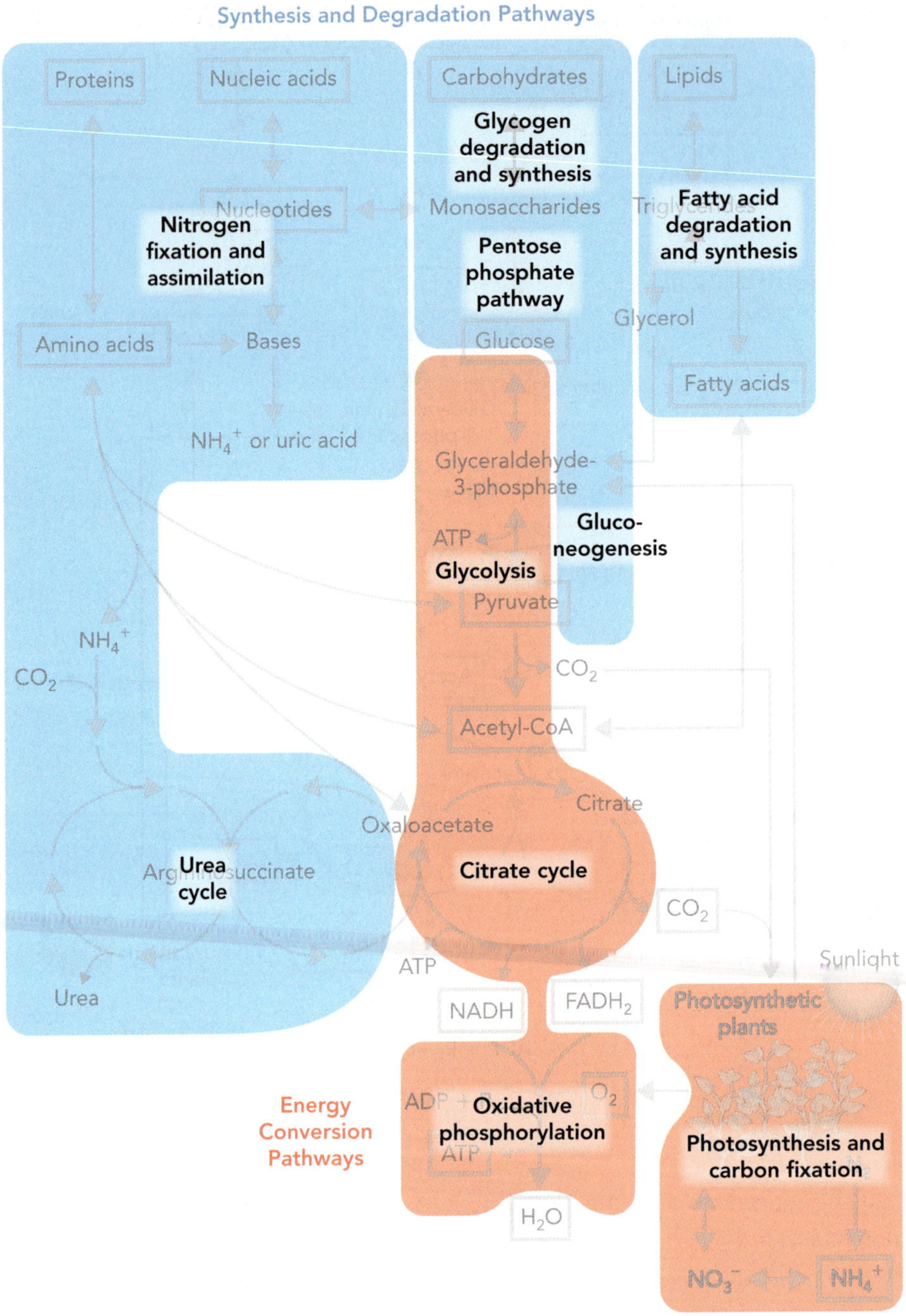

Figure 9.4 The basic metabolic map can be divided into two discrete groups of pathways, corresponding to energy conversion pathways (red) and metabolite synthesis and degradation pathways (blue). The ten sets of major metabolic pathways described in Part 3 (red) and Part 4 (blue) of the book are highlighted in bold.

To understand how flux is controlled, recall from Chapter 2 that the overall change in free energy ΔG of the chemical reaction $A \rightarrow B$ in living cells is equal to the sum of the biochemical standard free energy change $\Delta G^{\circ\prime}$ and the product of a value representing the gas constant, temperature, and actual concentrations of substrates and products:

$$\Delta G = \Delta G^{\circ\prime} + RT \ln Q \qquad (2.14)$$

Using the value of the gas constant R as 8.314×10^{-3} kJ/mol K and the temperature T in kelvins (K) gives units for free energy in kilojoules per mole (kJ/mol). As described in Chapter 2, the biochemical standard free energy change ($\Delta G^{\circ\prime}$) value for a reaction is experimentally determined by measuring the free energy change when the reaction is started with 1 M each of reactants and products (and standard conditions of pH, temperature, and pressure), then allowing the reaction to proceed to

Table 9.1 TEN MAJOR SETS OF METABOLIC PATHWAYS COLLECTIVELY FOUND IN PLANTS AND ANIMALS (SEE FIGURE 9.4)

Set	Major pathways in plants and animals	Chapter
	Energy conversion pathways	
1	Glycolysis	9
2	Citrate cycle	10
3	Oxidative phosphorylation	11
4[a]	Photosynthesis and carbon fixation	12
	Synthesis and degradation pathways	
5	Pentose phosphate pathway	14
6	Gluconeogenesis	14
7[b]	Glycogen degradation and synthesis	14
8	Fatty acid degradation and synthesis	16
9[c]	Nitrogen fixation and assimilation	17
10[d]	Urea cycle	17

[a]Pathways found only in plants.

[b]Animals store glucose as glycogen; plants store glucose as starch.

[c]Nitrogen fixation occurs in some bacteria, while nitrogen assimilation occurs in bacteria and plants.

[d]The urea cycle is found only in terrestrial animals.

equilibrium. By measuring the equilibrium concentrations of reactants and products, we can determine $\Delta G^{\circ\prime}$:

$$\Delta G^{\circ\prime} = -RT \ln \frac{[\mathrm{B}]_{\mathrm{eq}}}{[\mathrm{A}]_{\mathrm{eq}}}$$

or

$$\Delta G^{\circ\prime} = -RT \ln K_{\mathrm{eq}} \tag{2.12}$$

If the concentration of product at equilibrium is much greater than the concentration of substrate, $[\mathrm{B}]_{\mathrm{eq}} >> [\mathrm{A}]_{\mathrm{eq}}$, then the reaction $\mathrm{A} \rightarrow \mathrm{B}$ is highly favorable under standard conditions, leading to the accumulation of B. Because the natural log (ln) of a number greater than 1 is positive, the calculated $\Delta G^{\circ\prime}$ using Equation 2.12 will be negative. However, if the equilibrium concentration of substrate is much greater than that of product, $[\mathrm{A}]_{\mathrm{eq}} >> [\mathrm{B}]_{\mathrm{eq}}$, then $\Delta G^{\circ\prime}$ will be positive (the natural log of a number less than 1 is negative). A large negative $\Delta G^{\circ\prime}$ indicates that under standard conditions, the reaction proceeds from left to right (A is converted to B), whereas a large positive $\Delta G^{\circ\prime}$ indicates that the reaction proceeds from right to left (B is converted to A). Regardless of directionality, a large $\Delta G^{\circ\prime}$ means that a significant difference in free energy exists between the reaction components, and that the energy released during the drive toward equilibrium can potentially be used in energy conversion reactions.

Metabolic pathways function by using highly specific enzymes to harness the free energy released in small increments. Often, metabolic reactions with $\Delta G^{\circ\prime} << 0$ are used as a driving force to make reactions within a linked metabolic pathway proceed through use of shared intermediates (the product of reaction 1 is the substrate for reaction 2). The $\Delta G^{\circ\prime}$ of a coupled reaction is the sum of the $\Delta G^{\circ\prime}$ values for each individual reaction. For example, suppose the conversion of A to B is unfavorable ($\Delta G^{\circ\prime} > 0$), but the conversion of B to C is very favorable ($\Delta G^{\circ\prime} << 0$). By coupling these reactions, the conversion of A to B could occur because the net reaction (A converting to C) is favorable:

$$\begin{array}{ll} A \rightleftharpoons B & \Delta G^{\circ\prime} = +4 \text{ kJ/mol} \\ B \rightleftharpoons C & \Delta G^{\circ\prime} = -10 \text{ kJ/mol} \\ \hline A \rightleftharpoons C & \Delta G^{\circ\prime} = -6 \text{ kJ/mol} \end{array}$$

A more specific example involves the free energy released from ATP hydrolysis, which is relatively large ($\Delta G^{\circ\prime} = -30.5$ kJ/mol) and is frequently used in coupled reactions in metabolism. In fact, the first step in glycolysis is catalyzed by the enzyme hexokinase, which uses ATP hydrolysis to drive glucose phosphorylation in the following coupled reaction:

$$\begin{array}{ll} \text{Glucose} + P_i \rightleftharpoons \text{Glucose-6-phosphate} + H_2O & \Delta G^{\circ\prime} = +13.8 \text{ kJ/mol} \\ \text{ATP} + H_2O \rightleftharpoons \text{ADP} + P_i & \Delta G^{\circ\prime} = -30.5 \text{ kJ/mol} \\ \hline \text{Glucose} + \text{ATP} \rightleftharpoons \text{Glucose-6-phosphate} + \text{ADP} & \Delta G^{\circ\prime} = -16.7 \text{ kJ/mol} \end{array}$$

Coupled reactions can also be used to capture free energy in the form of ATP. This occurs in the last step in glycolysis in a reaction catalyzed by the enzyme pyruvate kinase, which converts phosphoenolpyruvate to pyruvate. In this case, the $\Delta G^{\circ\prime}$ of pyruvate formation is high enough to drive ATP synthesis:

$$\begin{array}{ll} \text{Phosphoenolpyruvate} + H_2O \rightleftharpoons \text{Pyruvate} + P_i & \Delta G^{\circ\prime} = -62.2 \text{ kJ/mol} \\ \text{ADP} + P_i \rightleftharpoons \text{ATP} + H_2O & \Delta G^{\circ\prime} = +30.5 \text{ kJ/mol} \\ \hline \text{Phosphoenolpyruvate} + \text{ADP} \rightleftharpoons \text{Pyruvate} + \text{ATP} & \Delta G^{\circ\prime} = -31.7 \text{ kJ/mol} \end{array}$$

How do we calculate metabolic flux, considering that most reactions are not linked directly to ATP hydrolysis, and substrates and products are not present at equal concentrations inside the cell? The answer comes from calculating the overall change in free energy (ΔG) using the **mass action ratio (Q)**, which is defined as the ratio of the product and substrate concentrations under actual conditions in the cell:

$$\Delta G = \Delta G^{\circ\prime} + RT \ln \frac{[B]_{\text{actual}}}{[A]_{\text{actual}}}$$

$$\Delta G = \Delta G^{\circ\prime} + RT \ln Q \qquad (2.14)$$

Individual metabolic reactions are most often not at equilibrium. The overall ΔG for a metabolic pathway indicates the driving force to reach equilibrium under cellular conditions. These metabolic reactions are not at equilibrium in a living organism; this only happens when the organism dies and reaches equilibrium with the surrounding environment. With that in mind, let's see how the actual free energy change ΔG of a reaction is affected by the concentrations of substrates and products in the cell.

We will use another glycolytic reaction as an example. In this case, the standard free energy change of the phosphoglucoisomerase reaction, in which glucose-6-phosphate (glucose-6-P) is converted to fructose-6-phosphate (fructose-6-P), is positive ($\Delta G^{\circ\prime} = +1.7$ kJ/mol) and unfavorable under standard conditions. However, because fructose-6-P can be converted to fructose-1,6-bisphosphate (fructose-1,6-BP) by the enzyme phosphofructokinase-1 (an ATP-coupled reaction with a $\Delta G^{\circ\prime} = -12.5$ kJ/mol), the concentration of fructose-6-P can be low enough under actual conditions in the cell to drive the reaction in the forward direction ($\Delta G = -2.9$ kJ/mol):

$$\text{Glucose-6-P} \rightleftharpoons \text{Fructose-6-P} \qquad \Delta G^{\circ\prime} = +1.7 \text{ kJ/mol}$$

$$[\text{Glucose-6-P}]_{\text{actual}} = 8.3 \times 10^{-5}\text{ M}$$

$$[\text{Fructose-6-P}]_{\text{actual}} = 1.4 \times 10^{-5}\text{ M}$$

$$\Delta G = \Delta G^{\circ\prime} + RT \ln Q$$

$$\Delta G = +1.7 \text{ kJ/mol} + (0.00831 \text{ kJ/mol K})(310 \text{ K}) \ln \frac{1.4 \times 10^{-5}\text{ M}}{8.3 \times 10^{-5}\text{M}}$$

$$\Delta G = +1.7 \text{ kJ/mol} + (-4.6 \text{ kJ/mol})$$

$$\Delta G = -2.9 \text{ kJ/mol}$$

Through reaction coupling, an unfavorable reaction may be driven forward by a favorable reaction. For example, when the phosphoglucoisomerase reaction is combined with the phosphofructokinase-1 reaction under standard conditions (1 M starting concentrations of glucose-6-P, fructose-6-P, and fructose-1,6-BP), the large negative standard free energy change for the phosphofructokinase-1 ATP-coupled reaction drives the phosphoglucomutase reaction in the forward direction:

Reactions in the glycolytic pathway:

$$\text{Glucose-6-P} \xrightleftharpoons{\text{Phosphoglucoisomerase}} \text{Fructose-6-P} + \text{ATP} \xrightleftharpoons{\text{Phosphofructokinase-1}} \text{Fructose-1,6-BP} + \text{ADP}$$

Combined reactions:

$$\begin{array}{ll}
\text{Glucose-6-P} \rightleftharpoons \text{Fructose-6-P} & \Delta G^{\circ\prime} = \ \ +1.7 \text{ kJ/mol} \\
\text{Fructose-6-P} + \text{P}_i \rightleftharpoons \text{Fructose-1,6-BP} & \Delta G^{\circ\prime} = +16.3 \text{ kJ/mol} \\
\text{ATP} \rightleftharpoons \text{ADP} + \text{P}_i & \Delta G^{\circ\prime} = -30.5 \text{ kJ/mol} \\
\hline
\text{Glucose-6-P} + \text{ATP} \rightleftharpoons \text{Fructose-1,6-BP} + \text{ADP} & \Delta G^{\circ\prime} = -12.5 \text{ kJ/mol}
\end{array}$$

The key to coupled reactions is the use of shared intermediates. For example, fructose-6-P in the glycolytic pathway is the shared intermediate in the reactions catalyzed by phosphoglucoisomerase and phosphofructokinase-1. One reason why shared intermediates are used so effectively in coupled reactions is that some metabolic enzymes are components of large multiprotein complexes, as illustrated in **Figure 9.5**. These types of close physical interactions limit product diffusion and function to "channel" shared intermediates from one enzyme to the next. Two types of data support the concept of metabolite channeling. One is the finding that some glycolytic enzymes

Figure 9.5 Protein complexes are likely to play an important role in metabolism by increasing the efficiency of coupled reactions through substrate channeling. Three types of complexes have been proposed to promote substrate channeling between metabolic enzymes (E): **a.** cytosolic protein complexes; **b.** membrane-bound protein complexes; **c.** protein complexes formed by binding to a shared scaffold protein. The labels E1, E2, and so forth, indicate proteins with separate enzymatic activities.

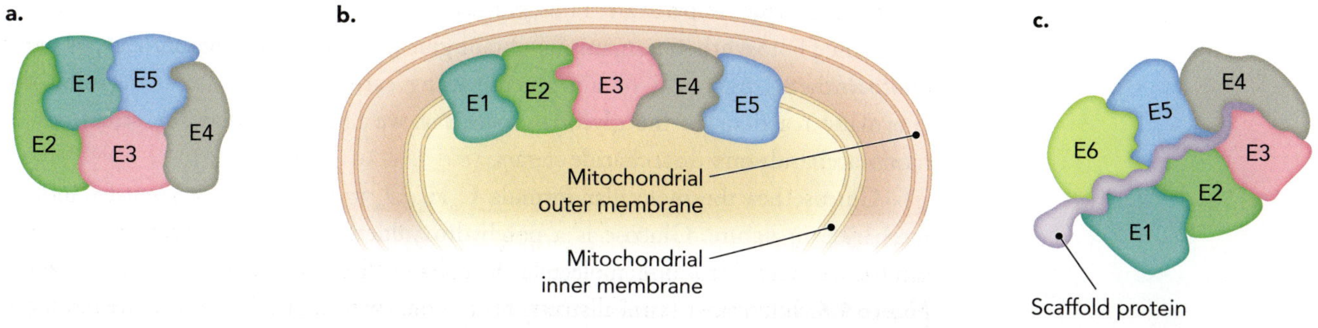

associate with each other in solution through high-affinity noncovalent interactions, suggesting that such complexes could form inside the cell. The other evidence comes from kinetic data, which indicate that rates of coupled reactions in protein aggregates *in vitro* closely resemble *in vivo* reaction rates; that is, these reactions are much faster than reactions in dilute solutions, in which protein aggregates do not form.

Although evidence for soluble cytosolic protein complexes containing two or more enzymes in a pathway is somewhat limited, it is clear that closely associated inner mitochondrial membrane-bound enzymes do play an important role in metabolism (Figure 9.5b). As we will see later, this is especially true in the citrate cycle pathway (see Chapter 10) and the electron transport system (see Chapter 11). It is also possible that scaffold proteins could serve as docking sites for several linked metabolic enzymes in a pathway. Such metabolic protein scaffolds would serve to increase the local concentration of metabolites and, moreover, could regulate enzyme activity through conformational changes induced by scaffold binding.

concept integration 9.1

How can changes in substrate and product concentrations for a single enzymatic reaction alter flux through multiple metabolic pathways?

The actual change in free energy ΔG for a reaction is the sum of the change in standard free energy $\Delta G°'$ and $RT \ln Q$, in which Q is the mass action ratio defined by $[\text{product}]_{\text{actual}}/[\text{substrate}]_{\text{actual}}$. In coupled reactions, as occur in metabolic pathways, products of one reaction are often metabolized by a linked reaction. Because the reactions do not reach equilibrium under normal cellular conditions, and $\Delta G°'$ and RT are constants that do not change, the continual loss of product results in a more negative ΔG (natural logarithm of a mass action ratio less than 1 is a negative number). The same thing happens when a large amount of substrate enters into the cell—for example, after a carbohydrate-rich meal—in that $[\text{substrate}]_{\text{actual}} >> [\text{product}]_{\text{actual}}$ and the mass action ratio is again less than 1, giving rise to a more negative ΔG. Changes in substrate and product concentrations for reactions in a single metabolic pathway affect flux through other metabolic pathways because metabolites are often shared between interconnected pathways.

9.2 Structures of Simple Sugars

The word *glycolysis* is derived from the Greek *glykys*, meaning "sweet," and *lysis*, which means "to split or break." Glycolysis is basically the splitting of one molecule of glucose into two molecules of pyruvate. Thus, glucose is the first reactant in the glycolytic pathway. Non-photosynthetic organisms obtain glucose from their environment, whereas photosynthetic organisms obtain glucose using carbon fixation. To understand the enzymatic reactions in the glycolytic pathway, we need to review the structures of monosaccharide and disaccharide sugars.

Glucose has the molecular formula $C_6H_{12}O_6$ and is the most plentiful monosaccharide in nature. Glucose is a polyhydroxyaldehyde, whereas fructose, a monosaccharide with the same molecular formula as glucose, is a polyhydroxyketone. **Figure 9.6** shows the chemical structures of glucose and fructose. The figure includes

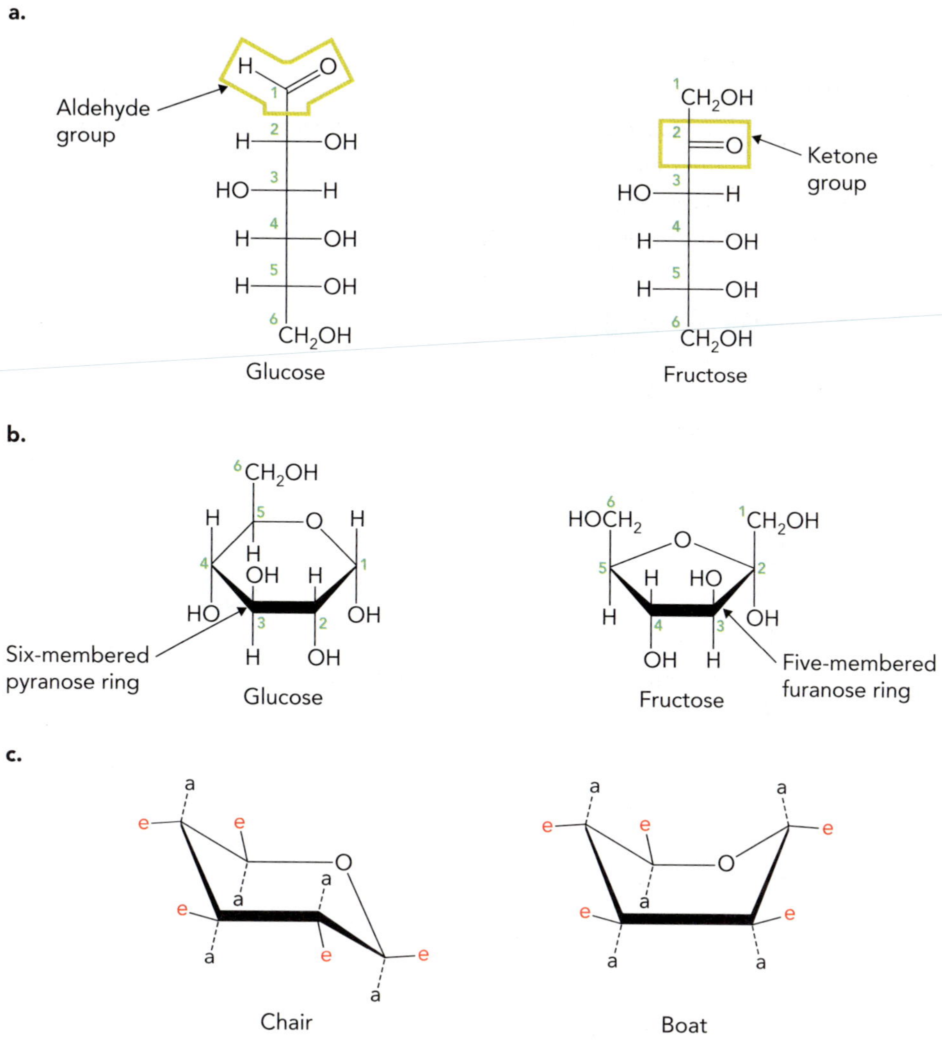

Figure 9.6 Glucose and fructose are both simple sugars with six carbon atoms. **a.** Glucose and fructose shown in linear conformations as Fischer projections. **b.** Glucose and fructose shown in cyclic conformations in Haworth perspective. **c.** Pyranose rings such as glucose are not actually planar, but rather take on one of two conformations referred to as the "chair" or "boat" conformation. Axial bonds (a) and equatorial bonds (e) are shown for each conformation.

both Fischer projections representing the linear forms of monosaccharides and Haworth perspectives illustrating the cyclic forms of monosaccharides. (We will see shortly that monosaccharides exist in an equilibrium between the linear and cyclic forms.) Glucose can also be drawn using a conformational formula in which glucose is represented by either the "chair" or "boat" conformations, reflecting the nonplanar structure of pyranose rings.

Simple sugars are sweet to the taste and are found naturally in many types of fruits and vegetables. They are also used commercially as additives to enhance the flavor of processed foods and beverages. The sensation of sweetness is the result of ligand activation of G protein–coupled receptor signaling in the taste cells of the tongue, leading to neuronal signaling from the tongue to the brain (see Figure 8.22). The G protein–coupled receptors in these cells bind sugars with differential affinities on the basis of their chemical structure, as demonstrated by human taste tests that measure the relative sweetness of various compounds (**Figure 9.7**). For example, fructose—which is found in high concentrations in many types of fruit and is added as a sweetener to processed foods—is perceived by the human tongue to be about three times sweeter than glucose and about two times sweeter than the disaccharide sucrose (table sugar). For comparison, Figure 9.7 also shows the structure of the artificial sweetener sucralose, which is a chlorinated sucrose molecule that is marketed under the brand name Splenda. Sucralose is currently the sweetest compound commercially available and is an amazing 600 times sweeter than sucrose because of its differential

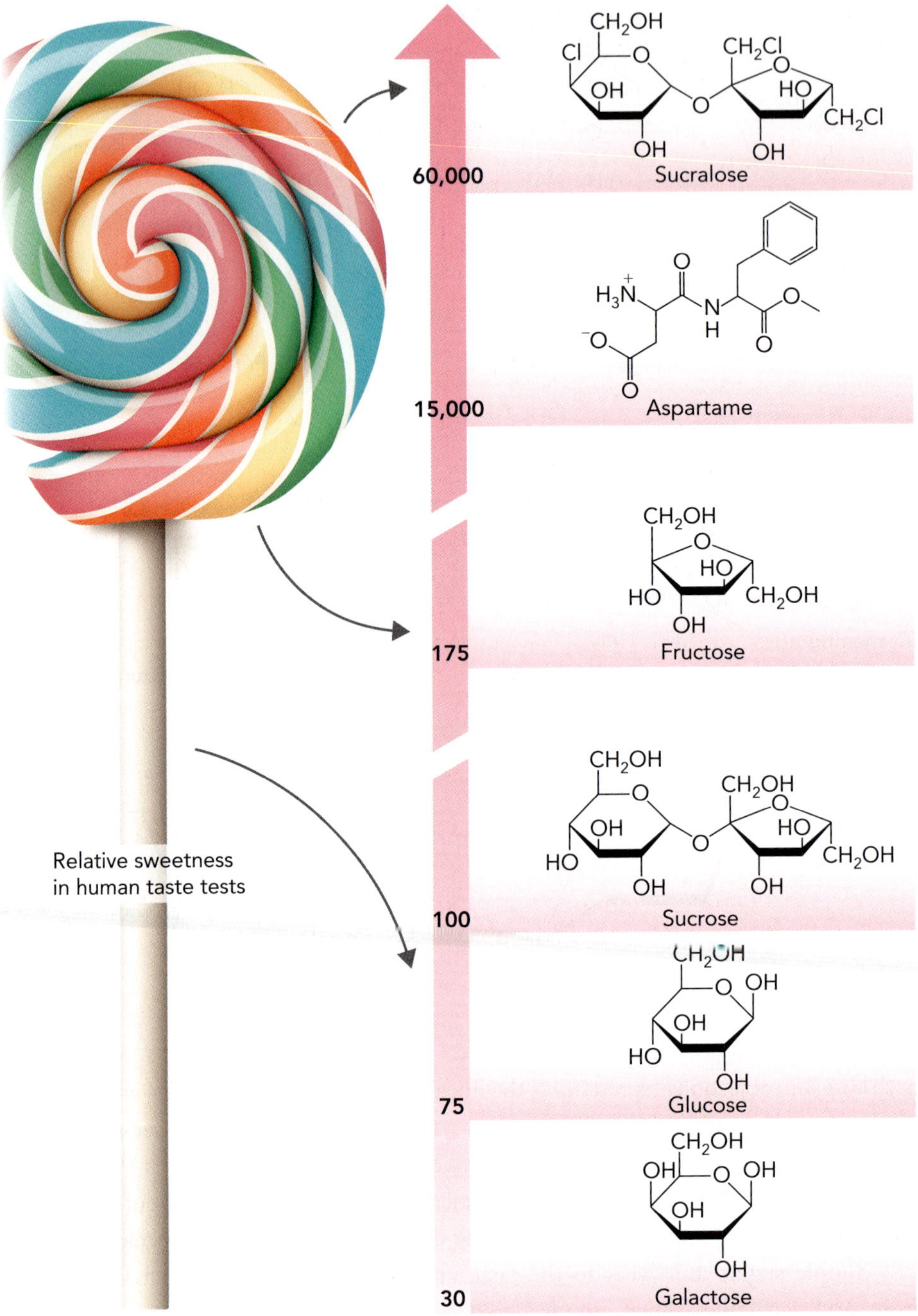

Figure 9.7 The relative sweetness of compounds can be determined from human taste tests. On the basis of a sweetness scale using arbitrary units, galactose has the lowest rating of "sweetness" in human taste tests, whereas sucralose (marketed as Splenda) is ~2,000 times sweeter. Because sucralose cannot be metabolized, it provides a sweet taste to foods without adding Calories (kilocalories). Aspartame (NutraSweet) can be metabolized, but because very little of it needs to be ingested to achieve a sweet taste sensation, the Calories it adds under normal circumstances are negligible.

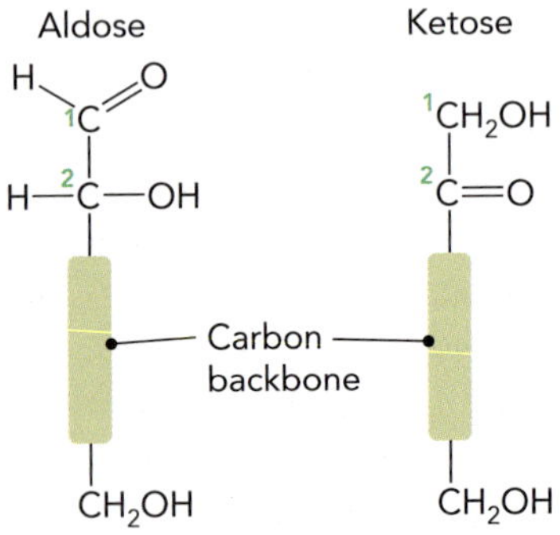

Figure 9.8 The generic structures of aldose and ketose sugars are shown here. When the carbonyl carbon (C=O) is the first carbon in the chain, it is an aldose sugar. Ketose sugars have the carbonyl carbon in the second position.

binding to taste receptor proteins on the tongue. Another artificial sweetener shown in Figure 9.7, aspartame, marketed as NutraSweet, is not a carbohydrate at all, but rather a dipeptide derivative of aspartate and phenylalanine.

Monosaccharides

Monosaccharides have either an aldehyde group at the end of the molecule, such as glucose, or a ketone group on the second carbon, as in fructose. Aldehyde-containing monosaccharides are called **aldoses**, and ketone-containing monosaccharides are called **ketoses**. All monosaccharides have a CH_2OH group on the other end of the carbon chain from the aldehyde or ketone group. Also, each of the carbons in the middle has an OH group and functions as a chiral center. **Figure 9.8** shows the generic structures of an aldose sugar and a ketose sugar, illustrating that the carbonyl carbon (C=O) is either at the end of the carbon chain (aldose) or at the second position (ketose). Monosaccharides can be categorized by the number of carbons in the chain; for example, an aldose with three carbons is a triose, with four carbons is a tetrose, with five carbons is a pentose, and with six carbons is a hexose.

The smallest monosaccharide is glyceraldehyde, a triose sugar with one chiral center. A carbon **chiral center** is an atom with four different functional groups. Chiral compounds lack a plane of symmetry and exist as two optical isomers, also called enantiomers. Enantiomers exist in nature as either right-handed (D form) or left-handed (L form) isomers and differentially rotate polarized light. The structures of D-glyceraldehyde and L-glyceraldehyde are shown in **Figure 9.9**, where you can see that these two isomers of glyceraldehyde are mirror images of each other. By convention, when the hydroxyl group in the chiral carbon is on the right side in a Fischer projection, it is the **D isomer**, and when it is on the left side, it is the **L isomer**.

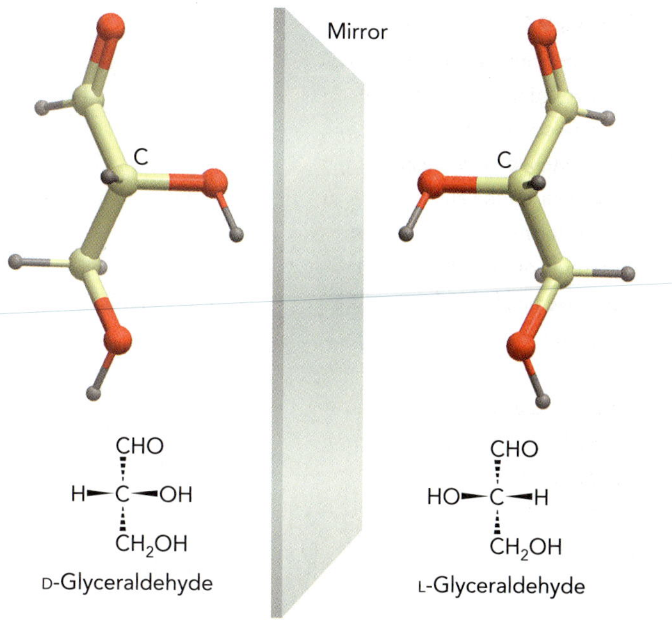

Figure 9.9 A chiral carbon (C) has four different functional groups, leading to mirror-image isomers (enantiomers). For example, D and L isomers of the triose sugar glyceraldehyde are mirror images of each other. The perspective projection uses solid wedges to indicate bonds that are coming out of the plane of the paper and hatched wedges to indicate bonds going behind the plane of the paper.

Most monosaccharides in living organisms are of the D conformation; however, L isomers do exist in nature, although they are usually conjugated to other molecules such as proteins or lipids. For monosaccharides that have multiple chiral centers, the D or L assignment refers to the chiral carbon farthest away from the carbonyl carbon. For example, the chiral carbons at C-5 in glucose and fructose as shown in Figure 9.6 are both in the D conformation, as seen in the Fischer projections.

Another structural feature of monosaccharides relates to the position of the hydroxyl group on the central carbon. Two monosaccharides that differ in the position of the hydroxyl group around only one carbon atom are called **epimers**. As shown in **Figure 9.10**, D-glucose and D-mannose are aldose epimers because of hydroxyl rotation around C-2, whereas D-ribose and D-xylose are aldose epimers because of hydroxyl rotation around C-3.

Monosaccharides of five, six, or seven carbons are often more stable in aqueous solution as cyclic structures than they are as open chains. Cyclic monosaccharides form spontaneously through a covalent linkage of the carbonyl carbon with a hydroxyl group in the carbon backbone. If this bond is the result of a reaction between an alcohol group and the aldehyde group of an aldose sugar, it forms a

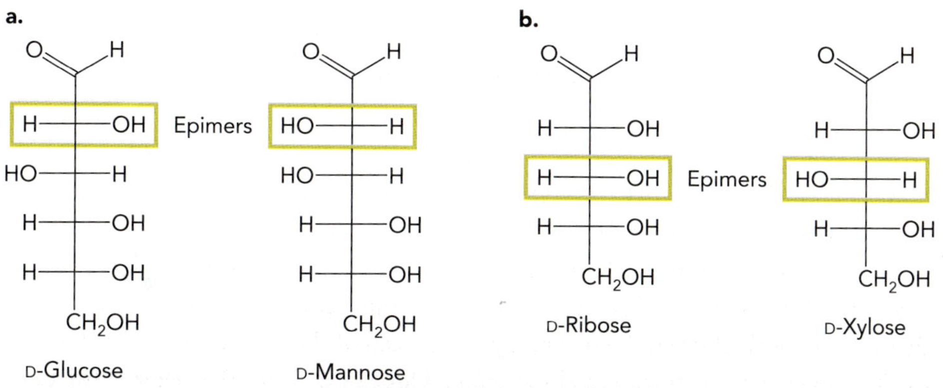

Figure 9.10 Epimers differ only by the position of the hydroxyl group around one of the carbon atoms. **a.** D-Glucose and D-mannose are C_6 monosaccharide epimers that differ only at C-2. **b.** D-Ribose and D-xylose are C_5 monosaccharide epimers that differ only at C-3.

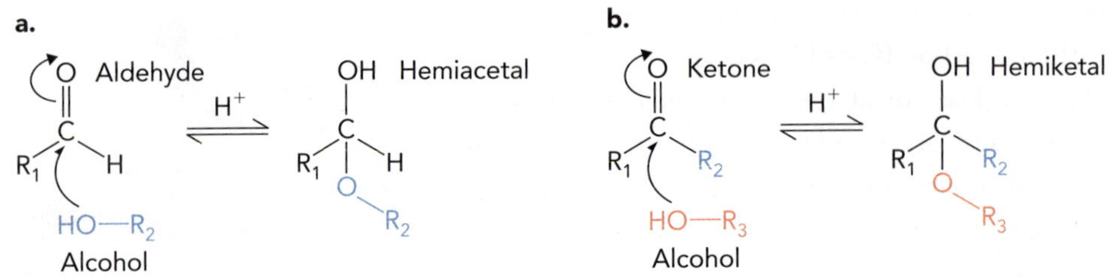

Figure 9.11 Formation of a hemiacetal and hemiketal. **a.** The reaction between an aldehyde and an alcohol forms a hemiacetal. **b.** The reaction between a ketone and an alcohol forms a hemiketal. The relevant R-groups and oxygen atoms are highlighted.

hemiacetal; a bond between an alcohol group and the ketone group of a ketose sugar forms a **hemiketal**. **Figure 9.11** shows the two general reactions that lead to the formation of a hemiacetal or a hemiketal. **Figure 9.12** illustrates the cyclization reaction that occurs when the C-5 hydroxyl group of D-glucose attacks the carbon atom of the C-1 aldehyde group to form a cyclic hemiacetal. In this conformation, C-1 of D-glucose becomes a new chiral center. Thus, cyclic forms of glucose exist either as β-D-glucose with the hydroxyl group at C-1 on the same side of the ring as CH_2OH or as α-D-glucose with the hydroxyl group on the opposite side of the ring as the CH_2OH. In aqueous solution, an equilibrium is established between the α-D-glucose and β-D-glucose conformations at a ratio of about 40:60 for α-D-glucose to β-D-glucose. A very small amount of the monosaccharide is found in the open-chain conformation (less than 0.05%).

The hemiacetal C-1 of cyclic D-glucose is called an **anomeric carbon**, and β-D-glucose and α-D-glucose are referred to as **anomers** because they differ only at the anomeric carbon. Cyclic conformations of hexose sugars are called **pyranoses** because the six-membered ring is similar to a pyran compound. Therefore, the two cyclic forms of glucose are sometimes called α-D-glucopyranose and β-D-glucopyranose (Figure 9.12). Ketoses such as fructose can also form cyclic structures, but because the carbonyl is in the C-2 position of the open chain, the ring that forms contains only five carbons. These sugars are called **furanoses** because they resemble a furan. Thus, cyclic conformations of fructose are called α-D-fructofuranose and β-D-fructofuranose. Note that because there is less strain in a six-membered ring than in a five-membered ring, pyranose rings are much more stable in solution than are furanose rings.

The normal concentration of glucose in human blood is ~3.5 to 5.5 mM (60–100 mg/dL), which is maintained by the hormones insulin and glucagon, as described

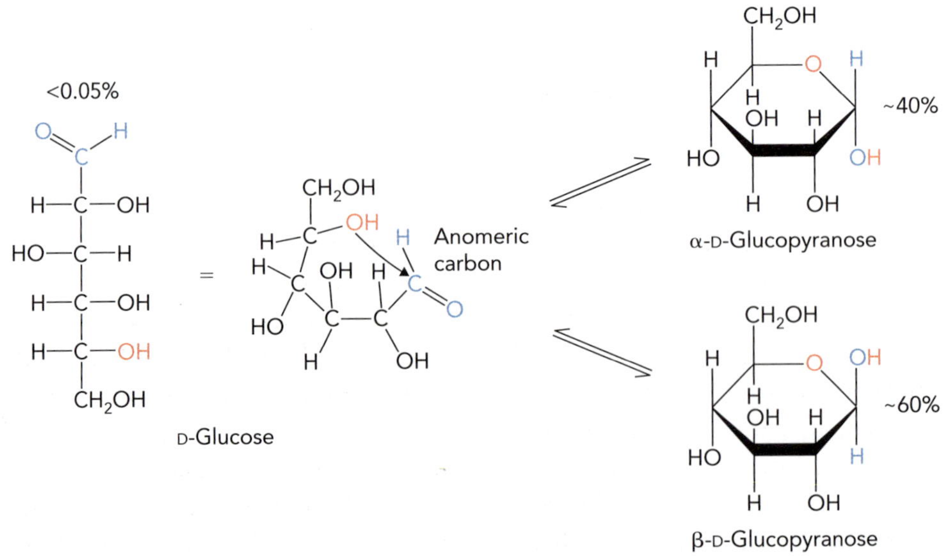

Figure 9.12 Cyclization reaction of linear glucose to form α-D-glucose and β-D-glucose. Cyclic hexoses resemble pyranose rings, giving rise to the names α-D-glucopyranose and β-D-glucopyranose.

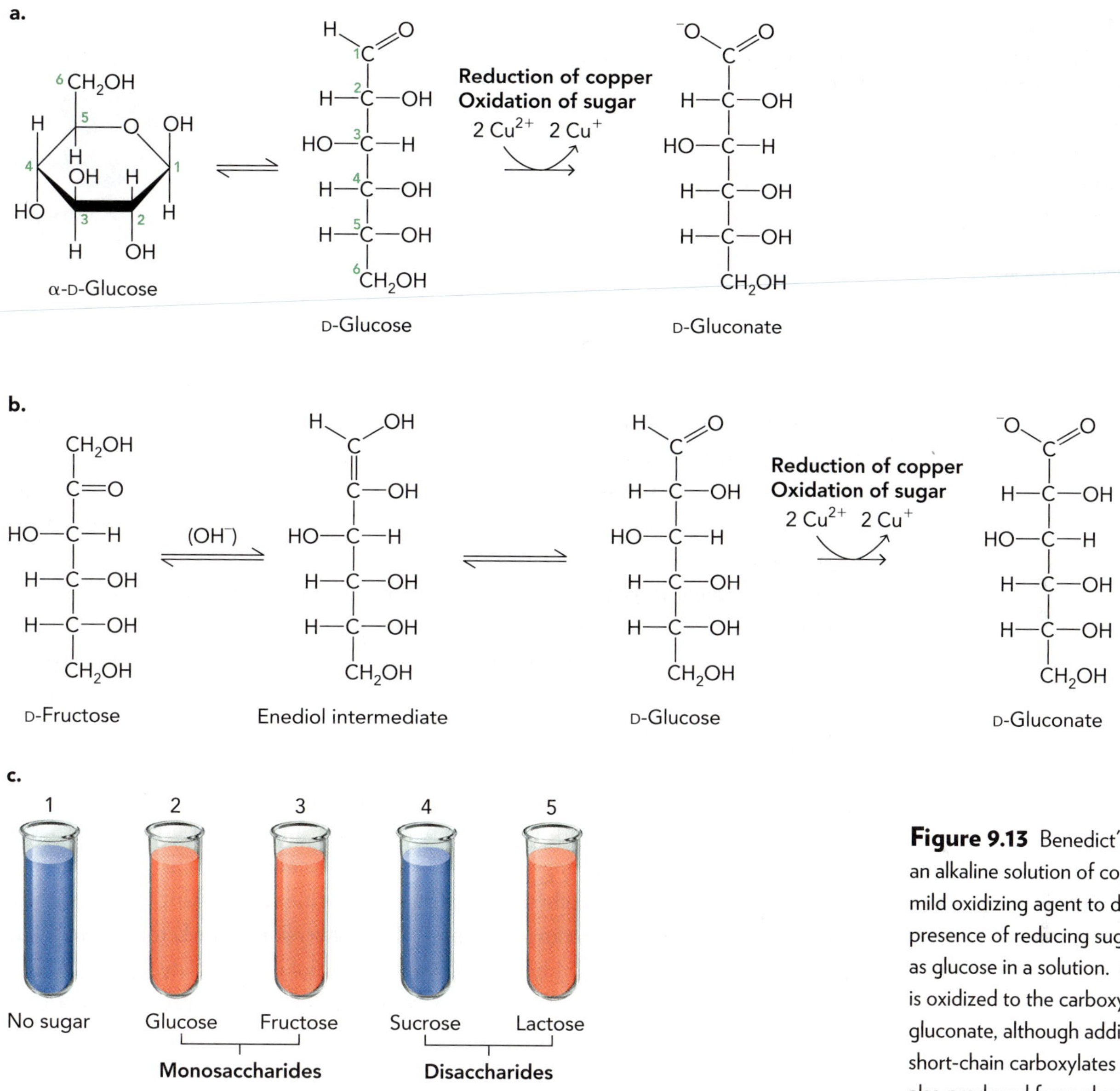

Figure 9.13 Benedict's test uses an alkaline solution of copper as a mild oxidizing agent to detect the presence of reducing sugars such as glucose in a solution. **a.** Glucose is oxidized to the carboxylic acid gluconate, although additional short-chain carboxylates are also produced from glucose in this reaction. The equilibrium between the cyclic and open-chain forms of glucose is pulled to the right as a result of glucose oxidation. **b.** Fructose, a ketose, gives a positive result in Benedict's test because the level of OH^- in the copper solution first converts fructose to glucose. **c.** Results of Benedict's test using five different solutions containing either (1) no sugar, (2) glucose, (3) fructose, (4) sucrose, or (5) lactose. The reduction of Cu^{2+} to Cu^+ changes the color of heated solutions with reducing sugars from blue to red, as a result of the formation of a copper oxide precipitate.

later in the chapter. Glucose (also called dextrose or blood glucose) levels are elevated in the blood of people with diabetes due to defects in insulin signaling (see Chapter 19). Glucose concentrations in the blood above 7.2 mM (130 mg/dL) are considered a diagnostic indicator of diabetes. Elevated blood glucose levels cause numerous symptoms, including dehydration, blurry vision, and fatigue. Fortunately, most patients with diabetes can control their blood glucose levels with insulin injections and diet, but they need to monitor glucose concentrations in the blood carefully to do this.

How are blood glucose levels measured? One way to do this is by taking advantage of the ability of aldose sugars in the open-chain conformation to be oxidized to carboxylic acids in a redox reaction with copper ($Cu^{2+} \rightarrow Cu^+$). As shown in **Figure 9.13**, the level of glucose in blood or urine (excess glucose in the urine is an indication of high blood glucose levels) can be seen in the test solution as a color change from blue (Cu^{2+}) to red (Cu^+). This reaction is called **Benedict's test**, and it qualitatively measures the amount of glucose in a solution on the basis of the amount of red copper oxide precipitate. Note that even though only the open-chain form of glucose reduces

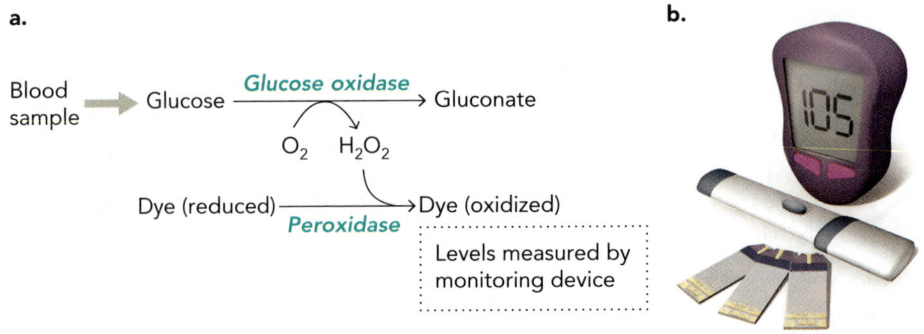

Figure 9.14 The glucose oxidase/peroxidase enzymatic reaction measures blood glucose levels more accurately than Benedict's test because it is highly specific for glucose and is much more sensitive. **a.** This assay is based on measuring the amount of oxidized dye in the reaction sample as a function of H_2O_2 levels produced by glucose oxidase. **b.** Personal glucose-monitoring devices based on the oxidase/peroxidase reaction are now commonly used by diabetic patients.

Cu^{2+} in this redox reaction, the oxidation of the aldehyde (D-glucose) to the carboxylic acid (D-gluconate) shifts more of the cyclic form of D-glucose to the open-chain form, so that the reaction eventually goes to completion.

Carbohydrates that react with oxidizing agents such as Cu^{2+} are called **reducing sugars** (glucose, galactose, and lactose), whereas sugars that cannot reduce Cu^{2+} are called **nonreducing sugars** (sucrose and trehalose). Although fructose is a ketose sugar and should not react with Cu^{2+} in Benedict's test, the assay includes sufficient amounts of base (OH^-) to convert some of the fructose to glucose, and thereby gives a positive result. The nonreducing sugars sucrose and trehalose are disaccharides that cannot be converted to an open-chain conformation and are therefore not substrates for the $Cu^{2+} \rightarrow Cu^{+}$ redox reaction.

The qualitative color change of Cu^{2+} reduction by glucose is not very accurate, nor is this reaction sensitive enough to detect small changes in blood or urine glucose levels. What's more, other reducing sugars besides glucose can reduce Cu^{2+}; for example, galactose and lactose (a disaccharide present in the urine of pregnant women). Even ketose sugars such as fructose can be oxidized, after conversion of the ketone in the alkaline Benedict's solution to an aldehyde through an enediol intermediate (Figure 9.13).

To solve these problems, biochemists have developed an improved blood glucose test that relies on the specificity and sensitivity of an enzymatic reaction. As shown in **Figure 9.14**, this enzyme-based test uses glucose oxidase to first convert glucose to gluconate and hydrogen peroxide (H_2O_2). The enzyme peroxidase then catalyzes a redox reaction in the presence of H_2O_2 that changes a colorless reduced dye into a colored oxidized product. The level of oxidized dye in the sample is measured using a blood glucose–monitoring instrument, which is now available for diabetic patients to use conveniently at home (Figure 9.14). The development of a home-based glucose-monitoring device for diabetic patients is another great example of how enzymes have been put to work as biological catalysts to improve the sensitivity and specificity of diagnostic tests.

Disaccharides

Disaccharide sugars are formed by a condensation reaction between two monosaccharides. The covalent linkage is called an **O-glycosidic bond** and represents the formation of an acetal from a hemiacetal and an alcohol. As shown in **Figure 9.15**, the disaccharide maltose is a degradation product of starch, a glucose polymer present in plants. Maltose is produced by the digestive enzyme α-amylase and consists of two glucose molecules linked together by a glycosidic bond between the anomeric C-1 (hemiacetal)

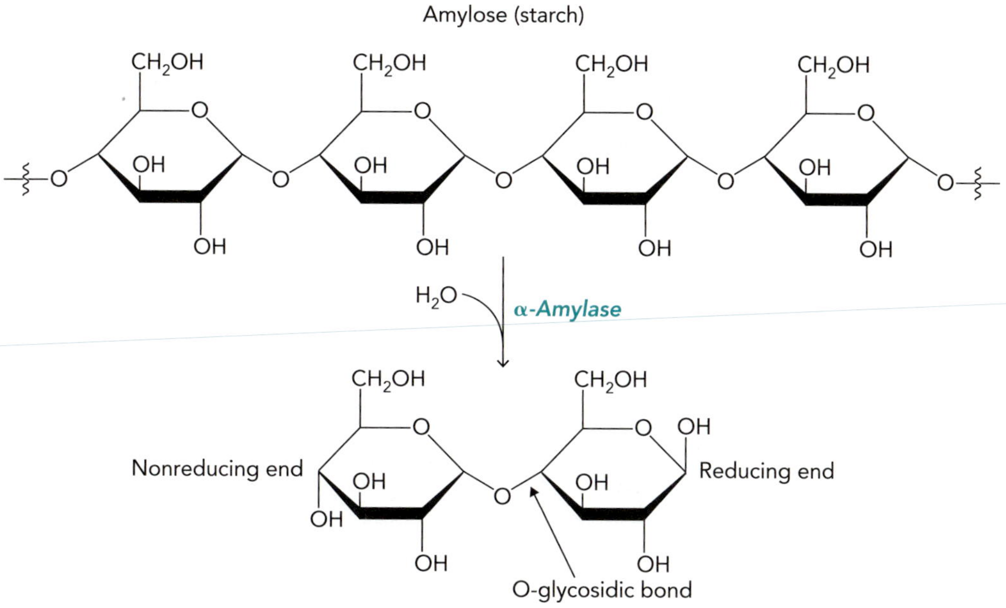

Maltose
Glc(α1→4)Glc

Figure 9.15 Maltose, a disaccharide derived from starch, consists of two glucose (Glc) units covalently attached by an α–1,4 linkage at the O-glycosidic bond. The Haworth perspective of the maltose structure shows the α–1,4 linkage between the two glucose units. The anomeric C-1 of the glucose unit on the right can be in either the α or β configuration (it is shown here in the β configuration). It functions as the reducing end of the molecule because it can convert to the open-chain conformation and participate in the $Cu^{2+} \rightarrow Cu^{+}$ redox reaction. In contrast, the glucose unit on the left represents the nonreducing end because C-1 is constrained within the glycosidic bond.

of one glucose and the C-4 (alcohol) of a second glucose. Maltose can be hydrolyzed by weak acid to release the two glucose units.

The glycosidic bond in maltose is called an α-1,4 linkage because the anomeric carbon is in the α conformation. The glucose molecule on the right retains the hemiacetal structure at its anomeric C-1 and can convert to the aldehyde open-chain form and reduce Cu^{2+}, thus designating this glucose as the reducing end of the disaccharide molecule. In contrast, the glucose on the left represents the nonreducing end because C-1 is part of the α-1,4 linkage and cannot form the open-chain structure needed to participate in the redox reaction. Because maltose contains one reducing end, it is called a reducing sugar.

Disaccharides can contain different monosaccharide units connected through α or β glycosidic bonds involving ring carbons. Therefore, it is convenient to name disaccharides using a descriptive nomenclature. Using standard conventions, the disaccharide is named by first listing the nonreducing monosaccharide on the left, followed by the glycosidic linkage between the two monosaccharides, and then the monosaccharide on the right. (If the linkage is between two anomeric carbons, one sugar is arbitrarily listed first, and a double-headed arrow is used). With this shorthand nomenclature, we can describe maltose as a Glc(α1→4)Glc disaccharide, in which the abbreviation "Glc" is used for glucose.

Figure 9.16 shows the structures of three common disaccharides found in nature:

1. *Lactose*, also called milk sugar, contains a β(1→4) glycosidic bond linking a galactose (Gal) to a glucose, forming Gal(β1→4)Glc;
2. *Sucrose*, made in plants and used as table sugar in its crystalline form, contains fructose (Fru) linked to glucose through the two anomeric carbons, forming Glc(α1↔β2)Fru; and
3. *Trehalose*, a glucose disaccharide made in insects, contains a glycosidic bond between the two anomeric carbons to form the disaccharide Glc(α1↔α1)Glc.

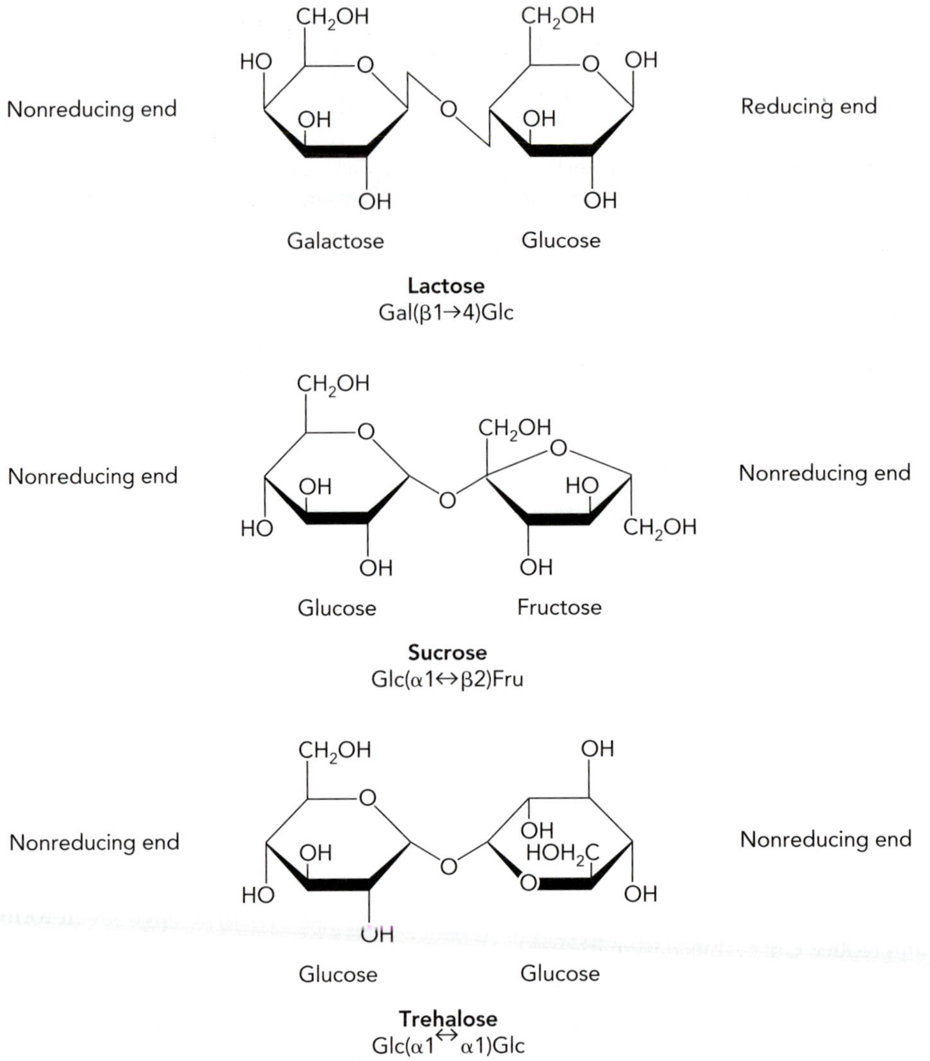

Figure 9.16 Lactose contains an anomeric carbon at C-1 that can form an aldehyde group in the open-chain conformation and function as a reductant, making it a reducing sugar. In contrast, sucrose and trehalose are nonreducing sugars because both C-1 anomeric carbons of the monosaccharide units are covalently linked by the glycosidic bond.

Of these three disaccharides, only lactose is a reducing sugar, because like maltose, it contains a free anomeric carbon that can interconvert the hemiacetal to an aldehyde (see Figure 9.13). Both sucrose and trehalose are nonreducing sugars because they lack a reducing end.

concept integration 9.2

Do disaccharides contain a reducing end and interconvert between the open and ring conformations like most monosaccharides?

Disaccharides can have either one or zero reducing ends. Lactose is an example of a disaccharide with one reducing end. Sucrose and trehalose are examples of disaccharides without reducing ends, as the anomeric carbons of both monomers are linked in a glycosidic bond. Opening of the ring form of a sugar requires the participation of the anomeric carbon. Thus, because of the linkage of the anomeric carbons of the monomers, disaccharides such as sucrose and trehalose are stabilized in the ring conformation and do not convert to the open-chain form.

9.3 Glycolysis Generates ATP under Anaerobic Conditions

As with all the metabolic pathways we will examine, we first need to see where glycolysis fits into our metabolic map. **Figure 9.17** shows glycolysis at the top of a set of interconnected pathways that include the citrate cycle and oxidative phosphorylation. Together with glycolysis, these pathways are responsible for the complete oxidation of glucose to CO_2 and H_2O by the following reaction:

$$\text{Glucose } (C_6H_{12}O_6) + 6\ O_2 \rightleftharpoons 6\ CO_2 + 6\ H_2O \qquad \Delta G^{\circ\prime} = -2{,}840 \text{ kJ/mol}$$

$$\Delta G = -2{,}938 \text{ kJ/mol}$$

Glycolysis, or the glycolytic pathway, is considered one of the core metabolic pathways in nature for three primary reasons:

1. Glycolytic enzymes are highly conserved among all living organisms, suggesting it is an ancient pathway.
2. Glycolysis is the primary pathway for ATP generation under anaerobic conditions and in cells lacking mitochondria, such as erythrocytes.
3. Metabolites of glycolysis are precursors for a large number of interdependent pathways, including mitochondrial ATP synthesis.

Glycolysis consists of 10 enzymatic reactions, which were elucidated in the early 1900s by chemists who were studying fermentation in brewer's yeast. This work followed the pioneering experiments of Eduard Buchner, who showed in the late 1890s that yeast cell extracts contained everything required for fermentation in a test tube (see Section 1.1). Today, we know the molecular structures of all intermediates in the glycolytic pathway, along with detailed structures of all 10 enzymes. In fact, mechanisms for most of the enzymatic reactions have been worked out, a few of which we will examine here.

Let's begin by answering the four questions about metabolic pathways as they pertain to glycolysis.

1. *What does glycolysis accomplish for the cell?* Glycolysis generates a small amount of ATP, which is critical under anaerobic conditions, and generates pyruvate, the precursor to acetyl-CoA, lactate, and ethanol.
2. *What is the overall net reaction of glycolysis?*

$$\text{Glucose} + 2\ NAD^+ + 2\ ADP + 2\ P_i \rightleftharpoons$$
$$2\ \text{Pyruvate} + 2\ NADH + 2\ H^+ + 2\ ATP + 2\ H_2O$$

3. *What are the key enzymes in glycolysis?* **Hexokinase** catalyzes the first step in glycolysis. The activity of hexokinase is feedback inhibited by glucose-6-P.

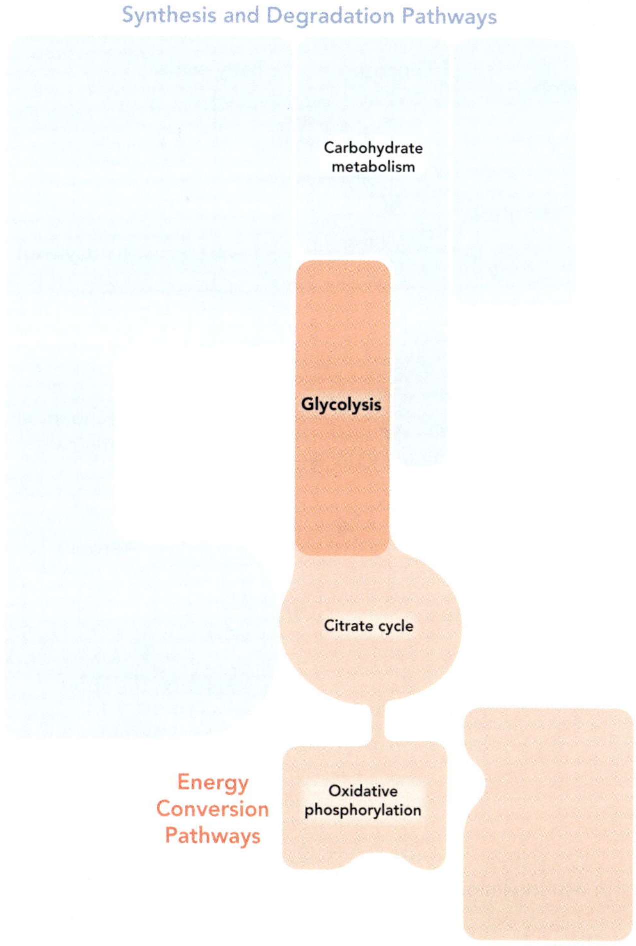

Figure 9.17 Glycolysis is the entry pathway for glucose catabolism. Although it is an anaerobic process, it generates metabolites for aerobic energy conversion by the citrate cycle and oxidative phosphorylation.

Figure 9.18 Glycolysis is an anaerobic pathway that occurs in the cytosol and produces 2 ATP for every 1 glucose molecule that is converted to 2 pyruvate molecules. In contrast, the oxidation of pyruvate to CO_2 and H_2O occurs in the mitochondrial matrix and depends on aerobic conditions because O_2 is the final electron acceptor in the electron transport chain. Note that acetyl-CoA derived from fatty acid oxidation is another major chemical energy source for ATP production in mitochondria.

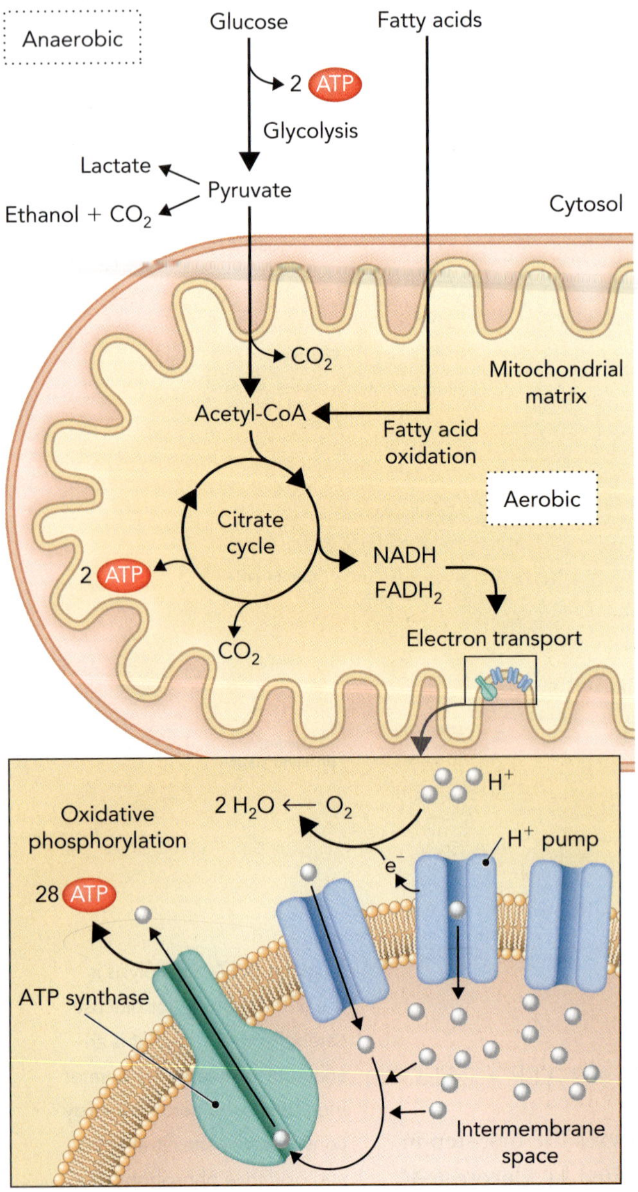

Phosphofructokinase-1, a highly regulated enzyme, is allosterically activated by AMP (low energy charge) and fructose-2,6-bisphosphate (produced when fructose-6-P is abundant). It is inhibited by ATP (high energy charge) and citrate (citrate cycle intermediate). **Pyruvate kinase** catalyzes the final reaction in glycolysis. It is activated by AMP and fructose-1,6-bisphosphate (feed forward) and is inhibited by ATP and acetyl-CoA (excess energy source).

4. *What is an example of this pathway in everyday biochemistry?* Glycolysis helps maintain appropriate blood glucose levels. A deficiency in the hexokinase-related enzyme glucokinase leads to a rare form of diabetes called maturity-onset diabetes of the young (MODY2). This disease is caused by the inability of liver and pancreatic cells to phosphorylate glucose and trap it inside cells. The result is abnormally high blood glucose levels.

The Glycolytic Pathway Consists of 10 Enzymatic Reactions

The conversion of one molecule of glucose to two molecules of pyruvate by the glycolytic pathway occurs in the cytosol and generates two molecules of ATP. As shown in **Figure 9.18**, pyruvate is transported into the mitochondrial matrix, where it is oxidized to CO_2 and H_2O under aerobic conditions, contributing to the generation of an additional 30 molecules of ATP. Thus, a total of 32 molecules of ATP can be derived from oxidation of one glucose molecule.

Because the conversion of glucose to pyruvate by glycolysis in the cytosol does not require oxygen, this metabolic pathway can operate under anaerobic conditions, which has advantages and disadvantages. The advantage is that a small amount of chemical energy in the form of 2 ATP can be derived from glucose metabolism under conditions when O_2 is limiting. For example, glycolysis can provide ATP to sustain muscle contraction during intense exercise. However, a yield of 2 ATP from glycolysis alone under anaerobic conditions, compared to a total of 32 ATP that can be generated from the complete oxidation of glucose under aerobic conditions, provides a yield of only 6% (2/32). Therefore, it is really the oxidation of pyruvate (and acetyl-CoA) in the mitochondria that generates the majority of ATP for most cells in an organism. In fact, as we will see in Chapter 16, in many animals stored fats in the form of fatty acids are actually the major source of chemical energy, which is made available through fatty acid oxidation and acetyl-CoA metabolism in the mitochondria (Figure 9.18).

Figure 9.19 shows the molecular structures of glucose and pyruvate. The six carbons and six oxygens present in glucose are stoichiometrically conserved by glycolysis in the two molecules of pyruvate that are produced. The 10 enzymatic reactions of glycolysis involve primarily bond rearrangements brought about by enzymes that catalyze phosphoryl transfer reactions, isomerizations, an aldol cleavage, an oxidation, and a dehydration. No net loss of carbon or oxygen atoms occurs. Because of the requirement for ATP hydrolysis in the initial reactions, followed by ATP synthesis in the later reactions, glycolysis is divided into two stages:

Stage 1: ATP investment (reactions 1–5)

Stage 2: ATP earnings (reactions 6–10)

Before we examine the biochemistry of each enzymatic reaction and look at the chemical structures of the substrates and products, let's take a quick survey of the glycolytic pathway to introduce the enzymes, metabolites, and ATP-dependent steps.

The input of chemical energy in the form of ATP hydrolysis in stage 1 of glycolysis is required to generate phosphorylated compounds that are metabolized in stage 2 (**Figure 9.20**). The enzyme hexokinase, which is present in all human cells, phosphorylates glucose using ATP to form glucose-6-phosphate (glucose-6-P). (The related enzyme glucokinase, which is expressed in liver and pancreatic cells, catalyzes this same reaction.) Glucose-6-P is isomerized by phosphoglucoisomerase in reaction 2 to form fructose-6-phosphate (fructose-6-P). In turn, fructose-6-P is a substrate for the phosphoryl transfer reaction catalyzed by phosphofructokinase-1 in reaction 3 to generate fructose-1,6-bisphosphate (fructose-1,6-BP). In reaction 4, the enzyme aldolase catalyzes an aldol cleavage reaction that splits the six-carbon fructose-1,6-BP intermediate into two phosphorylated products: glyceraldehyde-3-phosphate (glyceraldehyde-3-P) and dihydroxyacetone phosphate (dihydroxyacetone-P, or DHAP), each with three carbons. In the last reaction of stage 1, dihydroxyacetone-P is isomerized by the enzyme triose phosphate isomerase, a TIM barrel protein (see Figure 4.50), to form a second molecule of glyceraldehyde-3-P.

Stage 2 of glycolysis (Figure 9.20) begins with an oxidation reaction (reaction 6) catalyzed by the enzyme glyceraldehyde-3-phosphate dehydrogenase (glyceraldehyde-3-P dehydrogenase, or GAPDH). Reaction 6 utilizes the coenzyme NAD^+ to remove two electrons from glyceraldehyde-3-P, leading to the formation of 1,3-bisphosphoglycerate. Reaction 7 is catalyzed by the enzyme phosphoglycerate kinase, which converts 1,3-bisphosphoglycerate to 3-phosphoglycerate and represents the first of two ATP generating steps in glycolysis. This reaction is an example of **substrate-level phosphorylation**, in which ATP is produced directly by transfer of a phosphoryl group to ADP from a donor with high phosphoryl transfer energy. ATP synthesis by substrate-level phosphorylation is mechanistically distinct from oxidative phosphorylation (see Chapter 11) and photophosphorylation (see Chapter 12), the two major ATP synthesizing reactions in cells that require the enzyme ATP synthase. Because two pyruvate molecules are produced for every glucose, the phosphoglycerate kinase step provides the ATP payback in stage 2, which replaces the 2 ATP molecules invested in stage 1. In reaction 8 of glycolysis, phosphoglycerate mutase isomerizes 3-phosphoglycerate to form 2-phosphoglycerate, the substrate for a dehydration reaction catalyzed by enolase in reaction 9 to generate phosphoenolpyruvate. In the final reaction of glycolysis, a second substrate-level phosphorylation takes place when the enzyme pyruvate kinase converts phosphoenolpyruvate to pyruvate. In the process, this step provides the ATP earnings in glycolysis by the formation of two net ATP molecules.

One way to understand how a pathway is regulated is to examine the free energy changes ($\Delta G^{\circ\prime}$ and ΔG) that take place in each reaction. This helps to identify steps that are critical in driving the overall pathway toward product formation. The free energy

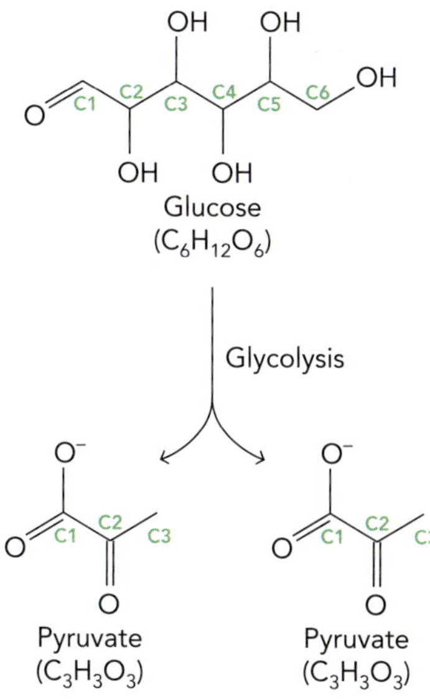

Figure 9.19 The glycolytic pathway cleaves glucose ($C_6H_{12}O_6$) into two molecules of pyruvate ($C_3H_3O_3$). The open conformation of glucose is shown here to emphasize the relationship between the carbon backbones of glucose and pyruvate.

▶ ANIMATION

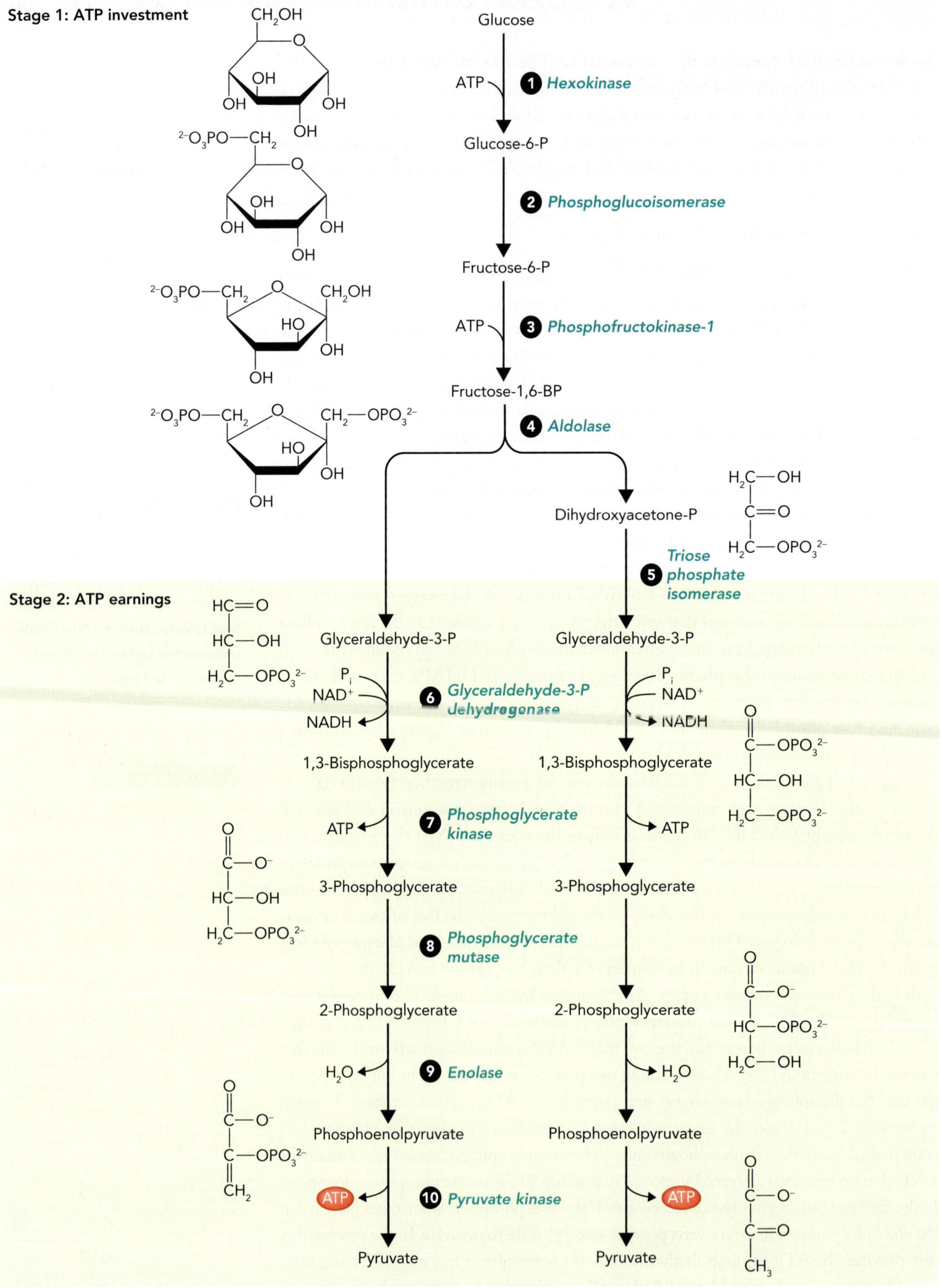

Figure 9.20 The glycolytic pathway consists of 10 enzymatic reactions that can be divided into two stages: ATP investment and ATP earnings. The 10th reaction in the glycolytic pathway is catalyzed by the enzyme pyruvate kinase and serves to generate the net 2 ATP derived from the metabolism of one molecule of glucose.

Table 9.2 LIST OF KNOWN $\Delta G°'$ VALUES FOR THE 10 GLYCOLYTIC REACTIONS AND THE ESTIMATED ΔG VALUES

	Reaction	Enzyme	$\Delta G°'$ (kJ/mol)	ΔG (kJ/mol)[a]
1	Glucose + **ATP** → Glucose-6-phosphate + ADP	Hexokinase	−16.7	−33.9
2	Glucose-6-phosphate → Fructose-6-phosphate	Phosphoglucoisomerase	+1.7	−2.9
3	Fructose-6-phosphate + **ATP** → Fructose-1,6-bisphosphate + ADP	Phosphofructokinase-1	−14.2	−18.8
4	Fructose-1,6-bisphosphate → Dihydroxyacetone phosphate + Glyceraldehyde-3-phosphate	Aldolase	+23.8	−0.4
5	Dihydroxyacetone phosphate → Glyceraldehyde-3-phosphate	Triose phosphate isomerase	+7.5	+2.4
6	Glyceraldehyde-3-phosphate + P_i + NAD^+ → 1,3-Bisphosphoglycerate + **NADH** + H^+	Glyceraldehyde-3-phosphate dehydrogenase	+6.3	−1.3
7	1,3-Bisphosphoglycerate + ADP → 3-Phosphoglycerate + **ATP**	Phosphoglycerate kinase	−18.8	−3.4
8	3-Phosphoglycerate → 2-Phosphoglycerate	Phosphoglycerate mutase	+4.6	+1.1
9	2-Phosphoglycerate → Phosphoenolpyruvate + H_2O	Enolase	+1.7	−1.0
10	Phosphoenolpyruvate + ADP → Pyruvate + **ATP**	Pyruvate kinase	−31.4	−26.8

[a]The ΔG values were calculated using Equation 2.14 and the metabolite concentrations given in Table 9.3 for erythrocytes under steady-state conditions at 37 °C (310 K).

changes for the 10 glycolytic reactions are listed in **Table 9.2**, including both the $\Delta G°'$ of each reaction, which is measured in the laboratory under standard conditions, and the calculated ΔG value, obtained using Equation 2.14 ($\Delta G = \Delta G°' + RT \ln Q$) and metabolite concentrations in erythrocytes under steady-state conditions (**Table 9.3**). You can see in Table 9.2 that the reactions catalyzed by the enzymes hexokinase, phosphofructokinase-1, and pyruvate kinase have large negative ΔG values and can therefore be considered irreversible reactions under physiologic conditions. As shown in **Figure 9.21**, the other reactions have small free energy changes, which means they are near equilibrium and can be reversed depending on metabolite concentrations.

Stage 1 of the Glycolytic Pathway: ATP Investment

Stage 1 of glycolysis includes five enzymatic reactions, which accomplish two tasks. First, using ATP as the phosphate donor, stage 1 reactions create phosphorylated compounds that are negatively charged and cannot diffuse out of the cell. These phosphorylated metabolites are specific substrates for glycolytic enzymes and are the precursors to the high-energy compounds 1,3-bisphosphoglycerate and phosphoenolpyruvate, which are used in stage 2 to generate ATP by

Table 9.3 CONCENTRATIONS OF GLYCOLYTIC METABOLITES IN ERYTHROCYTES UNDER STEADY-STATE CONDITIONS AT 37 °C

Metabolite	Concentration (mM)
Glucose	5
Glucose-6-phosphate	0.083
Fructose-6-phosphate	0.014
Fructose-1,6-bisphosphate	0.031
Dihydroxyacetone phosphate	0.138
Glyceraldehyde-3-phosphate	0.019
1,3-Bisphosphoglycerate	0.004
3-Phosphoglycerate	0.118
2-Phosphoglycerate	0.030
Phosphoenolpyruvate	0.023
Pyruvate	0.051
ATP	1.85
ADP	0.138
P_i	1

BASED ON DATA FROM S. MINAKAMI AND H. YOSHIKAWA (1965). THERMODYNAMIC CONSIDERATIONS ON ERYTHROCYTE GLYCOLYSIS. *BIOCHEMICAL AND BIOPHYSICAL RESEARCH COMMUNICATIONS, 18*, 345–349.

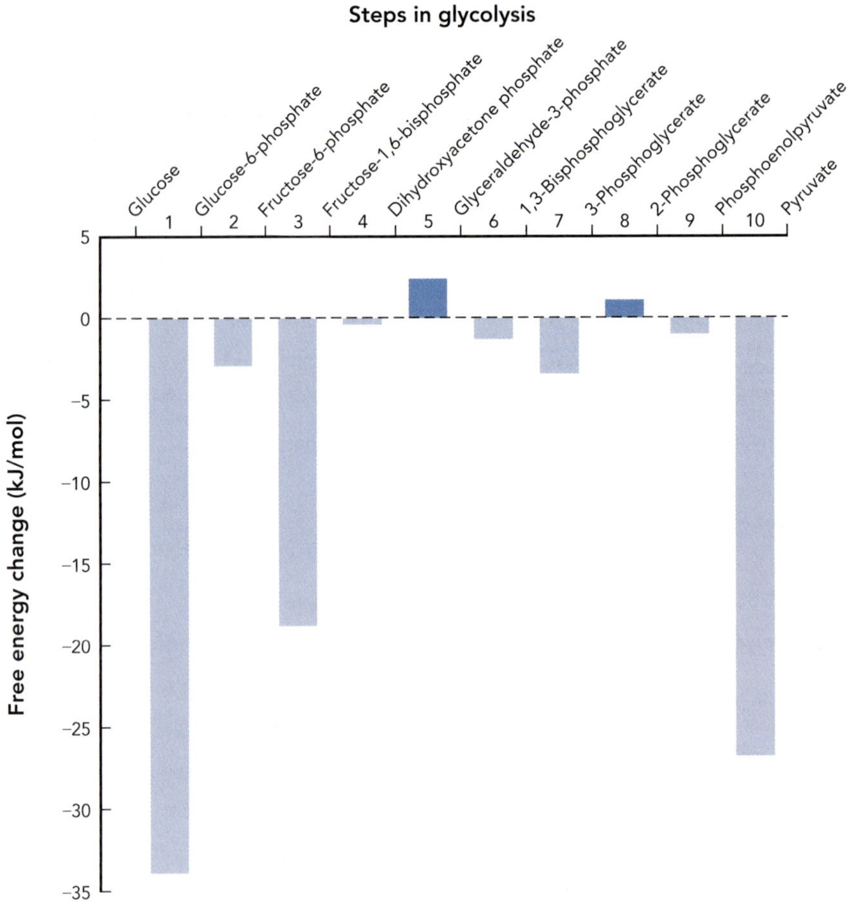

Figure 9.21 Free energy changes in the glycolytic pathway. The major reactants and products for each reaction are shown, along with the calculated free energy changes for each reaction (numbered 1–10). Three reactions (1, 3, and 10) have large negative free energy changes (which appear as large "steps" in this plot) and are considered irreversible under cellular conditions.

substrate-level phosphorylation. Second, the aldolase reaction in step 4 (reaction 4) splits the six-carbon fructose-1,6-BP compound into two halves, creating glyceraldehyde-3-P and dihydroxyacetone-P. The dihydroxyacetone-P is isomerized to form a second molecule of glyceraldehyde-3-P. We now look more closely at each step in the pathway to examine the chemistry of the reaction and what is known about the enzymes involved.

Reaction 1: Phosphorylation of Glucose by Hexokinase or Glucokinase The first reaction in glycolysis (**Figure 9.22**) serves to activate glucose for catabolism by attaching a phosphoryl group to the hydroxyl group on C-6, generating glucose-6-P. This is the first of two ATP investment steps in stage 1 of glycolysis, using the free energy released from ATP hydrolysis to drive the phosphoryl transfer reaction. As with all kinase reactions, the ATP substrate for the reaction is in a complex with the divalent cation Mg^{2+}, which functions to shield negative charges in the ATP molecule. In fact, it is the ATP–Mg^{2+} complex, not ATP alone, that is the high-affinity substrate for most kinase reactions.

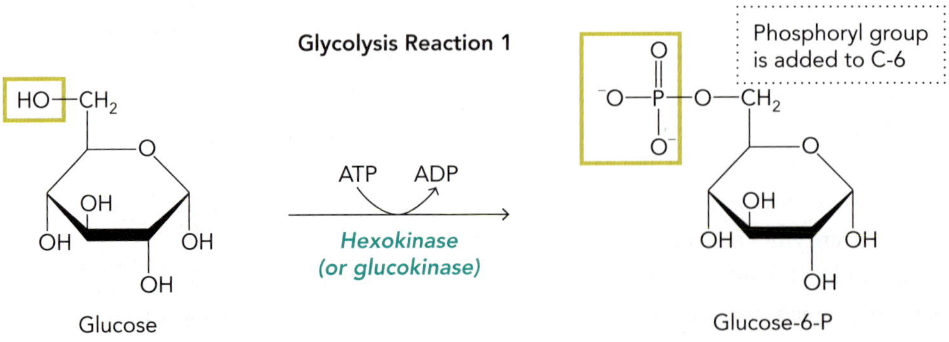

Figure 9.22 Glycolytic pathway reaction 1 is catalyzed by the enzymes hexokinase or glucokinase, generating glucose-6-P in a coupled reaction involving ATP hydrolysis.

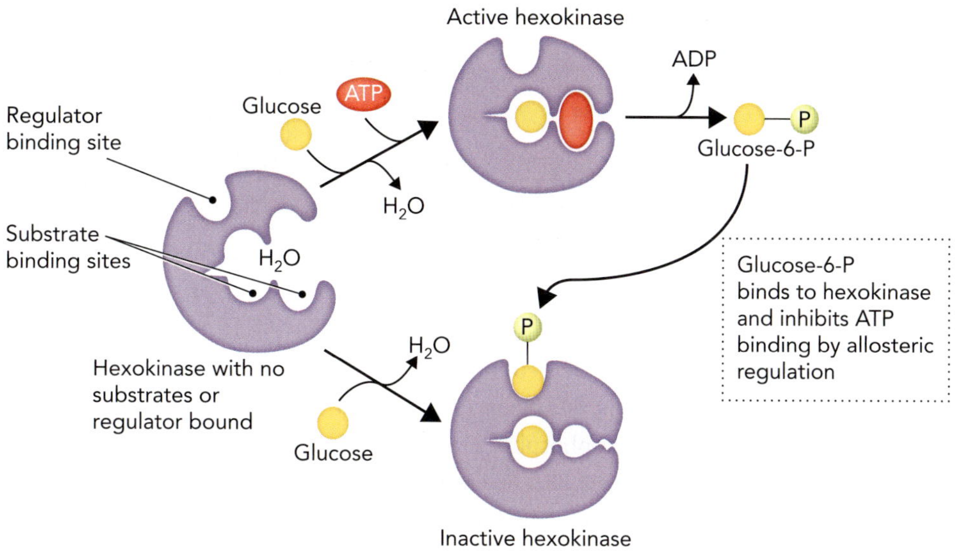

Figure 9.23 Hexokinase binds the substrates glucose and ATP by using an induced-fit mechanism, which excludes H_2O from the active site and facilitates phosphoryl transfer from ATP to glucose. Glucose-6-P is an inhibitor of hexokinase activity that binds to a regulatory site on the enzyme; it prevents phosphoryl transfer without altering glucose binding.

Two enzymes catalyze this phosphorylation reaction: hexokinase, which is found in all cells, and glucokinase, which is present only in liver and pancreatic cells. Hexokinase has a broad range of substrate specificities and also phosphorylates mannose and fructose. In contrast, glucokinase is highly specific for glucose. Hexokinase activity is inhibited by the product of the reaction, glucose-6-P, which accumulates in cells when flux through the glycolytic pathway is restricted. As described later in this chapter, glucokinase has a much lower affinity for glucose and is not feedback inhibited by glucose-6-P. These enzymatic properties of glucokinase, along with its selective expression in specific cell types, facilitate its function as a metabolic "sensor" of high blood glucose levels.

Hexokinase binds glucose through an induced-fit mechanism that excludes H_2O from the enzyme active site (refer to Figure 7.12) and brings the phosphoryl group of ATP into close proximity with C-6 of glucose. The molecular structures of yeast hexokinase in the presence and absence of glucose suggest that two domains of the enzyme act like jaws that clamp down on the substrate through a large conformational change (see Figure 7.3). **Figure 9.23** illustrates the induced-fit mechanism of hexokinase and also shows that glucose-6-P inhibition of hexokinase activity is mediated by binding of glucose-6-P to an allosteric effector site in the N-terminal domain of the protein.

Reaction 2: Isomerization of Glucose-6-P to Fructose-6-P by Phosphoglucoisomerase Phosphoglucoisomerase (also called phosphoglucose isomerase or phosphohexose isomerase) interconverts an aldose (glucose-6-P) and a ketose (fructose-6-P) through a multistep pathway that involves opening and closing of the ring structure (**Figure 9.24**). Because the reaction has a small free energy change ($\Delta G = -2.9$ kJ/mol), the reaction is readily reversible and is controlled by the levels of glucose-6-P and fructose-6-P in the cell.

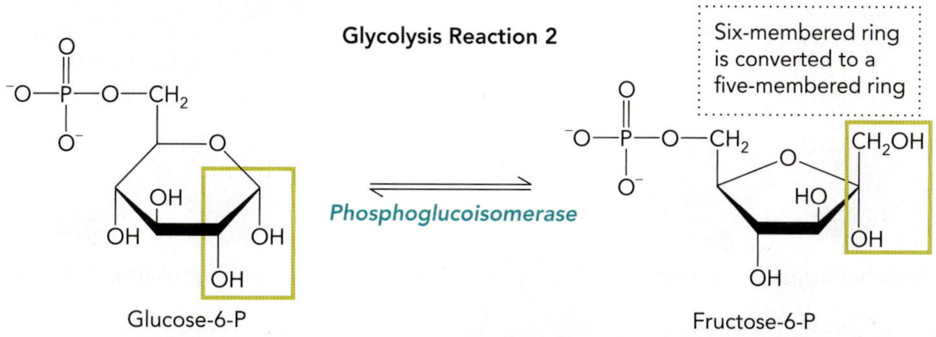

Figure 9.24 Glycolytic pathway reaction 2 is catalyzed by the enzyme phosphoglucoisomerase.

Figure 9.25 Glycolytic pathway reaction 3 is catalyzed by the enzyme phosphofructokinase-1. This enzyme is allosterically regulated and serves as a major control point in the glycolytic pathway.

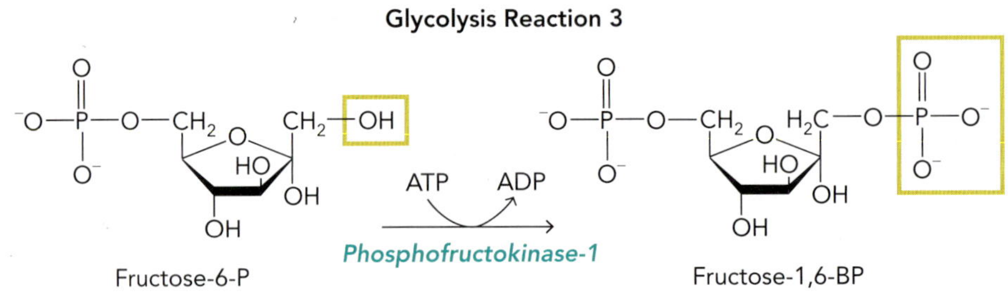

Reaction 3: Phosphorylation of Fructose-6-P to Fructose-1,6-BP by Phosphofructokinase-1 Reaction 3 is the second ATP investment reaction in glycolysis. It couples ATP hydrolysis with a phosphoryl transfer reaction catalyzed by the enzyme phosphofructokinase-1 (**Figure 9.25**). The reaction is essentially irreversible with a large negative free energy change ($\Delta G = -18.8$ kJ/mol). It also serves as a major regulatory site in the pathway through allosteric control of phosphofructokinase-1 activity. Phosphorylation of fructose-6-P generates fructose-1,6-BP, which forms two different triose phosphates when cleaved in reaction 4. Note that a bisphosphate compound contains two phosphates on different carbon atoms (for example, C-1 and C-6), whereas a diphosphate compound, such as ADP, contains two phosphates covalently linked to each other.

Phosphofructokinase-1 activity is allosterically controlled by the energy charge of the cell (see Equation 2.16) in such a way that increased AMP and ADP concentrations stimulate enzyme activity, whereas high ATP concentrations inhibit phosphofructokinase-1 activity. This regulation ensures that if the cell's energy needs are sufficient (high ATP), the glycolytic pathway does not proceed to produce more ATP. All three effectors—AMP, ADP, and ATP—compete for binding to the same allosteric site on the enzyme, making AMP and ADP positive allosteric effectors and ATP a negative allosteric effector. **Figure 9.26** shows space-filling models of the tetrameric phosphofructokinase-1 complex from the bacterium *Bacillus stearothermophilus* in both the active and inactive conformations. The tight packing of individual subunits is a common feature of allosteric enzymes because it facilitates effector-induced conformational changes across the entire complex. Details of the regulation of phosphofructokinase-1 are discussed later in this chapter (see Section 9.4).

Reaction 4: Cleavage of Fructose-1,6-BP into Glyceraldehyde-3-P and Dihydroxyacetone-P by Aldolase The splitting of fructose-1,6-BP into the triose phosphates glyceraldehyde-3-P and dihydroxyacetone-P is the reaction that puts the *lysis*

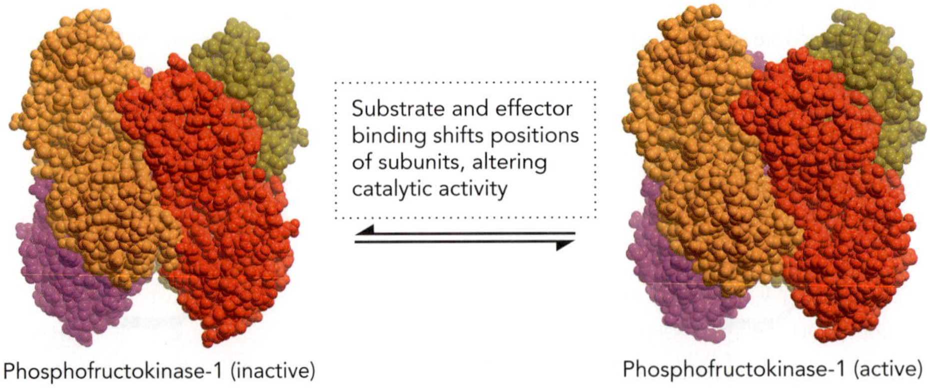

Figure 9.26 Phosphofructokinase-1 functions as a homotetramer in which each subunit contains a regulatory site and an active site. All four subunits share multiple contact points, which facilitates allosteric control by ATP, ADP, and AMP in response to the energy status of the cell. Binding of substrates and effectors influences the overall conformations of the subunits, which alter the structure of the active sites to modulate catalytic activity. The conformational changes between the active and inactive conformations are subtle in this enzyme. BASED ON PDB FILES 4PFK (ACTIVE) AND 6PFK (INACTIVE).

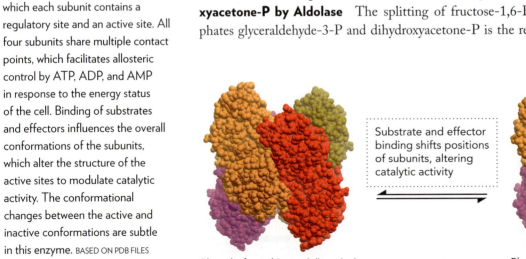

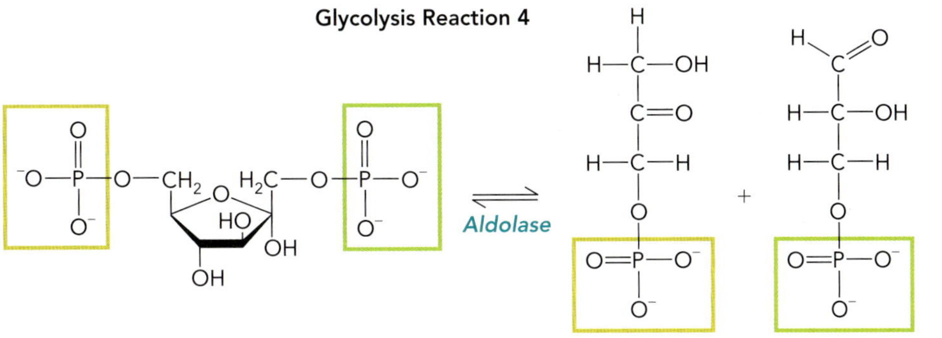

Fructose-1,6-BP Dihydroxyacetone-P Glyceraldehyde-3-P

Figure 9.27 Glycolytic pathway reaction 4 is catalyzed by the enzyme aldolase, which cleaves the six-carbon substrate fructose-1,6-BP between C-3 and C-4, forming the three-carbon products dihydroxyacetone-P and glyceraldehyde-3-P. Note that the phosphoryl groups on C-1 and C-6 of fructose-1,6-BP become the phosphates on the designated C-3 atoms of the two products.

in glycolysis (*lysis* means "splitting"), as shown in **Figure 9.27**. The enzyme responsible for this cleavage reaction between C-3 and C-4 in fructose-1,6-BP is aldolase (also called fructose bisphosphate aldolase). In the context of the glycolytic pathway, aldolase performs the reverse of an aldol condensation. As you can see in **Figure 9.28**, the mechanism of cleavage by aldolase involves the formation of a covalent enzyme–substrate complex through the generation of a Schiff base requiring a lysine residue in the enzyme active site. The five-step reaction can be summarized as follows:

1. The open-chain form of fructose-1,6-BP binds to the aldolase active site.
2. A lysine residue in the active site performs a nucleophilic attack on the ketose carbon, generating a covalently bound protonated Schiff base intermediate.
3. Base abstraction by an active site carboxyl group leads to C—C (aldol) cleavage between C-3 and C-4 of fructose, and a covalent enamine intermediate is formed. The first product, glyceraldehyde-3-P, is released.
4. Isomerization leads to formation of a second protonated Schiff base intermediate.
5. The Schiff base intermediate is hydrolyzed, releasing the second product, dihydroxyacetone-P.

The aldolase reaction illustrates the important difference between the $\Delta G^{\circ\prime}$ and ΔG values shown in Table 9.2. Under standard conditions, aldol condensation is favored, as the standard free energy for the cleavage reaction is highly positive ($\Delta G^{\circ\prime} = +23.8$ kJ/mol). In the cell, however, the cellular concentrations of metabolites result in a mass action ratio that favors the cleavage reaction during glycolysis ($\Delta G = -0.4$ kJ/mol).

Reaction 5: Isomerization of Dihydroxyacetone-P to Glyceraldehyde-3-P by Triose Phosphate Isomerase The production of dihydroxyacetone-P in reaction 4 creates a slight problem because glyceraldehyde-3-P, not dihydroxyacetone-P, is the substrate for reaction 6 in the glycolytic pathway. This dilemma is solved by the enzyme triose phosphate isomerase, which converts the ketose dihydroxyacetone-P to the aldose glyceraldehyde-3-P. This isomerization reaction is similar to reaction 2 (conversion of glucose-6-P to fructose-6-P), albeit in reverse (this time we need an aldose formed from a ketose). **Figure 9.29** (p. 457) shows that the reaction involves formation of an enediol intermediate.

Reaction 5 completes stage 1 of glycolysis. At the expense of investing 2 ATP, this stage produces two molecules of glyceraldehyde-3-P for every molecule of glucose that is phosphorylated in reaction 1.

Figure 9.28 The five-step reaction mechanism of fructose-1,6-bisphosphate cleavage by the enzyme aldolase in reaction 4 of glycolysis.

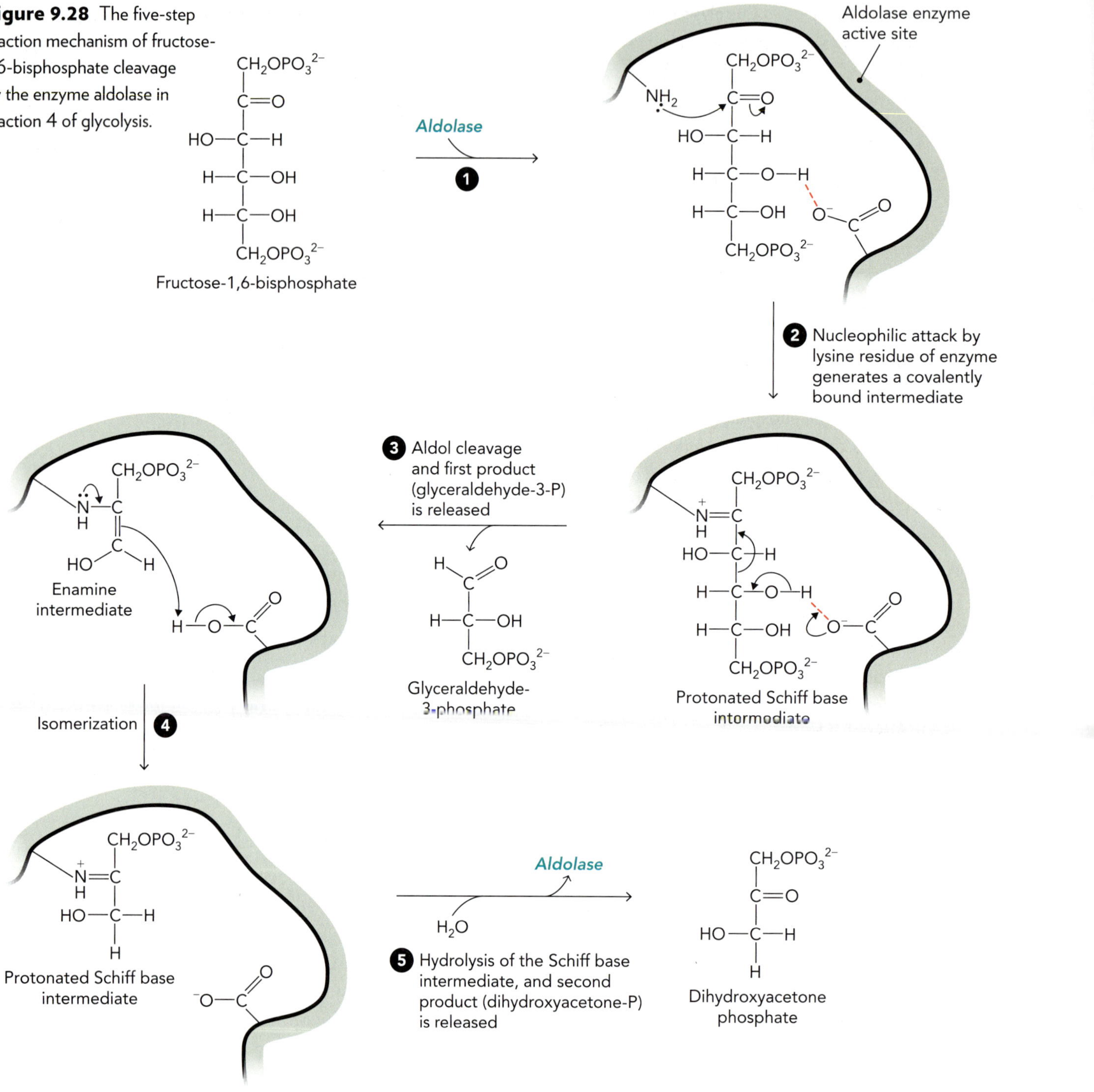

Stage 2 of the Glycolytic Pathway: ATP Earnings

As a group, the stage 2 reactions have four key features:

1. Two molecules of glyceraldehyde-3-P enter stage 2 for every molecule of glucose that is metabolized in stage 1.
2. Two substrate-level phosphorylation reactions, catalyzed by the enzymes phosphoglycerate kinase and pyruvate kinase, generate a total of 4 ATP per glucose metabolized. Considering the investment of 2 ATP in stage 1, this gives a net yield of 2 ATP per glucose.

3. An oxidation reaction catalyzed by glyceraldehyde-3-P dehydrogenase generates 2 NADH molecules. These molecules can be shuttled into the mitochondria to produce more ATP by oxidative phosphorylation. (Under anaerobic conditions, they are oxidized by lactate dehydrogenase in the reaction that forms lactate.) Most important, the glyceraldehyde-3-P dehydrogenase reaction requires a constant supply of NAD^+ to maintain flux through glycolysis. It is therefore linked to reactions downstream of pyruvate formation.
4. Reaction 10 is an irreversible reaction and ensures that pyruvate is the end point of glycolysis. This step helps maintain a distinction between the catabolic and anabolic pathways so that cells don't waste energy by undoing what the pathway has just accomplished. In gluconeogenesis (an anabolic pathway that synthesizes glucose for export—the reverse pathway of glycolysis), two separate enzymatic reactions are used to "reverse" this reaction. As described in Chapter 14, these reactions are energetically expensive because they require the hydrolysis of two high-energy phosphoryl compounds (ATP and GTP) to generate phosphoenolpyruvate from pyruvate to initiate gluconeogenesis.

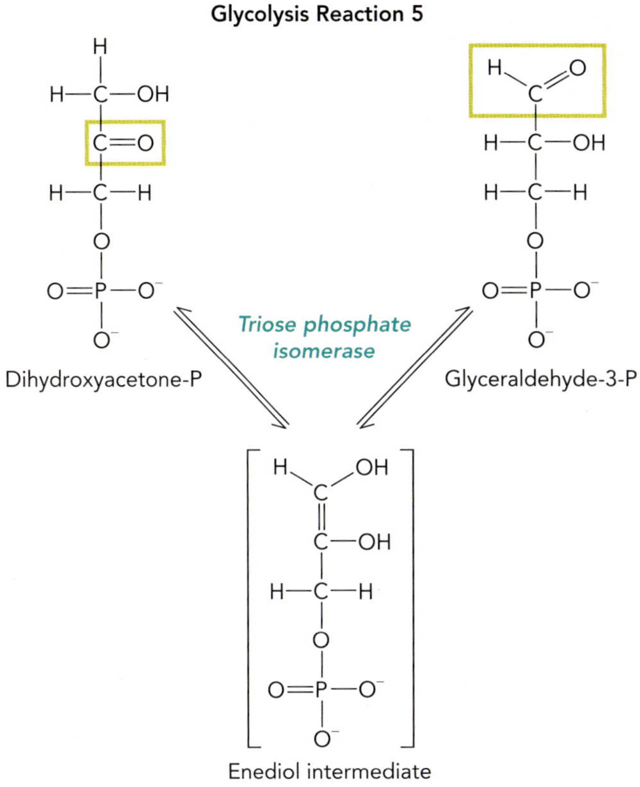

Figure 9.29 Glycolytic pathway reaction 5 is catalyzed by the enzyme triose phosphate isomerase through the formation of an enediol intermediate.

Let's now examine the biochemistry of each stage 2 reaction individually.

Reaction 6: Oxidation and Phosphorylation of Glyceraldehyde-3-P to Form 1,3-Bisphosphoglycerate by Glyceraldehyde-3-P Dehydrogenase The glyceraldehyde-3-P dehydrogenase reaction is a critical step in glycolysis because it generates a molecule with high phosphoryl transfer potential: 1,3-bisphosphoglycerate. The enzyme achieves this reaction by coupling the favorable reduction of the coenzyme NAD^+ with the oxidation of glyceraldehyde-3-P. The energy released from oxidation of glyceradehyde-3-P is used to drive a phosphoryl group transfer reaction, using inorganic phosphate (P_i) to produce 1,3-bisphosphoglycerate (**Figure 9.30**). Because

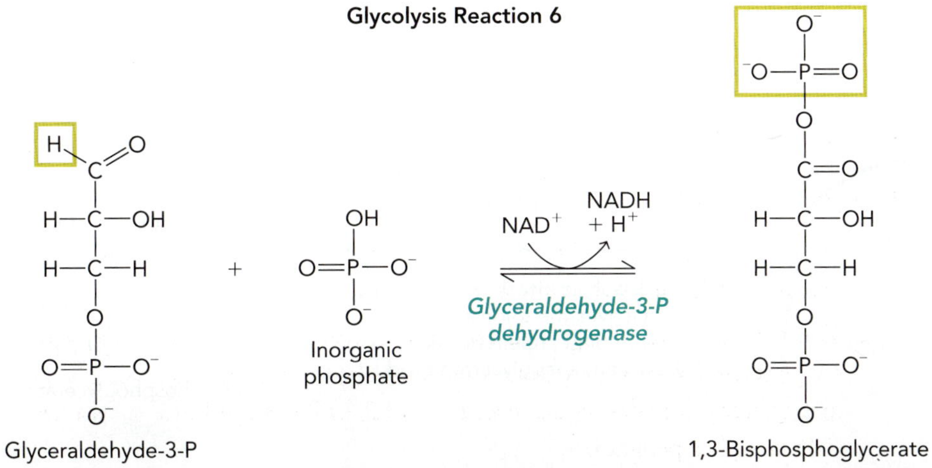

Figure 9.30 Glycolytic pathway reaction 6 is catalyzed by the enzyme glyceraldehyde-3-P dehydrogenase, which couples substrate oxidation to phosphorylation in a reaction requiring inorganic phosphate (P_i) and the coenzyme NAD^+.

NAD^+ is required for the oxidation step in this reaction, NAD^+ must be continually replenished within the cytosol to maintain flux through glycolysis. As discussed later in the chapter, this is accomplished aerobically in the mitochondrial matrix by the electron transport chain or anaerobically in the cytosol by the enzyme lactate dehydrogenase, which converts pyruvate to lactate.

The five-step enzyme mechanism for glyceraldehyde-3-P dehydrogenase has been worked out and is illustrated in **Figure 9.31**. In step 1, glyceraldehyde-3-P binds to the enzyme active site, which contains a bound NAD^+ coenzyme molecule that functions as an electron acceptor. In step 2, glyceraldehyde-3-P reacts with the sulfhydryl group of a cysteine residue in the enzyme active site, forming a thiohemiacetal intermediate. In step 3, the aldehyde group is then dehydrogenated by an oxidation reaction, in which a hydride ion ($:H^-$) is transferred from the enzyme-bound glyceraldehyde-3-P to the nicotinamide ring of NAD^+, producing $NADH + H^+$. This oxidation reaction results in the formation of an enzyme–substrate complex, called an acyl thioester, with a large

Figure 9.31 The formation of 1,3-bisphosphoglycerate from glyceraldehyde-3-P by the enzyme glyceraldehyde-3-P dehydrogenase in reaction 6 of glycolysis involves an NAD^+ coenzyme.

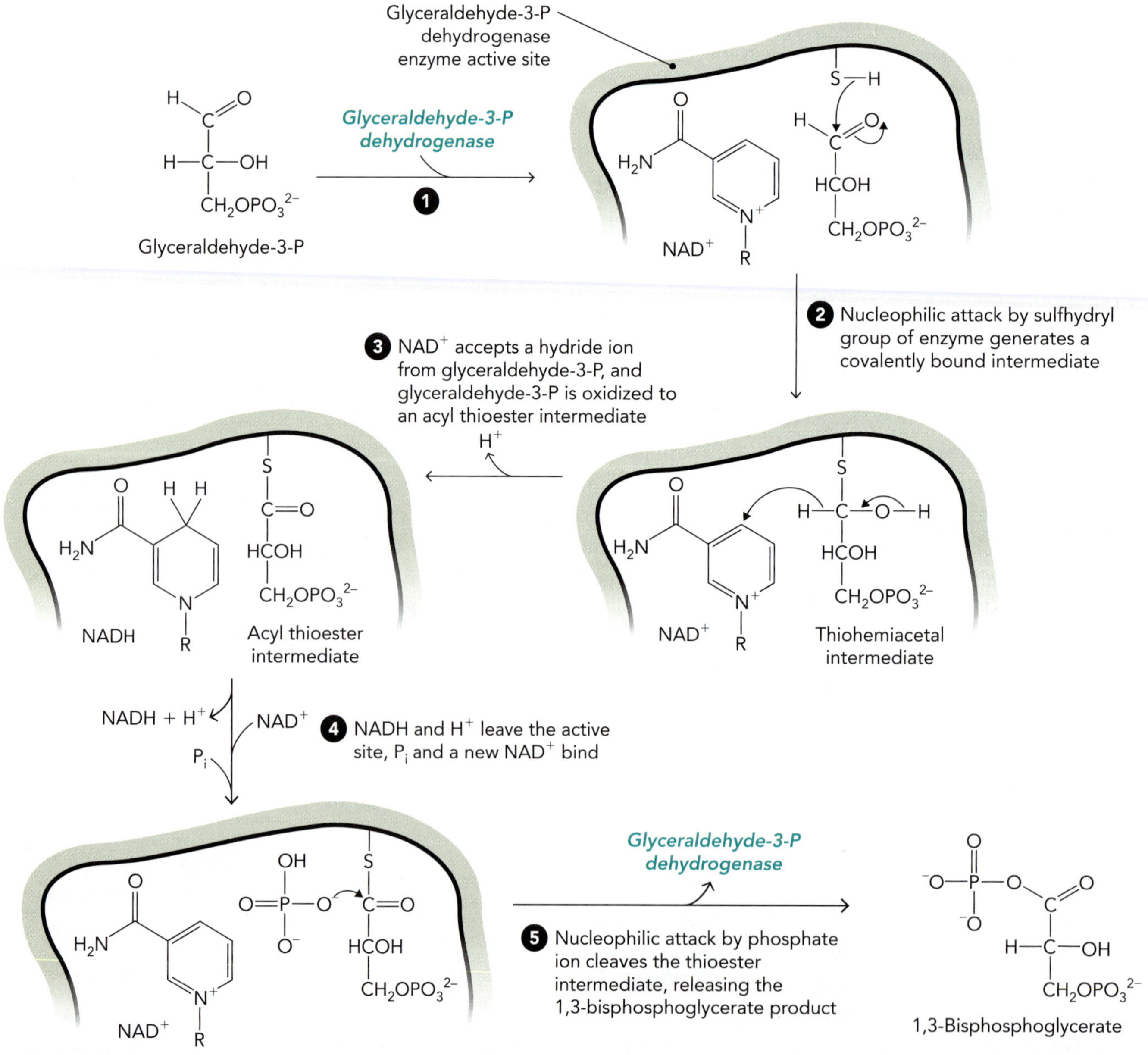

free energy of hydrolysis. In step 4, the reduced coenzyme NADH leaves the active site and is replaced by a new NAD^+ molecule. In step 5, the high-energy acyl thioester is then subject to nucleophilic attack by an incoming phosphate ion, which leads to product release (1,3-bisphosphoglycerate) and regeneration of the active enzyme.

As shown in **Table 9.4**, formation of 1,3-bisphosphoglycerate by this coupled oxidation–phosphorylation reaction generates a metabolite with a standard free energy of hydrolysis of $\Delta G°' = -49.4$ kJ/mol. This value is even more favorable than that for the phosphate hydrolysis of ATP ($\Delta G°' = -30.5$ kJ/mol). This difference in free energy changes is harnessed by the enzyme phosphoglycerate kinase in reaction 7, driving a coupled reaction that phosphorylates ADP to generate ATP by substrate-level phosphorylation. In that reaction, 1,3-bisphosphoglycerate serves as the phosphoryl group donor.

What accounts for the large favorable free energy change for phosphate hydrolysis for some compounds? A comparison of the reactants and products in the hydrolysis reactions provides some answers. For phosphate hydrolysis of 1,3-bisphosphoglycerate, the product 3-phosphoglycerate has two equally probable resonance forms of the carboxylate group. This resonance stabilization does not occur in 1,3-bisphosphoglycerate, and therefore the hydrolyzed products are favored. Table 9.4 also shows that phosphoenolpyruvate, another glycolytic intermediate, has a standard free energy change of phosphate hydrolysis of $\Delta G°' = -61.9$ kJ/mol, which exceeds that of both 1,3-bisphosphoglycerate and ATP. For phosphate hydrolysis of phosphoenolpyruvate, the pyruvate product can exist in either of two tautomeric forms, either an enol or a keto form, which provides significant stabilization. Tautomerization is not possible in phosphoenolpyruvate, so the forward hydrolysis reaction is greatly favored. Phosphoenolpyruvate provides the energy in reaction 10 to generate a molecule of ATP by substrate-level phosphorylation. Note in Table 9.4 that the standard free energy changes of phosphate hydrolysis for both glucose-6-P ($\Delta G°' = -13.8$ kJ/mol) and fructose-6-P ($\Delta G°' = -15.9$ kJ/mol) are less favorable than the phosphate hydrolysis of ATP ($\Delta G°' = -30.5$ kJ/mol), which explains why ATP investment is required in stage 1 of glycolysis.

Table 9.4 STANDARD FREE ENERGY CHANGES OF PHOSPHATE HYDROLYSIS FOR SELECTED BIOMOLECULES

Molecule	$\Delta G°'$ (kJ/mol)
Phosphoenolpyruvate	−61.9
1,3-Bisphosphoglycerate	−49.4
Creatine phosphate	−43.0
ADP → AMP + P_i	−32.8
ATP → ADP + P_i	−30.5
Glucose-1-phosphate	−20.9
2-Phosphoglycerate	−17.6
Fructose-6-phosphate	−15.9
Glucose-6-phosphate	−13.8
Glycerol-3-phosphate	−9.2

Reaction 7: Substrate-Level Phosphorylation to Generate ATP in the Conversion of 1,3-Bisphosphoglycerate to 3-Phosphoglycerate by Phosphoglycerate Kinase Phosphoglycerate kinase catalyzes the payback reaction in glycolysis because it replaces the 2 ATP that were used in stage 1 to prime the glycolytic pathway. As shown in **Figure 9.32**, the high phosphoryl transfer energy present in the substrate is used to phosphorylate ADP to form ATP by the mechanism of substrate-level phosphorylation, leading to the conversion of 1,3-bisphosphoglycerate to 3-phosphoglycerate. Recall that two molecules of 1,3-bisphosphoglycerate are formed from every molecule of glucose; therefore, this reaction occurs twice and generates 2 ATP per glucose metabolized (see Figure 9.20).

The molecular structure of phosphoglycerate kinase is similar to that of hexokinase, in that it has two lobes, each of which binds a substrate molecule ($ADP–Mg^{2+}$ for hexokinase or 1,3-bisphosphoglycerate for phosphoglycerate kinase). Substrate binding leads to a large

Glycolysis Reaction 7

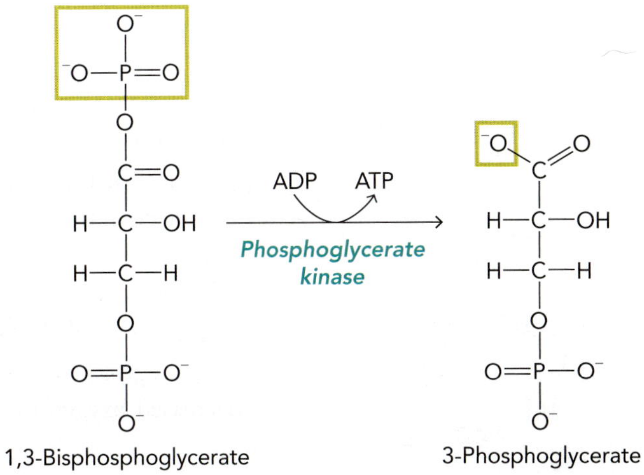

Figure 9.32 Glycolytic pathway reaction 7 is catalyzed by the enzyme phosphoglycerate kinase. This enzyme captures the high phosphoryl transfer energy available in 1,3-bisphosphoglycerate in a substrate-level phosphorylation reaction, generating ATP and 3-phosphoglycerate.

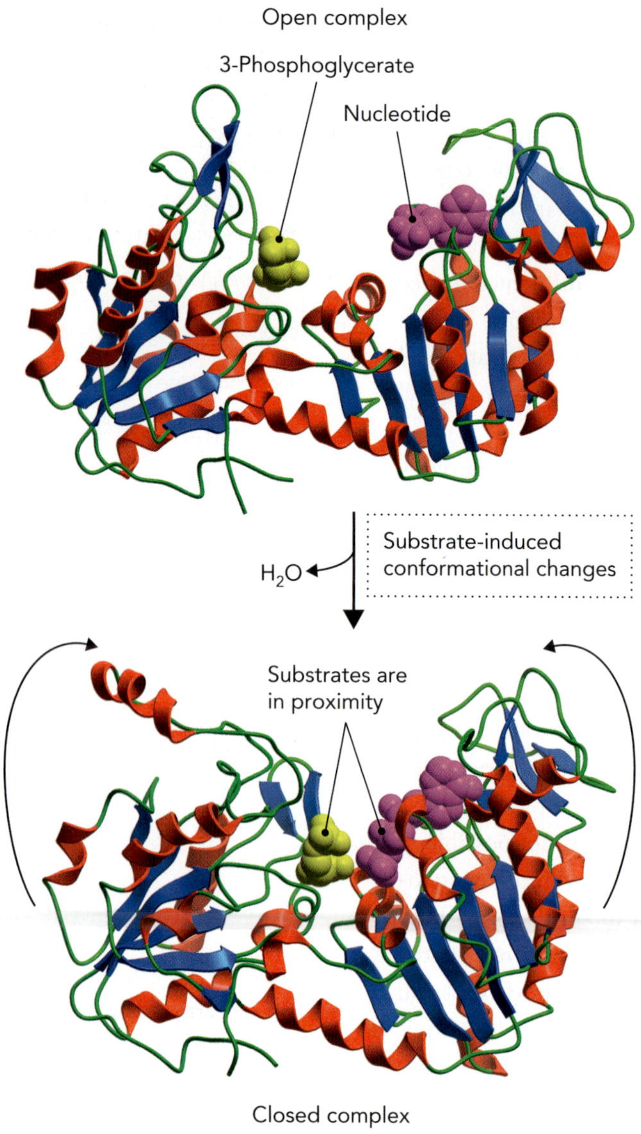

Figure 9.33 Substrate binding to phosphoglycerate kinase induces large movements in two domains of the protein, which brings the two substrates together for catalysis. The substrates 3-phosphoglycerate (yellow) and ADP or AMP (purple) are shown in space-filling style. BASED ON PDB FILES 1HDI (OPEN) AND 13PK (CLOSED).

conformational change in the enzyme that brings the substrates close together and excludes H_2O from the active site. The structures of phosphoglycerate kinase in the open and closed conformations are shown in **Figure 9.33**. As described in Chapter 7, the induced-fit mechanism of substrate binding maximizes accessibility to the enzyme active site without sacrificing the requirement that a hydrophobic environment be formed in the closed conformation.

Note that the actual changes in free energy for reactions 6 and 7 are close to zero ($\Delta G = -1.3$ kJ/mol; $\Delta G = -3.4$ kJ/mol), and therefore both reactions are in fact reversible inside the cell. In contrast, three reactions in glycolysis—catalyzed by the enzymes hexokinase, phosphofructokinase-1, and pyruvate kinase—are all irreversible because of large unfavorable changes in actual free energy (see Table 9.2). Small differences in metabolite concentrations will not change the sign of the free energy change to make these reactions favorable in the reverse direction. In the gluconeogenic pathway, which synthesizes glucose from noncarbohydrate sources (see Chapter 14), these reactions require different enzymes in order for them to proceed in the "reverse" direction. Throughout our study of metabolism, we will encounter numerous examples of reversible and irreversible reactions in opposing catabolic and anabolic pathways.

Before leaving reaction 7, we need to take a look at an important side reaction in erythrocytes. This reaction converts 1,3-bisphosphoglycerate to 2,3-bisphosphoglycerate (2,3-BPG), the allosteric regulator of hemoglobin (see Chapter 6). As shown in **Figure 9.34**, the enzymes bisphosphoglycerate mutase and 2,3-bisphosphoglycerate phosphatase interconvert 1,3-bisphosphoglycerate and 2,3-BPG in erythrocytes. The result is a metabolic link between glycolytic flux and oxygen transport.

This interdependence between 1,3-bisphosphoglycerate and 2,3-BPG explains why individuals with defects in glycolytic enzymes have altered oxygen-transport capabilities; it is due to increased or decreased levels of 2,3-BPG. If glycolytic enzymes corresponding to reactions upstream of 1,3-bisphosphoglycerate formation are defective (for example, hexokinase), then 2,3-BPG levels are reduced in erythrocytes. This shifts the oxygen saturation curve to the left (increased oxygen binding because of higher levels of the R conformation of hemoglobin; see Figure 6.17). In contrast, individuals with defects in glycolytic enzymes downstream of 1,3-bisphosphoglycerate contain higher levels of 2,3-BPG in erythrocytes. This results in decreased oxygen binding capability (the oxygen saturation curve is shifted to the

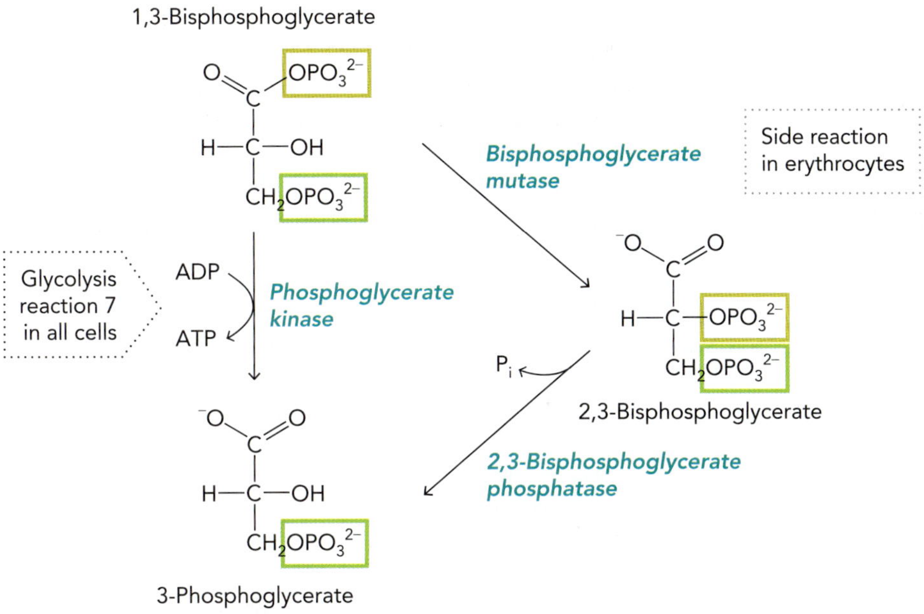

Figure 9.34 In addition to phosphoglycerate kinase, erythrocytes also contain the enzymes bisphosphoglycerate mutase and 2,3-bisphosphoglycerate phosphatase, which interconvert the molecules 1,3-bisphosphoglycerate, 2,3-bisphosphoglycerate, and 3-phosphoglycerate.

right because of high levels of hemoglobin in the T conformation). This is a good example of how biochemical defects at the enzyme level are often found to underlie seemingly unrelated physiologic conditions.

Reaction 8: Phosphoryl Shift in 3-Phosphoglycerate to Form 2-Phosphoglycerate by Phosphoglycerate Mutase The purpose of reaction 8 is to generate a compound, 2-phosphoglycerate, that can be converted to phosphoenolpyruvate in the next reaction, in preparation for a second substrate-level phosphorylation in step 10 that generates ATP earnings. The phosphoglycerate mutase reaction is shown in **Figure 9.35**.

The mechanism of this reversible reaction ($\Delta G = +1.1$ kJ/mol; see Table 9.2) is illustrated in **Figure 9.36**. You can see that the reaction requires a phosphoryl transfer from a phosphorylated histidine residue (His-P) located in the enzyme active site. In step 1 of the reaction, the substrate 3-phosphoglycerate binds to the enzyme active site and is phosphorylated on C-2 by a transfer reaction involving the His-P group. This step generates the short-lived intermediate 2,3-BPG. This mechanism involves the formation of noncovalent substrate interactions (denoted as substrate • enzyme), which are distinct from covalent substrate–enzyme complexes, as appeared in the glyceraldehyde-3-P dehydrogenase reaction (see Figure 9.31). In step 2 of the mechanism, the phosphate on C-3 is transferred to the histidine residue of the enzyme. This regenerates His-P, leading to the release of 2-phosphoglycerate and the binding of a new molecule of 3-phosphoglycerate in step 3. Note that the 2,3-BPG formed in step 1 can diffuse out of the active site, resulting in dephosphorylated enzyme (the histidine residue transferred its phosphate to the substrate). When this happens, trace amounts of 2,3-BPG in the cell can rebind to phosphorylate the histidine residue and activate the enzyme.

Figure 9.35 Glycolytic pathway reaction 8 is an isomerization reaction catalyzed by the enzyme phosphoglycerate mutase.

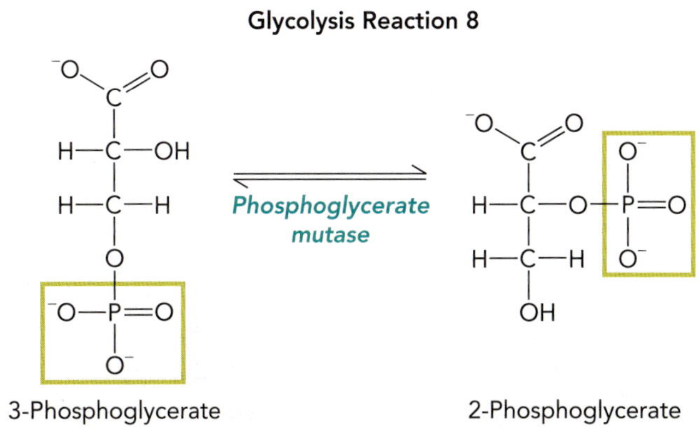

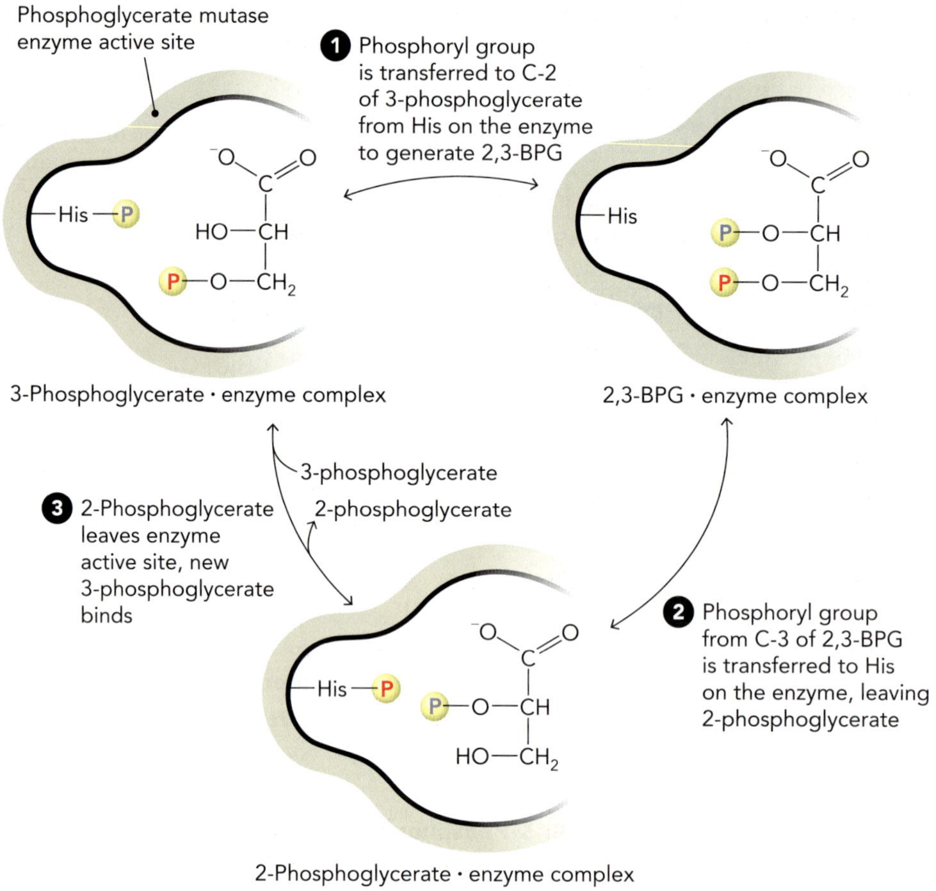

Phosphoglycerate mutase enzyme active site
1 Phosphoryl group is transferred to C-2 of 3-phosphoglycerate from His on the enzyme to generate 2,3-BPG
3-Phosphoglycerate · enzyme complex
2,3-BPG · enzyme complex
3-phosphoglycerate
2-phosphoglycerate
3 2-Phosphoglycerate leaves enzyme active site, new 3-phosphoglycerate binds
2 Phosphoryl group from C-3 of 2,3-BPG is transferred to His on the enzyme, leaving 2-phosphoglycerate
2-Phosphoglycerate · enzyme complex

Figure 9.36 The phosphoglycerate mutase reaction involves a phosphohistidine residue in the enzyme active site and the formation of noncovalent substrate · enzyme complexes in reaction 8 of glycolysis.

Reaction 9: Dehydration of 2-Phosphoglycerate to Form Phosphoenolpyruvate by Enolase In this penultimate step in glycolysis, a dehydration reaction catalyzed by the enzyme enolase converts 2-phosphoglycerate (a molecule with only moderate phosphoryl transfer potential) to phosphoenolpyruvate (which as we have already seen in Table 9.4 has extremely high phosphoryl transfer potential). **Figure 9.37** shows the enolase reaction, where it is observed that removal of water (dehydration) converts the phosphate ester present in 2-phosphoglycerate to an enol phosphate in phosphoenolpyruvate. The mechanism for this reaction was discussed in Chapter 7 as an example of general acid–base and metal ion–induced catalysis (see Figure 7.31). The arrangement of the phosphoryl group in phosphoenolpyruvate makes it much easier to transfer the phosphoryl group to ADP in the pyruvate kinase reaction (reaction 10). In this sense, enolase has the same function as glyceraldehyde-3-P dehydrogenase (reaction 6), generating a compound with high phosphoryl transfer potential in preparation for a substrate-level phosphorylation reaction.

It is noteworthy that the change in standard free energy for this reaction is relatively small ($\Delta G^{\circ\prime} = +1.7$ kJ/mol; see Table 9.2), meaning that the overall metabolic energy available from 2-phosphoglycerate or phosphoenolpyruvate is similar. However, when enolase converts 2-phosphoglycerate to phosphoenolpyruvate, it traps the phosphoryl group in an unstable enol form. This results in a dramatic increase in the phosphoryl transfer potential

Figure 9.37 Glycolytic pathway reaction 9 is a dehydration catalyzed by the enzyme enolase, generating phosphoenolpyruvate from 2-phosphoglycerate.

Glycolysis Reaction 9

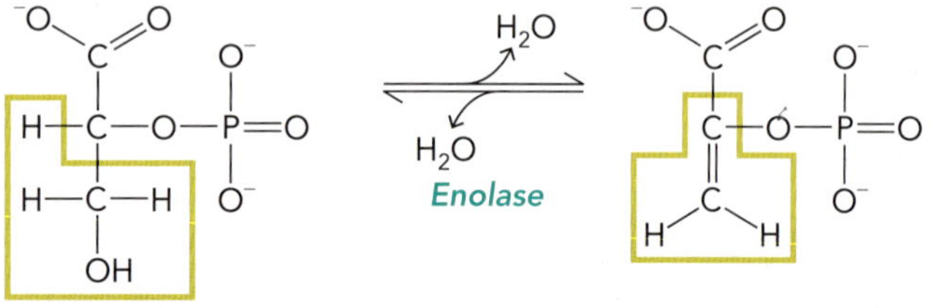

H_2O
H_2O
Enolase

2-Phosphoglycerate Phosphoenolpyruvate

of the triose sugar. Indeed, the standard free energy change for phosphate hydrolysis of 2-phosphoglycerate is $\Delta G^{\circ\prime} = -17.6$ kJ/mol, whereas for phosphoenolpyruvate it is an incredible $\Delta G^{\circ\prime} = -61.9$ kJ/mol (see Table 9.4).

Reaction 10: Substrate-Level Phosphorylation to Generate ATP in the Conversion of Phosphoenolpyruvate to Pyruvate by Pyruvate Kinase The phosphoglycerate kinase reaction (reaction 7) took care of the ATP payback step by replacing the 2 ATP molecules used up in stage 1 of glycolysis, so in reaction 10 we see where the net ATP earnings come from. In this reaction, the high phosphoryl transfer potential of phosphoenolpyruvate is used by the enzyme pyruvate kinase to generate pyruvate—the end product of glycolysis—and a net of 2 ATP for every glucose molecule entering the pathway (see Figure 9.20). This is the second of two substrate-level phosphorylation reactions in glycolysis that couple energy released from phosphate hydrolysis ($\Delta G^{\circ\prime} = -61.9$ kJ/mol) to energy consumed by ATP synthesis ($\Delta G^{\circ\prime} = +30.5$ kJ/mol), as shown in **Figure 9.38**. Unlike phosphoenolpyruvate, pyruvate is a stable compound in cells. It is used by many other metabolic pathways, as we will describe in the last section of the chapter.

Glycolysis Reaction 10

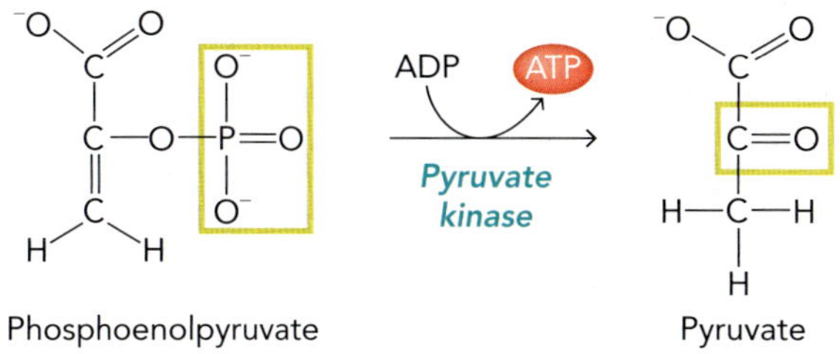

Figure 9.38 Glycolytic pathway reaction 10, catalyzed by the enzyme pyruvate kinase, is the second of two substrate-level phosphorylation reactions in glycolysis. This reaction takes advantage of the high phosphoryl transfer potential of phosphoenolpyruvate (see Table 9.4) to generate ATP and pyruvate, the end product of glycolysis.

concept integration 9.3

What is substrate-level phosphorylation and how is it used in the glycolytic pathway to generate a net 2 ATP under anaerobic conditions?

Substrate-level phosphorylation refers to phosphoryl transfer reactions that generate ATP through mechanisms that are independent of the enzyme ATP synthase, which is present in mitochondria and chloroplasts (see Chapter 11). In substrate-level phosphorylation, a phosphoryl group on a donor molecule with high phosphoryl transfer energy is transferred to ADP to form ATP. The enzymes phosphoglycerate kinase and pyruvate kinase catalyze the two substrate-level phosphorylation reactions in stage 2 of glycolysis. In both of these reactions, phosphoryl groups from substrates with high standard free energy changes for phosphate hydrolysis are transferred to ADP, generating ATP. Because two molecules of glyceraldehyde-3-P are generated for every molecule of glucose that enters the glycolytic pathway, together these two substrate-level phosphorylation reactions yield 4 ATP per glucose. However, when taking into account the 2 ATP that are invested in stage 1 of the glycolytic pathway to generate glucose-6-P and fructose-1,6-BP, the net ATP yield of glycolysis is 2 ATP per glucose under anaerobic conditions.

9.4 Regulation of the Glycolytic Pathway

We saw in Chapter 7 that enzyme activity is controlled by both substrate availability and levels of catalytic activity. The same factors control flux through all metabolic pathways. Reactions that operate at ΔG values close to zero are reversible and are characterized by enzymes that operate at full capacity. For these reactions, the direction of metabolic flux primarily depends on substrate availability. (Note that

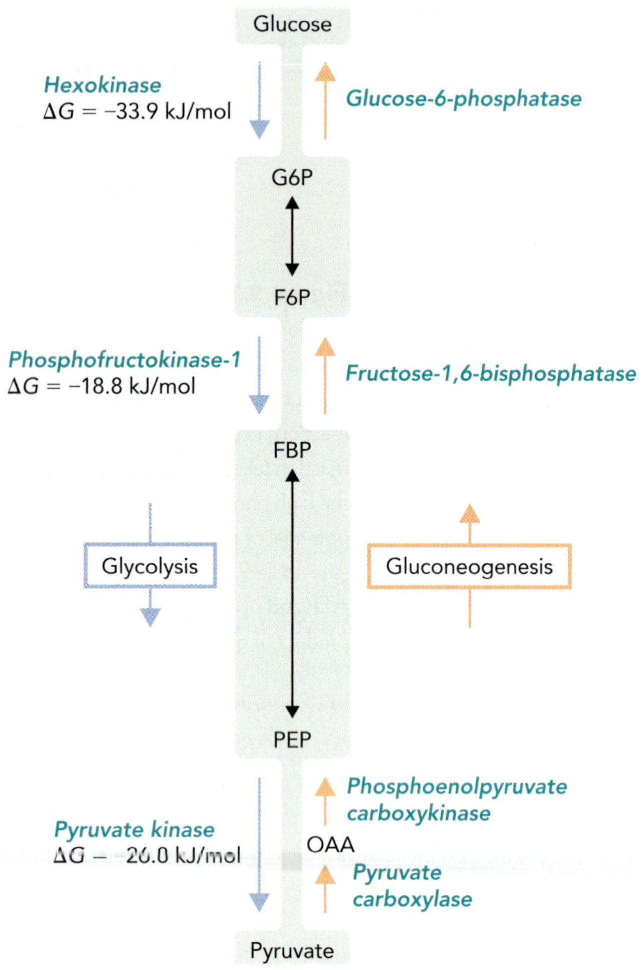

Figure 9.39 Metabolic flux through opposing pathways such as glycolysis and gluconeogenesis is controlled by rate-limiting enzymes that are highly regulated. Reactions that are readily reversible depending on metabolite concentrations are shown as "wide" regions where the flow through the pathway is unrestricted. The "bottlenecks" occur at the irreversible steps. In the bottlenecks, the opposing pathways of glycolysis and gluconeogenesis utilize different enzymes to accomplish these reactions in the appropriate directions. G6P = glucose-6-phosphate; F6P = fructose-6-phosphate; FBP = fructose-1,6-bisphosphate; PEP = phosphoenolpyruvate; OAA = oxaloacetate.

a 10-fold difference in substrate and product concentrations contributes ~5.9 kJ/mol to ΔG at 37 °C.) Table 9.2 lists seven reactions of this type in glycolysis. In contrast, reactions with highly negative ΔG values are functionally irreversible because of physiologic metabolite concentrations and are usually subject to enzymatic control. The three enzymes in glycolysis that fit this description are hexokinase, phosphofructokinase-1, and pyruvate kinase (see Table 9.2).

Irreversible reactions in metabolic pathways are called **rate-limiting** steps because the level of enzyme activity can be regulated to be low even when substrate levels are high. Rate-limiting enzymes in metabolic pathways serve as regulated "valves" that are opened or closed in response to cellular conditions. As illustrated in **Figure 9.39**, the reverse of glucose degradation via glycolysis (an ATP-generating catabolic pathway) is glucose synthesis via gluconeogenesis (an ATP-requiring anabolic pathway). Glycolysis and gluconeogenesis utilize the same enzymes for the reversible steps, whereas irreversible steps are rate limiting and require pathway-specific enzymes to control metabolic flux.

In this section, we examine three ways in which substrate availability and enzyme activity levels control flux through the glycolytic pathway. (We describe regulation of the gluconeogenic pathway in Chapter 14.) We first look at the enzyme glucokinase, which, like hexokinase, phosphorylates glucose to form glucose-6-P. Glucokinase has a unique role in liver and pancreatic cells as a molecular sensor of high blood glucose levels. We then examine the structure of phosphofructokinase-1 and describe how its function is allosterically regulated in response to changes in the energy charge of the cell. (Recall from Chapter 2 that the energy charge refers to the ratio of ATP, ADP, and AMP concentrations in the cell, which reflects the energy available from phosphoanhydride hydrolysis to do work.) We conclude with a discussion of the inflow and outflow of glycolytic intermediates, which affect metabolic flux through the pathway by altering the concentrations of reaction substrates.

Although we only consider short-term regulation of glycolytic enzymes here, several genes encoding glycolytic enzymes are also regulated at the transcriptional level in response to a high carbohydrate diet. These include phosphofructokinase-1, glyceraldehyde-3-P dehydrogenase, and pyruvate kinase. Transcriptional regulation of representative metabolic pathways in bacteria (lactose) and yeast (galactose) is described in Chapter 23.

Glucokinase Is a Molecular Sensor of High Glucose Levels

Four hexokinase genes have been identified in humans (hexokinase I–IV), all of which encode enzymes that are capable of converting glucose to glucose-6-P using the phosphoryl transfer energy of ATP hydrolysis (step 1 of glycolysis). We have already described one of these enzymes: Hexokinase I, often referred to simply as hexokinase, has a high affinity for substrate (K_m for glucose is ~0.1 mM), is expressed in all tissues, phosphorylates a variety of hexose sugars, and is inhibited by the product of the reaction, glucose-6-P (see Figure 9.23). In contrast, hexokinase IV, also known as glucokinase, has a low affinity for substrate (K_m for glucose is ~10 mM), is highly specific for glucose, is expressed primarily in liver and pancreatic cells, and is not inhibited by glucose-6-P. These differences in tissue expression and glucose affinity between hexokinase and glucokinase play important roles in controlling blood glucose levels, which ultimately control rates of glycolytic flux in all cells by limiting substrate availability.

As suggested by the different K_m values of hexokinase and glucokinase for glucose, substrate saturation curves for these two enzymes look markedly different (**Figure 9.40**). Because normal blood glucose levels are maintained at ~5 mM, significant levels of glucose phosphorylation by glucokinase only occur when blood glucose levels are much higher, such as after consuming a carbohydrate-rich meal. Moreover, because glucokinase is not inhibited by glucose-6-P, it is able to continue functioning even if flux through glycolysis cannot keep up with product formation.

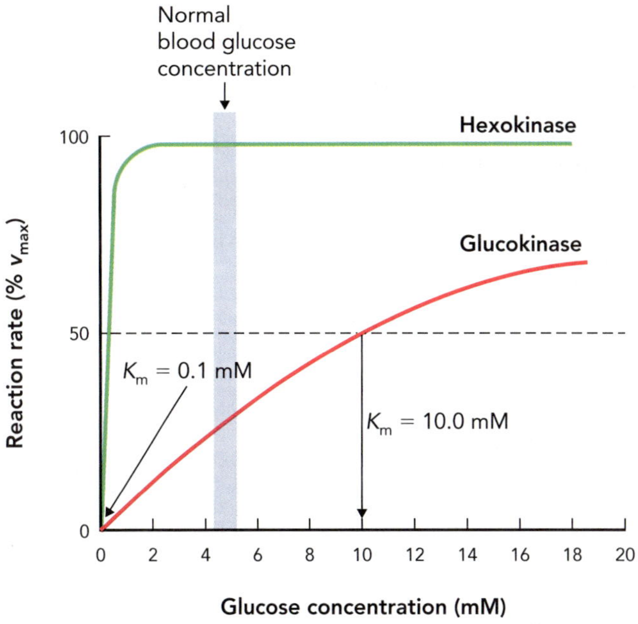

Figure 9.40 Different glucose binding affinities of hexokinase and glucokinase ensure that liver cells sequester glucose (by phosphorylating it to trap it within the cell) when it is very abundant.

The role of glucokinase in liver cells is to trap the extra glucose made available from the diet so that it can be stored as glycogen for an energy source later (for example, before breakfast; see Figure 9.1). Glucose transporters, which move glucose across cell membranes, do not transport phosphorylated glucose. This means that phosphorylation by glucokinase effectively retains glucose within the cell. Glucose-6-P is readily converted to glucose-1-phosphate (glucose-1-P) in liver cells by the enzyme phosphoglucomutase, which is the first step in glycogen synthesis (see Chapter 12). By being active in liver cells only when glucose concentrations exceed normal limits (greater than 5 mM), glucokinase ensures that the liver is the major sink for dietary glucose, efficiently removing glucose from the blood to help restore normal blood glucose concentrations.

Another important function of glucokinase is to act as a glucose sensor and stimulate insulin release from **pancreatic β cells** when blood glucose levels are elevated. Expression of glucokinase is increased in response to increased glucose import mediated by glucose transporter type 2 (GLUT2) protein. As shown in **Figure 9.41**, this results in increased flux through glycolysis and net ATP synthesis (step 2). The increase in ATP levels causes inhibition of ATP-sensitive K^+ channels (step 3), membrane depolarization (step 4), and activation of voltage-gated Ca^{2+} channels (step 5). In the final step, intracellular Ca^{2+} triggers fusion of insulin-containing vesicles with the plasma membrane (step 6), causing subsequent release of insulin into the blood (step 7).

Allosteric Control of Phosphofructokinase-1 Activity

Phosphofructokinase-1 (PFK-1) is a well-characterized enzyme in the glycolytic pathway because of its vital role in controlling flux through the pathway. **Figure 9.42** illustrates that AMP and ADP, which are indicators of low energy charge in the cell, and fructose-2,6-bisphosphate (fructose-2,6-BP), a metabolite that controls carbohydrate metabolism (see Chapter 14), function as allosteric activators of PFK-1.

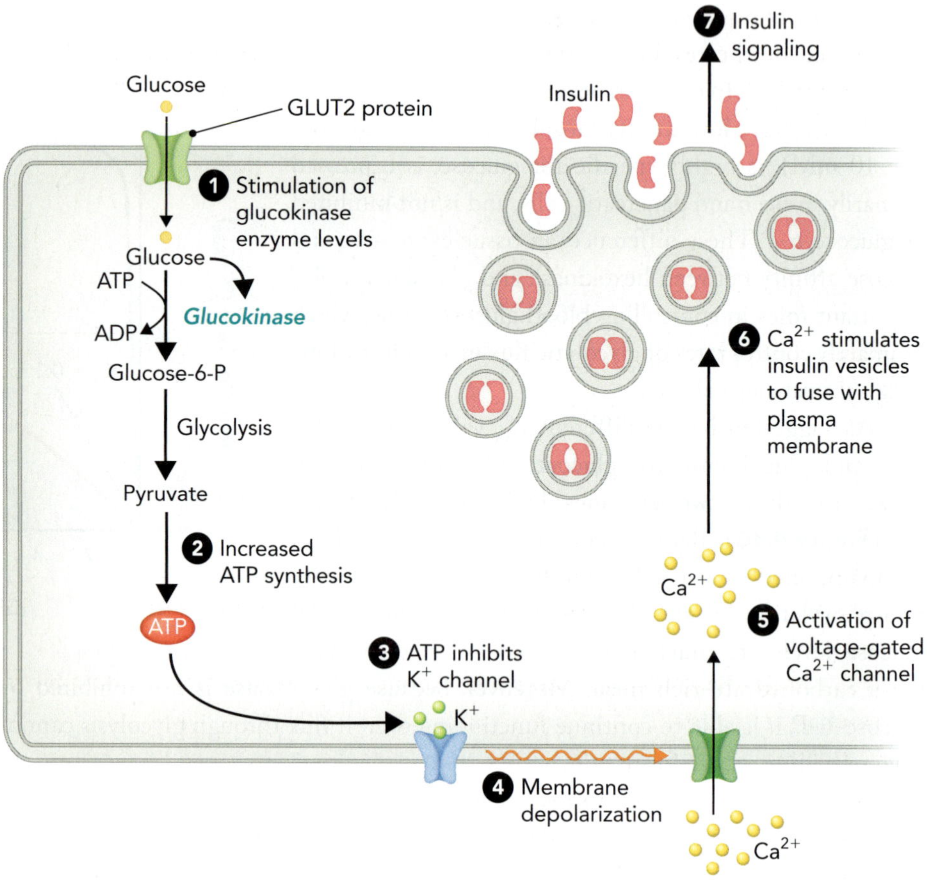

Figure 9.41 Glucokinase functions as a glucose sensor by inducing insulin release from pancreatic β cells in response to elevated levels of blood glucose.

In contrast, elevated levels of ATP, an indicator of high energy charge in the cell, and of citrate, which accumulates when the citrate cycle is inhibited, function as allosteric inhibitors of PFK-1. We showed the molecular structure of a PFK-1 tetramer earlier to emphasize the mechanism of allosteric control (see Figure 9.26). As you can see in **Figure 9.43**, the allosteric binding site for AMP, ADP, and ATP lies in close proximity to the substrate binding sites. Binding of the allosteric effectors changes the overall conformation of the PFK-1 subunits and alters activity at the enzyme active site.

As with other examples of allosteric regulation we have examined, PFK-1 exists in two conformations: the inactive T-state conformation with ATP bound to the allosteric effector site, and the active R-state conformation with AMP or ADP

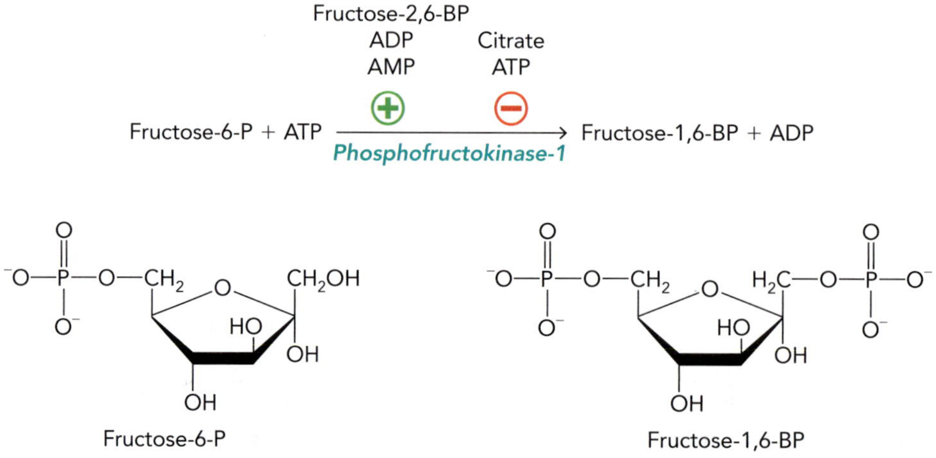

Figure 9.42 Allosteric activators and inhibitors of phosphofructokinase-1 enzyme activity regulate glycolytic flux through reaction 3 of the glycolytic pathway. Fructose-2,6-BP, ADP, and AMP function as activators. ATP and citrate function as inhibitors.

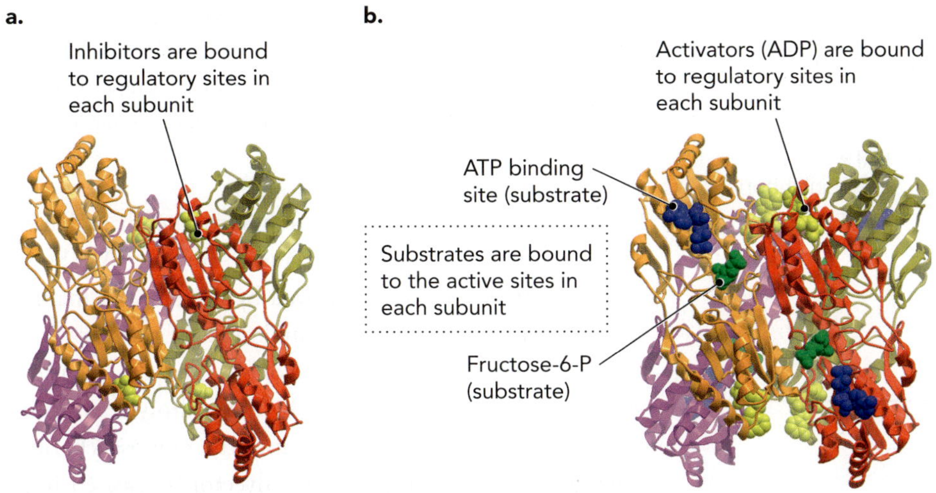

Figure 9.43 The molecular structures of PFK-1 heterotetramers show that the allosteric binding sites lie at the interface between subunits. Binding of allosteric effectors influences the positions of the subunits with respect to each other, which alters catalysis at the active site. **a.** The inactive conformation of PFK-1 is stabilized by binding of inhibitors (yellow space-filling representation) such as ATP or citrate. BASED ON PDB FILE 6PFK. **b.** The active conformation of PFK-1 is stabilized by ADP (yellow space-filling representation) binding to the regulatory site and substrate binding to the active site. BASED ON PDB FILE 4PFK. Note that space-filling models of these two molecular structures are shown in Figure 9.26.

bound to the same site (**Figure 9.44**). High ATP concentrations in the cell (high energy charge) increase the proportion of PFK-1 molecules in the inactive T-state conformation, which decreases the affinity of the enzyme active site for the substrate fructose-6-P. In contrast, high AMP and ADP concentrations (low energy charge) increase the proportion of PFK-1 molecules in the active R-state conformation, which favors fructose-6-P binding. (Note that ATP is bound to the active site in both the T-state and R-state conformations.)

Figure 9.45 shows how the allosteric regulators ATP, ADP, and fructose-2,6-BP alter the PFK-1 reaction rate as a function of substrate concentration (fructose-6-P). You can see that the activity of PFK-1 is inhibited by high ATP concentrations (shifts the activity curve to the right), whereas PFK-1 activity is maximally induced in the presence of fructose-2,6-BP. Note that fructose-2,6-BP accumulates when fructose-6-P levels increase because of insufficient PFK-1 activity (fructose-2,6-BP functions as an indirect feed-forward regulator). Elevated ADP levels also cause an increase in PFK-1 activity by preventing ATP from binding to the allosteric effector site.

Supply and Demand of Glycolytic Intermediates

Liver and muscle cells obtain glucose for the glycolytic pathway not only from dietary glucose, but also from glycogen, which is a polymeric storage form of glucose (see Chapter 13). Plants store polymeric glucose in starch granules, and animals that eat

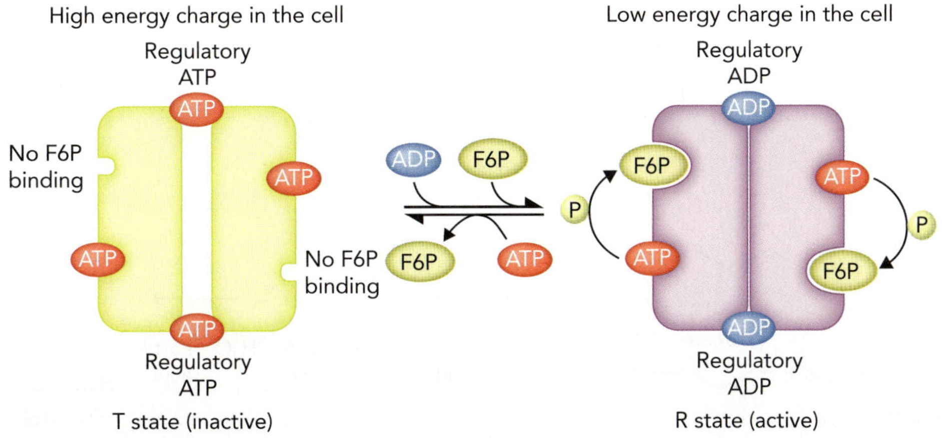

Figure 9.44 ADP and ATP shift the equilibrium between the R (active) and T (inactive) states of PFK-1. (Although PFK-1 functions as a tetramer, it is illustrated here as a dimer for clarity.) When the energy charge of the cell is high, ATP (red) binds to the effector site and stabilizes the T-state conformation. In the presence of elevated levels of ADP (low energy charge in the cell), the R-state conformation is stabilized and the affinity for fructose-6-P (F6P) is increased, leading to product formation.

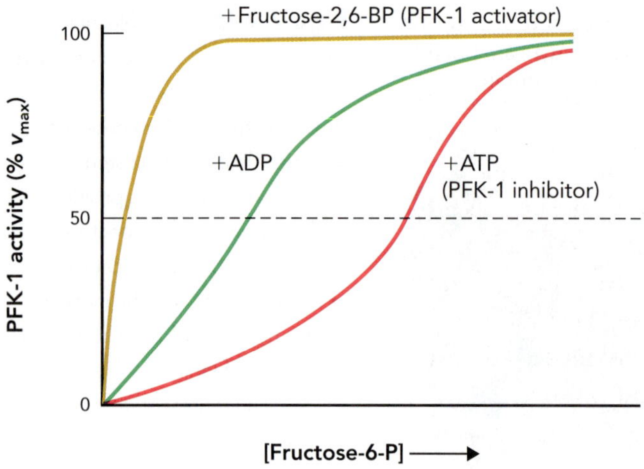

Figure 9.45 Allosteric regulators affect PFK-1 activity, as shown here in this plot of enzyme activity versus fructose-6-P substrate concentration. The enzyme activity curves represent the relative effect of each regulator when present at maximum levels.

plants break down dietary starch into the disaccharide maltose using the salivary enzyme α-amylase. As shown in **Figure 9.46**, maltose is cleaved by the enzyme maltase in the intestine to produce two molecules of glucose, which can be metabolized during glycolysis. The disaccharides sucrose and lactose are cleaved by the hydrolytic enzymes sucrase and lactase, respectively, into the monosaccharides glucose, fructose, and galactose. Glycerol is released from triglycerides after hydrolytic cleavage of fatty acids by lipase enzymes (see Chapter 16).

The conversion of lactose to galactose and glucose by the enzyme lactase may be familiar to you if you are lactose intolerant (lactose sensitive). People with this condition experience considerable discomfort from associated intestinal

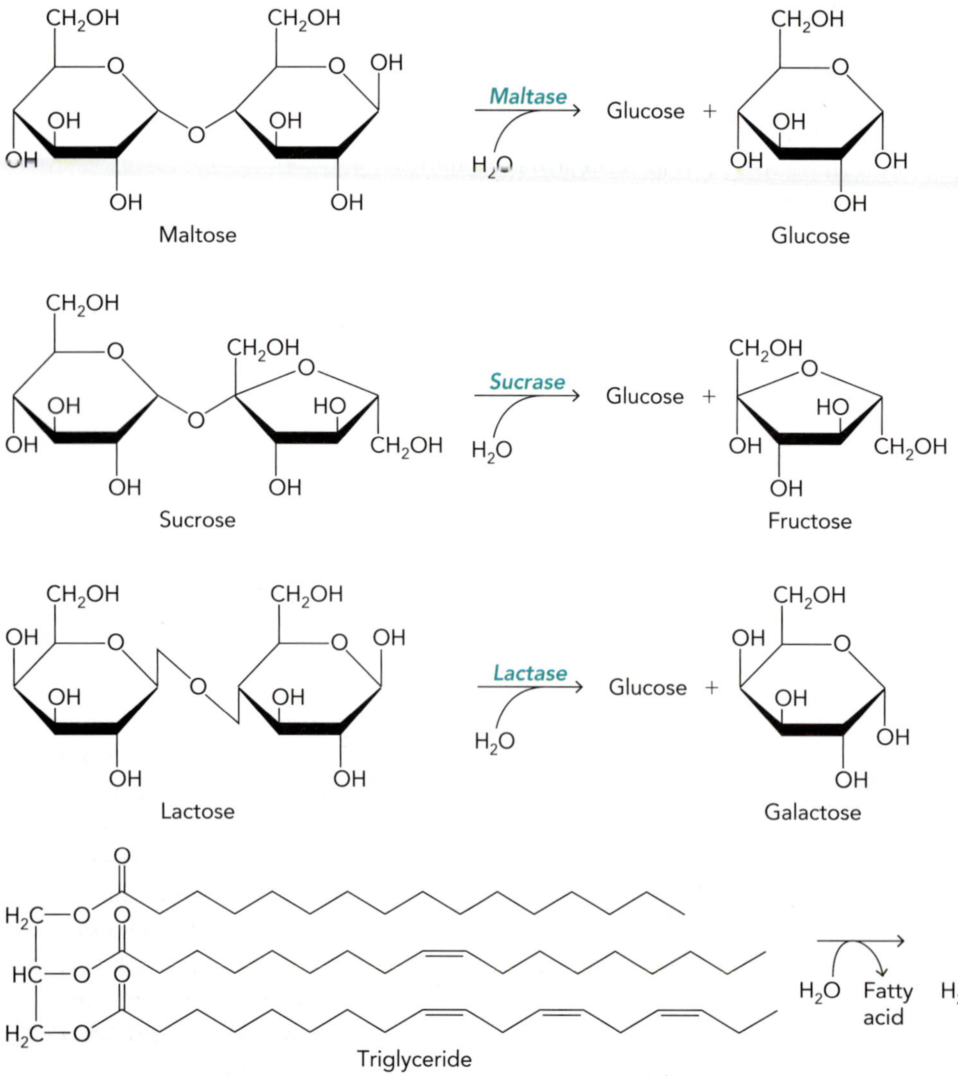

Figure 9.46 The disaccharides maltose, sucrose, and lactose are cleaved by hydrolytic enzymes in intestinal epithelial cells to produce the monosaccharides glucose, fructose, and galactose. Triglycerides are degraded by hydrolytic lipases to produce fatty acids and glycerol.

symptoms (excessive flatulence and diarrhea) if they eat lactose-containing foods such as dairy products. The human gene for lactase is expressed at high levels in infants to aid in the digestion of lactose in breast milk; however, lactase expression normally declines in adults, with the notable exception of people of Scandinavian descent. The explanation is that until modern times, most human populations did not include dairy products in their diets. Over evolutionary time, the need for lactase gene expression in adults was lost. In contrast, Scandinavians have a long history of dairy farming and maintained lactose in their diets well into adulthood.

The intestinal symptoms of **lactose intolerance** are caused by the activity of naturally occurring anaerobic bacteria in the human intestine. These bacteria, from the genus *Lactobacillus*, ferment undigested lactose to lactate, producing hydrogen (H_2) and methane (CH_4) gases as side products. Diarrhea becomes a problem if the amount of unhydrolyzed lactose is so high that it osmotically increases water flow into the intestine.

The simplest way to prevent the gastrointestinal symptoms resulting from lactose intolerance is to avoid food products containing lactose; for example, by eating soy milk products rather than dairy products. Biotechnology has provided a way to have your ice cream and eat it, too, through the industrial production of purified lactase enzyme (**Figure 9.47**). Commercial lactase can be ingested in pill form shortly before eating and, in most cases, can dramatically reduce the severity of symptoms. Lactase can also be added in liquid form to milk products, which hydrolyzes most of the lactose to glucose and galactose prior to consumption.

Fructose, galactose, and glycerol enter the glycolytic pathway through a variety of routes, many of which require additional enzymatic reactions, as shown in **Figure 9.48**. Glycerol, for example, enters glycolysis through a two-step reaction requiring the enzymes glycerol kinase and glycerol phosphate dehydrogenase to form the glycolytic intermediate dihydroxyacetone-P. Some metabolites enter glycolysis through a single phosphorylation reaction, such as fructose in muscle cells, which is converted to fructose-6-P by the enzyme hexokinase. Fructose metabolism in liver cells, however,

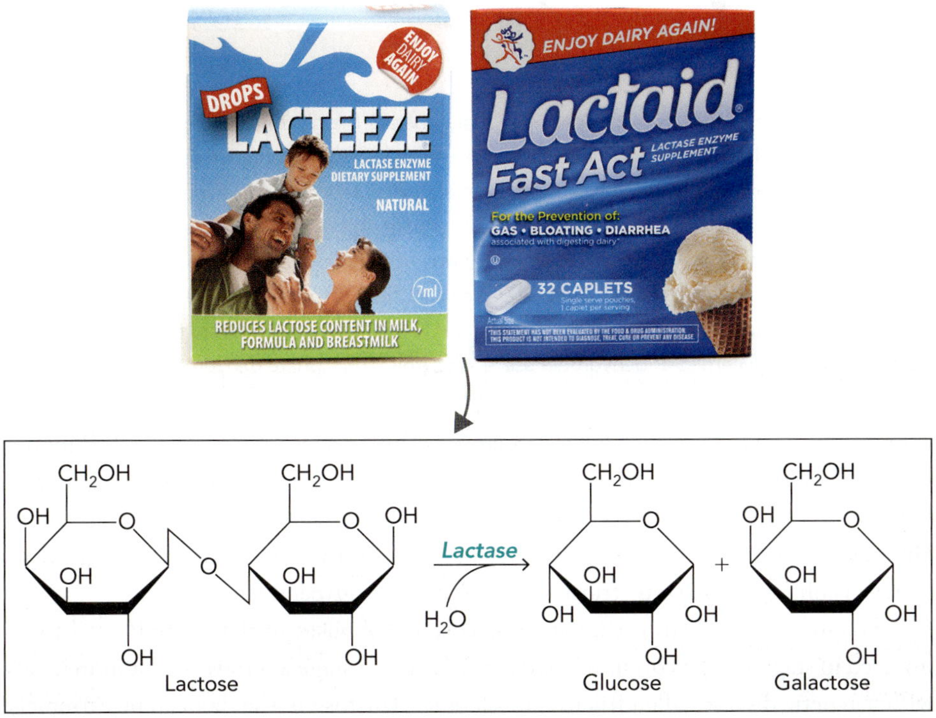

Figure 9.47 Purified lactase enzyme is available commercially and can be taken as a pill at mealtime or added to liquids prior to consumption. The lactase enzyme in these products hydrolyzes lactose and alleviates the gastrointestinal symptoms associated with lactose intolerance.
PHOTOS: THE PHOTO WORKS.

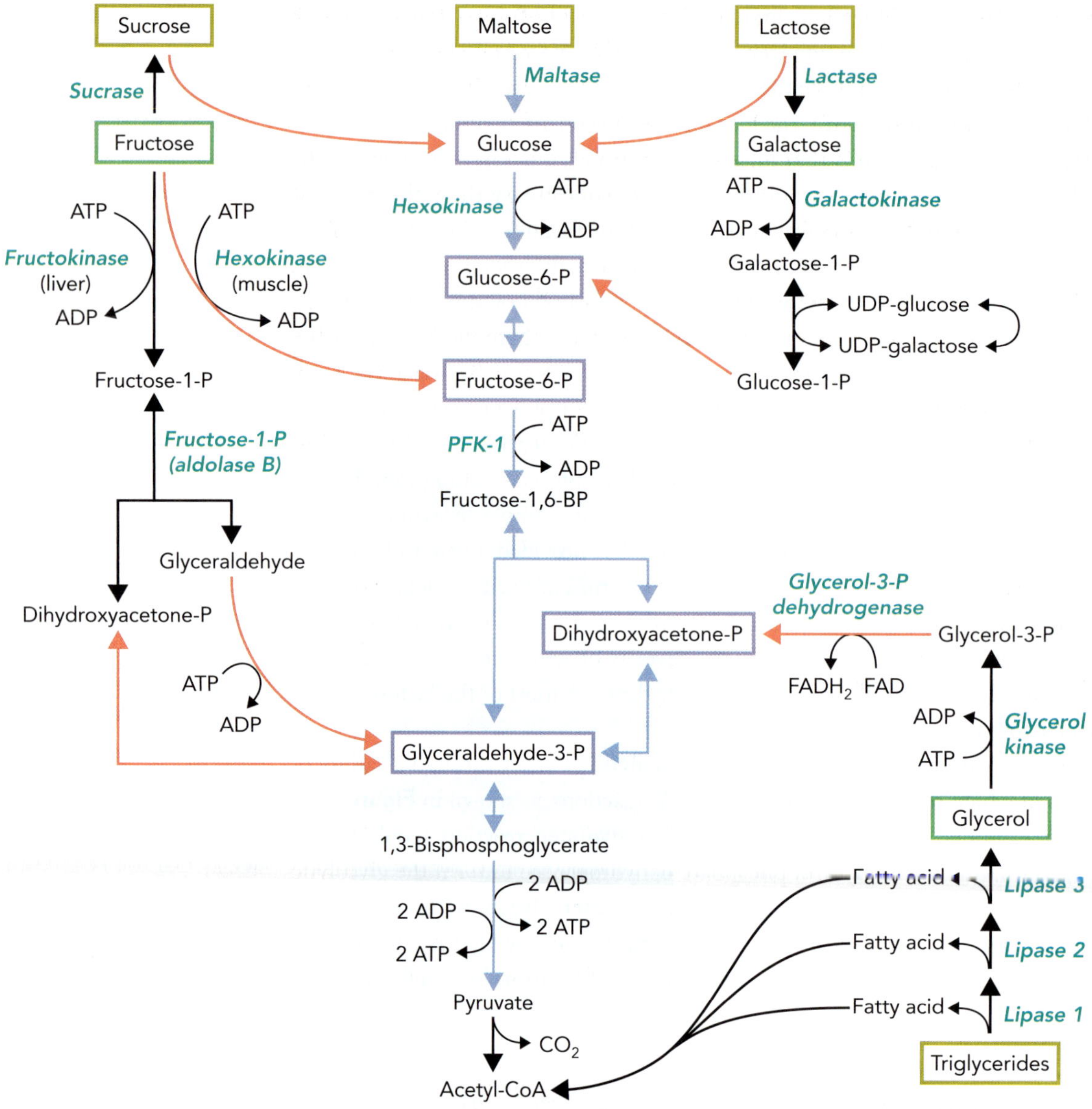

Figure 9.48 Metabolism of fructose, galactose, and glycerol involves feeder pathways that generate the glycolytic intermediates glucose-6-P, fructose-6-P, dihydroxyacetone-P, and glyceraldehyde-3-P. Increased substrate availability from alternative carbohydrate sources stimulates metabolic flux through the pathway. Red arrows indicate positions where metabolites feed into the glycolytic pathway, which is indicated by blue arrows. Glycolytic pathway metabolites produced from other pathways are boxed in blue. Dietary sources of carbohydrates and triglycerides are boxed in yellow green, which can be broken down into simpler molecules (boxed in green), which can eventually feed into the glycolytic pathway. Note that the 2 net ATP shown in the ATP earnings stage from glyceraldehyde-3-P to pyruvate refers to the metabolism of hexose sugars, whereas only 1 net ATP can be produced from the C_3 metabolite glycerol. Glycerol-3-P = glycerol-3-phosphate; glycerol-3-P dehydrogenase = glycerol-3-phosphate dehydrogenase.

is more complicated, in that dietary fructose is first converted to fructose-1-P by the enzyme fructokinase. Fructose-1-P is cleaved by fructose-1-P aldolase to generate dihydroxyacetone-P and glyceraldehyde. The dihydroxyacetone-P is isomerized to glyceraldehyde-3-P by the glycolytic enzyme triose phosphate isomerase, and glyceraldehyde is phosphorylated by triose kinase to produce glyceraldehyde-3-P.

Fructose metabolism is similar to glucose metabolism in that 2 ATP need to be invested in stage 1 of glycolysis to yield a net 2 ATP in stage 2. However, for individuals with a genetic disease called fructose intolerance, fructose in the diet can be extremely toxic. **Fructose intolerance** is due to deficiencies in the enzyme fructose-1-P aldolase, also called aldolase B. People with fructose intolerance cannot eat foods containing fructose because it leads to the buildup of fructose-1-P, which has no alternative metabolic fate in the absence of aldolase B. Normally, P_i turnover in cells is required to synthesize large amounts of ATP continually by oxidative phosphorylation, but the accumulation of fructose-1-P acts as a P_i "sink"—it ties up available phosphate in the liver that would normally be recycled by ATP hydrolysis. Under these conditions, liver cells are quickly depleted of ATP, causing ATP-dependent cation membrane pumps to shut down. This causes cell lysis and liver damage.

As shown in **Figure 9.49**, entry of galactose into glycolysis requires four enzymatic reactions; these begin with galactokinase, which phosphorylates galactose to form galactose-1-phosphate (galactose-1-P). Then, the enzyme galactose-1-P uridylyltransferase produces glucose-1-P and uridine diphosphate-galactose (UDP-galactose) from galactose-1-P and UDP-glucose. Of the two products produced,

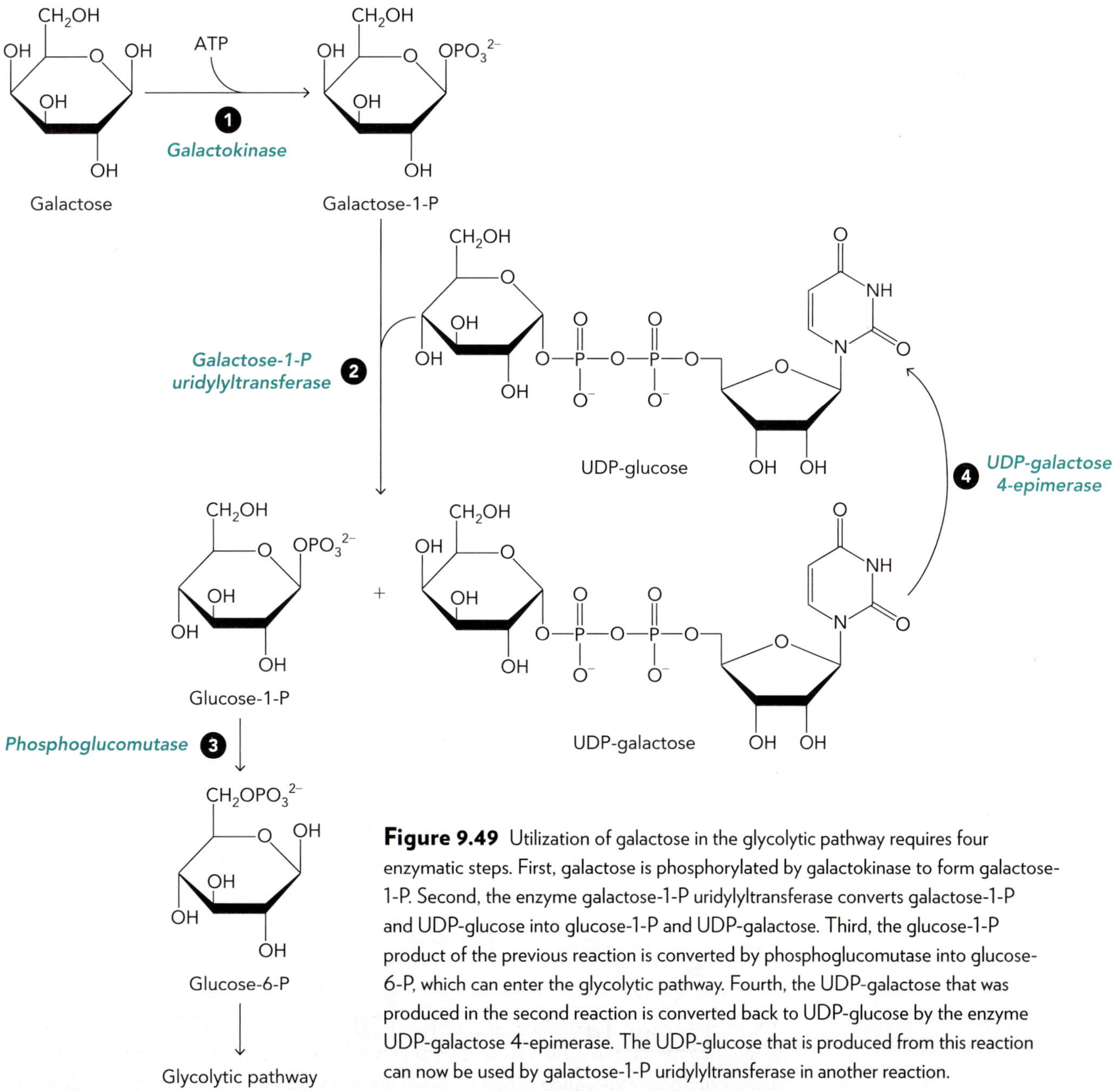

Figure 9.49 Utilization of galactose in the glycolytic pathway requires four enzymatic steps. First, galactose is phosphorylated by galactokinase to form galactose-1-P. Second, the enzyme galactose-1-P uridylyltransferase converts galactose-1-P and UDP-glucose into glucose-1-P and UDP-galactose. Third, the glucose-1-P product of the previous reaction is converted by phosphoglucomutase into glucose-6-P, which can enter the glycolytic pathway. Fourth, the UDP-galactose that was produced in the second reaction is converted back to UDP-glucose by the enzyme UDP-galactose 4-epimerase. The UDP-glucose that is produced from this reaction can now be used by galactose-1-P uridylyltransferase in another reaction.

glucose-1-P enters the glycolytic pathway by isomerization to glucose-6-P by phosphoglucomutase, which is the same enzyme that converts glucose-1-P units released by glycogen phosphorylase to glucose-6-P. The other product that is produced by galactose-1-P uridylyltransferase, UDP-galactose, is converted back to UDP-glucose by the enzyme UDP-galactose 4-epimerase. Now, this UDP-glucose can be used by galactose-1-P uridylyltransferase to form another molecule of glucose-1-P, which can enter the glycolytic pathway after conversion to glucose-6-P.

Defects in the enzyme galactose-1-P uridylyltransferase cause the hereditary disease **galactosemia**, which is characterized by developmental disabilities and liver damage in infants that obtain their nourishment from milk. The excess galactose accumulates in the blood and can be identified in the urine. If diagnosed in time, babies can be switched to a galactose-free diet, and most of the symptoms of galactose toxicity can be reversed.

Glycolytic intermediates serve important roles in anabolic pathways by providing carbon skeletons for biosynthesis (**Figure 9.50**). (We describe this in more detail in Chapter 14.) For example, pentose phosphates derived from glucose-6-P are required for nucleotide synthesis and the production of NADPH by the pentose phosphate pathway.

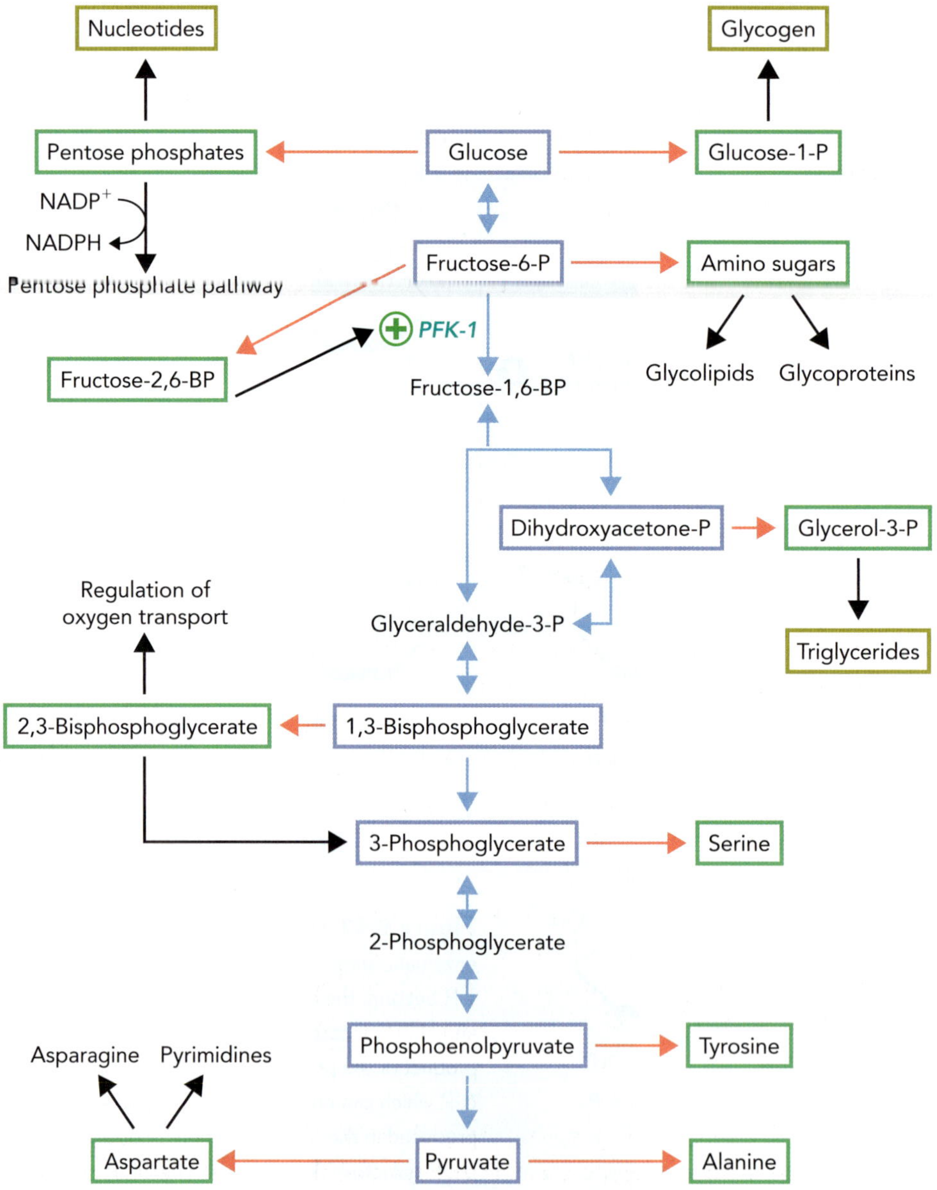

Figure 9.50 Glycolytic intermediates provide carbon backbones for numerous anabolic pathways and several regulatory molecules. Outflow of metabolites from glycolysis decreases substrate availability and thereby inhibits pathway flux. The glycolytic metabolites that are converted for use in other pathways are boxed in blue and are shown leaving the glycolytic pathway via red arrows. PFK-1 activity is stimulated by fructose-2,6-BP to produce fructose-1,6,-BP.

Moreover, glucose-6-P and dihydroxyacetone-P serve as building blocks in anabolic pathways when flux through glycolysis is reduced because of a high energy charge in the cell. Glucose-6-P is converted to glucose-1-P in liver and muscle cells for glycogen synthesis, and dihydroxyacetone-P is used to make glycerol phosphate, which is required in adipocytes to synthesize triglycerides. Figure 9.50 also shows that two important regulatory molecules—2,3-bisphosphoglycerate and fructose-2,6-bisphosphate—are derived from glycolytic intermediates through one-step phosphorylation reactions.

concept integration 9.4

Why is allosteric regulation of phosphofructokinase-1 activity so effective in controlling flux through the glycolytic pathway?

Three of the 10 glycolytic reactions are essentially irreversible under cellular conditions ($\Delta G << 0$) and are subject to regulation: the reactions catalyzed by hexokinase (and glucokinase), PFK-1, and pyruvate kinase. PFK-1 functions as a multi-subunit protein complex, which is allosterically controlled by metabolites that signal the energy charge of the cell. Specifically, AMP, ADP, and fructose-2,6-BP signal low energy charge and allosterically activate PFK-1 activity, whereas ATP and citrate signal high energy charge in the cell and allosterically inhibit PFK-1 activity. Allosteric control of enzyme activity permits dramatic increases or decreases in activity over narrow ranges of substrate concentration because of cooperative interactions between protein subunits. Because PFK-1 catalyzes an irreversible reaction in the glycolytic pathway ($\Delta G << 0$) and is subject to allosteric control, small changes in its activity have major effects on glycolytic flux.

9.5 Metabolic Fate of Pyruvate

Pyruvate, the final product of glycolysis, is metabolized in one of three ways, depending on the organism and the availability of oxygen (**Figure 9.51**). Under aerobic conditions, the majority of pyruvate is metabolized in the mitochondria to CO_2 and H_2O by the citrate cycle and electron transport chain. However, under anaerobic conditions, such as occurs in muscle cells during strenuous exercise, pyruvate is converted to lactate by the enzyme lactate dehydrogenase. A variety of microorganisms also convert pyruvate to lactate under anaerobic conditions; for example, *Lactococcus lactis*, a strain of bacteria used in the dairy industry to produce foods such as yogurt and cheese. A third fate of pyruvate occurs in some microorganisms, such as the yeast *Saccharomyces cerevisiae*, which convert pyruvate to CO_2 and ethanol using the enzymes pyruvate decarboxylase and alcohol dehydrogenase (see Figure 1.2).

A critical function of pyruvate metabolism in all three pathways is to regenerate the oxidized form of the

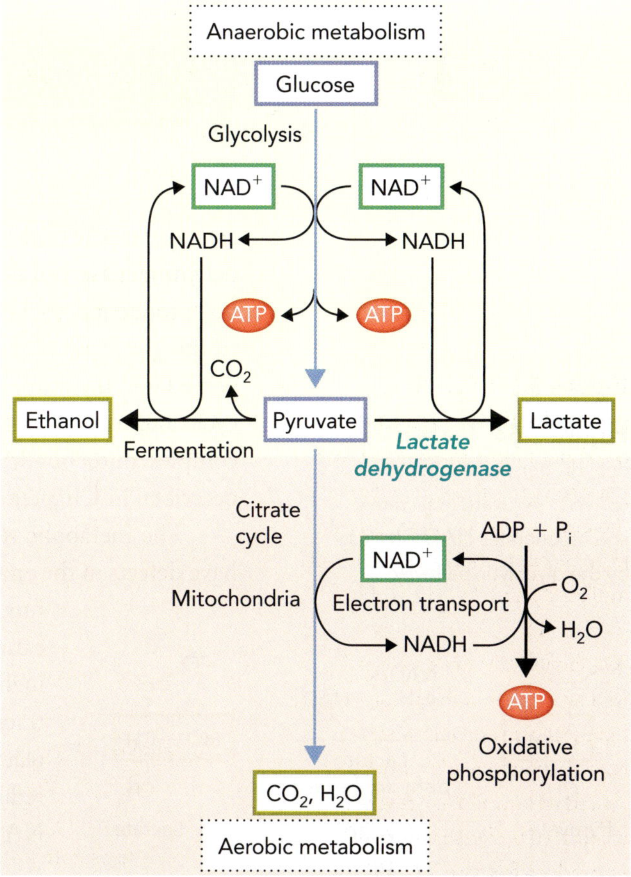

Figure 9.51 Pyruvate has three metabolic fates, depending on the type of organism and the availability of oxygen. Under aerobic conditions, pyruvate is metabolized in the mitochondria to produce CO_2 and H_2O. However, when oxygen is limited, pyruvate is converted to lactate or ethanol (fermentation).

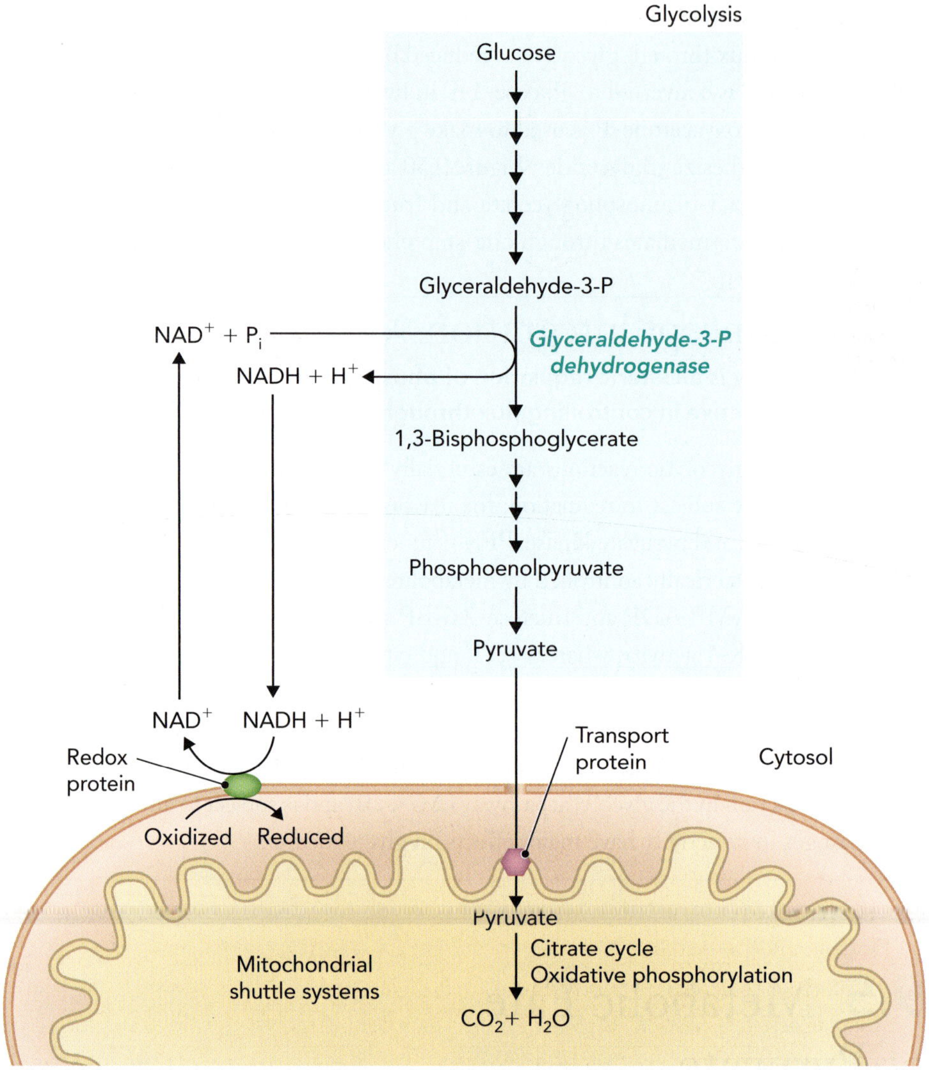

Figure 9.52 Regeneration of NAD^+ to maintain glycolytic flux through the glyceraldehyde-3-P dehydrogenase reaction requires the use of mitochondrial shuttle systems under aerobic conditions. This leads to the complete oxidation of pyruvate to CO_2 and H_2O by the citrate cycle and oxidative phosphorylation. The multiple short arrows represent individual enzymatic steps.

coenzyme NAD^+, which is required to maintain flux through the glyceraldehyde-3-P dehydrogenase reaction in glycolysis (Figure 9.51). In the case of lactate and ethanol production, NAD^+ is regenerated in the cytoplasm and made readily available to the glycolytic pathway. However, as shown in **Figure 9.52**, when pyruvate is transported into the mitochondrial matrix and fully oxidized to CO_2 and H_2O by the citrate cycle and oxidative phosphorylation, NAD^+ regeneration in the cytoplasm requires a complex mitochondrial shuttle system. This system is linked to the citrate cycle, as described in Chapter 10.

The metabolic requirement for NAD^+ in glycolysis can be seen in people who have defects in the enzyme lactate dehydrogenase. This recessive genetic disease (loss of function) is called **lactate dehydrogenase deficiency**. These patients cannot maintain sustained periods of intense exercise because of an inability to use glycolysis to produce ATP needed for muscle contraction under anaerobic conditions. The lactate dehydrogenase reaction is shown in **Figure 9.53**, where you can see that pyruvate reduction to lactate is coupled to the oxidation of NADH to produce NAD^+. When lactate dehydrogenase is not fully functional, NADH

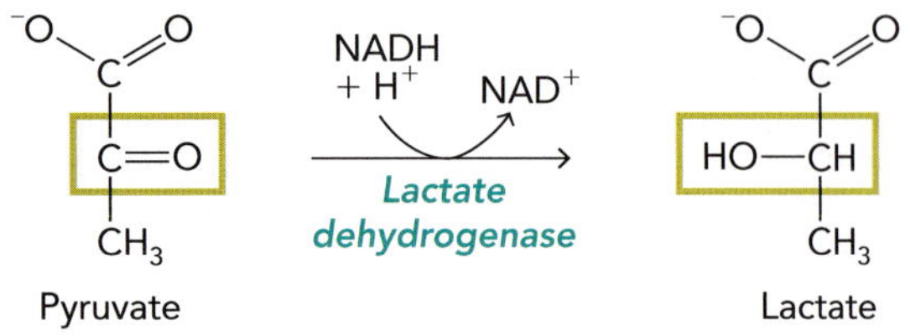

Figure 9.53 The lactate dehydrogenase reaction converts pyruvate to lactate and regenerates NAD^+ for use by the glycolytic pathway.

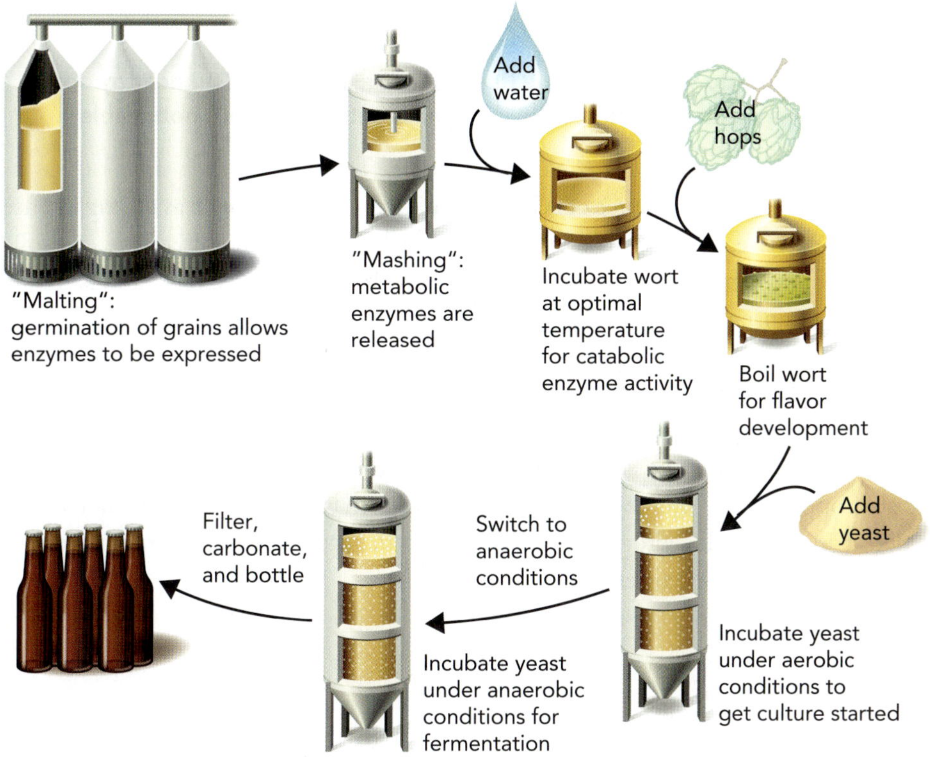

Figure 9.54 Carbonated alcoholic beer is made by yeast fermentation of sugars in barley. Yeast metabolize the simple sugars produced by enzymatic degradation of germinated barley seeds, generating pyruvate using the glycolytic pathway.

oxidation to regenerate NAD^+ does not occur at a high enough rate to sustain glycolysis. This causes the muscle cells to run out of ATP quickly, leading to fatigue and even muscle damage if anaerobic conditions persist.

Brewer's yeast, *Saccharomyces cerevisiae*, has been used for hundreds of years in the production of beer by taking advantage of the fact that anaerobic conditions induce yeast to rely on glycolysis to generate ATP. In the process, they produce pyruvate, which is metabolized by alcoholic fermentation to regenerate the necessary NAD^+. As shown in **Figure 9.54**, the source of carbohydrate used in the production of beer comes from cereal grains such as barley, which are first allowed to germinate during the malting process. This induces the expression of degradative enzymes such as α-amylase and maltase. The next step is mashing, which crushes the germinated barley to release metabolic enzymes, before adding water to produce a mixture called wort. Incubating the wort at an optimum temperature activates catabolic enzymes, which degrade carbohydrates and proteins in the mixture to generate metabolites for the glycolytic pathway. The undigested cellular debris is removed, and the wort is boiled with various ingredients, such as hops, to give the beer a distinct flavor. Once the wort has cooled, a live yeast culture is added under aerobic conditions to permit active growth of a starter culture for a short period of time. The yeast culture is then switched to anaerobic conditions, and fermentation of the remaining sugars takes place, producing ethanol and CO_2. In commercial beer production, the beer is aged, filtered, and carbonated (injection of CO_2) before bottling. Microbreweries, however, often use a second fermentation step at the time of bottling to carbonate the filtered beer by introducing a small amount of sugar and fresh yeast into the beer.

concept integration 9.5

How do you explain the finding that ~1 of every 500 healthy adults has reduced lactate dehydrogenase activity without noticeable symptoms?

Glycolysis is the primary source of ATP when oxygen is limiting, such as during intense strenuous exercise, and also in erythrocytes, which lack mitochondria. The cytosolic enzyme lactate dehydrogenase regenerates NAD^+ by oxidizing NADH and is highly active under anaerobic conditions because of the buildup of pyruvate, which cannot be metabolized to acetyl-CoA. Lactate dehydrogenase deficiency is a recessive genetic disease, and heterozygous carriers for lactate dehydrogenase mutations must be fairly common in the population. Because most of the pyruvate generated by glycolysis under normal aerobic conditions is completely oxidized to CO_2 and H_2O in mitochondria, most people with a heterozygous lactate dehydrogenase mutation would not notice a defect in lactate dehydrogenase unless they attempted intense anaerobic exercise, which is likely a rare occurrence.

chapter summary

9.1 Overview of Metabolism

- Catabolic pathways degrade macromolecules and nutrients to capture energy in the form of ATP and reduction potential. Anabolic pathways use this energy to synthesize biomolecules for the cell.
- Metabolic flux refers to the rate at which metabolites are interconverted in catabolic and anabolic pathways. It is determined by (1) availability of substrates and (2) amount of enzyme activity (enzyme level and catalytic activity).
- Metabolic pathways are highly interdependent and tightly controlled. Flux is always occurring in metabolic pathways, even in opposing catabolic and anabolic pathways, such as glycolysis and gluconeogenesis.
- Metabolism is hierarchical, consisting of macromolecules (proteins, nucleic acids, carbohydrates, and lipids), primary metabolites (amino acids, nucleotides, fatty acids, glucose, pyruvate, and acetyl-CoA), and small molecules (NH_4^+, CO_2, NADH, $FADH_2$, O_2, ATP, and H_2O).
- The spontaneity of metabolic reactions is determined by the change in actual free energy ΔG, which is calculated by taking into consideration the actual concentrations of substrates and products in the cell. Reactions with $\Delta G < 0$ are spontaneous in the forward direction, even if $\Delta G^{\circ\prime} > 0$ (nonspontaneous) for the same reaction.
- Coupled reactions in a pathway provide a mechanism to overcome unfavorable individual reactions if the total sum of all individual standard free energy changes is less than zero. Coupled reactions in cells often involve use of the phosphoryl transfer energy available in ATP ($\Delta G^{\circ\prime} = -30.5$ kJ/mol).
- Protein complexes containing multiple enzymes for a related set of pathway reactions provide one way to facilitate coupled reactions, as they decrease loss of shared intermediates due to diffusion. Some protein complexes are built around scaffold proteins that function as organizing centers.

9.2 Structures of Simple Sugars

- Simple sugars are metabolites that feed into the glycolytic pathway. They include aldose sugars, such as glucose, and ketose sugars, such as fructose.
- Aldose sugars in the open-chain conformation can be oxidized to carboxylic acids in a redox reaction with copper ($Cu^{2+} \rightarrow Cu^+$), a reaction known as Benedict's test. Sugars that react with Cu^{2+} are called reducing sugars (for example, glucose, galactose, and lactose), whereas unreactive sugars are nonreducing.
- Disaccharides contain two simple sugars covalently linked by an O-glycosidic bond, which can be either an α or β glycosidic bond. Maltose contains two glucose molecules linked by an α(1→4) glycosidic bond, and lactose contains glucose and galactose molecules linked by a β(1→4) glycosidic bond.

9.3 Glycolysis Generates ATP under Anaerobic Conditions

- Glycolysis is considered a core metabolic pathway for three reasons: (1) glycolytic enzymes are highly conserved; (2) glycolysis is the primary pathway for ATP generation under anaerobic conditions; and (3) glycolytic intermediates are shared metabolites in a variety of interconnected pathways.

- Glycolysis converts one molecule of glucose to two molecules of pyruvate with no loss of carbon or oxygen atoms. It generates two net ATP molecules and two NADH molecules for every glucose molecule that is metabolized.
- The glycolytic pathway consists of 10 enzymatic reactions: five reactions in stage 1, the ATP investment phase, and five reactions in stage 2, the ATP earnings phase. Because two molecules of glyceraldehyde-3-P are generated for every molecule of glucose that is metabolized, all of the reactions in stage 2 occur twice for each glucose molecule that enters the pathway.
- Three glycolytic enzymes (hexokinase, phosphofructokinase-1, pyruvate kinase) catalyze highly favorable reactions ($\Delta G << 0$) in glycolysis and are regulated to control flux through the pathway. The other seven enzymes catalyze readily reversible reactions that are shared with the gluconeogenic pathway.
- Substrate-level phosphorylation reactions generate ATP by direct transfer of a phosphoryl group from a donor to ADP. Phosphoryl donors in substrate-level phosphorylation reactions have standard free energy changes of phosphate hydrolysis that are more negative than that of ATP hydrolysis.
- Reaction 10 in the glycolytic pathway is catalyzed by the enzyme pyruvate kinase, which uses substrate-level phosphorylation to transfer the phosphoryl group from phosphoenolpyruvate to ADP. Because this reaction occurs twice for every glucose molecule that enters the glycolytic pathway, a net of 2 ATP is generated.

9.4 Regulation of the Glycolytic Pathway

- Glucokinase is an isozyme of hexokinase that converts glucose to glucose-6-P in pancreatic and liver cells. However, unlike hexokinase, glucokinase has a very low affinity for glucose (100 times lower) and is not feedback inhibited by glucose-6-P, which results in its activity being controlled by physiologic glucose levels.
- Increased blood glucose levels stimulate glucokinase activity in pancreatic β cells. This leads to increased flux through the glycolytic pathway, elevated ATP levels, and ultimately Ca^{2+}-mediated stimulation of insulin release from the cells. This glucose-sensing function of glucokinase is crucial to maintaining safe homeostatic levels of glucose in the blood.
- Phosphofructokinase-1 (PFK-1) is allosterically regulated by metabolites in the cell that signal changes in the energy charge and flux through the glycolytic and citrate cycle pathways. PFK-1 is activated by AMP, ADP, and fructose-2,6,-bisphosphate and is inhibited by ATP and citrate.
- ADP, AMP, and ATP bind to an allosteric effector site located at the interface of two PFK-1 subunits. ADP and AMP bind with high affinity to the active R-state conformation and stabilize it, whereas ATP binds to the same site when the protein complex is in the inactive T-state conformation, leading to inhibition of PFK-1 activity.
- In addition to glucose, fructose and galactose can also be converted to glycolytic intermediates and used to generate ATP. The disaccharide sucrose contains a glucose molecule linked to fructose and is hydrolyzed by the enzyme sucrase, whereas lactose contains a glucose molecule linked to galactose and is hydrolyzed by the enzyme lactase.
- Lactose intolerance is a natural condition in adult humans in which the level of lactase in the intestine decreases with age, resulting in gastrointestinal problems when large quantities of dairy products containing lactose are ingested. Intestinal anaerobic bacteria ferment the undigested lactose to produce H_2 and CH_4 gases.
- Under high energy charge conditions, glycolytic intermediates provide metabolites for anabolic pathways. For example, glucose-6-P is converted to pentose phosphates and used for nucleotide biosynthesis or production of NADPH by the pentose phosphate pathway. Several C_3 metabolites derived from stage 2 glycolytic reactions provide precursor molecules for amino acid biosynthesis.

9.5 Metabolic Fate of Pyruvate

- Pyruvate is metabolized under aerobic conditions in the mitochondria to acetyl-CoA and ultimately to CO_2 and H_2O by the citrate cycle and electron transport chain, generating the bulk of ATP derived from glucose metabolism.
- Under anaerobic conditions, such as in exercising muscle or in microorganisms when O_2 levels in the environment are low, pyruvate is reduced by the enzyme lactate dehydrogenase to produce lactate. The yeast *Saccharomyces cerevisiae* uses alcoholic fermentation under anaerobic conditions to convert pyruvate to CO_2 and ethanol.
- A critical function of pyruvate metabolism is to replenish NAD^+ levels in the cytoplasm in order to maintain flux through the glyceraldehyde-3-P dehydrogenase reaction. Under anaerobic conditions, this is done by the enzymes lactate dehydrogenase and alcohol dehydrogenase, whereas under aerobic conditions, mitochondrial shuttle systems in eukaryotic cells are required.

biochemical terms

(in order of appearance in text)
metabolism (p. 430)
glycolysis (p. 430)
KEGG (Kyoto Encyclopedia of Genes and Genomes) (p. 432)
mass action ratio (Q) (p. 436)
aldose (p. 440)
ketose (p. 440)
chiral center (p. 441)
D isomer (p. 441)
L isomer (p. 441)
epimers (p. 441)
hemiacetal (p. 442)
hemiketal (p. 442)
anomeric carbon (p. 442)
anomers (p. 442)
pyranose (p. 442)
furanose (p. 442)
Benedict's test (p. 443)
reducing sugar (p. 444)

nonreducing sugar (p. 444)
O-glycosidic bond (p. 444)
hexokinase (p. 447)
phosphofructokinase-1 (p. 448)
pyruvate kinase (p. 448)
substrate-level phosphorylation (p. 449)
rate limiting (p. 464)
pancreatic β cells (p. 465)
lactose intolerance (p. 469)
fructose intolerance (p. 470)
galactosemia (p. 472)
lactate dehydrogenase deficiency (p. 474)

review questions

1. Describe the functional role of catabolic and anabolic pathways in metabolism.
2. What is meant by the term *metabolic flux*, and how is it regulated?
3. Describe how glucagon and insulin signaling alter metabolic flux through glucose metabolizing pathways before and after breakfast.
4. List the key macromolecules, primary metabolites, and small molecules in metabolism.
5. How can a chemical reaction that is energetically unfavorable under standard conditions ($\Delta G^{\circ\prime} > 0$) occur in a cell?
6. Give three reasons why glycolysis is considered a core metabolic pathway.
7. Draw the chemical structures of glucose and fructose in the open-chain and closed-ring conformations. Which of these simple sugars is an aldose and which is a ketose?
8. Write the net reaction for the glycolytic pathway and describe the primary functions of stage 1 and stage 2.
9. Which reactions in the glycolytic pathway are substrate-level phosphorylations?
10. Compare and contrast the enzymatic properties of hexokinase and glucokinase.
11. Describe allosteric regulation of phosphofructokinase-1.
12. Briefly explain how the abundant dietary disaccharides maltose, lactose, and sucrose are metabolized by the glycolytic pathway.
13. What are the three metabolic fates of pyruvate? Include in your answer the conditions dictating which of these three metabolic fates is most likely to occur.

challenge problems

1. Describe what happens to the metabolic flux of glucose before and after eating breakfast in terms of glycogen and pyruvate levels in liver cells and the two peptide hormones that control blood glucose levels.
2. Explain why fructose is oxidized by Cu^{2+} in Benedict's test, even though this assay should only detect aldose sugars.
3. Is substrate-level phosphorylation restricted to aerobic metabolism? Explain.
4. Use the changes in standard free energy listed in Table 9.2 to calculate the ΔG value for the conversion of 2-phosphoglycerate to phosphoenolpyruvate by the enzyme enolase at 25 °C when the concentration of 2-phosphoglycerate is 10 times higher than the concentration of phosphoenolpyruvate. In the absence of enolase, is the ΔG value for this reaction higher, lower, or the same?
5. Calculate the $\Delta G^{\circ\prime}$ value at 25 °C for the aldolase reaction, which converts fructose-1,6-bisphosphate to glyceraldehyde-3-phosphate and dihydroxyacetone phosphate, given that the equilibrium constant (K_{eq}) for this reaction is 6.8×10^{-5}. What is the effect on the K_{eq} of adding twice as much aldolase to the reaction? Explain.
6. The 10 reactions in glycolysis demonstrate that all of the carbon and oxygen atoms in the substrate glucose are conserved in the product pyruvate.
   a. Draw the chemical structures of the glycolytic substrate glucose in a ring conformation (α-D-glucopyranose) and the two pyruvate products, and use them in the net reaction for the glycolytic pathway. Number all of the carbons in the two structures.
   b. Write out the two glycolytic reactions that account for the loss of six hydrogen atoms in the two pyruvate products relative to the substrate glucose.
7. Glycolysis converts glucose to pyruvate by the following stoichiometry:

$$\text{Glucose} + 2\,\text{ADP} + 2\,\text{P}_i + 2\,\text{NAD}^+ \rightarrow 2\,\text{Pyruvate} + 2\,\text{ATP} + 2\,\text{NADH} + 2\,\text{H}^+ + 2\,\text{H}_2\text{O}$$

Gluconeogenesis converts pyruvate to glucose by the following stoichiometry:

$$2\,\text{Pyruvate} + 4\,\text{ATP} + 2\,\text{GTP} + 2\,\text{NADH} + 2\,\text{H}^+ + 6\,\text{H}_2\text{O} \rightarrow \text{Glucose} + 4\,\text{ADP} + 2\,\text{GDP} + 6\,\text{P}_i + 2\,\text{NAD}^+$$

What is the net reaction when the glycolytic and gluconeogenic pathways are metabolically active at the same time? What does this tell us about the role of enzyme regulation in glycolysis and gluconeogenesis?
8. What key differences between hexokinase and glucokinase ensure that glucose is properly apportioned into muscle and liver cells, given their distinct physiologic roles?

9. Explain why it makes sense that the enzyme pyruvate kinase is activated by fructose-1,6-bisphosphate when blood glucose levels are high.
10. An individual with only 50% of the normal level of PFK-1 enzyme activity was found to be lacking one copy of the PKF-1 gene. After testing more than 100 people who were genetically related to this individual, it was found that a few also had only one PFK-1 gene copy, whereas all others had two copies of the normal PFK-1 gene. What is the simplest explanation for why no individuals were found who lacked both gene copies of PFK-1?
11. A muscle biopsy from an individual who was incapable of carrying out prolonged, intense exercise contained a severe deficiency in the glycolytic enzyme phosphoglycerate mutase, which converts 3-phosphoglycerate to 2-phosphoglycerate.
    a. What is the explanation for the inability of this individual to exercise intensely?
    b. Does this individual suffer from lactic acid buildup in the muscle? Explain.
12. What is the biochemical basis for liver damage in individuals with genetically inherited fructose intolerance? What are some examples of foods and beverages that need to be avoided by these individuals?
13. Infants express the enzyme lactase in their intestines to metabolize lactose. What are the two products of the lactase reaction? How many pyruvate molecules can be generated from the metabolism of one molecule of lactose? How many ATP molecules can be generated from lactose under anaerobic and aerobic conditions?
14. The yeast *Saccharomyces cerevisiae* is used in the production of beer because of its ability to shift readily from aerobic respiration, in which it converts glucose to CO_2 and H_2O, to anaerobic respiration, in which it converts glucose to C_2H_6O and CO_2.
    a. Explain why in the production of beer, the yeast are first grown under aerobic conditions, and then shifted to anaerobic conditions. Why not just grow them under anaerobic conditions the entire time?
    b. Why is it necessary to add sugar and fresh yeast to fermented beer just before bottling to ensure that the stored product is carbonated? Isn't there sufficient carbonation prior to bottling to provide the carbonation in stored beer?

smartwork5

If your instructor assigns homework with Smartwork5, access it here: digital.wwnorton.com/biochem.

suggested reading

Books and Reviews

Bouteldja, N., and Timson, D. J. (2010). The biochemical basis of hereditary fructose intolerance. *Journal of Inherited Metabolic Disease*, *33*, 105–112.

Campbell, A. K., Waud, J. P., and Matthews, S. B. (2009). The molecular basis of lactose intolerance. *Science Progress*, *92*, 241–287.

Gibson, D. M., and Harris, R. A. (2002). *Metabolic regulation in mammals*. London, England: Taylor & Francis.

Kaelin, W. G., Jr., and Thompson, C. B. (2010). Q&A: Cancer: clues from cell metabolism. *Nature*, *465*, 562–564.

Lodolo, E. J., Kock, J. L., Axcell, B. C., and Brooks, M. (2008). The yeast *Saccharomyces cerevisiae*—the main character in beer brewing. *FEMS Yeast Research*, *8*, 1018–1036.

Matschinsky, F. M. (1996). Banting Lecture 1995. A lesson in metabolic regulation inspired by the glucokinase glucose sensor paradigm. *Diabetes*, *45*, 223–241.

Nicholls, D. G., and Ferguson, S. J. (2002). *Bioenergetics3*. San Diego, CA: Academic Press.

Pawson, C. T., and Scott, J. D. (2010). Signal integration through blending, bolstering and bifurcating of intracellular information. *Nature Structural and Molecular Biology*, *17*, 653–658.

Primary Literature

Bernstein, B. E., Michels, P. A., and Hol, W. G. (1997). Synergistic effects of substrate-induced conformational changes in phosphoglycerate kinase activation. *Nature*, *385*, 275–278.

Frank, G. K., Oberndorfer, T. A., Simmons, A. N., Paulus, M. P., Fudge, J. L., Yang, T. T., and Kaye, W. H. (2008). Sucrose activates human taste pathways differently from artificial sweetener. *Neuroimage*, *39*, 1559–1569.

Maekawa, M., Kanda, S., Sudo, K., and Kanno, T. (1984). Estimation of the gene frequency of lactate dehydrogenase subunit deficiencies. *American Journal of Human Genetics*, *36*, 1204–1214.

Schirmer, T., and Evans, P. R. (1990). Structural basis of the allosteric behaviour of phosphofructokinase. *Nature*, *343*, 140–145.

Schleis, T. G. (2007). Interference of maltose, icodextrin, galactose, or xylose with some blood glucose monitoring systems. *Pharmacotherapy*, *27*, 1313–1321.

Citric acid can be isolated from lemons or, in much greater quantities, from a fermentative organism. A combination of high levels of the enzyme citrate synthase in the fungus *Aspergillus niger* and inhibition of other citrate cycle reactions under specialized culture conditions leads to overproduction and export of citrate into the culture medium. Citric acid is used commercially for a wide variety of purposes.

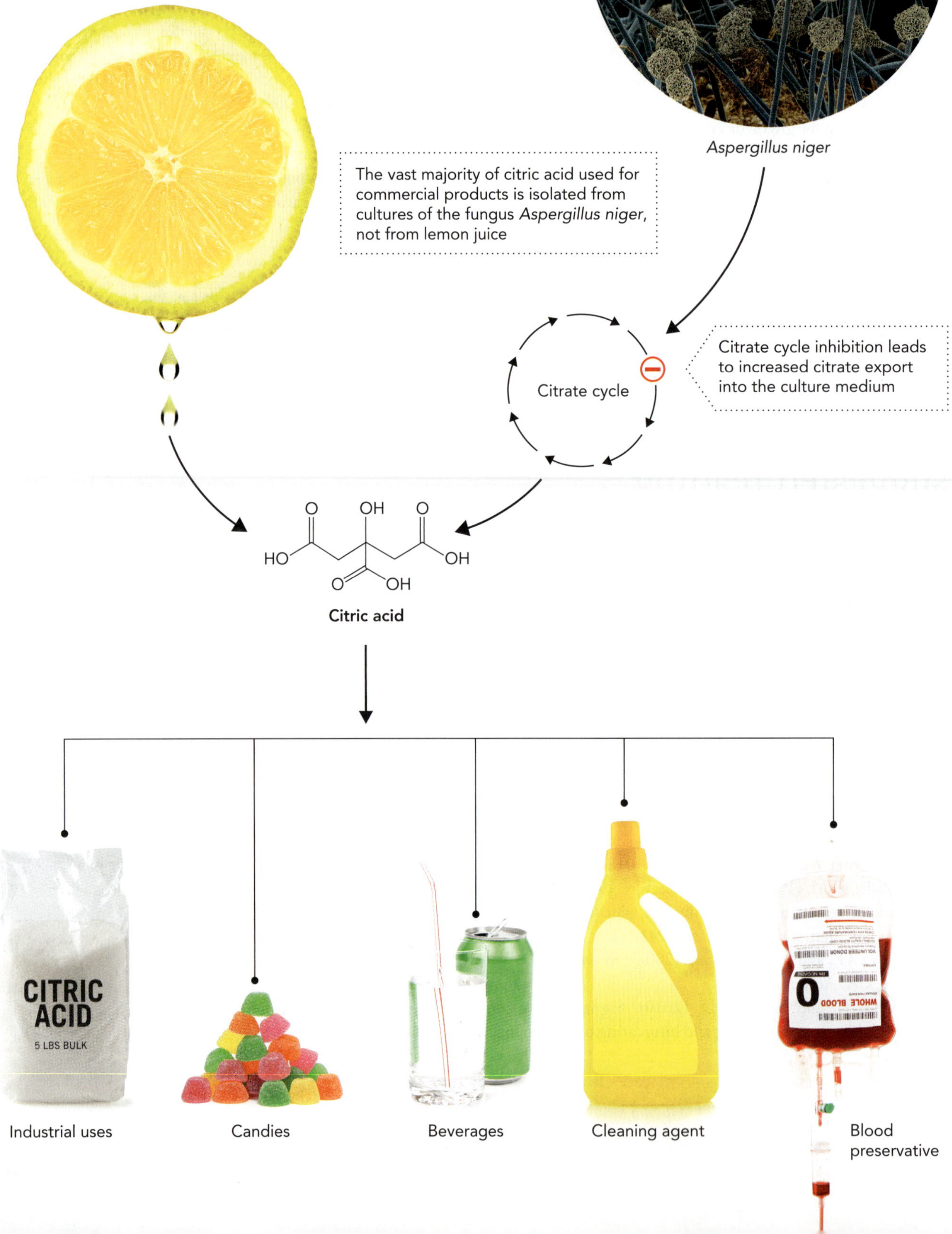

10

The Citrate Cycle

The first reaction of the citrate cycle is catalyzed by the enzyme citrate synthase. The product of this reaction is citrate, which is the conjugate base of citric acid. This tricarboxylic acid is used commercially to add flavor to beverages, to preserve processed foods, and for other commercial purposes. Historically, citric acid was isolated from lemons, but biotechnology now produces more than 500,000 metric tons of citric acid annually using fermentation methods that exploit citrate cycle reactions in the fungal mold *Aspergillus niger*.

Chemically, the distinctive feature of citric acid is its three carboxylic acid groups, which are ionized at physiologic pH. This makes it a good chelating agent (that is, a compound that binds to metals), as well as a flavoring agent.

CHAPTER OUTLINE

CREDITS: LEMON SLICE AND JUICE: FOTOEARL/SHUTTERSTOCK AND GALUSHKO SERGEY/SHUTTERSTOCK; *ASPERGILLUS NIGER* MOLD: DAVID SCHARF/SCIENCE SOURCE; BAG OF CITRIC ACID: EDGLIUS/SHUTTERSTOCK; CANDIES: NITO/SHUTTERSTOCK; BEVERAGES: CSP_DEVON/AGEFOTOSTOCK; CLEANING AGENT: VALIO84SL/ISTOCKPHOTO/GETTY IMAGES PLUS; BLOOD: DIETER MEYRL/ISTOCKPHOTO/GETTY IMAGES PLUS.

We have seen that glycolysis is an anaerobic process that takes place in the cytosol, producing some ATP as it converts glucose to pyruvate. We also discussed how pyruvate can be converted anaerobically into lactate or into CO_2 and ethanol, depending on the organism. However, for eukaryotes, the most important fate of pyruvate is its aerobic conversion to CO_2 and H_2O in the mitochondria, a process carried out by the citrate cycle (this chapter) and the electron transport system (Chapter 11). This process releases far more energy than anaerobic metabolism can produce, which enables cells to generate large amounts of ATP in a reaction catalyzed by the mitochondrial enzyme ATP synthase.

The citrate cycle is not simply an extension of glycolysis. Rather, it initiates the production of energy from fatty acids and proteins (amino acids), as well as from carbohydrates (glucose), linking several metabolic pathways. The oxidation of these metabolic fuels to produce ATP depends on enzyme-catalyzed oxidation–reduction reactions (redox reactions) in the citrate cycle. These reactions use the oxidized forms of the coenzymes nicotinamide adenine dinucleotide (NAD^+) and flavin adenine dinucleotide (FAD) as electron carriers, producing the reduced forms NADH and $FADH_2$.

The **citrate cycle** is considered the "hub" of cellular metabolism for three primary reasons:

1. It is central to aerobic metabolism and ATP production by generating the bulk of NADH and $FADH_2$, which are oxidized by the electron transport system to generate ATP by oxidative phosphorylation.
2. It links the oxidation of various metabolic fuels (carbohydrates, fatty acids, and proteins) to ATP synthesis through shared intermediates.
3. It provides metabolites for numerous biosynthetic pathways (**Figure 10.1**).

We begin this chapter by looking at what the citrate cycle accomplishes for the cell and reviewing the bioenergetics of redox reactions, which are used throughout the citrate cycle. This is followed by a description of the mechanism and regulation of the pyruvate dehydrogenase reaction, which uses five different enzyme cofactors to convert pyruvate (the product of glycolysis) to acetyl-CoA (the substrate for the first citrate cycle reaction). We next describe the eight reactions of the citrate cycle itself and how these reactions are regulated under normal physiologic conditions. Finally, we look at how citrate cycle intermediates are shared with other metabolic pathways in the cell, including anaplerotic reactions, which replenish citrate cycle intermediates.

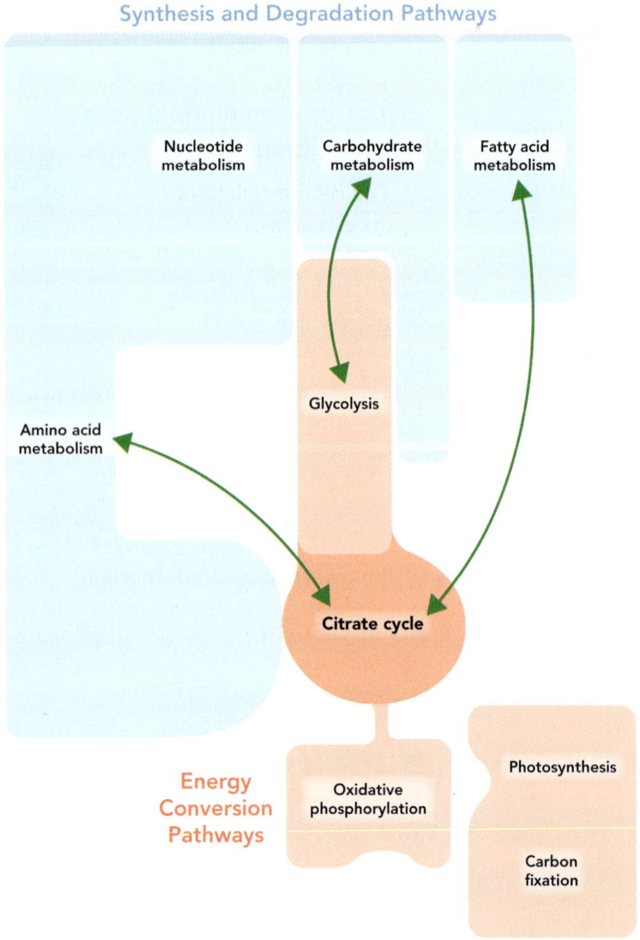

Figure 10.1 The citrate cycle functions as the hub of cellular metabolism by connecting both catabolic and anabolic pathways (arrows).

10.1 The Citrate Cycle Captures Energy Using Redox Reactions

In Chapter 2, we introduced the concept of coupled oxidation–reduction reactions in biochemical pathways as a source of energy for ATP synthesis in the mitochondria (see Figure 2.4). As you will see in this chapter, the citrate cycle is at the heart of this energy conversion process.

Overview of the Citrate Cycle

In eukaryotic cells, all of the enzymes in the citrate cycle and electron transport system reside in the mitochondrial inner membrane and matrix. The mitochondrial matrix is where pyruvate is converted to acetyl-CoA by the enzyme pyruvate dehydrogenase. (Prokaryotic cells do not contain any metabolically active organelles.) As illustrated in **Figure 10.2**, pyruvate and acetyl-CoA can be derived—aside from glycolysis—from amino acid catabolism (see Chapter 17). Also, acetyl-CoA is produced by fatty acid oxidation (see Chapter 16) as a major energy source in most organisms.

The primary function of the citrate cycle is to oxidize acetyl-CoA. In this process, each turn of the citrate cycle transfers four pairs of electrons ($8\,e^-$) from citrate cycle intermediates to NAD^+ and FAD, generating 3 NADH and 1 $FADH_2$ (**Figure 10.3**). The citrate cycle can be thought of as a "metabolic engine," in which the *fuel* is acetyl-CoA, the *exhaust* is CO_2, and the *work* performed is the transfer of electrons, using a series of linked redox reactions. The citrate cycle also contains one substrate-level phosphorylation reaction, which generates 1 GTP for every turn of the cycle. This GTP is used by the enzyme nucleoside diphosphate kinase to phosphorylate ADP to make ATP.

As described in Chapter 11, the ATP currency exchange for redox energy in the mitochondria, as a function of oxidative phosphorylation, is the generation of ~5 ATP for every 2 NADH that are oxidized. This gives an ATP currency exchange ratio of ~2.5 ATP per NADH. In contrast, oxidation of 2 $FADH_2$ molecules results in only ~3 molecules of ATP (~1.5 ATP per $FADH_2$), because of differences in where the electrons from these two molecules enter the electron transport system. On the basis of this ATP currency exchange ratio and the one substrate-level phosphorylation reaction, each turn of the cycle produces ~10 ATP for every acetyl-CoA molecule that is oxidized.

Figure 10.3 also illustrates that the oxidation of NADH and $FADH_2$ by the electron transport system is coupled to the reduction of O_2 to form H_2O. Regeneration of NAD^+ and FAD inside the mitochondrial matrix is required to maintain flux through the citrate cycle (four of the enzymes require NAD^+ or FAD as coenzymes). Thus, this metabolic engine depends on a constant supply of O_2, just like a combustion engine. The reliance on redox reactions in the electron transport system to maintain metabolic flux in the

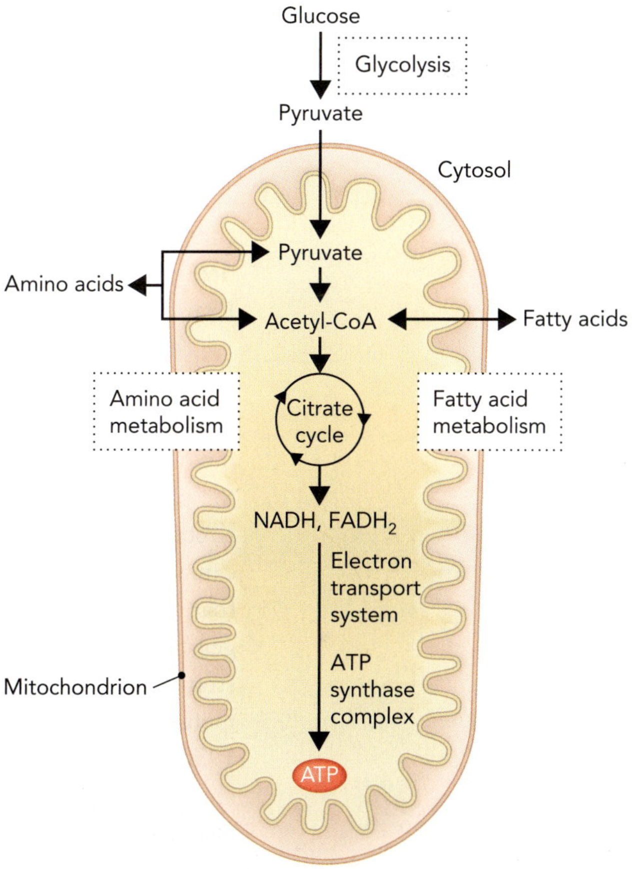

Figure 10.2 Citrate cycle reactions in eukaryotic cells take place in the mitochondrial matrix. They are linked to the electron transport system and the ATP synthase complex by the shared intermediates NADH and $FADH_2$. Many of the reactions involved in amino acid and fatty acid metabolism are also contained within the mitochondrial matrix, as described in later chapters.

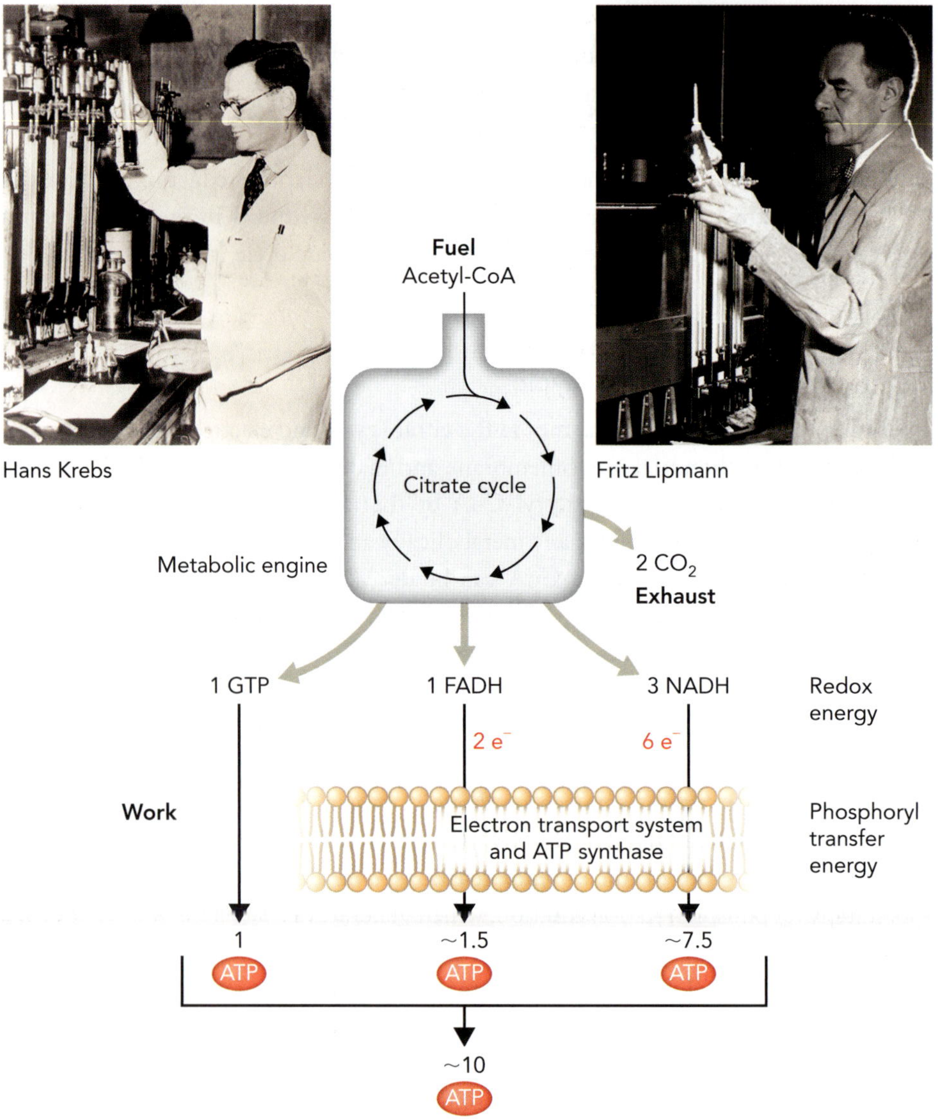

Figure 10.3 The citrate cycle is a metabolic engine in which all eight of the cycle intermediates are continually replenished, maintaining a smooth-running energy conversion process. The citrate cycle oxidizes acetyl-CoA to produce GTP, $FADH_2$, and NADH. The electron transport system converts the redox energy stored in NADH and $FADH_2$ into a proton gradient through a series of redox reactions, which donate the electrons to O_2 to produce H_2O. The ATP synthase uses the proton gradient to synthesize ATP from ADP and P_i. Hans Krebs and Fritz Lipmann shared the 1953 Nobel Prize in Physiology or Medicine for their discovery of the role of the citrate cycle in aerobic metabolism. H. KREBS: BETTMANN/CORBIS; F. LIPMANN: SCIENCE SOURCE.

citrate cycle is similar to the way in which the glyceraldehyde-3-P dehydrogenase reaction in glycolysis requires the regeneration of NAD^+ under anaerobic conditions (see Figure 9.51).

Hans Krebs, a biochemist who fled Nazi Germany for England in 1933, first described the citrate cycle in 1937. His discovery followed earlier experiments by others that had shown that minced muscle tissues (intact cells in buffered solution) rapidly oxidized organic acids such as citrate, fumarate, malate, and succinate in the presence of oxygen, resulting in the production of CO_2. Two observations gave Krebs the idea that these reactions might be part of a metabolic cycle. The first came from the work of Hungarian biochemist Albert Szent-Györgyi, who found that addition of malate or oxaloacetate to muscle cells stimulated reduction of oxygen to H_2O at much higher rates than would be needed to oxidize the added substrate completely. The second key observation came from Krebs's own work on urea metabolism, conducted with the German biochemist Kurt Henseleit in the early 1930s. Krebs and Henseleit discovered that urea is synthesized from amino acids and ammonia in a cyclic pathway, involving both cytosolic and mitochondrial enzymes (see Chapter 17).

Elucidation of the urea cycle gave Krebs the idea to test specifically to see if the generation of oxaloacetate from malate and the formation of citrate from pyruvate were linked reactions in a cyclic pathway. He eventually pieced the citrate cycle together using the compound malonate, which inhibits the cycle enzyme succinate dehydrogenase and causes the accumulation of cycle intermediates. Krebs and Fritz Lipmann, who discovered the role of acetyl-CoA in metabolism, shared the 1953 Nobel Prize in Physiology or Medicine for their work on this important energy-converting pathway (Figure 10.3b). As an honor to Krebs, the citrate cycle is sometimes called the **Krebs cycle**. It is also referred to as the **tricarboxylic acid (TCA) cycle**, as citrate is a tricarboxylate. We refer to this important metabolic cycle as the *citrate cycle* because citrate, the conjugate base of citric acid, is the first product of the pathway and the predominant species at physiologic pH (the pK_a values of the three carboxylate groups are 3.1, 4.7, and 6.4).

Two features of the citrate cycle distinguish it from linear metabolic pathways such as glycolysis. First, oxaloacetate is both the substrate for the first reaction (catalyzed by citrate synthase) and the product of the last reaction (catalyzed by malate dehydrogenase). This means that oxaloacetate must be regenerated from citrate after each turn of the cycle. Therefore, when just one citrate cycle intermediate at a time is added to cells, as in the early experiments of Szent-Györgyi and Krebs, the concentrations of all eight intermediates increase. These increased concentrations boost the capacity of the citrate cycle engine, leading to elevated acetyl-CoA consumption and CO_2 production.

A second important feature of the citrate cycle is that the enzymes pyruvate dehydrogenase and pyruvate carboxylase control the flow of acetyl-CoA and oxaloacetate, respectively, into the citrate cycle. Although these two enzymes are not counted among the eight citrate-cycle enzymes, we will discuss them in the context of the citrate cycle because of their important auxiliary role. As you will see later in Section 10.4, the pyruvate dehydrogenase reaction is inhibited when ATP and NADH levels are elevated, signaling that the energy charge in the cell is high and that excess acetyl-CoA should be diverted to fatty acid synthesis. This relationship between carbohydrate and fatty acid metabolism is the biochemical basis for low-carbohydrate diets, because if you limit carbohydrates in the diet, the body responds by burning fat through stimulating fatty acid metabolism to produce acetyl-CoA for the citrate cycle (see Chapter 19).

Now that we have a basic overview of the citrate cycle, we are ready to answer the four metabolic pathway questions.

1. *What does the citrate cycle accomplish for the cell?* The citrate cycle transfers eight electrons from acetyl-CoA to the coenzymes NAD^+ and FAD to form 3 NADH and 1 $FADH_2$. These molecules are oxidized by the electron transport system, producing ATP by the process of oxidative phosphorylation. The citrate cycle generates 2 CO_2 as waste products and uses substrate-level phosphorylation to generate 1 GTP, which is converted to ATP by nucleoside diphosphate kinase. The citrate cycle also supplies metabolic intermediates for gluconeogenesis and for amino acid and porphyrin biosynthesis.
2. *What is the overall net reaction of the citrate cycle?*

$$\text{Acetyl-CoA} + 3\ NAD^+ + \text{FAD} + \text{GDP} + P_i + 2\ H_2O \rightarrow$$
$$\text{CoA} + 2\ CO_2 + 3\ \text{NADH} + 3\ H^+ + FADH_2 + \text{GTP}$$

3. *What are the key enzymes in the citrate cycle?* Although not a citrate cycle enzyme, **pyruvate dehydrogenase** is critical to regulate the flux of acetyl-CoA through the cycle. This multi-subunit enzyme complex requires five coenzymes, is activated by NAD^+, CoA, and Ca^{2+} (in muscle cells), and is inhibited by acetyl-CoA, ATP, and NADH. **Citrate synthase** catalyzes the first reaction in the pathway. It is inhibited by citrate, succinyl-coenzyme A (succinyl-CoA), NADH, and ATP. **Isocitrate dehydrogenase** catalyzes the oxidative decarboxylation of isocitrate by transferring two electrons to NAD^+ to form NADH, releasing CO_2 in the process. It is activated by ADP and Ca^{2+} and inhibited by NADH and ATP. **α-Ketoglutarate dehydrogenase** is functionally similar to pyruvate dehydrogenase in that it is a multi-subunit complex, requires the same five coenzymes, and catalyzes an oxidative decarboxylation reaction that produces CO_2, NADH, and succinyl-CoA. It is activated by Ca^{2+} and AMP and is inhibited by NADH, succinyl-CoA, and ATP.
4. *What is an example of the citrate cycle in everyday biochemistry?* Fluoroacetate is naturally found in poisonous plants and is the active ingredient in Compound 1080, which has been used by ranchers to kill coyotes and foxes. Cells convert fluoroacetate to fluorocitrate, which is a potent inhibitor of the citrate cycle enzyme mitochondrial aconitase and results in cell death.

Redox Reactions Involve the Loss and Gain of Electrons

Many redox reactions in metabolism are a form of energy conversion involving the transfer of electron pairs from organic substrates to carrier molecules. The energy available from redox reactions is due to differences in the electron affinity of two compounds and is an inherent property of each molecule based on molecular structure. Electrons do not exist free in solution, so they must be passed from one compound to another in a coupled redox reaction. Coupled redox reactions consist of two half-reactions: (1) an **oxidation reaction** (loss of electrons) and (2) a **reduction reaction** (gain of electrons). Compounds that accept electrons are called **oxidants** and are reduced in the reaction; compounds that donate electrons are called **reductants** and are oxidized in the reaction.

Redox reactions in biochemistry rarely involve molecular oxygen (O_2) directly, but rather are characterized by the loss and gain of electrons from carbon. The terminology of biochemical redox reactions is the same as that used in inorganic chemistry. Specifically, each half-reaction consists of a **conjugate redox pair** represented by a molecule with and without an electron (e^-). For example, Fe^{2+}/Fe^{3+} is a conjugate redox pair, in which the ferrous ion (Fe^{2+}) is the reductant that loses an e^- during oxidation to generate the ferric ion (Fe^{3+}):

$$\underset{\text{Reductant}}{Fe^{2+}} \rightleftharpoons \underset{\text{Oxidant}}{Fe^{3+}} + e^-$$

Similarly, the cuprous ion (Cu^+) is the reductant that can be oxidized to form the oxidant, cupric ion (Cu^{2+}), plus an e^- in the reaction:

$$\underset{\text{Reductant}}{Cu^{+}} \rightleftharpoons \underset{\text{Oxidant}}{Cu^{2+}} + e^-$$

The two conjugate redox pairs in these reactions are Fe^{2+}/Fe^{3+} and Cu^+/Cu^{2+}. We can combine these two half-reactions into a coupled redox reaction by reversing the direction of the Fe^{3+} reduction reaction, so that the e^- functions as the common intermediate shared by the Cu^+ oxidation and Fe^{3+} reduction half-reactions:

$$Fe^{3+} + e^- \rightleftharpoons Fe^{2+} \text{ (reduction of } Fe^{2+}\text{)}$$
$$Cu^+ \rightleftharpoons Cu^{2+} + e^- \text{ (oxidation of } Cu^+\text{)}$$
$$\overline{Fe^{3+} + Cu^+ \rightleftharpoons Fe^{2+} + Cu^{2+} \text{ (coupled redox reaction)}}$$

Enzymes that catalyze redox reactions involving Fe^{2+}/Fe^{3+} and Cu^+/Cu^{2+} are components of the electron transport system, as described in Chapter 11.

Unlike the reduction of Fe^{3+} by Cu^+, which simply involves the transfer of one e^-, redox reactions in the citrate cycle involve the transfer of electron pairs ($2e^-$) to the electron carrier molecules NAD^+ and FAD. The reduction of NAD^+ to NADH involves the reductant giving up two hydrogen atoms, one in the form of a hydride ion ($:H^-$), which contains $2e^-$ and 1 H^+ proton, and the other as a proton (H^+) (see Figure 10.5). The products of the reaction are NADH and the release of a proton (H^+) into solution:

$$NAD^+ + :H^- + H^+ \rightleftharpoons NADH + H^+$$

Though this reaction is a hydride ion transfer, it is sometimes written as

$$NAD^+ + 2e^- + 2H^+ \rightleftharpoons NADH + H^+$$

In contrast, FAD is reduced by sequential addition of one hydrogen ($1e^-$ and 1 H^+) at a time, to give the fully reduced $FADH_2$ product:

$$FAD + 1e^- + 1H^+ \rightleftharpoons FADH^\bullet$$
$$FADH^\bullet + 1e^- + H^+ \rightleftharpoons FADH_2$$

Oxidations can also involve a direct combination with oxygen, which oxidizes the carbon by pulling e^- toward the more electronegative O atom. Enzymes that catalyze biochemical redox reactions are strictly called **oxidoreductases**; however, because most oxidation reactions involve the loss of one or more hydrogen atoms, they are often called **dehydrogenases**. We will look at the reduction of the coenzymes NAD^+ and FAD by dehydrogenases in more detail later in the chapter.

Free Energy Changes Can Be Calculated from Reduction Potential Differences

The two primary energy-conversion reactions in metabolism are (1) phosphoryl transfers involving ATP and (2) redox reactions that use the electron carriers NAD^+/NADH and FAD/$FADH_2$. We can compare the relative strengths of energy conversion reactions by comparing their changes in standard free energy.

As we discussed in Chapter 2, the change in biochemical standard free energy of a reaction ($\Delta G^{\circ\prime}$) is a measure of the spontaneity of the reaction. Measured in kilojoules per mole, it reflects the tendency of compound A to be converted to compound B ($A \rightarrow B$) when starting with a 1 M concentration of reactants and products. In redox reactions, we use the **biochemical standard reduction potential ($E^{\circ\prime}$)**, measured in volts (V), to represent the electron affinity of a given conjugate redox pair. Analogous to biochemical standard conditions that define the Gibbs free energy, $G^{\circ\prime}$ (25 °C, pH 7, and 1 M initial concentration of substrates and products), biochemical standard reduction potential refers to $E^{\circ\prime}$ under the same conditions.

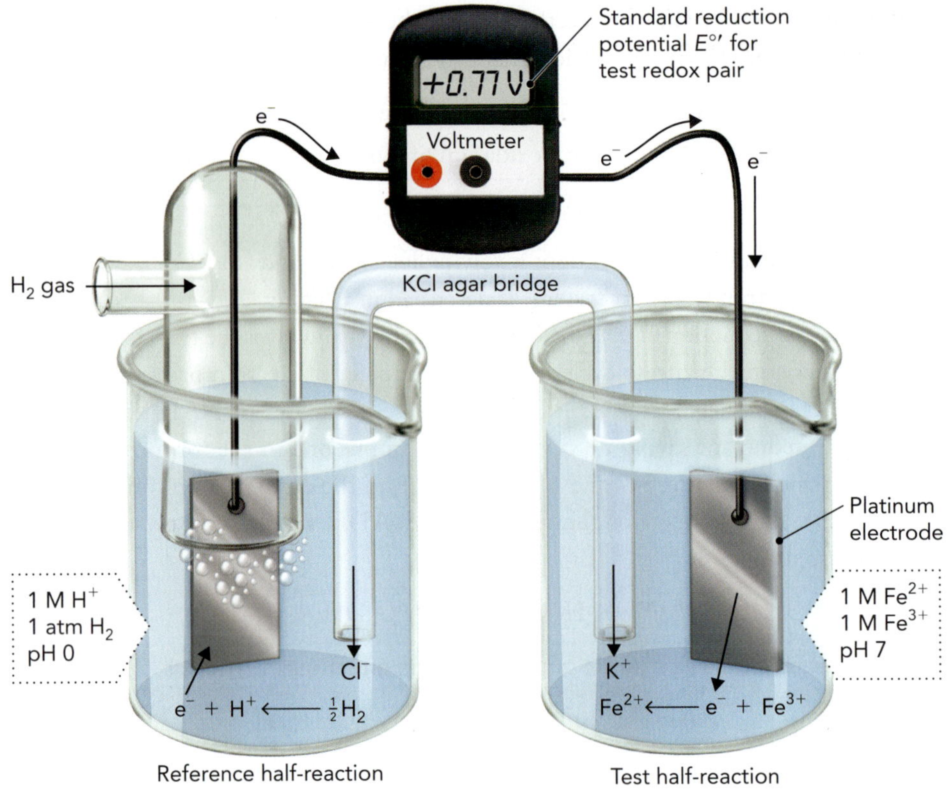

Figure 10.4 An electrochemical cell is used to measure the reduction potential of a test conjugate redox pair compared to that of a reference conjugate redox pair. In this example, the electrons (e^-) move through the wire from the hydrogen half-cell toward the Fe^{2+}/Fe^{3+} test half-cell because Fe^{3+} has a higher affinity for electrons than that of H^+. The potassium chloride (KCl) agar bridge allows counterion movement to maintain electrical neutrality.

Figure 10.4 illustrates how $E°'$ values can be determined in the laboratory using an apparatus called an **electrochemical cell.** The cell measures the relative e^- affinity of a test redox pair compared to that of the reference hydrogen half-reaction ($2\ H^+ + 2\ e^- \rightleftharpoons H_2$). For these measurements, two half-cells are connected by a voltmeter and use platinum electrodes to measure the movement of electrons from one half-cell to the other. Depending on the relative electron affinity of the test oxidant compared to that of H^+, the electrons either move from the hydrogen half-cell toward the test half-cell or from the test half-cell toward the hydrogen half-cell. The two half-cells are also connected by an agar bridge containing potassium chloride, which permits counterion movement to balance the charge.

The $E°'$ of oxidants with a higher affinity for electrons than that of H^+ are recorded as positive $E°'$ values ($E°' > 0$). Oxidants with a lower affinity for electrons than that of H^+ are recorded as negative $E°'$ values ($E°' < 0$). In the example shown in Figure 10.4, the test half-cell contains a 1 M concentration each of Fe^{3+} and Fe^{2+} (at pH 7), and the reference half-cell contains 1 M H^+ (pH 0) in equilibrium with H_2 gas. Under conditions of 25 °C and 1 atm pressure, the hydrogen half-cell is assigned the arbitrary $E°'$ value of 0.00, and the voltmeter registers an $E°'$ value of +0.77 V. This means that the electrons flow from the reference cell toward the Fe^{3+}/Fe^{2+} cell and that Fe^{3+} has a higher affinity for electrons than that of H^+.

Table 10.1 lists measured $E°'$ values for several conjugate redox pairs in biochemical reactions. You can see that O_2 is the most potent oxidant in the table, with an $E°'$ value of +0.82 V. Therefore, O_2 readily accepts electrons from all other conjugate redox pairs shown. In contrast, ferredoxin (a protein that contains Fe^{3+}) is the weakest oxidant but the most potent reductant. It can donate electrons to all other conjugate redox pairs in the table. By convention, standard reduction potentials are expressed as half-reactions written in the direction of a reduction

Table 10.1 STANDARD REDUCTION POTENTIALS ($E°'$) OF REPRESENTATIVE TEST HALF-REACTIONS

Half-reaction	$E°'$ (V)	
$\frac{1}{2}O_2 + 2\,H^+ + 2\,e^- \rightarrow H_2O$	0.82	
$Fe^{3+} + e^- \rightarrow Fe^{2+}$	0.77	
$NO_3^- + 2\,H^+ + 2\,e^- \rightarrow NO_2^- + H_2O$	0.42	Stronger oxidants
Cytochrome a_3 (Fe^{3+}) + $e^- \rightarrow$ Cytochrome a_3 (Fe^{2+})	0.35	
Cytochrome c (Fe^{3+}) + $e^- \rightarrow$ Cytochrome c (Fe^{2+})	0.25	
Ubiquinone + 2 H^+ + 2 $e^- \rightarrow$ Ubiquinol + H_2	0.05	
Fumarate + 2 H^+ + 2 $e^- \rightarrow$ Succinate	0.03	
$2\,H^+ + 2\,e^- \rightarrow H_2$ (at standard conditions, pH 0)	0.00	
Oxaloacetate + 2 H^+ + 2 $e^- \rightarrow$ Malate	−0.17	
Pyruvate + 2 H^+ + 2 $e^- \rightarrow$ Lactate	−0.19	
FAD + 2 H^+ + 2 $e^- \rightarrow$ $FADH_2$ (free FAD, enzyme-bound FAD is •0.0 V)	−0.22	
$NAD^+ + H^+ + 2\,e^- \rightarrow$ NADH	−0.32	Stronger reductants
$NADP^+ + H^+ + 2\,e^- \rightarrow$ NADPH	−0.32	
α-Ketoglutarate + CO_2 + 2 H^+ + 2 $e^- \rightarrow$ Isocitrate	−0.38	
$2\,H^+ + 2\,e^- \rightarrow H_2$ (at pH 7)	−0.41	
Ferredoxin (Fe^{3+}) + $e^- \rightarrow$ Ferredoxin (Fe^{2+})	−0.43	

Note: Values were obtained at 25 °C and pH 7 using the hydrogen half-cell (1 atm H_2 and 1 M H^+, pH 0) as the reference.

reaction. Note that when the test half-cell contains H^+ at 10^{-7} M (pH 7), and the standard half-cell contains H^+ at 1 M (pH 0), the $E°'$of hydrogen is measured as −0.41 V, meaning that electrons tend to move from the pH 7 electrode to the pH 0 electrode.

The amount of energy available from a coupled redox reaction is directly related to the difference between two reduction potentials, denoted as $\Delta E°'$. By convention, the $\Delta E°'$of a coupled redox reaction is determined by subtracting the $E°'$of the reductant (e^- donor) from the $E°'$of the oxidant (e^- acceptor) using the following equation:

$$\Delta E°' = (E°'_{e^-\,\text{acceptor}}) - (E°'_{e^-\,\text{donor}}) \tag{10.1}$$

Moreover, the $\Delta E°'$ for a coupled redox reaction is proportional to the change in free energy $\Delta G°'$, as described by the equation

$$\Delta G°' = -nF\Delta E°' \tag{10.2}$$

where n is the number of electrons transferred in the reaction, and F is the Faraday constant (96.48 kJ/V mol). You can see from this equation that when the difference in reduction potentials for a coupled redox reaction is positive ($\Delta E°' > 0$), then the reaction is favorable, as $\Delta G°'$ is negative. From the definition of $\Delta E°'$, this means that for a coupled redox reaction to be favorable, the reduction potential of the e^- acceptor needs to be more positive than that of the e^- donor.

To see how $\Delta G°'$ and $\Delta E°'$ are related, we can use the biochemical standard reduction potentials ($E°'$) in Table 10.1 to calculate the change in biochemical standard free energy ($\Delta G°'$) for the isocitrate dehydrogenase reaction of the citrate cycle:

$$\text{Isocitrate} + \text{NAD}^+ \rightleftharpoons \alpha\text{-Ketoglutarate} + \text{CO}_2 + \text{NADH} + \text{H}^+$$

In a spontaneous coupled redox reaction such as this, electrons flow from the reductant in the conjugate redox pair with the lower $E°'$ value (more negative) toward the oxidant in the conjugate redox pair with the higher $E°'$ value (less negative). Maintaining convention by writing all half-reactions in the direction of reductions, and using the $E°'$ values from Table 10.1, the two half-reactions for the isocitrate dehydrogenase reaction are

$$\alpha\text{-Ketoglutarate} + \text{CO}_2 + 2\,\text{e}^- + 2\,\text{H}^+ \rightarrow \text{Isocitrate} \qquad (E°' = -0.38\ \text{V})$$

$$\text{NAD}^+ + 2\,\text{e}^- + 2\,\text{H}^+ \rightarrow \text{NADH} + \text{H}^+ \qquad (E°' = -0.32\ \text{V})$$

For this coupled redox reaction, in which isocitrate is the reductant (e^- donor) and NAD^+ is the oxidant (e^- acceptor), we can calculate $\Delta E°'$ using the equation

$$\begin{aligned}\Delta E°' &= (E°'_{\text{e}^-\text{ acceptor}}) - (E°'_{\text{e}^-\text{ donor}})\\ &= (E°'_{\text{NAD}^+}) - (E°'_{\text{isocitrate}})\\ &= (-0.32\ \text{V}) - (-0.38\ \text{V}) = +0.06\ \text{V}\end{aligned}$$

Then, we can convert this $\Delta E°'$ value to $\Delta G°'$ using the relationship

$$\begin{aligned}\Delta G°' &= -nF\Delta E°'\\ &= -(2)(96.48\ \text{kJ/V mol})(+0.06\ \text{V})\\ &= -11.6\ \text{kJ/mol}\end{aligned}$$

This calculation shows that the conversion of isocitrate to α-ketoglutarate, reaction 3 in the citrate cycle, is favorable ($\Delta G°' < 0$) under standard biochemical conditions.

Biochemical conditions inside the mitochondrial matrix are not under standard conditions, so in order to calculate the actual reduction potentials for conjugate redox pairs, we need to take into account the concentration of the oxidant (e^- acceptor) and reductant (e^- donor) using an equation described by Walther Nernst in 1881:

$$E = E°' + \frac{RT}{nF}\ln\frac{[\text{e}^-\ \text{acceptor}]}{[\text{e}^-\ \text{donor}]} \tag{10.3}$$

In the **Nernst equation**, R is the gas constant (8.314×10^{-3} kJ/mol K), T is the absolute temperature in kelvin (K), n is the number of electrons transferred, and F is the Faraday constant (96.48 kJ/V mol). We can use the Nernst equation to calculate the actual reduction potential at 298 K for the NAD^+/NADH conjugate redox pair using a ratio of 20:1 for intramitochondrial NAD^+-to-NADH concentrations:

$$\begin{aligned}E &= E°' + \frac{RT}{nF}\ln\frac{[\text{e}^-\ \text{acceptor}]}{[\text{e}^-\ \text{donor}]}\\ &= -0.32\ \text{V} + \frac{(8.314 \times 10^{-3}\ \text{kJ/mol K})(298\ \text{K})}{2(96.48\ \text{kJ/V mol})}\ln 20\\ &= -0.32\ \text{V} + (0.013\ \text{V})(3) = -0.28\ \text{V}\end{aligned}$$

Because the actual reduction potential E value (−0.28 V) is more positive than the biochemical standard reduction potential $E°'$ (−0.32 V), it means that NAD^+ reduction is even more favorable inside the mitochondria due to the high NAD^+-to-NADH ratio.

concept integration 10.1

What does the $\Delta E°'$ of a redox reaction tell you about the spontaneity of the reaction, and how does this affect the order of multiple redox reactions within a metabolic pathway?

The difference in standard reduction potential ($\Delta E°'$) for a coupled redox reaction represents the tendency of the reductant in one conjugate redox pair to donate electrons to the oxidant of the other conjugate redox pair. Because the $\Delta E°'$ value of a redox reaction is proportional to the change in standard free energy ($\Delta G°'$) as described by the equation $\Delta G°' = -nF\Delta E°'$, redox reactions with positive $\Delta E°'$ values ($\Delta G°' < 0$) will proceed spontaneously while reactions with negative $\Delta E°'$ values ($\Delta G°' > 0$) will not. For electrons to move from a reductant to an oxidant, the standard reduction potential of the oxidant species must be more positive than that of the reductant species. Therefore, in a linked metabolic pathway, the oxidants in subsequent reactions must have progressively higher (more positive) standard reduction potentials $E°'$ than those of the reductants (reduced oxidants) in previous reactions. Linked reactions in a pathway are therefore assembled in such a way as to incrementally increase the standard reduction potential of each product so that chemical work can be obtained from as many reactions as possible.

10.2 Pyruvate Dehydrogenase Converts Pyruvate to Acetyl-CoA

Pyruvate must be transported from the cytosol into the mitochondrial matrix before it can serve as a source of reducing power for the cell (used in the citrate cycle) or as a precursor for glucose synthesis (used in gluconeogenesis). Pyruvate transport is accomplished by a transmembrane transporter protein. When the energy charge in the cell is low, pyruvate is metabolized in the mitochondria by the multi-subunit enzyme complex pyruvate dehydrogenase. This enzyme catalyzes the oxidative decarboxylation of pyruvate to form CO_2 and acetyl-CoA using a five-step reaction mechanism that requires three distinct enzymes and five different coenzymes. We begin this section by briefly looking at the biochemistry of the five coenzymes required for the pyruvate dehydrogenase reaction. Then, we examine the enzymatic mechanism of pyruvate dehydrogenase catalysis and the regulation of pyruvate dehydrogenase activity by phosphorylation.

Five Coenzymes Are Required for the Pyruvate Dehydrogenase Reaction

Generally, enzymes use amino acid side chains to catalyze chemical reactions in the active site of the protein (see Chapter 7). However, amino acids have relatively limited chemistry that they can accomplish. Therefore, enzyme cofactors and coenzymes are often involved in mediating enzyme catalysis (see Tables 7.1 and 7.2), especially in redox reactions, where they function as electron carriers. Because a dietary deficiency in coenzymes required for metabolic reactions can cause a variety of diseases, many of these compounds were first identified as *vitamins* by biochemists. As detailed in this subsection, the five coenzymes required for the pyruvate dehydrogenase reaction are NAD^+, FAD, CoA, thiamine pyrophosphate (TPP), and α-lipoic acid (lipoamide).

a.

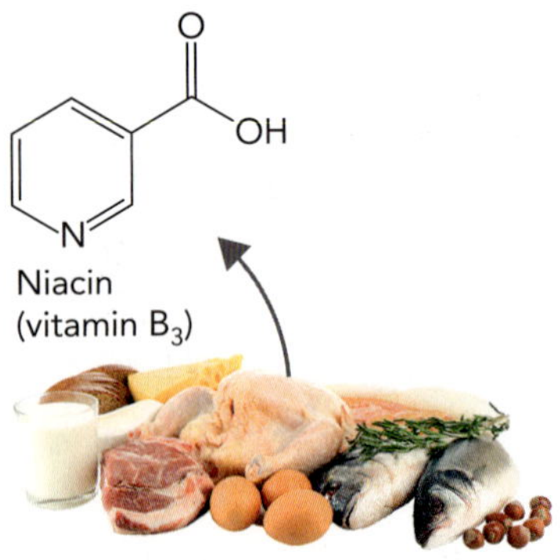

b.

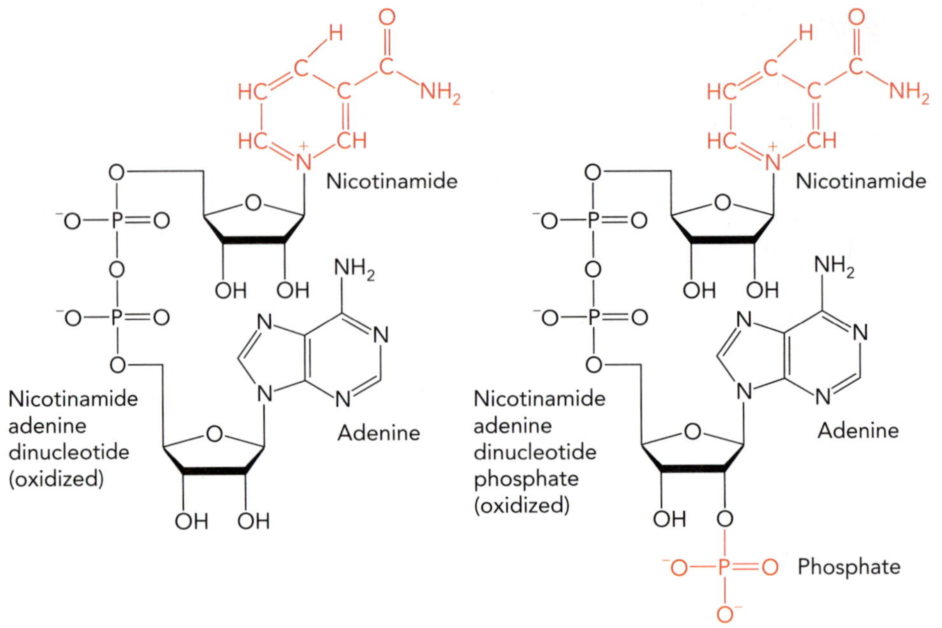

c.

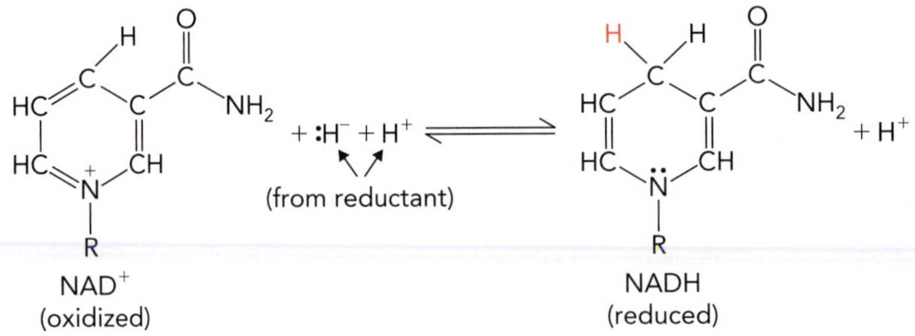

Figure 10.5 Niacin (vitamin B_3) is required to synthesize the electron carrier NAD^+. **a.** Structure of the vitamin niacin, along with representative foods containing high levels of this vitamin. PHOTO: AFRICA STUDIO/SHUTTERSTOCK. **b.** NAD^+ and $NADP^+$ are very similar in overall structure; however, $NADP^+$ has a phosphoryl group on the 2′-position of the ribose sugar. The nicotinamide ring, where the oxidation and reduction reactions take place, is identical in both molecules. **c.** Redox reactions using the conjugate redox pair NAD^+/NADH involve the loss of two hydrogens from the reductant in the form of a hydride ion ($:H^-$), which is transferred to NAD^+ to generate NADH, and a free H^+, which is released into solution.

Nicotinamide adenine dinucleotide (NAD^+) is derived from the water-soluble vitamin **niacin**, which is also called nicotinic acid, or vitamin B_3. Nutritional studies show that adults require 15–20 mg of niacin per day, most of which is obtained from poultry, fish, vegetables, nuts, and dairy products (**Figure 10.5a**). NAD^+, and its phosphorylated form $NADP^+$ (**Figure 10.5b**), are involved in at least 200 redox reactions in the cell, each of which is characterized by the transfer of 2 e^- in the form of a hydride ion ($:H^-$) transfer. Catabolic redox reactions primarily use the conjugate redox pair NAD^+/NADH, whereas anabolic reactions use $NADP^+$/NADPH. The structure of the oxidized and reduced forms of NAD^+ and NADH are shown in **Figure 10.5c**. Note that the "+" charge does not refer to the overall charge of the NAD molecule, but only to the charge on the ring N in the oxidized state.

Severe niacin deficiency causes the disease **pellagra**, which was first described in Europe in the early 1700s among peasants who relied on cultivated corn as their primary source of nutrition. The most common symptoms of pellagra are development of a skin rash (*pellagra* is Italian for "rough skin"), diarrhea, and neurologic problems, including depression and memory loss. Although it was initially thought that pellagra was caused by an infectious agent in contaminated corn, nutritional studies by physician Joseph Goldberger in the early 1900s showed that it was due to insufficient levels of a vitamin that was missing in a corn-rich diet (**Figure 10.6a**). Later, the biochemist Conrad Elvehjem showed that the missing vitamin was niacin. Surprisingly, niacin is

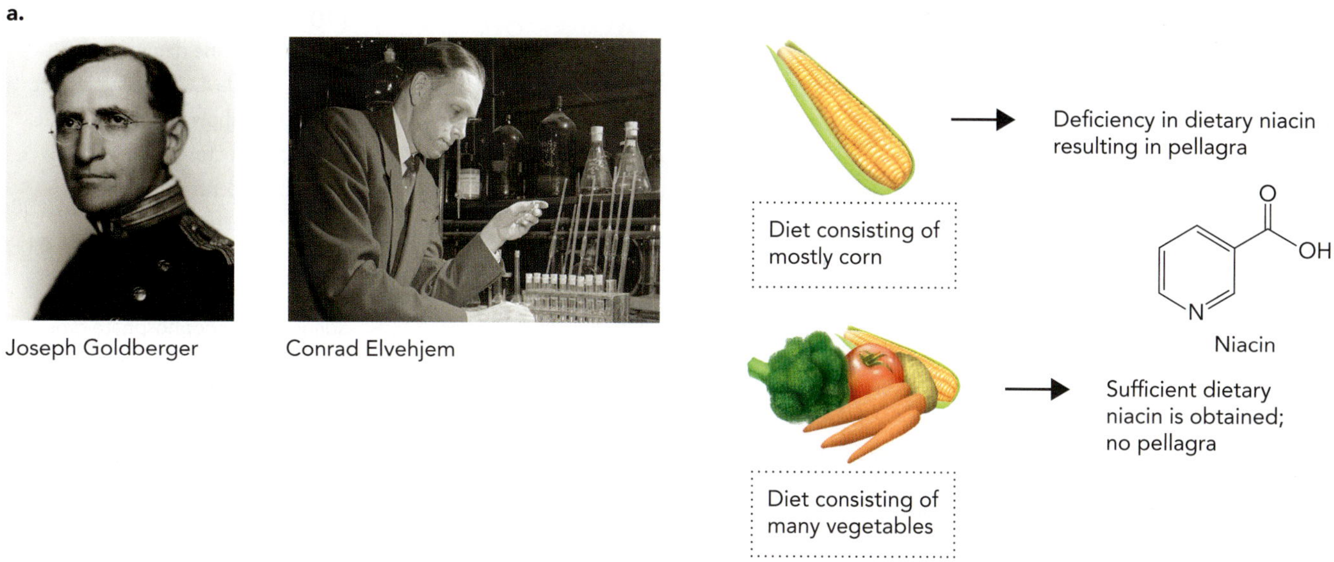
a.
Joseph Goldberger
Conrad Elvehjem
Diet consisting of mostly corn
Deficiency in dietary niacin resulting in pellagra
O
OH
N
Niacin
Diet consisting of many vegetables
Sufficient dietary niacin is obtained; no pellagra

b.
Soak corn in CaO (lime)
Cook the cornmeal
O
OH
N
Lime-soaked corn releases protein-bound niacin upon heating

Figure 10.6 Corn must be prepared properly in order to release bound niacin and prevent the vitamin-B_3-deficiency disease pellagra. **a.** Joseph Goldberger was the first to show that the disease pellagra is due to a dietary deficiency, and Conrad Elvehjem discovered that the missing vitamin is niacin. J. GOLDBERGER: PUBLIC HEALTH IMAGE LIBRARY/CDC; C. ELVEHJEM: UNIVERSITY OF WISCONSIN–MADISON ARCHIVES. **b.** The preparation of cornmeal in Mexican culture includes treating the corn with lime (calcium oxide), which releases the niacin upon cooking and makes corn tortillas a good source of vitamin B_3. PHOTO: GIL GIULIO/HEMIS/CORBIS.

actually present in corn, but in a protein-bound form (NAD^+/NADH) that drastically reduces its absorption in the intestine. One of the puzzling observations in cultures with corn-rich diets was that although pellagra was common in Europe, it was rare in Mexico. After the discovery of niacin, it was shown that corn used for tortillas in Mexican cooking is traditionally soaked in lime solution (calcium oxide), which releases niacin from its protein-bound form upon heating (**Figure 10.6b**).

Flavin adenine dinucleotide (FAD) is derived from the water-soluble vitamin **riboflavin**, which is also called vitamin B_2. Riboflavin is the precursor to FAD and flavin mononucleotide (FMN), both of which are tightly associated with enzymes that catalyze redox reactions. FAD is a coenzyme in the pyruvate dehydrogenase complex and is also covalently bound to a histidine residue in the citrate cycle enzyme succinate dehydrogenase. FAD is reduced to $FADH_2$ by the transfer of two electrons in the form of hydrogen-atom transfers. Unlike NAD^+, FAD can accept one electron at a time and form a partially reduced intermediate called a semiquinone ($FADH^{\bullet}$), as shown in **Figure 10.7**. $FADH_2$ functions as a reductant in the pyruvate dehydrogenase complex by transferring two electrons to NAD^+ in the final step of the reaction to generate NADH + H^+.

More than 50 metabolic enzymes have been characterized that contain FAD or FMN moieties in their active sites, all of which are called **flavoproteins**. Riboflavin deficiency is rare because only small amounts are required in the diet (~1 mg/day).

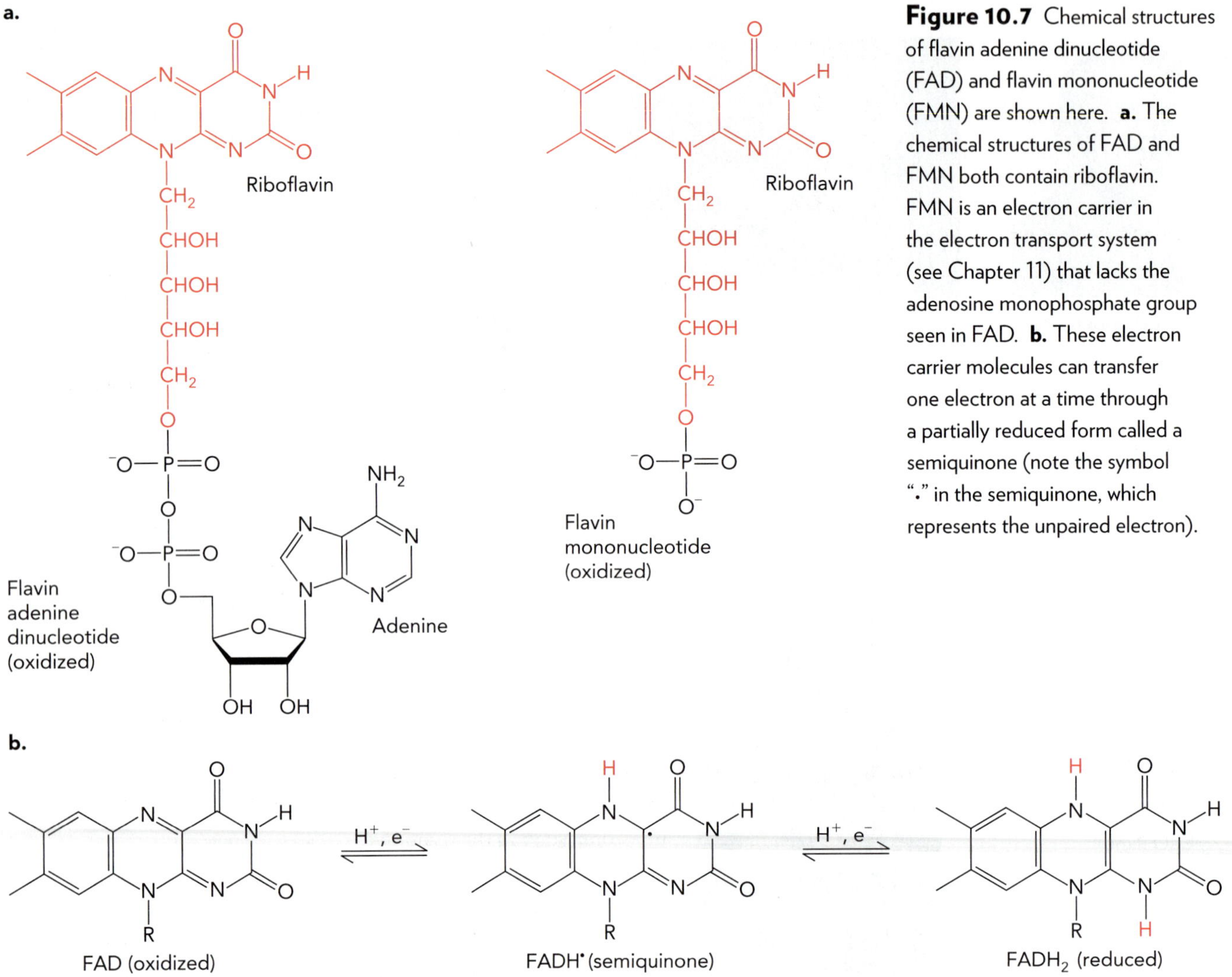

Figure 10.7 Chemical structures of flavin adenine dinucleotide (FAD) and flavin mononucleotide (FMN) are shown here. **a.** The chemical structures of FAD and FMN both contain riboflavin. FMN is an electron carrier in the electron transport system (see Chapter 11) that lacks the adenosine monophosphate group seen in FAD. **b.** These electron carrier molecules can transfer one electron at a time through a partially reduced form called a semiquinone (note the symbol "•" in the semiquinone, which represents the unpaired electron).

However, when it does occur because of malnutrition, it results in a symptom called cheilosis, characterized by swelling and cracked lips. Some foods that have been found to be high in riboflavin include dairy products (milk, cheese, eggs), almonds, and asparagus. Riboflavin, like several other vitamins, is destroyed by light, which is one reason why milk is no longer stored in clear containers.

Coenzyme A (CoA) is derived from the water-soluble vitamin **pantothenic acid,** which is also called vitamin B_5. Coenzyme A is absolutely essential for life because it is required for energy conversion by the citrate cycle and is a cofactor in biosynthetic pathways that produce fatty acids, acetylcholine, heme, and cholesterol. The primary role of CoA is to function as a carrier molecule for acetate units in the form of acetyl-CoA.

Figure 10.8 shows the structures of CoA and acetyl-CoA, which are reactants and products, respectively, in the pyruvate dehydrogenase reaction as described shortly. They consist of a central pantothenic acid unit that is linked on one side to a functional β-mercaptoethylamine group derived from cysteine and on the other side to 3′-phosphoadenosine diphosphate. The transported acetate unit of acetyl-CoA is covalently attached to CoA through an activated thioester bond, which has a high

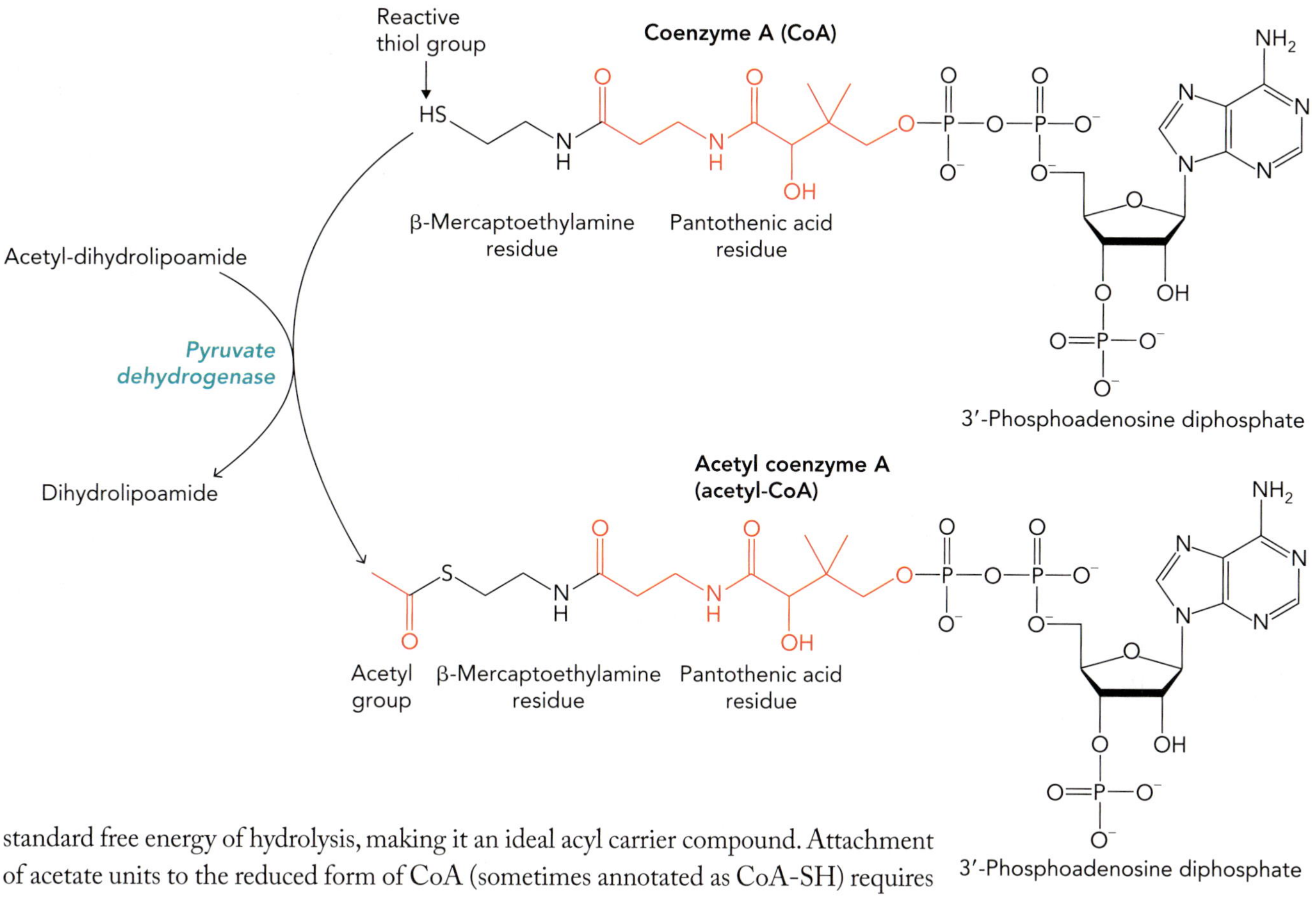

Figure 10.8 Coenzyme A is a common acyl carrier compound. The pantothenic acid moiety is linked to β-mercaptoethylamine and 3′-phosphoadenosine diphosphate. The reactive thiol group of CoA is indicated. In acetyl-CoA, the acetyl group is linked to the β-mercaptoethylamine group through a thioester bond. CoA and acetyl-CoA are reactants and products, respectively, in the pyruvate dehydrogenase reaction.

standard free energy of hydrolysis, making it an ideal acyl carrier compound. Attachment of acetate units to the reduced form of CoA (sometimes annotated as CoA-SH) requires reactions with sufficient energy to drive the formation of this bond, such as the reactions catalyzed by pyruvate dehydrogenase and α-ketoglutarate dehydrogenase.

Like many other coenzymes, CoA does not readily cross cell membranes. Therefore, dietary CoA is degraded by enzymes in the gut to yield pantothenic acid, which is absorbed and transported to tissues through the circulatory system. Once inside cells, pantothenic acid is converted back to CoA through a series of phosphorylation reactions and the addition of cysteine and ATP (**Figure 10.9**). The average daily intake of pantothenic acid in healthy adults is ~5 mg, which can be obtained easily from a balanced diet. Foods high in pantothenic acid include chicken, yogurt, and avocados.

Thiamine pyrophosphate (TPP) is derived from the water-soluble vitamin **thiamine**, which is also called vitamin B_1. TPP is an enzyme cofactor in relatively few metabolic reactions, two of which are the pyruvate dehydrogenase and α-ketoglutarate dehydrogenase reactions in the citrate cycle. The structures of thiamine and TPP are shown in **Figure 10.10**. You can see that a carbon atom on the thiazole ring is the functional component of the coenzyme involved in aldehyde transfer. Thiamine is absorbed in the gut and transported to tissues, where it is phosphorylated by the enzyme thiamine pyrophosphokinase in the presence of ATP to form TPP and AMP.

Thiamine deficiency was first described in Chinese literature more than 4,000 years ago and is the cause of **beriberi**, a disease characterized by anorexia, cardiovascular problems, and neurologic symptoms. Beriberi has been found in populations that rely on white polished rice as a primary source of nutrition (milling rice removes the bran, which contains thiamine). Beriberi has also been observed in chronic alcoholics, who have poor nutrition and obtain most of their calories for energy conversion from

Figure 10.9 Dietary pantothenic acid readily crosses cell membranes. Inside cells, it is used to synthesize CoA in a series of reactions requiring ATP, cysteine, and cytidine triphosphate (CTP).

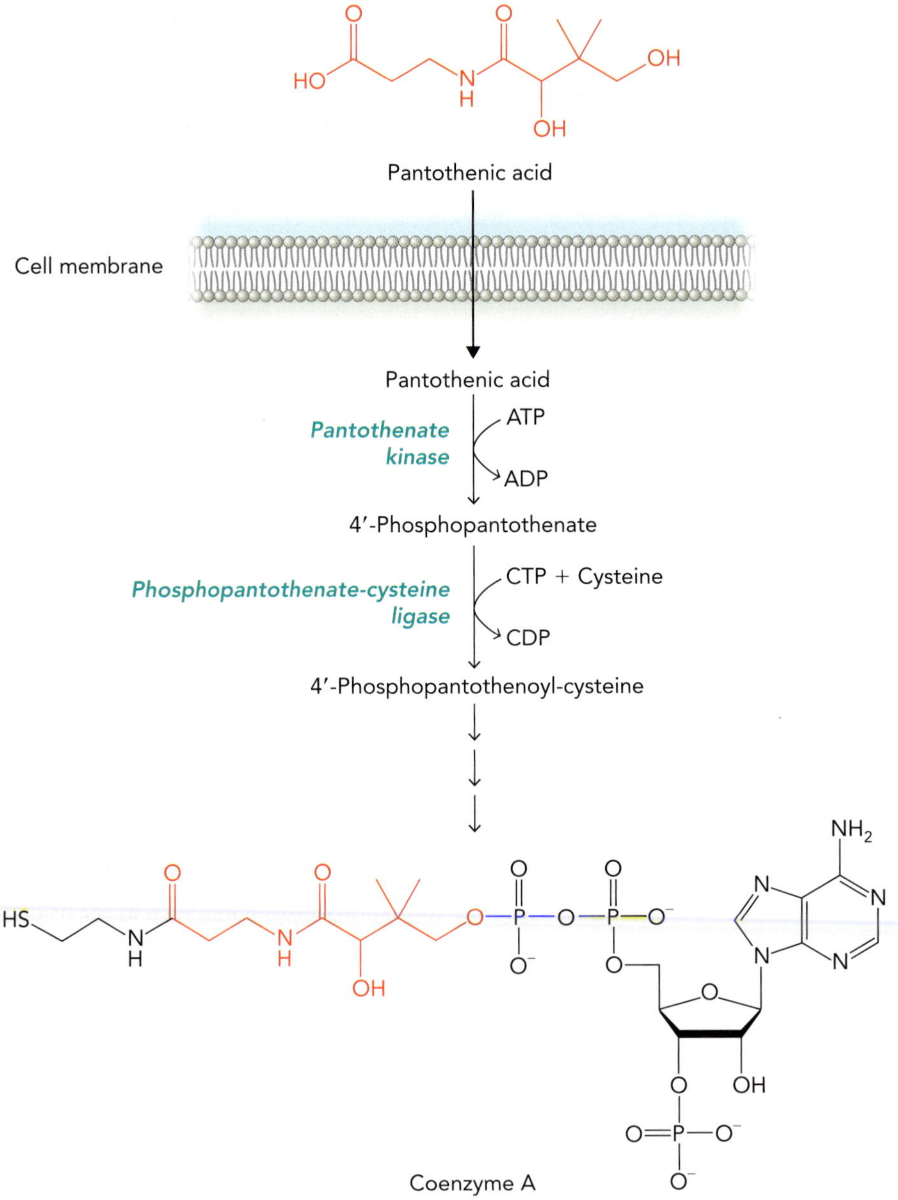

alcohol. Another dietary cause of beriberi is diets rich in foods that contain the enzyme thiaminase, which degrades thiamine during digestion (**Figure 10.11**). Raw fish contains thiaminase, as do African silkworms—a favorite food in some Nigerian cultures. Cooking these foods destroys the thiaminase and alleviates the symptoms of beriberi without substantially lowering the nutritional value. The recommended daily allowance of thiamine for adults is 1–2 mg/day, which can come from cooked lentils or brown rice or, more commonly in developed countries, from thiamine-fortified grains and cereals.

Figure 10.10 Dietary thiamine is phosphorylated by thiamine pyrophosphokinase inside cells, using ATP as the source of pyrophosphate. This generates the active coenzyme thiamine pyrophosphate. The carbon atom in the thiazole ring that is the attachment site for activated aldehydes is indicated with an arrow.

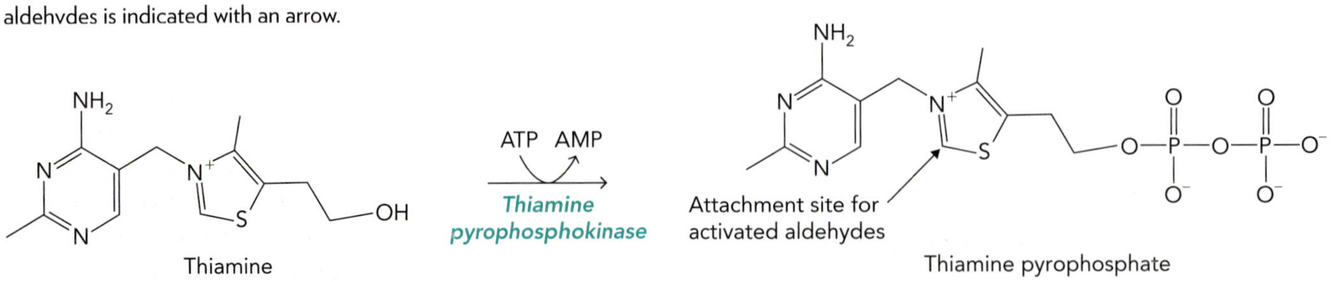

a.

b.

Figure 10.11 Thiamine deficiency leads to lack of the coenzyme thiamine pyrophosphate, resulting in the human disease beriberi. **a.** Thiamine deficiency has been found in populations that rely on white polished rice. Milling rice removes the bran, which contains thiamine. ZKRUGER/ISTOCK/GETTY IMAGES PLUS. **b.** Beriberi can also be caused by diets rich in uncooked fish or silkworms, both of which contain high levels of the thiamine-degrading enzyme thiaminase. RAW FISH: CYRIL HOU/SHUTTERSTOCK; SILKWORMS: RAYPHOTOGRAPHER/SHUTTERSTOCK.

α-Lipoic acid (lipoamide) is a coenzyme synthesized in plants and animals as a 6,8-dithiooctanoic acid. The role of α-lipoic acid in metabolic reactions is to provide a reactive disulfide that can participate in redox reactions within the enzyme active site. In the pyruvate dehydrogenase protein complex, the naturally occurring form of α-lipoic acid contains a covalent linkage to a lysine ε amino group on the E2 subunit of the complex (dihydrolipoyl acetyltransferase), forming a lipoamide. The lipoamide contains a reactive thiol, which accepts an acetyl group from hydroxyethyl-TPP in the multistep pyruvate dehydrogenase reaction to form acetyl-dihydrolipoamide (**Figure 10.12**). The acetyl group from this molecule is then transferred to CoA to generate acetyl-CoA, which is used to form citrate in the first reaction step of the citrate cycle. The long hydrocarbon chain bridging α-lipoic acid and the lysine residue in the enzyme provides a flexible extension to the reactive thiol group.

α-Lipoic acid is not considered a vitamin because it is synthesized at measurable levels in humans. It is, however, often promoted as a nutritional supplement because of its potential to function as an antioxidant in its reduced form. Foods with high levels of α-lipoic acid include broccoli, liver, spinach, and tomato.

The Pyruvate Dehydrogenase Complex Is a Metabolic Machine

The conversion of pyruvate to acetyl-CoA by the pyruvate dehydrogenase complex is the necessary transition from glycolysis to the citrate cycle. The reaction is an oxidative decarboxylation that represents another classic example of protein structure and function.

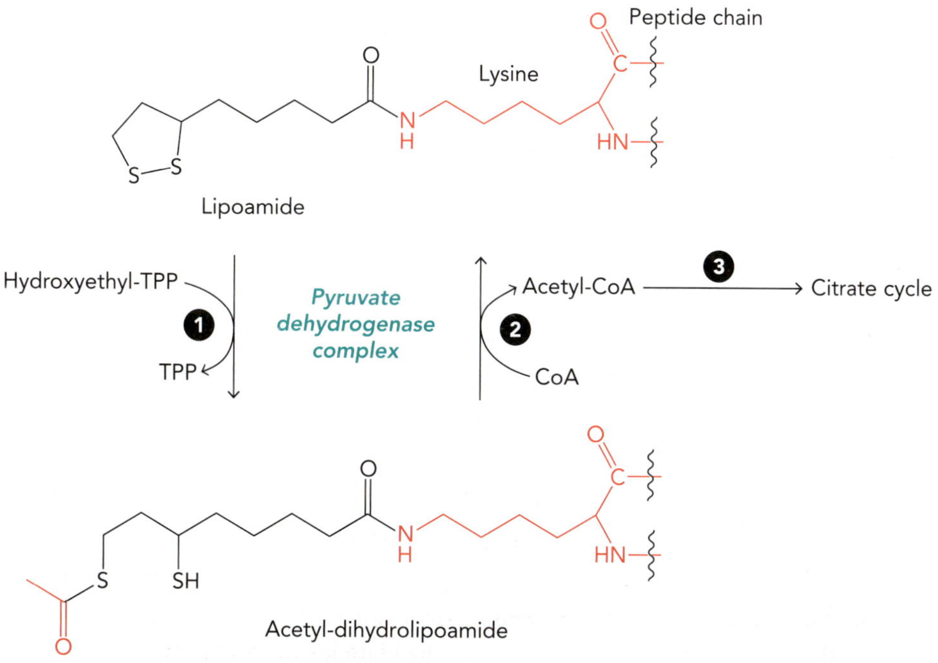

Figure 10.12 α-Lipoic acid is attached to lysine ε amino groups of proteins, such as the E2 enzyme subunit of the pyruvate dehydrogenase complex, to form a lipoamide. (1) Hydroxyethyl-TPP transfers an acetyl group to the reactive thiol group of lipoamide, forming acetyl-dihydrolipoamide. (2) The acetyl group is transferred from acetyl-dihydrolipoamide to reduced coenzyme A, generating the final product acetyl-CoA (see also Figure 7.10), which is (3) used in the citrate cycle.

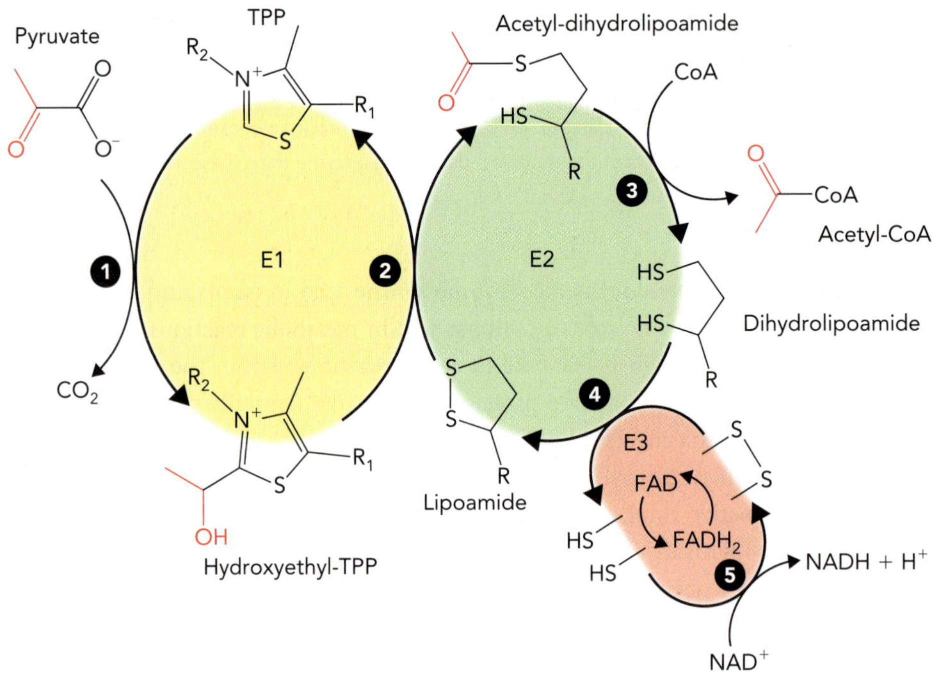

Figure 10.13 The pyruvate dehydrogenase reaction produces acetyl-CoA. This diagram shows the interdependence of each reaction step on coenzymes that are associated with the E1 (TPP), E2 (lipoamide), and E3 (FAD) subunits of the complex. The five numbered reaction steps are described in the text.

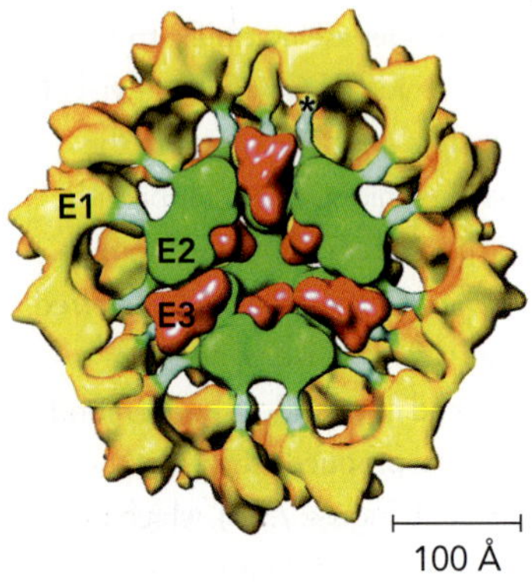

Figure 10.14 Bovine pyruvate dehydrogenase, shown here in cross section, forms a large spherical complex. The E1 (yellow), E2 (green), and E3 (red) pyruvate dehydrogenase complex subunits are labeled, and the linker regions connecting E2 to E1 are shaded blue. The lipoamide domain and its polypeptide tether (ball and chain) are not shown in this model. This three-dimensional reconstruction of the bovine pyruvate dehydrogenase complex was generated by Lester Reed and colleagues at the University of Texas, Austin, using a combination of biochemical, crystallographic, and electron microscopic methods. Z. H. ZHOU ET AL. (2001). THE REMARKABLE STRUCTURAL AND FUNCTIONAL ORGANIZATION OF THE EUKARYOTIC PYRUVATE DEHYDROGENASE COMPLEXES. *PROCEEDINGS OF THE NATIONAL ACADEMY OF SCIENCES USA, 98*(26), 14802–14807. PMID: 11752427 [PUBMED–INDEXED FOR MEDLINE] PMCID: PMC64939. © 2001 NATIONAL ACADEMY OF SCIENCES, USA.

The eukaryotic pyruvate dehydrogenase complex contains more than 90 subunits, which consist of three different catalytic enzymes: (1) E1, pyruvate dehydrogenase; (2) E2, dihydrolipoyl acetyltransferase; and (3) E3, dihydrolipoyl dehydrogenase. These three enzymes work together as a metabolic machine to carry out the following net reaction:

$$\text{Pyruvate} + \text{CoA} + \text{NAD}^+ \rightleftharpoons \text{Acetyl-CoA} + \text{CO}_2 + \text{NADH} + \text{H}^+$$
$$\Delta G^{\circ\prime} = -33.4 \text{ kJ/mol}$$

Coenzymes perform a critical role in the pyruvate dehydrogenase complex by providing a platform for the catalytic reactions. Three of the coenzymes are covalently linked to enzyme subunits: TPP is attached to the E1 subunit (pyruvate dehydrogenase), lipoamide is the functional component of the E2 subunit (dihydrolipoyl acetyltransferase), and FAD is covalently bound to the E3 subunit (dihydrolipoyl dehydrogenase). The two other coenzymes, CoA and NAD^+, are transiently associated with the E2 and E3 complexes, respectively.

Mechanism of the Pyruvate Dehydrogenase Reaction The pyruvate dehydrogenase reaction can be broken down into five distinct catalytic steps. Steps 1, 2, and 3 lead to the formation of acetyl-CoA, and steps 4 and 5 regenerate the oxidized form of lipoamide and reduce NAD^+ to NADH. The five reaction steps and the contributions of E1, E2, and E3 are illustrated in **Figure 10.13** and summarized below.

1. The E1 subunit (pyruvate dehydrogenase) binds pyruvate and catalyzes a decarboxylation reaction, resulting in the formation of hydroxyethyl-TPP and the subsequent release of CO_2.
2. The hydroxyethyl-TPP moiety of E1 reacts with the disulfide of the lipoamide group on the N-terminal domain of the E2 subunit (dihydrolipoyl acetyltransferase). This generates acetyl-dihydrolipoamide through a thioester bond.
3. The E2 acetyl-dihydrolipoamide group carries the acetyl group from the E1 catalytic site across a gap in the complex to the E2 catalytic

site, where it reacts with CoA to yield acetyl-CoA and fully reduced dihydrolipoamide.

4. The dihydrolipoamide group then swings over to the E3 subunit (dihydrolipoyl dehydrogenase), where it is reoxidized to the disulfide by a transfer of 2 e^- and 2 H^+ to a disulfide contained on the E3 subunit, forming a reduced dithiol. The reduced dithiol on the E3 subunit is reoxidized by transferring 2 e^- and 2 H^+ to the E3-linked FAD moiety, transiently forming E3-$FADH_2$.
5. The E3-$FADH_2$ coenzyme intermediate is reoxidized in a coupled redox reaction that transfers 2 e^- to NAD^+, producing NADH + H^+.

As shown in **Figure 10.14**, the E1, E2, and E3 subunits of mammalian pyruvate dehydrogenase are packed together in a huge sphere, ~400 Å in diameter, with a combined molecular weight of ~7,800 kDa. The pyruvate dehydrogenase complex consists of an inner core made up of tightly packed E2 subunits in close contact with the E3 subunits. The E1 subunits form an outer shell that is ~50 Å away from the E2 core. The E2 core of mammalian pyruvate dehydrogenase complexes is a pentagonal dodecahedron consisting of 60 E2 subunits arranged as trimers. They are connected by polypeptide linkers to an outer shell of 22 E1 subunits. Researchers have suggested that six E3 subunits stick into the E2 core at symmetrical positions within the complex (Figure 10.14). The approximate stoichiometry of the subunits (E1 to E2 to E3, 22:60:6) is consistent with the presence of ~60 acetyl-CoA synthesis sites in the pyruvate dehydrogenase complex.

The Lipoamide Ball and Chain **Figure 10.15** shows that the lipoamide moiety of the E2 subunit is attached near the end of a 221-amino-acid amino-terminal segment of the

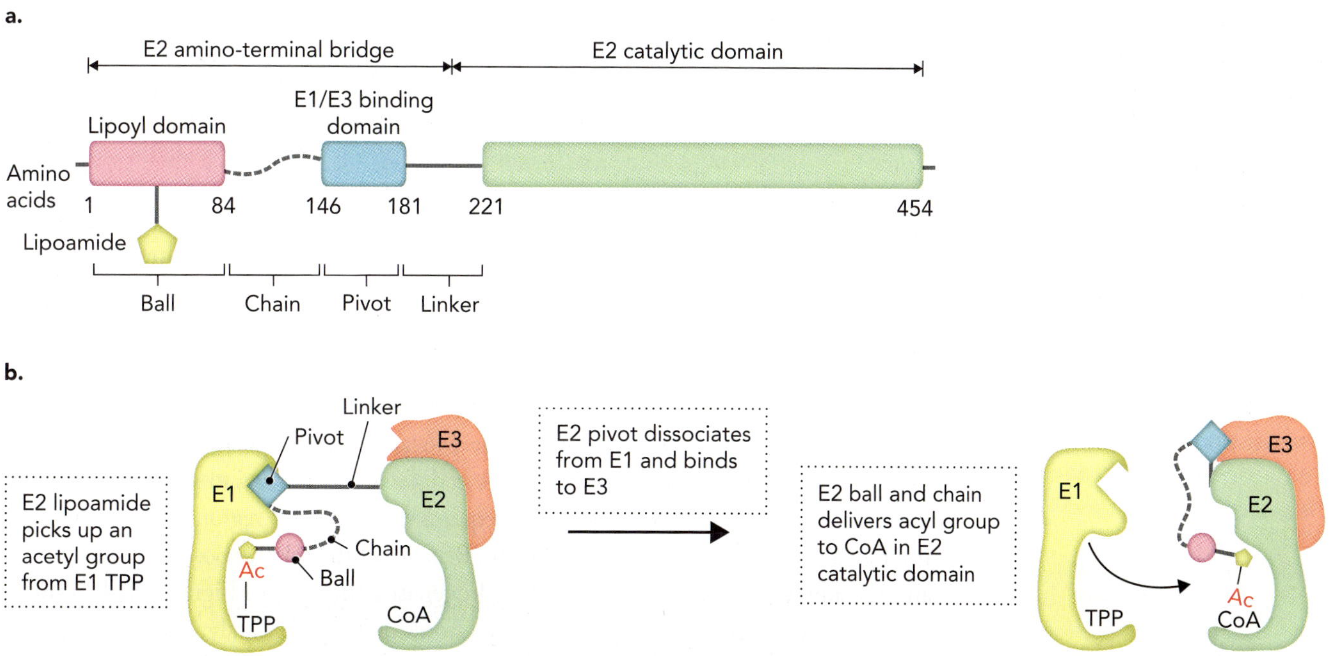

Figure 10.15 The N-terminal segment of the yeast E2 subunit (dihydrolipoyl acetyltransferase) of pyruvate dehydrogenase contains the subunit linker, a pivot, and the lipoamide moiety, attached to a polypeptide tether. Together, these pieces function as a ball and chain. **a.** Functional map of the 454-amino-acid yeast E2 subunit (dihydrolipoyl acetyltransferase), showing the location of the lipoyl domain (ball), polypeptide tether (chain), E1 binding domain (pivot), E1–E2 subunit linker, and the E2 catalytic domain. **b.** The E2 ball and chain moves the lipoamide domain between the E1 and E2 catalytic sites. The result is the transfer of an acyl group (Ac) from thiamine pyrophosphate (TPP) in E1 to the CoA substrate in the E2 catalytic site.

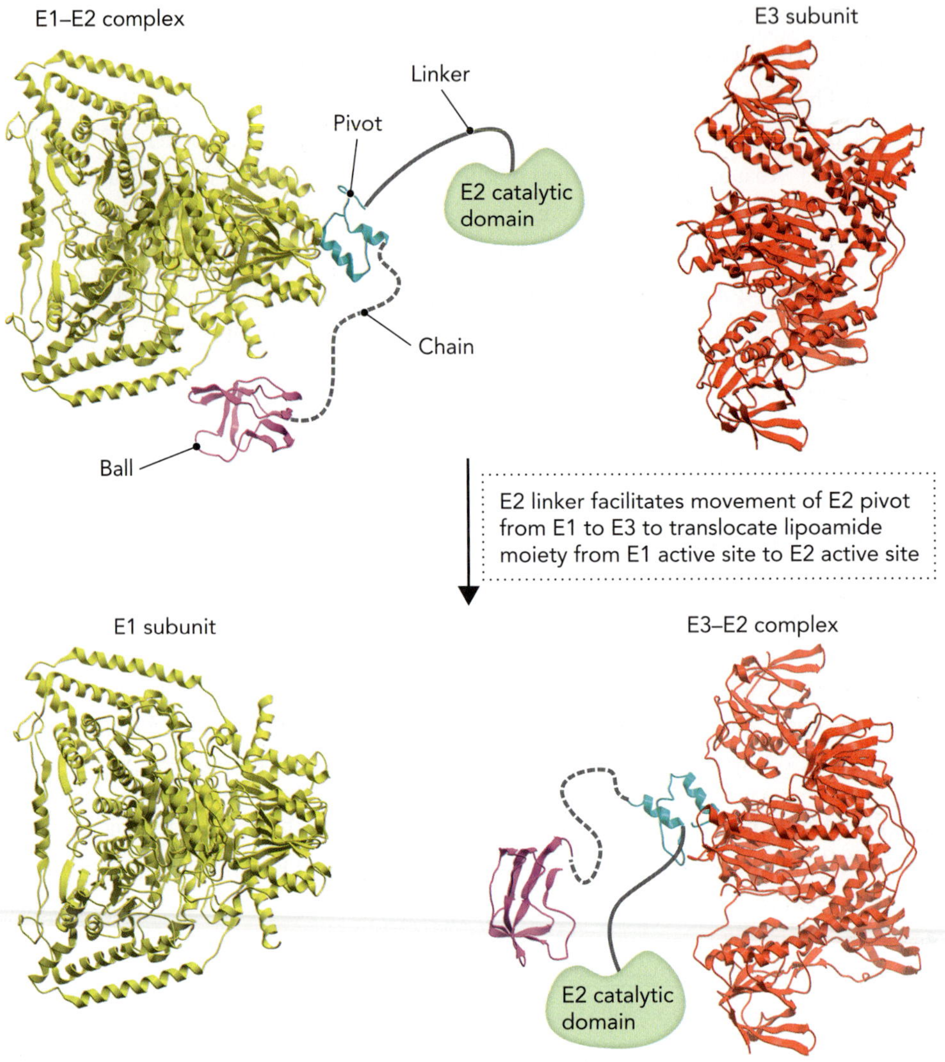

Figure 10.16 These structural models illustrate the movements of the ball-and-chain regions of the E2 subunit. Long, flexible polypeptide linkers (chain and linker, shown as gray dashed and solid lines, respectively) connect the lipoyl domain (ball), E1/E3 binding domain (pivot), and catalytic domain of the E2 subunit. These linker regions allow significant movements of the lipoyl and E1/E3 binding domains with respect to the E2 catalytic domain and the E1 (yellow) and E3 (red) subunits. In both representations, the position of the lipoyl domain is not known, but has been placed close to either the E1 or E2 subunits for illustrative purposes. BASED ON PDB FILES 1W85 (E1/E2 BINDING DOMAIN COMPLEX), 1EBD (E3/E2 BINDING DOMAIN COMPLEX), AND 1LAC (LIPOYL DOMAIN OF E2 SUBUNIT).

protein, which functions as a bridge between the E1 and E3 subunits (see the blue region in the pyruvate dehydrogenase complex in Figure 10.14). This E2 amino-terminal bridge consists of four functional components: (1) the *ball*, the lipoamide moiety covalently linked to the 84-amino-acid E2 lipoyl domain; (2) the *chain*, the 62-amino-acid flexible segment of E2; (3) the *pivot*, the 35-amino-acid E1/E3 binding domain; and (4) the *linker*, the 40-amino-acid segment needed to reorient the pivot between the E1 and E3 subunits. As illustrated in Figure 10.15, the 146-amino-acid lipoamide "ball and chain" swings across a 50-Å gap between the E1 and E2 catalytic sites to mediate the acetyl group transfer from TPP to CoA, which requires the E2 "pivot" region to dissociate from the E1 subunit and bind to the E3 subunit. Because of the relative positioning of the E2 and E1 subunits within the pyruvate dehydrogenase complex (see Figure 10.14), a single E2 subunit can obtain hydroxyethyl groups from multiple E1 catalytic sites and deliver them to the E2 subunit. **Figure 10.16** illustrates the pyruvate dehydrogenase ball-and-chain model using molecular structures of protein domains that have been characterized within the E1, E2, and E3 subunits.

Arsenic Is an Environmental Inhibitor of Pyruvate Dehydrogenase From Figures 10.14 and 10.15, it is clear that the lipoamide group is the workhorse in this catalytic machine. Also, without a fully functional pyruvate dehydrogenase complex,

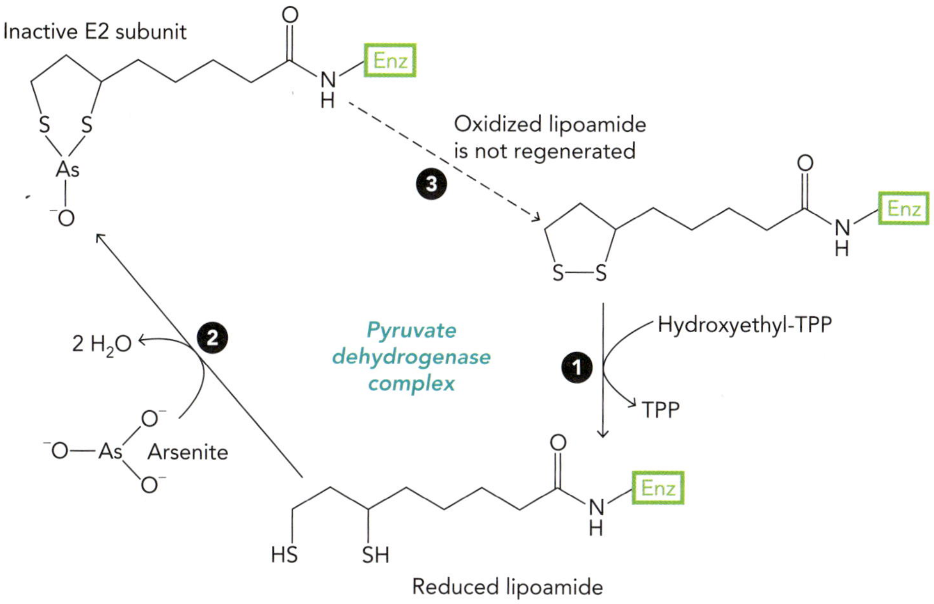

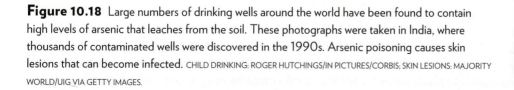

Figure 10.17 Arsenic poisoning occurs when covalent modification of lipoamide coenzymes by arsenite blocks enzyme activity. The E2 subunit (dihydrolipoyl acetyltransferase) is an example of a lipoamide-containing enzyme that is inhibited by arsenite. (1) The lipoamide in E2 is reduced during the enzyme reaction cycle. (2) Arsenite forms an adduct with the reduced lipoamide, resulting in inactivation of E2. (3) The oxidized lipoamide species cannot be regenerated, blocking further enzyme activity.

the link between glycolysis and the citrate cycle would be broken. A naturally occurring inhibitor of lipoamide coenzyme function is the element arsenic (As), which, in the form of arsenite (AsO_3^{3-}), creates bidentate adducts on dihydrolipoamide, as shown in **Figure 10.17**. Ingesting arsenite can lead to an untimely death by irreversibly blocking the catalytic activity of lipoamide-containing enzymes—in particular, the pyruvate dehydrogenase and α-ketoglutarate dehydrogenase complexes. Chronic arsenic poisoning can come from environmental sources, such as arsenic-contaminated drinking water or household paints, and results in the appearance of ulcerous skin lesions and an increased risk of a variety of cancers.

It is thought that arsenic poisoning was used routinely to kill off kings and queens in the Middle Ages and may even have been involved in the death of the exiled French emperor Napoleon Bonaparte. A more recent and tragic example is that of accidental arsenic poisoning of thousands of people in Bangladesh and India. Since the 1990s, it has been documented that millions of people in Bangladesh and India have been chronically exposed to toxic levels of arsenic in contaminated drinking water that was obtained from shallow, hand-pumped wells (**Figure 10.18**). During the 1970s and 1980s, UNICEF and other relief organizations helped drill thousands of wells in small Indian and Bangladeshi villages in a humanitarian effort to circumvent public water supplies that had become biologically contaminated. However, years later, large numbers of villagers in the Ganges Delta region developed skin lesions and cancers, and tests showed that the well water contained toxic levels of arsenic. Massive efforts were undertaken to close down contaminated wells and to develop purification systems to reduce the arsenic to safe levels in other water supplies. Arsenic-contaminated drinking water has also been found in Southeast Asia and South America, usually near areas that have been extensively mined.

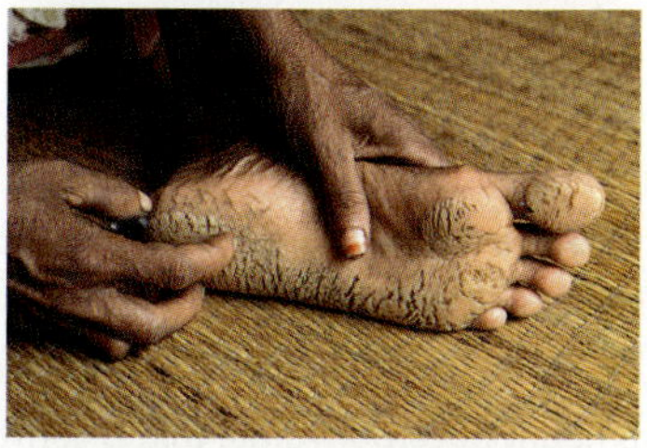

Figure 10.18 Large numbers of drinking wells around the world have been found to contain high levels of arsenic that leaches from the soil. These photographs were taken in India, where thousands of contaminated wells were discovered in the 1990s. Arsenic poisoning causes skin lesions that can become infected. CHILD DRINKING: ROGER HUTCHINGS/IN PICTURES/CORBIS; SKIN LESIONS: MAJORITY WORLD/UIG VIA GETTY IMAGES.

Pyruvate Dehydrogenase Activity Is Regulated by Allostery and Phosphorylation

Acetyl-CoA has only two metabolic fates in the cell (**Figure 10.19**):

1. It can be metabolized by the citrate cycle, ultimately converting redox energy to ATP by oxidative phosphorylation.
2. It can be used to store metabolic energy in the form of fatty acids, which are transported to adipose tissue (fat cells) as triglycerides.

Because the pyruvate dehydrogenase reaction is essentially irreversible, as physiologic concentrations of reactants and products will not be able to overcome the standard free energy change ($\Delta G^{\circ\prime} = -33.4$ kJ/mol), production of acetyl-CoA by the pyruvate dehydrogenase reaction is tightly controlled to coordinate the energy needs of the cell with the production of acetyl-CoA. This is especially important in animals, which lack the necessary enzymes to convert fats to carbohydrates, and therefore cannot reuse acetyl-CoA for glucose production when carbohydrate levels are low. Because of this, the pyruvate dehydrogenase complex is fully active in animal cells only when carbohydrate sources are plentiful.

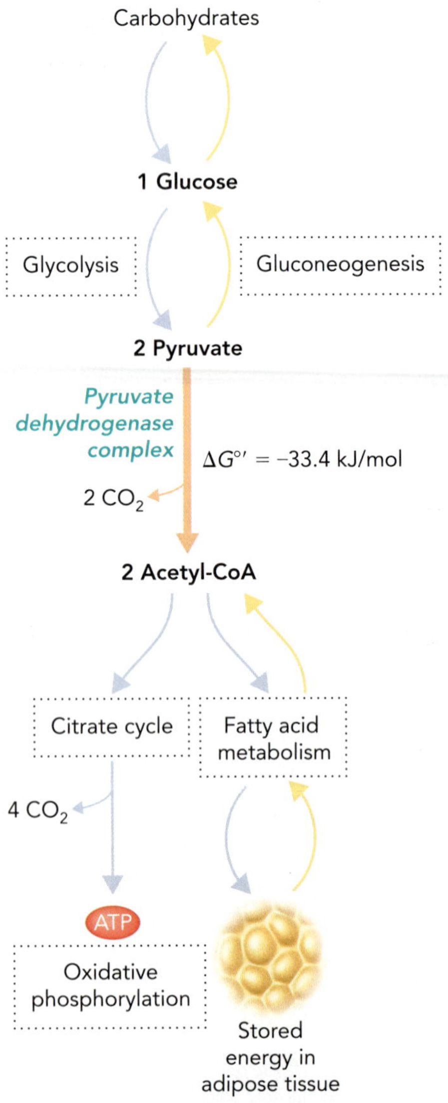

Figure 10.19 The pyruvate dehydrogenase reaction is essentially irreversible and plays a central role in metabolism by controlling the amount of carbohydrate that is converted to ATP or stored as fatty acids in adipose tissue.

The catalytic activity of the pyruvate dehydrogenase complex is regulated by both allosteric control and covalent modification. When NADH-to-NAD^+ ratios are high in the mitochondrial matrix, signaling that ATP synthesis by oxidative phosphorylation is slowing down, NADH competes with NAD^+ for binding to E3 (dihydrolipoyl dehydrogenase). NADH thereby blocks the last step in the reaction (see step 5 in Figure 10.13). Similarly, high acetyl-CoA levels compete with CoA binding to E2 (dihydrolipoyl acetyltransferase), preventing new rounds of pyruvate decarboxylation.

A more sensitive regulatory mechanism to control the rate of pyruvate catabolism is that of serine phosphorylation/dephosphorylation of the E1 pyruvate dehydrogenase subunit. This activity is regulated by the enzymes pyruvate dehydrogenase kinase and pyruvate dehydrogenase phosphatase-1, respectively. As shown in **Figure 10.20**, when the E1 subunit is phosphorylated by pyruvate dehydrogenase kinase, the catalytic activity of pyruvate dehydrogenase is decreased, and generation of the hydroxyethyl-TPP intermediate is blocked.

Several positive and negative effectors regulate pyruvate dehydrogenase kinase and pyruvate dehydrogenase phosphatase-1 activity in response to the anabolic and catabolic needs of the cell. Specifically, increased levels of NADH, acetyl-CoA, and ATP indicate that the energy charge in the cell is high, so it is time to switch from mostly catabolic pathways to anabolic pathways in order to synthesize macromolecules. Therefore, these metabolites function as activators of pyruvate dehydrogenase kinase activity, resulting in serine phosphorylation and inhibition of pyruvate dehydrogenase activity (Figure 10.20). In contrast, elevated levels of NAD^+, CoA, ADP, and Ca^{2+} reflect a negative energy charge in the cell, indicating the need to shift metabolic flux toward catabolic pathways through inhibition of pyruvate dehydrogenase kinase activity. Moreover, increased Ca^{2+} levels activate the pyruvate dehydrogenase phosphatase-1 enzyme, accelerating the rate at which pyruvate dehydrogenase is activated by dephosphorylation. Calcium regulation of pyruvate dehydrogenase kinase and pyruvate dehydrogenase phosphatase-1 activity is important in muscle cells, where muscle contraction is initiated by Ca^{2+} release and the increased demand for ATP.

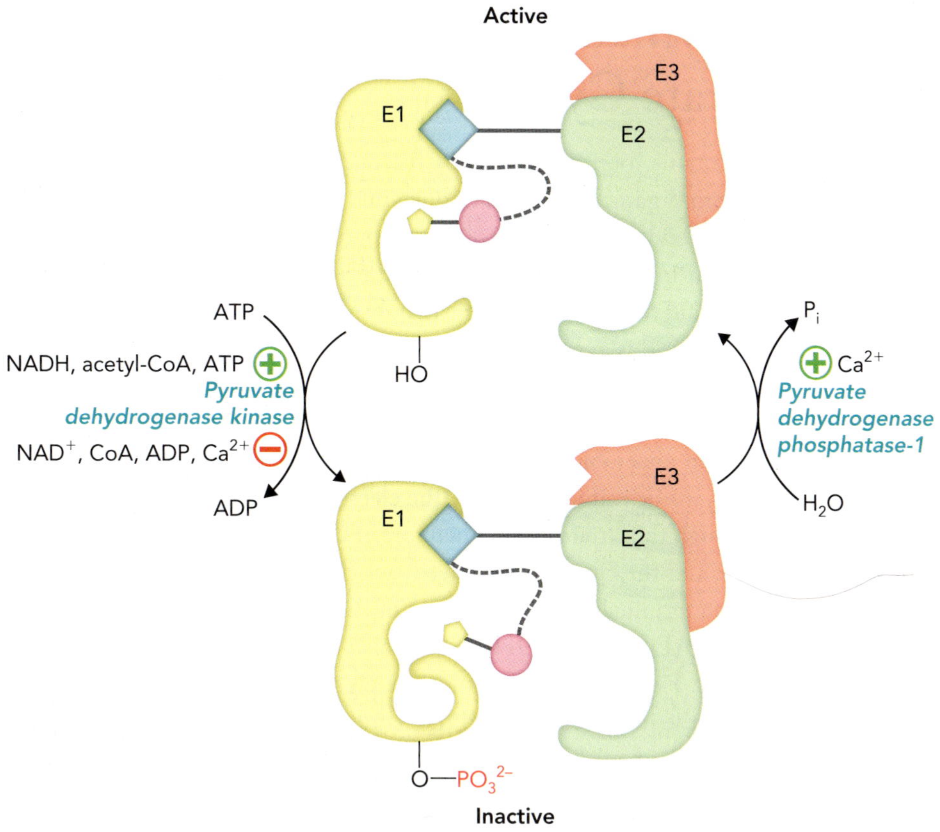

Figure 10.20 The activity of the pyruvate dehydrogenase complex is inhibited by phosphorylation of a serine residue on the E1 subunit. Allosteric activation of the enzyme pyruvate dehydrogenase kinase by NADH, acetyl-CoA, or ATP leads to phosphorylation of E1 and inhibition of the complex. In contrast, NAD^+, CoA, ADP, and Ca^{2+} allosterically inhibit pyruvate dehydrogenase kinase activity. The enzyme pyruvate dehydrogenase phosphatase-1 is activated by Ca^{2+} and functions to activate the pyruvate dehydrogenase complex by dephosphorylating the serine residue on the E1 subunit.

concept integration 10.2

What properties of coenzymes explain why they were initially discovered as vitamins, and what is their specific role in the pyruvate dehydrogenase reaction?

Coenzymes are biomolecules that provide additional functional groups to enzyme active sites and participate in catalytic mechanisms. They tend to be large, complex biomolecules that are not synthesized by all organisms because the process is energy demanding. Plants and many microorganisms devote some amount of their available energy toward synthesizing coenzymes; however, across evolutionary time most animals have lost the ability to synthesize many of these biomolecules because they obtain them from their diet. A *vitamin* is defined as a required nutrient that alleviates medical symptoms brought on by an unbalanced diet or unusual lifestyle. Because coenzymes are required for key metabolic enzymes, their presence or absence in the diet can be documented clinically and fit the definition of vitamin. The pyruvate dehydrogenase complex consists of multiple copies of three protein subunits (E1, E2, E3) and requires five coenzymes (TPP, lipoamide, CoA, FAD, and NAD^+). The coenzyme TPP is a component of the E1 subunit and functions in a decarboxylation reaction to convert pyruvate to CO_2 and hydroxyethyl-TPP. The two-carbon group on hydroxyethyl-TPP is transferred to the lipoamide moiety on the E2 subunit, and then acetyl-dihydrolipoamide passes the acetyl group to CoA to form acetyl-CoA, a substrate for the citrate cycle. The coenzyme FAD is linked covalently to the E3 subunit and is responsible for reoxidizing the dihydrolipoamide group in a reaction that ultimately passes 2 e^- as a hydride ion (:H^+) to the coenzyme NAD^+ to form NADH + H^+.

10.3 Enzymatic Reactions of the Citrate Cycle

The citrate cycle contains four coupled redox reactions, which serve to transfer four pairs of electrons from organic substrates (reductants) to the electron carriers NAD^+ and FAD (oxidants). Oxidative decarboxylation of pyruvate by the pyruvate dehydrogenase complex results in the transfer of a fifth pair of electrons to NAD^+, yielding a total of 4 NADH and 1 $FADH_2$ in the catabolism of pyruvate to CO_2. In addition, one molecule of GTP is generated by a substrate-level phosphorylation reaction catalyzed by the enzyme succinyl-CoA synthetase. It is important to note that the citrate cycle depends on the presence of oxygen (aerobic pathway), unlike glycolysis, which is considered an anaerobic pathway. The reason is that continual oxidation of NADH and $FADH_2$ inside mitochondria by the electron transport system is required to maintain flux through the citrate cycle. Therefore, if O_2 is limiting, then NADH and $FADH_2$ accumulate, and the citrate cycle is feedback inhibited. As we will describe in Chapter 11, O_2 is the terminal electron acceptor in the electron transport system and continually oxidizes NADH and $FADH_2$ to replenish mitochondrial pools of NAD^+ and FAD, respectively. **Figure 10.21** illustrates the eight citrate cycle reactions, along with the structures of the intermediates and the names of the citrate cycle enzymes.

Before we look individually at each of the eight enzymatic reactions, let's review the bioenergetics of the cycle, as shown in **Table 10.2**. Note that only free energy changes under standard conditions ($\Delta G°'$) are given, as it is not possible to obtain accurate measurements of ΔG values for citrate cycle intermediates in eukaryotic cells. The reason is that it is difficult to measure metabolite concentrations inside the mitochondrial matrix without contamination by the cytosolic components.

We can see from Table 10.2 that three of the reactions are highly favorable (the reactions catalyzed by citrate synthase, isocitrate dehydrogenase, and α-ketoglutarate dehydrogenase) and are probably even more favorable under actual conditions inside cells ($\Delta G << 0$). As described later, these are the same enzymes that function as key regulators of citrate cycle flux. With the exception of reaction 8, which is catalyzed by malate dehydrogenase ($\Delta G°' = +29.3$ kJ/mol), the other reactions have $\Delta G°'$ values close to zero and are likely to be readily reversible under cellular conditions.

Table 10.2 THE EIGHT REACTIONS OF THE CITRATE CYCLE, THE ENZYMES THAT CATALYZE THE REACTIONS, AND THE STANDARD FREE ENERGY CHANGES ($\Delta G°'$) FOR THE REACTIONS

Reaction	Enzyme	$\Delta G°'$ (kJ/mol)
1. Acetyl-CoA + Oxaloacetate + $H_2O \rightarrow$ CoA + Citrate	Citrate synthase	−31.4
2. Citrate $\rightleftharpoons$ Isocitrate	Aconitase	+6.7
3. Isocitrate + $NAD^+ \rightarrow$ α-Ketoglutarate + NADH + CO_2 + H^+	Isocitrate dehydrogenase	−20.9
4. α-Ketoglutarate + CoA + $NAD^+ \rightarrow$ Succinyl-CoA + NADH + CO_2 + H^+	α-Ketoglutarate dehydrogenase	−33.5
5. Succinyl-CoA + GDP + $P_i \rightleftharpoons$ Succinate + GTP + CoA	Succinyl-CoA synthetase	−2.9
6. Succinate + FAD $\rightleftharpoons$ Fumarate + $FADH_2$	Succinate dehydrogenase	+0.4
7. Fumarate + $H_2O \rightleftharpoons$ Malate	Fumarase	−3.8
8. Malate + $NAD^+ \rightleftharpoons$ Oxaloacetate + NADH + H^+	Malate dehydrogenase	+29.3

▶ ANIMATION

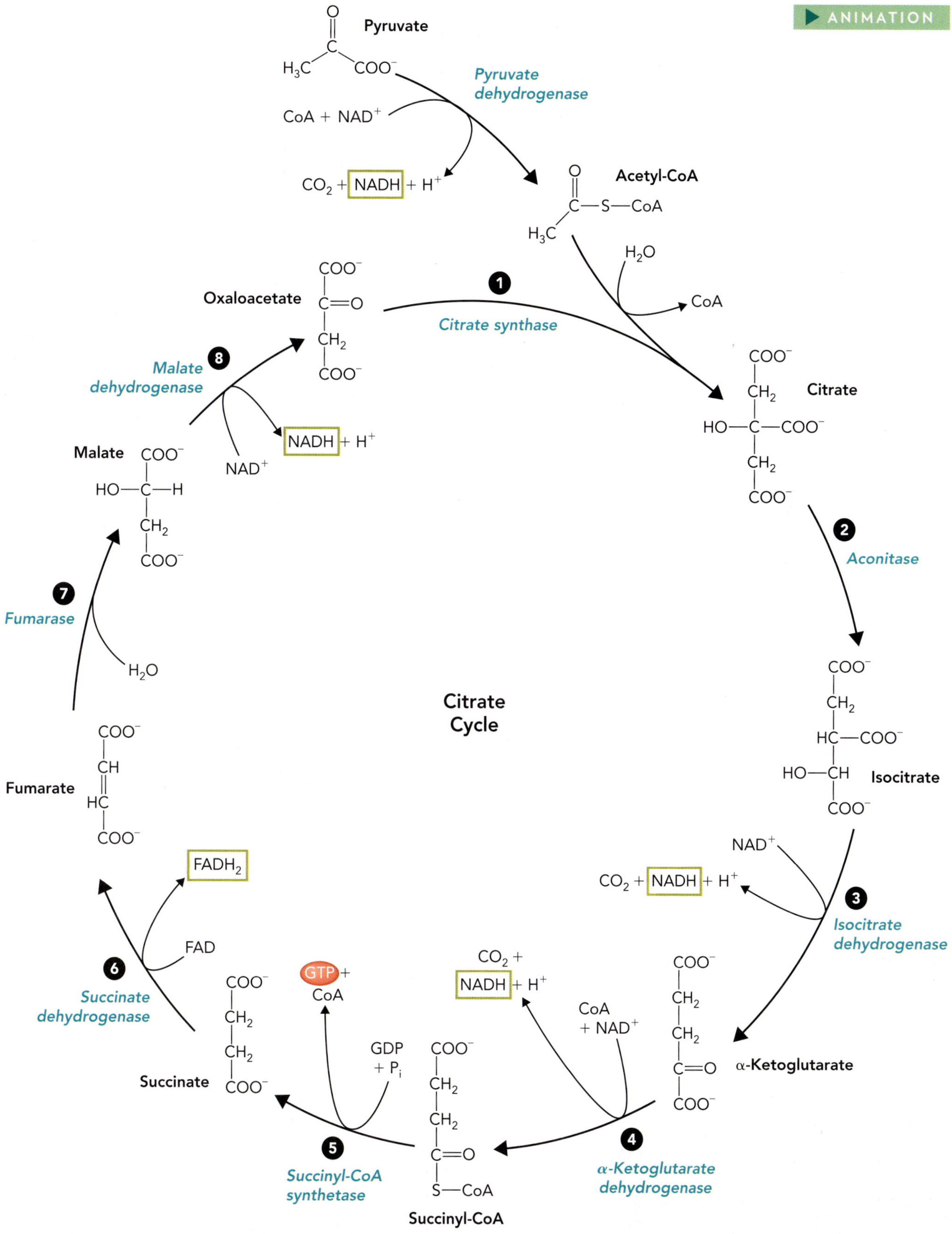

Figure 10.21 The eight enzymatic reactions in the citrate cycle.

Oxaloacetate concentrations are kept very low in the cell by its conversion to citrate in reaction 1, which allows reaction 8 to proceed in the forward direction even though the standard free energy change is large and positive (Figure 10.21).

The first four reactions of the citrate cycle include a condensation reaction that forms citrate from oxaloacetate and acetyl-CoA (reaction 1); an isomerization reaction (reaction 2); and two consecutive oxidative decarboxylation reactions, which release two CO_2 and generate two of the three NADH molecules (reactions 3 and 4). In the next four reactions, a thioester bond with high phosphoryl transfer energy is cleaved to generate GTP by substrate-level phosphorylation (reaction 5); FAD is reduced to $FADH_2$ (reaction 6); and a hydration reaction (reaction 7) sets up the final redox reaction to regenerate oxaloacetate and reduce the final NAD^+ to NADH (reaction 8).

The Eight Reactions of the Citrate Cycle

Reaction 1: Condensation of Oxaloacetate and Acetyl-CoA by Citrate Synthase to Form Citrate The purpose of the first reaction in the citrate cycle is to commit the acetate unit of acetyl-CoA to oxidative decarboxylation (**Figure 10.22**). The two originating carbons from acetyl-CoA that enter this round of the citrate cycle are highlighted in red in Figure 10.22 to distinguish them from carbons derived from oxaloacetate. On the basis of ^{14}C radiolabeling experiments, it was shown that these two carbons from acetyl-CoA are not lost as 2 CO_2 in the first round of the citrate cycle and are instead first incorporated into oxaloacetate. Analysis of ^{14}C-labeled carbons in the citrate cycle demonstrated the importance of regenerating metabolic intermediates in a cyclic pathway in order to maintain metabolic flux.

The citrate synthase reaction follows an ordered mechanism in which oxaloacetate binds first, inducing a conformational change in the enzyme that facilitates acetyl-CoA binding. Formation of the transient intermediate, citryl-CoA, is followed by a rapid hydrolysis reaction, which releases CoA (which is in the reduced form; see Figure 10.8) and citrate, as shown in **Figure 10.23**. The highly favorable change in standard free energy of this reaction ($\Delta G°' = -31.4$ kJ/mol) is due to hydrolysis of the thioester bond in citryl-CoA, which drives the reaction forward.

One of the interesting structural properties of citrate synthase is the large conformational change that accompanies oxaloacetate binding. The open conformation of the citrate synthase dimer promotes oxaloacetate binding, whereas the closed conformation creates the binding site for acetyl-CoA. It is thought that this large conformational change provides an environment for acetyl-CoA that promotes citryl-CoA formation rather than nonproductive hydrolysis of acetyl-CoA (side reaction). Structural studies of citrate synthase have shown that formation of

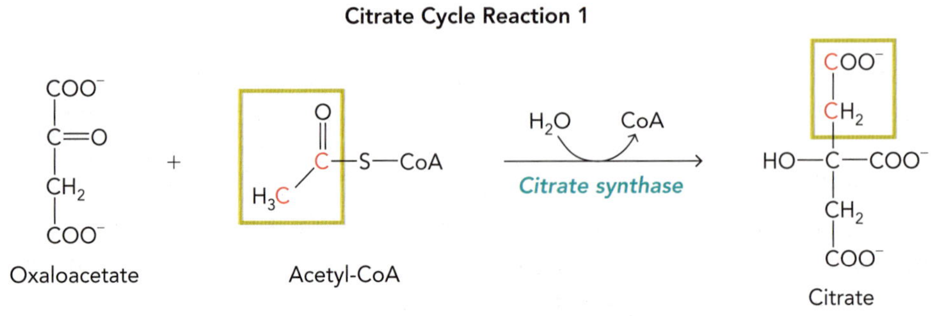

Figure 10.22 Citrate cycle reaction 1 is a highly favorable condensation reaction catalyzed by the enzyme citrate synthase ($\Delta G°' = -31.4$ kJ/mol). The two carbons shown in red are derived from the incoming acetyl-CoA.

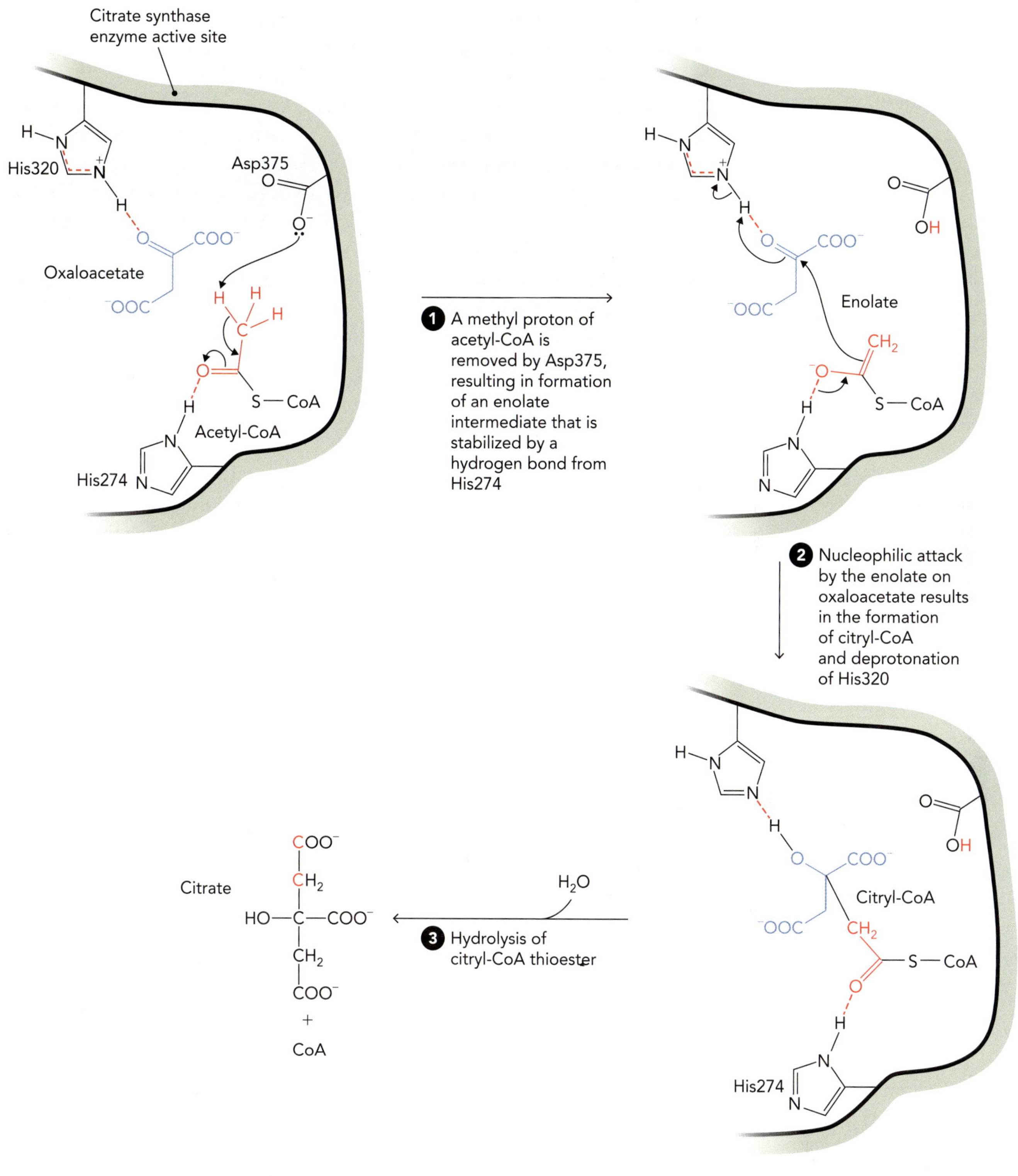
Citrate synthase
enzyme active site
His320
Asp375
Oxaloacetate
Acetyl-CoA
S—CoA
His274
1 A methyl proton of acetyl-CoA is removed by Asp375, resulting in formation of an enolate intermediate that is stabilized by a hydrogen bond from His274
Enolate
2 Nucleophilic attack by the enolate on oxaloacetate results in the formation of citryl-CoA and deprotonation of His320
Citryl-CoA
His274
H_2O
3 Hydrolysis of citryl-CoA thioester
Citrate
+
CoA

Figure 10.23 The citrate synthase reaction mechanism proceeds through an enolate intermediate.

citryl-CoA induces a second conformational change that promotes thioester cleavage and thereby releases citrate and CoA. **Figure 10.24** shows the molecular structure of a single subunit of citrate synthase enzyme in the two conformations: the open conformation with citrate bound to the enzyme active site, and the closed conformation with both citrate and acetyl-CoA bound in the active site. As with other enzymes we have examined, citrate and acetyl-CoA binding to citrate synthase is a classic example of induced-fit substrate binding.

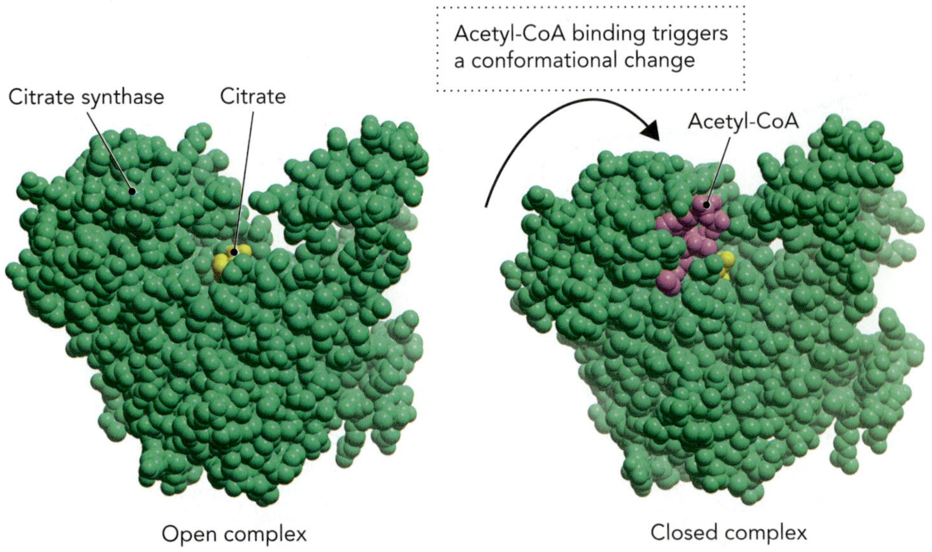

Figure 10.24 Citrate synthase undergoes a conformational change upon acetyl-CoA binding. BASED ON PDB FILES 1CTS (OPEN) AND 2CTS (CLOSED).

Reaction 2: Isomerization of Citrate by Aconitase to Form Isocitrate This reversible isomerization reaction involves a two-step mechanism. In the first step, the intermediate *cis*-aconitate is formed by a dehydration reaction requiring the participation of an iron–sulfur cluster (4 Fe–4 S) in the enzyme active site (**Figure 10.25**). In the second step, water is added back (hydration) to convert the double bond in *cis*-aconitate to a single bond with a hydroxyl group on the terminal carbon.

The iron–sulfur cluster in aconitase is unusual in that it serves to coordinate the OH group during its removal in the first step, and then to facilitate the subsequent hydration reaction. Most iron–sulfur clusters are found in enzymes that catalyze redox reactions; for example, the 4 Fe–4 S cluster in the NADH-Q reductase protein of the electron transport system (see Chapter 11).

Aconitase is one of the targets of the toxic compound **fluorocitrate**, which is derived from fluoroacetate, a naturally occurring poison found in some native Australian and African plants. When leaves of the Australian gidgee tree are mistakenly eaten by livestock, the fluoroacetate in the digested plant is converted to fluoroacetyl-CoA by the enzyme acetyl-CoA synthase. As shown in **Figure 10.26**, citrate synthase converts fluoroacetyl-CoA to the deadly compound fluorocitrate, which is a potent inhibitor of aconitase and also blocks citrate transport across the mitochondrial membrane. Plants

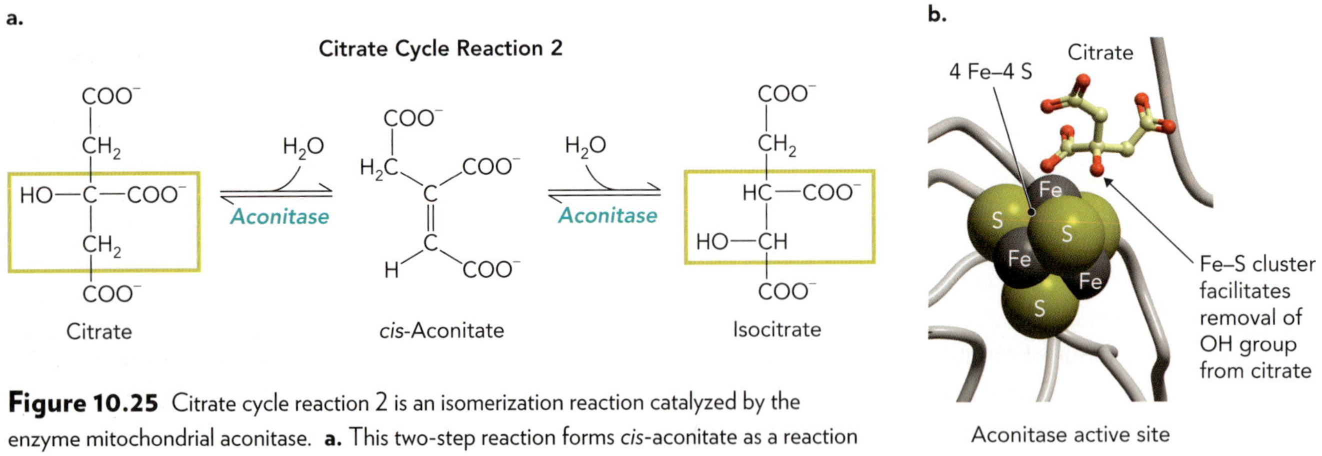

Figure 10.25 Citrate cycle reaction 2 is an isomerization reaction catalyzed by the enzyme mitochondrial aconitase. **a.** This two-step reaction forms *cis*-aconitate as a reaction intermediate. **b.** A 4 Fe–4 S cluster is used to facilitate removal of an OH group from citrate, as shown in this view of the enzyme active site. BASED ON PDB FILE 1C96.

a.

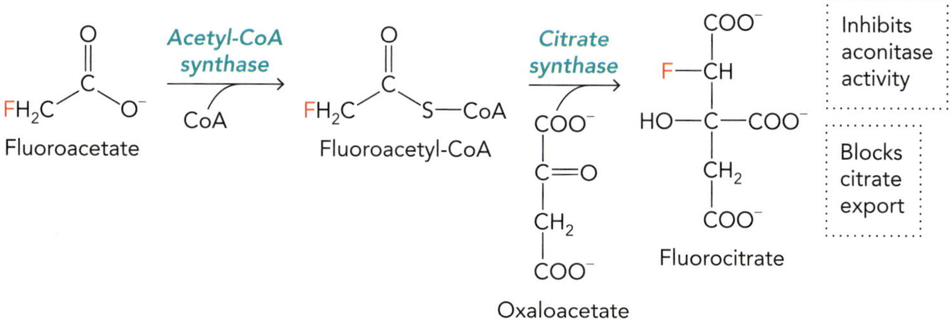

b.

c.

Figure 10.26 Fluoroacetate is a toxic compound that is converted to fluorocitrate, which inhibits aconitase activity and blocks citrate export from mitochondria. **a.** Reaction pathway converting fluoroacetate to fluorocitrate. **b.** The poisonous Australian gidgee tree (*Acacia georginae*) contains high levels of fluoroacetate. AUSCAPE/UIG/GETTY IMAGES. **c.** Sodium fluoroacetate is manufactured as Compound 1080 and is a controversial poison used to control populations of rodents, non-native mammals, coyotes, and foxes. This warning sign was posted on a tree after application of Compound 1080. TERRY WHITTAKER/FLPA IMAGE BROKER/NEWSCOM.

containing fluoroacetate are so deadly that Australian sheepherders have reported finding sheep with their heads still in the bush they were feeding on when they died.

Sodium fluoroacetate, manufactured under the trade name Compound 1080, is used as a highly lethal poison that kills most animals at a dose of only 2 mg/kg body weight. New Zealand, a country with no native land mammals, is the largest user of Compound 1080. Farmers apply it in attempts to eliminate non-native populations of possums, stoats, rats, and rabbits. Ranchers also use Compound 1080 to control coyote and fox populations.

There are actually two forms of aconitase: the mitochondrial form that catalyzes reaction 2 in the citrate cycle, as described here, and a cytosolic form that generates isocitrate for other metabolic pathways. Surprisingly, cytosolic aconitase also binds to double-stranded regions of mRNA transcripts that encode proteins involved in iron homeostasis (**Figure 10.27**). Studies have shown that when cytosolic aconitase binds to specific

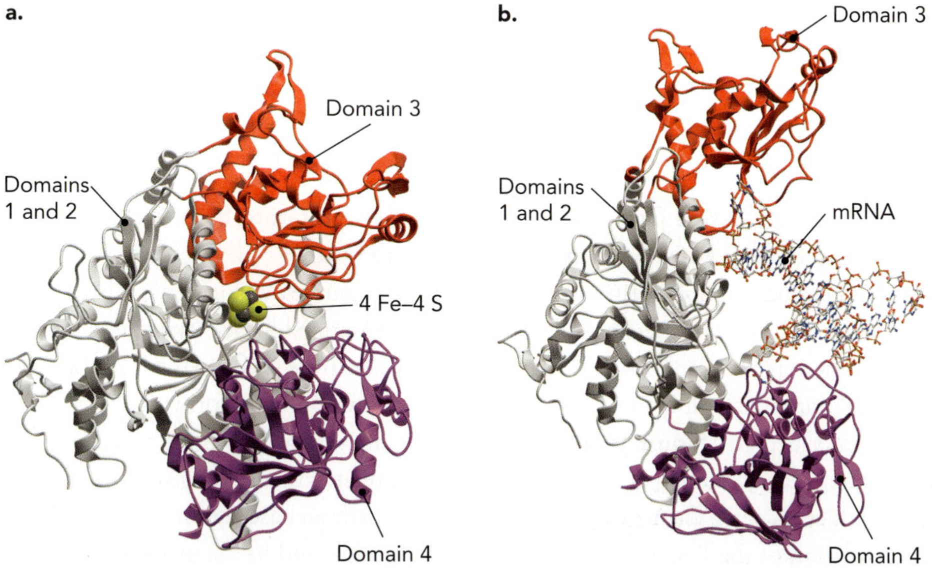

Figure 10.27 Cytosolic aconitase functions as both an enzyme that converts citrate to isocitrate, using an iron–sulfur cluster in the active site, and as an RNA binding protein, regulating the synthesis of ferritin and transferrin receptor proteins. **a.** Molecular structure of human cytosolic aconitase in the conformation that catalyzes the formation of isocitrate, showing the location of the four protein domains and the 4 Fe–4 S cluster in the enzyme active site. BASED ON PDB FILE 2B3Y. **b.** Molecular structure of the same four domains of the rabbit cytosolic aconitase bound to ferritin-encoding RNA in the conformation that regulates expression of iron-responsive genes. BASED ON PDB FILE 2IPY.

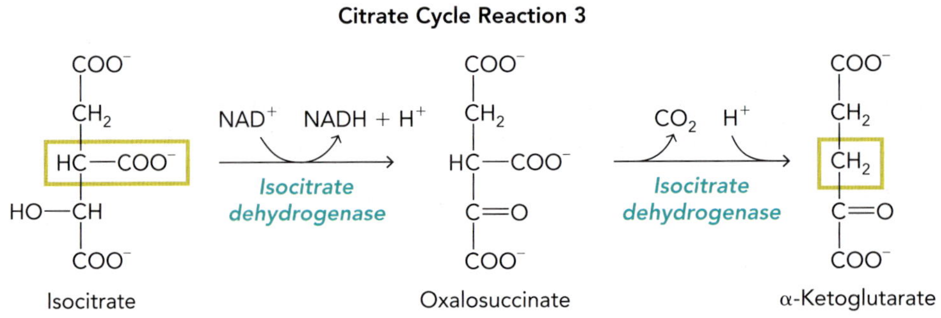

Figure 10.28 Citrate cycle reaction 3 is an **oxidative decarboxylation** reaction catalyzed by the enzyme isocitrate dehydrogenase. Oxalosuccinate and NADH are formed in the oxidation step of the reaction, followed by decarboxylation of oxalosuccinate to form α-ketoglutarate.

sequences in the 5′ end of the mRNA encoding ferritin, an iron storage protein, it inhibits ferritin protein synthesis. However, cytosolic aconitase binding to similar sequences in the 3′ end of the transferrin receptor mRNA promotes protein synthesis, thereby increasing iron import into cells. Researchers who first isolated the protein responsible for controlling the synthesis of ferritin and transferrin receptor called it iron regulatory protein-1 on the basis of its function in iron homeostasis and RNA binding properties. Not until the coding sequence of iron regulatory protein-1 was deduced using cDNA cloning techniques did it become clear that iron regulatory protein-1 and cytosolic aconitase were in fact the same protein. Another example of a dual-function protein is cytochrome c, which was first identified as an electron carrier in the electron transport system and later found to be a key signaling protein in the caspase-mediated apoptotic pathway.

Reaction 3: Oxidative Decarboxylation of Isocitrate by Isocitrate Dehydrogenase to Form α-Ketoglutarate, CO_2, and NADH Reaction 3 is the first of two decarboxylation steps in the citrate cycle. It is also the first reaction to generate NADH, which is used for energy conversion reactions in the electron transport system. As shown in **Figure 10.28**, NAD^+ triggers an oxidation reaction that generates the transient intermediate oxalosuccinate, which is then decarboxylated to form the product α-ketoglutarate (also known as 2-oxoglutarate). As described later, radioactive tracer experiments using ^{14}C show that the carbon in the CO_2 product of this reaction is derived from oxaloacetate rather than from the incoming acetyl-CoA.

The isocitrate dehydrogenase reaction is favorable under standard conditions ($\Delta G^{\circ\prime} = -20.9$ kJ/mol) and is likely irreversible under normal cellular conditions. The isocitrate dehydrogenase reaction is considered to be the rate-limiting step in the cycle and, as such, is highly regulated by ADP and Ca^{2+} (positive effectors) and ATP and NADH (negative effectors). We look at regulation of isocitrate dehydrogenase activity in the context of the entire citrate cycle later in the chapter. Note that α-ketoglutarate is one of the key shared metabolites between the citrate cycle and other pathways in the cell, primarily as a substrate and product in amino acid metabolism.

Reaction 4: Oxidative Decarboxylation of α-Ketoglutarate by α-Ketoglutarate Dehydrogenase to Form Succinyl-CoA, CO_2, and NADH In the second oxidative decarboxylation reaction of the citrate cycle, α-ketoglutarate dehydrogenase uses essentially the same catalytic mechanism as that already described for the pyruvate dehydrogenase reaction (see Figure 10.13). In the first step, isocitrate binds to the E1 subunit of α-ketoglutarate dehydrogenase, followed by decarboxylation and formation of a TPP-linked intermediate (**Figure 10.29**). The intermediate is then transferred to the lipoamide group of an E2 dihydrolipoyl succinyltransferase enzyme, leading to formation of the product succinyl-CoA. An E3 dihydrolipoyl dehydrogenase subunit then catalyzes the oxidation of the lipoamide group on the E2 subunit and in the process transfers a

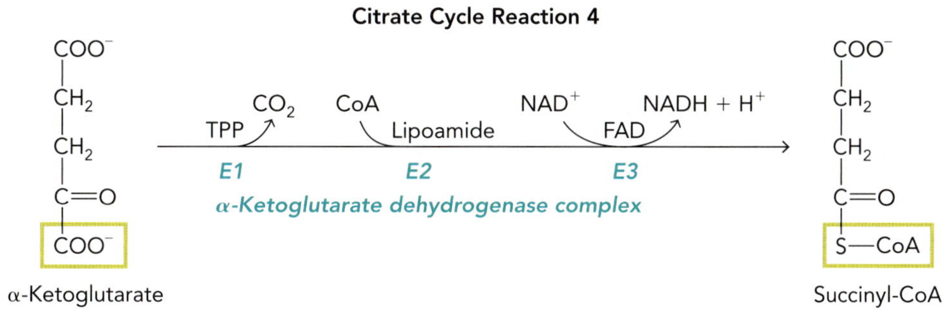

Figure 10.29 Citrate cycle reaction 4 is an oxidative decarboxylation reaction catalyzed by the enzyme α-ketoglutarate dehydrogenase, which converts α-ketoglutarate to succinyl-CoA. This reaction is mechanistically identical to the pyruvate dehydrogenase reaction in that it uses the same coenzymes and requires three distinct enzyme subunits (E1, α-ketoglutarate dehydrogenase; E2, dihydrolipoyl succinyltransferase; E3, dihydrolipoyl dehydrogenase). The carbon in the CO_2 product of this reaction is derived from oxaloacetate, not acetyl-CoA, in the most recent round of the citrate cycle (see Figure 10.22).

pair of electrons from $FADH_2$ to NADH. Similar to the pyruvate dehydrogenase reaction described earlier, the α-ketoglutarate dehydrogenase reaction has a large change in standard free energy ($\Delta G^{\circ\prime} = -33.5$ kJ/mol) and is allosterically regulated in response to changes in the energy charge of the cell.

Reaction 5: Conversion of Succinyl-CoA to Succinate by Succinyl-CoA Synthetase in a Substrate-Level Phosphorylation Reaction That Generates GTP The available free energy in the thioester bond of succinyl-CoA ($\Delta G^{\circ\prime} \approx -34$ kJ/mol) is used in the succinyl-CoA synthetase reaction to carry out a phosphoryl transfer ($\Delta G^{\circ\prime} = +30.5$ kJ/mol) leading to the production of GTP and succinate (**Figure 10.30**). Although the net standard free energy change for this coupled phosphoryl transfer reaction is still favorable ($\Delta G^{\circ\prime} = -2.9$ kJ/mol), the reaction mechanism requires that several different high-energy intermediates be formed along the way to ensure that the energy gained by cleavage of the thioester bond in succinyl-CoA is not lost before it can be used to generate GTP.

Figure 10.31 illustrates the proposed reaction mechanism for the succinyl-CoA synthetase reaction. In the first step, inorganic phosphate (P_i) replaces CoA to generate succinyl-phosphate, which retains the high energy of hydrolysis present in the succinyl-CoA substrate. Next, the phosphoryl group is transferred to a nearby His residue to create phosphohistidine, which then passes the phosphoryl group to GDP (or ADP) in the final step.

The mechanism for the succinyl-CoA synthetase reaction has been compared to the childhood game "pass the hot potato" because the high potential energy available in the thioester bond of succinyl-CoA must be converted to two separate high-energy intermediates (succinyl-phosphate and phosphohistidine) before it can be used to phosphorylate GDP. Note that even in cells that predominately generate GTP by the succinyl-CoA synthetase reaction, the enzyme nucleoside diphosphate kinase interconverts GTP and ATP using a reversible phosphoryl transfer reaction:

$$\text{GTP} + \text{ADP} \underset{}{\overset{\text{Nucleoside diphosphate kinase}}{\rightleftharpoons}} \text{GDP} + \text{ATP} \qquad \Delta G^{\circ\prime} = 0 \text{ kJ/mol}$$

Figure 10.30 Citrate cycle reaction 5 is catalyzed by the enzyme succinyl-CoA synthetase, which converts succinyl-CoA to succinate. The reaction involves a substrate-level phosphorylation reaction that generates GTP. (Some isoforms of succinyl-CoA synthetase produce ATP instead.)

Reaction 6: Oxidation of Succinate by Succinate Dehydrogenase to Form Fumarate Reaction 6 directly links the citrate cycle to the electron transport system through the redox conjugate pair $FAD/FADH_2$. This pair is covalently linked to the enzyme succinate dehydrogenase, a protein associated with the inner mitochondrial membrane. The succinate dehydrogenase reaction was critical in Hans Krebs's elucidation of the citrate

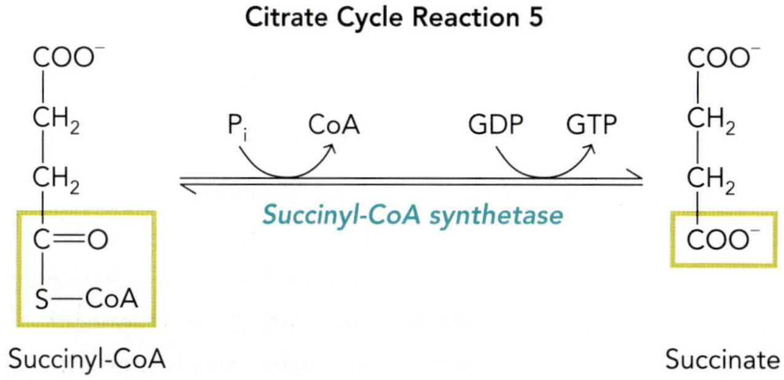

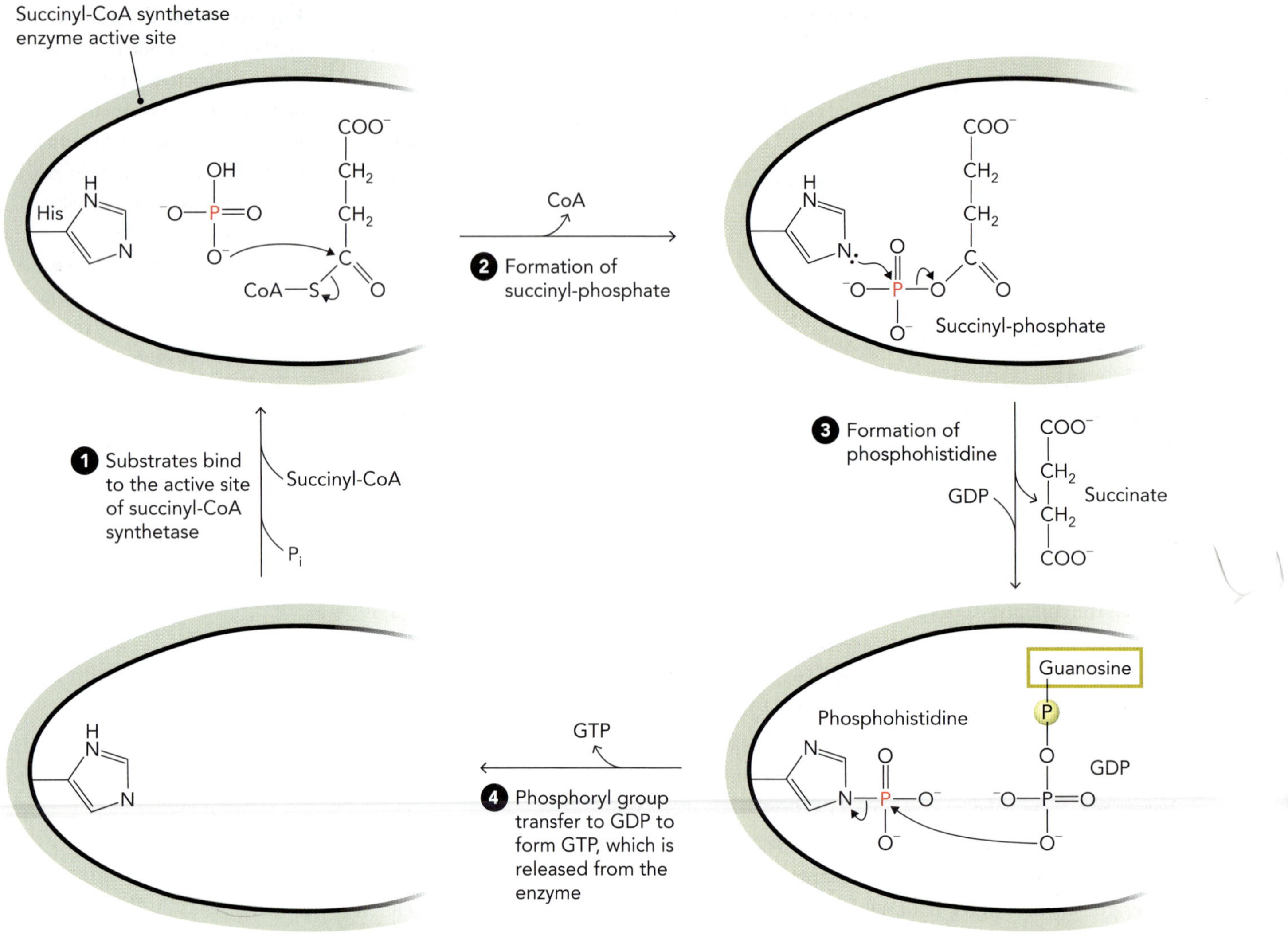

Figure 10.31 The succinyl-CoA synthetase reaction involves the formation succinyl-phosphate and then phosphohistidine in the enzyme active site. Depending on the nucleotide specificity of the β subunit, the phosphate can be donated to GDP or ADP to generate GTP or ATP, respectively.

cycle because he found that by inhibiting succinate dehydrogenase with malonate, a reversible inhibitor of the enzyme (see Figure 7.43), he could measure an increase in the accumulation of all five cycle intermediates that precede succinate oxidation (see Figure 10.21).

The succinate dehydrogenase reaction is shown in **Figure 10.32**. You can see that succinate oxidation results in the transfer of an electron pair (along with two protons) to the FAD moiety. Later on, these two electrons pass from $FADH_2$ to the electron carrier coenzyme Q in complex II of the electron transport system.

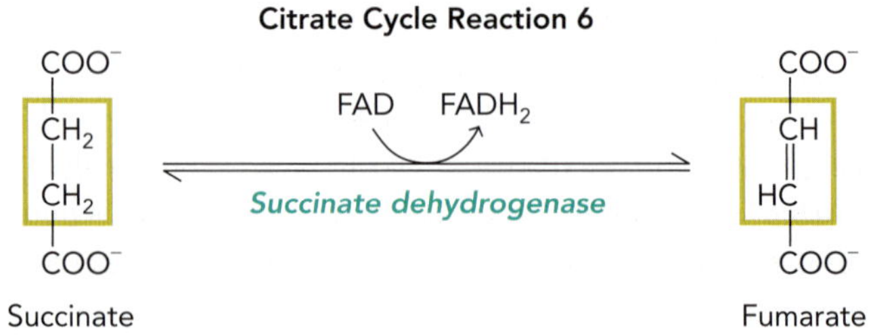

Figure 10.32 Succinate dehydrogenase catalyzes a redox reaction that couples succinate oxidation to FAD reduction, generating fumarate and $FADH_2$. The FAD coenzyme is covalently linked to succinate dehydrogenase and constitutes a component of complex II in the electron transport system (see Chapter 11).

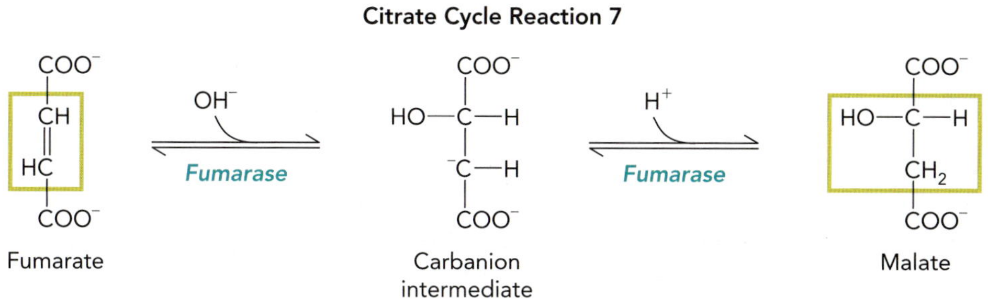

Figure 10.33 Citrate cycle reaction 7 is a hydration reaction catalyzed by the enzyme fumarase, which converts fumarate to malate. The proposed reaction mechanism involves the formation of a carbanion intermediate by hydroxylation, followed by a protonation step.

Reaction 7: Hydration of Fumarate by Fumarase to Form Malate Fumarase is a highly stereospecific enzyme that catalyzes the **hydration** of the C═C double bond in fumarate to generate the L isomer of malate. This reaction is shown in **Figure 10.33** and proceeds through the formation of a carbanion intermediate.

Reaction 8: Oxidation of Malate by Malate Dehydrogenase to Form Oxaloacetate The final reaction in the citrate cycle is catalyzed by the enzyme malate dehydrogenase, which oxidizes the hydroxyl group of malate to form oxaloacetate. In the process, malate dehydrogenase reduces NAD^+ to form the third and final NADH molecule generated by the citrate cycle (**Figure 10.34**). We will see the malate dehydrogenase reaction again in Chapter 11 when we describe the malate–aspartate shuttle that is responsible for transporting oxidized and reduced metabolites across the inner mitochondrial membrane (see Figure 11.36).

When Krebs first described the citrate cycle in 1937, it was not feasible to follow the fate of individual carbon atoms throughout the cycle, and therefore he could only propose how he thought each intermediate was formed. However, with the development of radioactive tracer techniques—in particular, ^{14}C labeling of citrate cycle metabolites—a more detailed picture of the eight citrate cycle reactions became possible. As shown in **Figure 10.35**, we can now trace the fate of the two incoming carbon atoms from acetyl-CoA and identify which carbons from citrate are lost as CO_2. Within a single round of the eight

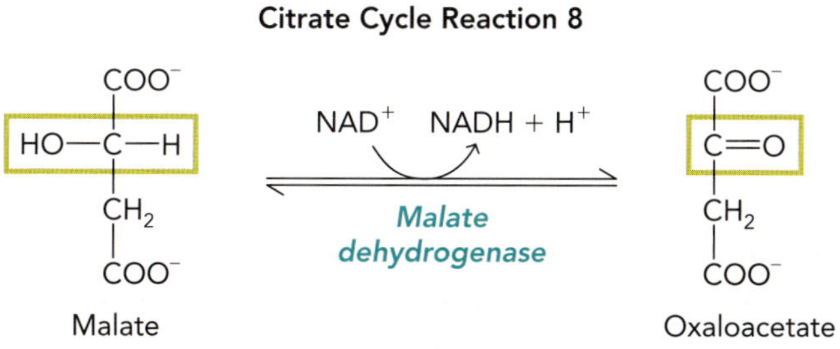

Figure 10.34 Citrate cycle reaction 8 is a redox reaction catalyzed by the enzyme malate dehydrogenase, which couples malate oxidation to NAD^+ reduction, generating oxaloacetate and NADH. Note that the two carbons that originated from acetyl-CoA in reaction 1 of this cycle are incorporated into oxaloacetate.

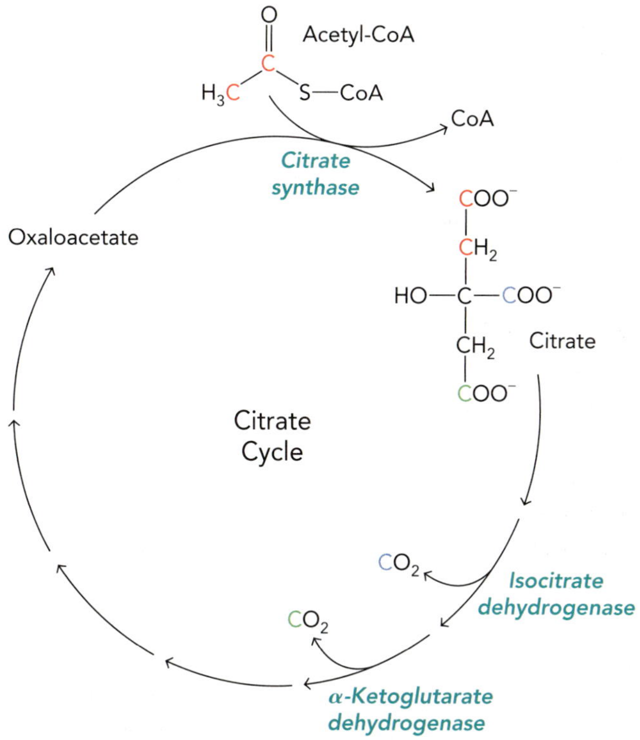

Figure 10.35 Carbon-14 radiolabeling experiments show that the carbons entering the citrate cycle from acetyl-CoA are not the carbons that are lost as CO_2 in that same round of the citrate cycle. Further rounds of the citrate cycle are needed to eliminate these carbons as CO_2.

reactions of the citrate cycle, the carbons incorporated from acetyl-CoA are not the same carbons that are eliminated as CO_2 in this round. This observation demonstrates the importance of regenerating metabolic intermediates in cyclic pathways.

concept integration 10.3

Which reactions catalyze the oxidative decarboxylation of pyruvate and acetyl-CoA to generate 3 CO_2 and 3 NADH? Explain.

Pyruvate is oxidatively decarboxylated by the enzyme pyruvate dehydrogenase in the presence of coenzyme A and NAD^+ to generate acetyl-CoA, CO_2, and NADH + H^+. The two carbons from pyruvate that are present in acetyl-CoA are added to oxaloacetate to produce citrate, which is then isomerized to isocitrate. Isocitrate is oxidatively decarboxylated by the enzyme isocitrate dehydrogenase in the presence of NAD^+ to generate α-ketoglutarate, CO_2, and NADH + H^+. The third enzyme required is α-ketoglutarate dehydrogenase, which uses NAD^+ and coenzyme A to generate succinyl-CoA, CO_2, and NADH.

10.4 Regulation of the Citrate Cycle

The end product of the citrate cycle, oxaloacetate, is the substrate for the first reaction. Therefore, flux through the pathway can be controlled by resetting the level of available substrate after each turn of the cycle. In addition to substrate availability, two other regulatory mechanisms operate in the citrate cycle: product inhibition and feedback control of key enzymes.

As shown in **Figure 10.36**, the three main control points within the citrate cycle are regulation of the enzymes citrate synthase, isocitrate dehydrogenase, and α-ketoglutarate dehydrogenase. Pyruvate dehydrogenase and pyruvate carboxylase—the two enzymes that control the amount of oxaloacetate and acetyl-CoA available for the citrate synthase reaction—are also regulated. Pyruvate dehydrogenase activity is regulated by metabolites that signal the energy charge in the cell and function to control the activity of pyruvate dehydrogenase kinase and pyruvate dehydrogenase phosphatase-1 (see Figure 10.20). Pyruvate carboxylase activity is allosterically activated directly by acetyl-CoA, leading to increased production of oxaloacetate. As described in the next section, this control mechanism makes sense metabolically because when acetyl-CoA levels are high—for example, when fatty acid oxidation is the primary source of acetyl-CoA—then oxaloacetate levels become limiting.

In vitro studies have shown that citrate synthase, isocitrate dehydrogenase, and α-ketoglutarate dehydrogenase are all inhibited by high NADH-to-NAD^+ ratios, which signal that the electron transport system is not working at full efficiency because of elevated ratios of ATP to ADP + AMP. All three enzymes are also inhibited by NADH and ATP, whereas citrate synthase and α-ketoglutarate dehydrogenase are both inhibited by succinyl-CoA as well. Citrate synthase is also potently inhibited by citrate, which is exported out of the mitochondria when its levels are too high. As shown in Figure 10.36, increased ADP levels activate both citrate synthase and isocitrate dehydrogenase by relieving the allosteric inhibition of ATP. Lastly, Ca^{2+} stimulates the activity of pyruvate dehydrogenase, isocitrate dehydrogenase, and α-ketoglutarate dehydrogenase, a finding that is consistent with the requirement for a continual supply of ATP during muscle contraction.

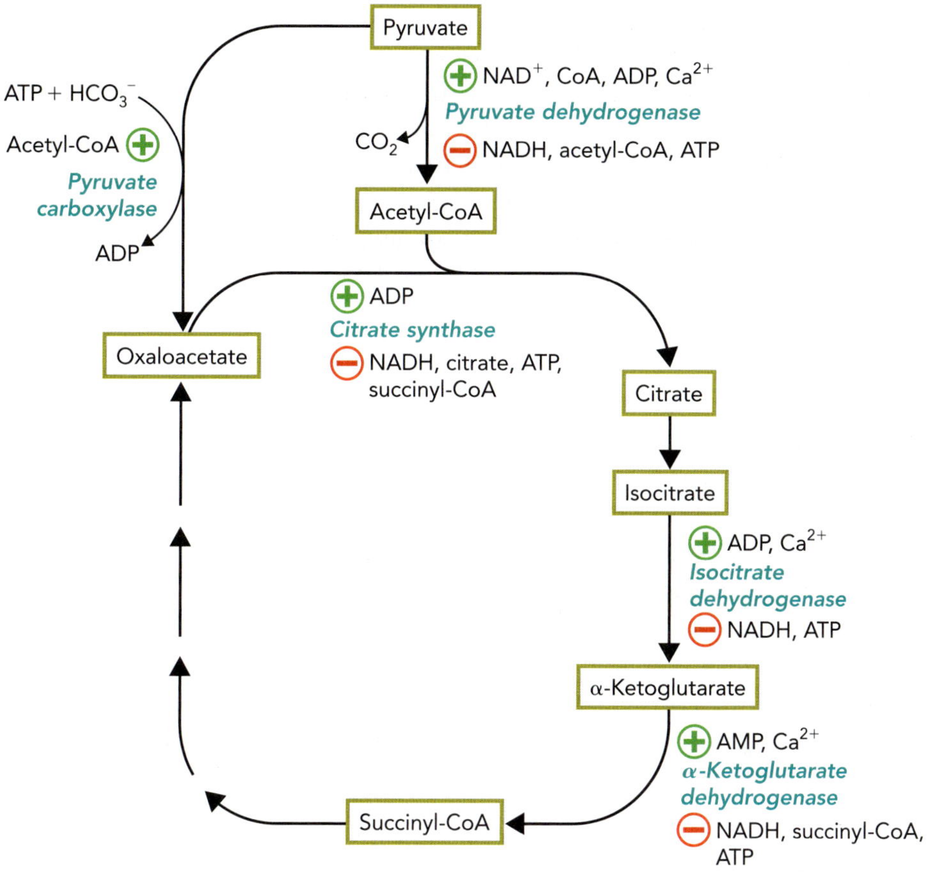

Figure 10.36 Regulation of key enzymes controls metabolic flux through the citrate cycle. Enzyme activators are denoted by the green icon and enzyme inhibitors by the red icon. The substrates and products of the pyruvate dehydrogenase reaction and the other regulated citrate cycle enzymes are shown in Figure 10.21, and the substrates and products of the pyruvate carboxylase reaction are shown later in Figure 10.40.

Citric acid is used in a wide variety of commercial applications, including preservation of processed foods, as a reactive acid in polymer chemistry, and as a stabilizer in pharmaceutical medicines (see the chapter-opening figure). Since the early 1900s, citric acid has been produced in large quantities for commercial purposes. Most citric acid at that time was isolated from concentrated lemon extract, which was sold by Italian cartels that controlled vast lemon groves throughout Italy. However, because the amount of citric acid that could be obtained from processing lemons could not keep up with global demand—especially after Italian lemon crops were nearly destroyed during World War I—alternative sources of citric acid were urgently sought.

The breakthrough came in 1917 when the American food chemist James Currie discovered that large amounts of citric acid in the form of citrate could be isolated from the mold *Aspergillus niger* when it was grown in a medium containing sucrose, salts, and iron. Currie teamed up with the chemical company Pfizer to develop proprietary methods to isolate large amounts of citrate from liquid *Aspergillus* cultures. Using optimized fermentation conditions, which take advantage of genetically selected *Aspergillus* strains, modern biotechnology methods can convert 140 g/L of sucrose into an amazing 115 g/L of citrate over a 10-day period (**Figure 10.37**). These closely guarded fermentation methods produce nearly 500,000 metric tons of citric acid annually by exploiting the citrate synthase reaction.

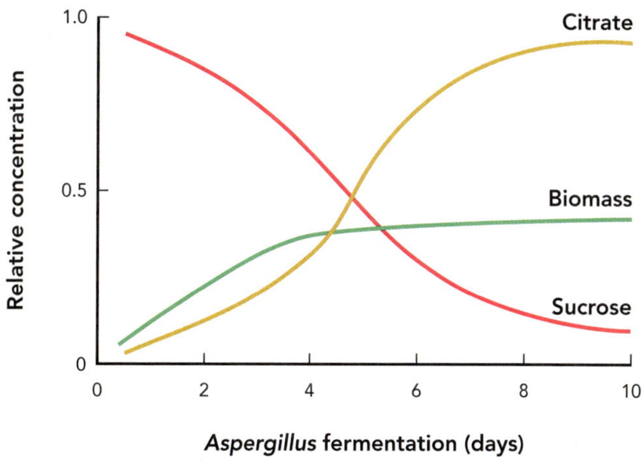

Figure 10.37 Most commercial citric acid is derived from a fermentation process using the mold *Aspergillus*, which efficiently converts sucrose to citrate, the conjugate base of citric acid. The schematic graph shows the relative conversion of sucrose to citrate by *Aspergillus* over a 10-day period of fermentation. Note that citrate production continues to rise even after a steady-state level of *Aspergillus* cell mass (biomass) has been reached.

How can citrate be produced at such high levels in *Aspergillus*, considering the feedback inhibitory pathways of the citrate cycle? The answer is twofold. First, by optimizing cell growth conditions, and by selecting for *Aspergillus* strains with high capacity for citrate export, citrate does not accumulate in the mitochondrial matrix and inhibit the citrate synthase reaction (**Figure 10.38**). Second, these same culture conditions result in inhibition of cytosolic citrate lyase, which decreases the rate of cytosolic citrate cleavage to generate oxaloacetate and acetyl-CoA.

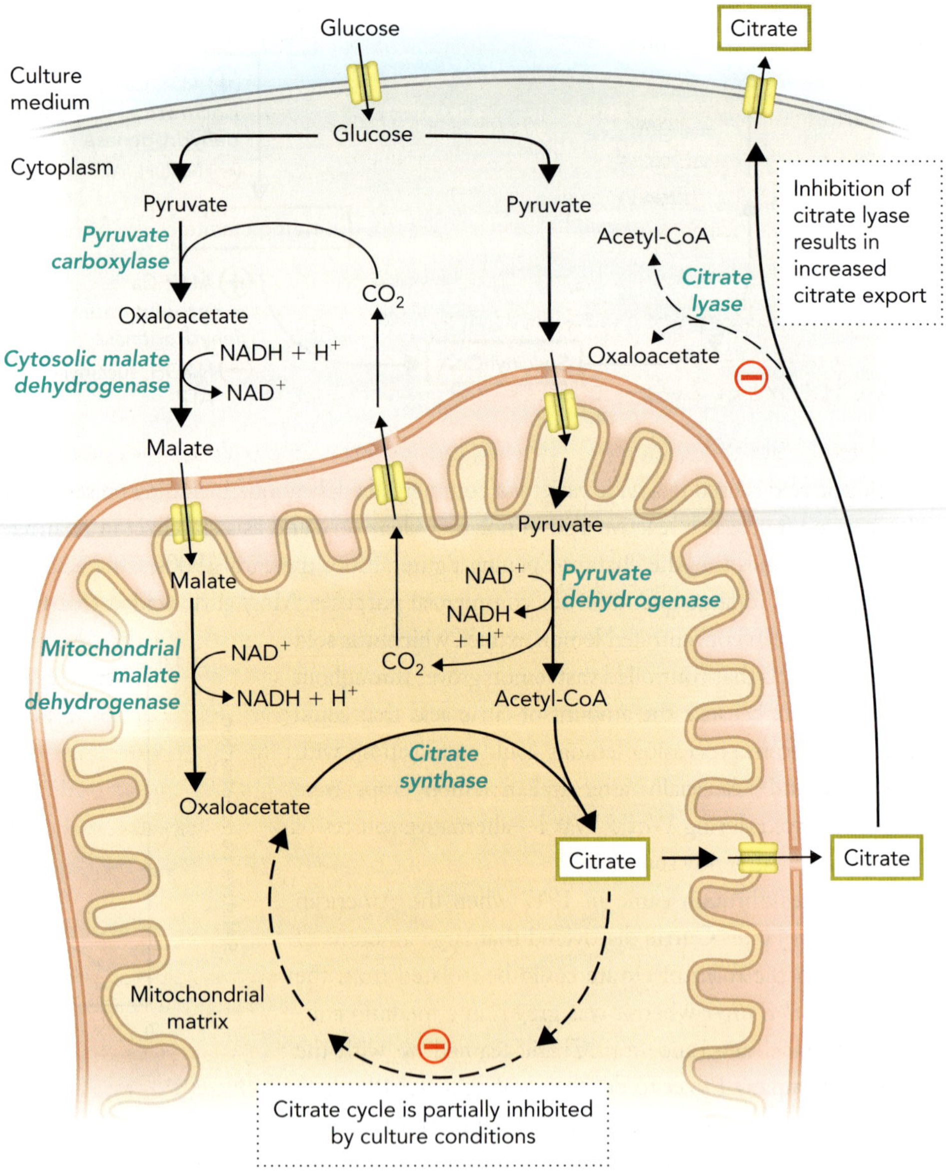

Figure 10.38 Metabolic pathways involved in citrate production by *Aspergillus* fermentation. Specific transport proteins (yellow cylinders) are required to move biomolecules across the mitochondrial and plasma membranes. Reactions that are decreased in *Aspergillus* compared to most other organisms are shown with dotted lines.

concept integration 10.4

In addition to substrate availability, regulation of enzyme activity in the citrate cycle is also achieved by product inhibition and feedback inhibition. Explain how these inhibitory strategies are effective in regulating flux through the citrate cycle.

Product inhibition provides a direct response to the reaction that is being catalyzed. For example, in the reaction catalyzed by citrate synthase, the product citrate inhibits the reaction. This provides immediate and direct feedback that citrate is already present at sufficient levels, and that citrate synthase does not need to produce more citrate. The three key regulated enzymes in the citrate cycle—citrate synthase, isocitrate dehydrogenase, and α-ketoglutarate dehydrogenase—are all inhibited by ATP. Regulation by ATP ensures that these enzymes have low activity under conditions where there is already a sufficiency of energy, and thus needless reactions in the citrate cycle are not wasted. Because ATP is not a direct product of any of these reactions, these are examples of feedback inhibition. This regulatory strategy is useful for preventing the earlier reactions in the pathway from beginning when the product from the end of the pathway, in this case ATP, is not needed.

10.5 Metabolism of Citrate Cycle Intermediates

We noted earlier that the citrate cycle not only generates the bulk of redox power for energy conversion by oxidizing acetyl-CoA to NADH and $FADH_2$, but it also provides biosynthetic precursors for numerous other pathways. The citrate cycle is therefore considered an **amphibolic pathway** because it functions in both catabolic (oxidation of acetyl-CoA) and anabolic (production of precursors for fatty acid, amino acid, and heme synthesis) pathways. In this section, we first identify the citrate cycle intermediates that are shared with other pathways and then look at how levels of citrate cycle intermediates are maintained by **anaplerotic reactions**, which are reactions that replenish citrate cycle intermediates that have been shunted to other metabolic pathways. (The word *anaplerotic* is derived from the Greek *ana* ("up") and *plerotikos* ("to fill")—the pyruvate carboxylase reaction "fills up" the citrate cycle by providing oxaloacetate.)

Citrate Cycle Intermediates Are Shared by Other Pathways

Five citrate cycle intermediates serve as biosynthetic precursors in other pathways: citrate, α-ketoglutarate, succinyl-CoA, malate, and oxaloacetate. As shown in **Figure 10.39**, citrate is exported from the mitochondria to the cytosol when the citrate cycle is feedback inhibited by ATP and NADH. Once in the cytosol, citrate is cleaved by the enzyme citrate lyase to release acetyl-CoA and oxaloacetate (this reaction is inhibited in *Aspergillus niger* strains; see Figure 10.38). Acetyl-CoA is used in the cytosol for fatty acid and cholesterol biosynthesis, while oxaloacetate can be converted to malate by cytosolic malate dehydrogenase and shuttled back into the mitochondria to complete the acetyl-CoA circuit. Alternatively, oxaloacetate also serves as a substrate for the enzyme phosphoenolpyruvate carboxykinase, which generates phosphoenolpyruvate for gluconeogenesis (see Chapter 14).

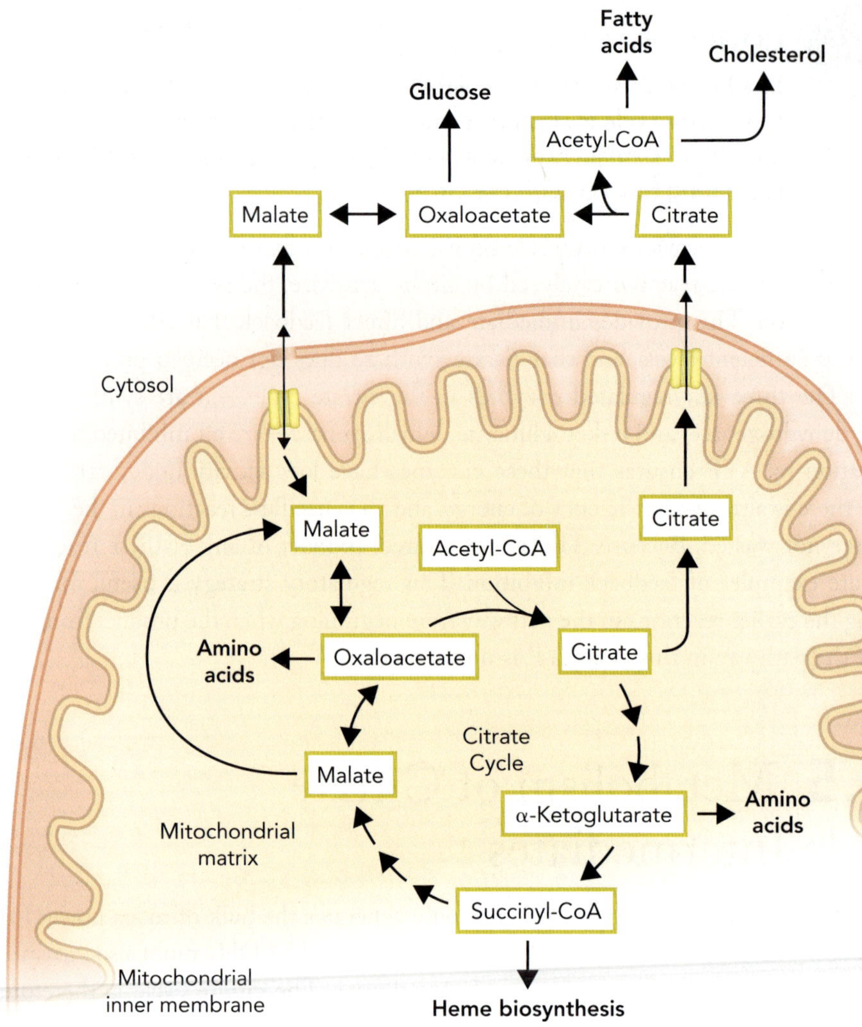

Figure 10.39 Citrate cycle intermediates serve as precursors for several biosynthetic pathways. Oxaloacetate and α-ketoglutarate provide carbon skeletons for amino acids, and succinyl-CoA is a precursor for heme biosynthesis. Citrate and malate are exported to the cytosol, where they are used to generate acetyl-CoA and oxaloacetate for fatty acid, cholesterol, and glucose synthesis.

The carbon backbones of the amino acids aspartate and glutamate are derived from oxaloacetate and α-ketoglutarate, respectively, through simple transamination reactions. These two amino acids are precursors for several other amino acids and for nucleotides. The citrate cycle also provides the essential component for heme biosynthesis, δ-aminolevulinic acid, which is synthesized in the mitochondria from succinyl-CoA and glycine by the enzyme δ-aminolevulinic acid synthase. Figure 10.39 also illustrates that malate can be siphoned off of the citrate cycle when demand arises in liver cells for glucose synthesis through the gluconeogenic pathway; for example, during starvation, when liver glycogen stores have been depleted. Malate export to the cytosol is used for oxaloacetate synthesis, which is converted to phosphoenolpyruvate by the enzyme phosphoenolpyruvate carboxykinase. We describe all of these citrate cycle-linked metabolic pathways in Part 4 of this book.

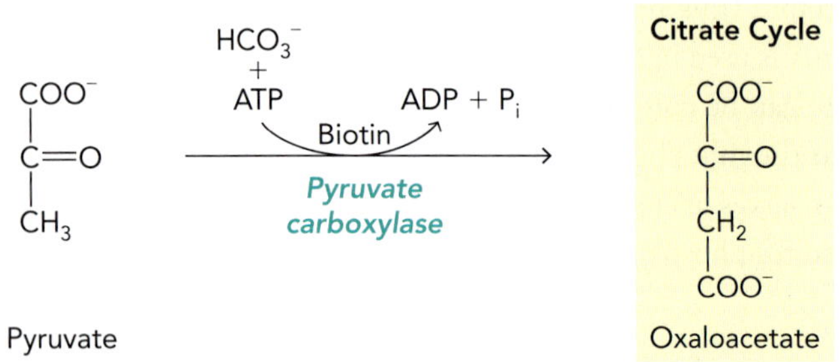

Figure 10.40 The pyruvate carboxylase reaction is ATP dependent and requires the coenzyme biotin to function as a carboxyl group carrier. Bicarbonate (HCO_3^-) is the predominant form of carbon dioxide at pH 7.

Pyruvate Carboxylase Catalyzes the Primary Anaplerotic Reaction

The enzyme pyruvate carboxylase has a crucial role in balancing the input of oxaloacetate with that of acetyl-CoA in the citrate cycle by providing a way to convert pyruvate into oxaloacetate using an ATP-dependent carboxylation reaction. For example, when flux through the citrate cycle is decreased because of loss of cycle intermediates to anabolic

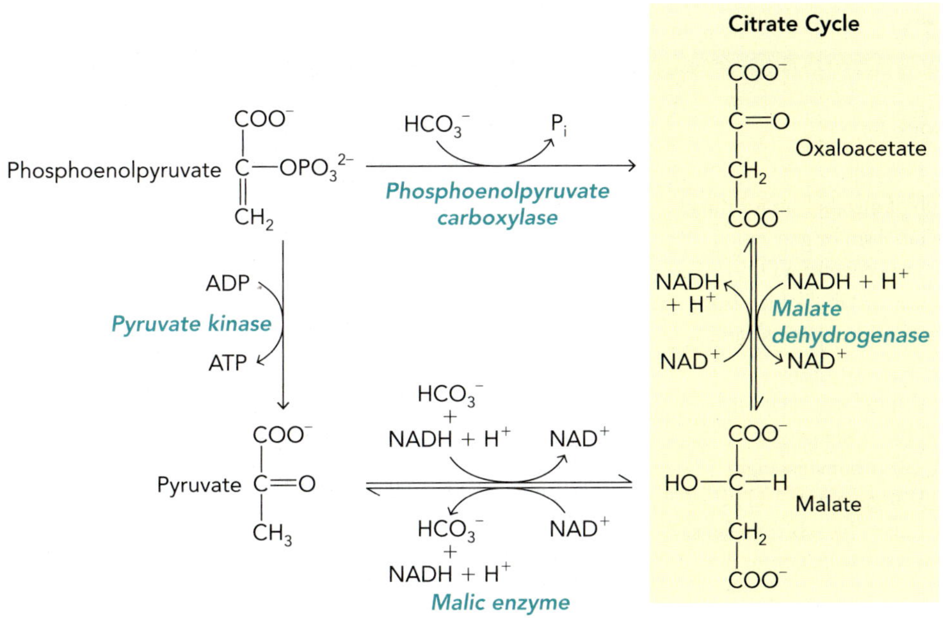

Figure 10.41
Phosphoenolpyruvate carboxylase and malic enzyme catalyze anaplerotic reactions that replenish the citrate cycle intermediates oxaloacetate and malate, respectively, in a variety of organisms. Note that the glycolytic enzyme pyruvate kinase converts phosphoenolpyruvate to pyruvate, and malate dehydrogenase interconverts malate and oxaloacetate.

pathways, acetyl-CoA accumulates in the mitochondrial matrix and allosterically activates pyruvate carboxylase (see Figure 10.36).

Pyruvate carboxylase uses the coenzyme biotin to catalyze an ATP-dependent reaction (**Figure 10.40**). Biotin, like many other coenzymes, is a vitamin; however, biotin deficiencies are rarely observed in humans because it is present at adequate levels in most foods (breads, cooked eggs, most vegetables), and daily requirements are very low (50 μg/day). We examine the biotin-dependent mechanism of the pyruvate carboxylase reaction in detail in Chapter 14 when we describe enzymes in the gluconeogenic pathway (see Figure 14.43).

The pyruvate carboxylase reaction is considered an anaplerotic reaction because it replenishes oxaloacetate and restores flux through the pathway. Two other anaplerotic reactions for the citrate cycle are shown in **Figure 10.41**. Plants, yeast, and bacteria contain the anaplerotic enzyme phosphoenolpyruvate carboxylase, which uses energy available from phosphate hydrolysis to drive a reaction generating oxaloacetate from phosphoenolpyruvate and bicarbonate. Malic enzyme, which is found in both prokaryotic and eukaryotic cells, can function as an anaplerotic enzyme by catalyzing the carboxylation of pyruvate to form malate in a reaction requiring the coenzyme NADH (or NADPH in some versions of the enzyme). Malic enzyme normally functions in plants as a component of the carbon fixation pathway (see Chapter 12), whereas in animal cells it plays a key role in fatty acid biosynthesis (see Chapter 16). These two anaplerotic reactions use the glycolytic and gluconeogenic metabolites phosphoenolpyruvate and pyruvate as substrates.

concept integration 10.5

What are the citrate cycle metabolites that are shared with other pathways, and what are the specific pathways?

Citrate, α-ketoglutarate, succinyl-CoA, malate, and oxaloacetate are all citrate cycle metabolites that are shared with other pathways. Citrate and malate may be exported to the cytosol, where they can be used in fatty acid and cholesterol synthesis. The carbon skeletons of oxaloacetate and α-ketoglutarate can be used to produce amino acids. Succinyl-CoA can be used as a precursor for heme biosynthesis.

chapter summary

10.1 The Citrate Cycle Captures Energy Using Redox Reactions

- The citrate cycle is considered the hub of metabolism for three reasons: (1) it is central to aerobic metabolism by generating the bulk of NADH and $FADH_2$ used to generate ATP by oxidative phosphorylation; (2) it oxidizes metabolic fuels from a variety of sources (carbohydrates, fatty acids, proteins); and (3) it provides metabolites for biosynthetic pathways.
- The citrate cycle accomplishes four things with each turn of the cycle: (1) it decarboxylates citrate to remove two carbons contributed by acetyl-CoA; (2) it transfers eight electrons to 3 NAD^+ and 1 FAD; (3) it generates 1 GTP (converted to ATP by diphosphate kinase) by substrate-level phosphorylation; and (4) it regenerates oxaloacetate so the cycle can start again.
- The citrate cycle can be thought of as a metabolic engine in which the fuel is acetyl-CoA, the exhaust is CO_2, and the work performed is the transfer of electrons using a series of linked redox reactions.
- The citrate cycle is also called the Krebs cycle in honor of Hans Krebs, who discovered it; alternatively, it is also called the citric acid cycle or the tricarboxylic acid (TCA) cycle. The descriptive term *citrate* cycle reflects that all three carboxylate groups on citrate are deprotonated at physiologic pH—it is not an acid in cells.
- Oxidation reactions are a loss of electrons, and reduction reactions are a gain of electrons. The energy available from redox reactions is due to differences in the electron affinity of reactants and is based on molecular structure.
- Coupled redox reactions consist of two half-reactions containing a conjugate redox pair. Compounds that accept electrons are called oxidants and are reduced in the reaction, whereas compounds that donate electrons are called reductants and are oxidized in the reaction.
- The reduction of NAD^+ to NADH involves the transfer of a hydride ion ($:H^-$), which contains 2 e^- and 1 H^+, and the release of a proton (H^+) into solution. FAD is reduced by sequential addition of one hydrogen (1 e^- and 1 H^+) at a time.
- The biochemical standard reduction potential ($E^{\circ\prime}$), measured in volts (V), represents the electron affinity of a given conjugate redox pair. The $E^{\circ\prime}$ of oxidants with a higher affinity for electrons than that of H^+ have positive $E^{\circ\prime}$ values ($E^{\circ\prime} > 0$); oxidants with a lower affinity for electrons than that of H^+ have negative $E^{\circ\prime}$ values ($E^{\circ\prime} < 0$).
- The amount of energy available from a coupled redox reaction is directly related to the difference between two reduction potentials and is defined by the term $\Delta E^{\circ\prime}$. The $\Delta E^{\circ\prime}$ of a coupled redox reaction is determined by subtracting the $E^{\circ\prime}$ of the reductant (e^- donor) from the $E^{\circ\prime}$ of the oxidant (e^- acceptor).
- The $\Delta E^{\circ\prime}$ for a coupled redox reaction is proportional to the standard free energy change $\Delta G^{\circ\prime}$ as described by the equation $\Delta G^{\circ\prime} = -nF\Delta E^{\circ\prime}$.
- To calculate the actual reduction potentials for conjugate redox pairs, the concentrations of the oxidant (e^- acceptor) and reductant (e^- donor) need to be accounted for using the Nernst equation.

10.2 Pyruvate Dehydrogenase Converts Pyruvate to Acetyl-CoA

- Pyruvate is oxidatively decarboxylated in the mitochondrial matrix by the multi-subunit pyruvate dehydrogenase complex, which uses a five-step reaction mechanism that requires three distinct enzymes and five different coenzymes.
- The five coenzymes used by the pyruvate dehydrogenase complex are NAD^+, FAD, CoA, TPP, and α-lipoic acid (lipoamide).
- Severe niacin deficiency causes the disease pellagra, which leads to insufficient levels of NAD^+/NADH for metabolic reactions. Corn-rich diets can lead to pellagra if the cornmeal is not prepared properly to release the protein-bound niacin.
- Thiamine, also called vitamin B_1, is the precursor to TPP, an important coenzyme in the pyruvate dehydrogenase and α-ketoglutarate dehydrogenase reactions. Thiamine deficiency causes the human disease beriberi.
- The eukaryotic pyruvate dehydrogenase complex contains three enzymes: E1, pyruvate dehydrogenase; E2, dihydrolipoyl acetyltransferase; and E3, dihydrolipoyl dehydrogenase.
- The pyruvate dehydrogenase reaction can be broken down into five distinct catalytic steps: steps 1, 2, and 3 lead to the formation of acetyl-CoA, and steps 4 and 5 regenerate the oxidized form of lipoamide and reduce NAD^+ to NADH + H^+.
- Pyruvate dehydrogenase activity is regulated by allosteric control in response to energy charge, NADH to NAD^+ ratios, and CoA to acetyl-CoA ratios. It is also regulated by serine phosphorylation, which is mediated by kinase and phosphatase enzymes that are themselves regulated by energy charge metabolites.

10.3 Enzymatic Reactions of the Citrate Cycle

- The citrate cycle consists of eight enzymatic reactions: four of these are coupled redox reactions that transfer electrons to NAD^+ and FAD; one is a substrate-level phosphorylation reaction generating GTP (converted to ATP by diphosphate kinase); and the other three reactions include two isomerizations and one hydration.
- Citrate synthase catalyzes the first reaction in the pathway, which is the condensation of oxaloacetate and acetyl-CoA to form citrate.

- Isocitrate dehydrogenase catalyzes the oxidative decarboxylation of isocitrate by transferring two electrons to NAD^+ to form NADH, releasing CO_2 in the process.
- α-Ketoglutarate dehydrogenase is functionally similar to pyruvate dehydrogenase and requires the same five coenzymes. It catalyzes an oxidative decarboxylation reaction that produces CO_2, NADH, and succinyl-CoA.
- Cytosolic aconitase is a dual-function protein that generates isocitrate for various metabolic pathways. It also binds specific sequences in mRNAs encoding iron-metabolizing proteins.
- Aconitase is a protein that contains 4 Fe–4 S and is inhibited by fluorocitrate. Citrate synthase converts fluoroacetyl-CoA to fluorocitrate, which is the lethal ingredient in the animal poison Compound 1080.
- GTP (or ATP) is generated by a substrate-level phosphorylation reaction catalyzed by the enzyme succinyl-CoA synthetase. GTP is readily converted to ATP in cells by the enzyme nucleoside diphosphate kinase.
- Carbon-14 labeling experiments using citrate cycle metabolites show that the two CO_2 molecules that are released per round of the citrate cycle are not derived from the acetyl-CoA molecule that entered in the same round of the cycle.
- Citrate, a flavoring for food and a preservative for medicines, is produced commercially using a proprietary biotechnology process based on *Aspergillus niger* fermentation of sucrose to generate citrate, which is exported to the culture medium and collected.

10.4 Regulation of the Citrate Cycle

- The three main control points for regulation of the citrate cycle are the reactions catalyzed by citrate synthase, isocitrate dehydrogenase, and α-ketoglutarate dehydrogenase.
- Regulation of citrate cycle reactions is accomplished by several mechanisms, including substrate availability, product inhibition, and feedback inhibition.
- Citrate synthase is inhibited by citrate, succinyl-CoA, NADH, and ATP; inhibition by ATP is reversed by ADP.
- Isocitrate dehydrogenase is activated by ADP and Ca^{2+} and inhibited by NADH and ATP.
- α-Ketoglutarate dehydrogenase is activated by Ca^{2+} and AMP and is inhibited by NADH, succinyl-CoA, and ATP.

10.5 Metabolism of Citrate Cycle Intermediates

- The citrate cycle provides biosynthetic precursors for several metabolic pathways; it is considered an amphibolic pathway because it functions in both catabolism and anabolism.
- Excess citrate in mitochondria is exported to the cytosol, where it is cleaved by the enzyme citrate lyase to release acetyl-CoA and oxaloacetate. Cytosolic acetyl-CoA is used for fatty acid and cholesterol biosynthesis, whereas oxaloacetate is used to generate phosphoenolpyruvate for gluconeogenesis.
- Oxaloacetate and α-ketoglutarate are metabolic precursors to aspartate and glutamate, respectively, and succinyl-CoA is a precursor in heme biosynthesis. Malate can be used as a source of carbon in the gluconeogenic pathway.
- The enzyme pyruvate carboxylase balances the input of oxaloacetate and acetyl-CoA into the citrate cycle by converting pyruvate into oxaloacetate using an ATP-dependent carboxylation reaction that is stimulated by acetyl-CoA.
- The pyruvate carboxylase reaction is called an anaplerotic reaction, meaning to "fill up," because it replenishes citrate cycle intermediates by providing oxaloacetate. Phosphoenolpyruvate carboxykinase and phosphoenolpyruvate carboxylase also supply oxaloacetate to the citrate cycle, and malic enzyme supplies malate.

biochemical terms

(in order of appearance in text)
citrate cycle (p. 482)
Krebs cycle (p. 485)
tricarboxylic acid (TCA) cycle (p. 485)
pyruvate dehydrogenase (p. 486)
citrate synthase (p. 486)
isocitrate dehydrogenase (p. 486)
α-ketoglutarate dehydrogenase (p. 486)
oxidation reaction (p. 486)
reduction reaction (p. 486)
oxidant (p. 486)
reductant (p. 486)
conjugate redox pair (p. 486)
oxidoreductase (p. 487)
dehydrogenase (p. 487)
biochemical standard reduction potential ($E°'$) (p. 487)
electrochemical cell (p. 488)
Nernst equation (p. 490)
niacin (p. 492)
pellagra (p. 492)
riboflavin (p. 493)
flavoprotein (p. 493)
coenzyme A (CoA) (p. 494)
pantothenic acid (p. 494)
thiamine (p. 495)
beriberi (p. 495)
α-lipoic acid (lipoamide) (p. 497)
fluorocitrate (p. 508)
oxidative decarboxylation (p. 510)
hydration (p. 513)
amphibolic pathway (p. 517)
anaplerotic reaction (p. 517)

review questions

1. What do lemons and the mold *Aspergillus niger* have in common?
2. Name four things the citrate cycle accomplishes.
3. Why is citrate cycle a more accurate name for this pathway than tricarboxylic acid (TCA) cycle? Why not just call it the Krebs cycle?

4. How is the change in standard reduction potential ($\Delta E°'$) determined for a redox reaction, and how is it related to the change in standard free energy $\Delta G°'$ for the same reaction?
5. Write the net reaction for pyruvate dehydrogenase and describe the nutritional origin of the five coenzymes required for pyruvate dehydrogenase function.
6. Describe the five steps in the pyruvate dehydrogenase reaction. Include in your answer the name of the pyruvate dehydrogenase enzyme subunit that participates in each step.
7. Describe control of pyruvate dehydrogenase activity by allostery and covalent modification.
8. Write the net reaction for the citrate cycle. Briefly describe the eight chemical reactions in the citrate cycle and include the names of the relevant enzymes.
9. What are the three primary control points in the citrate cycle with regard to energy charge in the cell?
10. Why is the citrate cycle considered an amphibolic pathway? Give an example of an anaplerotic reaction that provides substrates to the citrate cycle.

challenge problems

1. Use the $E°'$ values in Table 10.1 to calculate the $\Delta G°'$ values for the citrate cycle reactions catalyzed by malate dehydrogenase and succinate dehydrogenase. Are your calculated $\Delta G°'$ values within ~1 kJ/mol of the $\Delta G°'$ values found in Table 10.2? If not, what might be an explanation for the difference? What do you estimate the ΔG to be for this reaction? Explain.
2. The addition of ^{14}C-pyruvate to a preparation of mitochondria leads instantaneously to $^{14}CO_2$ production. However, adding $^{14}CO_2$ to the mitochondria does not result in ^{14}C-pyruvate. Why not?
3. Three of the five cofactors used in the pyruvate dehydrogenase complex are covalently attached to the enzyme complex. Name these three cofactors and describe their chemical functions in converting pyruvate to the acetyl group in acetyl-CoA.
4. Beriberi is a nutritional deficiency disorder that causes debilitating neurologic symptoms. Answer the following questions about this preventable condition.
   a. Why do individuals with beriberi have a large amount of pyruvate in their blood after eating a high-carbohydrate meal?
   b. Why would a diet rich in raw fish and African silkworms, both of which contain the enzyme thiaminase, lead to symptoms of beriberi, whereas if these foods are cooked, beriberi does not occur?
   c. A patient exhibiting all the symptoms of beriberi is placed on a thiamine-enriched diet; however, the symptoms do not disappear. Genetic screening suggests a defect in pyruvate dehydrogenase phosphatase. Briefly explain why the genetic defect caused beriberi symptoms and why thiamine supplementation had no effect.
5. The structure of citrate is shown here, with the C atoms numbered 1–6. Assume that carbon atoms 1 and 2 in this numbering scheme came from acetyl-CoA.

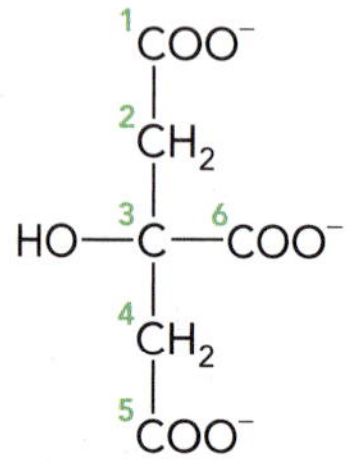

   a. Which carbon of citrate is the first carbon to be given off as CO_2 in the citrate cycle?
   b. What enzyme catalyzes that first decarboxylation?
   c. Which four carbons of the original six are found in oxaloacetate at the end of the first turn of the cycle?
6. Early biochemists used an *in vitro* system containing acetyl-CoA with radioactively labeled carbon to show that the citrate cycle was indeed a cycle.
   a. How did the pattern of ^{14}C-labeled cycle intermediates demonstrate that this was a cyclic pathway and not a linear pathway?
   b. Explain why no radioactive carbon was found in oxaloacetate when inorganic phosphate (P_i) was removed from the *in vitro* system prior to adding ^{14}C-acetyl-CoA.
7. A biochemist set up an *in vitro* system that measured oxygen consumption under conditions in which glucose was completely metabolized. She found that the addition of pyruvate to this glucose metabolizing system had little effect on the amount of oxygen consumed, whereas addition of fumarate resulted in a very large increase in oxygen consumption. Explain this observation.
8. A student set up an *in vitro* respiration system using cell extracts in which the rate of carbohydrate metabolism could be measured by monitoring the conversion of radioactive glucose to CO_2. The student found that the addition of citrate to this system led to a rapid decrease in the level of glucose; however, the addition of acetyl-CoA had little effect on the rate of glucose metabolism. Explain this observation.
9. On the basis of the nine mitochondrial reactions that are required to convert 1 pyruvate $\rightarrow$ 3 CO_2 under aerobic conditions and the ATP currency exchange ratios of ~5 ATP per 2 NADH and ~3 ATP per 2 $FADH_2$ that result from oxidative phosphorylation, calculate the total number of molecules of ATP that are generated for every molecule of pyruvate that is metabolized to 3 CO_2.
10. Pyruvate carboxylase is a mitochondrial enzyme that converts pyruvate to oxaloacetate. Explain why a mutation in this enzyme that blocks its ability to be allosterically activated by acetyl-CoA results in lower rates of energy conversion via the citrate cycle.

11. The liver mitochondrial enzyme pyruvate carboxylase catalyzes the reaction

$$\text{Pyruvate} + CO_2 + \text{ATP} + H_2O \rightarrow \text{Oxaloacetate} + \text{ADP} + P_i + H^+$$

What is most of the oxaloacetate converted into under the following conditions:

a. Blood glucose levels are low and the energy charge is high.

b. Mitochondrial acetyl-CoA levels are high and energy charge is low.

12. Citrate moves in and out of mitochondria via a specific transport protein. How would an inhibitor of this transport system affect the coordinate regulation of glycolysis and the citrate cycle?

13. Defects in the citrate cycle are rare but have been described. The data in the table that follows were collected from a blood analysis for a 1-month-old child. Which citrate cycle enzyme is defective in this patient? Explain.

Compound	Patient (mmol/L)	Normal control (mmol/L)
Lactate	25.5 ± 1.2	1.5 ± 0.3
Fumarate	7.1 ± 0.5	<0.1
Succinate	10.0 ± 1.0	0.5 ± 0.2
Malate	<0.1	4.2 ± 0.4
Citrate	5.5 ± 1.5	150 ± 25

smartwork5

If your instructor assigns homework with Smartwork5, access it here: digital.wwnorton.com/biochem.

suggested reading

Books and Reviews

Beinert, H., and Kennedy, M. C. (1993). Aconitase, a two-faced protein: enzyme and iron regulatory factor. *FASEB Journal*, 7, 1442–1449.

Buchanan, J. M. (2002). Biochemistry during the life and times of Hans Krebs and Fritz Lipmann. *Journal of Biological Chemistry*, *277*, 33531–33536.

Caussy, D., and Priest, N. D. (2008). Introduction to arsenic contamination and health risk assessment with special reference to Bangladesh. *Reviews of Environmental Contamination and Toxicology*, *197*, 1–15.

Earl, J. W., and McCleary, B. V. (1994). Mystery of the poisoned expedition. *Nature*, *368*, 683–684.

Fruton, J. S. (1999). *Proteins, enzymes, genes: the interplay of chemistry and biology*. New Haven, CT: Yale University Press.

Gibson, D. M., and Harris, R. A. (2002). *Metabolic regulation in mammals*. London, England: Taylor & Francis.

Max, B., Salgado, J. M., Rodriquez, N., Cortes, S., Converti, A., and Dominguez, J. M. (2010). Biotechnological production of citric acid. *Brazilian Journal of Microbiology*, *41*, 862–875.

Merrill, A. H., Jr., Bowman, B. B., and Preusch, P. C. (2000). Mechanistic aspects of vitamin and coenzyme utilization and function: a symposium in recognition of the distinguished career of Donald B. McCormick. *Journal of Nutrition*, *130*, 321S–322S.

Nicholls, D. G., and Ferguson, S. J. (2002). *Bioenergetics 3*. San Diego, CA: Academic Press.

Perham, R. N. (2000). Swinging arms and swinging domains in multifunctional enzymes: catalytic machines for multistep reactions. *Annual Review of Biochemistry*, *69*, 961–1004.

Quash, G., Fournet, G., and Reichert, U. (2003). Anaplerotic reactions in tumour proliferation and apoptosis. *Biochemical Pharmacology*, *66*, 365–370.

Swank, R. L. (1997). A prospective discussion of past international nutrition catastrophes—indications for the future. *Nutrition*, *13*, 344–348.

Zhou, Z. H., McCarthy, D. B., O'Connor, C. M., Reed, L. J., and Stoops, J. K. (2001). The remarkable structural and functional organization of the eukaryotic pyruvate dehydrogenase complexes. *Proceedings of the National Academy of Sciences USA*, *98*, 14802–14807.

Primary Literature

Beasley, M., Fisher, P., O'Connor, C., and Eason, C. (2009). Sodium fluoroacetate (1080): assessment of occupational exposures and selection of a provisional biological exposure index. *New Zealand Medical Journal*, *122*, 79–91.

Hoagland, M. B., and Novelli, G. D. (1954). Biosynthesis of coenzyme A from phospho-pantetheine and of pantetheine from pantothenate. *Journal of Biological Chemistry*, *207*, 767–773.

Leangon, S., Maddox, I., and Brooks, J. (2000). A proposed biochemical mechanism for citric acid accumulation by *Aspergillis niger* Yang No. 2 growing in solid state fermentation. *World Journal of Microbiology & Biotechnology*, *16*, 271–275.

Quash, G., Fournet, G., and Reichert, U. (2003). Anaplerotic reactions in tumour proliferation and apoptosis. *Biochemical Pharmacology*, *66*, 365–370.

Remington, S., Wiegand, G., and Huber, R. (1982). Crystallographic refinement and atomic models of two different forms of citrate synthase at 2.7 and 1.7 Å resolution. *Journal of Molecular Biology*, *158*, 111–152.

Sharma, N., Okere, I. C., Brunengraber, D. Z., McElfresh, T. A., King, K. L., Sterk, J. P., Huang, H., Chandler, M. P., and Stanley, W. C. (2005). Regulation of pyruvate dehydrogenase activity and citric acid cycle intermediates during high cardiac power generation. *Journal of Physiology*, *562*, 593–603.

Hibernating bear

When oxidative phosphorylation is "short-circuited," cells burn more fat than usual to produce enough ATP

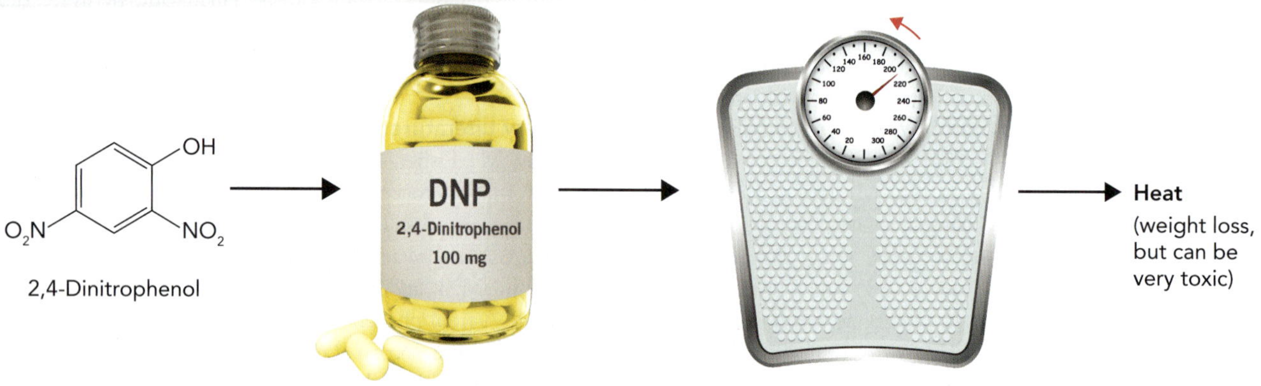

Diet pills

Hibernating animals rely on uncoupling proteins in brown adipose tissue to provide a mechanism that warms their tissues by using fatty acid degradation to convert chemical energy to thermal energy. 2,4-Dinitrophenol has been used as a diet pill because it works in a similar way to stimulate fatty acid degradation in adipose cells.

11

Oxidative Phosphorylation

◀ Hibernating organisms, such as bears, do not need much energy during their long winter sleep, but they must maintain a body temperature above the freezing point. In the brown adipose tissue of the hibernating bear, the membrane transporter protein called uncoupling protein 1 (UCP1) allows protons to flow across the inner mitochondrial membrane, which dissipates the proton gradient and decreases ATP synthesis. This results in increased fatty acid degradation to provide sufficient levels of ATP. The net result of this increase in fatty acid degradation in the hibernating bear is increased thermogenesis, which keeps the animal warm during the long winter.

Back in the 1930s, pharmaceutical companies began marketing the compound 2,4-dinitrophenol as a diet pill because it works in the same way as uncoupling proteins; namely, it dissipates the proton gradient and leads to increased rates of fatty acid degradation as a means to supply sufficient levels of ATP. However, ingestion of 2,4-dinitrophenol is dangerous because it accumulates in cell membranes, where its continued effect leads to liver damage, hyperthermia, and even death.

CHAPTER OUTLINE

Energy conversion processes at the molecular level sustain life on Earth. We noted in several places in Chapter 10 that the electron carriers produced in the citrate cycle lead to the production of energy through oxidative phosphorylation. In this chapter, we describe how this energy production occurs—the payoff for the molecular changes described in the preceding two chapters.

We focus on mitochondrial energy-conversion processes, which operate through a fundamental principle called chemiosmosis. The **chemiosmotic theory**, proposed by Peter Mitchell in 1961, is considered one of the great unifying principles in biology because it describes energy conversion in essentially all organisms. However, unlike the discovery of the DNA code by James Watson and Francis Crick in 1953—one of those "*Eureka!*" moments in science—Mitchell's chemiosmotic theory was ridiculed for years because it was such a radical departure from the biochemical mindset of the day.

We begin this chapter with an overview of the chemiosmotic theory. Then we examine in some detail what is currently known about the redox reactions and protein complexes that constitute the electron transport system in mitochondria, where these energy conversions take place. This is followed by a description of the ATP synthase complex and the transport systems that ferry biomolecules across the inner mitochondrial membrane. We end the chapter by looking at how the cell regulates oxidative phosphorylation.

11.1 The Chemiosmotic Theory

ATP synthesis in mitochondria is accomplished by establishing a proton (H^+) gradient across the mitochondrial inner membrane. This proton gradient harbors potential energy, which can be used to synthesize ATP. The movement of protons across the membrane down their concentration gradient to generate ATP is called **chemiosmosis**. The name *chemiosmosis* reflects the fact that this process is similar to osmosis, in which small molecules move across a membrane from regions of high concentration to regions of low concentration.

The proton gradient is established by the outward pumping of H^+ from the mitochondrial matrix by three large protein complexes and is powered by redox energy. Then, ATP synthesis is accomplished by the inward flow of H^+ (down the concentration gradient) through the membrane-bound ATP synthase complex. This same chemiosmotic process generates the bulk of ATP used by bacterial cells, which are ancestors of eukaryotic mitochondria. Chemiosmosis is also used to synthesize ATP in plant chloroplasts, which utilize sunlight energy rather than redox energy to establish a H^+ gradient across the thylakoid membrane (see Chapter 12). **Figure 11.1** shows where oxidative phosphorylation is located within our simplified metabolic map.

Glycolysis and the citrate cycle together provide only a small amount of the total ATP that can be obtained from glucose oxidation (4/32, or ~12%). This happens because most of the energy captured by glucose oxidation is not available for ATP synthesis until NADH and $FADH_2$ are oxidized by the electron transport system.

The term **electron transport system** (or electron transport chain) is used to describe the combined redox reactions that occur sequentially in a set of protein complexes embedded in the inner mitochondrial membrane. The net result of these redox reactions is the coupled oxidation of NADH and $FADH_2$ with the reduction of molecular oxygen to form NAD^+ and H_2O. As discussed later in this chapter, the difference in reduction potential ($\Delta E^{\circ\prime}$) between the NAD^+/NADH conjugate redox pair ($E^{\circ\prime} = -0.32$ V) and that of the $\frac{1}{2}\ O_2/H_2O$ conjugate redox pair ($E^{\circ\prime} = +0.82$ V) provides a large amount of free energy for ATP synthesis.

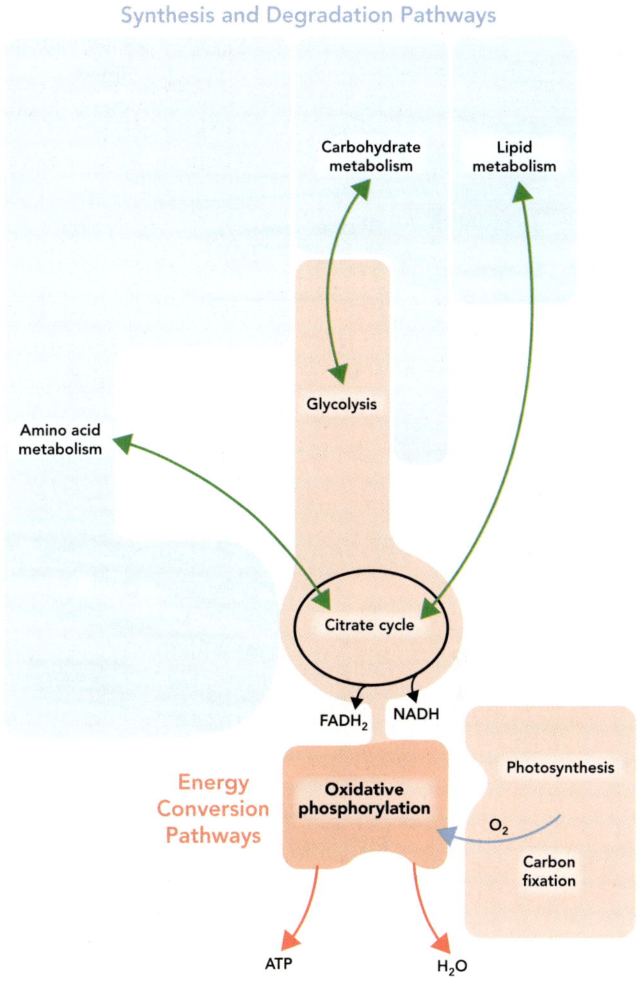

Figure 11.1 Oxidative phosphorylation represents the final step in the non-photosynthetic energy conversion pathways. The synthesis and degradation pathways are indicated with blue, whereas the energy conversion pathways are indicated with red. Overall, the non-photosynthetic energy conversion pathways catabolize carbon-based fuels (carbohydrates and lipids) to reduce O_2 and generate H_2O and ATP.

Redox Energy Drives Mitochondrial ATP Synthesis

The basic idea of chemiosmosis is that energy from redox reactions or light is coupled to electron transfer in membrane-bound proteins that use this energy to translocate protons across the membrane. These proteins thereby establish a proton gradient, which is a source of electrochemical energy. The chemical and electrical gradients across the membrane are favorable for the protons to flow back across the membrane through the ATP synthase complex, establishing a **proton circuit**. The ATP synthase complex couples this proton flow with production of ATP.

The ATP synthase complex was originally described as an ATP hydrolyzing enzyme, and in fact it can function as a reversible motor by hydrolyzing ATP under *in vitro* conditions, as described later in the chapter. However, under normal conditions in the cell where ATP is in constant demand, protons flow through the ATP synthase complex down their electrochemical gradient into the mitochondrial matrix, and thereby drive ATP synthesis rather than ATP hydrolysis.

Figure 11.2 shows a proton circuit across a mitochondrial membrane and another across a chloroplast thylakoid membrane, in response to redox or light energy, respectively. Note that directional H^+ pumping results in both a chemical gradient across the membrane, represented by ΔpH, and an electrical gradient due to the separation of charge, which can be measured as a membrane potential, $\Delta\psi$. The separation of charge is due to the buildup of protons (H^+) on one side of the membrane and accumulation of negative charges (OH^-) on the other side of the membrane. Figure 11.2 also shows that the relative inward/outward direction of proton movement is reversed in mitochondria compared to that in chloroplasts. Therefore, in order to keep track of the orientation of the proton gradient in a chemiosmotic system, we label the positive (P) and negative (N) sides of the membrane, with protons located on the positive side denoted as $H^+{}_P$ and those on the negative side as $H^+{}_N$.

An important concept of chemiosmosis is that the proton circuit behaves like an electrical circuit, in that potential energy is converted to work as a result of movement of charge. As shown in **Figure 11.3** (p. 529), the "resistor" in the proton circuit is the ATP synthase complex, which is analogous to an electric motor that performs work in an electrical circuit. The electrochemical proton gradient functions as a "capacitor" that temporarily stores energy generated by the "battery," which is the electron transport

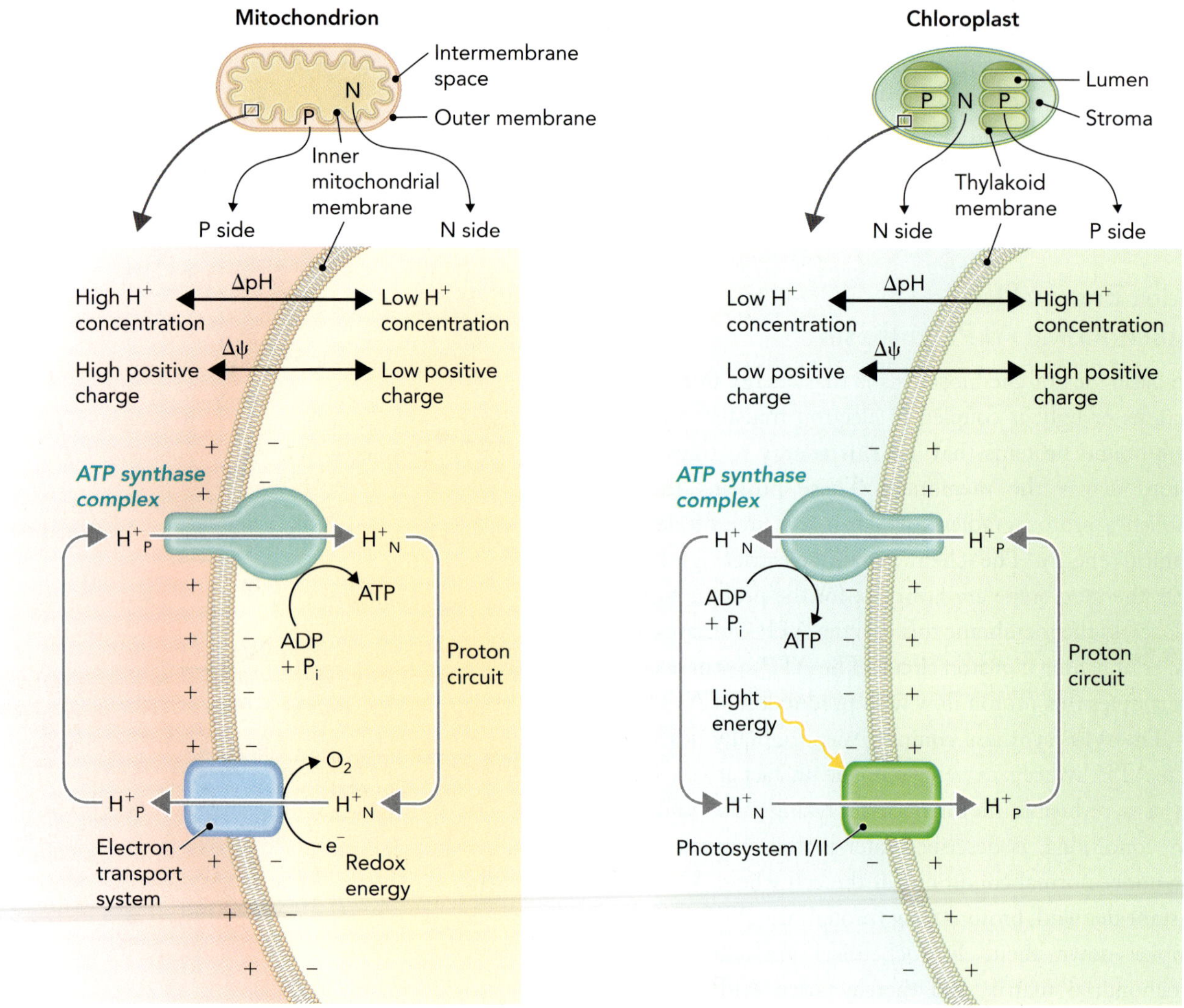

Figure 11.2 The chemiosmotic theory states that a proton circuit, consisting of a chemical gradient (ΔpH) and a membrane potential (Δψ), provides energy for ATP synthesis in both mitochondria and chloroplasts. The proton-impermeable membrane in a mitochondrion is the inner mitochondrial membrane, and in a chloroplast it is the thylakoid membrane. The two sides of the membrane are labeled "N" for negative charge and "P" for positive charge arising from proton translocation. Transmembrane proteins in the electron transport system (mitochondria) and photosystems I and II (chloroplasts) give rise to a proton gradient across the membrane, whereas ATP is generated in response to H^+ flowing back through the ATP synthase complex to complete the proton circuit.

▶ ANIMATION

system. Moreover, the proton circuit can be "shorted-out" or "uncoupled," so that useful work (ATP synthesis) no longer takes place. Proton circuit uncouplers—such as uncoupling proteins or the chemical compound 2,4-dinitrophenol, which allow H^+ leakage across the membrane without production of ATP—convert redox energy to heat. Figure 11.3 also shows that if ATP synthase complex activity is inhibited by, for example, the antibiotic oligomycin (analogous to a burned-out motor), proton flow shuts down and oxidative phosphorylation cannot occur. This eventually leads to a very large ΔpH and Δψ, causing inhibition of the electron transport system and cell death.

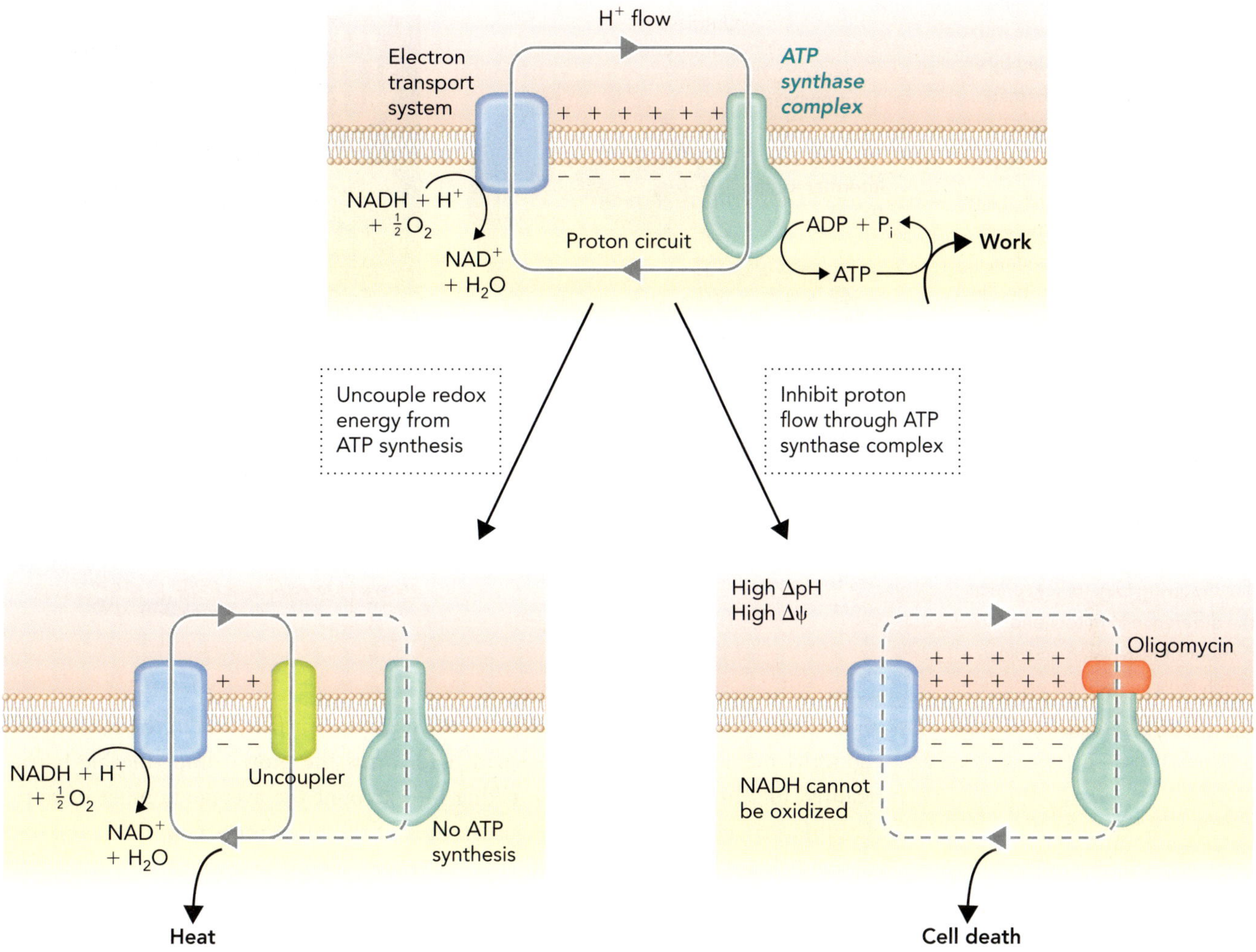

Figure 11.3 The chemiosmotic proton circuit is analogous to an electrical circuit. In this illustration, proton flow represents the electric current, the electron transport system functions as a battery, the proton gradient acts as a capacitor for energy storage, and the ATP synthase complex functions as a resistor, much like an electric motor. Uncouplers, such as uncoupling proteins or 2,4-dinitrophenol, "short-circuit" the proton flow so that energy is converted to heat rather than to ATP synthesis. In contrast, compounds that block the proton circuit (such as oligomycin) shut down energy conversion processes, leading to cell death.

Cell biological techniques based on electron tomography have revealed fine-structure details of mitochondria. As shown in **Figure 11.4**, the primary structural features of a mitochondrion are its outer mitochondrial membrane, an intermembrane space, and an inner mitochondrial membrane, which is highly invaginated to form internal structures called **cristae** (singular = crista). The aqueous interior of a mitochondrion is called the **mitochondrial matrix**, which is a protein-rich compartment containing enzymes required for the citrate cycle, the electron transport system, oxidative phosphorylation, and fatty acid degradation. The ATP synthase complexes found in mitochondria are embedded in the inner mitochondrial membrane and appear as knobs when viewed by electron microscopy. A single mitochondrion is about 0.4 μm in diameter and 2.5 μm in length.

The number of mitochondria per cell depends on the energy requirements of the cell. For example, slow-twitch muscle cells contain more mitochondria because they require aerobic energy conversion for long periods of time, whereas fast-twitch muscle cells have fewer mitochondria. The density of mitochondria in muscles contributes to the relative difference in coloration of meat in a turkey, for example. Dark meat found in the leg, thigh, and wing muscles is slow-twitch muscle and has high mitochondrial content, which is needed for sustained energy while running. Turkey breast is fast-twitch muscle and is white because it has low mitochondrial content, owing to the fact that turkeys grown on commercial farms rarely fly for more than a few feet (Figure 11.4c).

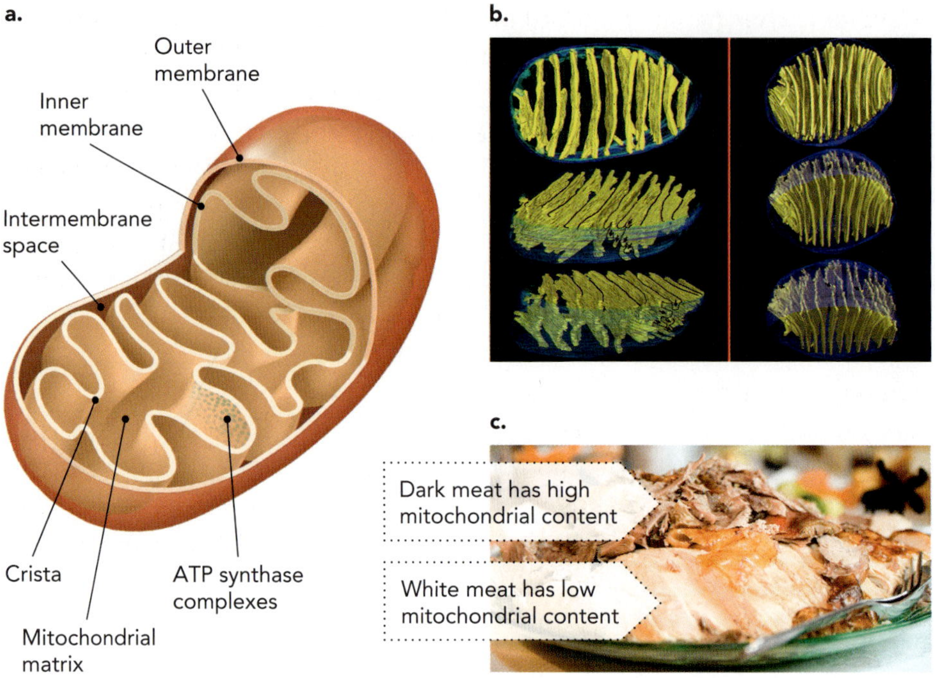

Figure 11.4 Mitochondria contain an inner membrane with a large surface area bounded by an outer membrane. **a.** Schematic drawing of a mitochondrion, showing the relative orientations of the outer mitochondrial membrane, the intermembrane space, the inner mitochondrial membrane, and the mitochondrial matrix. **b.** Reconstructed images, using electron tomography, of a single mitochondrion. G. A. PERKINS ET AL. (1997). ELECTRON TOMOGRAPHY OF LARGE, MULTICOMPONENT BIOLOGICAL STRUCTURES. *JOURNAL OF STRUCTURAL BIOLOGY*, 120. 219–227. **c.** The white meat and dark meat of a Thanksgiving Day turkey reflects the difference in mitochondrial content of fast-twitch and slow-twitch muscle, respectively. BRUCE BOULTON.CO.UK/ALAMY STOCK PHOTO.

A critical feature of mitochondria is the extensive surface area of the inner mitochondrial membrane, which is ~5 μm^2 and forms the proton-impermeable barrier required for chemiosmosis. When taking into account the average number of mitochondria per cell (10^1 to 10^3), the surface area of the inner mitochondrial membrane can easily be as much as 5,000 μm^2 per cell. Extrapolating this calculation to include the number of cells in the human body (10^{13} to 10^{14}), the surface area of the inner mitochondrial membrane is estimated to be 185,000 m^2 for an average-sized human—a total surface area equivalent to 26 full-sized soccer fields.

Each mitochondrion contains 5–10 copies of the circular mitochondrial DNA genome (the human mitochondrial genome is approximately 16,500 nucleotides long). This genome encodes some of the proteins and RNA molecules required for mitochondrial function. All other genes required for mitochondrial function are encoded by the nuclear genome (see Chapter 3). Mitochondria are inherited only from the female egg because sperm cells lose their mitochondria-rich tails upon fertilization. Therefore, as described later in the chapter, inherited mitochondrial gene mutations that affect the electron transport system are passed on through the mother.

Early research using purified mitochondria showed by fractionation methods and biochemical assays that five protein complexes contained within the inner mitochondrial membrane are required for oxidative phosphorylation. On the basis of the substrates and products of *in vitro* redox reactions containing the molecules NADH, coenzyme Q, succinate, cytochrome c, and O_2, four of the fractions were found to contain separate protein components of the electron transport system. These components were named

- complex I: NADH–ubiquinone oxidoreductase (NADH dehydrogenase)
- complex II: succinate dehydrogenase
- complex III: ubiquinone–cytochrome c oxidoreductase
- complex IV: cytochrome c oxidase

The fifth enzyme fraction contained the **ATP synthase complex**, which consists of a large, multi-subunit protein complex. One component of the ATP synthase complex was shown to be a membrane-bound "stalk" protein subcomplex (F_o) and the other component a spherical "head" protein subcomplex encoding the catalytic subunit (F_1), hence it is sometimes called the F_1F_o ATP synthase complex. As shown in **Figure 11.5**, researchers were able to order the four electron transport system complexes along an enzyme-linked redox pathway on the basis of the known reduction potentials ($E°'$) of conjugate redox pairs and the results from experiments using specific redox reaction inhibitors such as rotenone, antimycin A, and cyanide.

Metabolic fuel in the form of NADH and $FADH_2$ feeds into the electron transport system from the citrate cycle and fatty acid oxidation pathways. Pairs of electrons ($2\ e^-$) are donated by NADH and $FADH_2$ to complexes I and II, respectively, and flow through the electron transport system until they are used to reduce oxygen to form water ($\frac{1}{2}\ O_2 + 2\ e^- + 2\ H^+ \rightarrow H_2O$). The two mobile electron carriers in this series of reactions are **coenzyme Q (Q**; also called **ubiquinone**) and **cytochrome c**, which transfer electrons between various complexes.

The complexes are depicted in Figure 11.5 as proton translocators, but before Mitchell proposed his chemiosmotic theory, the only agreed-on connection between the mitochondrial redox reactions and ATP synthesis was that three "regions" of the system (complexes I, III, IV) were required to convert redox energy into ATP. It is now

Figure 11.5 The four protein complexes of the electron transport system (I–IV), cytochrome c (Cyt c), and the ATP synthase complex carry out the process of oxidative phosphorylation. Porin proteins provide channels for small molecules to diffuse across the outer membrane of mitochondria. Translocase proteins shuttle ATP, ADP, and P_i across the otherwise impermeable inner mitochondrial membrane. When starting with electrons from NADH that enter through complex I, a total of 10 H^+ are translocated by the electron transport system. Four H^+ reenter the matrix for every ATP that is synthesized (3 H^+ through the ATP synthase complex and 1 H^+ through the phosphate translocase).

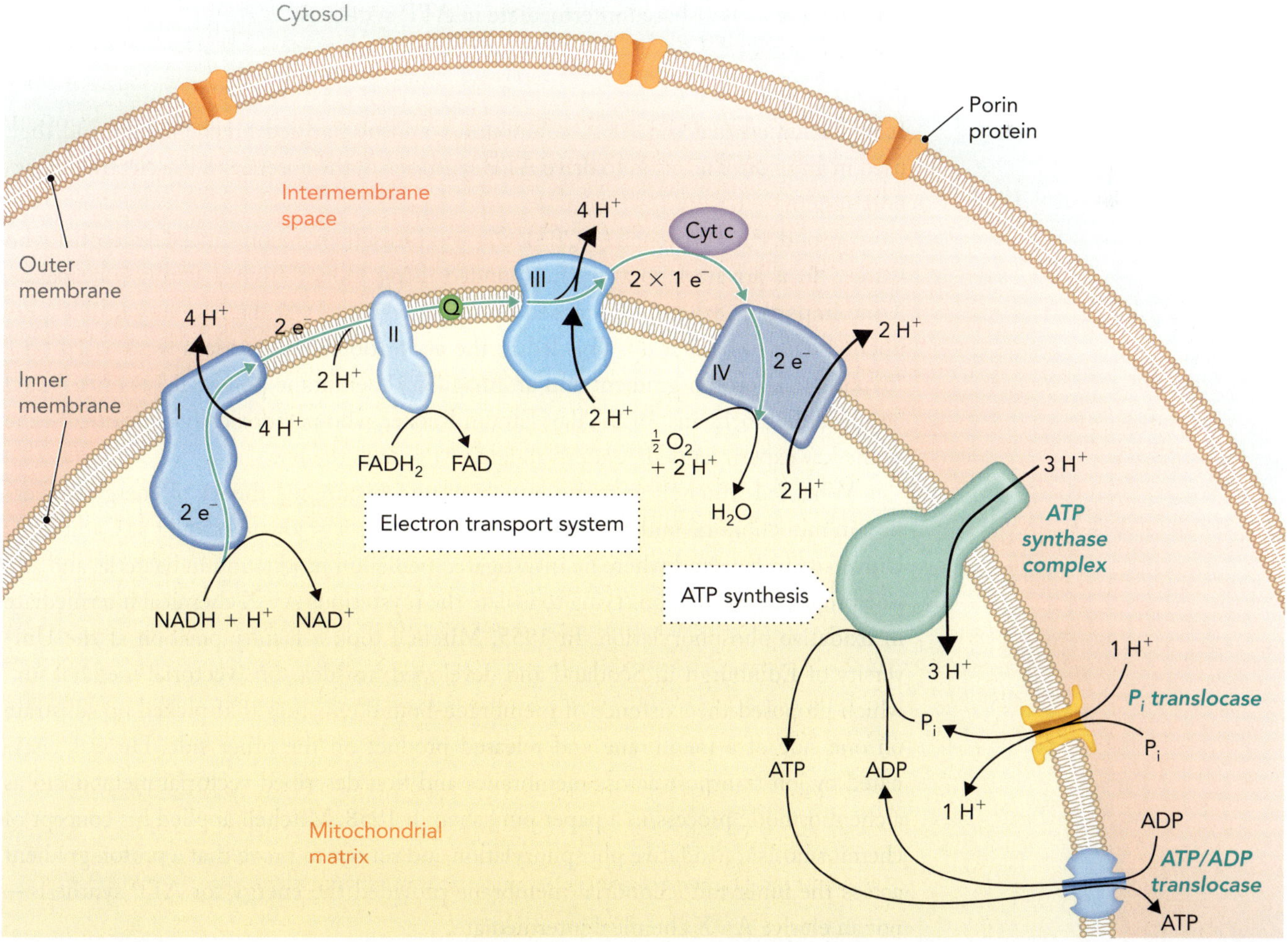

generally agreed that the stoichiometry of proton translocation starting with NADH is 4 H^+ in complex I, 4 H^+ in complex III, and 2 H^+ in complex IV (only 6 H^+ are translocated per $FADH_2$, as $FADH_2$ bypasses complex I as described later). The stoichiometry of proton influx back into the mitochondrial matrix is thought to be 3 H^+ per ATP and 1 H^+ per P_i, indicating that 4 H^+ are needed for every ATP synthesized, as P_i is a required substrate in the ATP synthase complex reaction. As described later in the chapter, the ATP/ADP translocase exchanges ATP for ADP, as shown in Figure 11.5.

Peter Mitchell and the Ox Phos Wars

The idea that an electrochemical proton gradient provides the power to activate an ATP-generating machine was not intuitive to most biochemists in 1961, when Peter Mitchell first proposed the chemiosmotic hypothesis. Indeed, it took more than 20 years for the theory to gain wide acceptance, and it did not appear in biochemistry textbooks as the primary energy-conversion process in mitochondria and chloroplasts until the mid-1980s, almost 10 years after Mitchell received the 1978 Nobel Prize in Chemistry.

There were two primary reasons for the extended resistance to the chemiosmotic hypothesis. First, a dominating concept in biochemistry at the time was that oxidative phosphorylation must involve an intermediate with a high-energy chemical bond called "squiggle X," denoted ~X (the "~" represents an unstable high-energy chemical bond). Supposedly, this intermediate was produced by the electron transport system and functioned as a shared intermediate in ATP synthesis:

$$A{\sim}X + P_i + ADP \rightleftharpoons A + X + ATP$$

It was assumed that the A~X compound could be purified biochemically and then used in an *in vitro* reaction to drive ATP synthesis in the absence of the electron transport system.

Second, Peter Mitchell (1920–1992) was himself an unconventional scientist who worked in a private laboratory he founded using money from a family inheritance. More important, when he proposed his hypothesis in 1961, he had very little experimental data to back it up. This led to the accusation that his ideas were based on a "hypothetical proton gradient and an imaginary membrane potential," as once stated by a leading biochemist of the day, Efraim Racker, who paradoxically performed some of the key experiments that later validated Mitchell's theory.

What led Mitchell to propose such an unconventional theory of energy conversion in mitochondria and chloroplasts? Mitchell was a graduate student at Cambridge University in England where he investigated penicillin production in bacteria, and as a side project he worked on trying to isolate the mysterious A~X chemical intermediate in oxidative phosphorylation. In 1955, Mitchell took a faculty position at the University of Edinburgh in Scotland and developed his ideas on "vectorial metabolism," which proposed the existence of membrane-bound enzymes that picked up substrate on one side of a membrane and released product on the other side. He was fascinated by ion transport across membranes and first described vectorial metabolism as a chemiosmotic process in a paper published in 1958. Mitchell applied his concept of chemiosmosis to oxidative phosphorylation and set out to prove that a proton gradient across the inner mitochondrial membrane provided the energy for ATP synthesis—not an elusive A~X chemical intermediate.

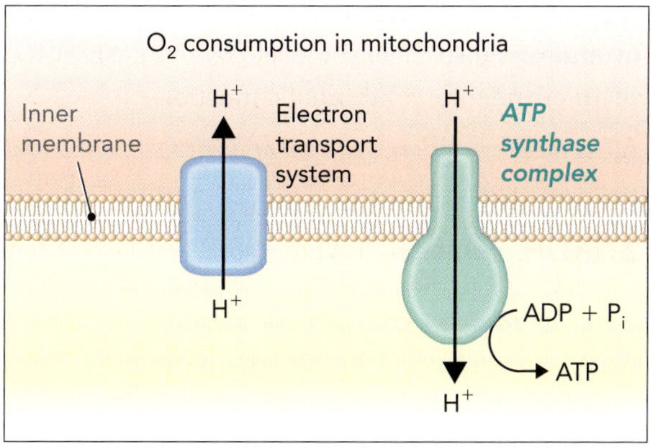

Figure 11.6 Peter D. Mitchell proposed the chemiosmotic theory of ATP synthesis through work conducted largely at his private laboratory at the Glynn Research Institute. P. D. MITCHELL: AP PHOTO; THE GLYNN RESEARCH INSTITUTE: MARTIN & CO. TRURO.

Using previously published data and his own ideas about vectorial metabolism as a basis for his theory, Mitchell correctly predicted that protons should be translocated across H^+-impermeable membranes in association with oxygen consumption in mitochondria and in response to light in chloroplasts. He also predicted that ATP synthesis should function as a reversible pump, in such a way that ATP hydrolysis should pump protons out of the mitochondrial matrix, and that biomolecules required for other mitochondrial pathways must be shuttled in and out of the mitochondria by specific carrier proteins. Mitchell and his research assistant, Jennifer Moyle, eventually provided data to support these predictions based primarily on experiments conducted in his private laboratory at a manor house in North Cornwall, England (**Figure 11.6**). The Glynn Research Institute was established by Peter Mitchell in the early 1960s with a research staff of less than 20, and it remained a private research institution for almost 30 years. (Mitchell's uncle was Sir Godfrey Mitchell, who owned George Wimpy and Company Limited, the largest construction company in England at the time.) The Glynn Research Institute gave Mitchell the freedom to test his theory without depending on conventional grant funding.

Not until the early 1970s was sufficient experimental evidence published to convince Mitchell's critics that his chemiosmotic hypothesis was probably correct. It was argued for years that a key experiment was needed to show that an electrochemical proton gradient could link directly with an electron transport system and provide the free energy needed for oxidative phosphorylation (ATP synthesis). The necessary data were published in 1973 by Efraim Racker and Walter Stoeckenius. They used reconstituted membrane vesicles containing a light-driven proton pump from bacteria (bacteriorhodopsin) and ATP synthase complexes isolated from bovine heart mitochondria (**Figure 11.7**). In the presence of light, the bacteriorhodopsin protein translocated protons into the vesicle interior, resulting in the establishment of an electrochemical proton gradient across the proton-impermeable membrane. When the protons flowed down the **electrochemical gradient**, they exited through the ATP synthase complexes and generated ATP on the outside of the vesicle, exactly as predicted by Mitchell's hypothesis.

Let's now answer our four key pathway questions regarding oxidative phosphorylation:

1. *What does oxidative phosphorylation accomplish for the cell?* It generates ATP derived from oxidation of metabolic fuels, accounting for 28 of 32 ATP (88%) obtained from glucose catabolism in most cells (4 ATP are generated by substrate-level phosphorylation reactions in the glycolytic pathway and citrate

cycle). The tissue-specific expression of UCP1 in brown adipose tissue of mammals short-circuits the electron transport system, producing heat for thermoregulation of metabolic functions.

2. *What is the overall net reaction of NADH oxidation by the oxidative phosphorylation pathway?*

$$2\ NADH + 2\ H^+ + 5\ ADP + 5\ P_i + O_2 \rightarrow 2\ NAD^+ + 5\ ATP + 2\ H_2O$$

3. *What are the key enzymes in the oxidative phosphorylation pathway?* The ATP synthase complex is the enzyme responsible for converting proton-motive force (energy available from the electrochemical proton gradient) into net ATP synthesis through a series of proton-driven conformational changes. NADH dehydrogenase, also called complex I or **NADH–ubiquinone oxidoreductase**, is the enzyme that catalyzes the first redox reaction in the electron transport system, in which NADH oxidation is coupled to flavin mononucleotide (FMN) reduction and translocates 4 H^+ into the intermembrane space. **Ubiquinone–cytochrome c oxidoreductase**, also called complex III, translocates 4 H^+ across the membrane via the Q cycle. It also has the important role of facilitating electron transfer from a two-electron carrier (QH_2) to cytochrome c, a mobile

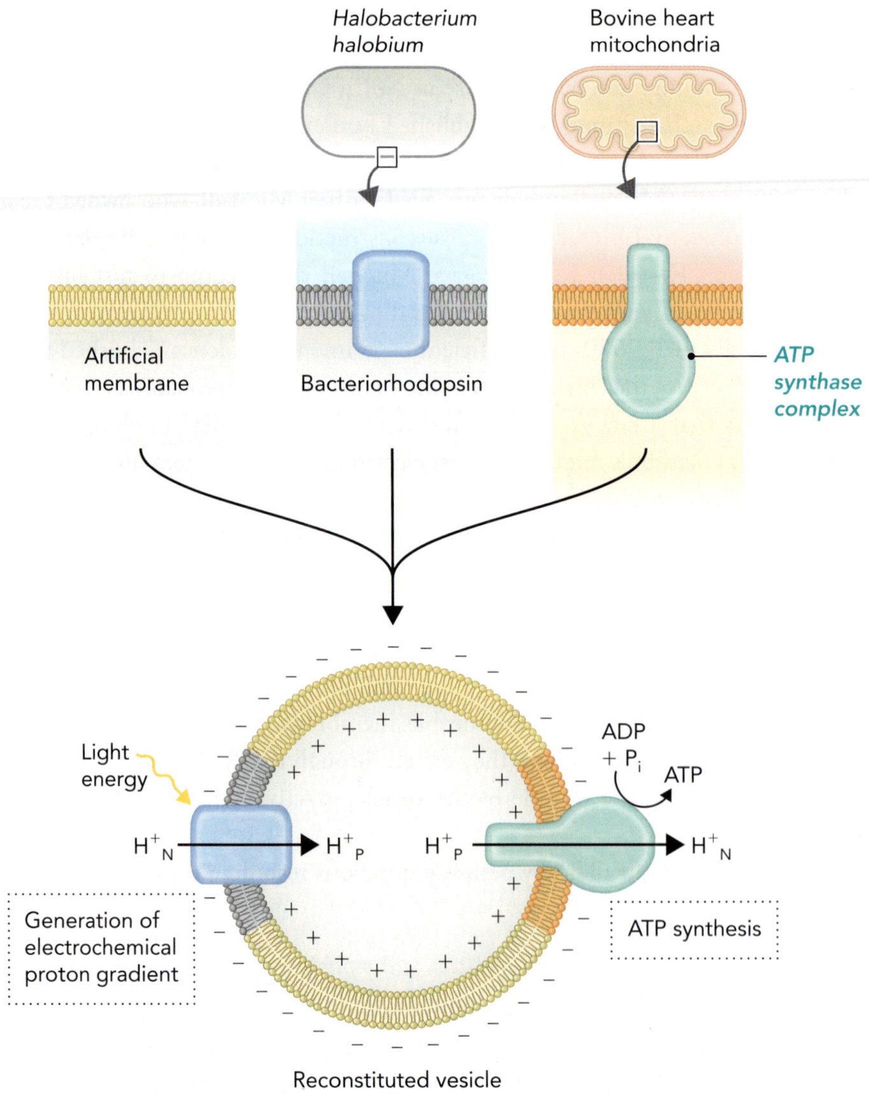

Figure 11.7 Light-activated ATP synthesis in reconstituted vesicles provided compelling evidence that Mitchell's chemiosmotic hypothesis was correct. The vesicles contained an artificial membrane, bacteriorhodopsin from *Halobacterium halobium*, and ATP synthase complexes from bovine heart mitochondria.

protein carrier that transfers one electron at a time to complex IV. **Cytochrome c oxidase**, also called complex IV, translocates 2 H^+ into the intermembrane space and catalyzes the last redox reaction in the electron transport system, in which cytochrome a_3 oxidation is coupled to the reduction of molecular oxygen to form water ($\frac{1}{2}O_2 + 2e^- + 2H^+ \rightarrow H_2O$).

4. *What are examples of oxidative phosphorylation in everyday biochemistry?* Consider hydrogen cyanide gas, which is the lethal compound produced in prison gas chambers when sodium cyanide crystals are dropped into sulfuric acid. Cyanide binds to the heme group in cytochrome a_3 of complex IV and blocks the electron transport system by preventing the reduction of oxygen to form H_2O. In other words, blocking the electron transport system leads to a quick death.

concept integration 11.1

What is the chemiosmotic theory, and how is a proton circuit involved?

The chemiosmotic theory states that energy captured by coupled redox reactions in the electron transport system is used to translocate protons across the inner mitochondrial membrane, creating a proton gradient. This proton gradient is a source of electrochemical energy due to both the chemical gradient and membrane potential ($\Delta pH + \Delta\psi$). The electrochemical gradient is used to drive ATP synthesis by the movement of protons through the ATP synthase complex. A proton circuit is established when protons are translocated into the mitochondrial intermembrane space by the electron transport system then flow back into the mitochondrial matrix through the ATP synthase complex.

11.2 The Mitochondrial Electron Transport System

The two metabolic pathways discussed so far take place in the cytosol (glycolysis) or the mitochondrial matrix (citrate cycle). The electron transport system involves several protein complexes associated with the inner membranes of mitochondria or chloroplasts.

The Mitochondrial Electron Transport System Is a Series of Coupled Redox Reactions

Oxidation of NADH by the mitochondrial electron transport system takes place on the matrix side of the inner mitochondrial membrane beginning with complex I (NADH–ubiquinone oxidoreductase). As shown in **Figure 11.8a**, the donation of 2 e^- (in the form of a hydride ion, $:H^-$) from NADH initiates a series of sequential redox reactions involving as many as 20 discrete electron carriers, culminating in the reduction of molecular oxygen to form water. A pair of electrons can also enter the electron transport system through oxidation of $FADH_2$ molecules (**Figure 11.8b**). These molecules are covalently attached to enzymes either associated with the cytosolic side of the inner mitochondrial membrane (mitochondrial glycerol-3-phosphate dehydrogenase) or that belong to metabolic pathways located within the mitochondrial matrix (succinate

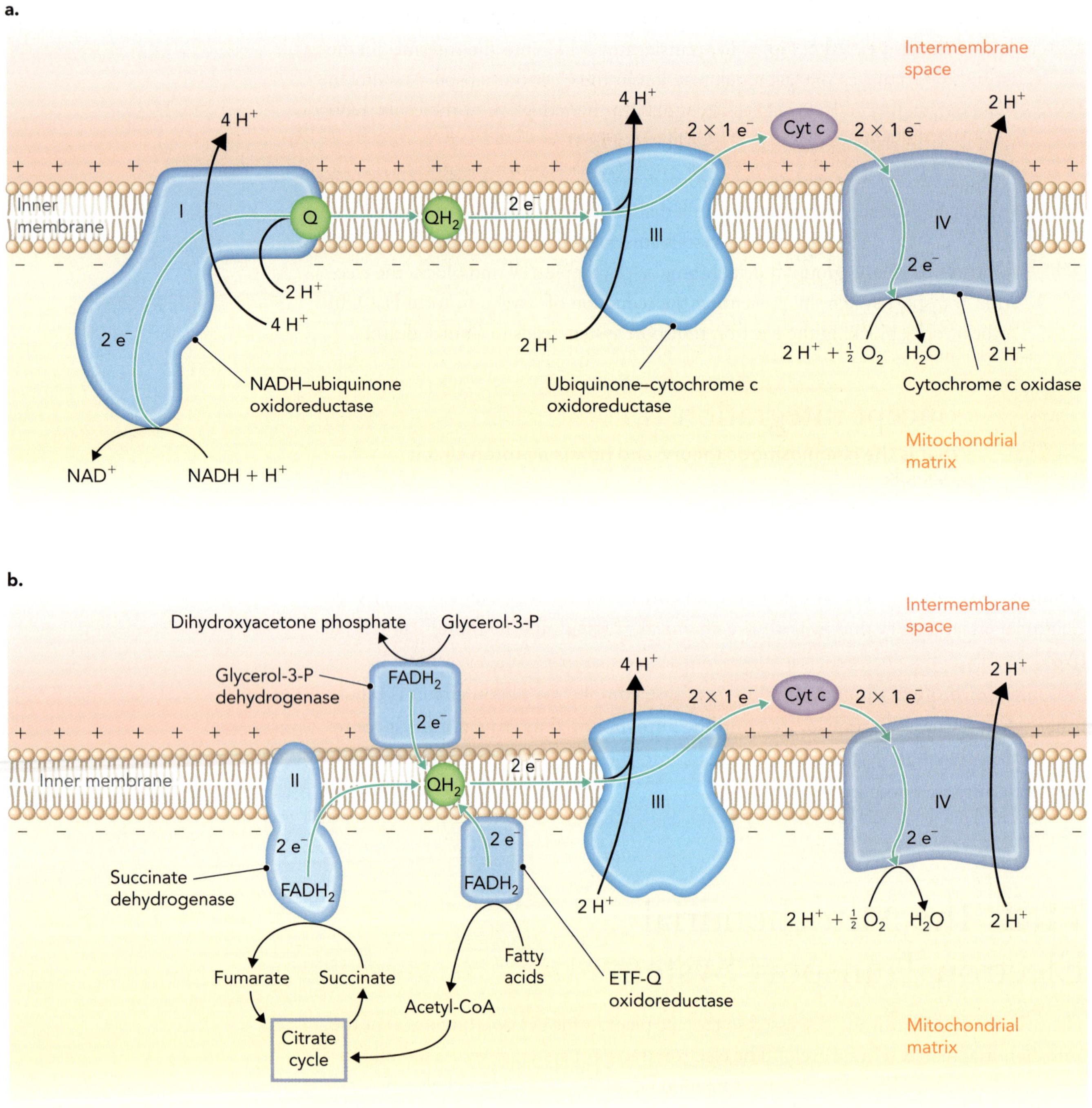

Figure 11.8 Electron pairs (2 e$^-$) from NADH and $FADH_2$ flow through the electron transport system. **a.** Electrons from NADH enter the electron transport system at complex I, then flow to coenzyme Q, complex III, and complex IV. A total of 10 H^+ are concomitantly translocated. **b.** Electron pairs are derived from $FADH_2$ oxidation at complex II (succinate dehydrogenase), from ETF-Q oxidoreductase of the fatty acid oxidation pathway, or from mitochondrial glycerol-3-phosphate dehydrogenase, which is part of the glycerol-3-phosphate shuttle. A total of 6 H^+ are concomitantly translocated when electrons are derived from $FADH_2$.

dehydrogenase [complex II] and electron-transferring flavoprotein–ubiquinone oxidoreductase [ETF-Q oxidoreductase]).

The flow of electrons through the electron transport system is facilitated by the sequential arrangement of electron carriers. These carriers are primarily proteins with Fe-containing prosthetic groups, which have the property of increasing reduction potentials ($E°'$), from a low of −0.32 V for NAD^+/NADH to a high of +0.82 V for $\frac{1}{2}$ O_2/H_2O. It is important to note that electron flow can only continue if each of the electron carriers is able to be reoxidized by giving up its electrons to the next carrier in the "chain." Any disruption of these reversible redox reactions brings the entire electron transport system to a grinding halt.

Because the electron transport system and ATP synthesis through the ATP synthase complex are coupled processes (see Figure 11.5), inhibition of electron transfer in the electron transport system results in loss of ATP synthesis by the ATP synthase complex. Some of the electron transport system inhibitors that block redox coupling are rotenone (a compound isolated from plant roots that was used by Amazon tribesmen to make poisonous arrows); the deadly gases hydrogen cyanide and carbon monoxide; and antimycin A (an antibiotic produced by *Streptomyces* bacteria that has been used as a pesticide to control non-native fish populations in streams and lakes).

The four functional components of the electron transport system that will be described in the next subsection are:

1. Three *large multi-subunit protein complexes, I, III, and IV*, which transverse the inner mitochondrial membrane and translocate protons across it.
2. *Coenzyme Q* (Q; also called ubiquinone), a small hydrophobic electron carrier that diffuses laterally within the membrane to donate electrons to complex III.
3. Three membrane-associated *FAD-containing enzymes*—succinate dehydrogenase (complex II); ETF-Q oxidoreductase; and glycerol-3-phosphate dehydrogenase—which pick up electrons from linked metabolic pathways and donate them to coenzyme Q.
4. *Cytochrome c*, a small protein that associates with the cytosolic side of the membrane and carries electrons one at a time from complex III to complex IV.

Table 11.1 lists the standard biochemical reduction potentials ($E°'$) for the key electron carriers in the electron transport system. You can see that for the most part, the reduction potentials gradually increase with each reaction in the electron transport system. It is important to point out, however, that the actual reduction potential (E) of the individual carriers depends on their *in vivo* concentrations, which are not necessarily equivalent molar ratios. Biochemical fractionation studies of mitochondrial membranes have shown that complex IV is the most abundant complex, with about twice as much protein as complex III. Complexes I and II are the least abundant of the major electron transport proteins. Coenzyme Q and cytochrome c are both very abundant, which is consistent with their functional role as mobile electron carriers.

How is the energy released by redox reactions used to translocate protons into the intermembrane space? The answer is not yet completely understood, but it is thought to involve the following:

1. A *redox loop mechanism* in which a separation of H^+ and e^- occurs on opposite sides of the membrane. The Q cycle in complex III uses this mechanism to translocate protons across the membrane.

2. A redox-driven *proton pump*, which is dependent on conformational changes in the protein complex. Protons move across the membrane as a result of protein conformational changes that alter pK_a values of amino acid functional groups located on the inner and outer faces of the membrane. Both complexes I and IV have properties that are consistent with such a mechanism.

It is likely that for some electron transport proteins, both a redox loop mechanism and proton pump are involved in net proton movement across the membrane. **Figure 11.9** illustrates how these two mechanisms could be operating.

Protein Components of the Electron Transport System

The electron transport system consists of large protein complexes containing numerous prosthetic groups that function as electron carriers. **Table 11.2** lists the four membrane-bound complexes (I–IV) that function within the inner mitochondrial membrane, as well as cytochrome c, which shuttles electrons from complex III to complex IV (see Figure 11.8). Here we describe each complex in more detail.

Table 11.1 STANDARD REDUCTION POTENTIALS ($E°'$) FOR SELECTED ELECTRON CARRIERS IN THE ELECTRON TRANSPORT SYSTEM

Electron carriers	$E°'$ (V)
$NAD^+ + H^+ + 2\,e^- \rightarrow NADH$	−0.32
Complex I (NADH–ubiquinone oxidoreductase)	
Fe–S (N-1b)	−0.25
Fe–S (N-3,4)	−0.24
Fe–S (N-5,6)	−0.27
Complex II (succinate dehydrogenase)	
$FAD + 2\,H^+ + 2\,e^- \rightarrow FADH_2$ (enzyme bound)	−0.04
Fe–S (S-1)	−0.03
Cytochrome b_{560}	−0.08
Coenzyme Q + $2\,H^+ + 2\,e^-$	+0.04
Complex III (ubiquinone–cytochrome c oxidoreductase)	
Cytochrome b_H	+0.03
Cytochrome b_L	−0.03
Fe–S	+0.28
Cytochrome c_1	+0.21
Cytochrome c (Cyt c)	+0.23
Complex IV (cytochrome c oxidase)	
Cytochrome a	+0.21
Cu_A	+0.24
Cytochrome a_3	+0.38
$\frac{1}{2}O_2 + 2\,H^+ + 2\,e^- \rightarrow H_2O$	+0.82

Direction of e^- flow

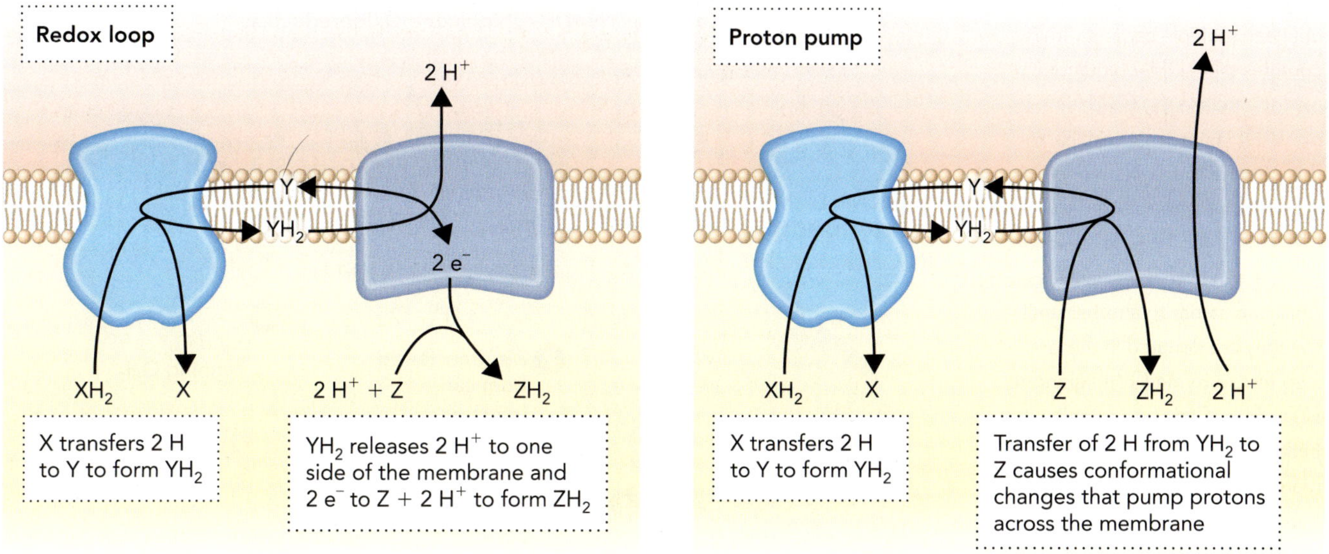

Figure 11.9 Proton translocation across an impermeable membrane can occur by a redox loop mechanism or by a proton pump. In the redox loop mechanism, the H^+ and e^- are separated to opposite sides of the membrane. The proton pump involves protein conformational changes that facilitate proton movement by altering pK_a values of amino acid functional groups. Both mechanisms are likely to function in the electron transport system.

Complex I: NADH–Ubiquinone Oxidoreductase Complex I is the largest of the four protein complexes in the mitochondrial electron transport system. It consists of 15–40 polypeptide chains, depending on the species. Electron microscopy contour mapping studies reveal that the mitochondrial complex I is shaped like a sideways "L," as illustrated in **Figure 11.10**. Complex I contains a covalently bound flavin mononucleotide (FMN), which accepts two electrons from NADH. The electron acceptor group of FMN is shown in **Figure 11.11**, where you can see that the structure of the isoalloxazine ring is identical to that of FAD (see Figure 10.7). Similarly to FAD, FMN can accept one or two electrons, going through a semiquinone intermediate step after a single reduction.

Table 11.2 PROTEIN COMPONENTS AND PROSTHETIC GROUPS IN COMPLEXES I–IV AND CYTOCHROME C OF THE ELECTRON TRANSPORT SYSTEM

Protein complex	Mass (kDa)	Subunits	Prosthetic group
I NADH–ubiquinone oxidoreductase	850	42	FMN, Fe–S
II Succinate dehydrogenase	140	5	FAD, Fe–S
III Ubiquinone–cytochrome c oxidoreductase	250	11	Hemes b_H, b_L, c_1, Fe–S
Cytochrome c	13	1	Heme c
IV Cytochrome c oxidase	160	13	Heme a, a_3, Cu_A, Cu_B

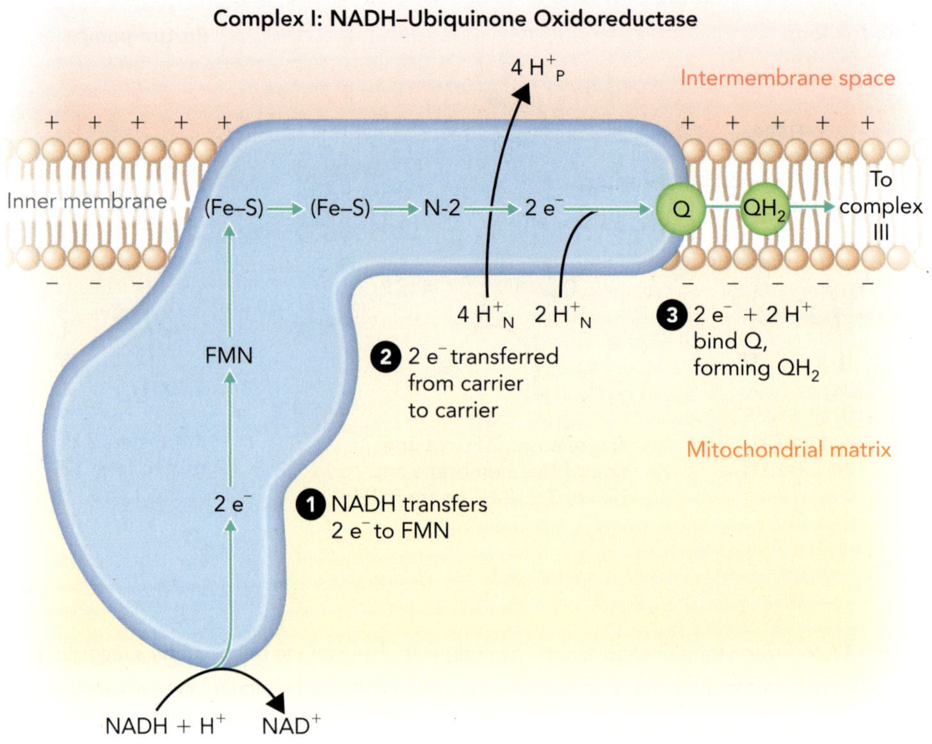

Figure 11.10 Electrons flow between multiple carriers within complex I: NADH–ubiquinone oxidoreductase. (1) Oxidation of NADH in the matrix releases 2 e^- (in the form of a hydride ion), which are transferred to FMN in a coupled redox reaction. (2) Electrons are then transferred from one carrier to another until (3) they are donated in the last step to coenzyme Q (ubiquinone; Q) to form QH_2 (ubiquinol). In the process, 4 H^+ from the matrix side of the membrane are translocated across the membrane by complex I, and 2 e^- and 2 H^+ are used to reduce coenzyme Q.

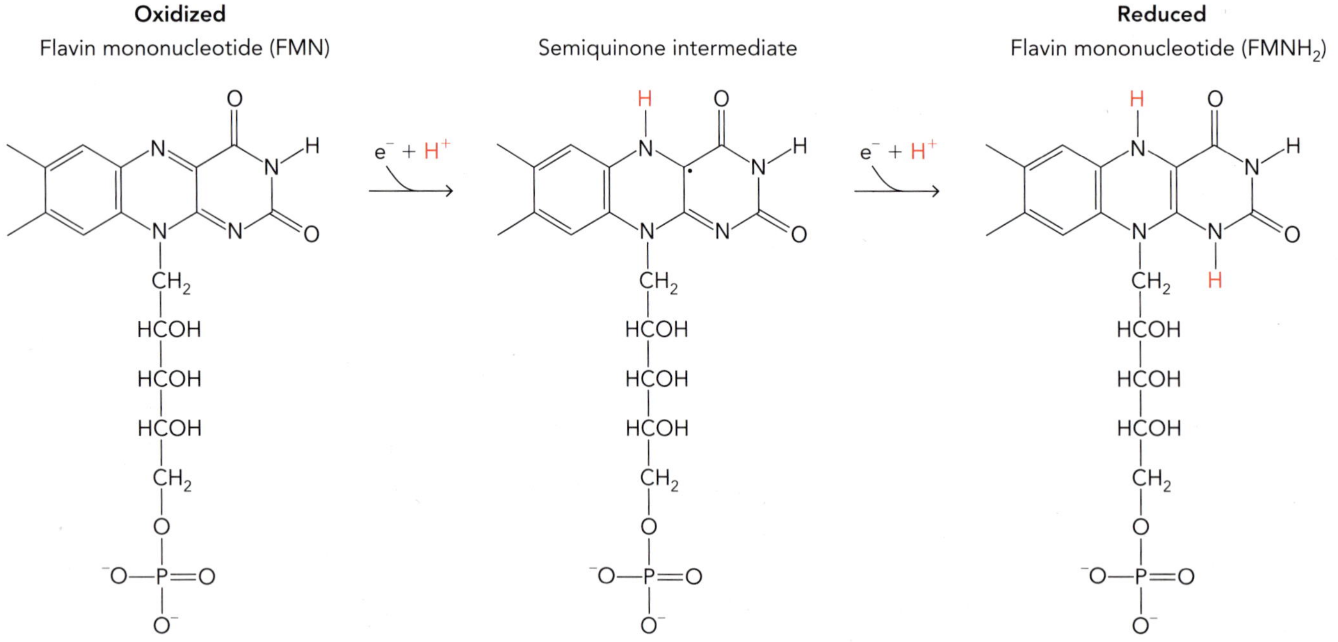

Figure 11.11 Flavin mononucleotide can accept electrons one at a time. Reduction by one electron forms the semiquinone intermediate, and the reduction by a second electron leads to the fully reduced species ($FMNH_2$).

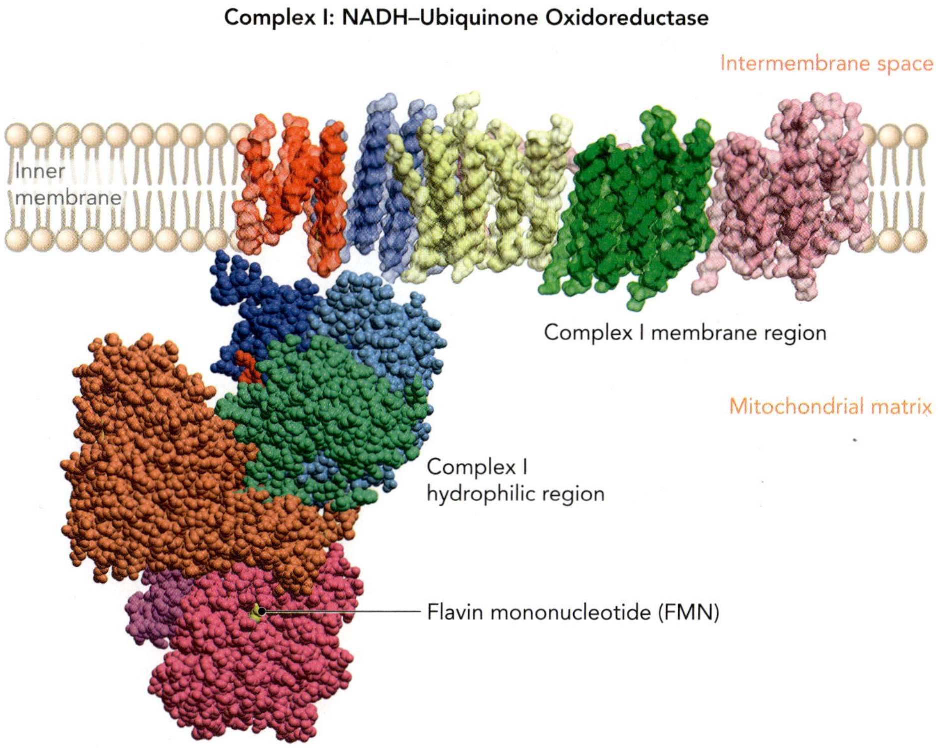

Figure 11.12 Complex I from the bacterium *Thermus thermophilus* contains both membrane-embedded and hydrophilic regions. The structures of the seven-subunit membrane region and eight-subunit hydrophilic region were determined separately and are shown here in close proximity to approximate the L-shaped structure observed by electron microscopy. Two subunits in the membrane region and one subunit in the hydrophilic region cannot be seen in this rendering of the composite structure. BASED ON PDB FILE 3M9S.

Figure 11.12 shows that the structure of complex I from *Thermus thermophilus* consists of a membrane region composed of seven subunits, containing a total of 63 transmembrane α helices. The membrane region is attached to the eight-subunit hydrophilic region that faces toward the mitochondrial matrix, where it catalyzes the oxidation of NADH. The FMN coenzyme is covalently attached to one of the subunits of the NADH–ubiquinone oxidoreductase.

In addition to FMN, complex I also contains seven or more **iron–sulfur (Fe–S) centers**, which carry one electron at a time from one end of the complex to the other. (The poison rotenone blocks electron transfer within complex I by preventing a redox reaction between two Fe–S centers.) **Figure 11.13** shows the structures of a 2 Fe–2 S and a 4 Fe–4 S center, both of which are present in complex I. The Fe–S centers are coordinated to cysteine residues in the protein.

The iron atom is reduced by 1 e^- transfers as a function of the redox reaction ($Fe^{3+} \rightarrow Fe^{2+}$). The reduction potentials ($E°'$) of 2 Fe–2 S and 4 Fe–4 S clusters within iron–sulfur proteins can vary from −0.5 V to +0.4 V, depending on the microenvironment of the protein. These reduction potentials explain how electrons can be transferred from one region of complex I to the other, based on the physical arrangement of neighboring Fe–S clusters with increasing reduction potentials (higher affinity for electrons).

The transfer of electrons through the various protein subunits of complex I involves numerous coupled redox reactions, with the net result being the oxidation of NADH

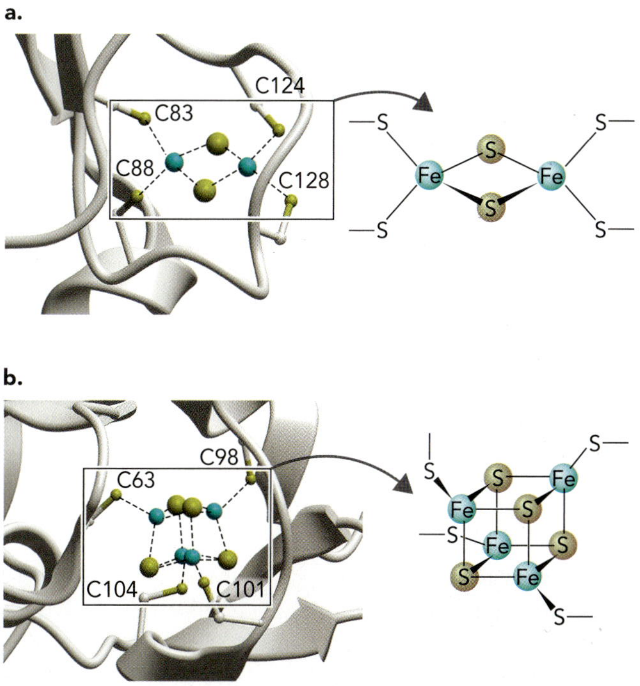

Figure 11.13 Complex I contains two types of Fe–S clusters, coordinated through cysteine residues in the protein subunits. **a.** The Nqo2 subunit contains a 2 Fe–2 S cluster. **b.** The Nqo9 subunit contains a 4 Fe–4 S cluster. BASED ON PDB FILES 3M9C AND 3I9V.

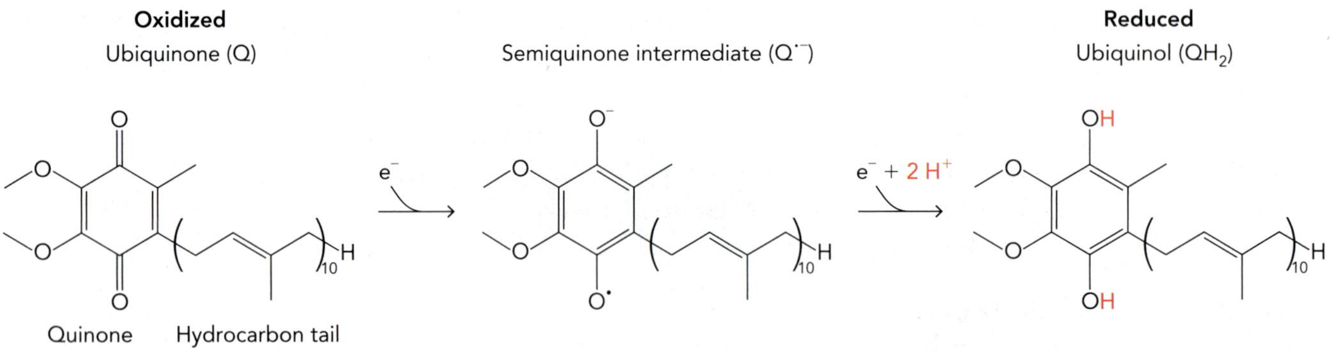

Figure 11.14 The structures of oxidized coenzyme Q (ubiquinone), the semiquinone intermediate ($Q^{\cdot -}$), and reduced coenzyme QH_2 (ubiquinol) are shown here. The hydrocarbon tail of human coenzyme Q contains 10 isoprenoid units, which led to the name CoQ_{10}.

and reduction of coenzyme Q (ubiquinone; Q) to form QH_2 (ubiquinol). In the process, 4 H^+ are translocated from the matrix side of the membrane (N; negative side) to the intermembrane space (P; positive side) as described by the equation

$$NADH + Q + 5\,H^+_N \rightarrow NAD^+ + QH_2 + 4\,H^+_P$$

Complex I passes the two electrons obtained from NADH to coenzyme Q, which has three critical roles in the electron transport system:

1. Coenzyme Q serves as a mobile electron carrier that transports electrons laterally in the membrane from complex I to complex III.
2. Coenzyme Q is the entry point into the electron transport system for electron pairs (2 e^-) obtained from the citrate cycle, fatty acid oxidation, and the enzyme glycerol-3-phosphate dehydrogenase (see Figure 11.8).
3. Coenzyme Q has the important task of converting a 2 e^- transport system in complexes I and II into a 1 e^- transport system in complex III, which passes electrons one at a time to the mobile carrier protein cytochrome c. This conversion process (2 $e^- \rightarrow 1\,e^- + 1\,e^-$) is accomplished by the Q cycle, as we will describe shortly.

As shown in **Figure 11.14**, coenzyme Q contains a quinone derivative attached to a long hydrophobic tail and can accept or donate one or two electrons at a time

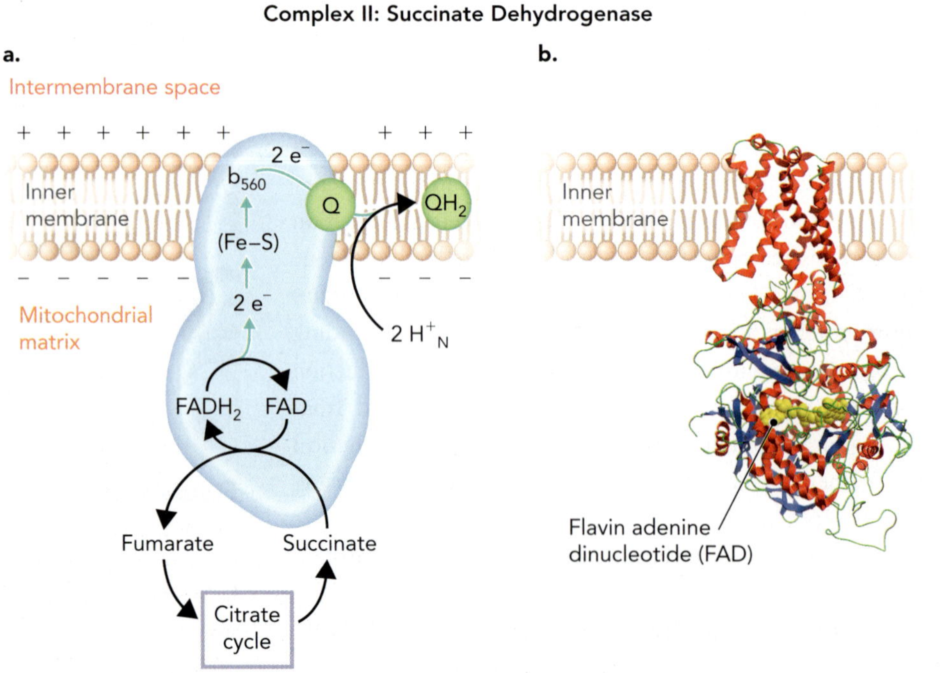

Figure 11.15 In the citrate cycle, oxidation of succinate to form fumarate by the membrane-bound enzyme succinate dehydrogenase (complex II) transfers a pair of electrons to FAD. **a.** Oxidation of $FADH_2$ within complex II transfers the electrons through a series of Fe–S clusters and cytochrome b_{560}. In the final redox reaction, 2 e^- + 2 H^+_N are used to reduce Q to form QH_2. No protons are translocated across the inner mitochondrial membrane by complex II. **b.** Molecular structure of complex II, isolated from porcine heart muscle mitochondria, bound to FAD. BASED ON PDB FILE 1ZOY.

by forming a semiquinone intermediate denoted as $Q^{\bullet -}$. The hydrophobic tail consists of repeating isoprenoid units (five-carbon building blocks also used in cholesterol biosynthesis) that vary in number, depending on the organism. The length of the hydrophobic tail in human coenzyme Q is 10 isoprenoid units (50 carbons), which gives it the name coenzyme Q_{10}, or CoQ_{10}. Although not approved by the U.S. Food and Drug Administration (FDA) to treat any human diseases, unregulated CoQ_{10} is sold as an over-the-counter nutritional supplement.

Complex II: Succinate Dehydrogenase Succinate dehydrogenase is a multi-subunit, membrane-bound enzyme that catalyzes an oxidation reaction converting succinate to fumarate in a coupled redox reaction involving the coenzyme FAD. This same enzyme also participates in the citrate cycle (see Chapter 10). The electron pair extracted from succinate through oxidation by succinate dehydrogenase is transferred to FAD and then used to reduce Q to generate QH_2, as shown in **Figure 11.15**. In addition to FAD, succinate dehydrogenase also contains three iron–sulfur centers (one 4 Fe–4 S and two 2 Fe–2 S) and a b-type heme (located in the protein cytochrome b_{560}). Heme groups are porphyrin rings with a central iron atom (Fe^{2+} or Fe^{3+}), which we saw as the prosthetic group that binds oxygen in the hemoglobin protein (see Chapter 6).

Complex III: Ubiquinone–Cytochrome c Oxidoreductase Complex III serves as the docking site for QH_2 and transfers electrons—one at a time—through a 2 Fe–2 S center, two b-type hemes (b_L and b_H), and one c-type heme (c_1). Like the other complexes in the electron transport system, complex III consists of multiple protein subunits. In its functional state, complex III consists of 11 different protein subunits, which form a dimeric complex for a total of 22 subunits (see Table 11.2).

Docking of the electron carrier protein cytochrome c to the intermembrane side of complex III leads to the transfer of an electron from the c_1 heme to the heme center of cytochrome c. **Figure 11.16** shows a diagram of the complex III monomer, illustrating the relative positions of the electron carriers. We also show the two distinct binding sites for ubiquinone, called Q_P and Q_N, which play a crucial role in diverting one electron

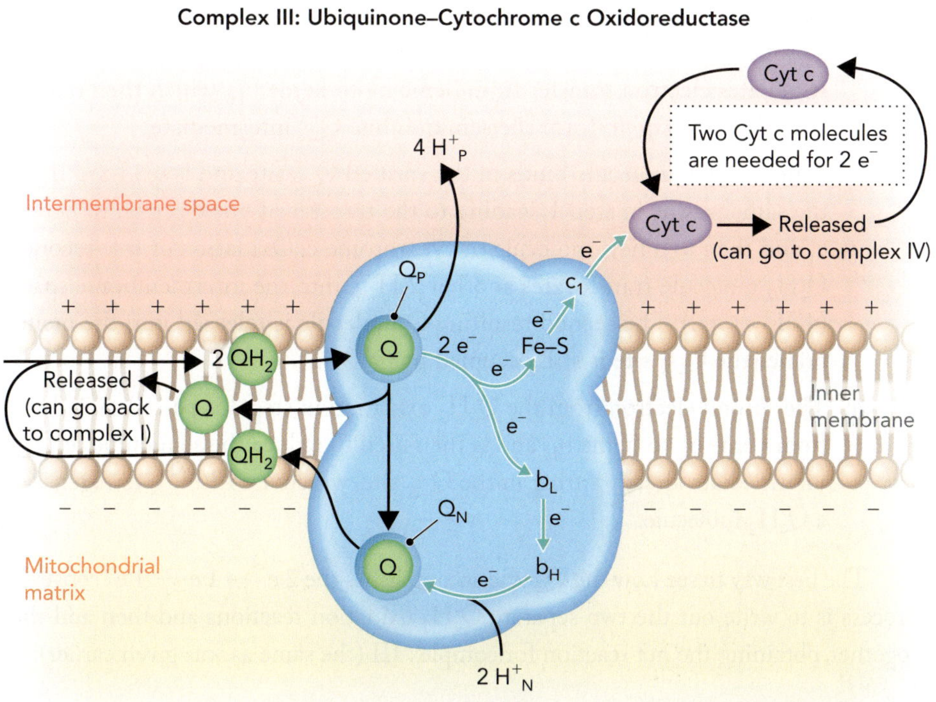

Figure 11.16 Complex III has two Q binding sites (Q_P and Q_N) and several prosthetic groups that function as electron carriers (Fe–S cluster and hemes b_L, b_H, and c_1). The oxidation of QH_2 to produce Q and two molecules of reduced cytochrome c (Cyt c) requires the Q cycle reactions illustrated in Figure 11.18.

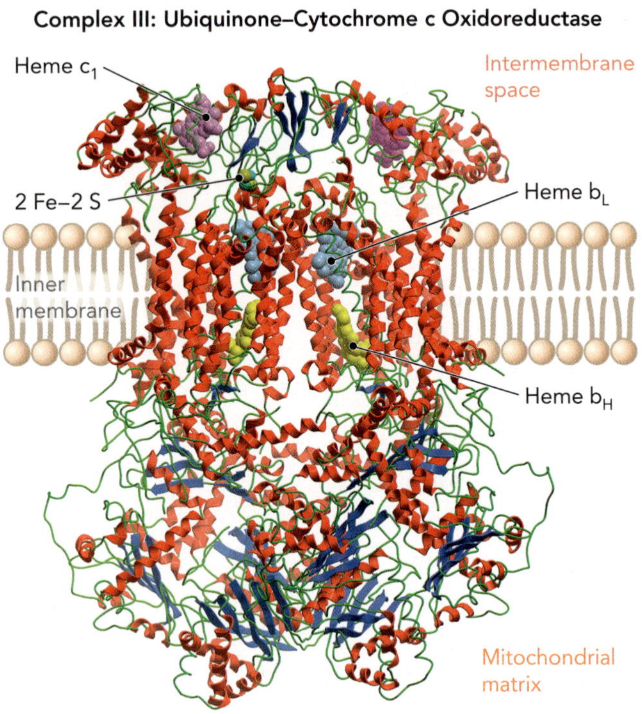

Figure 11.17 The structure of the complex III (ubiquinone–cytochrome c oxidoreductase) dimer from bovine heart muscle is shown here. The six heme groups (three hemes from each monomer, c_1, b_L, b_H) and a 2 Fe–2 S cluster in the complex III dimer are shown as space-filling models. BASED ON PDB FILE 1BGY.

at a time to cytochrome c via the Q cycle. The terms Q_P and Q_N refer to the proximity of the sites to the positive (intermembrane space) and negative (matrix) sides of the membrane.

Figure 11.17 shows the molecular structure of a complex III dimer from bovine heart muscle. You can see that the Fe–S center and c_1 heme are located in a region of the protein complex that is accessible to the cytochrome c carrier protein (see Figure 11.8). The b_L and b_H heme groups are packed closely together in the membrane-spanning region of the complex, where they interact with ubiquinol in a redox reaction that transfers electrons from QH_2 to complex III.

The net reaction of QH_2 oxidation by complex III can be written as

$$QH_2 + 2\,H^+{}_N + 2\text{ Cytochrome c (oxidized)} \rightarrow Q + 4\,H^+{}_P + 2\text{ Cytochrome c (reduced)}$$

The reaction shows that 4 H^+ are translocated to the intermembrane space (P side) with the reduction of two cytochrome c molecules.

How does the Q cycle convert a two-electron transport process into 2 one-electron transfers? The answer is by a four-step process involving the oxidation of two QH_2 molecules, which is summarized here and in **Figure 11.18**:

1. Oxidation of QH_2 at the Q_P site results in the transfer of one electron to the 2 Fe–2 S center. This electron is then transferred to the c_1 heme and then passed off to cytochrome c. The second electron derived from QH_2 oxidation is transferred to the b_L heme, which stores it temporarily. The oxidation of QH_2 in this first step contributes 2 $H^+{}_P$ to the intermembrane space.
2. The oxidized Q molecule moves from the Q_P site to the Q_N site through a proposed substrate channel within the protein complex. This movement stimulates electron transfer from heme b_L to heme b_H, which then reduces Q in the Q_N site to form the semiquinone $Q^{\bullet -}$ intermediate.
3. A new QH_2 molecule binds in the vacated Q_P site and is oxidized in the same way as in step 1, leading to the transfer of one electron to heme c_1 and then to a new molecule of cytochrome c. Oxidation of this second QH_2 molecule translocates another 2 $H^+{}_P$ into the intermembrane space (4 $H^+{}_P$ total). The second resulting Q molecule is released into the membrane (the Q_N site is still occupied by $Q^{\bullet -}$).
4. The second electron from the QH_2 oxidation in step 3 is passed directly from heme b_L to heme b_H and is then used to reduce the semiquinone $Q^{\bullet -}$ intermediate already sitting in the Q_N site, which uses 2 $H^+{}_N$ to regenerate a QH_2 molecule.

The best way to see how the Q cycle accomplishes the $2\,e^- \rightarrow 1\,e^- + 1\,e^-$ conversion process is to write out the two separate QH_2 oxidation reactions and then add them together, obtaining the net reaction for complex III (the same as was given earlier):

$$\mathbf{QH_2} + \text{Cytochrome c (oxidized)} \rightarrow \mathbf{Q^{\cdot-}} + 2\,H^+{}_P + \text{Cytochrome c (reduced)}$$

$$QH_2 + \mathbf{Q^{\cdot-}} + 2\,H^+{}_N + \text{Cytochrome c (oxidized)} \rightarrow Q + \mathbf{QH_2} + 2\,H^+{}_P + \text{Cytochrome c (reduced)}$$

$$QH_2 + 2\,H^+{}_N + 2\,\text{Cytochrome c (oxidized)} \rightarrow Q + 4\,H^+{}_P + 2\,\text{Cytochrome c (reduced)}$$

Cytochrome c Cytochrome c is a small protein of ~13 kDa that associates with the cytosolic side of the inner mitochondrial membrane. It is responsible for transporting one electron from complex III to complex IV, using an iron-containing heme prosthetic group. Oxidized cytochrome c contains ferric iron (Fe^{3+}) in the heme group, and

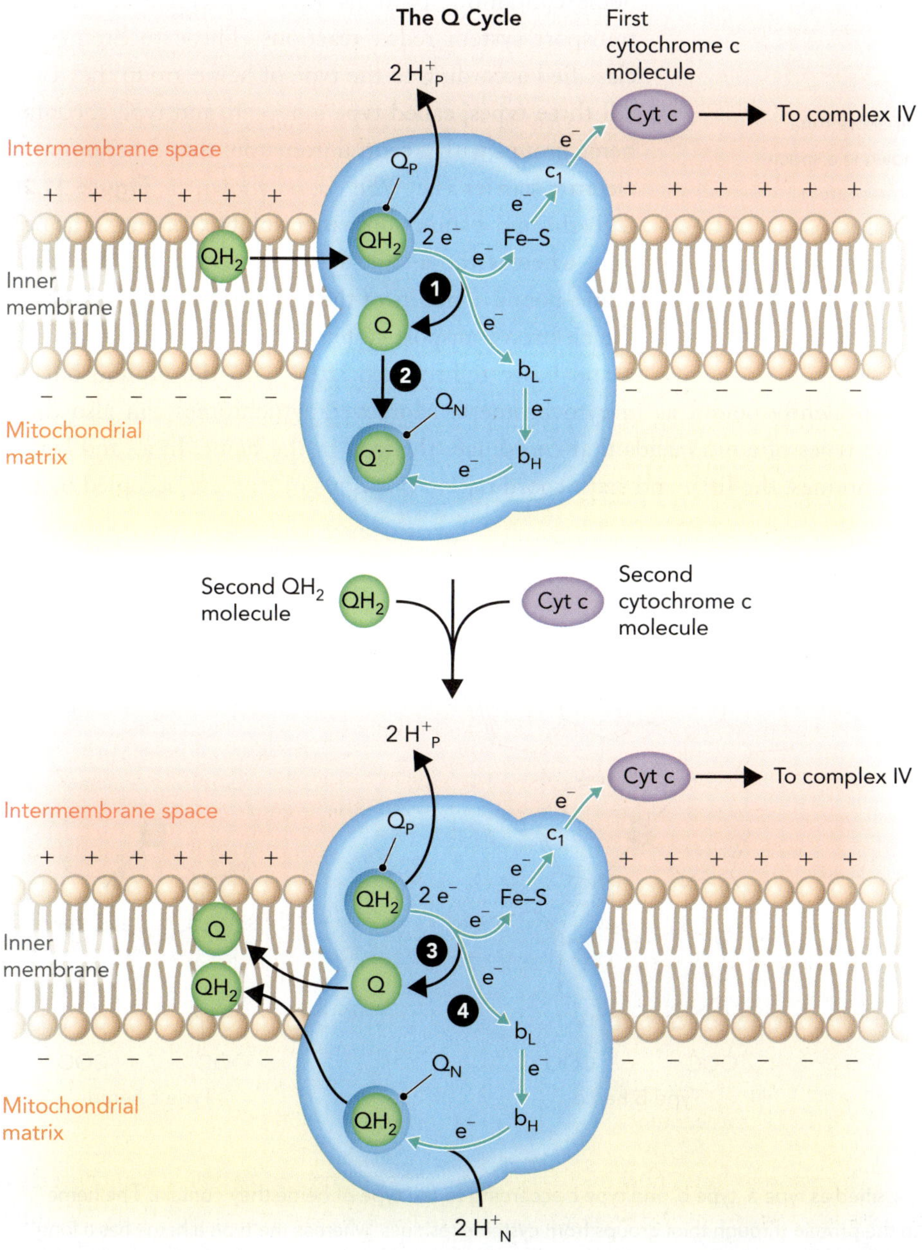

Figure 11.18 The Q cycle requires four steps, which together oxidize two QH_2 molecules (one is regenerated); translocate four protons ($H^+{}_P$) to the intermembrane space; and reduce two cytochrome c (Cyt c) molecules, which then transport 2 e^- to complex IV, one at a time. The four steps are described in the text.

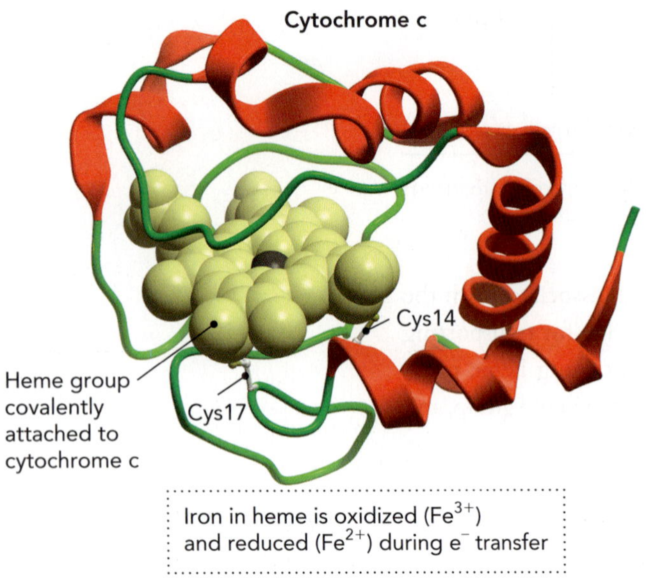

Figure 11.19 The molecular structure of albacore tuna cytochrome c shows that the heme group (shown in a space-filling model) sits in a hydrophobic pocket of the protein, surrounded by three α helices. In this protein, the heme is attached to the cytochrome c protein through the amino acid residues Cys14 and Cys17. BASED ON PDB FILE 3CYT.

reduced cytochrome c contains ferrous iron (Fe^{2+}). The heme group of cytochrome c is covalently attached to the protein through cysteine residues and lies within a pocket surrounded by three α helices. The molecular structure of a typical cytochrome c protein is shown in **Figure 11.19**.

Cytochrome c is a highly conserved protein in nature, with relatively few differences in the amino acid sequence of related species and nearly identical molecular structures across all species. The high level of cytochrome c structural conservation in nature reflects its critical function not only in the electron transport system as described in this chapter, but also as an activator of caspase enzymes in stress-induced apoptotic pathways (see Chapter 8).

Cytochrome proteins are membrane-associated, heme-containing proteins that participate in electron transport system redox reactions. The cytochromes are classified according to the type of heme group they carry. All three types, called type a, type b, and type c, contain heme groups with a central iron atom that functions as an electron carrier (Fe^{3+}/Fe^{2+}), as shown in **Figure 11.20**. Cytochrome c functions as a soluble, monomeric protein, whereas the a- and b-type cytochromes are membrane-bound subunits of the protein complexes that form the electron transport system. The hemes of cytochromes a and b are tightly associated with the protein but are not covalently bound as in cytochrome c. Moreover, cytochromes can also differ in the types of amino acids that coordinate the iron in the heme. In a- and b-type cytochromes, the fifth and sixth coordination positions of iron are occupied by two

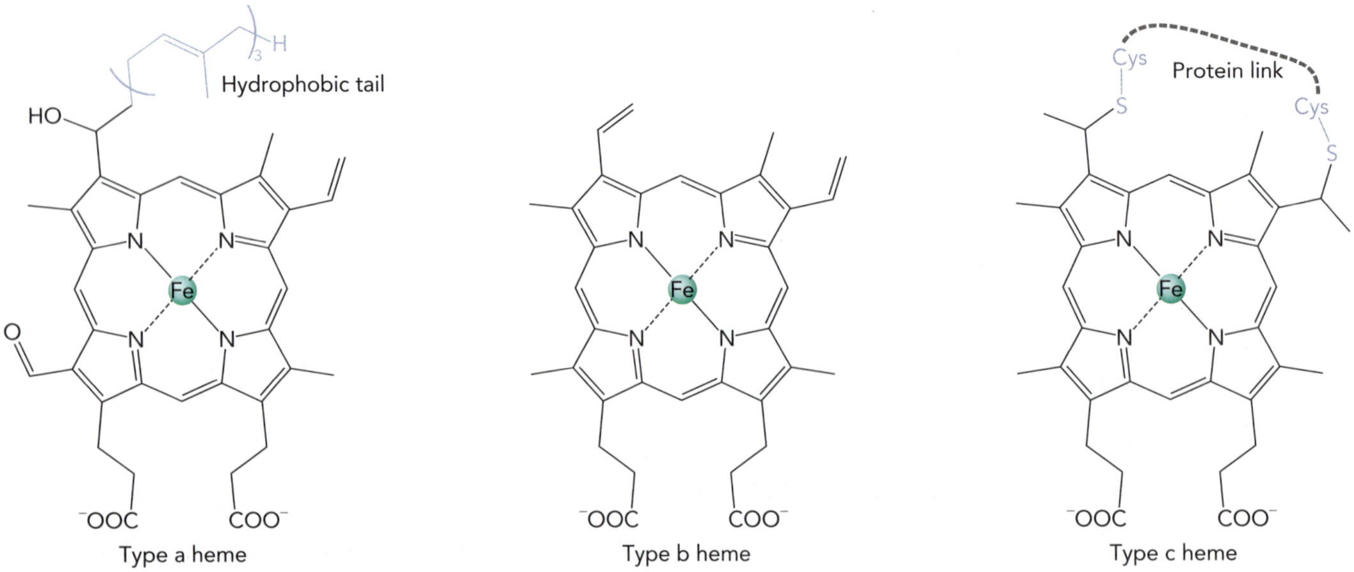

Figure 11.20 Cytochrome proteins are classified as type a, type b, and type c according to the type of heme they contain. The heme group of cytochrome c is covalently linked to the protein through thiol groups from cysteine residues, whereas the type a heme has a long hydrophobic tail.

histidine residues, whereas the heme iron in cytochrome c is coordinated by a histidine residue on one side of the heme and a methionine residue on the other side (through the sulfur atom).

Because of differences in the chemical structures of the hemes and in their environments within the cytochrome proteins, each type of cytochrome has a distinct infrared (IR) absorption spectrum. For example, the absorption spectrum of reduced cytochrome c shows three peaks at 418 nm (γ), 534 nm (β), and 554 nm (α), whereas the absorption spectrum of reduced cytochrome b has peaks at 429 nm (γ), 532 nm (β), and 563 nm (α). Cytochromes are sometimes given names that correspond to the wavelength of the α absorption peak of the reduced species. Thus, cytochrome b_L in complex III is also called cytochrome b_{566}, and cytochrome b_H is called cytochrome b_{562}. **Figure 11.21** shows that absorption spectroscopy can be used to distinguish between the oxidized and reduced states of cytochrome c.

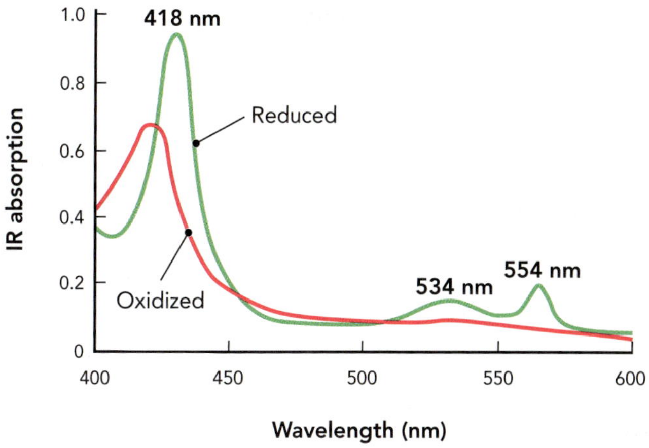

Figure 11.21 The oxidized and reduced forms of cytochrome c have different infrared absorption spectra.

Complex IV: Cytochrome c Oxidase Complex IV accepts electrons one at a time from cytochrome c and donates them to oxygen to form water. In the process, two net H^+ are translocated across the membrane, driven by conformational changes in complex IV proteins.

The mitochondrial complex IV is a homodimer containing two copper centers (Cu_A and Cu_B) and two heme groups: one in cytochrome a and one in cytochrome a_3. As illustrated in **Figure 11.22**, cytochrome c binds to complex IV near Cu_A, which accepts the electron in a coupled redox reaction leading to oxidation of the heme group in cytochrome c ($Fe^{2+} \rightarrow Fe^{3+}$). The reduced Cu_A passes the electron to the iron atom in the heme of cytochrome a, which then transfers it to cytochrome a_3. In the last step, the electron is passed to Cu_B, which donates it to oxygen to form water. The molecular structure of the complex IV homodimer is illustrated in **Figure 11.23**, showing where the two copper centers and two hemes in each monomeric subunit are located. The central core of the homodimer consists of more than 40 transmembrane α helices.

The redox reactions catalyzed by complex IV result in the uptake of 4 H^+ from the matrix and the translocation of 2 H^+ into the intermembrane space for every 2 e^- that pass through complex IV:

$$2\ \text{Cytochrome c (reduced)} + 4\ H^+_N + \tfrac{1}{2}\ O_2 \rightarrow 2\ \text{Cytochrome c (oxidized)} + 2\ H^+_P + H_2O$$

Note, however, that a total of 4 e^- need to be transferred from 4 cytochrome c to Cu_B in order to accomplish the

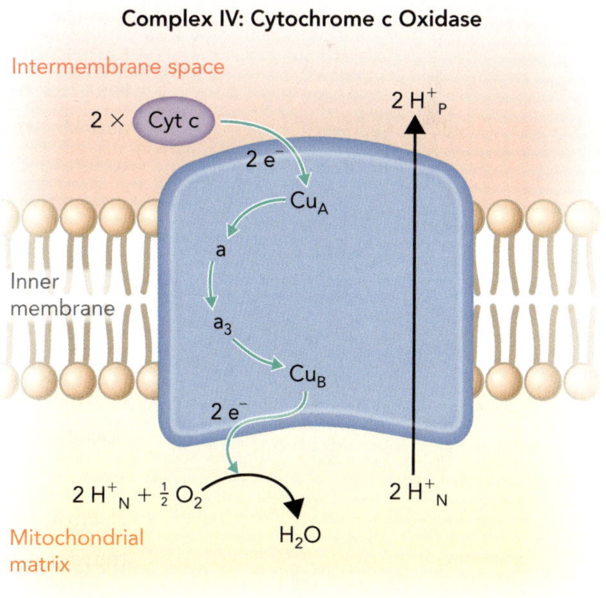

Figure 11.22 Oxidation of cytochrome c and subsequent electron transfer through one monomer of the complex IV homodimer, culminating in the reduction of oxygen to form water. A total of four H^+ are involved in the complex IV reactions, in which 2 H^+ are translocated into the intermembrane space and 2 H^+ are used to form H_2O.

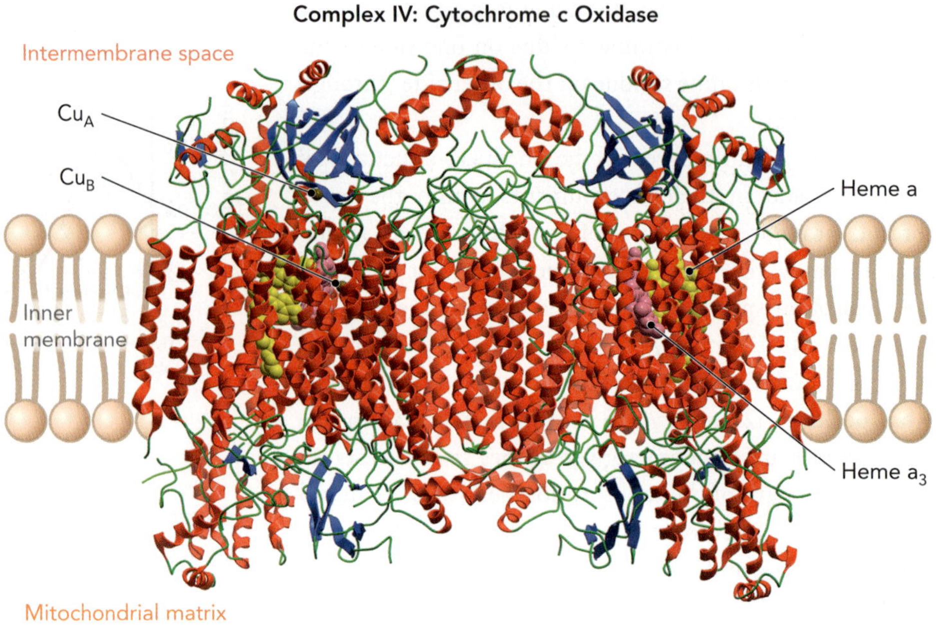

Figure 11.23 This molecular structure of a complex IV dimer shows the position of the copper centers and heme groups within the transmembrane α helices. BASED ON PDB FILE 3AG1.

reduction of 1 oxygen molecule to 2 water molecules ($O_2 \rightarrow 2\ H_2O$). Figure 11.22 shows the reduction $\frac{1}{2}\ O_2 \rightarrow H_2O$ for accounting purposes.

Now that we have looked closely at individual components of the electron transport system, let's reconstruct it using protein molecular structures, as shown in **Figure 11.24**. We emphasize three key elements of this energy conversion pathway.

1. Complexes I–IV all contain transmembrane regions that embed the proteins in the inner mitochondrial membrane, as well as functional domains that face toward the mitochondrial matrix. Moreover, because complexes III and IV need to transport electrons to and from cytochrome c, they contain functional domains that protrude into the intermembrane space, where they interact with this electron carrier protein.
2. The oxidation of NADH starting with complex I results in the translocation of 10 H^+_N across the inner mitochondrial membrane, whereas the oxidation of Q and $FADH_2$ starting with complex II leads to the translocation of only 6 H^+_N across the membrane. The difference in 4 H^+_N net translocated protons between these two oxidation reactions is due to the fact that complex II does not translocate any protons.
3. Both coenzyme Q and cytochrome c transport only 1 e^- at a time through the electron transport system and therefore must make two trips to complete the transfer of a pair of electrons from NADH or $FADH_2$ to $\frac{1}{2}\ O_2$ to form H_2O.

Bioenergetics of Proton-Motive Force

The term **proton-motive force** was coined by Peter Mitchell to describe the contributions of ΔpH and $\Delta\psi$ to the stored energy in the electrochemical proton gradient. (Note that like an electromotive force, which inspired this term, a proton-motive force is not a force, but a form of energy.) Because we can measure the ΔpH and $\Delta\psi$ values across the mitochondrial inner membrane experimentally, we can calculate how efficiently the electron transport system converts redox energy to a proton-motive force.

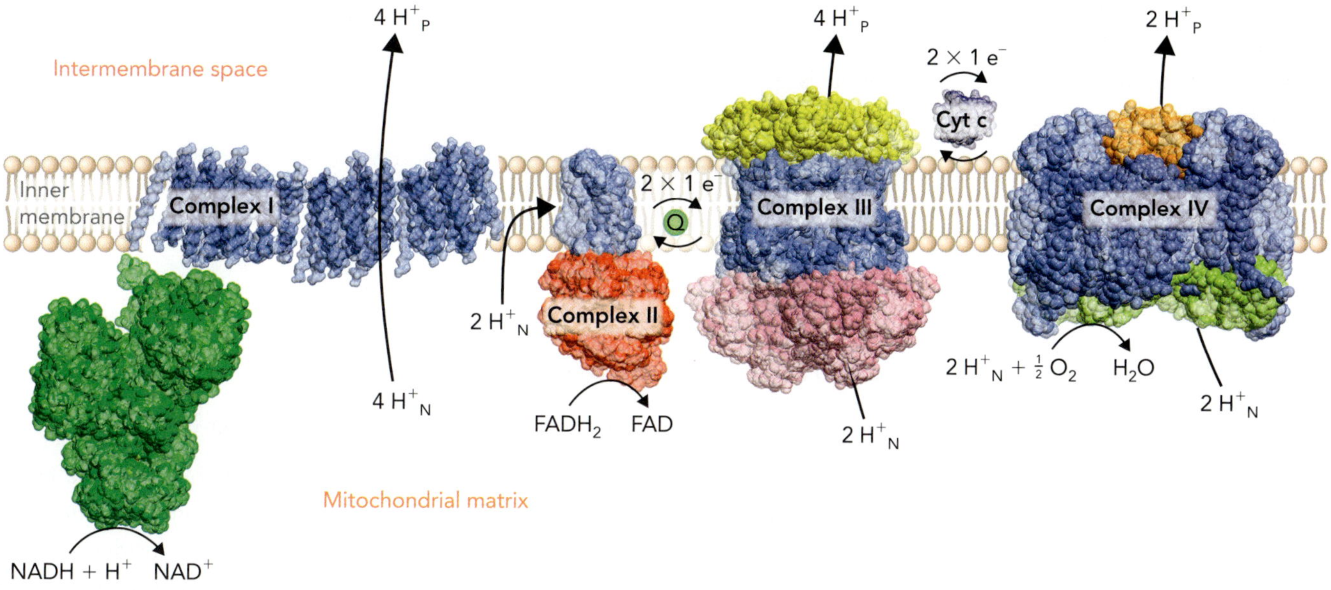

Figure 11.24 A reconstruction of the electron transport system is shown using the molecular structures of each protein component. BASED ON PDB FILES 3M9C AND 3I9V (COMPLEX I), 1ZOY (COMPLEX II), 1BGY (COMPLEX III), 3CYT (CYT C), AND 3AG1 (COMPLEX IV).

As described in Chapter 10, the change in standard free energy ($\Delta G°'$) of a coupled redox reaction can be calculated from the standard reduction potentials ($E°'$) of the corresponding conjugate redox pairs using Equation 10.2:

$$\Delta G°' = -nF\Delta E°'$$

where n is the number of electrons transferred in the reaction (two in the case of NADH oxidation), and F is the Faraday constant (96.48 kJ/V mol). As listed in Table 11.1, the $E°'$ value of the NAD^+/NADH conjugate redox pair is −0.32 V, and that of the $\frac{1}{2}$ O_2/H_2O conjugate redox pair is +0.82 V. Using these values, we can calculate the $\Delta G°'$ for translocation of 10 H^+ across the membrane by first determining $\Delta E°'$ with Equation 10.1:

$$\begin{aligned}\Delta E°' &= (E°'_{e^-\text{ acceptor}}) - (E°'_{e^-\text{ donor}})\\ &= (E°'_{1/2\,O_2}) - (E°'_{NADH})\\ &= (+0.82\text{ V}) - (-0.32\text{ V})\\ &= 1.14\text{ V}\end{aligned}$$

From this value for $\Delta E°'$, we can find $\Delta G°'$:

$$\begin{aligned}\Delta G°' &= -nF\Delta E°'\\ &= -2(96.48\text{ kJ/V mol})(1.14\text{ V})\\ &= -220\text{ kJ/mol of NADH oxidized}\end{aligned}$$

To determine the free energy available from the proton-motive force across the mitochondrial membrane, we need to take into account both the difference in H^+ concentration (ΔpH) and the membrane potential ($\Delta\psi$) using the following equation:

$$\Delta G = RT\ln(C_2/C_1) + ZF\Delta\psi \qquad (11.1)$$

In this equation, C_2/C_1 is the concentration ratio of the ion that is moving (H^+), Z is the absolute charge on the ion (+1), $\Delta\psi$ is the membrane potential (measured in volts), R is the gas constant (8.314×10^{-3} kJ/mol K), and T is the absolute temperature in kelvin (K). The ΔpH value for actively respiring mitochondria has been shown to be about 0.75 pH units, and the membrane potential across the inner mitochondrial membrane is approximately −0.15 V. Using these values, we can now calculate the free energy of the proton-motive force at 37 °C (310 K) as follows:

$$\begin{aligned}\Delta G &= RT\ln(C_2/C_1) + ZF\Delta\psi \\ &= (8.314 \times 10^{-3}\ \text{kJ/mol K})(310\ \text{K})(\ln 0.75) + (96.48\ \text{kJ/V mol})(-0.15\ \text{V}) \\ &= -0.74\ \text{kJ/mol} + -14.48\ \text{kJ/mol} \\ &= -15.21\ \text{kJ/mol}\end{aligned}$$

Because 10 $H^+{}_P$ are available from each NADH, we can calculate

$$\Delta G = -152.1\ \text{kJ/mol per } 10\ H^+{}_P = -152.1\ \text{kJ/mol per NADH}$$

This calculation shows us two things. First, most of the free energy available from the electrochemical proton gradient (proton-motive force) in mitochondria is derived from $\Delta\psi$ (−14.48 kJ/mol), not the ΔpH (−0.74 kJ/mol). Second, considering that −152.1 kJ/mol of energy is actually available for ATP synthesis, out of the theoretical total of 220 kJ/mol, energy conversion by the electron transport system is ~70% efficient (the ratio of ~150/220 = ~0.7). The energy difference of ~70 kJ/mol (220 − 150 = 70) is lost as heat and contributes to thermogenesis. The calculation of 70% efficiency assumes that the $\Delta G°'$ for NADH oxidation is equal to ΔG, which is probably an underestimation (that is, ΔG is likely to be more negative than −220 kJ/mol).

A second way to look at the overall efficiency of energy conversion in the mitochondria is to combine the electron transport system with ATP synthesis and compare the actual number of ATP synthesized (per NADH oxidized) to the theoretical yield of ATP. Using the ΔG value for proton-motive force, the energy derived from the oxidation of NADH is ~150 kJ/mol (see earlier). The estimated ΔG value for ATP synthesis inside the mitochondrial matrix is ~40 kJ/mol, based on the concentrations of ATP, ADP, and P_i. Thus, the theoretical yield per NADH oxidized is ~3.8 ATP (150 kJ/mol divided by 40 kJ/mol). However, the actual yield is closer to ~2.5 ATP per NADH, therefore the efficiency of energy conversion calculated with ΔG values is ~66% (2.5 ATP/3.8 ATP = 0.66), which is similar to the ~70% efficiency calculation based on proton-motive force.

To understand the mechanics of how proton-motive force is converted into the energy required for ATP synthesis, we now turn our attention to the structure and function of the ATP synthase complex.

concept integration 11.2

What is the key role of coenzyme Q in the electron transport system?

Coenzyme Q is a hydrophobic molecule that functions as a mobile electron carrier in the inner mitochondrial membrane, transporting a pair of electrons (2 e^-) in its reduced form of QH_2. The 2 e^- are obtained from either complex I after the oxidation of NADH or from membrane-bound $FADH_2$-containing enzymes, such as succinate dehydrogenase or glycerol-3-phosphate dehydrogenase. The reduced QH_2 donates the 2 e^- to complex III, which then transfers them one at a time to a second electron carrier called cytochrome c. An important function of coenzyme Q is to use the Q cycle to convert the two-electron transport system employed by complex I and II into the one-electron transport system used by cytochrome c and complex IV.

11.3 Structure and Function of the ATP Synthase Complex

The ATP synthase complex was originally called complex V and was considered part of the electron transport system. Later, it was purified as an ATP hydrolyzing enzyme using a cell-free assay system. It is now known to be one of the most essential protein machines found in all cells.

To explain how proton flow through the complex drives ATP synthesis, we first describe the arrangement of polypeptides within the structure. Then, we present a model that explains how protein conformational changes brought about by proton-motive force controls ATP synthesis by this enzyme.

Structural Organization of the ATP Synthase Complex

The mitochondrial ATP synthase complex consists of two large structural components: F_1, which encodes the catalytic activity, and F_o, which functions as a proton channel crossing the inner mitochondrial membrane. The subscript "o" in F_o, usually written in lowercase to avoid confusion with zero ("0"), refers to the early finding that the antibiotic **oligomycin** inhibits the activity of the F_o proton channel. As mentioned earlier, the ATP synthase complex is sometimes referred to as the F_1F_o ATP synthase complex; however, in this text it will simply be called the ATP synthase complex.

Most ATP synthase complexes that have been characterized from different organisms contain at least 20 polypeptide chains, associated with either the membrane-bound F_o component or the catalytic F_1 component. For example, the yeast mitochondrial F_1 component consists of three α subunits, three β subunits, and one each of γ, δ, and ε subunits, The β subunits contain the three catalytic sites located in the enzyme. The isolated F_o component contains one a subunit, two b subunits, and 10 c subunits, as well as the d, h, and OSCP subunits. The bacterial ATP synthase complex lacks some of the eukaryotic subunits, but overall is similar to the yeast enzyme. **Figure 11.25** compares the subunit arrangements of the F_1 and F_o components of the bacterial and

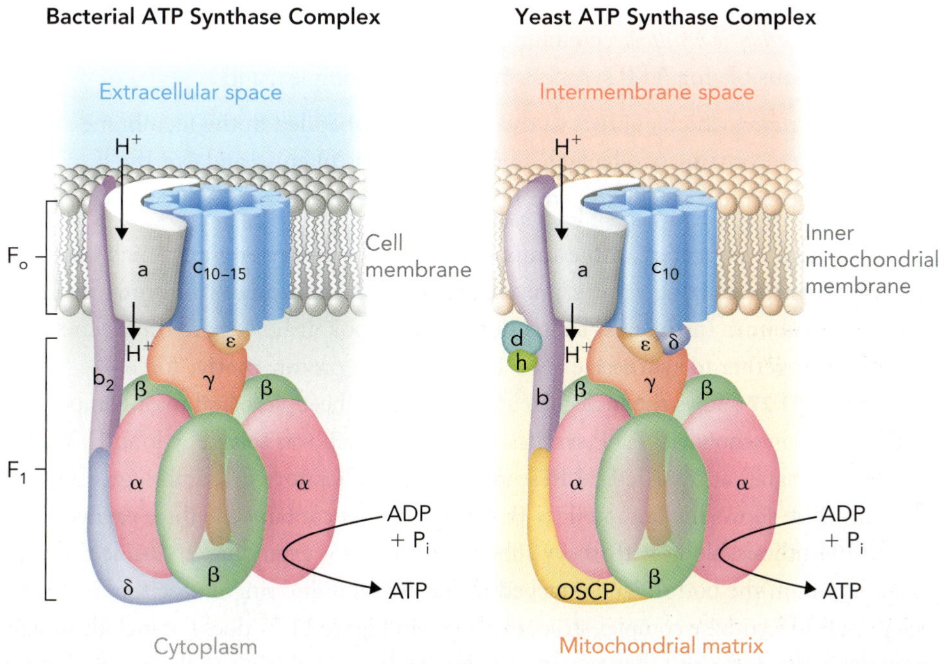

Figure 11.25 Models of the bacterial and yeast ATP synthase complexes, showing the various subunits and their relative orientation in the membrane. Each ATP synthase complex has three α and three β subunits, though the third α subunit is at the back of this drawing and is not visible here. The bacterial ATP synthase complex contains a subunit called ε that associates with the F_1 component. The δ subunit of the yeast ATP synthase complex is homologous to the bacterial ε subunit. The OSCP subunit in the yeast ATP synthase complex is the oligomycin-sensitivity-conferring protein.

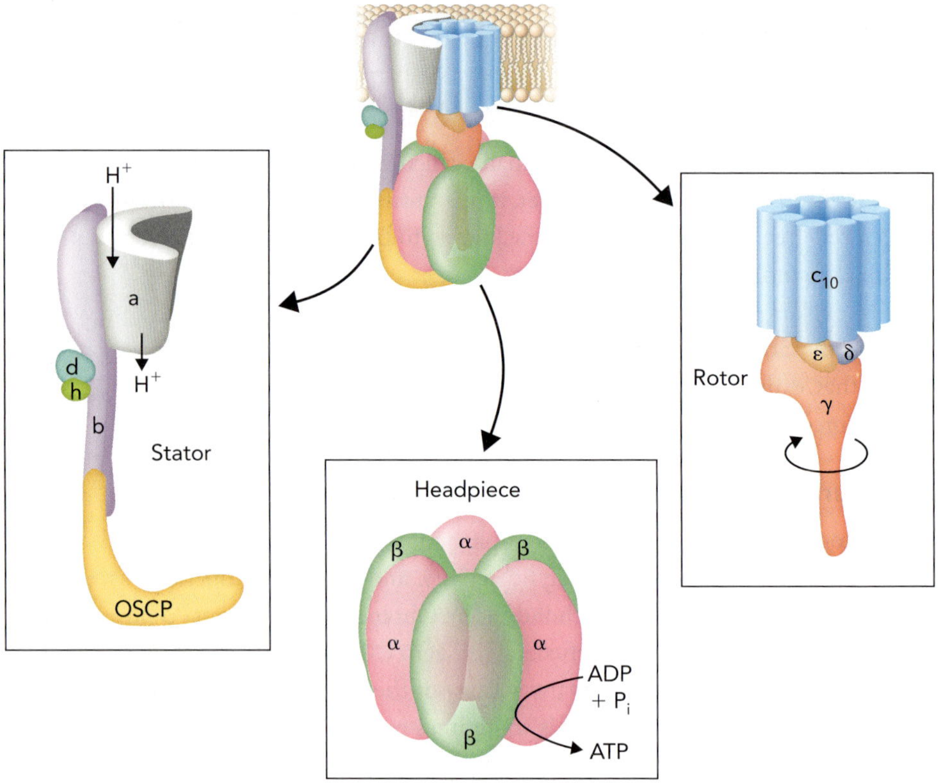

Figure 11.26 The ATP synthase complex consists of three functional components (a rotor, a headpiece, and a stator) that function together to coordinate proton flow with ATP synthesis. The stator holds the headpiece in place so that it does not turn with the rotor. The β subunits of the headpiece contain the catalytic sites for ATP synthesis, and the rotor is responsible for translating proton-motive force into protein conformational changes in the headpiece. The subunits for the yeast ATP synthase complex are illustrated here, but the same three functional components are found in all ATP synthase complexes.

yeast ATP synthase complexes. Note that some equivalent subunits are given different names in different organisms. For example, the δ subunit in yeast is equivalent to the bacterial ε subunit. The number of c subunits in the F_o component can vary between organisms, but also potentially within an organism depending on conditions.

It is useful to organize the ATP synthase complex into three functional units to visualize the working parts of this protein machine, as illustrated in **Figure 11.26**. These three functional units in the yeast ATP synthase complex can be described as

1. the *rotor*, made up of γ, δ, and ε subunits and the c-subunit ring, which function as a single unit and rotate as protons enter and exit the ring;
2. the *catalytic headpiece*, containing the hexameric $\alpha_3\beta_3$ unit, which is responsible for ATP synthesis in the intact complex; and
3. the *stator*, which consists of the a subunit embedded in the membrane and which contains two half-channels for protons to enter and exit the F_o component and a stabilizing arm made up of the b, d, h, and OSCP subunits.

The term *stator* refers to the immobile parts of a rotary engine, which describes the role of the stabilizing arm in preventing the $\alpha_3\beta_3$ headpiece from rotating along with the γ subunit. In the next subsection, we will look at how these three functional units work together to synthesize ATP in response to proton-motive force.

Figure 11.27 shows the molecular structure of the headpiece and rotor components of the yeast mitochondrial ATP synthase complex, which was solved by John E. Walker and his colleagues at the Medical Research Council laboratory in Cambridge, England. This structure shows the hexameric $\alpha_3\beta_3$ headpiece that contains the three catalytic sites for ATP synthesis, along with the γ subunit of the rotor extending up inside the $\alpha_3\beta_3$ headpiece from the bottom and attached to the 10-subunit c ring on the top. Note that the yeast ATP synthase complex structure shown in Figure 11.27 does not include protein subunits in the stator, and therefore it is not known how similar it is to the bacterial stator.

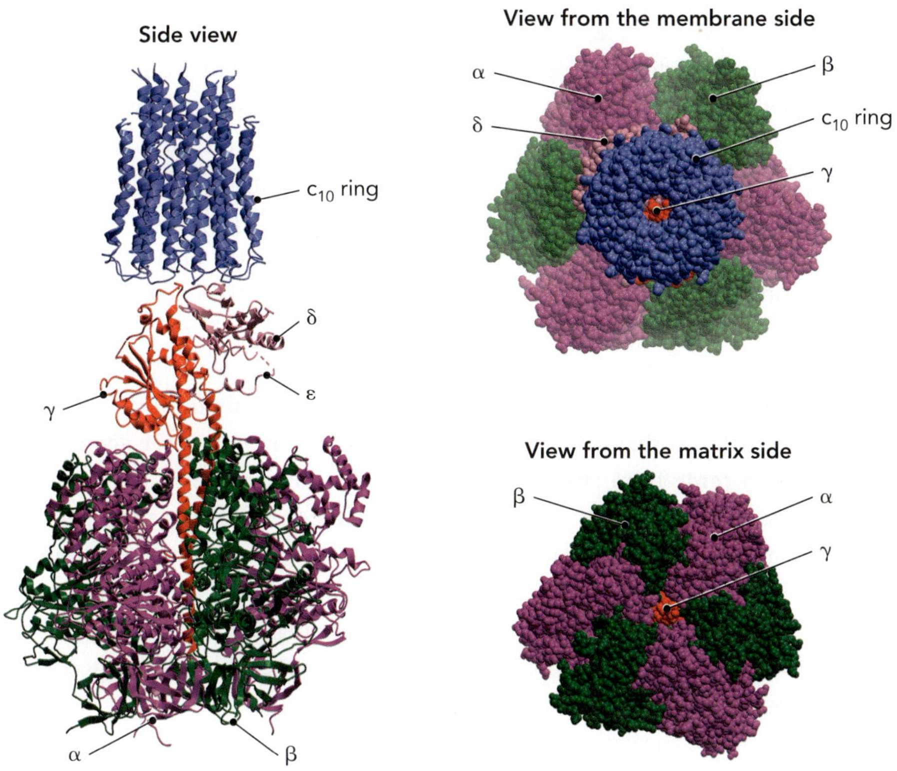

Figure 11.27 The molecular structure of the headpiece and rotor of the yeast mitochondrial ATP synthase complex is shown. Note that the δ subunit of the yeast enzyme, which is a component of the rotor, is equivalent to the bacterial ε subunit and should not be confused with the δ subunit of the bacterial stator component (see Figure 11.25). BASED ON PDB FILE 2XOK.

▶ ANIMATION

The flow of protons down their electrochemical gradient drives the formation of ATP from ADP and P_i at three catalytic sites in the F_1 component. **Figure 11.28a** shows one of the three catalytic sites, located on the outer edge of each β subunit in the yeast ATP synthase complex. In this structure, the β subunit contains a nonhydrolyzable analog of ATP called phosphoaminophosphonic acid–adenylate ester (ANP) in the catalytic site, which stabilizes the enzyme in a single conformation. Half of the Nobel Prize in Chemistry in 1997 was shared between Walker and Dr. Paul D. Boyer, who described how the ATP synthase complex catalyzes ATP synthesis is response to proton translocation (**Figure 11.28b**). The other half of the Nobel Prize in Chemistry that year was shared with Jens C. Skou, who discovered the Na^+–K^+ ATPase ion transporter (see Figure 6.47).

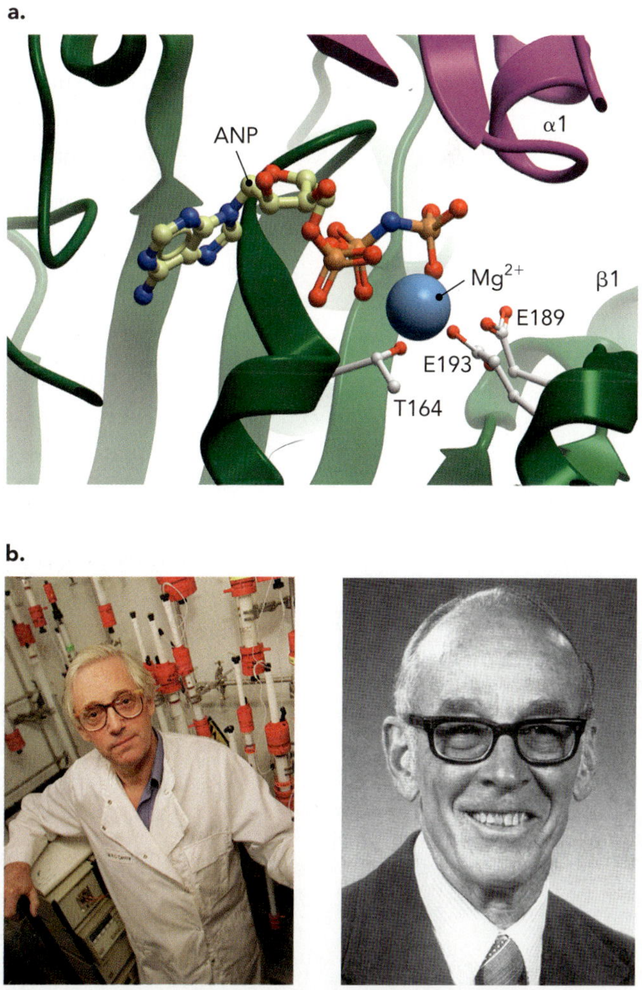

Figure 11.28 The catalytic site in the ATP synthase complex is located in the β subunit, near the interface with the α subunit. **a.** Phosphoaminophosphonic acid–adenylate ester (ANP) is a nonhydrolyzable analog of ATP that was used to trap the enzyme in a single conformation. Mg^{2+} ion is required for the enzyme reaction and is shown in the active site, coordinated with three residues in the β subunit. BASED ON PDB FILE 2XOK. **b.** John E. Walker and Paul D. Boyer (left to right) shared half of the 1997 Nobel Prize in Chemistry for their studies of the structure and function of the ATP synthase complex. J. E. WALKER: SION TOUHIG/SYGMA/CORBIS; P. D. BOYER: AP PHOTO.

Proton Flow through F_o Alters the Conformation of F_1 Subunits

Walker's structural studies revealed that differences in β-subunit conformations, as determined by specific interactions with the γ subunit, were associated with differential affinities for ATP in the catalytic sites. It was found that of the three catalytic sites present in F_1, one β subunit contained ATP (actually the analog ANP in the crystal structure), one contained ADP + P_i, and the third site usually lacked a substrate molecule. This discovery by Walker, combined with enzyme kinetic analysis of the ATP synthase complex carried out by Boyer, provided essential clues to understanding the molecular basis of proton-driven ATP synthesis.

Even before the molecular structure of the F_1 component was known, nucleotide binding studies revealed that it was the affinity of the β subunit for ATP, not the rate of ATP synthesis, that was altered by proton flow through the F_o component. This conclusion came from studies showing that in the absence of proton-motive force, the dissociation constant (K_d) of the F_1 headpiece for ATP was about 10^{-12} M, whereas when the electrochemical proton gradient was supplied, the K_d was found to be 10^{-6} M—a million-fold increase in K_d. In addition, radioactive exchange experiments, using purified F_1 and isotopically labeled P_i (using the isotope ^{18}O), showed that the ATP → ADP + P_i reaction was in fact readily reversible within the enzyme active site, confirming that proton-motive force was required for the release of ATP, not its formation. On the basis of these results and of what was known about the subunit composition of the F_1 component, Boyer proposed the **binding change mechanism** of ATP synthesis to explain how conformational changes in β subunits could control ATP production.

The binding change mechanism incorporates three basic principles (**Figure 11.29**):

1. The γ subunit directly contacts all three β subunits; however, each of these interactions is distinct, giving rise to three different β-subunit conformations.

2. The ATP binding affinities of the three β-subunit conformations are defined as L, loose; T, tight; and O, open. ADP and P_i bind to the L conformation, whereas ATP binds tightly to the T conformation. ATP is released from the enzyme when the β subunit is in the O conformation.

3. As protons flow through F_o, the γ subunit rotates counterclockwise during ATP synthesis (looking at F_1 from the matrix side), such that with each 120° rotation, the β subunits sequentially undergo conformational changes from L → T → O → L.

The binding change mechanism model predicts that one full rotation of the γ subunit should generate 3 ATP, as each β subunit will have cycled once through the T state. Moreover, the β-subunit conformations should be interdependent, so that when one subunit is in the T conformation, the other two subunits must be in the O and L conformations. As illustrated in Figure 11.29, which is drawn as if one is looking at F_1 from the mitochondrial matrix side, the cycle begins with ADP + P_i binding to the β1 subunit in the O conformation ($\beta1_O$), with ADP + P_i already in the active site of the β2 subunit in the L conformation ($\beta2_L$). The β3 subunit is in the T conformation ($\beta3_T$) and contains the tightly bound product of the reaction, ATP, still in the active site. With the input of sufficient proton-motive force (~3 H^+_P flowing through F_o to become ~3 H^+_N), the rotor turns 120° in the counterclockwise direction when looking

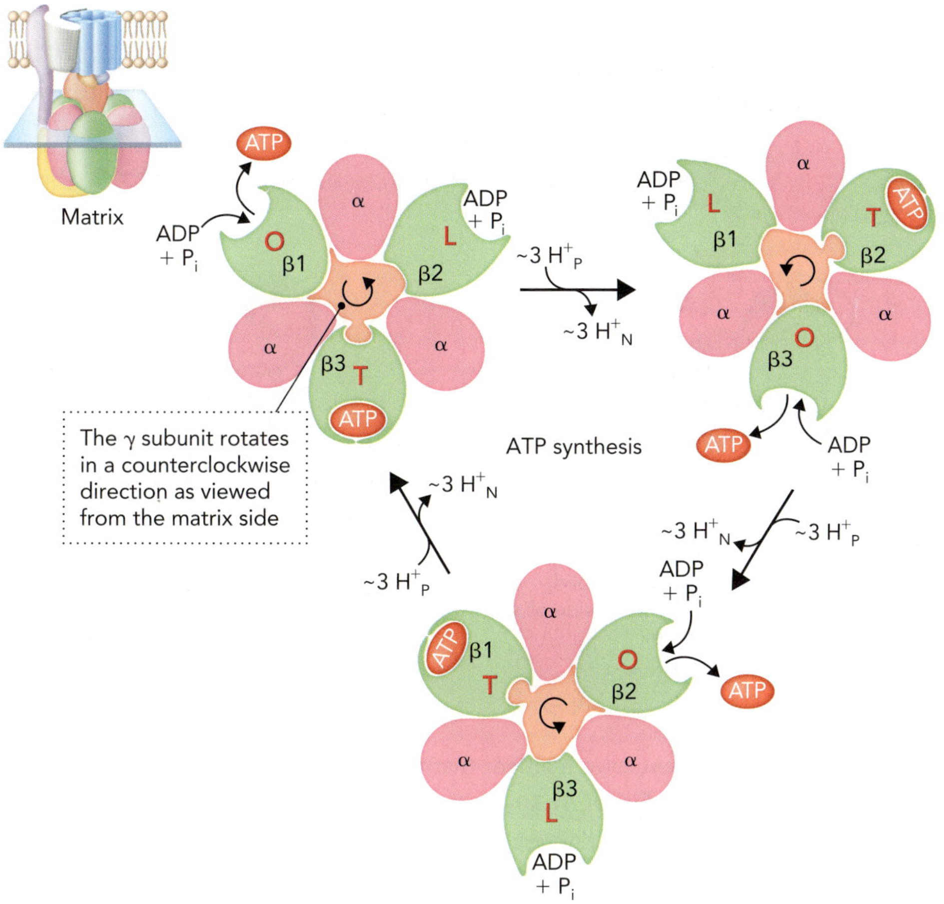

Figure 11.29 Diagram of the binding change mechanism of ATP synthesis. This diagram is drawn such that one is looking down at F_1 from the mitochondrial matrix side, and therefore the γ subunit (red-orange) rotates in the counterclockwise direction to alter the conformation of the β subunits (green). The protein contacts between the γ subunit and each β subunit are unique, such that when one β subunit is in the L conformation, the other two must be in the T and O conformations. Note that the sequential pattern of conformational changes in a single β subunit are L → T → O → L, and each β subunit conformation has a different nucleotide binding affinity. The number of H^+ required for each 120° turn of the γ subunit depends on the number of subunits in the c ring; the yeast mitochondrial ATP synthase complex appears to require ~3 H^+ for each ATP synthesized.

at F_1 from the matrix side. The rotation of the asymmetric γ subunit induces conformational changes in each of the β subunits following the sequence L → T → O → L.

After each 120° rotation, three things happen: (1) the conformation of $\beta3_T$ becomes $\beta3_O$, which releases one ATP; (2) $\beta2_L$ converts to $\beta2_T$, shifting the equilibrium toward ATP formation; and (3) $\beta1_O$ converts to $\beta1_L$, favoring ADP + P_i binding. This same sequence of events repeats two more times until the γ subunit has completed a 360° rotation, producing 3 ATP as a result of ~9 $H^+{}_P$ flowing down the electrochemical gradient through F_o.

Biochemical studies had shown that H^+ flow through F_o is not directly involved in the catalytic mechanism, and, moreover, that the F_1 headpiece functions as an ATP hydrolyzing enzyme (ATPase) in the absence of F_o. Therefore, Boyer's model predicts that ATP hydrolysis by the F_1 headpiece should reverse the direction of the γ subunit rotor; that is, rotation in a clockwise direction when looking at F_1 from the matrix side. This means that the conformational changes at a single β subunit should follow the sequential pattern O → T → L → O during ATP hydrolysis. To test this idea, Masasuke Yoshida of the Tokyo Institute of Technology and Kazuhiko Kinosita, Jr., of Keio University used recombinant DNA methods to modify the α, β, and γ subunits of the *E. coli* F_1 component to build a synthetic molecular motor. Using a technique called site-directed mutagenesis, they altered the genes encoding these subunits to add a short peptide that could immobilize the β subunit on a surface and to insert a single cysteine codon in the gene for the γ subunit. This cysteine residue was positioned in the γ subunit at a location near the normal c-ring binding site. The sulfhydryl group of this cysteine was used as a chemical attachment site for streptavidin, which was

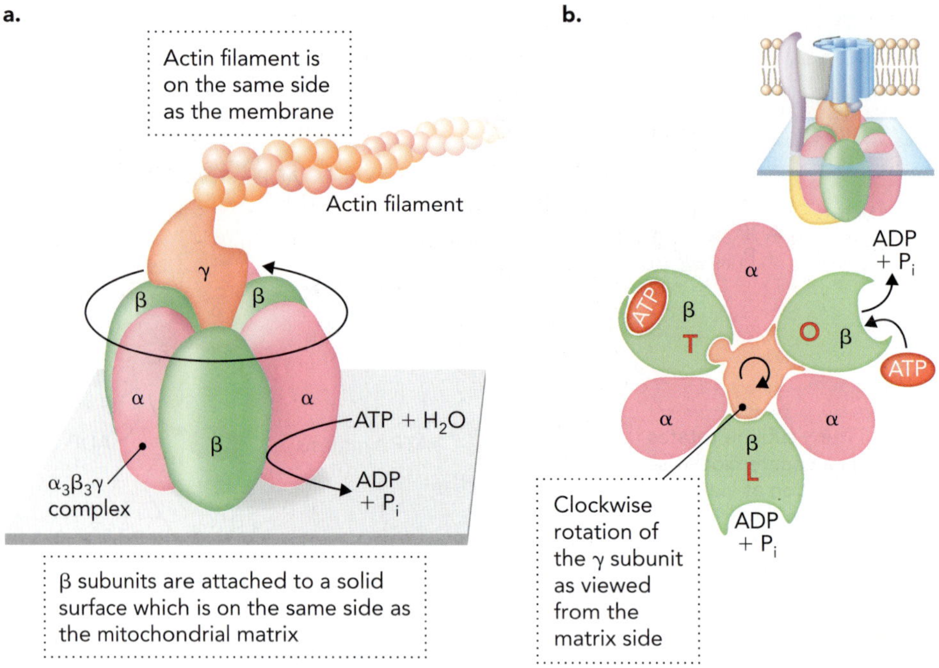

Figure 11.30 An ATP synthase complex molecular motor runs in reverse under conditions that favor ATP hydrolysis. **a.** The design of the Yoshida and Kinosita experiment, which was the first to demonstrate rotational movement in the ATPase. **b.** Using the same binding change mechanism model shown in Figure 11.29, it can be seen that ATP hydrolysis drives the γ-subunit rotation in the clockwise direction when viewing F_1 from the matrix side. ATP hydrolysis results in the conformational sequence O → T → L → O for a given β subunit, which is opposite that of ATP synthesis.

then linked to a fluorescent actin filament that was large enough to be observed with a microscope. As shown in **Figure 11.30**, the actin "propeller" allowed them to monitor rotation of the γ subunit using fluorescence microscopy.

When Yoshida and Kinosita added reaction buffer containing Mg^{2+} and ATP, they observed that the actin propeller rotated in the counterclockwise direction when viewed from the intermembrane side looking at F_o, which would be clockwise if looking at F_1 from the matrix side; that is, the opposite direction predicted for ATP synthesis. This observation confirmed γ-subunit rotation and Boyer's binding change mechanism model. The motion was not a smooth rotation, but rather a stepwise ratchet motion, consisting of three 120° rotations for each complete cycle. Under optimal conditions, Yoshida and Kinosita determined that the speed of the actin propeller was as high as 130 revolutions per second. In follow-up experiments, the same research group showed that ATP could be synthesized by forcing the γ subunit to turn in the clockwise direction when viewed from the intermembrane side looking at F_o, which would be counterclockwise if looking at F_1 from the matrix side. They accomplished this feat using a magnetic bead attached to the streptavidin linker on the γ subunit and a set of special electromagnets positioned in a circle around the chamber, as illustrated in **Figure 11.31**. They demonstrated the utility of their nanomotor by showing that ATP synthesis or ATP hydrolysis could be catalyzed by the $\alpha_3\beta_3$ headpiece simply as a function of the rotational direction of the γ subunit imposed by the electromagnets. Here, because ATP synthesis is independent of an electrochemical proton gradient, these elegant experiments demonstrated unequivocally that rotational movement of the γ subunit drives ATP synthesis, as predicted by Boyer's binding change mechanism model.

This brings us to the final mechanistic question: How does proton movement through the c-subunit ring of F_o cause rotation of the γ subunit? **Figure 11.32** (p. 558) shows a proposed model for the F_o "rotary engine" based on structural analysis of the yeast mitochondrial c-subunit ring, which contains 10 identical subunits. This "two-channel" model proposes that the a subunit contains a pair of proton channels, each of which provides access to the c-subunit ring from either the P or N side of the membrane. It is important to note that neither of the channels transverses the entire membrane, so they function only as half-channels. Because the concentration of H^+ on the P side is higher than it is on the N side, an $H^+{}_P$ will readily enter the half-channel in the a subunit, where it then comes in contact with a negatively charged Asp59 residue in the nearby c subunit. Protonation neutralizes the charge on the Asp59, allowing the c-subunit ring to rotate into the hydrophobic membrane in the counterclockwise direction (looking at F_1 from the matrix side). With this ~36° (360°per 10 c subunits) rotation, a proton bonded to Asp59 in a different c subunit, farther around the ring, gains access to the second half-channel in the a subunit and exits the channel because of the low $H^+{}_N$ concentration in the matrix. This model for the F_o rotor is similar to a carousel ride at a carnival, in which the carousel is the c-subunit ring, and the entrance and exit lines for the carousel are the two different proton channels in the a subunit. As illustrated in Figure 11.32, once an H^+ enters the a subunit from the P side, it must ride the c-subunit carousel once around until it is able to exit on the other side.

The two-channel model uses the electrochemical proton gradient as the driving force for rotation, which makes sense as there is always a long line of $H^+{}_P$ waiting to enter the c-subunit carousel and "take a seat" on the charged Asp59 residue. However, the model also presents a stoichiometry problem because the intuitive one-to-one relationship between an H^+ and an individual c subunit means that it should require 10 H^+ to turn the c-subunit ring a full 360°. Therefore, a strict interpretation of Boyer's binding change mechanism model would mean that a nonstoichiometric number of 3.3 H^+ crosses the membrane for each 120° rotation of the γ subunit.

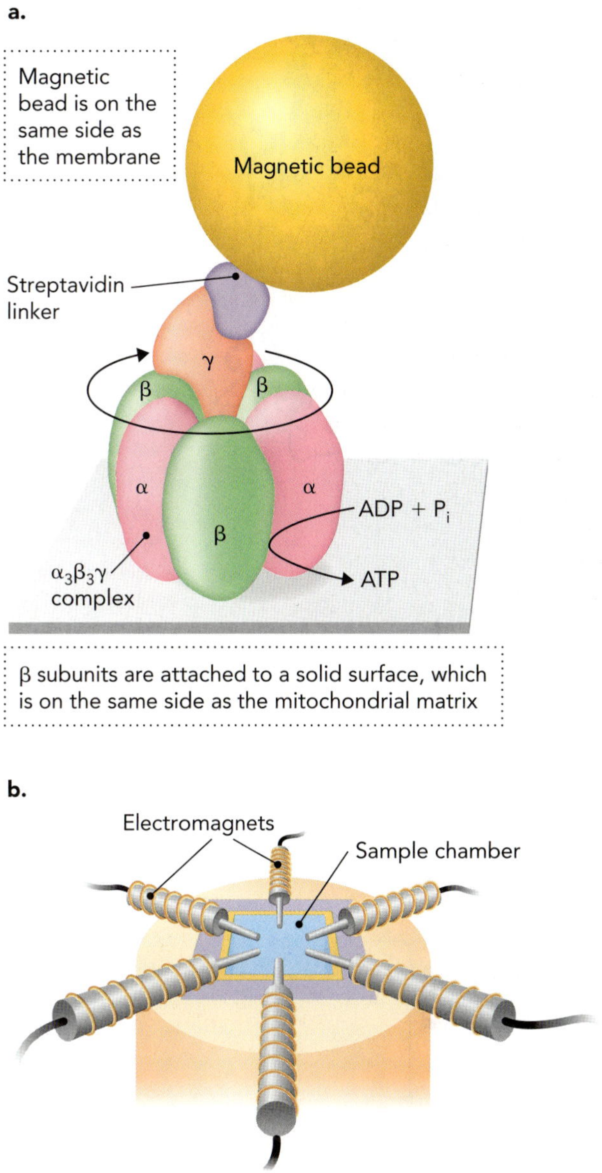

Figure 11.31 The F_1 component of the ATP synthase complex can be used as a nanomotor to drive ATP synthesis in the absence of an electrochemical proton gradient. **a.** In place of the actin filament shown in Figure 11.30, a magnetic bead was attached to the γ subunit via streptavidin. **b.** Six electromagnets in a circular arrangement around the experimental chamber were used to force rotation of the magnetic bead to drive ATP synthesis in the absence of a proton gradient.

There is likely to be structural flexibility in the c-subunit ring and the γ-subunit rotor, such that the degree of rotation of the c ring is occasionally different from the degree of rotation of the γ-subunit rotor. For example, most of the time, 3 H^+ entering and exiting the c ring lead to a 120° rotation of the c ring and γ-subunit rotor. However, about one-third of the time, 4 H^+ need to cross the membrane through the c ring in order to complete a 120° rotation of the γ-subunit rotor, resulting in 10 H^+ crossing the membrane for each full 360° rotation. A similar gear ratio difference has also been found in some bacterial flagellar motors, where the number of rotor subunits does not coincide stoichiometrically with the number of transported protons required for each 360° rotation. This explanation makes sense when considering the energetics of proton translocation. The membrane potential itself can change with the metabolic state of an organism, with the consequence that the energy available per proton varies proportionally. When the membrane potential is low, more protons are needed

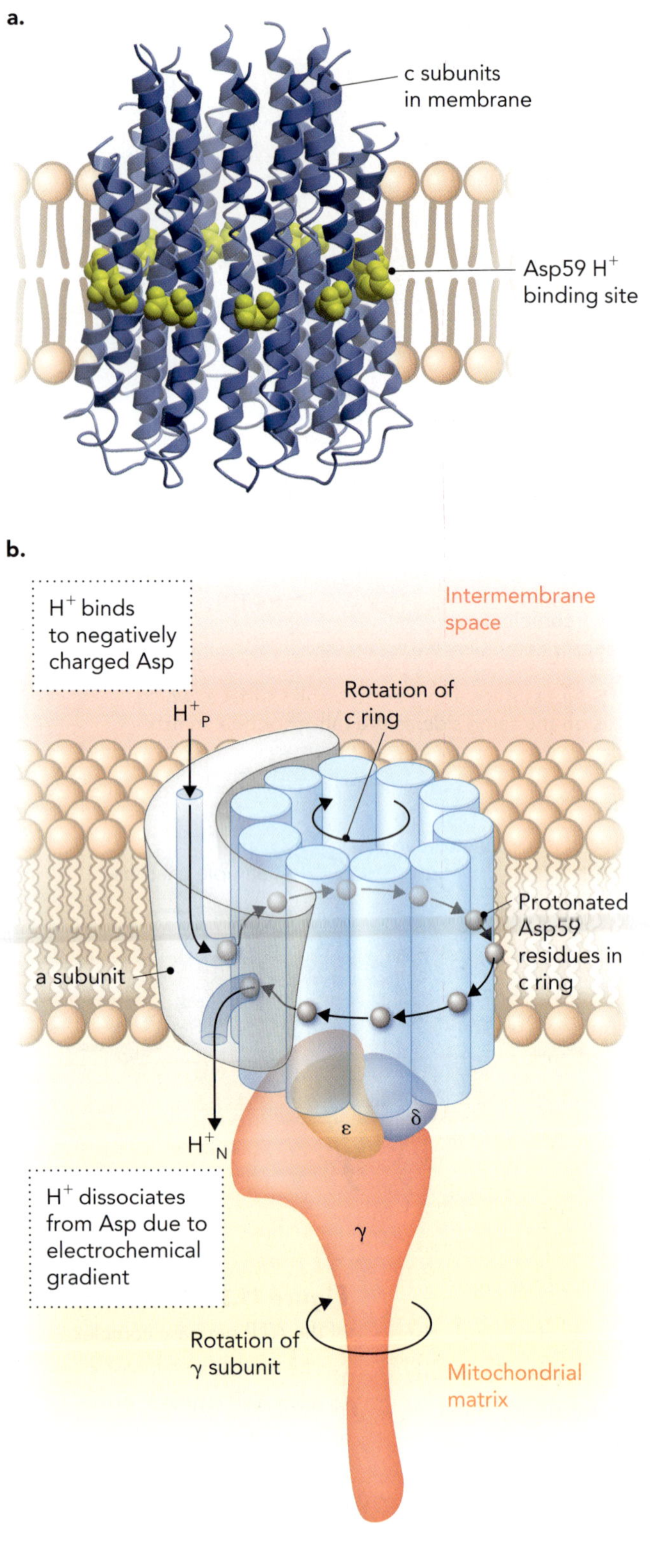

Figure 11.32 The two-channel model proposes that protons drive rotation of the c ring in response to the electrochemical proton gradient, through alternate protonation and deprotonation of a charged Asp residue in each c subunit. **a.** Location of Asp59 in one of the two α helices present in each c subunit in the yeast mitochondrial F_o component. BASED ON PDB FILE 2XOK. **b.** The electrochemical proton gradient drives rotation of the c ring as protons move from the intermembrane space into the mitochondrial matrix through two half-channels in the a subunit.

to produce the energy necessary to drive the synthesis of ATP. If the enzyme could only function with a fixed stoichiometry, it would be unable to adapt to changes in the membrane potential.

concept integration 11.3

How does proton flow through the ATP synthase complex drive ATP synthesis?

Proton flow through the ATP synthase complex drives ATP synthesis by inducing conformational changes in the subunits of the complex, which affects the affinity of the enzyme for ATP or ADP and P_i. The bacterial ATP synthase complex can be broken down into three functional units that work together to harness the energy of proton-motive force for the purpose of ATP synthesis: (1) the rotor, containing the c-subunit ring and the γ subunit, which turn together in a counterclockwise direction (looking at F_1 from the matrix side) as protons enter and exit the c-subunit ring; (2) the catalytic headpiece, containing the β subunits; and (3) the stator, consisting of the a subunit embedded in the membrane and a stabilizing arm, which prevents the catalytic headpiece from turning along with the rotor. The binding change mechanism proposed by Paul Boyer posits that rotation of the γ subunit induces three different conformations in the catalytic β subunits, called loose (L), tight (T), and open (O), which control the binding of ADP + P_i in the L conformation, the formation of ATP in the T conformation, and the release of ATP in the O conformation. According to the binding change mechanism, with every ~3 H^+ that pass through the c ring of the rotor, the γ subunit rotates 120° and the β subunits cycle through the L, T, and O conformations, producing 3 ATP for every complete 360° turn of the rotor.

11.4 Transport Systems in Mitochondria

A key element of the chemiosmotic theory is that in order to establish the proton gradient, the inner mitochondrial membrane must be impermeable to ions. Therefore, biomolecules required for the electron transport system and oxidative phosphorylation must be transported, or "shuttled," back and forth across the membrane by specialized proteins. In this section, we first describe the import of ADP and P_i and export of ATP across the inner mitochondrial membrane by a pair of translocase proteins. Then, we describe two shuttle systems: the malate–aspartate shuttle in liver cells and the glycerol-3-phosphate shuttle in muscle cells. These systems operate by transferring 2 e^- from cytosolic NADH to carrier molecules, which are then oxidized by the electron transport system. Because these two shuttle systems link cytosolic NADH to different components of the electron transport system (complex I or Q), the number of protons translocated across the membrane, and therefore the ATP yields, are not the same.

Transport of ATP, ADP, and P_i across the Mitochondrial Membrane

Most ATP-consuming reactions take place in the cytoplasm; however, the cell synthesizes most ATP in the mitochondrial matrix. The impermeability of the inner mitochondrial membrane therefore creates a problem for the ATP synthase reaction. Not only does newly synthesized ATP need to be exported from the matrix to replenish cytosolic ATP pools, but also ADP and P_i need to be continually imported into the matrix to maintain high rates of ATP synthesis.

The import of ADP and P_i into the mitochondrial matrix is accomplished by two translocase proteins located in the inner mitochondrial membrane. One is the **ATP/ADP translocase**, also called the adenine nucleotide translocase, which exports one ATP for every ADP that is imported. The second translocase protein is the **phosphate translocase**, which translocates one P_i and one H^+ into the matrix by an electrically neutral import mechanism.

The ATP/ADP translocase is a classic antiporter membrane transport protein, as first described in Chapter 6 (see Figure 6.44), because for every ATP molecule it exports out of the matrix, it imports an ADP molecule. **Figure 11.33** shows that the ATP/ADP translocase is a dimer that binds ATP and ADP with varying affinities, depending on the protein conformation. This "rocking bananas" mechanism is based on the finding that inhibitors differentially block binding of cytosolic ADP and matrix ATP, indicating that the dimeric protein exists in two conformations. These conformations are called the C state (cytosolic) and the M state (mitochondrial or matrix).

The plant toxin atractyloside, which can be isolated from the Mediterranean thistle, *Atractylis gummifera*, binds to the C-state conformation of mitochondrial ATP/ADP translocase and blocks ADP binding. In contrast, the poison bongkrekic acid, produced by bacteria sometimes present in fermented coconut, binds to the M-state conformation and inhibits ATP in the matrix from binding to the translocase. It is important to note that antiport translocation of ADP and ATP leads to a charge differential of -1 in the direction of the cytosol because under physiologic

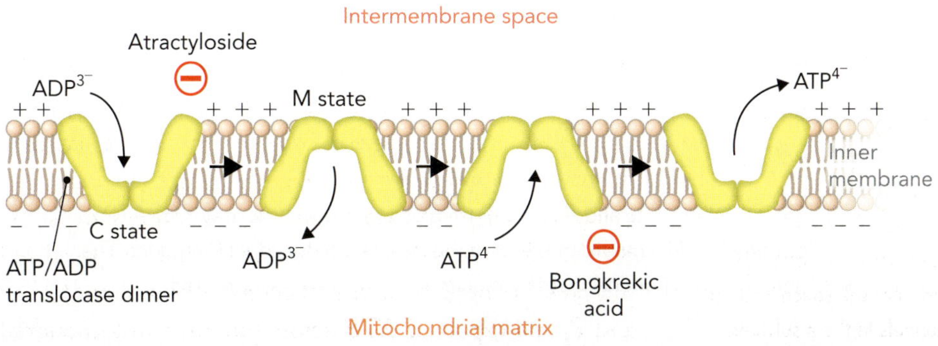

Figure 11.33 The mechanism of mitochondrial ATP/ADP translocase involves alternating protein conformations that exchange an ADP for an ATP. The poisons atractyloside and bongkrekic acid differentially block the C state and M state, respectively.

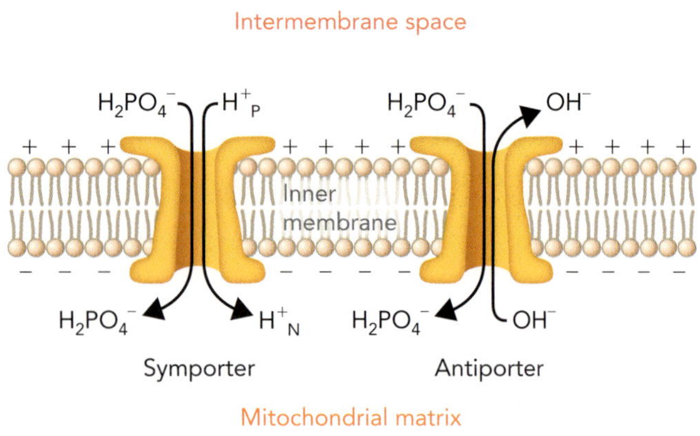

Figure 11.34 Phosphate translocase can function either as a symporter that translocates $H_2PO_4^-$ (P_i) and H^+_P across the membrane together or as an antiporter that exchanges $H_2PO_4^-$ for OH^-. In both cases, the movement of $H_2PO_4^-$ into the mitochondrial matrix is electrically neutral.

conditions, the phosphate groups of ADP contribute to a −3 charge (ADP^{3-}) in the molecule, but ATP has a −4 charge (ATP^{4-}). In this regard, the electron transport system provides energy for the ATP/ADP translocase by establishing a membrane potential that favors export of a net negative charge (ATP^{4-} for ADP^{3-}) in the direction of the more positively charged intermembrane space.

In contrast to the large conformational changes thought to occur with the mitochondrial ATP/ADP translocase, the phosphate translocase functions like a channel, which can have either symporter or antiporter functions. As shown in **Figure 11.34**, when the negatively charged P_i ion ($H_2PO_4^-$) accompanies the positively charged H^+ across the inner mitochondrial membrane in response to the proton gradient, the translocase acts as a symporter because both molecules are translocated in the same direction. This is an electrically neutral translocation because the two charges cancel each other out. The phosphate translocase can also function as an antiporter by exchanging a hydroxyl ion (OH^-), present at high levels in the matrix, for a cytosolic $H_2PO_4^-$ ion. However, for the purposes of tracking proton movement across the membrane, we will only consider the symporter role of the phosphate translocase and count $1\ H^+_P \rightarrow 1\ H^+_N$ for every P_i that is imported into the matrix.

Figure 11.35 summarizes the relationship between proton-motive force and the ATP/ADP and phosphate translocases in providing substrates for the ATP synthase complex and exporting ATP to the cytosol. Considering the rotary movement in the c ring of the ATP synthase complex (3 H^+ per ATP synthesized), you can see why $4\ H^+_P \rightarrow 4\ H^+_N$ are required for every ATP synthesized in the matrix.

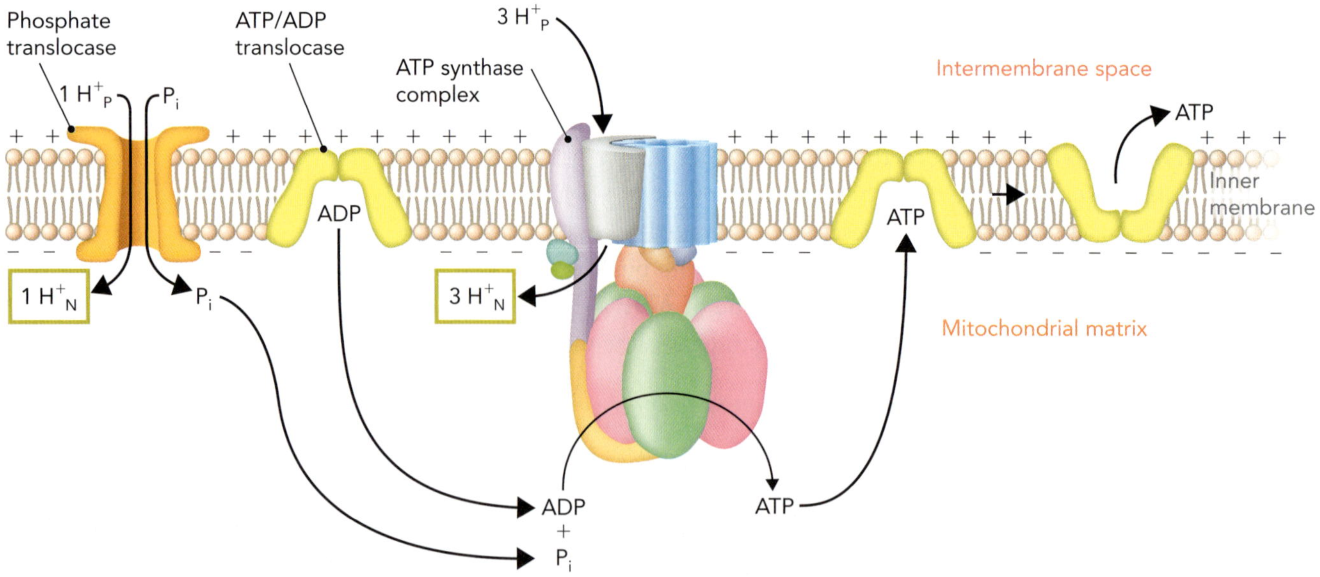

Figure 11.35 The phosphate and ATP/ADP translocases provide substrates for ATP synthesis in the mitochondrial matrix and remove the newly synthesized ATP product. The flow of 4 H^+ into the mitochondrial matrix is required for every ATP synthesized.

Cytosolic NADH Transfers Electrons to the Matrix via Shuttle Systems

Numerous dehydrogenase reactions in the cytosol generate NADH, including the reaction catalyzed by the glycolytic enzyme glyceraldehyde-3-phosphate dehydrogenase. However, cytosolic NADH cannot cross the inner mitochondrial membrane. Instead, an electron pair ($2\ e^-$) from NADH is transferred from the cytosol to the matrix, using two different shuttle systems.

The **malate–aspartate shuttle**, which functions as a reversible pathway, is the primary shuttle in liver cells (**Figure 11.36**). The key enzymes in this shuttle pathway are cytosolic malate dehydrogenase, which reduces oxaloacetate to malate, and mitochondrial malate dehydrogenase, the citrate cycle enzyme that oxidizes malate to form oxaloacetate. The inner mitochondrial membrane contains an antiporter protein that exchanges a malate molecule for α-ketoglutarate, which allows the electron pair to be transferred to a mitochondrial NAD^+ molecule to form NADH. Oxidation of this NADH by the electron transport system is then used to generate ~2.5 ATP by oxidative phosphorylation.

The malate–aspartate shuttle in liver cells has four key steps:

1. Reduction of oxaloacetate in the cytosol with $NADH + H^+$ by the enzyme cytosolic malate dehydrogenase to form malate and NAD^+.
2. Malate is transported into the mitochondrial matrix and oxidized by mitochondrial malate dehydrogenase with NAD^+ to form oxaloacetate and $NADH + H^+$.

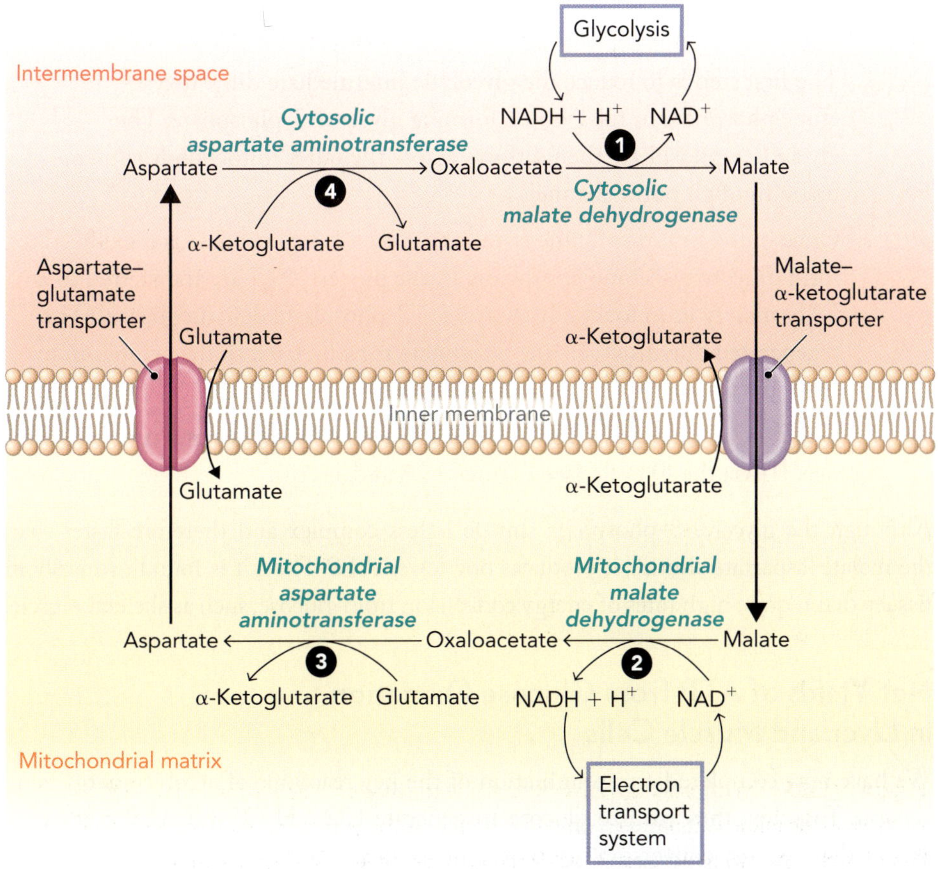

Figure 11.36 The malate–aspartate shuttle moves a pair of electrons from cytosolic NADH to mitochondrial NAD^+ using malate, which is transported across the inner mitochondrial membrane. The numbered steps 1–4 are described in the text.

3. Transamination of oxaloacetate by mitochondrial aspartate aminotransferase and glutamate to form α-ketoglutarate and aspartate, which is shuttled across the membrane.
4. Deamination of aspartate in the cytosol by cytosolic aspartate aminotransferase and α-ketoglutarate to form oxaloacetate and glutamate.

Because all of the reactions in the malate–aspartate shuttle are fully reversible, the direction of the reducing equivalents, into or out of the mitochondrial matrix, is driven by the cytosolic ratio of NADH and NAD^+. Another important function of the malate–aspartate shuttle is to ensure that under aerobic conditions, when oxidative phosphorylation is fully active, the supply of NAD^+ for the glyceraldehyde-3-phosphate dehydrogenase reaction is maintained in order to maximize flux through the glycolytic pathway. Replenishing NAD^+ for glycolysis under anaerobic conditions is the job of lactate dehydrogenase or alcohol dehydrogenase (see Figure 9.51).

Unlike liver cells, muscle cells use the **glycerol-3-phosphate shuttle** to deliver electrons from NADH to the mitochondrial matrix using the coenzyme FAD. This pathway differs from the malate–aspartate shuttle in that the electron pair extracted from cytosolic NADH enters the electron transport system through coenzyme Q, rather than at the point of complex I. This is important because the electron pair derived from cytosolic NADH under these conditions produces only ~1.5 ATP per NADH because complex I is bypassed, and therefore four fewer H^+ are translocated across the membrane (see Figure 11.8b). As shown in **Figure 11.37**, the glycerol-3-phosphate shuttle consists primarily of two isozymes of glycerol-3-phosphate dehydrogenase, which function together to transfer 2 e^- from NADH in the cytosol to an enzyme-bound FAD molecule in the inner mitochondrial membrane.

The glycerol-3-phosphate shuttle can be described in three main steps:

1. The first step is to reduce the glycolytic intermediate dihydroxyacetone phosphate in the cytosol, forming glycerol-3-phosphate. The glycerol-3-phosphate then diffuses across the outer mitochondrial membrane through porin channels.
2. Once in the intermembrane space, the glycerol-3-phosphate is reoxidized to form dihydroxyacetone phosphate. In the process, 2 e^- are transferred to an FAD moiety in mitochondrial glycerol-3-phosphate dehydrogenase. (The regenerated dihydroxyacetone phosphate recycles back to the cytoplasm by other porin channels.)
3. The 2 e^- are then passed to Q, which transfers them one at a time to complex III via the Q cycle (see Figures 11.8 and 11.18).

Although the glycerol-3-phosphate shuttle is less complex and therefore faster than the malate–aspartate shuttle, it produces one fewer ATP. Thus, it is found primarily in tissues that require high rates of energy conversion from glucose, such as skeletal muscle.

Net Yields of ATP from Glucose Oxidation in Liver and Muscle Cells

We have now completed our examination of the key reactions of aerobic energy conversion. Together, they oxidize glucose to generate CO_2, H_2O, and ATP. Let's now review these energy conversion reactions and see how 1 glucose molecule can be used

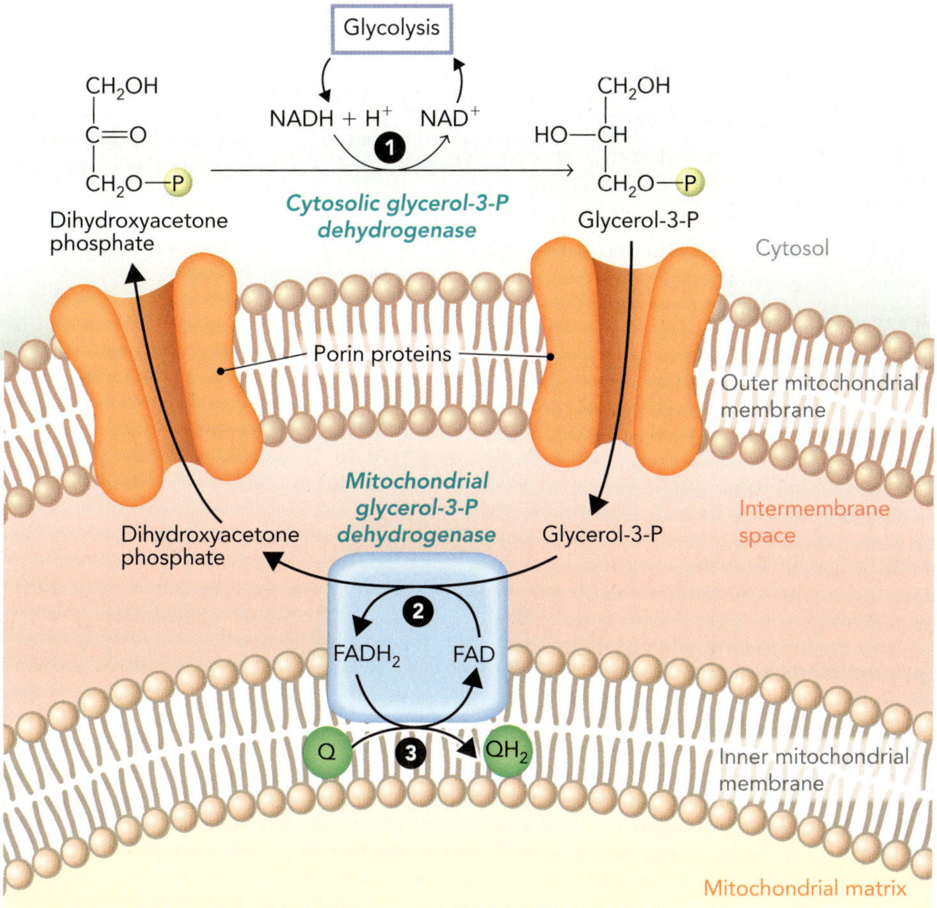

Figure 11.37 The glycerol-3-phosphate shuttle is active in muscle, where rapid conversion of redox energy from cytosolic NADH to ATP synthesis by oxidative phosphorylation is desirable.

to generate 32 ATP in liver cells via the malate–aspartate shuttle or 30 ATP in muscle cells via the glycerol-3-phosphate shuttle. As shown in **Table 11.3**, glycolysis produces 2 net ATP and 2 NADH, with another 2 NADH generated by the pyruvate dehydrogenase reaction when pyruvate is converted to acetyl-CoA inside the mitochondrial matrix.

As listed in Table 11.3, from 2 acetyl-CoA molecules the pyruvate dehydrogenase reaction produces 2 NADH and the citrate cycle generates 6 NADH, 2 $FADH_2$, and 2 ATP (GTP). This brings the total of mitochondrial NADH to 8, with another 2 NADH from the cytosol, for a grand total of 10 NADH. Considering that 3 H^+_P are required to synthesize 1 ATP by the ATP synthase complex (see Figure 11.29), and 1 H^+_P is needed to transport each negatively charged P_i molecule into the matrix (see Figure 11.34), we can now see where the ATP currency exchange ratios of ~2.5 ATP per NADH and ~1.5 ATP per $FADH_2$ come from. Specifically, because the oxidation of NADH translocates 10 H^+ out, and 4 H^+ flow in to synthesize each ATP, one NADH oxidation results in the synthesis of 2.5 ATP (10 H^+_{out}/4 H^+_{in}). Because only 6 H^+ are translocated out in the oxidation of $FADH_2$ and 4 H^+ are still needed to flow in to synthesize each ATP, one $FADH_2$ oxidation results in the synthesis of 1.5 ATP (6 H^+_{out}/4 H^+_{in}).

Now we can use these ATP currency exchange ratios to calculate ATP yields from oxidizing the 10 NADH generated by glucose oxidation. The result is 25 ATP in liver cells, but only 23 ATP in muscle cells. This 2 ATP difference is due to the

Table 11.3 SUMMARY OF ATP PRODUCTION FROM GLUCOSE OXIDATION IN LIVER AND MUSCLE CELLS, WHICH DIFFER IN THE USE OF NADH SHUTTLE SYSTEMS

	ATP yield per glucose	
Pathway	Malate–aspartate shuttle (liver)	Glycerol-3-phosphate shuttle (muscle)
Glycolysis (cytosol)		
Glucose phosphorylation (hexokinase or glucokinase)	−1	−1
Fructose-6-phosphate phosphorylation (phosphofructokinase-1)	−1	−1
Two substrate-level phosphorylations (phosphoglycerate kinase)	+2	+2
Two substrate-level phosphorylations (pyruvate kinase)	+2	+2
Reduction of 2 NAD^+ to form 2 NADH (glyceraldehyde-3-phosphate dehydrogenase)		
Pyruvate dehydrogenase reaction and citrate cycle (mitochondria)		
2 Pyruvate → acetyl-CoA (mitochondria)		
Reduction of 2 NAD^+ to form 2 NADH (pyruvate dehydrogenase)		
Two substrate-level phosphorylations (succinyl-CoA synthetase)	+2	+2
Reduction of 2 NAD^+ to form 2 NADH (isocitrate dehydrogenase)		
Reduction of 2 NAD^+ to form 2 NADH (α-ketoglutarate dehydrogenase)		
Reduction of 2 FAD to form 2 $FADH_2$ (succinate dehydrogenase)		
Reduction of 2 NAD^+ to form 2 NADH (malate dehydrogenase)		
Oxidative phosphorylation (mitochondria)		
Liver cells: oxidation of 2 NADH (malate–aspartate shuttle)	+5	
Muscle cells: oxidation of 2 NADH (glycerol-3-phosphate shuttle)		+3
Oxidation of 8 NADH produced in mitochondria	+20	+20
Oxidation of 2 $FADH_2$ produced in mitochondria	+3	+3
Net yield	**32**	**30**

Note: NADH highlighted in red are derived from the glycolytic pathway in the cytosol, whereas NADH and $FADH_2$ highlighted in blue are derived from the pyruvate dehydrogenase reaction and citrate cycle in the mitochondrial matrix.

two cytosolic NADH in muscle cells that donate their electrons to the FAD moiety of glycerol-3-phosphate dehydrogenase. We can now add in the 3 ATP derived from oxidation of the 2 $FADH_2$ produced by the succinate dehydrogenase reaction in the citrate cycle and the 4 ATP generated by substrate-level phosphorylation in glycolysis (2 ATP) and the citrate cycle (2 GTP [2 ATP equivalents]).

Because it is awkward to think about chemical reactions using noninteger values, we can simply double the amount of NADH or $FADH_2$ that is oxidized by the electron transport system and write the net equations for these reactions:

$$\mathbf{2\,NADH} + 22\,H^+{}_N + O_2 \rightarrow 2\,NAD^+ + 20\,H^+{}_P + 2\,H_2O$$

$$20\,H^+{}_P + 5\,ADP + 5\,P_i \rightarrow 20\,H^+{}_N + \mathbf{5\,ATP}$$

$$\mathbf{2\,NADH} + 2\,H^+ + 5\,ADP + 5\,P_i + O_2 \rightarrow 2\,NAD^+ + \mathbf{5\,ATP} + 2\,H_2O$$

$$\mathbf{2\,FADH_2} + 12\,H^+{}_N + O_2 \rightarrow 2\,FAD + 12\,H^+{}_P + 2\,H_2O$$

$$12\,H^+{}_P + 3\,ADP + 3\,P_i \rightarrow 12\,H^+{}_N + \mathbf{3\,ATP}$$

$$\mathbf{2\,FADH_2} + 2\,H^+ + 3\,ADP + 3\,P_i + O_2 \rightarrow 2\,FAD + \mathbf{3\,ATP} + 2\,H_2O$$

Now that we know how ATP is synthesized in the mitochondrial matrix by ATP synthase, let's look at how the process of oxidative phosphorylation is regulated.

concept integration 11.4

Why does glucose oxidation in muscle cells produce two fewer ATP than in liver cells?

Complete glucose oxidation in muscle cells produces two fewer ATP than in liver cells because of the difference in NADH shuttle systems used in these two tissue types. Liver cells use the malate–aspartate shuttle, which results in the net transfer of the electron pair from cytosolic NADH to mitochondrial NADH through the combined activity of cytosolic malate dehydrogenase and mitochondrial malate dehydrogenase, which interconvert malate and oxaloacetate. In contrast, muscle cells use the glycerol-3-phosphate shuttle, which transfers an electron pair from cytosolic NADH to FAD through a coupled reaction mechanism involving two isoforms of the enzyme glycerol-3-phosphate dehydrogenase. Because oxidation of the $FADH_2$ molecule in glycerol-3-phosphate dehydrogenase by coenzyme Q bypasses complex I in the electron transport system, the two cytosolic NADH molecules produced by glycolysis in muscle cells yield two fewer ATP molecules than that produced in liver cells (30 ATP instead of 32 ATP).

11.5 Regulation of Oxidative Phosphorylation

Oxidative phosphorylation is regulated by the cell's need for ATP. This need is indicated by the energy charge of the cell as a function of ATP concentration, ADP concentration, and AMP concentration, as described in Chapter 2. When ATP is being consumed at high rates to support metabolic functions, the level of ADP in the cell rises as a result of increased ATP hydrolysis. The cellular concentration of ADP is, in fact, a key control factor of oxidative phosphorylation because it determines not only the rate of ATP synthesis but also the rate of NADH oxidation and reduction of O_2 to form H_2O. As illustrated in **Figure 11.38**, the regulatory function of ADP and ATP in controlling aerobic respiration extends to the citrate cycle and glycolysis, both of which are activated by a low energy charge. In addition, the ratio of NADH to NAD^+ in the mitochondrial matrix controls multiple steps in the citrate cycle, which in turn determines the flow of electrons through the electron transport system and, ultimately, the rates of ATP synthesis.

Inhibitors of the Electron Transport System and ATP Synthesis

The role of the electrochemical proton gradient in linking substrate oxidation to ATP synthesis can be demonstrated by experiments using isolated mitochondria. The mitochondria are suspended in buffer containing O_2, but lacking ADP + P_i and an

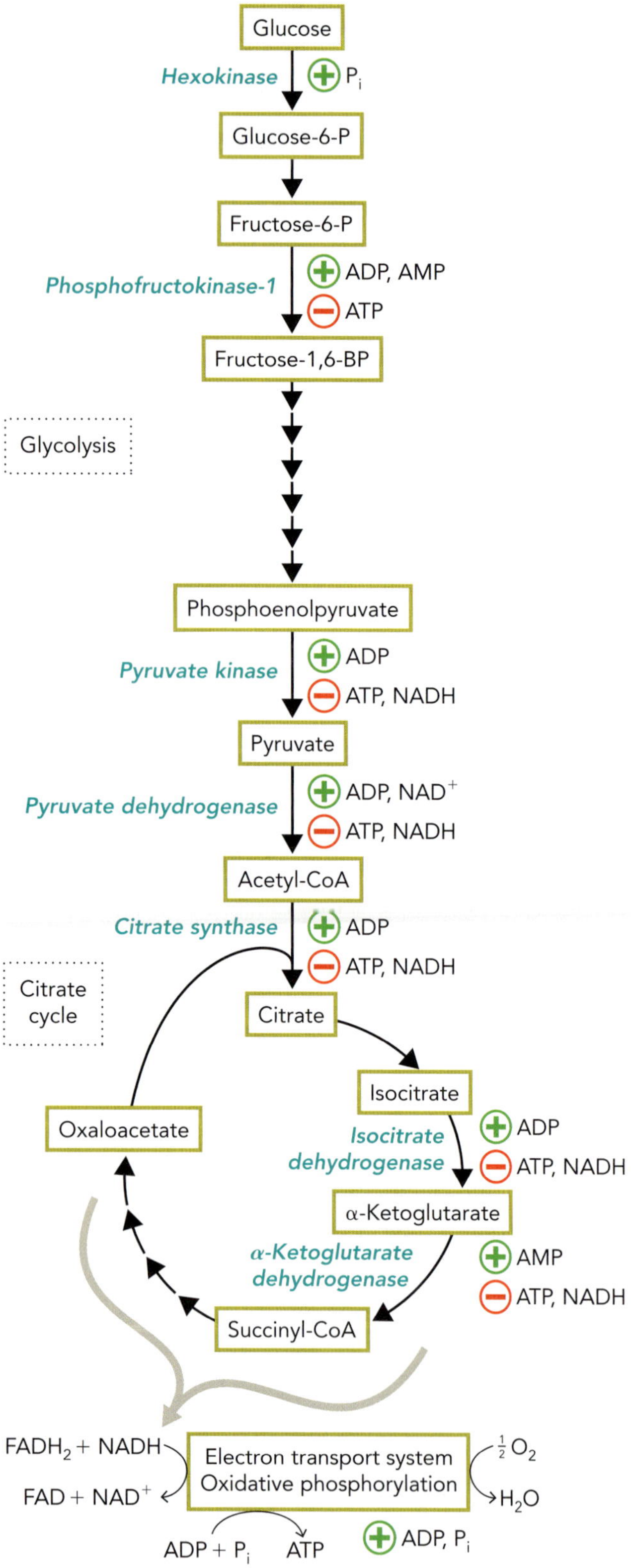

Figure 11.38 Numerous enzymes in glycolysis and the citrate cycle are regulated by intracellular concentrations of ATP, ADP, AMP, P_i, NAD^+, and NADH. Note that ADP + P_i stimulate both the rate of ATP synthesis and electron flow through the electron transport system, as these two processes are linked by the electrochemical proton gradient.

oxidizable substrate such as succinate, which donates a pair of electrons to FAD in complex II of the electron transport system. As shown in **Figure 11.39**, when ADP + P_i is added to a mitochondrial suspension lacking succinate, O_2 reduction (as measured by O_2 consumption) and ATP synthesis increase only slightly over time. However, when succinate is added to the mitochondrial suspension containing ADP + P_i, then the rates of O_2 reduction and ATP synthesis increase dramatically, until substrates become limiting. Conversely, if succinate is added first to the mitochondrial suspension, the addition of ADP + P_i is required to initiate high rates of O_2 reduction and ATP synthesis. These experiments demonstrate that substrates for both the electron transport system and ATP synthesis are required to maintain a high rate of oxidative phosphorylation in mitochondria.

What happens to the rates of O_2 reduction and ATP synthesis when we add inhibitors of the electron transport system or of the ATP synthase complex to the mitochondrial suspension? The answer is that both processes shut down, as predicted by the chemiosmotic theory. Specifically, in intact mitochondria, substrate oxidation by the electron transport system and ATP synthesis by the ATP synthase complex require an operational proton "circuit" (see Figure 11.3). For example, when cyanide is added to a mitochondrial suspension, it blocks the flow of electrons through complex IV. This results in the inhibition of the electron transport system because electron carriers in the rest of the chain cannot be reoxidized (all the electron carriers are stuck in the reduced state). This situation is illustrated in **Figure 11.40a**, where it is observed that the rate of O_2 consumption decreases rapidly after the addition of cyanide, which in turn inhibits ATP synthesis due to loss of proton-motive force.

Similarly, adding oligomycin to the mitochondrial suspension, which blocks proton flow through the ATP synthase complex, not only inhibits ATP synthesis, as would be expected, but also leads to a decrease in O_2 consumption. This happens because the energy required to move protons across the membrane (against the electrochemical proton gradient) is greater than the energy released from the combined coupled redox reactions of the electron transport system. To demonstrate that this is the case, **Figure 11.40b** shows that the rate of O_2 consumption increases in mitochondrial suspensions containing oligomycin when a chemical uncoupler such as **2,4-dinitrophenol** is added. Note that 2,4-dinitrophenol has no effect on O_2 consumption or ATP synthesis in mitochondrial suspensions containing cyanide (Figure 11.40a) because the electron

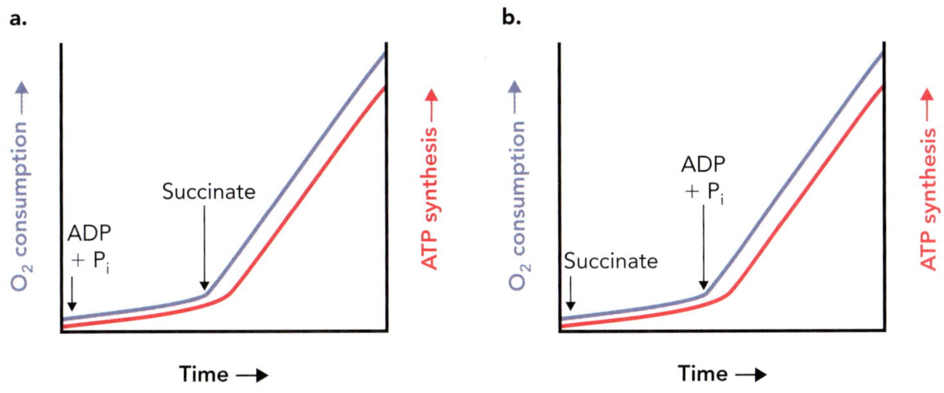

Figure 11.39 Oxidative phosphorylation depends on substrates for both the electron transport system and ATP synthesis. **a.** Addition of succinate to mitochondrial suspensions containing ADP + P_i stimulates O_2 consumption (blue line) and ATP synthesis (red line) by donating an electron pair to FAD in complex II, activating the electron transport system. **b.** Addition of ADP + P_i to mitochondrial suspensions containing succinate stimulates O_2 consumption (blue line) and ATP synthesis (red line) because the electron transport system depends on ongoing ATP synthesis (addition of ADP + P_i) in order to complete the proton circuit. In both experiments, O_2 consumption and ATP synthesis continue at a high rate until the substrates are depleted. BASED ON A. L. LEHNINGER, D. L. NELSON, AND M. M. COX (2013). *PRINCIPLES OF BIOCHEMISTRY* (6TH ED., P. 749). NEW YORK: W. H. FREEMAN.

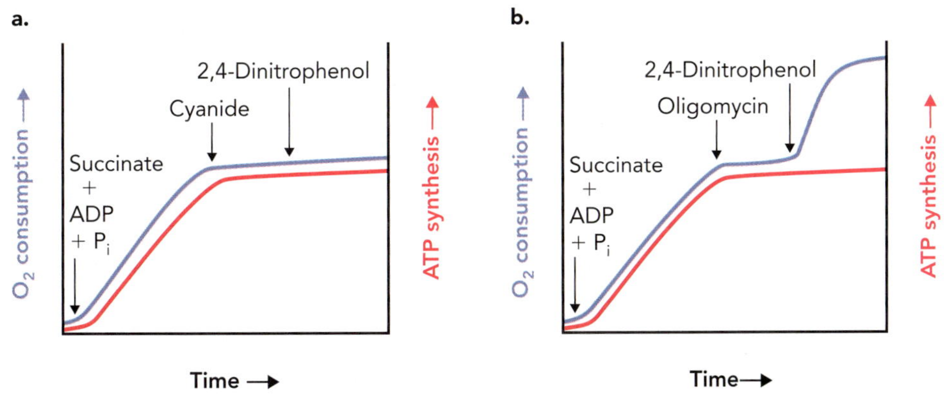

Figure 11.40 Effect of inhibitors on O_2 consumption and ATP synthesis in mitochondrial suspensions. **a.** Electron transport inhibitors such as cyanide block both O_2 consumption (blue line) and ATP synthesis (red line), as predicted by the chemiosmotic theory. Addition of an uncoupling agent such as 2,4-dinitrophenol to the sample has no effect because the electron transport system is blocked. **b.** Addition of the ATP synthase complex inhibitor oligomycin to mitochondrial suspensions shuts down the electron transport system (blocks O_2 consumption; blue line) by disrupting the proton circuit. Addition of 2,4-dinitrophenol to oligomycin-containing samples creates a "short-circuit" that permits the electron transport system to operate even in the absence of a functional ATP synthase complex. BASED ON A. L. LEHNINGER, D. L. NELSON, AND M. M. COX (2013). *PRINCIPLES OF BIOCHEMISTRY* (6TH ED., P. 749). NEW YORK: W. H. FREEMAN.

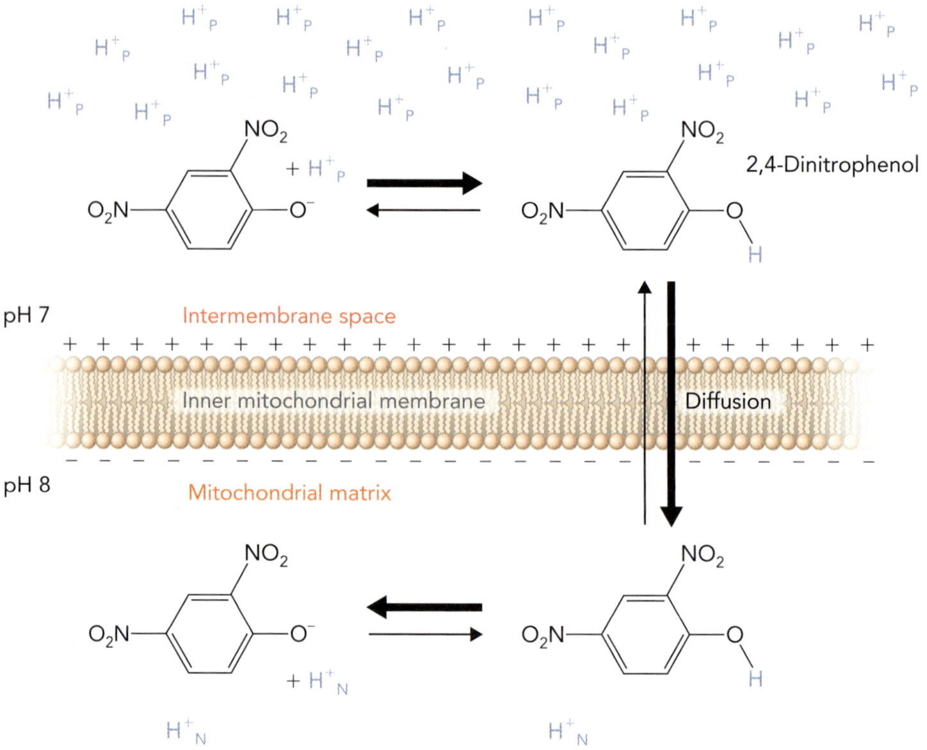

Figure 11.41 2,4-Dinitrophenol is a chemical uncoupler that carries protons across the membrane in response to an electrochemical gradient.

transport system is no longer operational, and the electrochemical proton gradient is already dissipated.

2,4-Dinitrophenol is a hydrophobic molecule that can easily diffuse across the inner mitochondrial membrane. In the process, it can carry protons one at a time from the intermitochondrial space (high H^+ concentration) to the matrix (low H^+ concentration) (**Figure 11.41**). This proton-transporting function explains why addition of 2,4-dinitrophenol to mitochondrial suspensions containing the ATP synthase complex inhibitor oligomycin stimulates O_2 consumption, as 2,4-dinitrophenol lowers the energy barrier to proton translocation. Energy normally derived from cellular respiration is wasted as heat. Because 2,4-dinitrophenol functions as a chemical uncoupler in the electron transport system, it is highly toxic, due to decreased efficiency of mitochondrial respiration and ATP production.

When liver and muscle cells are exposed to 2,4-dinitrophenol, the body attempts to offset the drop in energy charge by activating fatty acid degradation, which increases the supply of acetyl-CoA to the citrate cycle and stimulates oxidative phosphorylation. Because of these biochemical properties, 2,4-dinitrophenol was marketed in pill form in the 1930s as a "fat burner" and potent diet aid. However, it was soon realized that 2,4-dinitrophenol accumulates in cell membranes and is poisonous because individuals who ingest it have a difficult time maintaining energy homeostasis, resulting in liver damage, renal failure, extreme hyperthermia, and even death. By 1938, 2,4-dinitrophenol was banned as a human consumer product in the United States, and it is still illegal as a weight-loss product in many countries. However, despite the well-known toxicity of 2,4-dinitrophenol to humans and its chemical labeling as a poison, it has recently become widely available on the Internet through foreign chemical suppliers. Several deaths have recently been attributed to 2,4-dinitrophenol poisoning in people who ingested it in pill or powder form after purchasing it online as a "miracle" diet aid.

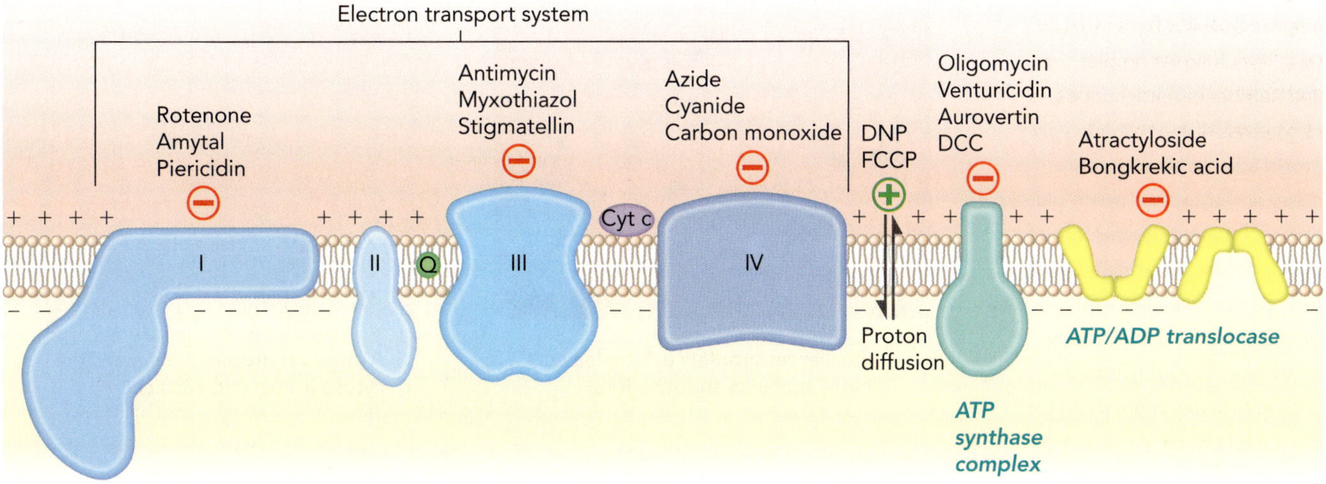

Figure 11.42 Summary of inhibitors that disrupt oxidative phosphorylation by targeting various components of the mitochondrial inner membrane. The chemical uncouplers 2,4-dinitrophenol (DNP) and carbonyl cyanide 4-(trifluoromethoxy) phenylhydrazone (FCCP) activate proton diffusion down the electrochemical proton gradient, which decreases the efficiency of oxidative phosphorylation. DCC = dicyclohexylcarbodiimide.

Figure 11.42 summarizes a variety of synthetic and naturally occurring inhibitors that have been shown to disrupt oxidative phosphorylation either by directly blocking the activity of the ATP synthase complex (oligomycin, aurovertin, venturicidin, dicyclohexylcarbodiimide) or by inhibiting the electron transport system (rotenone, amytal, piericidin, antimycin, myxothiazol, stigmatellin, cyanide, carbon monoxide, azide). Two other types of inhibitors that indirectly shut down oxidative phosphorylation are the chemical uncouplers [2,4-dinitrophenol, carbonyl cyanide 4-(trifluoromethoxy) phenylhydrazone] and inhibitors of the ATP/ADP translocase (atractyloside, bongkrekic acid), which shut off the supply of ADP required for ongoing ATP synthesis.

Uncoupling Proteins Mediate Biochemical Thermogenesis

Not all inhibitors of oxidative phosphorylation are detrimental. In fact, the **uncoupling protein 1 (UCP1)**, also called **thermogenin**, is necessary to control thermogenesis in animals. Cell-specific expression of the UCP1 protein leads to heat production under aerobic conditions by allowing protons to cross the inner mitochondrial membrane and thereby uncouple electron transport from ATP synthesis just as 2,4-dinitrophenol does (**Figure 11.43**).

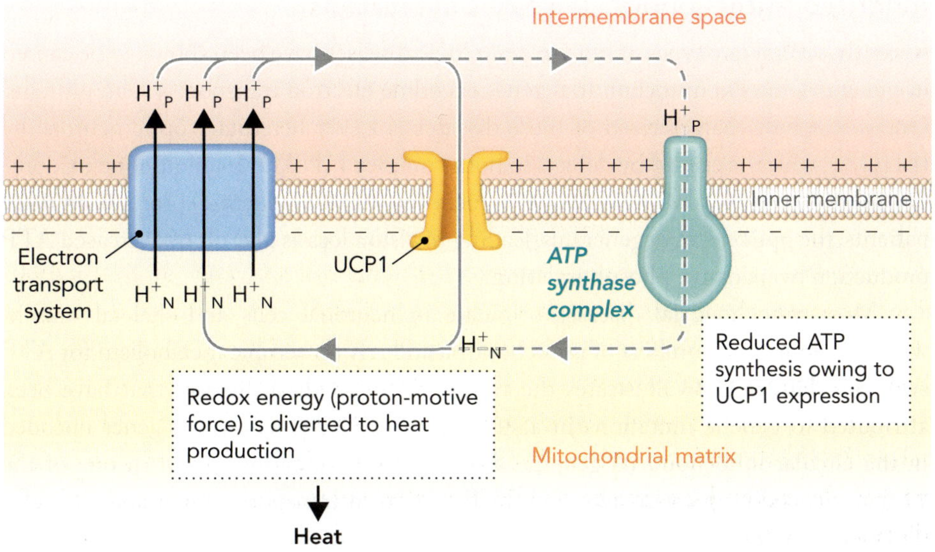

Figure 11.43 Increased expression of uncoupling protein 1 (UCP1) in brown adipose tissue short-circuits the electrochemical proton gradient and diverts redox energy to heat production.

Brown adipose tissue

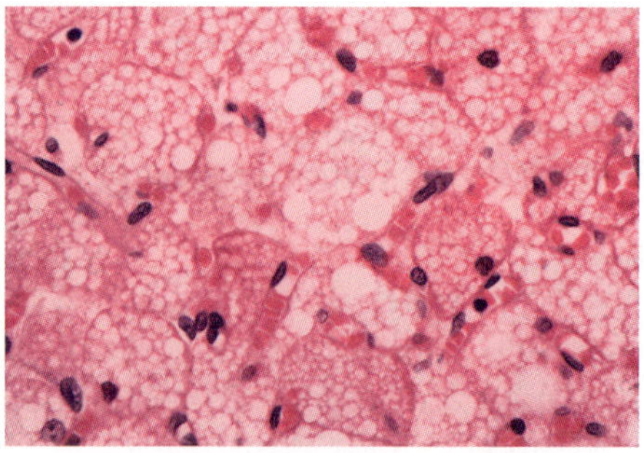

Smaller fat droplets (white) leave more space for mitochondria

White adipose tissue

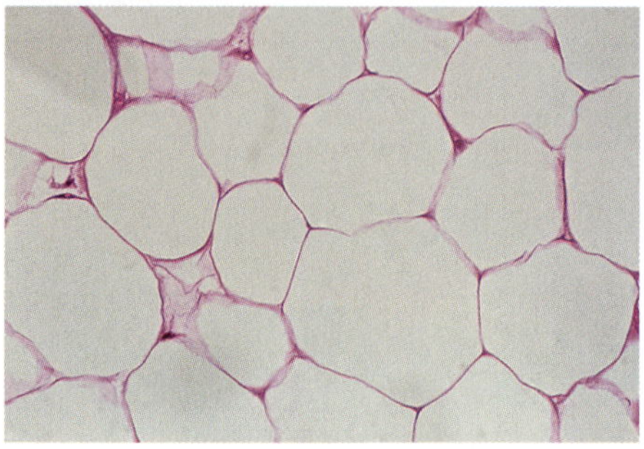

A large fat droplet (white) fills the cytosolic space in each cell

Figure 11.44 Brown and white adipose tissues differ in several ways, including fat droplet size and number of mitochondria. High levels of mitochondria give brown adipose tissue its dark color, similar to slow-twitch muscle fibers. Note that the majority of cytosolic space in white adipose tissue is filled with large fat droplets, whereas cells in brown adipose tissue have numerous smaller droplets, leaving more room for the mitochondria. BROWN ADIPOSE TISSUE: BIOPHOTO ASSOCIATES/SCIENCE SOURCE/GETTY IMAGES; WHITE ADIPOSE TISSUE: ED RECHKE/GETTY IMAGES.

The UCP1 protein is expressed at high levels in special fat cells called brown adipose tissue; these cells contain large numbers of mitochondria to increase the output of heat by the electron transport system. The brown color in these special fat cells is due to cytochromes in mitochondrial proteins, which is the same reason why slow-twitch muscles are observed as the dark meat of a turkey (see Figure 11.4c). **Figure 11.44** shows the difference between brown and white adipose tissues, which are characterized by the size of the lipid droplets inside the cell and the number of mitochondria.

Mammals, including humans, use metabolic heat produced by UCP1 uncoupling in brown adipose tissue to stay warm during the first few months after birth. As much as 5% of the total body fat of human infants at birth consists of brown adipose tissue, which gradually declines during childhood. Many types of hibernating animals—such as mice, squirrels, and bears—also rely on UCP1 protein in brown adipose tissue for thermogenesis (**Figure 11.45**). During the final weeks of hibernation, norepinephrine (noradrenaline) levels increase and activate UCP1 expression in brown adipose tissue at the base of the neck, which warms the blood traveling to the brain and signals the end of a long winter nap. This type of thermoregulation in brown adipose tissue is called nonshivering thermogenesis because it does not involve rapid muscle contraction to generate heat (shivering thermogenesis).

Figure 11.45 The hazel dormouse, an endangered species in the United Kingdom, relies on hibernation and brown adipose tissue to survive cold winters. Dormice eat acorns and are often found with stored acorns in their hibernation burrows, which can be a meter belowground. An acorn diet provides high-calorie oils that are used as a source of energy for thermogenesis during hibernation. ROGER TIDMAN/CORBIS.

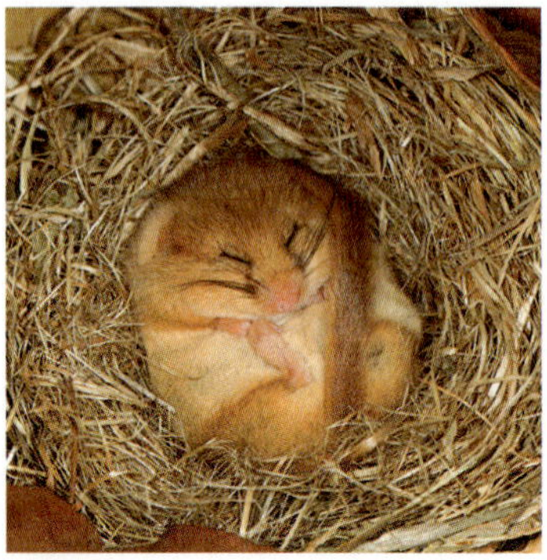

Inherited Mitochondrial Diseases in Humans

Recently, several rare types of human neurologic diseases have been shown to be caused by genetic defects in mitochondrial genes encoding electron transport system proteins. One of the best characterized of these diseases is Leber hereditary optic neuropathy (LHON), which is caused by defects in both complex I (NADH–ubiquinone oxidoreductase) and complex III (ubiquinone–cytochrome c oxidoreductase) proteins. In these patients, the optic nerve degenerates, leading to vision loss as a result of decreased ATP production by oxidative phosphorylation.

Most mitochondrial diseases originate in neuronal cells and skeletal muscle, which contain large numbers of mitochondria and rely on aerobic metabolism for ATP synthesis. **Figure 11.46** illustrates the various mitochondrial diseases that have been attributed to genetic mutations in a subset of the 37 mitochondrial genes encoded in the circular mitochondrial genome. Most of these mutations occur in one of the 12 genes encoding proteins required for the electron transport system and the ATP synthase complex.

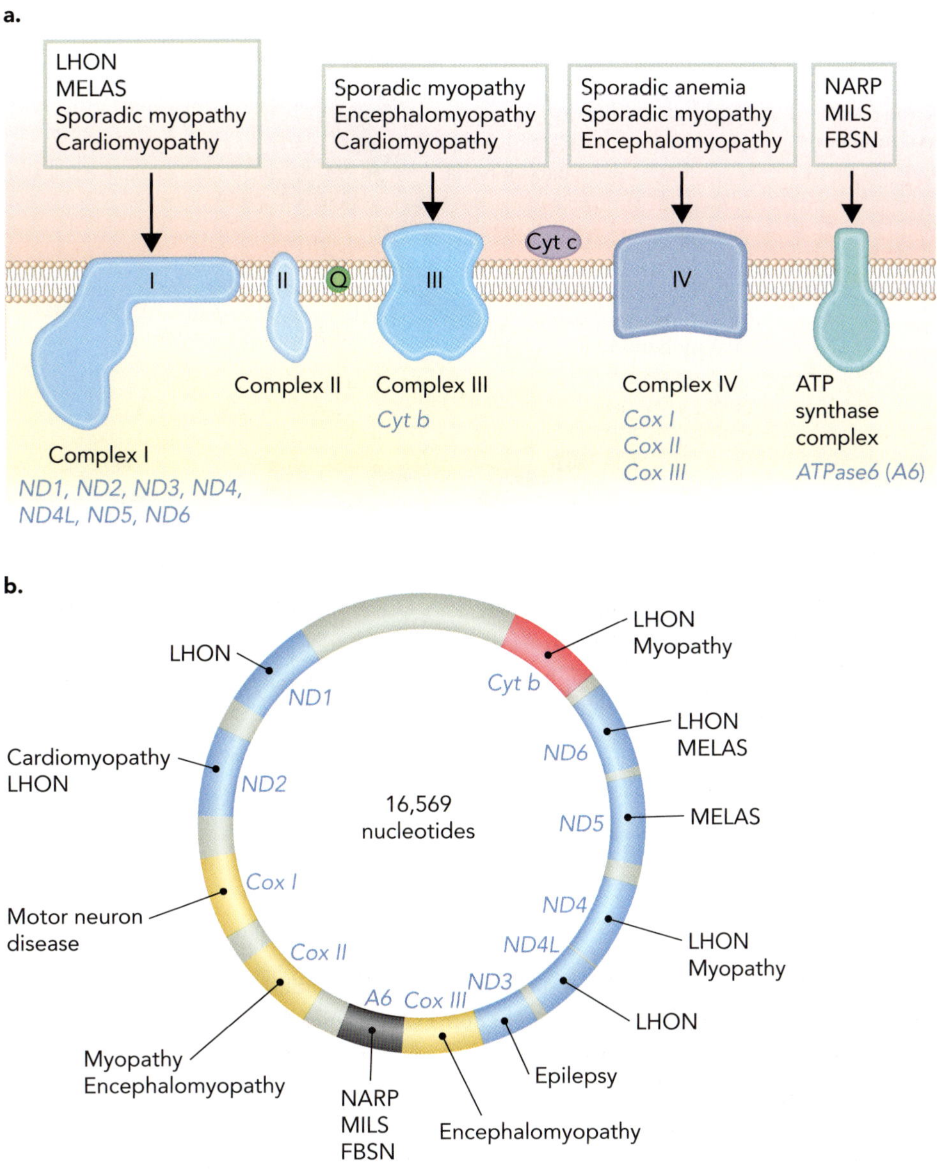

Figure 11.46 Numerous human mitochondrial diseases are associated with specific protein-coding genes in the mitochondrial genome. **a.** Diseases associated with complexes I, III, and IV of the electron transport system and the human A6 subunit of the ATP synthase complex are listed in the boxes. The affected genes in the individual complexes are listed in blue. **b.** Schematic diagram of the circular double-stranded human mitochondrial genome, highlighting the location of 12 genes named in blue that have been shown to be mutated in a variety of mitochondrial diseases. LHON = Leber hereditary optic neuropathy; MELAS = mitochondrial encephalomyopathy, lactic acidosis, and stroke-like episodes; NARP = neuropathy, ataxia, and retinitis pigmentosa; MILS = maternally inherited Leigh syndrome; FBSN = familial bilateral striatal necrosis.

Mitochondrial gene mutations are unusual in that they are passed through the mother, as mitochondria are derived from the egg cell, not the sperm cell. As shown in **Figure 11.47a**, this means that both males and females can be afflicted with the disease, but only females can pass the mitochondrial gene mutations to the next generation. Moreover, because a cell contains 10–1,000 mitochondria, the severity of the disease depends on the ratio of normal to mutant mitochondria, which could be different depending on random segregation of mitochondria in the early egg cell (**Figure 11.47b**). In addition, many mitochondrial diseases worsen with age, due in part to unequal partitioning of mutant mitochondria during somatic cell division (mitotic segregation).

Note that some mitochondrial diseases are caused by mutations in nuclear genes encoding proteins that are imported into the mitochondria by special targeting sequences in the N terminus of the polypeptide chain (see Chapter 22). These types of genetic mutations follow Mendelian inheritance patterns (derived from either the maternal or paternal genome) and affect all mitochondria in the cell.

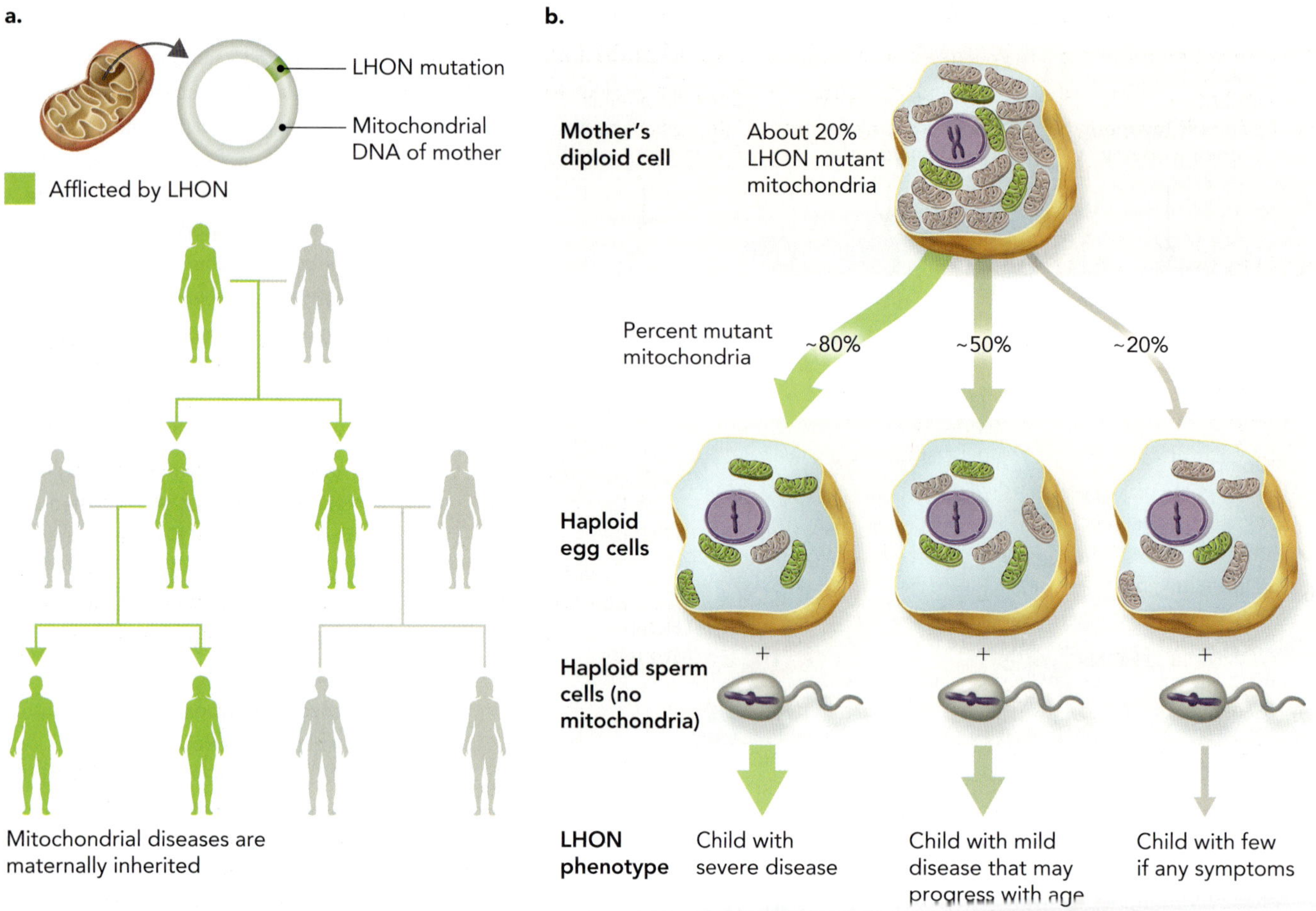

Figure 11.47 Mitochondrial diseases such as LHON are maternally inherited. **a.** Pedigree of the mitochondrial disease LHON, showing that both males and females can be afflicted, but only females can pass the LHON mutation onto their offspring. **b.** Depending on stochastic events during egg cell generation, different numbers of mutated mitochondria can be passed to haploid egg cells during meiosis. Because sperm do not donate mitochondria to the fertilized egg, offspring have disease symptoms that reflect the ratio of mutant to normal mitochondria in the egg cell.

concept integration 11.5

Why do electron transport uncouplers such as 2,4-dinitrophenol and UCP1 cause high rates of fatty acid degradation? Why is 2,4-dinitrophenol toxic to the body but UCP1 is not?

Rates of NADH oxidation by the electron transport system depend on both the availability of NADH generated by acetyl-CoA oxidation in the citrate cycle and the amount of energy required to translocate protons against the electrochemical gradient across the inner mitochondrial membrane. When ATP levels are high in the cell, feedback inhibition at multiple points in the energy conversion pathways normally decreases NADH production. However, in the presence of uncouplers such as 2,4-dinitrophenol and uncoupling proteins, ATP levels decrease because protons bypass the ATP synthase complex and reenter the mitochondrial matrix. Because this bypass mechanism causes a low energy charge in the cell, the body thinks it is starving and uses stored metabolic fuel from fat reserves to try and raise ATP levels. The futile cycle ends when either metabolic fuel reserves are exhausted or the uncoupler is inactivated. The reason 2,4-dinitrophenol is toxic is because it accumulates in mitochondrial membranes and has a slow turnover rate. In contrast, uncoupling proteins are subject to regulation at the level of RNA and protein synthesis and to protein degradation.

chapter summary

11.1 The Chemiosmotic Theory

- The chemiosmotic theory refers to the conversion of energy from redox reactions or light into potential energy in the form of an electrochemical proton gradient across a proton-impermeable organelle membrane.
- The proton circuit consists of both a chemical gradient (ΔpH) and a membrane potential ($\Delta\psi$), which together constitute proton-motive force and provide the necessary energy for ATP synthesis. Proteins in the inner mitochondrial membrane allow the passage of protons, though the membrane is otherwise largely impermeable to protons and other ions.
- The electron transport system consists of four protein complexes (complexes I–IV) embedded in the inner mitochondrial membrane and two mobile electron carriers (coenzyme Q and cytochrome c).
- Peter Mitchell correctly predicted that chemiosmosis is a process that couples redox and light energy to ATP synthesis without the need for a high-energy phosphate intermediate (A~X).
- Proof for the chemiosmotic theory came from experiments with a reconstituted artificial membrane system, showing that the protein bacteriorhodopsin is able to capture light energy and convert it into a proton gradient, which generates ATP using a membrane-embedded ATP synthase complex from bovine heart mitochondria.

11.2 The Mitochondrial Electron Transport System

- The donation of 2 e^- from NADH to the electron acceptor in complex I initiates a series of redox reactions through the rest of the electron transport system, resulting in the reduction of molecular oxygen to form water and translocation of 10 H^+ across the inner mitochondrial membrane.
- Flow of electrons through the electron transport system is facilitated by the sequential arrangement of electron carriers, primarily proteins with Fe-containing prosthetic groups that are arranged in order of increasing reduction potentials ($E°'$).
- Proton translocation across the inner mitochondrial membrane occurs by two mechanisms: (1) a redox loop in which a separation of the H^+ and e^- occurs on opposite sides of the membrane; and (2) redox-driven conformational changes, which alter pK_a values of functional groups located on the inner and outer faces of the membrane.
- Complex I (NADH–ubiquinone oxidoreductase) accepts two electrons from NADH in the form of a hydride ion ($:H^-$) and, through a series of redox reactions involving iron–sulfur centers (Fe–S), reduces Q to form QH_2. In the process, 4 H^+ are translocated from the matrix side of the membrane (N; negative side) to the intermembrane space (P; positive side).
- Complex II (succinate dehydrogenase) is a citrate cycle enzyme that also functions as a redox enzyme in the electron transport system. The $FADH_2$ moiety in complex II donates 2 e^- to Q to form QH_2 without translocating any protons across the inner mitochondrial membrane.
- Glycerol-3-phosphate dehydrogenase and ETF-Q oxidoreductase also donate 2 e^- from $FADH_2$ to Q and function as components of other energy-converting pathways.
- Complex III (ubiquinone–cytochrome c oxidoreductase) is the docking site for QH_2 and transfers electrons one at a time to the heme center of the mobile electron carrier protein cytochrome c. In the process, it translocates 4 H^+ across the inner mitochondrial membrane.
- The conversion of a two-electron carrier (QH_2) to a one-electron carrier (cytochrome c) within complex III requires a four-step reaction called the Q cycle, which uses two unique binding sites, Q_P and Q_N, to oxidize one net QH_2 and transfer 2 e^- to 2 cytochrome c carriers.
- Cytochrome c is localized to the intermembrane space and is responsible for transporting 1 e^- from complex III to complex IV using an iron-containing heme prosthetic group. Cytochrome c is a highly conserved protein in nature and has a critical role in both the electron transport system and the intrinsic apoptotic pathway.
- Complex IV (cytochrome c oxidase) accepts electrons one at a time from cytochrome c and donates them to oxygen to form water. In the process, it translocates 2 H^+ across the inner mitochondrial membrane.
- In total, when starting with 2 e^- from NADH, 10 H^+ are translocated from the matrix to the intermembrane space by the electron transport system: 4 H^+ in complex I, 4 H^+ in complex III, and 2 H^+ in complex IV (complex II does not translocate any protons).
- Using $\Delta G°'$ and ΔG values for NADH oxidation, the electron transport system is calculated to be ~70% efficient, with the other 30% lost as heat. The combined energy conversion efficiency of the electron transport system and the ATP synthase complex is ~66% based on actual and theoretical ATP yields.

11.3 Structure and Function of the ATP Synthase Complex

- The mitochondrial ATP synthase complex consists of numerous protein subunits distributed among two large structural components: one called F_1, which contains the catalytic activity, and the other called F_o, which functions as the proton channel crossing the inner mitochondrial membrane.
- The ATP synthase complex contains three functional units: (1) the *rotor* rotates as protons enter and exit the c ring; (2) the *catalytic headpiece* is responsible for ATP synthesis in the intact complex; and (3) the *stator* contains two half-channels for protons to enter and exit the c ring.
- The catalytic headpiece consists of an $\alpha_3\beta_3$ hexamer, with each β subunit containing a catalytic site for ATP synthesis. The catalytic sites have three occupancy states: T, tight with ATP bound; L, loose with ADP + P_i; and O, open with no reactants or products bound. Each state depends on the β-subunit conformation.
- The binding change mechanism model explains how conformational changes in β subunits control ATP production. The rate-limiting step is release of the newly formed ATP when the β-subunit conformation goes from T → O as a function of γ-subunit rotation in the rotor.
- Proton flow through F_o rotates the γ subunit such that with each 120° rotation, the β subunits sequentially undergo a conformational change from L → T → O → L. A 360° rotation of the γ subunit produces ~3 ATP as a result of ~9 H^+_P binding to the c ring and crossing the inner mitochondrial membrane back into the matrix.
- Structural flexibility in the c-subunit ring and the γ-subunit rotor likely accounts for the noninteger number of protons that was experimentally measured for ATP synthesis (~3.3 H^+ per ATP).
- Protons cross the membrane by entering and exiting through half-channels in the a subunit, where they come in contact with negatively charged Asp59 residues in the c subunit. Because of polarity differences between protonated and unprotonated Asp59 residues with respect to the hydrophobic membrane, the binding of 1 H^+ to Asp59 in one c subunit results in c-ring rotation of ~36°.

11.4 Transport Systems in Mitochondria

- A key element of the chemiosmotic theory is that the membrane must be impermeable to the free diffusion of ions to establish the proton gradient. Therefore, biomolecules must be shuttled in both directions across the mitochondrial and chloroplast membranes by transporter proteins.
- The ATP/ADP translocase is an antiporter that exports 1 ATP for every 1 ADP that is imported. Similarly, phosphate translocase translocates 1 P_i and 1 H^+ into the matrix by an electrically neutral import mechanism when functioning as a symporter or exchanges 1 P_i for 1 OH^- when it functions as an antiporter.
- NADH cannot cross the inner mitochondrial membrane. Instead, it must be oxidized in the cytosol to donate electrons via shuttle systems to NAD^+ inside the mitochondrial matrix. One shuttle is the malate–aspartate shuttle, which functions in liver cells; the other is the glycerol-3-phosphate shuttle, which functions in muscle cells.
- The malate–aspartate shuttle reduces oxaloacetate in the cytosol with NADH to generate NAD^+ and malate, which is shuttled into the matrix and oxidized to produce NADH and oxaloacetate. In contrast, the glycerol-3-phosphate shuttle oxidizes cytosolic NADH to yield $FADH_2$, which donates the electron pair directly to Q.
- The ATP currency exchange ratio is based on (1) the number of H^+ translocated across the inner mitochondrial membrane after NADH (10 H^+) or $FADH_2$ (6 H^+) oxidation by the electron transport system, and (2) the number of H^+ that must flow through the ATP synthase complex to generate each ATP (3 H^+ for each 120° rotation plus 1 H^+ for the P_i translocase = 4 H^+ per ATP). Therefore, ~2.5 ATP per NADH are generated (10 H^+/4 H^+ = 2.5), and ~1.5 ATP per $FADH_2$ are generated (6 H^+/4 H^+ = 1.5).
- The net ATP yield from oxidizing one molecule of glucose in liver cells is 32 molecules of ATP, considering the ATP exchange ratio of ~2.5 ATP per NADH and ~1.5 ATP per $FADH_2$ and use of the malate–aspartate shuttle. Oxidizing one molecule of glucose in muscle cells yields 30 molecules of ATP, as the 2 NADH generated by the glycolytic pathway are used to produce 2 $FADH_2$ by the glycerol-3-phosphate shuttle.

11.5 Regulation of Oxidative Phosphorylation

- When flux through catabolic pathways is high, it leads to increased levels of ATP and NADH. In contrast, when flux through anabolic pathways is high, ATP is consumed, and levels of ADP, AMP, and P_i increase. ATP and NADH are allosteric inhibitors of enzymes in energy-converting pathways, whereas ADP, AMP, and P_i are allosteric activators.
- Three classes of inhibitors decrease rates of mitochondrial ATP synthesis: (1) inhibitors of the electron transport system; (2) uncouplers that allow protons to cross the mitochondrial membrane without going through the ATP synthase complex; and (3) inhibitors of the ATP synthase complex and ATP/ADP translocase.
- Mitochondrial diseases originate in neuronal cells and skeletal muscle, which contain large numbers of mitochondria and rely on aerobic metabolism for ATP synthesis. Many of these diseases are due to mutations in mitochondrial genes, which are inherited maternally because mitochondria are derived from egg cells.

biochemical terms

(in order of appearance in text)

chemiosmotic theory (p. 526)
chemiosmosis (p. 526)
electron transport system (p. 526)
proton circuit (p. 527)
cristae (p. 529)
mitochondrial matrix (p. 529)
ATP synthase complex (p. 531)
coenzyme Q (Q) (p. 531)
ubiquinone (p. 531)
cytochrome c (p. 531)
electrochemical gradient (p. 533)
NADH–ubiquinone oxidoreductase (p. 534)
ubiquinone–cytochrome c oxidoreductase (p. 534)
cytochrome c oxidase (p. 535)
iron–sulfur (Fe–S) centers (p. 541)
proton-motive force (p. 548)
oligomycin (p. 551)
binding change mechanism (p. 554)
ATP/ADP translocase (p. 559)
phosphate translocase (p. 559)
malate–aspartate shuttle (p. 561)
glycerol-3-phosphate shuttle (p. 562)
2,4-dinitrophenol (p. 566)
uncoupling protein 1 (UCP1) (p. 569)
thermogenin (p. 569)

review questions

1. Who first proposed the chemiosmotic theory, and how does it explain the role of proton-motive force in ATP synthesis?
2. Describe the membrane reconstitution experiment, which provided convincing evidence that the chemiosmotic theory was likely to be correct.
3. Describe the two mechanisms by which protons are thought to be translocated across the inner mitochondrial membrane by electron transport system protein complexes.
4. Describe the structure and function of the six electron carriers in the mitochondrial electron transport system.
5. What are the functions of the three structural components of the mitochondrial ATP synthase complex?
6. Describe the binding change mechanism model with regard to the functional role of the γ subunit in mediating the synthesis of ~3 ATP for each 360° rotation.
7. Name two mitochondrial membrane translocase proteins, and describe their functions in mediating ongoing ATP synthesis by the ATP synthase complex.
8. Compare and contrast the functions of the malate–aspartate and glycerol-3-phosphate mitochondrial shuttle systems.
9. What is meant by the ATP currency exchange ratio? Why does the oxidation of mitochondrial $FADH_2$ generate one less ATP than oxidation of mitochondrial NADH?
10. Describe three classes of inhibitors known to decrease rates of ATP synthesis and give examples of each.

challenge problems

1. What might explain why a dramatic increase in cytosolic levels of cytochrome c functions as an initiating signal for cell death (apoptosis) in eukaryotes?
2. Why does the ATP exchange ratio in mitochondria count 4 H^+ translocated for every ATP synthesized, when only 3 H^+ need to cross the inner mitochondrial membrane to generate 1 ATP?
3. The energy required for mitochondrial ATP synthesis comes from the oxidation of NADH and $FADH_2$, which generates mitochondrial proton-motive force. The proton-motive force consists of a concentration gradient (ΔpH) and a membrane potential ($\Delta\psi$). Calculate ΔV across the inner mitochondrial membrane given that the ΔG needed to transport 1 H^+ across the same membrane at 25 °C from the matrix side (in) to the intermembrane space (out) is 21.8 kJ/mol, and the $[H^+]_{out} / [H^+]_{in}$ ratio is 25.
4. An ATP synthase complex has been characterized that contains a c ring with 12 identical c subunits. Experiments have shown that for this ATP synthase, 1 ATP is synthesized for every 4 H^+ that move through the complex into the matrix. Explain this 1:4 ratio in terms of the structure and function of the γ and β subunits of the ATP synthase complex.
5. Answer the following questions about an ATP synthase complex in which the counterclockwise rotation of the γ subunit was associated with ATP synthesis when the $\alpha_3\beta_3$ headpiece was viewed from the matrix side of the inner mitochondrial membrane.
   a. What was the direction of headpiece rotation (observed from the same relative orientation) when the ATP synthase complexes were placed in a cell-free buffer system that lacked intact membranes but contained ATP? Explain.

b. Would the levels of ATP in the system increase or decrease during a 30-minute incubation period? Explain.

6. Cyanide, oligomycin, and 2,4-dinitrophenol are all inhibitors of oxidative phosphorylation in mitochondria. Answer the following questions about these potent inhibitors.

a. Explain why adding cyanide to an active *in vitro* suspension of mitochondria blocks ATP synthesis. What happens to the rate of ATP synthesis when 2,4-dinitrophenol is added to this mitochondrial suspension after it was treated with cyanide? Explain.

b. Explain why the rate of oxygen consumption decreases in an *in vitro* suspension of mitochondria when oligomycin is added. What happens to the rate of oxygen consumption in this oligomycin-inhibited system after adding 2,4-dinitrophenol? Explain.

7. On the basis of the mitochondrial electron transport and ATP synthase inhibitors shown in Figure 11.42, fill in the following table, listing the effect of inhibitors on the rates (decrease activity, increase activity, no activity) of oxygen consumption (electron transport system activity) and ATP synthesis in a mitochondrial suspension containing all of the metabolites needed to reduce oxygen ($4\ H^+ + O_2 \rightarrow 2\ H_2O$) and synthesize ATP. Explain your reasoning.

Inhibitor(s) added	Electron transport system activity?	ATP synthesis?	Explain
Myxothiazol			
FCCP			
Venturicidin			
Venturicidin + FCCP			

8. Why is it advantageous for hibernating animals to have thermogenin in the mitochondria of special fat cells that surround arteries bringing blood to the brain?

9. Explain why different mitochondrial shuttle systems in liver and muscle cells result in the generation of 5 ATP and 3 ATP, respectively, by oxidative phosphorylation in the mitochondria for every 2 NADH that are produced in the cytosol by glycolysis.

10. When radioactive ^{3}H-NADH is added to a cell extract containing mitochondria, radioactivity quickly appears in the mitochondrial matrix. However, when radioactive ^{14}C-NADH is added to the same cell extract, no radioactivity is found in the mitochondrial matrix. Explain this observation.

11. On the basis of the mitochondrial reactions required to convert one molecule of pyruvate into three molecules of CO_2 under aerobic conditions (refer to Chapter 10 and see the figure that follows) and the ATP currency exchange ratios for NADH and $FADH_2$, fill in the blanks to calculate the total number of ATP that are generated in mitochondria for every 2 pyruvate that are metabolized to 6 CO_2.

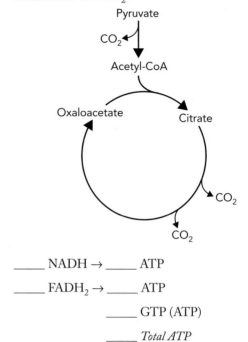

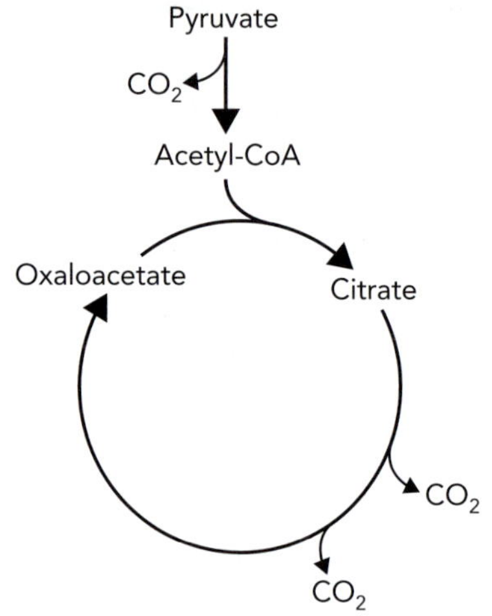

_____ NADH → _____ ATP

_____ $FADH_2$ → _____ ATP

_____ GTP (ATP)

_____ *Total ATP*

12. A pressurized hand sprayer is an airtight canister that has a nozzle to spray the liquid contents, a pump handle, and a pressure-release valve (see diagram). Using this pressurized sprayer as an analogy for a mitochondrion, match up the sprayer components on the left with the most appropriate mitochondrial component on the right to complete the analogy. Explain your reasoning.

a. Canister shell	**1.** Uncoupler (DNP or UCP1)
b. Spray nozzle	**2.** Reductants (NADH or $FADH_2$)
c. Pump handle	**3.** ATP synthase complex
d. Release valve	**4.** Inner mitochondrial membrane

suggested reading

Books and Reviews

Argyropoulos, G., and Harper, M. E. (2002). Uncoupling proteins and thermoregulation. *Journal of Applied Physiology*, *92*, 2187–2198.

Bartlett, J., Brunner, M., and Gough, K. (2010). Deliberate poisoning with dinitrophenol (DNP): an unlicensed weight loss pill. *Emergency Medicine Journal*, *27*, 159–160.

Bournat, J. C., and Brown, C. W. (2010). Mitochondrial dysfunction in obesity. *Current Opinion in Endocrinology, Diabetes, and Obesity*, *17*, 446–452.

Boyer, P. D. (1997). The ATP synthase—a splendid molecular machine. *Annual Review of Biochemistry*, *66*, 717–749.

Capaldi, R. A., and Aggeler, R. (2002). Mechanism of the F_1F_0-type ATP synthase, a biological rotary motor. *Trends in Biochemical Sciences*, *27*, 154–160.

DiMauro, S., and Schon, E. A. (2003). Mitochondrial respiratory-chain diseases. *New England Journal of Medicine*, *348*, 2656–2668.

Enerback, S. (2010). Human brown adipose tissue. *Cell Metabolism*, *11*, 248–252.

Gibson, F. (2000). The introduction of *Escherichia coli* and biochemical genetics to the study of oxidative phosphorylation. *Trends in Biochemical Sciences*, *25*, 342–344.

Hinkle, P. C. (2005). P/O ratios of mitochondrial oxidative phosphorylation. *Biochimica et Biophysica Acta*, *1706*, 1–11.

Man, P. Y., Turnbull, D. M., and Chinnery, P. F. (2002). Leber hereditary optic neuropathy. *Journal of Medical Genetics*, *39*, 162–169.

Nakamoto, R. K., Baylis, J. A., and Al-Shawi, M. K. (2008). The rotary mechanism of the ATP synthase. *Archives of Biochemistry and Biophysics*, *476*, 43–50.

Nicholls, D. G., and Ferguson, S. J. (2013). *Bioenergetics 4*. London, England: Academic Press.

Perkins, G. A., Sun, M. G., and Frey, T. G. (2009). Correlated light and electron microscopy/electron tomography of mitochondria in situ. *Methods in Enzymology*, *456*, 29–52.

Prebble, J. (2002). Peter Mitchell and the ox phos wars. *Trends in Biochemical Sciences*, *27*, 209–212.

Senior, A. E., and Weber, J. (2004). Happy motoring with ATP synthase. *Nature Structural & Molecular Biology*, *11*, 110–112.

Smeitink, J. A., Zeviani, M., Turnbull, D. M., and Jacobs, H. T. (2006). Mitochondrial medicine: a metabolic perspective on the pathology of oxidative phosphorylation disorders. *Cell Metabolism*, *3*, 9–13.

Storey, K. B. (2010). Out cold: biochemical regulation of mammalian hibernation—a mini-review. *Gerontology*, *56*, 220–230.

Primary Literature

Diez, M., Zimmermann, B., Borsch, M., Konig, M., Schweinberger, E., Steigmiller, S., Reuter, R., Felekyan, S., Kudryavtsev, V., Seidel, C. A., and Graber, P. (2004). Proton-powered subunit rotation in single membrane-bound F_oF_1-ATP synthase. *Nature Structural & Molecular Biology*, *11*, 135–141.

Efremov, R. G., Baradaran, R., and Sazanov, L. A. (2010). The architecture of respiratory complex I. *Nature*, *465*, 441–445.

Fillingame, R. H., Angevine, C. M., and Dmitriev, O. Y. (2003). Mechanics of coupling proton movements to c-ring rotation in ATP synthase. *FEBS Letters*, *555*, 29–34.

Itoh, H., Takahashi, A., Adachi, K., Noji, H., Yasuda, R., Yoshida, M., and Kinosita, K. (2004). Mechanically driven ATP synthesis by F_1-ATPase. *Nature*, *427*, 465–468.

Iwata, S., Lee, J. W., Okada, K., Lee, J. K., Iwata, M., Rasmussen, B., Link, T. A., Ramaswamy, S., and Jap, B. K. (1998). Complete structure of the 11-subunit bovine mitochondrial cytochrome bc_1 complex. *Science*, *281*, 64–71.

Miranda, E. J., McIntyre, I. M., Parker, D. R., Gary, R. D., and Logan, B. K. (2006). Two deaths attributed to the use of 2,4-dinitrophenol. *Journal of Analytical Toxicology*, *30*, 219–222.

Muramoto, K., Ohta, K., Shinzawa-Itoh, K., Kanda, K., Taniguchi, M., Nabekura, H., Yamashita, E., Tsukihara, T., and Yoshikawa, S. (2010). Bovine cytochrome c oxidase structures enable O_2 reduction with minimization of reactive oxygens and provide a proton-pumping gate. *Proceedings of the National Academy of Sciences USA*, *107*, 7740–7745.

Noji, H., Yasuda, R., Yoshida, M., and Kinosita, K., Jr. (1997). Direct observation of the rotation of F_1-ATPase. *Nature*, *386*, 299–302.

Rastogi, V. K., and Girvin, M. E. (1999). Structural changes linked to proton translocation by subunit *c* of the ATP synthase. *Nature*, *402*, 263–268.

Stock, D., Leslie, A. G., and Walker, J. E. (1999). Molecular architecture of the rotary motor in ATP synthase. *Science*, *286*, 1700–1705.

Sun, F., Huo, X., Zhai, Y., Wang, A., Xu, J., Su, D., Bartlam, M., and Rao, Z. (2005). Crystal structure of mitochondrial respiratory membrane protein complex II. *Cell*, *121*, 1043–1057.

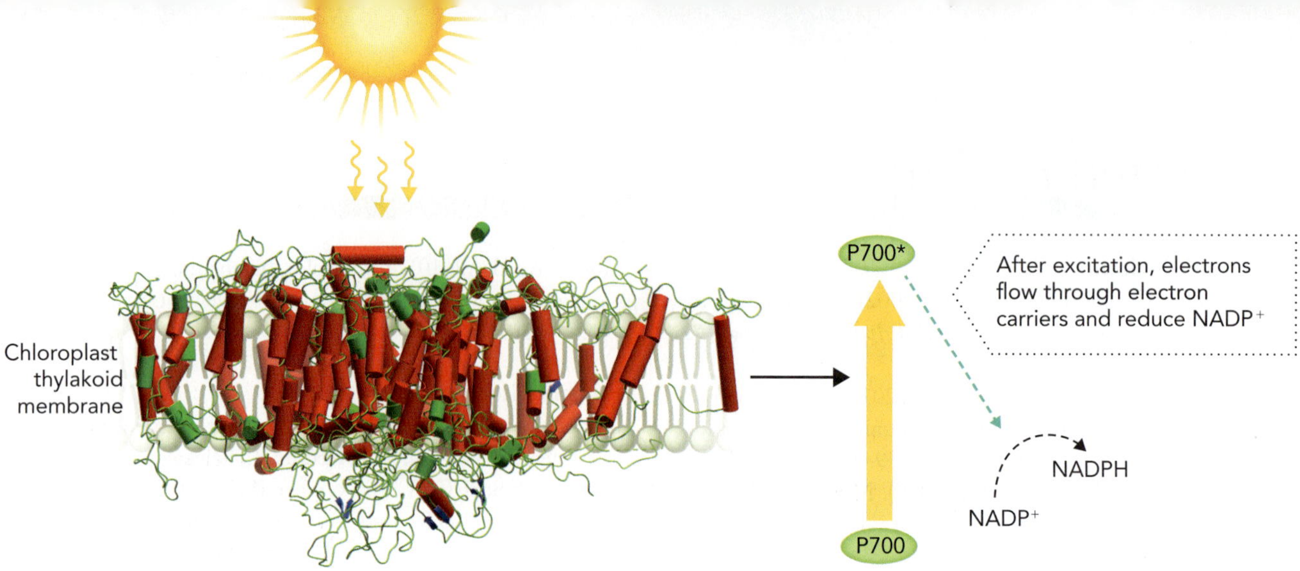

Photosystem I (PSI) reaction center

Healthy plant growth

Aerial spraying with paraquat

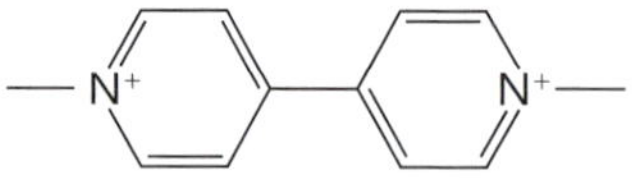

Paraquat

Paraquat is an inhibitor of redox reactions in PSI

Magazine cover from the 1980s

Paraquat is an herbicide that blocks electron transport in the photosystem I protein complex, leading to plant cell death (P700 is a chlorophyll reaction center). Paraquat was used in the 1980s to spray illegal marijuana plants, which did not always stop growers from harvesting the contaminated plants and selling them.

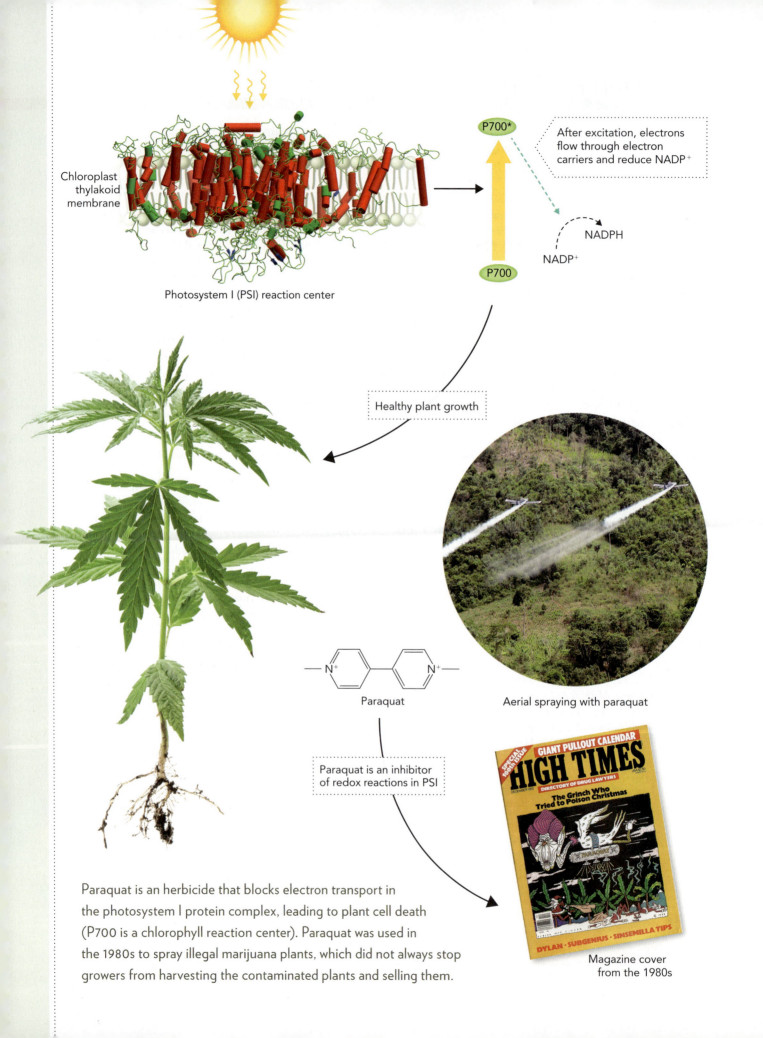

12

Photosynthesis

◀ Photosystem I (PSI) is a key component of the photosynthetic electron transport system in plants. It functions to transfer electrons from chlorophyll to an electron acceptor in a reaction called photooxidation. Subsequent redox reactions culminate in the reduction of $NADP^+$ to generate NADPH, which is required for CO_2 fixation by the Calvin cycle. Paraquat is an herbicide that prevents reduction of $NADP^+$ by accepting electrons from intermediate reductants in PSI. It also generates reactive oxygen species that are toxic to cells, thus killing the plant. Paraquat was used extensively in the 1980s as an herbicide in North and South America, primarily to destroy illegal marijuana crops by aerial spraying. In some cases, the herbicide-sprayed marijuana crops were harvested and sold in the United States, which led to a public outcry over fears that paraquat was causing human health problems as a result of people smoking paraquat-contaminated plant matter.

CHAPTER OUTLINE

We have observed how prokaryotic and eukaryotic cells use the metabolic pathways of glycolysis, the citrate cycle, and the electron transport system to extract energy from glucose. Plant cells also use glucose as a source of chemical energy by these same pathways, but the glucose is metabolized primarily at night when energy from sunlight is not available. In contrast to animal cells, eukaryotic plant cells have chloroplasts, which are organelles that convert light energy into chemical energy by the process of photosynthesis for the purpose of synthesizing carbohydrates from CO_2 and H_2O. Plants then use the carbohydrates they produce as a source of chemical energy as needed. As described in Chapter 2, plants are self-sufficient, making them autotrophs, whereas nonphotosynthetic organisms, including humans are heterotrophs because they obtain chemical energy directly or indirectly from plants.

Chloroplasts harvest light energy using protein molecules that include chlorophyll as a component; these protein molecules are embedded in chloroplast membranes and feed electrons into a series of coupled redox reactions called the photosynthetic electron transport system. In nature, sunlight is the energy input for photosynthesis. Light energy is converted to both chemical energy in the form of ATP and redox energy that is captured in the electron carrier NADPH. The energy in ATP and NADPH, obtained from photosynthetic reactions, is used by the Calvin cycle in chloroplasts to convert atmospheric CO_2 into the three-carbon molecule glyceraldehyde-3-phosphate, which is used to synthesize hexose sugars. Rubisco (ribulose-1,5-bisphosphate carboxylase/oxygenase), the key enzyme in the Calvin cycle, is arguably the most important enzyme on Earth, considering that plant-derived sugars are used directly or indirectly by nearly all nonphotosynthetic organisms. Not surprisingly, rubisco is the most abundant enzyme on our planet.

In this chapter, we first present an overview of photosynthesis and the assimilation of carbon dioxide into hexose sugars and then describe the photosynthetic electron transport system and the critical reaction of photooxidation. This is followed by a description of the chloroplast ATP synthase complex and of a specialized ATP generating pathway called cyclic photophosphorylation, which controls ATP-to-NADPH ratios in chloroplasts. We next examine enzymatic reactions in the Calvin cycle, which are dependent on a rate-limiting reaction catalyzed by the enzyme rubisco. This discussion includes a presentation of biochemical strategies that some plants use to survive in harsh conditions (the C_4 and CAM pathways), in which high temperatures limit the efficiency of carbon fixation by rubisco. Finally, we ask a question: How do germinating plant seeds get the energy they need for cell division and growth prior to the development of light-harvesting leaves? The answer is by using the glyoxylate cycle, which converts seed oils into carbohydrate precursors without loss of carbon in the form of carbon dioxide.

12.1 Plants Harvest Energy from Sunlight

In plants, the **photosynthetic electron transport system** and the **Calvin cycle** work together to convert light energy into chemical energy (ATP, NADPH, and triose phosphate), which in the process oxidizes H_2O to form O_2. The photosynthetic electron transport system is often referred to as the *light reactions* of photosynthesis, whereas the Calvin cycle has been called the *dark reactions*. However, as you will see later in this chapter, the term *dark reactions* can be misleading because the Calvin cycle is most active in the light, when ATP and NADPH levels are high. In the dark, plants depend on newly synthesized carbohydrates as a metabolic fuel for mitochondrial respiration, just like any other aerobic organism. **Figure 12.1** shows where these two processes fit into our overall metabolic map.

Overview of Photosynthesis and Carbon Fixation

The combined reactions of the photosynthetic electron transport system and the Calvin cycle result in the oxidation of H_2O, the production of O_2, and carbon fixation, which is the incorporation of atmospheric CO_2 into an organic compound. The organic molecules produced by the Calvin cycle are triose phosphates that can be used to produce hexose sugars. Because six CO_2 molecules are required to produce one 6-carbon glucose molecule, we can write the combined reaction that produces glucose as

$$6\ H_2O + 6\ CO_2 \xrightarrow{\text{Light energy}} C_6H_{12}O_6 + 6\ O_2$$

The change in standard free energy ($\Delta G°'$) for this reaction is +2,868 kJ/mol, which is outweighed by the potential energy stored in the products of photosynthetic electron transport; namely, ATP and NADPH. **Figure 12.2** illustrates the relationship between plants (autotrophs) and animals (heterotrophs) in the exchange of CO_2, H_2O, O_2, and carbohydrates.

It has been estimated that each year more than 10% of the available CO_2 in the atmosphere is converted to carbohydrate, with the majority of this fixed CO_2 being returned to the atmosphere as a result of aerobic respiration by microorganisms, plants, and animals. Carbon fixation accounts for an annual yield of 10^{11} tons of fixed carbon, representing as much as 10^{18} kJ of available energy. In addition, the photosynthetic electron transport system produces 10^{16} mol of O_2 per year as a result of H_2O oxidation.

Although photosynthetic processes remove large amounts of CO_2 from our atmosphere, the level of CO_2 is steadily increasing because of the burning of fossil fuels. Since the early 1800s, the concentration of atmospheric CO_2 has increased from 200 parts per million (ppm) to more than 400 ppm today. This excess CO_2 contributes to climate change by increasing the **greenhouse effect**, in which re-radiated heat from Earth cannot readily escape. Decreased global production of CO_2 and preservation of photosynthetic capacity in tropical rain forests are important strategies to reduce global warming and climate change due to the increased greenhouse effect.

The role of photosynthesis on Earth is not limited to the daily recycling of CO_2 and O_2 but is in fact responsible for how life evolved. The generation of O_2 over the past several billion years has provided the oxygen-rich atmosphere that supports aerobic respiration. Indeed, the emergence of O_2-producing photosynthetic organisms on Earth ~3 billion years ago represented a major turning point in evolution. This

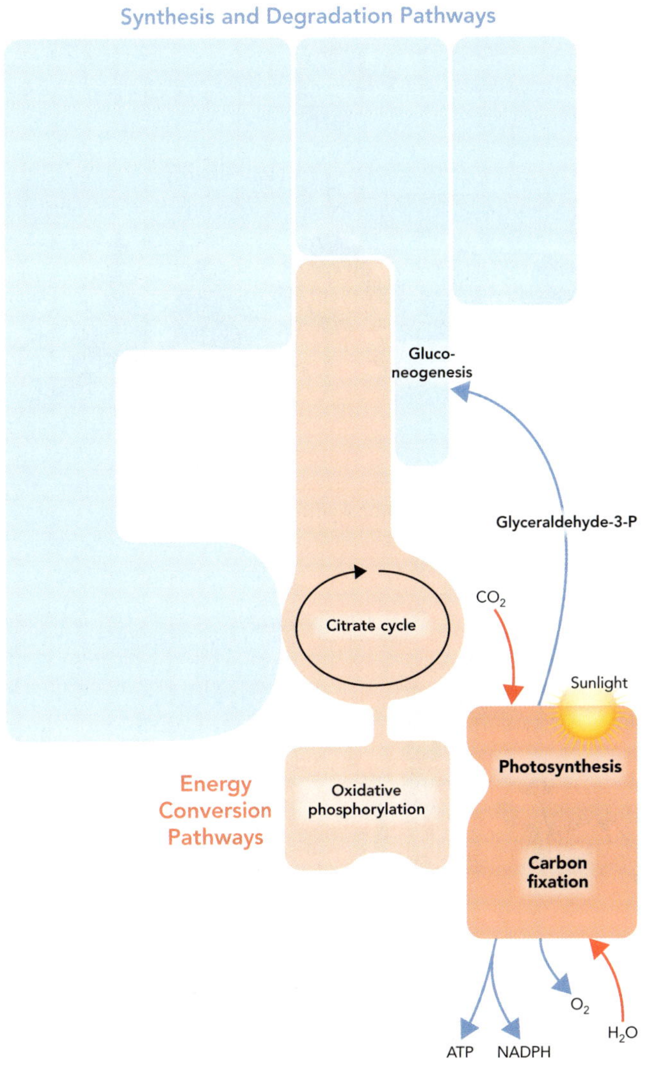

Figure 12.1 The photosynthetic electron transport system and carbon fixation (Calvin cycle) reactions together convert sunlight energy, CO_2, and H_2O into chemical energy (ATP, NADPH), molecular oxygen, and the triose phosphate glyceraldehyde-3-phosphate. These reactions are essential to life on Earth because they provide metabolic fuel in the form of carbohydrate (glyceraldehyde-3-phosphate) to nonphotosynthetic organisms, such as ourselves. Glyceraldehyde-3-phosphate feeds into the anabolic gluconeogenesis pathway (see Chapter 14), where it is used as a precursor for glucose synthesis.

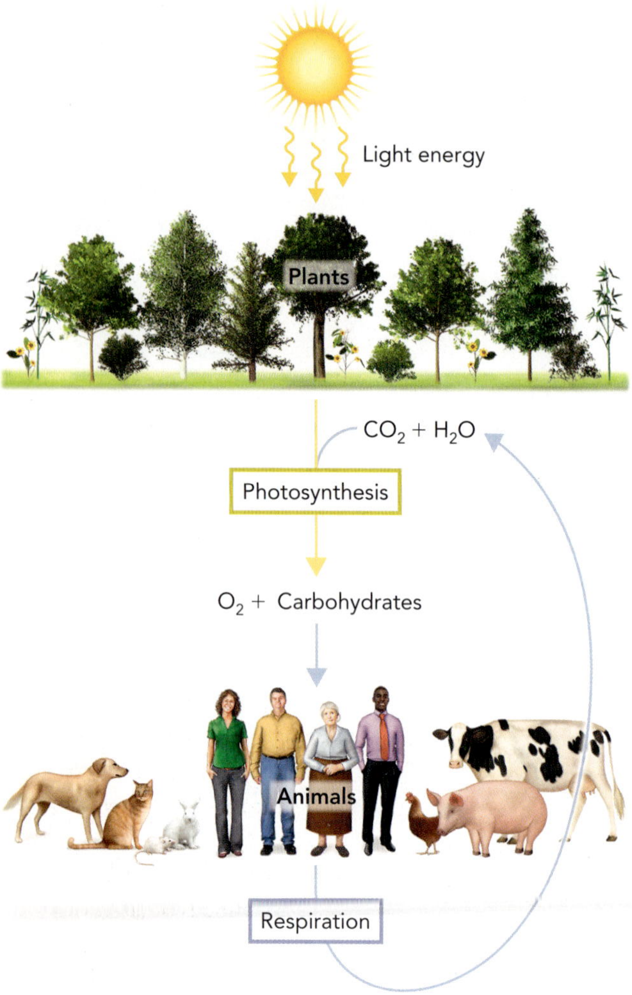

Figure 12.2 Plants obtain carbon from CO_2 by using light energy available during the day to generate oxygen from water and to synthesize carbohydrates, which are used at night as a source of chemical energy by the process of respiration. Animals use the oxygen and carbohydrates produced by photosynthetic organisms for respiration at all times; that is, animals are absolutely dependent on plants for life.

development not only put anaerobic organisms at a disadvantage (O_2 was a poisonous gas) but also provided a selective pressure on aerobic organisms to first adapt to and then exploit this changing environment. This history of life on Earth can be seen in the fossil record, which shows that the evolution of photosynthetic aquatic organisms, such as cyanobacteria and algae, preceded rapid increases in atmospheric O_2 levels (**Figure 12.3**). Land plants ultimately took advantage of this available O_2 for aerobic respiration in the dark, which allowed them to diversify and invade a variety of ecological niches.

For centuries, scientists have investigated how plants use sunlight and water to survive. A key finding was made in the 1770s by Joseph Priestly, an Englishman with training in chemistry and theology who showed that in the presence of light, plants produce oxygen. He went on to show that for a short period of time, animal respiration can be supported by a living plant in an airtight container. A modern version of the Priestly experiment was attempted in the early 1990s by a private company called Space Biospheres Ventures, which built an enclosed

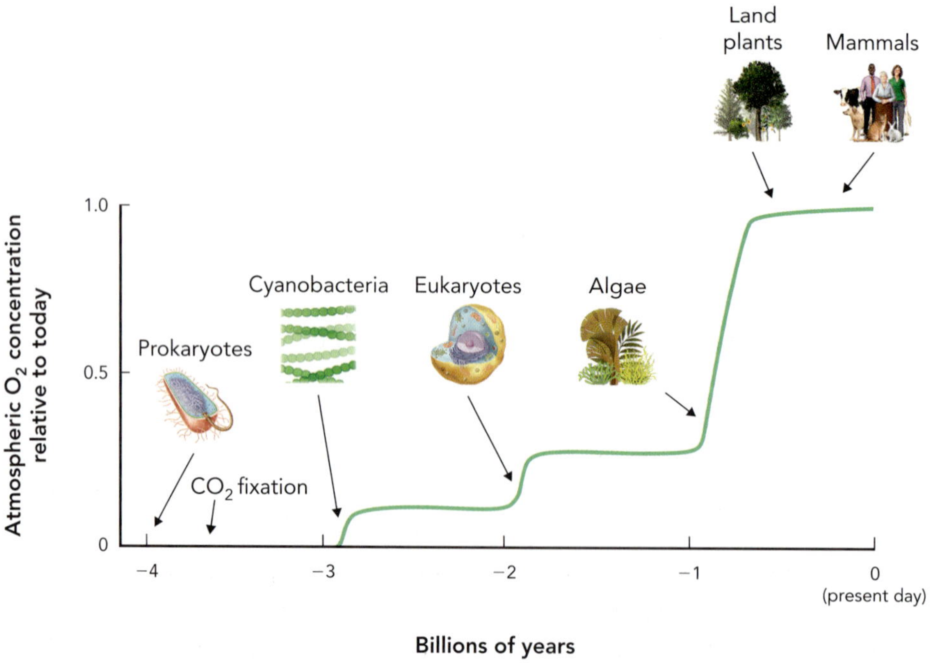

Figure 12.3 The emergence of photosynthetic organisms on Earth led to significant increases in the level of O_2 in the atmosphere and directly affected the evolution of multicellular organisms that use aerobic respiration for energy conversion. The ability to fix CO_2 into organic compounds was an earlier evolutionary trait.

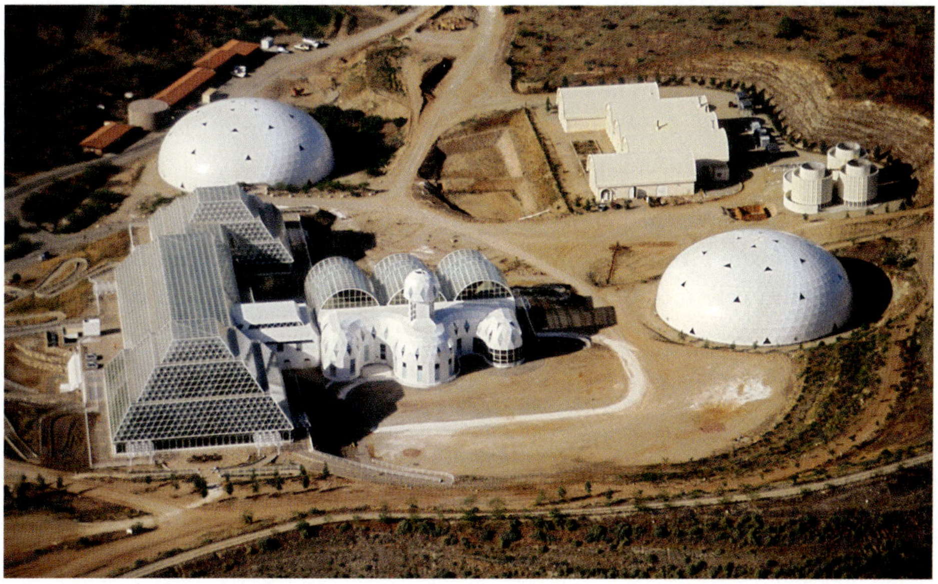

The inhabitants of the Biosphere 2 facility in 1991

Figure 12.4 Biosphere 2 is a sealed glass terrarium located outside of Tucson, Arizona. In 1991, Biosphere 2 was used for the first human experiment designed to test the idea that autotrophs and heterotrophs could survive within a sealed environment. Inside the Biosphere 2 facility are several different ecosystems, as well as agricultural areas and living quarters. The Biosphere 2 facility is now being used by the University of Arizona as an environmental laboratory and educational training center. FACILITY: JOE SOHM/THE IMAGE WORKS; ENTRANCE SIGN: IAN DAGNALL/ALAMY; INHABITANTS: REUTERS.

3.1-acre ecosystem in the desert outside of Tucson, Arizona. Biosphere 2, as it is called, was home to eight scientists who hoped to survive in this airtight terrarium for 2 years based solely on photosynthesis as the primary energy source for food production and for CO_2–O_2 gas exchange (**Figure 12.4**). Within a few months, however, CO_2 levels inside Biosphere 2 began to rise, and the experiment had to be interrupted several times to remove excess CO_2. Although not as successful as originally planned, the grand scale of the Biosphere 2 experiment serves as a reminder that photosynthetic organisms hold the key to survival on Earth.

Photosynthetic organisms use five steps to convert solar energy into chemical energy and precursor molecules for synthesis of carbohydrate sugars. The sugars are used as metabolic fuel at night when light is unavailable. **Figure 12.5** illustrates the five steps:

1. Photons are absorbed by chlorophyll molecules, which are components of photosystem II (PSII), a large protein complex embedded in chloroplast membranes. Light absorption by chlorophyll changes its reduction potential, so that it is able to donate one electron to a nearby electron acceptor molecule for every photon that is absorbed. The absorption of 4 photons by chlorophyll leads to the donation of 4 e^-, all of which are replaced in chlorophyll by the oxidation of 2 H_2O to form O_2. The 4 H^+ that are released from oxidation of 2 H_2O contribute to the proton-motive force.
2. Electron transport via carrier molecules (plastoquinone [PQ], cytochrome b_6f, and plastocyanin [PC]) in the photosynthetic electron transport system results in the establishment of a proton gradient across the membrane. Eight H^+ are translocated for every 4 e^- donated by 2 H_2O molecules.
3. Photon absorption by photosystem I continues the series of electron transfer events, resulting in the generation of reduced NADPH.
4. The proton-motive force is used to drive ATP synthesis by the chloroplast ATP synthase complex, with a stoichiometry of 12 H^+ generating 3 ATP (8 H^+ being translocated and 4 H^+ obtained from the oxidation of 2 H_2O).

5. The ATP and NADPH molecules generated by the photosynthetic electron transport system are used by enzymes in the Calvin cycle to drive carbon fixation reactions. The Calvin cycle uses 3 CO_2 molecules to generate 1 molecule of glyceraldehyde-3-phosphate, a precursor for glucose synthesis by the gluconeogenic pathway (see Chapter 14).

The membrane shown in Figure 12.5 represents the thylakoid membrane inside plant chloroplasts. However, similar photosynthetic reaction complexes are also present in the cell membrane of **cyanobacteria**, a prokaryotic photosynthetic microorganism.

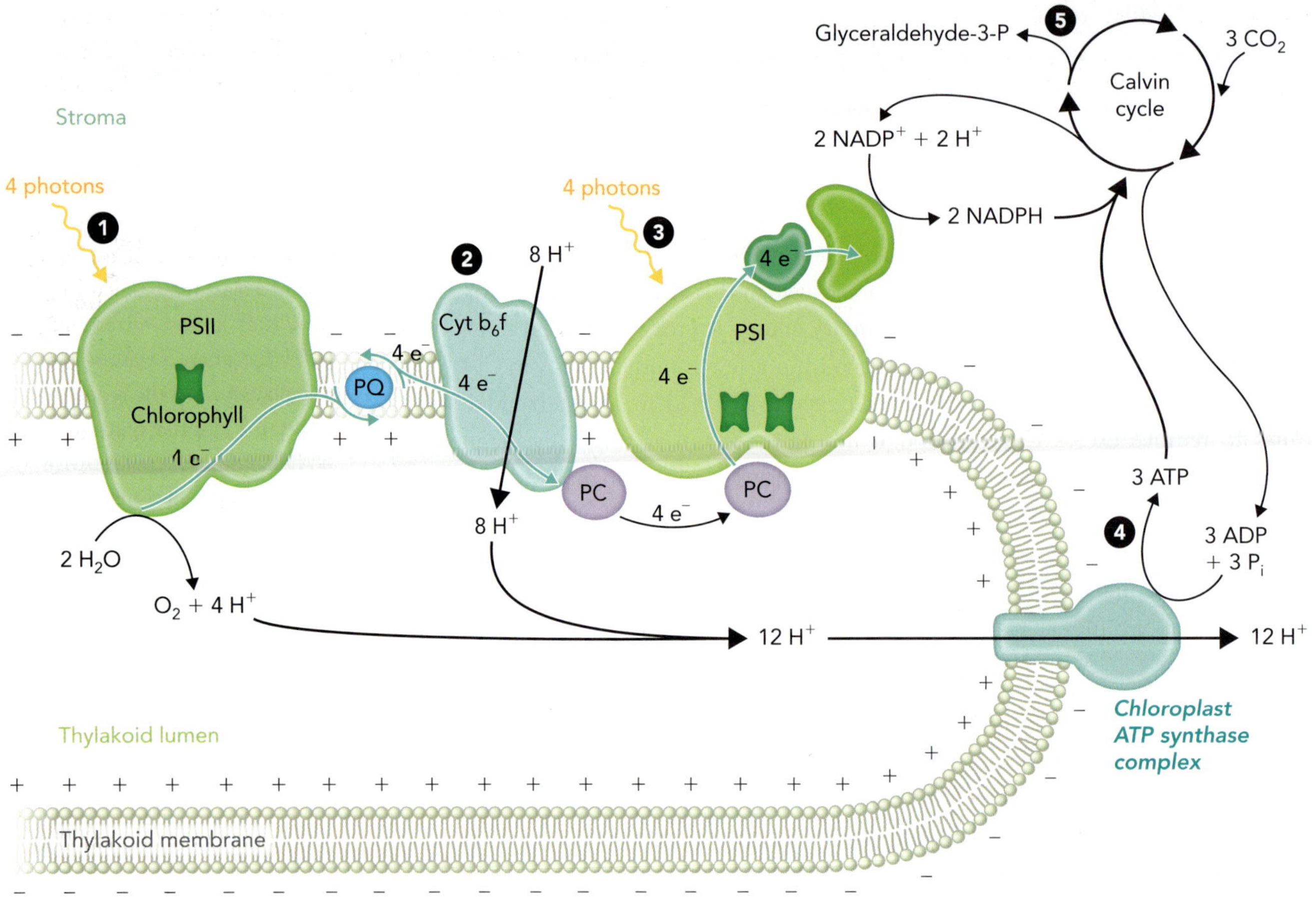

Figure 12.5 An overview of photosynthetic electron transport and photophosphorylation is shown here. (1) Photon absorption by the PSII reaction center complex leads to activation of the photosynthetic electron transport system and generation of O_2. Electrons pass through the system one at a time, but 4 e^- are generated from the oxidation of 2 water molecules. (2) Electrons flow through the photosynthetic electron transport system, which includes the electron carriers plastoquinone (PQ), plastocyanin (PC), and cytochrome b_6f (Cyt b_6f). Electron flow through cytochrome b_6f results in the buildup of a chemiosmotic proton gradient across the thylakoid membrane. (3) Photon absorption by the PSI reaction center complex continues the series of electron transfer reactions. The final electron acceptor in this pathway is $NADP^+$, which is reduced to form NADPH. (4) In response to proton flow from the thylakoid lumen into the stroma, the process of photophosphorylation occurs, in which ATP synthesis is catalyzed by the chloroplast ATP synthase complex. (5) CO_2 fixation by the Calvin cycle leads to the formation of glyceraldehyde-3-phosphate (glyceraldehyde-3-P), a triose phosphate used to make hexose sugars. Energy conversion from ATP hydrolysis and NADPH oxidation is used to drive the Calvin cycle reactions.

Structure and Function of Chloroplasts

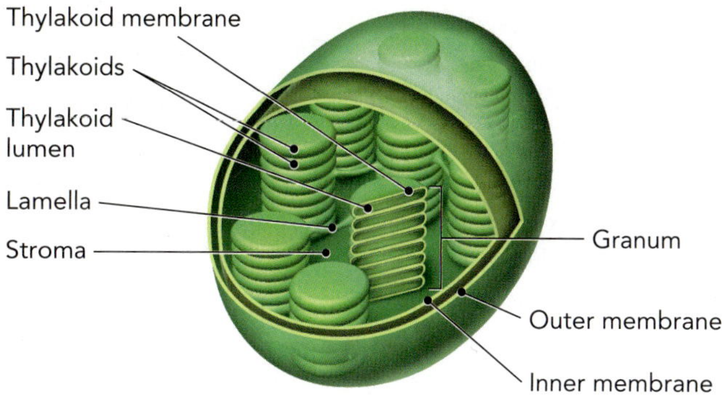

Figure 12.6 The anatomy of a plant chloroplast is shown here. The thylakoid disks are folded regions of the thylakoid membrane where proteins in the photosynthetic electron transport system are located. The aqueous compartment outside of the thylakoid membrane, but inside the inner chloroplast membrane, is called the stroma, which contains all of the Calvin cycle enzymes. The thylakoid membrane can be found in a stack of thylakoid disks called a granum or in an unstacked region called the lamella.

The photosynthetic machinery in eukaryotic plant cells is contained within organelles called chloroplasts, which number ~10 to 500 per cell. Each chloroplast is about 2 μm in diameter and ~5 μm long. Chloroplasts, like mitochondria, contain their own DNA and carry out protein synthesis within the organelle to make proteins required for chloroplast function. The remainder of the proteins needed for photosynthesis are encoded in the nuclear genome and imported into the chloroplast. Both chloroplasts and mitochondria are thought to represent an endosymbiotic relationship, which was established when a eukaryotic cell engulfed a bacterium. In the case of plant cells, the photosynthetic bacterium was most likely related to cyanobacteria, which have a photosynthetic electron transport system resembling that of chloroplasts.

Figure 12.6 shows the key physical features of chloroplasts. Chloroplasts contain three membranes. The outer membrane is permeable to most metabolites, whereas the inner membrane and the **thylakoid membrane** are essentially impermeable and form a barrier between two compartments within the chloroplast. These compartments are the **thylakoid lumen**, which is a continuous aqueous chamber inside the thylakoid membrane, and the **stroma**, which is the aqueous phase outside of the thylakoid membrane. Similar to the inner mitochondrial membrane, the thylakoid membrane contains all of the protein complexes that constitute the electron transport system and is the location of the ATP synthase complex. Thus, the thylakoid membrane is a physical barrier that restricts the free flow of ions, allowing the formation of the electrochemical proton gradient needed for ATP synthesis (see Figure 12.5).

Two other chloroplast structures shown in Figure 12.6 are the **granum**, which is the name given to a stack of thylakoid structures (actually highly invaginated thylakoid membranes), and the **lamella** (plural = lamellae), which is an unstacked region of the thylakoid membrane. Distinct protein complexes in the photosynthetic electron transport system are distributed differently in grana and lamellae. As we will discuss later, the relative distribution of photosynthetic electron transport proteins in these chloroplast structures can change, depending on the requirements of the Calvin cycle for ATP and NADPH.

As shown in **Figure 12.7**, light activation of the chloroplast electron transport system, located in the thylakoid membrane, results in proton translocation into the thylakoid lumen. The oxidation of H_2O takes place in the thylakoid lumen and generates protons, which contribute to the differential proton gradient across the thylakoid membrane. The ΔpH across the thylakoid membrane (pH 5 inside the lumen relative to pH 8 in the stroma) causes protons to flow out through the chloroplast ATP synthase complex, leading to ATP synthesis in the stroma.

The final electron acceptor in the photosynthetic electron transport system is $NADP^+$, which accepts a pair of electrons to produce NADPH in the stroma. The Calvin cycle enzymes located in the stroma use energy from ATP hydrolysis and NADPH oxidation to convert CO_2 into glyceraldehyde-3-phosphate. Most of the glyceraldehyde-3-phosphate is used to regenerate Calvin cycle intermediates, with the rest either used to synthesize starch in the stroma as an energy reserve or transported to the cytosol and converted to hexose sugars for export to other plant tissues.

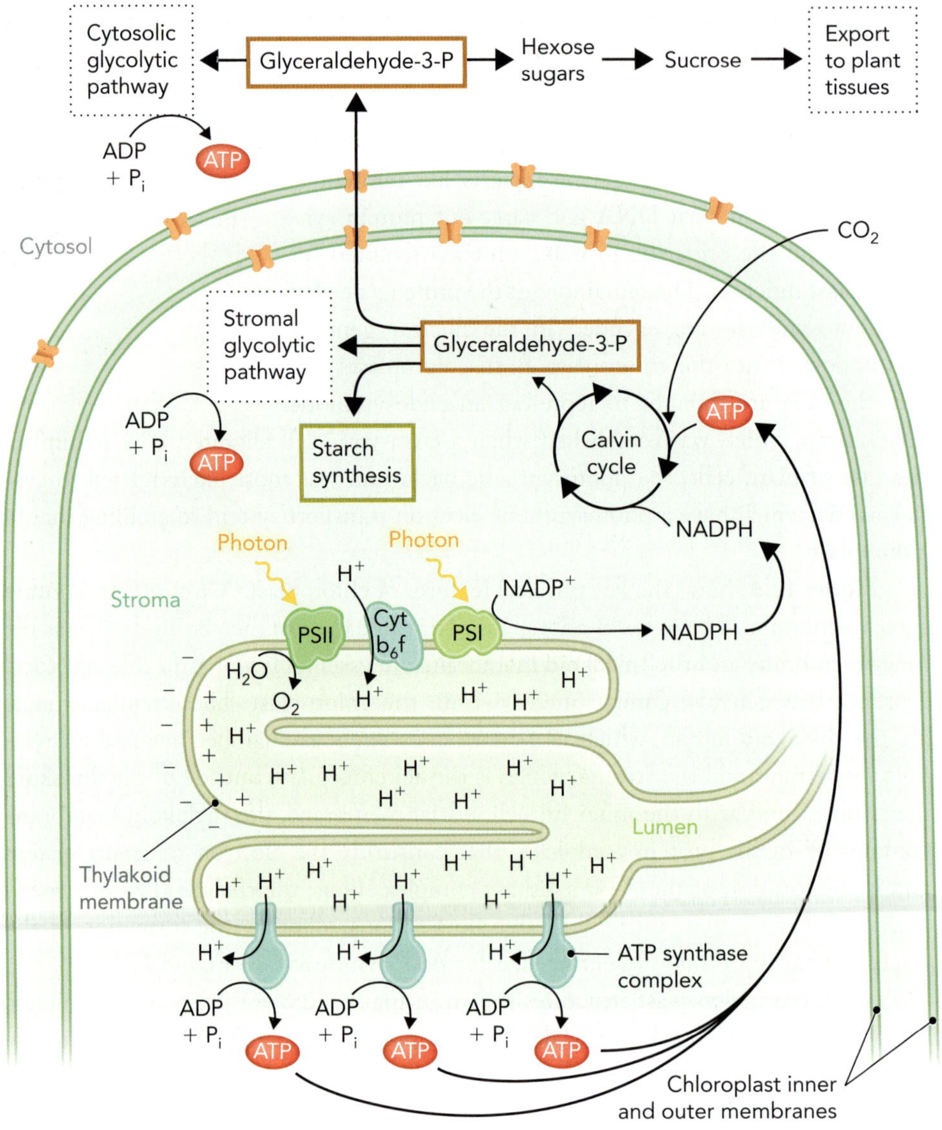

Figure 12.7 Photosynthetic electron transport leads to proton translocation into the thylakoid lumen and the production of NADPH in the stroma. The ATP synthase complex is oriented outward from the lumen, so that ATP synthesis occurs in the stromal compartment. Calvin cycle enzymes located in the stroma use NADPH and ATP to produce glyceraldehyde-3-phosphate from CO_2. This triose phosphate can either be used in the stroma or transported to the cytosol.

Glyceraldehyde-3-phosphate can also be metabolized by the glycolytic pathway in the stroma or cytosol to produce ATP for cellular processes.

Before describing these processes in detail, let's answer the four metabolic pathway questions that relate to the photosynthetic electron transport system and the Calvin cycle. Together, these two pathways are called photosynthesis.

1. *What does photosynthesis accomplish for the cell?* The photosynthetic electron transport system converts light energy into redox energy, which is used to generate ATP by chemiosmosis and reduce $NADP^+$ to form NADPH. Calvin cycle enzymes use energy available from ATP and NADPH to reduce CO_2 to form glyceraldehyde-3-phosphate, a triose phosphate that can be used to synthesize glucose. Photosynthetic cells use the triose phosphates produced by the Calvin cycle as a chemical energy source for mitochondrial respiration, which is independent of light. Photosynthetic organisms are autotrophs because they derive energy from light rather than from organic materials.

2. *What are the overall net reactions of the photosynthetic electron transport system and the Calvin cycle?* For the photosynthetic electron transport system (production of ATP and O_2):

$$2\ H_2O + 8\ \text{Photons} + 2\ NADP^+ + {\sim}3\ ADP + {\sim}3\ P_i \rightarrow O_2 + 2\ NADPH + {\sim}3\ ATP$$

For Calvin cycle reactions (net production of 1 triose phosphate from 3 CO_2):

$$3\ CO_2 + 6\ NADPH + 9\ ATP + 6\ H_2O \rightarrow \text{Glyceraldehyde-3-phosphate} + 6\ NADP^+ + 9\ ADP + 9\ P_i$$

3. *What are the key enzymes in the photosynthetic electron transport system and the Calvin cycle?* There are three protein components of the photosynthetic electron transport system. These three protein complexes are required for the oxidation of H_2O and reduction of $NADP^+$: **photosystem II** (P680 reaction center), **cytochrome b_6f** (proton translocation), and **photosystem I** (P700 reaction center). Chloroplast ATP synthase is the enzyme responsible for the process of photophosphorylation, which converts proton-motive force (energy available from the electrochemical proton gradient established by the photosynthetic electron transport system) into net ATP synthesis through a series of proton-driven conformational changes. This enzyme is very similar to mitochondrial ATP synthase in both structure and function. Ribulose-1,5-bisphosphate carboxylase/oxygenase (**rubisco**) is responsible for CO_2 fixation in the first step of the Calvin cycle. Rubisco activity is at a maximum in the light, when stromal pH is ~8 and Mg^{2+} levels are elevated because of proton translocation (proton influx to the thylakoid lumen is accompanied by Mg^{2+} efflux to the stroma).

4. *What is an example of the photosynthetic electron transport system and Calvin cycle in everyday biochemistry?* A sealed jar containing pond water, aquatic plants, and small organisms (such as ghost shrimp or snails) can create a tabletop ecosystem. The animals contribute organic waste to the environment; bacteria in the pond water break down this waste. The released CO_2 is fixed by the plants using light energy from photosynthesis, creating food and oxygen for the animals. When carefully set up with an appropriate balance of the components, these closed systems are self-sustainable for years (**Figure 12.8**).

Figure 12.8 An airtight glass carboy containing a spiderwort plant (*Tradescanthia*), soil, and a small amount of water was sealed in 1972 and forgotten for many years in a house in England. David Latimer, the amateur gardener who set up the terrarium, displayed it 40 years later as a perfect example of a plant's autotrophic life cycle in which carbon, oxygen, and hydrogen are continually recycled through metabolic pathways. P. YEOMANS/ BOURNEMOUTH NEWS AND PICTURE SERVICE.

concept integration 12.1

What biochemical role did photosynthetic organisms play in the evolution of other life-forms on Earth?

Photosynthesis uses light energy to oxidize H_2O and form O_2, which is used by nonphotosynthetic organisms as the ultimate electron acceptor in the electron transport system. Prior to the expansion of plants as the dominant life-form on Earth, the level of O_2 in the atmosphere was too low to support aerobic metabolism. However, as plants began to take over the planet, O_2 levels began to rise, and it provided aerobic organisms with a source of metabolic fuel (eating of plants), as well as a distinct advantage over anaerobic nonphotosynthetic organisms, for which O_2 is toxic. Moreover, aerobic metabolism also allowed plants to evolve and adapt to new environments as they could switch between photosynthesis and aerobic metabolism depending on light intensity.

12.2 Energy Conversion by Photosystems I and II

Light emanating from the Sun is the result of nuclear fusion reactions, which release energy in the form of photons. When this light energy reaches Earth, it is absorbed by special molecules called **chromophores**, which are contained within membrane proteins of photosynthetic organisms. The most important chromophore is called **chlorophyll**.

In this section, we first discuss how photon absorption by chlorophyll molecules causes an electron to be lifted to a higher energy level. With higher energy, an electron has the potential to be donated to a nearby oxidant molecule, resulting in charge separation. We then examine in some detail how each of the three protein complexes (PSII, cytochrome b_6f, and PSI) in the photosynthetic electron transport system uses redox energy to drive chemiosmosis and generate NADPH.

Chlorophyll Molecules Convert Light Energy to Redox Energy

Photosynthetic chromophore molecules absorb light in the visible range of the electromagnetic spectrum, which includes wavelengths of 400–700 nm. Given its frequency, the amount of energy in a photon can be calculated by the Planck relation:

$$E = h\nu = \frac{hc}{\lambda} \tag{12.1}$$

where h is Planck's constant (6.63×10^{-34} J s), ν is the frequency of the photon (s^{-1}), c is the speed of light (2.99×10^8 m/s), and λ is the wavelength of the light in meters. Using the Planck relation, we can calculate that a mole of photons, called an einstein (6.02×10^{23} photons), with a wavelength of 400 nm has ~300 kJ of energy, whereas a mole of photons of 700 nm has ~170 kJ of energy. (Note that shorter wavelengths [high frequency] have more energy than longer wavelengths [lower frequency].) As will be described later, for each electron that passes through the photosynthetic electron transport system of plants, two photons are required: one is absorbed by photosystem II, and one is absorbed by photosystem I. Because four electrons are required to oxidize 2 H_2O molecules to form O_2 and 2 NADPH (a pair of electrons is needed to reduce each $NADP^+$), the PSII–PSI photosynthetic electron transport system in plants involves the absorption of 8 photons in each complete cycle (see Figure 12.5).

To understand how light energy is converted to redox energy, we first need to consider what happens when a chlorophyll molecule (Chl) absorbs a photon. As shown in **Figure 12.9**, absorption of a photon of the appropriate wavelength excites an electron from its ground state and lifts it to a higher orbital, where it is said to be in an excited state, sometimes denoted with an asterisk (Chl*). At this point, three outcomes are possible.

1. The chlorophyll electron could return to its ground state and through a process called **resonance energy transfer** pass the absorbed energy to a nearby chlorophyll molecule of lower but overlapping energy. This energy transfer process is very important in photosynthesis because it serves to "harvest" light energy, allowing one chlorophyll molecule to absorb a photon and then pass the energy along to a nearby molecule.
2. The absorbed light energy could be lost as heat or fluorescence when the electron returns to the ground state. This represents a futile reaction, as none of the light energy is used to perform useful work in the plant.

3. After absorption of a photon, the chlorophyll molecule can transfer one electron to a nearby acceptor molecule of higher reduction potential (more positive $E°'$), resulting in oxidation of the chlorophyll molecule (loss of electron) and reduction of the acceptor molecule (gain of electron). This process is called photooxidation. This photooxidation reaction, or energy transduction, takes place in protein complexes called **reaction centers.** This outcome is the key to energy conversion in photosynthesis because the reaction results in photoinduced charge separation.

In the chloroplast PSII reaction center, there are two chlorophyll molecules—sometimes referred to as the "special pair"—that after photon absorption pass an excited electron to a nearby acceptor molecule. The electron acceptor is a molecule called pheophytin, which becomes negatively charged, denoted by $^{\bullet}Pheo^{-}$. Once the photooxidation reaction occurs, the reduced acceptor molecule (now negatively charged) donates an electron to another acceptor molecule of higher reduction potential, thereby activating the photosynthetic electron transport system. The oxidized chlorophyll molecule of the special pair (now positively charged; Chl^{+}) returns to the ground state by accepting an electron through a coupled redox reaction involving the oxidation of H_2O.

Chlorophylls absorb maximum amounts of light in the blue (~450 nm) and red (~700 nm) range of the visible spectrum, which is why they appear green. (When blue and red light are subtracted from white sunlight, the light left over and reflected to the eye is green.) To understand why chlorophyll molecules are ideal chromophores,

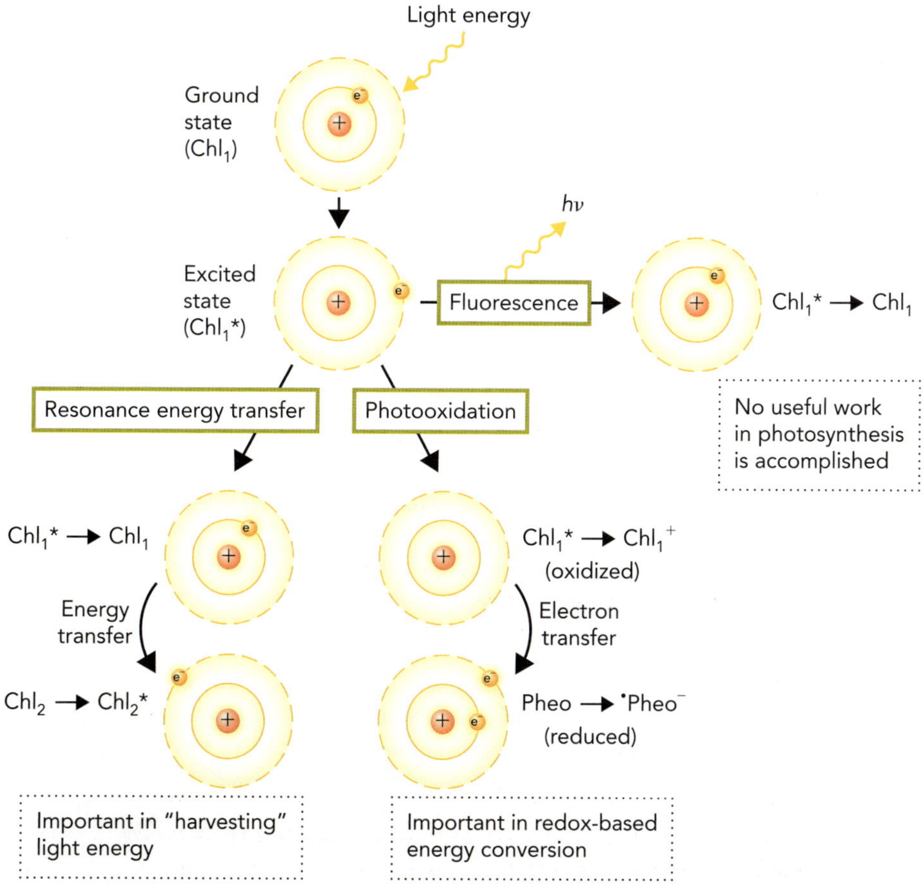

Figure 12.9 Photon absorption by chlorophyll molecules results in resonance energy transfer, fluorescence, or photooxidation. In resonance energy transfer, also called exciton transfer, the energy released by the excited chlorophyll molecule returning to the ground state ($Chl_1^* \rightarrow Chl_1$) is transferred to a nearby chlorophyll molecule, which causes it to be in the excited state ($Chl_2 \rightarrow Chl_2^*$). In photooxidation, an electron from the excited chlorophyll molecule is passed directly to a nearby acceptor molecule of higher (more positive) reduction potential, such as pheophytin (Pheo). This separation of charge creates a positively charged chlorophyll molecule ($Chl_1^* \rightarrow Chl^+$) and a negatively charged acceptor molecule (Pheo → $^{\bullet}Pheo^{-}$). Fluorescence does not result in useful work being performed by the plant as a result of light absorption.

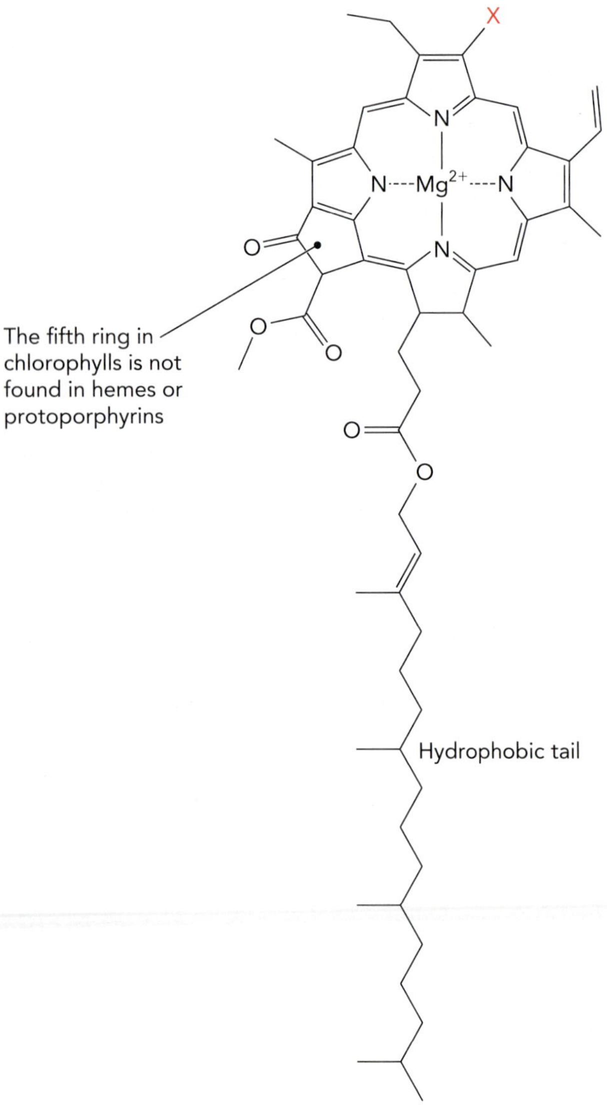

Figure 12.10 The light-harvesting pigments chlorophyll a and chlorophyll b consist of an Mg^{2+} ion–containing porphyrin ring and a hydrophobic phytol side chain. Chemical differences between chlorophylls a and b result in distinct light-absorbing properties. Chlorophyll a, X = CH_3; chlorophyll b, X = CHO.

we need to examine their chemical structure. As shown in **Figure 12.10**, chlorophyll molecules are polycyclic planar structures that resemble heme. Rather than having an Fe^{2+} ion in the middle of the porphyrin ring, as in the heme group of hemoglobin, chlorophyll molecules have an Mg^{2+} ion coordinated to the ring nitrogens. Chlorophylls also have an extended hydrophobic side chain, which anchors the chromophore to various pigment proteins.

The alternating single and double bonds of the chlorophyll polyene structure are essential to its light-absorbing properties because of delocalized electrons above and below the heterocyclic ring. These electrons have energy differences with the next higher molecular orbital that correspond to the energy of photons in the visible range. This is important because photon absorption is an all-or-none phenomenon—the photon energy must be exactly equal to the energy required to lift the electron to the next higher orbital. The two chlorophylls found in plants, chlorophyll a and chlorophyll b, differ only slightly in the structure of the heterocyclic ring (a CH_3 side group in chlorophyll a compared to a CHO group in chlorophyll b). This difference, however, is sufficient to alter the light-absorbing properties of these two chromophores in the blue and red ranges of the spectrum.

Photosynthetic organisms contain several other chromophores as well, which are capable of absorbing light at wavelengths in the green and yellow ranges of the visible spectrum. As shown in **Figure 12.11**, these

Figure 12.11 The accessory pigments β-carotene, phycocyanobilin, and phycoerythrobilin have extended conjugated double-bond systems to allow light absorption.

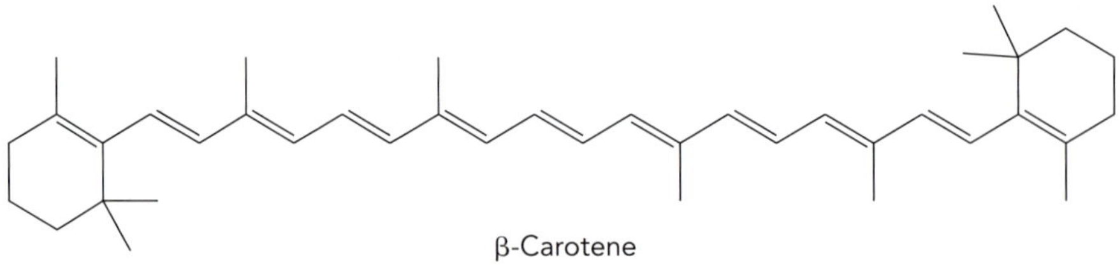

β-Carotene

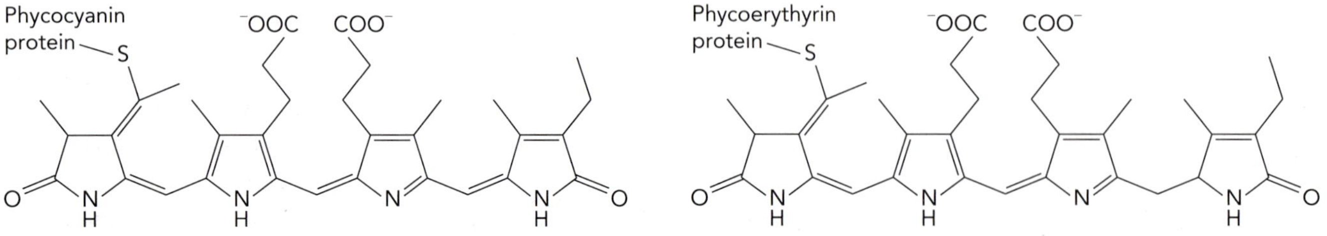

Phycocyanobilin

Phycoerythrobilin

pigments include β-carotene, an accessory pigment in plants; and phycocyanobilin and phycoerythrobilin, which are phycobilin-type chromophores found respectively in the phycocyanin and phycoerythrin proteins in red algae. **Figure 12.12** summarizes the absorption spectra of five major chromophores in proteins, showing that together they cover the entire solar spectrum. Green plants contain large amounts of chlorophylls a and b and, in most cases, the orange pigment β-carotene, whereas the phycobilin chromophores—phycocyanobilin and phycoerythrobilin—provide an evolutionary advantage to marine plants and phytoplankton that absorb light filtering through the upper layers of the ocean (**Figure 12.13**).

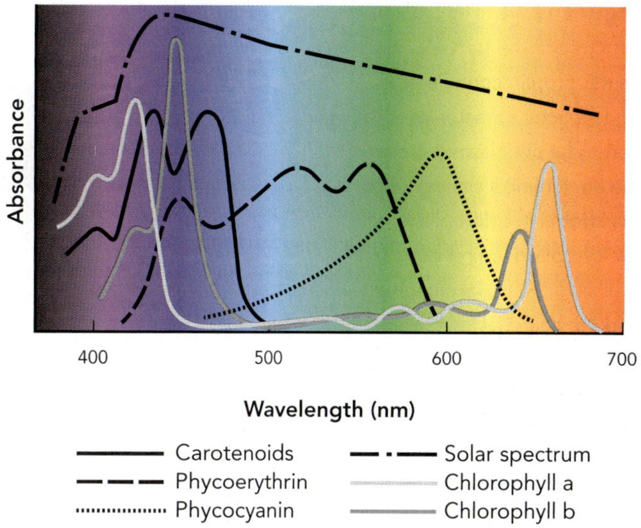

Figure 12.12 Photosynthetic pigments absorb light in the visible range of the electromagnetic spectrum. Chlorophylls a and b have two absorbance peaks at opposite ends of the solar spectrum. Carotenoids, such as β-carotene, absorb in the blue range, which is why they appear orange (reflected light). The phycobilins (phycoerythrobilin and phycocyanobilin) in phycoerythrin and phycocyanin absorb light in the middle of the solar spectrum and are primarily found in marine photosynthetic organisms.

How are these light-absorbing pigments arranged in photosynthetic membranes to maximize light absorption and photooxidation? To answer this question, let's examine the molecular structure of an individual pigment-containing protein and see where the chlorophyll molecules are located relative to the photosynthetic membrane.

Figure 12.14 shows the structure of a photosynthetic protein in the reaction center complex of the nonsulfur purple bacterium *Rhodobacter sphaeroides*. Each monomer of this chromophore protein consists of three polypeptide chains, two molecules of chlorophyll, and two types of electron carriers (pheophytin and ubiquinone). The chlorophyll molecules are anchored within the chromophore protein in such a way that each is surrounded by a unique chemical environment. Similar to the way an enzyme active site alters the chemical properties of a reaction substrate, the electrochemical properties of chlorophyll molecules are altered by the amino acid side groups of the protein.

Because of their unique chemical properties, chlorophyll molecules in reaction center complexes are responsible for photooxidation. However, most chlorophyll molecules in photosynthetic membranes function as light-harvesting antennae instead and are associated with chromophore proteins that participate in energy transfer reactions rather than photooxidation. The proteins containing these chromophore molecules are called **light-harvesting complexes (LHCs)** and are the most abundant proteins

a.

b.

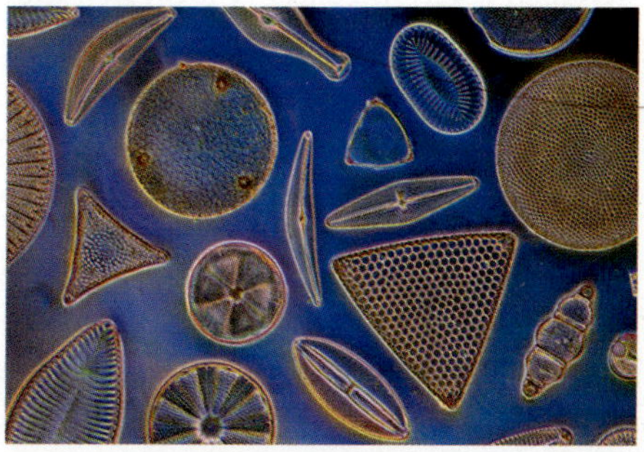

Figure 12.13 Marine plants and phytoplankton absorb light through water and must contain the optimal mixture of pigments to maximize rates of photosynthesis. **a.** Kelp forests along the Pacific Coast of California contain large numbers of *Macrocystis pyrifera*, which is one of the fastest-growing plants on Earth, able to grow up to 40 meters per year. ETHAN DANIELS/SHUTTERSTOCK. **b.** Phytoplankton are photosynthetic single-cell organisms that live in marine or freshwater environments and account for up to 50% of all photosynthesis on Earth. SCENICS & SCIENCE/ALAMY.

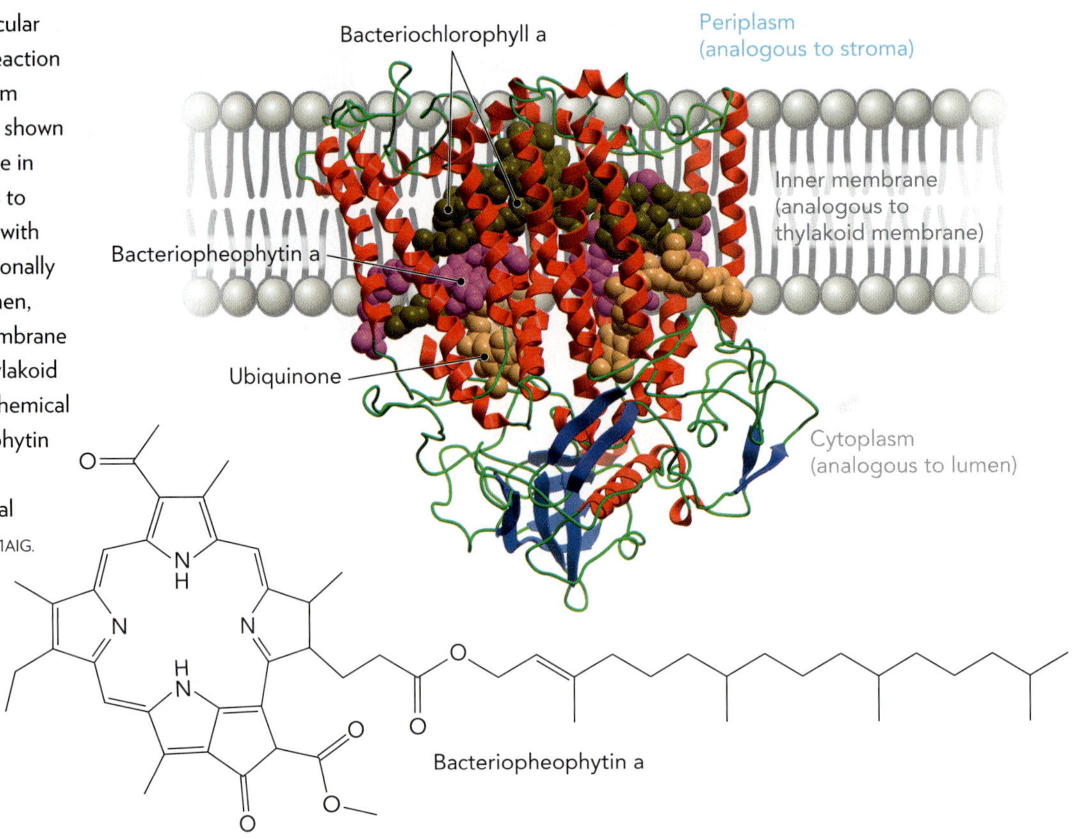

Figure 12.14 The molecular structure of the bacterial reaction center protein complex from *Rhodobacter sphaeroides* is shown here. The periplasmic space in the bacterium is analogous to the stroma in chloroplasts, with the cytoplasm being functionally similar to the thylakoid lumen, and the bacterial inner membrane being analogous to the thylakoid membrane. Shown is the chemical structure of bacteriopheophytin a, which is a chlorophyll molecule lacking the central Mg^{2+} ion. BASED ON PDB FILE 1AIG.

in the thylakoid membrane of chloroplasts. As shown in **Figure 12.15**, two types of light-harvesting complexes have been characterized, LHC I and LHC II, both of which consist of repeating α-helical transmembrane polypeptide segments containing large numbers of chlorophyll molecules and other accessory pigments.

The reaction center contains a special pair of chlorophyll molecules that accept light energy from the LHC II chlorophyll molecules arrayed around this reaction center. This process promotes excitation of an electron in one of the special pair of chlorophyll molecules in the reaction center. The different chemical environments of the chlorophyll molecules in the LHC I and reaction center protein complexes result in a higher reduction potential in the reaction center chlorophyll. Thus, energy transfer readily occurs from an LHC I chlorophyll to a reaction center chlorophyll. In most plants, LHC II greatly outnumbers LHC I and serves as the major light-gathering antenna in the photosynthetic membrane. In this way, once light energy is transferred from a chlorophyll molecule in an LHC II to a nearby chlorophyll in an LHC I, the reduced reaction center chlorophyll molecule is able to trap light energy.

The important role of chlorophyll antennae in capturing light energy for photosynthesis was first suggested by experiments done in 1932 by Robert Emerson and William Arnold. Using isolated spinach chloroplasts under conditions of saturating light, they found that one O_2 molecule was generated for every 2,400 chlorophyll molecules. Because 8 photons are required to form each O_2, and O_2 generation only occurs in reaction center complexes, this means that roughly 300 energy transfer events (2,400/8) in light-harvesting complexes occur for every one photooxidation reaction in a reaction center complex. **Figure 12.16** illustrates how energy transfer

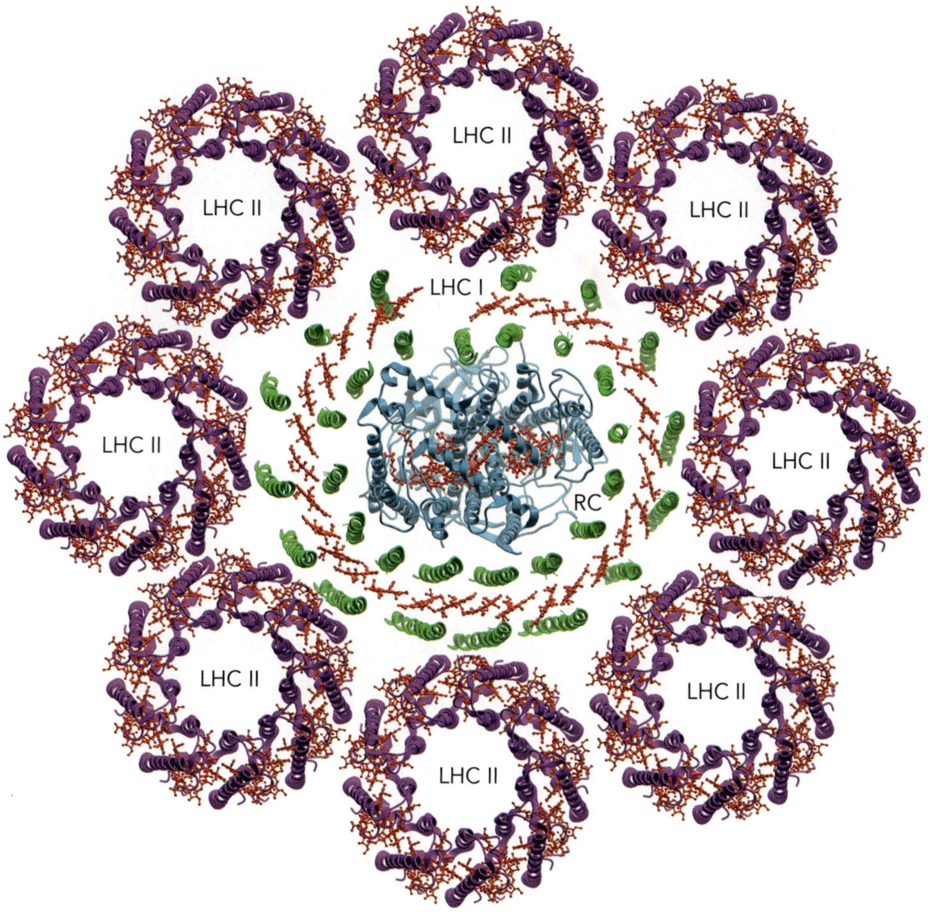

Figure 12.15 This model of light-harvesting complexes in the membrane of purple photosynthetic bacteria is shown in an orientation perpendicular to the membrane. This arrangement facilitates resonance-energy transfer between LHC II and LHC I, culminating in excitation of an electron in chlorophyll at the reaction center. The protein component of the reaction center (RC) is shown in cyan ribbon representation. The proteins of LHC II are shown as purple ribbons. The proteins of LHC I, which form a ring around the reaction center, are shown as green ribbons. The bacteriochlorophyll and bacteriopheophytin components are shown as red sticks. BASED ON PDB FILES 1PYH (RC AND LHC I) AND 1NKZ (LHC II).

reactions between LHC II and LHC I complexes in the chloroplast thylakoid membrane function as highly efficient "solar panels" to capture light energy for photooxidation in reaction center complexes.

A large surface area of leaves is one way to increase light absorption for photosynthesis. For instance, one of the largest leaves on Earth is that of the giant Amazon water

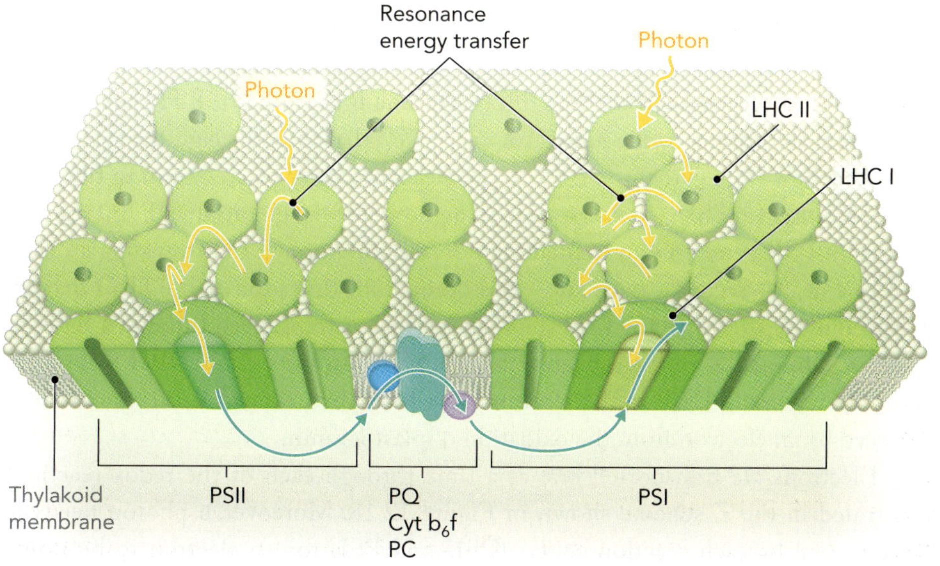

Figure 12.16 LHC II functions as a solar panel that harvests light energy for PSII and PSI reaction centers in the thylakoid membrane. A series of resonance-energy transfer reactions passes the energy along to LHC I surrounding the reaction centers, where photooxidation takes place.

a. b.

Figure 12.17 Plants can use a few large leaves or a large number of small leaves to capture the light they need for photosynthesis. **a.** Giant Amazon water lily *Victoria amazonica*. SONGSAK PANDET/SHUTTERSTOCK. **b.** Pine needles of *Pinus ponderosa*. IMAGES BY MARIA/SHUTTERSTOCK.

lily, which can be up to 6 feet in diameter (**Figure 12.17**). Another strategy is to greatly increase the number of leaves. The ponderosa pine needle, which is also a leaf, provides pine trees with a way to maintain high rates of photosynthesis in cold climates; however, each tree needs tens of thousands of pine needles to make up for their small surface area.

The Z Scheme of Photosynthetic Electron Transport

Plants and cyanobacteria contain two types of reaction center complexes: the photosystem II (PSII) reaction center complex, which absorbs light energy at 680 nm; and the photosystem I (PSI) reaction center complex, which absorbs light energy at 700 nm. These two complexes are functionally linked by three electron carriers called **plastoquinone (PQ)**, cytochrome b_6f (Cyt b_6f), and **plastocyanin (PC)**. Together, these elements constitute the so-called **Z scheme**.

In the Z scheme, shown in **Figure 12.18**, photooxidation releases an electron that can be used for chemical work by a series of coupled redox reactions. Just as we saw in the mitochondrial electron transport system, the physical proximity of electron carriers in the photosynthetic electron transport system, with increasing reduction potentials along the chain, results in electron flow.

The Z scheme begins with (1) absorption of a photon by the chlorophyll P680 reaction center (in PSII), which generates an excited-state P680*. (2) Transfer of an electron to the acceptor molecule **pheophytin (Pheo)** generates oxidized $P680^+$. (An electron resulting from the oxidation of H_2O through the O_2-evolving complex [OEC] reduces the $P680^+$ molecule to return it to the ground state P680.) (3) The electron on Pheo passes through the plastoquinone electron carriers (PQ_A and PQ_B). When electrons flow through cytochrome b_6f, a total of 8 H^+ are translocated from the stroma to the lumen for every 4 e^- that flow through the system. (4) Electrons flow from cytochrome b_6f to the electron carrier plastocyanin to the P700 reaction center (in PSI). (5) Absorption of a photon by the chlorophyll P700 reaction center (in PSI) generates an excited-state P700* chlorophyll that transfers an electron to a nearby acceptor molecule—this occurs four times for each 2 H_2O that are oxidized in step 1. (6) In the final steps, an electron is transferred through a series of carriers, culminating in the reduction of 2 $NADP^+$ to form 2 NADPH (four electrons are required). Each of the electrons lost from the oxidized $P700^+$ molecule is replaced by an electron from the oxidation of plastocyanin.

Electrons are transported one at a time through each of the redox reactions illustrated in the Z scheme shown in Figure 12.18. Moreover, a photon needs to be absorbed by each reaction center (PSII and PSI) for an electron to be transported through the Z scheme from H_2O to $NADP^+$. Therefore, a total of eight

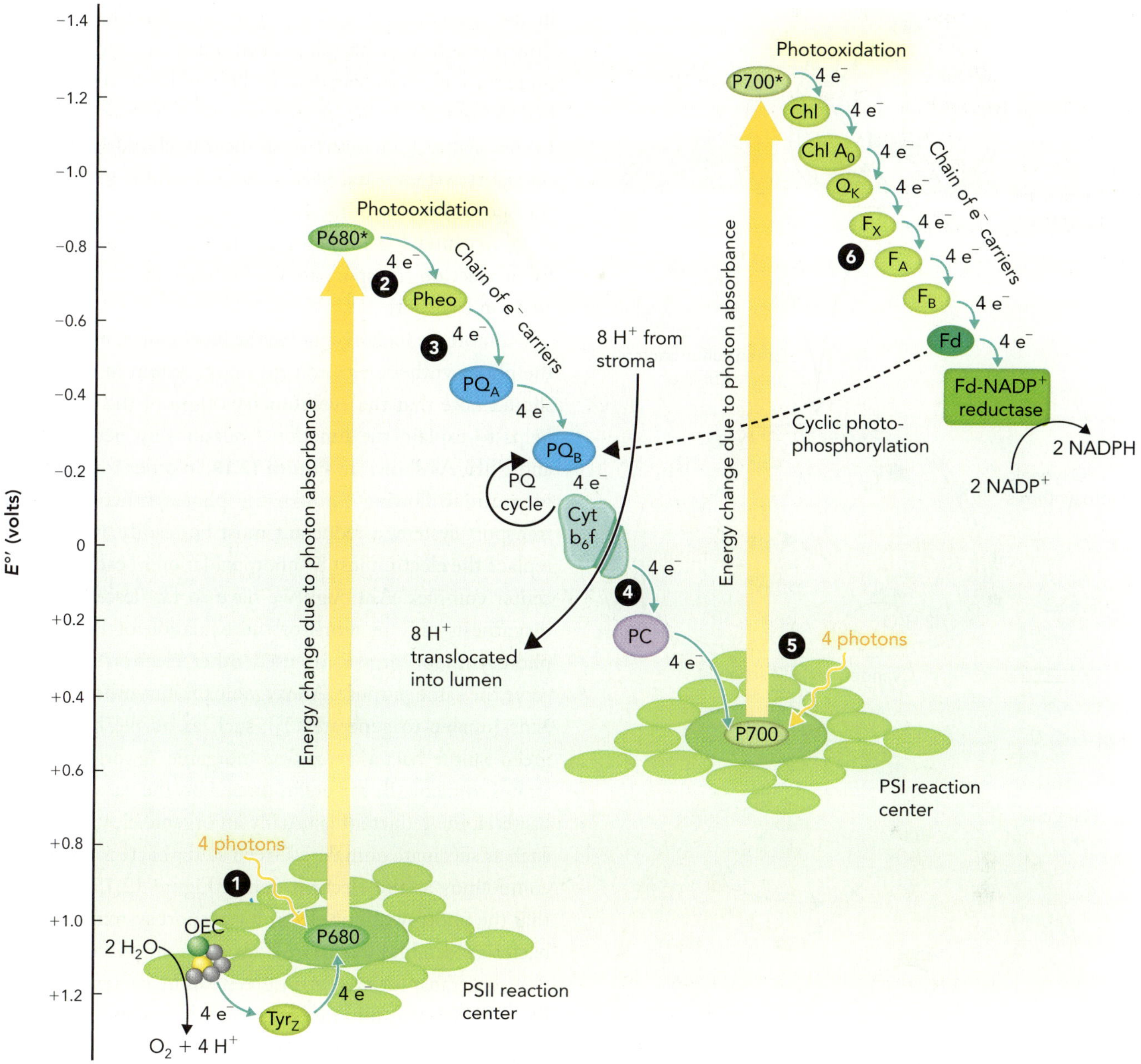

photons are required to transport the four electrons obtained from the oxidation of 2 H_2O. Note that in addition to the electron path through the Z scheme from H_2O to $NADP^+$, chloroplasts also contain an alternative electron path (dotted line in Figure 12.18) through the PSI reaction center. This process is called cyclic photophosphorylation and generates proton-motive force, and therefore ATP, in the absence of H_2O oxidation and NADPH. This process is described in more detail later in the chapter.

The photosynthetic electron transport system in plants consists of a series of photosystems, each requiring an input of energy from light absorption at PSII and PSI reaction center complexes to initiate electron flow. In each photosystem, photon absorption by the PSII reaction center complex results in electron flow from H_2O to plastocyanin, providing energy to translocate H^+ across the thylakoid membrane (see Figure 12.18). A second photon is absorbed by the PSI reaction center, which provides the energy necessary to initiate another series of redox reactions culminating

Figure 12.18 The Z scheme of electron flow in the photosynthetic electron transport system in plant chloroplasts is shown here. The redox energy made available by light absorption at reaction center complexes is used to establish a proton gradient across the thylakoid membrane and to reduce $NADP^+$. Note that Q_K is an electron carrier called phyloquinone, and F_X, F_A, and F_B are three protein-bound 4 Fe–4 S centers, in which F_B donates electrons to ferredoxin (Fd). See the text for details of the six steps shown here.

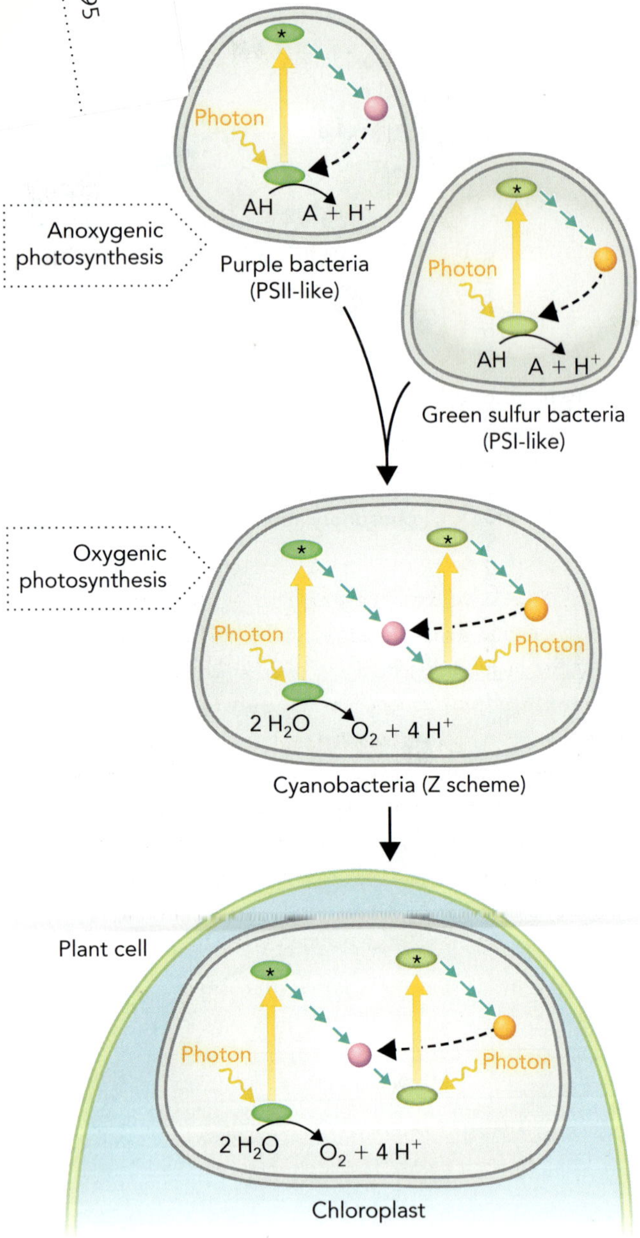

Figure 12.19 Components of the photosynthetic electron transport systems in bacteria resemble those in plant chloroplasts, suggesting an evolutionary relationship. The photosynthetic electron transport systems in purple bacteria and green sulfur bacteria use anoxygenic pathways to oxidize organic or inorganic substrates rather than H_2O. Cyanobacteria and plant chloroplasts are similar in that they both use oxygenic photosynthetic pathways (oxidize H_2O to generate O_2). The dotted lines represent cyclic electron paths that do not require substrate oxidation. The spheres represent ancestral gene products that gave rise to the eukaryotic Cyt b_6f and ferredoxin proteins, AH and A are the conjugate base and conjugate acids in a redox pair, and the asterisk (*) represents the excited state of the photosynthetic reaction center.

in the reduction of $NADP^+$ to form NADPH. Electron flow through these protein-linked redox reactions involves numerous electron carriers, including Fe–S centers, the hydrophobic molecule plastoquinone—which is analogous to coenzyme Q (ubiquinone) in the mitochondrial electron transport system—and plastocyanin, a soluble protein that transports 1 e^- at a time from cytochrome b_6f to the PSI reaction center. Plastocyanin has the same job in photosynthetic electron transport as cytochrome c in mitochondrial electron transport.

Before we look in detail at the biochemical reactions of the photosynthetic electron transport system in plants, we should note that the evolutionary origin of the Z scheme helps to explain the functional relationship between PSI and PSII. As shown in **Figure 12.19**, in order for electrons to continue flowing through the photosynthetic electron transport system, a reductant must be readily available to replace the electron lost by photooxidation in each reaction center complex. Although we have so far described photosynthesis only in terms of the oxidation of H_2O, early photosynthetic organisms used other electron donors to serve the same purpose. Anoxygenic photosynthetic organisms (unable to generate O_2), such as purple bacteria or green sulfur bacteria, oxidize inorganic or organic substrates available in the environment. In the case of purple bacteria, the reductant is usually an organic electron donor such as succinate or malate. Green sulfur bacteria use sulfur compounds as the electron donor. Figure 12.19 indicates that the photosynthetic electron transport system in purple bacteria resembles that of PSII, whereas the photosynthetic electron transport system in green sulfur bacteria is most like PSI. Plants and cyanobacteria (also called blue-green algae) are oxygenic photosynthetic organisms that contain both the PSII and PSI photosystems.

Protein Components of the Photosynthetic Electron Transport System

Figure 12.20 shows the path of electrons through the PSII reaction center. In PSII, absorption of a 680-nm photon results in excitation of a chlorophyll molecule to form P680*, which then transfers its electron to the electron acceptor pheophytin. This photooxidation event generates a positively charged chlorophyll P680 molecule ($P680^+$) and a negatively charged electron acceptor molecule (Pheo → $^\bullet Pheo^-$). Reduced pheophytin donates an electron to a plastoquinone molecule (PQ_A), which subsequently donates an electron by way of a nonheme iron to a second plastoquinone molecule (PQ_B). (These plastoquinone molecules are chemically identical but are distinct in their location within the photosynthetic electron transport system.) Once PQ_B acquires two electrons from PQ_A and 2 H^+ from the stroma compartment, it becomes fully reduced (PQ_BH_2). In this form, the molecule, now known as plastoquinol, translocates through the thylakoid membrane, where it binds to the cytochrome b_6f complex. This complex donates electrons one at time, using the **PQ cycle**, which is analogous to the Q cycle mechanism described in Chapter 11 (see Figure 11.18).

Electron loss from P680* creates a powerful oxidant in $P680^+$, which extracts an electron from H_2O through a series of reactions involving a manganese cluster (Mn_4). The subsequent reduction of $P680^+$ by H_2O regenerates P680 for another round of photooxidation. Taken together, in order to generate 1 O_2 from 2 H_2O, four photons must be absorbed by the PSII reaction center, leading to the transfer of 4 e^- to 2 PQ_B and the removal of 4 H^+ from the stromal compartment, to form 2 PQ_BH_2.

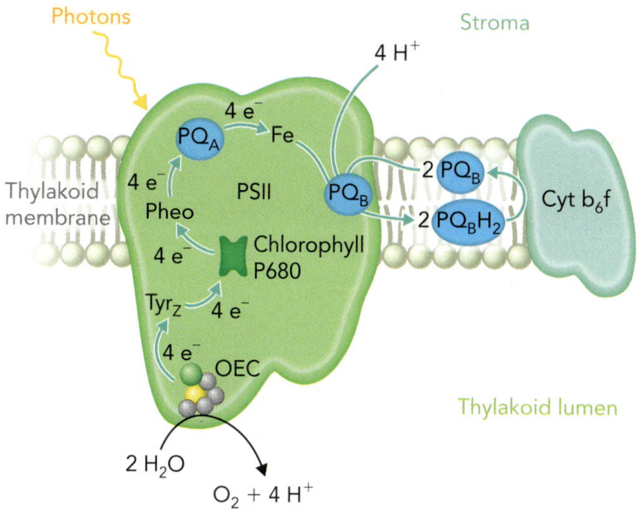

Figure 12.20 Electrons move between electron carriers in the PSII reaction center. Electrons extracted from H_2O in the Mn-containing O_2-evolving complex (OEC) replace electrons lost when chlorophyll P680* is photooxidized. These electrons follow a path through the protein complex from pheophytin to the plastoquinones PQ_A and PQ_B. The reduced PQ_BH_2 molecule diffuses through the thylakoid membrane and delivers electrons two at a time to the cytochrome b_6f complex. Though electrons are moved through PSII one at a time, 4 e^- are shown as being transferred, as a total of 4 e^- are needed to reduce 2 PQ_B to 2 PQ_BH_2.

Photosystem II Oxidizes H_2O to Generate O_2 **Figure 12.21** shows the molecular structure of a monomer of the dimeric PSII protein complex from the thermophilic cyanobacterium *Thermosynechococcus elongatus*. This structure contains 20 protein subunits, 35 chlorophyll a molecules,

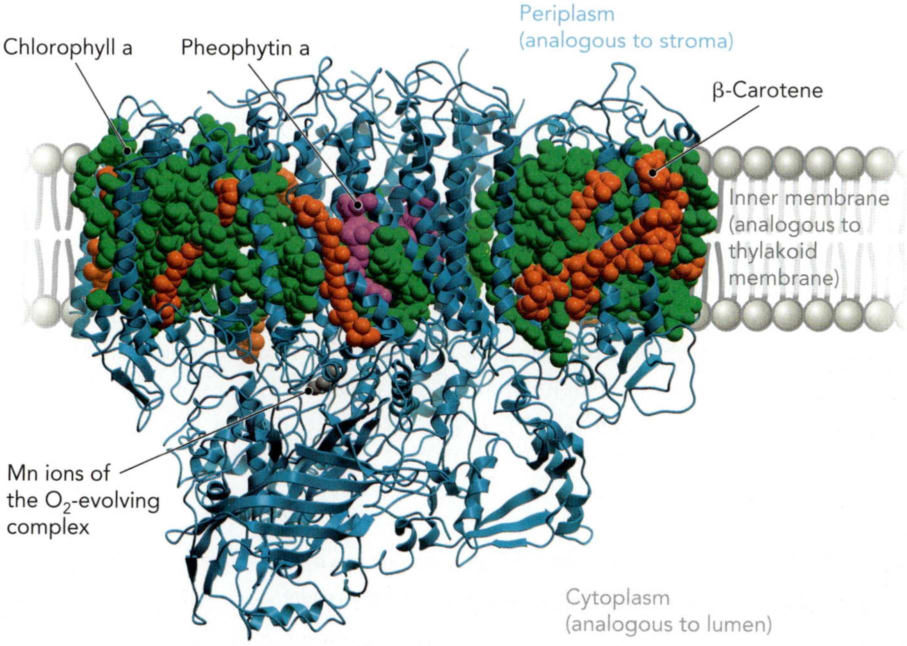

Figure 12.21 The molecular structure of a PSII monomer from the cyanobacteria *Thermosynechococcus elongatus* is shown. The protein subunits are shown in ribbon style, with the light-absorbing molecules (chlorophyll a and β-carotene) and electron transport molecules (pheophytin a and Mn ions of the O_2-evolving complex) shown in space-filling style. BASED ON PDB FILE 3BZ1.

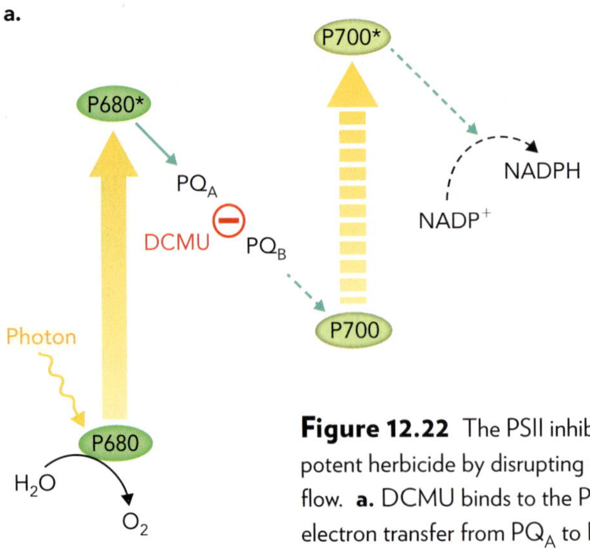

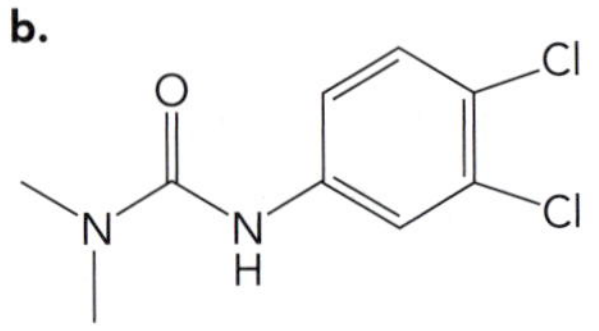

DCMU
3-(3,4-Dichlorophenyl)-1,
1-dimethylurea

Figure 12.22 The PSII inhibitor DCMU acts as a potent herbicide by disrupting the photosynthetic electron flow. **a.** DCMU binds to the PQ$_B$ site in PSII, thereby blocking electron transfer from PQ$_A$ to PQ$_B$. **b.** Structure of DCMU.

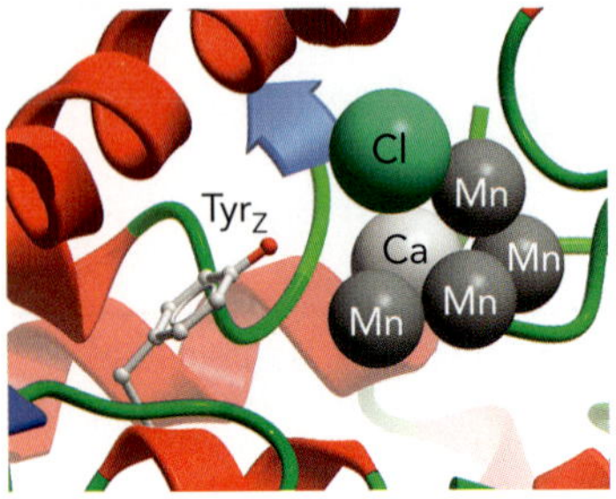

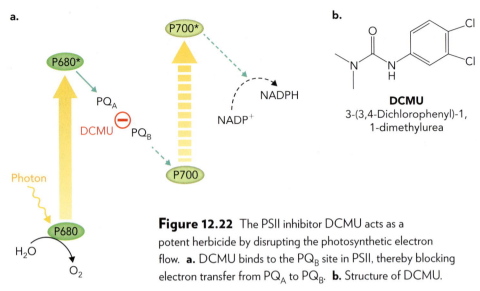

Figure 12.23 Arrangement of Mn atoms within the OEC of the cyanobacterial PSII shown in Figure 12.21. The electrons extracted from the oxidation of H_2O within the OEC are transferred to a tyrosine residue called Tyr$_Z$ in a nearby protein subunit. BASED ON PDB FILE 3BZ1.

12 β-carotene molecules, a single Mn_4 cluster in the O_2-evolving complex, and two molecules of the electron acceptor pheophytin a. Although not seen in this structure, the PSII complex also contains the two plastoquinones PQ$_A$ and PQ$_B$. Chemicals that disrupt electron transport within PSII are potent herbicides. One example is [3-(3,4-dichlorophenyl)-1,1-dimethylurea (DCMU; **Figure 12.22**), which blocks electron transfer from PQ$_A$ to PQ$_B$ by binding to the PQ$_B$ site in the PSII complex.

Because chlorophyll P680$^+$ requires only one electron to be reduced, how is this reaction linked to the oxidation of two H_2O molecules, which require the loss of four electrons to generate one O_2 molecule? The answer is that these four electrons are temporarily stored in an **O_2-evolving complex (OEC)**, consisting of four manganese ions, a calcium ion, a chloride ion, and a reactive tyrosine residue called Tyr$_Z$ (**Figure 12.23**). Each time chlorophyll P680 loses an electron because of photooxidation, an electron from the OEC is donated to Tyr$_Z$$^+$, which transfers the electron to P680$^+$ to return it to the ground state (**Figure 12.24**). Because manganese can exist in a variety of oxidation states (Mn^{2+} up to Mn^{5+}), it has been proposed that the Mn_4 cluster serves as an electron reservoir that feeds electrons one at a time to Tyr$_Z$.

Evidence for how this OEC functions came from an experiment in which it was shown that four photons are absorbed sequentially by P680 in the PSII complex of *T. elongatus* for each O_2 that is generated (**Figure 12.25**). In this experiment, short flashes of light were used to illuminate dark-adapted cyanobacteria, and oxygen production was measured in the solution. It was found that after the 3rd, 7th, and 11th flashes of light, there was a spike in the amount of O_2 generated. These results suggested that O_2 is not released from the complex until four

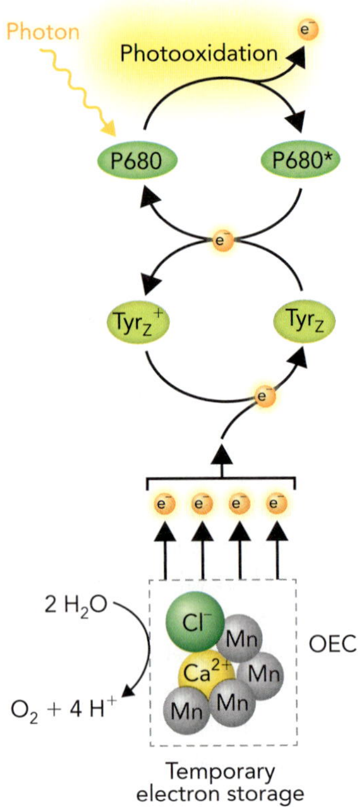

Figure 12.24 The Mn_4 cluster in the OEC donates one electron at a time to Tyr$_Z$, which then reduces P680$^+$ after each photooxidation event.

electrons are extracted from 2 H_2O, even though one electron at a time is passed to the $P680^+$ cation after each photooxidation event. Note that as the first O_2 is released after only three flashes, the resting state of the Mn_4 center must have already obtained one electron from H_2O.

Electrons are transported from PSII to the cytochrome b_6f complex by plastoquinol (PQ_BH_2), which has the same functional role as ubiquinol (QH_2) in the mitochondrial electron transport system. In mitochondria, the two electrons in QH_2 are transferred to two molecules of cytochrome c, whereas in chloroplasts, the two electrons in PQ_BH_2 are donated to two molecules of plastocyanin (see Figure 12.5). Plastocyanin is a small, soluble, Cu-containing protein located on the thylakoid lumen side of the membrane. It shuttles electrons one at a time from the cytochrome b_6f complex to the PSI reaction center. The PQ cycle in the cytochrome b_6f complex operates in much the same way as we observed in mitochondrial complex III (see Figure 11.18); that is, it sequentially oxidizes two molecules of PQ_BH_2, each time transporting one electron to plastocyanin and using the other electron to regenerate PQ_BH_2 from PQ_B to replenish the plastoquinone–plastoquinol pool. The structures of plastoquinone and plastoquinol are shown in **Figure 12.26**, where you can see that the long hydrocarbon tail serves to anchor PQ molecules in the membrane. Note that electrons are added one at a time to plastoquinone, forming the semiquinone intermediate in a reaction similar to that of coenzyme Q (see Figure 11.14).

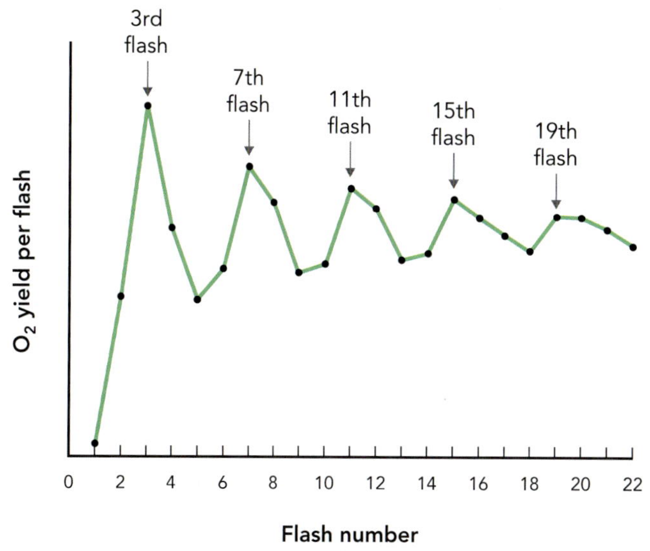

Figure 12.25 Four flashes of light are required to generate each molecule of O_2 from dark-adapted cyanobacteria. Oxygen yields were measured per flash in an experiment designed to determine how a four-electron oxidation reaction ($2\ H_2O \rightarrow O_2 + 4\ H^+ + 4\ e^-$) relates to amount of light absorbed. These results indicate that the resting state of the Mn_4 center has already obtained one electron from H_2O.

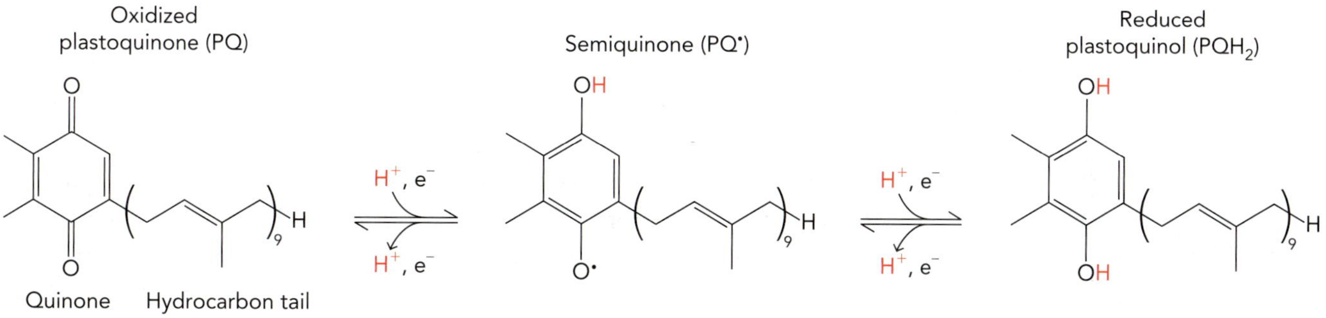

Figure 12.26 The structures of oxidized plastoquinone (PQ), the semiquinone intermediate ($PQ^•$), and reduced plastoquinol (PQH_2) contain a quinone molecule linked to a hydrocarbon tail.

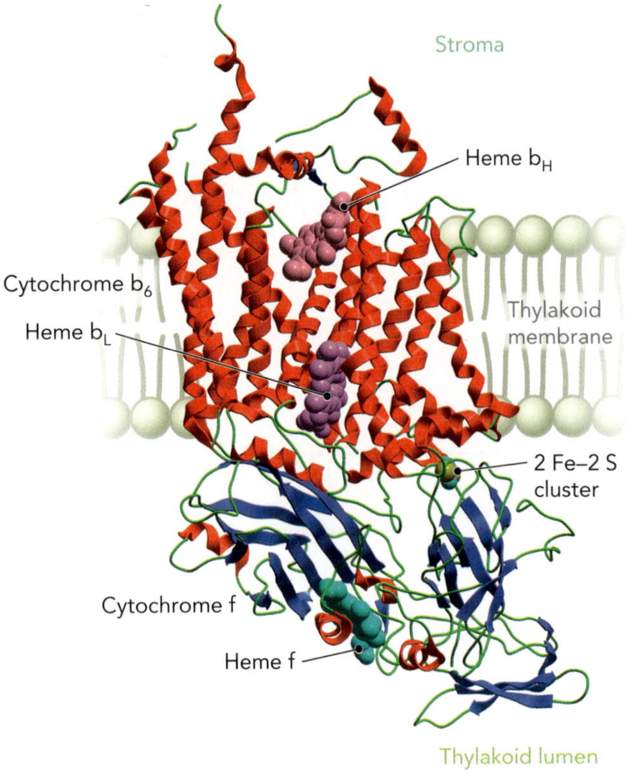

Figure 12.27 The cytochrome b_6f complex from the single-cell alga *Chlamydomonas reinhardtii* contains three heme groups and one 2 Fe–2 S cluster. BASED ON PDB FILE 1Q90.

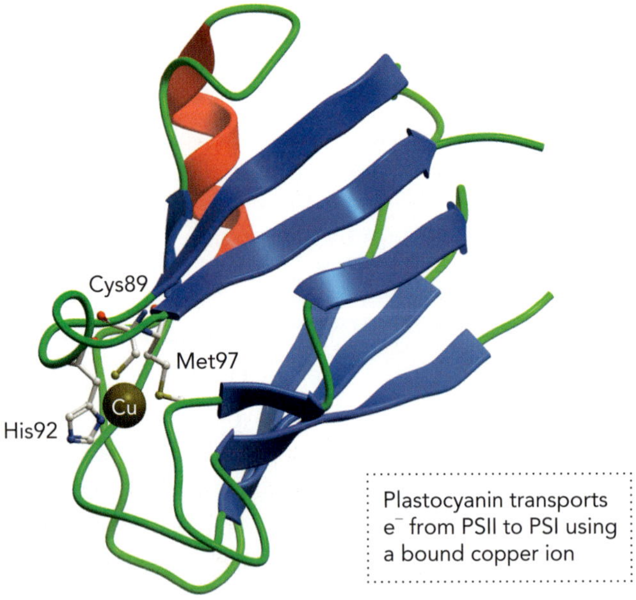

Figure 12.28 The structure of plastocyanin from the cyanobacterium *Anabaena variabilis* is shown here. Plastocyanin transports electrons one at a time from PSII to PSI. The three amino acid residues that coordinate the Cu^{2+}/Cu^+ ion within the protein are labeled. BASED ON PDB FILE 2GIM.

The Cytochrome b_6f Complex Translocates Protons across the Membrane The molecular structure of the cytochrome b_6f complex from the single-cell alga *Chlamydomonas reinhardtii* is shown in **Figure 12.27**. Cytochrome b_6f consists of a b-type cytochrome protein with b_H and b_L heme groups, which play a critical role in the PQ cycle; an iron–sulfur (2 Fe–2 S) cluster; and cytochrome f. The function of cytochrome b_6f is to donate electrons one at a time to the electron carrier protein plastocyanin, which transports electrons on the lumenal side of the membrane to PSI, using a redox reaction involving a Cu^{2+}/Cu^+ ion (**Figure 12.28**).

As summarized in **Figure 12.29**, the first half of the Z scheme consists of photon absorption by the chlorophyll P680 molecule in the PSII reaction center, which leads to the oxidation of two H_2O molecules and the production of four molecules of reduced plastocyanin (Cu^+). This segment of the photosynthetic electron transport system results in a net gain of 12 H^+ on the lumenal side of the thylakoid membrane: 8 H^+ are translocated when 4 PQ_BH_2 molecules are oxidized by the PQ cycle, and 4 H^+ are generated when 2 H_2O are oxidized to 1 O_2.

Photosystem I Generates NADPH for Carbohydrate Synthesis The PSI reaction center functions to transfer the electrons from plastocyanin to $NADP^+$ using the energy derived from photon absorption by a pair of chlorophyll P700 molecules contained within the large protein complex. As shown in **Figure 12.30**, photooxidation of either one of these chlorophyll P700 molecules increases their oxidation potential. This results in electron transfer through two nearby chlorophyll molecules, denoted as Chl and Chl A_0. These molecules pass the electron to **phylloquinone**, also called Q_K. Because electrons can travel either of two parallel paths within PSI, the complex actually contains two Chl chlorophylls, two Chl A_0 chlorophylls, and two phylloquinones. The molecular structure of the PSI complex from pea plants is shown in **Figure 12.31**. This large monomeric protein complex contains 18 protein subunits, 173 chlorophyll a molecules (most of which function as antenna chlorophylls), 15 β-carotene molecules, 2 phylloquinones, and 3 4 Fe–4 S clusters.

Similar to the role of H_2O oxidation in reducing the $P680^+$ cation in photosystem II, plastocyanin reduces the $P700^+$ cation to return it to the ground state in photosystem I. Oxidation of the phylloquinones involves a linked set of three protein-bound 4 Fe–4 S centers, called

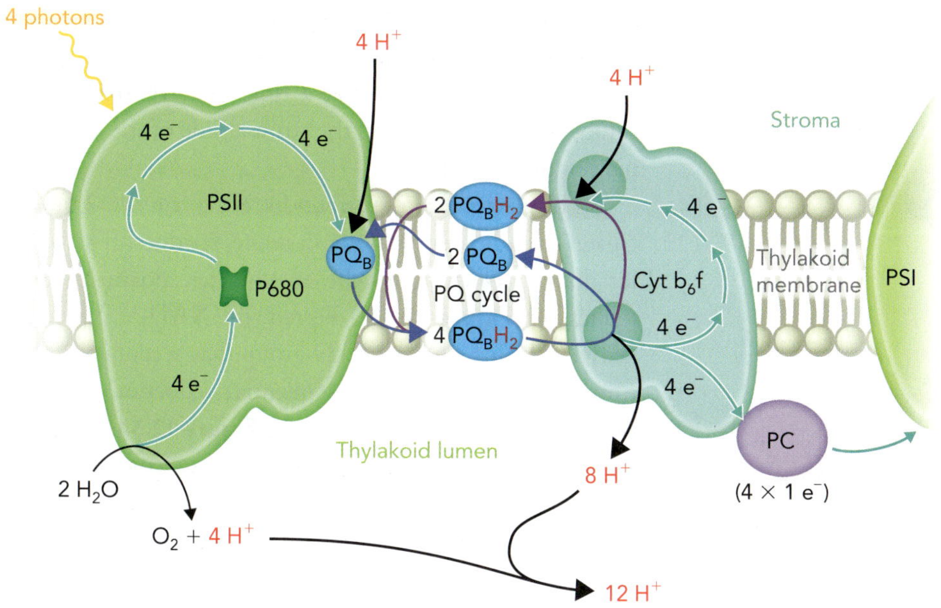

Figure 12.29 In the PQ cycle, 8 H^+ are translocated by the combined reactions in PSII and cytochrome b_6f. An additional 4 H^+ are generated by the oxidation of 2 H_2O to form O_2, leading to a net gain of 12 H^+ in the thylakoid lumen.

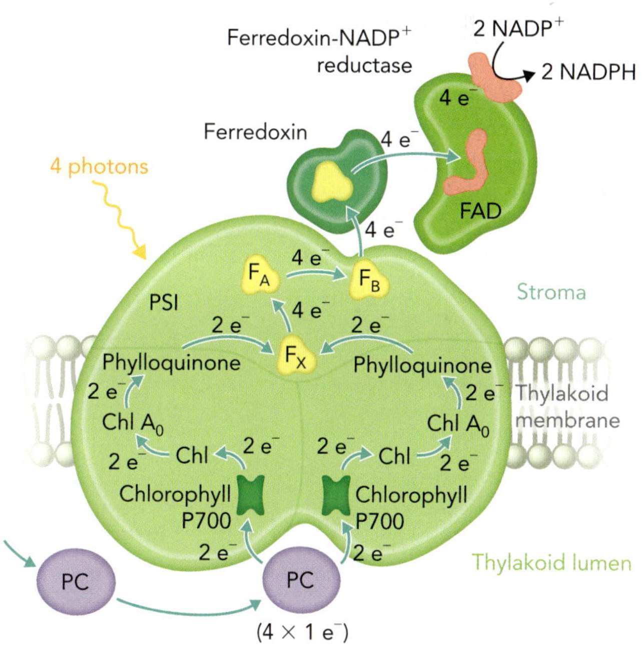

Figure 12.30 Photooxidation of chlorophyll P700 in the PSI complex results in electron flow through parallel pathways, culminating in the reduction of $NADP^+$ by the FAD-containing enzyme ferredoxin-$NADP^+$ reductase. Electrons from ferredoxin are used to reduce the FAD moiety in ferredoxin-$NADP^+$ reductase, which then reduces $NADP^+$ to form NADPH in the stroma. Plastocyanin replaces the electron lost from chlorophyll P700 after photooxidation. Only one electron at a time flows through each of the two parallel pathways; however, the numbers shown refer to the fact that a total of 4 e^- are needed to reduce 2 $NADP^+$ to 2 NADPH.

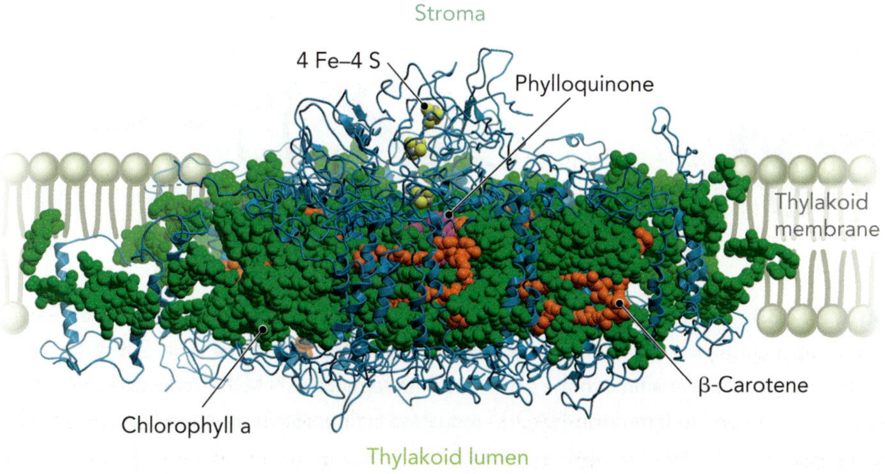

Figure 12.31 The structure of the chloroplast PSI complex in pea plants is shown here with the locations of the light-absorbing chlorophyll a (green) and β-carotene (orange) molecules labeled. In addition, the locations of the two phylloquinones (pink; only one is visible from this orientation) and three 4 Fe–4 S clusters (yellow) are shown. BASED ON PDB FILE 3LW5.

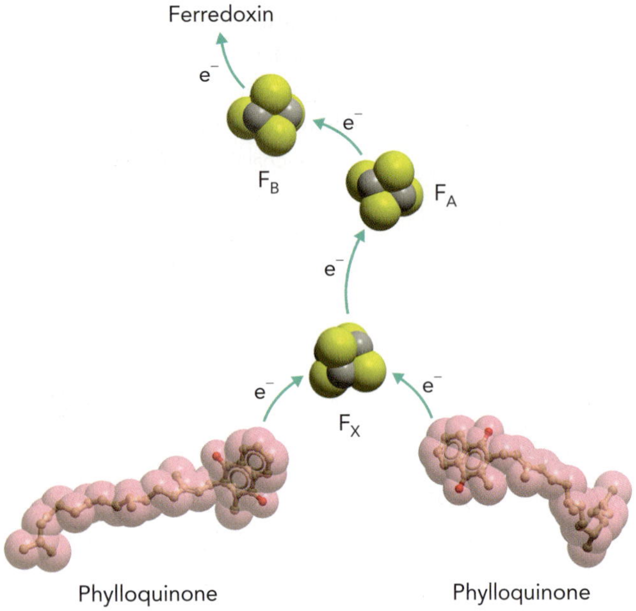

Figure 12.32 A molecular model showing the structural orientation of the two phylloquinones and three 4 Fe–4 S clusters in the chloroplast PSI complex. An electron is donated to F_X from either one of the two phylloquinones in PSI, and then F_X donates it to F_A, which transfers it to F_B. BASED ON PDB FILE 3LW5.

F_X, F_A, and F_B (**Figure 12.32**). The F_B 4 Fe–4 S cluster is the final electron carrier in PSI and is responsible for transferring electrons one at a time to **ferredoxin (Fd)**, a soluble protein located on the stromal side of the thylakoid membrane.

The electron carrier group in ferredoxin is a 2 Fe–2 S cluster coordinated to four cysteine residues. This cluster donates electrons to the enzyme ferredoxin-NADP$^+$ reductase (**Figure 12.33**). Ferredoxin-NADP$^+$ reductase (FdR) contains a bound FAD group that accepts one electron at a time from two sequential reduction reactions involving ferredoxin, as shown in **Figure 12.34**. The fully reduced $FADH_2$ moiety is then able to pass two electrons to NADP$^+$ in the form of a hydride ion, generating NADPH and thereby completing the photosynthetic electron transport system. Note that as an H^+ from the stromal side is used to produce NADPH, this reaction contributes to the ΔpH across the thylakoid membrane. The NADPH produced by ferredoxin-NADP$^+$ reductase is used by the Calvin cycle to generate triose phosphates after CO_2 fixation, as described later in the chapter.

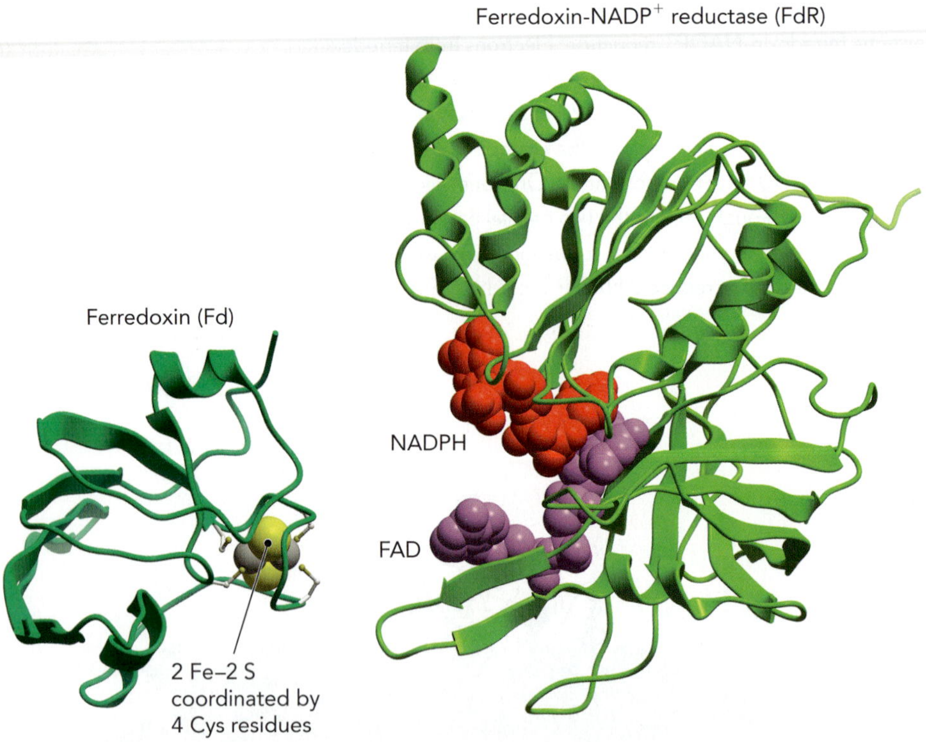

Figure 12.33 The structures of spinach chloroplast ferredoxin and pea chloroplast ferredoxin-NADP$^+$ reductase proteins are shown here, indicating the location of the 2 Fe–2 S cluster, which donates electrons one at a time to the FAD moiety in ferredoxin-NADP$^+$ reductase. A molecule of NADPH is bound to ferredoxin-NADP$^+$ reductase in this protein structure. BASED ON PDB FILES 1A70 (FERREDOXIN) AND 1QFZ (FERREDOXIN-NADP$^+$ REDUCTASE).

Paraquat is a potent herbicide that kills plants by accepting electrons from PSI and donating them to O_2. This electron capture not only blocks NADPH production but also produces superoxide anion ($O_2^{\bullet-}$) and hydrogen peroxide (H_2O_2), which are highly toxic to cells (**Figure 12.35**). Paraquat was used in the 1980s as an aerial herbicide to destroy fields of marijuana and coca plants in North and South America. The use of paraquat for this purpose was suspended when it was found that growers were harvesting the sprayed plants before they wilted and processing them for distribution. Smoking of paraquat-contaminated marijuana was thought to cause widespread lung damage through superoxide-mediated destruction of cell membranes.

Now that we have examined each of the protein components in photosynthetic electron transport, let's put them all together and summarize their contributions to this energy conversion process. As shown in **Figure 12.36**, the oxidation of 2 H_2O molecules within the thylakoid lumen requires the absorption of 8 photons (four each at PSII and PSI) and results in the transfer of 4 e^- through the system. This photon-induced redox energy results in the

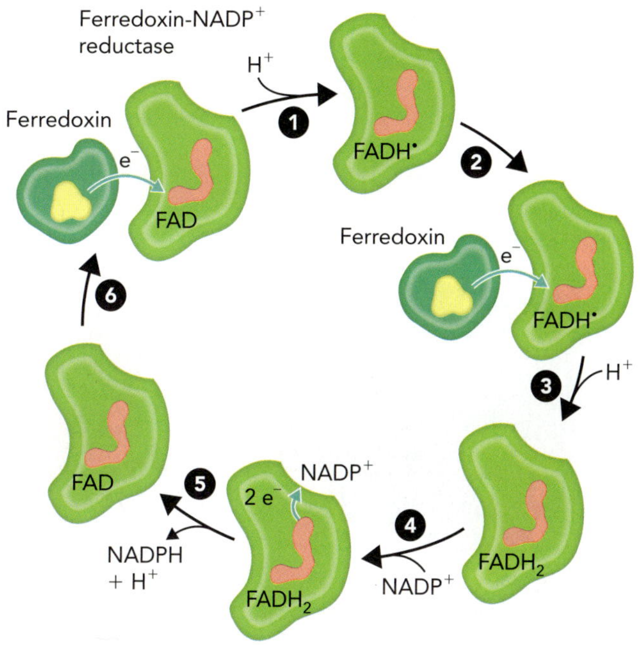

Figure 12.34 Ferredoxin-NADP$^+$ reductase contains an FAD moiety that converts a one-electron reduction reaction by ferredoxin into a two-electron reduction of NADP$^+$ through a semiquinone intermediate (FADH$^\bullet$). The net uptake of one H$^+$ from the stroma to reduce NADP$^+$ contributes to the ΔpH across the thylakoid membrane.

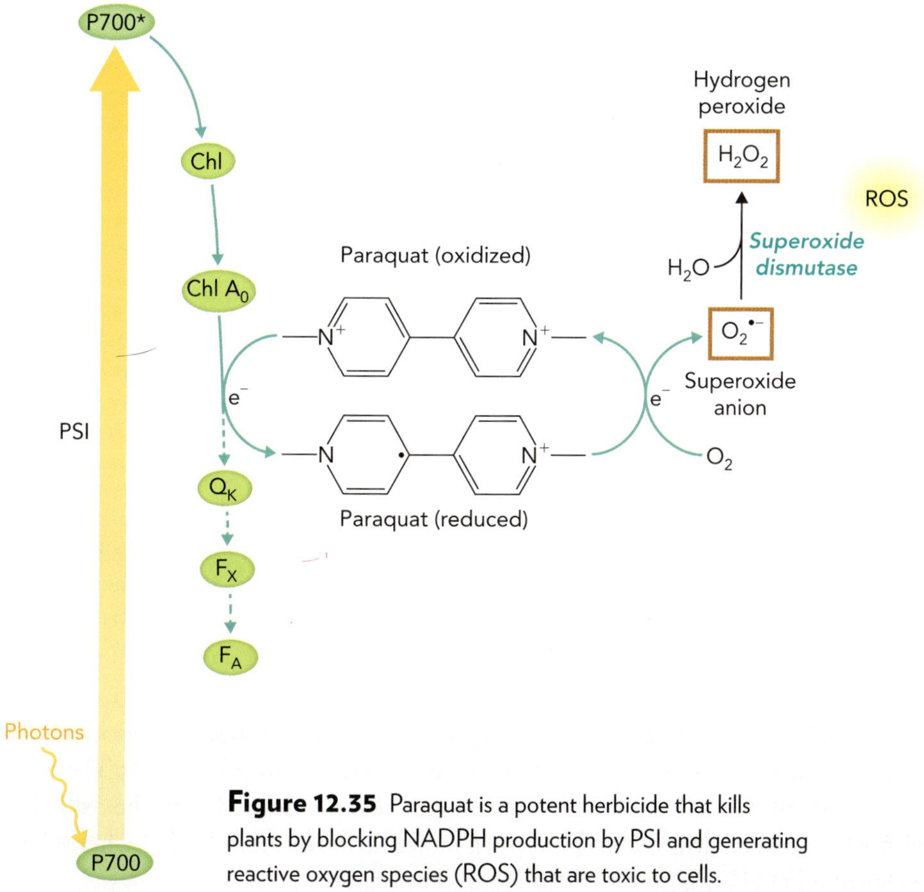

Figure 12.35 Paraquat is a potent herbicide that kills plants by blocking NADPH production by PSI and generating reactive oxygen species (ROS) that are toxic to cells.

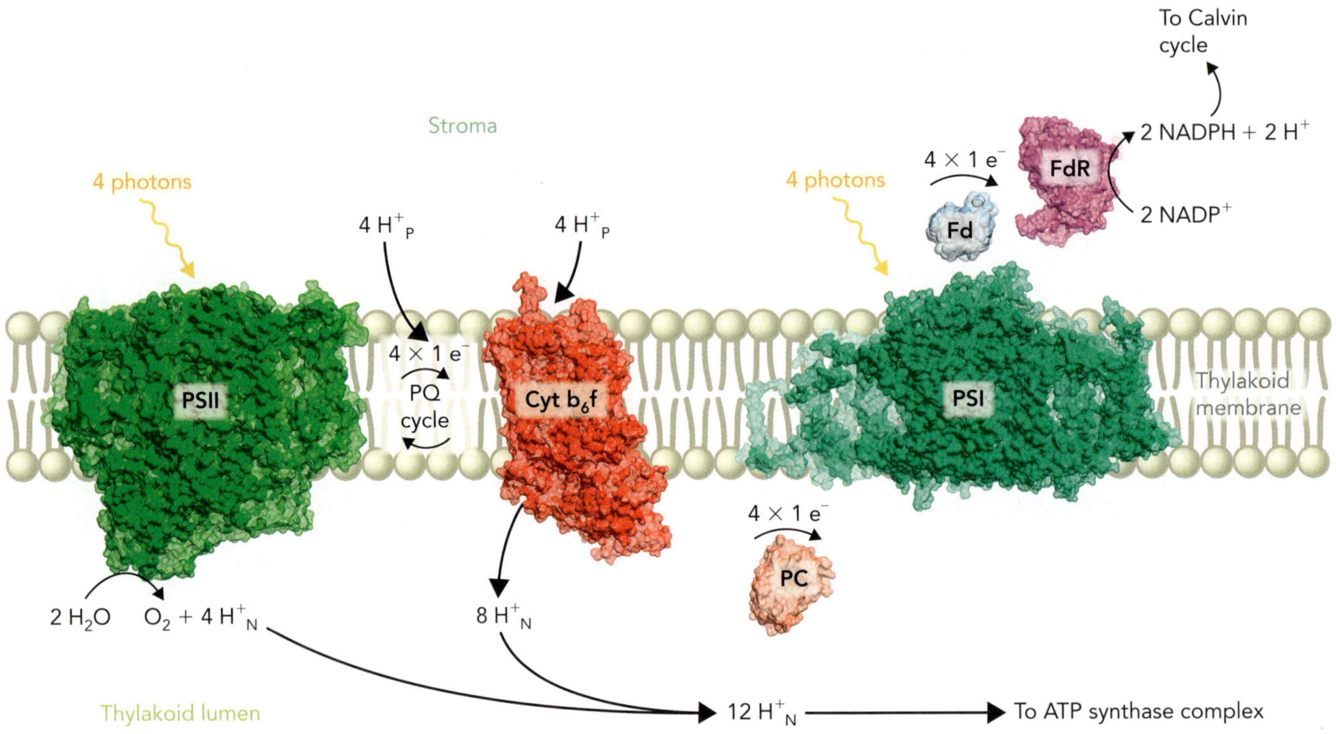

Figure 12.36 Protein complexes and electron carriers in the photosynthetic electron transport system using molecular structures that have been elucidated for all six of the protein components. Oxidation of 2 H_2O generates 4 e^-, which are transported through the system by coupled redox reactions. This leads to the production of 2 NADPH and a chemiosmotic gradient across the thylakoid membrane. PROTEIN SURFACE STRUCTURES ARE BASED ON PDB FILES LISTED IN FIGURES 12.21, 12.27, 12.28, 12.31, AND 12.33.

net accumulation of 12 H^+ in the stroma relative to the lumen, with the production of 1 O_2 and 2 NADPH, as shown by the equation

$$2\ H_2O + 8\ \text{Photons} + 12\ {H^+}_{N\ (\text{stroma})} + 2\ NADP^+ \rightarrow O_2 + 12\ {H^+}_{P\ (\text{lumen})} + 2\ NADPH + 2\ H^+$$

concept integration 12.2

How do photosynthetic reaction centers convert light energy into redox energy?

Photosynthetic reaction centers convert light energy into redox energy through photon absorption by chlorophyll molecules in the PSII and PSI reaction center complexes. This provides energy to lift a chlorophyll electron to a higher orbital, where it is said to be in an excited state. This excited electron is donated to a nearby acceptor molecule (pheophytin in the case of PSII or another chlorophyll molecule in PSI), leading to a separation of charge. The reduced acceptor molecule (now negatively charged) donates the electron to another acceptor molecule of lower reduction potential, thereby activating the photosynthetic electron transport system. The reaction center chlorophyll in PSII is reduced by an electron derived from the oxidation of H_2O, thereby returning the chlorophyll molecule to the ground state, whereas the reaction center chlorophyll in PSI is reduced by plastocyanin.

12.3 Photophosphorylation Generates ATP

We have seen many parallels between the electron transport system in mitochondria and the photosynthetic electron transport system in chloroplasts. Oxidative phosphorylation in mitochondria also has a number of similarities to the mechanism of ATP synthesis in photosynthetic organisms. In chloroplasts, these processes are together called photophosphorylation because the absorption of light energy (photon absorption) is coupled to a phosphoryl transfer reaction catalyzed by the ATP synthase complex in chloroplasts.

Structural studies of the chloroplast ATP synthase complex have shown that like the mitochondrial ATP synthase complex, it consists of two components: a membrane-bound F_o complex called CF_o and an F_1 catalytic complex called CF_1. Moreover, it has been shown that the same binding-change mechanism of mitochondrial ATP synthesis proposed by Paul Boyer also explains ATP production in chloroplasts. We first describe studies linking the electrochemical proton gradient across the thylakoid membrane to ATP synthesis. Then, we look at how cyclic photosynthetic electron transport through the PSI complex generates ATP without the reduction of $NADP^+$.

Proton-Motive Force Provides Energy for Photophosphorylation

One of the key biochemical experiments supporting Peter Mitchell's chemiosmotic theory was performed by Andre Jagendorf and Ernesto Uribe in 1966. They wanted to test directly if a proton gradient could support ATP synthesis in the absence of redox energy or light. For these experiments, they suspended isolated chloroplasts in an acidic solution at pH 4 in the dark and allowed the chloroplasts to equilibrate with the solution. As illustrated in **Figure 12.37**, the chloroplasts were subsequently resuspended in an alkali solution at pH 8, and ADP and ^{32}P (radioactive phosphorus) were then added to initiate proton flow through the chloroplast ATP synthase complex. Within just 15 seconds, a significant amount of radioactive ATP was synthesized in the chloroplast stroma. These results were consistent with Mitchell's chemiosmotic theory and demonstrated that an electrochemical proton gradient was sufficient to drive ATP synthesis, even in the absence of a functional energy-conversion system (redox or photon absorption).

Experiments have shown that the chloroplast proton circuit involves the net accumulation of ~12 H^+ inside the thylakoid lumen, relative to the stroma, for every O_2 produced by the oxidation of 2 H_2O. In addition, ~3 ATP appear to be synthesized per O_2 generated, suggesting that ~4 H^+ are transported through the chloroplast ATP synthase complex for every ATP synthesized (12 H^+/3 ATP). This represents one additional H^+ compared to H^+ transport through the mitochondrial ATP

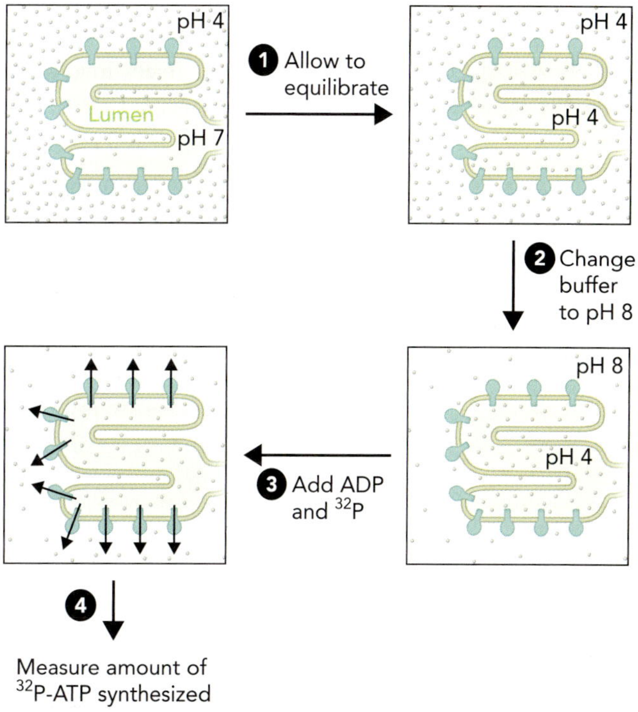

Figure 12.37 An artificial proton gradient is sufficient to drive ATP synthesis in isolated chloroplasts in the absence of light. For these experiments, chloroplasts were suspended in low-pH buffer in the dark and allowed to equilibrate (proton leakage across the thylakoid membrane occurs at a low rate). Upon resuspension of the chloroplasts in high-pH buffer, followed by the addition of ADP and ^{32}P, proton flow through the chloroplast ATP synthase complex resulted in the production of ^{32}P-ATP in the stroma. The results of this experiment were consistent with Peter Mitchell's chemiosmotic theory.

synthase complex (~3 H^+ per ATP). This difference could be due either to uncertainties in the experimental measurements or to differences in the "gear ratio" of the chloroplast ATP synthase complex, which has 14 subunits in the c ring rather than 10, as found in the c ring of the yeast mitochondrial ATP synthase complex (see Figure 11.32).

Cyclic Photophosphorylation Controls ATP-to-NADPH Ratios

Until now, we have described photosynthetic electron transport as a linked redox pathway in which electrons from H_2O are passed along to $NADP^+$ through the Z scheme. However, as shown in **Figure 12.38**, an alternative path exists for electrons in the PSI reaction center that pass through ferredoxin. Rather than ferredoxin donating electrons to $NADP^+$, the electrons can be passed instead to the plastoquinone pool (PQ_B) and fed into cytochrome b_6f. As electrons pass through cytochrome b_6f, protons are translocated across the membrane. In this alternative pathway, photon absorption by chlorophyll P700 initiates a cyclic electron transport system that moves the electron from plastocyanin though PSI, ferredoxin, the plastoquinone pool, cytochrome b_6f, and back to plastocyanin. This process is named **cyclic photophosphorylation** for the cyclic movement of electrons resulting in proton translocation across the membrane, which can be used to drive ATP synthesis. Cyclic photophosphorylation does not produce O_2 or NADPH from this path of electron flow.

Cyclic photophosphorylation in plants is primarily active when electron flow to $NADP^+$ is rate limiting, resulting in ferredoxin-mediated reduction of PQ_B as an alternative electron acceptor. It is not known precisely how ferredoxin reduces PQ_B, but it likely involves a ferredoxin–plastoquinone reductase enzyme that has yet to be identified. This cyclic pathway is a good way to control ATP-to-NADPH ratios because if NADPH ratios get too high ($NADP^+$ is rate limiting), then electrons are shunted off to plastoquinone. This stimulates ATP synthesis and restores the ratio of 3 ATP to 2 NADPH required for the Calvin cycle.

One way to stimulate cyclic photosynthetic electron transfer in chloroplasts is to increase the rate at which energy is transferred from the light-harvesting complexes (LHC I and LHC II) to the chlorophyll P700 reaction center (PSI). This is made possible by the uneven distribution of PSI and PSII complexes in the thylakoid membrane. PSII complexes are primarily associated with stacked thylakoid membranes (grana), whereas PSI and ATP synthase are more highly concentrated in the unstacked lamellar regions of the membrane. Cytochrome b_6f and molecules in the plastoquinone pool are distributed

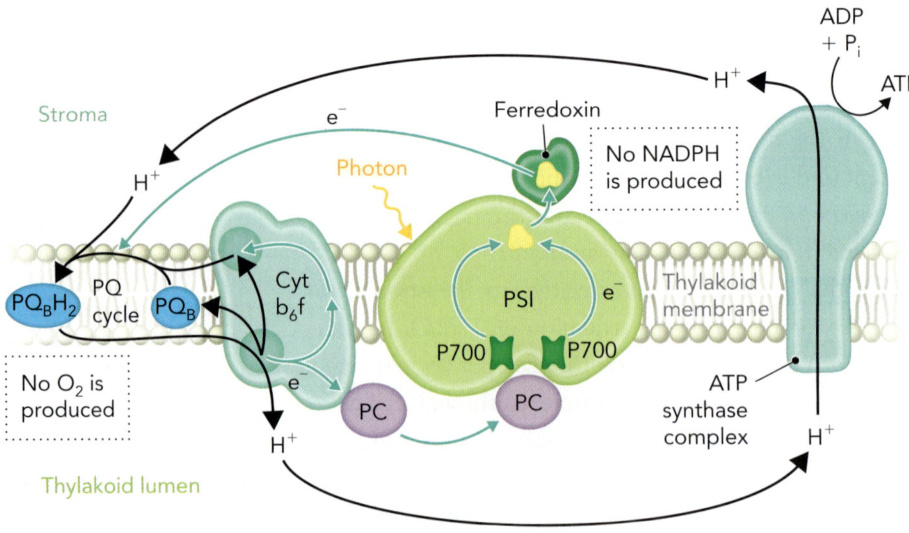

Figure 12.38 Cyclic photophosphorylation between PSI and cytochrome b_6f generates an electrochemical proton gradient independent of H_2O oxidation (O_2 is not produced) and $NADP^+$ reduction. When $NADP^+$ levels are limiting (high NADPH in the stroma), photooxidation of chloroplast P700 leads to electron flow between ferredoxin and the plastoquinone pool (PQ cycle), resulting in proton translocation by cytochrome b_6f and ATP synthesis.

throughout the grana and lamellae. Having PSII and PSI complexes localized to different regions of the convoluted thylakoid membrane makes sense for three reasons:

1. Localization of PSI in the lamellae increases accessibility to $NADP^+$ molecules present in the stroma. Moreover, ATP synthesis in this location provides ATP for the Calvin cycle, which is a stromal pathway.
2. If PSII and PSI complexes were too close together, then photon absorption by PSII would result in energy transfer to PSI, rather than a photooxidation event within PSII, because the energy required to excite chlorophyll P700 is less than that required for chlorophyll P680.
3. Separation of PSII and PSI complexes allows for the control of light absorption by these two complexes through a signaling mechanism involving the phosphorylation and dephosphorylation of LHC II.

When chlorophyll P680 is maximally excited, electron transfer through PSII results in elevated levels of PQ_BH_2 relative to PQ_B. Without sufficient photon absorption by chlorophyll P700, electron transfer through the Z scheme is limited because plastocyanin is not readily oxidized. As shown in **Figure 12.39**, under these

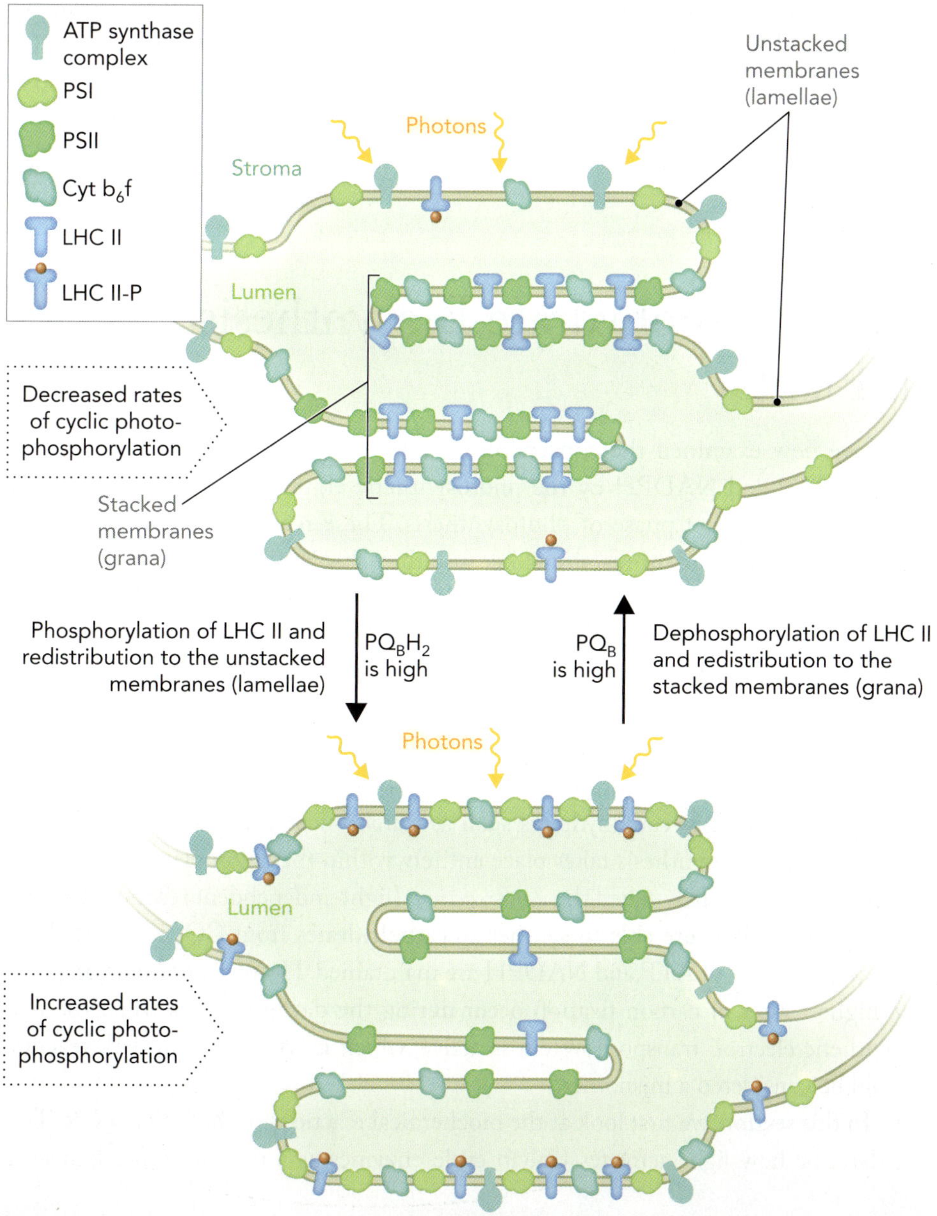

Figure 12.39 Redistribution of LHC II between the stacked and unstacked regions of the thylakoid membrane controls the rate of photooxidation at PSII and PSI reaction centers. When PQ_BH_2 levels are high, an LHC II kinase is activated, resulting in LHC II phosphorylation and movement of LHC II-P from the stacked to unstacked regions of the thylakoid membrane. This causes increased rates of PSI photooxidation and cyclic photophosphorylation.

conditions (high PQ_BH_2-to-PQ_B ratio), LHC II is phosphorylated by a PQ_BH_2-activated LHC II kinase to produce LHC II-P. Now, LHC II-P has a high binding affinity for PSI complexes, so elevated levels of PQ_BH_2 result in the redistribution of LHC II molecules to the lamellae, where they can harvest photons for PSI. Increased levels of photooxidation by PSI complexes not only stimulates electron flow through PSII by oxidizing plastocyanin but also leads to higher levels of cyclic photophosphorylation when $NADP^+$ levels become limiting.

concept integration 12.3

What conditions in the chloroplast stroma stimulate cyclic photophosphorylation?

Cyclic photophosphorylation is stimulated when $NADP^+$ levels in the stroma are low due to high levels of NADPH. This can happen when ATP levels are limiting, causing reduced flux through the Calvin cycle. Under these conditions, photon absorption by chlorophyll P700 in the PSI reaction center complex leads to increased electron flow through the cytochrome b_6f complex as a result of ferredoxin donating an electron back to the plastoquinone pool rather than to ferredoxin-$NADP^+$ reductase (see Figure 12.30). This PSI-dependent pathway stimulates proton translocation and ATP synthesis, thereby restoring the ratio of 3 ATP to 2 NADPH needed to convert CO_2 to carbohydrate by the Calvin cycle pathway.

12.4 Carbohydrate Biosynthesis in Plants

We have now examined the conversion of light energy to chemical energy in the form of ATP and NADPH by the photosynthetic electron transport system. This is considered the first phase of photosynthesis. These reactions are called the *light reactions* (light-dependent reactions) because they absolutely require light. However, in terms of energy conversion within the biosphere, heterotrophic organisms need plants to convert energy available from ATP and NADPH into metabolic fuel. This occurs through a second set of biochemical reactions called the Calvin cycle, which is a carbon fixation pathway involving CO_2.

The Calvin cycle is named after Melvin Calvin, who discovered it along with Andrew Benson and James Bassham in the early 1950s. The Calvin cycle converts atmospheric CO_2 into carbohydrates such as starch, sucrose, and cellulose. This second phase of photosynthesis takes place entirely within the chloroplast stroma. These reactions were originally called the *dark reactions* (light-independent reactions) because isolated chloroplasts are able to synthesize carbohydrates from CO_2 in the dark—as long as high levels of ATP and NADPH are maintained. However, normally in plants, the highest rates of carbon fixation occur during the daylight hours when the photosynthetic electron transport system is active, which is why the term dark reactions should be considered a misnomer.

In this section, we first look at the biochemical reactions of the Calvin cycle. Then we describe how light activates Calvin cycle enzymes as a regulatory mechanism to capitalize on the available ATP and NADPH produced when photosynthetic electron transport is stimulated. We conclude by studying two examples in nature in which CO_2 is first captured in the form of four-carbon (C_4) compounds. These compounds are then metabolized to release CO_2 under conditions that enhance the efficiency of the Calvin cycle enzyme rubisco.

Carbon Fixation by the Calvin Cycle

Figure 12.40 summarizes the main features of the Calvin cycle. It generates the metabolites 3-phosphoglycerate, glyceraldehyde-3-phosphate, and dihydroxyacetone phosphate, all of which may be used to synthesize the hexose phosphates fructose-1,6-bisphosphate and fructose-6-phosphate. Hexose phosphates are converted to (1) sucrose for transport to other plant tissues; (2) starch for energy stores within the cell; and (3) cellulose for cell wall synthesis. As discussed shortly, hexose phosphates are also used to generate pentose phosphates such as ribose-5-phosphate,

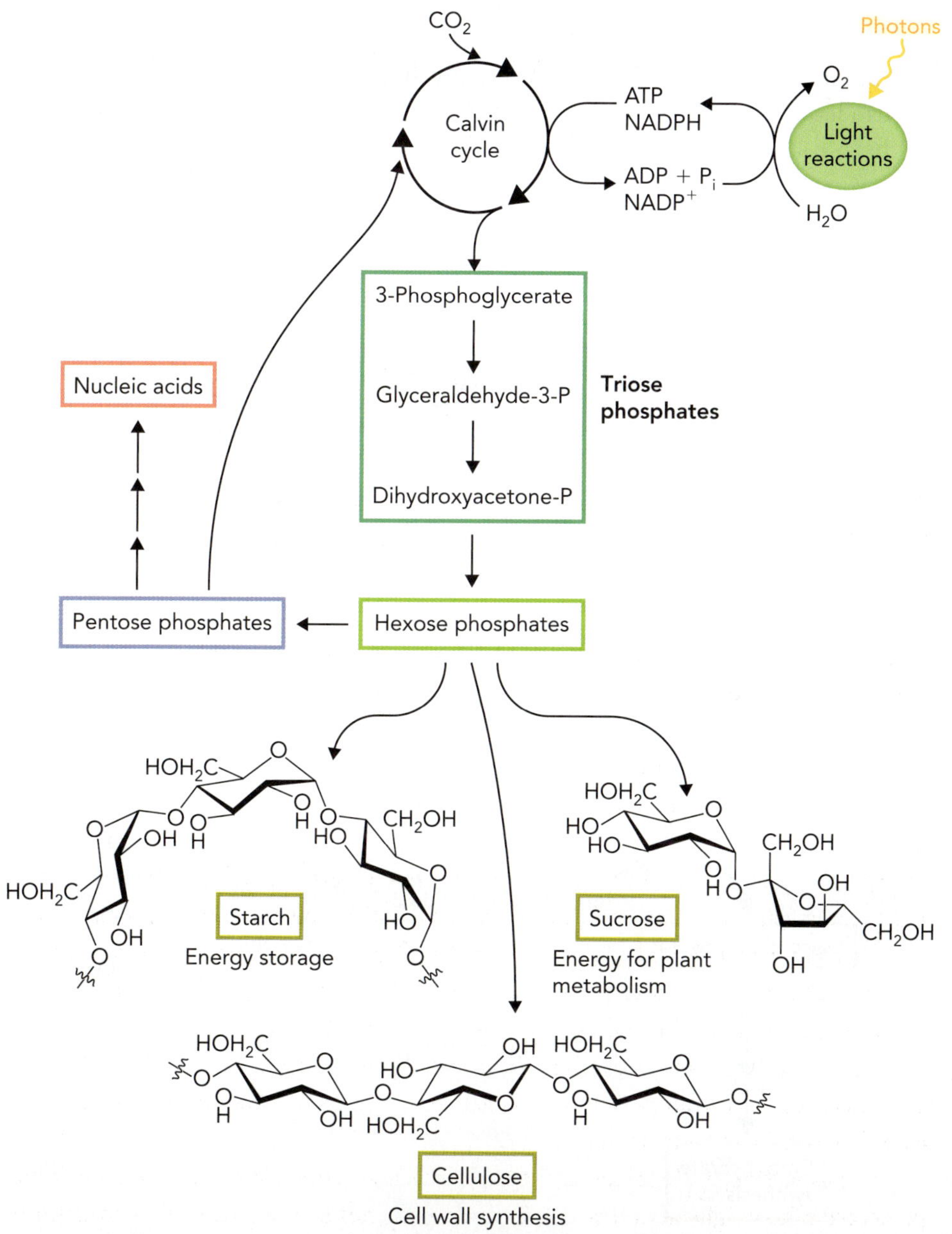

Figure 12.40 The Calvin cycle uses chemical energy in the form of ATP and NADPH to convert CO_2 into triose phosphates, which are used to produce hexose phosphates, the building blocks for the three primary carbohydrates in plant cells: starch, sucrose, and cellulose. Pentose phosphates are Calvin cycle intermediates and also form the carbohydrate backbone for DNA and RNA.

which is used to synthesize nucleic acids, and ribulose-1,5-bisphosphate, a Calvin cycle intermediate that needs to be regenerated in order for the cycle to continue.

The basic scheme of the Calvin cycle is illustrated in **Figure 12.41**. You can see that three molecules of CO_2 must enter the cycle to form one net C_3 molecule, glyceraldehyde-3-phosphate. The carbon fixation reaction is catalyzed by the enzyme ribulose-1,5-bisphosphate carboxylase/oxygenase, which is called rubisco for short.

- In *stage 1*, the key reaction catalyzed by rubisco combines ribulose-1,5-bisphosphate (RuBP), a five-carbon (C_5) compound, with CO_2 to form a transient six-carbon (C_6) intermediate, which is rapidly cleaved to two molecules of the C_3 compound 3-phosphoglycerate.
- In *stage 2*, the 3-phosphoglycerate molecule is reduced to glyceraldehyde-3-phosphate at the expense of ATP and NADPH.
- In *stage 3*, glyceraldehyde-3-phosphate is used to resynthesize ribulose-1,5-bisphosphate. The series of regeneration reactions that produce ribulose-1,5-bisphosphate are called "carbon shuffle" reactions. The role of ribulose-1,5-bisphosphate in the Calvin cycle is analogous to that of oxaloacetate in the citrate cycle; that is, to function as the acceptor molecule that initiates each new round. Alternatively, glyceraldehyde-3-phosphate can leave the Calvin cycle before stage 3 and be used in the net synthesis of hexose sugars.

▶ ANIMATION

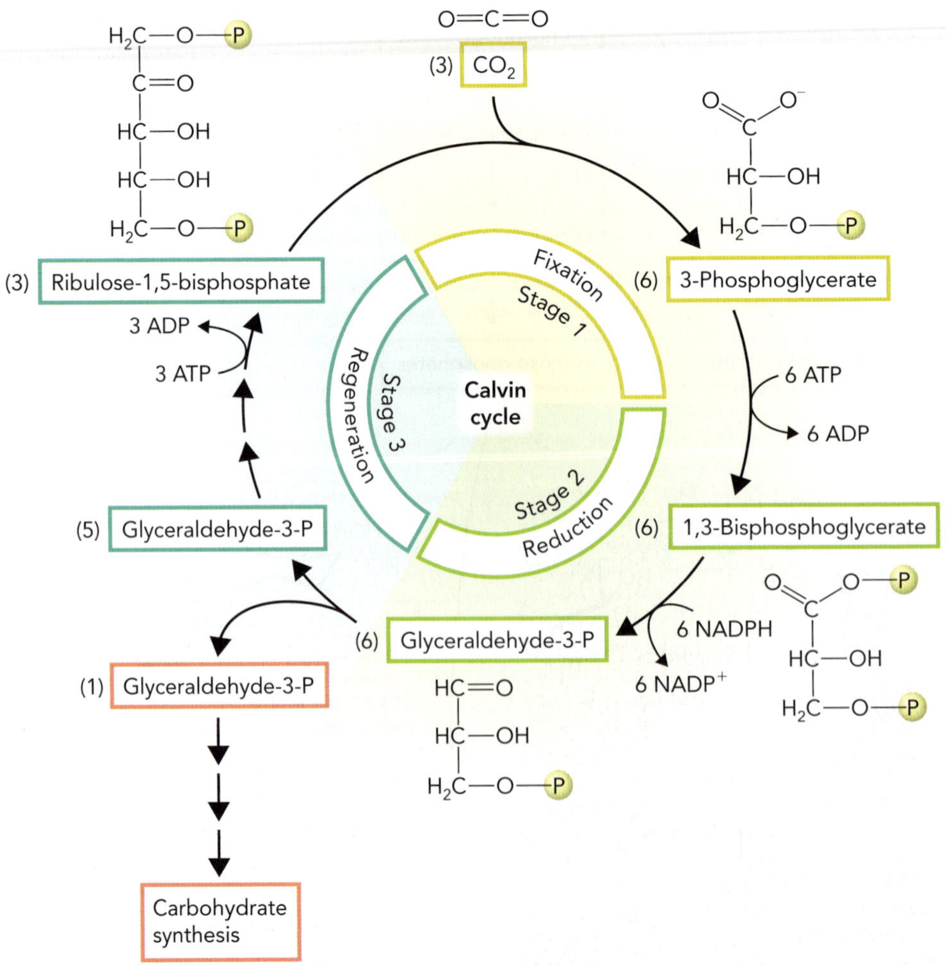

Figure 12.41 A summary of the three stages of the Calvin cycle is shown here. The numbers in parentheses next to each compound name represent the stoichiometry of molecules involved in fixation of three CO_2 molecules to produce one glyceraldehyde-3-phosphate and regenerate five ribulose-1,5-bisphosphate starting molecules.

The net Calvin cycle reactions fixing three CO_2 molecules and producing one glyceraldehyde-3-phosphate molecule can be written as

$$3\ CO_2 + 3\ \text{RuBP} + 6\ \text{NADPH} + 9\ \text{ATP} + 6\ H_2O \rightarrow$$
$$1\ \text{Glyceraldehyde-3-P} + 3\ \text{RuBP} + 6\ \text{NADP}^+ + 9\ \text{ADP} + 9\ P_i$$

If we look at the fate of just the carbons in the Calvin cycle reactions, we see that three CO_2 and three C_5 ribulose-1,5-bisphosphate molecules are consumed, forming one net C_3 compound (glyceraldehyde-3-phosphate) and regenerating three C_5 ribulose-1,5-bisphosphate molecules:

$$3\ C_1 + 3\ C_5 \rightarrow 1\ C_3 + 3\ C_5$$

Let's now examine each of these three stages in more detail to see how the 13 enzymes of the Calvin cycle accomplish the most important energy conversion process on Earth—turning atmospheric CO_2 into metabolic fuel.

Stage 1: Fixation of CO_2 into 3-Phosphoglycerate by Rubisco When Calvin began his experiments in the 1940s, it was known that CO_2 in plant chloroplasts was converted to hexose phosphates; however, it was not clear how this was accomplished. To identify the metabolic intermediates in this process, Calvin and his colleagues used radioactive labeling with $^{14}CO_2$ to follow carbon fixation in photosynthetic algae cells grown in culture (**Figure 12.42**). They found that within a few seconds of adding $^{14}CO_2$ to the culture, the cells accumulated ^{14}C-labeled 3-phosphoglycerate, suggesting that this was the first product of the carboxylation reaction. Although it was initially thought that the formation of this C_3 compound involved carboxylation of a C_2 precursor, additional $^{14}CO_2$ labeling experiments revealed that in fact, the C_5 compound RuBP was the CO_2 acceptor molecule. Calvin's elegant experiments were carried out at the University of California at Berkeley and earned him the 1961 Nobel Prize in Chemistry.

The work of Calvin and others led to the identification and characterization of rubisco as the key enzyme in the carbon fixation reaction. As shown in **Figure 12.43**, the rubisco reaction can be broken down into five basic steps. (1) The formation of an enediolate intermediate of ribulose-1,5-bisphosphate requires the participation of a carbamoylated lysine residue (Lys-CO_2) in the enzyme active site. (2) Carboxylation then occurs by nucleophilic attack on the CO_2 to form an unstable C_6 intermediate (2′-carboxy-3-keto-D-arabinitol-1,5-bisphosphate). (3) This intermediate is hydrated.

a. b.

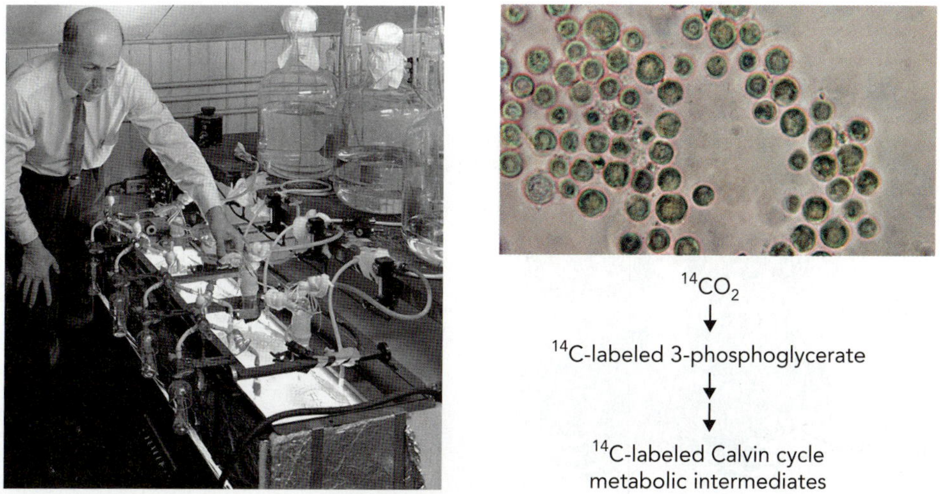

Figure 12.42 Melvin Calvin (1911–1997) discovered the cyclic pathway required for CO_2 fixation using single-cell algal cells and a ^{14}C radioactive labeling assay. **a.** Calvin in his laboratory at the University of California, Berkeley, in the 1960s. RANDSC/ALAMY [SIC]. **b.** *Chlorella pyrenoidosa*, the photosynthetic algal species Calvin used for his experiments. *CHLORELLA PYRENOIDOSA*: DE AGOSTINI PICTURE LIBRARY/SCIENCE SOURCE.

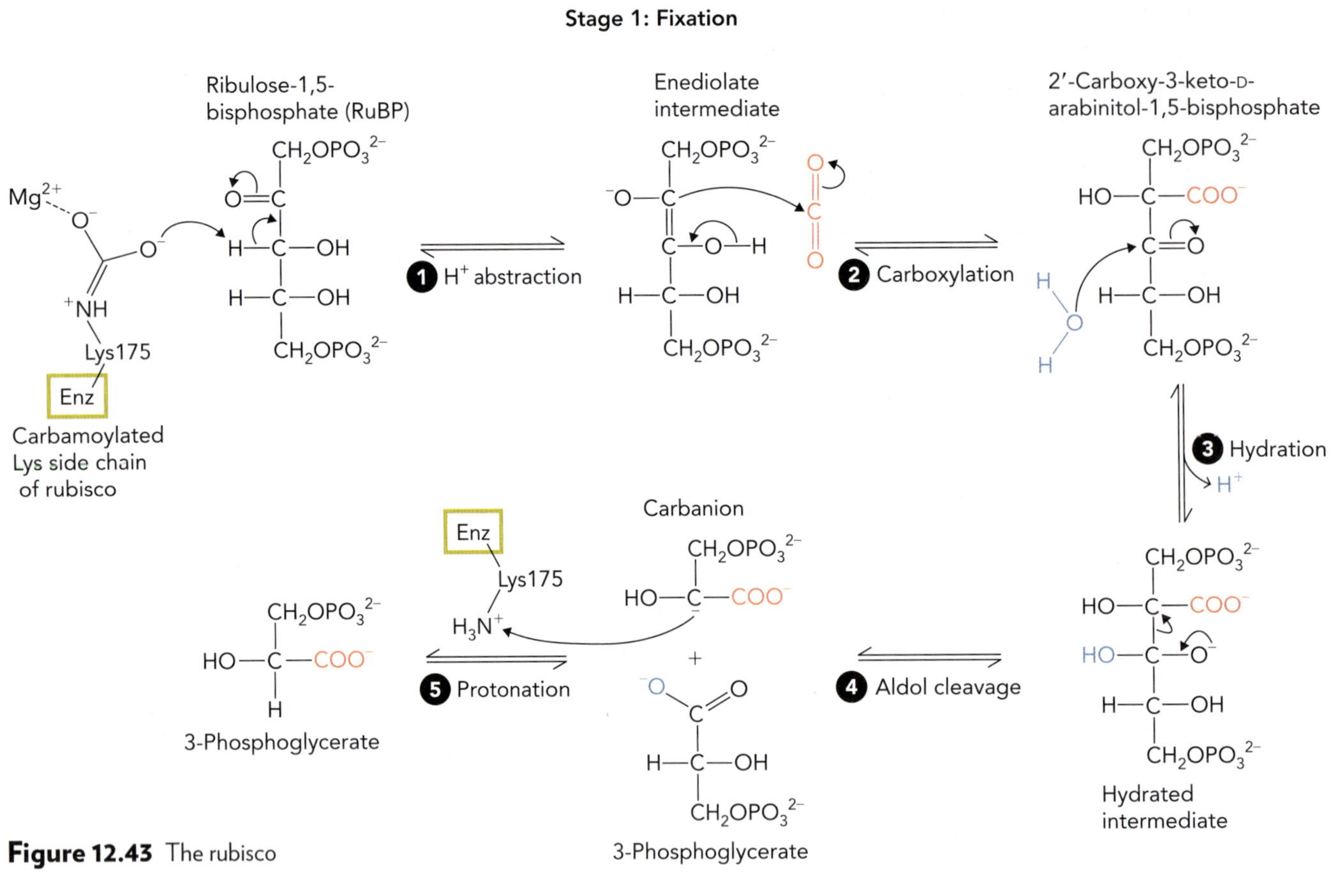

Figure 12.43 The rubisco reaction mechanism consists of five steps, involving the formation of a C_6 intermediate (2′-carboxy-3-keto-D-arabinitol-1,5-bisphosphate) that is split into two molecules of 3-phosphoglycerate by aldol cleavage. The abbreviation Enz here refers to the enzyme rubisco.

(4) Aldol cleavage leads to the formation of one molecule of 3-phosphoglycerate, which is released, and a C_3 carbanion intermediate within the active site. This carbanion harbors the newly carboxylated group (the ^{14}C-labeled carboxyl group in Calvin's experiments). (5) In the final step of the reaction, protonation of the carbanion by a nearby lysine residue (Lys175) releases the second molecule of 3-phosphoglycerate. The rubisco reaction is very exergonic ($\Delta G°' = -35.1$ kJ/mol), with the aldol cleavage step being a major contributor to the favorable change in free energy.

Rubisco is a multi-subunit enzyme consisting of eight large catalytic subunits (L chain) at the core, surrounded by eight smaller subunits (S chain), which function to stabilize the complex and enhance enzyme activity. The molecular structure of the spinach plant rubisco enzyme is shown in **Figure 12.44**. Because of its low catalytic efficiency—a mere three carboxylation reactions per second ($k_{cat} = \sim 3$/s)—chloroplasts

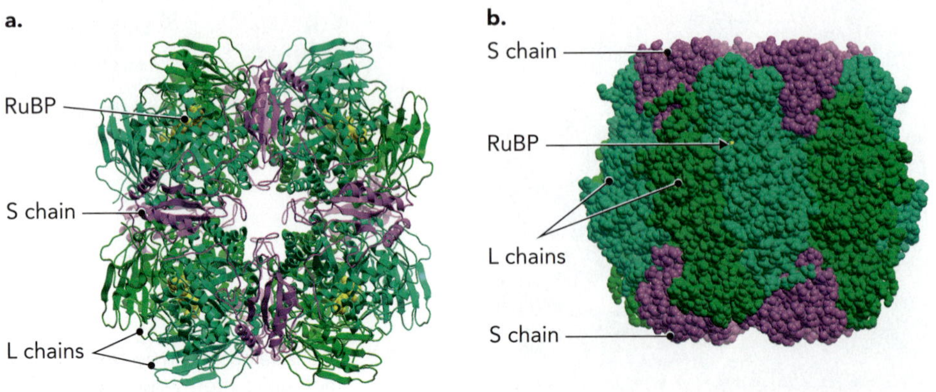

Figure 12.44 The structure of the rubisco enzyme complex that was isolated from spinach plant chloroplasts is shown. The complex consists of eight identical large subunits (L) and eight small subunits (S), arranged in layers of four subunits each. The catalytic sites are in the large subunits of the enzyme, shown here with the RuBP substrates bound to the eight independent active sites. **a.** Top view in ribbon style. **b.** Side view in space-filling style. BASED ON PDB FILE 1RCX.

contain large amounts of the enzyme in order to meet the demands of the cell. In fact, 50% of the total protein in a spinach leaf is rubisco, and the concentration of rubisco in chloroplast stroma is a staggering 250 mg/mL. Considering that rubisco plays a central role in all photosynthetic autotrophic organisms on Earth, of which ~85% are photosynthetic plants and microorganisms that inhabit the oceans, rubisco is arguably the most abundant enzyme, and perhaps the most abundant protein, on this planet.

Structural and biochemical analysis of the rubisco active site revealed that a divalent cation, most often Mg^{2+} in nature, is absolutely critical to enzyme activity. The Mg^{2+} ion stabilizes ribulose-1,5-bisphosphate binding and facilitates nucleophilic attack on the incoming CO_2 by the C_5 enediolate intermediate in step 2 of the reaction mechanism. As shown in **Figure 12.45**, three amino acids surrounding the active site (Asp203, Glu204, Lys201) serve to stabilize the Mg^{2+} ion, including the carbamoylated lysine residue that also has a catalytic function in the reaction. Formation of the carbamoylated lysine is absolutely required for rubisco activity. Formation involves a nonenzymatic carboxylation reaction (not the CO_2 that combines with ribulose-1,5-bisphosphate) that is facilitated by the regulatory enzyme rubisco activase. Rubisco activase is an ATP-dependent enzyme that removes ribulose-1,5-bisphosphate from the active site in order to expose the ε amino group of Lys201. Carbamoylation of Lys201 by CO_2 is favored under these conditions, which leads to binding of Mg^{2+} and enzyme activation.

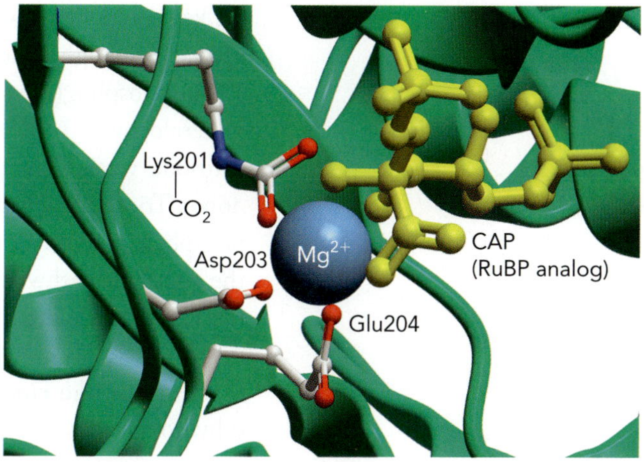

Figure 12.45 The structure of the enzyme active site of rubisco from rice chloroplasts is shown. The Mg^{2+} ion is coordinated by the ribulose-1,5-bisphosphate (RuBP) analog 2-carboxyarabinitol-1,5-diphosphate (CAP) and three highly conserved amino acids. The Lys201 residue in this protein structure is shown in the carbamoylated form. BASED ON PDB FILE 1WDD.

Stage 2: Reduction of 3-Phosphoglycerate to Form Hexose Sugars The 3-phosphoglycerate product of the rubisco reaction is converted to glyceraldehyde-3-phosphate by two isozymes of phosphoglycerate kinase and glyceraldehyde-3-phosphate dehydrogenase, as shown in **Figure 12.46**. The cytosolic forms of these two enzymes

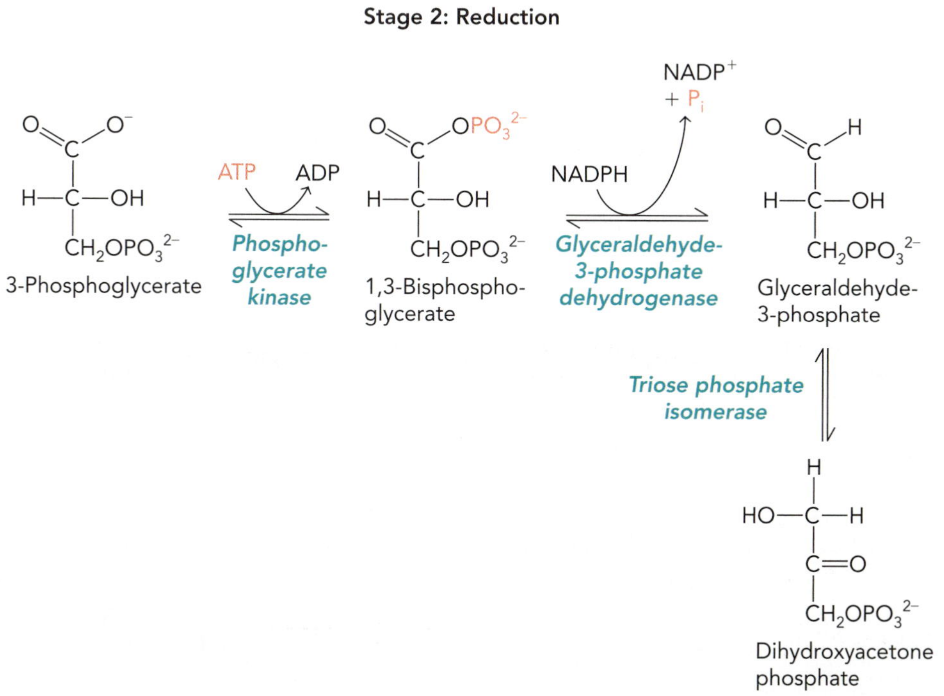

Figure 12.46 Reaction scheme showing the conversion of 3-phosphoglycerate to glyceraldehyde-3-phosphate in the reduction phase (stage 2) of the Calvin cycle. A phosphoryl transferase reaction, catalyzed by the enzyme phosphoglycerate kinase, converts 3-phosphoglycerate to 1,3-bisphosphoglycerate. This molecule is then reduced to glyceraldehyde-3-phosphate by glyceraldehyde-3-phosphate dehydrogenase. Triose phosphate isomerase interconverts glyceraldehyde-3-phosphate and dihydroxyacetone phosphate.

are key components of both the glycolytic and gluconeogenic pathways. In the Calvin cycle, phosphoglycerate kinase uses ATP in a phosphoryl transfer reaction to generate 1,3-bisphosphoglycerate, which is then reduced to glyceraldehyde-3-phosphate by glyceraldehyde-3-phosphate dehydrogenase. This stromal enzyme is similar to the cytosolic version except it uses NADPH instead of NADH as the electron-pair donor. Triose phosphate isomerase catalyzes the interconversion of glyceraldehyde-3-phosphate and dihydroxyacetone phosphate. (We have previously seen this reaction in the glycolytic pathway; see Figure 9.29.) Five triose phosphates formed in this stage of the Calvin cycle are needed to regenerate three ribulose-1,5-bisphosphate molecules in stage 3.

As shown in **Figure 12.47**, the triose phosphates glyceraldehyde-3-phosphate and dihydroxyacetone phosphate produced by the Calvin cycle are useful for the generation

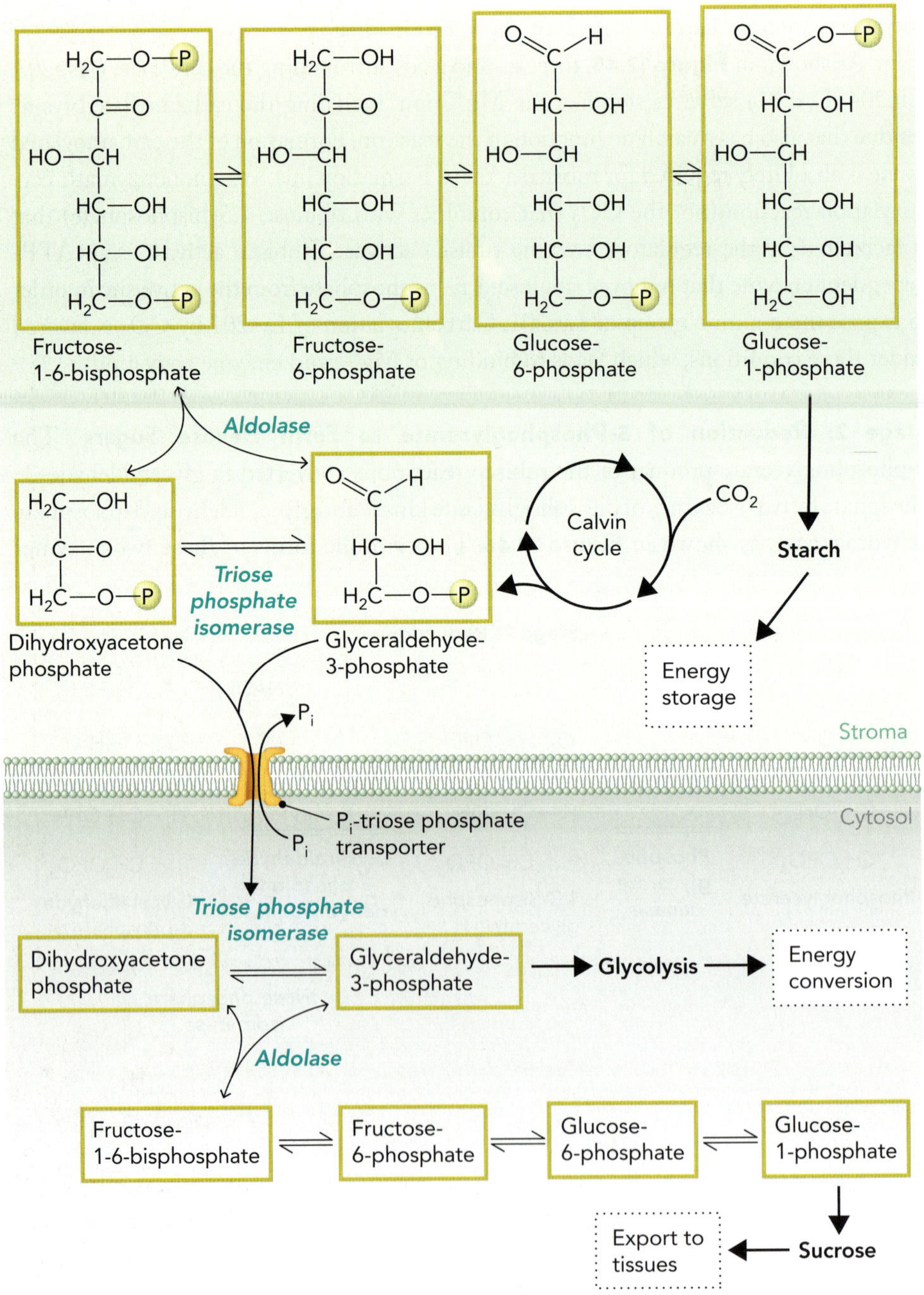

Figure 12.47 The triose phosphates glyceraldehyde-3-phosphate and dihydroxyacetone phosphate are building blocks for hexose sugars in the stroma and cytosol. Glyceraldehyde-3-phosphate and dihydroxyacetone phosphate are used in the stroma to synthesize starch for energy storage (not all reactants and products are shown). Alternatively, these triose phosphates are exported to the cytosol, where they are used to synthesize sucrose for export to plant tissues or catabolized by glycolysis for energy conversion.

of hexose sugars. Fructose-1,6-bisphosphate can be produced by a condensation reaction catalyzed by the enzyme aldolase. This leads to starch synthesis in the chloroplast and sucrose synthesis in the cytosol through interconversion of the hexose phosphates fructose-6-phosphate, glucose-6-phosphate, and glucose-1-phosphate. Aldolase catalyzes this same condensation reaction in the gluconeogenic pathway (see Chapter 14). Because the chloroplast membrane is not permeable to triose phosphates, glyceraldehyde-3-phosphate and dihydroxyacetone phosphate must be exported to the cytosol through a P_i-triose phosphate antiporter, which also serves to import P_i into the stroma. The P_i-triose phosphate antiporter is critical in chloroplasts because 9 ATP are consumed by Calvin cycle reactions in the fixation of 3 CO_2 molecules, but only 8 P_i are released into the stroma because one phosphate is used in the phosphoglycerate kinase reaction to form 1,3-bisphosphoglycerate. Therefore, if the P_i pool inside the stroma is not kept in balance with ADP levels, the rate of ATP synthesis by photophosphorylation will decrease because of limiting amounts of P_i. Note that at night, when the energy charge in the cell is low, cytosolic glyceraldehyde-3-phosphate and dihydroxyacetone phosphate are used by the glycolytic pathway to generate pyruvate for aerobic respiration in the mitochondria.

Stage 3: Regeneration of Ribulose-1,5-Bisphosphate In this final stage of the Calvin cycle, ribulose-1,5-bisphosphate must be regenerated in order to provide a substrate for the rubisco reaction in stage 1. A series of enzymatic reactions are needed to convert five C_3 molecules (glyceraldehyde-3-phosphate or dihydroxyacetone phosphate) into three C_5 molecules (ribulose-1,5-bisphosphate). Two of the primary enzymes in this carbon shuffle are transketolase and transaldolase, which are involved in interconverting C_3, C_4, C_6, and C_7 molecules.

The transketolase reaction transfers a C_2 group from a ketose to an aldose and requires thiamine pyrophosphate as a coenzyme in the reaction. In one case, transketolase transfers the C_2 unit (CO—CH_2OH) from a C_6 molecule (fructose-6-phosphate) to a C_3 molecule (glyceraldehyde-3-phosphate), forming C_4 (erythrose-4-phosphate) and C_5 (xylulose-5-phosphate) molecules. In a second transketolase reaction, a C_2 unit is transferred from a C_7 molecule (sedoheptulose-7-phosphate) to a C_3 molecule (glyceraldehyde-3-phosphate), forming two C_5 molecules (ribose-5-phosphate and xylulose-5-phosphate).

As we have already seen in stage 2 (see Figure 12.47), aldolase catalyzes the formation of a C_6 molecule (fructose-1,6-bisphosphate) from two C_3 molecules (glyceraldehyde-3-phosphate and dihydroxyacetone phosphate); here in stage 3, transaldolase generates a C_7 molecule (sedoheptulose-1,7-bisphosphate) from C_3 (dihydroxyacetone phosphate) and C_4 (erythrose-4-phosphate) molecules.

Figure 12.48 summarizes the first of these carbon shuffle reactions, showing how five C_3 molecules are converted to three C_5 molecules, consisting of two xylulose-5-phosphate and one ribose-5-phosphate. In the final steps of stage 3, shown in **Figure 12.49** (p. 617), the two xylulose-5-phosphate and one ribose-5-phosphate molecules are first converted to three ribulose-5-phosphate molecules by the enzymes ribulose-5-phosphate epimerase and ribose-5-phosphate isomerase, respectively. Then the enzyme ribulose-5-phosphate kinase catalyzes a phosphoryl transfer involving three ATP to generate the final three molecules of ribulose-1,5-bisphosphate. Note that in these carbon shuffle reactions, the net number of carbons (15) has not changed:

$$5\ C_3 \rightarrow 3\ C_5$$

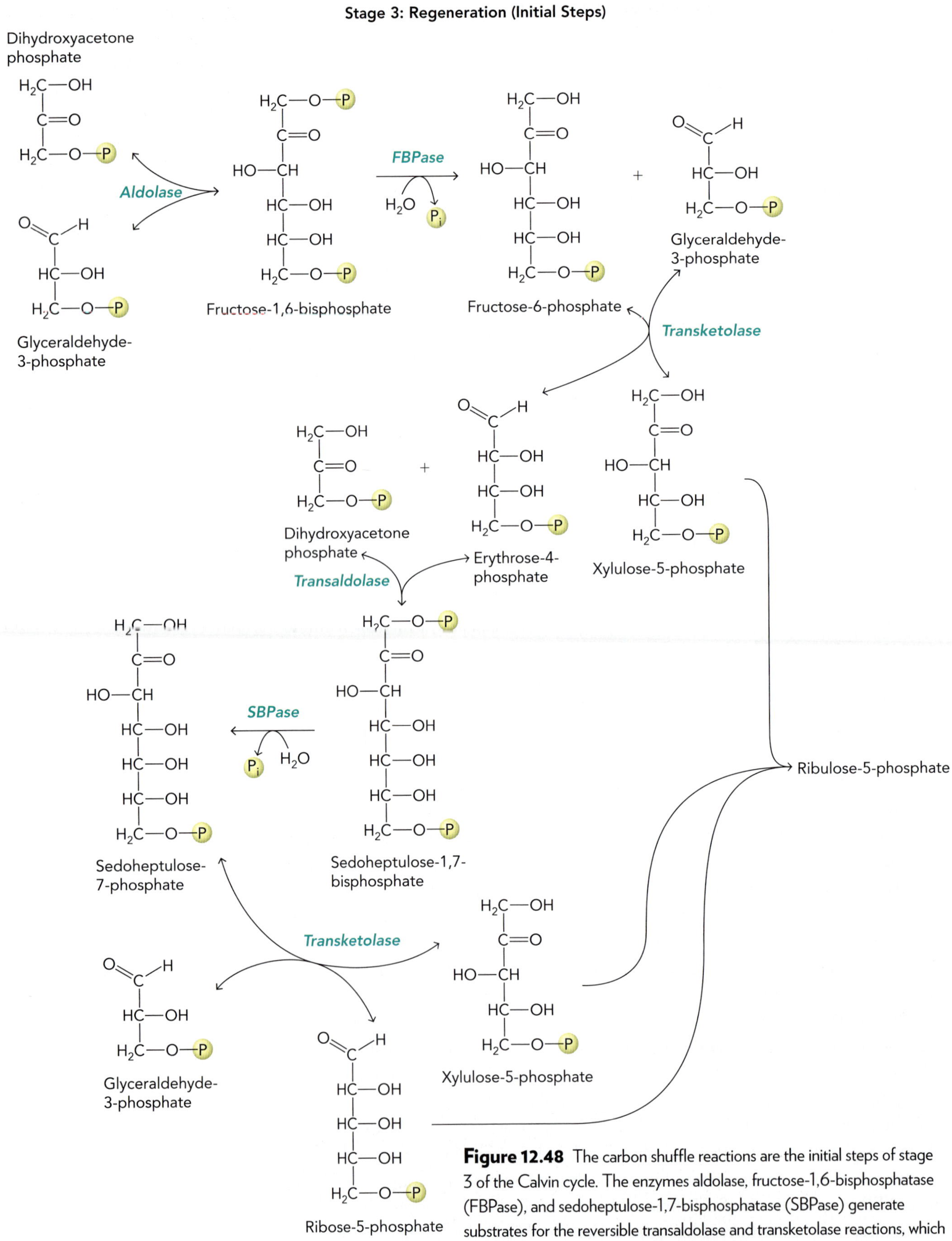

Figure 12.48 The carbon shuffle reactions are the initial steps of stage 3 of the Calvin cycle. The enzymes aldolase, fructose-1,6-bisphosphatase (FBPase), and sedoheptulose-1,7-bisphosphatase (SBPase) generate substrates for the reversible transaldolase and transketolase reactions, which interconvert C_3, C_4, C_6, and C_7 molecules. The three C_5 molecules, two xylulose-5-phosphate and one ribose-5-phosphate, are converted to three ribulose-5-phosphate molecules as shown in Figure 12.49.

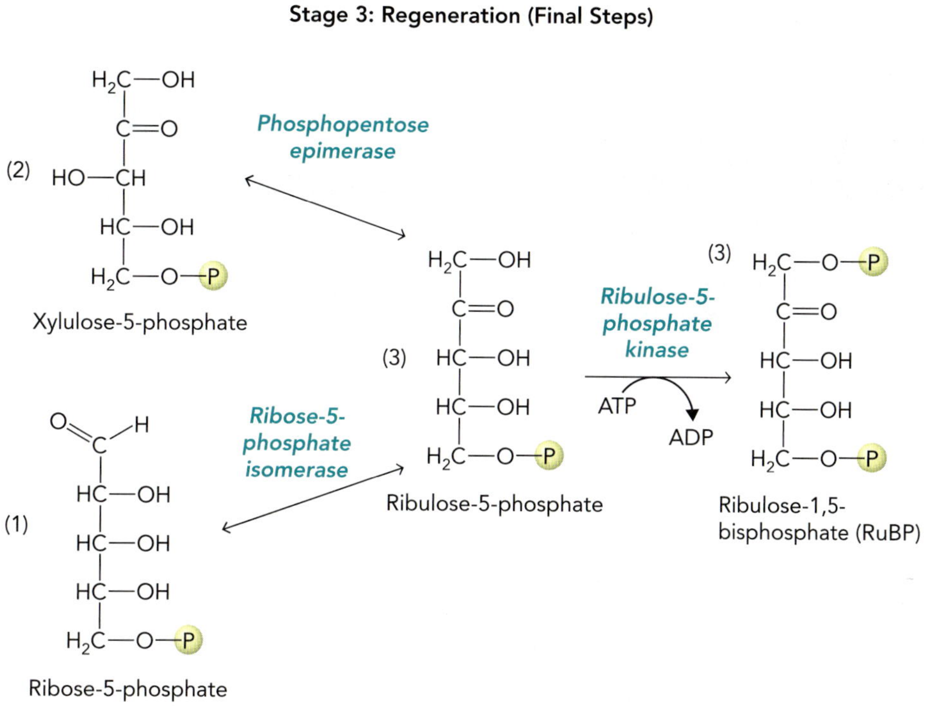

Figure 12.49 Conversion of two molecules of xylulose-5-phosphate and one molecule of ribose-5-phosphate to three molecules of ribulose-5-phosphate occurs in stage 3 of the Calvin cycle. The three molecules of ribulose-5-phosphate are then phosphorylated to form three ribulose-1,5-bisphosphate molecules.

The Activity of Calvin Cycle Enzymes Is Controlled by Light

At night, plant cells rely on glycolysis and mitochondrial aerobic respiration to generate ATP for cellular processes. They also rely on the pentose phosphate pathway for production of NADPH, which is used for many biosynthetic pathways (described in Chapter 14). The chloroplast stroma contains all the enzymes required for glycolysis and the pentose phosphate pathway in addition to the Calvin cycle enzymes.

Starch is stored in the chloroplast stroma during the day, when the Calvin cycle is active, and provides metabolic fuel for these pathways at night. Starch is similar to glycogen in animals in that they both provide carbohydrate substrates for energy, to enable aerobic respiration at times when other energy sources are not available. Because photophosphorylation and NADPH production by the photosynthetic electron transport system is shut down in the dark, it is crucial that the Calvin cycle be active only in the light. Otherwise, if glycolysis, the pentose phosphate pathway, and the Calvin cycle were all active at the same time, then simultaneous starch degradation and carbohydrate biosynthesis would result in decreased pools of ATP and NADPH in the stroma.

Three mechanisms control the activity of Calvin cycle enzymes in response to light intensity:

1. The stromal compartment contains several light-regulated enzyme inhibitors, which are present at higher levels in the dark. One of these inhibitors is called CA1P (2-carboxyarabinitol-1-phosphate), a transition-state analog that blocks rubisco activity. Another is CP12, which inhibits the activities of ribulose-5-phosphate kinase and glyceraldehyde-3-phosphate dehydrogenase by forming an inactive complex.
2. The Calvin cycle enzymes rubisco and fructose-1,6-bisphosphatase are activated by elevated pH and by increased Mg^{2+} concentrations in the stroma.

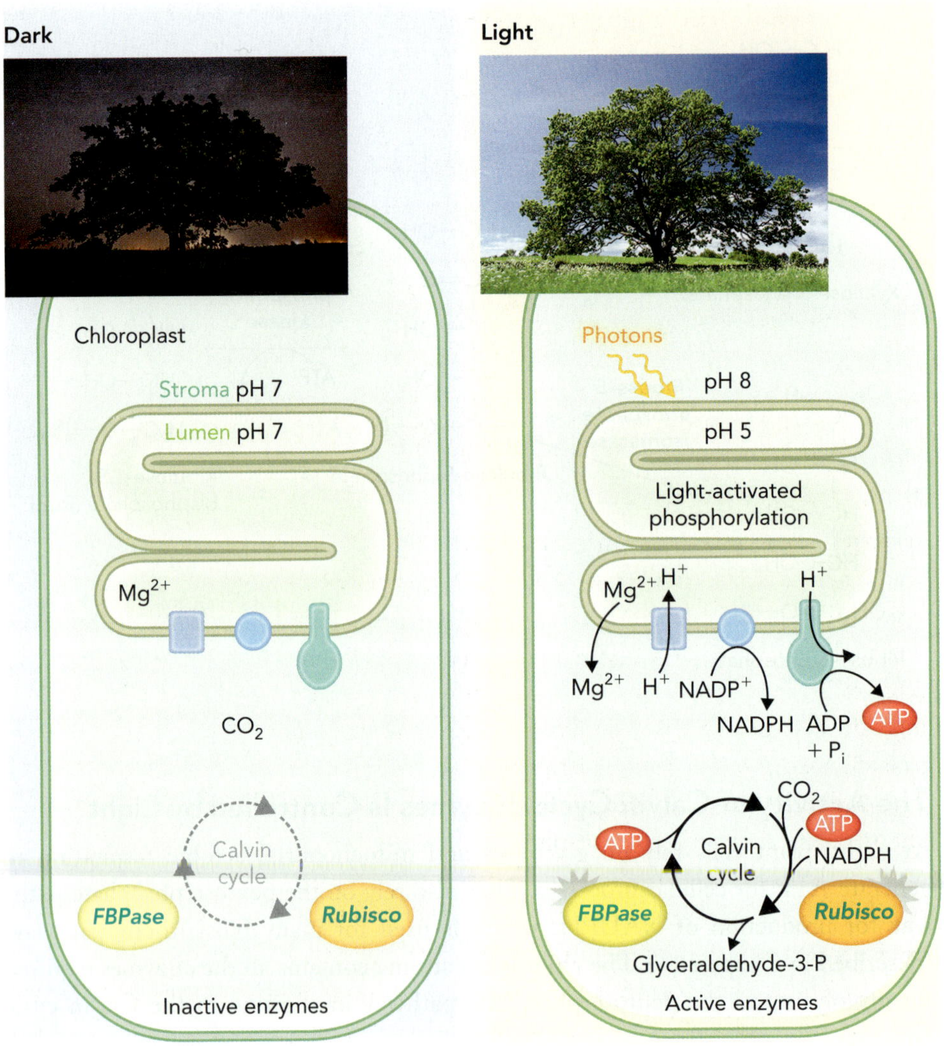

Figure 12.50 The Calvin cycle enzymes rubisco and fructose-1,6-bisphosphatase (FBPase) are stimulated during daylight by elevated pH and Mg^{2+} levels in the stroma. These elevated levels result from activation of the photosynthetic electron transport system in the thylakoid lumen.

NIGHT SKY TREE: ALEXUSSK/SHUTTERSTOCK; DAYLIGHT TREE: VISUALL2/SHUTTERSTOCK.

Light activation of the photosynthetic electron transport system causes stromal pH to increase from pH 7 to pH 8 as a result of proton translocation into the thylakoid lumen. As shown in **Figure 12.50**, this influx of H^+ into the lumen causes an efflux of Mg^{2+} to the stroma to balance the charge (the thylakoid membrane, unlike the mitochondrial inner membrane, is permeable to Mg^{2+} ions).

3. Some Calvin cycle enzymes are controlled by thioredoxin-mediated activation of Calvin cycle enzymes through reduction of disulfide bridges. **Thioredoxin** is a small protein of 12 kDa that is found throughout nature. It functions as a redox protein that can interconvert disulfide bridges and sulfhydryls in cysteine residues of target proteins. In the dark, fructose-1,6-bisphosphatase, sedoheptulose-1,7-bisphosphatase, ribulose-1,5-bisphosphate kinase, and glyceraldehyde-3-phosphate dehydrogenase all contain disulfide bridges that inhibit enzyme activity. Light activation of the photosynthetic electron transport system leads to reduction of thioredoxin present in the stroma, which in turn reduces disulfide bridges in these enzymes, resulting in their activation. As shown in **Figure 12.51**, photon absorption by the PSI reaction center leads

to increased levels of reduced ferredoxin, which passes electrons one at a time to the enzyme ferredoxin-thioredoxin reductase when $NADP^+$ becomes limiting. Reduced thioredoxin then uses these electrons to reduce the disulfide bridges in target enzymes. As long as reduced thioredoxin is present in the stroma, these Calvin cycle enzymes are maintained in the active state; however, when the Sun goes down, spontaneous oxidation leads to their inactivation.

The C_4 and CAM Pathways Reduce Photorespiration in Hot Climates

Rubisco, as its name implies (ribulose-1,5-bisphosphate carboxylase/oxygenase), also catalyzes an oxygenase reaction. As shown in **Figure 12.52**, this reaction combines ribulose-1,5-bisphosphate with O_2 to generate one molecule of 3-phosphoglycerate (C_3) and one molecule of 2-phosphoglycolate (C_2). The rubisco oxygenase reaction is similar to the rubisco carboxylation reaction (see Figure 12.43), except that in this case, the product 2-phosphoglycolate provides no obvious metabolic benefit to the cell. In fact, 2-phosphoglycolate is metabolized by the **glycolate pathway**, which generates CO_2 and requires a significant amount of cellular energy to salvage the C_2 group for reincorporation into the Calvin cycle.

As shown in **Figure 12.53**, the glycolate pathway in plants (also called the C_2 pathway) first converts 2-phosphoglycolate to glycolate in the chloroplast stroma. Glycolate is then exported to plant cell organelles called peroxisomes, where it is oxidized to form glyoxylate. Transamination of glyoxylate in peroxisomes produces glycine, which is then translocated to mitochondria. Here it is metabolized by the enzyme glycine decarboxylase to produce one serine from two glycines, releasing CO_2 and NH_3 in the process. In the next step, the serine goes back to the peroxisomes, which catalyze reactions that form hydroxypyruvate from serine, converting it to glycerate. The glycerate is exported back to the chloroplasts, where it is phosphorylated in the stromal compartment by glycerate kinase to produce 3-phosphoglycerate, a Calvin cycle substrate.

Collectively, oxygenation of ribulose-1,5-bisphosphate and metabolism of 2-phosphoglycolate by the glycolate pathway is called **photorespiration** because O_2 is consumed and CO_2 is released. However, unlike mitochondrial respiration, photorespiration requires energy input and therefore is considered a wasteful pathway in photosynthetic cells.

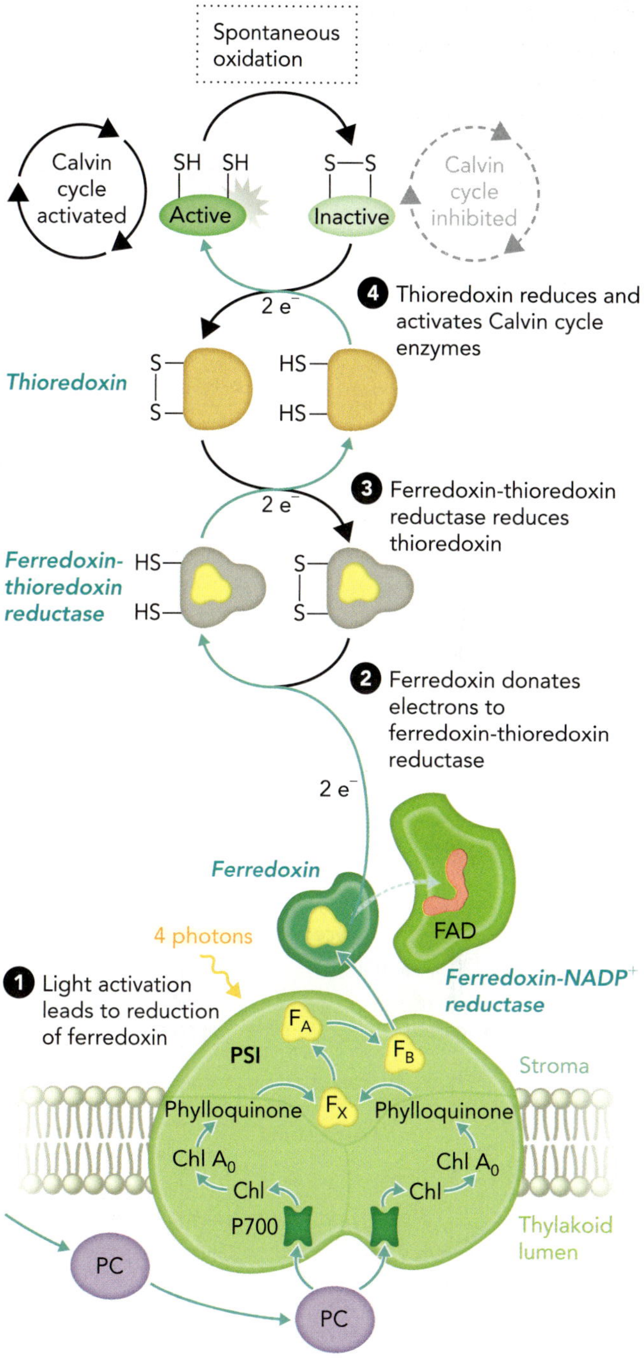

Figure 12.51 The activity of some Calvin cycle enzymes is controlled by a thioredoxin-mediated reduction reaction. When ferredoxin is in the reduced state, as a result of light activation of PSI and low $NADP^+$ levels in the stroma due to excess NADPH, ferredoxin can donate electrons to the soluble electron-carrier protein ferredoxin-thioredoxin reductase. Through a disulfide interchange reaction, ferredoxin-thioredoxin reductase reduces thioredoxin, which in turn reduces and activates the Calvin cycle enzymes fructose-1,6-bisphosphatase, sedoheptulose-1,7-bisphosphatase, ribulose-1,5-bisphosphate kinase, and glyceraldehyde-3-phosphate dehydrogenase.

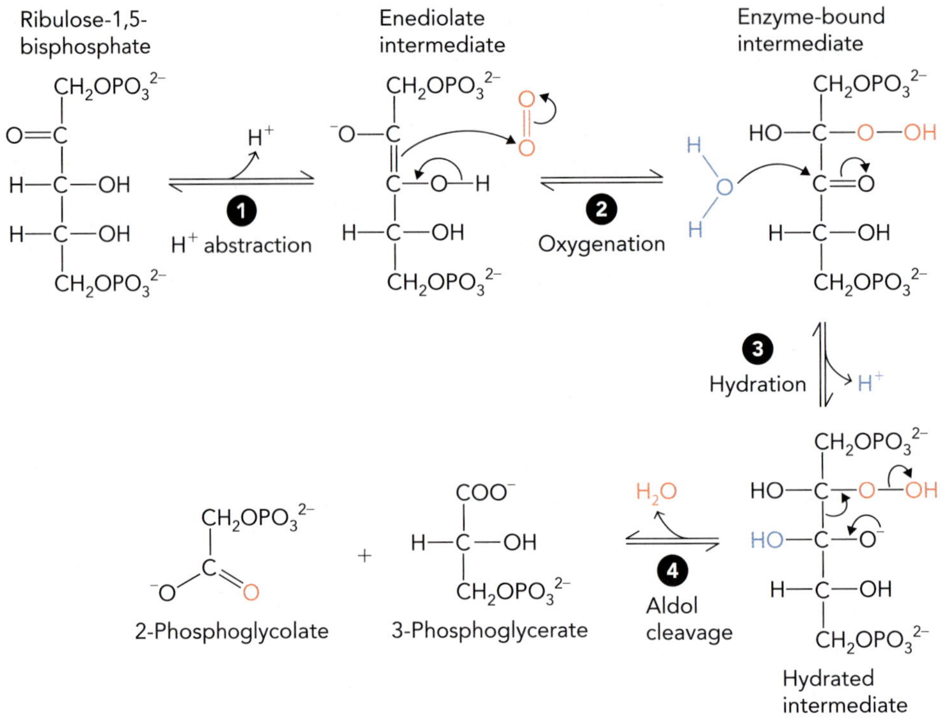

Figure 12.52 The products of the rubisco oxygenase reaction are 3-phosphoglycerate and 2-phosphoglycolate.

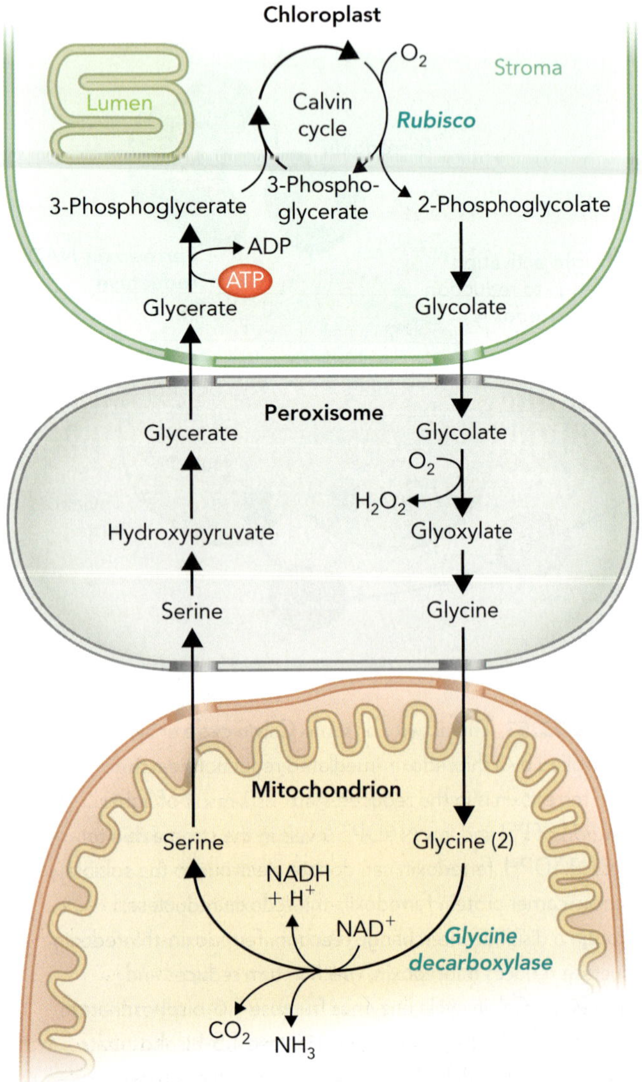

Molecular O_2 binds to the rubisco active site with an affinity that is ~30 times lower than that of CO_2 (K_d of ~10 μM for CO_2 compared to a K_d of ~300 μM for O_2). However, the concentration of O_2 in air is about 500 times higher than that of CO_2 (20% for O_2 compared to 0.04% for CO_2), leading to a significant concentration of O_2 in the chloroplast stroma. Indeed, it is estimated that ribulose-1,5-bisphosphate oxygenation occurs 20% to 30% of the time under normal conditions and can be as high as 50% when the ratio of O_2 to CO_2 rises in the chloroplast stroma. The two environmental conditions that increase O_2-to-CO_2 ratios in the stroma are (1) decreased CO_2 levels due to high rates of carbon fixation in intense light, and (2) increased temperature. Increased temperature leads to higher rates of photorespiration because rubisco is less specific for CO_2 than for O_2 at temperatures above 35 °C. Also,

Figure 12.53 The glycolate pathway involves enzymes in three cellular compartments. The 2-phosphoglycolate produced by rubisco in the chloroplast stroma is converted to glycolate, which is exported to peroxisomes. Glycolate is converted to glycine in peroxisomes and then exported to mitochondria, where two molecules of glycine are converted to one molecule of serine by the enzyme glycine decarboxylase. Serine is exported back to the peroxisomes and converted to glycerate, which returns to the chloroplasts. Here it is phosphorylated to produce 3-phosphoglycerate at the expense of ATP hydrolysis and carbon loss from CO_2.

O_2 is more soluble than CO_2 at higher temperatures, resulting in a higher O_2-to-CO_2 ratio of dissolved gases. High rates of photorespiration inhibit plant growth because of less carbon fixation.

Why does rubisco catalyze both a productive carboxylation reaction and a seemingly wasteful oxygenation reaction? One explanation is that rubisco is a very old enzyme, having evolved in bacteria before the emergence of cyanobacteria, at a time when CO_2 levels were much higher than they are today. Oxygenation by rubisco was presumably infrequent and therefore did not present a selective disadvantage. However, as photosynthetic organisms began to dominate the planet, O_2 levels rose as a direct result of increased photosynthesis, leading to higher rates of photorespiration. Plants in hot, sunny climates are especially susceptible to photorespiration because of high O_2-to-CO_2 ratios in these conditions.

In the 1960s, Marshall Hatch and Roger Slack, plant biochemists at the Colonial Sugar Refining Company in Brisbane, Australia, used $^{14}CO_2$ labeling experiments to determine what the initial products were in the carbon fixation reactions of sugarcane plants. To their surprise, they found that malate was more quickly labeled with ^{14}C than was 3-phosphoglycerate. Follow-up work showed that plants such as sugarcane and corn and weeds such as crabgrass thrive under high-temperature conditions by having very low levels of photorespiration. They found that rather than accumulating mutations in rubisco that increase the efficiency of carbon fixation, these plants evolved a way to increase the concentration of CO_2 inside the stroma dramatically, thereby preventing significant O_2 binding to the rubisco active site. The mechanism involves the carboxylation of phosphoenolpyruvate by the enzyme phosphoenolpyruvate carboxylase to form oxaloacetate, a four-carbon (C_4) intermediate that serves as a transient CO_2 carrier molecule.

Two variations of the "Hatch–Slack pathway" have been characterized, both of which separate CO_2 uptake from carbon fixation in the Calvin cycle (**Figure 12.54**). In the case of the **C_4 pathway**, which is found in tropical plants such as sugarcane, one cell type is required for CO_2 uptake, and a second cell type is required for rubisco-mediated carboxylation. In contrast, plants that use the **CAM pathway**, such as desert succulents

Sugarcane plants use the C_4 pathway
Saguaro cacti use the CAM pathway

Figure 12.54 Sugarcane plants grow in a hot and humid tropical climate and use the C_4 pathway for CO_2 fixation. Saguaro cacti, which grow in a hot and very dry desert climate, use the CAM pathway. SUGARCANE PLANTS: OLGA KHOROSHUNOVA/SHUTTERSTOCK; SAGUARO CACTI: ANTON FOLTIN/SHUTTERSTOCK.

▶ ANIMATION

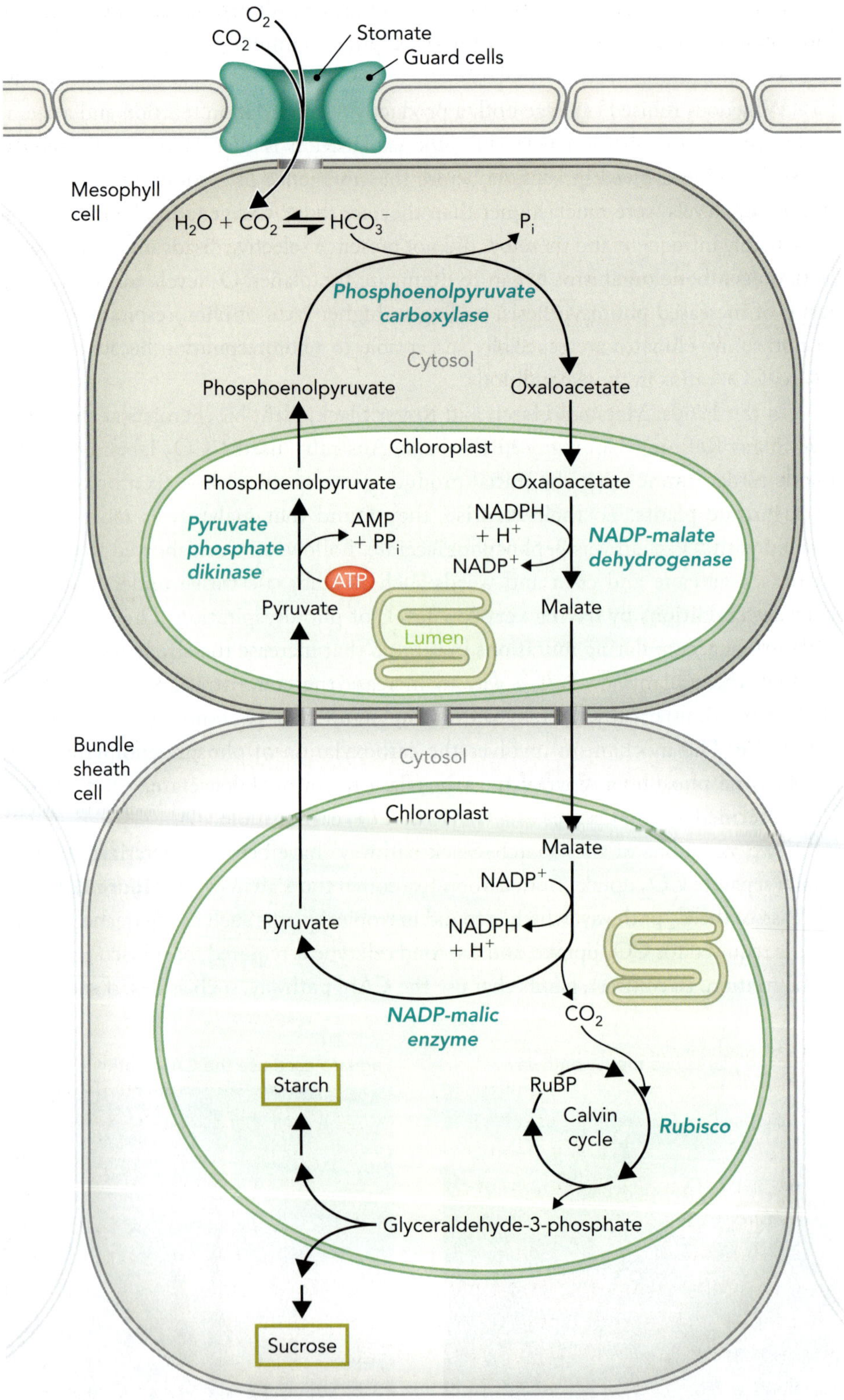

Figure 12.55 The C_4 pathway captures CO_2 in the form of oxaloacetate in mesophyll cells and then delivers it to bundle sheath cells as malate, another C_4 intermediate. Four enzymes are required for the C_4 pathway: phosphoenolpyruvate carboxylase, NADP-malate dehydrogenase, NADP-malic enzyme, and pyruvate phosphate dikinase. The stomate structure in leaves allows gas exchange by altering the shape of guard cells that surround the opening.

including the giant saguaro cactus, capture CO_2 at night when transpiration rates are low and then fix the stored CO_2 during the day.

To understand the functional differences between the C_4 pathway and the CAM pathway, it is important to know how plants obtain CO_2 from the environment. Gas exchange in land plants occurs through a special cell structure called a **stomate**, consisting of two guard cells that change shape to open and close in response to physiologic cues. Sugarcane grows in a moist, hot climate and the stomata (plural for stomate) open during the day to capture CO_2 when photosynthetic electron transport rates are high due to intense sunlight. In contrast, saguaro cacti live in a dry, hot climate and open their stomata only at night. Therefore, they must rely on the captured CO_2 to drive the Calvin cycle reactions during the day, when sunlight is available and the stomata are closed.

The C_4 pathway in sugarcane is shown in **Figure 12.55**. You can see that **mesophyll cells** are responsible for CO_2 capture in the form of oxaloacetate, whereas interior **bundle sheath cells** (farther away from atmospheric O_2) use CO_2 released from malate to carry out the Calvin cycle reactions. Malate functions in the C_4 pathway as the CO_2 transporter molecule from the mesophyll cells into the bundle sheath cells, whereas pyruvate serves to recycle the carbons back to the mesophyll cell. In addition, although chloroplasts in the mesophyll cells are normal, the bundle sheath cell chloroplasts lack PSII reaction centers and therefore cannot oxidize H_2O to generate O_2. This division of labor between the two cell types essentially eliminates the oxygenase reaction in rubisco and thereby blocks photorespiration.

In the first reaction of the C_4 pathway, phosphoenolpyruvate carboxylase catalyzes a reaction combining HCO_3^- with phosphoenolpyruvate to form oxaloacetate. The product oxaloacetate is then reduced to malate by the enzyme NADP-malate dehydrogenase. (Some other C_4 plants transaminate oxaloacetate to form aspartate, which functions as the CO_2 intermediate.) Malate is then transported to the bundle sheath cells through special protein-lined channels connecting the cells. Within the bundle sheath cells, malate is oxidized and decarboxylated in the chloroplast by NADP-malic enzyme, forming CO_2 and pyruvate. The released CO_2 is incorporated into ribulose-1,5-bisphosphate by rubisco, and the resulting 3-phosphoglycerate is metabolized by the Calvin cycle. In sugarcane, pyruvate is transported back to the mesophyll cells, where it is phosphorylated by the enzyme pyruvate phosphate dikinase to yield phosphoenolpyruvate. Because the ATP hydrolysis in this reaction produces AMP and PP_i, it consumes the equivalent of two high-energy phosphate bonds because two ATP are required to convert AMP to ATP (the PP_i is rapidly hydrolyzed to 2 P_i by the enzyme pyrophosphatase).

The additional input of energy required to store the CO_2 temporarily would seem to put C_4 plants at a disadvantage. However, the metabolic cost is more than compensated for by the increased carboxylation efficiency of rubisco in C_4 plants at temperatures above 30 °C. As shown in **Figure 12.56**, this growth advantage of C_4 plants at high temperatures is evident in the heat of summer, when crabgrass—a C_4 plant—is able to invade areas of turfgrass, which is growth-inhibited by high rates of photorespiration. However, at more moderate temperatures in the spring, when photorespiration rates in turfgrass are low, the higher energy cost of the C_4 pathway in the crabgrass is a disadvantage, and the turfgrass is able to prevail.

The CAM pathway was first discovered in succulent plants of the Crassulaceae family, and its full name is the Crassulacean acid metabolism (CAM) pathway. Just like the C_4 pathway in sugarcane and crabgrass, the CAM pathway functions to

a. Summer temperatures above 30 °C

b. Spring temperatures below 20 °C

Figure 12.56 C_4 plants have a growth advantage over C_3 plants in hot climates because of reduced rates of photorespiration; however, C_3 plants have a growth advantage over C_4 plants at cooler temperatures. **a.** Crabgrass is a C_4 plant that outgrows turfgrass in the summer. WALLY EBERHART/VISUALS UNLIMITED/CORBIS. **b.** Turfgrass grows faster than crabgrass in cooler climates because of the reduced energy requirements of the normal Calvin cycle pathway. FERRY/SHUTTERSTOCK.

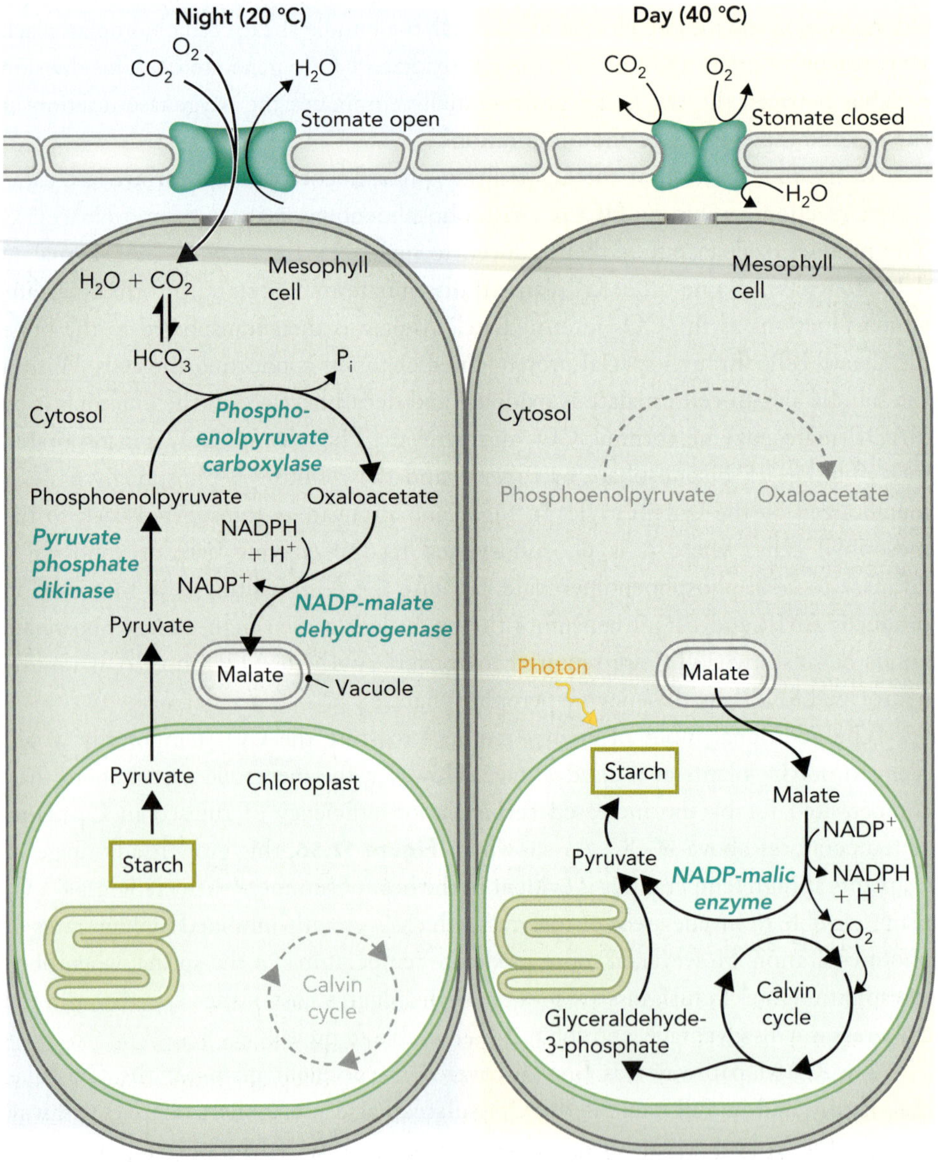

Figure 12.57 The CAM pathway traps CO_2 in the form of malate during the night, when the Calvin cycle is inactive and transpiration rates are low. During the day, malate is decarboxylated in the chloroplast and starch is produced.

concentrate CO_2 levels in the chloroplast stroma to limit the oxygenase activity of rubisco. The main difference is that instead of using two cell types spatially to prevent atmospheric O_2 from binding to rubisco as in C_4 plants, CAM plants use a temporal separation to accomplish the same goal.

As shown in **Figure 12.57**, during the night when the stomata are open in CAM plants, CO_2 is captured by the mesophyll cells and incorporated into oxaloacetate by phosphoenolpyruvate carboxylase. The oxaloacetate product is then reduced by the enzyme NADP-malate dehydrogenase to form malate. The malate is stored inside vacuoles overnight until sunlight activates the Calvin cycle enzymes the next morning. At that time, the malate is oxidized and decarboxylated by NADP-malic enzyme to generate CO_2 and pyruvate. Unlike in C_4 plants, in which the carbon skeleton of pyruvate is recycled directly to phosphoenolpyruvate, the pyruvate in CAM plants is stored as starch during the day. At night, the starch granules are degraded, releasing pyruvate to be converted into phosphoenolpyruvate for the phosphoenolpyruvate carboxylase reaction to capture CO_2 when the stomata are open. Photorespiration is kept to a minimum during the day when the stomata are closed and the CO_2 levels in the chloroplast are high because of the decarboxylation of malate. Moreover, by keeping the stomata closed, water loss by transpiration (evaporation) is inhibited at times of the day when temperatures can reach 40 °C in hot, dry climates like the desert.

concept integration 12.4

How do sugarcane plants and saguaro cacti avoid the wasteful side effects of photorespiration?

Sugarcane plants and saguaro cacti separate the process of gas exchange through the stroma from the carbon dioxide fixation reaction mediated by the Calvin cycle enzyme rubisco. In addition to carbon dioxide fixation, rubisco also catalyzes an oxygenase reaction that combines ribulose-1,5-bisphosphate with O_2 to generate one molecule of 3-phosphoglycerate and one molecule of 2-phosphoglycolate. This wasteful side reaction, called photorespiration, not only diminishes net conversion of CO_2 to carbohydrate but also requires that the glycolate pathway consume ATP to recover the two carbons from 2-phosphoglycolate. Sugarcane plants use the C_4 pathway to reduce photorespiration by separating gas exchange and CO_2 fixation into two different cell types (mesophyll and bundle sheath cells). In contrast, saguaro cacti use the CAM pathway—which limits photorespiration by capturing CO_2 at night and then recovering it from the C_4 intermediate in the same cells during the day—when it is too hot to open the stomata for gas exchange.

12.5 The Glyoxylate Cycle Converts Lipids into Carbohydrates

We have seen that in order for a plant to convert sunlight energy into chemical energy in the form of ATP, NADPH, and triose phosphates, chloroplast-containing structures such as leaves function as solar panels. Most germinated seedlings, however, require several days before leaves are able to break through the soil and initiate photosynthesis. Therefore, an alternative source of energy is needed to support plant growth during this time.

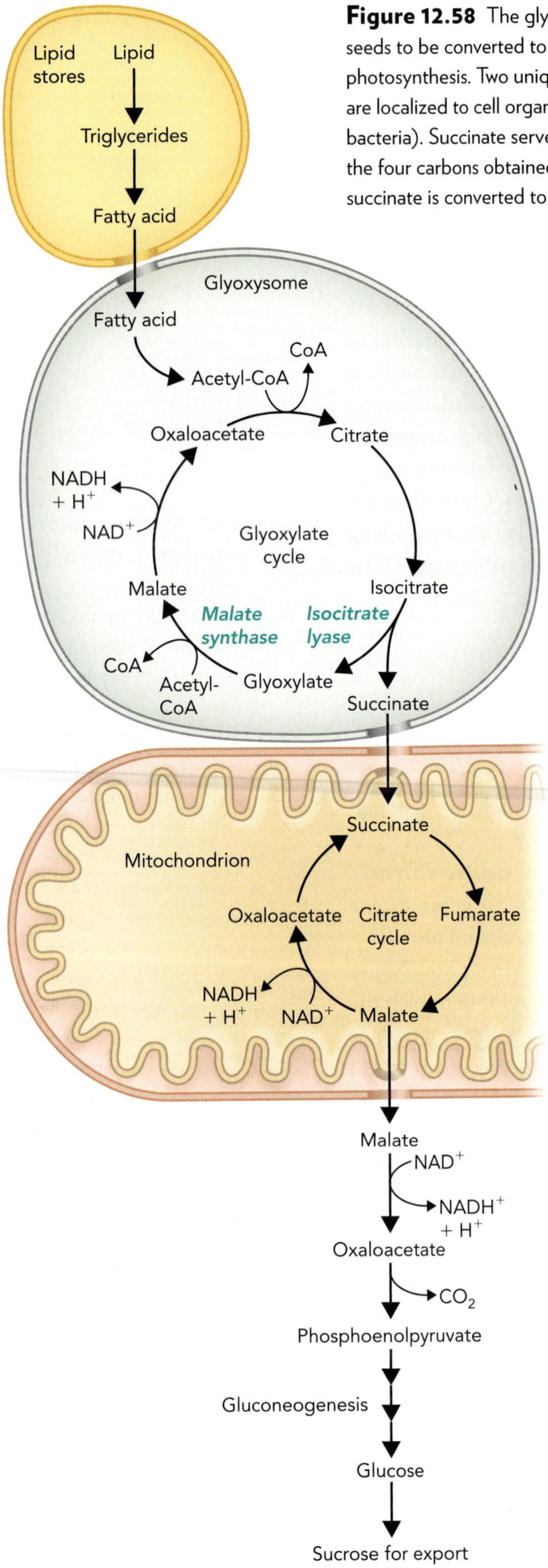

Figure 12.58 The glyoxylate cycle in plants provides a mechanism for fats stored in seeds to be converted to sucrose for export to developing plant tissues prior to the onset of photosynthesis. Two unique glyoxylate cycle enzymes, isocitrate lyase and malate synthase, are localized to cell organelles called glyoxysomes in plant cells (and are also present in bacteria). Succinate serves as the central metabolic intermediate in this pathway by transporting the four carbons obtained from two acetyl-CoA molecules to the mitochondria, where succinate is converted to malate for use in the citrate cycle or for export to the cytosol.

In many plants, metabolic energy is derived from lipid stores in the seed, which serve as a ready source of acetyl-CoA for aerobic respiration in the germ cell; however, this stored energy in the germ cell must also be transported to the growing stem and leaves. This requires, in turn, that hydrophobic fatty acids be converted into soluble carbohydrates such as sucrose. To solve this problem, the **glyoxylate cycle** is a metabolic pathway that converts two acetyl-CoA (C_2) molecules into one molecule of succinate (C_4), which serves as a carbon source for glucose biosynthesis. Animal genomes do not encode glyoxylate cycle enzymes and are not able to convert acetyl-CoA into glucose. The glyoxylate cycle was first discovered in bacteria and then later was shown to be present in plants by Harry Beevers and colleagues at Purdue University in Indiana.

As shown in **Figure 12.58**, the conversion of acetyl-CoA to succinate in plant cells takes place in special organelles called **glyoxysomes**. These contain two glyoxylate cycle enzymes absent from animal cells: isocitrate lyase and malate synthase. Fatty acids derived from lipid stores in the germ cell are converted to acetyl-CoA in glyoxysomes by enzymes in the fatty acid degradation pathway, for which isozymes are also present in mitochondria. Glyoxysomal citrate synthase then combines acetyl-CoA with oxaloacetate to form citrate, which is converted to isocitrate by aconitase, using the same reaction mechanisms as in the citrate cycle. Unlike the citrate cycle, however, isocitrate is not decarboxylated, but instead is cleaved by isocitrate lyase to form succinate and glyoxylate. Glyoxylate is then used to form malate by combining with a second molecule of acetyl-CoA in a reaction catalyzed by malate synthase. In the last step of the cycle, malate dehydrogenase oxidizes malate to form oxaloacetate and NADH.

The net reaction of the glyoxylate cycle can be written as

$$2\ \text{Acetyl-CoA} + \text{NAD}^+ + 2\ \text{H}_2\text{O} \rightarrow \text{Succinate} + 2\ \text{CoA} + \text{NADH} + \text{H}^+$$

Once succinate has been formed, it is transported from the glyoxysome to the mitochondria, where it is converted to fumarate by the citrate cycle enzyme succinate

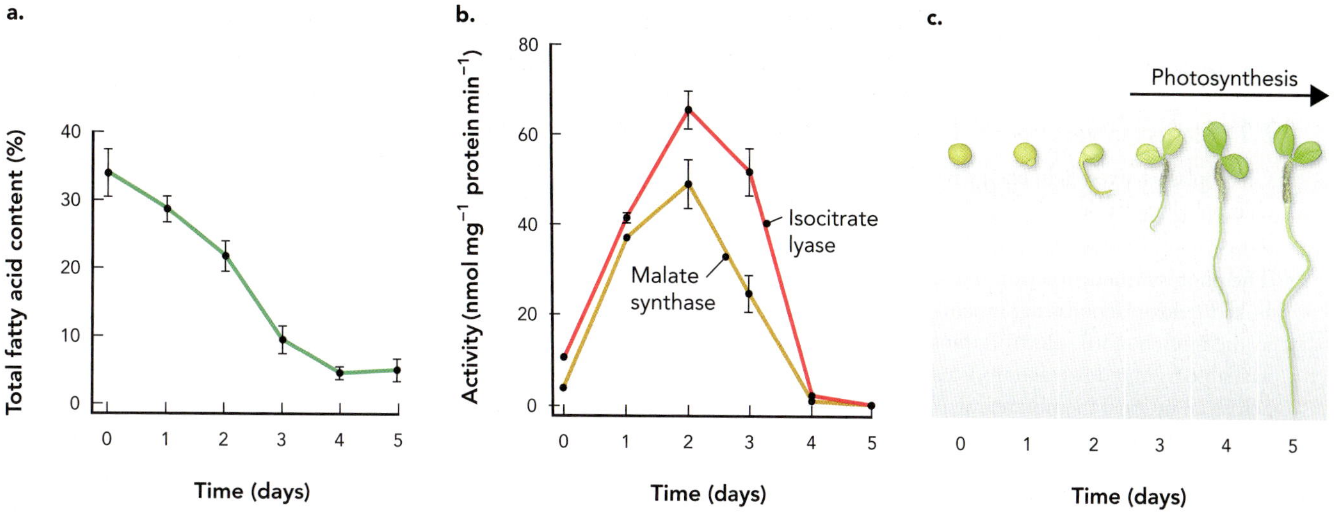

Figure 12.59 Lipid stores are converted to carbohydrates in germinated seeds by using the glyoxylate cycle to provide metabolic fuel to the plant before leaves are capable of photosynthesis. **a.** Fatty acid levels in seeds decrease rapidly during the first 3 days after wetting to induce germination (imbibition), when the glyoxylate cycle is fully active. **b.** The activity of two key glyoxylate cycle enzymes, isocitrate lyase and malate synthase, are highest in seedlings prior to the onset of photosynthesis in new leaves. **c.** Schematic drawing of a growing *Arabidopsis thaliana* seedling, showing the appearance of leaves at 3 days postimbibition.

BASED ON P. J. EASTMOND AND I. A. GRAHAM (2001). RE-EXAMINING THE ROLE OF THE GLYOXYLATE CYCLE IN OILSEEDS. *TRENDS IN PLANT SCIENCE*, 6(2), 72–78. © 2001 WITH PERMISSION FROM ELSEVIER.

dehydrogenase. Fumarate is then hydrated by fumarase to form malate. This product malate can either be converted to oxaloacetate by mitochondrial malate dehydrogenase or exported to the cytosol, where it is converted to oxaloacetate by cytosolic malate dehydrogenase. Cytosolic oxaloacetate is converted to phosphoenolpyruvate in a reaction catalyzed by the enzyme phosphoenolpyruvate carboxykinase, which leads to the formation of glucose by enzymes in the gluconeogenic pathway.

The glyoxylate cycle is not present in every plant tissue, and, in fact, it is most active in seedlings for only a few days after germination, when acetyl-CoA from lipid stores is used for sucrose synthesis. After that, photosynthetic activity in the newly formed leaves provides sucrose to the plant. As shown in **Figure 12.59**, the peak enzyme activity of isocitrate lyase and malate synthase in the plant *Arabidopsis thaliana* occurs at day 2, which is prior to the onset of photosynthesis and coincides with a rapid decline in the fatty acid content of the seedling.

concept integration 12.5

Why is it important that plants, but not animals, are able to convert lipids to carbohydrates?

The enzymes isocitrate lyase and malate synthase make it possible to convert lipids to carbohydrates in plants through the reactions of the glyoxylate cycle. Because newly germinated seedlings require several days to develop photosynthetic tissues (leaves) and capture light energy from the Sun, the glyoxylate cycle provides a mechanism, using stored lipids, to generate the carbohydrates needed for the citrate cycle. In developing plant root and stem cells, succinate is synthesized in organelles called glyoxysomes and exported to mitochondria, where it is converted to malate. Malate is then shuttled to the cytosol, where it is used to make oxaloacetate, a source of carbon for glucose synthesis by the gluconeogenic pathway. Seeds store lipids rather than carbohydrates because lipids have higher energy content and are not hydrated, which facilitates seed survival in dry climates. In contrast, carbohydrates are readily available to animal cells during all stages of development, either from nutrients stored in an egg or through maternal nutrients in the case of gestational animals, and therefore animal cells do not need the glyoxylate cycle.

chapter summary

12.1 Plants Harvest Energy from Sunlight

- Chloroplasts harvest light energy from the Sun using chlorophyll molecules, which absorb photons to excite electrons. The excited electrons are transferred to nearby carrier molecules.
- The photosynthetic electron transport system consists of linked redox reactions that function in much the same way as the mitochondrial electron transport system. Energy available from redox reactions is used to translocate protons across an internal membrane for ATP synthesis and to generate the reductant NADPH.
- ATP and NADPH produced by the photosynthetic electron transport system are used in the Calvin cycle to fix CO_2 in the form of triose phosphates. The triose phosphates can be used to form hexose sugars, which the plant uses as a source of chemical energy for mitochondrial electron transport.
- Photosynthetic organisms are considered autotrophs because they use light to supply all of their energy needs during the day and night. In contrast, heterotrophs are nonphotosynthetic organisms that depend on autotrophs for chemical energy.
- Photosynthesis and CO_2 fixation consume H_2O and generate O_2 and phosphorylated carbohydrates, whereas aerobic respiration consumes carbohydrate and O_2 and generates H_2O and CO_2.
- Chloroplasts contain three membranes: (1) a permeable outer membrane; (2) an impermeable inner membrane, which surrounds the stromal compartment; and (3) a proton-impermeable thylakoid membrane, which contains the ATP synthase complex and proteins in the photosynthetic electron transport system.

12.2 Energy Conversion by Photosystems I and II

- Nuclear fusion reactions occurring in the Sun release energy in the form of light, which reaches Earth and is absorbed by protein-embedded chromophores in plants. Different chromophores—for example, chlorophylls, β-carotene, and phycoerythrobilin—possess different light-absorbing properties.
- Light absorption by plant pigments excites an electron from the ground state to a higher orbital called the excited state. The electron can then (1) return to the ground state and transfer the absorbed energy to a nearby chlorophyll molecule (resonance energy transfer); (2) be transferred to a nearby acceptor molecule of higher reduction potential and thereby be reduced (photooxidation); or (3) return to the ground state and the absorbed energy is lost (fluorescence).
- Many light-harvesting protein complexes in a plant leaf absorb light and transfer the energy to nearby chlorophyll molecules, which eventually pass the energy to one of two reaction centers (PSII or PSI) where photooxidation takes place. About 300 energy transfer events occur for every one photooxidation.
- The PSII reaction center absorbs light at 680 nm, whereas the PSI reaction center absorbs light at 700 nm. The PSII and PSI reaction centers are linked by electron carriers: plastoquinone (PQ), which shuttles an electron from PSII to cytochrome b_6f; and plastocyanin (PC), which shuttles an electron from cytochrome b_6f to PSI.
- Redox reactions in photosynthesis constitute the Z scheme, which requires the absorption of 4 photons at PSII and 4 photons at PSI in order to transfer 4 electrons by photooxidation. In the process, water is fully oxidized to generate oxygen in the reaction $2\ H_2O \rightarrow O_2 + 4\ H^+ + 4\ e^-$.
- Electron transfer between PSII and cytochrome b_6f is mediated by plastoquinone through the PQ cycle mechanism, which is analogous to ubiquinol transfer of electrons from complexes I or II to complex III in mitochondria. Similarly, electron transfer between cytochrome b_6f and PSI is mediated by plastocyanin, which is a mobile electron carrier protein analogous to cytochrome c in mitochondria.
- A total of 12 H^+ accumulate inside the thylakoid lumen for every O_2 produced by the oxidation of 2 H_2O. This includes the translocation of protons from the stroma to the thylakoid lumen by the PQ cycle (4 H^+) and proton translocation by cytochrome b_6f (4 H^+), as well as the release of 4 H^+ by the oxidation of 2 H_2O.
- The 4 photons absorbed by PSI are used to reduce 2 $NADP^+$ to generate 2 NADPH in the stroma through a series of redox reactions requiring the electron carrier ferredoxin and the enzyme ferrodoxin-$NADP^+$ reductase.
- The net increase of 12 H^+ in the thylakoid from light absorption at PSII is used to generate ATP by the chloroplast ATP synthase complex; these 12 H^+ along with the 2 NADPH generated by light absorption at PSI are used for CO_2 fixation in the stroma by the Calvin cycle reactions.

12.3 Photophosphorylation Generates ATP

- The mechanism of ATP synthesis in photosynthetic organisms is called photophosphorylation because light absorption is linked to ATP synthesis through the chemiosmotic gradient produced by the photosynthetic electron transport system.
- The chloroplast ATP synthase complex is structurally similar to the yeast mitochondrial ATP synthase complex; however, 4 H^+ cross through the chloroplast ATP synthase CF_o complex for every 1 ATP that is generated.
- Cyclic photosynthetic electron transport bypasses electron transfer to ferredoxin-$NADP^+$ reductase; instead, ferredoxin transfers the electron to plastoquinone and recirculates through the electron transport system. The net result is higher rates of ATP synthesis and lower rates of NADPH generation. This is a good way to control ATP-to-NADPH ratios for biosynthesis.
- Differential localization of PSII and PSI reaction center complexes within the thylakoid membrane helps to control rates of cyclic photophosphorylation by increasing the frequency of energy transfer from the light-harvesting

complexes to PSI rather than PSII. The PSI and ATP synthase complexes are concentrated in unstacked lamellar regions of the thylakoid membranes, whereas PSII complexes are mostly localized to the stacked membranes (grana). Redistribution of PSII and PSI complexes between these two regions is controlled by phosphorylation.

12.4 Carbohydrate Biosynthesis in Plants

- The Calvin cycle is named after Melvin Calvin, who discovered it. Photosynthetic organisms use this metabolic pathway to assimilate CO_2 into triose phosphates. The Calvin cycle is useful because the triose phosphates can be converted into hexose sugars, which then serve as a source of metabolic energy. The Calvin cycle reactions occur entirely in the chloroplast stroma.
- Although reactions in the Calvin cycle have been called the dark reactions, this is a misnomer because the Calvin cycle does not run during the night due to limiting amounts of ATP and NADPH, which are only available during the daylight as a function of photosynthetic electron transport and photophosphorylation.
- The Calvin cycle consists of three stages. Stage 1: CO_2 fixation by the enzyme ribulose-1,5-bisphosphate carboxylase/oxygenase, also called rubisco. Stage 2: reduction of 3-phosphoglycerate to form glyceraldehyde-3-phosphate. Stage 3: regeneration of ribulose-1,5-bisphosphate to continue the cycle back at stage 1.
- To produce one net glyceraldehyde-3-phosphate molecule and regenerate the ribulose-1,5-bisphosphate starting material, the Calvin cycle requires 3 CO_2, 6 NADPH, 9 ATP, and 6 H_2O.
- Rubisco catalyzes an energetically favorable reaction ($\Delta G°' = -35$ kJ/mol), in which ribulose-1,5-bisphosphate is first carboxylated and then cleaved into two molecules of 3-phosphoglycerate. Rubisco consists of eight large catalytic subunits and eight small stabilizing subunits arranged in a ring structure. Because 85% of the organisms on Earth are photosynthetic and require rubisco for CO_2 fixation by the Calvin cycle, rubisco is the most abundant enzyme on the planet.
- The activity of Calvin cycle enzymes are controlled by light in three ways: (1) enzyme inhibitor proteins, which are present at higher levels in the dark than in the light; (2) activation of Calvin cycle enzymes by elevated pH and increased Mg^{2+} levels in the stroma (this only occurs in the light, when photosynthetic electron transport is fully active); and (3) thioredoxin-mediated activation of Calvin cycle enzymes through reduction of disulfide bridges.
- Rubisco not only carboxylates ribulose-1,5-bisphosphate but also catalyzes an oxygenase reaction, leading to the generation of 2-phosphoglycolate. This molecule must be converted to 3-phosphoglycerate through the glycolate pathway, a complex set of reactions requiring an investment of 1 ATP. This oxygenase reaction is therefore wasteful, and plant cells try to inhibit it by minimizing O_2 levels.
- Because high temperatures differentially increase levels of soluble O_2 relative to CO_2, plants that grow in high-temperature climates have adapted by trapping CO_2 in the form of malate and thereby limiting the exposure of rubisco to elevated O_2. Malate is then decarboxylated, and the released CO_2 is fixed by rubisco. This adaptation is called the Hatch–Slack pathway, after the biochemists who discovered it.
- Two variations of the Hatch–Slack pathway have been described. (1) The C_4 pathway, in tropical plants such as sugarcane, uses two separate cell types: one for CO_2 uptake and the other for rubisco-mediated carboxylation. (2) The CAM pathway, found in desert succulents such as the giant saguaro cactus, captures CO_2 at night when transpiration rates are low.

12.5 The Glyoxylate Cycle Converts Lipids into Carbohydrates

- The glyoxylate cycle is a metabolic pathway in plants that converts acetyl-CoA into succinate, which serves as a carbon source for glucose biosynthesis. Animals do not have glyoxylate cycle enzymes and cannot convert acetyl-CoA into glucose.
- Isocitrate lyase and malate synthase are glyoxylate cycle enzymes that are localized to glyoxysomes in plant cells and are also present in bacteria. The succinate produced by the glyoxylate cycle is converted to malate, which is then exported to the cytosol and oxidized to oxaloacetate, a substrate for glucose synthesis by gluconeogenesis.

biochemical terms

(in order of appearance in text)

photosynthetic electron transport system (p. 580)
Calvin cycle (p. 580)
greenhouse effect (p. 581)
cyanobacteria (p. 584)
thylakoid membrane (p. 585)
thylakoid lumen (p. 585)
stroma (p. 585)
granum (p. 585)
lamella (p. 585)
photosystem II (p. 587)
cytochrome b_6f (p. 587)
photosystem I (p. 587)
rubisco (p. 587)
chromophore (p. 588)
chlorophyll (p. 588)
resonance energy transfer (p. 588)
reaction center (p. 589)
light-harvesting complex (LHC) (p. 591)
plastoquinone (PQ) (p. 594)
plastocyanin (PC) (p. 594)
Z scheme (p. 594)
pheophytin (Pheo) (p. 594)
PQ cycle (p. 597)
O_2-evolving complex (OEC) (p. 598)
phylloquinone (p. 600)
ferredoxin (Fd) (p. 602)
cyclic photophosphorylation (p. 606)
thioredoxin (p. 618)
glycolate pathway (p. 619)
photorespiration (p. 619)
C_4 pathway (p. 621)
CAM pathway (p. 621)
stomate (p. 623)
mesophyll cell (p. 623)
bundle sheath cell (p. 623)
glyoxylate cycle (p. 626)
glyoxysome (p. 626)

review questions

1. What is the functional relationship between photosynthesis and aerobic respiration, and how does this relate to the metabolic needs of autotrophs and heterotrophs?
2. Describe the structure and function of chloroplasts, and compare the role of chloroplast membranes to that of mitochondrial membranes.
3. Describe the process of photooxidation at PSII and PSI reaction centers. How is photon energy transferred from light-harvesting complexes to PSII and PSI reaction centers?
4. Write the net equation for the photosynthetic electron transport system, following the absorption of 8 photons and oxidation of 2 H_2O. Briefly describe the structure and function of the six key proteins required for this process.
5. In what way does cyclic photophosphorylation differ from standard photophosphorylation?
6. How does differential localization of PSII and PSI complexes within the thylakoid membrane control rates of cyclic photophosphorylation?
7. Why is it incorrect to refer to the Calvin cycle as the dark reactions of photosynthesis?
8. What key chemical reactions occur in each of the three stages of the Calvin cycle?
9. Describe three ways in which the activity of Calvin cycle enzymes is controlled by light.
10. Why is the oxygenase reaction of rubisco considered wasteful to plants?
11. What is meant by a *C_4 plant*, and what two evolutionary adaptations permit C_4 plants to outcompete normal plants (C_3 plants) in hot climates?
12. What is the biochemical function of the glyoxylate cycle in plants?

challenge problems

1. Chloroplasts are plant organelles that convert sunlight energy into chemical energy. Answer the following questions about this energy-converting process.
   a. What do photosystems I and II have in common with the mitochondrial electron transport chain?
   b. What two products of the photosynthetic electron transport system (in addition to H_2O) are required by the Calvin cycle?
   c. What molecule is the C_1 substrate in the Calvin cycle?
   d. Do plant cells have mitochondria? Explain your answer.
2. The diagram that follows contains labels that define energy-converting processes in mitochondria and chloroplasts. For each letter A–G, draw an oval around the label that best describes a chloroplast, and draw a box around the label that best describes a mitochondrion. There should be a total of seven ovals and seven boxes.

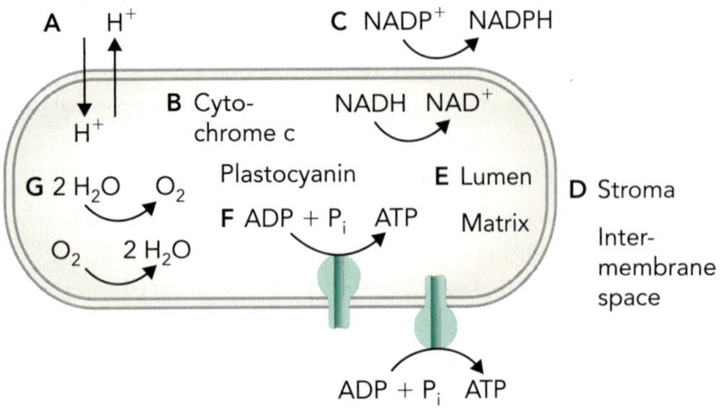

3. The cytochrome b_6f complex of the photosynthetic electron transport system is responsible for translocating 8 H^+ into the thylakoid space from the stroma in response to light. However, a total of 12 H^+ accumulate inside the thylakoid space as a result of photon absorption by PSII and PSI. What is the origin of the other 4 H^+?
4. Considering that the PSI complex alone does not translocate protons across the thylakoid membrane, why is photon absorption by PSI required to sustain proton translocation by the cytochrome b_6f complex in the presence of light?
5. Why are 8 photons required for the net reaction of the photosynthetic electron transport system (2 H_2O + 8 photons + 2 $NADP^+$ + 3 ADP + 3 $P_i \rightarrow O_2$ + 2 NADPH + 3 ATP)?
6. The key event in converting solar energy into redox energy is photooxidation of chlorophyll in the reaction centers. What must be the chemical property of the electron carrier molecule that *accepts* the electron from chlorophyll; that is, why is the electron transfer favorable?
7. What redox reaction ultimately replaces the transferred electron so that the chlorophyll molecule becomes reduced and can therefore undergo another round of photooxidation?
8. Paraquat is an effective herbicide that was used in the 1980s to control the growth of illegal crops in North and South America. Explain the redox mechanism by which paraquat blocks NADPH production.
9. Carbon dioxide is a C_1 substrate in the Calvin cycle reaction catalyzed by the enzyme rubisco. What are the one C_5 substrate and two C_3 products of this reaction? The mass of rubisco on Earth is higher than that of any other enzyme: Why?
10. The Calvin cycle is sometimes referred to as the dark reactions, even though light is required for flux to be

maintained. Describe the two primary mechanisms that regulate Calvin cycle enzymes so that they are most active in the light.

11. In terms of photorespiration and ATP requirements, what explains the observation that crabgrass, which uses the C_4 pathway of carbon fixation, has a growth advantage over turfgrass when temperatures are high and the O_2-to-CO_2 ratios are elevated because of increased O_2 solubility?

12. What are the two primary reasons why a suspension of chloroplasts is unable to synthesize glucose for an extended period of time when it is shifted from the light to the dark, even in the presence of the Calvin cycle substrates CO_2 and H_2O?

13. Light activation of PSI leads to the reduction of ferredoxin, which then reduces not only $NADP^+$ (to form NADPH) but also thioredoxin, which uses the electrons to reduce disulfide bridges in several Calvin cycle enzymes, leading to their activation. What turns off these Calvin cycle enzymes when the Sun goes down?

smartwork5

If your instructor assigns homework with Smartwork5, access it here: digital.wwnorton.com/biochem.

suggested reading

Books and Reviews

Amunts, A., and Nelson, N. (2009). Plant photosystem I design in the light of evolution. *Structure, 17*, 637–650.

Anderson, J. M., and Andersson, B. (1988). The dynamic photosynthetic membrane and regulation of solar energy conversion. *Trends in Biochemical Sciences, 13*, 351–355.

Barber, J. (2003). Photosystem II: the engine of life. *Quarterly Reviews of Biophysics, 36*, 71–89.

Barber, J. (2008). Photosynthetic generation of oxygen. *Philosophical Transactions of the Royal Society of London. Series B, Biological Sciences, 363*, 2665–2674.

Benson, A. A. (2010). Last days in the old radiation laboratory (ORL), Berkeley, California, 1954. *Photosynthesis Research, 105*, 209–212.

Blankenship, R. E. (2010). Early evolution of photosynthesis. *Plant Physiology, 154*, 434–438.

Croce, R., and van Amerongen, H. (2011). Light-harvesting and structural organization of Photosystem II: from individual complexes to thylakoid membrane. *Journal of Photochemistry and Photobiology. B, Biology, 104*, 142–153.

Eberhard, S., Finazzi, G., and Wollman, F. A. (2008). The dynamics of photosynthesis. *Annual Review of Genetics, 42*, 463–515.

Heathcote, P., Fyfe, P. K., and Jones, M. R. (2002). Reaction centres: the structure and evolution of biological solar power. *Trends in Biochemical Sciences, 27*, 79–87.

Marshall, E. (1982). Pot-spraying plan raises some smoke. *Science, 217*, 429.

Maurino, V. G., and Peterhansel, C. (2010). Photorespiration: current status and approaches for metabolic engineering. *Current Opinion in Plant Biology, 13*, 249–256.

Nelson, M., Allen, J. P., and Dempster, W. F. (1992). Biosphere 2: a prototype project for a permanent and evolving life system for Mars base. *Advances in Space Research, 12*, 211–217.

Penfield, S., Graham, S., and Graham, I. A. (2005). Storage reserve mobilization in germinating oilseeds: *Arabidopsis* as a model system. *Biochemical Society Transactions, 33*, 380–383.

Reski, R. (2009). Challenges to our current view on chloroplasts. *Biological Chemistry, 390*, 731–738.

Schneider, G., Lindqvist, Y., and Branden, C. I. (1992). RUBISCO: structure and mechanism. *Annual Review of Biophysics and Biomolecular Structure, 21*, 119–143.

Primary Literature

Amunts, A., Toporik, H., Borovikova, A., and Nelson, N. (2010). Structure determination and improved model of plant photosystem I. *Journal of Biological Chemistry, 285*, 3478–3486.

Barber, J. (2008). Crystal structure of the oxygen-evolving complex of photosystem II. *Inorganic Chemistry, 47*, 1700–1710.

Boussac, A., Rappaport, F., Carrier, P., Verbavatz, J. M., Gobin, R., Kirilovsky, D., Rutherford, A. W., and Sugiura, M. (2004). Biosynthetic Ca^{2+}/Sr^{2+} exchange in the photosystem II oxygen-evolving enzyme of *Thermosynechococcus elongatus*. *Journal of Biological Chemistry, 279*, 22809–22819.

Guskov, A., Kern, J., Gabdulkhakov, A., Broser, M., Zouni, A., and Saenger, W. (2009). Cyanobacterial photosystem II at 2.9-A resolution and the role of quinones, lipids, channels and chloride. *Nature Structural & Molecular Biology, 16*, 334–342.

Jagendorf, A. T., and Uribe, E. (1966). ATP formation caused by acid-base transition of spinach chloroplasts. *Proceedings of the National Academy of Sciences USA, 55*, 170–177.

Liu, Z., Yan, H., Wang, K., Kuang, T., Zhang, J., Gui, L., An, X., and Chang, W. (2004). Crystal structure of spinach major light-harvesting complex at 2.72 A resolution. *Nature, 428*, 287–292.

McConnell, I., Li, G., and Brudvig, G. W. (2010). Energy conversion in natural and artificial photosynthesis. *Chemistry & Biology, 17*, 434–447.

Melkozernov, A. N., Barber, J., and Blankenship, R. E. (2006). Light harvesting in photosystem I supercomplexes. *Biochemistry, 45*, 331–345.

Sener, M., Strumpfer, J., Timney, J. A., Freiberg, A., Hunter, C. N., and Schulten, K. (2010). Photosynthetic vesicle architecture and constraints on efficient energy harvesting. *Biophysical Journal, 99*, 67–75.

Slack, C. R., and Hatch, M. D. (1967). Comparative studies on the activity of carboxylases and other enzymes in relation to the new pathway of photosynthetic carbon dioxide fixation in tropical grasses. *Biochemical Journal, 103*, 660–665.

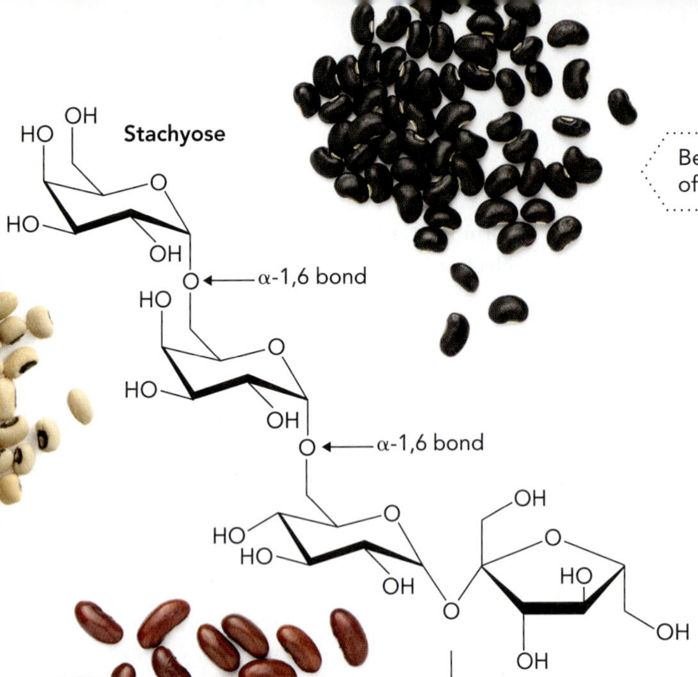

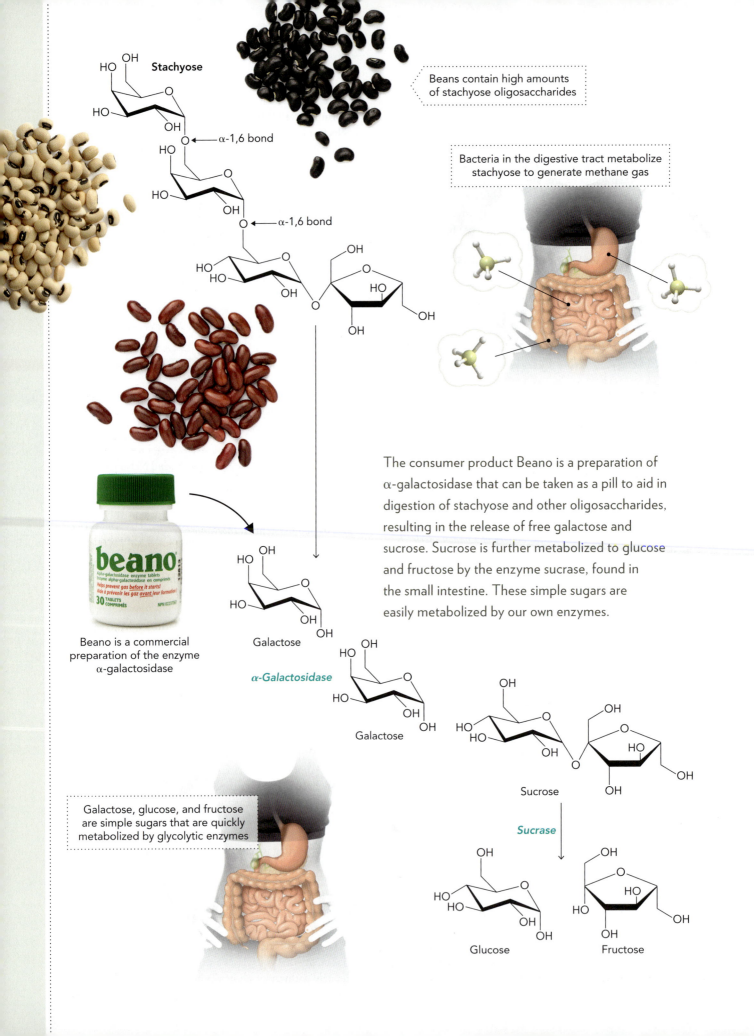

Beans contain high amounts of stachyose oligosaccharides

Bacteria in the digestive tract metabolize stachyose to generate methane gas

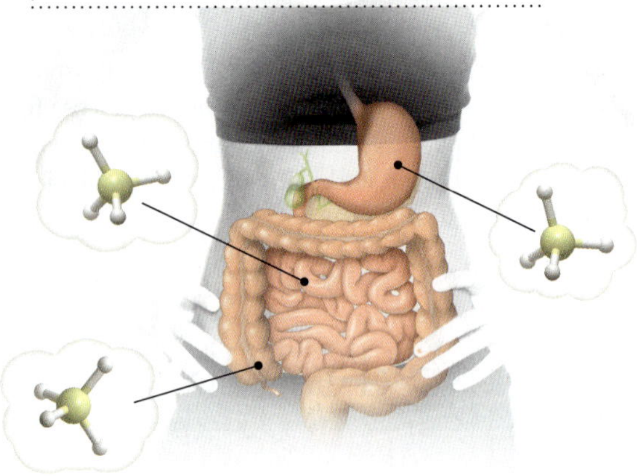

Beano is a commercial preparation of the enzyme α-galactosidase

The consumer product Beano is a preparation of α-galactosidase that can be taken as a pill to aid in digestion of stachyose and other oligosaccharides, resulting in the release of free galactose and sucrose. Sucrose is further metabolized to glucose and fructose by the enzyme sucrase, found in the small intestine. These simple sugars are easily metabolized by our own enzymes.

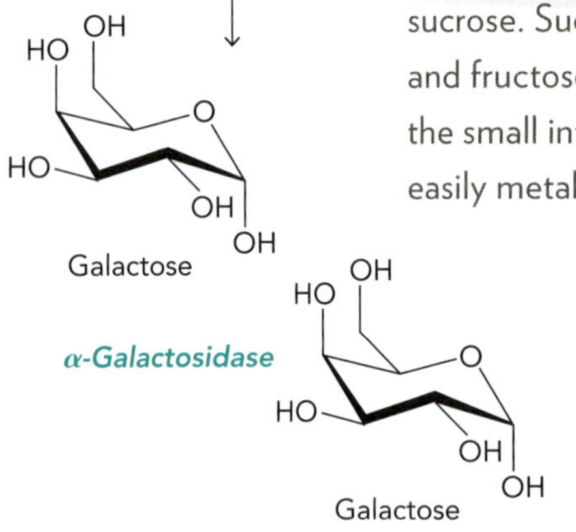

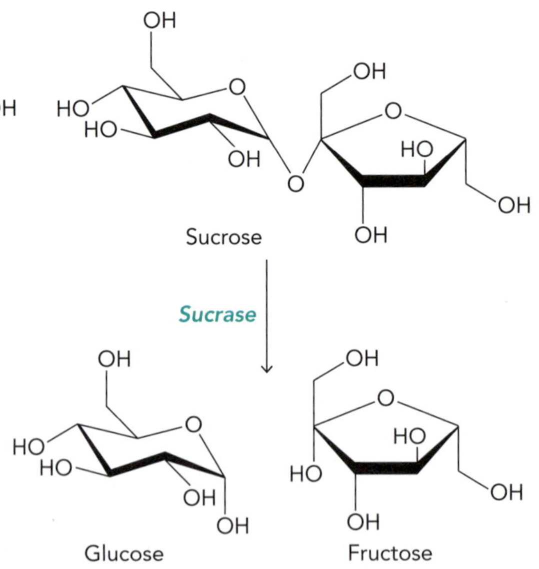

Galactose, glucose, and fructose are simple sugars that are quickly metabolized by glycolytic enzymes

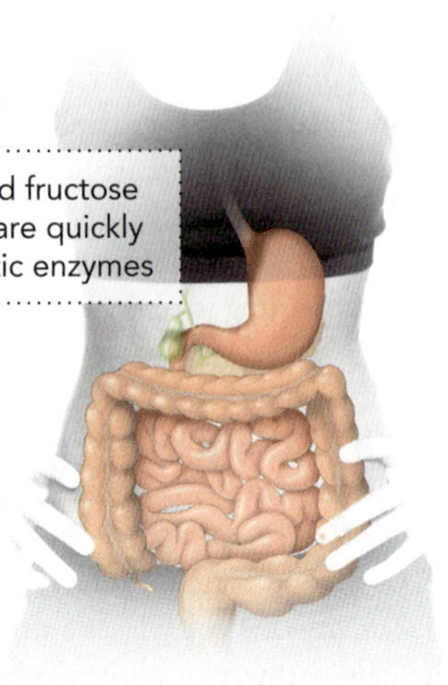

13

Carbohydrate Structure and Function

◀ Carbohydrates are important biological molecules that provide sources of metabolic energy in the form of dietary sugars; glucose storage in the form of glycogen (animals) and starch (plants); and structural components of cell walls and of the extracellular matrix in animals. Although many forms of carbohydrates in nature are polymers of glucose and galactose, most of these carbohydrate macromolecules cannot be used as a nutritional energy source by animals because they lack the required digestive enzymes.

For example, humans and nonruminating livestock do not have the α-galactosidase enzyme required to cleave the α-1,6 glycosidic bond present in stachyose, a raffinose-series oligosaccharide contained at high levels in beans. This can cause a problem when beans are consumed because the human and nonruminating livestock digestive tracts contain bacteria that express the α-galactosidase enzyme and are able to metabolize stachyose and other oligosaccharides. These bacterial fermentation processes generate methane gas and sulfur-containing compounds that cause bloat and flatulence.

CREDITS: BLACK BEANS: NUTTAPONG/SHUTTERSTOCK; WOMAN WITH HANDS ON HIPS: ALIN DRAGULIN/ALAMY; BEANO BOTTLE: HELEN SESSIONS/ALAMY STOCK PHOTO; KIDNEY BEANS: NUTTAPONG/SHUTTERSTOCK; BLACK-EYED PEAS: FLOORTJE/ISTOCK/GETTY IMAGES PLUS.

CHAPTER OUTLINE

In Part 3 of this book, we examined energy conversion pathways in cells and observed how sunlight absorbed by photosynthetic organisms is used as an energy source to drive the synthesis of carbohydrates in the form of starch and sucrose. These highly reduced, glucose-based compounds provide a readily available form of chemical energy for plants and animals. Here in Part 4, we explore biochemical pathways that control the synthesis and degradation of a variety of biomolecules. The energy required for biomolecular synthesis comes from both phosphoryl transfer energy available from ATP and redox energy provided by NADPH, the phosphorylated form of NADH. Part 4 also describes the degradation processes of biomolecules. These processes not only scavenge building blocks to support biosynthesis but also form a type of metabolic regulation by controlling the steady-state level of active biomolecules.

13.1 Carbohydrates: The Most Abundant Biomolecules in Nature

The word *carbohydrate* comes from the term *carbon hydrate*, which describes the empirical formula for carbohydrates, $(CH_2O)_n$, in which $n \geq 3$. Carbohydrates are also called **glycans**, which is a term frequently used in the context of **glycobiology**—an emerging area of modern biochemistry that focuses on the role of glycans in cell structure and function and in cell signaling. As shown in **Figure 13.1**, carbohydrates can be divided into three major groups on the basis of their structures.

1. *Simple sugars* consist of monosaccharides, disaccharides, and oligosaccharides, which often function as metabolic intermediates in energy conversion pathways.
2. *Polysaccharides* consist of either glucose homopolymers, such as cellulose, or disaccharide heteropolymers, such as chitin or heparan sulfate, in which one of the two sugars is a **hexosamine**—a monosaccharide containing an amino group.

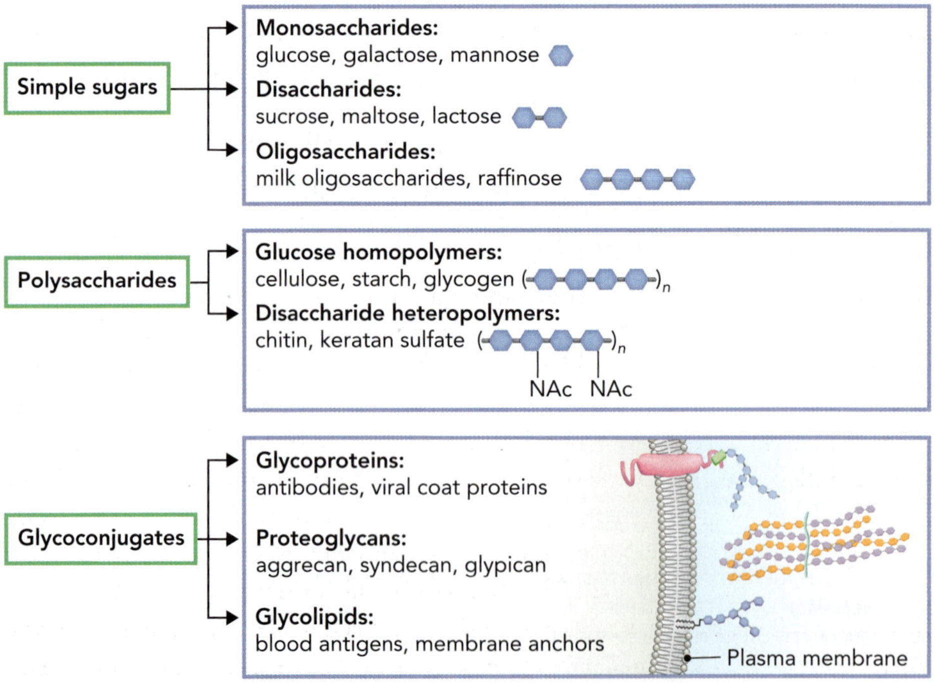

Figure 13.1 Carbohydrates, or glycans, can be divided into three major groups on the basis of their structures. Simple sugars are the building blocks of polysaccharides and glycoconjugates, both of which are large and often structurally complex biomolecules. NAc = *N*-acetyl.

3. **Glycoconjugates** are proteins or lipids with covalently attached glycans, which play a critical role in cellular communication.

Glycobiology: Study of Glycan Structure and Function

The polysaccharides cellulose and chitin account for most of the carbohydrates in the biosphere. This is due to the role of cellulose in forming plant cell walls and to chitin being the structural component of both invertebrate exoskeletons and the cell walls of fungi. Indeed, it has long been recognized that because of the large biomass of plants and insects on Earth, carbohydrates are the most abundant biomolecules in nature.

In addition to the structural importance that cellulose and chitin have in determining the physical properties of specialized cells, carbohydrates are also immensely important in biology because of their role as functional groups in glycoconjugates. It has been estimated that 50% of all human proteins are glycoproteins. Moreover, ~1% of the human genome encodes enzymes required for the synthesis and degradation of carbohydrates and glycoconjugates. These enzymes include **glycosyltransferases**, which covalently link glycans to proteins and lipids, and **glycosidases**, which remove glycans through hydrolysis reactions. Attesting to the importance of glycoconjugates in biology, enzymes that modify glycans are some of the most highly conserved proteins in eukaryotes.

As described later in this chapter, the high degree of variation among different glycan structures in glycoconjugates, which can be temporally and spatially modified in response to environmental changes, makes it difficult to decipher the chemical arrangement of sugar moieties relative to each other. But with recent advances in chemical separation techniques and the increased sensitivity of mass spectrometry, it has been possible to develop a number of useful methods for applications in glycobiology.

Figure 13.2a shows that at least 11 different monosaccharides are used as the building blocks of glycan groups in most glycoconjugates. These monosaccharides can be depicted as colored symbols based on notation developed by the **Consortium for Functional Glycomics (CFG)**. For example, the simple sugars glucose (Glc), galactose (Gal), and mannose (Man) are represented by blue, yellow, and green circles, respectively. The abundant hexosamines *N*-acetylgalactosamine (GalNAc), *N*-acetylglucosamine (GlcNAc), and *N*-acetylmuramic acid (MurNAc) are represented by colored squares (GalNAc and GlcNAc) and a colored hexagon (MurNAc). Less common glycan groups on glycoconjugates are xylose (Xyl), glucuronic acid (GlcA), fucose (Fuc), iduronic acid (IdoA), and *N*-acetylneuraminic acid (Neu5Ac), which is a form of sialic acid. As shown in **Figure 13.2b**, we can use the CFG notation to represent the structural arrangement of sugar residues in a glycan group linked to an asparagine amino acid residue in a glycoprotein.

The DNA code discovered by James Watson and Francis Crick in 1953 consists of G-C and A-T base pairs, which provide the molecular blueprint for generating an entire organism under appropriate developmental conditions. Similarly, the genetic code, elucidated in the 1960s by Marshall Nirenberg, Har Gobind Khorana, and Robert Holley, contains the information needed to direct the insertion of 20 different amino acids into a growing polypeptide chain, based on 61 triplet codons in mRNA transcripts (3 of the 64 triplet codons specify translation termination). However, an analogous predictive "sugar code," containing the information needed to specify the structure and function of glycan groups, has been difficult to identify for several reasons. First, unlike DNA and protein biosynthesis, monosaccharide addition to proteins and

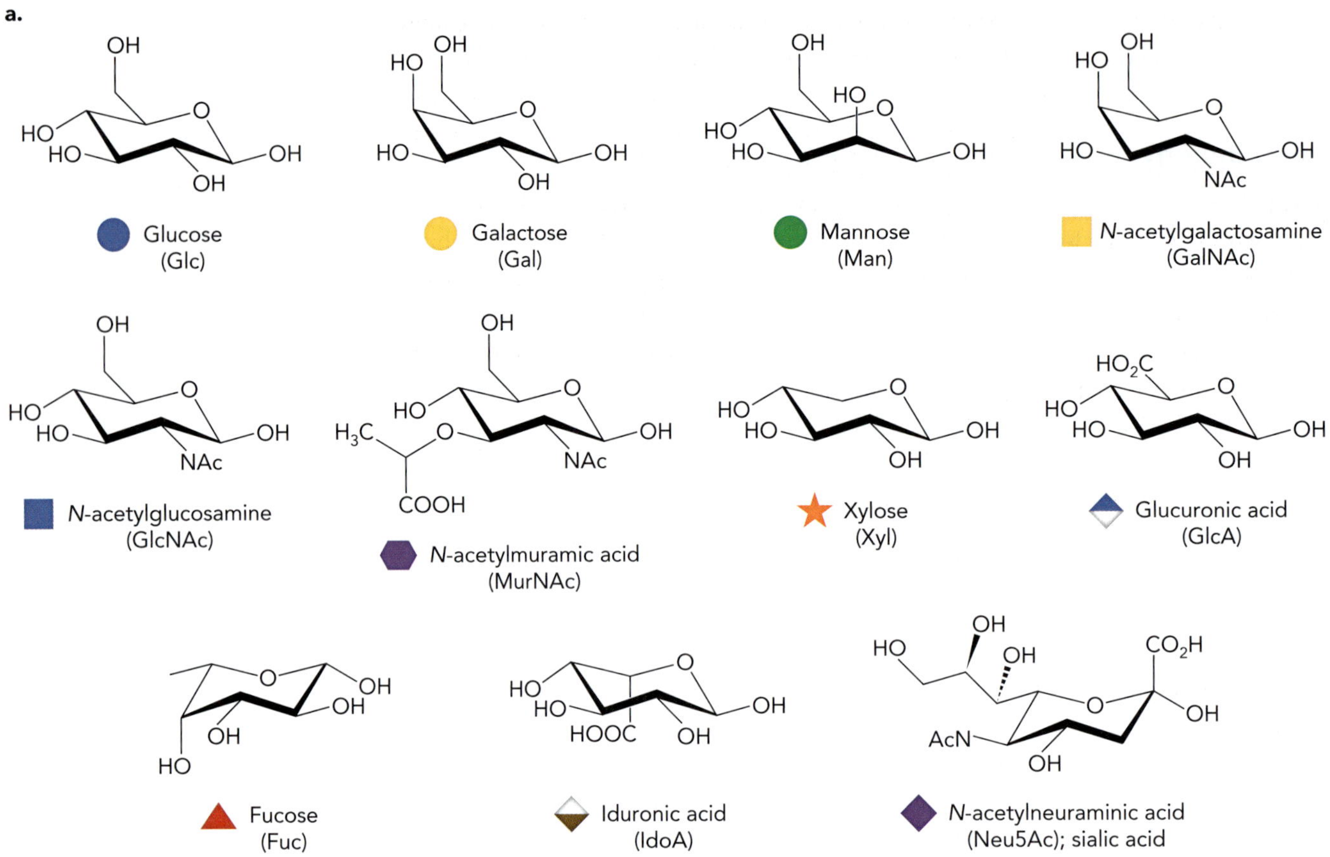

Figure 13.2 Glycoconjugates contain combinations of modified and unmodified monosaccharides, which are covalently attached to proteins and lipids as branched and unbranched structures. **a.** Structures of the most common monosaccharides found in glycoconjugates are shown with their corresponding symbols, as specified by the Consortium for Functional Glycomics (CFG). **b.** A representative branched glycan structure on an N-linked glycoprotein can be represented by chemical formulas, letter abbreviations, and CFG symbols.

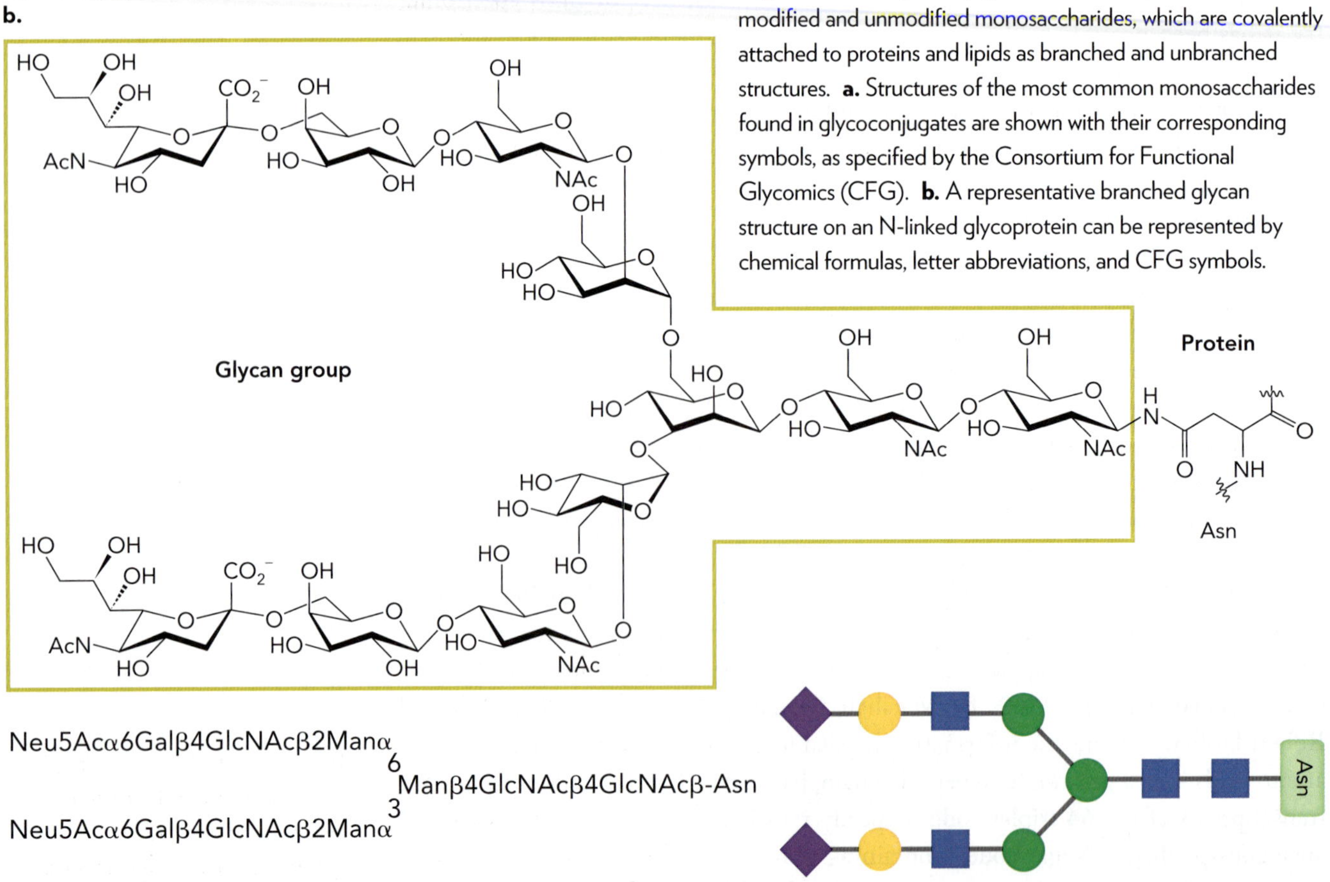

lipids by glycosyltransferase enzymes is not a template-directed process. This makes it difficult to predict glycan structures on the basis of substrate specificities of known glycosyltransferase enzymes. Second, glycan structures have been found to be similar, but not identical, among molecules of the same protein, owing to the effects of the cellular environment on glycan synthesis. Third, it is technically challenging to decipher complex glycan structures because of limitations in glycan analytical methods and instrument sensitivity.

Despite these difficulties, a short list of core "glycobiology principles" has been generated that describes the essentials of glycan structure and function in living cells. These principles are listed below and are illustrated in cellular processes in **Figure 13.3**.

1. *Glycan biochemistry.* Glycan chains are branched or linear carbohydrate structures consisting of modified (for example, acetylated or sulfated) and unmodified monosaccharides, which are covalently linked by glycosidic bonds in either of two conformations (α or β). Common glycan structures on glycoconjugates can be deduced by reiterative chemical and enzymatic cleavage using liquid chromatography and mass spectrometry.
2. *Glycan biosynthesis.* Monosaccharides are scavenged from the environment or synthesized *de novo* in the cell. They are covalently linked to proteins and lipids by unique glycosyltransferase enzymes that catalyze highly specific reactions, most often using nucleotide-activated monosaccharides. Eukaryotic glycosyltransferase enzymes function within the Golgi apparatus and endoplasmic reticulum (ER) as described in Chapter 22.
3. *Glycan diversity.* Cells contain a vast array of free and conjugated glycans, which can be found inside, attached to, and secreted from cells. Glycan structures are highly diverse, both intrinsically (variable between similar glycoconjugates) and extrinsically (variable between species). Because of the large number of chemically distinct monosaccharide building blocks used to generate glycan structures and the variety of possible linkage positions and conformations,

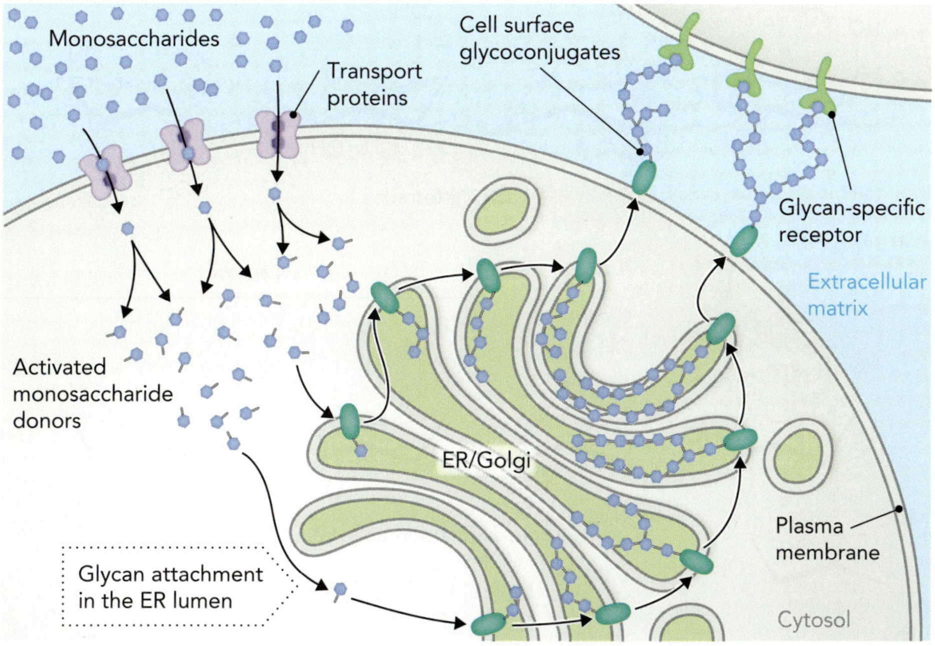

Figure 13.3 The core principles of glycobiology are illustrated by cellular processes that synthesize, secrete, and recognize glycoconjugates in eukaryotic cells.

glycan structures significantly increase the amount of biochemical information contained within nucleic acid and protein structures.

4. *Glycan recognition.* Specific binding proteins recognize and bind to glycans to elicit a biological response. One class of glycan binding proteins is the **lectins,** which are evolutionarily conserved. They possess high specificity but low affinity for any given glycan structure. Medical applications of glycobiology include development of reagents that disrupt lectin–glycan binding on the surfaces of host cells and microbial pathogens.

Related Oligosaccharides Are Derived from the Same Disaccharide

In addition to the representative monosaccharides and disaccharides described in Figure 13.1, oligosaccharides are the third type of simple sugar produced by cells. Oligosaccharides range in complexity from ~3 to ~20 branched and unbranched sugar residues. Biochemical characterization of oligosaccharides found in human breast milk and of the raffinose series of plant oligosaccharides indicates that related oligosaccharides are modifications of the same disaccharide. In the case of human milk oligosaccharides, the common disaccharide is lactose, whereas in raffinose-related oligosaccharides, the common disaccharide is sucrose.

The diversity of human milk oligosaccharides was discovered when glycobiology methods became sensitive enough to distinguish between highly related soluble carbohydrates present in human breast milk. By using liquid chromatography with mass spectrometry, researchers found that human milk contains well over 100 different lactose-derived oligosaccharides, with sizes ranging from 3 to 22 sugar residues. Notably, many of these milk oligosaccharides are not substrates for human digestive enzymes, but instead are degraded by bacterial flora in the infant's intestinal tract.

Figure 13.4 shows the chemical structure of two human milk oligosaccharides, lacto-*N*-tetraose and lacto-*N*-fucopentaose I, which are both derived from lactose.

Figure 13.4 Human milk oligosaccharides are derived from lactose. Only a few of the more than 100 oligosaccharides detected in breast milk have been biochemically characterized, two of which are lacto-*N*-tetraose and lacto-*N*-fucopentaose I. Lactose-derived oligosaccharides are the third most abundant component in human milk, after free lactose and neutral lipids. PHOTO: IN GREEN/SHUTTERSTOCK.

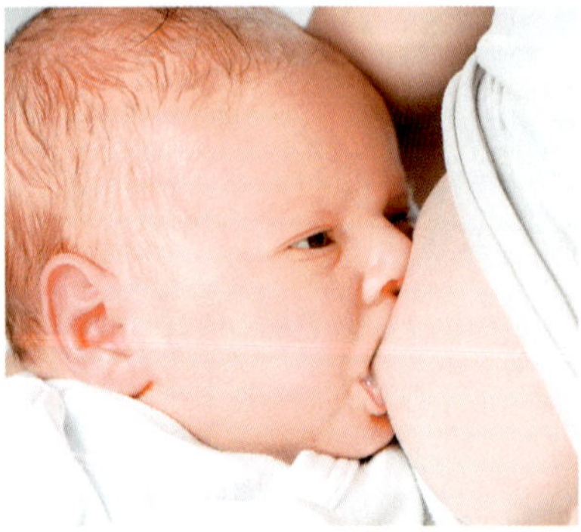

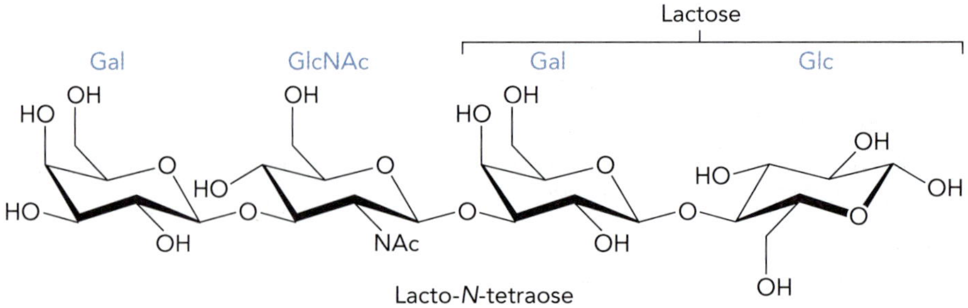

Lacto-*N*-tetraose

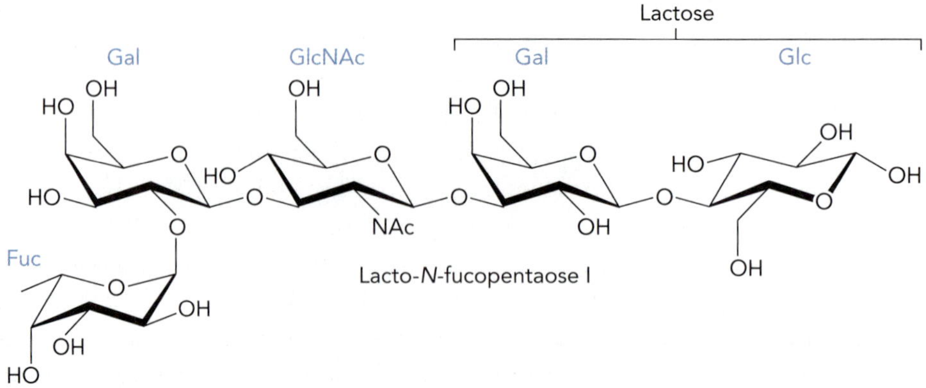

Lacto-*N*-fucopentaose I

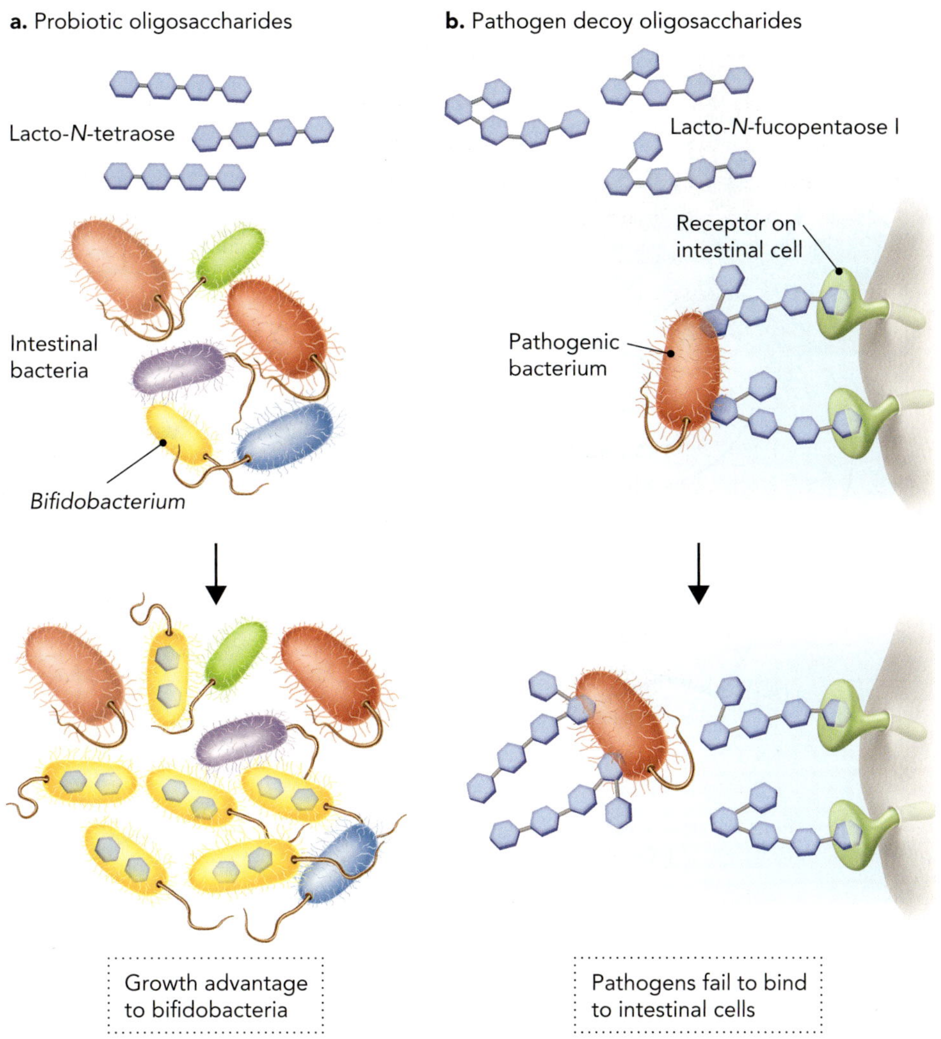

Figure 13.5 Human milk oligosaccharides are proposed to have several functions in the infant intestinal tract. **a.** Studies suggest that lacto-*N*-tetraose provides a probiotic growth advantage to bifidobacteria, which contain the glycosidases required to metabolize this oligosaccharide. Bifidobacteria are considered "good" bacteria in the human intestine because they aid the digestive process. **b.** Some milk oligosaccharides may function as soluble decoys that inhibit pathogenic bacteria from invading intestinal epithelial cells by providing a large number of competing glycan binding sites.

Studies suggest that lacto-*N*-tetraose functions as a probiotic in the infant intestine by supporting the growth of *Bifidobacterium* species, which contain the necessary enzymes to degrade and use this oligosaccharide as a source of metabolic fuel (**Figure 13.5a**). In contrast, lacto-*N*-fucopentaose I appears to protect human infants from microbial pathogens by functioning as a soluble glycan decoy. It prevents pathogen infection through binding to structurally related glycoconjugates on the surface of intestinal epithelial cells (**Figure 13.5b**).

Figure 13.6 shows three members of the raffinose series of plant oligosaccharides: raffinose, stachyose, and verbascose. These compounds are derived from sucrose and are found in high abundance in some types of vegetables. Humans and nonruminating animals such as pigs and poultry cannot digest these oligosaccharides because they lack the necessary α-galactosidase enzyme needed to hydrolyze the α-1,6 glycosidic bond. Eating foods high in raffinose-series oligosaccharides can lead to gastrointestinal discomfort (flatulence) because the undigested carbohydrates end up in the lower intestine, where bacteria—which do contain α-galactosidase—ferment the compounds to produce methane, carbon dioxide, and hydrogen gases.

The growth rates of pigs and chickens are often decreased by ingestion of raffinose-series oligosaccharides present in soybean-based feed because of the associated gastrointestinal problems they create. Therefore, their food is pretreated with a commercially produced α-galactosidase enzyme that first degrades the oligosaccharides present

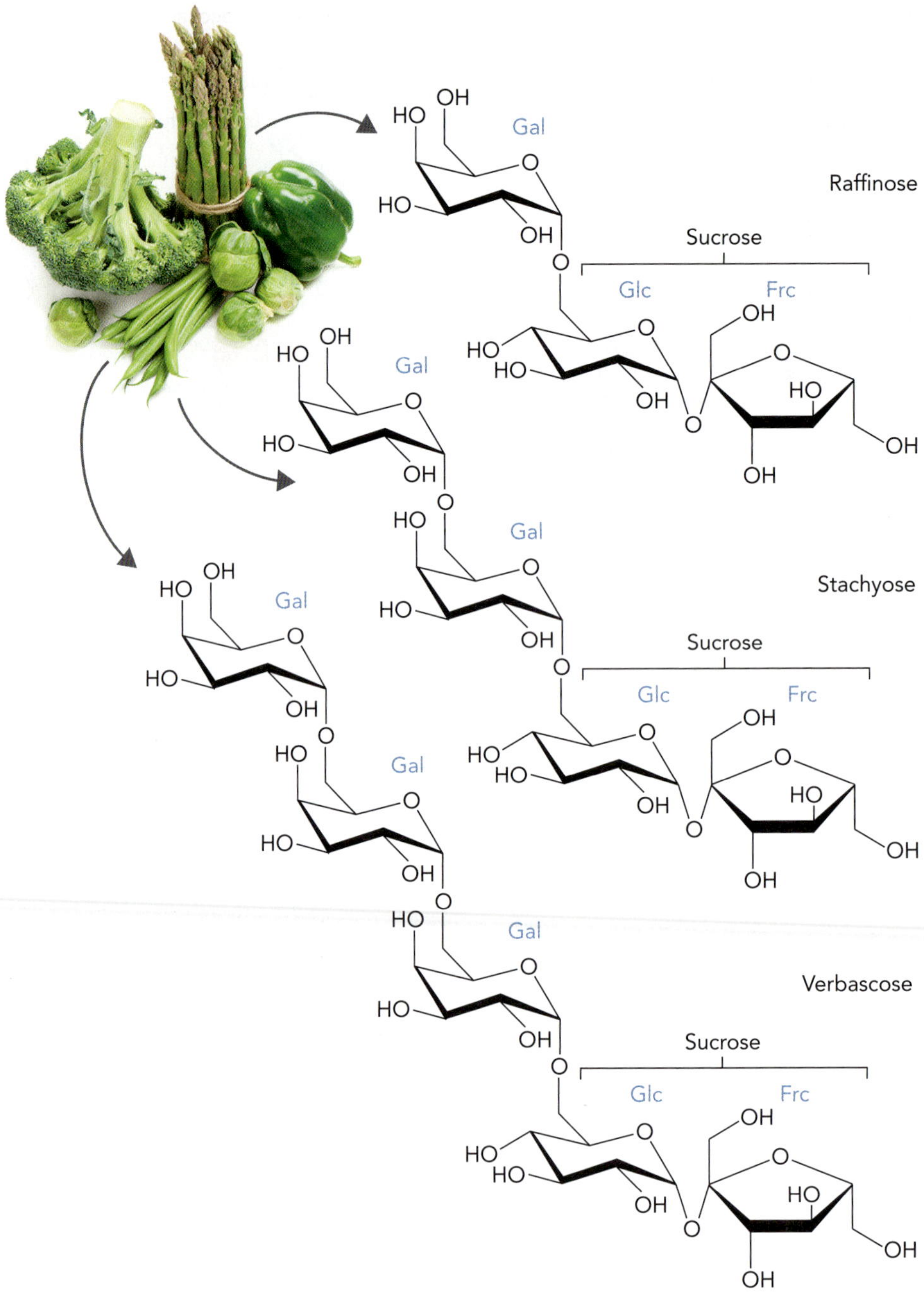

Figure 13.6 The oligosaccharides raffinose, stachyose, and verbascose, found in a variety of vegetables, differ in the number of galactose units attached to sucrose. PHOTO: IRYNA DENYSOVA/SHUTTERSTOCK.

in the feed. This α-galactosidase enzyme can be readily isolated from fungal cultures of *Aspergillus niger* that secrete the enzyme into the media, thus providing an efficient process for industrial scale-up (**Figure 13.7**). One form of the enzyme marketed as a dietary product for humans is called Beano.

Cellulose and Chitin Are Structural Carbohydrates

Cellulose is by far the most abundant carbohydrate on Earth because it is the core structural component of all plant cell walls. **Cellulose** is a homopolymeric molecule consisting of nearly a thousand repeating units of a disaccharide called **cellobiose**. These cellobiose units consist of two glucose residues linked by β-1,4 glycosidic bonds (**Figure 13.8**). Cellulose provides plants with a rigid cell wall consisting of hydrogen-bonded cellulose fibrils, which are held together by hemicellulose and pectin

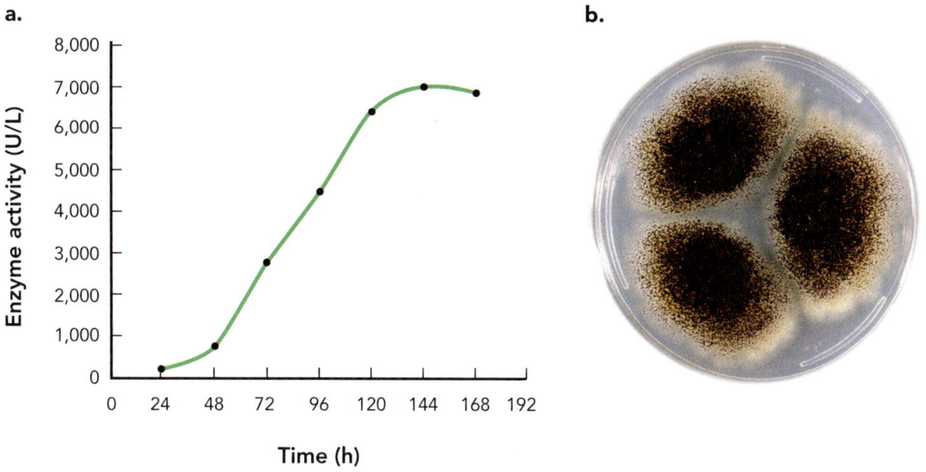

Figure 13.7 The α-galactosidase enzyme used for livestock feed preparation and sold commercially as Beano can be isolated from fungal cultures grown on soybean extracts. **a.** Secretion of α-galactosidase enzyme into the media of *Aspergillus niger* cultures as a function of time. BASED ON P. ALEKSIEVA, B. TCHORBANOV, AND L. NACHEVA (2010). HIGH-YIELD PRODUCTION OF ALPHA-GALACTOSIDASE EXCRETED FROM *PENICILLIUM CHRYSOGENUM* AND *ASPERGILLUS NIGER*. *BIOTECHNOLOGY & BIOTECHNOLOGICAL EQUIPMENT*, *24*(*1*), 1620–1623. **b.** Culture plate of *Aspergillus niger*. DR. DAVID ELLIS.

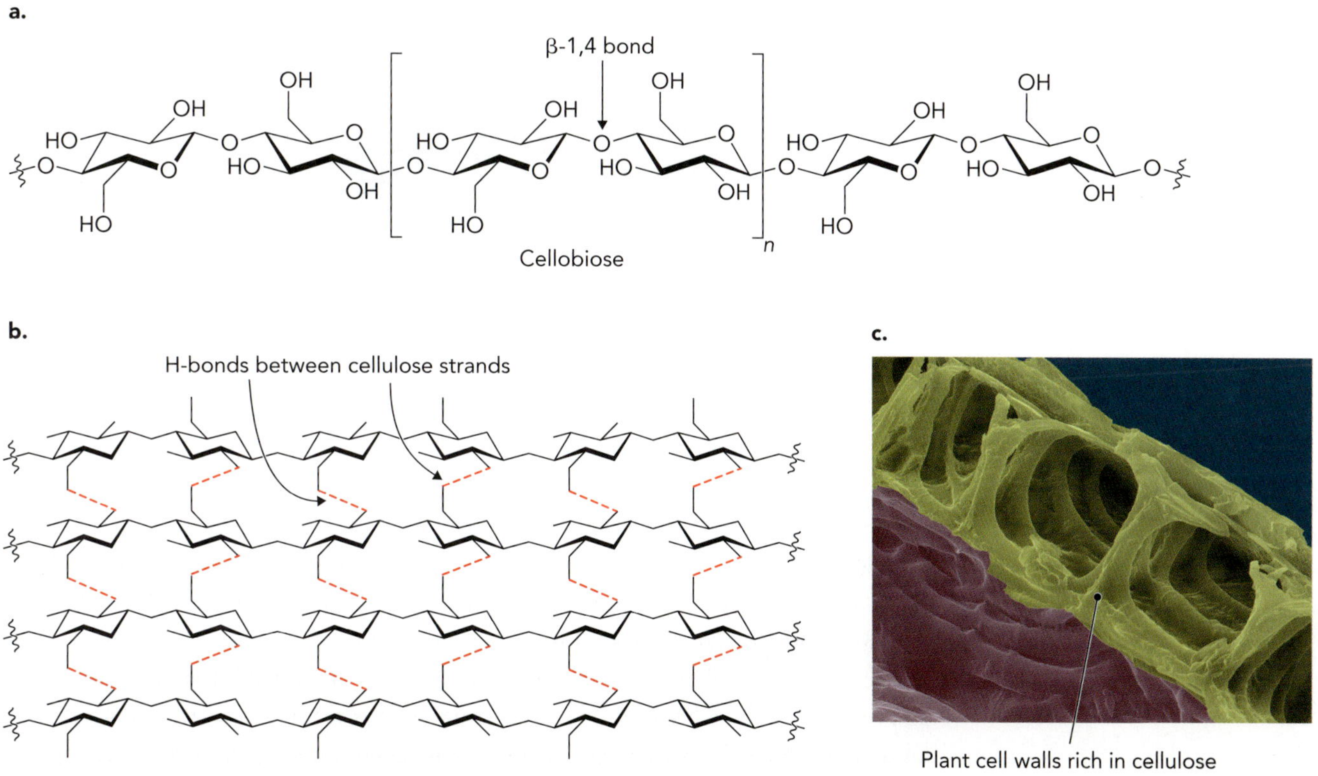

Figure 13.8 Cellulose is a glucose polysaccharide that provides strength to plant cell walls. **a.** Cellulose consists of the repeating disaccharide cellobiose, which consists of glucose residues linked by β-1,4 glycosidic bonds. **b.** Individual cellulose strands are extensively hydrogen bonded to build a polysaccharide fibril. **c.** Electron micrograph of cellulose-rich cell walls from the ragweed plant *Ambrosia psilostachya*. DR. DENNIS KUNKEL/VISUALS UNLIMITED.

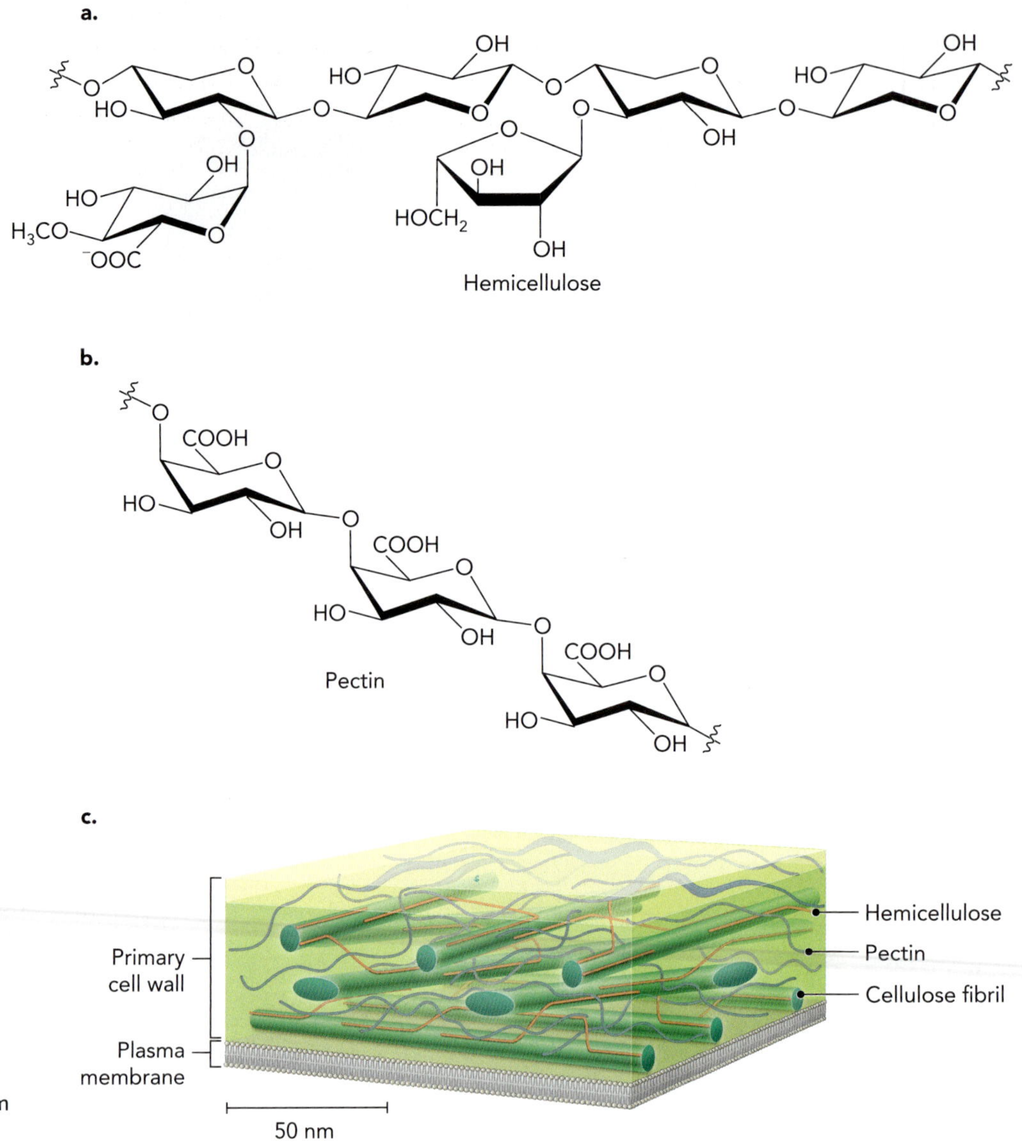

Figure 13.9 Plant cell walls consist of cellulose fibrils hydrogen bonded to the polysaccharides hemicellulose and pectin. **a.** Structure of the hemicellulose xylan isolated from softwood. Hemicelluloses are branched polysaccharides consisting of up to six different sugar residues. **b.** Pectin is a homopolymer of galacturonic acid. **c.** Cellulose fibrils, hemicellulose, and pectin are hydrogen bonded into a network of carbohydrate polymers to form plant cell walls.

polymers. **Hemicellulose** is a branched heteropolymer containing up to six different sugars, whereas **pectin** is a homopolymer of galacturonic acid (**Figure 13.9**).

Although cellulose consists entirely of glucose, which is a useful form of metabolic energy, most animals lack the enzyme **cellulase**, which is required to hydrolyze the β-1,4 glycosidic bonds between glucose residues. The enzymatic product of cellulase is **cellotetraose**, a tetrasaccharide consisting of four glucose residues (**Figure 13.10**). Plant material high in cellulose fiber is considered "roughage" in the diet because it passes through the digestive system without being degraded. Some animals have evolved symbiotic relationships with microorganisms that inhabit their digestive tracts and secrete cellulase. For example, ruminating herbivores such as cows and goats have an unusual stomach that permits them to regurgitate their food, thereby maximizing mechanical and enzymatic breakdown of cellulose with the help of bacteria.

Chitin is a linear hexosamine polysaccharide that forms the structural component of invertebrate exoskeletons in insects and crustaceans. Chitin is also a component of the cell walls in many types of fungi, including mushrooms. As illustrated in **Figure 13.11**, chitin consists of repeating GlcNAc hexosamine units linked by a β-1,4

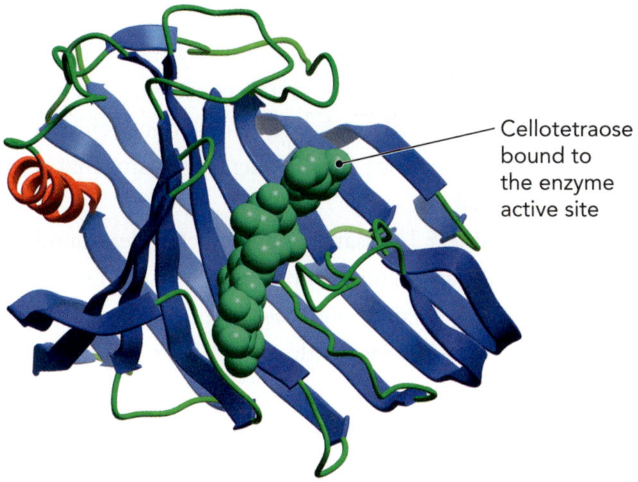

Figure 13.10 The structure of the enzyme cellulase that was isolated from the thermophilic bacterium *Thermotoga maritima* is shown. The oligosaccharide product of the cellulase reaction is cellotetraose, which is shown here bound to the enzyme active site.

BASED ON PDB FILE 3AMM.

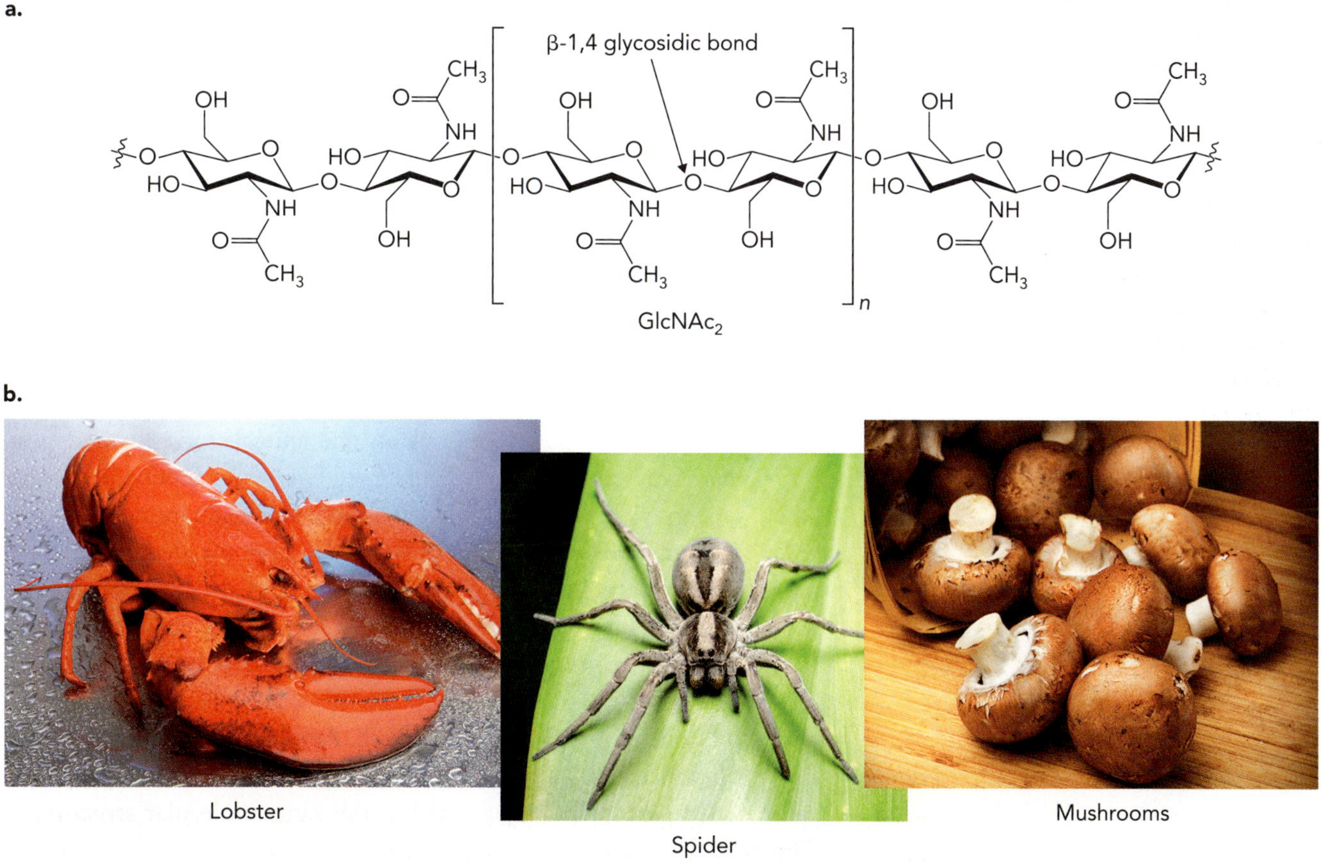

Figure 13.11 Chitin is formed from repeating GlcNAc units, which provides structural strength to the exoskeletons of many invertebrates. **a.** Chemical structure of chitin. **b.** The exoskeletons of a variety of invertebrates and fungi contain high amounts of chitin.

LOBSTER: AFRICA STUDIO/SHUTTERSTOCK; SPIDER: PATTARA PUTTIWONG/SHUTTERSTOCK; MUSHROOMS: KATALINKS/SHUTTERSTOCK.

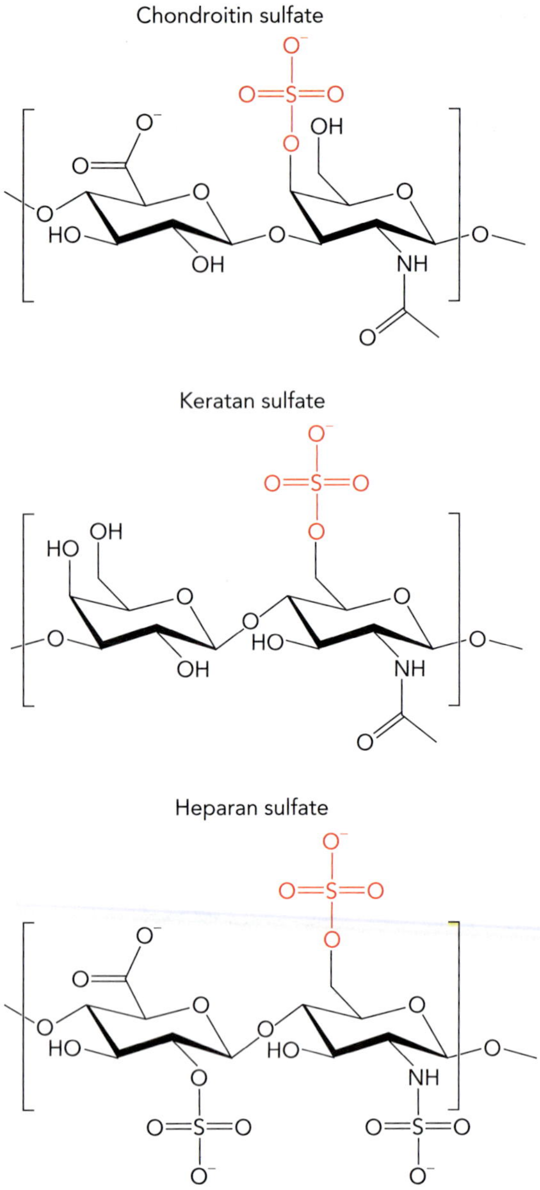

Figure 13.12 Repeating hexosamine disaccharide units in the glycosaminoglycans chondroitin sulfate, keratan sulfate, and heparan sulfate are sulfated in one or more positions.

glycosidic bond. Chitin, like cellulose, provides the organism with an excellent biomaterial for building a strong body frame by virtue of hydrogen-bonding contacts within and between polysaccharide strands. Moreover, because of the β-1,4 glycosidic bond, chitin can be used as a source of carbohydrate fuel by microorganisms that contain the enzyme **chitinase**.

Another class of linear hexosamine polysaccharides found in cells are **glycosaminoglycans**. Unlike chitin, however, glycosaminoglycans are covalently attached to proteins to form a class of glycoconjugates called **proteoglycans**, as described later. **Figure 13.12** shows representative repeating hexosamine disaccharides in the glycosaminoglycans **chondroitin sulfate**, **heparan sulfate**, and **keratan sulfate**. Glycosaminoglycans usually consist of ~20 to ~50 disaccharides in a single polysaccharide chain.

Glycosaminoglycans have important roles in biology, especially in the interstitial fluid between joints and tissues, where they help maintain a hydrated environment owing to their hygroscopic properties. In addition, glycosaminoglycans have a functional role in promoting structural organization of tissues. For example, keratan sulfate polymers function in the cornea to maintain the optimal arrangement of collagen fibrils to allow light to pass through the cornea. A genetic defect in the enzyme that modifies keratan, carbohydrate sulfotransferase 6 (CHST6), causes the disease **macular corneal dystrophy**. The associated loss of vision in patients with this disease is caused by disorganization of keratan sulfate proteoglycans due to reduced sulfation of the keratan disaccharide (**Figure 13.13**).

Starch and Glycogen Are Storage Forms of Glucose

Glucose homopolymers in plants and animals serve as short-term energy reserves to supplement stored lipids, which provide a higher energy yield per gram. The two most abundant forms of glucose polymers are starch, which plants use to store the glucose generated by photosynthesis during daylight hours, and glycogen, which many types of animals use to store dietary sources of glucose. Starch and glycogen contain linear or branched glucose polymers with α-1,4 and α-1,6 glycosidic bonds.

Plants actually synthesize two forms of starch. **Amylose**, a linear polysaccharide, contains ~100 glucose units linked by α-1,4 glycosidic bonds. **Amylopectin** is an α-1,6–branched glucose polymer that has the same molecular structure as glycogen (**Figure 13.14**, p. 646), though glycogen is more highly branched and generally contains more glucose units. Amylose, and presumably unbranched regions of amylopectin, can form stable left-handed helical structures as a result of intramolecular hydrogen bonding. Each turn of the helix contains six glucose molecules, which allows efficient packing of the glucose polymer within starch granules. The presence of α-1,6 bonds in amylopectin and glycogen creates branch points that greatly increase the number of free ends in the homopolymeric molecule. Unlike

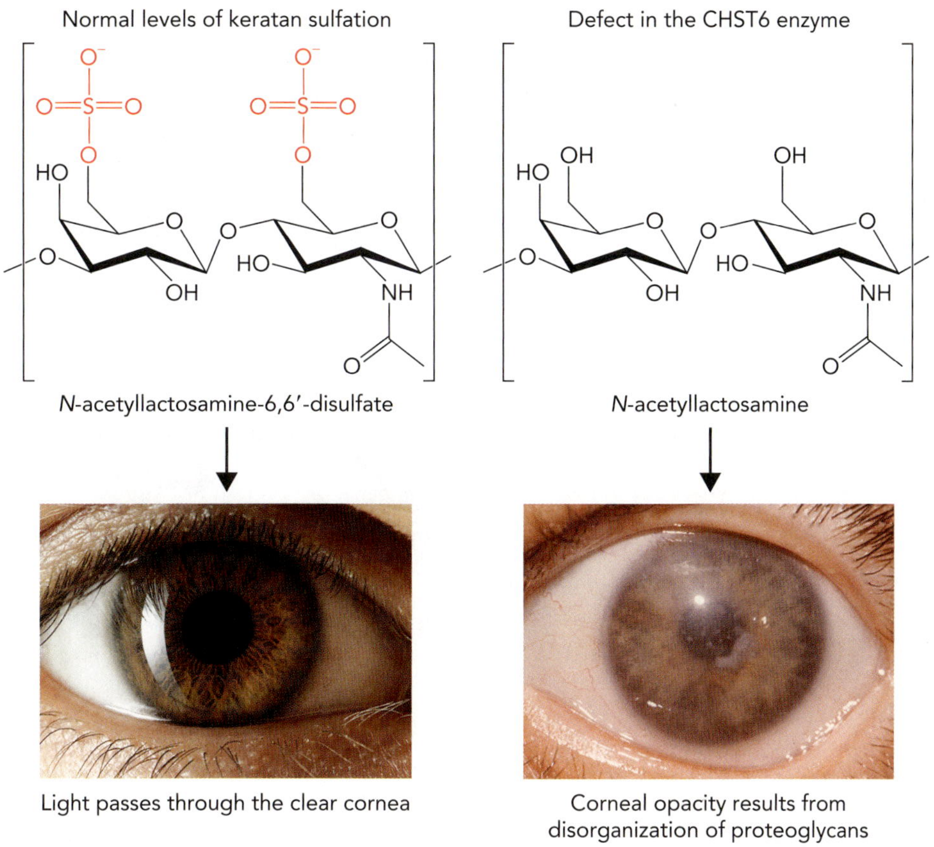

Figure 13.13 Keratan sulfate consists of the repeating disaccharide *N*-acetyllactosamine, which must be properly sulfated to provide structural support in the cornea of the eye. A rare genetic defect in the enzyme that sulfates keratan, CHST6, causes the disease macular corneal dystrophy, which is characterized by corneal opacity due to decreased keratan sulfation. CLEAR CORNEA: GERENME/ISTOCK/GETTY IMAGES PLUS; CORNEAL OPACITY: DR. HAROLD E. CROSS/USED WITH PERMISSION FROM THE AUTHOR OF THE HEREDITARY OCULAR DISEASE WEBSITE SPONSORED BY THE UNIVERSITY OF ARIZONA/UNIVERSITY OF ARIZONA COLLEGE OF MEDICINE. DEPT. OF OPHTHALMOLOGY AND VISION SCIENCE. © 2010–2015, ARIZONA BOARD OF REGENTS. ALL RIGHTS RESERVED.

cellulose and chitin, starch is an important dietary source of glucose for animals because it can readily be hydrolyzed by the enzyme α-amylase, which cleaves α-1,4 glycosidic bonds.

Amylopectin and glycogen both contain α-1,6 glycosidic bonds at branch points; however, the number of branch points in these two forms of stored glucose are quite different. As shown in **Figure 13.15** (p. 647), amylopectin contains an α-1,6 glycosidic bond about once every 15–30 glucose residues, and glycogen contains the same α-1,6 glycosidic bond about once every 8–12 glucose residues. Moreover, amylopectin contains a single glucose molecule at the reducing end, whereas glycogen contains a covalently linked dimeric protein called **glycogenin**, which functions as protein anchor. Amylopectin and glycogen structures are analogous to toy Koosh balls in which the number of fingers (nonreducing ends) is greater in glycogen than in amylopectin (Figure 13.15). Because glucose units can be added and removed from only the nonreducing ends of amylopectin and glycogen, the more branch points they have, the more ends that are available for glucose retrieval and storage. In this regard, glycogen is a more efficient storage form of glucose than amylopectin.

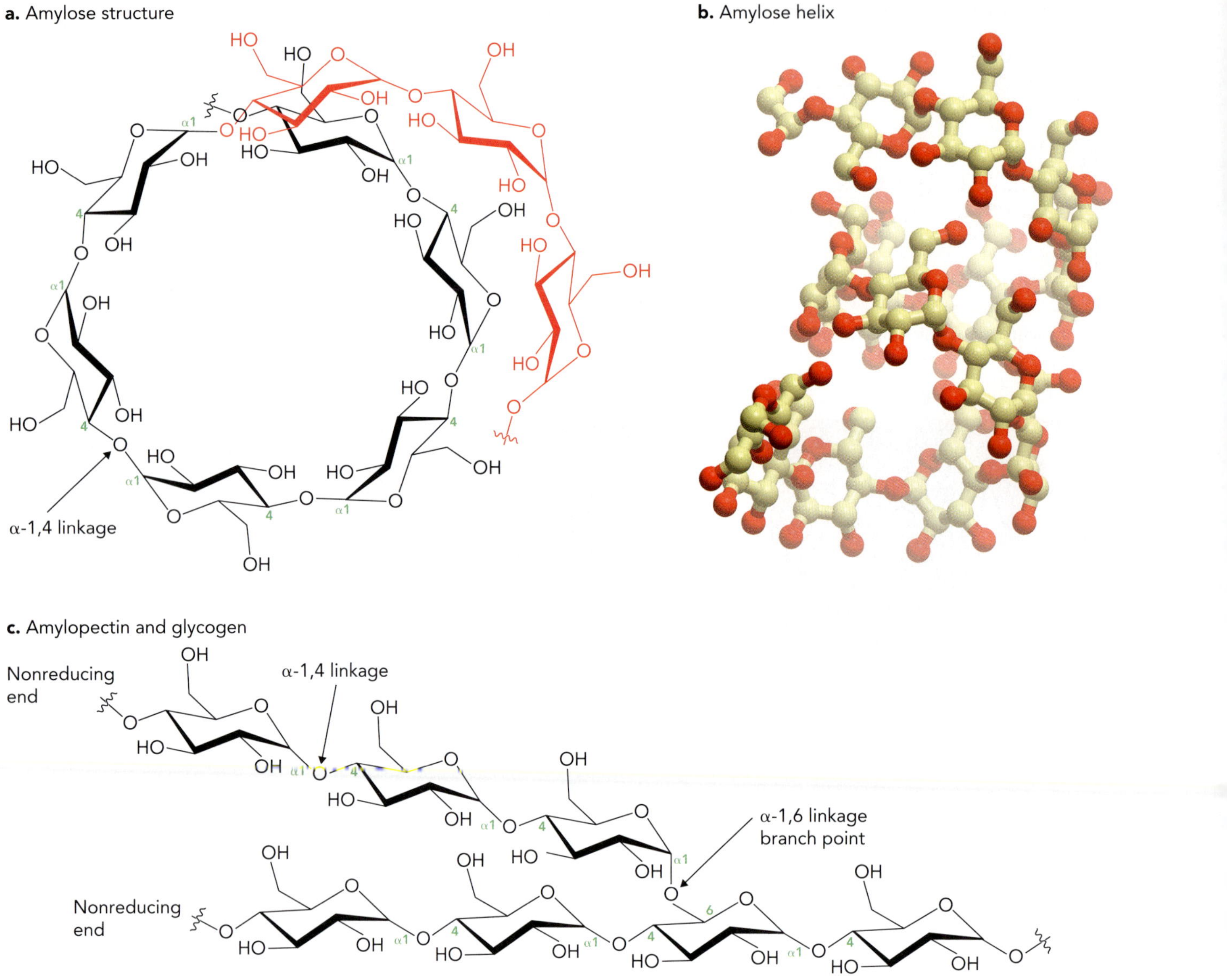

Figure 13.14 Starch, which consists of repeating units of glucose, exists as both linear and branched homopolymers. **a.** Amylose is a linear homopolymer of glucose with six glucose residues per turn. One turn is shown in black, and a portion of a second turn is shown in red. **b.** The amylose polymer forms a left-handed helix stabilized by intrastrand hydrogen bonds. BASED ON PDB FILE 1C58. **c.** Amylopectin and glycogen are branched forms of glucose homopolymers with α-1,6 linkages connecting linear strands of α-1,4–linked glucose units. Glycogen has more branching and more glucose units than amylopectin.

The glycogenin protein has a glucosyltransferase activity—an enzyme that adds glucose units to a glycan group—that links the reducing end of the initial glucose molecule to a tyrosine residue on the protein. As shown in **Figure 13.16** (p. 648), the glucose donor for this reaction is the nucleotide sugar uridine diphosphate-glucose (UDP-glucose), which is used to add up to eight α-1,4-linked glucose residues to the growing polymer. As described in Chapter 14, the nascent glycogen core molecule is extended by the enzymes glycogen synthase and glycogen-branching enzyme.

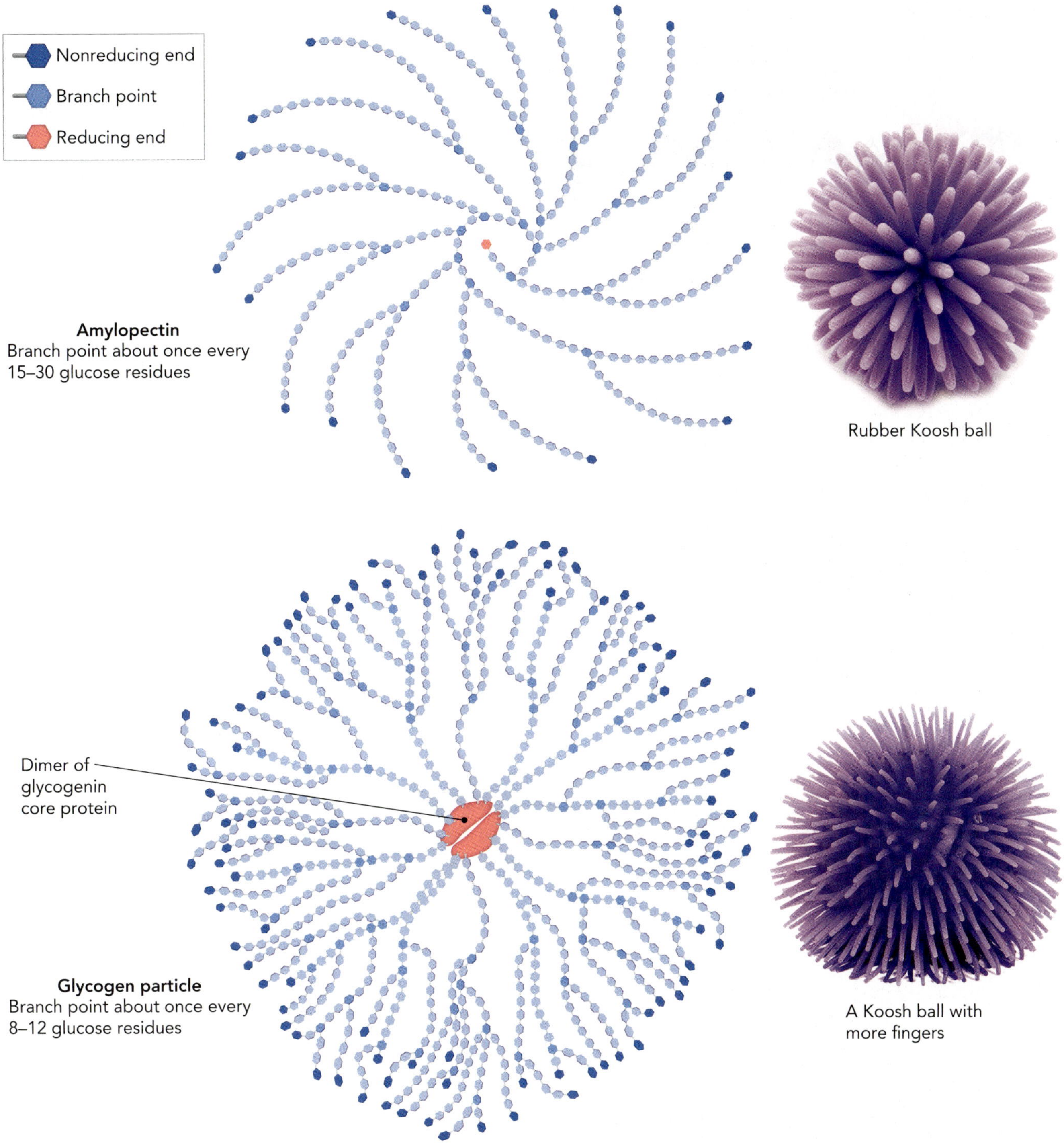

Figure 13.15 Macromolecular structures of amylopectin and glycogen. Amylopectin has a single reducing end (red) and α-1,6 branch points about once every 15–30 glucose residues. In contrast, glycogen has branch points about once every 8–12 glucose residues and a glycogenin homodimeric core protein (red) with two glucose molecules at the center. The increased number of α-1,6 branch points in glycogen compared to amylopectin results in a much larger number of nonreducing ends (indicated with dark-blue hexagons). A spherical toy Koosh ball is analogous to amylopectin and glycogen core particles with the difference being the number of fingers (nonreducing ends). KOOSH BALL, FEW FINGERS: KEITH BELL/SHUTTERSTOCK; KOOSH BALL, MANY FINGERS: STEPHEN BONK/SHUTTERSTOCK.

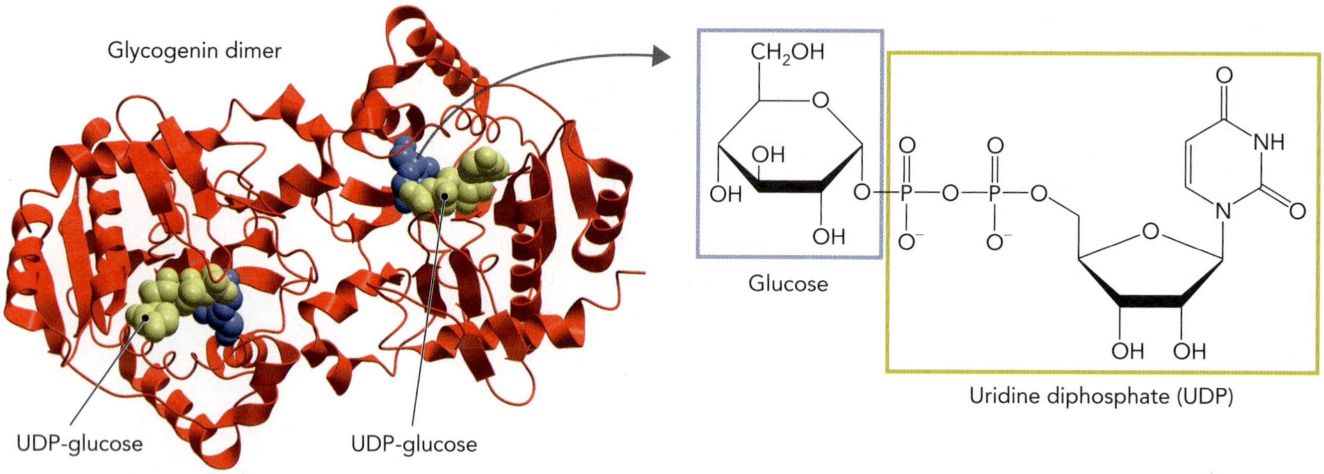

Figure 13.16 This molecular structure of the homodimeric rabbit glycogenin protein shows the nucleotide sugar UDP-glucose molecules bound to the enzyme active sites. BASED ON PDB FILE 3T7O.

concept integration 13.1

Describe three functions of carbohydrates that explain why they are the most abundant biomolecules in nature.

First, carbohydrates are the main structural component of cell walls in plants, bacteria, and the exoskeletons of many types of invertebrates. The carbohydrate in plant cell walls is cellulose, a homopolysaccharide of glucose, whereas the carbohydrate in the exoskeletons and in some types of fungi is chitin, a hexosamine polydisaccharide. Because plants (including marine algae) and insects constitute a huge biomass on Earth, carbohydrates are very abundant. Second, glucose is stored in branched polysaccharides called starch and glycogen, which are used for energy conversion reactions in organisms when sunlight and food, respectively, are scarce. For example, plants use light energy to convert CO_2 into glucose through the reactions of the Calvin cycle; the glucose is then stored as amylopectin for use at night when photosynthesis is inactive. Similarly, animals use excess glucose obtained from the diet to synthesize glycogen in the liver and muscle to fuel the glycolytic pathway between meals. Third, a large number of proteins and lipids in eukaryotic cells are glycoconjugates, containing covalently attached glycan groups. Glycoconjugates function in cell signaling, immunity, and the extracellular matrix in animals.

13.2 Important Biological Functions of Glycoconjugates

The two primary types of glycoconjugates in eukaryotic cells are **glycoproteins** and **glycolipids**. Glycan modification of proteins takes place within the lumen of the endoplasmic reticulum compartment of the cell, whereas glycolipids are primarily generated in the Golgi apparatus. We describe the structure and function of glycoproteins here in Chapter 13 and describe glycolipids in more detail in Chapter 15.

Glycoproteins are protein glycoconjugates in which the bulk of the macromolecule consists of protein. Some glycoproteins are transported by membrane vesicles to the plasma membrane for insertion or secretion. Others are targeted to cellular organelles such as the mitochondria or peroxisomes. The targeting information for protein sorting is contained with the N-terminal amino acid sequence, as described in Chapter 22.

Proteoglycans are a second type of protein glycoconjugate in which most of the macromolecule consists of carbohydrate units, with only a small contribution coming from protein (or short polypeptide segments). The human proteoglycans aggrecan and perlecan are located in the extracellular matrix and serve as the "gel" around tissues and in joints, whereas **peptidoglycans** are a type of proteoglycan found in bacterial cell walls. In this section, we describe the structure and function of representative glycoproteins, proteoglycans, and peptidoglycans, as well as several biomedical examples of these important glycoconjugates.

Glycoconjugates Function in Cell Signaling and Immunity

Glycan groups on glycoconjugates are recognized by glycan binding proteins, of which lectins are one of the best characterized. Two classes of glycoconjugate binding interactions have been characterized in humans: (1) *intrinsic* glycoconjugate binding of glycans to lectins within the same host cell (intracellular) or between host cells (intercellular), and (2) *extrinsic* glycoconjugate binding between glycans and lectins on human cells and pathogen cells (**Figure 13.17**). Intrinsic glycoconjugate interactions mediate cell recognition functions on immune cells and also facilitate cell attachment and cell migration. Extrinsic glycoconjugate interactions between immune cells and invading pathogens—such as bacteria, viruses, or fungi—protect host cells from infection by inducing immune responses that lead to the release of antibodies and hydrolytic enzymes that neutralize or kill the invading pathogen. However, pathogens evolve quickly both to avoid and to exploit these extrinsic glycoconjugate interactions. For example, pathogens can use host cell glycoconjugates as binding sites to aid in adhesion or can express their own glycoconjugates, which function as molecular mimics to promote host cell binding.

Figure 13.18 summarizes several types of intrinsic and extrinsic glycoconjugate interactions that have been characterized in humans. One example of an extrinsic glycoconjugate interaction is the binding of the *Escherichia coli* lectin protein FimH

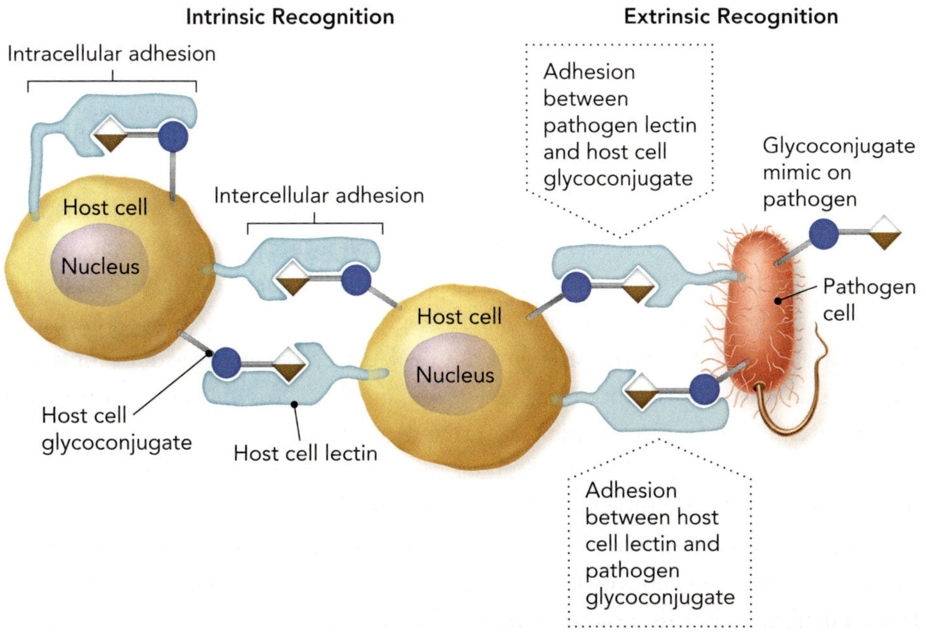

Figure 13.17 Glycoconjugate binding interactions can be characterized as intrinsic or extrinsic, depending on the cell origin of the glycoconjugate and lectin protein. Intrinsic glycoconjugate interactions occur within and between cells of the host organism; for example, mediation of self and non-self recognition in the immune system or promotion of cell migration in the neuronal system. In contrast, extrinsic glycoconjugate interactions occur between host cells and pathogen cells, which may involve adhesion or mimicry functions in the pathogen cell to promote host cell infection.

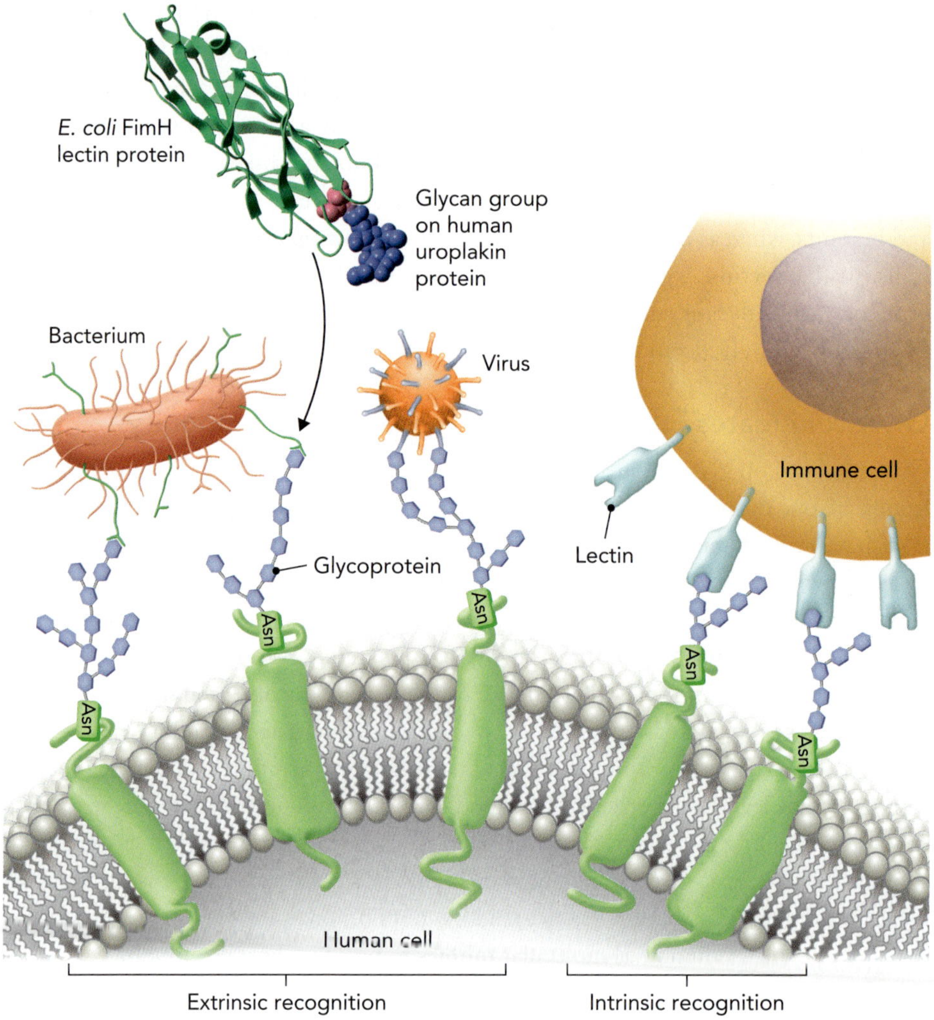

Figure 13.18 Glycoproteins have an important role in cell signaling and pathogen recognition. Immune cells communicate with each other by cell–cell interactions between glycoproteins and glycan binding proteins on the cell surface (intrinsic recognition). Two examples of extrinsic recognition, where pathogens are binding to host cells through glycoconjugate interactions, are shown here. In one example, a virus is binding to a host cell. In the other example, the *E. coli* lectin protein FimH is binding to a glycan group on the human glycoprotein uroplakin (inset molecular structure). BASED ON PDB FILE 2VCO.

to the mannose glycan group of the human uroplakin protein, which is expressed on the surface of human bladder cells. Because this binding interaction can lead to urinary tract infections, studies are under way to identify molecules that mimic glycan groups on human bladder cells and can function as inhibitors of bacterial cell adhesion.

As shown in **Figure 13.19**, carbohydrate linkage to glycoproteins occurs through either the amide nitrogen atom in the side chain of asparagine, leading to the generation of **N-linked oligosaccharides**, or through the oxygen atom of the side chain of serine or threonine residues, resulting in **O-linked oligosaccharides**. The most common N-glycosidic bond in glycoproteins is between asparagine and GlcNAc, although not all asparagine residues in proteins can be N-glycosylated. Studies have shown that N-glycosylating enzymes most often attach GlcNAc to asparagine residues contained within the amino acid sequence Asn-X-Ser/Thr, in which X is any amino acid except proline. In contrast, O-glycosylation of serine or threonine residues in glycoproteins does not seem to require a preferred amino acid recognition sequence, and the most common monosaccharide used to create the O-glycosidic bond is GalNAc (Figure 13.19). O-linked glycoproteins are also called **mucins**, which function in the human body to protect epithelial cells in the intestinal, urinary, and respiratory tracts from physical damage and pathogen infections.

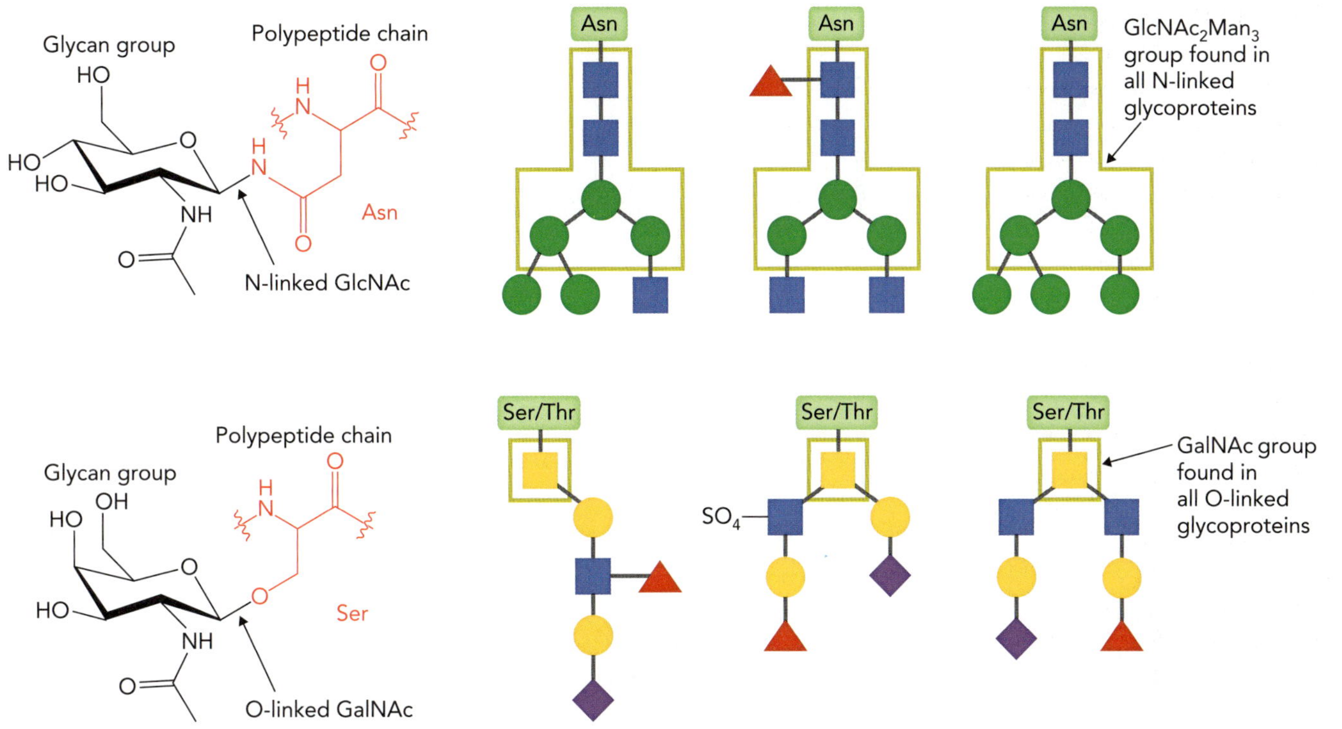

Figure 13.19 The structures shown are representative glycan groups in N-linked and O-linked glycoproteins. The N-linked glycan groups are attached to asparagine residues in the protein using a GlcNAc monosaccharide, whereas O-linked glycan groups are attached to serine or threonine residues using a GalNAc monosaccharide. N-linked glycan groups have a core structure of $GlcNAc_2Man_3$ (outlined), and O-linked glycan groups are GalNAc (outlined), often followed by Gal or GlcNAc. The symbols for the monosaccharides are the same as in Figure 13.2.

Both N-linked and O-linked glycoproteins have been found to contain one of several core glycan structures, which serve as scaffolds for the addition of a variety of monosaccharides using different glycosidic bonds (α-1,4, α-1,3, β-1,6, β-1,3, and so forth). For example, N-linked glycoproteins all contain a five-sugar core glycan group consisting of $GlcNAc_2Man_3$, covalently linked as Man α-1,6 (Man α-1,3) Man β-1,4 GlcNAc β-1,4 GlcNAc β-1-Asn (Figure 13.19). The O-linked glycoprotein core is more variable, although the O-linked glycosidic bond is usually with GalNAc, which is then most often linked to either Gal or GlcNAc monosaccharides. **Figure 13.20** shows the molecular structure of a portion of the extracellular domain of the human CD2 protein, with its N-linked oligosaccharide ($GlcNAc_2Man_7$) and four different types of glycosidic bonds.

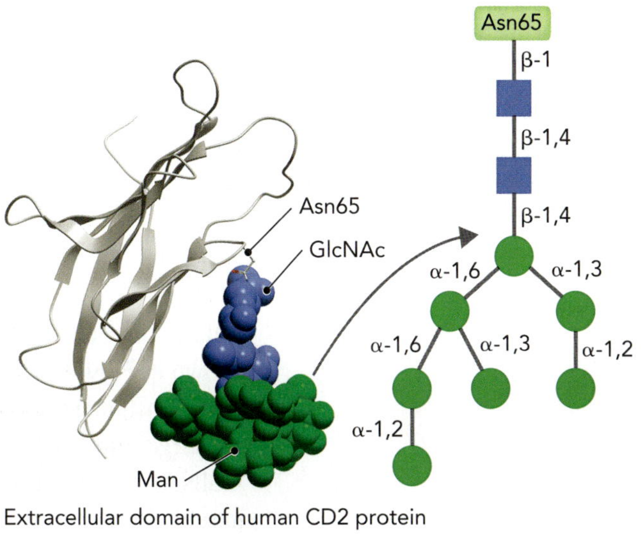

Figure 13.20 The molecular structure of a portion of the N-glycosylated extracellular domain of the human CD2 protein shows the arrangement of the nine sugar residues. BASED ON PDB FILE 1GYA.

ABO Human Blood Types Are Determined by Variant Glycosyltransferases

Protein glycosylation is highly specific and requires the activity of glycosyltransferase enzymes. The nucleotide sugar donor and glycan acceptor are substrates for these

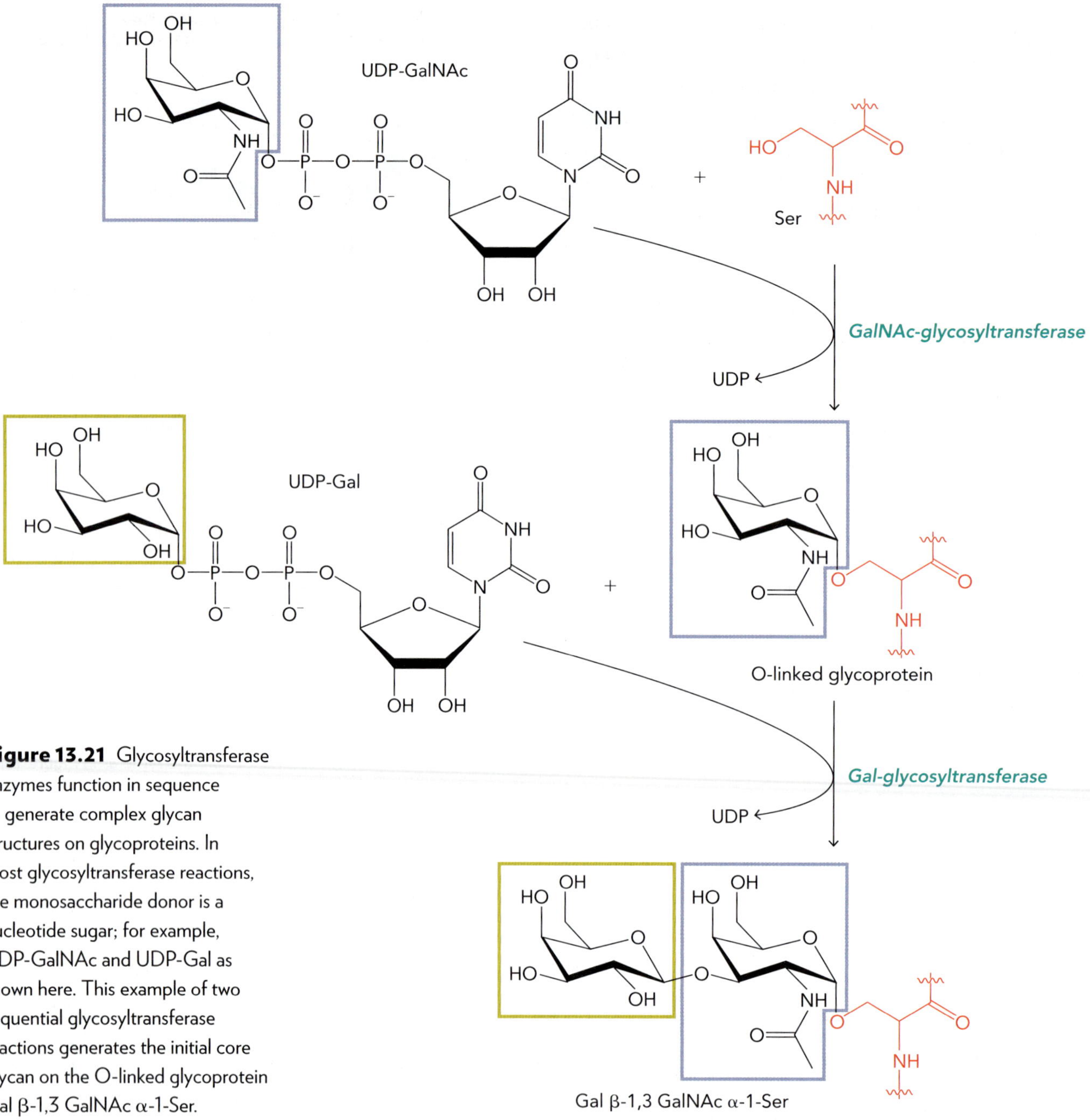

Figure 13.21 Glycosyltransferase enzymes function in sequence to generate complex glycan structures on glycoproteins. In most glycosyltransferase reactions, the monosaccharide donor is a nucleotide sugar; for example, UDP-GalNAc and UDP-Gal as shown here. This example of two sequential glycosyltransferase reactions generates the initial core glycan on the O-linked glycoprotein Gal β-1,3 GalNAc α-1-Ser.

linkage reactions, which proceed sequentially to generate functional glycan groups, as shown in **Figure 13.21**. Formation of an O-linked glycoprotein requires a specific glycosyltransferase enzyme that recognizes both UDP-GalNAc and a serine residue in the target protein as substrates in the reaction. After release of the products UDP and GalNAc α-1-Ser, the glycan group on the glycoprotein functions as the substrate in a second reaction, which is catalyzed by a different glycosyltransferase enzyme. This second enzyme uses UDP-Gal and GalNAc α-1-Ser as substrates to generate the products UDP and Gal β-1,3 GalNAc α-1-Ser.

Glycosyltransferase enzymes are themselves glycoproteins, and, not surprisingly, a large number are required to build the variety of glycan structures found in any one organism. Indeed, 250 glycosyltransferase-encoding genes have been identified in the human genome, representing ~1% of total protein-coding genes. Because glycosyltransferase enzymes function in the endoplasmic reticulum and Golgi compartments of the cell, they contain a membrane-spanning region targeting them to membranes (**Figure 13.22**). Although many of the glycosyltransferases remain embedded in membranes, others are proteolytically cleaved and secreted by cells.

Genetic differences in the expression and activity of glycosyltransferases account for immunologic incompatibility between individuals. One of the best examples of this is based on the biochemical properties of the **ABO blood groups** in humans. As illustrated in **Figure 13.23**, all three glycoconjugates in human blood types have a common *N*-acetyllactosamine disaccharide linked to glycoproteins or glycolipids on the surface of red blood cells. A typical human red blood cell has about 2×10^6 ABO glycoconjugate molecules attached to it, of which 75% is glycoproteins and 25% is glycolipids. The glycan groups on these glycoproteins and glycolipids are variable; however, they all contain a glycan subgroup called the O (or sometimes H) antigen (O-type blood). This subgroup functions as the attachment site for either a terminal GalNAc (A-type blood) or Gal (B-type blood) sugar residue.

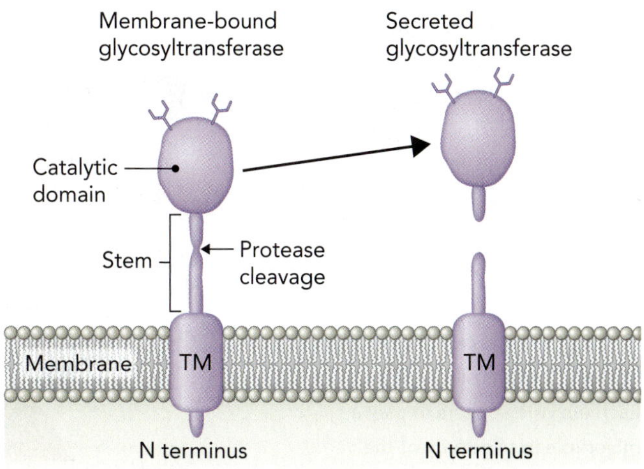

Figure 13.22 Most glycosyltransferase enzymes contain a single transmembrane region (TM), located at the N terminus, that targets the enzyme to the endoplasmic reticulum and Golgi membranes. Some glycosyltransferases are proteolytically cleaved in the stem region of the protein and secreted by cells.

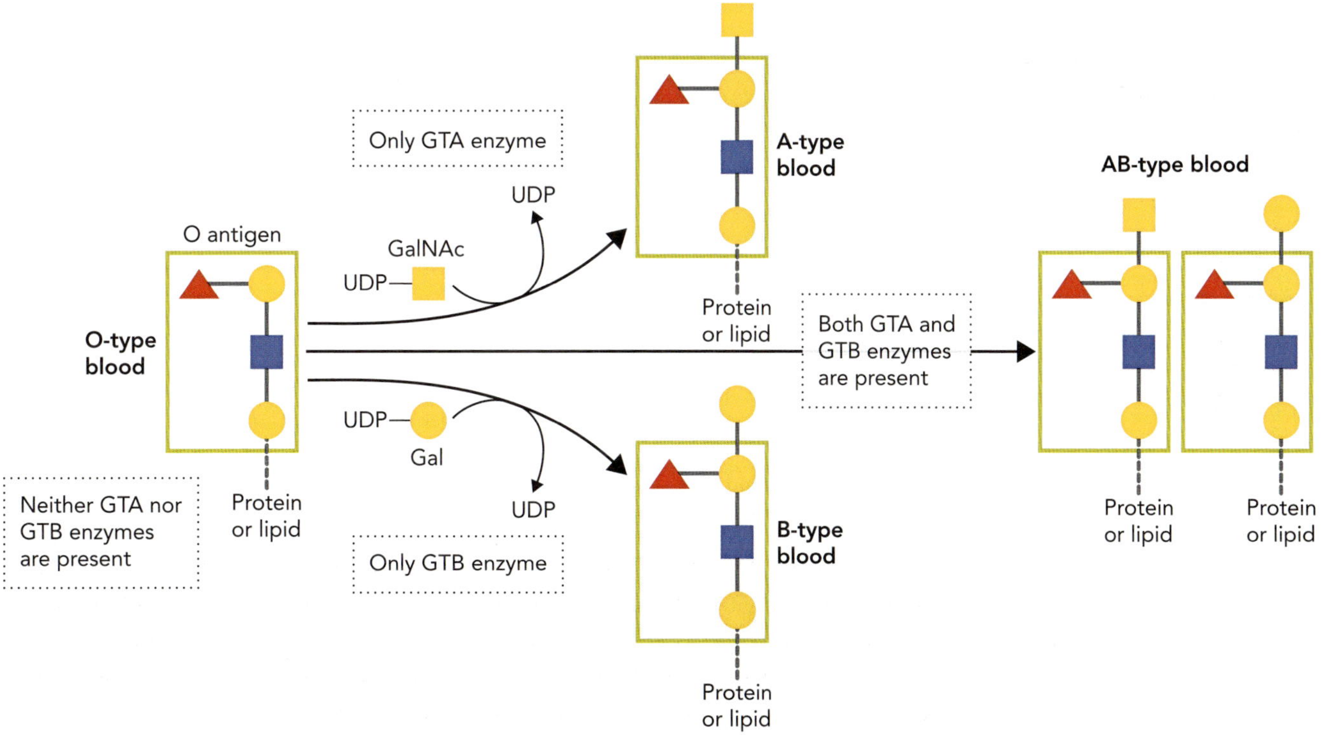

Figure 13.23 The human ABO blood groups are determined by expression of one, both, or neither of the two related enzymes α-1,3-*N*-acetylgalactosaminyltransferase (GTA) and α-1,3-galactosyltransferase (GTB). These enzymes attach either a GalNAc (GTA) or Gal (GTB) sugar residue to a glycan subgroup called the O antigen, which is present on glycoproteins and glycolipids on red blood cells. The symbols for the monosaccharides are the same as in Figure 13.2.

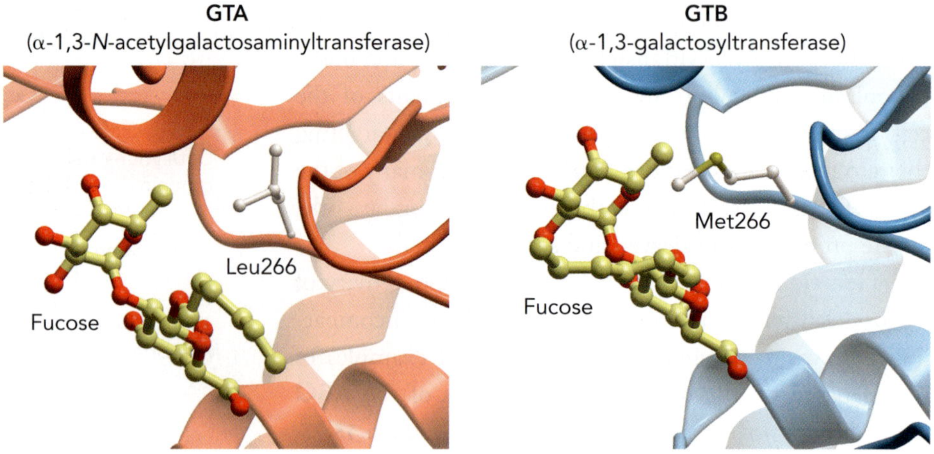

Figure 13.24 The molecular structures of the enzyme active sites are shown for the GTA and GTB enzymes. The two protein structures are shown with a fucose sugar in the enzyme active site, which represents a portion of the O-antigen glycan structure. The GTA and GTB enzymes differ by only four amino acids. The amino acid residue at position 266 of each enzyme appears to play a major role in selectivity of the nucleotide sugar substrate. BASED ON PDB FILES 1LZI (GTA) AND 1LZJ (GTB).

The glycosyltransferase enzymes that attach the GalNAc or Gal sugar residue are named α-1,3-*N*-acetylgalactosaminyltransferase (GTA) and α-1,3-galactosyltransferase (GTB); (Figure 13.23). These two evolutionarily related enzymes differ by only four amino acids, one of which is located at residue position 266 and appears to play a major role in the differential recognition of the GalNAc or Gal nucleotide sugars. The GTA enzyme has a leucine residue at position 266 and binds UDP-GalNAc as a substrate, whereas the GTB enzyme has a methionine residue in the same position and uses UDP-Gal as the nucleotide sugar donor (**Figure 13.24**). An individual expressing only the GTA enzyme has A-type blood, whereas an individual expressing only the GTB enzyme has B-type blood. If an individual expresses both the GTA and GTB enzymes, then that individual has AB-type blood, and if an individual fails to express functional GTA and GTB enzymes, then that individual has O-type blood.

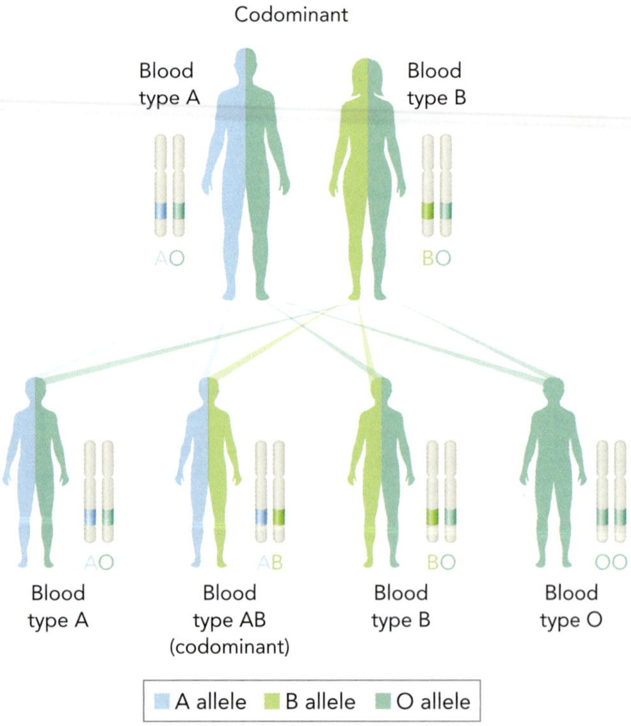

Figure 13.25 Human blood type is an example of a codominant inherited trait, as expression of both the GTA and GTB enzymes gives rise to the AB blood type. If one parent has A-type blood (GTA/–) and the other parent has B-type blood (GTB/–), then the probability of them having a child with any one of the four possible blood groups is 25%.

Inheritance of the A, B, AB, or O blood groups in humans is an example of a **codominant** genetic trait (**Figure 13.25**) because individuals inheriting one copy each of the GTA and GTB enzymes have red blood cells containing a mixture of A-type and B-type glycoconjugates (AB-type blood). The ABO blood groups are very important in blood transfusions because of the presence of antibodies in plasma (non-red-blood-cell component of blood) that recognize non-self glycoconjugate antigens. If blood plasma in the recipient contains antibodies recognizing non-self glycoconjugate antigens present on red blood cells from the donor, then the transfusion will be lethal to the patient due to red blood cell lysis (hemolysis), which can lead to blood clotting (agglutination).

As shown in **Figure 13.26**, individuals expressing only the GTA enzyme (A-type blood) have plasma antibodies that recognize the B-type glycoconjugate on transfused

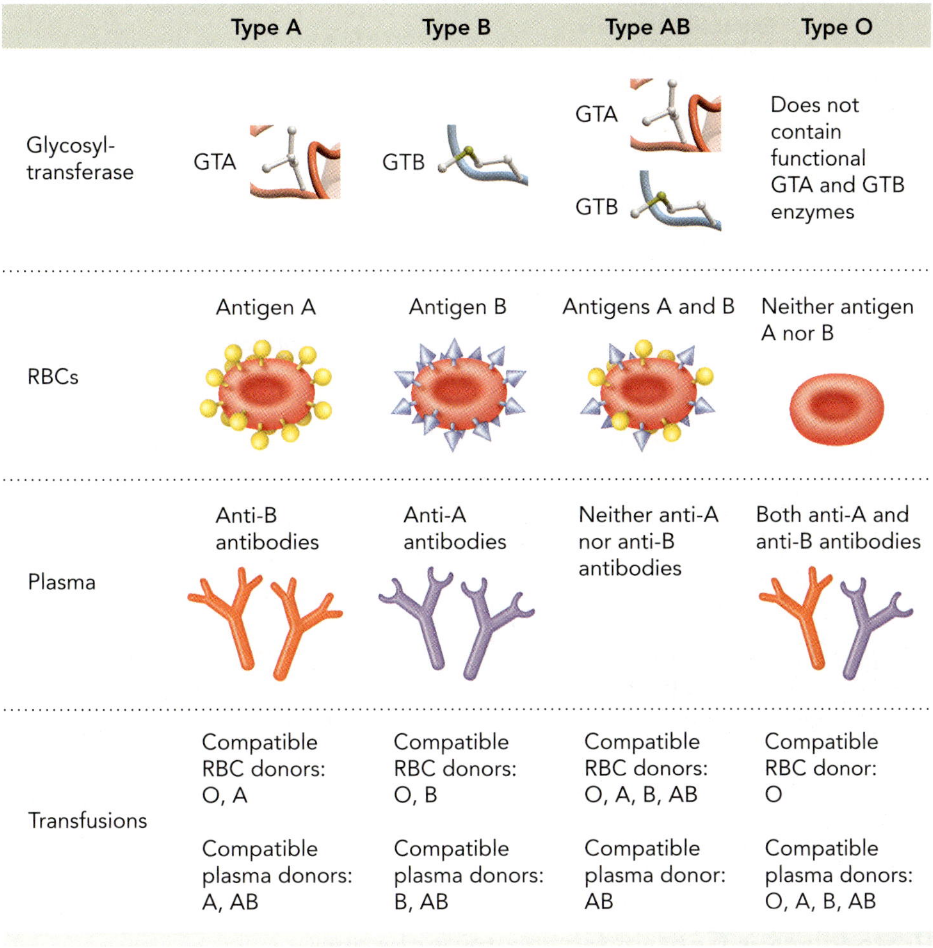

Type A
Type B
Type AB
Type O
Glycosyl-transferase
GTA
GTB
GTA
GTB
Does not contain functional GTA and GTB enzymes
RBCs
Antigen A
Antigen B
Antigens A and B
Neither antigen A nor B
Plasma
Anti-B antibodies
Anti-A antibodies
Neither anti-A nor anti-B antibodies
Both anti-A and anti-B antibodies
Transfusions
Compatible RBC donors: O, A
Compatible RBC donors: O, B
Compatible RBC donors: O, A, B, AB
Compatible RBC donor: O
Compatible plasma donors: A, AB
Compatible plasma donors: B, AB
Compatible plasma donor: AB
Compatible plasma donors: O, A, B, AB

Figure 13.26 Red blood cells (RBCs) and plasma need to be carefully matched in blood transfusions to avoid RBC hemolysis and the potentially lethal effects of blood clotting. Packed red blood cells lack antibodies present in plasma, which is separated from RBCs when blood is prepared for transfusion. Observe that O-type red blood cells can be used in transfusions with recipients having any of the four blood types, and AB-type individuals can receive packed red blood cells from donors with any blood type. O-type individuals can only receive blood transfusions from O-type donors. Plasma can also be donated, and in this case, the plasma of the donor must be compatible with the red blood cells of the recipient.

B-type red blood cells as non-self and initiate a life-threatening immune reaction. Likewise, individuals expressing only the GTB enzyme (B-type blood) have plasma antibodies that recognize the A-type glycoconjugate on A-type transfused red blood cells and initiate a similar life-threatening immune reaction. However, individuals with AB-type blood express both the GTA and GTB enzymes and contain red blood cells with both the A-type and B-type glycoconjugates, so neither are recognized as non-self in transfused red blood cells. Individuals with O-type blood lack functional copies of the GTA and GTB enzymes and therefore have plasma antibodies that recognize both the A-type and B-type glycoconjugates as non-self.

Figure 13.26 also shows that O-type individuals can donate packed red blood cells to A, B, AB, and O individuals because they do not express A or B glycoconjugate antigens. However, the O-type plasma must be separated and removed from the red blood cell preparation because it contains antibodies that recognize A and B glycoconjugates. As shown in Figure 13.26, AB-type individuals can accept packed red blood cells from any one donor type because their serum contains neither type of antibodies that recognize non-self and agglutination will not occur, whereas individuals with O-type blood can only accept packed red blood cells from donors with O-type blood. This same type of compatibility is required in plasma transfusions in that donated plasma must not contain antibodies that will recognize the red blood cells of the recipient (Figure 13.26).

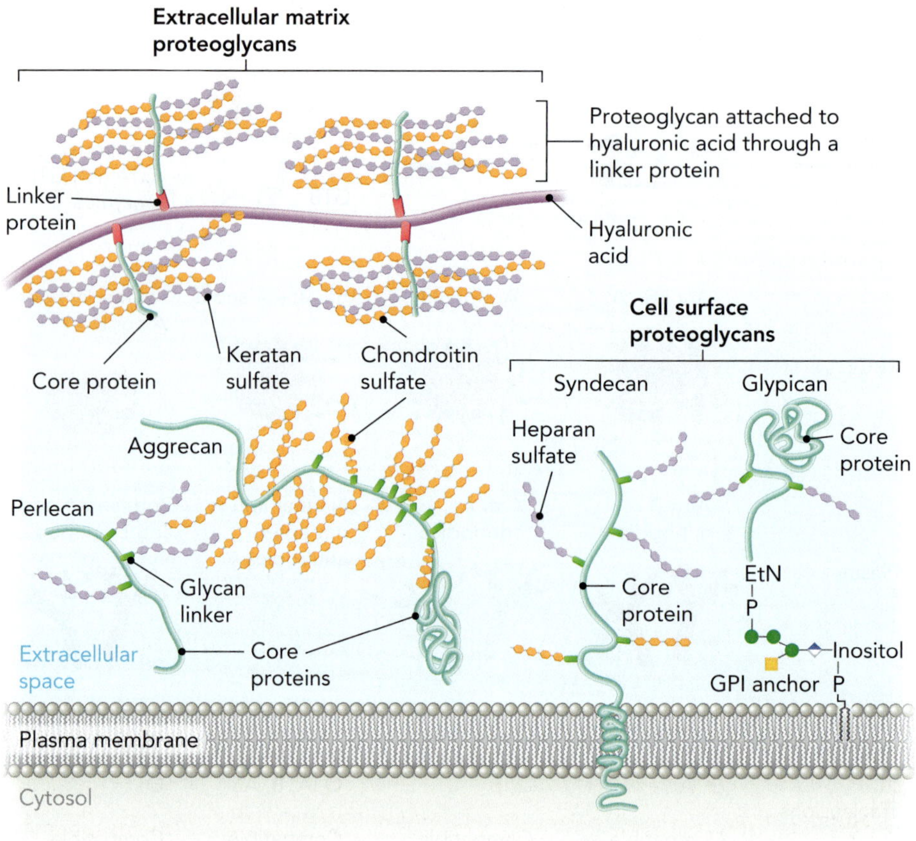

Figure 13.27 Proteoglycans are glycoproteins containing multiple long glycosaminoglycan chains covalently attached to a relatively small core protein. Proteoglycans primarily function in the extracellular space. They are either membrane-bound through a GPI anchor or a transmembrane α helix on the protein or are secreted into the extracellular matrix. Several representative proteoglycans containing the glycosaminoglycans heparan sulfate and/or chondroitin sulfate are illustrated. Large proteoglycan aggregates form in the extracellular matrix when linker proteins bind to the glycosaminoglycan hyaluronic acid. EtN = ethanolamine.

Proteoglycans Contain Glycosaminoglycans Attached to Core Proteins

Proteoglycans are a special type of glycoprotein containing covalently linked glycosaminoglycans, which function primarily in extracellular spaces such as the extracellular matrix. For example, proteoglycans are a major component of cartilage and play an important role in the interstitial space between tissues. Proteoglycans have protein binding activities, so they function in the extracellular matrix as regulators of signal transduction by sequestering cytokines and peptide hormones away from their cognate receptors. Proteoglycans associated with the fibers forming the basement membranes (which underlie the epithelium) facilitate cell migration during organismal development and regulate the permeability properties of some specialized basement membranes.

The glycoproteins involved in cell signaling have glycan groups that consist of ~10 sugar moieties each, covalently attached to a membrane protein of ~100 kDa. In contrast, proteoglycans contain multiple glycosaminoglycan chains of up to 100 sugar moieties, each attached to a small core protein of ~20 kDa. The most common glycosaminoglycans found within proteoglycans are the hexosamine polysaccharides chondroitin sulfate, keratan sulfate, and heparan sulfate, which were described earlier (see Figure 13.12).

Figure 13.27 illustrates the two major classes of proteoglycans in eukaryotes:

1. *cell surface proteoglycans*, which are anchored to the plasma membrane by either a glycophosphatidylinositol (GPI) anchor, as found in glypican, or a single transmembrane α helix of a protein, as in syndecan;
2. *extracellular matrix proteoglycans*, such as aggrecan and perlecan.

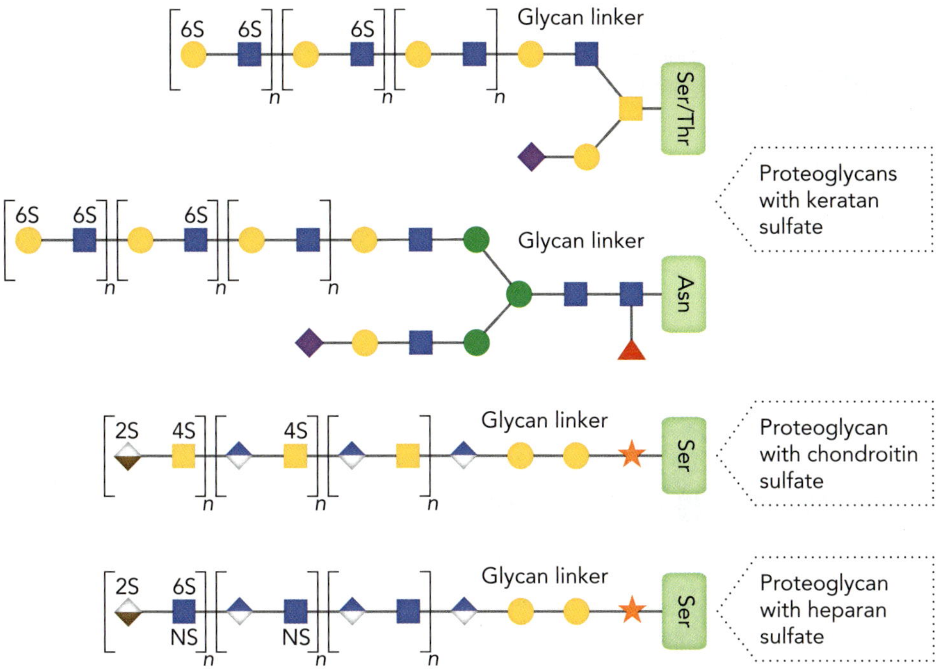

Figure 13.28 Many types of glycosaminoglycans are covalently attached to glycoproteins through O-linked or N-linked glycosidic bonds. Keratan sulfate can be N-linked or O-linked to the core protein, whereas chondroitin sulfate and heparan sulfate are attached to a glycan linker. The linker is O-linked to a serine residue in the core protein through a xylose monosaccharide. NS, 2S, 4S, and 6S refer to the atoms to which sulfate is attached. The polymers contain mixtures of nonsulfated, monosulfated, and polysulfated disaccharide units. The symbols for the monosaccharides are the same as in Figure 13.2.

Some proteoglycans contain only a single type of glycosaminoglycan; for instance, aggrecan has up to 100 chains of chondroitin sulfate attached to the core protein. Other proteoglycans, such as syndecan, contain both chondroitin sulfate and heparan sulfate glycosaminoglycan chains. Keratan sulfate is attached to the core protein through either N-linked or O-linked glycan core structures. Heparan sulfate and chondroitin sulfate are O-linked to a serine residue in the core protein through xylose (**Figure 13.28**). Very large aggregates of proteoglycans are formed in the extracellular matrix when they attach to the glycosaminoglycan **hyaluronic acid** through linker proteins associated with the proteoglycan. Hyaluronic acid is a large glycosaminoglycan polysaccharide consisting of alternating, nonsulfated D-glucuronic acid and GlcNAc residues. As shown in Figure 13.27, the proteoglycans attached to hyaluronic acid often contain the glycosaminoglycans chondroitin sulfate and keratan sulfate. It should be pointed out that the heavily sulfated glycosaminoglycan **heparin**—not to be confused with heparan sulfate—is a component of the proteoglycan serglycin produced by cells in the immune system. Although the physiologic function of heparin is unclear, commercial preparations of heparin are used clinically as a blood anticoagulant.

β-Lactam Antibiotics Target Peptidoglycan Synthesis

We mentioned earlier that peptidoglycans are found in bacterial cell walls, which are rigid, carbohydrate-rich oligopeptide structures. Peptidoglycans give bacteria their shape and serve as the physical boundary for bacterial membranes. The bacterial cell wall consists of multiple strands of hexosamine polysaccharide chains, each composed of repeating units of a β-1,4–linked disaccharide containing MurNAc and GlcNAc (**Figure 13.29a**). The repeating MurNAc-(β-1,4)-GlcNAc disaccharide forms polysaccharide chains, which are tethered together by linkages between short oligopeptides (**Figure 13.29b**). The β-1,4 glycosidic bond in the MurNAc-GlcNAc

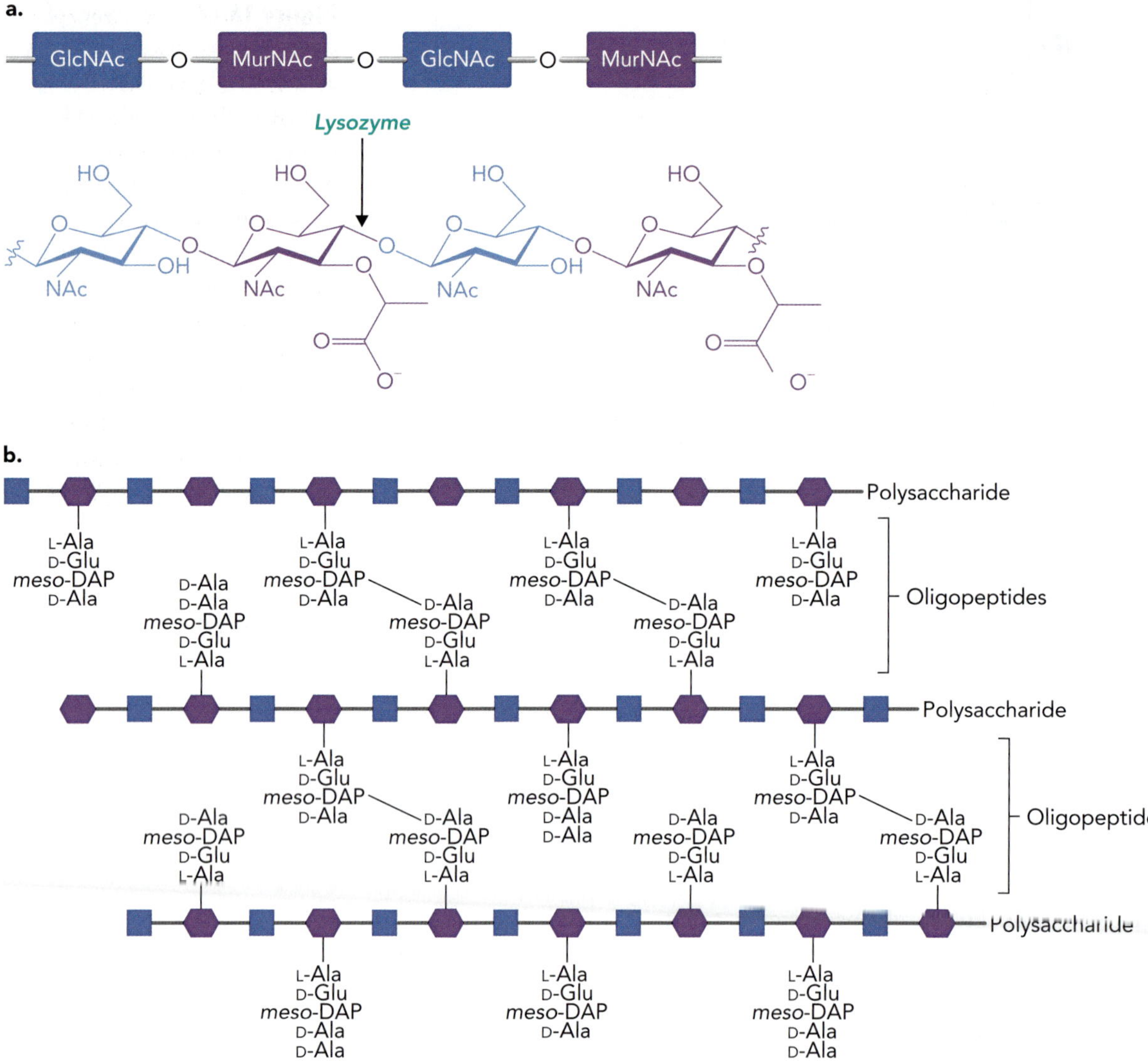

Figure 13.29 Bacterial cell walls contain peptidoglycans, which consist of hexosamine polysaccharide chains linked together by peptide bridges. **a.** The hexosamine polysaccharide contains repeating disaccharide MurNAc-(β-1,4)-GlcNAc units, which can be cleaved by the enzyme lysozyme. **b.** Adjacent hexosamine polysaccharide chains are covalently linked through oligopeptides, which are attached to MurNAc residues in each chain. The symbols for MurNAc and GlcNAc are the same as in Figure 13.2.

disaccharide is the cleavage site for the enzyme lysozyme, which degrades bacterial cell walls.

Unlike most naturally occurring proteins, the peptide linkers in bacterial peptidoglycans contain mixtures of D and L amino acids (**Figure 13.30a**). Although the MurNAc-GlcNAc polysaccharide chains are conserved between different types of bacteria, the precise peptide linkages are not all the same. For example, some bacteria have L-Lys as the third amino acid in the peptide, whereas others have *meso*-diaminopimelic acid (*meso*-DAP) in this position. (*meso*-DAP is an ε-carboxy derivative of lysine.) Moreover, the covalent linkage can be formed directly between oligopeptides in the apposing polysaccharide chain or involve a bridging oligopeptide such as pentaglycine (**Figure 13.30b**).

Bacteria can be classified into two broad groups on the basis of physical differences in their peptidoglycan layer (thick or thin) and whether or not they have an outer membrane, called a capsule, surrounding the peptidoglycan layer. Christian Gram, a Danish bacteriologist, demonstrated in 1884 that a histologic procedure could be used to distinguish these two types. **Gram-positive bacteria** have a thick peptidoglycan cell

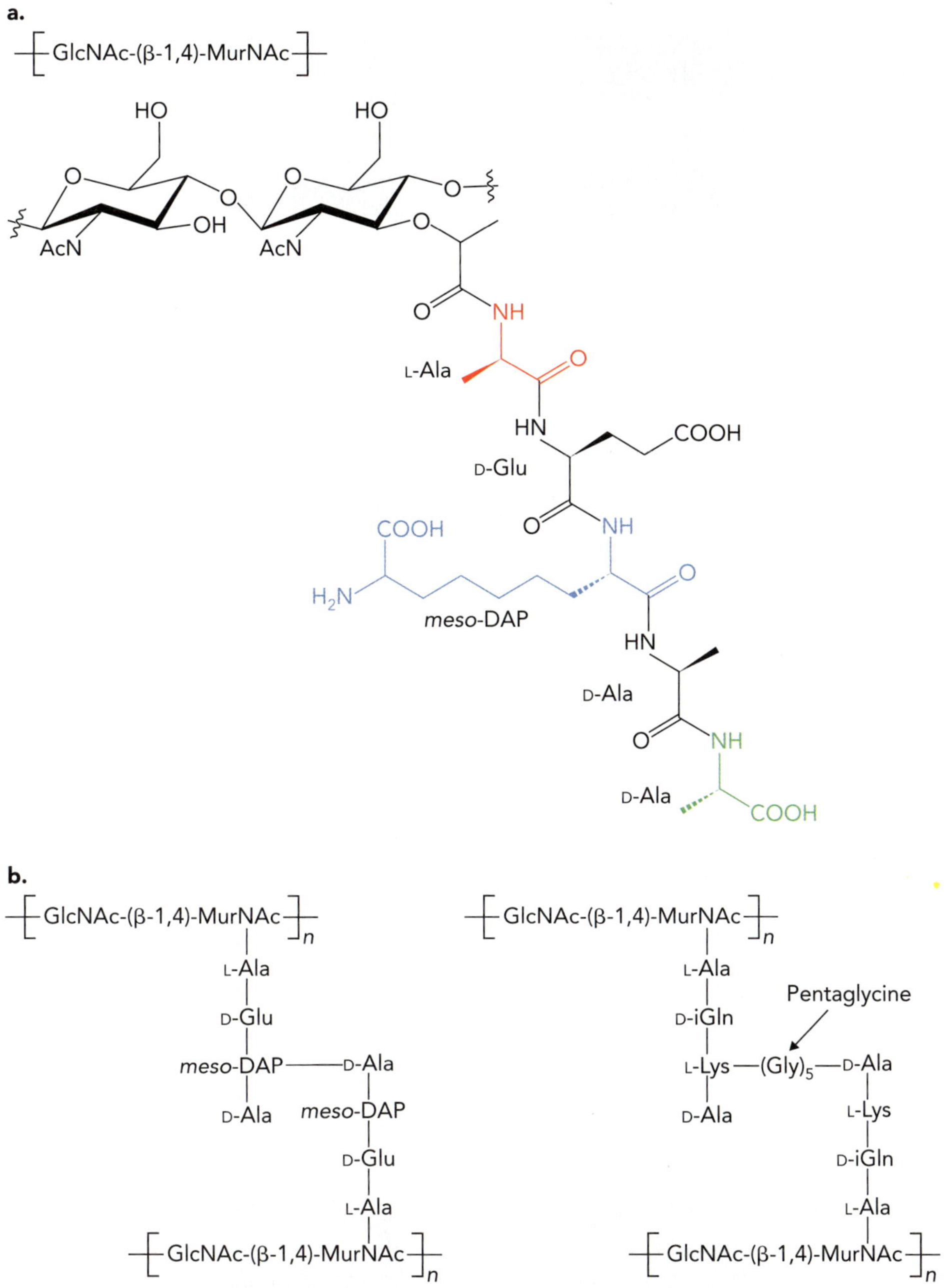

Figure 13.30 The oligopeptide linker in proteoglycans is variable among bacterial species. **a.** Structure of a pentapeptide linker showing the inclusion of D and L amino acids and the amino acid derivative *meso*-DAP. **b.** Peptide linkages in proteoglycans can be formed directly through the pentapeptides (the terminal D-Ala is removed during the cross-linking reaction) or involve an oligopeptide linker such as pentaglycine. iGln = isoglutamine.

wall (250 Å) with no outer membrane, whereas **Gram-negative bacteria** have a thin peptidoglycan cell wall (25 Å) surrounded by an outer membrane (**Figure 13.31**). The Gram test is done by staining bacteria with crystal violet dye, followed by treatment with an iodine solution, which functions as a mordant to aggregate the dye molecules into large particles. Upon washing the stained and iodine-treated bacteria with ethanol, the thick peptidoglycan wall collapses and traps the crystal violet dye inside Gram-positive bacteria. In contrast, Gram-negative bacteria do not retain the crystal violet dye after the ethanol wash because the capsule layer is removed and the peptidoglycan wall is too thin to trap the dye. Gram-negative bacteria are visualized by counterstaining with the red dye safranin, which has only a minimal effect on the color of Gram-positive bacterial cells.

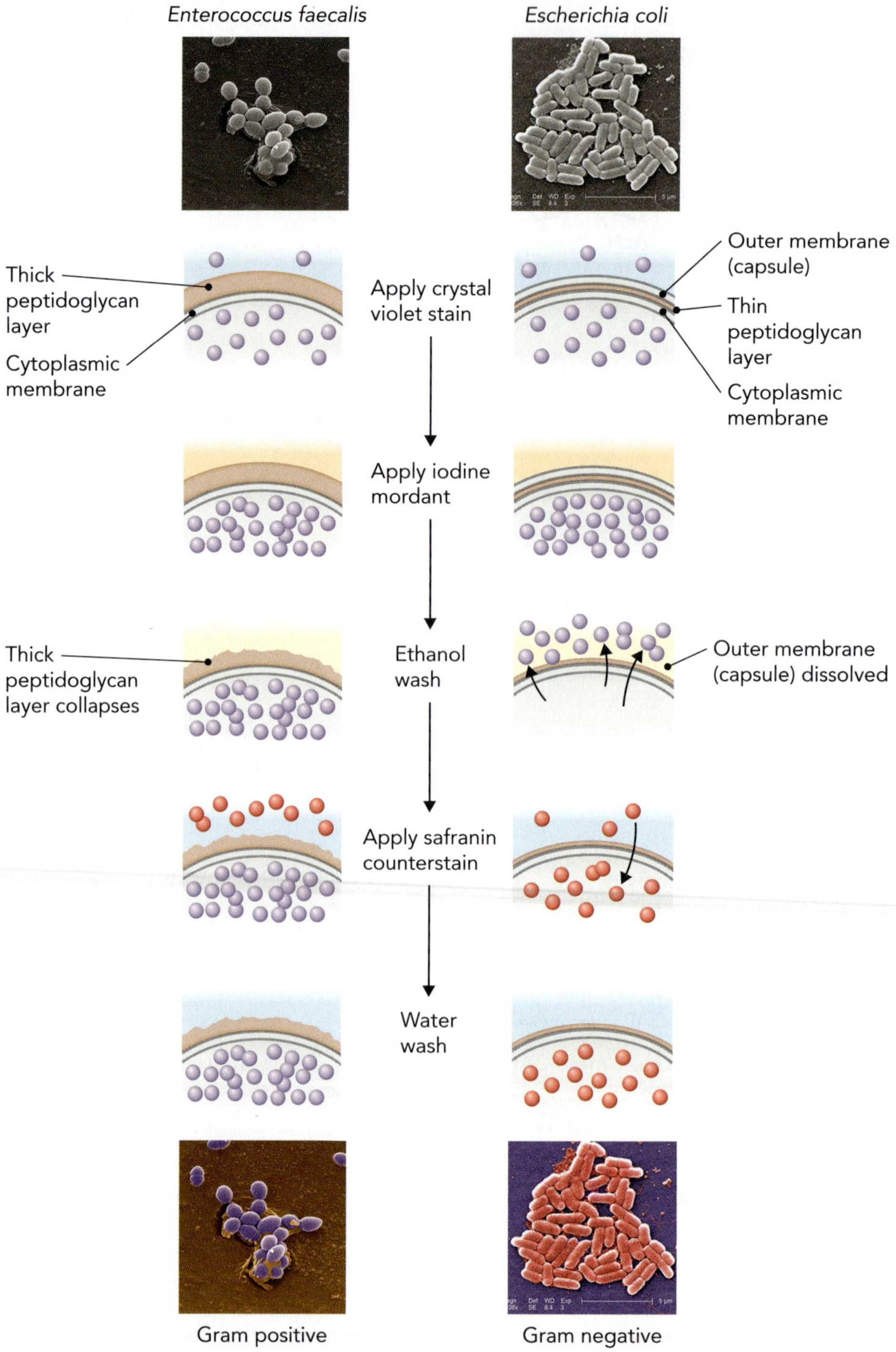

Figure 13.31 The Gram test can be used to distinguish between two types of peptidoglycan layers in bacterial cell walls. Gram-positive bacteria have a thick peptidoglycan layer that traps crystal violet dye after treatment with mordant (iodine solution) and ethanol. In contrast, Gram-negative bacteria have a thin peptidoglycan layer that allows the crystal violet dye to leak out during the ethanol wash. Counterstaining with red safranin dye is used to visualize Gram-negative bacteria. *Enterococcus faecalis* and *Escherichia coli* are representative examples of Gram-positive and Gram-negative bacterial species, respectively. ENTEROCOCCUS FAECALIS (BLACK AND WHITE) AND *ESCHERICHIA COLI* (BLACK AND WHITE; COLOR): CDC/JANICE HANEY CARR; *ENTEROCOCCUS FAECALIS* (COLOR): CDC/PETE WARDELL.

Figure 13.32 illustrates the key differences between the cell walls of Gram-positive and Gram-negative bacteria. Gram-positive bacteria lack an outer membrane and have a much thicker peptidoglycan layer. This thicker layer contains **lipoteichoic acid**, a negatively charged polymer of ribitol phosphate or glycerol phosphate. Lipoteichoic acid is thought to provide structural support to the cell wall of Gram-positive bacteria and influence the uptake of charged biomolecules. The outer membrane in Gram-negative bacteria is asymmetric with regard to glycolipids called **lipopolysaccharides (LPSs)**, which are localized exclusively to the outer leaf of the outer membrane. LPS, also called **endotoxin**, is a potent inflammatory agent in animals because immune cells

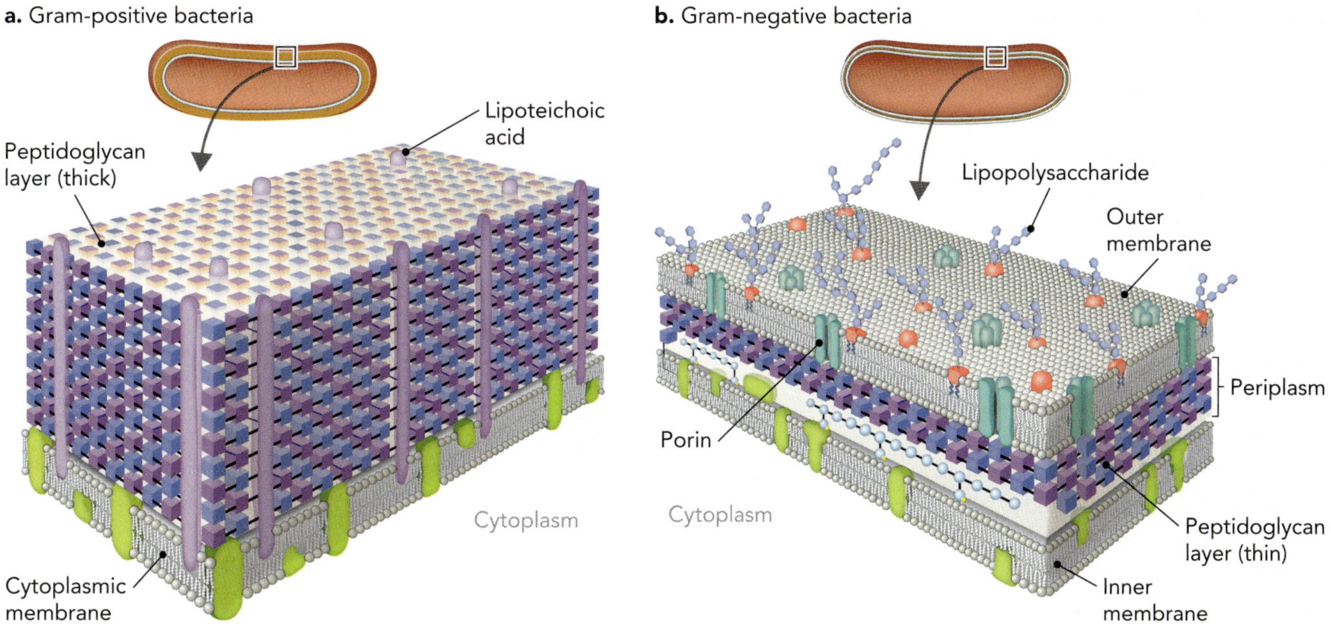

Figure 13.32 Structure of cell wall components for Gram-positive and Gram-negative bacteria are illustrated here to show the differences in organization of the peptidoglycan layers. **a.** Gram-positive bacteria, such as *Enterococcus faecalis*, lack an outer membrane and have a thick peptidoglycan layer studded with the negatively charged polymer lipoteichoic acid. **b.** Gram-negative bacteria, such as *Escherichia coli*, contain both an outer membrane and an inner membrane that surround a thin peptidoglycan layer. The space between the two membranes is called the periplasm. Lipopolysaccharides on the surface of Gram-negative bacteria induce a potent inflammatory response in animals by binding to lectin proteins on the surface of immune cells.

expressing lectins that bind LPS recognize this bacterial antigen as a non-self molecule. The outer membrane of Gram-negative bacteria also contains porin proteins, which were described in Chapter 6. These proteins function as channels to permit the exchange of small molecules between the periplasmic space and the extracellular environment.

The antibiotic **penicillin** blocks bacterial cell wall biosynthesis by inhibiting the enzyme **transpeptidase**, which is required to form the oligopeptide linkages between hexosamine polysaccharide chains in the peptidoglycan layer (**Figure 13.33**). Penicillin's mechanism of action involves the formation of a suicide inhibitor complex between a serine residue in transpeptidase and a carbonyl carbon in the β-lactam ring of penicillin. Without sufficient amounts of enzymatically active transpeptidase to support cell wall biosynthesis during cell division, the penicillin-treated bacteria die.

Penicillin was discovered in 1928 by the Scottish bacteriologist Alexander Fleming. Six years earlier, Fleming had found that lysozyme had antibacterial properties through its ability to cleave the glycoside bond in the MurNAc-(β-1,4)-GlcNAc disaccharide (see Figure 13.29). Although lysozyme itself was not suitable for development as a commercially available antibiotic, its discovery gave Fleming the idea that other antibacterial compounds might be present in nature that were just as potent. One day, while

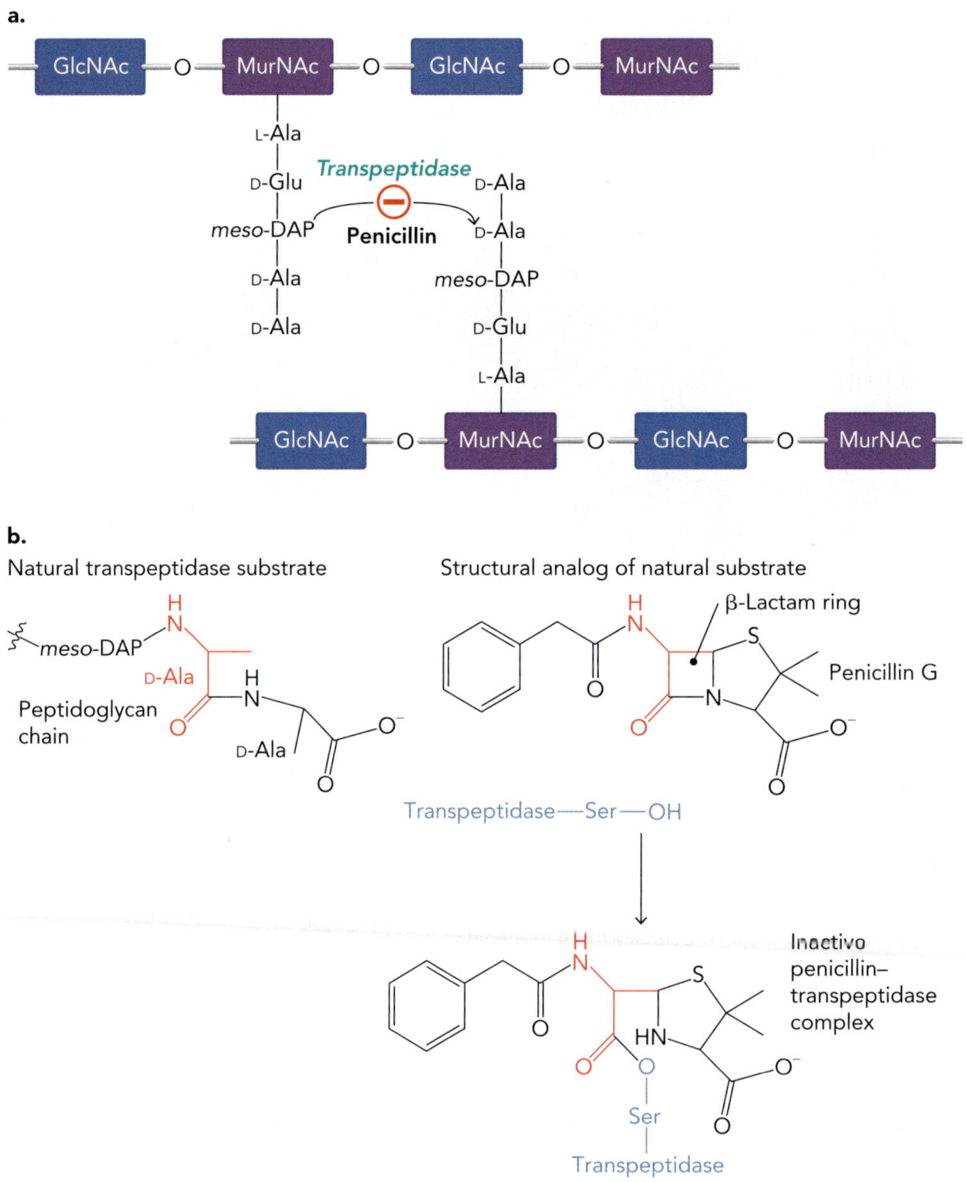

Figure 13.33 The antibiotic penicillin inhibits the bacterial enzyme transpeptidase. **a.** Cross-linking of hexosamine polysaccharide chains in peptidoglycans requires a transpeptidase enzyme, which first removes the terminal D-Ala residue from the oligopeptides and then generates the interstrand cross-link. Penicillin inhibits the transpeptidase cross-linking step and thereby blocks bacterial cell wall biosynthesis. **b.** Penicillin consists of a β-lactam ring fused to a thiazolidine ring and is structurally similar to the terminal dipeptide of the peptidoglycan. When penicillin binds to the transpeptidase enzyme active site, it initiates a reaction mechanism that opens the β-lactam ring and forms a covalent enzyme–substrate complex. Penicillin is an example of a "suicide" enzyme inhibitor because it forms an irreversible covalent bond with the enzyme during the catalytic reaction.

cleaning out a large pile of discarded agar plates used to grow the pathogenic bacteria *Staphylococcus aureus*, Fleming noticed that a mold was growing on one of the plates. What caught his eye about this particular plate was that all of the *S. aureus* bacteria had lysed in an area surrounding the mold, which was later identified as *Penicillium notatum* (**Figure 13.34**).

Fleming spent several years trying to purify penicillin from *P. notatum* cell extracts, but because of the chemical instability of penicillin, his efforts were unsuccessful. Finally, 10 years later, two Oxford chemists, Howard Florey and Ernst Chain, were able to isolate sufficient quantities of penicillin for clinical trials. In 1945, Fleming, Florey, and Chain shared the Nobel Prize in Physiology or Medicine for the development of penicillin, which was heralded as one of the miracle drugs of the 20th century.

Some bacteria are resistant to penicillin because they produce an enzyme called **β-lactamase**, which hydrolyzes the β-lactam ring in penicillin to inactivate it (the

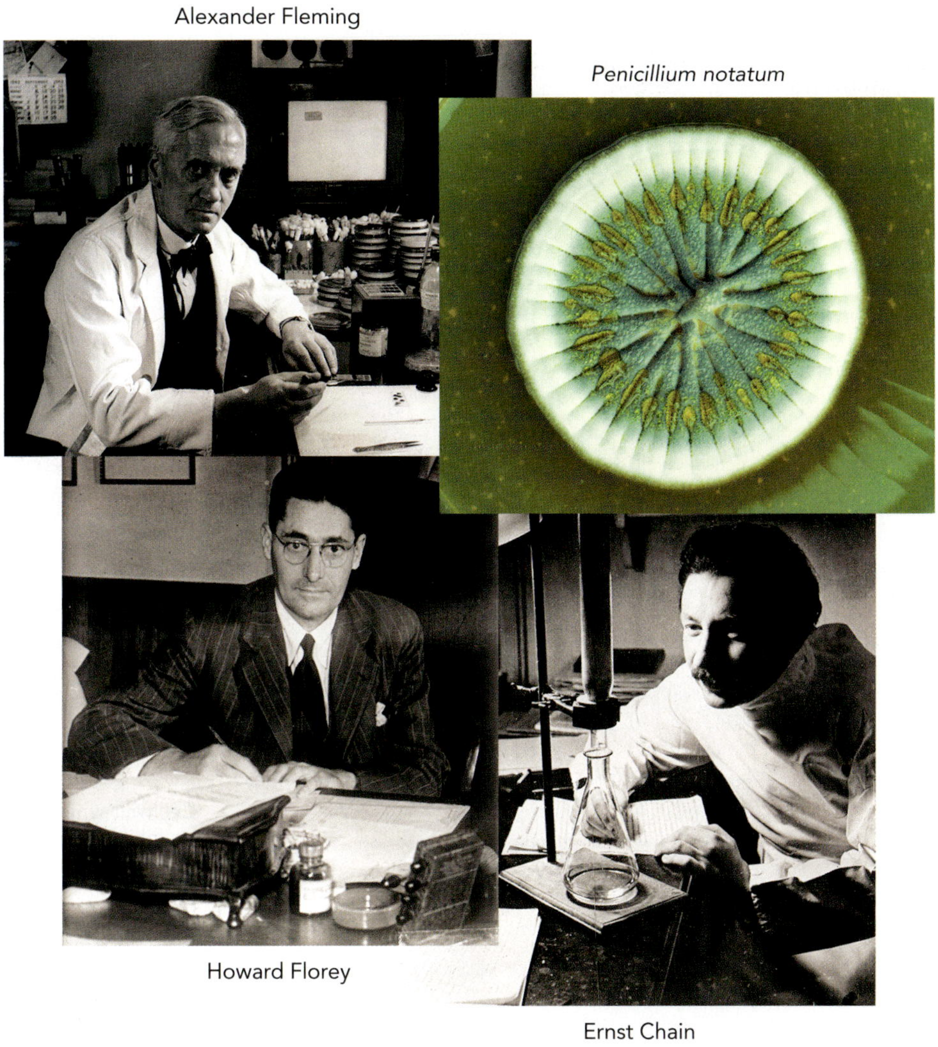

Figure 13.34 Alexander Fleming discovered penicillin in 1928 when he identified a compound in the mold *Penicillium notatum* that killed the pathogenic bacterium *Staphylococcus aureus*. Fleming, along with Howard Florey and Ernst Chain, who together developed a method to purify active penicillin on a commercial scale, shared the 1945 Nobel Prize in Physiology or Medicine.

hydrolyzed product is not a substrate for the transpeptidase). As β-lactamase is a secreted enzyme, it provides a protective shield around the bacterium. This type of penicillin resistance has been overcome by developing synthetic compounds such as **methicillin** that block transpeptidase activity without being substrates for β-lactamase (**Figure 13.35**). However, because of the widespread use of methicillin, particularly in hospitals where bacterial infections are treated aggressively, a methicillin-resistant strain of *Staphylococcus aureus* called MRSA has emerged (**Figure 13.36**, p. 665). One of the resistance mechanisms in MRSA strains is due to expression of a variant transpeptidase enzyme that does not bind methicillin.

DNA sequence analysis of one such methicillin-resistant transpeptidase gene revealed that it was not a mutant version of the *S. aureus* transpeptidase gene but instead

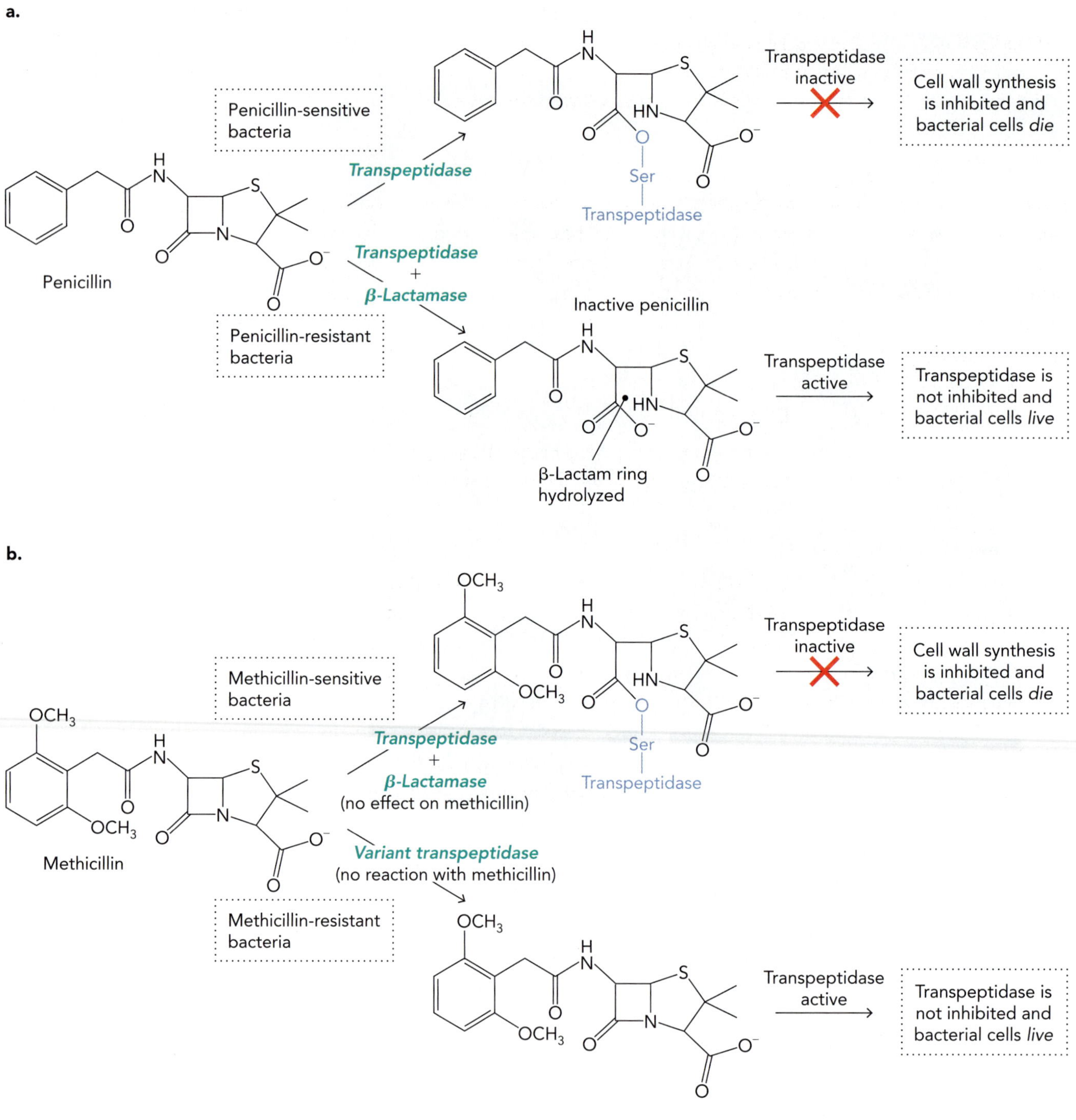

Figure 13.35 Bacteria have developed two mechanisms of resistance to β-lactam antibiotics. **a.** Penicillin-sensitive bacteria are killed by penicillin because of inactivation of transpeptidase activity and inhibition of cell wall biosynthesis. Penicillin-resistant bacteria express the enzyme β-lactamase, which inactivates penicillin by cleaving the β-lactam ring. **b.** Methicillin-resistant *S. aureus* bacterial strains have arisen that express a variant form of the transpeptidase enzyme that is not inhibited by the antibiotic.

Figure 13.36 A methicillin-resistant strain of *S. aureus* called MRSA is dangerous because few antibiotics are currently available to fight the infection. **a.** Photomicrograph of an MRSA strain. FRANK DELEO, NATIONAL INSTITUTE OF ALLERGY AND INFECTIOUS DISEASES (NIAID). **b.** MRSA infections can result in deep, pus-filled wounds that can be lethal if untreated. SCOTT CAMAZINE/ALAMY.

a. Methicillin-resistant *Staphylococcus aureus* (MRSA)

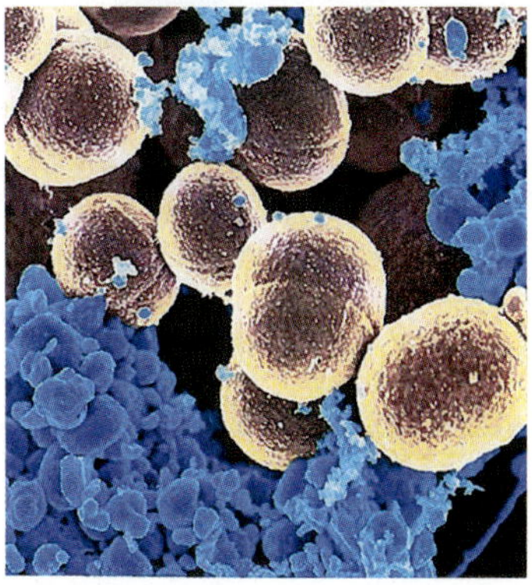

b. Example of an MRSA infection

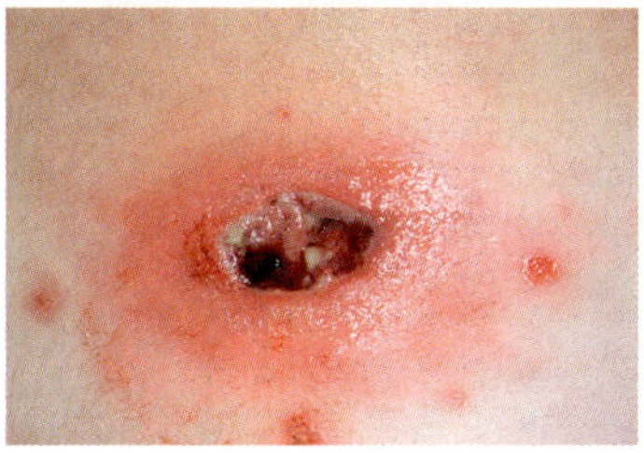

a transpeptidase gene obtained from another bacterial species through a process called **lateral gene transfer** (see Chapter 20). Acquired antibiotic resistance through lateral gene transfer is a serious concern because it demonstrates that genetic alterations in bacteria can occur rapidly, making it likely that resistance to other classes of antibiotics will also arise.

concept integration 13.2

What determines the structure of glycan groups on glycoproteins, and what contributes to glycan structural diversity within and between cell types in an organism?

Glycan groups are added to proteins and lipids by glycosyltransferase enzymes that recognize specific amino acids in proteins for the attachment of monosaccharides, using nucleotide sugars as donors. N-linked glycoproteins have GlcNAc attached to selected asparagine residues, whereas O-linked glycoproteins have GalNAc attached to selected serine or threonine residues. The substrates for glycosyltransferase reactions are both nucleotide sugars (containing one of more than 11 different monosaccharides) and the protein or growing glycan group. Because glycosyltransferases have distinct substrate recognition properties, the structure of glycan groups on glycoproteins is determined by the collection of glycosyltransferases in any given cell. The structural diversity of glycan groups on glycoproteins within the same cell is determined by subtle changes in the cellular environment over time. For example, differences in the bioavailability of nucleotide sugars or the activity of signaling cascades can modulate the activity of glycosyltransferase enzymes.

13.3 Biochemical Methods in Glycobiology

Glycoconjugates on the cell surface play a major role in cell recognition and signaling, especially with regard to human disease states. Therefore, much attention has been directed toward developing sensitive tools for glycobiology research. Biochemical analysis of glycoconjugates has primarily focused on two research objectives: (1) identification of glycan group structures on purified glycoproteins using liquid chromatography and mass spectrometry; and (2) applications of high-throughput, array-based screening assays to identify glycan binding interactions associated with cellular phenotypes.

As shown in **Figure 13.37**, cellular glycoconjugates can be characterized—most often glycoproteins—by either releasing the glycan group for structural analysis or incubating fluorescently labeled glycoproteins with lectin or antibody arrays to detect glycan functional groups.

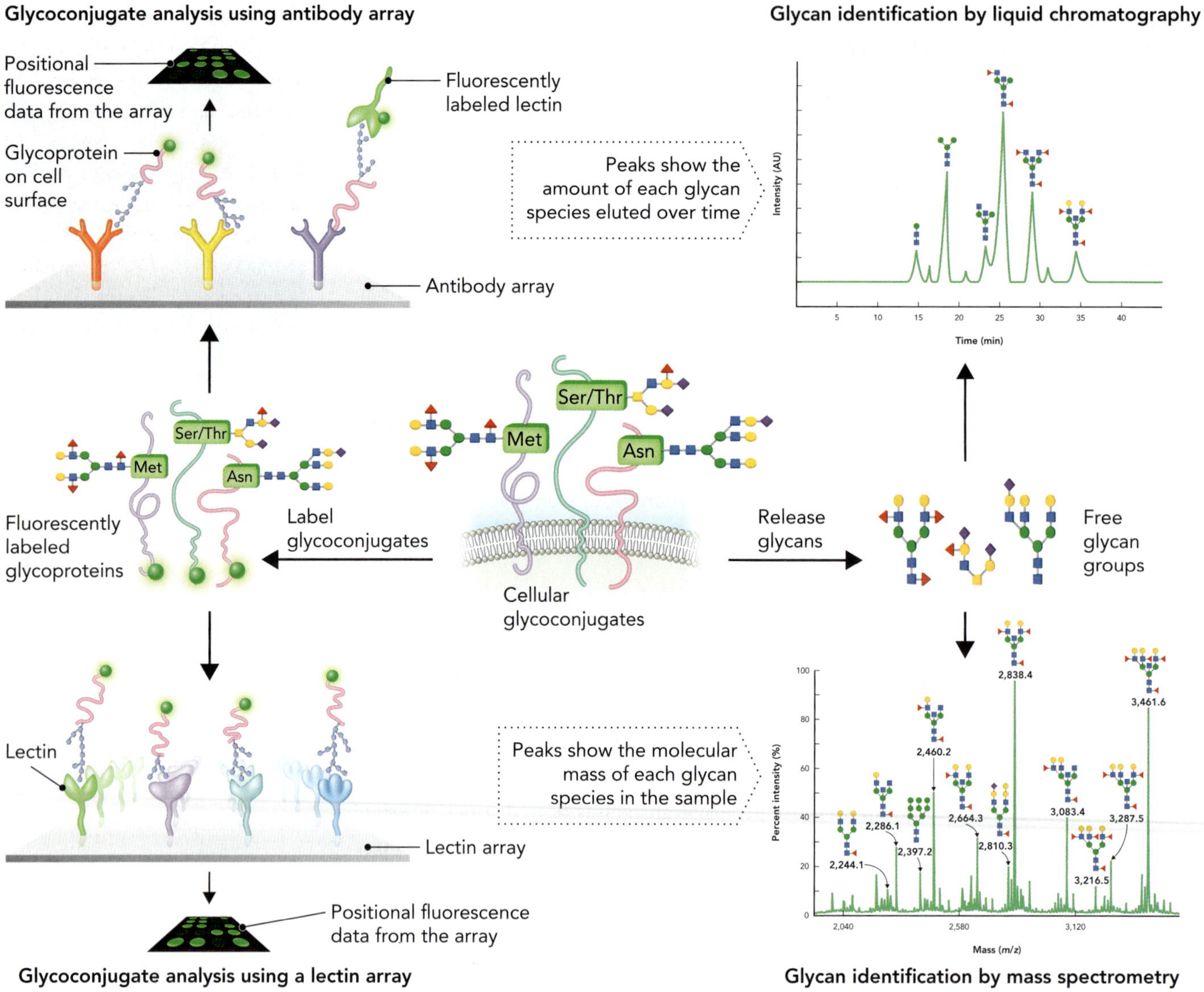

Figure 13.37 Cellular glycoproteins are biochemically characterized using either glycan structural analyses or high-throughput, protein-based arrays. Glycan groups released from glycoconjugates can be analyzed with liquid chromatography methods that separate glycan groups on the basis of size and chemistry or by mass spectrometry, which identifies common glycan groups using mass-to-charge ratios. Specific glycan binding interactions can be investigated using lectin and antibody arrays, which identify positive binding interactions using fluorescently labeled glycoproteins.

Glycan Determination by Chromatography and Mass Spectrometry

Structural characterization of glycans on glycoconjugates is technically challenging because of several factors not encountered in protein and nucleic acid structural analysis. These factors include multiple bonding arrangements between as many as 11 different sugars; presence of sugar stereoisomers with identical masses; and the fact that glycan structures can differ in subtle ways between identical classes of glycoconjugates. Two approaches are used to overcome these challenges: (1) a liquid chromatography–based sequencing strategy using specific glycosidase enzymes to generate related glycan structures with distinct

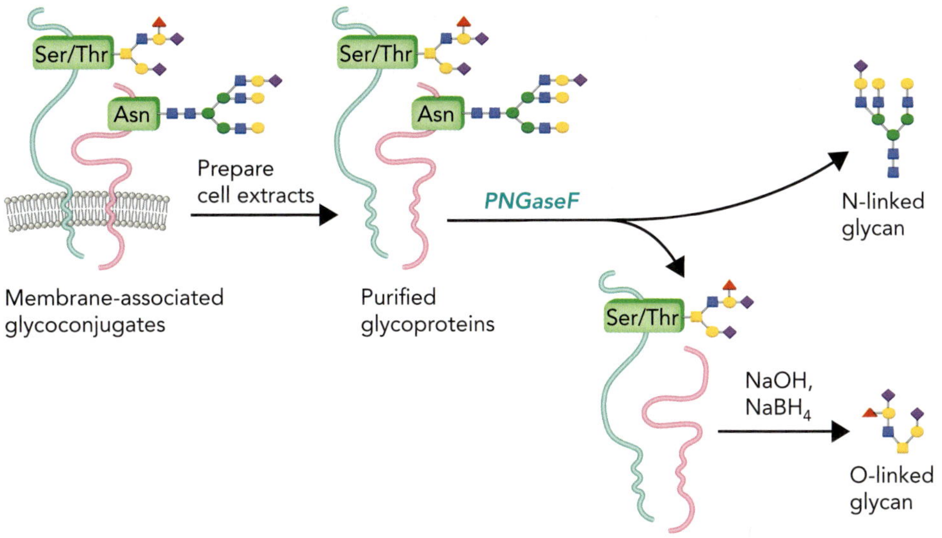

Figure 13.38 This flow scheme shows methods used for releasing glycan groups from membrane-associated glycoconjugates. N-linked glycan groups are removed from purified glycoproteins using enzymatic cleavage with the enzyme PNGaseF. O-linked glycans are released using chemical cleavage with NaOH and $NaBH_4$ (a β-elimination reaction).

elution profiles; and (2) highly accurate mass spectrometry instruments that are able to identify common glycan groups on the basis of predicted mass-to-charge ratios.

Before glycan groups on glycoproteins can be analyzed, they must first be separated from the protein moiety using a cleavage reaction. As shown in **Figure 13.38**, purified glycoproteins are treated with the enzyme peptide *N*-glycosidase F (PNGaseF), a glycosidase specific for N-linked glycan groups. Physical separation results in a glycan fraction containing the N-linked glycans and a glycoprotein fraction containing the released N-linked proteins as well as all O-linked glycoproteins. Although several O-glycosidases have recently been discovered, it is still common to use chemical cleavage of the O-glycosidic bond based on a β-elimination reaction using NaOH and $NaBH_4$ (Figure 13.38). Under optimal cleavage conditions, the purified N-linked and O-linked glycan groups remain essentially intact.

The free glycans are then subjected to structural analysis by either liquid chromatography or mass spectrometry, depending on the objective and instrument availability. Liquid chromatography, in combination with glycosylase cleavage, provides information about the arrangement of sugars in a highly purified glycan fraction using standard HPLC equipment. In contrast, mass spectrometry, which requires sophisticated mass analyzers, identifies common glycans present in a mixed glycan sample based on predicted mass-to-charge ratios. Ideally, both methods would be used to provide corroborative evidence for the identification of specific glycan groups.

Glycan sequencing of purified glycan groups is carried out using sugar-specific glycosylases to generate related glycan structures, which are then separated by HPLC. The sensitivity of this technique can be in the femtomole range when using fluorescently labeled glycans, as shown in **Figure 13.39a**. The fluorescent dye 2-aminobenzamide (2-AB) is covalently attached to acyclic glycan groups through a Schiff base intermediate. (Note that in solution, cyclic and acyclic glycan structures are present at equilibrium.) The intermediate is reduced with sodium cyanoborohydride to generate the fluorescently labeled glycan. The labeled glycans are subjected to stepwise treatment with sugar-specific glycosylases to yield related glycan structures that are detected by fluorescence analysis. For example, neuraminidases cleave nonreducing terminal sialic acids (Neu5Ac), whereas β-galactosidase cleaves nonreducing terminal galactose residues (**Figure 13.39b**).

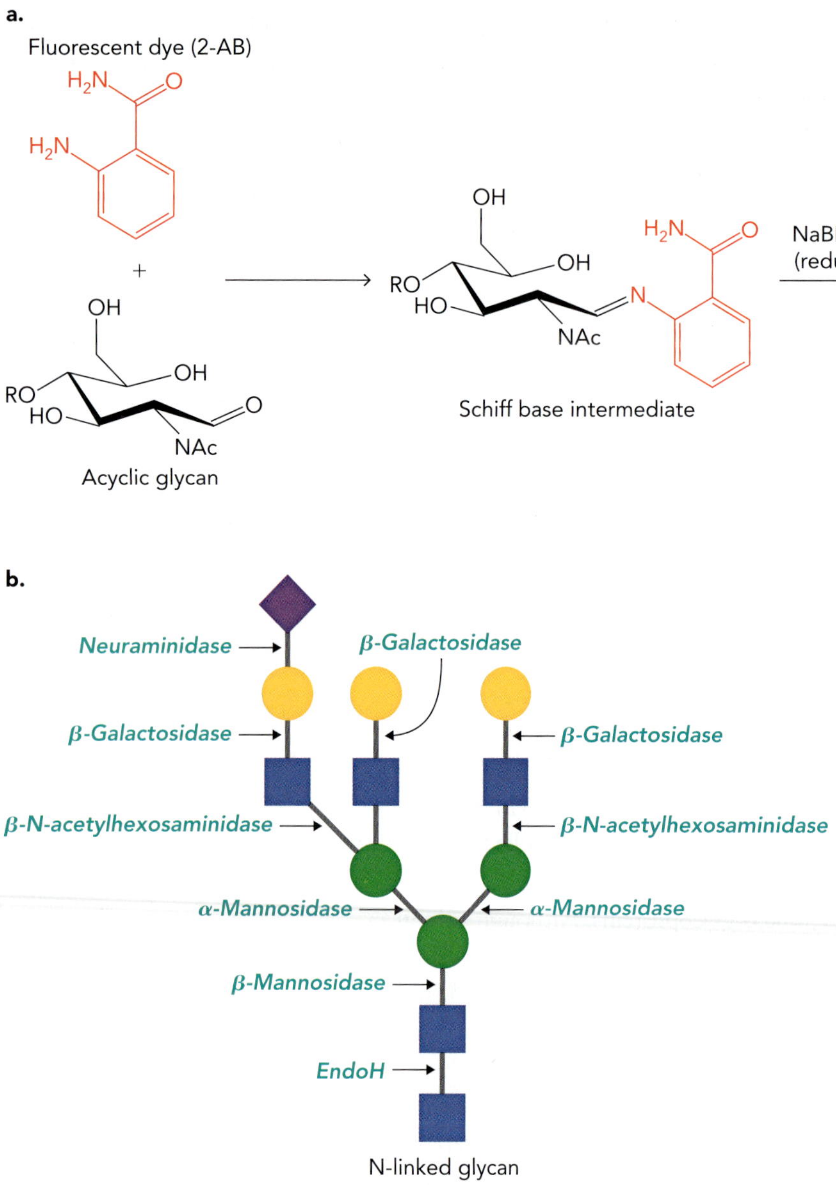

Figure 13.39 Glycans analyzed by liquid chromatography are fluorescently labeled and then sequentially treated with sugar-specific glycosylases that cleave nonreducing terminal sugar residues. **a.** Glycan groups are covalently labeled at free reducing ends on sugars with the fluorescent dye 2-AB to increase the sensitivity of detection by liquid chromatography. **b.** Glycosylase enzymes are used to cleave nonreducing terminal sugar residues at specific sites for use in glycan structural analyses. EndoH = endoglycosidase H.

Liquid chromatography separates molecules on the basis of size and chemical properties. This results in differential elution rates from solid matrices as a function of molecular interactions under various buffer conditions. In a process similar to protein separation by column chromatography (see Chapter 5), glycans are also differentially eluted from column matrices. Elution of fluorescently labeled glycans from the HPLC column can be measured either in minutes or in glucose units (GU), which are elution profiles based on glucose polymer standards of known lengths. As shown in **Figure 13.40**, five specific glycosylases can be used to generate six glycan samples, including the unreacted starting material. On the basis of the known specificities of the enzymes and the observed elution profiles of each sample relative to the glucose standards, this method can be used to deduce a glycan structure.

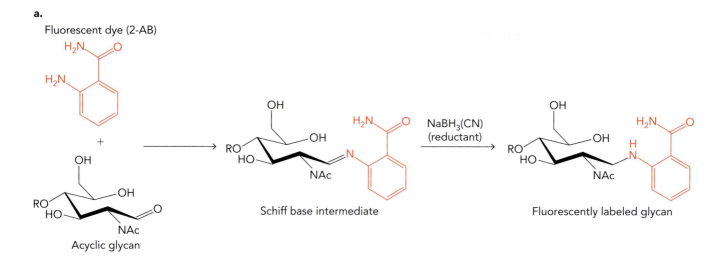

Mass spectrometry is a second method used for analyzing glycans released from glycoproteins, either as the primary analytical method or as a complementary technique to augment elution data derived from HPLC. In most cases, the glycan needs to be derivatized to enhance ionization efficiency; for example, by using permethylation, which adds CH_3 groups at random locations in the molecule. Glycan identification is often performed using matrix-assisted laser desorption/ionization (MALDI) in a time-of-flight (TOF) mass spectrometer to calculate mass-to-charge ratios (see Figure 5.23). For example, MALDI-TOF mass spectrometry can be used to determine the composition of a mixture of N-linked derived glycans from a glycoprotein sample that was treated with PNGaseF. As shown in **Figure 13.41**, by comparing the observed masses obtained from MALDI-TOF mass spectrometry to predicted masses of common N-linked glycan groups, specific peaks in the spectrum can be assigned to glycan groups.

Use of High-Throughput Arrays for Glycoconjugate Analysis

One of the most significant breakthroughs in glycobiology research has been the application of array technology to analyze glycan binding interactions with lectin proteins. Similar to nucleic acid arrays (see Figure 3.59), lectin and glycan arrays provide a platform to screen large numbers of samples for specific binding interactions using fluorescently labeled molecules. Two basic types of arrays have been developed for screening purposes: (1) protein arrays containing covalently attached lectin proteins or antibodies for the detection of labeled glycoproteins in experimental samples; and (2) glycan arrays containing covalently attached glycoproteins with intact glycan groups (or chemically synthesized glycan groups) for the detection of labeled lectin proteins or antibodies in experimental samples. The primary objective of these array studies is usually to quantitate the level of specific glycoproteins in an experimental sample on the basis of known glycan–lectin interactions. As new lectins are discovered, however, it should be possible to use lectin arrays to discover previously unknown glycoproteins that are identified by lectin-specific glycan interactions.

As shown in **Figure 13.42**, a lectin array contains a grid of lectin proteins that are covalently attached to a solid support. By incubating the lectin array with a fluorescently labeled biological sample containing glycoconjugates, such as cell surface glycoproteins, glycan–lectin binding interactions can be identified and quantitated on the basis of the fluorescence readout of the array.

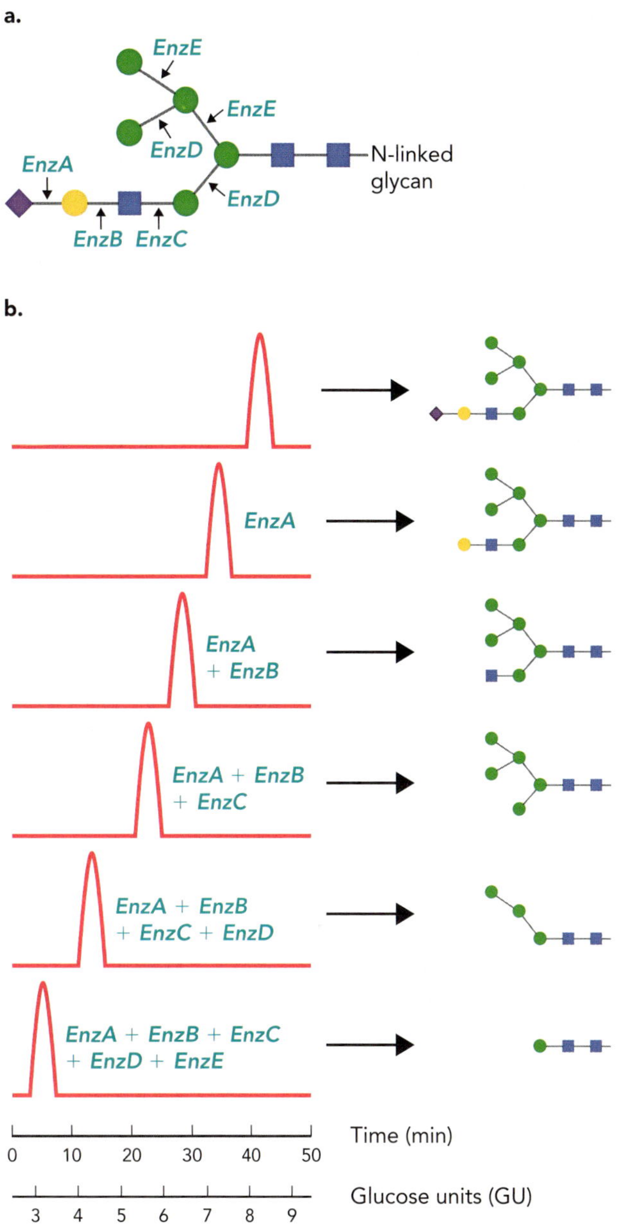

Figure 13.40 HPLC methods can be used to deduce the structure of a purified glycan group. **a.** Hypothetical cleavage sites of an N-linked glycan by sugar-specific glycosylases. **b.** Elution profiles of the fluorescently labeled glycan on an HPLC column after each successive cleavage reaction. Under the conditions shown here, smaller glycan groups elute first from the column.

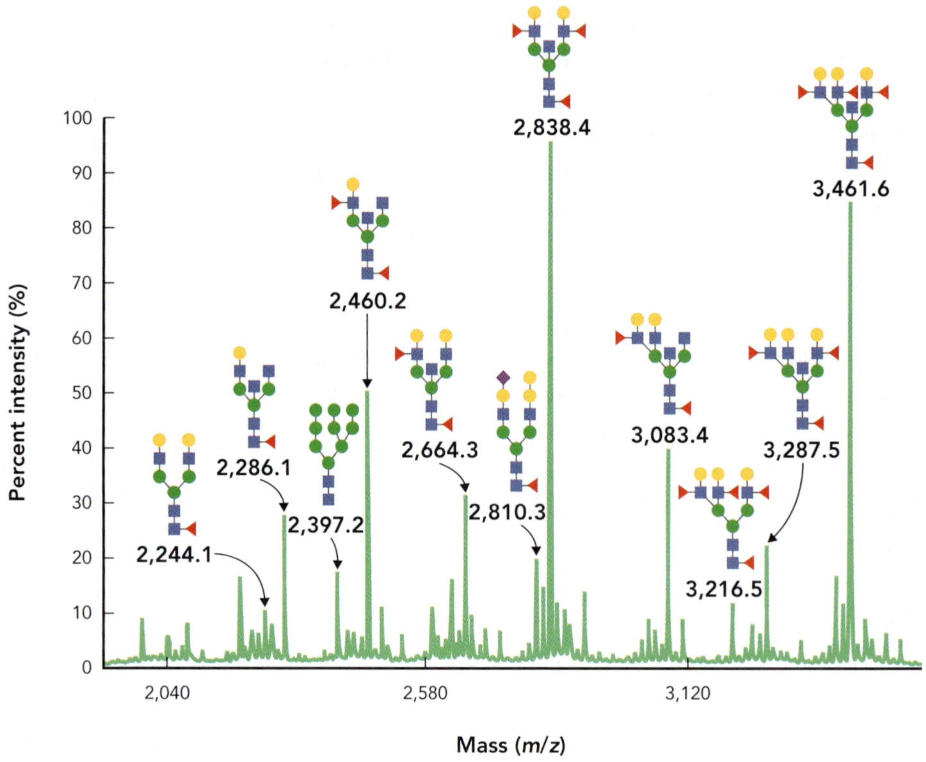

Figure 13.41 Mass spectrometry can be used to identify N-linked glycans in a mixed sample by comparing observed and predicted masses of common glycan groups.

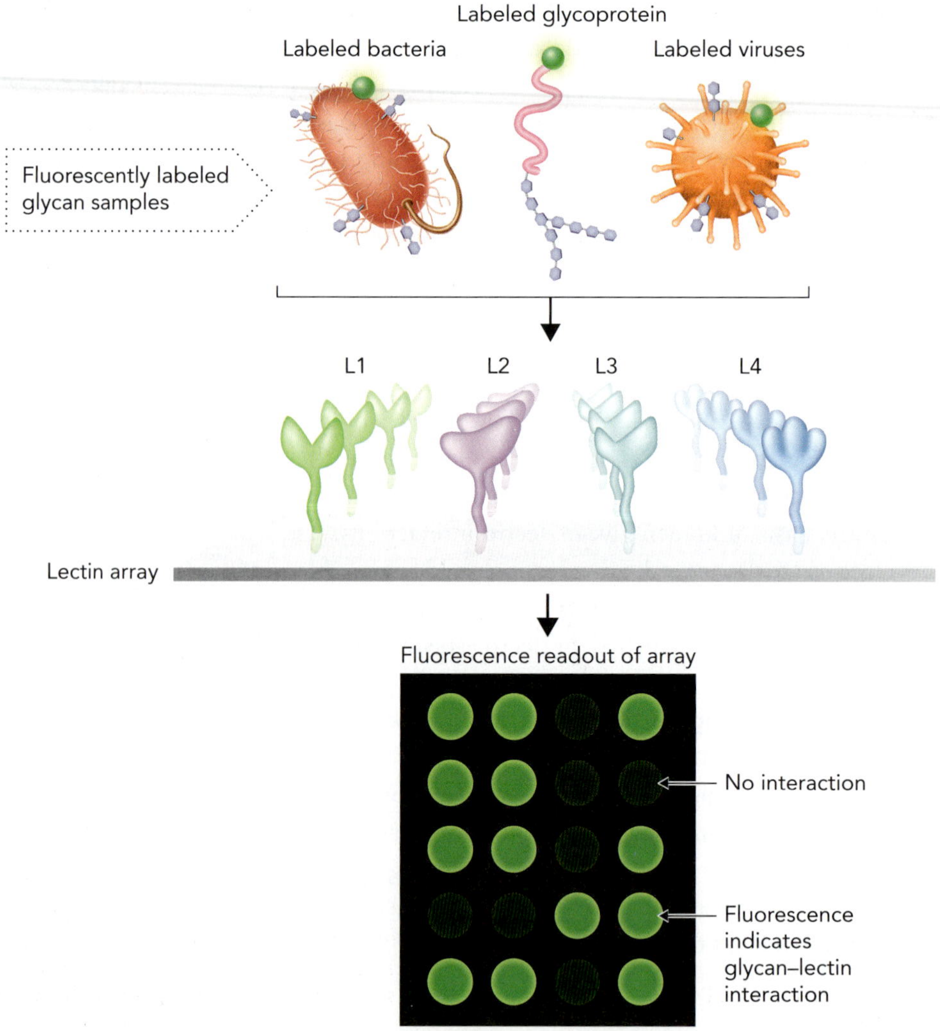

Figure 13.42 Lectin arrays contain hundreds of different, covalently linked lectin proteins on a solid surface. Each different lectin protein is at a different position in the array. The lectins can be tested for glycan binding using fluorescently labeled experimental samples of glycoproteins, viruses, and bacteria. Laser scanning of the array after incubation with the experimental sample generates a pattern of positive and negative signals, which are interpreted on the basis of the known location of specific lectins (L1, L2, L3, and so forth).

Another type of protein array used in glycobiology is one in which antibodies are covalently attached to the solid surface. As shown in **Figure 13.43**, antibodies can either bind specifically to glycan groups or bind to the protein moiety of well-characterized glycoproteins. When glycan binding antibodies are used to detect fluorescently labeled glycoproteins in an experimental sample, the antibody array functions in much the same way as a lectin array. Antibodies that recognize the protein component of a glycoprotein, rather than the glycan groups, can be used to investigate glycan modifications on glycoproteins when combined with fluorescently labeled lectins of known specificity.

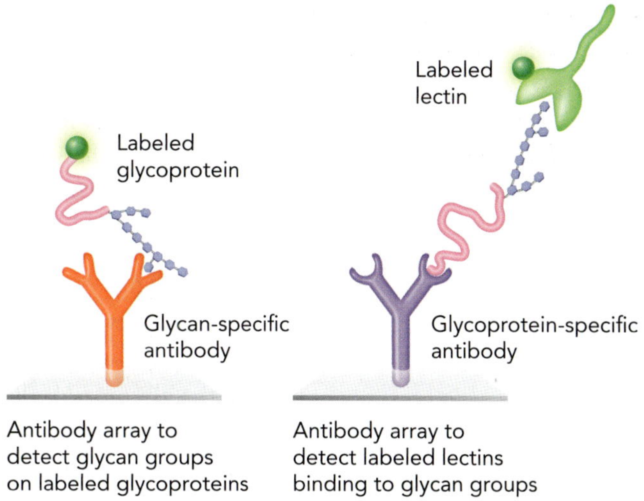

Figure 13.43 Antibody arrays in glycobiology research are used to detect glycan groups or glycoproteins in experimental samples. Glycan groups can be detected by glycan-specific antibody arrays using fluorescently labeled glycoproteins or by using protein-specific antibody arrays in combination with fluorescently labeled lectin proteins.

Glycan arrays are used to analyze the cellular **glycome** (complete set of cellular glycan groups) under various conditions using fractionated cell extracts. They are also used to detect lectins on the surface of pathogenic and non-pathogenic bacteria. As shown in **Figure 13.44**, glycan arrays can be constructed using HPLC fractions of cell extracts, which are covalently linked to solid surfaces through protein-specific linkages. This permits the associated glycans to be recognized by fluorescently labeled lectins or glycan-specific antibodies. This type of glycan array can be used to qualitatively compare fractionated cell extracts from different sources, such as mutant and wild-type cells, to detect differences in their respective glycomes. If the objective is to identify glycan binding proteins in an experimental sample, then the glycan array is constructed using chemically synthesized glycan groups of known structure. For example, if the pathogenicity of two strains of closely related bacteria is related to differences in their cell surface lectins, then a synthetic glycan array could provide structural information about the glycan binding properties of the bacterial lectins. A synthetic glycan array is also useful for investigating the glycan specificity of purified lectins and monoclonal antibodies based on their affinities for different glycan groups on the array.

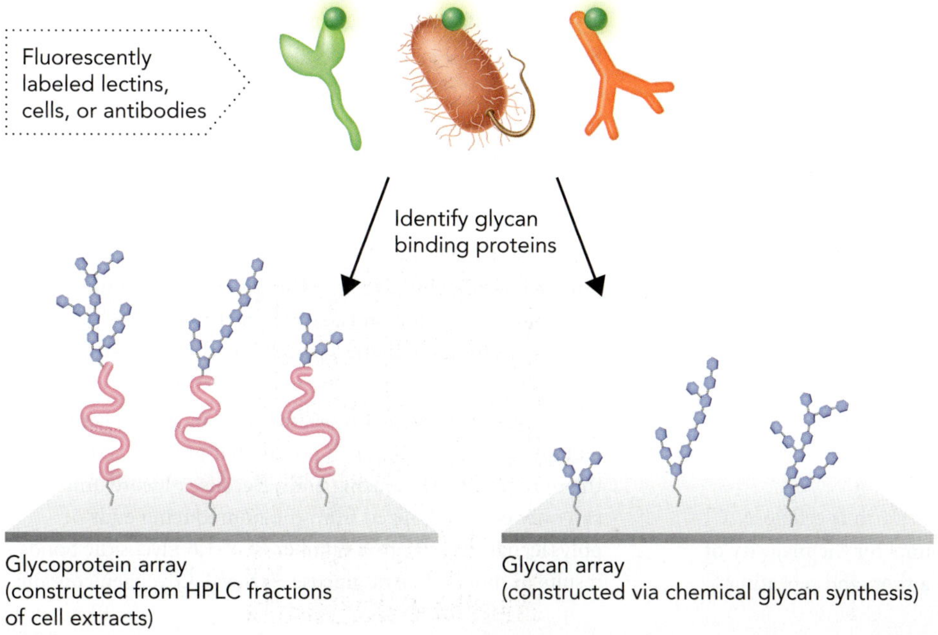

Figure 13.44 Glycan arrays are used to analyze cellular glycomes and to characterize glycan binding properties of lectins and antibodies. For glycome studies using fluorescently labeled lectins or antibodies, glycoprotein arrays can be constructed using HPLC fractions of cell extracts. Arrays made with chemically synthesized glycans are used to investigate glycan binding properties by comparing the affinity of lectins or glycan-specific antibodies with structurally related glycan groups contained on the array.

concept integration 13.3

How could lectin arrays and mass spectrometry be used together to identify glycan groups on the surface of cancer cells that distinguish malignant cancer (metastatic) from benign cancer (confined to the tumor)?

Lectin arrays can be used first to identify differences in the glycome of metastatic and benign cancer cells. To minimize differences between individuals, the glycoprotein samples used in the matched lectin arrays need to come from primary and secondary sites of cancer growth in the same individual. Once reproducible differences in the glycomes of the two cell types are verified, then isolation techniques would be used to enrich for the altered glycans. These could then be identified by HPLC and mass spectrometry using methods developed for glycobiology research. Depending on the nature of the glycan structural differences between the two cell populations, it may be possible to identify changes in the expression or activity of specific glycosyltransferase and glycosidase enzymes that explain the observed differences in metastatic properties.

chapter summary

13.1 Carbohydrates: The Most Abundant Biomolecules in Nature

- Carbohydrates, also called glycans, can be divided into three major groups: (1) *simple sugars* consist of monosaccharides, disaccharides, and oligosaccharides; (2) *polysaccharides* consist of glucose homopolymers or disaccharide heteropolymers in which one of the two sugars is a hexosamine; and (3) *glycoconjugates* of proteins or lipids with covalently attached glycans.
- Glycoconjugates contain combinations of modified and unmodified monosaccharides, which are covalently attached to proteins and lipids as branched and unbranched structures. Deciphering glycoconjugate structures is difficult.
- Glycosyltransferases covalently link glycans to proteins and lipids, whereas glycosidases remove glycans through hydrolysis reactions. Enzymes that modify glycans are some of the most highly conserved proteins in eukaryotes.
- Oligosaccharides range in size from ~3 to ~20 branched and unbranched sugar residues. Human milk contains lactose-derived oligosaccharides, which can function as probiotics or as soluble glycan decoys. Eating foods high in sucrose-based raffinose oligosaccharides can cause digestion problems because mammals, but not intestinal bacteria, lack the enzyme α-galactosidase.
- The most abundant carbohydrate on Earth is cellulose, a homopolymeric molecule that accounts for the majority of plant cell walls. It consists of up to a thousand repeating glucose disaccharides called cellobiose, which are linked by β-1,4 glycosidic bonds. Most organisms cannot metabolize cellobiose because they lack the enzyme cellulase, a β-1,4 glycosidase.
- Chitin is a linear hexosamine polysaccharide consisting of repeating GlcNAc units linked by β-1,4 glycosidic bonds. Chitin provides many types of insects and crustaceans with an excellent biomaterial for building a strong exoskeleton.
- Glycosaminoglycans are another type of linear hexosamine polysaccharide. Unlike chitin, however, glycosaminoglycans are covalently attached to proteins to form a class of glycoconjugates called proteoglycans.
- Starch and glycogen are glucose homopolymers used as quick sources of metabolic energy in plants and animals, respectively. Starch is generated during daylight hours by plants, using energy from photosynthesis, whereas glycogen stores in animals are synthesized from carbohydrates and proteins in the diet.
- Plants synthesize two forms of starch. Amylose is a linear polysaccharide containing ~100 glucose units linked by α-1,4 glycosidic bonds. Amylopectin is an α-1,6–branched glucose polymer, which has the same structure as glycogen.
- The presence of α-1,6 glycosidic bonds in amylopectin and glycogen creates branch points that greatly increase the number of nonreducing ends. Because glucose units can only be metabolized from the nonreducing ends of polysaccharides, a higher number of α-1,6 glycosidic bonds results in more efficient glucose retrieval. Glycogen contains

about three times as many α-1,6 glycosidic bonds per 100 glucose residues than amylopectin has.

13.2 Important Biological Functions of Glycoconjugates

- Glycan modification of proteins takes place within the lumen of the endoplasmic reticulum compartment of the cell, whereas glycolipids are primarily generated in the Golgi apparatus.
- Proteoglycans are protein glycoconjugates that consist mostly of carbohydrate with only a small protein component. Peptidoglycans are proteoglycans that are found in bacterial cell walls and contain peptide linkers of D and L amino acids between adjacent polysaccharide strands.
- Lectins are glycan binding proteins that mediate two classes of glycoconjugate binding interactions: (1) *intrinsic* glycoconjugate binding between glycans and lectins on human cells, and (2) *extrinsic* glycoconjugate binding between glycans and lectins on human cells and pathogen cells.
- Glycan attachments to glycoproteins occur through either the amide nitrogen atom of asparagine, leading to the generation of N-linked oligosaccharides, or through the oxygen atom of serine or threonine residues, resulting in O-linked oligosaccharides.
- The most common N-glycosidic bond in glycoproteins is between asparagine and GlcNAc, whereas the most common monosaccharide used to create the O-glycosidic bond is GalNAc.
- Glycosyltransferases use nucleotide sugars as the carbohydrate donor and sequentially add sugar residues to extend the glycan group. A large number of glycosyltransferases are required to build the variety of glycan structures found in any one organism.
- Genetic differences in the expression and activity of glycosyltransferases accounts for immunologic incompatibility between individuals. For example, the ABO blood groups in humans are determined by the presence or absence of variant glycosyltransferases.
- The GTA (A-type blood) and GTB (B-type blood) enzyme variants differ by only a few amino acids, yet they account for the differential addition of GalNAc or Gal to the O-antigen blood group glycan on erythrocyte membrane proteins and lipids.
- Individuals that inherit only one functional copy of the GTA gene have A-type blood, whereas individuals that inherit only one functional copy of the GTB gene have B-type blood. Individuals that inherit one copy each of the GTA and GTB genes have AB-type blood, an example of codominant inheritance. Individuals lacking functional copies of both the GTA and GTB genes have O-type blood.
- These ABO blood types are very important in blood transfusions because of the presence of antibodies in plasma that bind to glycoconjugate antigens. If blood plasma in the recipient contains antibodies that bind to glycoconjugate antigens present on red blood cells from the donor, then the transfusion will be lethal to the patient due to massive clotting.
- Proteoglycans contain multiple glycosaminoglycan chains attached to a small core protein of ~20 kDa. The most common glycosaminoglycans found in proteoglycans are the hexosamine polysaccharides chondroitin sulfate, keratan sulfate, and heparan sulfate.
- Proteoglycans are a major component of cartilage and play an important role in the interstitial space between tissues. Heparin is a sulfated glycosaminoglycan related to heparan sulfate that functions as an anticoagulant by binding to and inhibiting plasma proteins required for blood clotting.
- Bacterial cell walls are made of peptidoglycans, which are proteoglycans consisting of multiple strands of hexosamine polysaccharide chains. The chains consist of repeating units of a β-1,4–linked MurNAc-GlcNAc disaccharide, which are tethered together by linkages between short oligopeptides.
- Bacteria can be divided into two broad groups on the basis of physical differences in the peptidoglycan layer (thick or thin) and whether or not they have an outer membrane. Gram-positive bacteria have a thick peptidoglycan cell wall and no outer membrane, whereas Gram-negative bacteria have a thin peptidoglycan cell wall surrounded by an outer membrane.
- The Gram test is done by staining bacteria with crystal violet dye, which is trapped inside the thick peptidoglycan wall of Gram-positive bacteria after an ethanol wash. In contrast, Gram-negative bacteria do not retain the crystal violet dye after the ethanol wash because the peptidoglycan wall is too thin to trap the dye.
- The antibiotic penicillin blocks bacterial cell wall biosynthesis by inhibiting the enzyme transpeptidase, which is required to form the oligopeptide linkages between hexosamine polysaccharide chains. Without transpeptidase to support cell wall biosynthesis during cell division, the penicillin-treated bacteria die.
- Some bacteria are resistant to penicillin because they produce an enzyme called β-lactamase, which hydrolyzes the β-lactam ring in penicillin to inactivate it. This type of penicillin resistance has been overcome by methicillin, an antibiotic that blocks transpeptidase activity without being a substrate for β-lactamase.
- MRSA is a methicillin-resistant strain of *Staphylococcus aureus* that expresses a variant transpeptidase enzyme that does not bind methicillin. The methicillin-resistant transpeptidase gene was obtained from another bacterial species through a process called lateral gene transfer.

13.3 Biochemical Methods in Glycobiology

- Two research objectives of glycobiology research are (1) identification of glycan group structures on purified glycoproteins using liquid chromatography and mass spectrometry and (2) applications of high-throughput, array-based screening assays to identify biologically relevant glycan binding interactions.

- Structural characterization of glycans on glycoconjugates is technically challenging because of variable bonding arrangements between many different sugars, the presence of sugar stereoisomers with identical masses, and the fact that glycan structures can differ in subtle ways between identical classes of glycoconjugates.
- High-performance liquid chromatography (HPLC), in combination with glycosylase cleavage, can provide information about the arrangement of sugars in a highly purified glycan fraction using standard HPLC equipment. In contrast, mass spectrometry, which requires mass analyzers, identifies common glycans present in a mixed glycan sample based on predicted mass-to-charge ratios.
- The fluorescent dye 2-aminobenzamide (2-AB) is covalently attached to glycan groups prior to analysis of glycan structures by HPLC. The glycans labeled with 2-AB are subjected to stepwise treatment with sugar-specific glycosylases to yield related glycan structures that can be separated and identified by HPLC.
- Mass spectrometry is used either as the primary analytical method for glycan characterization or as a complementary technique to augment elution data derived from HPLC. By comparing the observed masses obtained from mass spectrometry to predicted masses of common N-linked glycan groups, specific peaks in the spectrum can be assigned to glycan groups.
- Glycan identification by mass spectrometry can be performed using matrix-assisted laser desorption/ionization (MALDI) and time of flight (TOF) to calculate mass-to-charge ratios.
- Lectin and glycan arrays provide platforms to screen large numbers of samples for specific binding interactions using fluorescently labeled molecules. Two basic types of arrays have been developed for screening purposes: (1) protein arrays, which contain covalently attached lectin proteins or antibodies; and (2) glycan arrays, which contain covalently attached glycoproteins with intact glycan groups or chemically synthesized glycan groups.
- Lectin arrays contain a grid of lectin proteins that are covalently attached to a solid support and used to detect glycan groups in an experimental sample. By using sample material that is fluorescently labeled, a pattern of binding interactions can be obtained indicating the presence or absence of glycans in the sample.
- Antibody arrays are also useful tools in glycobiology research. The antibodies can either bind to specific glycan groups or bind to the protein moiety of glycoproteins. When glycan binding antibodies are used to detect fluorescently labeled glycoproteins in an experimental sample, the antibody array functions in much the same way as a lectin array. Antibodies that recognize the protein component of a glycoprotein can be used to investigate glycan modifications on glycoproteins.
- Glycan arrays are used to analyze the cellular glycome under various conditions using fractionated cell extracts or to detect lectins on the surface of pathogenic and nonpathogenic bacteria. Glycan arrays can be constructed using HPLC fractions of cell extracts containing glycoprotein mixtures that are covalently linked to solid surfaces or using chemically synthesized glycans of known structure.

biochemical terms

(in order of appearance in text)
glycan (p. 634)
glycobiology (p. 634)
hexosamine (p. 634)
glycoconjugate (p. 635)
glycosyltransferase (p. 635)
glycosidase (p. 635)
Consortium for Functional Glycomics (CFG) (p. 635)
lectin (p. 638)
cellulose (p. 640)
cellobiose (p. 640)
hemicellulose (p. 642)
pectin (p. 642)
cellulase (p. 642)
cellotetraose (p. 642)
chitin (p. 642)
chitinase (p. 644)
glycosaminoglycan (p. 644)
proteoglycan (p. 644)
chondroitin sulfate (p. 644)
heparan sulfate (p. 644)
keratan sulfate (p. 644)
macular corneal dystrophy (p. 644)
amylose (p. 644)
amylopectin (p. 644)
glycogenin (p. 645)
glycoprotein (p. 648)
glycolipid (p. 648)
peptidoglycan (p. 649)
N-linked oligosaccharide (p. 650)
O-linked oligosaccharide (p. 650)
mucin (p. 650)
ABO blood groups (p. 653)
codominant (p. 654)
hyaluronic acid (p. 657)
heparin (p. 657)
Gram-positive bacteria (p. 658)
Gram-negative bacteria (p. 659)
lipoteichoic acid (p. 660)
lipopolysaccharide (LPS) (p. 660)
endotoxin (p. 660)
penicillin (p. 661)
transpeptidase (p. 661)
β-lactamase (p. 662)
methicillin (p. 663)
lateral gene transfer (p. 665)
glycome (p. 671)

review questions

1. Describe the structure and function of the three major glycan groups in nature.
2. Oligosaccharides often consist of an extended glycan disaccharide. Describe two examples of oligosaccharides that differ in their core disaccharide.
3. Describe the organization of abundant polysaccharides in plant cell walls, and explain how this organization contributes to the structural strength of plant tissues.
4. Starch and glycogen are storage forms of glucose for plants and animals, respectively. Compare and contrast these two abundant polysaccharides in nature.
5. What are lectins, and how do they function in the human immune system?
6. Describe two types of glycan linkages to glycoproteins.
7. What is the biochemical basis for A, B, O, and AB blood groups in humans?
8. How do β-lactam antibiotics such as penicillin kill bacterial cells, and what explains bacterial resistance to the antibiotic methicillin?
9. What are the two primary objectives of glycobiology research?
10. Why is structural characterization of biological glycans on glycoconjugates so challenging, and what two biochemical approaches have been developed to address these challenges?
11. Describe how lectin, antibody, and glycan arrays are used in glycobiology research.

challenge problems

1. Paper is made from cellulose fibers. Explain why paper loses its shape and planar strength when it is soaked in water but not when it is soaked in oil.
2. Glycan groups of membrane glycoproteins and glycolipids serve as high-specificity binding sites for cell recognition proteins. Which has higher structural complexity: a mixture of tripeptides generated from Thr, Glu, and Ala or a mixture of trisaccharides generated from D-xylose, D-glucose, and D-galactose? Explain your answer using numerical examples to illustrate the extent of structural variety in the mixtures.
3. Pig feed can contain vegetables with a high content of raffinose-series oligosaccharides.
   a. Explain why farmers premix this feed with commercial-grade α-galactosidase before feeding the pigs, as a means to increase weight gain.
   b. What causes flatulence in humans (and pigs) who consume vegetables such as broccoli, cabbage, and soybeans, but not when consuming potatoes, squash, and corn?
4. The number of branch points (α-1,6 glycosidic bonds) in amylopectin can be calculated using a chemical modification protocol based on extensive methylation, followed by hydrolysis, reduction, and acetylation. Because the hydrolysis products differ with regard to the position of methyl and acetyl groups in the released glucose molecules, it is possible to use this method to determine the number of α-1,6 glycosidic bonds and hence the number of branch points and nonreducing ends in the sample.
   a. Assuming methylation of glycogen at all available OH groups converts them to OCH_3 groups, label the amylopectin molecule that follows (which contains a single terminal α-1,6 glucose) with the correct glucose residues (A, B, C, D) based on the glucose products shown below the molecule.
   b. If a 0.5-g sample of amylopectin was found to have 25 mg of 2,3-dimethylglucose, calculate the ratio of glucose molecules with α-1,4 glycosidic bonds compared to those with α-1,6 glycosidic bonds. Show your work and include units. Note the molecular mass of a glucose residue in glycogen is 162 g/mol, and the molecular mass of 2,3-dimethylglucose is 208 g/mol.

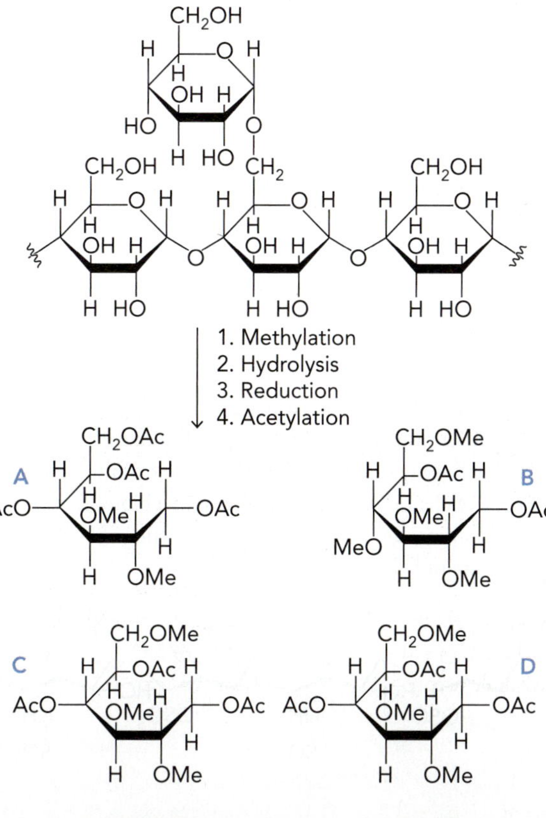

c. Calculate the approximate number of nonreducing ends in this 0.5-g sample of amylopectin.

5. Researchers identified a secreted glycoprotein enzyme in human serum and wanted to know if the glycan group is required for the normal function of the enzyme. To do this, they synthesized the glycoprotein in an *in vitro* system in such a way that they were able to purify the enzyme with and without the covalently attached glycan group. *In vitro* enzyme assays using a fluorogenic substrate showed that the enzyme has similar activity in the glycosylated and nonglycosylated forms, leading them to conclude that the glycan group is not required for normal biological function of this enzyme in cells. Give three reasons why this conclusion could be wrong.

6. The human ABO blood group antigens are glycoproteins and glycolipids on the surfaces of red blood cells. The antigens contain one of three different structural glycans or a mixture of two types. Cross-reacting antibodies present in serum from heterologous blood transfusions can cause blood agglutination (massive clotting) if compatible blood types are not used.

a. Describe the structural differences in the glycan groups of the human A, B, AB, and O blood group antigens.

b. What is the molecular explanation for the inheritance patterns of the A, B, AB, and O blood groups in humans?

c. Complete the following table by listing all possible paternal and maternal genotype variants of glycosyltransferase enzymes that could give rise to the A, B, AB, and O blood groups.

Blood group	Allele 1 (paternal or maternal)	Allele 2 (paternal or maternal)
A blood group		
B blood group		
AB blood group		
O blood group		

d. Which blood group is the most useful at blood bank donation centers? Explain. Which blood group is the best to have if you are in need of a transfusion? Explain.

7. Heparan sulfate and heparin are related glycosaminoglycans that differ primarily in the number of sulfate groups (heparin is more highly sulfated). Heparin has one of the highest negative charge densities ever found in a biomolecule and is used commercially to prevent blood clotting. The molecular structure of heparin is shown here.

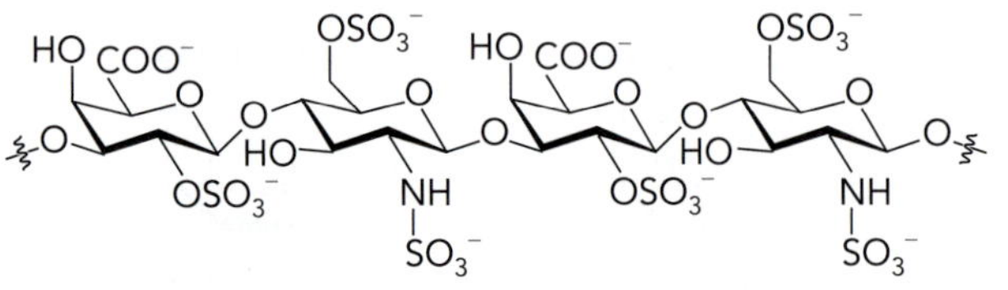

a. What accounts for the anticoagulant activity of commercially prepared heparin?

b. Fatal accidental heparin overdoses have tragically occurred in hospitals because of poor labeling on heparin bottles and human error. How does infusion with protamine sulfate, a cationic protein, counteract the effects of a heparin overdose?

8. Penicillin is an antibiotic that was discovered almost by accident when Alexander Fleming forgot to clean out old bacterial plates. Years earlier, Fleming had discovered that the enzyme lysozyme has antibacterial properties, but it was not amenable to commercial production as an antibiotic because lysozyme is chemically unstable.

a. What did Fleming notice on the bacterial plate that gave him the idea to look for the presence of an antibiotic? Why did he call it "penicillin"?

b. What is the biochemical target of penicillin in bacteria, and why is penicillin lethal to bacteria?

c. Overuse of penicillin led to the emergence of penicillin-resistant bacteria. What is the most common mechanism of penicillin resistance in bacteria, and how did pharmaceutical companies solve this problem?

d. Why is methicillin-resistant *Staphylococcus aureus* (MRSA) called the superbug? What is the molecular mechanism of antibiotic resistance in MRSA, and what might be a drug development strategy to overcome this resistance to treat MRSA infections?

9. Describe two methods to determine the structure of an N-linked glycan group of a soluble serum glycoprotein that was purified using an antibody affinity column with high binding capacity for the protein moiety of the glycoprotein.

10. Urinary tract infections are caused by pathogenic strains of *E. coli* bacteria containing a cell surface lectin called FimH. This bacterial lectin binds to a glycan group on the human membrane protein uroplakin, which is expressed on urinary tract epithelial cells.

a. Describe how a synthetic glycan array could be used to identify the glycan binding site of FimH on uroplakin.

b. Assuming the specific glycan group was identified, how could this information be used to develop glycan mimetics to treat urinary tract infections?

suggested reading

Books and Reviews

Anstee, D. J. (2009). Red cell genotyping and the future of pretransfusion testing. *Blood, 114*, 248–256.

Bode, L. (2006). Recent advances on structure, metabolism, and function of human milk oligosaccharides. *Journal of Nutrition, 136*, 2127–2130.

French, G. L. (2010). The continuing crisis in antibiotic resistance. *International Journal of Antimicrobial Agents, 36(suppl. 3)*, S3–7.

Ghazarian, H., Idoni, B., and Oppenheimer, S. B. (2011). A glycobiology review: carbohydrates, lectins and implications in cancer therapeutics. *Acta Histochemica, 113*, 236–247.

Hsu, K. L., and Mahal, L. K. (2009). Sweet tasting chips: microarray-based analysis of glycans. *Current Opinion in Chemical Biology, 13*, 427–432.

Lomako, J., Lomako, W. M., and Whelan, W. J. (2004). Glycogenin: the primer for mammalian and yeast glycogen synthesis. *Biochimica et Biophysica Acta, 1673*, 45–55.

Marino, K., Bones, J., Kattla, J. J., and Rudd, P. M. (2010). A systematic approach to protein glycosylation analysis: a path through the maze. *Nature Chemical Biology, 6*, 713–723.

Pabst, M., and Altmann, F. (2011). Glycan analysis by modern instrumental methods. *Proteomics, 11*, 631–643.

Pilobello, K. T., and Mahal, L. K. (2007). Deciphering the glycocode: the complexity and analytical challenge of glycomics. *Current Opinion in Chemical Biology, 11*, 300–305.

Rakus, J. F., and Mahal, L. K. (2011). New technologies for glycomic analysis: toward a systematic understanding of the glycome. *Annual Review of Analytical Chemistry (Palo Alto, Calif.), 4*, 367–392.

Varki, A., Cummings, R. D, Esko, J. D, Freeze, H. H., Stanley, P., Bertozzi, C. R., Wart, G. W., and Etzler, M. E. (2009). *Essentials of Glycobiology*. Cold Spring Harbor, NY: Cold Spring Harbor Press.

Primary Literature

Akama, T. O., Nishida, K., Nakayama, J., Watanabe, H., Ozaki, K., Nakamura, T., Dota, A., Kawasaki, S., Inoue, Y., Maeda, N., et al. (2000). Macular corneal dystrophy type I and type II are caused by distinct mutations in a new sulphotransferase gene. *Nature Genetics, 26*, 237–241.

Cheng, Y. S., Ko, T. P., Wu, T. H., Ma, Y., Huang, C. H., Lai, H. L., Wang, A. H., Liu, J. R., and Guo, R. T. (2011). Crystal structure and substrate-binding mode of cellulase 12A from *Thermotoga maritima*. *Proteins, 79*, 1193–1204.

Graham, K. K., Kerley, M. S., Firman, J. D., and Allee, G. L. (2002). The effect of enzyme treatment of soybean meal on oligosaccharide disappearance and chick growth performance. *Poultry Science, 81*, 1014–1019.

Marcus, S. L., Polakowski, R., Seto, N. O., Leinala, E., Borisova, S., Blancher, A., Roubinet, F., Evans, S. V., and Palcic, M. (2003). A single point mutation reverses the donor specificity of human blood group B-synthesizing galactosyltransferase. *Journal of Biological Chemistry, 278*, 12403–12405.

Patenaude, S. I., Seto, N. O., Borisova, S. N., Szpacenko, A., Marcus, S. L., Palcic, M. M., and Evans, S. V. (2002). The structural basis for specificity in human ABO(H) blood group biosynthesis. *Nature Structural Biology, 9*, 685–690.

Wellens, A., Garofalo, C., Nguyen, H., Van Gerven, N., Slattegard, R., Hernalsteens, J. P., Wyns, L., Oscarson, S., De Greve, H., Hultgren, S., and Bouckaert, J. (2008). Intervening with urinary tract infections using anti-adhesives based on the crystal structure of the FimH-oligomannose-3 complex. *PLoS ONE, 3*, e2040.

The pentose phosphate pathway enzyme glucose-6-phosphate dehydrogenase (G6PD) produces NADPH, which reduces glutathione and protects red blood cells against oxidative stress.

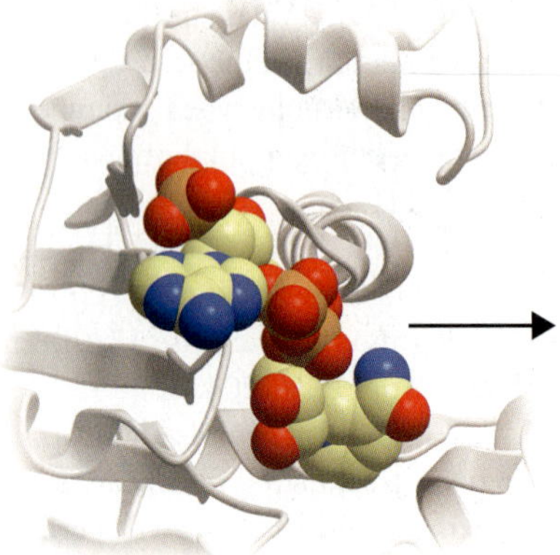

The molecular structure of glucose-6-phosphate dehydrogenase is shown here with $NADP^+$ bound to the active site

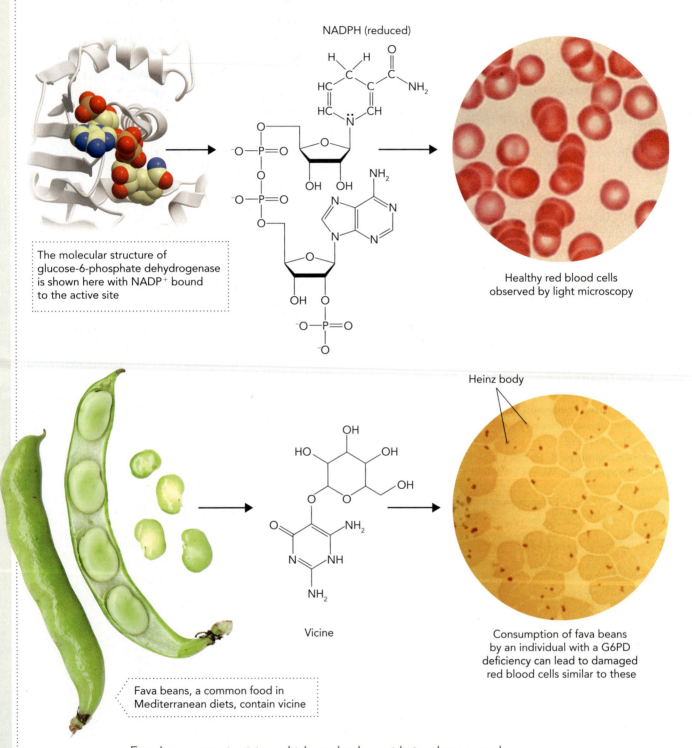

NADPH (reduced)

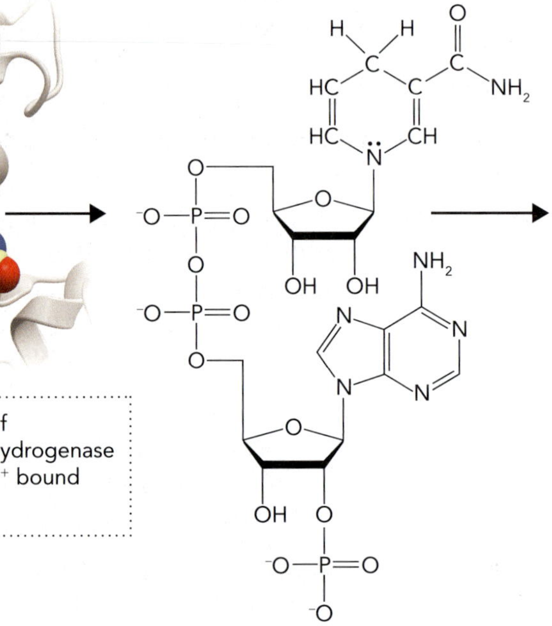

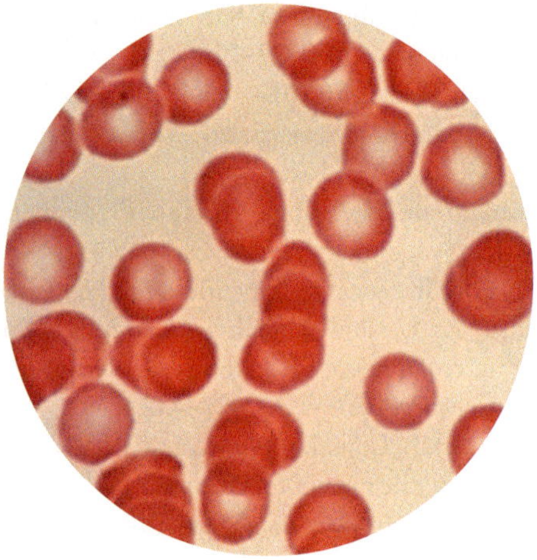

Healthy red blood cells observed by light microscopy

Fava beans, a common food in Mediterranean diets, contain vicine

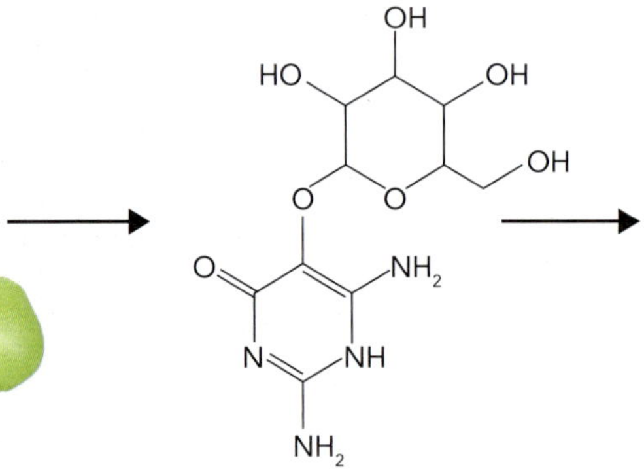

Vicine

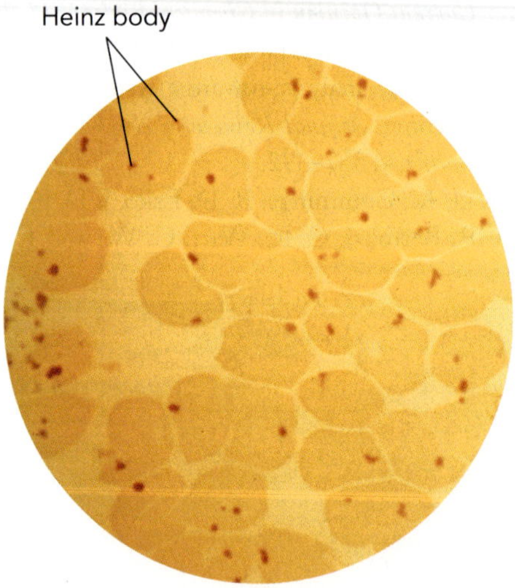

Consumption of fava beans by an individual with a G6PD deficiency can lead to damaged red blood cells similar to these

Fava beans contain vicine, which can lead to oxidative damage and hemolytic anemia in people with a deficiency in G6PD enzyme activity. Damaged red blood cells containing denatured hemoglobin can be identified by the presence of Heinz bodies in stained cells.

14

Carbohydrate Metabolism

Fava beans are a staple food in Mediterranean diets; however, they contain a glycoside compound called vicine, which generates free oxygen radicals and H_2O_2 when it is metabolized in red blood cells. The enzyme glucose-6-phosphate dehydrogenase (G6PD), which is part of the pentose phosphate pathway and generates NADPH for cells, helps protect cells against oxidative damage caused by compounds such as vicine. Individuals with a genetic defect in the glucose-6-phosphate dehydrogenase gene (*G6PD*) get sick from eating fava beans because these individuals are unable to produce high levels of NADPH to counteract the toxic effects of vicine in red blood cells. The name given to this diet-induced physiologic condition is favism, which in extreme cases can lead to hemolytic anemia (red blood cell death). Deficiency in the glucose-6-phosphate dehydrogenase enzyme represents the most common human genetic variant linked to metabolic disease, with an estimated 400 million people worldwide carrying *G6PD* mutations.

CHAPTER OUTLINE

In this chapter, we complete our study of carbohydrate metabolism by discussing how the cell metabolizes glucose-6-phosphate to generate NADPH by the pentose phosphate pathway, synthesizes glucose using the gluconeogenic pathway, and stores glucose in the form of glycogen as a short-term energy source to maintain metabolic homeostasis.

We begin this chapter by describing the pentose phosphate pathway, which is similar in many respects to the Calvin cycle (see Chapter 12), particularly with regard to the involvement of enzymes that mediate a series of "carbon shuffle" reactions. Next, we examine the gluconeogenic pathway, which is responsible for biosynthesis of glucose in the liver and kidneys for export to other tissues. The word *gluconeogenesis* combines Greek roots for the words generation (*genesis*), new (*neo*), and glucose (*gluco*). Finally, we describe how glycogen—a storage form of glucose in animal cells—is synthesized and degraded by enzymes that are regulated by phosphorylation cascades in response to hormone signaling. In particular, we look at how the peptide hormones epinephrine, glucagon, and insulin activate membrane receptors on the surface of liver and muscle cells, in turn activating downstream signals controlling key enzymes in glycogen metabolism.

The pentose phosphate pathway, gluconeogenesis, and glycogen degradation and synthesis all use glucose-6-phosphate (or its isomer glucose-1-phosphate) as a metabolic intermediate and represent the three major carbohydrate metabolizing pathways in animal cells (**Figure 14.1**).

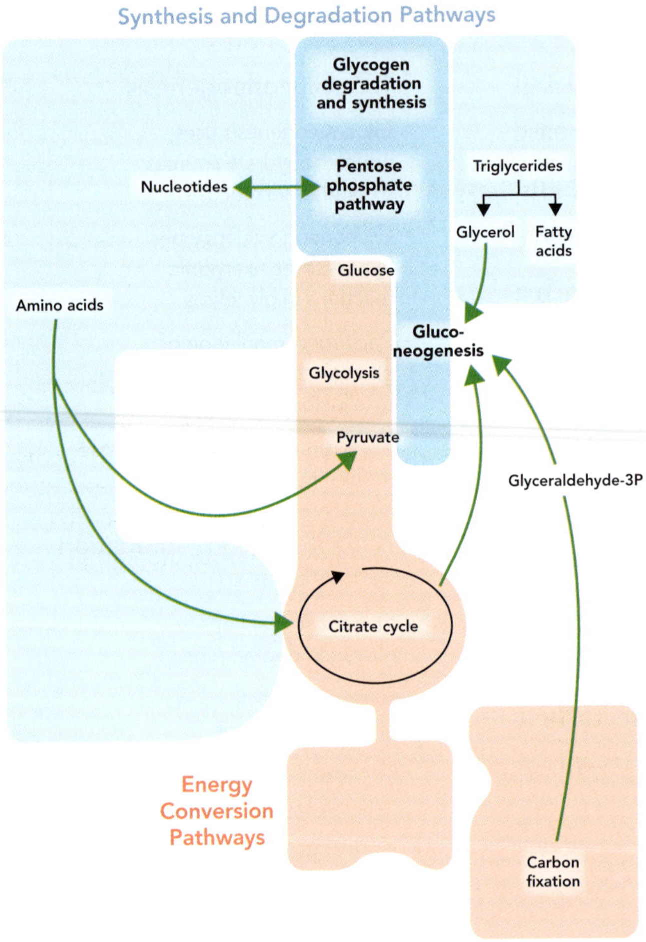

Figure 14.1 The three primary pathways for anabolic carbohydrate metabolism in nonphotosynthetic organisms are the pentose phosphate pathway, gluconeogenesis, and glycogen degradation and synthesis. Metabolism of ribose sugars in the pentose phosphate pathway is used to generate NADPH and to provide the carbohydrate component of nucleotides. The major sources of carbon in gluconeogenesis are amino acids and glycerol in animals and glyceraldehyde-3-phosphate in plants. As described in Chapter 9, glycolysis and gluconeogenesis are opposing pathways that metabolize glucose and pyruvate (see Figure 9.39).

14.1 The Pentose Phosphate Pathway

Some metabolic pathways we have studied occur in specialized organelles, such as the mitochondria or chloroplasts. Others occur partly in the cytoplasm and partly in organelles. The **pentose phosphate pathway** occurs entirely within the cytoplasm. It is also known as the hexose monophosphate shunt or the **phosphogluconate pathway**. The most important function of the pentose phosphate pathway is to reduce two molecules of $NADP^+$ to NADPH for each molecule of glucose-6-phosphate (glucose-6-P) that is oxidatively decarboxylated to ribulose-5-phosphate (ribulose-5-P). The pentose phosphate pathway is also responsible for producing ribose-5-phosphate (ribose-5-P) from glucose-6-P. This is very important because ribose-5-P is the carbohydrate component of nucleotides (precursors to DNA and RNA) and of the coenzymes ATP, NAD^+, $NADP^+$, FAD, and acetyl-CoA.

The coenzyme nicotinamide adenine dinucleotide phosphate (NADPH) functions as a strong reductant

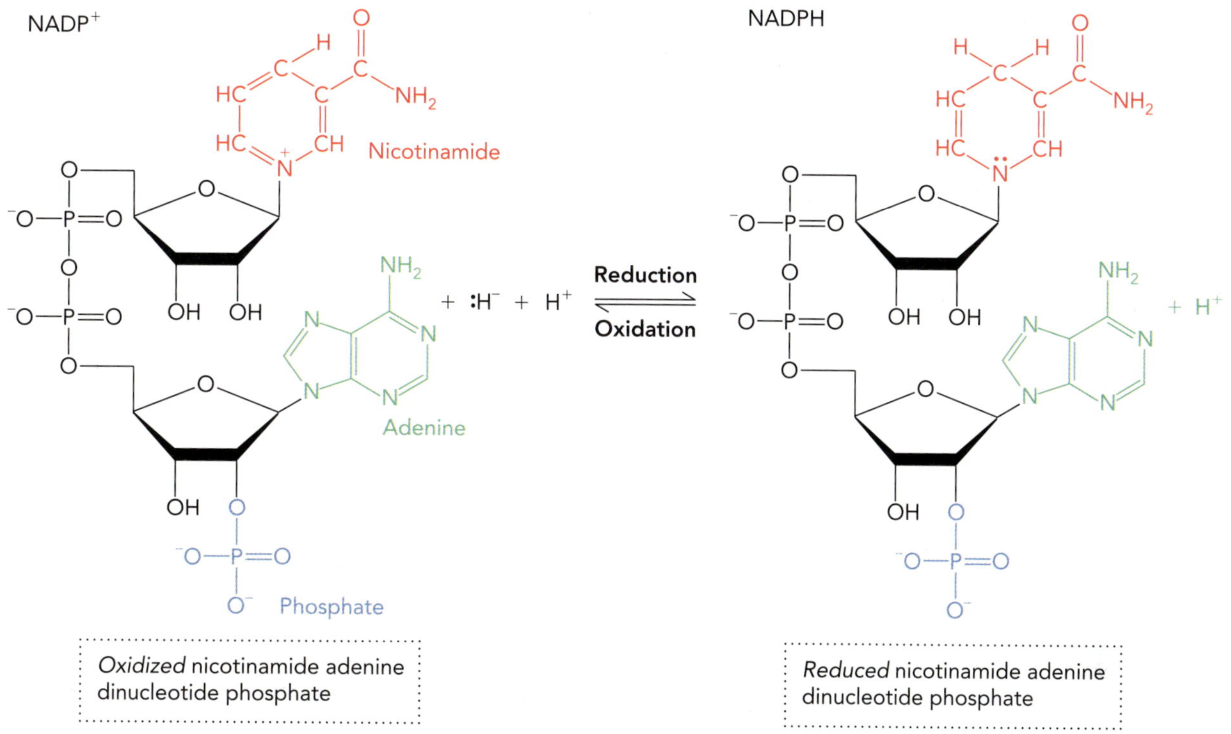

Figure 14.2 The structures of $NADP^+$ and NADPH are shown here. Reduction of $NADP^+$ involves the transfer of two electrons in the form of a hydride ion ($:H^-$) and the release of a proton (H^+).

(electron donor) in both biosynthetic pathways and in detoxification reactions, which neutralize reactive oxygen species. As shown in **Figure 14.2**, NADPH is structurally related to NADH, differing only in the presence of a phosphate group on the ribose sugar of the adenine nucleotide (see Figure 10.5). Similar to redox reactions involving the redox pair NAD^+/NADH, reduction of $NADP^+$ to NADPH involves the transfer of two electrons in the form of a hydride ion ($:H^-$) and release of a proton (H^+). Enzymes in anabolic pathways have evolved to use NADPH as the primary reductant, whereas enzymes in catabolic pathways use NAD^+ as the primary oxidant. This allows for separation of anabolic and catabolic pathways and prevents futile cycling. The distinct roles of these two related coenzymes in metabolism is indicated by the very different steady-state concentrations of the conjugate redox pairs. In liver cells, the ratio of NAD^+-to-NADH concentrations is close to 1,000; however, the ratio of $NADP^+$-to-NADPH concentrations is 0.01.

The pentose phosphate pathway can be divided into two phases: The *oxidative phase* generates NADPH, and the *nonoxidative phase* interconverts C_3, C_4, C_5, C_6, and C_7 sugar phosphates, using many of the same carbon shuffle reactions we saw in the Calvin cycle (see Figure 12.49). **Figure 14.3** provides an overview of the pentose phosphate pathway, illustrating the function of the oxidative and nonoxidative phases in producing NADPH and ribose-5-P, respectively. Flux through the oxidative and nonoxidative phases of the pentose phosphate pathway is regulated to meet three distinct metabolic states of the cell:

1. If *increased NADPH is required* for biosynthetic pathways or to provide reducing power for detoxification, then fructose-6-phosphate (fructose-6-P) and glyceraldehyde-3-phosphate (glyceraldehyde-3-P) are used to resynthesize glucose-6-P, thereby maintaining flux through the oxidative phase of the pathway.

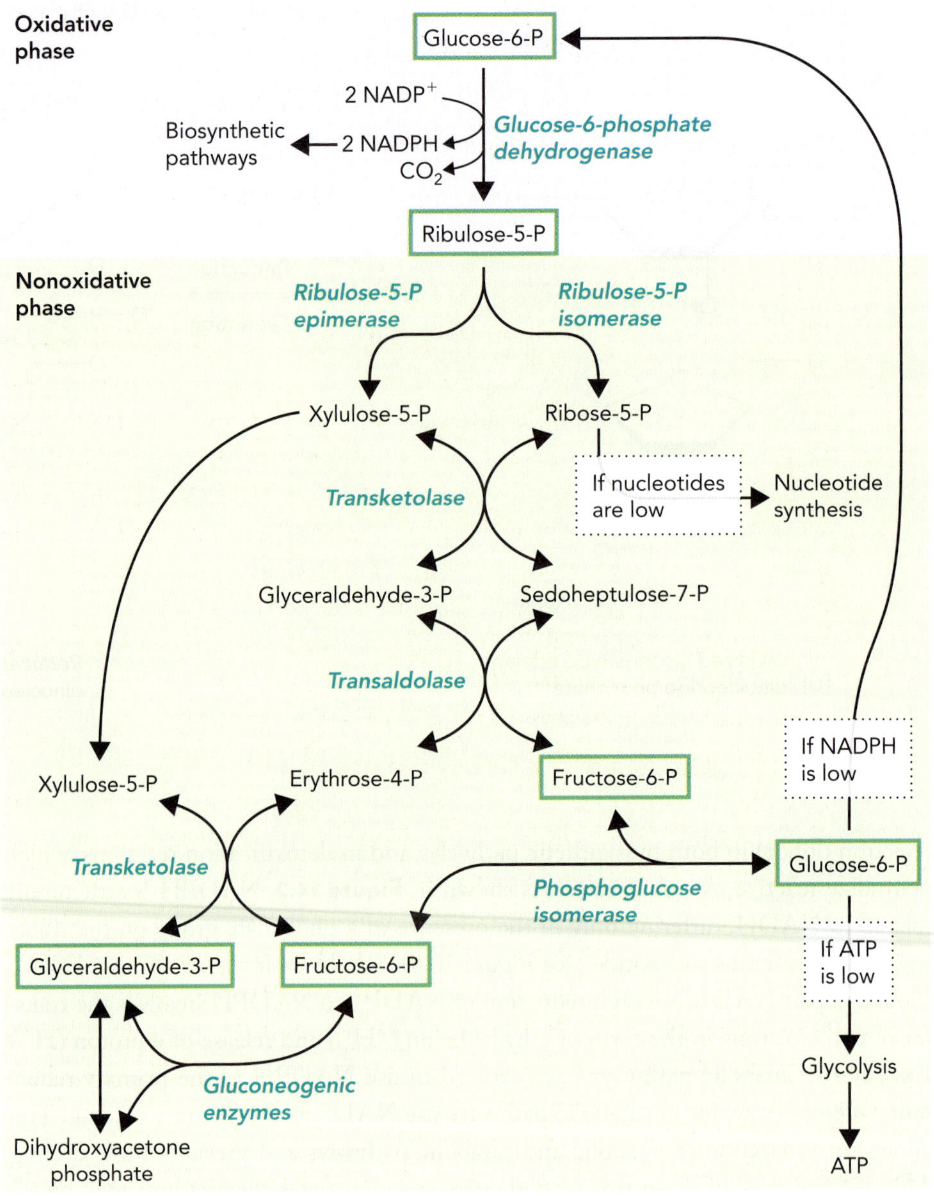

Figure 14.3 The pentose phosphate pathway generates NADPH for biosynthetic pathways and ribose-5-P for nucleotide synthesis. Flux through the pentose phosphate pathway is modulated in response to the metabolic needs of the cell. The conversion of glucose-6-P to ribulose-5-P is the oxidative phase of the pathway, whereas the conversion of ribulose-5-P to glyceraldehyde-3-phosphate (glyceraldehyde-3-P) and fructose-6-phosphate (fructose-6-P) is the nonoxidative phase of the pathway. The key enzymes in the pentose phosphate pathway are glucose-6-phosphate dehydrogenase, transketolase, and transaldolase.

2. If *nucleotide pools need to be replenished* because of high rates of DNA and RNA synthesis, then the bulk of ribulose-5-P is converted to ribose-5-P, stimulating nucleotide biosynthesis.
3. If *ATP levels in the cell are low*, then the enzyme glucose-6-phosphate dehydrogenase (G6PD) is inhibited, which shuts down the pentose phosphate pathway so that glucose-6-P can be metabolized directly by the glycolytic pathway.

Before we look at the details of the pentose phosphate pathway, we need to answer our four metabolic pathway questions:

1. *What does the pentose phosphate pathway accomplish for the cell?* The oxidative phase generates NADPH, which is required for many biosynthetic pathways and for detoxification of reactive oxygen species. The nonoxidative phase interconverts C_3, C_4, C_5, C_6, and C_7 monosaccharides to produce ribose-5-P for nucleotide synthesis. It also regenerates glucose-6-P to maintain NADPH production by the oxidative phase.

2. *What is the overall net reaction of the pentose phosphate pathway?* Oxidative phase:

$$6 \text{ Glucose-6-P} + 12 \text{ NADP}^+ + 6 \text{ H}_2\text{O} \rightarrow 6 \text{ Ribulose-5-P} + 12 \text{ NADPH} + 12 \text{ H}^+ + 6 \text{ CO}_2$$

Nonoxidative phase:

$$6 \text{ Ribulose-5-P} \rightarrow 4 \text{ Fructose-6-P} + 2 \text{ Glyceraldehyde-3-P}$$

Net reaction:

$$6 \text{ Glucose-6-P} + 12 \text{ NADP}^+ + 6 \text{ H}_2\text{O} \rightarrow 4 \text{ Fructose-6-P} + 2 \text{ Glyceraldehyde-3-P} + 12 \text{ NADPH} + 12 \text{ H}^+ + 6 \text{ CO}_2$$

3. *What are the key enzymes in the pentose phosphate pathway?* The enzyme **glucose-6-phosphate dehydrogenase (G6PD)** catalyzes the first reaction in the pathway, which converts glucose-6-P to 6-phosphogluconolactone. This reaction is the commitment step in the pathway and is feedback inhibited by NADPH. Defects in G6PD reduce NADPH levels and leave cells vulnerable to damage from reactive oxygen species. The enzymes **transketolase** and **transaldolase** together catalyze the reversible carbon shuffle reactions of the nonoxidative phase of the pathway. Transketolase catalyzes the transfer of C_2 units between sugars, whereas transaldolase catalyzes the transfer of C_3 units. These are the same enzymes used in the Calvin cycle to regenerate ribulose-5-P from glyceraldehyde-3-P.

4. *What are examples of the pentose phosphate pathway in everyday biochemistry?* Glucose-6-phosphate dehydrogenase deficiency is the most common enzyme deficiency in the world and affects more than 400 million people. A 90% decrease in enzyme activity results in the inability of red blood cells to produce enough NADPH to protect the cells from reactive oxygen species, which are generated, for example, by antimalarial drugs and noxious compounds in fava beans. This points to the importance of having a properly functioning pentose phosphate pathway.

Enzymatic Reactions in the Oxidative Phase

The oxidative phase of the pentose phosphate pathway includes three enzymatic reactions (**Figure 14.4**):

1. The first reaction is catalyzed by the enzyme G6PD. This irreversible reaction ($\Delta G°' = -17.6$ kJ/mol) represents the commitment step in the pathway because the product 6-phosphogluconolactone has no other metabolic fate. The oxidation of glucose-6-P by G6PD is coupled to the reduction of $NADP^+$, resulting in the formation of one molecule of NADPH.

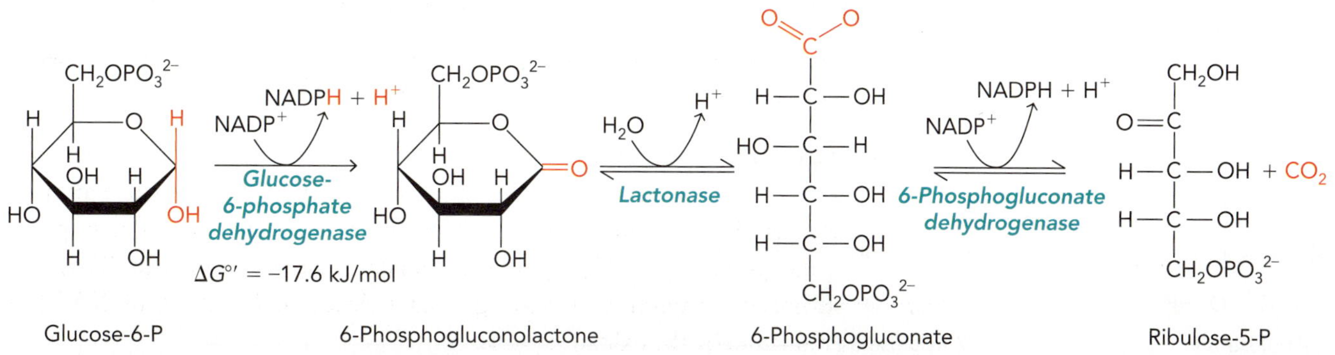

Figure 14.4 The reactions in the oxidative phase of the pentose phosphate pathway convert glucose-6-P to ribulose-5-P, resulting in the production of 2 NADPH and 1 CO_2. The enzyme G6PD catalyzes an irreversible reaction that represents the commitment step in the pathway.

2. In the second enzymatic reaction, 6-phosphogluconolactone is hydrolyzed by lactonase to produce the open-chain monosaccharide 6-phosphogluconate.
3. In the third enzymatic reaction, 6-phosphogluconate is oxidized and decarboxylated in a reaction catalyzed by 6-phosphogluconate dehydrogenase. This reaction generates ribulose-5-P and the second molecule of NADPH.

Enzymatic Reactions in the Nonoxidative Phase

In cells that require high levels of NADPH for biosynthetic reactions, the ribulose-5-P produced in the oxidative phase needs to be converted back into glucose-6-P. This conversion works to maintain flux through the G6PD reaction. The reaction involves enzymes in the nonoxidative phase of the pentose phosphate pathway.

Figure 14.5 The first steps of the nonoxidative phase of the pentose phosphate pathway are shown here. **a.** In this part of the pentose phosphate pathway, 6 ribulose-5-P are converted into 4 fructose-6-P and 2 glyceraldehyde-3-P using carbon shuffle reactions that are similar to those in the Calvin cycle (see Figure 12.49). **b.** Reactions catalyzed by the enzymes transketolase and transaldolase.

a.

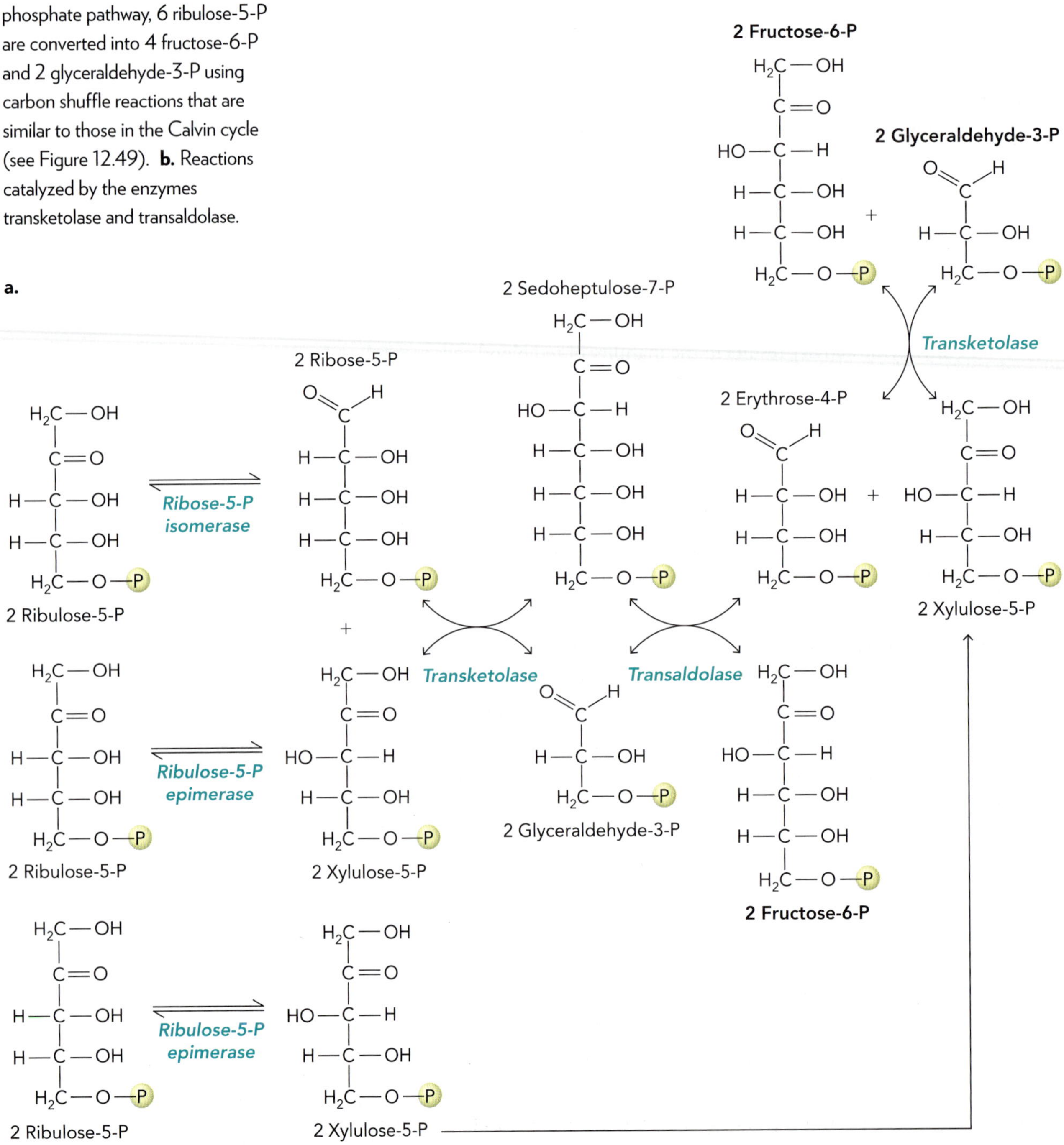

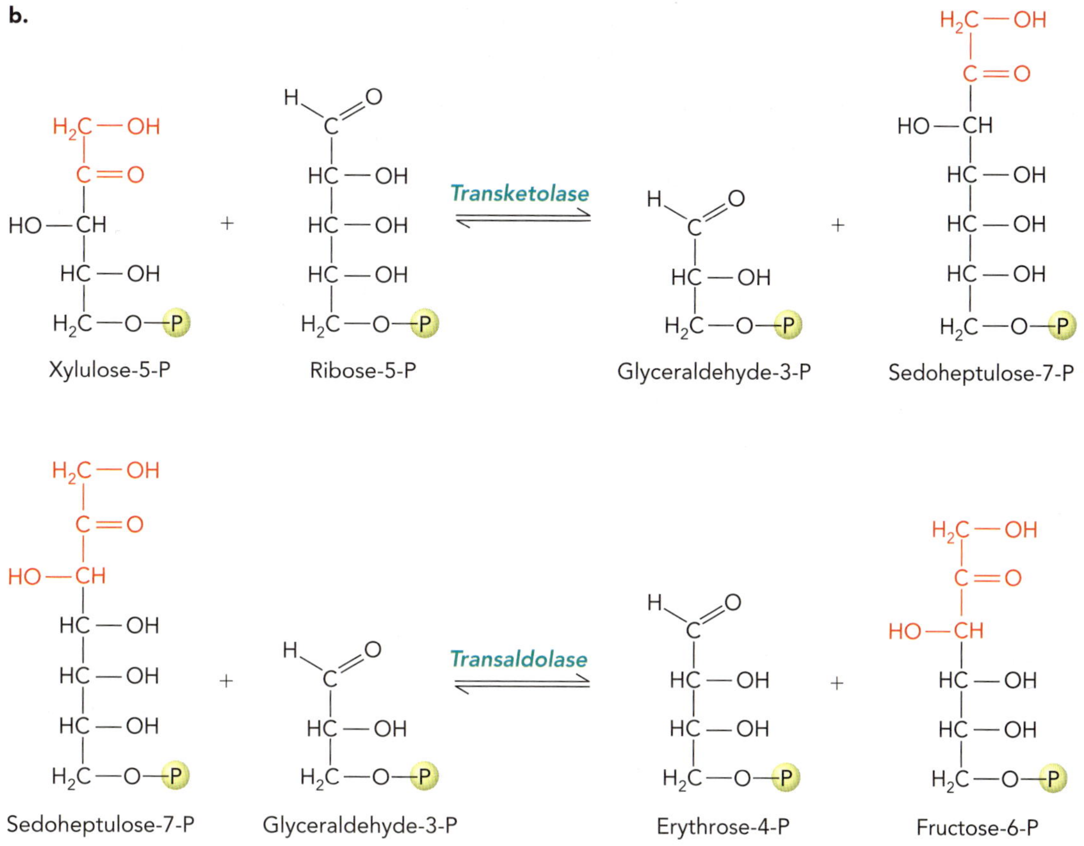

Figure 14.5
Continued

In the first set of reactions, two molecules of ribulose-5-P are converted into two molecules of ribose-5-P by the enzyme ribose-5-P isomerase. Then, four molecules of ribulose-5-P are converted into four molecules of xylulose-5-P by the enzyme ribulose-5-P epimerase (**Figure 14.5a**). These six C_5 molecules (two ribose-5-P and four xylulose-5-P) are converted into four molecules of fructose-6-P and two molecules of glyceraldehyde-3-P by the enzymes transketolase and transaldolase (**Figure 14.5b**). The net reaction of these carbon shuffle reactions in the nonoxidative phase is the conversion of six ribulose-5-P molecules into four molecules of fructose-6-P and two molecules of glyceraldehyde-3-P. As shown in **Figure 14.6**, the fructose-6-P and glyceraldehyde-3-P generated in the nonoxidative phase of the pentose phosphate pathway are used to regenerate five molecules of glucose-6-P to maintain flux through the oxidative phase. As discussed in Section 14.2, enzymes in the gluconeogenic pathway are responsible for regenerating these glucose-6-P molecules.

We can now add up the net reactions from the oxidative phase, the nonoxidative phase, and glucose-6-P regeneration by gluconeogenesis, giving us an overall net reaction for the production of NADPH by the pentose phosphate pathway:

Oxidative phase:

$$6\ \text{Glucose-6-P} + 12\ \text{NADP}^+ + 6\ \text{H}_2\text{O} \rightarrow 6\ \text{Ribulose-5-P} + 12\ \text{NADPH} + 12\ \text{H}^+ + 6\ \text{CO}_2$$

Nonoxidative phase (first part):

$$6\ \text{Ribulose-5-P} \rightarrow 4\ \text{Fructose-6-P} + 2\ \text{Glyceraldehyde-3-P}$$

Nonoxidative phase (second part, catalyzed by gluconeogenic enzymes):

$$4\ \text{Fructose-6-P} + 2\ \text{Glyceraldehyde-3-P} \rightarrow 5\ \text{Glucose-6-P}$$

Overall net reaction:

$$\text{Glucose-6-P} + 12\ \text{NADP}^+ + 6\ \text{H}_2\text{O} \rightarrow 12\ \text{NADPH} + 12\ \text{H}^+ + 6\ \text{CO}_2$$

a.

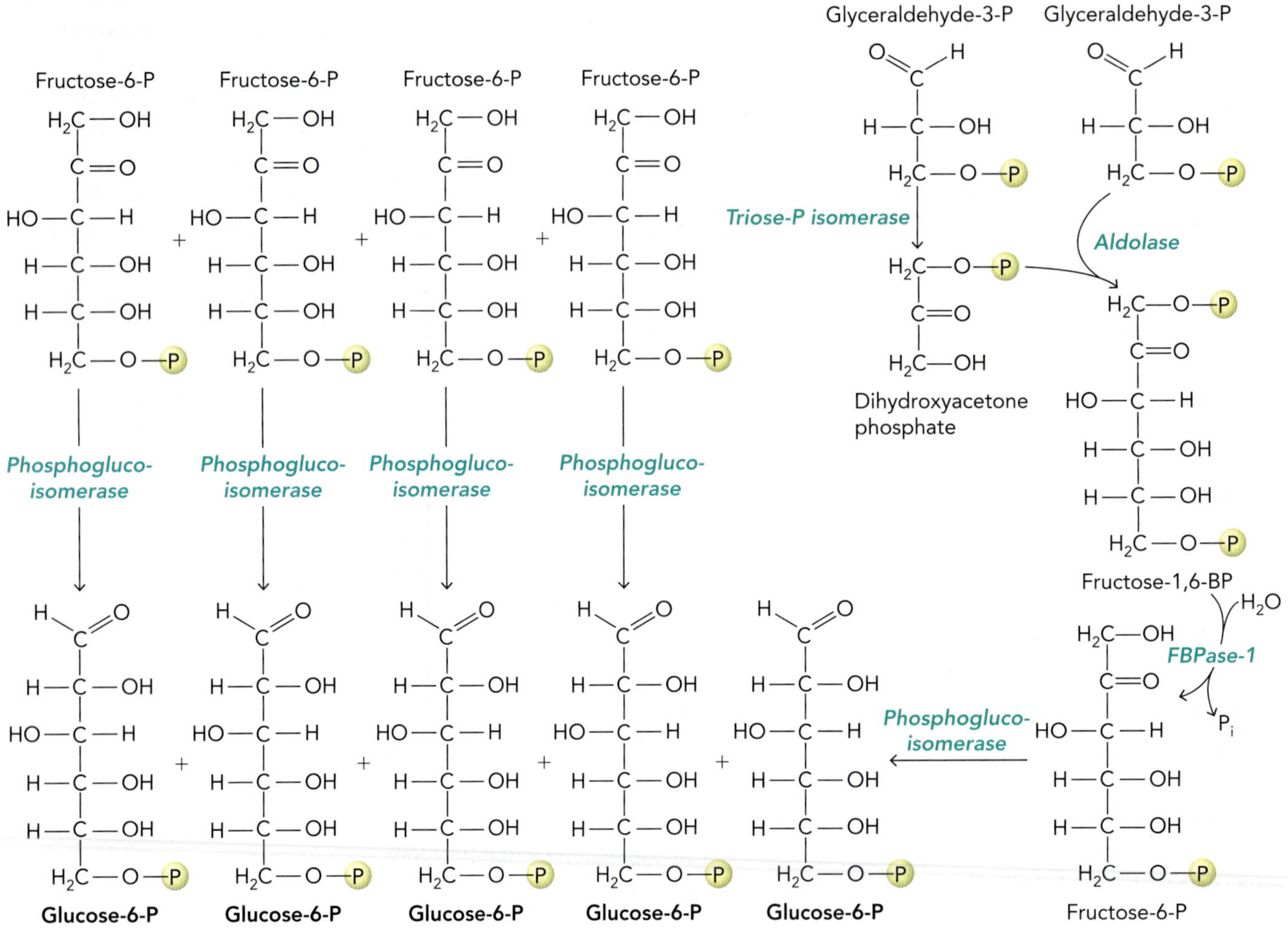

b.

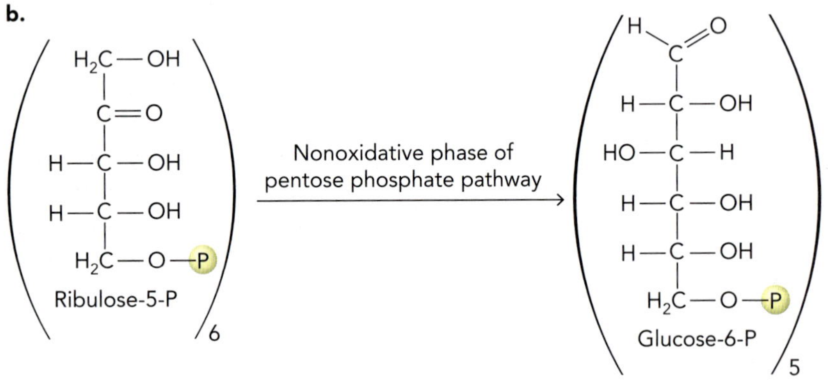

Figure 14.6 The final steps in the nonoxidative phase of the pentose phosphate pathway are shown here. **a.** Glucose-6-P is regenerated from fructose-6-P and glyceraldehyde-3-P by enzymes in the gluconeogenic pathway. The enzymes phosphoglucoisomerase, triose phosphate isomerase (triose-P isomerase), and aldolase are components of both the glycolytic pathway and the gluconeogenic pathway, whereas fructose-1,6-bisphosphatase-1 (FBPase-1) is unique to the gluconeogenic pathway. Fructose-1,6-BP = fructose-1,6-bisphosphate. **b.** Overall, the nonoxidative phase of the pentose phosphate pathway converts 6 ribulose-5-P to 5 glucose-6-P.

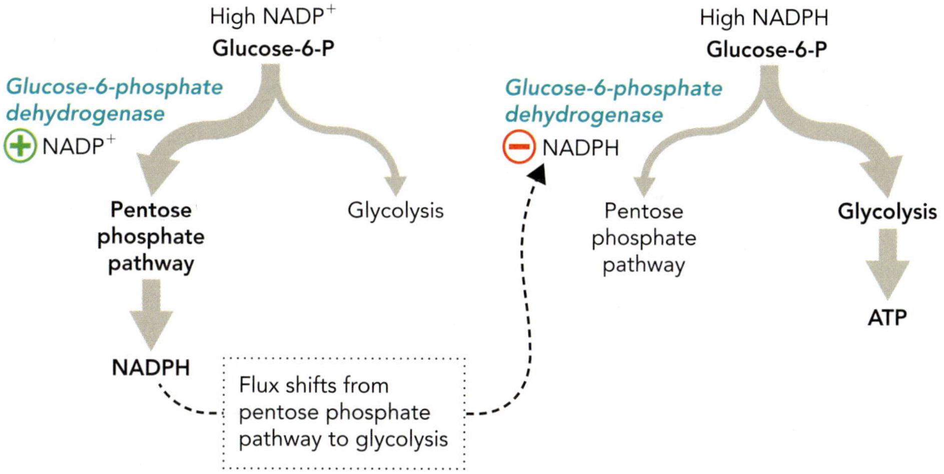

Figure 14.7 Glucose-6-phosphate dehydrogenase activity is allosterically regulated by $NADP^+$-to-NADPH concentration ratios in the cell. When $NADP^+$ levels are high, G6PD activity is allosterically activated. This causes most of the glucose-6-P to be metabolized by the pentose phosphate pathway, leading to increased production of NADPH. Because NADPH competes with $NADP^+$ for binding to G6PD, high NADPH levels result in lower enzyme activity and increased flux of glucose-6-P through glycolysis.

Considering that glucose-6-P is a substrate for both glycolysis and the pentose phosphate pathway, what controls the overall metabolic flux through these two pathways? The answer is the $NADP^+$-to-NADPH concentration ratio in the cytosol, which acts as a rheostat to regulate the activity of G6PD. As shown in **Figure 14.7**, when the rates of NADPH-dependent biosynthetic reactions are high in the cytosol, then the $NADP^+$-to-NADPH concentration ratio increases. This leads to allosteric activation of G6PD activity by $NADP^+$. This enzyme activation, in turn, increases flux through the pentose phosphate pathway to produce more NADPH. When the level of NADPH rises in the cell, it competes with $NADP^+$ for binding to G6PD, thereby reducing the activity of the enzyme. This results in decreased flux through the pentose phosphate pathway. The available glucose-6-P is then metabolized by the glycolytic pathway as a source of energy for the production of ATP. This makes sense because biosynthetic pathways require ATP, and when ATP levels drop (low energy charge in the cell), the demand for NADPH also diminishes, causing glucose-6-P to be shunted away from the pentose phosphate pathway and toward glycolysis.

Glucose-6-Phosphate Dehydrogenase Deficiency in Humans

In addition to its role in generating NADPH for biosynthetic pathways (primarily fatty acid and cholesterol biosynthesis in liver cells), the pentose phosphate pathway is also responsible for maintaining high levels of NADPH in red blood cells (erythrocytes). These cells use NADPH as a reductant in the glutathione reductase reaction (**Figure 14.8**). **Glutathione** is a tripeptide (γ-glutamylcysteinylglycine) that has a free

Figure 14.8 NADPH is required as a coenzyme in the glutathione reductase reaction to convert one molecule of oxidized glutathione (GSSG) to two molecules of reduced glutathione (GSH), which is a highly reactive electron donor in coupled redox reactions.

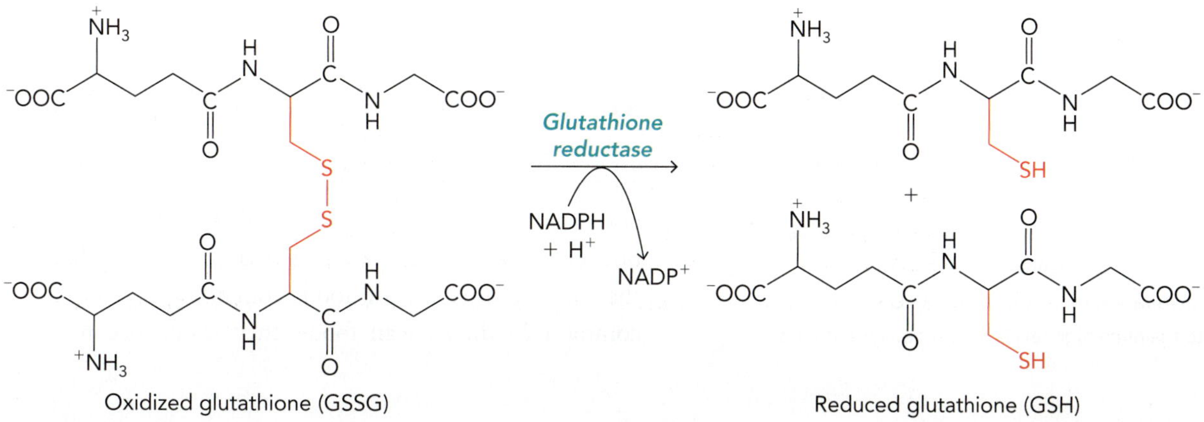

sulfhydryl group in the reduced form, which functions as an electron donor in a variety of coupled redox reactions in the cell. In erythrocytes, electrons from glutathione are used to keep cysteine residues in hemoglobin in the reduced state. They are also used for reducing harmful reactive oxygen species and hydroxyl free radicals that damage proteins and lipids through oxidation-induced cleavage reactions.

As shown in **Figure 14.9**, the enzyme glutathione peroxidase catalyzes a reduction reaction that neutralizes the damaging effects of H_2O_2, which accumulates in cells as a consequence of aerobic respiration and exposure to toxic compounds. This reduction reaction depends on the cell having sufficient levels of reduced glutathione, which in turn relies on the NADPH-dependent glutathione reductase reaction. Because G6PD is required to generate NADPH in the oxidative phase of the pentose phosphate pathway, defects in G6PD decrease the ability of cells to cope with oxidizing agents.

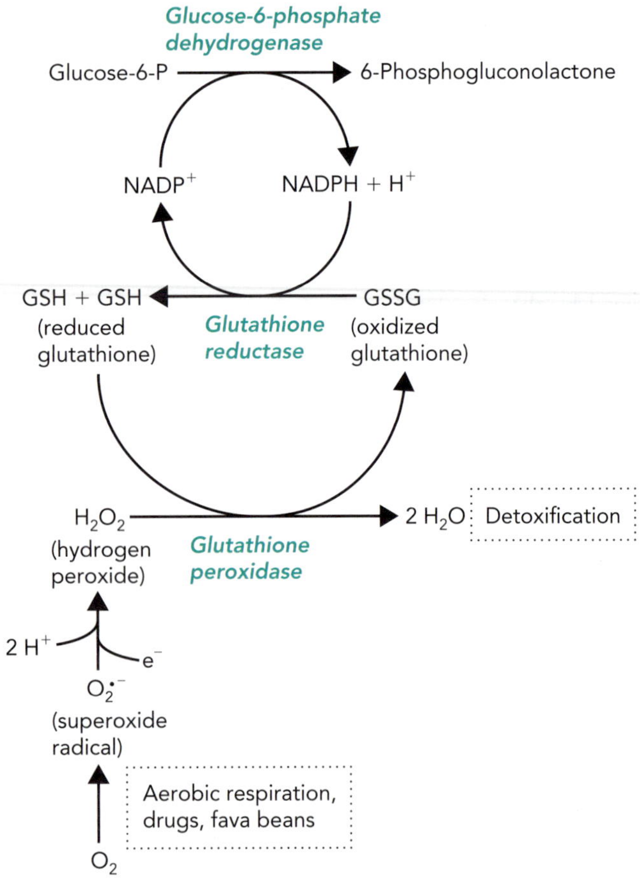

Figure 14.9 The buildup of harmful oxygen radicals in cells is prevented by detoxification reactions that require reduced glutathione and the enzyme glutathione peroxidase. The G6PD reaction in the pentose phosphate pathway is required to generate sufficient levels of NADPH in erythrocytes to maintain high levels of reduced glutathione.

The discovery in the mid-1950s of G6PD deficiency came as a result of observations made 30 years earlier when it was noticed that the antimalarial drug **primaquine** induced acute hemolytic anemia (red blood cell lysis) in a small percentage of patients who were taking the drug as a prophylactic antimalarial. The biochemical basis for this drug-induced illness was found to be lower-than-normal levels of NADPH in erythrocytes due to a G6PD deficiency. People with G6PD deficiency cannot tolerate primaquine because their erythrocytes do not contain enough reduced glutathione (GSH) to detoxify the reactive oxygen species produced by the antimalarial compound. In fact, the reason primaquine works as an antimalarial drug is that productive infection by the mosquito-borne *Plasmodium* pathogen is inhibited in erythrocytes by oxidative stress. Notably, it has been found that people who inherit the G6PD mutation have a lower incidence of malarial infection. The explanation is that reduced levels of NADPH, and the associated increase in oxidative stress in erythrocytes (coming from normal biochemical processes in the cell), creates a hostile environment for the malarial pathogen. This is analogous to how the hemoglobin S gene defect (HbS; see Figure 6.33), which causes sickle cell anemia, affords protection against malaria owing to the reduced ability of the pathogen to infect HbS-containing cells.

The finding that people with G6PD deficiency are for the most part asymptomatic (show no signs of illness), but can become gravely ill when given primaquine, led to the realization that another mysterious illness called **favism** was also caused by the same enzyme defect. As far back as the 6th century B.C., in the time of Pythagoras, it was observed that if certain people ate foods containing fava beans—a common Mediterranean food—they would become very

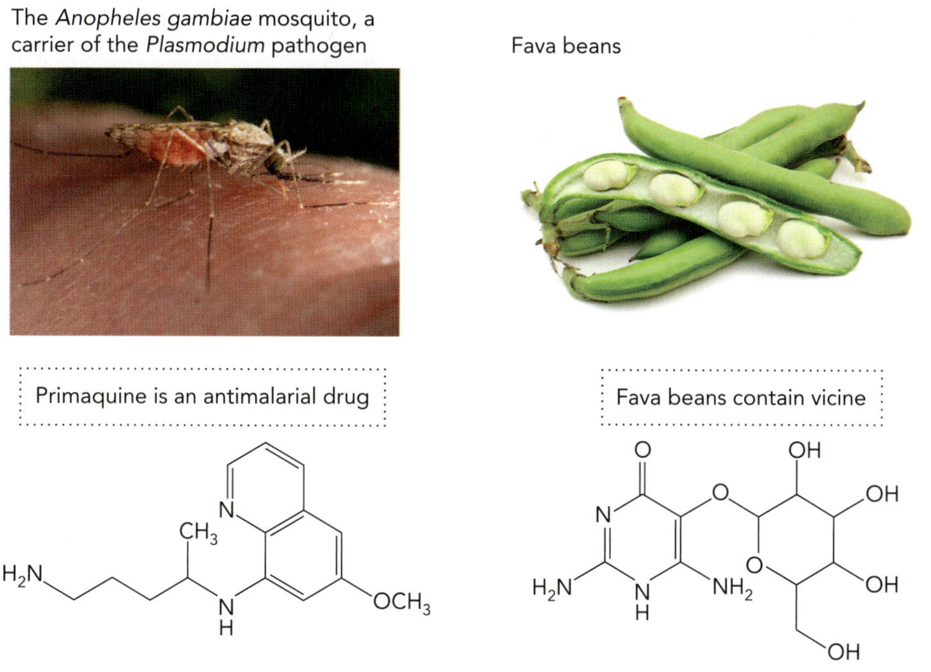

Figure 14.10 Individuals with defects in G6PD enzyme activity are less able to detoxify oxygen radicals that result from the metabolism of primaquine, an antimalarial drug, or vicine, a glycoside contained in fava beans. *ANOPHELES GAMBIAE* MOSQUITO: CDC/JAMES GATHANY; FAVA BEANS: BINH THANH BUI/SHUTTERSTOCK.

sick. It is now known that the same acute hemolytic anemia seen in individuals with G6PD deficiency who are treated with primaquine (or any other drug that induces oxidative stress in erythrocytes) also explains the symptoms of favism. One of the active compounds in fava beans is called **vicine,** a toxic glycoside that induces oxidative stress in erythrocytes (**Figure 14.10**).

concept integration 14.1

Explain why boys in Africa with a G6PD deficiency have a lower incidence of severe malaria than almost all other children in the same village.

A productive infection by the mosquito-borne malarial pathogen *Plasmodium* requires that erythrocytes have high levels of NADPH to control oxidative stress, which would otherwise inhibit the *Plasmodium* life cycle. Individuals with a G6PD deficiency have naturally reduced levels of NADPH in their erythrocytes and are more resistant to malarial infection than normal individuals because of higher intracellular levels of reactive oxygen species. This observation is consistent with the effects of the antimalarial drug primaquine on individuals with a G6PD deficiency—primaquine is known to induce oxidative stress in erythrocytes. Because the G6PD gene is carried on the X chromosome, G6PD deficiency is more prevalent in males than in females because only a single copy of the mutant G6PD gene need be inherited. Therefore, malaria resistance in boys with G6PD deficiency is more common than in most other children under the same conditions of exposure to malaria-infected mosquitoes.

14.2 Gluconeogenesis

We have seen that glucose is the primary chemical energy source for most nonphotosynthetic organisms (and for plants at night) and therefore must be readily available at all times. When dietary sources of glucose are insufficient and glucose stores have been depleted, cells synthesize glucose from noncarbohydrate compounds by a series of cytosolic reactions in an anabolic pathway called **gluconeogenesis.** Most animal tissues use a combination of glucose and fatty acids as a source of metabolic energy, but the

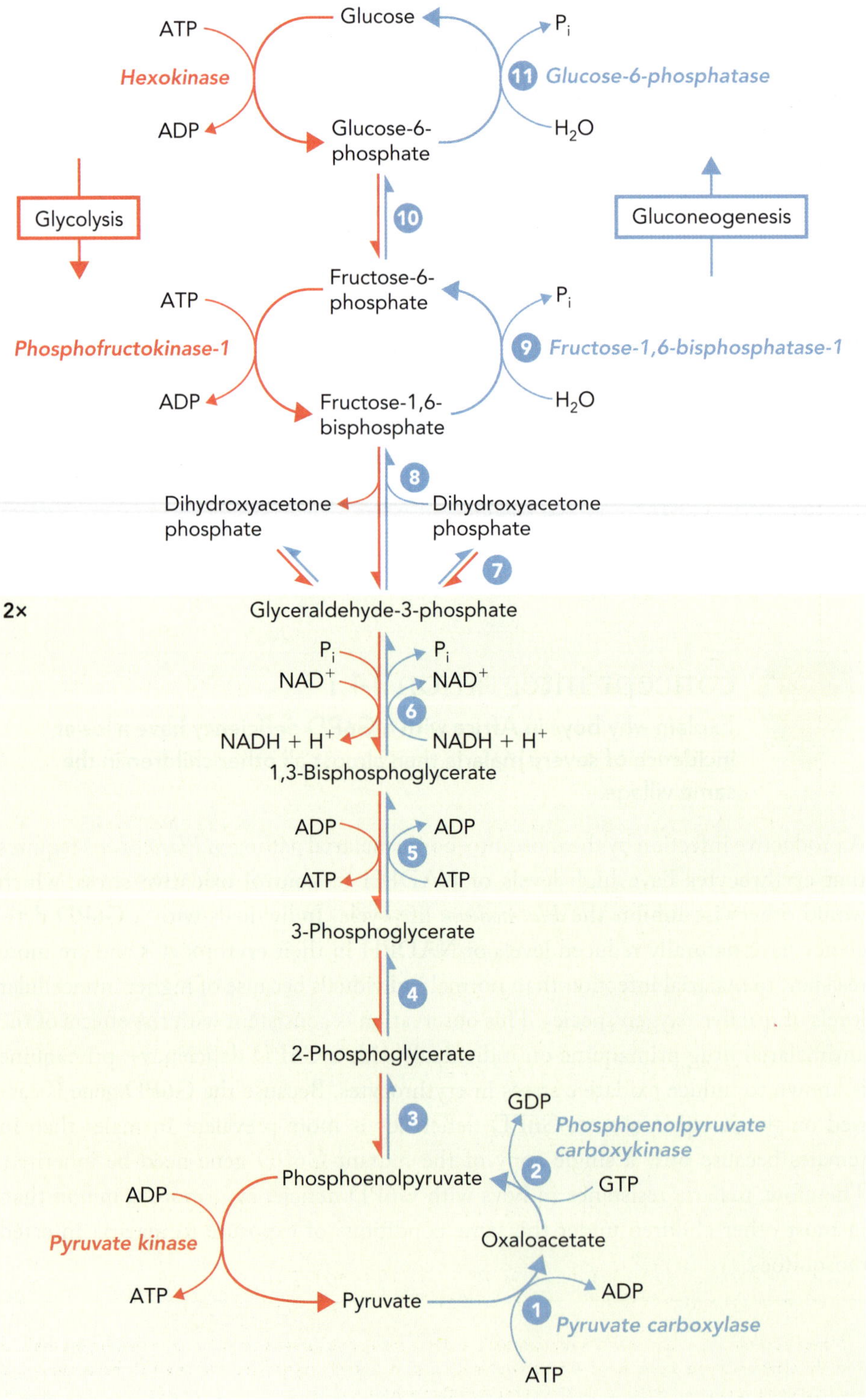

Figure 14.11 Three exergonic reactions in glycolysis (shown with enzyme names in red) are bypassed by four enzymes in the gluconeogenic pathway (shown with enzyme names in blue). The first bypass reaction involves the conversion of pyruvate to phosphoenolpyruvate, which requires pyruvate carboxylase, a mitochondrial enzyme, and phosphoenolpyruvate carboxykinase. The second bypass reaction is catalyzed by the enzyme fructose-1,6-bisphosphatase-1 and the third by glucose-6-phosphatase. Molar equivalents of the reactions shaded in green are doubled; that is, 2 pyruvate → 2 glyceraldehyde-3-P. The number labels refer to each enzymatic step, as shown in Table 14.1.

brain is totally dependent on glucose for energy under normal conditions. In humans, much of the glucose used by the brain and erythrocytes comes from gluconeogenesis occurring in liver and kidney cells when glycogen reserves are depleted.

Gluconeogenesis Uses Noncarbohydrate Sources to Synthesize Glucose

Gluconeogenesis and glycolysis share seven enzymes, each of which catalyzes a reversible reaction. The direction of each reaction is a response to the relative concentrations of shared metabolites (**Figure 14.11**). The two pathways differ significantly, however, by having pathway-specific enzymes that catalyze reactions at each of three metabolic "valves" responsible for controlling pathway flux. The three glycolytic-specific enzymes are hexokinase, phosphofructokinase-1 (PFK-1), and pyruvate kinase. The three enzymatic reactions catalyzed by these enzymes have standard free energies that are highly favorable for glycolysis ($\Delta G^{\circ\prime} \ll 0$ kJ/mol) but unfavorable ($\Delta G^{\circ\prime} \gg 0$ kJ/mol) for gluconeogenesis. Therefore, the conversion of pyruvate to glucose by the gluconeogenic pathway cannot simply be the reverse of glycolysis. Instead, it requires four separate enzymes that have evolved to catalyze "bypass" reactions and avoid the glycolysis-favored reactions. **Table 14.1** lists the 11 reactions in the gluconeogenic pathway and

Table 14.1 COMPARISON OF ENZYMES IN THE GLUCONEOGENIC AND GLYCOLYTIC PATHWAYS

	Reaction	Gluconeogenic enzyme	Glycolytic enzyme	Reaction type
	1. Pyruvate + CO_2 + ATP + $H_2O \rightarrow$ Oxaloacetate + ADP + P_i + 2 H^+	Pyruvate carboxylase	Pyruvate kinase	Bypass reaction(s)
	2. Oxaloacetate + GTP $\rightarrow$ Phosphoenolpyruvate + CO_2 + GDP	Phosphoenolpyruvate carboxykinase		
	3. Phosphoenolpyruvate + $H_2O \rightleftharpoons$ 2-Phosphoglycerate	Enolase	Enolase	Shared reactions
2×	4. 2-Phosphoglycerate $\rightleftharpoons$ 3-Phosphoglycerate	Phosphoglycerate mutase	Phosphoglycerate mutase	
	5. 3-Phosphoglycerate + ATP $\rightleftharpoons$ 1,3-Bisphosphoglycerate + ADP	Phosphoglycerate kinase	Phosphoglycerate kinase	
	6. 1,3-Bisphosphoglycerate + NADH + $H^+ \rightleftharpoons$ Glyceraldehyde-3-P + ADP	Glyceraldehyde-3-phosphate dehydrogenase	Glyceraldehyde-3-phosphate dehydrogenase	
	7. Glyceraldehyde-3-P $\rightleftharpoons$ Dihydroxyacetone phosphate	Triose phosphate isomerase	Triose phosphate isomerase	
	8. Glyceraldehyde-3-P + Dihydroxyacetone phosphate $\rightleftharpoons$ Fructose-1,6-BP	Aldolase	Aldolase	
	9. Fructose-1,6-BP + $H_2O \rightarrow$ Fructose-6-P + P_i	Fructose-1,6-bisphosphatase-1	Phosphofructokinase-1	Bypass reaction
	10. Fructose-6-P $\rightleftharpoons$ Glucose-6-P	Phosphoglucose isomerase	Phosphoglucose isomerase	Shared reaction
	11. Glucose-6-P + $H_2O \rightarrow$ Glucose + P_i	Glucose-6-phosphatase	Hexokinase	Bypass reaction
	Net: 2 Pyruvate + 2 NADH + 4 ATP + 2 GTP + 6 $H_2O \rightarrow$ Glucose + 2 NAD^+ + 2 H^+ + 4 ADP + 2 GDP + 6 P_i			

Note: Reactions 1, 2, 9, and 11 are highly exergonic and referred to as bypass reactions because they bypass the analogous reverse reactions in glycolysis. Seven of the eleven enzymes in gluconeogenesis are shared enzymes with glycolysis. The "2×" refers to the fact there are 2 mol of each reactant in this portion of the gluconeogenic pathway, which is not shown in the written reactions.

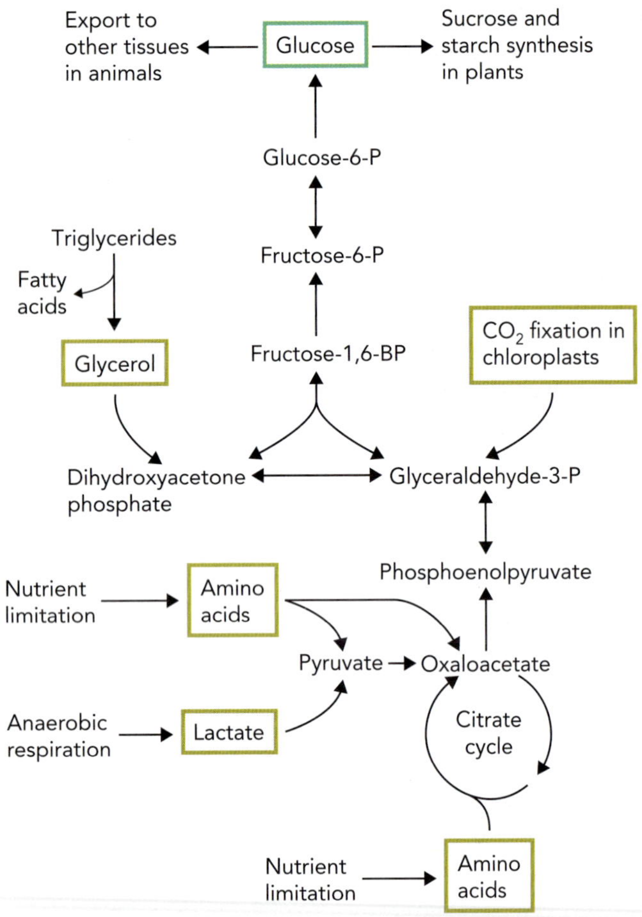

Figure 14.12 The four major sources of carbon for glucose synthesis by the gluconeogenic pathway are lactate, amino acids, glycerol, and CO_2 fixation (in plants). Gluconeogenesis is most active in animals when glycogen stores in the liver and muscle have been depleted. Triglycerides are continually converted into glycerol, which can enter the gluconeogenic pathway after conversion to dihydroxyacetone phosphate. Nutrient limitation increases the conversion of amino acids into pyruvate or citrate cycle metabolites. Anaerobic respiration increases the pool of lactate, which can be converted into pyruvate for entry into the gluconeogenic pathway. Plants use gluconeogenesis to synthesize glucose, which is used to produce sucrose and starch for energy storage.

compares them to the same or different enzymes catalyzing analogous reactions in the glycolytic pathway.

Two of the gluconeogenic-specific enzymes are pyruvate carboxylase and phosphoenolpyruvate carboxykinase. Together they bypass the pyruvate kinase reaction by converting pyruvate to phosphoenolpyruvate through the formation of oxaloacetate. These two reactions require the input of energy in the form of ATP and GTP hydrolysis and take place in two different cellular compartments. The second bypass is a reaction catalyzed by fructose-1,6-bisphosphatase-1 (FBPase-1), which converts fructose-1,6-bisphosphate to fructose-6-phosphate using energy from phosphate hydrolysis to drive the reaction. This enzyme is highly regulated by the allosteric effectors fructose-2,6-bisphosphate (fructose-2,6-BP) and AMP, both of which also regulate the activity of the corresponding glycolytic enzyme phosphofructokinase-1. The third bypass reaction in gluconeogenesis is catalyzed by the enzyme glucose-6-phosphatase. This enzyme converts glucose-6-phosphate to glucose, again using energy available from phosphate hydrolysis. Glucose-6-phosphatase is also an important enzyme in liver glycogen metabolism because it is required to dephosphorylate glucose-6-phosphate before glucose can be exported to other tissues. Muscle cells do not contain glucose-6-phosphatase, which ensures that muscle cells use the glucose-6-phosphate as a source of energy for muscle contraction. We will describe each of these bypass reactions in more detail in the next subsection.

As shown in **Figure 14.12**, when flux through the gluconeogenic pathway in the liver is stimulated to produce glucose for export to other tissues, the carbon required for glucose synthesis comes from lactate and amino acids that are converted into pyruvate or from the generation of oxaloacetate directly from amino acids. Carbon for gluconeogenesis can also be obtained from glycerol, which is one of the products of triglyceride catabolism (see Chapter 16). Glycerol can be converted into dihydroxyacetone phosphate through a two-step reaction involving the enzymes glycerol kinase and glycerol-3-phosphate dehydrogenase (see Figure 9.48). Finally, plants use the Calvin cycle to convert CO_2 into glyceraldehyde-3-P, which then enters the gluconeogenic pathway to produce glucose for synthesis of sucrose and starch.

With this overview of gluconeogenesis, we are now ready to answer our four key questions about this important biosynthetic pathway:

1. *What does gluconeogenesis accomplish for the organism?* The liver and kidneys use gluconeogenesis to generate glucose from noncarbohydrate sources (lactate, amino acids, glycerol) for export to other tissues that depend on glucose for energy—primarily the brain and erythrocytes. Plants use the gluconeogenic pathway to convert glyceraldehyde-3-P (the product of the Calvin cycle) into glucose, which is used to make sucrose and starch.
2. *What is the overall net reaction of gluconeogenesis?*

$$2\ \text{Pyruvate} + 2\ \text{NADH} + 4\ \text{ATP} + 2\ \text{GTP} + 6\ H_2O \rightarrow \text{Glucose} + 2\ \text{NAD}^+ + 2\ H^+ + 4\ \text{ADP} + 2\ \text{GDP} + 6\ P_i$$

3. *What are the key enzymes in gluconeogenesis?* The mitochondrial enzyme **pyruvate carboxylase** catalyzes a carboxylation reaction converting pyruvate to oxaloacetate using a reaction mechanism involving a biotinyl "swinging arm" and ATP hydrolysis. The enzyme **phosphoenolpyruvate carboxykinase**, localized to either the mitochondrial matrix or the cytosol (or both in the case of human liver cells), catalyzes a phosphoryl transfer reaction that converts oxaloacetate to phosphoenolpyruvate using the energy released by decarboxylation and GTP hydrolysis. **Fructose-1,6-bisphosphatase-1** catalyzes the dephosphorylation of fructose-1,6-BP to form fructose-6-P. Fructose-1,6-bisphosphatase-1 is inhibited by the allosteric regulators fructose-2,6-BP and AMP, which are also allosteric activators of PFK-1. **Glucose-6-phosphatase**, located in liver and kidney cells (not present in muscle cells), catalyzes the dephosphorylation of glucose-6-P to form glucose, which can be exported out of the cell. Glucose-6-phosphatase is located in the lumen of the endoplasmic reticulum.

4. *What are examples of gluconeogenesis in everyday biochemistry?* Athletes who exercise intensely for short periods of time, such as in a sprint race, build up large amounts of lactate in their muscles as a result of anaerobic glycolysis. These athletes do a "warming down" period of continual movement under aerobic conditions after a race to increase circulation and remove lactate from the muscle. The lactate is transported to the liver, where it is converted to glucose by the gluconeogenic pathway and shipped back to the muscles to replenish glycogen. This overall process is called the Cori cycle.

Gluconeogenic Enzymes Bypass Three Exergonic Reactions in Glycolysis

The enzymes pyruvate carboxylase and phosphoenolpyruvate carboxykinase together catalyze the first bypass reaction in gluconeogenesis, which converts pyruvate to phosphoenolpyruvate. As shown in **Figure 14.13**, pyruvate carboxylase uses phosphoryl transfer energy available in ATP to drive a carboxylation reaction that converts pyruvate to oxaloacetate. The oxaloacetate, which functions as an activated intermediate, is then decarboxylated by phosphoenolpyruvate carboxykinase and converted to phosphoenolpyruvate, using GTP as the phosphate donor. Pyruvate carboxylase is a mitochondrial enzyme that is allosterically activated by acetyl-CoA (Figure 14.13). Recall from Chapter 10 that pyruvate carboxylase uses pyruvate to replenish oxaloacetate in the citrate cycle, maintaining high rates of metabolic flux when acetyl-CoA levels are elevated (see Figure 10.40). Because gluconeogenesis is an anabolic pathway requiring a net investment of 6 ATP equivalents (4 ATP and 2 GTP) for each glucose synthesized, it is not surprising that both pyruvate

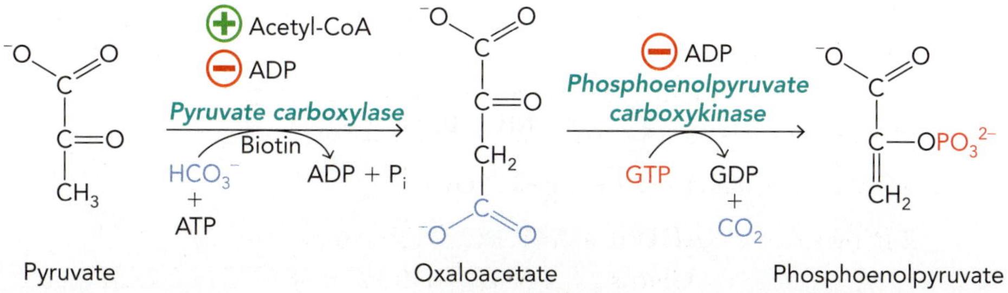

Figure 14.13 The conversion of pyruvate to phosphoenolypyruvate requires the two gluconeogenic enzymes pyruvate carboxylase and phosphoenolypyruvate carboxykinase. Both enzymes are allosterically inhibited by ADP, whereas pyruvate carboxylase is allosterically activated by acetyl-CoA.

carboxylase and phosphoenolpyruvate carboxykinase are allosterically inhibited by ADP (Figure 14.13). As described shortly, humans have both a cytosolic phosphoenolpyruvate carboxykinase and a mitochondrial phosphoenolpyruvate carboxykinase, which are differentially active in the gluconeogenic pathway, depending on the metabolic state of the cell.

Pyruvate carboxylase requires the cofactor **biotin** to function as a carboxyl group carrier. The biotin moiety is covalently linked to the ε amino group on a lysine residue in the enzyme active site, forming a biotinyllysine residue (**Figure 14.14**). Biotin contains four CH_2 groups, which—along with the four CH_2 groups in the lysine residue of the enzyme and the amide linkage—create a biotinylated "swinging arm" that is 14 Å long. The swinging arm in pyruvate carboxylase is not quite as long as the 50 Å lipoamide "ball and chain" in the pyruvate dehydrogenase complex, which swings between two enzyme subunits (see Figure 10.16). The swinging arm, however, is long enough to first position the biotin group in one region of the enzyme active site, where it is carboxylated in a reaction involving ATP hydrolysis. The arm then swings the carboxybiotin group across to the other side of the active site, where the CO_2 is released and subjected to nucleophilic attack by a pyruvate enolate intermediate, resulting in the formation of oxaloacetate. We will see a

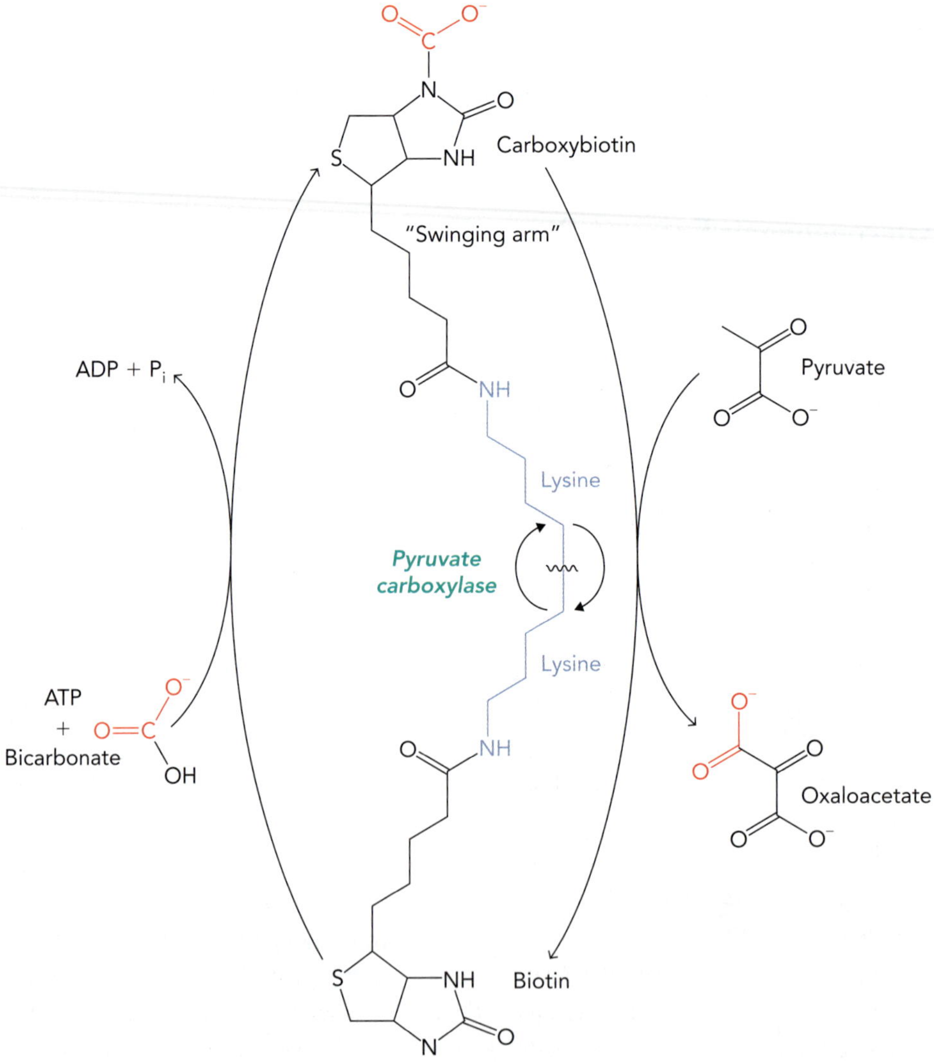

Figure 14.14 The biotinylated lysine group in pyruvate carboxylase functions as a swinging arm in the reaction to carry a carboxyl group from one region of the active site to another. Phosphotransfer energy from ATP is required in the first step of this reaction to carboxylate the biotin group of the enzyme.

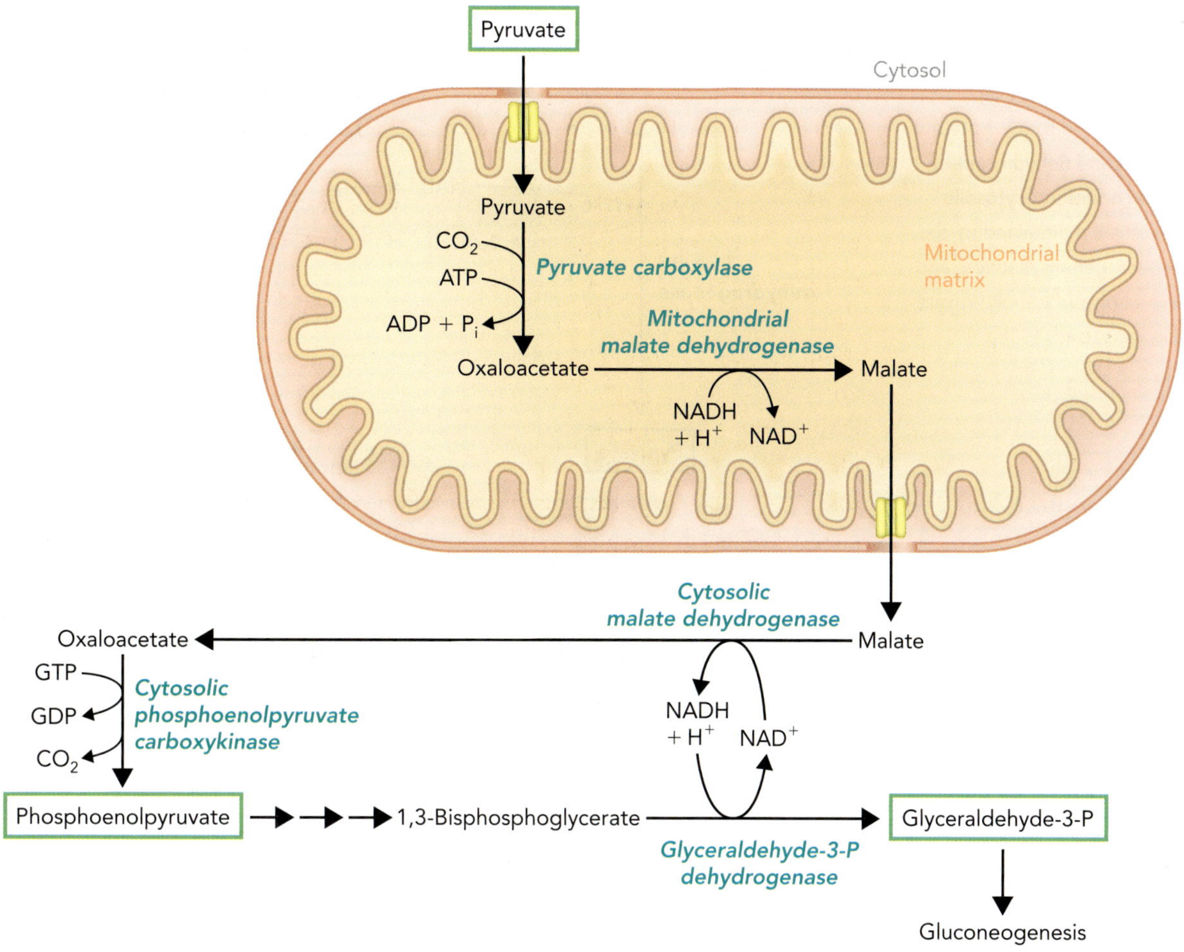

Figure 14.15 Malate transport from the mitochondrial matrix to the cytosol is used in human cells to increase cytosolic NADH levels under conditions when gluconeogenesis is favored. The net result of using cytosolic phosphoenolpyruvate carboxykinase to generate phosphoenolpyruvate is to supply NADH to the glyceraldehyde-3-phosphate dehydrogenase reaction and maintain flux through the gluconeogenic pathway.

similar biotin-mediated reaction mechanism in Chapter 16, when we describe the acetyl-CoA carboxylase reaction.

To maintain flux through the glyceraldehyde-3-phosphate dehydrogenase reaction in the gluconeogenic pathway, NADH equivalents need to be moved from the mitochondrial matrix to the cytosol. In humans, this is done by transporting malate from the mitochondrial matrix to the cytosol where it is converted to oxaloacetate and used by cytosolic phosphoenolpyruvate carboxykinase to generate phosphoenolpyruvate (**Figure 14.15**). Two isozymes of malate dehydrogenase, which catalyze opposing reactions in the mitochondrion and cytosol, are involved in this process. Oxidation of cytosolic malate by malate dehydrogenase results in the generation of NADH in the cytosol, where it is used by glyceraldehyde-3-phosphate dehydrogenase. This is the opposite of what happens when high rates of glycolysis are favored and cytosolic NAD^+ needs to be regenerated to maintain flux through the glyceraldehyde-3-phosphate dehydrogenase reaction (see Figure 9.52).

Figure 14.16 shows an alternative pathway from pyruvate to phosphoenolpyruvate that functions in humans when lactate builds up due to anaerobic metabolism in muscle cells. In this case, oxidation of lactate by lactate dehydrogenase in the cytosol generates pyruvate and the necessary NADH for the glyceraldehyde-3-phosphate dehydrogenase reaction, without requiring malate transport to the cytosol. The oxaloacetate produced in the mitochondria by pyruvate carboxylase can then be converted

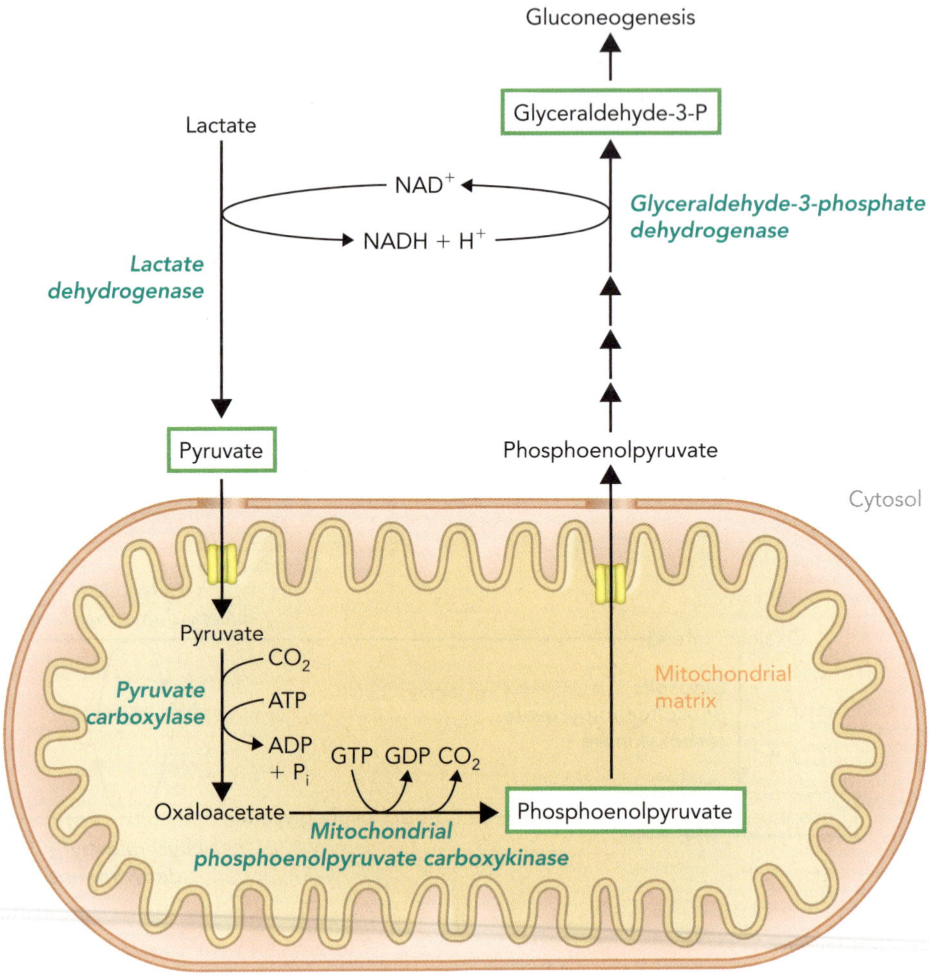

Figure 14.16 During vigorous exercise, lactate produced in muscle cells is transported to the liver, where it is converted to pyruvate by the enzyme lactate dehydrogenase. Under these conditions, cytosolic NADH levels are maintained by the lactate dehydrogenase reaction, and mitochondrial phosphoenolpyruvate carboxykinase is used.

to phosphoenolpyruvate by mitochondrial phosphoenolpyruvate carboxykinase. The phosphoenolpyruvate produced in the mitochondria is shuttled by a specific transport system to the cytosol, where it enters gluconeogenesis and is converted to glyceraldehyde-3-P. The use of mitochondrial phosphoenolpyruvate carboxykinase to facilitate the net conversion of two molecules of lactate to one molecule of glucose by gluconeogenesis in the liver is a way for the body to replenish muscle glycogen stores after vigorous exercise. This process, the Cori cycle, is described in more detail later in the chapter.

In glycolysis, the conversion of fructose-6-P to fructose-1,6-BP by the enzyme PFK-1 is considered the commitment step in the pathway and, as such, is highly regulated by allosteric effectors. The opposite reaction in gluconeogenesis is catalyzed by the enzyme FBPase-1, which represents the second bypass reaction in gluconeogenesis. The activities of PFK-1 and FBPase-1 are regulated by the allosteric effectors AMP and fructose-2,6-BP in a reciprocal manner (**Figure 14.17**). For example, when energy charge in the cell is low, AMP levels are high, leading to activation of PFK-1 (increased flux through glycolysis) and inhibition of FBPase-1 (decreased flux through gluconeogenesis). This makes sense because the pyruvate generated by glycolysis can then be used in the energy conversion pathways to replenish ATP, while at the same time glucose synthesis is shut down, resulting in a buildup of pyruvate. The allosteric regulator fructose-2,6-BP is an even more potent regulator of these two enzymes than AMP, as their activities are dramatically altered when fructose-2,6-BP levels increase. (The

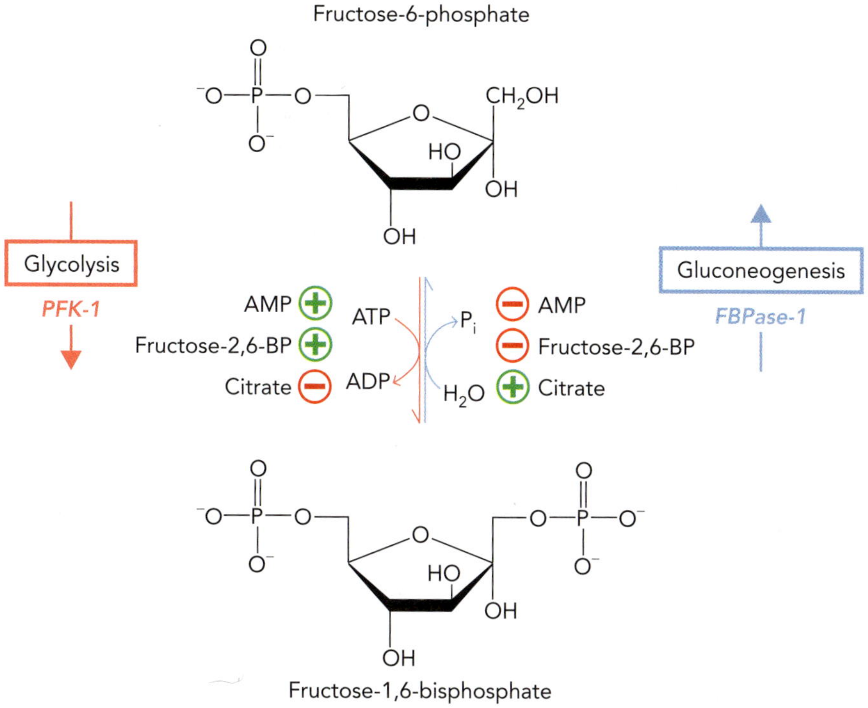

Figure 14.17 PFK-1 and FBPase-1 are reciprocally regulated by AMP, fructose-2,6-BP, and citrate in response to energy needs of the cell. AMP and fructose-2,6-BP allosterically activate PFK-1 and inhibit FBPase-1 in response to a low energy charge in the cell. In contrast, excess citrate in the cytosol signals decreased flux through the citrate cycle, in which case, PFK-1 activity is inhibited and FBPase-1 activity is stimulated.

reciprocal regulation of glycolysis and gluconeogenesis by fructose-2,6-BP is described in detail in the next subsection.)

A third metabolite that reciprocally regulates PFK-1 and FBPase-1 is citrate; however, citrate activates FBPase-1 and inhibits PFK-1. Increased levels of citrate in the cytosol indicate that the citrate cycle is backed up, causing excess citrate to be shuttled out of the mitochondrial matrix. By inhibiting PFK-1 and activating FBPase-1, citrate is able to redirect pyruvate away from the pyruvate dehydrogenase reaction and toward pyruvate carboxylase, thereby decreasing flux through the citrate cycle.

The third and final bypass reaction in gluconeogenesis is the dephosphorylation of glucose-6-P by glucose-6-phosphatase. This reaction generates free glucose that can be exported to the blood. The opposite reaction in glycolysis is the phosphorylation of glucose by hexokinase (see Figure 14.11). The catalytic activities of glucose-6-phosphatase and hexokinase are not reciprocally regulated by allosteric effectors. Instead, the two enzymes are located in different cellular compartments, which prevents futile cycling between glucose and glucose-6-P at the expense of ATP hydrolysis. As shown in **Figure 14.18**, glucose-6-phosphatase is localized to the lumen of the endoplasmic reticulum (ER) in liver cells, whereas hexokinase is in the cytosol. Glucose-6-P is transported into the ER lumen if cytosolic levels are high, which happens when flux through gluconeogenesis is favored over glycolysis or when glycogen degradation is generating large amounts of glucose-6-P. The glucose and P_i produced by glucose-6-phosphatase are exported out of the ER by separate transporter proteins, and then glucose is exported to the blood by the glucose transporter type 2 (GLUT2; see Figure 14.18). Glucose-6-phosphatase is present in liver and kidney cells, which have high rates of gluconeogenesis, but it is absent from muscle cells. The absence of glucose-6-phosphatase in muscle cells prevents glucose export and thereby maintains a readily available supply of glucose for energy conversion pathways in these cells.

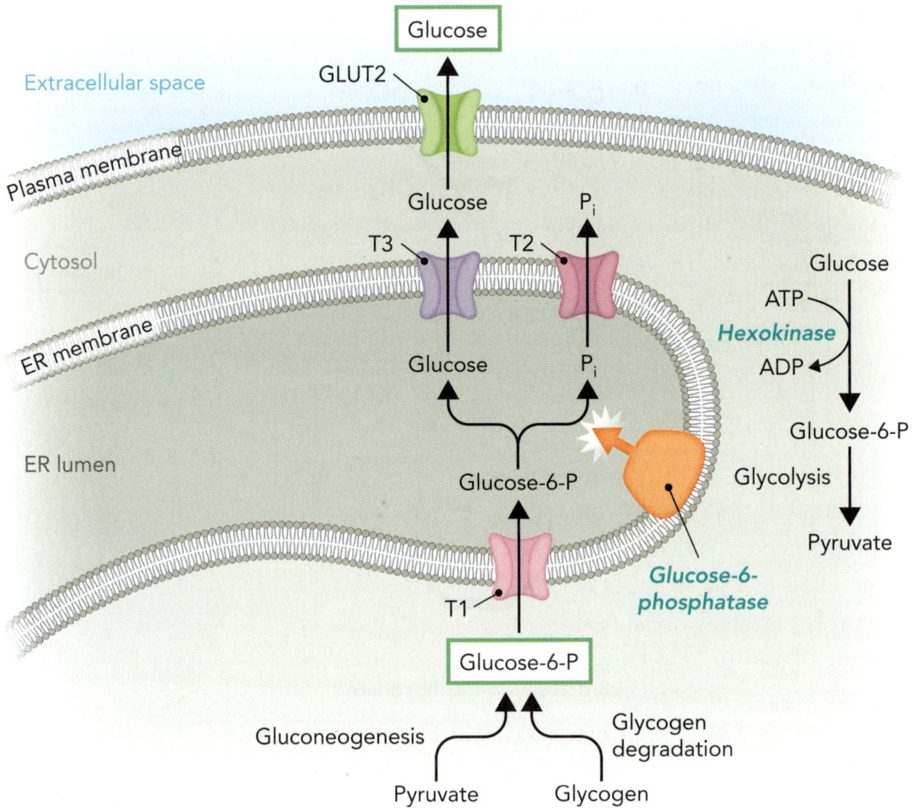

Figure 14.18 Hexokinase is a cytoplasmic enzyme, and glucose-6-phosphatase is localized to the lumen of the ER. Three different transporter proteins in the ER membrane (T1, T2, T3) translocate glucose-6-P, P_i, and glucose, respectively. The GLUT2 glucose transport protein in the plasma membrane of liver cells permits glucose exchange between the inside and outside of the cell in response to blood glucose concentrations.

Reciprocal Regulation of Gluconeogenesis and Glycolysis by Allosteric Effectors

Metabolic flux through gluconeogenesis and glycolysis is controlled by allosteric regulation of several key enzymes in each of the two pathways. As shown in **Figure 14.19**, the activities of PFK-1 and FBPase-1 are differentially regulated by energy charge in the cell. Specifically, high energy charge (high ATP levels) inhibits PFK-1 activity, whereas low energy charge (high AMP levels) inhibits FBPase-1 activity. PFK-1 and FBPase-1 are also reciprocally regulated by citrate and **fructose-2,6-bisphosphate (fructose-2,6-BP)**. Figure 14.19 also shows that pyruvate kinase is inhibited by ATP and alanine, which decreases flux through glycolysis, whereas acetyl-CoA stimulates pyruvate carboxylase activity, resulting in increased flux through gluconeogenesis. Note that although hexokinase and glucose-6-phosphatase are not subject to these same types of allosteric control mechanisms, their physical separation into two cellular compartments provides a means to control the activity of glucose-6-phosphatase when glucose-6-P levels are elevated (see Figure 14.18).

Reciprocal regulation of PFK-1 and FBPase-1 in response to AMP and citrate levels indicates metabolite control of glycolysis and gluconeogenesis as a function of energy charge in the cell. However, the activities of PFK-1 and FBPase-1 are also regulated in liver cells by glucagon and insulin signaling as a means to control blood glucose levels. The intracellular regulator that transmits the hormone signal in liver cells is fructose-2,6-BP. This potent effector molecule is structurally related to fructose-6-P and fructose-1,6-BP but is not a metabolic intermediate in either the glycolytic or gluconeogenic pathways. As shown in **Figure 14.20**, the apparent affinity of PFK-1 for its substrate fructose-6-P is 25 times higher when fructose-2,6-BP levels are elevated. In contrast, the apparent affinity of FBPase-1

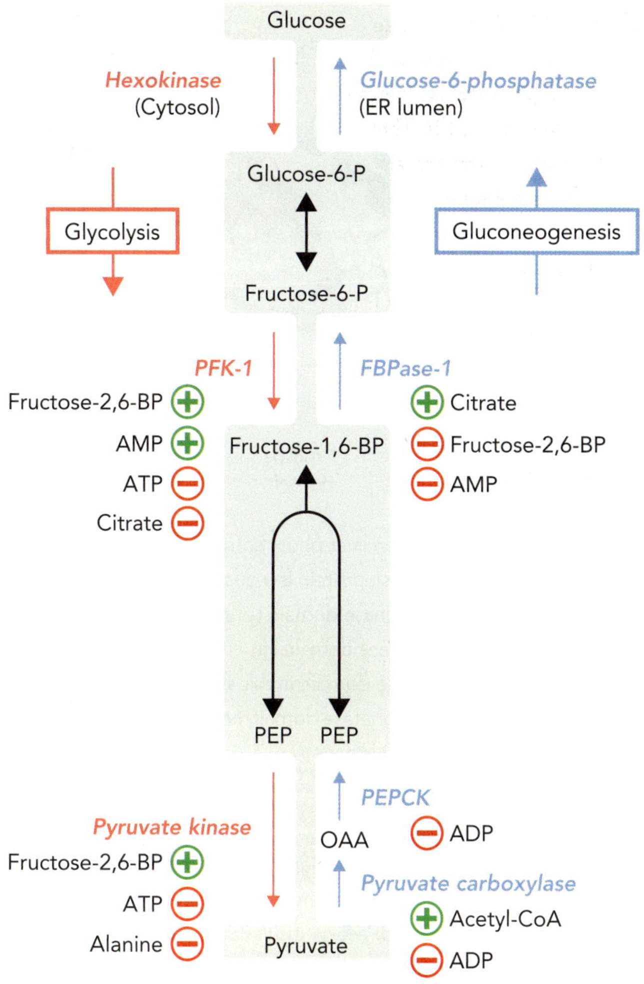

Figure 14.19 The metabolic control points in gluconeogenesis and glycolysis are summarized here. A low energy charge in the cell (high AMP and ADP levels) increases flux through the glycolytic pathway through stimulation of PFK-1 activity and inhibition of the gluconeogenic enzymes pyruvate carboxylase, phosphoenolpyruvate carboxykinase (PEPCK), and FBPase-1. In contrast, when metabolites in the citrate cycle accumulate, excess citrate and acetyl-CoA stimulate gluconeogenesis by activating pyruvate carboxylase and FBPase-1. Citrate and ATP also inhibit PFK-1 and pyruvate kinase to decrease flux through the glycolytic pathway. OAA = oxaloacetate; PEP = phosphoenolpyruvate.

for its substrate fructose-1,6-BP is 15 times lower under these same conditions. The net effect is that high levels of fructose-2,6-BP stimulate flux through the glycolytic pathway and inhibit flux through the gluconeogenic pathway. If you look at the shape of the enzyme activity curves in Figure 14.20 when fructose-2,6-BP levels are low, you can see that the opposite is also true; namely, decreased fructose-2,6-BP levels increase flux through the gluconeogenic pathway and at the same time decrease flux through the glycolytic pathway.

The amount of fructose-2,6-BP in the cell is regulated by hormone signaling as a means of mediating flux through the glycolytic and gluconeogenic pathways. Insulin, which signals high levels of blood glucose, increases production

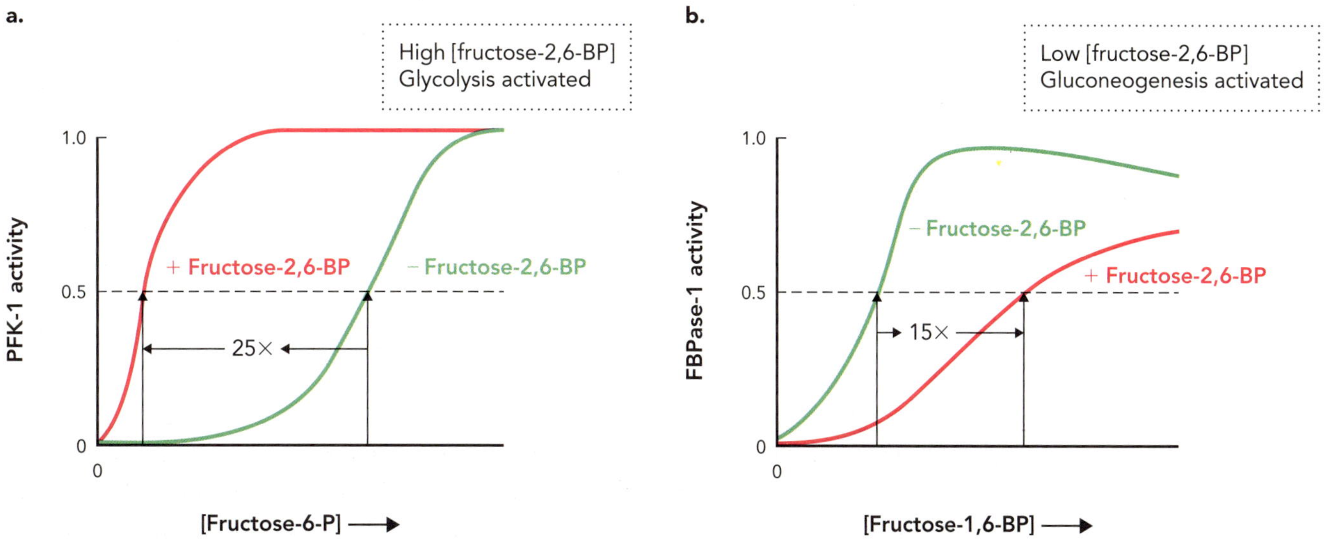

Figure 14.20 Fructose-2,6-BP is a potent allosteric regulator of PFK-1 and FBPase-1 activity. **a.** Activity of PFK-1 increases in the presence of fructose-2,6-BP, as shown by the curves of enzyme activity versus substrate concentration in the presence (red) and absence (green) of fructose-2,6-BP. In the presence of fructose-2,6-BP, half-maximal activity (dashed line) of PFK-1 is achieved at 25-fold lower substrate concentrations, reflecting the increase in apparent affinity of PFK-1 for substrate. **b.** In contrast, addition of fructose-2,6-BP to enzyme reactions containing FBPase-1 decreases its apparent affinity for fructose-1,6-BP by 15-fold, thereby inhibiting the FBPase-1 enzyme at low substrate concentrations.

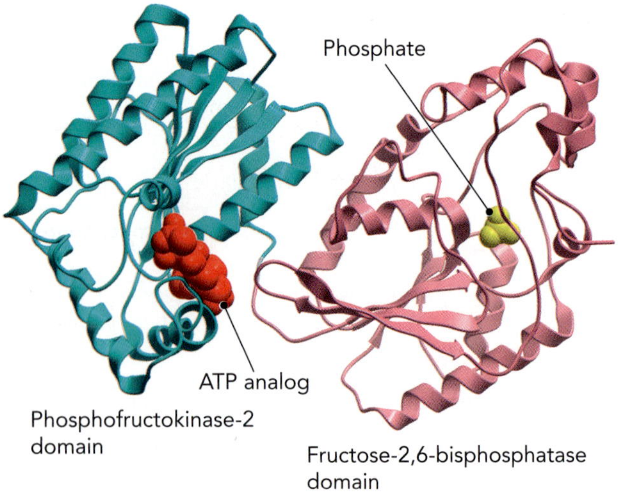

Figure 14.21 The human liver phosphofructokinase-2/fructose-2,6-bisphosphatase protein is a dual-function enzyme containing an N-terminal kinase domain (phosphofructokinase-2) and a C-terminal phosphatase domain (fructose-2,6-bisphosphatase). The kinase domain in this structure contains an ATP analog, and the phosphatase domain has a phosphate bound to the catalytic site. BASED ON PDB FILE 1K6M.

of fructose-2,6-BP levels. Conversely, glucagon signaling indicates low levels of blood glucose and decreases levels of fructose-2,6-BP. The mechanism by which insulin and glucagon control fructose-2,6-BP levels in liver cells is through modulating the activity of a dual-function enzyme called **phosphofructokinase-2/fructose-2,6-bisphosphatase (PFK-2/FBPase-2)**. As shown in **Figure 14.21**, this ~50-kDa enzyme includes two functional domains, each with its own catalytic activity: (1) a kinase activity that phosphorylates fructose-6-P to form fructose-2,6-BP; and (2) a phosphatase activity that dephosphorylates fructose-2,6-BP to form fructose-6-P. The kinase activity is localized to the N-terminal region and is called phosphofructokinase-2 (PFK-2), whereas the phosphatase activity is in the C-terminal half of the protein and is called fructose-2,6-bisphosphatase (FBPase-2).

The dual activities of PFK-2/FBPase-2 are controlled by phosphorylation, which in turn determines the intracellular level of fructose-2,6-BP. As illustrated in **Figure 14.22**, in the presence of insulin, the Ser32 residue on PFK-2/FBPase-2 is dephosphorylated by the enzyme protein phosphatase 1. This results in stimulation of the PFK-2 activity. However, when glucagon signaling is stimulated by low blood glucose levels, protein kinase A phosphorylates Ser32, which stimulates FBPase-2 activity. The effect of these two PFK-2/FBPase-2 activity states on fructose-2,6-BP production is that the PFK-2 domain phosphorylates fructose-6-P to generate fructose-2,6-BP, whereas the FBPase-2 domain dephosphorylates fructose-2,6-BP to generate fructose-6-P. As described earlier, fructose-2,6-BP reciprocally regulates PFK-1 and FBPase-1 activity (see Figure 14.20), and therefore, hormonal

Figure 14.22 Hormonal regulation of glycolysis and gluconeogenesis depends on the phosphorylation state of the dual-function enzyme PFK-2/FBPase-2. Insulin signaling stimulates flux through the glycolytic pathway by increasing levels of fructose-2,6-BP through dephosphorylation of Ser32 in the N-terminal domain and activation of the PFK-2 catalytic activity. Glucagon signaling stimulates flux through the gluconeogenic pathway by decreasing fructose-2,6-BP levels through phosphorylation and activation of the FBPase-2 catalytic activity. The asterisk denotes the active domain.

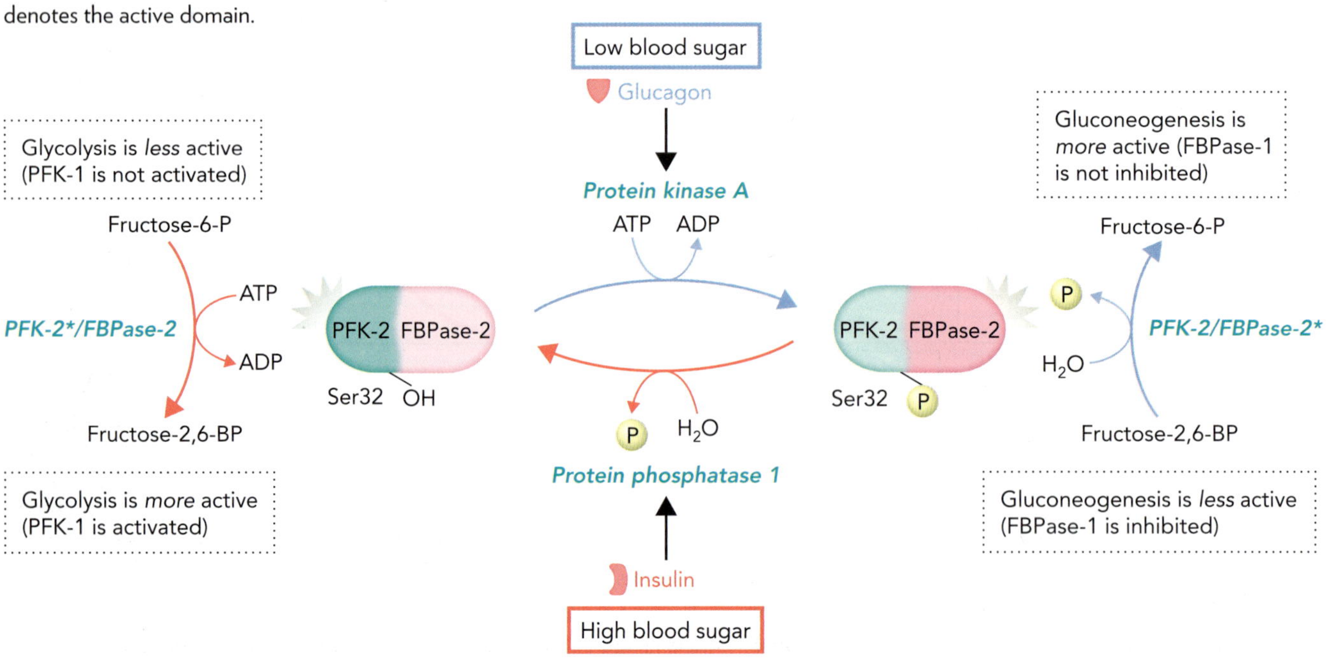

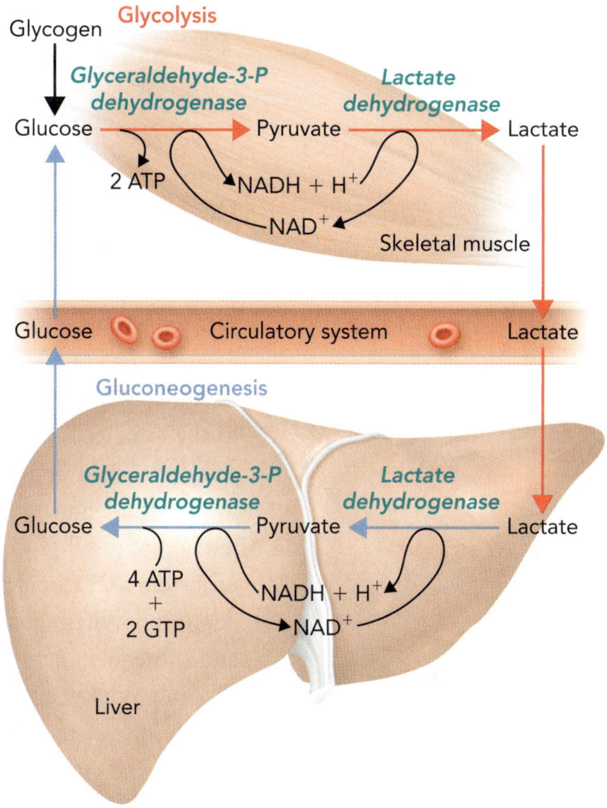

Figure 14.23 The Cori cycle provides a way for lactate produced by anaerobic glycolysis in the muscle cells to be converted to glucose through gluconeogenesis in liver cells, at a cost of net 4 ATP equivalents. The NAD^+/NADH coenzyme linkage between the glyceraldehyde-3-phosphate dehydrogenase and lactate dehydrogenase enzymes is maintained because these two reactions are reversible.

control of fructose-2,6-BP production in liver cells directly affects metabolic flux through the glycolytic and gluconeogenic pathways.

The Cori Cycle Provides Glucose to Muscle Cells during Exercise

As we have just seen, regulating metabolic flux through glycolysis and gluconeogenesis in the same cell requires reciprocal regulation of opposing enzymes to avoid futile cycling (that is, ATP hydrolysis in the absence of net chemical work). However, in one situation, having both pathways active at the same time, but in different cells, can be quite advantageous. The organism manages this feat through the Cori cycle.

The **Cori cycle** provides a mechanism to convert lactate produced by anaerobic glycolysis in muscle cells to glucose using the gluconeogenic pathway in liver cells (**Figure 14.23**). This glucose is then transported back to muscle cells to maintain anaerobic glycolysis and to replenish glycogen stores. Although it costs four high-energy phosphate bonds to run the Cori cycle (the difference between 2 ATP produced by anaerobic glycolysis and 4 ATP + 2 GTP consumed by gluconeogenesis), the benefit to the organism is that glycogen stores in the muscle can be quickly replenished after prolonged exercise. Most important, the reduction of pyruvate to lactate in muscle cells by the lactate dehydrogenase reaction regenerates NAD^+ for glycolysis, whereas the oxidation of lactate to pyruvate by lactate dehydrogenase in liver cells provides the NADH needed for gluconeogenesis.

The Cori cycle was elucidated by Carl and Gerty Cori in 1929 while both were in the Biochemistry Department at Washington University in St. Louis. Considered two of the most influential carbohydrate biochemists in the 20th century, they were awarded part of the 1947 Nobel Prize in Physiology or Medicine. Amazingly, six biochemists who were affiliated with the Cori laboratory at Washington University over the years went on to win Nobel Prizes for research they performed as independent investigators: Arthur Kornberg, Severo Ochoa, Luis F. Leloir, Edwin G. Krebs, Earl W. Sutherland, and Christian de Duve (**Figure 14.24**).

Gerty Cori and Carl Cori
1947 Nobel Prize
in Physiology or Medicine

Severo Ochoa and Arthur Kornberg
1959 Nobel Prize
in Physiology or Medicine

Luis F. Leloir
1970 Nobel Prize
in Chemistry

Earl W. Sutherland
1971 Nobel Prize
in Physiology
or Medicine

Christian de Duve
1974 Nobel Prize
in Physiology
or Medicine

Edwin G. Krebs
1992 Nobel Prize
in Physiology
or Medicine

Cori lab members whose subsequent work led to Nobel Prizes of their own

Figure 14.24 Carl and Gerty Cori were awarded the Nobel Prize for their elucidation of the link between lactate and glucose metabolism in muscle and liver tissue. Six scientists who were associated with the Cori laboratory at Washington University in St. Louis during the mid-1900s went on to win Nobel Prizes for their own research. G. CORI: SCIENCE SOURCE; C. CORI: NATIONAL LIBRARY OF MEDICINE/SCIENCE SOURCE; S. OCHOA: AP PHOTO; A. KORNBERG: AP PHOTO; L. LELOIR: PICTORIAL PRESS LTD/ALAMY; E. SUTHERLAND: SCIENCE SOURCE; C. DE DUVE: AFP/GETTY IMAGES; E. KREBS: AP PHOTO/MATT TODD.

concept integration 14.2

What is the metabolic logic for reciprocal regulation of liver phosphofructokinase-1 and fructose-1,6-bisphosphatase-1 activity by AMP and insulin?

High levels of AMP in the cell signal low energy charge and allosterically activate the glycolytic enzyme phosphofructokinase-1. This serves to stimulate flux through the glycolytic pathway and generate both ATP by substrate-level phosphorylation and pyruvate for aerobic respiration. Under these low-energy-charge conditions, flux through the gluconeogenic pathway is inhibited by AMP to prevent a net loss of 4 ATP as a result of futile cycling, which occurs when glycolysis and gluconeogenesis are both activated in the same cell. Reciprocal regulation of phosphofructokinase-1 and fructose-1,6-bisphosphatase-1 activity by insulin is another example of metabolic control; however, in this case, regulation is at the physiologic level in response to elevated blood glucose levels. Increased glucose in the blood after a carbohydrate-rich meal results in release of insulin and stimulation of protein phosphatase 1 activity in liver cells. This leads to dephosphorylation of the phosphofructokinase-2/fructose-2,6-bisphosphatase enzyme and subsequent activation of its kinase activity, which phosphorylates fructose-6-P on C-2 to produce the allosteric regulator fructose-2,6-BP. Binding of fructose-2,6-BP to the glycolytic enzyme phosphofructokinase-1 stimulates its activity and increases flux through the glycolytic pathway, while at the same time, fructose-2,6-BP binds to and inhibits the activity of the gluconeogenic enzyme fructose-1,6-bisphosphatase-1.

14.3 Glycogen Degradation and Synthesis

We saw in Chapter 13 that the storage form of glucose in most animals is glycogen, a large, highly branched polysaccharide consisting of glucose units joined by α-1,4 and α-1,6 glycosidic bonds (see Figure 13.14). Glycogen degradation and synthesis occurs in the cytosol, with the substrate for these reactions being the free ends of the branched polymer (nonreducing ends). The large number of branch points in glycogen results in the generation of multiple nonreducing ends, which provide a highly efficient mechanism either to release glucose from glycogen to meet energy needs or to rebuild glycogen particles when excess dietary glucose is available.

The three key reactions required for reversible degradation and synthesis of glycogen are catalyzed by glycogen phosphorylase, glycogen synthase, and the glycogen branching and debranching enzymes. **Figure 14.25** presents an overview of glycogen degradation and synthesis in animals by highlighting the function and regulation of these glycogen metabolizing enzymes:

1. *Glycogen phosphorylase* releases glucose-1-phosphate (glucose-1-P) from glycogen in a phosphorolysis reaction involving inorganic phosphate (P_i) and cleavage of the α-1,4 glycosidic bond. The glucose-1-P molecules are converted to glucose-6-P, which is either used for glycolysis in muscle cells or dephosphorylated in liver cells so that glucose can be exported to other tissues. The activity of glycogen phosphorylase in liver cells is activated by epinephrine and glucagon signaling.

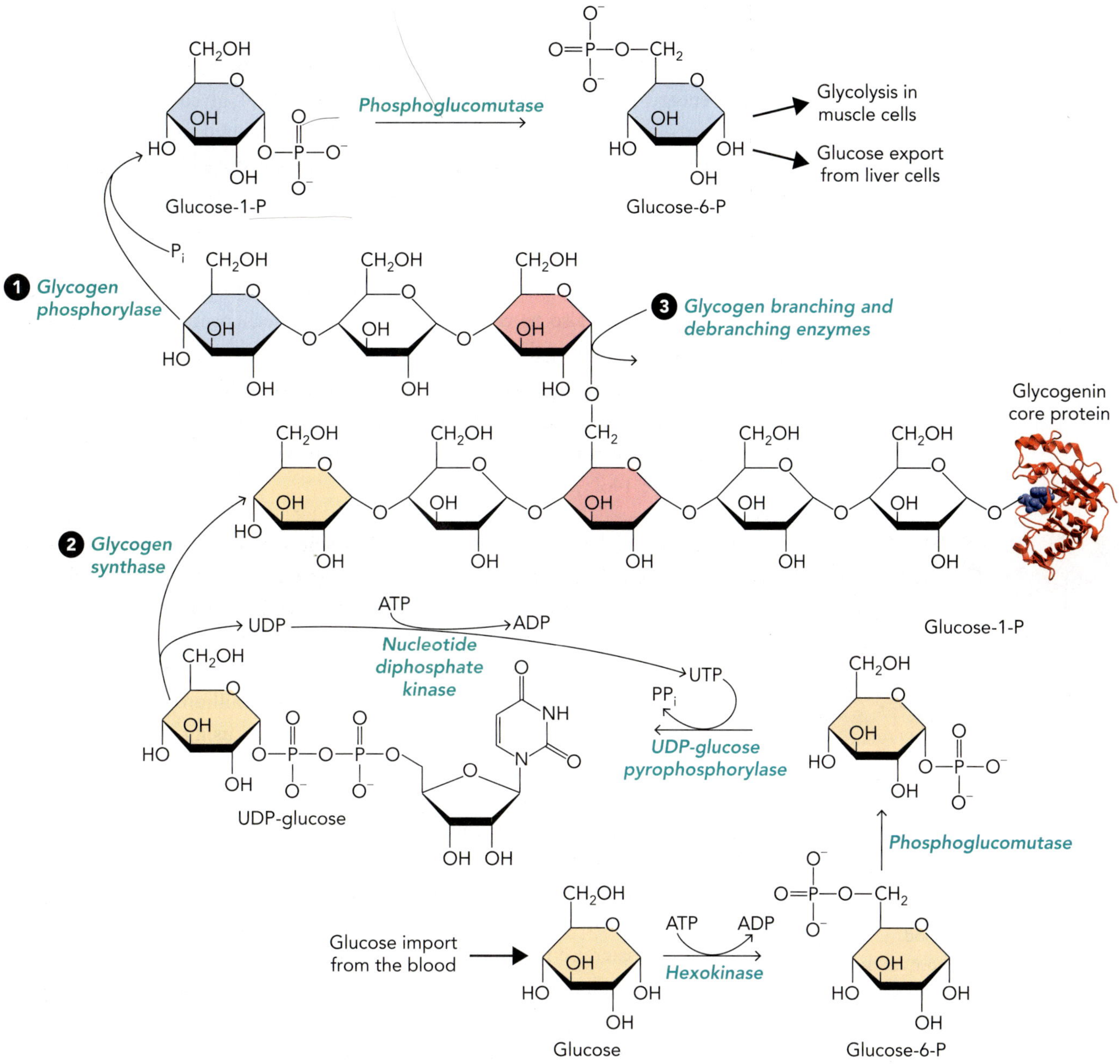

Figure 14.25 Glycogen degradation and synthesis involves three key reactions catalyzed by (1) glycogen phosphorylase, (2) glycogen synthase, and (3) the glycogen branching and debranching enzymes. Glycogen phosphorylase removes glucose from the nonreducing ends of glycogen through a phosphorolysis reaction that generates glucose-1-P. The enzyme phosphoglucomutase converts glucose-1-P to glucose-6-P, which is either used by muscle cells in glycolysis or dephosphorylated in liver cells and exported. The glycogen synthase enzyme catalyzes a reaction that links glucose to nonreducing ends of glycogen, using UDP-glucose as the substrate. Generation of UDP-glucose requires the enzymes UDP-glucose pyrophosphorylase and nucleotide diphosphate kinase. The branching and debranching enzymes are responsible for modifying the glycogen polymer to maximize the number of reducing ends available during glycogen degradation and synthesis. Hormonal signaling is the primary regulatory mechanism in controlling glycogen degradation (epinephrine, glucagon) and synthesis (insulin).

2. *Glycogen synthase* is activated by insulin signaling and is responsible for adding glucose to the nonreducing ends of glycogen in a reaction involving uridine diphosphate-glucose (UDP-glucose). Glycogen synthase uses the bond energy available in UDP-glucose to form α-1,4 glycosidic bonds at the growing nonreducing ends. ATP hydrolysis is required to regenerate uridine triphosphate (UTP) for subsequent rounds of glucose addition, therefore glycogen synthesis essentially requires 1 ATP per glucose residue added. As first discussed in Chapter 8, hormone signaling through glucagon, epinephrine, and insulin results in reversible phosphorylation of glycogen phosphorylase and glycogen synthase. The activities of glycogen phosphorylase and glycogen synthase can also be controlled by allosteric mechanisms in response to the metabolic state of the cell.
3. *Glycogen branching and debranching enzymes* ensure that glycogen phosphorylase and glycogen synthase have access to the maximum number of nonreducing ends for the cleavage or formation of α-1,4 glycosidic bonds (Figure 14.25).

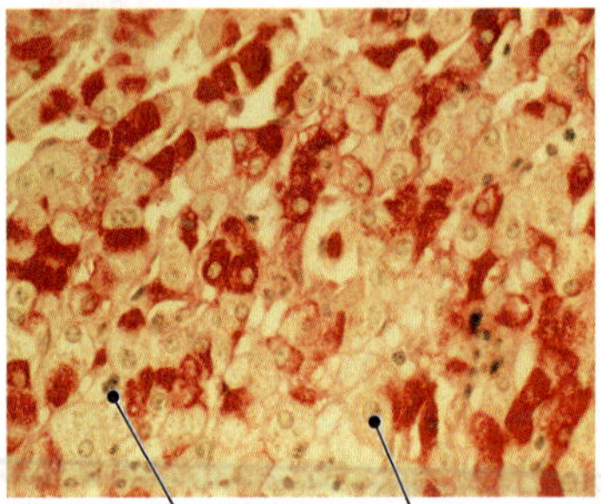

Figure 14.26 Liver cell section stained by the PAS method identifies glycogen deposits as darkly staining areas. SPL/SCIENCE SOURCE.

Although small amounts of glycogen are synthesized in many animal cell types, only the liver and skeletal muscle accumulate large amounts of glycogen. Glycogen core complexes consist of glycogenin protein and ~50,000 glucose molecules, with α-1,6 branches about every 8–12 residues, creating ~2,000 nonreducing ends (see Figure 13.15). Twenty to forty glycogen core complexes form **glycogen particles** inside liver and muscle cells, which contain more than a million glucose molecules. Glycogen particles can be visualized in liver cell sections using the periodic acid–Schiff (PAS) staining method, which identifies glycogen deposits as darkly staining regions (**Figure 14.26**). Liver glycogen can constitute up to 10% of a liver cell by weight and is used as a source of glucose for export to other tissues when dietary glucose is limiting (between meals). In contrast, the sole purpose of glycogen in muscle cells is to generate glucose-6-P for use as a chemical energy source in anaerobic and aerobic glycolysis. Muscle glycogen is 1% to 2% of muscle tissue by weight and can sustain vigorous exercise for up to an hour. Because muscle cells do not contain the enzyme glucose-6-phosphatase, all of the glucose-6-P that is made available from glycogen degradation stays inside the cell.

In the next two subsections we examine in more detail the enzymatic reactions that mediate glycogen degradation and synthesis in animals. This is followed by a description of several glycogen storage diseases in humans that result from defects in the enzymes that control glycogen metabolism. Before we get started, let's answer the four metabolic questions that summarize glycogen metabolism:

1. *What purpose does glycogen degradation and synthesis serve in animals?* Liver glycogen is used as a short-term energy source for the organism by providing a means to store and release glucose in response to blood glucose levels. Liver cells do not use this glucose for their own energy needs. (Fatty acids released from adipose tissue provide chemical energy to liver cells.) Muscle glycogen provides a readily available source of glucose during exercise to support anaerobic and aerobic energy-conversion pathways within muscle cells. Muscle cells lack the enzyme glucose-6-phosphatase and therefore cannot release glucose into the blood.

2. *What are the net reactions of glycogen degradation and synthesis?*

   Glycogen degradation:

   $$\text{Glycogen}_{n \text{ units of glucose}} + \text{P}_\text{i} \rightarrow \text{Glycogen}_{n-1 \text{ units of glucose}} + \text{Glucose-1-phosphate}$$

   Glycogen synthesis:

   $$\text{Glycogen}_{n \text{ units of glucose}} + \text{Glucose-1-phosphate} + \text{ATP} + \text{H}_2\text{O} \rightarrow \text{Glycogen}_{n+1 \text{ units of glucose}} + \text{ADP} + 2\ \text{P}_\text{i}$$

3. *What are the key enzymes in glycogen degradation and synthesis?* **Glycogen phosphorylase** catalyzes the phosphorolysis reaction that uses P_i to remove one glucose at a time from nonreducing ends of glycogen, resulting in the formation of glucose-1-P. Liver and muscle glycogen phosphorylase are isozymes (two different genes) that are both activated by phosphorylation but have distinct responses to allosteric effectors. **Glycogen synthase** catalyzes the addition of glucose residues to nonreducing ends of glycogen using UDP-glucose as the glucose donor. Glycogen synthase activity is inhibited by phosphorylation; binding of the allosteric activator glucose-6-P promotes dephosphorylation and enzyme activation. The two **branching and debranching enzymes** are responsible for adding glucose residues to (branching) or removing glucose residues from (debranching) the glycogen complex through the formation or cleavage of α-1,6 glycosidic bonds.
4. *What are examples of glycogen degradation and synthesis in everyday biochemistry?* The performance of elite endurance athletes (for example, marathon runners, professional bicycle riders) can benefit from a dietary regimen of carbohydrate "loading" prior to competition. Recent studies indicate that a short period of intense exercise to deplete muscle glycogen stores, followed by ingestion of 10 g/kg body mass of high-carbohydrate foods, can lead to a doubling of muscle glycogen in 24 hours. Carbohydrate loading regimens can result in a buildup of stored muscle glycogen that is sometimes higher than what can be obtained by simply following a high-carbohydrate diet.

Enzymatic Reactions in Glycogen Degradation

Glycogen degradation is initiated by glycogen phosphorylase, a homodimeric enzyme that catalyzes a phosphorolysis cleavage reaction of the α-1,4 glycosidic bond at the nonreducing ends of the glycogen molecule. Pyridoxal phosphate (PLP), which is attached to the enzyme as a coenzyme, functions as an acid–base catalyst that donates a proton to inorganic phosphate. The inorganic phosphate is then deprotonated by the oxygen at the glycosidic linkage, breaking the chain and releasing glucose-1-P as the product (**Figure 14.27**).

Glycogen phosphorylase is a processive enzyme, which means it stays attached to the glycogen substrate and continues to cleave α-1,4 glycosidic bonds and release glucose-1-P until the enzyme gets too close to an α-1,6 branch point. The coenzyme pyridoxal phosphate (derived from vitamin B_6) is covalently bound to the glycogen

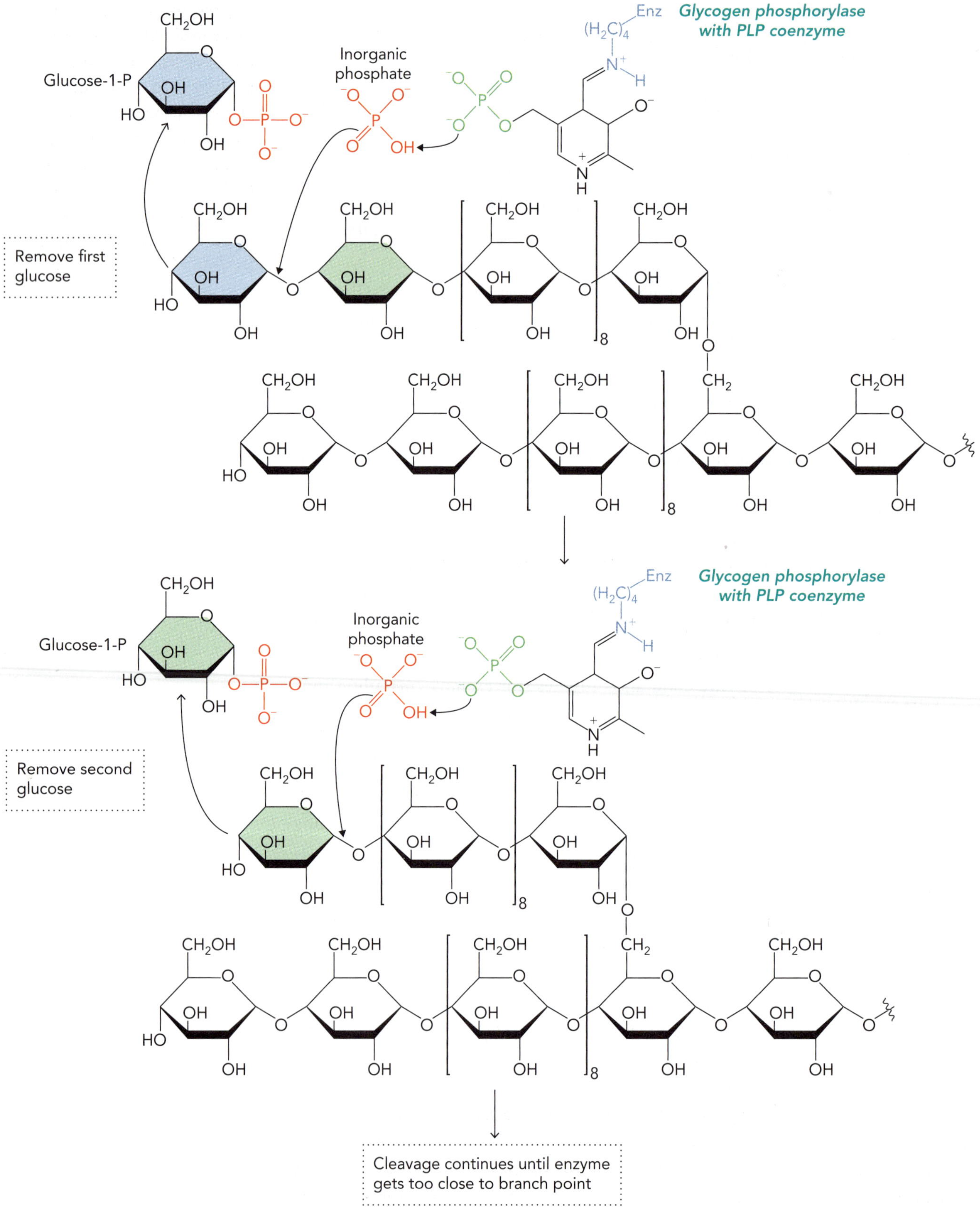

Figure 14.27 Glycogen phosphorylase catalyzes a phosphorolysis reaction using the enzyme-bound coenzyme PLP to generate the product glucose-1-P. Glycogen phosphorylase will processively cleave from nonreducing ends of glycogen until it is too close to an α-1,6 branch point.

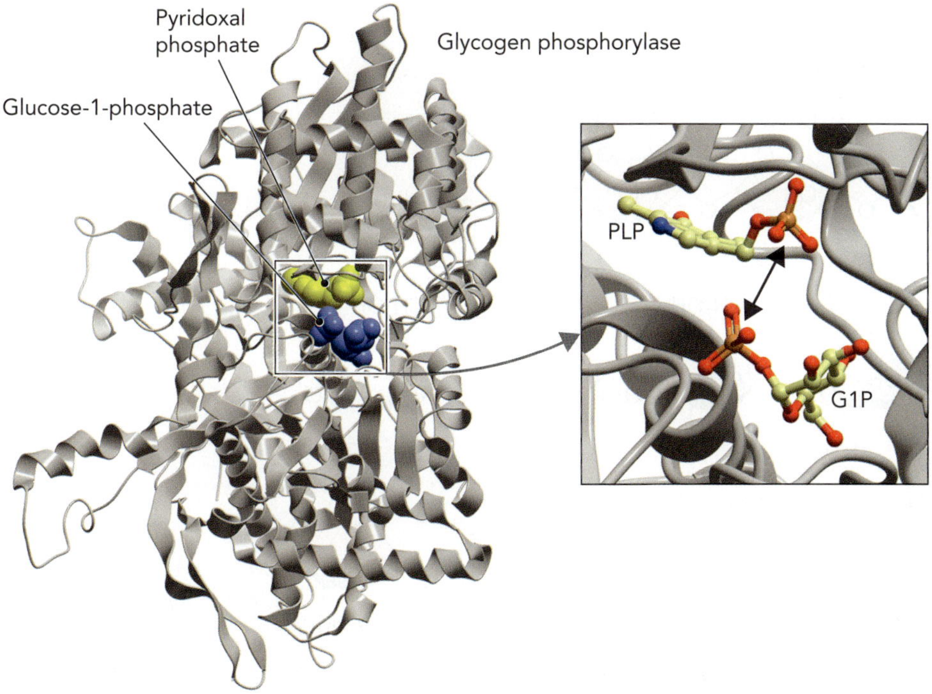

Figure 14.28 This molecular structure of rabbit muscle glycogen phosphorylase shows pyridoxal phosphate and the product glucose-1-phosphate bound in the enzyme active site. The phosphate groups of pyridoxal phosphate (PLP) and the product glucose-1-phosphate (G1P) are nearby in the active site, allowing proton transfer to occur between these groups. BASED ON PDB FILE 3GPB.

phosphorylase enzyme through a lysine residue (**Figure 14.28**). Although the standard free energy change for this phosphorolysis reaction is positive ($\Delta G^{\circ\prime} = +3.1$ kJ/mol), making the reaction unfavorable under standard conditions, the actual change in free energy is favorable ($\Delta G = -6$ kJ/mol) because of the high concentration of P_i relative to glucose-1-P inside the cell (ratio of close to 100:1).

Glycogen phosphorylase removes glucose units from the nonreducing end until it reaches within four glucose units of an α-1,6 branch point. As shown in **Figure 14.29**, at this stage, the glycogen debranching enzyme recognizes the partially degraded branch structure and remodels the substrate in a two-step reaction. First, the debranching enzyme transfers three glucose units to the nearest nonreducing end, generating a new substrate for glycogen phosphorylase. In the next step, the bifunctional debranching enzyme cleaves the α-1,6 glycosidic bond to release free glucose. In liver cells, the glucose is directly exported to the blood, and in muscle cells, it is phosphorylated by hexokinase as a first step in the glycolytic pathway. Because α-1,6 branch points occur about every 10 glucose residues in glycogen, complete degradation releases about 90% glucose-1-P and 10% glucose molecules.

The glucose-1-P product of the glycogen phosphorylase reaction is not an intermediate in glycolysis, and liver cells do not contain a "glucose-1-P-phosphatase activity" that could generate glucose. Therefore, the next reaction in glycogen degradation is the conversion of glucose-1-P to glucose-6-P by the enzyme phosphoglucomutase. This reaction is similar to the phosphoglycerate mutase reaction in glycolysis, in that the enzyme first donates a phosphate group to the substrate to generate an intermediate bisphosphate compound. Then, the bisphosphate compound is dephosphorylated to regenerate the phosphoenzyme and release the product. In the case of phosphoglycerate mutase, the phosphoryl transfer group is phosphohistidine, whereas the phosphoryl transfer group in phosphoglucomutase is a phosphoserine (see Figure 9.36).

As shown in **Figure 14.30** (p. 709), in the first step of this reversible reaction, the phosphoserine in the enzyme active site donates its phosphate to C-6 of glucose-1-P, forming glucose-1,6-bisphosphate (glucose-1,6-BP). In the next step, the phosphate group on C-1 is transferred to the serine residue to regenerate phosphoserine and release

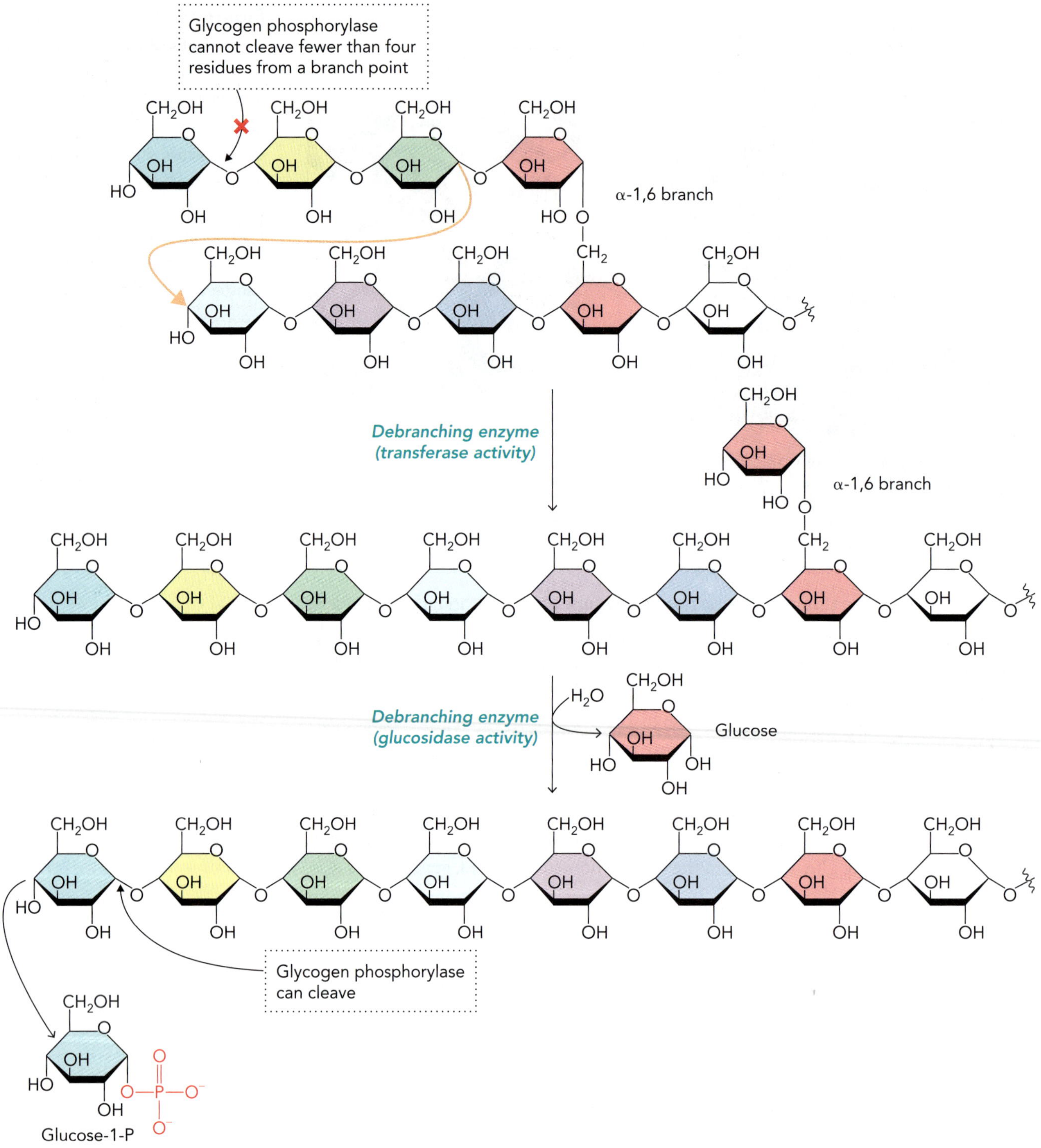

Figure 14.29 The dual-function glycogen debranching enzyme encodes both a glycosyltransferase activity and an α-1,6 glycosidase activity. Glycogen phosphorylase degrades glycogen from the nonreducing end until reaching within four residues of a branch point. Then, the glycogen debranching enzyme transfers three glucose residues to a nearby nonreducing end and subsequently cleaves the α-1,6 bond. Note that whereas the product of the glycogen phosphorylase reaction is glucose-1-P, the debranching enzyme releases free glucose.

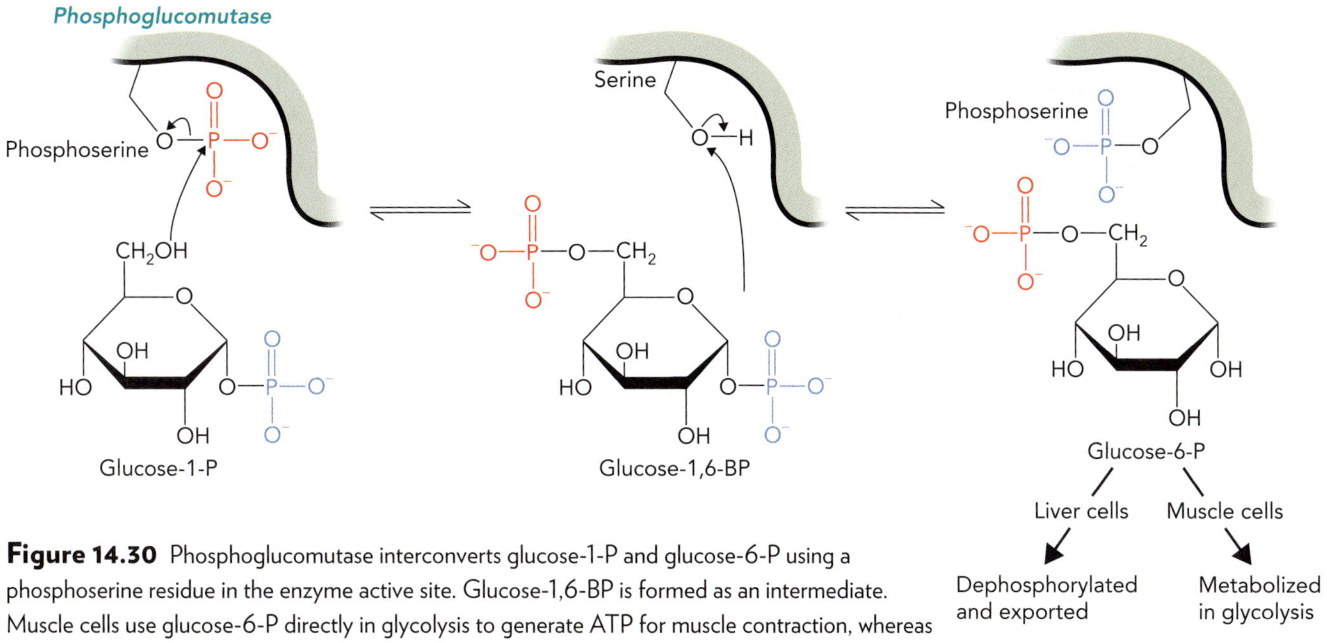

Figure 14.30 Phosphoglucomutase interconverts glucose-1-P and glucose-6-P using a phosphoserine residue in the enzyme active site. Glucose-1,6-BP is formed as an intermediate. Muscle cells use glucose-6-P directly in glycolysis to generate ATP for muscle contraction, whereas liver cells contain the enzyme glucose-6-phosphatase, which dephosphorylates the sugar and exports the glucose to other tissues.

glucose-6-P. In liver cells, glucose-6-P is dephosphorylated by glucose-6-phosphatase to generate glucose for export, whereas in muscle cells—which lack glucose-6-phosphatase—the glucose-6-P is used as a source of chemical energy in glycolysis.

Glycogen phosphorylase exists in two conformations: the R state, which is more active, and the T state, which is inactive. The equilibrium between the R state and the T state is shifted by both covalent modification and allosteric regulation. As shown in **Figure 14.31**, phosphorylation of Ser14 on the enzyme shifts the equilibrium toward the active R state, whereas dephosphorylation of Ser14 shifts the equilibrium back to the inactive T state. The enzyme responsible for phosphorylating glycogen phosphorylase is phosphorylase kinase, which is a downstream target of glucagon and epinephrine signaling in liver cells. Epinephrine signaling in muscle cells also results in phosphorylation of phosphorylase kinase (muscle cells do not contain glucagon receptors). Both the liver and muscle isozymes of glycogen phosphorylase are dephosphorylated by the enzyme protein phosphatase 1. As we describe later, the activity of protein phosphatase 1 is controlled by hormone signaling in liver and muscle cells.

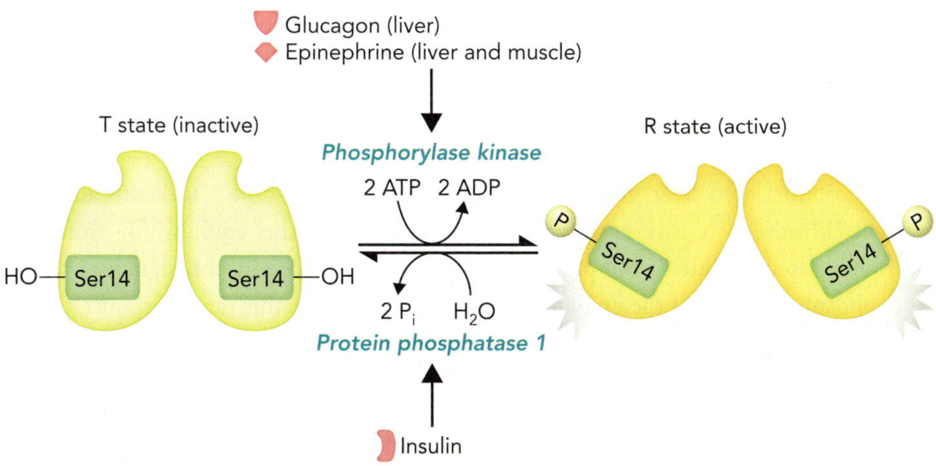

Figure 14.31 The activity of the glycogen phosphorylase dimer is stimulated by phosphorylation, which shifts the equilibrium from the T-state conformation (inactive) to the R-state conformation (active). The enzymes phosphorylase kinase and protein phosphatase 1 control the phosphorylation state of glycogen phosphorylase in response to hormone signaling.

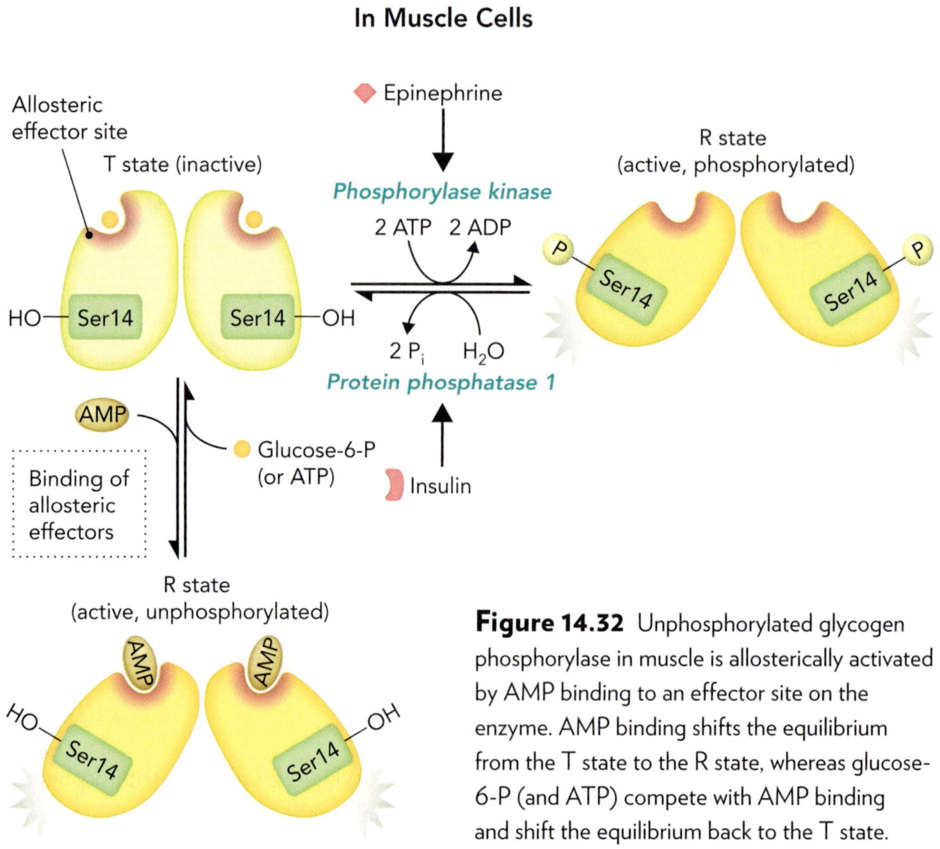

Figure 14.32 Unphosphorylated glycogen phosphorylase in muscle is allosterically activated by AMP binding to an effector site on the enzyme. AMP binding shifts the equilibrium from the T state to the R state, whereas glucose-6-P (and ATP) compete with AMP binding and shift the equilibrium back to the T state.

Glycogen degradation is a primary source of glucose for ATP production by anaerobic glycolysis during muscle contraction. To meet these energy needs quickly when ATP stores are depleted, unphosphorylated glycogen phosphorylase in muscle can be shifted from the T state to the R state by binding of the allosteric effector AMP (**Figure 14.32**). This makes sense because muscle contraction requires ATP hydrolysis, resulting in increased levels of AMP in the cell. These high AMP levels can be reached soon after continual muscle contraction occurs (for example, shortly after starting to exercise) and therefore signal glycogen degradation well before epinephrine-induced phosphorylation takes place. Importantly, glucose-6-P functions as an allosteric inhibitor of unphosphorylated muscle glycogen phosphorylase by competing with AMP for binding to the allosteric site, thereby pushing the equilibrium back to the T state as soon as energy sources become available. The interplay of acute allosteric control of muscle glycogen phosphorylase activity by energy sensing metabolites (AMP, ATP, glucose) and delayed hormonal control by epinephrine and insulin provides a temporal response to muscle contraction that spans minutes (allosteric control) to hours (hormonal control).

Phosphorylated liver glycogen phosphorylase is also subject to acute allosteric control; however, in this case, glucose binding shifts the equilibrium from the phosphorylated R state to the phosphorylated T state before insulin signaling results in dephosphorylation of the enzyme (**Figure 14.33**). Again, this rapid response to elevated glucose levels provides a mechanism to shut off glycogen degradation quickly when glucose is available, and this occurs well before insulin signaling is fully activated. Studies have shown that the T-state conformation of phosphorylated liver glycogen

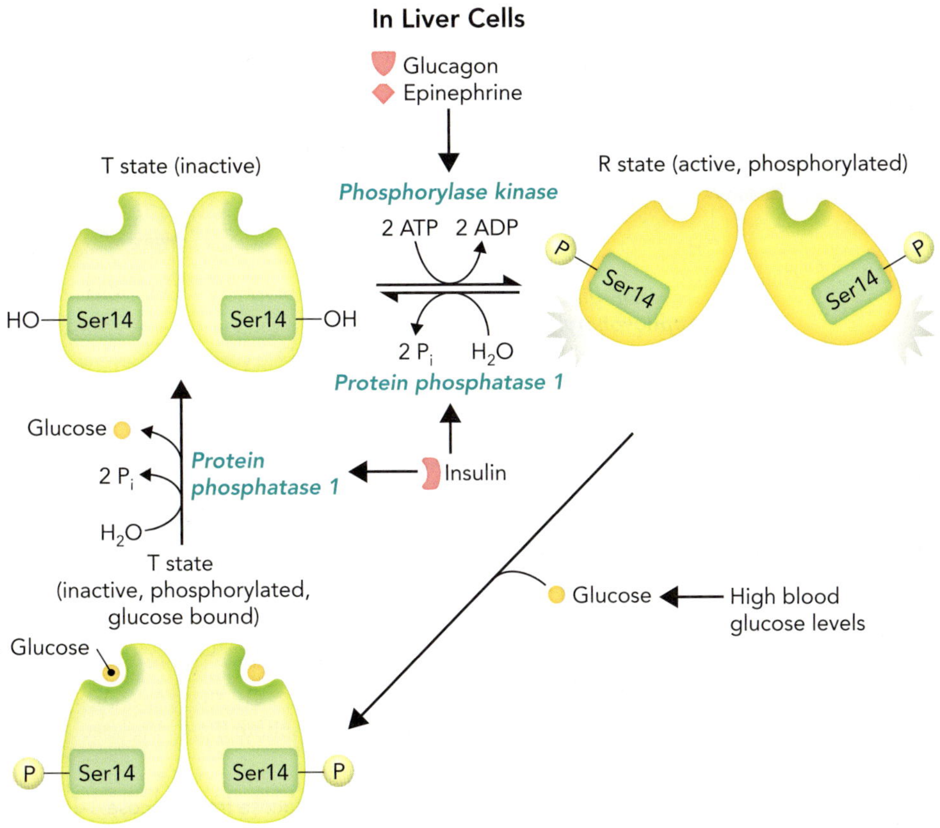

Figure 14.33 Glucose binding to an effector site (shaded in green) of phosphorylated liver glycogen phosphorylase shifts the equilibrium from the R state to the T state, which promotes dephosphorylation by protein phosphatase 1 in the presence of insulin. Phosphorylated liver glycogen phosphorylase functions as a glucose sensor that is rapidly inhibited by elevated serum glucose levels as a means to decrease the rate of glycogen degradation.

phosphorylase is a better substrate for dephosphorylation by protein phosphatase 1 than is the R-state conformation. Therefore, in the presence of high glucose levels, insulin signaling activates protein phosphatase 1, which results in dephosphorylation of both the R-state and T-state conformations to inhibit liver glycogen degradation. Because the GLUT2 transporter in liver cells permits free glucose exchange between the inside and outside of the cell, glucose-mediated dephosphorylation of liver glycogen phosphorylase serves as an acute biochemical sensing mechanism of high blood glucose levels.

We will observe how this mode of two-stage metabolic regulation combining rapidly acting allosteric effectors, such as AMP and ATP, with delayed hormonal signaling by glucagon (or epinephrine) and insulin also controls fatty acid metabolism when we examine regulation of the enzyme acetyl-CoA carboxylase in Chapter 16.

Enzymatic Reactions in Glycogen Synthesis

The addition of glucose units to the nonreducing ends of glycogen by the enzyme glycogen synthase requires the synthesis of an activated form of glucose called uridine diphosphate-glucose (UDP-glucose). Activated sugar nucleotides such as UDP-glucose serve a special role in metabolic pathways by providing high-energy compounds that are destined for specific reactions.

To initiate glycogen synthesis, glucose-6-P must first be converted to glucose-1-P by phosphoglucomutase (a reversible reaction; see Figure 14.30). The enzyme UDP-glucose pyrophosphorylase then catalyzes a reaction involving the attack of a phosphoryl oxygen from glucose-1-P on the α phosphate of uridine triphosphate (UTP). As shown in **Figure 14.34**, this reaction leads to the formation of UDP-glucose and

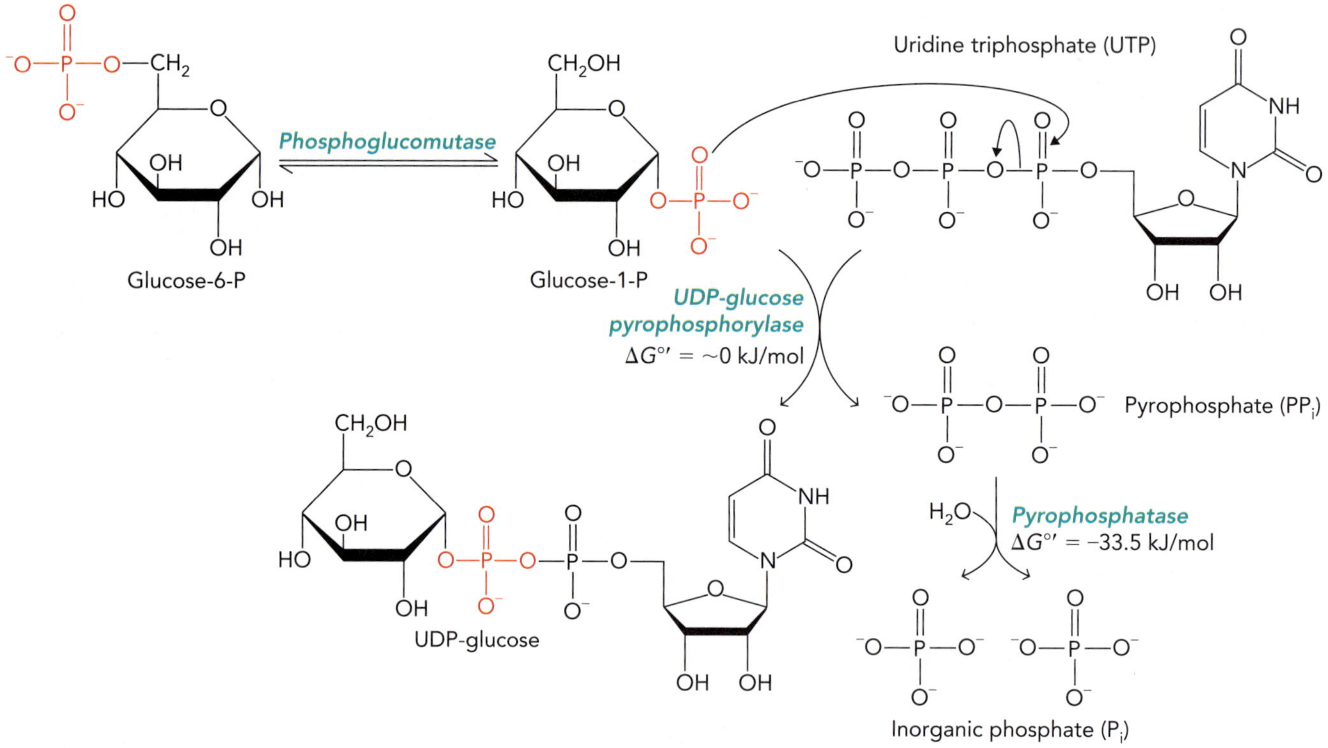

Figure 14.34 The enzyme UDP-glucose pyrophosphorylase catalyzes a reaction that generates UDP-glucose from glucose-1-P and UTP. The $\Delta G^{\circ\prime}$ of the UDP-glucose pyrophosphorylase reaction is close to zero; however, the rapid hydrolysis of pyrophosphate by the ubiquitous enzyme pyrophosphatase shifts the equilibrium of the reaction in favor of UDP-glucose formation.

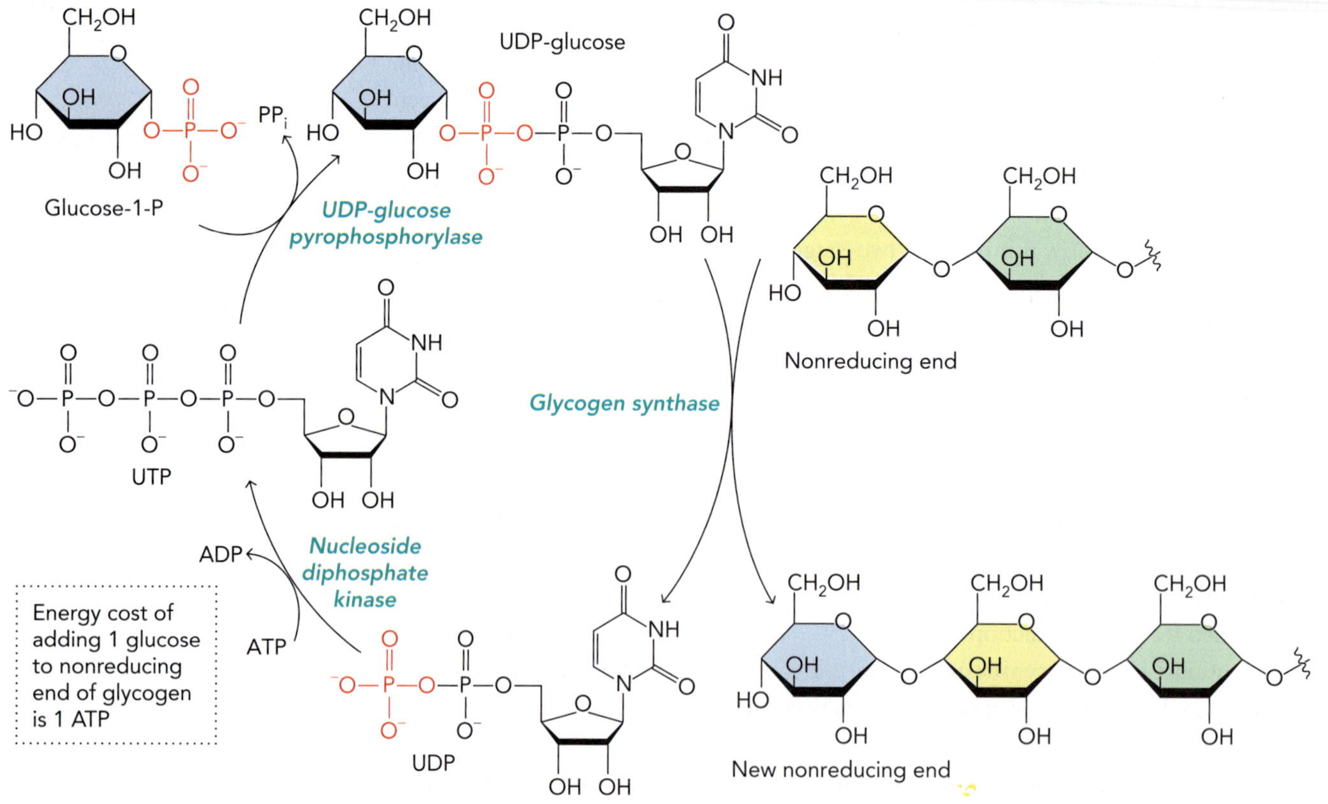

Figure 14.35 Glycogen synthase catalyzes a reaction using UDP-glucose to add glucose residues one at a time to the nonreducing ends of glycogen. UTP is regenerated from UDP by the enzyme nucleoside diphosphate kinase, which uses ATP as a phosphoryl donor.

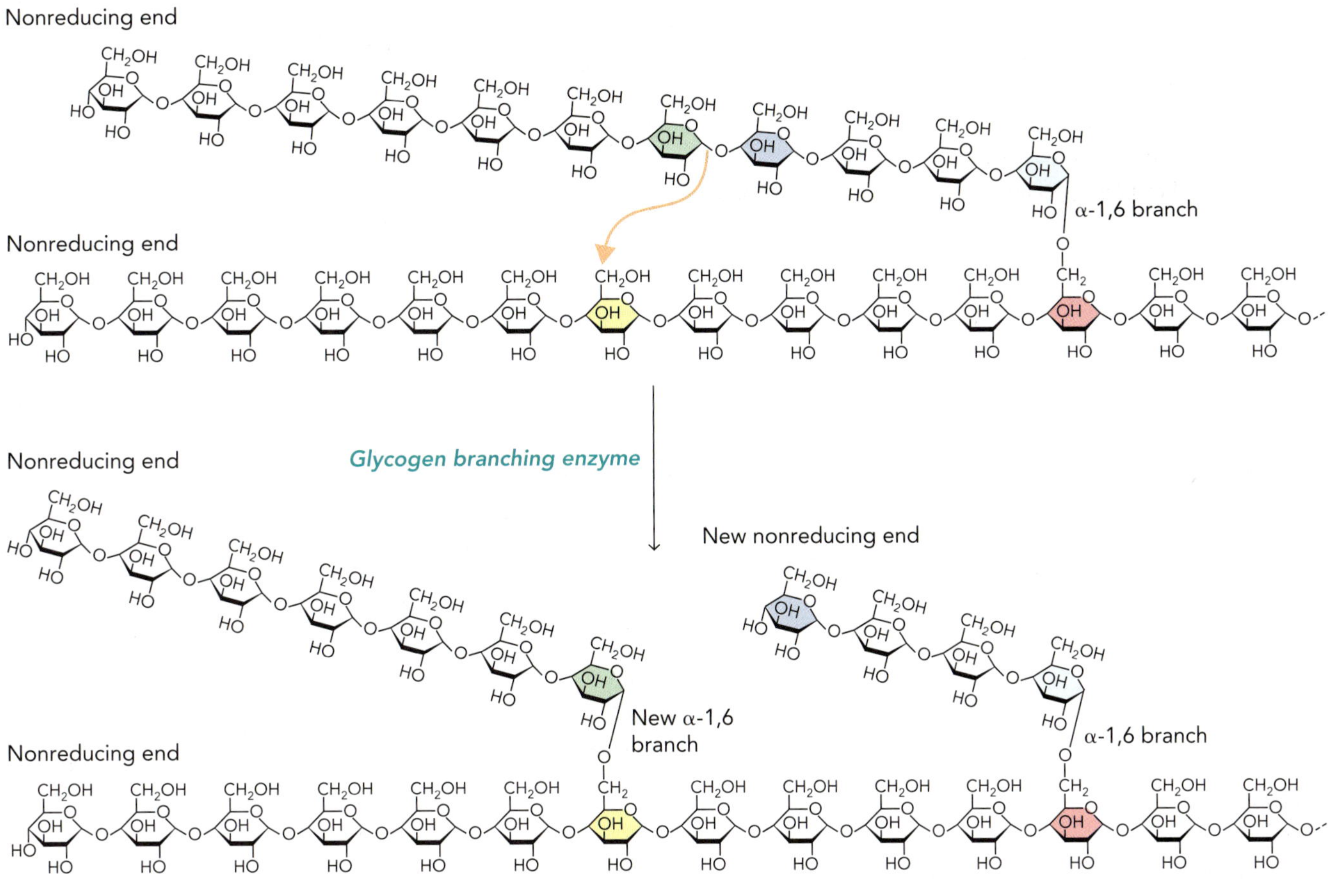

Figure 14.36 Glycogen branching enzyme creates a new α-1,6 branch point and a new nonreducing end by transferring seven glucose residues from the end of a growing chain to an internal point no closer than four glucose residues from another branch point.

the release of pyrophosphate (PP_i). Although the standard free energy change of the UDP-glucose pyrophosphorylase reaction is close to zero, the rapid hydrolysis of PP_i by pyrophosphatase ($\Delta G^{\circ\prime} = -33.5$ kJ/mol) shifts the equilibrium in favor of UDP-glucose formation.

Figure 14.35 shows how glycogen synthase uses UDP-glucose as a substrate to catalyze a glycosyltransferase reaction, which adds glucose to C-4 of the terminal glucose at the nonreducing end of a glycogen chain. This reaction leads to the formation of an α-1,4 glycosidic bond and extends the chain by one glucose unit, which becomes the new nonreducing end. To add another glucose molecule to the glycogen chain, UTP must be regenerated from UDP in a reaction involving ATP and the enzyme nucleoside diphosphate kinase. This investment of 1 ATP to regenerate UTP is the energy cost for each glucose molecule added to glycogen.

To maximize the number of nonreducing ends, α-1,6 branch points need to be introduced into the glycogen chain about every 8–12 glucose residues. As shown in **Figure 14.36**, this reaction is catalyzed by the glycogen branching enzyme, which transfers seven glucose residues from the end of one glycogen chain to an internal position in a nearby chain. The positioning of this new α-1,6 linkage needs to be at least four glucose residues away from the nearest branch point. The net result is the creation of a new nonreducing end, which not only facilitates the rapid addition and removal of glucose molecules from glycogen but also increases the solubility of glycogen.

It is clear that glycogen synthase can add glucose molecules to an extended glycogen chain, but what enzyme catalyzes the first α-1,4 linkage between two glucose molecules to build the initial glycogen chain? The answer is that the glycogen core protein, glycogenin, is both an anchor for the glycogen particle and the priming

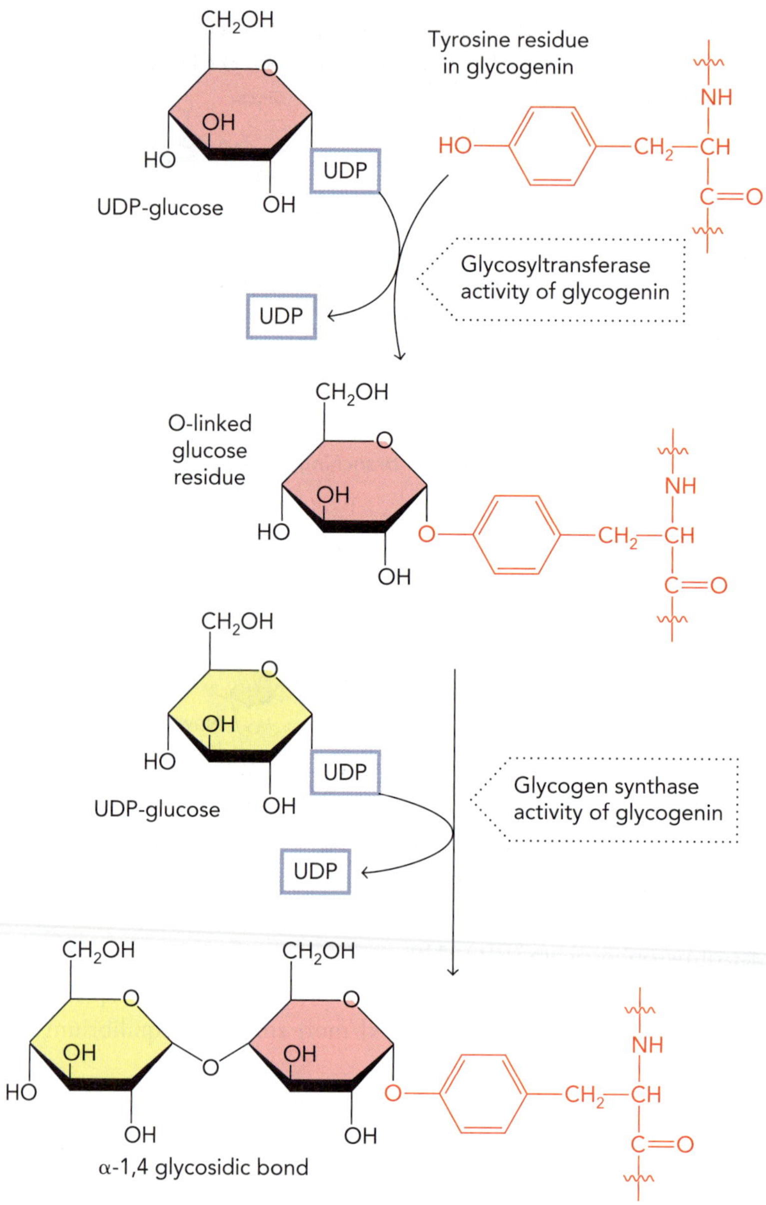

Figure 14.37 Glycogenin is both an anchor protein for the glycogen core particle and an enzyme that catalyzes the glycosyltransferase and synthesis reactions needed to generate the initial glycogen chain. Glucose derived from UDP-glucose is first attached to a tyrosine residue in glycogenin through an O-linked glycosidic bond that is formed via glycogenin's glycosyltransferase activity. Then, a second glucose is attached via glycogenin's intrinsic glycogen synthase activity, which forms an α-1,4 glycosidic bond. This glycogen synthase reaction is repeated five more times to yield an O-linked glucose septasaccharide.

enzyme. We first introduced glycogenin in Chapter 13, where we compared the macromolecular structures of amylopectin and glycogen (see Figure 13.15). As shown in **Figure 14.37**, glycogenin has two catalytic activities: (1) a glycosyltransferase activity that uses UDP-glucose as a donor to generate an O-linked glycosidic bond between glucose and a tyrosine residue in the protein; and (2) a glycogen synthase activity, which extends the glycogen chain out to seven glucose residues. The resulting septasaccharide glucose primer is a substrate for glycogen synthase and glycogen branching enzyme, which complete the process of adding up to ~50,000 glucose units to the glycogen core complex (**Figure 14.38**).

The catalytic activity of glycogen synthase is primarily controlled by reversible phosphorylation. However, unlike glycogen phosphorylase, glycogen synthase is inhibited by phosphorylation and activated by dephosphorylation. As shown in **Figure 14.39**, serine phosphorylation of glycogen synthase shifts the conformational equilibrium toward the inactive T state, whereas dephosphorylation by serine phosphatases, such as protein phosphatase 1, shifts the conformational equilibrium to the

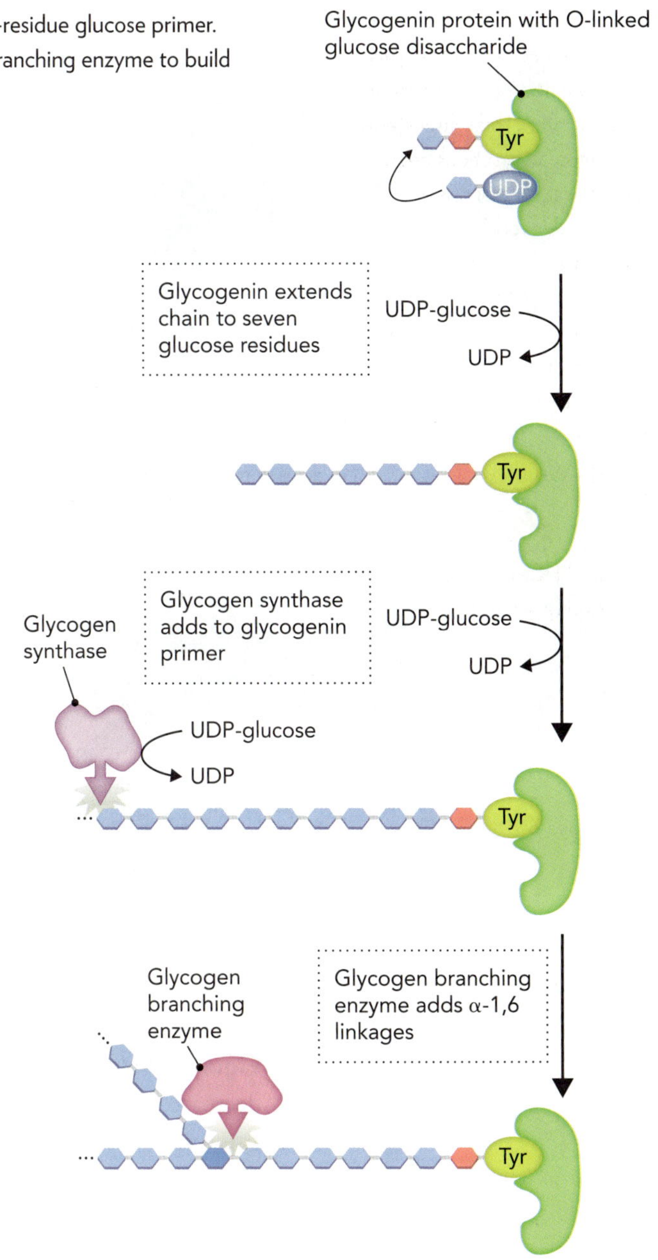

Figure 14.38 Autocatalytic reactions by glycogenin build the seven-residue glucose primer. This in turn is used as a substrate by glycogen synthase and glycogen branching enzyme to build the glycogen core particle.

active R state. Several inhibitory kinases have been identified that phosphorylate glycogen synthase, including protein kinase A, protein kinase C, glycogen synthase kinase 3, and Ca^{2+}/calmodulin kinase. As described in more detail in the next subsection, glycogen synthase is under the same hormonal control as glycogen phosphorylase, except in this case, glucagon and epinephrine signaling inhibit glycogen synthase activity. Figure 14.39 indicates that insulin signaling stimulates protein phosphatase 1 activity, which leads to dephosphorylation and activation of the enzyme.

Note that dephosphorylation of glycogen synthase is enhanced by transient binding of the allosteric effector glucose-6-P. Protein structure analysis of the yeast glycogen synthase enzyme in the absence and presence of glucose-6-P reveals that glycogen synthase protein dimers undergo a large conformational change in the presence of glucose-6-P (**Figure 14.40**). It is thought that this conformational change induced by glucose-6-P exposes key phosphoserine residues to dephosphorylation by protein phosphatase 1.

Hormonal Regulation of Glycogen Metabolism

Reciprocal regulation of glycogen metabolism by allosteric control and hormone signaling is a good example of how intracellular signaling pathways control biochemical processes through reversible phosphorylation. As shown in

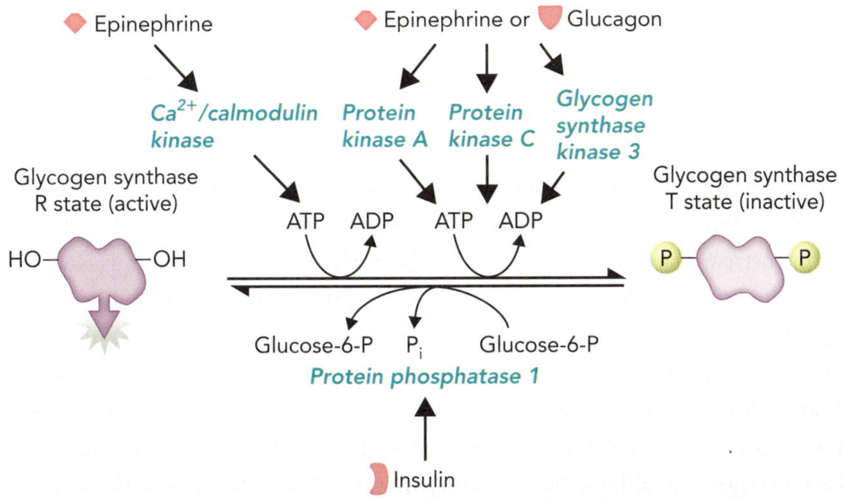

Figure 14.39 Glycogen synthase activity is hormonally controlled by reversible phosphorylation. The active R-state conformation of glycogen synthase is unphosphorylated, whereas phosphorylation by any one of several kinases converts it to the inactive T-state conformation. Epinephrine and glucagon signaling lead to phosphorylation and inactivation of glycogen synthase. In contrast, insulin signaling activates glycogen synthase by activating protein phosphatase 1, which is associated with transient binding of the allosteric effector glucose-6-P.

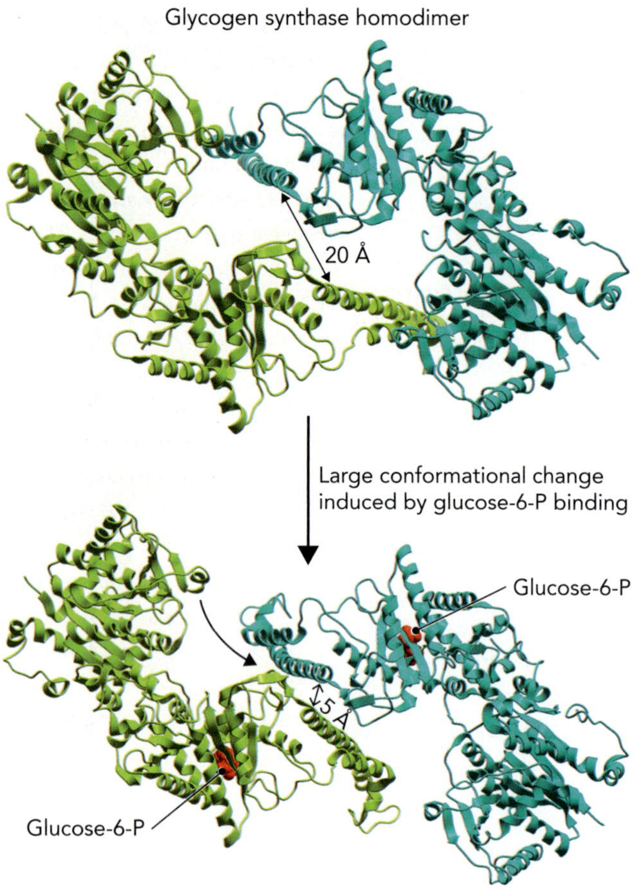

Figure 14.40 Binding of the allosteric effector glucose-6-P to glycogen synthase induces a large conformational change in this homodimeric enzyme. BASED ON PDB FILES 3O3C (TOP) AND 3NBO (BOTTOM).

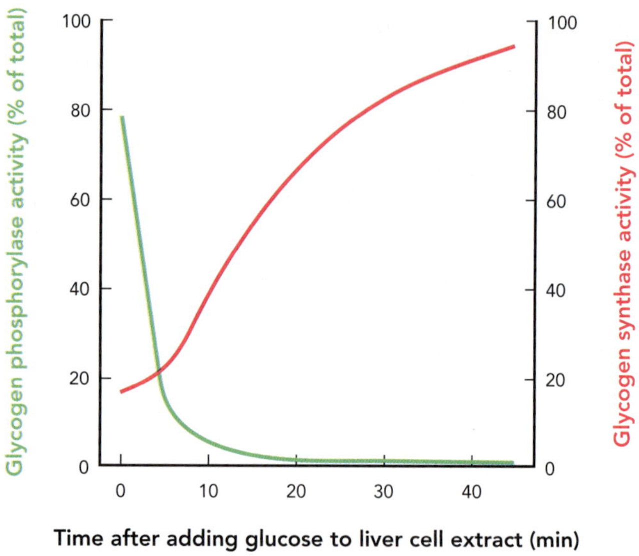

Figure 14.41 Glucose addition to liver cell extracts results in rapid allosteric inhibition of glycogen phosphorylase activity and, at the same time, allosteric activation of glycogen synthase activity. This mechanism of reciprocal regulation occurs on a timescale of minutes and reflects a sudden increase in blood glucose levels in response to ingesting a carbohydrate-rich meal. BASED ON J. MASSILLON ET AL. (1995). DEMONSTRATION OF A GLYCOGEN/GLUCOSE 1-PHOSPHATE CYCLE IN HEPATOCYTES FROM FASTED RATS. SELECTIVE INACTIVATION OF PHOSPHORYLASE BY 2-DEOXY-2-FLUORO-ALPHA-D-GLUCOPYRANOSYL FLUORIDE. *JOURNAL OF BIOLOGICAL CHEMISTRY, 270,* 19351–19356.

Figure 14.41, when glucose is added to a freshly prepared liver cell extract, a rapid decrease occurs in glycogen phosphorylase activity, which is coincident with a steady rise in the amount of glycogen synthase activity. This mechanism of reciprocal regulation occurs on a timescale of minutes and is the result of allosteric inhibition of glycogen phosphorylase activity by glucose (see Figure 14.33) and allosteric activation of glycogen synthase activity by glucose-6-P (see Figure 14.39). We now examine more closely how glucagon and insulin regulate these same two enzymes on a timescale of several hours, which is needed for homeostatic control of blood glucose levels before and after a meal.

Figure 14.42 summarizes the phosphorylation status and enzymatic activity of glycogen phosphorylase and glycogen synthase in liver cells as a function of glucagon and insulin signaling. When blood glucose levels are low, glucagon signaling initiates a phosphorylation cascade via protein kinase A, which directly phosphorylates and inhibits glycogen synthase. Glycogen phosphorylase activation by glucagon signaling involves a downstream signaling protein called phosphorylase kinase, which is phosphorylated by protein kinase A. Activated phosphorylase kinase then phosphorylates glycogen phosphorylase to initiate glycogen degradation. The net result of glucagon signaling in liver cells is increased rates of glycogen degradation and glucose export.

In contrast, high blood glucose levels stimulate insulin signaling, which initiates a dephosphorylation cascade via protein phosphatase 1 that activates glycogen synthase and inhibits glycogen phosphorylase, resulting in net glycogen synthesis and glucose

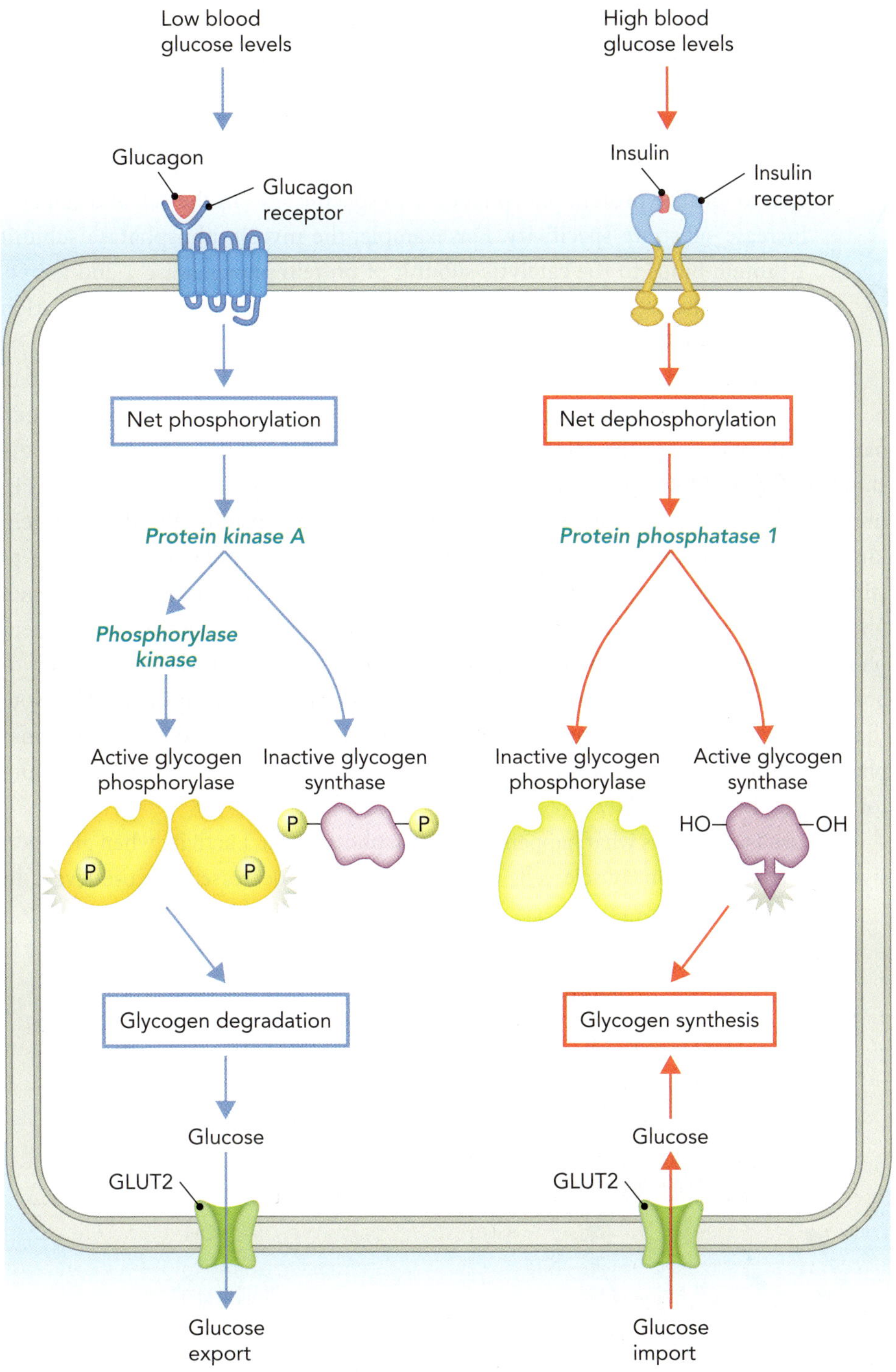

Figure 14.42 Delayed hormonal regulation of glycogen metabolism in liver cells is mediated by phosphorylation and dephosphorylation of glycogen phosphorylase and glycogen synthase, respectively. Glucagon signaling stimulates protein kinase A activity, leading to net phosphorylation of glycogen phosphorylase (via the enzyme phosphorylase kinase) and glycogen synthase, which results in increased glycogen degradation and glucose release. Insulin signaling stimulates protein phosphatase 1 activity and promotes net dephosphorylation of glycogen phosphorylase and glycogen synthase, resulting in increased glycogen synthesis and glucose import.

import. The same general pattern of phosphorylation and dephosphorylation is seen in the glucagon and insulin signaling pathways in muscle cells, with the exception that glucose is not exported from muscle cells because glucose-6-phosphatase is lacking (muscle cells use glucose released from glycogen degradation for their own energy needs).

In Chapter 8, we described the molecular mechanism by which cyclic AMP stimulates protein kinase A activity (see Figure 8.26), as well as how the insulin receptor activates downstream signaling pathways through insulin receptor substrate

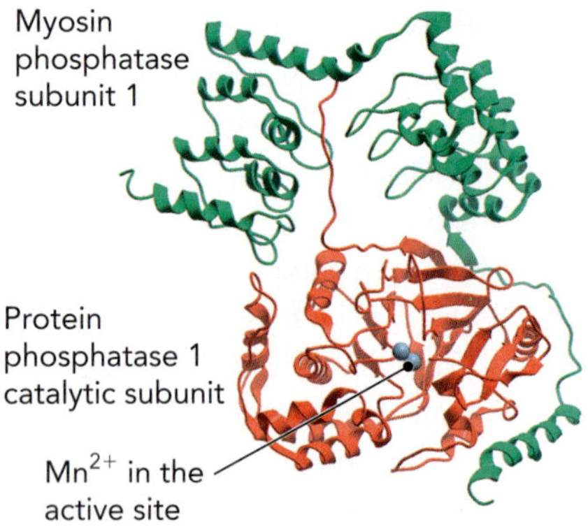

Figure 14.43 The molecular structure of chicken protein phosphatase 1 bound to the muscle targeting protein myosin phosphatase subunit 1 is shown here. Two manganese ions (Mn^{2+}) are bound to the protein phosphatase 1 enzyme active site. Myosin phosphatase subunit 1 interacts with protein phosphatase 1 over a large intermolecular surface, helping to shape the active site of protein phosphatase 1 and increase substrate specificity.
BASED ON PDB FILE 1S70.

proteins (see Figure 8.46). However, we still need to explain how **protein phosphatase 1** activity is stimulated by insulin signaling. Protein phosphatase 1 is a key signaling molecule in several pathways and has been shown to interact with as many as 50 regulatory proteins in mammals. One class of protein phosphatase 1 regulatory proteins is targeting proteins, which function to localize protein phosphatase 1 to specific sites in the cell and can also increase substrate specificity. For example, the myosin phosphatase subunit 1 protein binds to the catalytic subunit of protein phosphatase 1 and targets the phosphatase to myosin fibers, where it functions in muscle relaxation (**Figure 14.43**).

One of the best-characterized protein phosphatase 1 targeting proteins in muscle cells is called G_M, which localizes protein phosphatase 1 to glycogen particles (there is a corresponding G_L protein in liver cells with the same function). As shown in **Figure 14.44**, insulin signaling through insulin receptor substrate proteins in muscle cells activates a downstream signaling protein called insulin-stimulated protein kinase 1. This protein phosphorylates a serine residue on G_M called site 1, leading to stimulation of protein phosphatase 1 activity. Because glycogen synthase and glycogen phosphorylase are tightly associated with glycogen core particles, activation of protein phosphatase 1 by insulin-stimulated protein kinase 1 results in dephosphorylation of both enzymes and net stimulation of glycogen synthesis (see also Figure 14.42). Note that when glucose levels are high and insulin signaling is stimulated, a protein phosphatase 1 binding protein called Inhibitor 1 is unphosphorylated and has low affinity for protein phosphatase 1.

Figure 14.45 shows what happens to protein phosphatase 1 activity when glucagon signaling is activated in muscle cells under conditions of decreased glucose levels. In

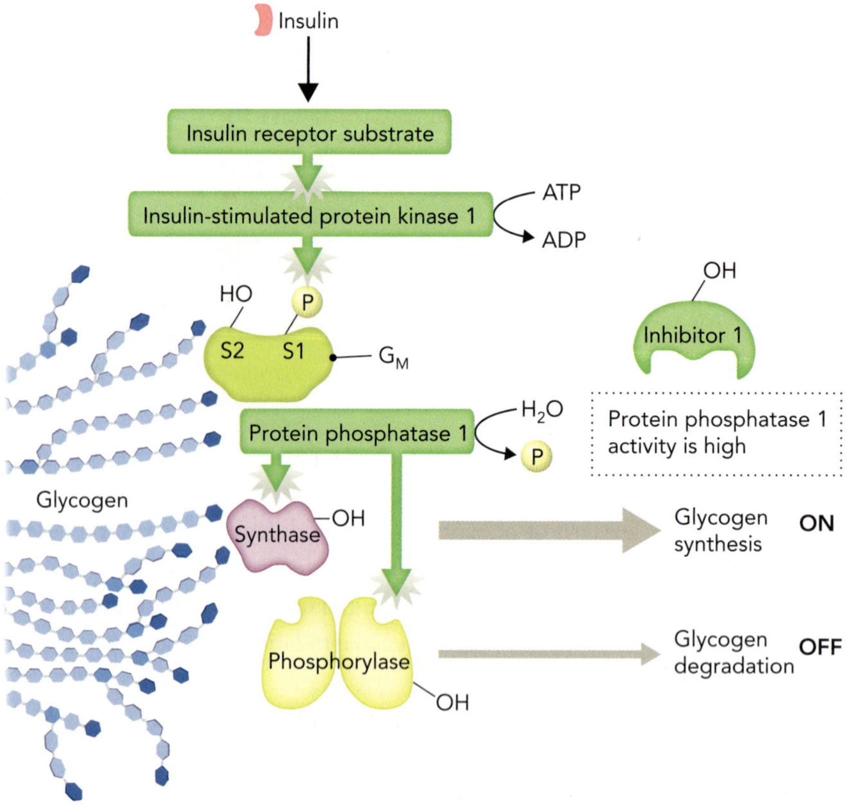

Figure 14.44 Insulin receptor activation in muscle cells stimulates signaling through insulin receptor substrate proteins and insulin-stimulated protein kinase 1. This leads to phosphorylation of the protein phosphatase 1 targeting protein G_M on the site 1 (S1) serine. Under these conditions, protein phosphatase 1 is activated and both glycogen synthase and glycogen phosphorylase are dephosphorylated, resulting in a net increase in glycogen synthesis. Inhibitor 1 is a protein phosphatase 1 binding protein with low affinity for protein phosphatase 1 in the absence of phosphorylation. S2 = site 2.

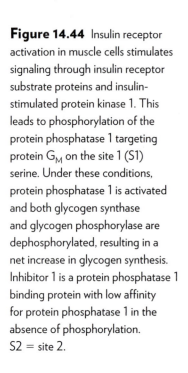

this case, glucagon signaling results in protein kinase A–mediated phosphorylation of the G_M subunit on the site 2 serine, which stimulates dissociation of protein phosphatase 1 from G_M. Protein kinase A also phosphorylates a serine residue on Inh-1, resulting in increased binding affinity for protein phosphatase 1. The net result is that glycogen degradation is stimulated because glycogen phosphorylase remains phosphorylated and active, whereas glycogen synthase becomes inactive (see also Figure 14.42).

Human Glycogen Storage Diseases

Several human diseases have been identified that affect glycogen metabolism. In many cases, disease symptoms include liver dysfunction due to excess glycogen; muscle defects (myopathy); fasting-induced hypoglycemia (low blood glucose levels); and, in the most severe diseases, death at an early age. **Figure 14.46** illustrates enzymes in glycogen metabolism that have been found to be defective in many of the human glycogen storage diseases that have been characterized. **Table 14.2** lists the primary organs affected by these enzyme deficiencies and the disease symptoms.

Although glycogen storage diseases are considered rare, six of these diseases have been well characterized and are briefly discussed in the paragraphs that follow. Note that von Gierke disease, Hers disease, and Pompe disease are the most common of these six diseases and occur in about 1 in 50,000 to 100,000 births.

von Gierke disease is due to a deficiency in the enzyme glucose-6-phosphatase. This causes a buildup of glycogen in the liver because glucose-6-P accumulates and activates glycogen synthase. As glucose-6-P cannot be dephosphorylated and released into the blood, the patient requires a nearly continual supply of dietary glucose.

Hers disease is a defect in the liver enzyme glycogen phosphorylase. It also leads to accumulation of glycogen in the liver and has many of the same clinical symptoms as von Gierke disease (see Table 14.2).

Pompe disease is due to a deficiency in lysosomal α-1,4-glucosidase, also called acid maltase. Lysosomes function as recycling centers in the cell and normally degrade excess muscle glycogen into glucose for the energy-converting reactions of the glycolytic pathway. Pompe disease was described in 1932 by Dutch pathologist Johann Pompe, who recognized that lysosomes in the affected patients accumulated large amounts of glycogen. However, Henri Hers—the same Belgian pathologist who described the symptoms of liver glycogen phosphorylase deficiency (Hers disease)—discovered in 1965 that Pompe disease was due to a lysosomal α-1,4-glucosidase deficiency.

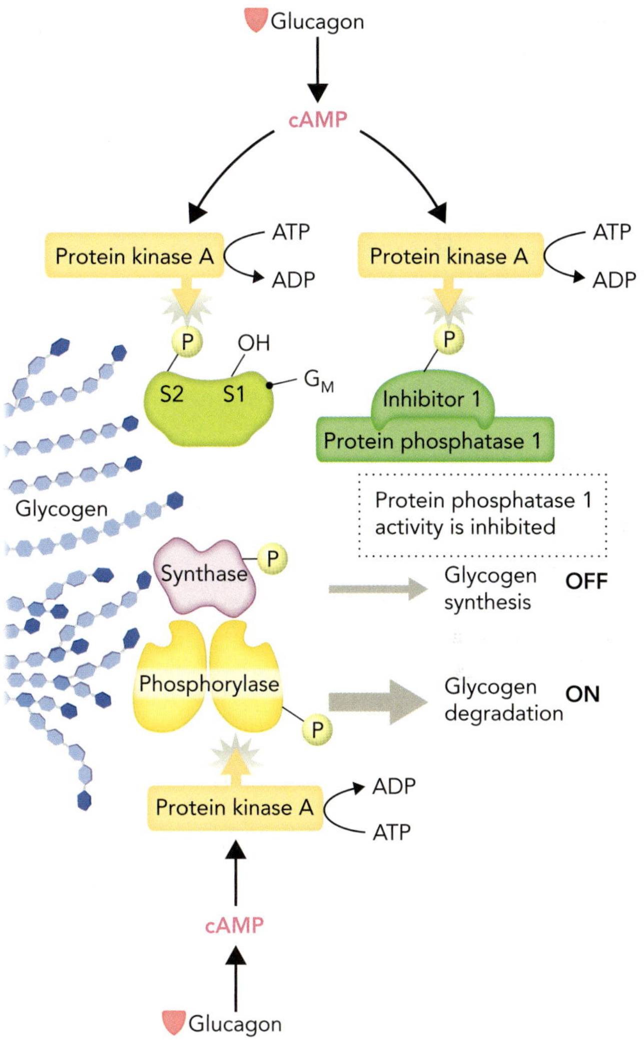

Figure 14.45 Glucagon receptor signaling in muscle cells activates glycogen degradation by stimulating protein kinase A signaling through the second messenger cyclic AMP (cAMP). Protein kinase A phosphorylates glycogen phosphorylase and glycogen synthase, which leads to a net increase in glycogen degradation. Protein kinase A also phosphorylates the site 2 (S2) serine in the G_M subunit, causing protein phosphatase 1 to dissociate from the glycogen particle. Protein phosphatase 1 is maintained in the inactive state through high-affinity binding to Inhibitor 1 protein, which is also phosphorylated by protein kinase A. S1 = site 1.

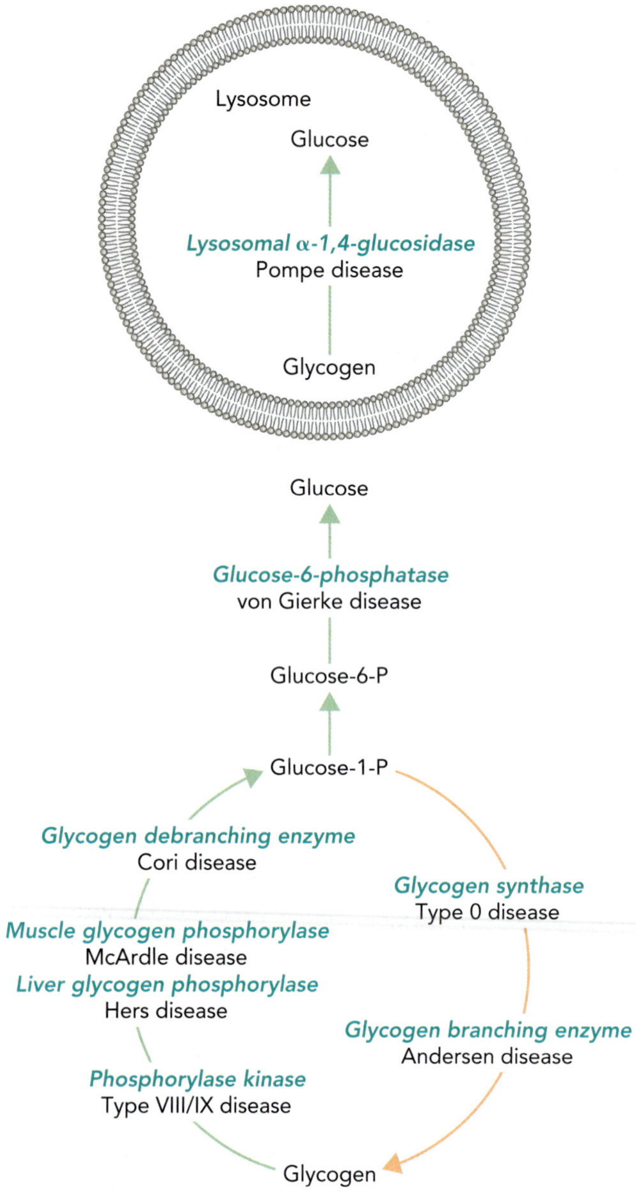

Figure 14.46 Human glycogen storage diseases have been shown to result from defects in key enzymes required for glycogen metabolism. See Table 14.2 and the text for details.

Gerty Cori discovered that patients suffering from **McArdle disease** harbor defects in muscle glycogen phosphorylase. These individuals suffer from exercise-induced cramps and muscle pain due to their inability to degrade muscle glycogen. Cori found that liver glycogen phosphorylase in patients with McArdle disease is normal, and that muscle glycogen phosphorylase in patients with Hers disease is also normal. Therefore, Cori proposed that the human genome contains two distinct genes encoding glycogen phosphorylase enzymes.

Cori disease, also described by Gerty Cori, results from defects in the glycogen debranching enzyme. The structures of both liver and muscle glycogen in these individuals are found to contain short outer chains in the glycogen particle that cannot be removed, thus blocking complete degradation.

The corresponding enzyme deficiency in glycogen branching enzyme is known as **Andersen disease**, which results in the synthesis of large unbranched glycogen molecules that are insoluble and cause the immune system to attack and destroy liver cells. This is one of the most severe glycogen storage diseases because liver damage occurs early in life, causing problems with normal physical and mental development. Unlike other glycogen storage diseases that can be treated to some extent with dietary glucose, the only treatment for Andersen disease is a liver transplant.

Table 14.2 ENZYME DEFICIENCIES AND DISEASE SYMPTOMS OF THE MOST COMMONLY DIAGNOSED HUMAN GLYCOGEN STORAGE DISEASES

Enzyme deficiency	Disease name	Organ affected	Disease symptoms
Glucose-6-phosphatase	von Gierke	Liver	Increased amount of glycogen stores; liver enlargement; kidney failure; hypoglycemia
Liver glycogen phosphorylase	Hers	Liver	Increased amount of liver glycogen; hypoglycemia
Muscle glycogen phosphorylase	McArdle	Muscle	Moderately increased amount of muscle glycogen; exercise-induced muscle cramps
Glycogen synthase	Type 0	Liver	Hypoglycemia; ketosis; failure to thrive
Glycogen debranching enzyme	Cori	Liver and muscle	Enlarged liver; mild hypoglycemia
Glycogen branching enzyme	Andersen	Liver and muscle	Enlarged liver; failure to thrive
Lysosomal α-1,4-glucosidase	Pompe	All organs	Heart failure in infantile form, muscle defects in juvenile form
Phosphorylase kinase	Type VIII/IX	Liver	Enlarged liver

The two other glycogen storage diseases listed in Table 14.2 are quite rare. One is type 0 disease, which is due to defects in glycogen synthase, causing chronic hypoglycemia and the constant need for dietary glucose to survive. The other disease, known as type VIII/IX disease, is due to a defect in the enzyme phosphorylase kinase. It causes the same type of liver dysfunction as Hers disease because of an inability to degrade glycogen in response to glucagon and epinephrine signaling (see Figure 14.42).

The treatments for most of these glycogen storage diseases are dependent on the severity of the symptoms and are of only moderate success in many cases. In the case of Pompe disease, however, it is now possible to use enzyme replacement therapy to treat these patients, which involves weekly infusions with recombinant lysosomal α-1,4-glucosidase protein. Although infusion of this recombinant enzyme does not cure the disease, studies have shown that it can slow the progression of disease symptoms in patients with juvenile and adult-onset Pompe disease. **Figure 14.47** shows muscle biopsies from a Pompe disease patient who had been treated weekly with recombinant lysosomal α-1,4-glucosidase protein for 1 year. In this particular case, the patient showed a remarkable decrease in the amount

a.

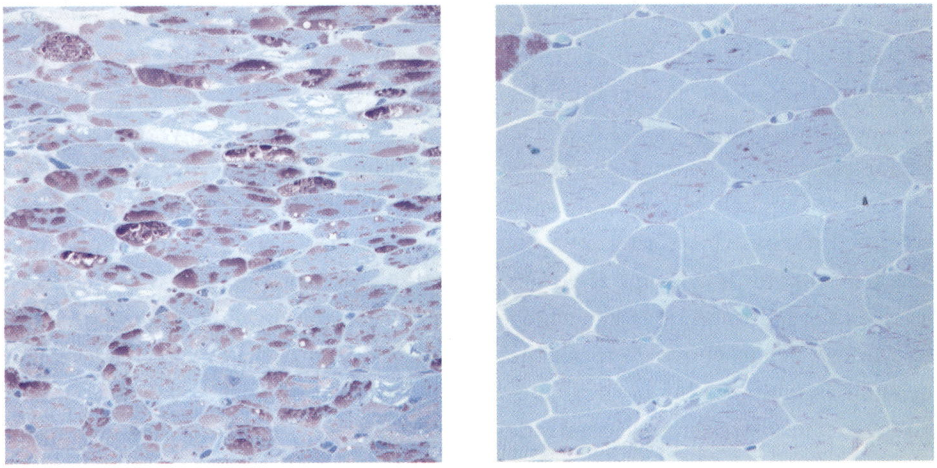

Excess lysosomal glycogen is present before treatment

Glycogen is reduced after treatment with α-1,4-glucosidase

b.

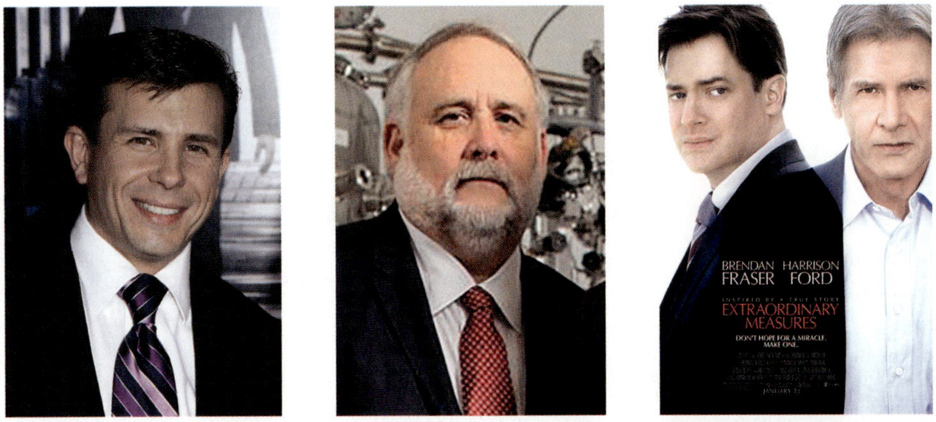

John Crowley

Dr. William Canfield

Extraordinary Measures portrayed the collaboration of Crowley and Canfield

Figure 14.47 Enzyme replacement therapy using recombinant lysosomal α-1,4-glucosidase protein has been used successfully to treat Pompe disease. **a.** Muscle biopsies of a Pompe disease patient before and after 52 weeks of treatment. The darkly stained material in the left panel is excess glycogen in lysosomal compartments, which was present prior to enzyme replacement therapy. REPRODUCED WITH PERMISSION OF THE AUTHOR. B. L. THURBERG ET AL. (2006). CHARACTERIZATION OF PRE- AND POST-TREATMENT PATHOLOGY AFTER ENZYME REPLACEMENT THERAPY FOR POMPE DISEASE. *LABORATORY INVESTIGATION, 86*, 1208–1220. DOI:10.1038/LABINVEST.3700484; PUBLISHED ONLINE 30 OCTOBER 2006. **b.** The collaboration between Dr. William Canfield and John Crowley to find a treatment for Pompe disease was portrayed in the 2010 movie *Extraordinary Measures*. J. CROWLEY: ALLSTAR PICTURE LIBRARY/ALAMY; W. CANFIELD: COURTESY CYTOVANCE BIOLOGICS; *EXTRAORDINARY MEASURES*: CBS FILMS/ COURTESY EVERETT COLLECTION.

of accumulated glycogen. William Canfield, a glycobiologist, and John Crowley, a biotechnology entrepreneur, worked together at their start-up company Novazyme to develop a clinical protocol to treat Pompe disease using enzyme replacement therapy, which was the inspiration for the movie *Extraordinary Measures*. Although an improved treatment for Pompe disease was later developed using a different preparation of α-1,4-glucosidase than that pioneered by Canfield and Crowley, the efforts of Canfield and Crowley demonstrated that protein-based drugs, also called **biologics**, could be used to treat human disease.

concept integration 14.3

Why does it make physiologic sense for muscle glycogen phosphorylase activity to be regulated by both metabolite allosteric control and hormone-dependent phosphorylation?

Muscle glycogen phosphorylase is allosterically activated by AMP, which signals low energy charge in the cell. High AMP levels also indicate a need for glycogen degradation and release of glucose substrate for ATP generation to support muscle contraction. Both ATP and glucose-6-P are allosteric inhibitors of muscle glycogen phosphorylase activity and signal a ready supply of chemical energy without the need for glycogen degradation. Both types of allosteric regulation occur rapidly on a timescale of seconds in response to sudden changes in AMP, ATP, and glucose-6-P levels. Allosteric control by metabolites provides a highly efficient means to control rates of glycogen degradation in response to the immediate energy needs of muscle cells. In contrast, hormonal regulation of muscle glycogen phosphorylase activity by glucagon and epinephrine is a delayed response (occurring on a timescale of hours), resulting in phosphorylation and activation of the enzyme after neuronal and physiologic inputs at the organismal level. Similarly, insulin signaling, which inhibits muscle glycogen phosphorylase activity through dephosphorylation, is also a delayed response at the organismal level and depends on multiple physiologic inputs. Taken together, allosteric regulation of muscle glycogen phosphorylase activity provides a rapid-response control mechanism to modulate muscle glucose levels, whereas hormonal signaling requires input from multiple stimuli at the organismal level and provides a longer-term effect on enzyme activity through covalent modifications.

chapter summary

14.1 The Pentose Phosphate Pathway

- The most important functions of the pentose phosphate pathway are to reduce two molecules of $NADP^+$ to NADPH for each molecule of glucose-6-P and to generate ribose-5-P for nucleotide and coenzyme biosynthesis.
- NADPH functions as a strong reductant in anabolic pathways and in detoxification reactions that neutralize reactive oxygen species. NAD^+ is primarily used as an oxidant in catabolic pathways.
- The pentose phosphate pathway is divided into the oxidative phase, which generates NADPH, and the nonoxidative phase, which interconverts sugar phosphates to regenerate glucose-6-P using transketolase and transaldolase enzymes.
- Flux through the pentose phosphate pathway is regulated to meet three distinct metabolic states of the cell: (1) a need for NADPH; (2) a need to replenish nucleotide pools; and (3) a need to generate ATP from glucose-6-P.

- Glucose-6-phosphate dehydrogenase (G6PD) catalyzes the commitment step in the pentose phosphate pathway and is highly regulated by the $NADP^+$-to-NADPH concentration ratio to control flux through the pathway ($NADP^+$ stimulates enzyme activity to increase NADPH levels in the cell).
- Mutations in the human G6PD gene cause the most common enzyme deficiency in the world. It results in reduced intracellular levels of NADPH, which is required to reduce glutathione and protect cells from compounds that can induce oxidative stress, such as primaquine (antimalarial drug) and vicine (from fava beans).

14.2 Gluconeogenesis

- Gluconeogenesis synthesizes glucose from noncarbohydrate sources when dietary glucose is limiting and glucose stores have been depleted. Much of the glucose used by the brain and erythrocytes in humans comes from gluconeogenesis occurring in liver and kidney cells.
- Gluconeogenesis and glycolysis share seven enzymes, catalyzing reversible reactions in both pathways. The four solely gluconeogenic enzymes are pyruvate carboxylase, phosphoenolpyruvate carboxykinase, fructose-1,6-bisphosphatase-1 (FBPase-1), and glucose-6-phosphatase, which bypass three exergonic reactions in glycolysis (catalyzed by hexokinase, phosphofructokinase-1, and pyruvate kinase).
- Pyruvate carboxylase and phosphoenolpyruvate carboxykinase catalyze the pyruvate kinase bypass reactions by converting pyruvate to phosphoenolpyruvate, using phosphoryl transfer energy from ATP (pyruvate carboxylase reaction) and GTP (phosphoenolpyruvate carboxykinase reaction).
- FBPase-1 catalyzes the phosphofructokinase-1 (PFK-1) bypass reaction by converting fructose-1,6-BP to fructose-6-P. FBPase-1 and PFK-1 are reciprocally regulated by the allosteric effectors AMP, fructose-2,6-BP, and citrate in response to energy charge and hormonal signaling.
- The hexokinase reaction in glycolysis is bypassed by the gluconeogenic enzyme glucose-6-phosphatase, which is localized to the ER lumen and thus physically isolated from the hexokinase reaction in the cytosol.
- Hormonal control of PFK-1 and FBPase-1 is mediated in liver cells by fructose-2,6-BP. Fructose-2,6-BP is structurally related to fructose-6-P and fructose-1,6-BP but is not a metabolic intermediate in either the glycolytic or gluconeogenic pathways.
- The apparent affinity of PFK-1 for fructose-6-P is stimulated 25-fold by fructose-2,6-BP levels, whereas fructose-2,6-BP decreases the apparent affinity of FBPase-1 for fructose-1,6-BP under the same conditions. The net effect is that high levels of fructose-2,6-BP stimulate flux through the glycolytic pathway and inhibit flux through the gluconeogenic pathway.
- The level of fructose-2,6-BP in liver cells is controlled by the dual-function enzyme phosphofructokinase-2/fructose-2,6-bisphosphatase (PFK-2/FBPase-2), which is hormonally regulated by phosphorylation. Insulin signaling stimulates dephosphorylation of PFK-2/FBPase-2, leading to higher levels of fructose-2,6-BP and activation of flux through the glycolytic pathway; glucagon signaling decreases fructose-2,6-BP levels and stimulates gluconeogenesis.
- The Cori cycle uses gluconeogenesis in liver cells to convert muscle-derived lactate into glucose, thereby replenishing glucose levels and supporting muscle contraction. The energy cost of running gluconeogenesis (liver) and glycolysis (muscle) at the same time is net 4 ATP equivalents (the difference between 2 ATP produced by anaerobic glycolysis and 4 ATP + 2 GTP consumed by gluconeogenesis).

14.3 Glycogen Degradation and Synthesis

- Glycogen is a storage form of glucose in animals and consists of branched homopolysaccharides linked by α-1,4 and α-1,6 glycosidic bonds. Glycogen degradation and synthesis occur in the cytosol, with the substrate for these reactions being the free ends (nonreducing ends) of the branched polymer.
- The three key enzymes required for reversible degradation and synthesis of glycogen are glycogen phosphorylase, glycogen synthase, and the glycogen branching and debranching enzymes.
- Glycogen phosphorylase catalyzes a reaction that releases glucose-1-P from glycogen in a phosphorolysis reaction involving inorganic phosphate and cleavage of the α-1,4 glycosidic bond. Glucose-1-P is converted to glucose-6-P, which is either used for glycolysis in muscle cells or dephosphorylated in liver cells and exported to other tissues. Glycogen phosphorylase activity is stimulated by glucagon and epinephrine signaling.
- Glycogen synthase is activated by insulin signaling and adds glucose to the nonreducing ends in a reaction involving UDP-glucose. Glycogen synthase uses the bond energy available in UDP-glucose to form α-1,4 glycosidic bonds at the nonreducing ends of the glycogen particle.
- Branching and debranching enzymes modify glycogen complexes to facilitate glycogen degradation (debranching) and glycogen synthesis (branching) through the cleavage and formation of α-1,6 glycosidic bonds. Both enzymes also catalyze reactions involving the transfer of glucose oligosaccharides between branched and unbranched regions of the molecule.
- Glycogen degradation and synthesis require the enzyme phosphoglucomutase, which interconverts glucose-1-P and glucose-6-P through the formation of a bisphosphorylated enzyme intermediate.
- Glycogen phosphorylase exists in the active R-state conformation and the inactive T-state conformation. The equilibrium between the R state and T state is shifted by covalent modification (phosphorylation/dephosphorylation) and metabolite allosteric regulation.

- The enzyme phosphorylase kinase phosphorylates glycogen phosphorylase and is a downstream target of glucagon and epinephrine signaling in liver cells.
- Allosteric activation of unphosphorylated muscle glycogen phosphorylase by AMP binding induces the R-state conformation of the enzyme, whereas ATP and glucose-6-P are allosteric inhibitors of unphosphorylated muscle glycogen phosphorylase by competing with AMP for binding to the allosteric site.
- Phosphorylated liver glycogen phosphorylase is allosterically inhibited by glucose binding, which shifts the equilibrium from the R state to the T state, and facilitates dephosphorylation of the enzyme in response to insulin signaling.
- Glycogen synthase adds glucose to the nonreducing ends of glycogen using the nucleotide sugar UDP-glucose. The enzyme UDP-glucose pyrophosphorylase catalyzes a reaction involving the attack of a phosphoryl oxygen from glucose-1-P on the α phosphate of UTP to generate UDP-glucose.
- Glycogenin is both an anchor protein for the glycogen core particle and an enzyme that catalyzes the glycosyltransferase and synthesis reactions needed to generate the initial glycogen chain.
- Glucagon signaling stimulates protein kinase A activity, leading to net phosphorylation of glycogen phosphorylase and glycogen synthase, which results in increased glycogen degradation and glucose release.
- Insulin signaling stimulates protein phosphatase 1 activity and promotes net dephosphorylation of glycogen phosphorylase and glycogen synthase, which results in increased glycogen synthesis and glucose import.
- von Gierke disease, Hers disease, and Pompe disease are the most prevalent of the six best-characterized human glycogen storage diseases. von Gierke disease is due to a deficiency in the enzyme glucose-6-phosphatase; Hers disease is a defect in the liver enzyme glycogen phosphorylase; and Pompe disease is due to a deficiency in lysosomal α-1,4-glucosidase, also called acid maltase.
- Three less common human glycogen storage diseases are McArdle disease (defect in muscle glycogen phosphorylase), Cori disease (defect in glycogen debranching enzyme), and Andersen disease (defect in glycogen branching enzyme).

biochemical terms

(in order of appearance in text)
pentose phosphate pathway (p. 680)
phosphogluconate pathway (p. 680)
glucose-6-phosphate dehydrogenase (G6PD) (p. 683)
transketolase (p. 683)
transaldolase (p. 683)
glutathione (p. 687)
primaquine (p. 688)
favism (p. 688)
vicine (p. 689)
gluconeogenesis (p. 690)
pyruvate carboxylase (p. 693)
phosphoenolpyruvate carboxykinase (p. 693)
fructose-1,6-bisphosphatase-1 (p. 693)
glucose-6-phosphatase (p. 693)
biotin (p. 694)
fructose-2,6-bisphosphate (fructose-2,6-BP) (p. 698)
phosphofructokinase-2/fructose-2,6-bisphosphatase (PFK-2/FBPase-2) (p. 700)
Cori cycle (p. 701)
glycogen particles (p. 704)
glycogen phosphorylase (p. 705)
glycogen synthase (p. 705)
branching and debranching enzymes (p. 705)
protein phosphatase 1 (p. 718)
von Gierke disease (p. 719)
Hers disease (p. 719)
Pompe disease (p. 719)
McArdle disease (p. 720)
Cori disease (p. 720)
Andersen disease (p. 720)
biologics (p. 722)

review questions

1. What are the chemical and functional differences between NADH and NADPH?
2. What are the three metabolic conditions that determine metabolic flux through the pentose phosphate pathway?
3. Write an equation for the rate-limiting reaction in the pentose phosphate pathway. What enzyme catalyzes this reaction, and what is the $\Delta G^{\circ\prime}$ value in the direction written?
4. How many molecules of NADPH are generated by the oxidative phase of the pentose phosphate pathway using six molecules of glucose-6-P as substrates? How many molecules of glucose-6-P are regenerated in the nonoxidative phase of the pentose phosphate pathway under these same conditions? What accounts for the loss of six carbons in these reactions?
5. What is the primary mechanism by which metabolic flux is regulated in the pentose phosphate pathway?
6. Why are reduced glutathione levels in red blood cells dependent on the activity of the enzyme glucose-6-phosphate dehydrogenase? Why are individuals with a glucose-6-phosphate dehydrogenase deficiency sensitive to prophylactic antimalarials such as primaquine, but at the same time more resistant to developing malaria even without primaquine treatment?
7. The conversion of pyruvate to phosphoenolpyruvate in the gluconeogenic pathway requires phosphoryl transfer energy

in reactions catalyzed by the enzymes pyruvate carboxylase (ATP dependent) and phosphoenolpyruvate carboxykinase (GTP dependent). Why is this pair of reactions counted as a cost of 4 ATP equivalents to convert pyruvate to phosphoenolpyruvate when counting up the number of ATP needed to generate one molecule of glucose?

8. Pyruvate carboxylase is often considered a gluconeogenic enzyme; however, it also plays an important role in an energy-converting pathway under conditions of low energy charge and high acetyl-CoA levels in mitochondria. What is the name of this pathway?
9. What is the metabolic logic of reciprocal regulation of the glycolytic and gluconeogenic pathways by citrate?
10. What is the functional role of fructose-2,6-bisphosphate (fructose-2,6-BP) in the gluconeogenic pathway? Is it a metabolite?
11. Is phosphofructokinase-2/fructose-2,6-bisphosphatase one protein with two catalytic activities or two protein subunits each encoding a single catalytic activity? What regulates the two catalytic activities of phosphofructokinase-2/fructose-2,6-bisphosphatase?
12. Explain why the Cori cycle has a net cost of 4 ATP equivalents per glucose to the organism.
13. Glycogen contains an α-1,6-glycosidic bond about once every 10 glucose residues, thereby creating a branch point and a corresponding nonreducing end for the removal and addition of glucose molecules. If a glycogen particle contains a total of 50,000 glucose residues, how many nonreducing ends are most likely to be found: ~25,000 ends, ~2,500 ends, or ~250 ends? Explain your answer.
14. The $\Delta G^{\circ\prime}$ of the glycogen phosphorylase reaction is +3.1 kJ/mol, whereas the ΔG under physiologic conditions is –6 kJ/mol. What is likely to account for this difference of ~9 kJ/mol between the $\Delta G^{\circ\prime}$ and ΔG values?
15. What is the function of glucose-6-phosphatase in liver and muscle cells?
16. The product of the glycogen phosphorylase reaction is glucose-1-P. Is there a difference in glycolytic ATP yield comparing the yield from the metabolism of glucose-1-P derived from glycogen degradation with the yield from the metabolism of dietary glucose? Explain.
17. Explain the metabolic logic of glucagon and insulin regulation of glycogen metabolism.
18. Defects in essentially every enzyme required for human glycogen metabolism have been identified and are collectively called glycogen storage diseases. Explain why Andersen disease, caused by a defect in glycogen branching enzyme, is fatal, whereas Cori disease, caused by a defect in glycogen debranching enzyme, only manifests in mild hypoglycemia.

challenge problems

1. Defects in the pentose phosphate pathway enzyme glucose-6-phosphate dehydrogenase represent the most common enzyme deficiency in humans.
   a. What explains the observation that individuals born with a deficiency in the enzyme glucose-6-phosphate dehydrogenase become clinically anemic if they have a diet rich in fava beans?
   b. Why is it thought that increased resistance to malaria is linked to glucose-6-phosphate dehydrogenase deficiencies?
2. Explain why 3 net ATP are generated in glycolysis using glucose-1-P derived from glycogen degradation, whereas dietary glucose only produces 2 net ATP by glycolysis.
3. What are the three potential substrate cycles (futile cycles between a reactant and its immediate enzymatic product) between the glycolytic and gluconeogenic pathways, and how is each substrate cycle prevented?
4. An individual with chronic hypoglycemia was suspected of having a defect in one of the enzymes unique to gluconeogenesis. To identify the defective enzyme, tissue samples from a normal liver were compared to samples from the patient's liver biopsy, using a biochemical assay that measures glucose production from glycerol or malate.

   It was found that incubation with glycerol produced normal amounts of glucose in both the control and biopsied liver samples; however, incubation with malate did not lead to glucose production in the liver biopsy, even though it did lead to glucose production in the control liver sample.

   On the basis of these observations, which of the four unique gluconeogenic enzymes is most likely defective in this individual? Hint: See Figure 9.48 to review glycerol entry into the gluconeogenic pathway and Figure 10.39 to review the transport of citrate cycle metabolites in and out of mitochondria.
5. Consuming large amounts of alcohol when blood glucose levels are low can lead to hypoglycemia due to low rates of gluconeogenesis. Considering that ethanol is oxidized to acetaldehyde by the enzyme alcohol dehydrogenase, how do you explain the inhibitory effect of alcohol consumption on gluconeogenesis and the subsequent hypoglycemia?

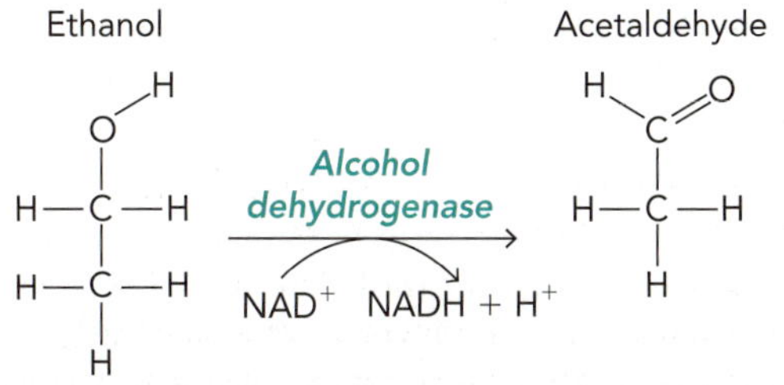

6. The Cori cycle provides a mechanism to replenish glycogen in muscle cells after anaerobic exercise by converting lactate, derived from muscle glycogen degradation, into glucose in liver cells, which is then exported back to the muscle cells.
   a. How many *net* ATP are generated in muscle cells *per glucose-1-P* generated by the glycogen phosphorylase reaction under these anaerobic conditions; that is, how many ATP are generated by reactions converting 1 glucose-1-P → 2 lactate? Explain.
   b. Once lactate is converted to glucose in liver cells by gluconeogenesis and then returned to the muscle to rebuild glycogen stores, how many ATP equivalents are required to incorporate *each of these glucose* molecules into glycogen; that is, 1 glucose + $\text{glycogen}_{(n)} \rightarrow \text{glycogen}_{(n+1)}$? Explain.
   c. Considering the ATP yields and ATP investments involved in removing glucose from glycogen in muscle cells, converting lactate to glucose in liver cells, and then resynthesizing glycogen using this glucose, what is the net ATP investment of glycogen metabolism with regard to running the Cori cycle?
7. What is the metabolic logic that explains why liver cells contain the enzyme glucose-6-phosphatase but muscle cells lack this enzyme?
8. Explain why individuals with von Gierke disease, which is a lack of glucose-6-phosphatase in the liver, accumulate large amounts of glycogen in the liver. Why would these same individuals release a small amount of glucose into the blood after injection with a high dose of glucagon?
9. McArdle disease is a glycogen metabolism disorder resulting from defects in the enzyme muscle glycogen phosphorylase. Explain why a defect in this enzyme makes it very difficult for these individuals to run up a flight of stairs.
10. Why does it make sense that muscle and liver glycogen phosphorylase are differentially regulated by glucose and AMP?
11. The intracellular signaling enzyme protein kinase A is activated by cyclic AMP (cAMP) binding, after upstream receptor activation through glucagon and epinephrine signaling.
    a. What is the downstream effect of protein kinase A activation on glycogen synthase and glycogen phosphorylase activities?
    b. Do these protein kinase A effects on glycogen synthase and glycogen phosphorylase increase or decrease the amount of stored glycogen in the body?
    c. Why does it make sense that glucagon and epinephrine have the same effect on glycogen metabolism, which is the exact opposite effect of insulin on glycogen metabolism?
12. The glucagon signaling pathway in liver cells (shown in the following illustration) activates glycogen degradation through a series of amplification steps, including (1) stoichiometric binding of glucagon to the glucagon receptor, (2) $G_{s\alpha}$ stimulation of adenylate cyclase (AC), (3) cAMP activation of protein kinase A (PKA), (4) protein kinase A activation of phosphorylase kinase (PhK), (5) phosphorylase kinase activation of glycogen phosphorylase (Ph), and (6) glycogen degradation and release of glucose. Note that the enzyme cAMP phosphodiesterase (cAMP PDE) cleaves cAMP to terminate glucagon signaling.

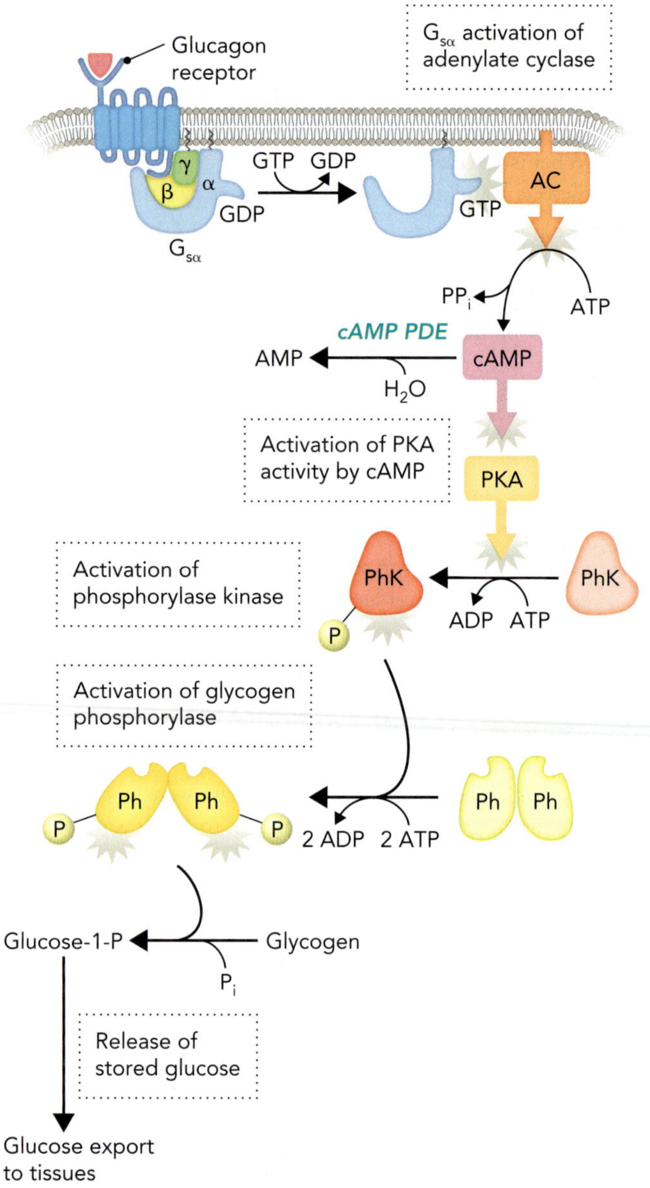

   a. Which enzymes in the cascade are regulated by covalent modification?
   b. Calculate the theoretical number of molecules of glucose-1-phosphate generated per second from glycogen as a result of the interaction of one molecule of glucagon with its receptor on a liver cell membrane using the following assumptions: (i) Ten $G_{s\alpha}$–GTP are formed for each molecule of glucagon bound to a receptor molecule. (ii) None of the downstream enzymes are rate limiting for the upstream glucagon signal. (iii) Adenylate cyclase, protein kinase A, phosphorylase kinase, and glycogen phosphorylase

have a catalytic turnover rate of 1,000 per second. (iv) cAMP phosphodiesterase is equimolar to adenylate cyclase and has a catalytic turnover rate of 100 per second. (v) Two cAMP are required to activate one protein kinase A catalytic subunit (see Figure 8.26).

smartwork5

If your instructor assigns homework with Smartwork5, access it here: digital.wwnorton.com/biochem.

suggested reading

Books and Reviews

Byrne, B. J., Falk, D. J., Pacak, C. A., Nayak, S., Herzog, R. W., Elder, M. E., Collins, S. W., Conlon, T. J., Clement, N., Cleaver, B. D., et al. (2011). Pompe disease gene therapy. *Human Molecular Genetics*, *20*, R61–68.

Cohen, P. T. (2002). Protein phosphatase 1—targeted in many directions. *Journal of Cell Science*, *115*, 241–256.

Frank, J. E. (2005). Diagnosis and management of G6PD deficiency. *American Family Physician*, *72*, 1277–1282.

Kornberg, A. (2001). Remembering our teachers. *Journal of Biological Chemistry*, *276*, 3–11.

Moslemi, A. R., Lindberg, C., Nilsson, J., Tajsharghi, H., Andersson, B., and Oldfors, A. (2010). Glycogenin-1 deficiency and inactivated priming of glycogen synthesis. *New England Journal of Medicine*, *362*, 1203–1210.

Ozen, H. (2007). Glycogen storage diseases: new perspectives. *World Journal of Gastroenterology*, *13*, 2541–2553.

Primary Literature

Baskaran, S., Roach, P. J., DePaoli-Roach, A. A., and Hurley, T. D. (2010). Structural basis for glucose-6-phosphate activation of glycogen synthase. *Proceedings of the National Academy of Sciences USA*, *107*, 17563–17568.

Jain, H., Beriwal, S., and Singh, S. (2002). Alcohol induced ketoacidosis, severe hypoglycemia and irreversible encephalopathy. *Medical Science Monitor*, *8*, CS77–79.

Kotaka, M., Gover, S., Vandeputte-Rutten, L., Au, S. W., Lam, V. M., and Adams, M. J. (2005). Structural studies of glucose-6-phosphate and $NADP^+$ binding to human glucose-6-phosphate dehydrogenase. *Acta Crystallographica. Section D, Biological Crystallography*, *61*, 495–504.

Lee, Y. H., Li, Y., Uyeda, K., and Hasemann, C. A. (2003). Tissue-specific structure/function differentiation of the liver isoform of 6-phosphofructo-2-kinase/fructose-2,6-bisphosphatase. *Journal of Biological Chemistry*, *278*, 523–530.

Louicharoen, C., Patin, E., Paul, R., Nuchprayoon, I., Witoonpanich, B., Peerapittayamongkol, C., Casademont, I., Sura, T., Laird, N. M., Singhasivanon, P., et al. (2009). Positively selected G6PD-Mahidol mutation reduces *Plasmodium vivax* density in Southeast Asians. *Science*, *326*, 1546–1549.

Massillon, D., Bollen, M., De Wulf, H., Overloop, K., Vanstapel, F., Van Hecke, P., and Stalmans, W. (1995). Demonstration of a glycogen/glucose 1-phosphate cycle in hepatocytes from fasted rats. Selective inactivation of phosphorylase by 2-deoxy-2-fluoro-alpha-D-glucopyranosyl fluoride. *Journal of Biological Chemistry*, *270*, 19351–19356.

Peters, A. L., and Van Noorden, C. J. (2009). Glucose-6-phosphate dehydrogenase deficiency and malaria: cytochemical detection of heterozygous G6PD deficiency in women. *Journal of Histochemistry and Cytochemistry*, *57*, 1003–1011.

Martin, J. L., Johnson, L. N., and Withers, S. G. (1990). Comparison of the binding of glucose and glucose-1-phosphate derivatives to T-state glycogen phosphorylase b. *Biochemistry*, *29*, 10745–10757.

Terrak, M., Kerff, F., Langsetmo, K., Tao, T., and Dominguez, R. (2004). Structural basis of protein phosphatase 1 regulation. *Nature*, *429*, 780–784.

Thurberg, B. L., Lynch Maloney, C., Vaccaro, C., Afonso, K., Tsai, A. C., Bossen, E., Kishnani, P. S., and O'Callaghan, M. (2006). Characterization of pre- and post-treatment pathology after enzyme replacement therapy for Pompe disease. *Laboratory Investigation*, *86*, 1208–1220.

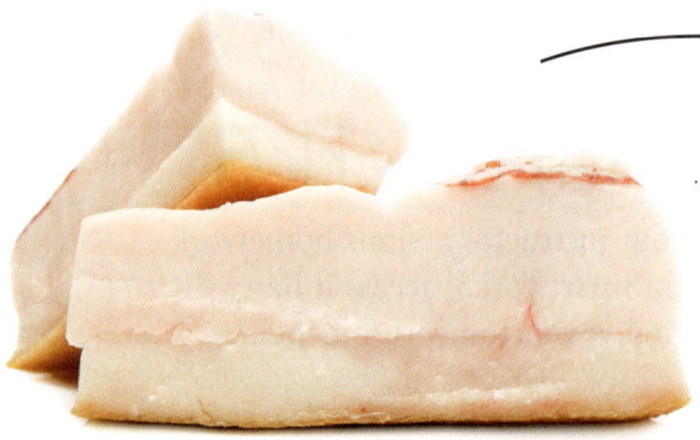

Pork lard

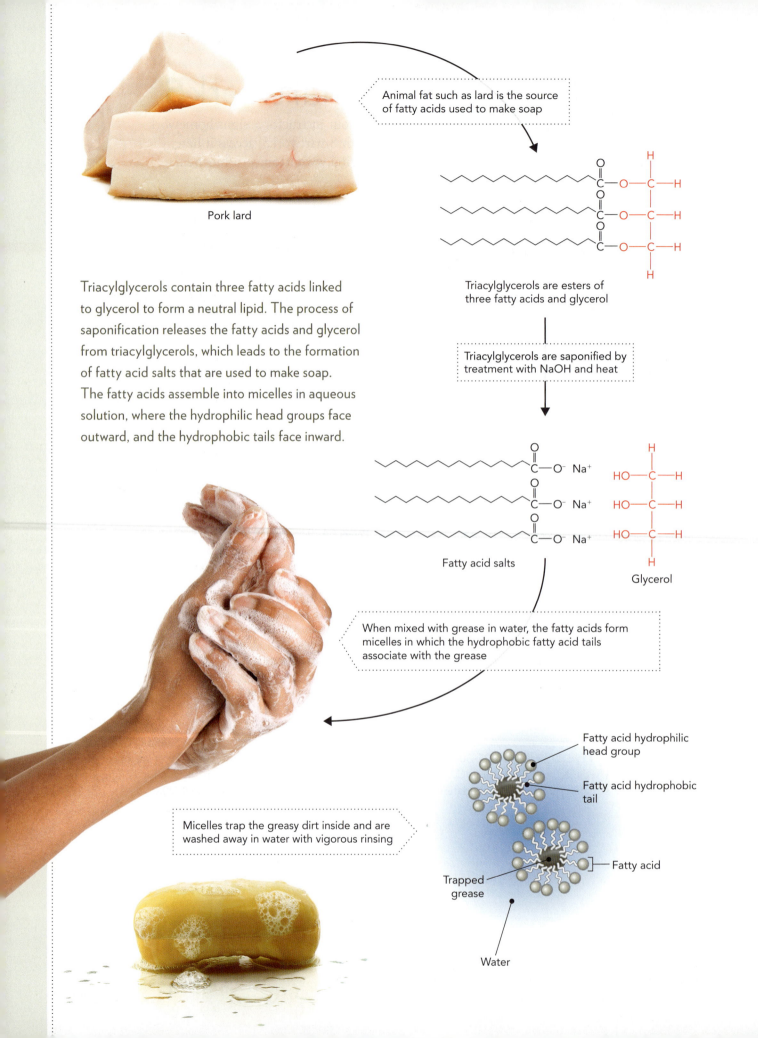

Animal fat such as lard is the source of fatty acids used to make soap

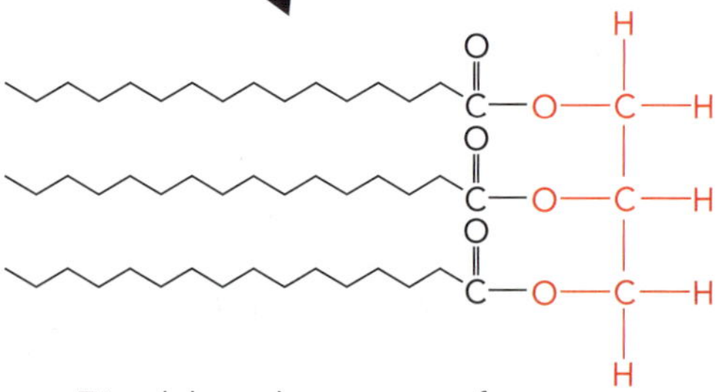

Triacylglycerols are esters of three fatty acids and glycerol

Triacylglycerols contain three fatty acids linked to glycerol to form a neutral lipid. The process of saponification releases the fatty acids and glycerol from triacylglycerols, which leads to the formation of fatty acid salts that are used to make soap. The fatty acids assemble into micelles in aqueous solution, where the hydrophilic head groups face outward, and the hydrophobic tails face inward.

Triacylglycerols are saponified by treatment with NaOH and heat

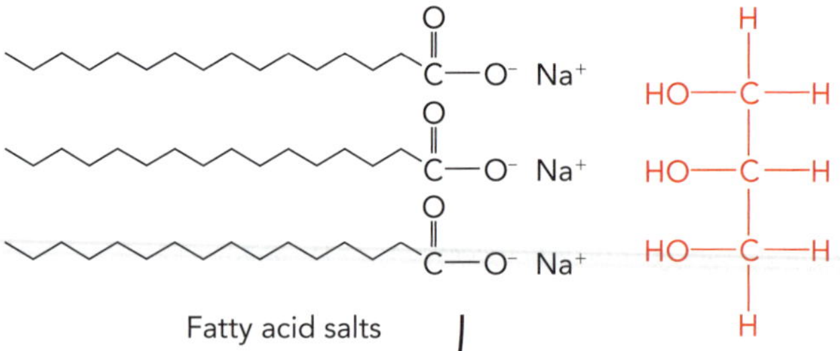

Fatty acid salts

Glycerol

When mixed with grease in water, the fatty acids form micelles in which the hydrophobic fatty acid tails associate with the grease

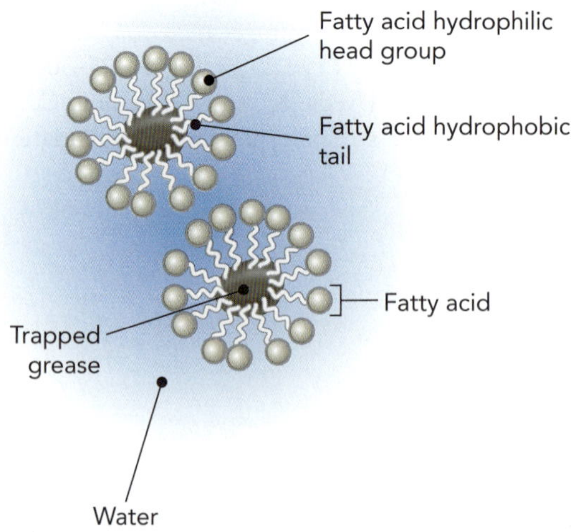

Micelles trap the greasy dirt inside and are washed away in water with vigorous rinsing

15

Lipid Structure and Function

Many people don't realize that animal fat is used to make most soap, which is interesting because we use soap to remove animal fat from our hands and clothing. Soap is made by subjecting animal lard to saponification, in which heat and a strong alkali (NaOH or KOH) release fatty acids in the form of sodium or potassium salts from triacylglycerols. When mixed with water, the fatty acids in the soap function as amphipathic molecules that form micelles. With sufficient agitation, these micelles trap greasy food particles released from hands or clothing and are washed away during the rinsing phase. Free fatty acids inside cells would also act as soap and disrupt biological membranes, but fatty acids inside cells are chemically neutralized by linkage to glycerol or are biologically sequestered in the cell by binding to fatty acid carrier proteins.

CREDITS: PORK LARD: MARTINA_L/SHUTTERSTOCK; SOAPY HAND: BORGES SAMUEL/ ALAMY STOCK PHOTO; WET SOAP: BIG PANTS PRODUCTION/SHUTTERSTOCK.

CHAPTER OUTLINE

15.1 Many Lipids Are Made from Fatty Acids

- Structures of the most common fatty acids
- Biological waxes have a variety of functions
- Structure and nonmetabolic uses of triacylglycerols

15.2 Triacylglycerols Are Energy Storage Lipids

- Dietary triacylglycerols are transported by chylomicrons
- Triacylglycerols synthesized in the liver are packaged in VLDL particles
- Adipocytes cleave stored triacylglycerols and release free fatty acids

15.3 Cell Membranes Contain Three Major Types of Lipids

- Cell membranes have distinct lipid and protein compositions
- Glycerophospholipids are the most abundant membrane lipids
- Sphingolipids contain one fatty acid linked to sphingosine
- Cholesterol is a rigid, four-ring molecule in plasma membranes

15.4 Lipids Function in Cell Signaling

- Cholesterol derivatives regulate the activity of nuclear receptor proteins
- Eicosanoids are derived from arachidonate

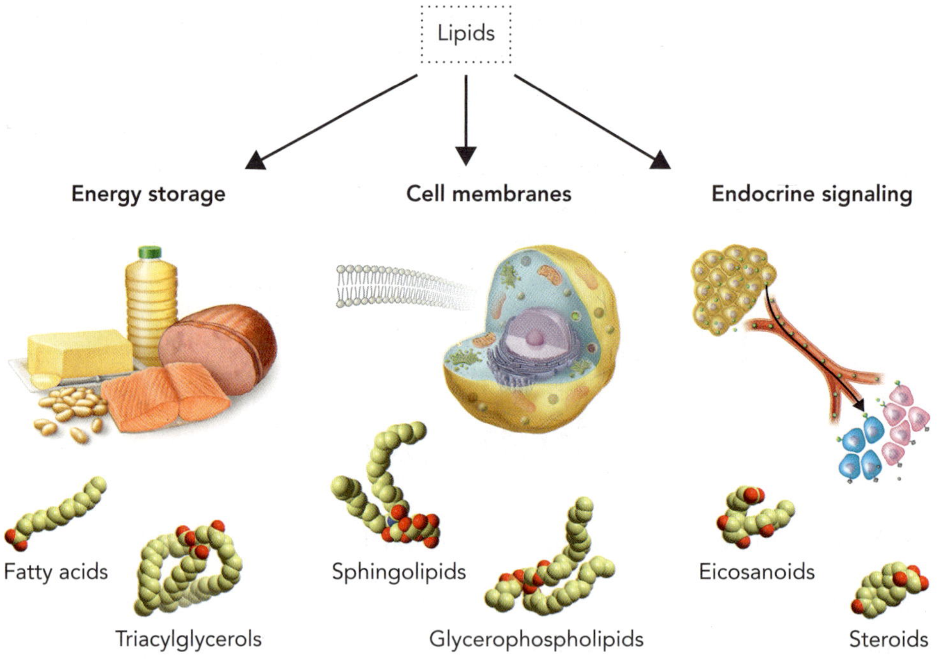

Figure 15.1 Lipids serve three important roles in biology. Fatty acids and triacylglycerols serve as energy storage molecules and are either obtained directly from the diet or synthesized from carbohydrates. Glycerophospholipids and sphingolipids form the hydrophobic barriers of cell membranes, including both the plasma membrane and endomembranes. Steroids and eicosanoids function in the endocrine system and activate a variety of signaling pathways. BASED ON PDB FILES 1E7I (STEARATE), 1V54 (TRIACYLGLYCEROL), 1GZP (GANGLIOSIDE), 1T27 (PHOSPHATIDYLCHOLINE), 1RYO (PROSTAGLANDIN), AND 2V95 (CORTISOL).

We briefly introduced membrane lipids in Chapter 2 as hydrophobic biomolecules that form a barrier between the cell and its surrounding environment and also partition the cytoplasm and nucleus into intracellular compartments. In this chapter, we examine membrane lipids in more detail and describe the structure and function of energy storage lipids and lipid signaling molecules (**Figure 15.1**). In Chapter 16, we continue our discussion of lipids by examining the metabolic pathways responsible for synthesizing and degrading fatty acids and cholesterol.

Lipids are a diverse class of biomolecules with a wide variety of structures. Many types of biomolecular complexes contain lipids; for example, lipoproteins and glycerophospholipids. A key characteristic of lipids is their hydrophobic nature, which is usually determined by hydrocarbon chains of various lengths (fatty acids) or fused cyclohexane rings (steroids). Because most lipids in biological systems are derived from fatty acids, we begin the chapter with a description of the structure and function of the most common saturated and unsaturated fatty acids.

15.1 Many Lipids Are Made from Fatty Acids

Fatty acids consist of a carboxylic acid head group and a hydrocarbon chain of 4–36 carbons, which forms a hydrophobic tail. The pK_a of the carboxyl group in free fatty acids is ~4.8, and therefore at pH 7, the carboxyl group is ionized and in the form of a conjugate base. As with other biomolecules we have described, this means at pH 7, the name of a free fatty acid ends with the suffix "-ate" as in *palmitate,* rather than *palmitic acid* (the conjugate acid that exists below pH 4.8). The same pH 7 standard applies to naming the *citrate cycle* (rather than *citric acid cycle*), as the citrate cycle reactions take place at ~pH 7 under standard physiologic conditions.

The most abundant fatty acids in nature are unbranched chains of 12–20 carbons. (Most fatty acids have an even number of carbons because they are synthesized from two-carbon units, as we discuss in Chapter 16.) If they contain fully reduced methylene groups (CH_2), they are called **saturated fatty acids**; if the hydrocarbon chains have one or more carbon–carbon double bonds (C=C bonds), they are called **unsaturated fatty acids**. Unsaturated fatty acids with one C=C bond are called **monounsaturated fatty acids**, and those with more than one C=C bond are called **polyunsaturated fatty acids**.

In general, cells more often use long-chain saturated fatty acids for energy storage because of their high redox potential (as they have many electrons to donate), whereas polyunsaturated fatty acids are usually components of membranes. Fatty acids covalently bound to proteins can be used to tether a protein to a biological membrane. Fatty-acylated cell signaling proteins, such as heterotrimeric G proteins and Ras, contain covalently attached myristate or palmitate fatty acid moieties, which serve to anchor the proteins to the cytosolic face of the plasma membrane (see Figure 8.20).

Structures of the Most Common Fatty Acids

Chemical structures of the most common saturated, unsaturated, and essential fatty acids are shown in **Figure 15.2**. Myristate, palmitate, and stearate are saturated fatty acids, whereas palmitoleate and oleate are monounsaturated fatty acids. The

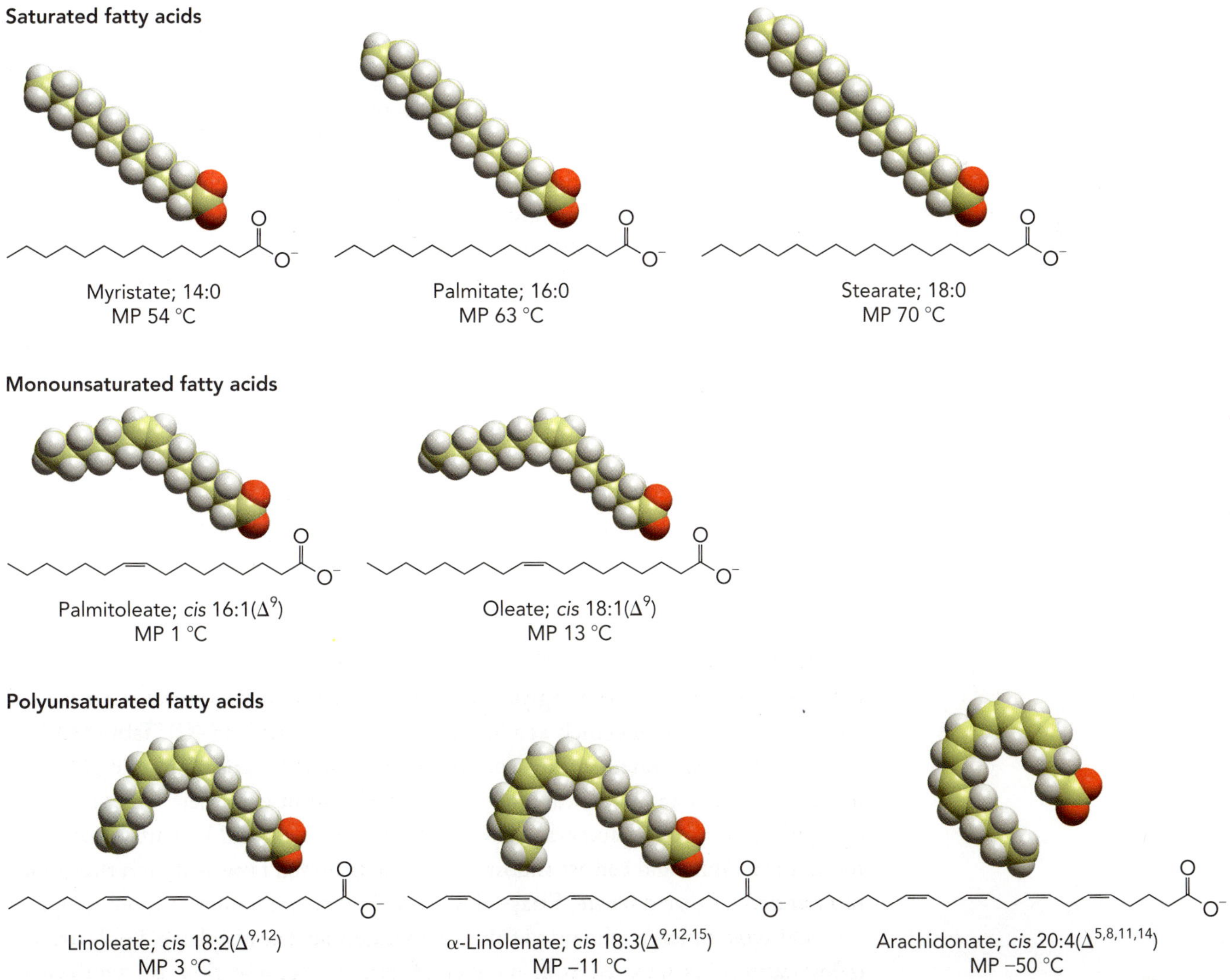

Figure 15.2 The structures and melting points of several common saturated and unsaturated fatty acids are shown. The Corey–Pauling–Koltun (CPK) models and chemical structures are shown for each fatty acid, along with the common name and C=C bond position and configuration (cis in all cases). Also listed is the melting point (MP) at which each fatty acid changes from a solid to a liquid and is at equilibrium between these states. The structures of unsaturated fatty acids in the cis configuration are shown here as unbent molecules for convenience. The CPK representations were made using a ChemDraw conversion method that generates PDB files.

polyunsaturated fatty acids linoleate, α-linolenate, and arachidonate are **essential fatty acids** in humans, which must be obtained from dietary sources because humans lack the enzymes required to synthesize them. The common names of fatty acids reflect how or where the fatty acid was discovered. For example, palmitate was first isolated from palm oil, and linoleate was isolated from flax seed (*linon* is Greek for "flax").

Figure 15.2 uses a common nomenclature for fatty acids, denoting the number of carbon atoms and carbon–carbon double bonds in the molecule separated by a colon. For example, palmitate is a fully saturated C_{16} fatty acid (16:0), whereas oleate is a monounsaturated C_{18} fatty acid (18:1). The locations of the C═C bonds are shown in unsaturated fatty acids by the symbol "Δ," followed by the numbers of the first carbons of the double bonds. (The carbons are numbered beginning with the carboxyl carbon as C-1.) The terms "cis" or "trans" are added to indicate the configuration of the double bonds. Because rotation about the C═C bond is restricted, interconversion between the cis and trans configurations is limited. Most fatty acids in nature have C═C bonds in the cis configuration, such as oleate, a *cis* 18:1(Δ^{9}) fatty acid; linoleate, a *cis* 18:2($\Delta^{9,12}$) fatty acid; and α-linolenate, a *cis* 18:3($\Delta^{9,12,15}$) fatty acid. As shown in **Figure 15.3**, saturated fatty acids or those fatty acids containing C═C bonds in the trans configuration are largely straight, whereas a C═C bond in the cis configuration introduces a bend in the chain. The cis or trans configuration of the double bonds in fatty acids affects the packing together of the hydrocarbon chains and thus their melting points.

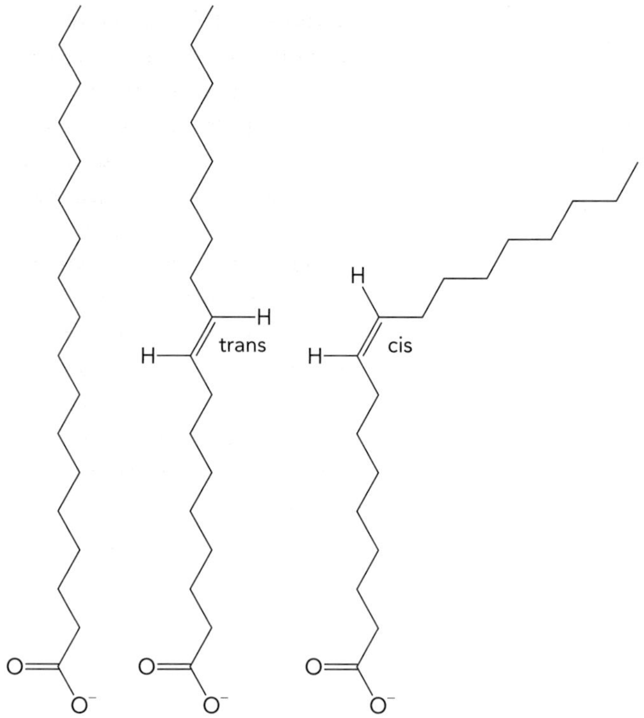

Figure 15.3 The degree of saturation and the configuration affect the structure of fatty acids. Saturated fatty acids have free rotation about every carbon–carbon single bond (C—C bond) and are largely linear. In unsaturated fatty acids, rotation is restricted about the carbon–carbon double bond (C═C bond). The cis configuration of the C═C bond introduces a bend in the fatty acid, whereas a trans C═C bond is more linear.

The number and configuration of C═C bonds in unsaturated fatty acids affect the **melting point (MP)** of fatty acid mixtures. The melting point is the temperature at which the fatty acid changes state from a solid to a liquid. At the melting point, fatty acids are in equilibrium between these states. Long-chain saturated fatty acids have higher melting points than long-chain unsaturated fatty acids because of differences in intermolecular interactions. Thus, lipid mixtures with large amounts of triacylglycerols that contain a high proportion of saturated fatty acids form solids at room temperature because of their higher melting point. In contrast, lipid mixtures with large amounts of triacylglycerols that contain a high proportion of unsaturated fatty acids are liquids under the same conditions. **Figure 15.4** shows that triacylglycerols that contain mostly saturated fatty acids (typical of animal fat) pack more tightly together and transition from a solid to a liquid state at a higher temperature (MP ~45 °C), whereas triacylglycerols that contain mostly unsaturated fatty acids (typical of vegetable oils) melt at a lower temperature (MP ~15 °C). **Table 15.1** lists the amounts of saturated and unsaturated fatty acids in a variety of biological lipid mixtures encountered in typical foods, which is consistent with their solid, semisolid, or liquid state at room temperature. For example, butterfat and lard are semisolid at room temperature and consist almost entirely of saturated fatty acids and the monounsaturated fatty acid oleate. Flaxseed oil and olive oil are liquids at room temperature and contain large amounts of the unsaturated fatty acids oleate, linoleate, and α-linolenate.

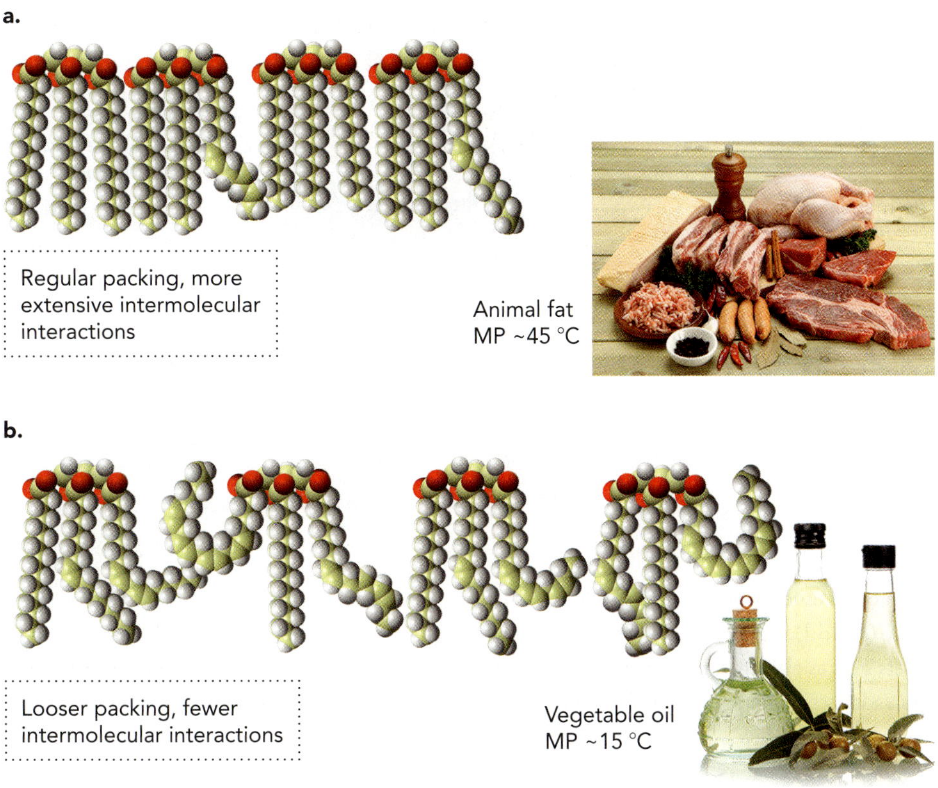

Figure 15.4 The degree of fatty acid saturation affects the melting point of lipids. **a.** Animal fats contain triacylglycerols with a high proportion of long-chain saturated fatty acids, which pack together to form solids at room temperature. PHOTO: DEEEPBLUE/ISTOCK/GETTY IMAGES PLUS. **b.** Vegetable oils contain triacylglycerols with mixtures of saturated and unsaturated fatty acids that disrupt tight packing, resulting in a low melting point. They are generally in the liquid state at room temperature. PHOTO: LABOKO/SHUTTERSTOCK.

It is possible to raise the melting point of lipid mixtures by converting unsaturated fatty acids into saturated fatty acids. This commercial process, called **hydrogenation**, heats oils in the presence of hydrogen gas to reduce C═C bonds. Partial hydrogenation of plant oils is used in the food industry to produce margarine, a semisolid lipid mixture. Besides providing an economical alternative to use of animal fats in food preparation (plant oils are less expensive to produce than animal fats), hydrogenated lipids have a longer shelf life than unsaturated fatty acids because they are more resistant to oxidation. Foods prepared with large amounts of unsaturated fatty acids can become rancid and have foul odors or tastes because of chemical changes in oxidized fatty acids.

Table 15.1 FATTY ACID COMPOSITION OF COMMON PLANT AND ANIMAL LIPID MIXTURES

	Saturated fatty acids			Unsaturated fatty acids			
Lipid	Myristate (%)	Palmitate (%)	Stearate (%)	Oleate (%)	Linoleate (%)	α-Linolenate (%)	Melting point (°C)
Fat							
Beef tallow	3	24	19	43	3	1	~45
Pork fat (lard)	2	26	14	44	10	ND	~40
Cow butterfat	11	27	12	29	2	1	~30
Oil							
Olive oil	ND	13	3	71	10	1	−6
Sunflower oil	ND	7	5	19	68	1	−17
Flaxseed oil	ND	3	7	21	16	53	−24

ND = not determined.

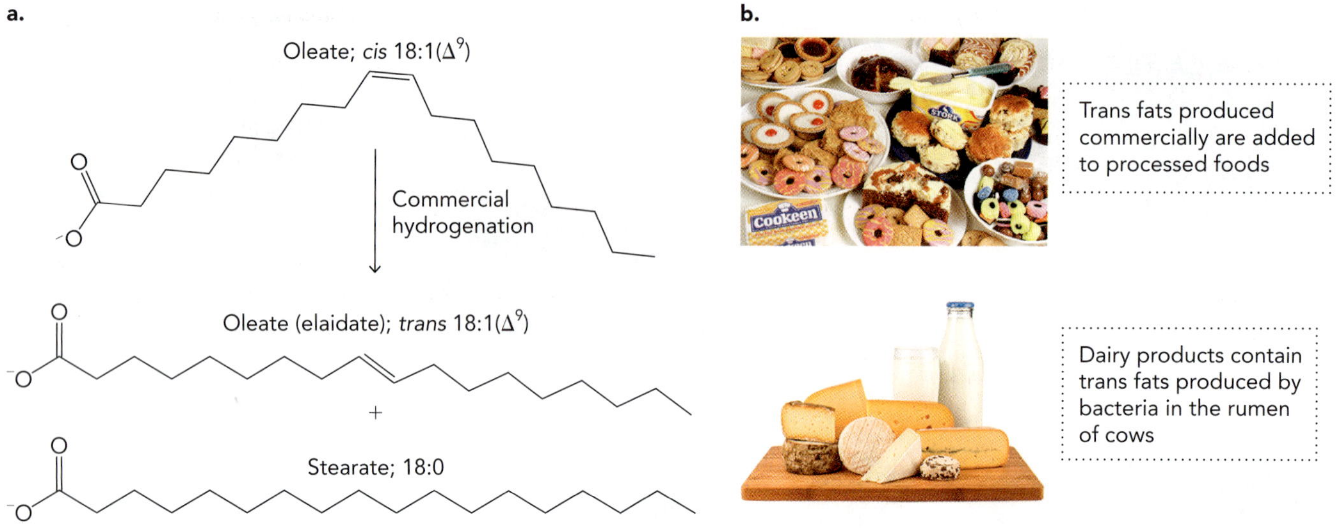

Figure 15.5 Hydrogenation is a commercial process used to reduce C═C bonds in unsaturated fatty acids to generate saturated fatty acids and trans fatty acids. **a.** Hydrogenation of oleate (*cis* 18:1) in plant oils, using heat and hydrogen gas, produces both stearate (18:0) and a form of oleate (*trans* 18:1) called elaidate. **b.** Trans fats such as elaidate are present in processed foods and in dairy products, the latter containing trans fats produced by bacteria in the rumen of cows. PROCESSED FOODS: CORDELIA MOLLOY/SCIENCE SOURCE; DAIRY PRODUCTS: MARIUS GRAF/ALAMY.

Figure 15.5a illustrates that commercial hydrogenation of oleate (*cis* 18:1) generates the fully saturated C_{18} fatty acid stearate (*cis* 18:0), as well as another form of oleate (*trans* 18:1), which is a **trans fat** called elaidate. Because a hydrogenated mixture of stearate and elaidate has a melting point of close to 50 °C, which is significantly higher than the melting point of oleate (13 °C), hydrogenation converts this plant oil into a semisolid lipid mixture. Naturally occurring trans fats are also present in dairy products as a result of bacterial metabolism in the rumen of cows (**Figure 15.5b**).

Commercial hydrogenation of plant oils was once considered a breakthrough in food preparation and preservation, but a number of nutrition studies have shown that consumption of trans fats in processed foods is associated with increased rates of cardiovascular disease. Indeed, plant-derived trans fats have been found to pose some of the same health risks as saturated animal fats. As shown in **Figure 15.6**, "Nutrition Facts" labels in the United States must include the amount of trans fat per serving of the food. Following the lead of New York City in 2006, numerous municipalities have passed laws against the use of trans fats in restaurants (Figure 15.6).

Carbons in fatty acids are usually numbered from the carboxylic acid end, with the carboxyl carbon being C-1 (**Figure 15.7**). However, the positions of C═C bonds in polyunsaturated fatty acids are often identified using the methyl carbon, which is called the omega carbon (ω-carbon). For example, linoleate is an **ω-6 fatty acid** and α-linolenate is an **ω-3 fatty acid** because of where the first C═C bonds are located relative to the ω-carbon. Linoleic acid and α-linolenic acid are both essential fatty acids for humans, meaning that these fatty acids must be obtained from the diet. As shown

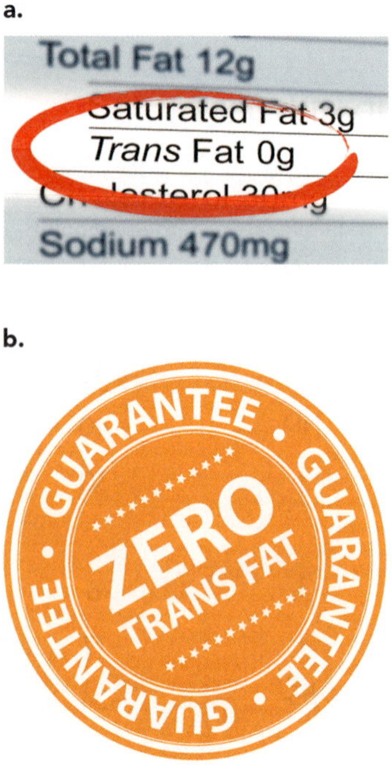

Figure 15.6 Consumption of large amounts of trans fats is associated with increased risks of cardiovascular disease. **a.** It is now required by law that the amount of trans fat per serving contained in foods be listed on the "Nutrition Facts" label. JONATHAN VASATA/SHUTTERSTOCK. **b.** Some cities, such as New York City, have made it illegal for restaurants to cook with trans fats. DEIMOSZ/SHUTTERSTOCK.

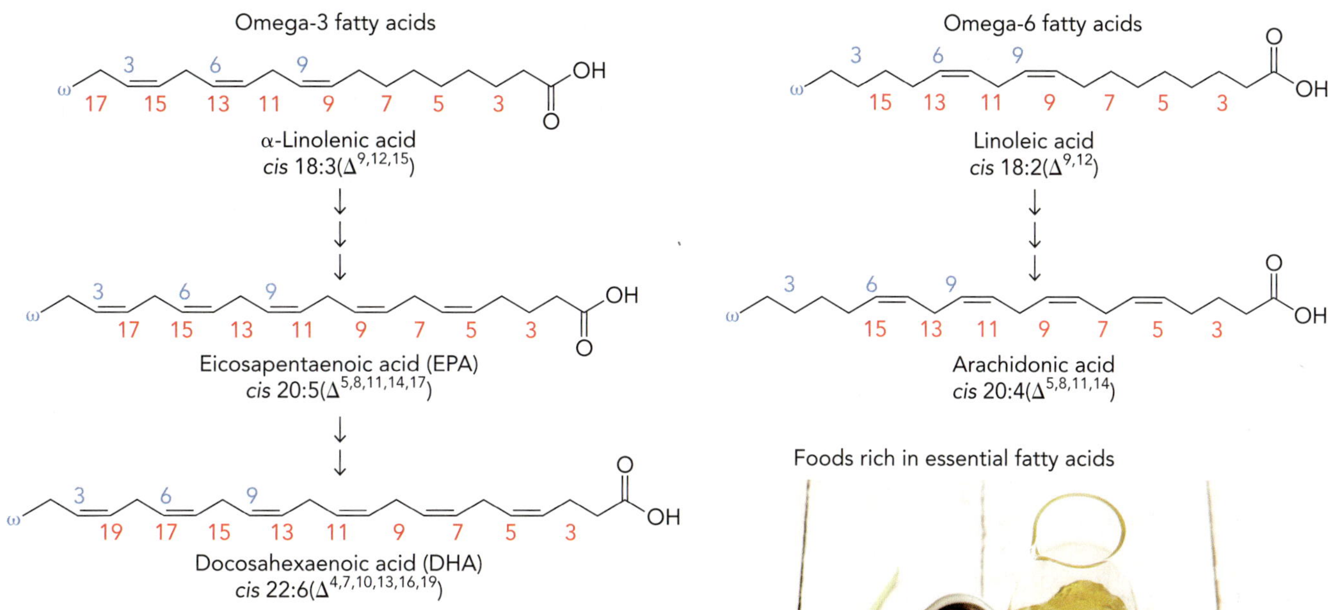

Figure 15.7 Linoleic acid and α-linolenic acid are essential fatty acids, which must be obtained from the diet. Humans have the enzymes needed to convert the ω-6 fatty acid linoleic acid into arachidonic acid, whereas the ω-3 fatty acid α-linolenic acid is converted into eicosapentaenoic acid and docosahexaenoic acid. Foods rich in ω-3 fatty acids include many types of fish, flaxseed oil, and nuts. Red numbers represent the numbering from the C-1 carboxylic carbon; blue numbers represent the numbering from the ω-carbon (methyl carbon). The structures of unsaturated fatty acids in the cis configuration are shown here as unbent molecules for convenience. PHOTO: BON APPETIT/ALAMY.

in Figure 15.7, linoleic acid and α-linolenic acid are converted in the body into three other essential fatty acids: the ω-6 fatty acid arachidonic acid and the ω-3 fatty acids **eicosapentaenoic acid (EPA)** and **docosahexaenoic acid (DHA)**. Studies have shown that diets containing foods rich in the ω-3 fatty acids EPA and DHA, such as fish, flaxseed oil, and nuts, are associated with reduced incidence of cardiovascular disease. The mechanistic basis for this observation is not completely understood; however, it may be due to binding of ω-3 fatty acid derivatives to peroxisome proliferator–activated receptors. As described in Chapter 19, peroxisome proliferator–activated receptors are a class of nuclear receptors that regulate the expression of lipid metabolizing enzymes in response to ligand binding.

Lipids are chemically very different from proteins and carbohydrates. Specifically, lipids are amphipathic, though largely hydrophobic biomolecules. In contrast, proteins and carbohydrates, for the most part, are hydrophilic biomolecules. Therefore, the methods used to isolate and characterize lipids depend on techniques in lipid biochemistry, including those of the emerging field of **lipidomics**, in which a collection of organic chemistry protocols has been modified for use with biological samples.

As shown in **Figure 15.8**, the first step in lipid purification requires an organic extraction of a homogenized biological sample using a mixture of chloroform and methanol. The extract is then washed with water to give an aqueous phase containing most of the carbohydrates, with lipids partitioned into the organic phase. Cell debris and much of the cellular protein is trapped in the interface between the aqueous and organic phases. The organic phase is then applied to a silicic acid column, which functions as an absorbent to permit the differential elution of lipids using successive washes with chloroform, acetone, and methanol. Depending on the objective and available instrumentation, the fractionated lipids are then further characterized by thin layer chromatography, gas phase chromatography, or mass spectrometry. Chemical properties that differentially affect lipid purification are the hydrocarbon chain length, the presence of polar head groups, and the number of C=C bonds.

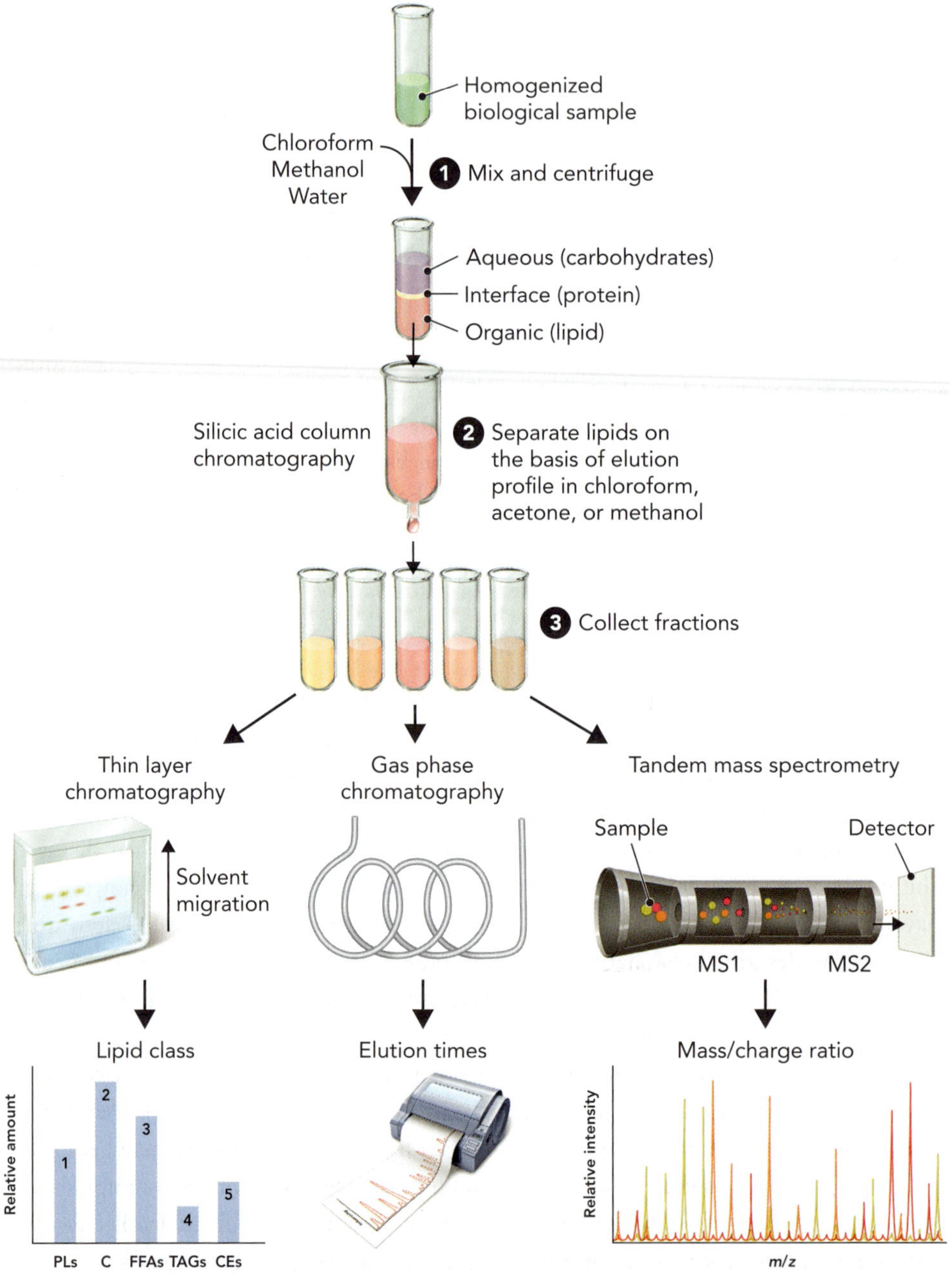

Figure 15.8 Cellular lipids are isolated and characterized using organic extraction methods developed for use in the field of lipidomics. Initially, a mixture of chloroform and methanol is used to isolate lipids from a cell extract. The lipids partitioned to the organic phase can be separated using silicic acid column chromatography and then characterized by chromatography or mass spectrometry. For example, thin layer chromatography can be used to separate phospholipids (PLs), cholesterol (C), free fatty acids (FFAs), triacylglycerols (TAGs), and cholesterol esters (CEs) on the basis of polarity.

a.

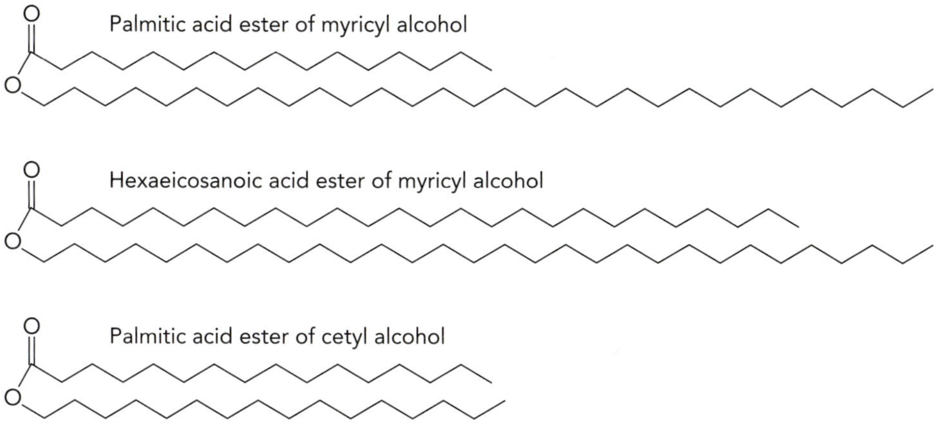

b.

Figure 15.9 Beeswax consists of 70% long-chain saturated wax esters and is solid at room temperature because of the high melting point of long-chain saturated wax esters. **a.** The three most abundant wax esters in beeswax consist of saturated C_{16} or C_{30} fatty alcohols linked to C_{16} or C_{26} fatty acids. **b.** Beeswax is secreted by bees and used to make honeycombs, but it also makes excellent candles. BEES AND HONEYCOMB: ALEKSANDRA PIKALOVA/SHUTTERSTOCK; BEESWAX CANDLES: ART DIRECTORS & TRIP/ALAMY.

Biological Waxes Have a Variety of Functions

Waxes are an abundant type of lipid in nature. They consist of long-chain fatty alcohols linked to long-chain fatty acids (**Figure 15.9**). The chemical structure of the lipid components of waxes can vary considerably but generally consists of various fatty alcohols containing hydrocarbon chains C_{16} to C_{30} in length, which are esterified to C_{16} or C_{26} fatty acids. The melting points of long-chain saturated wax esters are usually higher than ambient temperature and are therefore solids in their biological context. For example, beeswax has a melting point of 63 °C and is solid under most conditions. As shown in Figure 15.9, 70% of beeswax consists of three types of wax esters, all of which are fully saturated. This is important to the bees because they use the beeswax they secrete to construct a honeycomb, which is a solid support to hold the liquid honey they use for food. When beekeepers harvest honey, they also extract the beeswax and sell it as a high-quality candle wax.

Not all wax esters are solid at ambient temperature, owing to cis C=C bonds in the hydrocarbon backbone. These double bonds prevent tight packing and lower the melting point. An example of a liquid wax is jojoba oil, which is harvested from seeds produced by the desert plant *Simmondsia chinensis*. As shown in **Figure 15.10**, 90% of jojoba oil consists of wax esters in which the fatty acid moiety is unsaturated at the ω-9 position, resulting in a melting point of 7 °C. Oil produced from processing of jojoba seeds has many commercial uses in cosmetics and is also a potential source of alternative hydrocarbon fuel.

Marine copepods and sperm whales also contain large amounts of wax esters. It is thought that these compounds control buoyancy at different depths in the ocean. In the case of sperm whales, a large cavity in the head contains more than 1,000 L of oil, consisting of 75% wax esters and 25% triacylglycerols, with a melting point of ~30 °C. Although the primary purpose of this large, oil-filled cavity is for sonar navigation, it may also play a role in controlling the buoyancy of this large mammal when it dives into deep water to hunt for food. It is probable that whales can control the density of the oil by modulating their body temperature using thermoregulation and thereby regulate their buoyancy in the water. By allowing the oil to cool below 30 °C, it will become solid (more dense) and thereby decrease a whale's buoyancy, making it easier for the whale to dive into deep water where pressure will also contribute to the density of the oil. In

Figure 15.10 Some wax esters are liquid at ambient temperatures because of unsaturated C═C bonds and resulting low melting points. **a.** Jojoba oil from seeds of the plant *Simmondsia chinensis* consists almost entirely of wax esters containing long monounsaturated hydrocarbon chains, with a C═C bond at the ω-9 position. PHOTO: EMOTIVE IMAGES/AGEFOTOSTOCK. **b.** Sperm whales and some marine copepods store large amounts of wax esters with melting points that are close to the temperature of ocean water. It has been proposed that these organisms use the wax esters for buoyancy (low density at elevated temperature) and ballast (high density at decreased temperature and increased pressure) to maneuver effortlessly up and down, respectively, in the ocean. SPERM WHALE: REINHARD DIRSCHERL/GETTY IMAGES; MARINE COPEPOD: D. W. POND AND G. A. TARLING (2011). PHASE TRANSITIONS OF WAX ESTERS ADJUST BUOYANCY IN DIAPAUSING *CALANOIDES ACUTUS*. *LIMNOLOGY AND OCEANOGRAPHY*, *56*(4), 1310–1318. DOI: 10.4319/LO.2011.56.4.1310.

contrast, by heating the oil above 30 °C using thermoregulation, the oil will liquefy (less dense) and increase a whale's buoyancy, which makes it easier for the whale to rise to the ocean surface.

There are no direct data supporting this idea that sperm whales can control their buoyancy using thermoregulation of the stored oil, but scientists have studied ocean buoyancy in the marine copepod *Calanoides acutus*, which is found in the Antarctic Ocean. These tiny crustaceans feed on diatoms near the ocean surface in the summer but enter a dormant metabolic state in the winter and sink 1,500 meters or more below the surface ice. Research has showed that the *C. acutus* copepods do this by storing large amounts of wax esters in their bodies as they feed in the summer. This provides buoyancy to the organism because of the low density of the wax lipids at ocean surface temperatures. However, as winter approaches and the water cools, the wax density increases and begins to function as ballast, much like a weight belt on a scuba diver. As the copepod sinks into cooler water with higher pressure, the density of the wax esters continues to increase until the copepod reaches a depth where it has neutral buoyancy.

Structure and Nonmetabolic Uses of Triacylglycerols

The primary biological function of triacylglycerols is in energy storage, as we will describe in Section 15.2. The chemical properties of triacylglycerols, however, have also made them useful in other ways in everyday life.

The structure of triacylglycerols is illustrated in **Figure 15.11**. Observe that three fatty acids are esterified to glycerol, which makes them neutral lipids because the polar

a.

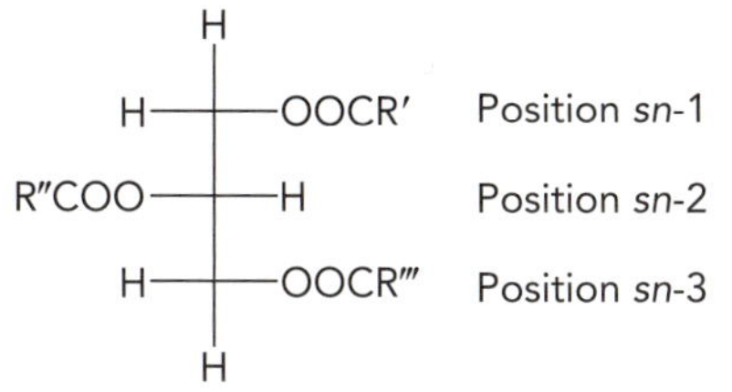

b.

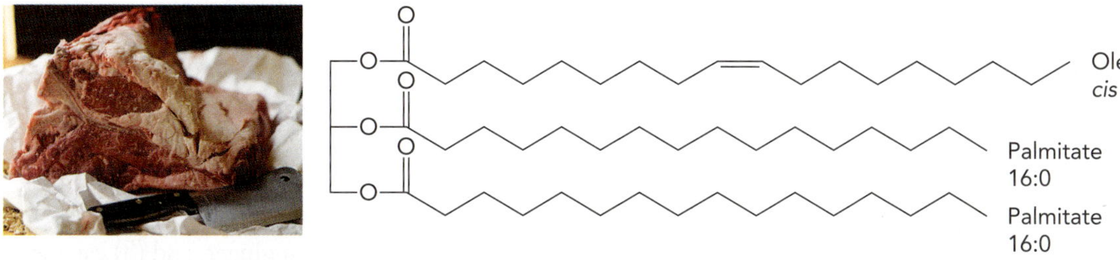

c.

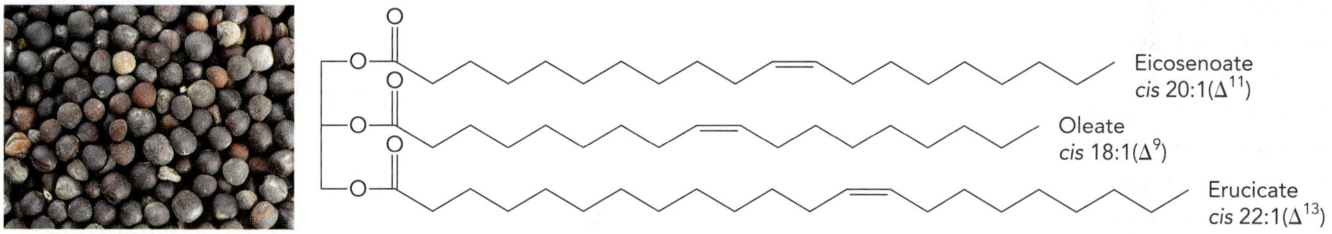

Figure 15.11 Triacylglycerols contain three fatty acids esterified to glycerol. **a.** In a Fischer projection of a triacylglycerol, the middle carbon is designated the *sn*-2 position using the stereospecific numbering system, whereas the top and bottom carbons are named the *sn*-1 and *sn*-3 positions, respectively. **b.** Beef fat contains triacylglycerols with mostly saturated long-chain fatty acids, which makes them solid at room temperature. PHOTO: JUPITERIMAGES/GETTY IMAGES. **c.** Triacylglycerols in plant seeds contain long-chain unsaturated fatty acids and are oils at room temperature. Rapeseed oil is also called canola oil. The structures of unsaturated fatty acids in the cis configuration are shown here as unbent molecules for convenience. PHOTO: DAVID DOHNAL/SHUTTERSTOCK.

carboxyl groups are neutralized. Using a Fischer projection, by convention the middle hydroxyl group attached to C-2 has its fatty acid drawn to the left, whereas the fatty acids attached to the hydroxyl groups on C-1 and C-3 are drawn to the right.

Because the three fatty acids usually differ in length and degree of saturation, triacylglycerols are asymmetric, and stereospecific numbering (*sn*) has been adapted for use in naming them. As shown in Figure 15.11a, the *sn*-1 position is at the top of the Fischer projection, with the *sn*-2 position in the middle and the *sn*-3 position at the bottom. This *sn* system is used to characterize triacylglycerols found in nature, which were initially described by the propensity of certain fatty acids to be found at the *sn*-1, *sn*-2, or *sn*-3 positions as a function of origin. For example, triacylglycerols isolated from beef fat often contain oleate in the *sn*-1 position, almost always have palmitate in the *sn*-2 position, and have either oleate or palmitate in the *sn*-3 position (Figure 15.11b). In contrast, triacylglycerols found in plant seeds contain much higher levels of unsaturated fatty acids, often in all three *sn* positions, as seen in an abundant triacylglycerol from the rapeseed *Brassica napus* (Figure 15.11c), which is used to produce rapeseed oil.

For centuries, the chemical properties of triacylglycerols have been exploited for two aspects of our daily lives: cooking and cleaning. For example, spices are often mixed with olive oil or butter to prepare a variety of cooked foods because of the flavor-enhancing properties of the fat. Moreover, adding spices to fatty meats such as lamb or beef results in a much more intense flavor than that from adding the same spices to lean meats such as chicken breasts. The reason that foods with high triacylglycerol content are so flavorful is that some common spices contain hydrophobic molecules that bind to taste receptor proteins on the tongue when they are solubilized in cooking oils and butter. Three distinctive

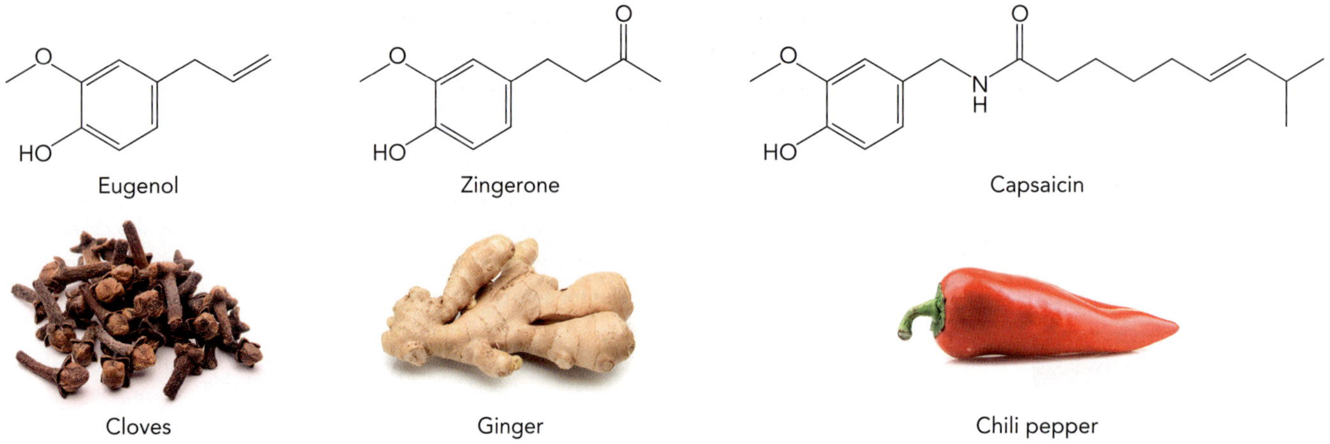

Figure 15.12 Many types of flavor molecules are hydrophobic and highly soluble in lipids used for cooking, such as butter or oil. Three examples of hydrophobic flavor molecules are eugenol, found in cloves; zingerone, found in ginger roots; and capsaicin, found in chili peppers. CLOVES: OLEKSANDR GULA/SHUTTERSTOCK; GINGER: JIANG HONGYAN/SHUTTERSTOCK; CHILI PEPPER: NENOV BROTHERS IMAGES/SHUTTERSTOCK.

hydrophobic "flavor" molecules are (1) eugenol, the active ingredient in bay leaves and cloves; (2) zingerone, which is found in high concentrations in ginger; and (3) capsaicin, the compound responsible for the hot flavor of chili peppers (**Figure 15.12**). The flavors associated with these hydrophobic molecules are the result of their function as ligands of G protein–coupled receptors present on tongue cells (see Figure 8.22c). Activation of the lingual G protein–coupled receptors by these molecules relays flavor signals to the brain through sensory neurons.

Another example of how the biochemistry of triacylglycerols has been exploited is the use of animal fat to make soap for cleaning clothes. The process of making soap from fat is called **saponification**, as illustrated in **Figure 15.13a**. Triacylglycerols present in animal fat are insoluble in water; however, during saponification, the amphipathic fatty acid molecules are released from triacylglycerol by boiling the fat in a strong alkali solution. This is done industrially with a concentrated NaOH solution, but it can also be done by leaching wood ashes to make an alkaline mixture called **lye**, which contains high levels of KOH and NaOH. Soap, representing the fatty acid sodium or potassium salts formed by this process, is readily ionized in water, leading to the formation of fatty acid micelles. Upon vigorous agitation—an important step in washing clothes—the micelles temporarily break apart and interact with grease (oily dirt) in the clothes. They then re-form into larger micelles that trap the hydrophobic "dirt" particles (**Figure 15.13b**). During the rinsing stage, the fatty acid micelles are washed away, taking the greasy dirt with them.

Note that washing clothes in water that contains a high mineral content (Mg^{2+} and Ca^{2+}), called hard water, causes the formation of a precipitate called soap scum that decreases the effectiveness of the washing process. In contrast, washing clothes in soft water, which has been treated to remove minerals, uses less soap because more of it remains in the solution as micelles that can interact with greasy dirt particles.

concept integration 15.1

Describe the chemical properties of trans fats that could help to explain why plant-derived trans fats and animal fats have similar detrimental health effects.

Hydrogenation of plant-derived fats results in the production of trans fats, which occur naturally in low abundance in plants. These trans fats, whether from an animal or plant-derived source, have similar chemical properties that result in overall less membrane fluidity. This is due to the configuration of atoms around a C=C bond, which affects the positions of the neighboring atoms. In a hydrocarbon chain, a cis C=C bond introduces a bend, whereas the hydrocarbon chain is much more linear in the trans configuration. The geometry of the hydrocarbon chains affects how tightly they pack, and thus the fluidity of the membrane containing these fatty acids. The bends in cis fatty acids result in less extensive interactions (less packing) between the fatty acid chains, lower melting points, and more fluid membranes. When the trans configuration is introduced, membranes containing these trans fats are more tightly packed, have higher melting points, and are less fluid.

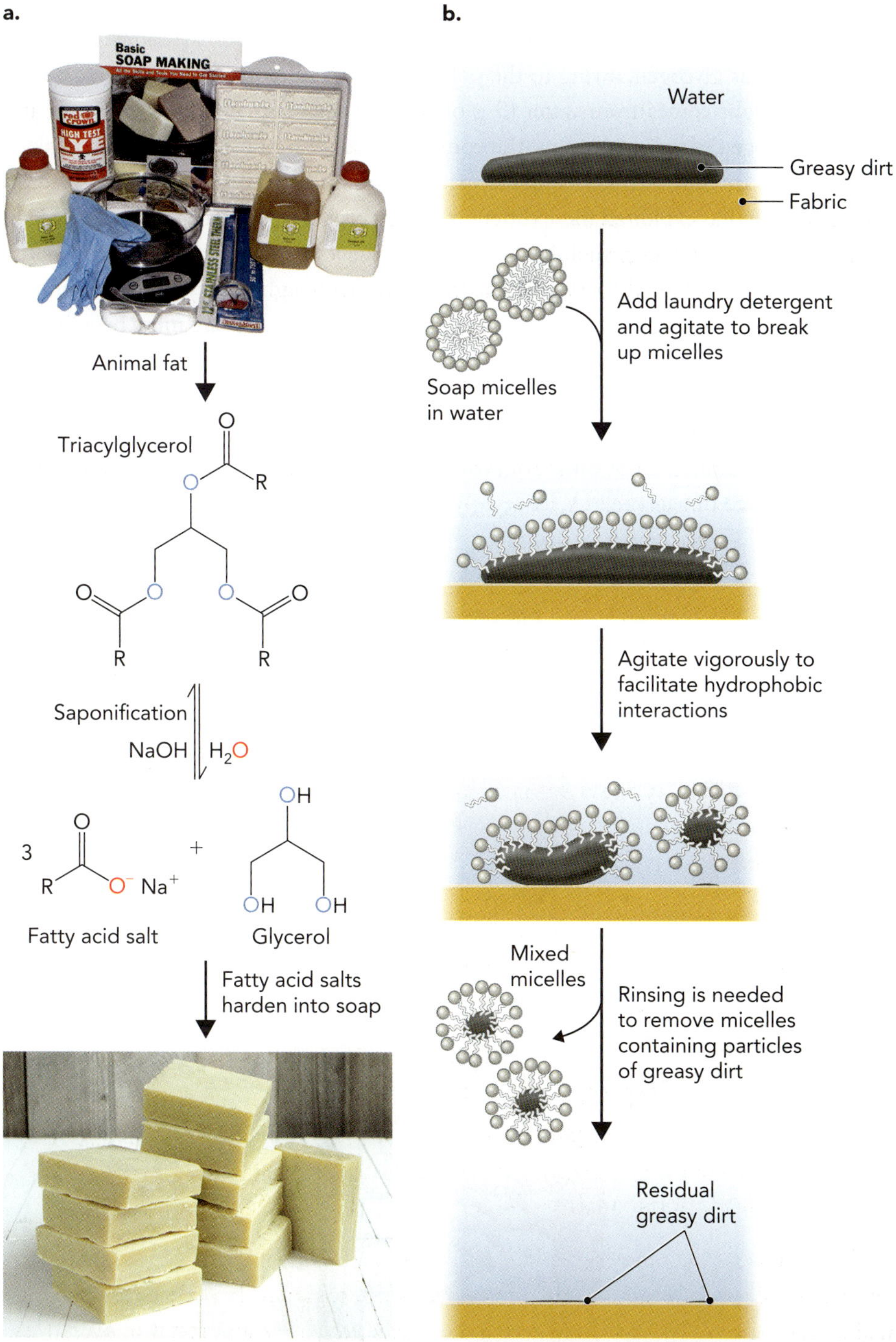

Figure 15.13 Soap is made from hydrolyzed triacylglycerols and works to remove grease from clothes by forming micelles. **a.** Basic ingredients for homemade soap include animal fat (lard), other oils such as coconut oil and olive oil, commercial lye (sodium hydroxide), and herbal ingredients to provide texture and fragrance. Fatty acids are released from triacylglycerols present in animal fat by heating in the presence of strong alkali to produce fatty acid salts, which harden into soap. SOAP-MAKING KIT: COURTESY BRUSHY MOUNTAIN BEE FARM; HOMEMADE SOAP: IMAGES72/SHUTTERSTOCK. **b.** Agitation of the soap together with dirty clothes results in the formation of mixed micelles containing fatty acids and hydrophobic dirt particles. Rinsing separates the micelles in the dirty water from the clean clothes.

15.2 Triacylglycerols Are Energy Storage Lipids

The primary biological function of triacylglycerols is energy storage. In animals, triacylglycerols are stored in fat cells called **adipocytes**. In plants, triacylglycerol storage is in the seeds, providing an oxidizable energy source for the developing embryo after germination.

There are two reasons why triacylglycerols are the major form of stored energy, rather than the glucose polymer glycogen. First, fatty acids are at a higher reduction state than glucose and therefore yield more energy for the same number of carbons upon oxidation in the mitochondrial matrix. Second, the hydrophobic nature of triacylglycerols means that they are not solvated by water and therefore have less mass for the same amount of stored energy as glycogen, owing to the polar properties of glucose. In glycogen, ~65% of its wet weight comes from water (3 g of wet glycogen contains ~2 g of water and ~1 g of glycogen). In energy terms, the complete oxidation of hydrated glycogen yields about 6 kJ/g. In contrast, the complete oxidation of triacylglycerols yields close to 38 kJ/g. Evolution has exploited these different energy properties between fat and glycogen (and mechanisms of energy mobilization) to optimize long-term and short-term energy storage. In humans, stored triacylglycerols can provide enough energy to sustain human life for almost 3 months without food. In contrast, glycogen stores are more beneficial in the short term by providing readily accessible chemical energy for rapid bursts of muscle activity and for maintaining blood glucose levels between meals.

As shown in **Figure 15.14**, triacylglycerols are either obtained from the diet—primarily from animal fat or nuts—or synthesized in the liver, using glucose and amino acids as a source of acetyl-CoA. Triacylglycerols are transported through the blood as

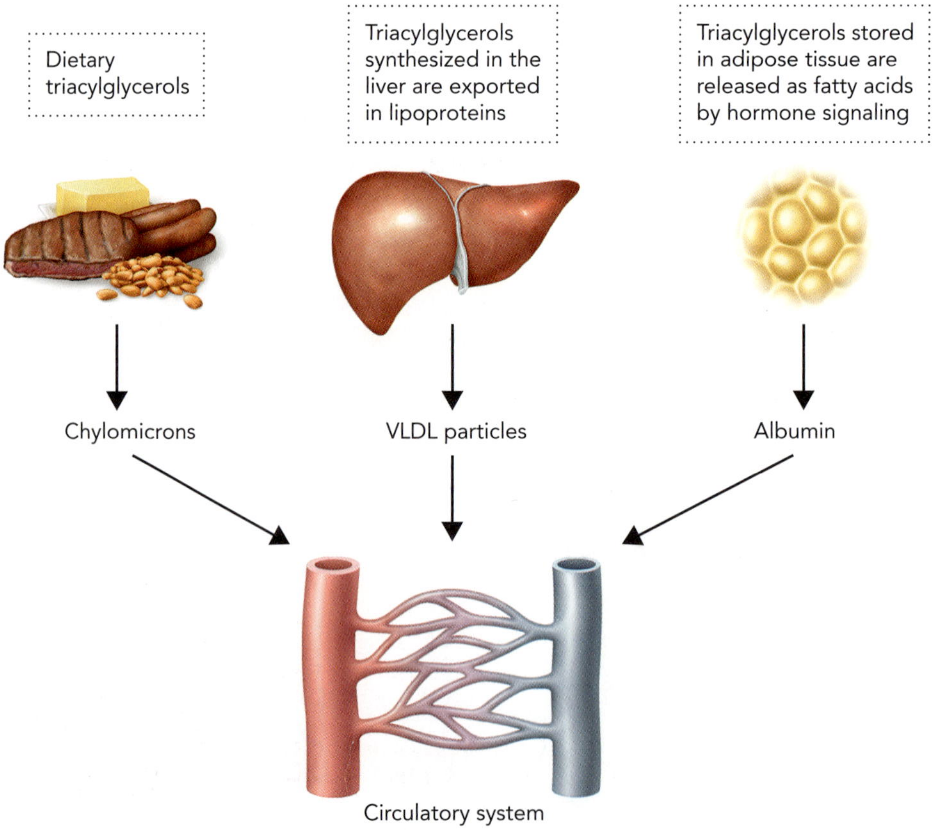

Figure 15.14 Triacylglycerols are the primary form of energy storage in animals and are derived from dietary sources, liver biosynthesis, and adipocytes. Dietary triacylglycerols are transported by large lipoprotein complexes called chylomicrons. Triacylglycerols synthesized in the liver are transported by VLDL particles. Triacylglycerols stored in adipocytes are released as free fatty acids in response to hormone signaling and are transported by the carrier protein albumin.

components of **lipoprotein** complexes. As described shortly, lipoproteins consist of a protein-rich phospholipid monolayer surrounding a hydrophobic core containing triacylglycerols and cholesterol esters. Lipoprotein complexes can be very large, such as chylomicrons produced by intestinal epithelial cells, or small, such as the very-low-density lipoprotein (VLDL) particles exported from liver cells. Although adipose tissue is the primary storage depot for triacylglycerols in animals, adipocytes do not export stored triacylglycerols in the form of lipoprotein complexes. Instead they hydrolyze triacylglycerols and release free fatty acids and glycerol. Free fatty acids released from adipose tissue are transported throughout the body by a carrier protein called **albumin**, which protects cell membranes from the detergent effect of amphipathic fatty acids. Let's now look at these three triacylglycerol transport processes in more detail.

Dietary Triacylglycerols Are Transported by Chylomicrons

Much of the triacylglycerol stored in adipose tissue originates from dietary lipids. Fats that enter the small intestine from the stomach are insoluble and must be emulsified by bile acids, which are secreted by the bile duct and function as detergents to promote micelle formation. **Lipases** are water-soluble enzymes in the small intestine that hydrolyze the acyl ester bonds in triacylglycerols to liberate free fatty acids, which then pass through the membrane on the lumenal side of intestinal epithelial cells (**Figure 15.15**).

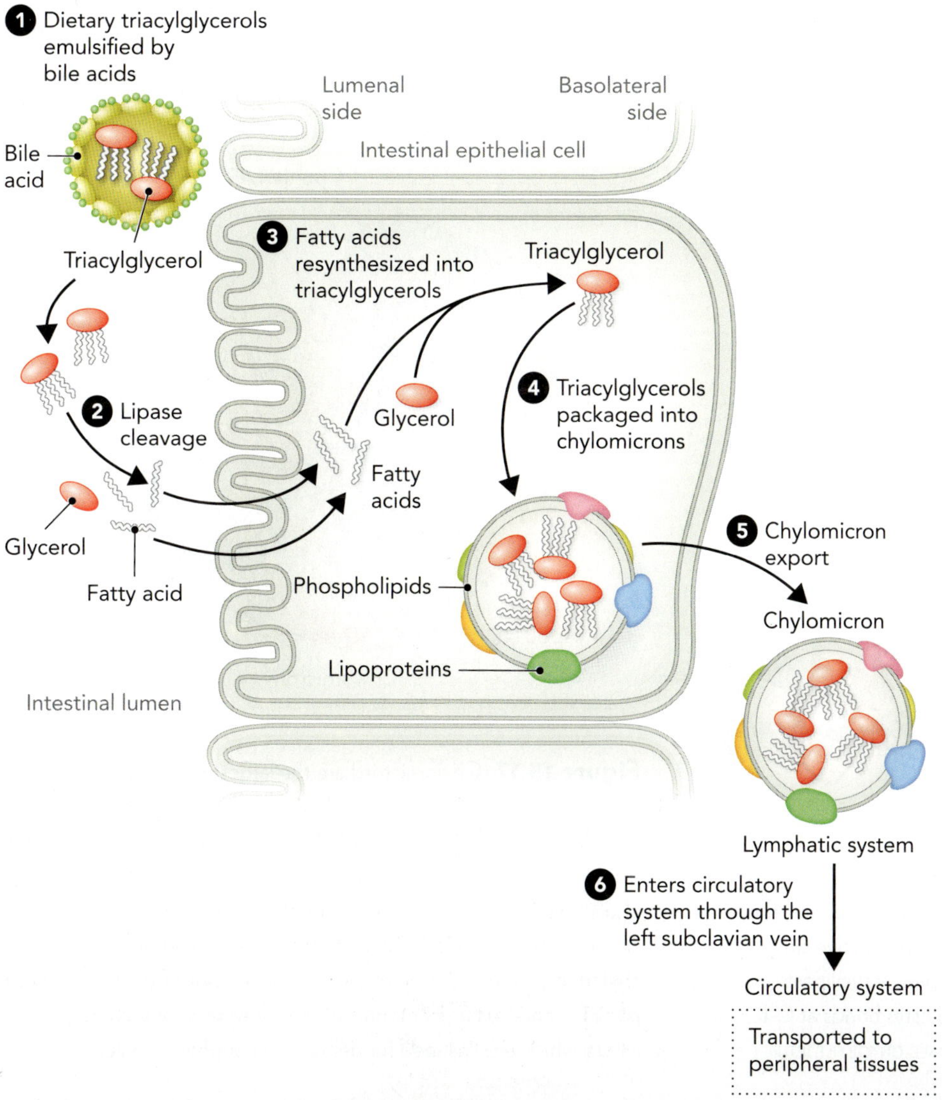

Figure 15.15 Dietary triacylglycerols are metabolized by enzymes in the intestine before being exported to the circulatory system as components of lipoproteins. The absorption and transport of dietary triacylglycerols can be broken down into six steps: (1) emulsification of triacylglycerols by bile acids; (2) hydrolysis of triacylglycerols by intestinal lipases to generate free fatty acids; (3) resynthesis of triacylglycerols inside intestinal epithelial cells; (4) packaging of triacylglycerols into large lipoprotein particles called chylomicrons; (5) export of chylomicrons to the lymphatic system; and (6) entry of chylomicrons into the circulatory system through the left subclavian vein.

Pancreatic lipase is one such lipase, which is secreted by the pancreas in response to feeding. It cleaves the ester bonds at C-1 and C-3 to release two free fatty acids and monoacylglycerol. The structure of porcine pancreatic lipase with a triacylglycerol substrate in the enzyme active site is shown in **Figure 15.16**. Observe that the lipase is complexed with a regulatory protein subunit called **colipase**. The function of colipase is to bind to the bile acids and promote pancreatic lipase cleavage of the C-1 and C-3 acyl ester bonds. The intestinal tract also contains other lipase enzymes that cleave the acyl bond at C-2 to generate glycerol and the third free fatty acid.

Free fatty acids are absorbed by the intestinal epithelial cells, where they are converted back into triacylglycerols and packaged into very large lipoprotein particles called **chylomicrons**. The chylomicrons transport the triacylglycerols to adipose tissue for storage and to muscle cells for energy conversion processes. As shown in Figure 15.15, chylomicrons are released from the basolateral surface of intestinal cells (the side opposite the intestinal lumen) into the lymphatic system and then enter the circulatory system through the left subclavian vein.

As shown in **Figure 15.17**, chylomicrons contain a phospholipid and cholesterol monolayer on the outside of the particle. This layer is studded with membrane proteins called apolipoproteins, of which there are at least six on chylomicrons (apoA-I, apoA-IV, apoB-48, apoC-II, apoC-III, apoE). **Apolipoproteins** function to promote lipoprotein particle formation in endomembrane systems of cells and also serve as molecular tags for lipoprotein metabolism in the body. For example, the apoC-II apolipoprotein on chylomicrons binds to and activates an enzyme on the surface of endothelial cells called lipoprotein lipase, which cleaves the acyl ester bond and releases free fatty acids and

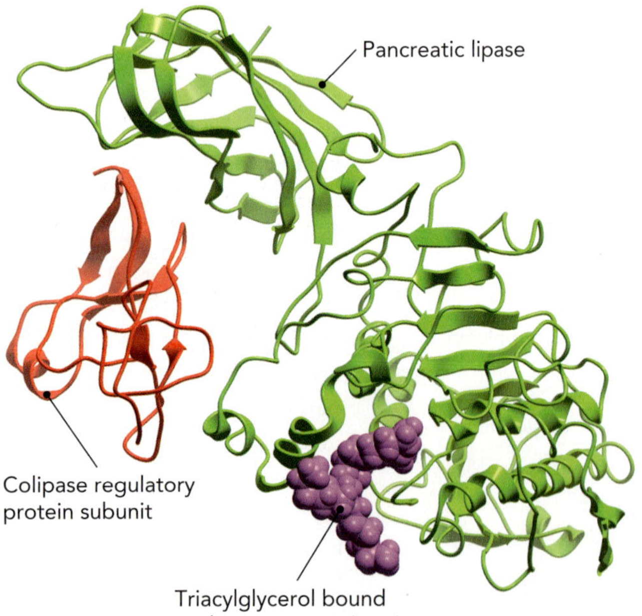

Figure 15.16 This molecular structure of porcine pancreatic lipase shows triacylglycerol bound to the enzyme active site. The colipase subunit facilitates binding of pancreatic lipase to bile acids and facilitates cleavage of the ester bonds at C-1 and C-3 of the triacylglycerol substrate. BASED ON PDB FILE 1ETH.

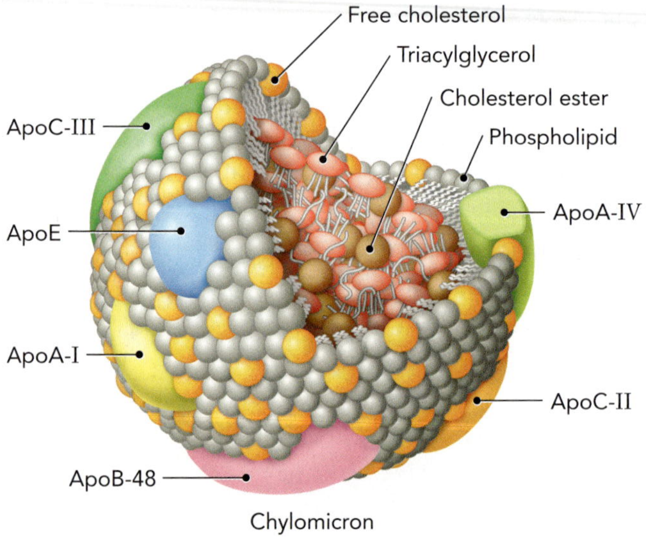

Figure 15.17 Chylomicrons are the largest of all lipoprotein particles. They consist of a monolayer phospholipid membrane containing free cholesterol. This phospholipid membrane is derived from the endomembrane system of intestinal epithelial cells. The phospholipid membrane is embedded with up to six different apolipoproteins, some of which function as binding sites for chylomicron metabolizing enzymes. The interior of the chylomicron particle contains large amounts of triacylglycerols and cholesterol esters, which are destined for delivery to peripheral tissues.

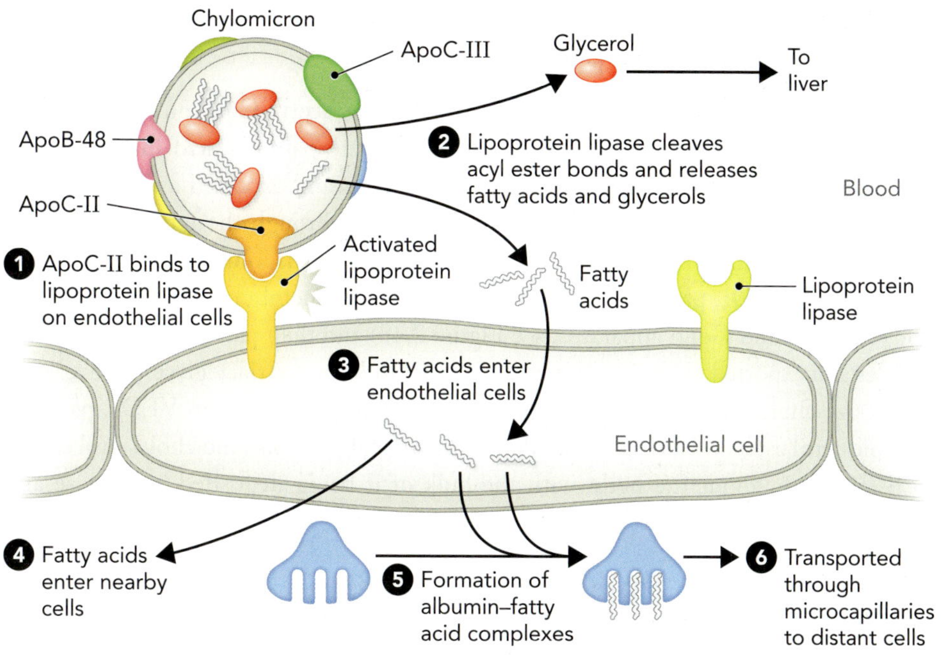

Figure 15.18 Apolipoprotein C-II on the surface of chylomicrons binds to and activates lipoprotein lipase on endothelial cells, leading to the release of fatty acids and glycerol. Fatty acids diffuse into the endothelial cells and then either enter nearby adipose and muscle cells or are transported by serum albumin protein to other tissues. The glycerol produced by lipoprotein lipase returns to the liver, where it is converted to dihydroxyacetone phosphate.

glycerol (**Figure 15.18**). The fatty acids diffuse into nearby adipose and muscle cells, whereas the glycerol travels through the blood to the liver. There it is converted to dihydroxyacetone phosphate by the enzymes glycerol kinase and glycerol-3-phosphate dehydrogenase (see Figure 9.48). As described in Chapter 16, chylomicron remnants, depleted of triacylglycerols but still containing cholesterol esters, are recycled to the liver, where they are repackaged with newly synthesized triacylglycerols and released into the blood as VLDL particles.

Figure 15.19 shows the molecular structure of human serum albumin, with seven fatty acids bound to distinct sites in the protein. This particular structure of an albumin–fatty acid complex contains oleic acid (monounsaturated fatty acid) and arachidonic acid (polyunsaturated fatty acid) in the binding sites. Albumin also binds large amounts of saturated fatty acids such as palmitic acid and stearic acid, as well as steroid and thyroid hormones. The fatty acids are sequestered within the interior of this transporter protein. Figure 15.19 shows that the fatty acid binding site in albumin contains tyrosine and phenylalanine amino acids that form a hydrophobic pocket.

Figure 15.19 Seven fatty acids are bound to distinct sites of human serum albumin. The fatty acids are sequestered within the interior of this transport protein in hydrophobic binding pockets formed from tyrosine and phenylalanine. BASED ON PDB FILE 1GNJ.

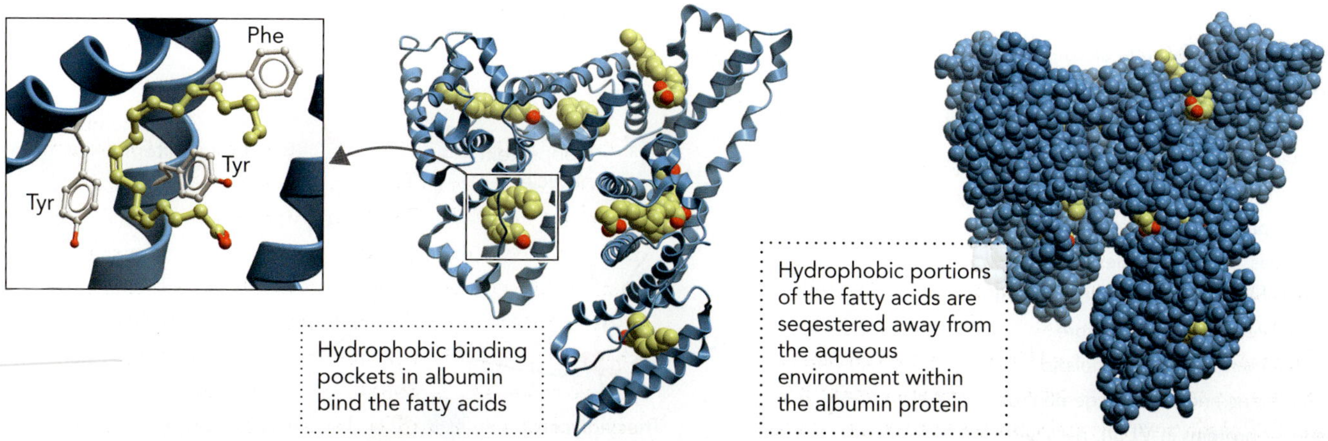

Triacylglycerols Synthesized in the Liver Are Packaged in VLDL Particles

The biosynthesis of triacylglycerols in animals uses acetyl-CoA produced by the degradation of carbohydrates and proteins to generate palmitic acid in the cytosol. The palmitic acid is then converted to triacylglycerols and exported as VLDL particles (**Figure 15.20**). As described in Chapter 16, flux through the fatty acid biosynthetic pathway in liver cells increases when the citrate cycle is inhibited by high energy charge. Liver VLDL particles are assembled in the endomembrane system (endoplasmic reticulum and Golgi apparatus; see Figure 2.48) and then packaged into secretory vesicles and exported to the circulatory system. In addition to triacylglycerols, liver cells synthesize cholesterol from acetyl-CoA, which is also packaged into VLDL particles as cholesterol esters. VLDL particles deliver the triacylglycerols (and cholesterol esters) to tissues throughout the body, with the bulk of it being deposited in adipose and muscle tissues for energy storage and energy conversion, respectively.

Adipocytes Cleave Stored Triacylglycerols and Release Free Fatty Acids

Fatty acids are released by endothelial cells after lipoprotein lipase cleavage of triacylglycerols in chylomicrons and VLDL particles. These fatty acids enter adipocytes, where they are combined with glycerol to regenerate triacylglycerols. The triacylglycerols are stored in large **lipid droplets** that fill nearly the entire volume of

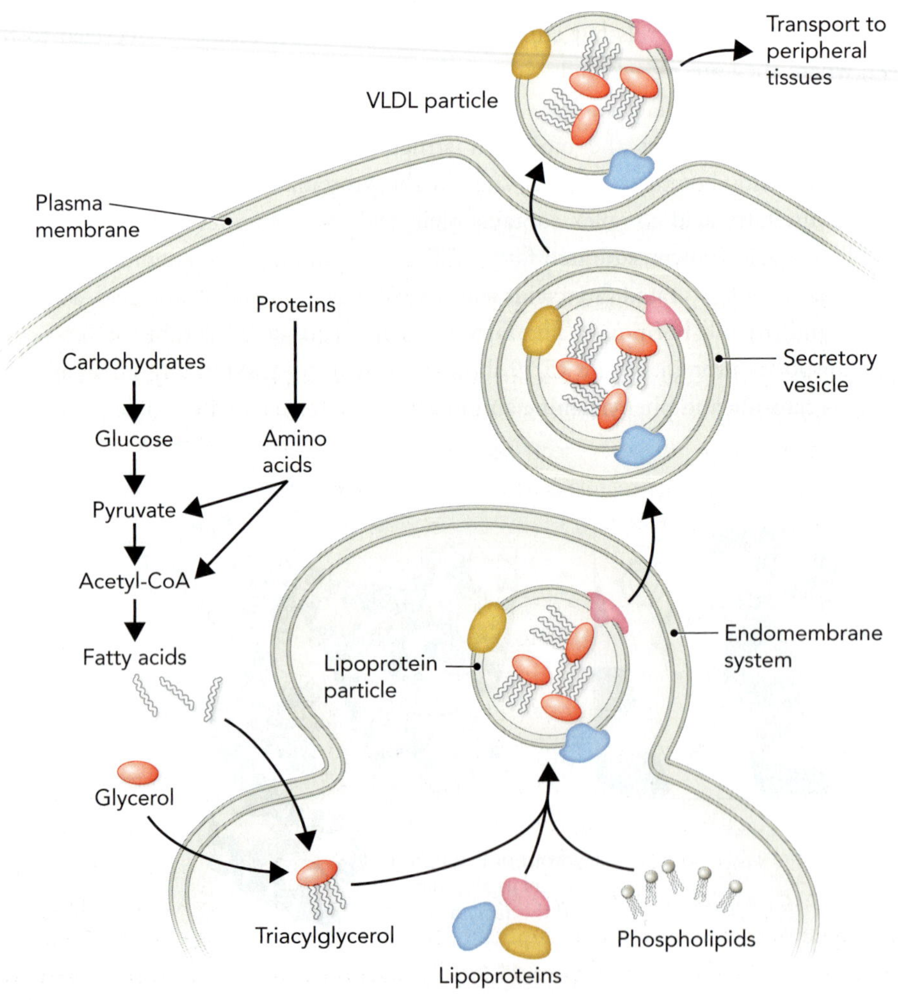

Figure 15.20 Fatty acids synthesized in the liver from acetyl-CoA in the cytosol are used to generate triacylglycerols in the endomembrane system before being packaged into lipoprotein particles. VLDL particles are one type of lipoprotein particle secreted from liver cells. The structure of the monolayer phospholipid membrane and the arrangement of apolipoproteins in VLDL particles are similar to those of chylomicrons.

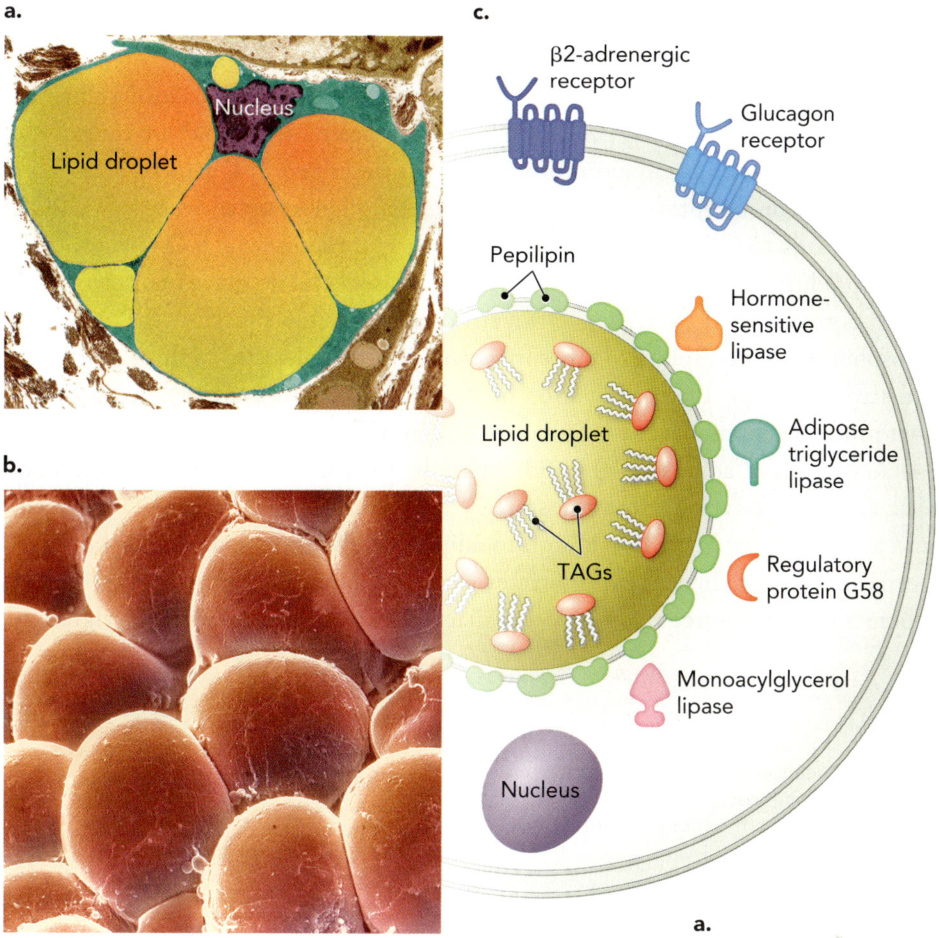

Figure 15.21 Adipocytes store triacylglycerols. **a.** Lipid droplets can occupy nearly the entire volume of an adipocyte. STEVE GSCHMEISSNER/SCIENCE SOURCE. **b.** This colorized scanning electron micrograph shows the size of adipocytes varies depending on the amount of stored triacylglycerol. DR. FRED HOSSLER/VISUALS UNLIMITED, INC. **c.** Lipid droplets are surrounded by a phospholipid monolayer containing perilipin protein, which prevents hydrolysis of stored triacylglycerols (TAGs) by the three major adipocyte lipases: adipose triglyceride lipase, hormone-sensitive lipase, and monoacylglycerol lipase. The regulatory protein G58 controls the lipase activity of adipose triglyceride lipase in response to epinephrine and glucagon signaling.

adipocytes (**Figure 15.21**). Lipid droplets are surrounded by a phospholipid monolayer containing large amounts of a protein called **perilipin,** which functions to prevent spurious degradation of triacylglycerols by lipase enzymes.

The three major lipases in human adipocytes are adipose triglyceride lipase, hormone-sensitive lipase, and monoacylglycerol lipase. Hormone signaling through the β2-adrenergic (epinephrine) and glucagon receptors in adipocytes stimulates lipase-mediated release of fatty acids in response to imminent danger (epinephrine) or lack of food (glucagon).

The diameter of an adipocyte cell is determined by the amount of stored triacylglycerols in the lipid droplet and is directly related to the proportion of an individual's weight consisting of body fat. The total percentage of body fat can be reduced by restricting caloric intake and by exercise, which shrinks the size of lipid droplets in adipocytes without decreasing the number of adipocytes. Because the total number of adipocytes does not change with dieting, a surgical procedure called liposuction was developed to physically remove adipocytes from unwanted fat deposits. As shown in **Figure 15.22**, liposuction can

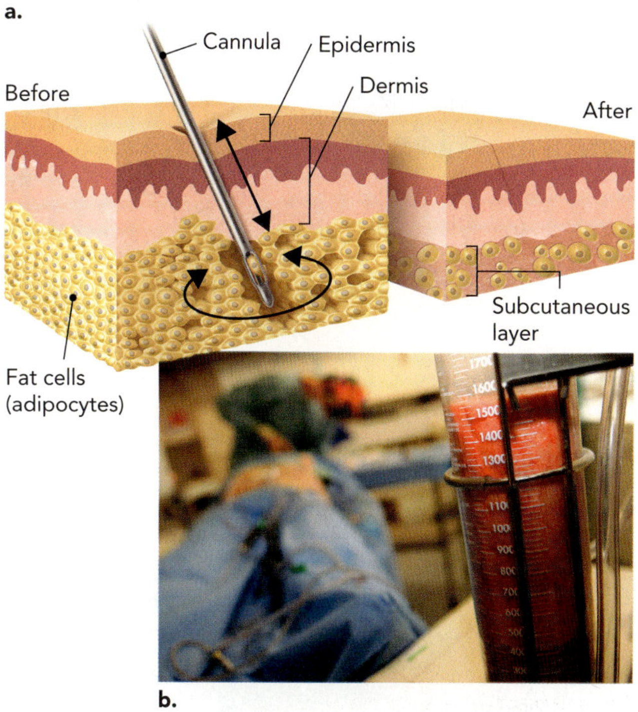

Figure 15.22 Liposuction is a surgical procedure that removes adipocytes from fat deposits in the body. **a.** Adipocytes are removed from subdermal layers of the body in the abdominal and thigh regions using a cannula under vacuum pressure. **b.** The fat tissue removed by liposuction is collected and discarded. MATTHEW TOTTON/ALAMY STOCK PHOTO.

remove up to 5 kg of fat cells from specific regions of the body at any one time and is considered cosmetic surgery.

Figure 15.23 illustrates how glucagon signaling initiates a cascade of events that results in the release of fatty acids into the bloodstream, where they are transported to peripheral tissues by albumin protein. (1) As described in Chapter 8, activation of the glucagon receptor—a member of the G protein–coupled receptor (GPCR) family—leads to GDP–GTP exchange in the $G_{s\alpha}$ protein, which stimulates cyclic AMP production by the enzyme adenylate cyclase. (2) The glucagon downstream signal is transmitted by cyclic AMP activation of protein kinase A (PKA), which phosphorylates perilipin on the surface of lipid droplets, as well as hormone-sensitive lipase. (3) Phosphorylation of perilipin leads to a conformational change that promotes binding of the regulatory protein G58 to adipose triglyceride lipase. (4) Activation of adipose triglyceride lipase by G58 initiates a hydrolysis reaction that cleaves a fatty acid from the stored triacylglycerols to generate diacylglycerol and a free fatty acid. (5) In addition to adipose triglyceride lipase, phosphorylated hormone-sensitive lipase also associates with perilipin on the lipid droplet surface, where it produces a free fatty acid from diacylglycerol to produce monoacylglycerol.

Figure 15.23 Glucagon signaling in adipocytes stimulates release of free fatty acids (FFAs) from triacylglycerols stored in lipid droplets as described in the text. Free fatty acids released by lipase cleavage are sequestered by human adipocyte fatty acid binding protein 4 and then exported to the circulatory system, where they are transported to peripheral tissues by the carrier protein albumin. AC = adenylate cyclase; DAG = diacylglycerol; MAG = monoacylglycerol; PKA = protein kinase A.

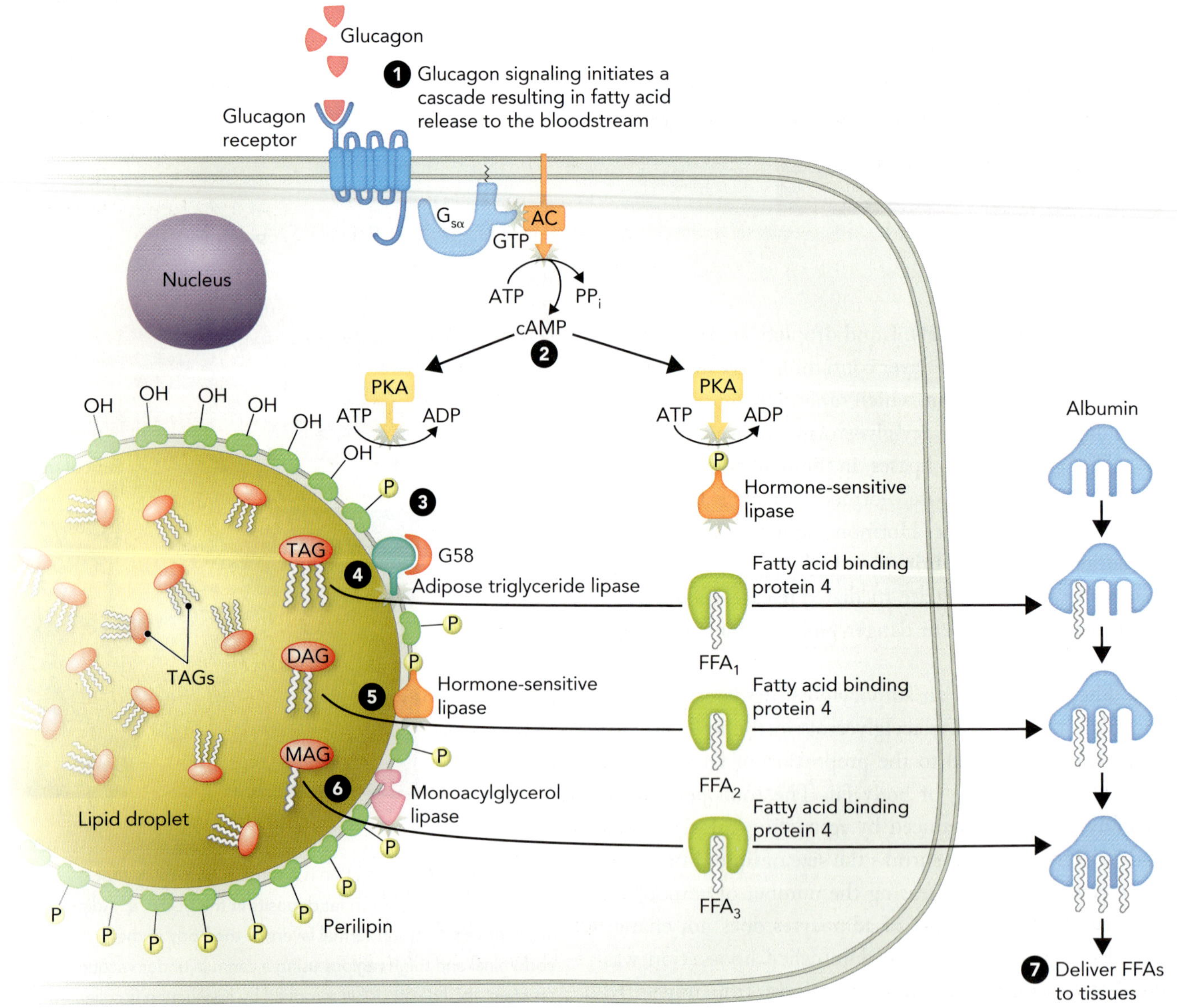

(6) The adipocyte enzyme monoacylglycerol lipase then generates a free fatty acid and glycerol from monoacylglycerol. (7) Lastly, albumin transports the free fatty acids through the circulatory system to tissues.

Because free fatty acids inside cells can act as a detergent and dissolve membranes owing to their amphipathic properties (see Figure 15.13), the released fatty acids are sequestered by fatty acid binding proteins in the cytosol. As shown in **Figure 15.24**, the human adipocyte fatty acid binding protein 4 has a flexible loop containing a Phe residue that functions as a lid, which opens and closes to provide access to the interior hydrophobic binding pocket. Note that epinephrine signaling in adipocytes through the β2-adrenergic receptor initiates the same downstream events as those of the glucagon receptor signaling pathway illustrated in Figure 15.23.

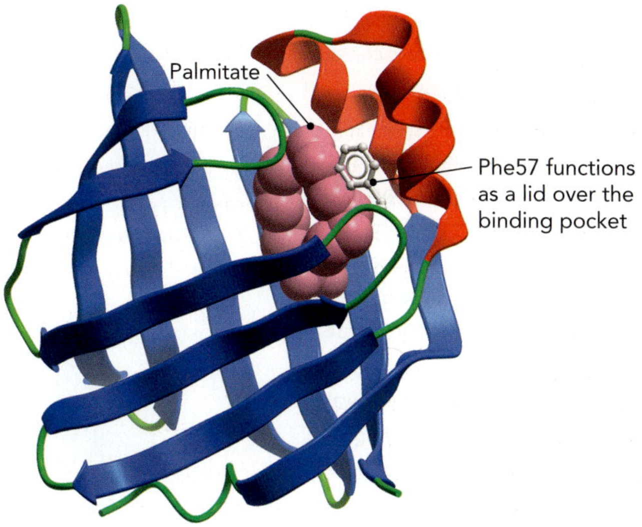

Figure 15.24 This molecular structure of fatty acid binding protein 4 from human adipocytes shows palmitate bound in the interior hydrophobic pocket. It has been proposed that the loop containing the residue Phe57 functions as a lid to regulate access to the fatty acid binding site. BASED ON PDB FILE 2HNX.

concept integration 15.2

Explain why cellular fatty acids are never really "free fatty acids," but rather are covalently linked to glycerol to form triacylglycerols or bound to carrier proteins such as fatty acid binding protein 4 and albumin.

Free fatty acids are amphipathic molecules containing a polar carboxyl group on one end and a long hydrocarbon chain with hydrophobic properties on the other end. This amphipathic property of fatty acids would dissolve cell membranes if not neutralized. Therefore, the primary storage form of fatty acids in organisms is triacylglycerols, in which the polar carboxyl group is neutralized by the covalent linkage to glycerol. Similarly, when triacylglycerols are hydrolyzed by lipase enzymes to release fatty acids, the fatty acids are quickly sequestered by intracellular binding proteins. These proteins deliver the fatty acids to the plasma membrane, where they are exported and carried to peripheral tissues by the fatty acid transport protein albumin.

15.3 Cell Membranes Contain Three Major Types of Lipids

As described in Chapter 2, cell membranes are responsible for protecting living cells from the environment by providing a hydrophobic barrier. The plasma membrane separates the outside of the cell from the inside of the cell and is studded with numerous proteins that function as metabolite transporters or signal transducers. The plasma membrane also provides a structural framework to maintain cell shape.

Analysis of plasma membranes in eukaryotic cells shows that by weight, most membranes contain ~50% protein and ~50% lipid, although the plasma membranes of some cell types contain about twice as much protein as lipid. For example, rat liver cell plasma membranes contain ~65% protein and ~35% lipid, owing to the high number of transmembrane receptor and transport proteins embedded in liver cell

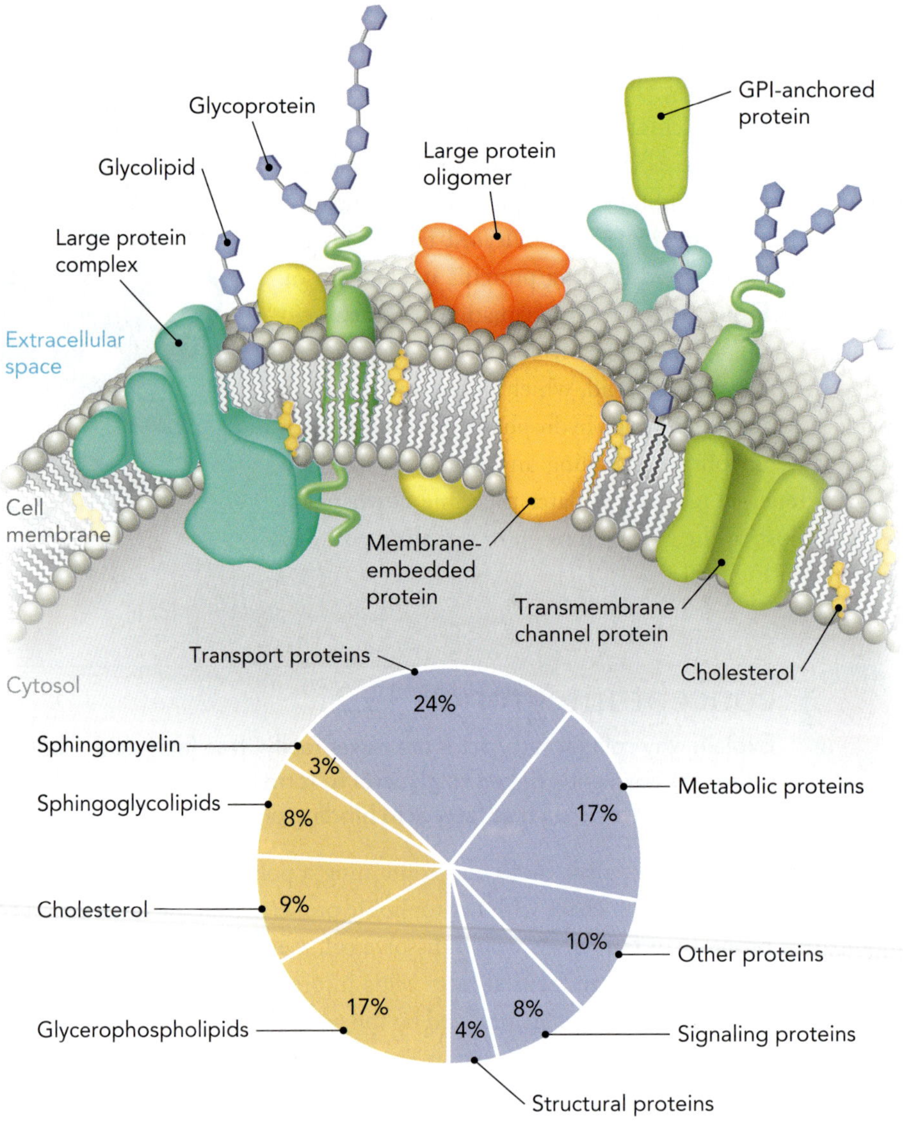

Figure 15.25 The plasma membrane of rat liver cells contains both proteins and lipids. The percentage of membrane proteins (blue) in each functional group was calculated using oligopeptide abundance as determined by mass spectrometry (R. CAO ET AL. 2008. *JOURNAL OF PROTEOME RESEARCH, 7,* 535–545). The percentage of lipid types (tan) within the membrane lipid fraction is based on mass and was determined by zonal centrifugation and thin layer chromatography (R. C. PFLEGER ET AL. 1968. *BIOCHEMISTRY, 7,* 2826–2833). The illustration at the top shows the various types of membrane proteins, including those modified with glycans or glycophosphatidylinositol (GPI), found in or associated with animal cell plasma membranes, as well as the basic organization of the lipid bilayer containing cholesterol.

plasma membranes relative to that in other cell types. **Figure 15.25** shows the relative distribution of major protein and lipid types in rat liver plasma membranes on the basis of proteomic and lipidomic studies, respectively. The protein fraction contains a large number of proteins involved in transport, metabolism, and signaling, whereas the lipid fraction consists of three major types of membrane lipids: **glycerophospholipids**, **sphingolipids**, and **cholesterol** (**Figure 15.26**). The relative ratio of these membrane lipids in rat liver cell plasma membranes is similar to that in other animal cell plasma membranes, although rat liver plasma membranes have higher amounts of protein than that in most cells.

Glycerophospholipids contain two fatty acids and a phosphate group attached to glycerol (see Figure 15.26). Representative glycerophospholipids present in eukaryotic cell membranes are **phosphatidylcholine**, **phosphatidylserine**, **phosphatidylethanolamine**, and **phosphatidylinositol**. Sphingolipids are derived from the biomolecule sphingosine (which is synthesized by linking serine to the carboxyl group of palmitate) and one fatty acid. The two most common types of sphingolipids are the **sphingophospholipids**, represented by sphingomyelin, and the sphingoglycolipids, which are called **cerebrosides** (monosaccharide glycan moiety) or **gangliosides**

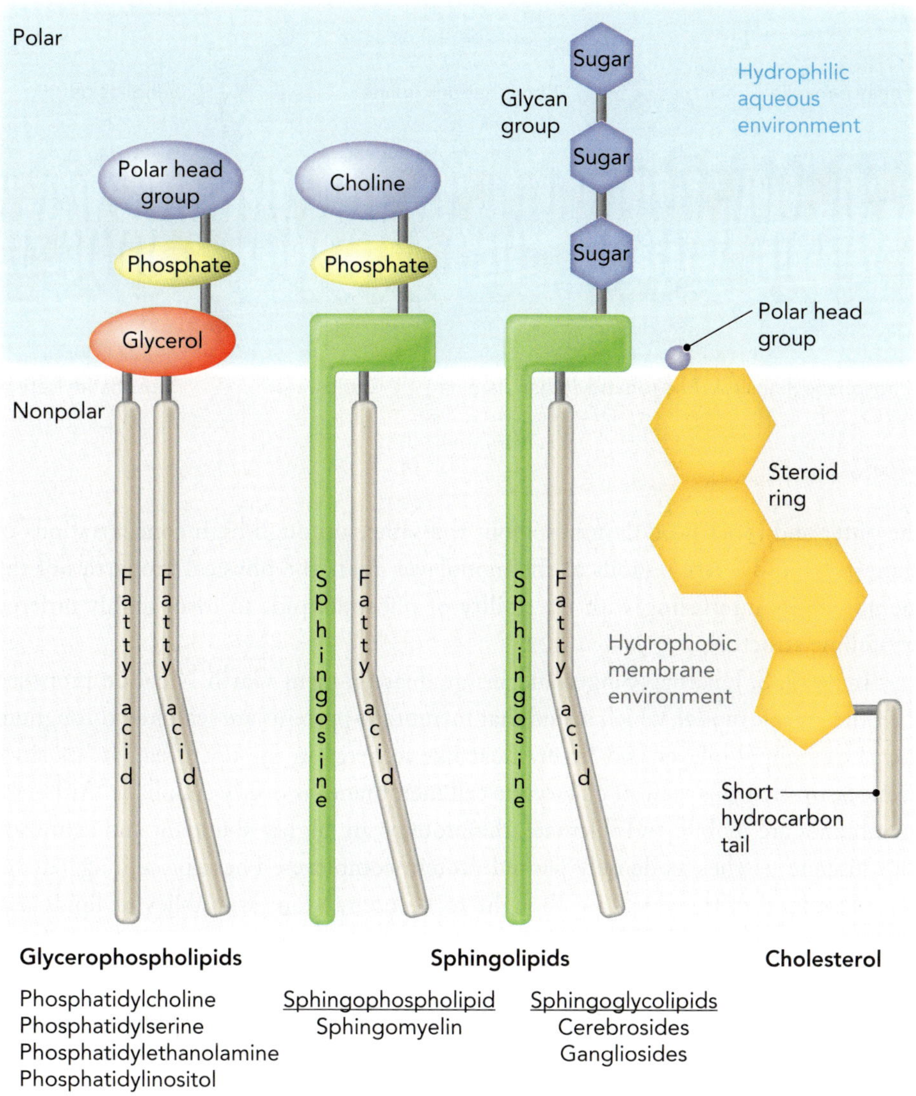

Figure 15.26 The three major types of membrane lipids are glycerophospholipids, sphingolipids, and cholesterol. The polar groups orient toward the aqueous environment, whereas the nonpolar tails of fatty acids, sphingosine, and cholesterol orient toward the hydrophobic center of the lipid bilayer. This illustration shows only a membrane monolayer. Sphingomyelin is also considered a phospholipid and is present at high levels in neuronal cell membranes.

(polysaccharide glycan moiety). Because sphingomyelin is a phospholipid, it is sometimes included with glycerophospholipids as an abundant membrane phospholipid, especially in neuronal cells. Lastly, cholesterol makes up 25% to 40% of the lipids in plasma membranes. It contains a rigid, four-ring steroid structure that disrupts packing of glycerophospholipids and sphingolipids. This affects the fluid properties of the membrane, as we now discuss.

Cell Membranes Have Distinct Lipid and Protein Compositions

As illustrated in **Figure 15.27**, the distribution of various glycerophospholipids and sphingolipids in the lipid bilayer of biological membranes is not uniform. The outer monolayer of the plasma membrane of human erythrocytes contains mostly phosphatidylcholine and the sphingomyelin and ganglioside sphingolipids. The inner monolayer contains mostly phosphatidylethanolamine, phosphatidylserine, and phosphatidylinositol (all glycerophospholipids). The uneven distribution of membrane lipids among the two monolayers reflects their distinct roles in determining the physical properties of the membrane (fluidity, charge, and thickness) and as mediators of signaling pathways. In contrast, the total cholesterol content of

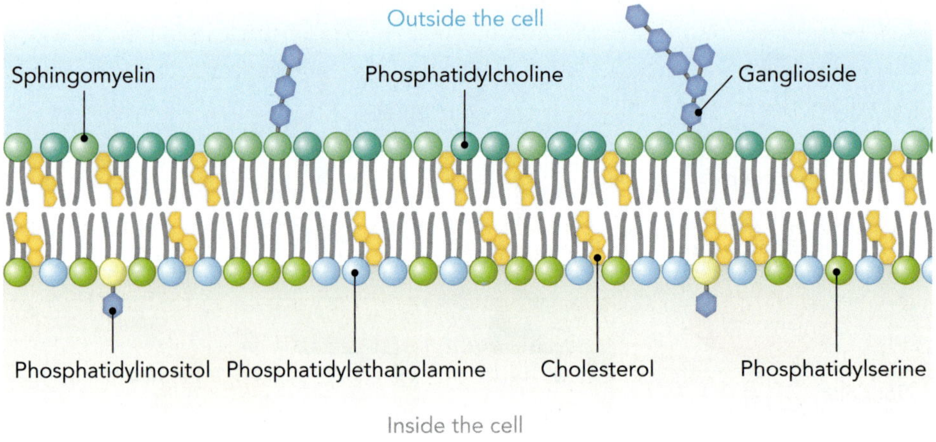

Figure 15.27 The two monolayers of a biological membrane contain different sets of membrane lipids. The outer monolayer of the plasma membrane in human erythrocytes contains mostly phosphatidylcholine and sphingolipids (sphingomyelin and gangliosides). In contrast, the inner monolayer consists almost entirely of the glycerophospholipids phosphatidylinositol, phosphatidylethanolamine, and phosphatidylserine. The overall cholesterol content of the inner and outer monolayers of the erythrocyte plasma membrane is similar.

the outer and inner monolayers is about the same, although high concentrations of cholesterol in discrete regions of the monolayer alters the physical properties of the membrane by interfering with the ability of phospholipids to form highly ordered crystalline structures.

In 1972, S. Jonathan Singer and his graduate student Garth Nicolson proposed the **fluid mosaic model**, which stated that membrane proteins are scattered throughout the phospholipid bilayer and "freely float like icebergs in the open sea." It has since been shown that this view of eukaryotic cell membranes is overly simplistic: Although membranes are mobile and fluctuate, the proteins are highly abundant and aggregate into discrete patches as densely packed protein complexes. These protein aggregates are called **lipid rafts**, which are thought to be nonrandom assemblies of lipids and proteins that serve as nucleating centers within the membrane to coordinate cell signaling processes, membrane trafficking, and neurotransmission (**Figure 15.28**). Lipid rafts have been shown to contain large transmembrane protein complexes, many of which are receptors for extracellular signals, as well as elevated levels of cholesterol and sphingolipids compared to other areas of the membrane. Lipid rafts may form in cell

Figure 15.28 Lipid rafts are thought to be discrete membrane regions that contain high concentrations of cholesterol (shown in yellow) and an aggregation of transmembrane receptor proteins and glycoproteins involved in cell signaling. Because of the way cholesterol interacts with the fatty acid side chains of membrane lipids, the membrane is thicker in the region of the lipid raft.

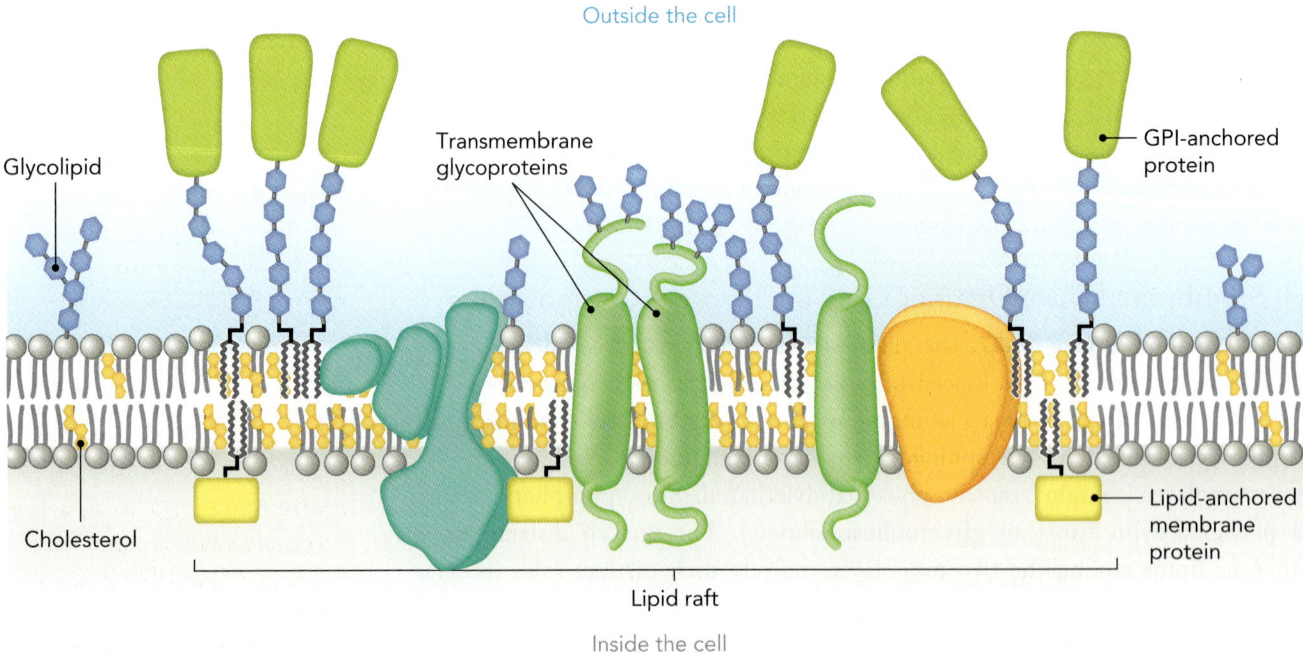

membranes as a result of ligand binding to the extracellular domains of transmembrane receptor proteins and GPI-anchored proteins. In some cell types, lipid rafts appear to be assembled by anchoring the cytoplasmic tails of transmembrane proteins to specific sites within the cytoskeleton.

Glycerophospholipids Are the Most Abundant Membrane Lipids

The first major type of membrane lipids is the glycerophospholipids, and the simplest glycerophospholipid is **phosphatidate**. Phosphatidate serves as the precursor to the more common glycerophospholipids, which are named according to the functional group attached to phosphate and include phosphatidylserine, phosphatidylethanolamine, phosphatidylcholine, and phosphatidylinositol. As shown in **Figure 15.29**, glycerophospholipids are triesters of glycerol-3-phosphate that have a polar head group in an ester linkage with the phosphate group. Many glycerophospholipids contain one saturated fatty acid, such as palmitate or stearate, and one monounsaturated fatty acid, which is usually oleate. The distribution of glycerophospholipids within the two monolayers (leaflets) of the plasma membrane differs depending on the cell type.

Glycerophospholipids are important sources of fatty acid–derived signaling molecules, which are released from membrane lipids by hydrolytic cleavage reactions catalyzed by phospholipase enzymes. As described in Chapter 8, phospholipase C cleaves the ester bond at C-3 in phosphatidylinositol-4,5-bisphosphate (PIP_2), which generates the signaling molecules inositol-1,4,5-trisphosphate (IP_3) and diacylglycerol (see Figure 8.10).

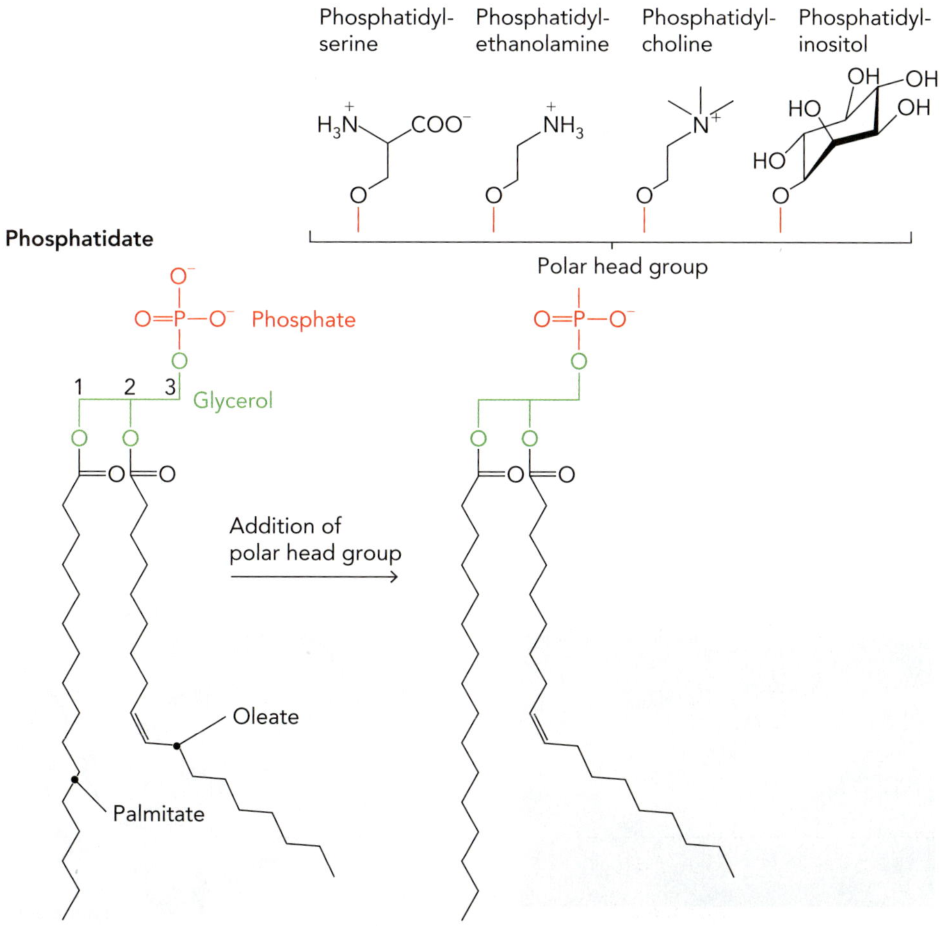

Figure 15.29 The most abundant glycerophospholipids in eukaryotic membranes are phosphatidylserine, phosphatidylethanolamine, phosphatidylcholine, and phosphatidylinositol. The substrate for glycerophospholipid synthesis is phosphatidate. Note that the fatty acids palmitate and oleate, shown here in ester linkage with C-1 (*sn*-1) and C-2 (*sn*-2), are two of the more common fatty acids found in glycerophospholipids.

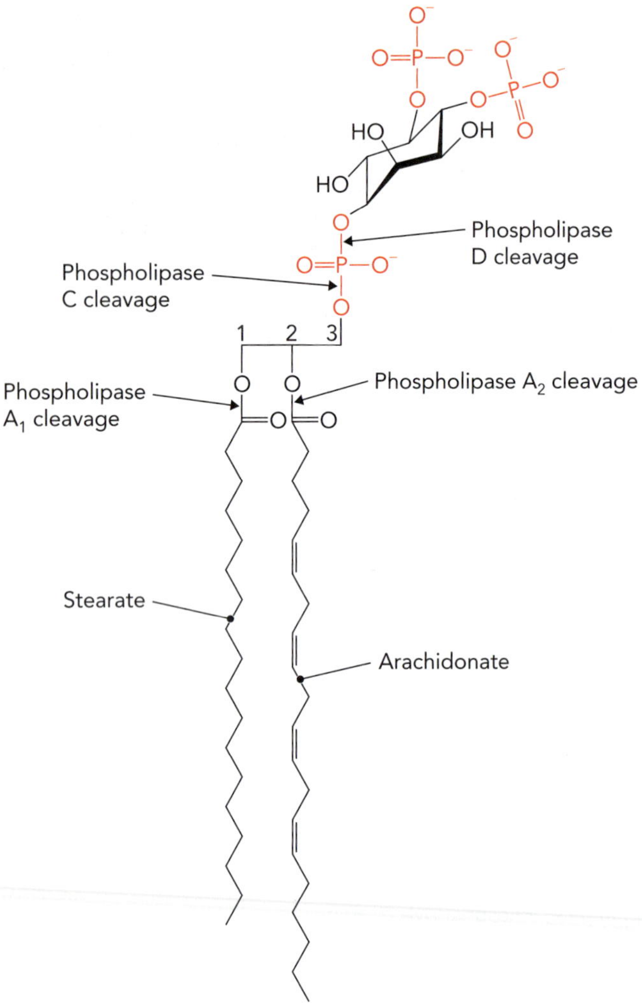

Phosphatidylinositol-4,5-bisphosphate (PIP_2)

Figure 15.30 Phospholipase enzymes cleave glycerophospholipids such as PIP_2 by catalyzing a hydrolysis reaction (see also Figure 8.10). In this example, cleavage by phospholipase A_2 generates arachidonate, an important mediator of inflammatory signaling, whereas cleavage by phospholipase A_1 produces stearate.

As illustrated in **Figure 15.30**, three other phospholipases have also been shown to cleave glycerophospholipids such as PIP_2. Phospholipase D cleaves the phosphodiester bond on the sugar side of the phosphate in PIP_2 and releases phosphatidate and inositol-4,5-bisphosphate (IP_2). The phospholipases A_1 and A_2 cleave at C-1 and C-2, respectively, which releases the fatty acids.

Snake venoms often contain secreted phospholipase A_2 enzymes, which cleave glycerophospholipids. This leads to tissue damage from membrane breakdown and the detergent effect of the free fatty acids. Without treatment, death can occur from massive internal bleeding within a few days. Several snake venom phospholipase A_2 enzymes have been biochemically characterized to identify inhibitors that could be useful in treating snakebites. **Figure 15.31** shows the molecular structure of a phospholipase A_2 enzyme from the deadly saw-scaled viper *Echis omanensis*. The enzyme active site is identified in this protein structure by the location of a bound molecule of lauric acid, a saturated C_{12} fatty acid. Similar to other secreted enzymes that need to retain structural stability in saliva or blood, the 122-amino-acid saw-scaled viper phospholipase A_2 enzyme has seven disulfide bonds.

Sphingolipids Contain One Fatty Acid Linked to Sphingosine

The second major type of membrane lipids is the sphingolipids, which are derivatives of **sphingosine**—a long-chain amino alcohol synthesized from palmitate and serine. Fatty acylation of sphingosine generates **ceramide**, the metabolic starting point for sphingophospholipids and sphingoglycolipids. As shown in **Figure 15.32**, the addition of phosphocholine to ceramide generates sphingomyelin, whereas the addition of glycan groups to ceramide generates cerebrosides and gangliosides.

Gangliosides play important roles in cell recognition and membrane function in several vital organs, including the nervous system. For example, the human ABO blood

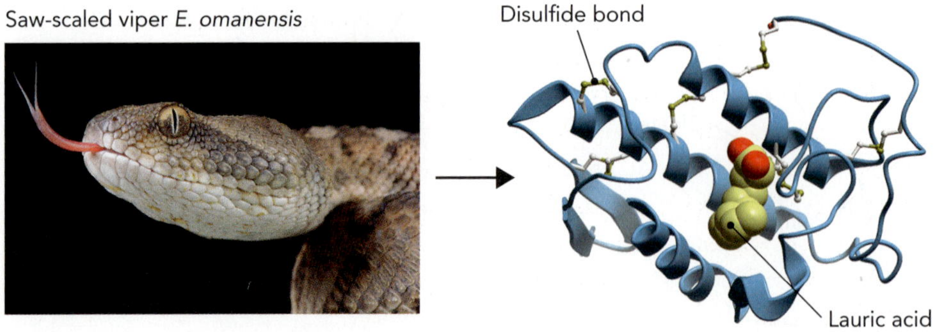

Figure 15.31 Most snake venoms contain a phospholipase A_2 enzyme that causes tissue damage by releasing free fatty acids from glycerophospholipids, which function as biological detergents. The phospholipase A_2 from the saw-scaled viper *E. omanensis* is shown here with a fatty acid bound to the enzyme active site and seven disulfide bonds. SAW-SCALED VIPER: REPTILES4ALL/SHUTTERSTOCK; PHOSPHOLIPASE A_2 ENZYME: BASED ON PDB FILE 2QHD.

Figure 15.32 The chemical structures of four types of sphingolipids are shown here. Ceramide is the core structure of all sphingolipids and consists of sphingosine and a fatty acid. Sphingomyelin is a sphingophospholipid, whereas cerebroside is a monoglycosylated ceramide. Ganglioside is a ceramide with an oligosaccharide moiety. AcN = *N*-acetyl group.

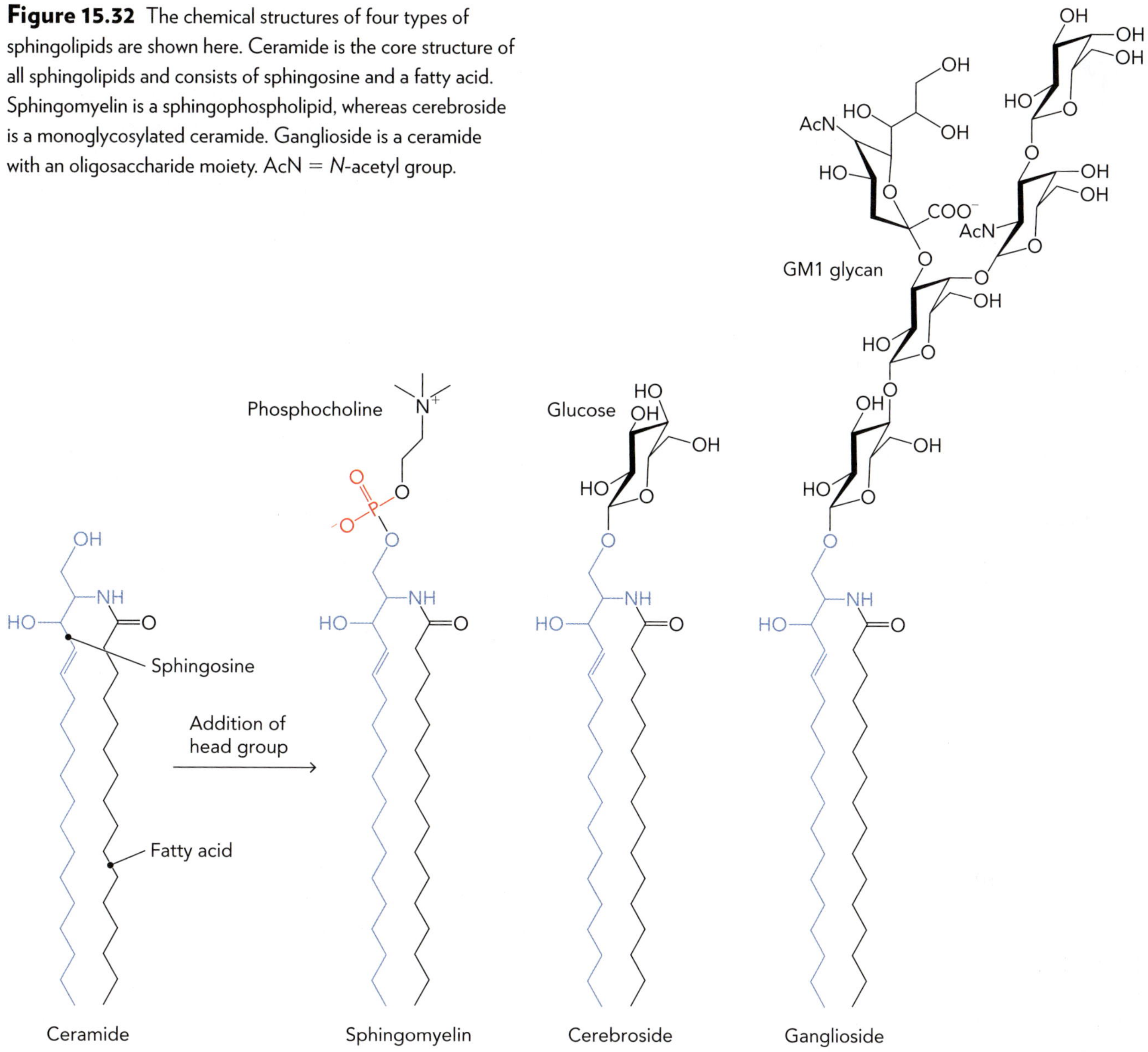

groups, described in Chapter 13 as components of glycoproteins (see Figure 13.23), are also present on erythrocytes in the form of ABO gangliosides, which are generated by the same two variant glycosyltransferase enzymes (GTA and GTB).

Defects in sphingolipid metabolism have been linked to several human lipid disorders, including **Tay–Sachs disease**, **Fabry disease**, and **Niemann–Pick disease**. In these three hereditary diseases, defects in enzymes responsible for sphingolipid degradation result in the buildup of metabolic precursors that lead to severe neuronal defects and death. As shown in **Figure 15.33a**, the ganglioside GM_2 is normally modified by the enzyme hexosaminidase A to remove a terminal GalNAc moiety and form the ganglioside GM_3. However, in patients with Tay–Sachs disease, hexosaminidase A is defective, causing a buildup of the GM_2 ganglioside in the spleen and brain. This leads to developmental delays, mental retardation, and death. Importantly, it is not the lack of the product GM_3 that causes the disease symptoms, but rather excess levels of the GM_2 substrate.

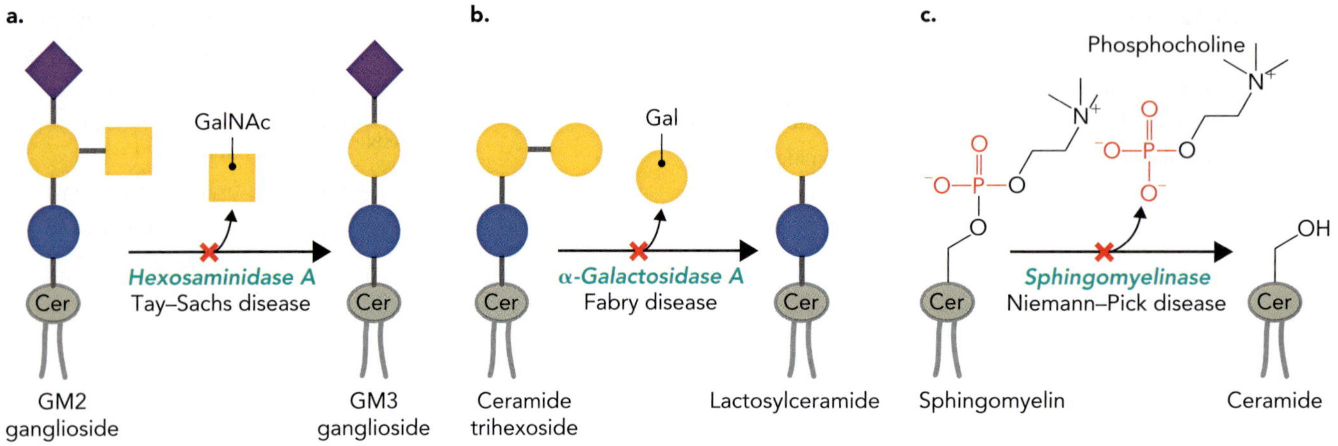

Figure 15.33 Defects in sphingolipid metabolizing enzymes cause several human neurologic diseases. **a.** Tay–Sachs disease is caused by a defect in the enzyme hexosaminidase A, leading to a buildup of the GM2 ganglioside. **b.** Fabry disease is caused by defects in α-galactosidase A and the accumulation of ceramide trihexoside. **c.** Niemann–Pick disease results from a defect in sphingomyelinase, which causes elevated levels of sphingomyelin. Symbols for the glycan groups are the same as those used in Figure 13.2. Cer = ceramide.

Similarly, Fabry disease (**Figure 15.33b**), which results from defects in the enzyme α-galactosidase A, and Niemann–Pick disease (**Figure 15.33c**), which is due to a defect in the enzyme sphingomyelinase, are both symptomatic because of accumulation of metabolic precursors. Although there are no cures at the present time for lipid metabolizing diseases, patients with Fabry disease, who often survive into early adulthood, do respond to enzyme replacement therapy. This treatment requires that the patient receive periodic infusions of recombinant human α-galactosidase A protein produced by biotechnology methods. Someday, it may be possible to use gene replacement therapy as a long-term treatment for certain hereditary diseases such as Fabry disease.

Cholesterol Is a Rigid, Four-Ring Molecule in Plasma Membranes

The third major type of lipid in animal cell membranes is cholesterol. Cholesterol is a C_{27} tetracyclic molecule consisting of a planar ring system with a polar hydroxyl group on one end and a hydrophobic alkyl side chain on the other end. Mammals primarily synthesize cholesterol *de novo* from mevalonate in the liver (see Chapter 16), but they can also obtain cholesterol from their diet. Although plants do not synthesize cholesterol, they do produce a related sterol compound called **stigmasterol**. As shown in **Figure 15.34**, stigmasterol is structurally similar to cholesterol. Some herbivorous insects, such as the tobacco hornworm *Manduca sexta*, are able to convert stigmasterol and other plant sterols into cholesterol, which they require for membrane biosynthesis during metamorphosis. Bacteria cannot synthesize sterols, although some bacterial species are able to incorporate cholesterol into their membranes.

The rigid structure of the nonpolar planar rings in cholesterol affects membrane fluidity. Therefore, the concentration of cholesterol in cell membranes modulates membrane functions such as the formation of lipid rafts (see Figure 15.28). The amount of cholesterol in cell membranes varies from 25% to 40% of lipids in the plasma membrane to ~10% of lipids in endoplasmic reticulum membranes. At physiologic temperatures, high local concentrations of cholesterol prevent lateral movement of

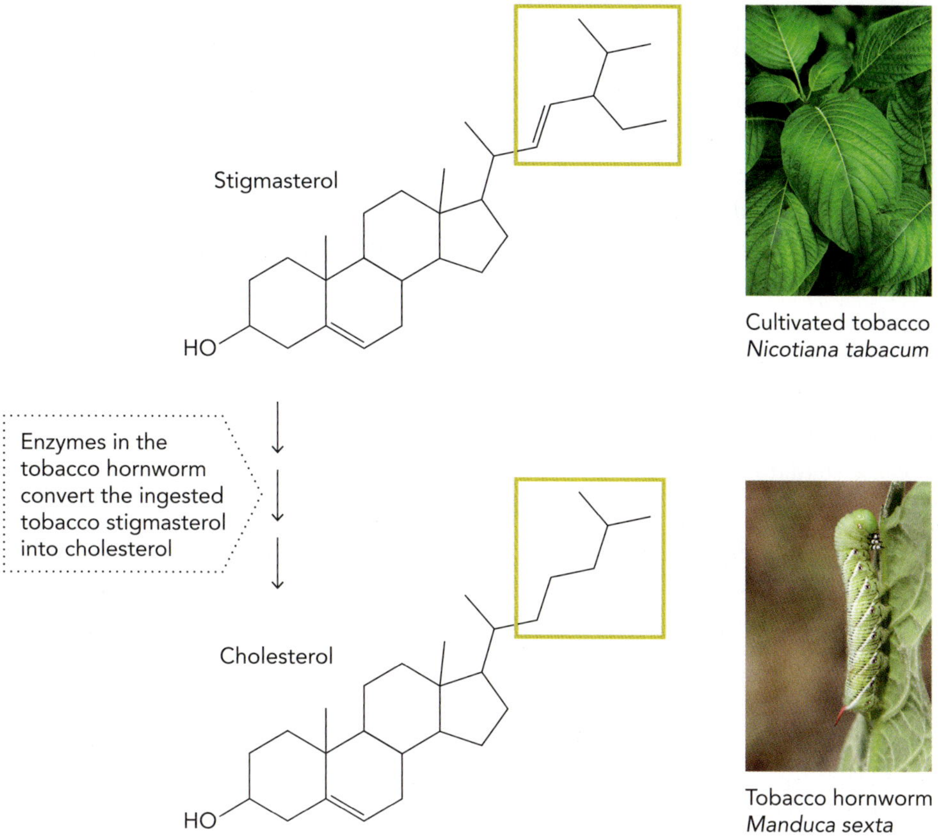

Figure 15.34 Stigmasterol is a plant sterol that insects such as the tobacco hornworm convert into cholesterol for use in membrane biosynthesis. CULTIVATED TOBACCO: ROBERTO A. SANCHEZ/GETTY IMAGES; TOBACCO HORNWORM: EDDIE MCGRIFF, UNIVERSITY OF GEORGIA, BUGWOOD .ORG/CC BY 3.0 ATTRIBUTION.

phospholipids and thereby decrease membrane fluidity in that region. But at low temperatures with the same local concentrations, cholesterol maintains membrane fluidity through its disruption of fatty acid packing.

concept integration 15.3

With regard to the structure and function of membrane lipids, what might explain the differential distribution of sphingomyelin and phosphatidylinositol to the outer and inner leaflets of the plasma membrane, respectively?

The hydrophobic fatty acid–derived tails of sphingomyelin and phosphatidylinositol are derived from ceramide and phosphatidate, respectively, which often contain distinct saturated and unsaturated hydrocarbon chains. Similarly, the polar head groups of these two membrane lipids are chemically quite different. For example, sphingomyelin, which is found at high concentrations in the outer leaflet of neuronal cell membranes, contains a phosphocholine group attached to ceramide, which may be important in facilitating neuronal cell recognition or migration. In contrast, phosphatidylinositol contains a glucose-derived inositol group, which serves as a binding site for intracellular signaling proteins and as a source of intracellular second messengers after phospholipase cleavage. Therefore, because chemical differences in sphingomyelin and phosphatidylinositol likely determine distinct functions in the extracellular space and cytosol, respectively, membrane modulating enzymes such as flippases (see Figure 2.46) must be able to recognize and selectively redistribute specific types of membrane lipids.

15.4 Lipids Function in Cell Signaling

Lipids constitute a truly diverse set of biological molecules. In addition to their roles in energy storage and membrane structure, they are also important signaling molecules. As described in this section, cholesterol is the precursor to steroid hormones (mineralocorticoids, glucocorticoids, progesterones, androgens, and estrogens) and to vitamin D. Another group of lipid signaling molecules are derived from the ω-6 fatty acid arachidonate, a polyunsaturated fatty acid that is converted into immune regulatory molecules called **eicosanoids**. The best-characterized eicosanoid signaling molecules are prostaglandins, prostacyclins, thromboxanes, and leukotrienes.

The hydrophobic property of lipids contributes to their function as high-affinity stereospecific ligands that bind to hydrophobic pockets in receptor proteins. Moreover, because steroid hormones and eicosanoids are not stored in appreciable amounts, it is convenient (and efficient) for them to be rapidly synthesized from preexisting lipids in the cell. Note that the diet-derived ω-3 fatty acids EPA and DHA are also thought to have signaling properties because of their low-affinity binding to peroxisome proliferator–activated receptors, which is described in more detail in Chapter 19.

Cholesterol Derivatives Regulate the Activity of Nuclear Receptor Proteins

Cholesterol is the metabolic precursor to steroid hormones and **bile acids** (**Figure 15.35**). Steroid hormones are ligands for nuclear receptor proteins, which mediate hormone signals by altering the expression of specific genes (see

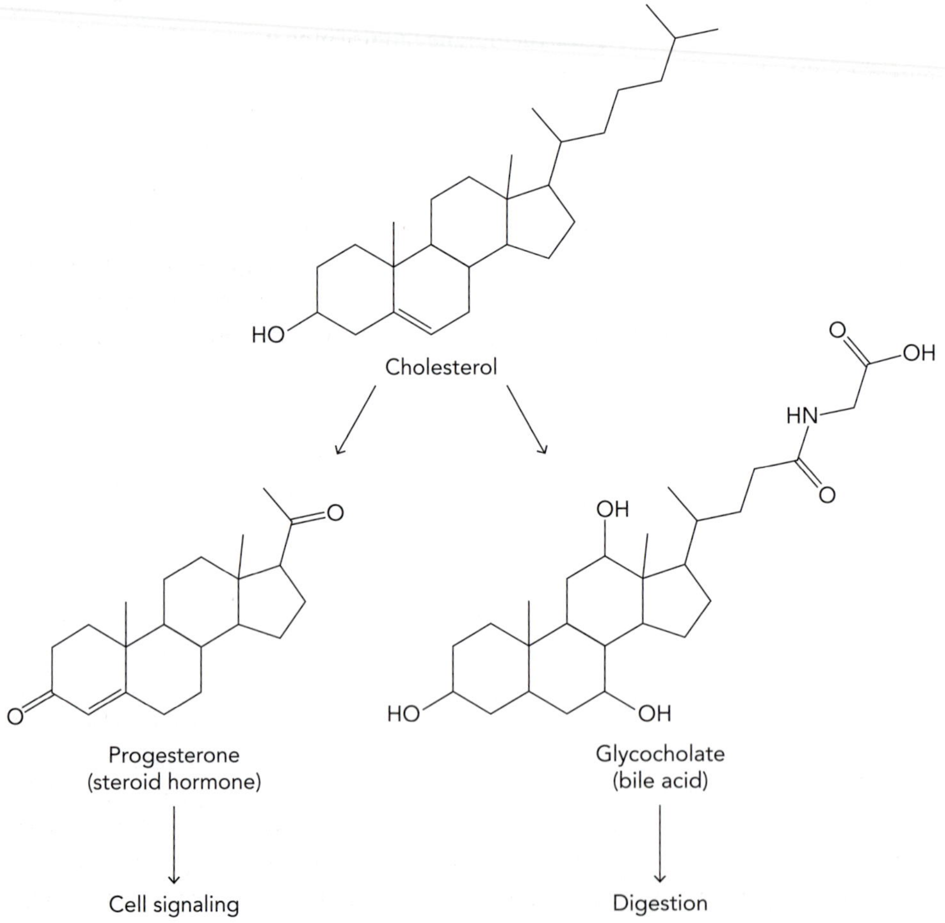

Figure 15.35 Cholesterol is the key metabolic precursor in steroid hormone and bile acid biosynthesis.

Figure 8.60). Bile acids such as **glycocholate** are synthesized in the liver and secreted into the intestine, where they function as emulsifying agents to aid in the absorption of dietary lipids.

Note that although cholesterol has many important functions in cells, excess levels of circulating cholesterol in humans contributes to cardiovascular disease. Cholesterol is transported through the body by lipoproteins and accumulates in atherosclerotic plaques that form in blood vessels and restrict blood flow. Effective medical treatment for high cholesterol levels involves the use of pharmaceutical drugs that block both cholesterol synthesis in the liver and dietary absorption.

Steroid hormones are potent signaling molecules that have critical roles in cell development, reproductive biology, and organismal physiology. Because of these wide-ranging functions, synthetic steroids are exploited as pharmacological agents to treat everything from skin rashes and asthma (glucocorticoids) to cancer (antagonists of estrogens and androgens). Steroid hormones are derived from modifications of cholesterol through a series of tissue-specific, enzyme-catalyzed reactions collectively called **steroidogenesis** (**Figure 15.36**). Many of the enzymes involved in steroidogenesis are hydroxylases (also called mixed-function oxidases), of which one type is the cytochrome **P450 monooxygenases** (oxidoreductases). The primary precursor in steroidogenesis is pregnenolone, which is generated from cholesterol after removal of the side chain at C-17 of the D ring. This C_{21} sterol gives rise to progesterone, which can be converted to cortisol, androstenedione, and corticosterone. Cortisol is a glucocorticoid hormone that regulates liver metabolism (gluconeogenesis), as well as a variety of cell functions in the immune system and the brain. Androstenedione is an androgenic hormone that gives rise to testosterone, which is then converted to either estradiol or dihydrotestosterone. Estradiol is the major reproductive hormone in females, and dihydrotestosterone is the high-affinity form of testosterone in males. Lastly, corticosterone can be converted to aldosterone, a mineralocorticoid that regulates ion transport in the kidney.

Mineralocorticoids and glucocorticoids are synthesized in the adrenal glands, which are located next to the kidneys. The three other major sites of steroid synthesis are the ovaries in females (estrogen), the testes in males (testosterone), and the corpus luteum in pregnant females (progesterone). The adrenal glands also synthesize androgens, which is how females acquire testosterone for estradiol production. **Table 15.2** (p. 761) summarizes the physiologic functions of the major classes of steroid hormones.

Many steroid hormones are chemically modified by enzymes belonging to the P450 family. P450 enzymes are membrane-bound mitochondrial proteins containing a Fe^{2+}/Fe^{3+} heme group that absorbs light at 450 nm, which is why they are called P450 enzymes (**Figure 15.37**, p. 761). The P450 enzymes are highly conserved and found in essentially all organisms, from bacteria to humans, owing to their critical role in chemical detoxification. In fact, the potency of pharmaceutical drugs is often affected as a result of modification by liver P450 enzymes. This can lead to inactivation of otherwise potent drugs or, in some cases, drug activation through the so-called first-pass effect. It is also well documented that some natural compounds are converted to carcinogens as a result of modification by P450 enzymes.

The modification of steroids by P450 enzymes primarily alters side groups on the cholesterol rings, which ensures that steroid receptor proteins are only activated by their cognate steroid hormone. Subtle differences in the side groups of steroid hormones account for their distinct agonist (activating) or antagonist (inhibiting)

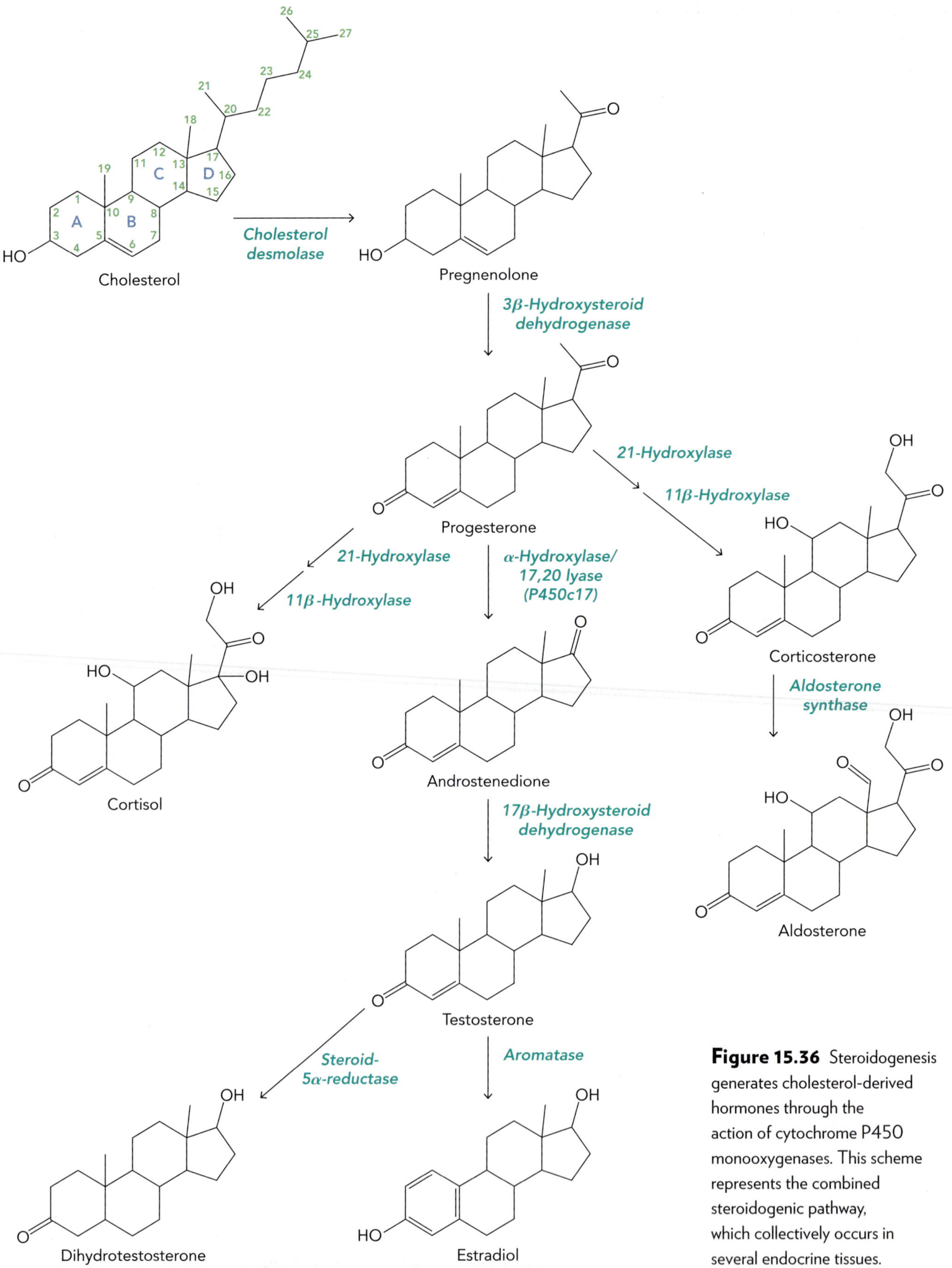

Figure 15.36 Steroidogenesis generates cholesterol-derived hormones through the action of cytochrome P450 monooxygenases. This scheme represents the combined steroidogenic pathway, which collectively occurs in several endocrine tissues.

Table 15.2 FUNCTIONS OF THE MAJOR CLASSES OF STEROID HORMONES IN MAMMALS

Steroid	Site of synthesis	Physiologic functions
Progesterone	Corpus luteum	Menstruation, development of mammary tissue
Cortisol	Adrenal cortex	Liver metabolism, immune functions, adaptation to stress
Aldosterone	Adrenal cortex	Ion transport in the kidneys, blood pressure regulation
Testosterone	Testes, adrenal cortex	Development of male reproductive organs
Estradiol	Ovaries	Development of female reproductive organs

functions. As shown in **Figure 15.38**, nandrolone is a synthetic androgen receptor agonist that is very similar in structure and function to the naturally occurring androgen receptor agonist dihydrotestosterone. Nandrolone has been used by bodybuilders to increase muscle mass because it has a selective effect on muscle cells. In contrast to structurally related steroid analogs with agonist or antagonist activities, bicalutamide (Casodex) is a nonsteroidal androgen that is used to treat metastatic prostate cancer because of its androgen receptor antagonist activity.

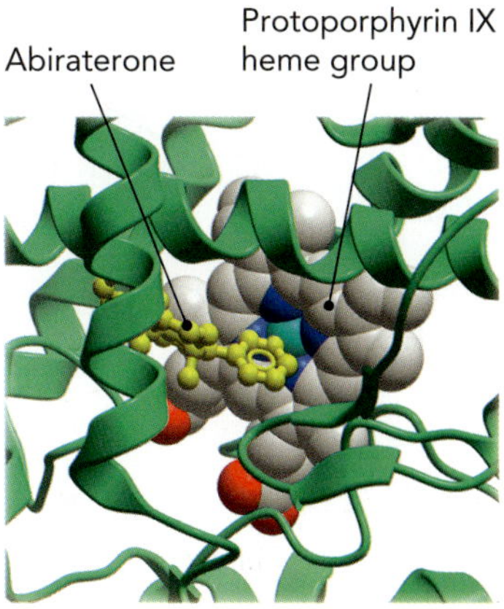

Figure 15.37 The molecular structure of the human enzyme cytochrome P450c17 is shown here with an enzyme inhibitor bound to the heme-containing active site. This enzyme is required in the steroidogenesis pathway (see Figure 15.36). The cytochrome P450c17 inhibitor abiraterone is being developed as a possible prostate cancer drug. BASED ON PDB FILE 3RUK.

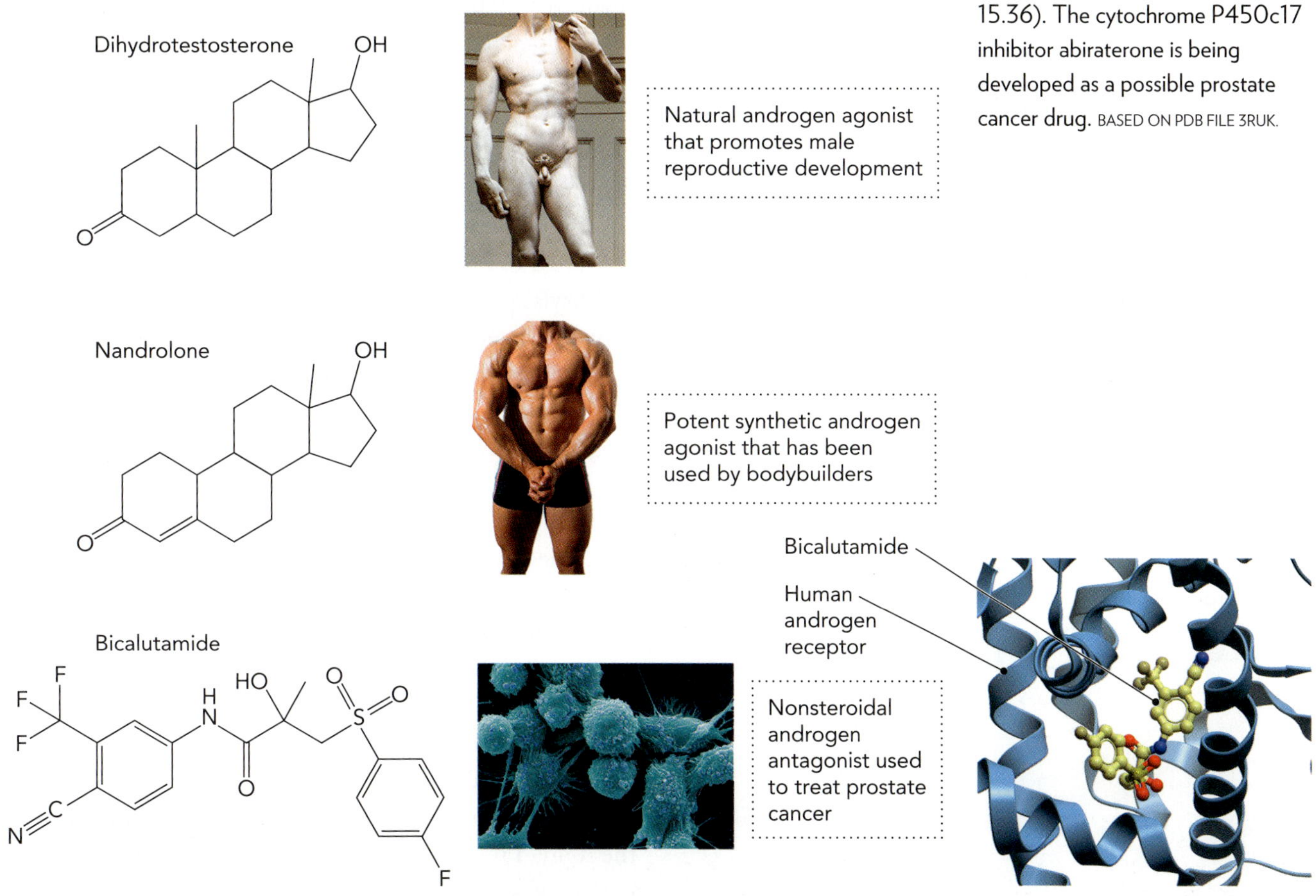

Figure 15.38 Modification of chemical side groups on steroid hormones alters their biological activity by affecting the structure and function of ligand-bound nuclear receptor complexes. Nandrolone, which is a synthetic androgen receptor agonist, is similar in molecular structure to the natural androgen steroid hormone dihydrotestosterone. Bicalutamide is a nonsteroidal androgen receptor antagonist that binds to the same site in the human androgen receptor protein as the natural ligand dihydrotestosterone. *DAVID* BY MICHELANGELO: MURATART/SHUTTERSTOCK; BODYBUILDER: WAVEBREAKMEDIA/SHUTTERSTOCK; PROSTATE CANCER CELLS: DR. GOPAL MURTI/VISUALS UNLIMITED/CORBIS; BICALUTAMIDE BOUND TO ANDROGEN RECEPTOR: BASED ON PDB FILE 1Z95.

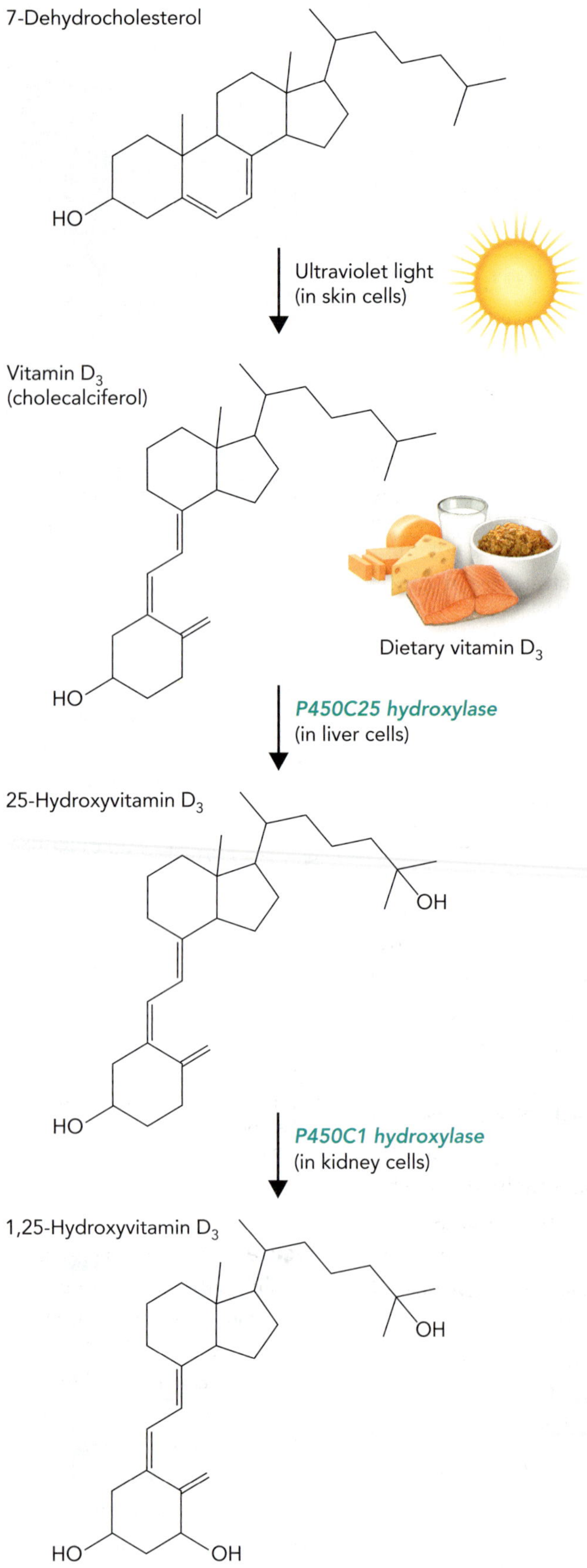

Figure 15.39 Sunlight converts 7-dehydrocholesterol into vitamin D_3, which is then hydroxylated to generate the biologically active form, 1,25-hydroxyvitamin D_3.

Vitamin D is another cholesterol-derived lipid signaling molecule. In the early 1920s, vitamin D was discovered as a nutritional supplement that could be added to the diet to treat the debilitating bone disease **rickets**, which is caused by insufficient calcium deposition. Adolph Windaus discovered in the mid-1920s that vitamin D is a derivative of 7-hydroxycholesterol and that humans synthesize sufficient quantities of a vitamin D precursor in skin cells after exposure to ultraviolet light. Windaus received the Nobel Prize in Chemistry in 1928 for his work on sterols and their connections to vitamins.

Few foods are naturally rich in vitamin D, but one of these is oily fish, which is why cod liver oil was often given to children as a dietary supplement to prevent rickets. In many countries, vitamin D has been added to processed foods, including pasteurized milk, orange juice, and breakfast cereals. Importantly, more than 90% of the vitamin D in the human body comes from sunlight conversion of 7-dehydrocholesterol. Studies show that it only requires ~10 minutes of sunlight exposure per day during the midday hours to produce sufficient levels of vitamin D in the skin.

The pathway for vitamin D synthesis is shown in **Figure 15.39**. You can see that 7-dehydrocholesterol is converted to **vitamin D_3** (cholecalciferol) in skin cells after absorption of energy from ultraviolet light. Vitamin D_3 is converted to 25-hydroxyvitamin D_3 in the liver by a P450 hydroxylase enzyme specific for C-25. Then, the product is hydroxylated in the kidney by a second P450 enzymatic reaction, leading to the formation of 1,25-hydroxyvitamin D_3. The 1,25-hydroxyvitamin D_3 compound is the physiologically active form that binds to the vitamin D receptor and regulates gene expression in target tissues. The vitamin D receptor is expressed in a variety of cell types, including bone, the small intestine, colon, thymocytes, and the β islet cells of the pancreas.

Windaus's discovery that sunlight was needed to convert 7-dehydrocholesterol to vitamin D_3 explained why rickets was found in children who lived in smoggy, industrialized urban centers, such as London at the beginning of the 1900s, where sunlight exposure was minimal. Indeed, it is thought that Charles Dickens's portrayal of the Tiny Tim character in his book *A Christmas Carol* is that of a child with rickets who was undernourished and had difficulty walking (**Figure 15.40**). Notably, even in some of the sunniest countries, vitamin D_3 deficiency can still occur if dietary sources of vitamin D_3 are not available. For example, rickets can occur in Middle Eastern countries where people cover themselves from head to toe with robes and scarves or in cases where people rarely go outside because of harsh weather or frailty.

Eicosanoids Are Derived from Arachidonate

Eicosanoids are a group of signaling molecules derived from C_{20} polyunsaturated fatty acids such as arachidonate. They are released from the membrane by phospholipases and modified by mitochondrial enzymes. Unlike steroid hormones, which function in tissues located far from the gland of origin (endocrine) and which have half-lives of hours, eicosanoids are produced by cells at their sites of action (paracrine) and have half-lives of only a few minutes. The four major classes of arachidonate-derived eicosanoids are prostaglandins, prostacyclins, thromboxanes, and leukotrienes.

Prostaglandins were discovered in the 1930s by the Swedish biochemist Ulf von Euler, who identified compounds in human semen that regulate blood flow when injected into animals. He called them prostaglandins because he thought they were made in the prostate gland. The prostaglandins Euler identified in semen are actually synthesized in the seminal vesicles, and in fact, prostaglandins are made in almost all tissues of both males and females.

A large number of prostaglandins have been isolated and functionally characterized over the past 50 years. It is now known that these potent and fast-acting lipid signaling molecules regulate a variety of cellular processes through activation of G protein–coupled receptors and second messenger signaling cascades. For example, prostaglandins play a primary role in regulating blood flow by modulating smooth muscle contraction and relaxation. This versatile class of eicosanoids also stimulate inflammatory responses (tissue swelling, pain, and fever), control ion transport, modulate the secretion of proteoglycans (which protect the stomach lining from the effects of low pH), and activate uterine contraction during birth.

Prostacyclins have been shown to control platelet aggregation and blood clot formation and to stimulate vasodilation. The other two classes of eicosanoids, thromboxanes and leukotrienes, were discovered in the 1970s by Bengt Samuelsson, a biochemist at the Karolinska Institute in Sweden. Thromboxanes regulate blood vessel constriction, and leukotrienes act as inflammatory mediators that also regulate smooth muscle contraction.

The eicosanoid synthetic pathway is illustrated in **Figure 15.41**. You can see that activation of phospholipase A_2 by the $G_{\beta\gamma}$ subunits of a G protein–coupled receptor results in the release of arachidonate from C-2 of membrane phospholipids. Arachidonate is converted into leukotrienes by lipoxygenase enzymes or is used to generate prostaglandin H_2, which is a precursor to other prostaglandins, thromboxanes, and prostacyclins. Cyclooxygenation of arachidonate by the enzymes cyclooxygenase-1 and cyclooxygenase-2 (COX-1 and COX-2; also called prostaglandin H_2 synthase-1 and prostaglandin H_2 synthase-2) requires two activities present in the cyclooxygenase enzymes. One is an oxygenase that uses 2 O_2 to oxygenate arachidonate, resulting in the formation of a five-carbon ring. The other activity, a peroxidase, removes one of these oxygens. Notably, the COX-1 and COX-2 enzymes are differentially inhibited by anti-inflammatory drugs.

Many of the eicosanoid signaling molecules derived from prostaglandin H_2 exacerbate the inflammatory response, and therefore drugs that inhibit prostaglandin H_2 synthesis function as anti-inflammatory agents. Because glucocorticoids also have anti-inflammatory properties, inhibitors of prostaglandin H_2 synthesis are called **nonsteroidal anti-inflammatory drugs (NSAIDs)** to distinguish them from

Tiny Tim in Charles Dickens's *A Christmas Carol* would be a typical child with rickets caused by vitamin D_3 deficiency

Even in sunny locations, women wearing burqas need dietary vitamin D_3 to protect against rickets

Figure 15.40 Lack of adequate sunlight and diets poor in vitamin D are the most common causes of the bone disease rickets. Tiny Tim in Charles Dickens's book *A Christmas Carol* is thought to be a young boy in London afflicted with rickets. Rickets still occurs in areas of the world where people do not get adequate exposure to the Sun and have dietary deficiencies. A CHRISTMAS CAROL: MARY EVANS PICTURE LIBRARY/ ALAMY STOCK PHOTO; WOMEN WEARING BURQAS: SHAH MARAI/AFP/GETTY IMAGES.

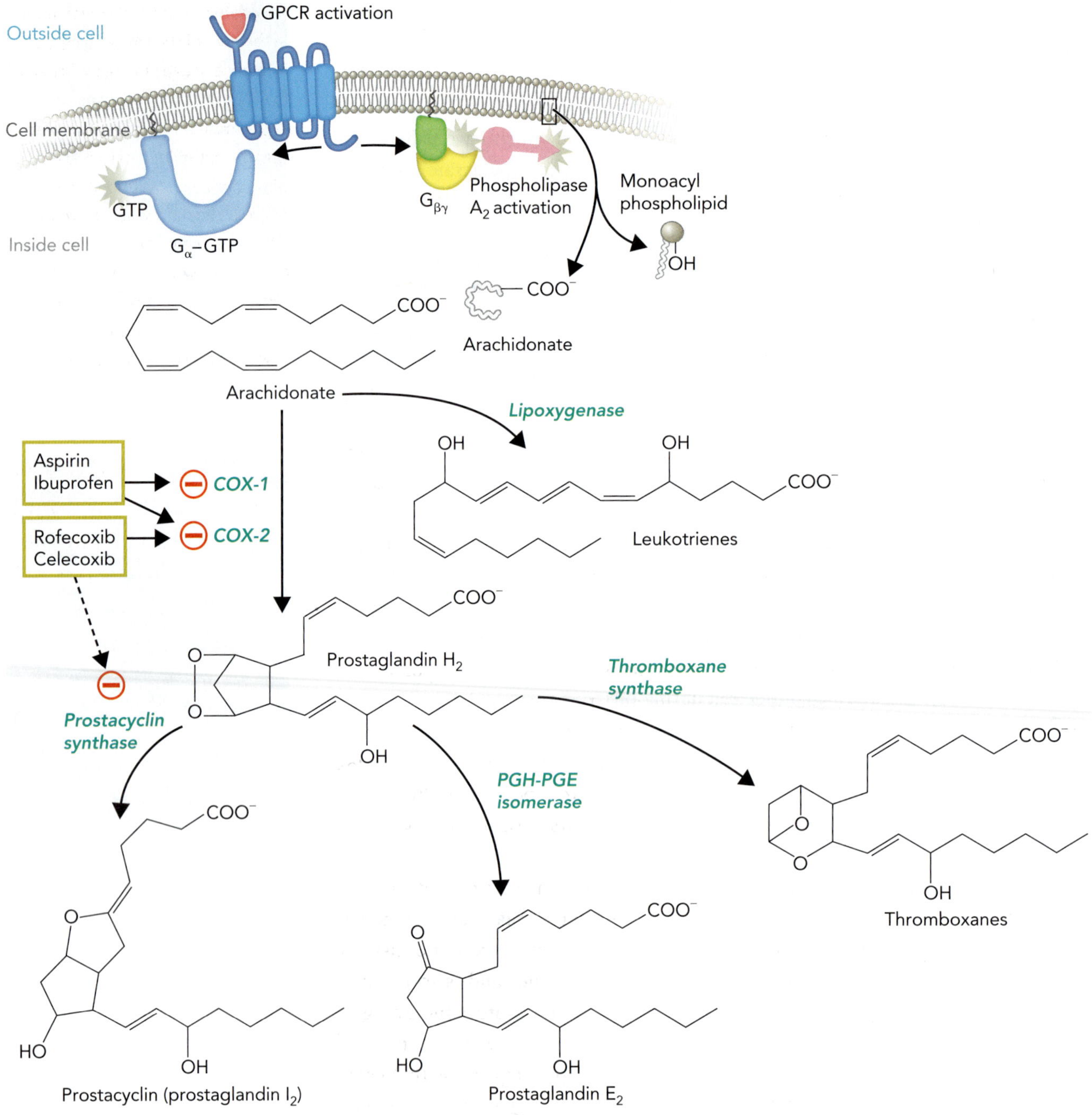

Figure 15.41 The four major classes of eicosanoids are all derived from the polyunsaturated fatty acid arachidonate. Activation of G protein–coupled receptors (GPCRs) leads to stimulation of phospholipase A_2 activity through the $G_{\beta\gamma}$ subunits, thereby releasing arachidonate from membrane phospholipids. Arachidonate is the precursor to prostaglandins, prostacyclins, leukotrienes, and thromboxanes. The cyclooxygenase enzymes, COX-1 and COX-2, are differentially inhibited by the anti-inflammatory drugs aspirin, ibuprofen, rofecoxib, and celecoxib. Prostacyclin is also known as prostaglandin I_2. PGH-PGE isomerase = prostaglandin H/prostaglandin E isomerase.

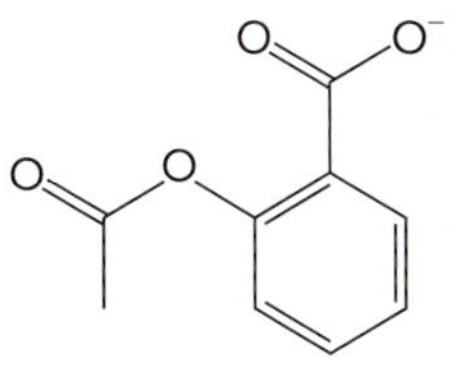

Acetylsalicylate

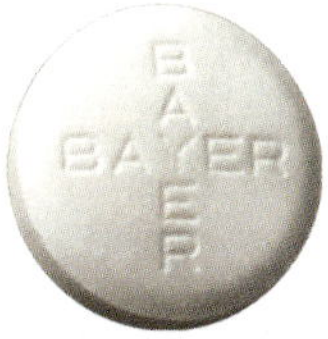

Bayer aspirin pill

Salicylate is produced in the bark of the white willow tree, *Salix alba*

John Vane discovered the mechanism of action of aspirin

Figure 15.42 Acetylsalicylate is the chemical name of Bayer's anti-inflammatory drug named aspirin. Salicylate is produced at high levels in the bark of the white willow tree and is the active ingredient in herbal teas used to treat fever and pain. Biochemist John Vane discovered that the anti-inflammatory action of aspirin is due to its inhibition of prostaglandin synthesis by the COX-2 enzyme. BAYER ASPIRIN PILL: CLYNT GARNHAM MEDICAL/ALAMY STOCK PHOTO; WHITE WILLOW TREE: PREMIUM STOCK PHOTOGRAPHY GMBH/ALAMY; J. VANE: WELLCOME LIBRARY, LONDON.

glucocorticoids. The first NSAID to be characterized was salicylate, a compound present in the bark of the white willow tree (*Salix alba*) that reduces inflammation and fever by inhibiting prostaglandin H_2 synthesis. Salicylate was the active ingredient in tea prepared by Hippocrates in ancient Greece using willow tree leaves, which he described as a potent herbal medicine for treating joint pain and fever. The German drug company Bayer isolated and commercialized a modified salicylate compound they called **aspirin** (acetylsalicylate), which could be taken orally as a pill to reduce inflammation (**Figure 15.42**). It wasn't until 1971—when British biochemist John Vane demonstrated that salicylate's anti-inflammatory action was in fact due to its inhibitory effect on COX-2—that a molecular explanation was presented for the potency of Hippocrates' tea. This key discovery was recognized by the 1982 Nobel Prize in Physiology or Medicine, which Vane shared with Sune Bergstrom and Bengt Samuelsson who together discovered and characterized prostaglandins and related biologically active substances.

The human COX-1 and COX-2 enzymes are biochemically related proteins in terms of overall three-dimensional structure and their similar catalytic activities. However, the physiologic functions of these two isozymes are quite distinct. For example, COX-1 is constitutively expressed in most tissues and is involved in producing prostaglandins that stimulate mucin secretion and protect the lining of the stomach from low pH. In contrast, COX-2 expression is specifically induced by inflammatory signals and is responsible for producing prostaglandins that cause the swelling, pain, and fever associated with inflammation (**Figure 15.43**). Aspirin inactivates both COX-1 and COX-2 enzymes by acetylating a serine residue in the enzyme active sites. Therefore, although aspirin offers relief from inflammatory responses mediated by COX-2, which are the cause of pain and fever, it also blocks COX-1 protection of the stomach lining, giving rise to bleeding ulcers in some individuals. The same undesirable gastrointestinal side effect is also true of the NSAIDs ibuprofen (Advil), flurbiprofen (Ansaid), and naproxen (Aleve), which function as competitive inhibitors of COX-1 and COX-2.

Once it was realized that COX-2 was the relevant biological target of NSAIDs and that COX-1 inhibition contributed to side effects associated with stomach bleeding, pharmaceutical companies sought to develop inhibitors selective for COX-2.

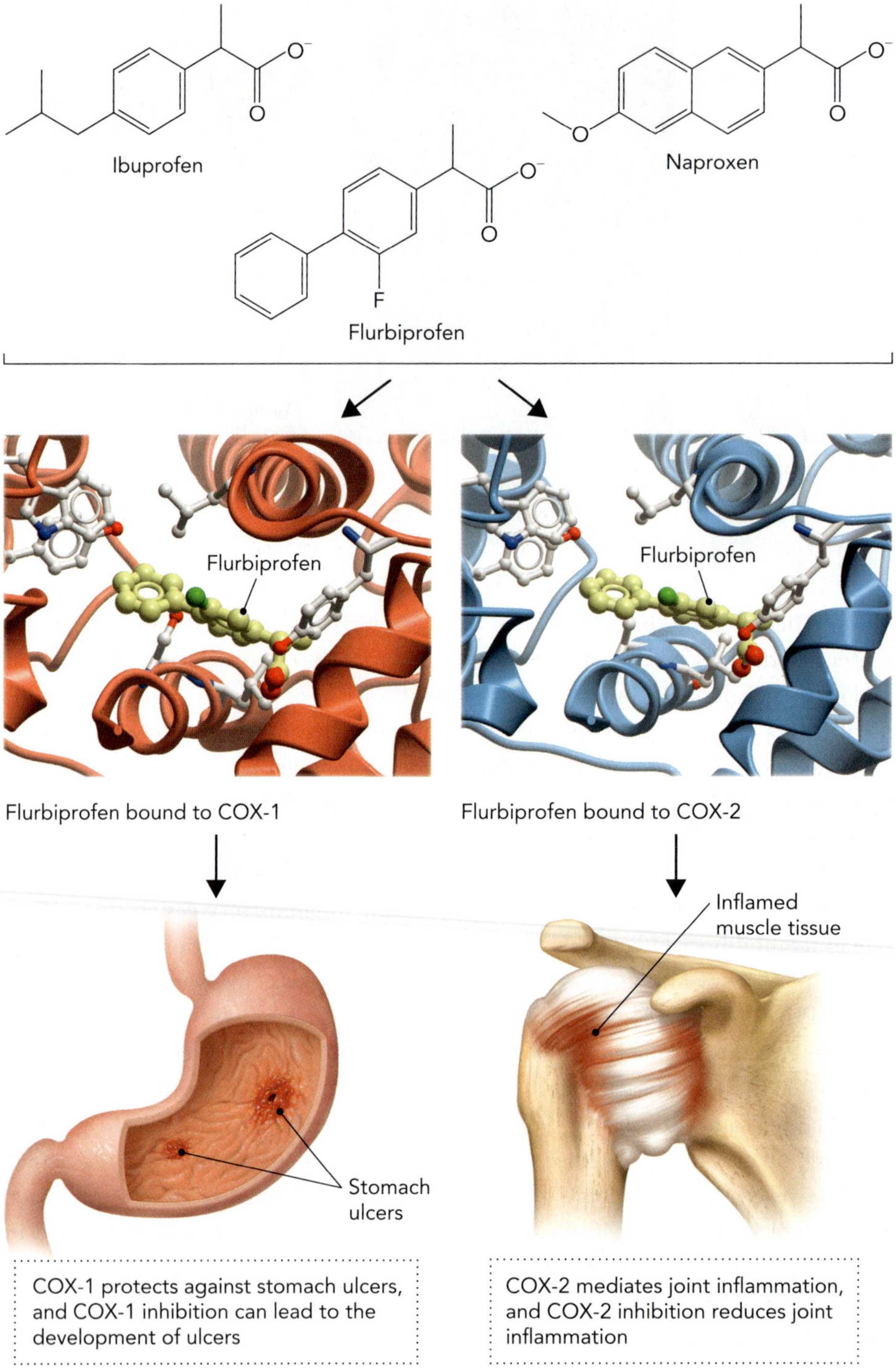

Figure 15.43 Ibuprofen, flurbiprofen, and naproxen are pharmaceutical NSAIDs that inhibit both COX-1 and COX-2 enzymes. Although COX-2 inhibition by these NSAIDs relieves symptoms associated with inflammation and fever, COX-1 inhibition can also cause stomach bleeding, which is an undesirable side effect. The highlighted amino acid side chains in the enzyme active site of COX-1 are L352, Y355, Y385, W387, S530, and L531, all of which are conserved in COX-2. BASED ON PDB FILES 3N8Z (MOUSE COX-1) AND 3PGH (MOUSE COX-2).

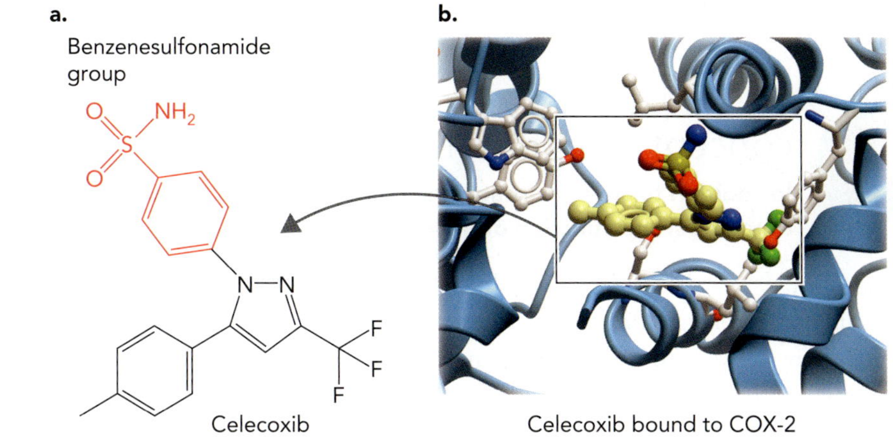

Figure 15.44 The selective COX-2 inhibitor celecoxib binds tightly to the COX-2 enzyme active site but with low affinity to catalytically active COX-1. **a.** The chemical structure of celecoxib, with the bulky benzenesulfonamide side group highlighted. **b.** The molecular structure of mouse COX-2 with celecoxib bound to the enzyme active site. The highlighted amino acid side chains in the enzyme active site are the same as those in Figure 15.43. BASED ON PDB FILE 3LN1.

The first two COX-2 inhibitors to reach the market were celecoxib (Celebrex) and rofecoxib (Vioxx), both of which exploit the difference in the size and geometry of the COX-1 and COX-2 active sites to target COX-2 selectively. As shown in **Figure 15.44**, celecoxib has a large benzenesulfonamide group that can fit in the COX-2 enzyme active site but not in the smaller COX-1 enzyme active site. Indeed, by comparing the molecular structure of the nonselective COX-1/COX-2 inhibitor flurbiprofen bound to the active site of mouse COX-2 (see Figure 15.43) with that of the selective COX-2 inhibitor celecoxib bound to the same mouse COX-2 enzyme (see Figure 15.44), it is clear that the bulky benzenesulfonamide side group of celecoxib completely fills the binding pocket. **Figure 15.45** illustrates how the differences in overall size of nonselective COX-1/COX-2 inhibitors and selective COX-2 inhibitors accounts for the biological activities of these two anti-inflammatory drugs. The nonselective COX inhibitors are small enough to bind to the active sites of both COX-1 and COX-2, whereas the larger, selective COX-2 inhibitors only bind with high affinity to the COX-2 enzyme.

Although selective COX-2 inhibitors do not have the side effects associated with COX-1 inhibition, long-term use can lead to an increased risk of cardiovascular disease. This clinical observation led to the removal of rofecoxib from the market and to the addition of warnings to celecoxib packaging. The explanation for this cardiovascular side effect appears to be low-level inhibition of prostacyclin synthase, which leads to lower levels of prostacyclin, an eicosanoid hormone that controls clot formation and platelet aggregation. Therefore, by partially inhibiting the activity of prostacyclin synthase, a serious side effect of some COX-2 inhibitors is increased platelet aggregation

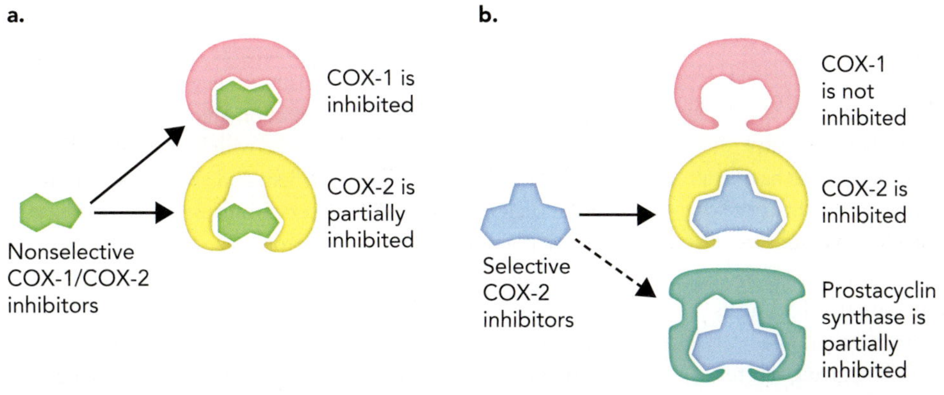

Figure 15.45 Cyclooxygenase inhibitors may have differential activity toward COX-1 and COX-2. **a.** Nonselective COX-1/COX-2 inhibitors such as aspirin and ibuprofen are small molecules that can bind to the active site of both COX-1 and COX-2. **b.** Selective COX-2 inhibitors such as celecoxib and rofecoxib are larger molecules that bind with high affinity to COX-2 but not COX-1. The cardiovascular problems associated with COX-2 inhibitors are due to their partial inhibition of prostacyclin synthase activity and reduced levels of prostacyclin.

and the formation of life-threatening blood clots. More selective COX-2 inhibitors are now under development to eliminate this inhibitory effect on prostacyclin synthase.

concept integration 15.4

In addition to energy storage and membrane structure, what is the third important function of biological lipids?

Biological lipids also function in signaling. Lipids are metabolic precursors to two important classes of signaling molecules: steroid hormones and eicosanoids. Steroid hormones are derived from cholesterol and function as ligands for nuclear receptor proteins, which function as transcription factors and regulate gene expression. In addition to the classic steroid hormones, which include progesterone, estrogen, androgen, cortisol (a glucocorticoid), and aldosterone (a mineralocorticoid), vitamin D is also a cholesterol-derived signaling molecule that binds to and activates a nuclear receptor protein. Eicosanoids, such as prostaglandins, prostacyclins, thromboxanes, and leukotrienes, are short-lived signaling molecules derived from arachidonate that bind to G protein–coupled receptors and stimulate an inflammatory response. Nonsteroidal anti-inflammatory drugs are inhibitors of the COX-1 and COX-2 enzymes, which convert arachidonate to the major prostaglandin, prostaglandin H_2.

chapter summary

15.1 Many Lipids Are Made from Fatty Acids

- Lipids serve three important roles in biology: (1) fatty acids and triacylglycerols serve as energy storage molecules; (2) glycerophospholipids and sphingolipids are major components of hydrophobic cell membranes; and (3) steroids and eicosanoids function in the endocrine system and activate receptor-mediated signaling pathways.
- The most abundant fatty acids in nature are unbranched hydrocarbon chains that contain either reduced methylene groups (CH_2), called saturated fatty acids, or oxidized C═C bonds, called unsaturated fatty acids. Fatty acids with multiple C═C bonds are called polyunsaturated fatty acids.
- The common names of fatty acids indicate how or where the fatty acid was discovered. For example, palmitate was first isolated from palm oil and linoleate was isolated from flax seed. Humans require three essential fatty acids in their diet: the polyunsaturated fatty acids linoleate, α-linolenate, and arachidonate.
- The number and configuration of C═C bonds in unsaturated fatty acids affect the melting points of fatty acid mixtures, with long-chain saturated fatty acids having a higher melting point than long-chain unsaturated fatty acids because of differences in intermolecular interactions.
- The melting points of lipid mixtures can be increased by converting unsaturated fatty acids into saturated fatty acids using a commercial process called hydrogenation. The hydrogenation process is not 100% efficient, giving rise to both fully saturated fatty acids and fatty acids containing C═C bonds with the trans configuration. Consumption of large amounts of trans fats is associated with increased risk of cardiovascular disease.
- Carbons in fatty acids are numbered from the carboxylic acid end, with the carboxyl carbon being C-1. Any C═C bonds present are denoted as superscript numerals associated with the symbol "Δ." For example, palmitate is a saturated C_{16} fatty acid (16:0), whereas α-linolenate is an unsaturated C_{18} fatty acid with three cis C═C bonds at C-9, C-12, and C-15, written as *cis* 18:2($\Delta^{9,12,15}$).
- Fatty acids can also be numbered from the methyl carbon (ω-carbon). Polyunsaturated fatty acids with the most distal C═C bond from the carboxyl group positioned three carbons away from the ω-carbon are called ω-3 fatty acids, such as eicosapentaenoic acid (EPA) and docosahexaenoic acid (DHA). The same nomenclature can be used to describe the ω-6 fatty acids linoleate and arachidonate.
- Methods used to isolate and characterize lipids depend on techniques in lipid biochemistry, including those of lipidomics, in which a collection of organic chemistry protocols has been modified for use with biological samples. Fractionated cellular lipids are identified and characterized using such methods as thin layer chromatography, gas phase chromatography, and mass spectrometry.
- Waxes are an abundant lipid found in nature, consisting of long-chain fatty alcohols linked to long-chain fatty acids to generate wax esters. The melting points of most waxes, such as beeswax, are higher than ambient temperature, making them solids in their biological context.

- Some wax esters, like those found in jojoba beans and sperm whale oil, are liquids under physiologic conditions, owing to the incorporation of unsaturated fatty acids.
- Triacylglycerols consist of three fatty acids esterified to glycerol, which makes them neutral lipids because the polar carboxyl groups are neutralized. In a Fischer projection, the middle carbon is designated the *sn*-2 position, using the stereospecific numbering system, whereas the top and bottom carbons are designated the *sn*-1 and *sn*-3 positions, respectively.

15.2 Triacylglycerols Are Energy Storage Lipids

- The primary function of triacylglycerols is energy storage. In animals, triacylglycerols are stored in fat cells called adipocytes; in plants, triacylglycerols are stored in the seeds and provide an oxidizable energy source for the developing embryo after germination.
- The two reasons why triacylglycerols are the primary form of energy storage instead of glucose-derived polymers are (1) fatty acids are at a higher reduction state than glucose and therefore yield more energy (electrons for redox reactions) for the same number of carbons upon oxidation; and (2) triacylglycerols are not solvated by water and therefore have less mass for the same amount of stored energy.
- Triacylglycerols are either obtained from the diet—primarily from digesting animal fat and nuts in the small intestine—or synthesized in the liver, using glucose and amino acids as a source of acetyl-CoA.
- Triacylglycerols are transported through the blood as components of lipoprotein complexes, which can be very large chylomicrons produced in intestinal epithelial cells or small very-low-density lipoprotein (VLDL) particles synthesized and exported by liver cells.
- Adipocytes hydrolyze stored triacylglycerols in response to hormone signaling (glucagon and epinephrine) and release free fatty acids and glycerol into the circulatory system, where the fatty acids are transported throughout the body by a carrier protein called albumin.
- Dietary triacylglycerols are hydrolyzed in the small intestine by lipase enzymes that release the three fatty acids and glycerol. Free fatty acids and glycerol enter the intestinal epithelial cells, where they are resynthesized into triacylglycerols and then exported to the lymphatic system as components of chylomicron lipoproteins.
- The biosynthesis of triacylglycerols in animals uses acetyl-CoA produced by the degradation of carbohydrates and proteins to generate palmitic acid in the cytosol, which is then converted to triacylglycerols and exported as VLDL particles.
- Lipoprotein particles contain proteins on the surface that facilitate fatty acid delivery to peripheral tissues through binding and activation of lipoprotein lipase on the surface of endothelial cells. The free fatty acids diffuse into nearby adipose and muscle cells, whereas the glycerol travels through the blood to the liver.

15.3 Cell Membranes Contain Three Major Types of Lipids

- The outer monolayer of the plasma membrane contains mostly phosphatidylcholine and sphingolipids, whereas the inner monolayer contains mostly glycerophospholipids. This difference is thought to reflect the distinct functional properties of the outer and inner monolayers of the plasma membrane.
- The total cholesterol content of the outer and inner monolayers is similar. High concentrations of cholesterol in discrete regions of the plasma membrane, such as in lipid rafts, alters the physical properties of the membrane by interfering with the ability of phospholipids to form highly ordered crystalline structures.
- The most abundant lipids in membranes are phospholipids—including both glycerophospholipids and sphingomyelin—followed by cholesterol and sphingoglycolipids. Glycerophospholipids represent about half of all membrane lipids in eukaryotic cell membranes, with sphingolipids and cholesterol making up the other half.
- Glycerophospholipids are derived from phosphatidate and are represented by the membrane phospholipids phosphatidylcholine, phosphatidylserine, phosphatidylethanolamine, and phosphatidylinositol.
- Many types of snake venom contain phospholipase enzymes that hydrolyze glycerophospholipids to release free fatty acids and destroy plasma cell membranes. If a snakebite is left untreated, death may occur from massive internal bleeding.
- Sphingolipids are derived from the biomolecule sphingosine (which is synthesized by linking serine to the carboxyl group of palmitate) and one fatty acid. There are two types of sphingolipids: the sphingophospholipids, represented by sphingomyelin, and the sphingoglycolipids, which are called cerebrosides and gangliosides.
- Defects in sphingolipid metabolism lead to three related hereditary disorders: Tay–Sachs disease, Fabry disease, and Niemann–Pick disease. They are all caused by the buildup of metabolic precursors, which leads to severe neuronal defects and death.
- Cholesterol contains a rigid, four-ring structure that disrupts the packing of glycerophospholipids and sphingolipids when inserted into membranes. Cholesterol also contributes to the formation of lipid rafts in the plasma membrane of eukaryotic cells.

15.4 Lipids Function in Cell Signaling

- Cholesterol is the precursor to steroid hormones (mineralocorticoids, glucocorticoids, progesterones, androgens, and estrogens) and vitamin D. Eicosanoids are derived from the ω-6 fatty acid arachidonate, a polyunsaturated fatty acid that is converted into prostaglandins, prostacyclins, thromboxanes, and leukotrienes.

- Steroid hormones are potent signaling molecules with critical roles in cell development, reproductive biology, and organismal physiology. Steroid signaling is mediated by steroid binding to nuclear receptor proteins that function as homodimeric transcription factors.
- Many of the enzymes involved in steroid biosynthesis (steroidogenesis) are hydroxylases, of which one type is the cytochrome P450 monooxygenases. Steroidogenesis begins with removal of the cholesterol side chain attached to the D ring to generate pregnenolone, the biosynthetic precursor to all animal steroids.
- Mineralocorticoids and glucocorticoids are synthesized in the adrenal glands, whereas estrogen is synthesized in the female ovaries, testosterone in the male testes, and progesterone in the corpus luteum in pregnant females. The adrenal glands also synthesize androgens, which is how females acquire testosterone for estradiol production.
- Vitamin D is another cholesterol-derived lipid signaling molecule, which was discovered as a nutritional supplement that could be added to the diet to treat the debilitating bone disease rickets. Only ~10% of vitamin D in the human body is acquired from the diet; the rest comes from sunlight conversion of 7-dehydrocholesterol in skin cells.
- Eicosanoids are a group of signaling molecules derived from C_{20} polyunsaturated fatty acids such as arachidonate, which are released from the membrane by phospholipases and modified by mitochondrial enzymes.
- Eicosanoids are produced by cells at their sites of action (paracrine) and have half-lives of only a few minutes. The four major classes of arachidonate-derived eicosanoids are prostaglandins, prostacyclins, thromboxanes, and leukotrienes, which mediate cell signaling by activating G protein–coupled receptors on target cells.
- The synthesis of prostaglandins, prostacyclins, and thromboxanes requires cyclooxygenation of arachidonate by either cyclooxygenase-1 (COX-1) or cyclooxygenase-2 (COX-2) to generate the precursor prostaglandin H_2. The enzyme lipoxygenase generates leukotrienes directly from arachidonate.
- Many of the eicosanoid signaling molecules derived from prostaglandin H_2 exacerbate the inflammatory response, and therefore drugs that inhibit prostaglandin H_2 synthesis function as anti-inflammatory agents. The most commonly used inhibitors of prostaglandin H_2 synthesis are called nonsteroidal anti-inflammatory drugs (NSAIDs) to distinguish them from glucocorticoids. These include aspirin (acetylsalicylate), ibuprofen, and naproxen.
- Human COX-1 and COX-2 enzymes have similar catalytic activities; however, they have distinct physiologic functions. For example, COX-1 produces prostaglandins that stimulate mucin secretion and protect the lining of the stomach from low pH, whereas COX-2 produces prostaglandins that cause the swelling, pain, and fever associated with inflammation.
- Aspirin, ibuprofen, and naproxen are considered nonselective COX-1/COX-2 inhibitors because they bind to both COX-1 and COX-2 enzymes, which leads to reduced inflammation through inhibition of COX-2 but increased risk of stomach bleeding through inhibition of COX-1.
- Selective COX-2 inhibitors such as celecoxib (Celebrex) and rofecoxib (Vioxx) have improved anti-inflammatory properties with fewer side effects due to stomach bleeding because they bind to and inhibit COX-2 without affecting COX-1. However, it was discovered that some selective COX-2 inhibitors also partially inhibit the activity of prostacyclin synthase, leading to increased blood clotting.

biochemical terms

(in order of appearance in text)

saturated fatty acid (p. 731)
unsaturated fatty acid (p. 731)
monounsaturated fatty acid (p. 731)
polyunsaturated fatty acid (p. 731)
essential fatty acid (p. 732)
melting point (MP) (p. 732)
hydrogenation (p. 733)
trans fat (p. 734)
ω-6 fatty acid (p. 734)
ω-3 fatty acid (p. 734)
eicosapentaenoic acid (EPA) (p. 735)
docosahexaenoic acid (DHA) (p. 735)
lipidomics (p. 735)
waxes (p. 737)
saponification (p. 740)
lye (p. 740)
adipocyte (p. 742)
lipoprotein (p. 743)
albumin (p. 743)
lipase (p. 743)
colipase (p. 744)
chylomicron (p. 744)
apolipoprotein (p. 744)
lipid droplet (p. 746)
perilipin (p. 747)
glycerophospholipid (p. 750)
sphingolipid (p. 750)
cholesterol (p. 750)
phosphatidylcholine (p. 750)
phosphatidylserine (p. 750)
phosphatidylethanolamine (p. 750)
phosphatidylinositol (p. 750)
sphingophospholipid (p. 750)
cerebroside (p. 750)
ganglioside (p. 750)
fluid mosaic model (p. 752)
lipid raft (p. 752)
phosphatidate (p. 753)
sphingosine (p. 754)
ceramide (p. 754)
Tay–Sachs disease (p. 755)
Fabry disease (p. 755)
Niemann–Pick disease (p. 755)
stigmasterol (p. 756)
eicosanoid (p. 758)
bile acid (p. 758)
glycocholate (p. 759)
steroidogenesis (p. 759)
P450 monooxygenases (p. 759)
rickets (p. 762)
vitamin D_3 (p. 762)
prostaglandins (p. 763)
prostacyclins (p. 763)
nonsteroidal anti-inflammatory drug (NSAID) (p. 763)
aspirin (p. 765)

review questions

1. What are the three major roles of lipids in biology? Provide examples of the lipids in these roles.
2. What are the molecular formulas for the following saturated and unsaturated fatty acids: *cis* 18:1(Δ^9), 18:0, *cis* 16:1(Δ^9), 16:0, *cis* 18:2($\Delta^{9,12}$), and *trans* 18:1(Δ^9)? What are the common names of these fatty acids at pH 3.0 and at physiologic pH 7.2?
3. Which has a higher stored energy potential per gram, glycogen or triglycerides? Explain.
4. What are the three major lipid components of beeswax, and how do their structures contribute to the high melting point of beeswax (63 °C) compared to that of jojoba oil (7 °C)?
5. Name two glycerophospholipids found in eukaryotic cell membranes and two sphingolipids. What fraction of the lipid in most plasma membranes is cholesterol, and how does cholesterol content affect membrane fluidity?
6. In addition to serving as a hydrophobic barrier in cell membranes, what is another important biological function of glycerophospholipids? Provide a specific example.
7. What is the precursor biomolecule of all sphingolipids, and what polar head group is attached to this precursor to form the abundant membrane lipid sphingomyelin?
8. What accounts for the fact that glycan structures of blood group ABO gangliosides are identical to the blood group ABO glycoproteins described in Chapter 13?
9. Name the five steroid hormones derived from cholesterol and briefly describe their metabolic functions in humans.
10. What is the physiologic difference between a drug that functions as a steroid agonist and one that functions as a steroid antagonist? Give a specific example of each type of drug.
11. What is the active form of vitamin D, and how is it synthesized in humans?
12. What four major classes of human eicosanoids are derived from arachidonate, and what are representative biological functions of each?

challenge problems

1. An average middle-aged man weighing 90 kg (200 lb) contains 15% body fat stored in adipose tissue.
   a. Calculate the amount of energy stored as fat in this man in kilojoules, assuming that the energy yield from fat is 37 kJ/g.
   b. If this same amount of energy were stored in hydrated glycogen rather than fat (hydrated glycogen has an energy yield of 6 kJ/g), how much would this man weigh in kilograms and in pounds?
2. Saponification releases free fatty acids from triacylglycerols by heating animal fat in an alkaline solution such as KOH.
   a. Calculate the average molecular mass of triacylglycerols in 1 kg of pure beef fat (tallow) if it requires 193.2 g of KOH for complete saponification.
   b. Calculate the average molecular mass of a triacylglycerol in this beef fat sample if the amount of oleate were twice the amount of palmitate and the amount of palmitate were equal to the amount of stearate.
   c. Are these calculations consistent with the measured fatty acid composition of beef tallow? Explain.
3. Describe how the export of lipids from liver cells (hepatocytes) and fat cells (adipocytes) differs with regard to fatty acids and triacylglycerols.
4. You have isolated two new species of bacteria. One of them grows best at 20 °C, and the other one grows best at 40 °C. Which bacterium would you expect to have a higher ratio of saturated to unsaturated fatty acids in its membrane phospholipids, and why?
5. The plasma membrane of a unicellular eukaryotic organism was found to have the same relative distribution of membrane lipids as the plasma membrane of rat liver cells. Calculate the number of moles of each of the four major types of membrane lipids in a 650-mg dried sample prepared by lipid extraction of the membrane fraction from this organism. Use the table that follows first to calculate the average molecular masses of glycerophospholipids, sphingoglycolipids, sphingomyelin, and cholesterol, and then use these values to calculate the number of moles of each in the 650-mg sample on the basis of the relative distributions in rat liver plasma membranes (see Figure 15.25).

Membrane lipid	Molecular formula of representative lipid	Molecular mass of representative lipid (g/mol)
Cerebroside	$C_{48}H_{93}NO_8$	782
Cholesterol	$C_{27}H_{46}O$	387
Ganglioside (GM1)	$C_{77}H_{139}N_3O_{31}$	1,603
Phosphatidylcholine	$C_{42}H_{82}NO_8P$	760
Phosphatidylethanolamine	$C_{41}H_{78}NO_8P$	744
Phosphatidylinositol	$C_{47}H_{83}O_{13}P$	887
Phosphatidylserine	$C_{13}H_{24}NO_{10}P$	812
Sphingomyelin	$C_{41}H_{83}N_2O_6P$	731

6. Some microorganisms in the Archaea evolutionary branch thrive in extreme environments of high temperature and low pH. Their membrane lipids consist of the tetraether compound shown here.

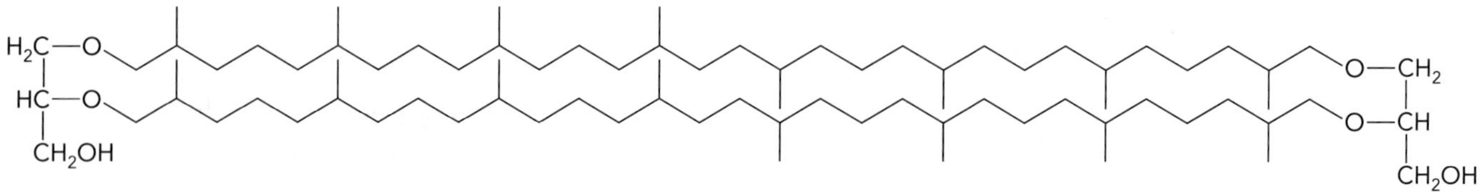

   a. How could a membrane lipid bilayer be constructed from this tetraether?
   b. Give two reasons why this archaeal compound provides thermal and chemical stability to the membrane. Briefly justify your answer.

7. Phospholipases hydrolyze membrane glycerophospholipids and release a variety of fatty acids and other signaling molecules, depending on the specific membrane lipid and the cleavage site. Use the structure of phosphatidylinositol-4,5-bisphosphate (PIP_2) shown in Figure 15.30 to answer the following questions.
   a. Draw the chemical structures of the two cleavage products derived from phospholipase C cleavage, and describe the downstream effect of these products on glycogen metabolism in liver cells.
   b. The bite of a saw-scaled viper injects soluble phospholipase A_2 into the wound and causes acute massive inflammation. What is the signaling molecule released by phospholipase A_2 cleavage, and why does this lead to inflammation?

8. Tay–Sachs disease is a fatal neurologic disease caused by a recessive genetic mutation on chromosome 15. There is no medical treatment for the disease, and children diagnosed with Tay–Sachs disease rarely live beyond 5 years of age.
   a. What is the biochemical basis for Tay–Sachs disease?
   b. What is the probability of a child having Tay–Sachs disease after inheriting two copies of the Tay–Sachs mutation (homozygous) if both parents are heterozygous carriers of the mutation? What is the probability of passing on the Tay–Sachs mutation to children of parents in which one is a heterozygous carrier and the other is homozygous normal?
   c. Why do you think it is rare to have children in developed countries diagnosed with Tay–Sachs disease today compared to 50 years ago?

9. Taurocholate, like glycocholate, is a bile acid synthesized in the liver and secreted into the intestinal lumen from the gall bladder in response to feeding.
   a. What is the function of taurocholate in digestion?
   b. Compare the chemical structures of taurocholate and glycocholate, and describe how the structures of these two bile acids determine their functions. Identify the polar and nonpolar regions of each bile acid.
   c. Bile soap is a natural product made from bovine gall bladder secretions mixed with soap. Would bile soap be more effective at removing red wine stains from clothing or removing stains caused by Italian salad dressing? Explain.

10. Defects in steroidogenic enzymes can lead to prenatal abnormalities in the development of reproductive organs, a pediatric condition known as ambiguous gender. For example, the genetic disease congenital adrenal hyperplasia (CAH) is caused by a deficiency in the enzyme 21-hydroxylase, which leads to masculinization of females. Similarly, defects in the enzyme 5α-reductase results in the genetic disorder 5α-reductase deficiency, which can result in feminization of males.
   a. What explains the masculinization of females with CAH, whereas males with CAH have normal reproductive development? Why will both females and males with CAH develop kidney disorders if not treated with hormonal therapy?
   b. Why are males born with a deficiency in 5α-reductase feminized at birth, but can become masculinized during the course of puberty? Why are females born with a deficiency in 5α-reductase asymptomatic?

11. Finasteride is an inhibitor of the enzyme 5α-reductase. It is used to treat enlargement of the prostate in men, a condition called benign prostatic hypertrophy, and is marketed for this use as the pharmaceutical drug Proscar. Moreover, finasteride is also marketed as the pharmaceutical drug Propecia for use to treat male pattern baldness. Explain why a balding man with benign prostatic hypertrophy may be able to take oral finasteride and treat both conditions at the same time; that is, why does inhibiting the activity of 5α-reductase block prostate cancer cell growth and induce hair growth?

12. The COX-2 inhibitors celecoxib (Celebrex) and rofecoxib (Vioxx) are anti-inflammatory drugs that have a reduced risk of stomach bleeding compared to that associated with aspirin and other nonsteroidal anti-inflammatory drugs (NSAIDs). Rofecoxib, and to a lesser degree celecoxib, also partially inhibit the activity of prostacyclin synthase.

a. What is the biochemical basis for celecoxib's reduced risk of stomach bleeding and its selective inhibition of COX-2 compared to NSAIDs?

b. Why would increased risk of blood clots be associated with rofecoxib treatment?

smartwork5

If your instructor assigns homework with Smartwork5, access it here: digital.wwnorton.com/biochem.

suggested reading

Books and Reviews

Ascherio, A., Katan, M. B., Zock, P. L., Stampfer, M. J., and Willett, W. C. (1999). Trans fatty acids and coronary heart disease. *New England Journal of Medicine*, *340*, 1994–1998.

Barros, R. P., and Gustafsson, J. A. (2011). Estrogen receptors and the metabolic network. *Cell Metabolism*, *14*, 289–299.

Fahy, E., Subramaniam, S., Brown, H. A., Glass, C. K., Merrill, A. H., Jr., Murphy, R. C., Raetz, C. R., Russell, D. W., Seyama, Y., Shaw, W., et al. (2005). A comprehensive classification system for lipids. *Journal of Lipid Research*, *46*, 839–861.

Funk, C. D. (2001). Prostaglandins and leukotrienes: advances in eicosanoid biology. *Science*, *294*, 1871–1875.

Greene, E. R., Huang, S., Serhan, C. N., and Panigrahy, D. (2011). Regulation of inflammation in cancer by eicosanoids. *Prostaglandins & Other Lipid Mediators*, *96*, 27–36.

Henry, H. L. (2011). Regulation of vitamin D metabolism. *Best Practice & Research. Clinical Endocrinology & Metabolism*, *25*, 531–541.

Imperato-McGinley, J., and Zhu, Y. S. (2002). Androgens and male physiology: the syndrome of 5α-reductase-2 deficiency. *Molecular and Cellular Endocrinology*, *198*, 51–59.

New, M. I. (1998). Inborn errors of steroidogenesis. *Steroids*, *63*, 238–242.

Remig, V., Franklin, B., Margolis, S., Kostas, G., Nece, T., and Street, J. C. (2010). Trans fats in America: a review of their use, consumption, health implications, and regulation. *Journal of the American Dietetic Association*, *110*, 585–592.

Seo, T., Blaner, W. S., and Deckelbaum, R. J. (2005). Omega-3 fatty acids: molecular approaches to optimal biological outcomes. *Current Opinion in Lipidology*, *16*, 11–18.

Stratakis, C. A., and Bossis, I. (2004). Genetics of the adrenal gland. *Reviews in Endocrine & Metabolic Disorders*, *5*, 53–68.

Svoboda, J. A., Kaplanis, J. N., Robbins, W. E., and Thompson, M. J. (1975). Recent developments in insect steroid metabolism. *Annual Review of Entomology*, *20*, 205–220.

Vance, D. E., and Vance, J. E. (Eds.). (2008). *Biochemistry of lipids, lipoproteins and membranes* (5th ed.). Amsterdam: Elsevier Publishing.

Wang, X., Baek, S. J., and Eling, T. (2011). COX inhibitors directly alter gene expression: role in cancer prevention? *Cancer Metastasis Reviews*, *30*, 641–657.

Primary Literature

Bohl, C. E., Gao, W., Miller, D. D., and Dalton, J. T. (2005). Structural basis for antagonism and resistance of bicalutamide in prostate cancer. *Proceedings of the National Academy of Sciences USA*, *102*, 6201–6206.

Cao, R., He, Q., Zhou, J., Liu, Z., Wang, X., Chen, P., Xie, J., and Liang, S. (2008). High-throughput analysis of rat liver plasma membrane proteome by a nonelectrophoretic in-gel tryptic digestion coupled with mass spectrometry identification. *Journal of Proteome Research*, *7*, 535–545.

Loll, P. J., Picot, D., and Garavito, R. M. (1995). The structural basis of aspirin activity inferred from the crystal structure of inactivated prostaglandin H_2 synthase. *Nature Structural Biology*, *2*, 637–643.

Marr, E., Tardie, M., Carty, M., Brown Phillips, T., Wang, I. K., Soeller, W., Qiu, X., and Karam, G. (2006). Expression, purification, crystallization and structure of human adipocyte lipid-binding protein (aP2). *Acta Crystallographica. Section F, Structural Biology and Crystallization Communications*, *62*, 1058–1060.

Miwa, T. K. (1971). Jojoba oil wax esters and derived fatty acids and alcohols: gas chromatographic analysis. *Journal of the American Oil Chemists' Society*, *48*, 259–264.

Pfleger, R. C., Anderson, N. G., and Snyder, F. (1968). Lipid class and fatty acid composition of rat liver plasma membranes isolated by zonal centrifugation. *Biochemistry*, *7*, 2826–2833.

Pond, D. W., and Tarling, G. A. (2011). Phase transitions of wax esters adjust buoyancy in diapausing *Calanoides acutus*. *Limnology and Oceanography*, *56*, 1310–1318.

Wang, J. L., Limburg, D., Graneto, M. J., Springer, J., Hamper, J. R., Liao, S., Pawlitz, J. L., Kurumbail, R. G., Maziasz, T., Talley, J. J., et al. (2010). The novel benzopyran class of selective cyclooxygenase-2 inhibitors. Part 2: the second clinical candidate having a shorter and favorable human half-life. *Bioorganic & Medicinal Chemistry Letters*, *20*, 7159–7163.

High-carbohydrate foods

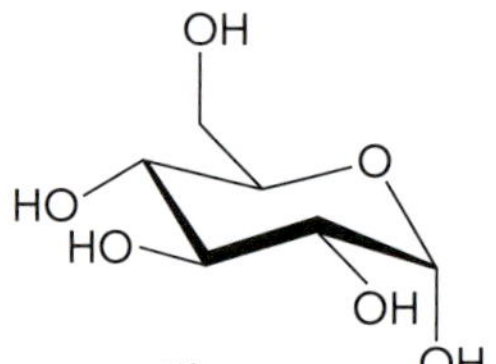

Glucose

Expanding waistlines are due in part to consumption of high-calorie, carbohydrate-rich foods, which can become a lifelong eating habit. Carbohydrate excess can lead to stimulation of the fatty acid synthesis pathway, which produces fatty acids. The molecular structure of a dimer of the multifunctional enzyme fatty acid synthase is shown here.

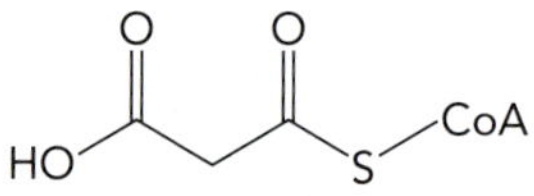

Malonyl-CoA

Substrate for fatty acid synthase

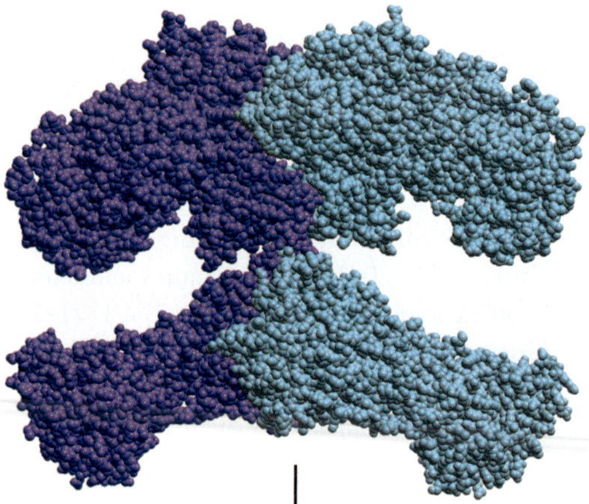

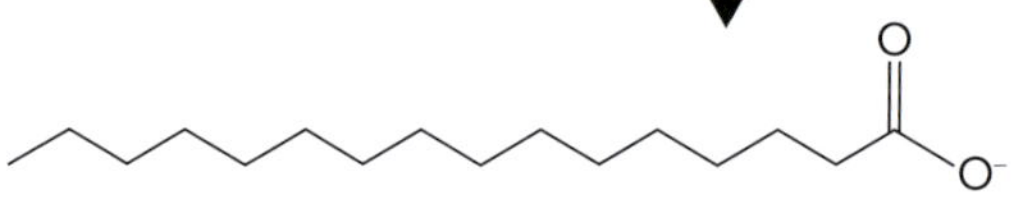

Palmitate

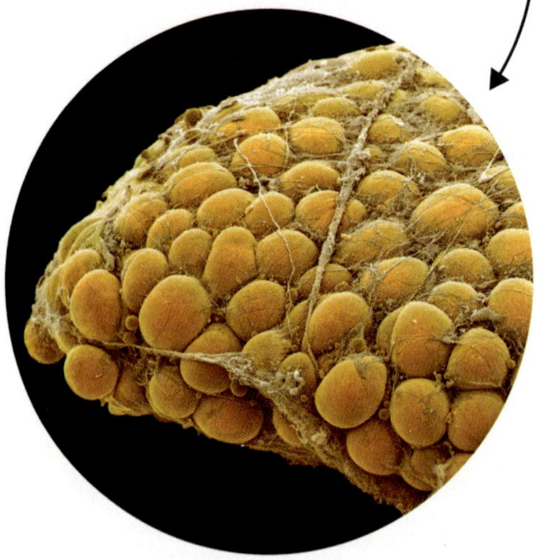

Adipose tissue

Weight gain with risk of diabetes and cardiovascular disease

16

Lipid Metabolism

◀ High-carbohydrate snacks such as cookies, soda, and potato chips are often consumed in large quantities, leading to excess levels of malonyl-CoA, the substrate for the enzyme fatty acid synthase. Synthesis of triacylglycerols from palmitate, a product of the fatty acid synthase reaction, results in increased stored fat in adipose tissue. Consumption of large amounts of carbohydrate-rich foods and beverages, even if they are low in fat, can lead to an increased risk of obesity and diabetes. The metabolic pathways converting nonfat carbohydrates to fats are described in this chapter.

CHAPTER OUTLINE

CREDITS: RED CUP: EVGENY KARANDAEV/SHUTTERSTOCK; POTATO CHIPS: TIGER IMAGES/SHUTTERSTOCK; FATTY ACID SYNTHASE: BASED ON PDB FILE 2UVB; ADIPOSE TISSUE: STEVE GSCHMEISSNER/SCIENCE SOURCE; CARDBOARD TAGS: ANDREY_KUZMIN/SHUTTERSTOCK.

In Chapter 15, we described the structure and function of fatty acids, triacylglycerols, membrane lipids, steroid hormones, and eicosanoids. Here in Chapter 16, we focus on the degradation and synthesis pathways of fatty acids, triacylglycerols, and cholesterol. In addition to describing the enzymes catalyzing these metabolic reactions, we also examine the regulation of lipid metabolism.

Figure 16.1 illustrates where fatty acid degradation and synthesis fits into our metabolic map. Observe that acetyl-CoA serves as the primary link between fatty acid metabolism and the energy conversion pathways. Acetyl-CoA also serves as the precursor to cholesterol, which is a component of biological membranes and the substrate for bile acid and steroid hormone synthesis. Understanding the biochemical mechanisms that control human lipid metabolism has been critical to the identification of genetic and environmental factors that contribute to the growing problem of obesity in industrialized countries.

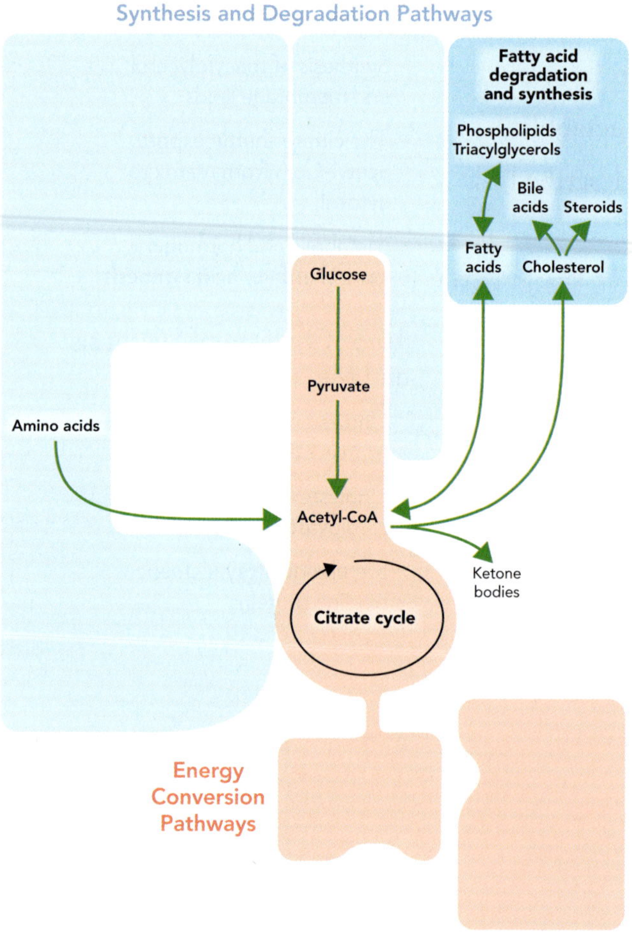

Figure 16.1 Acetyl-CoA is a key metabolite in cells, linking energy conversion pathways, amino acid metabolism, and lipid metabolism. Fatty acids and cholesterol, derived from acetyl-CoA, are the precursors to triacylglycerols, membrane lipids, steroids, and bile acids.

16.1 Fatty Acid Oxidation and Ketogenesis

The regulated degradation of fatty acids and the subsequent oxidation of acetyl-CoA by the citrate cycle reactions are essential processes for maintenance of metabolic homeostasis in terrestrial animals, which use lipids as their primary energy reserve. Moreover, replenishment of these lipid energy reserves using dietary nutrients as a carbon source for lipid biosynthesis is critical to long-term survival. Because lipid degradation and biosynthetic pathways are complementary in many ways, it is useful to answer the four metabolic questions as they relate to lipid metabolism, emphasizing key concepts in both fatty acid oxidation and fatty acid synthesis.

1. *What purposes do fatty acid oxidation and fatty acid synthesis serve in animals?* Fatty acid oxidation in mitochondria is responsible for providing energy to cells when glucose levels are low. Most humans store enough triacylglycerols in their adipose tissue to supply energy to the body for ~3 months during starvation. Liver and adipose cells convert excess acetyl-CoA into fatty acids that can be stored or exported as triacylglycerols.

2. *What are the* net *reactions of fatty acid degradation and synthesis for the typical C16 fatty acid palmitate?*

Fatty acid oxidation:

$$\text{Palmitate} + 7\ \text{NAD}^+ + 7\ \text{FAD} + 8\ \text{CoA} + 7\ \text{H}_2\text{O} + \text{ATP} \rightarrow$$
$$8\ \text{Acetyl-CoA} + 7\ \text{NADH} + 7\ \text{FADH}_2 + \text{AMP} + 2\ \text{P}_\text{i} + 7\ \text{H}^+$$

Fatty acid synthesis:

$$8\ \text{Acetyl-CoA} + 7\ \text{ATP} + 14\ \text{NADPH} + 14\ \text{H}^+ \rightarrow$$
$$\text{Palmitate} + 8\ \text{CoA} + 7\ \text{ADP} + 7\ \text{P}_\text{i} + 14\ \text{NADP}^+ + 6\ \text{H}_2\text{O}$$

3. *What are the key enzymes in fatty acid metabolism?* **Fatty acyl-CoA synthetase** catalyzes the "priming" reaction in fatty acid metabolism, converting free fatty acids in the cytosol into fatty acyl-CoA using the energy available from ATP and pyrophosphate hydrolysis. **Carnitine acyltransferase I** catalyzes the commitment step in fatty acid oxidation, which links fatty acyl-CoA molecules to carnitine so they can be transported across the inner mitochondrial membrane. The activity of carnitine acyltransferase I is inhibited by malonyl-CoA, which is the product of the acetyl-CoA carboxylase reaction. **Acetyl-CoA carboxylase** catalyzes the commitment step in fatty acid synthesis using acetyl-CoA to form the C_3 compound malonyl-CoA. The activity of acetyl-CoA carboxylase is regulated by both reversible phosphorylation and allosteric mechanisms. **Fatty acid synthase,** is responsible for catalyzing a series of reactions that sequentially adds C_2 units to a growing fatty acid chain.

4. *What are examples of fatty acid metabolism in everyday biochemistry? Fatty acid oxidation*: The kangaroo rat and the camel are two examples of animals that survive in desert environments for long periods of time without drinking water. They accomplish this by generating H_2O internally through the complete oxidation of fatty acids. The kangaroo rat obtains fatty acids from oils in seeds, whereas camels store triacylglycerols in the adipose tissue of their humps. *Fatty acid synthesis*: A variety of foods are prominently advertised as "nonfat," even though they can contain a high calorie count coming from carbohydrates. Eating nonfat, high-calorie foods in excess of energy needs activates the fatty acid synthesis pathway, resulting in the conversion of acetyl-CoA to fatty acids, which are stored as triacylglycerols.

The Fatty Acid β-Oxidation Pathway in Mitochondria

The mitochondrial degradation of fatty acids in animal cells was first described by Eugene Kennedy and Albert Lehninger in the late 1940s. Subsequent work showed that fatty acids need to be activated by coenzyme A on the cytosolic side of the outer mitochondrial membrane and then transported into the mitochondrial matrix by a specific carrier system. A set of three enzymes called fatty acyl-CoA synthetases, which differ in their specificity for short-, medium-, and long-chain fatty acids, catalyze the formation of fatty acyl-CoA derivatives using an ATP-coupled reaction mechanism illustrated in **Figure 16.2**.

In the first step of this reaction, fatty acyl-CoA synthetase catalyzes the adenylation of a fatty acid to form the enzyme-bound intermediate fatty acyl-adenylate. This involves an attack by the carboxylate ion of the fatty acid on the α phosphate of ATP and release of pyrophosphate (PP_i). In the second step of the reaction, the fatty acyl-adenylate intermediate is attacked by the thiol group of CoA, forming the thioester fatty acyl-CoA product and releasing AMP. This two-step reaction is energetically

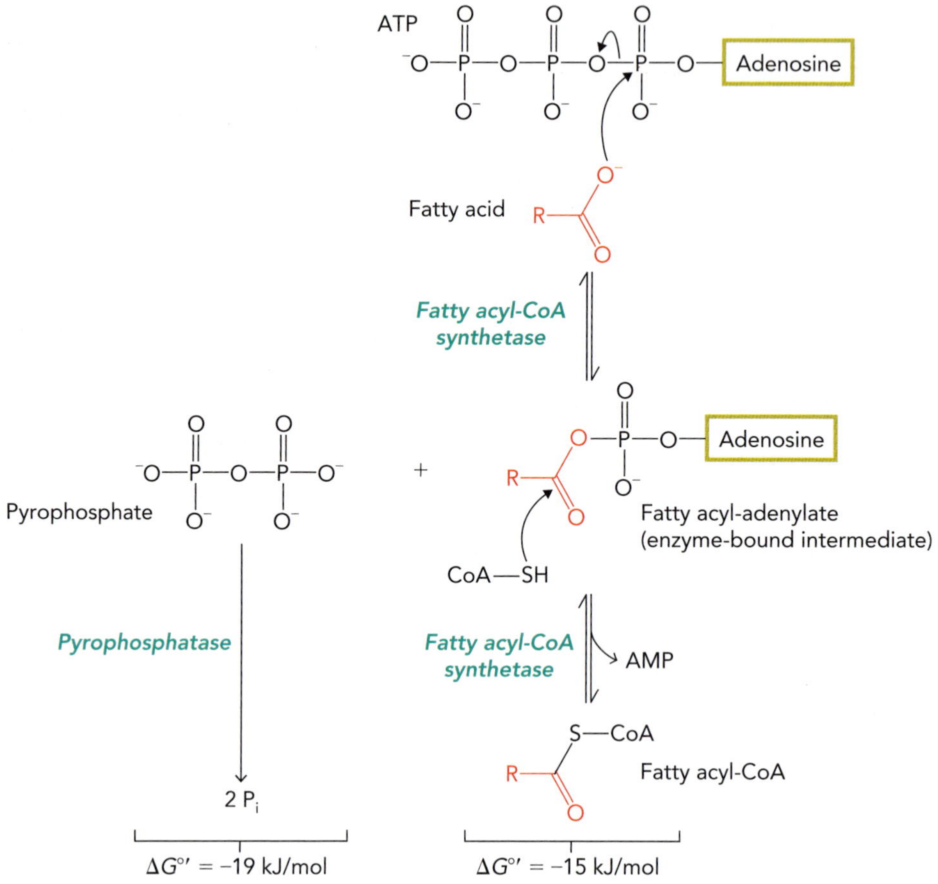

Figure 16.2 Formation of fatty acyl-CoA by fatty acyl-CoA synthetase requires a two-step reaction involving formation of an enzyme-bound adenylate intermediate and subsequent hydrolysis of pyrophosphate. The combined reactions catalyzed by fatty acyl-CoA synthetase and pyrophosphatase are highly exergonic ($\Delta G°' = -34$ kJ/mol). P_i = inorganic phosphate.

favorable, with $\Delta G°' = -15$ kJ/mol. As with other reactions we have seen involving PP_i, the rapid hydrolysis of PP_i by the enzyme pyrophosphatase pulls the reaction to the right, giving an overall change in standard free energy of $\Delta G°' = -34$ kJ/mol. We count this ATP-coupled reaction as an energy investment of 2 ATP because 2 ATP are required to phosphorylate the AMP product and regenerate ATP using the enzymes adenylate kinase and nucleoside diphosphate kinase. As shown in the chemical equations that follow, adenylate kinase uses one ATP to phosphorylate the input AMP (boldface) and generate an ADP product (boldface). In the second reaction, nucleoside diphosphate kinase uses a second ATP to phosphorylate this ADP (boldface) and generate the final ATP product (boldface). The net reaction therefore requires 2 ATP to regenerate an ATP from the input AMP.

$$\text{Adenylate kinase:} \quad \textbf{AMP} + \text{ATP} \rightleftharpoons \textbf{ADP} + \text{ADP}$$

$$\text{Nucleoside diphosphate kinase:} \quad \textbf{ADP} + \text{ATP} \rightleftharpoons \textbf{ATP} + \text{ADP}$$

$$\text{Net reaction:} \quad \textbf{AMP} + 2\,\text{ATP} \rightleftharpoons \textbf{ATP} + 2\,\text{ADP}$$

Fatty acyl-CoA has either of two fates, depending on the energy charge of the cell. If the energy charge is low, then fatty acyl-CoA is imported into the mitochondrial matrix by the **carnitine transport cycle** and degraded by the fatty acid oxidation reactions to yield acetyl-CoA, $FADH_2$, and NADH. However, if the energy charge is high and fatty acid synthesis is favored, then mitochondrial import of fatty acyl-CoA is inhibited by the fatty acid precursor malonyl-CoA.

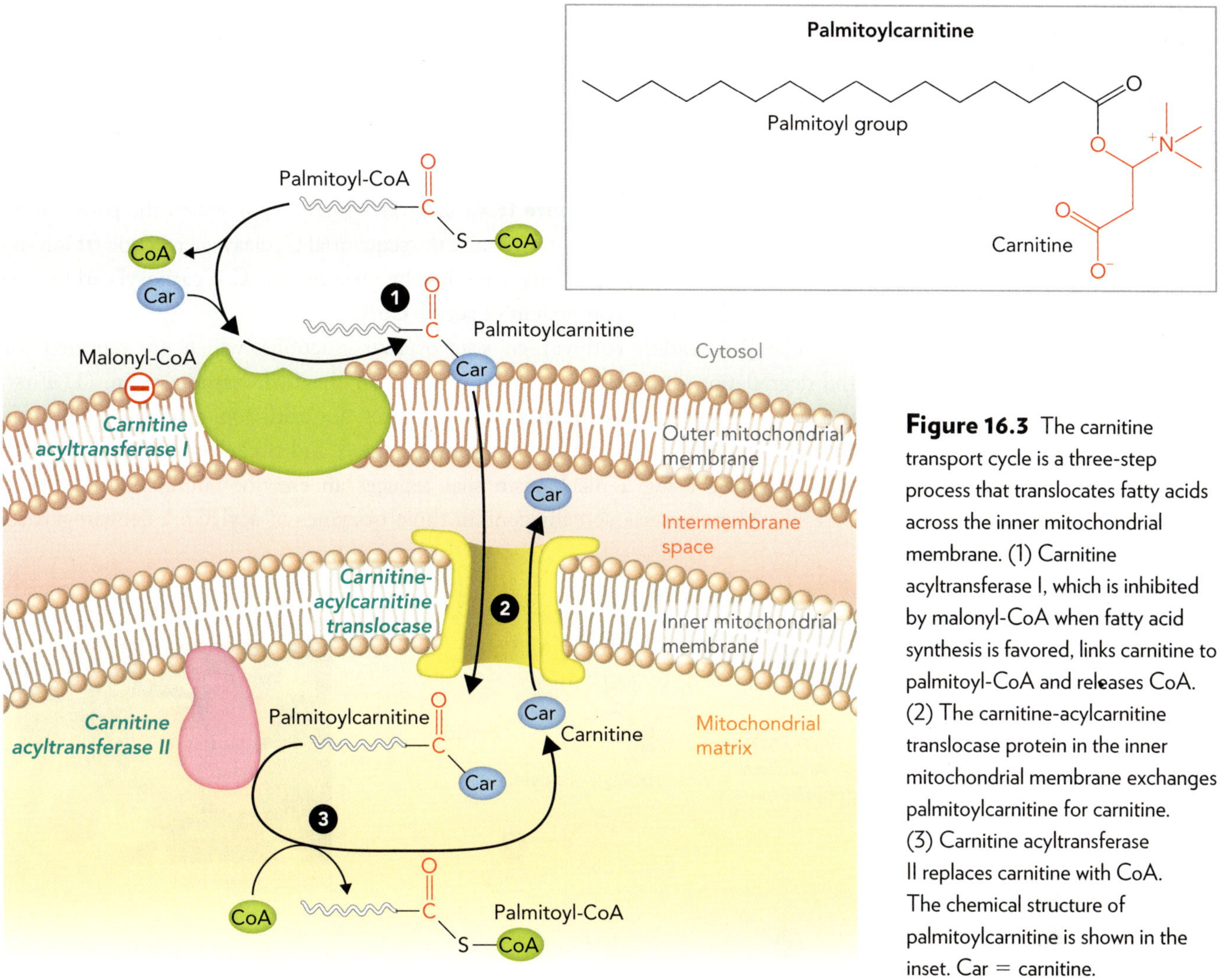

Figure 16.3 The carnitine transport cycle is a three-step process that translocates fatty acids across the inner mitochondrial membrane. (1) Carnitine acyltransferase I, which is inhibited by malonyl-CoA when fatty acid synthesis is favored, links carnitine to palmitoyl-CoA and releases CoA. (2) The carnitine-acylcarnitine translocase protein in the inner mitochondrial membrane exchanges palmitoylcarnitine for carnitine. (3) Carnitine acyltransferase II replaces carnitine with CoA. The chemical structure of palmitoylcarnitine is shown in the inset. Car = carnitine.

▶ ANIMATION

Figure 16.3 illustrates the carnitine transport cycle starting with palmitoyl-CoA. The cycle involves the function of three proteins and a molecular tag called carnitine. In the first reaction, carnitine acyltransferase I, which is located in the outer mitochondrial membrane, replaces the CoA moiety with carnitine to form palmitoylcarnitine. In the second reaction, the carnitine transport protein, called carnitine-acylcarnitine translocase, exchanges a palmitoylcarnitine molecule for a carnitine. The third reaction involves the conversion of palmitoylcarnitine back to palmitoyl-CoA in a reaction catalyzed by carnitine acyltransferase II, thereby releasing the carnitine so that it can be shuttled back across the inner mitochondrial membrane by the carnitine-acylcarnitine translocase protein.

The carnitine transport cycle serves two important functions in regulating cellular metabolism. First, it provides a mechanism to control the flux of fatty acids either into the degradative pathway inside the mitochondrial matrix or toward the synthesis of triacylglycerols and membrane lipids in the cytosol. This regulatory decision is controlled by malonyl-CoA, which inhibits the activity of carnitine acyltransferase I and prevents the import of fatty acyl-CoA molecules into the mitochondria when lipid synthesis is favored. Second, the carnitine transport cycle, as well as the citrate shuttle system that we describe later, functions to maintain separate pools of coenzyme A, which are involved in distinct processes. Cytosolic coenzyme A is used primarily for anabolic pathways, such as fatty acid synthesis, whereas mitochondrial coenzyme A is used for catabolic pathways, involving the degradation of pyruvate, fatty acids, and selected amino acids.

Once the electron-rich fatty acids are moved into the mitochondrial matrix, their value as high-energy metabolites is fully utilized to generate a substantial amount of ATP. This energy conversion process of fatty acid → ATP involves oxidation of fatty acids by sequential degradation of C_2 units, resulting in the production of $FADH_2$, NADH, and acetyl-CoA. **Figure 16.4a** illustrates this process, called the **β-oxidation pathway** of fatty acid degradation because the sequential C_2 cleavage reaction (thiolysis) occurs at the β carbon of the fatty acid, thereby releasing the C-1 carboxyl carbon and α carbon as the acetate component of acetyl-CoA.

The β-oxidation pathway consists of four reactions, which are repeated for the degradation of each two-carbon segment of the hydrocarbon chain. (1) First, the enzyme acyl-CoA dehydrogenase catalyzes an oxidation reaction. This reaction introduces a trans C=C bond between the α and β carbons of the fatty acyl-CoA molecule using a mechanism that reduces an enzyme-bound FAD to form $FADH_2$. Mitochondria actually contain three isozymes of acyl-CoA dehydrogenase,

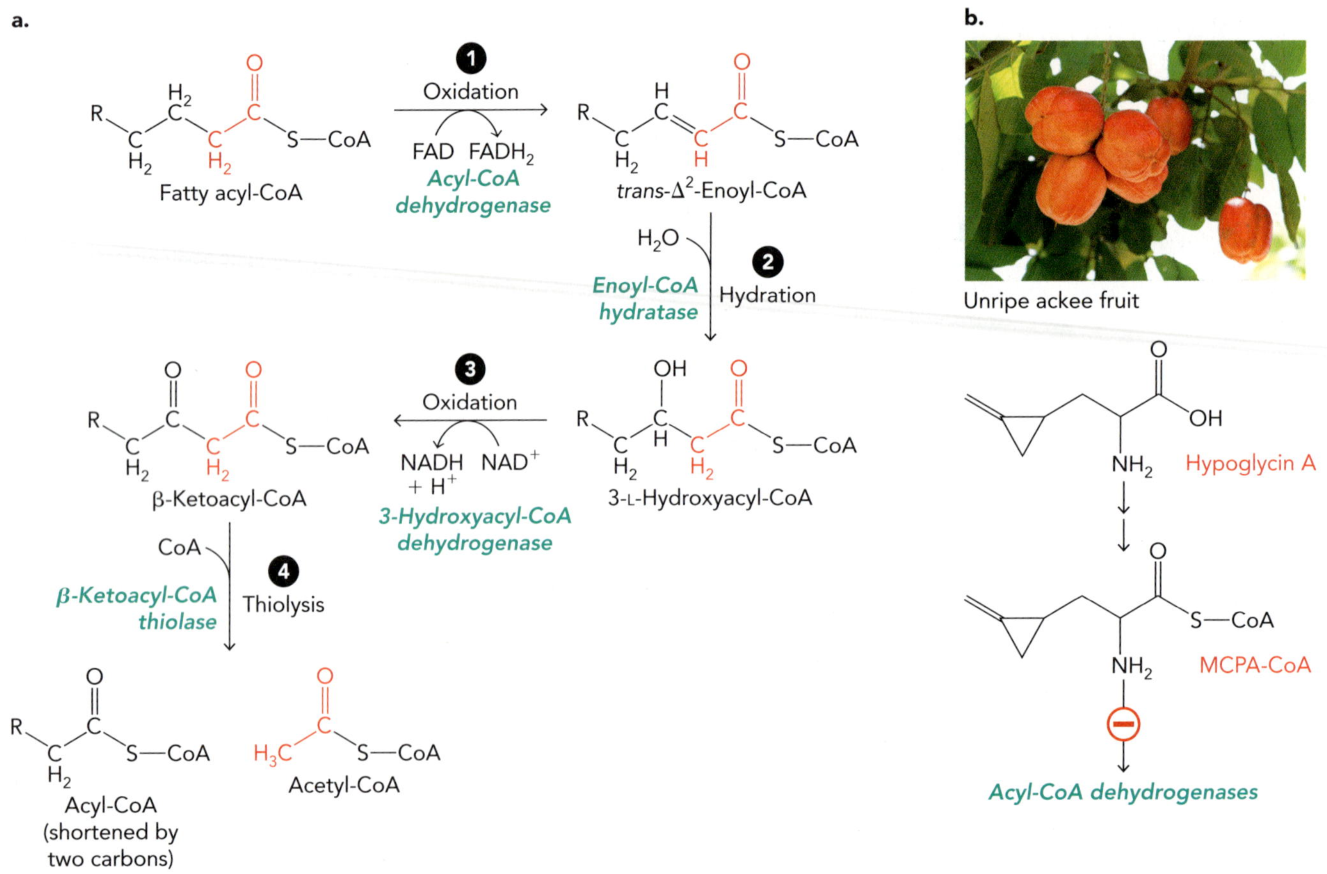

Figure 16.4 The β-oxidation pathway consists of four reaction steps, one of which is inhibited by the compound hypoglycin A found in unripe ackee fruit. **a.** (1) Oxidation of the fatty acyl-CoA substrate using an enzyme-linked FAD moiety forms a trans C=C bond between the α and β carbons. (2) Hydration across the C=C bond adds a hydroxyl group to the β carbon. (3) A second oxidation reaction, this time using NAD^+, removes two electrons and generates a C=O at the β carbon. (4) Thiolase-mediated cleavage at the β carbon releases acetyl-CoA and an acyl-CoA product that is two carbons shorter than the substrate. **b.** Hypoglycin A is an unusual amino acid produced at 10^4-fold higher levels in unripe ackee fruit than in ripe ackee fruit. After consumption of unripe ackee fruit, hypoglycin A is metabolized to a toxic intermediate called methylenecyclopropylacetic acid (MCPA), which is then is attached to CoA (MCPA-CoA). The MCPA-CoA inhibits liver mitochondrial acyl-CoA dehydrogenases and is the cause of Jamaican vomiting sickness. PHOTO: STEVEN FRAME/ALAMY.

which differ in their specificity for hydrocarbon chains of different lengths. These enzymes are referred to as long-chain (C_{12} to C_{18}), medium-chain (C_4 to C_{14}), and short-chain (C_4 to C_8) acyl-CoA dehydrogenases. (2) The second reaction in the β-oxidation pathway is a hydration step catalyzed by the enzyme enoyl-CoA hydratase, which adds H_2O across the C=C bond to convert *trans*-Δ^2-enoyl-CoA to 3-L-hydroxyacyl-CoA. (3) The third reaction is another oxidation step in which the enzyme 3-hydroxyacyl-CoA dehydrogenase removes an electron pair from the substrate and donates it to NAD^+ to form NADH. (4) Finally, coenzyme A is used in a thiolysis reaction catalyzed by the enzyme β-ketoacyl-CoA thiolase, which releases a molecule of acetyl-CoA. In the process, this reaction results in the formation of a fatty acyl-CoA product that is two carbons shorter than the starting substrate.

These four reactions together convert palmitoyl-CoA (C_{16}) into myristoyl-CoA (C_{14}) and in the process generate 1 $FADH_2$, 1 NADH, and 1 acetyl-CoA. The myristoyl-CoA product becomes the substrate for another round of β oxidation, resulting in the production of one more molecule each of $FADH_2$, NADH, and acetyl-CoA. Now you can see why three isozymes of acyl-CoA dehydrogenase exist: After sequential rounds of β oxidation, the progressively shorter fatty acyl-CoA substrate is no longer a high-affinity substrate for acyl-CoA dehydrogenase isozymes that recognize long-chain fatty acids. Notably, the compound **hypoglycin A**, which is found at high levels in unripe ackee fruit (**Figure 16.4b**), is a potent inhibitor of acyl-CoA dehydrogenases. The dietary condition known as Jamaican vomiting sickness results from eating unripe ackee fruit. The symptoms of this dietary ailment are due to acute inhibition of several metabolic processes in liver cells, including reduced transport of long-chain fatty acids into the mitochondria and decreased fatty acid oxidation.

As illustrated in **Figure 16.5**, the complete oxidation of palmitoyl-CoA (C_{16}) requires seven rounds of the β-oxidation pathway to convert one molecule of palmitoyl-CoA into eight molecules of acetyl-CoA. The net reaction can be written as

$$\text{Palmitoyl-CoA} + 7\ \text{CoA} + 7\ \text{FAD} + 7\ \text{NAD}^+ + 7\ \text{H}_2\text{O} \rightarrow 8\ \text{Acetyl-CoA} + 7\ \text{FADH}_2 + 7\ \text{NADH} + 7\ \text{H}^+$$

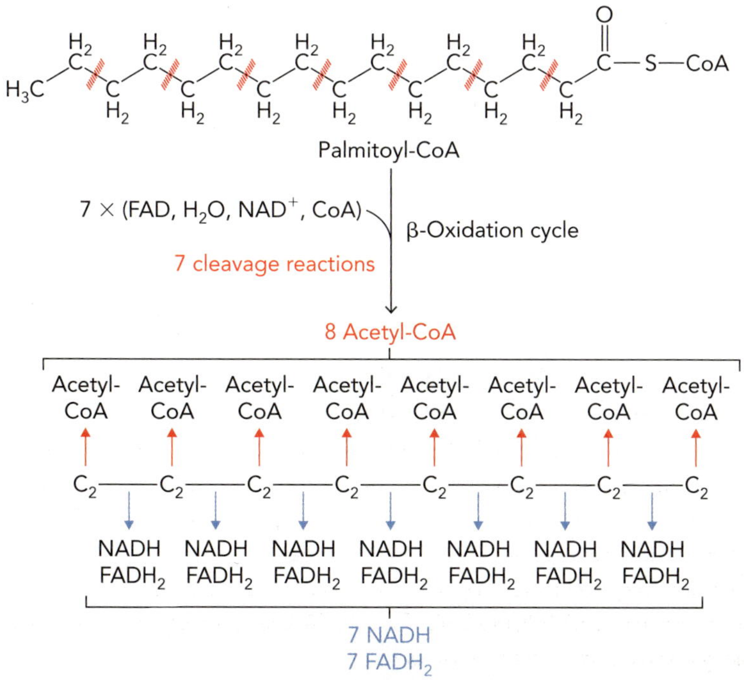

Figure 16.5 The complete oxidation of palmitoyl-CoA requires seven cycles of the four-step β-oxidation cycle to yield 8 acetyl-CoA + 7 $FADH_2$ + 7 NADH.

The first three steps in the β-oxidation pathway (oxidation, hydration, oxidation) are very similar to the three reactions in the citrate cycle that convert succinate to oxaloacetate (see reactions 6, 7, and 8 in Figure 10.21). Not only is the chemistry similar in this three-step series, but also the FAD-dependent oxidation step in both pathways involves enzymes that are tightly associated with the matrix side of the inner mitochondrial membrane. These two FAD-bound dehydrogenases pass an electron pair to ubiquinone, the mobile electron carrier in the electron transport system.

In the case of the citrate cycle enzyme succinate dehydrogenase—also referred to as complex II—the $FADH_2$ is oxidized by an Fe–S pair in the enzyme, which then passes the electrons on to coenzyme Q (ubiquinone) to form reduced coenzyme Q (ubiquinol). In the case of the β-oxidation pathway, the acyl-CoA dehydrogenase enzyme in step 1 (see Figure 16.4) donates the electrons to a protein called electron-transferring flavoprotein (ETF), which in turn passes them to the enzyme ETF-Q oxidoreductase (electron-transferring flavoprotein–ubiquinone oxidoreductase). As shown in **Figure 16.6**, it is the Fe–S center of ETF-Q oxidoreductase that donates the electrons to coenzyme Q. Because coenzyme Q functions downstream of complex I in the electron transport system, each electron pair originating from the acyl-CoA dehydrogenase step in β oxidation gives rise to ~1.5 ATP as a result of oxidative phosphorylation. This ATP currency exchange ratio of 1.5 ATP per $FADH_2$ is the same as that for the $FADH_2$ oxidation by succinate dehydrogenase (citrate cycle) and glycerol-3-phosphate dehydrogenase (mitochondrial shuttle) described in Chapter 11.

We are now ready to calculate the net ATP yield from the degradation of palmitoyl-CoA using the β-oxidation pathway. As shown in **Table 16.1**, after seven rounds of β oxidation, palmitoyl-CoA yields 8 acetyl-CoA, 7 NADH, and 7 $FADH_2$. The oxidation of acetyl-CoA by the citrate cycle then generates 24 NADH, 8 $FADH_2$, and 8 GTP (ATP). The combined reactions of the electron transport system and oxidative phosphorylation convert these 31 NADH into

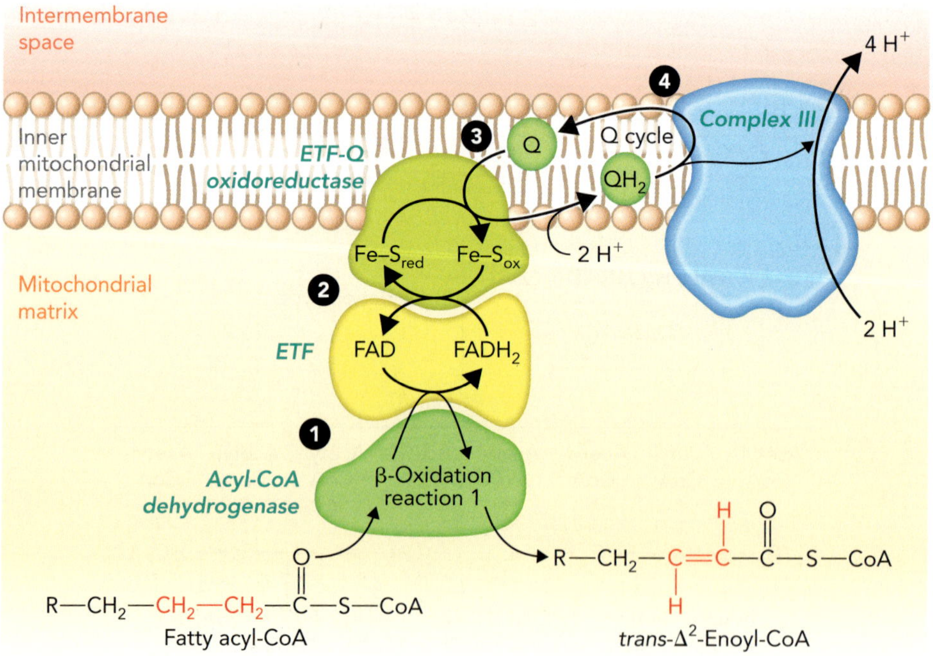

Figure 16.6 Beta oxidation results in electron transfer. (1) Acyl-CoA dehydrogenase donates the electrons from the first fatty acid oxidation reaction to an enzyme-bound FAD molecule in the ETF. (2) This electron pair is then passed to an Fe–S center in the ETF-Q oxidoreductase enzyme. (3) The electrons are passed from the Fe-S center to coenzyme Q (Q), generating reduced coenzyme Q (QH_2). (4) Reduced coenzyme Q is oxidized by complex III as part of the Q cycle (see Figure 11.8).

~77.5 ATP (31 × ~2.5 ATP), and the 15 $FADH_2$ are converted into ~22.5 ATP (15 × ~1.5 ATP). When adding in the 8 ATP derived from 8 GTP generated in the citrate cycle, the total ATP yield from palmitoyl-CoA is 108 ATP. It is important to note that if one is calculating net ATP yields beginning with palmitate rather than beginning with palmitoyl-CoA, it is necessary to subtract 2 ATP equivalents required for the fatty acyl-CoA activation (ATP → AMP + PP_i) of palmitate to generate palmitoyl-CoA (see Figure 16.2). In this case, the net ATP yield from palmitate is 106 ATP (108 ATP – 2 ATP = 106 ATP).

Table 16.1 ATP YIELD FROM THE COMPLETE OXIDATION OF PALMITOYL-COA

β oxidation of palmitoyl-CoA	Citrate cycle	ATP generated by oxidative phosphorylation
7 NADH →		17.5 ATP
7 $FADH_2$ →		10.5 ATP
	24 NADH →	60 ATP
8 Acetyl-CoA →	8 $FADH_2$ →	12 ATP
	8 GTP →	8 ATP
	Total	**108 ATP per palmitoyl-CoA**

Note: Values for the ATP currency exchange ratio are 2.5 ATP per NADH and 1.5 ATP per $FADH_2$.

To appreciate the increased energy yield from fatty acid oxidation compared to that of glucose oxidation, consider that the oxidation in muscle cells of one molecule of stearoyl-CoA, which contains a fully saturated C_{18} fatty acid, yields 122 ATP, using the same bookkeeping methods shown in Table 16.1. Thus, the net ATP yield from stearate accounting for the fatty acyl-CoA activation step is 120 ATP. In contrast, the complete oxidation of three molecules of glucose (3 × C_6) generates only 90 ATP (see Table 11.3). The increased energy yield of 33% (120/90 = 133%) for stearate compared to that for glucose is due to the increased number of electrons available to the electron transport system. When you take into account the fact that glucose in glycogen is much more hydrated than fatty acids in triacylglycerols, it is easy to see why evolution exploited the energy storage properties of lipids over carbohydrates.

In addition to the payout of ATP that comes from fatty acid oxidation, another benefit of β oxidation is the generation of molecular H_2O as a result of combined reactions in the electron transport system (oxidation of NADH and $FADH_2$ reduces O_2 to form 2 H_2O), and ATP synthase complex (formation of ATP from ADP + P_i produces H_2O). Indeed, when accounting for all biochemical reactions required to completely oxidize 1 mol of palmitate ~130 mol of H_2O are produced, which is essential to animals that live in dry climates, such as the desert kangaroo rat and the Arabian camel (**Figure 16.7**). Large animals that hibernate over the winter, such as the grizzly bear, also take advantage of fatty acid oxidation in order to replace H_2O that is lost by respiration.

Figure 16.7 The Arabian camel, desert kangaroo rat, and hibernating grizzly bear survive long periods of time without drinking H_2O by oxidizing stored triacylglycerols, which generates ~130 mol of H_2O per mole of palmitate. ARABIAN CAMEL: DAVID STEELE/SHUTTERSTOCK; DESERT KANGAROO RAT: ANTHONY MERCIECA/SCIENCE SOURCE/GETTY IMAGES; HIBERNATING GRIZZLY BEAR: STOUFFER PRODUCTIONS/AGEFOTOSTOCK.

Auxiliary Pathways for Fatty Acid Oxidation

Not all fatty acids used for energy conversion by the β-oxidation pathway are even-numbered and fully saturated fatty acids such as palmitate (16:0) or stearate (18:0). For example, oleate is an abundant dietary monounsaturated fatty acid with a cis C═C bond between C-9 and C-10 [*cis* 18:1(Δ^9) fatty acid], which needs to be removed before complete oxidation can occur. Starting with oleoyl-CoA as shown in **Figure 16.8**, after three rounds of β oxidation to remove three acetyl-CoA units, the cis C═C bond is removed and a trans C═C bond is added at the adjacent position by the enzyme Δ^3,Δ^2-enoyl-CoA isomerase. The product of this reaction, *trans*-Δ^2-dodecenoyl-CoA, is then hydrated, oxidized, and subjected to thiolysis to release an acetyl-CoA. The resulting decanoyl-CoA product undergoes four additional rounds of β oxidation, releasing another five acetyl-CoA units. Degradation of polyunsaturated fatty acids such as linoleate, an all-*cis* 18:2($\Delta^{9,12}$) fatty acid, is more complex. As shown in **Figure 16.9**, degradation of *cis*-Δ^9,*cis*-Δ^{12}-linoleoyl-CoA requires two isomerization reactions (catalyzed by Δ^3,Δ^2-enoyl-CoA isomerase and enoyl-CoA isomerase), as well as an NADPH-dependent reduction reaction catalyzed by the enzyme 2,4-dienoyl-CoA reductase.

Bovine milk contains high amounts of odd-numbered, long-chain fatty acids, one of which is the long-chain saturated fatty acid tricosanoate (23:0). As shown in **Figure 16.10**, the final product of odd-numbered fatty acid oxidation is the C_3 compound propionyl-CoA. The first step in oxidation of odd-numbered fatty acids is catalyzed by

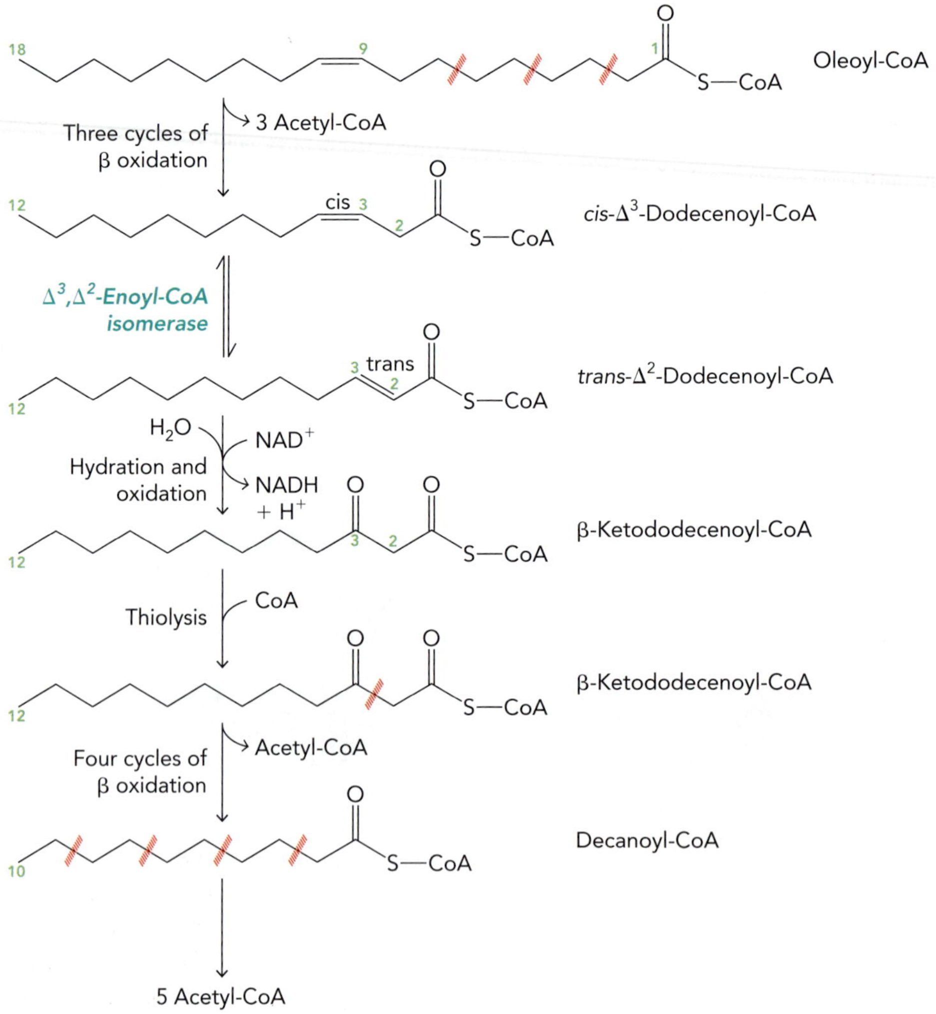

Figure 16.8 Degradation of the monounsaturated molecule oleoyl-CoA requires additional reactions to account for the C═C bond.

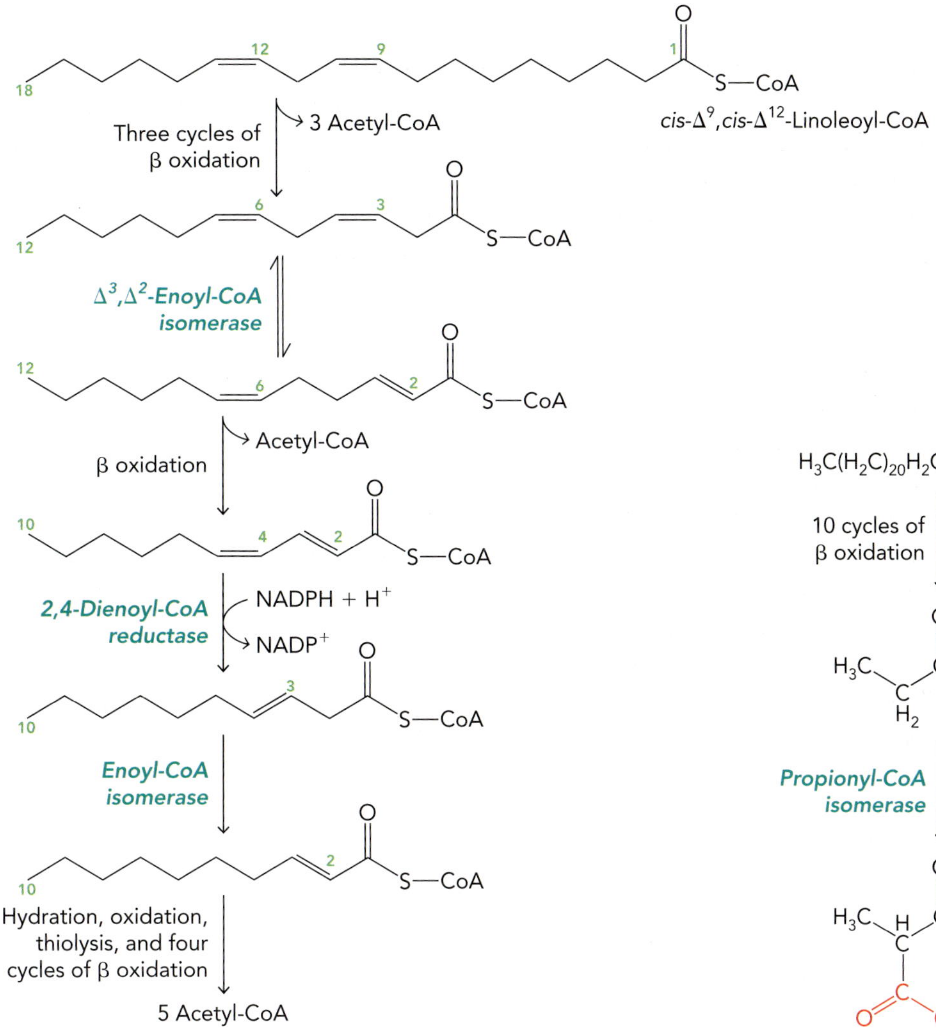

Figure 16.9 Degradation of the polyunsaturated molecule *cis*-Δ^9,*cis*-Δ^{12}-linoleoyl-CoA.

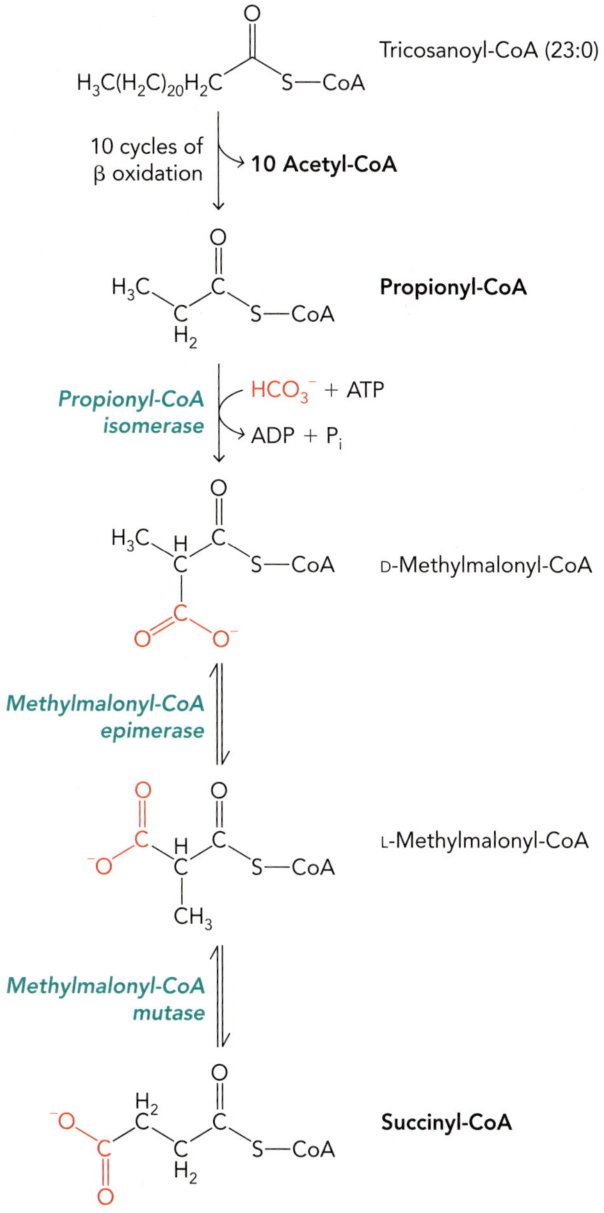

Figure 16.10 Degradation of the odd-numbered C_{23} tricosanoyl-CoA generates 10 C_2 acetyl-CoA molecules and one molecule of C_3 propionyl-CoA. Three enzymatic steps are required to convert propionyl-CoA to the citrate cycle intermediate succinyl-CoA.

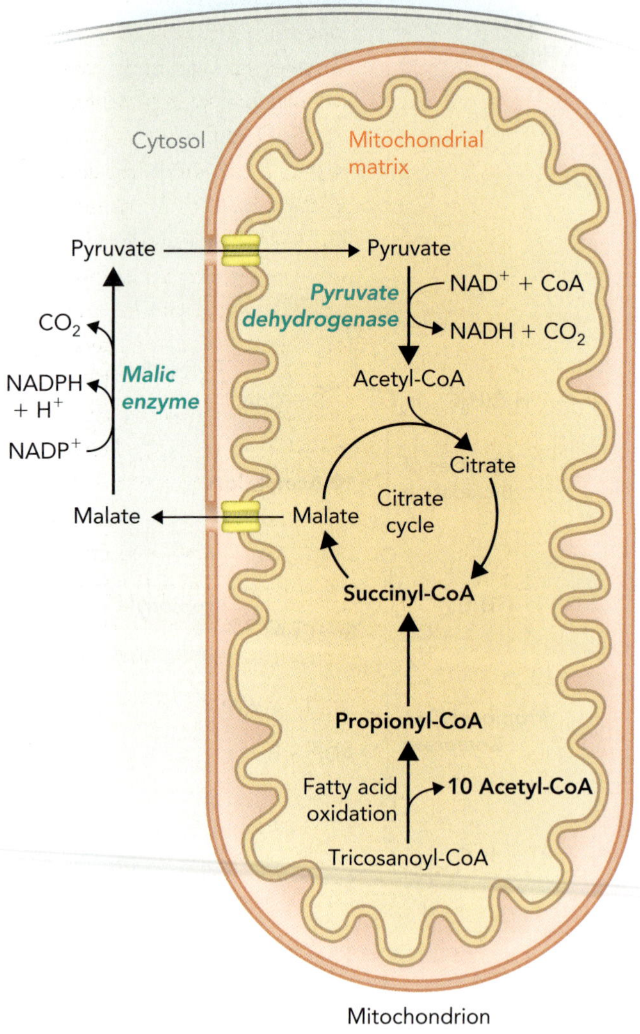

Figure 16.11 In order for the succinyl-CoA produced by odd-chain fatty acid oxidation of tricosanoyl-CoA to be used by energy conversion reactions, it must first be converted to malate, which is then decarboxylated in the cytosol by malic enzyme to yield pyruvate. Transport of pyruvate into the mitochondrial matrix under conditions of low energy charge leads to a net yield of one acetyl-CoA for each succinyl-CoA that enters the citrate cycle from odd-chain fatty acid oxidation. The total yield of acetyl-CoA from tricosanoyl-CoA oxidation is therefore 10 + 1 = 11 acetyl-CoA.

the biotin-containing enzyme propionyl-CoA carboxylase, which uses ATP to generate a carboxybiotin intermediate. The product of this reaction is D-methylmalonyl-CoA, which is then converted to the L isomer by the enzyme methylmalonyl-CoA epimerase. L-methylmalonyl-CoA is then rearranged by methylmalonyl-CoA mutase to generate succinyl-CoA, an intermediate in the citrate cycle.

In order for the carbons originating from propionyl-CoA to be recovered for energy conversion processes, succinyl-CoA must first be converted to malate by citrate cycle enzymes (**Figure 16.11**). Citrate cycle anaplerotic reactions do not result in a net energy yield because the products, such as succinyl-CoA, are themselves citrate cycle intermediates (see Chapter 10). Therefore, in order for the carbons of succinyl-CoA to yield energy upon oxidation, they must enter the citrate cycle in the form of acetyl-CoA. This is done by exporting the newly synthesized malate to the cytosol, where malic enzyme catalyzes an oxidative decarboxylation reaction to yield pyruvate and CO_2. If the cell requires energy in the form of ATP, then the pyruvate is imported into the mitochondria, where pyruvate dehydrogenase converts it to acetyl-CoA. In this scheme, complete oxidation of a C_{23} odd-numbered, saturated fatty acyl-CoA yields 11 acetyl-CoA and releases 1 CO_2.

In addition to mitochondrial fatty acid oxidation, animal cells also degrade fatty acids in organelles called peroxisomes, which are found in almost all cell types. Plants and eukaryotic microorganisms carry out fatty acid oxidation only in peroxisomes. The enzymatic steps in the peroxisomal fatty acid oxidation pathway are similar to β oxidation in mitochondria; however, utilization of acetyl-CoA, NADH, and $FADH_2$ in energy conversion processes does not occur because citrate cycle and oxidative phosphorylation enzymes are lacking.

Animal cell peroxisomes selectively degrade saturated, very-long-chain fatty acids such as eicosanoate (20:0), docosanoate (22:0), tetracosanoate (24:0), and hexacosanoate (26:0). As shown in **Figure 16.12**, acetyl-CoA generated by liver peroxisomal fatty acid oxidation is exported to the cytosol by a peroxisomal acylcarnitine transporter, where it is used to synthesize precursor metabolites for cholesterol and bile acid synthesis. One important difference between the mitochondrial and peroxisomal fatty acid degradation pathways is that the $FADH_2$ produced in the peroxisomal acyl-CoA dehydrogenase reaction donates its electron pair to H_2O to generate hydrogen peroxide (H_2O_2). Hydrogen peroxide is toxic and is converted to H_2O and ½ O_2 by the enzyme catalase. In mitochondrial β oxidation, the pair of electrons from $FADH_2$ is used to reduce an Fe–S center in ETF-Q oxidoreductase, which then transfers the electron pair to coenzyme Q in the electron transport system (see Figure 16.6). Electron micrographs of liver peroxisomes often show the presence of a proteinaceous crystalline structure called the core, which contains large amounts of peroxisomal enzymes (Figure 16.12, inset).

Two severe human metabolic diseases have been attributed to peroxisomal defects. One is **X-linked adrenoleukodystrophy (X-ALD)**, which is caused by a defect in a peroxisomal protein called adrenoleukodystrophy protein (ALDP) (see Figure 16.12). This

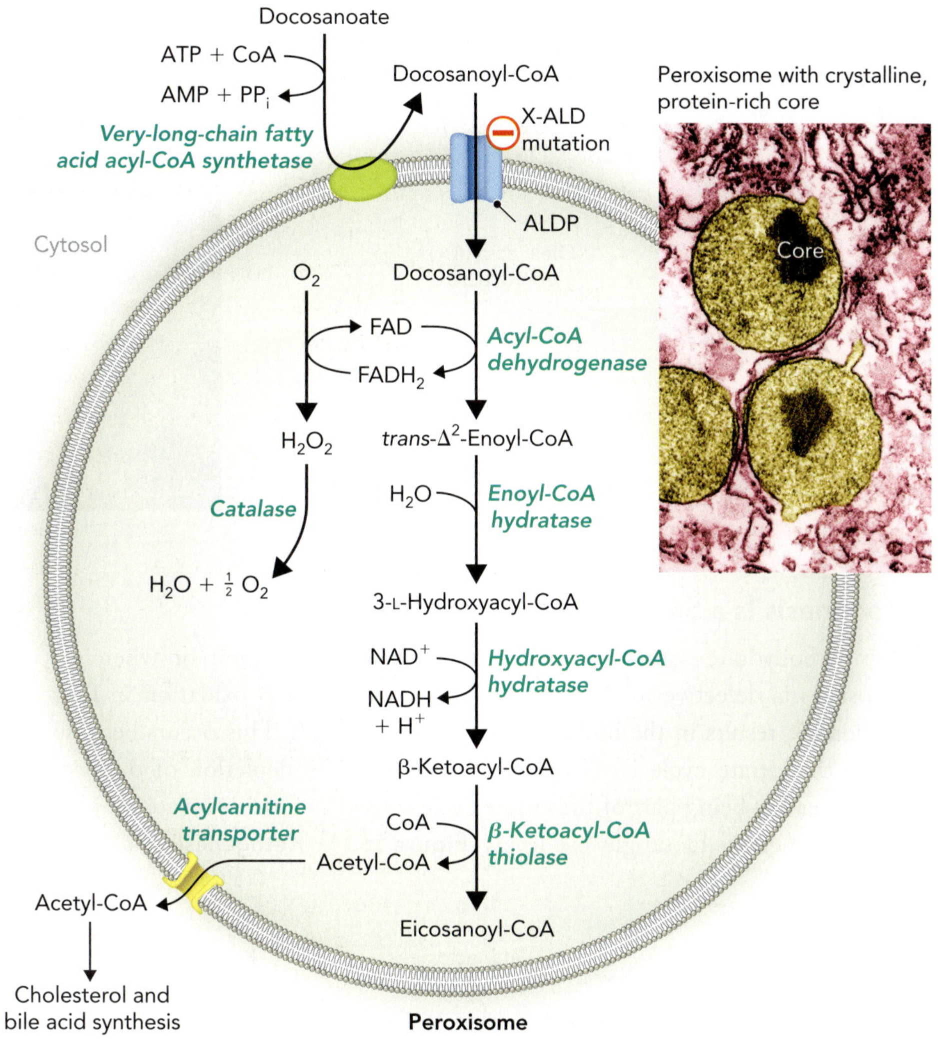

Figure 16.12 Peroxisomal fatty acid oxidation degrades very-long-chain fatty acids using specific enzymes that catalyze the same reactions present in mitochondria. The FAD/$FADH_2$ redox pair in acyl-CoA dehydrogenase donates its electron pair to O_2, generating H_2O_2, which is neutralized by peroxisomal catalase. The colorized electron micrograph of a peroxisome shows a dense region corresponding to the crystalline, protein-rich core. The peroxisomal deficiency disease X-linked adrenoleukodystrophy (X-ALD) is caused by mutations in the peroxisome transporting protein adrenoleukodystrophy protein (ALDP). PHOTO: FAWCETT & FRIEND/VISUALS UNLIMITED/CORBIS.

protein belongs to the ATP binding cassette transporter superfamily and is encoded by the ABCD1 gene. The defect in adrenoleukodystrophy protein blocks peroxisomal import of saturated very-long-chain fatty acids, which accumulate in the cytosol before being exported to the blood, where they destroy neuronal cell myelin sheaths. Because the defective gene in X-ALD is encoded on the X chromosome, this often-fatal disease primarily affects boys, who suffer from progressive neuronal degeneration.

One of the experimental treatments for X-ALD is ingestion of a 4:1 mixture of glycerol derivatives of the monounsaturated fatty acids oleate and erucidate, which is prepared from olive and rapeseed oils (**Figure 16.13**). This oil mixture is considered a homeopathic treatment for X-ALD and was given the name "Lorenzo's oil," which was popularized by a 1992 movie about a young X-ALD patient named Lorenzo Odone. On the basis of the observation that ingestion of Lorenzo's oil is most effective if given to presymptomatic young patients, it is thought that its mode of action is to decrease production of very-long-chain fatty acids and thereby limit myelin sheath damage in the early stages of the disease.

A second peroxisomal metabolic defect is **Zellweger syndrome**, which is caused by a complete lack of peroxisomes, resulting in death within the first year of birth. The rapid onset of lethal symptoms in infants with Zellweger syndrome attests to the critical role of peroxisomes in cellular metabolism.

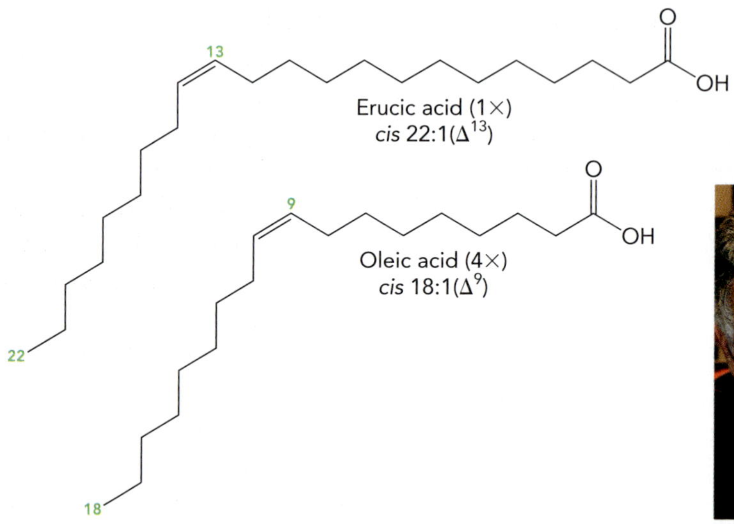

Figure 16.13 Lorenzo's oil is a 4:1 mixture of glycerated oleic acid and erucic acid used as a homeopathic treatment for X-ALD. Here Augusto Odone holds a bottle of pharmaceutical-grade Lorenzo's oil. PHOTO: RAPHAEL GALLARDE/GAMMA-RAPHO.

Ketogenesis Is a Salvage Pathway for Acetyl-CoA

When carbohydrate sources are limited because of starvation or when glucose homeostasis is defective in the case of diabetes, ongoing β oxidation in liver cell mitochondria results in the buildup of excess acetyl-CoA. This occurs because flux through the citrate cycle is diminished because of the depletion of oxaloacetate, which instead of being part of the citrate cycle is used to form pyruvate for gluconeogenesis to increase blood glucose levels (**Figure 16.14**). **Ketogenesis** is a process in

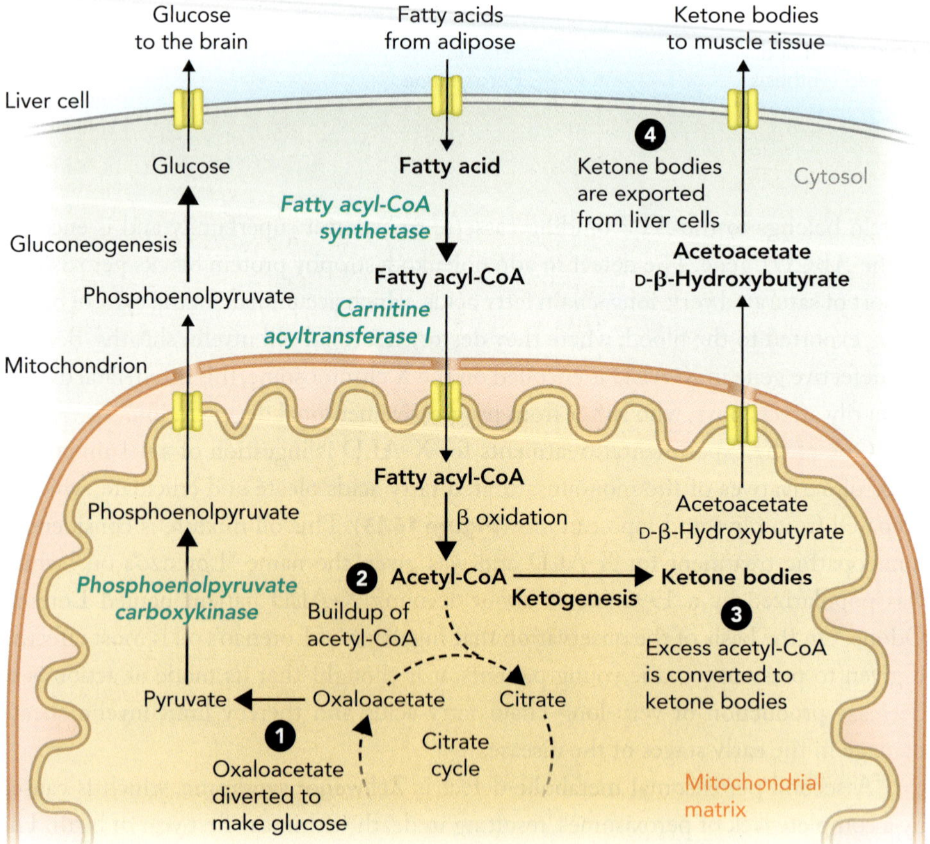

Figure 16.14 Ketogenesis salvages acetyl-CoA from liver mitochondria and converts it to acetoacetate and D-β-hydroxybutyrate, which are exported to skeletal and cardiac muscle where they are used for energy conversion. Flux through the ketogenic pathway is increased when glucose levels are low inside liver cells due to starvation or diabetes. Acetyl-CoA builds up under these conditions because oxaloacetate is used to make phosphoenolpyruvate for the gluconeogenic pathway, thereby decreasing flux through the citrate cycle.

liver cell mitochondria that takes the excess acetyl-CoA and converts it to acetoacetate and D-β-hydroxybutyrate. These two energy-rich compounds are called **ketone bodies** for historical reasons (they were originally thought to be liver precipitates, but they are actually quite soluble). Acetoacetate and D-β-hydroxybutyrate are exported from the liver to provide metabolic energy to muscle tissue.

Three mitochondrial enzymes are required to convert two acetyl-CoA molecules into acetoacetate, which is then reduced to form D-β-hydroxybutyrate (**Figure 16.15**). In the first step, the enzyme β-ketoacyl-CoA thiolase catalyzes a reaction that condenses two molecules of acetyl-CoA to form acetoacetyl-CoA. This is the same enzyme that releases one molecule of acetyl-CoA in reaction 4 of the β-oxidation pathway (see Figure 16.4); however, in this case, the reaction is driven toward condensation by the high concentration of acetyl-CoA in the mitochondria under ketogenic conditions. In the next step, the enzyme 3-hydroxy-3-methylglutaryl-CoA (HMG-CoA) synthase adds another acetyl-CoA group to form the intermediate HMG-CoA. This is converted to acetoacetate by the enzyme HMG-CoA lyase after removal of one of the two initiating acetyl-CoA groups. The acetoacetate can be exported directly or further metabolized to D-β-hydroxybutyrate by the enzyme

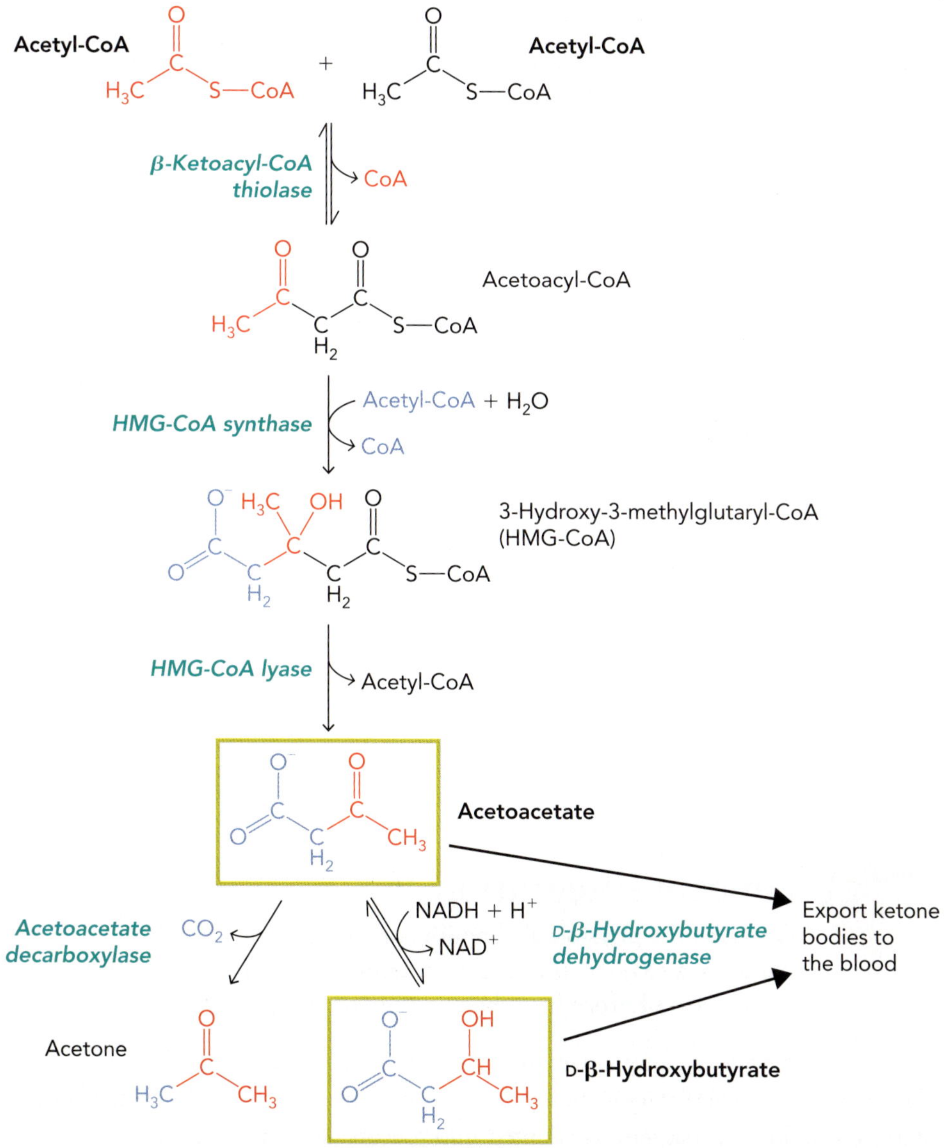

Figure 16.15 The ketone bodies acetoacetate and D-β-hydroxybutyrate are formed from two molecules of acetyl-CoA in a three-step reaction pathway. Ketone bodies produced in the liver are exported into the blood.

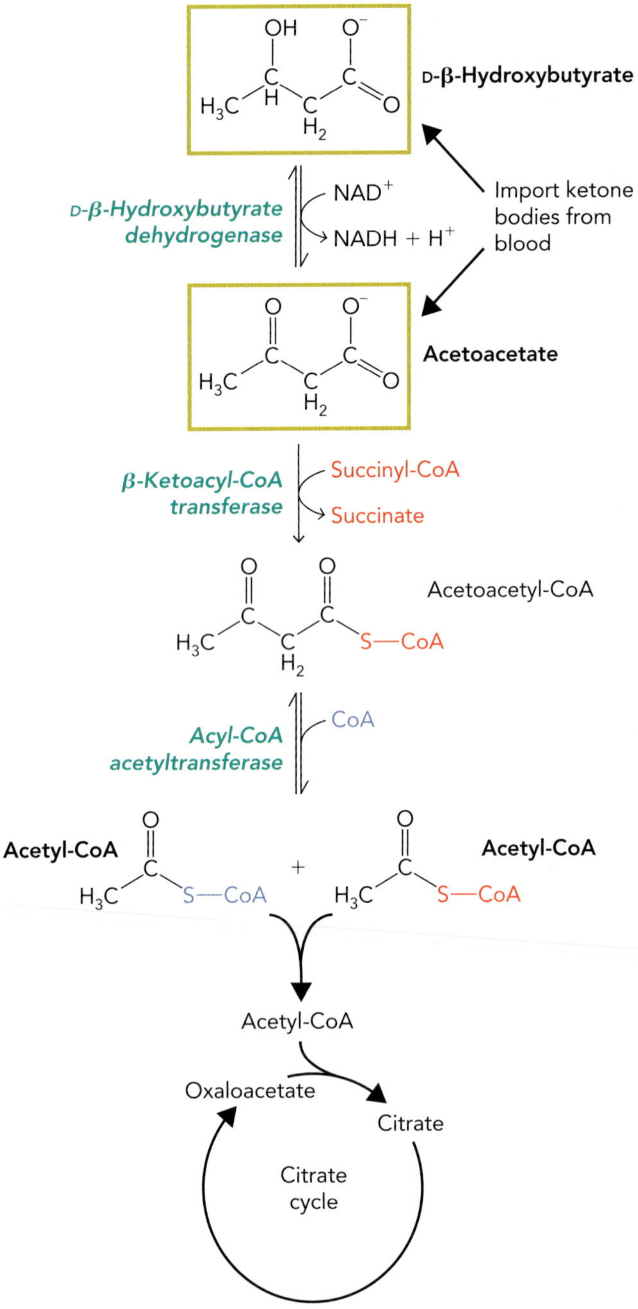

Figure 16.16 Skeletal and cardiac muscle cells use D-β-hydroxybutyrate and acetoacetate as a source of acetyl-CoA for energy conversion reactions in the citrate cycle.

D-β-hydroxybutyrate dehydrogenase. Note that acetoacetate can be converted to acetone plus CO_2 by spontaneous or enzymatic decarboxylation before liver cell export.

Acetoacetate and D-β-hydroxybutyrate are exported from the liver and used by other tissues, such as skeletal and heart muscle, to generate acetyl-CoA for energy conversion reactions. Even the brain, which prefers glucose as an energy source, can adapt to using ketone bodies as chemical energy during times of extreme starvation. In this energy conversion process, shown in **Figure 16.16**, D-β-hydroxybutyrate is first converted to acetoacetate by a reversal of the D-β-hydroxybutyrate dehydrogenase reaction. Then, the enzyme β-ketoacyl-CoA transferase uses an acetyl-CoA group obtained from succinyl-CoA to generate acetoacetyl-CoA. In the last step of this recovery process, acyl-CoA acetyltransferase cleaves acetoacetyl-CoA to generate two acetyl-CoA molecules, which are then metabolized by the citrate cycle. It is important to note that β-ketoacyl-CoA transferase is not present in liver cells, thus preventing this tissue from using the ketone bodies for its own energy needs.

Although ketogenesis is an important survival mechanism that maintains high rates of fatty acid oxidation when carbohydrate stores are depleted, it can also lead to pathologic conditions if acetoacetate and D-β-hydroxybutyrate levels in the blood become too high. **Ketoacidosis** is a condition caused by low blood pH, which can occur when ketogenesis produces more acetoacetate and D-β-hydroxybutyrate than can be used by the peripheral tissues. In patients with undiagnosed diabetes, elevated concentrations of acetoacetate and D-β-hydroxybutyrate in the blood and urine can be several orders of magnitude higher than normal, causing nausea, vomiting, and stomach pain. Moreover, these individuals also have high levels of acetone in their blood (which can be detected on their breath as a fruity odor) and can be delirious or even comatose due to acute hypoglycemia. There have been reports of fatalities of incarcerated individuals with severe ketoacidosis who were mistakenly arrested for public drunkenness and put in isolation for 24 hours.

concept integration 16.1

Describe the physiologic benefits of ketogenesis. Why is it dangerous to give insulin to an emergency room patient with severe ketoacidosis before first checking the patient's blood glucose levels?

Ketogenesis provides an alternative pathway to utilize excess acetyl-CoA in cardiac and skeletal muscle when dietary carbohydrates and protein are limiting. With the body in starvation mode, oxaloacetate is shunted into gluconeogenesis, and flux through the citrate cycle is restricted. The demand to oxidize stored fatty acids as a source of metabolic energy results in a buildup of acetyl-CoA in liver mitochondria. Ketogenesis allows for this valuable fuel to be converted to D-β-hydroxybutyrate and acetoacetate, which are then exported to the blood where they can be used as a source of acetyl-CoA by other tissues in the body. The reason it is critical to first check blood glucose levels in emergency-room patients presenting with severe ketoacidosis is that this condition can be caused by an insulin overdose and yet lead to the same symptoms as those of undiagnosed diabetes; that is, decreased glucose uptake into tissues. The way to distinguish between these two possible causes of severe ketoacidosis is to measure blood glucose levels. If blood glucose levels are extremely high (hyperglycemic), then the ketoacidosis is probably due to the onset of diabetes, and administration of insulin is the proper treatment. However, if glucose levels are low (hypoglycemic), then the ketoacidosis is probably the result of dietary carbohydrate deficiency, and infusion with a high glucose solution is warranted. Giving insulin to a patient with ketoacidosis and hypoglycemia would decrease the patient's blood glucose levels even lower and lead to a life-threatening response.

16.2 Synthesis of Fatty Acids and Triacylglycerols

We noted in Chapter 15 that triacylglycerols stored in adipocytes are derived from both dietary lipids and *de novo* fatty acid synthesis in liver cells (hepatocytes) and adipose tissue. The liver synthesizes triacylglycerols from fatty acids when glucose levels are high and the amount of acetyl-CoA produced exceeds the energy requirements of the cell.

Carbon substrates for fatty acid synthesis are primarily derived from dietary carbohydrates in the form of glucose, which is used to generate acetyl-CoA and glycerol for the production of triacylglycerols. The carbon flow from glucose to fatty acids is illustrated in **Figure 16.17**. Observe that glucose import is the starting point, which is then followed by the conversion of glucose to fructose-1,6-bisphosphate and the production of dihydroxyacetone phosphate and glyceraldehyde-3-phosphate. The dihydroxyacetone phosphate is used to make glycerol-3-phosphate, and the glyceraldehyde-3-phosphate is converted to pyruvate. The pyruvate is then oxidatively decarboxylated by the mitochondrial enzyme pyruvate dehydrogenase to form acetyl-CoA (see Chapter 9 for more details).

Citrate synthase, the first enzyme in the citrate cycle, combines oxaloacetate and acetyl-CoA to generate citrate, which is shuttled to the cytosol. Here it is cleaved by the enzyme **citrate lyase** to generate acetyl-CoA and oxaloacetate. The cytosolic acetyl-CoA is converted to malonyl-CoA by the enzyme acetyl-CoA carboxylase; the malonyl-CoA then serves as the building block for fatty acid synthesis by fatty acid synthase. In liver cells, fatty acids and glycerol are combined to form triacylglycerols, which are packaged into very-low-density lipoprotein (VLDL) particles. The VLDL particles are exported to the circulatory system where most of the triacylglycerol is taken up by adipose tissue and stored in lipid droplets. This series of biochemical processes and enzymatic reactions linking glycolysis, the citrate shuttle, and fatty acid synthesis explains why consuming excess calories in the form of carbohydrates, such as nonfat bagels, can result in increased amounts of stored fat in adipose tissue.

▶ ANIMATION

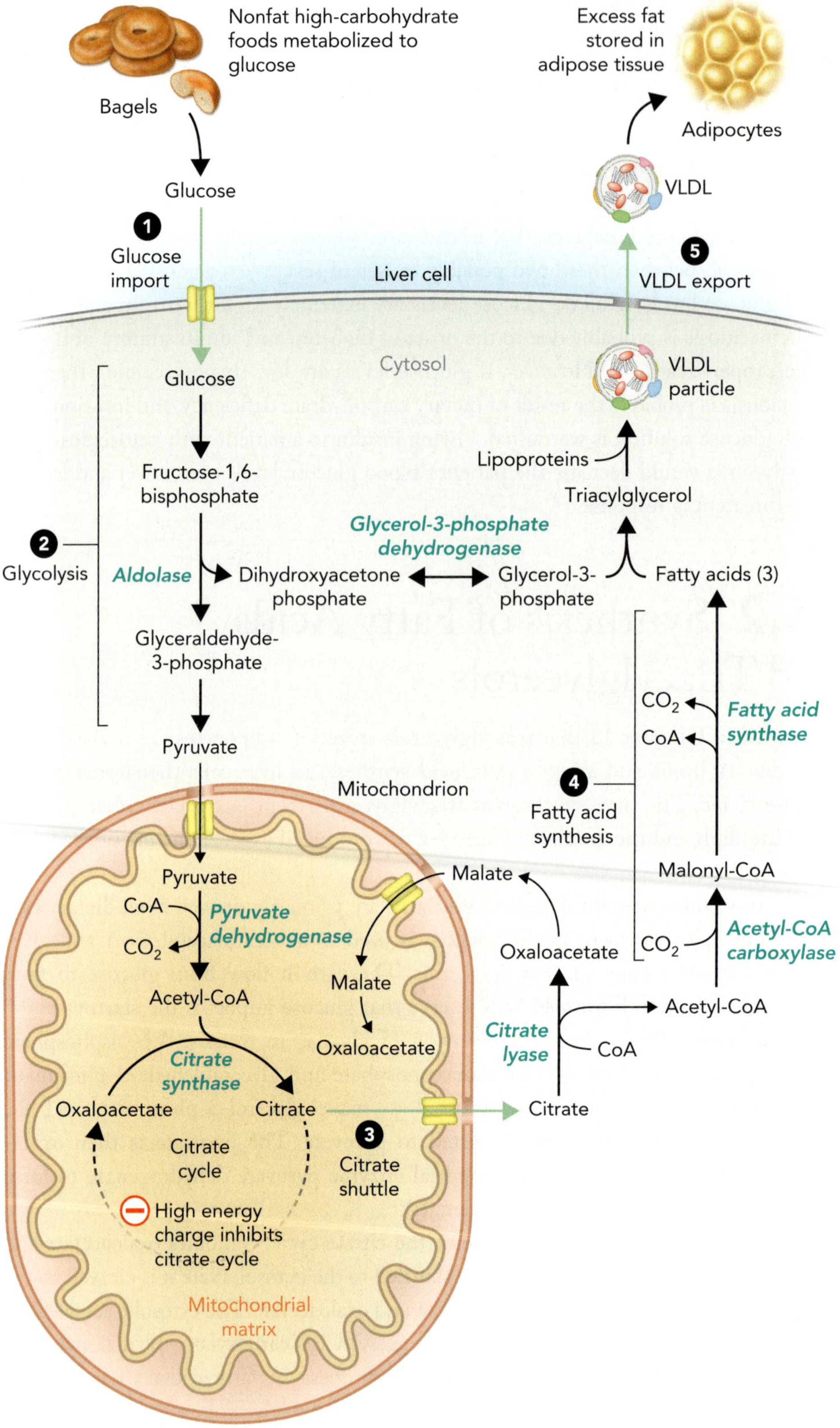

Figure 16.17 Excess glucose from eating too many nonfat high-carbohydrate foods is converted to triacylglycerol in hepatocytes and exported to adipose tissue as VLDL particles. Under conditions of excess carbohydrates in the liver, flux through the citrate cycle is inhibited by high energy charge in the cell (high ATP, low ADP), thereby stimulating citrate export from the mitochondrial matrix.

Fatty Acid Synthase Is a Multifunctional Enzyme

The fatty acid degradation and synthesis pathways are complementary in much the same way that glycolysis and gluconeogenesis are complementary (see Table 14.1). As shown in **Figure 16.18**, both fatty acid degradation and synthesis require a four-step reaction cycle, and each pathway involves the removal or addition of C_2 units attached to coenzyme A. There are also important differences between fatty acid degradation and synthesis in eukaryotes that prevent futile cycling between the two pathways. For example, fatty acid degradation occurs in the mitochondrial matrix and uses FAD and NAD^+ as oxidants, whereas fatty acid synthesis occurs in the cytosol and depends on NADPH serving as a reductant. Moreover, eukaryotic fatty acid degradation requires multiple enzymes for each reaction cycle and uses coenzyme A as the acetyl group anchor. In contrast, once malonyl-CoA is formed from acetyl-CoA and CO_2 by the enzyme acetyl-CoA carboxylase, the reaction cycle in fatty acid synthesis is catalyzed by a single multifunctional enzyme called fatty acid synthase and uses **acyl carrier protein (ACP)** as the hydrocarbon anchor. Lastly, the rate-limiting step in fatty acid degradation is carnitine-mediated transport into the mitochondrial matrix, whereas the rate-limiting step in fatty acid synthesis is generation of the reaction cycle substrate malonyl-CoA.

Fatty acid synthesis is accomplished by repeated reaction cycles within fatty acid synthase, each cycle adding two carbons at a time to the growing fatty acid chain.

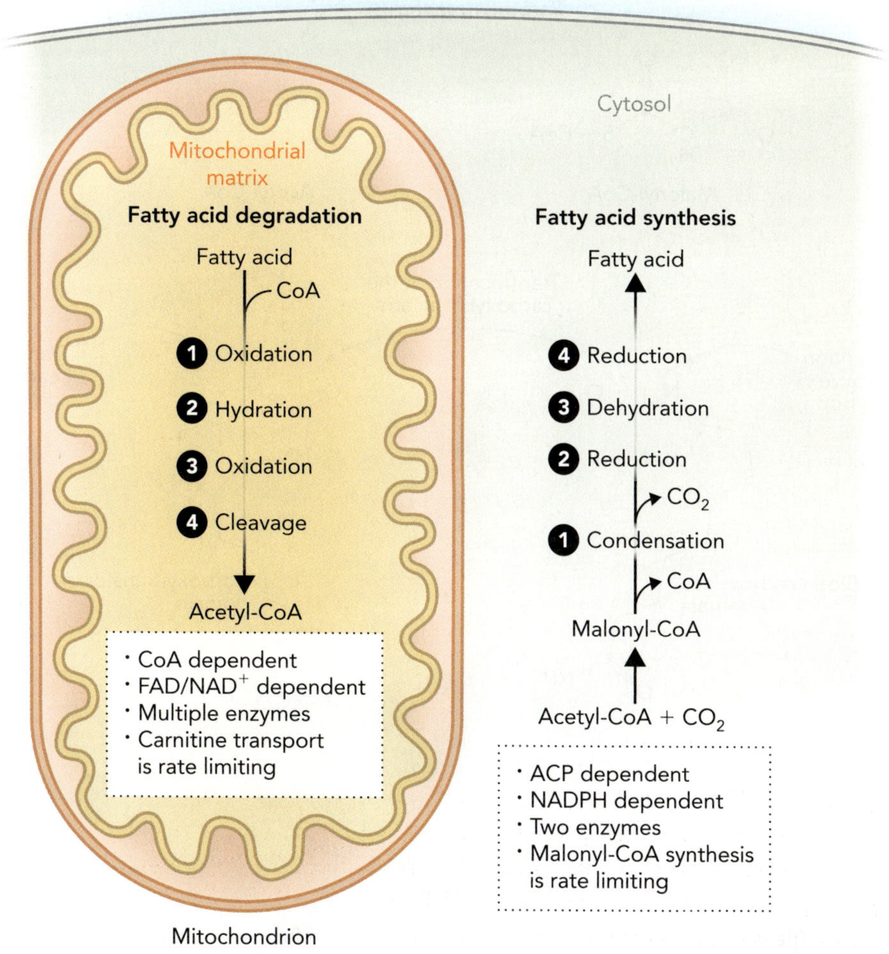

Figure 16.18 The eukaryotic fatty acid degradation and synthesis pathways are complementary with respect to substrates and products; however, there are important differences in these two pathways to avoid futile cycling.

Although acetyl-CoA is used directly as the anchoring primer in the first cycle of fatty acid synthesis, all subsequent cycles derive the two carbons that lengthen the chain from malonyl-CoA. This activated substrate is produced by the rate-limiting enzyme acetyl-CoA carboxylase, which is a biotin-dependent enzyme similar to pyruvate carboxylase. As shown in **Figure 16.19**, acetyl-CoA carboxylase carries out three activities, functioning as a biotin carboxylase, a biotin carrier, and a carboxyltransferase. Similar to fatty acid synthase, the prokaryotic acetyl-CoA carboxylases contain these three functions in a multi-subunit protein complex, whereas the eukaryotic acetyl-CoA carboxylases encode all three functions in a single multifunctional protein.

The first part of the acetyl-CoA carboxylase reaction produces malonyl-CoA. It starts when the biotin carboxylase active site catalyzes an ATP-dependent reaction in which bicarbonate (HCO_3^-) is used to form carboxyphosphate, which is then dephosphorylated to drive the formation of carboxybiotin. The carboxybiotin arm is attached to

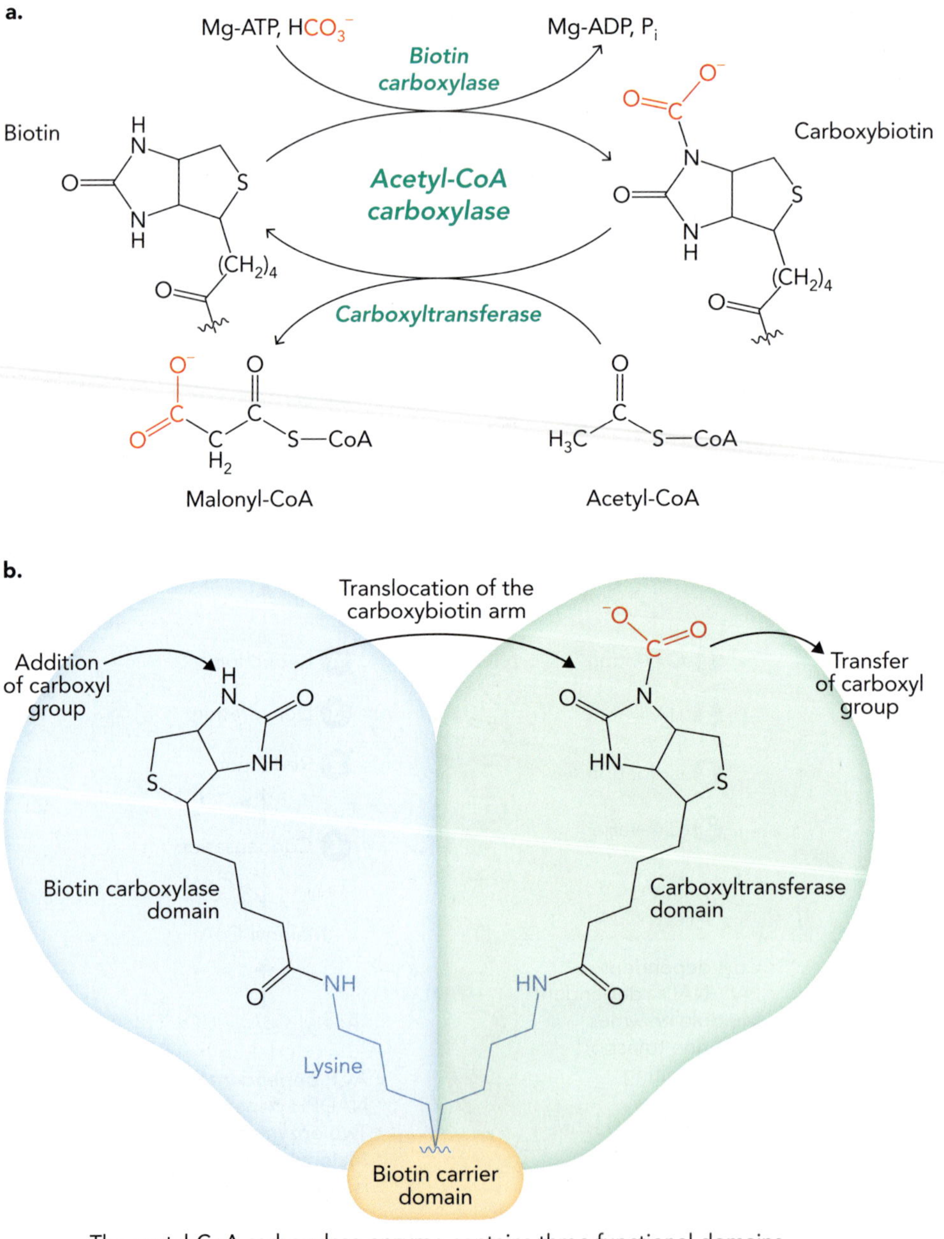

Figure 16.19 The production of malonyl-CoA by acetyl-CoA carboxylase involves a two-step ATP-dependent reaction that carries a carboxyl group from the biotin carboxylase active site to the carboxyltransferase active site using a carboxybiotin arm attached to the biotin carrier protein domain. **a.** The acetyl-CoA carboxylase reaction. **b.** Domain structure of acetyl-CoA carboxylase showing the three functional domains in eukaryotes, which are encoded on three separate polypeptides in prokaryotes.

a lysine residue in the biotin carrier domain, and once carboxybiotin is formed, it translocates 7 Å within the enzyme to the carboxyltransferase active site. In the second step of the reaction, the carboxyl group from carboxybiotin is attached to acetyl-CoA to form the reaction product malonyl-CoA.

Figure 16.20 shows the molecular structure of the biotin carrier protein subunit in a bacterial acetyl-CoA carboxylase enzyme. The attachment of biotin to the lysine side chain acts as a flexible tether to allow movement of the biotin group between different protein subunits. As you will see shortly, the use of a nonprotein tether to create a flexible extension is a strategy also used by the ACP subunit of fatty acid synthase.

Figure 16.21 shows the four conserved chemical reactions of the enzyme fatty acid synthase, which are required in each cycle of the fatty acid synthesis pathway. All of the carbons in this pathway are directly or indirectly derived from acetyl-CoA. Observe in Figure 16.21 that the first reaction in the fatty acid synthesis cycle is a condensation reaction catalyzed by the β-ketoacyl-ACP synthase (KS) subunit.

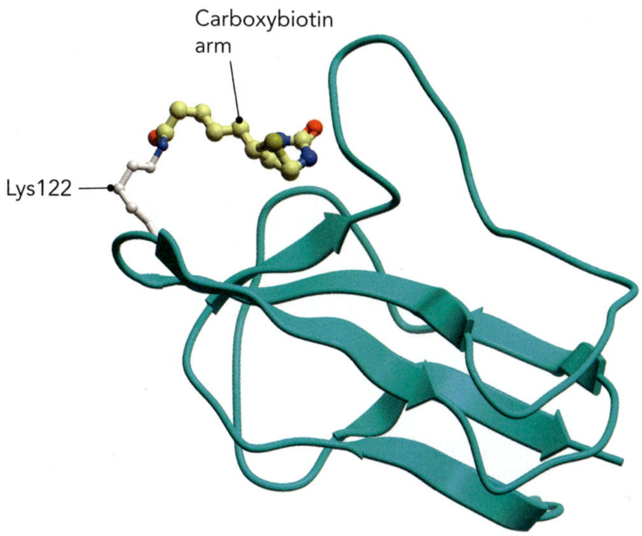

Figure 16.20 The biotin carrier domain of acetyl-CoA carboxylase contains a biotin group covalently attached to a lysine residue, as shown here in the molecular structure of the *E. coli* enzyme. BASED ON PDB FILE 1BDO.

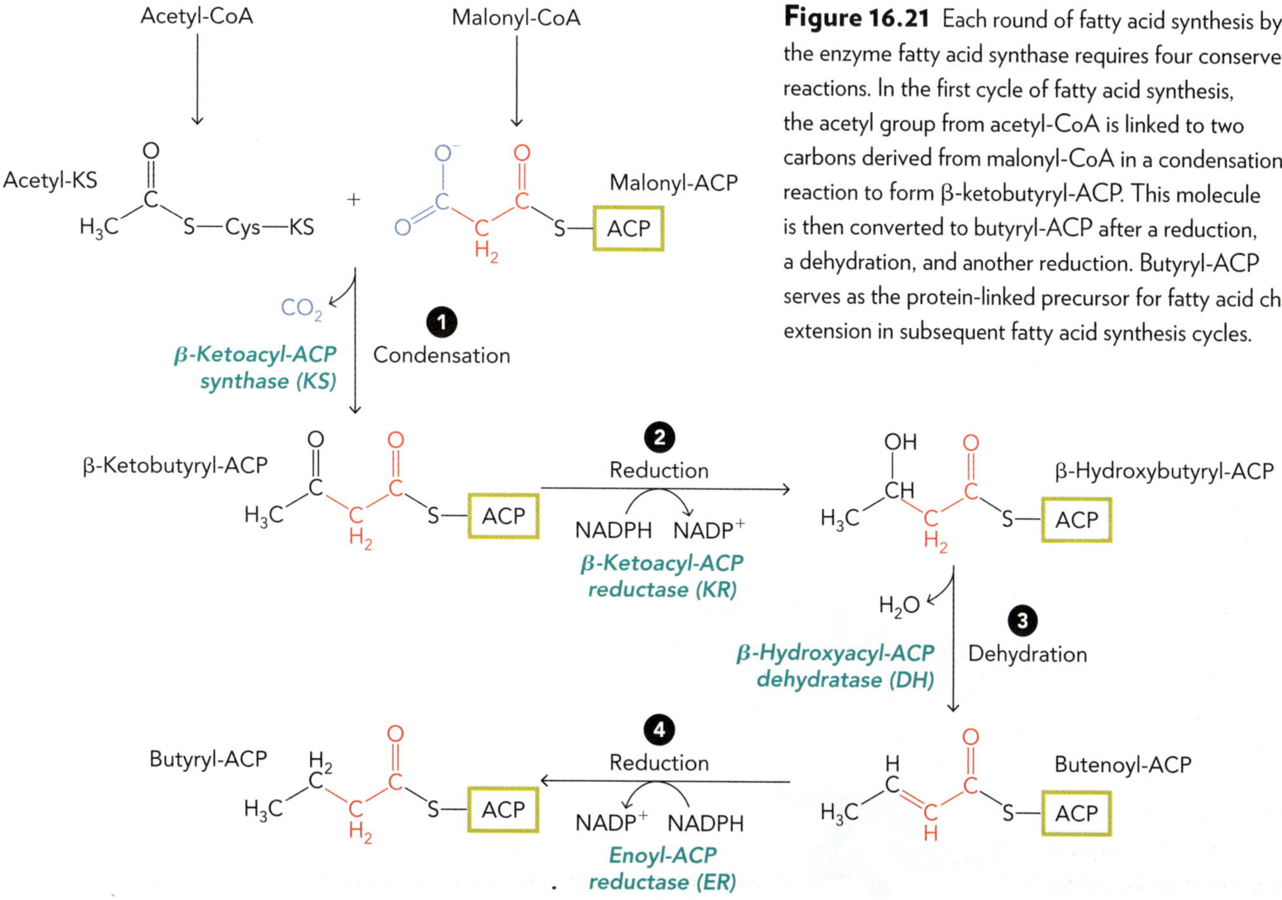

Figure 16.21 Each round of fatty acid synthesis by the enzyme fatty acid synthase requires four conserved reactions. In the first cycle of fatty acid synthesis, the acetyl group from acetyl-CoA is linked to two carbons derived from malonyl-CoA in a condensation reaction to form β-ketobutyryl-ACP. This molecule is then converted to butyryl-ACP after a reduction, a dehydration, and another reduction. Butyryl-ACP serves as the protein-linked precursor for fatty acid chain extension in subsequent fatty acid synthesis cycles.

This reaction involves decarboxylation of malonyl-ACP and formation of β-keto-butyryl-ACP after transfer of the cysteine-linked acetyl group from KS to ACP. In the second reaction, the β-ketoacyl-ACP reductase (KR) subunit of fatty acid synthase catalyzes a reduction using NADPH as the electron donor to generate β-hydroxybutyryl-ACP. This compound is then dehydrated in the third step by the β-hydroxyacyl-ACP dehydratase (DH) subunit of fatty acid synthase, producing butenoyl-ACP. In the final step, butenoyl-ACP is reduced by NADPH in a reaction catalyzed by the enoyl-ACP reductase (ER) subunit of fatty acid synthase to generate butyryl-ACP. The butyryl-ACP product is used in the next cycle of fatty acid synthesis, when the condensation reaction incorporates another two carbons derived from malonyl-CoA.

The ACP contains a nonprotein tether—functionally similar to the tether in acetyl-CoA carboxylase—that has sufficient flexibility to position the growing fatty acid substrate in the different subunits of fatty acid synthase. In this case, the enzyme-linked coenzyme that serves as a tether is phosphopantetheine, which is derived from vitamin B_5 (pantothenate) and also a component of coenzyme A (**Figure 16.22**). This

a.

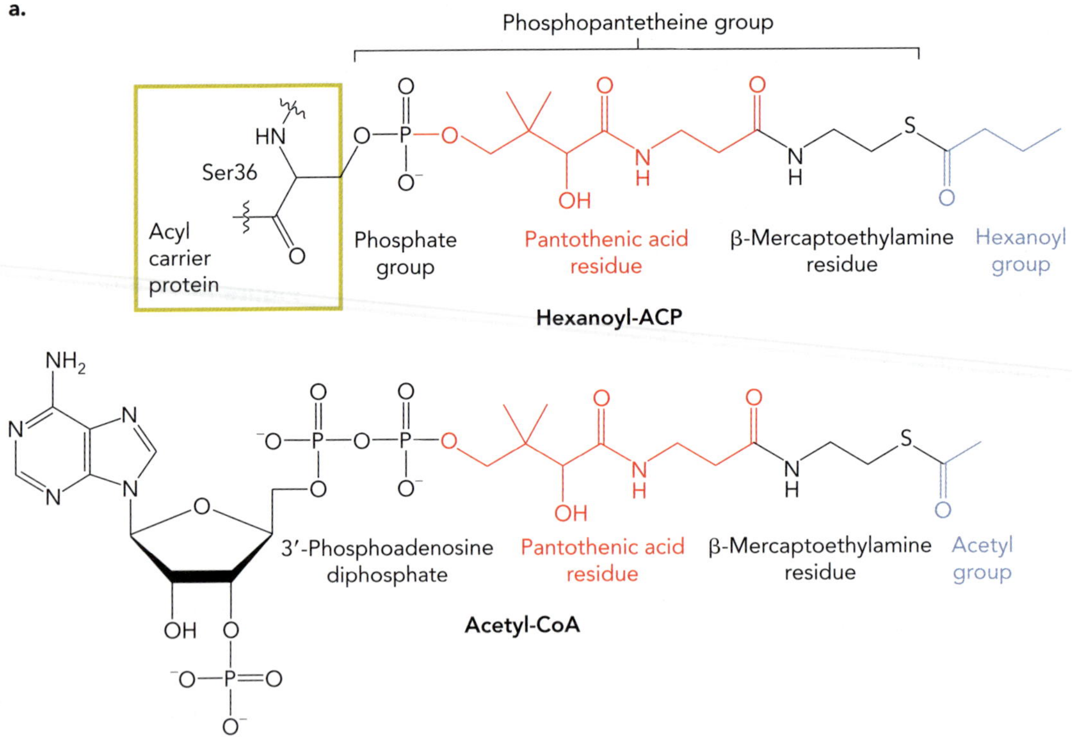

b.

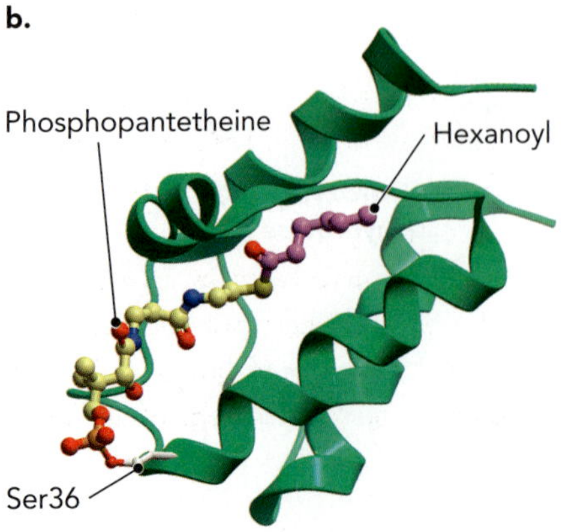

Figure 16.22 Acyl carrier protein functions as the attachment site for the growing hydrocarbon chain within the fatty acid synthase complex. **a.** Chemical structures of the phosphopantetheine groups in ACP and coenzyme A. **b.** Structure of *E. coli* ACP, showing the phosphopantetheine group attached to Ser36 in the protein and a C_6 hexanoyl hydrocarbon chain (red) linked to the sulfhydryl group. BASED ON PDB FILE 2FAC.

phosphopantetheine in ACP provides a sulfhydryl attachment site for the growing hydrocarbon chain in the fatty acid synthase complex.

Figure 16.23 shows the mammalian fatty acid synthase enzyme. It contains ~2,500 amino acids (~275 kDa), with seven protein functions encoded in a single polypeptide chain. On the basis of the molecular structure of the fatty acid synthase enzyme isolated from *Sus scrofa* (pig) shown in Figure 16.23b, it is observed that the enzyme is a dimeric protein with each monomer containing two functional lobes. The upper lobe of the fatty acid synthase complex contains the three modifying functions in the fatty acid synthesis cycle (KR, DH, and ER). The lower lobe contains the two condensing functions (KS and MAT). The ACP and palmitoyl thioesterase (TE) subunits, which are not considered components of the five enzymatic functions required for fatty acid synthesis, are not resolved in the structure shown here. As shown in Figure 16.23c, the growing fatty acid chain linked to the phosphopantetheine group on ACP sequentially interacts with all five enzyme activities contained in the modifying and condensing lobes (MAT, KS, KR, DH, ER). Once palmitoyl-ACP is generated, the phosphopantetheine group swings around and binds to the TE domain, which catalyzes a hydrolysis reaction to release palmitate as the final product.

The first step in the fatty acid synthesis pathway requires malonyl/acetyl-CoA ACP transacylase (MAT) to catalyze the priming reaction, which initially transfers the acetyl group from acetyl-CoA to the sulfhydryl group in ACP (**Figure 16.24**). This acetyl group primer is then linked to a cysteine residue in the KS domain to free up the phosphopantetheine group on ACP for the first malonyl group. The acetyl-CoA priming reaction on KS occurs only once for each palmitate product, as all subsequent reaction cycles use malonyl-CoA as the donor to produce malonyl-ACP. As seen in Figure 16.24, the MAT domain in the fatty acid synthase complex also catalyzes this malonyl transfer reaction that generates malonyl-ACP. With substrates attached to both the KS domain (acetyl-S-Cys) and ACP (malonyl-ACP), the four-step reaction sequence is initiated: (1) condensation catalyzed by KS; (2) reduction catalyzed by KR; (3) dehydration catalyzed by DH; and (4) reduction catalyzed by ER. In the final reaction of each synthesis cycle, the fully reduced product on ACP is translocated to the Cys residue on the KS subunit. The four-reaction synthesis cycle starts again when a new malonyl group is attached to ACP. As shown in **Figure 16.25** (p. 799), eight

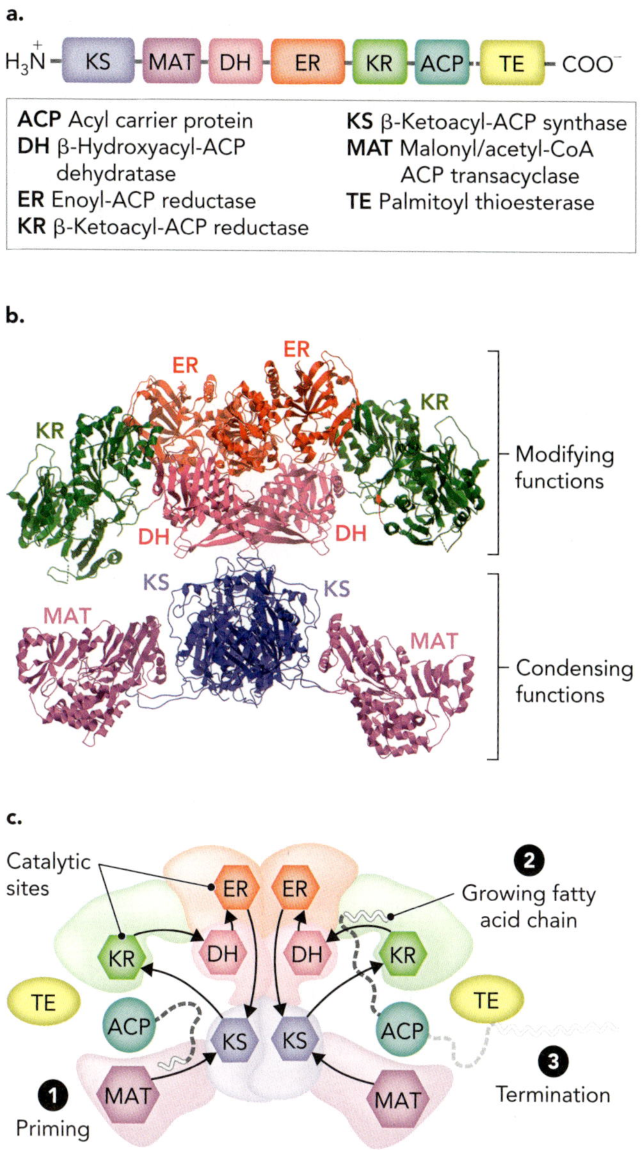

Figure 16.23 The mammalian fatty acid synthase complex has multiple functional domains. **a.** All five of the enzymatic functions (MAT, KS, KR, DH, ER) required for fatty acid synthesis, along with the two specialized domains (ACP, TE), are encoded on a single polypeptide of ~2,500 amino acid residues in mammals. Domain abbreviations are listed. **b.** Molecular structure of the dimeric fatty acid synthase from *Sus scrofa* (pig). The five enzymatic domains in this dimeric structure are labeled and color coded to match the domains in panel **a**. Note that the ACP and TE domains were not resolved in this structure; however, they are probably located between the modifying and condensing functional lobes. BASED ON PDB FILE 2VZ8. **c.** Proposed model of sequential interactions between the substrate on the phosphopantetheine group on ACP and the five enzymatic domains prior to cleavage of the final product by TE.

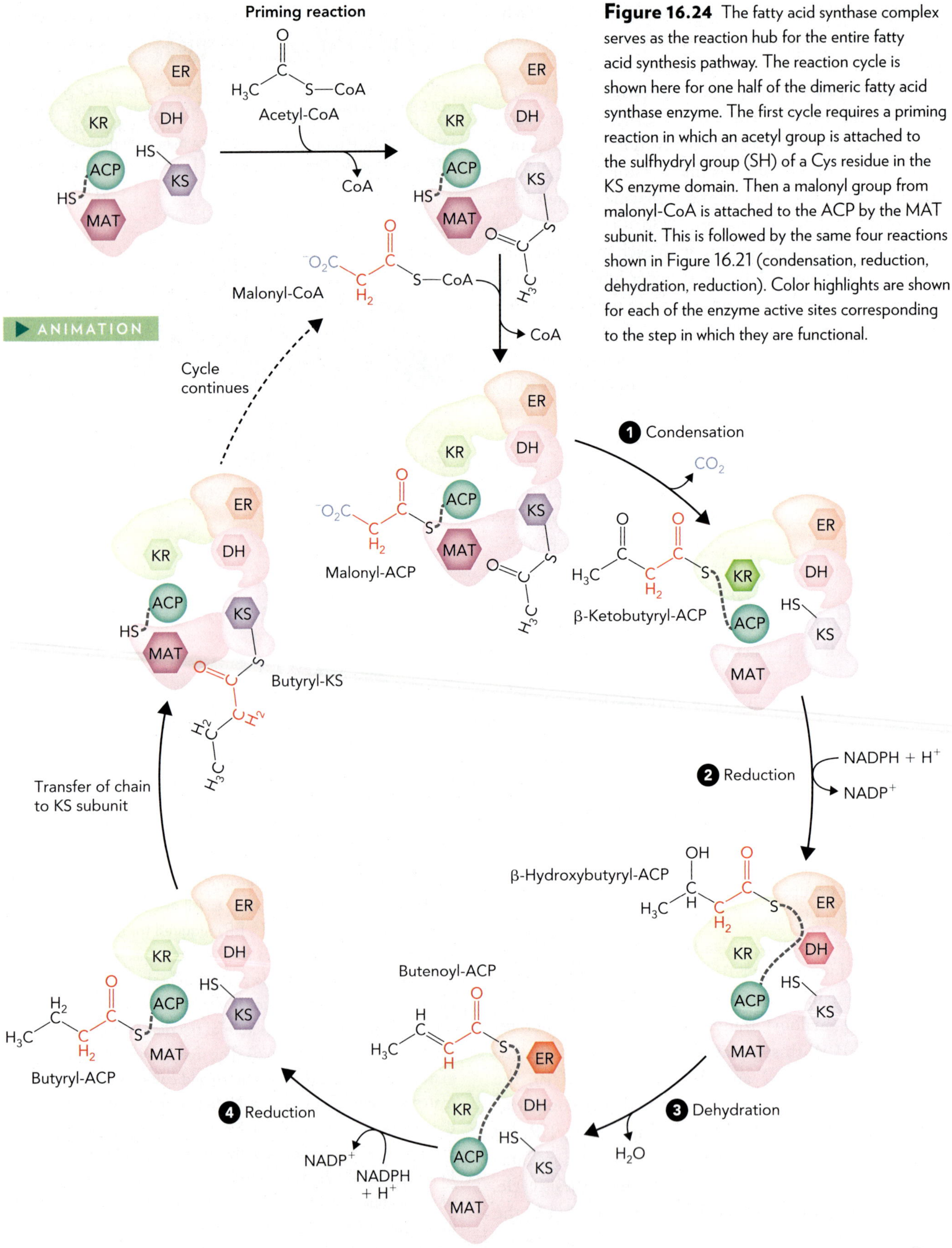

Figure 16.24 The fatty acid synthase complex serves as the reaction hub for the entire fatty acid synthesis pathway. The reaction cycle is shown here for one half of the dimeric fatty acid synthase enzyme. The first cycle requires a priming reaction in which an acetyl group is attached to the sulfhydryl group (SH) of a Cys residue in the KS enzyme domain. Then a malonyl group from malonyl-CoA is attached to the ACP by the MAT subunit. This is followed by the same four reactions shown in Figure 16.21 (condensation, reduction, dehydration, reduction). Color highlights are shown for each of the enzyme active sites corresponding to the step in which they are functional.

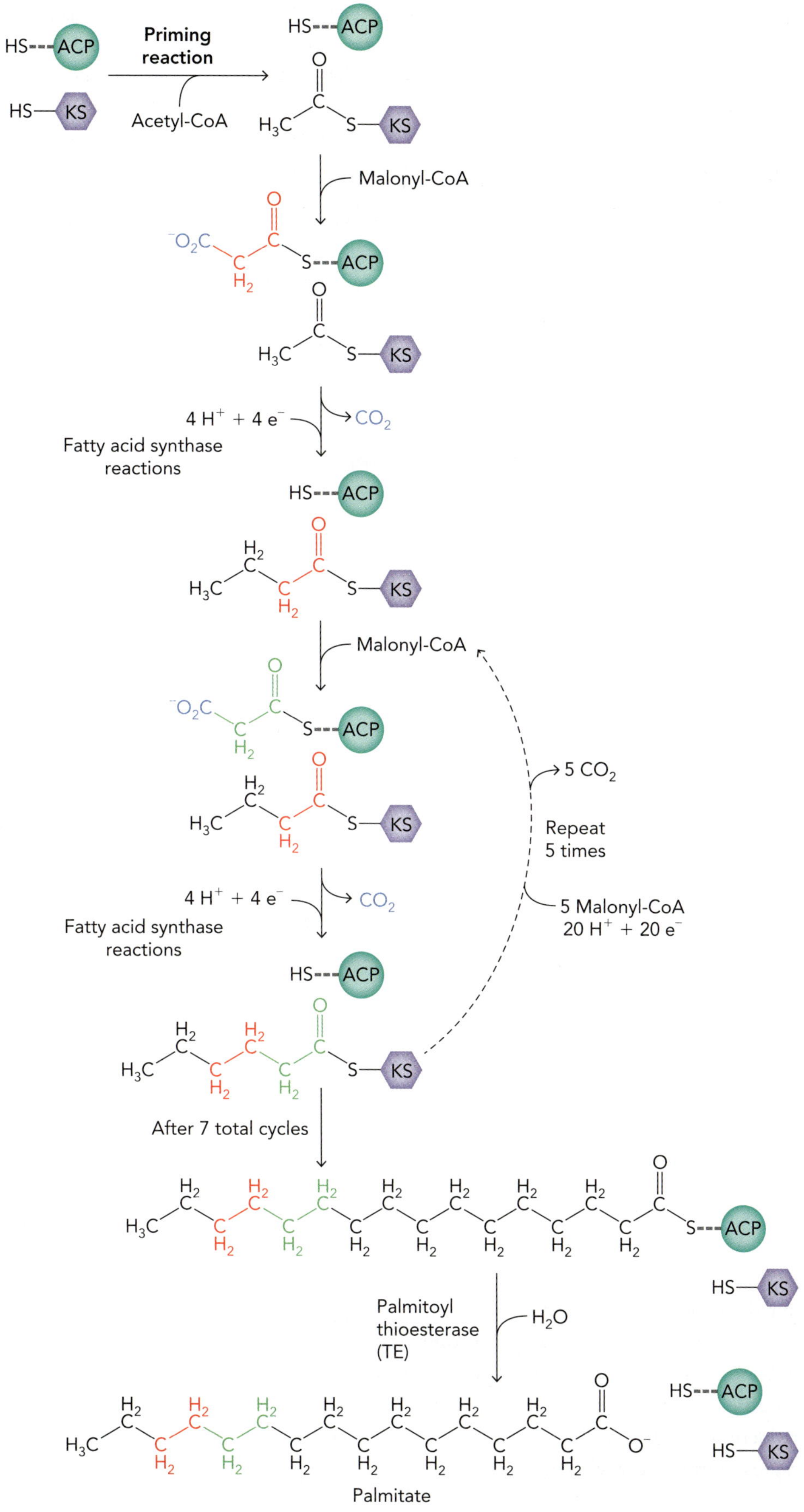

Figure 16.25 An overview of the fatty acid synthesis cycle resulting in the production of palmitate is shown here. Each cycle of the fatty acid synthase reactions elongates the chain by two carbons. Seven rounds of the cycle are required to produce C_{16} palmitate. Each round of the cycle uses 4 H^+ and 4 e^- (from NADPH and H^+) and releases 1 CO_2. In the final round of the cycle, the chain is not transferred to KS, but is instead hydrolyzed from ACP, releasing palmitate.

acetyl/malonyl-CoA molecules and seven turns of the fatty acid synthesis cycle are needed to generate the C_{16} pathway product palmitate $[CH_3(CH_2)_{14}CO_2^-]$. The final product of seven reaction cycles is palmitoyl-ACP, which is released as palmitate from ACP after hydrolysis by TE.

We can now calculate the ATP and NADPH requirements for synthesizing one molecule of the C_{16} fatty acid palmitate from eight molecules of the C_2 metabolite acetyl-CoA. We begin by forming seven molecules of malonyl-CoA using the acetyl-CoA carboxylase reaction:

$$7 \text{ Acetyl-CoA} + 7\ CO_2 + 7 \text{ ATP} \rightarrow 7 \text{ Malonyl-CoA} + 7 \text{ ADP} + 7\ P_i$$

We then use these seven malonyl-CoA molecules for seven turns of the reaction cycle, beginning with the priming of fatty acid synthase by one molecule of acetyl-CoA:

$$1 \text{ Acetyl-CoA} + 7 \text{ Malonyl-CoA} + 14 \text{ NADPH} + 14\ H^+ \rightarrow \text{Palmitate} + 7\ CO_2 + 8 \text{ CoA} + 14 \text{ NADP}^+ + 6\ H_2O$$

The net reaction can then be written as

$$8 \text{ Acetyl-CoA} + 7 \text{ ATP} + 14 \text{ NADPH} + 14\ H^+ \rightarrow \text{Palmitate} + 8 \text{ CoA} + 7 \text{ ADP} + 7\ P_i + 14 \text{ NADP}^+ + 6\ H_2O$$

The 14 NADPH molecules required to synthesize one molecule of palmitate come primarily from the pentose phosphate pathway (see Chapter 14), although some NADPH is also generated by reactions in the citrate shuttle, as described later. Note that although 7 H_2O are produced by the dehydration reaction during the seven rounds of synthesis, one H_2O molecule is required for the final TE-mediated cleavage reaction releasing palmitate from ACP (see Figure 16.25).

Elongation and Desaturation of Palmitate

The primary product of the fatty acid synthase reactions in most cells is palmitate. This molecule is then used as a substrate for both elongation and desaturation reactions to produce a variety of saturated and unsaturated fatty acids.

Enzymes that elongate palmitate exist in both the mitochondria and as membrane components of the endoplasmic reticulum (ER). The ER elongation reactions are most active in animal cells and use the same types of reactions used by fatty acid synthase to add C_2 units to a fatty acyl substrate. The primary difference between the ER elongation enzymes and fatty acid synthase is that the coenzyme A moiety of malonyl-CoA functions as the carrier molecule rather than ACP.

As shown in **Figure 16.26**, palmitoyl-CoA is formed from palmitate in a reaction requiring the enzyme fatty acyl-CoA synthetase, which we described earlier in our discussion of fatty acid oxidation (see Figure 16.2). Palmitoyl-CoA and malonyl-CoA are then linked through a condensation reaction catalyzed by one of several ER-localized fatty acyl elongation enzymes. The product of this reaction is then modified by the same three enzyme activities used in fatty acid synthesis; namely, a reduction, a dehydration, and a second reduction. In the final step, coenzyme A is removed from stearoyl-CoA to produce stearate.

The desaturating enzymes in animal cells are membrane-bound ER proteins that use molecular oxygen (O_2) as the oxidant. These desaturating enzymes are called **mixed-function oxidases** because they use O_2 in the reaction mechanism. However, as seen in **Figure 16.27**, the O_2 is not incorporated into the fatty acid product, but rather serves only as a strong oxidant to strip 2 e^- from the fatty acid substrate. In this example, stearoyl-CoA, 18:0, is desaturated at C-9 to form oleoyl-CoA, *cis* 18:1(Δ^9). This desaturation reaction requires three proteins embedded in the endoplasmic reticulum membrane: cytochrome b_5 reductase, cytochrome b_5, and stearoyl-CoA 9-desaturase. Together, these coupled redox reactions link the oxidation of a fatty acyl-CoA and NADH with the reduction of O_2 to form 2 H_2O.

Animal cells lack desaturating enzymes that add C=C bonds distal to C-9. This explains why the polyunsaturated fatty acids linoleate, 18:2($\Delta^{9,12}$), and α-linolenate, 18:3($\Delta^{9,12,15}$), are essential fatty acids for humans and must be obtained in the diet (see Figure 15.7).

Synthesis of Triacylglycerol and Membrane Lipids

Fatty acids are the building blocks of triacylglycerols and also the hydrocarbon components of the most common membrane lipids; namely, glycerophospholipids and sphingolipids. Both triacylglycerols and glycerophospholipids are derived from diacylglycerol-3-phosphate, also called phosphatidic acid. Sphingolipids, however, are derived from ceramide, as described in Chapter 15.

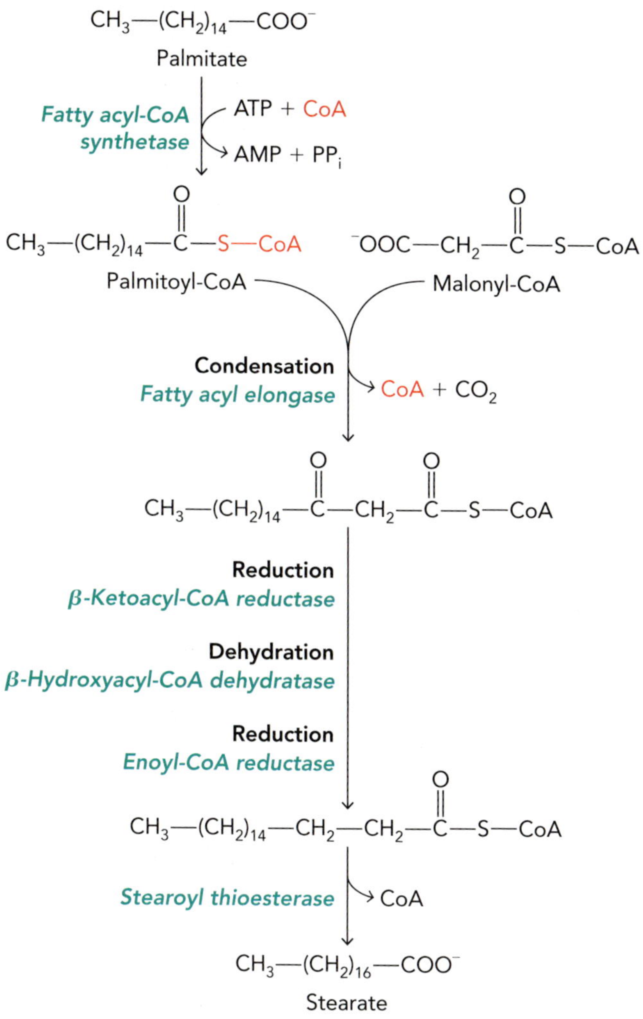

Figure 16.26 Elongase enzymes in the endoplasmic reticulum add C_2 units to palmitoyl-CoA, generating long-chain fatty acids.

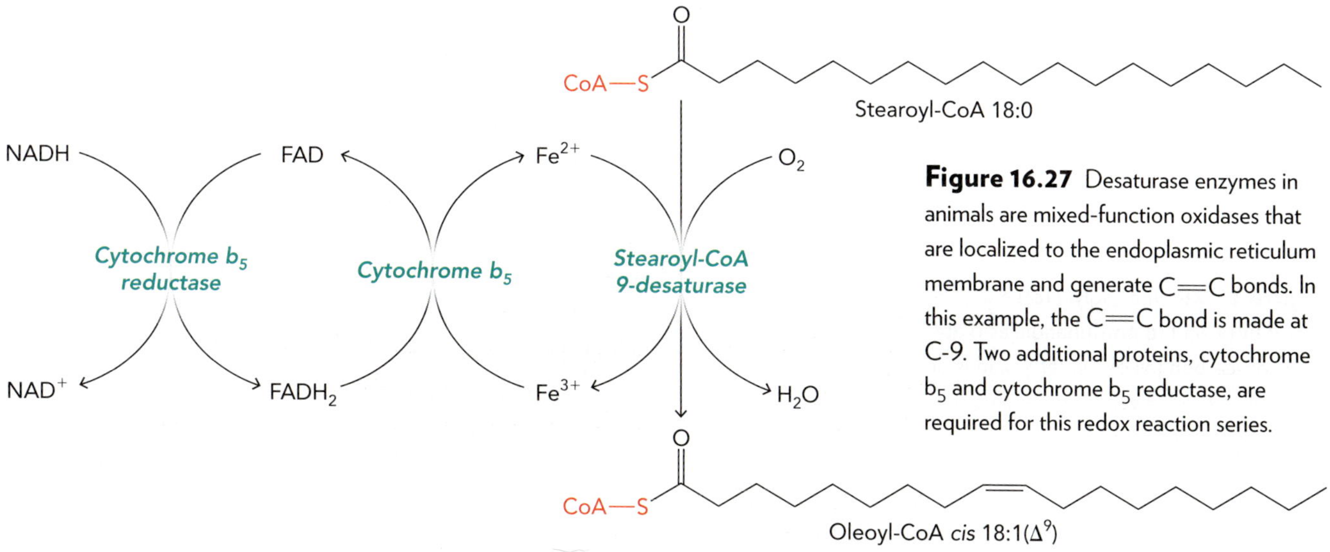

Figure 16.27 Desaturase enzymes in animals are mixed-function oxidases that are localized to the endoplasmic reticulum membrane and generate C=C bonds. In this example, the C=C bond is made at C-9. Two additional proteins, cytochrome b_5 and cytochrome b_5 reductase, are required for this redox reaction series.

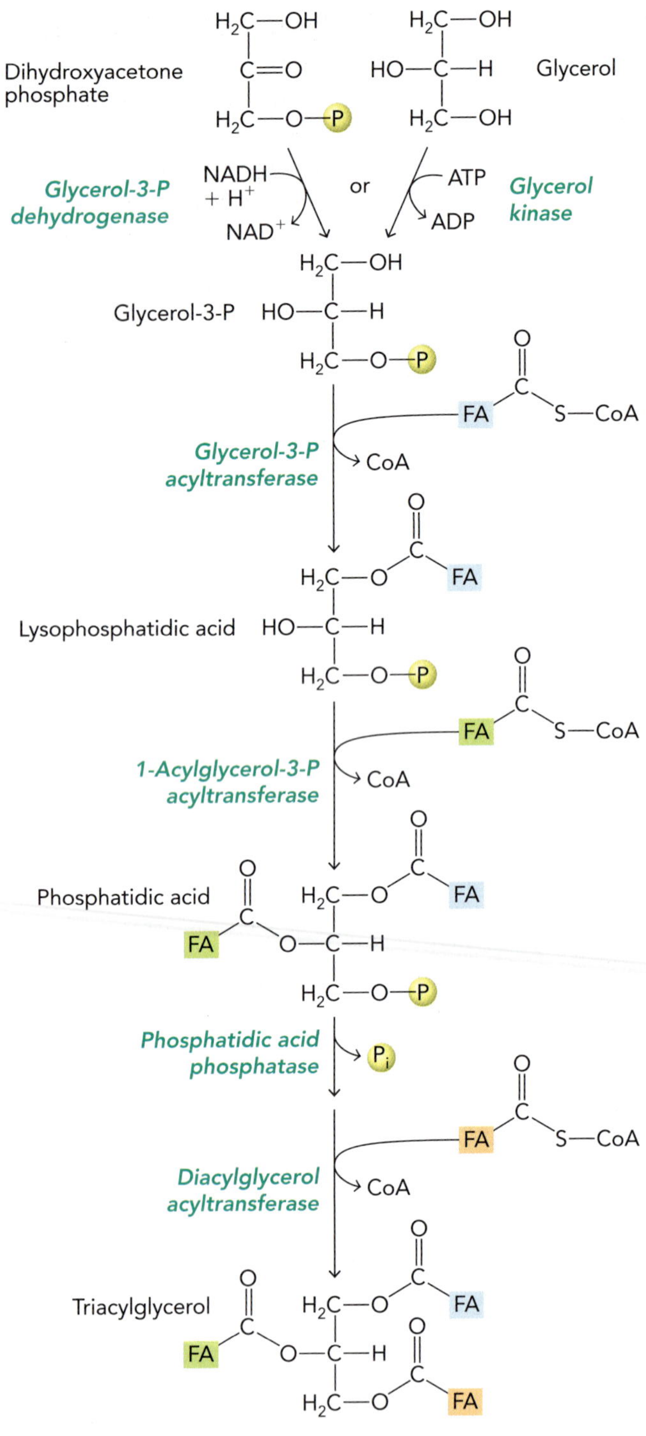

Figure 16.28 Triacylglycerols are generated by attaching a third fatty acid to phosphatidic acid at C-3. For simplicity, the hydrocarbon tail of the fatty acid is denoted by "FA." Glycerol-3-P can be generated from reduction of dihydroxyacetone phosphate or phosphorylation of glycerol.

As shown in **Figure 16.28**, triacylglycerols are generated from phosphatidic acid, which contains two fatty acids linked to C-1 and C-2 of glycerol-3-phosphate (glycerol-3-P). Phosphatidic acid is formed using fatty acyl-CoA as the substrate and the enzymes glycerol-3-P acyltransferase and 1-acylglycerol-3-P acyltransferase. The glycerol-3-P used to make phosphatidic acid is derived from either reduction of dihydroxyacetone phosphate by the enzyme glycerol-3-P dehydrogenase or by phosphorylation of glycerol by the enzyme glycerol kinase. Triacylglycerols are formed from phosphatidic acid by first dephosphorylating C-3 with the enzyme phosphatidic acid phosphatase, and then the enzyme diacylglycerol acyltransferase adds the third fatty acid to glycerol, yielding a triacylglycerol.

Membrane lipids are synthesized by enzymes associated with the smooth endoplasmic reticulum and must be transported to various membrane targets in the cell. **Figure 16.29** shows the synthesis of three common glycerophospholipids from phosphatidic acid (phosphatidylserine, phosphatidylethanolamine, and phosphatidylinositol). In the first step, the enzyme cytidine diphosphate (CDP)-diacylglycerol synthase activates phosphatidic acid by addition of cytidine monophosphate to the phosphate group on C-3 in a reaction that uses cytidine triphosphate and releases pyrophosphate. This nucleotide activation of phosphatidic acid is analogous to glucose activation by uridine triphosphate in the synthesis of glycogen (see Figure 14.35).

In the formation of phosphatidylserine, cytidine monophosphate is replaced by serine through a reaction catalyzed by the enzyme phosphatidylserine synthase. Decarboxylation of phosphatidylserine by the enzyme phosphatidylserine decarboxylase generates the cationic glycerophospholipid phosphatidylethanolamine. CDP-diacylglycerol is also a substrate for the enzyme phosphatidylinositol synthase, which replaces the nucleotide with inositol to generate the membrane lipid phosphatidylinositol. The hydroxyl groups on C-4 and C-5 of the sugar are phosphorylated by phosphatidylinositol kinases to generate phosphatidylinositol-4,5-bisphosphate (PIP_2), which functions in several signal transduction pathways.

As with glycerophospholipids, sphingolipids are also synthesized by enzymes associated with the smooth

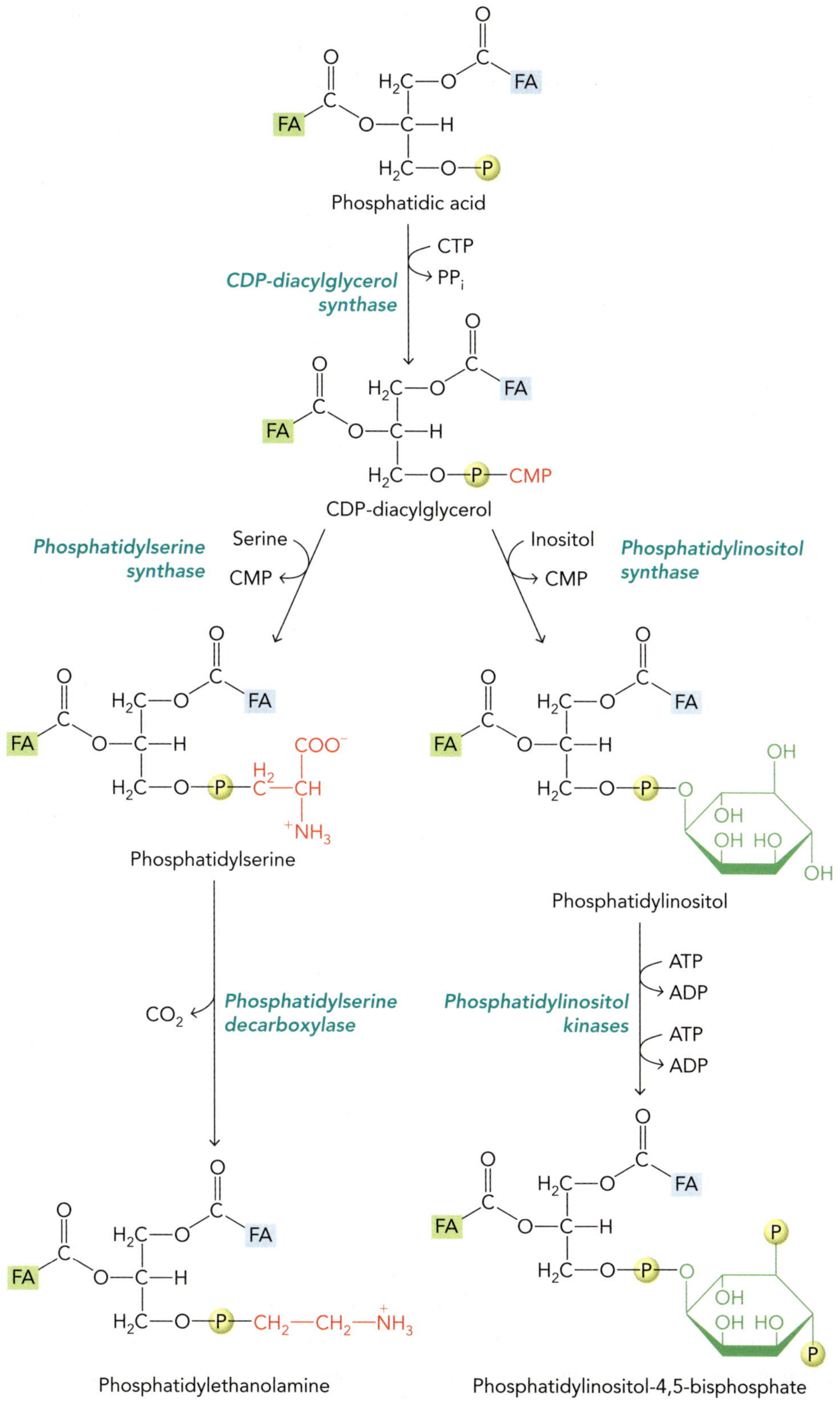

Figure 16.29 Phosphatidylserine, phosphatidylethanolamine, and phosphatidylinositol are synthesized from phosphatidic acid. Phosphorylation of phosphatidylinositol by phosphatidylinositol kinases generates phosphatidylinositol-4,5-bisphosphate, which is an important membrane lipid involved in cell signaling. CMP = cytidine-5′-monophosphate; CTP = cytidine triphosphate.

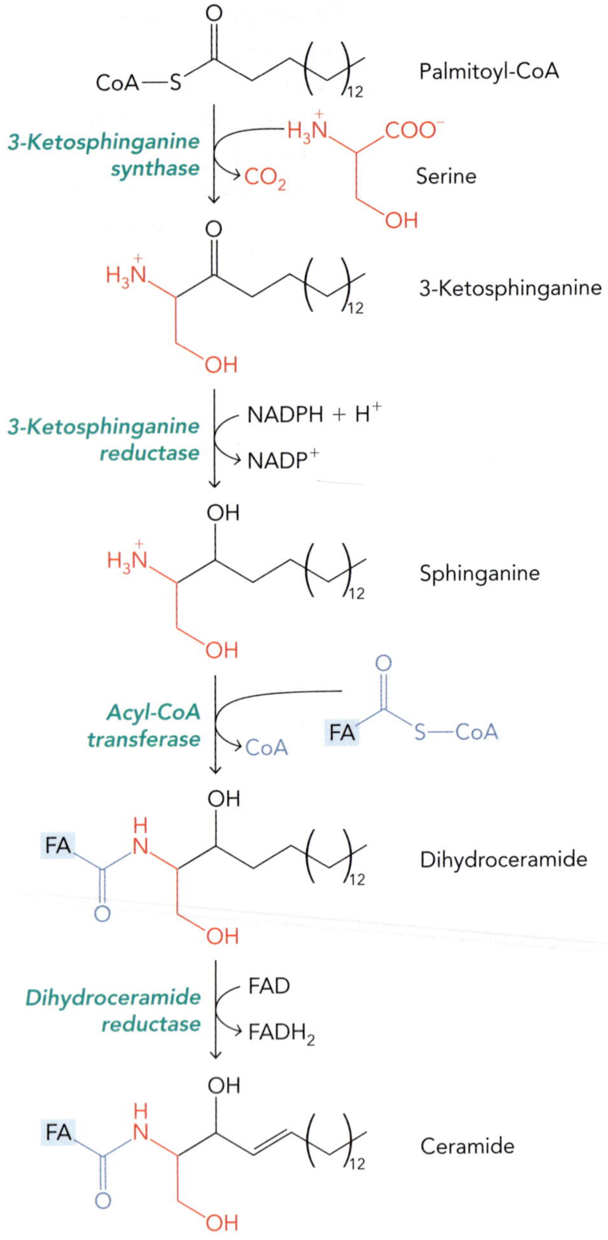

Figure 16.30 Ceramide is a fatty acylated form of sphinganine and is derived from palmitoyl-CoA and the amino acid serine.

endoplasmic reticulum. **Figure 16.30** shows an example of a sphingolipid synthesis pathway. First, the enzyme 3-ketosphinganine synthase catalyzes a reaction in which palmitoyl-CoA combines with the amino acid serine to form 3-ketosphinganine. Second, this sphingolipid intermediate is then reduced by NADPH to generate sphinganine in a reaction catalyzed by the enzyme 3-ketosphinganine reductase. Third, an acyl-CoA transferase enzyme adds a fatty acid to sphinganine through an amide linkage, creating *N*-acylsphinganine (dihydroceramide). Finally, dihydroceramide is oxidized by the enzyme dihydroceramide reductase in the presence of FAD to produce ceramide.

As shown in **Figure 16.31**, the enzyme sphingomyelin synthase can attach a phosphocholine head group derived from phosphatidylcholine to ceramide. This reaction leads to the production of sphingomyelin. Ceramide can also be modified through the addition of monosaccharides, such as glucose, by the enzyme ceramide glucosyltransferase, generating the membrane lipid glucocerebroside. The enzyme lactosylceramide synthase catalyzes a modifying reaction that adds galactose to glucocerebroside. This reaction generates lactosylceramide, which is a precursor for gangliosides. As described in Chapter 15, sphingomyelin is a sphingophospholipid, whereas cerebrosides and gangliosides are sphingoglycolipids (see Figure 15.26).

The Citrate Shuttle Exports Acetyl-CoA from Matrix to Cytosol

The majority of acetyl-CoA used for fatty acid synthesis in the cytosol is derived from the pyruvate dehydrogenase reaction, which takes place in the mitochondrial matrix (see Figure 16.17). However, because mitochondria do not contain an "acetyl-CoA transporter," the acetate units are exported to the cytosol via the **citrate shuttle**. As shown in **Figure 16.32** (p. 806), the citrate shuttle provides a mechanism to stimulate fatty acid synthesis in the cytosol when acetyl-CoA accumulates in the mitochondrial matrix. This buildup of mitochondrial acetyl-CoA occurs when glucose levels are high and the citrate cycle is feedback-inhibited by a high energy charge in the cell.

Once in the cytosol, the citrate is cleaved by citrate lyase in an ATP-dependent reaction to generate cytosolic acetyl-CoA and oxaloacetate. The acetyl-CoA is used for fatty acid synthesis, and the oxaloacetate is converted to malate by cytosolic malate dehydrogenase. The malate is then either transported to the mitochondrial matrix, where it is reduced by mitochondrial malate dehydrogenase, or it is oxidized and decarboxylated by malic enzyme in the cytosol to form pyruvate, NADPH, and CO_2. The

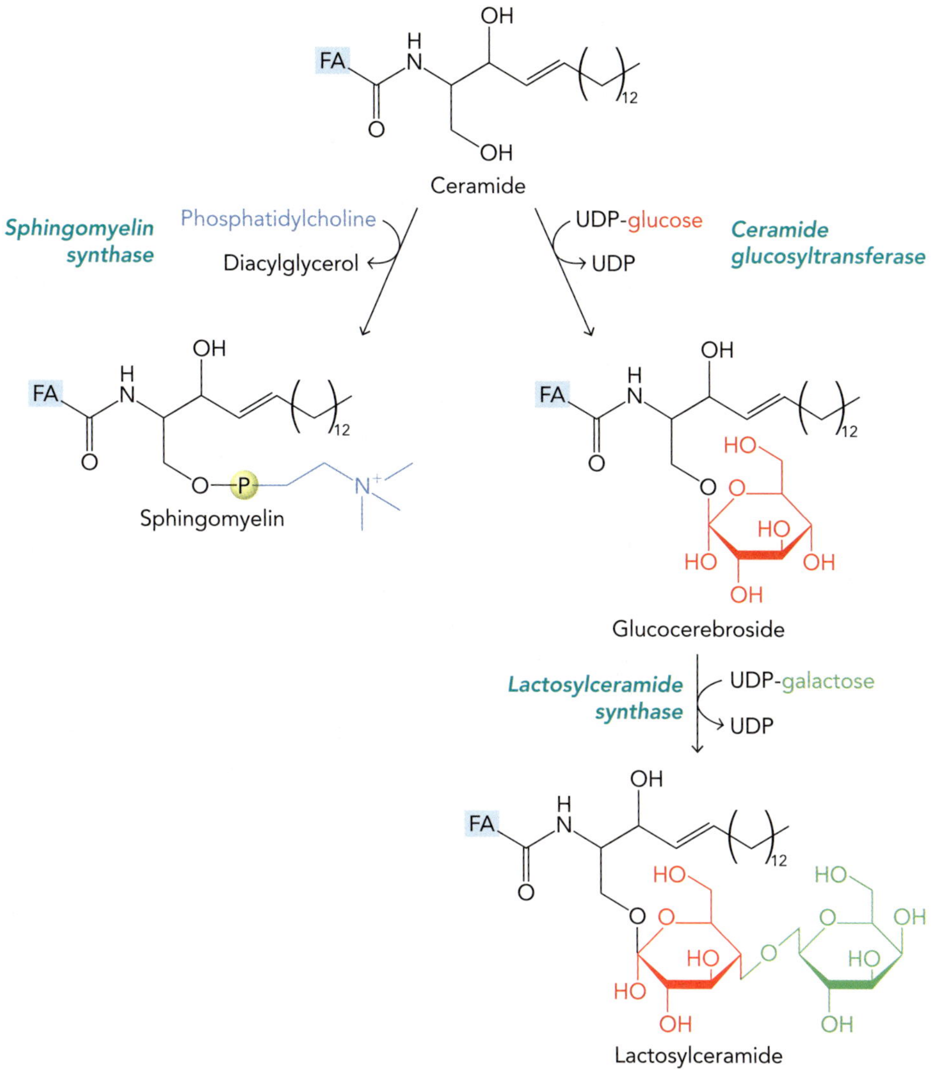

Figure 16.31 The two major classes of sphingolipids are sphingophospholipids, such as sphingomyelin, and sphingoglycolipids, which include glucocerebroside and lactosylceramide. They are all derived from ceramide.

pair of electrons used to oxidize $NADP^+$ in this reaction comes from the NADH-mediated reduction of oxaloacetate to form malate. Because flux through the glycolytic pathway is increased when glucose levels are elevated, much of this cytosolic NADH is generated by the glyceraldehyde-3-phosphate dehydrogenase step. Importantly, the production of cytosolic NADPH by malic enzyme provides additional reducing equivalents for fatty acid synthesis and supplements the NADPH generated by the pentose phosphate pathway. The pyruvate formed by malic enzyme is transported back into the mitochondrial matrix, where it is carboxylated by pyruvate carboxylase to form oxaloacetate. Note that although malic enzyme generates NADPH for fatty acid synthesis, the pyruvate carboxylase reaction consumes an extra ATP, making this a slightly more energetically expensive route back to oxaloacetate.

Metabolic and Hormonal Control of Fatty Acid Synthesis

The primary control point for regulating flux through the fatty acid biosynthetic pathway is the modulation of acetyl-CoA carboxylase activity. As with many of the key regulated enzymes in metabolism, the activity of acetyl-CoA carboxylase is controlled by both allosteric mechanisms (metabolic control) and covalent modification (hormonal control).

Figure 16.32 Under conditions when the citrate cycle is inhibited by a high energy charge in the cell, the citrate shuttle provides a mechanism to transport excess acetyl units from the mitochondria to the cytosol for fatty acid synthesis. The oxidation of malate to pyruvate by malic enzyme generates NADPH, which is used in the fatty acid synthesis pathway.

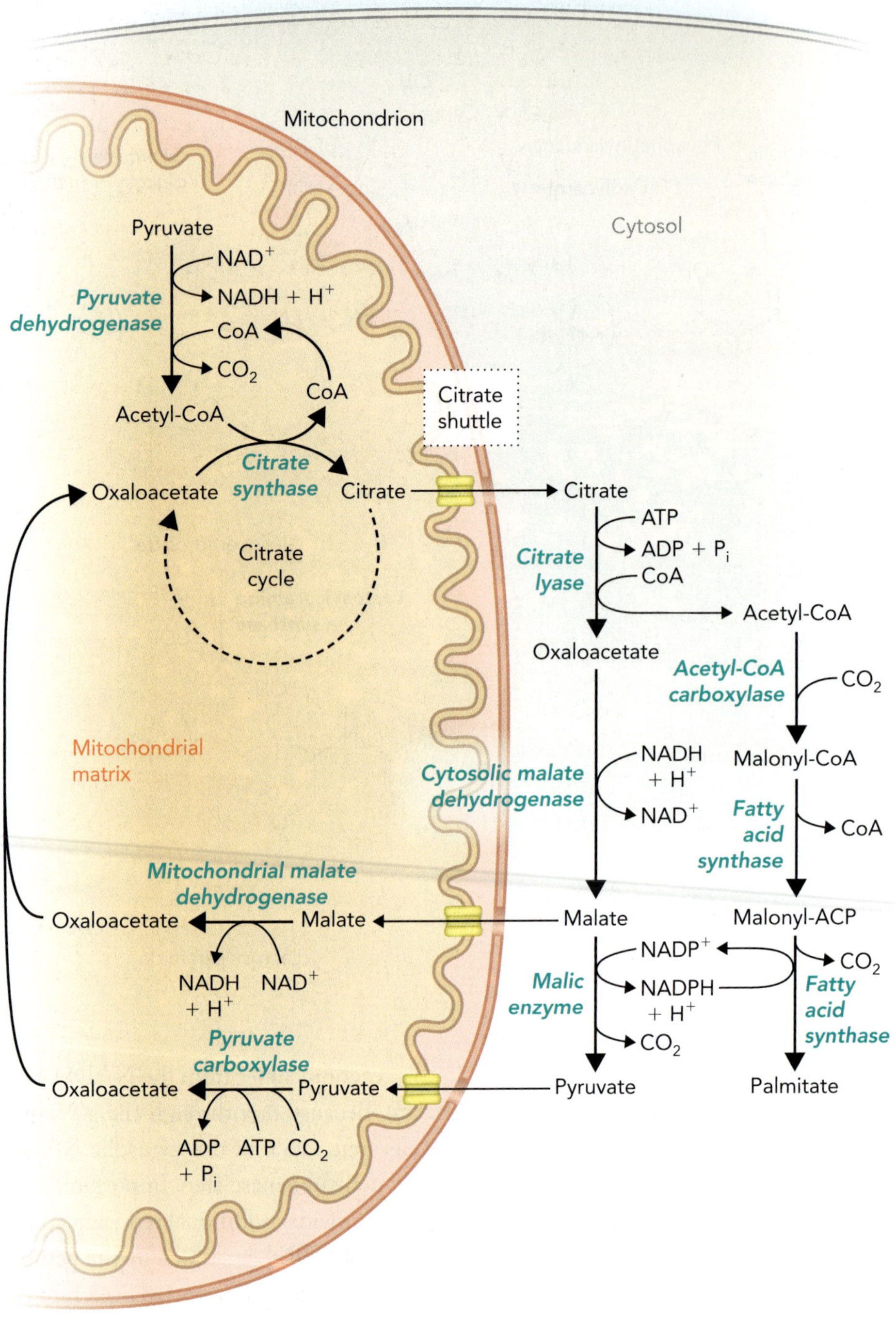

Figure 16.33a illustrates that acetyl-CoA carboxylase is most active when it is in a homopolymeric form. Citrate and palmitoyl-CoA are metabolites that bind to allosteric sites on the enzyme, stimulating polymerization or depolymerization, respectively. Rapid allosteric regulation of acetyl-CoA carboxylase activity makes sense, because when cytosolic citrate levels rise, it means that the citrate is being shuttled out of the mitochondria owing to citrate cycle inhibition, and therefore fatty acid synthesis is favored. In contrast, when palmitoyl-CoA levels begin to rise in the cytosol, it serves as a feedback inhibitor to decrease flux through the fatty acid synthesis pathway as the cell's need for fatty acid synthesis have been met.

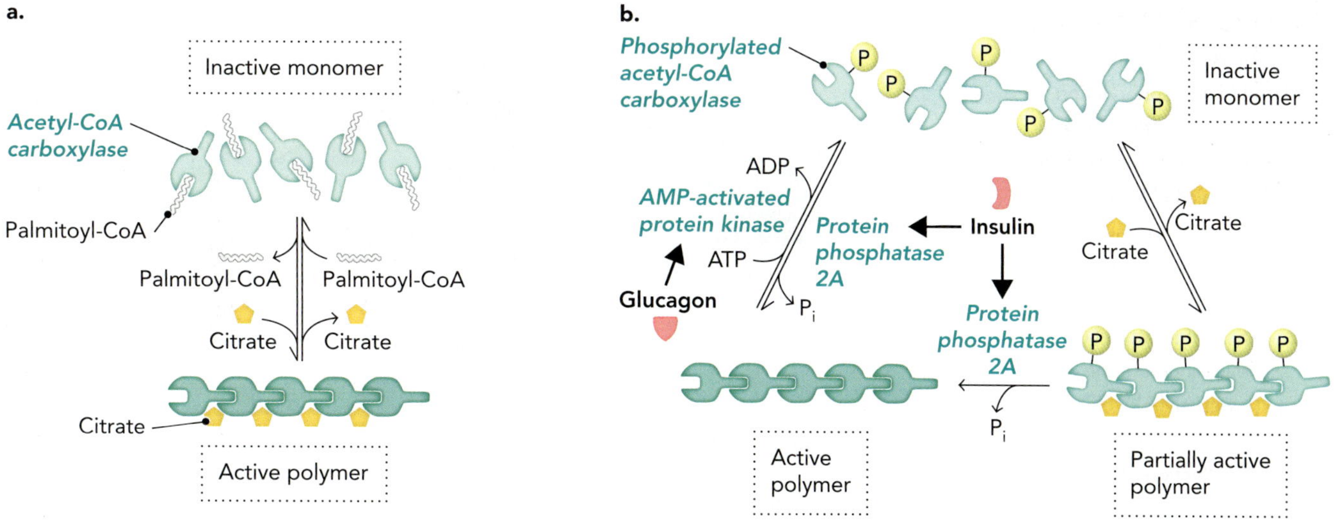

Figure 16.33 Acetyl-CoA carboxylase activity is stimulated by protein polymerization. **a.** Rapid metabolic regulation of acetyl-CoA carboxylase activity is mediated by citrate and palmitoyl-CoA, which are allosteric regulators that bind to the enzyme and alter the equilibrium between polymerization (active) and depolymerization (inactive). **b.** Delayed hormonal signaling by insulin stimulates dephosphorylation and polymerization of acetyl-CoA carboxylase through activation of protein phosphatase 2A. In contrast, glucagon stimulates phosphorylation and depolymerization of acetyl-CoA carboxylase through activation of AMP-activated protein kinase. Citrate binding to phosphorylated acetyl-CoA carboxylase partially activates the enzyme by stimulating polymerization in the absence of insulin signaling.

Hormone signaling provides longer-term regulation of acetyl-CoA carboxylase activity by shifting the equilibrium between the polymeric and monomeric forms as a function of phosphorylation state. As shown in **Figure 16.33b**, when the enzyme is phosphorylated it exists primarily in the monomeric form, whereas dephosphorylation favors polymer formation and leads to enzyme activation. Insulin activates acetyl-CoA carboxylase activity by stimulating dephosphorylation through protein phosphatase 2A. In contrast, glucagon and epinephrine signaling activate the enzyme AMP-activated protein kinase (AMPK), which phosphorylates acetyl-CoA carboxylase and shifts the equilibrium to the inactive monomeric form. Insulin signaling is activated by high serum glucose levels, and therefore activation of acetyl-CoA carboxylase activity ensures that excess glucose is converted to fatty acids for long-term energy storage. Similarly, when serum glucose levels are low, glucagon inhibits acetyl-CoA carboxylase activity to spare glucose for other purposes. This same type of short-term metabolic regulation of rate-limiting enzymes by allosteric effectors and longer-term regulation by insulin and glucagon modulation of enzyme phosphorylation was described in Chapter 14 when we examined regulation of the enzyme glycogen phosphorylase.

The regulatory enzyme **AMP-activated protein kinase (AMPK)** is activated by low energy charge in the cell (high levels of AMP) and is an important metabolic sensor that controls the activity of many key enzymes in both anabolic and catabolic pathways. The activity of AMPK is regulated by AMP binding, which induces a conformational change in the protein, facilitating its phosphorylation at a regulatory threonine residue (Thr172). AMPK is phosphorylated and activated by a variety of kinases, including the LKB1 catalytic subunit of the AMPK kinase protein complex.

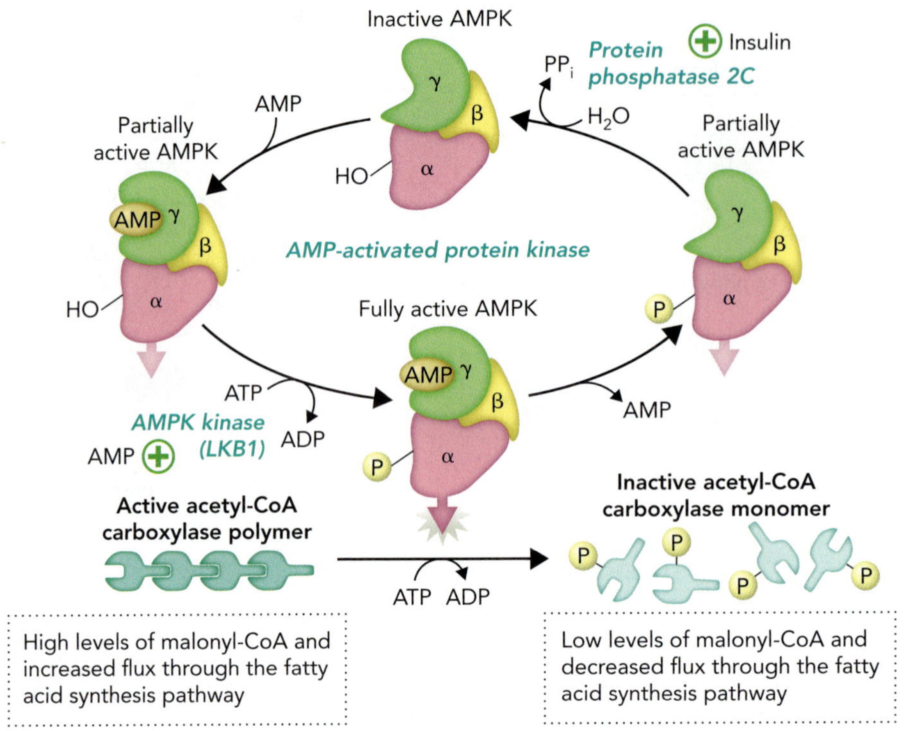

Figure 16.34 AMP-activated protein kinase activity is regulated by phosphorylation and dephosphorylation. Low energy charge results in activation of AMPK, which in turn phosphorylates and inactivates acetyl-CoA carboxylase to decrease flux through the fatty acid synthesis pathway. High glucose levels increase the energy charge (lower AMP levels) and stimulate insulin signaling, which leads to inactivation of AMPK and increased flux through the fatty acid synthesis pathway. Two enzymes that control AMPK activity through phosphorylation and dephosphorylation are AMPK kinase and protein phosphatase 2C, respectively.

When AMP levels increase as a function of decreased glucose and low energy charge, AMP binds to both AMPK and AMPK kinase (**Figure 16.34**). This leads to the phosphorylation and depolymerization of acetyl-CoA carboxylase, which decreases rates of fatty acid synthesis, thereby providing acetyl-CoA for energy conversion reactions in the citrate cycle. In contrast, when glucose levels are high, AMP levels decrease and insulin signaling stimulates protein phosphatase 2C, which dephosphorylates AMPK. This leads to increased levels of dephosphorylated acetyl-CoA carboxylase in the active polymer conformation and elevated flux through the fatty acid synthesis pathway. The central role of AMPK in integrating multiple metabolic pathways is described in Chapter 19.

In addition to hormonal regulation of acetyl-CoA carboxylase activity, which plays a major role in controlling flux through the fatty acid synthesis pathway (see Figures 16.33 and 16.34), flux through the fatty acid synthesis and degradation pathways is also allosterically controlled by metabolites. As shown in **Figure 16.35**, three regulatory mechanisms prevent futile cycling between anabolic and catabolic pathways:

1. Excess acetyl-CoA in the mitochondria signals feedback inhibition of the citrate cycle under conditions of high energy charge, leading to citrate export to the cytosol where it stimulates depolymerization and activation of acetyl-CoA carboxylase activity.
2. High levels of malonyl-CoA inhibit carnitine acyltransferase I activity, which prevents mitochondrial import and degradation of fatty acyl-CoA molecules at the same time that fatty acid synthesis is favored.
3. When fatty acyl-CoA levels are high owing to dietary influx of fatty acids or decreased rates of very-low-density lipoprotein export (liver cells), acetyl-CoA carboxylase activity is inhibited by fatty acyl-CoA binding (stimulates enzyme depolymerization).

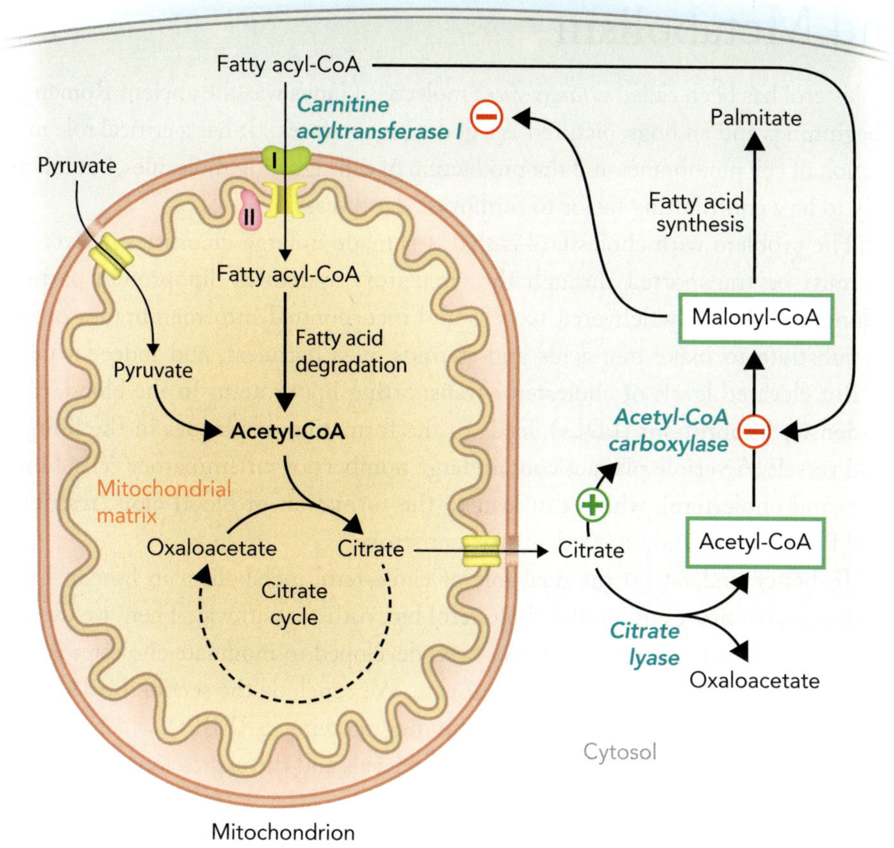

Figure 16.35 Flux through fatty acid synthesis and degradation pathways is controlled by three biochemical mechanisms: citrate activation of acetyl-CoA carboxylase activity; inhibition of carnitine acyltransferase I activity by malonyl-CoA; and inhibition of acetyl-CoA carboxylase activity by fatty acyl-CoA.

concept integration 16.2

What metabolic and hormonal signals control flux through the fatty acid synthesis pathway?

The primary control point for fatty acid synthesis is acetyl-CoA carboxylase, which catalyzes the commitment step of fatty acid synthesis. Allosteric control of acetyl-CoA carboxylase activity by citrate and palmitoyl-CoA binding reflects intracellular changes in metabolic flux, whereas phosphorylation control of acetyl-CoA carboxylase activity by insulin, glucagon, and epinephrine signaling occurs in response to metabolic changes in the whole organism. The activity of acetyl-CoA carboxylase is controlled by its polymerization state, with the active form being a polymer of identical subunits and the inactive state being monomeric. When the cell has excess glucose, citrate is transported out of the mitochondria to the cytosol. To ensure that acetyl-CoA carboxylase is in the active conformation, citrate binds to an allosteric site on the enzyme that promotes polymerization. In addition, high levels of serum glucose stimulate insulin signaling, which activates the enzyme protein phosphatase 2A, resulting in dephosphorylation and polymerization of acetyl-CoA carboxylase subunits. Inhibition of acetyl-CoA carboxylase occurs when fatty acyl-CoA molecules such as palmitoyl-CoA allosterically bind to acetyl-CoA and cause depolymerization. Lastly, when the energy charge in the cell is low (high AMP and low ATP), acetyl-CoA carboxylase is phosphorylated by the enzyme AMP kinase, resulting in depolymerization and inactivation.

16.3 Cholesterol Synthesis and Metabolism

Cholesterol has been called a *Janus-faced* molecule. (Janus was the ancient Roman god of beginnings and endings, pictured as a god with two faces.) It has a critical role in the function of cell membranes and the production of cell signaling molecules, but it is also likely to be a contributing factor to cardiovascular disease.

The problem with cholesterol is that it is made in large quantities in liver cells and must be transported through the circulatory system by lipoprotein particles. As long as it is safely delivered to cells and incorporated into membranes or used as a substrate to make bile acids and steroids, it is harmless, and indeed vital for life. But elevated levels of cholesterol-transporting lipoproteins in the blood, called **low-density lipoproteins (LDLs)**, leads to the formation of plaques in the lining of blood vessels. Arteriole plaques contain large numbers of inflammatory cells, fibrous tissue, and cholesterol, which can lead to the formation of blood clots, restricting blood flow and resulting in cardiac arrest or stroke.

To better understand the dual role of cholesterol metabolism in human health and disease, we first examine the cholesterol biosynthetic pathway. Then, we describe several pharmaceutical drugs that have been developed to modulate cholesterol levels and decrease the risk of cardiovascular disease. We conclude the section by describing the function of sterol regulatory element binding proteins (SREBPs), which provide a functional link between membrane cholesterol levels and the expression of cholesterol metabolizing genes.

Cholesterol Is Synthesized from Acetyl-CoA

Humans synthesize as much as 1 g of cholesterol per day and absorb ~0.3 g per day from their diet. Cholesterol synthesis occurs in all cells, but the largest amount is made in the liver. Because of this *de novo* cholesterol biosynthetic pathway, cholesterol is not an essential lipid in the human diet.

Figure 16.36 gives an overview of the cholesterol biosynthetic pathway, which takes place in the cytosol and consists of four distinct stages. (1) In stage 1, three molecules of acetyl-CoA are used to make **mevalonate**, a C_6 compound that is the product of a reaction catalyzed by the highly regulated enzyme **HMG-CoA reductase**. (2) In stage 2, mevalonate is phosphorylated and decarboxylated to form the activated C_5 isoprenoid intermediate **isopentenyl diphosphate**. This important metabolite is an intermediate in several other biosynthetic reactions, including those of plant chlorophylls and plant hormones (gibberellic acid), some vitamins (A, E, K), and the quinones (coenzyme Q and plastoquinone). (3) In stage 3, three molecules of isopentenyl diphosphate are combined to form farnesyl diphosphate (C_{15}). Then, two molecules of farnesyl diphosphate are used to generate squalene, a C_{30} cholesterol precursor. Farnesyl diphosphate is also used as a lipid anchor in some membrane-associated proteins, such as heterotrimeric G proteins involved in membrane receptor cell signaling. (4) In the fourth and final stage of cholesterol biosynthesis, squalene cyclizes to form a four-ringed molecule, which is then modified to remove three methyl groups and generate the four-ringed sterol molecule cholesterol (C_{27}).

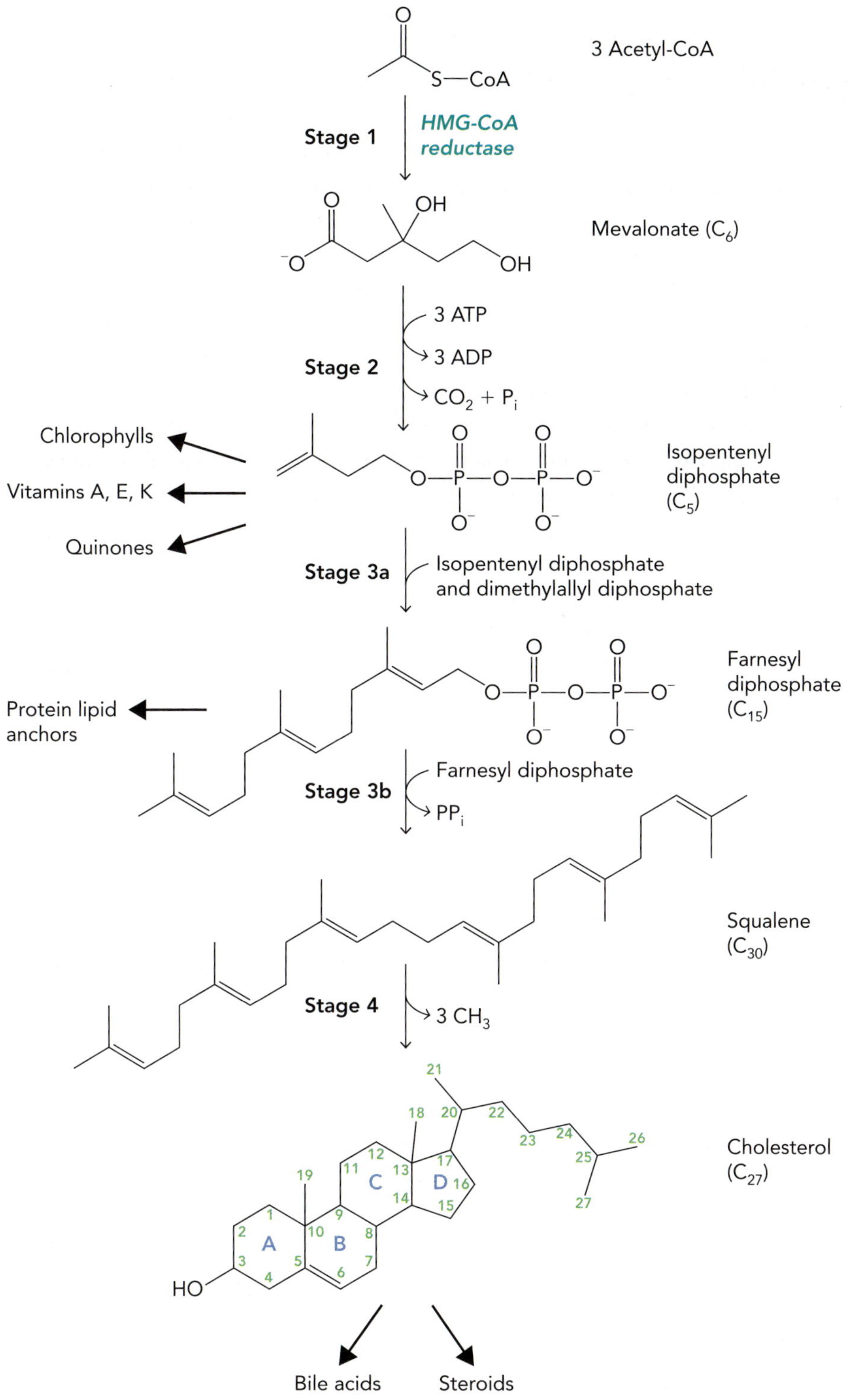

Figure 16.36 An overview of the cholesterol biosynthetic pathway is shown here. Several intermediates in the cholesterol biosynthetic pathway are precursors for other biomolecules, such as chlorophylls, some lipid-soluble vitamins, and quinones. Importantly, cholesterol is the precursor to bile acids and steroid hormones. The four rings of cholesterol (C_{27}) are labeled A, B, C, and D as shown. The four distinct stages in the cholesterol biosynthetic pathway are described in the text.

Let's now look more closely at the four stages of cholesterol biosynthesis.

Stage 1: Generation of Mevalonate from Acetyl-CoA In the first reaction of *de novo* cholesterol synthesis, two molecules of acetyl-CoA are condensed to form acetoacetyl-CoA through a reaction catalyzed by the enzyme thiolase

(**Figure 16.37**). A third acetyl-CoA is added to generate the C_6 compound 3-hydroxy-3-methylglutaryl-CoA (HMG-CoA) by the action of HMG-CoA synthase. In what turns out to be the rate-limiting step in the cholesterol biosynthetic pathway, the enzyme HMG-CoA reductase converts HMG-CoA to mevalonate in a reduction reaction that uses two molecules of NADPH and releases coenzyme A. **Statin drugs,** which are used to lower the risk of cardiovascular disease, bind to and inhibit HMG-CoA reductase activity, thereby decreasing flux through the cholesterol biosynthetic pathway.

Stage 2: Conversion of Mevalonate to Isopentenyl Diphosphate and Dimethylallyl Diphosphate Mevalonate is activated by the addition of two phosphoryl groups donated from ATP to generate 5-pyrophosphomevalonate (**Figure 16.38**). The enzyme pyrophosphomevalonate decarboxylase then catalyzes an ATP-dependent reaction that removes the terminal carboxyl group, generating isopentenyl diphosphate, which is readily isomerized to form dimethylallyl diphosphate. This is a typical example of an isoprene unit, which is a common building block of many natural products. In this case,

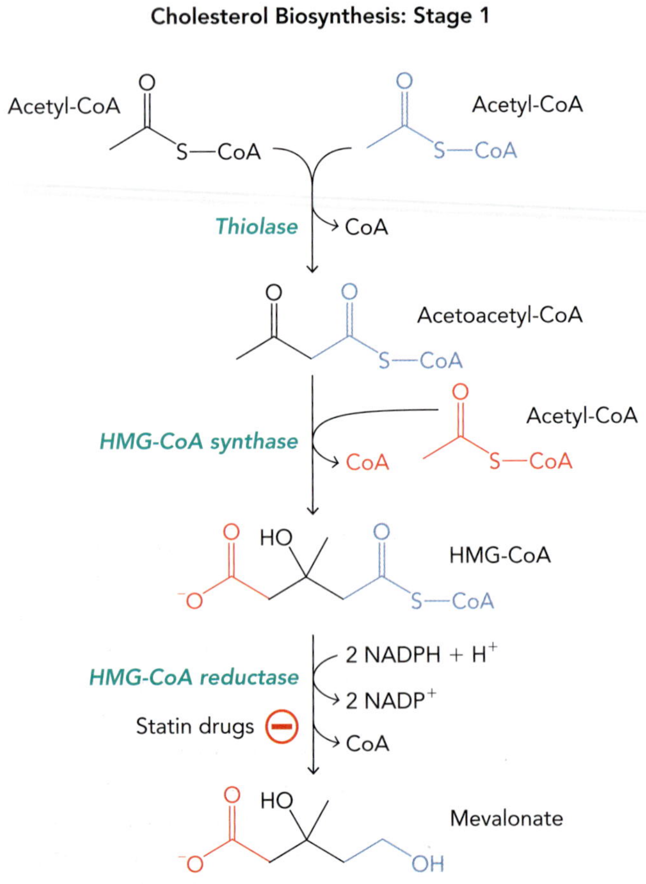

Figure 16.37 In stage 1 of the cholesterol biosynthetic pathway, mevalonate (C_6) is formed from three molecules of acetyl-CoA. The rate-limiting enzyme in the cholesterol biosynthetic pathway is HMG-CoA reductase, which is the target of statin drugs used to treat cardiovascular disease.

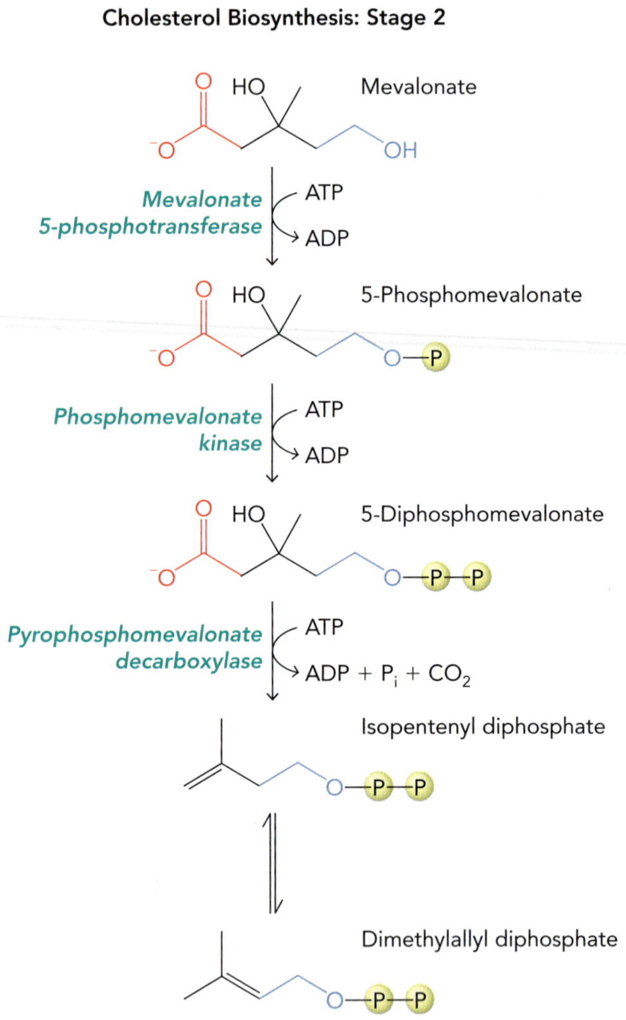

Figure 16.38 In stage 2 of the cholesterol biosynthetic pathway, phosphorylation and decarboxylation of mevalonate leads to the generation of isopentenyl diphosphate, which is isomerized to dimethylallyl diphosphate.

dimethylallyl diphosphate is an isoprene unit with a double bond and two phosphoryl groups attached.

Stage 3: Formation of Squalene from Four Isopentenyl Diphosphates and Two Dimethylallyl Diphosphates The next set of reactions makes use of the C_5 isoprene units generated in stage 2 to build the C_{30} carbon scaffold of squalene. In the first of two stage 3a reactions, prenyltransferase catalyzes a condensation reaction in which isopentenyl diphosphate and dimethylallyl diphosphate condense in a head-to-tail fashion (phosphoryl groups being the head) to form the C_{10} compound geranyl diphosphate (**Figure 16.39**). In the second stage 3a reaction, prenyltransferase adds an isopentenyl diphosphate to the head of geranyl diphosphate, generating the C_{15} intermediate farnesyl diphosphate. In the stage 3b reaction, two molecules of farnesyl diphosphate are linked

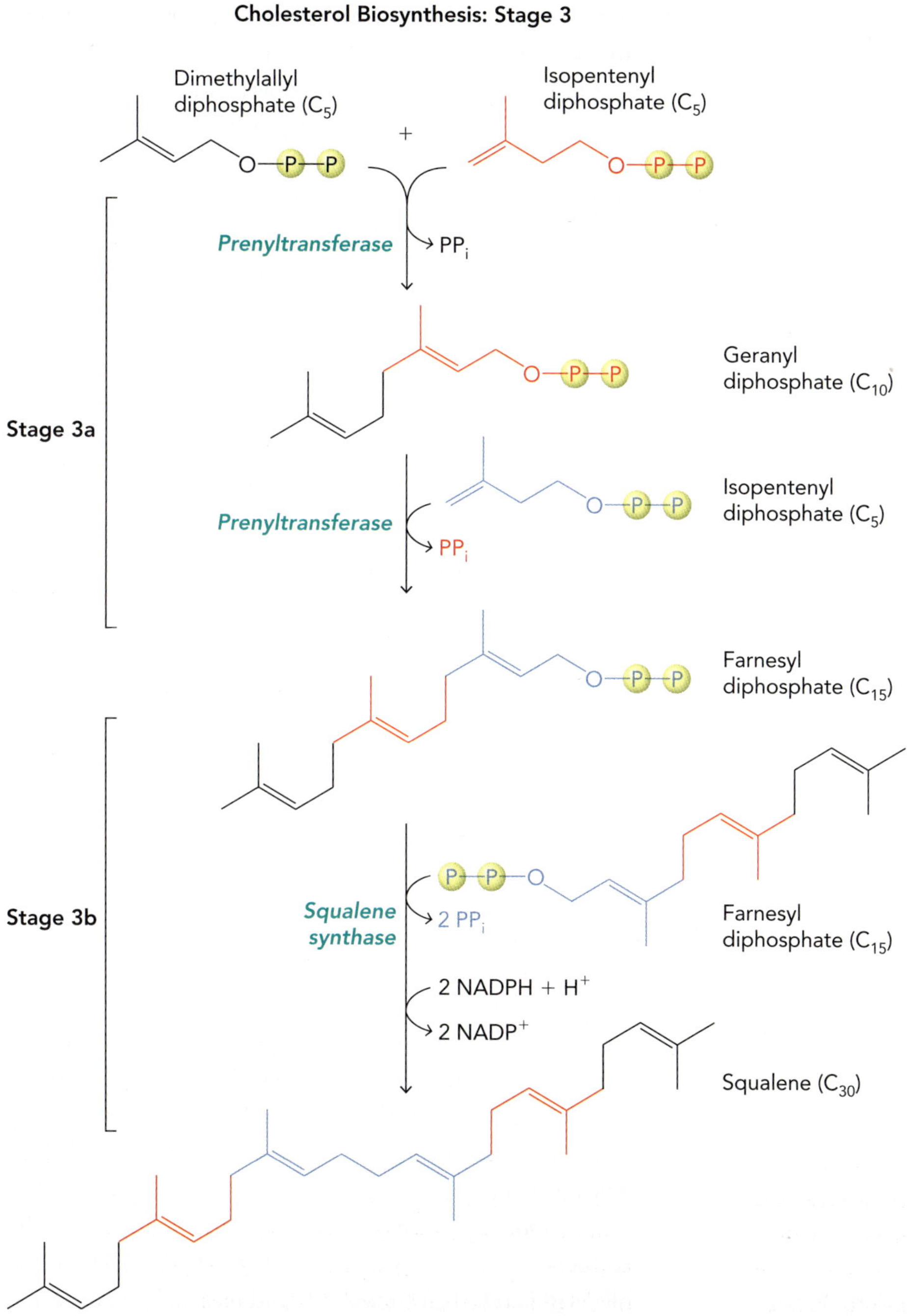

Figure 16.39 Reactions in stage 3 of the cholesterol biosynthetic pathway can be divided into two steps. In the stage 3a reactions, geranyl diphosphate is formed by a head-to-tail linkage of dimethylallyl diphosphate and isopentenyl diphosphate. Addition of another isopentenyl diphosphate to the head of geranyl diphosphate leads to the formation of farnesyl diphosphate. In the stage 3b reaction, two farnesyl diphosphate molecules are linked head-to-head to release two molecules of pyrophosphate and form the C_{30} compound squalene.

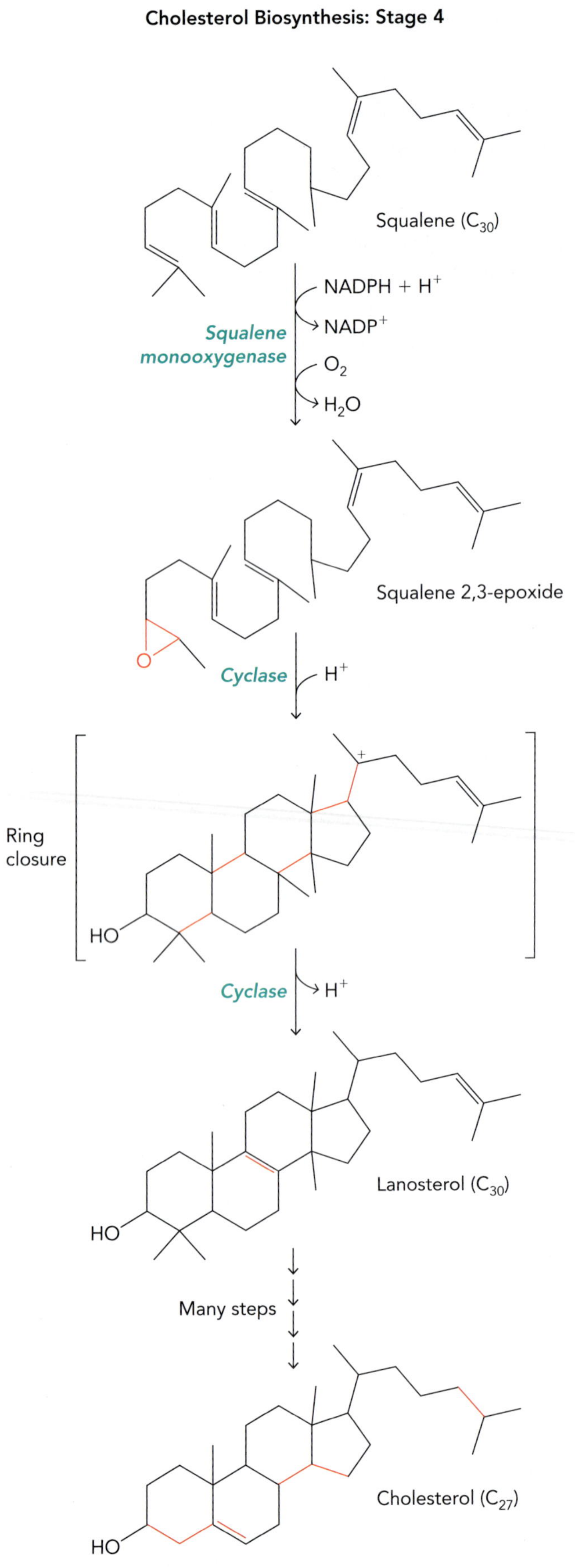

Figure 16.40 Reactions in stage 4 of the cholesterol biosynthetic pathway. In animals, cyclase enzymes catalyze bond rearrangements that convert squalene 2,3-epoxide into the four-ringed sterol molecule lanosterol. The conversion of lanosterol to cholesterol requires 19 additional steps and the removal of three methyl groups.

in a head-to-head arrangement by the enzyme squalene synthase, forming squalene (C_{30}) in a reduction reaction using NADPH as the electron donor. Hydrolysis of the two pyrophosphate products by the enzyme pyrophosphatase helps drive this reaction toward squalene formation.

Stage 4: Cyclization of Squalene and Lanosterol Modification to Form Cholesterol The conversion of a C_{30} hydrocarbon chain into a cyclic cholesterol molecule involves a complex set of reactions. It begins with the oxygenation of squalene by the enzyme squalene monooxygenase (**Figure 16.40**). The product of this reaction, squalene-2,3-epoxide, is then subjected to an amazing cyclization process in animal cells that is catalyzed by enzymes called cyclases. Cyclases promote the formation of the four-ringed cholesterol precursor lanosterol. Lanosterol is converted to cholesterol by a series of 19 reactions (not shown in Figure 16.40) that involve additional bond rearrangements and the removal of three methyl groups to generate the C_{27} product.

Plants also synthesize four-ringed sterol molecules (such as sitosterol and stigmasterol) from squalene, using a different set of reactions. Notably, although plant sterols are present at high levels in some types of foods (cilantro, soybeans, avocados), they are not readily absorbed by intestinal epithelial cells in humans. Some studies have suggested that diets rich in plant steroids can reduce the dietary uptake of cholesterol, resulting in lower serum LDL levels and decreased risk of cardiovascular disease.

As illustrated in **Figure 16.41**, cholesterol synthesized in the liver has three primary functions. If it is esterified with a fatty acid by the enzyme acyl-CoA cholesterol acyltransferase to make cholesterol esters, it is either (1) stored in intracellular lipid droplets or (2) packaged into lipoproteins and exported to the circulatory system. Alternatively, liver cholesterol can be converted into bile acids, which are (3) secreted into the small intestine from the bile duct during digestion.

Bile acids, which exist as bile salts in the unprotonated form, are amphipathic molecules stored in the gall bladder. The most abundant bile acid in humans is **cholate**, which is further modified to make the more water-soluble bile salts taurocholate and glycocholate (Figure 16.41). Most of the bile acid is returned to the liver and reused; however, some

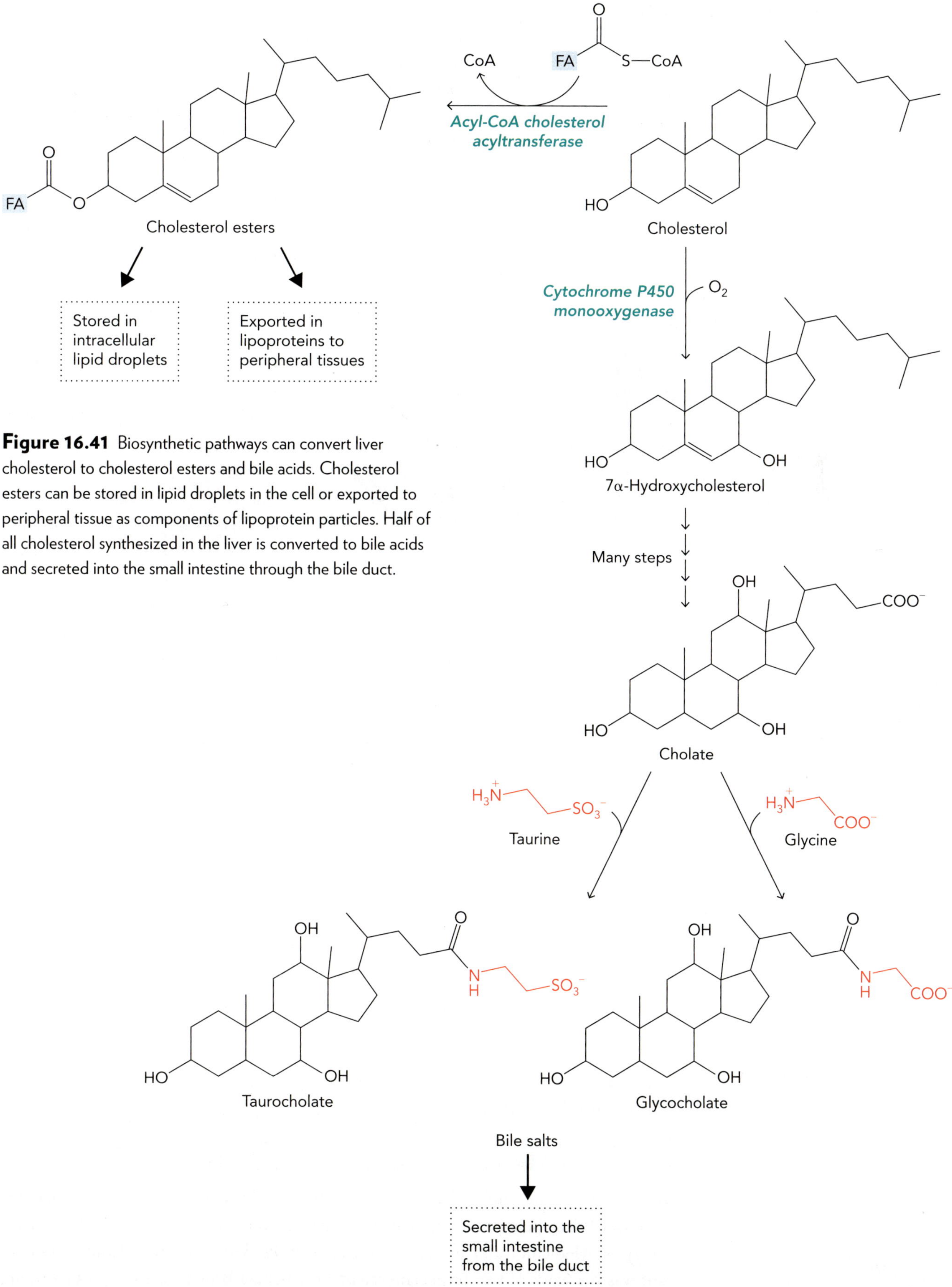

Figure 16.41 Biosynthetic pathways can convert liver cholesterol to cholesterol esters and bile acids. Cholesterol esters can be stored in lipid droplets in the cell or exported to peripheral tissue as components of lipoprotein particles. Half of all cholesterol synthesized in the liver is converted to bile acids and secreted into the small intestine through the bile duct.

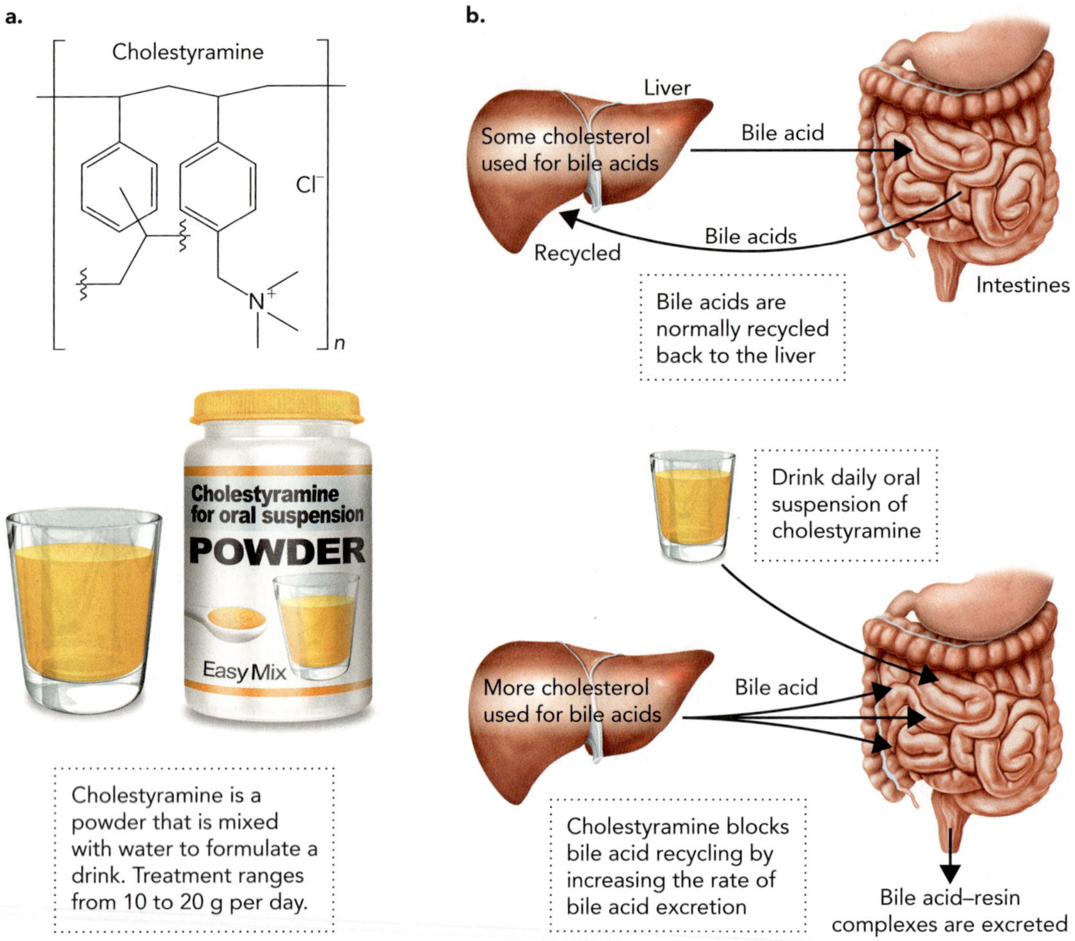

Figure 16.42 Bile acid binding resins are used to lower serum cholesterol levels by increasing the rate of bile acid excretion, which shunts cholesterol away from lipoprotein particle export. **a.** The bile acid binding resin cholestyramine is an insoluble mixed ionic polymer that binds to bile acids with high affinity. The bond coming from the center of the ring indicates that the substituent can be attached to variable positions in the ring. **b.** Drinking an oral suspension of insoluble resins results in the excretion of bile acid–resin complexes, triggering the production of more bile acids from cholesterol esters and thus lowering the total cholesterol pool.

of it is excreted as waste, which provides the only mechanism to rid the body of excess cholesterol. In fact, increasing bile acid excretion by ingesting insoluble resins that bind bile acids with high affinity is one of the available treatments to reduce overall cholesterol levels in the body (**Figure 16.42**). Because the excreted bile acids are not recycled, the liver shunts more of its available cholesterol into bile acid synthesis, thereby decreasing the size of the cholesterol ester pool. This treatment requires that patients ingest large amounts of resin to be effective in reducing serum cholesterol levels (as much as 10–20 g of resin each day).

Cholesterol Metabolism and Cardiovascular Disease

Elucidation of the cholesterol biosynthetic pathway was a heroic effort, which led to Nobel Prizes being awarded to Konrad Bloch and Feodor Lynen in 1964 (Physiology or Medicine) and to John Cornforth in 1975 (Chemistry). Equally important was the work of two physician–scientists, Michael Brown and Joseph Goldstein, who together uncovered the link between cholesterol homeostasis and cardiovascular disease (**Figure 16.43**). Brown and Goldstein were awarded the 1985 Nobel Prize in

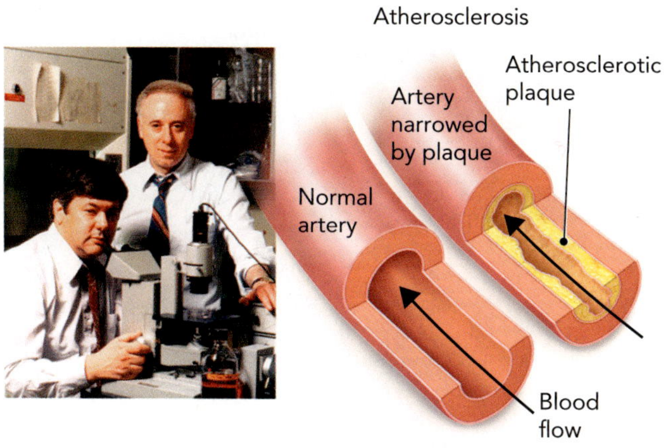

Figure 16.43 Michael Brown (left) and Joseph Goldstein (right) elucidated the link between cholesterol homeostasis and atherosclerosis and were recipients of the 1985 Nobel Prize in Physiology or Medicine.
PHOTO: COURTESY OF MICHAEL BROWN AND JOSEPH GOLDSTEIN.

Physiology or Medicine and pioneered the use of statin drugs to decrease the risk of cardiovascular disease in susceptible individuals.

As described in Chapter 15, triacylglycerols and cholesterol are transported through the circulatory system as components of plasma lipoprotein particles. These membrane-bound vesicles contain a hydrophobic core and one or more proteins on the surface, called apolipoproteins. Plasma lipoproteins have different functions in the body, depending on the type of apolipoproteins they contain and the lipid cargo they carry. **Table 16.2** lists the five major classes of human lipoproteins, ranging in size from 10 nm to 1 μm in diameter.

All lipoproteins consist of a phospholipid monolayer containing cholesterol and one or more apolipoproteins on the surface. These serve as signaling molecules that activate enzymes or bind to cell surface receptors (**Figure 16.44**). Depending on the size of the lipoprotein and the ratio of protein to lipid, lipoproteins have different densities, as determined by density gradient centrifugation. Chylomicrons, the largest of the lipoproteins, contain mostly triacylglycerols in their hydrophobic core and a low amount of protein relative to lipid, thereby giving them the lowest density. In contrast, **high-density lipoproteins (HDLs)** have a high percentage of protein compared to lipid and have the highest density. The three other major lipoprotein classes are very-low-density lipoproteins (VLDLs); intermediate-density lipoproteins (IDLs), also called VLDL remnants; and low-density lipoproteins (LDLs).

As illustrated in **Figure 16.45**, steady-state levels of circulating cholesterol are determined by the balance of cholesterol input (diet and *de novo* biosynthesis), cholesterol recycling (returning tissue cholesterol to the liver), and cholesterol output (loss of bile acids by excretion). The sizes of lipoproteins decrease as triacylglycerol and cholesterol cargo is transferred from the lipoprotein particle to endothelial cells lining the circulatory system. For example, chylomicron remnants are smaller chylomicrons that have unloaded most of their lipid cargo in the form of fatty acids. Moreover, IDL and LDL particles are VLDL particles containing proportionally fewer triacylglycerols owing to delivery

Table 16.2 MAJOR CLASSES OF LIPOPROTEINS IN HUMAN SERUM

Feature	Chylomicron	Very-low-density lipoprotein	Intermediate-density lipoprotein	Low-density lipoprotein	High-density lipoprotein
Diameter (nm)	100–1,000	30–80	25–35	18–25	5–12
Density (g/cm^3)	<0.95	<1.006	<1.006–1.019	<1.019–1.063	<1.063–1.210
Percent protein (%)	2	10	20	25	55
Percent triacylglycerol (%)	85	60	22	10	5
Percent cholesterol ester (%)	5	15	30	40	12
Major apolipoproteins	apoC-II apoC-III apoB-48 apoE	apoC-II apoC-III apoB-100 apoE	apoC-II apoC-III apoB-100 apoE	apoB-100	apoC-II apoC-III apoA-I apoA-II apoD

Note: apo = apolipoprotein.

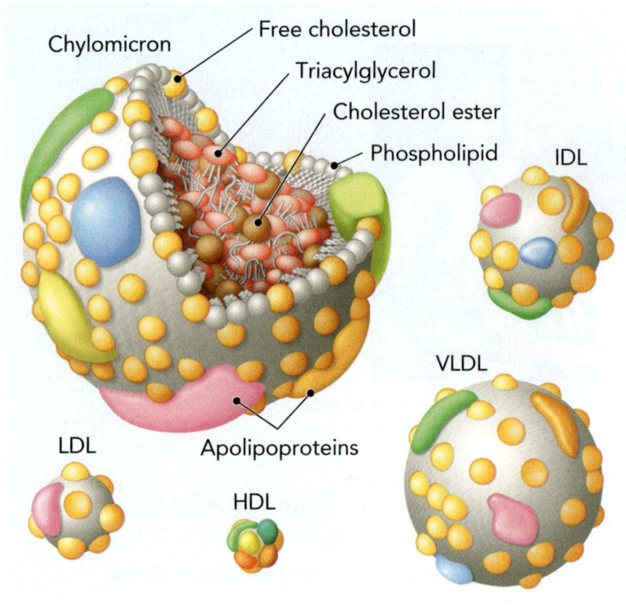

Figure 16.44 Lipoprotein particles contain cholesterol esters and triacylglycerols stored inside a phospholipid monolayer studded with unesterified cholesterol. Human serum contains five major classes of lipoproteins, which vary in size and cargo components as listed in Table 16.2.

of this valuable energy source to adipose and muscle tissues. Note from Table 16.2 that chylomicrons (and chylomicron remnants) uniquely contain apolipoprotein B-48 on their surface, whereas VLDL, IDL, and LDL particles are all identified as liver-derived lipoprotein particles because they contain apolipoprotein B-100. HDL particles are identified by the distinct apolipoproteins apoA-I, apoA-II, and apoD. HDL particles are also synthesized in the liver and function to return tissue cholesterol back to the liver through the process of reverse cholesterol transport (see Figure 16.45 and the discussion that follows).

Cholesterol deposits in blood vessels are considered one of the major causes of cardiovascular disease, and therefore control of cholesterol homeostasis is an important regulatory process. One of the important findings from large clinical studies designed to identify causes of cardiovascular disease is the association between levels of certain lipoproteins in serum and the lifetime risk of developing heart disease. For example, high amounts of LDL particles in the blood after a 12-hour fast are an indicator of elevated risk of cardiovascular disease, whereas high levels of HDL particles are associated with decreased risk of disease. These data have led to LDL particles being called "bad cholesterol" and HDL particles being called "good cholesterol." The biochemical

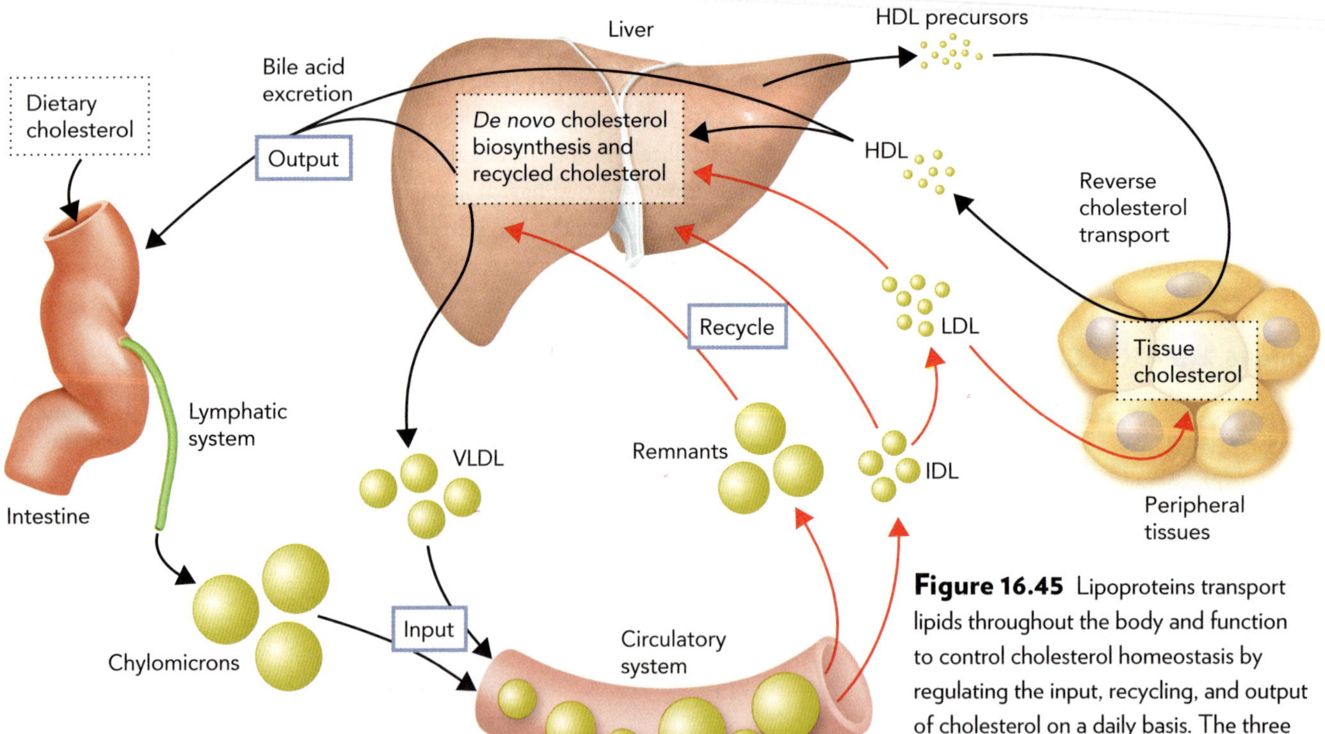

Figure 16.45 Lipoproteins transport lipids throughout the body and function to control cholesterol homeostasis by regulating the input, recycling, and output of cholesterol on a daily basis. The three sources of cholesterol in lipoproteins are dietary cholesterol; cholesterol biosynthesized in the liver; and cholesterol stored in peripheral tissues (see Table 16.2).

explanation for why LDL is bad and HDL is good is not totally understood, but it is related to differences in both cholesterol content and the function of lipoproteins contained in LDL and HDL particles (see Table 16.2). Think of LDL particles as delivering excess cholesterol from the liver to peripheral tissues (and arteries). In contrast, HDL particles scavenge excess cholesterol from the peripheral tissues (and arteries) and transport it back to the liver, where it can be converted to bile acid and excreted (see Figure 16.45).

Figure 16.46 illustrates how HDL particles function in reverse cholesterol transport. They first remove cholesterol from peripheral tissues through apoA-I activation of an enzyme in plasma called serum lecithin–cholesterol acyltransferase. Membrane-associated cholesterol is esterified by lecithin–cholesterol acyltransferase in a reaction involving the phospholipid phosphatidylcholine, also called lecithin, which generates lysolecithin as well as cholesterol esters that are incorporated into the HDL particle (**Figure 16.47**). The cholesterol obtained from peripheral tissues is taken back to the liver, where it is stored or converted to bile acids and excreted.

Cardiovascular disease is responsible for up to half of all deaths in some countries, including the United States. The main problem is arterial blockage, a condition called atherosclerosis, which leads to fatal heart attacks and strokes. As illustrated in **Figure 16.48**, **atherosclerosis** is characterized by the buildup of fibrous tissue in arterial walls. This fibrous tissue contains, among other things, large deposits of cholesterol-rich lipids. Atherosclerotic plaques are the result of infiltration of LDL particles into the space below the

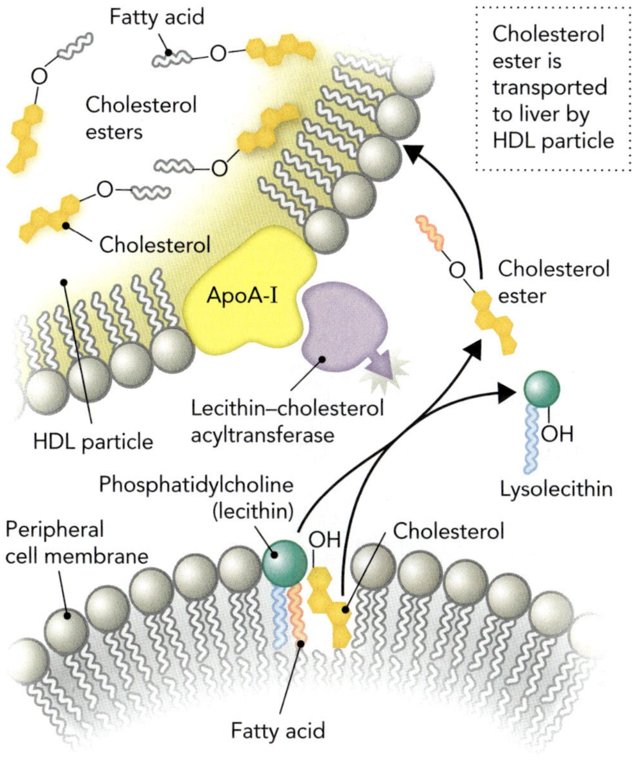

Figure 16.46 Apolipoprotein A-I (apoA-I) on the surface of HDL particles activates the enzyme lecithin–cholesterol acyltransferase. Lecithin–cholesterol acyltransferase esterifies cholesterol in peripheral cell membranes with a fatty acid tail derived from phosphatidylcholine (lecithin). The esterified cholesterol is taken up into HDL particles and can be returned to the liver.

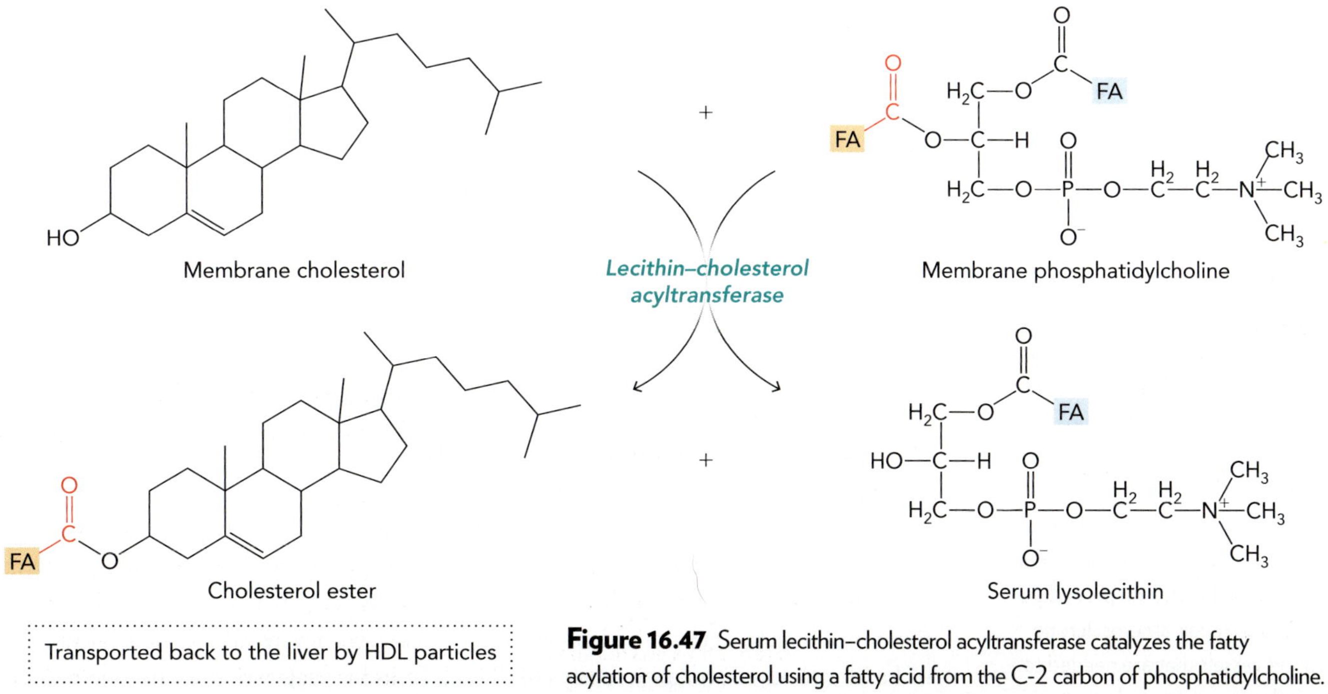

Figure 16.47 Serum lecithin–cholesterol acyltransferase catalyzes the fatty acylation of cholesterol using a fatty acid from the C-2 carbon of phosphatidylcholine.

Figure 16.48 Atherosclerotic plaques form when injury occurs to the endothelial cell lining, allowing LDL particles to invade and initiate an immune response. The plaques contain cholesterol and fat deposits, macrophage foam cells, and fibrous material.

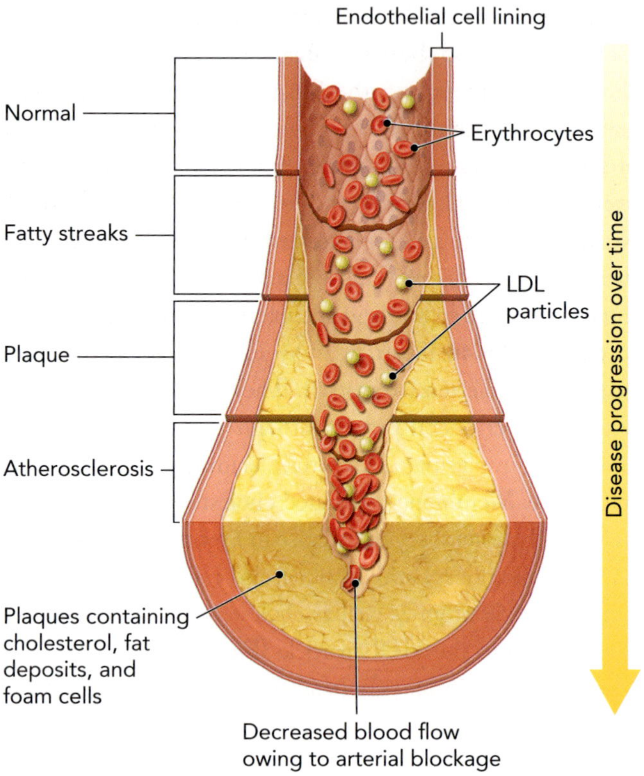

a.

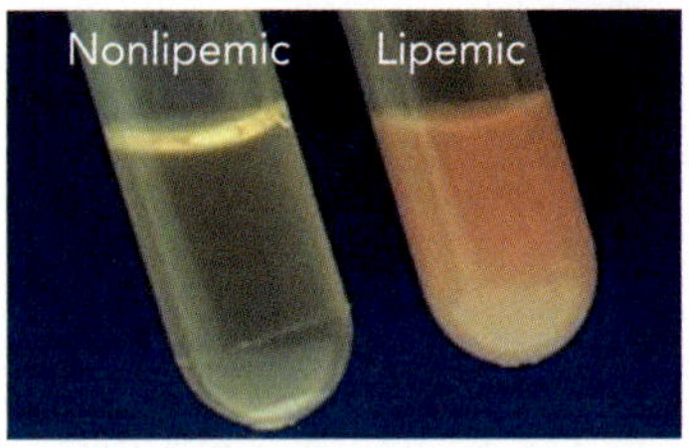

b.

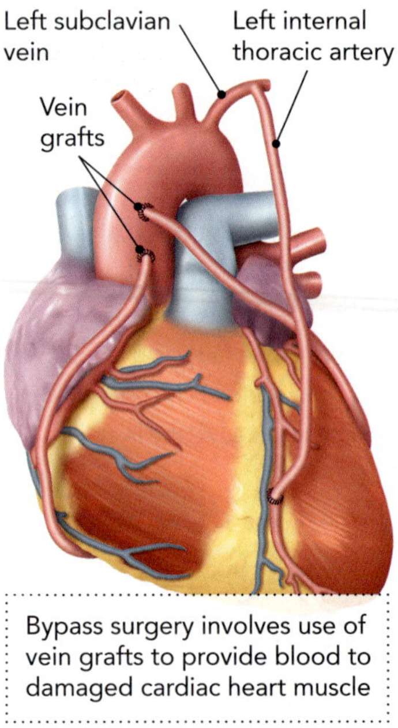

Figure 16.49 High-fat diets are associated with increased risk of cardiovascular disease. **a.** Diets containing high amounts of saturated animal fats are associated with elevated serum lipids in the plasma fraction of centrifuged whole blood as seen by the opaque color. RICHARD A. BOWEN, PHD. **b.** The coronary artery bypass graft is the most common type of surgery on the human heart. The procedure is used to repair atherosclerotic blood vessels that are needed to supply oxygen to cardiac muscle.

endothelial lining of blood vessels. Oxidized phospholipids in the LDL particle initiate an immune response, which triggers the local accumulation of inflammatory cells—a type of macrophage cell called a foam cell. Degradation of the LDL particles causes the cholesterol to be released and deposited in the plaque. The site of inflammation grows as more LDL particles attach to the plaque, leading to smooth muscle cell proliferation and calcification. If the plaque ruptures, a fibrous blood clot can form that blocks blood flow. If the clot occurs in a coronary or carotid artery, it can cause a heart attack or stroke, respectively. High blood pressure, also referred to as hypertension, increases the risk of arterial damage and plaque formation.

Diets containing high amounts of saturated animal fat lead to elevated levels of serum LDL, which contributes to the formation of atherosclerotic plaques. Studies have shown an association between diets containing high amounts of saturated fats and the risk of cardiovascular disease. A simple blood test can be used to show elevated levels of lipid in serum (lipemia), which can be caused by a high-fat diet. The plasma layer of lipemic serum is cloudy compared to that of normal serum (**Figure 16.49a**). When arterial blockages occur in blood vessels responsible for delivering oxygen to cardiac muscle, it can lead to a massive heart attack, or **myocardial infarction.** As shown in **Figure 16.49b**, the most common open-heart surgery in the world is a **coronary artery bypass graft**. This type of open-heart surgery replaces up to four coronary arteries with sections of veins taken from other parts of the body to physically bypass the arterial blockage caused by atherosclerotic plaques.

The connection between high serum LDL levels and atherosclerosis was made by Brown and Goldstein when they discovered that a genetic disorder called **familial hypercholesterolemia (FH)** was due to mutations in the LDL receptor gene. Individuals who have mutations in both copies of the LDL receptor gene (FH homozygous) have no LDL receptors on the cell surface, and consequently their serum LDL levels are five times higher than normal. These individuals have severe atherosclerosis and die of heart attacks or strokes in early childhood. In individuals who contain one mutant LDL receptor gene and one normal copy of the gene (FH heterozygous), their serum

LDL levels are two to three times higher than normal, and these patients often have their first heart attack before the age of 30.

Using a combination of biochemical and molecular biological approaches and cells isolated directly from FH patients, the Brown and Goldstein research teams deciphered the biochemical processes by which LDL receptors control serum LDL levels, as illustrated in **Figure 16.50**. This groundbreaking work led to two major discoveries. First, the researchers found that LDL binding to the LDL receptor through apoB-100 occurs at specific sites in the plasma membrane (called clathrin-coated pits). This binding initiates an endocytic process that degrades the LDL particle and returns the LDL receptor to the cell surface. The cholesterol esters are either released from lysosomes

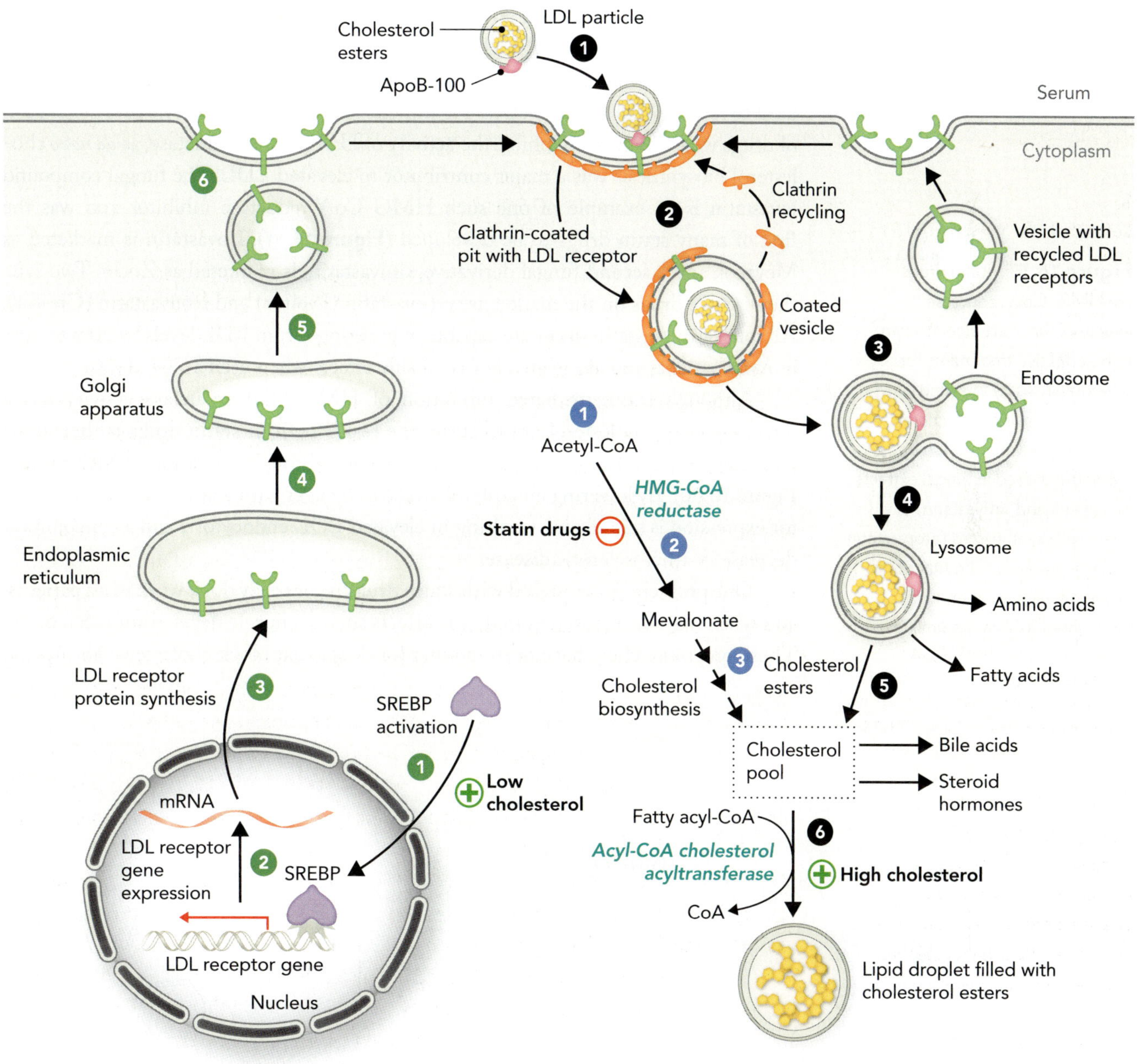

Figure 16.50 Three processes control cholesterol homeostasis in liver cells. The first is endocytosis of LDL particles by LDL receptors, which involves recycling of LDL receptors and recovery of the cholesterol cargo (black numbered circles). The second is *de novo* cholesterol biosynthesis by liver enzymes (blue numbered circles). This process is subject to inhibition by statin drugs that target the enzyme HMG-CoA reductase. The third is stimulation of LDL receptor gene expression by low intracellular cholesterol levels through activation of the transcription factor SREBP (green numbered circles). This process leads to increased levels of LDL receptors on the cell surface and higher rates of LDL endocytosis.

and stored in cholesterol droplets or are used to synthesize bile acids or steroids. The intracellular pool of cholesterol is determined by both cholesterol ester uptake from LDL particles and *de novo* cholesterol biosynthesis.

A second major discovery made by the Brown and Goldstein research teams was that low cholesterol levels stimulate HMG-CoA reductase activity (enzyme synthesis and activity). This activity induces expression of the LDL receptor gene through proteolytic activation of sterol regulatory element binding proteins (SREBPs), which are sequestered in the endoplasmic reticulum. As described in the next subsection, SREBPs are transcription factors that regulate the expression of numerous lipid metabolizing genes.

On the basis of the observation that high serum LDL levels in homozygous and heterozygous FH patients was positively correlated with severe atherosclerosis, it was proposed that age-dependent atherosclerosis in normal people might also be due to elevated serum LDL levels. Numerous clinical studies confirmed this prediction, and within a short period of time pharmaceutical companies developed drugs that could be used to lower serum LDL levels as a treatment for atherosclerosis. The first generation of drugs were designed to inhibit the activity of HMG-CoA reductase, as *de novo* cholesterol biosynthesis was a major contributor to elevated LDL. The fungal compound lovastatin is an example of one such HMG-CoA reductase inhibitor and was the first of many statin drugs to be developed (**Figure 16.51**). Lovastatin is marketed as Mevacor, and a second fungal derivative, simvastatin, is marketed as Zocor. Two synthetic statin drugs on the market are atorvastatin (Lipitor) and rosuvastatin (Crestor). All four of these statin drugs are capable of lowering serum LDL levels by 20% to 40% in most patients and are relatively free of side effects when used at low doses.

Although statin-mediated inhibition of HMG-CoA reductase activity has a direct effect on cholesterol homeostasis, the real benefit of statin drugs is their indirect effect of stimulating LDL receptor expression through activation of SREBP (see Figure 16.50). By lowering cholesterol levels in the liver with statin drugs, LDL receptor expression is stimulated, resulting in elevated LDL endocytosis and a concomitant decrease in atherosclerotic disease.

One problem encountered with statin drugs is that they don't work in all patients, and when high doses are required, side effects such as muscle degeneration can occur. Therefore, researchers have been looking for drugs that block cholesterol absorption

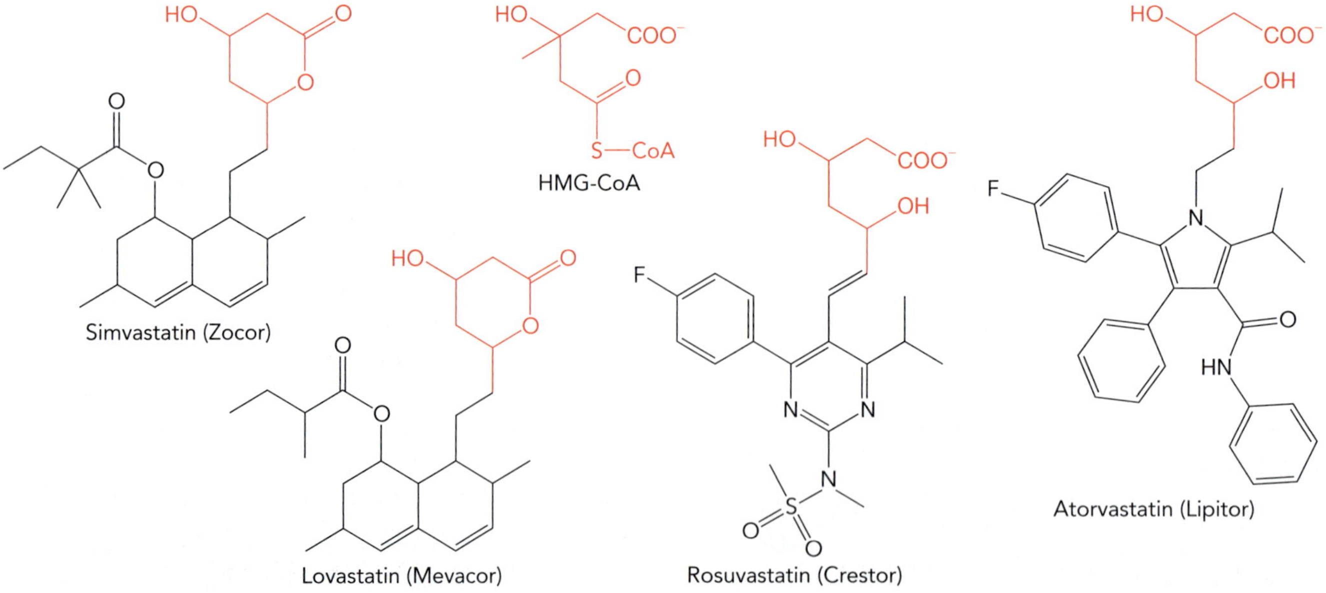

Figure 16.51 Statin drugs are HMG-CoA reductase inhibitors that have been shown to be effective treatments for atherosclerosis by lowering serum LDL levels. The chemical structures shown are statin drugs either discovered in fungal extracts (lovastatin and simvastatin) or chemically synthesized (atorvastatin and rosuvastatin). The functional group on each compound is highlighted to show its similarity to the substrate HMG-CoA.

in the intestine, as this would provide another means to decrease serum LDL levels. The first drug of this type is called ezetimibe (Zetia), which was found to block a cholesterol transporter in the small intestine known as Niemann–Pick C1-Like 1 (NPC1L1) protein (**Figure 16.52**). Studies have shown that ezetimibe treatment reduces serum LDL levels up to 20% by inhibiting cholesterol transport from the intestine to the liver. Like statins, ezetimibe also increases HDL levels by about 5%.

Because ezetimibe lowers cholesterol levels by interfering with intestinal absorption, and statin drugs lower cholesterol levels by inhibiting *de novo* biosynthesis in the liver and increasing LDL receptor expression, a combination drug called Vytorin was developed, which contained both ezetimibe and the statin drug simvastatin (**Figure 16.53**). Basic biochemical research on the function of HMG-CoA reductase (inhibited by simvastatin) and the NPC1L1 protein (inhibited by ezetimibe) suggested that Vytorin would be more effective in decreasing the risk of cardiovascular disease than either ezetimibe or

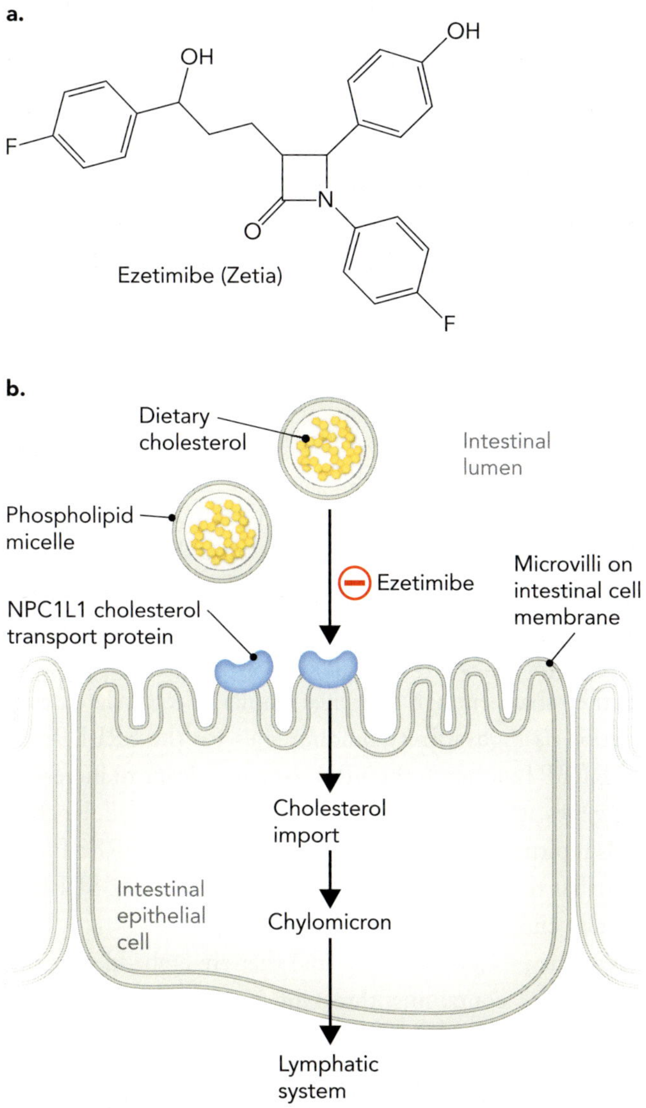

Figure 16.52 Ezetimibe blocks the import of dietary cholesterol by inhibiting the transport function of the NPC1L1 protein. **a.** Chemical structure of ezetimibe. **b.** Ezetimibe blocks the cholesterol import function of NPC1L1, which is expressed on the lumenal side of intestinal epithelial cells.

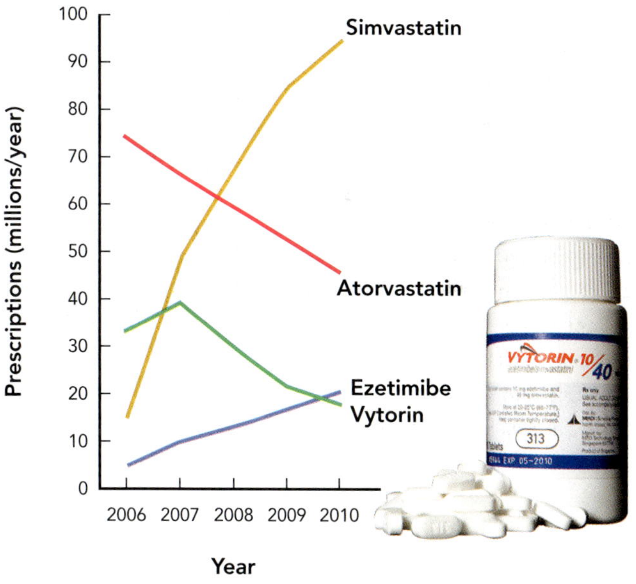

Figure 16.53 Vytorin is a combination drug containing both simvastatin and ezetimibe and was developed to decrease serum LDL levels and lower the risk of cardiovascular disease. It is sold with 10 mg of ezetimibe and different amounts of simvastatin; for example, Vytorin 10/40 contains 10 mg of ezetimibe and 40 mg of simvastatin. The number of Vytorin prescriptions dropped significantly after clinical tests showed it was no more effective than simvastatin alone. Atorvastatin (Lipitor) prescriptions also decreased during this same time period because it was more expensive than the generic drug simvastatin. Ezetimibe prescriptions increased during this time period as it is an alternative for patients who cannot tolerate statins. PHOTO: JB REED/BLOOMBERG VIA GETTY IMAGES.

simvastatin alone, but a number of clinical studies using large numbers of patients indicated this was not the case. Indeed, the most surprising result was that although Vytorin decreased total LDL levels below those measured in patients treated with only ezetimibe or simvastatin alone, cardiovascular disease risks were the same in all three patient groups. Because simvastatin is available as a generic drug and is much less expensive than Vytorin, the number of Vytorin prescriptions issued by doctors decreased significantly after the studies were published, whereas the number of simvastatin prescriptions rose sharply (Figure 16.53). The Vytorin story provides a good example of how basic biochemical research does not always translate into clinical success, owing to physiologic and pharmacological factors that can be measured only in large patient studies.

Sterol Regulatory Element Binding Proteins

Flux through lipid metabolizing pathways is regulated in response to three cellular demands: (1) the changing energy needs of the cell; (2) the requirement for membrane components in rapidly dividing cells; and (3) the need to synthesize cholesterol derivatives (bile acids and steroids). Under each of these conditions, a subset of functionally related enzymes is needed to synthesize and degrade a variety of lipid molecules.

Rather than maintaining adequate levels of these specialized enzymes in the cell at all times, many of the genes encoding lipid metabolizing enzymes are under transcriptional control. The best way to coordinate and control the expression of multiple genes is to regulate the activity of a single transcription factor protein, which binds to specific DNA sequences located in functionally related genes. One example of this principle in metabolism is the control of cholesterol biosynthesis and lipogenesis (lipid synthesis) by the **sterol regulatory element binding proteins (SREBPs)**.

The Brown and Goldstein labs discovered the SREBPs in 1993 on the basis of the ability of SREBPs to bind to specific DNA sequences, called **sterol regulatory elements (SREs)**, located in the proximal promoter of the LDL receptor gene. **Figure 16.54** shows a dimer of an SREBP bound to the SRE sequence in the LDL receptor gene. SREBPs share a common DNA binding domain motif with other helix–loop–helix transcription factors. SRE sequences similar to those found in the LDL receptor gene were soon discovered in the regulatory regions of the HMG-CoA synthase and HMG-CoA reductase genes (**Figure 16.55**). Although the SRE sequences in these three genes are not identical, they share enough sequence similarity to a related core SRE sequence that they function as high-affinity binding sites for the SREBPs. In cells depleted of cholesterol, SREBP binding to the SRE sequences leads to increased expression of SRE-containing genes, which results in higher levels of intracellular cholesterol. Once cholesterol levels return to normal, the transcription of the SREBP target genes decreases, indicating that the transcriptional regulatory activity of SREBPs must be controlled by intracellular cholesterol levels.

The SREBPs are large proteins (~1,100 amino acids) that are embedded in the ER membrane as inactive precursors containing three functional domains: (1) an N-terminal domain, encoding the DNA binding and gene regulatory functions; (2) a membrane anchoring domain, consisting of two transmembrane α helices joined by a 30-amino-acid loop protruding into the ER lumen; and (3) a C-terminal regulatory domain, which interacts with a cholesterol-sensing protein embedded in the ER membrane. Cleavage of precursor SREBPs by proteolytic enzymes in the Golgi apparatus releases the N-terminal transcriptional regulatory domain, which then enters the nucleus and binds to SRE sequences in numerous genes.

Three SREBPs have been characterized in animals: SREBP-1a, SREBP-1c, and SREBP-2, all of which contain the same three functional domains (**Figure 16.56**). The

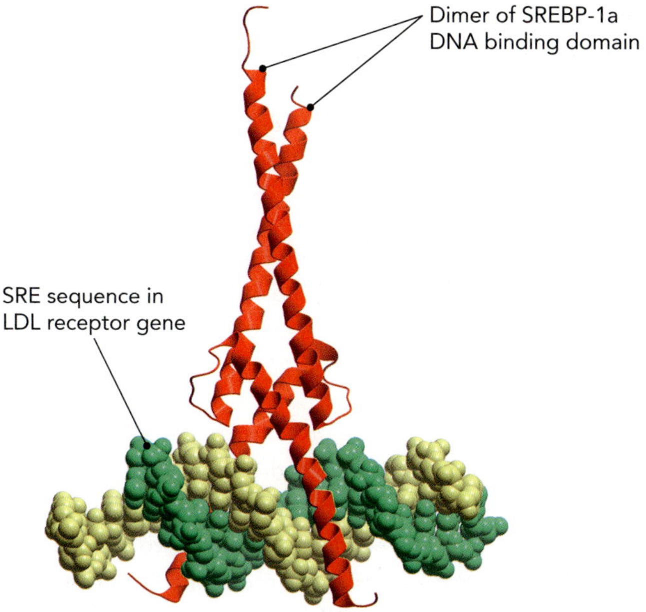

Figure 16.54 The DNA binding domain of a human SREBP-1a dimer is shown here bound to the SRE sequence TCACCCCAC in the human LDL receptor gene. BASED ON PDB FILE 1AM9.

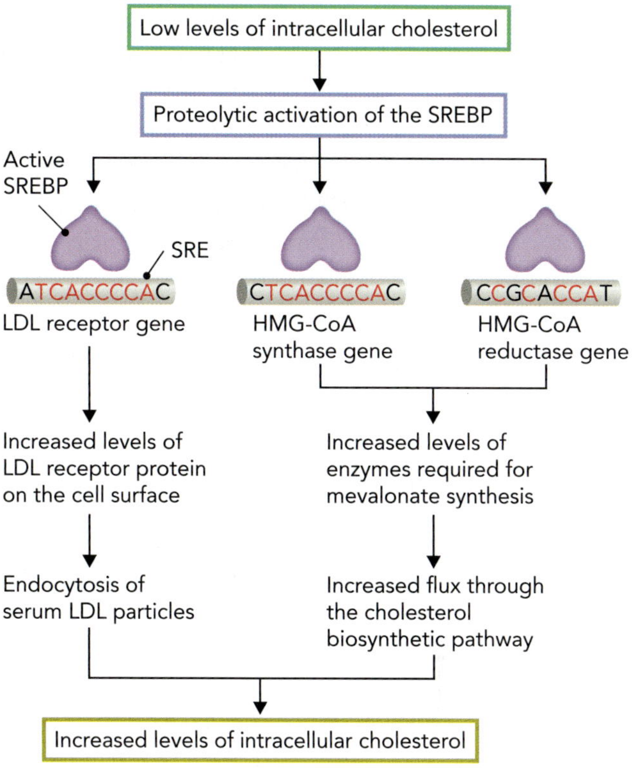

Figure 16.55 Low levels of intracellular cholesterol stimulate binding of SREBPs to SRE sequences located in the transcriptional control region of specific genes. Nucleotides in the shared core SRE sequence of these three genes are shown in red. The combined effect of SREBP activation of the LDL receptor gene, HMG-CoA synthase gene, and HMG-CoA reductase gene leads to increased levels of intracellular cholesterol.

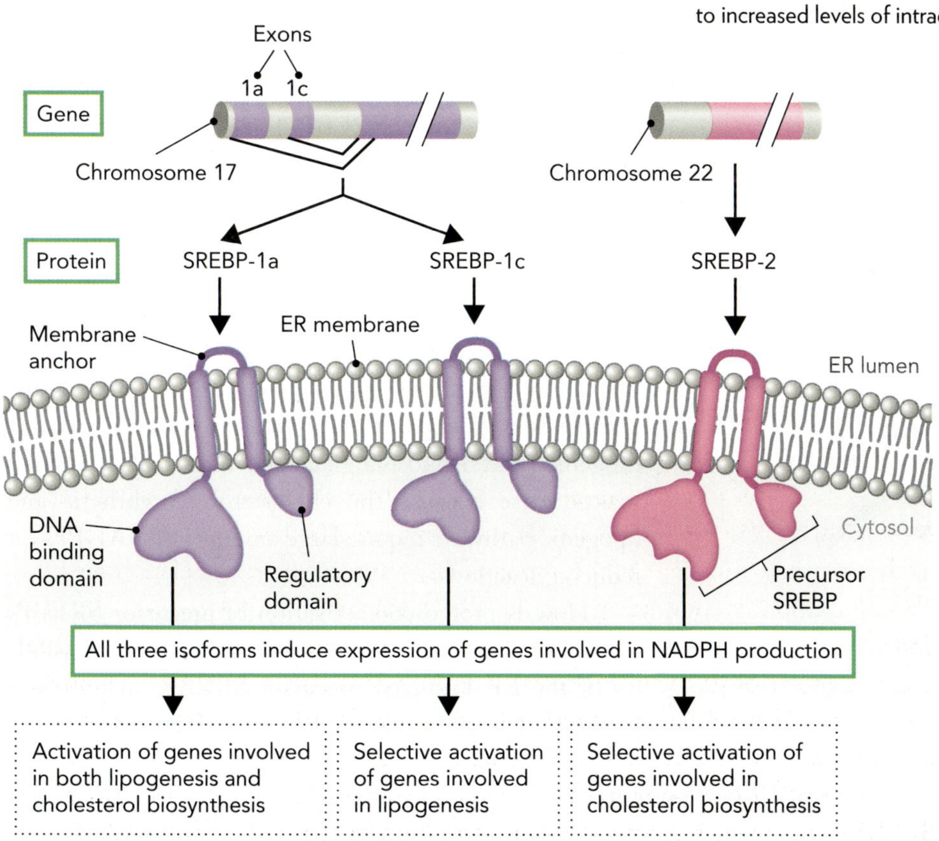

Figure 16.56 SREBPs are ER membrane–associated proteins consisting of a DNA binding domain, a membrane anchor domain, and a regulatory domain. Three SREBPs have been characterized: SREBP-1a and SREBP-1c, which are encoded by the same gene on chromosome 17 but differentially spliced; and SREBP-2, which is encoded by a gene on chromosome 22.

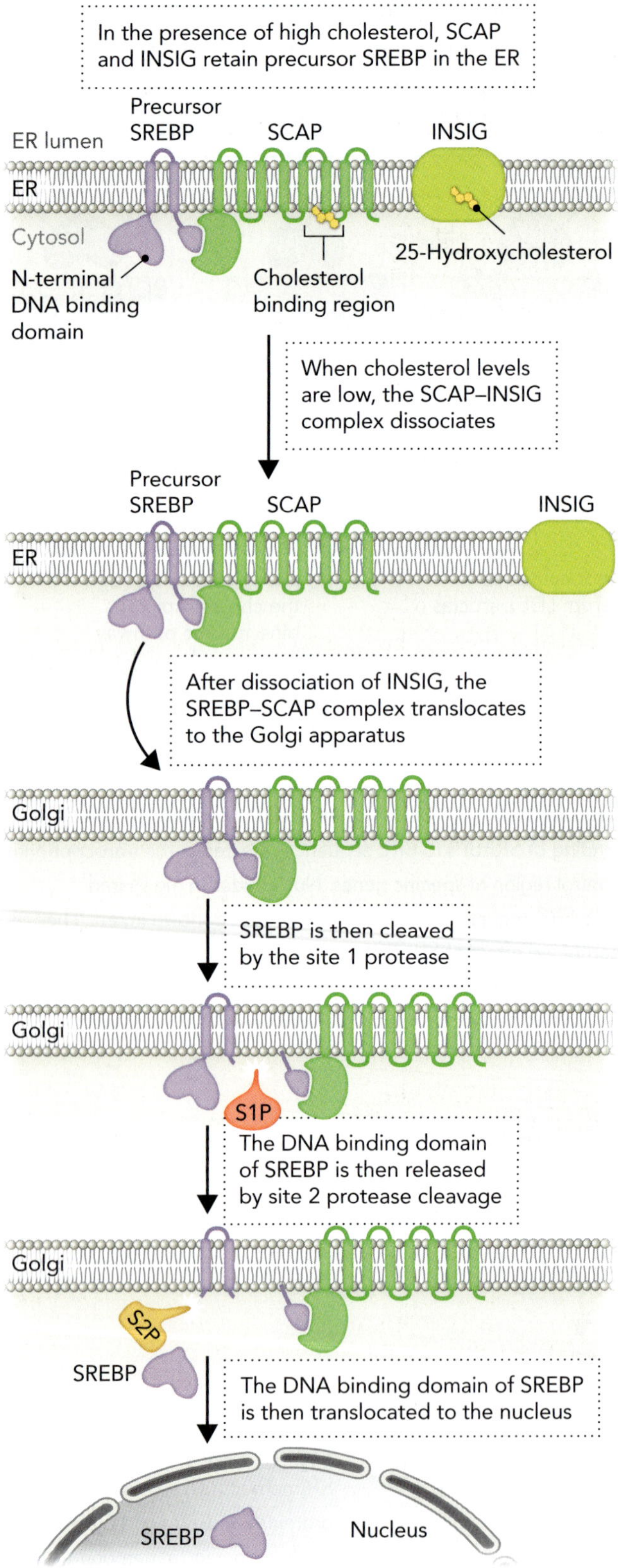

Figure 16.57 Precursor SREBPs are processed by proteolysis in the Golgi apparatus. Cholesterol binding to SCAP or 25-hydroxycholesterol binding to INSIG results in formation of a SREBP–SCAP–INSIG complex that maintains precursor SREBP in the ER. Activating signals lead to disruption of the inhibitory SCAP–INSIG complex, resulting in SCAP-mediated translocation of precursor SREBP to the Golgi apparatus. Sequential cleavage of SREBP proteins by the site 1 and site 2 proteases releases the N-terminal DNA binding domain, which is translocated to the nucleus.

SREBP-1a and SREBP-1c proteins are encoded by a single gene on chromosome 17 and are identical except for the length of the N-terminal domain, which is determined by alternative portions of the gene being used to form the mRNA. (This process, alternative exon splicing, is described in Chapter 21.) The SREBP-1a protein is expressed only in rapidly dividing cells, such as those found in the intestine and spleen. The SREBP-1c isoform is expressed at high levels in liver, adipose, and skeletal muscle cells, and its expression is induced by insulin signaling. The SREBP-2 protein is encoded by a second gene, located on chromosome 22, and is expressed constitutively in all tissues. The SREBP-1c and SREBP-2 proteins are the two major forms required for regulation of lipid metabolism in adult tissues.

Gene expression and DNA binding studies have shown that the SREBP-1c protein is primarily responsible for inducing the expression of genes required for lipogenesis. In contrast, the SREBP-2 protein selectively regulates the expression of genes in the cholesterol biosynthetic pathway. SREBP-1a regulates both sets of genes. This difference in gene specificity by SREBPs is due to differences in the amino acid sequence of the N-terminal domains, resulting in altered affinities for the SRE sequences present in these genes. As shown in Figure 16.56, all three SREBPs increase the expression of genes required for NADPH production, such as malic enzyme in the citrate shuttle and two enzymes in the pentose phosphate pathway (glucose-6-phosphate dehydrogenase and 6-phosphogluconate dehydrogenase). This makes sense because the cholesterol biosynthetic and lipogenic pathways require large amounts of NADPH for reducing reactions.

How is proteolytic activation of precursor SREBPs regulated? The answer is by controlling the accessibility of the ER-localized precursor SREBPs to proteases in the Golgi apparatus. As shown in **Figure 16.57**, low levels of intracellular cholesterol lead to proteolytic activation of precursor SREBPs. Precursor SREBPs remain anchored to the ER membrane in an inactive complex containing at least two other proteins: One is called SREBP cleavage-activating protein (SCAP), which interacts with the C-terminal regulatory domain of SREBP and is required to escort SREBP to the Golgi apparatus, where it is activated by proteolysis. The other protein in the complex is called insulin-induced gene (INSIG), which acts to inhibit SCAP function and thereby retain precursor SREBP in the ER. SCAP is a large membrane protein with eight transmembrane α helices, of which five constitute a "sterol-sensing" domain. Activating signals (low cholesterol levels or insulin) lead to a disruption of the INSIG–SCAP complex, allowing SCAP to escort precursor SREBP to the Golgi apparatus. Once in the Golgi apparatus, the loop in the precursor SREBP membrane domain is cleaved by the site 1 protease (S1P). Then, the N-terminal SREBP DNA binding domain is released into the cytosol by a second proteolytic cleavage catalyzed by the site 2 protease (S2P) and translocated into the nucleus.

Brown and Goldstein discovered that in addition to SREBP, other proteins are also stored as inactive precursors in the ER membrane and released by S1P and S2P cleavage in the Golgi apparatus. The general mechanism of protein activation by proteolytic cleavage is called **regulated intramembrane proteolysis**, and it has now been associated with numerous signaling pathways in bacteria, fruit flies, and humans. Examples include the production of epidermal growth factor receptor ligands during development and the processing of amyloid proteins, which can accumulate and cause neurologic symptoms associated with Alzheimer's disease.

concept integration 16.3

How do statin drugs reduce serum LDL levels and the risk of atherosclerosis?

Statin drugs block the *de novo* cholesterol biosynthetic pathway by inhibiting the activity of HMG-CoA reductase; this results in lower serum LDL levels through increase of LDL receptor gene expression by SREBP. Cholesterol is synthesized in the liver by a *de novo* biosynthetic pathway that is dependent on the activity of HMG-CoA reductase, the rate-limiting enzyme in the pathway. Statin drugs, such as lovastatin, inhibit cholesterol biosynthesis by blocking the activity of HMG-CoA reductase, leading to decreased levels of intracellular cholesterol in liver cells. This triggers proteolytic cleavage of the inactive precursor SREBP in the Golgi apparatus, thereby releasing the N-terminal domain, which translocates to the nucleus. The SREBP N-terminal domain encodes a transcription factor that regulates gene expression by binding to specific DNA sequences present in target genes, such as the LDL receptor gene. SREBP-mediated induction of LDL receptor gene expression leads to higher levels of LDL receptors on the cell surface and subsequent endocytosis of LDL particles, resulting in an overall reduction in serum LDL levels by 20% to 40%. Because high levels of serum LDL (bad cholesterol) are associated with atherosclerosis, statin drugs are an effective treatment for advanced cardiovascular disease.

chapter summary

16.1 Fatty Acid Oxidation and Ketogenesis

- Fatty acids destined for mitochondrial degradation are activated by coenzyme A on the cytosolic side of the outer mitochondrial membrane and then transported into the mitochondrial matrix by the carnitine transport cycle.
- Fatty acyl-CoA synthetases activate short-, medium-, and long-chain fatty acids in the cytosol using an ATP-dependent reaction involving formation of a fatty acyl-adenylate intermediate. The spontaneity of this reaction is enhanced by pyrophosphate hydrolysis.
- The carnitine transport cycle provides a mechanism to control the flux of fatty acids either into the degradative pathway inside the mitochondrial matrix or toward the synthesis of triacylglycerols and membrane lipids in the cytosol. It also serves to maintain separate pools of coenzyme A in the cytosol and mitochondrial matrix.
- Fatty acid degradation converts redox energy available in electron-rich fatty acids into ATP by producing acetyl-CoA, which is oxidized by the citrate cycle, as well as $FADH_2$ and NADH, which are oxidized by the electron transport system.
- The β-oxidation pathway of fatty acid degradation involves the sequential thiolytic cleavage of fatty acyl-CoA substrates at the β carbon of the fatty acid, releasing acetyl-CoA and a fatty acyl-CoA product that is two carbons (C_2) shorter.
- Each cycle of the β-oxidation pathway involves four reactions: (1) an oxidation catalyzed by acyl-CoA dehydrogenase, (2) a hydration catalyzed by enoyl-CoA hydratase, (3) an oxidation catalyzed by 3-L-hydroxyacyl-CoA dehydrogenase, and (4) a thiolysis catalyzed by β-ketoacyl-CoA thiolase.
- Hypoglycin A is a toxin present in unripe ackee fruit that is metabolized in liver mitochondria to produce methylenecyclopropylacetyl-CoA (MCPA-CoA), a potent inhibitor of acyl-CoA dehydrogenases and fatty acid oxidation.
- Degradation of the fully saturated C_{16} fatty acid palmitate by the β-oxidation pathway generates 8 acetyl-CoA, 7 $FADH_2$, and 7 NADH, which are then oxidized by the citrate cycle and electron transport system in the mitochondrial matrix. The overall yield is 106 net ATP per palmitate after subtracting the 2 ATP equivalents invested in the cytosolic acyl-CoA synthetase reaction.
- Oxidation of the fully saturated C_{18} fatty acid stearate yields 120 ATP compared to 90 ATP generated by the complete oxidation of three molecules of glucose having the same number of oxidizable carbons ($3 \times C_6$). Because glucose in glycogen is more hydrated than fatty acids in lipid stores, the difference in energy yield is even more pronounced when comparing a kilogram of lipid and a kilogram of hydrated glycogen.
- Another benefit of β oxidation is the generation of molecular H_2O as a result of combined reactions in the citrate cycle, electron transport system, and ATP synthase complex. These reactions generate 100 L of H_2O from the oxidation of 15 kg of palmitate.
- Oxidation of unsaturated fatty acids requires that cis C═C bonds be converted to trans C═C bonds by an isomerase reaction in order for enzymes in the β-oxidation pathway to completely degrade the unsaturated fatty acid to acetyl-CoA.
- Oxidation of odd-numbered fatty acids such as tricosanoate (23:0) results in the production of propionyl-CoA, which is converted to succinyl-CoA and used by the citrate cycle to generate malate. If the energy charge in the cell is low, the malate can be converted to pyruvate by cytosolic malic enzyme, and the pyruvate can be oxidized.
- Animal cell peroxisomes carry out β oxidation of saturated, very-long-chain fatty acids (VLCFAs) such as hexacosanoate (26:0). This process does not result in energy conversion because peroxisomes lack the necessary enzymes, but it does produce acetyl-CoA for cholesterol biosynthesis in the cytosol.
- X-linked adrenoleukodystrophy (X-ALD) is caused by a defect in the peroxisomal adrenoleukodystrophy protein (ALDP), which is required for peroxisomal import of saturated VLCFAs. Elevated levels of serum VLCFAs in X-ALD individuals causes destruction of neuronal myelin sheaths. Ingestion of a lipid mixture called Lorenzo's oil is a controversial homeopathic treatment for X-ALD.
- When carbohydrate sources are limited, ongoing β oxidation in liver cell mitochondria results in the buildup of excess acetyl-CoA. Ketogenesis is a process in liver cell mitochondria that takes this excess acetyl-CoA and converts it to acetoacetate and D-β-hydroxybutyrate, two energy-rich compounds known as ketone bodies. Spontaneous decarboxylation of acetoacetate generates acetone.
- Acetoacetate and D-β-hydroxybutyrate are exported from the liver and used by skeletal and heart muscle to generate acetyl-CoA for energy conversion reactions. Even the brain, which prefers glucose as an energy source, can adapt to using ketone bodies as chemical energy during times of extreme starvation.
- Ketoacidosis is a condition caused by low blood pH, which can occur when ketogenesis produces more acetoacetate and D-β-hydroxybutyrate than can be used by the peripheral tissues. Ketoacidosis is associated with hypoglycemia, delirium, nausea, vomiting, and a fruity odor on the breath from high acetone levels in the blood.

16.2 Synthesis of Fatty Acids and Triacylglycerols

- The liver synthesizes triacylglycerols from fatty acids when glucose levels are high and the amount of acetyl-CoA produced exceeds the energy requirements of the cell. Enzymatic reactions linking glycolysis, the citrate shuttle, and fatty acid synthesis efficiently convert excess carbohydrates into stored fat in adipose tissue.
- Two similarities between fatty acid degradation and fatty acid synthesis are that both require a four-step reaction cycle, and each pathway involves the removal or addition of C_2 units attached to coenzyme A.
- The distinguishing features of fatty acid degradation are (1) it occurs in the mitochondrial matrix; (2) it uses FAD and NAD^+ as oxidants; (3) it requires multiple enzymes for each reaction cycle; (4) it uses coenzyme A as the acetyl group anchor; and (5) the rate-limiting step is carnitine-mediated transport into the mitochondrial matrix.
- The distinguishing features of fatty acid synthesis are (1) it occurs in the cytosol; (2) it uses NADPH as a reductant; (3) it is catalyzed by a multifunctional enzyme; (4) it uses acyl carrier protein (ACP) as the hydrocarbon anchor; and (5) the rate-limiting step is generation of the reaction cycle substrate malonyl-CoA.
- Each cycle of the fatty acid synthesis pathway involves four reactions: (1) condensation catalyzed by β-ketoacyl-ACP synthase (KS); (2) reduction catalyzed by β-ketoacyl-ACP reductase (KR); (3) dehydration catalyzed by β-hydroxyacyl-ACP dehydratase (DH); and (4) reduction catalyzed by enoyl-ACP reductase (ER).
- Malonyl-CoA is the activated substrate in fatty acid synthesis and is produced by the rate-limiting enzyme acetyl-CoA carboxylase. This is a biotin-dependent enzyme that has three functional activities: a biotin carboxylase, a biotin carboxyl carrier, and a carboxyltransferase.
- The ACP subunit in fatty acid synthase requires an enzyme-linked phosphopantetheine coenzyme group derived from vitamin B_5 (pantothenate), which provides a sulfhydryl attachment site for the growing hydrocarbon chain.
- The first step in fatty acid synthesis requires malonyl/acetyl-CoA ACP transacylase (MAT) to catalyze the priming reaction, which transfers the acetyl group from acetyl-CoA to the coenzyme sulfhydryl group in ACP and then translocates it to a cysteine residue in the KS domain. The acetyl-CoA priming reaction occurs only once for each palmitate product. The other seven acetyl groups enter the cycle when MAT catalyzes a transfer reaction using malonyl-CoA to generate malonyl-ACP.
- With substrates attached to both the KS domain (acetyl-S-Cys) and ACP (malonyl-ACP), the four-step reaction sequence of condensation, reduction, dehydration, and reduction repeats seven times to yield palmitoyl-ACP, which is released as palmitate from ACP after hydrolysis by palmitoyl thioesterase (TE).
- The mammalian fatty acid synthase enzyme is a homodimer encoding all seven protein functions in each monomer subunit on a single polypeptide chain. The upper lobe of the fatty acid synthase complex contains the KR, DH, and ER catalytic sites responsible for the modifying functions, whereas the lower lobe contains the condensing functions encoded by the KS and MAT catalytic sites.
- The primary product of fatty acid synthesis is palmitate, which is used as a substrate for elongation and desaturation reactions localized to the endoplasmic reticulum (ER) in animal cells. The elongation reactions use coenzyme A as the carrier molecule rather than acyl carrier protein (ACP). The desaturating enzymes use molecular oxygen (O_2) as the oxidant and are called mixed-function oxidases.
- Triacylglycerols and glycerophospholipids are derived from diacylglycerol-3-phosphate, also called phosphatidic acid, whereas sphingolipids are derived from ceramide.
- Membrane lipids are synthesized by enzymes associated with the smooth endoplasmic reticulum and must be transported to various membrane targets in the cell.
- The most abundant membrane lipids in cells are glycerophospholipids, which include phosphatidylserine, phosphatidylethanolamine, and phosphatidylinositol. Phosphatidic acid is the building block for all three of these glycerophospholipids.
- Attachment of a phosphocholine head group to ceramide leads to the production of sphingomyelin. Ceramide can also be modified by the addition of monosaccharides, such as glucose, via the enzyme ceramide glucosyltransferase to generate the membrane lipid glucocerebroside.
- Sphingomyelin is a sphingophospholipid, whereas cerebrosides and gangliosides are sphingoglycolipids.
- The citrate shuttle provides a mechanism to stimulate fatty acid synthesis in the cytosol when acetyl-CoA accumulates in the mitochondrial matrix. Citrate export to the cytosol is balanced by malate and pyruvate import, thereby maintaining a steady supply of carbohydrate-derived C_2 acetate units for fatty acid synthesis.
- The primary control point for regulating flux through the fatty acid biosynthetic pathway is modulation of acetyl-CoA carboxylase activity. The activity of acetyl-CoA carboxylase is controlled by both allosteric mechanisms (metabolic control) and covalent modification (hormonal control).
- Metabolic regulation of acetyl-CoA carboxylase activity is mediated by citrate and palmitoyl-CoA, which are allosteric regulators that bind to the enzyme and alter the equilibrium between polymerization (active) and depolymerization (inactive).
- Insulin stimulates dephosphorylation and polymerization of acetyl-CoA carboxylase through activation of protein phosphatase 2A, whereas glucagon stimulates phosphorylation and depolymerization of acetyl-CoA carboxylase through activation of AMP-activated protein kinase (AMPK). Citrate binding to phosphorylated acetyl-CoA carboxylase partially activates the enzyme by stimulating polymerization in the absence of insulin signaling.

- Low energy charge activates AMPK through AMP binding, leading to AMPK-mediated phosphorylation and inactivation of acetyl-CoA carboxylase, thus decreasing flux through the fatty acid synthesis pathway. High glucose levels increase the energy charge and stimulate insulin signaling, which leads to inactivation of AMPK and subsequent increased flux through the fatty acid synthesis pathway.
- Three metabolic control mechanisms regulate flux through the fatty acid synthesis pathway: (1) citrate export from the mitochondrial matrix activates acetyl-CoA carboxylase activity; (2) malonyl-CoA inhibits carnitine acyltransferase I activity to prevent mitochondrial import of fatty acyl-CoA molecules; and (3) palmitoyl-CoA inhibits acetyl-CoA carboxylase activity to decrease fatty acid synthesis.

16.3 Cholesterol Synthesis and Metabolism

- Cholesterol has a critical role in the function of cell membranes and as a precursor of cell signaling molecules, but it is also a contributing factor to cardiovascular disease.
- Cholesterol synthesis occurs in all cells, with the largest amounts made in the liver. Because of this *de novo* cholesterol biosynthetic pathway, cholesterol is not an essential lipid in the diet.
- *De novo* cholesterol biosynthesis consists of four stages: (1) synthesis of mevalonate, a C_6 compound, which is generated by the rate-limiting enzyme HMG-CoA reductase; (2) conversion of mevalonate to isopentenyl diphosphate (C_5) and dimethylallyl diphosphate; (3) formation of squalene (C_{30}) from isopentenyl diphosphate and dimethylallyl diphosphate; and (4) squalene cyclization to form cholesterol (C_{27}).
- Liver cholesterol has three metabolic fates: (1) it is stored as cholesterol esters in lipid droplets; (2) it is packaged into lipoproteins and exported to the circulatory system; and (3) it is converted into bile acids, which aid in digestion of fatty foods.
- Most of the bile acid secreted into the small intestine is returned to the liver and reused; however, some of it is excreted as waste, which provides the only mechanism to rid the body of excess cholesterol. Ingestion of insoluble bile acid resins can be used to decrease serum cholesterol by depleting the body of bile acids and thereby shunting more cholesterol toward bile acid synthesis.
- Steady-state levels of circulating cholesterol are determined by the balance of cholesterol input (diet and *de novo* biosynthesis), cholesterol recycling (returning tissue cholesterol to the liver), and cholesterol output (excretion of bile acids).
- Triacylglycerols and cholesterol are transported through the circulatory system as components of plasma lipoprotein particles, which are membrane-bound vesicles containing a hydrophobic core and apolipoproteins. The sizes of lipoproteins decrease as their lipid cargo is transferred to endothelial cells during circulation.
- There are five major classes of lipoproteins, which differ in density depending on their size and the ratio of protein to lipid: (1) chylomicrons contain large amounts of dietary triacylglycerols obtained in the small intestine; (2) very-low-density lipoproteins (VLDLs) are synthesized in the liver and transport fatty acids in the form of triacylglycerols; (3) intermediate-density lipoproteins (IDLs) are VLDL remnants; (4) low-density lipoproteins (LDLs) transport large amounts of cholesterol esters and are known as "bad cholesterol"; and (5) high-density lipoproteins (HDLs) are cholesterol scavengers and are known as "good cholesterol."
- HDL particles function in reverse cholesterol transport by removing cholesterol from peripheral tissues through activation of serum lecithin–cholesterol acyltransferase, which generates cholesterol esters that are taken back to the liver by HDL.
- Cardiovascular disease is characterized by the buildup of fibrous tissue and cholesterol deposits in arterial walls, a condition called atherosclerosis. Atherosclerotic plaques can rupture, resulting in myocardial infarction (heart attack) or stroke (blockage of the carotid artery to the brain).
- In addition to *de novo* cholesterol biosynthesis, liver cholesterol is also obtained by endocytosis of serum LDL, which is initiated by LDL binding to LDL receptors on the cell surface. Low levels of intracellular liver cholesterol stimulate LDL receptor expression, leading to increased LDL endocytosis and decreased serum LDL levels.
- Inhibition of HMG-CoA reductase activity by statin drugs such as simvastatin decreases liver cholesterol and induces LDL receptor expression through proteolytic activation of sterol regulatory element binding proteins (SREBPs). Statin drugs therefore decrease the risk of cardiovascular disease by lowering serum LDL levels.
- Ezetimibe blocks cholesterol transport in the small intestine, which reduces serum LDL levels by decreasing uptake of dietary cholesterol. Vytorin is a combination simvastatin–ezetimibe drug that lowers serum LDL levels by inhibiting both cholesterol uptake and cholesterol biosynthesis, but the combination drug is not more effective at preventing cardiovascular disease than simvastatin alone.
- SREBPs are embedded in the ER membrane as inactive precursors containing three functional domains: (1) an N-terminal DNA binding domain; (2) a membrane-anchoring domain; and (3) a C-terminal regulatory domain, which interacts with a cholesterol-sensing protein embedded in the ER membrane.
- Cleavage of precursor SREBPs by proteolytic enzymes in the Golgi apparatus releases the N-terminal transcriptional regulatory domain, which then enters the nucleus. It induces expression of genes that modulate enzymes of the cholesterol biosynthetic pathway, such as malic enzyme, glucose-6-phosphate dehydrogenase, and 6-phosphogluconate dehydrogenase.
- Low cholesterol levels in liver cells disrupt the association of SREBPs with two SREBP regulatory proteins in the ER membrane called SCAP and INSIG. This results in proteolytic cleavage and release of the SREBP N-terminal DNA binding domain.

biochemical terms

(in order of appearance in text)
fatty acyl-CoA synthetase (p. 777)
carnitine acyltransferase I (p. 777)
acetyl-CoA carboxylase (p. 777)
fatty acid synthase (p. 777)
carnitine transport cycle (p. 778)
β-oxidation pathway (p. 780)
hypoglycin A (p. 781)
X-linked adrenoleukodystrophy (X-ALD) (p. 786)
Zellweger syndrome (p. 787)
ketogenesis (p. 788)
ketone bodies (p. 789)
ketoacidosis (p. 790)
citrate lyase (p. 791)
acyl carrier protein (ACP) (p. 793)
mixed-function oxidase (p. 801)
citrate shuttle (p. 804)
AMP-activated protein kinase (AMPK) (p. 807)
low-density lipoprotein (LDL) (p. 810)
mevalonate (p. 810)
HMG-CoA reductase (p. 810)
isopentenyl diphosphate (p. 810)
statin drug (p. 812)
cholate (p. 814)
high-density lipoprotein (HDL) (p. 817)
atherosclerosis (p. 819)
myocardial infarction (p. 820)
coronary artery bypass graft (p. 820)
familial hypercholesterolemia (FH) (p. 820)
sterol regulatory element binding protein (SREBP) (p. 824)
sterol regulatory element (SRE) (p. 824)
regulated intramembrane proteolysis (p. 827)

review questions

1. Why do fatty acids destined for degradation in the mitochondrial matrix need to first be activated by linkage to coenzyme A; that is, what purpose does it serve?
2. What are the four types of chemical reactions needed in the mitochondrial β-oxidation pathway to remove a C_2 acetyl group in the form of acetyl-CoA? Include the names of the mitochondrial enzymes that catalyze each reaction.
3. How many molecules of acetyl-CoA, $FADH_2$, and NADH are generated by the mitochondrial β-oxidation pathway from one molecule of stearoyl-CoA? How many net ATP are produced from cytosolic stearate by this same pathway? Show your work.
4. What is the biochemical basis for X-linked adrenoleukodystrophy (X-ALD)? What is the formulation of "Lorenzo's oil," and what is thought to be its mode of action?
5. What causes ketoacidosis, and what are its symptoms?
6. Compare and contrast the fatty acid degradation and fatty acid synthesis pathways in eukaryotic cells.
7. What are the four types of chemical reactions needed in the fatty acid synthesis pathway to add a C_2 acetyl group to the growing hydrocarbon chain? Include the names of the cytosolic enzyme subunits that catalyze each reaction.
8. Considering that malonyl-CoA is the source of C_2 acetyl groups during each cycle of fatty acid synthesis, why are only 7 malonyl-CoA required ($7 \times C_2 = C_{14}$) to synthesize palmitate (a C_{16} fatty acid)? Where does the other C_2 group come from?
9. What reactions in animal cells are required to generate long-chain saturated and unsaturated fatty acids from palmitate?
10. Sphingomyelin contains a phosphate in its head group, so is it a phospholipid or a sphingolipid? Where does the phosphate group come from in sphingomyelin?
11. Describe the pathway by which excess acetyl-CoA produced by the pyruvate dehydrogenase reaction in the mitochondrial matrix of liver cells is used to synthesize triacylglycerols that are exported into the blood as VLDL particles. Specifically, what key enzymes are required, and how is flux through this carbohydrate → lipid pathway maintained under conditions of high energy charge?
12. Describe the four stages of *de novo* cholesterol biosynthesis.
13. What are the primary metabolic fates of liver cholesterol?
14. Describe the three processes that determine the steady-state level of circulating cholesterol in humans.

challenge problems

1. A young child was admitted to a hospital emergency room several hours after consuming a meal containing large amounts of fat. A sample of her blood was found to have extremely high levels of triacylglycerols in the form of chylomicrons. Assuming she had a previously undiagnosed genetic disorder, what are the two most likely protein defects she could have?
2. Why do people with an enzyme deficiency in carnitine acyltransferase have severe hypoglycemia but do not manifest an increase in plasma ketones during starvation?

3. How is the transport of fat mobilized from adipose tissue to peripheral tissues different from the way that dietary fat is transported from the intestine to peripheral tissues?

4. Palmitate uniformly labeled with tritium (^{3}H) to a specific radioactivity of 3.0×10^8 counts per minute per micromole was added to a mitochondrial preparation that oxidizes ^{3}H-palmitate to acetyl-CoA. The experiment calls for hydrolysis of the recovered acetyl-CoA to acetate and determination of the specific radioactivity of the acetate in counts per minute per micromole (cpm/μmol). Assuming that the palmitate is oxidized by the β-oxidation pathway and that there is no unlabeled fatty acid to dilute the tritium, what would be the expected specific radioactivity of the acetate? Show your reasoning and calculations.

5. The compound 2-bromopalmitoyl-CoA inhibits the oxidation of palmitoyl-CoA added directly to isolated, intact mitochondria. In contrast, it has no effect on oxidation of palmitoylcarnitine. What is the likely site of inhibition by 2-bromopalmitoyl-CoA? Explain your conclusion.

6. Oxidation of fatty acids is a major source of water for many desert animals. How much water (in milliliters) is generated by the complete oxidation of 1 g of palmitate given that 130 mol of H_2O are generated per mole of oxidized palmitate?

7. Although animals cannot convert even-numbered fatty acids stored in adipose tissue into carbohydrates, small amounts of glucose can be made from oxidation of ingested odd-chain fatty acids. What are the three key metabolites needed for this to occur?

8. Explain why in humans, triglycerides can be used to produce glucose via the gluconeogenic pathway, but fatty acids cannot.

9. Explain the fact that when palmitate and radioactive CoA are added to a liver homogenate, the palmitoyl-CoA isolated from the cytoplasm is radioactive, but the palmitoyl-CoA isolated from the mitochondria is not radioactively labeled.

10. Calculate the net ATP yield from the complete oxidation of pentanoyl-CoA, the five-carbon acyl-CoA intermediate generated during degradation of odd-chain fatty acids. Use the standard energy currency ratios of 5 ATP/2 NADH and 3 ATP/2 $FADH_2$.

11. When acetyl-CoA produced by β oxidation in the liver exceeds the capacity of the citrate cycle to use acetyl-CoA, the excess acetyl-CoA is used to form acetoacetate and D-β-hydroxybutyrate. High rates of ketogenesis are observed in severe uncontrolled diabetes because the tissues cannot use glucose and must use fats for energy instead. Answer the following questions about ketogenesis.

- **a.** What problems would occur if the liver did not divert acetyl-CoA to ketones? How does diversion to ketone bodies solve the problem?
- **b.** Explain why prolonged aerobic exercise by an athlete on a low-carbohydrate diet can induce ketogenesis.
- **c.** What is the effect of a high-fat and low-carbohydrate diet on ketone body formation? Explain.

12. Fatty acid biosynthesis is an endergonic process. What processes are the thermodynamic driving forces for the following reactions?

- **a.** Acetyl-CoA carboxylase
- **b.** β-Ketoacyl-ACP synthase
- **c.** Conversion of a β-ketoacyl-ACP to a fully saturated acyl-ACP

13. Acetyl-CoA carboxylase is the principal regulation point in the biosynthesis of fatty acids. Addition of citrate raises the v_{max} of the enzyme as much as 10-fold, which exists in polymerized and monomeric forms that differ markedly in their activities.

- **a.** How is the allosteric regulation by citrate linked to the availability of the substrates for fatty acid biosynthesis?
- **b.** How is the activity of acetyl-CoA carboxylase covalently regulated by glucagon?

14. Describe the metabolic and regulatory roles of citrate in controlling fatty acid synthesis.

15. List three reasons why it makes metabolic sense that malonyl-CoA is the carbon donor in fatty acid synthesis rather than acetyl-CoA. Include in your answer the role of ATP and both the chemical and regulatory roles of malonyl-CoA.

16. Familial hypercholesterolemia is a hereditary disease resulting from defects in the LDL receptor. What two primary factors contribute to the high level of serum cholesterol found in patients with this disease?

17. Bassen–Kornzweig syndrome, also known as abetalipoproteinemia, is a rare disorder in which the ability to make apolipoprotein B (apoB) is lost.

- **a.** Describe the expected lipoprotein composition in the blood of patients with this disorder.
- **b.** Is cholesterol biosynthesis regulated up or down in these patients? Explain.
- **c.** Why do these patients accumulate fat in their livers on a high-carbohydrate diet?

18. Fat malabsorption can occur in several disorders, such as pancreatitis, liver disease, AIDS, or intestinal disease. Testing for fat malabsorption traditionally required collecting stools and measuring the fat content—an unpleasant task at best. It has been proposed that a simpler test would be to measure the amount of $^{13}CO_2$ expired in the breath after consuming a meal prepared with a ^{13}C-labeled, fatty acid–containing triacylglycerol.

- **a.** Explain how such a test would work.
- **b.** Can this test distinguish between a defect in pancreatic lipase and a lack of bile salts? Justify your answer.

suggested reading

Books and Reviews

Anderson, R. G. (2003). Joe Goldstein and Mike Brown: from cholesterol homeostasis to new paradigms in membrane biology. *Trends in Cell Biology*, *13*, 534–539.

Brand-Miller, J., and Buyken, A. E. (2012). The glycemic index issue. *Current Opinion in Lipidology*, *23*, 62–67.

Brown, M. S., and Goldstein, J. L. (2006). Biomedicine. Lowering LDL—not only how low, but how long? *Science*, *311*, 1721–1723.

Daniels, T. F., Killinger, K. M., Michal, J. J., Wright, R. W., Jr., and Jiang, Z. (2009). Lipoproteins, cholesterol homeostasis and cardiac health. *International Journal of Biological Sciences*, *5*, 474–488.

Eberle, D., Hegarty, B., Bossard, P., Ferre, P., and Foufelle, F. (2004). SREBP transcription factors: master regulators of lipid homeostasis. *Biochimie*, *86*, 839–848.

Goldstein, J. L., and Brown, M. S. (2009). The LDL receptor. *Arteriosclerosis, Thrombosis, and Vascular Biology*, *29*, 431–438.

Hansson, G. K. (2005). Inflammation, atherosclerosis, and coronary artery disease. *New England Journal of Medicine*, *352*, 1685–1695.

Jeon, T. I., and Osborne, T. F. (2012). SREBPs: metabolic integrators in physiology and metabolism. *Trends in Endocrinology and Metabolism*, *23*, 65–72.

Moser, H. W., Moser, A. B., Hollandsworth, K., Brereton, N. H., and Raymond, G. V. (2007). "Lorenzo's oil" therapy for X-linked adrenoleukodystrophy: rationale and current assessment of efficacy. *Journal of Molecular Neuroscience*, *33*, 105–113.

Schrader, M., and Fahimi, H. D. (2008). The peroxisome: still a mysterious organelle. *Histochemistry and Cell Biology*, *129*, 421–440.

Sherratt, H. S. A. (1986). Hypoglycin, the famous toxin in unripe Jamaican ackee fruit. *Trends in Pharmacological Sciences*, *7*, 186–191.

Smith, J. L., and Sherman, D. H. (2008). Biochemistry. An enzyme assembly line. *Science*, *321*, 1304–1305.

Tong, L. (2005). Acetyl-coenzyme A carboxylase: crucial metabolic enzyme and attractive target for drug discovery. *Cell and Molecular Life Sciences*, *62*, 1784–1803.

Primary Literature

Amemiya-Kudo, M., Shimano, H., Hasty, A. H., Yahagi, N., Yoshikawa, T., Matsuzaka, T., Okazaki, H., Tamura, Y., Iizuka, Y., Ohashi, K., et al. (2002). Transcriptional activities of nuclear SREBP-1a, -1c, and -2 to different target promoters of lipogenic and cholesterogenic genes. *Journal of Lipid Research*, *43*, 1220–1235.

Athappilly, F. K., and Hendrickson, W. A. (1995). Structure of the biotinyl domain of acetyl-coenzyme A carboxylase determined by MAD phasing. *Structure*, *3*, 1407–1419.

Gaussin, V., Hue, L., Stalmans, W., and Bollen, M. (1996). Activation of hepatic acetyl-CoA carboxylase by glutamate and Mg^{2+} is mediated by protein phosphatase-2A. *Biochemical Journal*, *316(Pt 1)*, 217–224.

Horton, J. D., Shah, N. A., Warrington, J. A., Anderson, N. N., Park, S. W., Brown, M. S., and Goldstein, J. L. (2003). Combined analysis of oligonucleotide microarray data from transgenic and knockout mice identifies direct SREBP target genes. *Proceedings of the National Academy of Sciences USA*, *100*, 12027–12032.

Ijioma, N., and Robinson, J. G. (2011). Lipid-lowering effects of ezetimibe and simvastatin in combination. *Expert Review of Cardiovascular Therapy*, *9*, 131–145.

Leibundgut, M., Maier, T., Jenni, S., and Ban, N. (2008). The multienzyme architecture of eukaryotic fatty acid synthases. *Current Opinion in Structural Biology*, *18*, 714–725.

Lomakin, I. B., Xiong, Y., and Steitz, T. A. (2007). The crystal structure of yeast fatty acid synthase, a cellular machine with eight active sites working together. *Cell*, *129*, 319–332.

Maier, T., Leibundgut, M., and Ban, N. (2008). The crystal structure of a mammalian fatty acid synthase. *Science*, *321*, 1315–1322.

Roujeinikova, A., Simon, W. J., Gilroy, J., Rice, D. W., Rafferty, J. B., and Slabas, A. R. (2007). Structural studies of fatty acyl-(acyl carrier protein) thioesters reveal a hydrophobic binding cavity that can expand to fit longer substrates. *Journal of Molecular Biology*, *365*, 135–145.

van Roermund, C. W., Hettema, E. H., van den Berg, M., Tabak, H. F., and Wanders, R. J. (1999). Molecular characterization of carnitine-dependent transport of acetyl-CoA from peroxisomes to mitochondria in *Saccharomyces cerevisiae* and identification of a plasma membrane carnitine transporter, Agp2p. *EMBO Journal*, *18*, 5843–5852.

Zhang, H., Yang, Z., Shen, Y., and Tong, L. (2003). Crystal structure of the carboxyltransferase domain of acetyl-coenzyme A carboxylase. *Science*, *299*, 2064–2067.

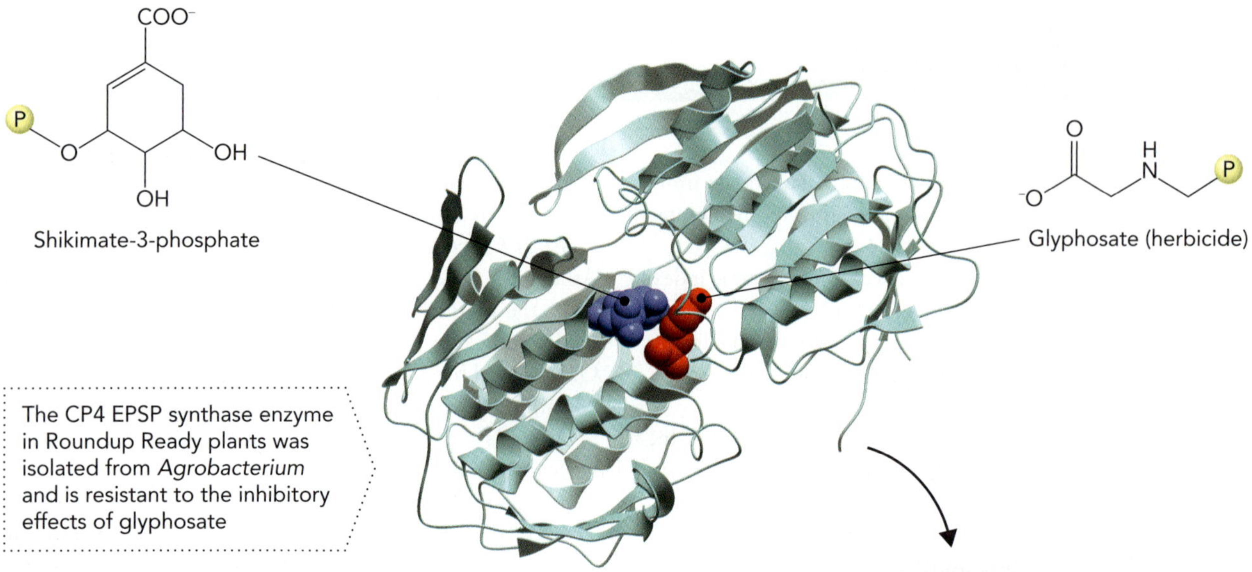

The CP4 EPSP synthase enzyme in Roundup Ready plants was isolated from *Agrobacterium* and is resistant to the inhibitory effects of glyphosate

genuity

ROUNDUP READY®
SUGARBEETS

Roundup Ready sugar beets treated with glyphosate show a 20% increase in crop yields compared to those of nontransgenic sugar beets treated with conventional herbicides and grown under the same conditions

In 2010, up to 95% of the sugar beet crops grown in the USA were Monsanto's Roundup Ready sugar beets, which accounts for nearly half of all sugar sold (the rest comes from sugarcane). The molecular structure of the glyphosate-resistant bacterial enzyme CP4 EPSP synthase is shown with both its substrate, shikimate-3-phosphate, and glyphosate bound to the active site.

17

Amino Acid Metabolism

◀ Glyphosate is the active compound in the herbicide Roundup. It inhibits the enzyme 5-enolpyruvylshikimate-3-phosphate (EPSP) synthase, which is required to synthesize the aromatic amino acids tryptophan, tyrosine, and phenylalanine. Plants sprayed with glyphosate stop growing and die because they cannot make these aromatic amino acids needed for protein synthesis. Transgenic crop plants that contain a gene encoding the glyphosate-resistant bacterial enzyme CP4 EPSP synthase are called Roundup Ready plants and do not die when treated with glyphosate. Studies have shown that Roundup Ready plants produce higher crop yields per acre because weeds are killed and thus are not able to compete for nutrients. Although increased crop yields are desirable, environmental concerns have arisen because of the widespread use of glyphosate and its detrimental effect on native plants. Moreover, glyphosate-resistant weeds are becoming common.

CHAPTER OUTLINE

CREDITS: CP4 EPSP SYNTHASE ENZYME: BASED ON PDB FILE 2GGA; LOGO: COURTESY MONSANTO; SUGAR BEET: BARCIN/ISTOCK/GETTY IMAGES PLUS; SUGAR ON SPOON: MAEXICO/SHUTTERSTOCK.

As discussed in Chapter 1, amino acids and nucleotides are among the fundamental building blocks of life, forming polymers that encode proteins and nucleic acids, respectively. Thus, the synthesis and degradation processes for these molecules are of vital importance to all organisms. Amino acids and nucleotides contain nitrogen, which makes their metabolism different from that in the pathways we have discussed to this point because nitrogen cannot readily be stored for use in anabolic pathways. This means nitrogen-containing compounds need to be continually interconverted by complex pathways to move the nitrogen around. To understand the metabolism of nitrogen-containing compounds, we describe the degradation and synthesis of amino acids here in Chapter 17, which includes the last two of the 10 major metabolic pathways that were introduced in Chapter 9 (see Table 9.1). These two pathways are (1) nitrogen fixation and assimilation and (2) the urea cycle. The metabolism of nucleotides is described in Chapter 18.

As shown in **Figure 17.1**, amino acid metabolism is tightly integrated with the glycolytic, gluconeogenic, and citrate cycle pathways, providing the carbon backbone for shared intermediates. In addition, amino acids are the source of nitrogen required for nucleotide base synthesis (purines and pyrimidines). Most important, whereas bacteria and plants can synthesize all 20 common amino acids, animals can only synthesize 10 amino acids. Therefore, animals must rely on dietary proteins as the source of the other 10 amino acids, which are commonly called *essential* amino acids.

We begin our study of amino acid metabolism by describing the enzymatic process of nitrogen fixation, in which specialized bacteria convert atmospheric nitrogen (N_2) into ammonia (NH_3) through a series of ATP-dependent redox reactions. Note that NH_3 is found as ammonium ion (NH_4^+) under physiologic conditions. In plants that have a symbiotic relationship with these bacteria, amino acids are synthesized by the bacteria and shared with the plants in exchange for carbon-based metabolites. For most other plants, free-living, nitrogen-fixing bacteria in the soil generate ammonium ion, which plants can use directly to synthesize their own amino acids. We then examine protein turnover and amino acid degradation inside cells and observe how the urea cycle provides a means for animals to remove excess ammonia from their bodies. This is followed by a description of essential and nonessential amino acids and a brief look at several representative amino acid biosynthetic pathways. We conclude the chapter by discussing how cells use amino acids as precursors for the synthesis of other nitrogen-containing biomolecules.

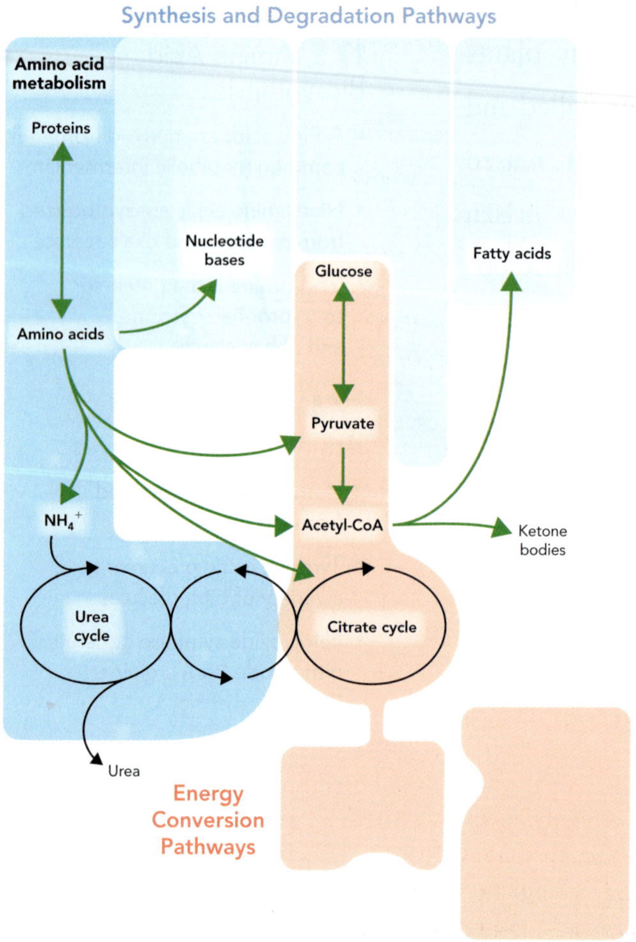

Figure 17.1 Amino acids are the building blocks of proteins and provide the nitrogen for nucleotide bases. The carbon backbone of amino acids can be used for energy conversion processes through metabolic links with glycolysis, the citrate cycle, and gluconeogenesis. Some amino acids are converted to acetyl-CoA, which is a precursor to ketone bodies and fatty acids.

17.1 Nitrogen Fixation and Assimilation

Nitrogen is the most abundant element in the air we breathe (air is 78% nitrogen) and is an essential component of amino acids and nucleotides. Nitrogen is also found in some carbohydrates (glucosamines) and lipids (sphingosines) and in the enzyme cofactors thiamine, NAD^+, and FAD. Nitrogen in biological compounds ultimately comes from the nitrogen gas (N_2) in Earth's atmosphere. But N_2 must first be reduced to NH_3 (or NH_4^+) by the process of **nitrogen fixation** or oxidized to nitrate (NO_3^-) by atmospheric lightning before nonbacterial organisms can use it.

Nitrogen fixation in nature is carried out by bacteria (for example, *Azotobacter*, *Klebsiella,* and *Rhizobium*) and also by species of cyanobacteria (blue-green algae), which live in both soil and aquatic environments. *Rhizobium* is an example of a nitrogen-fixing soil bacterium that lives symbiotically with leguminous plants, such as beans and alfalfa. It has an important role in agriculture by reducing the need for commercial fertilizers. Leguminous plants incorporate nitrogen from bacterially synthesized amino acids into their own amino acids, whereas nonleguminous plants incorporate NH_4^+ produced by nitrogen-fixing soil bacteria directly into amino acids in a process called **nitrogen assimilation.** As you will see, the amino acids glutamate and glutamine are the primary nitrogen carriers in cells. When animals eat plants, the glutamate and glutamine they ingest, along with other amino acids and nucleotides, provide the nitrogen needed to synthesize a variety of biomolecules. As animals are dependent on plants as their only source of nitrogen, the vital role of fixation and cycling of nitrogen by bacteria and plants mirrors the vital role that plants play in providing carbohydrates and O_2 for aerobic respiration in heterotrophic organisms (see Figure 2.2).

Before examining nitrogen metabolism in more detail, we need to answer our four pathway questions about nitrogen fixation and assimilation.

1. *What purpose does nitrogen fixation and assimilation serve in the biosphere?* Nitrogen fixation takes place in bacteria and is the primary process by which atmospheric N_2 gas is converted to ammonium (NH_4^+) and nitrogen oxides (NO_2^- and NO_3^-) in the biosphere. Two other nitrogen fixation processes exist: one is industrial (the Haber process) and the other atmospheric (lightning). Nitrogen assimilation is the process by which plants and bacteria incorporate nitrogen into organic compounds. Most often, NH_4^+ is incorporated into the amino acids glutamate and glutamine.

2. *What are the net reactions of nitrogen fixation and assimilation by plants and bacteria?* Nitrogen fixation in bacteria is mediated by an enzyme called the nitrogenase complex:

$$N_2 + 8\,H^+ + 8\,e^- + 16\,ATP + 16\,H_2O \rightarrow 2\,NH_3 + H_2 + 16\,ADP + 16\,P_i$$

Nitrogen assimilation in plants requires the enzymes glutamine synthetase and glutamate synthase:

$$\alpha\text{-Ketoglutarate} + NH_4^+ + ATP + NADPH + H^+ \rightarrow \text{Glutamate} + ADP + P_i + NADP^+$$

3. *What are the key enzymes in nitrogen fixation and assimilation in plants and bacteria?* The nitrogenase complex uses redox reactions coupled to ATP hydrolysis to convert N_2 gas into 2 NH_3. This enzyme has two functional components: the **Fe protein** contains the binding site for ATP and the 4 Fe–4 S redox

center, and the **MoFe protein** carries out the N_2 reduction reaction using an iron–molybdenum (Fe–Mo) redox center. The enzyme **glutamine synthetase**, which is found in all organisms, incorporates NH_4^+ into glutamate to form glutamine through an ATP-coupled redox reaction. The activity of glutamine synthetase is regulated by allosteric inhibitors and by covalent modification, which is mediated by adenylylation. The enzyme **glutamate synthase**, which is found in bacteria, plants, and some insects, works in concert with glutamine synthetase to replenish glutamate so that the glutamine synthetase reaction is not substrate limited. Glutamate synthase converts α-ketoglutarate and glutamine to two molecules of glutamate. **Glutamate dehydrogenase**, an enzyme also found in all organisms, interconverts glutamate, NH_4^+, and α-ketoglutarate in a redox reaction using NAD^+/NADH or $NADP^+$/NADPH. [We will indicate situations when either of these cofactors can be used by the notation $NAD(P)^+$ and NAD(P)H.] Under conditions of high NH_4^+ concentrations in nature (for example, applications of fertilizer to crop fields), glutamate dehydrogenase assimilates NH_4^+ into glutamate. In animals, however, glutamate dehydrogenase most often generates NH_4^+ from glutamate to initiate the process of nitrogen excretion as urea or uric acid.

4. *What are examples of nitrogen fixation and assimilation in everyday biochemistry?* Natural fertilizers are used in organic farming to reduce the dependence on industrial sources of nitrogen. The two most common sources of natural fertilizers are manure, if livestock are readily available, and crop rotation practices, in which leguminous plants, such as soybean or clover, and nonleguminous plants, such as corn and wheat, are planted in alternating seasons. After the growing season dedicated to leguminous plants is complete, the leguminous plants are plowed under the soil. Nitrogen contained in the plants is released into the soil and processed by soil bacteria to provide nitrogenous compounds for the nonleguminous corn and wheat plants to be planted the next season.

Nitrogen Fixation Reduces N_2 to form NH_3

To obtain nitrogen from the atmosphere for incorporation into biomolecules, a multi-subunit enzyme complex called **nitrogenase** must break the triple bond of the N_2 molecule. But this is not easily done, considering that the bond energy of N_2 is a staggering 930 kJ/mol. To overcome this high energy barrier, one of three processes is required:

1. *biological fixation* by bacteria, which reduces N_2 to NH_3 through an ATP-dependent process that is catalyzed by the nitrogenase complex;
2. *industrial fixation*, in which N_2 and H_2 gases are heated to ~500 °C under a pressure of ~250 atm (~350 kPa) to produce liquid ammonia; or
3. *atmospheric fixation* as a result of lightning, which breaks the triple bond in N_2 and allows nitrogen to combine with oxygen to form nitrogen oxides. These oxides dissolve in rain and fall to Earth.

It has been estimated that ~90% of the nitrogen incorporated into the biosphere (that is, all living organisms on Earth) comes from biological and industrial fixation, with the other 10% resulting from atmospheric fixation. Modern agricultural practices depend heavily on industrial fixation as a source of fertilizer, which accounts for most of the nitrogen humans obtain from eating plants and animals.

Figure 17.2 Industrial nitrogen fixation uses the Haber process, which mixes one part nitrogen with three parts hydrogen to make NH_3 gas that can be cooled and transported as liquid ammonia. F. HABER: ULLSTEIN BILD/ULLSTEIN BILD VIA GETTY IMAGES; C. BOSCH: SUEDDEUTSCHE ZEITUNG PHOTO/ALAMY STOCK PHOTO.

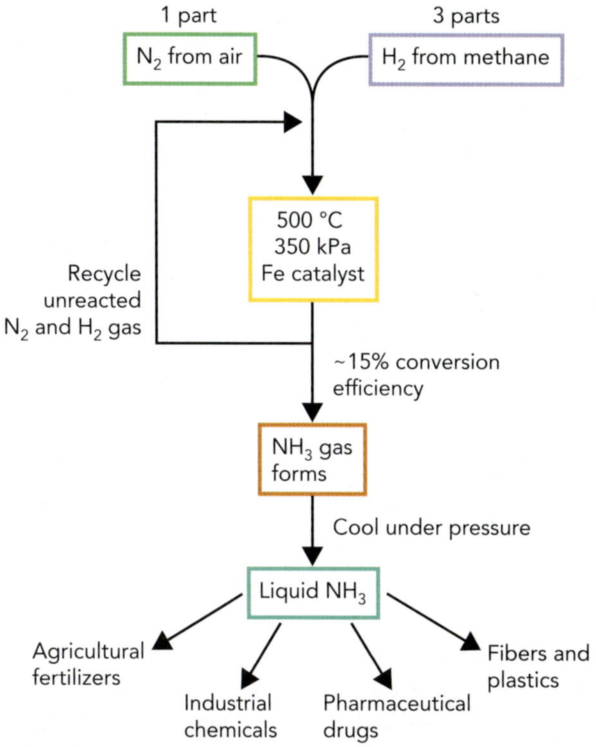

Fritz Haber (1868–1934)

Carl Bosch (1874–1940)

Figure 17.2 illustrates that industrial nitrogen fixation is based on the synthesis of liquid NH_3 from nitrogen and hydrogen gases using the **Haber process**, also known as the Haber–Bosch process. The process is named after the German physical chemist Fritz Haber, who first developed this method in 1909 with the help of the German industrial chemist Carl Bosch. Haber received the 1918 Nobel Prize in Chemistry for his development of industrial ammonia synthesis. Bosch was awarded the 1931 Nobel Prize in Chemistry for his work on chemical high-pressure methods.

The Haber process converts ~15% of the input nitrogen and hydrogen gases into NH_3 during each reaction cycle. The unreacted gases are recycled until a conversion rate of 98% is achieved. Note that the extreme conditions of temperature and pressure used in the Haber process must balance with the theoretical maximum yield of NH_3 formation to make the process cost-effective, hence economically feasible. Commonly used agricultural fertilizers produced with this industrial source of ammonia are ammonium sulfate, $(NH_4)_2SO_4$; ammonium phosphate, $(NH_4)_3PO_4$; and ammonium nitrate, NH_4NO_3. Other uses of liquid ammonia include the production of industrial chemicals (nitric acid, sodium carbonate, hydrogen cyanide), explosives, pharmaceutical drugs, synthetic fibers (nylon), and plastics.

Biological nitrogen fixation by bacteria requires the activity of nitrogenase, which consists of a dimer of two functional components (**Figure 17.3**). One of these

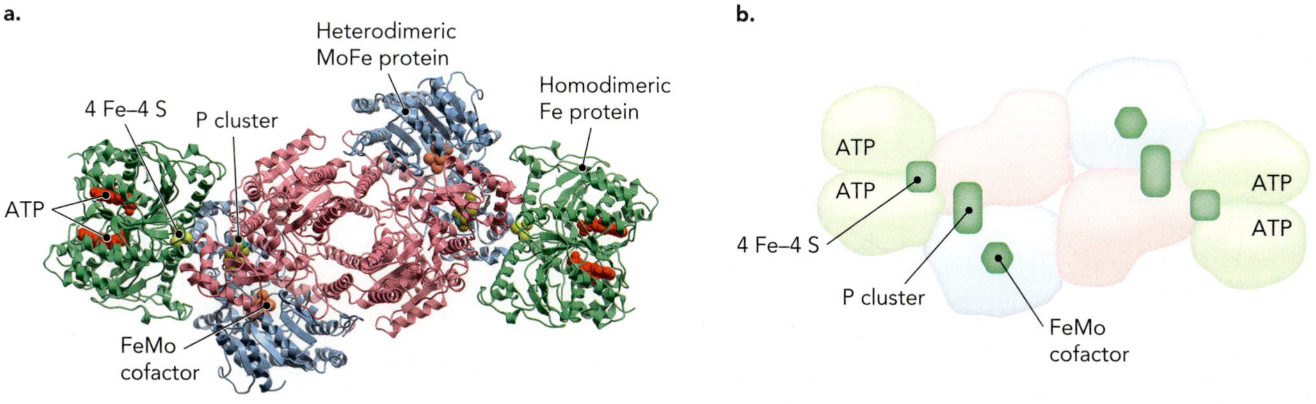

Figure 17.3 The structure of the nitrogenase complex from the soil bacterium *Azotobacter vinelandii* reveals eight polypeptides. **a.** The nitrogenase complex consists of two homodimeric Fe proteins (green) and two heterodimeric MoFe proteins (blue and pink), with a total of six redox centers. BASED ON PDB FILE 1N2C. **b.** Functional arrangement of the eight protein subunits within the *Azotobacter* nitrogenase complex.

functional components is the homodimeric Fe protein, and the other is the heterodimeric MoFe protein. Each subunit of the Fe protein contains a binding site for ATP and a single 4 Fe–4 S redox center connected to cysteine residues. The function of the Fe protein is to obtain electrons from ferredoxin (or flavodoxin) and pass them on to the MoFe protein, which then catalyzes the reduction of N_2 to generate 2 NH_3. Each MoFe protein contains two redox centers: the P cluster, containing an 8 Fe–7 S center, and an iron–molybdenum redox center called the FeMo cofactor (Mo–7 Fe–9 S), which is the site of N_2 reduction. The molecular structure of the nitrogenase complex from *Azotobacter vinelandii*, an aerobic soil bacterium, is shown in Figure 17.3, where it is observed that the two homodimeric Fe proteins and the two heterodimeric MoFe proteins are arranged in a head-to-head configuration.

Each round of nitrogen reduction involves the transfer of 1 e^- to the nitrogenase complex and the hydrolysis of 2 ATP (**Figure 17.4**). Because 6 e^- are required to generate 2 NH_3 from the reduction of N_2, we expect a total of 12 ATP to be invested. But H_2 is produced in a wasteful side reaction from 2 H^+ + 2 e^-, which decreases the efficiency of NH_3 production by nitrogenase. In most bacteria where it has been measured, ~25% of the reduction activity in the MoFe protein complex is diverted to H_2 production. Therefore, on average, 8 e^- and 16 ATP are required to generate 2 NH_3 from N_2.

As illustrated in Figure 17.4, the nitrogenase reaction cycle consists of six steps. In step 1, the exchange of 2 ADP for 2 ATP is accompanied by reduction of the 4 Fe–4 S redox center in the Fe protein. In step 2, the reduced Fe protein forms a complex

Figure 17.4 A schematic illustration is shown of the proposed six-step, ATP-dependent nitrogenase reaction in soil bacteria. Oxidized redox centers are shown in blue outline, and reduced redox centers are shown in red outline. Note that only one electron is transferred at a time, and therefore two cycles are needed at step 5 to convert N_2 to N_2H_2.

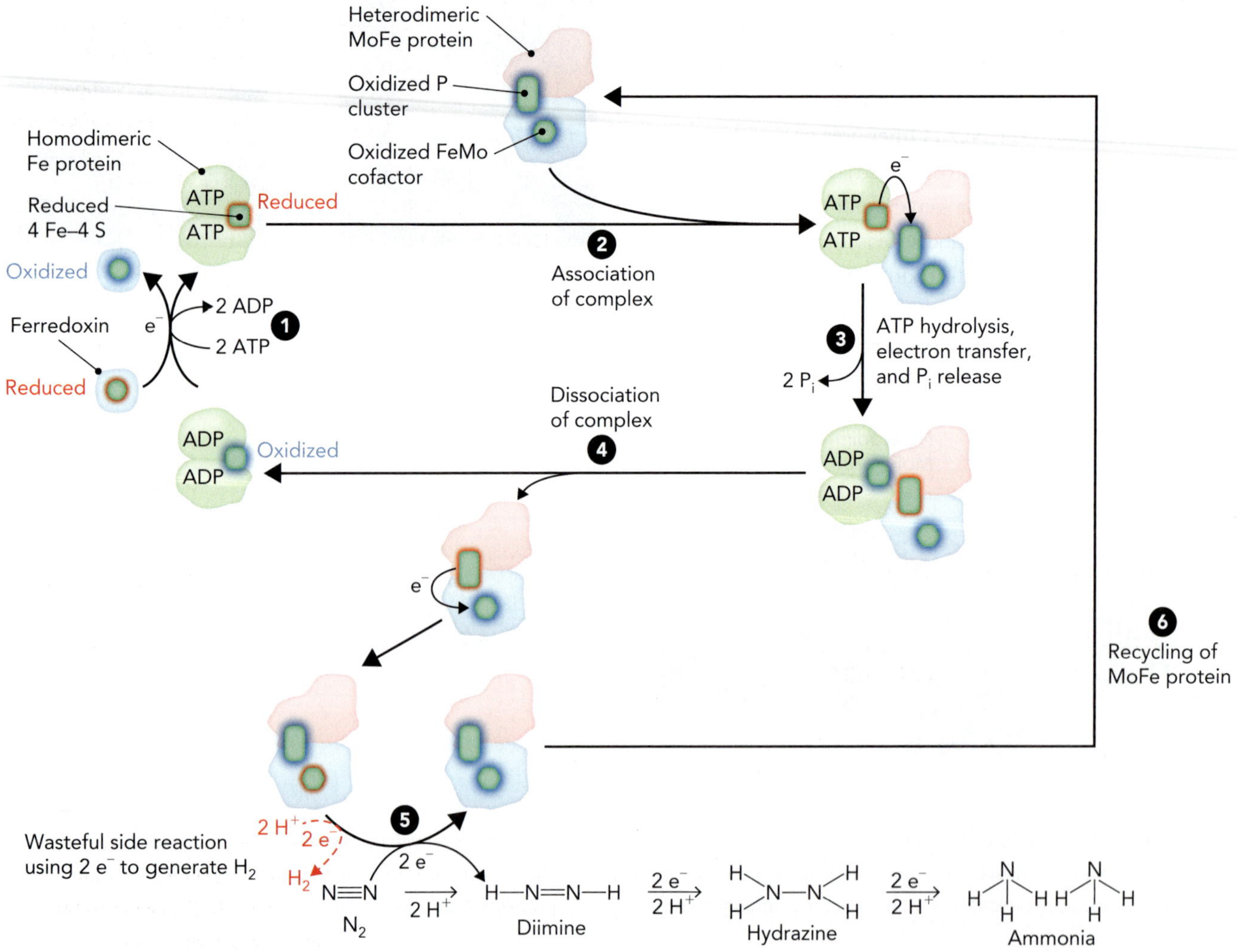

with the oxidized MoFe protein. In step 3, ATP is hydrolyzed, and the electron is transferred from the 4 Fe–4 S cluster in the Fe protein to the P cluster in the MoFe protein. In step 4, the ADP-bound Fe protein dissociates from the MoFe protein complex, and the electron in the P cluster is transferred to the FeMo cofactor, which uses the electron to reduce N_2 (or H^+). In step 5, after a second 1 e^- transfer to N_2, the product diimine (N_2H_2) is generated. In step 6, the oxidized MoFe protein is recycled to obtain another 1 e^- transfer to the FeMo cofactor to complete the reduction of N_2.

In addition to being energetically expensive—and somewhat wasteful by generating unneeded H_2—the nitrogenase reaction is inhibited by O_2. This means nitrogen-fixing bacteria need to either perform this reaction under anaerobic conditions or find a way to reduce O_2 levels locally within the cell. The free-living facultative aerobe *Klebsiella pneumoniae*, for example, only synthesizes the protein components of the nitrogenase complex when it is living in an anaerobic environment and nitrogen fixation is favorable. *Azotobacter vinelandii* uses a different strategy. This bacterium decreases local O_2 concentrations by increasing flux through the electron transport system to reduce O_2 to H_2O rapidly. These high rates of O_2 consumption are accomplished by partially uncoupling the electron transport system from oxidative phosphorylation without completely shutting down ATP synthesis.

A third mechanism to increase the efficiency of the nitrogenase reaction is used by nitrogen-fixing bacteria that live as plant **symbionts**. One of these bacterial species is *Sinorhizobium meliloti*, which invades the roots of leguminous plants through tubular structures called infection threads (**Figure 17.5**). The bacteria exit the infection thread and invade nearby root cells, causing the plant cells to divide and form root nodules harboring the nitrogen-fixing bacteria inside the cytoplasm. Once inside the plant cell, the bacterium loses its cell wall and becomes a **bacteroid**, containing an inner and outer membrane. The symbiotic relationship between *S. meliloti* and clover plants is a classic example of two organisms that find a way to exploit an ecological niche by sharing limited resources.

As shown in **Figure 17.6**, the plant provides fumarate and malate to the bacteroids, which use these citrate cycle metabolites as oxidizable energy sources to generate NADH. The NADH is used to generate ATP by the bacterial electron transport system and oxidative phosphorylation, thereby providing chemical energy to the

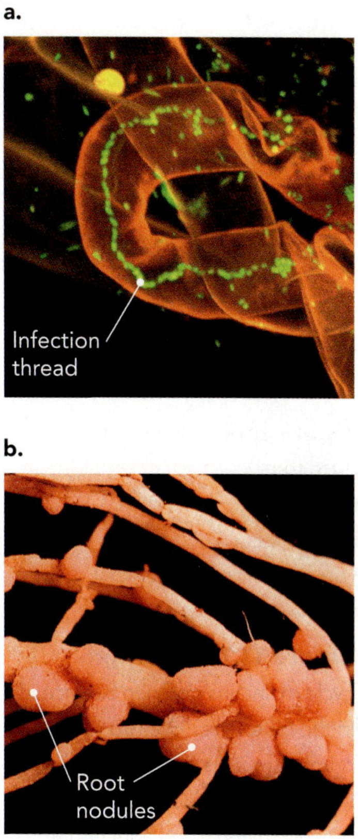

Figure 17.5 Nitrogen-fixing bacteria live symbiotically with leguminous plants. **a.** Nitrogen-fixing bacteria such as *Sinorhizobium meliloti* invade plant root hairs through infection threads. COURTESY OF THE SAMUEL ROBERTS NOBLE FOUNDATION (C. PISLARIU AND M. UDVARDI), ARDMORE, OKLAHOMA. **b.** Root nodules form as a result of plant cell proliferation. HUGH SPENCER/SCIENCE SOURCE/GETTY IMAGES.

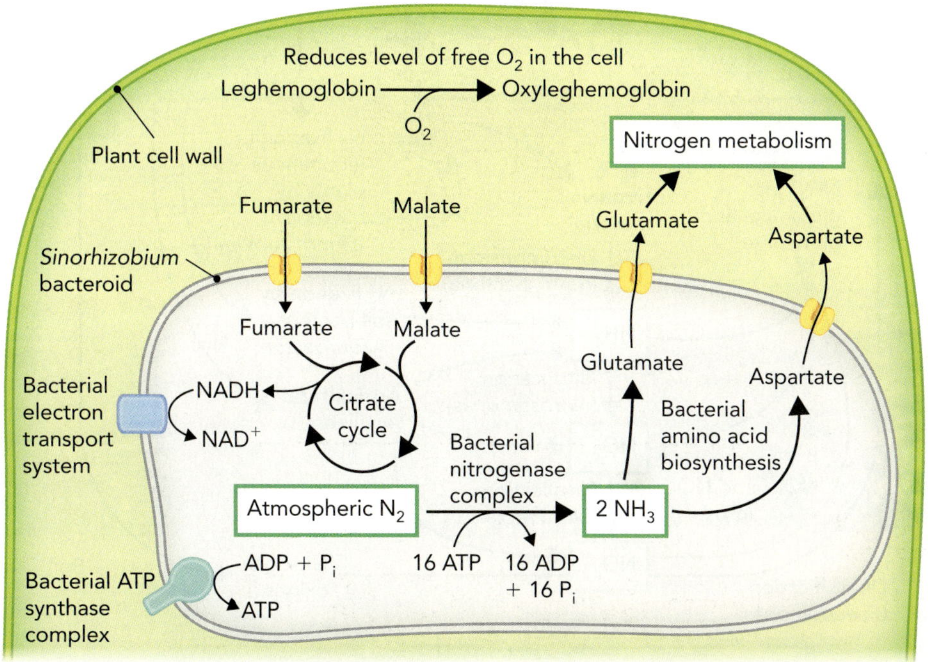

Figure 17.6 Infected plant cells benefit from the nitrogen-fixing activity of symbiotic bacteria, which synthesize amino acids (glutamate and aspartate) that are used by the plant cell as a source of nitrogen. In turn, the plant provides the bacteroid with oxidizable substrates (fumarate and malate) that can be used to generate NADH for ATP production by oxidative phosphorylation.

Sinorhizobium bacteroid. In turn, the nitrogenase complex of the bacteroid generates NH_3, which is used to synthesize amino acids such as glutamate and aspartate that the plant can use as sources of nitrogen. The plant also synthesizes a heme-containing protein called **leghemoglobin**, which sequesters O_2 away from the nitrogenase complex. Leghemoglobin has an extremely high affinity for O_2 and thereby decreases intracellular levels of O_2, which results in increased efficiency of the nitrogenase reaction. Although it was once thought that *Sinorhizobium* synthesized all of the heme needed for production of the plant leghemoglobin protein, it has been shown that many leguminous plants are able to synthesize their own heme in response to *Sinorhizobium* infection.

The decomposition of organic material is a major source of NH_4^+ in the soil. This process begins with the degradation of plant and animal parts by invertebrates (insects and worms); then, a variety of bacterial and fungal species complete the decomposition of nitrogen-containing compounds by releasing NH_4^+. Most of this NH_4^+, however, as well as the NH_4^+ produced by the free-living soil bacteria *Azotobacter* and *Klebsiella,* enters the biosphere in the form of nitrite (NO_2^-) and nitrate (NO_3^-), which are produced by bacteria in the soil through the process of **nitrification.** These bacteria use NH_4^+ as a source of electrons for energy conversion processes and convert most of the NH_4^+ in the soil to NO_2^- and NO_3^-. For example, bacteria in the genus *Nitrosomonas* oxidize NH_4^+ to NO_2^-, and *Nitrobacter* bacteria oxidize NO_2^- to form NO_3^-. Plants take up the NO_2^- and NO_3^- in the soil and convert it back to NH_4^+ using the enzymes **nitrite reductase** and **nitrate reductase.**

Nitrogen balance in the biosphere can be illustrated as a **nitrogen cycle** (**Figure 17.7**), incorporating the processes of nitrogen fixation, nitrification, and nitrate

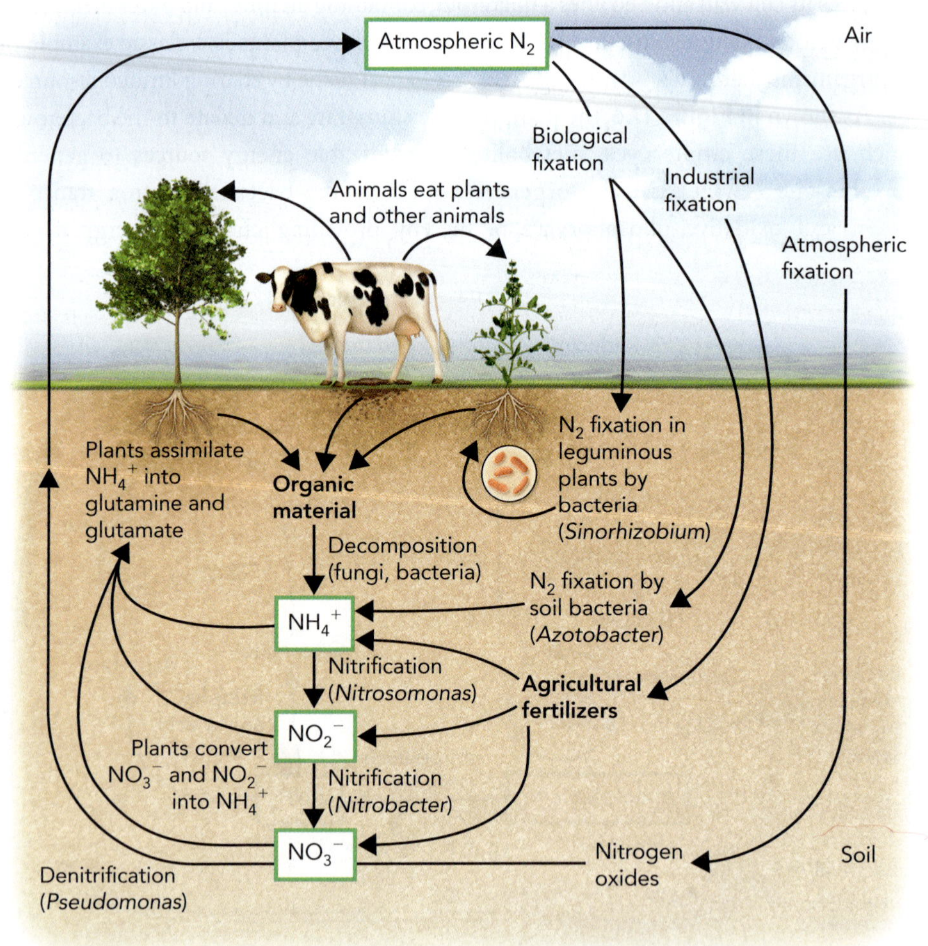

Figure 17.7 The nitrogen cycle maintains nitrogen balance in the biosphere. The two primary sources of nitrogen in the soil are the decomposition of organic material and agricultural fertilizers. Atmospheric nitrogen is converted to NH_4^+ by biological, industrial, and atmospheric fixation processes. Bacteria in the soil convert nitrogen to NH_4^+ either as symbionts with leguminous plants (for example, *Sinorhizobium*) or as free-living organisms (for example, *Azotobacter*). The NH_4^+ in the soil, derived from decomposition, free-living soil bacteria, and human-made fertilizers, is converted to NO_2^- and NO_3^- by soil bacteria that carry out the process of nitrification (for example, *Nitrosomonas* and *Nitrobacter*). Plant roots absorb nitrites and nitrates (NO_2^- and NO_3^-) in the soil and convert them back into NH_4^+ using nitrite and nitrate reductase enzymes. The assimilation of NH_4^+ into amino acids by plants provides a source of nitrogen for animals (directly or indirectly). The denitrification process carried out by bacteria that reduce nitrites and nitrates (for example, *Pseudomonas*) releases N_2 back into the atmosphere.

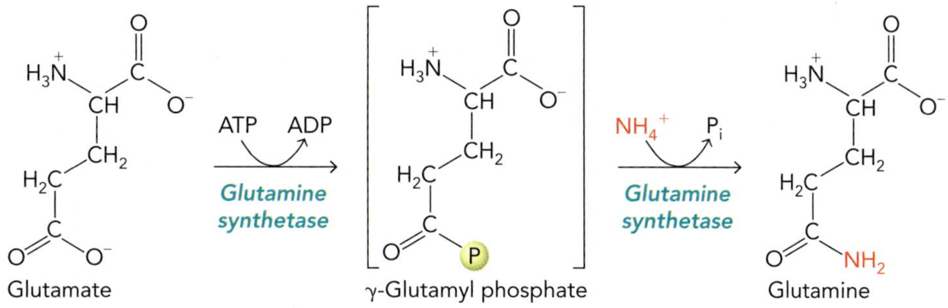

Figure 17.8 The glutamine synthetase reaction requires ATP hydrolysis and involves the formation of γ-glutamyl phosphate as a reaction intermediate.

reduction. It is important to note that the level of atmospheric N_2 is maintained by several species of *Pseudomonas* and *Bacillus* bacteria that reduce NO_3^- and NO_2^- to N_2 under anaerobic conditions by the process of **denitrification**. This form of "anaerobic respiration" results from the denitrifying bacteria using NO_3^- rather than O_2 as the final electron acceptor in the electron transport system.

Assimilation of Ammonia into Glutamate and Glutamine

Plants and bacteria use the available NH_4^+ generated by nitrogen fixation to synthesize the amino acids glutamate and glutamine, which are the primary sources of nitrogen for biosynthetic pathways in all organisms. Glutamate provides nitrogen for amino acid biosynthesis through the action of aminotransferase enzymes, which transfer the α amino group from glutamate to α-keto acids. In contrast, glutamine is the primary source of amino groups for the biosynthesis of a variety of biomolecules, including nucleotide bases, glucosamine-6-phosphate, carbamoyl phosphate, and the side chains of tryptophan and histidine. The incorporation of NH_4^+ into glutamate and glutamine is referred to as **ammonia assimilation** and is mediated by three enzymes: (1) glutamine synthetase, which is found in all organisms; (2) glutamate synthase, which is found in plants, bacteria, and some insects (mosquitoes and silkworms); and (3) glutamate dehydrogenase, which is found in all organisms but functional in ammonia assimilation only when NH_4^+ levels are very high.

Glutamine synthetase uses NH_4^+ in a reaction that converts glutamate to glutamine. This two-step reaction requires ATP and involves the formation of a phosphoryl intermediate, γ-glutamyl phosphate, in the first step (**Figure 17.8**). In the second step of this reaction, NH_4^+ replaces the phosphate group to form glutamine. This reaction is the primary entry point for NH_4^+ into biomolecules. But glutamine synthetase is also important in animal cells as a means to transport NH_4^+ from peripheral tissues to the liver, where it can be excreted in the form of urea.

Considering that glutamine synthetase is a primary NH_4^+ assimilation reaction in plants, how is the level of glutamate maintained to keep the reaction going? The answer is through the reductive amination of α-ketoglutarate by the enzyme glutamate synthase, which uses glutamine as the amino donor (**Figure 17.9**). The glutamate synthase

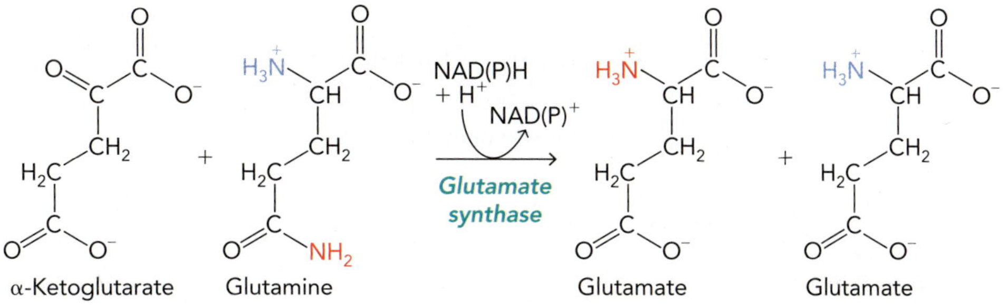

Figure 17.9 The glutamate synthase reaction transfers the amide nitrogen from glutamine to α-ketoglutarate, generating two molecules of glutamate.

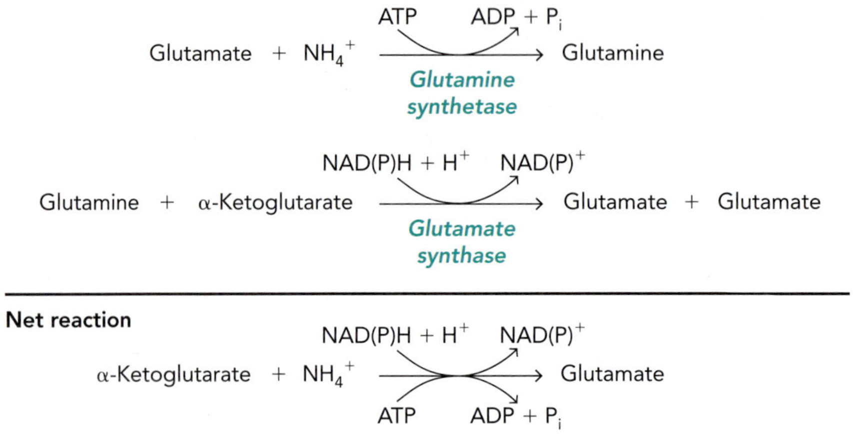

Figure 17.10 Under normal conditions of low NH_4^+ concentration, NH_4^+ assimilation into glutamate requires the combined reactions of glutamine synthetase and glutamate synthase in plants and bacteria.

reaction uses NAD(P)H as the reductant, catalyzing a reaction transferring the amide nitrogen from glutamine to α-ketoglutarate. This results in the production of two molecules of glutamate. Note that by combining the glutamate synthase reaction with the glutamine synthetase reaction, the net result is incorporation of NH_4^+ into glutamate at the expense of ATP hydrolysis and NAD(P)H oxidation (**Figure 17.10**). Because the glutamate synthase reaction occurs only in plants and bacteria, animals depend on these organisms for their source of nitrogen for biosynthetic pathways.

The third important enzyme in nitrogen assimilation is glutamate dehydrogenase, which catalyzes a reaction interconverting α-ketoglutarate and glutamate in the presence of NH_4^+ (**Figure 17.11**). It was once thought that this reaction was primarily responsible for nitrogen assimilation in plants because it should be able to incorporate NH_4^+ from nitrogen-fixing bacteria into glutamate. It was discovered, however, that the K_m of glutamate dehydrogenase for NH_4^+ is ~1 mM, and moreover, the $\Delta G°'$ of the reaction in the direction written is +30 kJ/mol. Therefore, it is more likely that glutamate dehydrogenase only plays a role in nitrogen assimilation in crop plants when nitrogenous fertilizers are applied to the fields in high concentrations. Indeed, as will be described later in the chapter, the glutamate dehydrogenase reaction in animals most often generates NH_4^+ for carbamoyl phosphate synthesis by deaminating glutamate in the highly favorable reverse reaction ($\Delta G°'$ = −30 kJ/mol).

Metabolite Regulation of Glutamine Synthetase Activity

Glutamine synthetase is best characterized in bacteria, where it exists as a large protein complex containing 12 identical subunits, as shown in **Figure 17.12**. The activity of glutamine synthetase is tightly controlled by a form of allosteric regulation known

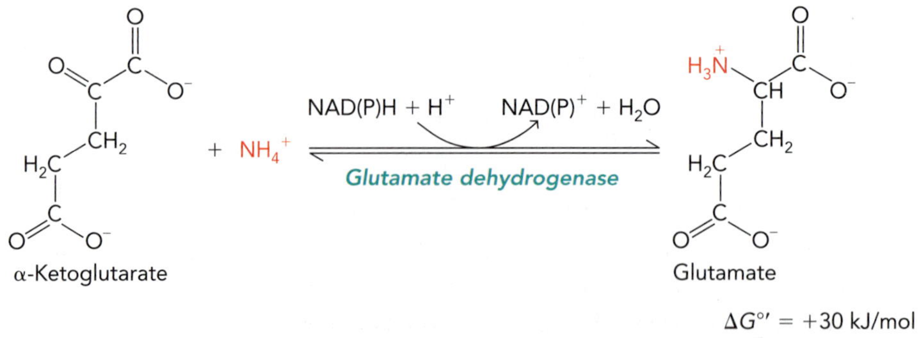

Figure 17.11 The glutamate dehydrogenase reaction can assimilate NH_4^+ into glutamate under conditions of very high NH_4^+ concentrations, as in heavily fertilized agricultural fields. Glutamate dehydrogenase can use either NADH or NADPH as the reductant.

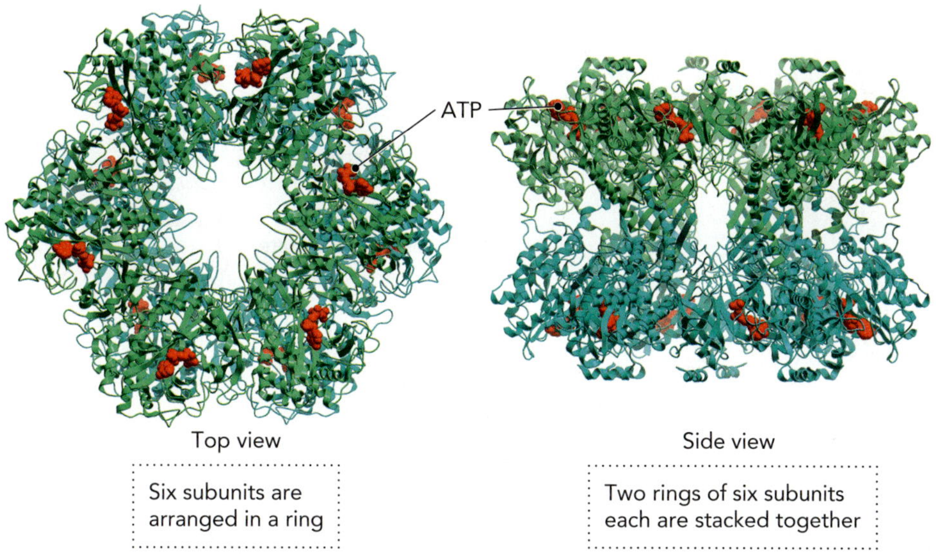

Figure 17.12 In the bacterial glutamine synthetase complex from *Salmonella typhimurium*, 12 identical subunits are arranged in a stack of two hexameric rings. An ATP substrate is bound to each of the 12 enzyme active sites (red molecules). BASED ON PDB FILE 1FPY.

as feedback inhibition and by an unusual type of covalent modification called adenylylation. This modification involves a phosphodiester linkage of adenosine monophosphate (AMP) to a tyrosine residue in glutamine synthetase. These two regulatory mechanisms function together to precisely control the activity of this rate-limiting enzyme in response to steady-state levels of glutamine and glutamine metabolites.

In addition to an active site in each of the 12 subunits that binds the substrates glutamate, ATP, and NH_4^+, the subunits also contain binding sites for negative allosteric regulators derived from glutamine, which inhibit enzyme activity. Cumulative binding of these metabolites to the glutamine synthetase complex results in synergistic inhibition of enzyme activity. As shown in **Figure 17.13**, allosteric inhibitors of the enzyme include glucosamine-6-phosphate, carbamoyl phosphate, AMP, cytidine triphosphate (CTP), and the amino acids histidine and tryptophan. Three other allosteric inhibitors are the amino acids alanine, glycine, and serine, which signal an overabundance of these amino acids commonly used in protein synthesis.

As described in Chapter 7 (see Figure 7.58), bacterial glutamine synthetase activity is inhibited by covalent modification through adenylylation of Tyr397 by the enzyme glutamine synthetase adenylyltransferase. Glutamine synthetase adenylyltransferase catalyzes both the adenylylation (inactivation) and deadenylylation (activation) of glutamine synthetase, depending on the activity of a regulatory subunit in the enzyme called PII (**Figure 17.14**). Similar to the control of glutamine synthetase activity by AMP modification of a tyrosine residue, regulatory activity of the glutamine synthetase adenylyltransferase PII subunit is also controlled by modification with a mononucleotide (nucleotidylylation). In this case, uridine monophosphate (UMP) modification (uridylylation) of glutamine synthetase adenylyltransferase on the Tyr51 residue of PII by the enzyme uridylyltransferase activates the deadenylylating activity of glutamine

Figure 17.13 Glutamine synthetase activity is regulated by allosteric inhibitors. At least six metabolites of glutamine are known to be negative allosteric regulators of glutamine synthetase activity, as are the amino acids alanine, glycine, and serine.

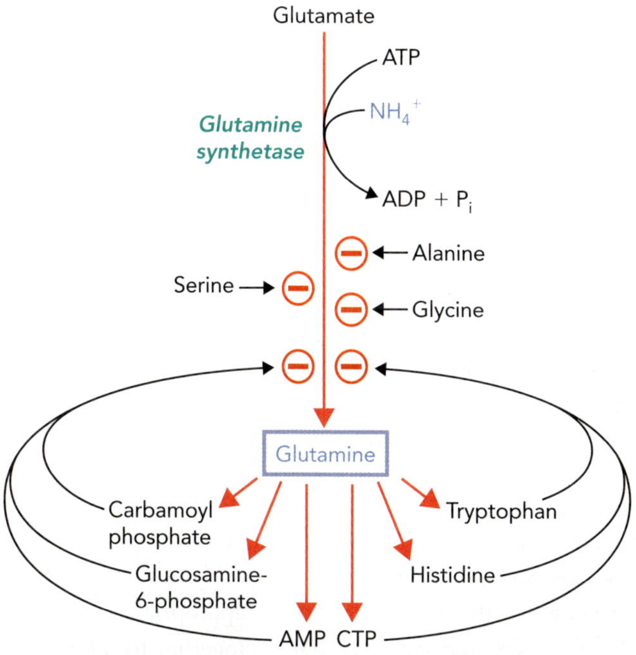

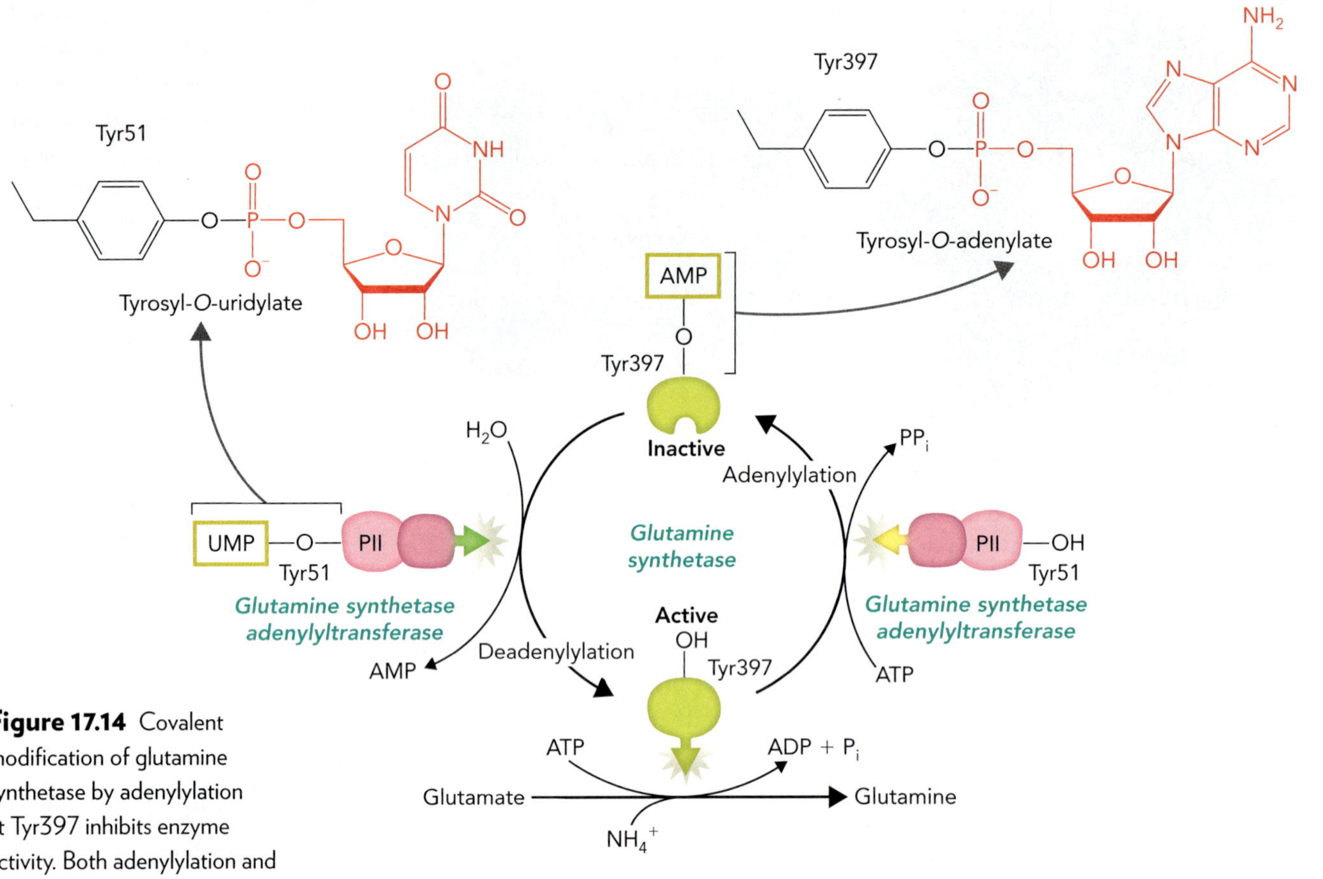

Figure 17.14 Covalent modification of glutamine synthetase by adenylylation at Tyr397 inhibits enzyme activity. Both adenylylation and deadenylylation of glutamine synthetase are mediated by the enzyme glutamine synthetase adenylyltransferase. In turn, the enzyme activity is controlled by uridylylation of Tyr51 in the PII regulatory subunit by uridylyltransferase (see Figure 7.59). The chemical structures of the adenylyl and uridylyl modifications on glutamine synthetase and glutamine synthetase adenylyltransferase, respectively, are shown in the insets.

synthetase adenylyltransferase, leading to activation of glutamine synthetase (see Figure 7.59). But when the uridylyl group is removed from the glutamine synthetase adenylyltransferase PII subunit by the same uridylyltransferase enzyme, then the adenylylating activity of glutamine synthetase adenylyltransferase is stimulated and glutamine synthetase is inhibited.

Aminotransferase Enzymes Play a Key Role in Amino Acid Metabolism

Aminotransferases (also known as transaminases) play an important role in amino acid degradation and synthesis, as well as in nitrogen transport between tissues. Aminotransferase reactions are close to equilibrium under standard conditions ($\Delta G^{\circ\prime} = {\sim}0$), which means the direction of the reaction under physiologic conditions is determined by substrate availability. **Figure 17.15** illustrates a typical aminotransferase reaction in which α-ketoglutarate serves as the acceptor molecule, resulting in the formation of glutamate and the α-keto acid of the deaminated amino acid. Two specific examples of aminotransferase reactions of clinical relevance are also shown: the aspartate aminotransferase reaction and the alanine aminotransferase reaction. Both aspartate aminotransferase and alanine aminotransferase are commonly measured as indicators of liver health.

Most animals contain at least 18 aminotransferase enzymes, each of which recognizes a different amino acid (aminotransferases specific for lysine and threonine are lacking). Because the reactions are reversible, the relative levels of some amino acids can be adjusted to reflect the protein synthesis needs of the cell. As long as the appropriate α-keto acids are available, two aminotransferase reactions can be linked together to interconvert amino acid pools. As shown below, by linking the aspartate aminotransferase and alanine aminotransferase reactions together (in opposite directions), aspartate can be converted to alanine through the common intermediates glutamate and α-ketoglutarate:

$$\textbf{Aspartate} + \alpha\text{-Ketoglutarate} \xrightleftharpoons{\text{Aspartate aminotransferase}} \text{Oxaloacetate} + \text{Glutamate}$$

$$\text{Glutamate} + \text{Pyruvate} \xrightleftharpoons{\text{Alanine aminotransferase}} \alpha\text{-Ketoglutarate} + \textbf{Alanine}$$

$$\textbf{Aspartate} + \text{Pyruvate} \rightleftharpoons \text{Oxaloacetate} + \textbf{Alanine}$$

In all of the aminotransferases that have been characterized, the catalytic mechanism involves a two-stage reaction in which the α amino group of the amino acid is first transferred to an enzyme-linked **pyridoxal phosphate (PLP)** group. Pyridoxal phosphate is a coenzyme derivative of vitamin B_6 (pyridoxine). This first stage results

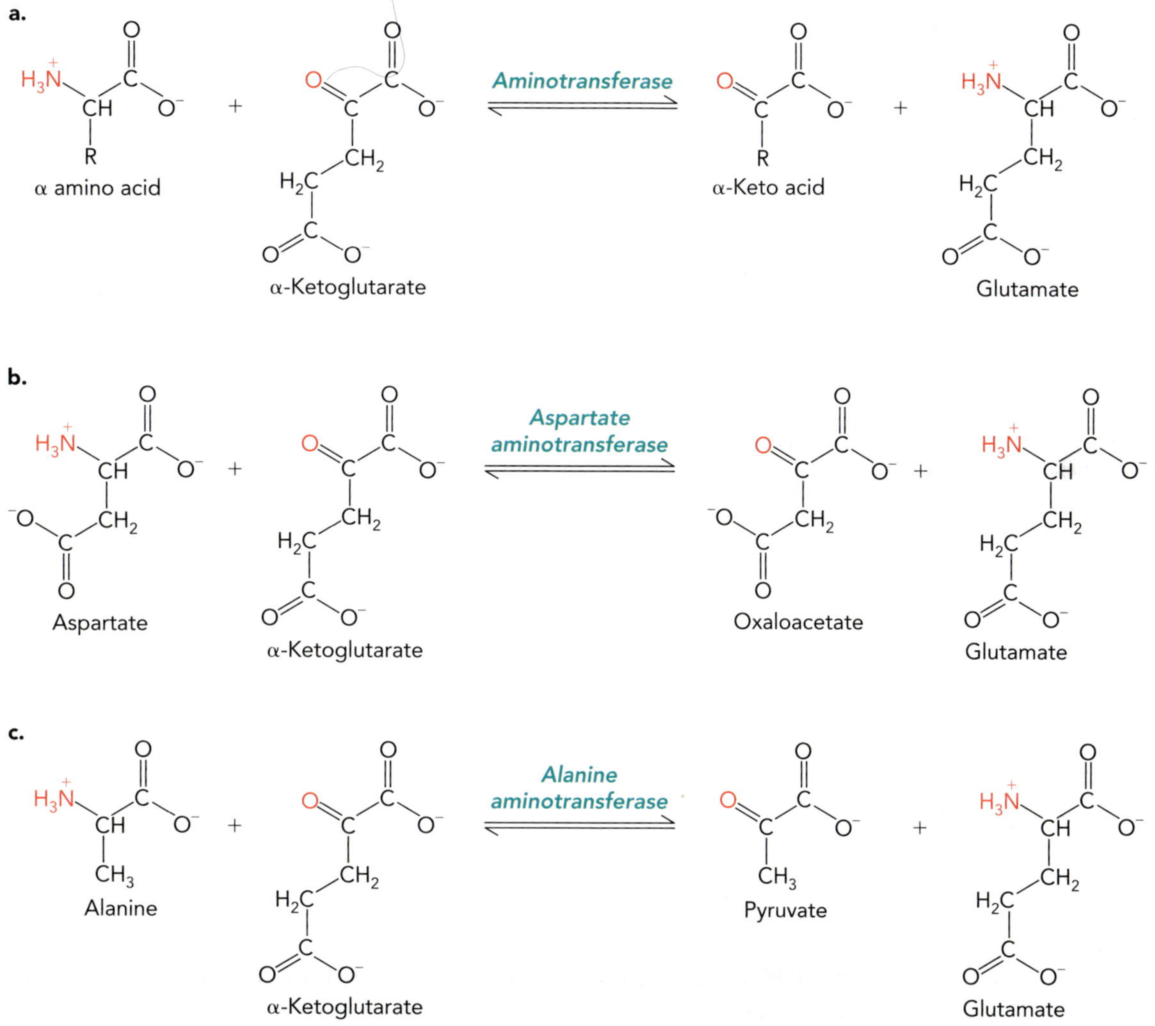

Figure 17.15 Aminotransferase enzymes catalyze reversible reactions that transfer the amino group of amino acids to α-keto acids. **a.** Aminotransferase reaction using α-ketoglutarate as the acceptor of the α amino group, resulting in the formation of glutamate and the corresponding α-keto acid. **b.** The α amino acid in the aspartate aminotransferase reaction is aspartate, and the α-keto acid is oxaloacetate. **c.** The α amino acid in the alanine aminotransferase reaction is alanine, and the α-keto acid is pyruvate. Note that the direction of aminotransferase reactions is determined by the relative concentrations of substrates and products.

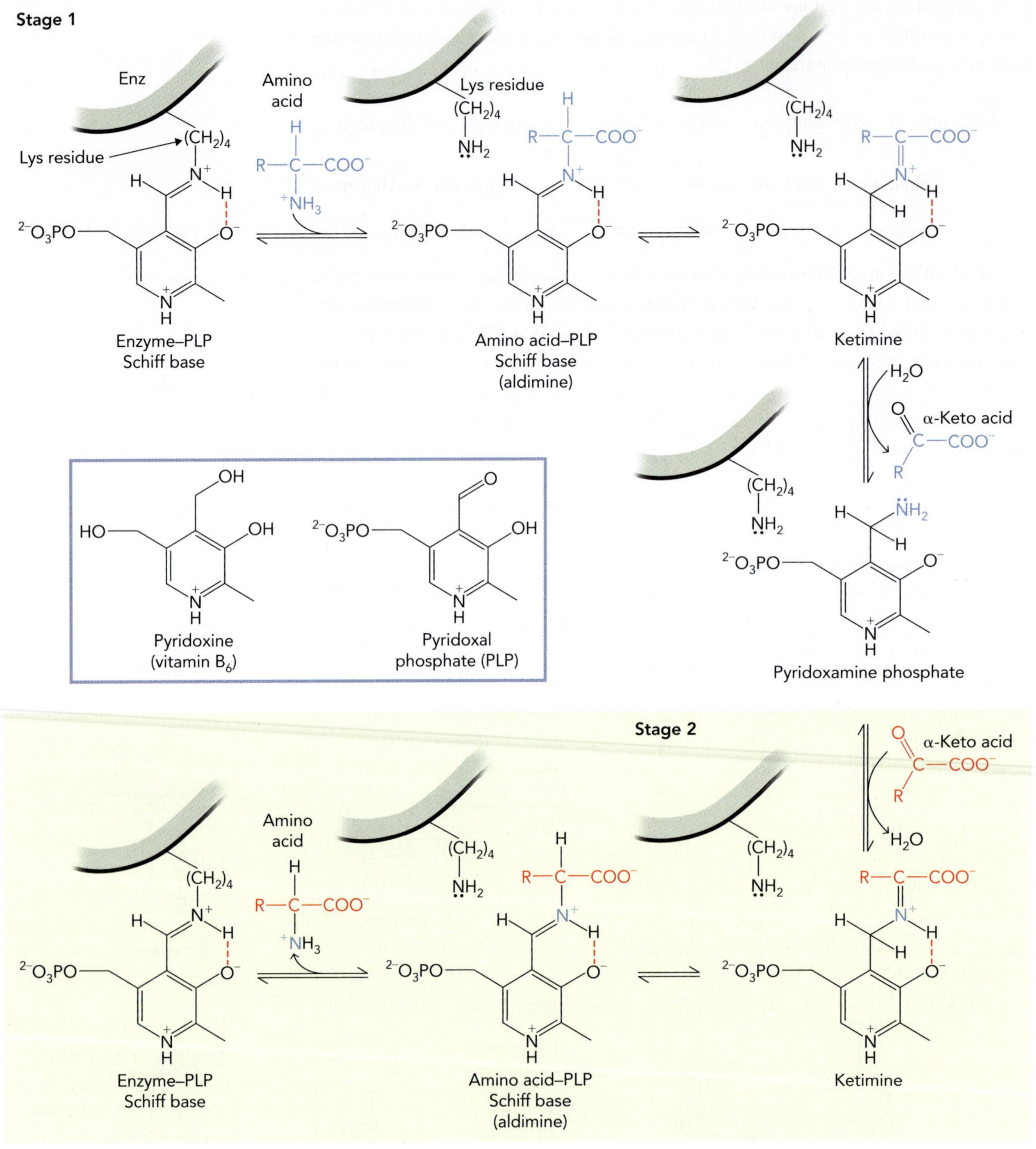

Figure 17.16 The aminotransferase reaction involves a two-stage ping-pong mechanism that requires the coenzyme PLP. In the first stage of the reaction, the α amino group of the amino acid substrate is transferred to the enzyme-bound PLP group to form pyridoxamine phosphate and an α-keto acid. In the second stage, this nitrogen is transferred to an incoming α-keto acid (α-ketoglutarate or oxaloacetate) to form the amino acid product (glutamate or aspartate). The inset shows the structure of pyridoxine (vitamin B_6) and PLP.

in the formation of pyridoxamine phosphate and release of the corresponding α-keto acid. In the second stage, the amino group from pyridoxamine phosphate is transferred to an incoming α-keto acid (α-ketoglutarate or oxaloacetate) to generate the amino acid product.

The aminotransferase reaction mechanism is a classic example of so-called ping-pong enzyme kinetics in which the first product leaves the active site before the second substrate enters. This two-stage mechanism is illustrated in **Figure 17.16**. Observe that in the absence of substrate, the aldehyde group on PLP is condensed with a lysine residue in the enzyme through a Schiff base linkage. When the amino acid substrate binds to the active site, the amino group undergoes a nucleophilic attack on the PLP moiety to form an amino acid–PLP Schiff base in place of the Schiff base formed between the ε amino group of the lysine residue and PLP. The resulting aldimine intermediate undergoes tautomerization, converting the amino acid–PLP Schiff base into an α-keto acid–pyridoxamine phosphate Schiff base (ketimine). The α-keto acid is then released by hydrolysis, leaving behind pyridoxamine phosphate in the enzyme active site.

In the second stage of the aminotransferase reaction, these same chemical reactions occur in the reverse order. First, pyridoxamine phosphate reacts with an incoming α-keto acid to form the amino acid–pyridoxamine phosphate Schiff base (ketimine) intermediate. The tautomerization reaction then converts the ketimine into the corresponding amino acid–PLP Schiff base (aldimine) intermediate. In the last step, the ε amino group of the lysine residue in the enzyme attacks the Schiff base to re-form the PLP–enzyme linkage and release the amino acid product. Note that the different aminotransferases bind a specific amino acid substrate in stage 1 of the reaction, but only transfer the amino group from pyridoxamine phosphate to either α-ketoglutarate or oxaloacetate in stage 2. This results in the funneling of the α amino nitrogen derived from amino acid degradation to only glutamate or aspartate, which are then used to form urea as described in the next section.

Aspartate aminotransferase and alanine aminotransferase are found at high levels in muscle and liver cells and are not normally found at appreciable levels in the blood. But heart or liver damage, which can occur as a result of a heart attack or liver degeneration due to cirrhosis or hepatitis C virus infection, leads to cell death and subsequent leakage of these enzymes into the blood. By measuring the level of aspartate aminotransferase and alanine aminotransferase (ALT) in serum, physicians can determine the extent of tissue damage. **Figure 17.17** shows a typical time course of hepatitis

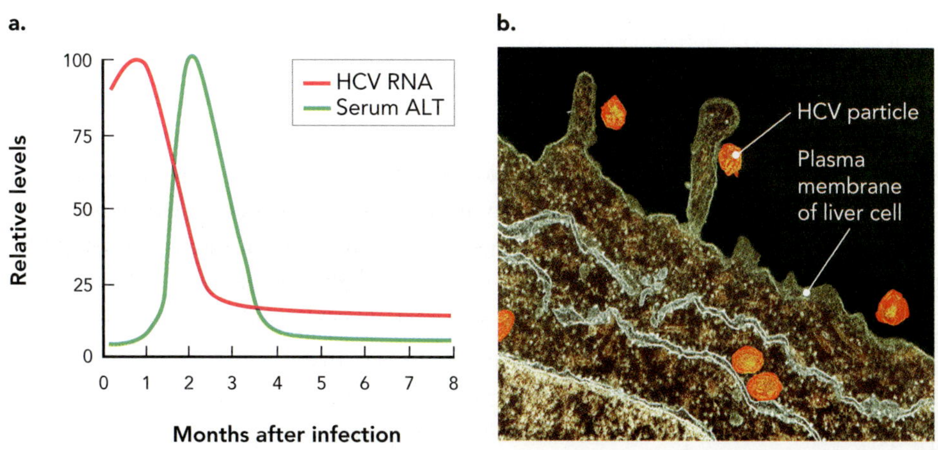

Figure 17.17 Elevated levels of alanine aminotransferase (ALT) in the blood are an indication of liver damage from an active hepatitis C virus (HCV) infection. **a.** Typical time course of HCV infection illustrated by HCV RNA levels. The associated increase in serum ALT levels indicates that liver damage has occurred. BASED ON DATA FROM UNIVERSITY OF WASHINGTON, HEPATITIS WEB STUDY. **b.** HCV infects human liver cells. THOMAS DEERINCK, NCMIR/SCIENCE SOURCE.

C virus (HCV) infection and the appearance of alanine aminotransferase protein in the serum, which is used as a biochemical marker of liver damage. Although many of the biochemical assays currently used to detect protein markers of disease are based on enzymatic or antibody methods, more sensitive techniques are being developed using mass spectrometry in the expanding field of clinical proteomics.

concept integration 17.1

Starting with the process of nitrogen fixation by bacteria, explain how atmospheric nitrogen is incorporated into the amino acid aspartate.

Atmospheric nitrogen (N_2) is converted into ammonia (NH_3) as a nitrogen source for plants by three distinct processes: (1) biological fixation by soil bacteria, (2) industrial fixation by the Haber process, and (3) atmospheric fixation as a result of lightning. All organisms require nitrogen for the synthesis of numerous biomolecules, the most abundant of which are amino acids and nucleotides. Biological fixation is carried out by two types of soil bacteria: free-living bacteria and symbiotic bacteria. Both of these bacteria use an ATP-dependent reaction mechanism catalyzed by the enzyme nitrogenase to reduce N_2 to NH_3 (which becomes ammonium ion, NH_4^+). Much of the NH_4^+ present in the soil is oxidized to nitrate (NO_3^-) and nitrite (NO_2^-) by nitrifying bacteria, so plants have nitrate and nitrite reductase enzymes to convert these compounds back into NH_4^+. The enzymes glutamine synthetase and glutamate synthase work together to assimilate the NH_4^+ into glutamine and glutamate, which are then used in aminotransferase reactions to add an amino group to α-keto acids, generating the corresponding α amino acid. One such enzyme is aspartate aminotransferase, which generates aspartate and α-ketoglutarate from glutamate and oxaloacetate in a reversible reaction. Under conditions of extremely high NH_4^+ concentrations in the soil, such as fertilized crop fields, the enzyme glutamate dehydrogenase assimilates NH_4^+ into glutamate, using α-ketoglutarate as the carbon donor.

17.2 Amino Acid Degradation

We saw in earlier chapters that glucose can be stored in the body as glycogen or converted to acetyl-CoA and stored as fatty acids. Unlike glucose, however, nitrogen cannot be stored in a useable form because NH_4^+ is toxic. Therefore, nitrogen lost as a result of protein and nucleic acid degradation must be replaced from the diet. When an individual is in **nitrogen balance**, it means that the daily intake of nitrogen, primarily in the form of protein, equals the amount of nitrogen lost by excretion in the feces and urine. A normal healthy adult needs about 60 g of protein per day to maintain nitrogen balance. In contrast, young children and pregnant women have a positive nitrogen balance because they accumulate nitrogen in the form of new protein, which is needed to support tissue growth. Negative nitrogen balance is a sign of disease or starvation and occurs in individuals with elevated rates of protein breakdown (loss of muscle tissue) or an inability to obtain sufficient amounts of amino acids in their diets.

To understand the biochemical processes that govern nitrogen balance in the body, we begin this section by describing the degradation of dietary proteins and the proteolysis of cellular proteins by the ubiquitin–proteasome pathway. The resulting amino acids are either recycled for synthesis of other proteins or deaminated so that the carbon skeletons can be used as metabolites in energy conversion pathways. We next examine the transport of NH_4^+ from peripheral tissues to the liver, where it is incorporated into urea and excreted to maintain nitrogen balance. Lastly, we take a look at representative amino acid degradation pathways and describe several metabolic disorders that result from defects in amino acid catabolism.

Dietary and Cellular Proteins Are Degraded into Amino Acids

Plants and bacteria have the necessary enzymes to synthesize all 20 amino acids. But, as we noted earlier, animals depend on protein in their diets to obtain the 10 essential amino acids they require for growth and development.

Protein digestion in humans takes place in the stomach and the small intestine, where proteases cleave the peptide bond to yield amino acids and small peptides. As illustrated in **Figure 17.18**, when food enters the stomach through the esophagus, it stimulates the release of **gastrin**, a small peptide hormone synthesized in the mucosal cells that line the stomach. Gastrin triggers the release of gastric juices containing hydrochloric acid from parietal cells and the secretion of pepsinogen from chief cells; pepsinogen is the inactive state, or zymogen, of the protease pepsin. The resulting increase in acidity (pH of ~2) in the stomach denatures the dietary proteins and kills most bacteria contained in the food. The low pH in the stomach also results in the activation of the protease by autocatalytic cleavage of pepsinogen to expose the protease active site, and now it is in the active form known as pepsin (see Figure 7.61). Pepsin is maximally active at a pH value of ~2 and preferentially cleaves polypeptide bonds on the amino-terminal side of the aromatic amino acids Phe, Trp, and Tyr.

The highly acidic slurry of digested food, called chyme, leaves the stomach by passing through the pyloric valve and into the duodenum, which is the first section

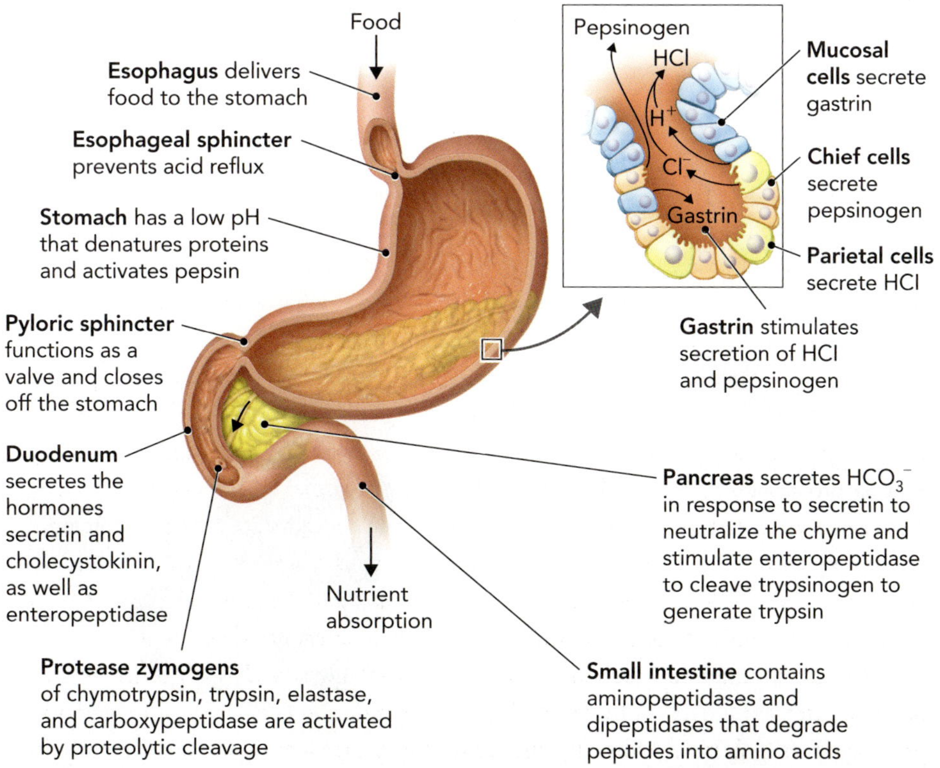
Food
Esophagus delivers food to the stomach
Esophageal sphincter prevents acid reflux
Stomach has a low pH that denatures proteins and activates pepsin
Pyloric sphincter functions as a valve and closes off the stomach
Duodenum secretes the hormones secretin and cholecystokinin, as well as enteropeptidase
Protease zymogens of chymotrypsin, trypsin, elastase, and carboxypeptidase are activated by proteolytic cleavage
Nutrient absorption
Pepsinogen
HCl
H+
Cl−
Gastrin
Mucosal cells secrete gastrin
Chief cells secrete pepsinogen
Parietal cells secrete HCl
Gastrin stimulates secretion of HCl and pepsinogen
Pancreas secretes HCO3− in response to secretin to neutralize the chyme and stimulate enteropeptidase to cleave trypsinogen to generate trypsin
Small intestine contains aminopeptidases and dipeptidases that degrade peptides into amino acids

Figure 17.18 Proteases play a central role in the digestion of dietary proteins. Food enters the stomach through the esophagus, and protein digestion begins in the stomach as a result of the acidic environment and pepsin activation. After leaving the stomach, the food bolus (chyme) stimulates secretion of duodenal hormones that induce secretion of protease zymogens from the pancreas. Once inside the small intestine, other protease zymogens are activated by cleavage, leading to further digestion of proteins in the meal to generate protein fragments (peptides). Aminopeptidases and dipeptidases in the intestinal mucosal cells degrade the peptides into single amino acids, which are then absorbed into the bloodstream and transported to the liver.

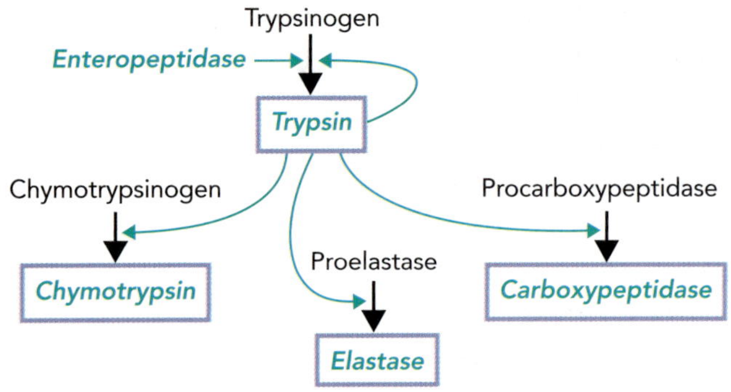

Figure 17.19 Enteropeptidase cleaves the zymogen trypsinogen, resulting in the active protease trypsin. Trypsin cleaves and activates more trypsinogen, as well as other zymogens. The net result of trypsin activation by enteropeptidase is a proteolytic cascade and initiation of protein degradation in the small intestine.

of the small intestine. This results in the release of two duodenal hormones: secretin and cholecystokinin. The duodenum also secretes **enteropeptidase**, a protease that specifically activates several proteolytic zymogens released from the pancreas.

The role of secretin is to stimulate the pancreas to release an alkaline fluid containing bicarbonate (HCO_3^-), which neutralizes the chyme. At a pH of ~7, enteropeptidase is then able to cleave a pancreatic zymogen called trypsinogen to form the endopeptidase trypsin. As shown in **Figure 17.19**, the activated trypsin enzyme then cleaves numerous pancreatic zymogens, including chymotrypsinogen, proelastase, and procarboxypeptidase, as well as trypsinogen itself to amplify the proteolytic cascade. The proteolytic cleavage of chymotrypsinogen by trypsin to generate the activating protease π-chymotrypsin was described in Chapter 7 (see Figure 7.62). The combined activity of the pancreatic proteases and the **aminopeptidases** and **dipeptidases**, which are located on the membrane of intestinal mucosal cells, generates peptides and amino acids. These are transported into intestinal epithelial cells before being exported to the blood.

Degradation of cellular proteins is another source of free amino acids and occurs continually in all cells. Most enzymes and structural proteins have half-lives of several hours, but numerous regulatory proteins, such as transcription factors and signaling proteins, are degraded within a few minutes of being synthesized. This ensures that upstream signaling and downstream signaling are coordinated. In addition, a cell's quality control of protein synthesis leads to the degradation of misfolded proteins and truncated proteins that occur as a result of aborted translation.

Most eukaryotic cellular proteins are degraded by one of two pathways: (1) an *ATP-independent* process that degrades proteins inside lysosomes, which are intracellular vesicles derived from Golgi membranes (**Figure 17.20a**); or (2) an *ATP-dependent*

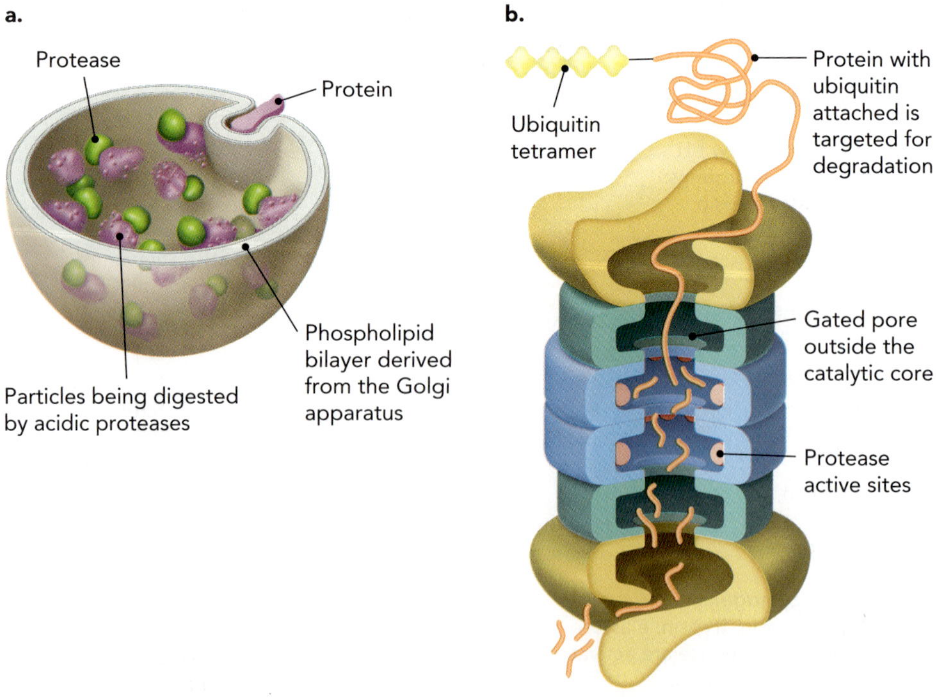

Figure 17.20 Degradation of eukaryotic cellular proteins is mediated by lysosomes and proteasomes. **a.** Lysosomes contain acidic proteases that nonselectively digest protein particles delivered to lysosomes by phagocytosis or endosomal fusion. **b.** Proteasomes degrade ubiquitinated proteins by using three different protease activities located within the central core.

process that degrades proteins containing a polymer of **ubiquitin** protein. This process occurs in a large, multi-subunit complex called a **proteasome** (**Figure 17.20b**).

Lysosomes are a low-pH (~5) compartment filled with digestive enzymes that function nonselectively. Protons are pumped into lysosomes by vacuolar ATPases, and the acidic pH aids in protein unfolding and provides an optimal environment for the activation of cysteine and aspartate proteases. The proteolytic enzymes within the lysosome are optimized to function at low pH; this ensures that if these enzymes leak out of the lysosome into the cytosol, their activity is low so they do not proteolyze cytosolic proteins. Many macromolecules are degraded within lysosomes, including large particles (viruses or bacteria) brought in by phagocytosis or receptor particles internalized by endocytosis. One example of lysosomal protein degradation is the proteolysis of lipoproteins contained in LDL particles that have been endocytosed after binding to LDL receptors on liver cells (see Figure 16.50).

In contrast to the nonselective activity of lysosomes, eukaryotic proteasomes selectively degrade ubiquitinated proteins delivered to a catalytic core. This core uses three distinct protease activities, which are characterized as chymotrypsin-like, trypsin-like, and a caspase-like peptidyl-glutamyl peptide-hydrolyzing enzyme. Archaebacteria also contain proteasomes, the best characterized of which is the *Thermoplasma acidophilum* proteasome. This complex contains an unusual serine-like protease that uses a threonine residue as the catalytic nucleophile. In addition, a selective ATP-dependent protein degradation pathway mediated by the Clp-chaperone protease family is found in bacteria. Structure–function analyses of the eukaryotic and archaeal proteasomes show that they are evolutionarily related to the bacterial Clp-chaperone proteases.

▶ ANIMATION

The proteasome consists of a 20S core particle (S is the Svedberg sedimentation unit) and two 19S regulatory particles, which serve as caps to regulate protein entry into and exit from the proteolytic core. The 19S complexes contain binding sites for ubiquitinated proteins and also encode ATP hydrolyzing enzymes that function in protein unfolding, which is required before the polypeptide can enter the internal chamber and be degraded. The intact proteasome, consisting of two regulatory subunits and the proteolytic core, has a sedimentation coefficient of 30S (26S proteasomes have only a single 19S subunit).

Figure 17.21 shows the protein structure of the yeast 30S proteasome. It can be seen that the 20S core particle consists of two α rings and two β rings, each

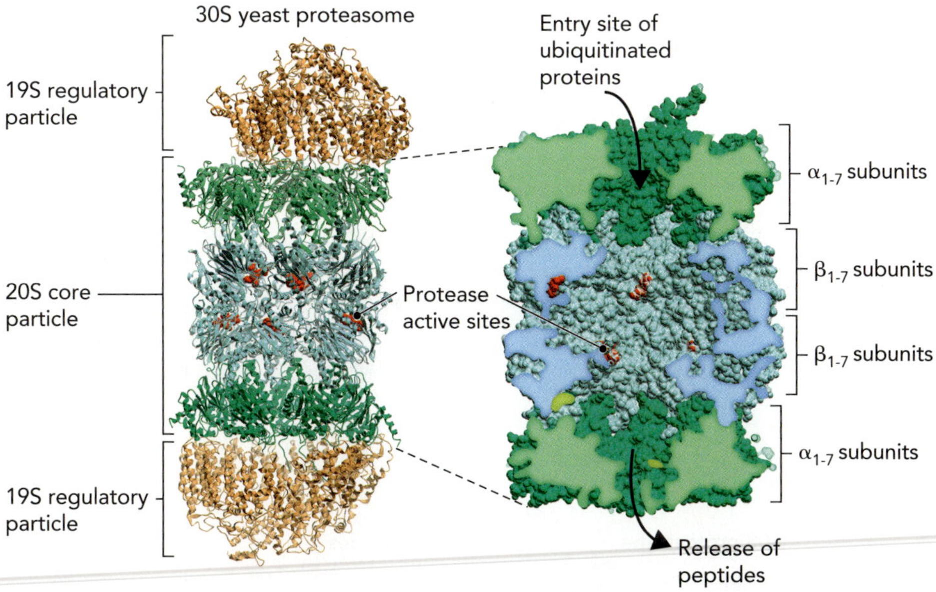

Figure 17.21 The molecular structure of the yeast 30S proteasome shows that it consists of two 19S regulatory complexes and a 20S core complex. The six protease active sites located near the N termini of the β_1, β_2, and β_5 subunits are highlighted in red. The cross section of the 20S core particle on the right shows the proteasome's internal chambers formed by the α and β rings. Four of the six protease active sites are highlighted in red in the cross section. BASED ON PDB FILE 3L5Q.

containing seven structurally related protein subunits to form an $\alpha_{1\text{-}7}\beta_{1\text{-}7}\beta_{1\text{-}7}\alpha_{1\text{-}7}$ cylindrical complex. The 19S regulatory complexes located at the two ends of the core complex function to regulate entry of ubiquitinated proteins and to release oligopeptides 3–15 amino acids long in an ATP-dependent reaction. Within the core particle are six protease active sites located near the N termini of the β_1, β_2, and β_5 subunits, all of which contain an N-terminal threonine residue. The location and function of each of the three types of proteases within the yeast proteasomal core particle were identified using a combination of mutational analysis and biochemical substrate cleavage assays.

Proteins targeted for proteasomal degradation must first be "tagged" on lysine residues by covalent linkage of ubiquitin through its carboxyl-terminal glycine residue. The tagging process requires three classes of ubiquitinating proteins, called E1, E2, and E3, which work together in an ATP-dependent pathway to attach ubiquitin to target proteins (**Figure 17.22**). Ubiquitinating proteins recognize either specific residues at the N terminus of the target protein or a structural property of the protein, such as a phosphorylated residue or abnormal conformation. Ubiquitinated proteins enter the proteasome one at a time, where the ubiquitin is removed and recycled. Then, the polypeptide is cleaved into small oligopeptides (6–10 amino acids long), which are released into the cytosol and degraded into individual amino acids.

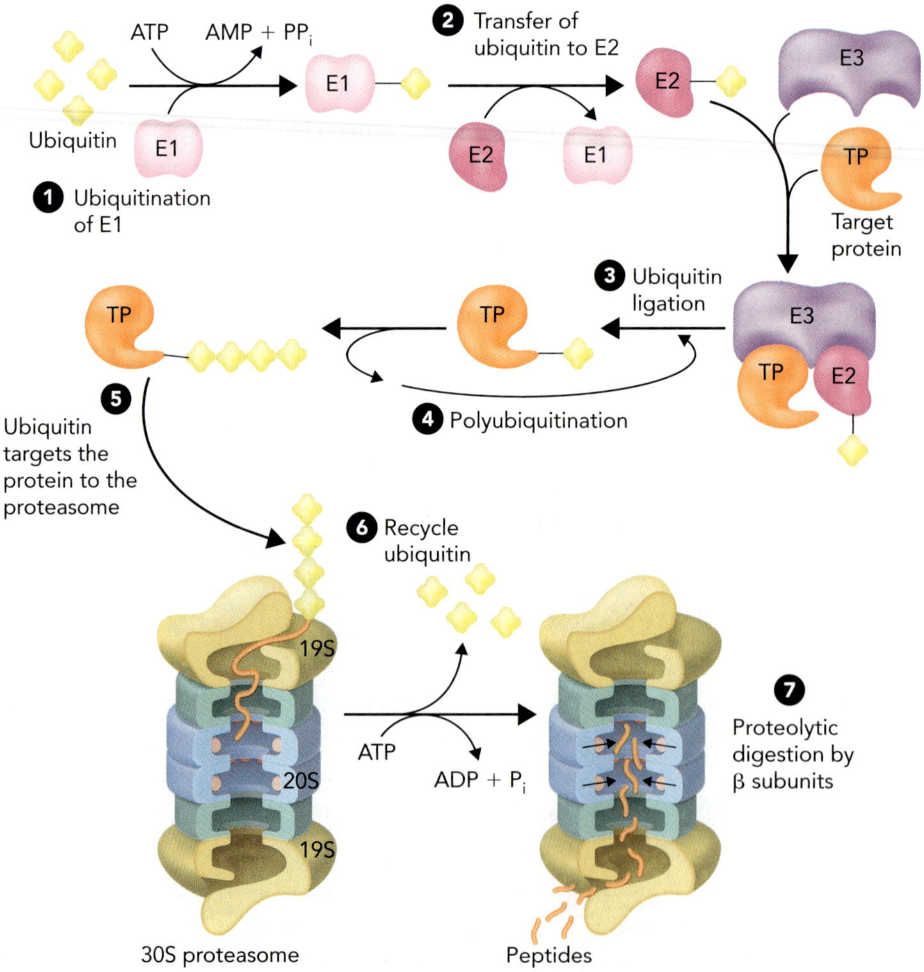

Figure 17.22 The ubiquitin–proteasome pathway in eukaryotic cells. Target proteins are destined for proteasomal degradation by the covalent attachment of ubiquitin subunits to lysine residues. Binding of the polyubiquitinated target protein to the 19S proteasomal complex initiates protein unfolding and hydrolysis of the ubiquitin subunits. The unfolded protein is then degraded into oligopeptides within the interior chamber. TP = target protein.

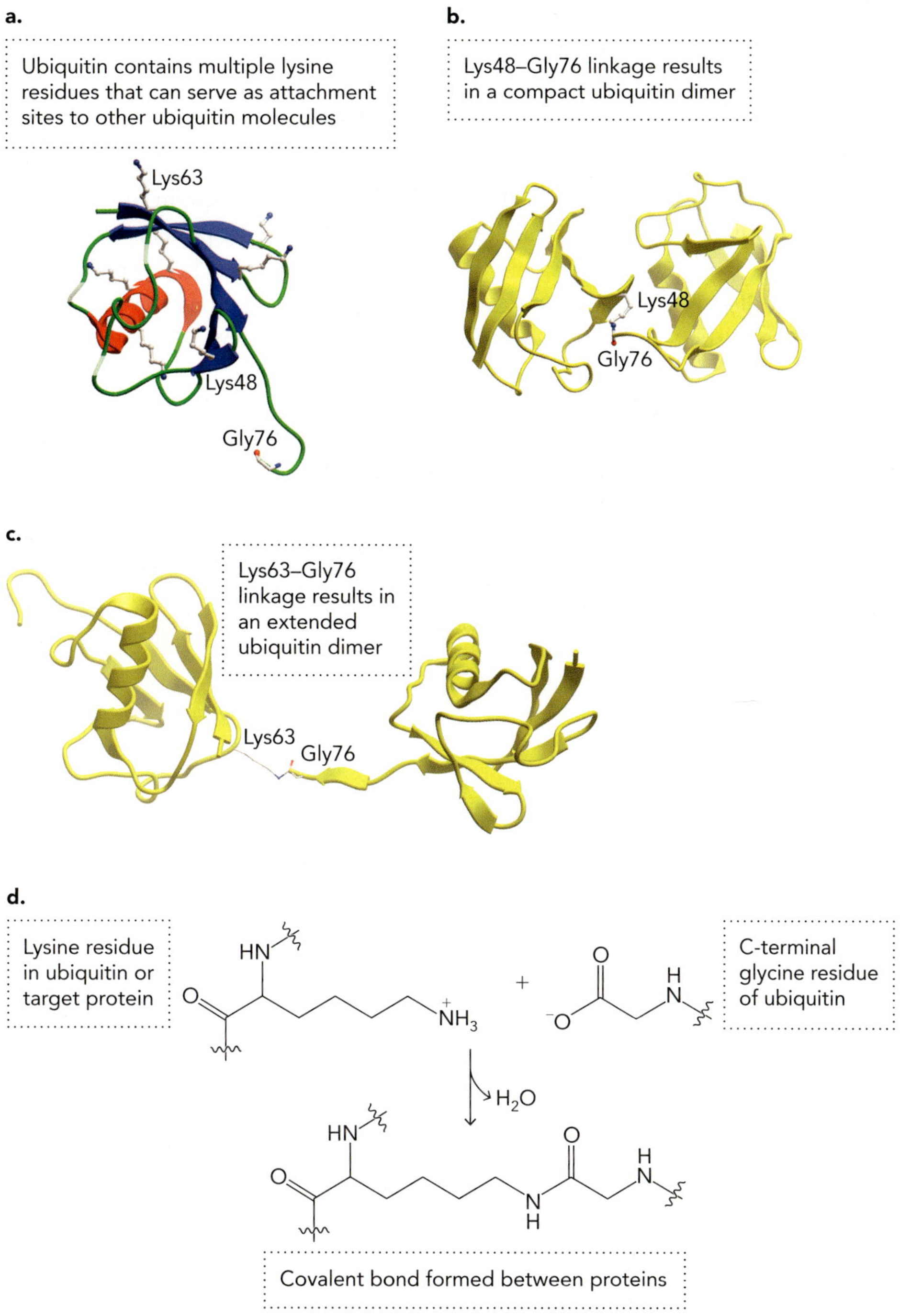

Figure 17.23 Most human polyubiquitinated proteins contain ubiquitin monomers linked through the C-terminal Gly76 residue on one monomer and either the Lys48 or Lys63 residue of the adjacent monomer. **a.** Molecular structure of human ubiquitin, showing the locations of all seven lysine residues in the protein as ball-and-stick representations. Two of these, Lys48 and Lys63, are the most common linkage sites to Gly76 in the adjacent ubiquitin monomer. BASED ON PDB FILE 2ZNV. **b.** Molecular structure of human ubiquitin dimers linked through Gly76–Lys48. BASED ON PDB FILE 1TBE. **c.** Molecular structure of human ubiquitin dimers linked through Gly76–Lys63. BASED ON PDB FILE 2ZNV. **d.** Chemical structure of the linkage between a lysine side chain and the C-terminal glycine of ubiquitin. The lysine side chain can be from either a target protein or another ubiquitin monomer.

As shown in **Figure 17.23**, human ubiquitin is a 76-amino-acid-long protein with seven lysine residues, all of which can serve as linking sites between ubiquitin monomers. The C-terminal glycine residue of ubiquitin (Gly76) serves as the attachment site to targeted proteins or to other ubiquitin monomers. The most common ubiquitin linkage in proteins with at least four ubiquitin subunits is between Gly76 and Lys48, which identifies proteins targeted for proteasomal degradation. In contrast, a Gly76–Lys63 ubiquitin linkage targets tetraubiquitinated proteins to the intracellular secretory pathway, which translocates proteins to the plasma membrane rather than to proteasomes. As you can see in Figure 17.23, a Gly76–Lys48 linkage generates a much more compact ubiquitin dimer than does a ubiquitin dimer containing a Gly76–Lys63 linkage. This structural difference likely plays a role in targeting polyubiquitinated proteins to distinct intracellular locations. Monoubiquitinated proteins have also been identified in cells, which seems to be a posttranslational modification used to control the function of gene regulatory proteins (see Chapter 23).

The three classes of proteins involved in ubiquitination are (1) E1 enzymes that attach ubiquitin to E2 enzymes; (2) E2 enzymes that attach ubiquitin to target proteins; and (3) E3 proteins that facilitate ubiquitination of target proteins by forming heterotrimeric complexes with E2 enzymes and target proteins. The E3 proteins are commonly called E3 **ubiquitin ligases**, of which ~500 are encoded in the human genome. In addition, the human genome encodes ~30 different E2 genes, but only two E1 genes. Considering that the formation of distinct E2–E3 complexes (from more than a thousand different combinations) ultimately determines which proteins

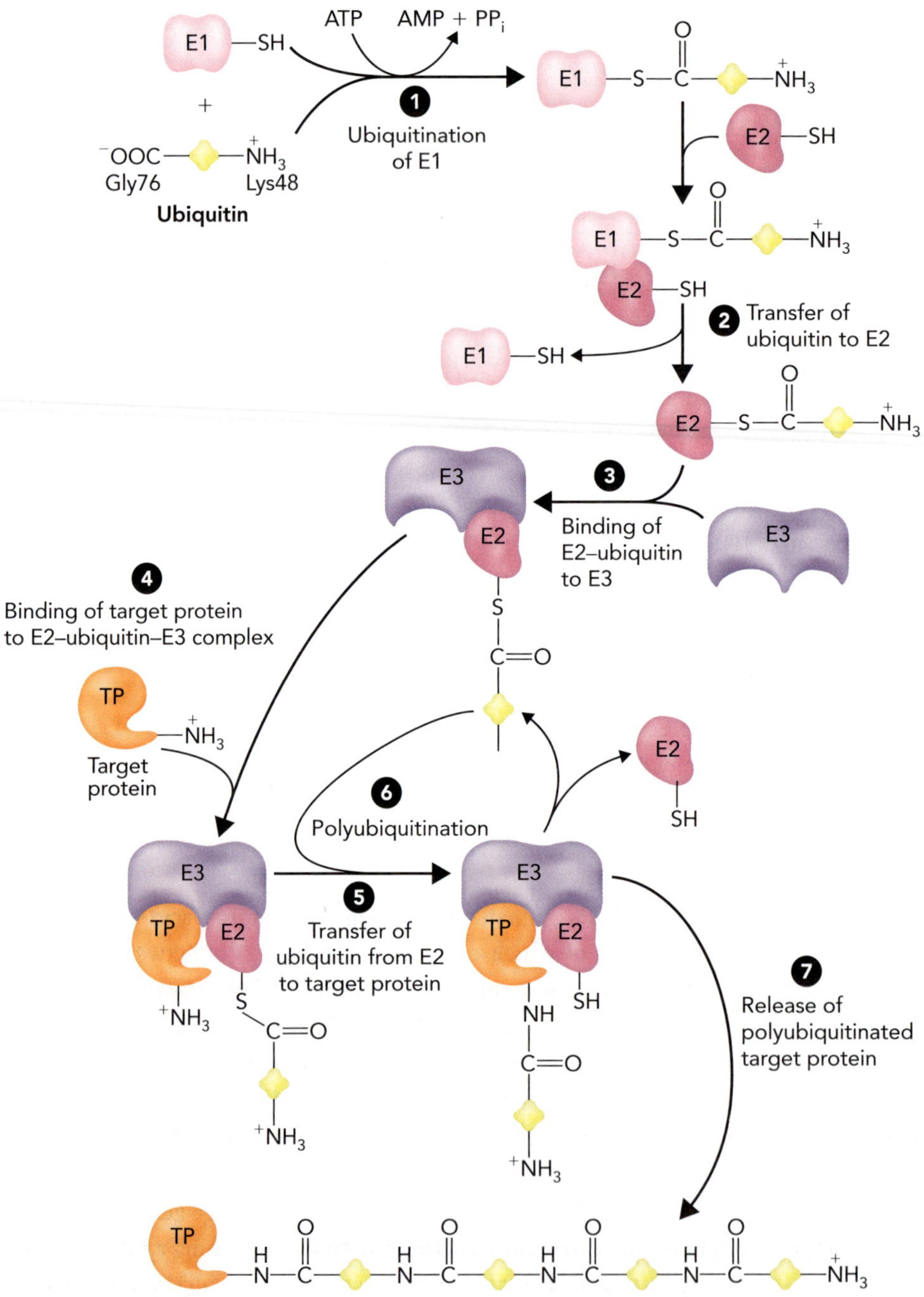

Figure 17.24 Attachment of ubiquitin to a target protein (TP) requires ubiquitin activation, ubiquitin ligation, and polyubiquitination. Ubiquitin activation is an ATP-dependent reaction that links the terminal carboxyl group of Gly76 of ubiquitin to a cysteine residue in the E1 enzyme. Transfer of this ubiquitin to a cysteine residue in E2 releases E1 and leads to the formation of an E2–E3 complex. Ubiquitination of a lysine residue on the target protein initiates the polyubiquitination process, which links at least four ubiquitin subunits together through a series of Gly76–Lys48 linkages.

are ubiquitinated, it makes sense to have multiple copies of E2- and E3-related genes. In contrast, because the function of E1 enzymes is to ubiquitinate E2 enzymes in an ATP-dependent reaction that does not involve target proteins, there is no need to encode more than a few E1 genes in the human genome.

The biochemical roles of E1, E2, and E3 proteins are illustrated in **Figure 17.24**. (1) In the first step, ubiquitin is covalently attached to a cysteine residue in the E1 active site through an ATP-dependent reaction, involving the carboxyl-terminal glycine residue of ubiquitin (Gly76). (2) In the second step, E1 associates with an E2 enzyme and transfers the ubiquitin to a cysteine residue in the E2 active site. (3) Each E2–ubiquitin complex then binds to an E3 enzyme to form an E2–ubiquitin–E3 complex. (4) This complex attaches ubiquitin to target proteins recognized by the complex. (5) Depending on which E2 and E3 proteins constitute the recognition complex, the ubiquitin is either first transferred from E2 to E3 before the target protein is ubiquitinated or the target protein is ubiquitinated directly by the E2 subunit in the complex. The E2 and E3 enzymes link ubiquitin to lysine residues in target proteins through the Gly76 carboxyl-terminal residue of ubiquitin. (6) After covalent attachment of one ubiquitin to the target protein, a minimum of three more ubiquitins must be attached by Gly76–Lys48 linkages before the protein is recognized by the 19S complex of the proteasome. (7) The sequential attachment of ubiquitin subunits occurs within the same E3–target protein complex through continual cycling of E2–ubiquitin moieties.

How is protein ubiquitination regulated? The two most common mechanisms are (1) biochemical changes in target proteins, such as N-terminal cleavage or phosphorylation, and (2) biochemical changes in E3 ligases, including activator protein binding or phosphorylation. As shown in **Figure 17.25a**, target proteins that lack the N-terminal methionine (Met) residue, but instead have a Phe, Leu, Asp, Lys, or Arg residue at the N terminus, are recognized and ubiquitinated by certain E2–E3 complexes, which follow the **N-end rule** of protein degradation. The N-end rule refers to the propensity of polypeptides with specific N-terminal amino acids to have short or long half-lives in cells as a result of ubiquitination and proteasomal degradation. Alexander Varshavsky and his colleagues at the California Institute of Technology described the N-end rule in 1986 and are credited with elucidating many of the molecular details of protein recognition in the ubiquitin–proteasome pathway.

In addition, some target proteins are ubiquitinated as a result of phosphorylation, which increases their binding affinity for E2–ubiquitin–E3 complexes (**Figure 17.25b**). An example of this type of regulatory mechanism is the phosphorylation, ubiquitination, and proteasomal degradation of IκBα protein, which regulates inflammatory signaling and apoptosis through the transcription factor NFκB (see Figure 8.54).

E2–ubiquitin–E3 complexes themselves can be regulated as a means to control ubiquitin-mediated degradation of target proteins. For example, phosphorylation of a protein complex containing an E3 ubiquitinating ligase (**Figure 17.25c**) leads to the degradation of regulatory proteins called cyclins, which control cell cycle progression. Degradation of specific proteins can also be initiated by the binding of regulatory molecules to E3 enzymes (or target proteins), thereby inducing a conformational change that promotes target protein binding and ubiquitination (**Figure 17.25d**).

The Urea Cycle Removes Toxic Ammonia from the Body

Cells cannot store amino acids that accumulate as a result of protein degradation; they must be either recycled for protein synthesis or deaminated in order to salvage their

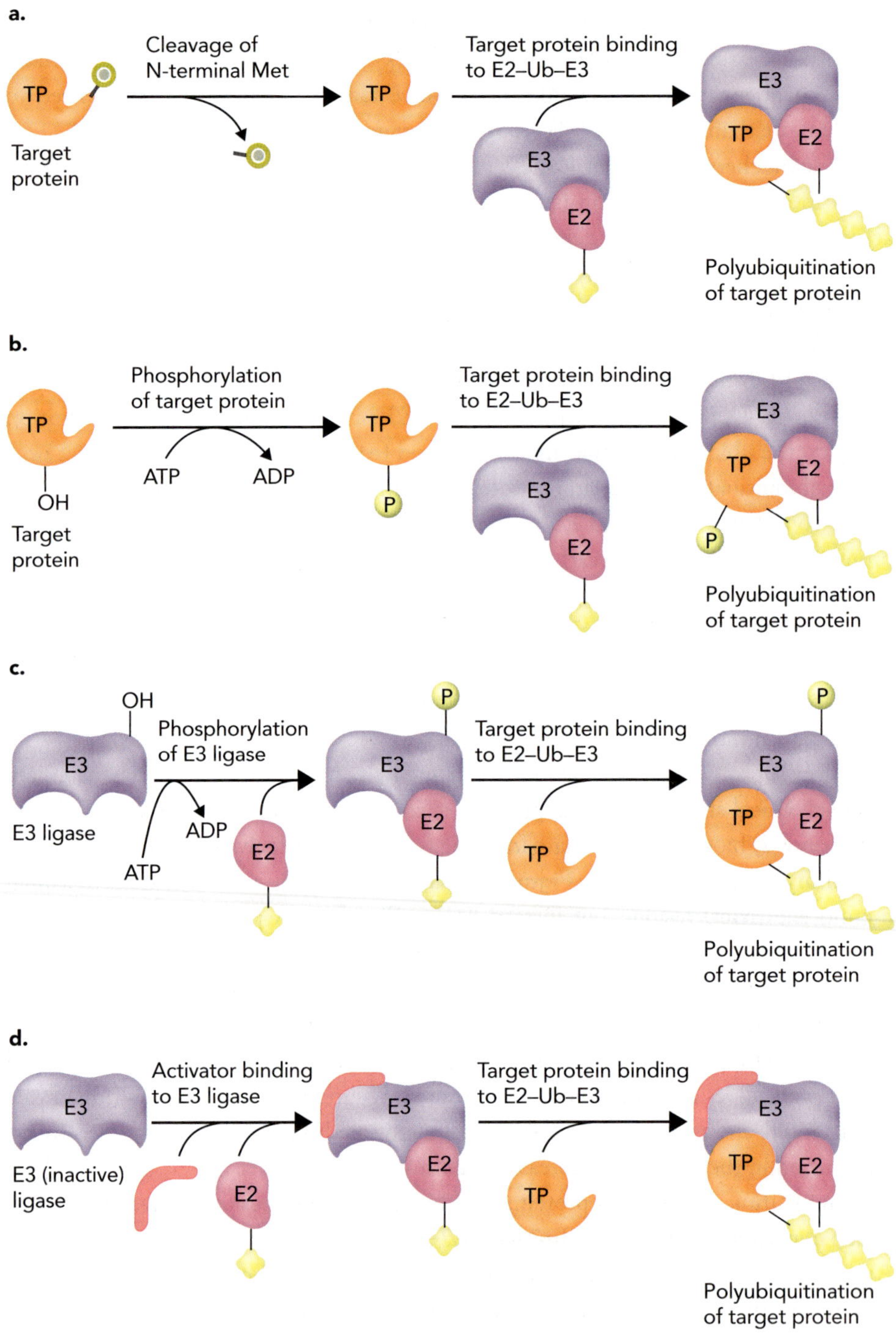

Figure 17.25 Ubiquitination of target proteins is a regulated process. **a.** Cleavage of the N terminus of a target protein can facilitate E2–Ub–E3 binding, as proposed by the N-end rule of protein degradation. **b.** Phosphorylation of the target protein can make it a better substrate for the E2–Ub–E3 complex. **c.** Phosphorylation of the E3 ligase can increase its binding affinity for E2 or for target proteins. **d.** Binding of an activator protein to E3 ligase can alter its binding affinity for E2 or for target proteins. TP = target protein; Ub = ubiquitin.

carbon skeletons for use in other pathways. As shown in **Figure 17.26**, deamination of amino acids results in generation of NH_4^+. This molecule, in turn, is used in the synthesis of other nitrogen-containing compounds or excreted in the form of urea in most animals. The remaining carbon skeletons are used as metabolites in energy conversion pathways. In this subsection, we describe how excess NH_4^+ is removed from the body; in the next subsection, we examine the fate of the carbon skeletons in representative pathways of amino acid degradation.

Glutamate and glutamine function as the primary nitrogen carriers in most organisms. In mammals, this nitrogen ends up in the liver, where it is converted to urea.

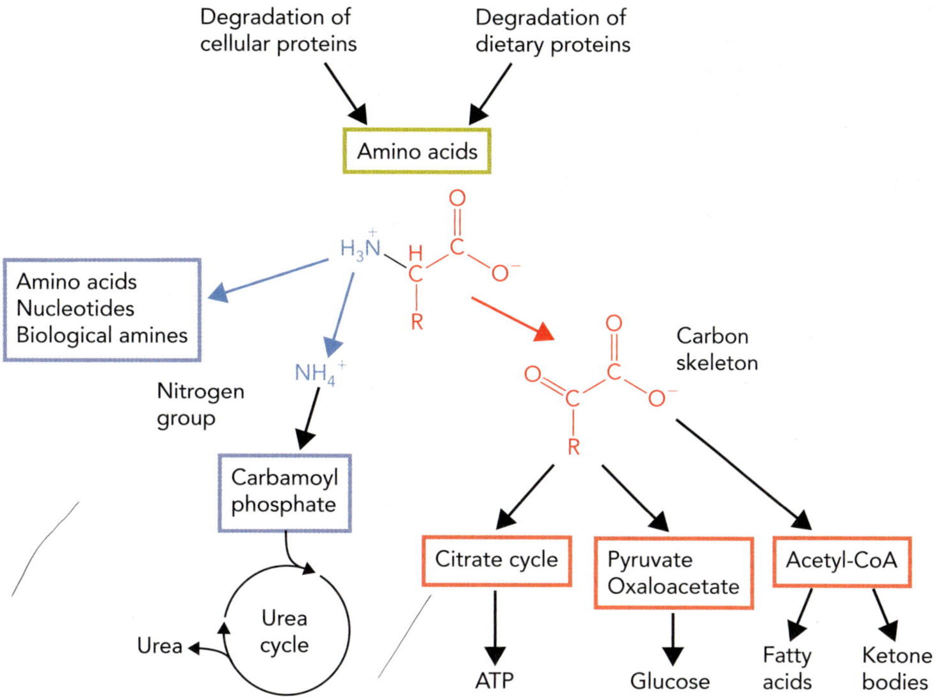

Figure 17.26 An overview of amino acid catabolism is shown here. In mammals, amino acids derived from the degradation of dietary proteins or the turnover of cellular proteins are deaminated, and the nitrogen is used for biosynthetic pathways or excreted as urea. The remaining carbon skeletons are metabolized by energy conversion pathways to generate ATP, glucose, fatty acids, or ketone bodies.

Figure 17.27 shows that the nitrogens in urea are derived from two sources: (1) the NH_4^+ released when glutamate or glutamine is deaminated; and (2) aspartate, which is formed when oxaloacetate is transaminated by aspartate aminotransferase. The carbon atom in urea comes from the CO_2 (HCO_3^-) produced in the mitochondrial matrix by the citrate cycle, and the oxygen atom is derived from H_2O in the final reaction of the cycle.

Rather than synthesizing urea, which is an energy-consuming process, fish and aquatic amphibians simply excrete NH_4^+ directly into water. Terrestrial amphibians excrete nitrogen as urea or, in some cases, as uric acid, which is not water soluble and is excreted by birds, reptiles, and insects that need to conserve water. Notably, the African lungfish, *Protopterus annectens,* has the ability to excrete NH_4^+ when it lives in water and to activate urea cycle enzymes when it lives on land (**Figure 17.28**). This metabolic switch is triggered by prolonged exposure to air, which stimulates a signaling cascade leading to increased expression of urea cycle enzymes.

Amino acids are transported to the liver, where the nitrogen is removed and used for urea synthesis. These amino acids have three sources: (1) amino acids derived from the digestion of dietary proteins; (2) the amino acid glutamine, which is generated from glutamate and NH_4^+ in peripheral tissues by glutamine synthetase; and (3) the amino acid alanine, which is formed by the alanine aminotransferase reaction as a way to remove excess nitrogen

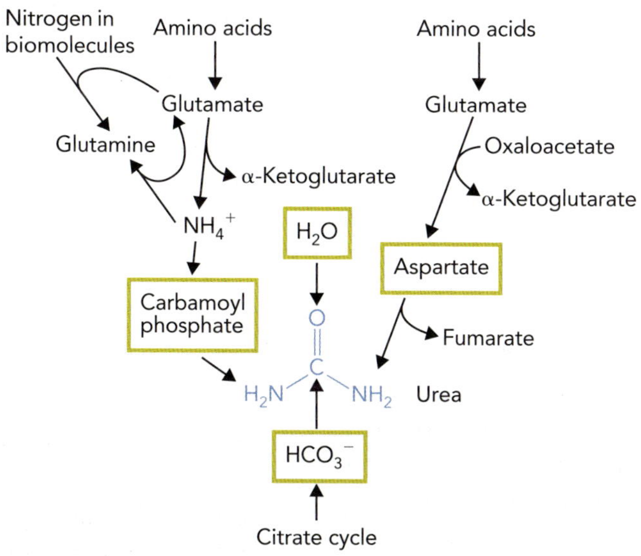

Figure 17.27 The two nitrogens in urea are derived from NH_4^+ and aspartate; the carbon is derived from bicarbonate (HCO_3^-) produced in the citrate cycle; and the oxygen is derived from H_2O.

Figure 17.28 The African lungfish, *Protopterus annectens,* excretes nitrogen as ammonia when living in water; however, when the fish lives on land, urea cycle enzymes are activated, and the ammonia is converted to urea. ROBERT S. MICHELSON/AGEFOTOSTOCK.

from exercising (or starving) skeletal muscle. As shown in **Figure 17.29**, dietary amino acids in the blood are taken up by the liver, where aminotransferase enzymes in the mitochondrial matrix transfer the amino group to α-ketoglutarate to form glutamate. Amino acids derived from the degradation of cellular proteins are also deaminated to generate glutamate. In liver cells, the glutamate is first imported into the mitochondrial matrix, where it is metabolized by the enzyme glutamate dehydrogenase to produce NH_4^+. This is used to make the urea cycle precursor **carbamoyl phosphate.** In addition, some of the glutamate is converted to aspartate by the aspartate aminotransferase reaction. This aspartate is fed into the urea cycle as the second source of nitrogen. Glutamine, which carries two excess nitrogen atoms to the liver from peripheral tissues, is deaminated by the enzyme glutaminase to generate NH_4^+ and glutamate. The NH_4^+

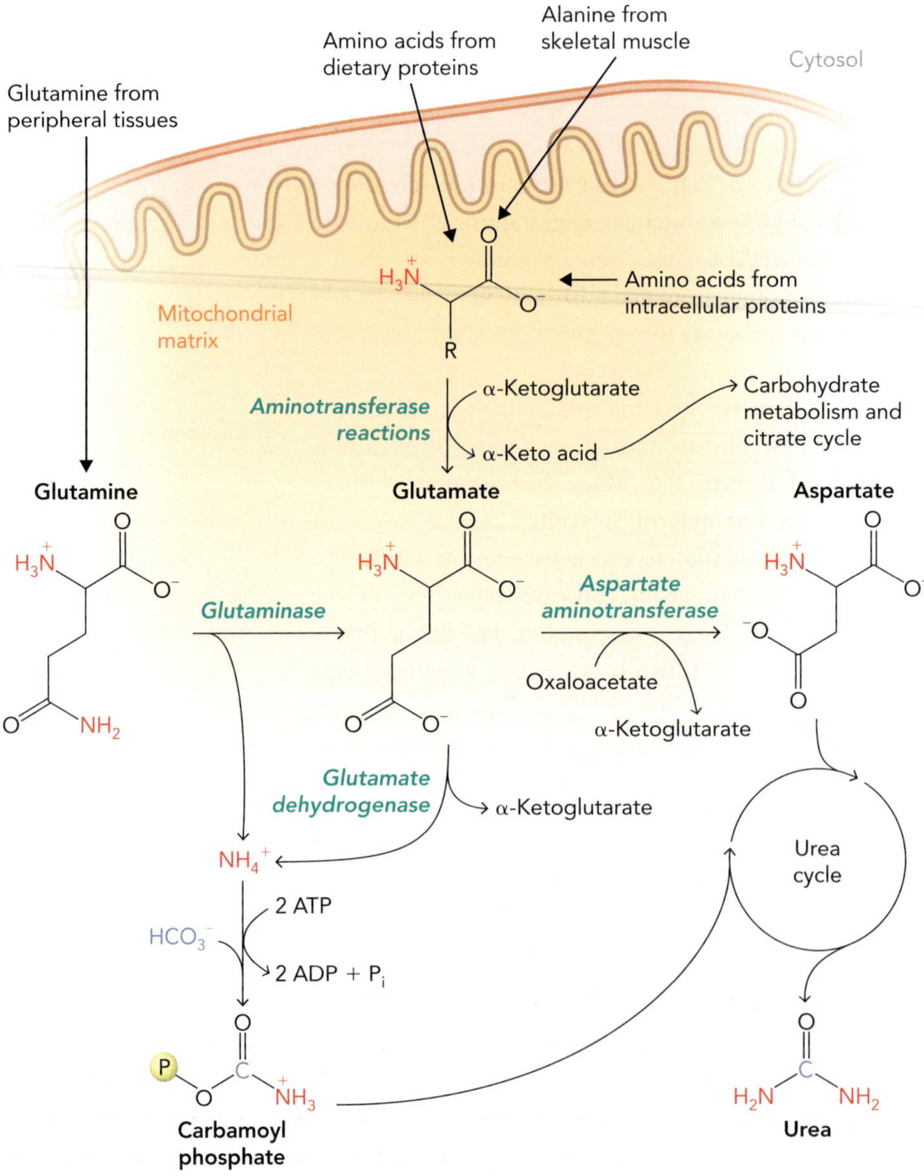

Figure 17.29 Amino acids are transported to the liver, where the nitrogen is used to make urea. Glutamine and glutamate give rise to NH_4^+, which is used to make carbamoyl phosphate. Glutamate is also used to make aspartate, which donates the second nitrogen to urea. Note that some of these reactions take place in the cytosol, whereas others are mitochondrial reactions.

is used to make carbamoyl phosphate directly, and the glutamate is deaminated by glutamate dehydrogenase to liberate a second molecule of NH_4^+ for carbamoyl phosphate synthesis.

During vigorous anaerobic exercise, glycogen degradation leads to the buildup of pyruvate, which can be oxidized to lactate and exported out to the liver as a component of the Cori cycle (see Figure 14.23). Because muscle protein is also degraded during exercise, the excess nitrogen coming from amino acid catabolism must be removed to avoid cell toxicity. As shown in **Figure 17.30**, the **alanine–glucose cycle** solves this problem by linking nitrogen metabolizing reactions in muscle and liver cells, using alanine as the nitrogen carrier. This is done by transferring the α amino group of an amino acid to α-ketoglutarate to form glutamate, which is then used as a substrate in the alanine aminotransferase reaction to convert pyruvate to alanine. The alanine is exported to the blood, where it is taken up by the liver and deaminated by alanine aminotransferase to regenerate glutamate and pyruvate. The resulting glutamate is metabolized by glutamate dehydrogenase to release NH_4^+ for urea synthesis, whereas the pyruvate is used to synthesize glucose via the gluconeogenic pathway. Glucose export to muscle cells completes the cycle and provides a renewable source of metabolic energy for continued muscle contraction or for replenishing glycogen stores.

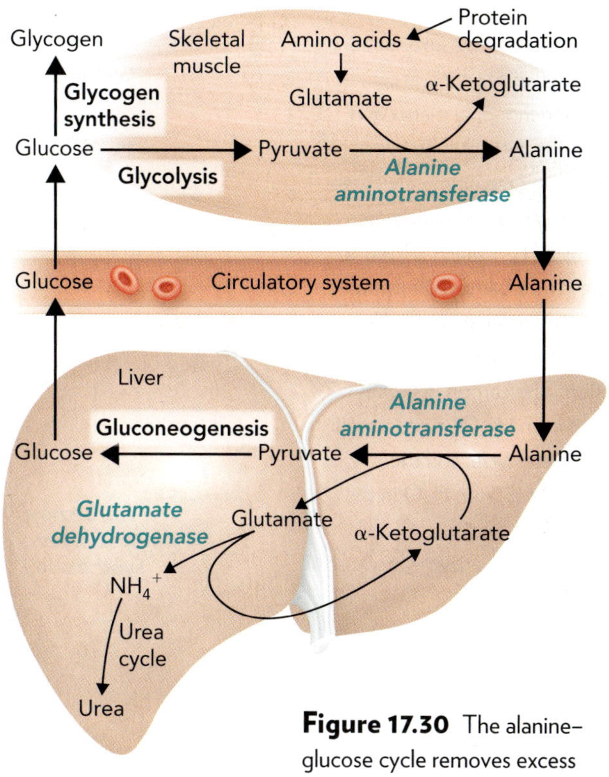

Figure 17.30 The alanine–glucose cycle removes excess nitrogen from muscle cells, using alanine as the carrier. The alanine aminotransferase reaction in liver cells removes the nitrogen from alanine to generate pyruvate and glutamate. The pyruvate is then used to produce glucose via gluconeogenesis, and the glucose is exported back to muscle cells, where it can be used as a source of energy for muscle contraction or converted to glycogen. Deamination of glutamate by glutamate dehydrogenase in the liver generates NH_4^+, which is used to make urea for nitrogen excretion.

Before we examine urea synthesis in detail, let's answer the four metabolic questions that pertain to the urea cycle.

1. *What does the urea cycle accomplish for the organism?* Urea synthesis provides an efficient mechanism to remove excess nitrogen from the body.
2. *What is the net reaction of the urea cycle?*

$$NH_4^+ + HCO_3^- + \text{Aspartate} + 3\ \text{ATP} \rightarrow \text{Urea} + \text{Fumarate} + 2\ \text{ADP} + 2\ P_i + \text{AMP} + PP_i$$

3. *What is the key regulated enzyme in urea synthesis?* The mitochondrial enzyme **carbamoyl phosphate synthetase I** catalyzes the commitment step in the urea cycle. Its activity is regulated by *N*-acetylglutamate in response to elevated levels of glutamate and arginine.
4. *What is an example of the urea cycle in everyday biochemistry?* A deficiency in the enzyme argininosuccinase inhibits flux through the urea cycle, causing hyperammonemia and neurologic symptoms. This metabolic disease can be treated with a low-protein diet supplemented with arginine, thereby resulting in argininosuccinate excretion as a substitute for urea. Normal functioning of the urea cycle maintains proper nitrogen balance in the body.

In humans, urea is synthesized in the liver and transported through the blood to the kidneys, where it is concentrated and excreted in urine. As shown in **Figure 17.31**, five enzymatic reactions are required for urea synthesis. Two of these occur inside mitochondria (catalyzed by carbamoyl phosphate synthetase I and ornithine transcarbamoylase),

Figure 17.31 The urea cycle consists of five enzymatic reactions, two of which occur in the mitochondrial matrix and three in the cytosol. The two nitrogens in urea ultimately come from glutamine and glutamate—either as NH_4^+, which is used to synthesize carbamoyl phosphate, or from aspartate produced by the aspartate aminotransferase reaction. The carbon in urea is derived from HCO_3^- generated in the citrate cycle, and the oxygen comes from H_2O in the final cleavage reaction, which is catalyzed by the enzyme arginase.

▶ ANIMATION

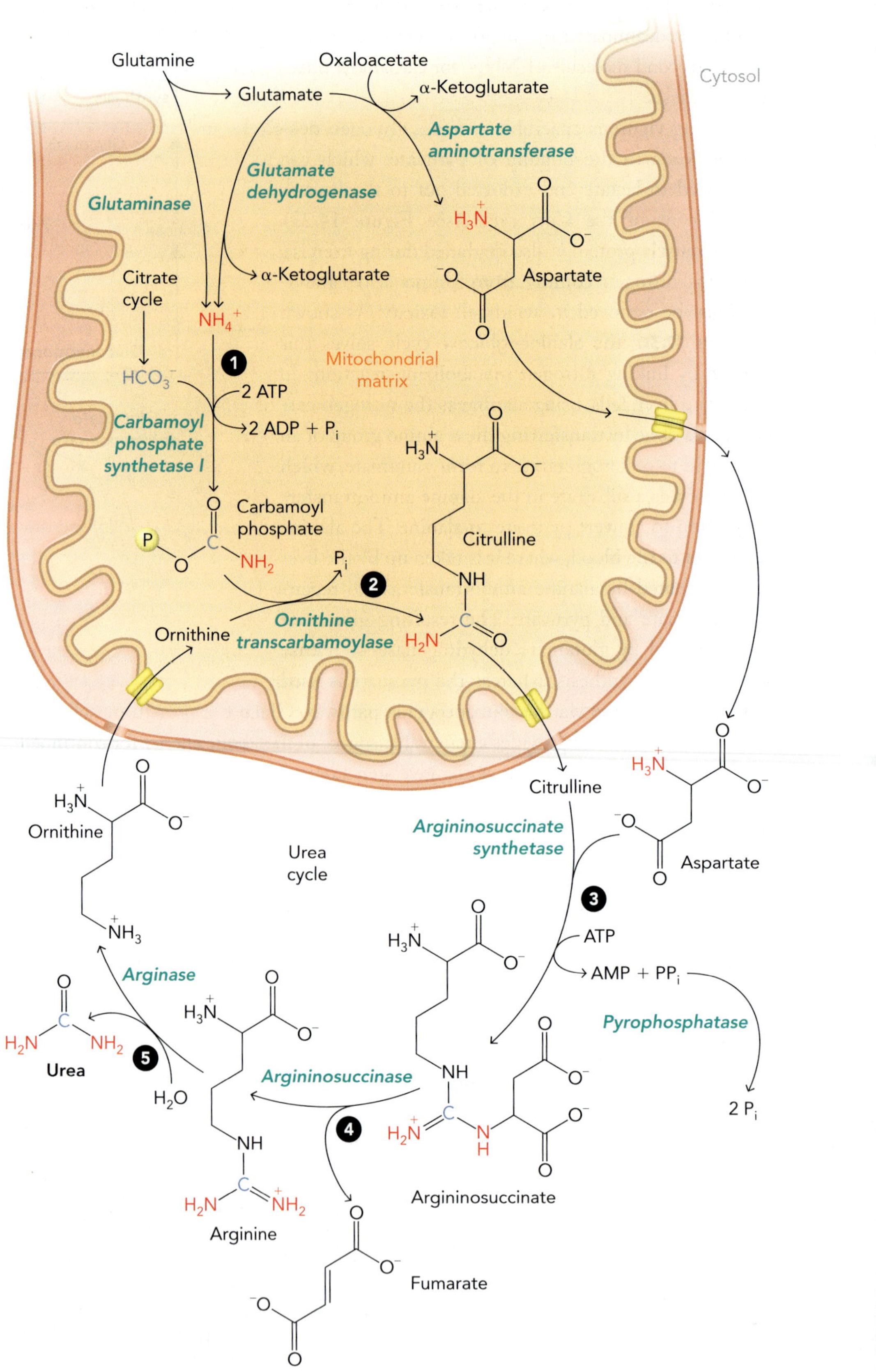

and three others take place in the cytosol (catalyzed by argininosuccinate synthetase, argininosuccinase, and arginase). Both the carbamoyl phosphate synthetase I and argininosuccinate synthetase reactions are ATP dependent, using a total of four high-energy phosphate bonds for every molecule of urea produced.

The urea cycle was discovered in 1932 by Hans Krebs and a medical student who worked in his lab, Kurt Henseleit. Krebs is the same biochemist who later described the citrate cycle and was corecipient of the 1953 Nobel Prize in Physiology or Medicine. In fact, elucidation of the urea cycle gave Krebs insights into how cyclic pathways work; he exploited this knowledge to unravel the complexities of the citrate cycle 5 years later.

The five key steps in the urea cycle are illustrated in Figure 17.31: (1) carbamoyl phosphate is formed in the mitochondrial matrix by the ATP-dependent enzyme carbamoyl phosphate synthetase I, using the substrates HCO_3^- generated in the citrate cycle and NH_4^+ derived from the deamination of glutamine and glutamate. (2) Carbamoyl phosphate is combined with ornithine to form citrulline in a mitochondrial reaction catalyzed by the enzyme ornithine transcarbamoylase. (3) The citrulline is then exported to the cytosol, where it is first activated by AMP before being converted to argininosuccinate when aspartate displaces the AMP. This reaction is catalyzed by the cytosolic enzyme argininosuccinate synthetase and results in the incorporation of a second nitrogen atom into the product. Note that cleavage of PP_i by pyrophosphatase means that this reaction consumes two high-energy phosphate bonds. (4) Argininosuccinate is cleaved in the next reaction by the enzyme argininosuccinase to yield fumarate and arginine, the latter containing both nitrogens. (5) Lastly, the enzyme arginase converts arginine to urea and ornithine to complete the cycle. Ornithine has the same role in the urea cycle as oxaloacetate does in the citrate cycle; namely, as both the product of the last reaction and the substrate of the first reaction.

By including the pyrophosphatase reaction, you can see that four high-energy phosphate bonds are required (4 ATP equivalents) for every molecule of urea that is synthesized. Moreover, the carbon backbone of aspartate gives rise to fumarate:

$$NH_4^+ + CO_2 + \text{Aspartate} + 3\text{ ATP} \rightarrow \text{Urea} + \text{Fumarate} + 2\text{ ADP} + \text{AMP} + 4\text{ P}_i$$

The enzyme carbamoyl phosphate synthetase I catalyzes the rate-limiting reaction in the urea cycle and, as such, is the key regulated step in the pathway. This mitochondrial enzyme is distinct from carbamoyl phosphate synthetase II, which is a cytosolic enzyme involved in pyrimidine biosynthesis (see Chapter 18). As shown in **Figure 17.32**, the three-step reaction catalyzed by carbamoyl phosphate synthetase I requires the hydrolysis of 2 ATP. In the first step, ATP activates HCO_3^- to form carboxyphosphate, which is then attacked by NH_3 in the second step to form carbamate. In the third step, ATP phosphorylates carbamate to form carbamoyl phosphate.

This key reaction in urea synthesis is allosterically regulated by *N*-acetylglutamate, a metabolite that signals high levels of glutamate in the cell (**Figure 17.33**). The enzyme *N*-acetylglutamate synthase catalyzes the formation of *N*-acetylglutamate from glutamate and acetyl-CoA and is activated by arginine, a urea cycle intermediate. The net result is that glutamate and arginine stimulate flux through the urea cycle by increasing the rate of carbamoyl phosphate synthesis. Note that other enzymes required for urea synthesis are regulated at the level of gene expression in response to the amount of protein in the diet and to glucocorticoid hormone signaling.

The urea cycle and citrate cycle are metabolically linked through the shared intermediate fumarate. Fumarate provides the carbon backbone for aspartate by providing

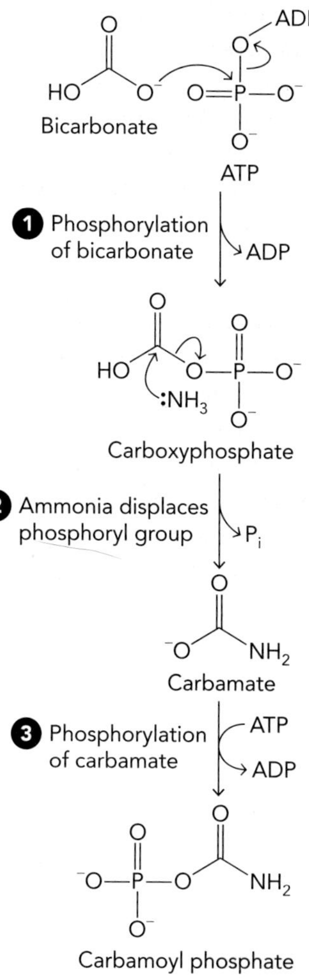

Figure 17.32 The reaction mechanism for carbamoyl phosphate synthetase I is shown here. In the first step of this reaction, bicarbonate (HCO_3^-) is phosphorylated by ATP to form carboxyphosphate, which is then attacked by $:NH_3$ in the second step to release the phosphate and generate carbamate. In the final step, a second ATP molecule is used to phosphorylate carbamate to form the product of the reaction, carbamoyl phosphate.

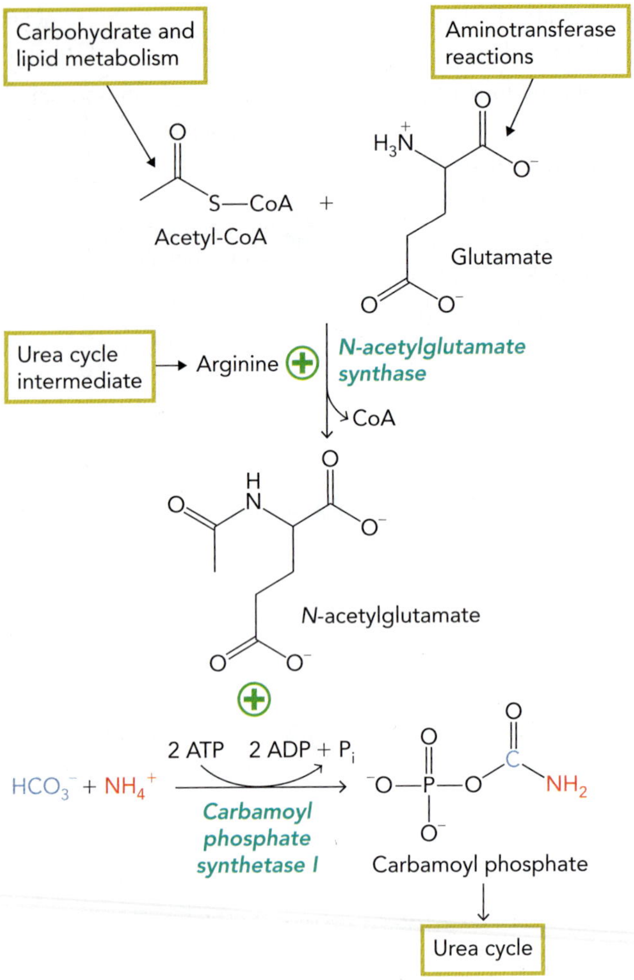

Figure 17.33 Carbamoyl phosphate synthetase I activity is stimulated by *N*-acetylglutamate. High levels of glutamate and acetyl-CoA in liver cells lead to the synthesis of *N*-acetylglutamate, which is an allosteric activator of carbamoyl phosphate synthetase I. Arginine is an allosteric activator of the enzyme *N*-acetylglutamate synthase.

the oxaloacetate needed in the aspartate aminotransferase reaction. As shown in **Figure 17.34**, the aspartate–argininosuccinate shunt converts fumarate, produced in the cytosol by the urea cycle, into malate that is used to make oxaloacetate in the citrate cycle. Oxaloacetate combines with glutamate to generate aspartate and α-ketoglutarate; then, the aspartate is transported back into the cytosol, where it is used as a substrate in the argininosuccinate synthetase reaction of the urea cycle.

In the simplest version of this bypass reaction, sometimes called the **Krebs bicycle** pathway, fumarate is converted to malate in the cytosol by an isozyme of fumarase. The malate can then be transported into the mitochondrial matrix using the malate–aspartate shuttle and converted to oxaloacetate by malate dehydrogenase. The resulting oxaloacetate is used as a substrate in the aspartate aminotransferase reaction to generate aspartate, which is transported to the cytosol, where it serves as a urea cycle substrate. Note that nitrogen from amino acid degradation enters the Krebs bicycle at two points: (1) deamination of glutamine and glutamate to generate NH_4^+ for the carbamoyl phosphate synthetase I reaction; and (2) transfer of the amino group from glutamate to oxaloacetate to generate aspartate in the aspartate aminotransferase reaction. Most important, the recycling of fumarate to generate oxaloacetate for the aspartate aminotransferase reaction produces NADH in the malate dehydrogenase reaction, which can be used by the electron transport system to generate 2.5 ATP. This net yield of ATP helps offset the energy cost of the urea cycle (4 ATP equivalents).

Inherited defects in many of the urea cycle enzymes have been observed clinically. Complete loss of a urea cycle enzyme causes death shortly after birth; however, a deficiency in urea cycle enzymes results in **hyperammonemia** (elevated ammonia levels in the blood). Most urea cycle disorders also lead to a buildup of glutamine and glutamate, which function as osmolytes (compounds affecting osmosis) that can cause brain swelling and associated neurologic symptoms. Fortunately, it is possible to treat some urea cycle disorders by restricting dietary protein as a means to limit nitrogen intake. In addition, by providing metabolic substrates that increase the biosynthesis of nitrogen-containing compounds that can be excreted, it is often possible to decrease the severity of hyperammonemia. As shown in **Figure 17.35**, one way to remove excess nitrogen in individuals with urea cycle deficiencies is to treat them with the compound phenylbutyrate. Phenylbutyrate is metabolized to the compound phenylacetylglutamine by the enzyme glutamine *N*-acetyltransferase and is excreted in the urine. This results in increased synthesis of glutamine from glutamate and NH_4^+ by the glutamate synthase reaction, thereby lowering NH_4^+ levels.

Another approach to treating urea cycle disorders is to provide the metabolites to maintain flux through the cycle and avoid substrate buildup. This strategy works in

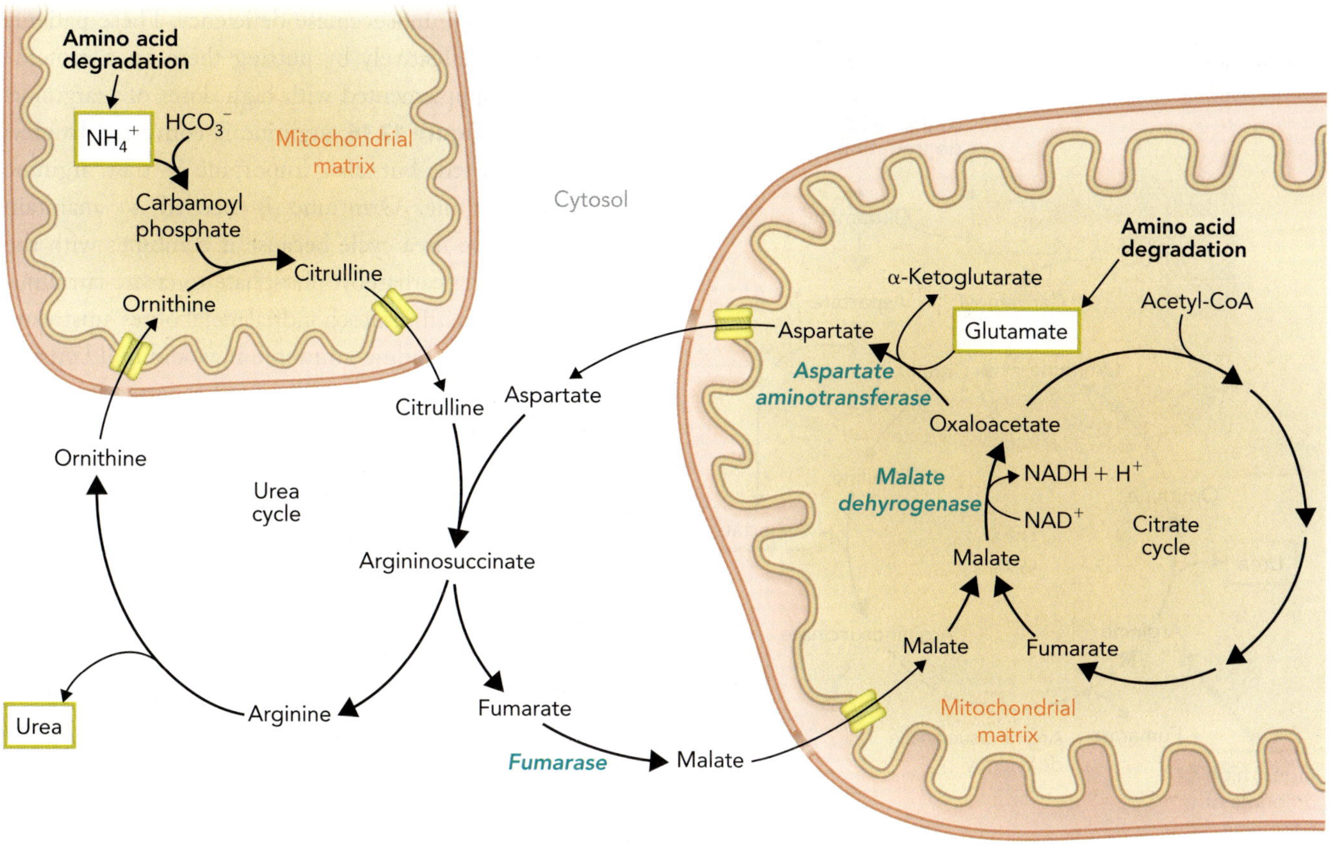

Figure 17.34 The aspartate–argininosuccinate shunt (Krebs bicycle) connects the urea cycle and the citrate cycle. Fumarate produced in the urea cycle can be used to make aspartate in the mitochondrial matrix, thereby linking the urea cycle to the citrate cycle. See text for details. The two primary entry points of nitrogen from amino acid degradation are shown.

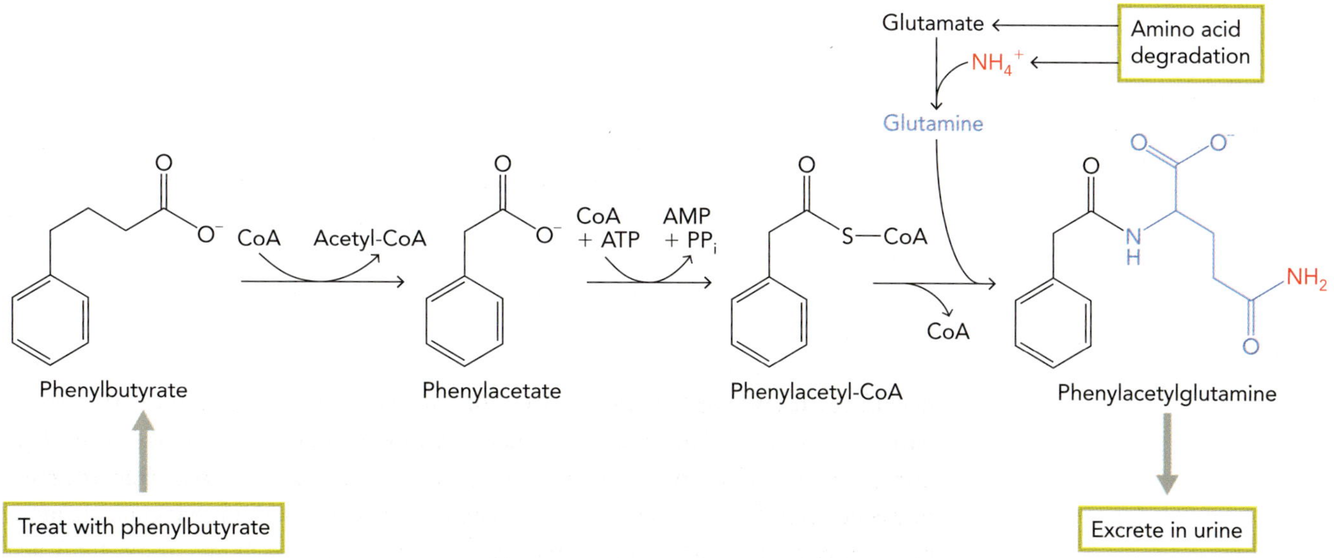

Figure 17.35 Some individuals with urea cycle deficiencies can be treated with sodium phenylbutyrate, which is metabolized to phenylacetylglutamine and excreted in the urine. This treatment strategy provides a mechanism to remove excess NH_4^+ and glutamine from the body.

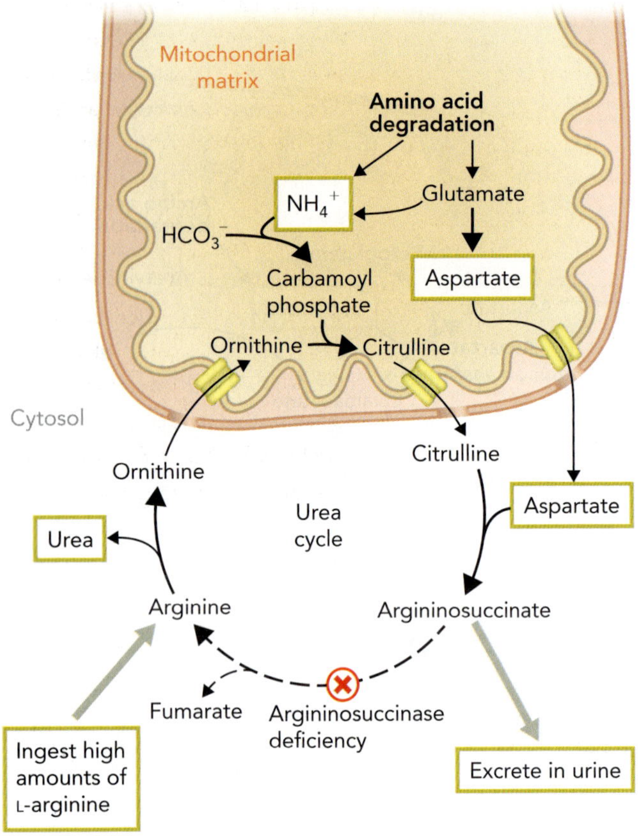

Figure 17.36 Patients with argininosuccinase deficiency can be treated with L-arginine. Conversion of arginine to urea and ornithine provides the ornithine needed to maintain flux through the remainder of the urea cycle. Ornithine combines with carbamoyl phosphate to generate citrulline and then argininosuccinate. Because argininosuccinate is water soluble, it is excreted in urine.

patients with argininosuccinase deficiency. These patients can be treated effectively by putting them on a protein-depleted diet supplemented with high doses of L-arginine. As shown in **Figure 17.36**, arginine is converted to urea, which is excreted, but also important is that arginine produces ornithine. Ornithine is needed to maintain flux through the urea cycle because it combines with the nitrogen-carrying carbamoyl phosphate to create citrulline. Without ornithine, the mitochondrial metabolites "upstream" of this reaction, including ammonium, would build up and cause ammonia toxicity. Citrulline and aspartate combine to generate argininosuccinate, which is soluble and can be excreted in the urine. Note that supplementing the diet with ornithine would also give the same result, but using arginine is more cost-effective.

Degradation of Glucogenic and Ketogenic Amino Acids

The carbon backbones of 11 of the 20 standard amino acids can be converted into pyruvate or acetyl-CoA, which can then be used for energy conversion by the citrate cycle and oxidative phosphorylation reactions. The other nine amino acids are converted to the citrate cycle intermediates α-ketoglutarate, fumarate, succinyl-CoA, and oxaloacetate, which can be used for glucose synthesis by conversion of oxaloacetate to phosphoenolpyruvate. Amino acid degradation pathways are somewhat complex, so it is convenient to think about them in terms of the metabolites they produce, and whether these metabolites are precursors to glucose or ketone bodies.

As shown in **Figure 17.37**, amino acids that give rise to pyruvate or any of the citrate cycle intermediates are called **glucogenic** amino acids because pyruvate and oxaloacetate are precursors in the gluconeogenic pathway (α-ketoglutarate, succinyl-CoA, and fumarate can be converted to oxaloacetate). In contrast, amino acids that are converted into acetyl-CoA or acetoacetyl-CoA are called **ketogenic** amino acids because they can give rise to ketone bodies (see Figure 16.41). Five amino acids provide carbon backbones that can be used for both glucose production (glucogenic) and the synthesis of acetoacetyl-CoA or acetyl-CoA (ketogenic), which means they are categorized as both glucogenic and ketogenic amino acids. As shown in Figure 17.37, these five amino acids are threonine, tryptophan, phenylalanine, tyrosine, and isoleucine.

Rather than describe all of the known amino acid degradation pathways in detail, we instead present an overview of three interconnected amino acid degradation pathways, which together account for 13 of the 20 amino acids. These three groups of pathways can be categorized as follows: *group 1* pathways, which degrade four glucogenic amino acids (alanine, cysteine, glycine, and serine) that generate pyruvate and two amino acids (threonine and tryptophan) that generate acetyl-CoA and acetoacetyl-CoA; *group 2* pathways, which degrade five glucogenic amino acids (arginine, histidine, glutamate, glutamine, and proline) that generate α-ketoglutarate; and *group 3* pathways, which convert the ketogenic amino acid phenylalanine to tyrosine, also a ketogenic amino acid, which is then degraded to acetoacetyl-CoA.

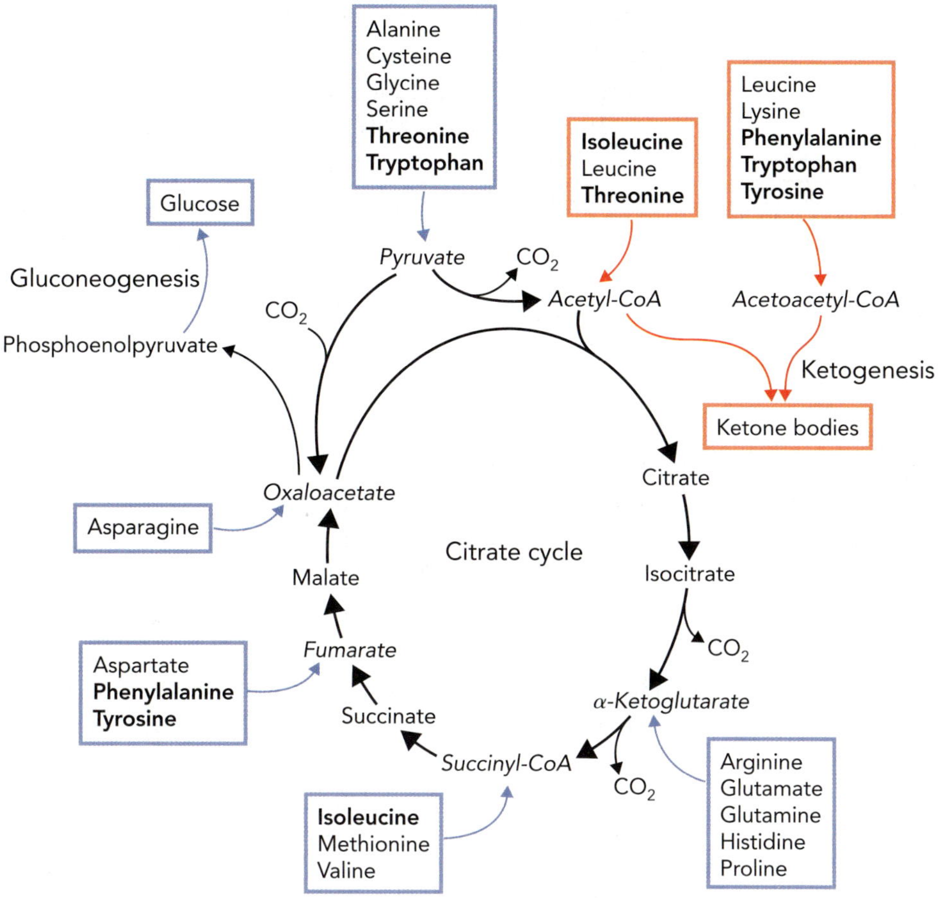

Figure 17.37 An overview of the glucogenic and ketogenic amino acid degradation pathways is shown. Glucogenic amino acids are amino acids with carbon skeletons that can be used to synthesize glucose, whereas ketogenic amino acids give rise to acetyl-CoA and acetoacetyl-CoA, which are precursors to ketone bodies. Five amino acids are considered both glucogenic and ketogenic and are shown in boldface. Italicized compounds are products of amino acid degradation pathways.

Group 1 amino acid degradation pathways are shown in **Figure 17.38**. This group includes the simplest conversion of an amino acid to pyruvate; namely, deamination of alanine to generate pyruvate in a reaction catalyzed by alanine aminotransferase. Another single-step conversion is that of serine to pyruvate by the pyridoxal phosphate–dependent enzyme serine dehydratase, which removes both the hydroxyl group and amino group through formation of a covalent enzyme intermediate. The amino acid cysteine is converted to pyruvate by a single *Escherichia coli* enzyme, whereas in animals, the enzyme cysteine dioxygenase first oxidizes the sulfhydryl on cysteine to form cysteine sulfinate, which is then deaminated by cysteine aminotransferase to form 3-sulfinylpyruvate. The sulfur group is then spontaneously cleaved to release SO_3^{2-} and pyruvate.

Glycine degradation can proceed by at least three different routes, as shown in Figure 17.38: (1) The enzyme serine hydroxymethyltransferase converts glycine to serine, which is then a substrate for the serine dehydratase reaction leading to the formation of pyruvate. The serine hydroxymethyltransferase reaction requires the coenzymes pyridoxal phosphate and tetrahydrofolate (N^5,N^{10}-methylenetetrahydrofolate), which are derived from vitamin B_6 and folate, respectively. As described in Chapter 18, tetrahydrofolate is a carbon donor used as a coenzyme in a reaction catalyzed by thymidylate synthase. (2) Glycine can also be degraded completely to CO_2 and NH_4^+ by the enzyme glycine synthase, which is the predominant fate of glycine in animal cells. (3) Alternatively, glycine can be converted to glyoxylate by the enzyme D-amino acid oxidase. Tryptophan degradation is more complex, consisting of multiple enzymatic reactions leading to the production of acetoacetyl-CoA and alanine, which is then converted to pyruvate. Lastly, threonine

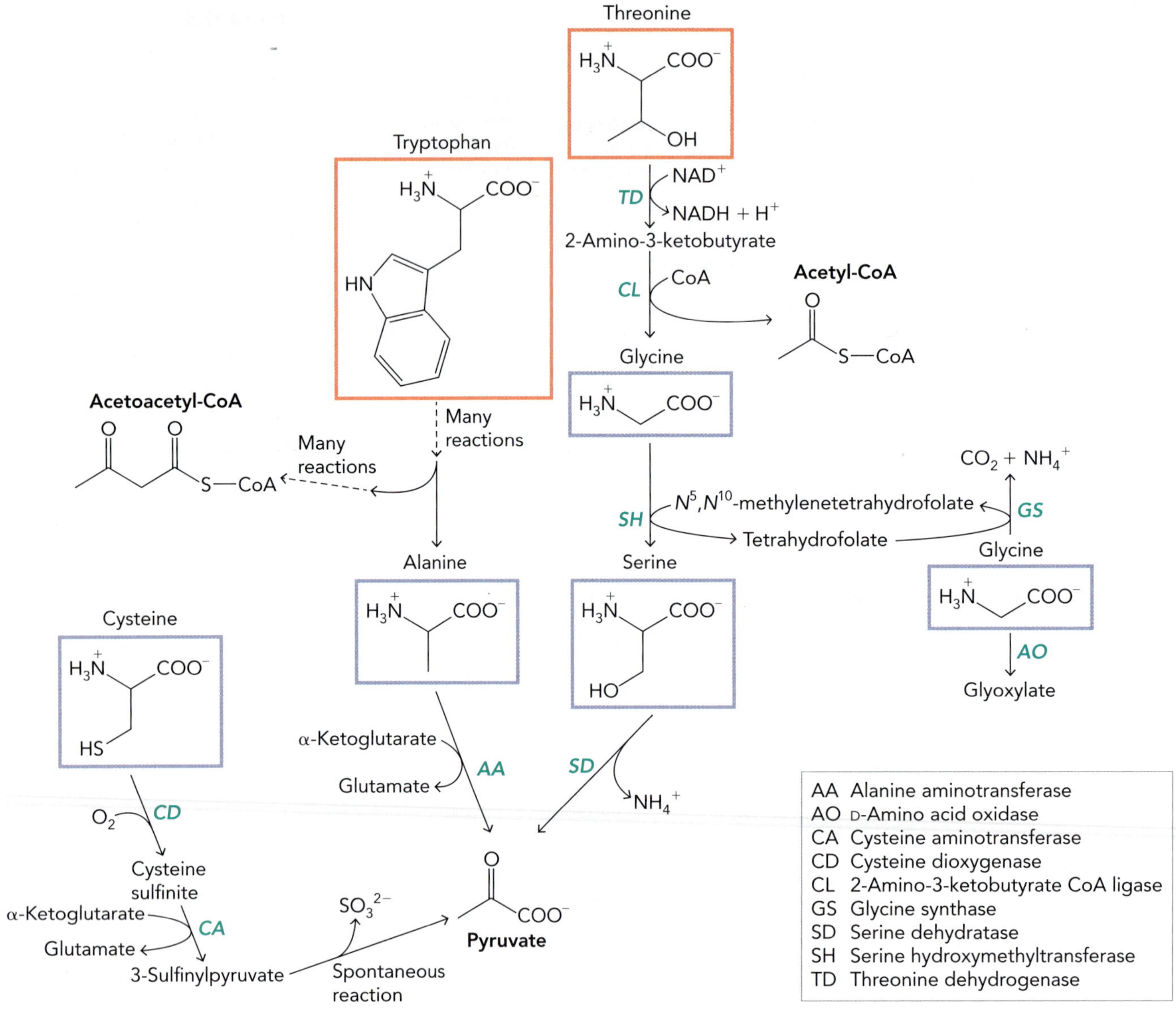

Figure 17.38 Group 1 amino acid degradation pathways. The glucogenic amino acids alanine, cysteine, glycine, and serine are deaminated and converted to pyruvate, whereas the amino acids threonine and tryptophan give rise to acetyl-CoA and acetoacetyl-CoA, respectively, though they also lead to pyruvate. Threonine and tryptophan can be considered both ketogenic and glucogenic amino acids, as parts of their carbon skeletons are used to generate pyruvate.

is converted to acetyl-CoA and glycine by the enzymes threonine dehydrogenase and 2-amino-3-ketobutyrate CoA ligase.

Group 2 amino acid degradation pathways are shown in **Figure 17.39**. Here it is observed that arginine, proline, histidine, and glutamine are all converted to glutamate, which is then deaminated by glutamate dehydrogenase to form α-ketoglutarate. The most straightforward conversion reaction in this series is that of glutamine deamination by glutaminase to generate NH_4^+ and glutamate, which was introduced earlier when describing entry points for NH_4^+ into the urea cycle (see Figure 17.29). Another enzyme we have already discussed is arginase, which cleaves arginine to form urea and ornithine in the final step of the urea cycle (see Figure 17.31). To recover the carbon skeleton from ornithine to make α-ketoglutarate, the amide group of ornithine is removed by ornithine δ-aminotransferase to form glutamate and glutamate-5-semialdehyde. The enzyme glutamate semialdehyde dehydrogenase then reduces glutamate-5-semialdehyde to form glutamate. Proline degradation also leads to the production of glutamate-5-semialdehyde, after reduction by the enzyme proline oxidase and a spontaneous hydration reaction. The most complex set of reactions in this series is that of histidine degradation, which requires four reactions. The first

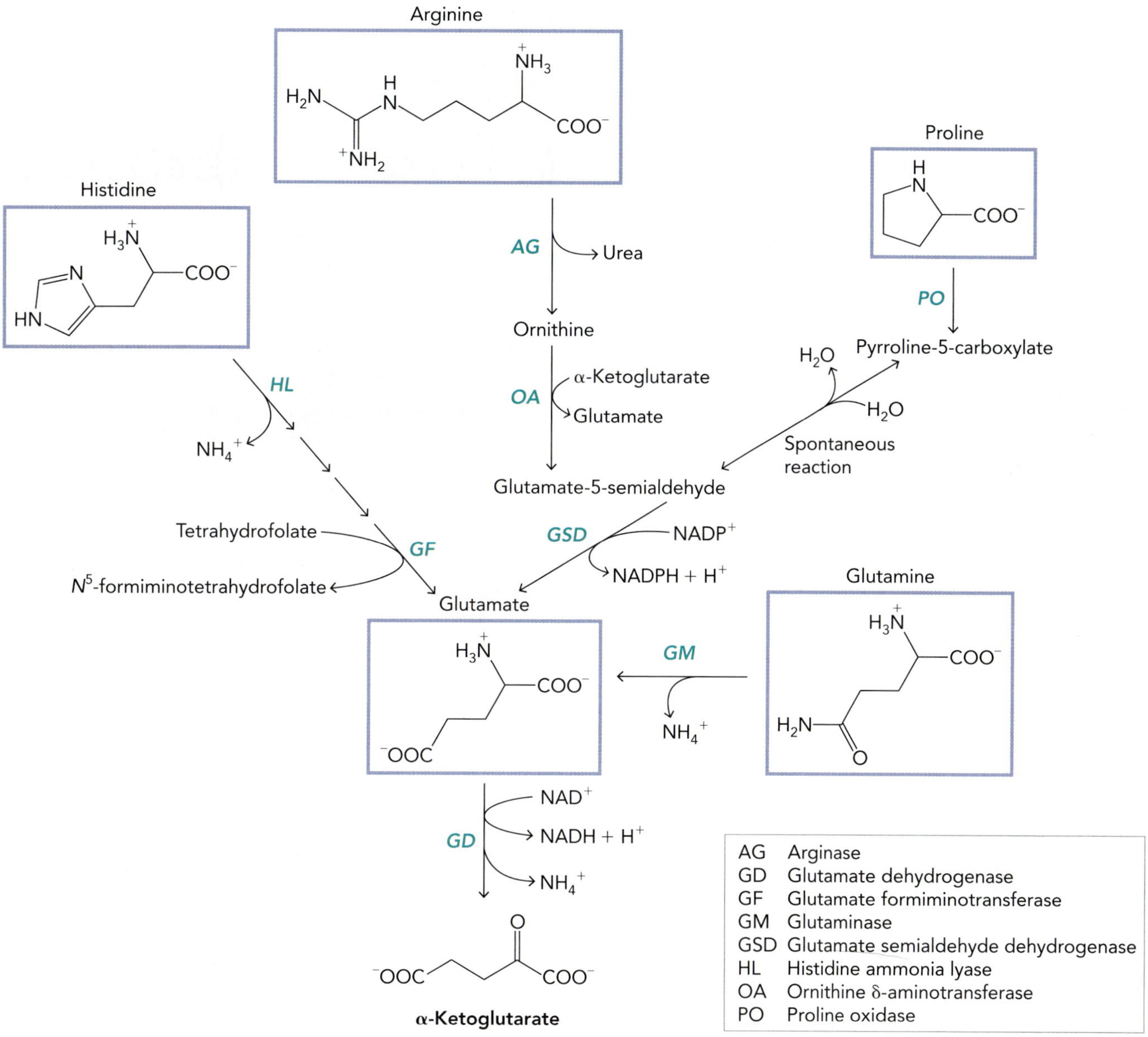

Figure 17.39 Group 2 amino acid degradation pathways are shown. Four glucogenic amino acids are converted to glutamate, which is then deaminated to form α-ketoglutarate.

reaction removes NH_4^+ in a reaction catalyzed by the enzyme histidine ammonia lyase. In the final reaction, glutamate formiminotransferase uses the coenzyme tetrahydrofolate to remove one carbon in order to generate glutamate.

Group 3 amino acid degradation pathways are shown in **Figure 17.40**. The first reaction converts the ketogenic amino acid phenylalanine into tyrosine, which is then metabolized to generate acetoacetyl-CoA. The conversion of phenylalanine to tyrosine by the enzyme phenylalanine hydroxylase serves as an initiating step in phenylalanine degradation and is also responsible for tyrosine production in animals. Animals lack the enzymes needed to synthesize phenylalanine *de novo*, but because they have the enzyme phenylalanine hydroxylase, they can convert dietary phenylalanine into tyrosine. Tyrosine is the metabolic precursor for the neurotransmitter dopamine and for skin pigments (melanins) and epinephrine.

Phenylalanine degradation was the first metabolic pathway to be linked to a human disease resulting from a single gene mutation. In 1902, Sir Archibald Garrod, a London physician, published a scientific paper describing high levels of homogentisate in the urine excreted by genetically related individuals with a particular

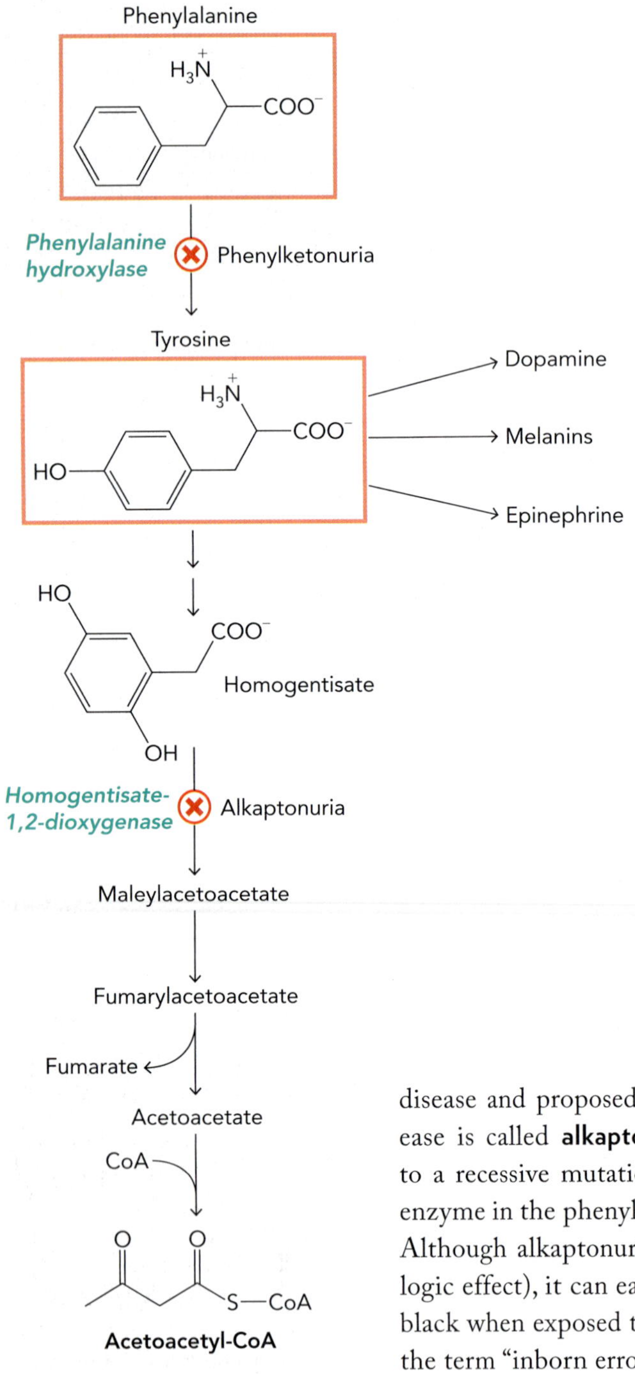

Figure 17.40 Group 3 amino acid degradation pathways are shown. The ketogenic amino acids phenylalanine and tyrosine are converted to acetoacetyl-CoA. The inherited metabolic diseases phenylketonuria and alkaptonuria are due to disruption of two key enzymes in this amino acid degradation pathway.

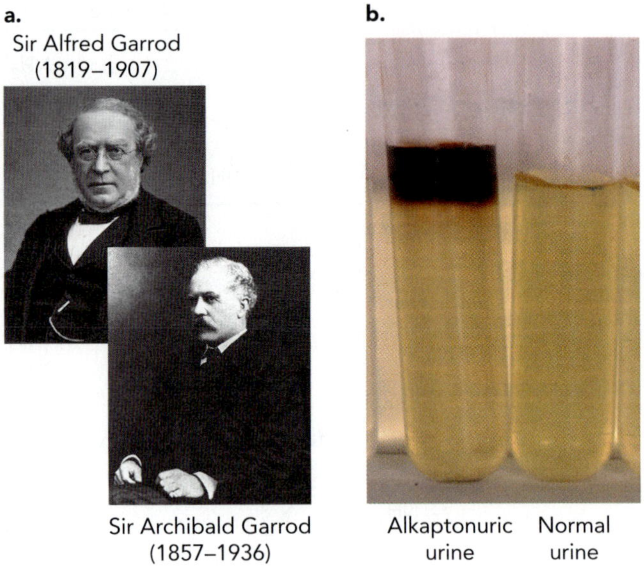

Figure 17.41 The link between genetic inheritance and metabolic disease was first made for the disease alkaptonuria, also called black urine disease. **a.** Sir Archibald Garrod, a British physician-scientist, described the first case of alkaptonuria. His father, Sir Alfred Garrod, also a physician-scientist, was the first to describe chronic joint inflammation. SIR ALFRED GARROD: WELLCOME LIBRARY, LONDON; SIR ARCHIBALD GARROD: © USED WITH PERMISSION FROM COLD SPRING HARBOR LABORATORY ARCHIVES. **b.** Alkaptonuria is diagnosed by the dark color of urine, which appears within 24 hours through oxidation of the accumulated metabolite homogentisate when exposed to air (see Figure 17.40). COURTESY ALKAPTONURIA SOCIETY.

disease and proposed that the disease was due to an enzyme deficiency. The disease is called **alkaptonuria**, or black urine disease. It was later found to be due to a recessive mutation in the gene encoding homogentisate-1,2-dioxygenase, an enzyme in the phenylalanine and tyrosine degradation pathways (see Figure 17.40). Although alkaptonuria is usually not debilitating (arthritis is the primary pathologic effect), it can easily be diagnosed because the urine of these individuals turns black when exposed to air as a result of oxidation of homogentisate. Garrod coined the term "inborn errors of metabolism" in 1908 to describe the link between genetics and metabolic disease and published a seminal book on this subject in 1923 (**Figure 17.41**). Garrod's approach to medical science was likely influenced by the work of his father, Sir Alfred Garrod, who was himself a well-known physician in London in the 1800s. Alfred Garrod is credited with discovering the link between uric acid and gout, which he reported in an 1848 article published in *Medical Chirurgical Transactions*. Like his son Archibald, Alfred also had a penchant for defining medical conditions: In 1859, he was the first to use the term "rheumatoid arthritis" as a description for chronic inflammation of the joints.

Another example of a genetically linked metabolic disorder is **phenylketonuria (PKU)**, which is due to defects in the enzyme phenylalanine hydroxylase. The disease occurs about once in every 15,000 births, making it one of the more common genetic metabolic disorders. Disease symptoms in untreated individuals include severe mental

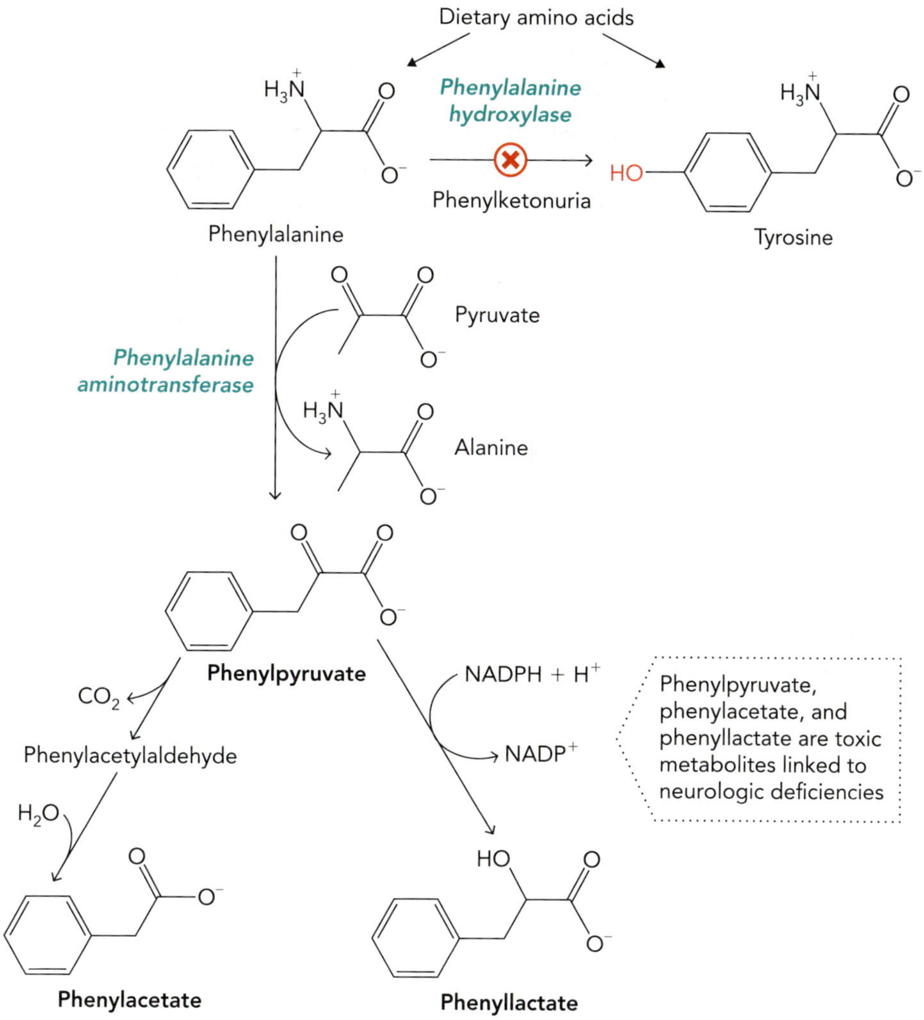

Figure 17.42 The neurologic symptoms associated with phenylketonuria are due to buildup of the toxic phenylalanine metabolites phenylpyruvate, phenylacetate, and phenyllactate in the brain. The primary treatment for phenylketonuria is to limit the amount of phenylalanine in the diet.

retardation, stunted growth, and dental problems. The clinical symptoms of phenylketonuria are caused by the accumulation of phenylalanine in the blood at a concentration 30–50 times higher than normal. This high level of phenylalanine leads to the production of phenylalanine metabolites such as phenylpyruvate, phenylacetate, and phenyllactate (**Figure 17.42**), all of which are associated with the observed neurologic and developmental problems.

The genetic cause of phenylketonuria was discovered in the 1930s by Dr. Ivar Asbjørn Følling of Norway, who made the connection between this metabolic disorder and a significant number of institutionalized mental patients (**Figure 17.43**). The American writer Pearl Buck described the effects of phenylketonuria on her daughter Carol in the essay *The Child Who Never Grew*, which was published in 1950. Buck's international fame (she had won both the Pulitzer Prize and the Nobel Prize) brought much-needed attention to the devastating—and avoidable—neurologic effects of phenylketonuria on children. With the introduction of a reliable phenylketonuria blood test for newborns

Ivar Asbjørn Følling (1888–1973)

Carol Buck (1920–1992), left, and Pearl Buck, (1892–1973), right

Figure 17.43 The genetic link between defects in phenylalanine metabolism and severe mental retardation was first described in 1934 by the Norwegian physician Dr. Ivar Asbjørn Følling. Pearl Buck brought attention to the tragedy of this treatable disease in her 1950 essay *The Child Who Never Grew*, which was first published in a women's magazine. IVAR ASBJØRN FØLLING: SCANPIX NORWAY/SIPA USA; CAROL AND PEARL BUCK: HULTON ARCHIVE/GETTY IMAGES.

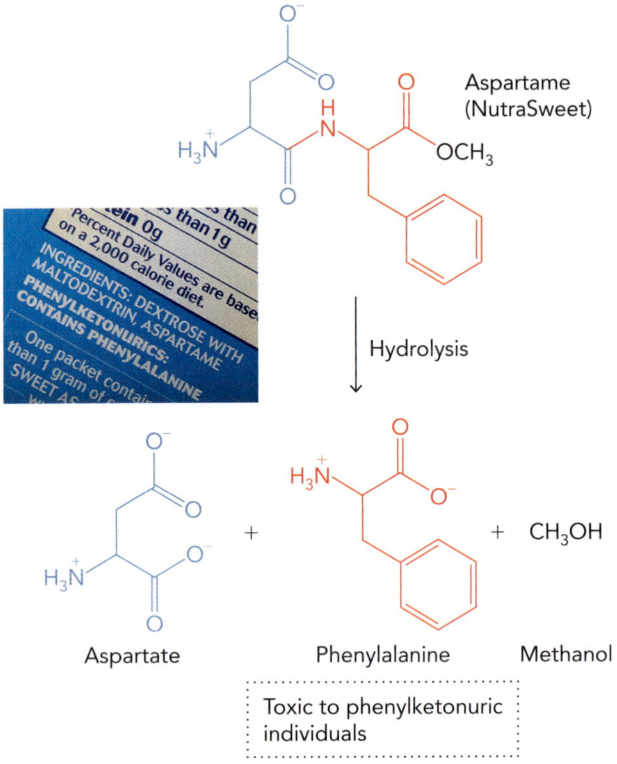

Figure 17.44 Phenylketonuric individuals must avoid foods containing the artificial sweetener aspartame because it is hydrolyzed in the stomach to phenylalanine and aspartate. NutraSweet, the commercial name for aspartame, is a popular zero-calorie sweetener found in a variety of processed foods and beverages. PHOTO: EDITORIAL IMAGE, LLC/ALAMY.

in the 1960s, dietary regimens were developed for infants that limited phenylalanine ingestion, thereby preventing the associated neurologic damage.

Because the symptoms of phenylketonuria are caused by excess phenylalanine and its metabolites, and because humans require phenylalanine in their diets (it is an essential amino acid), dietary treatment focuses on careful monitoring of phenylalanine intake to provide just enough for protein synthesis without causing phenylalanine accumulation. Phenylketonuric individuals also have to be careful to avoid processed foods and beverages containing the food additive **aspartame** (aspartyl-phenylalanine methyl ester) because it is hydrolyzed to aspartate and phenylalanine in the intestine (**Figure 17.44**).

Aspartame is an artificial sweetener that was discovered in 1965 by James Schlatter, a chemist at the pharmaceutical company G.D. Searle. Schlatter was synthesizing compounds to be tested as a treatment for gastric ulcers and, unknowingly, got one of the intermediate products on his finger while working at the bench. When he licked his finger later in the day to pick up a piece of paper, he noticed an intense sweet flavor and initially thought it came from his breakfast donut. He eventually figured out it was the aspartyl-phenylalanine methyl ester compound he had made. Twenty years later, NutraSweet became a billion-dollar-a-year product for the company. Aspartame binds to G protein–coupled receptors on taste cells in the tongue and is 150 times sweeter than sucrose in human taste tests (see Figure 9.7).

concept integration 17.2

Which two amino acids serve as the primary nitrogen donors in urea synthesis? Why isn't aspartate considered a primary nitrogen donor?

The two amino acids serving as primary nitrogen donors in urea synthesis are glutamine and glutamate. Urea is a waste product synthesized in terrestrial vertebrates (and some invertebrates) from the NH_4^+ released from glutamine and glutamate and from the nitrogen atom of aspartate, which is derived from glutamate in the aspartate aminotransferase reaction. The carbon atom in urea is derived from HCO_3^- produced by the citrate cycle (CO_2), and the oxygen atom comes from H_2O used to hydrolyze arginine. Glutamine serves as a nitrogen carrier in the body that transports NH_4^+ from the peripheral tissues to the liver, where it is deaminated by the enzyme glutaminase in the mitochondrial matrix to release NH_4^+ and regenerate glutamate. The enzyme glutamate dehydrogenase has the important job in the liver of releasing the NH_4^+ from glutamate and producing α-ketoglutarate. The free NH_4^+ is then combined with CO_2 to form the urea cycle substrate carbamoyl phosphate in a mitochondrial reaction catalyzed by the enzyme carbamoyl phosphate synthetase I. The second nitrogen in urea comes directly from aspartate; however, aspartate is not considered a primary nitrogen donor to urea because it is itself generated in the aspartate aminotransferase reaction from glutamate in a cytosolic reaction in which citrulline is converted to argininosuccinate by the addition of aspartate.

17.3 Amino Acid Biosynthesis

We now turn our attention to the metabolic pathways involved in amino acid biosynthesis. To get an overview of this process, we first look at the metabolic pathways used by plants and bacteria to synthesize nine amino acids, starting with the metabolic intermediates pyruvate and oxaloacetate (alanine, valine, leucine, isoleucine, aspartate, asparagine, methionine, threonine, and lysine). We then examine the biosynthetic pathways needed to convert phosphoenolpyruvate and the pentose phosphate pathway intermediate erythrose-4-phosphate into the aromatic amino acids tryptophan, phenylalanine, and tyrosine.

Amino Acids Are Derived from Common Metabolic Intermediates

The carbon skeletons of all 20 amino acid side chains are derived from just seven metabolic intermediates, which are found in three metabolic pathways (**Figure 17.45**):

1. Three glycolytic pathway intermediates: 3-phosphoglycerate, phosphoenolpyruvate, and pyruvate
2. Two pentose phosphate pathway intermediates: ribose-5-phosphate and erythrose-4-phosphate
3. Two citrate cycle intermediates: α-ketoglutarate and oxaloacetate

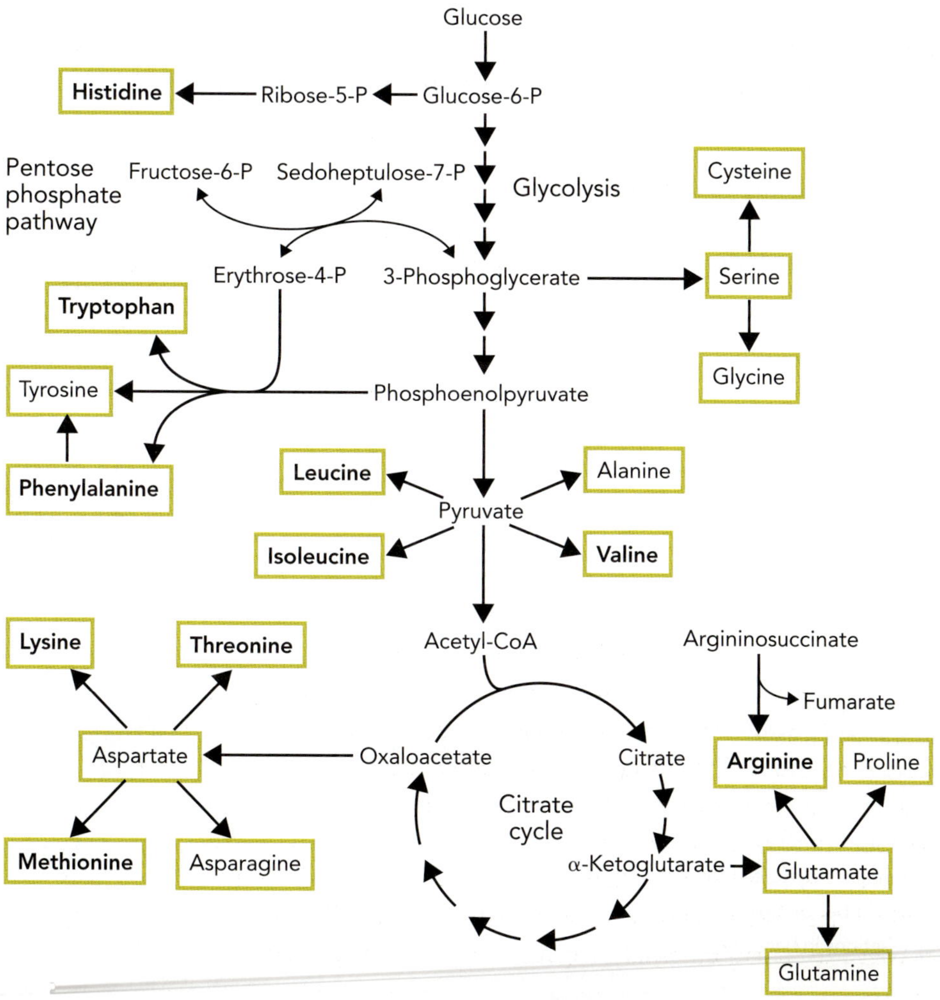

Figure 17.45 Plants and bacteria synthesize all 20 amino acids using seven different metabolic precursors derived from glycolysis, the pentose phosphate pathway, and the citrate cycle. Amino acids shown in boldface are essential amino acids for animals.

Table 17.1 LIST OF ESSENTIAL AND NONESSENTIAL AMINO ACIDS IN HUMANS

Essential amino acids	Nonessential amino acids
Arginine[a]	Alanine
Histidine	Asparagine
Isoleucine	Aspartate
Leucine	Cysteine
Lysine	Glutamate
Methionine	Glutamine
Phenylalanine	Glycine
Threonine	Proline
Tryptophan	Serine
Valine	Tyrosine[b]

[a]Mammals synthesize arginine from argininosuccinate in the urea cycle; however, most of it is cleaved to form urea and ornithine and therefore not available for protein synthesis.

[b]Tyrosine is generated from phenylalanine in animals that contain the enzyme phenylalanine hydroxylase, and therefore it is not considered an essential amino acid even though phenylalanine is essential.

Plants and bacteria are capable of synthesizing all 20 amino acids, but most animals synthesize only about half of the amino acids because they lack many of the required enzymes. This is thought to be a result of evolutionary adaptation in animals, which, because of their dependence on food for chemical energy, have diets rich in protein. Therefore, animals do not need to commit energy reserves to synthesizing amino acids that they can obtain in their diets.

As shown in **Table 17.1**, humans can synthesize only 10 of the 20 amino acids, which are called *nonessential* amino acids. In contrast, the other 10 amino acids, called *essential* amino acids, must be obtained from the diet. Alanine and aspartate are nonessential amino acids because humans can make them from pyruvate and oxaloacetate, respectively, using transamination reactions. Essential amino acids, such as tryptophan and methionine, must be obtained from the diet because humans lack the enzymes necessary to synthesize them *de novo*. Arginine is listed in Table 17.1 as an essential amino acid because humans need arginine in the diet to support rapid growth during childhood and during pregnancy. Arginine, however, is actually generated from argininosuccinate in the urea cycle (see Figure 17.31), which means that a small amount of this "essential" amino acid is made available for protein synthesis through this route. Tyrosine is listed in Table 17.1 as a nonessential amino acid because it is made in humans from the essential amino acid phenylalanine by the enzyme phenylalanine hydroxylase (see Figure 17.40). Therefore, humans can generate tyrosine as long as there is enough phenylalanine in the diet. But in fact, much of the tyrosine in the human body comes directly from dietary tyrosine.

In general, the structures of the essential amino acids are more complex than those of the nonessential amino acids, which is reflected in the number of enzymatic reactions required to synthesize the essential amino acids. As shown in **Figure 17.46**, the nonessential amino acids alanine, aspartate, and serine are synthesized by all organisms

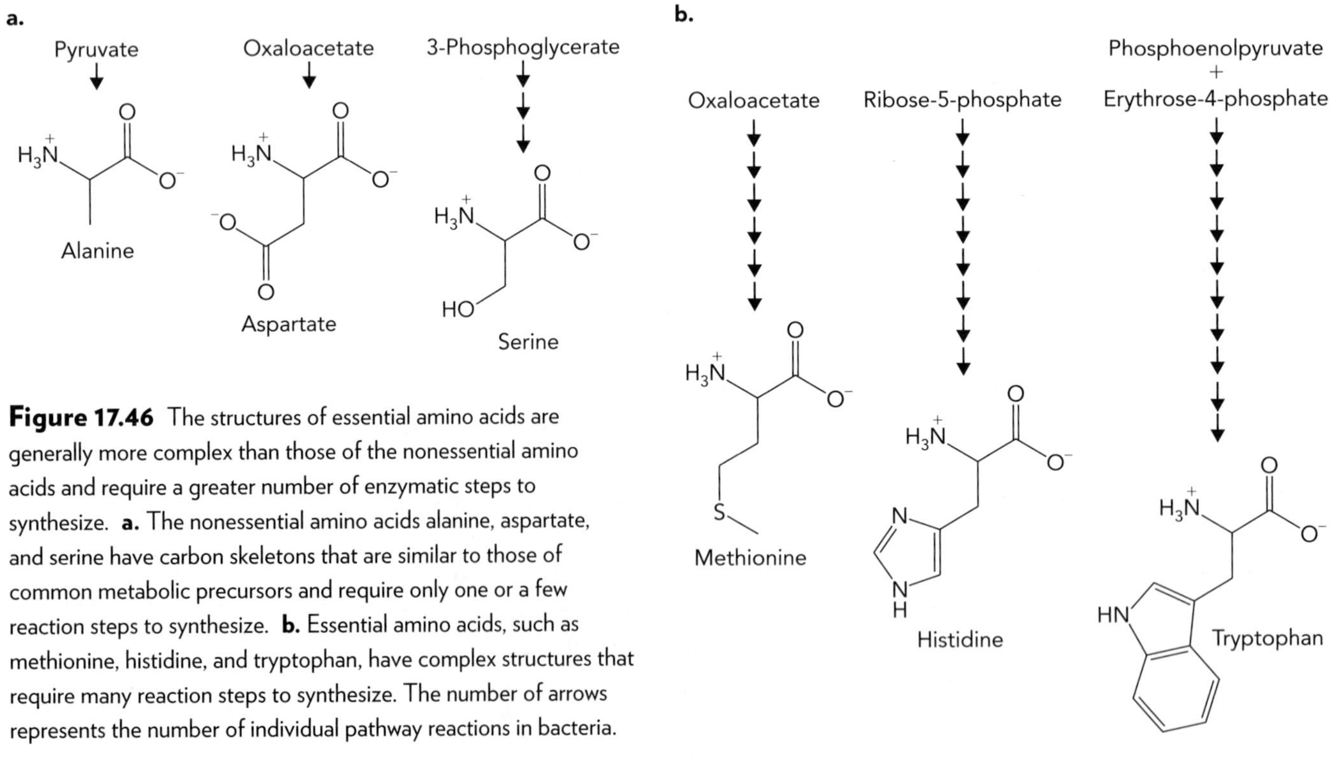

Figure 17.46 The structures of essential amino acids are generally more complex than those of the nonessential amino acids and require a greater number of enzymatic steps to synthesize. **a.** The nonessential amino acids alanine, aspartate, and serine have carbon skeletons that are similar to those of common metabolic precursors and require only one or a few reaction steps to synthesize. **b.** Essential amino acids, such as methionine, histidine, and tryptophan, have complex structures that require many reaction steps to synthesize. The number of arrows represents the number of individual pathway reactions in bacteria.

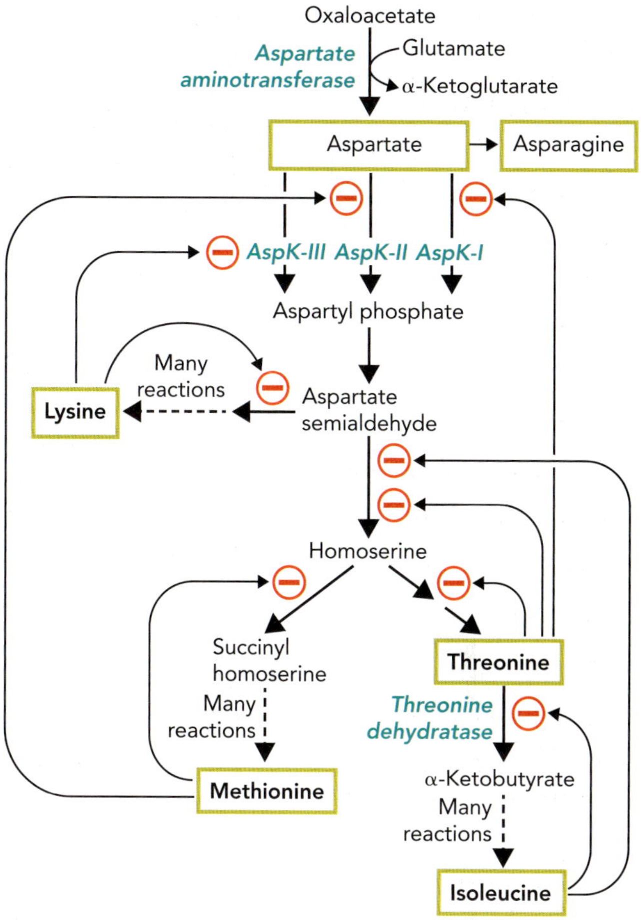

Figure 17.47 Feedback regulation is critical in controlling flux through amino acid biosynthetic pathways in plants and bacteria. Amino acids function as allosteric inhibitors that bind to regulatory sites in one or more upstream enzymes in the pathway. The conversion of aspartate to aspartyl phosphate is catalyzed by three isozymes of aspartokinase (shown here as AspK-I, AspK-II, and AspK-III), which are each inhibited by different amino acids. Another key enzyme in this regulated pathway is threonine dehydratase, which is inhibited by isoleucine and serves to link the oxaloacetate and pyruvate amino acid biosynthetic pathways. Essential amino acids in humans are shown in boldface.

using simple reaction pathways, whereas plants and bacteria synthesize the essential amino acids methionine, histidine, and tryptophan using multienzyme pathways.

Metabolic flux through amino acid biosynthetic pathways is tightly regulated to maintain a pool of amino acids that optimally supports protein synthesis. For example, lysine, alanine, valine, isoleucine, glycine, and glutamic acid are the most common amino acids in proteins, whereas cysteine, tyrosine, histidine, and methionine are relatively rare. As illustrated in **Figure 17.47**, the general principle of feedback inhibition plays a pivotal role in modulating flux through linked amino acid biosynthetic pathways. For example, in the bacterial pathway that converts oxaloacetate to aspartate—which is used to synthesize asparagine, lysine, methionine, threonine, and isoleucine—numerous reactions in the pathway are allosterically inhibited by the downstream amino acid products. At each control point, amino acid binding to the allosteric site on the enzyme leads to decreased metabolic flux through that reaction. If the control point is at a branch in the pathway, then feedback inhibition in only one side of the branch temporarily results in increased synthesis of the other product. In most cases, however, the same amino acid also inhibits synthesis of a shared intermediate, thereby eventually blocking synthesis of both products.

Threonine, for example, inhibits its own synthesis and the synthesis of isoleucine. It regulates the activities of homoserine dehydrogenase and homoserine kinase, which produce and break down homoserine. By inhibiting homoserine synthesis, it also affects the methionine branch of the pathway. Lastly, threonine inhibits the conversion of aspartate to aspartyl phosphate by the enzyme aspartokinase. Notably, *E. coli* contains three isozymes of aspartokinase, two of which are allosterically inhibited: one by threonine (aspartokinase I; AspK-I) and the other by lysine (aspartokinase III; AspK-III). The third isozyme, aspartokinase II (AspK-II), is inhibited at the transcriptional level by methionine. This mechanism of isozyme regulation by individual amino acids provides a metabolic "rheostat" that modulates flux through the entire pathway in response to the protein synthesis needs of the cell.

Nine Amino Acids Are Synthesized from Pyruvate and Oxaloacetate

Now that we have a global view of amino acid biosynthesis and its regulation, let's look at a few examples in more detail. As shown in **Figure 17.48**, the biosynthesis in *E. coli* of three nonessential amino acids in humans (alanine, aspartate, and asparagine) and six essential amino acids in humans (methionine, threonine, lysine, isoleucine, valine, and leucine) involves two interconnected pathways that use pyruvate and oxaloacetate as precursors. The metabolic intermediate α-ketobutyrate links the oxaloacetate and pyruvate pathways together, owing to the fact that the carbon skeleton of isoleucine is derived from both α-ketobutyrate and pyruvate. Although most organisms contain orthologous enzymes required for the synthesis of alanine, aspartate, and asparagine, the pathways leading to the synthesis of methionine, threonine, lysine, isoleucine, valine, and leucine vary considerably among plant and bacterial species.

Figure 17.49 illustrates the enzymatic reaction in *E. coli* that converts oxaloacetate to aspartate, along with the enzymatic reactions then used to synthesize asparagine, lysine, methionine, and threonine. Asparagine is made from aspartate in an ATP-dependent reaction catalyzed by the enzyme asparagine synthetase, which uses glutamine as the NH_3 donor. Phosphorylation of aspartate by one of three aspartokinase isozymes (see Figure 17.47) leads to the formation of aspartyl phosphate, which is then reduced and dephosphorylated by aspartate semialdehyde dehydrogenase. The lysine pathway then splits off from the methionine and threonine pathways, when aspartate semialdehyde is converted to either dihydrodipicolinate or homoserine. Lysine biosynthesis requires another five reactions, one of which is catalyzed by the enzyme succinyl diaminopimelate aminotransferase, which acquires the NH_3 group from glutamate. Finally, methionine and threonine are both formed from homoserine. For methionine, cysteine is the source of sulfur in a reaction catalyzed by cystathionine synthase, and the terminal CH_3 group comes from N^5-methyltetrahydrofolate in the methionine synthase reaction.

In the pyruvate side of these two linked pathways, alanine is formed by a transamination reaction catalyzed by alanine aminotransferase using the α amino

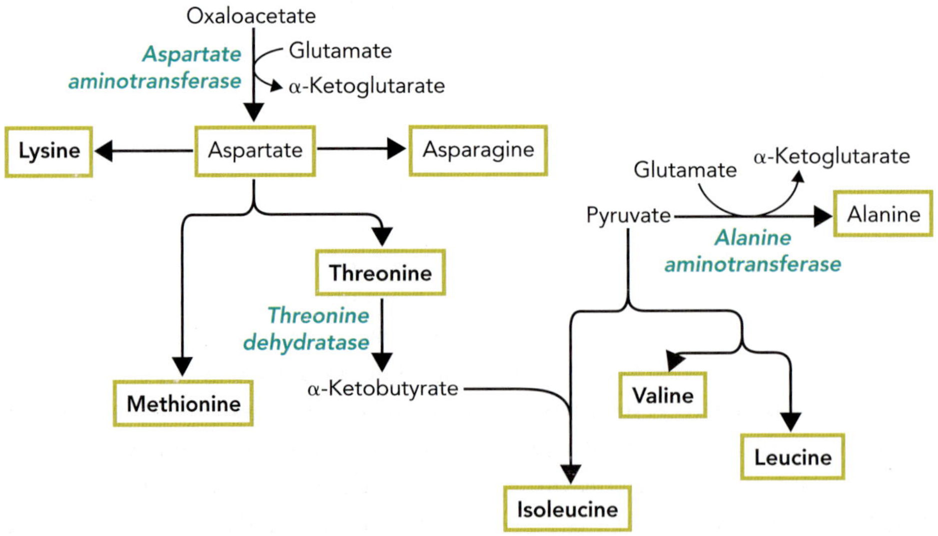

Figure 17.48 Nine amino acids are synthesized from oxaloacetate and pyruvate in *E. coli*. The oxaloacetate and pyruvate pathways are linked by α-ketobutyrate, which is a shared product in the two pathways. Essential amino acids in humans are shown in boldface.

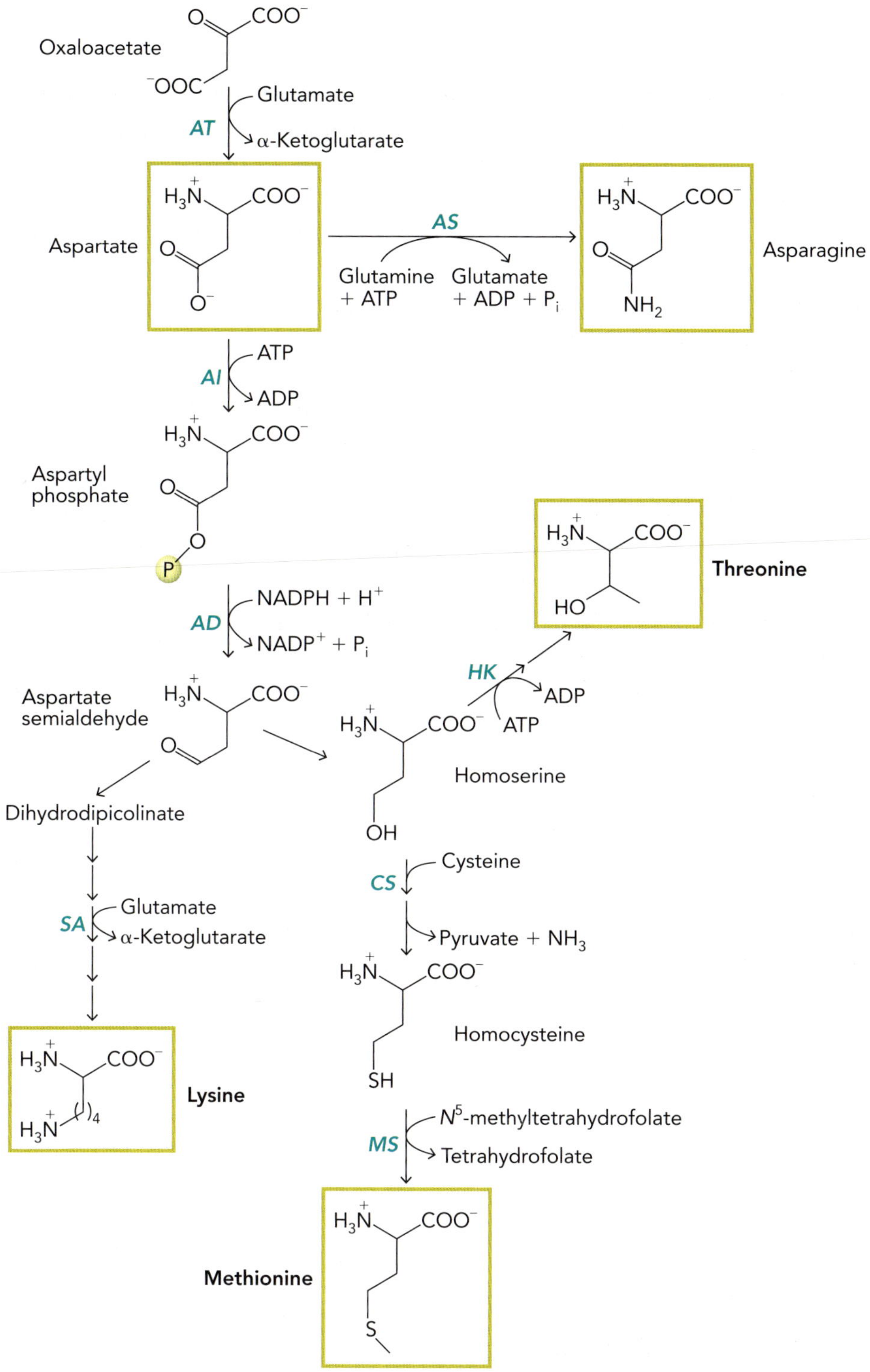

AD Aspartate semialdehyde dehydrogenase
AI Aspartokinase isozymes
AS Asparagine synthetase
AT Aspartate aminotransferase
CS Cystathionine synthase
HK Homoserine kinase
MS Methionine synthase
SA Succinyl diaminopimelate aminotransferase

Figure 17.49 Oxaloacetate is the metabolic precursor to aspartate, asparagine, lysine, threonine, and methionine. Essential amino acids in humans are shown in boldface.

group from glutamate. As shown in **Figure 17.50**, pyruvate is also the precursor to isoleucine, valine, and leucine. These syntheses begin with decarboxylation of pyruvate by the enzyme acetolactate synthase in a reaction requiring the coenzyme thiamine pyrophosphate (TPP) and forming the intermediate hydroxyethyl-TPP. This compound is then combined with α-ketobutyrate, a product of the threonine dehydratase reaction, to form α-aceto-α-hydroxybutyrate. This molecule is used

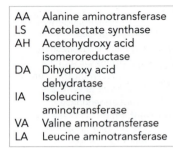

AA	Alanine aminotransferase
LS	Acetolactate synthase
AH	Acetohydroxy acid isomeroreductase
DA	Dihydroxy acid dehydratase
IA	Isoleucine aminotransferase
VA	Valine aminotransferase
LA	Leucine aminotransferase

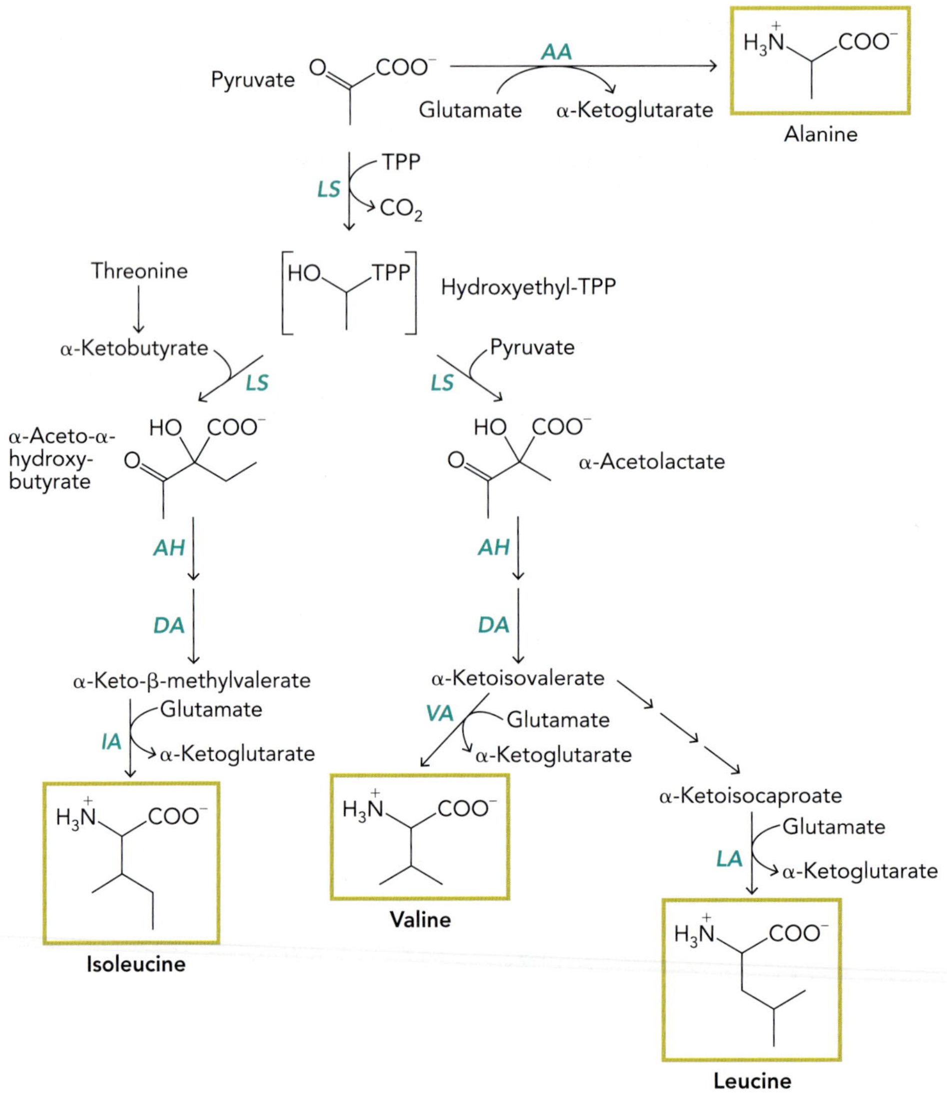

Figure 17.50 Pyruvate is the metabolic precursor to alanine, isoleucine, valine, and leucine. Note that α-ketobutyrate is derived from threonine in a reaction catalyzed by the enzyme threonine dehydratase. Essential amino acids in humans are shown in boldface.

to generate isoleucine in a series of three enzymatic reactions, the last of which is transamination by isoleucine aminotransferase. Valine is formed by combining the hydroxyethyl-TPP made in the first step with a second molecule of pyruvate to generate α-acetolactate. This reaction is also catalyzed by acetolactate synthase, the only difference being the use of pyruvate rather than α-ketobutyrate as the second substrate. The α-acetolactate is then converted to α-ketoisovalerate by two reactions that use the same enzymes as those in the isoleucine branch of the pathway. Transamination of α-ketoisovalerate by valine aminotransferase results in the formation of valine. In the other side of this branched pathway, α-ketoisovalerate is used to generate α-ketoisocaproate, which is then transaminated by leucine aminotransferase to form leucine.

Chorismate Is the Precursor to Tryptophan, Tyrosine, and Phenylalanine

Aromatic amino acids are synthesized in plants, fungi, and bacteria by the **shikimate pathway**. This pathway involves the condensation of phosphoenolpyruvate and erythrose-4-phosphate, followed by hydrocarbon ring formation.

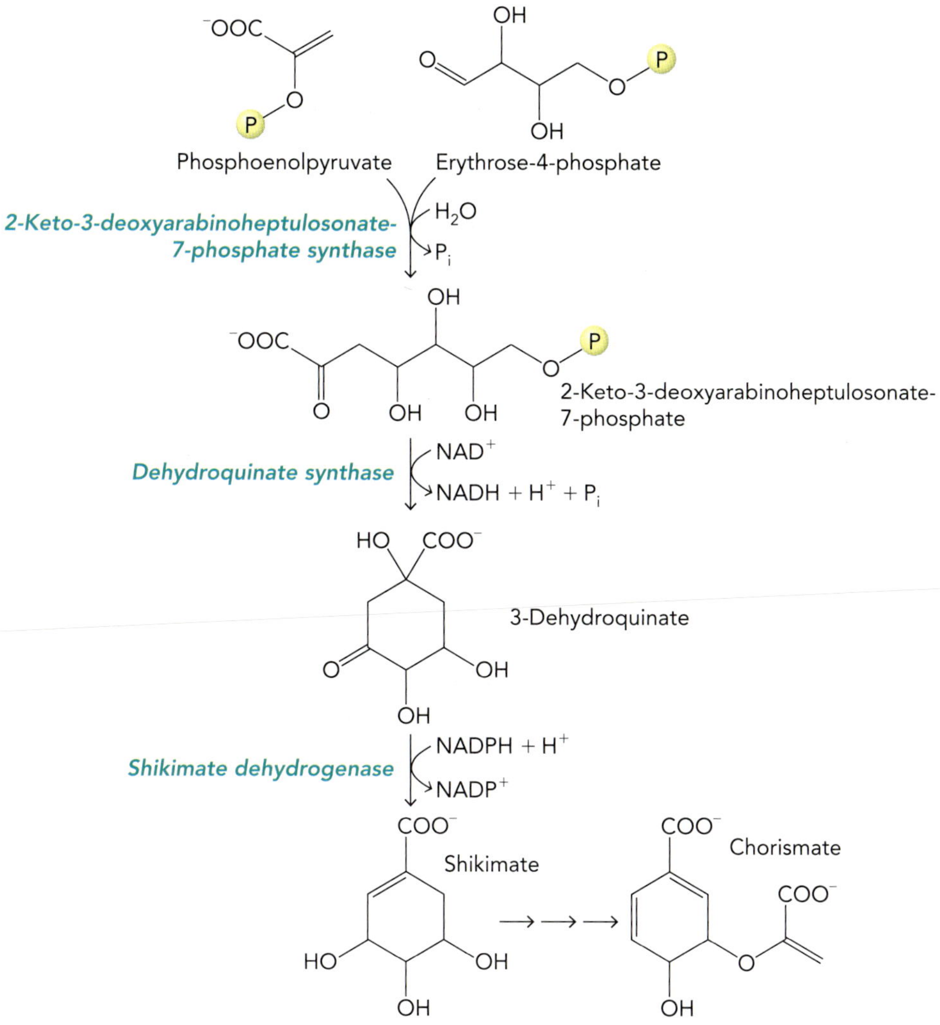

Figure 17.51 Plants, fungi, and bacteria synthesize chorismate from the metabolic precursors phosphoenolpyruvate and erythrose-4-phosphate.

In the first stage of this pathway, a C_{10} compound is formed called **chorismate**, which is the precursor to the three aromatic amino acids tryptophan, tyrosine, and phenylalanine. As shown in **Figure 17.51**, the C_7 compound 2-keto-3-deoxyarabinoheptulosonate-7-phosphate is generated by the enzyme 2-keto-3-deoxyarabinoheptulosonate-7-phosphate synthase in a condensation reaction releasing inorganic phosphate. Cyclization of 2-keto-3-deoxyarabinoheptulosonate-7-phosphate by dehydroquinate synthase results in the formation of 3-dehydroquinate. This molecule is then reduced by NADPH in a reaction catalyzed by shikimate dehydrogenase to form shikimate. Finally, conversion of shikimate to chorismate requires three reactions, two of which involve phosphoryl transfer.

Because animal cells do not contain the enzymes required for chorismate biosynthesis, enzyme inhibitors of this pathway have been developed for use as animal-safe herbicides. One of the most widely used herbicides in this class is **glyphosate**, the active ingredient in Roundup. As shown in **Figure 17.52**, glyphosate is a competitive inhibitor of the enzyme 5-enolpyruvylshikimate-3-phosphate (EPSP) synthase, which is required to convert shikimate-3-phosphate to 5-enolpyruvylshikimate-3-phosphate in one of the final three steps leading to chorismate. By spraying plants with glyphosate, EPSP synthase activity is inhibited, leading to insufficient levels of chorismate. The plants are then unable to synthesize tryptophan, tyrosine, and phenylalanine and die off.

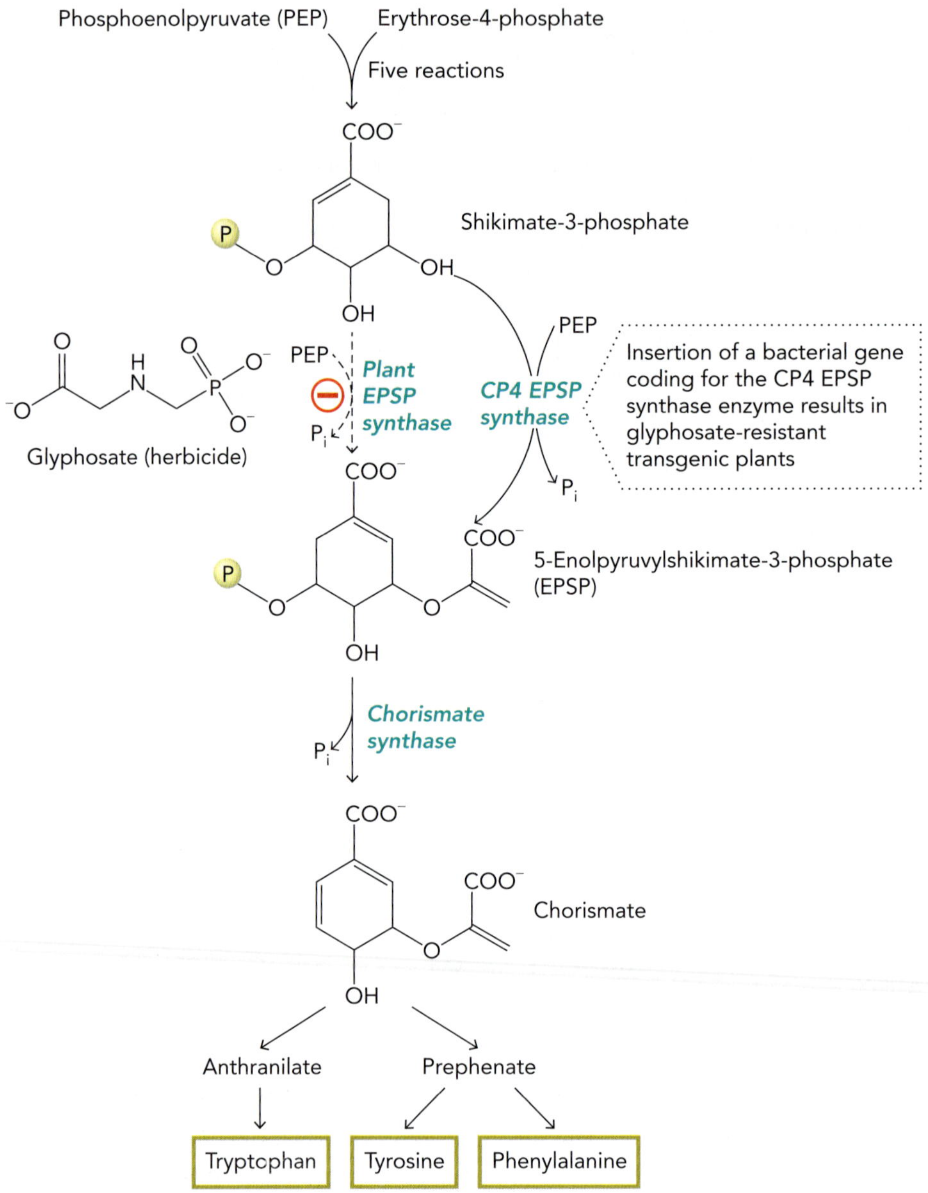

Figure 17.52 The herbicide glyphosate (Roundup) blocks tryptophan, tyrosine, and phenylalanine synthesis in plants by inhibiting the activity of EPSP synthase. The bacterial CP4 EPSP synthase enzyme is not inhibited by glyphosate, and therefore transgenic plants containing the gene for CP4 EPSP synthase are glyphosate resistant.

Because glyphosate was proved to be an animal-safe herbicide, plant scientists at the biotechnology company Monsanto reasoned that by developing glyphosate-resistant crop plants, it would be possible to spray fields with the Roundup herbicide to eliminate weeds without killing the crop plants. The first glyphosate-resistant crop plant developed was a strain of soybeans marketed as Roundup Ready soybeans. Farmers planting fields with Roundup Ready soybean seeds spray their crops with glyphosate throughout the growing season to kill weeds that compete with the crop plants for nutrients and water (**Figure 17.53**). By reducing weed growth through aerial spraying of glyphosate, it is possible to achieve significantly higher crop yields with improved quality of the crops.

The strategy used to develop Roundup Ready crop plants was to isolate an EPSP synthase gene from glyphosate-resistant bacteria (the CP4 strain of *Agrobacterium tumefaciens*) and insert it into the crop plants to confer glyphosate resistance. The CP4 EPSP synthase enzyme does not bind glyphosate but is still able to convert shikimate-3-phosphate to EPSP and thereby maintain flux through this amino acid biosynthetic pathway (see Figure 17.52). Monsanto's Roundup Ready crop plants are by far the most widely

Figure 17.53 Glyphosate-resistant soybean plants were the first major genetically modified organism (GMO) crop to be grown commercially. **a.** A field of Monsanto's Roundup Ready soybean plants. PAULO FRIDMAN/BLOOMBERG VIA GETTY IMAGES. **b.** Strains of glyphosate-resistant pigweed (*Amaranthus palmeri*) have emerged in Georgia and often grow alongside Roundup Ready soybeans. © ALAN CRESSLER.

used genetically modified organisms (GMOs) in commercial agriculture, which now includes soybeans, corn, sugar beets, cotton, and alfalfa. Although agricultural yields are higher using Roundup Ready seeds, the use of GMO crop plants in processed foods intended for human consumption is controversial, in part because strains of weeds have emerged that are glyphosate resistant (see Figure 17.53).

concept integration 17.3

What is the biochemical basis for the most common form of phenylketonuria?

Phenylketonuria is a genetically inherited metabolic disease caused by deficiencies in the enzyme phenylalanine hydroxylase, which is required to convert phenylalanine to tyrosine in animal cells. Newborn infants are tested for phenylketonuria at birth by measuring the level of phenylalanine in the blood. If they are found to be deficient in phenylalanine hydroxylase, they are immediately put on a phenylalanine-restricted diet. Aspartame (aspartyl-phenylalanine methyl ester) is an artificial sweetener that is hydrolyzed to aspartate and phenylalanine in the intestine, and therefore cannot be included in the diet of an individual with phenylketonuria. Notably, because tyrosine (the product of the phenylalanine hydroxylase reaction) is required for the synthesis of hair and skin pigments (melanins), infants with phenylketonuria have lightly colored hair and skin until they obtain sufficient amounts of tyrosine in their diets.

17.4 Biosynthesis of Amino Acid Derivatives

We noted earlier that the bulk of amino acids recovered from protein degradation is used to support ongoing protein synthesis in cells. The remainder is used as a source of nitrogen for the biosynthesis of hemes (hemoglobin and cytochromes), nucleotides (purines and pyrimidines), and signaling molecules (neurotransmitters, hormones, and nitric oxide). We begin this section by describing the synthesis and degradation of heme, which obtains all four of its iron-coordinating nitrogen atoms from the amino acid glycine. We then look at reactions that convert tyrosine to the signaling molecules dopamine, norepinephrine, and epinephrine and to melanin-derived pigments that give hair and skin their color. We conclude with a description of nitric oxide production from arginine by the enzyme nitric oxide synthase.

Heme Nitrogen Is Derived from Glycine

The prosthetic group of hemoglobin, myoglobin, and cytochromes is heme, a porphyrin ring containing iron that is derived from the amino acid glycine. The heme biosynthetic pathway is illustrated in **Figure 17.54**. The heme biosynthetic pathway requires enzymes that reside in both the mitochondrial matrix and the cytosol.

Figure 17.54 Glycine contributes the four nitrogen atoms to heme in a series of reactions that take place in both the mitochondrial matrix and the cytosol.

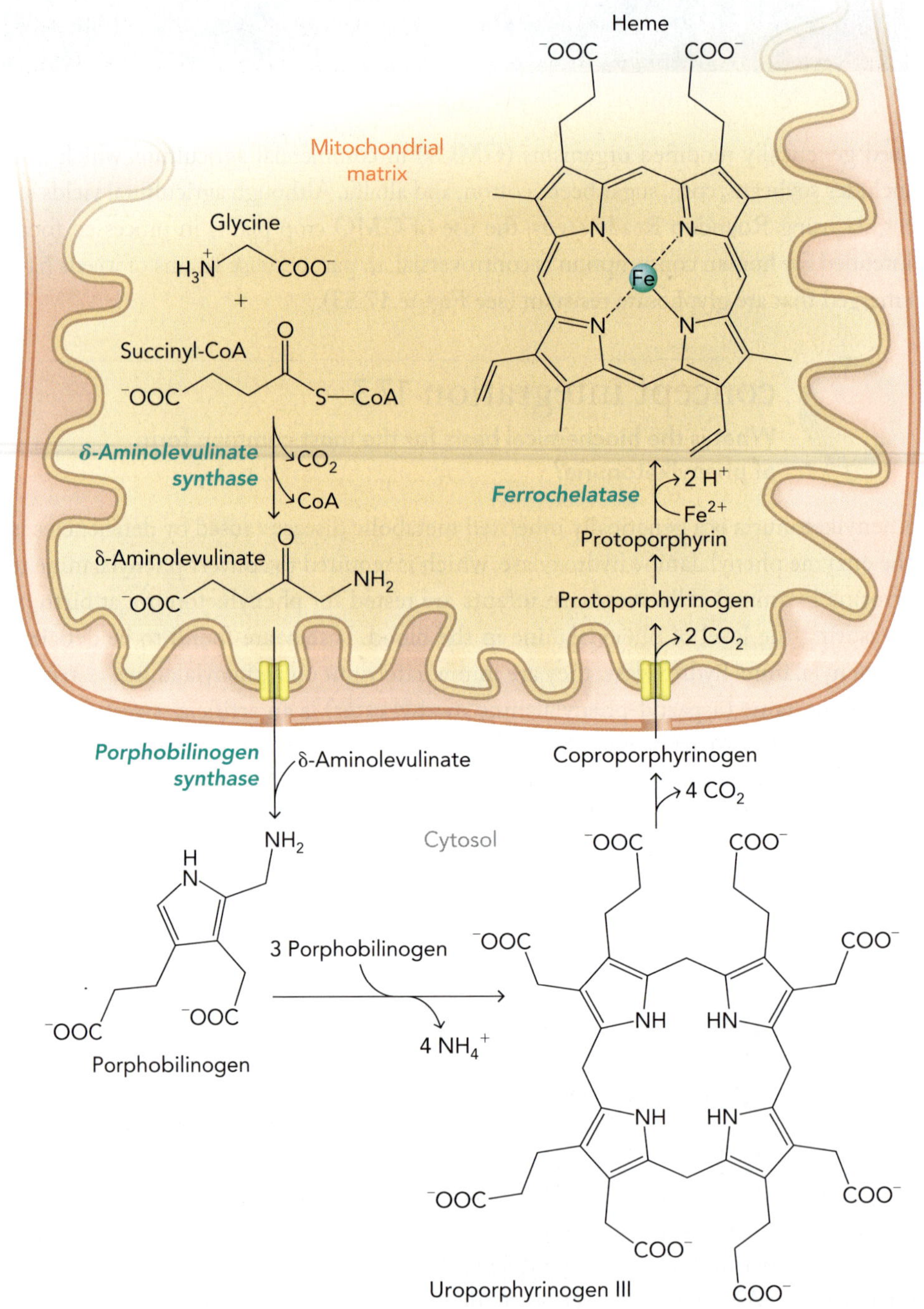

In the first reaction, the enzyme δ-aminolevulinate synthase catalyzes a reaction combining glycine and succinyl-CoA to form δ-aminolevulinate. This is then exported to the cytosol, where it condenses with another molecule of δ-aminolevulinate to generate porphobilinogen. In the subsequent reaction, four molecules of porphobilinogen are deaminated and combined to synthesize the heme precursor uroporphyrinogen III. In the next two reactions, six CO_2 are removed from uroporphyrinogen III to form protoporphyrinogen in the mitochondrial matrix. After an oxidation step that converts protoporphyrinogen to protoporphyrin, the enzyme ferrochelatase incorporates Fe^{2+} into the heme ring.

Heme biosynthesis takes place in erythrocyte precursors in the bone marrow to produce hemoglobin and in liver cells to provide heme for enzymes. Numerous metabolic diseases affecting heme biosynthesis have been linked directly to enzyme deficiencies in the heme biosynthetic pathway. These diseases are characterized by the accumulation of heme precursors in the blood and liver; they are collectively called **porphyrias** because they inhibit porphyrin ring synthesis. As shown in **Figure 17.55**, defects in all eight enzymes in the heme biosynthetic pathway have been identified, although some of these enzyme deficiencies are extremely rare.

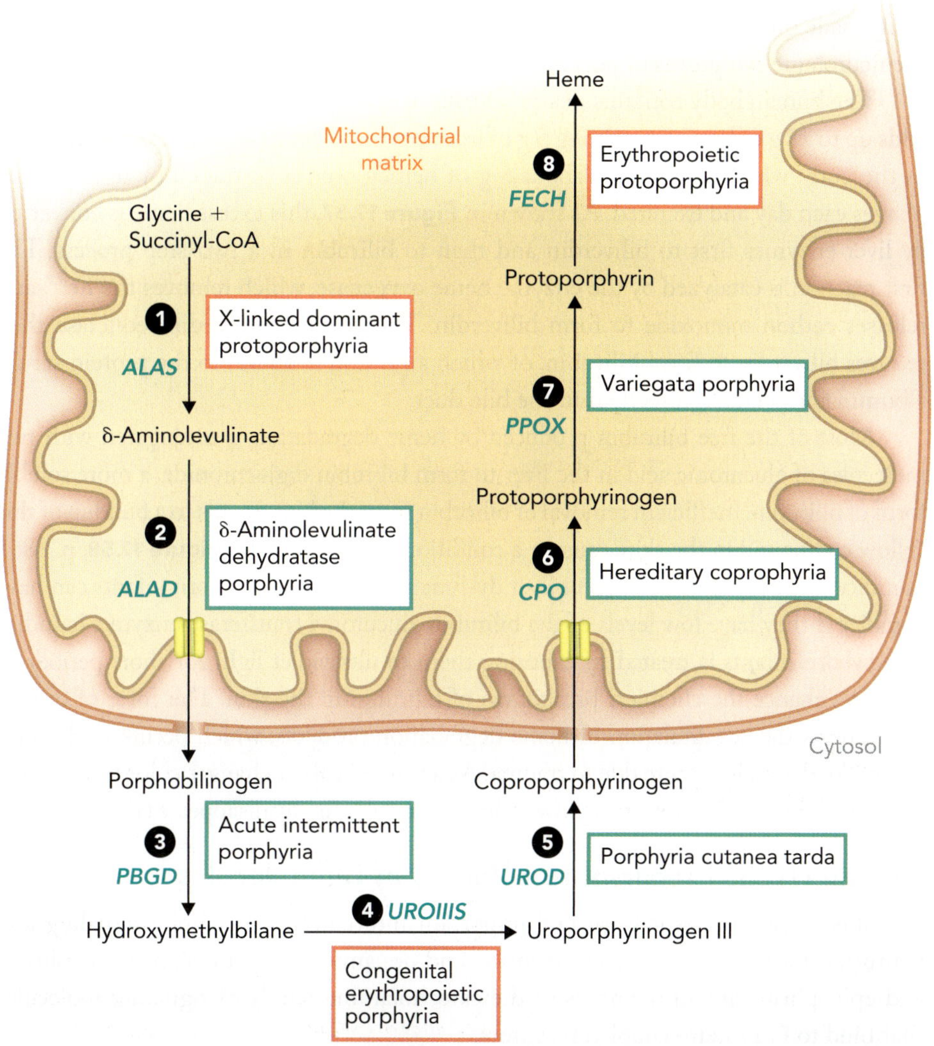

Figure 17.55 Porphyrias are a class of inherited metabolic diseases that inhibit heme biosynthesis and cause an accumulation of heme precursors in the blood and liver. Blue-green boxes identify porphyrias caused by defects in liver enzymes, whereas red-brown boxes identify porphyrias caused by enzyme defects in erythrocyte precursor cells. Abbreviations for enzymes that catalyze each of the eight reactions in humans are shown in the key.

ALAS	δ-Aminolevulinate synthase
ALAD	Porphobilinogen synthase
PBGD	Porphobilinogen deaminase
UROIIIS	Uroporphyrinogen III synthase
UROD	Uroporphyrinogen decarboxylase
CPO	Coproporphyrinogen oxidase
PPOX	Protoporphyrinogen oxidase
FECH	Ferrochelatase

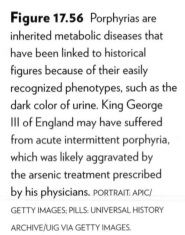

Figure 17.56 Porphyrias are inherited metabolic diseases that have been linked to historical figures because of their easily recognized phenotypes, such as the dark color of urine. King George III of England may have suffered from acute intermittent porphyria, which was likely aggravated by the arsenic treatment prescribed by his physicians. PORTRAIT: APIC/GETTY IMAGES; PILLS: UNIVERSAL HISTORY ARCHIVE/UIG VIA GETTY IMAGES.

One of the most common porphyrias is acute intermittent porphyria, which is due to dominant mutations in the porphobilinogen deaminase gene. Dominant mutations result in gain-of-function phenotypes, in which disease symptoms result when only one copy of the gene is defective. Unlike most dominant mutations, however, the severity of symptoms in individuals with acute intermittent porphyria are often mild unless physiologic triggers exacerbate the disease. For example, hormonal fluctuations, certain drugs, or dietary changes can trigger stomach pain and neurologic problems stemming from accumulation of the heme precursors δ-aminolevulinate and porphobilinogen in the blood and liver.

On the basis of these disease symptoms, it has been suggested that King George III of England, also known as Mad King George, suffered from acute intermittent porphyria during his time on the British throne from 1760 to 1820 (**Figure 17.56**). In fact, it has been speculated that acute intermittent porphyria may have affected the outcome of the American Revolution because King George was known to suffer from excruciating stomach pain and episodes of delirium, which could have been symptoms of the disease. Unfortunately, physicians at the time would routinely treat numerous diseases with arsenic, which likely made the king's symptoms worse because it is now known that arsenic aggravates the neurologic symptoms of porphyria.

The human body contains nearly 30 trillion erythrocytes in ~5 L of blood, which adds up to 75 g of heme containing 5 g of iron. Erythrocytes have a life span of 120 days in the body, which means that about 6 g of hemoglobin is degraded by liver macrophages each day and excreted. As shown in **Figure 17.57**, this excess heme is converted by liver enzymes first to biliverdin and then to bilirubin in a two-step process. The first reaction is catalyzed by the enzyme heme oxygenase, which removes the Fe^{2+} and releases carbon monoxide to form biliverdin. The enzyme biliverdin reductase then reduces biliverdin to form bilirubin, of which a portion is bound to the protein serum albumin and secreted directly into the bile duct.

Most of the free bilirubin produced by heme degradation is conjugated with two molecules of glucuronic acid in the liver to form bilirubin diglucuronide, a more soluble form of bilirubin. Inefficient removal of bilirubin from the blood leads to a buildup of this yellow compound in the skin, causing a condition called **jaundice** (**Figure 17.58**, p. 886). Jaundice is usually an indication of liver dysfunction; however, newborn infants can have jaundice if they have low levels of the bilirubin glucuronyl transferase enzyme. Jaundice in newborn infants is treated by exposing them to ultraviolet light for short periods of time to enhance the chemical breakdown of bilirubin in the skin. The color of bruises also reflects the accumulation of heme degradation products, which occurs in the skin when blood vessels rupture due to trauma. As a bruise heals, it changes color from purple (heme) to green (biliverdin) to yellow (bilirubin) until it finally disappears (Figure 17.58).

Tyrosine Is the Precursor to a Variety of Biomolecules

Tyrosine is the precursor to several important molecules in metabolic signaling and neurotransmission, including epinephrine and dopamine. Dopamine, norepinephrine, and epinephrine are all members of the catecholamine family of signaling molecules that bind to G protein–coupled receptors.

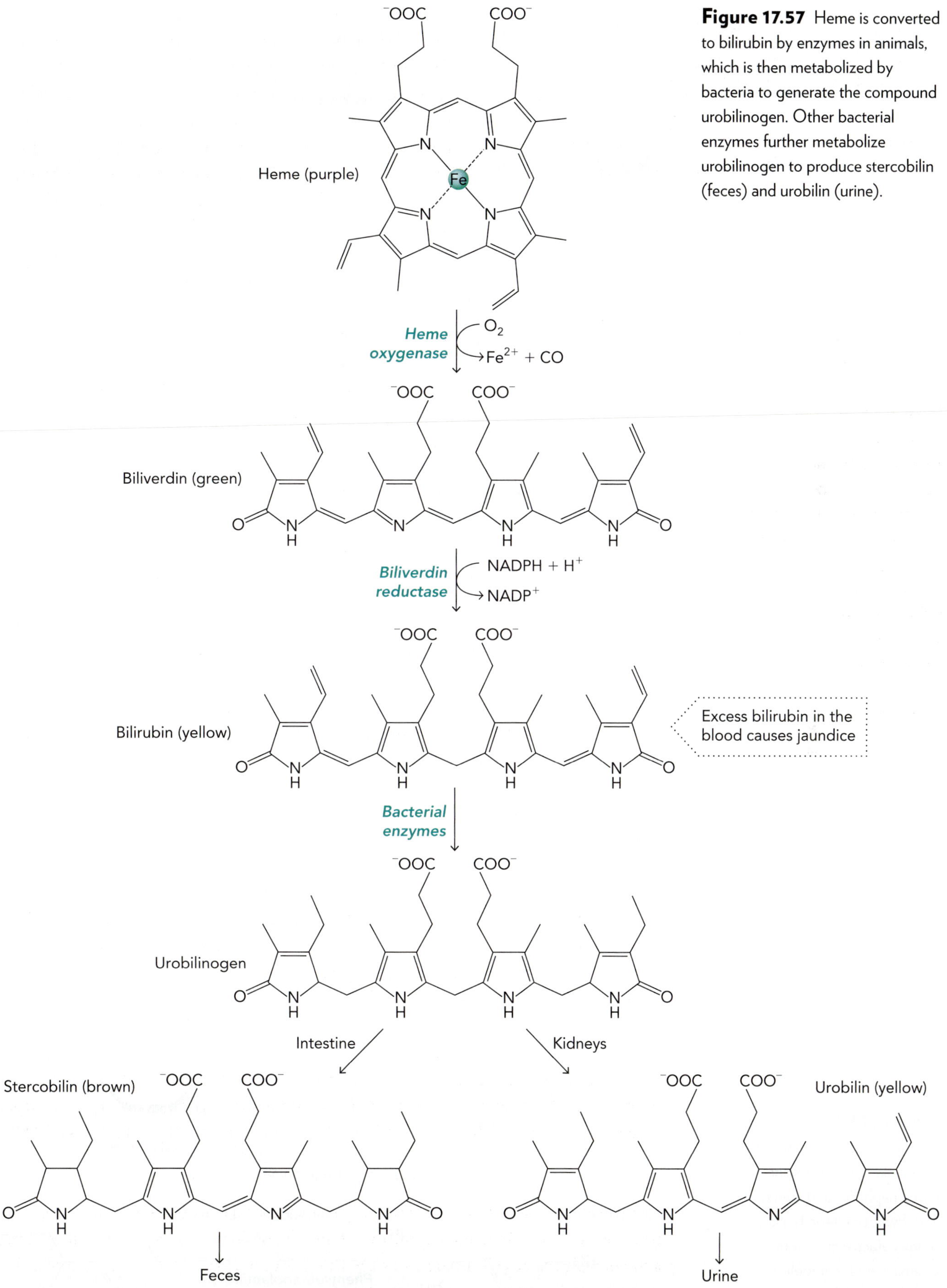

Figure 17.57 Heme is converted to bilirubin by enzymes in animals, which is then metabolized by bacteria to generate the compound urobilinogen. Other bacterial enzymes further metabolize urobilinogen to produce stercobilin (feces) and urobilin (urine).

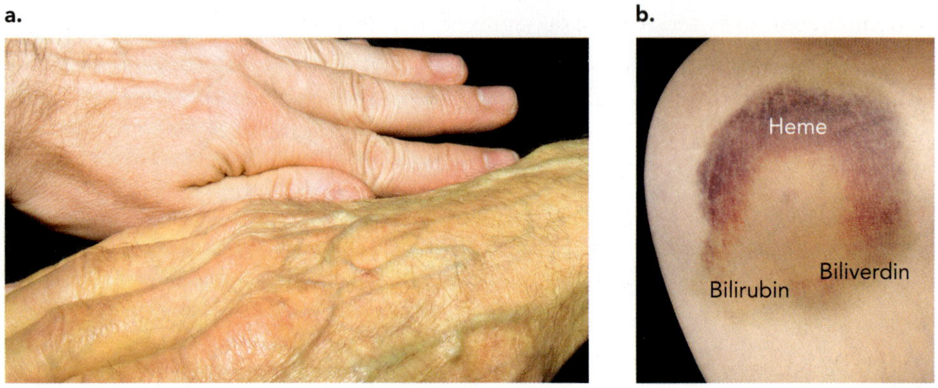

Figure 17.58 Acute changes in skin color are often the result of accumulated heme metabolites. **a.** Decreased efficiency of removing bilirubin from the blood causes the skin to turn yellow, a condition called jaundice. The hand in the foreground is jaundiced from the high levels of bilirubin. DR. P. MARAZZI/SCIENCE SOURCE. **b.** Bruises change color as they heal, going from dark purple (heme) to green (biliverdin) to yellow (bilirubin). This bruise is healing from the center outward. MARK VOLK/GETTY IMAGES.

The enzyme tyrosine hydroxylase oxidizes tyrosine in a reaction requiring the enzyme cofactor **tetrahydrobiopterin (BH_4)** to form dihydroxyphenylalanine (L-dopa), a metabolic precursor to dopamine and dopaquinone (**Figure 17.59**). In the dopamine branch of the pathway, dihydroxyphenylalanine is converted to dopamine by the enzyme aromatic amino acid decarboxylase. In the next reaction, dopamine is converted

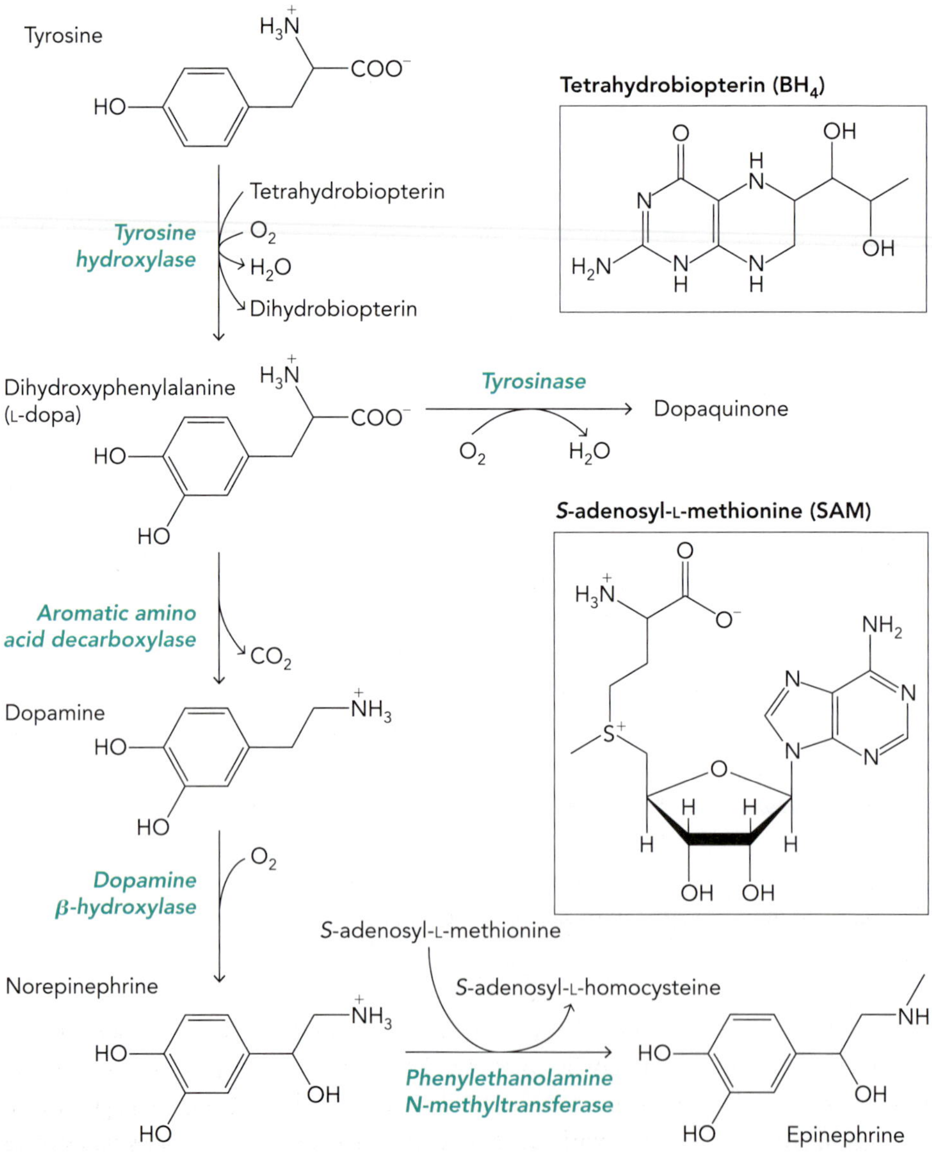

Figure 17.59 Tyrosine is the metabolic precursor to the catecholamines dopamine, norepinephrine, and epinephrine. Tetrahydrobiopterin is a key redox cofactor in a variety of amino acid metabolic pathways, and S-adenosyl-L-methionine is a methyl donor.

to norepinephrine by the enzyme dopamine β-hydroxylase, which leads to the production of epinephrine in a reaction requiring the enzyme phenylethanolamine *N*-methyltransferase. This methylation reaction uses *S*-adenosyl-L-methionine (SAM) as the methyl donor.

Dopamine is a potent neurotransmitter that is required for numerous brain functions. Patients with **Parkinson disease**, a debilitating neurologic syndrome, have less than 20% of the normal dopamine-producing cells in the substantia nigra region of the brain. It may be possible to treat Parkinson disease with stem cell therapy by implanting dopamine-producing cells into the affected area of the brain and thereby provide a better treatment with fewer side effects (see Chapter 23).

Tyrosine is also the precursor to pigment molecules called **melanins**, which are produced from dopaquinone as shown in **Figure 17.60**. The two primary types of melanins are eumelanins, which are dark pigments having a brown or black color, and pheomelanins, which have red or yellow color. The yellow color of pheomelanin pigments comes from the sulfur in cysteine, which is combined with dopaquinone.

Type 1 **albinism** is an autosomal recessive disease caused by mutations in the gene encoding tyrosinase. As shown in **Figure 17.61**, albinism occurs in a variety of organisms, including humans; such individuals are referred to as albinos. Individuals with severe albinism have higher rates of skin cancer, blindness, and bleeding disorders. Notably, individuals with the metabolic disease phenylketonuria, which is caused by a deficiency in the enzyme phenylalanine hydroxylase (see Figure 17.42), have light skin and hair at birth because phenylalanine is not readily converted to tyrosine. Phenylketonuric individuals are not albinos, however, and gain pigmentation over time as a result of tyrosine in their diets.

Melanocytes are skin cells located near the base of hair follicles that produce melanins. Depending on the ratio of eumelanin and pheomelanin pigments, a person can have dark hair or light hair, depending on the distribution of melanin-filled granules along the hair shaft. Beginning around the age of 35 years, melanin production in human melanocytes begins to shut down because the aging melanocytes are not replaced from a population of stem cells. Eventually, the amount of pigment in the hair decreases to the point where the hair appears gray. Gray hair can be colored by treating it with a mixture of hydrogen peroxide and an ammonia-based solution containing artificial pigments. The ammonia causes the hair follicles to swell, allowing the pigments to enter the hair shaft and undergo a chemical reaction with the hydrogen peroxide

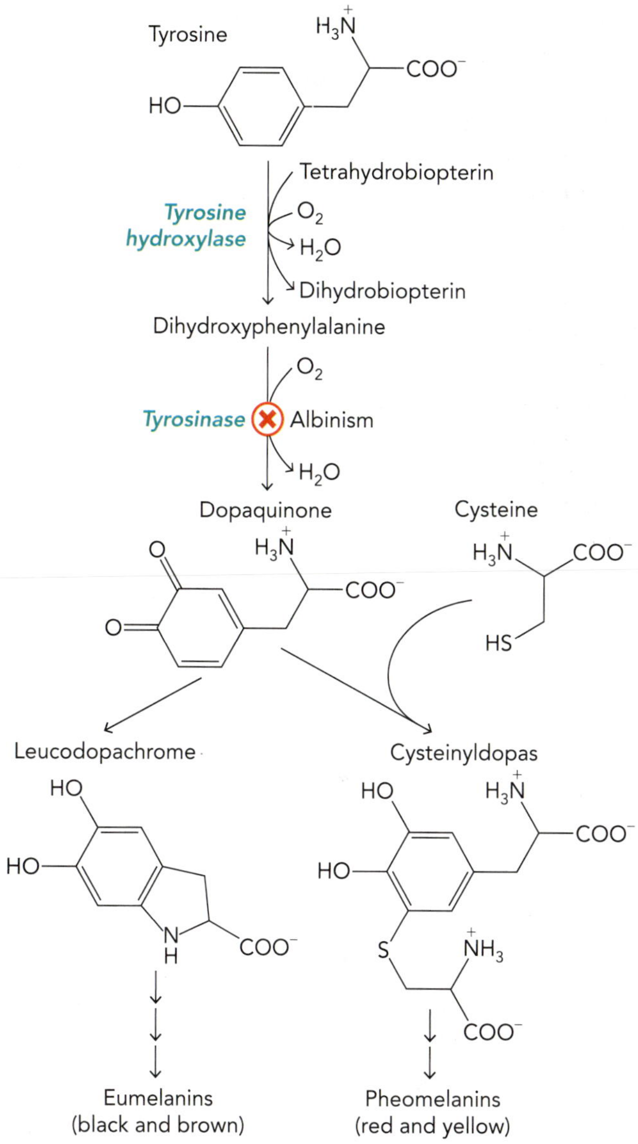

Figure 17.60 Tyrosine is the precursor to melanin pigments that give hair and skin their color. Mutations in the gene encoding the enzyme tyrosinase cause albinism, a condition in which individuals lack both eumelanin and pheomelanin pigments.

Figure 17.61 Albinism is rare in nature because affected animals lack the ability to effectively hide from predators. The photograph shows an albino baby kangaroo and its mother.
ROLAND WEIHRAUCH/DPA/CORBIS.

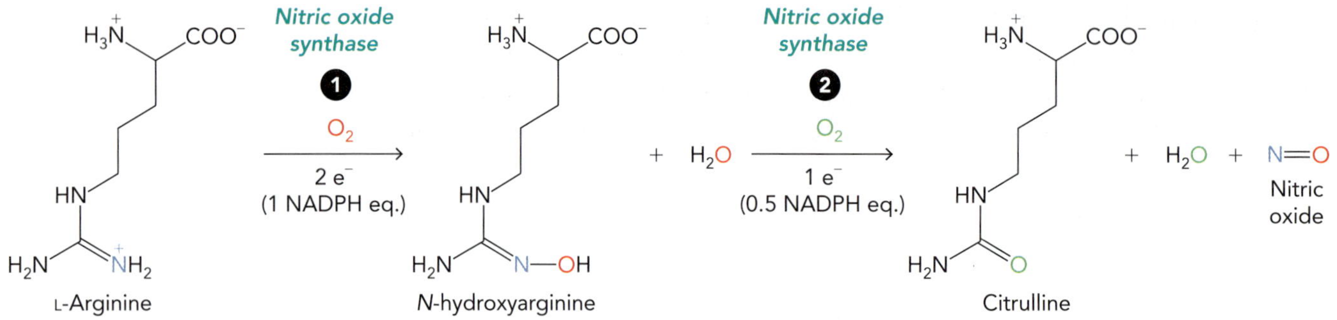

Figure 17.62 Nitric oxide synthase uses a two-step reaction, each step involving the input of O_2 and NADPH. eq. = equivalent.

Figure 17.63 The molecular structure of the dimeric oxygenase domain of bovine endothelial nitric oxide synthase is shown with tetrahydrobiopterin and arginine bound to the heme-containing active site. The heme groups in the dimeric protein are shown in red, tetrahydrobiopterin (BH_4) is shown in yellow, and arginine (Arg) is colored by atom. BASED ON PDB FILE 2NSE.

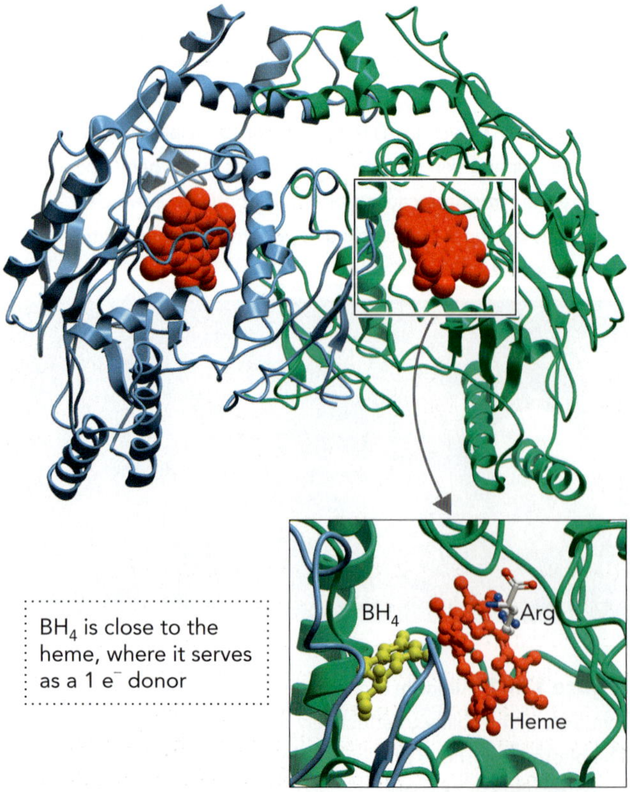

to permanently stain the hair. As new hair grows out from the hair follicle, however, it is easy to distinguish between the unpigmented gray hair and the artificially colored hair.

Nitric Oxide Synthase Generates Nitric Oxide from Arginine

One of the most surprising discoveries in the field of signal transduction over the past 30 years was the finding that "endothelial-derived relaxing factor"—a potent vasodilator in humans—is actually soluble nitric oxide (NO). During the 1980s and 1990s, a flurry of research uncovered the biochemical basis of nitric oxide signaling in a variety of biological systems, which led to the 1998 Nobel Prize in Physiology or Medicine for Robert Furchgott of SUNY Downstate Medical Center in Brooklyn; Louis Ignarro, then at Tulane University in New Orleans; and Ferid Murad at the University of Virginia. We first discussed the signaling properties of NO in Chapter 8 when we described the effect of nitric oxide on blood vessel vasodilation through a mechanism involving production of the second messenger cyclic GMP (cGMP; see Figure 8.7).

As shown in **Figure 17.62**, cellular nitric oxide is produced from arginine in a two-step oxidation reaction catalyzed by the enzyme nitric oxide synthase. The product of the first NADPH-dependent reaction is the stable intermediate *N*-hydroxyarginine, which is then cleaved by a second NADPH-dependent reaction to release citrulline and NO. As seen in the net reaction that follows, the nitric oxide synthase reaction requires 3 e^- donated by the equivalent of 1.5 NADPH molecules for each NO generated:

$$\text{L-Arginine} + 1.5\ \text{NADPH} + \text{H}^+ + 2\ \text{O}_2 \rightarrow \text{Citrulline} + \text{NO} + 1.5\ \text{NADP}^+$$

Three related forms of nitric oxide synthase have been characterized. The endothelial form is involved in vasodilation; a neuronal form is required for neuronal signaling in the brain; and an inducible form is present in immune cells. The dimeric structure of the bovine endothelial nitric oxide synthase oxygenase domain is shown in **Figure 17.63**. Each dimer contains a heme group and tetrahydrobiopterin, which together catalyze the oxygenase reactions required to produce NO and citrulline from arginine. The reductase domain of the enzyme, containing bound forms of the FAD and FMN coenzymes, is not shown in the figure.

What controls the activity of endothelial nitric oxide synthase? The answer is calcium–calmodulin binding to the endothelial nitric oxide synthase enzyme in response to receptor-mediated signaling through G protein–coupled receptors. As shown in **Figure 17.64**, neuronal input stimulates acetylcholine release from nerve cells

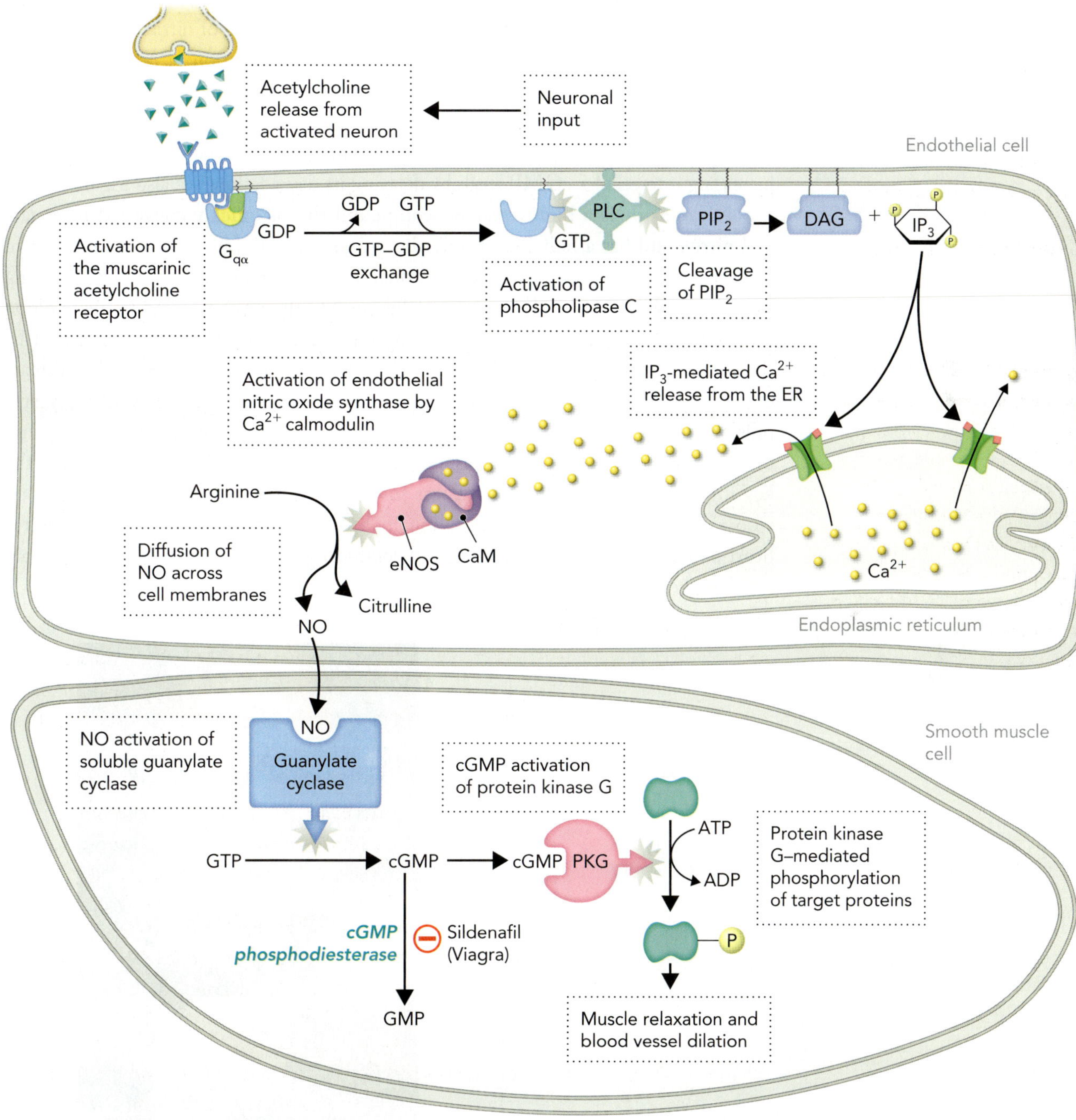

Figure 17.64 Nitric oxide production in endothelial cells by endothelial nitric oxide synthase (eNOS) leads to relaxation of smooth muscle cells and blood vessel dilation. Binding of acetylcholine to muscarinic acetylcholine receptors results in activation of phospholipase C (PLC) by $G_{q\alpha}$ signaling, which in turn leads to Ca^{2+} release from the endoplasmic reticulum (ER) in response to inositol-1,4,5-trisphosphate (IP_3) and stimulation of eNOS activity. Nitric oxide diffuses across the cell membranes and binds to the heme group in soluble guanylate cyclase, resulting in enzyme activation and production of the second messenger signaling molecule cGMP. Activation of protein kinase G (PKG) by cGMP leads to the phosphorylation of target proteins and relaxation of smooth muscle cells. CaM = calmodulin; DAG = diacylglycerol; eNOS = endothelial nitric oxide synthase; PIP_2 = phosphatidylinositol-4,5-bisphosphate.

and activation of $G_{q\alpha}$ signaling through the **muscarinic acetylcholine receptor**. This leads to stimulation of phospholipase C activity and cleavage of phosphatidylinositol-4,5-bisphosphate, producing the second messengers diacylglycerol and inositol-1,4,5-trisphosphate (IP_3). Binding of IP_3 to IP_3 receptors on the endoplasmic reticulum leads to release of internal Ca^{2+}, which binds to calmodulin and activates endothelial nitric oxide synthase activity. The nitric oxide produced from arginine diffuses out of the endothelial cells and into nearby smooth muscle cells. Binding of nitric oxide to soluble guanylate cyclase stimulates production of cGMP and activation of protein kinase G, which phosphorylates downstream target proteins, resulting in muscle relaxation. As illustrated in Figure 17.64, sildenafil (Viagra) maintains the vasodilated state in erectile tissue by inhibiting the activity of cGMP phosphodiesterase, which has the net result of increasing intracellular cGMP levels by delaying its degradation (see also Figure 8.7).

Nitric oxide is a highly reactive compound that is quickly converted to nitrate and nitrite and has a half-life of only about 5 seconds. Because Ca^{2+} stimulation of nitric oxide synthase activity after neuronal activation requires a complex signaling cascade, nitroglycerin has long been used as an NO donor compound to dilate blood vessels rapidly in patients suffering from angina (chest pain). Angina is caused by the constriction of blood vessels surrounding the heart and can precede a life-threatening heart attack. People at risk of a heart attack sometimes carry nitroglycerin pills or nitroglycerin mouth sprays, which can be ingested to release NO rapidly and dilate blood vessels surrounding the heart (**Figure 17.65**). Nitroglycerin is also the key ingredient in dynamite, which was invented by Alfred Nobel, the same person who established the Nobel Foundation in Sweden. Nitroglycerin formulations for heart patients are carefully prepared and chemically stabilized with cellulose to avoid accidental explosions.

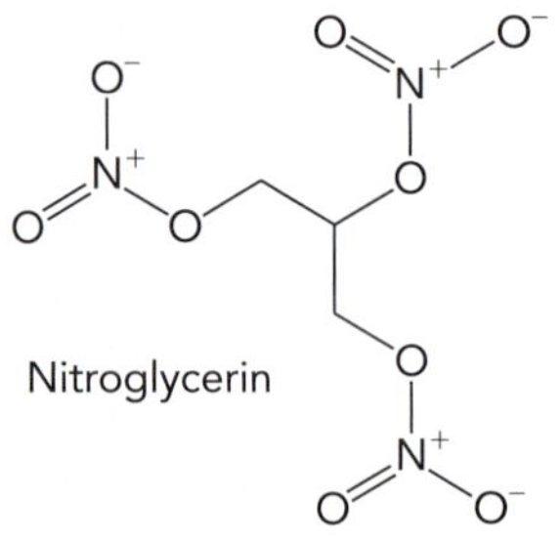

Figure 17.65 Nitroglycerin is used to treat angina caused by vasoconstriction of blood vessels that supply blood to the heart. Each molecule of nitroglycerin releases three molecules of NO and can be kept in pill or inhalant form as an emergency treatment. CREATISTA/SHUTTERSTOCK.

concept integration 17.4

Which four amino acids are required for the biosynthesis of heme, epinephrine, pheomelanins, and nitric oxide? Describe the function of each of these amino acid–derived biomolecules in humans.

Glycine is required for heme biosynthesis in the first reaction of the pathway, which is catalyzed by the enzyme δ-aminolevulinate synthase. Glycine provides all four nitrogen atoms to the heme ring. Heme biomolecules in humans are involved in oxygen transport as components of hemoglobin and in electron transfer in the electron transport system; for example, in cytochrome c oxidase (complex IV). Tyrosine is the amino acid precursor in the synthesis of epinephrine, an endocrine hormone, and in the synthesis of pheomelanin, a skin and hair pigment in humans. In addition to tyrosine, which provides nitrogen for pheomelanin synthesis, cysteine is also required, providing the sulfur atom to this pigment. Lastly, arginine is the amino acid substrate for nitric oxide production by the enzyme nitric oxide synthase. Nitric oxide (NO) is a soluble gas that functions in signal transduction pathways involving NO activation of the enzyme guanylate cyclase, which produces the second messenger cyclic GMP.

chapter summary

17.1 Nitrogen Fixation and Assimilation

- Biological nitrogen fixation is catalyzed by the enzyme nitrogenase, contained in microorganisms that live in soil and aquatic environments. Rhizobia are nitrogen-fixing soil bacteria that live symbiotically with root cells in leguminous plants; rhizobia produce NH_4^+, which the plant uses to generate glutamate and glutamine. Animals obtain nitrogen for amino acid synthesis and other biomolecules by eating plants and other animals.
- Industrial nitrogen fixation uses N_2 and H_2 gases under conditions of extreme temperature and pressure to produce NH_3 by the Haber process. Liquid ammonia generated by this industrial process is used to produce agricultural fertilizers, which are the major source of biological nitrogen in developed countries.
- Atmospheric nitrogen fixation occurs when energy from lightning combines N_2 with O_2 to form nitrogen oxides, which are dissolved in rain and fall to Earth. Nitrates produced by lightning are incorporated into plants that convert nitrates and nitrites into NH_3, which is used to synthesize glutamate and glutamine.
- Nitrogenase is a large protein complex in nitrogen-fixing bacteria that catalyzes an ATP-dependent redox reaction converting N_2 into 2 NH_3. Three rounds of nitrogen reduction sequentially convert $N_2 \rightarrow$ diimine $\rightarrow$ hydrazine $\rightarrow$ 2 NH_3; each reduction transfers 1 e^- to the nitrogenase complex with the hydrolysis of 2 ATP. Because 6 e^- are required to generate 2 NH_3 from the reduction of N_2, and H_2 is produced in a wasteful side reaction, a total of 8 e^- and 16 ATP are required.
- The nitrogenase reaction is inhibited by O_2, so microorganisms have evolved mechanisms to limit O_2 access to the enzyme active site: (1) *Klebsiella pneumoniae* only synthesizes the protein components of the nitrogenase complex when it is living in an anaerobic environment. (2) *Azotobacter vinelandii* decreases local O_2 concentrations by increasing flux through the electron transport system to rapidly reduce O_2 to H_2O. (3) *Rhizobium meliloti* invades the roots of leguminous plants, which synthesize a heme-containing protein called leghemoglobin that sequesters O_2 away from the nitrogenase complex.
- A major source of nitrogen in the soil is NH_4^+, which is produced by the degradation of plant and animal parts by invertebrates, bacteria, and fungi that live in the soil. Another major source of nitrogen in the soil comes from the process of nitrification, which bacteria use to convert NH_4^+ in the soil into NO_2^- (nitrite), and then into NO_3^- (nitrate).
- The nitrogen cycle refers to the recycling of soil NH_4^+, NO_2^-, and NO_3^- back to either plant roots or, through a process of bacterial denitrification, to atmospheric N_2.

- Plants and bacteria use the available NH_4^+ generated by nitrogen fixation to synthesize the amino acids glutamate and glutamine. Glutamate is the source of nitrogen in amino acid biosynthesis through the action of aminotransferase enzymes. Glutamine is the primary source of amino groups for the biosynthesis of nucleotide bases and carbamoyl phosphate, a metabolite in the urea cycle.
- The incorporation of NH_4^+ into glutamate and glutamine is called ammonia assimilation and is mediated by three enzymes: (1) glutamine synthetase, (2) glutamate synthase, and (3) glutamate dehydrogenase.
- The combined action of glutamine synthetase and glutamate synthase catalyzes a net reaction that generates glutamate from α-ketoglutarate and NH_4^+. Glutamate dehydrogenase catalyzes the same reaction, but only under conditions of very high NH_4^+ levels, such as those after application of agricultural fertilizers in fields ($\Delta G°' =$ +30 kJ/mol).
- Glutamine synthetase is regulated by both feedback inhibition and covalent attachment of AMP to a tyrosine residue in the enzyme. Adenylylation of glutamine synthetase increases its sensitivity to allosteric inhibitors, which include the metabolites carbamoyl phosphate, AMP, CTP, histidine, tryptophan, and serine.
- Aminotransferases play an important role in amino acid degradation and synthesis by transferring amino groups between amino acids and α-keto acids. For example, aspartate aminotransferase transfers the α amino group of aspartate to α-ketoglutarate to form glutamate and oxaloacetate. The aminotransferase reaction mechanism is a classic example of ping-pong enzyme kinetics, in which the first product of the reaction leaves the active site before the second substrate enters.

17.2 Amino Acid Degradation

- Nitrogen cannot be stored in the body in a useable form because NH_4^+ is toxic. Therefore, nitrogen lost as a result of protein and nucleic acid degradation must be replaced from the diet. When an individual is in nitrogen balance, it means that the daily intake of nitrogen equals the amount of nitrogen lost by excretion.
- Plants and bacteria synthesize all 20 amino acids; however, animals depend on protein in their diets to obtain the 10 essential amino acids they require for growth and development. Protein digestion in humans occurs in the stomach and small intestine, where proteases cleave proteins to yield amino acids and oligopeptides.
- Degradation of cellular proteins is another source of amino acids. Two mechanisms of intracellular protein degradation are ATP-independent lysosomal degradation and ATP-dependent proteasomal degradation of ubiquitin-tagged proteins. Lysosomal vesicles have an acidic lumen, which provides optimal conditions for protein unfolding and proteolytic degradation. Proteasomes degrade ubiquitinated proteins using three distinct protease activities located within the internal chamber of the catalytic core.
- Proteins destined for proteasomal degradation are tagged on lysine residues by covalent linkage of ubiquitin through its carboxyl-terminal glycine residue. The proteasome consists of a 20S core particle and two 19S regulatory particles, which together form the 30S proteasome. The 19S complexes serve as caps to regulate protein entry into and exit from the 20S proteolytic core.
- Ubiquitinating enzymes recognize specific residues at the N terminus of the target protein or a structural property of the protein, such as phosphorylation or misfolding. There are three classes of ubiquitinating enzymes: E1 enzymes attach ubiquitin to E2 enzymes; E2 enzymes conjugate ubiquitin to target proteins; and E3 enzymes recognize target proteins and facilitate ubiquitination by interacting directly with E2–ubiquitin and the target protein.
- Ubiquitin activation is ATP dependent and links the terminal carboxyl group of Gly76 in ubiquitin to a Cys residue in the E1 enzyme. Transfer of this ubiquitin to a Cys residue in E2 releases E1 and leads to the formation of an E2–ubiquitin–E3 complex. Ubiquitination of a Lys residue on target proteins initiates polyubiquitination, which links at least four ubiquitin subunits together through Gly76–Lys48 linkages.
- Target proteins having a Phe, Leu, Asp, Lys, or Arg residue at the N terminus instead of Met are recognized and ubiquitinated by certain E2–ubiquitin–E3 complexes that follow the "N-end rule" of protein degradation. The N-end rule refers to the propensity of polypeptides with specific N-terminal amino acids to have short half-lives in cells.
- Glutamate and glutamine function as the primary nitrogen carriers in most organisms. Mammals excrete excess nitrogen as urea, which is synthesized in the liver. Fish and aquatic amphibians excrete NH_4^+ directly into water, whereas terrestrial amphibians excrete nitrogen as urea or uric acid.
- The two nitrogens in urea are derived from (1) NH_4^+ released from the deamination of glutamate and glutamine and (2) incorporation of aspartate into the urea cycle intermediate argininosuccinate. The carbon atom in urea comes from CO_2 (HCO_3^-) produced in the citrate cycle, and the oxygen atom is derived from H_2O.
- In the first reaction of the urea cycle, HCO_3^- and NH_4^+ are used to synthesize carbamoyl phosphate, which is then combined with ornithine to form citrulline. Citrulline is exported to the cytosol and activated by AMP before being converted to argininosuccinate when aspartate displaces the AMP. Argininosuccinate is cleaved to yield fumarate and arginine, followed by arginase cleavage to produce urea.
- The urea cycle and citrate cycle are metabolically linked through the shared intermediate fumarate, which provides the carbon backbone for aspartate by supplying

the oxaloacetate needed in the aspartate aminotransferase reaction. The net result is that the amino group of glutamate is transferred to urea via a pathway shunt involving aspartate, fumarate, and oxaloacetate. These reactions are called the "Krebs bicycle," or aspartate–argininosuccinate shunt.

- Urea cycle enzyme deficiencies result in hyperammonemia and a buildup of glutamine and glutamate, which function as osmolytes that can cause brain swelling and associated neurologic symptoms. Some urea cycle disorders can be treated by restricting dietary protein or by providing an alternative path for nitrogen excretion.
- The carbon backbones of 11 amino acids can be converted into pyruvate or acetyl-CoA, which are used for energy conversion by the citrate cycle and oxidative phosphorylation reactions. The other nine amino acids are converted to α-ketoglutarate, fumarate, succinyl-CoA, and oxaloacetate for glucose synthesis.
- Amino acids that give rise to pyruvate or citrate cycle intermediates are called glucogenic because pyruvate and oxaloacetate are precursors in the gluconeogenic pathway. In contrast, amino acids converted into acetyl-CoA or acetoacetyl-CoA are called ketogenic amino acids because they can give rise to ketone bodies.
- Three interconnected amino acid degradation pathways cover 13 amino acids: (1) amino acids that directly (glycine, serine, alanine, and cysteine) or indirectly (threonine and tryptophan) give rise to pyruvate; (2) amino acids that generate α-ketoglutarate (arginine, histidine, proline, glutamate, and glutamine); and (3) the degradation of phenylalanine.
- Phenylalanine degradation was the first metabolic pathway to be linked to a human disease resulting from a single gene mutation. The disease is called alkaptonuria, or black urine disease, and is due to a recessive mutation in the gene encoding homogentisate-1,2-dioxygenase, an enzyme in the phenylalanine and tyrosine degradation pathways. Large amounts of homogentisate accumulate in the urine and cause the black color upon oxidation.
- Phenylketonuria is due to defects in the enzyme phenylalanine hydroxylase, which converts phenylalanine to tyrosine. Phenylketonuria symptoms, which include severe mental retardation, are caused by accumulation of phenylalanine metabolites (phenylpyruvate, phenylacetate, and phenyllactate). The primary treatment for phenylketonuria is to limit phenylalanine in the diet, beginning shortly after birth.

17.3 Amino Acid Biosynthesis

- The side chains of amino acids are derived from seven metabolic intermediates in three metabolic pathways: (1) the glycolytic pathway (3-phosphoglycerate, phosphoenolpyruvate, and pyruvate); (2) the pentose phosphate pathway (ribose-5-phosphate and erythrose-4-phosphate); and (3) the citrate cycle (α-ketoglutarate and oxaloacetate).
- Plants and bacteria synthesize all 20 amino acids, whereas animals synthesize only ~10 amino acids (nonessential amino acids), owing to evolutionary adaptation in that animals have diets rich in proteins containing the other ~10 amino acids (essential amino acids).
- The structures of the essential amino acids are more complex than those of the nonessential amino acids, which is reflected in the number of enzymatic reactions required to synthesize them. For example, alanine, serine, and aspartate are synthesized by all organisms using simple reaction pathways, whereas plants and bacteria synthesize tryptophan, histidine, and methionine by pathways requiring multiple enzymes.
- Regulation of metabolic flux through amino acid biosynthetic pathways is tightly controlled to maintain a pool of amino acids that optimally supports protein synthesis. The general principle of feedback inhibition plays a pivotal role in modulating flux through linked amino acid biosynthetic pathways.
- Biosynthesis of three nonessential amino acids (alanine, aspartate, and asparagine) and six essential amino acids (methionine, threonine, lysine, isoleucine, valine, and leucine) involves two interconnected pathways in bacteria that use pyruvate and oxaloacetate as precursors.
- Aromatic amino acids are synthesized in plants, fungi, and bacteria by the shikimate pathway, which uses the substrates phosphoenolpyruvate and erythrose-4-phosphate to generate chorismate, the metabolic precursor to tryptophan, tyrosine, and phenylalanine.
- Animal cells do not contain the enzymes required for chorismate biosynthesis, and therefore enzyme inhibitors of this pathway are used as animal-safe herbicides. For example, glyphosate (Roundup) inhibits the enzyme EPSP synthase, which is required to convert shikimate-3-phosphate to 5-enolpyruvylshikimate-3-phosphate (EPSP). Glyphosate-treated plants die because they are deficient in tryptophan, tyrosine, and phenylalanine.
- Roundup Ready soybeans contain a bacterial gene coding for the enzyme CP4 EPSP synthase, which does not bind glyphosate and thereby provides glyphosate resistance to the plant. Farmers growing Roundup Ready soybeans spray their crops with glyphosate to kill weeds that compete with the crop plants for nutrients and water.

17.4 Biosynthesis of Amino Acid Derivatives

- The iron porphyrin ring of hemoglobin, myoglobin, and the cytochromes is derived from the amino acid glycine and is synthesized by a complex pathway requiring eight enzymes localized to either the mitochondrial matrix or the cytosol.
- Heme biosynthesis takes place in erythrocyte precursors in the bone marrow to produce hemoglobin and in liver cells to provide heme for enzymes. Numerous metabolic diseases called porphyrias have been linked directly to enzymes in the heme biosynthetic pathway. One of the

most common porphyrias is acute intermittent porphyria, which is due to dominant mutations in the porphobilinogen deaminase gene.

- About 6 g of hemoglobin are degraded every day by liver enzymes. One of the products of hemoglobin degradation is bilirubin, a yellow metabolite that can accumulate in the blood and cause a condition called jaundice. Most bilirubin degradation products produced each day are excreted in the feces and urine.
- Tyrosine is the precursor to biomolecules required in metabolic signaling and neurotransmission (epinephrine and dopamine) and to pigments (eumelanins and pheomelanins).
- Two human diseases related to tyrosine metabolism are Parkinson disease, which is caused by a loss of dopamine-producing cells in the brain, and albinism, which results from genetic defects in the enzyme tyrosinase.
- Nitric oxide (NO) is a soluble gas that functions as a potent vasodilator. NO is produced from arginine in a two-step oxidation reaction catalyzed by the enzyme nitric oxide synthase. Humans have three nitric oxide synthase enzymes: (1) the endothelial form involved in vasodilation; (2) the neuronal form required for neuronal signaling in the brain; and (3) the inducible form that is present in immune cells.
- NO-mediated vasodilation in smooth muscle cells is induced by acetylcholine activation of calcium signaling through the muscarinic acetylcholine receptor in endothelial cells, which stimulates calmodulin-dependent endothelial nitric oxide synthase activity. Diffusion of NO into nearby smooth muscle cells activates soluble guanylate cyclase and production of cGMP, which binds to protein kinase G, leading to phosphorylation of downstream target proteins.

biochemical terms

(in order of appearance in text)

review questions

1. Describe three processes by which nitrogen gas (N_2) is incorporated into the biosphere.
2. What accounts for the energy requirement of 16 ATP to convert $N_2 \rightarrow 2\ NH_3$ by the nitrogenase complex?
3. What three mechanisms have evolved in nitrogen-fixing microorganisms to limit the inhibitory effects of O_2 on the nitrogenase reaction?
4. Describe how the nitrogen cycle works to redistribute nitrogen in Earth's biosphere.
5. How does the combined activity of glutamine synthetase and glutamate synthase assimilate NH_4^+ into amino acids? Why doesn't the glutamate dehydrogenase reaction contribute much to ammonia assimilation under normal physiologic conditions?
6. Describe the biochemical function of the E1, E2, and E3 enzymes in the ubiquitinating system. Why are there ~500 E3 genes in the human genome but only two E1 genes?
7. What is the metabolic function of the "Krebs bicycle" in nitrogen metabolism?
8. What is the difference between a glucogenic and a ketogenic amino acid? Which amino acids fit the description of both a glucogenic and a ketogenic amino acid?

9. What is meant by the term "inborn errors of metabolism"? Give three examples of inborn errors that are directly related to the metabolism of phenylalanine and tyrosine.
10. What is the functional definition of essential and nonessential amino acids in the human diet? How might this functional definition be explained in evolutionary terms?
11. Glyphosate (Roundup) is a very effective herbicide that is toxic to almost all types of plants.
    a. Why is glyphosate considered an animal-safe herbicide?
    b. Why does it take up to 7 days before there is any noticeable herbicidal effect of glyphosate?
    c. Why don't Roundup Ready soybean plants die when they are sprayed with glyphosate?
12. Phenylketonuria is caused by a recessive genetic mutation, whereas acute intermittent porphyria is caused by a dominant genetic mutation.
    a. What is the biochemical difference in terms of protein function that distinguishes most recessive mutations from dominant mutations?
    b. What is the probability that a father with phenylketonuria and a homozygous normal mother will have a child that is a phenylketonuria carrier?
    c. What is the probability that a heterozygous mother with acute intermittent porphyria will have a child with the disease if the father is disease free?

challenge problems

1. Bacteria can both fix and assimilate nitrogen, plants can only assimilate nitrogen, and animals can do neither. What is the main purpose of nitrogen fixation and assimilation?
2. The histogram that follows shows results from an experiment in which *Arabidopsis* plants were grown hydroponically in either NH_4NO_3 medium or nitrogen-free medium during a preincubation period. The plants were then switched to media containing either $^{15}NH_4^+$ or $^{15}NO_3^-$, supplied as $(NH_4)_2SO_4$ or KNO_3, respectively, and grown during the experimental period for 24 hours. The amount of ^{15}N influx into the plant was measured as a function of root mass in grams and time in hours.

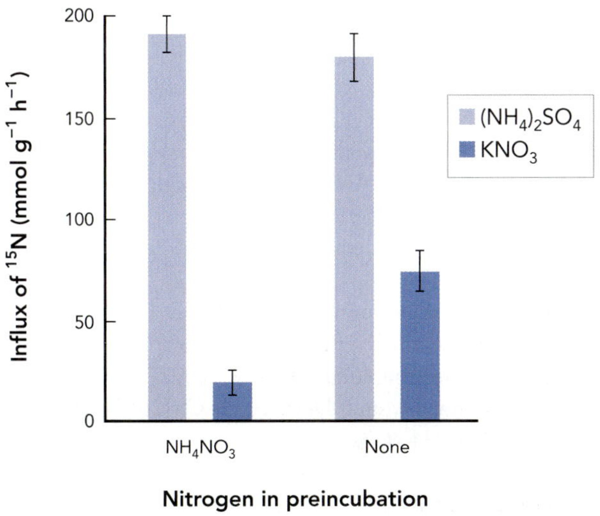

    a. Why is NH_4^+ preferred by the plant during the 24-hour experimental period, rather than NO_3^-, regardless of the preincubation media?
    b. What explains the large increase in NO_3^- utilization when the plants were preincubated in nitrogen-free medium compared to preincubation in NH_4NO_3 medium?
    c. What would you propose as the best form of nitrogen for use as fertilizer?
3. Why would you expect plant root nodules to contain high levels of glutamine synthetase?
4. Bacterial glutamine synthetase is subject to complex regulation. Describe two types of regulation of this enzyme, and give specific examples of each.
5. When ^{15}N-labeled aspartate is fed to animals, many ^{15}N-labeled amino acids appear in protein within a short time.
    a. Explain this observation and include in your answer the most likely enzyme reaction responsible for this finding.
    b. What is the enzyme cofactor in this reaction, and what type of chemical bond must be formed for the reaction to proceed?
6. Cats were fasted overnight and then given a single meal containing all amino acids except arginine. Within 2 hours, blood ammonia levels rose by 800%. A control group of cats was fasted and then fed the same meal, but with arginine included, and showed no change in blood ammonia levels.
    a. Fasting was required to obtain this difference in blood ammonia levels. Why?
    b. What caused the ammonia level to rise in the experimental group compared to the control group?
7. Why does it make metabolic sense that tadpoles (which live in water) have low levels of the enzyme arginase, but after their metamorphosis into frogs (which spend extended periods of time on land) their arginase levels increase dramatically?

8. A newborn infant with highly elevated levels of blood ammonia was diagnosed to have a urea cycle enzyme defect in either carbamoyl phosphate synthetase or arginase. The clinical lab was able to measure the concentrations of alanine, glutamine, and arginine in the blood, but enzyme assays were not available. How would knowledge of abnormal levels of alanine, glutamine, or arginine in the blood be useful to distinguish between an enzyme defect in carbamoyl phosphate synthetase or arginase?
9. How is excess intracellular ammonia removed from
   a. muscle tissue?
   b. liver tissue?
   c. all other tissues?
10. Individuals with enzyme deficiencies in the urea cycle have elevated blood ammonia levels, which can be life threatening. One useful therapy is to feed these individuals benzoic acid, which reacts with glycine to form hippuric acid. This compound can be excreted in the urine. Why does benzoic acid therapy lower blood ammonia levels in these individuals?
11. High-protein and low-carbohydrate diets involving daily ingestion of large amounts of protein hydrolysate (a mixture of amino acids and small oligopeptides) can be dangerous, although body weight often decreases quickly. Explain why most of the rapid weight loss in a high-protein diet is due to water loss rather than increased rates of fatty acid oxidation.
12. All 20 amino acids are found in the bloodstream, but alanine and glutamine are by far the most abundant. Explain this observation.
13. Kwashiorkor is a dietary deficiency disease characterized by decreased pigment in the skin and hair.
    a. Which major food group is likely missing from the diet? Justify your answer.
    b. What would you conclude if adding this missing food group to the diet had no effect on the skin and hair condition?
14. Individuals with the disease phenylketonuria lack the enzyme phenylalanine hydroxylase, which is required to convert phenylalanine to tyrosine. Albinism is a disease characterized by lack of skin pigments due to a deficiency in the enzyme tyrosinase.
    a. Explain why patients with phenylketonuria can be spared from many of the deleterious effects of the disease, whereas there is no feasible treatment for albinism.
    b. Explain why individuals with phenylketonuria are not complete albinos, even though they cannot synthesize tyrosine from phenylalanine.
15. What explains the observation that some forms of porphyria are associated with jaundice while others are not?

smartwork5

If your instructor assigns homework with Smartwork5, access it here: digital.wwnorton.com/biochem.

suggested reading

Books and Reviews

Batista, R., and Oliveira, M. M. (2009). Facts and fiction of genetically engineered food. *Trends in Biotechnology, 27*, 277–286.

Bird, I. M. (2011). Endothelial nitric oxide synthase activation and nitric oxide function: new light through old windows. *Journal of Endocrinology, 210*, 239–241.

Derbyshire, E. R., and Marletta, M. A. (2012). Structure and regulation of soluble guanylate cyclase. *Annual Review of Biochemistry, 81*, 533–559.

Dixon, R., and Kahn, D. (2004). Genetic regulation of biological nitrogen fixation. *Nature Reviews Microbiology, 2*, 621–631.

Dougan, D. A., Micevski, D., and Truscott, K. N. (2012). The N-end rule pathway: from recognition by N-recognins, to destruction by AAA+ proteases. *Biochimica et Biophysica Acta, 1823*, 83–91.

Elsasser, S., and Finley, D. (2005). Delivery of ubiquitinated substrates to protein-unfolding machines. *Nature Cell Biology, 7*, 742–749.

Finger, S., and Christ, S. E. (2004). Pearl S. Buck and phenylketonuria (PKU). *Journal of the History of the Neurosciences, 13*, 44–57.

Frankland-Searby, S., and Bhaumik, S. R. (2012). The 26S proteasome complex: an attractive target for cancer therapy. *Biochimica et Biophysica Acta, 1825*, 64–76.

Hershko, A., Ciechanover, A., and Varshavsky, A. (2000). Basic Medical Research Award. The ubiquitin system. *Nature Medicine, 6*, 1073–1081.

Pilcher, H. (2005). Microbiology: pipe dreams. *Nature, 437*, 1227–1228.

Pollegioni, L., Schonbrunn, E., and Siehl, D. (2011). Molecular basis of glyphosate resistance—different approaches through protein engineering. *FEBS Journal, 278*, 2753–2766.

Puy, H., Gouya, L., and Deybach, J. C. (2010). Porphyrias. *Lancet, 375*, 924–937.

Primary Literature

Fusetti, F., Erlandsen, H., Flatmark, T., and Stevens, R. C. (1998). Structure of tetrameric human phenylalanine hydroxylase and its implications for phenylketonuria. *Journal of Biological Chemistry, 273*, 16962–16967.

Gill, H. S., and Eisenberg, D. (2001). The crystal structure of phosphinothricin in the active site of glutamine synthetase illuminates the mechanism of enzymatic inhibition. *Biochemistry*, *40*, 1903–1912.

Meierhofer, D., Wang, X., Huang, L., and Kaiser, P. (2008). Quantitative analysis of global ubiquitination in HeLa cells by mass spectrometry. *Journal of Proteome Research*, *7*, 4566–4576.

Molenda, M. A., Sroa, N., Campbell, S. M., Bechtel, M. A., and Mitch Opremcak, E. (2010). Peroxide as a novel treatment for ecchymoses. *Journal of Clinical and Aesthetic Dermatology*, *3*, 36–38.

Raman, C. S., Li, H., Martasek, P., Kral, V., Masters, B. S., and Poulos, T. L. (1998). Crystal structure of constitutive endothelial nitric oxide synthase: a paradigm for pterin function involving a novel metal center. *Cell*, *95*, 939–950.

Sadre-Bazzaz, K., Whitby, F. G., Robinson, H., Formosa, T., and Hill, C. P. (2010). Structure of a Blm10 complex reveals common mechanisms for proteasome binding and gate opening. *Molecular Cell*, *37*, 728–735.

Schindelin, H., Kisker, C., Schlessman, J. L., Howard, J. B., and Rees, D. C. (1997). Structure of ADP·AlF_4^--stabilized nitrogenase complex and its implications for signal transduction. *Nature*, *387*, 370–376.

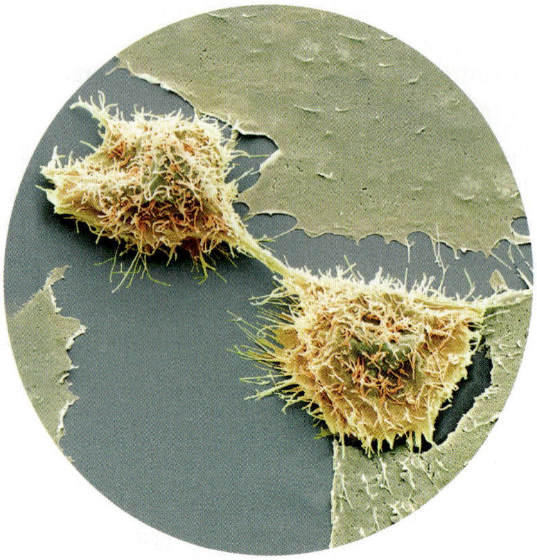

Human cancer cells have abnormally high rates of cell division and growth

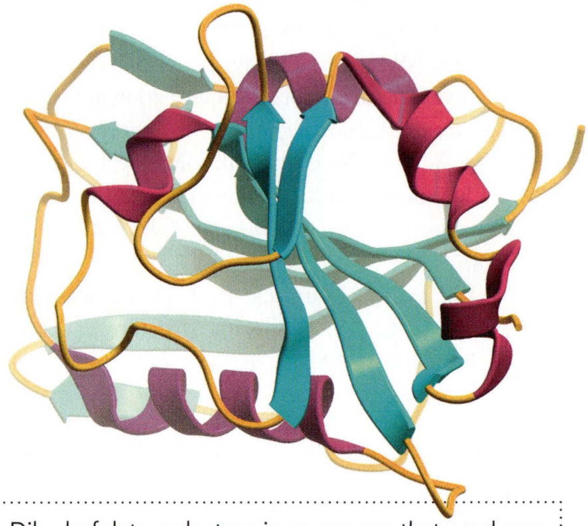

Dihydrofolate reductase is an enzyme that produces a coenzyme needed for the biosynthesis of thymine, a nucleotide base in DNA

Human cancer cells often acquire a Leu→Arg mutation at residue 22 (L22R) of dihydrofolate reductase. The biochemical effect of the L22R mutation in dihydrofolate reductase is a significant decrease in methotrexate binding affinity, leading to cancer cell proliferation.

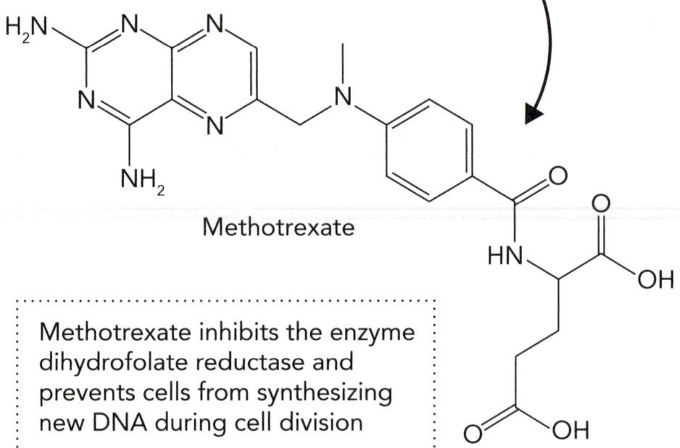

Methotrexate

Methotrexate inhibits the enzyme dihydrofolate reductase and prevents cells from synthesizing new DNA during cell division

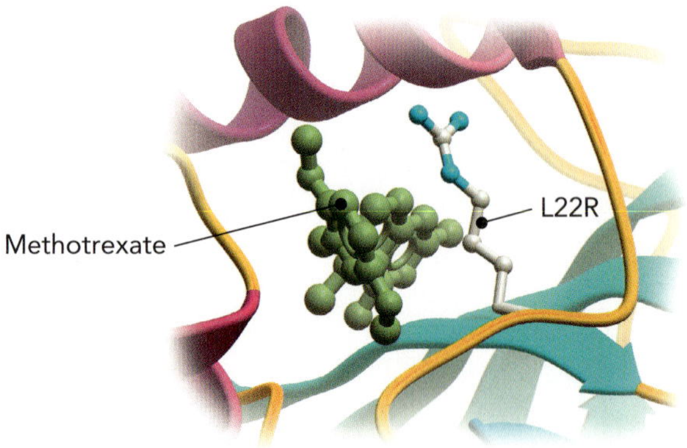

Location of the L22R mutation in the human dihydrofolate reductase enzyme active site

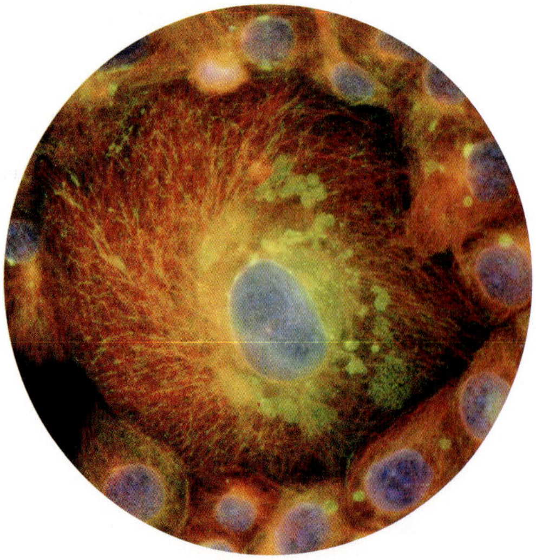

Cancer cells that become resistant to one cancer drug sometimes become resistant to many cancer drugs, such as the cancer cell shown here

18

Nucleotide Metabolism

◀ Inhibitors of thymidine synthesis can be used as anticancer drugs because they block DNA synthesis and decrease rates of cell division. One of the anticancer drugs that inhibits thymidine synthesis is methotrexate, which targets the enzyme dihydrofolate reductase. This enzyme is required to generate the thymidylate synthase substrate methylenetetrahydrofolate, a coenzyme required in the pathway to generate the nucleotide base thymine. By inhibiting methylenetetrahydrofolate synthesis, rapidly dividing cells cannot synthesize DNA quickly enough because they run out of the required nucleoside thymidine.

CHAPTER OUTLINE

CREDITS: HUMAN CANCER CELLS UNDERGOING CELL DIVISION: STEVE GSCHMEISSNER/SCIENCE SOURCE; DIHYDROFOLATE REDUCTASE AND LOCATION OF THE L22R MUTATION: BASED ON PDB FILES 1U71 AND 1U72; DRUG-RESISTANT HUMAN CANCER CELL: NANCY KEDERSHA/SCIENCE SOURCE.

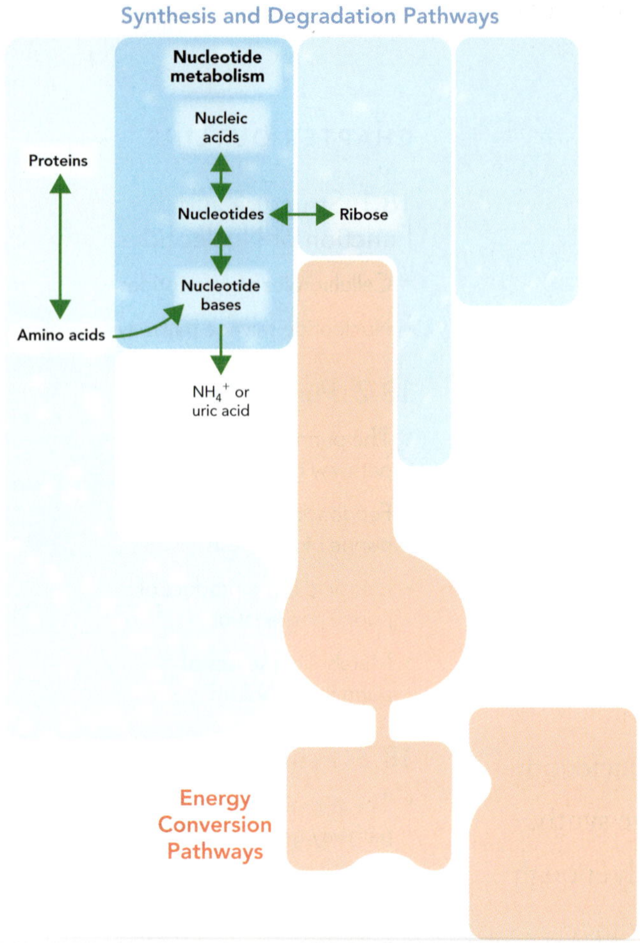

Figure 18.1 The nitrogen atoms of nucleotide bases are derived from amino acids, whereas the sugar backbones come from ribose. Thus, nucleotide metabolism intersects with both carbohydrate and amino acid metabolism.

Nucleotides are one of the fundamental types of biomolecules and participate in most reaction sequences in the cell. We will discuss them again in Part 5 of the book in the context of DNA and RNA structure and function. Here, we take a closer look at the chemical structures of nucleotides and describe the biosynthesis and degradation of purines and pyrimidines. We also describe the biochemical function and allosteric regulation of the enzyme ribonucleotide reductase, which is one of the most highly conserved enzymes in nature. This enzyme regulates a key step in DNA synthesis. Along the way, we examine the importance of anticancer drugs that target thymine biosynthesis and thereby inhibit DNA replication in rapidly dividing cancer cells (see the chapter-opening application).

Figure 18.1 shows where nucleotide metabolism fits into our metabolic map. The interdependence of nucleotides, amino acids, and carbohydrates is reflected in the chemical structures of purine and pyrimidine bases and the sugar ribose. Moreover, degradation of purines produces uric acid, whereas pyrimidine catabolism leads to the production of NH_4^+; both of these products need to be excreted to avoid nitrogen toxicity, as we described in Chapter 17.

18.1 Structure and Function of Nucleotides

Nucleotides participate in diverse cellular processes, and therefore it is vital that a sufficient supply of nucleotides be available. In this section, we will first describe the general roles of nucleotides in cells and then discuss salvage pathways whereby nucleotides are reused to minimize the energy expenditure needed to synthesise these complex molecules *de novo*.

Cellular Roles of Nucleotides

Nucleotides participate in four important biochemical processes in cells: energy conversion reactions, signal transduction pathways, coenzyme-dependent reactions, and genetic information storage and transfer. This is best exemplified (**Figure 18.2**) by the most abundant nucleotide in cells, adenosine-5′-triphosphate (ATP). Although we highlight only ATP here as an example, biomolecules containing any one of the five major nucleotide bases described in this chapter (adenine, guanine, cytosine, thymine, and uracil) all have important roles in biochemistry.

1. ATP plays a key role in energy conversion reactions. These reactions generally begin with ATP synthesis by the ATP synthase complex in response to an electrochemical proton gradient.
2. ATP derivatives are required for many signal transduction pathways. 3′,5′-Cyclic adenosine monophosphate (cyclic AMP, or cAMP) is

commonly used as a second messenger. But nucleotides also appear in phosphoryl transfer reactions catalyzed by receptor tyrosine kinases and downstream signaling kinases such as protein kinase A.

3. ATP-derived coenzymes, such as nicotinamide adenine dinucleotide (NAD^+), flavin adenine dinucleotide (FAD), and coenzyme A (CoA), provide essential chemical groups to numerous metabolic enzymes.
4. ATP is one of the essential biomolecules required for genetic information storage and transfer, which depend on hydrogen bonding between adenine and thymine nucleotide bases in duplex DNA or between adenine and uracil nucleotide bases in DNA–RNA and RNA–RNA hybrids (see Chapter 3).

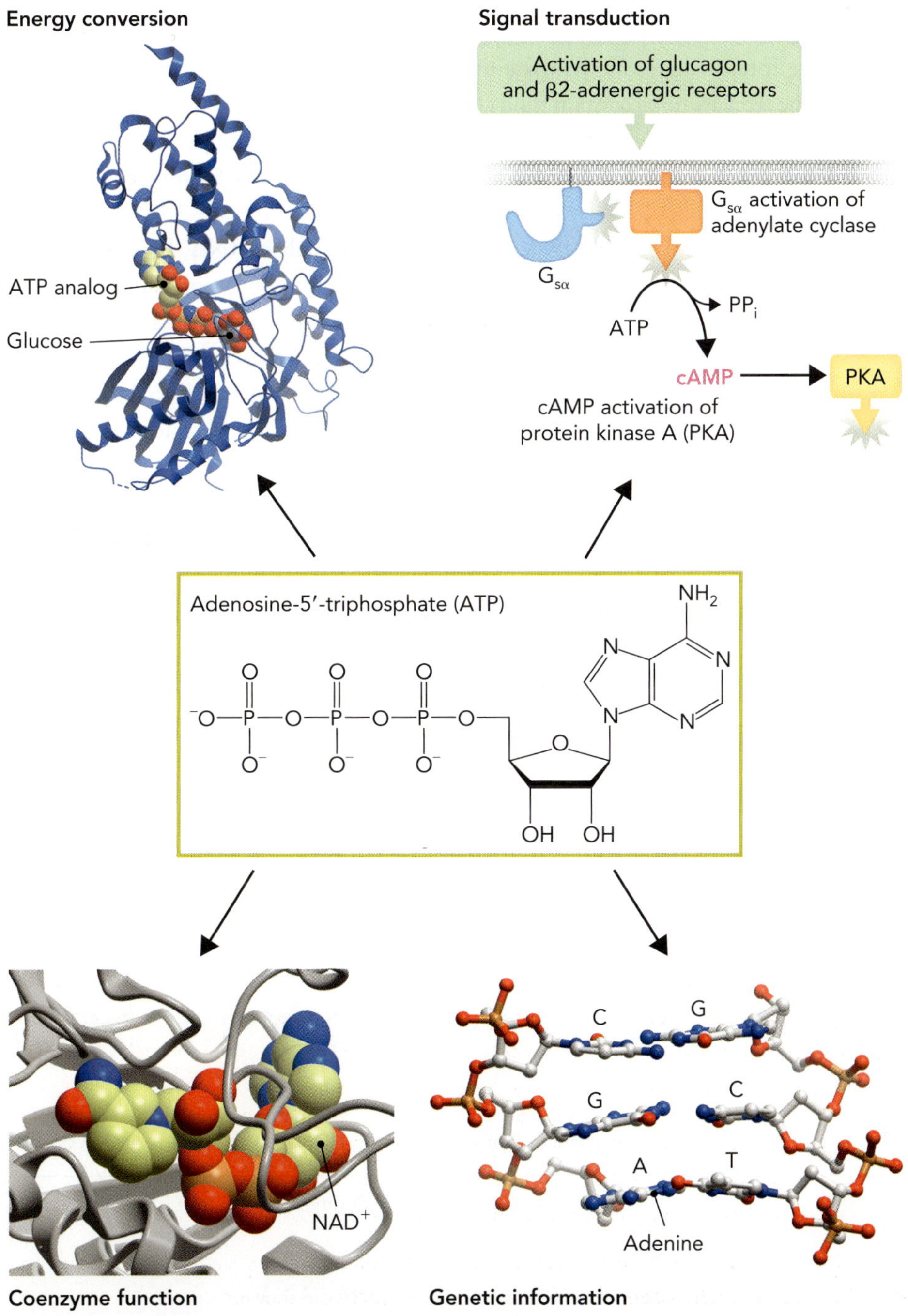

Figure 18.2 Nucleotides have key roles in four biochemical processes, as exemplified by the nucleotide ATP. These four processes are (1) energy conversion reactions (for example, glucokinase uses ATP to phosphorylate glucose to produce glucose-6-phosphate); (2) signal transduction (mediated by cyclic AMP in kinase-mediated phosphoryl transfer reactions); (3) coenzyme-dependent reactions such as those using nicotinamide adenine dinucleotide (NAD^+) shown here bound to lactate dehydrogenase; and (4) genetic information storage and transfer (based on nucleotide base pairing through adenine–thymine base pairs in DNA). BASED ON PDB FILES 3FGU (GLUKOKINASE), 1BNA (BASE PAIRS), AND 1T2C (LDH-NAD^+).

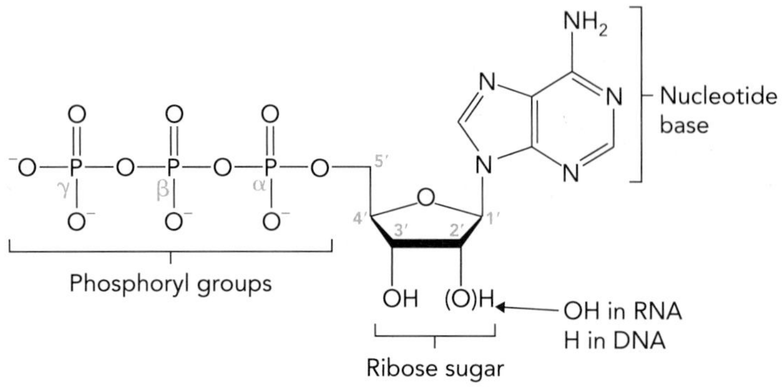

Figure 18.3 Nucleotides consist of a nucleotide base (purine or pyrimidine), a ribose sugar, and one or more phosphoryl groups linked to the ribose sugar. ATP, shown here, is an example of a nucleotide. Ribonucleotides contain a hydroxyl group on the C-2′ of ribose, whereas deoxyribonucleotides have an H atom.

As shown in **Figure 18.3**, nucleotides consist of a nucleotide base, which is either a purine or pyrimidine; a ribose ring; and one or more phosphoryl groups linked to the ribose. These phosphoryl groups are commonly denoted as α, β, and γ. Ribonucleotides contain hydroxyl groups at both the 3′-carbon and 2′-carbon of the ribose, whereas deoxyribonucleotides lack the hydroxyl group at the 2′-carbon. (Numbered carbons in the ribose ring are labeled as prime [′] to distinguish them from numbered atoms in the nucleotide base.)

Figure 18.4 shows the structures of the two major purine bases, **adenine (A)** and **guanine (G)**, along with the structures of the three major pyrimidine bases, **cytosine (C)**, **thymine (T)**, and **uracil (U)**. Other purine and pyrimidine bases that serve as key metabolite intermediates are **hypoxanthine** (purine) and **orotate** (pyrimidine), which we describe later. **Purines** contain nine atoms in the heterocyclic ring, of which the N-9 atom links to the ribose sugar in a standard nucleotide. **Pyrimidines** contain six atoms in the ring, and the ribose sugar covalently attaches to the N-1 atom in a standard nucleotide.

Thymine is found only in DNA, and uracil is found only in RNA. This arrangement helps to protect against DNA mutations arising from spontaneous cytosine deamination, which would generate uracil. As described in Chapter 3 (see Figure 3.22), if uracil were a DNA base that normally base paired with adenine (U-A), then a spontaneous cytosine deamination would not be recognized as DNA damage and would lead to the conversion of a G-C base pair into an A-T base pair after two rounds of DNA replication. However the use of thymine in place of uracil in DNA enables DNA repair enzymes to recognize uracil bases in DNA as spontaneous mutations and remove them.

In Chapter 3, we listed the names of the five common nucleotide bases and their derivatives (see Table 3.1). Recall that a nucleoside consists of a base and a sugar, whereas a nucleotide is a phosphorylated nucleoside. Using this nomenclature, ATP is named adenosine-5′-triphosphate, and AMP is adenosine-5′-monophosphate. All of the nucleoside monophosphates also have generic names that end in the suffix "-ylate"; for example, adenosine-5′-monophosphate is called adenylate, and cytosine-5′-monophosphate is called cytidylate. Deoxythymidine and deoxythymidylate are sometimes referred to as thymidine and thymidylate, respectively, as the thymine base is only found in deoxyribonucleotides.

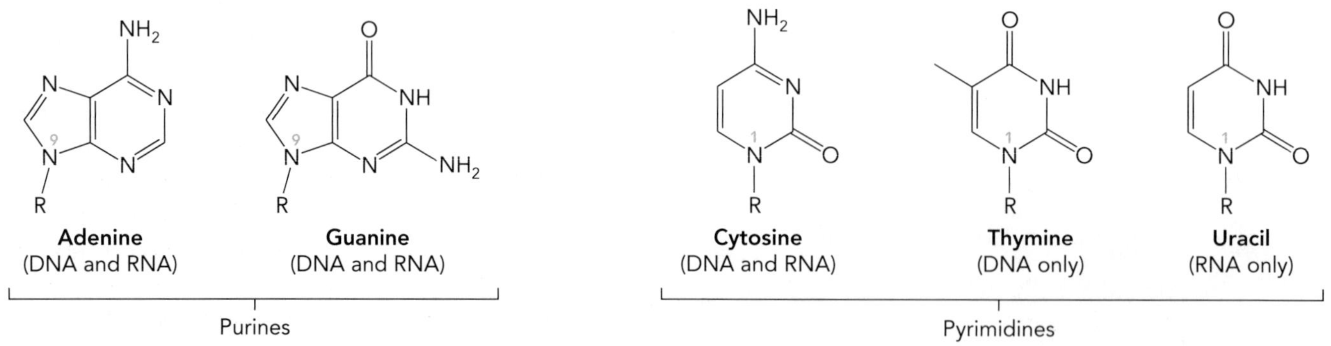

Figure 18.4 The chemical structures of the five nucleotide bases commonly found in RNA and DNA are shown.

Nucleotide Salvage Pathways

The *de novo* biosynthetic pathways for purines and pyrimidines are complex, requiring an energy investment of up to nine high-energy phosphate bonds in the case of adenylate synthesis. To avoid this energy expense when nucleotide bases are available from nucleic acid degradation, cells use **salvage pathways** to resynthesize nucleoside monophosphates from free bases. Although some nucleic acids are obtained in the diet and can be used as a source of nucleotide bases, most nucleic acid degradation takes place inside cells as a normal part of nucleic acid turnover. The majority of recycled nucleotide bases (except thymine) come from RNA, which serves as the transient copy of biochemical information stored in DNA. As shown in **Figure 18.5**, RNA degradation by endonuclease enzymes produces short oligonucleotides, which are cleaved into nucleoside monophosphates by phosphodiesterases. These nucleoside

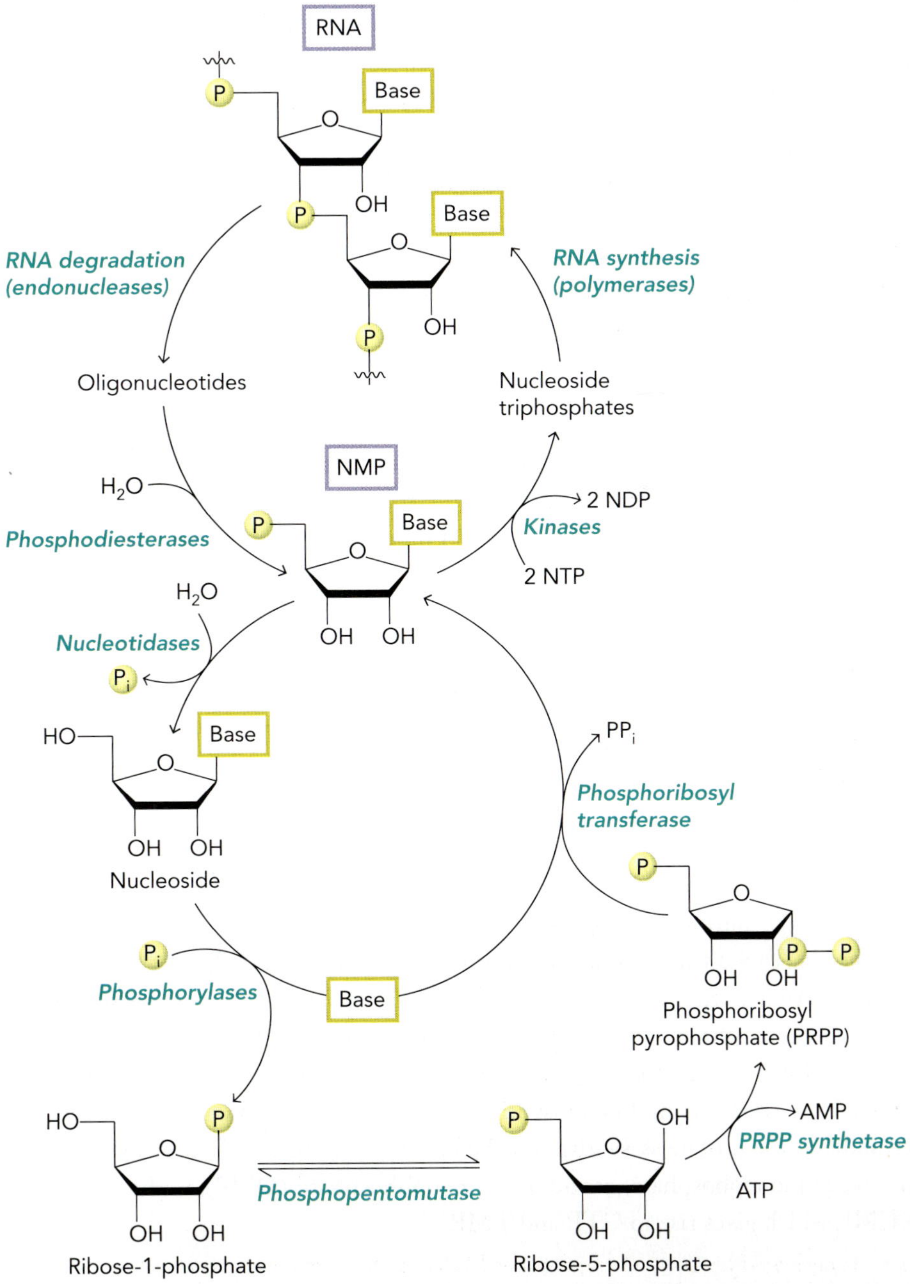

Figure 18.5 Nucleotide salvage pathways recycle bases and use less energy than *de novo* biosynthesis. Degradation of RNA produces oligonucleotides, which are cleaved by phosphodiesterases to yield nucleoside monophosphates (NMPs). These are used for nucleic acid synthesis or degraded to ribose-1-phosphate and free bases. Deoxyribonucleotides are salvaged by similar pathways. NDP = nucleoside diphosphate; NTP = nucleoside triphosphate.

monophosphates can be converted to nucleoside triphosphates by kinase enzymes and used for RNA synthesis. Alternatively, they can be degraded by enzymes called nucleotidases that remove the phosphoryl group from the ribose sugar.

Nucleosides are degraded into the free base and ribose-1-phosphate (ribose-1-P) by purine or pyrimidine **phosphorylase** enzymes. These enzymes use inorganic phosphate in a phosphorolytic cleavage reaction, similar to that of glycogen phosphorylase (see Figure 14.27). To recycle the free base, ribose-1-P must first be converted to ribose-5-phosphate (ribose-5-P) by the enzyme phosphopentomutase. Ribose-5-P is then phosphorylated by the enzyme phosphoribosyl pyrophosphate synthetase (PRPP synthetase) to produce **phosphoribosyl pyrophosphate (PRPP)**. Nucleoside monophosphates are regenerated from PRPP by phosphoribosyl transferase enzymes, which link free bases to PRPP. This phosphorylated ribose sugar is also required in the *de novo* purine and pyrimidine biosynthetic pathways. As discussed in the next section, defects in one of the phosphoribosyl transferase enzymes, called hypoxanthine-guanine phosphoribosyl transferase (HGPRT), causes the severe metabolic disease Lesch–Nyhan syndrome. Although not shown in Figure 18.5, deoxyribonucleotides are salvaged by similar salvage pathways.

concept integration 18.1

It is important to salvage nucleotide bases because they are energetically costly to synthesize by *de novo* pathways. The ribose sugar, however, is a metabolite in two other biochemical pathways. Describe these two pathways.

Ribose-5-P is an intermediate in the nonoxidative phase of the pentose phosphate pathway; it is the product of the ribulose-5-phosphate (ribulose-5-P) isomerase reaction that converts ribulose-5-P to ribose-5-P. When nucleotide pools are low, ribose-5-P is shunted from the pentose phosphate pathway to the nucleotide salvage pathway and *de novo* biosynthetic pathways. In plants and other photosynthetic organisms that use the Calvin cycle to fix CO_2 into hexose sugars, ribose-5-P is produced in the regeneration phase (stage 3) by the enzyme ribose-5-P isomerase. Ribose-5-P is also a reactant in the transketolase reaction. The enzyme ribose-5-P isomerase in plants interconverts ribulose-5-P and ribose-5-P in both the pentose phosphate pathway and the Calvin cycle.

18.2 Purine Metabolism

Biosynthesis of purines features some fundamental differences from biosynthesis of pyrimidines. To see this, consider the purine ribonucleotides AMP and GMP, which are synthesized from the common intermediate **inosine-5′-monophosphate (IMP)**. This nucleotide contains the purine base hypoxanthine, which is built directly on the PRPP scaffold using a stepwise reaction pathway. As shown in **Figure 18.6**, this process contrasts with pyrimidine biosynthesis, in which the six-membered heterocyclic pyrimidine base orotate is synthesized before it is attached to PRPP. The resulting nucleoside monophosphate, **orotidine-5′-monophosphate (OMP)**, is then converted to UMP, which gives rise to CTP and TMP.

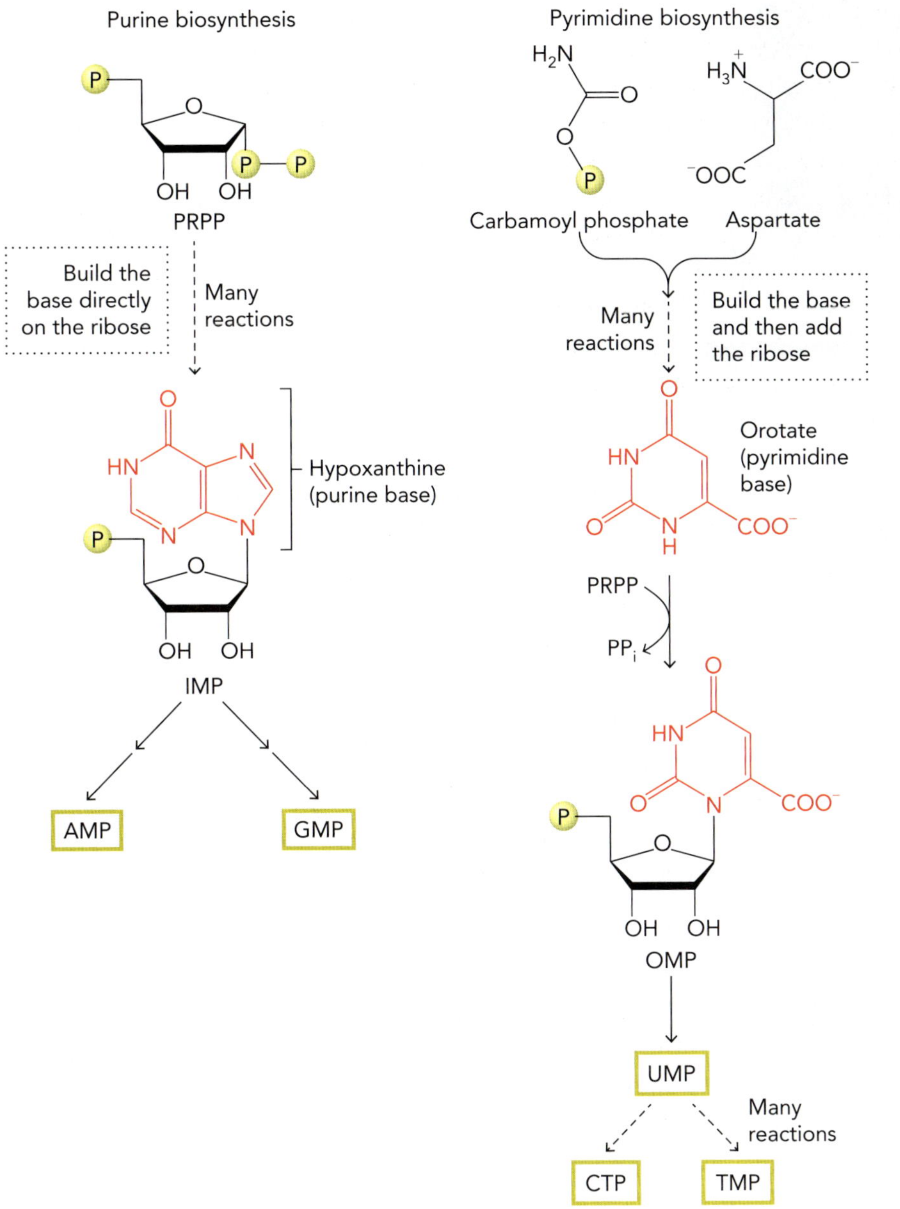

Figure 18.6 An overview is shown of the *de novo* biosynthetic pathways for purines and pyrimidines. Purines are synthesized directly on the ribose sugar PRPP to generate the common intermediate IMP, which contains the purine base hypoxanthine. Pyrimidine bases are synthesized from carbamoyl phosphate and aspartate to yield orotate, which is then linked to PRPP to generate OMP.

The Purine Biosynthetic Pathway Generates IMP

The purine biosynthetic pathway was initially characterized by John Buchanan, a biochemist at the University of Pennsylvania who used pigeons collected from nearby parks in his research. Buchanan fed the pigeons a variety of radioisotopically labeled compounds and then analyzed the uric acid they excreted as a waste product. Pigeons, like all birds, synthesize purines as a way to remove nitrogen waste generated by amino acid catabolism. By analyzing the position of the labeled atoms in excreted uric acid, Buchanan was able to identify the origin of all nine nonhydrogen atoms in the purine ring.

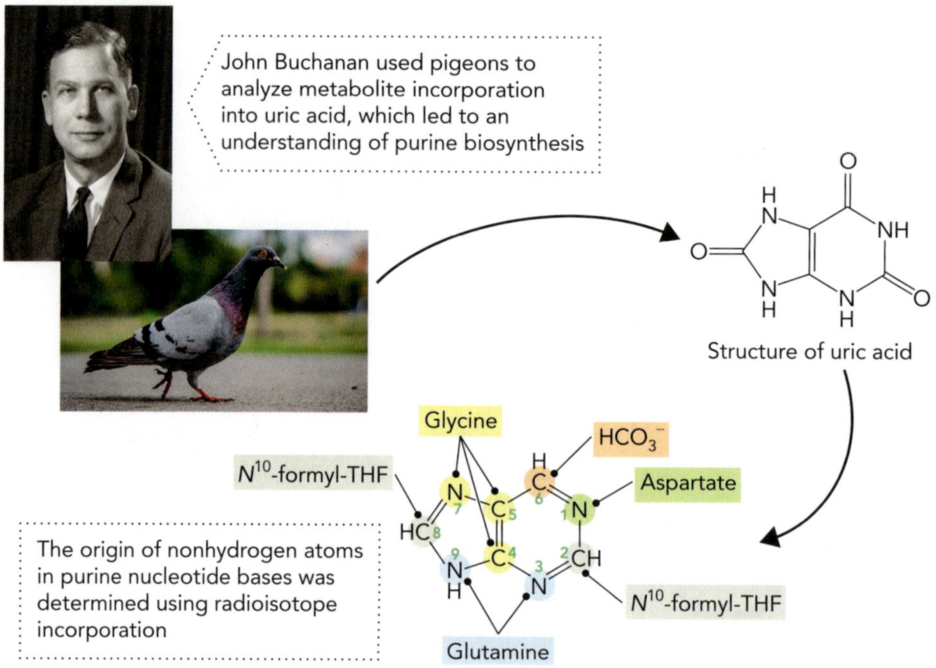

John Buchanan used pigeons to analyze metabolite incorporation into uric acid, which led to an understanding of purine biosynthesis
Structure of uric acid
Glycine
HCO_3^-
N^{10}-formyl-THF
Aspartate
The origin of nonhydrogen atoms in purine nucleotide bases was determined using radioisotope incorporation
N^{10}-formyl-THF
Glutamine

Figure 18.7 John Buchanan used pigeons in radioisotope labeling experiments to determine the origin of all nine nonhydrogen atoms in the purine ring by analyzing uric acid. The metabolite N^{10}-formyl-tetrahydrofolate is abbreviated as N^{10}-formyl-THF.

J. BUCHANAN: COURTESY MIT MUSEUM; PIGEON: OMIHAY/SHUTTERSTOCK.

As shown in **Figure 18.7**, Buchanan's experiments revealed that the four nitrogen atoms in purine bases are derived from aspartate (N-1), glycine (N-7), and two glutamines (N-3 and N-9). The five carbons come from glycine (C-4 and C-5), HCO_3^- (C-6), and two molecules of N^{10}-formyl-tetrahydrofolate (C-2 and C-8). Similar to some of the other metabolic pathways we have described, many of the enzymes in the eukaryotic purine biosynthetic pathway are multifunctional proteins, whereas the same catalytic activities in bacteria are encoded by separate polypeptides. We will first describe purine biosynthesis in *Escherichia coli* and will then present the case for a eukaryotic "purinosome" in human cells that may function as a purine biosynthetic machine.

The biosynthesis of IMP in *E. coli* can be broken down into two stages, each consisting of five well-characterized biochemical reactions. In the first stage, the five-membered imidazole ring of the purine base is built on PRPP in a series of reactions leading to the formation of 5-aminoimidazole ribonucleotide (AIR; **Figure 18.8**). The second ring of the purine molecule is then generated in stage 2 by five reactions that convert AIR to IMP (**Figure 18.9**, p. 908). Before synthesis of the purine ring can even start, a pyrophosphoryl group needs to be added to the C-1′ of ribose-5-P by the enzyme PRPP synthetase to form PRPP. Recall that ribose-5-P is a product of the pentose phosphate pathway and is redirected toward nucleotide biosynthesis when required for ongoing DNA and RNA synthesis (see Figure 14.3).

As shown in Figure 18.8, the amide group of glutamine replaces the pyrophosphoryl group on PRPP in reaction 1 of the purine biosynthetic pathway to form 5-phosphoribosylamine (5PRA). (This nitrogen will become N-9 of the purine ring.) This rate-limiting reaction is catalyzed by the next enzyme in the pathway, glutamine-PRPP amidotransferase. (This step is subject to feedback inhibition by AMP and GMP, as described later.) In reaction 2, glycine is added to the nitrogen in an ATP-dependent reaction catalyzed by glycinamide ribonucleotide (GAR) synthetase, which gives rise to the C-4, C-5, and N-7 atoms of the purine ring. Then, C-8 is added in reaction 3 by GAR transformylase, using N^{10}-formyl-tetrahydrofolate as the carbon donor to generate formylglycinamide ribonucleotide (FGAR). Reaction 4

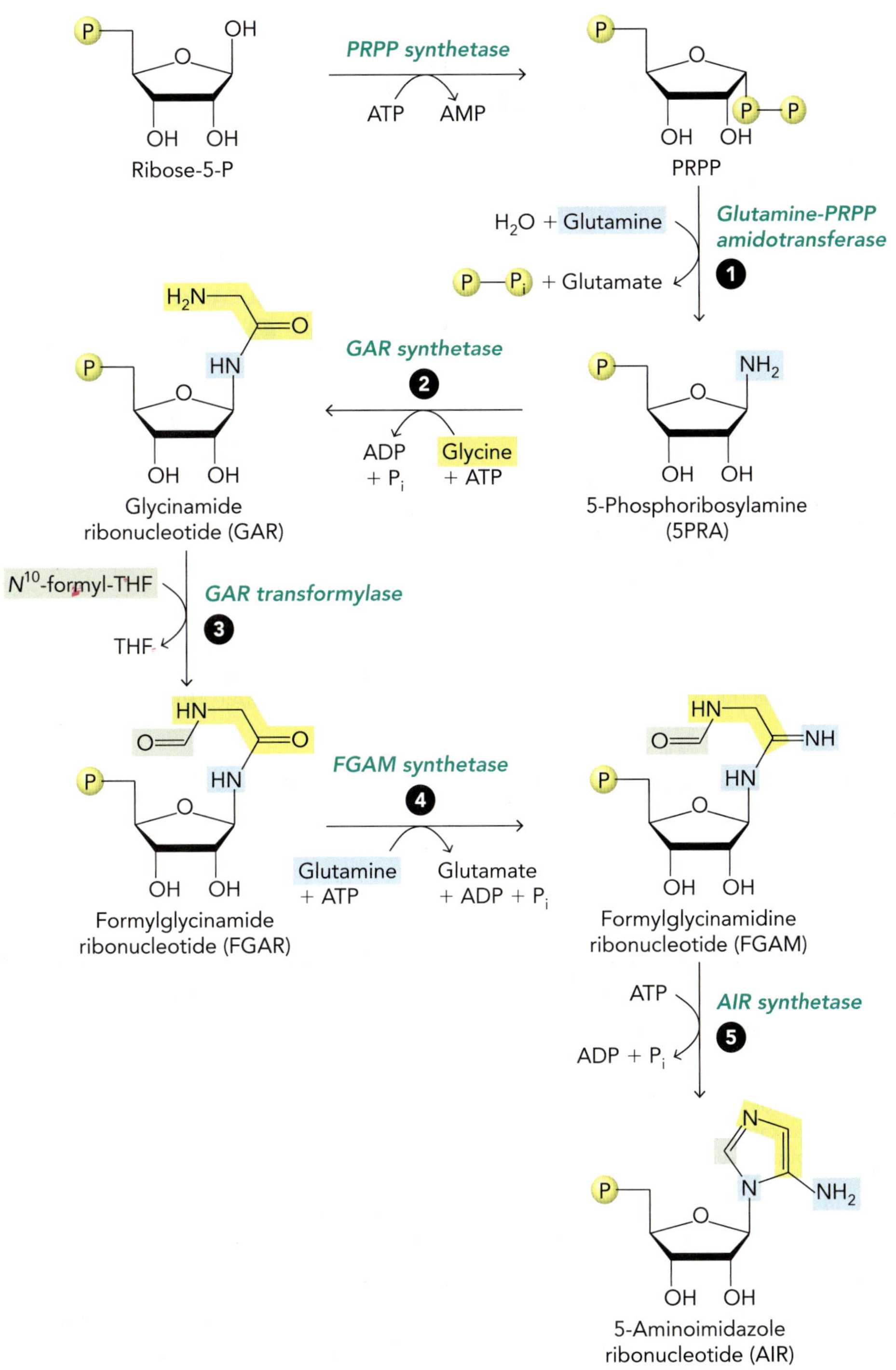

Figure 18.8 Stage 1 of the purine biosynthetic pathway in *E. coli* is shown. The purine ring is built onto the PRPP ribose sugar, beginning with addition of the first nitrogen obtained from glutamine. The final product of stage 1 is AIR, which contains six of the nine atoms that make up the purine ring. The formyl donor is N^{10}-formyl-tetrahydrofolate (N^{10}-formyl-THF).

is an ATP-dependent reaction in which another glutamine donates a nitrogen atom in a reaction catalyzed by the enzyme formylglycinamidine ribonucleotide (FGAM) synthetase. The imidazole ring of the purine base is then completed in reaction 5 by the enzyme 5-aminoimidazole ribonucleotide synthetase (AIR synthetase) using energy made available from ATP hydrolysis. Taken together, 5 ATP equivalents are needed in stage 1 to build AIR when starting with ribose-5-P, counting ATP → AMP as 2 ATP equivalents in the PRPP synthetase reaction.

Stage 2 of the purine biosynthetic pathway in *E. coli* begins with reaction 6, in which AIR is carboxylated by AIR carboxylase in an ATP-dependent reaction to add C-6 to the purine ring (Figure 18.9). The resulting product, carboxyaminoimidazole

Figure 18.9 Stage 2 of the purine biosynthetic pathway in *E. coli* is shown. The purine product of stage 2 is hypoxanthine, which is the nucleotide base of IMP.

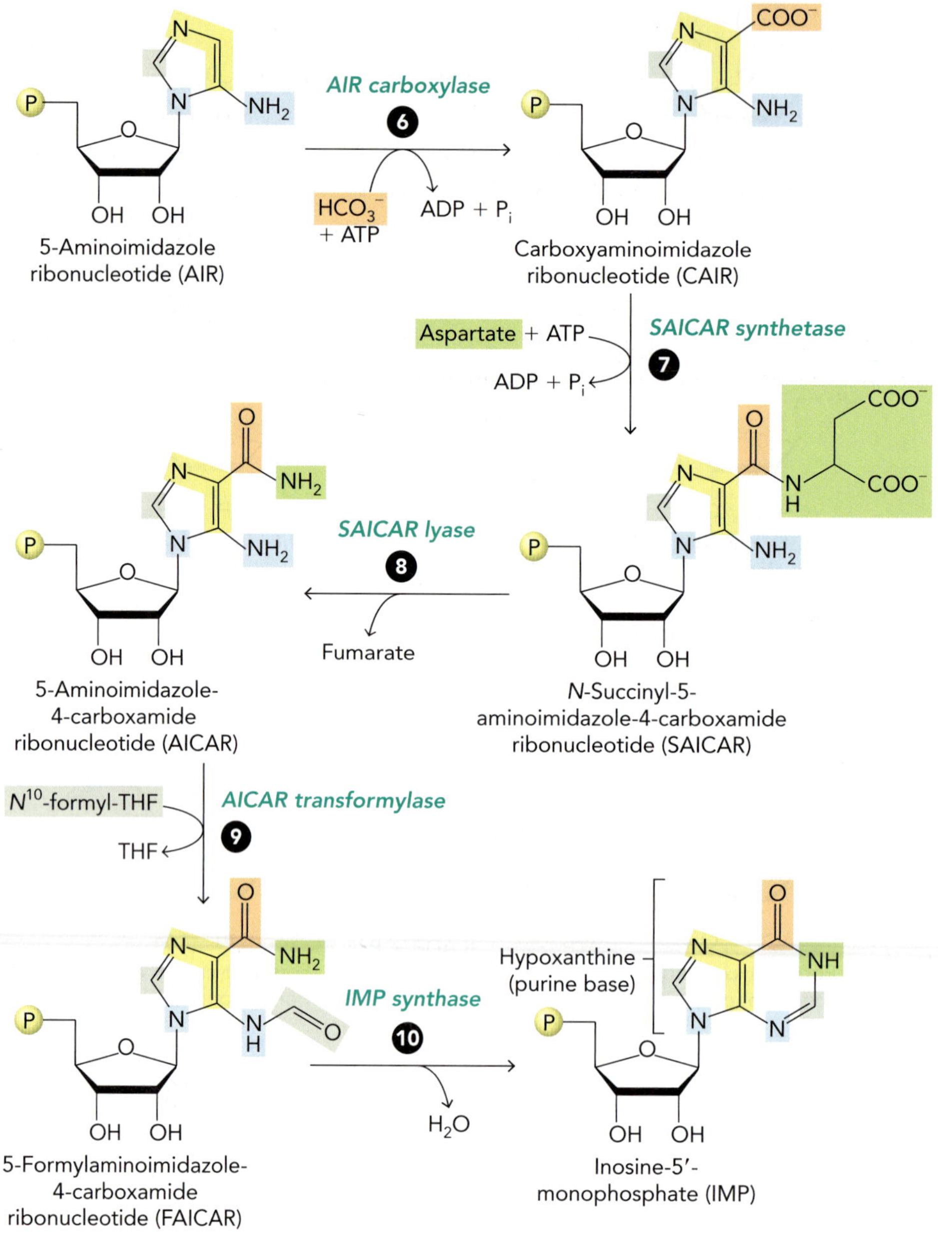

ribonucleotide (CAIR), is then converted to *N*-succinyl-5-aminoimidazole-4-carboxamide ribonucleotide (SAICAR) in reaction 7 by SAICAR synthetase through attachment of an aspartate to C-6. In a reaction similar to the argininosuccinase reaction in the urea cycle, SAICAR lyase cleaves SAICAR in reaction 8 to generate fumarate and 5-aminoimidazole-4-carboxamide ribonucleotide (AICAR). This reaction leaves the nitrogen originating from aspartate at the N-1 position on the growing purine ring. The final atom needed to form the purine ring is C-1, which is added in reaction 9 by the enzyme AICAR transformylase using N^{10}-formyl-tetrahydrofolate to generate 5-formylaminoimidazole-4-carboxamide ribonucleotide (FAICAR). The purine ring is then closed in reaction 10 to form the purine base hypoxanthine in a dehydration reaction catalyzed by IMP synthase, thereby yielding the nucleotide IMP. Considering that the synthesis of IMP from ribose-5-P requires the input of 8 ATP equivalents (5 ATP in stage 1 and 3 ATP in stage 2), it is easy to see why recycling purine bases using salvage pathways is so beneficial to cells (see Figure 18.5).

Researchers have shown that the enzymes required to convert PRPP to IMP in animal cells exist in a large protein complex called a **purinosome**, which functions as a multi-subunit purine-synthesizing machine. Evidence for the existence of purinosomes came from tissue culture experiments using human cancer cells grown in purine-depleted media. By expressing in these cells fluorescent recombinant proteins fused to the coding sequences of purine biosynthetic enzymes, it was possible to visualize the location of these purine biosynthetic enzymes in live cells.

Unlike *E. coli*, which requires 10 separate proteins for purine biosynthesis, human cells contain only six enzymes for purine biosynthesis. One of these is hTrifGART, a trifunctional enzyme encoding the catalytic activities found in the *E. coli* GAR synthetase (reaction 2), GAR transformylase (reaction 3), and AIR synthetase (reaction 5) enzymes (**Figure 18.10a**). As shown in **Figure 18.10b**, human cells expressing

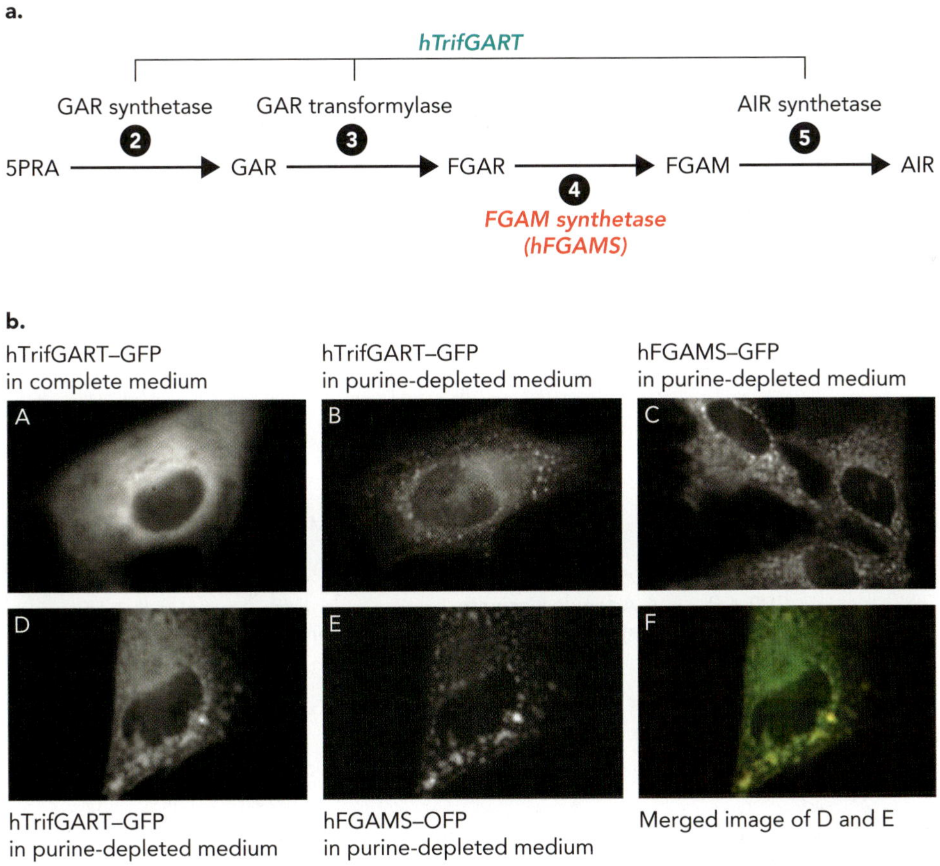

Figure 18.10 Purine biosynthetic enzymes colocalize in large multienzyme complexes when human cells are cultured in purine-depleted media. **a.** The human enzyme hTrifGART encodes the same three activities as the *E. coli* enzymes catalyzing reactions 2, 3, and 5 in the purine biosynthetic pathway shown in Figure 18.8. The human enzyme hFGAMS is the ortholog of the *E. coli* enzyme FGAM synthetase, which catalyzes reaction 4 (see Figure 18.8). **b.** The human purine biosynthetic enzymes hTrifGART and hFGAMS localize to discrete regions in cells cultured in purine-depleted media but not in the same cells grown in complete media containing purines. The punctate fluorescent patterns in cells grown in purine-depleted media are indicative of multienzyme complexes functioning as "purinosomes." Panels A–C: Representative HeLa cells expressing the fusion proteins hTrifGART–GFP or hFGAMS–GFP cultured in complete or purine-depleted media. Panels D–F: Fluorescence pattern seen in a single cell expressing both hTrifGART–GFP and hFGAMS–OFP cultured in purine-depleted media. S. AN ET AL. (2008). REVERSIBLE COMPARTMENTALIZATION OF DE NOVO PURINE BIOSYNTHETIC COMPLEXES IN LIVING CELLS. *SCIENCE*, *320*, 103–106. DOI: 10.1126/SCIENCE.1152241.

a green fluorescent protein (GFP) fused with the human multifunctional enzyme TrifGART (hTrifGART–GFP) display a punctate fluorescent pattern when cultured in purine-depleted media. Further support for the existence of purinosomes in human cells came from similar experiments done using a GFP fusion protein containing human FGAM synthetase (hFGAMS–GFP). As observed in Figure 18.10b, the hFGAMS–GFP fusion protein produced the same punctate staining pattern of purinosomes as did the hTrifGART–GFP fusion protein. Moreover, when the hFGAMS coding sequences were fused to orange fluorescent protein (OFP) to generate hFGAMS–OFP, both hTrifGART–GFP and hFGAMS–OFP were also found to colocalize to purinosomes. Multi-subunit complexes such as purinosomes can enhance metabolite flux in the open cellular environment where reaction efficiency may otherwise be affected by diffusion.

As shown in **Figure 18.11**, IMP is the precursor to both AMP and GMP, which are generated in parallel biosynthetic pathways. The nitrogen of AMP comes from aspartate in a reaction catalyzed by adenylosuccinate synthetase. This reaction uses GTP, rather than ATP to drive the reaction to convert IMP to adenylosuccinate. The enzyme adenylosuccinate lyase then cleaves adenylosuccinate, generating fumarate and AMP. In the other branch of this pathway, IMP is converted to GMP by oxidation of the

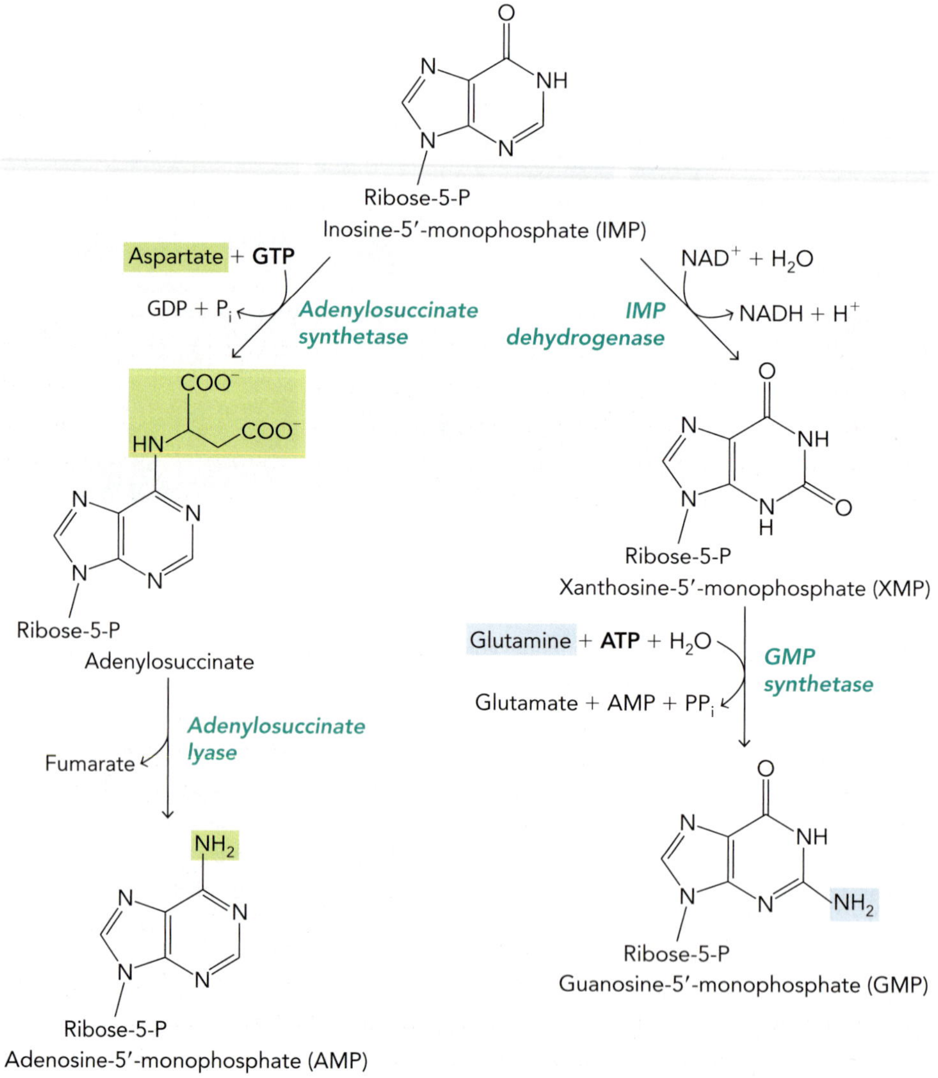

Figure 18.11 IMP is the precursor to both AMP and GMP. Note that GTP is used to drive the adenylosuccinate synthetase reaction in the AMP branch of the pathway, whereas ATP is required for the GMP branch.

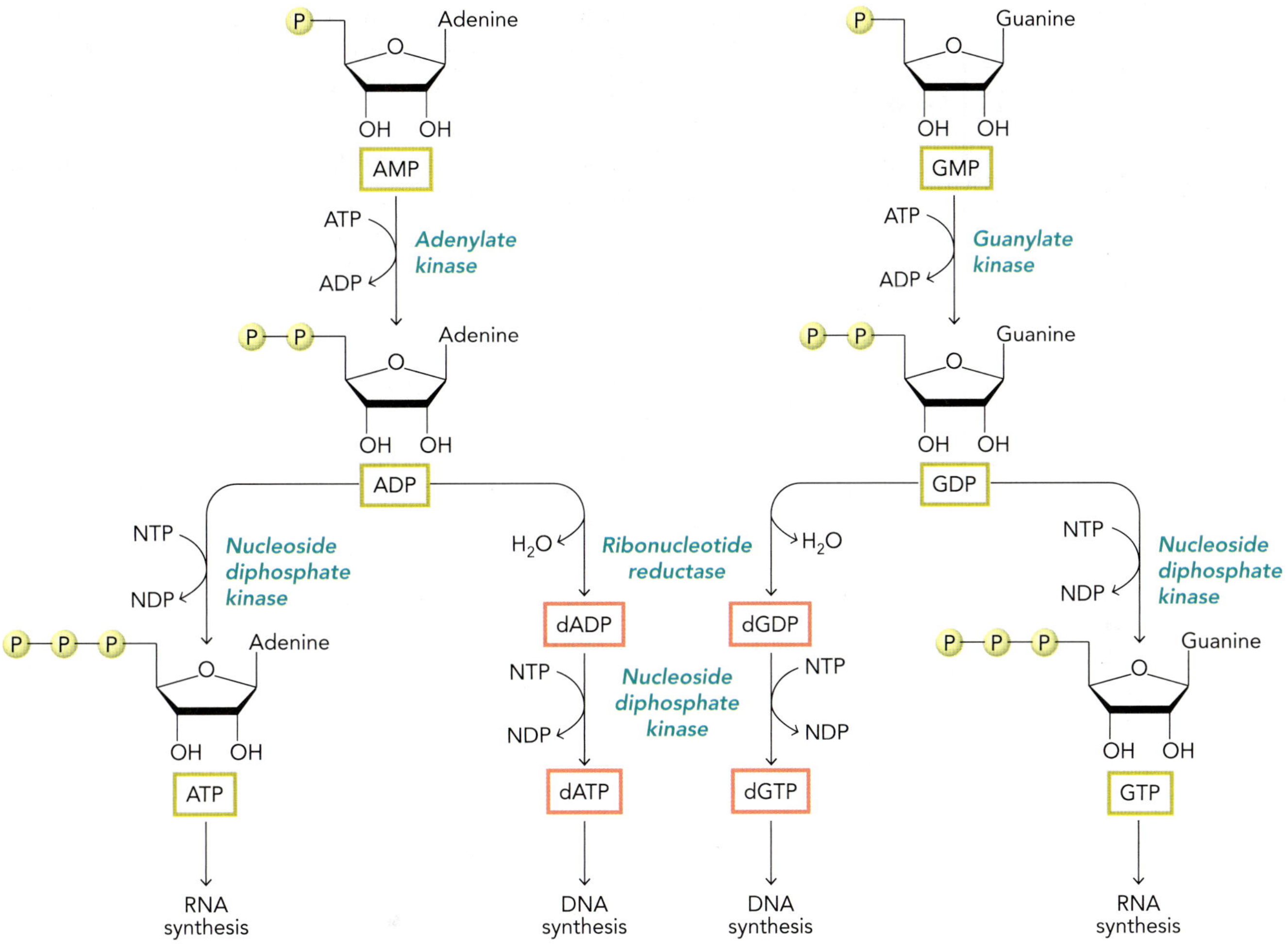

Figure 18.12 AMP and GMP are phosphorylated to generate nucleoside triphosphates for RNA and DNA synthesis. The enzyme nucleoside diphosphate kinase can use any nucleoside triphosphate as a phosphate donor and phosphorylates both nucleoside diphosphates and deoxynucleoside diphosphates. dATP = deoxyadenosine triphosphate; dGTP = deoxyguanosine triphosphate.

hypoxanthine moiety at C-2 using the enzyme IMP dehydrogenase; this reaction generates xanthosine-5′-monophosphate (XMP). The oxygen on C-2 is then replaced with an amino group from glutamine in an ATP-dependent reaction catalyzed by GMP synthetase to generate GMP. It is significant that GTP is required as the energy source to drive AMP synthesis, whereas ATP is needed for GMP synthesis. This type of substrate specificity in the two energy-requiring reactions of the branched purine biosynthetic pathway helps to balance AMP and GMP production, as we explain shortly.

In order for AMP and GMP to be useful in the cell, they must first be converted to their respective nucleoside triphosphates through a pair of phosphorylation reactions. As shown in **Figure 18.12**, ATP is used in a phosphoryl transfer reaction to convert AMP to ADP by the enzyme adenylate kinase. Similarly, GMP is converted to GDP by guanylate kinase, again using ATP as the source of phosphate. Nucleoside diphosphates can either be phosphorylated directly by nucleoside diphosphate kinase or first reduced by ribonucleotide reductase to form deoxyadenosine diphosphate (dADP) and deoxyguanosine diphosphate (dGDP).

It is important to note that nucleoside diphosphate kinase can use any nucleoside triphosphate as a source of phosphate, although ATP is the usual substrate because it is so abundant in cells. Moreover, nucleoside diphosphate kinase is not specific for ribonucleotides, as this same enzyme also phosphorylates deoxynucleoside diphosphates to generate deoxynucleoside triphosphates. Although nucleoside diphosphate kinase is able to phosphorylate any purine or pyrimidine nucleoside diphosphate, most of

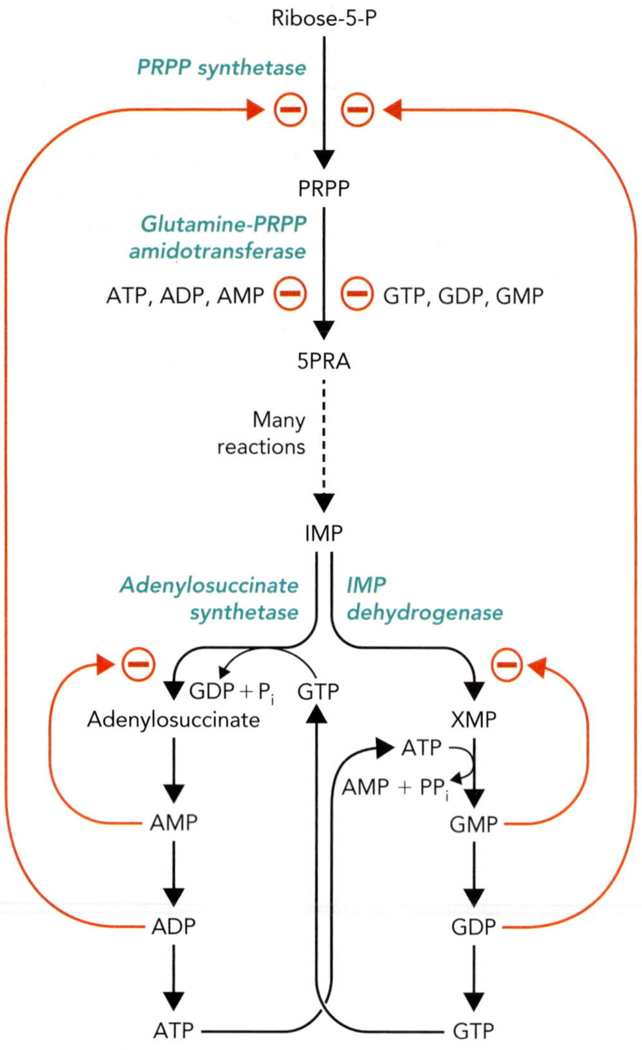

Figure 18.13 This diagram shows the regulation of the purine biosynthetic pathway in *E. coli* by feedback inhibition. Flux through the purine biosynthetic pathway is primarily controlled by inhibition of the PRPP synthetase and glutamine-PRPP amidotransferase reactions. The balance of AMP and GMP synthesis is controlled by both feedback inhibition of the individual branches in the pathway and by ATP and GTP cross-talk regulation. 5PRA = 5-phosphoribosylamine.

the ADP in the cell is phosphorylated by the two substrate-level phosphorylation reactions in glycolysis and by ATP synthase in the mitochondria or chloroplast.

Feedback Inhibition of Purine Biosynthesis

As described in Chapter 17, feedback inhibition is an important regulatory mechanism in amino acid biosynthesis because it ensures that amino acids are synthesized in the correct proportions needed to sustain protein synthesis. Feedback inhibition is also important for nucleic acid biosynthesis, which requires equivalent amounts of purine and pyrimidine nucleotides.

As shown in **Figure 18.13**, the purine biosynthetic pathway is regulated at multiple steps by feedback inhibition. The PRPP synthetase reaction is inhibited by both ADP and GDP; this feedback loop prevents the expenditure of 2 ATP equivalents to convert ribose-5-P into PRPP if it is not needed. The enzyme glutamine-PRPP amidotransferase is feedback inhibited by all three phosphorylated forms of adenine and guanine nucleotides. Note that AMP, ADP, and ATP bind to one regulatory site on the enzyme, whereas GMP, GDP, and GTP bind to a second regulatory site. This permits both independent and synergistic regulation of this key enzyme.

As shown in Figure 18.13, the individual branches of the purine biosynthetic pathway are also subject to feedback inhibition. For example, a buildup of AMP inhibits its own synthesis by binding to the enzyme adenylosuccinate synthetase. Similarly, GMP binds to and inhibits the activity of IMP dehydrogenase. This form of feedback inhibition leads to increased flux through the alternative branch in the pathway. Specifically, AMP inhibition of adenylosuccinate synthetase causes accumulation of IMP, leading to increased synthesis of XMP and flux through the guanylate synthesis pathway. Likewise, a shift in flux from GMP synthesis to AMP synthesis occurs when GMP builds up and inhibits the IMP dehydrogenase reaction. Moreover, because ATP is required for GMP synthesis, and GTP is the energy source for AMP synthesis (see Figure 18.11), this provides a form of regulatory "cross talk" in that there is a built-in balance for the production rates of AMP and GMP.

Uric Acid Is the Product of Purine Degradation

Excess purines from cellular or dietary nucleic acids are degraded into **uric acid**, which is then excreted directly or further metabolized to other nitrogen waste products. As shown in **Figure 18.14**, AMP, IMP, and GMP are all dephosphorylated by the enzyme 5′-nucleotidase to generate adenosine, inosine, and guanosine, respectively. AMP can also be deaminated by the enzyme AMP deaminase to generate IMP, which is then dephosphorylated to produce inosine. Inosine is cleaved by purine nucleoside phosphorylase to generate hypoxanthine and ribose-1-P. Hypoxanthine is oxidized in

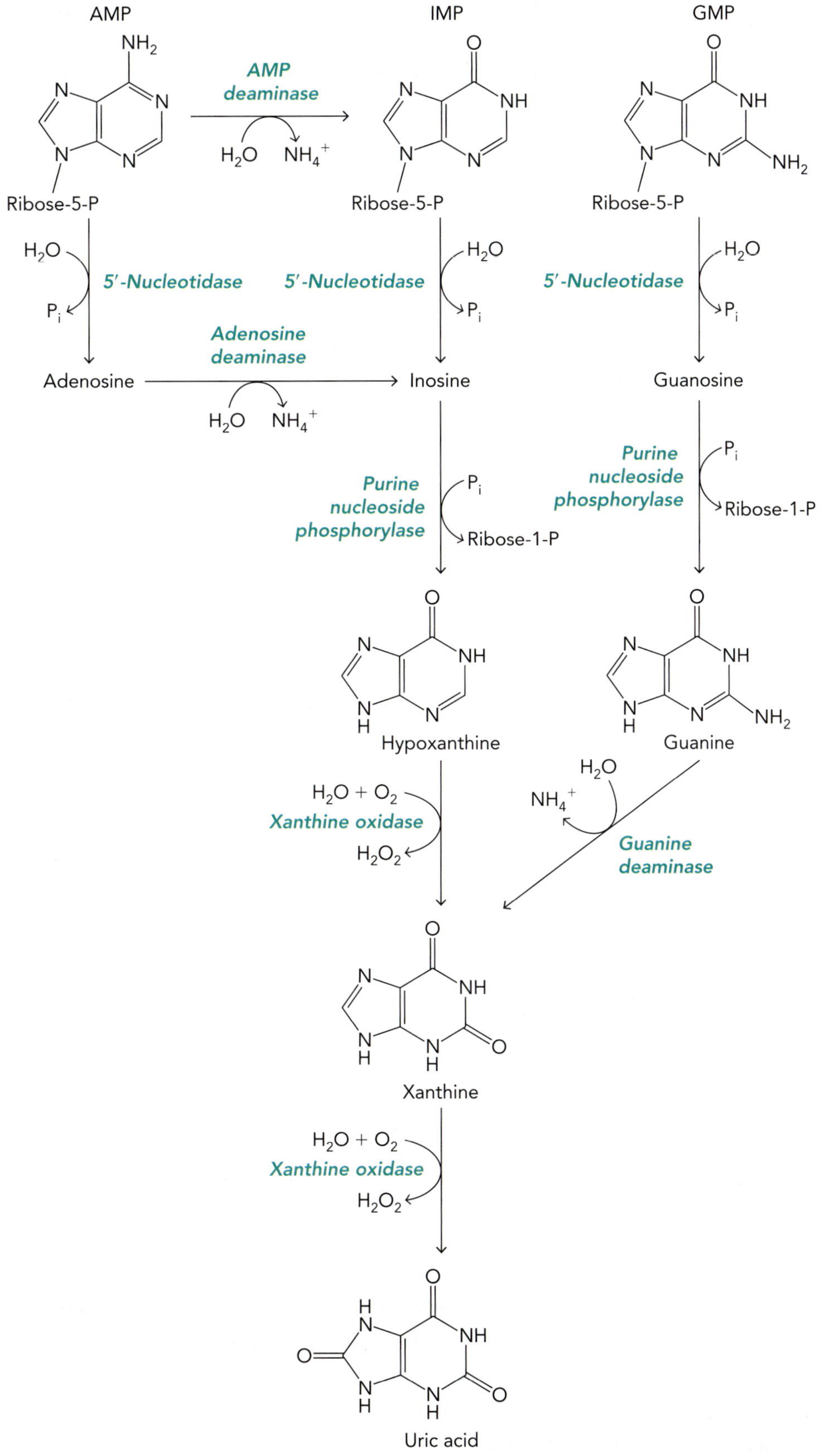

Figure 18.14 The purine degradation pathway involves removal of the ribose sugar and deamination to generate the free bases hypoxanthine or xanthine. The enzyme purine nucleoside phosphorylase cleaves inosine or guanosine to release the free base hypoxanthine or guanine, respectively, and ribose-1-P. Oxidation of hypoxanthine and xanthine by xanthine oxidase generates uric acid, which can be excreted as nitrogen waste.

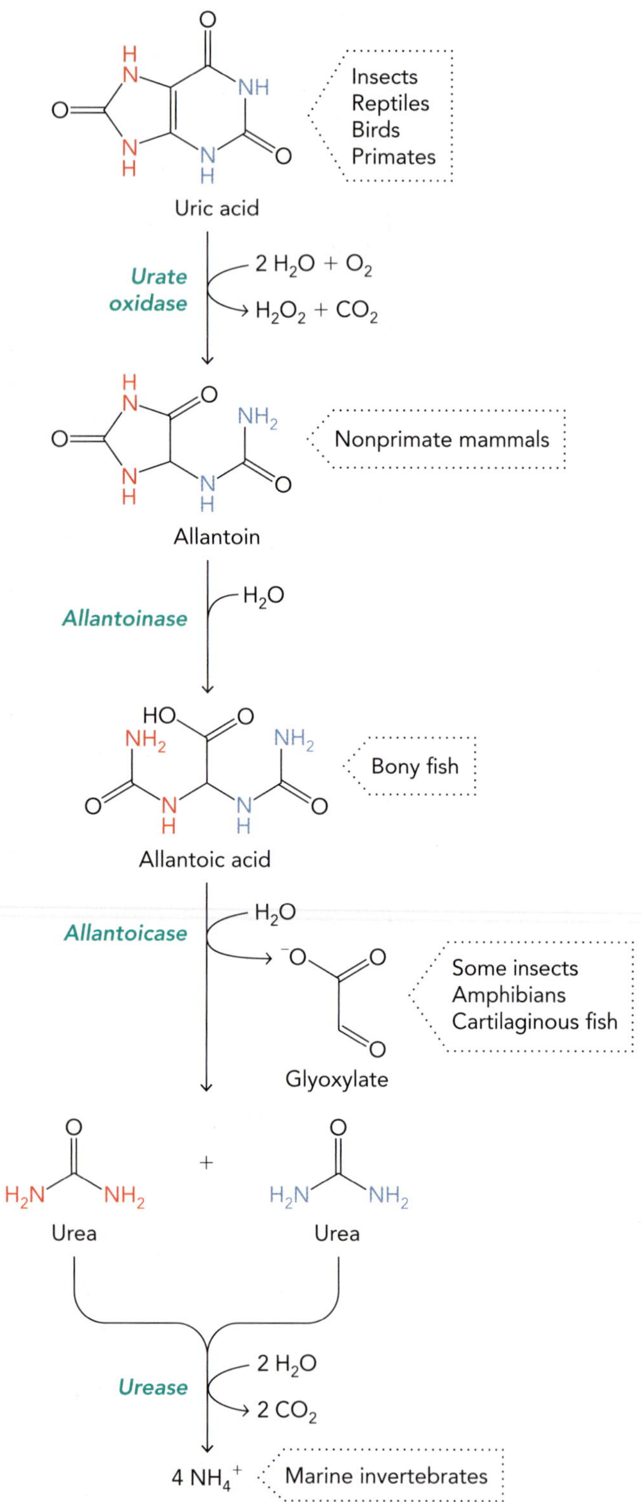

Figure 18.15 Depending on the organism, uric acid can be degraded into other nitrogen-containing compounds that are excreted. Although most mammals excrete excess nitrogen primarily as urea, other organisms convert nitrogen metabolites to uric acid that is either excreted directly or further metabolized.

two successive reactions by the enzyme xanthine oxidase to produce xanthine and uric acid, respectively. Guanosine degradation also gives rise to xanthine, after removal of the ribose sugar and deamination by guanine deaminase.

Urea excretion requires large amounts of water, whereas uric acid, which is essentially insoluble in water, is excreted as a dry paste by organisms that need to conserve water, including birds, reptiles, and insects. (Birds need to keep their weight to a minimum in order to fly; reptiles often live in a dry environment; and insects continually lose water through their cuticle [outer layer] by transpiration.) Primates, including humans, also excrete small amounts of uric acid, even though most of their nitrogen waste products are excreted as urea produced by the urea cycle. Indeed, because the four nitrogens in the purine ring come from the amino acids glutamine, aspartate, and glycine (see Figure 18.7), purine degradation is an effective way to remove excess nitrogen-derived amino acids. As shown in **Figure 18.15**, uric acid is further degraded in some animals to other nitrogen-containing waste products. For example, nonprimate mammals produce **allantoin**, whereas bony fish (teleosts) produce **allantoic acid**. Cartilaginous fish, such as sharks, as well as amphibians and some insects (mosquitoes and honeybees), convert allantoic acid into urea and glyoxylate. Marine invertebrates cleave urea using the enzyme urease to produce CO_2 and NH_4^+, which is freely excreted into water.

Metabolic Diseases of Purine Metabolism

Several metabolic diseases in humans have been associated with defects in purine degradation. The most common is **gout**, which is caused by the buildup of uric acid crystals (sodium urate) in the joints and kidneys (**Figure 18.16**). The big toe is a common joint affected by uric acid because of poor circulation in the foot and the frequency of blunt injury (for example, you stub your toe), which releases uric acid crystals into the synovial fluid. Gout has often been called the "disease of kings" because it can be associated with consumption of alcohol and meat, as was common for nobility in the time of King Henry VIII (A.D. ~1500), who most likely suffered from gout. Alcohol is thought to interfere with uric acid excretion, whereas meats contain high amounts of nucleic acids (RNA and DNA) that lead to increased uric acid production.

Gout is considered a metabolic disease because it can be caused by deregulation of the purine biosynthetic pathway, resulting in uric acid accumulation. As shown in

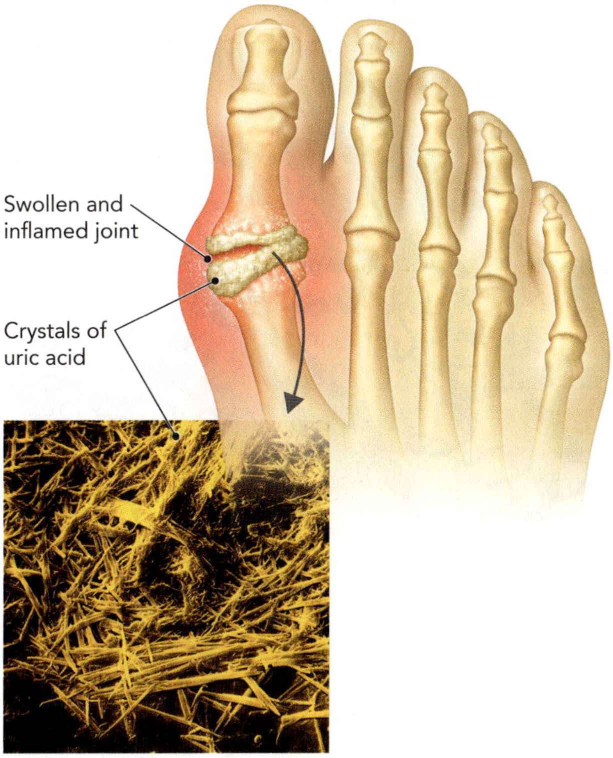

Figure 18.16 Gout is caused by a buildup of uric acid crystals in the joints and kidneys, which can be very painful. One of the most common symptoms of gout is accumulation of uric acid crystals in the synovial fluid around the joint of the big toe. The inset shows a photo of uric acid crystals isolated from a patient with gout and observed by polarized light microscopy. PHOTO: DR. GILBERT FAURE/SCIENCE SOURCE.

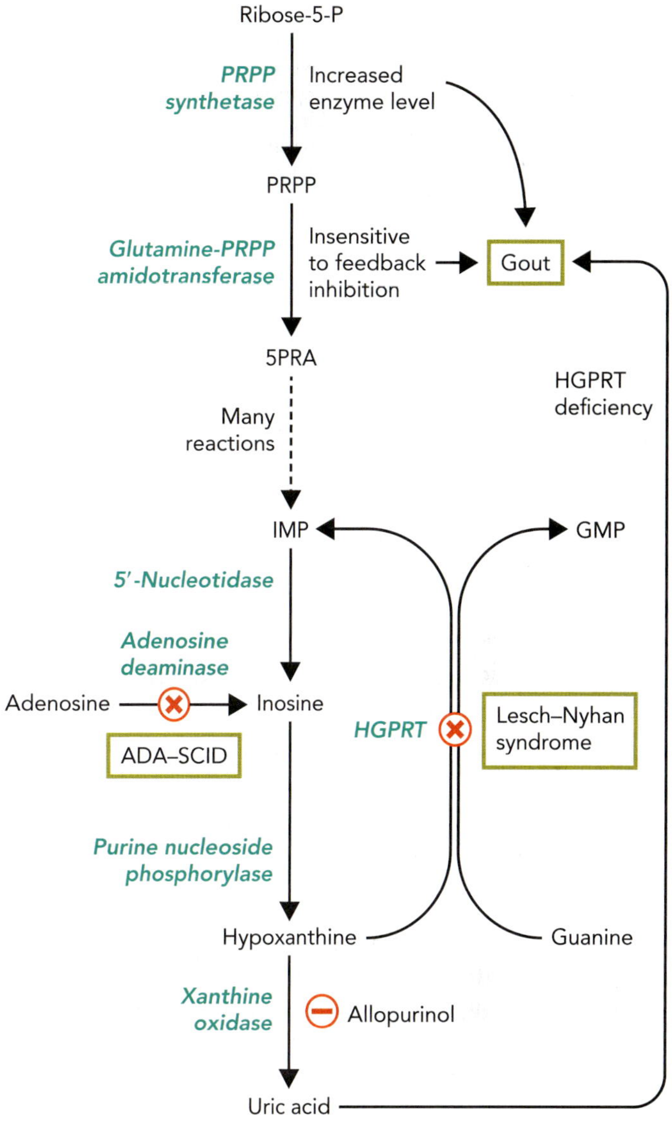

Figure 18.17 Defects in purine degradation have been linked to the human metabolic diseases gout, Lesch–Nyhan syndrome, and ADA–SCID, the latter also an immunodeficiency disease. Allopurinol treatment reduces gout symptoms by inhibiting the activity of xanthine oxidase, thereby decreasing rates of uric acid production.

Figure 18.17, gout has been linked to increased levels of PRPP synthetase, resulting in overproduction of IMP, and to defects in feedback inhibition of the enzyme glutamate-PRPP amidotransferase. Deficiencies in the salvage enzyme hypoxanthine-guanine phosphoribosyltransferase (HGPRT) is also associated with gout, as it leads to a buildup of hypoxanthine, which is converted to uric acid by xanthine oxidase. Patients with gout are usually treated by putting them on restricted diets that reduce the intake of purine-rich foods. In addition to diet, uric acid levels can be reduced by treatment with **allopurinol**, a competitive inhibitor of the enzyme xanthine oxidase.

Two other purine degradation diseases in humans, **Lesch–Nyhan syndrome** and **adenosine deaminase deficiency**, are caused by mutations in specific genes. As shown in Figure 18.17, Lesch–Nyhan syndrome is due to an HGPRT enzyme deficiency, whereas adenosine deaminase (ADA) deficiency leads to a **severe combined immunodeficiency disease (SCID)** called ADA–SCID. Lesch–Nyhan syndrome is a rare recessive genetic disease that is characterized by unusual neurologic symptoms, including severe anxiety and self-mutilation. The human HGPRT gene

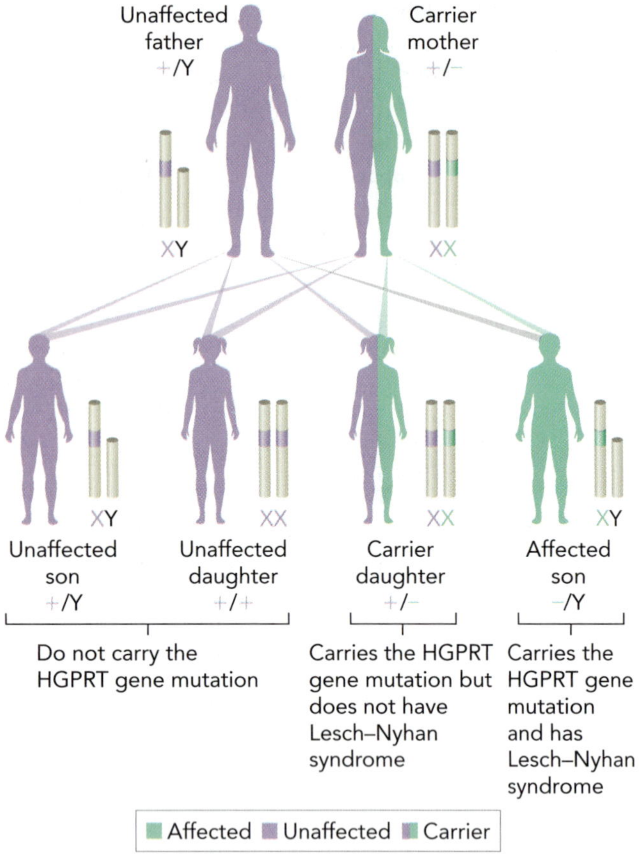

Figure 18.18 Lesch–Nyhan syndrome is an X-linked recessive genetic disease associated with neurologic disorders and gout. X-linked recessive genetic diseases such as Lesch–Nyhan syndrome affect males at a much higher frequency than females because males contain only a single X chromosome.

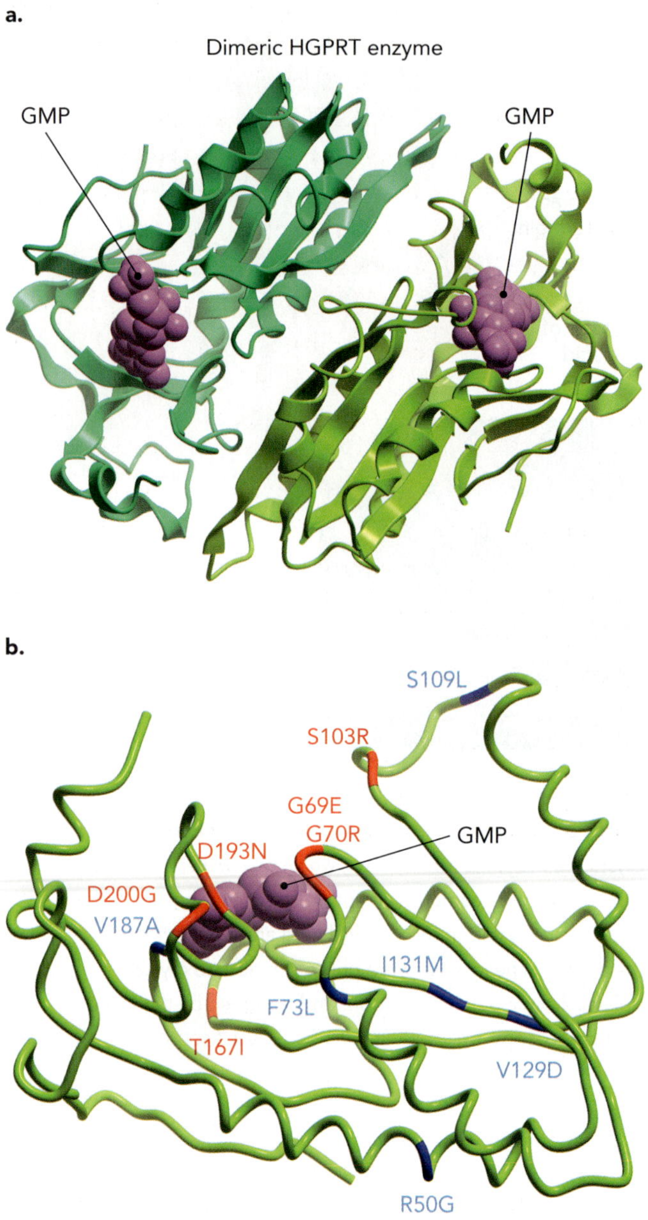

Figure 18.19 Mutations in the salvage enzyme HGPRT cause Lesch–Nyhan syndrome. **a.** Structure of an HGPRT dimer, showing GMP bound to the enzyme active sites. **b.** Two classes of mutations have been characterized in the human HGPRT gene: (1) mutations that decrease enzyme activity (red labels); and (2) mutations associated with decreased enzyme levels in the cell, owing to destabilization of the protein structure (blue labels). BASED ON PDB FILE 1HMP.

was genetically mapped to the X chromosome (Xq26.1) in the 1970s using low-resolution chromosome analysis methods. Later, in the 1990s, the gene was physically located on the X chromosome using high-throughput DNA sequencing. As shown in **Figure 18.18**, Lesch–Nyhan syndrome follows the inheritance pattern of an X-linked recessive genetic mutation. Unlike human genetic diseases caused by autosomal recessive mutations (see Figure 6.32), diseases caused by X-linked recessive mutations appear at a much higher frequency in males than in females because males contain only a single X chromosome. Indeed, 50% of sons born to a heterozygous mother carrying an HGRPT gene mutation on one X chromosome and a father having one normal X chromosome will have Lesch–Nyhan syndrome, whereas none of the daughters will be stricken with the disease.

The HGPRT enzyme is required to salvage hypoxanthine and guanine bases; however, in individuals with an HGPRT deficiency, the bases are converted to uric acid rather than being recycled. This explains the symptoms of gout in Lesch–Nyhan patients. The basis for the neurologic symptoms of Lesch–Nyhan syndrome are unclear but could be related to the dependence of brain cells on nucleotide salvage pathways. Without sufficient levels of nucleoside triphosphates to maintain DNA and RNA synthesis during childhood development, brain cell function would be compromised. Numerous mutations have been identified in the dimeric human HGPRT protein, all of which are linked to Lesch–Nyhan syndrome (**Figure 18.19**). In one class, the mutations are located near the enzyme active site and disrupt enzyme activity (G69E, G70R, S103R, T167I, D193N, D200G). A second class of mutations is those associated with decreased HGPRT enzyme levels in the cells as a result of protein destabilization and degradation (R50G, R73L, S109L, V129D, I131M, V187A).

ADA–SCID is a third human metabolic disease associated with purine degradation. It is caused by defects in the enzyme adenosine deaminase (see Figure 18.17). The primary symptom of ADA–SCID is a severely defective immune system resulting from low levels of functional B and T cells. The biochemical explanation for immunodeficiency in ADA–SCID patients is that defects in adenosine deaminase lead to adenosine accumulation. This excess adenosine goes into dATP production. Because dATP is an inhibitor of the enzyme ribonucleotide reductase, rapidly dividing B and T cells are starved of deoxyribonucleotides and fail to proliferate.

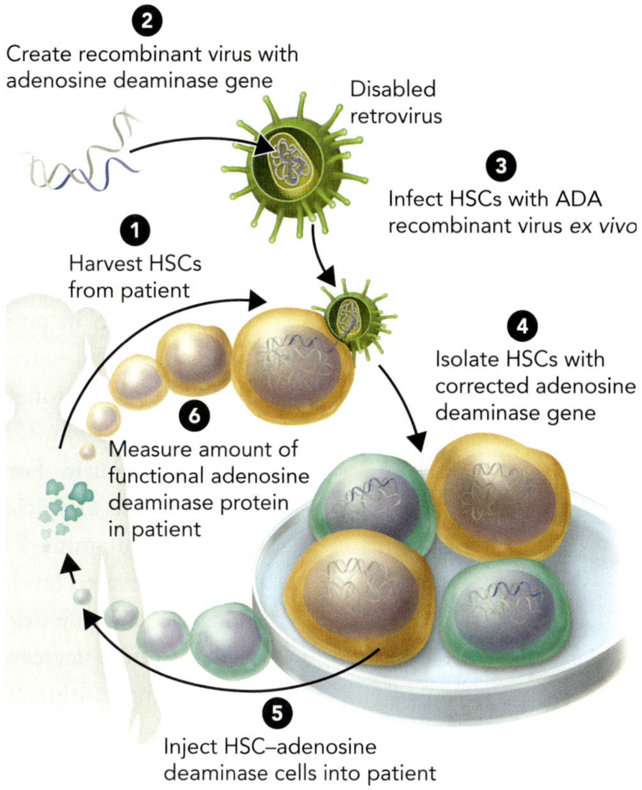
2
Create recombinant virus with adenosine deaminase gene
Disabled retrovirus
3
Infect HSCs with ADA recombinant virus *ex vivo*
1
Harvest HSCs from patient
4
Isolate HSCs with corrected adenosine deaminase gene
6
Measure amount of functional adenosine deaminase protein in patient
5
Inject HSC–adenosine deaminase cells into patient

Figure 18.20 Human gene therapy is an experimental treatment for ADA–SCID. A functional copy of the adenosine deaminase gene is inserted into a disabled, nonreplicating retrovirus, which is used to infect the patient's hematopoietic stem cells (HSCs). Cells containing the corrected adenosine deaminase gene are expanded in culture and then injected back into the ADA–SCID patient, where they continually produce functional adenosine deaminase protein.

Patients with ADA–SCID must initially be kept in isolation to avoid life-threatening infections and are given frequent intravenous infusions of immunoglobulins or recombinant adenosine deaminase enzyme. The only approved curative treatment for ADA–SCID is a bone marrow transplant, which usually requires a sibling match to avoid tissue rejection. An experimental treatment for ADA–SCID is human gene therapy using hematopoietic stem cells (HSCs). These cells are isolated from a patient and infected with a nonreplicating retrovirus that carries a functional copy of the adenosine deaminase gene (**Figure 18.20**). Healthy ADA-producing cells are identified in culture and then expanded in number and injected back into the patient. A successful treatment regimen using human gene therapy results in production of functional adenosine deaminase protein and the recovery of a disease-fighting immune system.

concept integration 18.2

Describe the biochemical basis for nutritional and genetic factors that contribute to gout. Why does allopurinol relieve gout symptoms regardless of the underlying cause?

Gout is caused by the accumulation of uric acid crystals in the joints and kidneys, which can be very painful when they impinge on sensory nerves. Normally, purines are degraded into uric acid, which is efficiently excreted; however, certain foods are naturally rich in purines, which leads to increased uric acid production and joint deposition. Moreover, excessive alcohol consumption has also been associated with gout symptoms, presumably because alcohol interferes with uric acid excretion in some way. Gout can also be caused by genetic deficiencies that affect uric acid metabolism. For example, increased levels of PRPP synthetase stimulate flux through the uric acid pathway, as does deregulation of feedback inhibition of the enzyme glutamine-PRPP amidotransferase. Lastly, defects in the enzyme HGPRT lead to elevated levels of hypoxanthine, which is then converted to uric acid by the enzyme xanthine oxidase. Allopurinol is a competitive inhibitor of xanthine oxidase and therefore decreases the rate of uric acid production regardless of the mechanism underlying xanthine accumulation.

18.3 Pyrimidine Metabolism

The *de novo* biosynthesis of pyrimidines requires fewer enzymatic reactions than that of purines, which is not surprising given that pyrimidines are less complex molecules than purines. As shown in **Figure 18.21**, the six atoms in the pyrimidine ring are derived from aspartate (C-1, C-4, C-5, C-6) and carbamoyl phosphate, which is itself generated from glutamine (N-3) and HCO_3^- (C-2). We first look at the six enzymatic reactions in the bacterial pyrimidine biosynthetic pathway leading to production of UMP, and then describe the conversion of UMP to UTP and CTP. Thymine, the third pyrimidine base, is generated from dUMP, which is discussed in Section 18.4 when we examine deoxyribonucleotide metabolism.

The Pyrimidine Biosynthetic Pathway Generates UMP

The pyrimidine biosynthetic pathway takes place in six steps: (1) Reaction 1 uses the enzyme carbamoyl phosphate synthetase II to generate carbamoyl phosphate from HCO_3^- and the amide group of glutamine (**Figure 18.22**). Note that this cytosolic enzyme is distinct from the mitochondrial enzyme carbamoyl phosphate synthetase I, which is a urea cycle enzyme (see Figure 17.31). (2) In reaction 2 of the pathway, aspartate transcarbamoylase (ATCase) combines carbamoyl phosphate with aspartate to form carbamoyl aspartate. (We described regulation of the *E. coli* ATCase enzyme by CTP and ATP in Chapter 7 as a classic example of allosteric control; see Figure 7.50.) (3) In reaction 3, a dehydration reaction catalyzed by dihydroorotase closes the pyrimidine ring to produce dihydroorotate. (4) Dihydroorotate is oxidized by dihydroorotate dehydrogenase in reaction 4 to yield the pyrimidine base orotate. (5) The attachment of orotate to ribose-5-P in reaction 5 is catalyzed by the enzyme orotate phosphoribosyl

Figure 18.21 The six atoms in the pyrimidine ring are all derived from either aspartate or carbamoyl phosphate, with carbamoyl phosphate being synthesized from glutamine and HCO_3^-.

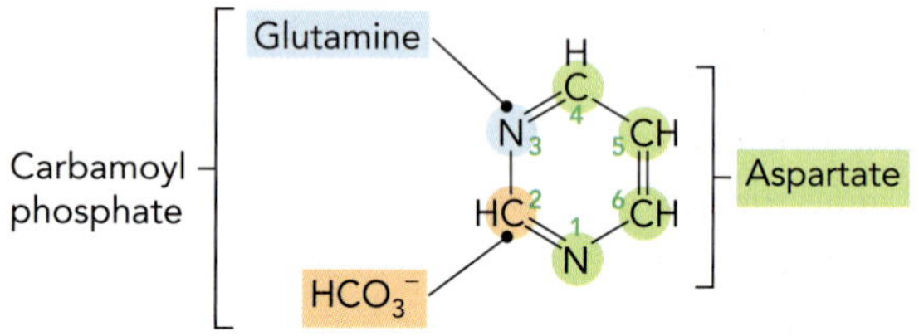

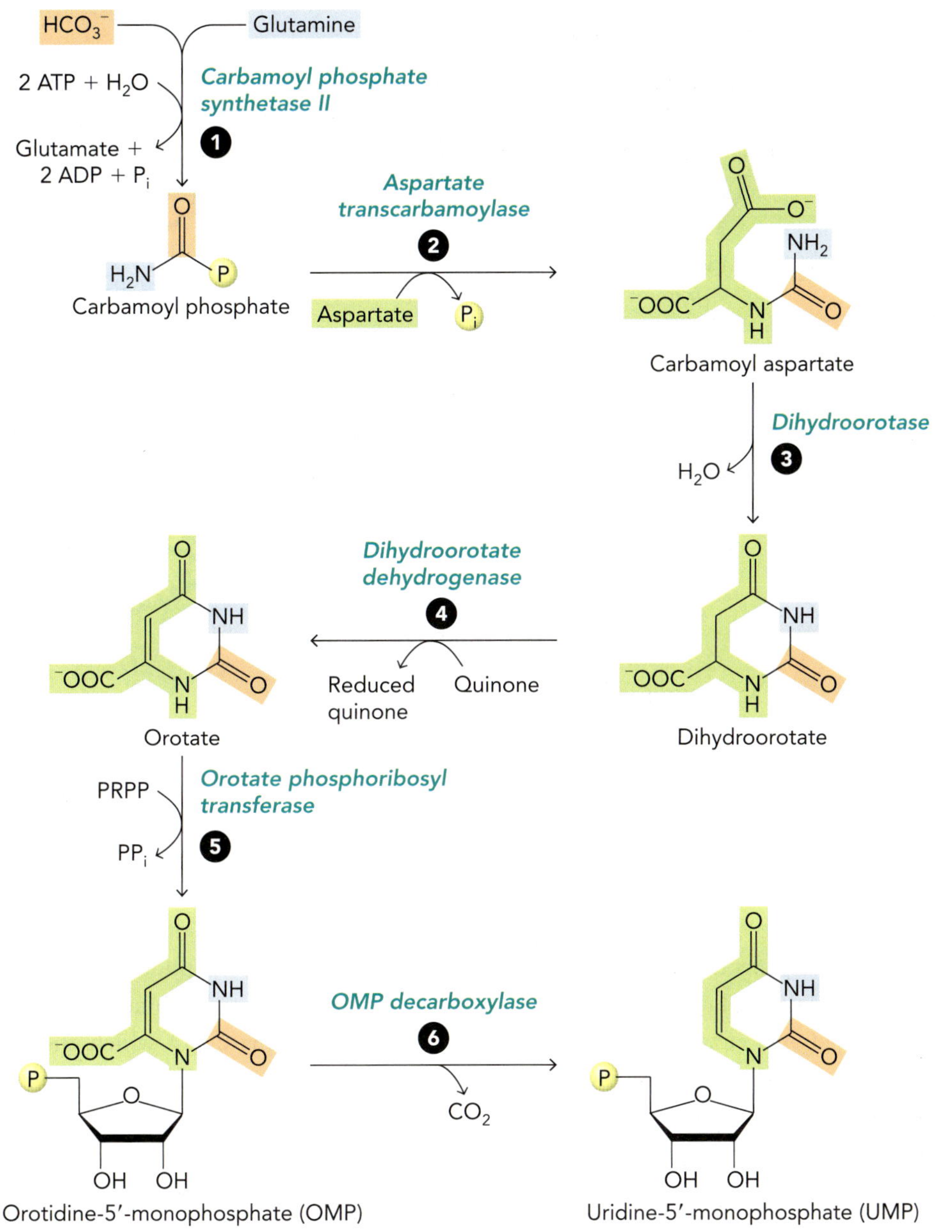

Figure 18.22 The pyrimidine biosynthetic pathway in *E. coli* consists of six reactions that use aspartate, glutamine, and HCO_3^- as substrates to generate the pyrimidine nucleotide UMP. The color coding of atoms in the pyrimidine ring indicates their origin from carbamoyl phosphate or aspartate.

transferase, leading to the formation of orotidine-5′-monophosphate (OMP). This reaction uses PRPP as the activated sugar and releases pyrophosphate, which is cleaved by pyrophosphatase to drive the reaction in the forward direction. Orotate phosphoribosyl transferase also functions as a salvage enzyme, recycling uracil and cytosine bases by reattaching them to the ribose sugar. (6) In the final reaction of the pathway, OMP is decarboxylated in reaction 6 by the enzyme OMP decarboxylase to generate the pyrimidine nucleotide UMP.

As shown in **Figure 18.23**, UMP is phosphorylated by a UMP kinase to generate uridine-5′-diphosphate (UDP). The UDP is then phosphorylated by nucleoside diphosphate kinase to produce UTP. Replacement of the oxygen on C-4 of the UTP pyrimidine ring with an amino group from glutamine generates CTP in a reaction catalyzed by the enzyme CTP synthetase. This amidation reaction uses glutamine as the source of nitrogen in most species; however, bacteria use NH_4^+ directly.

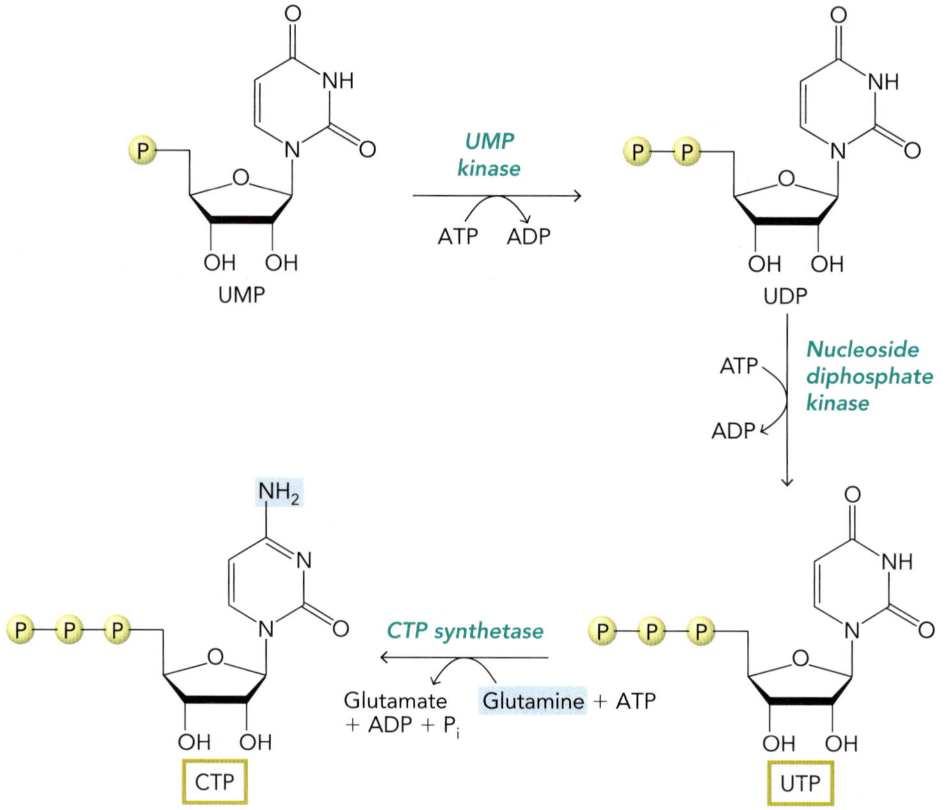

Figure 18.23 UMP is converted to UTP by sequential phosphorylation reactions. The UTP is then aminated by CTP synthetase to generate CTP.

Similar to the purine biosynthetic pathway, *E. coli* pyrimidine biosynthetic enzymes are encoded in the bacterial genome as separate polypeptides, whereas in most animals (including humans), several catalytic activities are combined into large multifunctional enzymes. One example is the human enzyme CAD, which encodes all three catalytic activities contained in the bacterial enzymes carbamoyl phosphate synthetase II (reaction 1), aspartate transcarbamoylase (reaction 2), and dihydroorotase (reaction 3). In addition, the bifunctional enzyme UMP synthase in animals encodes the catalytic activities corresponding to the *E. coli* enzymes orotate phosphoribosyl transferase (reaction 5) and OMP decarboxylase (reaction 6). Dihydroorotate dehydrogenase, the enzyme catalyzing reaction 4 in the pyrimidine biosynthetic pathway, is a single polypeptide in both bacterial and animal cells.

Allosteric Regulation of Pyrimidine Biosynthesis

Pyrimidine biosynthesis is regulated by both feedback inhibition and allosteric activation in bacterial and animal cells; however, the allosteric effectors and target enzymes are quite different in each case, as illustrated in **Figure 18.24**. The *E. coli* ATCase enzyme is controlled by heterotropic allosteric regulation and serves as the primary regulated enzyme in the bacterial pyrimidine biosynthetic pathway. As described in Chapter 7, ATP is an allosteric activator of ATCase, whereas CTP is an allosteric inhibitor. ATP and CTP bind to the same allosteric effector site in the regulatory subunit of the bacterial enzyme.

In contrast, pyrimidine biosynthesis in animal cells is regulated at both the first and last steps in the pathway. As shown in Figure 18.24, CTP, UTP, and UDP are all negative allosteric regulators of the carbamoyl phosphate synthetase II activity

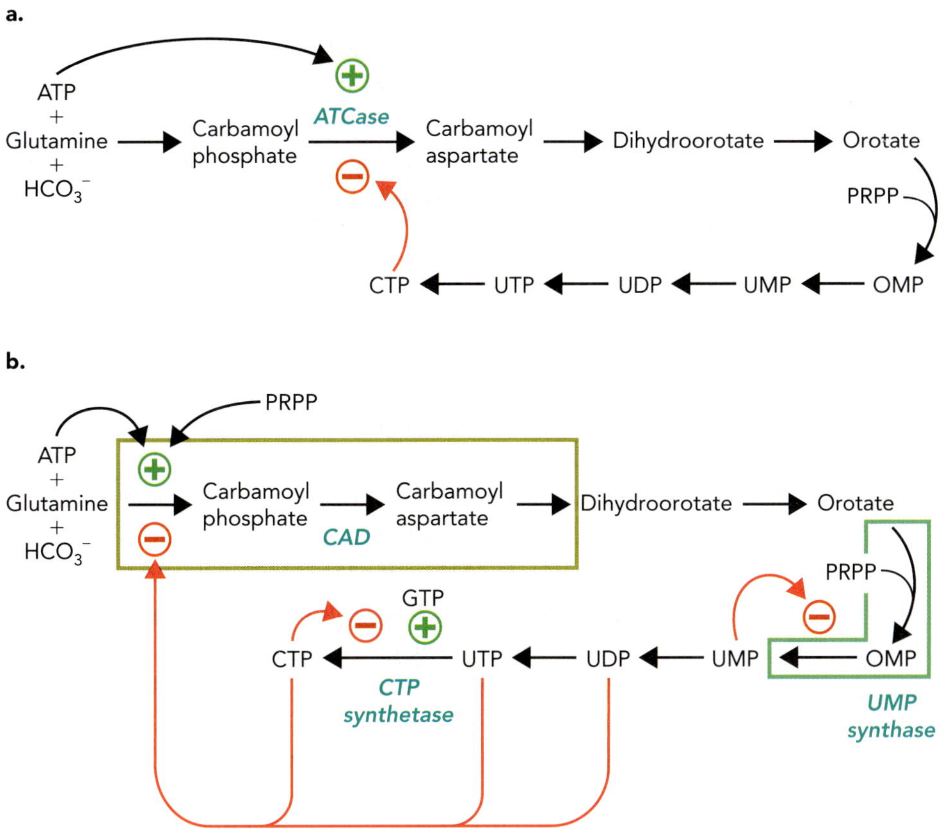

Figure 18.24 The pyrimidine biosynthetic pathways in *E. coli* and animal cells are controlled by positive and negative regulation. **a.** ATCase is the key regulated enzyme in the pyrimidine biosynthetic pathway in *E. coli* cells, being activated by ATP and inhibited by CTP. **b.** Flux through the pyrimidine biosynthetic pathway in animal cells is controlled by allosteric regulation of the CAD enzyme, UMP synthase, and CTP synthetase. The boxes show the corresponding enzymatic activities in *E. coli* that are encoded by the CAD and UMP synthase genes.

in the trifunctional CAD enzyme, whereas PRPP and ATP are allosteric activators. The bifunctional UMP synthase enzyme in animal cells is feedback inhibited by UMP, whereas CTP synthetase is feedback inhibited by CTP. The CTP synthetase enzyme in animal cells is also allosterically activated by GTP, which makes sense because this ensures that the GTP and CTP nucleotide pools are balanced to support ongoing DNA synthesis.

Pyrimidines Are Degraded by a Common Pathway

The pyrimidine nucleotides UMP, CMP, and deoxythymidine-5′-monophosphate (dTMP) are degraded by a common three-reaction pathway. This pathway converts uracil and thymine into β-alanine and β-aminoisobutyrate, respectively.

As shown in **Figure 18.25**, CMP, UMP, and dTMP are first dephosphorylated by 5′-nucleotidase. Then, the ribose sugar is removed by pyrimidine phosphorylases to generate the free bases (cytidine is first converted to uridine by cytidine deaminase). In the first reaction of the common degradation pathway, dihydropyrimidine dehydrogenase (sometimes abbreviated as DPD) catalyzes the reduction of uracil to dihydrouracil and of thymine to dihydrothymine. Dihydropyrimidine dehydrogenase is the rate-limiting enzyme in the pyrimidine degradation pathway. In the next reaction, dihydropyrimidinase catalyzes a hydrolytic cleavage that opens the pyrimidine ring, generating *N*-carbamoyl-β-alanine (uracil degradation) and *N*-carbamoyl-β-aminoisobutyrate (thymine degradation). In the final reaction, β-ureidopropionase converts *N*-carbamoyl-β-alanine to β-alanine, NH_4^+, and HCO_3^-, whereas *N*-carbamoyl-β-aminoisobutyrate is converted to β-aminoisobutyrate, NH_4^+, and HCO_3^-. The amino groups of β-alanine and β-aminoisobutyrate are transferred to α-ketoglutarate

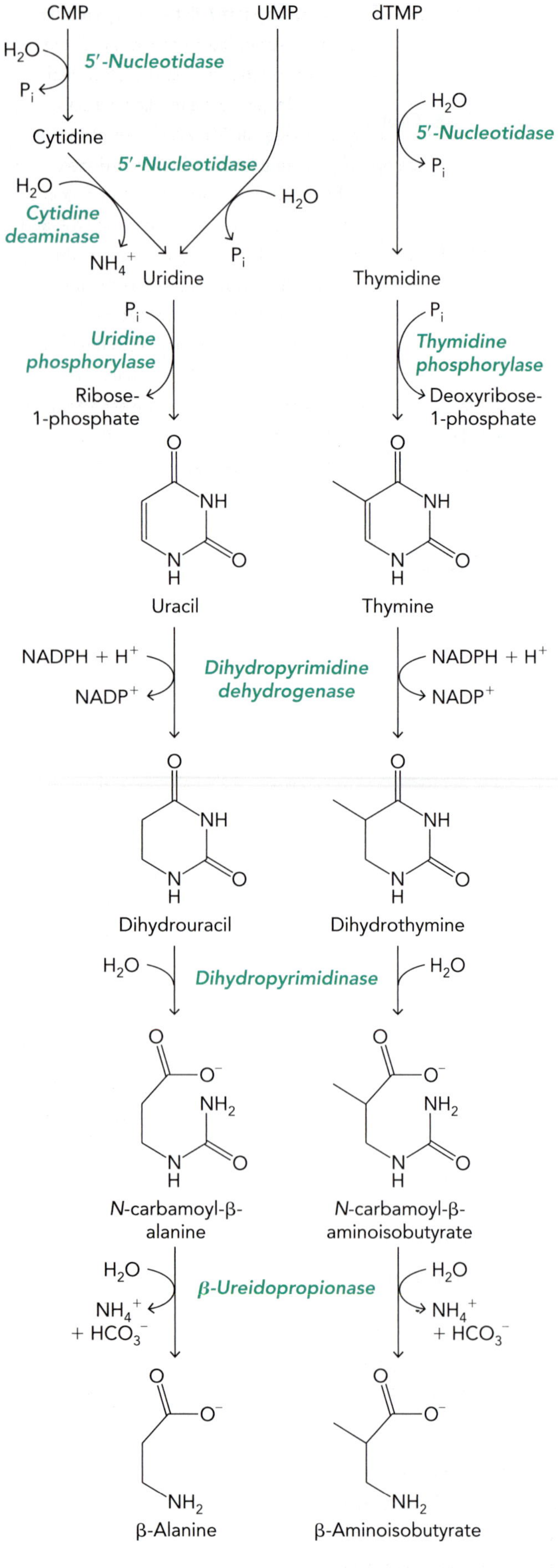

Figure 18.25 A common pathway degrades CMP, UMP, and dTMP into NH_4^+, HCO_3^-, and either β-alanine (uracil degradation) or β-aminoisobutyrate (thymine degradation).

by liver aminotransferase enzyme, generating glutamate and malonic semialdehyde and methylmalonic semialdehyde, respectively, which are converted to CoA derivatives and metabolized in other pathways.

A high prevalence of dihydropyrimidine dehydrogenase deficiencies in the human population (~5%) was discovered by accident when it was observed that some cancer patients cannot tolerate aggressive treatment with the anticancer drug **5-fluorouracil**. As described in the next section, 5-fluorouracil is a potent inhibitor of the thymidylate synthesis reaction, which is required to convert dUMP to dTMP and thereby provide deoxynucleotides for DNA synthesis. Although treatment with 5-fluorouracil can be quite effective in blocking DNA synthesis in rapidly dividing cancer cells, it also blocks DNA synthesis in normal cells, such as intestinal epithelial cells and hair follicle cells. After reports that several cancer patients died following 5-fluorouracil treatment, it was determined that the underlying cause of this toxic effect was defects in the enzyme dihydropyrimidine dehydrogenase. Indeed, much like the drug-induced hemolytic anemia observed in individuals with glucose-6-phosphate dehydrogenase deficiencies who take primaquine to prevent malaria (see Figure 14.10), 5-fluorouracil treatment can be more harmful to cancer patients with dihydropyrimidine dehydrogenase deficiencies than the cancer it is intended to treat.

The dimeric dihydropyrimidine dehydrogenase enzyme is shown in **Figure 18.26**. You can see that the NADPH binding site is nearly 50 Å away from the enzyme active site, which in this model contains a bound 5-fluorouracil molecule. A proposed electron transfer path consisting of four 4 Fe–4 S clusters is thought to connect the two functional sites in the enzyme. Pharmacokinetic studies have shown that up to 80% of the 5-fluorouracil used in chemotherapy is degraded by dihydropyrimidine dehydrogenase in the liver. Because of this, cancer patients are given high doses of 5-fluorouracil as a way to compensate for this metabolic loss. When individuals with a dihydropyrimidine dehydrogenase deficiency are unknowingly given these same high doses of 5-fluorouracil, dangerous tissue toxicity arises because most of the 5-fluorouracil remains in the circulatory system rather than being degraded. Cancer patients being considered for 5-fluorouracil treatment are now routinely screened for dihydropyrimidine dehydrogenase deficiencies.

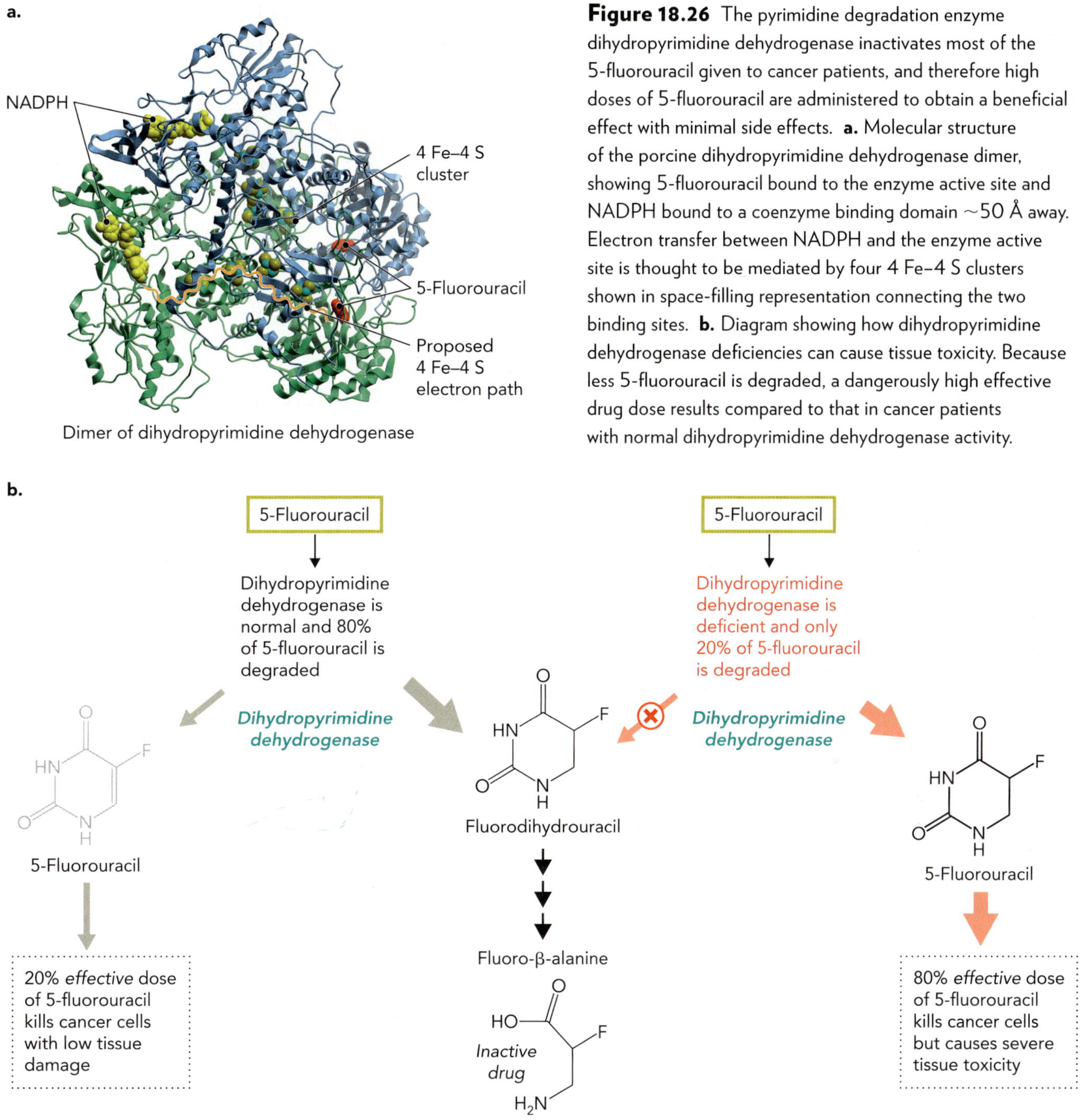

Figure 18.26 The pyrimidine degradation enzyme dihydropyrimidine dehydrogenase inactivates most of the 5-fluorouracil given to cancer patients, and therefore high doses of 5-fluorouracil are administered to obtain a beneficial effect with minimal side effects. **a.** Molecular structure of the porcine dihydropyrimidine dehydrogenase dimer, showing 5-fluorouracil bound to the enzyme active site and NADPH bound to a coenzyme binding domain ~50 Å away. Electron transfer between NADPH and the enzyme active site is thought to be mediated by four 4 Fe–4 S clusters shown in space-filling representation connecting the two binding sites. **b.** Diagram showing how dihydropyrimidine dehydrogenase deficiencies can cause tissue toxicity. Because less 5-fluorouracil is degraded, a dangerously high effective drug dose results compared to that in cancer patients with normal dihydropyrimidine dehydrogenase activity.

concept integration 18.3

Describe two trifunctional enzymes in humans that catalyze nucleotide biosynthetic reactions. What is the likely explanation for the existence of these multifunctional enzymes in animal cells but not in bacterial cells?

In the human purine biosynthetic pathway, the trifunctional enzyme hTrifGART includes three catalytic activities encoded in the orthologous *E. coli* enzymes GAR synthetase, GAR transformylase, and AIR synthetase. Similarly, in the human pyrimidine biosynthetic pathway, the trifunctional enzyme CAD includes three catalytic activities encoded in the orthologous *E. coli* enzymes carbamoyl phosphate synthetase II, aspartate transcarbamoylase, and dihydroorotase. A likely explanation for the existence of these multifunctional enzymes in humans is that gene fusions occurred over evolutionary time as a consequence of chromosomal recombination, which gave rise to more efficient enzymes. For example, CAD encodes the first three catalytic activities in the pyrimidine biosynthetic pathway, which use glutamine and HCO_3^- as the initiating substrates to produce the end product dihydroorotate, a closed pyrimidine ring. Because reactants are less likely to diffuse away in a multifunctional enzyme than in a protein complex consisting of individual subunits, the trifunctional enzymes may be more efficient.

18.4 Deoxyribonucleotide Metabolism

We now turn our attention to the synthesis of deoxyribonucleotides, the building blocks of DNA. As described earlier, DNA differs from RNA in two important ways. First, the ribose sugar in DNA lacks the hydroxyl group at C-2′. This prevents spontaneous cleavage of the phosphodiester backbone through formation of 2′,3′-cyclic phosphate, as occurs in RNA (see Figure 3.21). Second, thymine replaces uracil as the pyrimidine base that forms hydrogen bonds with the purine base adenine. This feature also has important evolutionary implications because it ensures that DNA repair mechanisms in the cell quickly identify and remove uracil bases resulting from spontaneous cytosine deamination.

Generation of Deoxyribonucleotides by Ribonucleotide Reductase

Deoxyribonucleotides are derived from ribonucleotides by a single reaction catalyzed by the enzyme **ribonucleotide reductase**. The reaction substrates are any of the four nucleoside 5′-diphosphates (GDP, ADP, CDP, UDP) and NADPH + H^+. As shown below, the products of the reaction are H_2O and the corresponding deoxynucleoside 5′-diphosphates (dGDP, dADP, dCDP, dUDP):

$$\text{NADPH} + \text{H}^+ + \text{Nucleoside 5}'\text{-diphosphate} \rightarrow \text{H}_2\text{O} + \text{Deoxynucleoside 5}'\text{-diphosphate} + \text{NADP}^+$$

Ribonucleotide reductase is a highly conserved enzyme consisting of an all-α-helical structure that is structurally similar in all organisms. This ancient enzyme likely played a pivotal role in converting an RNA world, in which RNA was the nucleic acid blueprint for life, into the present-day DNA world. The idea that RNA preceded DNA as the genetic basis of life was first proposed by Carl Woese in 1967. Support for the **RNA world** hypothesis came in the 1980s when Thomas Cech and Sidney Altman discovered that RNA molecules can function as biochemical catalysts, called ribozymes, that are chemically distinct from protein-based enzymes. It was later found that ribosomes, the protein-synthesizing machines in cells, are also ribozymes, which confirmed the central role of RNA in peptide bond formation.

As shown in **Figure 18.27**, the RNA world hypothesis proposes that RNA was the original blueprint for life, and that proteins were synthesized from amino acids using RNA molecules as both the template and the biochemical catalysts for protein synthesis. When one of these proteins evolved to become the enzyme ribonucleotide

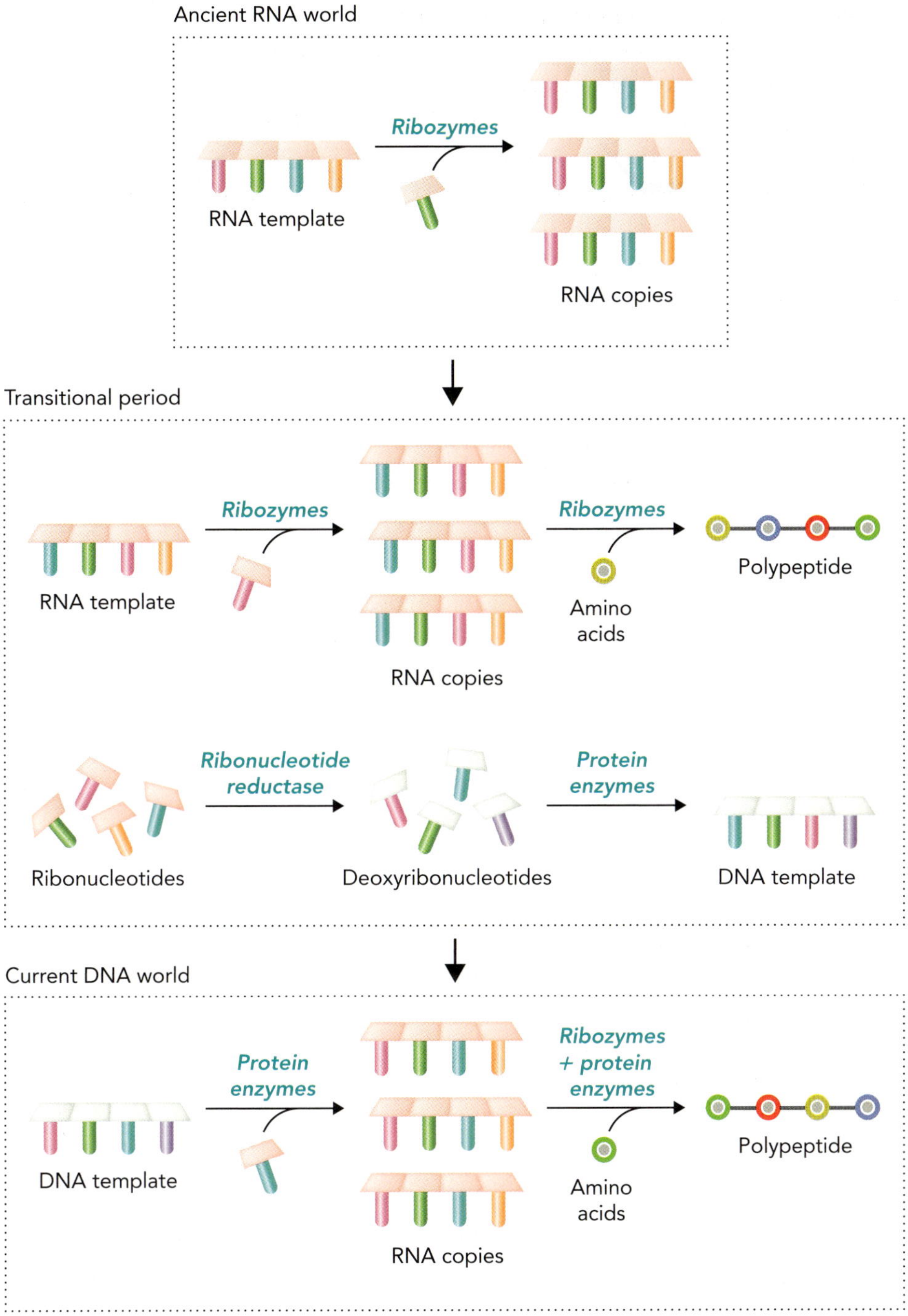

Figure 18.27 The RNA world hypothesis proposes that RNA was the first genetic molecule and that ribozymes were the biochemical catalysts for RNA and protein synthesis in cells. After the emergence of ribonucleotide reductase, deoxyribonucleotides were polymerized into DNA, which is better suited to function as genetic material. Note that even in our current DNA world, ribozymes play a key role in RNA metabolism (see Chapter 21) and protein synthesis (see Chapter 22).

reductase, it allowed for the transition from an RNA world to a DNA world in a relatively short period of time. The switch from ribonucleotides to deoxyribonucleotides as the building blocks of genetic information greatly enhanced evolutionary processes owing to the improved chemical properties of double-stranded DNA compared to those of single-stranded RNA (see Chapter 3).

The reduction of C-2′ on nucleoside diphosphates by ribonucleotide reductase requires the input of two electrons derived from NADPH. These electrons are used to reduce a pair of sulfhydryl groups in the enzyme active site. As shown in **Figure 18.28**, the reduction of these sulfhydryls in ribonucleotide reductase is not

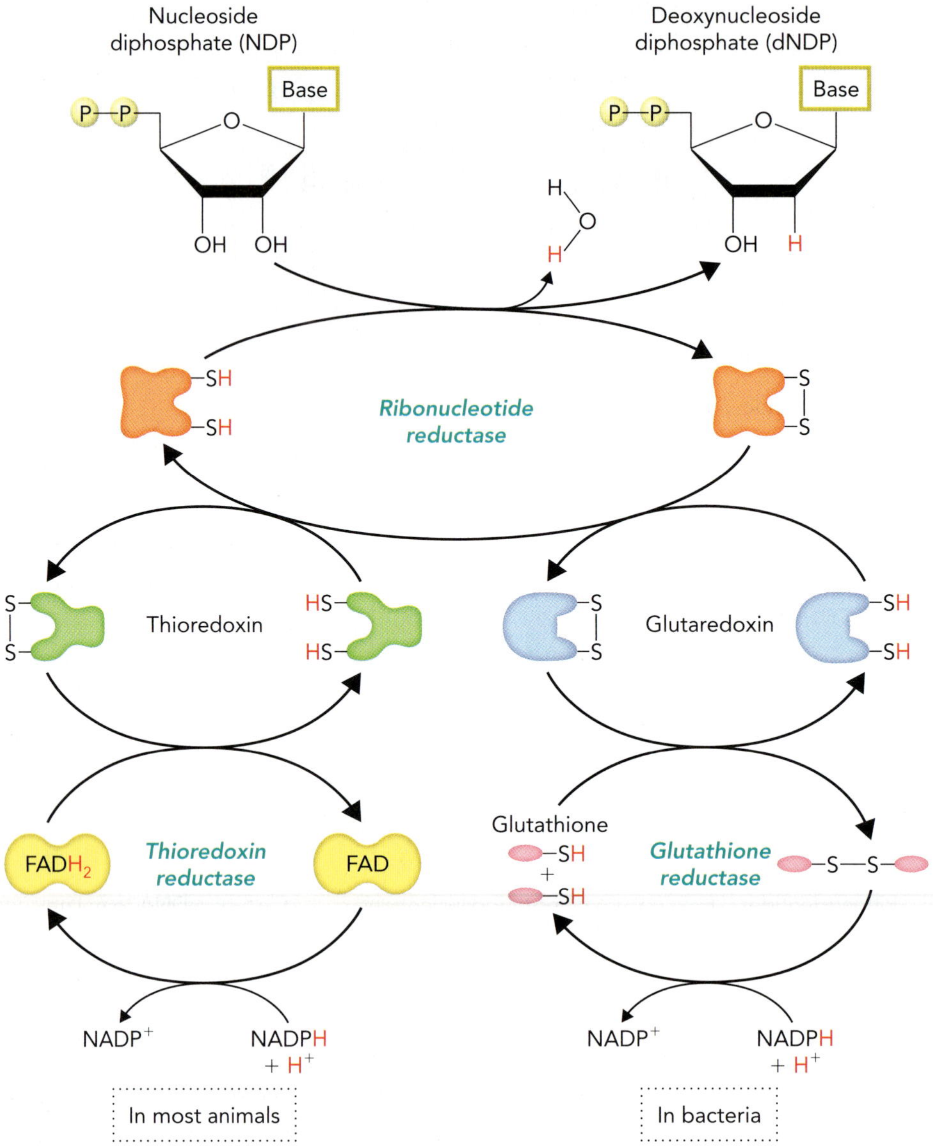

Figure 18.28 Two sulfhydryl groups in the ribonucleotide reductase active site need to be reduced to form deoxynucleoside diphosphates from nucleoside diphosphates. To regenerate the active form of ribonucleotide reductase, two electrons are extracted from NADPH through a series of redox reactions involving either thioredoxin or glutaredoxin. Note that a nucleoside diphosphate may also be called a ribonucleoside diphosphate.

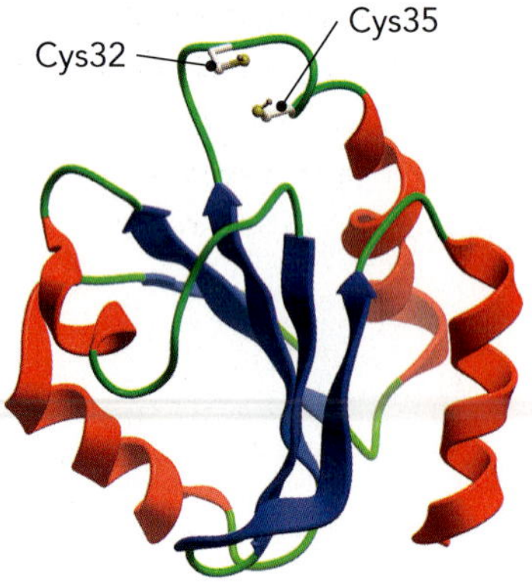

Reduced human thioredoxin

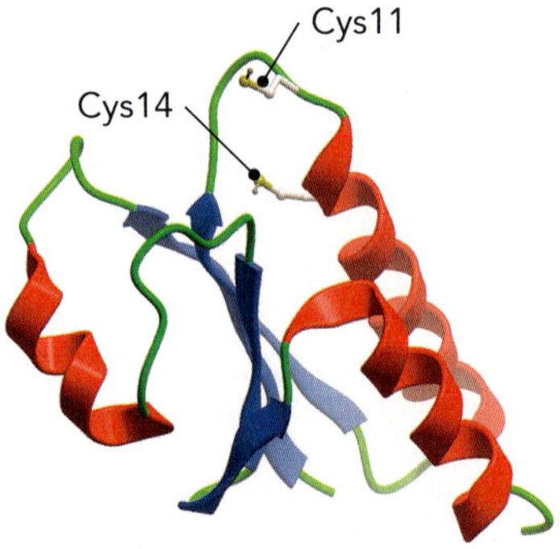

Reduced *E. coli* glutaredoxin

Figure 18.29 The molecular structures of reduced human thioredoxin and reduced *E. coli* glutaredoxin have been obtained by NMR spectroscopy. The two sulfhydryl-containing cysteines in the active site of each redox protein are labeled. BASED ON PDB FILES 3TRX (THIOREDOXIN) AND 1EGR (GLUTAREDOXIN).

done by NADPH directly, but rather through a redox circuit requiring intermediary proteins. In most organisms, the oxidized ribonucleotide reductase enzyme is reduced by the abundant cellular redox protein thioredoxin, which also regulates the activity of photosynthetic enzymes in response to light (see Figure 12.52). Thioredoxin is reduced by the enzyme thioredoxin reductase, an FAD-containing enzyme that is itself reduced by NADPH. Bacterial cells reduce ribonucleotide reductase by a different redox mechanism involving the protein **glutaredoxin**. Glutaredoxin is reduced by glutathione, the substrate for glutathione reductase (Figure 18.28). The molecular structures of human thioredoxin and *E. coli* glutaredoxin proteins are shown in **Figure 18.29**. Observe that the overall tertiary structures and relative positions of the active-site sulfhydryls are similar between the two proteins.

Ribonucleotide reductase contains two subunits, R1 and R2, that function as a tetrameric complex ($R1_2R2_2$). The catalytic mechanism in *E. coli* requires contributions from amino acids in both the R1 and R2 subunits, as well as a dinuclear Fe^{3+} iron center coordinated by an oxide ion (O_2^-). As illustrated in

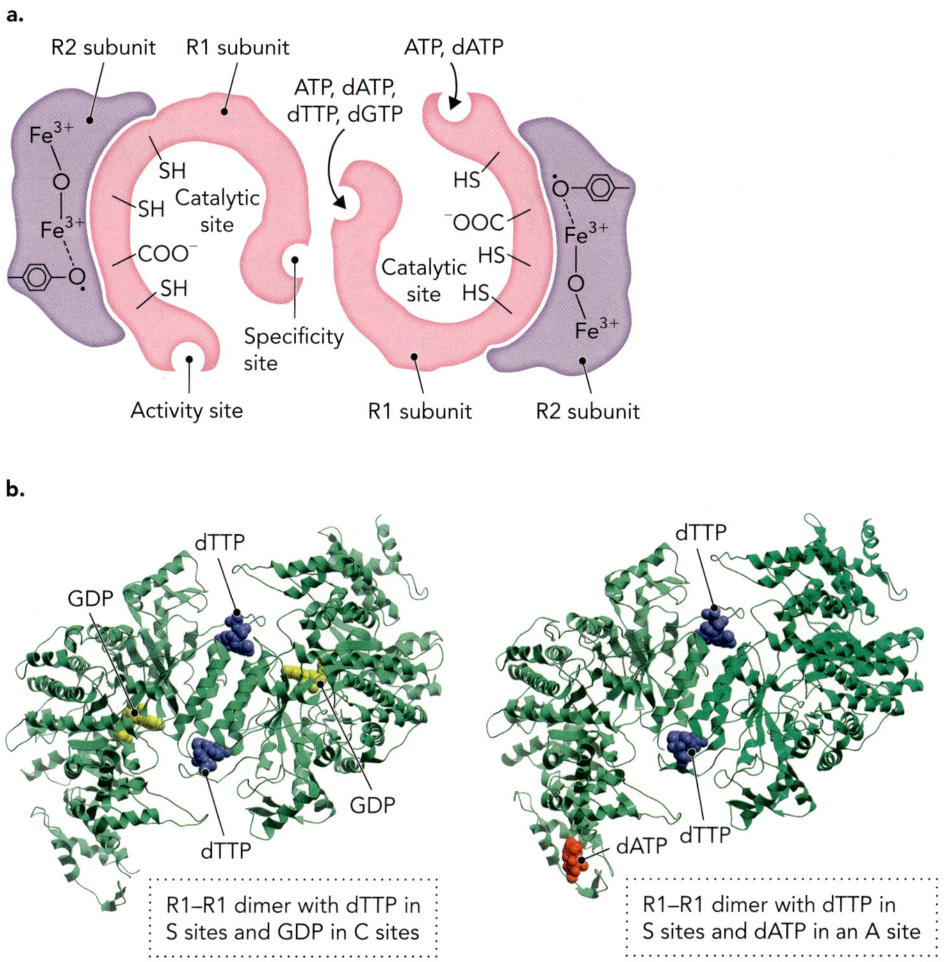

Figure 18.30 The ribonucleotide reductase enzyme consists of two R1 regulatory subunits and two R2 subunits. Each R2 subunit contains a dinuclear Fe^{3+} center. **a.** Schematic representation of the $R1_2R2_2$ tetrameric complex, showing the allosteric effectors that bind to the substrate specificity site (S site) and those that bind to the activity site (A site). The reduction reaction takes place in the catalytic site (C site) and requires three cysteine residues and a glutamate residue in the R1 subunit and the dinuclear Fe^{3+} center in the R2 subunit. **b.** Molecular structures of human ribonucleotide reductase R1 dimers with dTTP bound to the two S sites and either GDP bound to the C sites or dATP bound to one of the two A sites.
BASED ON PDB FILES 3HND AND 3HNF.

Figure 18.30a, the R1 subunit contains two types of allosteric effector sites: one that determines substrate specificity (S site) and another that controls enzyme activity (A site). We will look at the function of these allosteric sites in more detail shortly. The iron center in the R2 subunit creates a free radical at a nearby tyrosine residue, which in turn abstracts an electron from a cysteine residue in the R1 subunit to create a free radical in the enzyme catalytic site (C site). Two other cysteine residues and a glutamate residue in the R1 subunit also have crucial roles in the reduction reaction.

The ribonucleotide reductase enzyme in most eukaryotes and aerobic bacteria is an example of a class I ribonucleotide reductase. All such enzymes contain a dinuclear Fe^{3+} center in the R2 subunit. Other prokaryotes contain either a class II ribonucleotide reductase with a cobalt reactive center or a class III enzyme with a 4 Fe–4 S center. **Figure 18.30b** shows the molecular structure of a human R1 ribonucleotide reductase dimer with two dATP molecules bound at the regulatory sites.

The ribonucleotide reductase mechanism is unusual in that it depends on the formation of a free radical to catalyze the reaction. As shown in **Figure 18.31**, three cysteine residues in the R1 subunit play a role in the catalytic reaction. Two of these are directly involved in the reduction reaction at C-2′, and a third (Cys439) functions as a free radical group in the reaction mechanism. Also, an important glutamate residue (Glu441) in the R1 subunit stabilizes the substrate–enzyme interactions by forming a hydrogen bond with the hydroxyl group on C-3′.

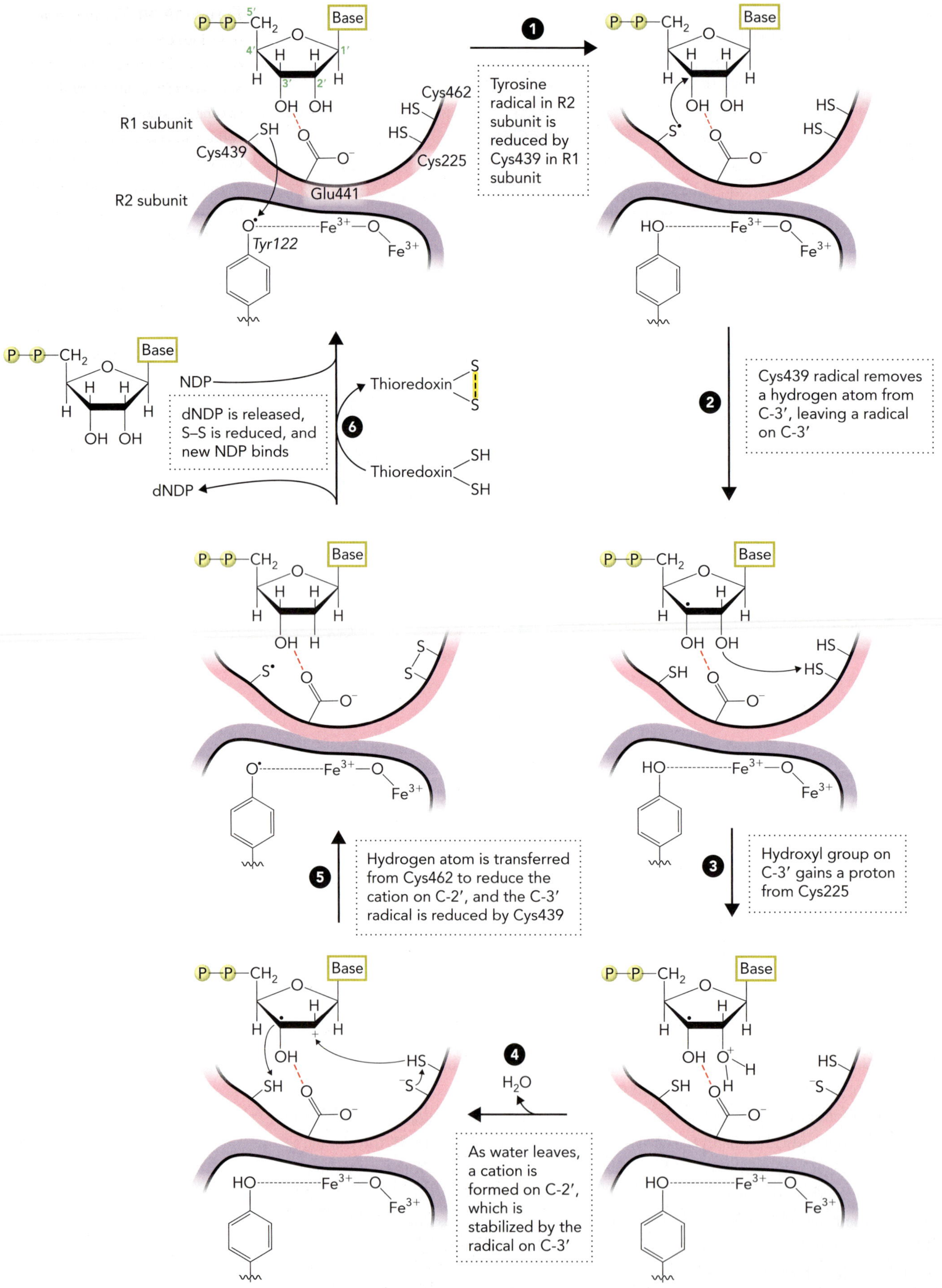

Figure 18.31 Proposed mechanism for the *E. coli* ribonucleotide reductase reaction. See text for details.

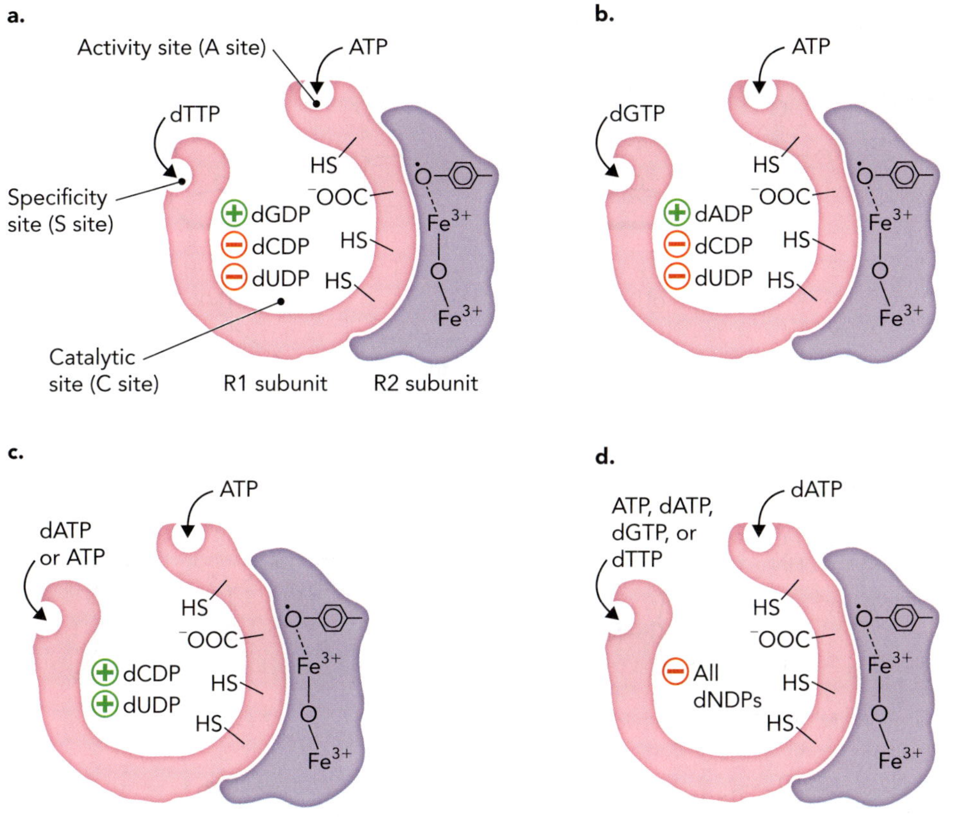

Figure 18.32 The pool size of various deoxynucleotides is controlled by effector binding to the specificity and activity sites of ribonucleotide reductase. **a–c.** Relative increases or decreases in the size of deoxynucleoside diphosphate pools are regulated by allosteric effector binding to the specificity sites, as long as ATP is bound to the activity site. **d.** When dATP is bound to the activity site, then ribonucleotide reductase activity is inhibited under all conditions.

In the proposed reaction mechanism for *E. coli* ribonucleotide reductase shown in Figure 18.31, in step 1 a tyrosyl free radical at Tyr122 abstracts a hydrogen atom from the Cys439 sulfhydryl, leading to the formation of a cysteine thiyl radical in the enzyme active site. In step 2, this thiyl free radical removes a hydrogen from C-3′ of the nucleoside diphosphate, generating a radical at the C-3′ position. In steps 3 and 4, the free radical on C-3′ promotes the removal of the hydroxyl group on C-2′ as a water molecule through an acid-catalyzed reaction involving the Cys225 sulfhydryl group. In step 5, the radical cation on C-2′ is reduced by the Cys462 sulfhydryl, creating the C-2′ deoxynucleotide and a Cys225–Cys462 disulfide bridge in the enzyme. In addition, the radical at the C-3′ position is reduced by retrieving the hydrogen atom back from the reduced Cys439 residue. Finally, in step 6, the deoxynucleoside diphosphate (dNDP) is released from the active site, and the Cys225–Cys462 disulfide group is reduced by thioredoxin or glutaredoxin to regenerate the enzyme active site.

How does the cell regulate the substrate specificity of ribonucleotide reductase to maintain equal amounts of all four deoxynucleoside triphosphates (dNTPs)? The answer is by allosteric regulation of enzyme specificity through binding of dTTP, dGTP, dATP, or ATP to the substrate specificity site in the R1 subunit. As shown in **Figure 18.32a**, when dTTP is bound to the specificity site, more dGDP is produced, whereas less dCDP and dUDP are generated. Similarly, binding of dGTP to the specificity site increases the pool of dADP in the cell but decreases the pools of dCDP and dUDP (**Figure 18.32b**). **Figure 18.32c** shows that when ATP or dATP is bound to the specificity site, then the levels of dCDP and dUDP are both increased.

In addition to substrate specificity, the overall activity of ribonucleotide reductase is regulated by an allosteric effector binding to the activity site. In this case, ATP binding to the activity site stimulates ribonucleotide reductase activity (Figure 18.32a–c),

Figure 18.33 Thymidylate synthase catalyzes a methylation reaction that converts deoxyuridine-5′-monophosphate (deoxyuridylate; dUMP) into deoxythymidine-5′-monophosphate (deoxythymidylate; dTMP) using the coenzyme N^5,N^{10}-methylenetetrahydrofolate as the methyl donor. Regeneration of N^5,N^{10}-methylenetetrahydrofolate requires the enzymes dihydrofolate reductase and serine hydroxymethyltransferase.

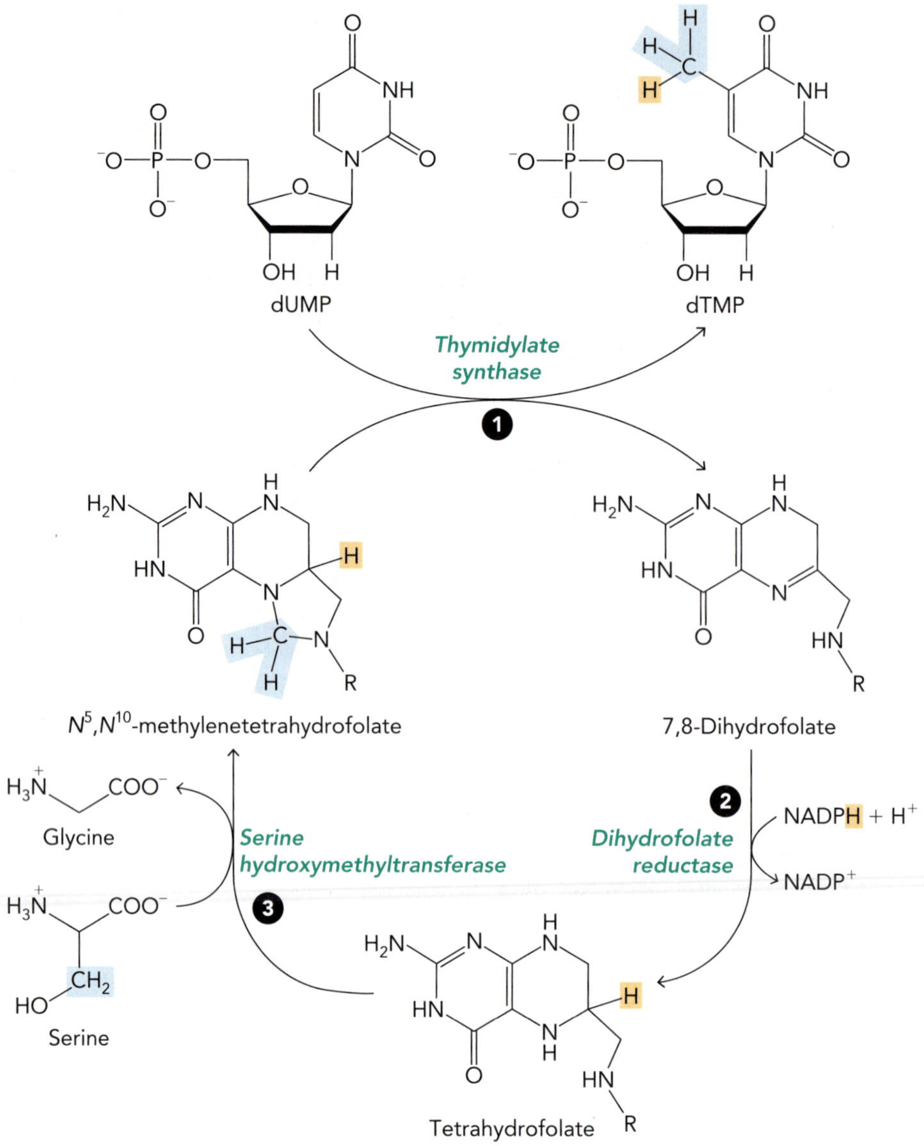

whereas dATP binding to this site inhibits enzyme activity (**Figure 18.32d**). Inhibition of ribonucleotide reductase activity by dATP binding to the activity site is the biochemical basis for severe combined immunodeficiency disease (SCID), which is the result of mutations in the adenine deaminase gene (see Figure 18.17).

Metabolism of Thymine Deoxyribonucleotides

Thymine is a pyrimidine base formed by methylation of uracil on C-5 in a reaction catalyzed by the enzyme thymidylate synthase. As shown in **Figure 18.33**, this enzyme reaction converts dUMP (deoxyuridylate) to dTMP (deoxythymidylate, or simply thymidylate; see Table 3.1) through a mechanism involving a C_1 transfer from the coenzyme N^5,N^{10}-methylenetetrahydrofolate. The resulting oxidization of N^5,N^{10}-methylenetetrahydrofolate produces **dihydrofolate**, which is then reduced by the enzyme dihydrofolate reductase (DHFR) to form **tetrahydrofolate**. To regenerate N^5,N^{10}-methylenetetrahydrofolate, the C_1 unit is replaced using serine in a reaction catalyzed by the enzyme serine hydroxymethyltransferase.

As shown in **Figure 18.34**, dUMP in *E. coli* cells is derived from one of three sources: (1) dephosphorylation of dUTP by the enzyme dUTP diphosphohydrolase;

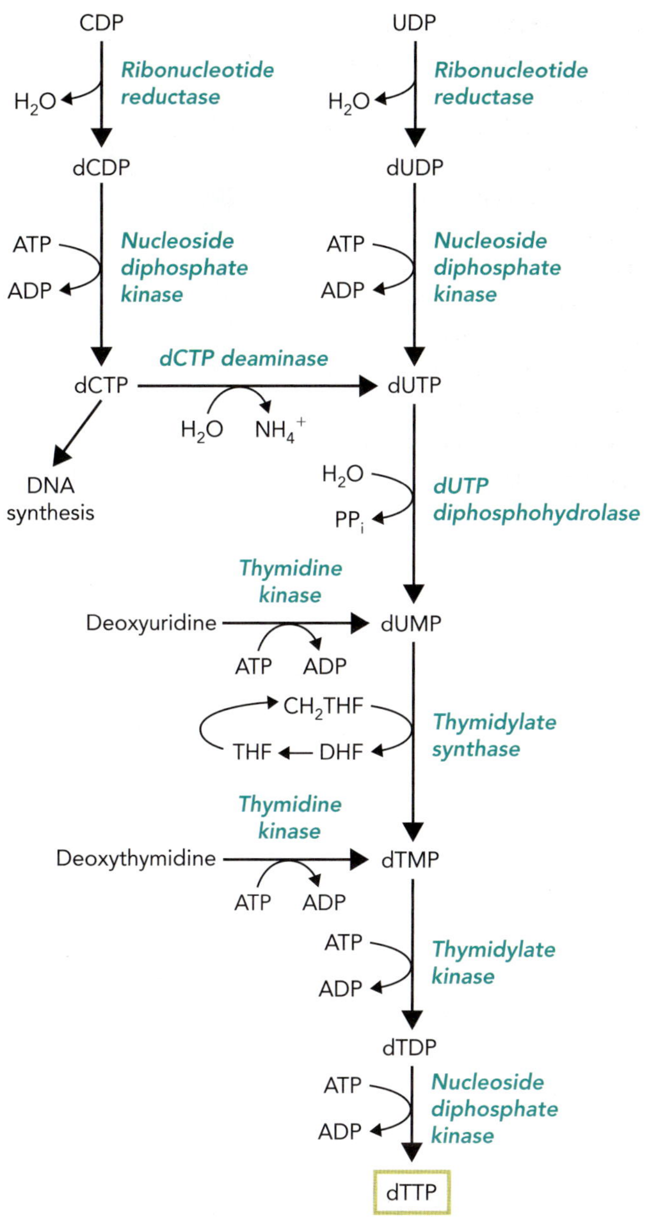

Figure 18.34 The three sources of dUMP for the thymidylate synthase reaction in *E. coli* are dCDP, dUDP, and deoxyuridine. CH_2THF = N^5,N^{10}-methylenetetrahydrofolate; DHF = dihydrofolate; THF = tetrahydrofolate.

(2) deamination of dCTP by dCTP deaminase to generate dUTP, which again is converted to dUMP by dUTP diphosphohydrolase; or (3) phosphorylation of deoxyuridine by the enzyme thymidine kinase to yield dUMP. Efficient dephosphorylation of dUTP by dUTP diphosphohydrolase is an important reaction in the cell because it prevents dUTP from accumulating and mistakenly being incorporated into DNA. Thymidine kinase is a salvage enzyme that can recognize both deoxyuridine and deoxythymidine (thymidine) as substrates. The dTMP product of the thymidylate synthase reaction is phosphorylated by thymidylate kinase to generate dTDP, which is then phosphorylated by nucleoside diphosphate kinase to yield dTTP, the substrate for DNA synthesis.

Inhibitors of Thymidylate Synthesis Are Effective Anticancer Drugs

Thymidylate synthesis is required in rapidly dividing cells that need to maintain high rates of DNA synthesis. Because cancer cells generally grow faster than most normal cells in the body, inhibitors of thymidylate synthesis have been developed as anticancer drugs to block DNA synthesis.

As shown in **Figure 18.35**, thymidylate synthesis can be disrupted by two mechanisms. The most direct route is to block the activity of thymidylate synthase using uracil-based compounds such as 5-fluorouracil or **5-fluorodeoxyuridine**.

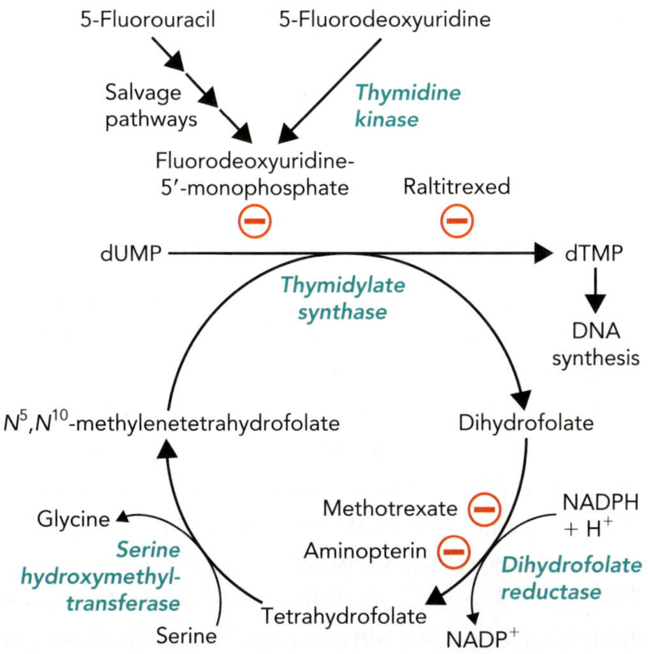

Figure 18.35 Inhibition of dTMP (thymidylate) synthesis is an effective anticancer therapy. The anticancer drugs 5-fluorouracil and 5-fluorodeoxyuridine are converted to the potent thymidylate synthase inhibitor 5-fluorodeoxyuridine-5′-monophosphate by cellular enzymes. Another option is raltitrexed, which is a folate analog that binds to and inhibits thymidylate synthase directly. Methotrexate and aminopterin are folate analogs that inhibit the enzyme dihydrofolate reductase, preventing regeneration of N^5,N^{10}-methylenetetrahydrofolate, which is required for the thymidylate synthase reaction.

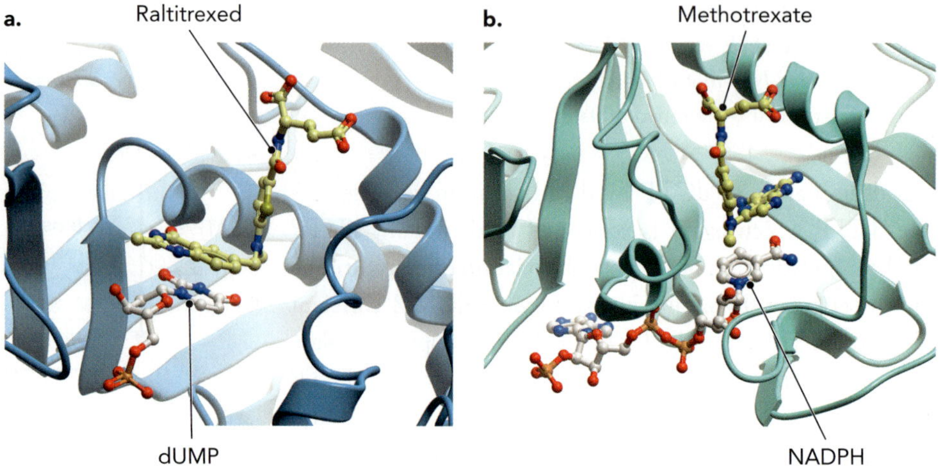

Figure 18.36 Raltitrexed and methotrexate are chemotherapeutic drugs used to inhibit DNA synthesis in rapidly dividing cells by decreasing pools of dTMP. **a.** Protein structure of rat thymidylate synthase showing the binding sites of raltitrexed and the enzyme substrate dUMP. BASED ON PDB FILE 2TSR. **b.** Protein structure of human dihydrofolate reductase showing the binding sites of methotrexate and the coenzyme NADPH. BASED ON PDB FILE 1U72.

These compounds are converted to the potent thymidylate synthase inhibitor 5-fluorodeoxyuridine-5′-monophosphate. Another option with this strategy is to treat patients with raltitrexed (Tomudex), a noncompetitive inhibitor (see Figure 7.50b) that binds to thymidylate synthase and decreases enzymatic activity (**Figure 18.36**). Raltitrexed is structurally related to folate compounds.

The second mechanism uses a class of anticancer drugs (also related to folate compounds) that inhibit dihydrofolate reductase (**Figure 18.37**). The dihydrofolate reductase inhibitors **methotrexate** and aminopterin bind to dihydrofolate reductase and prevent regeneration of N^5,N^{10}-methylenetetrahydrofolate, which indirectly inhibits flux through the thymidylate synthase reaction. High doses of methotrexate are used to treat cancer, whereas low doses are used to treat rheumatoid arthritis. Another dihydrofolate reductase inhibitor is trimethoprim, which has been used as a selective antibiotic that binds bacterial dihydrofolate reductase with an affinity that is 30,000 times higher than its affinity for human dihydrofolate reductase. A combination drug containing trimethoprim and sulfamethoxazole (a sulfonamide-based antibiotic) is an effective antibiotic treatment for common urinary tract infections.

One of the challenges in cancer therapy is that cancer cells can often become resistant to anticancer drugs. Three mechanisms of methotrexate drug resistance have been characterized in human cancer cells: (1) point mutations in the dihydrofolate reductase coding sequence that lower methotrexate binding affinity; (2) dihydrofolate reductase gene amplification to increase expression of the dihydrofolate reductase enzyme; and (3) overexpression of membrane transport proteins that rapidly export methotrexate from the cell.

1. As shown in **Figure 18.38** (p. 934), a Leu→Arg mutation at residue 22 (L22R) in the human dihydrofolate reductase protein places a large, positively charged side chain into the substrate binding pocket, which likely accounts for the ~1,000-fold decrease in methotrexate binding affinity.

2. In addition to dihydrofolate reductase mutations, human cancer cells have also been shown to acquire methotrexate resistance by a mechanism called **gene amplification**, which leads to the accumulation of hundreds of copies of the dihydrofolate reductase gene and high-level expression of the dihydrofolate reductase enzyme (**Figure 18.39**, p. 934). Dihydrofolate reductase gene amplification allows cancer cells to maintain sufficient levels of tetrahydrofolate to support the thymidylate synthase reaction even in the presence of methotrexate.

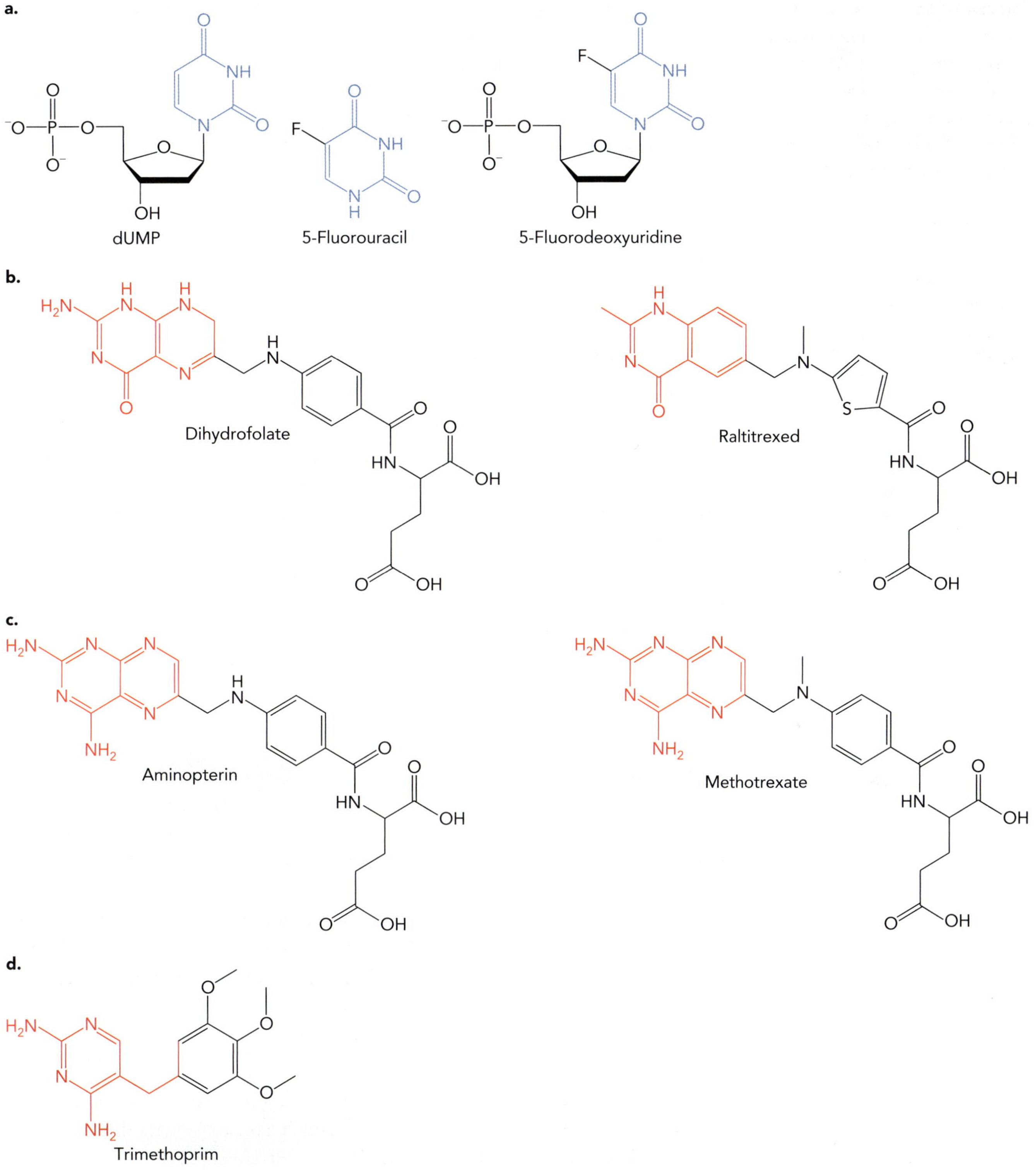

Figure 18.37 Shown are chemical structures of pharmaceutical drugs used to inhibit dTMP synthesis as a means to block DNA replication in rapidly dividing cells. **a.** The thymidylate synthase inhibitors 5-fluorouracil and 5-fluorodeoxyuridine are substrate analogs of dUMP. **b.** Raltitrexed is an inhibitor of thymidylate synthase and is a substrate analog of folate metabolites, such as dihydrofolate. **c.** Methotrexate and aminopterin are folate analogs that inhibit the enzyme activity of dihydrofolate reductase. **d.** Trimethoprim is a potent inhibitor of bacterial dihydrofolate reductase.

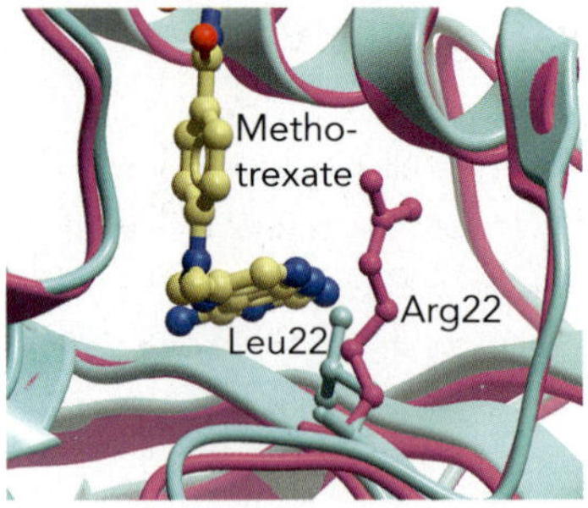

Figure 18.38 An overlay is shown of the molecular structures at the active site of wild-type human dihydrofolate reductase with that of the L22R methotrexate-resistant mutant. The decreased affinity of the L22R mutant for methotrexate is probably due to the positively charged side chain of arginine compared to leucine. Structure of the wild-type dihydrofolate reductase protein is shown in blue, and the L22R mutant is shown in magenta. BASED ON PDB FILES 1U72 (WILD-TYPE) AND 1U71 (L22R MUTANT).

3. The third mechanism of methotrexate resistance in human cancer cells is illustrated in **Figure 18.40**. In this case, cancer cells have increased expression of an ATP-dependent transporter protein called the **multidrug resistance** protein, also called P-glycoprotein. The multidrug resistance protein is normally present at low levels in cells and functions to export out of the cell xenobiotic compounds (foreign molecules) that may be harmful. Similar to the way cancer cells have been shown to amplify the dihydrofolate reductase gene to become methotrexate resistant, examples are known of multidrug resistance gene amplification as a mechanism of methotrexate resistance. Cancer cells with elevated levels of the multidrug resistance protein are able to escape the toxic effects of anticancer drugs by maintaining a low intracellular level. The multidrug resistance protein is a member of the ATP-binding cassette (ABC) transporters (see Figures 6.53 and 6.54).

In some types of cancer therapies, patients are treated with a "cocktail" of anticancer drugs. Each drug blocks a specific process in order to maximize chances of killing the cancer cells before drug resistance occurs. The first example of this was the combination of methotrexate or 5-fluorouracil to inhibit dTMP synthesis and **verapamil**, a multidrug resistance protein inhibitor that improves efficacy by preventing rapid drug efflux (**Figure 18.41**). More recently, recombinant antibodies have been developed that target membrane proteins and initiate a cell death response. Highly specific anticancer drugs such as imatinib (Gleevec), which inhibit oncogenic kinases, are also now available. By including all four of these inhibitors

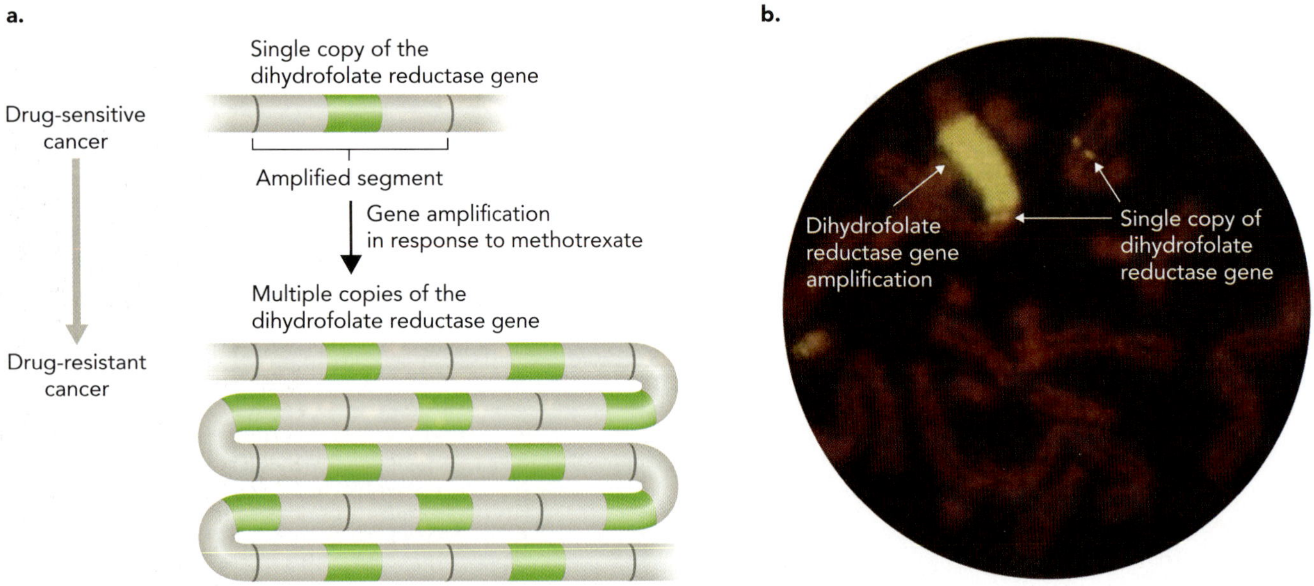

Figure 18.39 Cancer cells can become resistant to anticancer drugs by DNA recombination events that lead to gene amplification. **a.** Treatment of cells with the dihydrofolate reductase inhibitor methotrexate causes amplification of large segments of DNA surrounding the dihydrofolate reductase gene. **b.** Results of an *in situ* hybridization analysis of metaphase cells using a DNA probe against the dihydrofolate reductase gene. Chinese hamster ovary cells were selected for resistance to growth in methotrexate. This resistant cell line was estimated to contain ~150 copies of the dihydrofolate reductase gene on chromosome 2. B. J. TRASK AND J. L. HAMLIN (1989). EARLY DIHYDROFOLATE REDUCTASE GENE AMPLIFICATION EVENTS IN CHO CELLS USUALLY OCCUR ON THE SAME CHROMOSOME ARM AND LOCUS. *GENES & DEVELOPMENT, 3*, 1913–1925. DOI:10.1101/GAD.3.12A.1913. © COLD SPRING HARBOR LABORATORY PRESS.

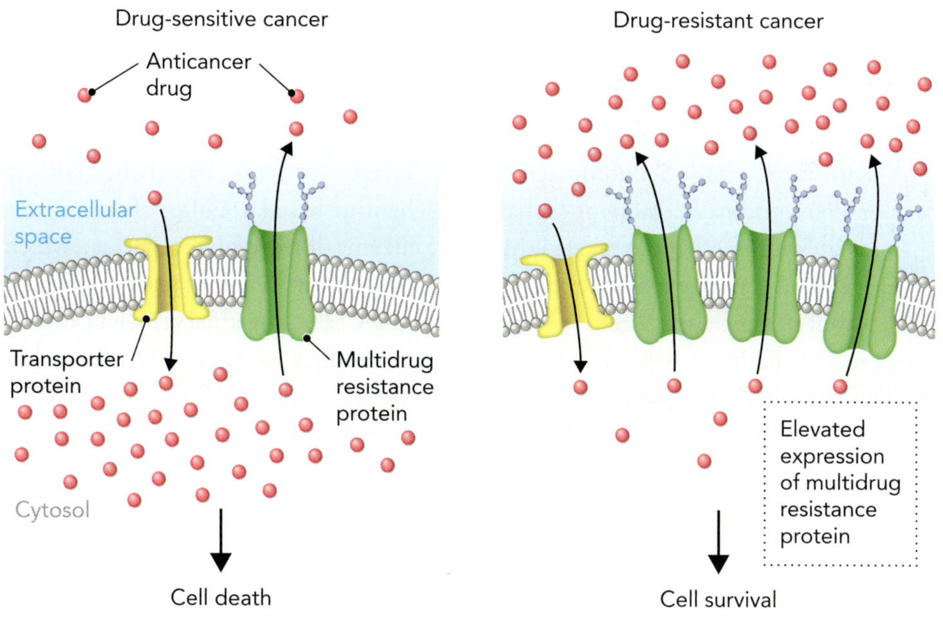

Figure 18.40 Elevated expression of multidrug resistance protein, also known as P-glycoprotein, results in drug resistance due to a rapid efflux of anticancer drugs. The multidrug resistance protein is an ATP-dependent membrane-spanning pump that transports xenobiotic compounds out of the cell.

in cancer therapy cocktails, the odds improve of killing off most of the cancer cells before they develop drug resistance.

This same multipronged pharmaceutical approach has also been shown to provide beneficial long-term treatment to patients infected with the human immunodeficiency virus (HIV). A combination therapy called highly active antiretroviral therapy (HAART) consists of at least three classes of HIV drugs: (1) inhibitors of viral replication; (2) inhibitors of viral proteases, required for protein maturation; and (3) inhibitors of the HIV integrase enzyme. Most studies show that HIV patients treated with HAART survive ~30% to 50% longer than those who are not treated for HIV infection.

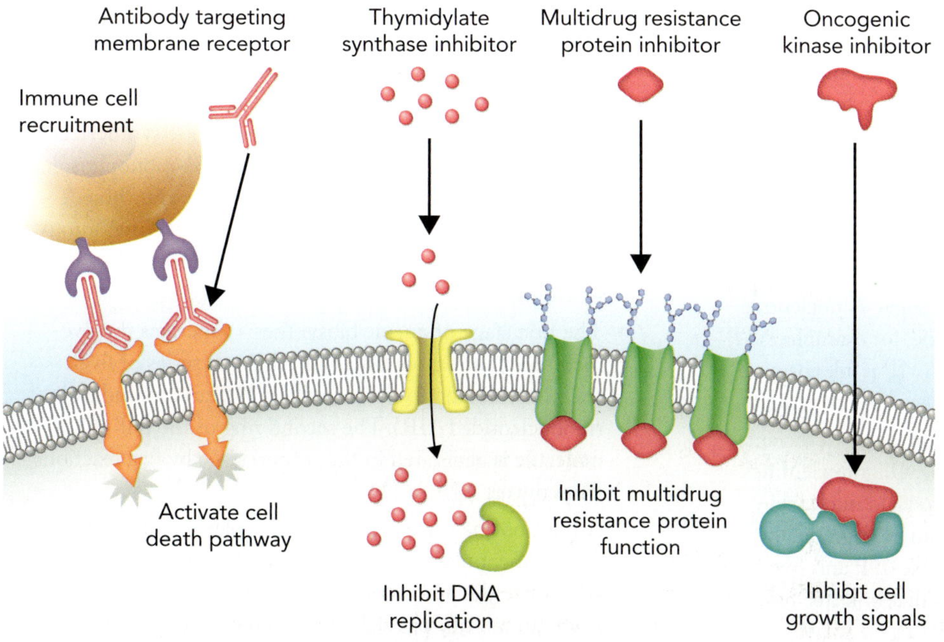

Figure 18.41 Combination anticancer drug therapy helps avoid the problem of drug resistance by targeting multiple proteins with the same treatment regimen.

concept integration 18.4

How do the anticancer drugs 5-fluorouracil and methotrexate block DNA synthesis?

The anticancer drug 5-fluorouracil blocks DNA synthesis by giving rise to 5-fluorodeoxyuridine-5′-monophosphate, a mechanism-based inhibitor of the enzyme thymidylate synthase. Methotrexate inhibits the enzyme dihydrofolate reductase, which is required to regenerate tetrahydrofolate for the thymidylate synthase reaction. Cancer cells, as well as some normal cells in the body (immune cells, intestinal cells, hair follicle cells), require high levels of thymine deoxyribonucleotides to support DNA synthesis. The enzyme thymidylate synthase converts dUMP into dTMP by a methylation reaction involving the coenzyme N^5,N^{10}-methylenetetrahydrofolate. Salvage pathway enzymes convert 5-fluorouracil into 5-fluorodeoxyuridine-5′-monophosphate, which binds to thymidylate synthase and irreversibly blocks its activity by forming a covalent enzyme intermediate with N^5,N^{10}-methylenetetrahydrofolate. Without sufficient levels of thymidylate, rapidly dividing cancer cells (as well as some other cell types in the body) cannot maintain the high rates of DNA synthesis needed to proliferate. Methotrexate also blocks thymidylate synthesis; however, in this case methotrexate inhibits the enzyme activity of dihydrofolate reductase, which prevents the regeneration of N^5,N^{10}-methylenetetrahydrofolate.

chapter summary

18.1 Structure and Function of Nucleotides

- Nucleotides participate in four important biochemical processes: (1) energy conversion reactions; (2) signal transduction pathways; (3) coenzyme-dependent reactions; and (4) genetic information storage and transfer.
- Purine bases contain nine atoms in the heterocyclic ring, whereas pyrimidine bases contain six atoms.
- The two common purine bases are adenine and guanine. The three common pyrimidine bases are cytosine, thymine, and uracil. Uracil is found only in RNA, and thymine is found only in DNA.
- Ribonucleotides contain hydroxyl groups at both C-3′ and C-2′ of the ribose, whereas deoxyribonucleotides lack the hydroxyl group at C-2′.
- A nucleoside consists of a base and sugar. A nucleotide refers to a phosphorylated nucleoside; for example, ATP is adenosine-5′-triphosphate, and AMP is adenosine-5′-monophosphate.
- Nucleotide salvage pathways are more energy efficient than *de novo* nucleotide biosynthesis. Oligonucleotides are cleaved by phosphodiesterases to yield nucleoside monophosphates (NMPs), which can be degraded by phosphorylase enzymes to yield ribose-1-P and free bases. NMPs are regenerated by enzymes that link free bases to phosphoribosyl pyrophosphate (PRPP).

18.2 Purine Metabolism

- Purine bases are synthesized directly on the ribose sugar, whereas pyrimidine bases are first synthesized as a closed ring before attaching to the ribose sugar.
- The four nitrogen atoms in purine bases are derived from aspartate (N-1), glycine (N-7), and two glutamines (N-3 and N-9). The five carbons come from glycine (C-4 and C-5), HCO_3^- (C-6), and two molecules of N^{10}-formyltetrahydrofolate (C-2 and C-8).
- The purine nucleotides AMP and GMP are synthesized from the common intermediate inosine-5′-monophosphate (IMP), which contains the purine base hypoxanthine. The biosynthesis of IMP in *E. coli* can be broken down into two stages, each consisting of five well-characterized biochemical reactions.
- The first stage of purine biosynthesis generates the five-membered imidazole ring on PRPP in a series of five reactions, leading to the formation of 5-aminoimidazole ribonucleotide (AIR). The second ring of the purine molecule is generated in the second stage by five reactions that convert AIR to IMP.
- A purinosome is a large protein complex in eukaryotic cells that contains all of the enzymatic activities needed to synthesize the purine ring. Evidence for purinosomes comes from human cancer cells grown in purine-depleted media

that express fluorescent fusion proteins linked to purine biosynthetic enzymes.

- IMP is the precursor to both AMP and GMP, which are generated in parallel biosynthetic pathways. Balanced regulation of metabolic flux through these two parallel pathways is mediated by GTP and ATP, which are required for AMP and GMP synthesis, respectively.
- AMP and GMP are phosphorylated to generate nucleoside triphosphates for RNA and DNA synthesis. The enzyme nucleoside diphosphate kinase can use any nucleoside triphosphate as a phosphate donor and phosphorylates both nucleoside diphosphates and deoxynucleoside diphosphates.
- Flux through the purine biosynthetic pathway is primarily controlled by inhibition of the PRPP synthetase and glutamine-PRPP amidotransferase reactions. The balance of AMP and GMP synthesis is controlled by both feedback inhibition of the individual branches in the pathway and by ATP and GTP cross-talk regulation.
- Excess AMP, GMP, and IMP nucleotides from cellular or dietary nucleic acids are dephosphorylated by the enzyme 5′-nucleotidase to generate adenosine, inosine, and guanosine, respectively, which are then further degraded into uric acid.
- Uric acid is essentially insoluble in water and is excreted as a dry paste by organisms that need to conserve water, such as birds, reptiles, and insects. Primates also excrete small amounts uric acid, but most of their nitrogen waste products are excreted as urea produced by the urea cycle. Uric acid is further degraded in some animals to other nitrogen-containing waste products.
- Gout is caused by the buildup of uric acid crystals (sodium urate) in the joints and kidneys. The big toe is a common joint affected by uric acid because of poor circulation in the foot and the frequency of blunt injury, which releases uric acid crystals into the synovial fluid.
- Dietary causes of gout include alcohol, which interferes with uric acid excretion, and the high amounts of nucleic acids contained in meats. Gout has also been linked to increased levels of PRPP synthetase, defects in feedback inhibition of glutamate-PRPP amidotransferase, and deficiencies in the salvage enzyme hypoxanthine-guanine phosphoribosyltransferase (HGPRT).
- Lesch–Nyhan syndrome is a rare recessive genetic disease caused by defects in HGPRT, which leads to a buildup of guanine and hypoxanthine and is characterized by unusual neurologic symptoms. Lesch–Nyhan syndrome follows the inheritance pattern of an X-linked recessive genetic mutation owing to the fact that the HGPRT gene is located on the X chromosome.
- Defects in the enzyme adenosine deaminase (ADA) cause ADA–SCID (SCID: severe combined immunodeficiency disease), leading to adenosine accumulation, which in turn shunts excess adenosine into dATP production. Because dATP is an inhibitor of the enzyme ribonucleotide reductase, rapidly dividing B and T cells in the immune system are starved of deoxyribonucleotides and fail to proliferate.

18.3 Pyrimidine Metabolism

- The six atoms in the pyrimidine ring are derived from just two precursor biomolecules: aspartate (C-1, C-4, C-5, C-6) and carbamoyl phosphate, which is generated from glutamine (N-3) and HCO_3^- (C-2).
- The pyrimidine biosynthetic pathway in *E. coli* consists of six reactions to generate the pyrimidine nucleotide UMP, which is converted to UTP by sequential phosphorylation reactions and then aminated by CTP synthetase to generate CTP.
- *E. coli* purine and pyrimidine biosynthetic enzymes are encoded in the bacterial genome as separate polypeptides, whereas in most animals, including humans, several catalytic activities required for purine and pyrimidine biosynthesis are combined into large multifunctional enzymes.
- Pyrimidine biosynthesis is regulated by both feedback inhibition and allosteric activation in bacteria and animal cells. Aspartate transcarbamoylase is the key regulated enzyme in the pyrimidine biosynthetic pathway in *E. coli* cells, being activated by ATP and inhibited by CTP. Flux through the pyrimidine biosynthetic pathway in animal cells is controlled by allosteric regulation of the CAD enzyme, UMP synthase, and CTP synthetase.
- The pyrimidine nucleotides UMP, CMP, and deoxythymidine-5′-phosphate (dTMP) are degraded by a common three-reaction pathway converting uracil and thymine into β-alanine and β-aminoisobutyrate, respectively.
- The first reaction in the purine degradation pathway is catalyzed by the rate-limiting enzyme dihydropyrimidine dehydrogenase, which converts uracil and thymine to dihydrouracil and dihydrothymine, respectively. Dihydropyrimidine dehydrogenase enzyme deficiencies are fairly common, being present in ~5% of the human population.
- Dihydropyrimidine dehydrogenase deficiency is the cause of drug toxicity in cancer patients being treated with 5-fluorouracil because the dose of 5-fluorouracil is too high, owing to its reduced degradation by dihydropyrimidine dehydrogenase in the liver. Cancer patients being considered for 5-fluorouracil treatment are now routinely screened for dihydropyrimidine dehydrogenase deficiencies.

18.4 Deoxyribonucleotide Metabolism

- Nucleoside 5′-diphosphates (GDP, ADP, CDP, UDP) are converted into the corresponding deoxynucleoside 5′-diphosphates (dGDP, dADP, dCDP, dUDP) by the enzyme ribonucleotide reductase using NADPH as a coenzyme.

- Ribonucleotide reductase is an ancient enzyme that is structurally similar across all species and likely played a pivotal role in converting the ancestral RNA world into the present-day DNA world.
- The RNA world hypothesis proposes that RNA was the first genetic molecule and that ribozymes were the biochemical catalysts for RNA and protein synthesis. After the emergence of ribonucleotide reductase, deoxyribonucleotides were polymerized into DNA, which is better suited to function as genetic material.
- The reduction of C-2′ on nucleoside diphosphates by nucleotide reductase requires the input of two electrons derived from NADPH that are used to reduce a pair of sulfhydryl groups in the enzyme active site. The reduction of these sulfhydryls in ribonucleotide reductase is not done by NADPH directly, but rather by a redox circuit requiring intermediary proteins (thioredoxin or glutaredoxin).
- Ribonucleotide reductase contains two subunits, R1 and R2, that function as a tetrameric complex ($R1_2R2_2$). The catalytic mechanism in *E. coli* is unusual in that it depends on formation of a free radical to catalyze the reaction and requires contributions from a dinuclear Fe^{3+} iron center that is coordinated by an oxide ion (O_2^-).
- Substrate specificity of ribonucleotide reductase is regulated by allosteric binding of dTTP, dGTP, dATP, or ATP to the substrate specificity site in the R1 subunit. The overall activity of ribonucleotide reductase is regulated by allosteric binding of ATP and dTTP to the regulatory site, such that ATP is an activator and dTTP an inhibitor.
- Thymine is a pyrimidine base formed by methylation of uracil on C-5 in a reaction catalyzed by the enzyme thymidylate synthase, which converts dUMP (deoxyuridylate) to dTMP (thymidylate) using the coenzyme N^5,N^{10}-methylenetetrahydrofolate.
- dUMP is derived from three sources in *E. coli* cells: (1) dephosphorylation of dUTP by the enzyme dUTP diphosphohydrolase; (2) deamination of dCTP by dCTP deaminase to generate dUTP, which is then converted to dUMP by dUTP diphosphohydrolase; or (3) phosphorylation of deoxyuridine by the enzyme thymidine kinase to yield dUMP.
- Thymidylate synthesis is required for rapidly dividing cells that need to maintain high rates of DNA synthesis. Because cancer cells generally grow faster than most normal cells in the body, inhibitors of thymidylate synthesis have been developed as anticancer agents to block DNA synthesis.
- Thymidylate synthesis can be disrupted by two mechanisms: (1) direct inhibition of thymidylate synthase by 5-fluorodeoxyuridine-5′-monophosphate or by folate analogs such as raltitrexed; or (2) indirect inhibition of thymidylate synthase using the folate analog methotrexate, which prevents regeneration of N^5,N^{10}-methylenetetrahydrofolate by inhibiting dihydrofolate reductase activity.
- Three mechanisms of methotrexate drug resistance have been characterized in human cancer cells: (1) point mutations in the dihydrofolate reductase coding sequence to lower methotrexate binding affinity; (2) dihydrofolate reductase gene amplification to increase expression of the dihydrofolate enzyme; and (3) overexpression of the multidrug resistance protein that rapidly exports methotrexate from the cell.
- In some types of cancer therapies, patients are treated with a "cocktail" of anticancer drugs, each of which blocks a specific process to maximize chances of killing the cancer cells before drug resistance occurs. For example, treatment might include thymidylate synthase inhibitors, multidrug resistance protein inhibitors, oncogenic kinase inhibitors, and antibodies that target membrane proteins and stimulate cell death pathways.

biochemical terms

(in order of appearance in text)

adenine (A) (p. 902)
guanine (G) (p. 902)
cytosine (C) (p. 902)
thymine (T) (p. 902)
uracil (U) (p. 902)
hypoxanthine (p. 902)
orotate (p. 902)
purine (p. 902)
pyrimidine (p. 902)
salvage pathway (p. 903)
phosphorylase (p. 904)
phosphoribosyl pyrophosphate (PRPP) (p. 904)
inosine-5′-monophosphate (IMP) (p. 904)
orotidine-5′-monophosphate (OMP) (p. 904)
purinosome (p. 908)
uric acid (p. 912)
allantoin (p. 914)
allantoic acid (p. 914)
gout (p. 914)
allopurinol (p. 915)
Lesch–Nyhan syndrome (p. 915)
adenosine deaminase deficiency (p. 915)
severe combined immunodeficiency disease (SCID) (p. 915)
5-fluorouracil (p. 922)
ribonucleotide reductase (p. 924)
RNA world (p. 924)
glutaredoxin (p. 926)
dihydrofolate (p. 930)
tetrahydrofolate (p. 930)
5-fluorodeoxyuridine (p. 931)
methotrexate (p. 932)
gene amplification (p. 932)
multidrug resistance (p. 934)
verapamil (p. 934)

review questions

1. Describe the four biochemical processes that nucleotides participate in. Give examples of nucleotides—other than those containing the nucleotide base adenine—that have a role in each of these processes.
2. What are the names of the following five chemical structures?

a.

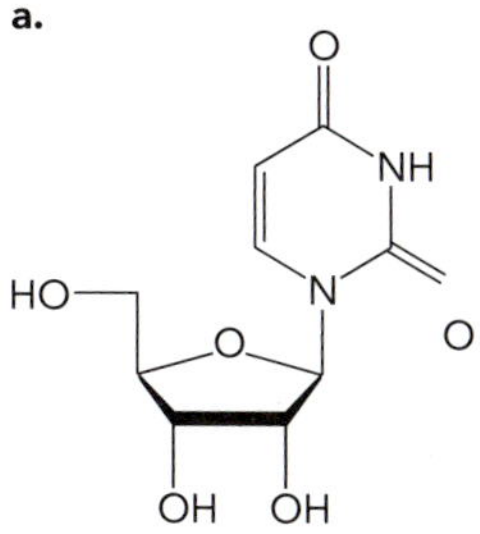

b.

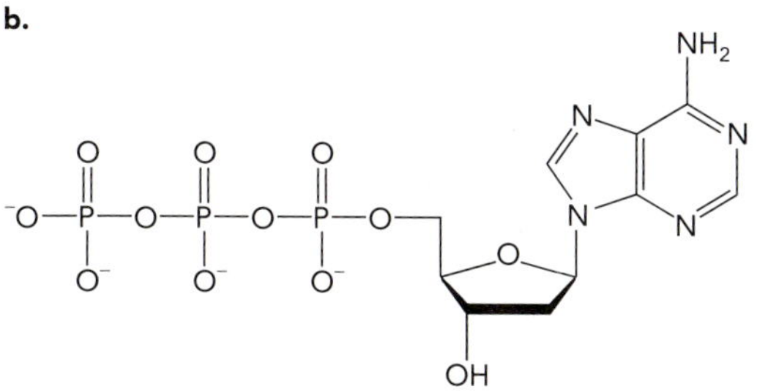

c.

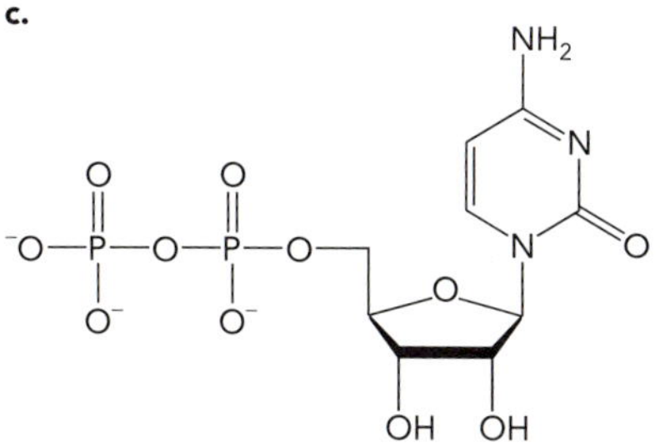

d.

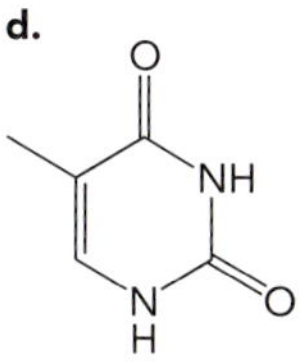

e.

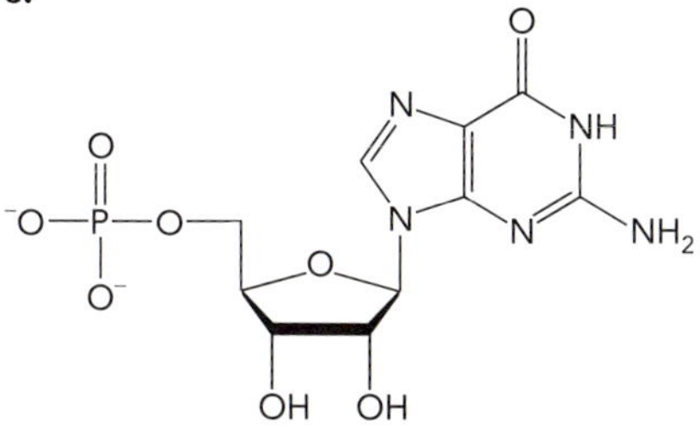

3. Nucleotide salvage pathways are an important recycling process in cells because they require less ATP than the *de novo* biosynthetic pathways use. What is the source of most nucleotide bases salvaged by recycling processes in cells?
4. Describe the primary difference between the *E. coli* purine and pyrimidine *de novo* biosynthetic pathways.
5. What metabolites contribute to the nine atoms in purine bases? What is the name of the metabolic precursor to adenylate and guanylate?
6. What is a purinosome, and what is the experimental evidence for the existence of purinosomes in cells? What biochemical methods could be used to "prove" the existence of purinosomes in cells under normal physiologic conditions?
7. What regulatory mechanisms ensure a balanced pool of guanine and adenine in cells?
8. What is the primary metabolic fate of excess purine nucleotides in humans, Siberian tigers, flamingos, rattlesnakes, tuna, great white sharks, salamanders, cockroaches, and lobsters?
9. What is the biochemical basis for ADA–SCID, and what are the treatment options for patients diagnosed with ADA–SCID?
10. Describe similarities and differences between the regulation of the *de novo* pyrimidine biosynthetic pathway in *E. coli* compared to that in humans.
11. How would a deficiency in the human enzyme dihydropyrimidine dehydrogenase cause 5-fluorouracil toxicity in cancer patients undergoing chemotherapy?
12. Describe the RNA world hypothesis. What is the best evidence to date that this view of early biochemical life on Earth may be accurate?
13. Describe the function and regulation of *E. coli* ribonucleotide reductase.
14. What are three sources of deoxyuridylate in bacterial cells?
15. Describe three mechanisms of methotrexate drug resistance in human cancer cells.

challenge problems

1. Phosphoribosyl pyrophosphate (PRPP) is a central intermediate in nucleotide metabolism. Name three *distinct* pathways in which PRPP is required for the production of a nucleoside monophosphate and describe how it is incorporated into the pathway.
2. Gout can be due to either excess production of purines by *de novo* synthesis (defect A) or inability to excrete excess dietary uric acid properly (defect B). Explain how feeding a patient ^{15}N-glycine, and determining the amount of ^{15}N

in their excreted uric acid, could be used to distinguish between defect A and defect B.

3. Why is *de novo* biosynthesis of purines markedly elevated in patients with a deficiency in hypoxanthine-guanine phosphoribosyl transferase (HGPRT)?

4. In a case study of Lesch–Nyhan syndrome dating from the early 1990s, an asymptomatic individual with a family history of the disease wanted to know if he or she carried the genetic defect before deciding on starting a family. Because this case study predated high-throughput DNA sequencing, a functional biochemical test was performed using fibroblast cell cultures made from a skin punch. Cells were incubated for 4 hours in culture media containing radioactive hypoxanthine, and radioactive DNA was measured as evidence of ongoing pyrimidine metabolism. It was found that incorporation of radioactive hypoxanthine into DNA was about half of what would be expected in an individual with normal pyrimidine metabolism.
   a. What enzyme is defective in Lesch–Nyhan syndrome, and why was the amount of radioactive DNA in this cell culture assay ~50% of normal?
   b. What is the genetic sex of the cell donor? Explain your answer.
   c. Why doesn't this individual show signs of Lesch–Nyhan syndrome?
   d. What advice would you give this individual regarding the probability that his or her child will develop Lesch–Nyhan syndrome?

5. How is the metabolism of AMP linked to energy metabolism in muscle?

6. The enzyme adenosine deaminase converts adenosine and deoxyadenosine to inosine and deoxyinosine, respectively, as shown in the reaction below.

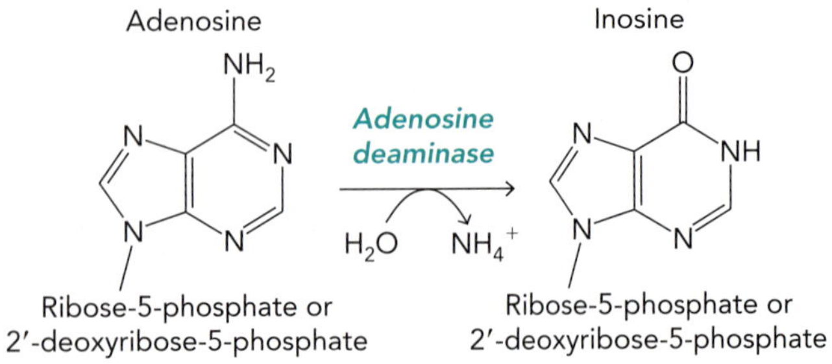

   a. What is the name of the human disease caused by an adenosine deaminase deficiency, and what are the clinical symptoms?
   b. How does a defect in adenosine deaminase lead to inhibition of DNA synthesis, and why does this have an effect only on certain types of human cells?

7. Individuals lacking orotate phosphoribosyl transferase excrete high levels of orotic acid in their urine and develop anemia (reduced numbers of red blood cells). When patients are fed uridine or cytidine, the anemia is reduced, and levels of orotic acid in the urine decline.
   a. Identify the metabolic pathway that requires orotate phosphoribosyl transferase, and write the reaction it catalyzes.
   b. Why does feeding patients uridine or cytidine help alleviate the symptoms of anemia and high levels of orotic acid in the urine?

8. Uracil is not found in DNA, so why does it make sense that ribonucleotide reductase converts UDP to dUDP?

9. The analog 5-bromodeoxyuridine can be phosphorylated in human cells and incorporated into DNA as a complementary base pair with adenine and thereby functions as a thymine analog. In the next round of DNA replication, however, 5-bromodeoxyuridine can form a base pair with guanine and function as a cytosine analog, resulting in the mutation of an A-T base pair into a G-C base pair. Explain why cells with defects in thymidine kinase function can grow in media containing high levels of 5-bromodeoxyuridine, whereas cells with normal thymidine kinase activity die under these conditions.

10. The hypoxanthine analog 6-mercaptopurine can be converted to 6-mercaptopurine-5′-monophosphate by the salvage pathway to generate a potent competitive inhibitor of adenylosuccinate synthetase and IMP dehydrogenase, which generate AMP and GMP, respectively, from IMP. Leukemia patients are sometimes treated with 6-mercaptopurine to decrease AMP and GMP production in rapidly dividing cancer cells. It has been found that the inhibitory effect of 6-mercaptopurine on leukemia cell growth is enhanced if the treatment includes allopurinol, a specific inhibitor of the purine degradation enzyme xanthine oxidase.
   a. Explain the improved efficacy of 6-mercaptopurine when combined with allopurinol.
   b. How would the 6-mercaptopurine treatment regimen need to be modified if the leukemia patient had a deficiency in xanthine oxidase? What other nucleotide metabolic disease would this be similar to, with regard to modifying a standard disease treatment on the basis of a patient's genetic profile?

11. Cultured cells can be prevented from initiating DNA replication by adding excess thymidine to the growth media. Propose a mechanism to account for this "thymidine block" on the basis of the specificity and activity of ribonucleotide reductase under different physiologic conditions.

12. Cells defective in thymidylate synthase are able to undergo cell division if methotrexate and thymidine are both provided; however, cells with fully functional thymidylate synthase cannot divide under these same culture conditions. What is the biochemical basis for this observation?

smartwork5

If your instructor assigns homework with Smartwork5, access it here: digital.wwnorton.com/biochem.

suggested reading

Books and Reviews

Berg, L. J. (2008). The "bubble boy" paradox: an answer that led to a question. *Journal of Immunology*, *181*, 5815–5816.

Buchanan, J. M. (1994). Aspects of nucleotide enzymology and biology. *Protein Science*, *3*, 2151–2157.

Hager, A. J., Pollard, J. D., and Szostak, J. W. (1996). Ribozymes: aiming at RNA replication and protein synthesis. *Chemistry & Biology*, *3*, 717–725.

Mondello, C., Smirnova, A., and Giulotto, E. (2010). Gene amplification, radiation sensitivity and DNA double-strand breaks. *Mutation Research*, *704*, 29–37.

Muller, U. F. (2006). Re-creating an RNA world. *Cellular and Molecular Life Sciences*, *63*, 1278–1293.

Noller, H. F. (2012). Evolution of protein synthesis from an RNA world. *Cold Spring Harbor Perspectives in Biology*, *4*, a003681.

Nyhan, W. L. (2005). Disorders of purine and pyrimidine metabolism. *Molecular Genetics and Metabolism*, *86*, 25–33.

Poole, A. M., Logan, D. T., and Sjoberg, B. M. (2002). The evolution of the ribonucleotide reductases: much ado about oxygen. *Journal of Molecular Evolution*, *55*, 180–196.

Stubbe, J. (1998). Ribonucleotide reductases in the twenty-first century. *Proceedings of the National Academy of Sciences USA*, *95*, 2723–2724.

Terkeltaub, R. (2010). Update on gout: new therapeutic strategies and options. *Nature Reviews Rheumatology*, *6*, 30–38.

van Kuilenburg, A. B., Meinsma, R., and van Gennip, A. H. (2004). Pyrimidine degradation defects and severe 5-fluorouracil toxicity. *Nucleosides, Nucleotides & Nucleic Acids*, *23*, 1371–1375.

Primary Literature

An, S., Kumar, R., Sheets, E. D., and Benkovic, S. J. (2008). Reversible compartmentalization of de novo purine biosynthetic complexes in living cells. *Science*, *320*, 103–106.

Bystroff, C., Oatley, S. J., and Kraut, J. (1990). Crystal structures of *Escherichia coli* dihydrofolate reductase: the $NADP^+$ holoenzyme and the folate-$NADP^+$ ternary complex. Substrate binding and a model for the transition state. *Biochemistry*, *29*, 3263–3277.

Cody, V., Luft, J. R., and Pangborn, W. (2005). Understanding the role of Leu22 variants in methotrexate resistance: comparison of wild-type and Leu22Arg variant mouse and human dihydrofolate reductase ternary crystal complexes with methotrexate and NADPH. *Acta Crystallographica. Section D, Biological Crystallography*, *61*, 147–155.

Dobritzsch, D., Schneider, G., Schnackerz, K. D., and Lindqvist, Y. (2001). Crystal structure of dihydropyrimidine dehydrogenase, a major determinant of the pharmacokinetics of the anti-cancer drug 5-fluorouracil. *EMBO Journal*, *20*, 650–660.

Eads, J. C., Scapin, G., Xu, Y., Grubmeyer, C., and Sacchettini, J. C. (1994). The crystal structure of human hypoxanthine-guanine phosphoribosyltransferase with bound GMP. *Cell*, *78*, 325–334.

Fairman, J. W., Wijerathna, S. R., Ahmad, M. F., Xu, H., Nakano, R., Jha, S., Prendergast, J., Welin, R. M., Flodin, S., Roos, A., et al. (2011). Structural basis for allosteric regulation of human ribonucleotide reductase by nucleotide-induced oligomerization. *Nature Structural & Molecular Biology*, *18*, 316–322.

Fang, C. T., Chang, Y. Y., Hsu, H. M., Twu, S. J., Chen, K. T., Lin, C. C., Huang, L. Y., Chen, M. Y., Hwang, J. S., Wang, J. D., and Chuang, C. Y. (2007). Life expectancy of patients with newly-diagnosed HIV infection in the era of highly active antiretroviral therapy. *QJM*, *100*, 97–105.

Field, M. S., Anderson, D. D., and Stover, P. J. (2011). *Mthfs* is an essential gene in mice and a component of the purinosome. *Frontiers in Genetics*, *2*, 36.

Gaspar, H. B., Cooray, S., Gilmour, K. C., Parsley, K. L., Zhang, F., Adams, S., Bjorkegren, E., Bayford, J., Brown, L., Davies, E. G., et al. (2011). Hematopoietic stem cell gene therapy for adenosine deaminase-deficient severe combined immunodeficiency leads to long-term immunological recovery and metabolic correction. *Science Translational Medicine*, *3*, 97ra80.

Morgan, E., Honig, G., and Nelson, D. J. (1981). Acute lymphocytic leukemia in a child with congenital xanthine oxidase deficiency: implications for therapy. *American Journal of Pediatric Hematology/Oncology*, *3*, 439–441.

Ragab, A. H., Gilkerson, E., and Myers, M. (1974). The effect of 6-mercaptopurine and allopurinol on granulopoiesis. *Cancer Research*, *34*, 2246–2249.

Scaraffia, P. Y., Tan, G., Isoe, J., Wysocki, V. H., Wells, M. A., and Miesfeld, R. L. (2008). Discovery of an alternate metabolic pathway for urea synthesis in adult *Aedes aegypti* mosquitoes. *Proceedings of the National Academy of Sciences USA*, *105*, 518–523.

Sotelo-Mundo, R. R., Ciesla, J., Dzik, J. M., Rode, W., Maley, F., Maley, G. F., Hardy, L. W., and Montfort, W. R. (1999). Crystal structures of rat thymidylate synthase inhibited by Tomudex, a potent anticancer drug. *Biochemistry*, *38*, 1087–1094.

Trask, B. J., and Hamlin, J. L. (1989). Early dihydrofolate reductase gene amplification events in CHO cells usually occur on the same chromosome arm and locus. *Genes & Development*, *3*, 1913–1925.

Yu, D. S., Sun, G. H., Ma, C. P., and Chang, S. Y. (1999). Cocktail modulator mixtures for overcoming multidrug resistance in renal cell carcinoma. *Urology*, *54*, 377–381.

Glucose uptake into tissues is regulated by insulin signaling

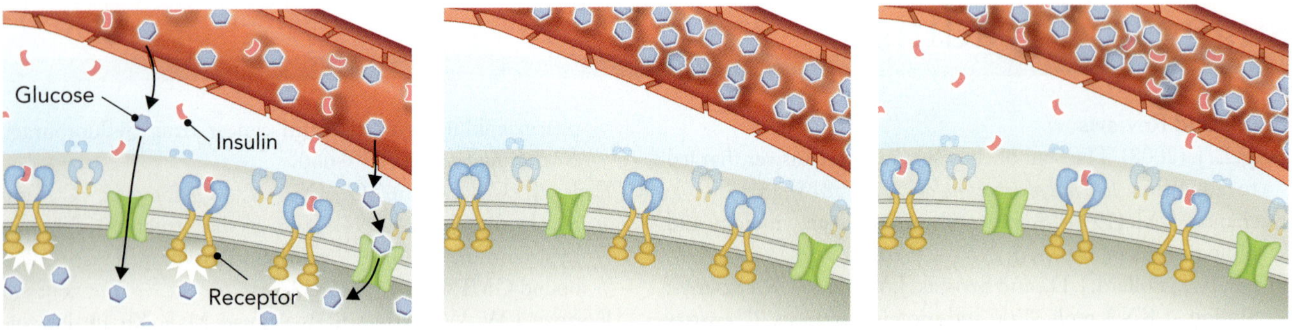

Normal Response
Insulin binding to insulin receptors activates downstream signaling pathways, leading to glucose uptake

Type 1 Diabetes
Insulin is not produced by the pancreas, and blood glucose levels are high

Type 2 Diabetes
Insulin receptors do not activate downstream signaling pathways, and blood glucose levels are high

Patients with type 2 diabetes need to measure blood sugar levels frequently with personal testing devices. To maintain blood sugar levels within a safe range, the patients can take rosiglitazone daily. This thiazolidinedione drug binds to and activates the PPARγ nuclear receptor, which is shown here with rosiglitazone bound to the ligand binding domain. Activated PPARγ increases insulin sensitivity in liver, skeletal, and adipose tissue.

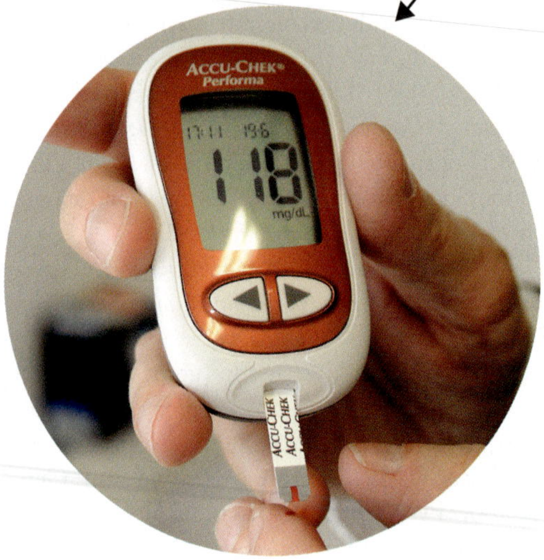

Type 2 diabetes patient measuring blood glucose levels

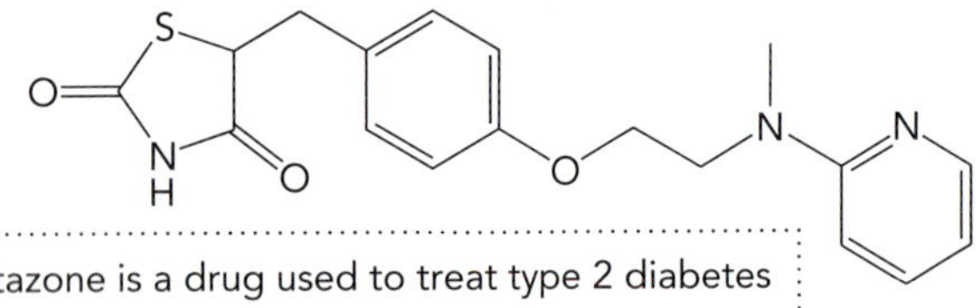

Rosiglitazone is a drug used to treat type 2 diabetes

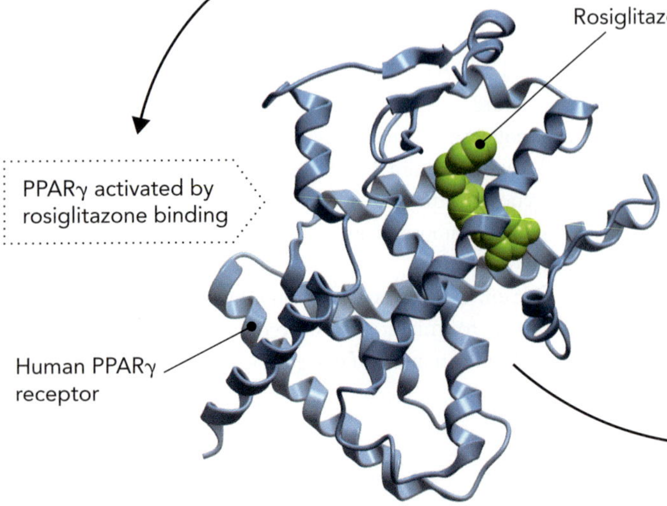

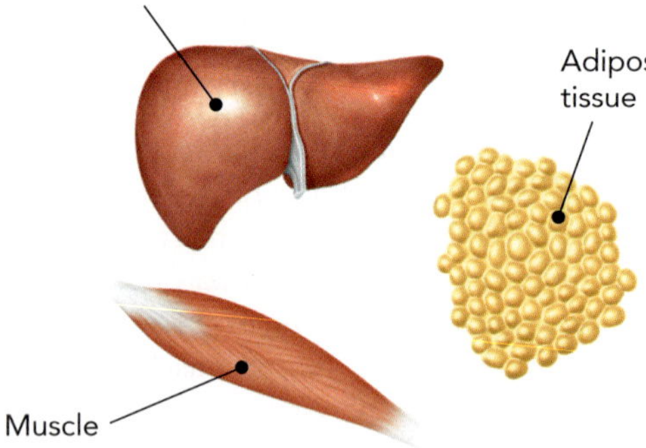

Increased insulin sensitivity in liver, muscle, and adipose tissue after rosiglitazone treatment

19

Metabolic Integration

◀ Insulin signaling stimulates blood glucose uptake into tissues that have functional insulin receptors. An inability to control blood glucose levels by insulin signaling is the molecular basis for the disease diabetes. Type 1 diabetes is characterized by lack of insulin production in the pancreas in response to increased glucose levels, whereas type 2 diabetes is characterized by defects in insulin receptor signaling despite the production of insulin and binding to insulin receptors, which is called insulin resistance. The effect of insulin resistance is high glucose levels in the blood, which can lead to heart disease, problems with eyesight, kidney failure, and poor circulation in extremities.

Thiazolidinediones have been used to treat patients with insulin-resistant type 2 diabetes because these drugs improve insulin sensitivity through modulation of glucose and fatty acid metabolism. Thiazolidinediones function as ligands for the peroxisome proliferator–activated receptor gamma (PPARγ) protein. Although thiazolidinedione-mediated activation of PPARγ improves insulin sensitivity, this treatment has been associated with an increased risk of cardiovascular disease in some patients.

CHAPTER OUTLINE

Multicellular organisms have evolved specialized cells and organs to maximize metabolic efficiency and to cope with environmental changes. Thus, a brain cell, a liver cell, a muscle cell, and a nerve cell are quite different. Yet they all have the same metabolic needs and carry out synthesis of new molecules and degradation of old molecules at the same time. In this chapter, we examine this concept, using humans as a model organism. We take advantage of what we have learned in previous chapters about intracellular metabolic pathways to now look more broadly at metabolic flux within the whole organism. We also describe the biochemical processes that mediate and integrate metabolism among the various pathways.

We begin by examining the metabolic roles of several major tissues and organs in humans. We pay special attention to key interorgan signaling pathways that coordinate metabolic flux when a person has just eaten or when no food has been eaten in some time. This is followed by a discussion of the role of genes and environment in determining energy balance within individuals. Recent discoveries have begun to shed light on metabolic control signals that regulate energy balance in humans. Some of these signals are mediated by peptide hormones, which are secreted from adipose tissue and the gastrointestinal tract in response to nutritional status. In many cases, these peptide hormones activate neuronal signaling pathways in the brain that affect metabolic flux and eating behavior. We end the chapter by looking at the biochemistry of diet and exercise in terms of metabolic integration, including a description of two intracellular signaling pathways that have been linked to muscle metabolism.

19.1 Metabolic Integration at the Physiologic Level

Back in Chapter 9, we showed a map of the major pathways in cells that convert the chemical energy in metabolic fuels into ATP (see Figure 9.3). In subsequent chapters, we went on to describe a variety of anabolic and catabolic pathways that metabolize carbohydrates, lipids, proteins, and nucleic acids. As shown in **Figure 19.1**, we can modify the metabolic map to emphasize how dietary proteins, carbohydrates, and lipids contribute to ATP production (nucleic acids have been left out because they contribute very little to energy metabolism). This revised metabolic map focuses on five energy conversion processes that are described in Parts 3 and 4 of the book: (1) carbohydrate metabolism (glycolysis and gluconeogenesis); (2) lipid metabolism (fatty acid oxidation and synthesis); (3) amino acid metabolism (degradation and synthesis); (4) the citrate cycle; and (5) oxidative phosphorylation.

The term **energy balance** relates energy input in the whole organism to energy expenditure. Positive or negative energy balance is determined by the energy content of the metabolic fuels ingested compared to the amount of energy expended through chemical reactions, physical exertion, and thermogenic processes. In the simplest case, energy balance exists when energy input equals energy expenditure on a daily basis.

Note that the relative proportions of carbohydrate, fat, and protein in the human diet need to be optimized to prevent metabolic disorders that can occur even under conditions of Caloric energy balance. For example, obtaining too many daily Calories from saturated fats can lead to cardiovascular disease, whereas excessive amounts of protein can cause nitrogen toxicity due to NH_4^+ overload.

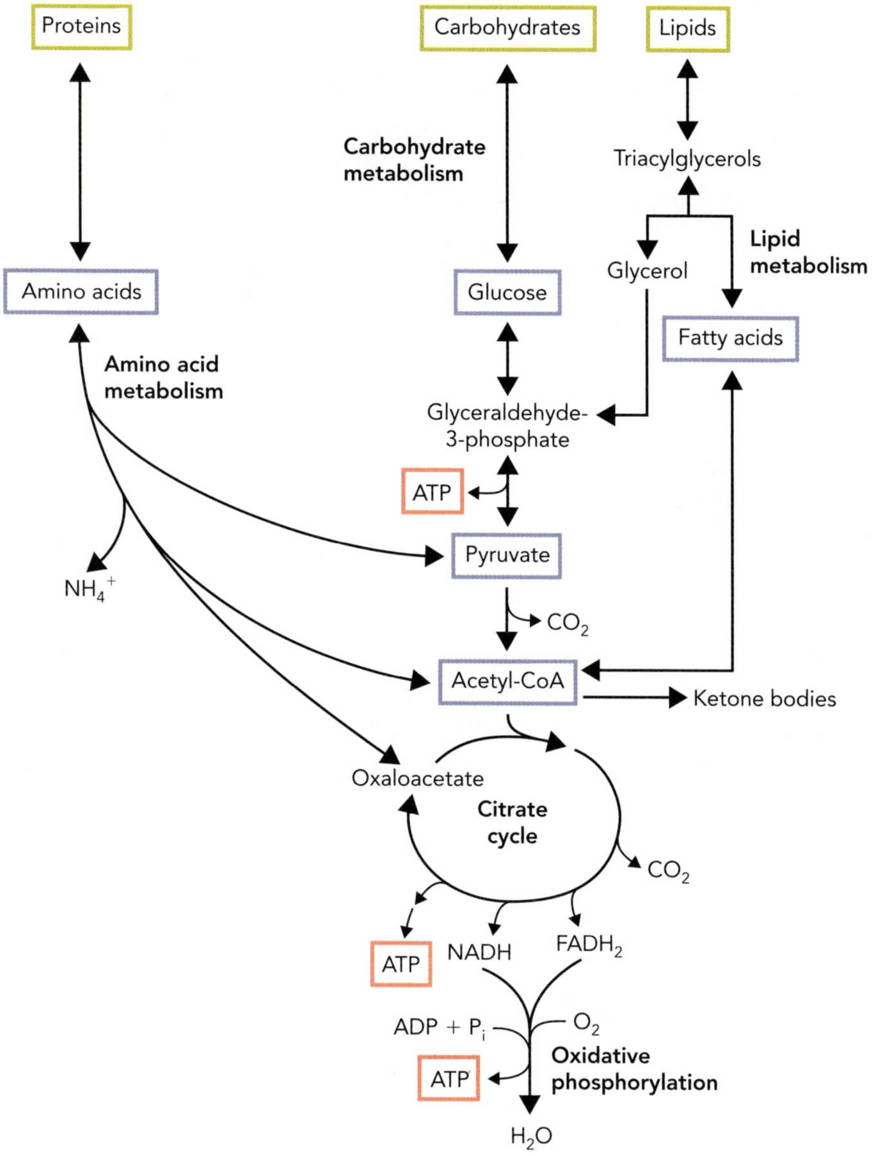

Figure 19.1 Proteins, carbohydrates, and lipids are the three major sources of metabolic fuel for the body. The key pathways required to convert these metabolic fuels into ATP are those involved in amino acid metabolism, carbohydrate metabolism, lipid metabolism, the citrate cycle, and oxidative phosphorylation. Primary metabolites in these pathways are amino acids, glucose, fatty acids, pyruvate, and acetyl-CoA. Waste products are NH_4^+, CO_2, and H_2O. See Figure 9.3 for a more complete metabolic map.

Specialized Metabolic Functions of Major Tissues and Organs

Figure 19.2 shows the locations and functions of the primary tissues and organs in the human body that play a direct role in metabolic flux. In addition to the liver, the muscles (skeletal and heart), adipose tissue, brain, kidneys (each described below) and several other organs play important supporting roles in metabolic integration.

One of these organs is the pancreas, a vital organ that secretes the hormones insulin and glucagon in response to changes in blood glucose levels. The pancreas also secretes digestive proteases (trypsin, chymotrypsin, and elastase) that degrade dietary proteins in the small intestine. The small intestine and large intestine are the two major tissue types in the gastrointestinal tract and function to absorb nutrients (small intestine) and water and electrolytes (large intestine). The stomach prepares food for the small intestine by producing an acidic food slurry called **chyme**. The chyme partially digests the food using chemical hydrolysis and also by activating the digestive protease pepsin. The stomach and small intestine were recently found also to secrete peptide hormones that control eating behaviors through neuronal signaling in the brain.

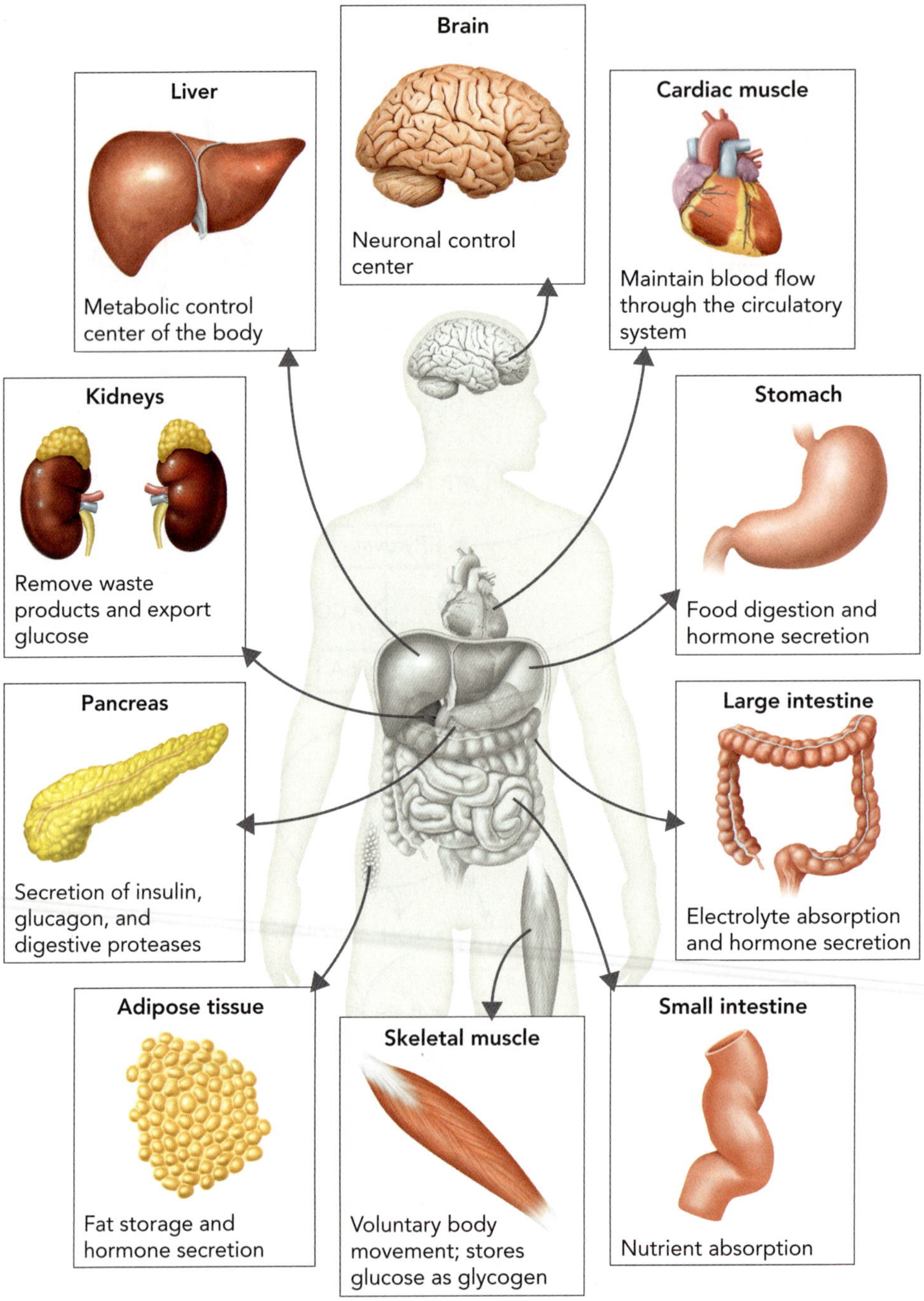

Figure 19.2 The metabolic functions are shown for major human organs and tissues.

Liver The liver serves as the central processing facility and metabolic hub of the human body. It determines what dietary nutrients and metabolic fuels are distributed to the peripheral (nonliver) tissues. The liver functions as a physiologic glucose regulator that removes excess glucose from the blood when carbohydrate levels are high (glucose influx). The liver also releases glucose from stored glycogen or as a product of gluconeogenesis when blood glucose levels are low (glucose efflux). Blood glucose regulation by the liver is controlled through the insulin and glucagon signaling pathways, which modulate metabolic flux through glycolysis, gluconeogenesis, and glycogen metabolism.

We have seen that dietary triacylglycerols are transported from the small intestine to peripheral tissues by chylomicrons that enter the lymphatic system (see Figure 15.15). But most nutrients absorbed in the small intestine are delivered directly to the liver via the **portal vein,** which is why the liver plays a key role in coordinating the distribution of dietary nutrients, as it is the first organ to inventory the contents of a meal. The

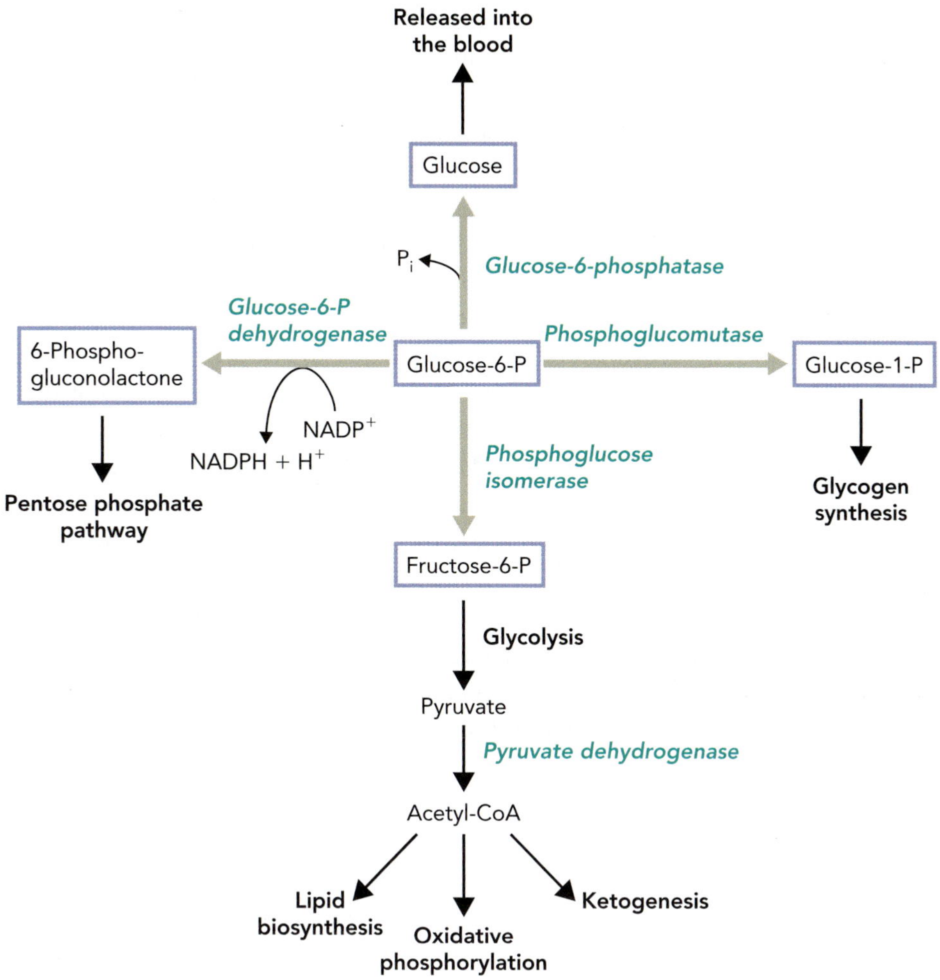

Figure 19.3 Glucose-6-phosphate in liver cells has several metabolic fates. Glucose-6-phosphate (glucose-6-P) is converted into four major products by liver enzymes: (1) glucose-1-phosphate (glucose-1-P) for use in glycogen synthesis; (2) glucose for release into the blood; (3) 6-phosphogluconolactone to generate NADPH by the pentose phosphate pathway; and (4) fructose-6-phosphate (fructose-6-P), which is used in the glycolytic pathway to produce pyruvate.

portal vein facilitates another liver function: that of inactivating toxins contained in the diet that would otherwise enter the circulatory system and cause cellular damage. Liver detoxification involves oxidation reactions by the P450 monooxygenase enzyme system (see Section 15.4). A large proportion of the dietary monosaccharides delivered by the portal vein are retained by the liver in the form of glucose-6-phosphate after phosphorylation of glucose by the enzymes hexokinase and glucokinase.

As shown in **Figure 19.3**, glucose-6-phosphate has at least four different fates, depending on the metabolic needs of the liver and the peripheral tissues. Most of the glucose-6-phosphate is used to synthesize liver glycogen after isomerization of glucose-6-phosphate to glucose-1-phosphate by the enzyme phosphoglucomutase. Glucose-6-phosphate can also be dephosphorylated in the liver by glucose-6-phosphatase and released into the blood to be used by other tissues, in particular the brain. If liver cells are in need of NADPH for biosynthetic reactions, then glucose-6-phosphate is converted to 6-phosphogluconolactone by glucose-6-phosphate dehydrogenase in the first reaction of the pentose phosphate pathway. Lastly, glucose-6-phosphate can be converted to fructose-6-phosphate by phosphoglucose isomerase and metabolized by the glycolytic pathway and the pyruvate dehydrogenase reaction to generate acetyl-CoA. The three primary metabolic fates of acetyl-CoA are lipid biosynthesis, ketogenesis, and oxidative phosphorylation.

Muscle The human body contains two types of muscle tissue that play a major role in metabolic integration: (1) **Skeletal muscle** uses different amounts of free fatty acids, glucose, or ketone bodies for metabolic fuel, depending on the physical movements required (rapid burst of activity or endurance activity). (2) **Cardiac muscle** uses mostly fatty acids and ketone bodies as metabolic fuel to sustain a steady heartbeat (averaging ~100,000 beats per day).

During the resting state, skeletal muscle uses fatty acids released from adipose tissue as a source of energy. The fatty acids are oxidized to generate acetyl-CoA, which is then used in the citrate cycle to produce reducing power (NADH and $FADH_2$) for oxidative phosphorylation. But when muscle contraction is required for a very short burst of activity—for example, serving a tennis ball to your opponent (2–3 seconds)—the exercising muscles make use of the intracellular ATP pool. If a more sustained level of muscle activity is needed, then additional ATP is quickly synthesized by the enzyme creatine kinase, using **phosphocreatine** as the phosphoryl group donor (**Figure 19.4**). The creatine kinase reaction is readily reversible and catalyzes the resynthesis of phosphocreatine when cellular ATP levels return to normal during muscle recovery.

Most of the stored glycogen in humans exists in muscle tissue that is spread throughout the body. But unlike the liver, which contains 10% glycogen by weight, individual muscle groups contain only ~1% glycogen by weight. Therefore, glycogen stores in the most active muscle groups become depleted after an hour of continual use, whereas liver glycogen maintains safe blood glucose levels for 12–18 hours. As muscle glycogen levels decline with continual use, muscle cells depend more on fatty acids released from adipose tissue and on ketone bodies produced in the liver to maintain the high rates of ATP synthesis needed for contraction. Muscle cells lack fatty acid synthase and glucose-6-phosphatase, which means that they cannot export fatty acids or glucose. Instead, they use these energy-rich compounds for muscle contraction. In this regard, muscle is truly a "selfish" tissue, using energy made available from other parts of the body for its own purpose of mechanical movement. Skeletal muscle can be used as an energy source for the body during times of starvation by providing amino acid substrates for liver and kidney gluconeogenesis.

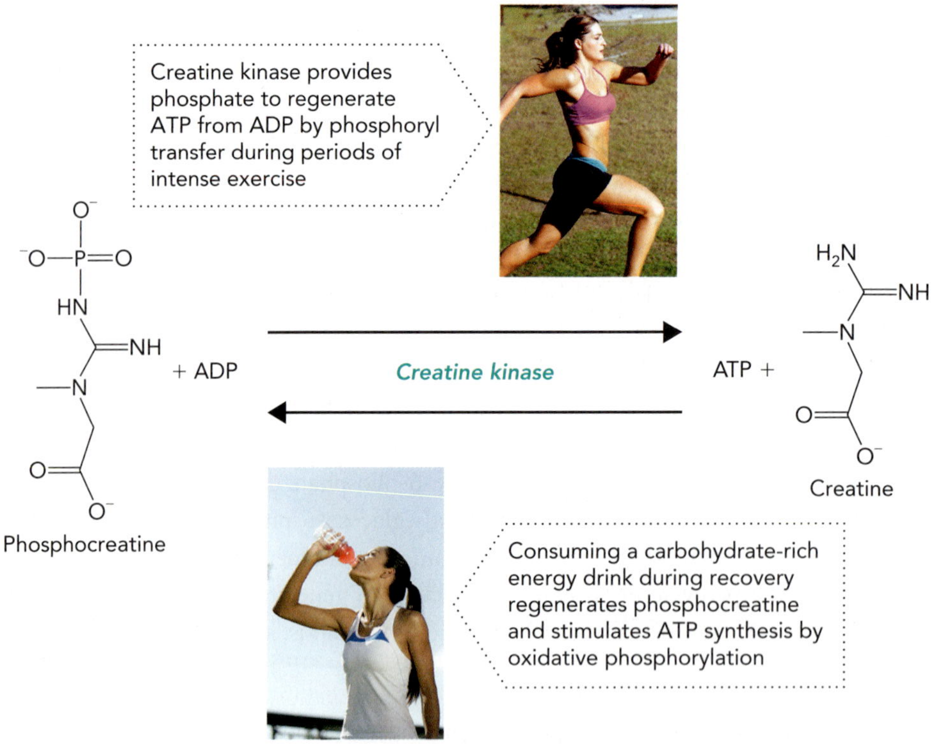
Creatine kinase provides phosphate to regenerate ATP from ADP by phosphoryl transfer during periods of intense exercise
+ ADP
Creatine kinase
ATP +
Phosphocreatine
Creatine
Consuming a carbohydrate-rich energy drink during recovery regenerates phosphocreatine and stimulates ATP synthesis by oxidative phosphorylation

Figure 19.4 Creatine kinase catalyzes a reversible phosphoryl transfer reaction that generates ATP in muscle cells as a readily available source of energy for muscle contraction. Phosphocreatine is resynthesized from ATP and creatine during the muscle recovery period, when ATP pools return to normal. INTENSE EXERCISE: CARL STEWART/ALAMY; MUSCLE RECOVERY: SANDRO DI CARLO DARSA/PHOTOALTO/GETTY IMAGES.

The energy requirements of cardiac muscle are quite different from those of skeletal muscle because the heart must be able to contract continually at the same rate for an entire lifetime. Therefore, the primary mode of energy conversion in cardiac muscle is aerobic respiration, using acetyl-CoA provided by fatty acid oxidation and ketone bodies. This high level of continual oxidative phosphorylation is supported by large numbers of mitochondria, which take up as much as 50% of the cytoplasmic space. Cardiac muscle does contain a small amount of phosphocreatine, but it does not store glycogen or fatty acids. Although the heart can use blood glucose for ATP generation, this is not a significant energy source for cardiac muscle cells, thereby ensuring that glucose is spared for the brain, which is a glucose-dependent tissue under normal conditions.

Adipose Tissue The function of adipose tissue was once thought to be limited to that of a fat depot storing fatty acids from adipocytes and releasing them in response to metabolic needs. It is now known, however, that adipose tissue is an endocrine organ that secretes peptide hormones called **adipokines** (adipocyte hormones) to regulate metabolic homeostasis. Adipokines are key regulators of metabolism and control important immunologic, neurologic, and developmental functions in the body.

Adipose tissue is widely distributed throughout the body, making up ~15% to 25% of an individual's mass, and it accounts for more than 500,000 kJ of stored energy. Fat stored in adipose tissue consists of two basic types: (1) **subcutaneous fat**, which is located just below the skin surface in the thighs, buttocks, arms, and face; and (2) **visceral fat** (sometimes called abdominal fat), which lies deep within the abdominal cavity and is known to secrete a variety of adipokines. Small amounts of fat deposits are also found near skeletal muscle, surrounding blood vessels, and in the mammary glands. The relative amounts of visceral and subcutaneous fat in an overweight individual can give rise to a so-called apple body type (more visceral fat), or a pear body type (more subcutaneous fat). Some studies suggest that the risk of cardiovascular disease may be higher in overweight individuals with high levels of visceral fat (apple shape) than in overweight individuals with high levels of subcutaneous fat (pear shape). It has been proposed that the molecular basis for this difference could be related to the secretion of adipokines from visceral fat but not from subcutaneous fat. Elevated levels of certain types of adipokines in the blood have been associated with increased risk of cardiovascular disease.

Most studies designed to calculate disease risks in overweight individuals rely on recorded height and weight data as a function of age, gender, and lifestyle. A single value called the **body mass index (BMI)** provides an estimation of total body fat in an average person. BMI values are derived by dividing a person's weight in kilograms by the square of his or her height in meters:

$$\text{Body mass index (BMI)} = \frac{\text{Weight (kg)}}{[\text{Height (m)}]^2}$$

It is generally accepted that a BMI value of less than 18.5 is considered underweight; values of 18.5–24.9 are within the normal weight range; values of 25–29.9 correspond to overweight; and BMI values ≥ 30 signify obese (**Figure 19.5**). Two other measurements used in studies to determine disease risks in overweight individuals are (1) the **waist-to-hip ratio**, which is based on the circumferences of an individual's waist and hips, and (2) **percent body fat**, which can be determined using whole-body bioelectrical impedance (fat impedes electrical conductance).

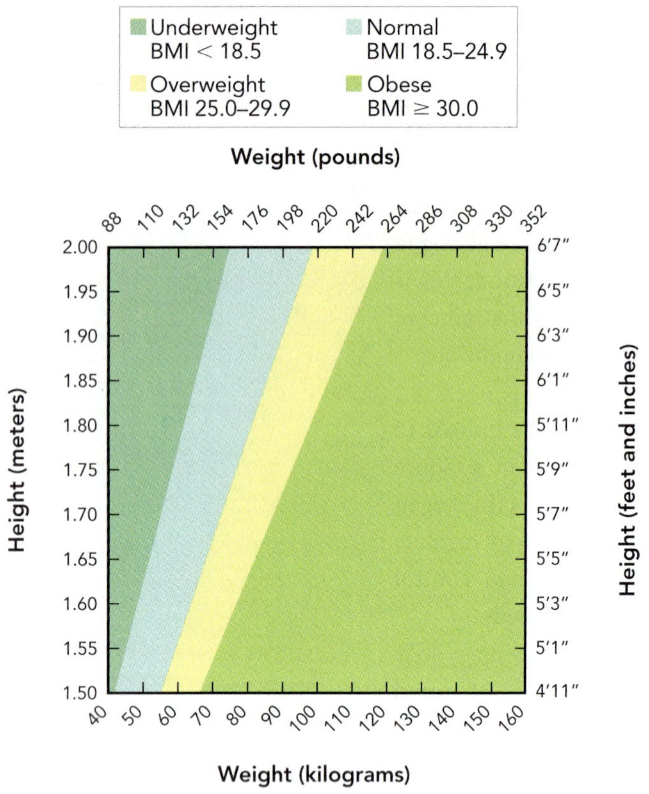

Figure 19.5 BMI is calculated from height and weight measurements, which can be used to estimate body fat, assuming that increased weight is due to increased adipose tissue. The ranges of BMI values corresponding to the terms *underweight, normal, overweight,* and *obese* are generally accepted based on the results of large clinical studies. This chart can be used to estimate one's BMI value; height and weight may be measured in either metric or imperial units.

Brain The brain is the control center of our bodies. It consists of about 100 billion nerve cells (neurons) that transmit electrical information along the neuronal axon using action potentials driven by changes in charge distribution across the plasma membrane. The key to these electrical impulses are ions that cross the membrane through channels controlled by neurotransmitter substances such as acetylcholine. These compounds function as signaling molecules between adjacent neurons (see Figure 8.14). The steady-state electrical charge across the membrane is maintained by ATP-dependent ion pumps, the most important of which is the Na^+–K^+ ATPase transporter protein (see Figure 6.46). On the basis of studies using ouabain to inhibit the Na^+–K^+ ATPase transporter, up to half of all the ATP generated in the brain goes toward keeping this critical ion pump fully active.

Studies have shown that about 20% of the oxygen consumed by the body is used for oxidative phosphorylation in the brain. Under normal conditions, the brain—unlike most other organs—depends exclusively on glucose, which provides the critical chemical energy needed for maintaining neuronal membrane potentials. Glucose is delivered to brain cells by microcapillaries that are surrounded by glial cells called **astrocytes**. These cells functionally define the **blood–brain barrier** (**Figure 19.6a**). Glucose metabolism by brain cells is the basis for **positron emission tomography (PET)** imaging, which uses the metabolic radioisotope 2-deoxy-2-(^{18}F)fluoro-D-glucose (^{18}F-deoxyglucose) as an indicator of high glycolytic activity (ATP production). **Figure 19.6b** shows how differences between ^{18}F-deoxyglucose metabolism in the brain of a healthy individual and that in the brain of an Alzheimer's patient can be visualized by PET imaging, indicating less glucose metabolism in the brain of the Alzheimer's patient.

The brain requires as much as 120 g of glucose each day, which accounts for 60% of the glucose used by the human body under normal conditions. The brain's dependence on glucose is illustrated by the feeling of light-headedness one experiences when blood glucose decreases from normal levels of ~4.5 mM (~80 mg/dL) to ~3.5 mM (~60 mg/dL), which can be brought on by lack of food or prolonged intense exercise. A more serious condition called **hypoglycemia** develops when blood glucose levels drop to 2.5 mM (45 mg/dL) as a result of fasting or excessive alcohol consumption. Symptoms of hypoglycemia include perspiring, mental confusion, and fainting. If glucose levels fall below ~2.2 mM (~40 mg/dL), then lethargy, coma, and death occur if the condition is not reversed.

Fatty acids cannot cross the blood–brain barrier because brain astrocytes lack the necessary enzymes to recover fatty acids contained within lipoprotein particles as triacylglycerols. But the energy-rich ketone bodies acetoacetate and D-β-hydroxybutyrate are able to enter the brain through astrocytes during prolonged starvation, when glucose levels are abnormally low. The brain adapts to using ketone bodies to supply the acetyl-CoA needed for ATP synthesis by oxidative phosphorylation.

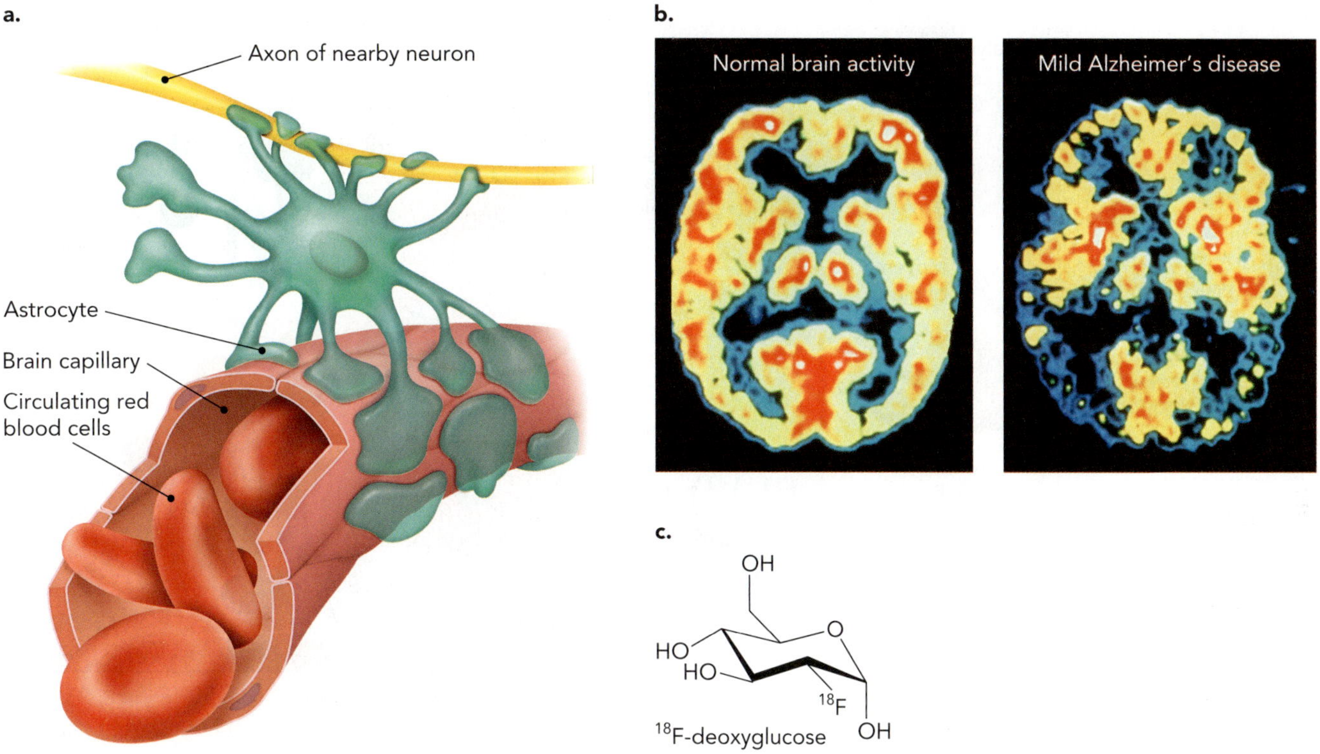

Figure 19.6 The brain depends on glucose as its sole energy source under normal physiologic conditions. **a.** Brain astrocytes make multiple contacts with the endothelial cells of microcapillaries. The astrocytes function to transport glucose metabolites to surrounding neurons for use in energy conversion reactions that generate ATP. **b.** Glucose use by brain cells is the molecular basis for PET imaging using ^{18}F-deoxyglucose. Areas of the brain showing bright yellow and red coloring signify elevated glucose metabolism and therefore high levels of brain activity. The brain of an Alzheimer's patient shows reduced brain activity using PET imaging. DR. ROBERT FRIEDLAND/SCIENCE SOURCE. **c.** Chemical structure of the metabolic radioisotope 2-deoxy-2-(^{18}F)fluoro-D-glucose (^{18}F-deoxyglucose).

Kidneys Humans have two kidneys, one located on either side of the abdominal cavity (see Figure 19.2). The kidneys are the only major organ system other than the liver that uses the gluconeogenic pathway to synthesize glucose for export to other tissues. Humans can survive with only one functioning kidney, making kidney transplants one of the few organ surgeries performed on live donor patients. Two healthy kidneys filter 6 L of human blood up to 30 times each day and remove 2 L of water containing concentrated levels of urea, NH_4^+, ketone bodies, and other soluble metabolites. To sustain this level of kidney function, a person needs to drink 2–3 L of water every day to replace the water lost by excretion, perspiration, and from evaporation when exhaling.

As shown in **Figure 19.7**, kidney function can be provided artificially using a procedure called **hemodialysis**, in which a dialysis machine filters blood from a patient with failing kidneys to remove waste products. This straightforward procedure is based on osmosis: A semipermeable membrane in the dialysis machine allows water, small molecules such as urea, and ions to diffuse across in response to a concentration gradient. Because the pore size of the semipermeable membrane is small, red blood cells and other large protein complexes in the blood do not diffuse across the membrane and are not removed in the buffered dialysate solution. Efficient diffusion rates across the semipermeable membrane are achieved by circulating the patient's blood in a direction opposite that of the circulation of the buffered dialysate solution, which is under negative pressure and contains physiologic concentrations of Na^+, K^+, Ca^{2+}, Mg^{2+}, HCO_3^-,

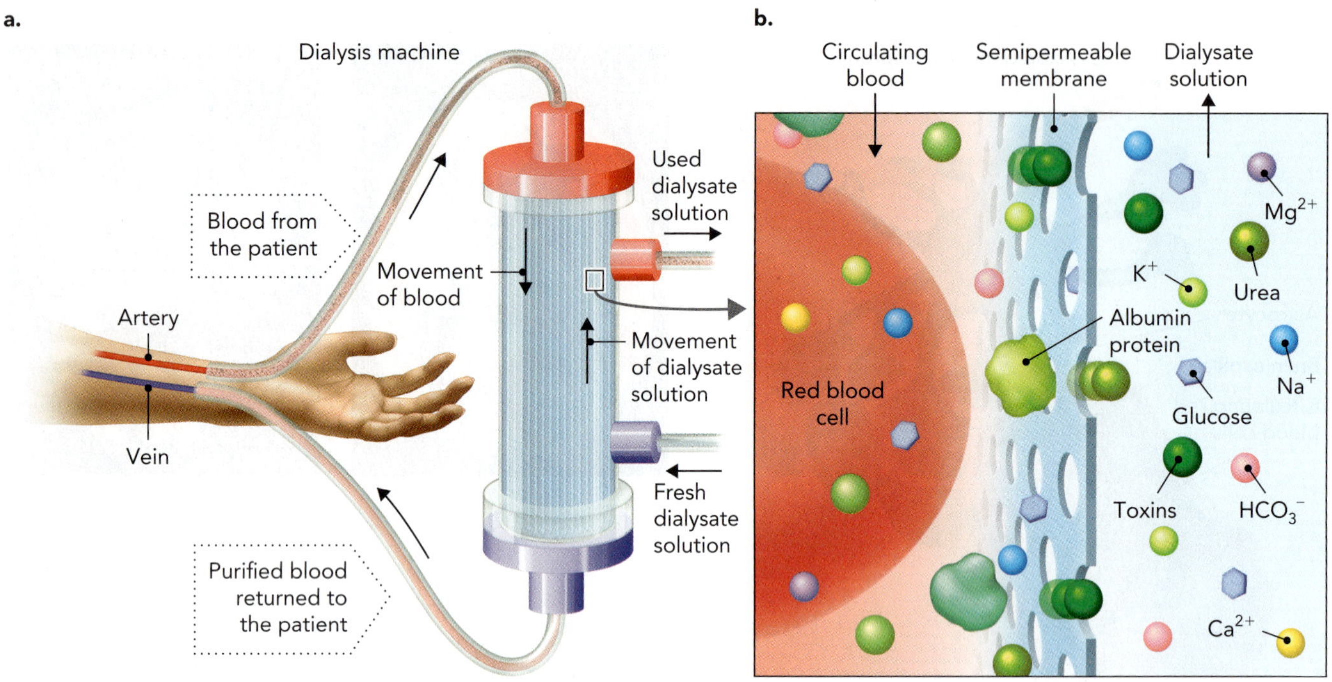

Figure 19.7 Hemodialysis is a procedure that removes waste products from the blood of patients with kidney disease using high diffusion rates across a semipermeable membrane. **a.** The patient's blood is circulated through the dialysis machine in a direction opposite that of the circulation of buffered dialysate solution to increase diffusion rates across the semipermeable membrane. The dialysate solution is continually replenished, and the purified blood is returned to the patient. **b.** Urea, toxins, and other small molecules can cross the semipermeable membrane from the blood into the buffered dialysate solution, which contains physiologic levels of ions and glucose. Red blood cells and proteins (not drawn to scale) are larger than the pore size of the membrane and therefore are retained in the circulating blood.

and glucose. Patients with severe kidney disease require hemodialysis up to five times a week for ~4 hours at a time.

Metabolite Flux between Tissues Optimizes Use of Stored Energy

Metabolic integration within the human body depends on the redistribution of metabolites, ions, and hormones by the circulatory system. This complex network consists of ~150,000 km of blood vessels (from major veins and arteries to micro-capillaries) that recycle 6 L of blood every minute throughout the body. The circulatory system links together the major tissues and organs of the body in such a way that biochemical pathways in different cells share metabolites, ensuring that the metabolic efficiency of the whole organism is greater than the sum of its parts. This process of maintaining optimal metabolite concentrations and managing chemical energy reserves in tissues is called **metabolic homeostasis**. This term describes steady-state conditions that apply to a wide variety of physiologic parameters. Metabolic homeostasis is affected by physical activity, psychological stress, timing and extent of feeding, and tissue dysfunction.

The liver is the control center of this metabolic network and plays a crucial role in regulating metabolite flux among tissues and organs under normal homeostatic conditions. The six primary functions together required to maintain metabolic homeostasis under normal conditions are illustrated in **Figure 19.8**:

1. The primary role of the liver in this metabolic network is to export glucose and triacylglycerols to the peripheral tissues for use as metabolic fuel.
2. The brain requires a constant input of glucose, one of the body's most precious metabolites. Though considered to be the most vital human organ, the brain is also an energy drain on the metabolic system.
3. Cardiac muscle uses fatty acids and ketone bodies for most of its energy needs but also uses small amounts of glucose.

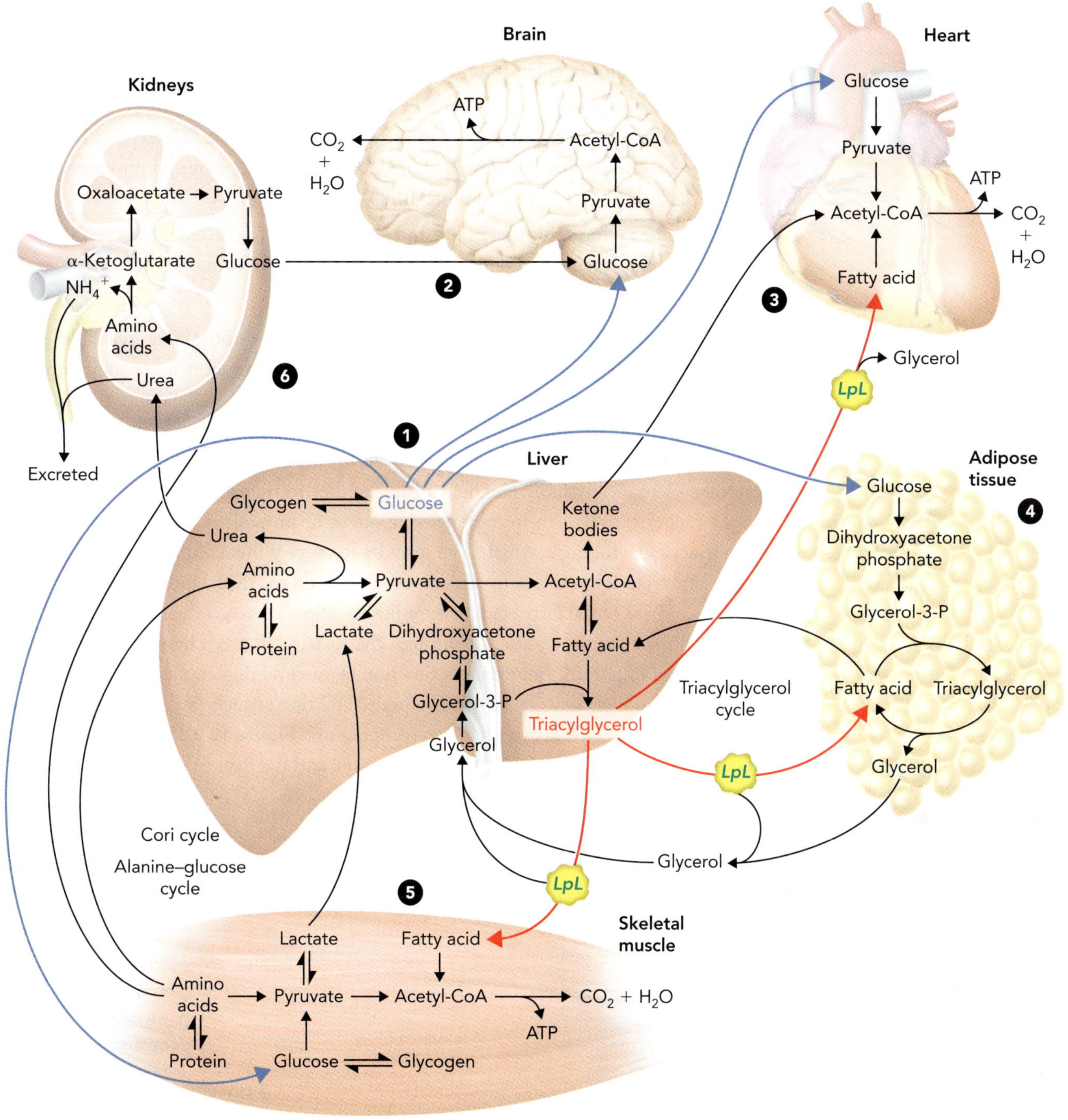

Figure 19.8 The flux of metabolites between major tissues in the human body under normal homeostatic conditions is shown. The liver serves as the metabolic hub of the body and coordinates the exchange of metabolites between major tissues and organs. Glucose and triacylglycerols are the primary export products of the liver under normal physiologic conditions and are used as sources of energy for oxidative phosphorylation in most tissues. Triacylglycerols in the blood are hydrolyzed by the enzyme lipoprotein lipase (LpL) on the surface of endothelial tissue to release free fatty acids and glycerol into nearby tissues as shown here. Amino acids, ketones, and lactate are usually found at relatively low levels in the blood, but they can be very important metabolites at times of limited food availability or intense physical activity. See the text for details about the six enumerated functions that are required to maintain metabolic homeostasis.

4. The exchange of fatty acids and triacylglycerols between the liver and adipose tissue is an ongoing process called the triacylglycerol cycle, which maintains circulation of high-energy fatty acids, as described shortly.
5. Skeletal muscle uses glucose and fatty acids derived from both the liver and dietary sources for ATP synthesis. In turn, skeletal muscle exports lactate back to the liver to complete the Cori cycle during times of prolonged physical exertion (see Figure 14.23).
6. The amino acids glutamine and alanine transport excess nitrogen obtained from protein degradation in the muscle to the liver and kidneys for excretion as nitrogen waste in the form of urea.

Metabolite exchange between tissues is critical to optimizing available energy stores at the physiologic level. For example, the brain requires a constant supply of glucose to ensure high-fidelity neuronal transmissions, and skeletal muscle must have enough glycogen to permit rapid muscle contraction in response to imminent danger or to obtain food. Similarly, adipose tissue must be able to control the release and storage of triacylglycerols obtained from the diet to manage this high-energy metabolic fuel effectively. An important component of physiologic energy homeostasis is the **triacylglycerol cycle**, which is an interorgan process that continually circulates fatty acids and triacylglycerols between adipose tissue and the liver. Under homeostatic conditions, ~75% of the fatty acids released from adipocytes into the blood is returned to adipose tissue as triacylglycerols through the systemic route. The triacylglycerol cycle provides an important homeostatic function by maintaining energy-rich fatty acids in circulation so that they can be used by peripheral tissues such as skeletal muscle.

The triacylglycerol cycle has two components (**Figure 19.9**): (1) the *systemic* component cycles fatty acids between adipose tissue and the liver in the form of fatty acids

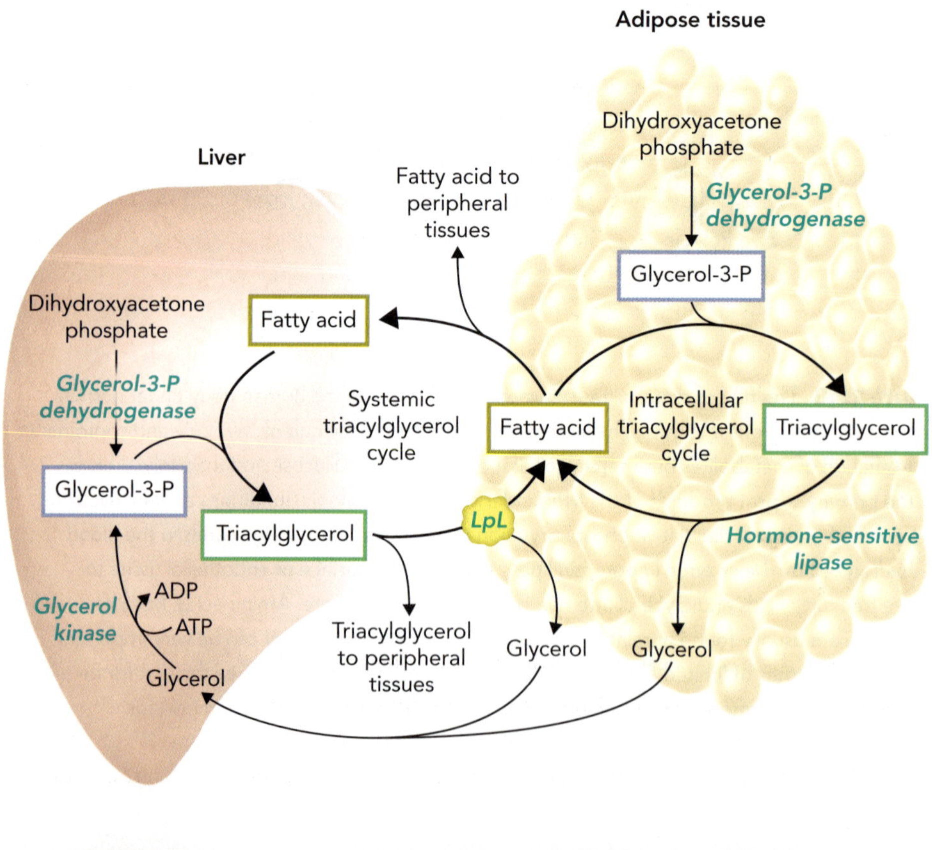

Figure 19.9 The triacylglycerol cycle exchanges triacylglycerols synthesized in the adipose tissue with those synthesized in liver cells, using the same pool of fatty acids. The systemic component of the triacylglycerol cycle exchanges free fatty acids and triacylglycerols between adipose tissue and liver cells, whereas the intracellular component of the cycle interconverts pools of free fatty acids and triacylglycerols in adipocytes. Most of the fatty acids and triacylglycerols are recycled back to the adipose tissue unless needed by the peripheral tissues for energy conversion reactions or membrane biosynthesis. LpL = lipoprotein lipase.

bound to albumin and triacylglycerols contained in lipoprotein particles; and (2) the *intracellular* component in adipocytes cycles fatty acids in the form of cytosolic free fatty acids and triacylglycerols stored in lipid droplets. Both the systemic and intracellular components of the triacylglycerol cycle depend on sufficient levels of glycerol-3-phosphate (glycerol-3-P) to resynthesize triacylglycerol from fatty acids and maintain flux through the cycle. The enzyme glycerol-3-phosphate dehydrogenase (glycerol-3-P dehydrogenase) generates glycerol-3-P from dihydroxyacetone phosphate in both hepatocytes and adipocytes. In addition, hepatocytes contain the enzyme glycerol kinase, which phosphorylates serum glycerol to produce glycerol-3-P.

Flux through the triacylglycerol cycle depends on availability of glycerol-3-P for triacylglycerol resynthesis in adipocytes and hepatocytes. Because adipocytes lack appreciable amounts of the enzyme glycerol kinase, they can only generate glycerol-3-P from dihydroxyacetone phosphate using glycerol-3-P dehydrogenase (see Figure 19.9). When dietary glucose is readily available, dihydroxyacetone phosphate in adipocytes and hepatocytes is derived from glycolysis and is used to make glycerol-3-P for triacylglycerol biosynthesis. When blood glucose levels are low, however, dihydroxyacetone phosphate is synthesized from the carbon backbones of amino acids and lactate via pyruvate using a pathway called **glyceroneogenesis**.

As shown in **Figure 19.10**, the glyceroneogenic pathway converts pyruvate to phosphoenolpyruvate using the gluconeogenic enzymes pyruvate carboxylase and malate dehydrogenase in mitochondria and the cytosolic enzymes malate dehydrogenase and phosphoenolpyruvate carboxykinase. By reversing several glycolytic reactions, phosphoenolpyruvate is then converted to glyceraldehyde-3-phosphate (glyceraldehyde-3-P), which is isomerized to dihydroxyacetone phosphate by triose phosphate isomerase. Because adipocytes do not contain the gluconeogenic enzymes fructose-1,6-bisphosphatase and glucose-6-phosphatase, the dihydroxyacetone phosphate generated by glyceroneogenesis in adipose tissue is used exclusively for glycerol-3-P synthesis.

Control of Metabolic Homeostasis by Signal Transduction

Metabolic homeostasis is regulated by physiologic inputs in response to fluctuating nutrient levels in the blood, as well as by neuronal inputs to the brain in response to environmental changes. In other words, human biochemical balance differs when one is hungry from when one has just eaten and also differs when one is hot from when one is cold. In this subsection, we examine physiologic control of metabolic homeostasis by describing biochemical responses to insulin and glucagon signaling and control of metabolic gene expression in response to signaling through the peroxisome proliferator–activated receptor (PPAR) nuclear receptors. In Chapter 8, we described the biochemical mechanisms by which the insulin receptor (see Figure 8.50), glucagon receptor (see Figure 8.28), and PPAR nuclear receptors (see Figure 8.60) mediate cellular responses. Neuronal inputs to the brain regulating metabolic homeostasis are described in Section 19.2. We include in this subsection a number of cross-references to figures in other chapters or to figures elsewhere in this chapter so that you can more easily see the integration between different metabolic pathways.

Insulin and Glucagon Control of Glucose Homeostasis Insulin and glucagon are considered the "yin and yang" of glucose homeostasis in that they have opposing functions in controlling metabolic homeostasis. Insulin release from the pancreas is associated with the fed state, making it the "I just ate" hormone. Consistent with metabolic

Figure 19.10 The glyceroneogenic pathway functions in adipocytes and hepatocytes. In this pathway, pyruvate is used to generate phosphoenolpyruvate, which is then converted to dihydroxyacetone phosphate. The dihydroxyacetone phosphate is then converted to glycerol-3-P by the enzyme glycerol-3-P dehydrogenase. Gluconeogenic reactions 3–5 are listed in Table 14.1. Fructose-1,6-BP = fructose-1,6-bisphosphate.

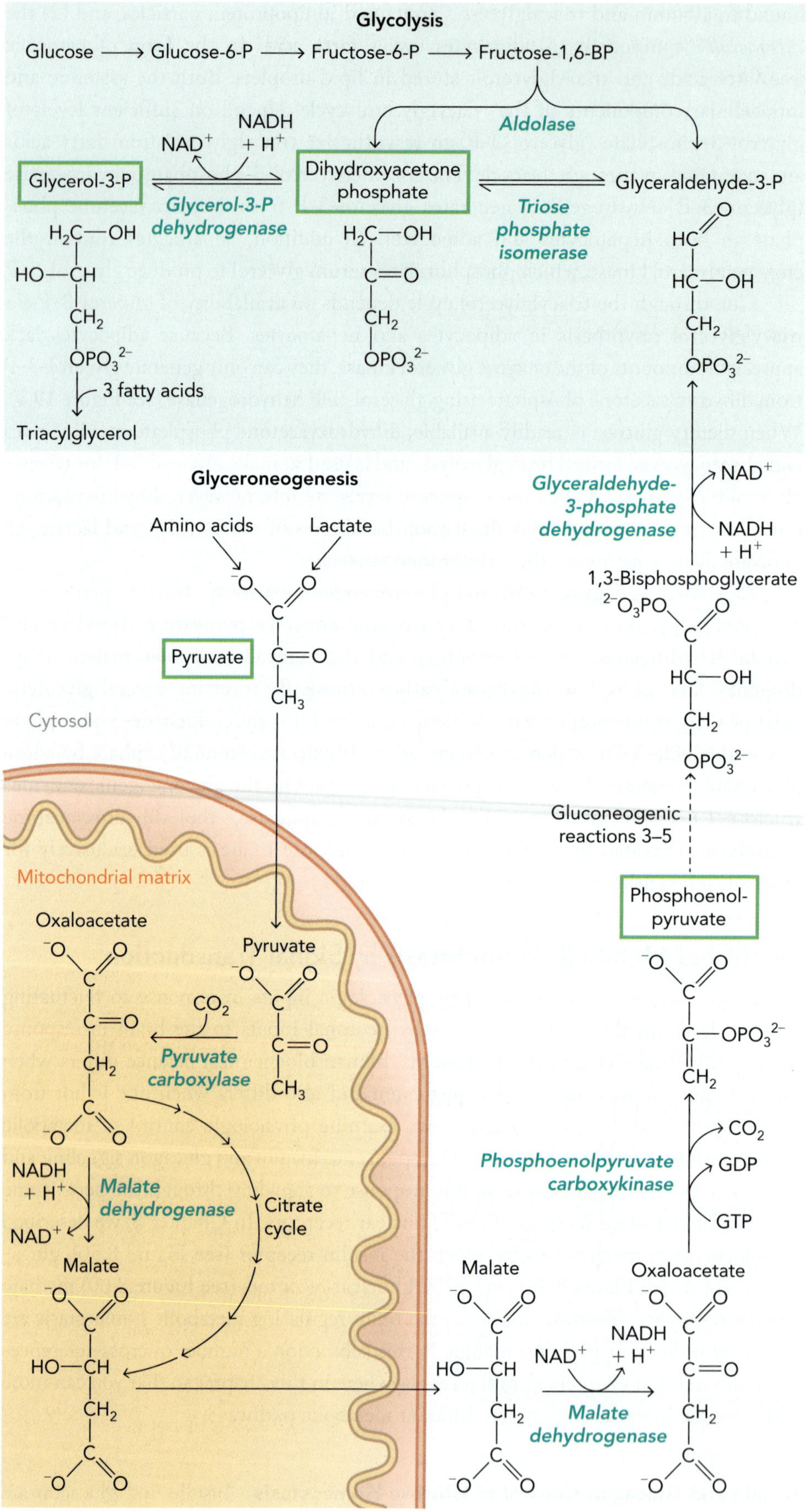

changes induced by eating, insulin stimulates glucose uptake, activates glycogen and fatty acid synthesis, and decreases appetite through neuronal signaling in the brain. In contrast, glucagon is the "I am hungry" hormone and accordingly stimulates gluconeogenesis, glycogen degradation, and fatty acid export from adipose tissue when food is not available.

Insulin and glucagon are synthesized as prohormones in regions of the pancreas called the **islets of Langerhans**, named after the German medical student Paul Langerhans, who first described these hormone-secreting cell clusters in 1869. The islets of Langerhans have three types of cells that produce peptide hormones (**Figure 19.11**). The first type, the β cells, make up the majority of cells in this region of the pancreas and are responsible for insulin secretion. The second type, called α cells, secrete glucagon. The third type, the δ cells, produce **somatostatin**, which is a paracrine hormone that functions locally to control the secretion of insulin, glucagon, and digestive proteases. In addition to hormone secretion by the cells in the islets of Langerhans, the pancreatic **acinar cells** secrete digestive proteases into the pancreatic duct, which empties into the duodenum of the small intestine.

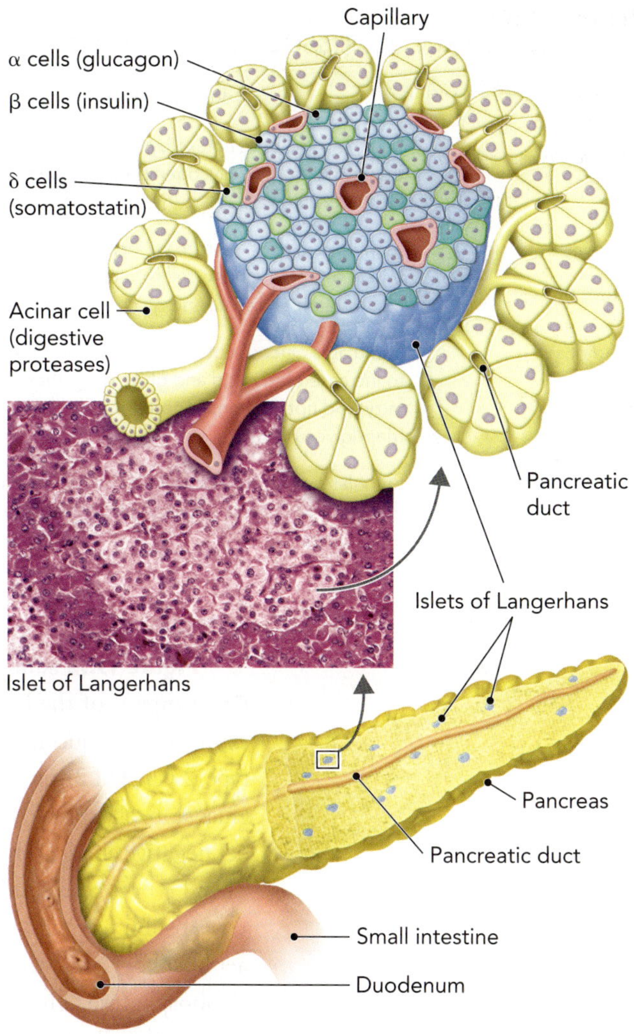

Figure 19.11 Insulin and glucagon are peptide hormones produced in the pancreas. Cells in regions of the pancreas called the islets of Langerhans secrete insulin (β cells) and glucagon (α cells) into the capillaries in response to blood glucose levels; the δ cells produce somatostatin, which is a paracrine hormone that controls secretory processes in the pancreas. Acinar cells synthesize digestive enzymes that are secreted into the pancreatic duct and delivered to the small intestine through the duodenum. PHOTO: DR. GLADDEN WILLIS/VISUALS UNLIMITED/CORBIS.

Elevated blood glucose levels stimulate the activity of glucokinase in pancreatic β cells, which leads to the release of insulin from intracellular vesicles (see Figure 9.41). Regulation of glucagon secretion from α cells is more complex and is thought to involve both stimulatory neuronal signals and inhibitory paracrine effects of insulin and somatostatin. Note that the pancreas secretes insulin and glucagon at the same time; however, the relative ratio of these two hormones is tightly regulated to achieve optimal blood glucose levels. When the balance is shifted toward increased insulin secretion, then blood glucose levels are reduced, whereas elevated levels of glucagon secretion result in increased blood glucose levels.

Insulin signaling stimulates glucose uptake in liver, skeletal muscle, and adipose tissue and activates fatty acid uptake and triacylglycerol storage in adipose tissue (**Table 19.1**). Glucose uptake in liver cells is primarily due to increased metabolic flux through glycolytic, glycogen synthesis, and triacylglycerol synthesis pathways. In addition to activating glycolysis and glycogen synthesis in skeletal muscle cells, insulin also stimulates translocation of the glucose transporter type 4 (GLUT4) protein to the plasma membrane. In adipose tissue, insulin signaling leads to GLUT4 translocation and increased rates of fatty acid uptake and triacylglycerol storage. Insulin stimulates neuronal signaling in the hypothalamus region of the brain, which controls eating behavior and energy expenditure.

Glucagon signaling in liver tissue stimulates glucose export as a result of increased rates of gluconeogenesis and glycogen degradation. In adipose tissue, glucagon activates triacylglycerol hydrolysis and fatty acid export. Skeletal muscle and brain cells lack appreciable levels of glucagon receptors and are considered to be glucagon insensitive.

Table 19.1 EFFECTS OF INSULIN AND GLUCAGON SIGNALING ON METABOLIC PATHWAYS IN MAJOR TISSUES OF THE HUMAN BODY

Tissue	Insulin signaling	Glucagon signaling
Liver	Stimulates glucose uptake by increasing metabolic flux through glycolytic, glycogen synthesis, and triacylglycerol synthesis pathways	Stimulates glucose export by increasing metabolic flux through gluconeogenesis and glycogen degradation pathways
Skeletal muscle	Stimulates glucose uptake by increasing the level of GLUT4 protein on the cell surface and also by increasing flux through glycolytic and glycogen synthesis pathways	No effect
Adipose	Stimulates glucose uptake by increasing GLUT4, leading to increased fatty acid and glycerol synthesis; activates fatty acid uptake from lipoprotein particles and promotes triacylglycerol storage	Stimulates fatty acid export by activating triacylglycerol hydrolysis at the surface of lipid droplets
Brain	Stimulates neuronal signaling in the hypothalamus that leads to decreased eating and increased energy expenditure	No effect

Figure 19.12 summarizes the primary responses of insulin signaling in four major target organs of the body. Activation of insulin signaling in liver cells leads to elevated expression of glucokinase and a net influx of glucose through the glucose transporter type 2 (GLUT2) protein. Glucokinase-mediated phosphorylation of glucose generates glucose-6-phosphate, which cannot be transported back through GLUT2 because of its net negative charge. The trapped glucose-6-phosphate is oxidized by glycolysis to replenish ATP or is converted to glucose-1-phosphate and used to synthesize liver glycogen.

Insulin signaling in liver cells also activates protein phosphatase 1, which dephosphorylates the dual-function enzyme phosphofructokinase-2/fructose-2,6-bisphosphatase. The result is activation of phosphofructokinase-1 through increased levels of the allosteric regulator fructose-2,6-bisphosphate (see Figure 14.20). Increased levels of fructose-2,6-bisphosphate, in turn, inhibit fructose-1,6-bisphosphatase, thereby decreasing metabolic flux through the gluconeogenic pathway.

Another downstream effect of protein phosphatase 1 activation in liver cells is a net increase in glycogen synthesis, which results from stimulation of glycogen synthase activity and inhibition of phosphorylase kinase and glycogen phosphorylase activities (see Figure 14.45). Lastly, the pyruvate dehydrogenase complex and protein phosphatase 2A are both stimulated by insulin signaling in liver cells. Together, this increased enzyme activity leads to increased fatty acid synthesis through production and utilization of acetyl-CoA. Specifically, protein phosphatase 2A activates acetyl-CoA carboxylase, which synthesizes malonyl-CoA from acetyl-CoA to provide substrate for the fatty acid synthase reaction (see Figure 16.33).

As shown in Figure 19.12, insulin signaling in skeletal muscle cells is similar to insulin signaling in liver cells in that the primary downstream effect is activation of protein phosphatase 1, leading to increased glucose influx and elevated rates of glycogen synthesis. Activity of the pyruvate dehydrogenase complex is also stimulated by insulin signaling in skeletal muscle cells. But three key differences occur with respect to insulin effects in these two cell types.

1. The primary glucose transporter in skeletal muscle cells (and adipose tissue) is GLUT4, which is present at low levels and translocated to the cell surface in response to insulin signaling. This is in contrast to GLUT2, the major glucose

transporter in liver and pancreatic β cells, which is present on the cell surface independent of hormone signaling.

2. Skeletal muscle cells do not contain the enzymes glucose-6-phosphatase and fatty acid synthase. Therefore, all of the glucose that enters muscle cells is used for metabolic fuel within this tissue; it cannot be exported or converted to fatty acids to be stored in adipose tissue.
3. The mass of skeletal muscle tissue in the body is ~30 times greater than the mass of the liver, and skeletal muscle uses glucose for its own energy needs or converts it to glycogen. Therefore, insulin stimulation of glucose uptake in muscle cells is the primary mechanism by which glucose levels are reduced in the blood after a high-carbohydrate meal.

The global effects of insulin signaling in adipose tissue leads to translocation of GLUT4 proteins to the cell surface, resulting in increased glucose influx into the cells. The glucose is used for fatty acid synthesis in adipocytes, following the same metabolic route as in liver cells (glycolysis, citrate shuttle, fatty acid synthesis). Glucose is also used for the production of glycerol for triacylglycerol biosynthesis (see Figure 19.10). To increase triacylglycerol storage in adipocytes after a meal, insulin stimulates the activity of lipoprotein lipase, which hydrolyzes fatty acids from chylomicrons and very-low-density lipoprotein (VLDL) particles to promote lipid uptake (see Figure 15.18). In addition, insulin signaling inhibits hormone-sensitive lipase to decrease fatty acid release from adipocytes and thereby alters flux through the triacylglycerol cycle to favor lipid storage in adipocytes.

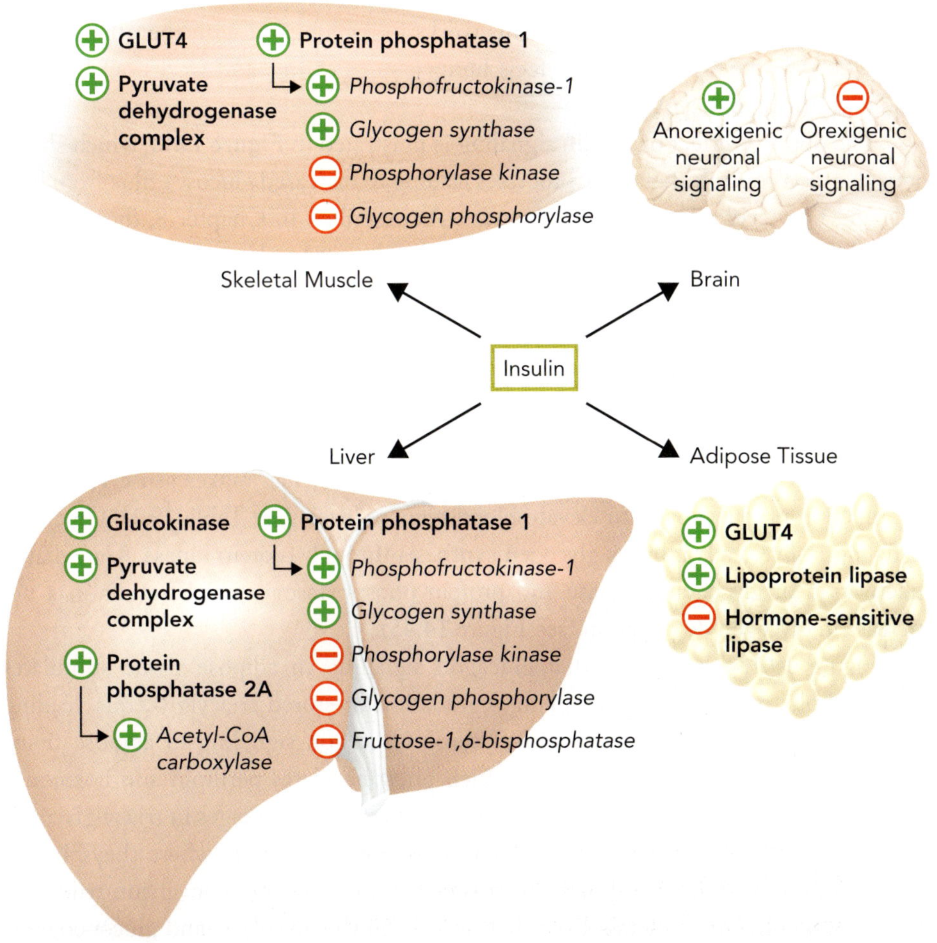

Figure 19.12 Insulin has effects in liver, skeletal muscle, adipose tissue, and the brain. Enzymes or proteins regulated directly by insulin signaling are shown in boldface (activated or inhibited), and enzymes or proteins affected by insulin regulation of protein phosphatase 1 or protein phosphatase 2A are shown in italics (activated or inhibited). Insulin signaling in the brain activates neuronal signals, regulating eating behaviors and energy expenditure (see Figure 19.25 later in this chapter).

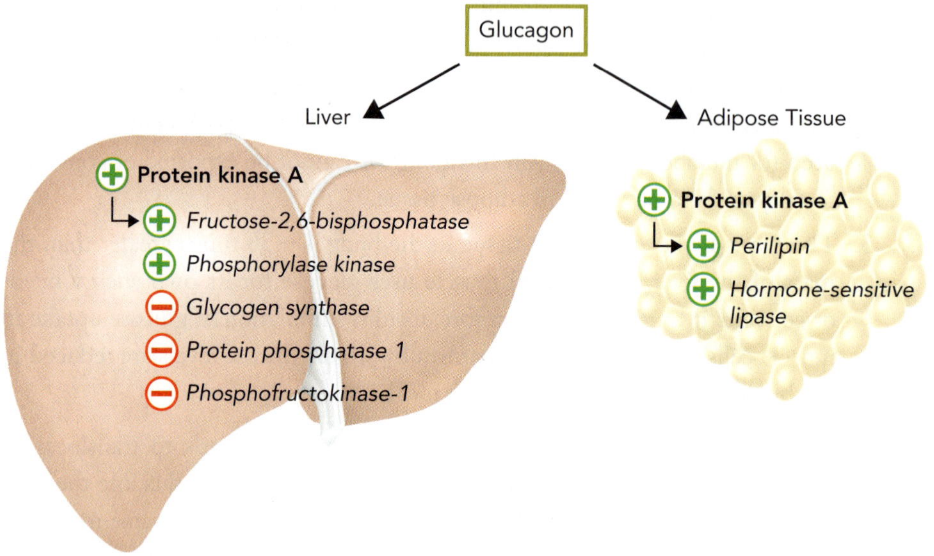

Figure 19.13 Glucagon stimulates signaling systems in liver and adipose tissue. Activation of glucagon receptors in liver and adipose tissue increases intracellular levels of cyclic AMP, which activates protein kinase A signaling. Enzymes or proteins regulated by protein kinase A are shown in italics (activated or inhibited). In liver cells, protein kinase A phosphorylates the dual-specificity enzyme phosphofructokinase-2/fructose-2,6-bisphosphatase, leading to increased flux through gluconeogenesis and decreased flux through glycolysis. Protein kinase A also phosphorylates glycogen synthase and phosphorylase kinase in liver cells, resulting in inhibition of glycogen synthesis and activation of glycogen degradation, respectively. Glucagon signaling in adipose tissue activates triacylglycerol hydrolysis through protein kinase A–mediated phosphorylation of perilipin and hormone-sensitive lipase.

The importance of insulin signaling in the brain has only recently been elucidated. It is now known that insulin regulates the release of neuropeptides in the hypothalamus. As indicated in Figure 19.12, insulin activates anorexigenic neurons in the brain, sending out the message "eat less and catabolize more." At the same time, insulin inhibits signaling through orexigenic neurons in the hypothalamus, which normally sends out the message "eat more and catabolize less." The net effect of insulin on the brain is to modify behavior and energy expenditure to provide feedback inhibition in response to eating a carbohydrate-rich meal. This works in conjunction with the other insulin signals so that the body no longer feels hungry, stops eating, and works on digesting food and storing nutrients.

The opposite effect of insulin signaling is illustrated in **Figure 19.13**, which shows how glucagon controls metabolic processes leading to increased glucose efflux from liver cells and fatty acid release from adipocytes. As described in Chapter 8, the glucagon receptor is a G protein–coupled receptor that stimulates adenylate cyclase activity, leading to the production of cyclic AMP and activation of protein kinase A (see Figure 8.28). As described in Chapter 14, the two major targets of protein kinase A in liver cells are phosphorylase kinase, which induces glycogen degradation, and phosphofructokinase-2/fructose-2,6-bisphosphatase, resulting in decreased fructose-2,6-bisphosphate levels. The net effect of lowering fructose-2,6-bisphosphate levels is increased flux through gluconeogenesis and decreased flux through glycolysis, which makes sense given that the function of glucagon is to elevate blood glucose levels (see Table 19.1). Glucagon signaling in liver cells inhibits glycogen synthesis through protein kinase A–mediated inactivation of glycogen synthase and protein phosphatase 1, which results in a net increase in glycogen degradation (see Figure 14.45).

Figure 19.13 also shows that glucagon signaling in adipose tissue stimulates the metabolic processes that result in degradation of stored triacylglycerols, leading to the release of free fatty acids and glycerol. Glucagon signaling in adipose tissue is also mediated by protein kinase A, which phosphorylates perilipin and hormone-sensitive lipase, resulting in hydrolysis of the three acyl ester linkages in triacylglycerols (see Figure 15.23). The released free fatty acids enter the blood, where they bind to serum albumin and are transported to tissues for use in energy conversion reactions. The released glycerol is metabolized by liver cells in the glycolytic and gluconeogenic

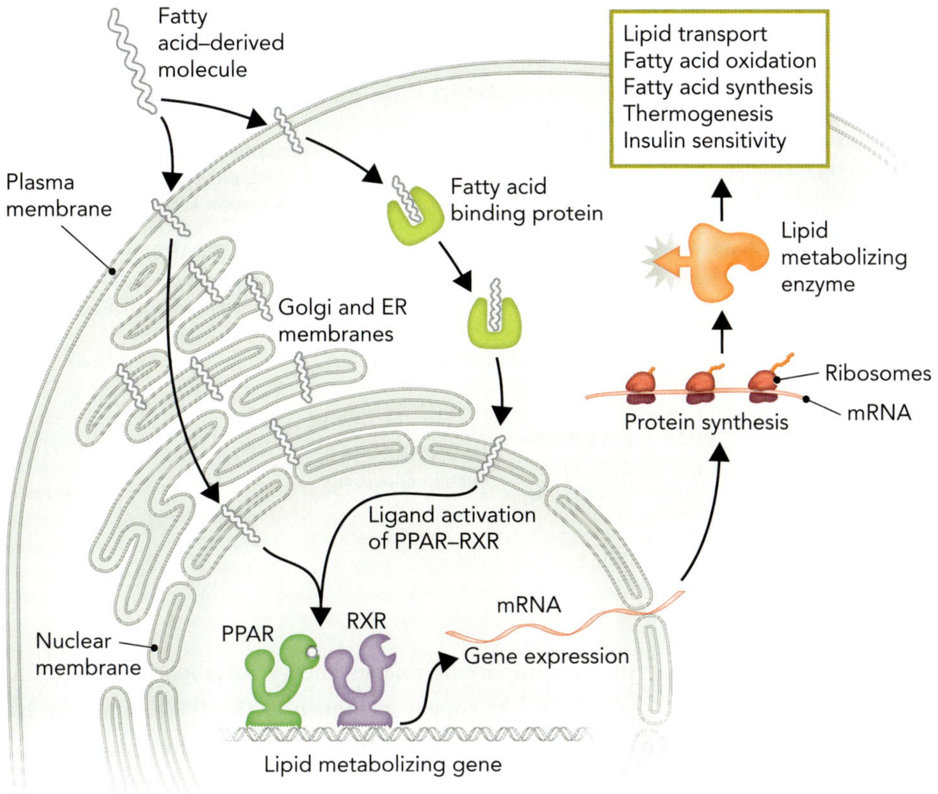

Figure 19.14 PPAR signaling mediates long-term control of lipid metabolism through regulation of gene expression. Fatty acid–derived metabolites are PPAR ligands that enter the nucleus through intracellular membranes or by binding to fatty acid binding proteins. Ligand-activated PPAR–retinoid X receptor (RXR) heterodimers bind to gene regulatory sequences and control the expression of lipid metabolizing genes (see Figure 8.60). The newly synthesized lipid metabolizing enzymes control a variety of cellular processes, depending on the function of the enzyme and the cell type.

pathways. The overall effect of glucagon signaling on the body when it is low on metabolic fuel is to mobilize energy reserves from glycogen and lipid stores and, at the same time, stimulate the desire to eat.

Peroxisome Proliferator–Activated Receptor Signaling Metabolic homeostasis is controlled not only by peptide hormone signaling through insulin and glucagon but also by the transcriptional regulatory activities of three peroxisome proliferator–activated receptor (PPAR) nuclear receptor proteins. First discovered in the early 1990s, the PPARα, PPARγ, and PPARδ nuclear receptor proteins are key molecules in controlling metabolic homeostasis in humans. But unlike the insulin and glucagon receptors, which rapidly activate intracellular phosphorylation signaling cascades, the PPARs regulate gene expression in response to the binding of low-affinity, fatty acid–derived nutrients such as polyunsaturated fatty acids and eicosanoids (**Figure 19.14**).

The ligand binding function of PPARs makes them ideal metabolic sensors of lipid homeostasis, resulting in long-term control of pathway flux by directly altering the steady-state levels of key proteins. The name *peroxisome proliferator–activated receptor* was originally coined to describe PPARα, which was discovered in rat hepatocytes as the mediator of peroxisome biogenesis in response to the drug **clofibrate**. Clofibrate-related drugs (fibrates) have been used to treat high serum cholesterol in humans, and although they do not cause peroxisome proliferation in human hepatocytes, the name peroxisome proliferator–activated receptor has been applied to all three of the PPAR family members: PPARα, PPARγ, and PPARδ.

PPARs have a variety of functions in lipid metabolism, ranging from regulation of lipid transport and mobilization to fatty acid oxidation and lipid synthesis. They also

Figure 19.15 Stimulatory effects of PPAR signaling are shown for liver, muscle, and adipose tissue. The primary role of PPARα in liver and skeletal muscle is to stimulate fatty acid oxidation, whereas PPARγ stimulates lipid synthesis and improves insulin sensitivity in liver and adipose tissue. PPARδ signaling in adipose tissue and skeletal muscle increases rates of fatty acid oxidation and thermogenesis. In liver cells, PPARδ stimulates flux through the pentose phosphate pathway.

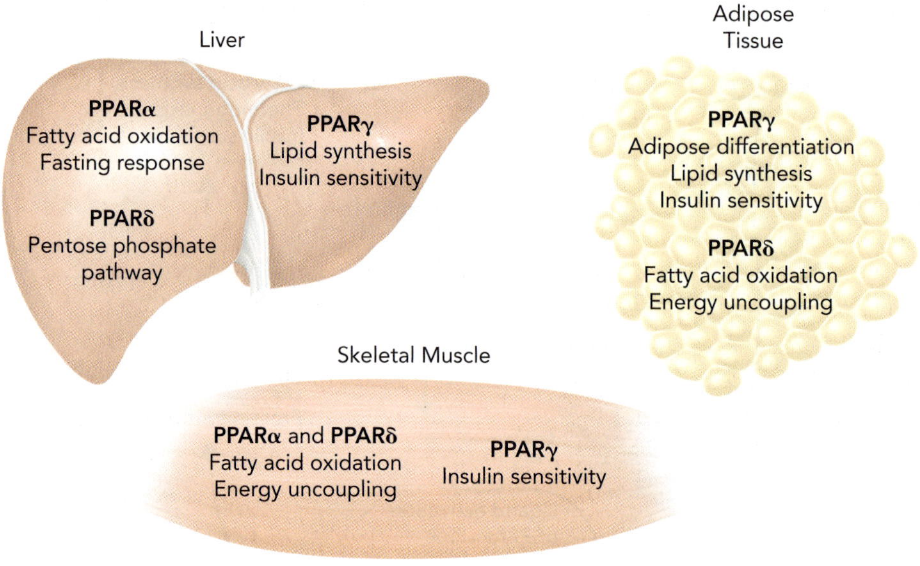

play an important role in energy metabolism and insulin sensitivity. As shown in **Figure 19.15**, PPAR signaling has different stimulatory effects on liver, adipose, and skeletal muscle metabolism. PPARγ and PPARδ are expressed in all three tissues (and in other tissues), whereas PPARα expression is restricted to liver and skeletal muscle. Notably, overnight fasting stimulates PPARα signaling in liver cells, which increases rates of fatty acid oxidation, resulting in elevated production and export of ketone bodies and glucose to the peripheral tissues. One of the most important functions of PPARγ is to control adipocyte differentiation and lipid synthesis in adipose tissue, but it also regulates insulin sensitivity in all three tissues. PPARγ is the primary therapeutic target of **thiazolidinediones**, which improve insulin sensitivity in type 2 diabetics by activating PPARγ target genes involved in lipid synthesis. Lastly, PPARδ signaling in adipose tissue and skeletal muscle leads to increased rates of fatty acid oxidation and thermogenesis, whereas in liver cells, PPARδ stimulates flux through the pentose phosphate pathway. Taken together, PPAR signaling regulates metabolic homeostasis at the physiologic level by controlling flux

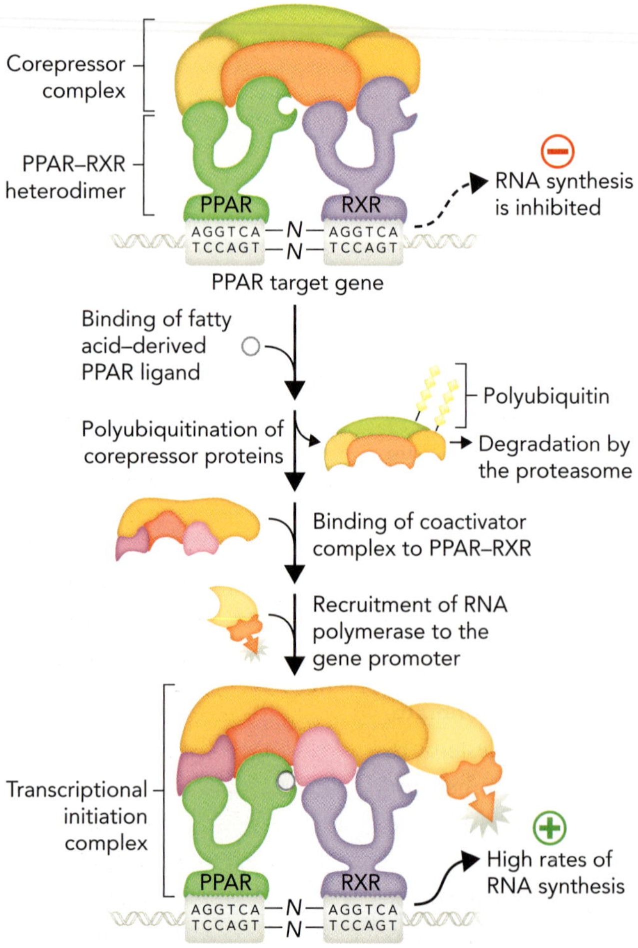

Figure 19.16 Ligand binding to the PPAR nuclear receptor leads to gene activation through a mechanism involving the assembly of a transcriptional initiation complex. Conformational changes in the PPAR–RXR heterodimer, triggered by ligand binding, stimulate polyubiquitination of the corepressor proteins and degradation by the proteasome. Coactivator proteins bind to the ligand-activated PPAR–RXR heterodimer and recruit RNA polymerase to the target gene promoter. *N* refers to a variable number of spacer nucleotides between adjacent DNA binding sites.

through anabolic and catabolic pathways in multiple tissues in response to lipid-based signaling molecules.

Activation of PPAR gene regulatory functions is illustrated in **Figure 19.16**. Observe that PPARs bind to direct repeats of DNA sequences related to the consensus binding site 5′-AGGTCA-3′. (Differences in target gene specificity between the highly similar PPARα, PPARγ, and PPARδ proteins are primarily a function of receptor-specific amino acids in the DNA binding domains. The amino acid sequences in this region are ~80% identical.) The DNA-bound PPAR–RXR heterodimer complexes are either inhibited or activated, which depends on ligand binding and binding of coregulatory (corepressor or coactivator) protein complexes. In the absence of PPAR ligands, corepressor proteins associate with the PPAR–RXR heterodimers, which prevents enzymes required for RNA synthesis from binding to the target gene promoter. When ligand concentrations increase inside the nucleus, ligand binding to PPAR leads to a conformational change in the receptor complex, facilitating ubiquitination and degradation of corepressor proteins. The corepressor complex is then replaced by a coactivator complex. An important function of coactivator proteins is to facilitate assembly of the transcriptional initiation complex, which contains the RNA polymerase enzyme. The fully assembled transcriptional initiation complex then activates expression of the PPAR target gene.

The PPARs represent an attractive class of protein targets for the development of pharmaceutical drugs for treating human metabolic disease. **Figure 19.17** shows chemical structures of three PPAR-selective agonists that have been developed. Gemfibrozil is a PPARα-selective fibrate used to treat high cholesterol; rosiglitazone is a PPARγ-selective agonist used to treat the symptoms of type 2 diabetes; and

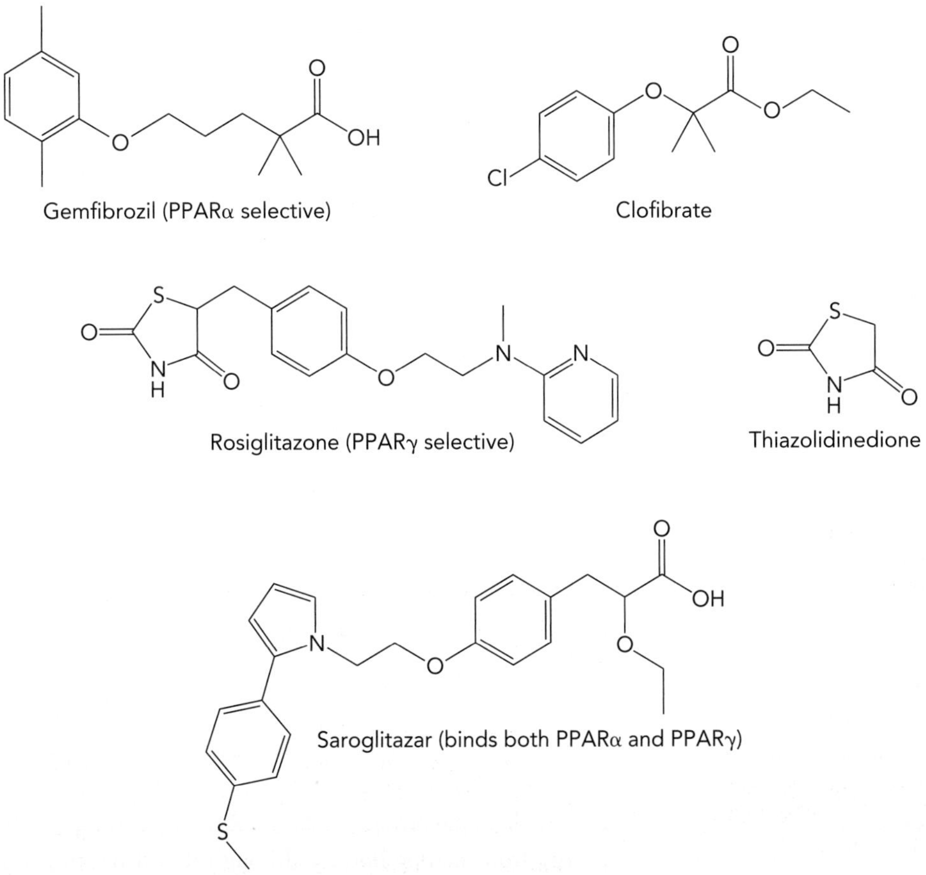

Figure 19.17 PPAR-selective agonists have been developed to treat a variety of metabolic disorders. Gemfibrozil is a clofibrate-related drug that binds selectively to PPARα and is used to lower serum cholesterol levels. Rosiglitazone binds with high affinity to PPARγ and belongs to the thiazolidinedione class of drugs, which are used to improve insulin sensitivity in patients with type 2 diabetes. Saroglitazar binds to both PPARα and PPARγ, and thereby has a dual mode of action in both lowering cholesterol and improving insulin sensitivity.

saroglitazar is a dual PPARα and PPARγ agonist that acts both to lower serum triglycerides and to improve insulin sensitivity. Saroglitazar is the first dual PPAR agonist approved for treatment of type 2 diabetes.

Mobilization of Metabolic Fuel during Starvation

Metabolic adaptations to food shortages have been preserved over evolutionary time to ensure survival during famine. The human body adapts to these near-starvation conditions by altering the flux of metabolites between various tissues to extend life as long as possible.

The primary metabolic challenge is to provide enough glucose for the brain to maintain normal neuronal cell functions. As mentioned earlier, the brain cannot use fatty acids for metabolic fuel because they cannot cross the blood–brain barrier. Erythrocytes (red blood cells) also depend on blood glucose for energy conversion because they lack mitochondria and therefore must derive all of their ATP from glycolysis (fatty acid oxidation is localized to the mitochondrial matrix).

With the onset of starvation, blood glucose levels are initially maintained by degradation of liver glycogen in response to glucagon signaling. This source of glucose is quickly depleted, however, resulting in a drop in blood glucose levels. To cope with the nutrient imbalance brought on by starvation and to maintain blood glucose levels (**Figure 19.18**), flux through metabolic pathways is altered in two important ways. First, the gluconeogenic pathway in the liver and kidneys is stimulated to generate glucose for brain cells and erythrocytes. The noncarbohydrate substrates for gluconeogenesis under these conditions are glycerol, alanine, glutamate, and lactate (see Figure 14.12). The glycerol is derived from triacylglycerol hydrolysis in adipose tissue, whereas alanine and lactate are produced in muscle cells by transamination reactions and anaerobic respiration, respectively. Glutamate, which is the preferred gluconeogenic precursor in kidney cells, is an abundant metabolite in the blood used to generate α-ketoglutarate, a citrate cycle precursor to oxaloacetate. The second way our bodies cope with the depletion of liver glycogen is to begin using fatty acids as the primary metabolic fuel in almost all tissues, which spares glucose for energy conversion reactions in brain cells and erythrocytes (Figure 19.18). The fatty acids are derived from triacylglycerol hydrolysis in adipose tissue after glucagon stimulation of protein kinase A–mediated signaling.

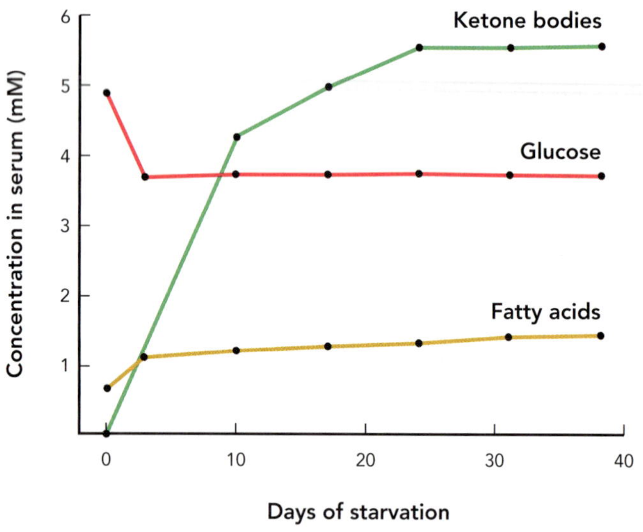

Figure 19.18 This graph shows the relative changes in concentration (millimolar) of glucose, fatty acids, and ketone bodies during 40 days of starvation. Blood glucose levels are maintained at ~3.5 mM by high rates of gluconeogenesis, coupled with a metabolic switch to fatty acid utilization by most tissues to spare glucose for brain cells and erythrocytes. High rates of fatty acid oxidation lead to a buildup of acetyl-CoA in the liver; the acetyl-CoA is converted to ketone bodies that are used by cardiac muscle and brain cells as sources of metabolic fuel. BASED ON DATA FROM F. G. CAHILL AND R. L. VEECH (2003). KETOACIDS? GOOD MEDICINE? *TRANSACTIONS OF THE AMERICAN CLINICAL AND CLIMATOLOGICAL ASSOCIATION, 114*, 149–163.

By the sixth day of starvation, high levels of acetyl-CoA produced by fatty acid oxidation in the liver lead to a dramatic increase in blood concentrations of acetoacetate and D-β-hydroxybutyrate (ketone bodies). The onset of ketogenesis is crucial because it provides a second energy source for the brain and heart, which use ketone bodies to generate acetyl-CoA. By the second week of starvation, the brain obtains up to half of its energy from ketone bodies, although it still depends on

Table 19.2 METABOLIC FUELS AVAILABLE TO A 70-KG MAN PRIOR TO FASTING

Type of fuel	Tissue	Mass (kg)	Energy (Calories)	Survival time[a]
Triacylglycerols	Adipose tissue	15	141,000	83 days
Proteins	Skeletal muscle	6	24,000	14 days
Glycogen	Skeletal muscle	0.15	600	8.4 hours
Glycogen	Liver	0.07	300	4.2 hours
Glucose, fatty acids, triacylglycerols	Circulatory system	0.023	100	1.4 hours
Total			166,000	~98 days

[a]Assuming a basal metabolism of 1,700 Calories per day.

glucose production in the liver and kidneys to provide the remaining energy for ATP synthesis. An important outcome of the utilization of ketone bodies as a partial replacement for glucose is that it delays the wholesale degradation of skeletal muscle protein, which would otherwise be required to generate gluconeogenic substrates from amino acids. By using muscle protein sparingly as a source of metabolic fuel, it increases the chances that the body will be strong enough to obtain food and thereby prevent death.

As shown in **Table 19.2**, an average-size man of 70 kg contains enough metabolic fuel to live ~98 days without food, assuming a minimal energy expenditure of 1,700 Calories per day. (Recall that in nutritional science, Calorie with a capital "C" is a kilocalorie [kcal], so 1 Calorie = 1 kcal = 4.184 kJ. In this chapter, we will refer to energy values in Calories, as these values are most commonly reported on nutrition labels and in diet and exercise plans.) By far, the bulk of stored metabolic fuel is in the form of triacylglycerols in adipose tissue, which is sufficient to prolong life for 3 months. Protein is the second most abundant stored fuel, but as described earlier, metabolic adaptations to starvation ensure that this form of energy is spared for as long as possible. Notably, an obese individual with three times as much body fat as a normal person could theoretically survive starvation for up to 8 months (249 days). But as discussed in the next section, chronic obesity actually shortens life span due to an increased incidence of type 2 diabetes and cardiovascular disease.

We can now revisit the metabolic map presented in Figure 19.8, showing metabolite flux between six major tissues in the human body under normal conditions, and observe how metabolite flux is altered by starvation. As shown in **Figure 19.19**, four major changes occur in metabolic flux, which together facilitate human survival for long periods of time without food. These changes are as follows:

1. *Increased release of fatty acids from adipose tissue.* After depletion of liver glycogen within the first 12–24 hours of starvation, triacylglycerol hydrolysis is stimulated in adipose tissue, resulting in release of fatty acids and glycerol into the blood. Moreover, glucose uptake by skeletal muscle is inhibited due to the low levels of insulin in the blood. This has the effect of shifting energy away from glucose utilization and toward fatty acid oxidation for most tissues.
2. *Increased gluconeogenesis in liver and kidney cells.* To keep blood glucose levels above ~3.5 mM, which is needed for brain cells and erythrocytes, flux through

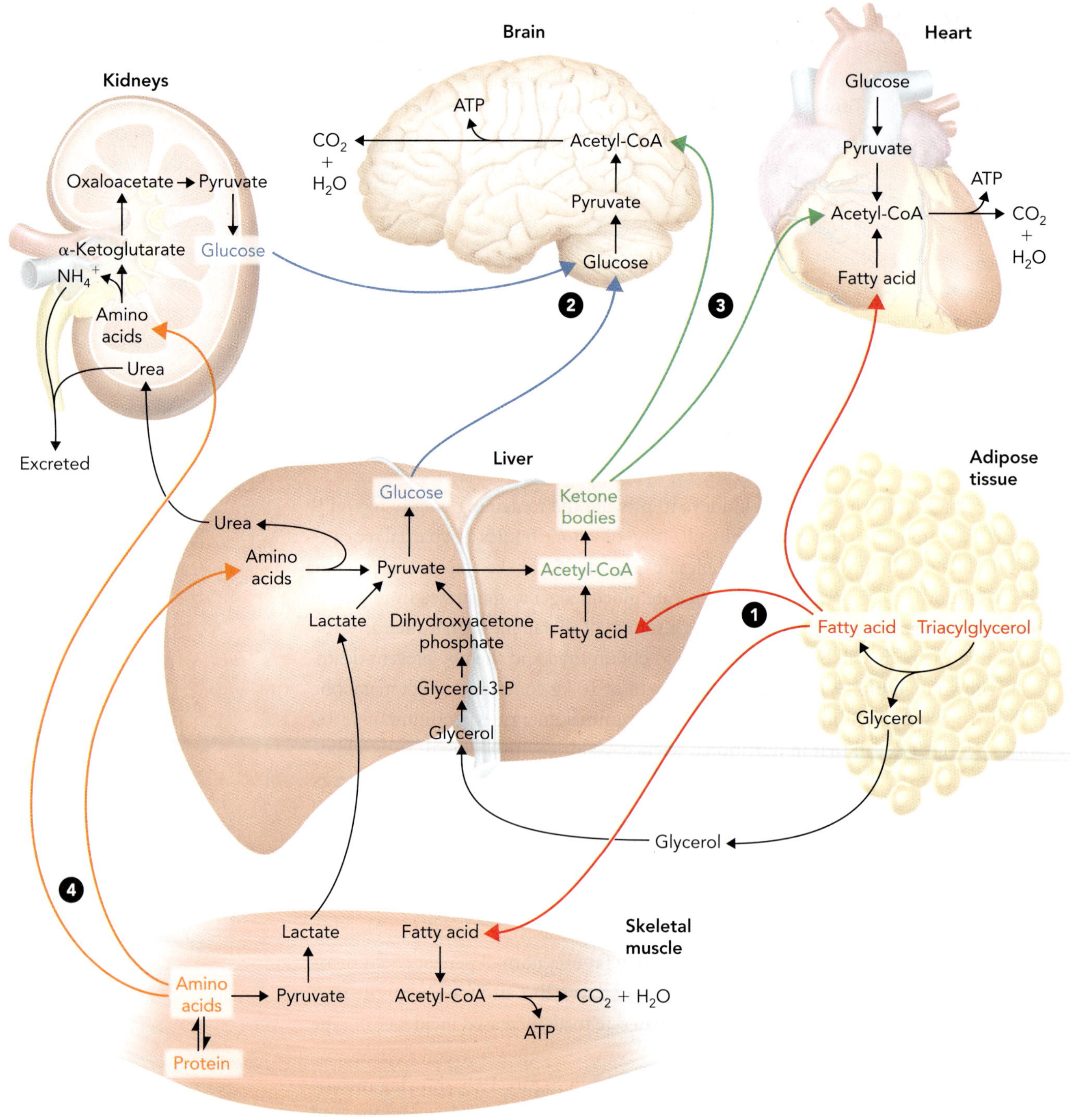

Figure 19.19 The flux of metabolites between major tissues and organs in the human body under starvation conditions is shown here (see also Figure 19.8). Once glycogen stores are depleted, adipose tissue and skeletal muscle are the primary sources of metabolic fuel during starvation. Fatty acids released from triacylglycerol hydrolysis in adipose tissue are transported to skeletal muscle and the heart by serum albumin protein. These fatty acids are used to generate acetyl-CoA for the citrate cycle and oxidative phosphorylation. Acetyl-CoA produced from fatty acids in the liver is used for the production of ketone bodies, which are an important energy source for the heart and brain during starvation. Amino acids derived from protein degradation in skeletal muscle provide the necessary carbon to make pyruvate or α-ketoglutarate. The α-ketoglutarate is used in the citrate cycle to make pyruvate, which is then used to produce glucose in liver cells by gluconeogenesis. See the text for details about the four enumerated metabolic responses to starvation.

the gluconeogenic pathway is increased in liver and kidney cells. The major substrates for glucose biosynthesis in the liver are glycerol, alanine, and lactate, which are all converted to pyruvate. Kidney cells primarily use glutamate from the blood to generate oxaloacetate.

3. *Increased ketogenesis in liver cells.* As a result of high rates of fatty acid oxidation in mitochondria, acetyl-CoA levels in the liver increase dramatically, giving rise to high levels of ketogenesis. This occurs because acetyl-CoA generated by fatty acid oxidation cannot be metabolized by the citrate cycle because of oxaloacetate being shunted toward gluconeogenesis. Under these conditions, the brain and heart can use significant amounts of ketone bodies as a source of acetyl-CoA for aerobic respiration.
4. *Protein degradation in skeletal muscle.* Muscle protein provides amino acids that serve as gluconeogenic substrates in the liver and kidneys. Although this is a good source of energy reserves (see Table 19.2), catabolism of skeletal muscle is delayed as long as possible to maintain mobility and enable the ongoing search for food. Once protein stores fall below 50% of prestarvation levels, life can no longer be sustained.

Metabolic adaptations to starvation are also found in untreated diabetics, who have high levels of blood glucose that cannot be used by tissues because of insulin resistance. In terms of metabolic homeostasis, there is little difference between not having enough carbohydrates because of a food shortage and not being able to metabolize blood glucose because of a defect in insulin signaling. Untreated diabetics have increased rates of ketogenesis in the liver as a result of elevated fatty acid oxidation and degradation of skeletal muscle proteins as sources of gluconeogenic precursors.

concept integration 19.1

How do the pancreas and liver work together to maintain blood glucose levels at ~4.5 mM?

The pancreas is an endocrine organ that produces insulin and glucagon, the two major peptide hormones involved in blood glucose regulation. Insulin and glucagon are synthesized in regions of the pancreas called the islets of Langerhans, with the β cells secreting insulin and the α cells secreting glucagon. Insulin signaling stimulates glucose uptake primarily into liver, skeletal muscle, and adipose tissue in response to elevated blood glucose levels.

In liver cells, glucose is used for glycogen synthesis or is converted to acetyl-CoA, which is a substrate for fatty acid synthesis. In contrast, glucagon secretion from the pancreas promotes glucose efflux from liver cells by activating glycogen degradation and gluconeogenesis.

19.2 Metabolic Energy Balance

Unhealthy lifestyles, characterized by consumption of high-Calorie foods and lack of daily exercise, are becoming increasingly common in the modern world. This fundamental change in how people live today compared to how they lived 100 years ago appears to be why an increasing number of children and adults in the United

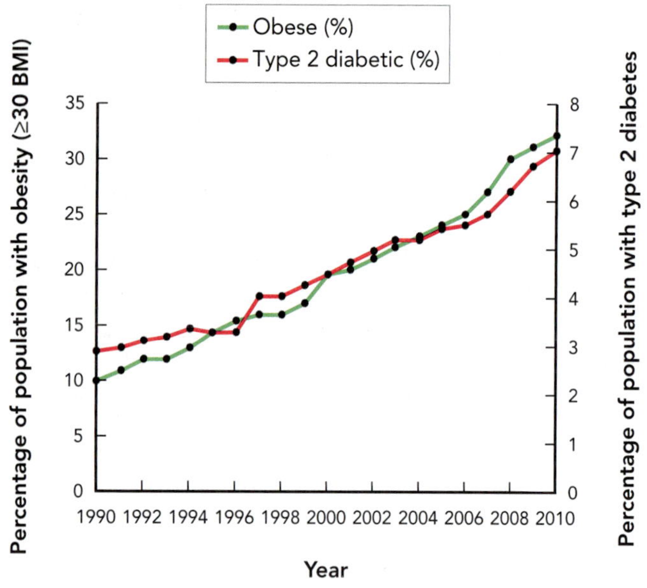

Figure 19.20 The prevalence of adult obesity and diabetes has significantly increased in the United States between 1990 and 2010. Normalized data were used to account for differences in the percentage of people in all age groups during any given year. DATA FROM THE CENTERS FOR DISEASE CONTROL AND PREVENTION, ATLANTA, GEORGIA.

States are defined as obese; that is, having a BMI value ≥30. Moreover, physicians have observed a concomitant rise in the incidence of insulin-resistant type 2 diabetes (**Figure 19.20**). Perhaps most alarmingly, the number of children diagnosed with type 2 diabetes has increased 10-fold in the past 20 years.

The coincident rise in rates for both obesity and type 2 diabetes is compelling evidence that the two conditions may be causally linked. We discuss the biochemical basis for this connection in this section, using metabolic energy balance as the common theme. Indeed, researchers have recently begun to uncover biochemical mechanisms that control energy balance in mammals, raising new hope that these discoveries can be used to stem what has been called the "diabesity epidemic." As many as 400,000 deaths per year in the United States are attributed to obesity-related disease, and more than 10% of the health-care budget in the United States is used to treat these patients. This gives great urgency to identifying the underlying biochemical causes of obesity and diabetes in order to develop a rational treatment strategy.

The Role of Genes and Environment in Energy Balance

The key to maintaining energy balance is having a daily intake of dietary nutrients that is equivalent to Caloric energy expenditure. Energy balance is defined as

$$\text{Calories consumed/day (input)} = \text{Calories expended/day (output)}$$

If Caloric input > output, then an individual has a positive energy imbalance. If Caloric input < output, then an individual has a negative energy imbalance.

For most adults, a balanced input and output occurs at ~1,500–2,000 Calories/day, depending on gender, body size, and physical activity. For the **macronutrients** carbohydrate, protein, and fat, the Caloric ratio of consumed Calories should be close to 2:1:1 (carbohydrate to protein to fat). As long as the Calories derived from dietary nutrients are completely used each day for chemical reactions in the body and for muscle contraction or in conversion to heat through thermogenesis, an individual's weight remains constant.

The concept of energy balance can be used to explain how energy input (food Calories) and energy expenditure (basal metabolism, physical activity, exercise) alter metabolic homeostasis. This can bring about a measurable weight gain or weight loss, as reflected in the amount of stored fat in the body (**Figure 19.21**). For example, consuming an extra ~115 Calories/day for a month (total of ~3,500 Calories) is a positive energy imbalance that would cause most people to add 1 pound of stored fat to their bodies. The best way to lose this pound of fat is to maintain a negative energy imbalance until 3,500 extra Calories of energy are expended. This could occur either by consuming 3,500 Calories fewer than needed or by increasing physical activity to burn off an extra 3,500 Calories. Note that an individual can be in energy balance and still be overweight. As illustrated in Figure 19.21, a period of positive energy imbalance leads to weight gain and an increase in BMI that is maintained unless a period of negative

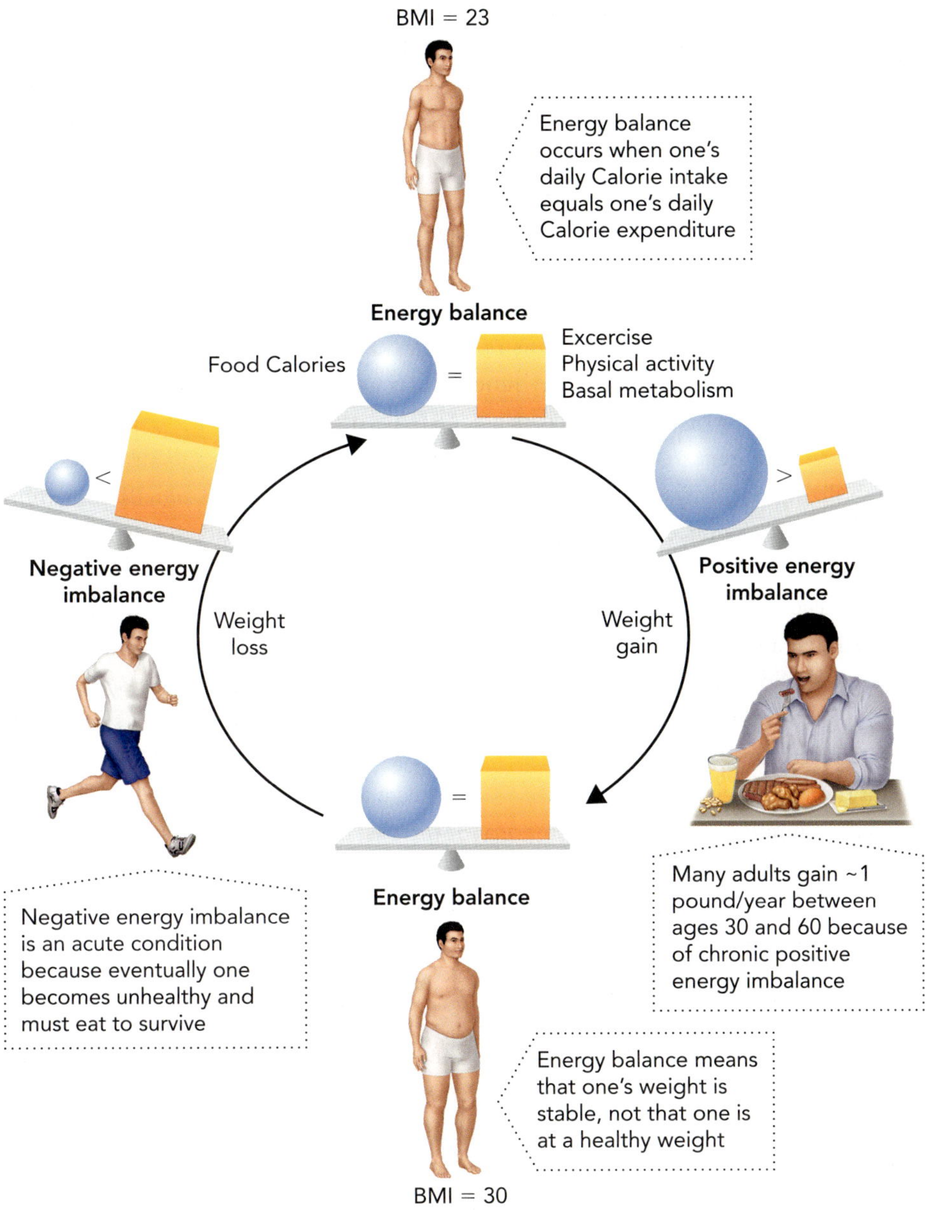
BMI = 23
Energy balance occurs when one's daily Calorie intake equals one's daily Calorie expenditure
Energy balance
Food Calories
=
Excercise
Physical activity
Basal metabolism
<
Negative energy imbalance
>
Positive energy imbalance
Weight loss
Weight gain
=
Energy balance
Negative energy imbalance is an acute condition because eventually one becomes unhealthy and must eat to survive
Many adults gain ~1 pound/year between ages 30 and 60 because of chronic positive energy imbalance
Energy balance means that one's weight is stable, not that one is at a healthy weight
BMI = 30

Figure 19.21 Energy balance is achieved by consuming and expending the same number of Calories on a daily basis. Under conditions of a positive energy imbalance, one gains weight, whereas if one is in a negative energy imbalance, one loses weight. Energy balance is not an indication of healthy weight: One can be in energy balance and still have a BMI outside the normal range.

energy imbalance is initiated. In fact, many overweight individuals maintain energy balance and often consume fewer Calories than someone of normal weight, but they remain overweight because they are less physically active and burn fewer Calories. It is not the amount of food a person eats that determines his or her weight, but rather how many Calories a person consumes relative to how many Calories he or she expends.

Nutritional studies have shown that most people gain weight slowly over a long period of time, and once they have gained the extra weight, they have a very difficult time losing it. A sedentary lifestyle and hormone-related changes in basal metabolic rates are two of the primary reasons most adults gain an average of 1–3 pounds/year from the ages of 30 to 60. Research suggests that a pattern of long-term positive energy imbalance may cause changes in the fat storage **set point** of the body, which is a term describing the average amount of adipose tissue the body maintains at physiologic homeostasis. Although the molecular basis for such a fat storage set point is not completely understood, it could explain why it is so difficult to reduce fat stores over a short period of time by dieting alone.

Research into the biochemistry of energy balance has recently provided clues as to why people are so different with regard to their ability to control their weight. As with many other biological processes, genetic inheritance influences who is at risk for being overweight, but environment (that is, quality and quantity of diet and exercise) determines who will actually be overweight. Two types of studies have led to this conclusion. In the first type, the tendency to be overweight for genetically identical twins who were separated at birth and raised in different homes was compared to that for genetically identical twins who were raised in the same household. It was found that twins adopted at birth and raised in separate households were just as likely to be lean, normal, or overweight as twins raised in the same household. Statistical analyses based on results such as these indicate that genetic inheritance can account for up to 40% to 70% of the risk factors associated with obesity. Many of these genetic risk factors are genes that regulate metabolite flux through carbohydrate and lipid metabolizing pathways, as well as genes that regulate neurologic signals controlling eating behaviors and energy expenditure in the form of thermogenesis (uncoupled mitochondrial ATP synthesis).

A second type of study investigating metabolism and energy balance clearly showed that although genes contribute to the risk of obesity and diabetes, dramatically different environmental factors do, in fact, have a major impact on disease incidence. In one such study, two populations of Pima Indians were shown to have significantly different rates of obesity and insulin-resistant type 2 diabetes (**Figure 19.22**). The two Pima Indian populations speak the same language and are genetically related but have lived in different parts of North America for the past 1,000 years. A National Institutes of Health (NIH) study conducted in the 1980s showed that Pima Indians of southern Arizona have rates of obesity and type 2 diabetes that are among the highest in the world, with 80% of the adults being overweight, obese, or diabetic. The Pima Indians living in northern Mexico, however, weigh on average ~60 pounds less than their Arizona relatives when accounting for gender and age. More important, the incidence of insulin-resistant type 2 diabetes in the Pima Indians of southern Arizona is nearly 10 times higher than it is in the Pima Indian population living in northern Mexico.

What accounts for this difference in rates of obesity and diabetes between the genetically similar Pima Indians of Arizona and Mexico? The answer appears to be a

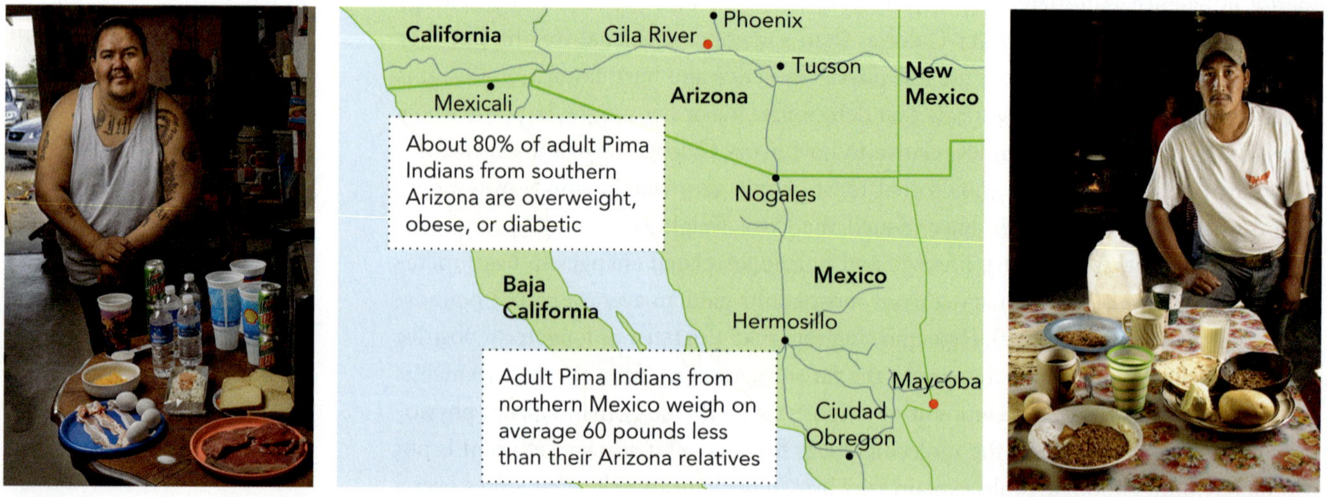

Figure 19.22 The Pima Indians of Arizona and Mexico are genetically related but have very different lifestyles and eating habits, which affects the rates of obesity and diabetes in the two populations. The Pima Indians of southern Arizona generally have a sedentary lifestyle and a high-Calorie diet. This leads to an incidence of diabetes that is nearly 10 times higher than that of the Pima Indians of Maycoba, Mexico, who spend most of their day working in agricultural fields and consume mostly unprocessed food. PHOTOS: PETER MENZEL PHOTOGRAPHY.

high genetic risk factor for obesity combined with lifestyles characterized by significant differences in daily Caloric intake and energy expenditure. In southern Arizona, the Pima Indians live on government reservation land that no longer has sufficient water to support the agricultural-based lifestyle of their past. This has led to changes in their daily routines, making them more similar to Arizona residents living in nearby Phoenix. The Pima Indians of this population have very little daily physical activity, and their diets consist of high-Calorie beverages and foods containing large amounts of saturated fats. In contrast, the Pima Indians living in northern Mexico have retained their physically demanding agricultural lifestyle of thousands of years, which requires that they expend large amounts of energy obtaining food that is naturally low in Calories and saturated fat. It appears that the more traditional Pima Indian lifestyle has allowed them to escape the fate of their Arizona relatives, despite their shared genetic inheritance, which is now known to include high risk factors for obesity and diabetes.

The **thrifty gene hypothesis**, proposed in the 1960s by James Neel, a University of Michigan physician and scientist, is consistent with the Pima Indian study and provides an explanation for the high rates of obesity and diabetes in developed countries. Neel, who had training in genetics and understood the role of natural selection in evolution, proposed that humans contain gene variants (*thrifty genes*) favoring individuals with a capacity to store extra fat during times of feast as a way to prolong survival during times of famine. Neel argued that although this genetic background was a good thing to have when humans depended on a hunter-and-gatherer lifestyle, these same thrifty gene variants become disease genes in a modern society, where diets high in saturated fats and carbohydrate-rich foods are combined with sedentary lifestyles. Because the rates of obesity and diabetes have increased significantly over the past 20 years (see Figure 19.20), with little change in the human gene pool over the past ~5,000 years, it should be possible to reverse obesity and diabetes trends by realigning our environment with a hunter-and-gatherer lifestyle. This is best done by altering our diets to include more whole grains, fruits, and vegetables and by including a regular amount of physical exercise in our daily routine.

Control of Energy Balance by Hormone Signaling in the Brain

A major breakthrough in obesity research came in 1994 when Jeffrey Friedman and his colleagues at Rockefeller University identified a gene mutation in a strain of obese mice called OB (obese) mice. A typical OB mouse contains up to three times more adipose tissue than normal mice and develops symptoms consistent with insulin-resistant type 2 diabetes. The Friedman lab discovered that a recessive mutation in the ob gene is responsible for the obesity phenotype seen in homozygous mice (*ob/ob* mice). Gene cloning and biochemical characterization of the ob gene revealed that it encodes an adipocyte peptide hormone, which they named **leptin** after the Greek word *leptos,* meaning "thin." As shown in **Figure 19.23**, when *ob/ob* mice were injected with purified recombinant leptin protein, the mice ate less food and lost weight. Moreover, the high serum levels of insulin and glucose in the *ob/ob* mice, which are characteristic symptoms of type 2 diabetes, were reduced by leptin injections. Taken together, these data suggested that the newly discovered leptin hormone may explain the connection between obesity and diabetes if variants in the human ob gene were common in the population.

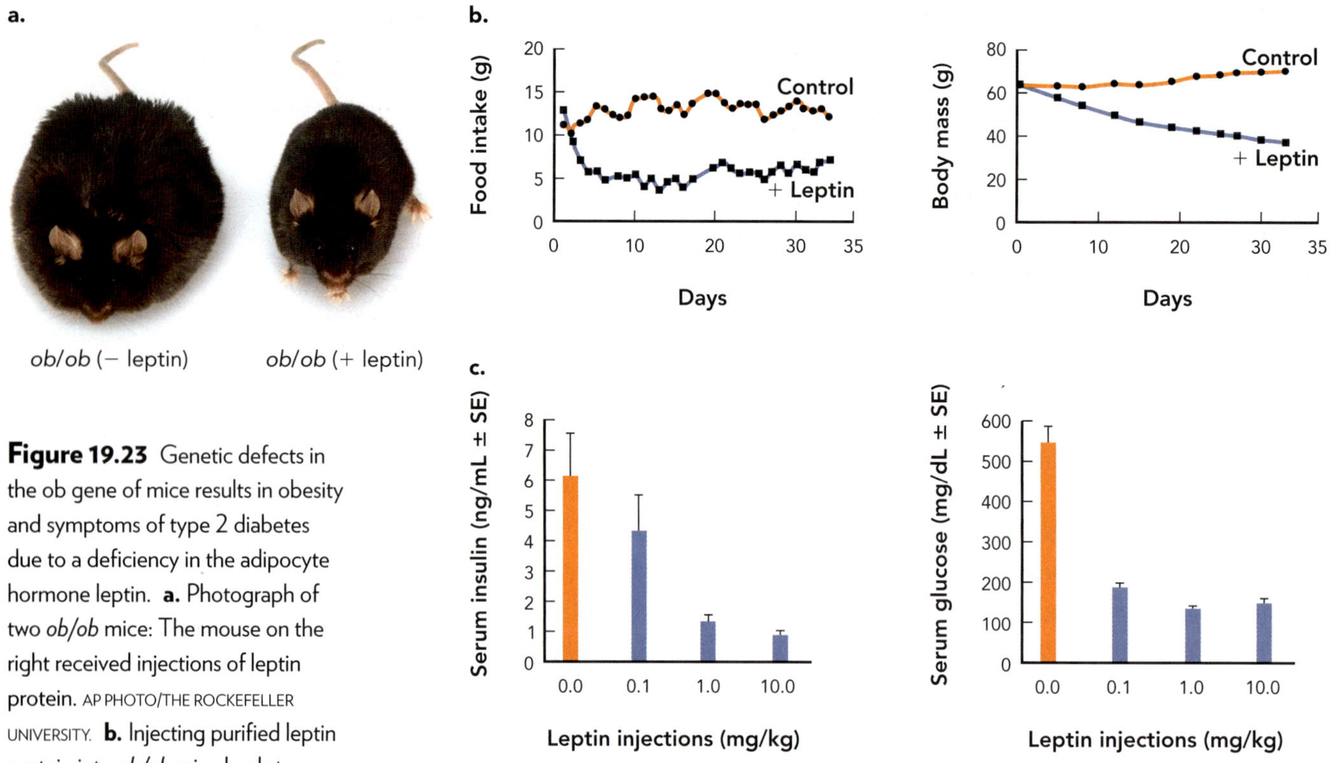

Figure 19.23 Genetic defects in the ob gene of mice results in obesity and symptoms of type 2 diabetes due to a deficiency in the adipocyte hormone leptin. **a.** Photograph of two *ob/ob* mice: The mouse on the right received injections of leptin protein. AP PHOTO/THE ROCKEFELLER UNIVERSITY. **b.** Injecting purified leptin protein into *ob/ob* mice leads to reduced food intake and body mass. BASED ON DATA FROM J. HALAAS ET AL. (1995). WEIGHT-REDUCING EFFECTS OF THE PLASMA PROTEIN ENCODED BY THE OBESE GENE. *SCIENCE*, *269*, 543. **c.** Elevated insulin and glucose levels in the serum of *ob/ob* mice are reversed by leptin injections. BASED ON DATA FROM M. A. PELLEYMOUNTER ET AL. (1995). EFFECTS OF THE OBESE GENE PRODUCT ON BODY WEIGHT REGULATION IN OB/OB MICE. *SCIENCE*, *269*, 540.

Encouraged by their breakthrough finding in OB mice, Friedman's lab next focused on finding the gene mutation responsible for a second strain of obese mice called DB (diabetic), which had very similar phenotypes to that of the OB mice. Using the same genetic mapping techniques, Friedman's lab reported in 1996 that the genetic defect in DB mice was in fact a mutation in the gene for the leptin receptor, which had been cloned a year earlier by Louis Tartaglia's group at a biotechnology company in Cambridge, Massachusetts. The phenotypes of OB and DB mice are very similar in that both are grossly overweight and have elevated serum levels of glucose and insulin due to insulin resistance (**Table 19.3**). In addition, whereas OB mice lack leptin in their serum, DB mice have leptin levels that are 35 times higher than those of normal mice. Although the identification of mutations in the leptin gene and the leptin receptor gene in the overweight OB and DB mice, respectively, was exciting, it was soon discovered that leptin deficiencies did not explain the majority of cases of obesity and diabetes in humans. Indeed, most obese individuals

Table 19.3 EFFECT OF THE *OB* AND *DB* MUTATIONS ON SERUM GLUCOSE AND INSULIN LEVELS IN MICE

Genotype	Leptin production	Leptin receptor	Body weight (g)	Serum glucose (mg/dL)	Serum insulin (ng/mL)	Serum leptin (ng/mL)
Wild type (+/+)	Yes	Yes	23.9	209.4	2.4	2.0
OB (*ob/ob*)	No	Yes	43.5	362.9	10.3	ND
DB (*db/db*)	Yes	No	41.4	489.9	21.3	73.1

Note: The symbol "+" refers to a normal copy of the genes encoding leptin (*ob*) and the leptin receptor (*db*). ND = not detected.

BASED ON S. KONSTANTINIDES ET AL. (2001). LEPTIN-DEPENDENT PLATELET AGGREGATION AND ARTERIAL THROMBOSIS SUGGESTS A MECHANISM FOR ATHEROTHROMBOTIC DISEASE IN OBESITY. *JOURNAL OF CLINICAL INVESTIGATION*, *108*, 1533–1540.

have very high circulating levels of leptin, and mutations in the human leptin or leptin receptor genes are extremely rare. Therefore, researchers focused their attention on how leptin signaling worked in humans to try and understand its functional role, if any, in energy balance.

Leptin circulates throughout the body and activates leptin signaling in a variety of tissues, including the brain, where it binds to leptin receptors in the **hypothalamus**. Leptin controls appetite and energy expenditure through a hierarchical neuronal signaling network involving first-order and second-order neurons. The net effect of leptin neuronal signaling in the brain is to reduce fat storage through a combination of decreased appetite and increased energy expenditure (by elevating basal metabolic rates). As illustrated in **Figure 19.24**, the hypothalamus has two types of first-order neurons that contain leptin receptors: the proopiomelanocortin (POMC) neurons and the neuropeptide Y/agouti-related peptide (NPY/AGRP) neurons. The POMC neurons secrete a neuropeptide called α-melanocyte stimulating hormone (α-MSH), whereas the NPY/AGRP neurons secrete the neuropeptides NPY (neuropeptide Y) and AGRP (agouti-related peptide).

Leptin receptor signaling in POMC neurons stimulates the activity of a membrane-associated signaling protein called Janus kinase 2 (JAK2), which leads to phosphorylation and activation of the transcription factor signal transducer and activator of transcription 3 (STAT3). STAT3 enters the nucleus in POMC neurons and induces POMC gene expression, producing the nascent POMC polypeptide, which is proteolytically processed to generate α-MSH. Secretion of α-MSH from POMC neurons leads to stimulation of neuronal signaling in second-order neurons, resulting in decreased fat storage. In contrast, leptin activation of JAK2 signaling in NPY/AGRP neurons stimulates the phosphoinositide-3 kinase (PI-3K) signaling pathway, which leads to transcriptional repression of the NPY and AGRP genes and decreased synthesis of the NPY and AGRP neuropeptides, respectively. Because NPY and AGRP signaling in second-order neurons functions to increase fat storage, leptin inhibition of NPY and AGRP production in NPY/AGRP neurons also leads to decreased fat storage.

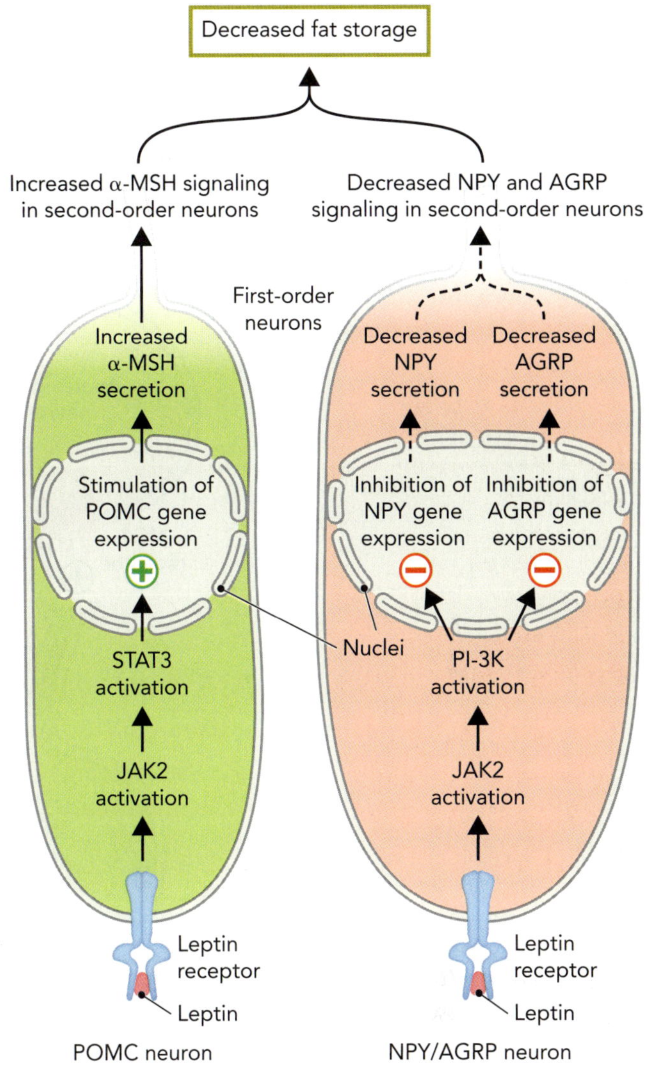

Figure 19.24 Leptin signaling pathways in first-order POMC and NPY/AGRP neurons leads to decreased fat storage. Leptin signaling in POMC neurons stimulates JAK2 activity, which phosphorylates the transcription factor STAT3. This stimulates expression of the POMC gene and production of the neuropeptide α-MSH. In contrast, JAK2 activation by leptin in NPY/AGRP neurons stimulates the PI-3K pathway, which leads to reduced expression of the NPY and AGRP genes and decreased signaling in second-order neurons. The net result of reciprocal leptin signaling is decreased fat storage.

The α-MSH, NPY, and AGRP neuropeptides secreted from first-order neurons bind to G protein–coupled receptors on two types of second-order neurons: the anorexigenic neurons and the orexigenic neurons. **Anorexigenic** (without appetite) neurons send out signals that decrease appetite and increase energy expenditure; these neurons are therefore responsible for reducing fat stores. The function of **orexigenic** (with appetite) neurons is to increase appetite and decrease energy expenditure, thereby increasing fat stores. As shown in **Figure 19.25**, α-MSH secretion from

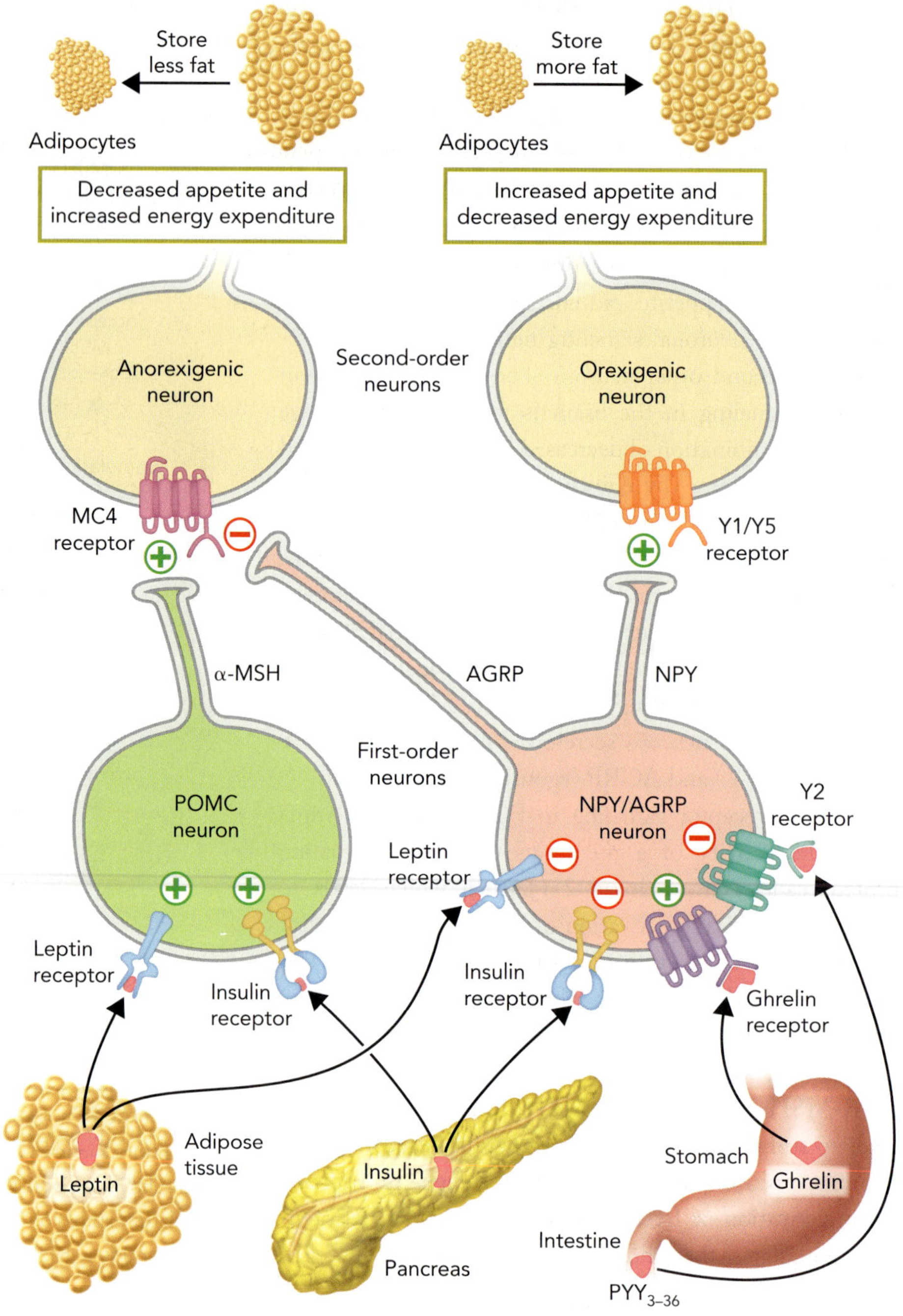

Figure 19.25 Neuropeptide signaling in the brain controls fat storage by regulating appetite and energy expenditure. Leptin receptor activation and insulin receptor activation have the same effect on neuronal signaling in POMC and NPY/AGRP neurons, where the activated receptors stimulate or inhibit signaling, respectively. Ghrelin, a peptide hormone synthesized in the stomach, stimulates orexigenic neuronal signaling by binding to ghrelin receptors on NPY/POMC neurons. $PYY_{3\text{-}36}$, a peptide hormone synthesized in the intestine, binds to Y2 receptors on NPY/AGRP neurons and inhibits orexigenic signaling. The neuropeptides α-MSH, AGRP, and NPY are secreted from first-order neurons and control neuronal signaling in second-order neurons. Adipocyte size is determined by the amount of lipid stored in cytosolic lipid droplets.

POMC neurons activates melanocortin 4 (MC4) receptors on anorexigenic neurons. This leads to stimulation of neuronal signals that decrease appetite and increase energy expenditure. In contrast, secretion of NPY from NPY/AGRP neurons activates Y1/Y5 receptors on orexigenic neurons, resulting in activation of neuronal signals that increase appetite and decrease energy expenditure. In addition, AGRP secretion from NPY/AGRP neurons interferes with signaling in anorexigenic neurons by competing with α-MSH for binding to MC4 receptors, making AGRP an anorexigenic antagonist. Because leptin binding to NPY/AGRP neurons inhibits NPY and AGRP secretion, the net effect of leptin signaling in the hypothalamus is decreased stimulation of orexigenic neurons by NPY and decreased inhibition of anorexigenic neurons by AGRP.

In addition to leptin signaling in the hypothalamus, three other hormones—insulin, ghrelin, and $PYY_{3\text{-}36}$—also signal through the same set of neurons (Figure 19.25). Insulin receptor activation in POMC and NPY/AGRP neurons leads to the same neuronal output as leptin signaling; namely, decreased appetite and increased energy expenditure. This makes sense because high levels of insulin in the blood signal that carbohydrates are plentiful, which leads to increased fatty acid synthesis in the liver. Ghrelin, a 28-amino-acid peptide synthesized at high levels in the stomach between meals, sends signals to the brain that it is time to eat. It functions through activation of ghrelin receptors on NPY/AGRP neurons, which leads to activation of orexigenic neuron signaling (NPY) and inhibition of anorexigenic neuron signaling (AGRP). The peptide hormone $PYY_{3\text{-}36}$, a 34-amino-acid peptide synthesized in the ileum and colon (large intestine), counters the effect of ghrelin by sending signals to the brain that one has had enough to eat. As illustrated in Figure 19.25, $PYY_{3\text{-}36}$ binds to Y2 receptors on NPY/AGRP neurons, where it inhibits neuropeptide release and orexigenic signaling. Although it was initially thought that food had to pass through the intestine physically in order for $PYY_{3\text{-}36}$ to be secreted, it has been discovered that $PYY_{3\text{-}36}$ release is triggered by neuronal inputs associated with eating behaviors.

The Metabolic Link between Obesity and Diabetes

A lifestyle of chronic positive energy imbalance can lead to a condition in humans called the **metabolic syndrome**. An individual is diagnosed with metabolic syndrome if he or she has three or more of the following symptoms:

1. Abdominal obesity (large amounts of visceral fat)
2. Insulin resistance (prediabetes)
3. Hypertension (high blood pressure)
4. Hyperlipidemia (high LDL and low HDL levels)
5. High risk for cardiovascular disease (blood protein profiles associated with atherosclerosis)

Although many of these disease states are associated with alterations in carbohydrate or lipid metabolism, the relationship between obesity and insulin-resistant **type 2 diabetes** is most directly coupled to a positive energy imbalance. Insulin-resistant type 2 diabetes is distinct from **type 1 diabetes**, which is due to insufficient insulin production by the pancreatic β cells. Type 2 diabetes is characterized by desensitization of insulin receptor signaling in muscle, liver, and adipose tissue. Once known as adult-onset diabetes

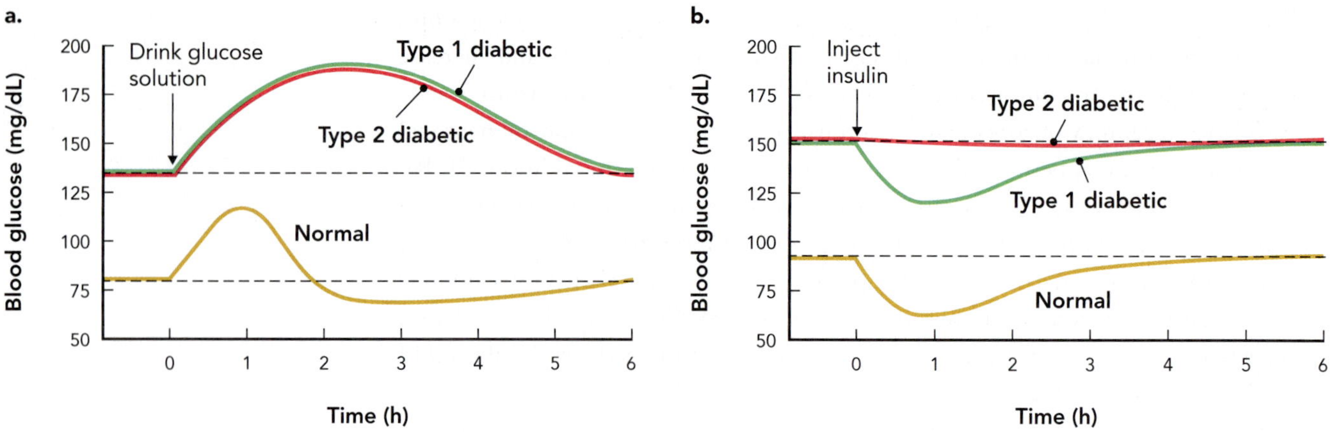

Figure 19.26 Blood glucose levels in patients with type 1 or type 2 diabetes after treatment with glucose or insulin are shown. **a.** Both type 1 and type 2 diabetics have elevated levels of blood glucose for a long period of time in a glucose tolerance test. **b.** Type 1 diabetics are insulin sensitive, whereas type 2 diabetics are insulin resistant, as seen by blood glucose levels after an insulin injection.

because of its prevalence in people over the age of 40, type 2 diabetes is not confined to this age group, as evidenced by the increasing number of obese children being diagnosed with insulin resistance.

As illustrated in **Figure 19.26a**, a **glucose tolerance test** shows that patients with type 1 or type 2 diabetes both have abnormal glucose clearance kinetics compared to that of normal individuals. In this test, an oral glucose solution or intravenous feeding of glucose is administered to the individual, and then blood glucose levels are measured over time. In addition to having elevated blood glucose levels before the test begins, both type 1 and type 2 diabetics are unable to clear the excess glucose from their blood within 2 hours. Moreover, blood glucose levels do not return to pretest levels for 6 hours. Because both type 1 and type 2 diabetics have defects in insulin signaling, neither is able to mount a normal physiologic response to high blood glucose levels. The best way to distinguish between defects in insulin production (type 1) and insulin sensitivity (type 2) is to administer insulin and monitor blood glucose levels over time. As seen in **Figure 19.26b**, only type 1 diabetics are able to show a significant response to insulin treatment by lowering blood glucose levels, which is why this form of diabetes can be controlled by daily insulin injections.

Factors That Can Lead to Type 2 Diabetes Several human gene mutations have been associated with an increased risk of type 2 diabetes, including mutations in signaling molecules that control glucose and lipid metabolism. Symptoms of type 2 diabetes are initially mild and can include frequent urination, thirst, and blurry vision, all of which are directly related to the osmotic effects of high glucose. But if elevated blood glucose levels and insulin resistance are not brought under control, diabetic patients can experience renal failure, blindness, and gangrene infections in the legs and feet due to poor circulation. Because of the deleterious effects of high glucose and free fatty acid levels on the cardiovascular system, most advanced-stage diabetics die of coronary artery disease brought on by hypertension and atherosclerosis.

Body weight is one of the best predictors that an individual may develop type 2 diabetes. Studies have shown that people who are obese before the age of 30 have significantly higher risks of developing type 2 diabetes by the age of 60, which is consistent with the fact that 80% of individuals with type 2 diabetes are obese. What is the biochemical basis for a causal link between obesity and type 2 diabetes? Although the answer is complex, two phenotypes in obese individuals are strongly associated with insulin-resistant type 2 diabetes: (1) elevated levels of free fatty acids in the serum, and (2) altered secretion of peptide hormones from adipose tissue.

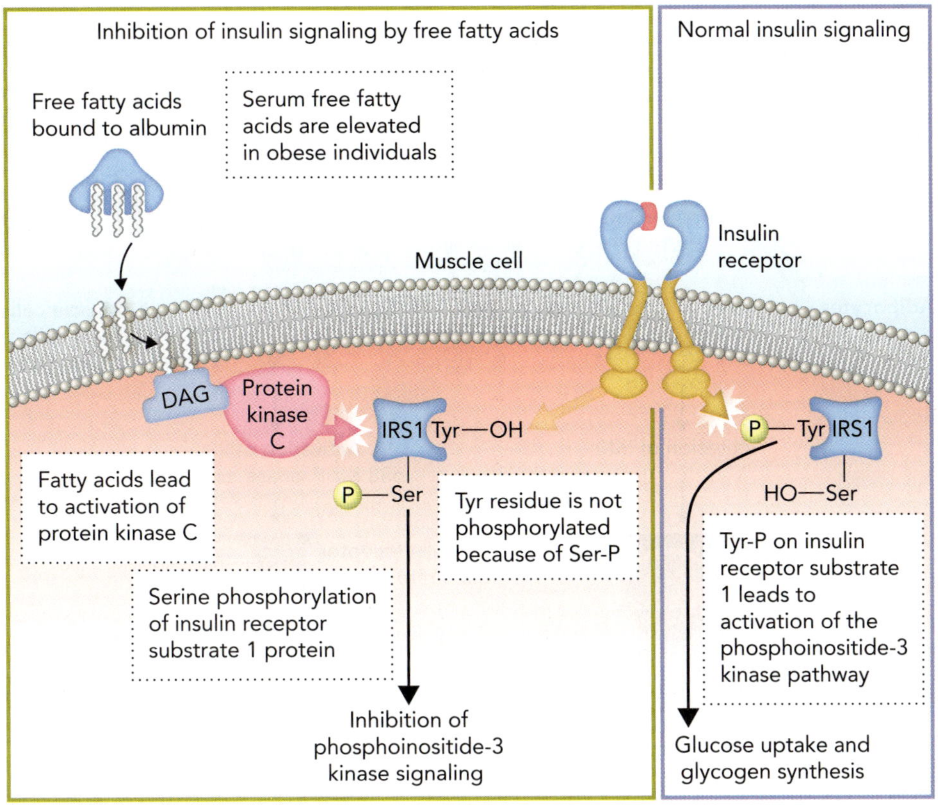

Figure 19.27 A proposed mechanism for how elevated serum levels of free fatty acids in obese individuals could promote insulin resistance in muscle cells through inhibition of the intracellular insulin signaling pathway is shown. In this model, increased levels of free fatty acids lead to the production of diacylglycerol (DAG) in the plasma membrane, which stimulates protein kinase C signaling. Phosphorylation of insulin receptor substrate 1 (IRS1) on serine residues by protein kinase C inhibits the normal phosphorylation of insulin receptor substrate 1 tyrosine residues by the insulin receptor, which is required for downstream phosphoinositide-3 kinase activation.

Figure 19.27 illustrates how elevated levels of free fatty acids in serum could interfere with insulin signaling in muscle cells by inducing phosphorylation of downstream insulin signaling proteins at inhibitory sites. In this proposed mechanism, activation of protein kinase C in skeletal muscle cells is the result of increased levels of fatty acyl-CoA derivatives and diacylglycerol. Activated protein kinase C phosphorylates the insulin receptor substrate 1 protein on serine residues, which is known to inhibit insulin receptor–mediated phosphorylation of tyrosine residues on insulin receptor substrate 1. The result is lack of insulin signaling leading to high glucose levels in the blood. Phosphotyrosines on insulin receptor substrate 1 function as protein docking sites and are required for downstream insulin signaling through the phosphoinositide-3 kinase pathway (see Figure 8.47).

Studies in mice and humans have shown that the production of numerous hormones in adipose tissue, including leptin, is affected by increased fat storage. The expression and secretion of two other hormones in adipose tissue is also affected by the amount of stored lipid. One of these is tumor necrosis factor-α (TNF-α), a well-characterized inflammatory cytokine that is produced at higher levels in adipocytes when lipid stores are high. The other hormone is **adiponectin**, which is expressed at lower levels in adipocytes under the same conditions. Although the

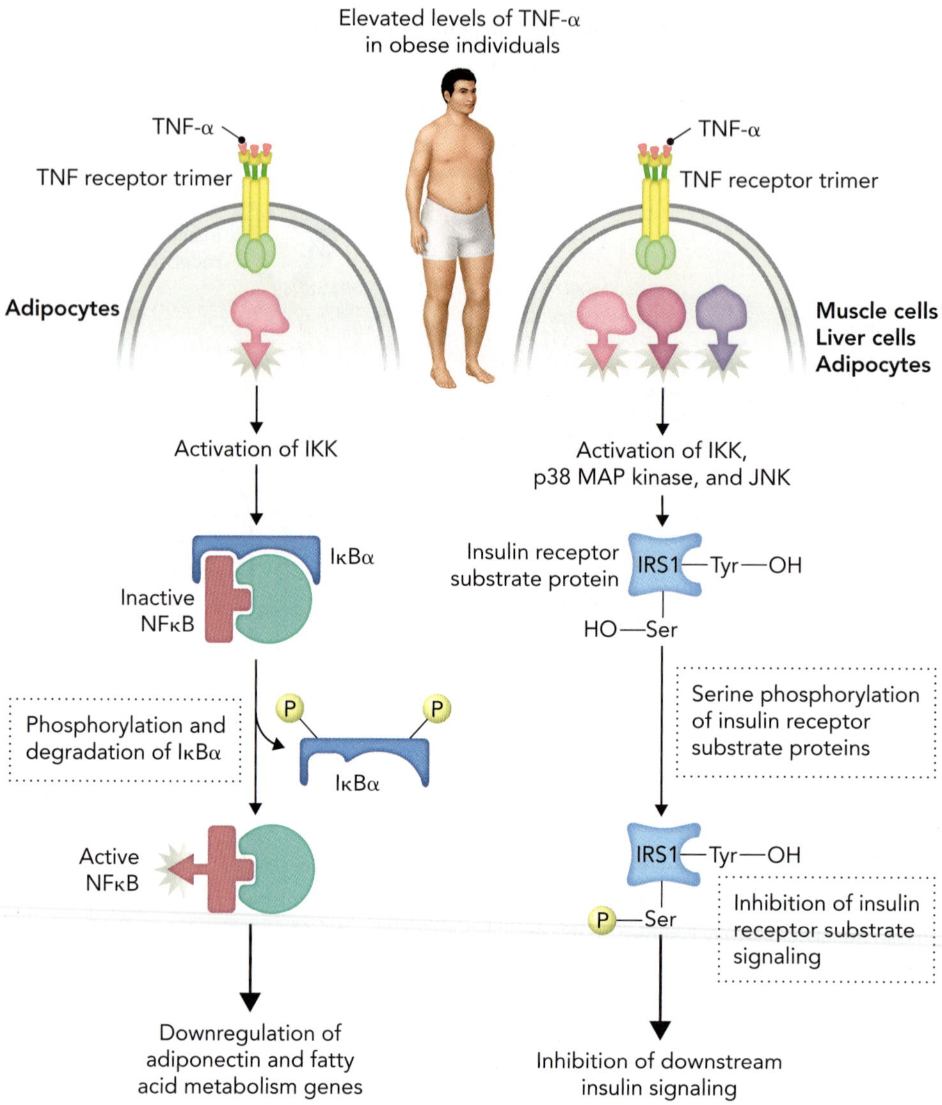

Figure 19.28 Elevated levels of TNF-α in the serum of obese individuals leads to downregulation of gene expression in adipose tissue and inhibition of insulin signaling in muscle, liver, and adipose tissue. In adipocytes, TNF-α signaling through NFκB inhibits expression of the adiponectin gene and genes involved in fatty acid uptake and storage. TNF-α signaling in muscle, liver, and adipose tissue also interferes with insulin signaling by inducing serine phosphorylation of insulin receptor substrate proteins.

connection between obesity and insulin-resistant type 2 diabetes is complex, the effect of altered levels of secretion of TNF-α and adiponectin from adipocytes in visceral fat is likely to be a contributing factor.

In addition to modulating apoptotic signaling pathways (see Figure 8.54), TNF-α also regulates several metabolic processes in muscle, liver, and adipose tissue. As shown in **Figure 19.28**, autocrine signaling by TNF-α in adipocytes leads to downregulation of genes involved in the uptake and storage of fatty acids. This contributes to increased levels of fatty acids in serum and interferes with insulin signaling in muscle cells (see Figure 19.27). TNF-α signaling in adipocytes also downregulates expression of adiponectin; this lowers insulin sensitivity in muscle and liver cells by reducing downstream signaling through AMP-activated protein kinase (AMPK).

Many of these effects on gene expression mediated by TNF-α require the transcription factor NFκB, a heterodimer protein that is held in an inactive cytoplasmic complex by the NFκB regulatory protein IκBα (inhibitor of κBα). TNF receptor signaling induces the activity of IκBα kinase (IKK), which phosphorylates IκBα, stimulating its degradation by proteasomes. Once this occurs, the activated NFκB heterodimer enters the nucleus, where it binds to DNA and regulates gene expression (see Figure 8.54).

Besides these transcriptional effects, TNF-α signaling also activates a phosphorylation cascade in muscle, liver, and adipose tissue that leads to serine phosphorylation of insulin receptor substrate proteins and decreased insulin receptor signaling. The serine kinases that have been implicated in insulin receptor substrate protein inactivation by TNF-α signaling are IKK, p38 mitogen-activated protein kinase (p38 MAP kinase), and Jun N-terminal kinase (JNK). TNF receptor activation of the IKK, p38 MAP kinase, and JNK serine kinases involves the formation of large protein complexes on the cytoplasmic side of the plasma membrane, which serve as docking sites for signaling proteins. Phosphorylation of insulin receptor substrate 1 on a serine residue leads to inhibition of downstream signaling (see Figure 19.27).

Adiponectin is a 247-amino-acid multimeric protein that is produced exclusively in adipose tissue and binds to transmembrane adiponectin receptors in a variety of tissues. Adiponectin hexamers consist of two trimers covalently linked by disulfide bonds. These hexamers then combine to form the active high-molecular-weight species containing 12–18 adiponectin subunits. As shown in **Figure 19.29**, binding of high-molecular-weight adiponectin oligomers to adiponectin receptors in muscle cells stimulates glucose uptake and fatty acid oxidation, which reduces blood glucose and free fatty acid levels. Serum levels of high-molecular-weight adiponectin are decreased in obese individuals, in part due to TNF-α suppression of adiponectin expression in adipocytes (see Figure 19.28).

Reduced AMPK signaling in obese individuals can contribute to insulin resistance in two ways. First, expression of the GLUT4 glucose transporter protein and its translocation to the plasma membrane is reduced in muscle cells when AMPK signaling is not fully activated by adiponectin. Second, reduced inhibition of muscle cell acetyl-CoA carboxylase by AMPK results in decreased rates of fatty acid oxidation, owing to higher levels of malonyl-CoA and inhibition of carnitine acyltransferase I activity. The net result of decreased fatty acid oxidation is increased levels of free fatty acids in the blood, which can induce insulin resistance by interfering with insulin receptor signaling (see Figure 19.27).

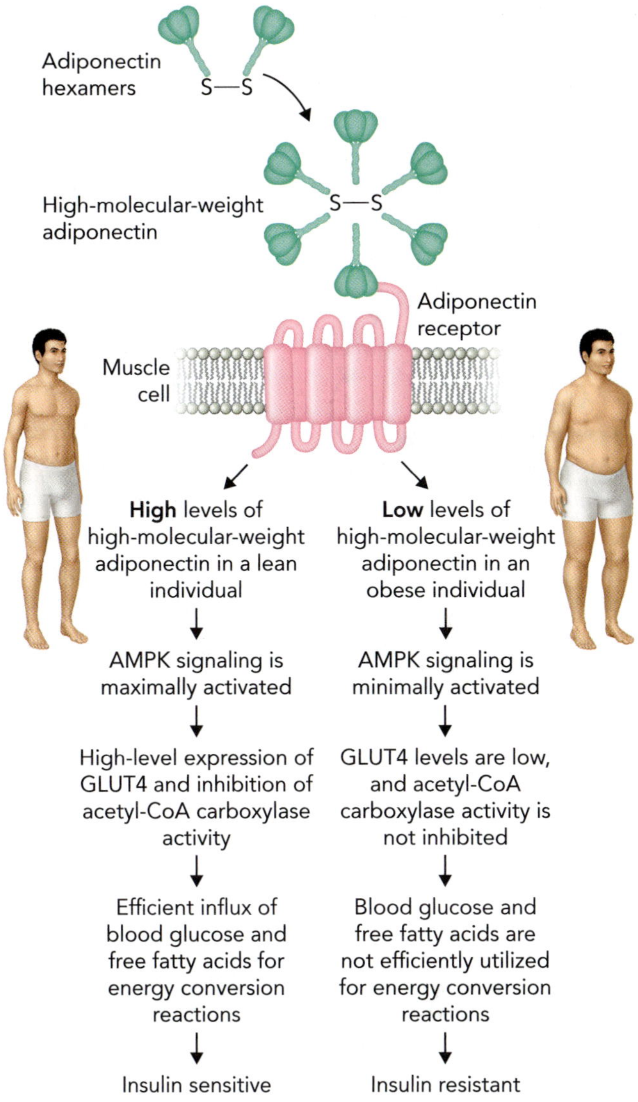

Figure 19.29 Reduced levels of high-molecular-weight serum adiponectin in obese individuals results in decreased activation of AMPK signaling in muscle cells and the development of insulin resistance. Adiponectin receptors contain seven transmembrane-spanning regions, but unlike G protein–coupled receptors, the N terminus of adiponectin receptors is in the cytoplasm and the C terminus is in the extracellular space.

Although many of the details still need to be worked out, research over the past 20 years has confirmed that both genes and environment contribute to the development of obesity and type 2 diabetes. As described earlier, increased rates of obesity may be related to the presence of thrifty genes, which function as metabolic disease genes in an environment consisting of an overabundance of high-Calorie food and a sedentary lifestyle (long-term positive energy imbalance). Although not all obese individuals develop symptoms associated with type 2 diabetes, many do. It often begins with a diagnosis of prediabetic symptoms characterized by elevated levels of blood glucose (**hyperglycemia**) and free fatty acids (**hyperlipidemia**). If this condition persists, it can lead to chronic insulin secretion, then to insulin resistance, and eventually to

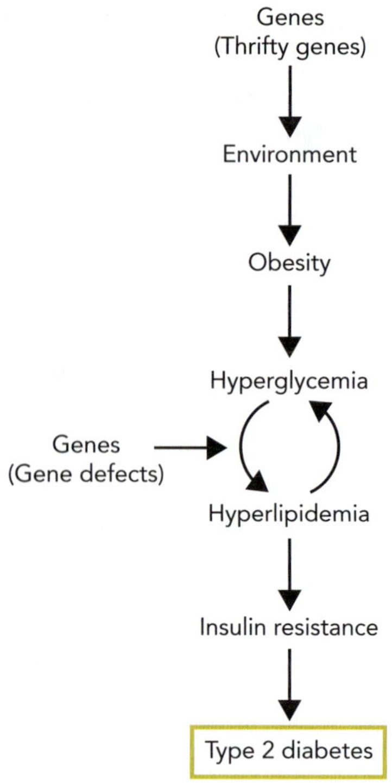

Figure 19.30 Type 2 diabetes often results from a state of prediabetes characterized by hyperglycemia and hyperlipidemia. In many individuals, chronic hyperglycemia and hyperlipidemia can lead to symptoms of insulin resistance and type 2 diabetes.

the clinical symptoms of type 2 diabetes (**Figure 19.30**). Type 2 diabetes can also be initiated independently of obesity as a consequence of genetic defects that alter insulin signaling pathways, which explains why not all obese individuals develop type 2 diabetes, and 20% of type 2 diabetics have never been overweight or obese.

Treatment of Type 2 Diabetes The key to successful treatment of type 2 diabetes is to lower blood glucose levels and increase insulin sensitivity before physiologic damage to the insulin signaling pathway occurs. Four major classes of drugs have been developed to treat type 2 diabetes: (1) α-glucosidase inhibitors (miglitol); (2) sulfonylurea drugs that inhibit the pancreatic ATP-dependent K^+ channel (glipizide); (3) drugs that stimulate the activity of AMPK (metformin); and (4) ligand agonists of the nuclear receptor PPARγ (thiazolidinediones).

The α-glucosidase inhibitors block carbohydrate degradation in the intestines by binding to the active sites of pancreatic α-amylase and membrane-bound intestinal α-glucosidase. The effect of these drugs is to lower glucose levels in the blood by decreasing intestinal absorption of carbohydrates in the diet. In contrast, sulfonylurea drugs, such as glipizide, increase insulin secretion from the pancreatic β cells by inhibiting the activity of the pancreatic ATP-dependent K^+ channel. This leads to membrane depolarization and fusion of insulin-containing vesicles with the plasma membrane, which then release insulin into the blood (see Figure 9.41). The use of α-glucosidase inhibitors and sulfonylurea drugs is limited to the treatment of patients with mild type 2 diabetes because the mechanisms of action of these drugs do not target the underlying cause of the disease; namely, insulin resistance.

Metformin is one of the most widely prescribed drugs for treating type 2 diabetes. It is a guanidine analog that improves insulin sensitivity in multiple tissues by elevating AMP levels, which activates AMPK signaling. As far back as the Middle Ages, metformin-related compounds such as isoamylene guanidine isolated from the French lilac, *Galega officinalis*, have been used to treat symptoms of diabetes (**Figure 19.31**). Metformin stimulation of AMPK signaling in skeletal muscle leads to increased glucose uptake and fatty acid oxidation, as well as mitochondrial biogenesis and increased rates of oxidative phosphorylation (**Figure 19.32**). In cardiac muscle, AMPK activation increases flux through the glycolytic pathway and also increases glucose uptake and fatty acid oxidation. One of the major mechanisms by which metformin reduces blood glucose levels is by AMPK-mediated inhibition of liver gluconeogenesis.

Thiazolidinediones are high-affinity ligands for PPAR nuclear receptor proteins. The insulin-sensitizing activity of thiazolidinediones is due to PPARγ-mediated regulation of adipocyte gene expression, which affects glucose and lipid metabolism in liver and muscle tissue. As shown in **Figure 19.33**, PPARγ induces the expression of fatty acid transport protein and lipoprotein lipase, which

French lilac, *Galega officinalis*

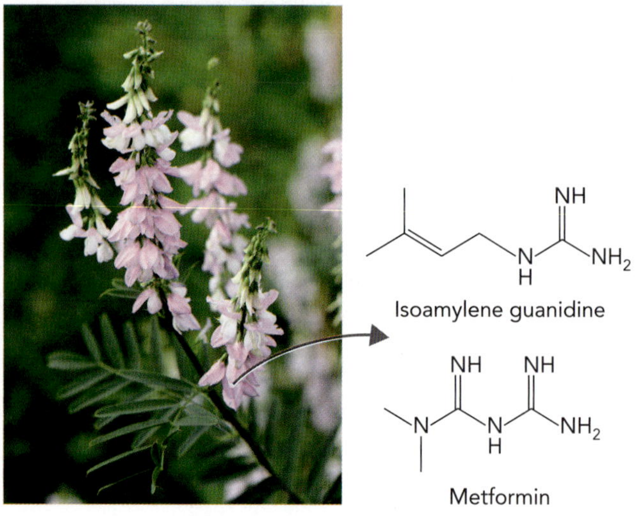

Figure 19.31 Metformin-related compounds such as isoamylene guanidine are present in the flower of the French lilac, *Galega officinalis*, and have been used for centuries to treat symptoms of diabetes. PHOTO: RAMI AAPASUO/ALAMY.

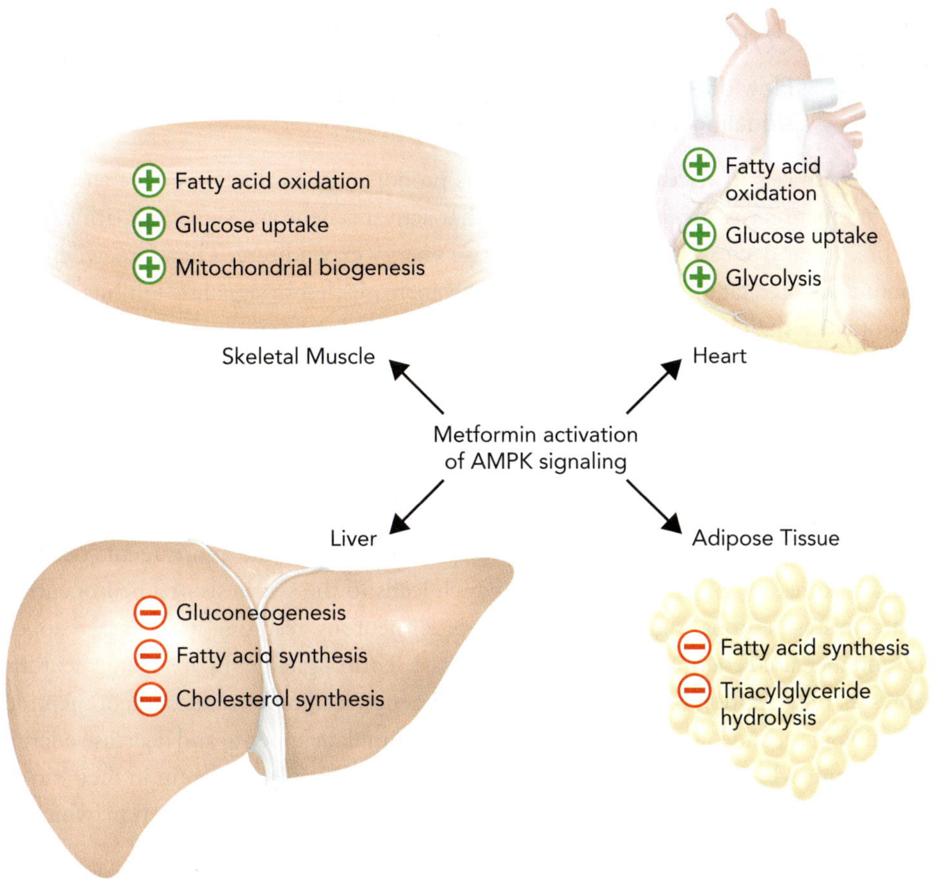

Figure 19.32 Metformin increases AMP levels and activates AMPK signaling in skeletal muscle, heart, liver, and adipose tissue.

increase the influx of fatty acids into adipocytes. In addition, PPARγ increases the expression of the phosphoenolypyruvate carboxykinase and glycerol kinase genes, resulting in activation of the triacylglycerol synthesis pathway through elevated levels of glycerol-3-phosphate. The other major function of thiazolidinediones in adipose tissue is to alter the expression of peptide hormones. This includes induction of adiponectin gene expression and downregulation of the inflammatory cytokines TNF-α, interleukin-6, and plasminogen activator inhibitor 1. Although thiazolidinedione treatment has been shown to improve insulin sensitivity in type 2 diabetics, the long-term use of these potent PPAR agonists is associated with significant side effects. The most serious side effects have occurred with the PPARγ agonists rosiglitazone and pioglitazone, which can include excessive weight gain and an increased risk of cardiovascular disease.

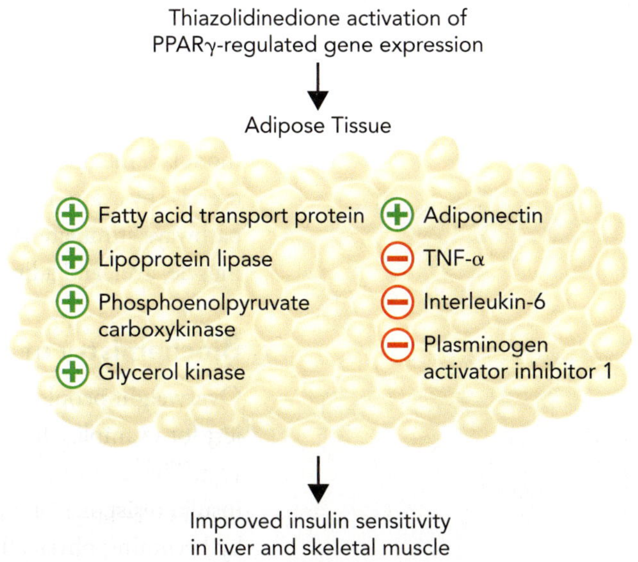

Figure 19.33 Thiazolidinedione activation of PPARγ-regulated gene expression in adipose tissue leads to increased insulin sensitivity in liver and skeletal muscle tissue. Two major effects of PPARγ activation on adipocytes are increased expression of target genes encoding proteins involved in uptake and storage of fatty acids, and altered expression of target genes encoding inflammatory cytokines.

concept integration 19.2

How does leptin help control energy balance through neuronal signaling in the brain?

Leptin synthesis in visceral adipose tissue is proportional to the amount of stored fat. Leptin functions as a metabolic regulator that activates neuronal signaling pathways in the brain that decrease appetite and increase energy expenditure. Therefore, when the amount of stored fat in adipose tissue suddenly increases, leptin signaling inhibits further fat storage to maintain energy balance. Similarly, if fat storage suddenly decreases, leptin levels decrease, and fat storage increases because it is not subject to leptin suppression. Leptin signaling in the brain involves activation of the leptin receptor in a subregion of the hypothalamus that contains anorexigenic neurons and orexigenic neurons. Anorexigenic neurons send signals that decrease appetite and increase energy expenditure to help reduce fat stores, whereas orexigenic neurons stimulate appetite and decrease energy expenditure to increase fat stores. Leptin signaling in the anorexigenic neurons induces expression of the POMC gene, which leads to the synthesis of a neuropeptide that activates neuronal signaling in second-order neurons, resulting in decreased appetite and increased energy expenditure. In contrast, leptin signaling in orexigenic neurons represses transcription of the NPY and AGRP genes, leading to reduced neuropeptide signaling in second-order neurons that normally stimulate appetite and decrease energy expenditure. Most obese individuals have high levels of circulating leptin in their blood and are leptin resistant, apparently due to chronic leptin receptor stimulation, which is similar to the cause of insulin resistance in type 2 diabetics.

19.3 Nutrition and Exercise

The three primary factors influencing metabolic homeostasis are genetic inheritance, nutrition, and exercise. You cannot change who your parents are, but you can control what you eat and how much you exercise. Maintaining a healthy weight significantly lowers the risk of type 2 diabetes, cardiovascular disease, and some types of cancer. Although there is no escaping the biochemical reality that energy balance determines body weight (see Figure 19.21), not all foods of equal Calories provide the same nutritional value (fiber, vitamins, essential amino acids). Moreover, as we discuss in this section, the adage "you are what you eat" may have biochemical implications beyond Calorie counting, in that the quantity and quality of macronutrients (carbohydrates, proteins, fats) appear to modulate metabolic flux under some conditions. We also present a biochemical view of exercise by examining how the AMPK and PPARγ coactivator-1α (PGC-1α) signaling pathways stimulate ATP synthesis in muscle cells by regulating metabolic pathways in muscle, liver, and adipose tissue in response to muscle contraction.

Biochemistry of Macronutrition and Dieting

Overweight people often make New Year's resolutions to lose weight by eating healthier; for example, they might aim to lose 10% of their body weight by the end of the year. The motivation could be to reverse the effects of obesity-related disease such as insulin resistance or a psychological need to improve self-esteem and instill confidence by becoming physically more attractive. On the basis of thermodynamic principles, a person eventually loses weight if he or she eats fewer Calories than expended, which is best accomplished by restricting the amount of high-Calorie foods in the diet and by exercising. Because eating fewer desserts and running around the neighborhood do not appeal to most people, a variety of weight-loss pills have been developed to promote negative energy imbalance in the short term.

One of the oldest drugs used to induce weight loss is **ephedrine**. This drug acts indirectly to stimulate adrenergic receptor signaling through enhanced release of noradrenaline, a catecholamine signaling molecule. Ephedrine is an alkaloid derived from plants in the genus *Ephedra* and is the active ingredient in the Chinese herbal medicine *ma huang*, which is prepared from the plant *Ephedra sinica*. Another class of drugs developed for weight control are those that modulate eating behavior; for example, **lorcaserin** (Belviq). Lorcaserin is a high-affinity receptor agonist for a class of serotonin receptors called 5-HT_{2C} that suppress appetite. Although lorcaserin provides only modest effects (most patients lose 5% to 10% of their body weight over 1 year), it was the first approved weight-loss pill that specifically targets neuronal control of food consumption and energy expenditure.

A more direct method to lose weight is to limit absorption of dietary fats in the small intestine by inhibiting the enzyme pancreatic lipase using the pharmaceutical drug **orlistat** (Xenical). The appeal of orlistat is that one can enjoy the flavor of high-fat food when it is eaten, but because pancreatic lipase is inhibited, the triglycerides contained in the food pass through the digestive tract without being metabolized and are excreted. A complementary approach is to replace fats in processed foods with a fat substitute that adds the same flavor to foods but is not a substrate for digestive lipases and is excreted before it can be metabolized. One such compound is the food additive **olestra**, which is a sucrose molecule containing up to eight covalently linked fatty acyl groups (**Figure 19.34**). As some Calorie-conscious consumers have found out, taking orlistat before eating a high-fat meal or ingesting large amounts of olestra-containing snack foods can cause unpleasant side effects, such as flatulence and loose stools due to undigested lipids in the colon.

The most common method to lose weight is to reduce Caloric intake by dieting. This can be as simple as counting Calories using the Weight Watchers plan, altering the macronutrient composition of a daily diet in an attempt to metabolically stimulate degradation of stored fats, or an approach that combines Calories and macronutrients into one number (also available through Weight Watchers). Three of the most popular macronutrient-based diets are the low-carbohydrate Atkins Diet; the high-protein, low-carbohydrate Zone Diet; and the low-fat, high-fiber Ornish Diet (**Figure 19.35**). Because these diets are promoted through the popular press they are often called "fad" diets. Notably, these diets were developed by medical doctors or research scientists and are loosely based on biochemical principles. The Atkins Diet limits carbohydrates, which stimulates fatty acid oxidation as a source of metabolic fuel. The Zone Diet proposes that limiting carbohydrates while increasing protein intake at every meal results in a more favorable insulin-to-glucagon ratio and produces eicosanoids that stimulate PPAR activity. The Ornish Diet is very low in fat and depends on fruits and vegetables for much of the daily Caloric intake. The high fiber content in this diet slows the absorption of food into the digestive system, and thus makes an individual feel fuller for a longer period of time.

Figure 19.34 Olestra is a zero-Calorie food substitute containing fatty acid side chains covalently linked to sucrose. The taste of olestra is similar to that of normal dietary fats such as triglycerides, though the bulky, irregular shape of this molecule prevents its breakdown by digestive lipases. FA represents the hydrocarbon chain of a fatty acid.

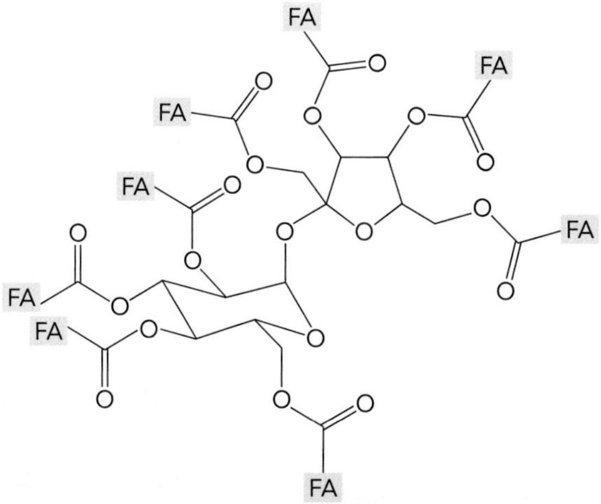

Olestra (fat substitute)

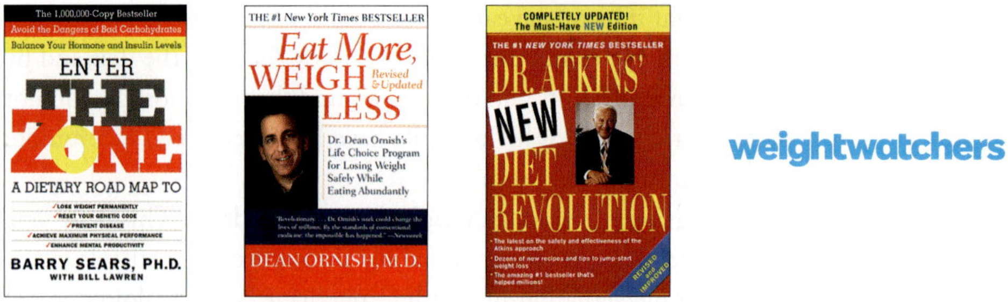

High protein, low carbohydrate | Low fat, high fiber | Low carbohydrate | Restricted Calories

Figure 19.35 Altering the macronutrient profile: a Calorie-restricted diet is a hallmark of many of the most popular fad diets. The Zone Diet is considered a high-protein, low-carbohydrate diet, the Ornish Diet is a low-fat vegetarian diet, and the Atkins Diet is a low-carbohydrate diet. The Weight Watchers diet is based on a regimen of Calorie counting using a Web-based personal diary to achieve short-term negative energy imbalance.

Table 19.4 RESULTS FROM A SCIENTIFIC STUDY COMPARING THE EFFECTS OF FOUR DIFFERENT WEIGHT-LOSS DIETS ON KEY METABOLIC PARAMETERS

Parameters	Atkins Diet (low carbohydrate)	Zone Diet (high protein)	Ornish Diet (low fat)	Weight Watchers (low Calorie)
Baseline				
Participants (*n*)	40	40	40	40
Age (y)	47	51	49	49
Weight (kg)	100	99	103	97
BMI	35	34	35	35
Waist (cm)	109	108	111	108
Cholesterol (mg/dL)	214	222	214	221
LDL/HDL ratio	3.2	3.1	3.2	3.3
12 months later				
Participants (*n*)	21	26	20	26
Weight (kg)	−3.9	−4.9	−6.6	−4.6
BMI	−1.4	−1.6	−2.3	−1.7
Waist (cm)	−4.7	−4.5	−4.3	−5.0
Cholesterol (mg/dL)	−8.1	−15.6	−21.5	−12.6
LDL/HDL ratio	−0.73	−0.61	−0.59	−1.07

Note: Mean values are shown.

BASED ON DATA FROM M. L. DANSINGER ET AL. (2005). COMPARISON OF THE ATKINS, ORNISH, WEIGHT WATCHERS, AND ZONE DIETS FOR WEIGHT LOSS AND HEART DISEASE RISK REDUCTION: A RANDOMIZED TRIAL. *JAMA, 293*, 43–53.

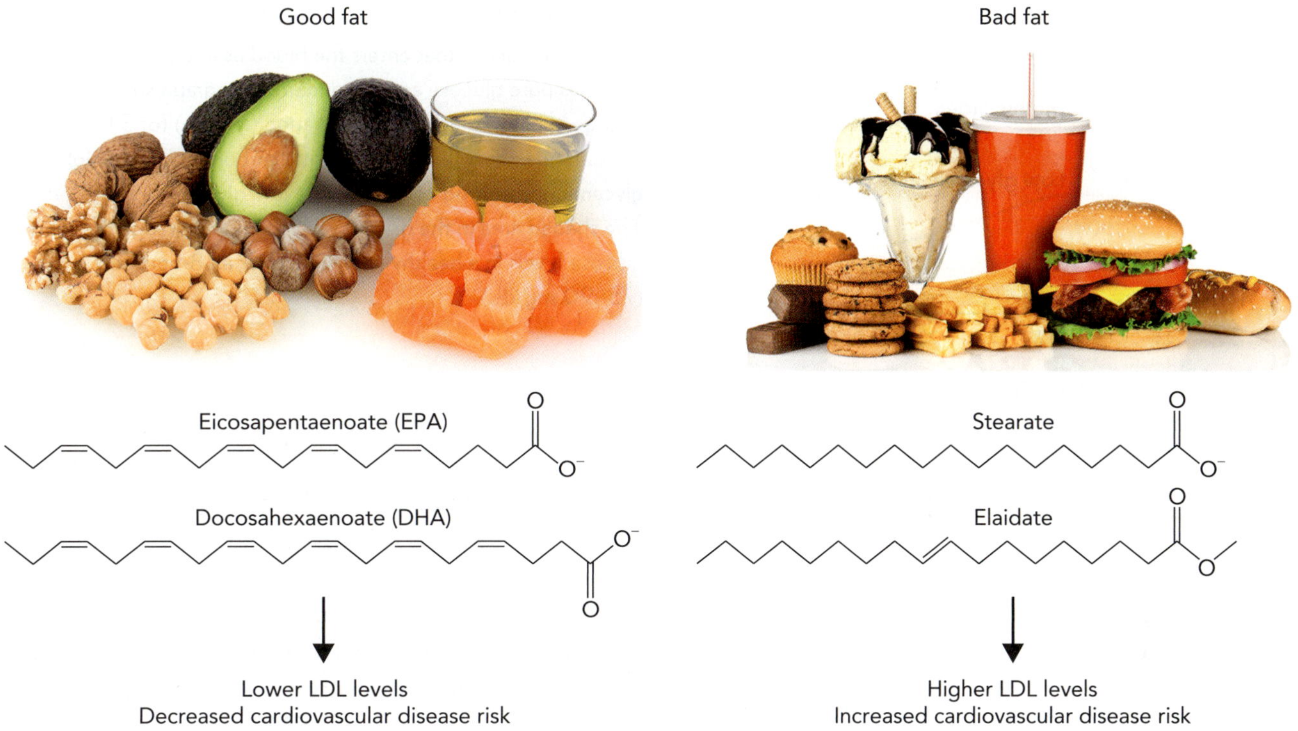

Figure 19.36 The dietary ω-3 polyunsaturated fatty acids eicosapentaenoate and docosahexaenoate are associated with decreased risk of cardiovascular disease, whereas saturated fatty acids (stearate) and trans fatty acids (elaidate) are associated with increased risk of cardiovascular disease. Both of these effects are linked to changes in serum LDL levels. GOOD FAT: TINA LARSSON/SHUTTERSTOCK; BAD FAT: FCAFOTODIGITAL/ISTOCK/GETTY IMAGES PLUS.

Although each of the four diets shown in Figure 19.35 is somewhat unique, little clinical evidence exists that any one works better than the others. Indeed, results from a randomized trial comparing the effectiveness of the Weight Watchers, Atkins, Zone, and Ornish diets in treating obese adults showed no significant differences in several key metrics. As shown in **Table 19.4**, not only was the average weight loss after a 12-month period nearly the same for participants, regardless of the diet regimen (~5% total body weight), but also the observed improvements in metabolic indicators of total cholesterol and the LDL-to-HDL ratios were very similar. Therefore, although the biochemistry behind these diets makes sense in terms of promoting fatty acid oxidation as a way to lose weight, the reality is that for most people, genes and environment play a big part in determining how successful the diet will be. It may be noteworthy that ~40% of the volunteers in each of the four study groups dropped out before the 12-month study was completed, pointing to the fact that eating behavior is a key component to weight control.

Two types of studies have indicated that macronutrient composition of daily food intake may be an important factor in maintaining energy balance and overall physiologic health. The first type of study focused on the contribution of dietary saturated and unsaturated fatty acids to long-term cardiovascular health, and the second type of study compared the effects of different forms of dietary carbohydrates on blood glucose levels within the first few hours after eating.

Consuming high amounts of saturated fatty acids and trans fatty acids on a regular basis has been shown to increase LDL levels in the blood, which is associated with a higher risk of cardiovascular disease. The long-chain saturated fatty acid stearate (18:0) is commonly found in animal products (see Table 15.1), whereas the C_{18} trans fatty acid elaidate [*trans* 18:1(Δ^9)] is contained in many types of snack foods prepared using hydrogenated vegetable oils (see Figure 15.5). In contrast, diets containing the ω-3 polyunsaturated fatty acids eicosapentaenoate [all-*cis* 20:5($\Delta^{5,8,11,14,17}$)] and docosahexaenoate [all-*cis* 22:6($\Delta^{4,7,10,13,16,19}$)] decrease serum LDL levels and lower the risk of cardiovascular disease (**Figure 19.36**). Therefore, because both types of fatty acids

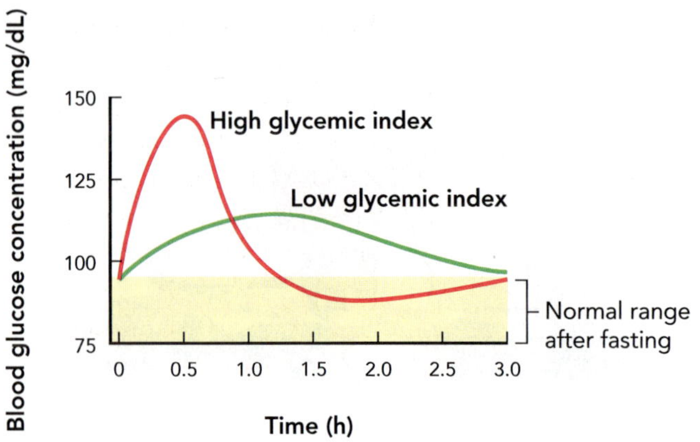

Figure 19.37 The glycemic index of foods is a measure of the amount of glucose that enters the blood as a function of time and uses pure glucose as the standard. The graph shows the change in blood glucose concentration (mg/dL) for 3 hours after consuming the same amount of total carbohydrate in a high-glycemic-index food compared to a low-glycemic-index food. The shaded area in the graph shows the normal range of blood glucose levels in a healthy individual after an 8-hour fast (75–90 mg/dL).

contain ~10 Calories/g (41.8 kJ/g), it is advisable for an individual to consume more of his or her daily fat Calories from foods such as fish, olive oil, and nuts (high in polyunsaturated fats), rather than from fried foods, pizza, and donuts (high in saturated and trans fatty acids). One explanation for the physiologic differences between saturated and polyunsaturated fatty acids may be related to the fact that some of these dietary fatty acids stimulate the transcriptional regulatory activity of PPAR nuclear receptors. As described earlier in this chapter, ligand activation of PPAR nuclear receptors has a dramatic effect on metabolic flux in multiple tissues because they function as regulators of metabolic gene expression (see Figure 19.16).

Consuming foods containing carbohydrates that lead to rapid and dramatic increases in glucose levels within 30 minutes of eating has been shown to increase rates of obesity compared to consuming foods with carbohydrates that cause only moderate increases in blood glucose levels over several hours. These studies compared different types of carbohydrates using a physiologic measurement of glucose metabolism called the glycemic index. The **glycemic index** of foods is a numerical value indicating how quickly glucose is released into the blood after eating different types of carbohydrate-containing foods relative to the rise in blood glucose levels after drinking a solution of pure glucose (**Figure 19.37**). The glycemic index can be calculated by dividing the area under the glucose curve obtained with the test food by the area under the glucose curve using pure glucose and then multiplying the ratio by 100. If the glycemic index is above 70, then the food is considered to have a high glycemic index and is associated with a rapid increase in blood glucose levels (**Table 19.5**). Foods with a glycemic index below 55 are considered to be healthy carbohydrates because their effects on blood glucose levels are less dramatic, resulting in a graded insulin response. Ingesting foods with a high glycemic index causes a sudden rise in blood glucose levels, which induces a rapid increase in insulin levels and stimulates conversion of this dietary glucose into fatty acids. Foods with a high glycemic index are associated with a hypoglycemic refractory period in which blood glucose levels drop below the normal fasting state (75–90 mg/dL), which can leave one feeling lethargic (see Figure 19.37). This steep drop in blood glucose levels between 1.5 and 2.5 hours after eating high-glycemic-index foods results from the spike in insulin release that occurs ~30 minutes after eating. In contrast, ingesting foods with a low glycemic index leads to a more gradual increase in blood glucose and insulin levels, resulting in more of the dietary glucose being used for energy conversion and glycogen synthesis rather than fatty acid synthesis.

Table 19.5 GLYCEMIC INDEX VALUES FOR REPRESENTATIVE FOODS

Food	Glycemic index
Broccoli	10
Peanuts	14
Low-fat yogurt	20
Cherries	22
Grapefruit	25
Skim milk	37
Lentils, boiled	41
Grapes	43
Spaghetti	44
Chocolate bar	49
Pound cake	54
Soda	72
Donut	76
Jelly beans	80
Corn chips	105
Pretzel	116

Note: The glycemic index value is a ratio of blood glucose levels with test food (area under the curve) divided by blood glucose levels using pure glucose (area under the curve). The ratio value is then multiplied by 100.

Eating a macronutrient-balanced diet is the key to maintaining metabolic homeostasis within a healthy range. But it can be difficult to determine what the best food choices should be, given that much of the nutritional information in the

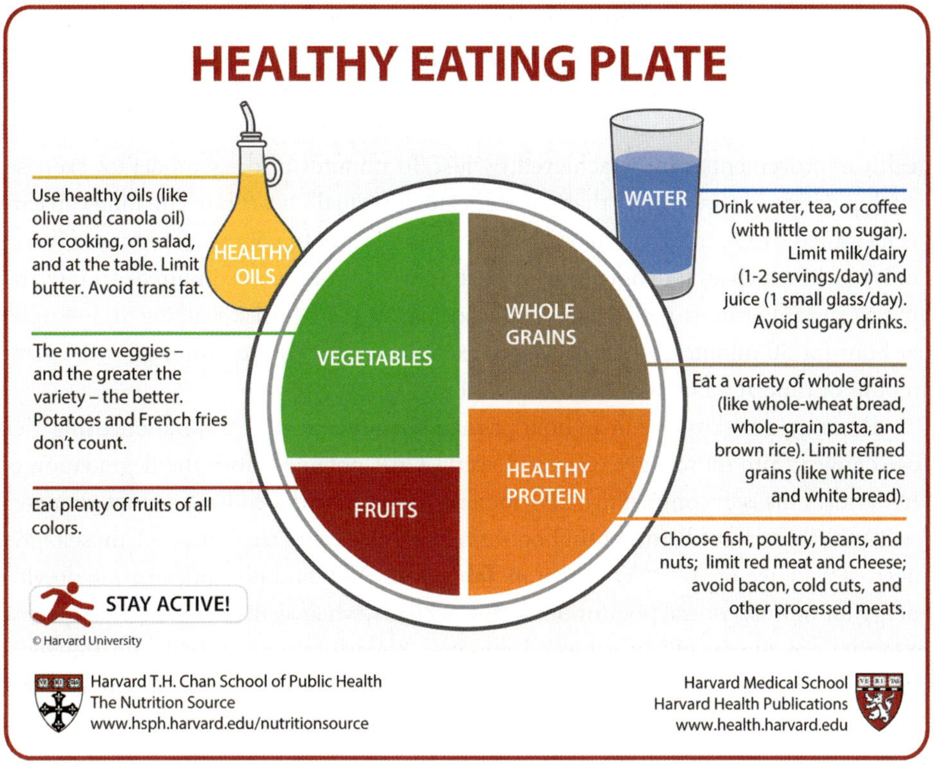

Figure 19.38 The "healthy eating plate" nutritional recommendations developed at the Harvard School of Public Health provide study-based guidelines to healthy living. AS PRINTED IN IMAGE AND © HARVARD UNIVERSITY. FOR MORE INFORMATION VISIT: WWW.HEALTH.HARVARD.EDU. NOTE: HARVARD HEALTH PUBLICATIONS DOES NOT ENDORSE ANY PRODUCTS OR MEDICAL PROCEDURES.

media and on packaging labels is provided by the commercial food industry. To help consumers make healthy food choices, the U.S. Department of Agriculture (USDA) has recently developed an education program based on the concept of a dinner plate that contains five food groups served in relative proportions. The USDA recommends that 50% of daily Calories be derived from fruits and vegetables, 30% from whole grains, and 20% from protein, with an optional serving of dairy in the form of nonfat milk or yogurt. As an extension of the USDA dinner plate, Walter Willett, a physician and epidemiologist at the Harvard School of Public Health, used data from large-scale human studies to develop what he calls a "healthy eating plate." As shown in **Figure 19.38**, the Harvard healthy eating plate is similar in many ways to the USDA's recommendations, in that half of the plate is filled with fruits and vegetables, with healthy grains and proteins making up the other half of the plate. A major difference, however, is an emphasis on water rather than milk as a daily beverage, and the inclusion of healthy cooking oils with low amounts of saturated and trans fats. The Harvard healthy eating plate campaign uses a prominent "stay active" message to remind consumers of the importance of healthy eating *and* daily exercise.

Metabolic Effects of Physical Exercise

Achieving energy balance requires that Caloric intake equals Caloric expenditure. Considering that an average-sized person of normal weight burns about 70 Calories/h when at rest under normal physiologic conditions (~1,700 Calories/d), that person would need to expend another 300 Calories/d through physical activity to maintain an energy balance on a 2,000 Calorie/d diet. As shown in **Table 19.6**, this could be done by combining several low-intensity activities, such as walking briskly (170 Calories for men), gardening (170 Calories for men), or even vacuuming (130 Calories for men). One could expend additional energy by engaging in vigorous exercises like jogging, swimming, cycling, or weight training, each of which burns 200–400 Calories in a 30-minute workout.

Table 19.6 ENERGY EXPENDITURES FOR COMMON ACTIVITIES AND SPORTS

Activity	Female (55 kg)	Male (80 kg)
Bicycling	250	350
Bowling	70	95
Gardening	120	170
Jogging	285	400
Shopping	75	105
Sleeping	60	70
Swimming laps	280	390
Tennis	200	280
Vacuuming	95	130
Walking (briskly)	120	170
Weight training	235	330

Note: Values are given in Calories for a 30-minute session.

Table 19.7 AVAILABLE ENERGY STORES FOR ATP PRODUCTION DURING INTENSE EXERCISE

Fuel source for ATP production	Time until fuel depleted
Muscle ATP	3 seconds
Muscle phosphocreatine	6 seconds
Muscle glycogen (anaerobic)	3 minutes
Liver glycogen	50 minutes
Muscle glycogen (aerobic)	1 hour 25 minutes
Adipose tissue fatty acids	>5 hours

In earlier times, the energy expenditure required to obtain food was sufficient to maintain energy balance, and therefore high levels of physical activity were a part of daily life. In modern cultures, however, food is plentiful, and many people do not need to do strenuous physical activity to make a living. Studies have shown that significant health improvements can be achieved by just 30 minutes of daily moderate exercise, which is defined as an activity that increases an individual's heart rate and lung capacity (as measured by O_2 capacity) to 60% of maximum levels. In contrast, an intense workout is one in which an individual exercises at 80% of his or her maximum heart rate and lung capacity. For most people, moderate exercise would be brisk walking at 3–4 miles per hour for 30 minutes, whereas intense exercise would require running for the same amount of time at a rate of 6–8 miles per hour.

Running a 10-km race in an hour primarily requires aerobic respiration, using glucose derived from muscle glycogen and acetyl-CoA obtained from the degradation of fatty acids. This is in contrast to a 200-meter sprint, which depends on a burst of energy for 20–30 seconds, utilizing phosphocreatine and anaerobic respiration of muscle glycogen as a source of ATP. As shown in **Table 19.7**, ATP and phosphocreatine provide energy for muscle contraction for only a few seconds, whereas the conversion of glucose to lactate can supply ATP for several minutes. Marathon runners who run the 26.2-mile race in under 4 hours use up most of their glycogen reserves within the first 60–90 minutes, and then must rely on fatty acid oxidation for ATP generation. If this metabolic crossover from muscle glycogen to fatty acids is too abrupt, it causes a physiologic condition marathon runners call "hitting the wall," in which ATP levels in the muscles suddenly decline. The reason for this is it takes almost three times longer to generate the same amount of ATP from fatty acids stored in the adipose tissue as it does using glucose released from muscle glycogen. To avoid this energy deficit, some marathon runners eat carbohydrate energy bars or drink liquids with added carbohydrate (sucrose, glucose, and fructose) during the race to sustain blood glucose levels. Another strategy is to develop a training regimen that adapts the body to a more gradual change in energy use during the race.

AMPK and PPARγ Coactivator-1α Signaling in Skeletal Muscle

The physiologic effects of exercise in humans have been studied for hundreds of years. But only recently have we begun to unravel the biochemical mechanisms that underlie the physiologic and metabolic changes brought on by exercise.

One of the most important mechanisms is exercise-induced activation of AMPK signaling, which alters metabolic flux through energy conversion pathways to increase ATP production. AMPK is a heterotrimeric serine/threonine kinase that is highly conserved in eukaryotes (**Figure 19.39**).

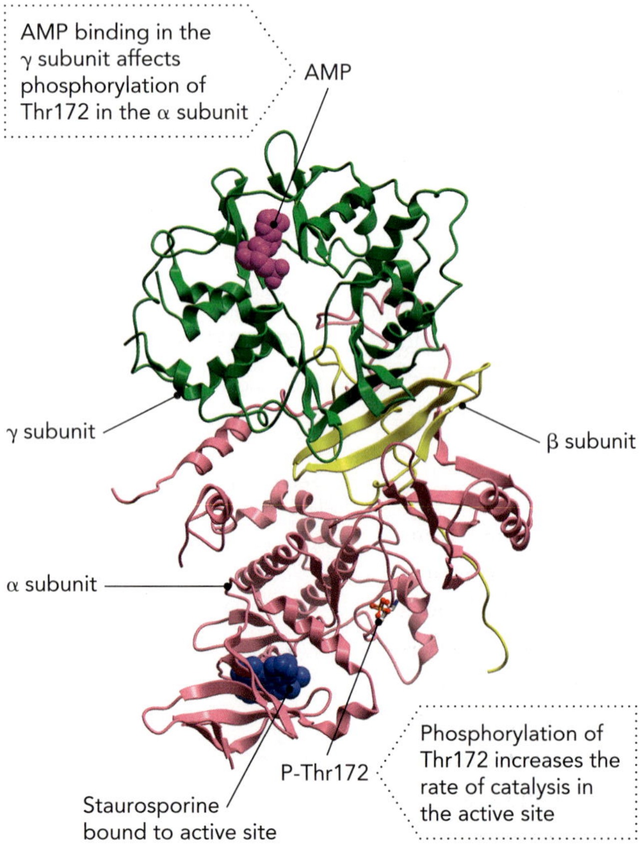

Figure 19.39 This structure of a mammalian AMPK heterotrimeric complex shows AMP bound to the regulatory γ subunit and the kinase inhibitor staurosporine bound to the active site in the catalytic α subunit. The location of phosphothreonine 172 (P-Thr172) in the α subunit is labeled. For clarity, only a small portion of the β subunit is included in this model of the protein complex. BASED ON PDB FILE 2Y94.

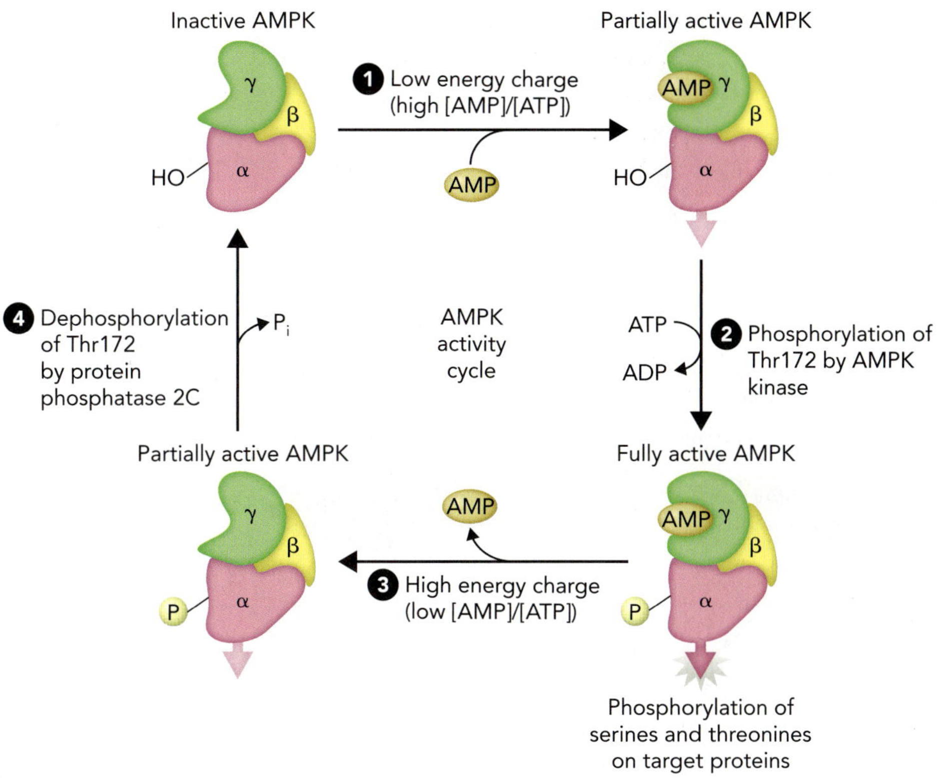

Figure 19.40 AMPK serine/threonine kinase activity is controlled by AMP binding to the regulatory γ subunit and phosphorylation of the catalytic α subunit. Under conditions of high AMP concentration, AMP binding to the AMPK γ subunit induces a conformational change that partially activates the enzyme. This conformational change facilitates phosphorylation of Thr172 on the α subunit by the regulatory protein AMPK kinase, which leads to full activation of AMPK activity. Phosphorylation of AMPK target proteins alters metabolic flux in numerous cellular pathways, resulting in increased rates of ATP synthesis and as a result decreased AMP concentration. Dissociation of AMP from the γ subunit stimulates dephosphorylation of AMPK by protein phosphatase 2C.

It consists of a catalytic α subunit; a regulatory γ subunit that binds an adenosine nucleotide; and a β subunit, which functions as a molecular scaffold. As shown in **Figure 19.40**, under conditions of low energy charge in muscle cells (high AMP concentration), AMP binding to the AMPK γ subunit leads to a conformational change in the heterotrimeric complex, facilitating phosphorylation of Thr172 in the α subunit by AMPK kinases. Although AMP binding to the γ subunit alone increases AMPK activity by ~10-fold, phosphorylation of AMPK on Thr172 stimulates AMPK activity by 100-fold.

A primary metabolic response to AMPK signaling is increased rates of ATP synthesis. This lowers AMP levels, resulting in AMP dissociation from the γ subunit and rapid dephosphorylation of the α subunit by protein phosphatase 2C. Note that the AMP binding site in the regulatory γ subunit also binds ATP and ADP and is referred to as the AXP site. The regulatory γ subunit senses the AMP-to-ATP concentration ratio in the cell on the basis of nucleotide occupancy of the AXP site: AMP binding shifts the conformational equilibrium toward the active conformation, and ATP binding shifts it toward the inactive conformation.

Figure 19.41 shows how AMPK-mediated phosphorylation of serine or threonine residues on metabolic target proteins in muscle cells leads to a net increase in ATP concentration by three mechanisms: (1) stimulation of flux through glycolysis; (2) stimulation of flux through fatty acid oxidation; and (3) increased oxidative phosphorylation. Some of these AMPK-mediated effects involve direct phosphorylation of enzymes that control metabolic flux (glycogen synthase, phosphofructokinase-2, acetyl-CoA carboxylase, hormone-sensitive lipase). Other effects involve phosphorylation of transcription factors that regulate the expression of metabolic genes, such as PPARγ coactivator-1α (PGC-1α) and cAMP response element binding protein. For example, Ser133 phosphorylation of the cAMP response element binding protein transcription factor leads to increased expression of hexokinase, whereas phosphorylation of PPARγ coactivator-1α on Ser538 and Thr177 stimulates myocyte-enhancer

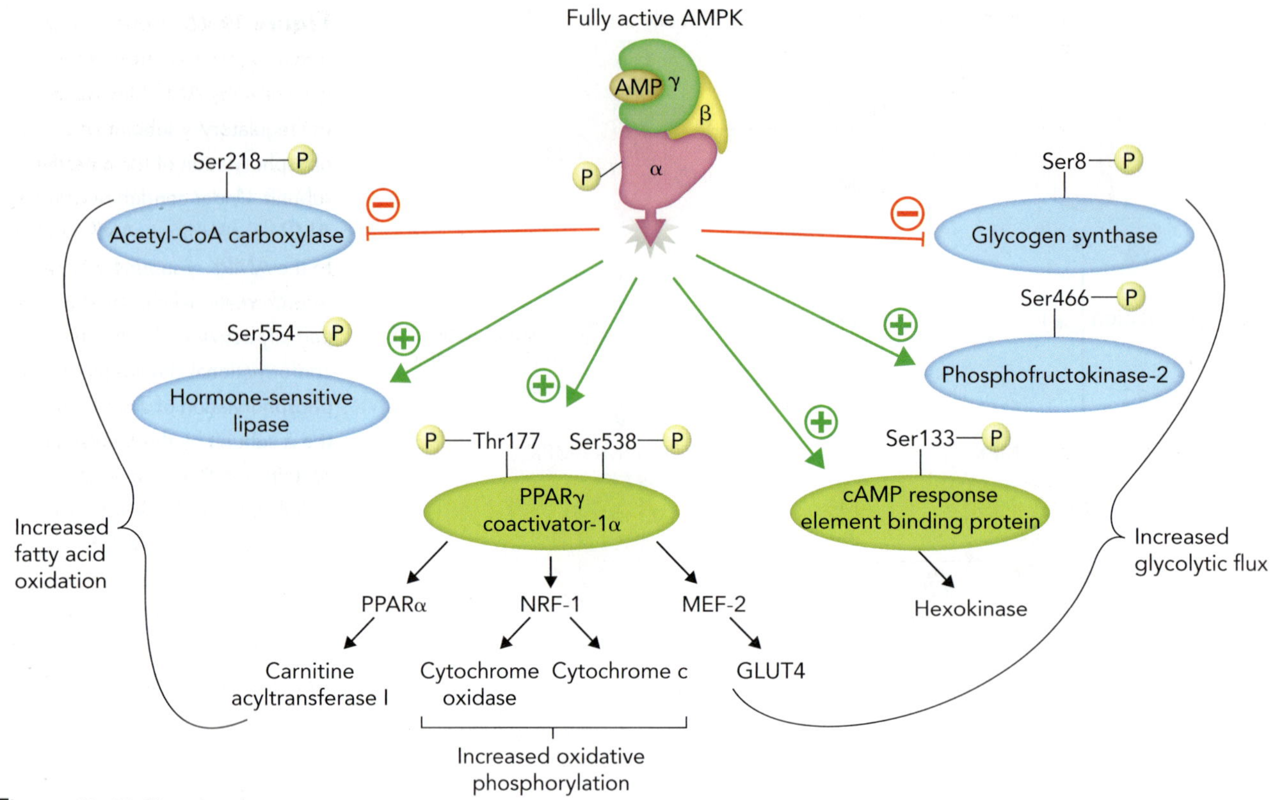

Figure 19.41 Phosphorylation of serine and threonine residues on AMPK target proteins in muscle cells leads to a net increase in ATP synthesis by three distinct mechanisms. AMPK-mediated phosphorylation of metabolic enzymes is shown by blue ovals, and phosphorylation of transcription factors is shown by green ovals. NRF-1 = nuclear respiratory factor 1.

factor 2 (MEF-2)-mediated transcription of the GLUT4 glucose transporter. As shown in Figure 19.41, AMPK-mediated phosphorylation of target proteins can be inhibitory or stimulatory. Thus, acetyl-CoA carboxylase activity is inhibited by AMPK phosphorylation of Ser218, whereas hormone-sensitive lipase activity is stimulated by phosphorylation of Ser554.

AMPK-mediated phosphorylation of the transcription factor PPARγ coactivator-1α is of particular importance in muscle cells because PPARγ coactivator-1α signaling has been linked to several exercise-induced responses. As shown in **Figure 19.42**, PPARγ coactivator-1α is a multifunctional adaptor protein that contains two types of

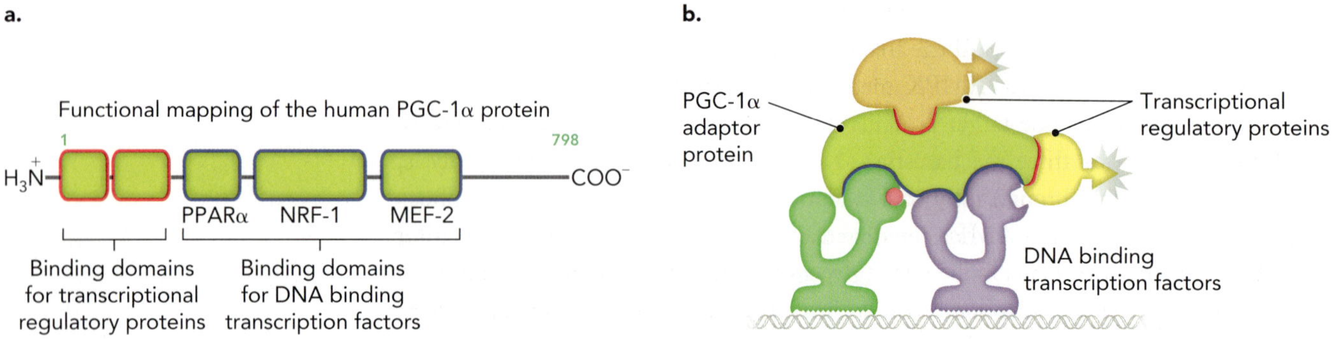

Figure 19.42 PPARγ coactivator-1α (PGC-1α) is a transcriptional coactivator that functions as an adaptor protein at target gene promoters. **a.** The molecular structure of the intact human PPARγ coactivator-1α protein is not yet known; however, functional mapping studies of the human PPARγ coactivator-1α gene have shown that it encodes a 798-amino-acid protein with multiple domains that interact with specific transcription factors. **b.** PPARγ coactivator-1α recruits transcriptional regulatory proteins to target genes by interacting with DNA-bound transcription factors.

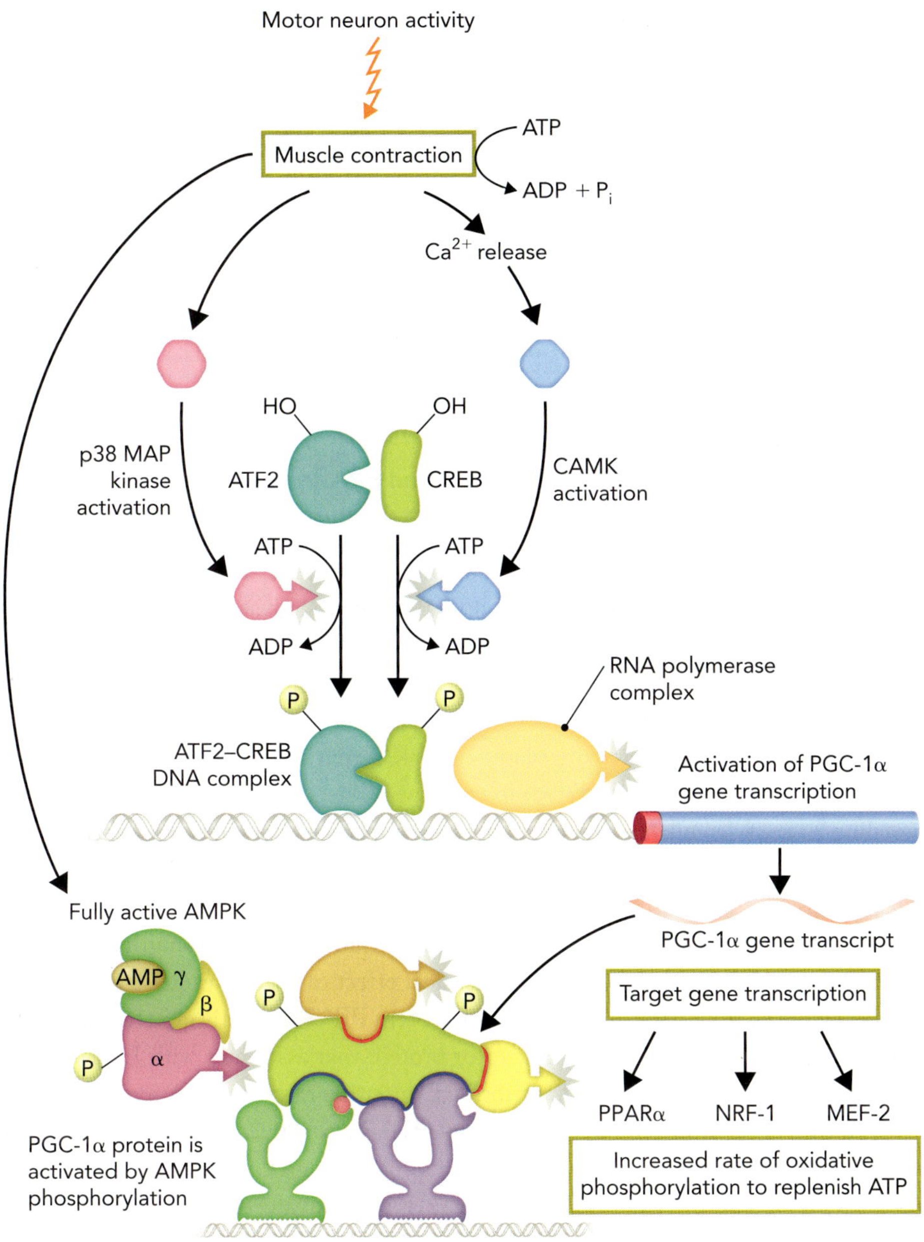

Figure 19.43 Proposed mechanism for the observed increase in PPARγ coactivator-1α (PGC-1α) gene expression in exercising muscle. In this model, muscle contraction stimulates the activity of AMPK, CAMK, and p38 MAP kinase, which phosphorylate downstream target proteins. CAMK and p38 MAP kinase phosphorylate and activate the transcription factors CREB and ATF2, respectively, whereas AMPK phosphorylates and activates PGC-1α protein.

protein interaction domains. One type of protein binding domain interacts with the transcription factors PPARα, NRF-1, and MEF-2, which bind DNA directly and regulate gene expression. The second type of protein binding domain on PPARγ coactivator-1α recruits histone deacetylases and RNA polymerase activator proteins, which modulate rates of transcriptional initiation without directly binding to DNA. Studies have shown that intense exercise stimulates the expression of PPARγ coactivator-1α in human skeletal muscle up to 10-fold, with the peak expression occurring 2 hours after completion of the exercise. This observation is important because muscle cells with higher levels of PPARγ coactivator-1α respond more quickly to AMPK-mediated phosphorylation and are better able to provide ATP for muscle contraction.

Figure 19.43 shows a signaling pathway that could explain how PPARγ coactivator-1α gene expression is induced in exercising muscle through two signaling pathways. One is a calcium-dependent signaling pathway through Ca^{2+}/calmodulin kinase (CAMK), leading to phosphorylation of the cAMP response element binding protein (CREB) transcription factor. A second signaling pathway is mediated by the downstream kinase p38 MAP kinase, which phosphorylates the activating transcription factor 2 (ATF2) transcription factor. Binding of the phosphorylated ATF2–CREB complex to DNA sequences located in the promoter region of the PPARγ coactivator-1α gene results in elevated levels of the PPARγ coactivator-1α protein. This is followed by phosphorylation of PPARγ coactivator-1α protein by AMPK and regulation of metabolic genes controlled by the transcription factors PPARα, NRF-1, and MEF-2. The induced expression of PPARγ coactivator-1α in contracting muscle leads to increased production of ATP and sustained muscle activity until energy reserves (carbohydrates and lipids) are exhausted.

concept integration 19.3

How does activation of AMPK and PPARγ coactivator-1α signaling in skeletal muscle function to sustain muscle contraction?

Regular exercise improves a person's metabolic profile and athletic performance, in part through activation of the AMPK and PPARγ coactivator-1α signaling pathways in skeletal muscle cells. Exercise activates AMPK signaling in skeletal muscle cells through an increase in the cellular AMP-to-ATP concentration ratio after ATP hydrolysis during muscle contraction. Phosphorylation of Thr172 on the α subunit of AMPK by the AMPK kinase stimulates AMPK activity, leading to phosphorylation of protein targets on serine or threonine residues. The short-term effects of AMPK signaling in muscle cells are increased glucose import, glycolytic flux, and fatty acid oxidation, all of which lead to increased levels of ATP and the ability to sustain muscle contractions. AMPK signaling in skeletal muscle cells also leads to changes in muscle cell differentiation and proliferation, which improve muscle strength and physical endurance. A second important biochemical effect of regular exercise is increased expression of the transcriptional coactivator protein PPARγ coactivator-1α. This metabolic regulator is a multifunctional adaptor protein that associates with DNA-bound transcription factors and boosts transcriptional initiation rates at target gene promoters by recruiting chromatin remodeling proteins. PPARγ coactivator-1α gene expression in exercising muscle is induced by calcium signaling through CAMK and by p38 MAP kinase. The long-term effects of elevated PPARγ coactivator-1α expression in skeletal muscle cells results from its coactivation of numerous transcription factors that together regulate the expression of genes involved in mitochondrial DNA replication, the electron transport system, oxidative phosphorylation, glucose uptake, fatty acid oxidation, and regulation of pyruvate dehydrogenase activity.

chapter summary

19.1 Metabolic Integration at the Physiologic Level

- The term *energy balance* relates energy input to energy expenditure in the whole organism. Positive and negative energy imbalances are determined by the energy content of the metabolic fuels ingested compared to the amount of energy expended through chemical reactions, physical exertion, and thermogenic processes.
- The liver is the metabolic hub of the human body, functioning as a physiologic glucose regulator to maintain safe blood glucose levels of 4.5 mM. The liver removes excess glucose from the blood when carbohydrate levels are high (glucose influx) and releases glucose from stored glycogen or as a product of gluconeogenesis when blood glucose levels are low (glucose efflux).

- Glucose-6-P is converted into four major products by liver enzymes: (1) glucose for release into the blood; (2) glucose-1-P for use in glycogen synthesis; (3) 6-phosphogluconolactone to generate NADPH by the pentose phosphate pathway; and (4) fructose-6-P, which is used in the glycolytic pathway to produce pyruvate.
- The human body contains two types of muscle tissue: (1) skeletal muscle, which uses different amounts of free fatty acids, glucose, or ketone bodies for metabolic fuel, depending on the physical movements; and (2) cardiac muscle, which uses fatty acids and ketone bodies as metabolic fuel to sustain a steady heartbeat.
- Adipose tissue functions as a fat depot that stores and releases fatty acids from adipocytes in response to metabolic needs. It also serves as an endocrine organ that secretes peptide hormones called adipokines to regulate metabolic homeostasis.
- Fat stored in adipose tissue consists of two basic types: (1) subcutaneous fat, which is located just below the skin surface in the thighs, buttocks, arms, and face; and (2) visceral fat, which lies deep within the abdominal cavity and is known to secrete a variety of adipokines.
- Three parameters have been developed to describe fat distribution in the human body: (1) body mass index (BMI) values are derived by dividing a person's weight in kilograms by the square of his or her height in meters; (2) the waist-to-hip ratio, which is based on the circumference of an individual's waist and hips; and (3) percent body fat, which can be determined using whole-body bioelectrical impedance (fat impedes electrical conductance).
- The brain is the neuronal control center of the human body. Glucose is delivered to brain cells by microcapillaries that are surrounded by cells called astrocytes, which functionally define the blood–brain barrier.
- Humans contain two kidneys that are capable of filtering 6 L of human blood up to 30 times each day and removing 2 L of water containing concentrated levels of urea, NH_4^+, ketone bodies, and other soluble metabolites.
- Metabolic homeostasis is the process of maintaining optimal metabolite concentrations and managing chemical energy reserves in tissues. Metabolic homeostasis is regulated by physiologic inputs in response to fluctuating nutrient levels in the blood and by neuronal inputs to the brain in response to environmental changes.
- The triacylglycerol cycle is an interorgan process that continually circulates fatty acids and triacylglycerols between adipose tissue and the liver to maintain energy-rich fatty acids in circulation so that they can be used by peripheral tissues.
- Insulin is the "I just ate" hormone, which stimulates glucose uptake, activates glycogen and fatty acid synthesis, and decreases appetite through neuronal signaling in the brain. In contrast, glucagon is the "I am hungry" hormone, which stimulates gluconeogenesis, glycogen degradation, and fatty acid export from adipose tissue.
- Insulin and glucagon are synthesized as prohormones in regions of the pancreas called the islets of Langerhans. The β cells secrete insulin, the α cells secrete glucagon, and the δ cells secrete somatostatin.
- Insulin signaling in liver, skeletal muscle, and adipose tissue stimulates glucose uptake and glycogen and lipid storage, whereas insulin signaling in the brain stimulates the anorexigenic neurons that decrease appetite and increase energy expenditure.
- Glucagon signaling in the liver and adipose tissue stimulates glycogen and triacylglycerol degradation. Skeletal muscle and brain cells do not express glucagon receptors.
- PPAR signaling is tissue specific and regulates lipid transport and mobilization, fatty acid oxidation, and lipid synthesis, which together affect energy metabolism and insulin sensitivity.
- PPARs are protein targets for a variety of pharmaceutical drugs developed to treat human metabolic disease. PPARγ is the primary therapeutic target of thiazolidinediones, which improve insulin sensitivity in type 2 diabetes.
- The human body adapts to starvation conditions by altering the flux of metabolites between various tissues with the primary metabolic objective being to supply the brain with glucose to maintain ATP-dependent ion pumps and ensure normal neuronal cell functions.
- The four major changes in metabolic flux under starvation conditions are (1) increased triacylglycerol hydrolysis in adipose tissue, (2) increased gluconeogenesis in liver and kidney cells, (3) increased ketogenesis in liver cells, and (4) protein degradation in skeletal muscle.

19.2 Metabolic Energy Balance

- A sedentary lifestyle and hormone-related changes in basal metabolic rates are two of the primary reasons most adults gain weight. Genetic inheritance influences who is at risk for being overweight, but the environment (quality and quantity of diet and exercise) determines who will actually become overweight.
- The thrifty gene hypothesis states that humans contain metabolic gene variants that provide protection against famine by maximizing fat storage during times of feast. These same gene variants contribute to the epidemic of obesity in countries where high-Calorie food is readily available.
- Mouse genetics led to the discovery of the leptin gene in obese (*ob/ob*) mice and the leptin receptor gene in diabetic (*db/db*) mice. Leptin is an adipokine hormone synthesized in adipose tissue at levels proportional to the amount of stored fat. Most obese humans have normal leptin signaling, unlike the OB and DB mice.
- Leptin circulates throughout the body and activates signal tranduction in a variety of tissues, including the hypothalamus region of the brain. Activation of leptin receptors decreases appetite and increases energy expenditure to reduce lipid stores.
- Leptin (and insulin) bind to first-order POMC and NPY/AGRP neurons that produce neuropeptides (α-MSH, NPY, and AGRP), which bind to their cognate receptors on

anorexigenic (eat less, metabolize more) and orexigenic (eat more, metabolize less) second-order neurons.

- The neuropeptides ghrelin and $PYY_{3\text{-}36}$ also signal through NPY/AGRP neurons in the hypothalamus to modulate energy balance. Ghrelin is synthesized in the stomach and sends signals to the brain that it is time to eat. $PYY_{3\text{-}36}$ is synthesized in the colon and counters the effect of ghrelin by sending signals that it is time to stop eating.
- A lifestyle of chronic positive energy imbalance can lead to a condition in humans called the metabolic syndrome, which is defined as the appearance of three or more of the following symptoms: (1) abdominal obesity, (2) insulin resistance, (3) hypertension, (4) hyperlipidemia, (5) high risk for cardiovascular disease.
- Insulin-resistant type 2 diabetes is characterized at initial diagnosis by high levels of circulating insulin and desensitization of insulin receptor signaling in muscle, liver, and adipose tissue. In contrast, type 1 diabetes is due to insufficient insulin production by the pancreatic β cells and is treatable with insulin injections.
- Unlike individuals with normal insulin signaling, type 1 and type 2 diabetics are unable to lower blood glucose levels within 2 hours. Type 1 and type 2 diabetes can be distinguished by an insulin sensitivity test, which shows a decrease in blood glucose levels for type 1 diabetics but not for type 2 diabetics.
- The biochemical basis for a causal link between obesity and type 2 diabetes is complex; however, two phenotypes in obese individuals are strongly associated with insulin-resistant type 2 diabetes: (1) elevated levels of free fatty acids in the serum, and (2) altered secretion of peptide hormones from adipose tissue.
- One model proposes that elevated fatty acids in the serum of obese individuals leads to insulin resistance in muscle cells through production of diacylglycerol, which stimulates protein kinase C signaling. Phosphorylation of insulin receptor substrate 1 on serine residues by protein kinase C inhibits the normal phosphorylation of insulin receptor substrate 1 tyrosine residues by the insulin receptor.
- A second model proposes that elevated expression of the inflammatory cytokine TNF-α in adipocytes leads to autocrine-mediated downregulation of adiponectin expression in adipocytes, which is associated with insulin resistance in skeletal muscle cells.
- Not all obese individuals develop type 2 diabetes, and moreover, not all type 2 diabetics are overweight or obese. This suggests that the link between obesity and type 2 diabetes likely involves metabolic gene variants that contribute to an elevated risk of developing insulin resistance, as seen in the Pima Indians.
- Four major classes of drugs have been developed to treat type 2 diabetes: (1) α-glucosidase inhibitors (miglitol); (2) sulfonylurea drugs that inhibit the pancreatic ATP-dependent K^+ channel (glipizide); (3) drugs that stimulate the activity of AMPK (metformin); and (4) ligand agonists of the nuclear receptor PPARγ (thiazolidinediones).

19.3 Nutrition and Exercise

- Three factors that affect metabolic homeostasis are genetic inheritance, nutrition, and exercise. Energy balance determines body weight; however, not all foods of equal Calories provide the same nutritional value.
- Four classes of compounds have been developed as weight-loss drugs: (1) ephedrine is a stimulant that increases basal metabolic rates; (2) lorcaserin targets neuronal signaling in the brain to control appetite; (3) orlistat inhibits the activity of pancreatic lipase in the small intestine; and (4) olestra is a zero-Calorie food substitute containing fatty acid side chains covalently linked to sucrose.
- The most common methods to lose weight are a Calorie-restricted diet using a formula-based approach, such as Weight Watchers, or by altering the macronutrient composition, such as the low-carbohydrate Atkins Diet, the high-protein, low-carbohydrate Zone Diet, and the low-fat Ornish Diet.
- Consuming high amounts of saturated fatty acids and trans fatty acids on a regular basis increases LDL levels in the blood, which is associated with a higher risk of cardiovascular disease. Dietary fatty acids function as ligands of PPAR nuclear receptors and may have differential effects on transcription of metabolic genes.
- The glycemic index of foods is a numerical value indicating how quickly glucose is released into the blood after eating different types of carbohydrate-containing foods. Carbohydrates with a low glycemic index cause only moderate increases in blood glucose levels over several hours.
- Exercise-induced activation of AMPK signaling alters metabolic flux through energy conversion pathways to increase ATP production in skeletal muscle cells. AMPK is a heterotrimeric serine/threonine kinase consisting of a catalytic α subunit, a regulatory γ subunit that binds AMP, and the β subunit, which functions as a molecular scaffold.
- AMP binding to the AMPK γ subunit leads to a conformational change facilitating Thr172 phosphorylation on the α subunit by AMPK kinases. AMP binding to the γ subunit induces AMPK activity by ~10-fold, whereas phosphorylation of AMPK on Thr172 stimulates activity by 100-fold.
- AMPK-mediated phosphorylation of serine or threonine residues on metabolic target proteins in muscle cells leads to a net increase in ATP concentration by three mechanisms: (1) stimulation of flux through glycolysis; (2) stimulation of flux through fatty acid oxidation; and (3) increased oxidative phosphorylation.
- The transcription factor PPARγ coactivator-1α is an AMPK target protein that has been linked to exercise-induced responses. Two signaling pathways induce PPARγ coactivator-1α gene expression in exercising muscle: (1) a calcium-dependent signaling pathway through CAMK and (2) activation of p38 MAP kinase.

biochemical terms

(in order of appearance in text)
energy balance (p. 944)
chyme (p. 945)
portal vein (p. 946)
skeletal muscle (p. 947)
cardiac muscle (p. 948)
phosphocreatine (p. 948)
adipokine (p. 949)
subcutaneous fat (p. 949)
visceral fat (p. 949)
body mass index (BMI) (p. 949)
waist-to-hip ratio (p. 949)
percent body fat (p. 949)
astrocyte (p. 950)
blood–brain barrier (p. 950)
positron emission tomography (PET) (p. 950)
hypoglycemia (p. 950)
hemodialysis (p. 951)
metabolic homeostasis (p. 952)
triacylglycerol cycle (p. 954)
glyceroneogenesis (p. 955)
islet of Langerhans (p. 957)
somatostatin (p. 957)
acinar cell (p. 957)
clofibrate (p. 961)
thiazolidinedione (p. 962)
macronutrient (p. 968)
set point (p. 969)
thrifty gene hypothesis (p. 971)
leptin (p. 971)
hypothalamus (p. 973)
anorexigenic (p. 973)
orexigenic (p. 973)
metabolic syndrome (p. 975)
type 975 diabetes (p. 975)
type 975 diabetes (p. 975)
glucose tolerance test (p. 976)
adiponectin (p. 977)
hyperglycemia (p. 979)
hyperlipidemia (p. 979)
metformin (p. 980)
ephedrine (p. 983)
lorcaserin (p. 983)
orlistat (p. 983)
olestra (p. 983)
glycemic index (p. 986)

review questions

1. How does the liver maintain safe blood glucose levels during a normal 24-hour day? What are the physiologic symptoms of low blood glucose levels?
2. What metabolite is central to liver metabolism? What are the four fates of this metabolite in liver cells, and how is metabolite flux among these pathways regulated?
3. What are the two primary types of muscle in the human body, and what are their primary metabolic fuels?
4. What are the two primary types of adipose tissue in the human body, and how do they differ anatomically and physiologically?
5. What three parameters are used to estimate the amount of stored fat in an individual? How might they be used to distinguish between individuals in a large retrospective clinical study designed to find associations between obesity and cardiovascular disease?
6. What is the function of the triacylglycerol cycle? How does glyceroneogenesis help to maintain flux through the triacylglycerol cycle?
7. Why is insulin called the "I just ate" hormone and glucagon the "I am hungry hormone"? Compare and contrast the endocrine functions of these two peptide hormones in terms of regulating metabolic homeostasis.
8. What are the physiologic ligands for PPAR nuclear receptors, and why does this make sense given the downstream targets of PPAR signaling pathways?
9. What are the two human tissues that must be supplied with metabolic energy to ensure survival during times of starvation? What are the four major metabolic adaptations that take place in humans deprived of food for at least 7 days?
10. Describe the thrifty gene hypothesis and give an example that supports it.
11. What is the biochemical basis for energy balance regulation by the peptide hormones leptin, insulin, ghrelin, and PYY_{3-36}?
12. What is the difference between type 1 and type 2 diabetes, and how are these two endocrine disorders diagnosed in patients?
13. What are the four major classes of drugs used to treat type 2 diabetes, and what are their biochemical functions?
14. Describe the mechanism of action of four types of weight-loss drugs. Include in your answer any known side effects of each drug. What advice would you give to someone who wanted a drug-free alternative to weight loss?
15. Describe three ways that AMPK signaling in contracting skeletal muscle cells increases ATP concentrations to sustain muscle activity during exercise.

challenge problems

1. Measurements show that 1 g of glucose contains 3.7 kcal of food energy. Moreover, the $\Delta G^{\circ\prime}$ for ATP synthesis is +30.5 kJ/mol. Use this information to answer the following questions. Show your work.
   a. Assuming 65% efficiency of converting glucose to ATP by oxidative phosphorylation, how many moles of ATP can be synthesized by consuming a large bag of potato chips (~30 chips) containing 56 g of carbohydrate in the form of starch?
   b. Using the information in Table 19.6, how long would a female need to garden and a male need to vacuum in order to burn off the Calories contained in the carbohydrate portion of this large bag of potato chips?

2. A variety of metabolic parameters affect homeostatic blood glucose levels. For each of the altered metabolic conditions listed below, predict if it would result in higher or lower blood glucose levels compared to that of the normal metabolic condition.
   a. Amino acids are not efficiently converted to keto acids during starvation.
   b. Glycogen debranching enzyme does not function during intense exercise.
   c. The glucagon receptor is always active, even in the absence of glucagon.
   d. The enzyme pyruvate carboxylase is expressed at abnormally high levels.
   e. Pancreatic α cells do not secrete functional glucagon.
3. In addition to stimulating uptake of excess blood glucose into adipose, insulin stimulates fatty acid synthesis and storage of triacylglycerols in adipose. On the basis of these insulin effects, why is it a bad idea to eat six sugar donuts before a 60-mile bike race to get "energy" for the 3-hour event?
4. Fasting studies performed in an animal model showed that rates of muscle protein hydrolysis change dramatically in the first 7 days. In the first 24 hours, the rates of protein hydrolysis were low, but then increased substantially for the next 3–4 days. By day 5, the rates of protein hydrolysis decreased to much lower levels for the duration of the fast. Given that there are no known regulatory mechanisms controlling muscle protein hydrolysis, how can you explain this pattern?
5. On the basis of the premise of the thrifty gene hypothesis, predict if the following hypothetical gene variants would be candidate thrifty genes. Explain your answer.
   a. Acetyl-CoA carboxylase that is insensitive to feedback inhibition by palmitoyl-CoA
   b. Insulin receptor that has an increased affinity for insulin
   c. High basal levels of uncoupling protein in liver cells
   d. A hyperactive lipoprotein lipase on the plasma membrane in adipose cells
   e. A hyperactive hormone-sensitive lipase on lipid droplets in adipose cells
6. One of the effects of leptin in nonadipose tissue is to upregulate the synthesis of uncoupler protein. How might this effect help prevent obesity?
7. Leptin is an adipocyte hormone that sends signals to the brain to "eat less" and "metabolize more." What is the explanation for why leptin injections cause weight loss in the strain of OB mutant mice, but leptin injections have no effect in the majority of obese people? Is this observation analogous to type 1 or type 2 diabetes? Explain.
8. On the basis of the function of ghrelin and PYY_{3-36}, explain why it is good advice to eat slowly if you are trying to lose weight.
9. The primary organ responsible for converting excess carbohydrate into fat is the liver. The observation that increased annual consumption of beverages containing high-fructose corn syrup in the United States between 1980 and 2000 overlaps with increased rates of obesity led to the proposal that high-fructose corn syrup consumption may be a causative factor. The composition of high-fructose corn syrup used in beverages is 55% fructose and 42% glucose.
   a. On the basis of the metabolic fate of fructose in liver cells (see Figure 9.48), provide a plausible biochemical explanation for why increased consumption of fructose could potentially lead to higher rates of obesity than would be expected if an equimolar amount of sucrose were the sweetener. The K_m of liver fructokinase for fructose is 0.5 mM, and the K_m of liver glucokinase for glucose is 10 mM.
   b. What percentage of the recommended 2,000 Calories/d is accounted for in a late-night snack consisting of a large bag of potato chips (56 g starch) and a 1-L soda containing 110 g of high-fructose corn syrup? Assume 4 Calories/g of carbohydrate (starch or high-fructose corn syrup).
10. Treatment of type 2 diabetes to lower blood glucose includes the use of sulfonylureas, which stimulate insulin secretion, and metformin, which inhibits gluconeogenesis. Explain the biochemical rationale behind the use of each drug.
11. What is the biochemical explanation for the high level of ketone bodies in untreated diabetics?
12. Of the three diets listed below, which would provide the most complete nutrition? Explain.
    1. Carbohydrate + protein
    2. Carbohydrate + fats
    3. Protein + fats
13. What explains the often-encountered side effect of "acetone breath" on dieters following a very strict low-carbohydrate diet such as the Atkins Diet?
14. What explains the dramatic weight loss associated with dinitrophenol pills? What are the possible dire consequences of using dinitrophenol pills to lose weight?
15. Why is it "healthier" to consume 200 Calories of a low-glycemic-index food, such as kidney beans, compared to 200 Calories of a high-glycemic-index food, such as jelly beans?
16. What type of data supports the finding that the fatty acids in avocados and cashews are "good fat," whereas the fatty acids in donuts and fried chicken are "bad fat?"
17. After short-term vigorous anaerobic exercise, you continue to breathe rapidly for some time. The oxygen consumed during this period is called the oxygen debt and is equal to the amount of oxygen that would have been consumed if the exercise had been completely aerobic. What is the oxygen debt used for in muscle and liver cells?

smartwork5

If your instructor assigns homework with Smartwork5, access it here: digital.wwnorton.com/biochem.

suggested reading

Books and Reviews

Beale, E. G., Hammer, R. E., Antoine, B., and Forest, C. (2002). Glyceroneogenesis comes of age. *FASEB Journal, 16*, 1695–1696.

Bensinger, S. J., and Tontonoz, P. (2008). Integration of metabolism and inflammation by lipid-activated nuclear receptors. *Nature, 454*, 470–477.

Canto, C., and Auwerx, J. (2010). AMP-activated protein kinase and its downstream transcriptional pathways. *Cellular and Molecular Life Sciences, 67*, 3407–3423.

Chakravarthy, M. V., and Booth, F. W. (2004). Eating, exercise, and "thrifty" genotypes: connecting the dots toward an evolutionary understanding of modern chronic diseases. *Journal of Applied Physiology, 96*, 3–10.

Farmer, S. R. (2009). Obesity: be cool, lose weight. *Nature, 458*, 839–840.

Fernandez-Marcos, P. J., and Auwerx, J. (2011). Regulation of PGC-1α, a nodal regulator of mitochondrial biogenesis. *American Journal of Clinical Nutrition, 93*, 884S–890S.

Gibbons, A. (2009). Human evolution. What's for dinner? Researchers seek our ancestors' answers. *Science, 326*, 1478–1479.

Handschin, C., and Spiegelman, B. M. (2008). The role of exercise and PGC1α in inflammation and chronic disease. *Nature, 454*, 463–469.

Heal, D. J., Gosden, J., and Smith, S. L. (2012). What is the prognosis for new centrally-acting anti-obesity drugs? *Neuropharmacology, 63*, 132–146.

Lindsay, R. S., and Bennett, P. H. (2001). Type 2 diabetes, the thrifty phenotype—an overview. *British Medical Bulletin, 60*, 21–32.

Mihaylova, M. M., and Shaw, R. J. (2011). The AMPK signalling pathway coordinates cell growth, autophagy and metabolism. *Nature Cell Biology, 13*, 1016–1023.

Neel, J. V. (1999). The "thrifty genotype" in 1998. *Nutrition Reviews, 57*, S2–9.

Nye, C., Kim, J., Kalhan, S. C., and Hanson, R. W. (2008). Reassessing triglyceride synthesis in adipose tissue. *Trends in Endocrinology and Metabolism, 19*, 356–361.

O'Rahilly, S. (2009). Human genetics illuminates the paths to metabolic disease. *Nature, 462*, 307–314.

Ogawa, W., and Kasuga, M. (2008). Cell signaling. Fat stress and liver resistance. *Science, 322*, 1483–1484.

Ruan, H., and Dong, L. Q. (2016). Adiponectin signaling and function in insulin target tissues. *Journal of Molecular Cell Biology, 8*, 101–109.

Sanchez, A. M., Candau, R. B., Csibi, A., Raibon, A., and Bernardi, H. (2012). The role of AMP-activated protein kinase in the coordination of skeletal muscle turnover and energy homeostasis. *American Journal of Physiology. Cell Physiology, 303*, C475–C485.

Taubes, G. (2009). Insulin resistance. Prosperity's plague. *Science, 325*, 256–260.

Primary Literature

Bostrom, P., Wu, J., Jedrychowski, M. P., Korde, A., Ye, L., Lo, J. C., Rasbach, K. A., Bostrom, E. A., Choi, J. H., Long, J. Z., et al. (2012). A PGC-1-α-dependent myokine that drives brown-fat-like development of white fat and thermogenesis. *Nature, 481*, 463–468.

Ebbeling, C. B., Swain, J. F., Feldman, H. A., Wong, W. W., Hachey, D. L., Garcia-Lago, E., and Ludwig, D. S. (2012). Effects of dietary composition on energy expenditure during weight-loss maintenance. *JAMA, 307*, 2627–2634.

Frankenfield, D. C., Rowe, W. A., Cooney, R. N., Smith, J. S., and Becker, D. (2001). Limits of body mass index to detect obesity and predict body composition. *Nutrition, 17*, 26–30.

Gibson, W. T., Farooqi, I. S., Moreau, M., DePaoli, A. M., Lawrence, E., O'Rahilly, S., and Trussell, R. A. (2004). Congenital leptin deficiency due to homozygosity for the Δ133G mutation: report of another case and evaluation of response to four years of leptin therapy. *Journal of Clinical Endocrinology and Metabolism, 89*, 4821–4826.

Gleyzer, N., Vercauteren, K., and Scarpulla, R. C. (2005). Control of mitochondrial transcription specificity factors (TFB1M and TFB2M) by nuclear respiratory factors (NRF-1 and NRF-2) and PGC-1 family coactivators. *Molecular and Cellular Biology, 25*, 1354–1366.

Halaas, J. L., Gajiwala, K. S., Maffei, M., Cohen, S. L., Chait, B. T., Rabinowitz, D., Lallone, R. L., Burley, S. K., and Friedman, J. M. (1995). Weight-reducing effects of the plasma protein encoded by the obese gene. *Science, 269*, 543–546.

Pelleymounter, M. A., Cullen, M. J., Baker, M. B., Hecht, R., Winters, D., Boone, T., and Collins, F. (1995). Effects of the obese gene product on body weight regulation in *ob/ob* mice. *Science, 269*, 540–543.

Pilegaard, H., Saltin, B., and Neufer, P. D. (2003). Exercise induces transient transcriptional activation of the PGC-1α gene in human skeletal muscle. *Journal of Physiology, 546*, 851–858.

Schulz, L. O., Bennett, P. H., Ravussin, E., Kidd, J. R., Kidd, K. K., Esparza, J., and Valencia, M. E. (2006). Effects of traditional and western environments on prevalence of type 2 diabetes in Pima Indians in Mexico and the U.S. *Diabetes Care, 29*, 1866–1871.

Shai, I., Schwarzfuchs, D., Henkin, Y., Shahar, D. R., Witkow, S., Greenberg, I., Golan, R., Fraser, D., Bolotin, A., Vardi, H., et al. (2008). Weight loss with a low-carbohydrate, Mediterranean, or low-fat diet. *New England Journal of Medicine, 359*, 229–241.

Wormser, D., Kaptoge, S., Di Angelantonio, E., Wood, A. M., Pennells, L., Thompson, A., Sarwar, N., Kizer, J. R., Lawlor, D. A., Nordestgaard, B. G., et al. (2011). Separate and combined associations of body-mass index and abdominal adiposity with cardiovascular disease: collaborative analysis of 58 prospective studies. *Lancet, 377*, 1085–1095.

Xiao, B., Sanders, M. J., Underwood, E., Heath, R., Mayer, F. V., Carmena, D., Jing, C., Walker, P. A., Eccleston, J. F., Haire, L. F., et al. (2011). Structure of mammalian AMPK and its regulation by ADP. *Nature, 472*, 230–233.

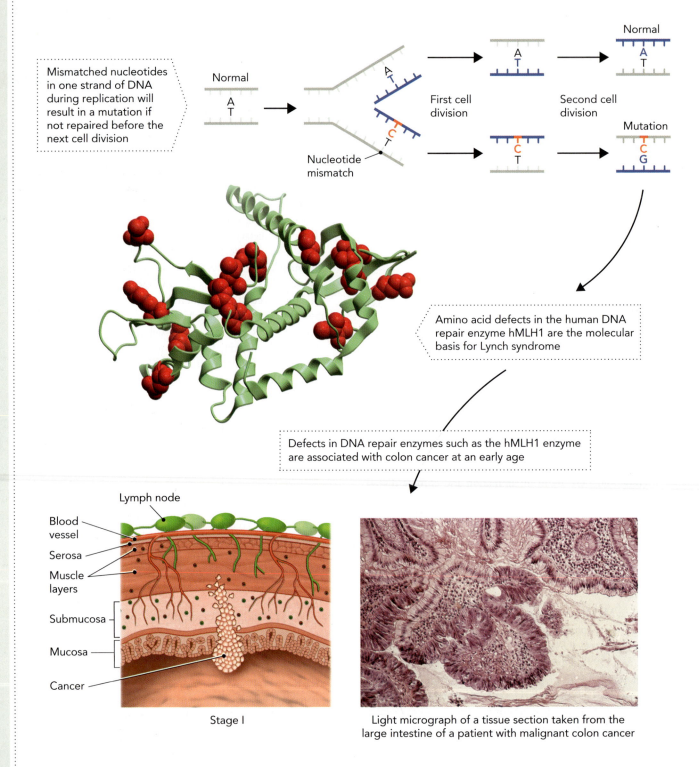

Mismatched nucleotides in one strand of DNA during replication will result in a mutation if not repaired before the next cell division

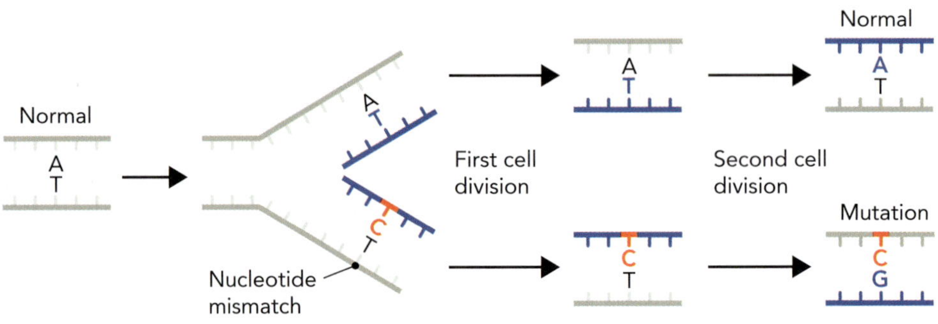

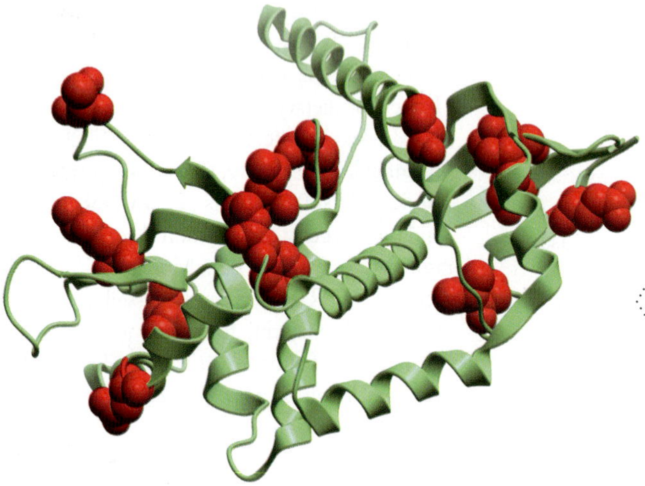

Amino acid defects in the human DNA repair enzyme hMLH1 are the molecular basis for Lynch syndrome

Defects in DNA repair enzymes such as the hMLH1 enzyme are associated with colon cancer at an early age

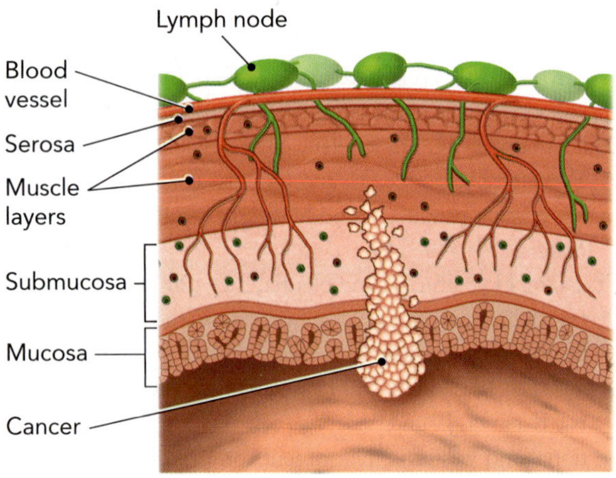

Stage I

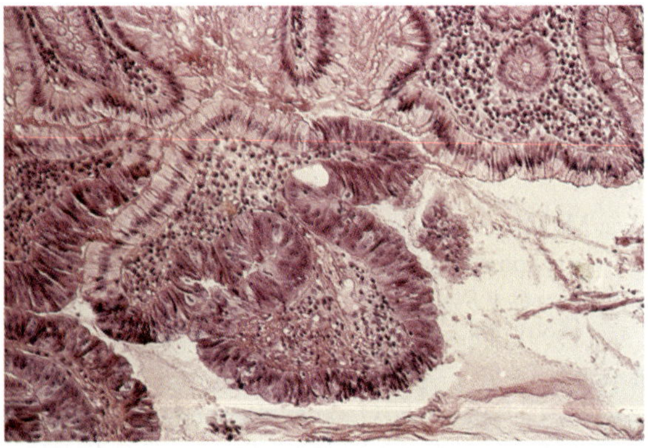

Light micrograph of a tissue section taken from the large intestine of a patient with malignant colon cancer

If a mismatched nucleotide arises during DNA replication, it must be repaired before the second cell division after it occurs or else it will become a fixed mutation. In the case shown here, an A-T base pair is mutated to a C-G base pair. Lynch syndrome (HNPCC) is caused by one of any number of autosomal dominant mutations in the gene encoding the human DNA repair enzyme hMLH1. Amino acid substitutions that have been demonstrated to cause HNPCC are shown here in space-filling style to distinguish them from all other amino acids in the enzyme, which are shown in ribbon style. Colon cancer begins as a single mutant cell, which then divides and eventually generates progeny cancer cells with additional mutations that cause cancer.

20

DNA Replication, Repair, and Recombination

◀ Maintaining the integrity of genomic DNA is an ongoing challenge for the cell's DNA repair enzymes. Nucleotide mismatches can occur during DNA replication that must be repaired before the next cell division. DNA is also subjected to mutations that result from spontaneous chemical changes or exposure to ultraviolet light. If the damaged nucleotides and bases are not repaired, they can lead to diseases such as cancer. If the damage occurs in reproductive cells, it can cause transmissible, or heritable, genetic mutations. It is easy to see why mutations in genes encoding DNA repair enzymes can be catastrophic.

One such example is the human disease known as hereditary non-polyposis colorectal cancer (HNPCC), or Lynch syndrome. Lynch syndrome is characterized by the appearance of cancer at an early age due to mutations in the genes encoding the DNA repair enzymes responsible for mismatch repair. In many cases, the first cancers to arise in these individuals are colon cancers owing to mutations occurring in the rapidly dividing intestinal epithelial cells.

CHAPTER OUTLINE

CREDITS: HMLH1: BASED ON PDB FILE 4E4W; LIGHT MICROGRAPH: CNRI/SCIENCE SOURCE.

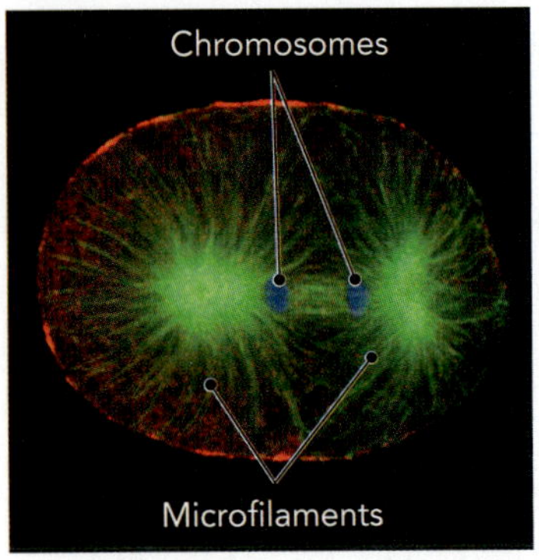

Figure 20.1 A photomicrograph of chromosomes segregating in a roundworm embryo provides an image of genome duplication at the cellular level. The chromosomes are stained blue. COURTESY BRUCE BOWERMAN, UNIVERSITY OF OREGON.

DNA is present in a fully functional form each time a new cell is generated by cell division. In fact, the synthesis of a new copy of DNA largely defines the generation of a new cell from an older one. In this chapter, we describe the processes of DNA replication, repair of DNA damage, and recombination of DNA.

DNA replication is the process by which a cell copies its entire genome and is a required element of cell division (**Figure 20.1**). To prevent mutations, the process must have high fidelity, ensuring that the DNA copy is the same as the original. The multitude of enzymes and other protein factors required for this process make this one of the most highly coordinated events within the cell.

With so many resources invested in keeping DNA from changing, it would seem that diversity of life is difficult to achieve, as mutations are what give rise to evolution. But other processes such as DNA recombination alter DNA in controlled and beneficial ways, giving rise to new species and explaining some of the differences between two children of the same parents. Another examble is DNA integration, which permits viruses to invade a host organism, become part of the host cell genome, and then produce viral progeny that invade nearby cells. Indeed, DNA metabolism is fascinating and has captivated the attention of biochemists since the structure of DNA was discovered more than six decades ago.

20.1 DNA Replication

DNA replication must proceed at a high rate in order to copy the entire genome in the time available for cell division. The result is one of nature's most elegant biochemical processes. In this section, we first discuss the details of the replication process itself and then describe how the initiation and termination stages operate to start and end replication.

Overview of Genome Duplication

Cell division requires the production of a complete copy of the DNA contained within that cell. In this way, the resulting daughter cells will each contain a complete copy of the genomic DNA. To understand how genome duplication occurs, several classic experiments were performed in the 1950s and 1960s to elucidate the mechanism of DNA replication.

DNA Replication Is Semiconservative One of the initial questions about DNA replication was the relationship of the initial DNA template to the DNA product. The three hypotheses that were proposed were conservative replication, dispersive replication, and semiconservative replication (**Figure 20.2**). In conservative replication, the template—the original duplex DNA molecule—remains intact; thus, after replication, one of the DNA molecules contains two newly synthesized strands, and the other is the original duplex DNA molecule. In dispersive replication, the original duplex DNA template is broken into fragments, giving products that contain a mixture of both newly synthesized DNA and original template DNA. In semiconservative replication, the original duplex DNA template is separated into single strands prior to replication, resulting in duplex daughter molecules that each contain one strand from the original template and one newly synthesized strand.

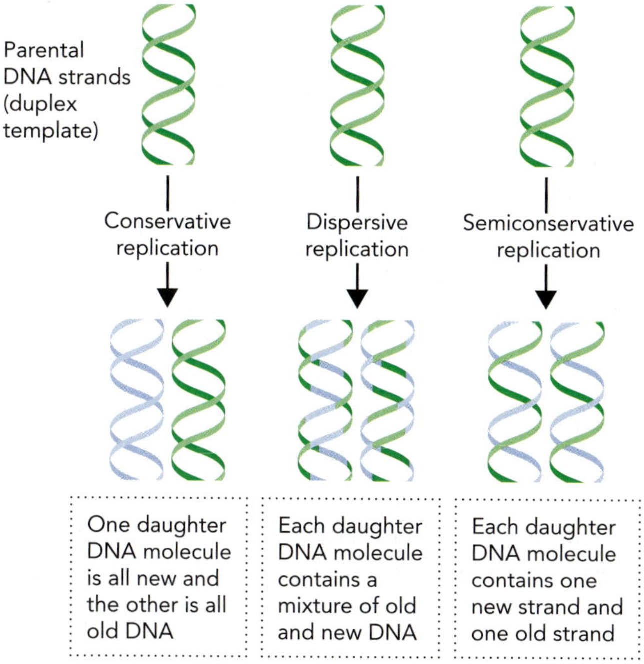

Figure 20.2 Three hypothesized mechanisms for DNA replication are shown. Conservative replication results in one duplex daughter molecule containing two newly synthesized DNA strands and the original duplex molecule. Dispersive replication results in portions of the original strands mixed throughout the daughter strands. Semiconservative replication results in duplex daughter molecules that contain one newly synthesized strand and one original template strand.

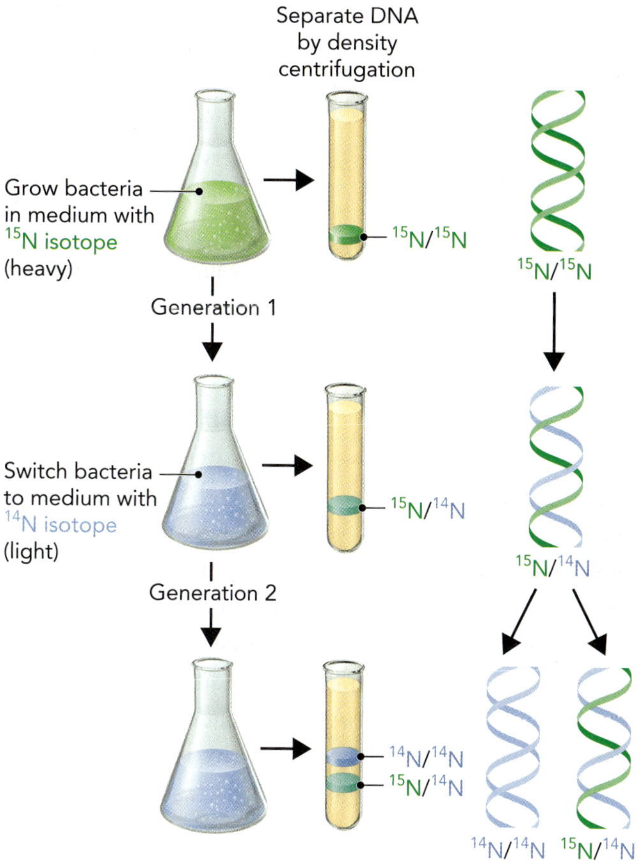

Figure 20.3 The Meselson–Stahl experiment provided biochemical evidence that DNA replication is semiconservative. *Escherichia coli* grown in ^{15}N-containing medium contain heavy DNA. When allowed to divide in ^{14}N-containing medium, a hybrid DNA is produced that contains one strand of heavy DNA and one strand of light DNA. A second round of division produces one hybrid DNA molecule and one light DNA molecule. This result could only occur through semiconservative replication.

When Watson and Crick published their structure of DNA in 1953, they specifically noted that the double-helical structure "suggests a possible copying mechanism"; however, any of the three replication hypotheses subsequently proposed by scientists seemed equally valid. It was not until a few years later (1958) that Matthew Meselson and Franklin Stahl conclusively demonstrated that **semiconservative replication** takes place (**Figure 20.3**). Their experiment was done by first growing bacteria in a medium containing nitrogen in only the ^{15}N isotopic form, which was incorporated into the newly synthesized DNA. DNA that contains ^{15}N ("heavy DNA") will be slightly more dense than DNA containing the usual ^{14}N ("light DNA"). DNA with different densities can be separated by centrifugation in cesium chloride and will accumulate as bands at various positions in the test tube. After several rounds of cell division, all of the bacteria contained DNA that was labeled with ^{15}N.

When the ^{15}N-containing medium was replaced with normal (^{14}N) medium, the first round of cell division after changing the medium produced a single band of DNA that was of intermediate density between the heavy DNA and light DNA. This ruled out conservative replication, which would have produced two different bands: one from the template containing nothing but heavy DNA and a second from the newly synthesized strand containing nothing but light DNA. But this result did not show that semiconservative replication occurred exclusively, as dispersive replication could give a similar result. A second round of cell division in ^{14}N-containing medium produced two bands after centrifugation: one identical to the band produced in the first round of cell division and a second band containing only light DNA. This result proved semiconservative replication because the template after the first round of replication must have been composed of one heavy strand and one light strand. The heavy template strand gave rise to the intermediate band in the second round of replication, and the light template strand gave rise to the light band. Had dispersive replication occurred, the product of the second round would have again been a single band that was less dense than the template but denser than the light DNA.

DNA Replication Is Bidirectional Understanding the relationship between the parent and daughter DNA strands was an important step in understanding how DNA replicates. But it was still largely unknown how the replication process begins and ends in a cell as a function of time. For example, where does replication initiate? This question is harder to answer than it seems, considering that most early research on DNA replication was performed using bacteria with circular chromosomes, which have no beginning or end. Clearly, there must be an **origin of replication** where replication begins, but how DNA replication proceeded from that point was not well understood. The easiest imaginable mechanism to produce DNA in a semiconservative manner would be to separate the two strands of DNA completely and replicate each strand. Single-stranded DNA, however, was known to be rather unstable. Thus, a mechanism that produced only localized areas of single-stranded DNA (ssDNA) would be the most likely; that is, DNA double helices would only partially unwind at certain regions.

Support for this hypothesis came in 1963 from the work of John Cairns, who proved that complete unwinding of the chromosome does not occur. Cairns grew *E. coli*, which contains a circular chromosome, in the presence of ^{3}H-thymidine, thus enabling him to visualize individual molecules of replicating DNA by use of autoradiography. Even individual molecules in the midst of replication could be seen (**Figure 20.4a**). The experiment showed that both single-stranded and double-stranded regions are found within a DNA molecule. Furthermore, Cairns showed that DNA replicates through the formation of a **replication fork**, which is the Y-shaped region of DNA

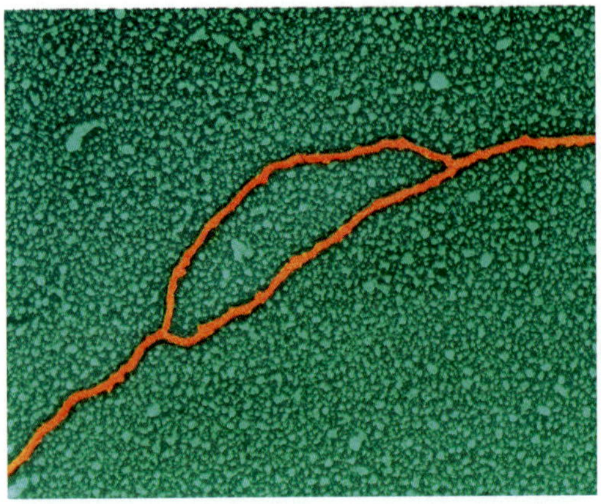

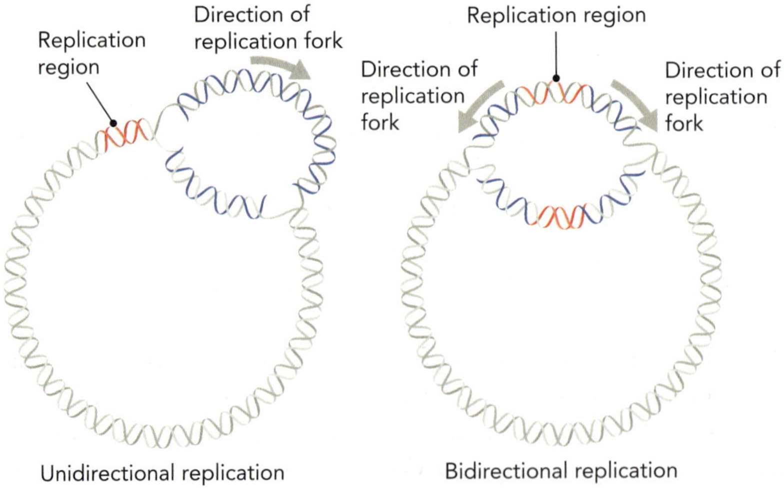

Figure 20.4 Unidirectional and bidirectional replication are two possible mechanisms for DNA replication from a single origin. **a.** This colorized electron micrograph shows a single DNA molecule partially separated during replication, resulting in both single-stranded and double-stranded regions. Electron microscopy is a more modern technique to visualize individual molecules than autoradiography, which was used by Cairns. DR. GOPA MURTI/SCIENCE SOURCE. **b.** Once replication is initiated, the replication fork could proceed in one direction or in both directions from the origin. It is now confirmed that replication is bidirectional from a single origin.

formed where the DNA duplex has been separated into two single strands and synthesis begins. With regard to replication of a circular bacterial chromosome, replication could possibly occur either unidirectionally (one replication fork) or bidirectionally (two replication forks) as shown in **Figure 20.4b**.

Further experimentation with actively replicating *E. coli* cells showed that replication occurs bidirectionally. Indeed, most genomic DNA is replicated bidirectionally, with prokaryotes having a single origin on a circular chromosome and two replication forks that meet at a point opposite the origin. The replicating circular chromosome is often referred to as a theta structure due to its similar shape to the Greek letter (θ). Eukaryotes, with their much larger chromosomes, have multiple origins of replication on each chromosome, each producing two replication forks, as described later in the chapter.

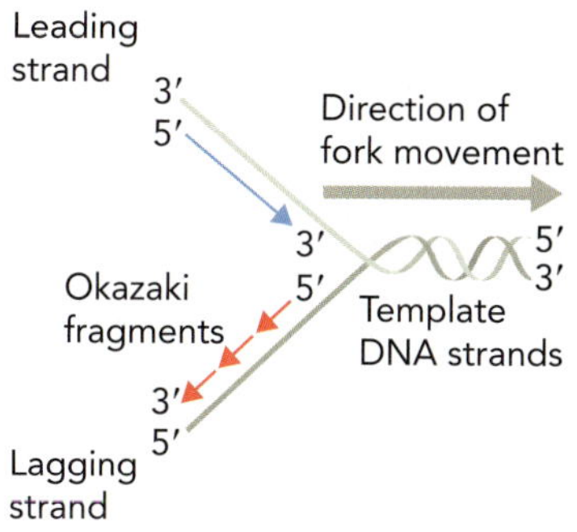

Figure 20.5 This simplified diagram shows the orientation of the leading and lagging strands at the replication fork. Synthesis of the lagging strand is performed in a discontinuous fashion, producing small fragments that must be joined at a later time. These fragments are known as Okazaki fragments. The leading strand is synthesized as one continuous segment of DNA.

Discovery of Okazaki Fragments It was found that the enzymes responsible for DNA polymerization all synthesize DNA in a 5′→3′ direction. Because of the antiparallel nature of DNA, this works fine for one strand of DNA, but the other strand cannot be synthesized in a 5′→3′ fashion without stopping the polymerization process. How could both strands be synthesized simultaneously? The answer came from Reiji Okazaki and his colleagues in 1966 who found that while one strand is synthesized continuously, the other is synthesized in fragments known as **Okazaki fragments** (**Figure 20.5**). The continuously synthesized strand is called the **leading strand**, while the strand containing the Okazaki fragments is called the **lagging strand** and is synthesized discontinuously. Figure 20.5 illustrates the relative orientation of the leading and lagging strands at the replication fork. Note that DNA synthesis occurs at the 3′ end of both the leading and lagging strands; however, the net movement of the replication fork is in one direction, as shown by the arrow in Figure 20.5.

Okazaki and his colleagues discovered **discontinuous synthesis** by performing a pulse-chase experiment, which is a method that allows for identification of the most recently synthesized DNA during the replication process. The researchers provided replicating cells with a small amount of ^{3}H-thymidine for short periods of time—the "pulse"—followed by isolation and separation of the DNA strands. Okazaki was able to show that one new strand of DNA was produced originally in fragments of ~2,000 nucleotides, while the other strand was produced in a continuous fashion. (In eukaryotes, these fragments are ~200 nucleotides in length.) This allows for both strands to be synthesized in a 5′→3′ direction, with the gaps between Okazaki fragments sealed later in the replication process by the enzyme DNA ligase.

Structure and Function of DNA Polymerases

The enzymes responsible for the bulk of DNA synthesis are known as **DNA polymerases**. These enzymes, particularly in prokaryotes, must be very efficient to allow the complete replication of the genome in a short period of time. Additionally, DNA polymerases must have a high degree of accuracy for incorporation of the correct nucleotide at each site. To understand how DNA polymerases function, we first need to describe the mechanism for polynucleotide synthesis.

Biochemistry of the DNA Synthesis Reaction DNA polymerases catalyze the addition of a deoxynucleotide to the 3′ end of the growing DNA molecule (**Figure 20.6**). Two divalent metal cations, generally Mg^{2+}, are bound in the active site and are essential for the reaction. One of these Mg^{2+} ions facilitates the deprotonation

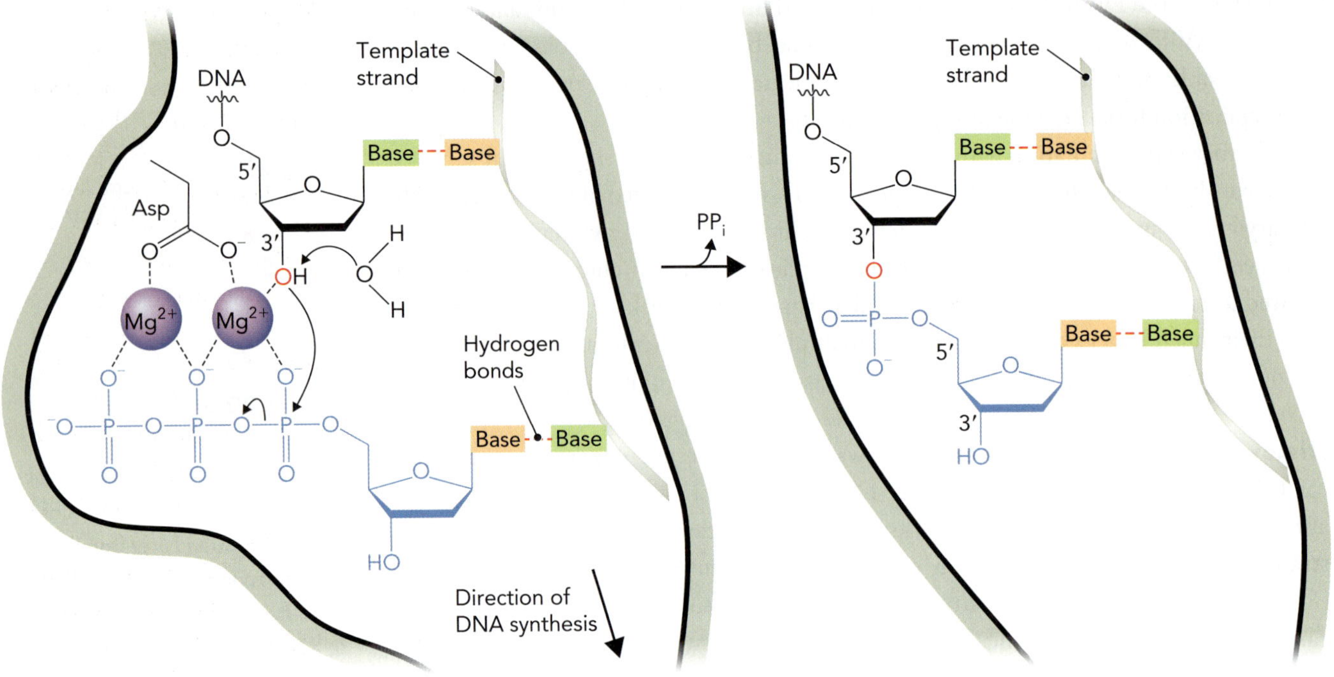

Figure 20.6 The DNA polymerase reaction adds a deoxynucleotide to the 3′ end of the growing DNA chain. The Mg^{2+} ions stabilize the negative charges on the deoxynucleotide and assist in deprotonation of the 3′-OH by a base (shown as a water molecule). The 3′-oxygen of the growing DNA strand serves as the nucleophile in this reaction, displacing the pyrophosphate from the deoxynucleoside triphosphate (dNTP) in the active site. The product is a DNA strand that has been extended by one nucleotide in the 3′ direction.

of the 3′-OH group by a base (shown in Figure 20.6 as a water molecule), creating a nucleophile that attacks at the α-phosphoryl group of the incoming deoxynucleotide. The other Mg^{2+} ion stabilizes the negative charges on the incoming deoxynucleotide, ultimately making the pyrophosphate a better leaving group. Subsequent hydrolysis of pyrophosphate to inorganic phosphate ions makes the reaction essentially irreversible. (We saw an example of how pyrophosphate hydrolysis can help drive a reaction forward in the glycogen synthesis reactions in Chapter 14.)

Because a 3′-hydroxyl group is required for the reaction, DNA polymerase cannot simply bind to a single-stranded template DNA molecule and start synthesis. The first 3′-hydroxyl used to initiate DNA synthesis is supplied by an oligonucleotide, usually RNA, which is bound to the template DNA and is known as a **primer**. Primers are synthesized by the enzyme primase, which creates one primer on the leading strand template and multiple primers on the lagging strand template. The free 3′-hydroxyl supplied by the primer, and later by the growing DNA strand, is one of the substrates for DNA polymerase. RNA primers must later be removed and replaced with DNA, a process that has a distinct set of enzymes that we will discuss later.

DNA polymerase must be able to accommodate all four deoxynucleotides; however, only one will form the correct base pairing with the template to give the correct configuration for the active site (**Figure 20.7**). Pyrimidine–pyrimidine and purine–purine mismatches are prevented largely on the basis of incorrect size, whereas other mismatches, such as A-C or G-T, are largely excluded from the active site because their molecular geometry is not well matched with the structure of the active site. Even with these mechanisms in place, mismatches do occur at a rate of about 1 base every 10^4 to 10^5 base pairs.

To duplicate a bacterial genome containing 10^6 to 10^7 base pairs, DNA polymerases must catalyze many nucleotide addition reactions before separating from the template DNA. Because the product of one round of polymerization becomes the substrate for the next round, dissociation of the enzyme from the DNA

a.

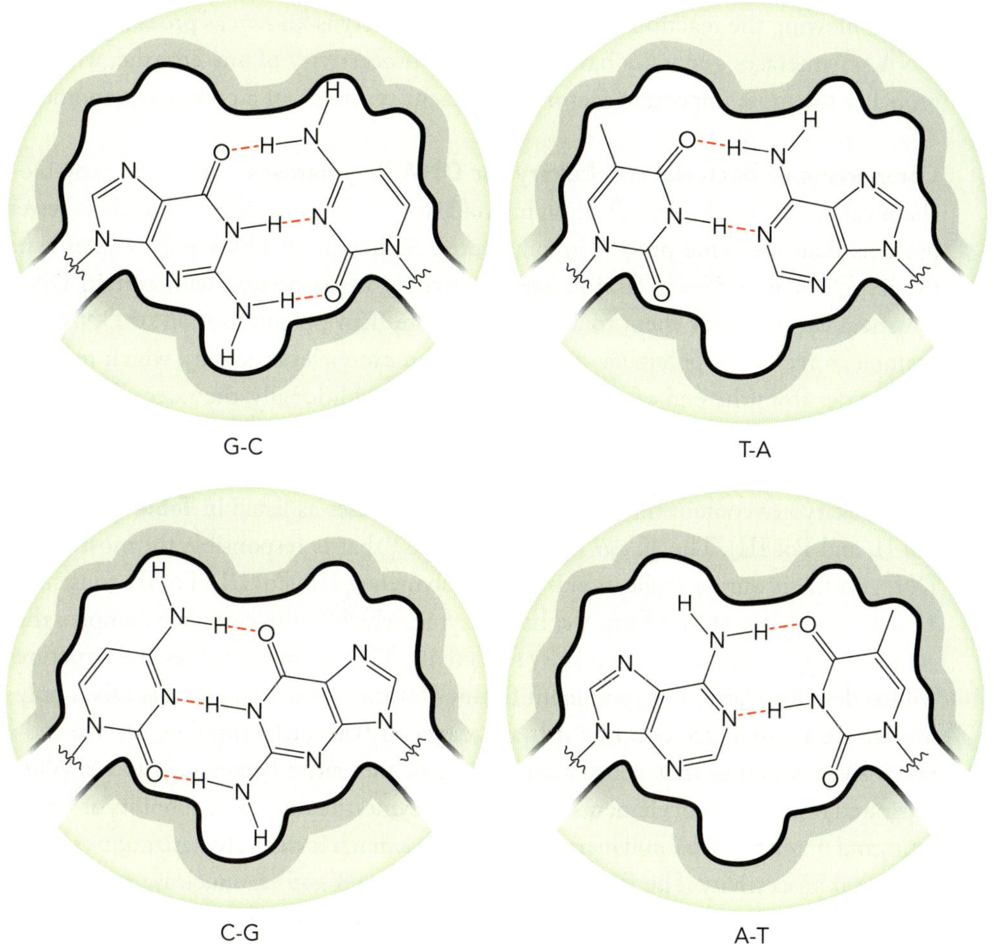

b.

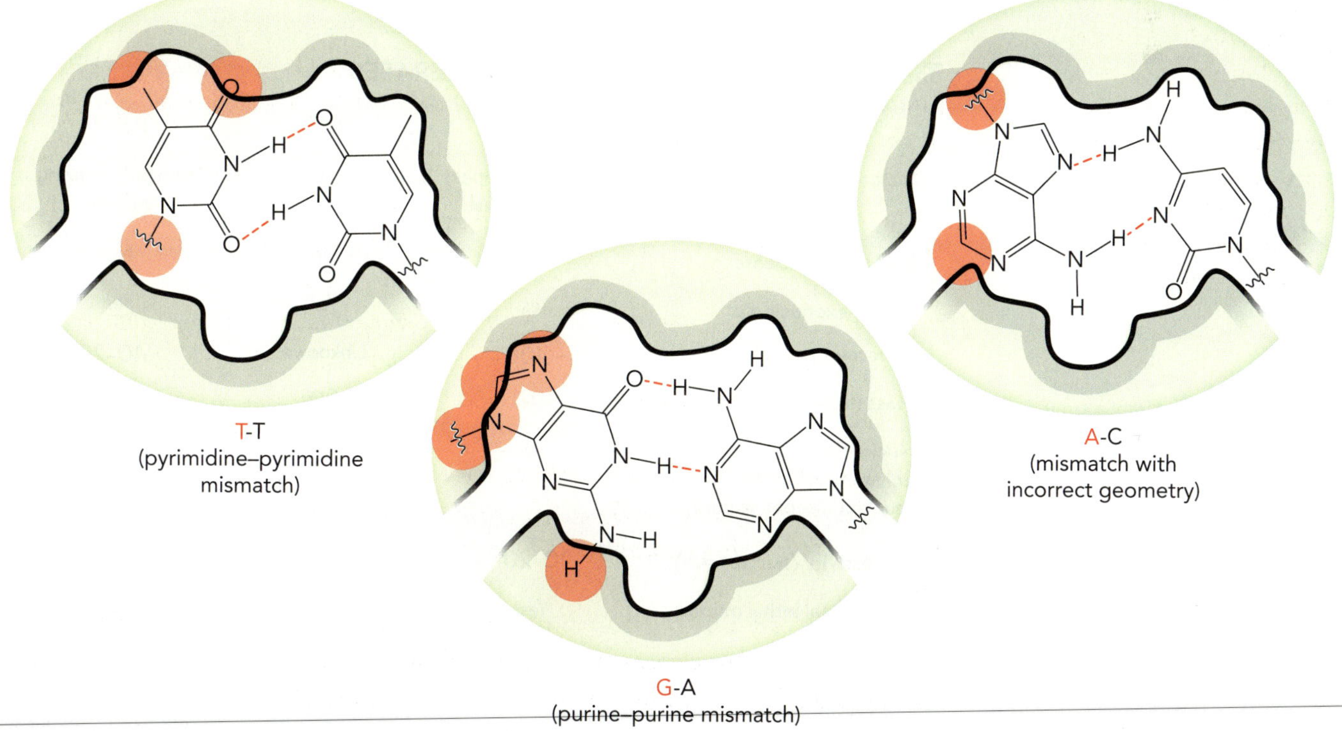

Figure 20.7 The shape of the DNA polymerase active site facilitates formation of the correct Watson–Crick base pairs. **a.** Correct base pairs fit well into the DNA polymerase active site. **b.** Mismatches do not fit well into the active site pocket and thus are largely excluded. Examples are shown of pyrimidine–pyrimidine (T-T) and purine–purine (G-A) mismatches, as well as an A-C mismatch. The position of attachment to the ribose C-1′ is indicated by the squiggly line. In each of these examples, when the nucleotide base on the right is correctly positioned within the active site, the base on the left will not properly fit into the active site. Significant clashes are noted by shaded red circles.

molecule would be a wasteful process. Instead, the enzyme remains bound to the DNA, allowing the reaction to proceed. This property is known as **processivity**, and DNA polymerases generally have the highest processivity of any enzyme, which is probably the most important factor for efficient synthesis of a complete genome.

Comparison of Bacterial and Eukaryotic DNA Polymerases As with many biochemical processes, DNA replication in prokaryotes is less complex and requires fewer proteins than the same process in eukaryotes. Although all DNA polymerases synthesize DNA in a 5′→3′ fashion, other activities are associated with certain DNA polymerases that allow them to accomplish more than polymerization of DNA. For example, some DNA polymerases also encode an exonuclease activity, which removes nucleotides from the end of a nucleic acid strand. Additionally, the rates of polymerization, the degree of processivity, and the number of DNA polymerase molecules per cell can vary tremendously from one polymerase to another.

Prokaryotes contain three major DNA polymerases as listed in **Table 20.1** (Pol I, Pol II, and Pol III). The *E. coli* DNA polymerase that is responsible for most of the DNA synthesis during replication is DNA polymerase III (Pol III). Pol III consists of 17 subunits that together form the holoenzyme, which is the complete complex that contains all the components necessary for activity. The subunit activities and structures will be described later. The prominent features for the holoenzyme of this DNA polymerase are a polymerization rate that approaches 1,000 nucleotides per second and a processivity function that permits replication of the entire bacterial genome. When separated from the other elements of the holoenzyme, the three core subunits of Pol III (α, ε, and θ) retain DNA polymerase activity but a much lower polymerization rate and decreased processivity. The Pol III holoenzyme has a 3′→5′ exonuclease activity that allows for proofreading during polymerization, resulting in an even greater decrease in the occurrence of mismatches. Because only two copies of the enzyme are needed for replication to begin (one at each replication fork), there are ~10–20 copies of this enzyme in each cell.

Table 20.1 MAJOR *E. COLI* DNA POLYMERASES

	Pol I	Pol II	Pol III
Major function(s)	Proofreading, repair, primer removal	Repair	Main polymerizing enzyme
5′→3′ exonuclease	Yes	No	No
3′→5′ exonuclease	Yes	Yes	Yes
Molecules per cell	~400	Unknown	~10–20
Mass (kDa)	103	90	167[a]
Nucleotide polymerization rate (nucleotides per second)	10	0.5	~1,000
Processivity (nucleotides)	10–20	1,500	Unlimited[b]
Conditional lethal mutant	Yes	No	Yes

[a]This represents the core polymerase that contains the α, ε, and θ subunits.

[b]For holoenzyme.

A second bacterial polymerase is DNA polymerase I (Pol I), which is involved in proofreading of newly synthesized DNA, DNA repair, and removal of RNA primers. To accomplish these functions, it has both 3′→5′ and 5′→3′ exonuclease activity. Although this was the first well-characterized DNA polymerase, its slow polymerase activity (~10–20 nucleotides per second) is not sufficient for timely duplication of the entire genome. Furthermore, it has low processivity, although because of its ability to remove primers and replace those sections with DNA, its role in lagging strand synthesis is vital for the cell. Each of the three distinct activities of Pol I—that is, polymerization, 3′→5′ exonuclease, and 5′→3′ exonuclease—has its own active site. Proteolytic cleavage of this 103-kDa enzyme results in a Klenow fragment that has polymerase and 3′→5′ exonuclease activity, whereas the 5′→3′ exonuclease activity is present on a smaller fragment of the original enzyme. (We discuss functions of the Klenow fragment in more detail later.)

A third bacterial polymerase, DNA polymerase II (Pol II), has a lower synthesis rate than that of Pol I. Conditional mutants that lack Pol I or Pol III cannot survive, but mutants lacking only Pol II are viable. The primary function of Pol II is thought to be DNA repair, which we discuss in Section 20.2.

Although there are 15 known eukaryotic polymerases, only three are responsible for the bulk of DNA replication (**Table 20.2**). The two principal enzymes for eukaryotic DNA replication are DNA polymerase δ (Pol δ) and DNA polymerase ε (Pol ε). Like *E. coli* Pol III, Pol δ has an extremely high processivity, but it has a much lower reaction rate (~40–50 nucleotides per second); however, because eukaryotic cells have multiple replication forks on each chromosome instead of the two replication forks found in prokaryotic cells, the entire yeast genome can be replicated in less than an hour. The high processivity of Pol δ is only observed in the presence of a protein called **proliferating cell nuclear antigen (PCNA)**, a protein that is structurally similar to one of the subunits of *E. coli* Pol III. Like its prokaryotic counterpart, Pol δ contains 3′→5′ exonuclease activity to enable proofreading.

Researchers have recently discovered that Pol δ is responsible for lagging strand synthesis, and Pol ε is responsible for leading strand synthesis. Unlike Pol δ, Pol ε does not have a requirement for PCNA, but it is generally found associated with PCNA at the replication fork. Also, Pol ε does have 3′→5′ exonuclease activity, so like Pol δ it is capable of proofreading. The combined activity of these two polymerases accomplishes the bulk of eukaryotic DNA synthesis.

DNA polymerase α (Pol α) is similar to bacterial Pol I in that it is most often associated with lagging strand synthesis. One of its subunits contains primase activity, and like Pol I it has very low processivity. However, Pol α has no 3′→5′ exonuclease activity.

Table 20.2 MAJOR EUKARYOTIC DNA POLYMERASES

	Pol δ	Pol ε	Pol α
Major function(s)	Lagging strand synthesis	Leading strand synthesis	Initial synthesis from primer, primase
3′→5′ exonuclease	Yes	Yes	No
Processivity (nucleotides)	High	High	Low
PCNA association	Yes	Yes	No

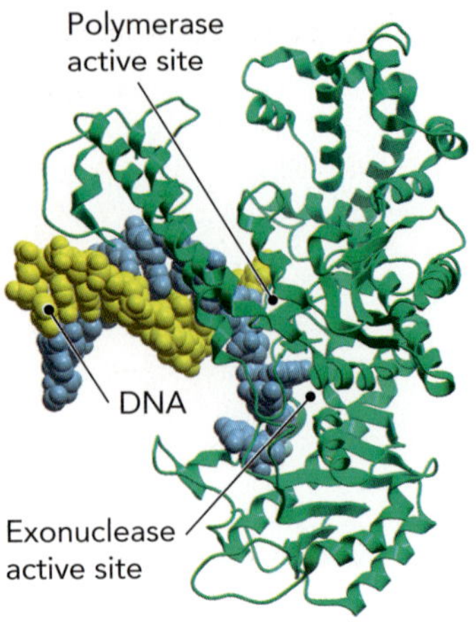

Figure 20.8 The structure of the *E. coli* DNA Pol I Klenow fragment bound to DNA is shown. The polymerase domain is at the top of the structure, and the exonuclease domain is located at the bottom. BASED ON PDB FILE KLN.

▶ ANIMATION

Other eukaryotic DNA polymerases include DNA polymerase γ, which is found in the mitochondria and is responsible for replication of the mitochondrial DNA. DNA polymerase β has very low processivity (it only adds one nucleotide before dissociating from the DNA) and shows increased presence after exposure to DNA-damaging events, so it is most likely involved in DNA repair.

Structure and Function of *E. coli* DNA Polymerases I and III Arthur Kornberg was the first to isolate an enzyme with DNA polymerase activity. He did this in 1957, only 4 years after Watson and Crick published the structure of DNA. The enzyme Kornberg isolated was *E. coli* DNA polymerase I (Pol I), which he found was composed of a single 103-kDa polypeptide containing 928 amino acid residues. As mentioned earlier, proteolytic cleavage of the 103-kDa polypeptide produces two fragments. One contains the 5′→3′ exonuclease activity, and the other is referred to as the **Klenow fragment.** The 5′→3′ exonuclease activity is used during DNA repair and will be discussed later. The Pol I Klenow fragment is a 68-kDa polypeptide that is very similar in function to Pol III. It contains two domains: the larger C-terminal portion contains the polymerase active site, and the smaller domain contains the active site for the 3′→5′ exonuclease activity.

As shown in **Figure 20.8**, the Pol I Klenow fragment shows a clear separation of the two domains, with a distance between the two active sites of ~25–35 Å. The polymerase active site contains at least three negatively charged amino acids that are complexed to divalent cations, usually Mg^{2+}. These cations improve the leaving-group potential of the pyrophosphate, thus improving the rate-limiting step of the reaction. Numerous positively charged amino acids are found near the active site, which are responsible for proper positioning of the negatively charged DNA substrate.

DNA polymerases do not play a direct role in assisting proper base pairing between the substrates, but rather base pairing is determined by the physical and chemical properties of the enzyme active site (see Figure 20.7). Although only proper Watson–Crick base pairs are complementary to the active site, base pair mismatches can occur. This mistake is corrected by the 3′→5′ exonuclease activity of Pol I. As shown in **Figure 20.9**, when a base pair mismatch occurs during replication, the growing DNA strand is flipped from the polymerase domain to the exonuclease domain, and the mismatched base is excised. The growing strand then flips back to the polymerase domain to continue the replication process.

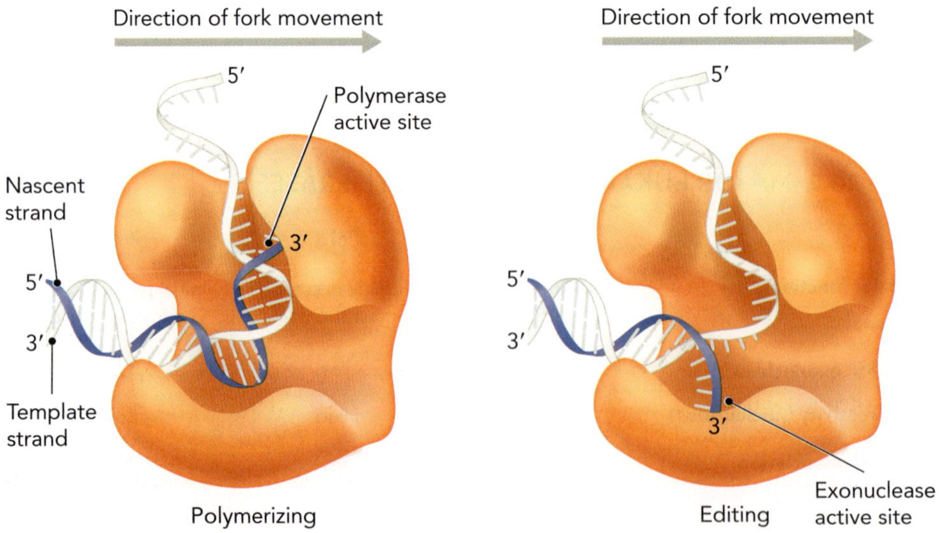

Figure 20.9 The Pol I Klenow fragment has both polymerizing and editing modes. In the polymerizing mode, nucleotides are added to the end of the growing chain in the polymerase active site. In editing mode, the newly added nucleotide is flipped out of the polymerase active site and into the exonuclease site, where it is excised from the growing DNA strand.

Although the Klenow fragment of Pol I is an excellent model of a bacterial DNA polymerase, Pol I is not responsible for the bulk of *E. coli* DNA synthesis. That role falls to Pol III, an enzyme whose polymerase and proofreading activities are found on separate subunits. The core Pol III enzyme consists of an α subunit that is responsible for polymerization, an ε subunit that is responsible for proofreading, and a θ subunit whose role is not entirely understood, but most likely aids the ε subunit. The Pol III holoenzyme consists of this core, plus at least seven other subunits. The interplay of these different subunits at the replication fork will be discussed later when we look at the replication fork in greater detail.

Structure and Function of Reverse Transcriptase A virus that uses RNA as its genetic material, such as human immunodeficiency virus (HIV), must have a mechanism for converting RNA to DNA before inserting its genome into the host DNA. The viral enzyme reverse transcriptase is responsible for this conversion and shares many similarities with an analogous region in the *E. coli* Pol I enzyme. As shown in **Figure 20.10**, the HIV reverse transcriptase (HIVRT) and *E. coli* Pol I enzymes both resemble the shape of a hand, in which the three protein domains are referred to as the fingers, the thumb, and the palm. Structure–function studies of numerous DNA polymerizing enzymes have shown that the catalytic residues are found within the β sheets of the palm region, whereas the finger and thumb regions form a tunnel through which the DNA passes.

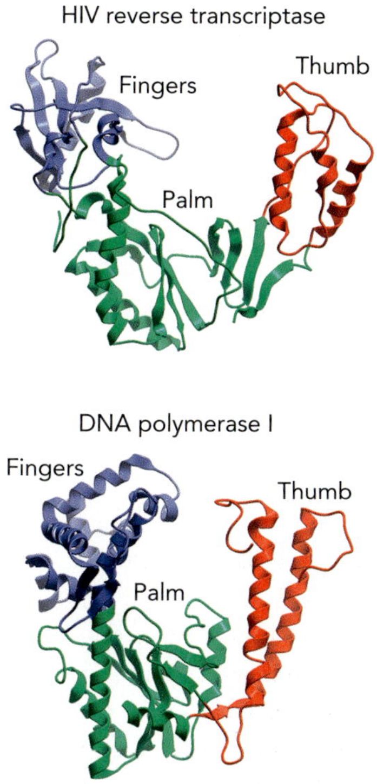

Figure 20.10 A structural comparison of HIV reverse transcriptase (p66 subunit) and the *E. coli* Pol I Klenow fragment is shown. The protein domains referred to as the fingers, thumb, and palm are labeled. Biochemical studies have shown that the DNA moves between the fingers and thumb, while the palm contains the catalytic residues. BASED ON PDB FILES 3HVT (HIVRT) AND 1KLN (KLENOW FRAGMENT).

The HIVRT enzyme is probably the best-studied reverse transcriptase, as inhibition of this enzyme has been critical in combating HIV infections. HIVRT has two separate enzymatic activities. In addition to the reverse transcriptase activity that can create DNA from an RNA template, the enzyme possesses RNA-cleaving activity (also called ribonuclease, or RNase, activity), which allows for degradation of the viral RNA genome after conversion to DNA. HIVRT is highly error-prone, as no proofreading function is associated with the enzyme. The typical error rate is approximately 1 in 2,000 nucleotides, with A-T mismatches representing the most prevalent error. This high error rate of about 10 errors per genome replication is responsible for the hypermutability of the virus.

As a reverse transcriptase, HIVRT first uses the single-stranded RNA genome as a template to produce a DNA–RNA hybrid. This is then converted into double-stranded DNA, again using HIVRT. Thus, HIVRT is able to use both DNA and RNA as a template. As with DNA polymerases, a free 3′-hydroxyl group is required for the polymerase reaction to take place. This hydroxyl group is provided by the 3′ end of the transfer RNA (tRNA) for Lys, which binds to the viral RNA with rather high affinity. After the production of the first DNA strand, the RNA is degraded by the RNase activity to provide a single-stranded template for the second round of DNA synthesis by HIVRT, ultimately producing double-stranded DNA. The *in vitro* synthesis of double-stranded complementary DNA (cDNA) from an RNA template using a reverse transcriptase enzyme is described in Chapter 3 (see Figure 3.53).

Structure and Function of Replication Fork Proteins

Many proteins are required at the replication fork to allow DNA polymerase to function properly. The complete complex that contains the enzymes and proteins required to replicate DNA is called the **replisome**. As with most biochemical processes, the events at the eukaryotic replication fork are more complex than those at the prokaryotic replication fork and as such are less well understood. For that reason, we focus primarily on the prokaryotic system.

Table 20.3 REPLICATION FORK PROTEINS IN *E. COLI*

Replication fork proteins	Function in DNA replication
DNA polymerase	Synthesizes nascent DNA on leading and lagging strands
DNA helicase	Unwinds the DNA double helix ahead of the replicating complex
Primase	Synthesizes RNA primers on the lagging strand
Single-stranded DNA binding proteins	Binds to single-stranded DNA to prevent reannealing of the DNA double helix
DNA gyrase	Relieves torsional stress in front of the fork (a type II topoisomerase)
β clamp	Prevents DNA polymerase III from dissociating from the DNA

Anatomy of a Prokaryotic Replication Fork Once the replication fork is formed, three events must take place to synthesize new DNA: (1) conversion of double-stranded DNA to single-stranded DNA; (2) addition of an RNA primer; and (3) extension of the RNA primer. To produce the single-stranded DNA templates, the enzymes **gyrase** (a topoisomerase) and **helicase** (also called **DnaB**) are found at the front of the replication fork. The active form of helicase is a hexamer that is able to encircle a single strand of DNA completely. The unwinding of DNA by helicase puts torsional strain upon DNA, requiring the action of gyrase, which moves in front of helicase, to relieve that strain. Helicase binds to only one of the DNA strands and moves in a 5′→3′ direction. Because single-stranded DNA will quickly form double-stranded DNA whenever possible, **single-stranded DNA binding proteins (SSBs)** bind to the single-stranded DNA after the helicase action to prevent reassociation of the two DNA strands. The enzyme **primase** (DnaG) associates with helicase to synthesize the primers necessary for lagging strand synthesis. The combination of helicase and primase is often referred to as the **primosome**. The primary contributors to the replication fork are summarized in **Table 20.3** and shown in **Figure 20.11**.

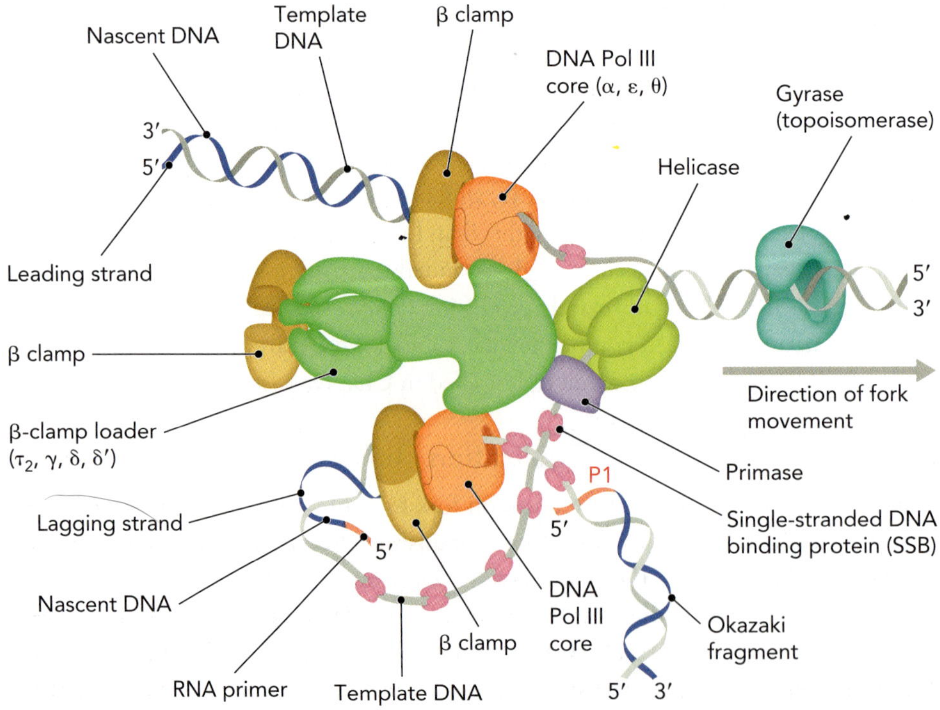

Figure 20.11 This schematic drawing shows the anatomy of the *E. coli* replication fork. The *E. coli* replication fork consists of Pol III as well as helicase, primase, gyrase, and single-stranded DNA binding proteins. This proposed arrangement of proteins within the *E. coli* replication fork accommodates 5′ to 3′ directional DNA synthesis of the leading and lagging strands. Tethering of both Pol III core complexes to helicase by τ subunits in the β-clamp loading complex ensures that leading and lagging strand synthesis occur at a similar rate. See the text for descriptions of the components. P1 is an RNA primer.

As illustrated in Figure 20.11, the Pol III complex is tethered to the helicase by the τ subunit of the polymerase complex. At the replication fork, Pol III consists of two cores, with each of the cores composed of α, ε, and θ subunits. Each core is associated with a **β clamp**—a structure consisting of two β subunits that are able to encircle the DNA, resulting in the drastically increased processivity associated with Pol III. A third β clamp is associated with Pol III by interactions with a γ subunit and a δ subunit, which together are responsible for loading the clamp onto the DNA, along with a δ subunit that opens the clamp during the loading process. This third β clamp is positioned for rapid attachment to the lagging strand when a new Okazaki fragment is started, ensuring that lagging strand synthesis proceeds at a rate similar to that of leading strand synthesis.

Function of the Replisome Proteins Now that we have introduced the major proteins associated with the replication fork, let's look at the functions of each in more detail. The first is DNA gyrase, which is located in front of the replication fork. Because DNA helicase unwinds the double-stranded DNA and adds positive supercoils ahead of the replication fork, gyrase functions to introduce negative supercoiling and thereby alleviate the strain added by helicase. The topoisomerase II activity of gyrase cleaves both strands of DNA in order to introduce the negative supercoils, and then in an ATP-dependent step, the two strands are repaired (see the topoisomerase II reaction cycle in Figure 3.20b). As shown in **Figure 20.12**, *E. coli* gyrase is a tetramer composed of two GyrA subunits, which are involved in DNA binding and cleavage, and two GyrB subunits, which are the sites of ATP hydrolysis.

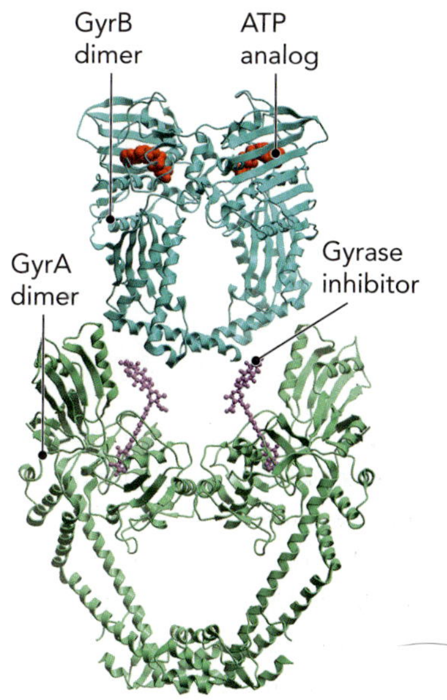

Figure 20.12 The structure of *E. coli* gyrase is shown. This composite figure shows the structure of GyrB dimer with ATP analogs bound to the nucleotide binding sites in each subunit (top) and a GyrA dimer with gyrase inhibitor molecules bound to the catalytic active sites (bottom). BASED ON PDB FILES 1EI1 (GYRB) AND 2Y3P (GYRA).

Helicase plays a very important role in DNA replication, but helicase activity is also involved in many processes that require single-stranded DNA, such as DNA repair and recombination. Thus, the cell contains multiple helicases. The particular helicase involved in *E. coli* DNA replication is called DnaB. Like all helicases, DnaB cannot itself start the separation of double-stranded DNA and instead requires a small section of single-stranded DNA before it can further separate the DNA strands. The initial unwinding of DNA is brought about by several factors, which are discussed when we look at initiation in greater detail. The true function of DnaB is thus the propagation of single-stranded DNA, not the initial formation of it.

Once helicase has generated single-stranded DNA, keeping it in this form to enable DNA synthesis to occur requires stabilization by SSBs. In *E. coli*, SSBs are tetramers, and binding to DNA often results in two tetramers associating with each other (see Figure 3.29). This occurs in a cooperative manner, so that the resulting octamer binds much more tightly to single-stranded DNA than does the tetramer. Structures larger than an octamer are possible and result in still stronger binding to single-stranded DNA. There are only about 300 SSB tetramer molecules in an *E. coli* cell, and therefore they are constantly recycled during the replication process. It is estimated that approximately 70 SSB tetramers are associated with each replication fork, with most associated with the lagging strand. The binding of SSBs to DNA takes place largely through interactions between nucleotide bases in DNA and aromatic amino acids in an SSB, primarily tryptophan and phenylalanine. Positively charged lysine residues in SSBs are also involved in DNA binding.

The RNA primer required for DNA synthesis at the replication fork is synthesized by the enzyme primase (DnaG) in *E. coli*. Primase makes an RNA primer using a DNA template, so it is similar in function to RNA polymerase. But unlike RNA polymerase, primase has very low processivity, producing a typical primer length of 10–12 nucleotides. Also, the rate of polymerization is about 1,000 times slower than that of *E. coli* RNA polymerase and 10,000 times slower than that of Pol I. Because primase is tightly coupled to helicase at the front of the replication fork (see Figure 20.11), the slow speed of primase might play a role in keeping the leading strand from moving too quickly in relation to the lagging strand. Because the DNA polymerase components on the leading strand template cannot pass the primosome, the rate of leading strand synthesis is ultimately controlled by primase as well.

In the absence of the β clamp, Pol III incorporates about 20 nucleotides per second and dissociates after addition of about 10 nucleotides. When Pol III is associated with the β clamp, however, the polymerization rate exceeds 750 nucleotides per second, and processivity exceeds 50,000 bases. The β clamp is a homodimer with the monomers having a crescent shape, so that the dimer forms a ring that completely encircles the double-stranded DNA. Addition and removal of the β-clamp dimer on double-stranded DNA requires the clamp loader protein. As shown in **Figure 20.13**, the clamp loader binds to one of the β monomers and, in an ATP-dependent mechanism, opens a gap at the interface between the two β monomers. The β clamp is distorted, allowing the DNA to enter or exit the now-open ring. The β clamp has a high affinity for primed template DNA when associated with the clamp loader, and the primed DNA enters the clamp through the opening between the β subunits. This stimulates ATP cleavage by the clamp loader protein, causing the clamp loader to dissociate from the β clamp, which then recloses.

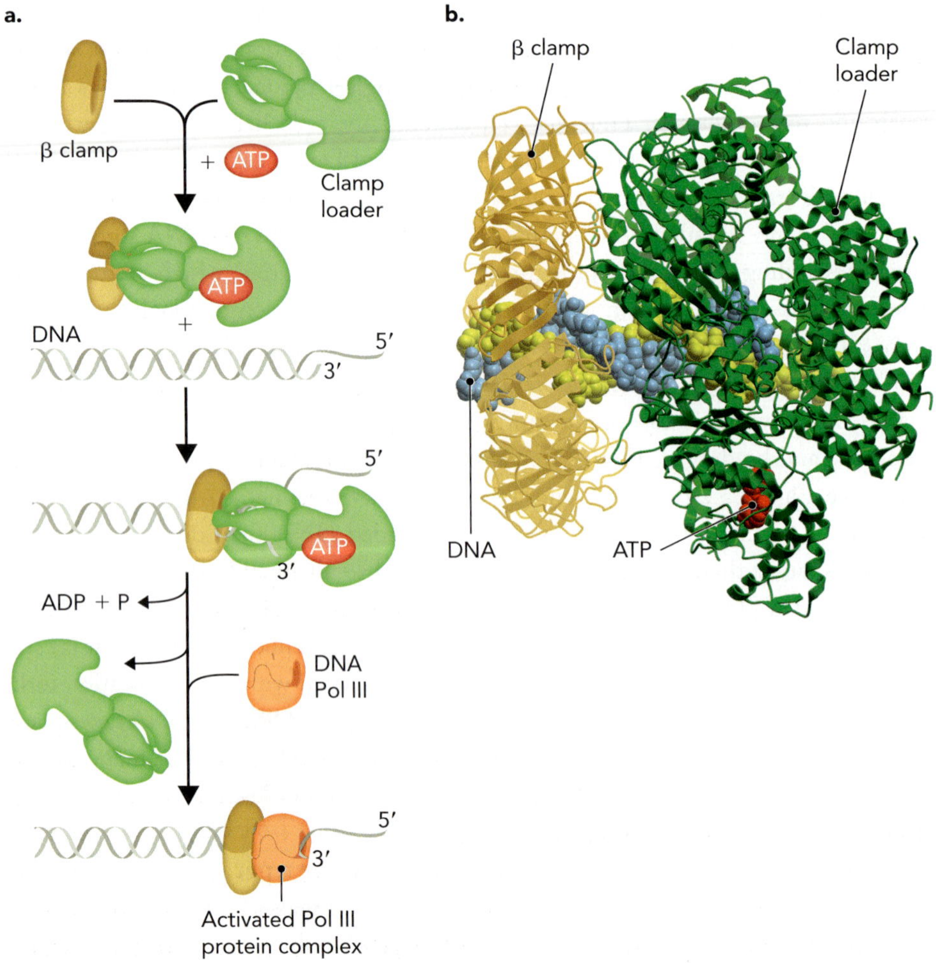

Figure 20.13 The β-clamp complex is used to keep the Pol III complex associated with DNA. **a.** The clamp loader complex is required to load the β clamp onto the DNA in an ATP-dependent manner. Once the β clamp is in position, the clamp loader complex dissociates, and the Pol III complex associates with the β clamp. **b.** Structure of the T4 bacteriophage β clamp bound to DNA and the clamp loader complex. The *E. coli* β clamp and clamp loader complex is similar, with the exception that the T4 β clamp has three subunits instead of two subunits. BASED ON PDB FILE 3U5Z.

Structure and Mechanism of DNA Helicases Helicases are important not only in separating double-stranded DNA but also in many aspects of RNA metabolism. Helicases are rather specific for the type of reaction they catalyze, so the cell requires a large variety of helicases to provide the necessary activities. Because of the large number of different structures and activities, helicases have been classified into six different superfamilies. Superfamilies 1 and 2 represent monomeric helicases, whereas superfamilies 3–6 are hexameric. The hexameric helicases are further divided by their directionality. DnaB, being a hexameric 5′→3′ helicase, is a member of superfamily 4.

As shown in **Figure 20.14**, hexameric helicases function by forming a barrel-like structure through which the nucleic acid passes, whether DNA or RNA. ATP provides the energy for translocation along the DNA, with the ATP binding pocket found at the interfaces between the monomers. To separate the two strands of DNA, a helicase must break the hydrogen bonds holding the helix together. In doing so, it must associate with the single-stranded DNA passing through the helicase to prevent the helix from re-forming. In the case of DnaB, 12 nucleotides of the single-stranded DNA are associated with the helicase at any given time, with each of the six monomers bound to two nucleotides.

Strand separation by DnaB occurs by passing one strand of DNA through the helicase while the other strand is forced outside the enzyme. Not all helicases work in this fashion; some work by routing the duplex nucleic acid into the helicase with the separate strands emerging from the enzyme. Strand separation by helicases can be thought of as a wedge driving apart the two strands of DNA. In order for the wedge to proceed, energy must be provided so the helicase can move stepwise along the DNA. Movement of each monomer requires ATP hydrolysis, and each movement of one monomer advances the helicase two nucleotides. Thus, the unzipping of DNA for replication is an energy-intensive process requiring one ATP for every two bases.

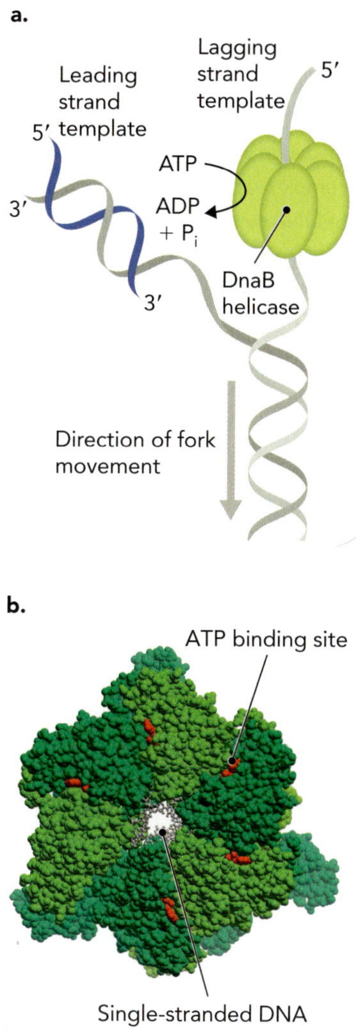

Figure 20.14 The *E. coli* replication helicase DnaB unwinds the DNA double helix ahead of the replication fork. **a.** Model depicting DnaB moving along the lagging strand template to unwind the helix using ATP hydrolysis. **b.** Structure of the *E. coli* DNA helicase bound to single-stranded DNA. The location of the ATP binding sites are identified by the nucleotides shown in red space-filling style. BASED ON PDB FILE 4ESV.

Coordinate Processivity Ensures Efficient Fork Movement How do proteins at the replication fork work together to synthesize both the leading and lagging strands simultaneously? As mentioned earlier, the two Pol III cores are linked together through the primosome, and therefore the rate of DNA synthesis of the leading strand cannot exceed the rate of synthesis of the lagging strand. Coordination of DNA synthesis on the leading and lagging strand templates is also facilitated by the τ subunits of the β-clamp loader protein, which bind to two Pol III core complexes (see Figure 20.11). Through the combined activity of the helicase and primase enzymes at the leading edge of the fork, the DNA helix is unwound, and the RNA primers are synthesized on the lagging strand template.

As shown in **Figure 20.15**, the "trombone model" of DNA synthesis at the fork proposes that the Pol III core on the lagging strand template alternates between bound and unbound forms. The Pol III core on the lagging strand template synthesizes an Okazaki fragment from the 3′ end of one RNA primer until it reaches the 5′ end of the RNA primer farther down the lagging strand template. This primer extension and DNA synthesis on the lagging strand template resembles the slide arm of a trombone as the loop is extended. Once the Pol III core releases the lagging strand template DNA strand, the DNA loop shrinks in size much like the retraction of the trombone arm. Reassociation of the Pol III core with the lagging strand template at the site of the next RNA primer starts the process over again with synthesis of the next Okazaki fragment. Behind the Pol III complex, the RNA primers are removed by Pol I, which

Figure 20.15 Coordinated DNA synthesis on the leading and lagging strand templates by two Pol III core complexes is mediated by the clamp loader complex. The "trombone model" of DNA synthesis proposes that the lagging strand DNA template is alternately bound and released by the Pol III core complex as each Okazaki fragment is synthesized.

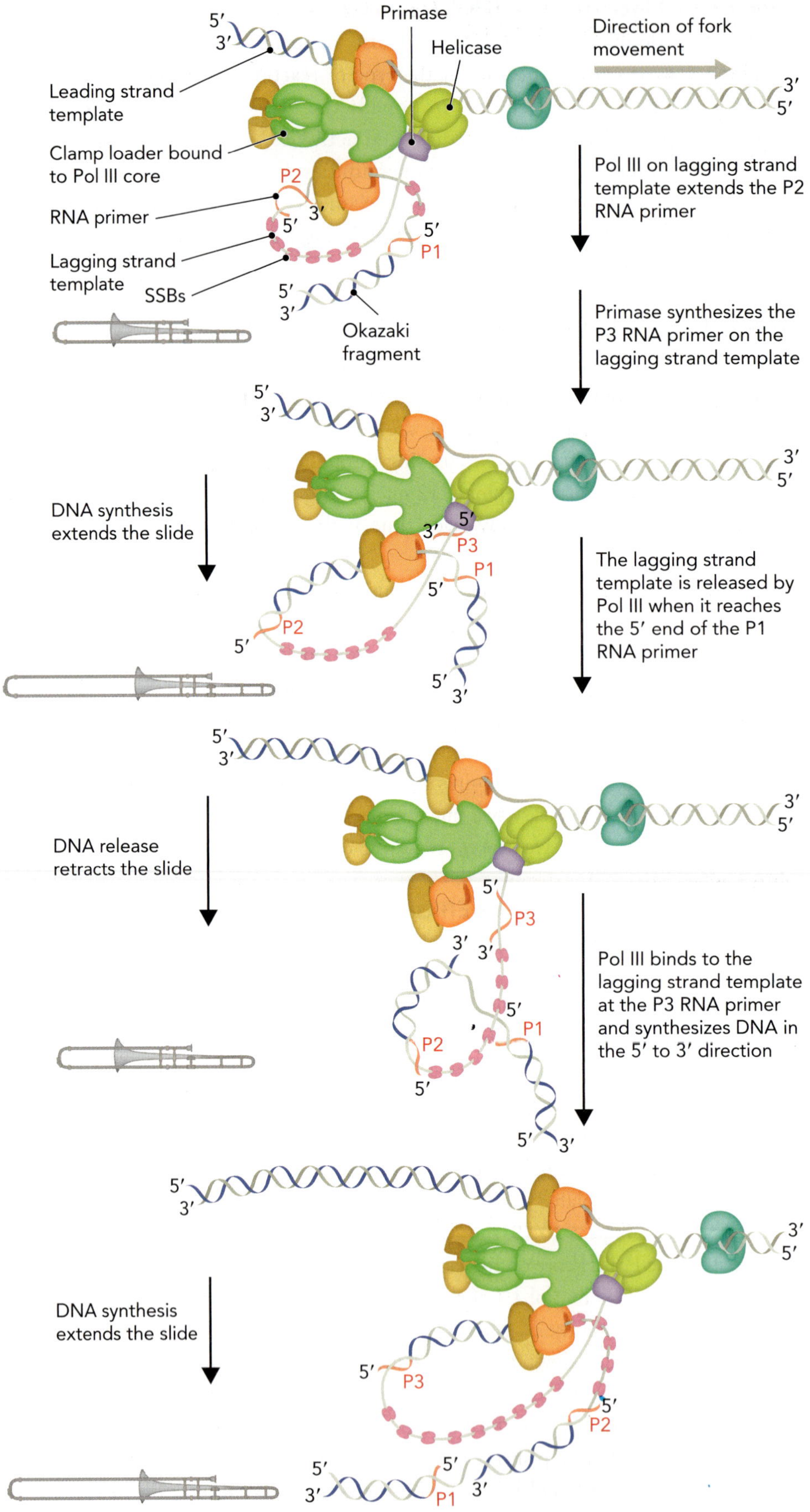

Table 20.4 PROKARYOTIC AND EUKARYOTIC REPLICATION FORK PROTEINS

Replication fork proteins	Prokaryotic	Eukaryotic
Core polymerases	α, ε, θ	α, δ, ε
Clamp	β	PCNA
Clamp loader	τ_2, γ, δ, δ′	RFC
Helicase	DnaB	MCM2-7
Primase	DnaG	Primase
Single-stranded DNA binding protein	SSB	RPA

also replaces the gaps with DNA. Pol I uses the 3′-hydroxyl group adjacent to the site of primer removal as the starting point for DNA synthesis. When Pol I reaches the other end of the primer site, however, it cannot form a phosphodiester bond with the existing 5′ end that was originally connected to the primer. The enzyme DNA ligase forms the requisite phosphodiester bond, thus sealing the nick in the DNA.

Overview of Eukaryotic Replication **Table 20.4** lists five protein complexes that have similar functions in prokaryotes and eukaryotes, with most prokaryotic protein names referring to genes in *E. coli* and the eukaryotic proteins corresponding to genes in yeast. Eukaryotic DNA replication most likely uses a similar "trombone" mechanism to coordinate leading strand and lagging strand DNA synthesis (see Figure 20.15); however, many more proteins appear to be required than in prokaryotic DNA replication. For example, eukaryotic DNA synthesis requires two different DNA polymerases: Pol ε for the leading strand and Pol δ for the lagging strand.

As shown in **Figure 20.16**, the primary helicase protein required for eukaryotic DNA replication is called the mini chromosome maintenance protein (MCM2-7) and

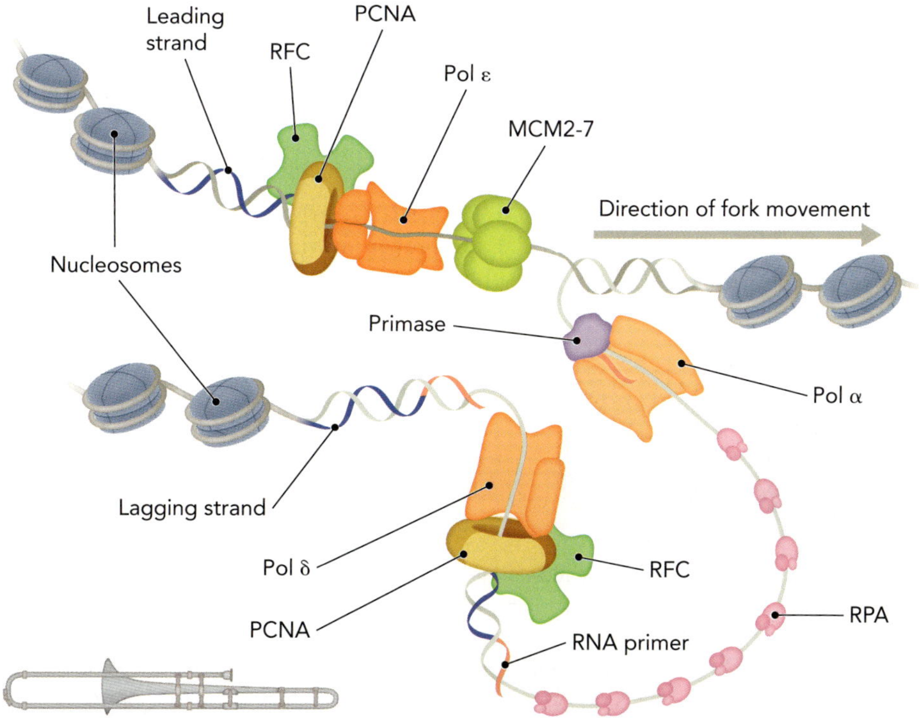

Figure 20.16 The eukaryotic replication fork contains five protein complexes that have analogous functions in prokaryotes. The eukaryotic and prokaryotic replication forks are largely similar, although two different DNA polymerases are required for leading strand (Pol ε) and lagging strand (Pol δ) synthesis. A third DNA polymerase (Pol α) is associated with primase. Moreover, unlike prokaryotic DNA, eukaryotic DNA is packaged into nucleosomes. The orientation depicted here occurs shortly after the lagging strand is released from the Pol ε/Pol δ complex held together by RFC proteins; that is, the trombone slide has been retracted (see Figure 20.15).

consists of six different subunits. Like the *E. coli* helicase, the MCM2-7 heterohexamer encircles one strand of DNA while the other strand routes around the helicase complex. In *E. coli*, the lagging strand template passes through the hexamer, whereas in eukaryotes, the leading strand template passes through the hexamer. The functional equivalent of prokaryotic SSBs in eukaryotes is the replication protein A (RPA) proteins. The eukaryotic RNA primase is a protein complex consisting of both the primase and a DNA polymerase called Pol α.

The functional equivalent in eukaryotes of the *E. coli* β-clamp protein is the proliferating cell nuclear antigen (PCNA) protein. The PCNA protein is essential for rapid replication, as it increases the processivity of the δ and ε polymerases by approximately 1,000-fold. Like the β clamp, it must be loaded on the single-stranded DNA by a clamp loading complex known as replication factor C (RFC). PCNA has also been shown to have a role in DNA repair, both during and after replication. An important difference between eukaryotic and prokaryotic replication is that eukaryotic DNA is packaged in histone-containing nucleosomes, which must be removed ahead of the replication fork and replaced in each of the daughter strands (Figure 20.16).

Initiation and Termination of DNA Replication

We have looked at the structure and the function of proteins that are required for DNA synthesis at the replication fork, and now we examine the biochemical processes required for initiation and termination of DNA replication. Initiation and termination of DNA replication is better understood in prokaryotes than in eukaryotes, so we start once again by describing these processes in *E. coli*.

The *E. coli* Origin of Replication Is Defined by Sequence As shown in **Figure 20.17**, initiation of DNA replication in *E. coli* occurs at a specific site on the circular genome called *oriC* (origin). To initiate replication, DNA at *oriC* must be separated to form a single-stranded region that allows binding of the primosome protein complex. A large number of proteins are involved in this process, with several of the steps involving ATP binding or hydrolysis. Two key elements of *oriC* include three 13-bp repeats and four 9-bp repeats, both of which are enriched in A-T base pairs. The 9-bp repeats serve as binding sites for the initiator protein DnaA, whereas the 13-bp repeats are the actual site where DNA unwinding begins.

Replication is initiated at *oriC* by the binding of about 20 subunits of DnaA to sequences very near the 9-bp repeats (**Figure 20.18**). The monomers form a right-handed helical filament that is crucial for proper unwinding—inability of DnaA oligomerization prevents initiation. In addition to DnaA, other proteins also bind to

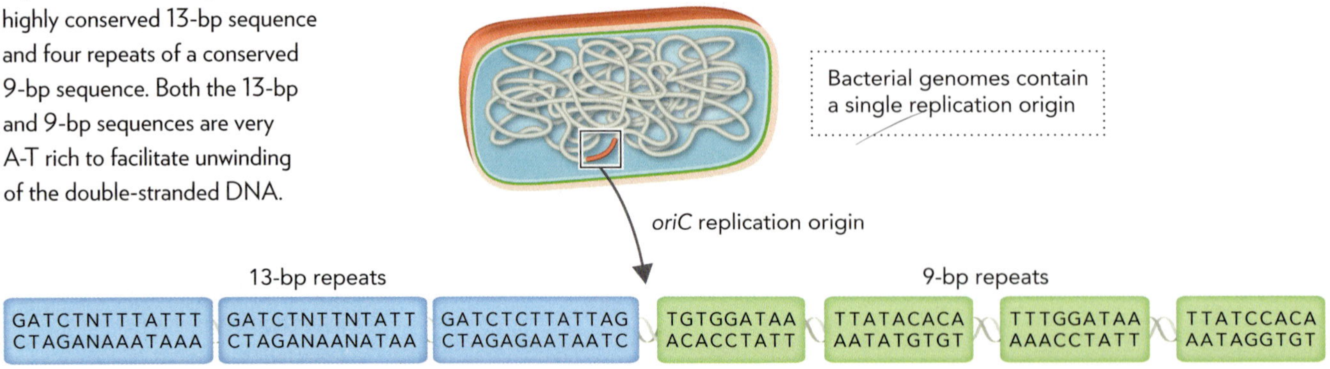

Figure 20.17 Prokaryotic replication is initiated at a single sequence in the bacterial genome call *oriC*, which contains two key elements: three repeats of a highly conserved 13-bp sequence and four repeats of a conserved 9-bp sequence. Both the 13-bp and 9-bp sequences are very A-T rich to facilitate unwinding of the double-stranded DNA.

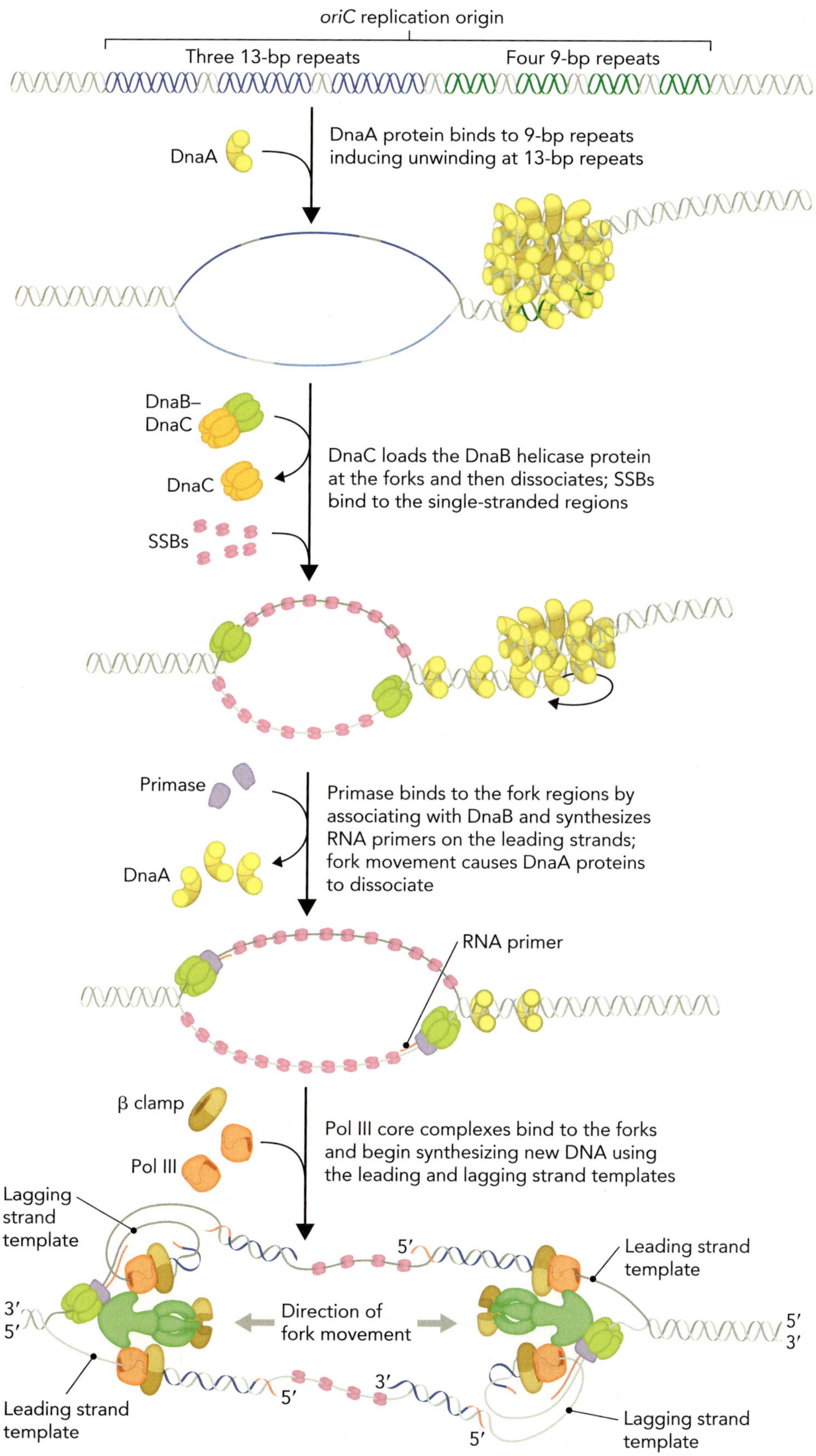

oriC replication origin
Three 13-bp repeats
Four 9-bp repeats
DnaA
DnaA protein binds to 9-bp repeats inducing unwinding at 13-bp repeats
DnaB–DnaC
DnaC
SSBs
DnaC loads the DnaB helicase protein at the forks and then dissociates; SSBs bind to the single-stranded regions
Primase
DnaA
Primase binds to the fork regions by associating with DnaB and synthesizes RNA primers on the leading strands; fork movement causes DnaA proteins to dissociate
RNA primer
β clamp
Pol III
Pol III core complexes bind to the forks and begin synthesizing new DNA using the leading and lagging strand templates
Lagging strand template
Leading strand template
3′
5′
Direction of fork movement
5′
3′
5′
3′
5′
5′
Leading strand template
Lagging strand template

Figure 20.18 Initiation of DNA replication at *oriC* requires assembly of a replication complex at each of the two forks. About 20 molecules of DnaA protein bind to *oriC*, which allows for unwinding of the double helix. This provides a site for primosome assembly. Binding of the DnaB–DnaC complex occurs after DNA unwinding. Association of primase is followed by addition of Pol III and transitioning of the initiation complex to a replicative complex.

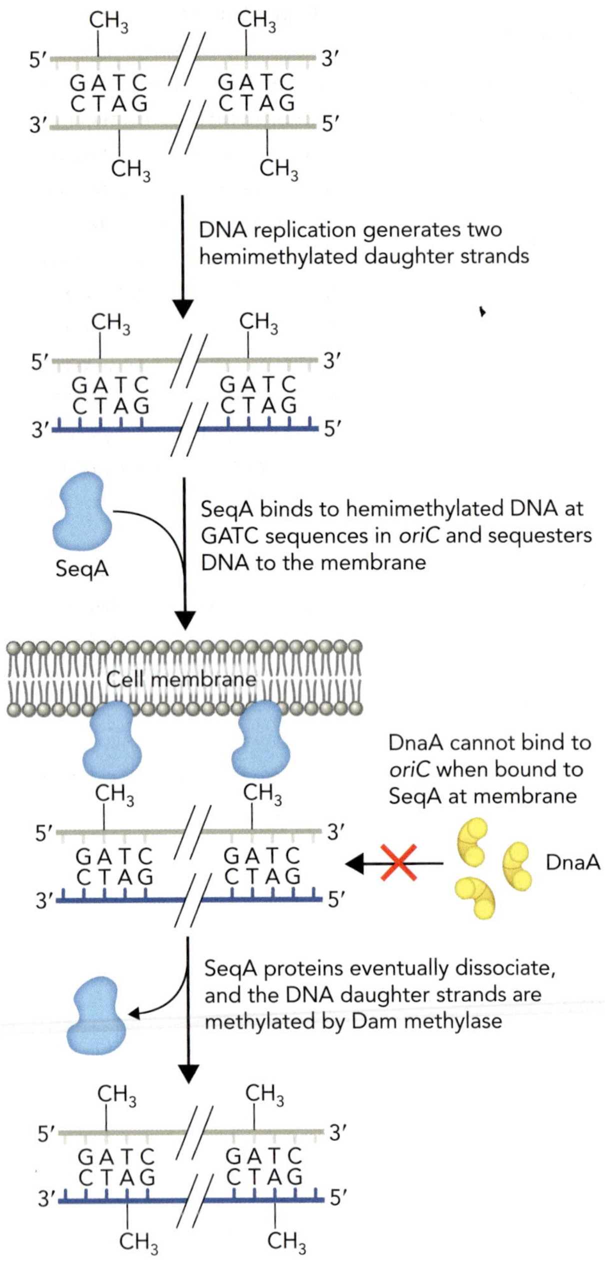

Figure 20.19 Binding of SeqA to hemimethylated DNA in *oriC*, followed by membrane association, prevents a second round of initiation from occurring.

sequences in *oriC* and, together with DnaA, induce bending of the DNA. This bending is instrumental in forming the "open complex," which is the region of single-stranded DNA at *oriC*. As shown in Figure 20.18, multiple copies of DnaA binding near the 9-bp repeats causes opening of the DNA helix in the region of A-T–rich 13-bp repeats.

Once the open complex is formed, DnaA facilities the binding of a complex consisting of the hexameric DnaB protein (helicase) and DnaC. Binding of the DnaB–DnaC complex to the *oriC* region further opens the left-hand 13-bp site and stimulates the release of DnaC. This is followed by binding of SSBs to form the pre-priming complex. This process is repeated to form the second replication fork. Once DnaB is loaded, association with gyrase and SSB allows further opening of the growing replication forks. Primase interacts once with the leading strand template and only intermittently with the lagging strand template, whenever synthesis of a primer is required. Interaction of Pol III via the β-clamp τ subunit promotes release from DnaA and transition from initiation to elongation. Release of DnaA also plays a role in regulating replication, ensuring that a second round of replication is not initiated until the first is well under way.

In addition to a sequence that allows for easy formation of single-stranded DNA, *oriC* contains 11 sites that are recognized by the enzyme **Dam methylase** (DNA adenine methylase). Methylation by Dam methylase is used to distinguish the original strand in newly synthesized DNA, as the newly synthesized strand does not have methyl groups present. This serves as a mechanism for repairing mismatches that occurred during polymerization, as the mismatch repair system uses the methylation to determine the template strand. For most of the chromosome, the hemimethylated state—in which one strand is methylated and the other is not—lasts about 2 minutes. At *oriC*, the hemimethylated state lasts substantially longer, generally about a third of the cell cycle. This is an important control mechanism in replication initiation at *oriC* because it apparently prevents a second round of initiation from occurring before the first round has been completed.

As shown in **Figure 20.19**, the protein SeqA binds to *oriC*, using the high concentration of hemimethylated adenines as recognition sites. Two SeqA tetramers bind to two different hemimethylated sites that are no more than about 30 bp away from each other. The SeqA–*oriC* complex becomes sequestered at the cell membrane, thus preventing DnaA protein from binding to the 9-bp repeats and initiating another round of replication. Eventually the SeqA–*oriC* complex is released from the membrane, SeqA dissociates from *oriC*, and Dam methylase methylates the *oriC* portion of the newly formed strand. The structure of a SeqA monomer bound to a DNA sequence containing an N^6-methyladenine residue within one of the two GATC sequences (hemimethylated) is shown in **Figure 20.20**. Structural analysis

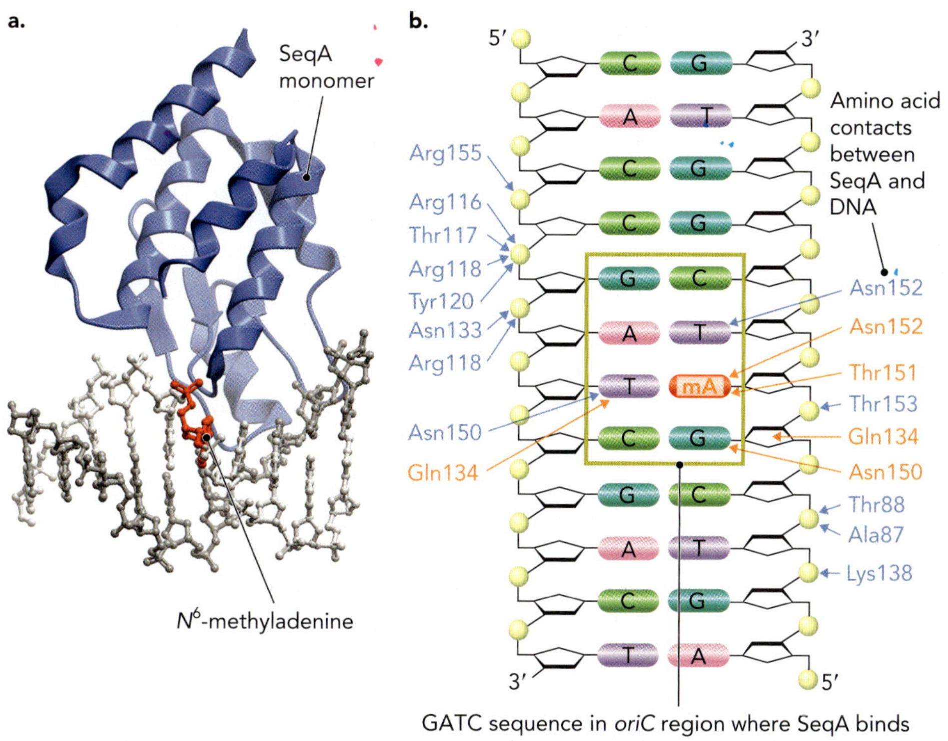

Figure 20.20 The structure of SeqA is shown bound to hemimethylated DNA containing an N^6-methyladenine residue (mA) within the GATC binding site. **a.** SeqA monomer bound to a double-stranded DNA segment containing the sequence 5′-AGTCCG(mA)TCGG-3′. BASED ON PDB FILE 1LRR. **b.** Contact sites between the SeqA protein and its DNA substrate based on structural analysis of the protein–DNA complex shown in panel **a.** Amino acids shown in blue make hydrogen bond contacts, and those shown in orange make van der Waals contacts. BASED ON DATA DESCRIBED IN A. GUARNE ET AL. (2002). INSIGHTS INTO NEGATIVE MODULATION OF *E. COLI* REPLICATION INITIATION FROM THE STRUCTURE OF SEQA-METHYLATED DNA COMPLEX. *NATURE STRUCTURAL BIOLOGY*, *11*, 839–843.

revealed that SeqA amino acid residues Thr151 and Asn152 make specific contacts with the N^6-methyladenine base, with additional base-specific contacts between the SeqA protein and nucleotide bases on both strands of the hemimethylated DNA.

Termination Proteins Resolve Colliding DNA Replication Forks in the *E. coli* Genome DNA synthesis in *E. coli* is initiated at *oriC* and proceeds bidirectionally until the DNA replication forks reach each other halfway around the genome at a sequence we call the termination region. As shown in **Figure 20.21**, resolution of

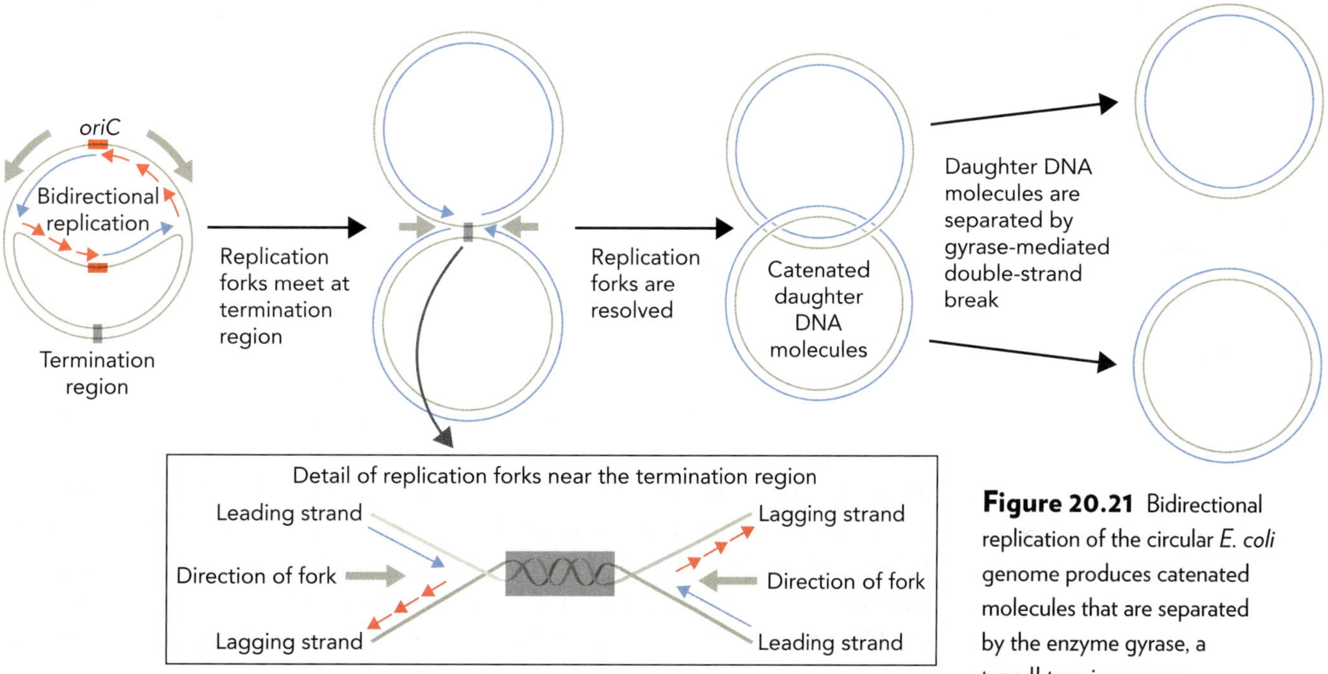

Figure 20.21 Bidirectional replication of the circular *E. coli* genome produces catenated molecules that are separated by the enzyme gyrase, a type II topoisomerase.

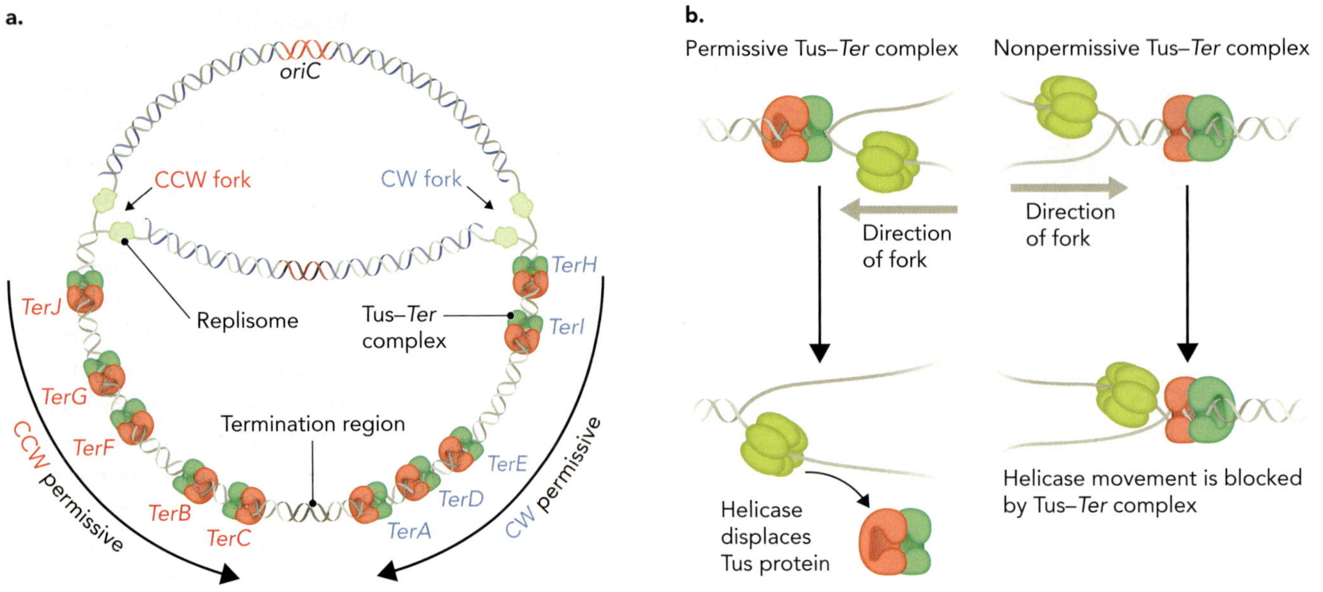

Figure 20.22 Function of Tus–*Ter* complexes in regulating replication termination in *E. coli*. **a.** The *E. coli* genome encodes 10 *Ter* sequences (*TerA*–*TerJ*), five of which are permissive for counterclockwise (CCW) replication fork movement, and five that are permissive for clockwise (CW) replication fork movement. Binding of the Tus protein to the *Ter* sequence is directional such that one side of the Tus protein is permissive for replication fork movement (green), whereas the opposite side of the Tus protein is not (red). **b.** When a replication fork encounters a permissive Tus–*Ter* complex (green side of Tus), the Tus protein dissociates from DNA and the replication fork continues. But when a replication fork encounters a nonpermissive Tus–*Ter* complex (red side of Tus), it is halted and cannot continue.

the two opposing replication forks produces two daughter DNA molecules, each containing one old strand of DNA and one nascent DNA strand. Because of the way the replication forks are resolved, the two DNA molecules are catenated as interlocking rings, in the same way two links in a chain are catenated. The final step in *E. coli* DNA synthesis is the separation of the two daughter molecules by the action of a type II topoisomerase (gyrase), which breaks and seals double-stranded DNA.

How is termination of replication controlled in *E. coli*, and does this always occur in the termination region? For example, if one fork is moving at a slower rate than the other or has stalled due to encountering DNA damage, then termination might occur at a random place on the chromosome. This issue is solved by having 10 copies of a 23-bp termination sequence called *Ter* (terminator), which are located opposite of *oriC* on the *E. coli* chromosome. As shown in **Figure 20.22**, these 10 sites are referred to as *TerA* through *TerJ* and serve as permissive or nonpermissive sites for passage of the replication fork, depending on the direction of movement. The mechanism by which *Ter* sites regulate termination of DNA replication is through the binding of Tus (terminus utilization substance) protein to *Ter* sequences. When helicase encounters a Tus–*Ter* complex oriented in the proper way to halt the replication fork, the helicase stops. This is made possible by "plugging" the opening of the helicase such that it cannot move any further along the DNA strand. Helicase

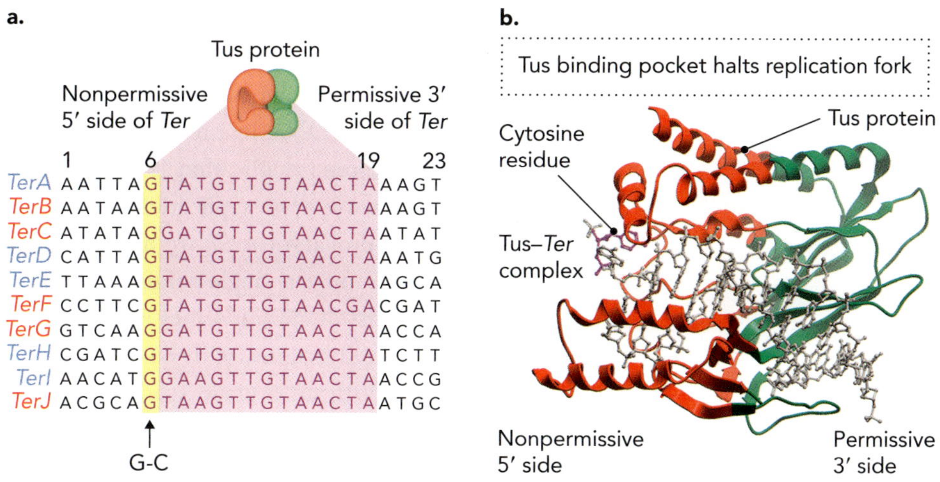

Figure 20.23 A conserved G-C base pair in each of the *Ter* sequences plays a functional role in determining whether the Tus–*Ter* complex is permissive or nonpermissive for replication fork movement. **a.** DNA sequences written 5′ to 3′ for the 10 *Ter* sequences encoded in the *E. coli* genome. The invariant G-C base pair at position 6 of the 23-bp *Ter* sequence is highlighted. **b.** Structure of the nonpermissive *E. coli* Tus–*Ter* complex, showing the conserved cytosine of the *Ter* sequence bound in a pocket on the Tus protein. BASED ON PDB FILE 2EWJ.

dissociates, and the remainder of the replisome soon dissociates as well, once DNA polymerase has finished synthesizing new DNA up to the Tus–*Ter* complex.

As shown in **Figure 20.23**, a conserved G-C base pair found in all *Ter* sequences plays a key role in blocking movement of the helicase. When Tus protein binds to a *Ter* sequence, the cytosine residue is flipped out of its normal base pair position and is buried in the Tus binding pocket. Replication forks moving in the nonpermissive direction push this cytosine residue farther into the binding pocket, thus increasing the affinity of the Tus–*Ter* interaction and halting the replication fork. In contrast, when the replication fork is moving in the permissive direction, the cytosine residue is extracted from the Tus binding pocket, thus allowing the replication fork to continue.

Initiation of DNA Replication in Eukaryotes Is Regulated by Timing of the Cell Cycle Many of the differences between prokaryotic and eukaryotic replication arise from the larger size of eukaryotic chromosomes, the packaging of eukaryotic DNA into nucleosomes (see Figure 3.28), and the slower rate of eukaryotic replication (**Table 20.5**). The DNA synthesis rate in eukaryotes is about 20-fold slower than in prokaryotes, as measured by both the number of nucleotides incorporated per second (~50 nucleotides/second in human cells compared to ~1,000 nucleotides/second in

Table 20.5 COMPARISON OF PROKARYOTIC AND EUKARYOTIC REPLICATION PROCESSES

DNA replication parameter	Bacteria	Human
Genome size (base pairs)	4.6×10^6	3.3×10^9
Number of chromosomes	1 chromosome	23 chromosome pairs
DNA is packaged into nucleosomes	No	Yes
Time to complete one round of genome duplication	~40 minutes	~8 hours
Number of replication origins	1	~10,000
Nucleotide polymerization rate (nucleotides/second)	~1,000	~50
Rate of replication fork movement (μm/min)	~20	~0.5–2

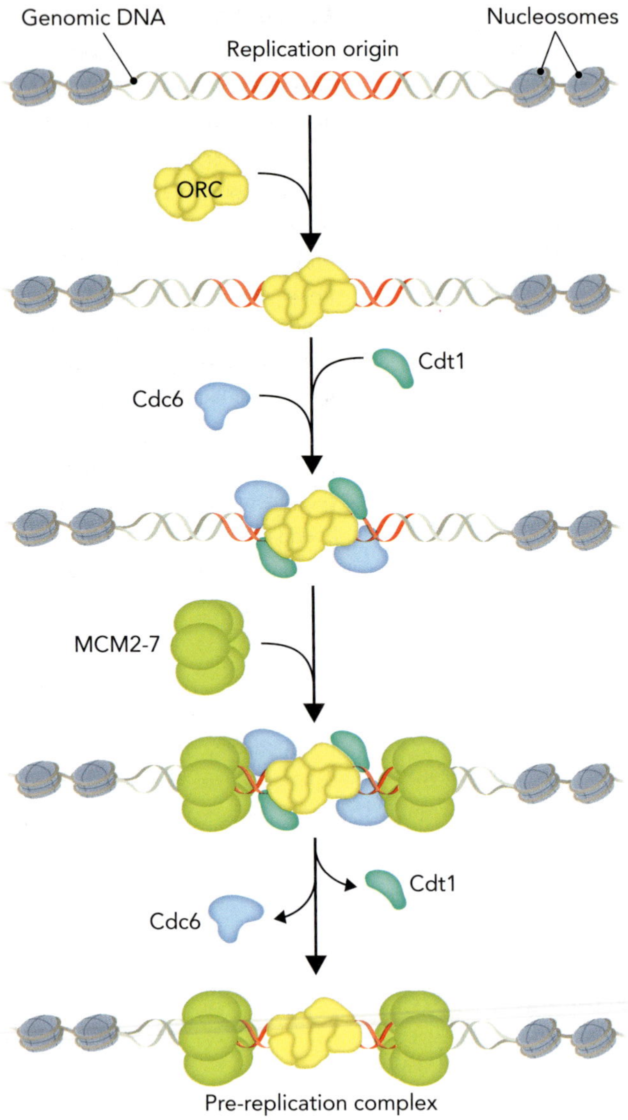

Figure 20.24 Binding of Cdc6, Cdt1, and two MCM2-7 helicases to the ORC forms the pre-replication complex (pre-RC). Each pre-RC is formed during the G1 phase, but activation only occurs at specific sites during the S phase. The pre-RC converts to a replisome progression complex (RPC) after the binding of additional proteins at the two replication forks.

bacterial cells) and the rate of replication fork progression (~0.5–20 μm/min in human cells compared to 20 μm/min in bacterial cells). The challenge of moving replication forks through nucleosomal DNA in a reasonable amount of time to complete genome duplication is overcome in eukaryotes by having many more replication origins per kilobase of DNA than prokaryotes. As we have seen, the *E. coli* genome encodes only a single origin of replication, whereas the 23 pairs of chromosomes in the human genome contain ~10,000 origins, with some estimates as high as ~30,000. With an average space of 30–300 kilobases (kb) between each origin, a typical replisome must replicate 15–150 kb, as opposed to the 2.3 million bp replicated by each *E. coli* replisome.

Replication in an *E. coli* cell often starts about halfway through the previous replication cycle, whereas eukaryotic replication does not occur nearly as often. Initiation of eukaryotic replication is a highly regulated event that occurs within the S phase of the cell cycle. In a typical human cell, the S phase lasts 6 to 8 hours, and although that might seem like more than enough time for a replisome to replicate even the larger 150-kb regions, not all origins are active at the same time. Some origins are replicated early on in S phase, while others are replicated later. Generally, areas of chromosomes that are transcriptionally active tend to be replicated early in S phase. This is highly cell-dependent, as genes that are transcribed in a liver cell are different from those transcribed in a muscle cell.

The key elements required for initiation of eukaryotic replication are the origin recognition complex (ORC) and MCM2-7 protein (**Figure 20.24**). The ORC consists of six protein subunits and plays a role very similar to that of DnaA in that it loads the MCM2-7 helicase onto the chromosome, a process known as DNA to replication licensing. The binding of ORC to DNA is followed by the recruitment of the licensing factors Cdc6 and Cdt1. Cdt1 enters, either by itself or already bound to MCM2-7, and associates with ORC. Closing of the helicase occurs with ATP hydrolysis by Cdc6, followed by dissociation of Ctd1 and Cdc6 from the complex. The complex of ORC and MCM2-7 bound at an origin sequence is referred to as the pre-replication complex (pre-RC). The pre-RC is formed early in the G1 phase but is not activated until S phase, when it converts to a replisome progression complex (RPC). Most likely, far more pre-RCs form than are actually used, as some might be removed prior to use due to transcription and others might not be used and only be activated in the event of a stalled replication fork (**Figure 20.25**).

The conversion of a pre-RC to an RPC at an active replication fork occurs at the transition of the G1 phase to the S phase of the cell cycle. Replication fork activation involves a variety of proteins, the most important of which are **cyclin-dependent protein kinases (CDKs)**, which are activated by regulatory proteins called **cyclins**.

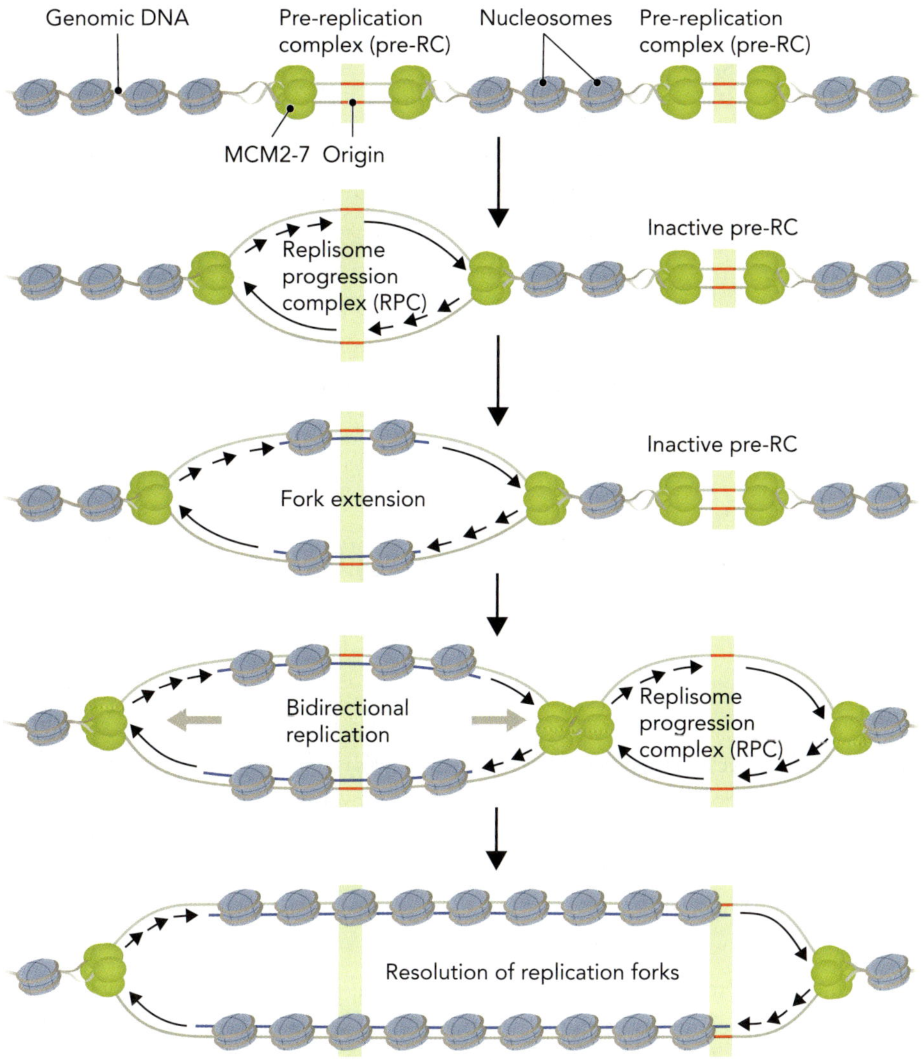

Figure 20.25 Eukaryotic DNA replication begins with formation of a pre-replication complex (pre-RC) and activation of bidirectional replication. Some pre-RCs are converted to a replisome progression complex (RPC) in S phase of the cell cycle, while others are converted later in the cell cycle or even disassembled before ever being converted. Replication bubbles in eukaryotic DNA eventually fuse to complete the replication process, which requires resolution of opposing replication forks. The displacement of nucleosomes in front of the replication fork and reorganization of nucleosomes behind the replication fork contribute to slow replication rates in eukaryotic cells. Additional proteins required for fork movement and DNA synthesis are not included in this figure.

A variety of CDKs exist, each activated by a different cyclin, and specific cyclins are produced at certain points within the cell cycle. For example, cyclin E, which activates CDK2, is largely present late in G1 phase and early in S phase (**Figure 20.26**). CDKs are thought to prevent reinitiation from occurring in S, G2, and M phases. Because of proteolytic destruction of cyclins, CDKs generally lack activity early in G1, allowing the ORC to localize at origins and recruit MCM2-7. During S phase, the increase in CDK activity blocks further production of pre-RCs while promoting the conversion of existing pre-RCs to RPCs. Although a variety of control mechanisms are involved in the initiation process, the cyclin–CDK mechanism is probably one of the most well studied and could offer significant therapeutic potential for diseases that are linked to cell cycle issues.

Telomerase Enzyme Resolves DNA Replication at the Ends of Eukaryotic Chromosomes The linear chromosomes of eukaryotes present a special challenge for replication at the ends of the chromosomes. Recall that primers are needed to begin DNA synthesis; however, at the end of a linear chromosome, there is no template on which to produce the primer. Thus, the ends of the chromosome would not be replicated, and the eukaryotic chromosomes would potentially be shortened each time

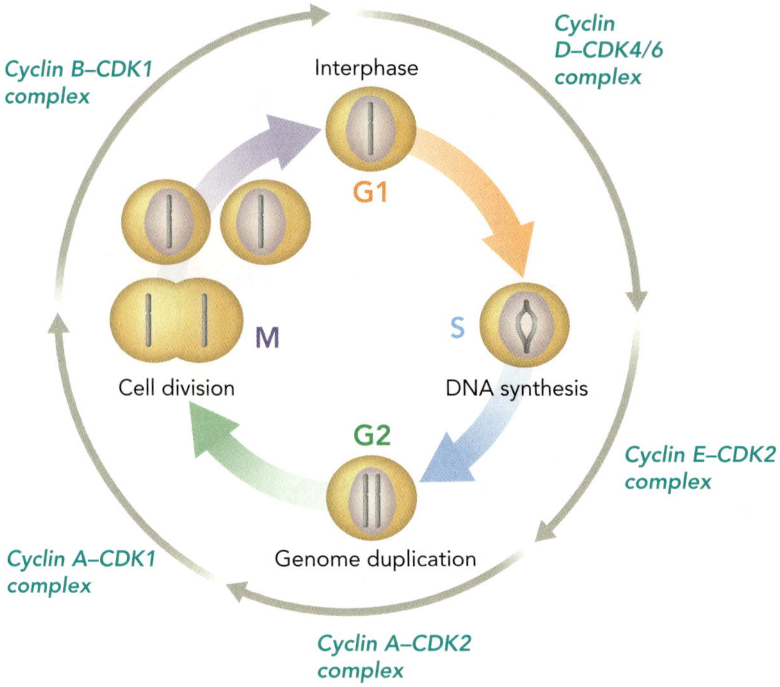

Figure 20.26 Cyclin-dependent protein kinases (CDKs) control eukaryotic DNA initiation events. CDK activation is required in multiple steps for conversion of the pre-RC to an active replication fork at RPCs. The presence of different cyclins at different times during the cell cycle controls initiation of DNA synthesis at origins.

DNA underwent replication. To solve this problem, the eukaryotic chromosome has structures called telomeres at the ends of each chromosome (see Figure 3.35). Telomeres contain tandem repeats that vary among species, but all contain a high percentage of G residues. For example, in the yeast *Saccharomyces cerevisiae*, the telomere contains between 60 and 100 copies of the sequence $TG_{2\text{-}3}(TG)_{1\text{-}6}$, whereas humans have as many as 8,000 repeats of TTAGGG.

The addition of telomeric DNA sequences to the ends of chromosomal DNA is accomplished by the enzyme **telomerase**. This ribonucleoprotein acts as a reverse transcriptase and synthesizes telomeric DNA from an RNA template. Human telomerase contains a 451-nucleotide RNA molecule with a complementary sequence that forms base pairs with the telomere repeat sequence. As shown in **Figure 20.27**, telomerase binds to the 3′ end of the DNA strand by base pairing of the DNA with the telomerase RNA, followed by reverse transcriptase action to synthesize new telomeric DNA in a 5′→3′ direction. Telomerase then repositions, using the newly synthesized telomeric DNA to once again pair bases and thus repeat the process. An RNA primer is synthesized by primase using the newly extended DNA strand as a template, and then DNA polymerase extends the RNA primer to fill in the missing DNA. The catalytic subunit of an insect telomerase bound to the template region of RNA and the complementary DNA is shown in **Figure 20.28a** (p. 1026). The telomerase enzyme and the mechanism of telomere extension werediscovered by Elizabeth Blackburn and Carol Grieder in 1984 when Greider was a graduate student in Blackburn's lab at the University of California, Berkeley. Blackburn and

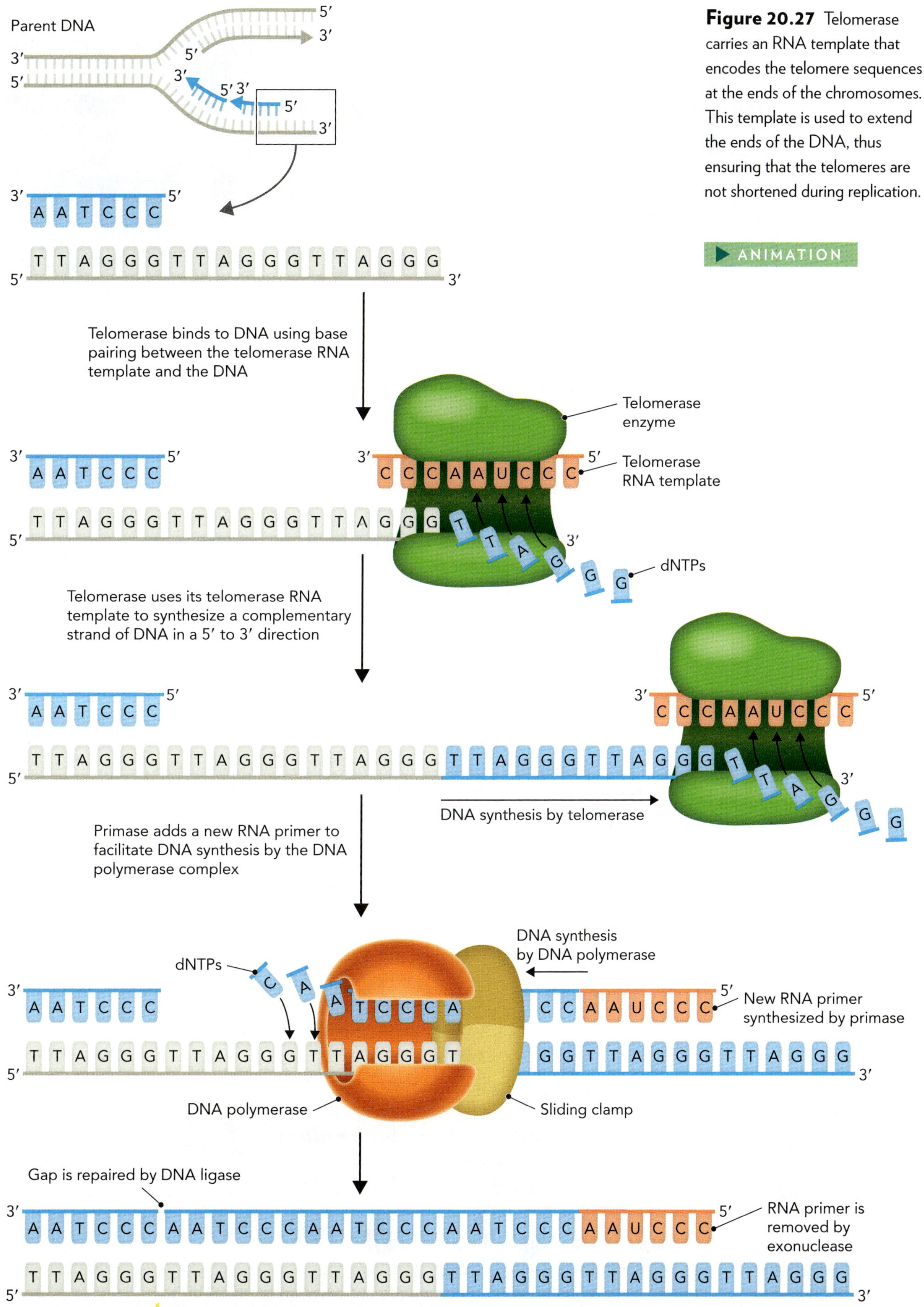

Figure 20.27 Telomerase carries an RNA template that encodes the telomere sequences at the ends of the chromosomes. This template is used to extend the ends of the DNA, thus ensuring that the telomeres are not shortened during replication.

ANIMATION

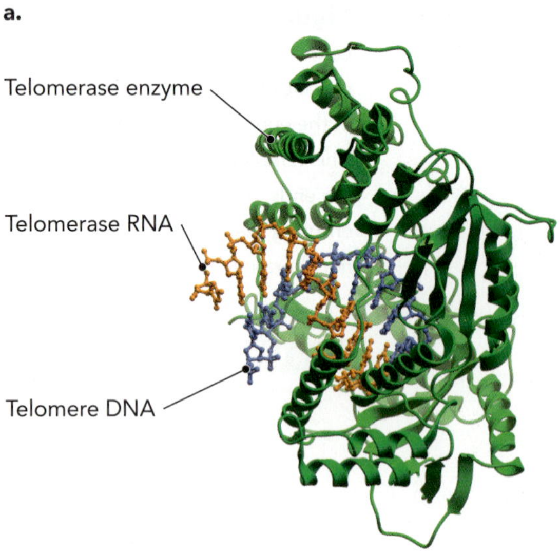

Figure 20.28 Telomerase adds DNA to the ends of eukaryotic chromosomes. **a.** Structure of the telomerase catalytic subunit from the beetle *Tribolium castaneum* bound to the RNA template and complementary DNA. The majority of the telomerase RNA component is not present in this structure. BASED ON PDB FILE 3KYL. **b.** Elizabeth Blackburn and Carol Greider discovered the enzyme telomerase in 1984 and were awarded a share of the 2009 Nobel Prize in Physiology or Medicine. E. BLACKBURN: ELISABETH FALL/FALLFOTO.COM/SIPA PRESS/NEWSCOM; C. GREIDER: COURTESY JOHNS HOPKINS UNIVERSITY.

Greider were awarded a share of the 2009 Nobel Prize in Physiology or Medicine for their discovery of telomerase (**Figure 20.28b**).

Telomerase activity is largely limited within human cells to undifferentiated embryonic stem cells, male germ cells, and activated lymphocytes. The lack of telomerase activity in differentiated cells controls the number of times a cell is able to divide, because without telomerase, the length of the telomeres decreases with each division (**Figure 20.29**). Telomere shortening is a major factor in aging, as cells cease to divide (referred to as cell senescence) when a minimum telomere size is reached. Elevated telomerase activity can result in the immortalization of cells, thus giving rise to tumorigenesis, as evidenced by the approximately 90% presence of telomerase activity in tumor samples. Although telomerase by itself does not make a cell cancerous, division

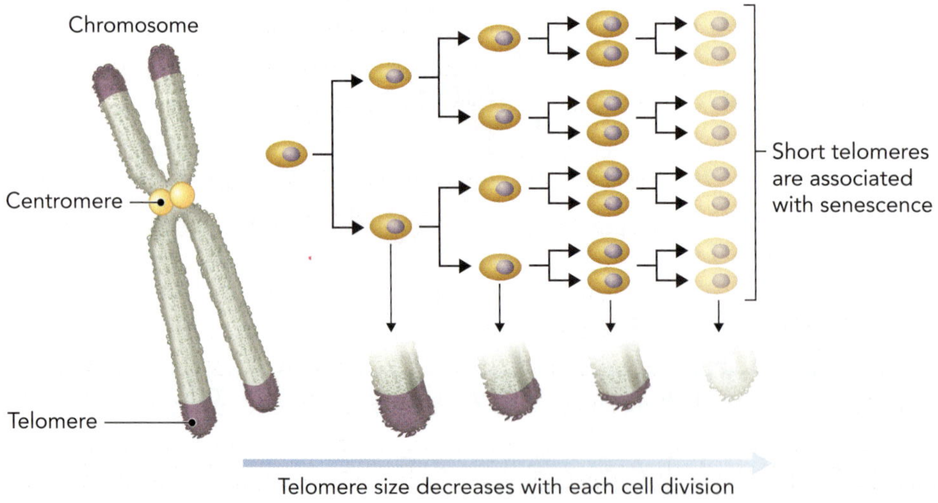

Figure 20.29 Decreased levels of telomerase are associated with telomere shortening. Once the telomeres have reached a minimum size, the cells stop dividing and become senescent. Telomere shortening is thought to play a role in aging.

of cells that have oncogenic mutations does lead to cancer. The combination of oncogenic mutation(s) with telomerase activity allows cells that have become cancerous to divide without restraint, a hallmark of cancerous growth.

concept integration 20.1

Why does it make sense that a protein complex carries out DNA replication rather than utilizing individual proteins that do not associate with each other?

Replicating DNA with a protein complex rather than individual proteins increases the efficiency of the duplication process. Because the polymerases that replicate the leading and lagging strand templates are associated with each other, these strands are produced at roughly the same time. Association of the helicase and primase with the polymerases also means that these processes are coordinated with replication.

20.2 DNA Damage and Repair

Most biological molecules, such as proteins and membrane lipids, can be completely replaced when damaged. But it is not possible for an organism to replace all of its DNA, and therefore chemical damage to DNA must be repaired immediately. Sometimes, the amount of DNA damage can overwhelm the repair capacity of a cell, resulting in conditions such as cancerous growth or cell death. In this section, we look at the cellular effects of DNA damage, the specific types of DNA damage that have been characterized, and the mechanisms that cells have evolved to repair DNA damage.

Unrepaired DNA Damage Leads to Genetic Mutations

If damage to DNA is beyond the capacity of the cell to repair or the damage is not discovered by the cellular DNA repair systems, then a genetic mutation will occur. Mutations are frequently associated with diseases such as cancer, which is often caused by mutations that occur during the lifetime of an individual and are not inherited (somatic cell mutations). In other cases, mutations occur in the germ-line cells and are inherited by the offspring of the organism. These inherited mutations could have no effect, a negative effect, or, in rare cases, a positive effect through the acquisition of an enhanced trait. Indeed, the evolutionary process is a result of DNA mutations that accumulate in the gene pool of a species over a long period of time.

Overview of DNA Damage and the Consequences The sequence of nucleotides within a protein-coding gene specifies the amino acid sequence of the protein product. Changing a single base within a gene can result in a change of the resulting protein. Mutations can occur anywhere in the DNA sequence and can affect gene function. DNA mutations can alter the protein-coding sequence, the fidelity of RNA splicing reactions, or regulation of gene transcription. **Figure 20.30** summarizes the major types of DNA damage described in Section 20.2.

One of the most common DNA mutations is substitution of one base for another during DNA replication. The deletion or insertion of a base can also occur, resulting in

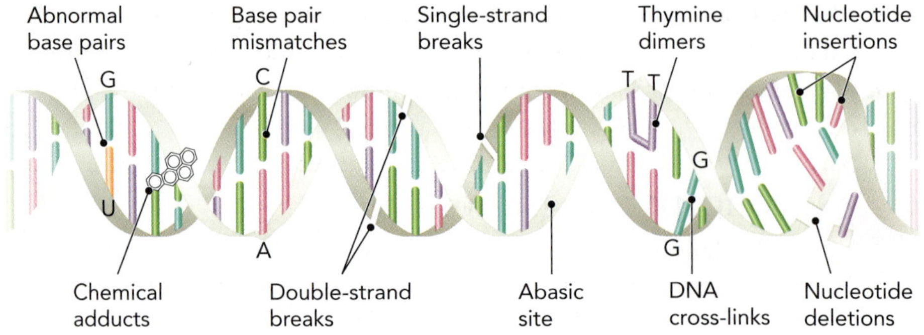

Figure 20.30 There are 10 major types of DNA damage in eukaryotic cells.

a shortened or lengthened nucleotide sequence. In all organisms, a substitution mutation is far more prevalent than a combination of insertions and deletions.

Other mutations can occur as a result of chemical alteration to nucleotide bases, resulting in a significant change to the DNA structure. These mutations are most likely the result of exposure to certain chemicals or various types of radiation. When such events occur immediately before replication, these types of damage can lead to a stalled replication fork. In bacteria, this generally leads to an event known as the SOS response, which can result in even greater amounts of mutation (see Chapter 23). Here, we examine the various kinds of DNA damage encountered by most cells and the various biochemical mechanisms that have evolved to repair the damage.

Most Cancers Are Caused by Somatic DNA Mutations Although mutations are the basis for the diversity of life on our planet, they are often associated with life-threatening events. **Somatic mutations** are those that occur after zygote formation and thus are not genetically inherited. Somatic mutations that result in decreased cell viability are most likely to be lost because they affect only one cell before it is able to divide and produce daughter cells. Some somatic mutations alter genes that are regulators of cell division, which can lead to uncontrolled cell growth and allow the cell to outcompete those cells around it. Such uncontrolled cell growth—resulting from one or more somatic mutations—is the hallmark of cancer (see Figure 8.42). Indeed, most cancer cells have several early mutations that affect cell division, which can lead to additional mutations that cause cancer cell metastasis (invasive cancer cells that leave the site of origin). As shown in **Figure 20.31** for colon cancer, mutations in both copies of the APC tumor suppressor gene (loss of function) initiate the cancer process

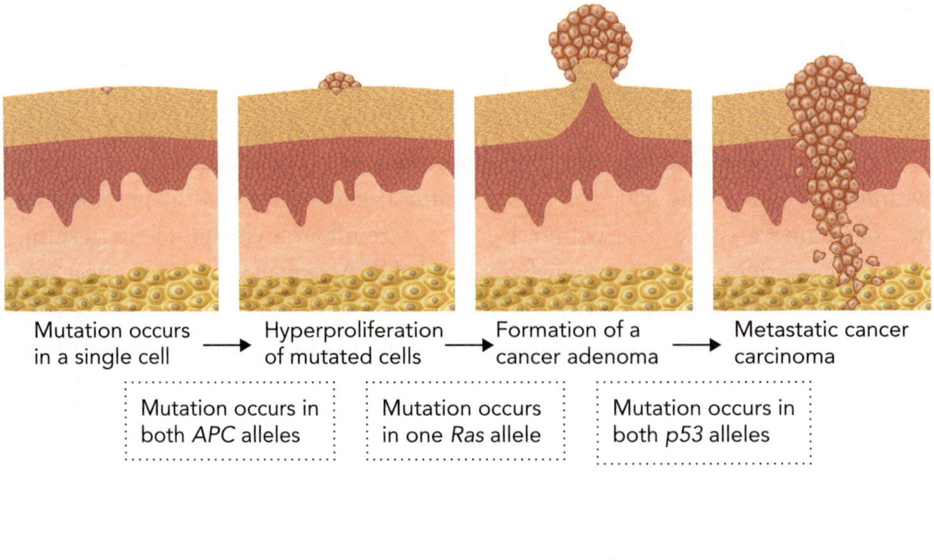

Figure 20.31 Cancer progression is a result of accumulated mutations.

in a single cell. This is followed by a mutation in one of the two Ras oncogenes (gain of function), and, eventually, mutations in both copies of the gene encoding the p53 tumor suppressor (loss of function), resulting in metastatic colon carcinoma. Exposure to certain environmental factors, such as cigarette smoke or UV radiation, can increase the number of somatic mutations, resulting in an increased incidence of cancer.

The ability to determine whether or not a given chemical can cause DNA damage can provide important information about the possible biological toxicity of the compound. Trying to determine exactly what damage might be done to DNA can be somewhat difficult, but testing for DNA damage that leads to mutations is relatively easy. The most widely used test of this nature is the **Ames test**, developed by Bruce Ames in the 1970s. The test determines whether or not a substance is mutagenic to a particular strain of bacteria, *Salmonella typhimurium*, which lacks the ability to produce histidine because of mutations in the genes for histidine synthesis (**Figure 20.32**). The bacteria are first exposed to the agent being tested. If the bacteria are able to grow in a medium that is not supplemented with the amino acid histidine (that is, they have

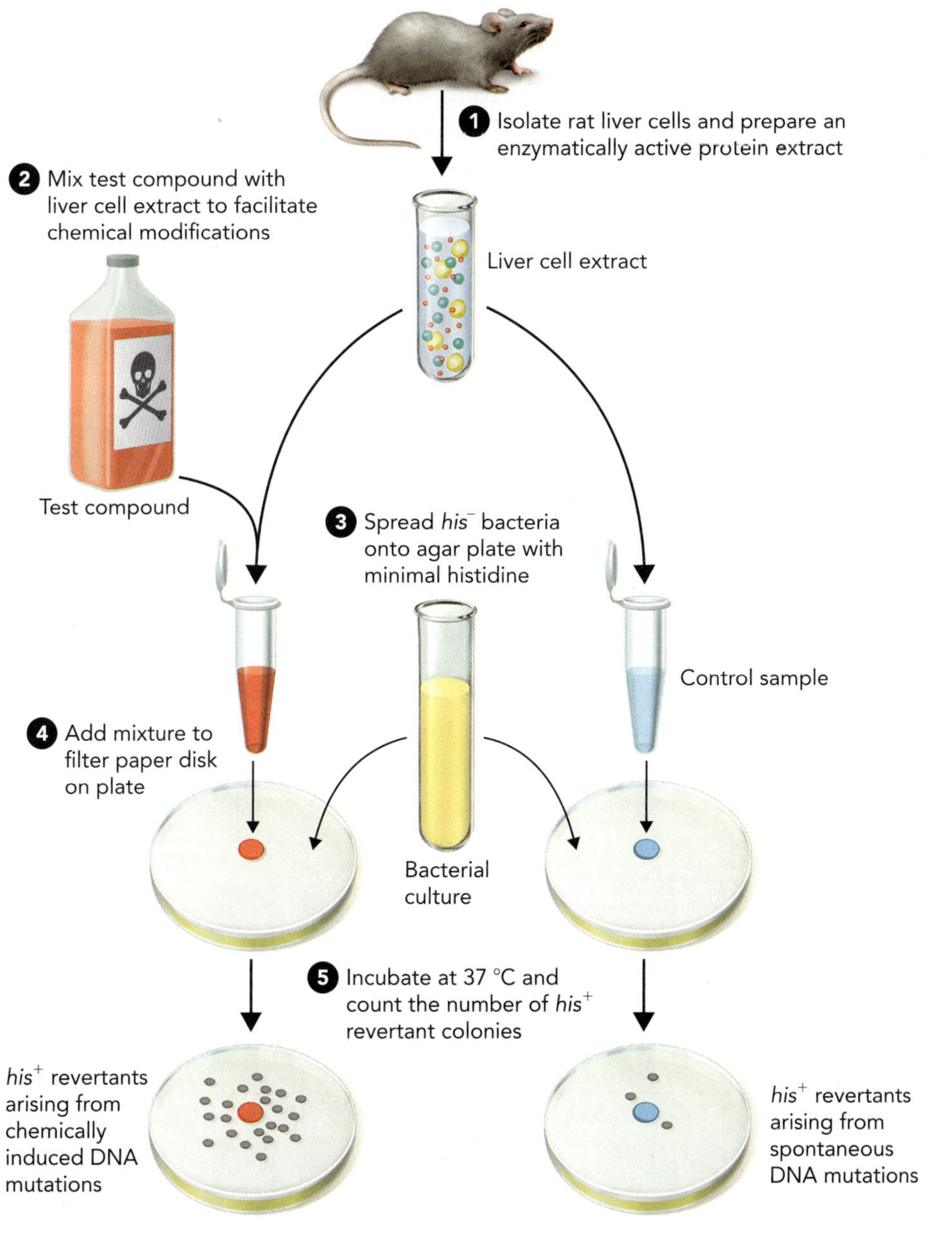
1 Isolate rat liver cells and prepare an enzymatically active protein extract
2 Mix test compound with liver cell extract to facilitate chemical modifications
Liver cell extract
Test compound
3 Spread his⁻ bacteria onto agar plate with minimal histidine
Control sample
4 Add mixture to filter paper disk on plate
Bacterial culture
5 Incubate at 37 °C and count the number of his⁺ revertant colonies
his⁺ revertants arising from chemically induced DNA mutations
his⁺ revertants arising from spontaneous DNA mutations

Figure 20.32 A suspected mutagen is analyzed in the Ames test by incubating it with *his*$^-$ *Salmonella typhimurium* and rat liver extract, which is then added to a medium containing a minimal amount of histidine. If bacteria are capable of growing to the point where colonies form, then the bacteria are producing histidine. This indicates that a back mutation has occurred and the agent is indeed mutagenic. A negative control is used to observe the rate of spontaneous reversion.

regained the ability to produce their own histidine), this indicates a back mutation has occurred as a result of exposure to the agent that allowed the histidine synthesis genes to regain their function.

Some substances are not necessarily mutagens when ingested by an organism but become mutagens when metabolized. For that reason, an extract of rat liver enzymes is often added to the Ames test to simulate metabolism of the substance by the organism. Generally, substances that are mutagens in bacteria have the potential to be carcinogens in humans. Of the chemical agents identified as mutagenic in bacteria by the Ames test during the past four decades, more than 75% have also been found to be carcinogenic in animals. Because of the ease and low cost of the Ames test, it is the most widely used initial test for chemical mutagenicity.

Biological and Chemical Causes of DNA Damage

Types of DNA Damage The major biological damage to DNA comes from substitution mutations arising from replication. The rate of mutation varies greatly from one organism to another, with humans having a relatively high rate of mutation at 10^{-4} to 10^{-6} mutations per haploid genome for a given gene. *E. coli* has a significantly lower replication-induced mutation rate of 10^{-7} to 10^{-8} mutations per cell division. Cellular metabolism can also result in DNA damage, as the production of reactive oxygen species (ROS) is a by-product of several metabolic pathways. ROS damage to DNA is rather ubiquitous, and we discuss its mechanism throughout our investigation of DNA damage.

One of the most common types of DNA damage is the spontaneous deamination of cytosine to uracil (**Figure 20.33**). Because uracil is not present in DNA, it is removed by glycosylase enzymes (see Figure 3.22), which generates an **abasic site**. Generally, the abasic site is repaired by the base excision repair pathway, which we describe later. But if an abasic site is present during DNA replication, it can stall replication. Some DNA polymerases are able to bypass an abasic site, often inserting a cytosine opposite the abasic site. As shown in **Figure 20.34**, if the abasic site is the result of cytosine deamination, the result is a C→G substitution in the template strand after repair of the abasic site. If replication occurs prior to removal of the uracil, adenine is added in the daughter strand instead of guanine, ultimately resulting in a C-G base pair converting to a T-A base pair.

Damage from environmental factors is generally more severe than the substitution mutations of replication. One of the best-characterized DNA damaging agents is UV radiation, resulting largely from exposure to the Sun. The lesions formed by UV damage are often referred to as **photoproducts**. As shown in **Figure 20.35**, the two most prevalent are cyclobutane pyrimidine dimers and (6-4)

Figure 20.33 The spontaneous deamination of cytosine produces uracil, which is cleaved by a glycosylase, resulting in an abasic nucleotide. Cytosine is shown in the tautomeric form in which hydrolytic attack likely takes place.

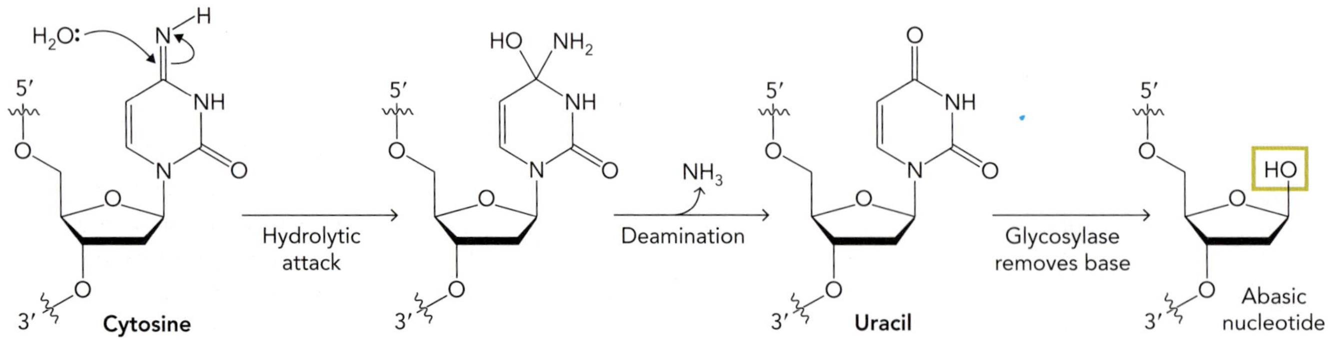

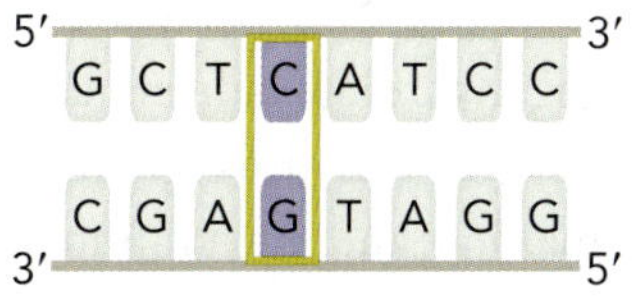

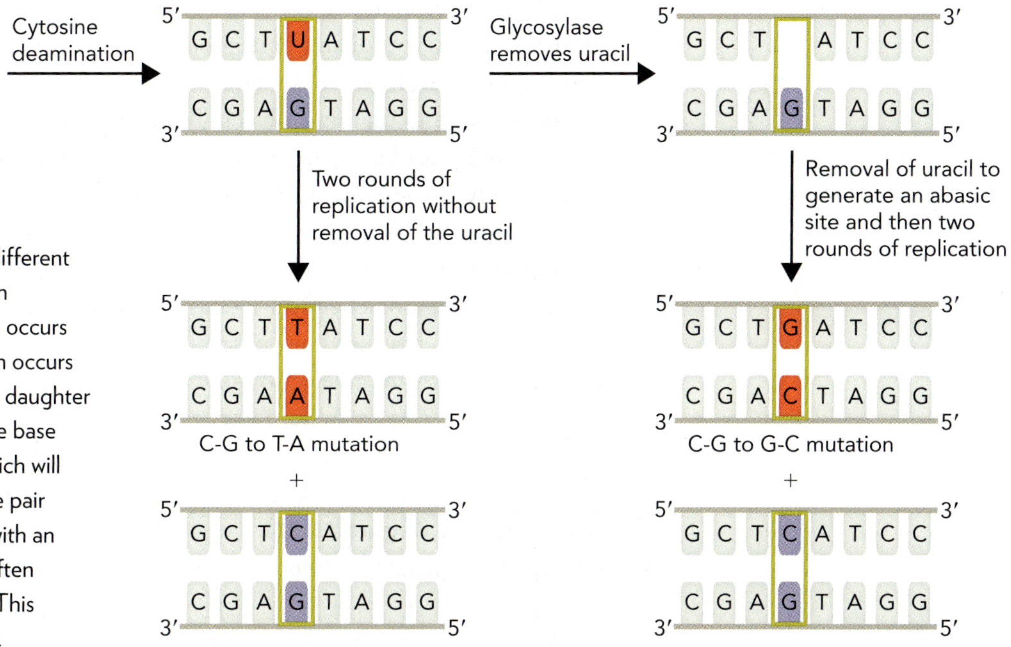

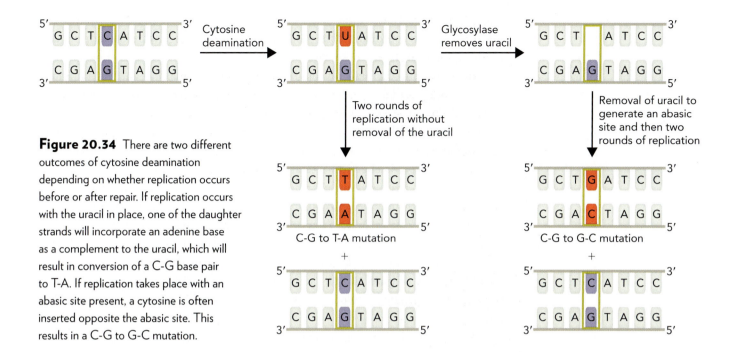

Figure 20.34 There are two different outcomes of cytosine deamination depending on whether replication occurs before or after repair. If replication occurs with the uracil in place, one of the daughter strands will incorporate an adenine base as a complement to the uracil, which will result in conversion of a C-G base pair to T-A. If replication takes place with an abasic site present, a cytosine is often inserted opposite the abasic site. This results in a C-G to G-C mutation.

pyrimidine–pyrimidone photoproducts. Both of these lesions form from adjacent pyrimidines, with two adjacent thymines giving rise to about 55% of the cyclobutane pyrimidine dimers. Adjacent thymines can also form a (6-4) pyrimidine–pyrimidone photoproduct; however, most of these photoproducts are actually formed by adjacent T-C sequences. The (6-4) pyrimidine–pyrimidone photoproducts result in a DNA molecule that does not adopt the proper conformation for replication, resulting in a stalled replication fork and possibly a double-strand break.

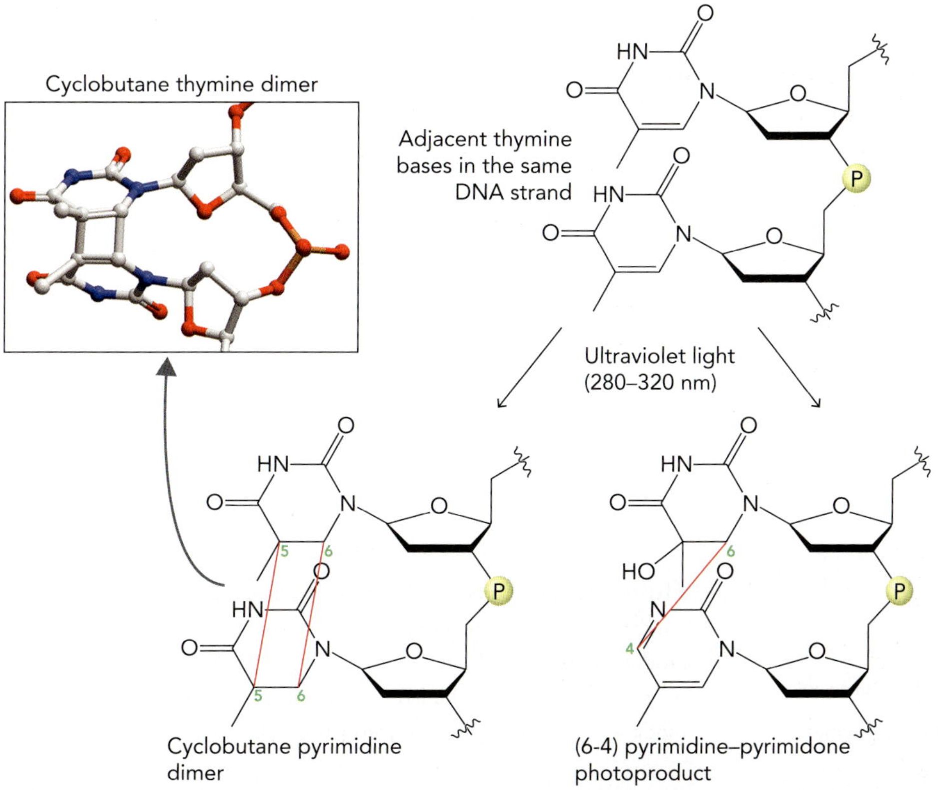

Figure 20.35 Adjacent pyrimidines often produce cyclobutane pyrimidine dimers or (6-4) pyrimidine–pyrimidone photoproducts after UV irradiation. BASED ON PDB FILE 1SM5.

UV light can also react with other molecules within the cell, particularly riboflavin and tryptophan. This produces ROS, including hydroxyl radicals and singlet oxygen. Both of these agents can convert guanine into 8-hydroxyguanine and thymine into thymine glycol. Because 8-hydroxyguanine and its tautomer, 8-oxoguanine, can base pair with adenine, the major result of its production is a G-C to T-A conversion. Thymine glycols often result in either single-strand or double-strand DNA breaks, with the extent of damage dependent upon the presence of other DNA lesions.

Another type of chemical alteration leading to mutations is DNA alkylation (**Figure 20.36a**). Because most of the heteroatoms of the bases are nucleophilic, they react rapidly with many electrophilic alkylating agents. The most common site of alkylation is N-7 of guanine. Alkylation of purines, particularly at N-7 and N-3, yields an unstable molecule that is subject to a deglycosylation reaction, creating an abasic site. Alternatively, alkylation at other sites, particularly O-6 of guanine and O-4 of thymine (**Figure 20.36b**), yields stable adducts with alternative base pairing patterns. The alkylated products can result in mutagenesis during subsequent DNA replication.

Most of the alkylating agents that attack DNA are produced by chemical reactions within cells, often as the result of ROS interactions with membrane lipids. The byproducts of the reaction can be electrophilic alkylating agents such as malondialdehyde and acrolein. These agents can react with a single guanine residue to produce ethenoguanine adducts, which severely alter the normal Watson–Crick base pairing and interfere with DNA synthesis. In addition, two adjacent guanine residues can be cross-linked by reaction of the ethenoguanine adduct with the exocyclic amine of the second guanine.

a.

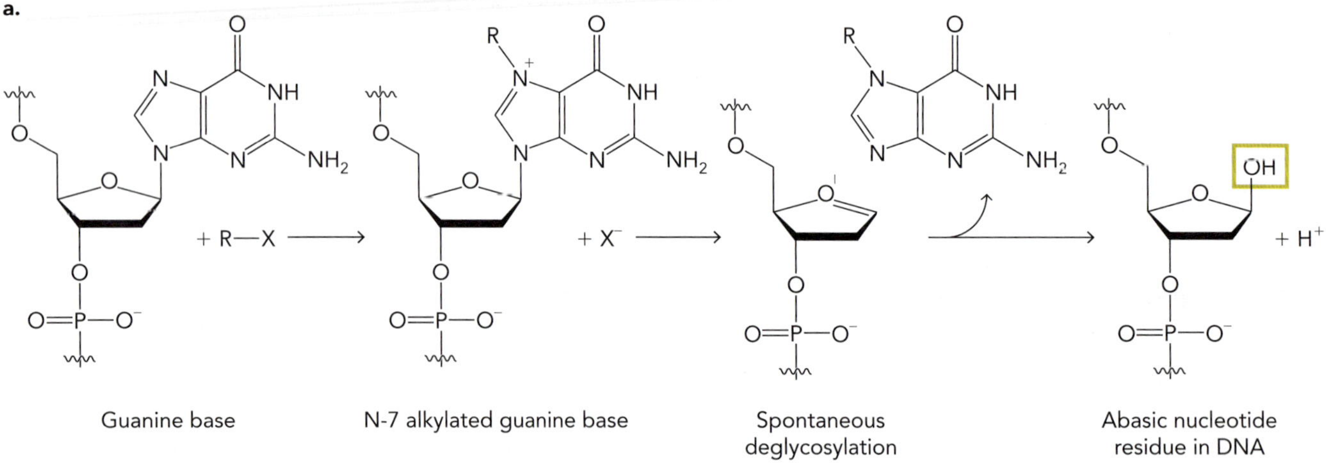

b.

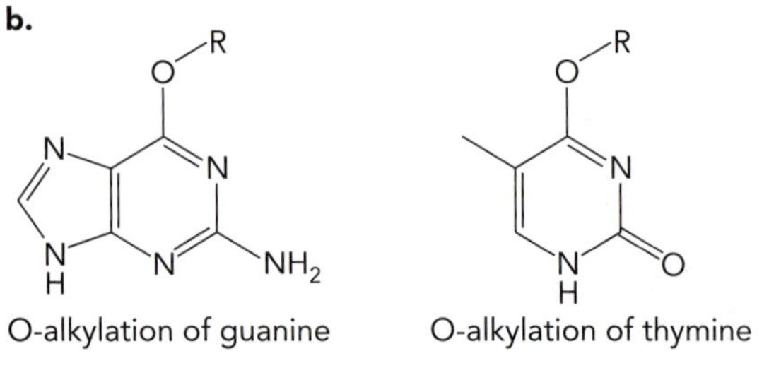

Figure 20.36 Common chemical modifications of DNA bases are shown. **a.** N-7 of guanine is commonly alkylated, with the product subject to spontaneous deglycosylation. **b.** O-alkylation products. O-alkylation of guanine and thymine produce modified bases with different hydrogen-bonding patterns.

This type of lesion changes the structure of DNA, making replication past the lesion difficult or impossible.

Effects of DNA Damage on Protein-Coding Sequences In order for the information stored in a gene to be converted into a protein, it must be converted into RNA. For this conversion to occur, RNA polymerase must be able to bind to single-stranded DNA and move along the template similar to that of DNA polymerase. Any lesion in DNA, such as a pyrimidine dimer or cross-linked guanine, contorts the DNA strand and can prevent RNA polymerase from proceeding through the lesion. In contrast, abasic sites in the DNA template often lead to the insertion of an adenine residue by RNA polymerase. Because any of the four nucleotides could have occupied the abasic site, the chances of the added adenine being the incorrect residue are essentially 75%.

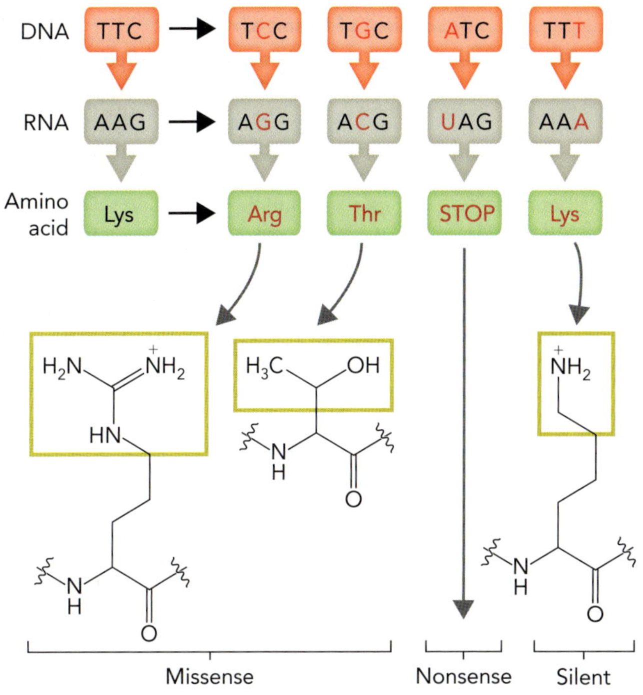

Figure 20.37 Single point mutations in DNA can lead to coding for a different amino acid (missense mutation); premature termination of protein synthesis (nonsense mutation); or no change in the resulting protein (silent mutation).

It takes three nucleotide bases in DNA to specify an amino acid codon in RNA. A DNA mutation in one of the three nucleotide bases in a codon might not result in a change of the amino acid sequence, which is called a silent mutation. As shown in **Figure 20.37**, alternatively a single nucleotide change could result in incorporation of a chemically similar or dissimilar amino acid; such mutations are called missense mutations. If the mutation results in the substitution of the original amino acid with one that is chemically similar, then it is called a conservative missense mutation, whereas if the substituted amino acid is chemically dissimilar, then it is called a nonconservative missense mutation. A nonsense mutation in DNA is one in which the nucleotide change converts an amino acid codon into a stop codon that terminates the protein synthesis process before the true end of the protein is reached (see Chapter 22).

DNA Repair Mechanisms

Although DNA repair mechanisms are largely conserved between prokaryotic and eukaryotic organisms, some differences have been identified. For example, in prokaryotes, the pyrimidine dimers formed as a result of UV damage are repaired by an enzyme called **photolyase**, an enzyme that repairs only one type of DNA damage. In eukaryotes, such dimers (as well as many other types of damage) are repaired by a process called **nucleotide excision repair (NER)**. This process requires at least 18 different protein complexes, which are the products of 30 different genes. Prokaryotes do possess the NER process and use it for other types of damage, but it is much simpler in prokaryotes, requiring only three proteins.

An important consideration in DNA repair mechanisms is that as long as the damage occurs on only one of the DNA strands, the other DNA strand can be used as the template for repair. This assumes that the undamaged strand can be identified and thus used as the repair template. Several of the repair mechanisms that we will discuss have mechanisms for determining which DNA strand needs to be used as the template.

▶ ANIMATION

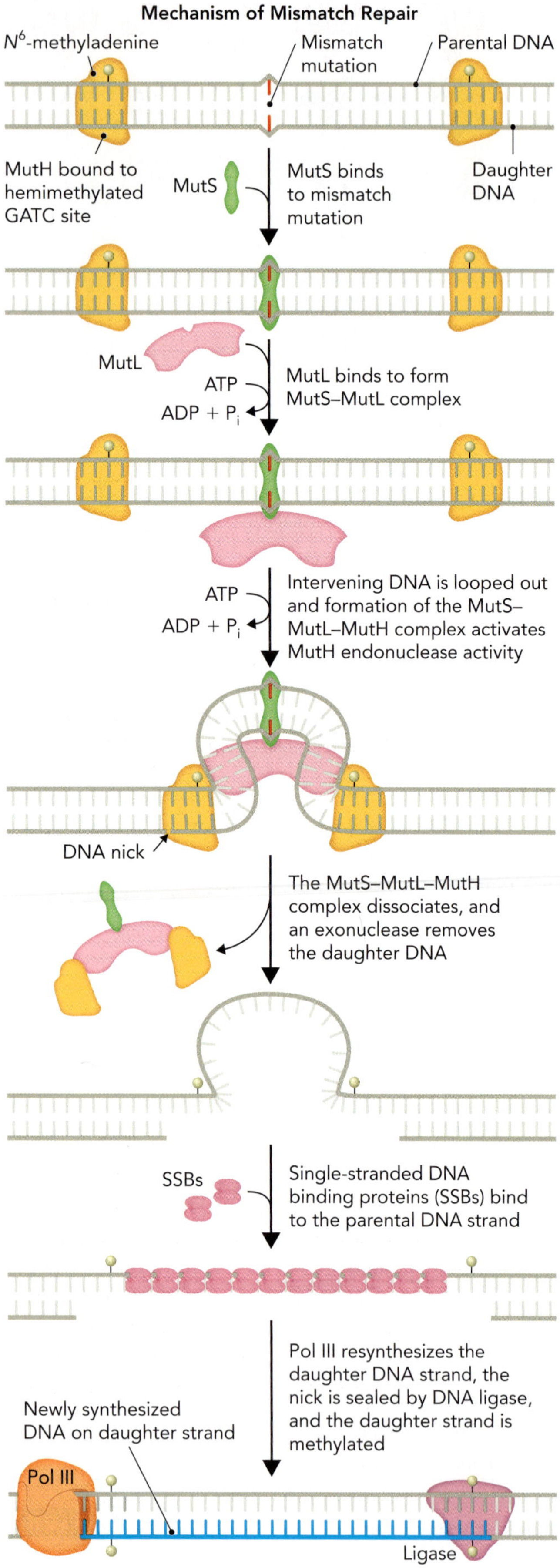

Figure 20.38 The *E. coli* MutS-MutL-MutH protein complex is responsible for mediating mismatch repair.

Mismatch Repair of Replication Errors The typical error rate of a DNA polymerase with proofreading activity is one error per 10^6 to 10^8 bp, indicating that one mismatch per replication is a fairly common occurrence. The error rate for the *E. coli* replication process, however, is approximately one error per 10^9 to 10^{10} bp, indicating that mismatch errors that are not removed by the polymerase proofreading system are fixed by the DNA **mismatch repair** system.

As described earlier, *E. coli* cells use an enzyme called Dam methylase to specifically methylate the N-6 of adenines in the sequence 5′-GATC-3′. Prior to replication, all of these adenines should be methylated, but immediately behind the replication fork, only the original template strands are methylated (see Figure 20.19). This hemimethylated state can be checked for errors because the unmethylated strand is recognized as being newly synthesized. In the case of a mismatch, the repair enzymes can recognize the original template by the presence of methylated adenines.

The process of mismatch repair in *E. coli* is initiated when a homodimer of the mismatch repair protein MutS recognizes and binds adjacent to a base pair mismatch (**Figure 20.38**). A second mismatch repair protein, MutL, forms a complex with MutS through an ATP-dependent association. Once this initiating complex of MutS and MutL has formed, the DNA is threaded through the complex in the direction of an adjacent methylated GATC sequence. The ATP-dependent movement of DNA through the MutS–MutL complex is similar to the β sliding clamp of polymerase III. As the DNA moves through the MutS–MutL complex, it forms a loop that contains the mismatch.

In the next step, the MutS–MutL complex associates with the third mismatch repair protein, MutH, which is bound at methylated GATC sequences in the parental DNA strand. Formation of the heterotrimeric mismatch repair protein complex (MutS–MutL–MutH) activates the endonuclease activity of MutH, which makes a single-strand cut in the unmethylated DNA daughter strand. The intervening DNA daughter strand between the mismatch and the methylated GATC sequence in the parental strand is removed by a helicase and an exonuclease. Depending on how far the nearest GATC site is from the mismatch, the number of nucleotides removed could be fairly large. Single-stranded DNA binding protein (SSB) protects the single-strand region of the parental DNA strand during this process and facilitates the binding of Pol III to resynthesize the DNA daughter strand. In the final step of mismatch repair, ligase seals the nick left behind by Pol III.

The eukaryotic mismatch repair process is more complex than that observed in *E. coli*. Mammals have homologs of both MutS and MutL proteins, and there are a number of eukaryotic MutS homologs (MSHs) of which three—MSH2, MSH3, and MSH6—participate in DNA mismatch repair. The homologs give rise to heterodimeric MutS complexes.

Homologs of MutL in eukaryotes include MutLα and MutLγ. Like the homologs of MutS, MutLα and MutLγ are heterodimers composed of MLH1 (MutL homolog 1) and either PMS2 (post-meiotic segregation protein 2) or MLH3, respectively. As with the MutL function in prokaryotic repair, the eukaryotic MutL heterodimers assist the MutS homologs. Unlike *E. coli* DNA, which is hemimethylated, eukaryotic systems do not have such a marker to differentiate the DNA strands, and the mechanism of differentiation is still unknown.

Base Excision Repair **Base excision repair** is responsible for removal and replacement of individual bases that are damaged by various chemical reactions, including damage by reactive oxygen species. The process is best described as removal of the base by cleavage of the N-glycosyl bond to create an abasic site, followed by removal of the deoxyribose-5′-phosphate, replacement of the gap with the correct nucleotide, and ligation of the nick. As shown in **Figure 20.39**, creation of an abasic site is catalyzed by DNA glycosylase enzymes. A variety of modified bases serve as substrates for DNA glycosylase, including uracil, 8-hydroxyguanine, and 7-methylguanine.

Spontaneous deamination of cytosine produces uracil and is one of the most common initiators of base excision repair. The major function of the *E. coli* uracil DNA glycosylase (UNG) is the removal of misincorporated uracil during DNA replication, which occurs when dUTP is substituted for dTTP. The first step in base excision repair involves an apurinic/apyrimidinic (AP) endonuclease (AP1) that cuts the DNA strand containing the abasic site on the 5′ side of the lesion to generate an abasic deoxyribose phosphate (dRP). One type of base excision repair is called *short-patch* repair, which involves removal and replacement of a single nucleotide in a reaction catalyzed by the enzymes DNA polymerase and DNA lyase. In vertebrates, this is carried out by a single enzyme, DNA polymerase β, which contains

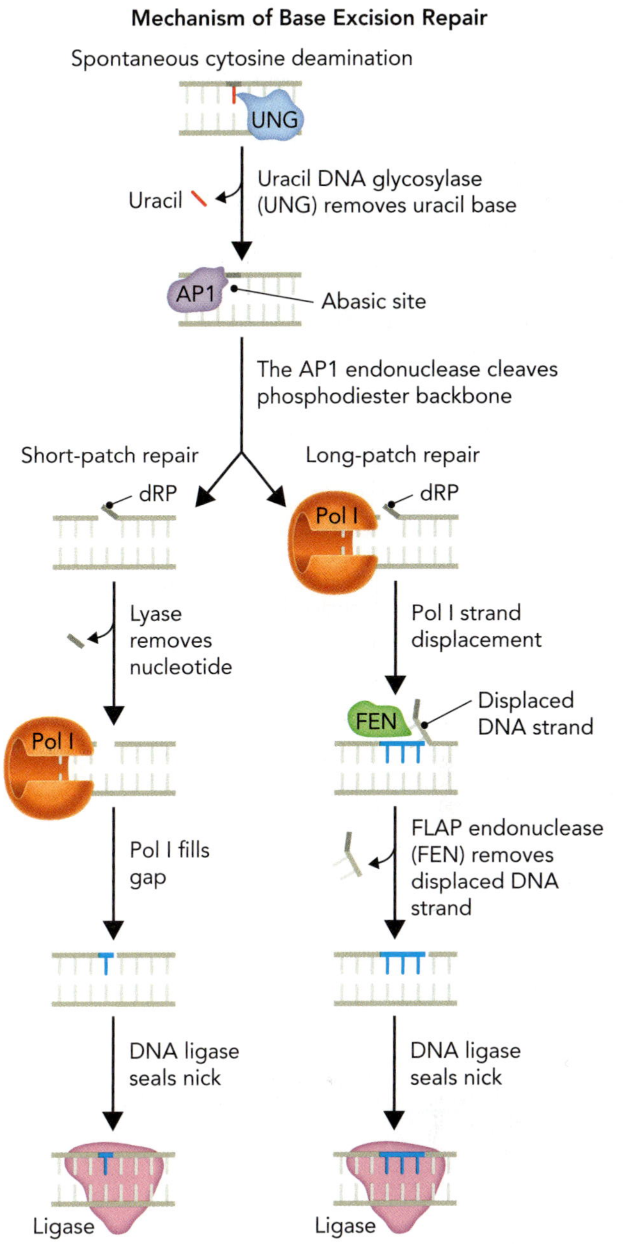

Figure 20.39 Base excision repair is mediated by short-patch and long-patch mechanisms. Short-patch base excision repair removes only one nucleotide, whereas long-patch base excision repair removes up to 10 nucleotides. This example shows creation of the abasic site after uracil DNA glycosylase (UNG)-mediated removal of uracil after cytosine deamination. The API endonuclease cleaves the phosphodiester backbone to generate an abasic deoxyribose phosphate (dRP), which is either removed by the enzyme lyase (short-patch repair) or displaced by Pol I as a result of DNA synthesis (long-patch repair).

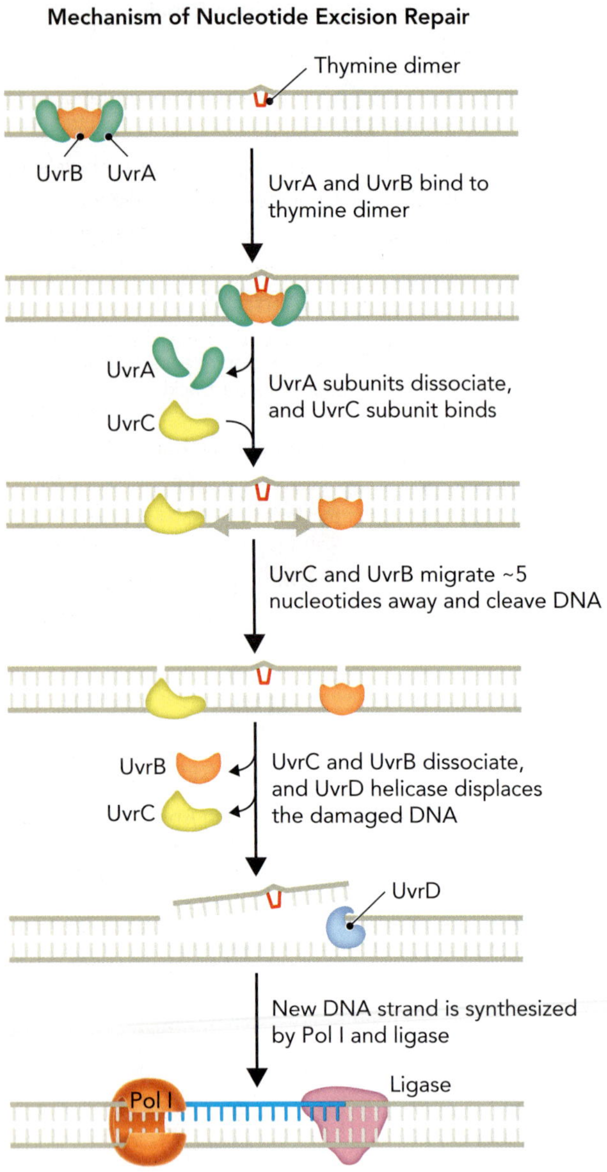

Figure 20.40 Nucleotide excision repair is initiated by recognition of the lesion by UvrAB complex. Nicks in the DNA are created by UvrB and UvrC. UvrD removes the polynucleotide containing the lesion, followed by gap repair using Pol I and ligase.

both the lyase function (removal of an abasic 5′-deoxyribose-5-phosphate) and the polymerase function (replaces the missing nucleotide using the template strand). In both prokaryotes and eukaryotes, DNA ligase seals the nick in the repaired strand to complete the process.

A second type of base excision repair is called *long-patch* repair, which involves the synthesis of up to 10 or more nucleotides using a strand displacement mechanism (see Figure 20.39). An enzyme called flap endonuclease (FEN) removes the displaced strand, leaving behind a single-strand nick that is sealed by DNA ligase. The selection of short-patch versus long-patch repair is not entirely understood, but most likely it is determined by the type of damage that occurred and the specific endonuclease that was used during the repair process.

Nucleotide Excision Repair Nucleotide excision repair is used for large lesions that distort the helical nature of DNA. These lesions include cyclobutane pyrimidine dimers, (6-4) pyrimidine–pyrimidone photoproducts, and other lesions that arise from cross-linking of bases. As shown in **Figure 20.40**, nucleotide excision repair in *E. coli* proceeds in a similar manner to base excision repair, with recognition of the damage, excision of a string of nucleotides surrounding the damage, filling of the resulting gap, and ligation of the nick. Detection of lesions is accomplished by an excinuclease complex that binds a region containing several nucleotides and hydrolyzes phosphodiester bonds on both sides of the lesion. The *E. coli* UvrABC excinuclease scans for errors in DNA, and once the lesion is recognized, UvrA dissociates, leaving UvrB bound to the lesion. Binding of UvrC induces UvrB to cut the DNA backbone four to five nucleotides away on the 3′ side of the lesion. UvrC also cuts the DNA eight nucleotides away on the 5′ side of the lesion. The UvrD helicase then removes the resulting 12- to 13-base section of DNA. Pol I fills the gap left by the damaged DNA, and ligase seals the nick to finish the process.

A similar process exists for nucleotide excision repair in eukaryotes, which is divided into two different types of repair: *global genomic* nucleotide excision repair and *transcription-coupled* nucleotide excision repair. The global genomic nucleotide excision repair is very similar to that observed in *E. coli,* as the protein complex that locates damage scans the chromosomes for lesions. Once a lesion is located, other proteins are recruited to the site, and repair is completed through the same steps as in *E. coli* nucleotide excision repair. Transcription-coupled nucleotide excision repair results in very rapid repair of lesions in transcribed genes by using RNA polymerase to scan the genes for lesions. When RNA polymerase stalls due to a lesion, many of the same proteins involved in global genomic nucleotide excision repair bind to the damaged DNA to initiate the repair process. The human disease xeroderma pigmentosum, a condition characterized by an abnormally high rate of skin cancer resulting from an inability to repair UV-induced lesions, is caused by mutations in proteins required for nucleotide excision repair.

Direct Repair of Damaged DNA Most organisms can remove lesions such as pyrimidine dimers by using nucleotide excision repair or methylated guanine residues by using base excision repair; however, two specialized **direct repair** pathways also exist for eliminating these specific lesions. One is the DNA photolyase system, which uses visible light to reverse the formation of cyclobutane dimers by UV light, and the other is dealkylation of O^6-methylguanine (O^6MeG) by the enzyme O^6-methylguanine-DNA methyltransferase (MGMT).

As shown in **Figure 20.41**, the enzyme DNA photolyase specifically binds to pyrimidine dimers and flips the dimer out of the helix and into the active site of the enzyme. The coenzyme methenyltetrahydrofolate (MTHF) serves as a chromophore, absorbing light in the near-UV to blue spectrum. The excited MTHF transfers energy

Figure 20.41 DNA damage caused by cyclobutane dimers is repaired by the enzyme photolyase using a six-step mechanism. (1) Absorption of light by the folate coenzyme (MTHF). (2) Energy transfer from MTHF* to $FADH^-$ to generate the excited state $^*FADH^-$. (3) Electron transfer from $^*FADH^-$ to the cyclobutane dimer substrate. (4) Radical reaction breaks the first bond. (5) Radical reaction breaks the second bond. (6) An electron is transferred to FADH, and the adjacent pyrimidines are restored to their normal structure.

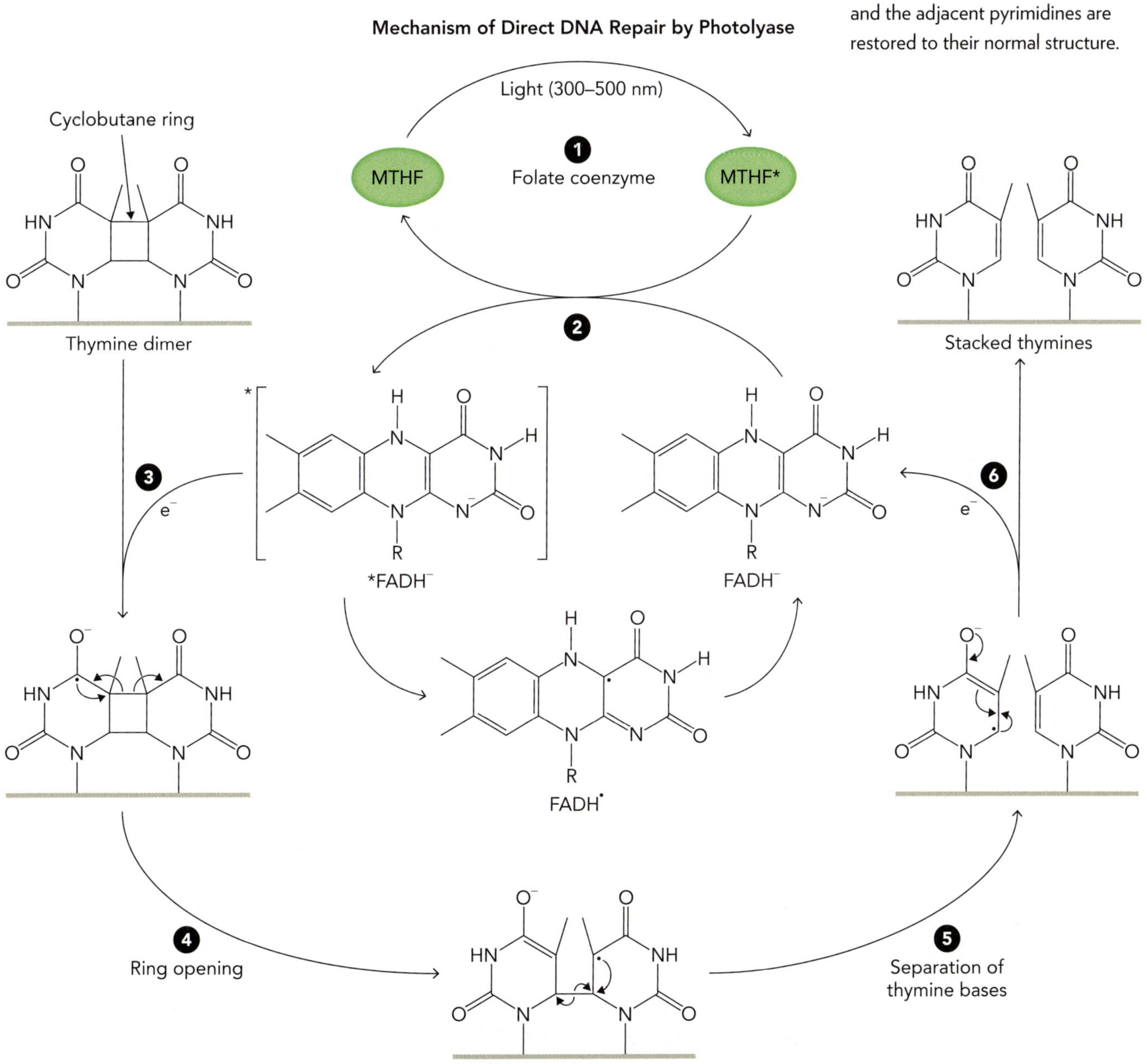

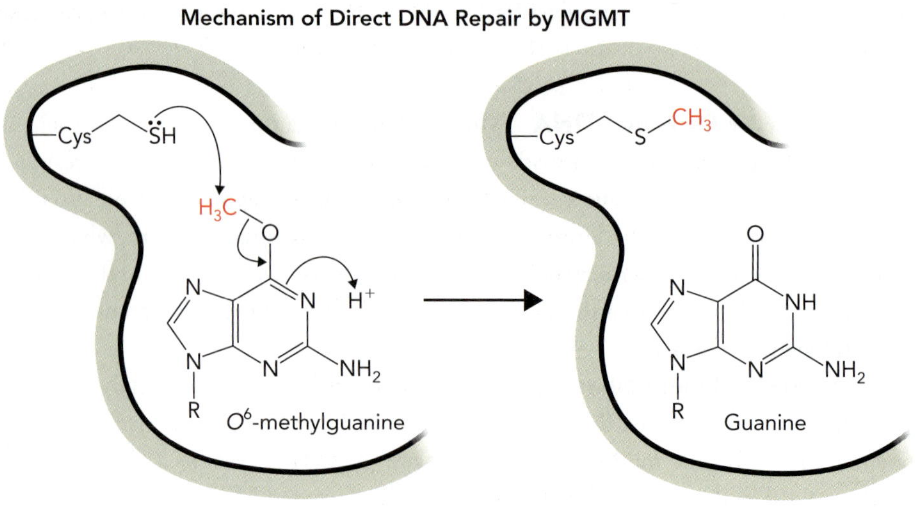

Figure 20.42 The MGMT enzyme uses a suicide mechanism to remove the CH_3 group form O^6-methylguanine. An active site cysteine residue displaces the methyl group to produce a homocysteine residue. This modification inactivates the protein, which is then degraded to metabolize the homocysteine.

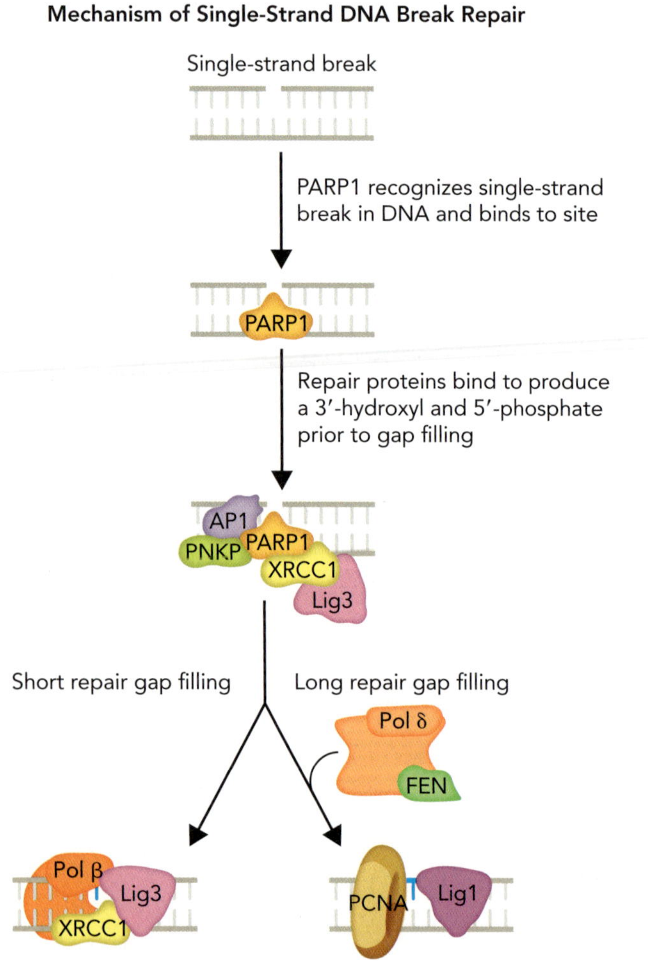

Figure 20.43 The process for repairing a single-strand DNA break in eukaryotes is very similar to base excision repair and nucleotide excision repair mechanisms. Different repair proteins are required for short repair and long repair. The gap remaining after short or long DNA repair is sealed by one of several DNA ligases (Lig1 or Lig3).

to a flavin coenzyme in the $FADH^-$ form, which then transfers an electron to the pyrimidine dimer. A radical-induced rearrangement occurs that breaks the bonds of the cyclobutane ring, with the unpaired electron transferred back to the flavin to re-form $FADH^-$.

Although most alkylation products rely on base excision repair, the MGMT enzyme is able to repair alkylated bases such as O^6MeG directly. MGMT is referred to as a "suicide enzyme" because during the reaction with O^6MeG, it becomes alkylated itself and cannot be regenerated to its active form. As shown in **Figure 20.42**, MGMT uses an active site cysteine to remove the methyl group from the guanine, with the cysteine residue being converted to 5-methylcysteine. The only way to convert the 5-methylcysteine residue back to cysteine is by digestion of the protein and metabolism of the free 5-methylcysteine.

Repair of Single-Strand and Double-Strand DNA Breaks In a typical cell, **single-strand DNA breaks** occur at the rate of tens of thousands per day, largely as a result of reactive oxygen species, alkylating agents, or damage to the deoxyribose. Many single-strand DNA breaks occur as an intermediate during base excision repair, which is a common method for removal of damage from reactive oxygen species and alkylation. When a single-strand DNA break occurs, most often the ends of the DNA must be modified to produce a 3′-hydroxyl and a 5′-phosphate. As shown in **Figure 20.43**, a single-strand DNA break in eukaryotic DNA is located by poly(ADP-ribose) polymerase 1 (PARP1), which signals the X-ray cross-complementing

(XRCC1) scaffold protein to bind to the site. XRCC1 enables a variety of other proteins that are responsible for the repair to bind to the site. This collection of enzymes includes AP endonuclease (AP1) and polynucleotide kinase (PNKP), which are responsible for forming the 3′-hydroxyl and 5′-phosphate. DNA polymerase β fills single nucleotide gaps using short-patch repair, followed by sealing with ligase. If a long-patch repair must be done, then DNA Pol δ and Pol ε and flap endonuclease (FEN) perform long repair.

Double-strand DNA breaks have two major pathways for repair: homologous recombination and nonhomologous end joining (**Figure 20.44**). (Homologous recombination is also used during meiosis and is discussed in Section 20.3 in connection with that process.) **Homologous recombination** can only be used during late S phase or G2 phase, as it requires the presence of an intact sister chromatid as a template for DNA polymerase. A large collection of proteins is required, including a recombinase known as Rad51 and the breast cancer susceptibility proteins BRCA1 and BRCA2. The ends of the DNA are processed by a nuclease complex to give a 3′-hydroxyl and 5′-phosphate. Rad51 binds to the double-strand DNA break as an oligomer; the size of the complex depends on the length of the overhangs created during end processing. Once this presynaptic complex is formed, the complex searches for and binds to the homologous chromosome. This initiates strand invasion, which is assisted by BRCA1 and BRCA2. Strand invasion provides the template for DNA polymerase to repair the missing nucleotides, and resolution of the sister chromatids completes the process.

Both BRCA (breast cancer susceptibility gene) proteins are tumor suppressors and are found in many cells. Mutations to their genes can give rise to a variety of cancers, most notably breast cancer, but also ovarian, colon, pancreatic, and prostate cancers. The discovery of the mutated genes came from analysis of breast cancer tissue, and their names were derived from this finding. Individuals with mutations of either BRCA1 or BRCA2 have an increased incidence of cancer due to a substantially decreased ability to repair double-strand DNA breaks.

Nonhomologous end joining is considered the simplest double-strand break repair mechanism and can be used during any portion of the cell cycle, although it is most active during G1. The “nonhomologous” portion of the name arises from the lack of requirement for a homologous template. As shown in **Figure 20.45**, the ends of the double-strand DNA break are bound by a heterodimer

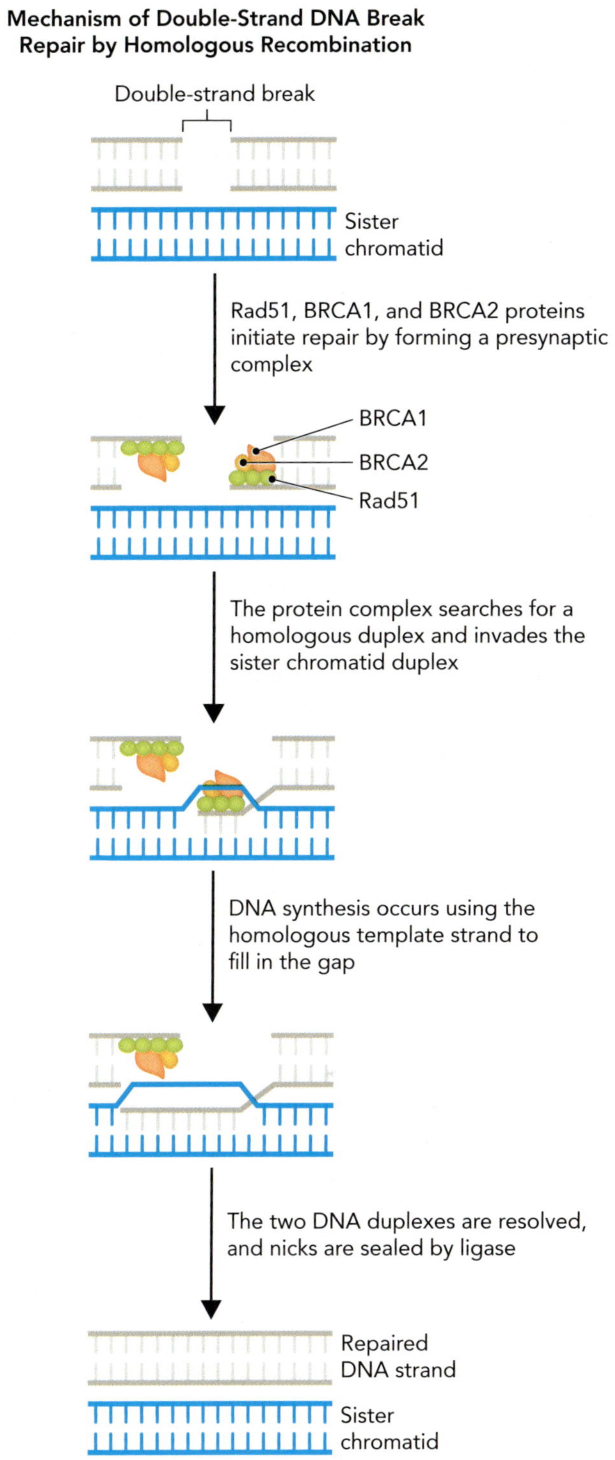

Figure 20.44 Double-strand DNA breaks in eukaryotes are repaired by homologous recombination.

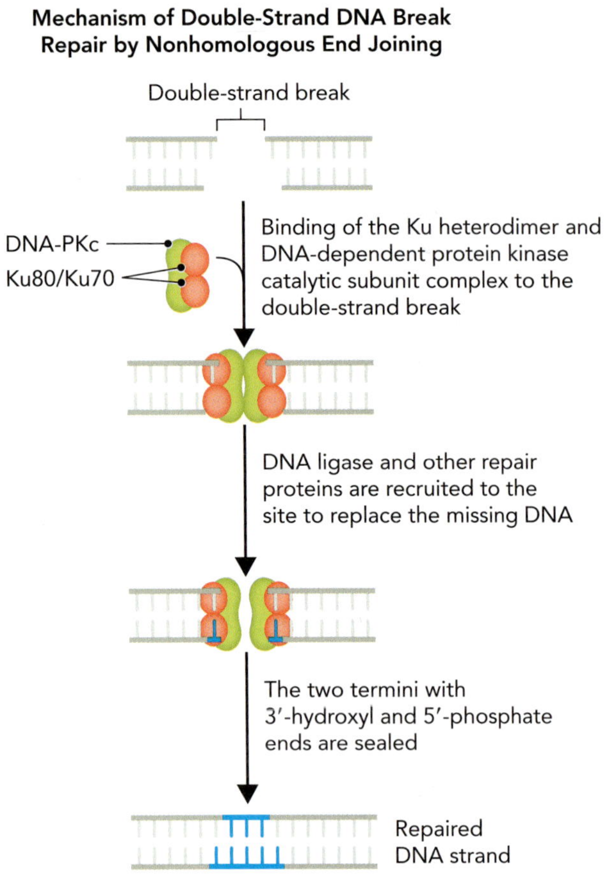

Figure 20.45 Double-strand DNA break repair by nonhomologous end joining does not require a DNA template strand.

of the Ku protein (Ku80/Ku70) and DNA-dependent protein kinase catalytic subunit (DNA-PKc). This Ku–DNA-PKc complex then recruits ligase and other repair proteins, which replace a small number of nucleotide bases on either side of the break and then rejoin the two DNA strands using 3′-hydroxyl and 5′-phosphoryl termini on each strand.

Lynch Syndrome Is a Human Genetic Deficiency in Mismatch Repair Enzyme Most human cancers are caused by somatic mutations occurring during an individual's lifetime, but some cancers are the result of inherited mutations. One of the most common types of inherited cancers is Lynch syndrome, which is also called hereditary non-polyposis colorectal cancer (HNPCC). Lynch syndrome is caused by autosomal dominant mutations in the genes encoding the human mismatch repair enzymes hMLH1 and hMSH2. Mutations in just one copy of either of these DNA repair enzyme genes lead to HNPCC at a relatively early age, which is not surprising given the critical role of mismatch repair in maintaining genome integrity.

Lynch syndrome is divided into two classifications, called Lynch syndrome I and Lynch syndrome II, which together account for more than 90% of all patients observed with HNPCC (**Figure 20.46**). Lynch syndrome I results primarily in colon cancer due to mutations in either the *hMSH2* or *hMLH1* genes. Lynch syndrome II results in a variety of other cancers, including endometrial, ovarian, and stomach cancer, and most often is associated with mutations in *hMLH1*. Often, individuals with HNPCC develop more than one type of cancer during their lives.

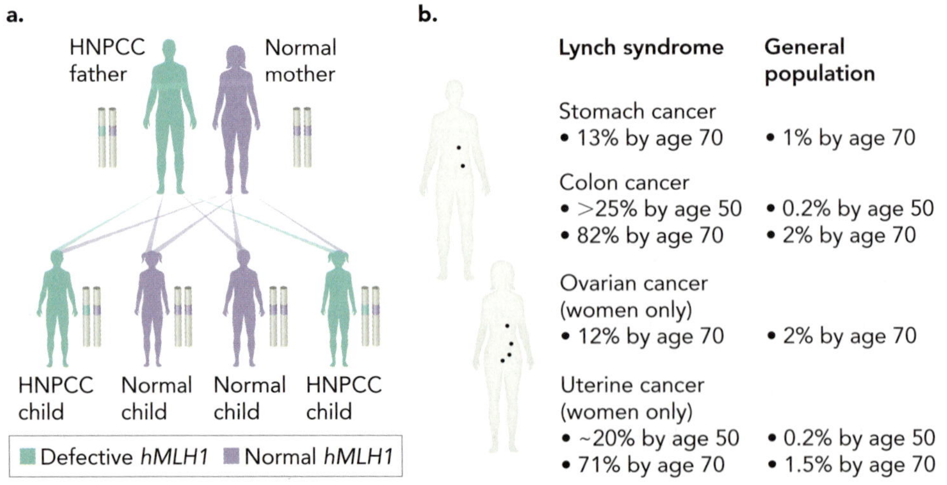

Figure 20.46 Lynch syndrome is due to inherited autosomal dominant mutations in the human mismatch repair genes *hMLH1* and *hMSH2*. **a.** Autosomal dominant mutations cause disease when only a single copy of the mutant gene is inherited. The molecular basis for dominant mutations can be a gain of function (unregulated enzyme activity) or loss of function (mutant protein inhibits activity of the normal protein). Autosomal dominant mutations are passed on to 50% of the offspring. **b.** In addition to colon cancer, individuals with Lynch syndrome also develop other types of cancers at a significantly higher rate than that in the normal population.

concept integration 20.2

How might a high metabolic rate in some individuals lead to an increased risk of cancer?

High metabolic rates increase the rate of oxygen use, specifically by the electron transport process. Increased use of oxygen has the potential to increase reactive oxygen species. This increases the probability of oxidative damage to DNA, thus increasing the mutation rate and with it the possibility of cancerous growth. Metabolic rates can be affected by diet and exercise, which may explain associations between the probability of some diseases and lifestyle choices. For example, high metabolic rates and unrepaired DNA damage could be contributing factors in cancer.

20.3 DNA Recombination

While the vast majority of nucleotide mutations in DNA are quickly repaired, it is the unrepaired mutations that add to genetic diversity and are indeed one of the driving forces in evolution. Another driver of genetic diversity is DNA rearrangement, which can occur within an organism's genome (in a directed or random way) or by transferring DNA from one organism to another. Genetic recombination is the biochemical process that mediates DNA rearrangements within and between organisms. In this section, we will look at several examples of genetic recombination, including the process of meiosis, the movement of viral genes to a host genome, and the rise of diversity in the immune system through recombination.

Homologous Recombination during Meiosis

We have seen that homologous recombination is an important method of double-strand DNA repair; however, more generally, homologous recombination moves genetic material from one place within the genome to another. The movement can be within a chromosome or from one chromosome to another. In eukaryotes, homologous recombination largely occurs during meiosis.

The highest rate of recombination in eukaryotes occurs during meiotic cell division. Sister chromatids generated during prophase of the first meiotic division are able to exchange DNA through homologous recombination, also called chromosomal crossover. This DNA crossover event permits exchange of genetic information between paternal and maternal copies of homologous chromosomes, which are very similar but not identical. The organization of genes on the chromosomes does not change; only the sequence of nucleotides within the genes changes. The products of DNA crossover events during meiosis are passed on to their progeny and thus contribute to genetic diversity. As shown in **Figure 20.47**, DNA crossover events occur during meiosis I, when four sister

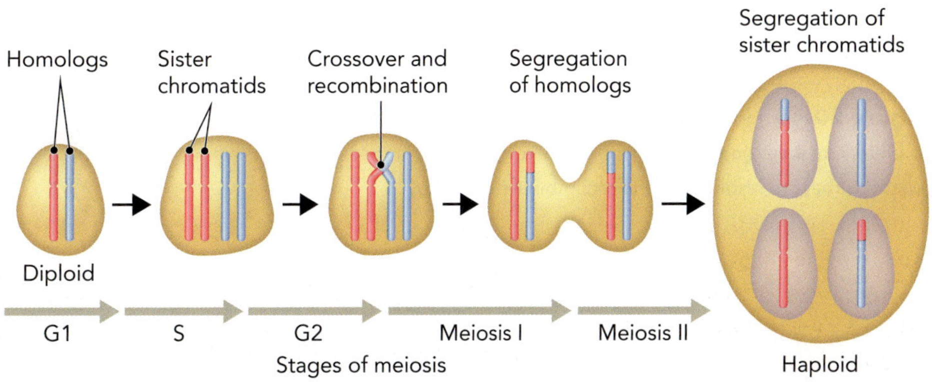

Figure 20.47 DNA recombination events during meiosis introduce genetic diversity. During meiosis I, crossover can occur between homologous chromosomes. In this case, the gametes produced in meiosis II show two gametes with parental genotypes and two gametes with recombinant genotypes.

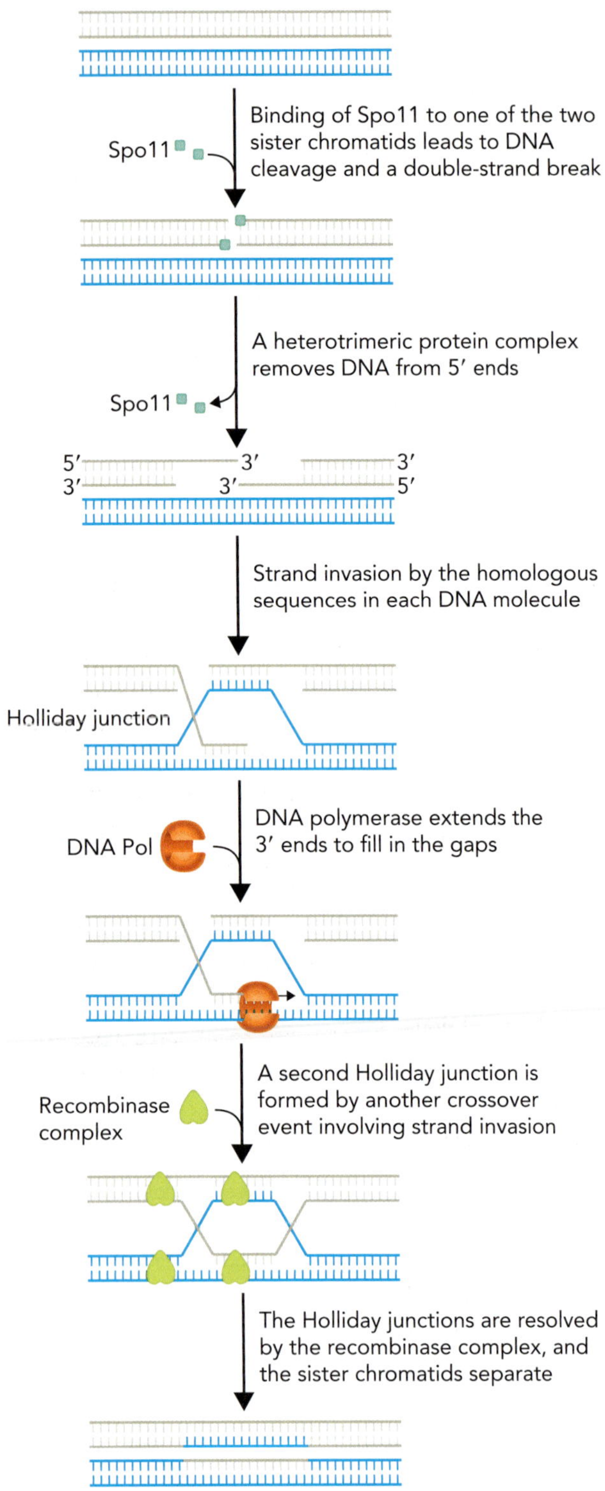

Figure 20.48 Homologous recombination leads to DNA crossover and the formation of two Holliday junctions that are resolved by the recombinase complex.

chromatids are present. During meiosis II, the gametes that are produced can have either entirely maternal or entirely paternal genetic material, resulting from no crossover, or they can contain a recombinant chromosome with a mixture of maternal and paternal genes.

Double-strand DNA breaks are required in meiosis for homologous recombination to occur. As shown in **Figure 20.48**, double-strand breaks in meiotic homologous recombination are initiated by the binding of two molecules of Spo11, one to each of the DNA strands. Cleavage of the strands and dissociation of Spo11 allows for binding by a heterotrimeric protein complex that removes DNA from the 5′ ends. This leaves 3′ single-stranded DNA overhangs that can cross over to bind homologous sequences in each DNA molecule. Extension of both 3′ ends by DNA polymerization replaces the DNA that was removed at the initiation of the process. Ligation of the ends forms a region where multiple DNA strands come together, called a **Holliday junction**. The Holliday junctions can be moved along the now-joined chromosomes by helicase activity in a process called branch migration. This can increase or decrease the size of the heteroduplex region. Ultimately, the Holliday junctions are resolved by the enzyme recombinase to separate the joined chromosomes.

The Holliday junction is a region of quadruplex DNA where four different DNA strands come together (**Figure 20.49**). The structure was proposed in 1964 by Robin Holliday, yet the details of how these junctions form and how they are resolved are still being investigated. The double-strand break recombination model just described produces a double Holliday junction, but other mechanisms exist for Holliday junction production, most of which result in only a single Holliday junction.

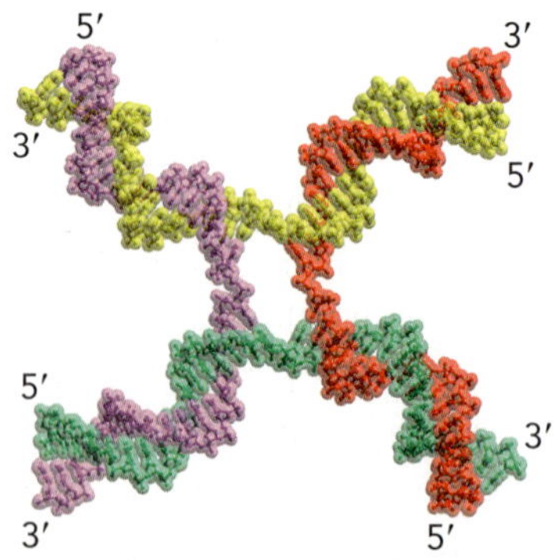

Figure 20.49 The structure of a Holliday junction is shown. BASED ON PDB FILE 3CRX.

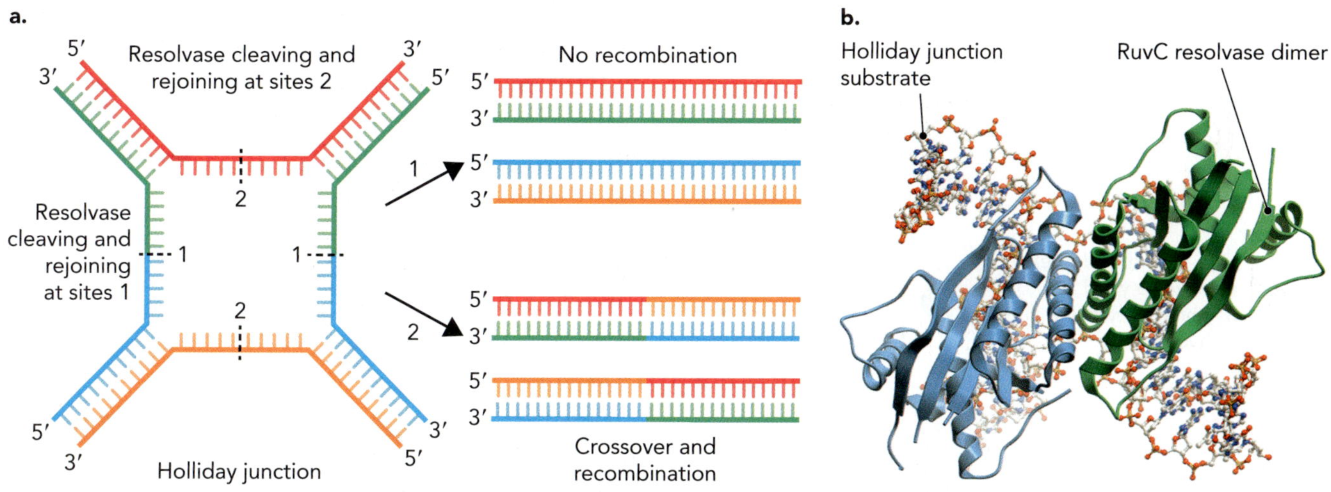

Figure 20.50 Holliday junction resolution can result in two different outcomes, only one of which results in DNA crossover. **a.** A resolvase can bind at either sites 1 or 2. If binding occurs such that cuts are made at sites 1 followed by ligation, then the product shows no crossover. If binding occurs such that cuts are made at sites 2 followed by ligation, crossover occurs. **b.** The bacterial resolvase RuvC binds as a dimer to a Holliday junction structure. BASED ON PDB FILE 4LDO.

Resolution of Holliday junctions is accomplished by resolvase enzymes, which are part of the recombinase complex. Many resolvases are dimeric (or dimers of dimers), such that they have two active sites for cleavage of two DNA strands without dissociation from the Holliday junction. The mechanism of binding and cleavage determines what type of product results. A single junction presents two possible ways to resolve, each giving a different product as shown in **Figure 20.50**.

The activity of one class of resolvase enzymes (such as the Cre recombinase) arises from tyrosine residues present in the active site. As shown in **Figure 20.51**, each strand of the sister chromatids is first cleaved to yield a 3′-phosphotyrosine and a 5′-hydroxyl. After strand swapping and covalent linkage of DNA strands through a nucleophilic attack by the 5′-hydroxyl on the phosphotyrosine, a Holliday junction is formed. This complex is then isomerized, and a second round of strand cleavage, strand invasion, and covalent linkage results in recombined sister chromatids. The cleavage of the junction is always performed on opposite sides of the junction, thus ensuring that the products are properly paired chromatids. Even in the non-crossover product, small changes to the DNA sequence can still occur due to the extension of the 3′ ends after strand invasion.

Integration and Transposition of Viral Genomes

For a virus to insert its genomic DNA into a host cell genome successfully, it must undergo a DNA integration event, which requires a viral enzyme called integrase. Successful viral integration also relies on DNA modifying enzymes in the host cell; for example, enzymes involved in host cell homologous recombination and DNA repair.

Bacteriophage λ Integration into the *E. coli* Genome Bacteriophages are viruses that invade bacteria. Bacteriophage λ is one of the most well-characterized bacteriophages. The integration of bacteriophage λ into the *E. coli* genome begins with the injection of phage DNA into the cell (**Figure 20.52**, p. 1045). The DNA is injected in a linear form and immediately circularizes within the cell, using the complementary overhangs on the ends of the DNA to form the circularstructure. The ends are then ligated by a host ligase that prepares the DNA for one of two potential activities: insertion into the host chromosome (lysogeny) or production of new bacteriophage λ particles (lysis). If the lytic cycle is activated, bacteriophage λ DNA is replicated and viral proteins are produced, followed by assembly of bacteriophage λ particles, lysis of the cell, and release of the virions. If the lysogenic cycle is induced, bacteriophage λ DNA is

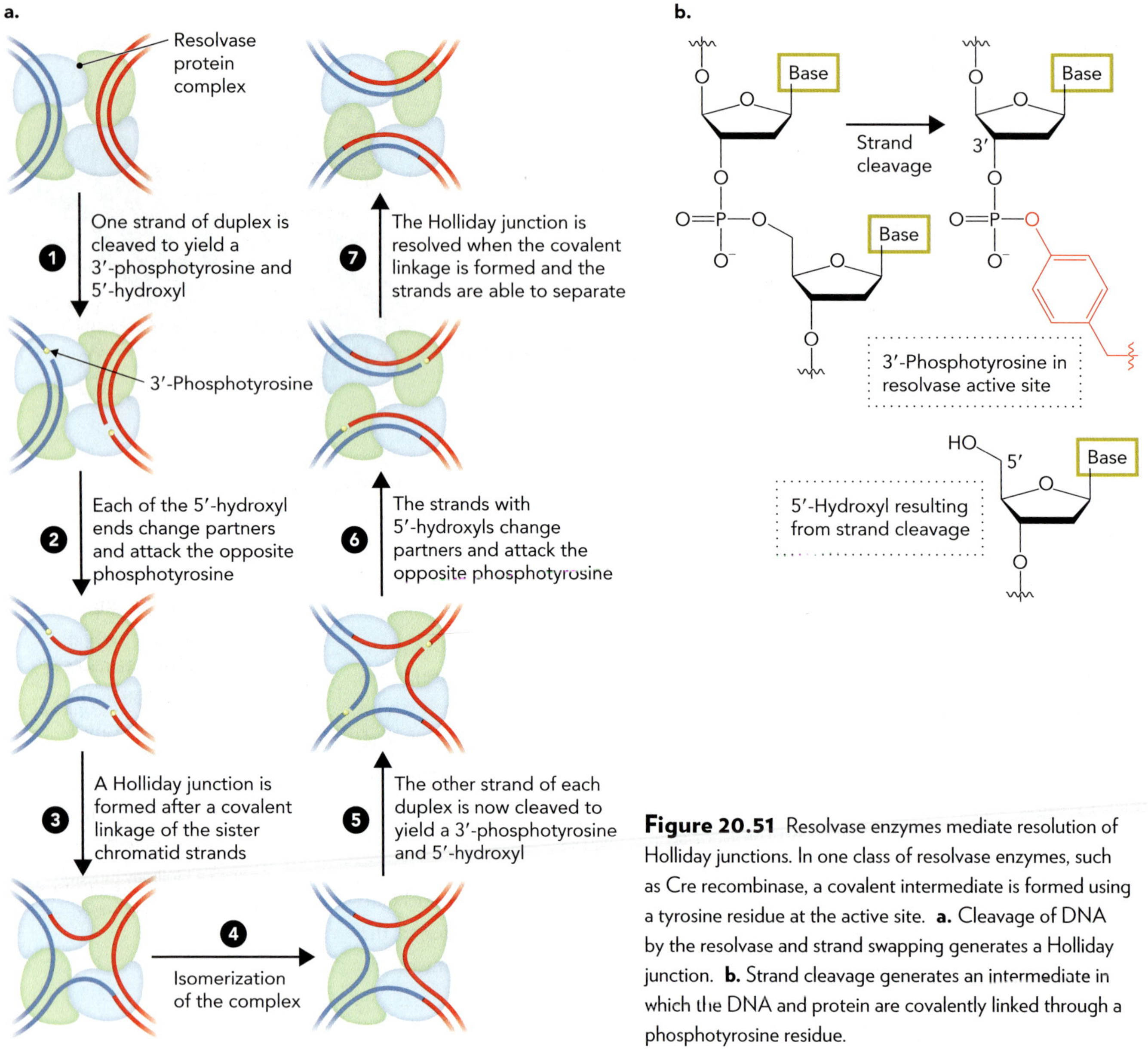
a.
Resolvase protein complex
1
One strand of duplex is cleaved to yield a 3′-phosphotyrosine and 5′-hydroxyl
3′-Phosphotyrosine
2
Each of the 5′-hydroxyl ends change partners and attack the opposite phosphotyrosine
3
A Holliday junction is formed after a covalent linkage of the sister chromatid strands
4
Isomerization of the complex
5
The other strand of each duplex is now cleaved to yield a 3′-phosphotyrosine and 5′-hydroxyl
6
The strands with 5′-hydroxyls change partners and attack the opposite phosphotyrosine
7
The Holliday junction is resolved when the covalent linkage is formed and the strands are able to separate
b.
Base
Strand cleavage
3′
3′-Phosphotyrosine in resolvase active site
HO
5′
5′-Hydroxyl resulting from strand cleavage

Figure 20.51 Resolvase enzymes mediate resolution of Holliday junctions. In one class of resolvase enzymes, such as Cre recombinase, a covalent intermediate is formed using a tyrosine residue at the active site. **a.** Cleavage of DNA by the resolvase and strand swapping generates a Holliday junction. **b.** Strand cleavage generates an intermediate in which the DNA and protein are covalently linked through a phosphotyrosine residue.

integrated into the host chromosome and replicated each time the cell divides. The decision of lysis versus lysogeny is influenced by the status of the bacterial cell; for example, lysis is activated by DNA damage caused by UV irradiation. In contrast, cells that have a poor nutrient supply often do not have the nutrient levels to support a lytic cycle, and thus lysogeny represents a way for the phage to wait for better conditions. Regulatory mechanisms governing the lysis–lysogeny decision in bacteriophage λ are discussed in Chapter 23.

As shown in **Figure 20.53**, bacteriophage λ DNA insertion into the *E. coli* genome begins with alignment of the bacterial and bacteriophage λ attachment sites, *att*B and *att*P, respectively. Both DNA molecules are cleaved, with the viral DNA looped such that both ends of the 48-kb genome are aligned at the bacterial insertion site. The bacterial protein integration host factor (IHF) is involved in this site-specific recombination event, forming a complex with the Int protein at the *att*P site. A Holliday junction is an intermediate in the process, resolving with the aid of Int to yield the modified bacterial chromosome with viral DNA inserted. Integration of bacteriophage

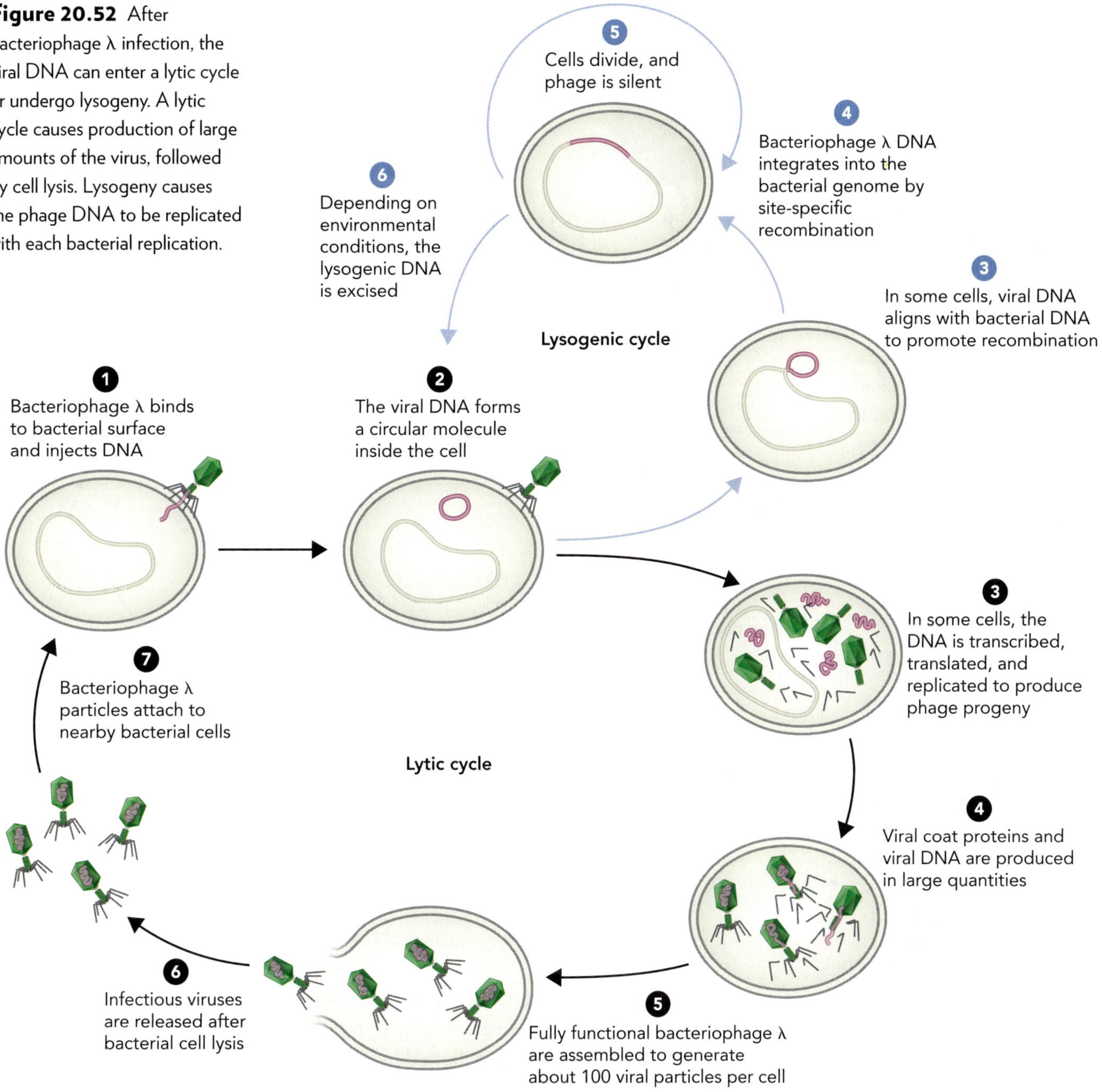
5
Cells divide, and phage is silent
4
Bacteriophage λ DNA integrates into the bacterial genome by site-specific recombination
6
Depending on environmental conditions, the lysogenic DNA is excised
3
In some cells, viral DNA aligns with bacterial DNA to promote recombination
Lysogenic cycle
1
Bacteriophage λ binds to bacterial surface and injects DNA
2
The viral DNA forms a circular molecule inside the cell
3
In some cells, the DNA is transcribed, translated, and replicated to produce phage progeny
7
Bacteriophage λ particles attach to nearby bacterial cells
Lytic cycle
4
Viral coat proteins and viral DNA are produced in large quantities
6
Infectious viruses are released after bacterial cell lysis
5
Fully functional bacteriophage λ are assembled to generate about 100 viral particles per cell

Figure 20.52 After bacteriophage λ infection, the viral DNA can enter a lytic cycle or undergo lysogeny. A lytic cycle causes production of large amounts of the virus, followed by cell lysis. Lysogeny causes the phage DNA to be replicated with each bacterial replication.

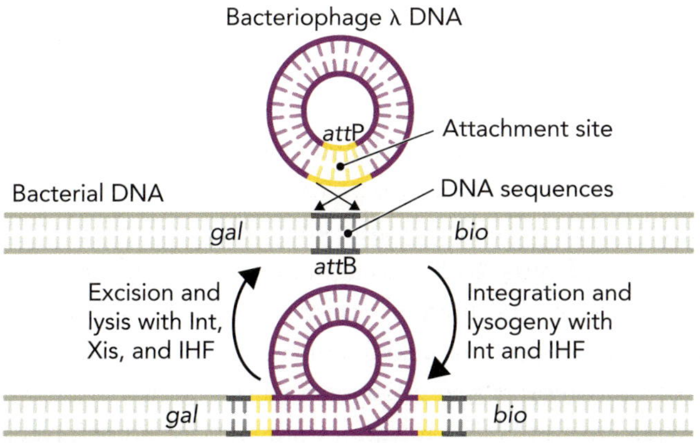
Bacteriophage λ DNA
attP
Attachment site
Bacterial DNA
DNA sequences
gal
bio
attB
Excision and lysis with Int, Xis, and IHF
Integration and lysogeny with Int and IHF
gal
bio

Figure 20.53 Integration of bacteriophage λ DNA into the *E. coli* chromosome requires a site-specific recombination event. DNA sequences in the *att*P site of the viral DNA align with the DNA sequences in the bacterial *att*B site, which is located between the galactose (*gal*) and biotin (*bio*) operons in the *E. coli* chromosome. Bacteriophage λ proteins Int and IHF aid in the recombination process. Excision from the *E. coli* chromosome requires the bacterial protein IHF and the bacteriophage λ Int and Xis proteins.

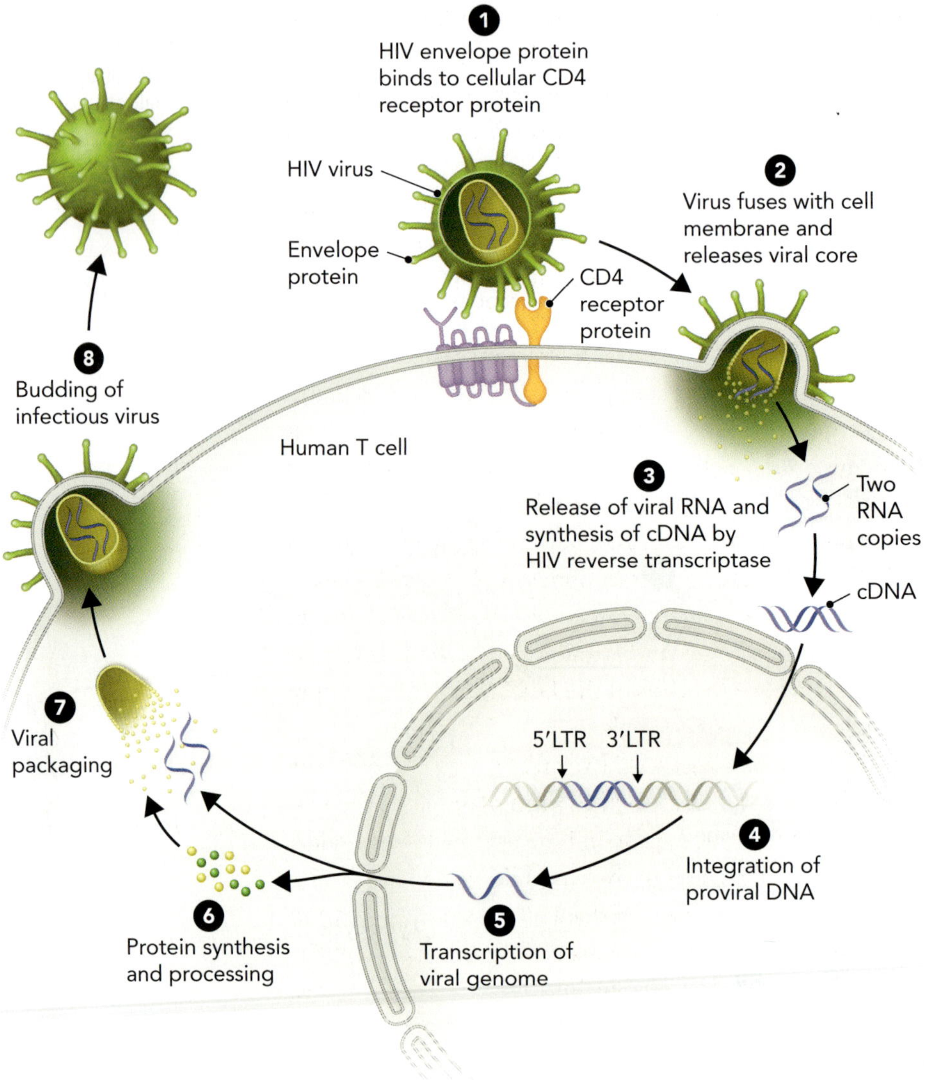

Figure 20.54 The HIV infection cycle requires conversion of the viral genome to cDNA that is then integrated into the host cell genome by a recombination event. After binding of the virus to the CD4 receptor and membrane fusion, the RNA genome is inserted into the host cell. Reverse transcriptase uses the single-stranded RNA to produce a cDNA–RNA hybrid. The RNase activity of reverse transcriptase degrades the RNA, then forms duplex DNA by polymerization of the complementary strand. LTR = long terminal repeat.

λ is reversible under conditions where the lytic cycle is supported. Expression of the bacteriophage λ Xis gene generates an enzyme that excises the viral genome from the bacterial chromosome. This is followed by production of virions and lysis of the cell.

HIV Retroviral Integration into the Human Genome The human immunodeficiency virus (HIV) is a retrovirus containing an RNA genome that is converted to DNA using the viral reverse transcriptase enzyme encoded in its genome. This DNA copy of the virus is then integrated into the human genome by nonhomologous recombination. Studies of the HIV infection cycle have identified at least 19 host cell proteins—including cytoplasmic, nuclear, and nuclear membrane proteins—as playing some role in the integration process.

The HIV infection life cycle begins with the virion binding to a receptor on the host cell membrane. As shown in **Figure 20.54**, fusion of the cell membrane and viral envelope allows for entry of the viral proteins and genome into the cell. The virion contains two copies of the RNA genome, held together at the ends of the RNA molecules by the presence of short repeats that allow base pairing between the two molecules. The RNA molecules are also bound to various proteins, including reverse transcriptase. The process of reverse transcription first produces a transient cDNA–RNA hybrid, then the RNA is degraded by the RNase activity of reverse transcriptase. A small portion of RNA is left to serve as the primer for polymerization of the second DNA strand, which ensues immediately after completion of the initial DNA strand. We described the biochemistry of *in vitro* cDNA synthesis by the enzyme reverse transcriptase in Chapter 3 (see Figure 3.53). The double-stranded viral DNA is then integrated into the host cell genome by a recombination reaction that is catalyzed by the HIV integrase enzyme. Transcription of the integrated viral genome generates viral mRNA, which is translated into viral proteins, as well as packaged directly into new HIV virions to produce infectious viral particles that bud off from the infected cell (Figure 20.54). Integrase binds to the DNA via the long terminal repeats and processes the ends to give an overhang of four nucleotides and a 3′-hydroxyl. Integrase remains bound to the ends of the DNA, providing the core of the pre-integration complex.

The mechanism of DNA viral integration into the host cell genome is illustrated in **Figure 20.55**. In the first step, the HIV integrase protein binds to the **long terminal repeat (LTR)** sequences in the viral DNA and uses 3′-end processing to generate an overhang of four nucleotides and a 3′-hydroxyl. Integrase remains bound to the ends of the DNA to facilitate integration into the host cell genome. After translocation of the viral DNA–integrase complex into the host cell nucleus, integrase recognizes the target DNA and binds to it. With the viral DNA looped around integrase so that the ends of the DNA are both bound to the enzyme, integrase cleaves the host DNA, providing overhangs that do not necessarily base pair with the viral DNA overhangs. Strand transfer occurs by moving the host DNA ends into place with the viral DNA ends, followed by dissociation of integrase. Because the overhangs of host and viral DNA do not correctly base pair, a DNA flap is created by the 5′ ends of the viral DNA. An endonuclease removes the unpaired 5′ ends of the viral DNA, leaving a gap that can be repaired by the host DNA gap repair system. DNA polymerase fills the gaps and the nick is sealed, using DNA ligase, to complete the integration.

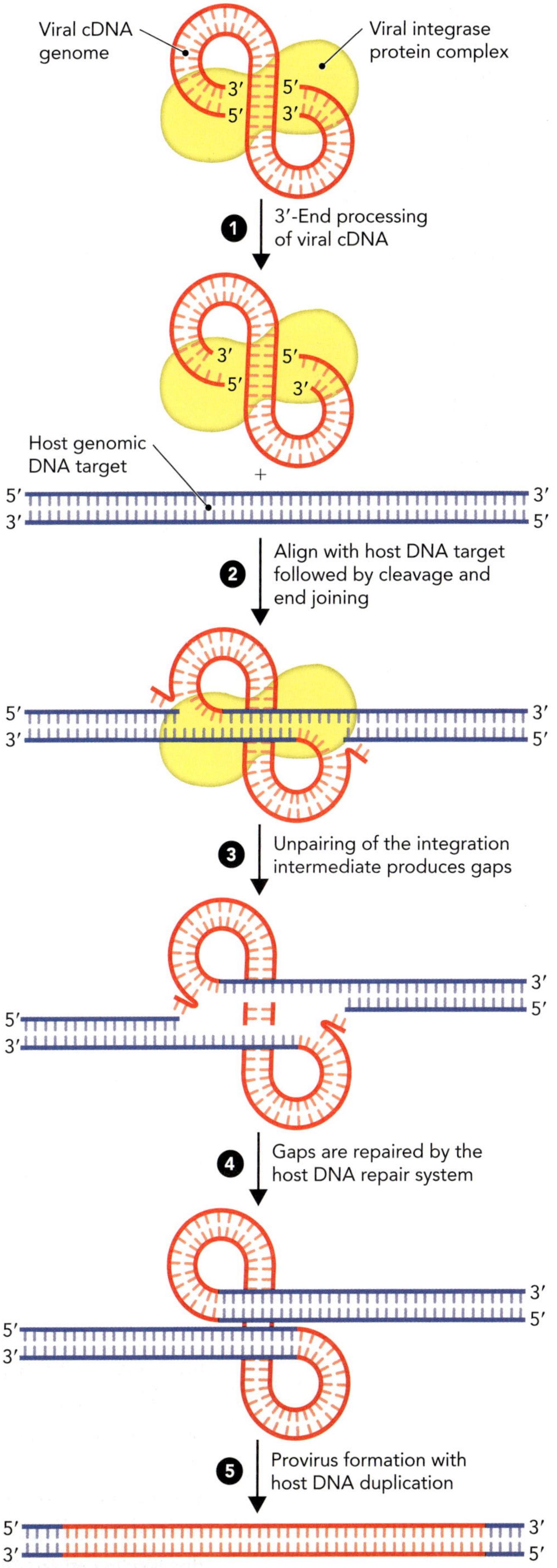

Figure 20.55 HIV integration into the host cell genome is initiated by integrase binding both ends of the viral DNA and removing several nucleotides to create a 5′ overhang. The host DNA is captured and cleaved to produce complementary ends. Integrase facilitates strand transfer to join the ends of viral and host DNA, followed by gap repair to complete the integration process.

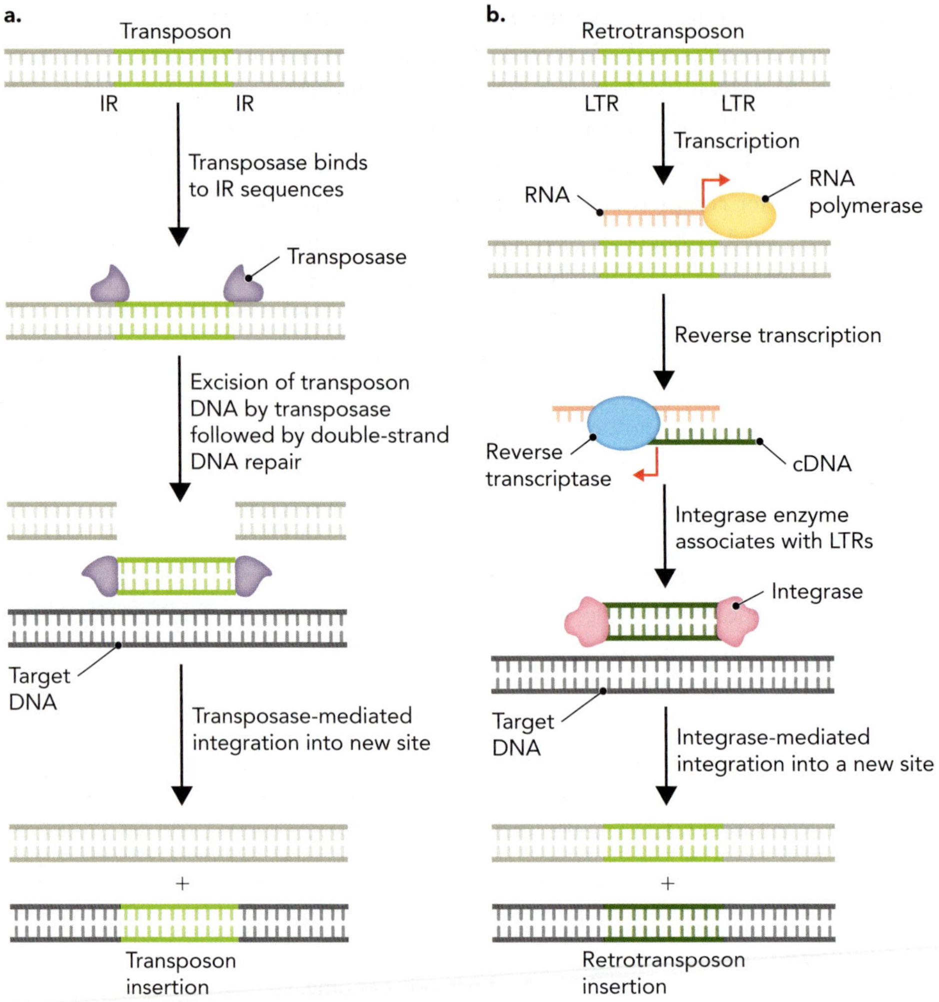

Figure 20.56 Transposition involves the movement of DNA sequences to new sites in DNA. **a.** Transposons move by a cut-and-paste mechanism mediated by transposase enzymes and inverted repeat (IR) sequences. **b.** Retrotransposons move through an RNA intermediate, which is transcribed into cDNA by the enzyme reverse transcriptase. Insertion requires an integrase enzyme and LTR sequences.

Transposon Integration and Transposition A **transposon** is a segment of DNA that can move from one region in the genome to another, either within the same chromosome or between two chromosomes. As shown in **Figure 20.56**, DNA transposons require a transposase protein for both removal of the transposon from its current site and insertion into a new site in the genome. DNA sequences at the ends of the transposon segment are arranged as inverted repeats (IRs) and function in the recombination process by promoting sequence alignment. The mechanism of transposase-mediated recombination is similar to that of viral integration, using a cross-over homologous recombination process similar to that of viral integrase (see Figure 20.55).

Genome analysis has shown that almost 50% of the human genome is composed of transposons, with 2% of these being DNA transposons and the remaining being **retrotransposons**. Retrotransposons are transcribed into RNA and moved into place after conversion into DNA by the action of reverse transcriptase. Retrotransposons are divided into two major classes: those that contain LTR sequences, and those that do not contain LTR sequences. LTR transposons are not capable of direct DNA movement (LTR retrotransposons use an RNA-mediated mechanism); thus, the only type of DNA transposon that is active in humans is the non-LTR retrotransposon.

Rearrangement of Immunoglobulin Genes

Similar to the importance of genetic recombination in meiosis to generate organismal diversity, genetic recombination of antibody genes within the immune system is required to generate antibody diversity and protection against pathogens. An antibody contains a heavy chain and a light chain; the heavy chain arises from variable (V), diverse (D), joining (J), and constant (C) segments, and the light chain arises from V, J, and C segments (see Figure 4.61). The recombination events that assemble the immunoglobulin genes for antibody production are often referred to as **V(D)J recombination**. Computational studies have shown that through a process called junctional diversification, immunoglobulin gene rearrangements can give rise to billions of different antibodies.

The ability of the human immune system to generate the large array of antibodies begins with the number of V, D, and J genes that can be recombined to give rise to the variable regions of antibodies. The light-chain immunoglobulin genes are of two types: the κ chain and the λ chain. The κ-chain variable genes include 40 V-segment genes and 5 J-segment genes, whereas the λ-chain variable genes include 30 V-segment genes and 5 J-segment genes. Because the variable region of each light chain is made from one V and one J segment, the number of different κ light chains that can be produced by recombination of V and J segments is about 200, and the number of λ light chains is about 150. The heavy-chain immunoglobulin genes include 51 V-segment genes, 27 D-segment genes, and 6 J-segment genes, which together can be recombined into more than 8,000 different heavy-chain variable regions.

As shown in **Figure 20.57**, these immunoglobulin gene segments are recombined during the process of immune development to generate a functional antibody consisting of two light-chain protein subunits and two heavy-chain protein subunits. By taking into account the total number of V, D, and J segments that recombine into light-chain and heavy-chain protein subunits, the number of different antibodies that can be assembled from the combination of these genes is approximately 2.9×10^6. The true diversity of antibodies, however, arises from the V(D)J recombination events, which have the potential to produce more than 10^9 unique antibodies.

Figure 20.57 Rearrangement of the V and J gene segments in the light chain genes and the V, D, and J gene segments in the heavy genes by DNA recombination generates the required antibody diversity needed for a functional immune system.

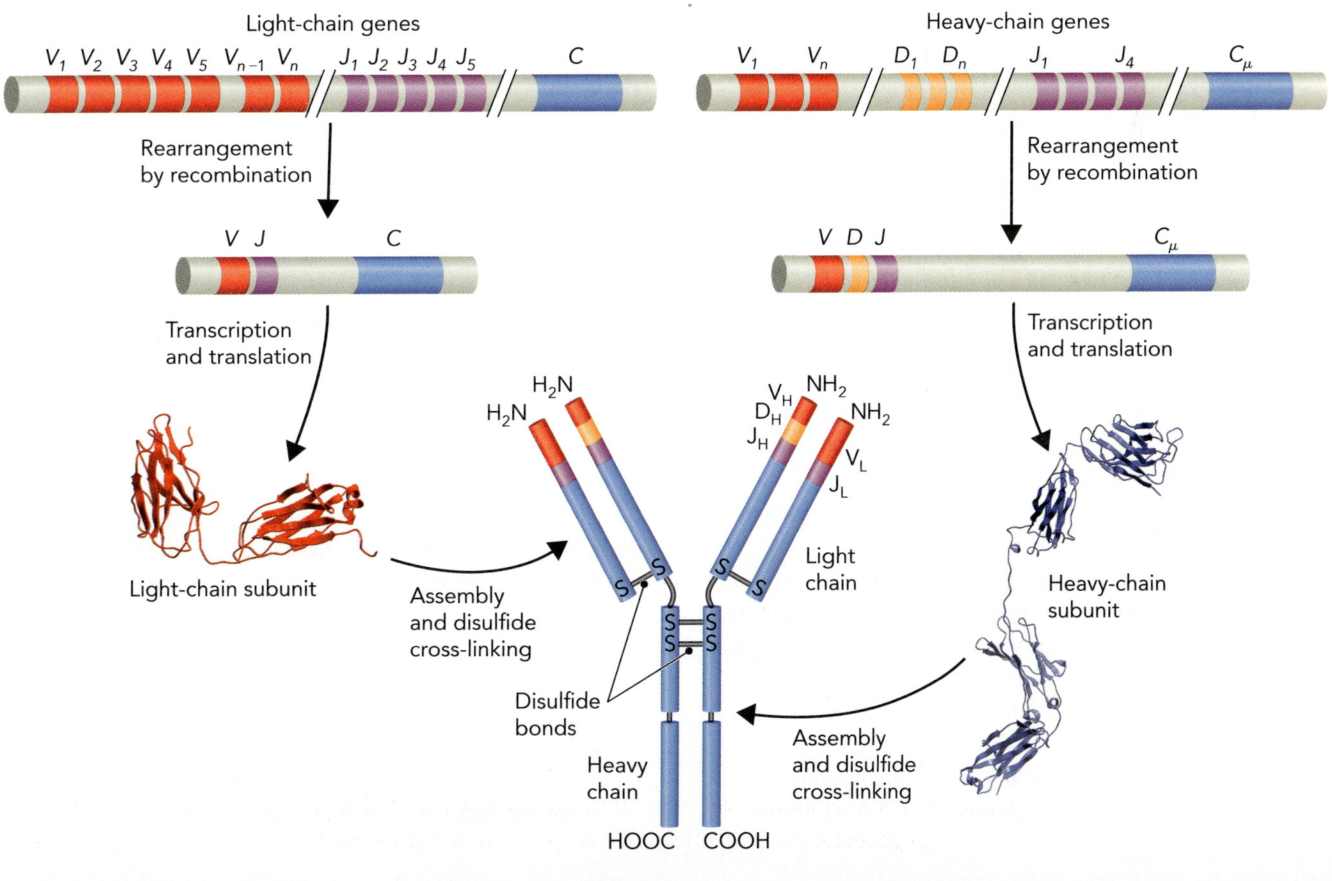

concept integration 20.3

Azidothymidine (AZT) is a nucleoside analog that lacks a 3′-hydroxyl and is used in treatment of HIV. On the basis of your knowledge of the HIV integration process, how does AZT work as an inhibitor? Where are protease inhibitors effective against HIV integration?

AZT is an analog of thymidine and is converted to AZT triphosphate in a cell. It then serves as a substrate for HIV reverse transcriptase and is incorporated into the growing DNA. Because the newly incorporated nucleotide analog lacks a 3′-hydroxyl group, more nucleotides cannot be added, so replication of DNA stops. Protease inhibitors prevent the proteolysis of the reverse transcriptase, which prevents optimal activity of the enzyme. Without proteolysis, the dissociation constant of reverse transcriptase is increased 10^4-fold, and its catalytic efficiency is reduced by about 50%. The combination of the two drugs has an additive effect in reducing the reverse transcriptase activity.

chapter summary

20.1 DNA Replication

- DNA replication is semiconservative, with each daughter molecule composed of one newly synthesized strand and one template strand. DNA replication begins at an origin and proceeds in both directions using two replication forks.
- Each replication fork produces one leading strand and one lagging strand. The leading strand is synthesized continuously, while the lagging strand is synthesized discontinuously, producing Okazaki fragments. The presence of multiple β clamps allows for rapid lagging strand synthesis.
- Many DNA polymerases exist. In prokaryotes, DNA polymerase III is the main polymerizing enzyme; in eukaryotes, DNA polymerase δ and DNA polymerase ε fulfill that role. Other polymerases are used specifically during removal of primers, DNA repair, or polymerization of non-nuclear DNA, such as mitochondrial DNA.
- An active prokaryotic replication fork requires topoisomerase, DNA helicase, primase, single-stranded DNA binding proteins, the core polymerase, and a spare β clamp for synthesis of the next Okazaki fragment. Binding of helicase is a critical part of the initiation process. Association of most of the elements at the replication fork allows for leading and lagging strands to be synthesized at a similar rate.
- High fidelity is needed to eliminate errors of replication. DNA polymerases discriminate by proper base pairing, as base pairs that are not Watson–Crick base pairs do not fit the geometry of the active site. Improperly inserted bases can be removed by the proofreading function of polymerases.
- High processivity is critical if entire chromosomes are to be replicated in the time allowed for DNA replication. Prokaryotic DNA polymerase III uses a β clamp to increase processivity. Eukaryotic DNA polymerases δ and ε use PCNA to achieve high processivity.
- Prokaryotic replication has one origin and one site of termination opposite the origin. Eukaryotic replication involves multiple origins on each chromosome, and origins are activated at different times in the cell cycle. Because of their linear nature, eukaryotic chromosomes have telomeres at the chromosome ends that shorten with each replication process unless the enzyme telomerase is present.

20.2 DNA Damage and Repair

- DNA damage can lead to mutations, depending on the severity of the damage. Common damaging agents include reactive oxygen species, UV light, alkylating agents, and errors of replication.
- Mutations to a single nucleotide can be silent or can cause changes to gene products through either single amino acid substitutions or production of a truncated protein. Mutations can lead to a variety of issues, including cancerous cell growth. Generally, many mutations must accumulate in a cell before it becomes cancerous.
- Cells have the ability to repair DNA, including mismatch repair, base excision repair, and nucleotide excision repair. Single-strand DNA breaks and double-strand DNA breaks also must be repaired.
- Some DNA repair occurs by highly specialized systems that deal with only one type of damage. Photolyase specifically reverts cyclobutane pyrimidine dimers to the original pyrimidines. O^6-methylguanine-DNA methyltransferase is the only mechanism for removal of O^6-methylguanine, a common alkylated form of guanine.

20.3 DNA Recombination

- Recombination moves genetic material from one place to another. Movement can be from one place in the genome to another or from one organism to another. Recombination gives rise to genetic diversity.
- During meiosis, homologous chromosomes can undergo crossover to produce chromosomes that contain a mix of maternal and paternal genes. This has the ability to drastically increase the genetic variety within gametes.
- Homologous recombination requires a double-strand break to allow the formation of a Holliday junction. Resolving the Holliday junction produces either a crossover product or reversion to the original chromosomes.
- Integration of a viral genome into host DNA is a recombination process involving a double-strand break, mediated by enzymes from both the virus and the host cell. Bacteriophage λ integrates its genome into a specific site on the *E. coli* chromosome, allowing for passive replication of the viral genome each time the *E. coli* chromosome is replicated. Alternatively, lysis of the cell can be initiated by production of a large number of phage virions.
- HIV uses reverse transcriptase to convert its single-stranded RNA genome into double-stranded DNA prior to insertion into the host genome. The viral integrase enzyme is required for integration, along with several host proteins associated with double-strand break repair.
- A significant portion of the human genome is composed of transposons that are capable of being moved from one place to another within the genome. Movement of a transposon can alter gene activity and cause increasing or decreasing gene expression or disruption of a gene if insertion occurs within the gene.
- Recombination of immunoglobulin genes has the capacity to generate billions of different antibodies through the process of V(D)J recombination. This results in a permanent change to the B-cell DNA and the production of only one specific antibody per cell.

biochemical terms

(in order of appearance in text)

semiconservative replication (p. 1001)
origin of replication (p. 1002)
replication fork (p. 1002)
Okazaki fragment (p. 1003)
leading strand (p. 1003)
lagging strand (p. 1003)
discontinuous synthesis (p. 1003)
DNA polymerase (p. 1003)
primer (p. 1004)
processivity (p. 1006)
proliferating cell nuclear antigen (PCNA) (p. 1007)
Klenow fragment (p. 1008)
replisome (p. 1009)
gyrase (p. 1010)
helicase (DnaB) (p. 1010)
single-stranded DNA binding protein (SSB) (p. 1010)
primase (p. 1010)
primosome (p. 1010)
β clamp (p. 1011)
Dam methylase (p. 1018)
cyclin-dependent protein kinase (CDK) (p. 1022)
cyclin (p. 1022)
telomerase (p. 1024)
somatic mutation (p. 1028)
Ames test (p. 1029)
abasic site (p. 1030)
photoproduct (p. 1030)
photolyase (p. 1033)
nucleotide excision repair (NER) (p. 1033)
mismatch repair (p. 1034)
base excision repair (p. 1035)
direct repair (p. 1037)
single-strand DNA break (p. 1038)
double-strand DNA break (p. 1039)
homologous recombination (p. 1039)
Holliday junction (p. 1042)
long terminal repeat (LTR) (p. 1047)
transposon (p. 1048)
retrotransposon (p. 1048)
V(D)J recombination (p. 1049)

review questions

1. What is meant by semiconservative replication? How is this different from conservative or dispersive replication?
2. What are Okazaki fragments?
3. What is the nucleophile in the reaction catalyzed by DNA polymerase? What is the significance of this in terms of initiating DNA synthesis?
4. What are the primary replicative polymerases in prokaryotic and eukaryotic cells?
5. Describe the mechanism by which HIV-1 reverse transcriptase (HIVRT) produces DNA for chromosomal integration.
6. Describe the typical major elements of a replication fork.
7. What are the major events that occur at *oriC* to allow initiation of DNA synthesis?
8. What is unique about the structure of telomerase that allows it to extend the ends of chromosomes?
9. Describe the Ames test and explain its importance.
10. What happens during cytosine deamination, and how is it repaired?
11. If unrepaired, what effect can cyclobutane pyrimidine dimers or cross-linked guanine residues have on cellular processes?
12. What are the three types of mutations that result from single point mutations? What is the potential severity of each?
13. Show the product of O^6-methylation of guanine, and explain how it is repaired.

14. What are the two roles of methylation by Dam methylase—one in the replication process, the other in repair?
15. What are the major differences between meiotic homologous recombination by double-strand break and homologous recombination for the repair of double-strand breaks caused by DNA damaging events?
16. Describe what occurs when bacteriophage λ invades an *E. coli* cell and establishes lysogeny. What would happen if the *E. coli* cell were exposed to a high dose of UV light?
17. Describe the events in V(D)J recombination and explain the importance of this process.

challenge problems

1. Calculate the number of Okazaki fragments produced during the replication of a single *E. coli* chromosome.
2. Explain the events that would need to occur at the replication fork if conservative replication occurred.
3. What would be the effect of a mutation in the HIV-1 reverse transcriptase gene that decreased the rate of mismatches?
4. What effect would an absence of Mg^{2+} have on the activity of DNA polymerase?
5. Approximately how many ATP are required by helicase for the replication of the *E. coli* chromosome?
6. What must the rate of nucleotide incorporation by primase be such that it completes a primer in exactly the same time that an Okazaki fragment is completed?
7. If the average Okazaki fragment in a yeast cell is 200 nucleotides, what is the maximum number of times that a yeast cell can replicate before its telomeres are gone?
8. An Ames test of a suspected mutagen was examined both before and after incubation with rat liver extract, giving the following results. What can you conclude about the suspected mutagen?

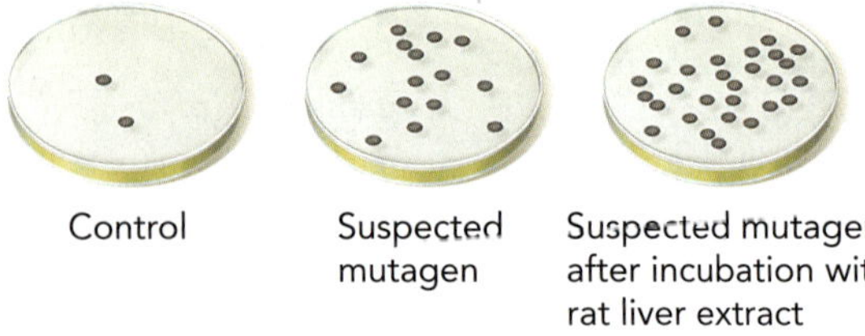

9. If the sequence 5′-AACGC-3′ were damaged by reactive oxygen species, what would be the most prevalent product, and what would be the result of replication (show both strands after replication)?
10. From the following sequence, locate the sites of potential photoproduct formation. Indicate what photoproducts will most likely form and what the potential effect on the DNA would be if left unrepaired.

    5′-ACGTCAGTTACGTACTGACGT

11. What would happen if a mutation in the λ phage Xis gene occurred such that the resulting protein was not functional?
12. What effect would an RNase inhibitor found in a human cell have on HIV? What problems might be associated with such an agent?

smartwork5

If your instructor assigns homework with Smartwork5, access it here: digital.wwnorton.com/biochem.

suggested reading

Books and Reviews

Bell, S. P., and Dutta, A. (2002). DNA replication in eukaryotic cells. *Annual Review of Biochemistry*, *71*, 333–374.

Benkovic, S. J., and Valentine, A. M. (2001). Replisome-mediated DNA replication. *Annual Review of Biochemistry*, *70*, 181–208.

Blackburn, E. H. (2005). Telomeres and telomerase: their mechanisms of action and the effects of altering their functions. *FEBS Letters*, *579*, 859–862.

Chase, J. W., and Williams, K. R. (1986). Single-stranded DNA binding proteins required for DNA replication. *Annual Review of Biochemistry*, *55*, 103–136.

Frick, D. N., and Richardson, C. C. (2001). DNA primases. *Annual Review of Biochemistry*, *70*, 39–80.

Gates, K. S. (2009). An overview of chemical processes that damage cellular DNA: spontaneous hydrolysis, alkylation, and reactions with radicals. *Chemical Research in Toxicology*, *22*, 1747–1760.

Gerton, J. L., and Hawley, R. S. (2005). Homologous chromosome interactions in meiosis: diversity amidst conservation. *Nature Reviews Genetics*, *6*, 477–487.

Heyer, W.-D., Ehmsen, K. T., and Solinger, J. A. (2003). Holliday junctions in the eukaryotic nucleus: resolution in sight? *Trends in Biochemical Sciences*, *28*, 548–557.

Hoeijmakers, J. H. (2001). Genome maintenance mechanisms for preventing cancer. *Nature*, *411*, 366–374.

Ikehata, H., and Ono, T. (2011). The mechanisms of UV mutagenesis. *Journal of Radiation Research*, *52*, 115–125.

Joyce, C. M., and Steitz, T. A. (1994). Function and structure relationships in DNA polymerases. *Annual Review of Biochemistry*, *63*, 777–822.

McHenry, C. S. (2011). DNA replicases from a bacterial perspective. *Annual Review of Biochemistry*, *80*, 403–436.

Ogawa, T., and Okazaki, T. (1980). Discontinuous DNA replication. *Annual Review of Biochemistry*, *49*, 421–457.

Sancar, A. (2003). Structure and function of DNA photolyase and cryptochrome blue-light photoreceptors. *Chemical Reviews*, *103*, 2203–2237.

Singleton, M. R., Dillingham, M. S., and Wigley, D. B. (2007). Structure and mechanism of helicases and nucleic acid translocases. *Annual Review of Biochemistry*, *76*, 23–50.

Smith, G. R. (1987). Mechanism and control of homologous recombination in *Escherichia coli*. *Annual Review of Genetics*, *21*, 179–201.

van Gent, D. C., Hoeijmakers, J. H., and Kanaar, R. (2001). Chromosomal stability and the DNA double-stranded break connection. *Nature Reviews Genetics*, *2*, 196–206.

Primary Literature

Beese, L. S., Derbyshire, V., and Steitz, T. A. (1993). Structure of DNA polymerase I Klenow fragment bound to duplex DNA. *Science*, *260*, 352–355.

Brino, L., Urzhumtsev, A. Mousli, M., Bronner, C., Mitschler, A., Oudet, P., and Moras, D. (2000). Dimerization of *Escherichia coli* DNA-gyrase B provides a structural mechanism for activating the ATPase catalytic center. *Journal of Biological Chemistry*, *275*, 9468–9475.

Casali, P., and Zan, H. (2004). Class switching and Myc translocation: how does DNA break? *Nature Immunology*, *5*, 1101–1103.

Cheng, K. C., Cahill, D. S., Kasai, H., Nishimura, S., and Loeb, L. A. (1992). 8-Hydroxyguanine, an abundant form of oxidation DNA damage causes G T and A C substitutions. *Journal of Biological Chemistry*, *267*, 166–172.

Coskun-Ari, F. F., and Hill, T. M. (1997). Sequence-specific Interactions in the Tus-Ter complex and the effect of base pair in substitutions on arrest of DNA replication in *Escherichia coli*. *Journal of Biological Chemistry*, *272*, 26448–26456.

Fu, Y. V., Yardimci, H., Long, D. T., Guainazzi, A., Bermudez, V. P., Hurwitz, J., Van Oijen, A., Schärer, O. D., and Walter, J. C. (2011). Selective bypass of a lagging strand roadblock by the eukaryotic replicative DNA helicase. *Cell*, *146*, 931–941.

Gorecka, K. M., Komorowska, W., and Nowotony, M. (2013). Crystal structure of RuvC resolvase in complex with Holliday junction substrate. *Nucleic Acids Research*, *41*, 9945–9955.

Guarne, A., Zhao, Q., Ghirlando, R., and Yang, W. (2002). Insights into negative modulation of *E. coli* replication initiation from the structure of SeqA-hemimethylated DNA complex. *Nature Structural Biology*, *11*, 839–843.

Gueneau, E., Dherin, C., Legrand, P., Tellier-Lebegue, C., Gilquin, B., Bonnesoeur, P., Londino, F., Quemener, C., Le Du, M. H., Marquez, J. A., et al. (2013). Structure of the MutL C-terminal domain reveals how MLH1 contributes to Pms1 endonuclease site. *Nature Structural & Molecular Biology*, *20*, 461.

Guo, F., Deshmukh, N. G., and van Duyne, G. D. (1997). Structure of Cre recombinase complexed with DNA in a site-specific recombination synapse. *Nature*, *389*, 40–46.

Heyer, W.-D. (2004). Recombination: Holliday junction resolution and crossover formation. *Current Biology*, *14*, R56–R58.

Ip, S. C. Y., Rass, U., Blanco, M. G., Flynn, H. R., Skehel, J. M., and West, S. C. (2008). Identification of Holliday junction resolvases from humans and yeast. *Nature*, *456*, 357–361.

Itsathitphaisarn, O., Wing, R. A., Eliason, W. K., Wang, J., and Steitz, T. A. (2012). The hexameric helicase DnaB adopts a nonplanar conformation during translocation. *Cell*, *151*, 267–277.

Mitchell, M., Gillis, A., Futahashi, M., Fugiwara, H., and Skordalakes, E. (2010). Structural basis for telomerase catalytic subunit TERT binding to RNA template and telomeric DNA. *Nature Structural Biology*, *17*, 513–519.

Mulcair, M. D., Schaeffer, P. M., Oakley, A. J., Cross, H. F., Neylon, C., Hill, T. M., and Dixon, N. E. (2006). A molecular mousetrap determines polarity of termination of DNA replication in *E. coli*. *Cell*, *125*, 1309–1319.

Nick McElhinny, S. A., Gordenin, D. A., Stith, C. M., Burgers, P. M. J., and Kunkel, T. A. (2008). Division of labor at the eukaryotic replication fork. *Molecular Cell*, *30*, 137–144.

Patel, P. H., Jacobo-Molina, A., Ding, J., Tantillo, C., Clark, A. D., Raag, R., Nanni, R. G., Hughes, S. H., and Arnold, E. (1995). Insights into DNA polymerization mechanisms from structure and function analysis of HIV-1 reverse transcriptase. *Biochemistry*, *34*, 5351–5363.

Raghavendra, N. K., Shkriabai, N., Graham, R. L., Hess, S., Kvaratskhelia, M., and Wu, L. (2010). Identification of host proteins associated with HIV-1 preintegration complexes isolated from infected CD4+ cells. *Retrovirology*, *7*, 66.

Rice, P. (2005). Resolving integral questions in site-specific recombination. *Nature Structural & Molecular Biology*, *12*, 641–643.

Stratmann, S. A., and van Oijen, A. M. (2014). DNA replication at the single-molecule level. *Chemical Society Reviews*, *43*, 1201–1220.

Normal view

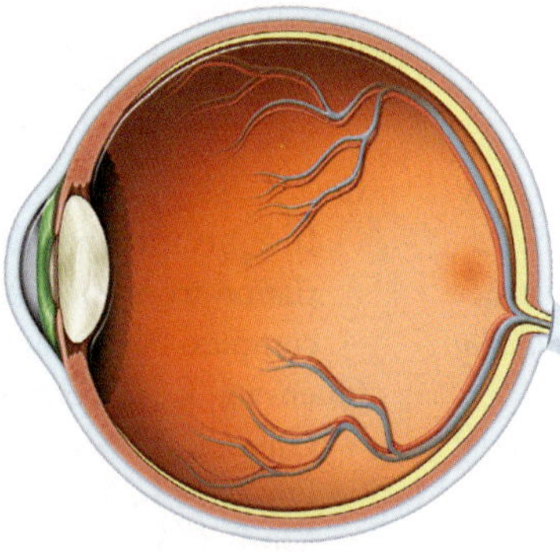

Normal retina

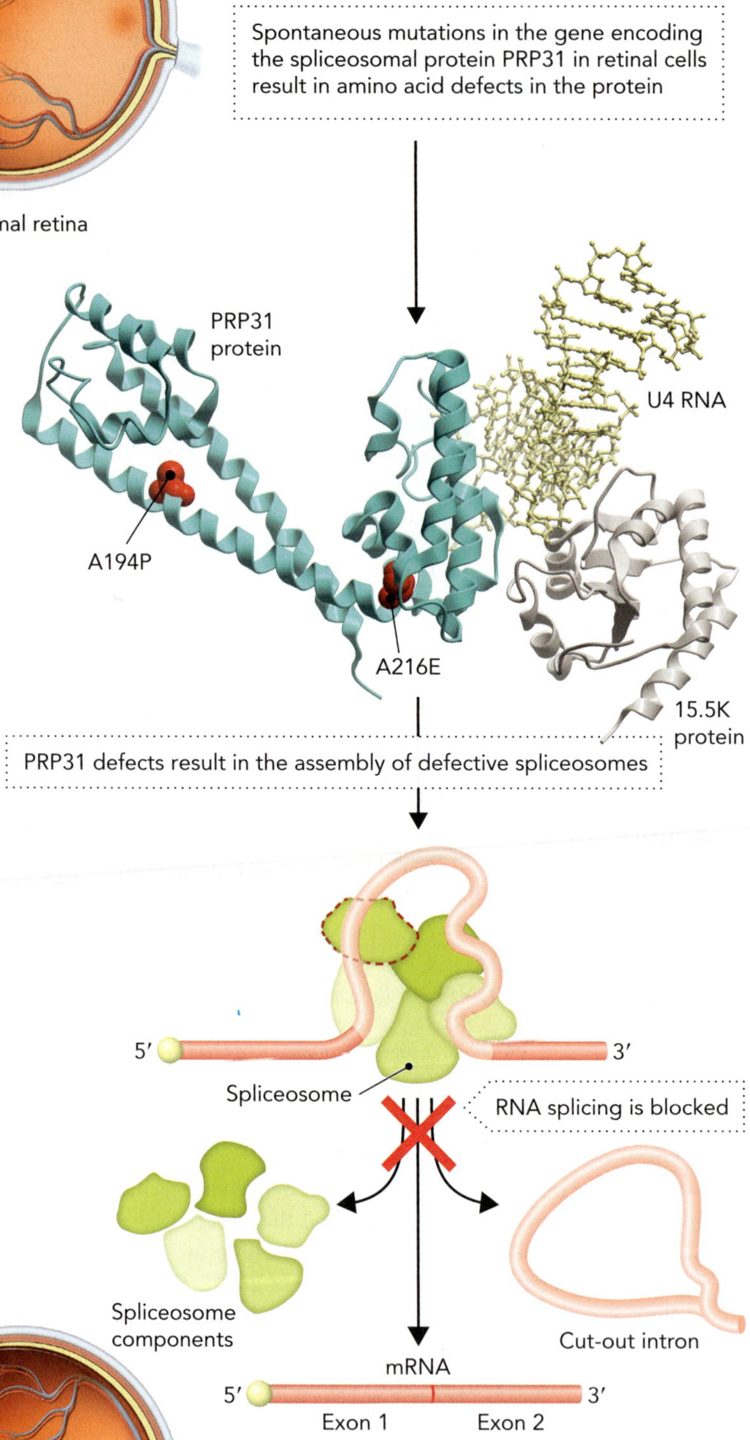
Spontaneous mutations in the gene encoding the spliceosomal protein PRP31 in retinal cells result in amino acid defects in the protein
PRP31 protein
U4 RNA
A194P
A216E
15.5K protein
PRP31 defects result in the assembly of defective spliceosomes
5′
3′
Spliceosome
RNA splicing is blocked
Spliceosome components
Cut-out intron
mRNA
5′
3′
Exon 1
Exon 2

Retinitis pigmentosa results from a defect in the assembly of spliceosomes, which are protein complexes that activate mRNA by splicing and removing introns. The top photograph represents normal vision. The reduced area in the bottom photograph illustrates the loss of peripheral vision that occurs with retinitis pigmentosa. The structure of the PRP31–U4–15.5K trimeric complex is shown on the right, with PRP31 depicted as a ribbon structure and two amino acid defects (A194P and A216E) identified by space-filling style. Both the A194P and A216E defects lead to incorrect assembly of spliceosomes due to the absence of the U4 snRNP (red dashed circle). This results in pigment deposits and cell degeneration.

View with retinitis pigmentosa

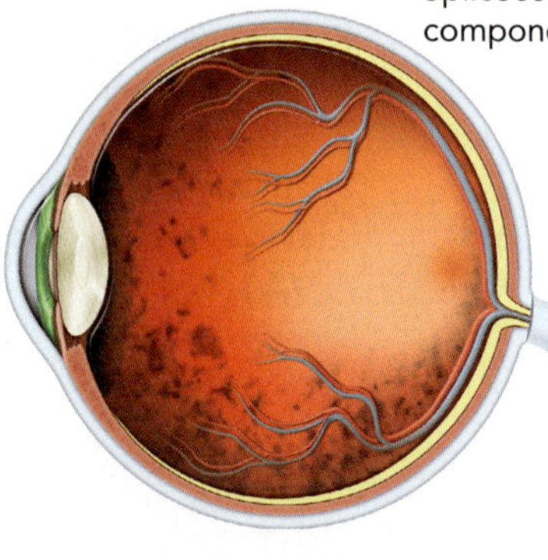
Pigment clumps accumulate in the retina causing vision loss

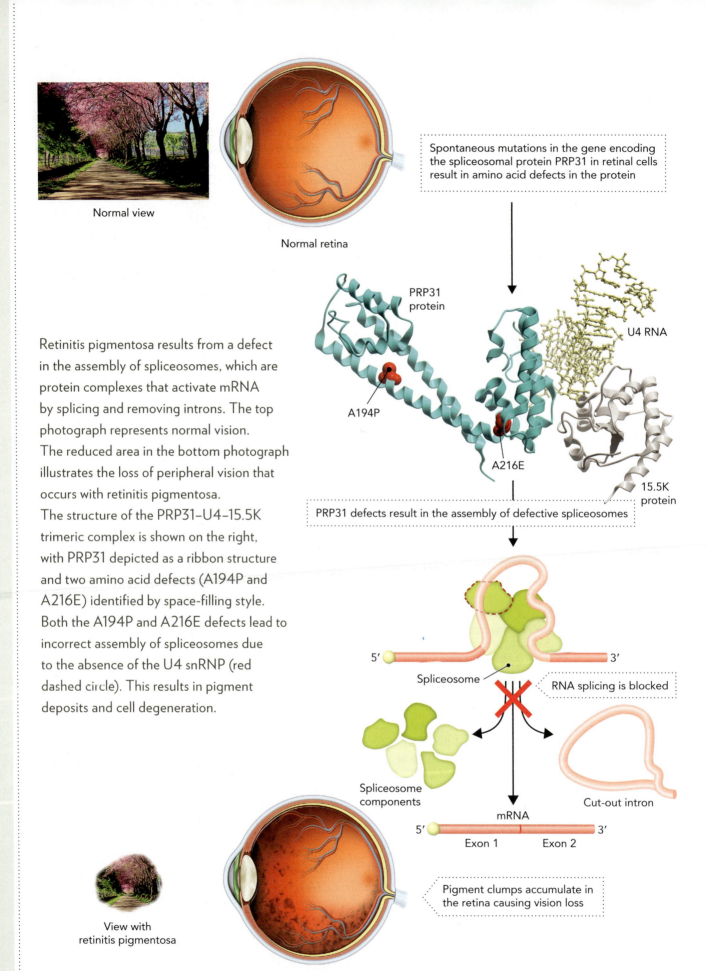

Retina with retinitis pigmentosa

21

RNA Synthesis, Processing, and Gene Silencing

◀ Retinitis pigmentosa is a degenerative disease of the retina characterized by loss of the light-sensing photoreceptor cells and black mottling of the retina. Several mutations associated with the development of retinitis pigmentosa have been found in factors that mediate splicing of mRNA. For example, missense mutations in the gene encoding the human protein PRP31 result in an autosomal dominant form of the disease. The PRP31 protein promotes assembly of the PRP31–U4–15.5K trimeric complex, which is essential for correct splicing of mRNA. The two PRP31 missense mutations, A194P and A216E, are believed to lead to interference with formation of the trimeric complex, thus preventing proper binding of the spliceosome to mRNA to facilitate intron removal. Loss of PRP31-mediated RNA splicing in photoreceptor cells leads to cell death and vision loss.

CHAPTER OUTLINE

CREDITS: CHERRY BLOSSOM PATHWAY: I LOVE PHOTO/SHUTTERSTOCK; PRP31-U4-15.5K TRIMERIC COMPLEX: BASED ON PDB FILE 3SIU.

In Chapter 3, we described the general structure of RNA and the role it plays in directing protein synthesis. In this chapter, we focus on the biochemistry of RNA synthesis and how the nascent RNA transcripts are processed into fully functional biomolecules. We first describe the biochemical properties of RNA as a versatile nucleic acid polymer that has many functions in the cell, including encoding the amino acid sequence of proteins, directly catalyzing chemical reactions, and regulating gene expression. Next we review the structure and function of the most abundant RNA molecules found in prokaryotic and eukaryotic cells. These are the RNA molecules required for protein synthesis: messenger RNA (mRNA), ribosomal RNA (rRNA), and transfer RNA (tRNA). Along with these three protein-synthesizing RNA molecules, which are found in all organisms, there is a diverse class of RNA molecules collectively referred to as **noncoding RNA (ncRNA)** found primarily in eukaryotes. As shown in **Table 21.1**, these ncRNA classes can be divided into three groups on the basis of length: short (miRNA, siRNA, piRNA), small (snRNA, snoRNA), and large (RNaseP, TERC, lncRNA). The structure and function of ncRNA molecules is only now beginning to be appreciated.

21.1 Structure and Function of RNA

By the early 1940s, it was clear to scientists that nucleic acids could be divided into two groups, DNA and RNA, on the basis of the structure of their ribose unit and the bases contained within each macromolecule. As described in Chapter 3, both RNA and DNA have the same backbone structure, with alternating phosphate and ribose units connected by a phosphodiester bond. In contrast to DNA, the ribose residue in RNA contains a 2′-OH that makes it more susceptible to autocleavage by base hydrolysis. The glycosidic bond connecting each nitrogenous base to a ribose is identical in both DNA and RNA, and both macromolecules contain the bases adenine, guanine, and cytosine. RNA contains the base uracil instead of thymine, which is found in DNA. As we will see in this chapter, RNA can contain both single-stranded and double-stranded regions that fold into a variety of structures and, in some cases, has catalytic functions.

Although RNA was initially discovered at the same time as DNA, it took many years before it was clear that there were different forms of RNA in the cell and that at least three of these forms were required for protein synthesis. Interest in RNA had taken time to develop, as scientists were primarily focused on understanding the structure and function of DNA. The sensitivity of RNA to autocleavage also meant that it often broke down in early preparations. Not until scientists began paying attention to the presence of **RNases**—enzymes that preferentially cleave RNA—were substantial amounts of RNA available for analysis.

By the late 1950s, it was generally accepted that RNA was the intermediate biomolecule connecting the genetic information encoded by DNA with the amino acid sequence of each protein; however, the confirmation of RNA as a "messenger" did not occur until scientists developed techniques that allowed them to isolate ribosomes from the rest of the cellular proteins. In 1960, François Jacob and Matthew Meselson confirmed that ribosomes were the site of protein synthesis and that they contained a significant RNA component. Although it was initially thought that the RNA component of ribosomes carried the genetic message, further experiments demonstrated that

Table 21.1 ABUNDANT RNA SPECIES IN PROKARYOTIC AND EUKARYOTIC CELLS

Types of RNA	Length (nt)[a]	Organisms	Function
Protein-synthesizing RNA			
Messenger RNA (mRNA)	~1,000–10,000	Prokaryotes, eukaryotes	Information transfer
Transfer RNA (tRNA)	~70–130	Prokaryotes, eukaryotes	Adaptor function
Ribosomal RNA (rRNA)	~120–4,300	Prokaryotes, eukaryotes	Peptidyl transfer reaction
Short noncoding RNA			
Micro RNA (miRNA)	~18–23	Eukaryotes	Translational regulation
Short interfering RNA (siRNA)	~18–23	Eukaryotes	Viral RNA degradation
Piwi-interacting RNA (piRNA)	~24–32	Eukaryotes	Genome stabilization
Small noncoding RNA			
Small nuclear RNA (snRNA)	~70–200	Eukaryotes	RNA splicing
Small nucleolar RNA (snoRNA)	~70–200	Eukaryotes	rRNA processing
Long noncoding RNA			
RNaseP RNA	~200–500	Prokaryotes, eukaryotes	Riboendonuclease
TERC RNA	~400–500	Eukaryotes	Telomerase reaction
Long ncRNA (lncRNA)	~200–1,000	Eukaryotes	Gene regulation

[a]nt = nucleotides.

this abundant RNA was identical in all ribosomes and therefore could not encode for cellular proteins.

RNA Is a Biochemical Polymer with Functional Diversity

RNA is a remarkably versatile biomolecule. Some RNA molecules are information carriers, like DNA, while others play structural or catalytic roles. Indeed, it is likely that RNA is the ancestral complex biomolecule upon which is built the biochemical framework of life on Earth. The RNA world hypothesis described in Chapter 18 was formulated in the 1980s when it was discovered that some RNAs have enzymatic properties and can function as a nucleic acid catalyst called a ribozyme (see Figure 18.27).

RNA is a biochemical polymer of four ribonucleotides containing the bases adenine, uracil, guanine, and cytosine. These bases can be chemically modified posttranscriptionally in a way that is analogous to posttranslational modification of proteins. Moreover, RNA is able to fold into a variety of tertiary structures made up of single-stranded and double-stranded regions and may have triplet and quadruplet interactions between its nucleotide bases (see Figure 3.26). These complex RNA structures afford enormous chemical functionality, which, along with cofactors that bind to RNA, can catalyze chemical reactions, provide scaffolds for multi-subunit protein complexes, and perform all functions required for protein synthesis in the form of mRNA, tRNA, and rRNA.

RNA is a highly dynamic biomolecule in three important ways:

1. RNA can undergo a rapid cycle of synthesis, functional interactions, and degradation, all of which can take place in the nucleus of a eukaryotic cell (**Figure 21.1a**). This temporal dynamic for RNA is in stark contrast to

proteins, which require RNA transcription, processing, and nuclear export of mRNA to the cytoplasm before protein synthesis can even occur.

2. RNA structure is altered by the binding of ligands, which can be large proteins or small metabolites or cofactors (**Figure 21.1b**). These molecules change interactions within the RNA molecule, which can promote the formation of alternative RNA tertiary structures. This gives rise to a different chemical functionality.
3. Base pairing between mRNA molecules and ncRNA molecules can modulate the process of protein synthesis. This dynamic interstrand RNA base pairing is transient, depending on the structure and abundance of the ncRNA molecules (**Figure 21.1c**).

An important feature of RNA is that the formation of RNA–RNA and RNA–DNA base pairs is not evolutionarily constrained in the same way protein–RNA or protein–DNA interactions are constrained. In contrast, a single nucleotide mutation in the protein-coding region of DNA can drastically alter the protein product by disrupting the translational reading frame or by encoding a different amino acid. Single RNA mutations often do not impart enough thermodynamic instability that the tertiary structure is lost. Indeed, the ability of RNA to maintain its biochemical function despite the accumulation of mutations supports the RNA world hypothesis (see Figure 18.27).

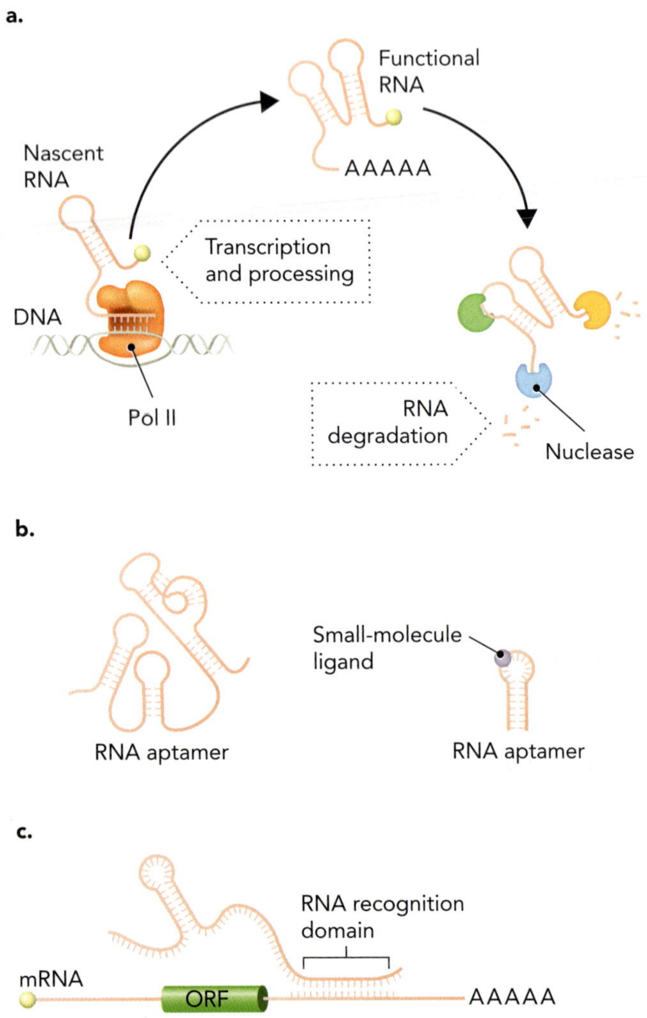

Figure 21.1 RNA is a dynamic biochemical polymer. **a.** RNA is synthesized and degraded within the nucleus, where it functions to control numerous biochemical processes. **b.** RNA aptamers are single-stranded molecules with well-defined tertiary structure. Their structure can be altered by the binding of protein and small-molecule ligands. Structural changes in RNA affect chemical functionality. **c.** RNA is able to participate in functional base-pairing interactions with RNA or DNA; for example, interactions between ncRNA and mRNA as shown here. ORF = open reading frame.

Protein-Synthesizing RNA Molecules: mRNA, tRNA, rRNA

Protein synthesis is very similar among vastly different organisms. Therefore, the three different types of RNA required to complete protein synthesis are well conserved in both structure and function.

As summarized in **Figure 21.2**, mRNA is a carrier of the genetic information, rRNA is the major constituent of ribosomes, and tRNA is the adaptor molecule that connects RNA synthesis to protein synthesis. The number of protein-coding genes varies widely between organisms, with humans containing ~20,000 protein-coding genes, and the *Escherichia coli* genome encoding 4,288 validated protein-coding genes. Most organisms contain multiple copies of rRNA genes to sustain the high levels of rRNA molecules needed to support protein synthesis. Human cells, for example, contain ~400 copies of the 45S rRNA gene, whereas *E. coli* has seven copies of the rRNA operon. Similarly, most organisms have many copies of tRNA genes, of the order 300–600. (Only about 30 tRNA genes need to be unique in order to cover the 61 codons specified by the genetic code in most organisms.) Consistent with the central dogma of molecular biology (see Figure 1.23), all three of these major RNA classes are required for the process of protein synthesis using the genetic information contained in DNA.

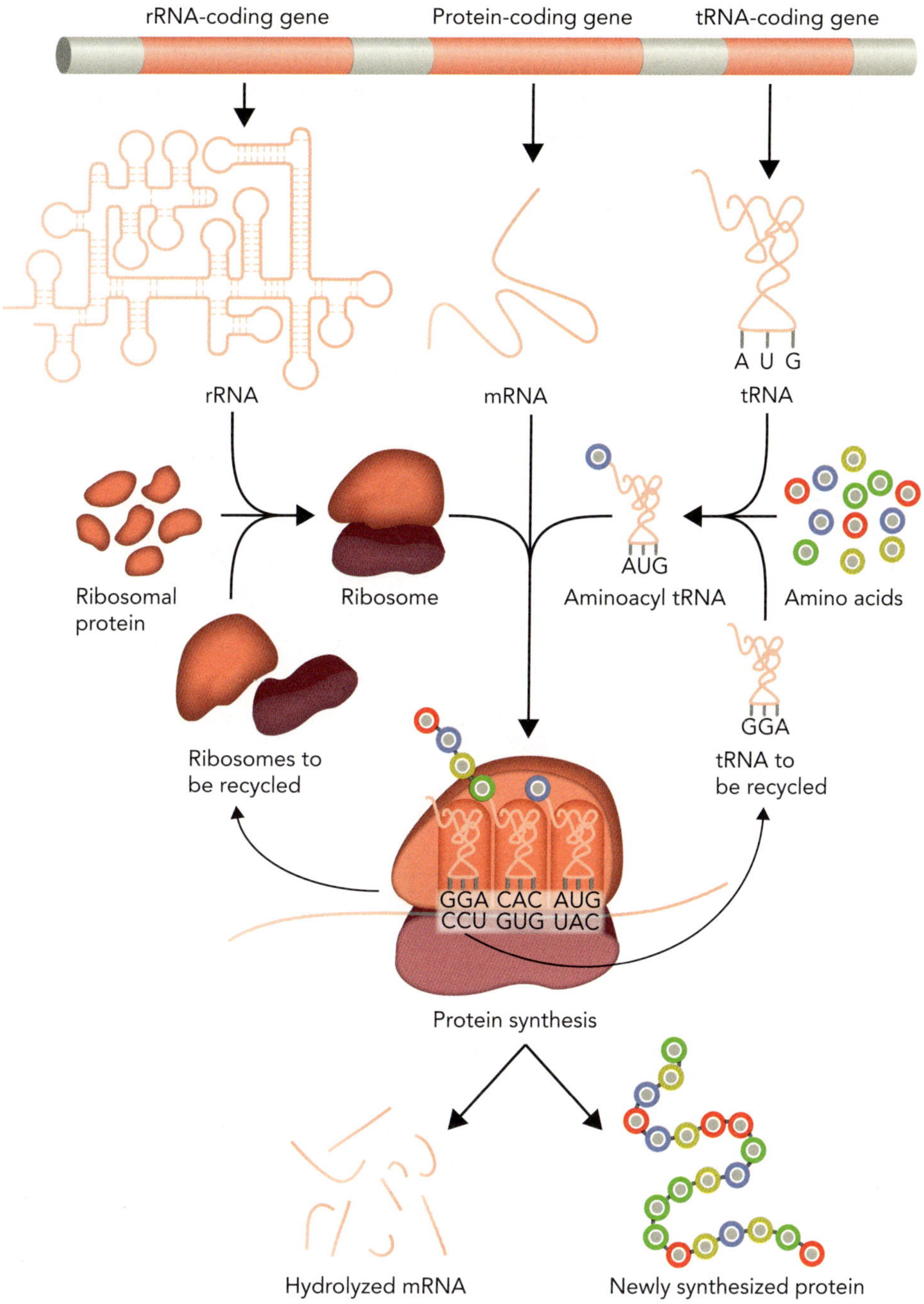

Figure 21.2 Protein synthesis requires three types of RNA molecules; mRNA, tRNA, and rRNA. Genes encoding rRNA provide the RNA component of ribosomes, which also require ribosomal proteins. Protein-coding genes direct the synthesis of mRNA, and tRNA genes produce the adaptor molecules to translate the genetic information in mRNA into polypeptide chains. Note that RNA is recycled as intact functional molecules (rRNA and tRNA) or degraded (mRNA).

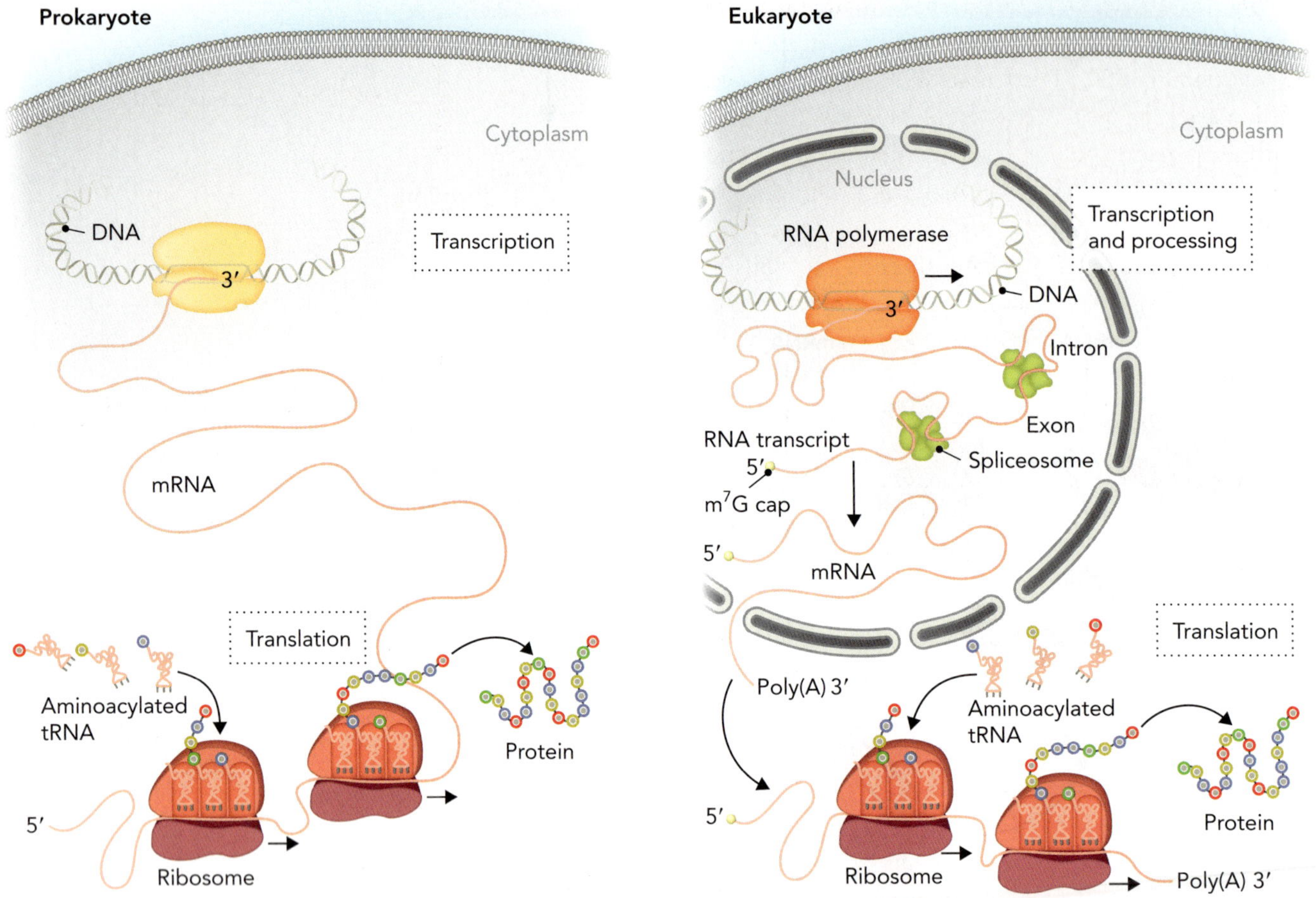

Figure 21.3 Transcription and translation in prokaryotes and eukaryotes is differentiated by compartmentalization. Prokaryotic protein synthesis occurs at the same time, and in the same cellular space, as RNA synthesis. In contrast, eukaryotic RNA synthesis and RNA processing events occur in the nucleus, whereas protein synthesis takes place in the cytoplasm. The spliceosomes are associated with the RNA polymerase, but this is not shown here for clarity.

Although protein synthesis occurs through the same general mechanism in prokaryotes and eukaryotes, prokaryotes lack a membrane-bound nucleus, which means that both transcription and translation occur in the same cellular compartment. As a result, mRNA is bound by ribosomes as soon as it is transcribed, and the processes of transcription and translation are tightly coupled. As shown in **Figure 21.3**, synthesis of prokaryotic tRNA and rRNA, along with assembly of the ribosomes, also occur in the cytoplasm. In eukaryotes, transcription occurs in the nucleus, and translation occurs in the cytoplasm. The physical separation of the two processes requires many additional factors to transport RNA from one compartment to the next as we describe later in the chapter. Eukaryotic cells also contain a subdomain within the nucleus known as the nucleolus, which is the site of rRNA transcription and the beginning stages of ribosome assembly.

Recall that in prokaryotes, mRNA can be monocistronic (encodes a single protein) or polycistronic (encodes multiple proteins on a single mRNA), whereas eukaryotic mRNA is most often monocistronic (see Figure 3.37). Unlike prokaryotic mRNA, the coding sequence of eukaryotic mRNA is often discontinuous due to the presence of intervening RNA sequences called introns, which interrupt the protein coding sequences called exons. Intronic sequences must be removed from the precursor mRNA by the process of RNA splicing to generate a continuous open reading frame consisting of exonic sequences (**Figure 21.4**).

In addition to RNA splicing, the physical separation of transcription and translation in eukaryotes requires that mRNA be chemically modified at both the 5′ and 3′

termini. The 5′ end of mature mRNA contains a 7-methylguanylate residue called the **7-methylguanylate cap (m^7G cap)**, whereas the 3′ terminus is extended by the addition of 50–200 adenine residues in a process called polyadenylation to generate a poly(A) tail. The 5′ m^7G cap and 3′ poly(A) tail facilitate the interaction between mRNA and the ribosome, increasing translational efficiency. Eukaryotic mRNA molecules also can undergo modifications to the bases that can affect mRNA stability, splicing, or even alter the coding sequence. We describe eukaryotic mRNA processing in more detail later in the chapter.

Although prokaryotic mRNA is not processed from a primary transcript in the same way as is eukaryotic mRNA, prokaryotic tRNA and rRNA molecules do require removal of spacer sequences to generate the mature forms of tRNA and rRNA. As shown in **Figure 21.5**, removal of this spacer RNA in prokaryotic tRNA and rRNA primary transcripts involves endoribonucleolytic cleavage by the ribozyme ribonuclease P (RNaseP) to release functional RNA molecules. Bacterial tRNA genes are located between rRNA genes and can be transcribed as components of a polycistronic RNA that is processed. Alternatively, tRNA can be encoded by primary tRNA transcripts that contain one to seven tRNAs surrounded by lengthy flanking sequences. We describe the processing of eukaryotic tRNA and rRNA genes later in the chapter.

Noncoding RNA Serves Important Functions in Eukaryotes

Eukaryotes encode multiple forms of noncoding RNA (ncRNA), which are classified into short, small, and long RNAs (see Table 21.1). **Figure 21.6** shows that the extent to which genomic sequences are transcribed

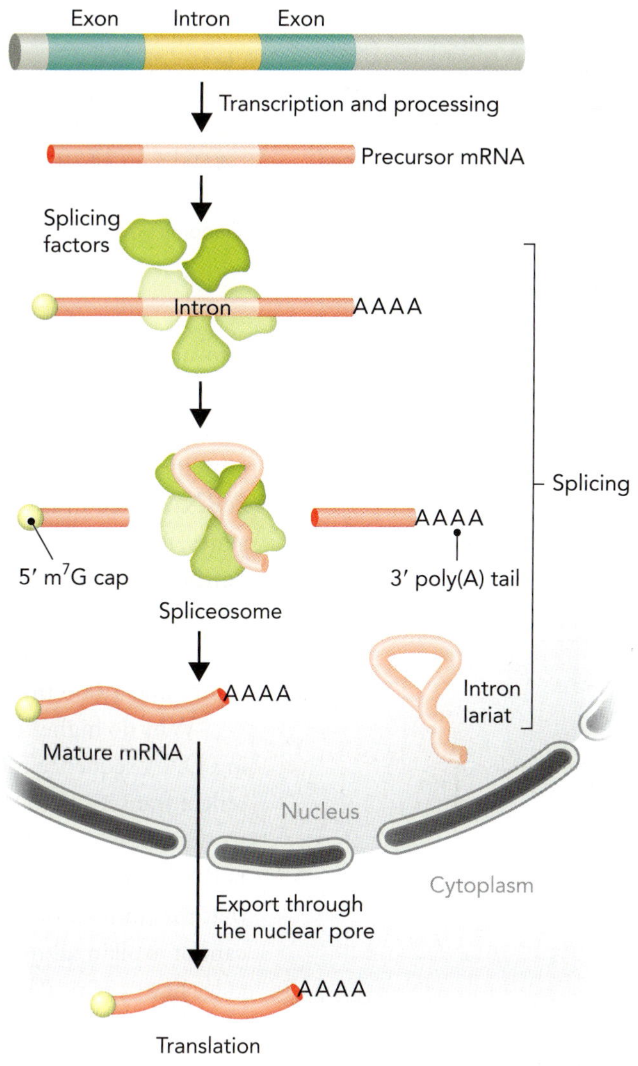

Figure 21.4 Eukaryotic mRNA is processed in the nucleus to remove intron sequences (splicing), protect the 5′ end with an m^7G cap, and facilitate mRNA translation by adding a poly(A) tail at the 3′ end.

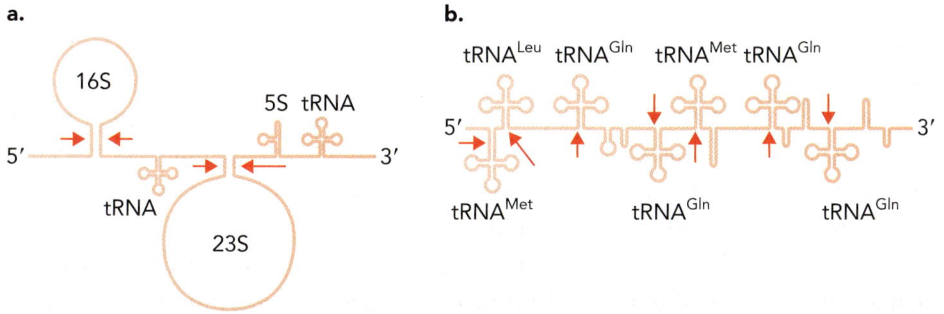

Figure 21.5 Prokaryotic tRNA and rRNA are encoded on polycistronic transcripts. **a.** An example of a prokaryotic primary transcript containing both tRNA and rRNA. **b.** An example of multiple tRNAs encoded in a single polycistronic transcript. Red arrows indicate RNaseP endonuclease cleavage sites.

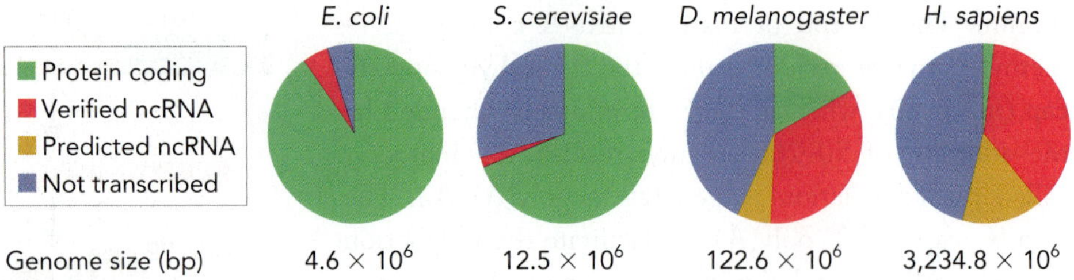

Figure 21.6 The fraction of genomic DNA corresponding to transcribed protein-coding genes is inversely related to genome size and to the amount of DNA transcribed into ncRNA. These values are estimates based on bioinformatic analysis of whole-genome sequencing and transcriptome characterization.

into ncRNA is proportionately higher in complex multicellular organisms than in bacteria such as *E. coli*. It is also inversely related to the fraction of the genome containing protein-coding genes. For example, 90% of the *E. coli* genome contains protein-coding genes, with very little of the *E. coli* genome accounting for transcribed ncRNA. In contrast, whereas only 2% of the human genome contains protein-coding genes, nearly 40% of the genomic sequences have been found to be transcribed into ncRNA, with another 15% predicted to be transcribed into ncRNA on the basis of bioinformatic analysis of cryptic promoter sequences. The amount of genomic DNA is an important difference between bacterial cells and human cells, with humans containing 1,000 times more DNA ($\sim 10^9$ bp) than *E. coli* (10^6 bp).

Why do higher organisms with large genomes transcribe so much of the noncoding DNA sequence into ncRNA molecules? For some years it was simply thought to reflect transcriptional noise due to aborted RNA synthesis events at cryptic promoter sequences, which are found scattered throughout the genome. Indeed, some of the ncRNA molecules identified in cells may in fact reflect transcriptional noise, given that the number of transcripts of any given ncRNA species is relatively low. But this cannot explain all ncRNA transcripts, as it is known that some small ncRNA molecules have a critical role in RNA processing, functioning as catalytic ribozymes. Other ncRNAs, such as the short ncRNA molecules miRNA and siRNA, mediate gene silencing. With the advent of high-throughput sequencing of RNA transcripts using the method of RNA-seq (see Figure 3.58), it was found that much more of the genome is transcribed into **long noncoding RNA (lncRNA)** molecules (>200 nucleotides) than was previously thought.

The first clues that some lncRNA molecules found in eukaryotic cells have specific functions came from genetic experiments in which DNA mutations that were biochemically linked to a cellular phenotype were shown to disrupt ncRNA expression. One of the best characterized of these lncRNA molecules is known as **X-inactive specific transcript (XIST)**, which regulates X-chromosome inactivation in females. As shown in **Figure 21.7**, the XIST gene is encoded on the human X chromosome, and expression of this lncRNA (17 kb in humans) leads to recruitment of proteins to only one of the two X chromosomes, resulting in chromosome inactivation. The stochastic mechanism giving rise to only one of the two X chromosomes undergoing X inactivation is related to a feedback regulatory loop involving a XIST antisense gene transcript called TSIX.

As shown in **Figure 21.8**, lncRNA sequences are found scattered throughout the human genome, as defined by genomic mapping of cDNA transcripts. For example, lncRNA sequences can be found in the same orientation and on the same DNA strand (sense) or can be derived from the opposite DNA strand (antisense). Other lncRNA sequences map to opposite strands but are separated by 1,000 bp or more (bidirectional) or are located entirely within introns between exon sequences of a protein-coding gene

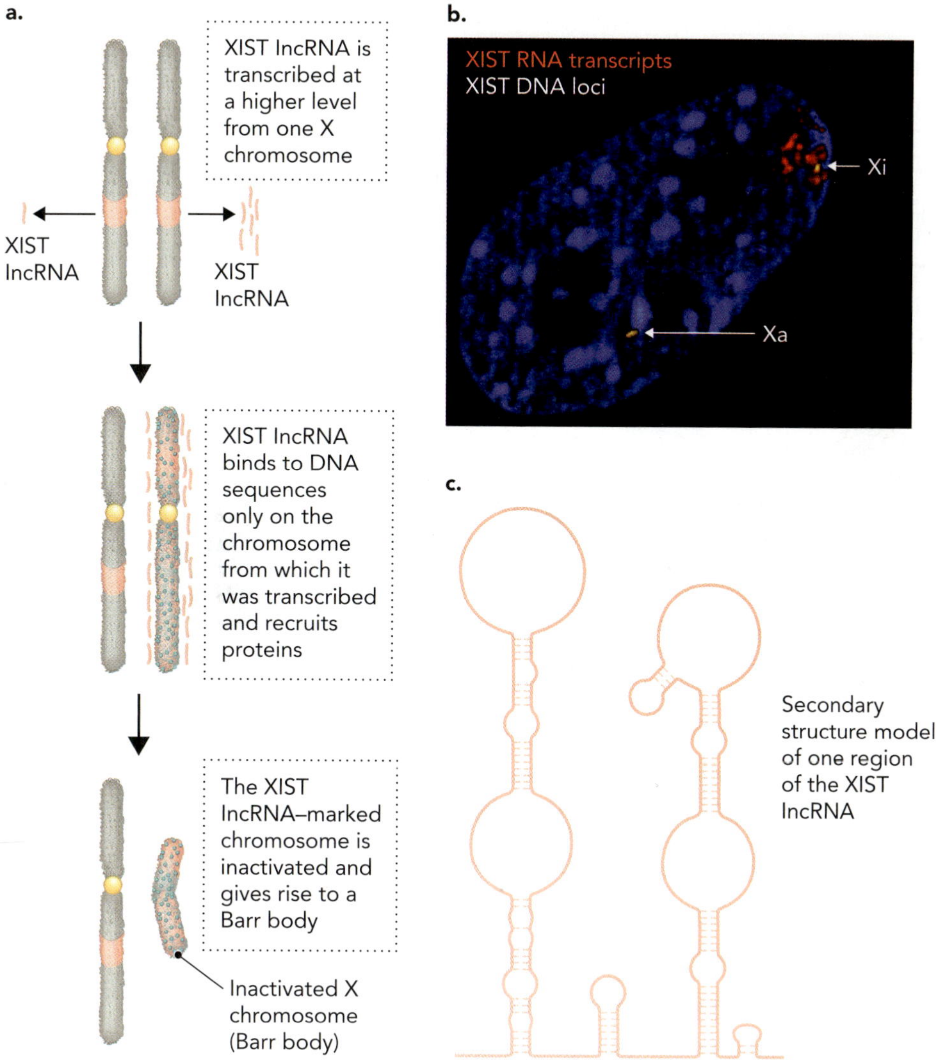
a.
XIST lncRNA is transcribed at a higher level from one X chromosome
XIST lncRNA
XIST lncRNA
XIST lncRNA binds to DNA sequences only on the chromosome from which it was transcribed and recruits proteins
The XIST lncRNA–marked chromosome is inactivated and gives rise to a Barr body
Inactivated X chromosome (Barr body)
b.
XIST RNA transcripts
XIST DNA loci
Xi
Xa
c.
Secondary structure model of one region of the XIST lncRNA

Figure 21.7 The XIST gene on the human X chromosome is transcribed into an lncRNA that regulates the process of X inactivation. **a.** The XIST gene is transcribed from both X chromosomes in females, but by a stochastic mechanism involving feedback regulation, only one of the two chromosomes produces enough XIST lncRNA to cause X inactivation. **b.** *In situ* hybridization with fluorescent nucleic acid probes identifies two X chromosomes in a cell, of which only one (Xi) is associated with XIST lncRNA. B. REINIUS ET AL. (2010). FEMALE-BIASED EXPRESSION OF LONG NON-CODING RNAS IN DOMAINS THAT ESCAPE X-INACTIVATION IN MOUSE. *BMC GENOMICS, 11*, 614. DOI: 10.1186/1471-2164-11-614. © REINIUS ET AL. 2010. **c.** Proposed secondary structure model of XIST lncRNA shows a variety of stem and loop structures.

(intronic). Lastly, some lncRNA sequences map to regions of DNA between separated protein-coding genes (intergenic). Most lncRNA sequences are likely to have arisen from mutations in protein-coding genes that maintain promoter sequences to direct transcription. Alternatively, mutations in noncoding DNA regions could have given rise to cryptic promoter sequences that convert a region of nontranscribed DNA into one that is transcribed.

With the discovery that up to ~50% of the human genome is transcribed into ncRNA (see Figure 21.6), and that much of this is lncRNA, the search began for examples other than XIST where lncRNA molecules were required for cellular function. The methods used in this search include selectively reducing the expression of specific lncRNA sequences and then looking for changes in cellular phenotypes. A more biochemical approach is based on isolating ribonucleoprotein complexes and characterizing the function of associated lncRNA molecules. As shown in **Figure 21.9**,

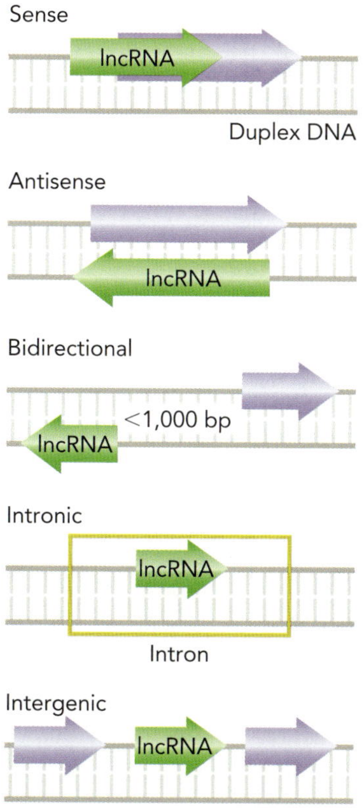
Sense
lncRNA
Duplex DNA
Antisense
lncRNA
Bidirectional
<1,000 bp
lncRNA
Intronic
lncRNA
Intron
Intergenic
lncRNA

Figure 21.8 A schematic map of lncRNA sequences in the genome is shown relative to known protein-coding genes. The purple arrows represent a protein-coding gene, and the green arrows represent lncRNAs.

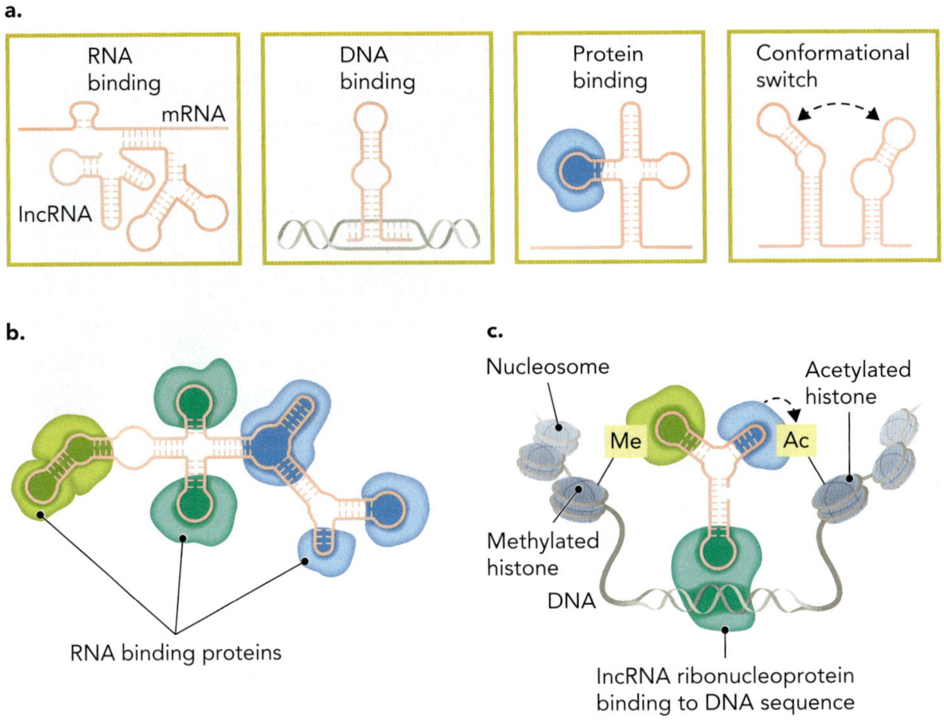

Figure 21.9 lncRNA has several proposed modes of action in mediating cellular functions. **a.** Studies investigating the biochemical properties of lncRNA have shown that lncRNAs are associated with other RNA and DNA molecules through sequence-specific base pairing, which could provide a regulatory function. Other modes of action include the formation of ribonucleoprotein complexes and undergoing conformational changes in response to signaling stimuli. **b.** lncRNA can function as a scaffold to facilitate the formation of large ribonucleoprotein complexes. **c.** lncRNA has been shown to mediate gene regulation by recruiting chromatin remodeling proteins to promoter sequences. Ac = acetyl; Me = methyl.

these studies have identified four functional mechanisms by which lncRNA molecules could regulate cellular processes: (1) base pairing between nucleotides in the lncRNA and target RNA, (2) base pairing between lncRNA and single-stranded regions of DNA, (3) formation of functional ribonucleoprotein complexes similar to ribosomes and spliceosomes, and (4) ligand-induced **riboswitches** that function in signaling pathways. (A riboswitch can be an mRNA that regulates its own translation in response to metabolite-induced structural changes.) Available data suggests that all four of these mechanisms are likely utilized by lncRNA molecules to mediate the assembly of large scaffolding complexes and sequence-specific gene regulation (Figure 21.9).

The list of specific cases where lncRNA molecules have been proved to be required for normal cellular functions is still relatively short, but many examples are known where changes in the pattern of lncRNA expression at the transcriptome level is associated with human disease states, such as cancer or neurologic dysfunction. Understanding the full role of this RNA species in regulating gene expression could provide many insights that will challenge our current models of RNA behavior.

concept integration 21.1

What contributes to the high percentage of ncRNA in organisms with large genomes compared to the percentage of ncRNA in organisms with smaller genomes?

The two contributing factors to the synthesis of ncRNA are transcriptional noise (that is, abortive transcripts) and regulatory functions of ncRNA that are not performed by proteins or other biomolecules. The higher percentage of ncRNA in higher organisms compared to that in lower organisms could be because large genomes are often found in complex multicellular organisms, which would give rise to a higher level of abortive transcripts owing to increased chance of a cryptic promoter. Moreover, higher eukaryotes have a greater need for a regulatory mechanism to control gene expression in response to environmental stimuli, and therefore a greater number of ncRNA species are needed to control the complexity.

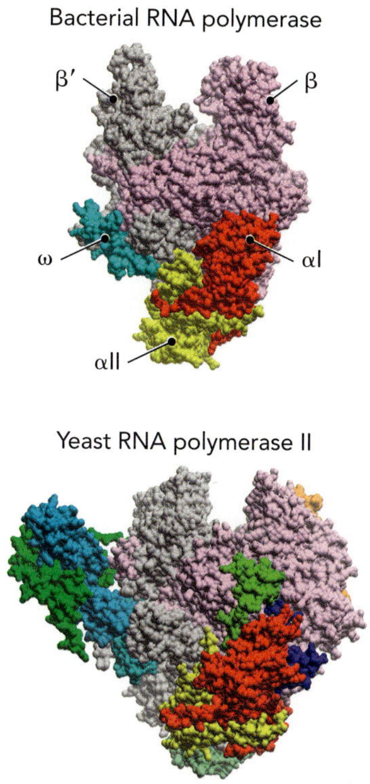

Figure 21.10 Molecular structures are shown for the RNA polymerase enzyme from the bacterial species *Thermus aquaticus* and for the RNA polymerase II enzyme from the yeast *Saccharomyces cerevisiae*. The five core subunits shared by prokaryotes and eukaryotes are colored the same. BASED ON PDB FILES 1HQM (BACTERIA) AND 1YIV (YEAST).

21.2 Biochemistry of RNA Synthesis

The hunt to understand the process of RNA synthesis gained its first major breakthrough in 1959 with the discovery by Samuel Weiss and Leonard Gladstone of a DNA-dependent RNA polymerase in extracts of mammalian cells. This was followed by the isolation of an RNA polymerase from bacterial species by several laboratories. Crystallization of the yeast RNA polymerase II and the RNA polymerase from the thermophilic bacterium *Thermus aquaticus* in 1999 revealed that the two enzymes shared a high degree of structural homology. To date, more than 1,700 crystal structures of RNA polymerase subunits are available in the Protein Data Bank. As shown in **Figure 21.10**, bacterial RNA polymerases consist of five subunits with the structure $\alpha_2\beta\beta'\omega$, whereas the yeast RNA polymerase contains these same five orthologous core subunits and another seven auxiliary protein subunits (**Figure 21.11**).

The α_2 dimer of the bacterial RNA polymerase is asymmetric and provides a scaffold for the holoenzyme assembly. *Holoenzyme* here refers to the minimal number of subunits required for catalytic activity without the addition of regulatory subunits, and the subunits can contain cofactors such as metal ions. Although the α subunits are primarily a structural component of the holoenzyme, the C-terminal domain of one of the subunits is also important for interactions with positive activators of transcription, such as the **cAMP receptor protein (CRP)**. The β and β′ subunits of the bacterial RNA polymerase holoenzyme form the catalytic center, with the ω subunit functioning to stabilize the interaction between one of the α subunits and the β′ subunit. As seen in Figure 21.10, the β and β′ subunits form a structure resembling a claw, which forms a channel that serves as the binding site for the single-stranded DNA template.

Eukaryotic RNA polymerases contain as many as 10–20 protein subunits depending on the organism, and the fully assembled RNA polymerase complex can range in size from 0.4 to 0.8 megadalton (MDa). As described shortly, three distinct eukaryotic

Bacteria	Yeast	
β′	Rpb1	Core RNA polymerase subunits in the holoenzyme
β	Rpb2	
αI	Rpb3	
αII	Rpb11	
ω	Rpb6	
	Rpb4	
	Rpb5	
	Rpb7	
	Rpb8	
	Rpb9	
	Rpb10	
	Rpb12	

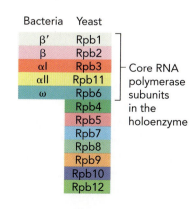

Figure 21.11 The names of RNA polymerase subunits in bacteria and yeast are shown to provide a comparison between the two holoenzymes. Core RNA polymerase subunits are aligned and indicated by the same color as used in Figure 21.10 for the corresponding polypeptides.

RNA polymerases have been characterized: RNA polymerase I, RNA polymerase II, and RNA polymerase III. The protein subunits found in these three types of eukaryotic RNA polymerases can be roughly divided into three categories: those homologous to the bacterial RNA polymerase subunits ($\alpha_2\beta\beta'\omega$), those common to all three RNA polymerases, and those that are specific to only one type of eukaryotic RNA polymerase. The bacterial homologues are similar in structure and function, whereas the remaining subunits are primarily involved in interactions with other eukaryotic transcription factors or in stabilizing the interaction between RNA polymerase and the DNA template. Although both bacterial and eukaryotic RNA polymerase holoenzymes are sufficient to catalyze the synthesis of RNA from a DNA template, additional transcription factors are needed to initiate and regulate transcription.

RNA Polymerase Is Recruited to Gene Promoter Sequences

The RNA polymerase holoenzyme has the ability to bind nonspecifically to DNA; however, it is the auxiliary RNA polymerase subunits that direct the holoenzyme to specific gene promoter sequences. In bacteria, the $\alpha_2\beta\beta'\omega$ holoenzyme is associated with a family of auxiliary proteins called **σ factors**, which are transcription factors that bind to specific DNA sequences known as the −35 and −10 boxes (**Figure 21.12**). The σ factor targets RNA polymerase to the promoter by decreasing the affinity of the protein for nonspecific DNA sequences by a factor of 10^4. In this way, only promoter region DNA is bound efficiently by the complete polymerase, ensuring that transcription begins at the correct location in the DNA template.

Bacteria contain a primary σ factor responsible for initiating transcription of genes required for essential biochemical processes, which are often called **housekeeping genes**. In addition, bacteria contain a variable number of related σ-factor proteins that activate transcription of specific genes in response to growth

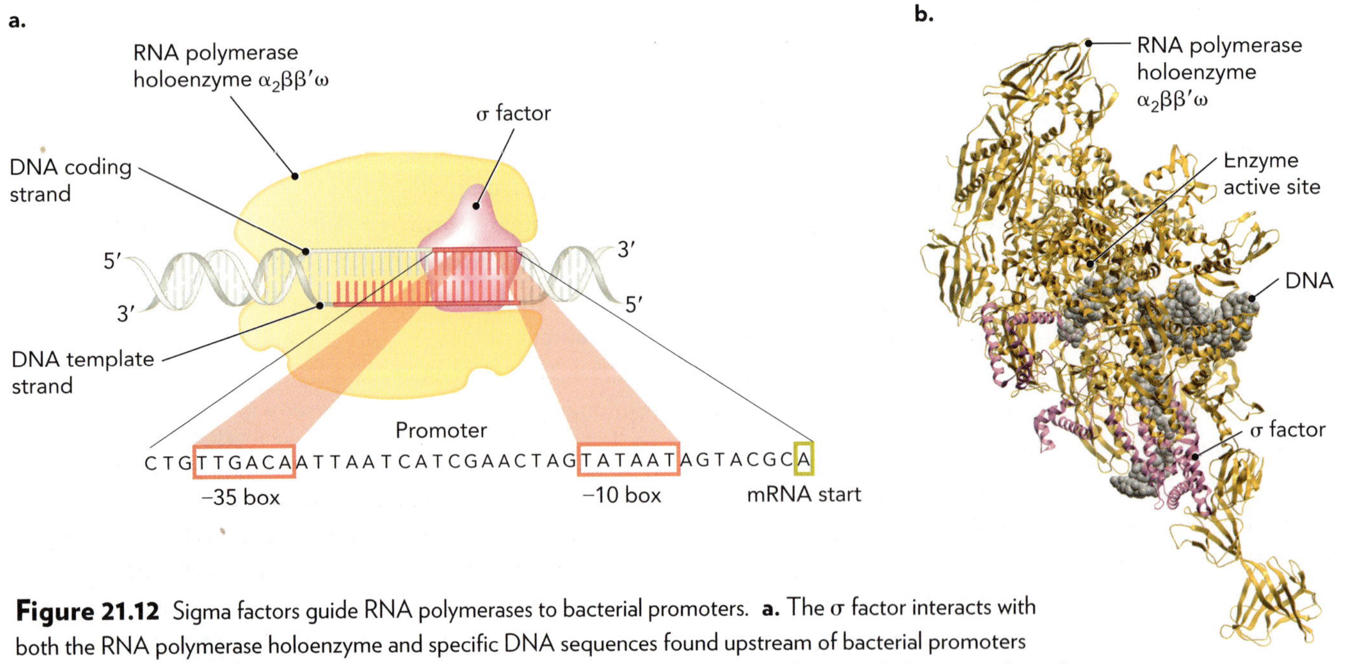

Figure 21.12 Sigma factors guide RNA polymerases to bacterial promoters. **a.** The σ factor interacts with both the RNA polymerase holoenzyme and specific DNA sequences found upstream of bacterial promoters called the −35 and −10 boxes. **b.** Structure of the *Thermus thermophilus* RNA polymerase holoenzyme bound to a bacterial gene promoter. The DNA strands, shown in space-filling representation, are partially separated for transcription. BASED ON PDB FILE 4Q4Z.

signals or stress conditions. The number of σ factors can be quite different among bacterial species. For example, the *E. coli* genome encodes seven different σ factors, whereas the bacterium *Streptomyces coelicolor* encodes more than 60 different σ factors in its genome. The number of alternative σ factors present in a specific bacterium is likely due to the living conditions of the organism. *E. coli*, for example, live in the relatively sheltered environment of the gut of other organisms and therefore are not subjected to dramatic changes in temperature or nutrient availability. This is in contrast to *S. coelicolor*, which live in the soil and must be able to adapt to several different environmental conditions.

Identification of gene promoter sequences is achieved in the laboratory using a technique called **DNase I footprinting.** This technique is based on the fact that DNase I, an endonuclease that cleaves the phosphodiester backbone of DNA on both strands, does not cleave DNA where it interacts with a protein. When a protein is bound to a region of DNA, it prevents the DNase I enzyme from accessing the DNA minor groove and therefore protects the DNA from cleavage. Although cleavage is random, DNase I preferentially cleaves phosphodiester bonds between adjacent purines and pyrimidines. Thus, a double-stranded DNA devoid of bound proteins will produce a random set of cleavage products of various lengths as long as the DNase I reaction is not allowed to go to completion (complete DNA digestion would yield short oligonucleotides).

As shown in **Figure 21.13**, the DNase I footprinting method can be used to map the location of protein binding sites on DNA. The DNA molecule is labeled at only one end of the duplex with radioactive phosphate (^{32}P). After partial digestion of the DNA with limiting amounts of DNase I for short periods of time, the resulting cleavage products are separated on the basis of nucleotide length using gel electrophoresis. The DNA fragments can be visualized using autoradiography, which detects decay of the ^{32}P isotope. In the reaction mixture that lacks the DNA binding protein, the DNA should be cleaved at every position. Therefore, gel electrophoresis should yield a "ladder" of radioactive DNA cleavage products differing in length by a single nucleotide. In the reaction mixture where the DNA sample is first incubated with the DNA binding protein, such as *E. coli* RNA polymerase containing the holoenzyme and the σ subunit, then some DNA sites cannot be cleaved by DNase I because they are blocked by the DNA binding protein. This absence of specific cleavage products gives rise to a so-called DNA footprint, which corresponds to the protein-protected DNA sequences.

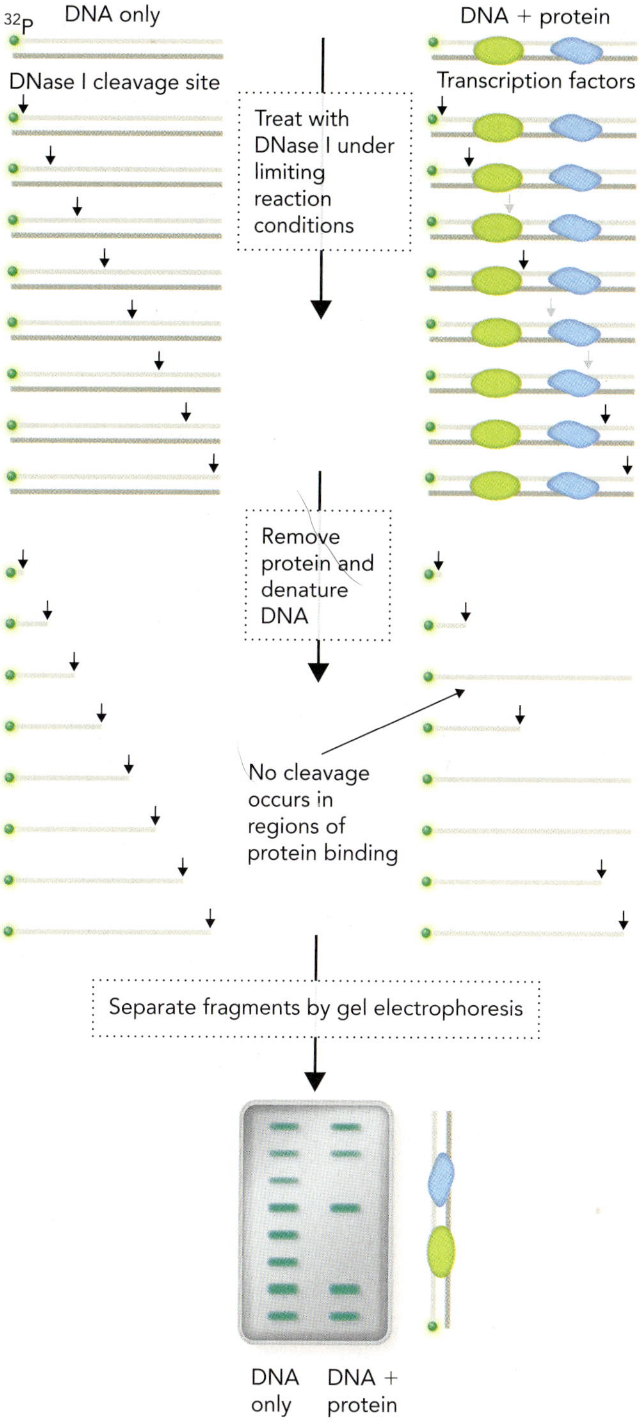

Figure 21.13 Protein binding sites are identified by DNase I footprinting. A schematic of the DNase I footprinting assay is shown with two transcription factors bound to DNA. The fragments resulting from random (and limited) DNase I cleavage are shown from both the control and protein-added assays. A schematic of an autoradiograph from a DNA footprinting experiment is shown at the bottom with protein binding sites identified.

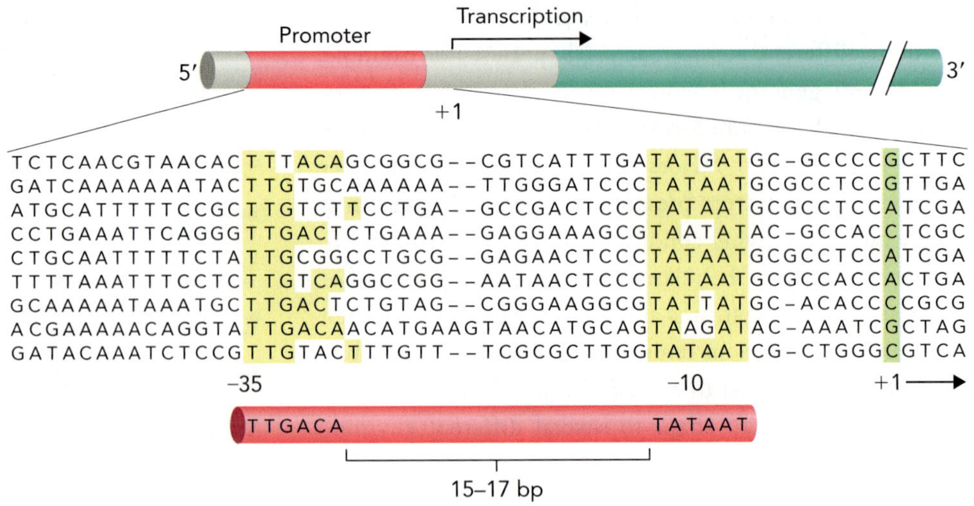

Figure 21.14 This schematic of a bacterial genomic region shows the promoter upstream of the transcribed region. An alignment of numerous *E. coli* promoters shows homology at positions −35 and −10 relative to the transcription start site.

DNase I footprinting experiments using bacterial promoter regions were initially used to identify two short sequences of DNA that were protected by *in vitro* binding of RNA polymerase in a complex with σ-factor protein. As shown in **Figure 21.14**, the most commonly protected sequences under these conditions were found to be located about ~10 nucleotides and ~35 nucleotides upstream of the transcriptional start site. Additional bioinformatic analysis of hundreds of prokaryotic promoter sequences were used to define consensus sequences for what are referred to as the −35 box (5′-TTGACA-3′) and the −10 box (5′-TATAAT-3′). Alignment of these different promoter sequences, however, made scientists aware that not all positions within the −35 and −10 boxes were equally conserved. These differences are due in part to the need for different σ factors to recognize different coding regions. Variations from the conserved sequence also indicate the strength of the promoter, with strong promoters having −35 and −10 boxes that differ only slightly from the consensus sequence, whereas weak promoters may contain several nucleotide changes. A strong promoter is one in which the rate of transcriptional initiation is higher than that of most promoters, whereas a weak promoter is one in which the rate of transcriptional initiation is lower than that of most promoters.

Similar DNase I footprinting assays and bioinformatic analyses have been used to identify conserved sequences in eukaryotic promoters. The organization of these promoter regions, however, is much more complex than that of bacteria. The complexity of eukaryotic promoters is due to the existence of three RNA polymerases (I, II, III), as well as the need for an increased level of gene regulation in these complex organisms. **Figure 21.15** illustrates the conserved DNA sequences found in gene promoters that are recognized by the three eukaryotic RNA polymerases. In most of these eukaryotic gene promoter regions, the conserved DNA binding sites for RNA polymerase protein complexes are located on the 5′ side of the +1 nucleotide in the primary RNA transcript (upstream). Additional conserved DNA binding sites are found at the start site of transcription, and some are located on the 3′ side of the +1 nucleotide (downstream).

Most eukaryotic promoters have multiple protein binding sites—some for components of the RNA polymerase complex, and others for transcription factors that function as regulatory proteins. A subset of these transcriptional regulatory factors function like the bacterial σ factor in that they recruit RNA polymerases to specific

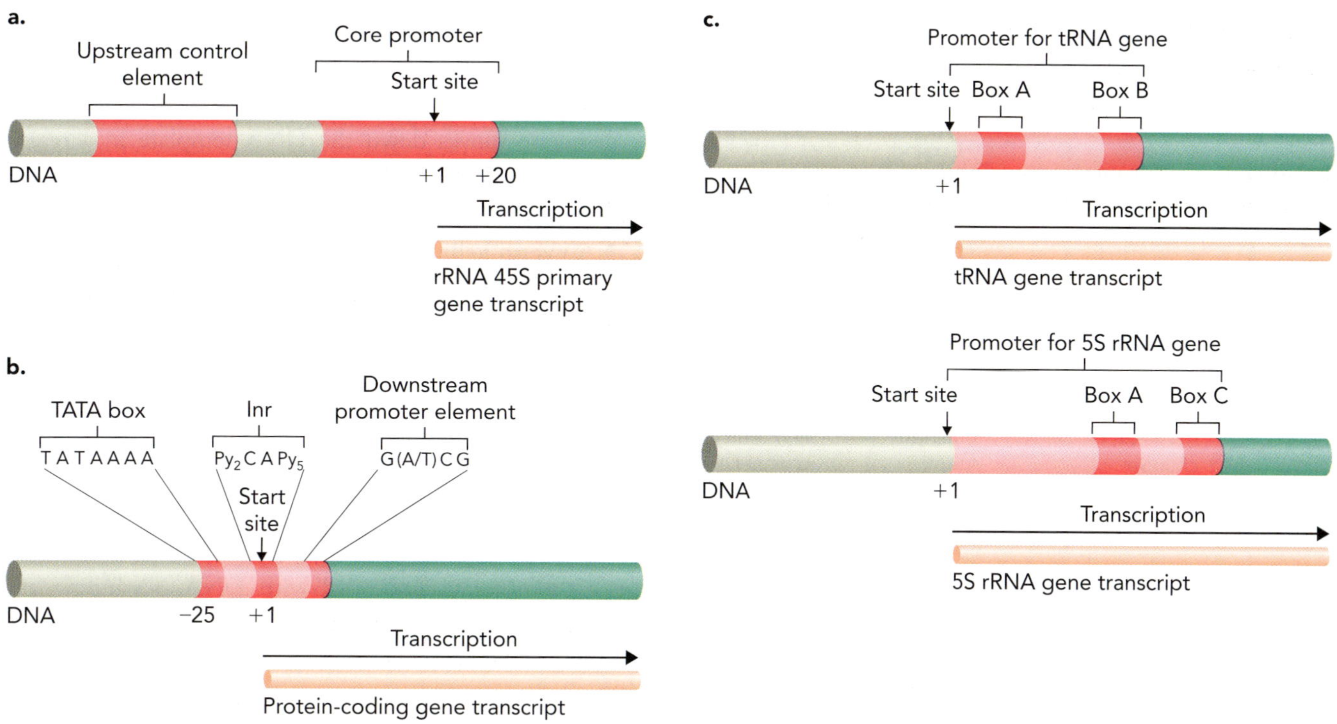

Figure 21.15 Eukaryotic promoters contain conserved DNA sequences that function to specifically recruit RNA polymerase I, II, or III protein complexes. **a.** The RNA polymerase I promoter region for a ribosomal RNA (rRNA) gene contains an upstream control element and a core element. **b.** RNA polymerase II promoter regions have an upstream TATA box sequence where the TATA binding protein binds. The initiator region (Inr) is located at the transcription start site, and the downstream promoter element is downstream of the start site for transcription. The abbreviations Py_2 and Py_5 refer to a tract of two or five pyrimidine residues (C or T), respectively. **c.** RNA polymerase III promoters, such as those for tRNA (top) and 5S rRNA (bottom), have box A and box B sequences located downstream of the transcription start site.

gene promoters in response to normal growth signals or environmental stress. One such transcription factor is called the **TATA binding protein (TBP)**, which binds to the sequence 5′-TATAAAA-3′ located ~30 nucleotides upstream of the transcriptional start site in gene promoters transcribed by RNA polymerase II. TBP is a single polypeptide chain that folds to produce two symmetric halves. These halves flank the DNA and generate a sharp bend as a result of two phenylalanine residues on each side that insert into the minor groove and force a kink into the DNA backbone (**Figure 21.16**). The strong interaction between TBP and the TATA box is influenced by a string of lysine and arginine residues that bind in the minor groove and by asparagine residues that make specific base contacts at the center.

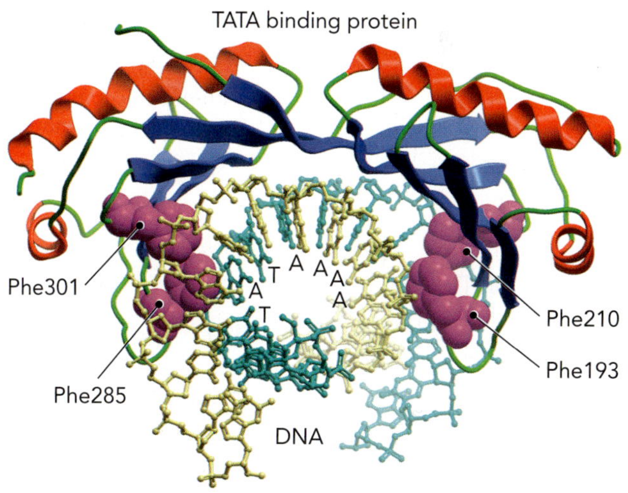

Figure 21.16 TBP binds to specific sequences in eukaryotic RNA polymerase II promoters. The single polypeptide chain of the human TBP has two nearly identical structural domains, each with a pair of Phe residues (shown in pink space-filling representation) that insert into the DNA double helix. The 5′-TATAAAA-3′ sequence is labeled. BASED ON PDB FILE 1CDW.

Proteins Required for RNA Synthesis in Prokaryotes

During RNA synthesis in all organisms, one strand of DNA is used as a template for synthesis of a complementary strand.

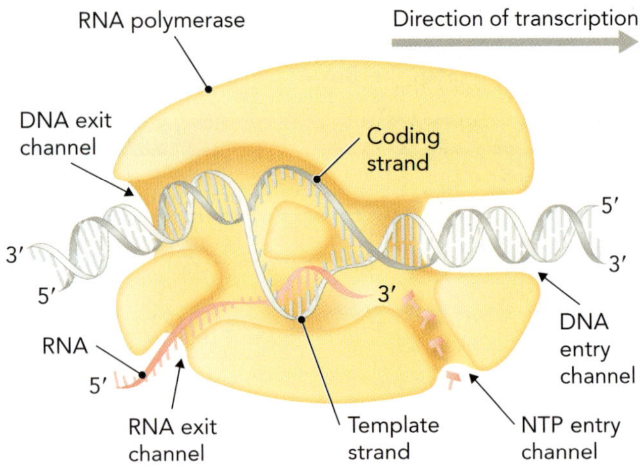

Figure 21.17 RNA synthesis uses the template strand to generate a transcript.

As shown in **Figure 21.17**, in order to access the template strand, the DNA must be unwound by RNA polymerase. As RNA synthesis takes place in the 5′ to 3′ direction, RNA polymerase moves down the DNA template in a region known as the transcription bubble, which contains the enzyme, a locally unwound DNA "bubble," and an RNA–DNA hybrid helix, usually about 8 base pairs in length.

The biochemistry of RNA synthesis has been most extensively characterized in *E. coli*, and therefore we will use this model system to describe the basic mechanisms of RNA synthesis. As shown in **Figure 21.18**, transcription of bacterial promoters is initiated by binding of the RNA polymerase complex in association with the σ-factor protein. Conversion of the RNA polymerase–DNA **closed complex** (inactive) to the **open complex** (active) at the gene promoter is associated with DNA unwinding and formation of the first phosphodiester bond in the nascent RNA transcript. As described in Chapter 20, initiation of

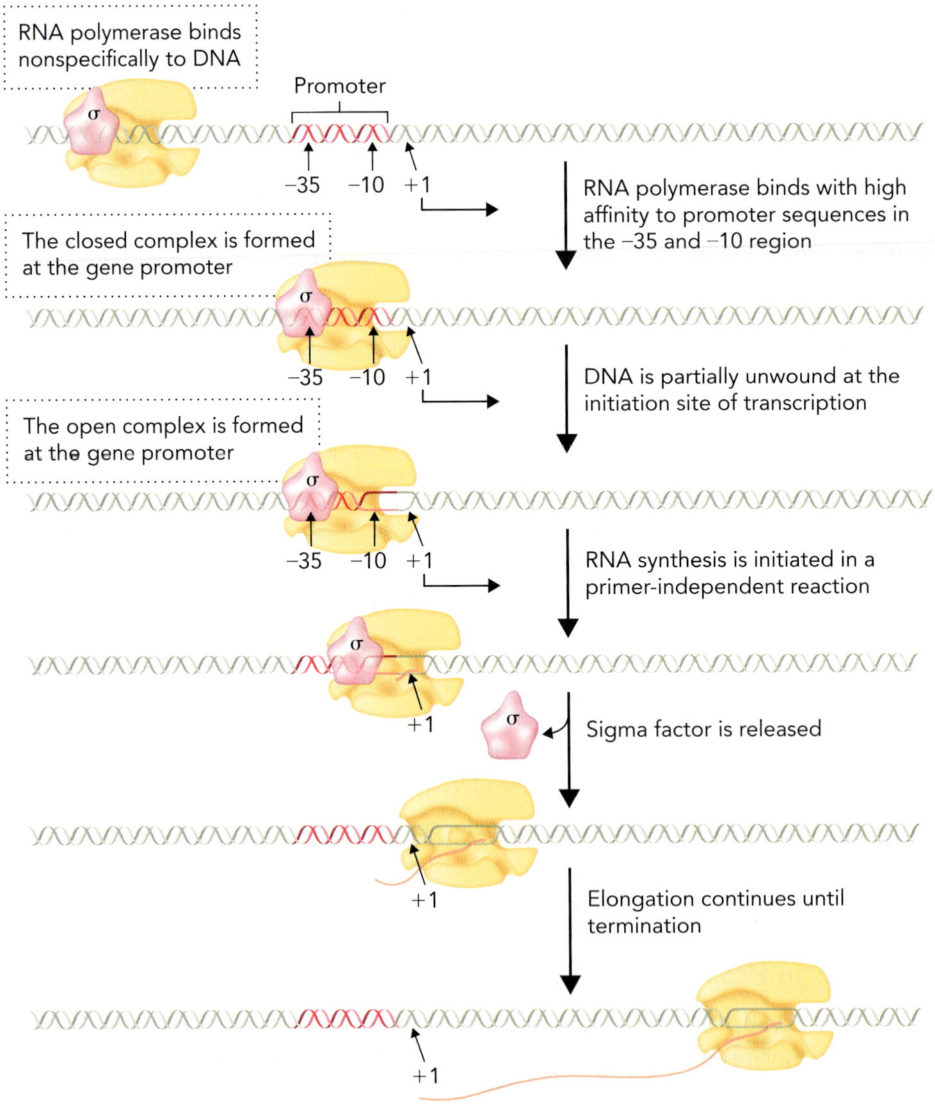

Figure 21.18 The process of bacterial RNA polymerase recruitment to the promoter by σ factor, transcription initiation, and elongation in *E. coli* is shown sequentially from top to bottom. At the top, the enzyme complex is shown scanning the DNA. Upon binding to the −35 and −10 boxes, the closed promoter complex transitions to an open complex, which initiates transcription and the release of the σ factor. Transcription elongation then proceeds, beginning at the +1 site.

RNA synthesis during the process of DNA replication by the enzyme primase does not require a primer; this is also true for RNA synthesis during the process of transcription. The elongation phase of RNA synthesis in *E. coli* is promoted by dissociation of the σ-factor protein from the RNA polymerase complex, which leads to increased rates of transcription and translocation of the transcription bubble down the DNA strand, with about 17 bp of DNA separated into single strands at a time. Average elongation rates are ~20 to 80 nucleotides per second. RNA synthesis is therefore slower than DNA synthesis, which averages more than 1,000 nucleotides per second. The low rate of RNA synthesis is due to the discontinuous nature of the elongation reaction, as it frequently pauses, stalls, and arrests. Frequent pausing of transcription is actually used as a mechanism of gene regulation.

As DNA enters the active site of the *E. coli* β′ subunit, the RNA–DNA hybrid rotates so that the 3′-OH of the growing RNA remains accessible to the catalytic site. The incoming NTP is stabilized through interactions with two Mg^{2+} ions that are coordinated by Asp residues in the polymerase. One of these Mg^{2+} ions interacts with the 3′-OH of the RNA chain, facilitating deprotonation of the oxygen. This mechanism is the same as that described for DNA synthesis, with the only difference being a deoxyribose sugar in the case of the DNA synthesis reaction catalyzed by DNA polymerase (see Figure 20.6). As the RNA chain grows, the DNA–RNA helix is forced to separate near the exit point of the active site, which allows re-formation of the DNA double helix and exit of the newly synthesized RNA chain. In bacteria, as with most prokaryotes, elongation continues until one of two mechanisms occurs to signal the end of synthesis.

The signals controlling transcriptional termination are equally important to those controlling transcriptional initiation, as failure to stop RNA synthesis efficiently would have significant consequences for the organism. In bacteria, transcriptional termination can occur as either a **Rho-dependent termination** or **Rho-independent termination** process. Rho is an ATP-dependent helicase that binds to the newly synthesized RNA chain at a C-rich sequence. This destabilizes the RNA–DNA helix, thus terminating transcription and dissociating the polymerase complex (**Figure 21.19**). Rho-dependent transcription therefore requires both cis and trans factors. By contrast, Rho-independent termination, also called intrinsic termination, does not require any trans-acting factors and depends only on the presence of certain sequences within the RNA. The termination signal in bacterial DNA consists of a GC-rich region followed by an A-rich region. The RNA synthesized from this region forms a stem–loop structure at the GC-rich region. The presence of the stem–loop causes the polymerase complex to pause, while the weak interaction between the dA–rU base pairs leads to a dissociation of the RNA–DNA duplex (Figure 21.19).

▶ ANIMATION

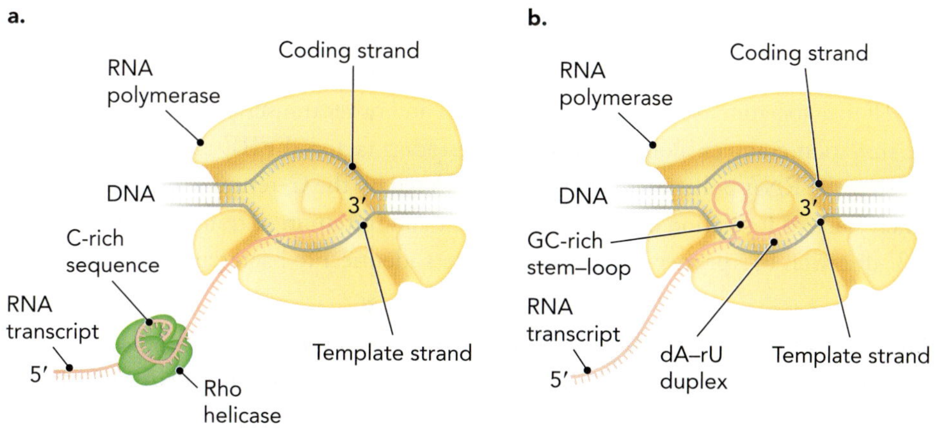

Figure 21.19 Transcription termination in prokaryotes is either Rho dependent or Rho independent. **a.** Rho-dependent termination is caused by binding of the Rho protein to a C-rich region in the newly synthesized RNA. **b.** Rho-independent (intrinsic) termination is triggered by synthesis of a GC-rich stem–loop followed by a sequence with multiple uridine residues. The GC stem–loop structure causes the polymerase to pause, and the relatively unstable dA–rU hybrid duplex dissociates. dA = deoxyadenylate; rU = uridylate.

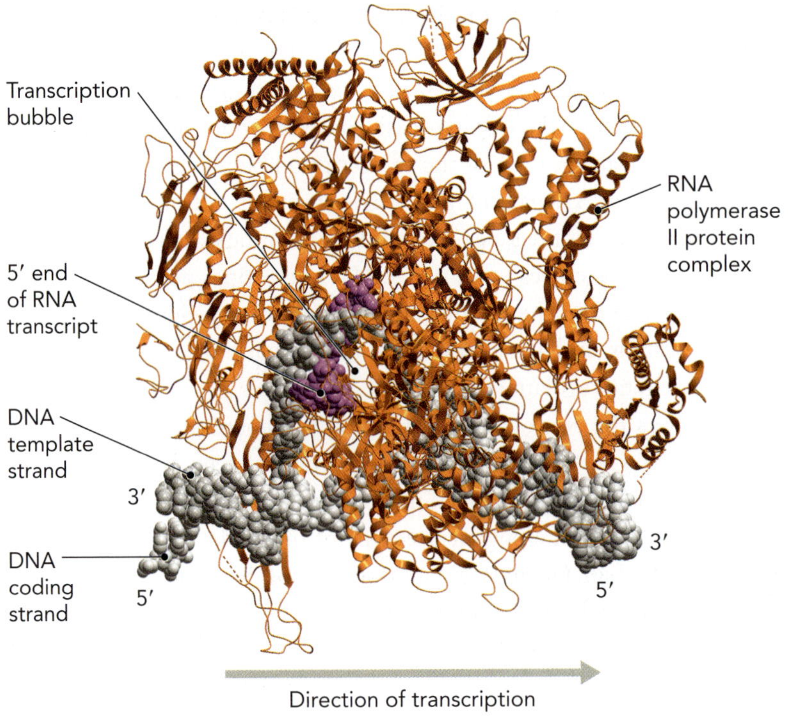

Figure 21.20 The structure of a yeast RNA polymerase II transcription bubble is shown. This structure has a DNA duplex of 38 base pairs with a transcription bubble containing an RNA strand of nine residues annealed to the template DNA. The yeast RNA polymerase II protein complex consists of 12 polypeptide subunits. The direction of transcription is oriented from left to right. BASED ON PDB FILE 5C4X.

Proteins Required for RNA Synthesis in Eukaryotes

The eukaryotic RNA polymerases (I, II, and III) differ in the associated protein components required for promoter recognition and transcriptional initiation. In this section, we focus only on the proteins required for transcription by the eukaryotic RNA polymerase II complex, which is the primary polymerase responsible for transcribing protein-coding genes leading to the synthesis of mRNA. The mechanism of RNA synthesis by the eukaryotic RNA polymerase II protein complex is very similar to that by the prokaryotic RNA polymerase in that the DNA helix forms a transcription bubble to facilitate RNA synthesis on the DNA template strand. **Figure 21.20** shows the structure of the yeast RNA polymerase II protein complex with a short RNA strand annealed to the DNA template strand in the middle of a transcription bubble.

As described earlier, the RNA polymerase II complex is recruited to gene promoter regions by the DNA binding protein TBP, which is functionally analogous to the bacterial σ-factor protein. As shown in **Figure 21.21**, TBP is a component of the transcription factor IID (TFIID) complex, which does not include the RNA polymerase II complex. In addition to TBP, TFIID contains numerous TBP-associated factors (TAFs), which are responsible for stabilizing the interaction with DNA and providing a platform for the assembly of additional transcription factors. After binding of the TFIID–TBP complex to the gene promoter, the transcription factors TFIIA and TFIIB bind to the growing transcriptional initiation complex. In the next step, TFIIF in association with RNA polymerase II and other transcription factors binds to the initiation complex along with the transcription factors TFIIE and TFIIH. In the last step, the RNA polymerase II is phosphorylated on the C-terminal domain (CTD), which leads to formation of the open complex and initiation of transcription in the 5′ to 3′ direction in a primer-independent process identical to bacterial RNA synthesis.

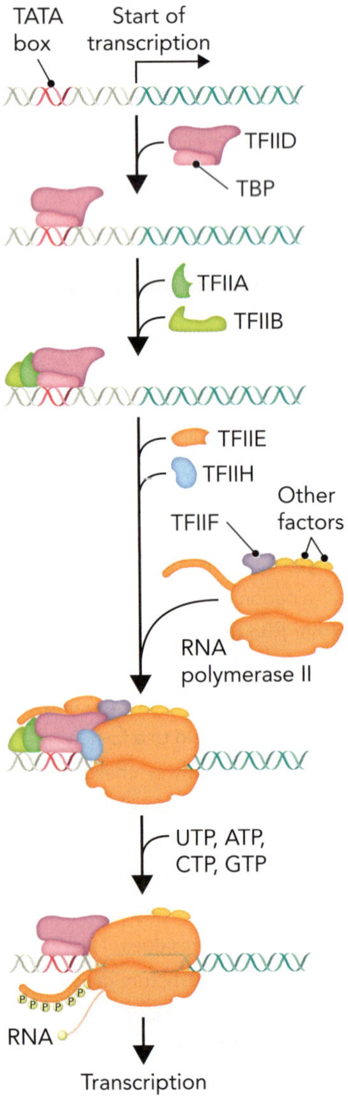

Figure 21.21 Sequential loading of transcription factor proteins to RNA polymerase II promoters is required to initiate formation of the open complex and the start of RNA synthesis.

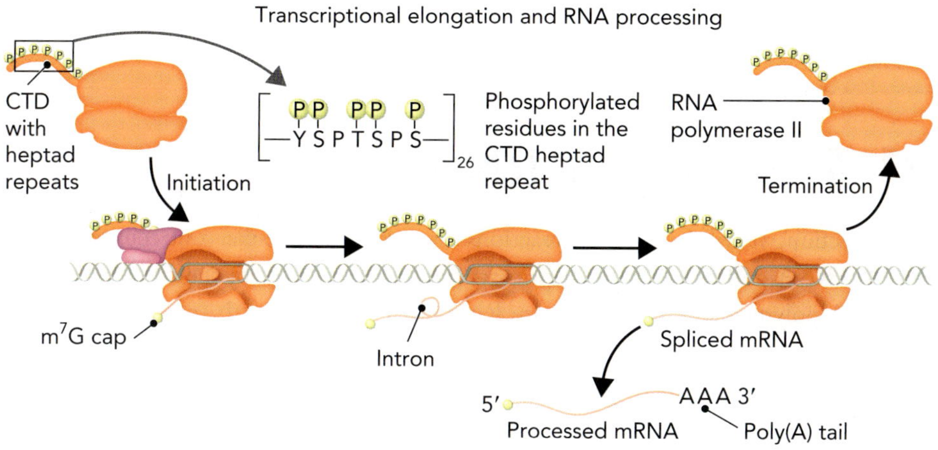

Figure 21.22 Transcription initiation, elongation, RNA processing, and termination is shown in RNA polymerase II–dependent transcription. Phosphorylation sites in the CTD heptad repeat sequence of yeast RNA polymerase II are shown.

An important point for understanding control of eukaryotic transcription is the functional role of the RNA polymerase II CTD in coordinating the biochemical processes of mRNA transcription and processing, which occur simultaneously. The CTD consists of multiple repeats of the heptapeptide sequence YSPTSPS, with the number of repeats varying from 26 in yeast to 52 in mammals. Hyperphosphorylation of the CTD on serine, tyrosine, and threonine residues is a key step in the transition of the enzyme complex from the initiation to the elongation stage (**Figure 21.22**). The CTD also serves as an assembly platform for a variety of factors involved in both transcription elongation and RNA processing. The interaction with these factors depends on both the overall phosphorylation state of the CTD and the phosphorylation of specific amino acid residues. The CTD remains hyperphosphorylated throughout elongation, however the specific sites of phosphorylation differ as a function of early or late events during elongation.

Eukaryotic transcription continues until signals within the DNA template work together with trans-acting factors to signal termination. Transcription termination in eukaryotes is a complicated process because it is coupled with processing of the 3′ end of mRNA. A functional poly(A) signal is generally required for efficient termination of mRNA synthesis and, in some cases, may be sufficient to lead to termination. But additional downstream elements may also be required in other genes, and RNA polymerase II often continues transcription for several hundreds to thousands of nucleotides after the poly(A) signal. The process of transcriptional termination and polyadenylation is described in more detail in the next section.

concept integration 21.2

What are the functions of RNA polymerase ancillary proteins with regard to promoter recognition and transcriptional initiation, and how do they differ between prokaryotes and eukaryotes?

In bacteria, for example, the addition of the σ factor to the holoenzyme significantly increases the affinity for DNA. In eukaryotes, TATA binding protein must first bind to DNA before other transcription factors and the RNA polymerase II holoenzyme are recruited. These additional factors are necessary as part of a regulatory measure. Because RNA polymerases must transcribe many different regions of DNA, their specificity for the interaction with DNA is very low; that is, they can theoretically bind any DNA sequence. Alone, this would not be a very good system for transcription because the polymerase would have no way of distinguishing coding regions from intergenic regions.

The additional transcription factors are necessary both to direct the polymerase to the promoter region and to ensure that genes are transcribed at the correct time.

21.3 Eukaryotic RNA Processing

In this section, we will focus on the reactions that take place in eukaryotic cells to convert mRNA, rRNA, and tRNA from the primary transcript produced by the different eukaryotic RNA polymerases into the mature RNAs that take part in the process of translation. We also examine the roles of the enzymes that catalyze these reactions and how the CTD of RNA polymerase II is responsible for coordinating many of these mRNA processing events. Finally, we look at how alternative processing of mRNA transcripts in eukaryotic cells provides the necessary cell-specific specialization that multicellular organisms depend on and how mutations in RNA processing can lead to disease.

Ribozymes Mediate RNA Cleavage and Splicing Reactions

Unlike many of the protein-mediated, enzyme-catalyzed reactions discussed in previous chapters, RNA processing reactions are mostly catalyzed by RNA molecules that function as ribozymes. The existence of ribozymes provides support for the RNA world hypothesis, which proposes that early forms of life on Earth relied on RNA to carry genetic information and catalyze reactions (see Figure 18.27). According to the RNA world hypothesis, the instability and poor catalytic properties of RNA led to the evolution of DNA and proteins, although ribozymes remain a significant part of RNA processing and of other cellular processes such as translation.

Ribozymes function as catalysts that react with substrate to mediate either intramolecular (cis) or intermolecular (trans) cleavage of the phosphodiester backbone. Because ribozymes are catalysts, they affect only the rate of the reaction, not the equilibrium. Turnover rates for ribozymes are measured the same way they are for protein enzymes (product released over time). As shown in **Figure 21.23**, ribozymes, like essentially all enzymes, are regenerated after the trans cleavage of target RNA. To understand the catalytic activity of RNA, we first need to examine ribozymes that lack protein components. **Figure 21.24** shows general examples of acid–base catalysis and two-metal ion catalysis, which are the two primary mechanisms proposed for ribozyme-mediated phosphodiester cleavage.

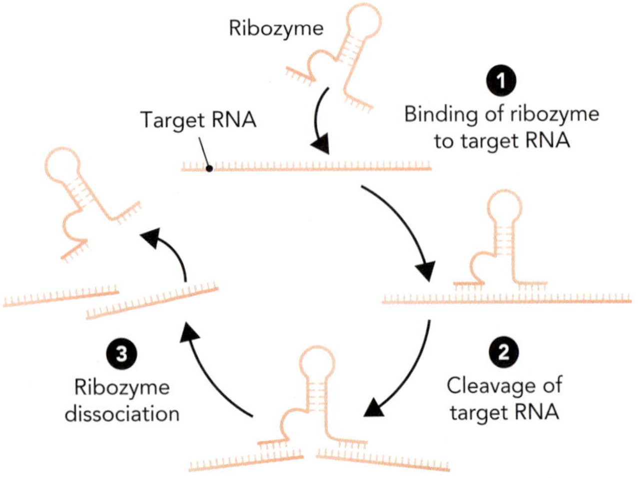

Figure 21.23 Ribozymes are RNA enzymes. They generally mediate intramolecular or intermolecular cleavage reactions of RNA substrates. An example of intermolecular cleavage is shown here.

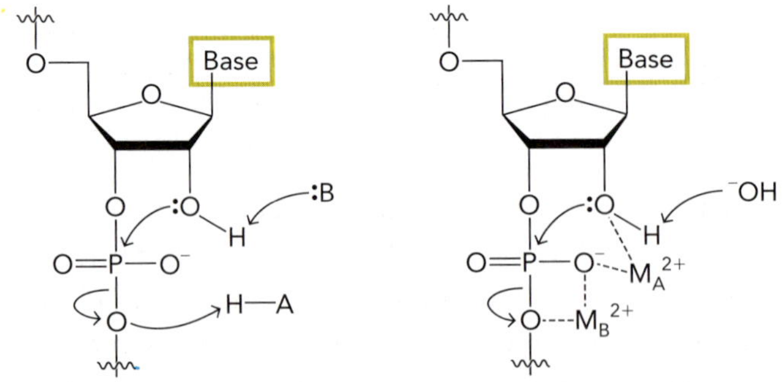

Figure 21.24 Ribozyme-mediated phosphodiester cleavage can be accomplished through acid–base catalysis and two-metal ion catalysis.

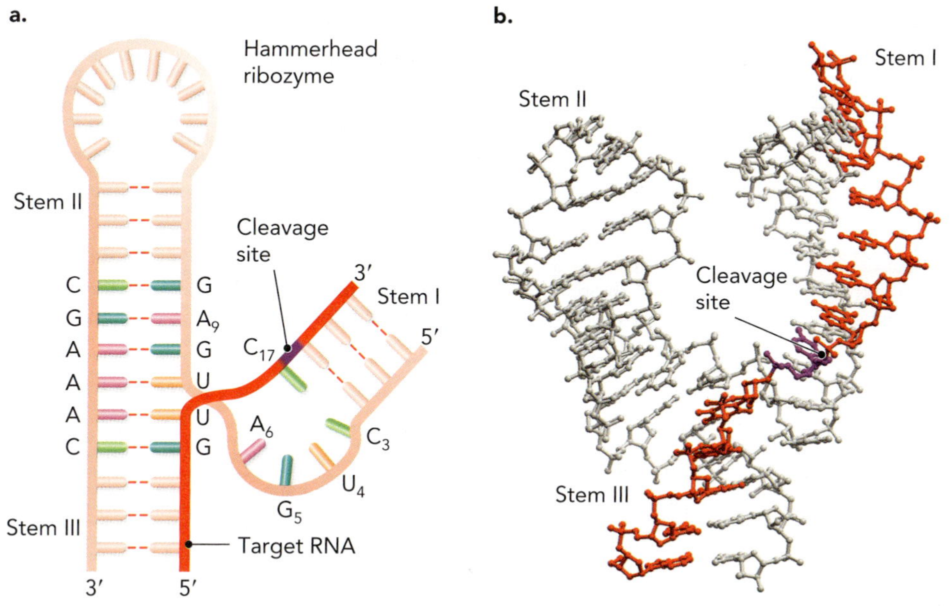

Figure 21.25 Hammerhead ribozymes contain a sequence that is capable of trans phosphodiester cleavage. **a.** Topological map of the minimal hammerhead ribozyme shown in panel **b**. This hammerhead ribozyme consists of two RNA components and performs trans cleavage at the site indicated. **b.** The molecular structure is shown of a full-length hammerhead ribozyme bound to a target RNA substrate. BASED ON PDB FILE 2OEU.

The **hammerhead ribozyme** is one of the best-characterized catalytic RNAs; it was discovered as a component of the satellite RNA of the tobacco ringspot virus. Hammerhead ribozymes, which range in size from 50 to 150 nucleotides, occur in three different natural forms. Ribozymes in the same family usually contain sequence variations with only a few conserved or invariant residues, which form the active site. The overall secondary and tertiary structures of ribozymes are conserved, though the sequences can vary. Nucleotide changes are tolerated as long as the catalytic site maintains the functional groups needed to recognize specific substrates and cleave the target RNA. In hammerhead ribozymes, the minimal requirements of the ribozyme are a 15-nucleotide sequence that folds into three base-paired stems, with the catalytic core formed through interactions between conserved nucleotides (**Figure 21.25**).

This ribozyme is interesting because it has the ability to self-cleave and to act on other substrates. The distinction between the cis (self-cleaving) and trans (cleaving of target RNA) activities of this ribozyme depends on the connection between the enzyme strand and the substrate strand of the RNA. Specifically, when both RNA strands are in the same molecule, the RNA self-cleaves. When the RNA strands are unconnected, the ribozyme is capable of binding other RNA transcripts and catalyzing multiple reactions.

Hammerhead ribozymes are not the catalysts responsible for removing intronic or intergenic sequences from eukaryotic mRNA, rRNA, or tRNA. Instead, these intervening RNA sequences are removed by an RNA-catalyzed reaction using either a self-splicing cis reaction, or a trans-splicing reaction mediated by ribonucleoprotein complexes called **spliceosomes**, which mediate the splicing of mRNA precursors. We first look at cis-acting self-splicing introns and then turn our attention to trans-acting spliceosomes.

The discovery of self-splicing introns was reported in the early 1980s by Thomas Cech and his co-workers at the University of Colorado in Boulder, who discovered that intronic RNA sequences in genes from the protozoan *Tetrahymena thermophila* function to catalyze an intrastrand cleavage reaction that removes the intron and ligates the exonic sequences (**Figure 21.26**). They called these catalytic RNA sequences **group I introns**, which are

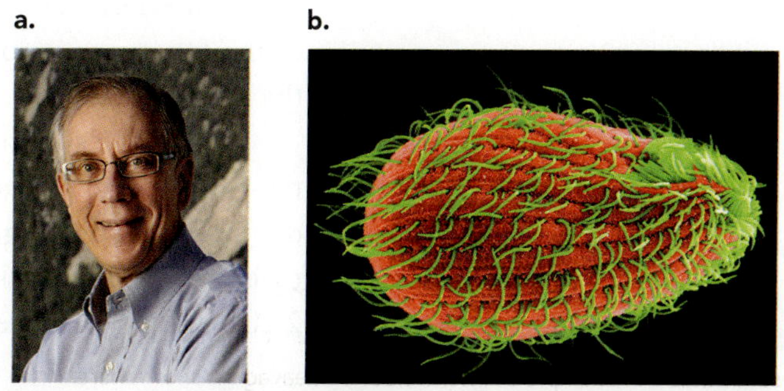

Figure 21.26 Thomas Cech shared the 1989 Nobel Prize in Chemistry for his discovery of RNA self-splicing in a protozoan organism. **a.** Thomas Cech in his lab at the University of Colorado. GLENN ASAKAWA/UNIVERSITY OF COLORADO. **b.** The protozoan *Tetrahymena thermophila* was the model organism used in the Cech lab that led to the discovery and characterization of eukaryotic RNA self-splicing. AARON J. BELL/SCIENCE SOURCE.

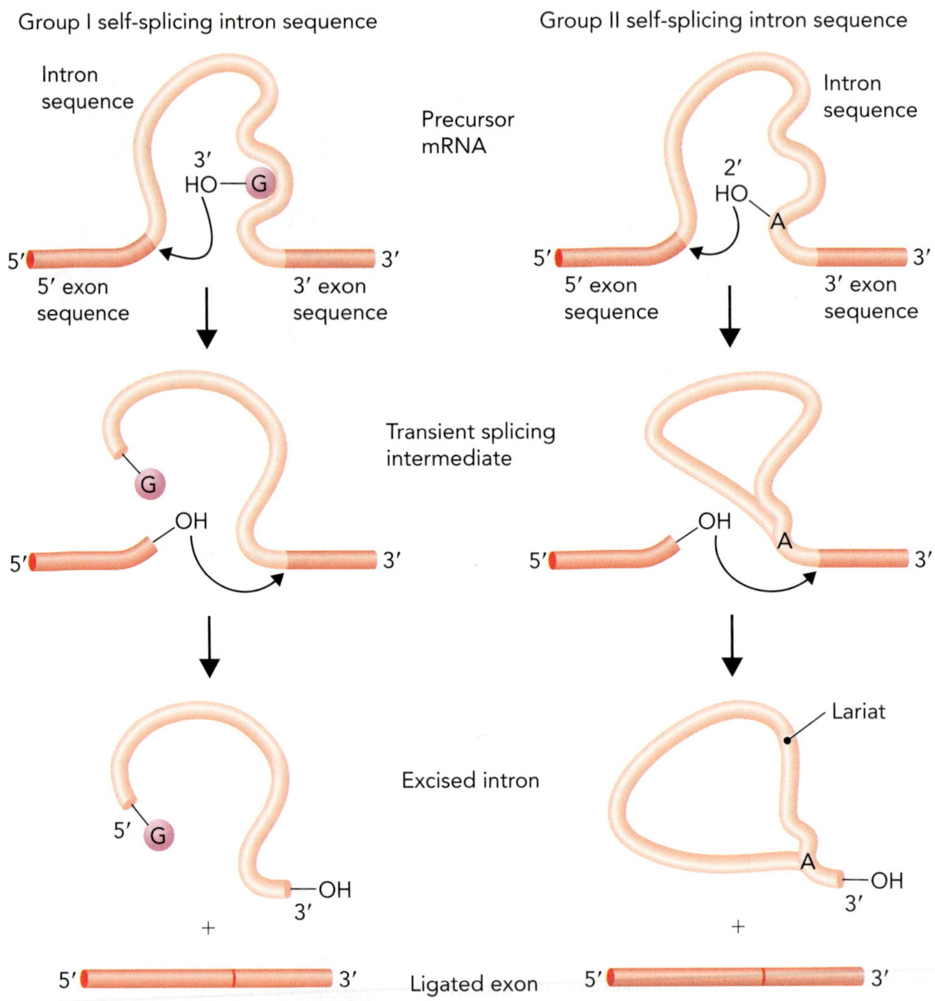

Figure 21.27 A comparison of group I and group II intron self-splicing is shown. In group I introns, the 3′-OH of an exogenous guanosine cofactor (G) participates in a transesterification reaction with the phosphoryl group at the 5′ end of the intron. This reaction is similar to the reaction between 5′-phosphoryl group and the 2′-OH of an adenosine (A) at the branch site in group II introns. For group I introns, the excised intron is linear, whereas group II introns have a covalently joined lariat structure after the excision reaction.

now known to be found in precursors to rRNA, tRNA, and mRNA. Group I introns are most prominent in fungal and plant mitochondrial RNAs but can also be found in both bacteria and other eukaryotes, although not in vertebrates.

Group I self-splicing introns have relatively little sequence similarity but share a series of short conserved sequence elements, including a U at the 5′ splice site and a G at the 3′ splice site. Tertiary interactions are very important for the formation of this active site because minimal base pairing occurs between cleavage sites at the 5′ and 3′ ends of the intron and the catalytic center. All RNA splicing events are a series of transesterification reactions in which the intron is removed and the exonic sequences are rejoined to generate a mature transcript. In group I introns, splicing is initiated by the binding of an essential guanosine cofactor. This cofactor initiates a nucleophilic attack on the phosphodiester bond at the 5′ splice site to begin the first transesterification reaction (**Figure 21.27**). Removal of the intron and joining of the two exons is completed as the new 3′-OH nucleophile of the exon attacks the 3′ splice site.

A second class of self-splicing introns is known as the **group II introns.** There are two notable differences between group I and group II introns. First, group II introns do not require an exogenous nucleoside to initiate the cleavage reaction. Second, the excised intron forms a lariat structure in which the 5′ end of the intron is linked to the 2′-OH of an adenosine residue near the 3′ end of the intron.

Structure and Function of Spliceosomes

Precursor mRNA transcripts in higher eukaryotes contain an average of ~10 exons, each ~150–200 nucleotides in length (encoding ~50–70 amino acids), while the introns separating the exons are often 10–100 times longer (1.5–15 kb in size). Removal of introns from precursor mRNA transcripts is mediated by the spliceosome complex, which catalyzes a trans RNA splicing reaction. As shown in **Figure 21.28**, the mechanism of spliceosome-mediated trans RNA splicing is similar to the self-splicing mechanism found in group II introns, involving an intronic adenine residue and the formation of a lariat structure to remove the intronic sequences (see Figure 21.27).

As shown in **Figure 21.29**, introns removed by spliceosomes from primary mRNA transcripts contain short conserved sequences at the 5′ (GU) and 3′ (AG) splice sites, as well as in the branch site. These are similar to the sequences observed in the group II intron. The branch site in the introns of primary mRNA transcripts is located 18–40 nucleotides upstream of the 3′ splice site and is often composed of a polypyrimidine tract. Eukaryotic spliceosomes recognize these splice-site target sequences in the precursor mRNA molecule and remove the intronic sequences by ribozyme-mediated lariat formation.

Eukaryotic spliceosomes are composed of small nuclear RNAs (snRNAs) and proteins, which together are called **small nuclear ribonucleoproteins (snRNPs)**. The snRNPs are named after the snRNA molecules contained within each RNA–protein complex; for example, the large spliceosome assembly contains U1, U2, U5, and U4/U6 snRNPs consisting of the corresponding snRNA molecules and their associated proteins. The U1, U2, U4, and U5 snRNAs are transcribed by RNA polymerase II as 1.5- to 2-kb transcripts that are trimmed at the 3′ end both in the nucleus and cytoplasm to a final length of less than 200 nt. Like other RNA polymerase II transcripts, these snRNA molecules undergo the same 5′ capping reaction as mRNA in the nucleus but are not spliced or polyadenylated. As shown in **Figure 21.30**, the capped pre-snRNA transcripts are exported from the nucleus. In the cytoplasm, each snRNA is assembled with a heptameric ring of proteins, called Sm proteins, to form the snRNP core complex. The snRNA is hypermethylated on the 5′ cap, which converts the cap structure from m^7G to

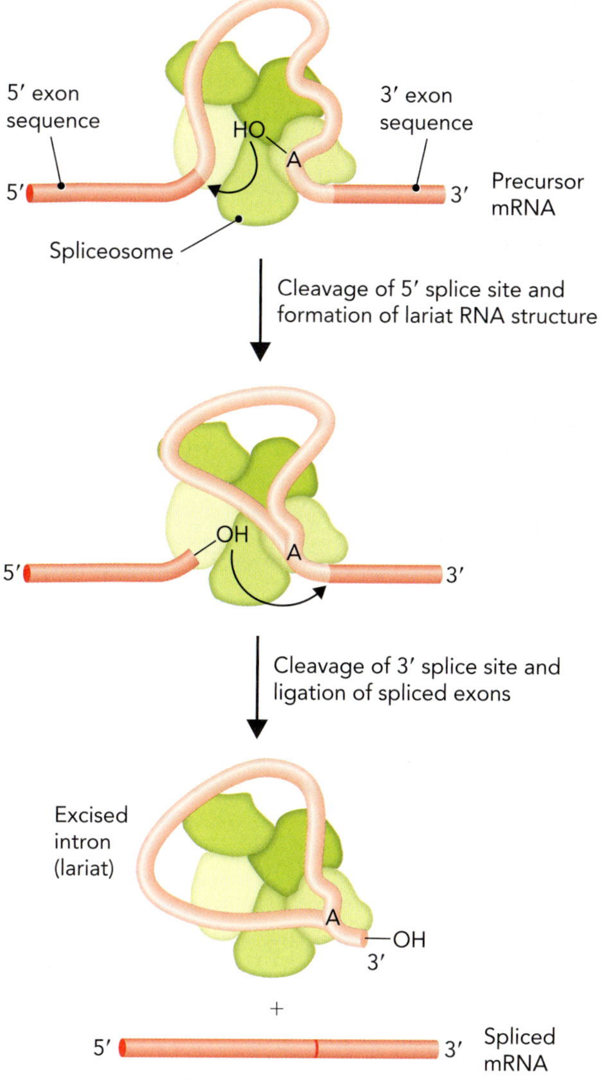

Figure 21.28 The mechanism of spliceosome-mediated trans splicing of precursor mRNA transcripts is similar to the mechanism of self-splicing of group II introns.

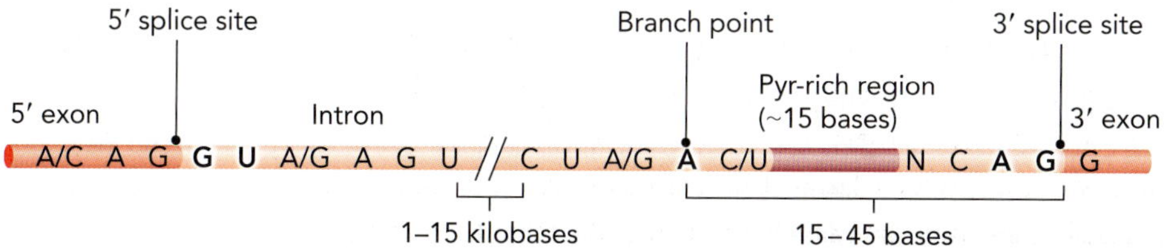

Figure 21.29 A schematic intron is shown with the 5′ splice site, branch point, and 3′ splice site labeled. The letters in boldface show the most highly conserved nucleotides in the precursor mRNA transcript at the 5′ end of the intron (GU), the branch point (A), and the 3′ end of the intron (AG).

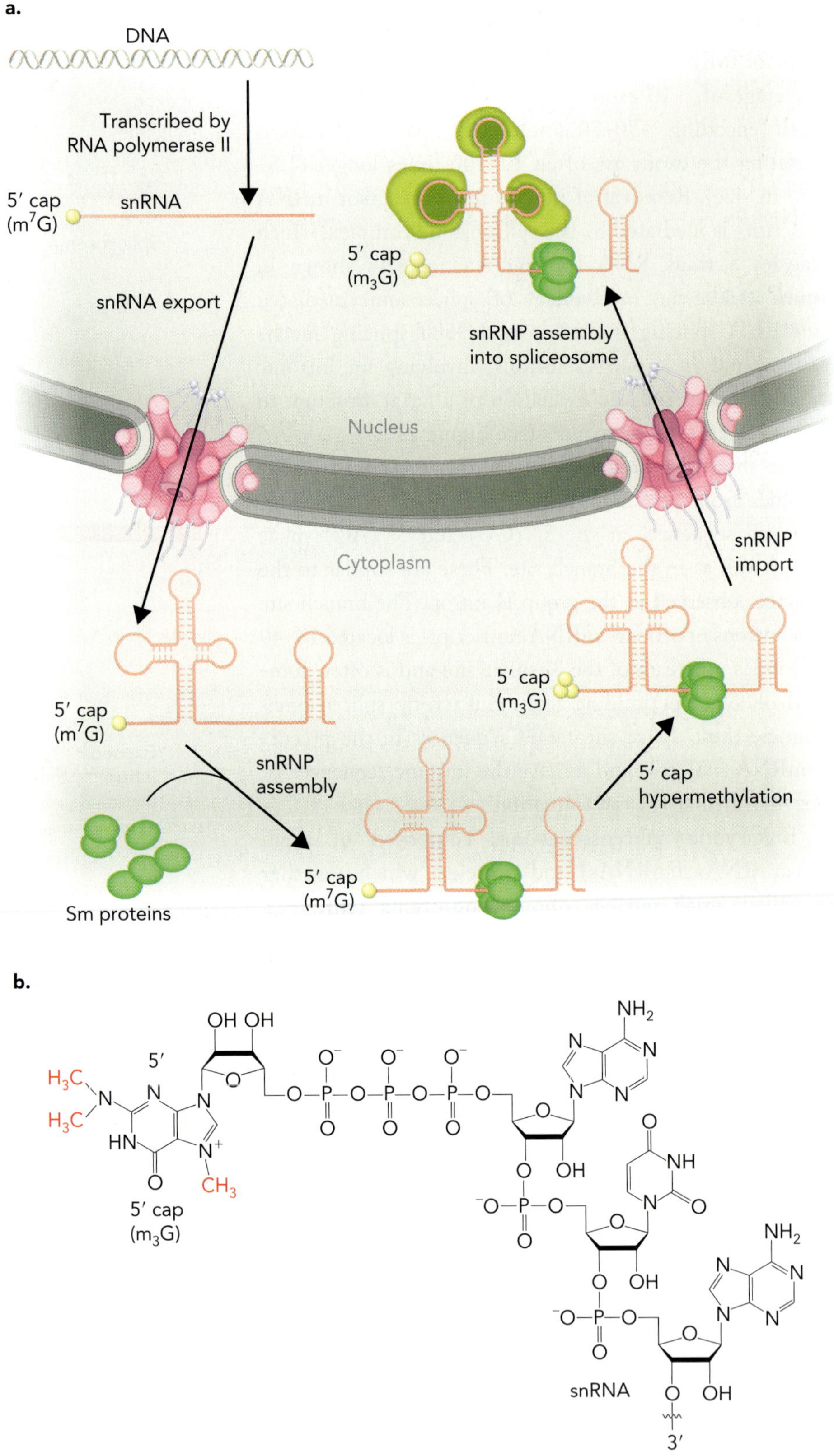

Figure 21.30 The spliceosome assembly pathway requires snRNA nuclear export and snRNP nuclear import steps. **a.** The primary snRNA transcript is capped at the 5′ end, exported to the cytoplasm, assembled into a core snRNP complex containing seven Sm proteins, and then imported back into the nucleus, where a functional spliceosome complex is generated. **b.** The structure of the hypermethylated snRNP m_3G cap is shown.

trimethylguanosine (m_3G). The snRNP complex is then transported back into the nucleus, where the snRNP is assembled into a spliceosome complex. By contrast, U6 snRNA is transcribed by RNA polymerase III, and its assembly into an snRNP complex occurs in the nucleus. **Figure 21.31** shows the structure of the human U4 snRNP complex, containing the heptameric Sm protein ring and a 68-nucleotide segment of U4 snRNA.

As shown in **Figure 21.32**, spliceosome-mediated splicing consists of two transesterification steps, which result in the release of an excised lariat intron and the correctly spliced exonic sequences. Splicing of precursor mRNA after this two-step transesterification reaction occurs through ordered interaction of the snRNAs and associated complex proteins with the RNA transcript.

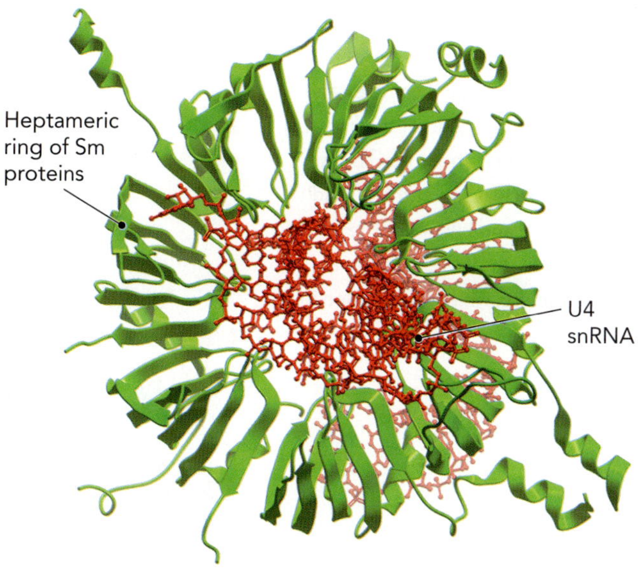

Figure 21.31 The structure of the human U4 snRNP complex is shown. BASED ON PDB FILE 2Y9C.

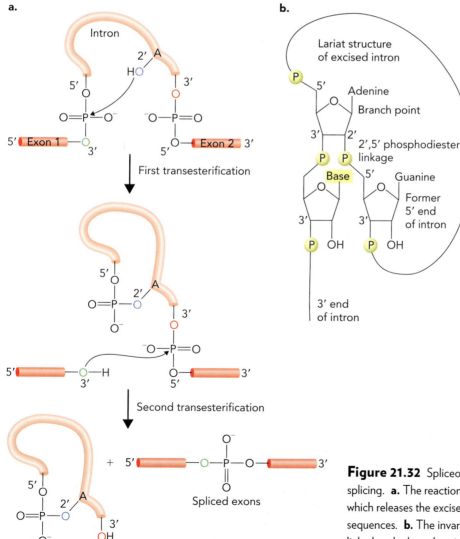

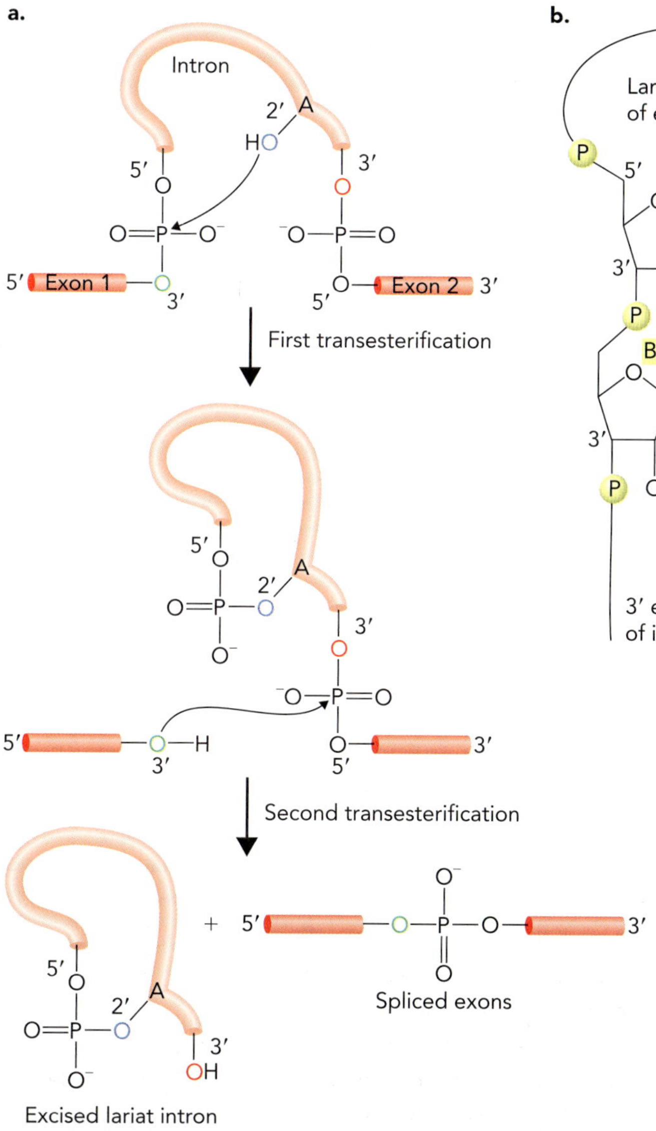

Figure 21.32 Spliceosome-mediated precursor mRNA splicing. **a.** The reaction consists of two transesterification steps, which releases the excised lariat intron and the spliced exonic sequences. **b.** The invariant G from the 5′ end of the intron is linked to the branch point A through a 2′,5′ phosphodiester bond, forming a lariat structure.

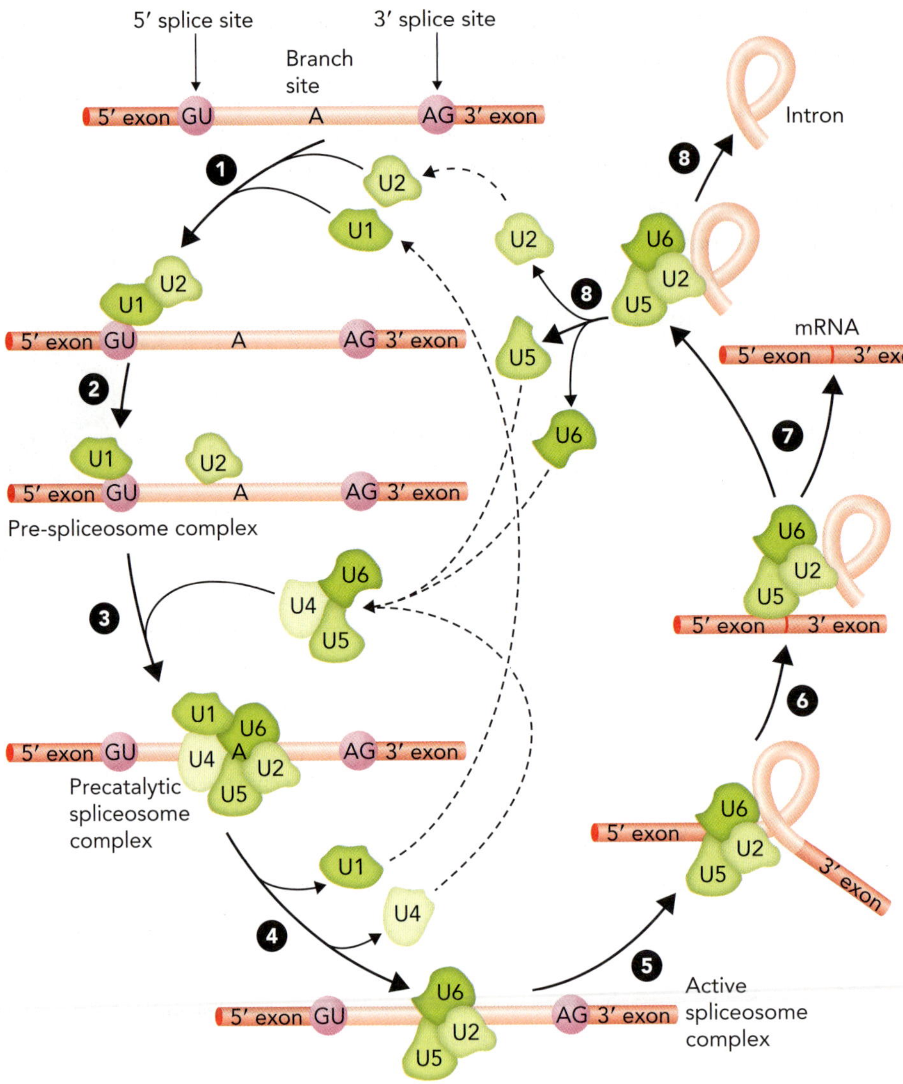

Figure 21.33 The spliceosome reaction cycle starts with the assembly of the active complex and cleavage at the 5′ end of the intron to form the lariat structure. This is followed by cleavage at the 3′ end of the intron and joining of the two exons. The intron is released as a lariat structure, which is degraded in the nucleus.

As shown in **Figure 21.33**, the first step in the pathway involves the recruiting of U1 and U2 snRNPs to the 5′ splice site. In step 2, U2 associates with the branch site, forming the pre-spliceosome complex. The interaction between the spliceosome and mRNA is mediated by the snRNAs in both U1 and U2 snRNPs. These U1 and U2 snRNA molecules base pair with the conserved sequences in the 5′ splice site and branch point, respectively, to direct binding of the remaining spliceosome complex. In step 3 of the splicing reaction, U4–U5–U6 form a tri-snRNP complex, which is then recruited to the transcript to form the precatalytic spliceosome complex, consisting of multiple intrastrand and interstrand base-pairing interactions between the snRNAs (**Figure 21.34**). Binding of the

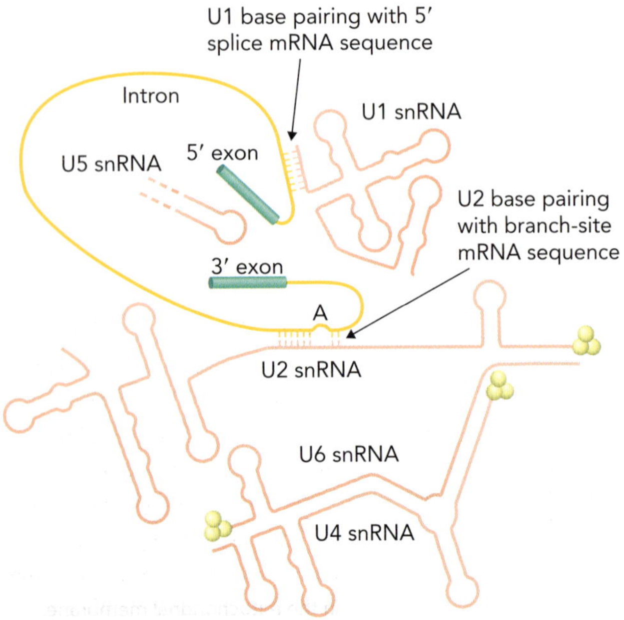

Figure 21.34 RNA–RNA interactions within the precatalytic spliceosome complex are shown. Multiple intrastrand and interstrand base pairs form between the snRNA molecules. The U1 snRNA directly base pairs with sequences in the 5′ donor site, whereas the U2 snRNA base pairs with residues in the branch site.

tri-snRNP to the precursor mRNA induces rearrangements in the RNA–RNA and RNA–proteins interactions and destabilizes the U1 and U4 snRNPs, thereby releasing them from the spliceosome in step 4 (Figure 21.33). The transesterification reaction takes place in steps 5 and 6 of the spliceosome pathway, releasing mRNA in step 7. In step 8, the U2–U5–U6 complex dissociates from the lariat intron to initiate another round of splicing.

Processing of Eukaryotic tRNA and rRNA Transcripts

Precursor mRNAs are not the only transcripts that require processing in eukaryotic cells before they are fully functional. Both precursor tRNA and rRNA molecules also require processing.

Processing of eukaryotic tRNA begins with cleavage of the primary transcript from the surrounding region or cleavage of a 5′ sequence of variable length by RNaseP, similar to prokaryotic tRNA processing (see Figure 21.5). As discussed in Section 21.1, the RNA transcripts fold into their conserved secondary structures prior to cleavage. This co-transcriptional folding is an important part of the process and often produces transient secondary structures that can change as the transcript lengthens. The presence of these transient structures, particularly in large molecules such as rRNA, may help narrow down the possible conformations and thereby increase the ability of the RNA to fold into the correct final structure.

As shown in **Figure 21.35**, yeast precursor tRNA is processed in the nucleus by removal of 5′-ribonucleotides, addition of the trinucleotide sequence CCA to the 3′ end, and chemical modification of some bases. Addition of a CCA trinucleotide to the 3′ end is catalyzed by the enzyme tRNA nucleotidyltransferase, which synthesizes the trinucleotide sequence using CTP and ATP as substrates. The enzyme uses the

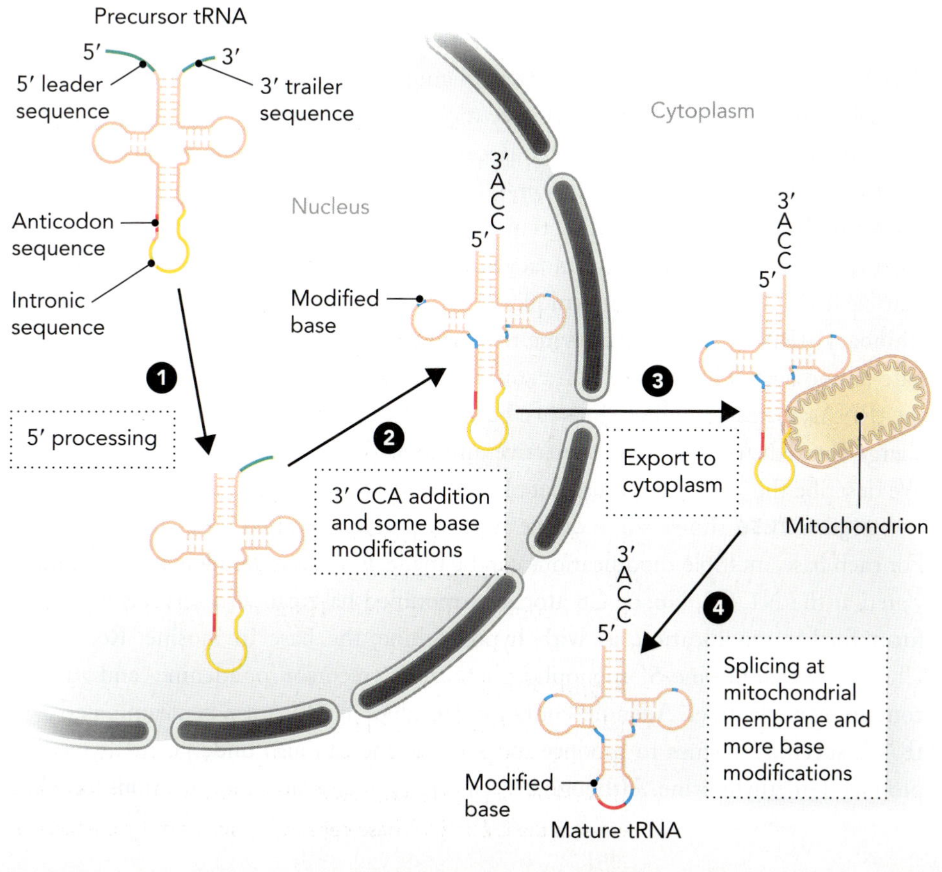

Figure 21.35 Yeast tRNA cleavage, splicing, and modification steps are shown. Cleavage of the 5′ leader sequence occurs first, but the remaining reactions do not necessarily occur in the order shown here. Moreover, not all tRNA precursors are modified by enzymes in the mitochondrial membrane.

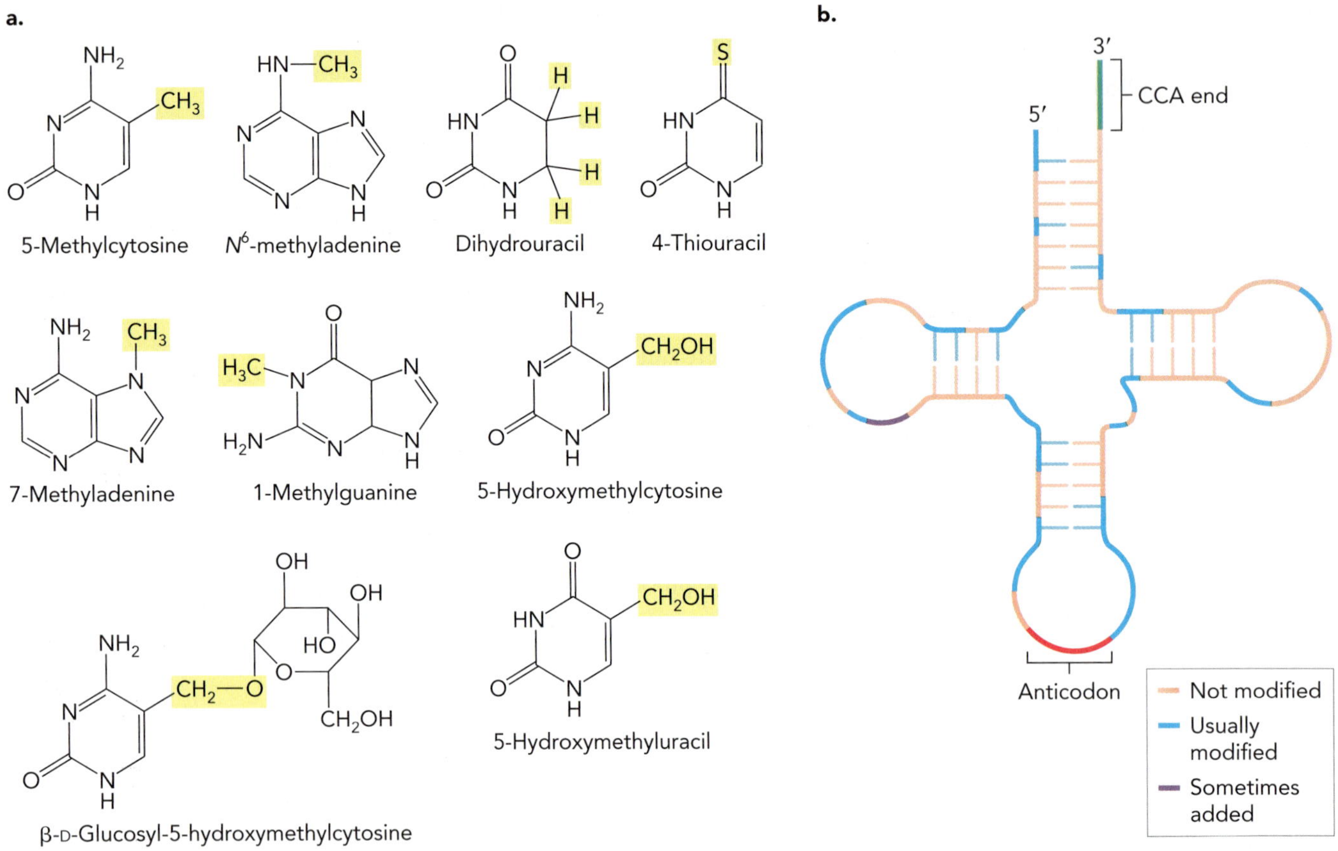

Figure 21.36 Common modifications in yeast tRNA molecules are shown. **a.** Chemical structures of the most common ribonucleotide base modifications in tRNA. **b.** Summary of residues in yeast tRNA molecules that have been found to be modified. The positions of invariant anticodon and CCA end sequences are also shown.

tertiary structure of tRNA, rather than any specific base recognition, as a guide to bind and locate the 3′ end. The tRNA is then exported to the cytoplasm, where it is further processed by enzymes on the mitochondrial membrane. Splicing and modification of eukaryotic tRNA molecules do not always occur in the same order, and in some cases tRNA is modified both before and after splicing.

More than 100 different RNA base modifications have been identified, with the majority of these occurring in tRNA. Because all tRNAs must have similar structures in order to interact with the ribosome, they also have a certain degree of sequence similarity. But for tRNAs to function in protein synthesis, they must be linked to specific amino acids to form aminoacylated tRNAs. Correct aminoacylation requires that each tRNA must contain unique features that allow it to be recognized by the aminoacyl-tRNA synthetase enzyme that will covalently join the tRNA with its cognate amino acid. This recognition cannot be based on structural differences between the tRNAs; instead, modifications of a base or ribose are one way that the enzymes that charge the tRNAs with their respective amino acids can differentiate between them. We describe this process in more detail in Chapter 22.

Figure 21.36 shows some of the most common modified bases found in RNA. For each base, multiple modifications can be made. For example, adenine can be methylated at the N1, C2, N6, or C8 atoms. A modified base can also serve as a substrate for a further modification, as with hypoxanthine, the base in inosine. Recall from Chapter 18 that inosine-5′-monophosphate is the precursor for adenine- and guanine-containing nucleotides. Adenine-containing residues in tRNA can be deaminated using tRNA-specific enzymes to produce inosine. Inosine can also undergo methylation to produce 1-methylinosine. Although a large variety of chemical modifications have been

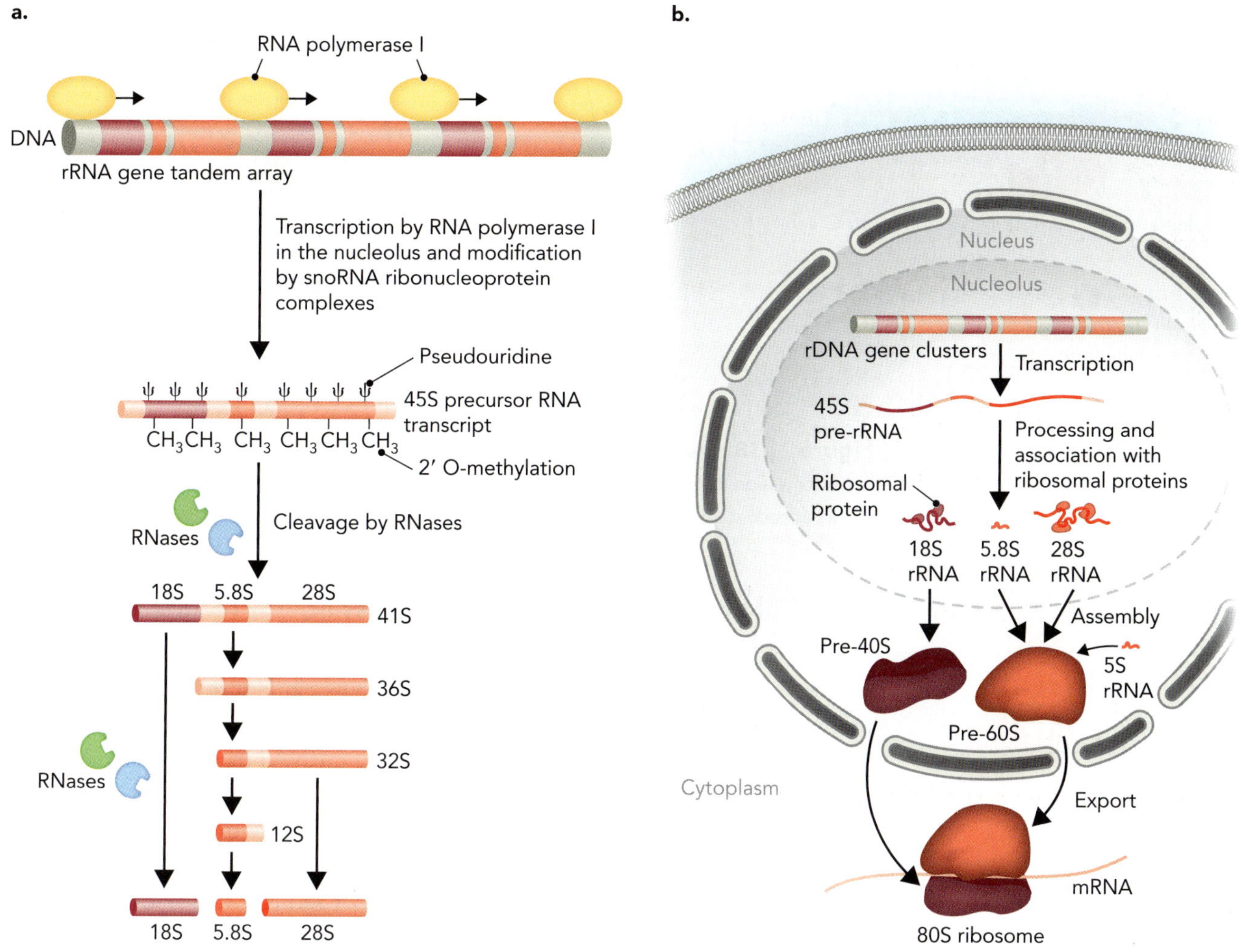

Figure 21.37 The steps required for eukaryotic precursor rRNA transcripts in the nucleolus and assembly of the mature 80S ribosome are shown. **a.** Modification of a precursor 45S rRNA transcript by snoRNA-guided complexes precedes cleavage by RNaseP-like ribozymes to yield the mature 28S, 18S, and 5.8S rRNA molecules. **b.** Ribosomal proteins are imported into the nucleolus and combine with rRNA molecules. Assembly of the pre-40S and pre-60S ribosomal complexes occurs in the nucleolus. The functional ribosome is formed in the cytoplasm during translation initiation.

characterized in tRNA, some modifications are conserved in all organisms. For example, a uracil in one of the loops is modified to pseudouridine (Ψ) in all tRNA species.

Eukaryotic precursor rRNA transcripts contain coding sequences for the 28S, 18S, and 5.8S rRNA molecules, which are all components of the 80S ribosomal complex. As shown in **Figure 21.37**, eukaryotic rRNA genes are arrayed in tandem in the genome and are transcribed in the nucleolus by a transcriptional initiation complex containing the enzyme RNA polymerase I. The 45S precursor transcript is modified by ribonucleoprotein complexes containing **small nucleolar RNA (snoRNA)** molecules (see Table 21.1). The snoRNA molecules form sequence-specific base-pairing complexes with the precursor rRNA transcript and function to guide RNA-modifying enzymes to specific sites to convert uridine to pseudouridine and to methylate other residues. The modified precursor rRNA is then cleaved by RNaseP-like ribozymes to successively process the transcript, eventually yielding the mature 28S, 18S, and 5.8S rRNA products. The 40S and 60S ribosomal subunits are assembled in the nucleolus through complex formation with more than 70 ribosomal proteins to generate the 80S ribosome.

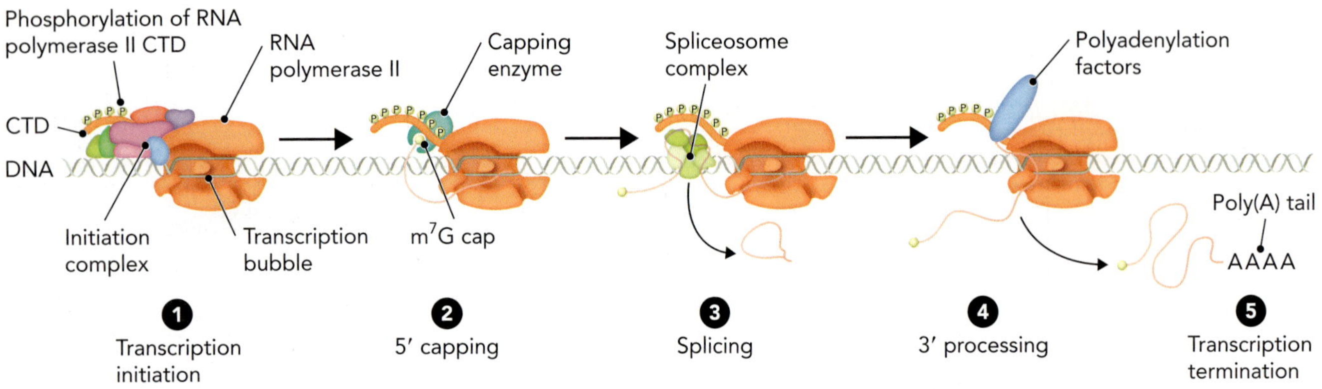

Figure 21.38 The CTD of RNA polymerase II binds mRNA capping, splicing, and polyadenylation factors. This overview of mRNA transcription indicates the mRNA processing that occurs at each stage.

RNA Polymerase II Coordinates Processing of Precursor mRNA

As with tRNA and rRNA, processing reactions in mRNA occur co-transcriptionally and begin very early in the elongation cycle. As shown in **Figure 21.38**, processing of precursor mRNA occurs during transcription and is coordinated by the CTD of RNA polymerase II, which serves as a binding domain for mRNA processing complexes such as the 5′ capping enzyme, guanine-N7 methyltransferase (**Figure 21.39**). Addition of the m^7G cap occurs soon after initiation, generally when the transcript has reached a length of 25–30 nucleotides. The CTD of RNA polymerase II becomes hyperphosphorylated by TFIIH at serine 5 (Ser5) and 7 (Ser7) in the conserved heptapeptide sequence (YSPTSPS) during the formation of the initiation complex (see Figure 21.22). This favors the association with the 5′ capping enzyme. Because the capping enzyme is recruited by the CTD, this reaction is unique to transcripts produced by RNA polymerase II, although it is not unique to mRNA.

All RNA polymerase II transcripts begin with a 5′-triphosphate on the first nucleotide. This is the template for the addition of a 7-methylguanosine in a three-part reaction to form a 5′ m^7G cap (**Figure 21.40**). First, the enzyme RNA triphosphatase removes the terminal phosphate on the 5′-nucleotide. Next, RNA guanylyltransferase transfers GMP from a molecule of GTP to the diphosphate end of the precursor mRNA. Finally, the enzyme guanine-N7 methyltransferase transfers a methyl group from *S*-adenosyl-L-methionine to the cap guanine to form a 7-methylguanosine with a 5′,5′ triphosphate linkage to the terminal nucleoside and release *S*-adenosyl-L-homocysteine. In some lower eukaryotes, the first two reactions are catalyzed by two separate enzymes, whereas in multicellular organisms a single bifunctional enzyme has both RNA triphosphatase and RNA guanylyltransferase activity.

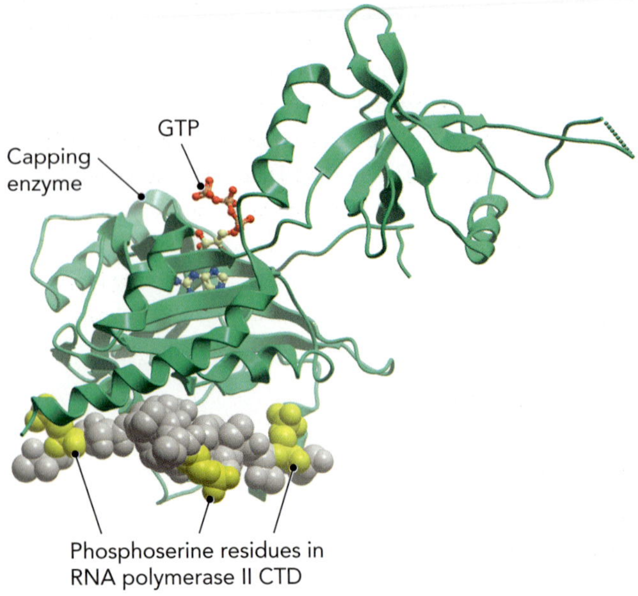

Figure 21.39 The structure of the yeast 5′ capping enzyme, guanine-N7 methyltransferase, is shown here bound to a 17-amino-acid segment of the RNA polymerase II CTD containing copies of the serine-rich YSPTSPS heptad repeat (shown in space-filling style). BASED ON PDB FILE 1P16.

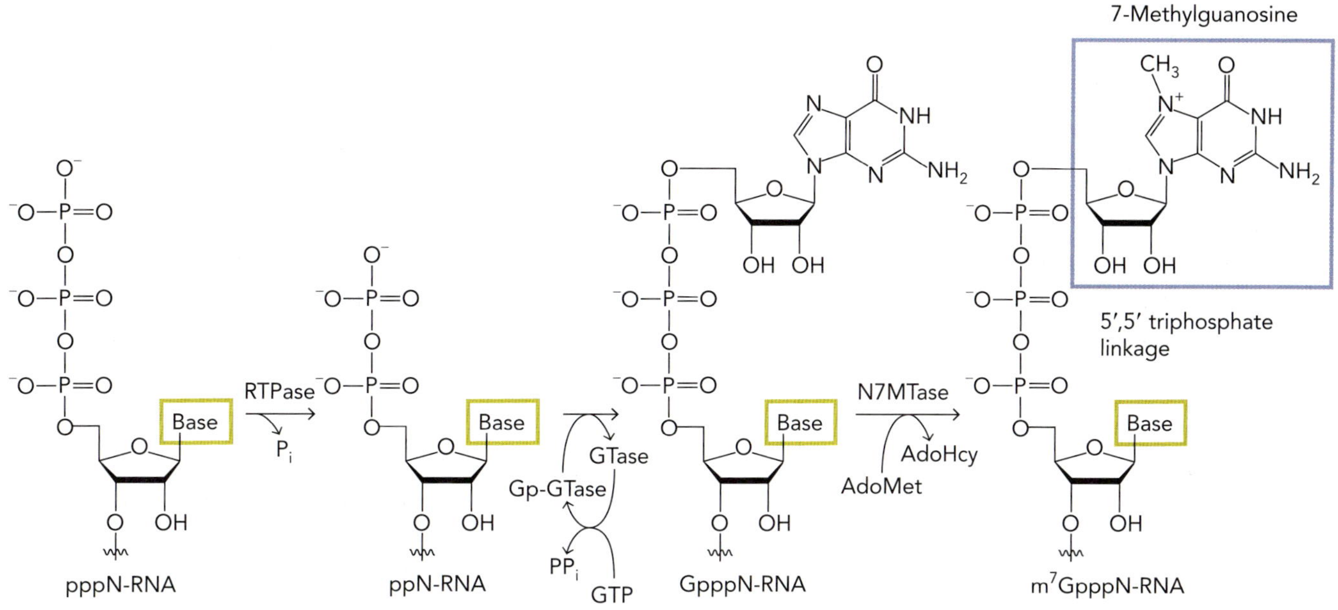

Figure 21.40 The three steps in the 5′ capping reaction on precursor mRNA are shown. In the first step, the enzyme RNA triphosphatase (RTPase) hydrolyzes the γ phosphate on the 5′ terminal nucleotide to generate diphosphate RNA (ppN-RNA; N is any ribonucleotide). In the second step, RNA guanylyltransferase (GTase) adds a GMP moiety to ppN-RNA to form GpppN-RNA. In the final step, guanine-N7 methyltransferase (N7MTase) transfers a methyl group from *S*-adenosyl-L-methionine (AdoMet) to the guanine to form 7-methyl-GpppN-RNA and release *S*-adenosyl-L-homocysteine (AdoHcy).

The 5′ m^7G cap protects mRNA from the actions of 5′ to 3′ exonucleases and is a binding site for factors that direct splicing, nuclear export, and efficient translation. After the capping reaction is completed, a cap-binding complex binds to 7-methylguanosine and initiates splicing of the first intron by stabilizing the interaction with components of the spliceosome complex. This complex formation is facilitated by hyperphosphorylation of serine residues on the RNA polymerase II CTD. Splicing by the spliceosome complex begins during elongation; however, the rate of transcription is too slow to allow all splicing to take place co-transcriptionally, so splicing must also occur posttranscriptionally.

Sequences near the 3′ end of precursor mRNA mediate transcription termination through binding of cleavage and polyadenylation factors. As shown in **Figure 21.41**, the precursor mRNA sequence that promotes polyadenylation is 5′-AAUAAA-3′, which is bound by a protein called the cleavage and polyadenylation specificity factor (CPSF). Moreover, the downstream G and U (G/U)-rich region (3′ side of the AAUAAA sequence) is bound by another regulatory protein called the cleavage stimulatory factor (CStF). Two additional cleavage factors, CFI and CFII, then bind to the complex, followed by binding of the enzyme **poly(A) polymerase**, which synthesizes the poly(A) tail. Cleavage cannot occur until poly(A) polymerase binds, linking polyadenylation with the cleavage event. As soon as cleavage occurs, poly(A) polymerase catalyzes the addition of ~10 adenine residues to the 3′-hydroxyl terminus of the precursor mRNA. The rate of polyadenylation by poly(A) polymerase increases when **poly(A) binding protein** binds to the short tract of adenine residues, which presumably increases the activity of the poly(A) polymerase enzyme.

With the physical separation of the transcription and translational machinery in eukaryotic cells comes the requirement to transport all RNA components from the site of synthesis—the nucleus or nucleolus—to the cytoplasm, which is the site of protein synthesis. Because most processing reactions take place in the nuclear compartment, many proteins involved in facilitating transport preferentially recognize the mature RNA species. All macromolecule transport between the nucleus and cytoplasm must occur through ~2,000 **nuclear pore complexes** on the nuclear membrane. Nuclear

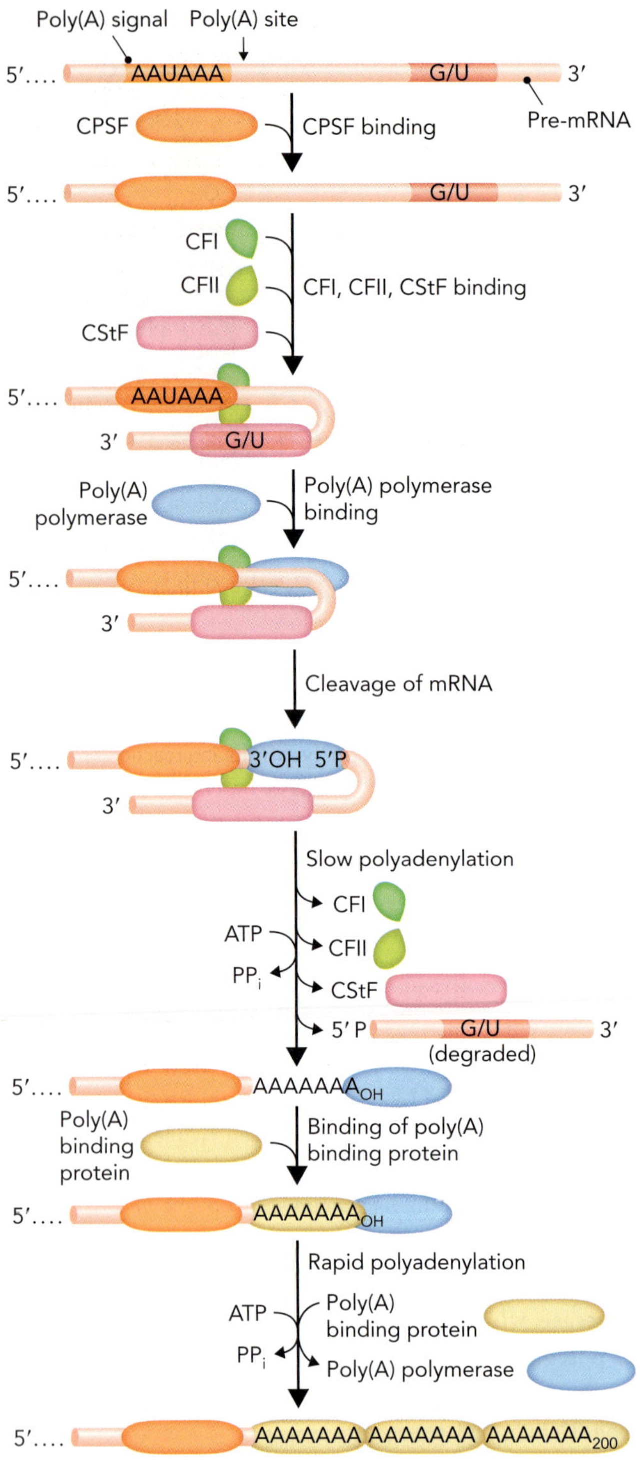

Figure 21.41 Cleavage and polyadenylation of eukaryotic mRNA is a coupled process. First, the RNA polymerase II elongation complex passes through the poly(A) signal to synthesize a precursor mRNA with the sequence AAUAAA. Then, the protein cleavage and polyadenylation specificity factor (CPSF) binds to initiate the termination process. This is followed by the binding of the G/U sequence recognition protein cleavage stimulatory factor (CStF) and two additional cleavage factors called CFI and CFII, which stimulates the binding of the enzyme poly(A) polymerase to the termination complex. The precursor mRNA is then cleaved to the 3′ side of the AAUAAA sequence in an ATP-dependent reaction, which leads to disassembly of the complex and the addition of ~10 adenine nucleotides to the 3′-hydroxyl on the terminus. In the final step, the stimulatory factor poly(A) binding protein binds to the short poly(A) tail and significantly increases the activity of poly(A) polymerase until ~200 adenine nucleotides are added to the terminus.

pore complexes are large complexes of up to 125 MDa that protrude on both sides of the membrane (**Figure 21.42**). RNA cannot pass through these complexes alone, but instead requires the assistance of one or more transport proteins that bind preferentially to the mature RNA structure. One of these proteins is called **Ras-related nuclear protein (Ran)**, which plays an important role in transporting RNA and proteins through the nuclear pore complex.

As shown in **Figure 21.43**, different classes of RNAs each require a distinct set of transport proteins, making this process very specific and ensuring that only the correct RNAs are transported into the cytoplasm. With the exception of mRNA, all other classes of RNA require the Ran protein for transport through the nuclear pore in a Ran-dependent manner. (Protein transport across the nuclear membrane also uses Ran; see Figure 23.22.) In many cases, proteins that bind as part of the processing reaction for each RNA species are also involved in transport. For example, the cap binding complex remains associated with the 5′ cap on RNA polymerase II transcripts. The cap binding complex is not essential for transport; rather, it helps signal the orientation of mRNA to facilitate transport through the nuclear pore complex in a 5′ to 3′ direction.

Messenger RNA Decay Is Mediated by 3′ Deadenylation and 5′ Decapping

RNA decay is a normal cellular process that can help eliminate incorrectly processed RNA and promote ribonucleotide turnover. Degradation of tRNA and rRNA occurs primarily as a stress response or to eliminate a defective transcript. In contrast, mRNA degradation appears to be regulated as a mechanism to control protein production.

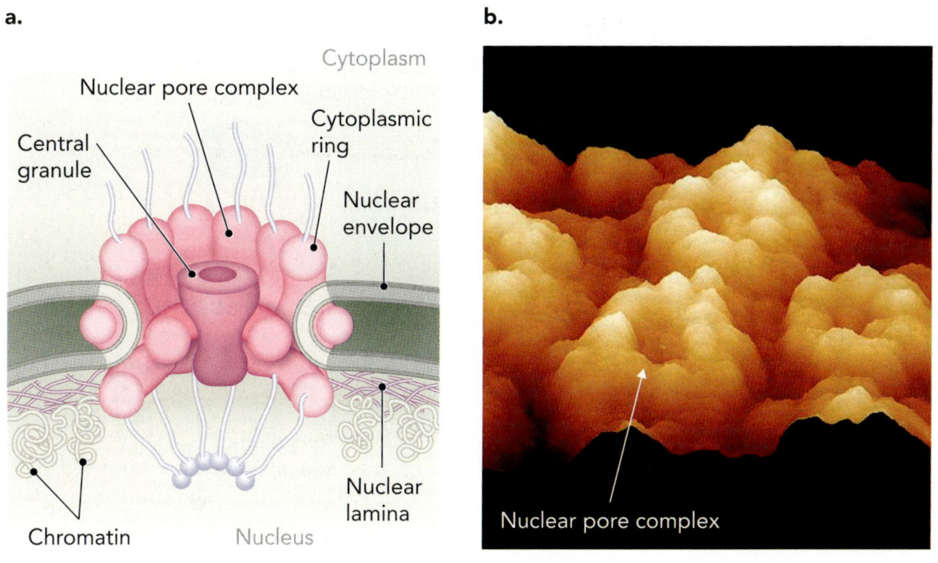

Figure 21.42 The nuclear pore complex transports proteins and ribonucleoprotein complexes across the nuclear membrane in eukaryotic cells. **a.** Schematic drawing showing the major nuclear pore complex components and their asymmetric distribution across the nuclear membrane. **b.** Atomic force microscopy image of an isolated yeast cell nucleus, showing the presence of nuclear pore complexes on the cytoplasmic side of the nuclear membrane. VICTOR SHAHIN, PROF. DR. H. OBERLEITHNER, UNIVERSITY HOSPITAL OF MUENSTER/SCIENCE PHOTO LIBRARY/SCIENCE SOURCE.

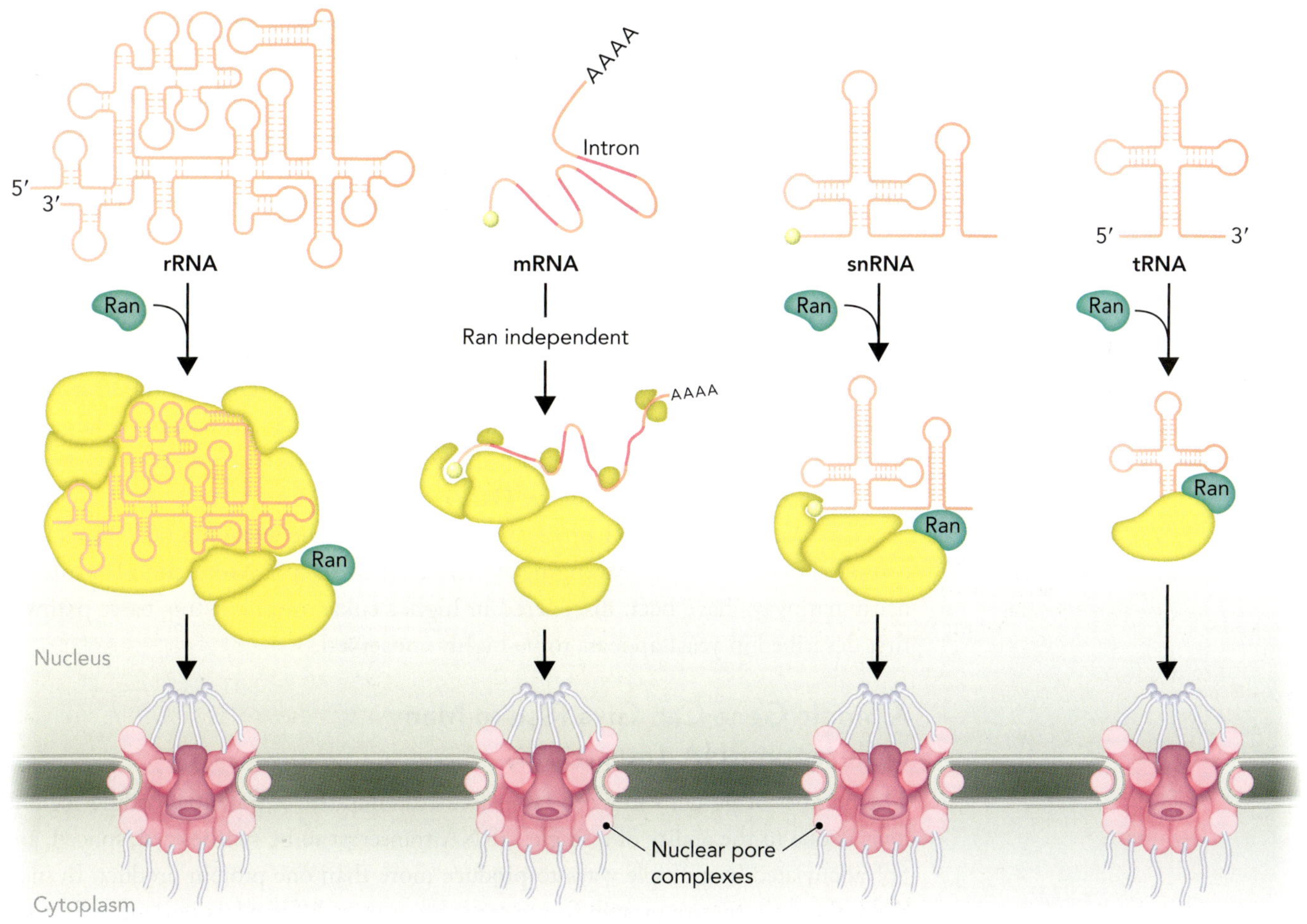

Figure 21.43 RNA processing and assembly of ribonucleoprotein complexes occurs prior to transport through the nuclear pore complex. Ran-dependent transport is required for all classes of RNA except mRNA, which is transported through the nuclear pore complex in a Ran–independent manner. Distinct ribonucleoprotein complexes are assembled to mediate transport of each RNA class.

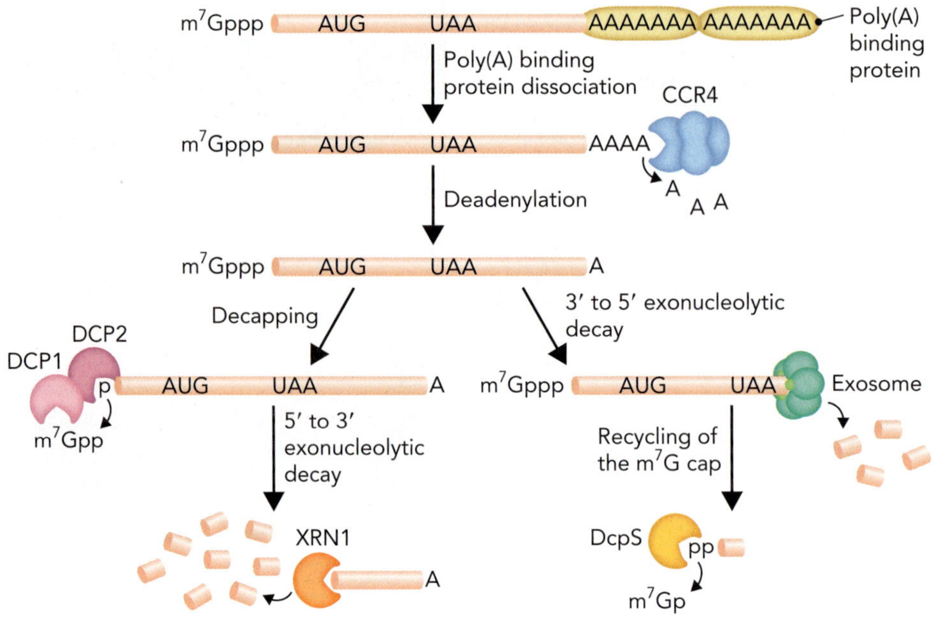

Figure 21.44 Post-transcriptional mRNA decay in yeast requires five major enzyme activities: deadenylation, decapping, 5′ to 3′ exonucleolytic decay, 3′ to 5′ exonucleolytic decay, and scavenging of the 5′ m^7G cap. After CCR4-mediated deadenylation, the unprotected mRNA can be degraded either from the 5′ end by the enzymes DCP1–DCP2 and XRN1 or from the 3′ end by the exosome and DcpS. Similar mRNA decay enzymes are found in most eukaryotes.

Eukaryotic mRNA decay can occur during transcription, as well as post-transcriptionally through several different mechanisms. The decay of mRNA in the cytoplasm can occur to decrease the expression of a toxic protein or to reduce the abundance of functional proteins by the half-life of their mRNAs. As shown in **Figure 21.44**, the primary pathway to degrade eukaryotic mRNA begins with removal of the poly(A) tail by a process called **deadenylation**. The major deadenylase activity in yeast is associated with a large multi-subunit protein complex containing the enzyme CCR4, which is a member of the *Exo*III endonuclease family. Shortening of the poly(A) tail by CCR4-mediated deadenylation prevents binding of poly(A) binding protein, which functions to protect the 3′ end of mRNA from 3′ to 5′ exonuclease decay. Poly(A) binding protein also plays a role in shielding the 5′ m^7G cap from 5′ to 3′ exonuclease decay, and therefore in the absence of poly(A) binding protein binding to the poly(A) tail, the 5′ m^7G cap is a target for degradation. As shown in Figure 21.44, once deadenylation occurs, two subsequent fates are possible: (1) decapping by DCP1 and DCP2 and 5′ to 3′ exonucleolytic decay by the enzyme XRN1 or (2) 3′ to 5′ decay by the **exosome** (an exonuclease complex) and m^7G cap scavenging by the enzyme DcpS. More complex mRNA decay pathways have been discovered in higher eukaryotes, but this basic pathway first described in yeast appears to be highly conserved.

A Single Gene Can Give Rise to Many Different mRNA Transcripts

The number of proteins in a cell far exceeds the number of potential coding regions. This is due to the ability of a single mRNA transcript to be synthesized, spliced, and polyadenylated in multiple ways to produce more than one protein product. In most cases, these alternative protein forms are expressed in different tissues and reflect the developmental lineage of that tissue or cell type.

As shown in **Figure 21.45**, the three primary mechanisms eukaryotic cells use to generate distinct mRNA products from the same gene are (1) alternative promoters,

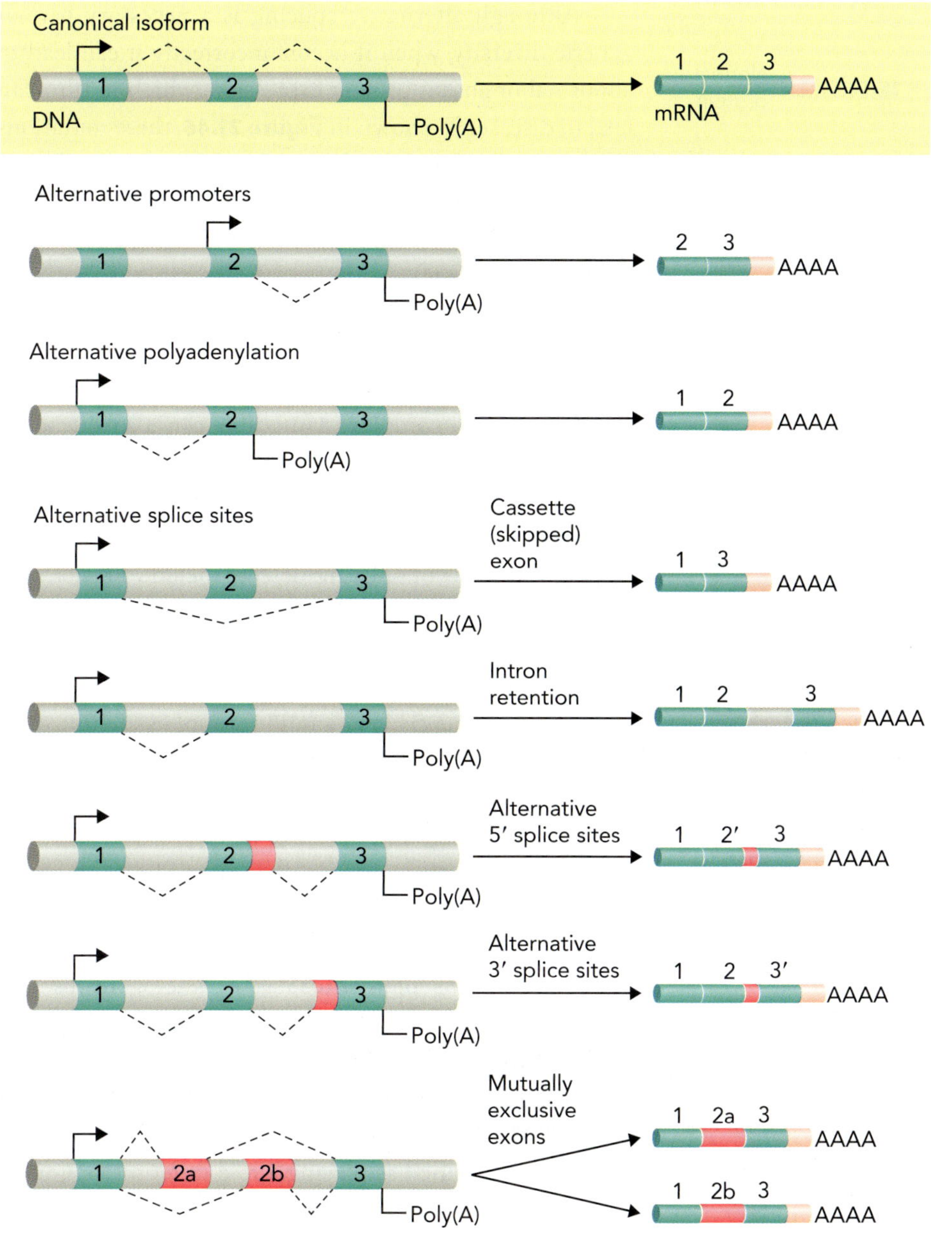

Figure 21.45 Protein diversity in eukaryotes reflects the use of alternative promoters and polyadenylation sites or, more commonly, alternative splicing. The canonical transcription, splicing, and polyadenylation pattern for this hypothetical gene with three exons is shown at the top, and the predicted coding sequence of the resulting mRNA transcript is shown at the right. Translation of each mRNA results in a unique protein.

(2) alternative polyadenylation, and (3) alternative splicing. Differential promoters generate distinct protein products if the downstream promoter eliminates the coding sequences in an upstream exon. Numerous examples have been found where an alternative downstream promoter is activated in a cell-specific manner by the binding of developmentally regulated transcription factors. Similarly, alternative polyadenylation alters the protein product if use of an upstream polyadenylation site excludes coding sequences in a downstream exon as a result of 3′ end processing. Alternative splicing can be accomplished through multiple mechanisms. Skipped exons is the most common mechanism of protein diversity and likely reflects gene evolution through the addition of exons encoding protein subdomains or activities. Intron retention also adds to protein diversity of protein-coding sequences. Lastly, use of alternative 5′ donor splice sites or 3′ acceptor splice sites is often associated with activation of cryptic splice sites or can be the result of DNA mutations.

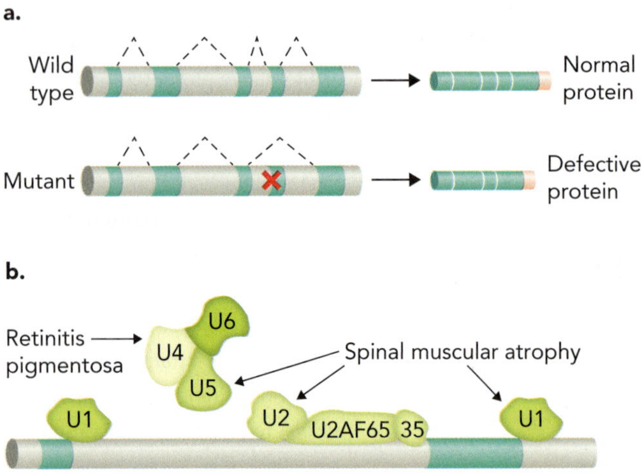

Figure 21.46 Splicing defects that cause disease can be due to defects in splicing recognition sequences (cis mutations) or in components of spliceosome complexes (trans mutations). **a.** The most common cis mutations in splicing that cause human disease are the loss of an exon-coding sequence generating proteins that lack a required exon. **b.** A rare number of trans mutations in the snRNA or protein components of spliceosomes have been identified in several human diseases. In the cases of retinitis pigmentosa and spinal muscular atrophy, the mutations affect the basal spliceosome machinery in specific cell types.

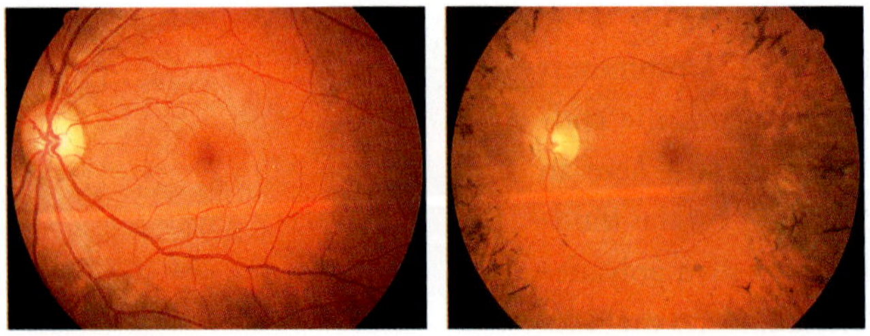

Figure 21.47 Retinitis pigmentosa causes degenerative changes in the retina that lead to total blindness. Shown are images of a normal retina and one affected by retinitis pigmentosa. The pigment buildup can be seen as the dark dots on the peripheral edge of the retina. NORMAL RETINA: LEFT-HANDED PHOTOGRAPHY/SHUTTERSTOCK; RETINITIS PIGMENTOSA: ISM/PHOTOTAKE.

Although alternative splicing is a necessary part of genetic diversity, when it occurs incorrectly it can lead to abnormal or absent protein expression, which in turn can lead to disease. As shown in **Figure 21.46**, these mutations can be grouped into two classes: those that disrupt the use of alternative splice sites (cis effects) and those that affect expression of spliceosome components or regulatory proteins (trans effects). Mutation of alternative splice sites can have multiple effects, although the most frequent is exon skipping and occurs as a result of a base change at one of the conserved regions that must be recognized by the spliceosome complex. Mutations can also introduce new splice sites or cause introns to be retained within the mature transcript, all of which result in an abnormal protein product. Some well-characterized diseases that result from these types of splicing defects include Frasier syndrome, which affects the development of the kidneys and genitalia; aggregation of the microtubule-associated protein tau into neuronal cytoplasmic inclusions, leading to progressive dementia; and a mild form of cystic fibrosis, a disease in which a transmembrane chloride channel in the secretory epithelium is disrupted, resulting in a buildup of mucus in the lungs and digestive defects.

Mutations that disrupt the assembly or function of the snRNPs are responsible for two well-characterized diseases: retinitis pigmentosa and spinal muscular atrophy. **Retinitis pigmentosa** is a heterogeneous disease affecting 1 in 4,000 individuals and is characterized by progressive retinal degeneration, night blindness, and loss of peripheral vision leading to total blindness (**Figure 21.47**). Although more than 30 different genes have been identified as playing a role in retinitis pigmentosa, autosomal dominant retinitis pigmentosa is linked to mutations in three genes that encode components of the spliceosome complex: *PRP8*, *PRP31*, and *PRP3*. The three proteins coded by these genes are all required for the formation of a stable U4–U6 complex and assembly of the U5–U4–U6 trimer. Each of these genes can contain more than one retinitis pigmentosa–linked mutation; for example, seven different mutations have been identified in *PRP31*, including insertions, deletions, and missense mutations that lead to a single amino acid change. Although the mutations in all three PRP genes should theoretically cause disruptions in splicing in all cells of the body, only the photoreceptor cells are affected in individuals with retinitis pigmentosa. One explanation for this cell-specific defect is the high level of splicing activity in retinal cells, which is needed to synthesize large amounts of rhodopsin protein as a result of cell renewal in the outer segments of the rod photoreceptor. Because these mutations are autosomal dominant, it is possible for individuals to have one normal copy of an affected PRP gene, which would lead to only a reduction in splicing activity.

concept integration 21.3

Explain why alternative splicing can be both an advantage and a disadvantage to an organism when it occurs in mRNA but would only have negative effects if it occurred in ncRNA.

Alternative splicing in mRNA can be an advantage because it increases the complexity of the proteome. By producing different proteins from a single transcript, phenotypic differences that enhance the survival of an organism can occur without modifying the genome. Alternative splicing in mRNA can be a disadvantage if one or more exons are skipped that encode critical domains of the protein or if the new splice site results in premature truncation of the protein. The activity of ncRNAs, however, is dependent entirely on their structure. Here, alternative splicing would result in a different secondary or tertiary structure, rendering the RNA inactive.

21.4 RNA-Mediated Gene Silencing

A major focus of RNA research in the 1990s was devoted to uncovering the function of ncRNAs; in particular, the short ncRNA species siRNA and miRNA (see Table 21.1). These short ncRNAs mediate a process of sequence-specific RNA interference called **gene silencing** in eukaryotic cells, which directly inhibits mRNA translation (siRNA) or gene expression (miRNA). Both of these short ncRNAs regulate normal cellular processes and likely evolved as defense mechanisms against RNA-containing pathogens (viruses) or to fine tune the regulation of RNA synthesis, processing, and degradation. We will examine how both of these pathways were discovered and their mechanisms. We also describe how these pathways have been exploited to increase our gene manipulation capabilities and how ncRNAs can be used as a diagnostic tool to detect certain diseases in humans.

The Discovery of RNA Interference

The discovery of gene silencing mediated by short ncRNA was the result of an experiment designed to achieve a completely different result. In 1990, Rich Jorgensen and Carolyn Napoli at the University of Arizona in Tucson were studying enzymes involved in generating the deep violet color observed in petunia flowers. Flavonoid pigments known as anthocyanins produce this color. The biosynthesis of anthocyanins begins with the condensation of malonyl-CoA and 4-coumaroyl-CoA by the enzyme chalcone synthase.

Jorgensen and Napoli inserted a second copy of the chalcone synthase gene into a plasmid, which was transformed into petunias to determine if the chalcone synthase enzyme was the rate-limiting step in anthocyanin biosynthesis. They expected to generate petunias with a darker violet color than that of the wild-type plants, owing to increased expression of the chalcone synthase gene due to the second copy. Unexpectedly, the petunias with an extra copy of the chalcone synthase gene contained white

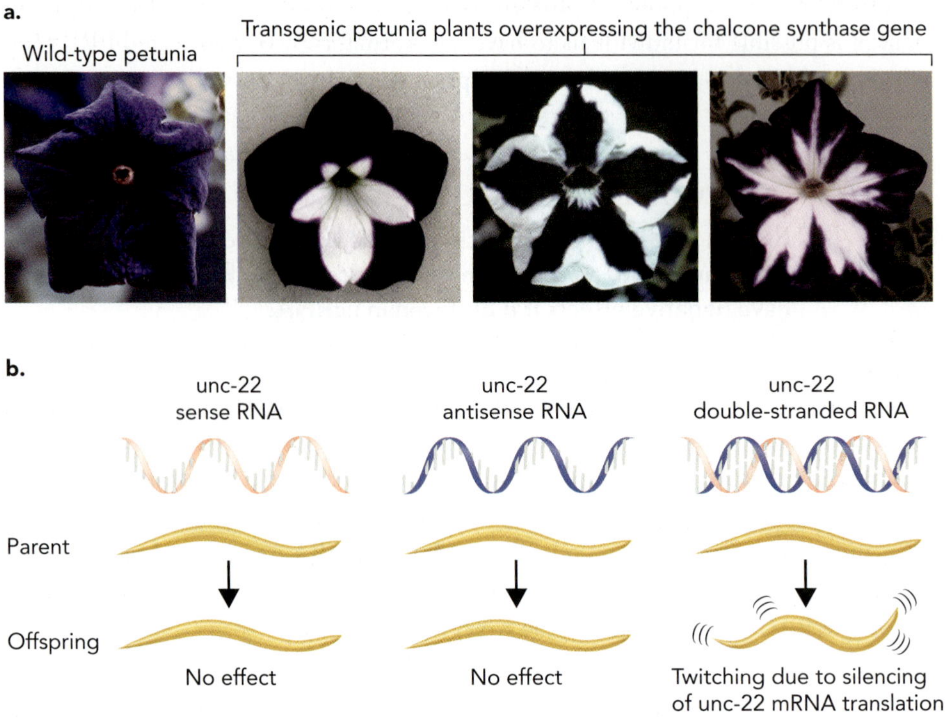

Figure 21.48 RNA interference was discovered in flowers and worms. **a.** White streaks in petonia flowers are caused by overexpression of the gene for chalcone synthase, the enzyme responsible for violet color. The introduction of the exogenous copy of the chalcone synthase gene interfered with expression of the endogenous chalcone synthase enzyme, resulting in loss of violet color in some cells. WILD-TYPE PETUNIA: R. A. JORGENSEN (1995). COSUPPRESSION, FLOWER COLOR PATTERNS, AND METASTABLE GENE EXPRESSION STATES. *SCIENCE, 268(5211)*, 686–691. © 1995 AMERICAN ASSOCIATION FOR THE ADVANCEMENT OF SCIENCE; TRANSGENIC PETUNIA PLANTS OVEREXPRESSING THE CHS GENE: (LEFT) Q. QUE, H. WANG, AND R. A. JORGENSEN (1998). DISTINCT PATTERNS OF PIGMENT SUPPRESSION ARE PRODUCED BY ALLELIC SENSE AND ANTISENSE CHALCONE SYNTHASE TRANSGENES IN PETUNIA FLOWERS. *THE PLANT JOURNAL, 13(3)*, 401–409. © 2002 JOHN WILEY AND SONS; (CENTER) C. NAPOLI, C. LEMIEUX, AND R. JORGENSEN (1990). INTRODUCTION OF A CHIMERIC CHALCONE SYNTHASE GENE INTO PETUNIA RESULTS IN REVERSIBLE CO-SUPPRESSION OF HOMOLOGOUS GENES IN TRANS. *THE PLANT CELL*, 2, 279–289. © 1990 AMERICAN SOCIETY OF PLANT PHYSIOLOGISTS; (RIGHT) COURTESY RICHARD JORGENSEN. **b.** A gene-silencing experiment performed in the worm *Caenorhabditis elegans* showed that injection of double-stranded RNA targeting the unc-22 gene, which encodes a myofilament protein, was much more effective at inhibiting the expression of endogenous unc-22 mRNA than either sense or antisense RNA alone.

streaks (**Figure 21.48**). When Jorgensen and Napoli investigated this result further, they discovered that the levels of chalcone synthase mRNA and protein in the plants with the extra copy of the gene were lower than in wild-type plants. From these results they deduced that the white color was due to destruction of chalcone synthase mRNA. They hypothesized that this color phenotype was the result of the introduced copy of the chalcone synthase gene somehow interfering with production of the endogenous chalcone synthase enzyme. They called the inhibitory effect **co-suppression** because expression of both the endogenous chalcone synthase gene and the chalcone synthase transgene was suppressed after insertion of the exogenous chalcone synthase DNA. The molecular mechanism of co-suppression was unclear at the time.

In 1992, Su Gou and Kenneth Kemphues documented the same type of co-suppression effect in the worm *Caenorhabditis elegans* that had been observed in flowers. The working model at the time was that co-suppression resulted from antisense RNA hybridizing to endogenous mRNA, which generated a double-stranded RNA species that could not be translated or was quickly degraded by riboendonucleases. Guo and Kemphues initially observed co-suppression using RNA with the same sequence as the DNA-coding strand (antisense RNA), as well as with RNA having the same sequence as the DNA template (sense RNA). This was a surprising result because only the antisense RNA should be able to form a double-strand RNA–RNA hybrid with endogenous mRNA, which has the same sequence as the DNA-coding strand (except uridine replaces thymine).

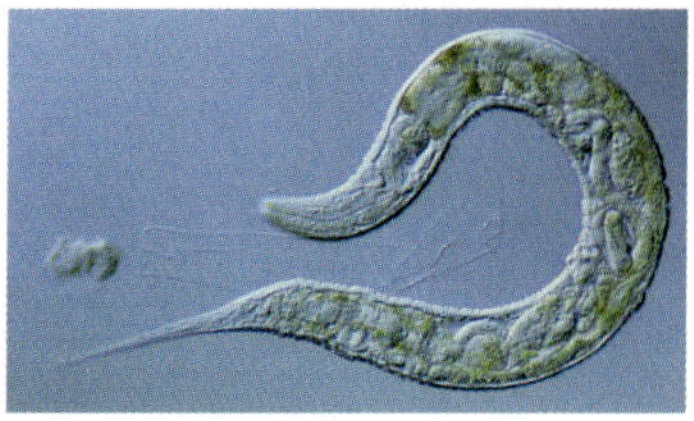

Caenorhabditis elegans

Figure 21.49 Craig Mello and Andrew Fire were awarded the 2006 Nobel Prize in Physiology or Medicine for their discovery of RNA interference in the nematode worm *C. elegans*. C. MELLO: MICHAEL FEIN/BLOOMBERG VIA GETTY IMAGES; A. FIRE: LINDA A. CICERO/STANFORD NEWS; *C. ELEGANS*: SINCLAIR STAMMERS/SCIENCE SOURCE.

In 1998, however, a paper coauthored by Andrew Fire of the Carnegie Institution for Science in Washington, D.C., and Craig Mello of the University of Massachusetts–Worcester reported that co-suppression of gene expression in *C. elegans* was actually mediated by dsRNA, not by the single-strand sense or antisense RNA, as originally proposed. They suggested that the reason Guo and Kemphues observed co-suppression under both conditions (adding only sense or antisense RNA) was that small amounts of single-stranded RNA were present in their preparations, and it led to the formation of double-stranded RNA that triggered co-suppression. As verification of their reasoning, Fire and Mello were able to show that double-stranded RNA was 10–100 times more effective in targeting mRNA transcribed from the unc-22 gene, which is required for controlling muscle activity, than either the sense or antisense strand of unc-22 RNA alone. Fire and Mello were awarded the 2006 Nobel Prize in Physiology or Medicine for their discovery of the underlying mechanism of RNA interference (RNAi) using the model organism *C. elegans* (**Figure 21.49**).

Using the work of Fire and Mello as a starting point, scientists began to explore the mechanism behind RNAi. It was known from their work that RNAi required double-stranded RNA as the initiator, but the steps between introduction of double-stranded RNA into cells and the eventual destruction of mRNA remained unclear. The ability of the double-stranded RNA to induce a systemic effect on gene expression, as well as its ability to be inherited, argued for the existence of a stable intermediate in the pathway. In addition, short (21–25 nt) RNA fragments, called short interfering RNA (siRNA), were detected after introducing double-stranded RNA into plant and animal cells. Characterization of siRNA revealed that it contains a $5'$-phosphoryl group, a $3'$-hydroxyl group, and two- to three-nucleotide $3'$ overhangs, much like the products of RNase III digestion. An RNase III–like enzyme in *Drosophila melanogaster* extracts was identified that cleaved double-stranded RNA into fragments similar to those produced in the RNAi pathway; this enzyme was given the name Dicer. Dicer is a versatile enzyme that cleaves any double-stranded RNA template regardless of the presence of noncanonical base pairing or minor deviations in duplex structure.

The interaction between siRNA and the target mRNA is mediated by a family of **RNA-induced silencing complexes (RISCs)**, which are composed of ribonucleoproteins that can be programmed to target almost any gene for silencing (**Figure 21.50**). Dicer cleaves double-stranded RNA to produce siRNA, which is then assembled into RISC. The siRNA and mRNA form duplex RNA between the complementary single-stranded siRNA and mRNA. Although the proteins involved in RISC assembly and gene silencing differ slightly between organisms, the basic

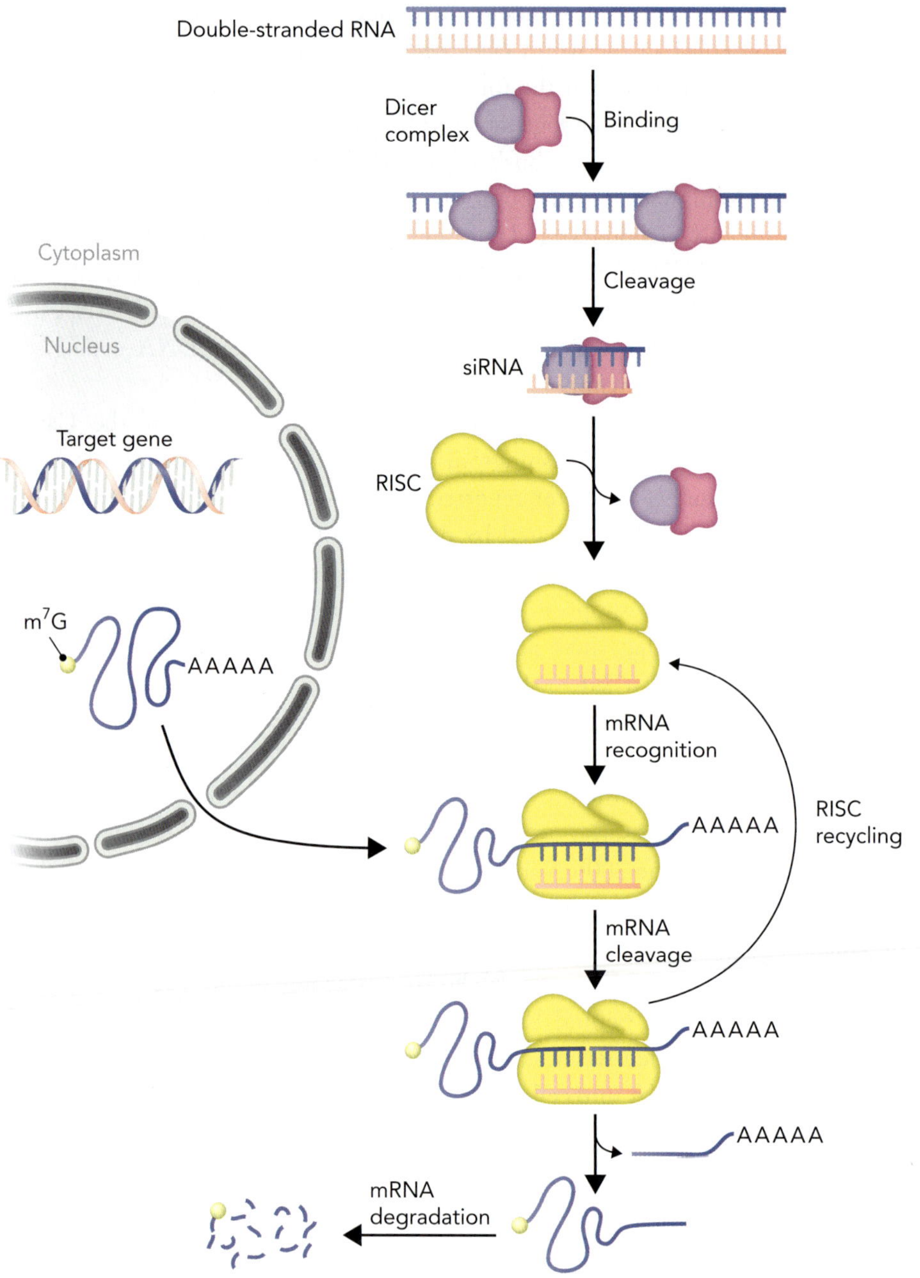

Figure 21.50 The RNAi pathway degrades target mRNA using siRNA generated from double-stranded RNA. In the example shown here, the double-stranded RNA is added exogenously, which could be from a double-stranded RNA virus or some other source of double-stranded RNA. (Endogenous sources of double-stranded RNA could be from viral replication.) The Dicer enzyme recognizes and binds to double-stranded RNA to generate siRNA. The RISC assembles the mRNA-degrading complex using single-stranded siRNA.

▶ ANIMATION

biochemical processes of siRNA production are similar. Importantly, the majority of siRNA molecules are a perfect match to their target, resulting in the destruction of the bound mRNA.

Biogenesis and Function of miRNA

The siRNA fragments that mediate RNAi are not the only example of noncoding RNAs that have an effect on gene expression. Micro RNA (miRNA) is another class of short ncRNAs that bind to mRNA and mediate gene silencing. Many similarities appear in the pathways induced by siRNA and miRNA, along with shared protein components; however, the initiating steps of the pathway are different, as is the mechanism of translational repression (**Figure 21.51**). Although siRNA is most often generated from exogenous double-stranded RNA, such as a virus or other pathogen, miRNA is encoded in the genome. Because of this difference, miRNA is often not a perfect complement to its

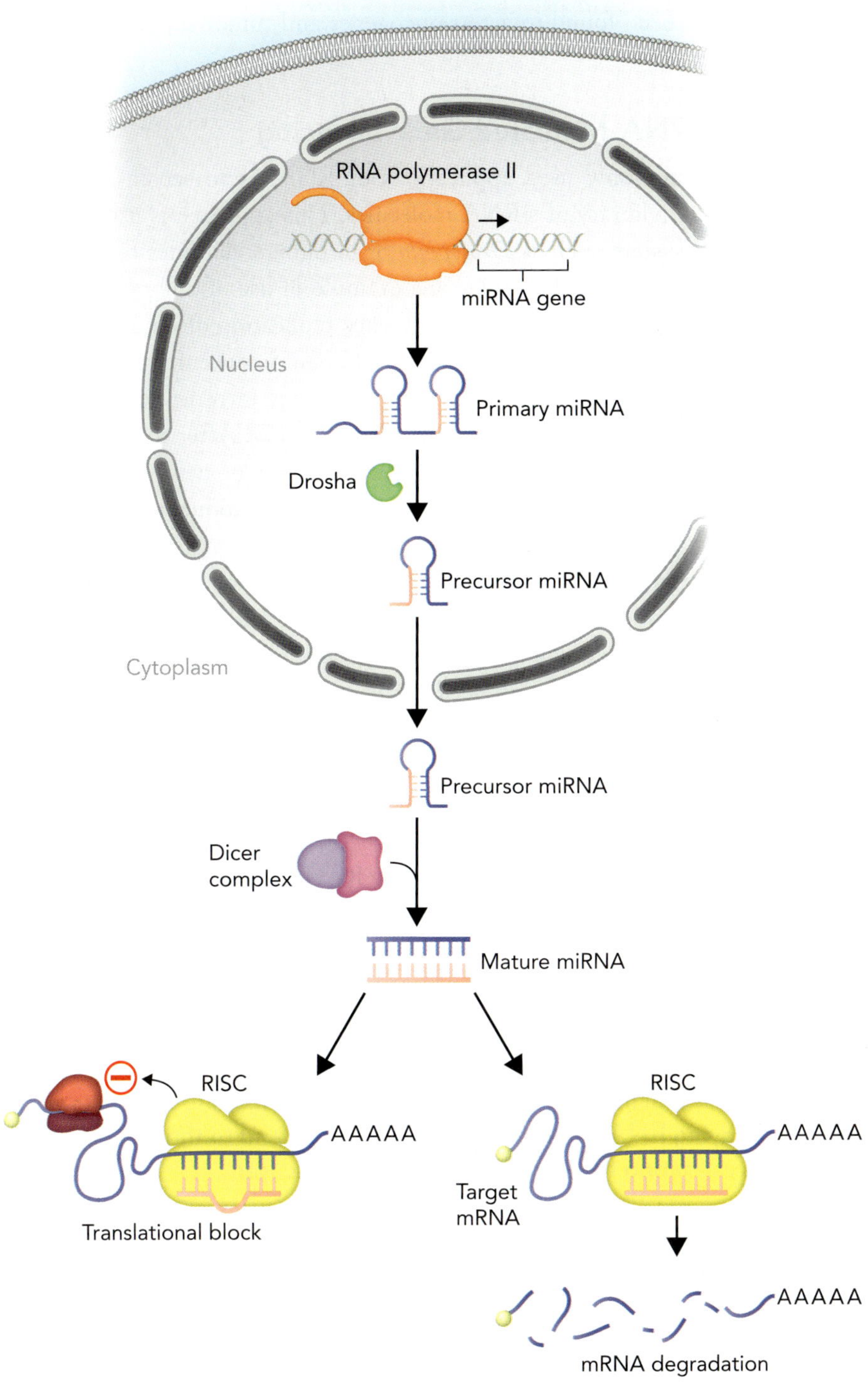

Figure 21.51 The miRNA pathways share many common features with the RNAi pathway. Expression of an miRNA gene by RNA polymerase II generates primary miRNA, which is processed by Drosha and exported to the cytoplasm. Drosha is an endoribonuclease that cleaves primary miRNA into precursor miRNA. The precursor miRNA is cleaved in the cytoplasm by Dicer, another endoribonuclease, to generate the mature miRNA product. It is this mature miRNA that mediates gene silencing of specific mRNA targets by translational inhibition of ribosome function or mRNA degradation.

target mRNA. Both pre-miRNA and mature miRNA contain a more complex secondary structure than siRNA, with hairpins and internal loops resulting from one or more unpaired regions. Unlike RNAi, which evolved as a protective mechanism, miRNA has been shown to be an essential part of normal cell development and regulation.

A large number of miRNAs are known to regulate a variety of cellular processes, such as development, differentiation, proliferation, and apoptosis. More than 1,400 miRNAs have been identified in humans, and these can target significantly more transcripts given the ability of a single miRNA to repress translation of up to several hundred mRNAs. Micro RNA plays an important role in regulating cellular

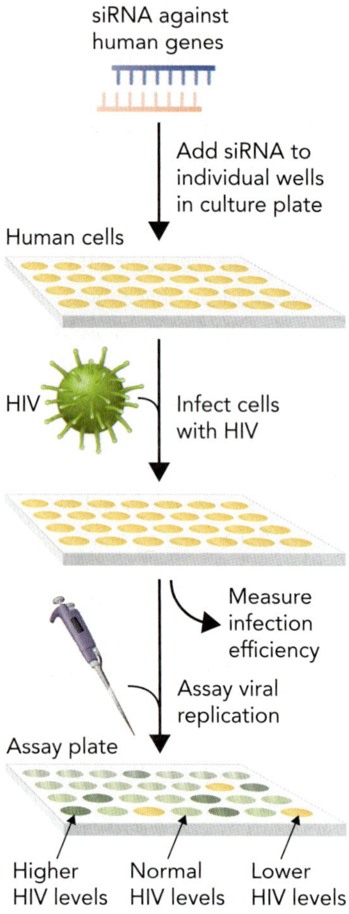

Figure 21.52 Applications of RNAi include the use of siRNA screens in human cell lines to identify genes affecting HIV replication. By incubating human cells in small-volume culture plates with a library of siRNA molecules that specifically decrease protein expression from thousands of human genes prior to HIV infection, it is possible to identify human genes required for viral replication (decreased HIV replication) and those that inhibit viral replication (increased HIV replication).

processes, and the aberrant expression of miRNA has been linked to diseases. In fact, many miRNAs have been found to act as oncogenes and tumor suppressors, giving them a role in cancer.

Applications of RNA-Mediated Gene Silencing

One of the most commonly used applications of RNAi is to screen for phenotypic responses using a library of siRNA molecules. This approach is used for basic research in a model system such as *Drosophila* or in human cell lines to investigate a disease phenotype. **Figure 21.52** shows one example of use of a library of siRNA molecules to screen for human genes that influence replication of human immunodeficiency virus (HIV). In this screening strategy, human tissue culture cells are first incubated with siRNA molecules targeting human genes in order to reduce their level of expression prior to infecting the cells with live HIV. After the normal time required for viral replication in infected cells, the amount of new virus produced is quantitated using a fluorescence assay. If a human gene produces a product that inhibits viral production, then reduced expression of that human gene would lead to elevated viral replication. In contrast, if a specific human gene product is required for viral production under normal conditions (the virus utilizes the human gene product), then siRNA knockdown of the human gene should lead to decreased viral replication. The level of viral infection needs to be monitored independently of measuring viral replication in order to distinguish between effects on viral infectivity and viral replication in normally infected cells. This experiment could be designed so that a predetermined set of human genes is targeted that are thought to be involved in HIV replication or it could be designed so that it is an unbiased search for human genes that are functionally identified as necessary for HIV replication. Such a global approach to gene silencing using siRNA is analogous to mutant screens done with bacteria or fruit flies more than 50 years ago.

concept integration 21.4

Why is oral ingestion of siRNA a topic of concern to scientists studying RNAi applications in GMO crops and to those investigating the use of siRNA to treat disease?

The ability of short RNA species to be ingested and retain their function is a topic of great interest to all studying potential applications of siRNA. A major concern in the generation of GMO crops by RNAi is that the siRNA used to modify gene expression in plants could survive the digestive tract and pass into the cells of humans or other animals that ingest these crops. Once in the cell, these siRNAs would form RISCs as any other siRNA or miRNA and could match, even imperfectly, with the host mRNA, leading to changes in gene expression. Although this is a disadvantage to the potential widespread use of RNAi to modify plants, it could be a benefit in the use of RNAi to treat some diseases. Although some cancer-targeting RNAi treatments would need to rely on a more focused delivery system so that the expression of the target gene would not be affected in normal cells, the ability to turn RNAi into a pill-based therapy could be advantageous for treating a number of human diseases. Not only does the use of an oral delivery system generally make the drugs cheaper to manufacture, but also it is easier for the patient to self-administer the treatment.

chapter summary

21.1 Structure and Function of RNA

- Prokaryotic and eukaryotic cells contain abundant protein-synthesizing RNAs (messenger RNA, ribosomal RNA, and transfer RNA), and eukaryotic cells also contain abundant noncoding RNA (ncRNA).
- Three classes of eukaryotic ncRNA have been characterized: short (miRNA, siRNA, piRNA), small (snRNA, snoRNA), and long (RNaseP, TERC, lncRNA).
- Eukaryotic mRNA is transcribed by RNA polymerase II; tRNA by RNA polymerase III; and rRNA by RNA polymerases I and III.
- Transcription and translation in prokaryotes are coupled processes, whereas these two processes in eukaryotes take place in the nucleus and cytoplasm, respectively.
- Precursor mRNA in eukaryotes is transcribed and processed in the nucleus, with 5′ capping, splicing, and polyadenylation being coordinated by functions associated with RNA polymerase II.
- Genes encoding rRNA are transcribed as a single unit that is then cleaved and processed; in prokaryotes, rRNA genes often flank a tRNA gene.
- The portion of the genome encoding protein genes is inversely proportional to the size of the genome and the amount of ncRNA; ~50% of the human genome sequence is found represented in ncRNA compared to 2% ncRNA in yeast.
- One of the best-studied examples of a functional lncRNA is the sequence XIST, which regulates X inactivation in females on only one of two X chromosomes.
- Four modes of action have been attributed to lncRNA: (1) base pairing between nucleotides in the lncRNA and target RNA, (2) base pairing between lncRNA and single-stranded regions of DNA, (3) formation of functional ribonucleoprotein complexes similar to ribosomes and spliceosomes, and (4) ligand-induced riboswitches that function in signaling pathways and gene expression.

21.2 Biochemistry of RNA Synthesis

- The bacterial RNA polymerase holoenzyme is composed of an α_2 dimer, β, β', and ω subunits. One of several σ transcription factors associates with the holoenzyme and directs binding of RNA polymerase to specific promoter regions.
- Bacterial promoter regions contain two conserved regions, called the −35 box and −10 box, which are responsible for σ-factor binding.
- DNase I footprinting is an experimental technique that can be used to identify regions of DNA bound by proteins, such as transcription factors or RNA polymerase.
- The eukaryotic TATA binding protein (TBP) transcription factor is responsible for recruiting all three RNA polymerases to promoter regions.
- RNA polymerase synthesizes a complement of the template strand of DNA using ATP, CTP, UTP, and GTP. The transcription bubble contains the enzyme, a locally unwound region of DNA, and an RNA–DNA hybrid helix of usually 8 bp.
- Bacterial transcriptional termination occurs as either a Rho-dependent or Rho-independent process.
- In RNA polymerase II transcription, the TFIID–TBP complex binds to the core promoter sequence, followed by the binding of TFIIA, TFIIB, TFIIF, TFIIE, and TFIIH and the RNA polymerase II enzyme.
- The CTD in eukaryotic RNA polymerase II is required for coordinating precursor mRNA processing, and its functions are regulated by phosphorylation and dephosphorylation on multiple repeats of the heptapeptide sequence YSPTSPS.
- RNA polymerase II transcriptional termination is coupled to processing of the 3′ end, which involves polyadenylation by the enzyme poly(A) polymerase.

21.3 Eukaryotic RNA Processing

- Ribozymes are enzymes that contain a catalytically active RNA component. Some ribozymes are composed only of RNA, whereas others contain both RNA and protein subunits.
- The hammerhead ribozyme is a catalytic RNA that can self-cleave as well as catalyze cleavage of other RNA molecules.
- RNaseP is a ribonuclease present in bacteria, archaea, and eukaryotes that generates the mature 5′ end of tRNAs and cleaves some precursor rRNA transcripts.
- Eukaryotic RNA splicing occurs when an intron is removed in a transesterification reaction, which is coupled to rejoining of the 5′ and 3′ ends to generate a processed transcript.
- Group I and group II self-splicing intron reactions do not require proteins, whereas spliceosome-mediated precursor mRNA splicing involves small nuclear ribonucleoproteins (snRNPs).
- Group II intron self-splicing and spliceosome-mediated splicing give rise to an excised lariat intron structure that is degraded.
- Eukaryotic precursor mRNA introns are flanked by short conserved sequences at the 5′ and 3′ splice sites. The branch site is located 18–40 nucleotides upstream of the 3′ splice site, and in higher eukaryotes it contains a polypyrimidine tract.
- Precursor tRNA transcripts undergo cleavage at the 5′ and 3′ ends, addition of a CCA trinucleotide sequence to the

3′ acceptor stem, and base modification. Many tRNAs also are spliced to remove an intron in the anticodon loop.

- Processing of precursor rRNA occurs co-transcriptionally in the nucleolus, along with assembly of ribosomal proteins on the nascent transcript. Base modification of eukaryotic rRNA is catalyzed by small nucleolar RNAs (snoRNAs).
- A 7-methylguanylate cap (m^7G cap) is added to the 5′ end of RNA polymerase II transcripts and protects the mRNA from degradation by 5′ to 3′ exonucleases. The m^7G cap also serves as a binding site for factors that direct splicing, nuclear export, and efficient translation.
- Transport of eukaryotic RNA from the nucleus to the cytoplasm requires associations with transport proteins, one of which is the protein Ras-related nuclear protein (Ran); export of mRNA from the nucleus is Ran independent.
- Targeted RNA decay is a normal cellular process that promotes mRNA turnover as a method of gene regulation. Decapping and deadenylation of the transcript are a component of most mRNA decay mechanisms.
- Alternative splicing of mRNA can increase genomic complexity but can also cause disease. DNA mutations can lead to a gain or loss of splice sites, resulting in alternative splicing and production of aberrant proteins.

21.4 RNA-Mediated Gene Silencing

- RNA interference (RNAi) refers to the process of gene silencing, a process that is mediated by long or short double-stranded RNA molecules that can form base pairs with a target RNA and direct its degradation or inhibit translation.
- Double-stranded RNA is cleaved into short double-stranded fragments called siRNA (21–25 bp) by the RNase III–like enzyme Dicer and loaded onto the RNA-induced silencing complex (RISC). The RISC binds to and catalyzes cleavage of a complementary target RNA, which could be mRNA or viral RNA.
- Micro RNAs (miRNAs) are short, untranslated RNAs that bind to mRNA and negatively regulate gene expression. They are encoded in the genome, transcribed by RNA polymerase II or III, and cleaved into siRNA by the Dicer enzyme.
- Applications of RNAi include use as a research tool to investigate gene function in model organisms, as a reagent for high-throughput screening strategies to identify functions of genes required for a specific process, and, perhaps someday, as a therapeutic agent to treat human disease.

biochemical terms

(in order of appearance in text)

noncoding RNA (ncRNA) (p. 1056)
RNase (p. 1056)
7-methylguanylate cap (m^7G cap) (p. 1061)
long noncoding RNA (lncRNA) (p. 1062)
X-inactive specific transcript (XIST) (p. 1062)
riboswitch (p. 1064)
cAMP receptor protein (CRP) (p. 1065)
σ factor (p. 1066)
housekeeping gene (p. 1066)
DNase I footprinting (p. 1067)
TATA binding protein (TBP) (p. 1069)
closed complex (p. 1070)
open complex (p. 1070)
Rho-dependent termination (p. 1071)
Rho-independent termination (p. 1071)
hammerhead ribozyme (p. 1075)
spliceosome (p. 1075)
group I introns (p. 1075)
group II introns (p. 1076)
small nuclear ribonucleoprotein (snRNP) (p. 1077)
small nucleolar RNA (snoRNA) (p. 1083)
poly(A) polymerase (p. 1085)
poly(A) binding protein (p. 1085)
nuclear pore complex (p. 1085)
Ras-related nuclear protein (Ran) (p. 1086)
deadenylation (p. 1088)
exosome (p. 1088)
retinitis pigmentosa (p. 1090)
gene silencing (p. 1091)
co-suppression (p. 1092)
RNA-induced silencing complex (RISC) (p. 1093)

review questions

1. What three forms of RNA are found in both prokaryotes and eukaryotes? Why are these three common to all organisms?
2. Why is it beneficial for organisms to have more than one copy of rRNA genes in the genome? Why do more complex organisms tend to have more copies of these genes?
3. The DNA sequence of a gene follows. If this is the sense strand, write the sequence of the mRNA transcript formed.

   CGCGGATCCTTGAATTCTAAATAAACCATTT
   ACCACCATGACC

4. What feature of rRNA makes co-transcriptional folding necessary for correct assembly of the ribosome? How do ribosomal proteins contribute to the formation of the conserved secondary structure of rRNA?
5. The following sequence is from a region of the M13 bacteriophage genome. Identify and label the promoter

elements that would be recognized by the bacterial RNA polymerase. Where would transcription begin?

CAGGCGATGATCAAATCTCCGTTGTACTTT
GTTTCGCGCGTTGGTATAATCGCTGGGGTCAA
GATGAGT

6. Why do bacteria contain multiple σ factors?
7. Why is Mg^{2+} essential for the activity of RNA polymerase?
8. Compare and contrast Rho-dependent and Rho-independent termination. What is a common feature in both mechanisms?
9. TATA-less promoters do not contain a recognizable TATA box but are still associated with TBP. What features are often found in these promoters?
10. Why is the phosphorylation state of the RNA polymerase II CTD an important component of transcription? Describe the changes that the CTD undergoes from initiation to termination.
11. Explain why decapping and deadenylation are common mechanisms involved in RNA decay.
12. How does the existence of ribozymes like the hammerhead ribozyme and group I introns provide support for the RNA world hypothesis?
13. How does the mechanism of group I introns differ from that of group II introns? Which is more similar to the spliceosome-catalyzed reaction?
14. Why does tRNA contain more types of base modifications than rRNA?
15. Explain how the spliceosome recognizes introns.
16. Describe two ways in which a point mutation in a gene can result in alternative splicing.
17. What are the differences in the RNAi and miRNA pathways?

challenge problems

1. A scientist wishes to express a eukaryotic protein in bacterial cells. The gene is cloned along with its promoter region and is inserted into a plasmid. After transforming the plasmid into bacterial cells, protein expression is initiated, but no protein is observed after the cells are lysed. Why? How could this problem be fixed?
2. In the experiment by Jacob and Meselson, bacteria were grown for several generations in a medium containing heavy isotopes ($^{15}NH_4Cl$ and ^{13}C-glucose). The bacteria used these isotopes to synthesize all cellular components, resulting in their incorporation into all of the bacterial nucleic acids and proteins. The bacteria were switched to a normal medium after infection with bacteriophage, and the ribosomes were isolated. Jacob and Meselson discovered that the isolated ribosomes contained the heavy isotopes. Why did this result indicate that ribosomes are not carriers of genetic information?
3. In bacterial transcription, the σ factor dissociates during elongation. How would the activity of the polymerase and transcription elongation be affected if the σ factor remained bound to the holoenzyme after initiation?
4. RNaseP can cleave both pre-tRNA and pre-rRNA sequences. If the primary sequences of these RNAs are not always similar, how the does the enzyme recognize and bind its substrates?
5. RNAi-based modification of amylopectin content in wheat occurs through decreased expression of starch-branching enzymes. If the siRNA used to knock down plant starch-branching enzymes was transmitted to humans, what enzyme might it affect, and why would this be harmful?
6. Many antibacterial agents work by inhibiting RNA polymerase. These agents are typically molecules that interact with various regions of the catalytic center to prevent DNA binding or transcript synthesis. Why are many of these considered "broad-spectrum antibiotics"; that is, antibacterial agents that act against a large number of different bacterial species? Why are these broad-spectrum antibiotics advantageous for both physicians and their patients? These agents, however, do not affect eukaryotic RNA Pol I, II, or III. What does this tell you about the sequence conservation between the bacterial and eukaryotic enzymes?
7. Describe the similarities and differences between snRNA and snoRNA, including cellular location and function.
8. Open the PDB file 3RTX of mouse guanylyltransferase (Mce1) in a molecular viewer and read the associated paper in *Molecular Cell* (2011) describing the crystal structure. What role does phosphorylation at Ser5 play in the interaction between mammalian RNA guanylyltransferase and the RNA polymerase II CTD? Is phosphorylation at this residue essential for the interaction? Why is Ser2 phosphorylation not required for efficient capping?
9. The review by Valadkhan [Valadkhan, S. (2010). Role of the snRNAs in the spliceosomal active site. *RNA Biology*, 7, 345–353] describes the catalytic activity of the spliceosome complex, and Figure 1 therein illustrates interactions between U2 and U6 snRNA. Describe these

interactions, and explain why they are necessary for correct positioning of the intron and for catalysis.

10. In bacteria, the same DNA strand is often a template for both replication and transcription. Why are codirectional collisions a common occurrence?

11. The paper by Balbo and Bohm [Balbo, P. B., and Bohm, A. (2007). Mechanism of poly(A) polymerase: structure of the enzyme-MgATP-RNA ternary complex and kinetic analysis. *Structure, 15,* 1117–1131] illustrates the interactions between poly(A) polymerase, the 3′ end of mRNA, and ATP. Figures 2 and 3 therein detail the interactions between the protein and the two substrates. Why is single-stranded DNA a poor substrate for this reaction?

12. Formation of the 5′ m^7G cap requires that the 5′-nucleotide be a triphosphate. Explain why only this nucleotide retains all three phosphoryl groups.

13. What is the primary difference between strong and weak bacterial promoters? How does transcription differ from these two types of promoters?

smartwork5

If your instructor assigns homework with Smartwork5, access it here: digital.wwnorton.com/biochem.

suggested reading

Books and Reviews

Agrawal, N., Dasaradhi, P. V., Mohmmed, A., Malhotra, P., Bhatnagar, R. K., and Mukhergee, S. K. (2003). RNA interference: biology, mechanism, and applications. *Microbiology and Molecular Biology Reviews, 67,* 657–685.

Brown, T. A. (2002). *Genomes.* Oxford, England: Wiley-Liss.

Chem, C. A., and Shyu, A. (2011). Mechanisms of deadenylation-dependent decay. *Wiley Interdisciplinary Reviews RNA, 2(2)*: 167–183.

Decker, C. J., and Parker, R. (2002). mRNA decay enzymes: decappers conserved between yeast and mammals. *Proceedings of the National Academy of Sciences USA, 99,* 12512.

Dong, H., Lei, J., Wen, Y., Ju, H., and Zhang, X. (2013). MicroRNA: function, detection, and bioanalysis. *Chemical Reviews, 113,* 6207–6233.

Fedor, M. J., and Williamson, J. R. (2005). The catalytic diversity of RNAs. *Nature Reviews Molecular Cell Biology, 6,* 399.

Geisler, S., and Coller, J. (2013). RNA in unexpected places: long noncoding RNA functions in diverse cellular contexts. *Nature Reviews Molecular Cell Biology, 14,* 699.

Kim, E., Goren, A., and Ast, G. (2007). Alternative splicing: current perspectives. *BioEssays, 30,* 38–47.

Kohler, A., and Hurt, E. (2007). Exporting RNA from the nucleus to the cytoplasm. *Nature Reviews Molecular Cell Biology, 8,* 761–773.

Mercer, T. R., and Mattick, J. S. (2013). Structure and function of long noncoding RNAs in epigenetic regulation. *Nature Structural and Molecular Biology, 20,* 300.

Simms, R. J., Belotserkovskaya, R., and Reinberg, D. (2004). Elongation by RNA polymerase II: the short and long of it. *Genes and Development, 18,* 2437–2468.

Scott, W., Horan, L. H., and Martick, M. (2013). The hammerhead ribozyme: structure, catalysis and gene regulation. *Progress in Molecular Biology and Translational Science, 120,* 1–23.

Thomas, M. C., and Chang, C. (2006). The general transcription machinery and general cofactors. *Critical Reviews in Biochemistry and Molecular Biology, 41,* 105–178.

Valadkhan, S. (2007). The spliceosome: a ribozyme at heart? *Biological Chemistry, 388,* 693–697.

Will, C. L., and Luhrmann, R. (2011). Spliceosome structure and function. *Cold Spring Harbor Perspectives in Biology, 3(7),* a003707.

Primary Literature

Brass, A. L., Dykxhoorn, D. M., Benita, Y., Yan, N., Engelman, A., Xavier, R. J., Lieberman, J., and Elledge, S. J. (2008). Identification of host proteins required for HIV infection through a functional genomic screen. *Science, 15,* 921.

Liu, S., Ghalei, H., Luhrmann, R., and Wahl, M. C. (2011). Structural basis for the dual U4 and U4atac snRNA-binding specificity of spliceosomal protein hPrp31. *RNA, 17,* 1655–1663.

Maenner, S., Blaud, M., Fouillen, L., Savoye, A., Marchand, V., Dubois, A., Sanglier-Cianférani, S., Van Dorsselaer, A., Clerc, P., Avner, P., et al. (2010). 2-D Structure of the A region of Xist RNA and its implication for PRC2 association. *PLoS Biology, 8,* e1000276.

Nikolov, D. B., and Burley, S. K. (1997). RNA polymerase II transcription initiation: a structural view. *Proceedings of the National Academy of Sciences USA, 94,* 15–22.

Pratt, A., and MacRae, I. J. (2009). The RNA-induced silencing complex: a versatile gene-silencing machine. *Journal of Biological Chemistry, 284,* 17897–17901.

Regina, A., Bird, A., Topping, D., Bowden, S., Freeman, J., Barsby, T., Kosar-Hashemi, B., Li, Z., Rhaman, S., and Morell, M. (2006). High-amylose wheat generated by RNA interference improves indices of large-bowel health in rats. *Proceedings of the National Academy of Sciences USA*, *103*, 3546–3551.

Salananha, R., Mohr, G., Belfort, M., and Lambowitz, A. M. (1993). Group I and group II introns. *FASEB Journal*, *7*, 15–24.

Tabernero, J., Shapiro, G. I., LoRusso, P. M., Cervantes, A., Schwartz, G. K., Weiss, G. J., Paz-Ares, L., Cho, D. C., Infante, J. R., Alsina, M., et al. (2013). First-in-human trial of an RNA interference therapeutic targeting VEGF and KSP in cancer patients with liver involvement. *Cancer Discovery*, *3*, 406–417.

Tanackovic, G., Ransijn, A., Thibault, P., Elela, S., Klinck, R., Berson, E. L., Chabot, B., and Rivolta, C. (2011). PRPF mutations are associated with generalized defects in spliceosome formation and pre-mRNA splicing in patients with retinitis pigmentosa. *Human Molecular Genetics*, *20*(*11*), 2116–2130.

Uncharacterized soil bacteria can be a rich source of new antibiotics, which are critically needed to treat antibiotic-resistant infections.

Samples can be obtained directly from the soil or from plant parts and debris

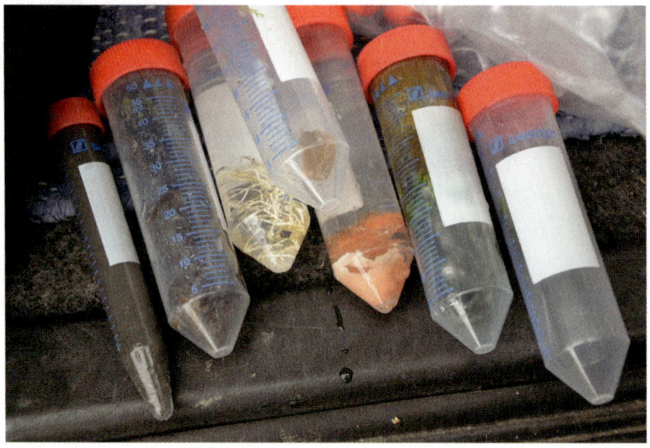

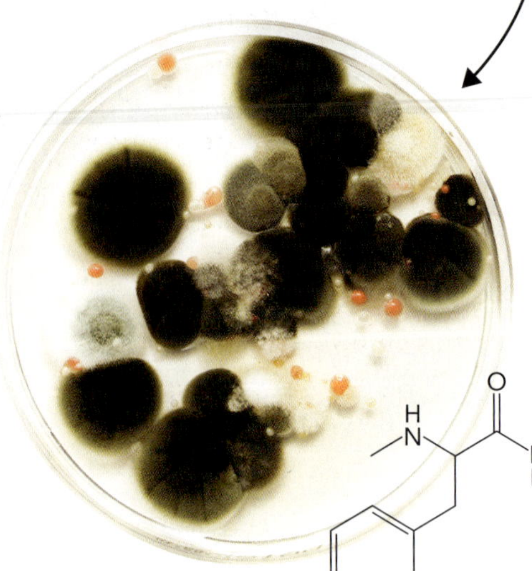

Culturing bacteria in the lab can be a challenging task for microbiologists

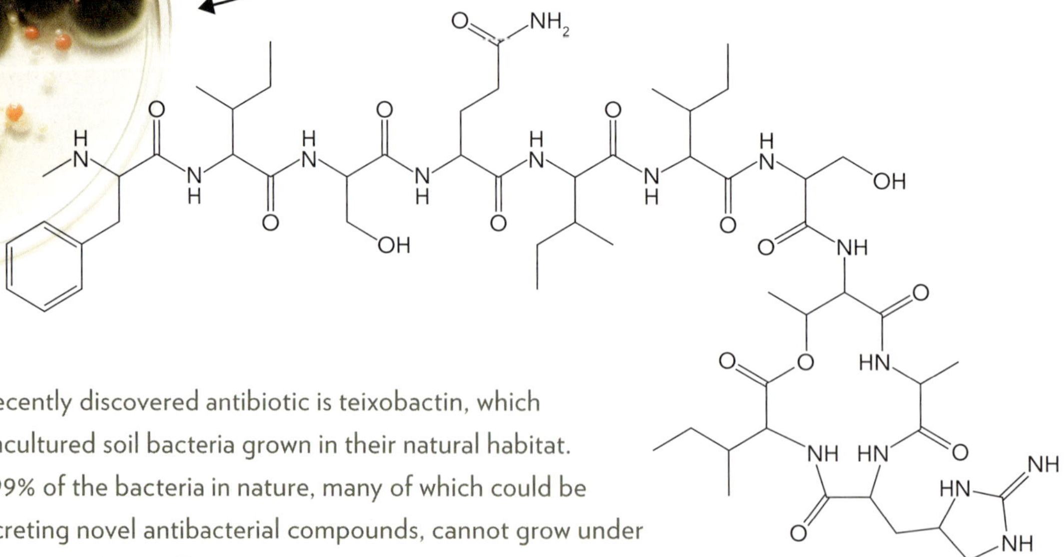

One example of a recently discovered antibiotic is teixobactin, which was isolated from uncultured soil bacteria grown in their natural habitat. It is estimated that 99% of the bacteria in nature, many of which could be synthesizing and secreting novel antibacterial compounds, cannot grow under conventional laboratory conditions. Teixobactin has been shown to inhibit cell wall synthesis in *Staphylococcus aureus* and *Mycobacterium tuberculosis* grown *in vitro* and *in vivo* without leading to detectable resistance.

22

Protein Synthesis, Posttranslational Modification, and Transport

◀ Overuse of antibiotics in modern society has led to the emergence of "superbugs" that are resistant to essentially all known antibiotics. The two major classes of antibiotics currently in use today are those that block bacterial cell wall synthesis, such as penicillin and methicillin (see Chapter 13), and those that block biochemical steps required for protein synthesis. As examples of the latter, streptomycin alters the structure of the bacterial 30S ribosome causing it to misread mRNA during translation, and chloramphenicol binds to the 50S ribosome and prevents peptide bond formation. We describe these classes of antibiotics in this chapter. Making chemical modifications to existing antibiotics is unlikely to produce truly novel antibiotics, and resistance to such new derivatives is inevitable. An alternative approach to discovery is to scour the earth looking for microorganisms that have never been tested to see if they secrete chemical compounds that kill, for instance, *Staphylococcus aureus* bacteria. This hunt has taken scientists to underwater volcanic vents off the coast of Iceland and to the most remote jungles, but even our own backyards could yield the next generation of antibiotics using new culturing methods.

CHAPTER OUTLINE

CREDITS: FOREST ASSORTMENT: ANNE DEMARINIS; MUD SPLATS: DIOGOPPR/SHUTTERSTOCK; RED SOIL: GALAPAGOSPHOTO/SHUTTERSTOCK; ICHIP DEVICE: PROF. SLAVA EPSTEIN, NORTHEASTERN UNIVERSITY; SAMPLES OF UNCULTURED SOIL BACTERIA GROWN IN THEIR NATURAL HABITAT: JENNIFER YANG/TORONTO STAR VIA GETTY IMAGES; SAMPLES OF CULTURES GROWN IN A LAB SETTING: GRECHA/SHUTTERSTOCK.

Figure 22.1 The RNA Tie Club in Cambridge, England, was dedicated to finding the link between nucleic acids and proteins. A few of the notable RNA Tie Club members are shown (from left to right: Francis Crick, Alexander Rich, Leslie E. Orgel, and James Watson). The woolen necktie, designed by George Gamow, contained an embroidered helix given to each RNA Tie Club member.

ALEXANDER RICH, COURTESY JOSIAH RICH.

We have observed that proteins are key biomolecules in essentially all biochemical processes and have a variety of functions essential to cell survival. In Chapter 4 we discussed the primary, secondary, tertiary, and quaternary structures of proteins, and in subsequent chapters we studied the functions of many different proteins and the importance of the relationship between structure and function.

We now turn to a closer examination of how the primary structure of a protein is determined by biochemical processes that convert molecular information in mRNA into a polypeptide sequence as dictated by the genetic code. We first explore how scientists came to understand and "crack" the genetic code. Next, we take a closer look at the process of translation and discuss how various factors work together to produce cellular proteins. Finally, we examine some of the functional modifications made to proteins during and after translation and how proteins are transported throughout the cell.

22.1 Deciphering the Genetic Code

In Chapter 3, we described how James Watson and Francis Crick elucidated the structure of the DNA double helix in 1953 based in part on the X-ray data of Rosalind Franklin. They next turned their attention to how the nucleotide sequence of DNA can result in a protein of a particular sequence. Soon after their elucidation of the DNA double helix, Watson and Crick received a letter from physicist George Gamow. Working from a mathematical perspective, Gamow proposed that a genetic sequence composed of four nucleotides (A, C, G, and T/U) and linked to the production of 20 amino acids would require at least three of these four nucleotides to serve as a code for each individual amino acid. Gamow was correct, but still the problem remained as to exactly how each series of three nucleotides, or codons, could be matched with an amino acid.

Gamow, Watson, and Crick, along with other notable scientists, joined together to form the RNA Tie Club in 1954. The goal of this club was to provide a sounding board for members to discuss their theories for the process of protein synthesis. The club was made up of 20 members, each representing an amino acid, and four honorary members representing nucleotides. **Figure 22.1** shows a few of the members, including Crick (tyrosine) and Watson (proline), wearing the woolen, embroidered tie given to each member.

The Molecular Adaptor Required for Protein Synthesis Is tRNA

Gamow, along with others, formulated various hypotheses about the translation of the genetic code. The majority of these ideas hinged on the existence of a direct physical interaction between nucleotides and amino acids, which Crick did not favor because of where these components were located in the cell and known nucleotide structures. Crick's dissatisfaction with these ideas led him to develop a novel hypothesis in which the physical connection between the DNA–RNA genetic code and amino acids was accomplished by an adaptor molecule. As shown in **Figure 22.2**, Crick's proposed

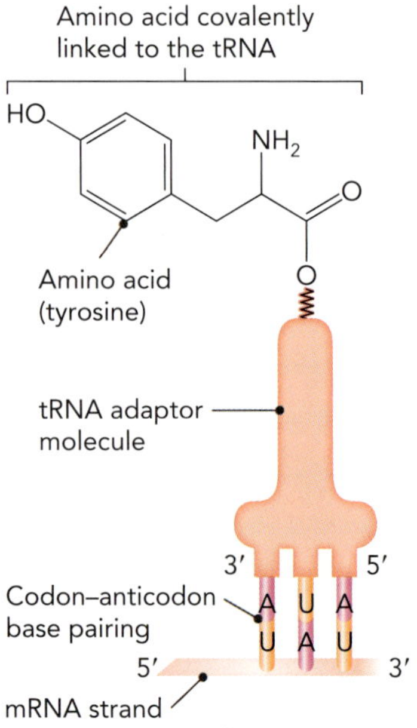

Figure 22.2 Crick's adaptor hypothesis proposed the link between RNA and protein. A general schematic is shown illustrating how tRNA functions as an adaptor molecule to connect the nucleotide sequence to a specific amino acid. The anticodon of the tRNA makes specific base pairs with the mRNA codon.

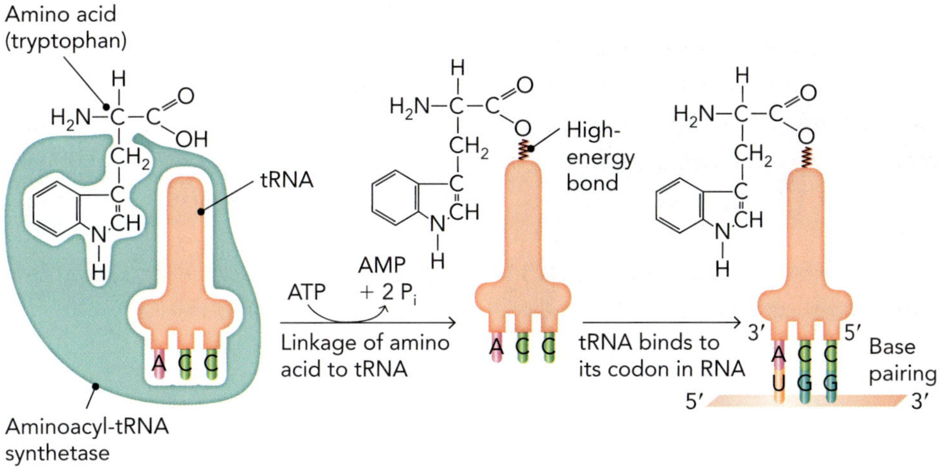

Figure 22.3 The function of aminoacyl-tRNA synthetase enzymes is to charge tRNA molecules with the correct amino acid. This process is covered in more detail later in the chapter.

adaptor molecule is now known to be transfer RNA (tRNA). The structure of tRNA was not discovered until the 1970s, but Crick's hypothesis initiated the hunt to isolate, sequence, and determine the secondary and tertiary structures of this molecule. We first introduced tRNA in Chapter 3 when we described that RNA contains modified nucleotide bases (see Figure 3.24).

Crick proposed that the existence of an adaptor molecule implied the need for an enzyme to catalyze the attachment of an amino acid to this molecule. **Figure 22.3** illustrates the process of aminoacylation, in which an enzyme known as an **aminoacyl-tRNA synthetase** attaches a specific amino acid to the 3′ acceptor stem of the cognate tRNA. The synthesis of an aminoacylated, or charged, tRNA is coupled with the hydrolysis of ATP. (We discuss aminoacyl-tRNA synthetases and their functions in more detail later in this chapter.)

Solving the Genetic Code Using Experimental Biochemistry

In addition to searching for the elusive adaptor molecule, scientists were also eager to decipher the genetic code in order to explain how the information in mRNA was encoded in the polypeptide sequence. The task was to assign each of the 64 possible nucleotide codons to each of the 20 known amino acids. (Four bases can be arranged in three positions, and $4 \times 4 \times 4 = 64$.) The first codon was serendipitously deciphered in the 1960s by Marshall Nirenberg and his postdoctoral fellow Heinrich Matthaei, who were testing a method for *in vitro* protein synthesis at the National Institutes of Health in Washington, D.C. (**Figure 22.4**). Nirenberg and Matthaei discovered that addition of ribosomal RNA (rRNA) increased the amount of protein synthesized *in vitro* by a mixture of lysed bacterial cells, which contained all of the biomolecules needed to synthesize proteins. In the course of these experiments, they added RNA containing only uridine (polyU RNA) to the system as a negative control to demonstrate that it was rRNA that uniquely stimulated protein synthesis. To their surprise, however, they found that addition of polyU RNA to the protein synthesis system not only stimulated protein synthesis even more than by addition of rRNA alone, but also the protein that was produced contained only phenylalanine.

This discovery confirmed the ability of RNA to serve as a template for protein synthesis and showed that phenylalanine was coded for by UUU. They next tested polyA and polyC RNA polymers using the bacterial cell-free extracts, each containing

a.

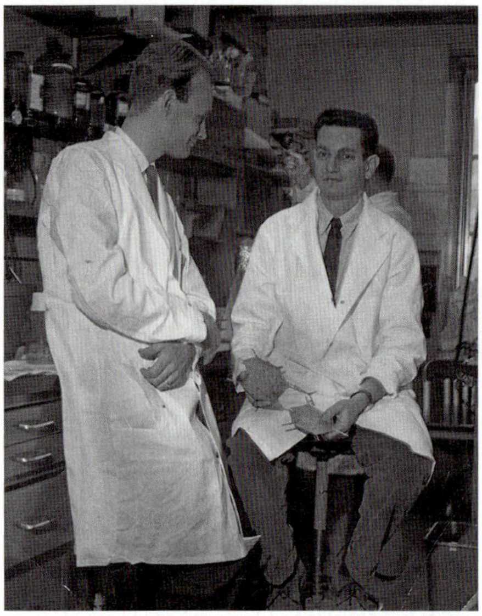

b.

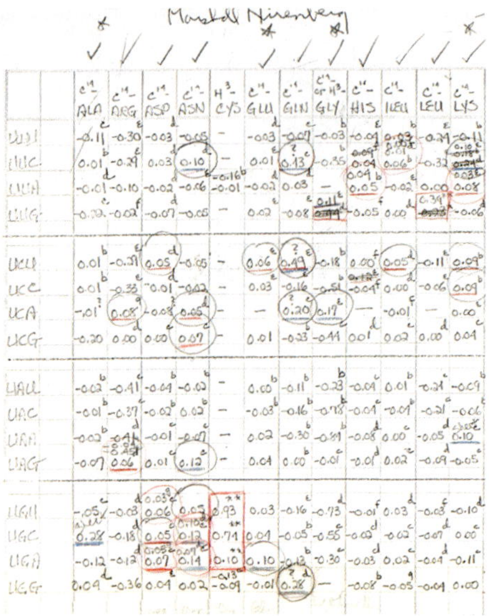

Figure 22.4 The genetic code was deciphered by biochemists using defined *in vitro* protein synthesis reactions. **a.** Marshall Nirenberg (right) with his postdoctoral fellow Heinrich Matthaei (left) in the lab at the National Institutes of Health in 1962. COURTESY OF THE NATIONAL INSTITUTES OF HEALTH. **b.** A page from one of Nirenberg's notebooks showing his work on elucidation of the genetic code. NATIONAL LIBRARY OF MEDICINE, NATIONAL INSTITUTES OF HEALTH.

only one of the 20 amino acids for potential incorporation into a polypeptide. As shown in **Figure 22.5**, they confirmed that polyU RNA only encoded polyphenylalanine, and moreover, that polyA RNA only encoded polylysine and polyC RNA only encoded polyproline. The cornerstone of the genetic code was now in place, and the race was on to identify the triplet codons specifying each of the 20 amino acids.

Because there are 64 possible nucleotide codons, but proteins consist of only 20 different amino acids, scientists were left to wonder if there were redundancies within the genetic code or if some codons did not code for any particular amino acid. This question could only be answered by testing all 64 possible codon variations, and scientists including Nirenberg, Philip Leder, and Har Gobind Khorana set out to achieve this goal. Their efforts were largely successful owing to Khorana's ability to synthesize DNA molecules with known sequences of repeating two, three, or four nucleotides. Khorana's DNA molecules were used to synthesize the RNA templates needed for protein synthesis. The polypeptides synthesized from these synthetic RNA templates were analyzed to determine their amino acid composition. For example, the repeating triplet nucleotide RNA sequence

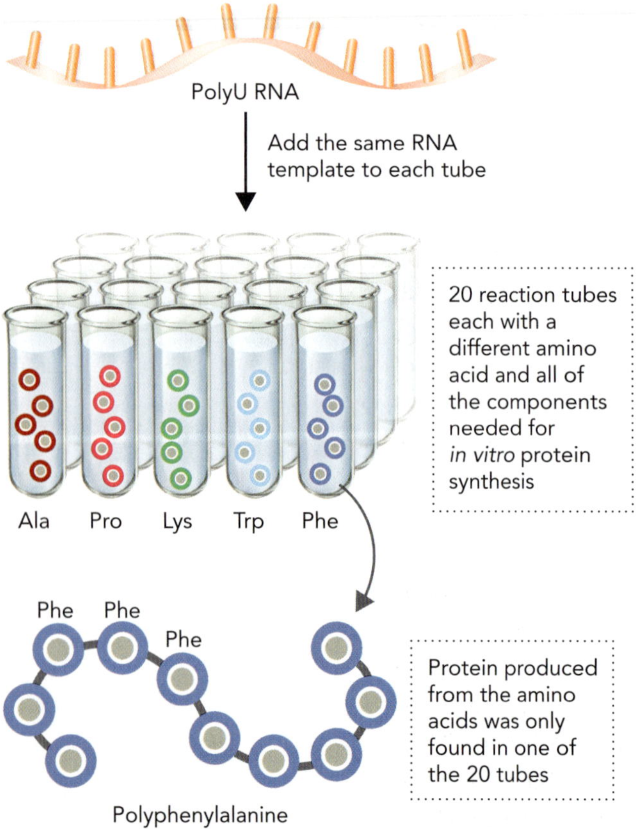

Figure 22.5 The Nirenberg–Matthaei experiment demonstrated that polyU RNA contains a triplet codon for the amino acid phenylalanine. Bacterial cell-free extracts containing all components necessary for protein synthesis were incubated with polyU RNA in 20 separate reactions, each with only one amino acid. The only protein that was produced was from the reaction containing phenylalanine; the other reactions contained only amino acids, not proteins, at the end of the experiment. This experiment was repeated with polyC and polyA RNA, leading to the identification of the codons for proline and lysine, respectively.

a.

b.

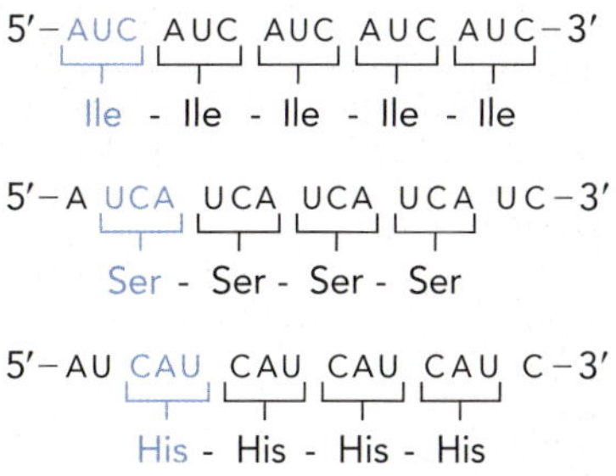

Figure 22.6 Khorana's synthesis of nucleotide polymers with defined sequences was a key component in the cracking of the genetic code. **a.** Har Gobind Khorana is shown in his laboratory at the University of Wisconsin in the 1960s. UNIVERSITY OF WISCONSIN-MADISON ARCHIVES, IMAGE #S10956. **b.** The repeating triplet nucleotide RNA sequence AUCAUCAUC directs the synthesis of repeating Ile, Ser, and His residues depending on the triplet register.

AUCAUCAUC contains three potential codons: AUC, UCA, and CAU. Translation of this RNA in an *in vitro* reaction resulted in the synthesis of polypeptides of repeating Ile, Ser, or His residues, depending on the triplet register, which is often referred to as the reading frame (**Figure 22.6**). This information was compared with the results from other related RNA sequences (for example, UUCUUC or ACACAC) to match codons with the appropriate amino acids.

Nirenberg and Leder designed experiments to match trinucleotide RNA sequences with their associated tRNA to further elucidate the genetic code. These short sequences can be bound by the ribosome and recognized by the cognate tRNA but are too short to allow for protein synthesis. Instead, the trinucleotides and tRNA remain bound to the ribosome and can be isolated as a complex by passing the reaction mixture through a filter membrane. In each experiment, a unique synthetic trinucleotide sequence was mixed with protein synthesis components. Then, amino acid mixtures, each containing a different ^{14}C-labeled amino acid, were added individually to the mixture before it was passed through a filter. As shown in **Figure 22.7**, if the trinucleotide does not code for the labeled amino acid on the aminoacylated tRNA, then the radioactive tRNA does not form a complex with the ribosome, and it passes through the filter. But if the anticodon of the radioactive aminoacylated tRNA is complementary to the triplet codon, then a complex is formed, and the radioactivity is retained on the filter. Using this assay in combination with the data resulting from Khorana's repeating nucleotide sequences (see Figure 22.6), it was possible to assign 61 of the 64 possible codons to specific amino acids, with the remaining three codons serving as termination signals, or "stop" codons, for protein synthesis (see Figure 4.22). In recognition of breaking the genetic code and discovery of the role of tRNA as the adaptor molecule, Nirenberg and Khorana, along with Robert Holley of Cornell University, were awarded the 1968 Nobel Prize in Physiology or Medicine.

Is the genetic code universal? A universal genetic code would imply that mRNA from any species could be translated by the protein synthesis components of any other species. In many cases, prokaryotic organisms can be used to express recombinant eukaryotic proteins in the laboratory. In some cases, however, problems can occur because of differences in the frequency of certain codons. For example, the triplet AAA is used to code for lysine three times more frequently than AAG in *E. coli*, whereas the two codons are used at almost the same frequency in humans. If the reduced codon usage is reflected in a lower expression of the corresponding tRNA,

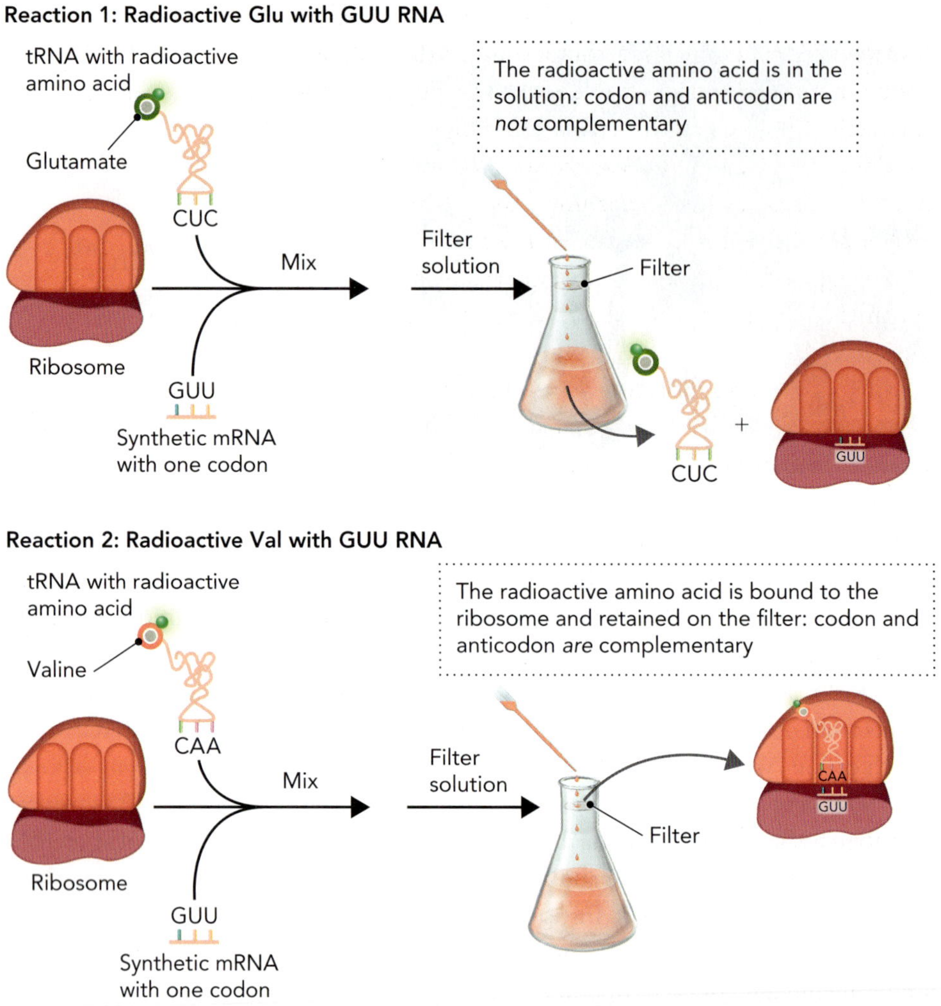

Figure 22.7 The Nirenberg–Leder experiment was used to assign triplet codons to specific amino acids on the basis of retention of radioactive aminoacylated tRNA molecules on a nitrocellulose filter that binds to proteins (such as ribosomal proteins). If the anticodon and codon were not complementary, then the radioactive aminoacyl tRNA passed through the filter and was found in the wash solution.

then the rate of protein synthesis can be affected. Different organisms also show preferences for stop codons, with UAA found most commonly in simple organisms such as *E. coli* and yeast, and UGA is found more frequently in complex organisms such as mice, rats, and humans.

In addition to the genetic code variations found in some species, differences also exist between the preferred codon usage of nuclear and mitochondrial DNA. For example, in humans two codons exist for methionine in the mitochondria (AUG and AUA), in contrast to the single methionine codon in nuclear genes. The differences in mitochondrial codon usage are due to the presence of alternative tRNA genes found within the mitochondrial genome.

The tRNA Wobble Position Explains Redundancy in the Genetic Code

At the most basic level, we can view the interaction between mRNA and tRNA as a series of three adjacent nucleotide base pairs formed in antiparallel fashion. But if the interaction were really this simple and followed the rules for Watson and Crick base pairing, then the number of codons and tRNA molecules would be identical. In fact, the number of tRNAs varies among organisms, with most species having 30–35 different tRNA genes, rather than the maximum number of 61 tRNA genes.

Yet even before it was realized that organisms contain less than 61 tRNA genes, a comparison of the different codons for a single amino acid led to the proposal that the interaction between tRNA and mRNA might not follow canonical G-C and A-U base pairing. For example, a G nucleotide base at the 5′-position (first position) in the tRNA anticodon can base pair with either its canonical C nucleotide base in the 3′-position (third position) of the mRNA or the noncanonical U nucleotide base in this position.

As shown in **Figure 22.8**, noncanonical base pairings between tRNA anticodon bases and mRNA codon bases can also occur with the purine nucleotide inosine (I), which is formed by the deamination of adenosine by the enzyme tRNA-specific adenosine deaminase. The four most common noncanonical base pairings between the 5′-position of the tRNA anticodon and the 3′-position of the mRNA codon are G-U, I-C, I-U, and I-A base pairs. The knowledge that different codons for a single amino acid often vary only at the third position of the codon, combined with the known anticodon sequences of several yeast tRNAs, led Francis Crick to propose the **wobble hypothesis**, which stated that the base pairing rules may be relaxed at the third position of the codon (the first position of the anticodon). By building models of the possible interactions between bases at this position, wobble base pairing became firmly established as the reason why ~35–50 tRNA genes are able to provide coverage of all 61 amino acid codons in the genetic code. **Figure 22.9** illustrates how noncanonical codon–anticodon base pairing can occur between the 5′-position of the tRNA anticodon (wobble position) and the 3′-position of the mRNA codon. Note that inosine is not found in mRNA (except infrequently by deamination of adenosine) and therefore is present only in the 5′-position of the tRNA anticodon.

As the genes for different tRNAs were identified, it became increasingly apparent that a tRNA may recognize different mRNA codons because of noncanonical base pairing in the wobble position. Moreover, there are also related tRNAs that may be charged with the same amino acid but have distinct anticodons. As shown in **Figure 22.10**, there are four codons for alanine (GCA, GCU, GCC, GCG), three of which are recognized by $tRNA^{Ala1}$, whereas a fourth is recognized by $tRNA^{Ala2}$. Another example is the leucine family of tRNAs. There are six codons for leucine, which are recognized by five tRNAs: two recognize a single codon each (CUG or UUG), while the remaining three each recognize

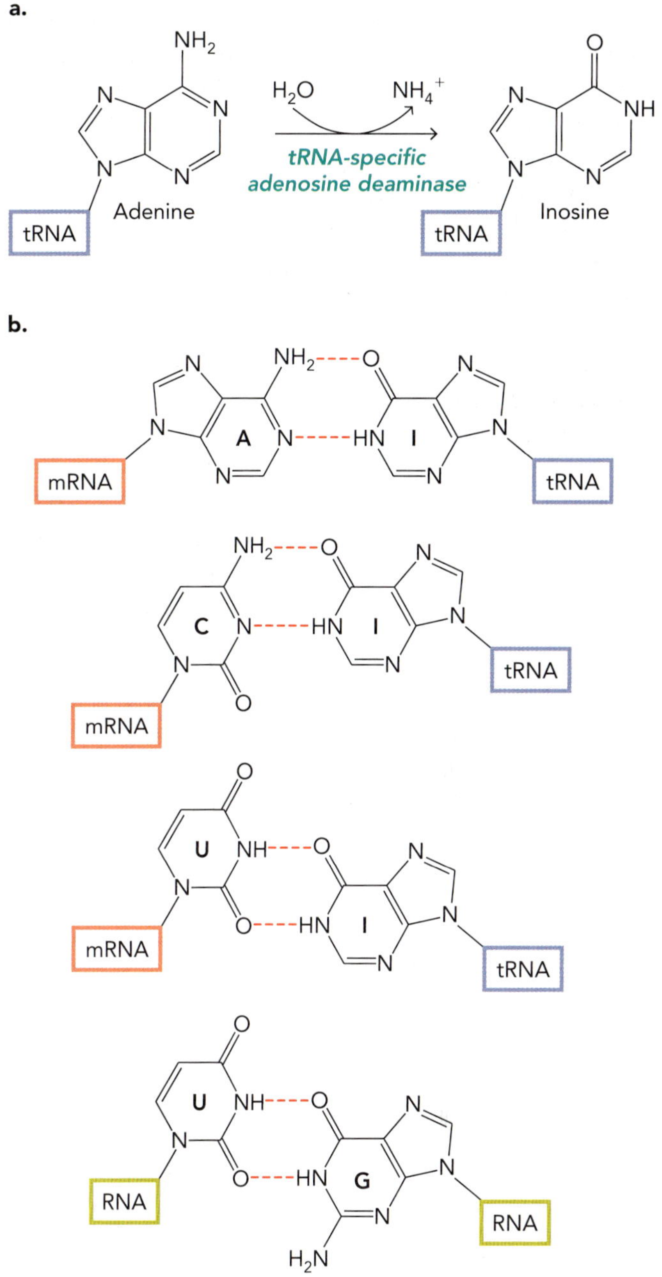

Figure 22.8 Structures of noncanonical base pairs between tRNA and mRNA are shown. **a.** Inosine in tRNA is formed by deamination of adenosine by the enzyme tRNA-specific adenosine deaminase. **b.** Common base pairs found between the tRNA anticodon and the mRNA codon, which include inosine (I) base paired with A, C, and U. Note that U-G base pairs occur in many types of RNA, including both tRNA and mRNA, whereas inosine is usually only found in tRNA anticodons.

a.

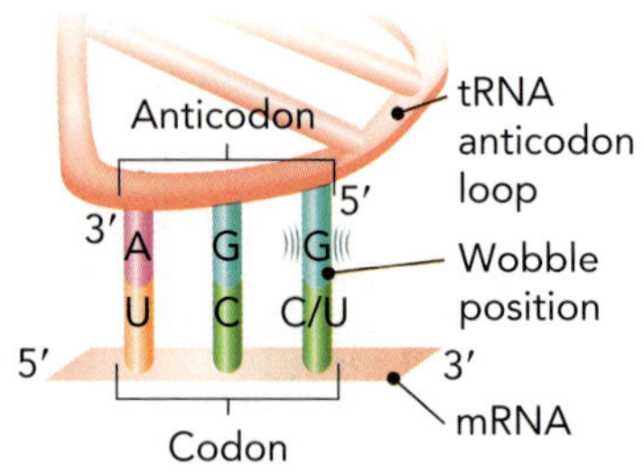

b.

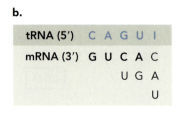

tRNA (5′)	C	A	G	U	I
mRNA (3′)	**G**	**U**	**C**	**A**	C
			U	G	A
					U

Figure 22.9 The wobble hypothesis proposes that noncanonical base pairing can occur between the 3′-position of the mRNA codon and the 5′-position of the tRNA anticodon. **a.** A G nucleotide base in the 5′-position of the tRNA anticodon can base pair with either a C nucleotide base or a U nucleotide base in the 3′-position of the mRNA codon. **b.** Summary table of the canonical (boldface) and noncanonical base pairings frequently observed between the wobble position of tRNA and mRNA. Other noncanonical base pairs are observed less commonly, and these are not shown.

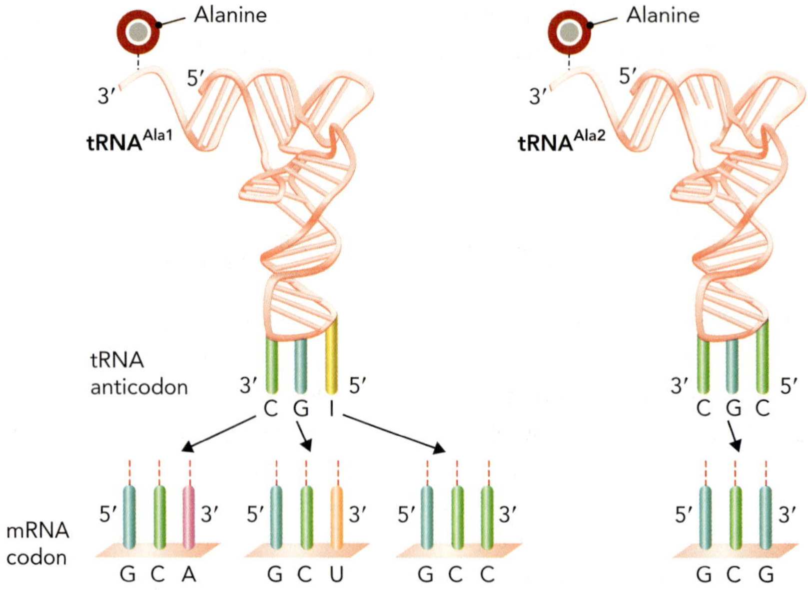

Figure 22.10 The amino acid alanine (Ala) is specified by four different codons, which requires two different tRNAs to function as adaptor molecules ($tRNA^{Ala1}$ and $tRNA^{Ala2}$). The minimum number of tRNAs required to specify each of the 20 amino acids is determined by the wobble position and varies between organisms.

two codons (CUU and CUC, CUA and CUG, or UUA and UUG). Depending on the organism, different tRNAs may be aminoacylated by the same or different aminoacyl-tRNA synthetases depending on how many ribonucleotides outside of the anticodon are distinct.

concept integration 22.1

Adenosine deaminase converts adenosine in tRNA to inosine, which is important for base pairing between the wobble position in tRNA and codons in mRNA. How would fidelity of translation be affected if adenosine in mRNA, rather than in tRNA, was converted to inosine?

If adenosine deaminase were to catalyze the conversion of an adenosine to inosine in mRNA, it could have a variety of effects on translation, including misincorporation of amino acids or premature truncation of the protein. Replacing the first or second nucleotide in the mRNA codon with an inosine would have a more significant effect than a change at the third position, as a change at either of the first two positions would likely result in the incorporation of a different amino acid. For example, if the ACU codon for Thr was mistakenly converted to ICU, then tRNAs specifying Pro (GGU), Ala (CGU), and Ser (UGU) tRNAs could all base pair with this aberrant ICU Thr codon. A change in the second codon position could have a similar effect, whereas a change at the third position would not affect the matching tRNA because of wobble base pairing. The specific modifications of the tRNA anticodon that convert adenosine residues to inosine provides the needed redundancy in the genetic code, whereas the same mRNA modifications would be deleterious because of altered amino acid sequences.

22.2 Biochemistry of mRNA Translation

In Chapter 21, we saw how mRNA is processed so that it can serve as a template for protein translation. In the preceding section, we explored the interaction between an mRNA codon and the matching anticodon on tRNA. In this section, we combine these processes to examine how ribosomes catalyze the peptide synthesis reaction by serving as the molecular mediator of mRNA translation. We first take a closer look at the structure of tRNA and the mechanism of aminoacyl-tRNA synthetases and how they are able to charge a tRNA with a specific amino acid. Next, we examine the reactions required for each round of peptide bond formation and the mechanism of translational termination. Lastly, we describe antibiotics that block protein synthesis and show how understanding the mechanism behind these antibiotics has led to a greater understanding of protein synthesis.

Transfer RNA Synthetases Provide a Second Genetic Code

The first full nucleotide sequence of a tRNA was determined in 1965 by Robert Holley's group at Cornell University (Holley was awarded a share of the 1968 Nobel Prize with Nirenberg and Khorana for this accomplishment). Since then, more than 4,000 tRNA genes from 300 organisms have been sequenced. As shown in **Figure 22.11**, tRNA consists of four structural components known as the acceptor stem, the TΨC arm, the anticodon arm, and the D arm. The acceptor stem is the site of aminoacylation, and the anticodon of the anticodon arm forms base pairs with the

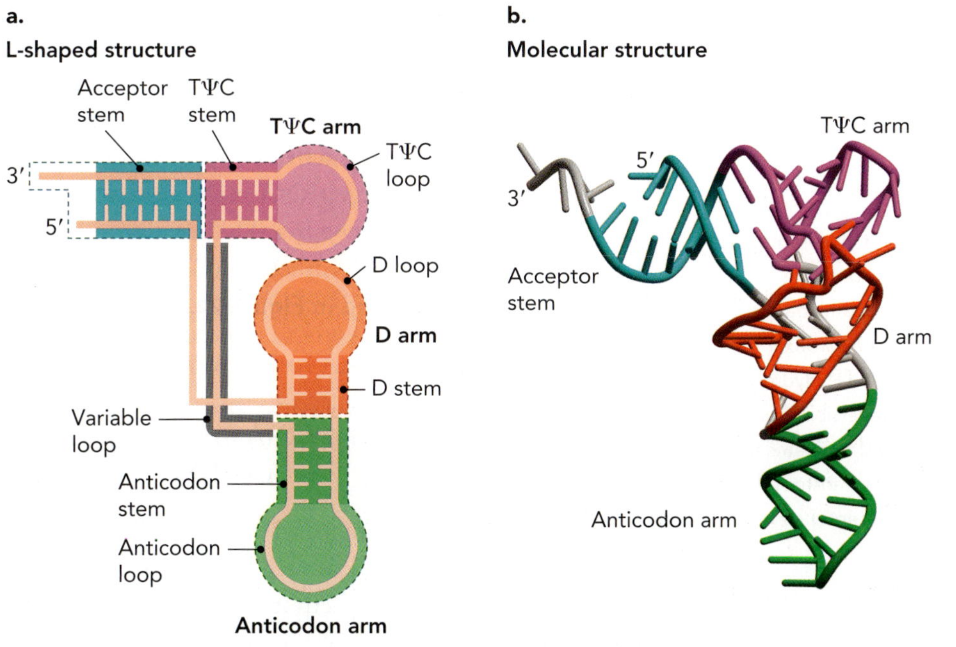

Figure 22.11 The structural organization of a tRNA molecule is shown. **a.** The tRNA L-shaped structure shows the interactions between the TΨC and D loops that take place when the tRNA is folded in three dimensions. **b.** The molecular structure of tRNA shows that the site of amino acid attachment on the acceptor stem and the anticodon are at opposite ends of the molecule. BASED ON PDB FILE 1EHZ.

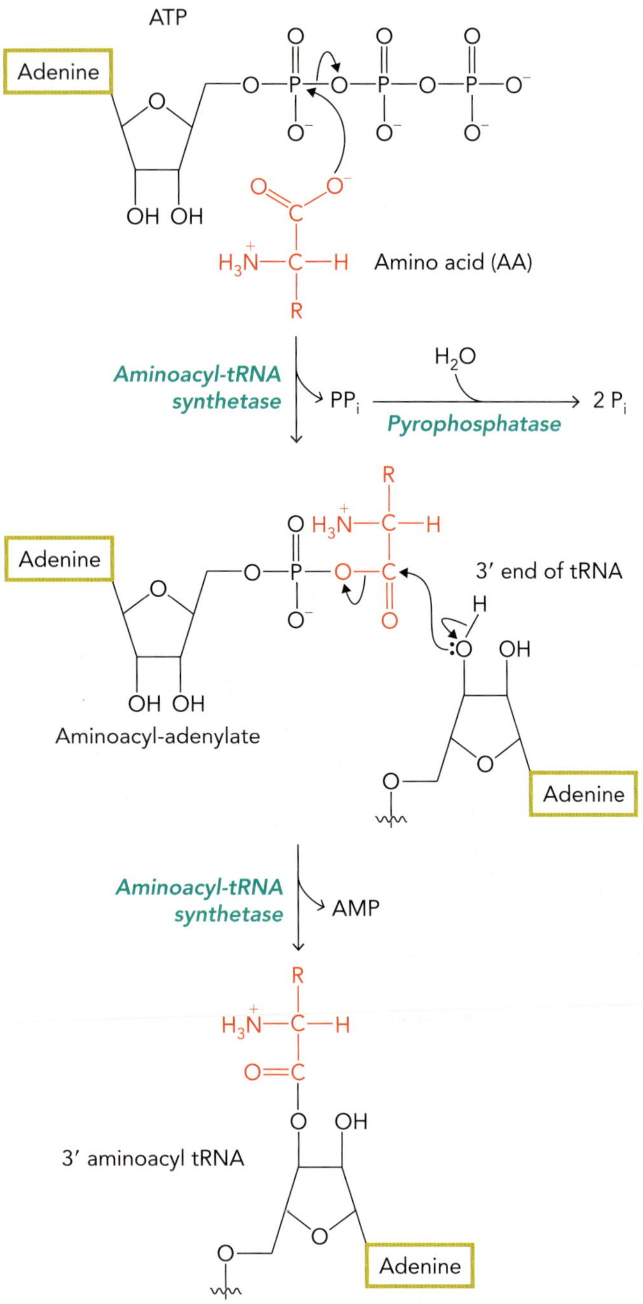

Figure 22.12 The two steps of the aminoacyl-tRNA synthetase charging reaction are shown here. In the first step, both the amino acid and ATP are bound to the enzyme. An aminoacyl-adenylate intermediate is then formed between the carboxyl group of the amino acid and the α phosphate of ATP. Hydrolysis of the released PP$_i$ makes the overall reaction irreversible. In the second step, cleavage of the phosphodiester bond facilitates the formation of aminoacyl tRNA as the amino acid is transferred to the 2′-hydroxyl or 3′-hydroxyl of the tRNA acceptor stem.

corresponding nucleotide bases in the mRNA. The TΨC arm is named for the thymidine–pseudouridine–cytidine trinucleotide sequence present in every tRNA, and the D arm contains an invariant dihydrouridine residue, which is found in all tRNA molecules within this region. Recall that these modified bases are generated during the processing of precursor tRNA molecules (see Figure 21.35). These tRNA structural components can be displayed in a two-dimensional "cloverleaf" formation or as an L shape, the latter showing the long-range interactions that are made between the loops when the tRNA adopts its three-dimensional structure. This L shape more accurately represents the molecular structure of tRNAs that have been determined by X-ray crystallography. The large amount of sequence data that exists regarding tRNAs has identified certain regions of the molecule that are invariant; that is, there are regions of the sequence that are identical in all species, whereas other regions can have a small or large amount of variation. The invariant and semiconserved regions are generally involved in tertiary structure formation, while the variable regions determine the identity of the tRNA.

As described earlier, tRNA serves as an adaptor molecule to translate nucleotides into amino acids, but it also provides a driving force for protein synthesis. The formation of a peptide bond between free amino acids is not thermodynamically favorable; however, the aminoacylated tRNA serves as an activated form of the amino acid, thus the addition of an amino acid (AA) from the tRNA to the growing peptide is thermodynamically favorable. The enzymes that produce aminoacylated tRNAs are known as aminoacyl-tRNA synthetases. There are 36 functional aminoacyl-tRNA synthetases in human cells: 16 of these function exclusively in the cytoplasm, another 17 are only active in mitochondria, and 3 are bifunctional and charge tRNA molecules in both the cytoplasm and the mitochondria. The reaction catalyzed by an aminoacyl-tRNA synthetase (E) occurs in two stages, as shown below and in **Figure 22.12**:

$$\text{Stage 1: E + ATP + AA} \rightarrow \text{E(AA–AMP) + PP}_i$$

$$\text{Stage 2: E(AA–AMP) + tRNA} \rightarrow \text{AA–tRNA + E + AMP}$$

In stage 1 of the aminoacyl-tRNA synthetase reaction, free energy from ATP is used to form an aminoacyl-adenylate. This reaction activates the amino acid by preserving the free energy of the phosphodiester bond in

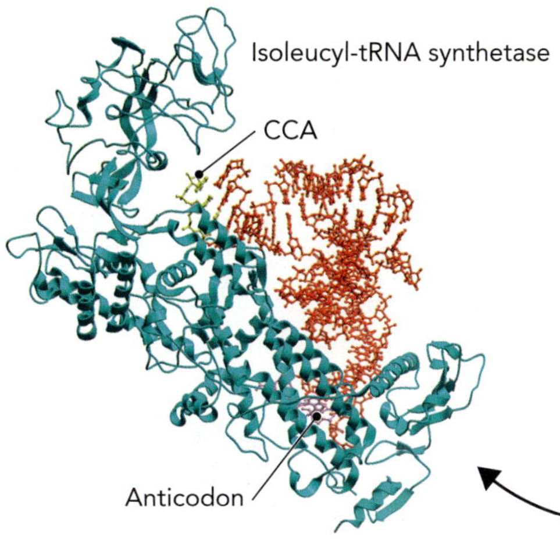

Two Classes of Aminoacyl-tRNA Synthetases	
Class I	**Class II**
Arg	Ala
Cys	Asn
Gln	Asp
Glu	Gly
Leu	His
Met	Lys
Trp	Phe
Tyr	Pro
Val	Ser
Ile	Thr

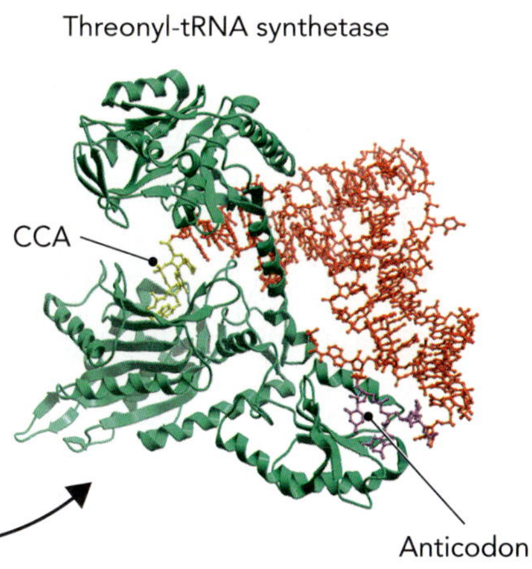

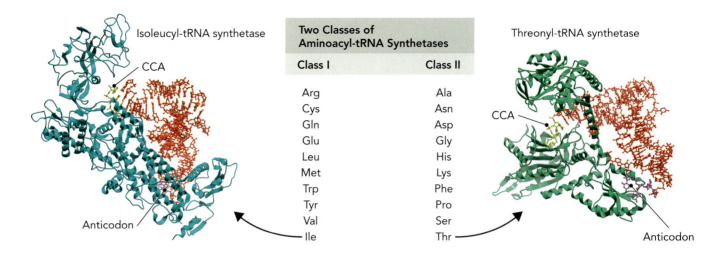

Figure 22.13 The overall structures of class I and class II aminoacyl-tRNA synthetases are distinct. The molecular structures of the *E. coli* isoleucyl-tRNA synthetase (class I) and the threonyl-tRNA synthetase (class II) are shown. BASED ON PDB FILES 1FFY AND 1QF6.

the aminoacyl-AMP intermediate (Figure 22.12). The amino acid is then transferred to the acceptor stem of tRNA in stage 2, driven by the cleavage of aminoacyl-AMP. The $\Delta G^{\circ\prime}$ of the two combined steps is near equilibrium, but as we have seen before in several metabolic reactions, hydrolysis of PP_i released after the first step of the aminoacylation reaction makes this process irreversible.

Although all aminoacyl-tRNA synthetases catalyze the same overall reaction, they exhibit a surprising amount of structural and sequence diversity. These synthetases can vary from monomeric to tetrameric and range in size from 51 kDa to more than 350 kDa. They can be divided into two classes on the basis of their active site configuration and the placement of the amino acid on the acceptor stem of tRNA. **Figure 22.13** shows the division of the synthetases into class I or class II. It is thought that each synthetase class originated from a single ancestral gene, which expanded through gene duplication to give 10 synthetases per class. All aminoacyl-tRNA synthetases contain a similar ATP binding domain; the designation of a synthetase as class I or II is based primarily on the structure of the active site. Class I enzymes have a Rossmann nucleotide binding fold structure in their active sites (see Figure 4.48) and initially link the aminoacyl group to their cognate tRNA on the 2′-hydroxyl group of the terminal adenosine in the acceptor stem prior to moving the aminoacyl group by a transesterification reaction to the 3′-hydroxyl group of the ribose. Class II enzymes have a catalytic core composed of a seven-stranded β-sheet structure flanked by α helices and directly charge the 3′-hydroxyl of the adenosine.

The interaction between an aminoacyl-tRNA synthetase and its cognate tRNA is mediated through a variety of noncovalent interactions. Class I aminoacyl-tRNA synthetases bind the D-loop side of the tRNA and interact with the acceptor stem in the minor groove, whereas class II enzymes recognize the opposite side of the tRNA and bind the acceptor stem in the major groove. At least one of the last three base pairs in the acceptor stem serves as a discriminatory site to promote optimal binding between the correct tRNA–enzyme pair. **Isoacceptor tRNAs**, which are different tRNAs for the same amino acid that bind alternate codons, must all be recognized by a single aminoacyl-tRNA synthetase and therefore have an invariant residue at this recognition position in the acceptor stem. Both enzyme classes also make contact with the anticodon loop, recognizing at least one and sometimes all of the anticodon bases.

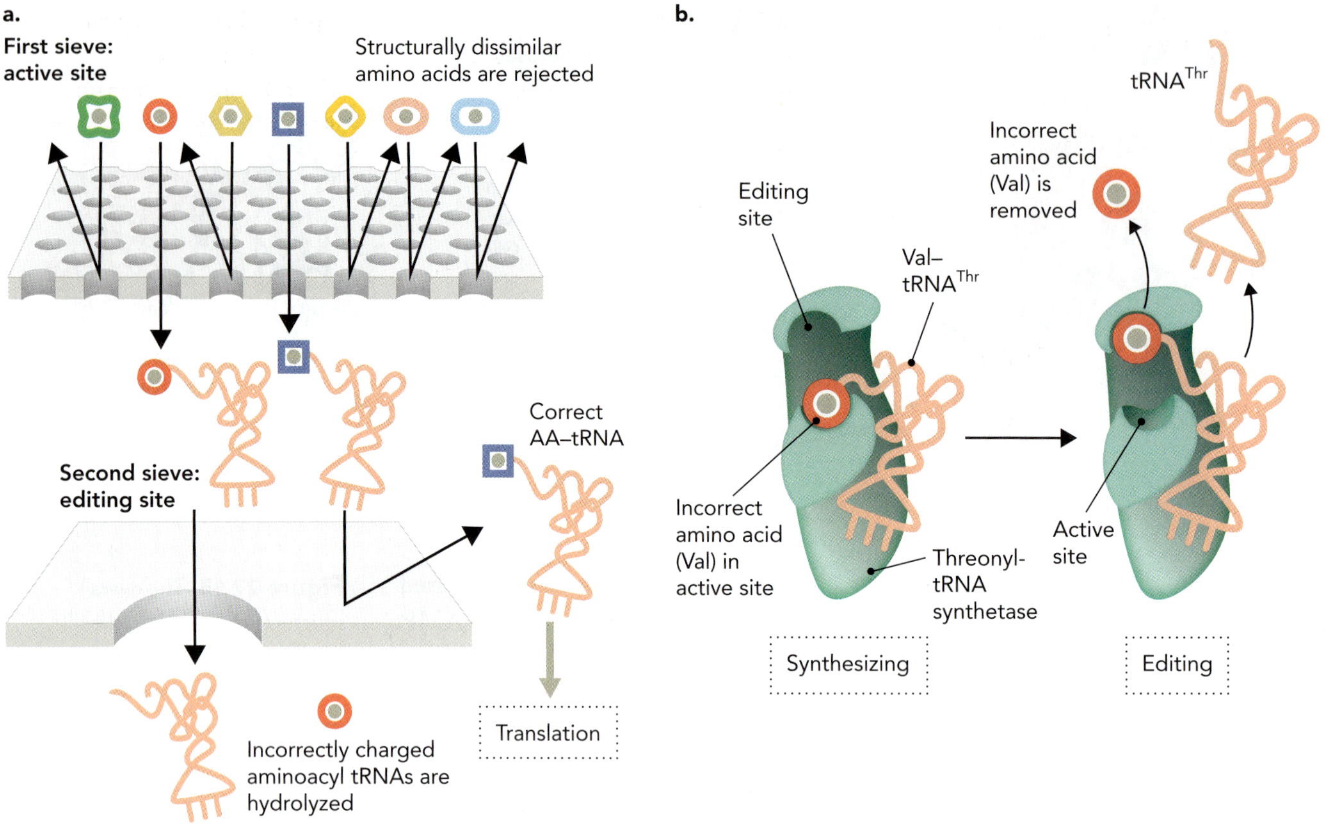

Figure 22.14 Proofreading and editing functions of tRNA synthetases are shown. **a.** Proofreading can be thought of as a double-sieve mechanism in which structurally related amino acids are charged to the tRNA after the first sieve, whereas structurally dissimilar amino acids are rejected from the sieve. In the second sieve, the correct aminoacyl tRNA is released from the enzyme, but the incorrect aminoacyl tRNA is first hydrolyzed and then released. **b.** Schematic showing that the active site of the enzyme and the editing site are physically distinct such that an incorrectly charged tRNA (Val–$tRNA^{Thr}$) is hydrolyzed in the editing site, and both the amino acid and tRNA are released.

Recognizing and binding the correct tRNA is only part of the mechanism that ensures these enzymes have an error rate of less than 10^{-4}; there are also two proofreading steps. As shown in **Figure 22.14**, this proofreading can be thought of as a double-sieve mechanism. In the first sieve, structurally similar amino acids are covalently linked to the cognate tRNA, whereas structurally dissimilar amino acids are rejected from the enzyme active site. In the second sieve, an incorrectly charged aminoacyl $tRNA^{AA}$ is hydrolyzed releasing the amino acid, whereas a correctly charged aminoacyl $tRNA^{AA}$ is not hydrolyzed and exits the editing site intact. The reason for the two-sieve mechanism is that the pretransfer proofreading cannot eliminate binding of all similar amino acids, and therefore incorrectly charged tRNAs must be eliminated by a posttransfer editing process. For example, the enzyme threonyl-tRNA synthetase functions to charge $tRNA^{Thr}$ with threonine; however, because the amino acids threonine and valine are similar in size, $tRNA^{Thr}$ is sometimes charged with valine. As shown in Figure 22.14b, a mischarged valine residue bound to the acceptor stem will pass into the editing site where the amino acid–tRNA bond is hydrolyzed, and the valine amino acid and the uncharged $tRNA^{Thr}$ are released. In contrast, Thr–$tRNA^{Thr}$ does not fit into this editing site, so threonine charging to $tRNA^{Thr}$ is excluded, and threonine is not hydrolyzed.

Ribosomes Are Protein Synthesis Machines

The codon AUG, which specifies the amino acid methionine, most often signals the start of the protein-coding sequence and is used to establish the reading frame for translation. An mRNA may have multiple initiator AUG codons in the 5′ region of the transcript, and the mechanism by which the correct AUG codon is identified in prokaryotes and eukaryotes is mechanistically different as will be described shortly.

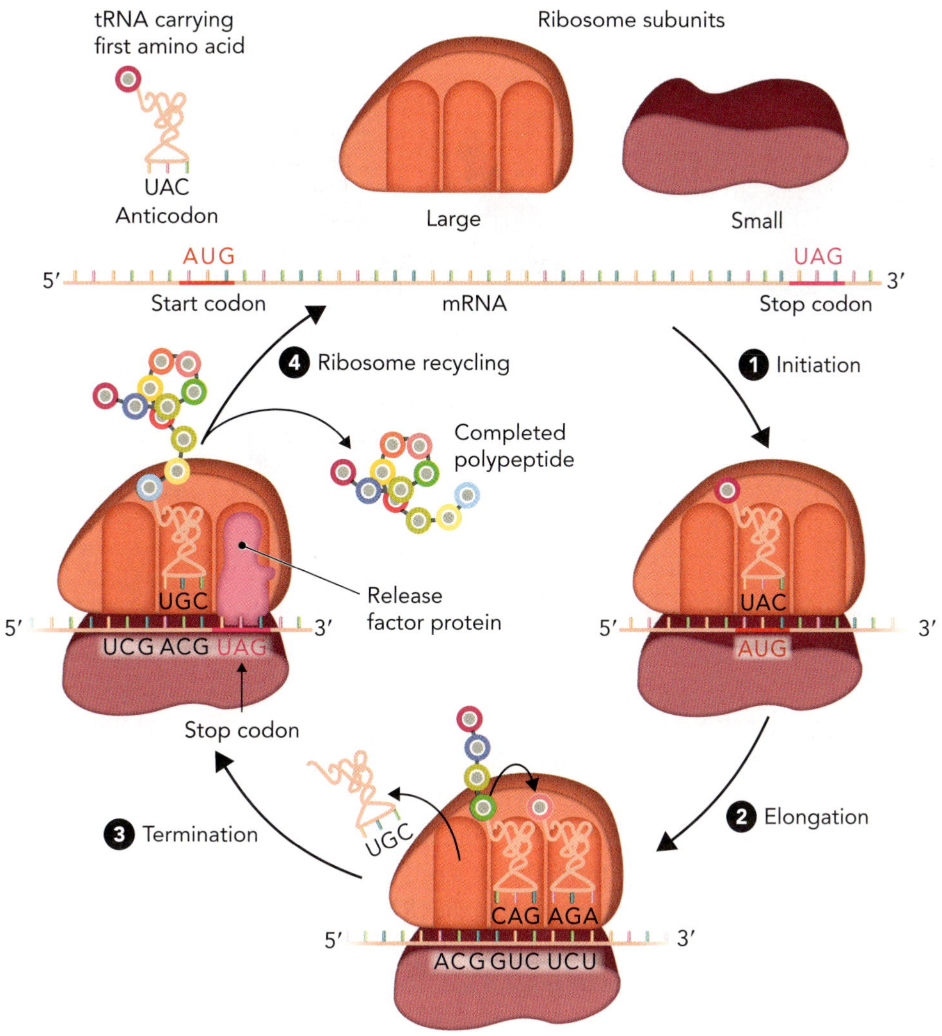

Figure 22.15 Polypeptide synthesis is composed of four key steps: initiation, elongation, termination, and ribosome recycling.

The methionine family of tRNAs contains two members that both recognize the AUG codon. One, an **initiator tRNA** charged with methionine (Met–$tRNA_i^{Met}$), is able to bind to the otherwise empty ribosome and match with the AUG codon. The second Met–$tRNA^{Met}$ is only able to bind after protein synthesis has been initiated. In prokaryotes, the methionine on the initiator tRNA is formylated to produce *N*-formylmethionine soon after aminoacylation (fMet–$tRNA_i^{Met}$). In eukaryotes, Met–$tRNA_i^{Met}$ undergoes modification on the ribosome.

An overview of protein synthesis is shown in **Figure 22.15**. Translation initiation in the 5′ region of the mRNA transcript involves binding of the small and large ribosomal subunits along with the initiator Met–$tRNA_i^{Met}$. Elongation of the polypeptide chain involves binding of aminoacylated tRNA molecules to provide the substrates for polypeptide synthesis within the ribozyme active site. Finally, when a stop codon is reached in the mRNA transcript, a release factor binds to the ribosome in place of a charged tRNA, and the polypeptide chain is released from the ribosome.

As we saw in Chapter 21, the ribosome is a two-subunit complex with each subunit containing both protein and RNA components. The large and small ribosomal subunits are brought together during initiation of translation. Within the ribosome, there are three tRNA binding sites that are occupied by a tRNA during

▶ ANIMATION

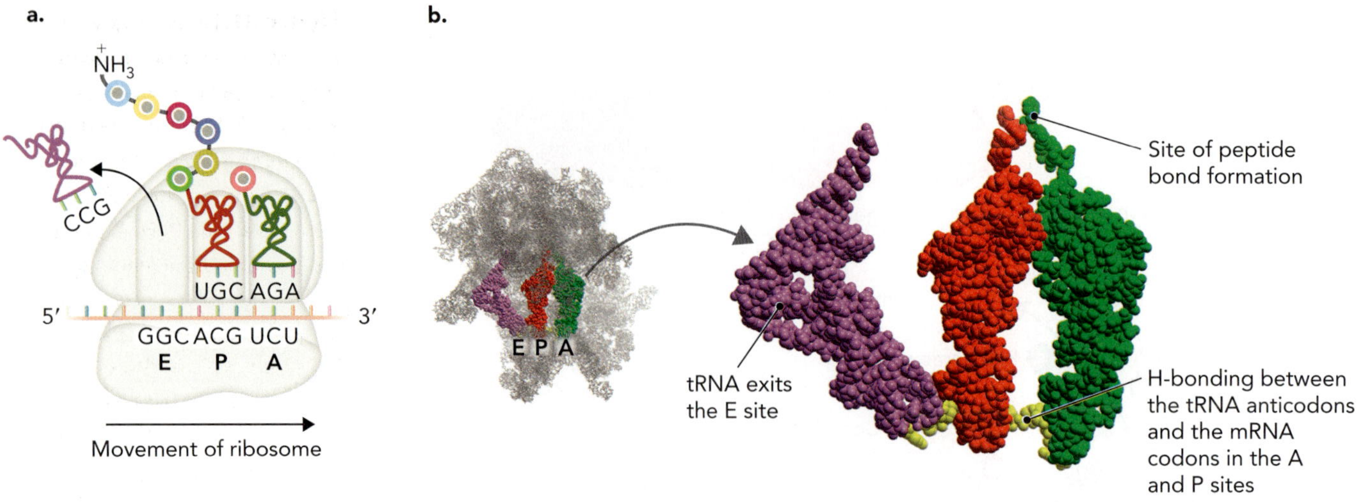

Figure 22.16 Locations of the E, P, and A sites in the ribosome complex are shown. **a.** Schematic drawing of a ribosome complex illustrating tRNA molecules occupying the exit (E), peptidyl (P), and aminoacyl (A) sites during the elongation phase of translation. The anticodons of the tRNAs in the A and P sites are base paired to the codons of the mRNA. **b.** Molecular structure of the 70S bacterial ribosome with tRNA molecules in the E, P, and A sites and a short mRNA transcript. This structure shows only the RNA components of the small and large subunits. The three tRNA molecules (purple, red, and green) and the mRNA transcript (yellow) are shown at bottom.
BASED ON PDB FILES 2WDK AND 2WDL.

the process of translation as shown in **Figure 22.16**. Two of the tRNA binding sites are positioned so that the anticodon of a charged tRNA is able to make contact with mRNA.

The initiator tRNA binds to the **peptidyl site (P site)**. The **aminoacyl site (A site)** is the initial binding site for all other charged tRNAs after Met–$tRNA_i^{Met}$. Binding of the second tRNA to the A site often involves recruitment of factors necessary for polypeptide elongation. As each new amino acid is bound to the A site, ribosomal RNA (rRNA) assists with the formation of a peptide bond between the last residue on the growing polypeptide chain (attached to the tRNA in the P site) and the next amino acid (attached to the tRNA in the A site). Formation of this peptide bond transfers the polypeptide chain from the P site–bound tRNA to the A-site tRNA. Elongation factors catalyze the translocation of tRNAs, which facilitates release of the uncharged tRNA from the exit site (E site) and frees the A site for a new charged tRNA to bind and continue synthesis.

Polypeptide Synthesis: Initiation, Elongation, Termination

Initiation of protein synthesis in prokaryotes and eukaryotes is mechanistically different, and therefore we need to look at each separately. The primary distinction is the mechanism by which the AUG initiation codon is identified by the small ribosomal subunit. There are also differences with regard to binding of translation initiation factors to the small ribosomal subunit and the role of ATP/GTP hydrolysis.

Translational Initiation A prokaryotic ribosome is composed of 30S and 50S subunits, which together form the fully functional 70S complex. The "S" used in the designation of each component refers to the sedimentation rate, or **Svedberg value**, which depends on the size and shape of the component. (Svedberg units are not additive, so the individual sedimentation rates of each subunit may not add up to the Svedberg value of the final complex.) The larger 50S subunit contains two ribosomal RNAs—5S and 23S rRNA—and the smaller 30S subunit contains a single 16S rRNA; both subunits also contain a number of proteins. As shown in **Figure 22.17**, initiation begins when three **initiation factors**, or **IFs** (IF1, IF2, IF3), bind to the 30S ribosome, followed by binding of the mRNA and fMet–$tRNA_i^{Met}$ to form the 30S initiation complex. The 50S ribosomal subunit then binds to form the functional prokaryotic 70S initiation complex.

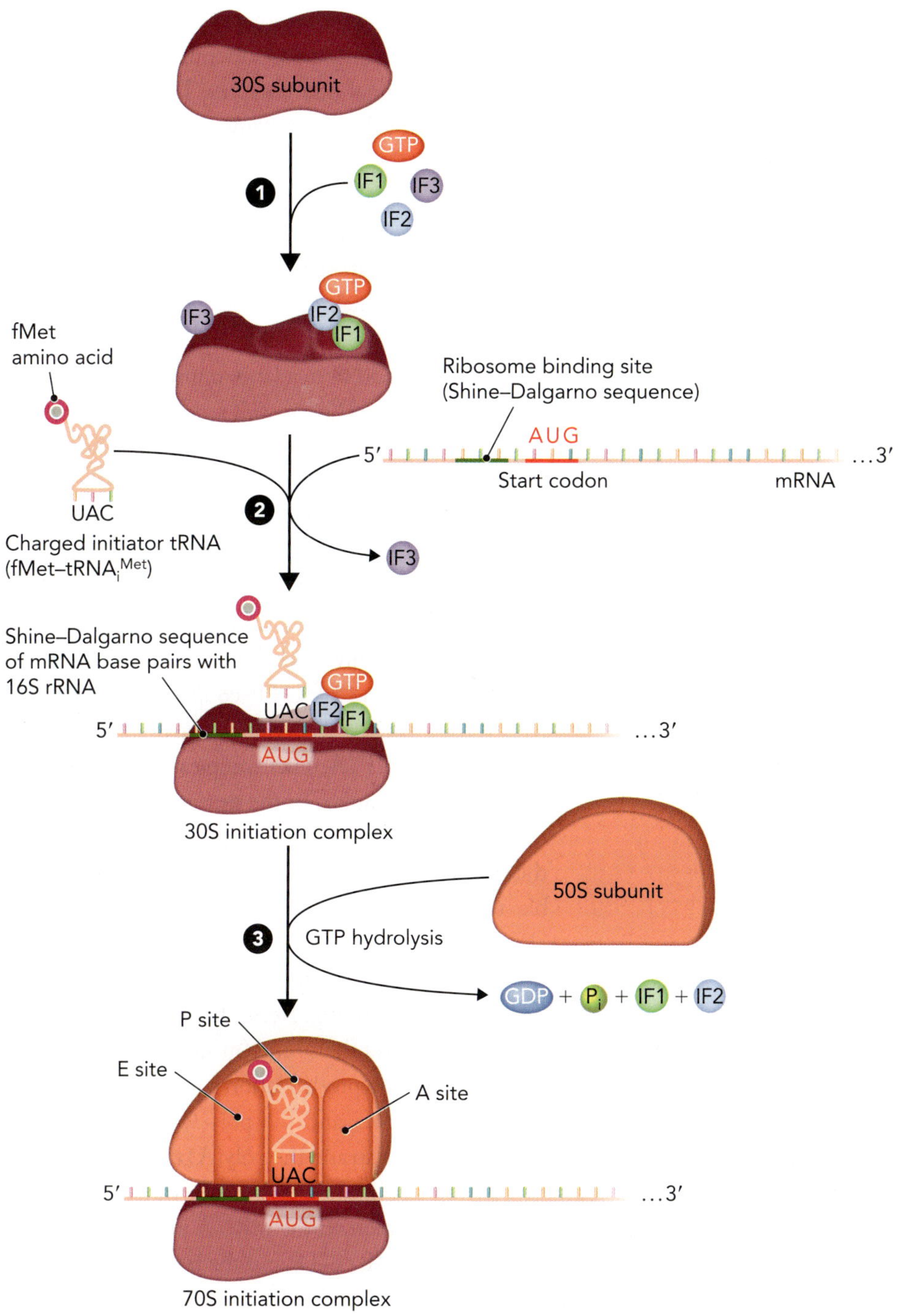

Figure 22.17 Formation of a prokaryotic translation initiation complex requires three steps. In step 1, GTP, IF1, IF2, and IF3 bind to the 30S ribosomal subunit. In step 2, the mRNA and fMet–tRNA$_i^{Met}$ bind to the complex with the Shine–Dalgarno sequence of mRNA aligning with complementary sequences in the 16S rRNA in the 30S ribosomal subunit, and IF3 is released. In step 3, hydrolysis of GTP facilitates binding of the 50S ribosomal subunit, with the release of IF1 and IF2 to form the functional 70S ribosomal initiation complex. The prokaryotic initiation complex contains fMet–tRNA$_i^{Met}$ bound in the P site, with the E site and A site empty until elongation begins.

In order for the initial AUG codon to be brought into the correct orientation below the P site, the ribosome must be able to identify the 5′ end of the mRNA transcript. A specific nucleotide sequence known as the **Shine–Dalgarno sequence** appears upstream of the first AUG to orient mRNA within the ribosome. **Figure 22.18** shows the interaction between the Shine–Dalgarno sequence and the 16S rRNA within the prokaryotic 30S ribosomal subunit. Base pairing between the 16S rRNA and the mRNA helps position the mRNA beneath the P site, allowing fMet–tRNA$_i^{Met}$ to bind along with initiation factors and the 50S ribosomal subunit. The Shine–Dalgarno sequence is a purine-rich region with a core consensus sequence of 5′-AGGAG-3′ (though other variations exist) and is found 5–13 base pairs upstream of the initiator AUG codon in the mRNA. The Shine–Dalgarno sequence is complementary to the 16S rRNA sequence 5′-CUCCU-3′.

a.

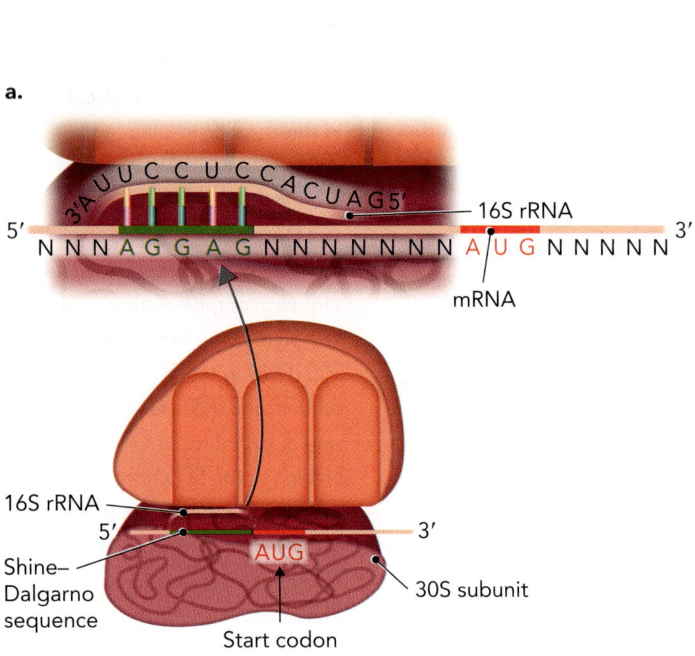

b.

	Shine–Dalgarno sequence / Start codon
rplE	5′–GUAAUUUGGAGUAGUACGAUGGCGAAA–3′
rplJ	5′–CAUCCAGGAGCAAAGCTAAUGGCUUUA–3′
trpL	5′–GUAAAAAGGGUAUCGACAAUGAAAGCA–3′
araB	5′–UUUGGAUGGAGUGAAACGAUGGCGAUU–3′
lacZ	5′–AAUACACAGGAAACAGCUAUGACCAUG–3′
dnaK	5′–UAUAGUGGAGACGUUUAGAUGGGUAAA–3′
galE	5′–AGCCUAAUGGAGCGAAUUAUGAGAGUU–3′

Figure 22.18 Prokaryotic 16S rRNA forms base pairs with the mRNA containing the Shine–Dalgarno sequence. **a.** The prokaryotic 30S ribosomal subunit is shown with a close-up of the association between 16S rRNA and the Shine–Dalgarno sequence. **b.** Examples of Shine–Dalgarno sequences found upstream of the initiator (AUG) codon in seven different *E. coli* genes.

The eukaryotic ribosomal complex is composed of 40S and 60S ribosomal subunits, which together form the functional 80S complex. The small eukaryotic ribosomal subunit contains only a single rRNA (18S), while the large ribosomal subunit contains three rRNAs (5S, 5.8S, and 28S). Formation of the eukaryotic translation initiation complex requires specific mRNA sequences at both the 5′ and 3′ ends of the mRNA, which direct the formation of RNA–protein complexes at the 5′ cap and poly(A) tail, respectively. In addition, a large number of eukaryotic initiation factors (eIFs), as well as GTP and ATP, are required to assemble the complete 80S initiation complex. Eukaryotic ribosomes scan the mRNA in a 5′ to 3′ direction from the 5′ cap structure until finding the first AUG and do not require a specific ribosome binding site such as a Shine–Dalgarno sequence.

As shown in **Figure 22.19**, eukaryotic translation initiation begins with the binding of a ternary complex consisting of GTP-coupled eIF2 and the Met–$tRNA_i^{Met}$ to the 40S ribosomal subunit along with eIF3 and eIF1A. This complex is called the 43S pre-initiation complex. In the next step, eIF3 binds to the eIF4G subunit of the 5′ cap binding complex on the mRNA in an ATP-dependent reaction to assemble the 48S pre-initiation complex along with the eIF4E cap binding protein and eIF4A, an RNA helicase required for unwinding of the mRNA to facilitate ribosome scanning. Recall from Chapter 21 that eukaryotic mRNA contains both a methylated cap at the 5′ end (m^7G) and a poly(A) tail at the 3′ end. As illustrated in Figure 22.19, this 48S pre-initiation complex also includes a direct protein–protein interaction between eIF4G and the poly(A) binding protein (PABP) to form a circularized mRNA with the 5′ cap and the poly(A) tail both bound in a single large RNA–protein complex. Assembly of this looped mRNA initiation complex promotes ribosomal scanning in the 5′ to 3′ direction on the mRNA until it reaches the first AUG codon. In the final step, addition of eIF5B • GTP and the large 60S ribosomal subunit results in GTP hydrolysis on eIF5B and eIF2, which completes assembly of the 80S ribosome on the mRNA to stimulate polypeptide synthesis.

Figure 22.19 illustrates the importance of specific mRNA sequences in the association with eIF4. Eukaryotic translation initiation is cap dependent, which means that this 5′ cap is an important feature not only to protect the mRNA during transport from the nucleus to the cytosol but also in the ribosome, where it serves to orient mRNA within eIF4. Once eIF4 has bound the methylated mRNA cap, any secondary structure downstream of the cap is unwound by the eIF4A helicase. This allows access to the initial AUG and prepares the remainder of the RNA for translation.

Although the 5′ cap and poly(A) tail assist with proper unwinding and orientation of mRNA within the ribosome, an additional nucleotide sequence appears to enhance binding of Met–$tRNA_i^{Met}$ to the appropriate AUG. In contrast to prokaryotic mRNA, eukaryotic mRNA can contain several hundred to a thousand base pairs

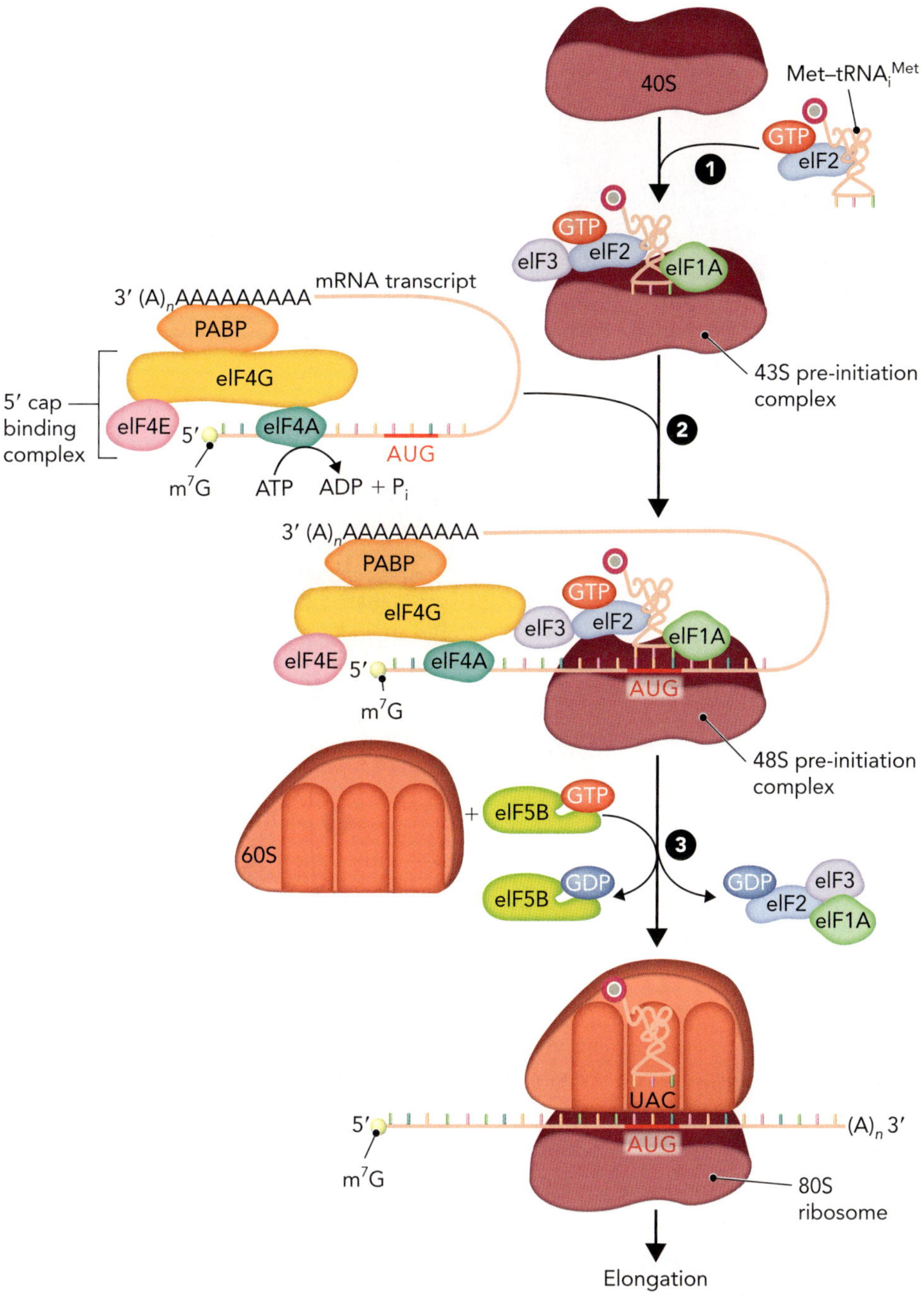

Figure 22.19 Assembly of the eukaryotic translation initiation complex is shown. Significant differences between prokaryotic and eukaryotic translation initiation complexes include the lack of a specific ribosome binding site in the mRNA in eukaryotes and the number of initiation factors required.

between the 5′ cap and the initiation AUG. The eukaryotic 80S ribosome scans in the 5′ to 3′ direction until reaching the first AUG, which often is embedded within the sequence CC(A/G)CC<u>AUG</u>G known as the **Kozak sequence**. The exact Kozak sequence can vary among eukaryotes, and, in fact, not all mRNA transcripts contain a Kozak sequence. This can be explained by the observation that it is the assembly of the pre-initiation complex consisting of proteins bound to both the 5′ cap and the poly(A) tail that determines where the ribosome begins scanning, rather than a base pairing interaction between rRNA and mRNA as it is in prokaryotes.

Translational Elongation In both prokaryotes and eukaryotes, formation of the functional ribosome complex completes the initiation phase of translation and leads to the elongation phase. The biochemical process of elongation is functionally similar in prokaryotes and eukaryotes, and therefore we will describe translational elongation

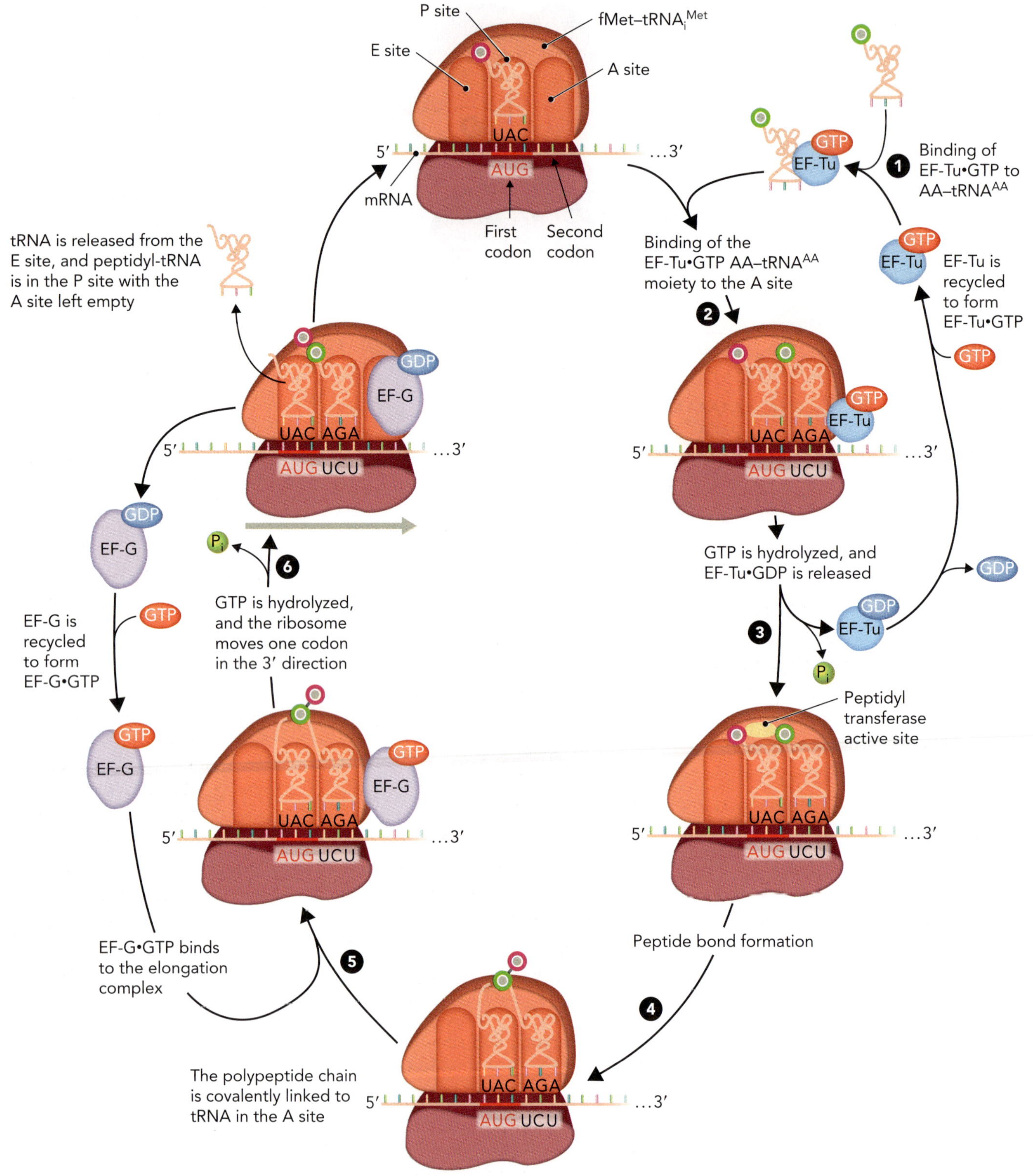

Figure 22.20 Translational elongation in prokaryotes and eukaryotes is similar. The six-step elongation cycle requires the hydrolysis of 2 GTP, one to promote binding of the AA–tRNA^{AA} to the A site, and one to facilitate translocation of the ribosome in the 3′ direction.

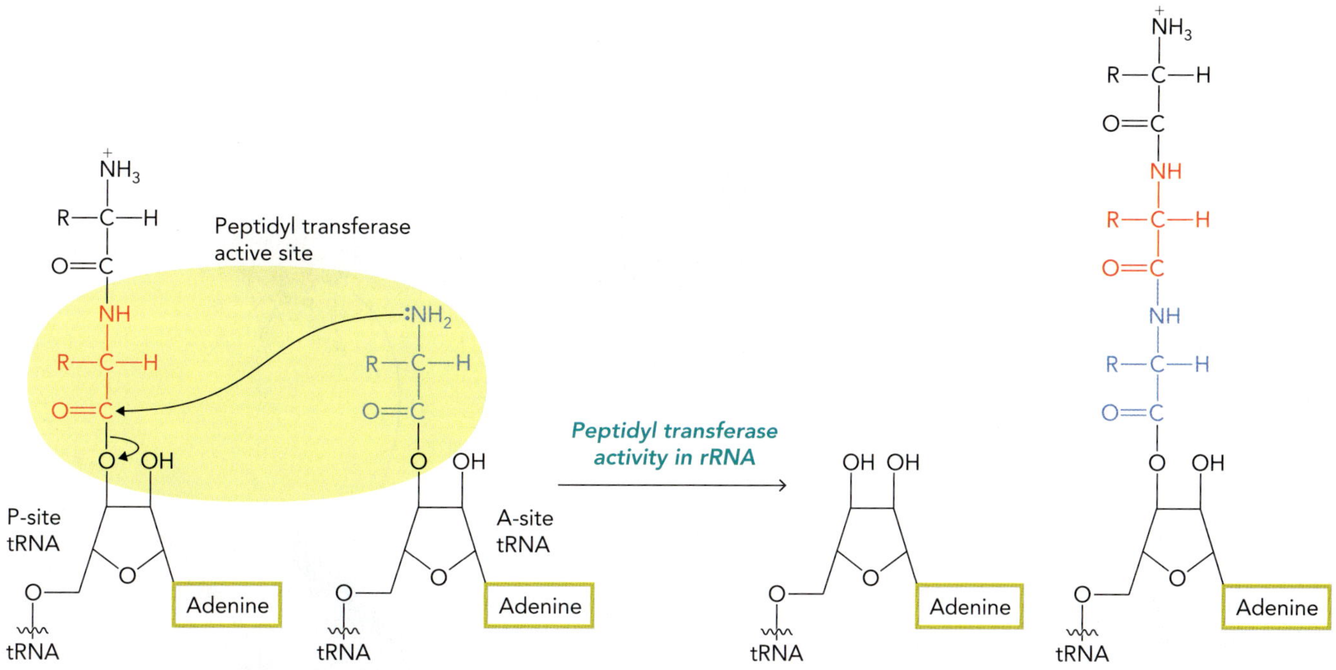

Figure 22.21 The peptidyl transferase reaction is catalyzed by the ribozyme activity of the 23S rRNA in the large ribosomal subunit. The nucleophilic amino group of the amino acid bound to the 3′ terminus of A-site tRNA attacks the electrophilic carbonyl carbon in the ester bond between the 3′ terminus of P-site tRNA and its bound amino acid. Rearrangement of the tetrahedral intermediate results in transfer of the formerly P-site amino acid (or polypeptide) to the amino acid on the A site–bound tRNA.

in the *E. coli* system. **Figure 22.20** shows the process of elongation and the role of two **elongation factors (EFs)**, EF-Tu and EF-G, in extending the polypeptide chain. In step 1, the charged tRNA is brought to the ribosome by EF-Tu bound to GTP. Interactions among the tRNA, mRNA, and rRNA ensure that the correct tRNA has been selected. Once the appropriate codon–anticodon base pairings have formed in the A site (step 2), EF-Tu•GTP is converted to EF-Tu•GDP + P_i, allowing release of EF-Tu (step 3). Mispairings between the codon and anticodon lead to the tRNA being released from the ribosome without hydrolysis of the GTP bound to EF-Tu. In step 4 of the elongation cycle, the peptide bond is formed in the peptidyl transferase active site of the ribosome between the amino acids bound to the tRNAs in the P and A sites, resulting in the growing peptide chain being covalently linked to the tRNA base paired to the A site. After peptide bond formation, EF-G•GTP binds to the elongation complex in step 5, followed by GTP hydrolysis in step 6 to promote translocation of the ribosome along the mRNA in the 3′ direction. This ribosome movement shifts the tRNAs from A to P sites and from P to E sites. The tRNA in the E site is no longer bound to the peptide and is ejected. The peptidyl-tRNA is in the P site, and the A site is empty and ready for the next AA–$tRNA^{AA}$. Note that the binding sites for EF-Tu and EF-G overlap, so only one of these proteins can be bound to the ribosome at any given time.

As described earlier, the peptidyl transferase reaction in polypeptide synthesis is catalyzed by the 23S rRNA in the prokaryotic large ribosomal subunit (28S rRNA in the eukaryotic ribosome), and thus the ribosome is a ribozyme. As shown in **Figure 22.21**, the amino group of the amino acid attached to the 3′ terminus of the tRNA in the A site attacks the carbonyl carbon of the peptidyl-tRNA in the P site, resulting in an extended A-site peptidyl-tRNA.

Translational Termination Termination is also similar in both prokaryotes and eukaryotes. The elongation process continues until one of the three stop codons enters the A site of the ribosome. The stop codon is recognized by a protein

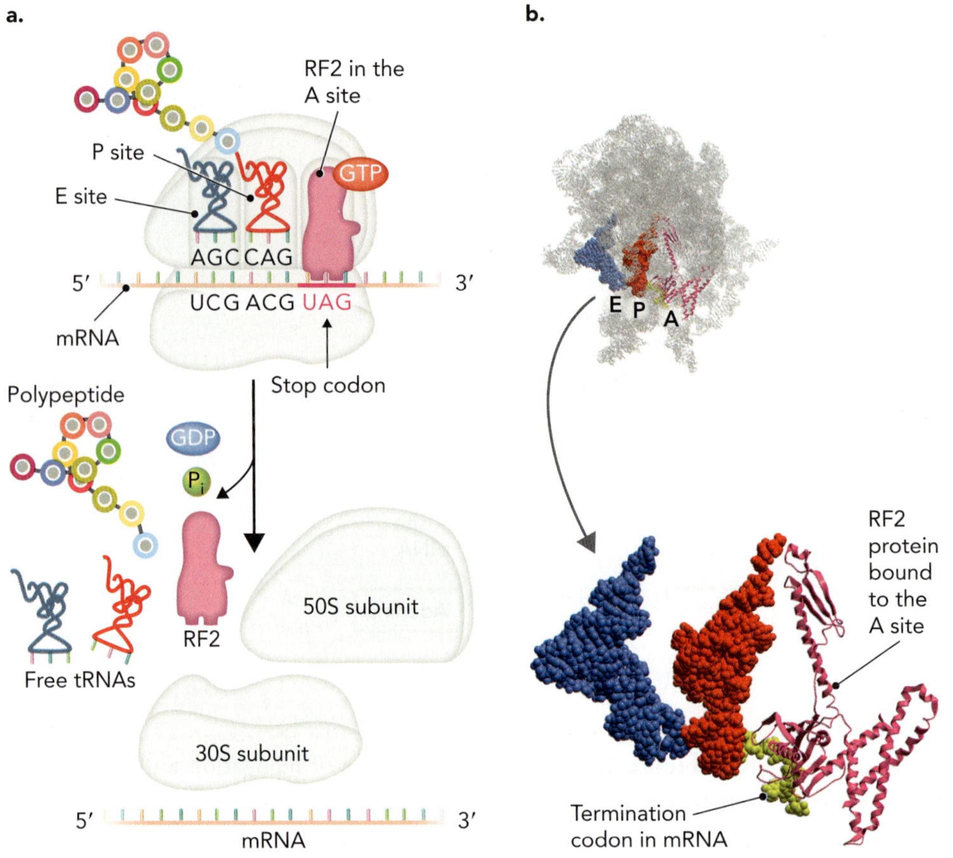

Figure 22.22 Release factors bind to the termination codon and lead to release of the polypeptide chain. **a.** Schematic showing the function of prokaryotic release factors. Binding of the associated release factor protein (release factor 2; RF2) to the stop codon in the mRNA leads to GTP hydrolysis and release of all bound translational components. **b.** Molecular structure of the 70S bacterial ribosome with tRNA molecules in the E and P sites, RF2 in the A site, and a short mRNA substrate containing the stop codon UAA. The structure shows only the RNA components of the small and large subunits.
BASED ON PDB FILES 3F1G AND 3F1H.

release factor (RF), of which there are three. Release factors hydrolyze GTP to promote disassembly of the ribosomal complex (**Figure 22.22**).

Some Antibiotics Target Bacterial Protein Synthesis

Many organisms produce small molecules that inhibit bacterial growth, as bacteria can compete for available nutrients and harm the organism in the process. Natural products with antibacterial properties were first discovered as secreted products of fungi or other microorganisms growing in culture and were named **antibiotics**. The first well-characterized antibiotic was penicillin, which was discovered by Alexander Fleming in 1928. Even before Fleming's discovery, however, Paul Ehrlich hypothesized in the early 1900s that it should be possible to synthesize antimicrobial compounds that would kill pathogenic organisms without harming the host. Ehrlich's lab discovered an arsenic-containing compound called Salvarsan (arsphenamine), which effectively inhibits the bacterium that causes syphilis. This drug became widely used to treat syphilis until it was replaced in the 1940s by penicillin. As shown in **Figure 22.23**, antibiotics were developed continually up through the 1970s; however, relatively few new antibiotics have been discovered in the past 30 years. The number of antibiotic-resistant strains of bacteria in hospitals has been steadily on the rise, and much research is under way to try to reverse this trend. Importantly, by understanding the mechanisms of action of antibiotics, researchers have been able to discover the key components of a number of biochemical processes, which in turn has led to further advances in the diagnosis and treatment of disease. For example, as shown in **Figure 22.24**, penicillin blocks bacterial cell wall synthesis, chloramphenicol interferes with bacterial protein synthesis, and coumermycin and oxolinic acid inhibit DNA gyrase thus blocking DNA replication.

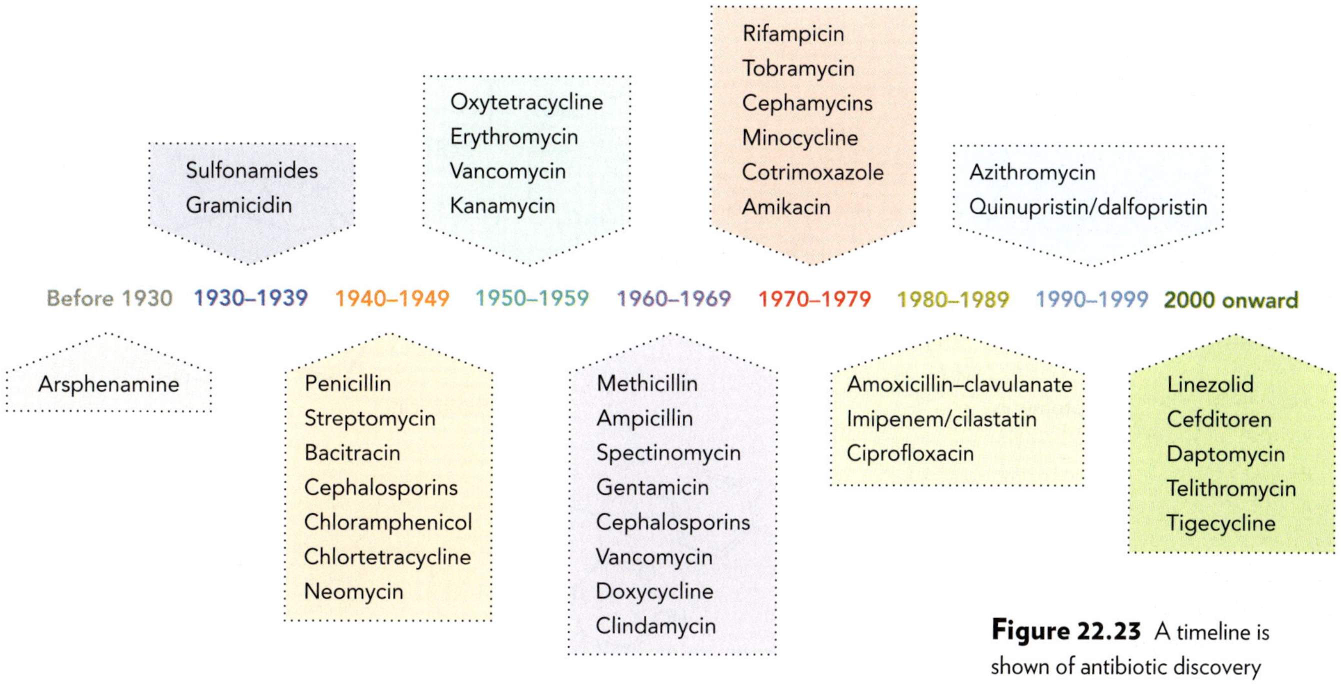

Figure 22.23 A timeline is shown of antibiotic discovery and introduction to the clinic.

Antibiotics that interfere with bacterial protein synthesis have been useful in deciphering the details of this biochemical process. **Figure 22.25** shows the structures of streptomycin, erythromycin, tetracycline, and chloramphenicol, which represent important examples of inhibitors of ribosomal function in prokaryotes, and the structure of puromycin, which affects peptidyl transfer in both prokaryotes and eukaryotes. Because these drugs inhibit ribosomal function in more than one specific organism, they are often referred to as broad-spectrum antibiotics and can be used to kill a number of different organisms.

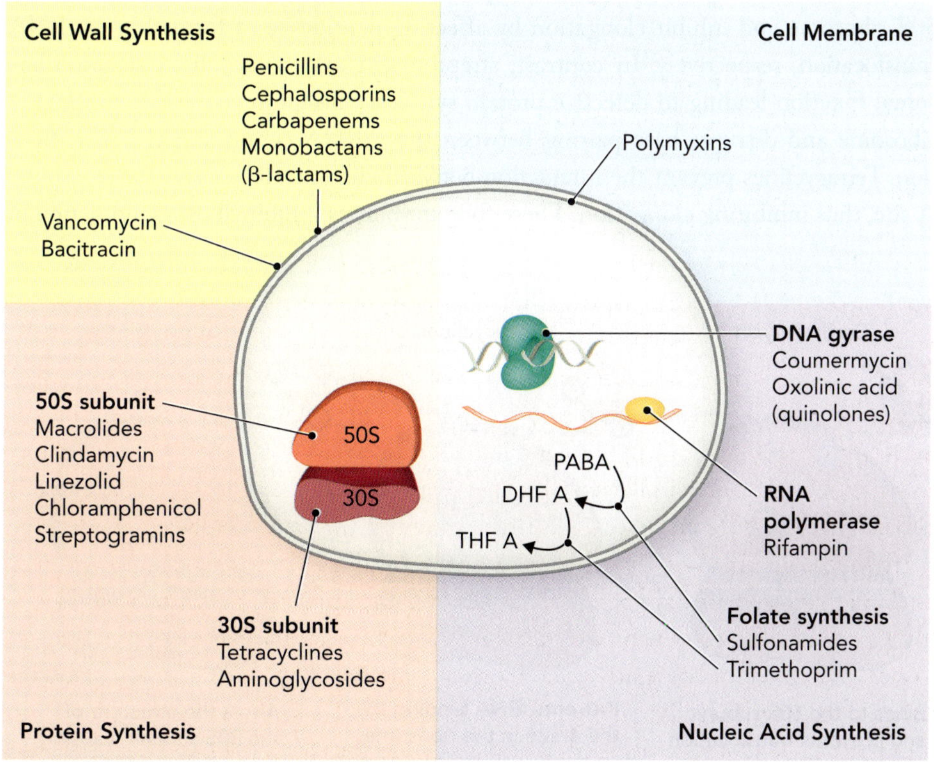

Figure 22.24 Major biochemical targets in bacteria are shown for four classes of antibiotics. DHF A = dihydrofolate A; PABA = *para*-aminobenzoic acid; THF A = tetrahydrofolate A.

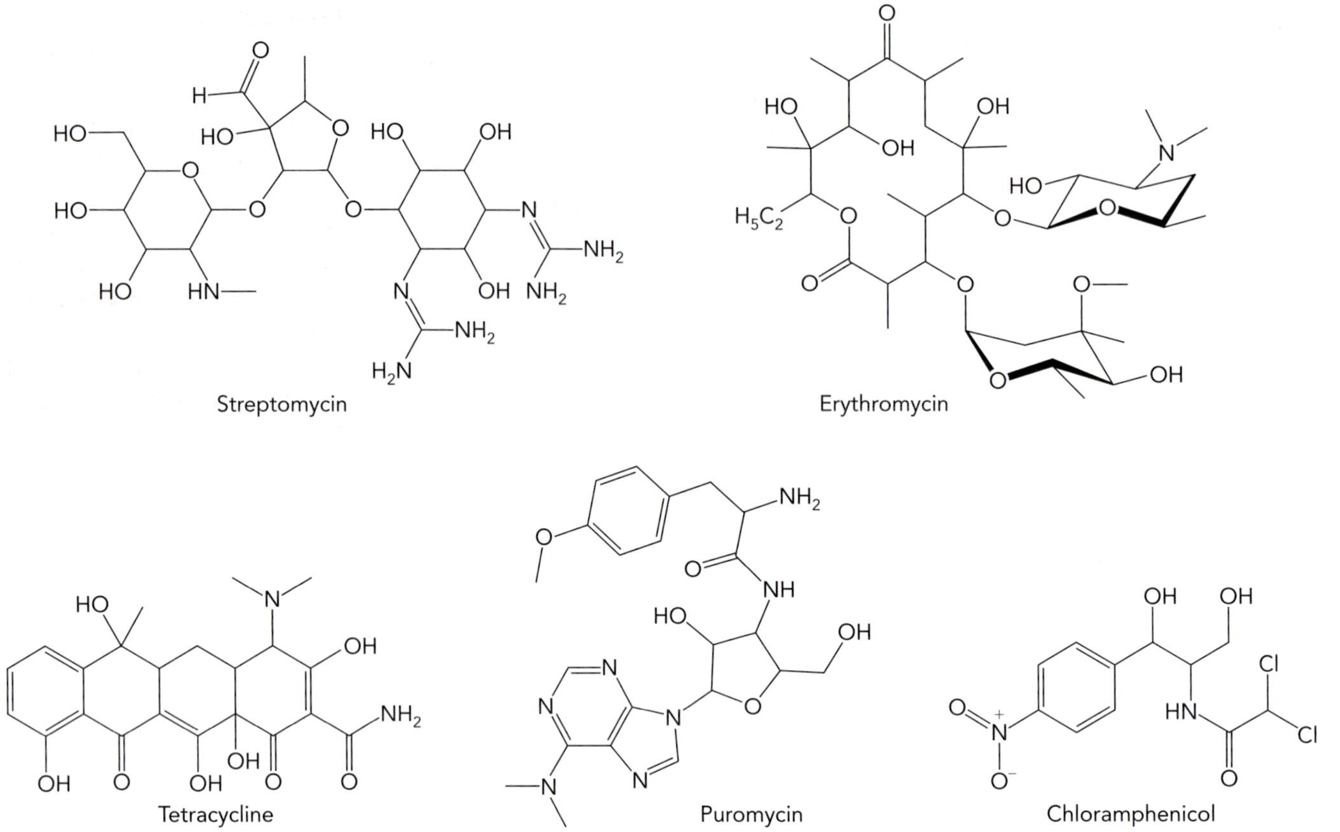

Figure 22.25 Shown are the chemical structures of four antibiotics that affect prokaryotic protein synthesis: streptomycin, erythromycin, tetracycline, and chloramphenicol. The structure of puromycin, which affects both prokaryotic and eukaryotic protein synthesis, is also shown.

As shown in **Figure 22.26**, chloramphenicol and erythromycin both bind to the 50S ribosome and inhibit elongation by affecting peptide bond formation or tRNA translocation, respectively. In contrast, streptomycin and tetracyclines impair ribosomal function leading to defective protein synthesis. Streptomycin binds to the 30S ribosome and disrupts base pairing between the mRNA codon and tRNA anticodon. Tetracyclines prevent the interaction between charged tRNA and the ribosomal A site, thus inhibiting elongation. These four antibiotics inhibit ribosomal function in

Figure 22.26 Examples are shown of four antibiotics that inhibit bacterial protein synthesis at distinct steps.

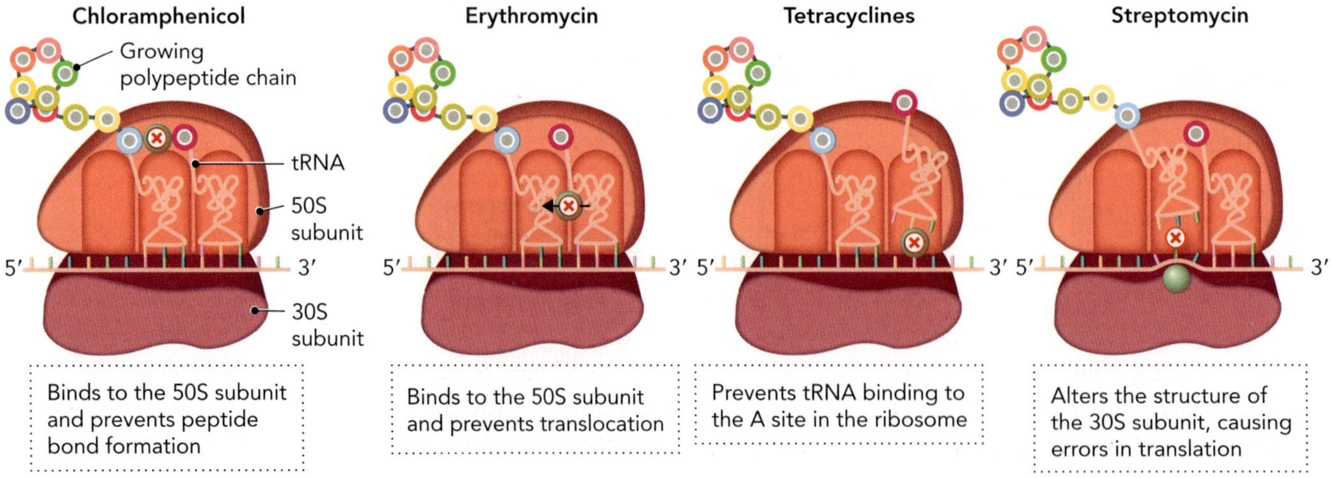

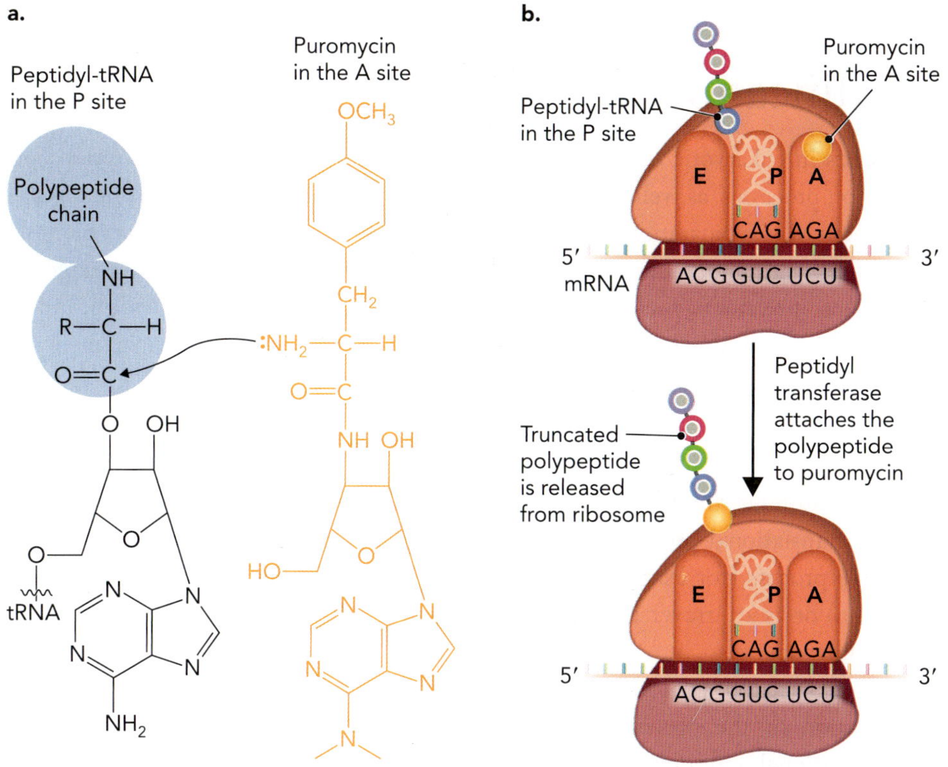

Figure 22.27 Puromycin functions as a peptidyl acceptor to terminate translation. **a.** The amino group of puromycin attacks the carbonyl carbon of the polypeptide chain in the P site. **b.** The orange circle represents puromycin as it binds in the A site. When a peptide bond is formed and the polypeptide transfers to the puromycin molecule, this compound is released from the ribosome, thus inhibiting further addition of amino acids.

bacteria but have little or no effect on general protein synthesis in eukaryotes because of structural differences in the prokaryotic and eukaryotic ribosomes.

For example, puromycin is structurally similar to the terminal aminoacyl-adenosine of tRNA and therefore can bind to the A site (**Figure 22.27**). Puromycin acts as an acceptor for the growing polypeptide chain and causes premature termination of translation. Puromycin was used to confirm that the 23S rRNA was a ribozyme and participated directly in the formation of the peptide bond between the growing polypeptide chain and the amino acid charged to the tRNA in the A site.

concept integration 22.2

Why is proofreading required during both aminoacylation and elongation on the ribosome to ensure the fidelity of protein synthesis? Can these processes work in combination to prevent translation errors?

Proofreading is required both before and during translation to ensure that the synthesis of proteins proceeds with as few errors as possible. Proofreading begins during the aminoacylation reaction. Most aminoacyl-tRNA synthetases have mechanisms to prevent binding of an incorrect amino acid and to hydrolyze incorrectly charged tRNAs. Together, the pre-editing and postediting functions result in a very low error rate for aminoacylation. But the ribosome must also have a proofreading function to prevent mismatched codon–anticodon interactions from resulting in the incorporation of an incorrect amino acid. Here, the elongation factors prevent peptide bond formation unless the appropriate codon–anticodon interaction is present, which triggers GTP hydrolysis. Once the elongation factor is in the GDP-bound form, it is released from the ribosome, allowing peptidyl transferase activity to occur. Proofreading on the ribosome, however, is not able to prevent binding of an incorrectly charged tRNA to the ribosome.

22.3 Posttranslational Modification of Proteins

Organelles in eukaryotic cells provide physical separation between transcription and translation processes, as well as between catabolic and anabolic pathways. The regulatory benefit of this separation, however, is balanced with the need to transport biomolecules across the membranes of these organelles. In previous chapters, we have explored transport as it relates to fuel metabolism, cell signaling, and energy production, and in this section we examine how proteins are targeted to specific cellular compartments as a function of protein synthesis and protein modification.

Covalent Attachment of Functional Groups to Proteins

For some proteins, transport begins before translation is complete; for others, this is a posttranslational process. As shown in **Figure 22.28**, transport can occur as a reversible process (for example, between the Golgi apparatus and the endoplasmic reticulum) or as an irreversible process (for example, from secretory granules to the extracellular space).

In many cases, protein transport is driven by the need for **posttranslational modification** of the protein. In other cases, modification itself drives transport. Posttranslational modifications involve the attachment of a functional group or

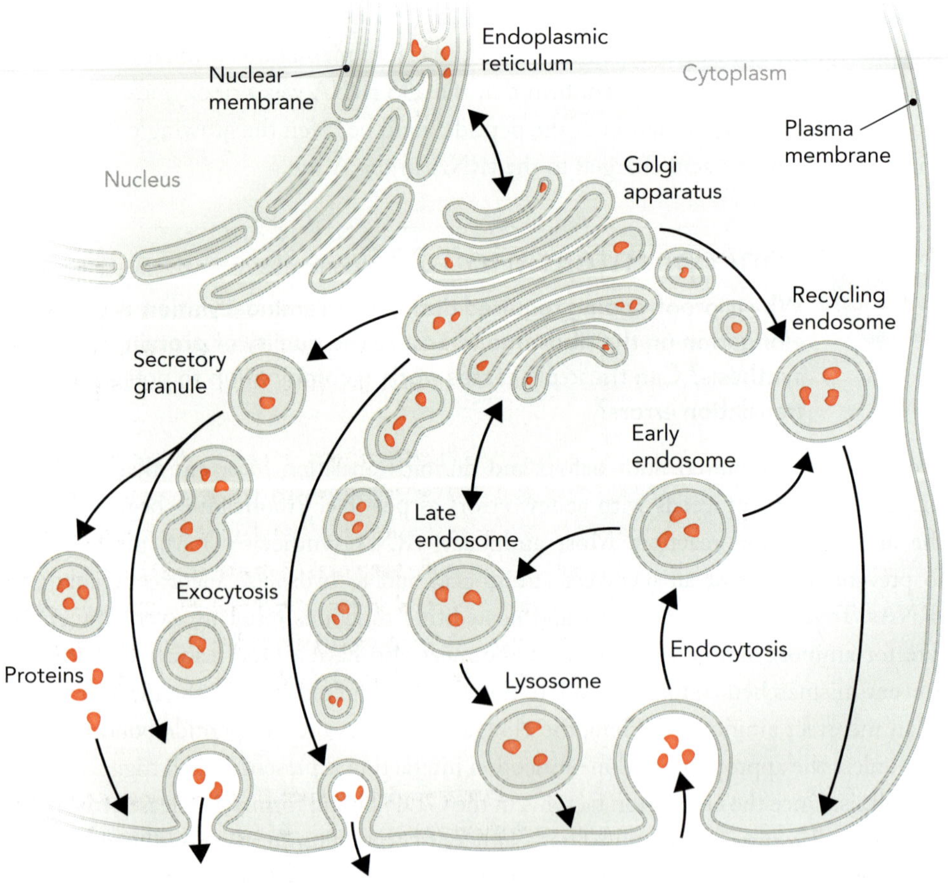

Figure 22.28 Protein transport between subcellular compartments and the plasma membrane. The direction of protein transport is indicated by the arrows.

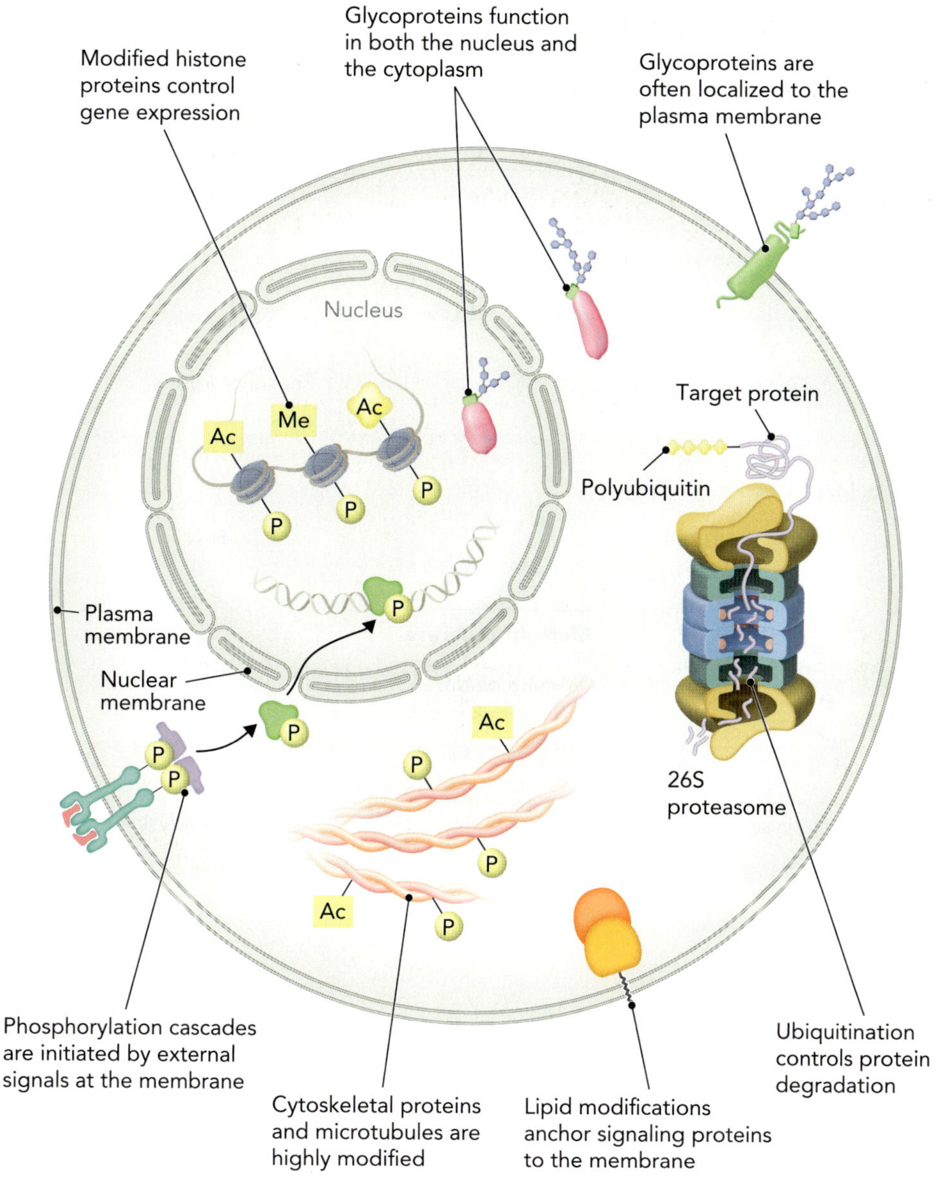

Figure 22.29 Many posttranslational modifications involve attachment of a functional group or macromolecule to the fully synthesized protein. Glycoproteins are produced by glycosylation. Lipid modifications include farnesylation and prenylation. Ac = acetyl group; Me = methyl group; P = phosphoryl group.

macromolecule to a specific amino acid side chain within a protein. As shown in **Figure 22.29**, the most common enzyme-mediated protein modifications in eukaryotic cells are phosphorylation, methylation, acetylation, ubiquitination, glycosylation, and lipid modification; some of these protein modifications also occur in prokaryotic cells. These posttranslational modifications are commonly reversible reactions requiring distinct enzymes for the forward and reverse reactions (**Figure 22.30**). Note that the ubiquitination of eukaryotic proteins destined for proteasomal degradation was described in Chapter 17 (see Figures 17.22–17.25), and the modification of proteins by carbohydrates and lipids is described in detail later in this section.

Ran-Mediated Nuclear Import and Export of Eukaryotic Proteins

Nuclear functions such as replication and transcription require a variety of proteins, all of which are synthesized in the cytosol. These proteins, along with histones, are imported to the nucleus to mediate DNA replication and transcription and then

Figure 22.30 Four common types of enzyme-mediated posttranslational modification of proteins are shown. **a.** Phosphorylation and dephosphorylation of amino acid residues by kinase and phosphatase enzymes, respectively. Tyrosine (shown), serine, and threonine (not shown) residues are common sites of phosphorylation in proteins. **b.** Methylation and demethylation of lysine residues by methyltransferase and amine oxidase enzymes, respectively. SAH = *S*-adenosyl-L-homocysteine; SAM = *S*-adenosyl-L-methionine. **c.** Acetylation and deacetylation of histone proteins by histone acetyltransferase (HAT) and histone deacetylase (HDAC) enzymes. **d.** Ubiquitination and deubiquitination of lysine residues by ubiquitin ligase and deubiquitinating isopeptidase enzymes.

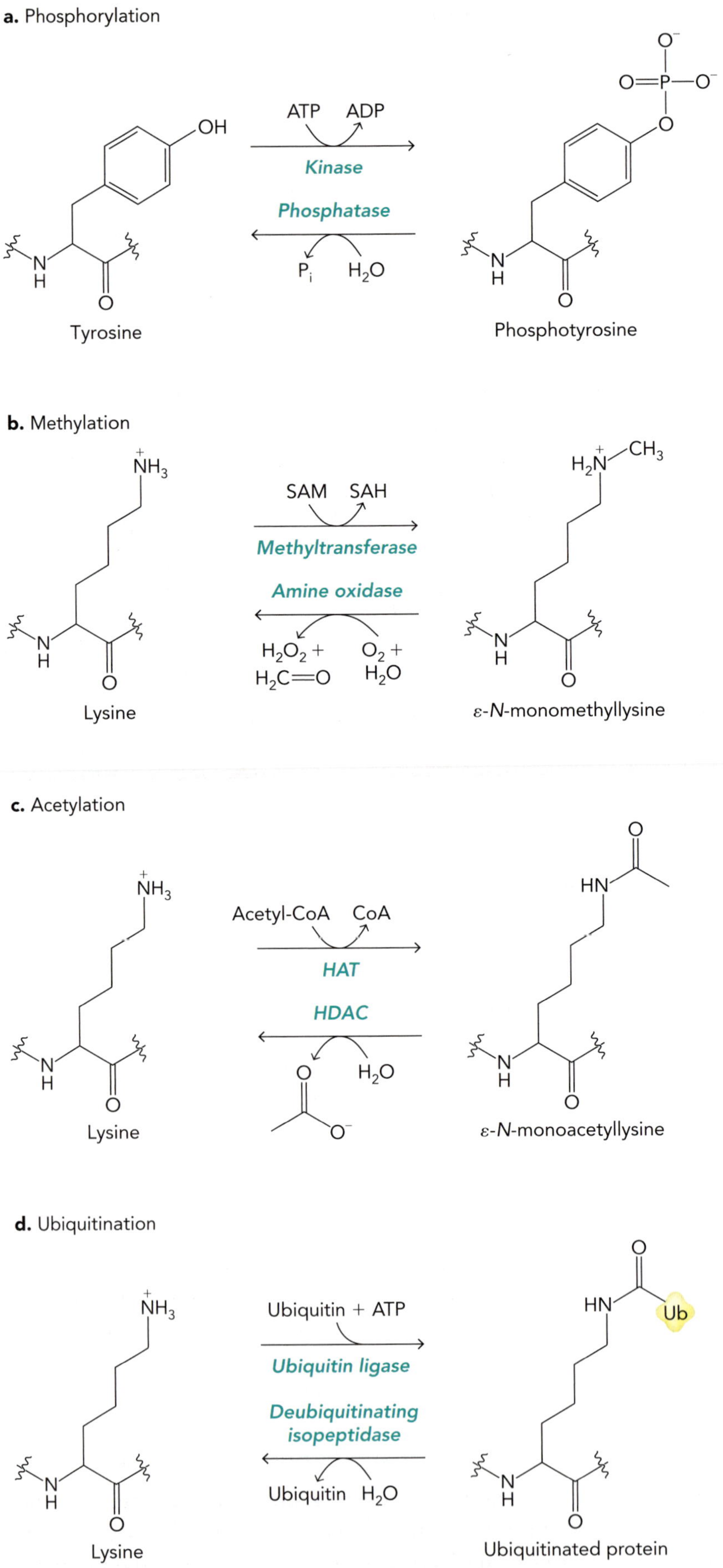

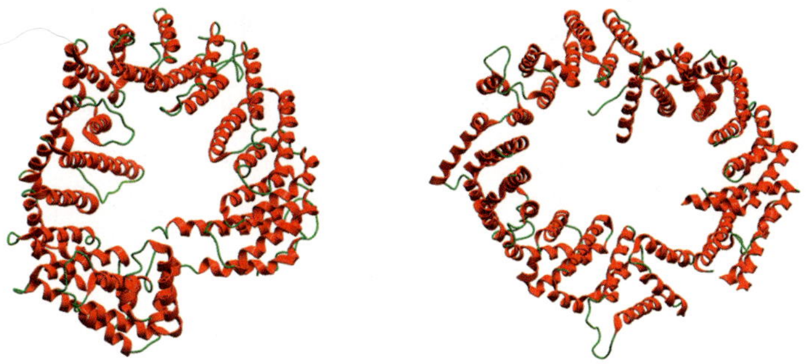

Figure 22.31 Importin and exportin proteins are structurally similar. Both importins and exportins bind directly to the cargo proteins that are being transported through the nuclear pore complex. BASED ON PDB FILES 2Q5D (IMPORTIN) AND 3GB8 (EXPORTIN).

exported to the cytosol where they are recycled. As described in Chapter 21, nuclear pore complexes allow diffusion of small molecules and selective transport of larger molecules such as DNA, RNA, and proteins across the nuclear envelope (see Figure 21.42). The number of pore complexes in the nuclear envelope increases with DNA replication to facilitate the entry of histones required for compaction of the newly synthesized DNA.

Because the nuclear pore complexes limit the entry of proteins into the nucleus, all proteins imported or exported across the nuclear envelope must contain a signal sequence, which is an amino acid sequence that interacts with components of the pore complex to allow for entry. The process of nuclear import requires a stretch of basic amino acids called a **nuclear localization signal (NLS)**. The NLS must be exposed on the surface of the protein in order to be recognized and bound by nuclear transport proteins. A family of proteins known as **importins** and **exportins** facilitate active transport through the nuclear pore complex (**Figure 22.31**). Importins and exportins have similar primary structures and differ mainly in their function, with importins binding to their target proteins in the cytosol and releasing them in the nucleus, whereas exportins follow the opposite path. Some exportins recognize proteins containing **nuclear export signal (NES)** sequences, while others recognize RNA sequences (see Figure 21.43).

As shown in **Figure 22.32**, Ran protein is required for protein transport through the nuclear pore complex just as it is for RNA transport (see Figure 21.43). The affinity of Ran for other proteins in the transport complexes changes depending on whether Ran is bound to GTP or GDP. In the case of importin, Ran–GTP transports unbound importin back into the cytoplasm after it delivers cargo to the nucleus, whereas an exportin–cargo complex is transported by Ran–GTP from the nucleus to the cytoplasm to deliver cargo. In both the import cycle and export cycle, the GTPase activity of Ran is activated by GTPase activating proteins (GAPs) in the cytoplasm to generate Ran–GDP and release the importin or exportin proteins. Ran–GDP is then recycled back into the nucleus where guanine nucleotide exchange factor (GEF) proteins stimulate replacement of GDP with GTP to regenerate activated Ran–GTP.

Co-translational Modification of Proteins in the Endoplasmic Reticulum

The nucleus is but one of many destinations for eukaryotic proteins. As shown in **Figure 22.33** (p. 1131), thesite of protein synthesis determines for the most part where a protein will end up in its final form. As with nuclear-localized proteins that are

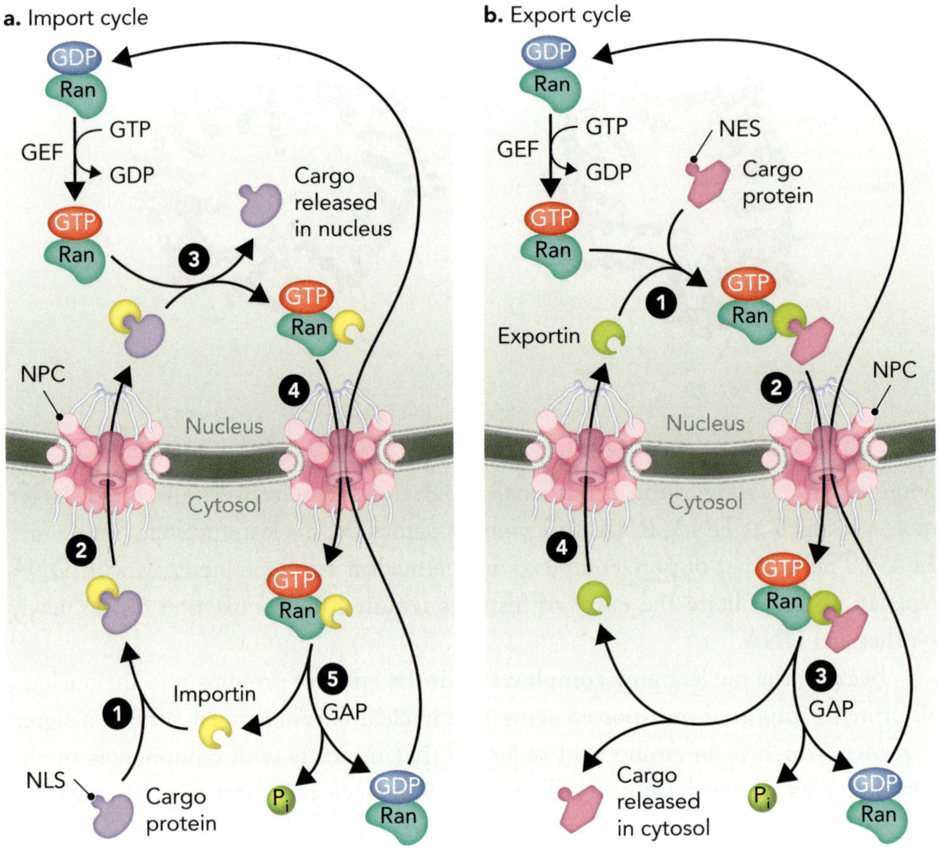

Figure 22.32 Ran-dependent nuclear import and export of protein and RNA cargo is mediated by importin and exportin proteins. **a.** (1) The nuclear import cycle is initiated in the cytoplasm by the binding of importin to protein cargo containing an NLS sequence. (2) The importin–cargo complex passes through the nuclear pore complex (NPC) where (3) importin binds to Ran–GTP to release the cargo protein. (4) The Ran–GTP–importin complex is then transported through the NPC to the cytoplasm where (5) importin is released after GAP-stimulated GTP hydrolysis. The Ran–GDP protein is recycled back into the nucleus where it is reactivated by GEF-mediated GTP–GDP exchange to regenerate Ran–GTP. **b.** The export cycle is similar to the import cycle except that (1) it is initiated by GEF-mediated formation of a Ran–GTP–exportin–cargo complex in the nucleus. (2) The complex is transported through the NPC to the cytoplasm, where (3) a GAP protein stimulates GTPase activity of Ran to dissociate the complex and form Ran–GDP. (4) Exportin is transported back into the nucleus.

synthesized on free ribosomes, proteins destined for mitochondria, peroxisomes, and chloroplasts (in photosynthesizing cells) include targeting sequences. But for proteins destined for the plasma membrane, where they are either secreted into the extracellular space or inserted into the membrane to function as receptor proteins, protein synthesis must take place on membranes, which in eukaryotic cells is the endoplasmic reticulum (ER) membrane. ER-synthesized proteins are also targeted to lysosomes via vesicle transport pathway endosomes.

Targeting of ribosomal complexes to the rough ER depends on the presence of a specific amino acid sequence at the N terminus of the nascent polypeptide known as the ER **signal peptide sequence**. These ER-targeting protein sequences were first discovered in the 1970s, during a comparison of protein synthesis in a cell-free system versus that in systems containing microsomes derived from the rough ER. Proteins synthesized in the cell-free system were slightly larger than those synthesized in the presence of the ER, indicating that some region of the protein had been cleaved during ER-directed translation. Further investigation into the amino acid sequences of these proteins led to the identification of N-terminal signal sequences. As shown in **Figure 22.34**, ER-targeted proteins contain signal peptide sequences with four functional regions: (1) an N-terminal segment, often containing one or more positively charged amino acids; (2) a segment of hydrophobic amino acids; (3) a C segment on the carboxyl-terminal side of the hydrophobic amino acids; and (4) a protease cleavage site. The confirmed signal peptide sequences of seven human ER-targeted proteins are shown in Figure 22.34 where it is observed that there are ~10–15 amino acids in the hydrophobic segment.

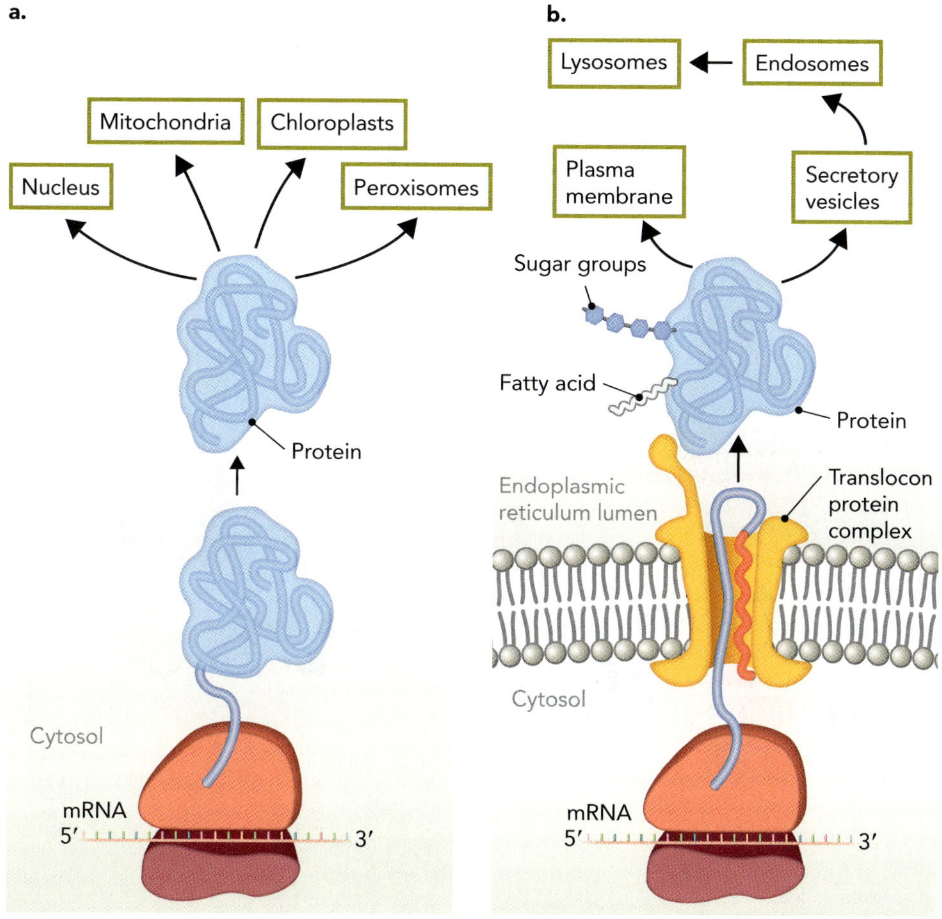

Figure 22.33 Subcellular localization of newly synthesized proteins is determined by whether or not the ribosome is associated with the ER membrane. **a.** Proteins synthesized on free ribosomes are directed to subcellular locations on the basis of targeting sequences, which specify the nucleus, mitochondria, peroxisomes, and chloroplasts. **b.** Proteins synthesized on the ER membrane are destined for the plasma membrane or for various membrane compartments in the cytosol. ER-synthesized proteins are often posttranslationally modified with covalently attached lipids and carbohydrates. The translocon protein complex is described in the text.

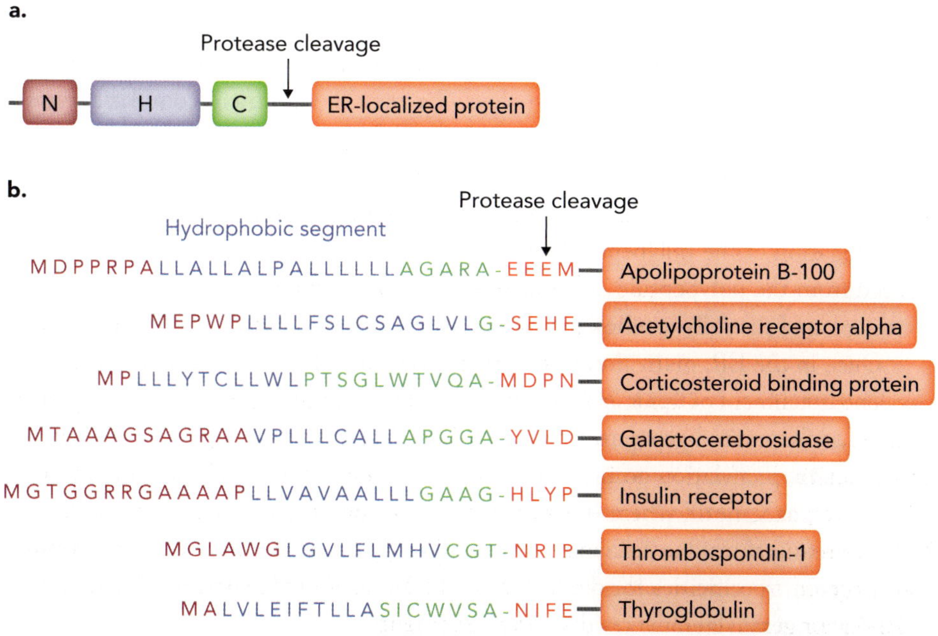

Figure 22.34 Signal peptide sequences are located at the amino terminus of ER-targeted proteins. **a.** The functional components of signal peptide sequences are the N-terminal segment (N), the hydrophobic segment (H) consisting of ~10–15 amino acids, and the C segment, which is on the C-terminal side of the hydrophobic segment and adjacent to the protease cleavage site. **b.** Signal peptide sequences of seven representative human proteins obtained from the Signal Peptide Database (www.signalpeptide.de).

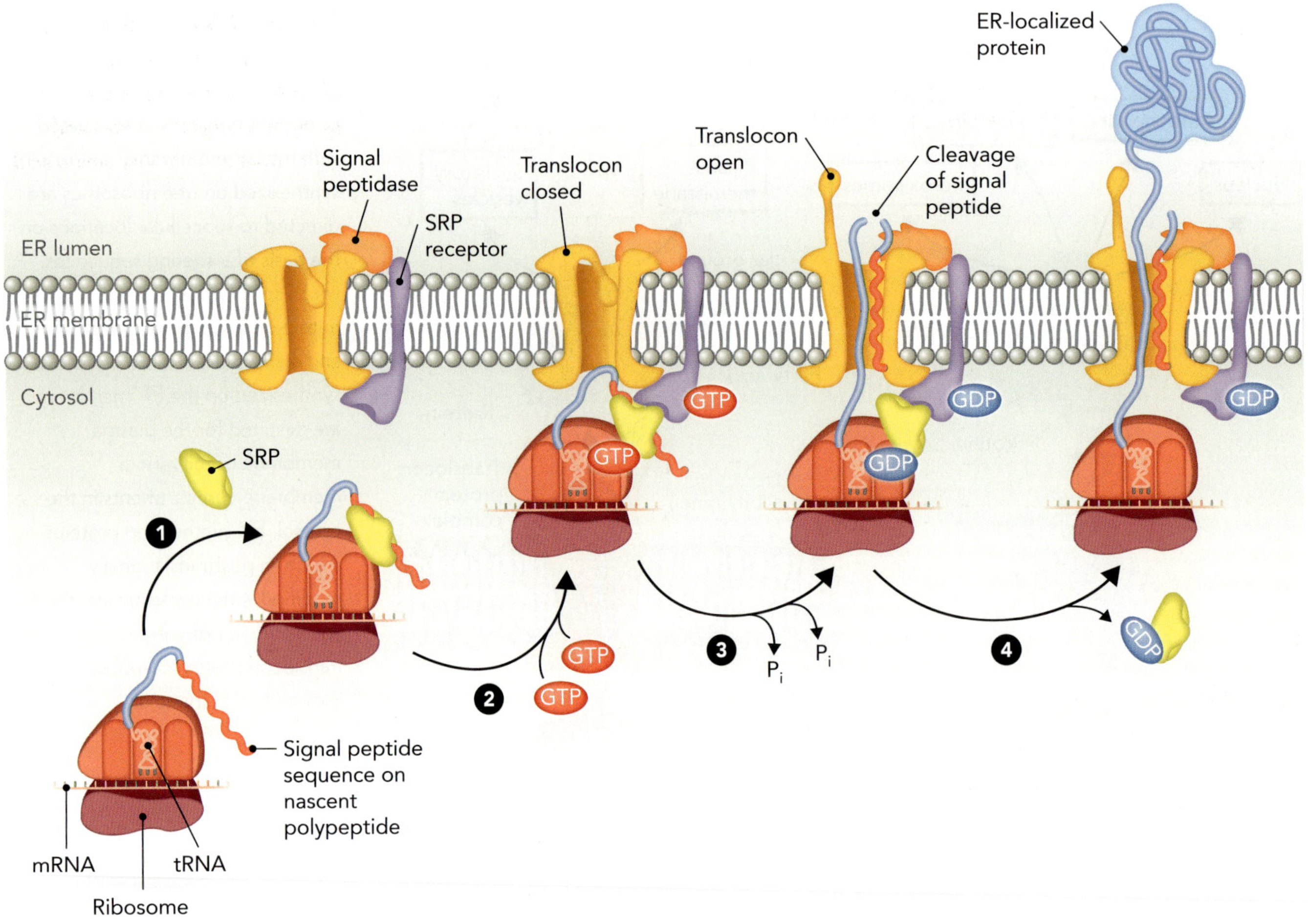

Figure 22.35 SRP binding to N-terminal sequences targets the ribosome to the SRP receptor. Step 1: SRP binds to the signal peptide sequence on the nascent polypeptide and halts ribosomal translocation. Step 2: GTP binding to SRP and the SRP receptor generates the translationally competent complex bound to the SRP receptor. Step 3: GTP hydrolysis stimulates protein synthesis, and the signal peptidase cleaves the peptide sequence when it enters the ER lumen. Step 4: SRP–GDP dissociates, and protein synthesis continues until the protein is released from the ribosome into the ER lumen.

Translation of a protein containing an ER signal peptide sequence begins on a free ribosome. As shown in **Figure 22.35**, once the N-terminal region has been translated, the ER signal sequence emerges from the ribosome and is bound by the **signal recognition particle (SRP)**. SRPs contain both a protein and an RNA component. Upon binding of SRP to the ER signal peptide sequence on the nascent polypeptide, the ribosome pauses on the mRNA to halt translation. The SRP–ribosome complex then binds to the SRP receptor on the ER membrane, and after GTP hydrolysis in both the SRP protein and the SRP receptor, the signal peptide sequence is inserted into the open translocon protein complex. The translocon protein complex traverses the ER membrane and provides a conduit for the nascent polypeptide chain to cross the membrane when the translocon is in the open conformation. GTP hydrolysis also leads to dissociation of the SRP–ribosome complex, which stimulates protein synthesis and elongation of the polypeptide into the ER lumen. The signal peptide sequence is cleaved from the polypeptide by a signal peptidase enzyme in the lumen. Translation continues, and the fully synthesized protein is released into the lumen of the ER.

Once in the ER, many proteins undergo a variety of modifications, including the covalent attachment of lipids and carbohydrates. For example, prenylation is the attachment of an isoprenoid group to a cysteine residue via a thioester linkage. As shown in **Figure 22.36**, prenylation occurs at the C terminus of a protein, which is marked by a conserved CaaX motif, where "C" is cysteine, "a" is any aliphatic amino acid, and "X" is the C-terminal residue. Enzymes that catalyze the addition of the isoprenoid group have different specificities for the C-terminal residue. Prenylated proteins have either a farnesyl or geranylgeranyl residue attached (Figure 22.36), resulting in a protein that is tightly associated with cell membranes.

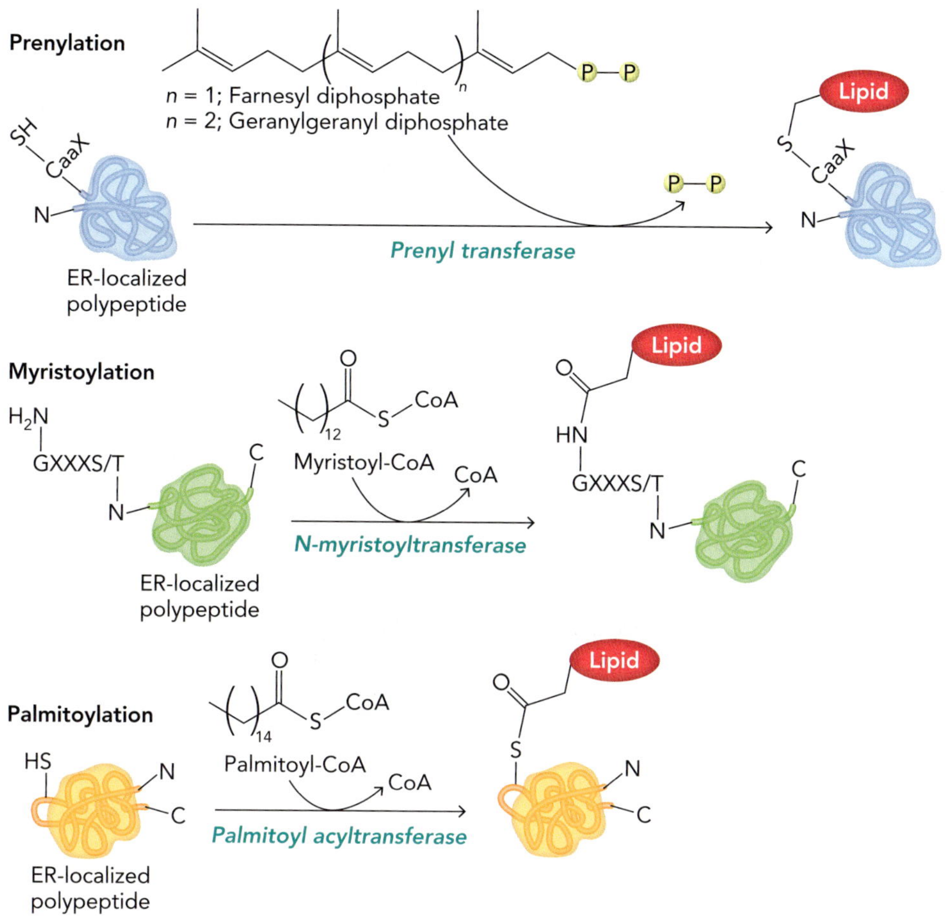

Figure 22.36 The three most common lipid modifications of ER-localized proteins are prenylation, myristoylation, and palmitoylation. X = any amino acid.

Palmitoylation and **myristoylation** are the attachment of a saturated 16-carbon or 14-carbon fatty acid, respectively, to the target protein. For myristoylation, the fatty acid myristoylate is transferred to an N-terminal glycine residue of the target protein, and attachment often occurs co-translationally as the N terminus of the protein emerges from the ribosome. Like prenylation, myristoylation increases the affinity of the target protein for a cell membrane; however, this attachment may be mediated by other regions of the protein or be made transient by conformational changes that hide the myristoylated residues within the interior of the protein.

Palmitoylation of proteins occurs on the sulfur atom of internal cysteine residues found anywhere in the protein. Fatty acids other than palmitate can also be attached at these amino acid residues, and the process is therefore often referred to as thioacylation. Palmitoylation is unique in that no specific sequence motif surrounds the cysteine residues that would allow for prediction of fatty acid attachment.

Attachment of a protein to a cell membrane often requires more than one lipid modification or the combination of a lipid modification and interaction between a hydrophobic region on the protein and the membrane. The type of attachment often dictates the region of the membrane that interacts with the protein. Myristoylation and prenylation lead to interactions with the cytoplasmic leaflet of the cell membrane, whereas palmitoylation occurs more often in transmembrane proteins and acts to target the protein within subdomains of the membrane.

As described in Chapter 13, proteins can also be modified by the addition of oligosaccharide chains. The proteins most often modified in this manner are those found

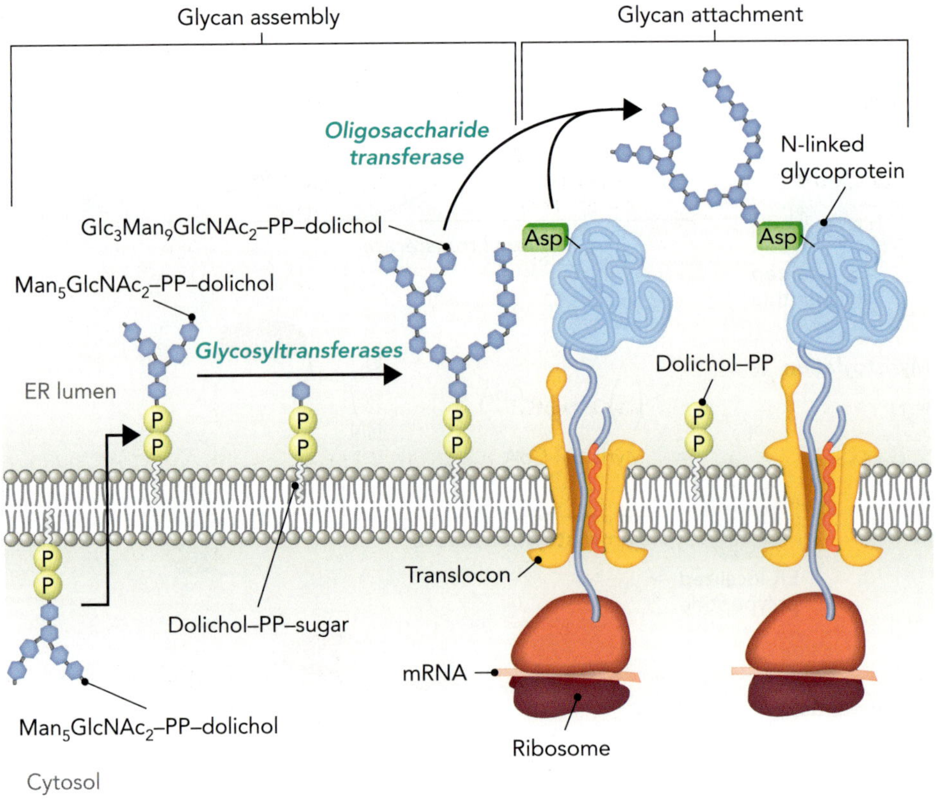

Figure 22.37 Some ER proteins are co-translationally glycosylated. The process of co-translational N-linked glycosylation consists of glycan assembly and glycan attachment.

within the plasma membrane or secreted from the cell. Oligosaccharide attachments can occur as N-linked or O-linked attachments (see Figure 13.19). N-linked oligosaccharides are covalently linked to the amide nitrogen of the side chain of asparagine and contain a standard core of five residues. O-linked oligosaccharides are attached to the hydroxyl group of serine and threonine and are generally short, containing only a few residues.

Although many glycosylation events occur after translation, N-linked glycosylations also occur during translation as the nascent polypeptide is emerging into the lumen of the ER. As shown in **Figure 22.37**, an oligosaccharide precursor is first assembled on the cytosolic side of the ER membrane. The oligosaccharide $Man_5GlcNAc_2$–PP–dolichol, which is composed of five mannose (Man) and two *N*-acetylglucosamine (GlcNAc) molecules, is linked to the polyisoprenoid lipid dolichol through a phosphoribosyl residue to form the dolichol–glycan precursor molecule. This dolichol–glycan is then flipped in the ER membrane so that the glycan is oriented inside the lumen of the ER. After modification of the precursor glycan by ER glycosyltransferase enzymes that add glucose and mannose residues to generate $Glc_3Man_9GlcNAc_2$–PP–dolichol, the assembled glycan is transferred to an asparagine residue on the polypeptide by an oligosaccharide transferase enzyme.

Glycan modification of ER-localized proteins can be exploited to add chemical tags to cellular proteins via a method called bioorthogonal labeling, which uses metabolic incorporation of azide (N_3)-containing carbohydrates to generate labeled biological macromolecules. These azide-containing glycoproteins can be identified by attaching a biochemical tag such as biotin or a fluorescent probe to the azide group using a phosphine-mediated Staudinger ligation reaction (**Figure 22.38**). The biotin tag can then be identified by a secondary reaction using a streptavidin fluorescent probe or an anti-biotin antibody. Alternatively, a reaction between the azide

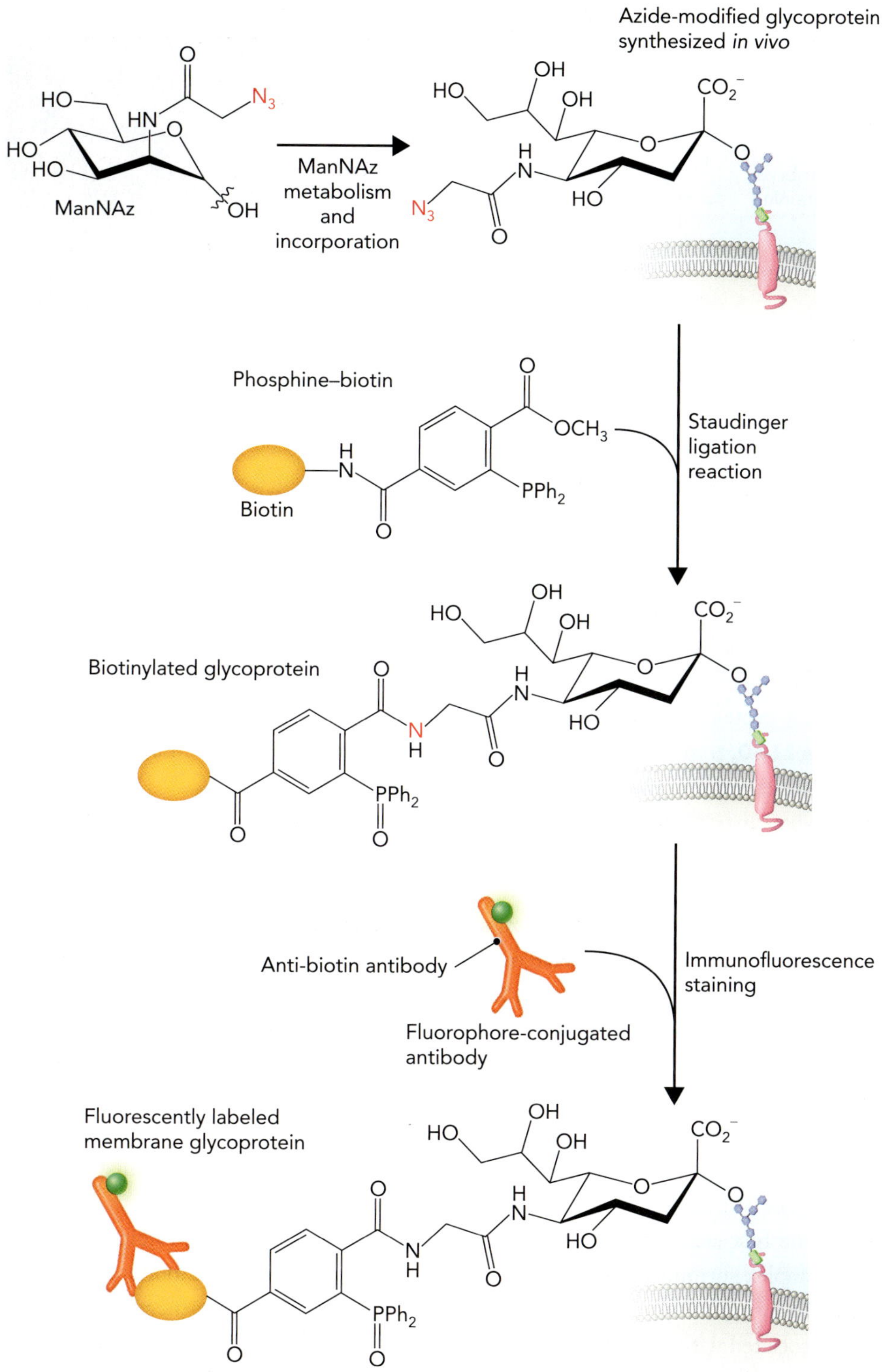

Figure 22.38 Glycoproteins can be chemically labeled using azide-modified sugars that are attached *in vivo* by glycosyltransferases in the cell. The Staudinger ligation reaction between an azide and a phosphine-containing compound can be used to add a biotin functional group that is recognized by a fluorescently labeled anti-biotin antibody. ManNAz = *N*-azidoacetylmannosamine; PPh_2 = phosphine.

and an internal alkyne known as the Huisgen cycloaddition reaction can be used to label the glycoprotein. As shown in **Figure 22.39**, if a fluorescently labeled difluorocyclooctyne (DIFO) compound is used in this reaction, then the glycan-modified biomolecule can be detected directly in tissue samples. Bioorthogonal labeling of glycoproteins under various metabolic conditions makes it possible to investigate regulation of cellular processes such as ER-localized glycan modification of proteins in mosquito tissues.

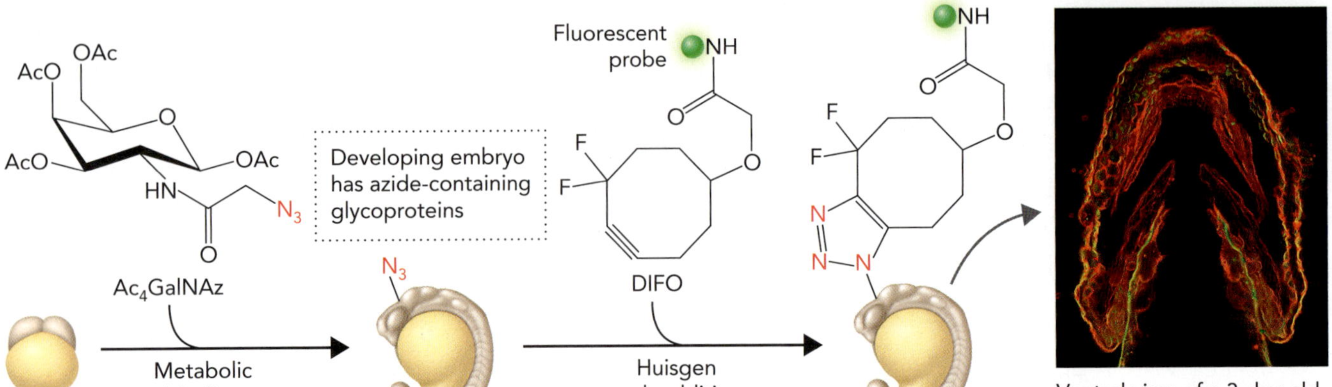

Figure 22.39 Bioorthogonal labeling of zebrafish glycoproteins in the embryo can be done by adding azide-labeled sugars that are incorporated into the glycans. The detection of these azide-glycans is accomplished by the addition of fluorescently labeled alkyne molecules such as difluorocyclooctyne (DIFO), which reacts with the azide via a Huisgen cycloaddition reaction. $Ac_4GalNAz$ = peracetylated *N*-azidoacetylgalactosamine. PHOTO: CAROLYN BERTOZZI/STANFORD UNIVERSITY.

Vesicle Transport Systems in Eukaryotic Cells

Many proteins with N-linked glycosylations must undergo further processing in the Golgi apparatus. From there, these and other proteins can travel to other organelles or to the cell membrane as part of a secretory pathway consisting of transport vesicles.

The outside or cytoplasmic surface of these vesicles is coated with one of three different proteins, depending on the membrane source of the vesicle. As shown in **Figure 22.40**, transport from the Golgi apparatus to the endoplasmic reticulum and within the Golgi apparatus is mediated by **cytoplasmic coat I (COPI)-coated vesicles**, whereas transport from the ER to the Golgi is mediated by **cytoplasmic coat II (COPII)-coated vesicles.** Transport from the trans-Golgi network or from the plasma membrane occurs in **clathrin**-coated vesicles. The protein coats form cage-like structures around the vesicle as it buds off of the donor membrane.

The overall process of vesicle transport with COPI-coated or COPII-coated vesicles is similar. As shown in **Figure 22.41**, a vesicle forms as a bud extending from the ER membrane. The interior of the vesicle contains proteins or other biomolecules to be transported, and the membrane that forms these vesicles often contains transmembrane proteins. Once the vesicle is released from the parent membrane, it travels to the appropriate target membrane and fuses with the membrane to release its contents within the organelle. Membrane-bound proteins in the vesicle now become part of the membrane of the target organelle.

The first step in vesicle formation is the recruitment of the proteins that form the complex surrounding the vesicles, or **coatomer proteins**, to the donor membrane, directed by a GTP binding protein. Once the protein has bound to the membrane and has associated with a molecule of GTP, it directs the assembly of other proteins, which polymerize and cause a mechanical modification of the membrane, leading to vesicle formation. Proteins and other molecules to be transported often associate with the donor membrane through membrane-bound receptors, although the complete mechanism that directs transport is not fully understood. When formation of the vesicle is complete, it buds from the donor membrane and is released into the cellular compartment. Just as the coatomer proteins differ based on the origin of the vesicle, the destination of the vesicle is dictated by the presence of transmembrane proteins that extend to the outside of the vesicle. These proteins are known as **soluble NSF attachment protein receptors (SNAREs)** and are divided into two groups: v-SNAREs found on vesicle membranes and t-SNAREs found on the target membrane (Figure 22.41). Different types of membranes contain variations of these SNARE proteins so that only the

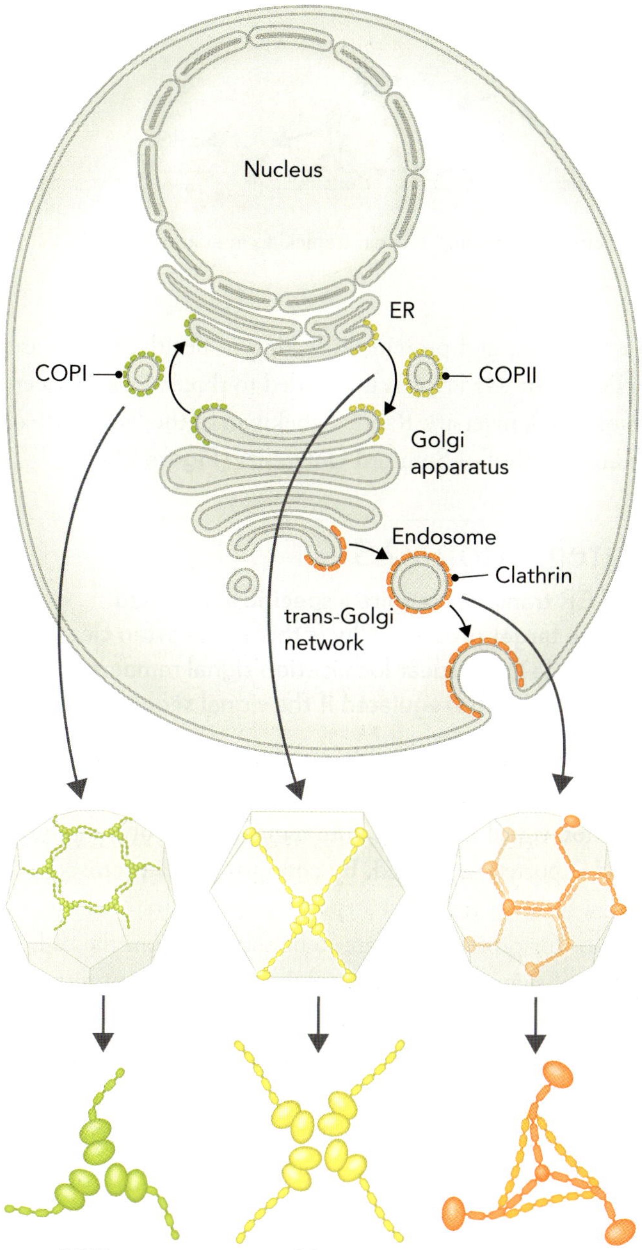

Figure 22.40 Intracellular vesicle transport in eukaryotic cells is mediated by COPI, COPII, and clathrin coat proteins that induce membrane budding in the ER, Golgi, and endosomes. The cages formed by these proteins are architecturally different from each other.

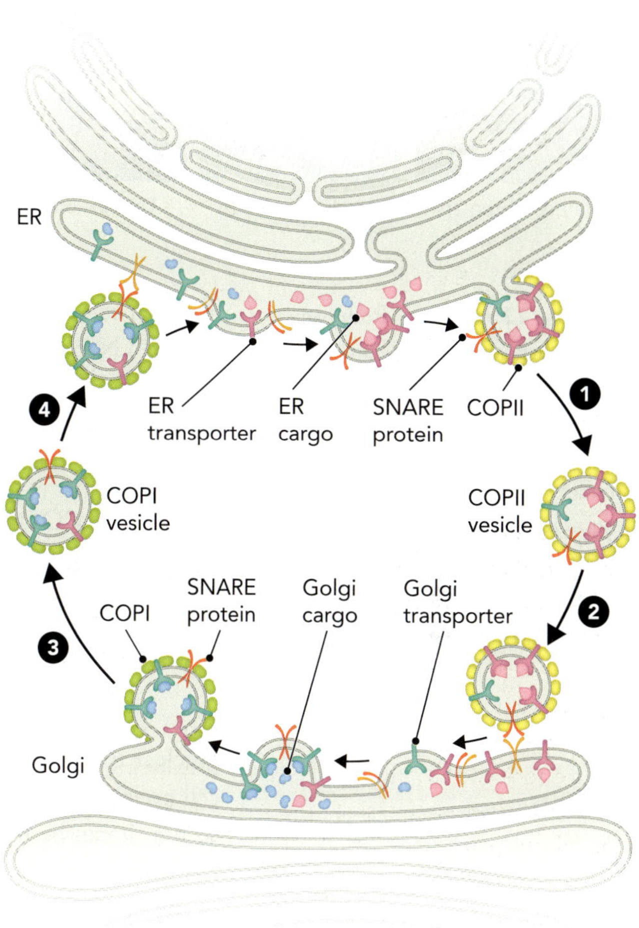

Figure 22.41 Protein sorting between the ER and Golgi requires COPI-coated and COPII-coated vesicles. Vesicles bud from their associated membranes and travel through the cytosol before fusing with a different organelle or the outer cell membrane. The functional role of SNARE proteins is to serve as vesicle receptors at the respective membrane surfaces. The cargo proteins and the cargo transporter proteins are partitioned differently between the ER and Golgi compartments. The starting point is shown here as the ER because most newly synthesized proteins destined for vesicle transport originate in the ER. COPII vesicles often transport proteins back to the ER.

correct association between matching v-SNAREs and t-SNAREs results in vesicle fusion to the target membrane. Formation of the coat is necessary for vesicle formation, but the presence of the coat proteins would inhibit fusion with the target membrane by blocking these transmembrane proteins. Therefore, the coat is released from the vesicle after it buds from the donor membrane, triggered by GTP hydrolysis.

Understanding the mechanisms of vesicle transport in eukaryotic cells required a combination of *in vitro* and *in vivo* biochemical assays designed to identify the proteins involved and the signals required for subcellular targeting. Because of the

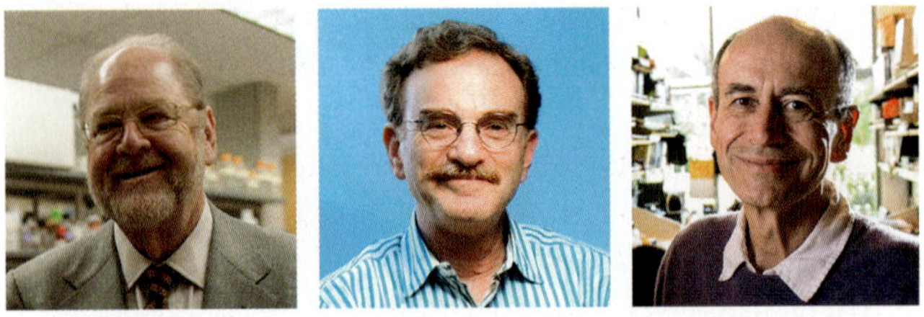

James Rothman Randy Schekman Thomas Südhof

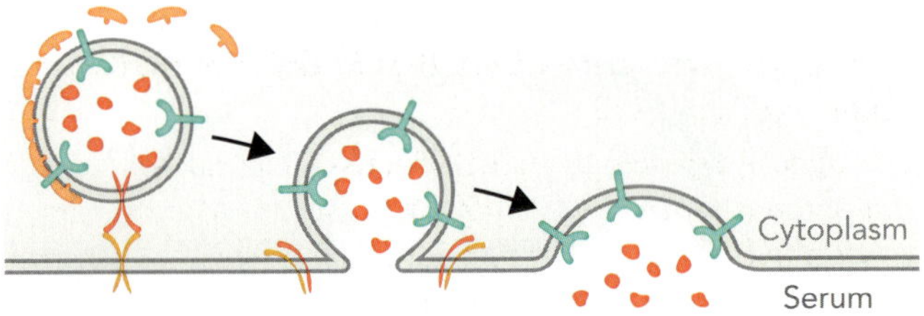

Vesicle transport and protein trafficking in eukaryotic cells

Figure 22.42 James Rothman, Randy Schekman, and Thomas Südhof shared the 2013 Nobel Prize in Physiology or Medicine for their contributions to our understanding of vesicle transport in eukaryotic cells. J. ROTHMAN: DOUGLAS HEALEY/POLARIS/NEWSCOM; R. SCHEKMAN: ZUMA PRESS, INC./ALAMY STOCK PHOTO; T. SÜDHOF: XU YONG/XINHUA PRESS/CORBIS.

importance of cellular vesicle transport and protein trafficking in health and disease, the 2013 Nobel Prize in Physiology or Medicine was awarded to three of the pioneers in this field, James Rothman of Yale University, Randy Schekman of the University of California, Berkeley, and Thomas Südhof of Stanford University (**Figure 22.42**).

concept integration 22.3

Both nuclear and ER transport require a specific amino acid sequence within the target protein. Why is this signal often cleaved from ER proteins while the nuclear localization signal remains intact? How can nuclear transport be regulated if the signal sequence is always present on the target protein?

Nuclear transport is bidirectional, with proteins in a constant flux depending on the cell cycle. The nuclear localization signal must therefore remain a part of the protein so that it can be imported and exported as needed. By contrast, protein transport to the ER is generally the first in a series of transport steps for proteins destined for one of the other cellular organelles or for proteins that are to be secreted from the cell. In this case, cleavage of the ER signal sequence ensures that the protein is committed to the secretory pathway. Because the nuclear localization signal is not cleaved, transport can be regulated instead by making the sequence more or less accessible to transport proteins. The nuclear localization signal can be modified by a posttranslational modification (that is, phosphorylation or acetylation) that inhibits interaction between the target protein and importin, thus excluding it from the nucleus.

chapter summary

22.1 Deciphering the Genetic Code

- tRNA is an adaptor molecule that connects three-nucleotide codons in mRNA with a specific amino acid.
- Nirenberg and Matthaei developed *in vitro* translation experiments to decipher the first three codons, UUU, AAA, and CCC.
- Khorana's synthesis of DNA molecules of known sequence, along with filter binding assays developed by Nirenberg and Leder, allowed most of the remaining codons to be decrypted.
- The standard genetic code has 64 triplet codons, 61 of which correspond with one or more of the 20 amino acids plus 3 codons that specify termination.
- Only tryptophan and methionine are encoded by a single codon each. The other 18 amino acids use from two to six codons.
- Noncanonical base pairings can occur between mRNA and tRNA at the wobble position, which is the first (5′) position of the tRNA anticodon and the third (3′) position of the mRNA codon. These alternative

base pairings mean that more than one codon may be recognized by a single tRNA, thus 61 different tRNAs are not needed in a cell.

- Inosine is produced by the deamination of adenosine and is a common modification in tRNA at the anticodon wobble position. The inosine base can form noncanonical base pairs with adenine, cytosine, and uracil.

22.2 Biochemistry of mRNA Translation

- tRNA molecules are 70–90 bp in length and contain many invariant and semiconserved residues. All tRNA molecules contain an invariant CCA sequence at the 3′ end of the acceptor stem, which is the site of amino acid attachment.
- tRNA folds into a cloverleaf secondary structure consisting of the D, TΨC, and anticodon arms and the acceptor stem. tRNA adopts an L-shaped tertiary structure.
- tRNAs are charged with amino acids by aminoacyl-tRNA synthetases. These enzymes catalyze a two-stage reaction in which the amino acid is first adenylated using ATP, and then the aminoacyl-adenylate is used to add the amino acid to the tRNA.
- Aminoacyl-tRNA synthetases are grouped into two classes depending on their structural differences, which determine whether they add the amino acid to the 2′-hydroxyl or 3′-hydroxyl of the adenosine at the terminus of the tRNA acceptor stem.
- Aminoacyl-tRNA synthetases have both pretransfer and posttransfer editing functions to prevent the incorrect charging of tRNA.
- The ribosome is the site of protein synthesis in prokaryotes and eukaryotes. It is composed of two subunits that contain both protein and RNA components.
- Within the ribosome, there are three tRNA binding sites and an mRNA binding site. During elongation, a charged tRNA binds first to the A site and translocates to the P site after peptide bond formation. After transfer of the polypeptide chain, the tRNA is released from the E site.
- Different mechanisms are used to determine the starting AUG codon in prokaryotes and eukaryotes. Prokaryotic mRNA transcripts contain a Shine–Dalgarno sequence that binds to the ribosome and positions the AUG codon in the P site. Eukaryotes scan the mRNA in a 5′ to 3′ direction until the first AUG codon is encountered, which may be within a sequence known as the Kozak sequence.
- During initiation, several initiation factors are involved in assembling the ribosome and bringing the Met–$tRNA_i^{Met}$ to the P site on the ribosome.
- Elongation factors bring charged tRNAs to the A site on the ribosome (EF-Tu) and catalyze translocation (EF-G) in GTP-dependent processes. Peptide bond formation is catalyzed by rRNA.
- Binding of a protein release factor to a termination codon in mRNA halts translation and leads to dissociation of the ribosomal complex.
- Many antibiotics inhibit bacterial translation without having a significant effect on the process in the host organism because of differences in the structures of prokaryotic and eukaryotic ribosomes. Antibiotics are also used by scientists to study the process of translation in both prokaryotic and eukaryotic cells.

22.3 Posttranslational Modification of Proteins

- Posttranslational modification can alter the structure or function of the target protein. Common modifications include phosphorylation, methylation, acetylation, glycosylation, and ubiquitination, as well as lipid modifications.
- Posttranslational modifications are often reversible, but the modification or removal of the modification is catalyzed by separate enzymes.
- Nuclear import and export utilize Ran proteins, importins, and exportins. Importins recognize nuclear localization signal sequences on the proteins that will be imported. Ran proteins cycle through GTP or GDP bound states, which alter their affinity for other proteins in the transport complexes.
- Protein translocation to the ER often occurs during synthesis. The signal recognition particle (SRP) binds to a signal sequence at the N terminus of the nascent polypeptide and directs the ribosome to the SRP receptor in the ER. As the remainder of the protein is synthesized, it directly enters the ER lumen.
- Proteins that must attach to a membrane as part of their function are posttranslationally modified in the ER by the attachment of one or more lipid residues, such as farnesyl, geranylgeranyl, palmitoyl, or myristoyl residues.
- N-linked and O-linked glycosylation can occur both in the ER and the cytosol. N-linked glycosylation often occurs co-translationally.
- Azide-containing carbohydrate modifications can be used to identify cellular proteins. This process is known as bioorthogonal labeling.
- Transport between the ER and other cellular organelles occurs through the formation of vesicles. A coatomer protein is first recruited to the donor membrane surface, which initiates vesicle formation. The type of protein coat is dictated by the donor membrane.
- Transport within the Golgi apparatus is mediated by COPI-coated vesicles, whereas transport from the ER to the Golgi is mediated by COPII-coated vesicles. Transport from the trans-Golgi or from the plasma membrane occurs in clathrin-coated vesicles.

biochemical terms

(in order of appearance in text)

aminoacyl-tRNA synthetase (p. 1105)
wobble hypothesis (p. 1109)
isoacceptor tRNA (p. 1113)
initiator tRNA (p. 1115)
peptidyl site (P site) (p. 1116)
aminoacyl site (A site) (p. 1116)
Svedberg value (p. 1116)
initiation factor (IF) (p. 1116)
Shine–Dalgarno sequence (p. 1117)
Kozak sequence (p. 1119)
elongation factor (EF) (p. 1121)
release factor (RF) (p. 1122)
antibiotics (p. 1122)
posttranslational modification (p. 1126)
nuclear localization signal (NLS) (p. 1129)
importin (p. 1129)
exportin (p. 1129)
nuclear export signal (NES) (p. 1129)
signal peptide sequence (p. 1130)
signal recognition particle (SRP) (p. 1132)
palmitoylation (p. 1133)
myristoylation (p. 1133)
cytoplasmic coat I (COPI)-coated vesicles (p. 1136)
cytoplasmic coat II (COPII)-coated vesicles (p. 1136)
clathrin (p. 1136)
coatomer proteins (p. 1136)
soluble NSF attachment protein receptor (SNARE) (p. 1136)

review questions

1. Why was Francis Crick dissatisfied with protein synthesis models that brought amino acids into direct contact with DNA or mRNA?
2. What are the three possible amino acids resulting from the translation of the RNA sequence CCAAA?
3. What is the wobble hypothesis, and how does it explain the degeneracy of the genetic code?
4. Describe the secondary and tertiary structure of tRNA.
5. Describe the basic aminoacylation reaction. How does this reaction differ in class I and class II aminoacyl-tRNA synthetases?
6. What are the functions of the three tRNA binding sites on the ribosome?
7. How does translation termination occur?
8. Why is puromycin not useful as an antibiotic to treat bacterial infections in humans?
9. How are proteins directed to the endoplasmic reticulum during protein synthesis?
10. Why do acetylation and methylation not occur simultaneously on the same lysine residue on histones?
11. What are the three types of coatomer proteins that assist with vesicle formation, and with what type of transport is each associated?

challenge problems

1. Why did George Gamow state that at least three nucleotides must be required to code for each amino acid?
2. Why was Khorana's method of nucleotide synthesis so essential to the discovery of the genetic code?
3. The mRNA sequence AUGCACAGU codes for the first three amino acids of a particular protein. Which nucleotides can be changed without modifying the amino acid sequence that will result after translation?
4. Why does the human mitochondrion have a slightly different genetic code than that found in the cytosol?
5. Why is it important that all tRNA molecules adopt a similar tertiary structure?
6. Why does the initiator tRNA bind in the P site rather than the A site of the ribosome?
7. Why do some bacterial mRNA transcripts contain more than one Shine–Dalgarno sequence?
8. When comparing the three phases of translation in prokaryotes and eukaryotes, why is initiation the most different?
9. Why must proteins be completely translated before nuclear transport can occur?
10. Why is the ribosome considered a ribozyme?
11. Why are multiple lipid modifications often found on membrane-associated proteins?

315-91 10/11

suggested reading

Books and Reviews

Aminov, R. I. (2010). A brief history of the antibiotic era: lessons learned and challenges for the future. *Frontiers in Microbiology*, *1*: 134–160.

Antonellis, A., and Green, E. D. (2008). The role of aminoacyl-tRNA synthetases in genetic diseases. *Annual Review of Genomics and Human Genetics*, *9*, 87–107.

Crick, F. H. (1966). Codon-anticodon pairing: the wobble hypothesis. *Journal of Molecular Biology*, *19*, 548–555.

Cooper, G. M. (2000). *The Cell: A Molecular Approach*. Sunderland, MA: Sinauer Associates.

Giege, R., and Frugier, M. (2000). Transfer RNA structure and identity. In Madame Curie Bioscience Database. Austin, TX: Landes Bioscience. Available at http://www.ncbi.nlm.nih.gov/books/NBK6236.

Hughson, F. M. (2010). Copy coats: COPI mimics clathrin and COPII. *Cell*, *142*, 19–21.

Südhof, T., and Rizo, J. (2011). Synaptic vesicle exocytosis. *Cold Spring Harbor Perspectives in Biology*, *2011*(*3*): a005637.

Tasaki, T., Sriram, S. M., Park, K. S., and Kwon, Y. T. (2012). The N-end rule pathway. *Annual Review of Biochemistry*, *81*, 261–289.

Primary Literature

Blanchard, S. C., Cooperman, B. S., and Wilson, D. N. (2010). Probing translation with small-molecule inhibitors. *Chemistry & Biology*, *17*(*6*), 633–645.

Dong, X., Biswas, A., Suel, K. E., Jackson, L. K., Martinez, R., Gu, H., and Cook, Y. M. (2009). Structural basis for leucine-rich nuclear export signal recognition by CRM1. *Nature*, *458*, 1136–1141.

Fu, J., Jiang, Q., and Zhang, C. (2010). Coordination of cell cycle events by Ran GTPase. *Nature Education*, *3*(*9*), 32.

Fukai, S., Nureki, O., Sekine, S., Shimada, A., Tao, J., Vassylyev, D. G., and Yokoyama, S. (2000). Structural basis for double-sieve discrimination of L-valine from L-isoleucine and L-threonine by the complex tRNA(Val) and valyl-tRNA synthetase. *Cell*, *103*, 793–803.

Goodenbour, J. M., and Pan, T. (2006). Diversity of tRNA genes in eukaryotes. *Nucleic Acids Research*, *34*(*21*), 6137–6146.

Ishitani, R., Nureki, O., Nameki, N., Okada, N., Nishimura, S., and Yokoyama, S. (2003). Alternative tertiary structure of tRNA for recognition by a posttranscriptional modification enzyme. *Cell*, *113*, 383–394.

Korostelev, A., Asahara, H., Lancaster, L., Laurberg, M., Hirschi, A., Zhu, J., Trakhanov, S., Scott, W. G., and Noller, H. F. (2008). Crystal structure of a translation termination complex formed with release factor RF2. *Proceedings of the National Academy of Sciences USA*, *105*, 19684–19689.

Kozak, M. (1986). Point mutations define a sequence flanking the AUG initiator codon that modulates translation by eukaryotic ribosomes. *Cell*, *44*, 283–292.

Moore, P. B., and Steitz, A. T. (2003). After the ribosome structure: how does peptidyl transferase work? *RNA*, *9*, 155–159.

Schimmel, P. (2008). Development of tRNA synthetases and connection to genetic code and disease. *Protein Sciences*, *17*(*10*), 1643–1652.

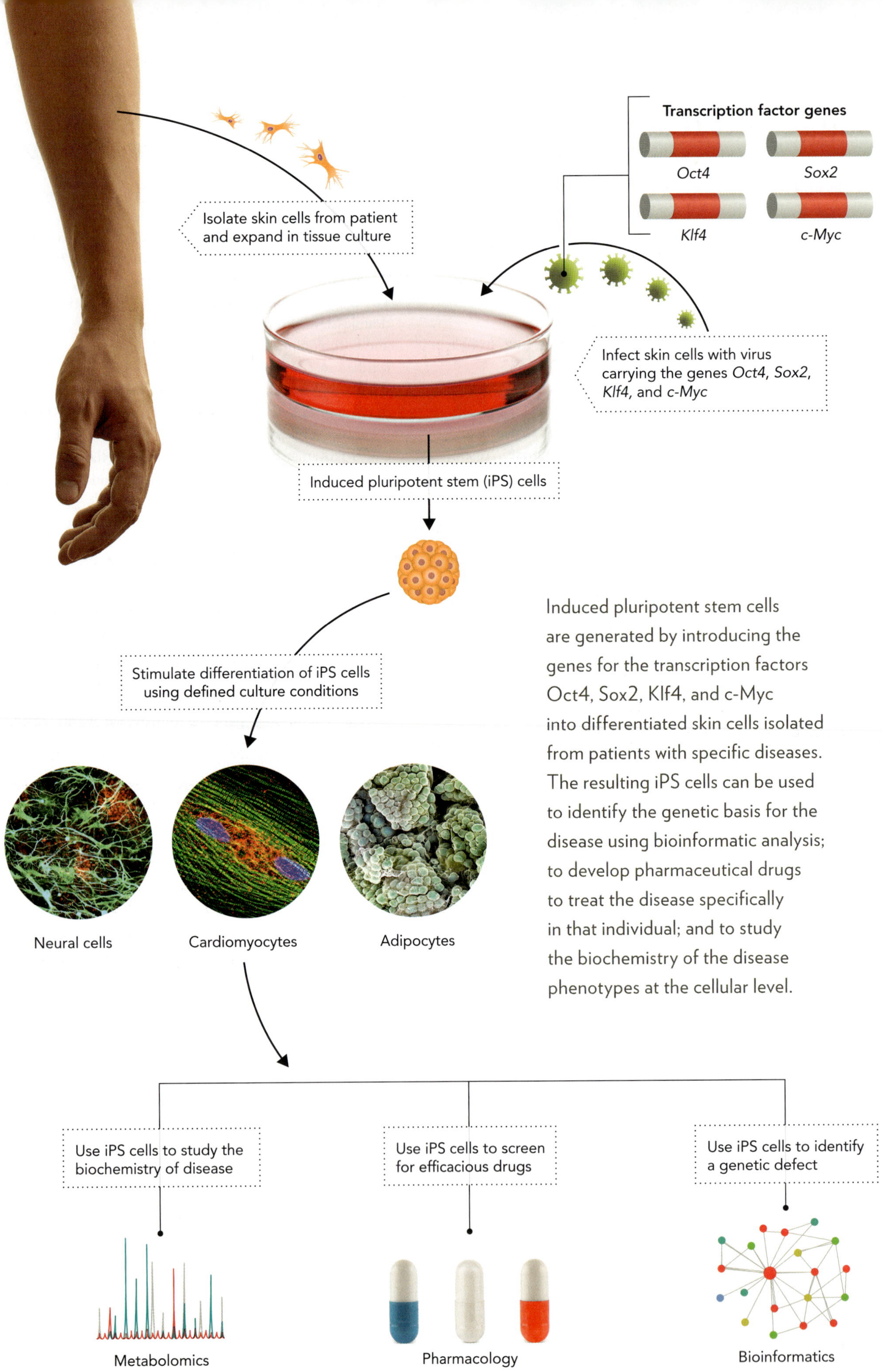

Induced pluripotent stem cells are generated by introducing the genes for the transcription factors Oct4, Sox2, Klf4, and c-Myc into differentiated skin cells isolated from patients with specific diseases. The resulting iPS cells can be used to identify the genetic basis for the disease using bioinformatic analysis; to develop pharmaceutical drugs to treat the disease specifically in that individual; and to study the biochemistry of the disease phenotypes at the cellular level.

23

Gene Regulation

Gene regulation is controlled by proteins called transcription factors, which determine the differentiated state of cells. Pluripotent stem cells such as embryonic stem cells have the potential to differentiate into any cell type, depending on external stimuli. Gene regulation studies in stem cells led to the discovery that a small number of specific transcription factors can revert differentiated cells into pluripotent stem cells. Using gene transfer techniques, methods were developed to generate induced pluripotent stem (iPS) cells that can be stimulated to differentiate into specific cell types and used for a variety of biochemical studies.

CHAPTER OUTLINE

CREDITS: ARM: DREAMBIG/SHUTTERSTOCK; PETRI DISH: PHOTOEVENT/ISTOCK/GETTY IMAGES PLUS; NEURAL CELLS: SILVIA RICCARDI/SCIENCE SOURCE; CARDIOMYOCYTES: ROGER J. BICK & BRIAN J. POINDEXTER/ UT-HOUSTON MEDICAL SCHOOL/SCIENCE SOURCE; ADIPOCYTES: STEVE GSCHMEISSNER/SCIENCE SOURCE; THREE PILLS: PIOTR_PABIJAN/SHUTTERSTOCK.

The human body contains several hundred different cell types, but most cell types contain the same complement of genes. Cell types exist largely because each type has its own pattern of gene expression. Red blood cells make hemoglobin; muscle cells make myosin, actin, and other components of muscle; certain immune cells make antibodies; pancreatic islet cells make digestive enzymes; and rod cells in the eye make rhodopsin. In each case, the particular proteins made by that cell type largely determine its function in the body.

How are the genes for these proteins selectively expressed? The answer is by highly regulated biochemical processes involving transcription factor proteins that function as activators or repressors of gene expression through direct interactions with DNA or with DNA binding proteins. To understand these processes, we first discuss several general principles of gene regulation. We then examine gene regulation in prokaryotes, where the mechanisms are relatively simple and well understood. Finally, we turn to gene regulation in eukaryotes, in which additional levels of control are needed because the DNA is packaged into protein–DNA complexes called nucleosomes. The structural arrangement of nucleosomes in eukaryotic chromatin affects the ability of transcription factors to bind DNA and regulate transcription.

Before beginning our discussion of gene regulatory mechanisms in prokaryotic and eukaryotic organisms, it is worth pointing out that biochemists typically dissect complex processes by analyzing the components and their interactions *in vitro*, using either crude extracts or, ideally, purified components. In the study of eukaryotic gene regulation, this task is difficult, in part because of the very large number of interacting components. Hence, much of the progress in this area has come from a complementary approach; namely, the use of genetics in several well-characterized model organisms.

A small group of organisms have received intensive study over the years, perhaps less for their intrinsic interest than for their ease of use in experiments. A major reason we still use model organisms is the great wealth of knowledge from past research, so that new information can be integrated with known results. Examples of model organisms are the bacterium *Escherichia coli*, the single-cell budding yeast *Saccharomyces cerevisiae*, and the multicellular fruit fly *Drosophila melanogaster*. These and other models have been chosen for a range of properties. They are easy to grow and maintain in the lab; most of them allow biochemical analysis of components; and genetic analysis is easy and well developed in most model organisms.

Using biochemistry and genetics together is a very powerful combination. Biochemistry often takes a divide-and-conquer approach, studying the properties of each component separately. Genetics has a much more "synthetic" approach because all the components of the cell are present, and changes in components by mutation can affect multiple processes. Thus, each method provides a different type of information. In addition, a common way of combining these approaches is to perform biochemical analysis of mutant components.

Model organisms are also useful because many processes are universal, or nearly so, across the entire range of biology. All cells have DNA as their genetic material, and the processes of DNA replication, transcription, and translation are chemically the same and occur in similar ways. Among eukaryotes, many processes in humans have their counterparts in yeast and flies. Our awareness of many processes in humans came about from studies in model organisms, and our understanding of some human diseases relied initially on such studies.

23.1 Principles of Gene Regulation

The overall process of gene expression involves a number of control points. As shown in **Figure 23.1**, eukaryotic DNA must first be made accessible to the RNA synthesizing machinery through a process known as chromatin remodeling. In both prokaryotes and eukaryotes, initiation of RNA synthesis is a key regulatory step, as is RNA degradation. Eukaryotic mRNA, but not prokaryotic mRNA, is processed (splicing, capping, and polyadenylation), whereas primary transcripts of rRNA and tRNA in prokaryotes and eukaryotes both require processing. In all organisms, mRNA transcripts are translated into protein, and the efficiency of this process can also be regulated. As described in Chapter 22, protein products of mRNA translation are often posttranslationally modified in a regulated fashion, and protein degradation in prokaryotes and eukaryotes is also a regulated process. In eukaryotes, proteins can be targeted to organelles in a regulated manner; for instance, proteins can be targeted to the nucleus by a nuclear localization signal, which can be masked or unmasked by a regulatory factor. Changes in any of these steps can affect how much of the functional protein is present in cells.

The primary control point for most gene regulation occurs at the initiation of transcription, and therefore our main focus in this chapter will be on regulation of this step. But before we describe specific examples of prokaryotic and eukaryotic gene regulation, we need to discuss several general principles governing gene regulatory processes.

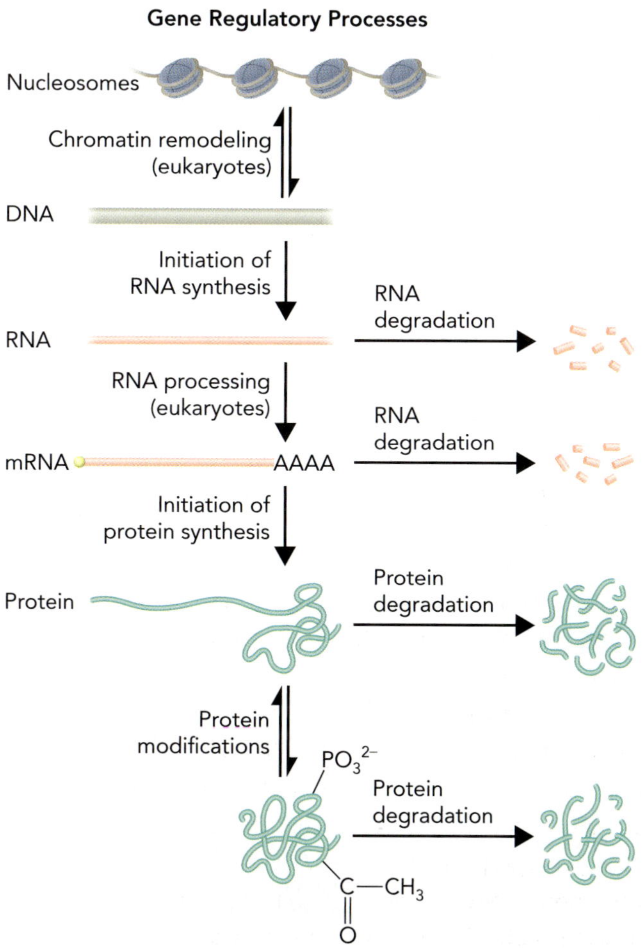

Figure 23.1 There are discrete control points in the expression of protein-coding genes that generate mRNA, some of which are specific for eukaryotes.

Specificity of Gene Regulation

Gene regulation involves a great deal of specificity. Regulatory systems generally operate on a small set of genes, not on the entire complement of genes in the cell. Accordingly, the cell needs mechanisms that operate specifically on the genes to be regulated. For simplicity, we will discuss regulatory proteins that bind to DNA.

Specificity comes about by the interaction of transcription factor proteins, often called **trans-acting factors**, which bind to particular regulatory sites on DNA termed **cis-acting sites** (**Figure 23.2**). These interactions can be highly specific; for example, the binding of the trans-acting σ factor to cis-acting prokaryotic promoter sequences or the binding of the trans-acting TATA binding protein (TBP) to cis-acting eukaryotic promoter sequences (see Chapter 21). Trans-acting factors can bind to specific DNA sequences located throughout the genome, whereas cis-acting sites affect only the DNA element to which they are physically linked. Note that while most trans-acting factors are proteins and most cis-acting sites are DNA sequences, some trans-acting factors are RNA molecules and some cis-acting sites are located on RNA.

Binding of specific DNA binding proteins to their cognate recognition sites involves two types of noncovalent interaction. One is a nonspecific component, in which the phosphate backbone of the DNA interacts with the protein, and thus is general to almost any DNA sequence. These nonspecific interactions are often electrostatic or, in some cases, hydrogen-bonding interactions. The other interaction type is specific, typically between the base pairs and amino acid side chains, allowing specific DNA sequences to be recognized by the protein. DNA binding proteins often have α helices that will bind in the major groove of DNA, and the side chains that project from these helices can interact with the DNA. The binding affinity of transcription factor proteins to DNA at specific sequences is 10^4- to 10^6-fold higher than at nonspecific sequences.

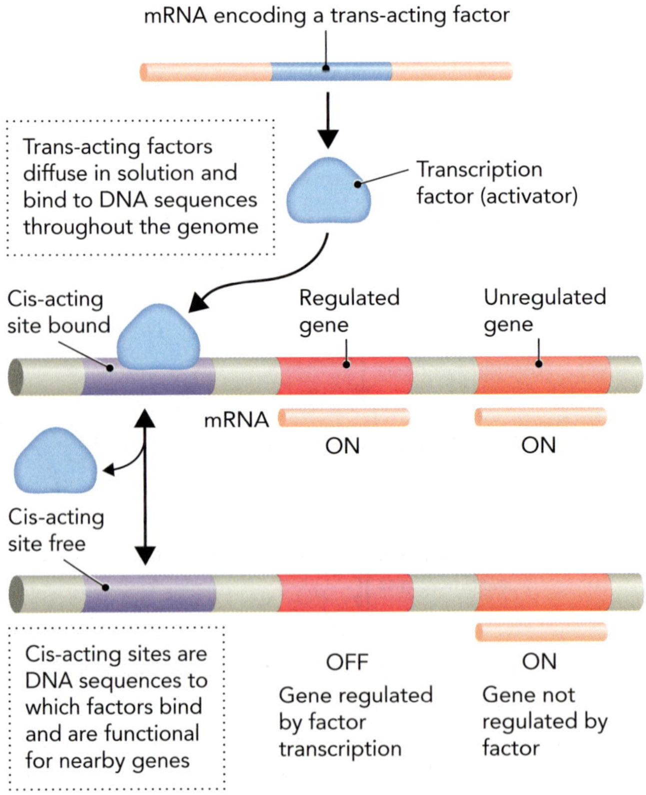

Figure 23.2 Cis-acting DNA sequences and trans-acting transcription factor proteins are the key components of gene regulatory processes. When the trans-acting transcription factor is made, it can diffuse in solution and interact with any DNA site that is accessible. This site is a cis-acting site, meaning that it only operates on the regulated gene to which it is physically linked. Another gene located nearby is not regulated by the presence of the factor.

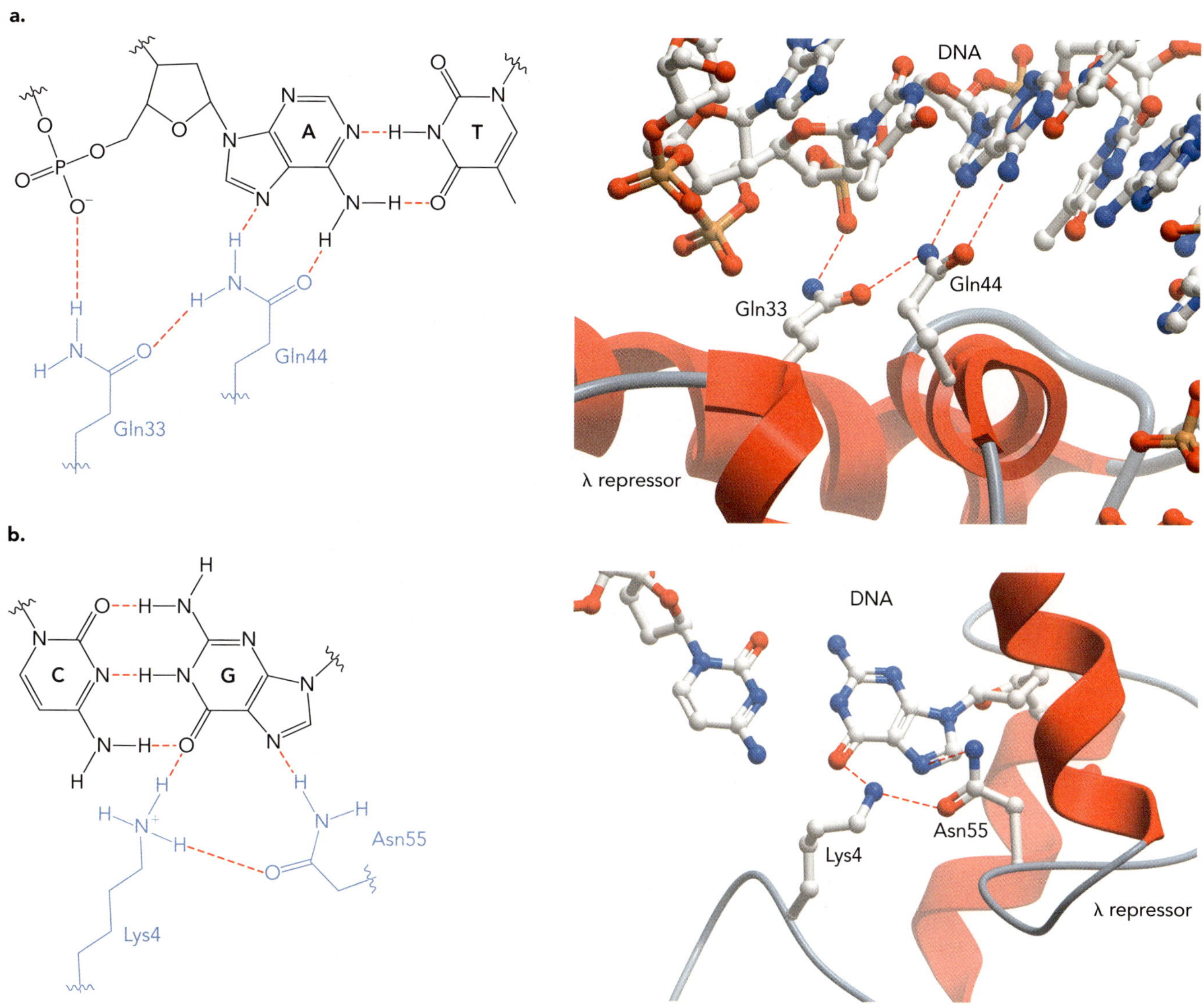

Figure 23.3 Specific binding interactions between the λ repressor protein and target DNA are shown. **a.** Gln33 and Gln44 make specific contacts with DNA in the region of an A-T base pair in the DNA. The amino acid Gln33 functions as a hydrogen-bond donor to a phosphate group in DNA, whereas Gln44 functions as both a hydrogen-bond acceptor and a hydrogen-bond donor with the same purine base. Gln33 and Gln44 also hydrogen bond with each other. **b.** Lys4 and Asn55 residues in the λ repressor protein function as hydrogen-bond donors to a guanine base in a C-G base pair and hydrogen bond with each other. (In panel **b**, only a single base pair in the ball-and-stick structure is shown for clarity.) BASED ON PDB FILE 1LMB.

The most common types of reversible binding interaction between proteins and DNA at specific sites are hydrogen bonds, with donors and acceptors coming from the nucleotide bases and amino acid side chains, and van der Waals interactions between hydrophobic groups. The amino acids Arg, Lys, Gln, and Asn are commonly found to make specific contact with the nucleotide bases. Water molecules can also play a role in mediating specific protein–DNA interactions, but more often water serves as a solvent at the binding interface. Intrinsic sequence-specific interactions between amino acids and nucleotide bases contribute to the specificity of binding, as does conformational variability, which enables DNA and the protein to have an overall shape complementarity. **Figure 23.3** illustrates two specific interactions between the λ repressor protein and its target DNA sequences that involve hydrogen bonding. In this case, Gln44

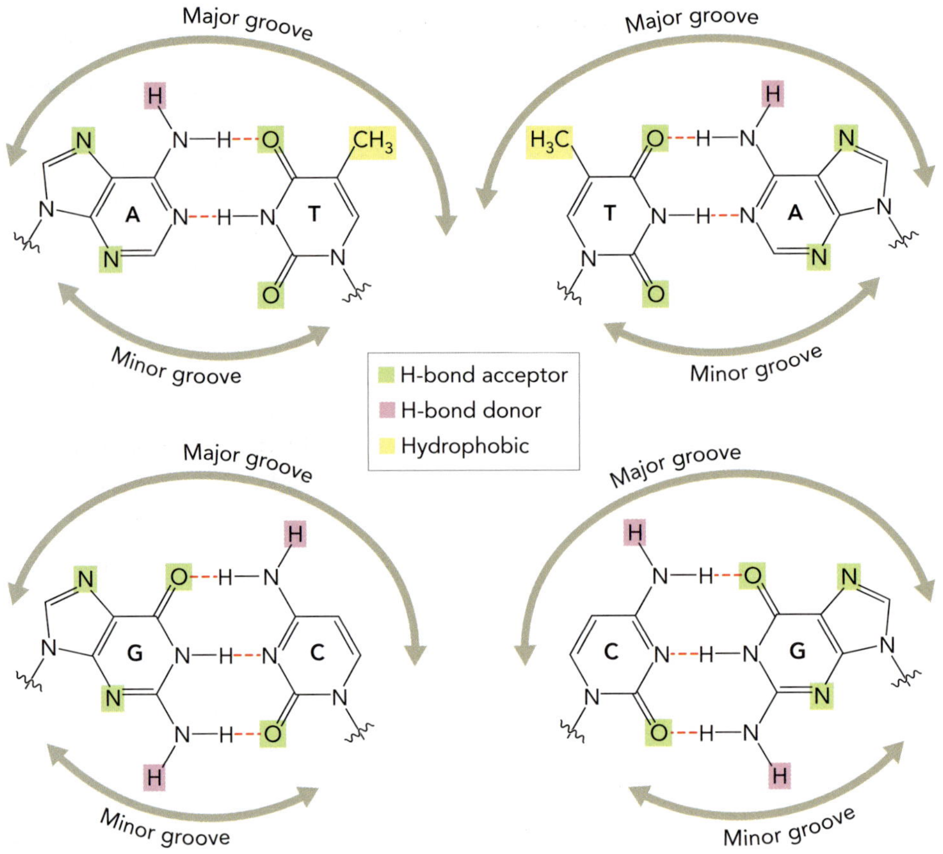

Figure 23.4 There are a variety of noncovalent protein interaction sites with base pairs in the major and minor grooves of the DNA double helix. Shown are the four base pairs, with the locations of hydrogen-bond donors and acceptors and hydrophobic methyl groups indicated.

forms hydrogen bonds with the adenine base of an A-T base pair and with a carbonyl oxygen on Gln33. The Gln33 residue also forms a hydrogen bond with the phosphate group of the same adenine base. Lys4 and Asn55 form hydrogen bonds with the guanine base of a C-G base pair and with each other. In both cases, the protein side chains contact the DNA bases through the major groove of the DNA.

The structural arrangements of the base pairs present distinctive patterns of hydrogen-bond donors and acceptors to the solvent. As shown in **Figure 23.4**, all four base pairs (considering A-T as different from T-A, and C-G from G-C) present distinctive patterns of potential hydrogen-bond donors and acceptors and hydrophobic groups in the major groove of DNA. Hence, each base pair provides a unique binding interface for a protein. For example, the hydrogen-bonding pattern for the A-T base pair in the major groove from the perspective of the DNA binding protein is acceptor–donor–acceptor–CH_3, whereas the pattern for a T-A base pair in the same orientation is CH_3–acceptor–donor–acceptor. Similarly for a G-C base pair the hydrogen-bonding pattern in the major groove is acceptor–acceptor–donor, but the pattern for a C-G base pair in the same orientation is donor–acceptor–acceptor. This is not the case in the minor groove of DNA, where the patterns of hydrogen-bond donors and acceptors are indistinguishable between A-T and T-A base pairs (acceptor–acceptor) and between G-C and C-G base pairs (acceptor–donor–acceptor).

Figure 23.5 illustrates this principle using a region of DNA that contains an A-T base pair adjacent to a T-A base pair. In this example, the pattern of hydrogen-bond donors, hydrogen-bond acceptors, and hydrophobic groups in the major groove is reversed for A-T (blue) and T-A (red) relative to the phosphate backbone. But when this same molecule of DNA is rotated 180° to show the minor groove on the opposite face, the

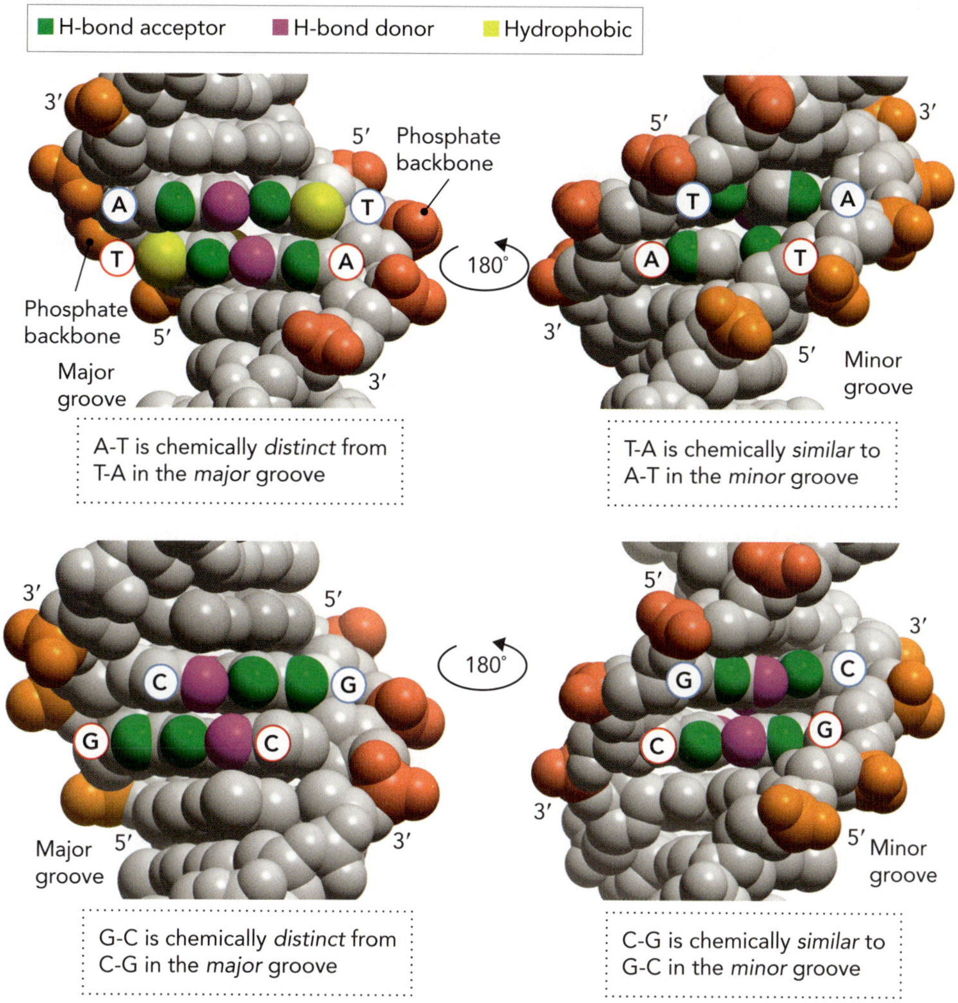

Figure 23.5 The chemical patterns of potential protein interaction sites for different base pair combinations are distinct in the major groove of DNA. However in the minor groove, the pattern of A-T base pairs is similar to that of T-A base pairs, and the pattern of C-G base pairs is similar to that of G-C base pairs. BASED ON PDB FILE 1GLU.

pattern of potential protein interaction sites is the same for A-T and T-A base pairs. The same holds true when comparing the potential protein interaction sites for a C-G base pair and a G-C base pair in the major groove and minor groove of the DNA double helix.

Specific recognition of DNA by a protein often includes direct contact with the nucleotide bases, usually through the DNA major groove. An α helix of a protein has an appropriate diameter to fit in the major groove of DNA, and thus α helices are a common feature in DNA binding proteins for making base-specific contacts. These α helices are usually found as part of a protein domain of less than 100 amino acid residues, which helps to position the α helix correctly within the major groove. Transcription factor proteins and their DNA binding domains are most often structural components of larger proteins: A transcription factor protein binds to DNA, and parts of the same protein form protein–protein interactions with other proteins in the transcription complex, such as RNA polymerase. A variety of specific cis-acting DNA binding sites are short-direct or inverted-repeat DNA sequences, which means that the corresponding proteins that bind to these sequences need to form dimeric protein complexes.

Figure 23.6 shows one of the most common DNA binding motifs found in transcription factor proteins: the **helix–turn–helix motif.** This protein motif is only ~20 amino acids in length and consists of two α helices connected by a short turn. In this example, the helix–turn–helix domain of the λ repressor protein is bound to its cognate DNA sequence in the bacteriophage λ genome. The second helix of the motif (called the recognition helix) is particularly important for making direct contacts with the bases

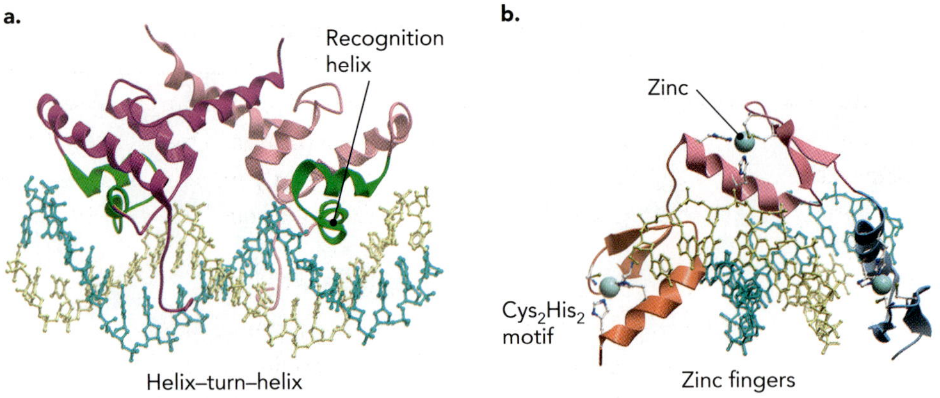

Figure 23.6 Two of the most common DNA binding motifs found in transcription factor proteins are the helix–turn–helix motif and the zinc finger motif. **a.** The DNA binding domain of the dimeric λ repressor protein contains a helix–turn–helix motif (green). The recognition α helix of the motif binds in the major groove of the DNA. BASED ON PDB FILE 3BDN. **b.** The Zif268 transcription factor contains three zinc finger motifs in tandem. The two cysteine and two histidine residues (Cys$_2$His$_2$ motif) are shown in each domain coordinated to a zinc ion. The α helix of the zinc finger motif binds in the major groove of the DNA helix. BASED ON PDB FILE 1ZAA.

in the major groove of the DNA helix (see Figure 23.3a, where Gln44 is part of the recognition helix). As we describe later in this chapter, both the *E. coli* lac repressor protein and the Trp repressor protein bind to specific DNA sequences using helix–turn–helix motifs. Figure 23.6 also shows the molecular structure of another common DNA binding motif found in transcription factor proteins: the zinc finger motif. There are few examples of zinc finger–containing proteins in prokaryotes; most transcription factors with the zinc finger DNA-binding motif are found in eukaryotic cells. The canonical zinc finger motif is ~30 amino acids in length and coordinates one zinc atom with two cysteine and two histidine residues (Cys$_2$His$_2$; see Figure 4.53). In the example shown here, the Zif268 transcription factor contains three zinc finger motifs that each coordinate a zinc atom using two cysteine and two histidine residues. The α helix of each zinc finger motif is situated in the major groove of the DNA.

Two other protein domain motifs found in eukaryotic transcription factor proteins are shown in **Figure 23.7**. One is called the **leucine zipper motif**, and the other is called the **helix–loop–helix motif**. Unlike the helix–turn–helix and zinc finger DNA-binding domain motifs described earlier, the leucine zipper and helix–loop–helix motifs refer to a protein–protein interface found in transcription factors containing these motifs. It can be seen that both motifs insert α helices within the major groove of DNA. The DNA

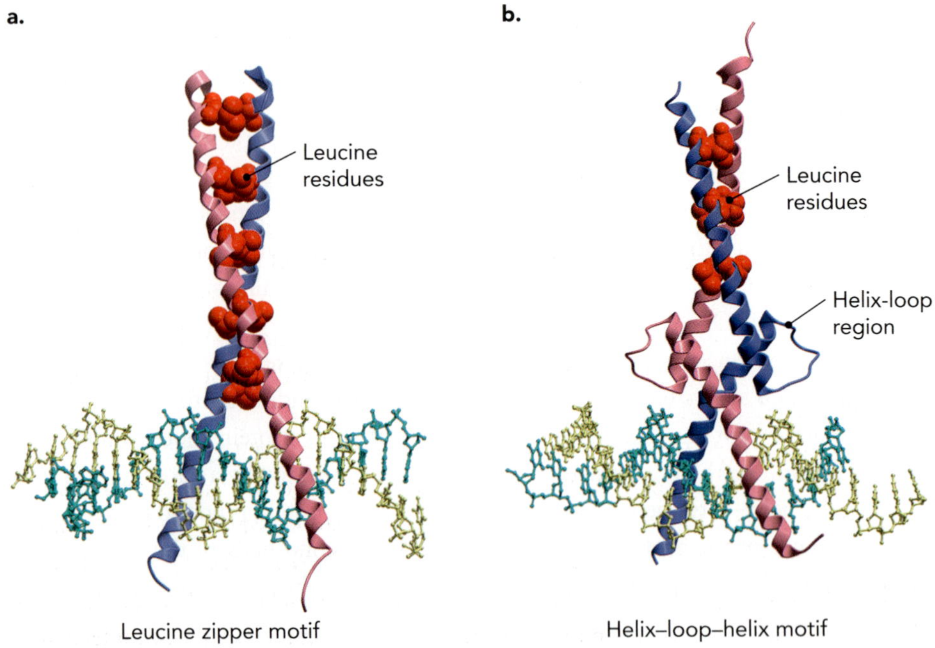

Figure 23.7 Dimeric transcription factors often contain one of two types of protein–protein interaction motifs, which are the leucine zipper or the helix–loop–helix motifs. **a.** The leucine zipper motif in the yeast GCN4 transcription factor contains leucines spaced seven residues apart to promote the formation of a hydrophobic core when the α helices associate in a coiled coil structure. BASED ON PDB FILE 1YSA. **b.** The helix–loop–helix motif of the SREBP-1A transcription factor also contains regularly spaced leucine residues at the dimer interface and is considered a basic helix–loop–helix leucine zipper transcription factor. BASED ON PDB FILE 1AM9.

binding regions of the α helices in these motifs contain many basic amino acid residues (Arg and Lys), which are required for specific DNA binding interactions. In the case of leucine zipper proteins, a leucine residue is found about every seventh position in the region of the protein required for dimerization. This spacing of the leucine residues facilitates direct interactions between similarly spaced leucine residues in the other protein subunit, resulting in the formation of a coiled coil protein structure. Helix–loop–helix proteins are characterized by a loop region between the DNA binding α helix and the dimerization α helix. Some helix–loop–helix proteins, like the one shown in Figure 23.7, contain leucine residues to promote dimerization at the protein interface and, because of the basic amino acid residues in the DNA binding domain, are called basic helix–loop–helix leucine zipper transcription factors.

The forces favoring specific protein–DNA interactions are complicated and not well understood. Much, if not most, of the energy involved is actually derived from the many nonspecific protein–DNA interactions. Another consideration is that entropy is often increased by the release of bound water molecules and cations when a protein–DNA complex forms. Importantly, the hydrogen bonds in the protein–DNA interface do not contribute greatly to the affinity of binding, but rather to the specificity of binding. This is because hydrogen bonds with water must be broken for these bonds to form. Note that a DNA base change within a cis-acting DNA sequence could reduce the affinity of protein binding to this sequence by up to 100-fold.

Before a specific protein–DNA complex is formed, the protein binds nonspecifically to DNA, mostly using interactions with the backbone phosphates, and slides up and down the DNA until a specific binding site is encountered. In this nonspecific binding mode, binding is relatively weak, in part because the protein is only loosely associated with the DNA. When a specific binding site is reached, the protein assumes a closer contact with the DNA and forms a specific complex.

For a binding site to be unique in the genome, or nearly so, it must be fairly large. Consider, for instance, the frequency of a 4-bp sequence in a genome of random sequence in which each of the four positions on one strand could be a G, A, T, or C nucleotide base. The frequency with which any one 4-bp sequence would occur is once every 256 base pairs. Therefore, a protein that recognizes a 4-bp sequence could bind to many such sites within a genome, given that most genomes contain 10^6 to 10^9 base pairs (see Figure 3.31). For a binding site to be unique in a genome the size of that of *E. coli* (4.7×10^6 bp), the binding site should contain about 12 specific positions. These positions need not be contiguous; they could be spaced out over a longer region, as is often observed. From this perspective then, it is useful to have a large number of interactions in the complex.

Many gene regulatory proteins bind to multiple sites on the genome. Typically, the affinity of the protein for the DNA differs somewhat from one site to another. Therefore, these slight sequence differences can have regulatory consequences. For instance, if a protein has weak affinity for a DNA site, it will only bind to the DNA when there are high levels of the regulatory protein. But other sites, for which the protein has high affinity, may be occupied even at low levels of the protein. Ultimately this means for the same protein at a given concentration, some DNA sites will be bound by the protein, while other sites are free.

An additional level of control over gene expression arises from control over the assembly of the regulatory proteins. As we will describe later, eukaryotic transcription involves several large, multi-subunit complexes that associate in various ways with DNA binding proteins and with other complexes. These complexes are held together by a large number of relatively weak interactions. Each individual interaction would not suffice to hold a complex together, but in combination these can result in strong and highly specific interactions.

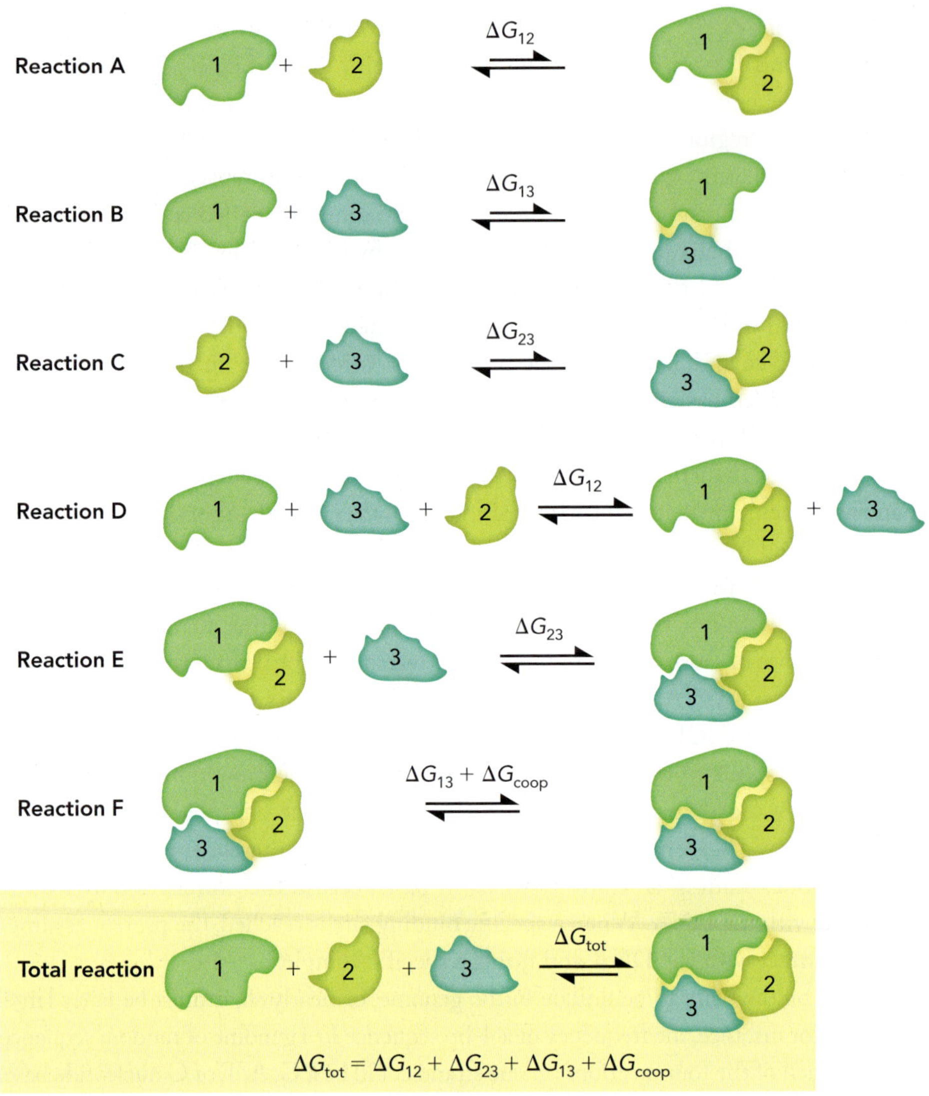

Figure 23.8 Weak interactions and cooperative DNA binding can be modeled using three protein components, each of which can interact weakly with the other two. When the third component binds to a binary complex, the same interactions form, but only small amounts of entropy are lost when this complex forms, so it is strongly favored. Note that the overall assembly process (reaction total) must have the same free energy change irrespective of the assembly pathway.

Consider the energetics of assembly of the heterotrimeric protein complex illustrated in **Figure 23.8**. Each protein can form a binary complex with either of the other two (reactions A–C), but each individual binary interaction is weak. In other words, the free energy difference between the free and bound states of each binary complex is small, and at equilibrium there is little dimer formation. One factor contributing to this free energy difference is the entropic cost of localizing the bound subunits, which can diffuse freely relative to the third subunit that is not bound to a subunit.

Now consider the assembly reactions D, E, and F in Figure 23.8, where protein subunits 1 and 2 form a binary complex (reaction D), which is then joined by subunit 3 (reaction E). Addition of this third component involves formation of interactions with both subunits 2 (reaction E) and 1 (reaction F), both of which contribute to stabilizing the complex. But in addition, once one of the two interactions (for example, between subunits 2 and 3) forms, formation of the remaining interaction (for example, between subunits 1 and 2) is far more favored because subunits 1 and 3 are juxtaposed. Therefore, most of the entropic cost of localizing subunit 3 was paid upon formation of an interaction between subunits 2 and 3. This model proposes that the interaction between subunits 1 and 3 within the complex (reaction F) is favored relative to their interaction in isolation (reaction B). As would be expected, the total change in free energy for the assembly of the trimeric complex is the same regardless of the order of assembly.

Overall, the tendency toward heterotrimerization, as governed by the total change in free energy, is much greater than the tendency toward formation of any of the three possible dimers. This is an example of **cooperativity**, or cooperative binding, and is often termed the chelate effect. Note that cooperative DNA binding can be visualized in this same example by considering component 2 as the DNA. In this case, reactions A and C show proteins 1 and 3 binding to their cognate site on DNA in the absence of the other protein. When both proteins and the DNA are present, the interaction between them allows the occupancy of one site to influence the occupancy of the other, and in practice may affect affinity by 100- to 1,000-fold.

A special case of this interaction occurs with DNA binding proteins, in which two molecules bind to adjacent sites. In cooperative DNA binding, binding of one molecule is more favored when another one is bound to a nearby site on the DNA. The two molecules can be the same factor or two different factors. In either case, a weak interaction between two proteins can lead to much tighter binding to DNA than is observed with each protein alone. It is likely that formation of such interfaces between proteins could arise relatively easily during evolution, as weak interactions usually involve only a small number of residues. Another consequence of cooperativity, often seen in eukaryotes, is that the DNA binding sites for each component can be smaller than would generally be seen for a single protein binding by itself; effectively, the binding site is composed of the two binding sites together.

Gene regulatory proteins usually have a modular organization; that is, they consist of multiple domains within a single polypeptide that act as autonomous structural and functional units. For example, in eukaryotic activator proteins, a separate domain from the DNA binding domain often carries out the activator function. Frequently, these domains can carry out their functions when artificially separated from the rest of the protein. For instance, the DNA binding domain often can bind to DNA independent of the rest of the protein. The X-ray crystal structures of DNA–protein complexes are commonly determined with separated domains because the intact protein may be so flexible that it will not form crystals or the crystals containing larger proteins give very poor resolution.

Basic Mechanisms of Gene Regulation

Single-celled organisms, such as prokaryotes and simple eukaryotes such as yeast, must be able to respond to changes in the environment (for example, sources of nutrition) that are often unpredictable. Therefore, the mechanisms controlling gene expression are usually reversible to cope with these environmental changes. In multicellular organisms, such as humans, the changes more often refer to the patterns of gene expression in different cell types. Thus, the regulatory decisions favoring expression of some genes and repression of others are often irreversible. For instance, the globin genes transcribed in red blood cells are permanently turned off in highly condensed chromatin in other cell types, usually in response to cell signaling—we previously discussed the example of steroid hormone. Later in this chapter, we discuss the biochemical process of embryonic development in the model organism *Drosophila*, in which entire programs of gene expression are contingent upon the sequential expression of transcription factors.

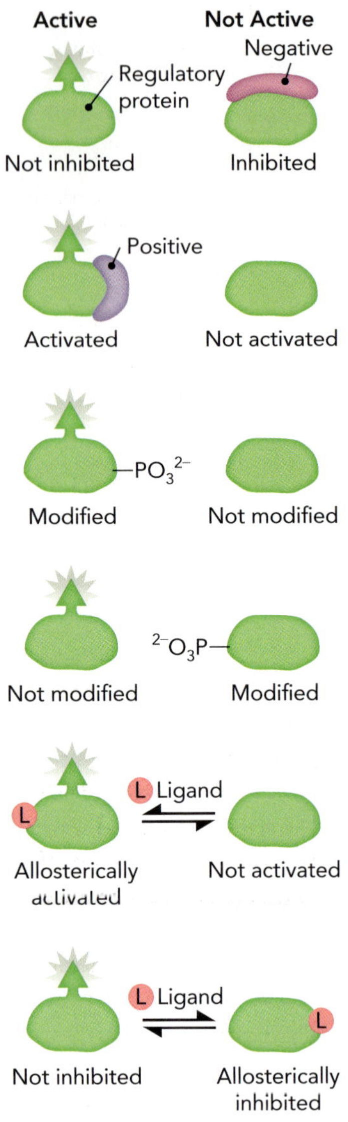

Figure 23.9 Several common mechanisms for modifying the activity of transcription factors are shown. Transcription factors can be regulated by the binding of proteins or ligands (L) that negatively or positively modulate their activity. Posttranslational regulation such as phosphorylation and allosteric control are often initiated by extracellular signals.

The cell uses several types of mechanisms for controlling the activity of regulatory proteins such as transcription factors. As shown in **Figure 23.9**, the steady-state level, or subcellular localization, of a transcription factor protein is subject to regulation. Moreover, the activity of a transcription factor can be affected by the presence or absence of protein binding partners, which can be either negative or positive modulators. Transcription factor activity can be regulated by a range of covalent and noncovalent posttranslational mechanisms; for example, covalent modification of proteins by phosphorylation of serine, threonine, and tyrosine amino acid residues or noncovalent modification by the binding of allosteric regulatory molecules functioning as ligands. Certainly, modulation of the activity of preexisting transcription factor proteins is much faster than regulation of the expression of the transcription factor gene itself, which is a theme that recurs throughout this chapter.

Controlling the synthesis or stability of a regulatory factor is more common in eukaryotes than in prokaryotes, although examples of this strategy are seen in prokaryotes as well. During the process of development in higher eukaryotes, numerous cell types are created, each with its own pattern of gene expression, and much of this is directed by sequential expression of transcription factors. In prokaryotes, posttranslational modulation of activity is more common. Several different mechanisms are known that modulate DNA binding. Many cases involve allosteric transitions between different conformations of the regulatory protein that can affect DNA binding affinity. In this mechanism, a ligand molecule, typically one that is related to the function of the genes being regulated, binds to the transcriptional regulatory protein. Binding of the ligand will cause conformational changes in the transcription factor that either decrease or increase the protein's affinity for DNA. In other cases, covalent modifications of the transcription factor can occur, such as phosphorylation or regulated degradation (see Figure 23.9).

In eukaryotes, control of transcription factor activity is most often by covalent modifications, with allosteric control of nuclear receptor proteins by ligand binding being the exception rather than the rule (see Figure 8.64). The covalent modifications of eukaryotic transcription factors often occur as a late event in a long signal-transduction pathway. In some cases, covalent protein modification results in dissociation of an inhibitory molecule, which leads to activation through a process known as derepression. Modification might also directly change the nuclear localization of the regulatory molecule.

Positive and Negative Gene Regulation We can distinguish two basic types of regulation, called positive control and negative control (**Figure 23.10**). For now, we restrict our discussion to prokaryotes, in which the mechanisms are relatively simple, in contrast to eukaryotes, where the presence of chromatin greatly complicates the regulatory process by introducing additional levels of control.

As shown in Figure 23.10a and b, access to prokaryotic promoters by RNA polymerase is blocked by the presence of a regulatory protein, termed a repressor (a trans-acting factor), which binds to a cis-acting sequence on DNA called an operator that is located at or very near the promoter. As noted earlier, for this to be considered regulation, some mechanism must exist for modulating the affinity of the repressor for the operator. Allosteric transitions are a common mechanism, especially in regulation of metabolic pathways. These can either favor binding to DNA or prevent it. In the case of ligand-regulated repression, the affinity of the repressor for the operator sequence in DNA is increased by the binding of a ligand (L) to the repressor (Figure 23.10a).

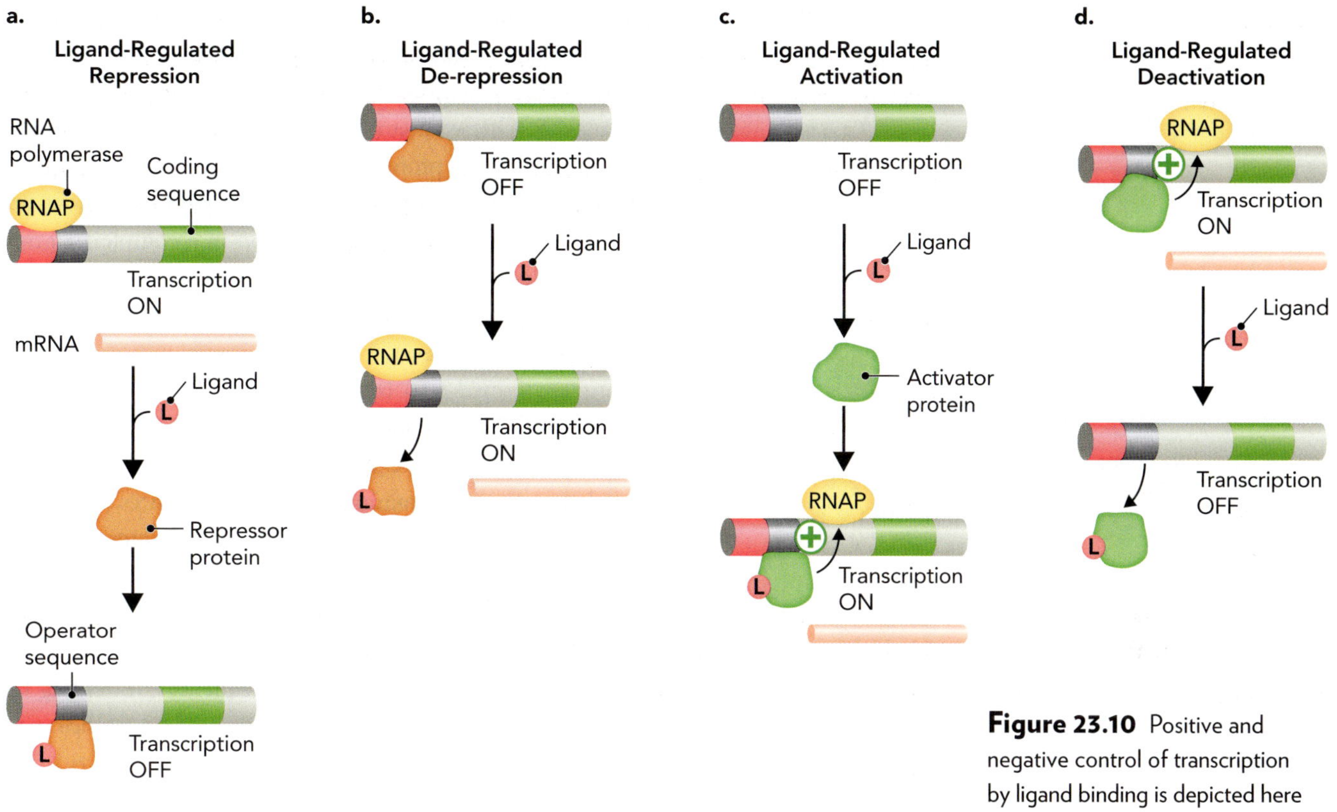

Figure 23.10 Positive and negative control of transcription by ligand binding is depicted here for prokaryotes. In negative control, expression of a gene is turned off by either (**a**) ligand-induced binding of a repressor protein to DNA or (**d**) ligand-induced dissociation of activator binding to DNA. In positive control, expression is turned on by either (**c**) ligand-induced binding of an activator protein or (**b**) ligand-induced dissociation of repressor binding to DNA. In these examples, the regulatory protein is present at all times, though its function is altered by its interaction with a small-molecule ligand (**L**). Interaction of the regulatory protein with **L** causes an allosteric change in the regulatory protein, denoted by a change of shape in the protein. This change affects the ability of the protein to bind to DNA. RNAP = RNA polymerase.

In an example of positive gene regulation as a result of ligand-regulated de-repression (Figure 23.10b), ligand binding to the repressor protein decreases the repressor's affinity for DNA, which leads to RNA polymerase binding and gene transcription. The DNA binding affinity of activator proteins can also be regulated by ligand binding. In this case, ligand-regulated activation occurs when binding of the activator protein to the operator sequence is increased by ligand binding, which in turn recruits RNA polymerase to the promoter through a protein–protein interaction (Figure 23.10c). Lastly, if ligand binding to the activator protein decreases the activator protein's affinity for the operator sequence on DNA, then ligand-regulated deactivation occurs as a result of dissociation of the activator protein from DNA and loss of RNA polymerase recruitment (Figure 23.10d).

Positive control occurs in cases where the gene promoters being regulated have low rates of transcriptional initiation in the absence of activator proteins. Such promoters are often called "weak" promoters, which are characterized biochemically as promoter sequences in which RNA polymerase either cannot bind tightly to the promoter or cannot readily undergo subsequent steps in the initiation process (conversion from the closed complex to the open complex; see Figure 21.18). Binding of an activator protein to a cis-acting DNA sequence facilitates one or more steps in initiation. Activator proteins increase the affinity of RNA polymerase for the promoter, or increase the rate of open complex formation, or both. Again, some mechanism must be present to modulate the activity of the activator protein itself, such as binding of a small-molecule ligand to facilitate or prevent its DNA binding. In other cases, synthesis of an activator protein depends on the particular conditions, so that when the activator is made, the target gene is expressed, whereas if the activator is not made, the target gene is not expressed. In such cases, the activator itself is often unstable and readily degraded, so that its continued action depends on continuing synthesis.

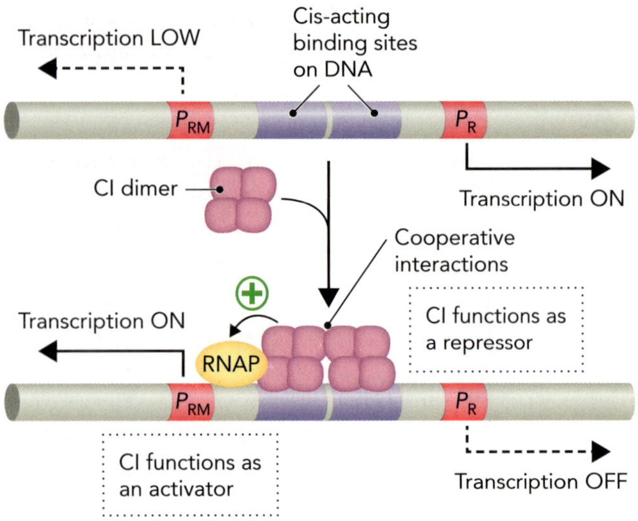

Figure 23.11 The function of a DNA binding protein depends on the location of its binding site, as illustrated by the properties of the bacteriophage λ regulatory protein CI. This region contains two promoters, P_{RM} and P_R.

The function of regulatory proteins often depends on the regulatory context in which they act. A given protein might serve as a repressor in one context and an activator in another. One well-studied case is the CI protein of bacteriophage λ. This protein is often called λ repressor protein, because it serves to repress viral genes. **Figure 23.11** shows an example of λ repressor protein action in a complex control region called the O_R region. When CI binds to two adjacent sites, it represses expression from a promoter termed P_R. At the same time, however, it activates expression from another promoter, termed P_{RM}, by making a close contact with the σ subunit of RNA polymerase. As illustrated in Figure 23.11, CI binds as a dimer to two DNA sequence elements in a regulatory region containing many cis-acting sites. When CI is absent, expression occurs from the P_R promoter; P_{RM} is a very weak promoter. When CI binds to the two cis-acting sites, it works both as an activator, stimulating expression from the P_{RM} promoter about 10-fold, and as a repressor, blocking RNA polymerase from binding to P_R. Hence, the context of a cis-acting site is important in determining the consequences of binding by a trans-acting factor; simply identifying the site by analyzing the DNA sequence may not suffice to indicate the function. In addition, the function of a protein at a site or its affinity can be modified by the presence of a second protein located nearby.

Negative and positive regulation are also used in eukaryotes, although as we will see, the mechanisms are generally different. Often, regulation is achieved by establishing open or closed states of chromatin. Even when the chromatin is relatively accessible, moving particular nucleosomes is often an important part of turning a gene on. For now, the important point is that positive and negative control play crucial roles in eukaryotic regulation as well as in prokaryotic regulation.

Gene Regulatory Circuits Cells use positive and negative control in a variety of ways. Controls can be connected in cells in simple or elaborate ways to build up networks of what are called regulatory circuits.

Perhaps the simplest type of circuit involves autoregulation, in which a regulatory protein acts to modify its own expression. This creates feedback, in which the output of the system is used as input at a later time. In *negative autoregulation* (**Figure 23.12**), a protein represses its own expression, giving rise to negative feedback, which acts to modulate fluctuations in the level of the regulator. For example, when repressor protein levels are high at the onset, negative autoregulation decreases the repressor protein's own expression until it reaches steady-state levels. In contrast, when repressor protein levels are low at the onset, negative autoregulation does not occur until repressor protein levels are high enough to affect transcription rates. A familiar example of negative feedback is a thermostat, in which the temperature of the room is used as input to control future output (turning the furnace on or off). In *positive autoregulation*, a protein activates its own expression (**Figure 23.13**). This confers positive feedback and is often used in regulatory circuits to drive regulatory decisions or switches toward a particular regulatory state.

More complex regulatory circuits can combine positive and negative regulation mechanisms in a wide variety of ways. The mechanisms by which complex circuits

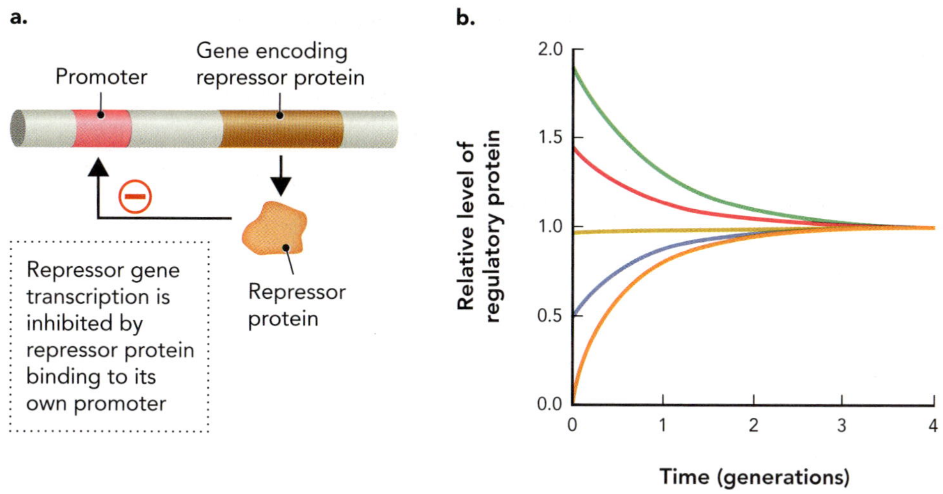

Figure 23.12 A model showing how negative autoregulation controls repressor protein levels. **a.** In a negative regulatory circuit, a repressor binds near its own promoter and represses its own transcription. **b.** The graph shows several time courses (time is in cell generations), starting from various initial values of the regulatory protein. In each case, because of negative autoregulation, the level of the regulatory protein reaches the same final value.

operate have been best studied in model organisms, such as yeast and *Drosophila*, in which genetic approaches allow identification and analysis of cis-acting sites and trans-acting factors. Indeed, much of the current knowledge of regulatory circuits in mammals is based on work with these simpler organisms.

During development of multicellular eukaryotes, complex programs of gene expression are established that eventually lead to different cell types. Often, the regulation comes from sequential cascades of regulatory factors, as in the example of *Drosophila* embryogenesis described later in the chapter. Some regulatory circuits lead to patterns in gene expression that are then inherited through multiple cell generations (**Figure 23.14**). These changes are often called epigenetic changes, which means that the pattern of gene expression is altered without a change in the DNA sequence. The resulting states are called **epigenetic states**. Examples are the stable patterns of gene expression seen in different cell types in multicellular organisms. Epigenetic states can be characterized as being maintained in trans or in cis, as in the case of DNA methylation, in which the state is maintained by a feature that only works in cis. For example, a modification such as DNA methylation may take place at one site, but not another, and this pattern is inherited. A classic example

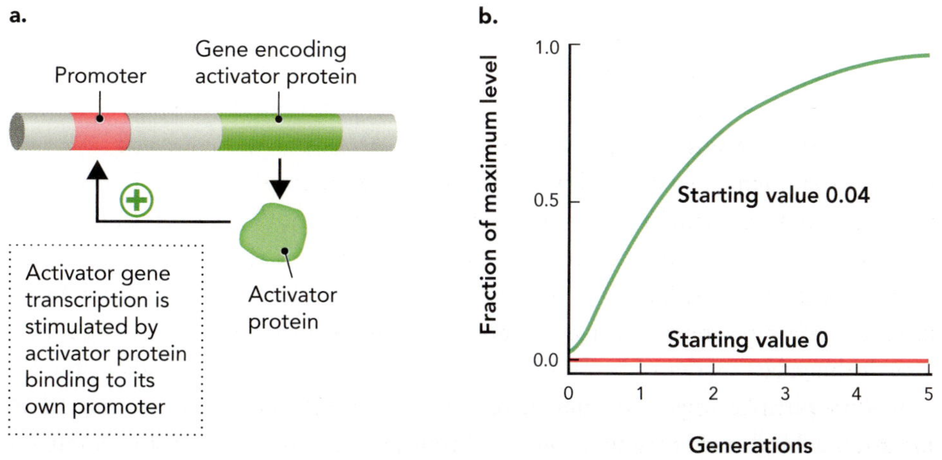

Figure 23.13 A model showing how positive autoregulation controls activator protein levels. **a.** Schematic of a simple positive autoregulatory loop. Expression of the promoter is zero in the absence of activator but at a maximum rate when activator is bound. **b.** The graph shows the time course (in cell generations) of cells with low or zero starting values. Starting with a range of values greater than zero results in maximum expression.

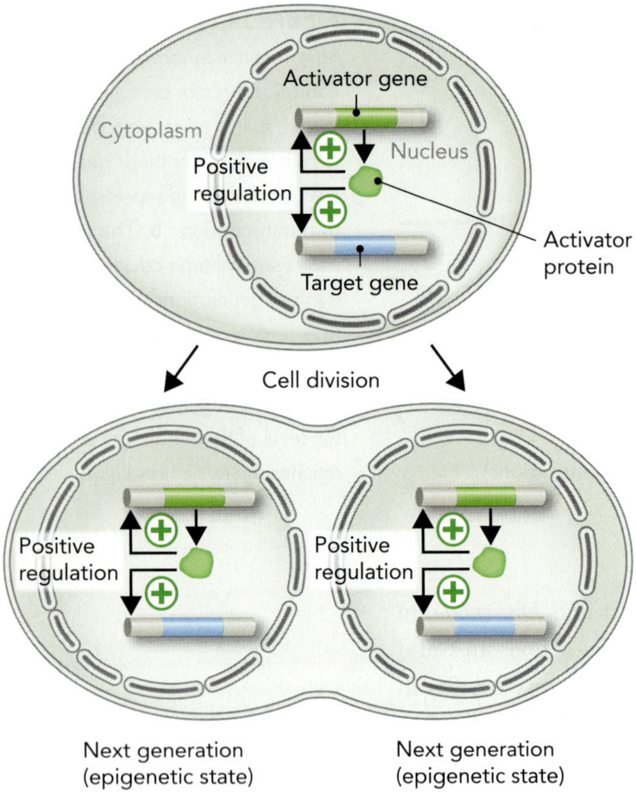

Figure 23.14 Epigenetic states are stable for multiple generations. As an example, a regulatory protein (green icon) is expressed and has two functions: It stimulates its own expression, and it stimulates expression of one or more target genes (blue). Self-activation ensures that the state continues. This condition may persist for many cell generations; only one cell division is shown.

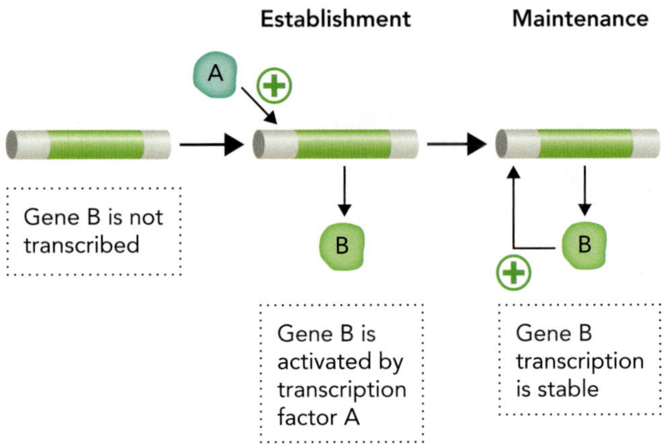

Figure 23.15 Epigenetic states are typically set up by one set of regulatory events and kept in a stable, ongoing state by a different set of events. In this example, the state is established by the presence of regulatory factor A. Once the state is set up, it is maintained by the presence and ongoing action of B stimulating its own expression; factor A is no longer needed.

is X-chromosome inactivation in female mammals: one of the two X chromosomes in a cell is active while the other is inactivated by long noncoding RNA (lncRNA; see Figure 21.7).

In general, the regulatory circuitry that sets up an epigenetic state can be divided into two phases (**Figure 23.15**): an *establishment phase*, in which a particular state of the circuitry is set up, and a *maintenance phase*, in which that state is perpetuated by the ongoing action of regulatory proteins or perpetuation of cis-acting marks. This phase usually operates in the absence of the regulators that were at work in the establishment phase.

Biochemical Applications That Exploit Gene Regulatory Processes

Understanding how regulatory circuits work has led to numerous applications, including the use of recombinant DNA. Parts of regulatory circuits can be assembled in novel fashions to achieve various goals.

Large amounts of a particular protein can be synthesized by fusing its gene to a promoter with cis-acting regulatory sequences, such as the *lac* promoter (described in detail in the next section). Protein over-production is very useful when large quantities of protein are need for biochemical analysis. As shown in **Figure 23.16**, the *lac* promoter is repressed by the **lac repressor**: If an inducer called isopropyl-β-D-1-thiogalactopyranoside (IPTG) is added to the cells, IPTG binds to the lac repressor, which causes it to dissociate from the DNA. The result is that RNA polymerase binds to the *lac* promoter, and transcription is turned on. Recombinant

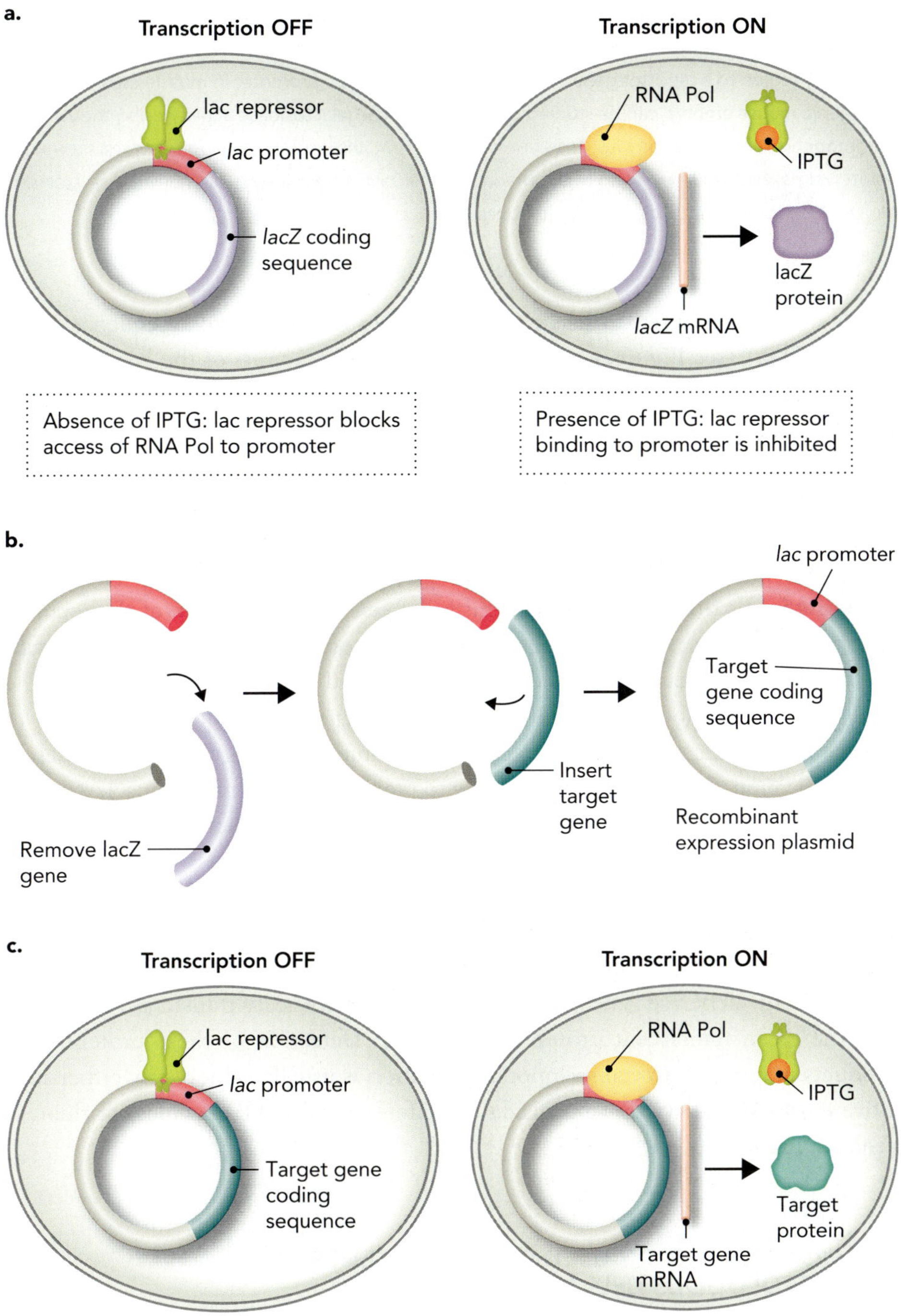

Figure 23.16 The *lac* promoter can be used to regulate the expression of a heterologous target gene using IPTG, an inhibitor of the lac repressor. **a.** Regulation of transcription by the *lac* promoter using IPTG to inhibit lac repressor binding to the *lac* operator sequences. **b.** Construction of a recombinant plasmid by replacing the naturally occurring lacZ gene located downstream of the *lac* promoter with a target gene sequence of interest. **c.** Cells containing the recombinant plasmid are first grown in the absence of IPTG to achieve high levels of cell growth. Upon addition of IPTG, the gene is activated (unrepressed), and large amounts of the gene product are produced.

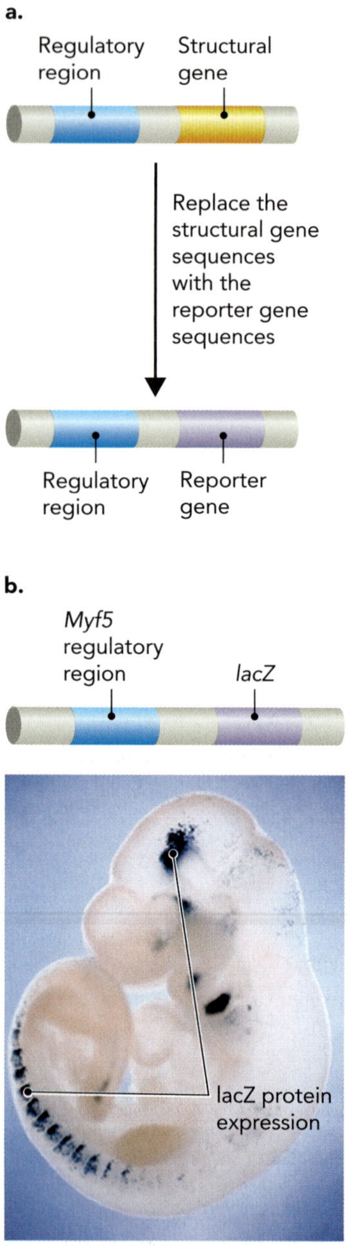

Figure 23.17 Reporter genes are used to understand gene regulation. **a.** A typical gene has a regulatory region and a structural gene. Recombinant DNA methods can be used to fuse a regulatory region to a reporter gene. **b.** Activation of the *Myf5* promoter in muscle cells of this transgenic mouse embryo can be studied by assaying for β-galactosidase activity using the X-gal chromogenic substrate. PHOTO: A. BUCHBERGER, N. NOMOKONOVA, AND H. H. ARNOLD (2003). MYF5 EXPRESSION IN SOMITES AND LIMB BUDS OF MOUSE EMBRYOS IS CONTROLLED BY TWO DISTINCT DISTAL ENHANCER ACTIVITIES. *DEVELOPMENT, 130*, 3297–3307. DOI: 10.1242/DEV.00557. © 2003 THE COMPANY OF BIOLOGISTS.

DNA methods are used to fuse the *lac* promoter to a target gene sequence using plasmid DNA (see Figure 3.52). To make abundant amounts of the gene-encoded target protein, IPTG is added to bacterial cells containing the recombinant plasmid, thereby leading to synthesis of the target protein (Figure 23.16).

Another application of gene regulation principles is to use **reporter genes**—genes whose products are easy to detect or assay—to simplify analysis of how regulatory regions behave. For example, recombinant DNA methods can be used to fuse a cis-acting regulatory region to a reporter gene. This hybrid gene is introduced into cells that do not contain this same gene, and expression of the reporter is tested in response to various conditions. The reporter genes are chosen because their products are easy to biochemically detect and analyze, in part because the genes are heterologous (not present) to the cell type being studied. As shown in **Figure 23.17**, using the bacterial lacZ gene to study gene regulatory processes in transgenic mouse embryos provides a way to monitor activity of the muscle-specific Myf5 gene by assaying for activity of β-galactosidase (product of the lacZ gene) using the X-gal chromogenic substrate (see Figure 3.51). Regions of the developing mouse embryo that activate the *Myf5* promoter can be identified by the presence of β-galactosidase activity, which is visualized as a blue stain. As shown in **Table 23.1**, in addition to *lacZ* encoding β-galactosidase, three other common reporter genes are the firefly luciferase gene (*luc*), the chloramphenicol acetyltransferase gene (*cat*), and the green fluorescent protein (GFP) gene. Detection of these protein gene products in response to transcriptional regulation is usually achieved by either sensitive enzymatic assays (β-galactosidase, luciferase, and CAT enzymes) or fluorescence in live cells (GFP).

Table 23.1 COMMONLY USED REPORTER GENES IN HETEROLOGOUS CELL SYSTEMS

Reporter gene	Protein product	Biochemical assay
lacZ	β-galactosidase	Indicator plates; enzyme assay of cell extracts; *in situ* colorimetric assay
luc	Luciferase	Enzyme assay of cell extracts; *in situ* fluorescence assay
cat	Chloramphenicol acetyltransferase (CAT)	Enzyme assay of cell extracts; genetic selection for chloramphenicol resistance
gfp	Green fluorescent protein	Autofluorescence in live cells

Note: The product of these genes can be detected biochemically using protein assays or by directly detecting the RNA transcript using polymerase chain reaction (PCR) or *in situ* RNA hybridization.

concept integration 23.1

How do gene regulatory circuits affect only the genes that are toberegulated and not all genes in the cell?

The basis for specificity is interaction of trans-acting gene regulatory proteins with cis-acting sites lying near the genes to be regulated. These interactions are highly specific, in that only one or a few binding sites are generally present or accessible in the genome. The specificity comes from interactions between amino acid side chains and the sides of base pairs in double-stranded DNA.

23.2 Mechanisms of Prokaryotic Gene Regulation

We have discussed some of the fundamental principles governing gene regulation at the level of genetic interactions. We now turn to analyzing the operation of regulatory circuits at the molecular and mechanistic level. We begin with prokaryotes, where the mechanisms are much better understood.

Regulation of the *E. coli lac* Operon

The French scientists François Jacob and Jacques Monod dissected the first prokaryotic regulatory system to be characterized at the molecular level using a combination of genetics and biochemistry. This classic study investigated the regulatory mechanisms controlling transcription of the *E. coli* ***lac*** **operon** (**Figure 23.18**). Their groundbreaking studies revealed the existence of operons, which are units of DNA containing multiple genes under control of a single promoter. In addition, they elucidated the concept of cis-acting regulatory sites on DNA, which are binding sites for trans-acting regulatory proteins. Jacob and Monod also provided crucial evidence that led to the discovery of mRNA as the transcript of gene sequences in DNA and of the role of allostery in protein structure and function. For this work, Jacob and Monod (along with André Lwoff) were awarded the Nobel Prize in Physiology or Medicine in 1965.

The function of the *lac* operon in *E. coli* is to provide the enzymes needed to utilize the disaccharide lactose as a nutritional source of galactose and glucose for energy conversion reactions (**Figure 23.19**). But lactose is not always present in the environment of *E. coli*, and it would be wasteful of energy to express the lactose metabolizing genes in the absence of lactose. Hence, the operon is regulated so that it is only expressed when lactose is present. As we will observe, an additional layer of control also exists: The operon is not expressed when glucose is also present, which make sense because glucose is the preferred carbohydrate source for growth.

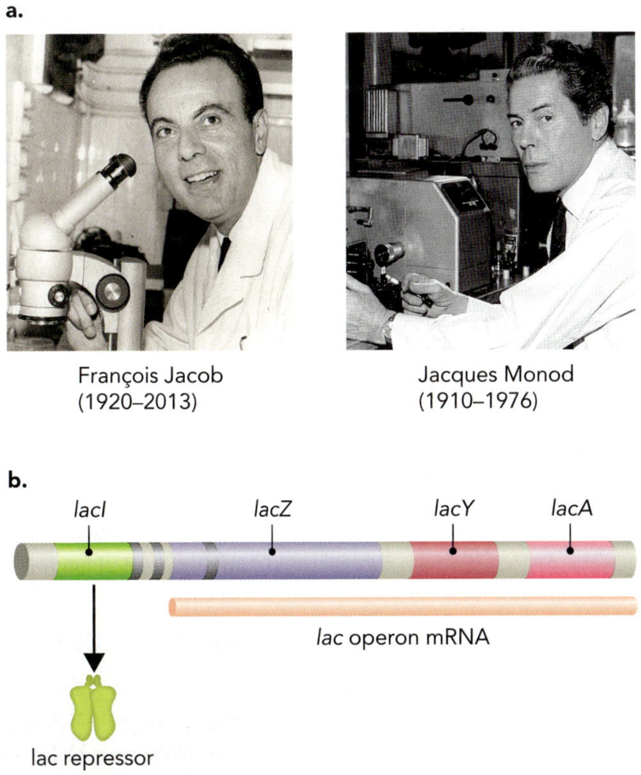

Figure 23.18 The work of François Jacob and Jacques Monod led to our understanding of gene regulation using the *E. coli lac* operon as a model system. **a.** François Jacob and Jacques Monod in their laboratories. F. JACOB: KEYSTONE/GETTY IMAGES; J. MONOD: KEYSTONE-FRANCE/GAMMA-KEYSTONE VIA GETTY IMAGES. **b.** Functional map of the *lac* operon.

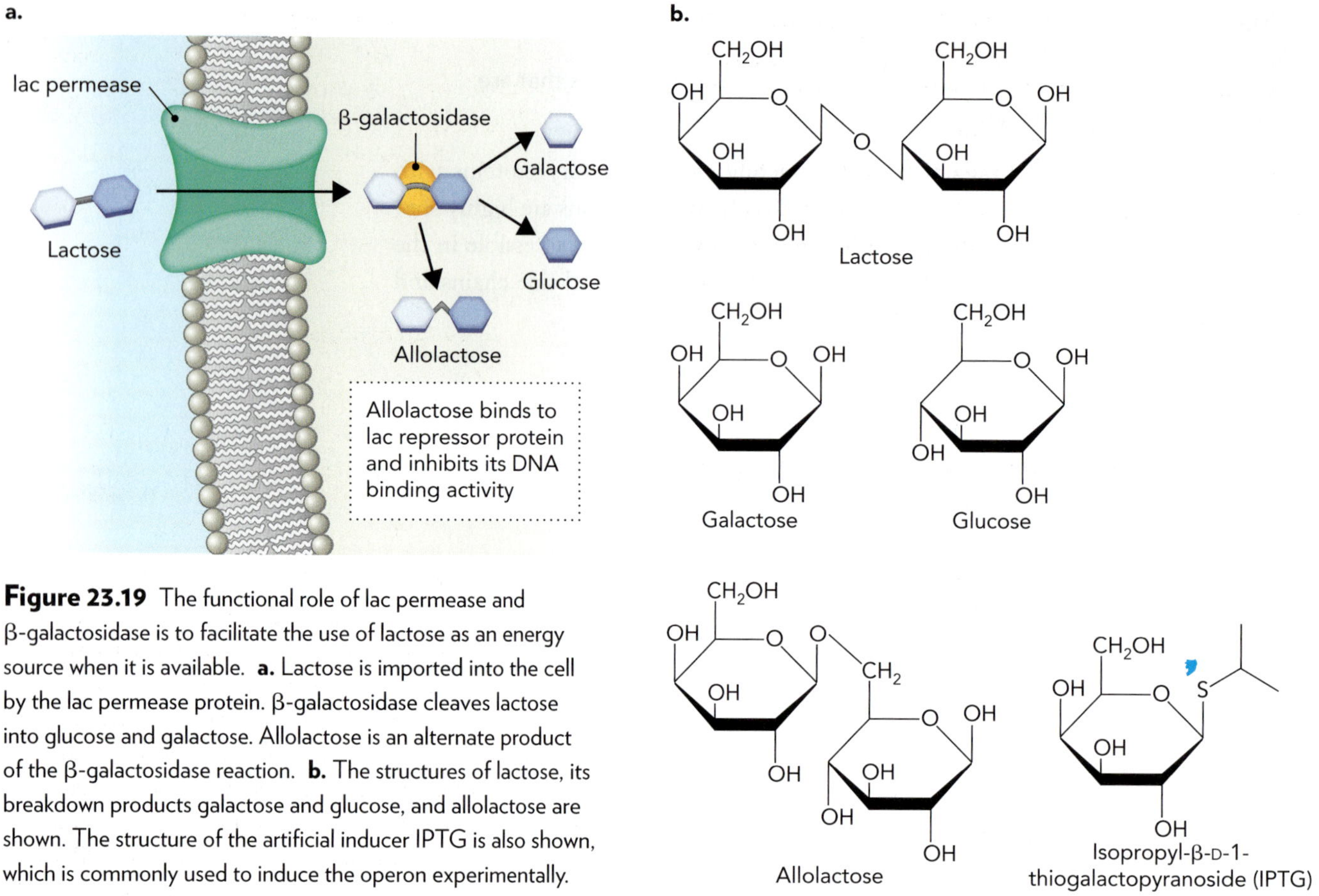

Figure 23.19 The functional role of lac permease and β-galactosidase is to facilitate the use of lactose as an energy source when it is available. **a.** Lactose is imported into the cell by the lac permease protein. β-galactosidase cleaves lactose into glucose and galactose. Allolactose is an alternate product of the β-galactosidase reaction. **b.** The structures of lactose, its breakdown products galactose and glucose, and allolactose are shown. The structure of the artificial inducer IPTG is also shown, which is commonly used to induce the operon experimentally.

▶ ANIMATION

Organization of the *lac* Operon The *lac* operon (**Figure 23.20**) consists of two parts: a regulatory region and the structural genes. The regulatory region contains cis-acting sites that control *lac* operon expression and regulation. These include the promoter, to which RNA polymerase binds to transcribe the operon; a primary binding site for the lac repressor, which blocks RNA polymerase binding; a binding site for a positive activator, cyclic AMP receptor protein (CRP), which stimulates expression; and two ancillary binding sites for the lac repressor, which together mediate complete repression. The gene for the lac repressor, *lacI*, lies adjacent to the operon, but this feature is not important for the function of the regulatory system. The lac repressor is a trans-acting factor and can work just as well to regulate *lac* operon expression even if its gene is located elsewhere.

Three structural genes are expressed from the operon: *lacZ, lacY,* and *lacA*. A single mRNA is made that encodes all three proteins. Only the first two gene products are required for breakdown of lactose into glucose and galactose. The lacZ gene encodes the enzyme β-galactosidase, which hydrolyzes lactose into glucose and galactose and also produces the metabolic *lac* operon inducer allolactose (see Figure 23.19); the lacY gene encodes lac permease, which transports lactose into the cell; and the lacA gene encodes an enzyme called lac transacetylase, which has an unknown function in lactose metabolism.

Metabolite Control of *lac* Operon Expression Expression of the *lac* operon involves two regulatory controls—one negative and one positive. First, it is repressed by the action of the lac repressor, which prevents binding of RNA polymerase to the promoter. When lactose is present, some of it is converted by β-galactosidase to a related molecule, **allolactose,** which binds to the lac repressor and causes it to dissociate from the operator. Thus, allolactose is the inducer of the operon. Second, the operon is activated by the binding of CRP when CRP is activated by cyclic AMP binding (CRP is also known as *catabolite gene activator protein*, or CAP). The CRP–cAMP complex binds to a site upstream of the promoter and stimulates the promoter by causing tighter binding of RNA polymerase. This action depends on the growth conditions; the levels of cAMP are high only in the absence of glucose as described later.

Accordingly, we can consider four different growth conditions, which differ by the absence or presence of glucose and lactose (**Figure 23.21**). Full expression of the genes in the *lac* operon occurs only in the absence of glucose and presence of lactose.

How does the level of glucose control the level of cAMP? Cells contain a sugar transport system called the phosphotransferase system (PTS), in which phosphate groups are transferred sequentially from phosphoenolpyruvate through a series of proteins and eventually to sugars, including glucose. As shown in **Figure 23.22**, in the presence of glucose the PTS is in dephosphorylated form, which binds to the enzyme adenylate cyclase and inhibits

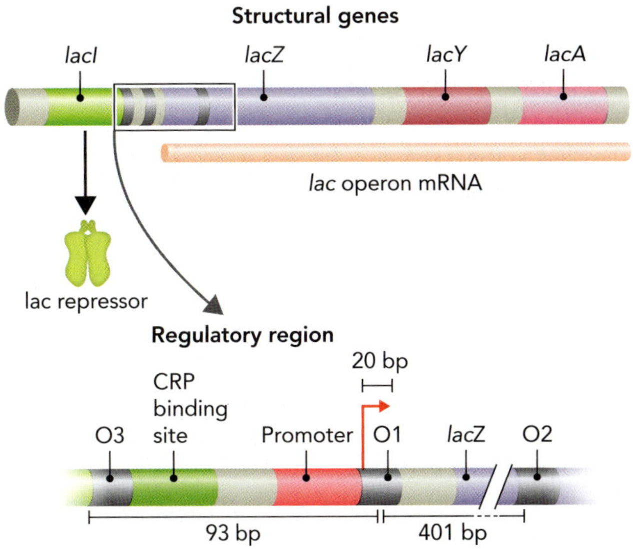

Figure 23.20 The *lac* operon of *E. coli* consists of a regulatory region plus three genes expressed via a single mRNA. The gene encoding the lac repressor, *lacI,* is located very close to the *lac* operon but is not considered part of the *lac* operon. An expanded view of the *lac* operon regulatory region shows the center-to-center spacing between the O1 operator and the O2 and O3 operator sequences, which are binding sites for the lac repressor protein. The arrow to the right of the promoter indicates the start point of transcription for the *lac* operon. The regulatory protein CRP is a trans-acting protein that induces *lac* operon expression.

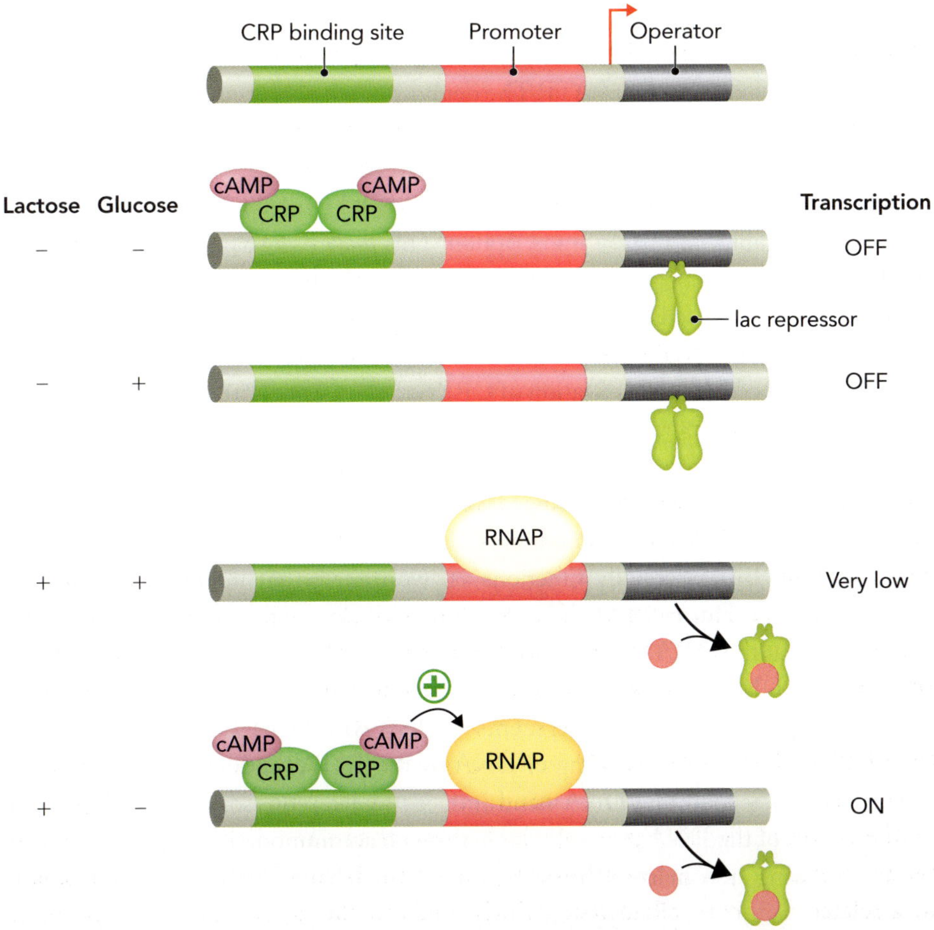

Figure 23.21 Effect of growth conditions on expression from the *lac* operon are shown. The locations of cis-acting sites are denoted at the top. For each of the four growth conditions shown, the binding of regulatory proteins is indicated. In the presence of glucose, CRP does not bind because cAMP levels are low. In the presence of lactose alone, lac repressor does not bind because allolactose binds the repressor, and the complex does not bind the operator. When both sugars are present, a very low level of expression occurs because the promoter is not completely silent without CRP–cAMP binding.

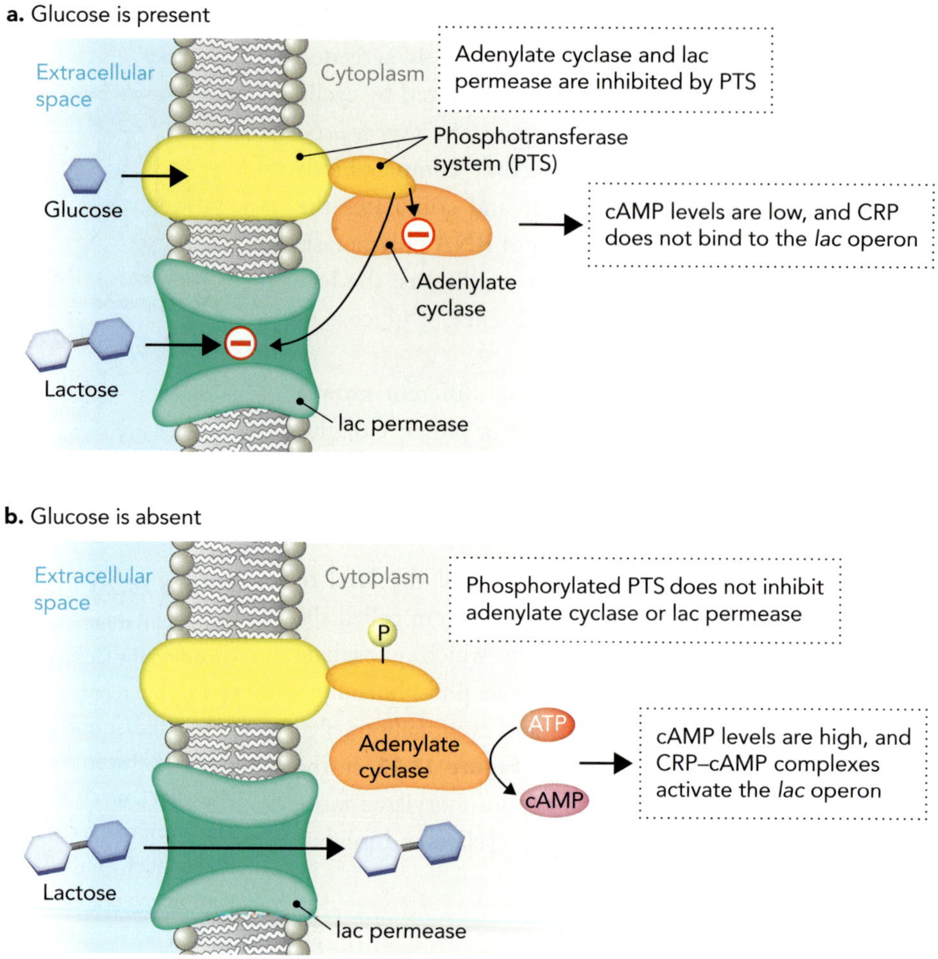

Figure 23.22 Inhibition of cAMP production by glucose transport. Glucose is transported into the cell by the phosphotransferase system (PTS). During transport, glucose is phosphorylated to generate glucose-6-phosphate (glucose-6-P). **a.** When glucose is present, the PTS is in the nonphosphorylated form, and both adenylate cyclase and lac permease functions are inhibited. **b.** In the absence of glucose, the PTS is phosphorylated, which stimulates adenylate cyclase and reverses the inhibition of lac permease. This raises the cAMP levels so that CRP can act as a regulatory protein and stimulate *lac* operon expression in the presence of lactose.

the production of cAMP from ATP. The dephosphorylated form of PTS also inhibits lac permease, so that energy is not expended transporting lactose into the cell when glucose is available. Because cAMP is required to activate CRP binding to DNA and thereby stimulate transcription of the *lac* operon (see Figure 23.21), glucose transport by the PTS inhibits *lac* operon transcription. This makes sense because glucose is the preferred sugar, and there is no need to synthesize enzymes required for lactose metabolism as long as glucose is available. When cells are depleted of glucose, however, then the PTS is phosphorylated, and this form stimulates adenylate cyclase, resulting in higher levels of cAMP and activation of CRP and the *lac* operon—as long as lactose is present and the lac repressor is not bound to the *lac* operon (see Figure 23.21).

Structure and Function of lac Repressor Protein Most specific DNA binding proteins in prokaryotes are homodimers. In contrast, the lac repressor is a homotetramer, though it essentially functions as two dimers that interact through their C-terminal ends (**Figure 23.23**). The structure of the lac repressor dimer–DNA complex shows that the DNA is bent ~40° in the complex. Such distortion of the DNA is often seen in DNA binding proteins to facilitate binding or assembly of multiprotein complexes by bringing different segments of DNA with DNA-bound proteins closer together; for example the bending of DNA by the TATA binding protein, TBP (see Figure 21.16). Second, the DNA bend induced by lac repressor binding causes the minor groove of the DNA, normally too narrow to accommodate an α helix, to widen greatly, allowing an α helix in the protein (the "hinge helix") to bind in that position.

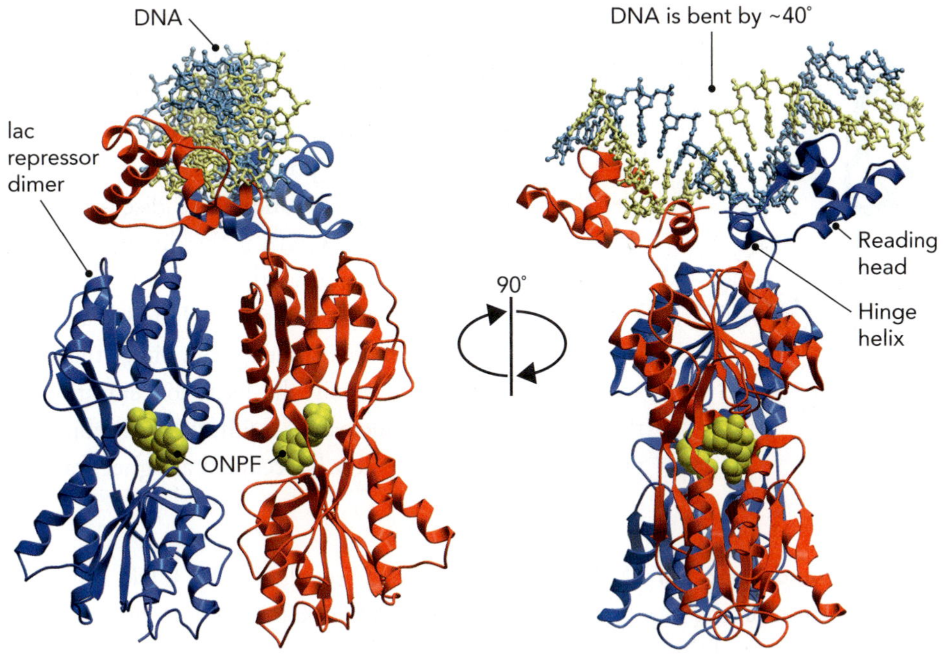

Figure 23.23 Structure of the lac repressor dimer is shown bound to DNA. In this structure, the repressor is missing the portion responsible for tetramer formation. The helix–turn–helix motif in the reading head of the lac repressor makes sequence-specific contacts in the major groove of the DNA. Orthonitrophenylfucoside (ONPF) binds to a cleft between two subdomains of the core. IPTG and allolactose, which weaken the interaction of lac repressor protein with DNA, would bind at a similar position as ONPF. BASED ON PDB FILE 1EFA.

Third, the portion of the protein that forms the hinge helix is unstructured when the protein is not bound to DNA; the helix forms upon binding. This feature is also seen in other DNA binding proteins. For example, bZIP proteins have α helices that bind in the major groove of DNA; these helices are unstructured when the protein is not bound to DNA.

The natural lac repressor is a tetramer—a dimer of dimers. Only one of the dimers contacts an operator, but the other dimer plays an important role in repressing the operon. In addition to the primary operator site for lac repressor binding called O1, which overlaps with the promoter sequence where RNA polymerase binds, the *lac* operon contains two additional binding sites for the lac repressor, called O2 and O3 (see Figure 23.20). Assays of gene expression show that the presence of the O2 and O3 sites in the *lac* operon makes transcriptional repression more efficient, presumably by increasing the lac repressor protein occupancy of the primary O1 site. As shown in **Figure 23.24**, the presence of the O1 and O3 binding sites facilitates DNA looping, which occurs when

Figure 23.24 Model for looping by lac repressor is shown. For simplicity, only O1 and the auxiliary operator O3 are shown. The lac repressor is depicted as a dimer of dimers. Loop formation proceeds in two steps. First, a lac repressor tetramer binds to one of the operators using one of the dimers. Second, when both operators are bound by a tetrameric lac repressor, a loop structure will be formed.

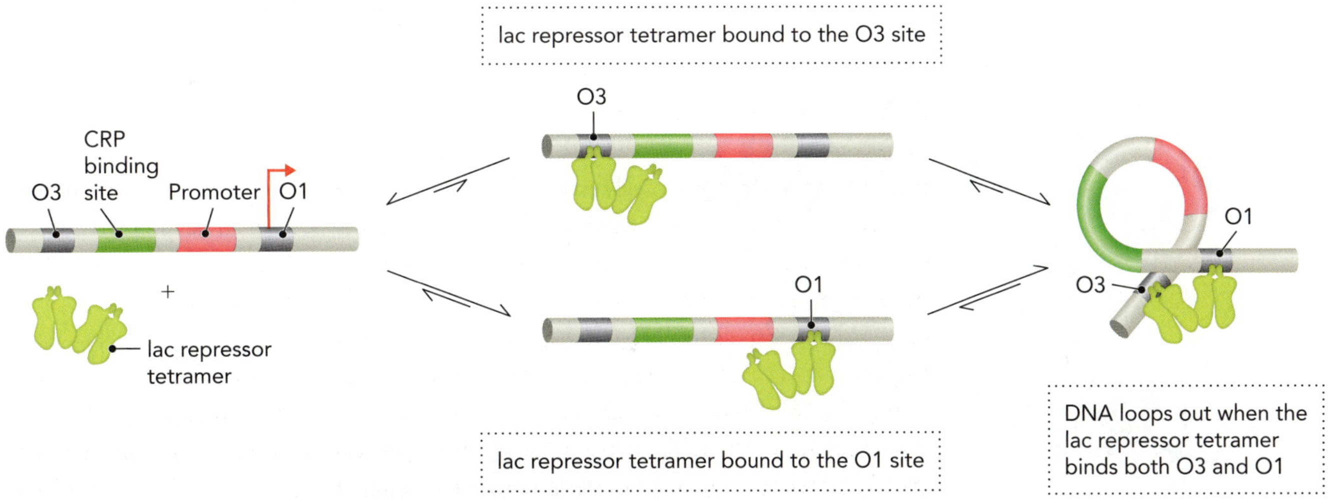

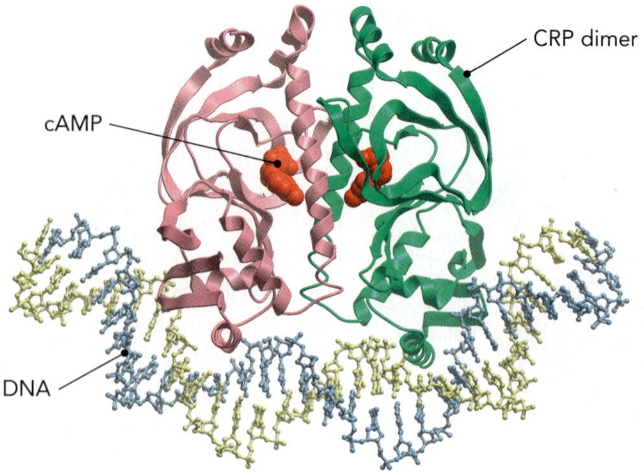

Figure 23.25 Complex between CRP–cAMP and DNA is shown. In this complex, one subunit of CRP contacts each half of the DNA binding site, which is symmetric in sequence. BASED ON PDB FILE 1CGP.

the second dimer of the lac tetramer binds to one of these sites. Because the local concentration of the second operator relative to the second dimer is very high, this reaction is strongly favored, forming the looped structure.

The binding of the lac repressor to DNA is an example of cooperativity in DNA binding, as discussed earlier (see Figure 23.8). In this case, if the dimer bound to the primary operator O1 transiently dissociates from the DNA, the fact that the tetramer is still anchored to another site near O1 makes it much easier for the dimer that dissociated to rebind. Effectively, the local concentration of the repressor is greatly increased, facilitating the rebinding process, so that the operator is occupied a higher fraction of the time.

Structure and Function of CRP Transcriptional regulation by CRP, the positive regulator of the *lac* operon, is complex. When CRP–cAMP binds DNA (**Figure 23.25**), the DNA is bent nearly 90°. A region of the CRP protein has been demonstrated to make a direct contact with the α subunit of RNA polymerase, specifically with the C-terminal domain of the α subunit (α-CTD). As shown in **Figure 23.26**, this domain of the α subunit also contacts DNA, often nonspecifically, and it interacts with many activator proteins. The contact surface between CRP and the α-CTD is small but contributes to cooperative binding.

CRP acts to regulate at least 150 operons across the *E. coli* genome. Some of these are involved in metabolism of other sugars, such as maltose, and these operons are also repressed if glucose is present. CRP is viewed as a global regulator, which integrates many facets of cell metabolism in response to glucose availability.

Regulation of the *E. coli* SOS Regulon

The **SOS regulatory system** regulates a set of genes involved in the response to DNA damage or inhibition of replication. Like the *lac* system, it involves two regulatory proteins. The negative regulator of this gene circuit, **LexA repressor**, turns down the expression of several dozen genes under normal growth conditions. The target genes of LexA repressor are collectively termed a **regulon**, as the genes are regulated in parallel but are not located in a single transcriptional unit, as in an operon. As shown in **Figure 23.27**, there are four distinct transcription states in the SOS regulatory circuit: (1) When the system is in the OFF state, the LexA protein acts to repress several dozen SOS genes in the SOS regulon. Genes in the SOS regulon function in a variety of roles in dealing with DNA

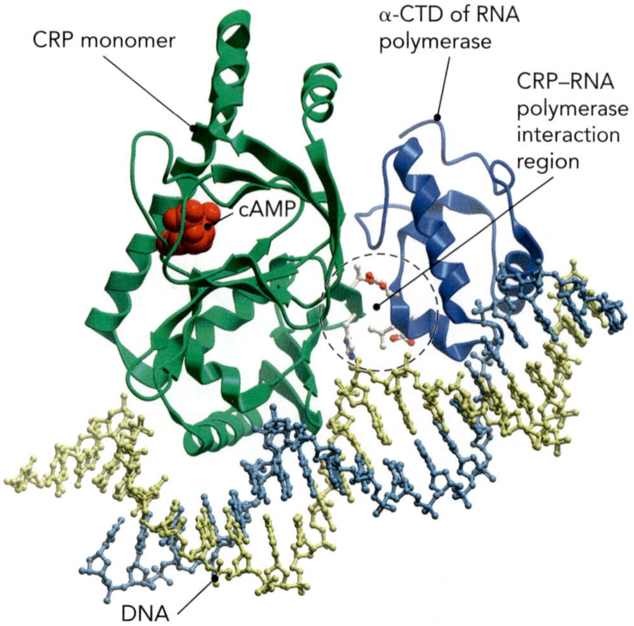

Figure 23.26 CRP makes direct contacts with the α-CTD of RNA polymerase. Only one subunit of the CRP dimer is shown. Amino acid residues in CRP (Thr158, His159) and in α-CTD (Glu286, Val287, Glu288) that make direct contact are highlighted by a circle. BASED ON PDB FILE 1LB2.

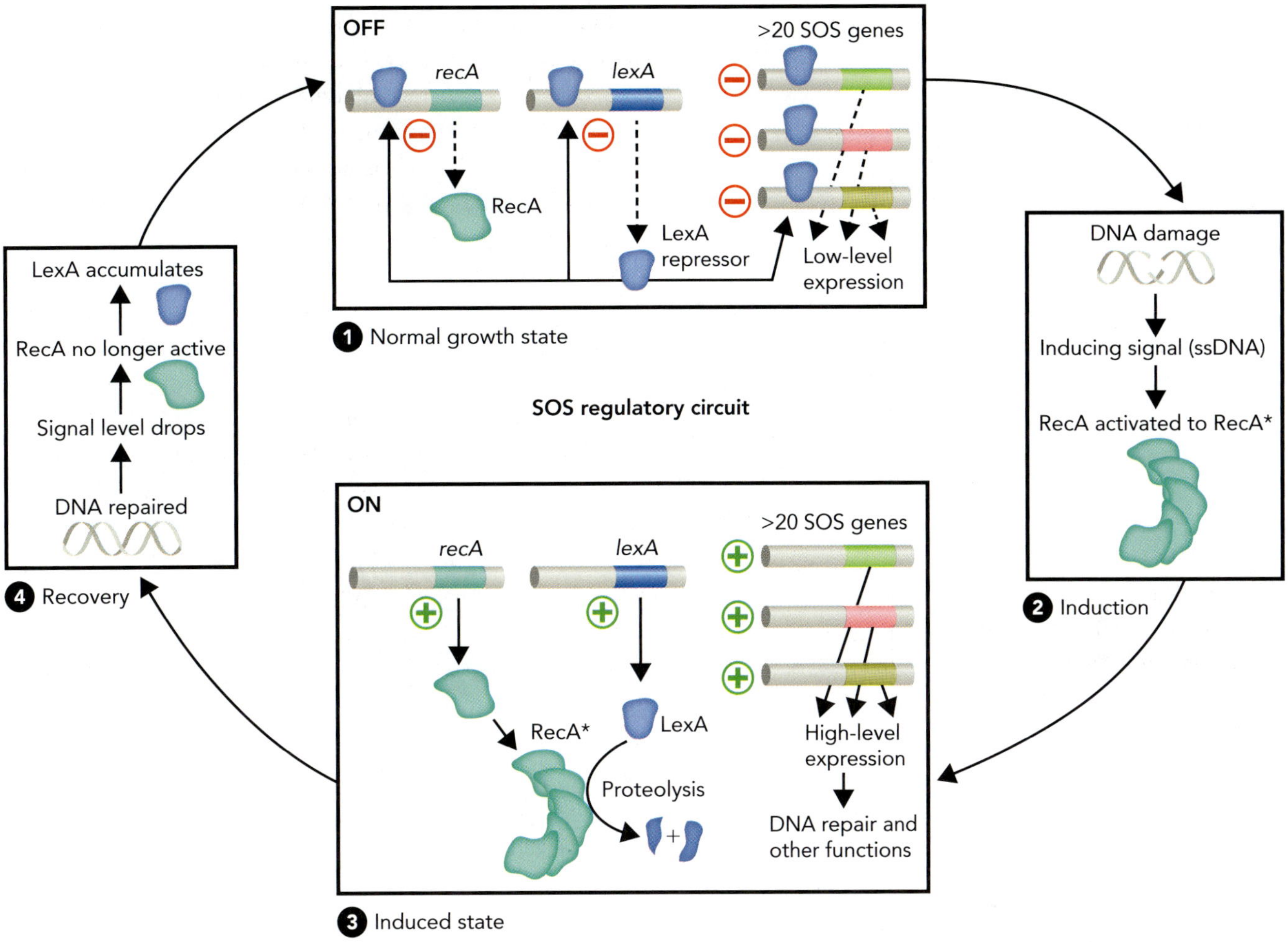

Figure 23.27 A model showing the SOS gene regulatory system in *E. coli*. See the text for details.

damage, including several directly involved in DNA repair. In this state, LexA also inhibits expression of its own gene (autorepression) and expression of the recA gene. (2) DNA damage leads to activation of the RecA protein, denoted RecA*. (3) RecA* stimulates LexA autocleavage, which reduces the amount of LexA repressor in the cell, thereby increasing the expression of the SOS regulon genes, *rec*A, and *lex*A as a result of de-repression. Because LexA protein is stimulated to undergo autocleavage by RecA*, it does not accumulate. (4) When DNA is repaired and replication can resume, RecA is no longer active. This enables LexA to accumulate, allowing a return to the OFF state.

The gene products in the SOS regulon act in various ways to counteract the effects of DNA damage. Some are directly involved in DNA repair, such as the products of the uvrA and uvrB genes (see Figure 20.40). Others, such as the products of the *umuDC* operon, are involved in mutagenesis in cases where damage cannot be repaired. The extent to which these genes are repressed varies from one gene to another. Some, such as *uvrA*, are only repressed a few-fold. This provides enough UvrA protein to cope with small amounts of sporadic DNA damage without turning on the entire regulon. Others, such as the *umuDC* operon, are almost completely turned off, as these gene products are harmful during normal growth. It is important to note that the level of repression is dictated by how tightly LexA binds to the various operators in the regulon. Weak binding sites lead to higher levels of basal expression, whereas strong binding sites afford more complete repression. One of the weakest binding sites is in

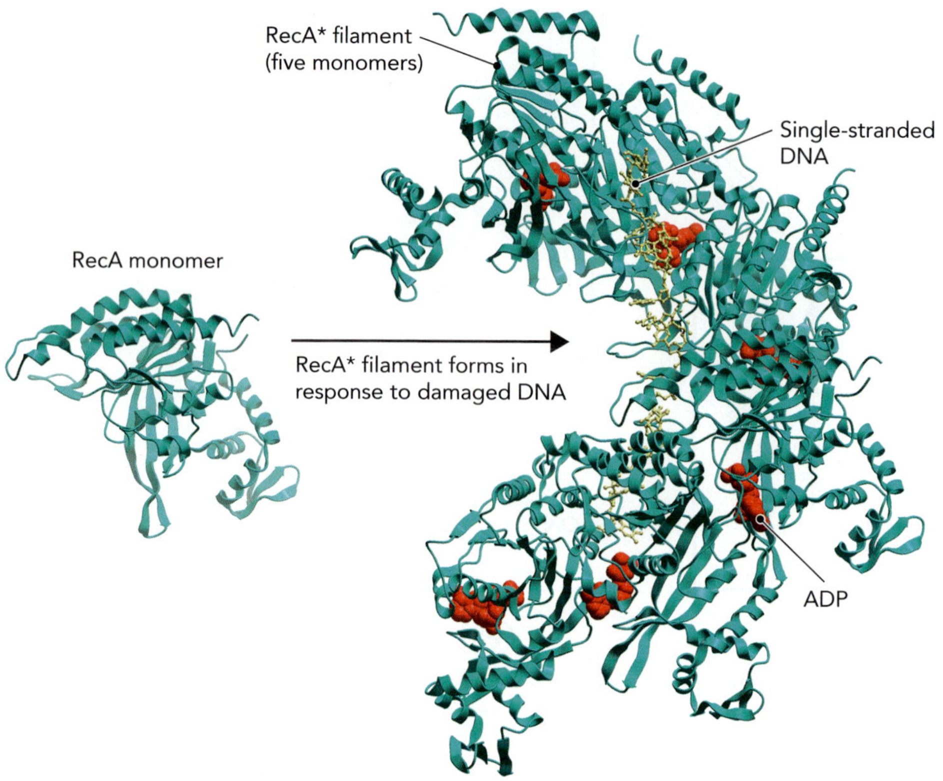

Figure 23.28 The monomer form of RecA is converted to the activated RecA* filament form by binding to single-stranded DNA. As seen in the macromolecular structure, RecA* forms a right-handed helical filament on single-stranded DNA in a complex containing ATP. The complex contains three nucleotides of DNA per RecA* subunit. BASED ON PDB FILE 1CMU.

the lexA gene, which makes sense: If LexA bound tightly to its own operator, it would not be able to build up to levels that allowed repression of weaker sites.

The SOS regulon is turned on by DNA damage. Replication forks often halt at the site of DNA damage, generating single-stranded DNA, which is a binding site for the second SOS regulatory protein called **RecA**. The RecA protein plays important roles in homologous recombination, catalyzing strand transfer reactions at an early stage of the recombination process. In the first stage of this process, RecA protein binds to single-stranded DNA, forming a helical filament (**Figure 23.28**). This same filament forms after DNA damage and acts in the SOS regulatory circuit as a positive regulatory molecule, termed RecA*, which functions to stimulate LexA-mediated cleavage of LexA; that is, LexA is cleaved in an autoproteolytic reaction that requires RecA*. But RecA* is not a typical protease with an active site, such as chymotrypsin; instead, it acts as a co-protease that stimulates the LexA-mediated autocleavage reaction. As shown in **Figure 23.29**, the C-terminal domain of LexA exists in two different protein conformations: (1) the noncleavable form in which the cleavage site is 20 Å from the active site, and (2) the cleavable form in which the cleavage site and active sites are juxtaposed. The binding of RecA* to LexA stabilizes the cleavable form, most likely by binding preferentially to this conformation.

Activation of the SOS system is reversible: Once the DNA damage is repaired, RecA is no longer activated, and LexA becomes stable. Because LexA regulates its own expression, the cell can recover more rapidly: When LexA levels are low, the lexA gene is expressed at high rates, allowing the LexA protein level to build up rapidly.

Figure 23.29 Regulation of LexA autodigestion is shown. These two structures depict the C-terminal domain of LexA. In yellow is the cleavage site region, which adopts a different position in the two forms. The cleavage site is red, and the active site residues are green. BASED ON PDB FILES 1JHE AND 1JHH.

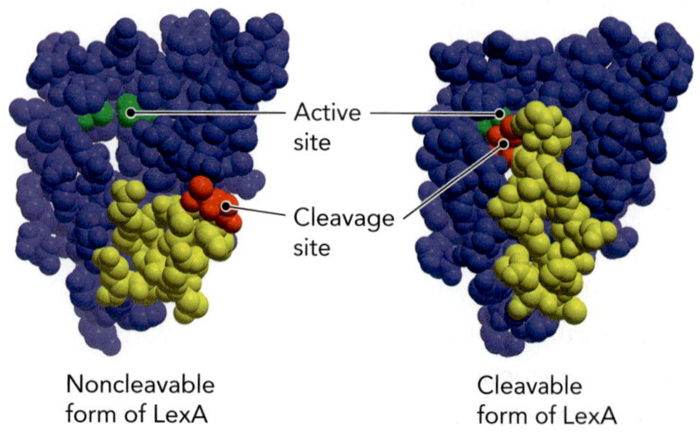

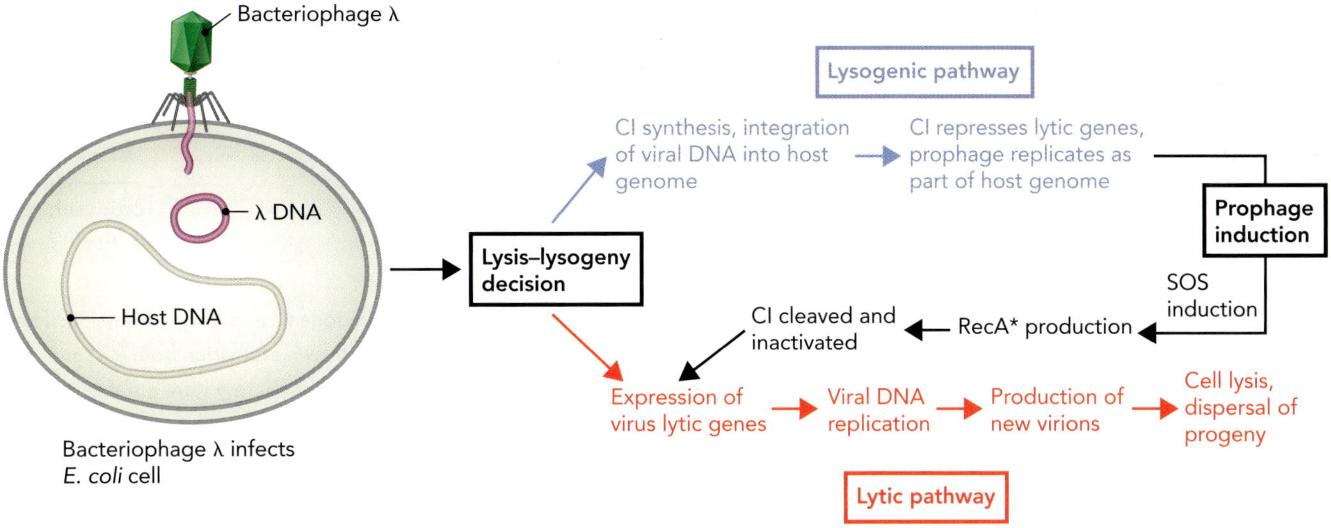

Figure 23.30 The life cycle of bacteriophage λ can be diagrammed as a gene regulatory circuit in which the lysis–lysogeny decision results in either the lytic pathway or the lysogenic pathway (see Figure 20.52). Conversion of the lysogenic pathway to the lytic pathway involves the SOS gene circuit and is controlled by RecA*-mediated cleavage and inactivation of bacteriophage λ CI regulatory protein (λ repressor protein), much like RecA*-mediated cleavage of the LexA repressor protein (see Figure 23.27).

Regulation of an Epigenetic Switch in Bacteriophage λ

The SOS system also stimulates the lytic pathway in *E. coli* cells infected with bacteriophage λ. To understand how the system works in this case, we first outline the gene regulatory circuit of this virus. As is typical of bacterial viruses, the mature virus consists of viral DNA surrounded by a protective protein coat, as well as an apparatus for attaching to a cell and injecting the DNA to initiate an infection.

As first described in Chapter 20, bacteriophage λ has two distinct lifestyles: one in which the bacteriophage integrates into the *E. coli* genome, called **lysogeny**, or a lifestyle in which it generates progeny viruses that kill the bacterial host upon cell **lysis** (see Figure 20.52). What gene regulatory circuits control this genetic switch between the two viral lifestyles? As shown in **Figure 23.30**, when bacteriophage λ infects an *E. coli* cell, it is faced with the lysis–lysogeny decision, which in biochemical terms is essentially a mechanistic response favoring expression of either genes required for the lytic pathway or genes required for the lysogenic pathway. The lysis–lysogeny decision at the biochemical level is dependent on environmental factors that favor one of the two lifestyle choices.

What prevents the prophage, which is the latent form of the bacteriophage in the lysogenic pathway, from expressing the lytic genes? The prophage makes a gene regulatory protein called CI, also called λ repressor (see Figure 23.11). The CI protein represses expression of the lytic genes and also regulates its own expression, acting both as an activator and, at higher concentrations, as a repressor. As shown in **Figure 23.31**, most of the regulatory circuit centers on a complex regulatory region of 100 bp that contains five cis-acting sites: two promoters (P_R and P_{RM}) and three binding sites for CI (O_R1, O_R2, and O_R3). When CI binds cooperatively to O_R1 and O_R2, it represses the lytic P_R promoter and activates the lysogenic P_{RM} promoter, which expresses the gene *cI* in the lysogenic pathway to produce more CI protein. Therefore, as long as CI protein is present, it autoactivates expression of its own promoter (P_{RM}) and prevents expression of the lytic promoter (P_R). This gene regulatory circuit can be considered a stable epigenetic state.

Figure 23.31 additionally shows that bacteriophage λ also makes a second repressor, called Cro, which is produced from the first gene in the P_R operon. Cro binds to the same sites in the O_R region as does CI, but it binds most tightly to O_R3. When Cro

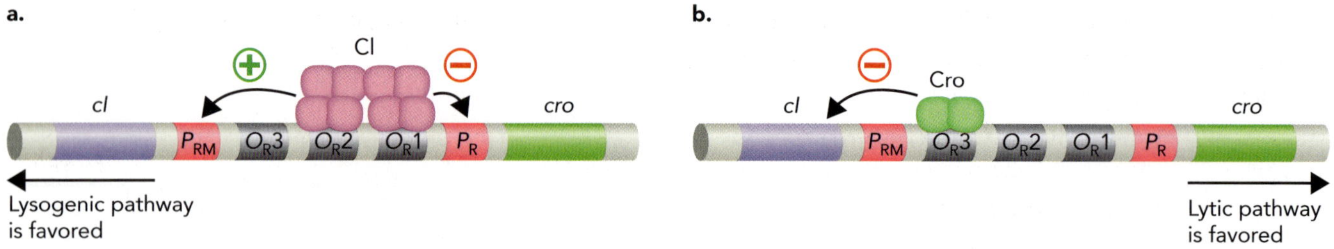

Figure 23.31 Transcriptional regulation is the molecular basis for the control of the bacteriophage λ lysogenic and lytic pathways. **a.** The lysogenic state is favored when CI binds tightly and cooperatively to sites O_R1 and O_R2, which activates its own expression from P_{RM} and represses P_R. **b.** Cro binds tightly to O_R3, repressing P_{RM} without affecting its own expression from P_R. This regulatory state is stable because Cro continues to be made, and CI is repressed.

is bound to O_R3, it represses P_{RM}, turning off expression of CI without affecting its own expression. Under these conditions, the lytic pathway is favored.

Although the lysogenic state of bacteriophage λ is highly stable, it can switch to the lytic state in the process of prophage induction. When the SOS system is triggered by DNA damage (see Figure 23.27), activation of RecA to generate RecA* leads to cleavage and inactivation of CI by the same self-cleavage mechanism that inactivates the host LexA protein (see Figure 23.29). The result is expression of the lytic genes, including *cro*, which shuts off further CI synthesis. The system then follows the lytic pathway. Here, bacteriophage λ has taken advantage of a signal that the cell is in trouble and may not survive.

Regulatory Mechanisms Governing the *trp* Operon

As described earlier, not all regulation occurs at the level of transcriptional initiation (see Figure 23.1). A striking example is seen in the ***trp* operon** of *E. coli*. This operon has a set of five genes that code for proteins involved in biosynthesis of the amino acid tryptophan (**Figure 23.32**). As with the *lac* operon, expression of these genes is regulated so that they are only turned on when tryptophan levels are low. Strikingly, two different mechanisms are used to control the *trp* operon. The first mechanism is by the action of a repressor protein (Trp repressor) that inhibits transcriptional initiation by RNA polymerase, whereas the second mechanism involves disruption of transcriptional elongation by RNA polymerase through a mechanism called **attenuation**. In addition to these gene regulatory mechanisms, the first enzyme in the pathway, anthranilate synthase, is allosterically inhibited by tryptophan. This feedback inhibition ensures that if tryptophan suddenly becomes available to the bacterial cell, it quickly blocks the tryptophan biosynthetic pathway at an early step.

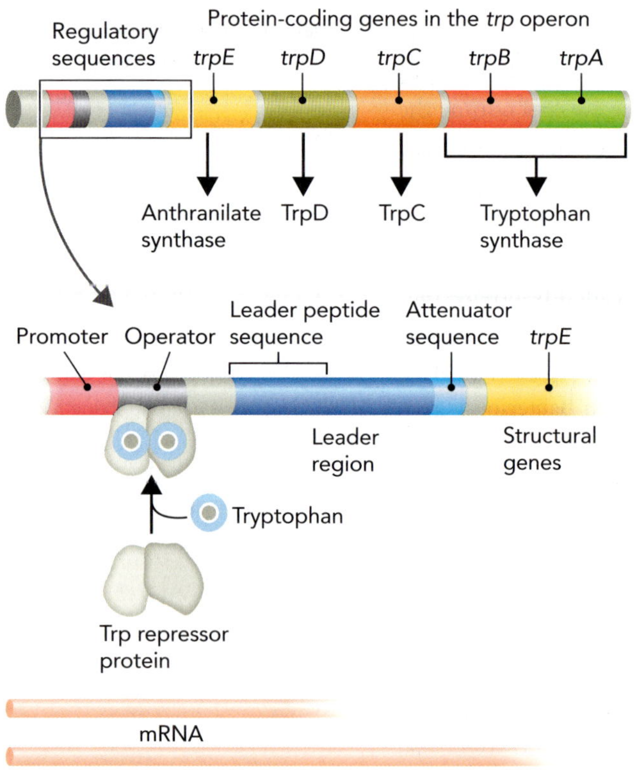

Figure 23.32 The *trp* operon of *E. coli* contains a gene regulatory region and the coding sequences for five enzymes in the tryptophan biosynthetic pathway. The gene regulatory region is expanded to show the location of the promoter sequence and the operator sequence to which Trp repressor binds when activated by tryptophan (T). A segment of the *trp* operon called the leader region encodes the leader peptide and the attenuator sequence, which regulates transcriptional elongation. Two mRNA transcripts are produced by the *trp* operon: a short mRNA of 140 nucleotides that terminates in the attenuator sequence, and a long mRNA of 9,300 nucleotides that encodes the five proteins.

Transcriptional initiation of the *trp* operon is regulated by the **Trp repressor**, which binds as a dimer to an operator sequence overlapping the *trp* operon promoter, but only in the presence of tryptophan (**Figure 23.33**). Under these conditions, transcription of the *trp* operon is repressed by the Trp repressor up to 80-fold compared to conditions when tryptophan is limiting. This makes sense because production of the tryptophan biosynthetic enzymes is a waste of metabolic energy if tryptophan is available to the bacterial cell from the environment.

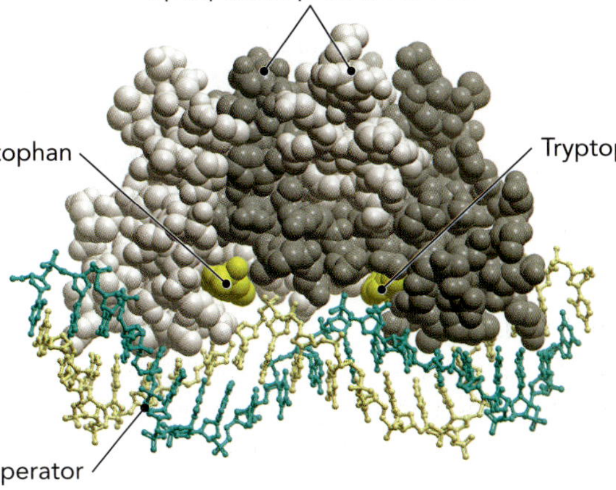

Figure 23.33 The *E. coli* Trp repressor protein dimer is shown bound to the *trp* operator sequence in the presence of tryptophan. BASED ON PDB FILE 1RCS.

Attenuation, the second regulatory mechanism controlling expression of the *trp* operon, functions only after the initiation of transcription. The molecular basis for transcriptional attenuation in the *trp* operon in *E. coli* is dependent on the fact that transcription and translation are coupled processes in bacteria. As described in Chapter 21, ribosome binding and initiation of protein synthesis in bacteria begins shortly after transcription (see Figure 21.3). Therefore, as soon as the 5′ end of the nascent mRNA transcript is available, ribosomes can bind and begin synthesizing the encoded polypeptide. Because of these coupled processes, factors that control the rate of protein synthesis by ribosomes can affect rates of RNA synthesis by RNA polymerase. The attenuation mechanism exploits this potential for regulation in two ways. First, the initial region of transcribed RNA can fold into alternative secondary structures called the termination and antitermination forms, which affect elongation rates of RNA polymerase by creating a transcriptional pause site. Second, the rate at which ribosomes are able to synthesize polypeptides is dependent on the availability of appropriately charged tRNA molecules to enter the A site of the ribosome. Therefore, if a specific amino acid pool is depleted, in this case the tryptophan pool, then the ribosome stalls, which can affect the formation of the alternative RNA secondary structures in the nascent mRNA transcript.

As shown in **Figure 23.34**, the leader sequence of the *trp* operon encodes four related DNA segments that can pair up in two different ways to form stem–loop structures when transcribed into the corresponding mRNA transcript. The partially complementary sequences are known as regions 1–4. Moreover, located on the 5′ side of region 1 is a series of triplet codons specifying tryptophan, which is critical to understanding how tryptophan levels in the cell control transcription of the *trp* operon. In one configuration of the stem–loop structure, the so-called termination form, pairing of region 3 with region 4 forms a stem–loop structure directing Rho-independent transcriptional termination facilitated by a run of uridine ribonucleotides (see Figure 21.19). An alternative stem–loop structure can form in the mRNA in which region 3 pairs with region 2. In this configuration, the RNA polymerase termination site formed by the region 3–4 stem–loop does not form, and transcription is allowed to ensue. The choice between forming the termination or antitermination stem–loop structures depends on whether region 2 is available to pair with region 3.

As shown in **Figure 23.35a** (p. 1173), if charged Trp–$tRNA^{Trp}$ levels are high in the bacterial cell, Trp is incorporated into the leader peptide, and the ribosome proceeds to the stop codon. The ribosome positioned here partially covers up region 2. Meanwhile, the RNA polymerase proceeds through region 3, but region 3 cannot pair with region 2. Once region 4 is transcribed, it pairs with region 3 to form the terminator, and the RNA polymerase terminates, having made only a short mRNA

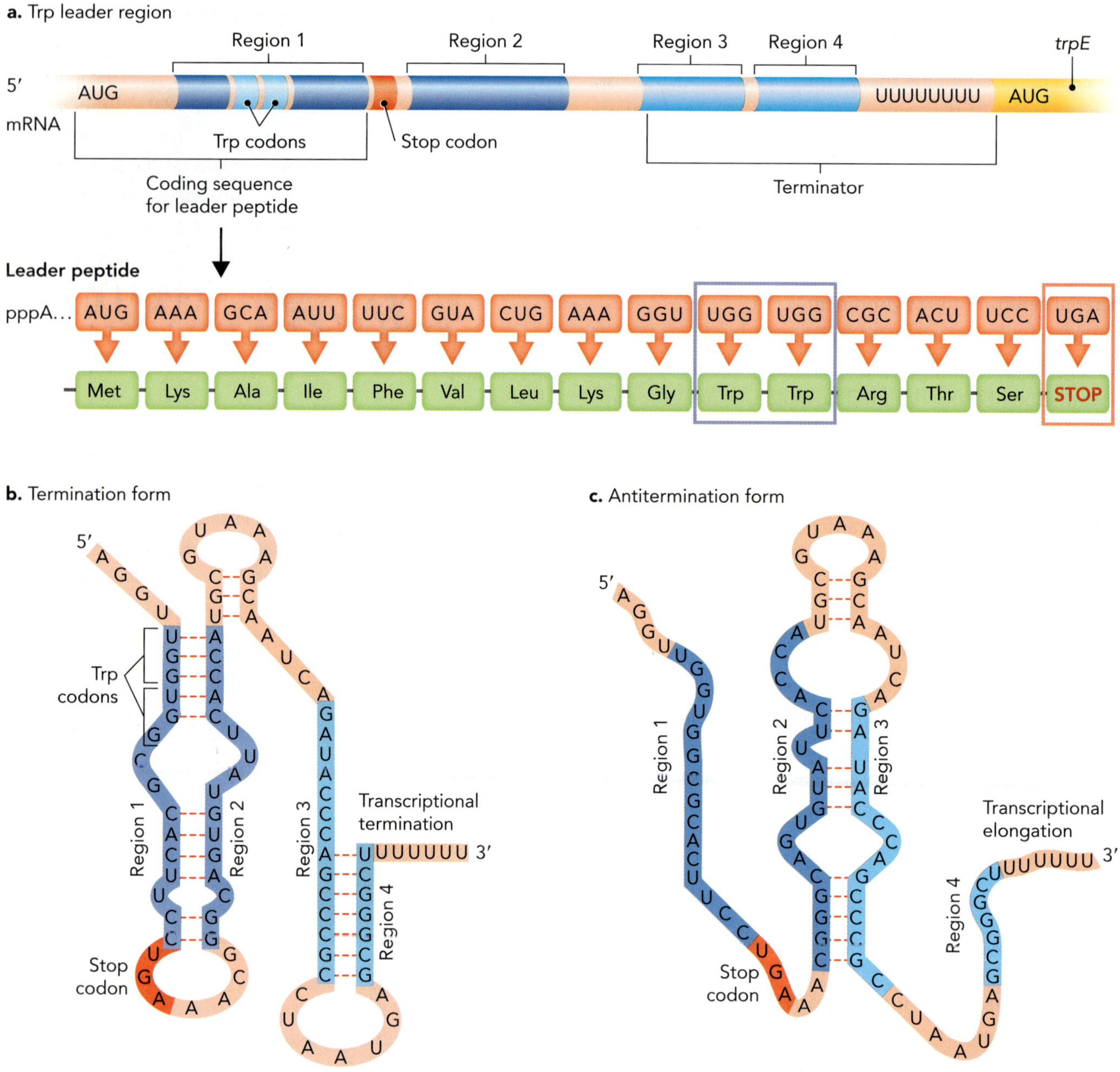

Figure 23.34 Alternative RNA stem–loop structures can form in the *E. coli trp* operon leader region. **a.** Functional map of the leader region showing the location of the coding sequence for the 14-amino-acid leader peptide, which contains two Trp codons, and the location of the four sequences that can form alternative stem–loop structures (regions 1–4). **b.** The termination form consists of a 3–4 stem–loop structure that stimulates Rho-independent transcriptional termination. **c.** The antitermination form is a 2–3 stem–loop structure, which permits transcriptional elongation by RNA polymerase.

transcript. But if Trp–tRNA^{Trp} levels are low in the bacterial cell (**Figure 23.35b**), then the ribosome stalls at the Trp codons, leaving region 2 uncovered. Once region 3 is transcribed, it can pair with the available region 2, forming the antiterminator stem–loop. Because region 3 is unavailable to pair with region 4, the RNA polymerase proceeds past the attenuator into the structural genes, and transcriptional elongation occurs.

The leader peptide itself has no role in biosynthesis of tryptophan; its only role is as a sensor of the Trp–tRNA^{Trp} levels. In addition, the ribosome involved in translating

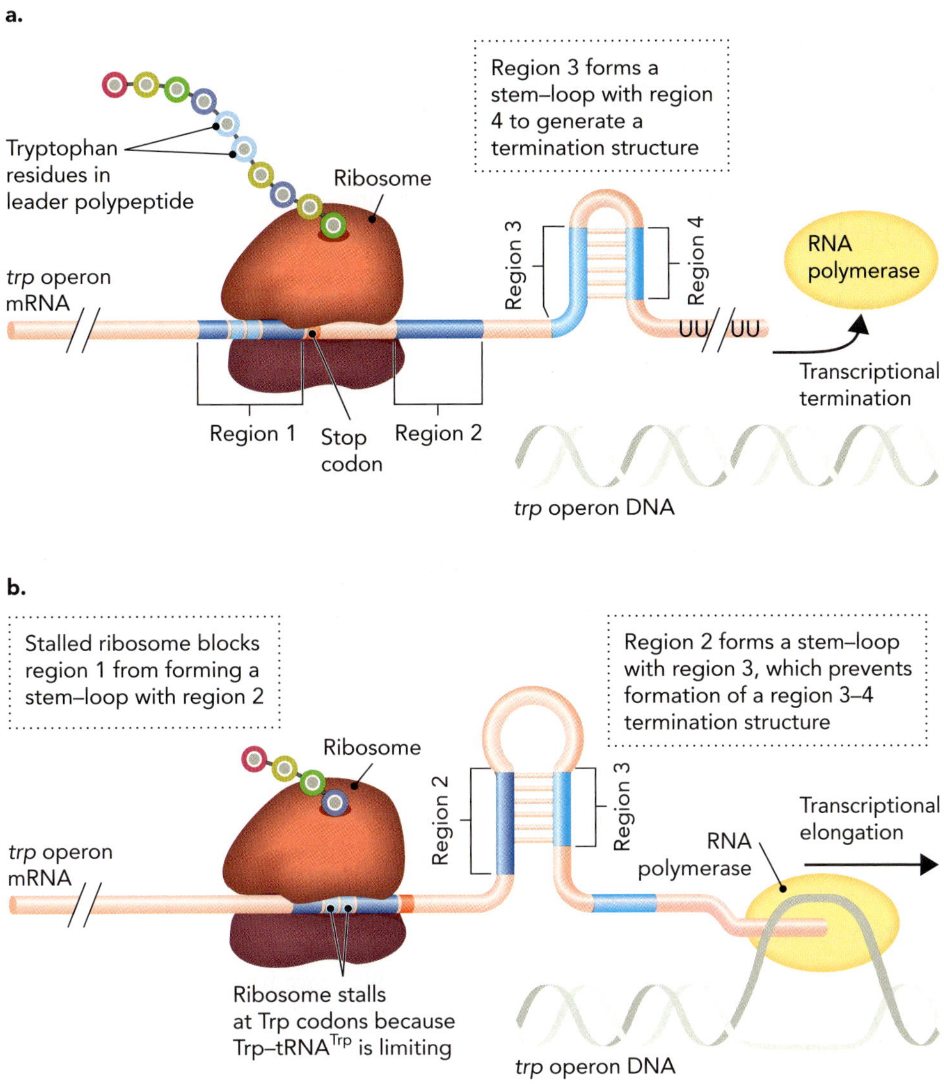
a.
Region 3 forms a stem–loop with region 4 to generate a termination structure
Tryptophan residues in leader polypeptide
Ribosome
trp operon mRNA
Region 3
Region 4
RNA polymerase
UU
UU
Transcriptional termination
Region 1
Stop codon
Region 2
trp operon DNA
b.
Stalled ribosome blocks region 1 from forming a stem–loop with region 2
Region 2 forms a stem–loop with region 3, which prevents formation of a region 3–4 termination structure
Ribosome
Region 2
Region 3
Transcriptional elongation
RNA polymerase
trp operon mRNA
Ribosome stalls at Trp codons because Trp–tRNA^Trp is limiting
trp operon DNA

Figure 23.35 Mechanism of attenuation of the *trp* operon in *E. coli* is shown. **a.** Transcriptional termination by RNA polymerase is favored when tryptophan levels are high because the abundance of charged Trp–$tRNA^{Trp}$ means the ribosome quickly synthesizes the leader polypeptide, and the termination structure generated by the region 3–4 stem–loop is allowed to form. **b.** Transcriptional elongation is favored when tryptophan levels are low because the ribosome stalls at the pair of Trp codons owing to low levels of Trp–$tRNA^{Trp}$ in the cell. The ribosome stalling prevents region 1 from pairing with region 2, which is then free to pair with region 3. Formation of the region 2–3 antitermination structure favors transcriptional elongation by preventing the formation of the region 3–4 termination structure.

the leader peptide is not involved in translation of the structural genes, which have their own ribosome binding sites on the mRNA. Why does *E. coli* have two different mechanisms for controlling transcription of the *trp* structural genes? One answer is that these mechanisms extend the range of expression of the operon: repression reduces the level about 80-fold, and attenuation another 6- to 8-fold. Attenuation is only a factor under conditions of extreme starvation for tryptophan, so having a second mechanism also allows the system to respond to a wider range of conditions.

concept integration 23.2

How do prokaryotic regulatory proteins affect transcription of targetgenes?

There are two general classes of regulatory proteins: repressors and activators. Repressors typically bind at or very near the promoter and block access of RNA polymerase to the promoter, preventing transcription from occurring. Activators typically bind somewhat upstream of the promoter and act to stimulate one or more steps in the formation of open complexes. They can enhance the binding of RNA polymerase and/or they can speed the transition from a closed complex to an open complex.

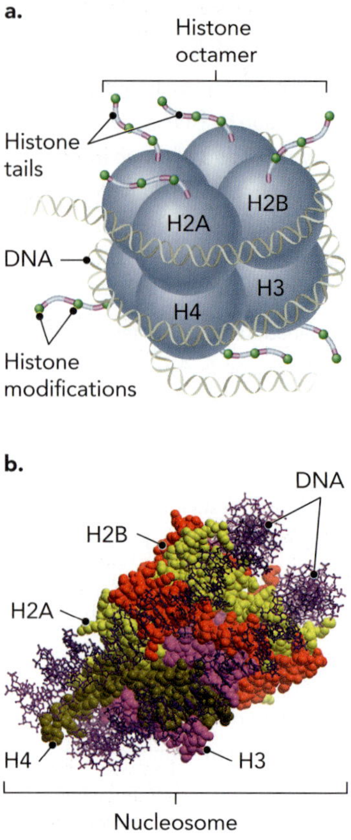

Figure 23.36 The basic building block of chromatin is the nucleosome. **a.** A nucleosome consists of a protein core containing two molecules each of four histones called the histone octamer and a 147-bp segment of DNA that wraps around this positively charged protein complex. **b.** Structure of a nucleosome containing human histone proteins.

BASED ON PDB FILE 3AFA.

23.3 Mechanisms of Eukaryotic Gene Regulation

Gene regulation in eukaryotes is vastly more complex than in prokaryotes. The main reason is that eukaryotic DNA is packaged into chromatin. The basic building block of chromatin is the nucleosome, in which about 147 bp of DNA are wrapped around a complex of eight histone molecules, called the histone octamer (**Figure 23.36**). This arrangement compacts the DNA about sevenfold. The histones carry amino-terminal tails that protrude from the surface of the structure; these are especially rich in positive charges and are one of the main sites for posttranslational modifications such as acetylation. As we saw in Chapter 3, chromatin has multiple levels of organization, in which nucleosomes are arranged in progressively more compact arrays (see Figure 3.32). Moreover, for a eukaryotic gene to be expressed, it must reside in a region of chromatin that is accessible to transcription factors and to other DNA binding proteins (**Figure 23.37**). This means that condensed chromatin, which is considered transcriptionally repressed, must be remodeled into decondensed chromatin consisting of widely spaced nucleosomes resembling "beads on a string." In addition, nucleosomes must be absent entirely from the DNA region surrounding the start site of transcription where RNA polymerase binds. Achieving a state of transcriptional competence is an active, energy-requiring process involving an incredibly large number of molecular players, which includes both proteins and small noncoding RNAs.

Eukaryotic Gene Regulation Is Most Often Transcriptional Activation

The presence of chromatin creates a fundamental difference between gene expression in prokaryotes and eukaryotes. In prokaryotes, the "ground state" of a gene is that it is available to be expressed. Some prokaryotic promoters are weak and require activators for expression, but the promoter is available by default. In contrast, in eukaryotes the ground state of a gene is that it is unavailable because it is masked by chromatin, and a complex sequence of events must ensue to allow transcription to occur. Therefore in most cases, regulation in eukaryotes is based on gene activation rather than gene

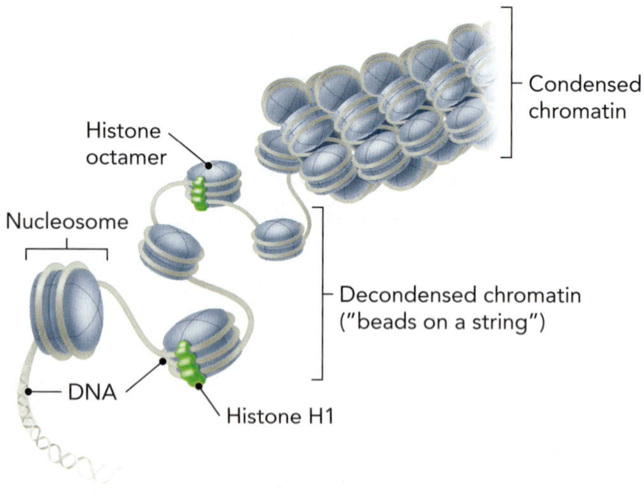

Figure 23.37 Chromatin can be found in two structural states: condensed and decondensed. In condensed chromatin, the nucleosomes associate into tightly packed structures involving interactions between the N-terminal tails of histone proteins. In decondensed chromatin, the DNA is extended with widely spaced nucleosomes.

repression. Eukaryotic transcription factors that function as activator proteins counteract the repressive effects of chromatin and help recruit components of the transcription machinery to promoters.

Most eukaryotic activators are specific DNA binding proteins, and they carry out their various roles by binding to cis-acting sites located in the vicinity of genes to be regulated. These cis-acting regulatory sites are most often in regions of decondensed chromatin and usually not covered by nucleosomes. Some cis-acting sites lie relatively close to the transcription start site for a promoter, and others lie much farther away, in regions termed **enhancers** (**Figure 23.38**). Both at enhancers and near promoters, chromatin must be modified to allow binding of other factors, so we begin by discussing how activators do this. Most activators have a modular organization, with separate DNA binding and transcriptional activation domains, in which each domain is structurally and functionally autonomous. Activator proteins stimulate chromatin decondensation and help assemble the transcription machinery by recruiting other proteins to the gene promoter region. In turn, these recruited proteins carry out a diverse range of functions. Activator protein complexes carry out two broad roles. First, they modify chromatin, often freeing other binding sites for activators. Second, they are part of the transcription machinery and help to assemble a pre-initiation complex at the promoter.

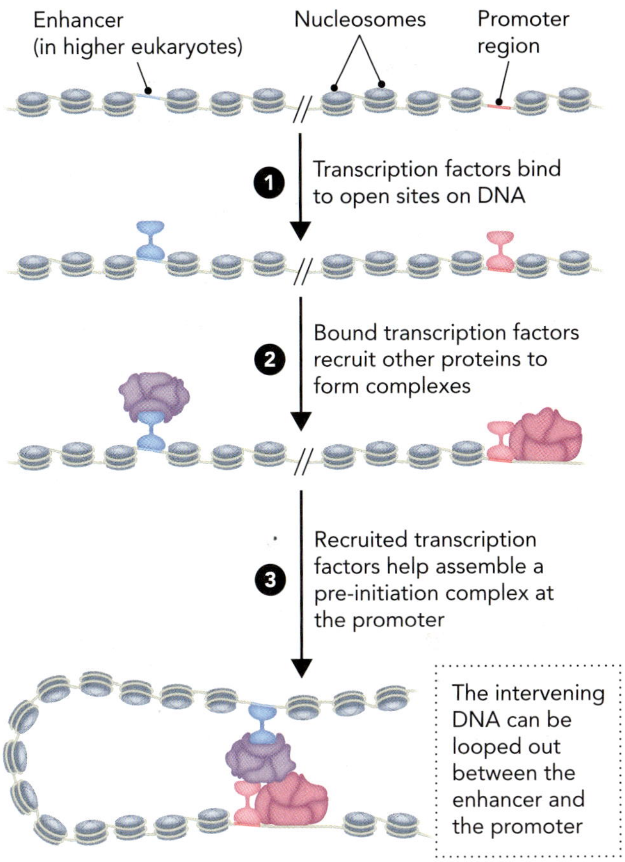

Figure 23.38 Transcription factors that function as activators recruit other transcription factors to DNA regulatory sites. Some activator binding sites are not covered with nucleosomes, allowing activators to bind.

Gene Activation Is Associated with Chromatin Modifications There are three broad classes of nucleosome modifications:

1. Histones in nucleosomes undergo a wide variety of posttranslational modifications on their amino acid side chains.
2. The positions of nucleosomes relative to the DNA sequence can be changed—they can be moved to new positions or even removed from the DNA. This can expose or hide cis-acting sites.
3. Some nucleosomes contain histone variants, in which a particular histone is replaced by a closely related protein. This can alter the properties of the nucleosome in ways that are not yet well understood.

As shown in **Figure 23.39**, all three of these functions are carried out by individual proteins or protein complexes that are recruited by transcription factors. We will focus on the first two types of modifications, which are better understood.

Posttranslational Histone Modifications One major modification to histones is lysine acetylation on particular residues. These residues lie in the **histone tails**—portions of histones that lie at their amino-terminal ends and protrude from the main body of the nucleosome (see Figure 23.36). Studies have shown that histone

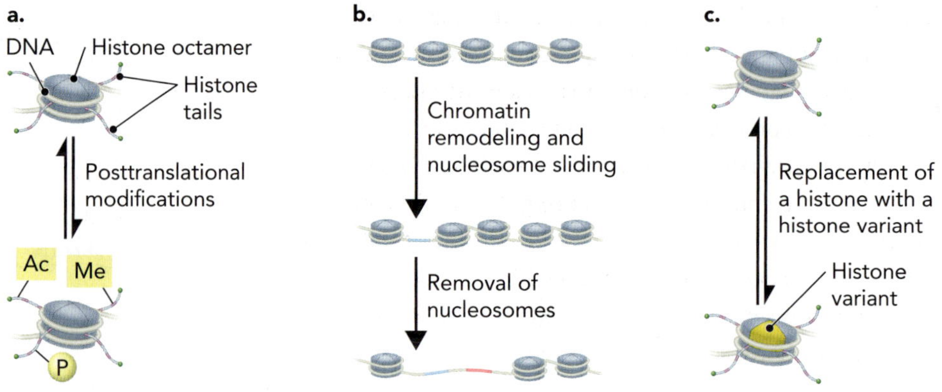

Figure 23.39 Three general types of modifications to nucleosomes affect gene expression. **a.** Posttranslational modification of histones. **b.** Nucleosome positioning relative to specific DNA sequences. **c.** Switching out of common histones in nucleosomes with histone variants. Ac = acetyl; Me = methyl; P = phosphoryl.

acetylation is associated with increases in transcriptional activity, whereas histone deacetylation is associated with repression of transcription.

How might histone acetylation on lysine residues lead to transcriptional activation? Consider that histone acetylation removes the positive charge from a lysine side chain. This has at least three biochemical consequences, each of which contributes to increased transcription. First, it weakens the interaction between the histone tail and the DNA by removing an electrostatic interaction and therefore destabilizes the nucleosome. Second, acetylation weakens interactions between nucleosomes, thereby disassembling the higher-order chromatin structures. Third, histone acetylation provides binding sites for additional transcription factor proteins. As shown in **Figure 23.40**, histone acetylation is carried out by enzymes called **histone acetyltransferases,** or **HATs** (Figure 23.40). These enzymes are recruited to cis-acting regulatory sequences by activator proteins. HAT enzymes are components of multi-subunit protein complexes that are recruited to gene promoters by trans-acting activator proteins that bind to cis-acting regulatory sequences. HAT enzymes can be multidomain proteins containing a catalytic domain and a separate protein interaction domain. Conversely, acetyl groups on histone lysine residues are removed by the action of enzymes called **histone deacetylases (HDACs)**, leading to loss of activation and chromatin decondensation.

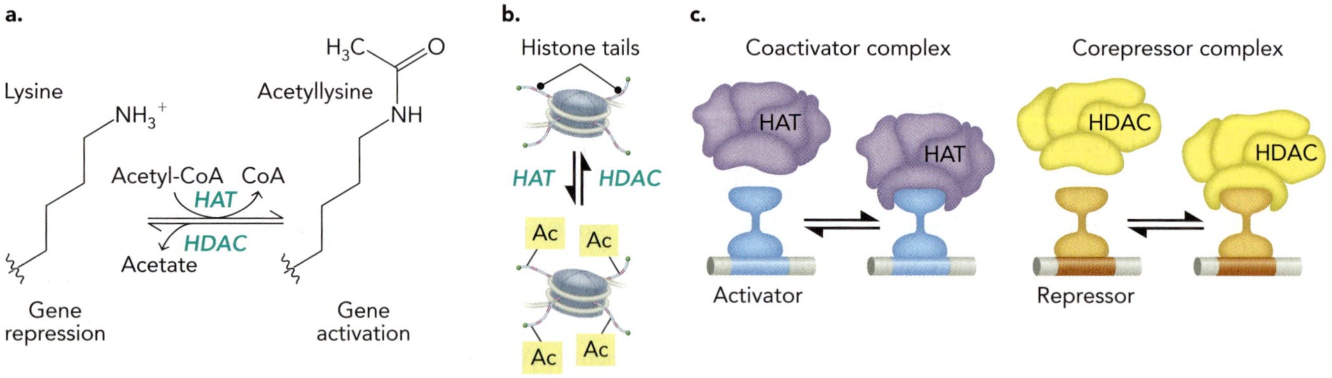

Figure 23.40 Histone acetylating and deacetylating reactions are chromatin-modifying processes. **a.** The chemical reactions catalyzed by histone acetyltransferase (HAT) enzymes and histone deacetylase (HDAC) enzymes are associated most often with gene activation and gene repression, respectively. The amino group in a lysine side chain in a histone is acetylated by HAT enzymes, neutralizing the positive charge. HDAC enzymes hydrolyze the amide bond, restoring the free amino group. **b.** Acetyl groups are added to N-terminal tails on one or more histones in a nucleosome. **c.** HAT and HDAC enzymes are typically components of large transcription factor complexes.

HDAC enzymes typically are found in large protein complexes and are recruited to promoters by the action of trans-acting repressor proteins bound at cis-acting regulatory sites. Acetylation creates an additional docking site for recruiting new factors. Acetylated histones interact with proteins containing **bromodomains**, whereas methylated histones interact with proteins containing **chromodomains** (**Figure 23.41**).

One striking example of histone modification appears in mammalian nuclear receptors, which were described in Chapter 8. Recall that these nuclear receptor proteins bind small lipophilic ligands and that many of these receptors are always bound at their respective cis-acting sites. For example, the thyroid hormone receptor forms a DNA-bound complex with the retinoid X receptor (RXR) protein. In the absence of ligand, this complex recruits a corepressor complex called the silencing mediator for RAR and TR (SMRT) complex (**Figure 23.42**). One component of the SMRT complex is an HDAC enzyme whose deacetylase activity is activated by binding to the complex. The HDAC enzyme then acts to remove acetyl groups from neighboring nucleosomes. When ligand binds to thyroid receptor, the structure of the receptor changes, and the corepressor can no longer bind. Instead, a coactivator complex binds, and this complex includes a HAT enzyme, leading to gene activation.

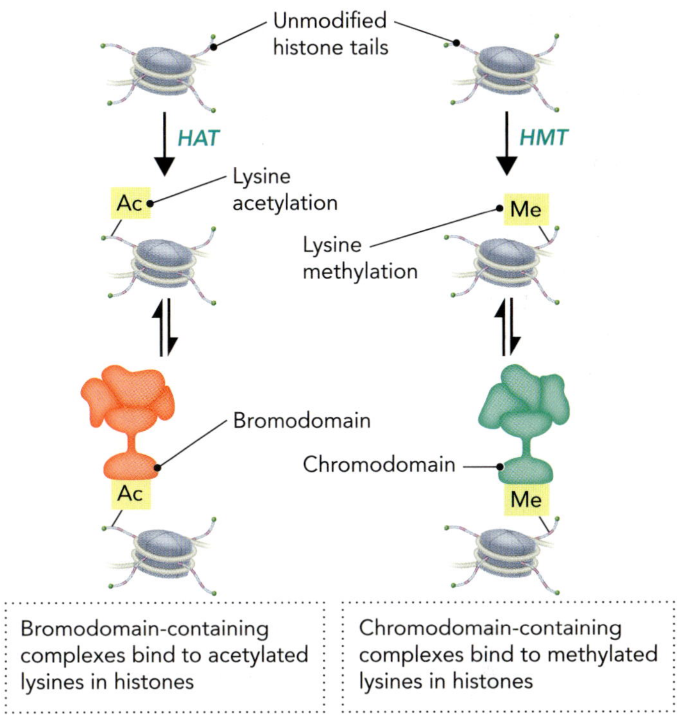

Figure 23.41 Lysines in the tails of unmodified nucleosomes are acetylated by HAT enzymes (left) or methylated by histone methyltransferase (HMT) enzymes (right). The acetylation or methylation modification serves as a binding site for bromodomain or chromodomain proteins, respectively.

In addition to acetylation, histone proteins in nucleosomes undergo a wide variety of other posttranslational modifications, including methylation of lysine residues and phosphorylation of serine residues. Histone methylation and phosphorylation have been shown to provide chromatin docking sites for additional transcription factor proteins. Lysine methylation differs from acetylation in several ways. First, methylation does not affect the positive charge on the lysine residue. Second, a lysine amino group can carry one, two, or three methyl groups, and in some cases the number of substituents affects the function of the modified group. Third, methylation is generally a longer-lasting modification than acetylation. Methylation and demethylation enzymes are highly specific for particular lysine residues in histone proteins. Methylation of some histone lysine residues is associated with gene activation, whereas methylation of other lysine residues is associated with gene repression.

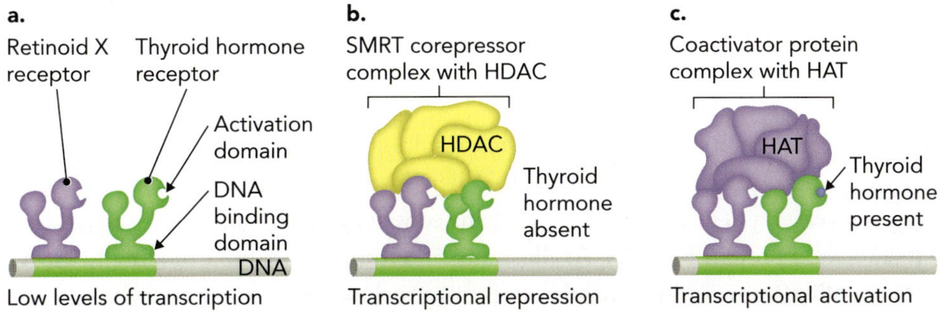

Figure 23.42 Nuclear receptors recruit HAT and HDAC enzymes to regulatory regions near promoters. **a.** Heterodimer of retinoid X receptor (RXR) and thyroid receptor (TR) bound to a DNA site. Both proteins have DNA binding domains and activation domains. **b.** In the absence of thyroid hormone, the RXR–TR heterodimer recruits a corepressor such as the SMRT complex, which contains an HDAC. **c.** When thyroid hormone binds to the TR activation domain, the structure of TR changes, leading to HAT enzyme recruitment to the complex, histone acetylation, and gene activation.

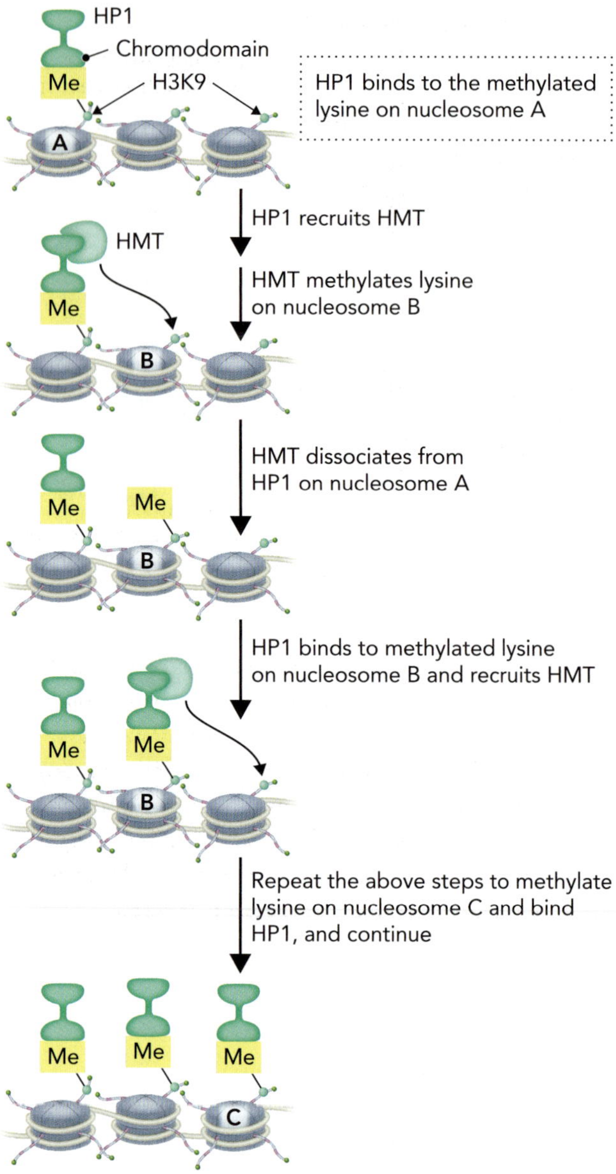

Figure 23.43 The sequential pathway shown is for converting a tract of nucleosomes into methylated forms. The process starts at the top with a single methylated nucleosome. HP1 then recruits an HMT that specifically methylates H3K9 on neighboring nucleosomes.

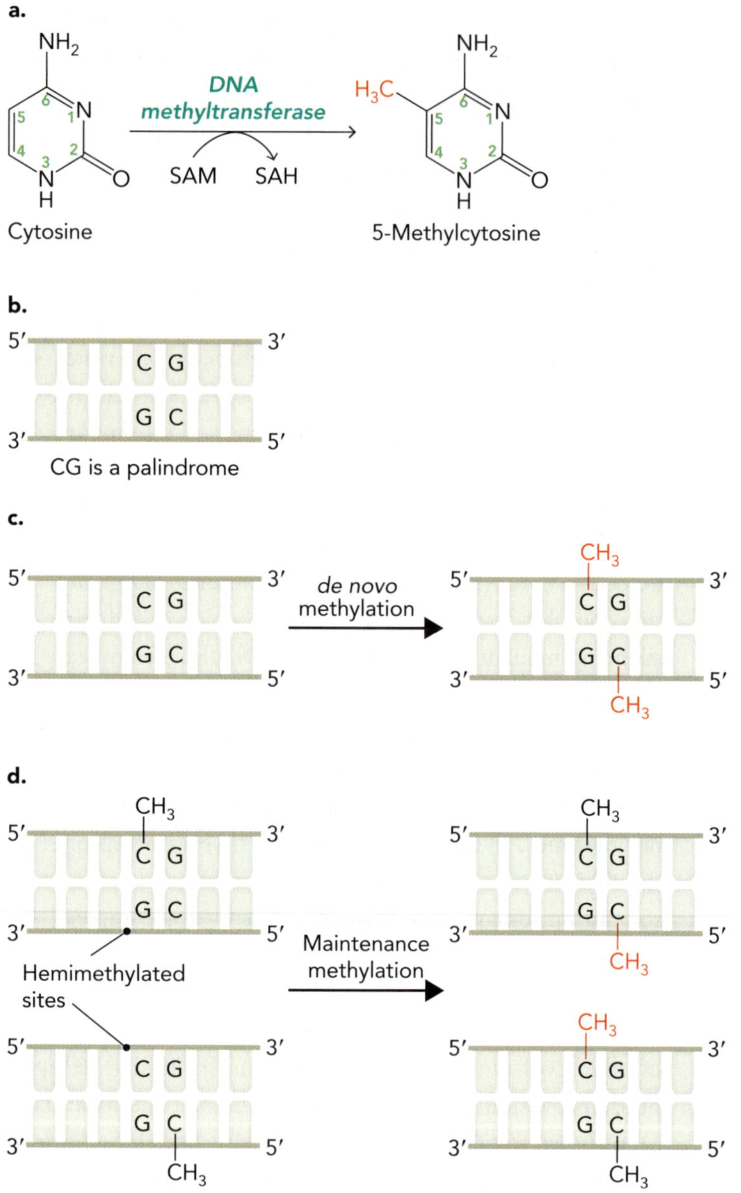

Figure 23.44 Two classes of methylation events and DNA methyltransferases exist. **a.** DNA methyltransferases transfer a methyl group from *S*-adenosyl-L-methionine (SAM) to cytosine. **b.** Most methylation in higher eukaryotes occurs at the palindromic sequence 5′-CG-3′. **c.** In *de novo* methylation, a 5′-CG-3′ sequence that is initially unmethylated on both strands becomes methylated. **d.** Maintenance methylation follows after DNA replication of a fully methylated site, which produces two progeny duplexes, each with a methyl group only on the parental strand (a hemimethylated site). A maintenance methyltransferase then adds a methyl group to each progeny strand. SAH = *S*-adenosyl-L-homocysteine.

A specific example of gene repression resulting from lysine methylation involves methylation of Lys9 in histone 3 (H3K9). When this residue is dimethylated or trimethylated, it serves as a docking site for proteins containing chromodomains such as heterochromatin protein 1 (HP1). As illustrated in **Figure 23.43**, HP1 binding to H3K9 recruits other transcriptional regulatory factors, including a histone methyltransferase that methylates H3K9 on neighboring nucleosomes. This creates a new binding site for HP1, which recruits another HMT, and the cycle repeats. The result is methylation and HP1 binding to a large tract of adjacent nucleosomes, which constitutes an epigenetic change that can continue for many cell generations.

Another type of chromatin modification that affects gene expression is methylation of the nucleotide base cytosine in regions of DNA near gene promoters. As shown in **Figure 23.44**, cytosine can be methylated on its 5-position to form 5-methylcytosine (5-MeC). DNA methylation occurs mainly on cytosine residues found in the palindromic sequence 5′-CG-3′. Both DNA strands are typically methylated at this site. Two types of enzymes catalyze the formation of 5-MeC. First is an enzyme that methylates nonmethylated CG sequences and is termed a *de novo* DNA cytosine methyltransferase. The second enzyme, a maintenance cytosine methyltransferase, works on a hemimethylated sequence, in which one DNA strand is methylated and the other is not. Hemimethylated DNA results from DNA replication of a fully methylated CG sequence. The maintenance cytosine methyltransferase perpetuates the preexisting pattern of methylation and represents another example of an epigenetic modification, as it does not change the DNA sequence itself.

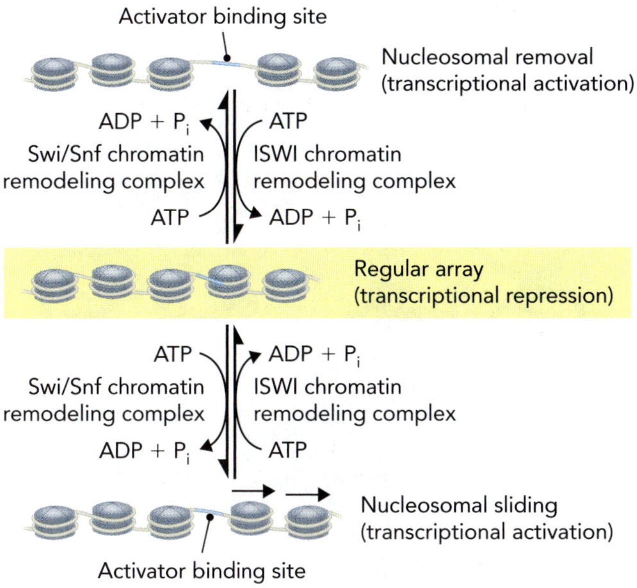

Figure 23.45 Chromatin remodeling complexes change the spatial relationships among nucleosomes. Action of the Swi/Snf protein complex shifts nucleosome positions along the chromatin or removes them altogether. This exposes cis-acting sequences that are binding sites for trans-acting transcriptional activator proteins. Other types of chromatin remodelers, such as the ISWI protein complex, function to establish equally spaced nucleosome arrays that contribute to gene inactivation.

DNA methylation does not directly cause transcriptional repression, but rather marks regions of the genome that are destined to be less transcriptionally active than unmethylated regions. Methylated CG sites in DNA appear to affect the recruitment of chromatin remodeling proteins that condense chromatin to silence gene transcription. Although DNA methylation is considered a heritable epigenetic marker, it can be reversed. Two mechanisms for demethylation are (1) inhibition of the maintenance methyltransferase, followed by multiple rounds of DNA replication during cell growth and division, and (2) oxidation of the methyl group to make 5-hydroxymethylcytosine, which is then removed by DNA repair mechanisms.

Nucleosome Positioning Another major class of chromatin modifications affecting gene activation is known as chromatin remodeling. Several large protein complexes (the chromatin remodeling machines) are involved in this process. These complexes interact with promoters and enhancers in different ways. In some cases, specific DNA binding proteins are required to form a larger protein–DNA complex; alternatively, preexisting histone modifications (such as acetylation or methylation) serve as docking sites. For example, some chromatin remodeling machines contain bromodomains, which facilitate binding to acetylated histones. These machines then use the energy available from ATP to alter the positions of nucleosomes in various ways.

Different chromatin remodeling machines have distinct effects on nucleosome positioning. As shown in **Figure 23.45**, one type is the **Swi/Snf complex** (switch/sucrose non-fermentable), which moves nucleosomes along the DNA or leads to their displacement. This has the effect of creating nucleosome-free regions near a promoter, which contributes to transcriptional activation. Another chromatin remodeling machine is

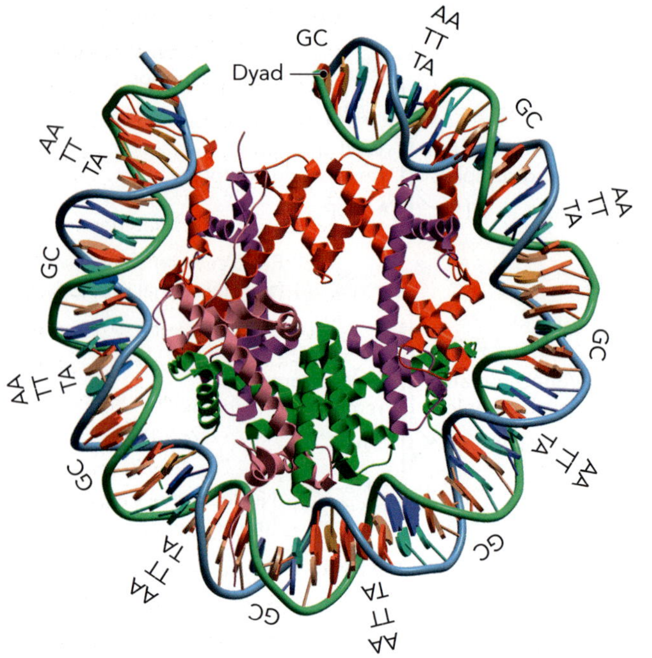

Figure 23.46 Alternating TA and GC sequences about every 10 nucleotides favors nucleosome positioning. About one turn of the DNA is represented, and the dyad axis of the DNA (the midway point of the DNA in the nucleosome) is indicated. The protein structure in the middle is part of the histone core. BASED ON PDB FILE 3AFA.

typified by the ISWI complex (imitation SWI), which functions to space nucleosomes at regular intervals along the chromatin and is associated with transcriptional repression.

The DNA sequence itself often contributes to the location of nucleosomes by the presence of nucleosome-positioning sequences. Some promoters, especially in yeast, contain DNA elements that prevent nucleosome binding. This is likely related to the propensity of particular dinucleotides to allow the sharp bending needed to form a nucleosome. As shown in **Figure 23.46**, some nucleotides favor bending when located properly with respect to the histones. For example, the dinucleotide sequences AA, TT, and TA in DNA are often found in regions of the nucleosome facing in toward the histone core, whereas GC dinucleotides are most often in regions of the DNA that face outward away from the histone core. Moreover, long tracts of A residues are unusually stiff and cannot be easily bent to wrap around the nucleosome and therefore are rarely found in regions of stable nucleosome positioning.

The first transcription factor protein complex that binds to an accessible cis-acting sequence near a gene promoter is called a **pioneer factor** because it sets in motion the subsequent chain of events. After binding of the pioneer transcription factor protein complex to the regulatory site on DNA, additional transcription factor complexes bind in the same region and initiate the process of transcriptional activation. As shown in **Figure 23.47**, at least three possibilities exist for how pioneer transcription factor binding can be initiated on cis-acting sites located in gene regulatory regions: (1) pioneer transcription factor complexes can bind to accessible cis-acting sites in the linker DNA regions between nucleosomes; (2) pioneer transcription factor complexes can bind to transiently exposed cis-acting sequences that

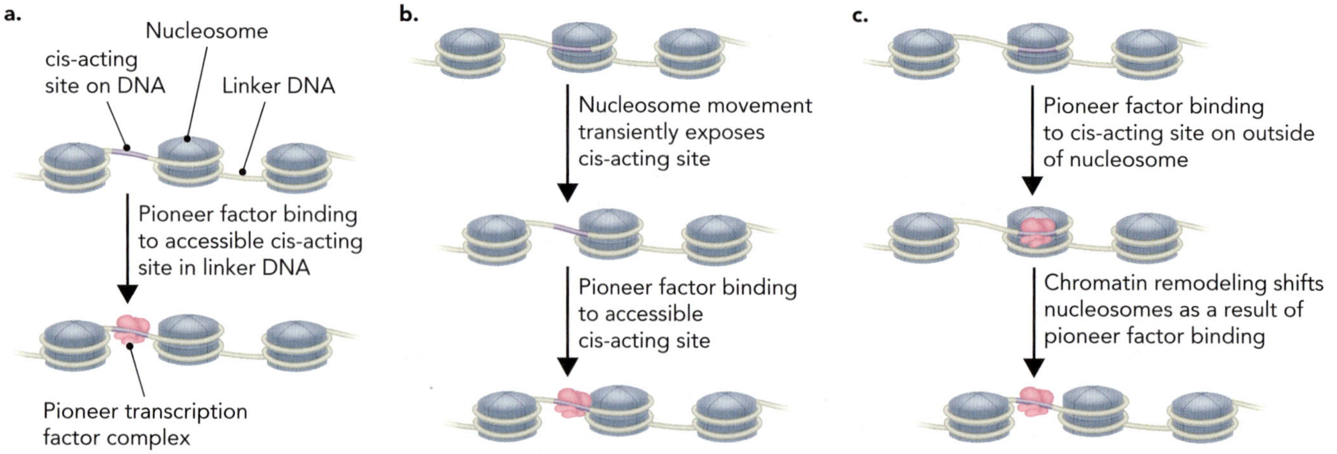

Figure 23.47 Pioneer factors are the first transcription factors to bind to a promoter or enhancer region, leading to changes that allow other factors to bind. Several mechanisms can enable their binding. **a.** The nucleosomes might be positioned so that the binding site is free. **b.** The DNA near the end of the portion wrapped around the core can dissociate briefly, exposing the site. If a factor binds to such a site, it prevents the DNA from reassociating with the core. **c.** Some pioneer transcription factors can bind to DNA that is wrapped in a nucleosome and initiate chromatin remodeling to expose the cis-acting site.

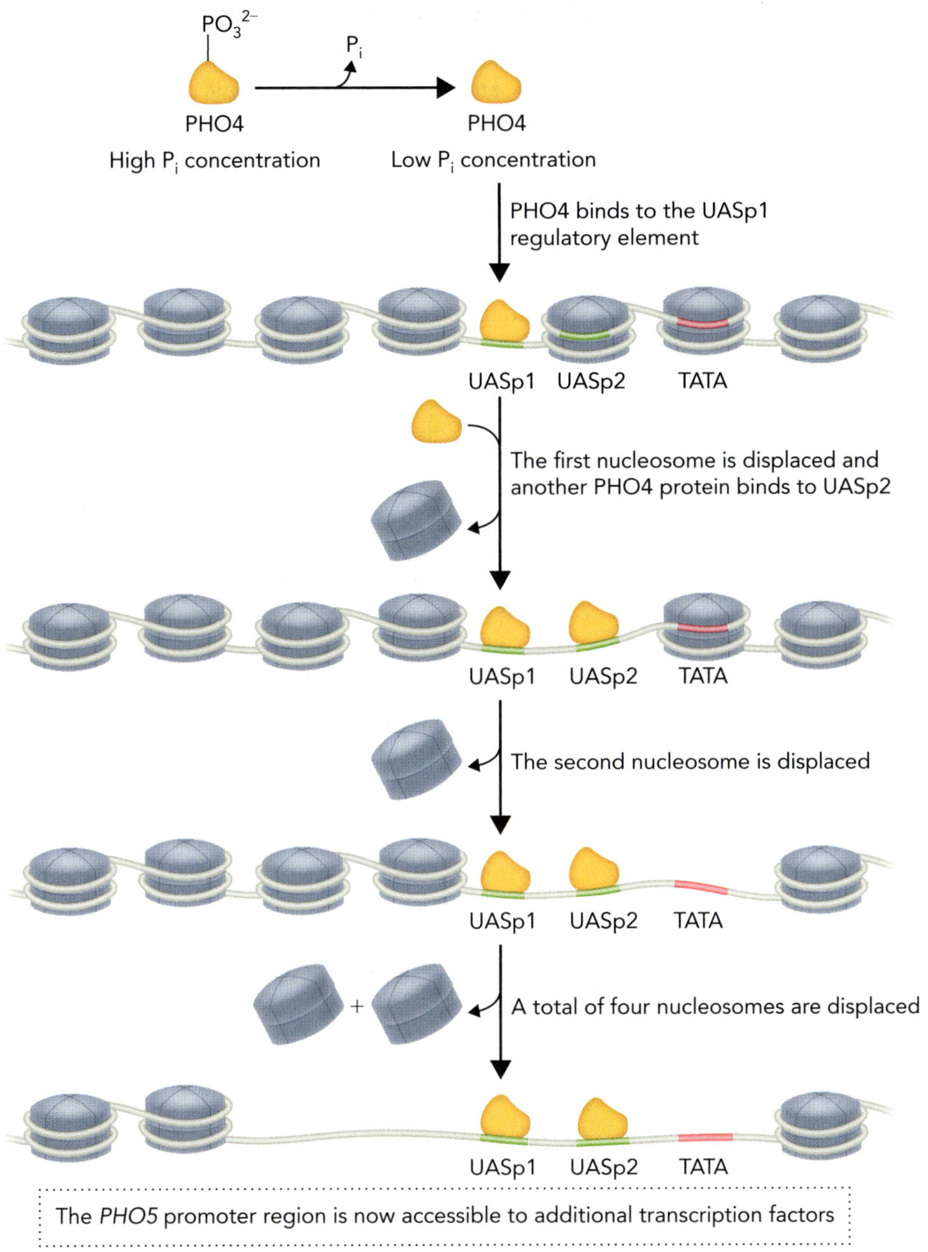

Figure 23.48 Transcription of the yeast PHO5 gene is regulated by PHO4-mediated removal of nucleosomes near the *PHO5* promoter. Under conditions of low phosphate levels, PHO4 is dephosphorylated and translocated to the nucleus where it binds to a regulatory element called UASp1. This leads to repositioning of the neighboring nucleosome, exposing a second PHO4 binding site, UASp2, which is then free to bind another PHO4 protein. Eventually, a total of four nucleosomes are removed from a region of the *PHO5* promoter, thereby establishing a nucleosome-free region to facilitate the binding of additional transcription factors.

are located near the linker DNA; or (3) pioneer transcription factor complexes can bind to cis-acting sites located on the outside of the nucleosome surface.

A well-studied example of nucleosome positioning comes from the promoter for the yeast PHO5 gene (**Figure 23.48**). This gene is turned on when the yeast cells are starved for phosphate. It codes for a secreted phosphatase and acts to release phosphate groups from phosphomonoesters in the environment. Hence, it only needs to be expressed when phosphate is limiting. The PHO5 gene promoter region contains binding sites for several transcription factors, including the transcription factor PHO4, which is itself regulated by phosphorylation. When phosphate becomes limiting, PHO4 is dephosphorylated and moves to the nucleus, allowing it to bind to specific sites on DNA near the *PHO5* promoter. PHO4 recruits chromatin modifiers, and their action results in removal of four nucleosomes, which then promotes activation of the *PHO5* promoter. But when phosphate is abundant, these four nucleosomes are positioned in the regulatory region for *PHO5* and block the binding transcription factors such as the TATA binding protein.

Activator Proteins Recruit Transcriptional Machinery to Gene Promoters Transcriptional activator proteins have two primary functions: (1) recruit chromatin modifiers to DNA regions near genes, and (2) recruit components of the general transcriptional machinery to gene promoter sequences. Recall from Chapter 21 that RNA polymerase II does not act by itself, unlike prokaryotic RNA polymerase, but rather enters the growing transcriptional pre-initiation complex at a late stage of its assembly (see Figure 21.21). Indeed, most of the general transcription factors, such as TFIID, TFIIB, and TFIIH, were initially identified by their ability to stimulate basal transcription of a relatively active promoter in an *in vitro* assay system using nucleosome-free DNA as a template. In these experiments, no transcriptional activator protein was required, and the complications due to chromatin were absent. On most promoters, transcriptional activator proteins are required, as well as a large multiprotein complex called the **mediator complex**. The mediator complex is required *in vivo* for expression of essentially all eukaryotic genes, and by this criterion, it is considered a general transcription factor complex.

Whether activator proteins are bound at enhancer sequences located at a far distance from the gene promoter or to cis-acting sequences very near to the gene promoter, transcriptional activator proteins interact with many of the general transcription factors to help recruit them to the growing pre-initiation complex. As shown in **Figure 23.49**, two of the key protein complexes recruited to the pre-initiation complex by activator proteins are the mediator complex and TFIID. Transcriptional activator proteins often make multiple contacts with these large protein complexes, and when two or more activator proteins are involved, they often contact different parts of the transcriptional machinery.

When a gene is active for transcription, it generally is transcribed multiple times. After RNA polymerase II initiates transcription and leaves the promoter, some portion of the transcription machinery is left behind at the promoter, forming a scaffold that most likely facilitates entry of another RNA polymerase II molecule for reinitiation

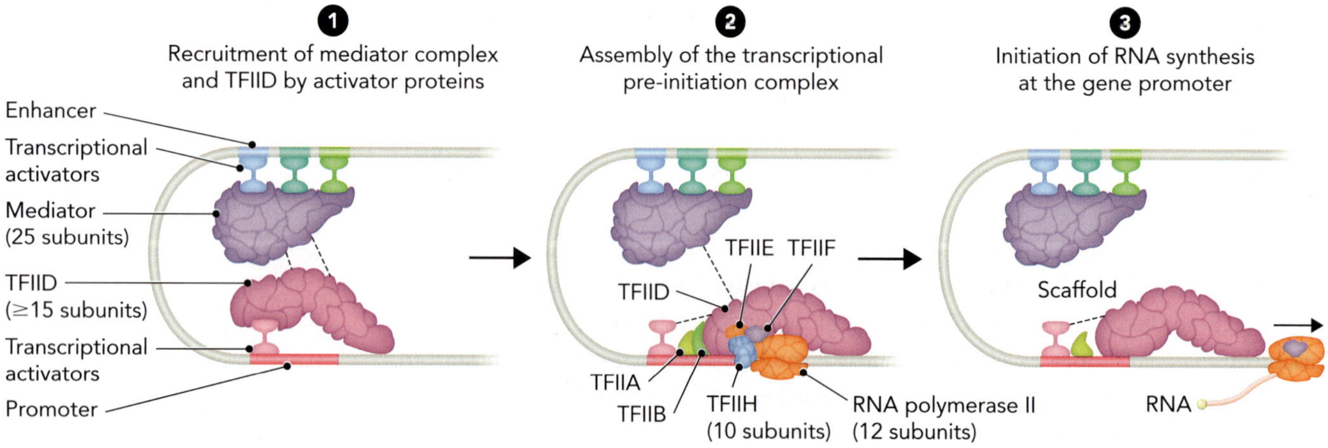

Figure 23.49 Transcriptional activator proteins recruit mediator and TFIID complexes to gene promoters to initiate transcription by RNA polymerase II. In step 1, activator proteins bind to sequences near gene promoters and recruit the TFIID complex, whereas activator proteins bound to sequences far upstream of the gene promoter (enhancer sequences) recruit mediator protein complexes. Nucleosomes are not depicted for simplicity, and heavy dashed lines represent interactions between different components. In step 2, the transcriptional pre-initiation complex forms, which consists of the mediator complex, RNA polymerase II, TFIID, and a number of other proteins such as TFIIA, TFIIB, TFIIF, TFIIE, and TFIIH. In step 3, after initiation of transcription some of the general transcription factors, along with the mediator, are left behind at the promoter, forming a scaffold. This facilitates assembly of a transcriptional pre-initiation complex for the next round of transcription.

(see Figure 23.49). The result is that subsequent rounds of transcription are more efficient because important parts of the complex do not need to be recruited anew. Hence, the presence of the scaffold bypasses some rate-limiting steps in the initial assembly of the transcriptional pre-initiation complex.

Because enhancers can work over long distances, other mechanisms need to operate to keep them specific for the genes they are designed to control. As shown in **Figure 23.50**, DNA sequence elements termed **insulator sequences** counteract the action of enhancers when located between an enhancer and a promoter. A current model for insulator function is that insulators divide chromosomes into discrete domains, perhaps by long-range looping, and that an enhancer can work only on a promoter that lies in the same domain.

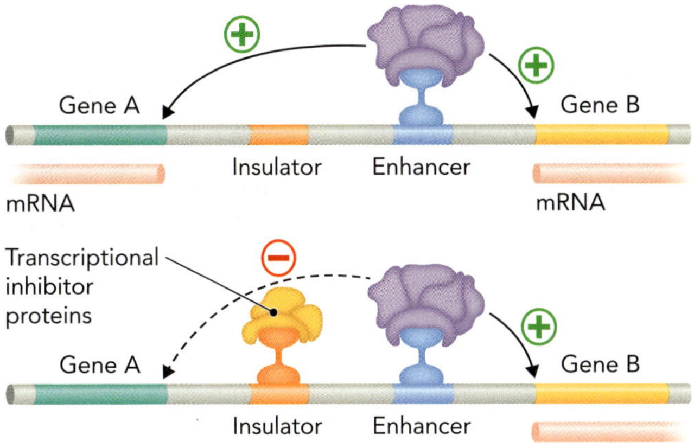

Figure 23.50 Insulator sequences are binding sites for transcriptional inhibitor protein complexes.

In summary, transcriptional regulation involves a vast network of interacting proteins and multiprotein complexes. Biochemical analysis can detect a multitude of interactions, but it is harder to tell which ones are actually important. The use of classical genetics and molecular genetic methods in model organisms complements a purely biochemical approach. Molecular genetic methods such as RNA interference and gene knockouts have been used to investigate mechanisms of cell-specific gene regulation.

Regulation of Galactose Metabolism in Yeast

We have already seen examples of two predominant ways of regulating the activities of transcription factors. First, the activities of existing transcription factor proteins are modulated, either by allosteric mechanisms or by posttranslational modifications. Second, the synthesis of transcription factor proteins can be regulated, often by the previous history of the cell. In general, single-celled organisms such as prokaryotes and yeast utilize allosteric mechanisms to regulate the activity of transcription factors, whereas multicellular organisms rely more on the synthesis and degradation of transcription factors to regulate gene expression. We describe examples of each strategy, starting with the yeast **GAL4 activator protein**.

Figure 23.51 The metabolic pathway for galactose catabolism in yeast is shown. The names of genes are shown in parentheses and are associated with the proteins they encode.

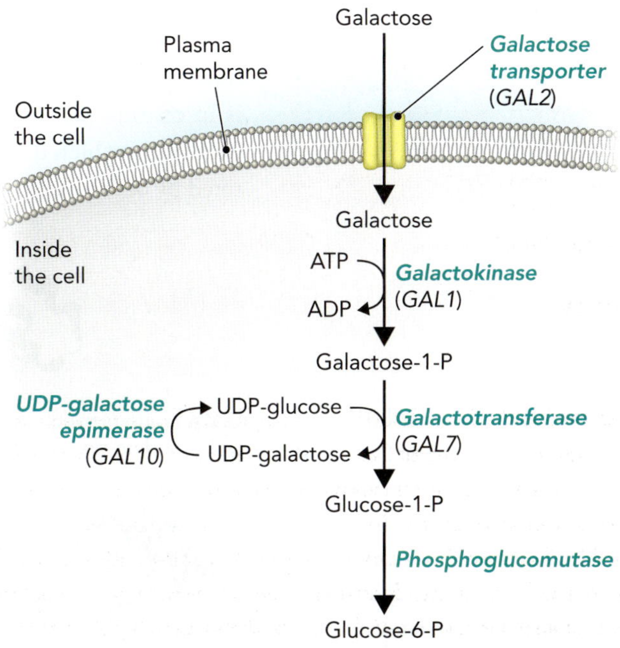

Similar to bacteria such as *E. coli*, yeast cells are exposed to a wide range of nutrients in the environment. Moreover, as with the *E. coli lac* operon, there is little reason to make catabolic enzymes if their substrates are not present. In yeast, the sugar galactose is converted in several steps to glucose-6-phosphate, which then enters the glycolytic pathway. There are at least four genes involved in this metabolic pathway, and all are transcriptionally regulated in response to galactose availability. As shown in **Figure 23.51**, these four yeast genes are named *GAL1*, *GAL2*, *GAL7*, and *GAL10*, each of which contributes to galactose catabolism.

The best characterized gene regulatory circuit in the yeast galactose catabolic pathway is the transcriptional regulation of the GAL1 and GAL10 genes by the interplay of three regulatory proteins. As shown in **Figure 23.52**, the GAL4 protein contains a DNA binding domain and a transcriptional activation domain that functions to recruit

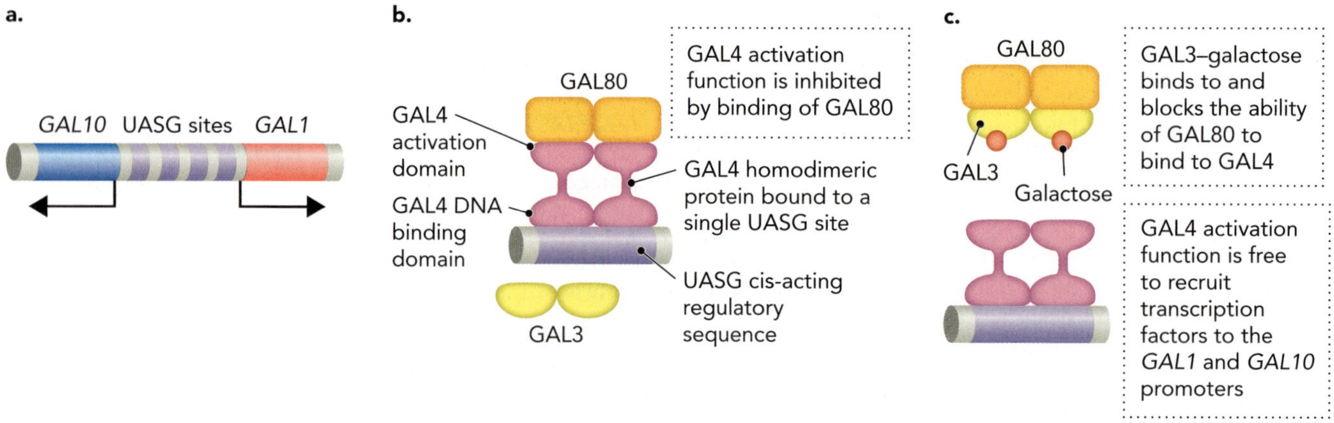

Figure 23.52 Transcriptional regulation of the GAL1 and GAL10 genes in yeast requires the GAL4, GAL80, and GAL3 proteins. **a.** Location of the cis-acting UASG sequence between the GAL1 and GAL10 genes. This site contains four binding sites for the homodimeric GAL4 transcription factor. **b.** The GAL80 allosteric regulatory protein binds to the GAL4 transcriptional activation domain in the absence of galactose and inhibits its ability to recruit transcription factors to the GAL1 and GAL10 gene promoters. Only a single GAL4 UASG binding site is shown. **c.** In the presence of galactose, the GAL3 allosteric regulatory protein undergoes a conformational change in response to binding galactose, which promotes its binding to GAL80. The GAL80–GAL3–galactose complex is unable to bind to the GAL4 transcriptional activation domain.

transcription factors to the GAL1 and GAL10 gene promoters. The GAL4 protein binds to four cis-acting regulatory sites, called the galactose upstream activator sequence (UASG), which are located in the intervening DNA segment between the opposing GAL1 and GAL10 genes. In the absence of galactose, the GAL80 regulatory protein binds to the GAL4 protein activation domain and inhibits the recruitment of transcription factors to the GAL1 and GAL10 gene promoters. Upon addition of galactose, a third protein, GAL3, binds galactose, which induces a protein conformational change in GAL3 that promotes its binding to the GAL80 protein. Once this happens, the GAL80 protein is no longer able to bind to and inhibit the GAL4 protein activation function. The result is GAL4-induced transcription of the yeast galactose genes and activation of the galactose catabolic pathway.

As shown in **Figure 23.53**, the GAL4 homodimeric protein binds to the UASG sequences using an unusual Zn_2Cys_6 binuclear cluster in which two short α helices bind to the major groove of DNA at sites that are separated by 14–16 nucleotides. This is in contrast to binding of the homodimeric glucocorticoid receptor DNA binding domain, in which the α helices of this Zn_2Cys_8 protein bind to DNA sequences separated by only 8–10 nucleotides.

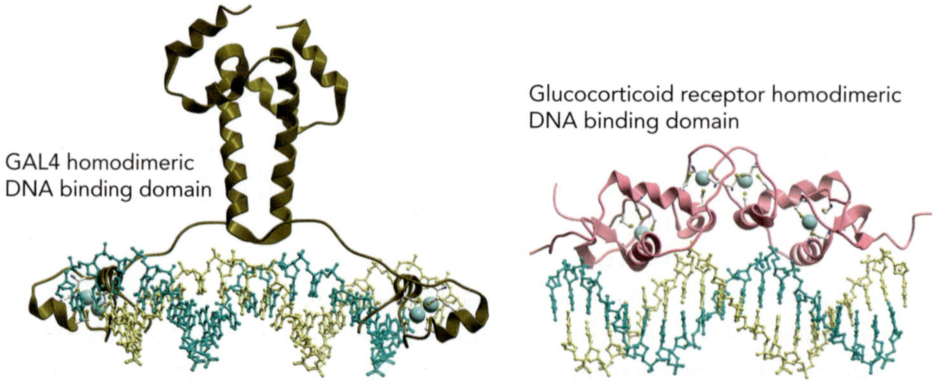

Figure 23.53 The GAL4 and glucorticoid receptor DNA binding domains both contain Zn-coordinated cysteine residues but bind to their cognate DNA sequences using distinct structures.
BASED ON PDB FILES 3COQ AND 1GLU.

Gene Expression Patterns in Developing *Drosophila* Embryos

We now turn to an example in which the control of target gene expression is dependent on transcriptional regulation of transcription factors themselves. A striking example of this is seen in the early stages of *Drosophila* development, a time at which the embryo is divided into a series of segments along an axis from the anterior end to the posterior end. A hallmark of this segmentation process can be visualized in the *Drosophila* embryo as a series of stripes that correspond to cells expressing distinct gene products. As shown in **Figure 23.54**, the locations of these stripes are determined by the concentrations of multiple transcriptional activators and repressors. A well-studied example is expression of the transcription factor gene, *even-skipped*, or *eve*, which is expressed in seven stripes along the axis of the *Drosophila* embryo (Figure 23.54a). For example, *eve* gene expression in the region of the embryo corresponding to *eve* stripe 2 is controlled by a dedicated enhancer sequence (*eve* stripe 2 enhancer). The **eve stripe 2 enhancer** contains multiple cis-acting binding sites for at least four transcription factor proteins: Bicoid, Hunchback, Giant, and Krüppel. The transcription factor proteins Bicoid and Hunchback are activators, whereas Giant and Krüppel proteins are transcriptional repressors. Expression of these four transcription factors is initiated at earlier stages of embryonic development. Most important, at the time of development when the *eve* stripe 2 enhancer is activated, the intracellular concentrations of each of these transcription factors differ as a function of cell position along the axis of the embryo. As seen in Figure 23.54b, cells close to the anterior end of the embryo express high levels of the Giant and Hunchback proteins, whereas Bicoid protein levels are relatively low, and Krüppel protein is not expressed at all. In contrast, near the posterior end of the embryo at this same time, Krüppel protein levels are high, Hunchback protein levels are lower, and Bicoid and Giant protein levels are extremely low. It is only in the region of the embryo corresponding to activation of the *eve* stripe 2 enhancer that expression of the activator proteins Hunchback and Bicoid is higher than that of the repressor proteins Giant and Krüppel.

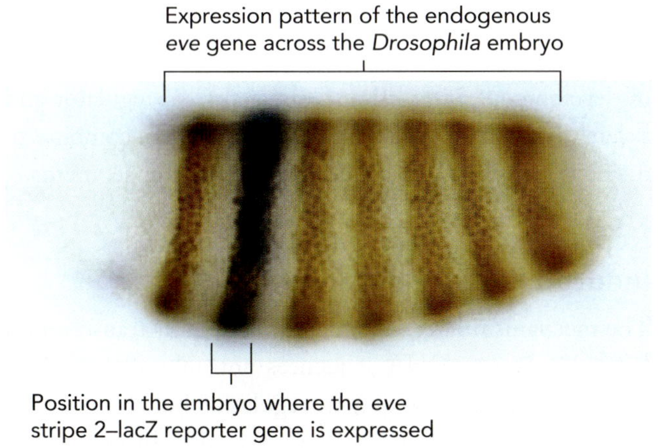

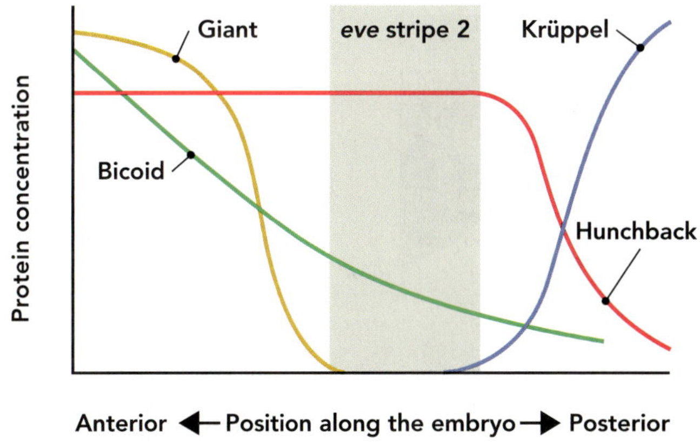

Figure 23.54 Expression and regulation of the even-skipped (*eve*) pair-rule gene in *Drosophila* is determined by the cellular concentration of four transcription factors. **a.** Stained *Drosophila* embryo showing the seven stripes of *eve* gene expression extending from the anterior end to the posterior end. This embryo contains a transgenic reporter gene in which the *eve* stripe 2 promoter is fused to the lacZ gene. The brown stripes show where the endogenous *eve* gene is expressed, and the blue stripe identifies cells in the embryo that activate only the *eve* stripe 2 enhancer in response to expression levels of Bicoid, Hunchback, Giant, and Krüppel. **b.** Spatial distribution and relative expression levels of the Bicoid, Hunchback, Krüppel, and Giant proteins across the embryo at this time in *Drosophila* development. M. LEVINE (2013). DEVELOPMENT: COMPUTING AWAY THE MAGIC? *ELIFE, 2*, E01135. © LEVINE.

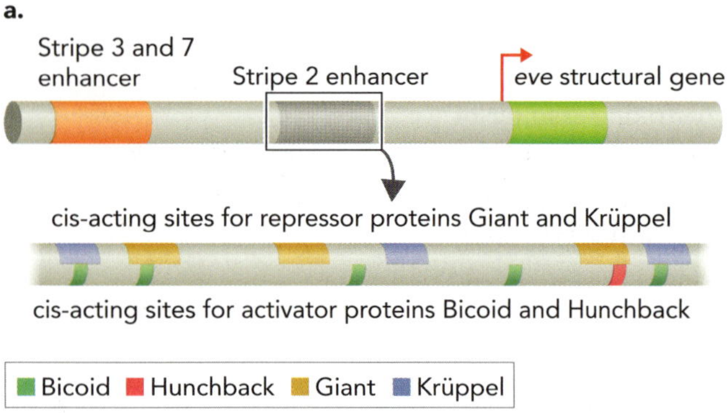

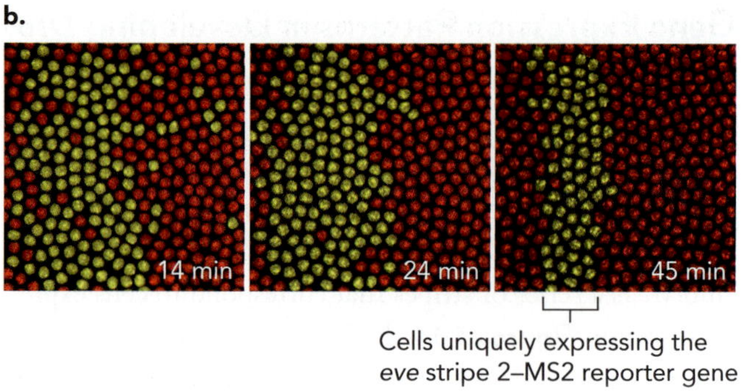

Figure 23.55 The combined occupancy of cis-acting binding sites by the Bicoid, Hunchback, Krüppel, and Giant transcription factors within the *eve* stripe 2 enhancer controls expression of the *eve* structural gene in specific cells. **a.** Expanded portion of the *eve* stripe 2 enhancer sequence shows the location of binding sites for the four regulatory proteins that specify *eve* gene expression. **b.** Time-lapse photography showing expression of an *eve* stripe 2 enhancer reporter gene as a function of transcription factor gradient formation over time. The MS2 reporter gene mRNA transcript is recognized by a fluorescently labeled MS2 protein. Live cell imaging using fluorescent microscopy allows visualization of reporter gene activation in specific cells. J. P. BOTHA ET AL. (2014). DYNAMIC REGULATION OF *EVE* STRIPE 2 EXPRESSION REVEALS TRANSCRIPTIONAL BURSTS IN LIVING *DROSOPHILA* EMBRYOS. *PROCEEDINGS OF THE NATIONAL ACADEMY OF SCIENCES USA, 111*(29), 10598–10603. DOI: 10.1073/PNAS.1410022111.

The conditions required to regulate expression of the *eve* gene precisely in the region of the embryo where its protein product is needed depends on the relative occupancy of the cognate cis-acting sequences for each of the seven *eve* stripes. Expression of the *eve* gene is controlled by binding of the Bicoid, Hunchback, Krüppel, and Giant transcription factors to their cognate DNA sequences within the *eve* stripe 2 enhancer region (**Figure 23.55a**). As shown in **Figure 23.55b**, this precise pattern of *eve* stripe 2 enhancer occupancy develops over time, so that eventually only a small number of cells express the *eve* structural gene in that region of the embryo at that time in development. The regulatory gene networks controlling expression of *eve* in cells corresponding to the other six *eve* stripes in the *Drosophila* embryo are also based on transcription factor gradients and cis-acting sequences. In the next stage of development, interactions among the *eve* regulator and several other factors, also expressed in stripes, combine to divide the embryo into smaller subdivisions or segments.

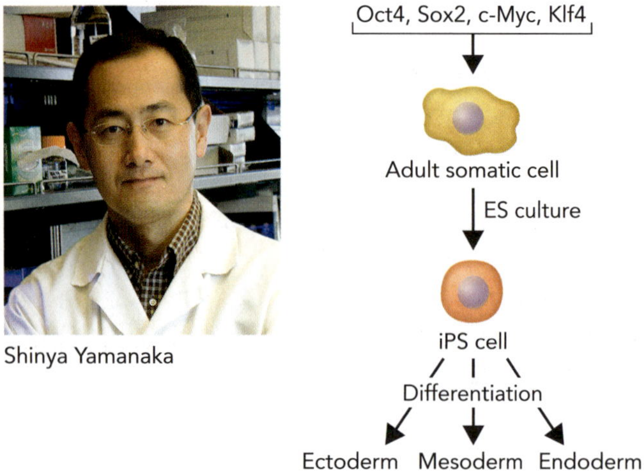

Figure 23.56 Shinya Yamanaka was awarded a portion of the 2012 Nobel Prize in Physiology or Medicine for his work on the transcriptional control of cell fate determination. ES = embryonic stem; iPS = induced pluripotent stem. PHOTO: RICK WOOD/MILWAUKEE JOURNAL-SENTINEL/RAPPORT SYNDICATION/NEWSCOM.

Reprogramming Gene Expression: Induced Pluripotent Stem Cells

The mechanism of cell fate determination by transcription factor binding to DNA sequences not only offers insight into basic mechanisms of gene regulation but also provides an opportunity to improve human health. In one of the most amazing feats of trial and error in modern biomedical science, Shinya Yamanaka of Kyoto University and his team discovered that introduction of DNA encoding the genes for four mammalian transcription factors (Oct4, Sox2, c-Myc, Klf4) into a differentiated cell was necessary and sufficient to cause induction of a pluripotent state (**Figure 23.56**). In other words, by simply forcing

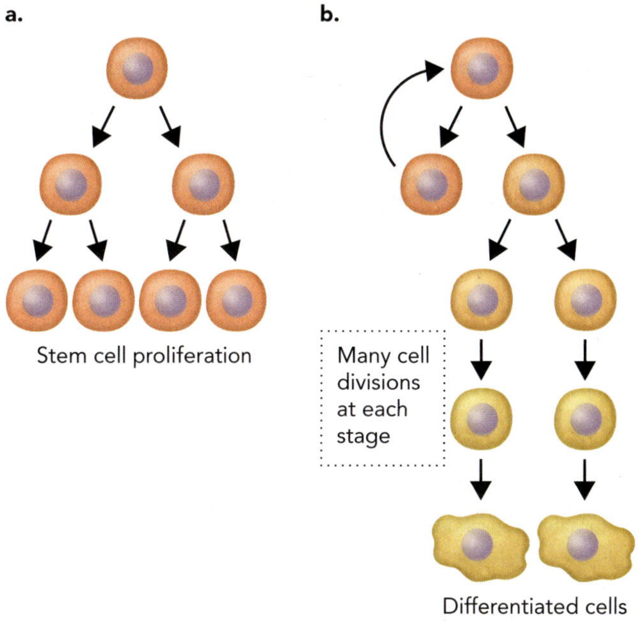

Figure 23.57 Stem cells have two properties: cell proliferation and cell differentiation. **a.** Stem cells can divide to give two cells like themselves (self-renewal). **b.** Stem cells can divide asymmetrically to give one stem cell and a cell that is at the start of a pathway to a differentiated cell type. These cells can continue dividing and developing through a series of stages toward the final differentiated state.

the expression of these four specific transcription factors in a fully differentiated cell, the genetic program of the cell reverted to a **pluripotent state**, meaning it reversed its developmental program to become a "predifferentiated" cell. Yamanaka and others were able to then redirect these artificially induced pluripotent stem cells into a variety of differentiated cell types by changing the growth media to stimulate specific cell differentiation programs. To understand the importance of this work, we first need to understand stem cell differentiation.

Stem cells are a special type of cell that can grow and give rise to more cells like themselves through a process called self-renewal. As shown in **Figure 23.57**, stem cells are able to differentiate into specialized cell types. There are two main types of stem cells: (1) **somatic stem cells**, also called adult stem cells, which continually replenish dying cells in differentiated tissues such the skin, intestine, and bone marrow; and (2) **embryonic stem (ES) cells**, which are found in early embryos, are undifferentiated, and can develop into all of the different cell types of the body during the course of development.

Although study of human ES cells can be very useful, both for understanding basic biological processes and for medical applications, the use of human ES cells is controversial because isolating them involves destruction of an embryo. One of the reasons Yamanaka's work on transcriptional control of the pluripotent state by a small number of transcription factors was so important is it offers an alternative way of obtaining stem cells that are very similar, if not equivalent, to ES cells. These dedifferentiated cells are called **induced pluripotent stem cells (iPS cells)**. The use of iPS cells avoids the ethical issues involved with ES cells and offers considerable advantages for medical applications, as we will discuss.

Work with mammalian cells is commonly done outside the organism using tissue culture methods (**Figure 23.58**). In this procedure, individual cells are isolated from an organism and then cultured in special media containing nutrients and growth factors that stimulate cell division. Exogenous DNA can be introduced into tissue culture cells, typically as free DNA or in the form of viruses, which can then be expressed to produce their encoded protein products.

The experiment done by Yamanaka and colleagues used tissue culture methods and introduction of DNA encoding four specific transcription factor genes; namely, *Oct4*,

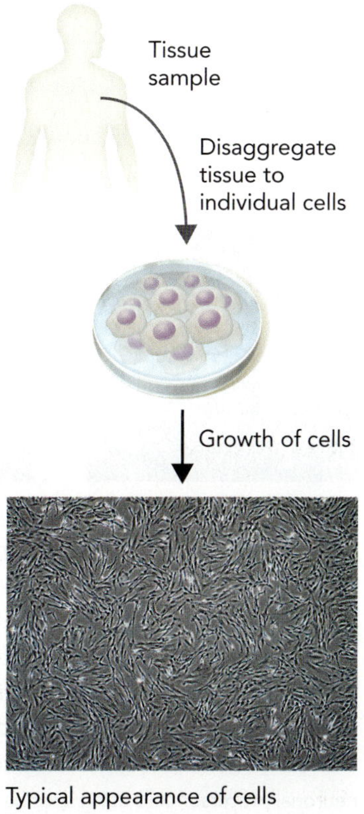

Figure 23.58 Human tissue samples are obtained from a biopsy and then treated to separate cells from one another before adding the cells to a culture dish containing a precise mixture of nutrients. Depending on the type of cells, they can sometimes be passaged for many generations in culture. PHOTO: DR. SHOUKHRAT MITALIPOV/OREGON HEALTH & SCIENCE UNIVERSITY.

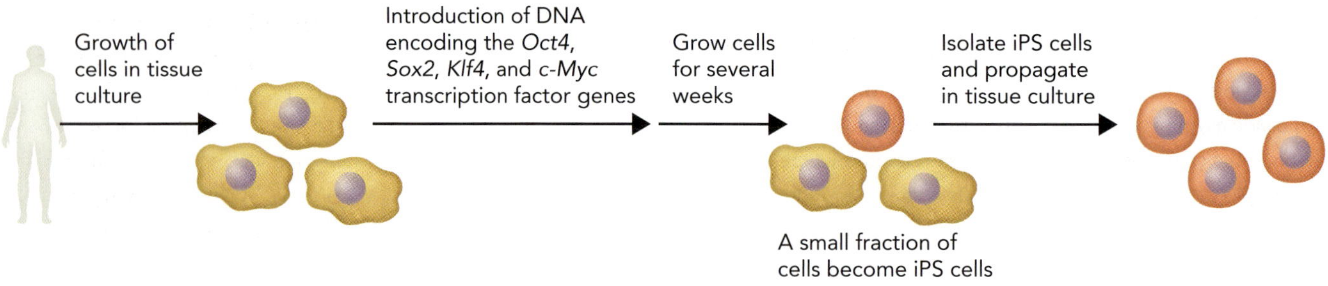

Figure 23.59 Reprogramming of differentiated cells to iPS cells is done by introduction of the four transcription factor genes *Oct4*, *Sox2*, *Klf4*, and *c-Myc*. A human tissue sample is grown in culture, and then viruses carrying the transcription factor genes are introduced into the cells. After continued growth for several weeks, a few cells become iPS cells. These iPS cells are then expanded in culture and further characterized.

Sox2, *Klf4*, and *c-Myc*. As shown in **Figure 23.59**, the combined expression of these four transcription factor genes in cultures containing differentiated cells resulted in the generation of iPS cells. The initial experiments used human fibroblast cells (cells that make up the connective tissue), and although the frequency of conversion from differentiated cells into iPS cells was low, the recovered iPS cells were found to be very similar to ES cells. As shown in **Figure 23.60**, colonies of iPS cells form in a culture of differentiated cells, which can then be isolated and induced to differentiate by additional treatments that specify cell fate. For example, iPS cells derived from human skin fibroblasts can be differentiated into neuronal cells under the appropriate cell culture conditions.

Induced pluripotency of differentiated cells by transcription factor expression is remarkable for several reasons. First, the differentiated state is highly stable, being maintained in an epigenetic state by several different mechanisms. The success of Yamanaka and colleagues using just four transcription factor genes to generate iPS cells from differentiated cells suggests that these epigenetic mechanisms can be reversed, given the appropriate treatment. Second, because a range of cell types can be converted to iPS cells, the pathway for conversion is general, not specific to a particular cell type. Third, in addition to transfecting DNA encoding Oct4, Sox2, Klf4, and c-Myc into the differentiated cells, induction of the pluripotent state requires that several endogenous genes be transcriptionally activated, even though they are located in highly condensed chromatin.

Pluripotent cells have a very stable epigenetic state, which is controlled in part by the transcription factors Oct4 and Sox2, as well as a transcription factor called Nanog. Oct4 and Sox2 function as a heterodimer when bound to DNA (**Figure 23.61**),

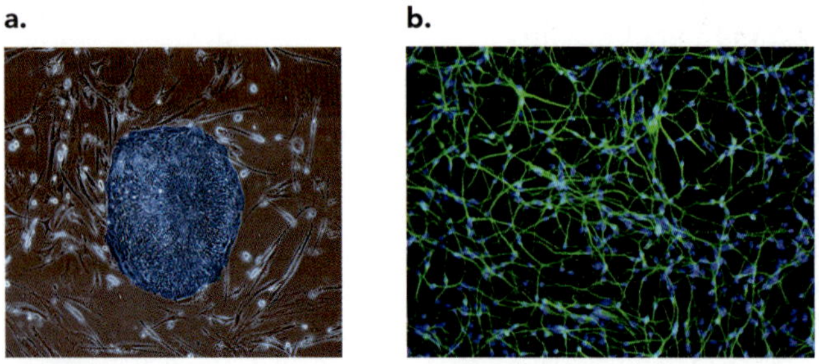

Figure 23.60 Isolation and differentiation of iPS cells using tissue culture methods are shown. **a.** Human iPS cells appear as distinct colonies in a culture of differentiated fibroblast cells 3–4 weeks after DNA transfection with the four transcription factor genes. The frequency of iPS cell formation from fibroblast cells is less than 1% of the DNA-treated cells. SCIENCE SOURCE/COLORIZATION BY MARY MARTIN. **b.** One example of iPS cell differentiation is the conversion of iPS cells to neuronal precursor cells using specific culture conditions. Blue identifies cell nuclei, and green highlights neuronal processes. © MTI-GLOBALSTEM.

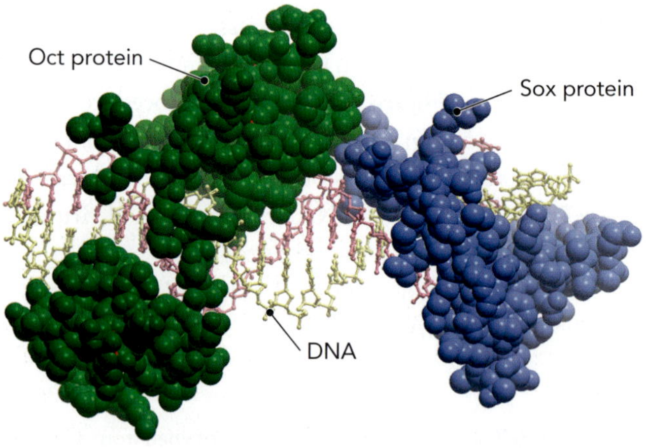

Figure 23.61 Structure of the mouse Oct1–Sox2 heterodimer protein bound to DNA is shown. The mouse Oct1 protein is structurally similar to the human Oct4 protein. BASED ON PDB FILE 1GTO.

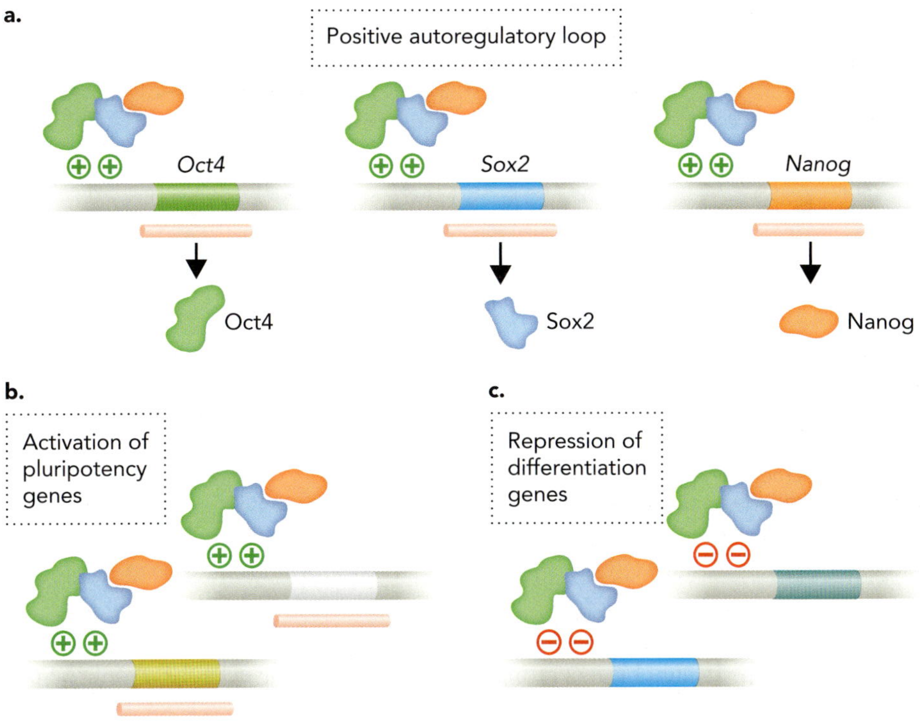

Figure 23.62 A model proposing the three transcriptional regulatory mechanisms that control the expression of the Oct4, Sox2, and Nanog transcription factors and their downstream targets. This model assumes a low level of Nanog protein is present in cells before introduction of the Oct4 and Sox2 genes. **a.** The Oct4–Sox2 heterodimer and Nanog proteins bind to enhancers for each of the three genes, activating their continued expression in a positive feedback loop. **b, c.** Oct4, Sox2, and Nanog transcription factors activate the expression of pluripotency genes and inhibit the expression of differentiation genes.

which explains why DNA encoding both of these genes is required for generation of iPS cells (Oct4 or Sox2 alone is insufficient). As shown in **Figure 23.62**, the Oct4–Sox2 heterodimeric complex, along with the transcription factor Nanog, function in a positive autoregulatory loop to reinforce transcription of each of the three corresponding genes. This positive feedback loop provides sufficient amounts of each of the three transcription factors to then activate a set of pluripotency genes and repress a set of differentiation genes, thereby initiating the conversion of a differentiated cell into a pluripotent cell. This is a surprising finding considering that most of the Oct4–Sox2 target genes are in highly condensed chromatin, including the Oct4 and Sox2 genes themselves.

The development of methods to generate iPS cells from differentiated cells holds great promise for medical applications, especially in the field of personalized medicine because such methods offer the possibility of obtaining cells from a patient for manipulation and reintroducing them into the same patient. This avoids a major concern of tissue transplants from other donors, which is that the donor tissue is often rejected by the immune system of the recipient.

Applications of iPS cell technology are in their very early stages, and many steps in the pathways to such applications are poorly understood. Broadly speaking, however, two main approaches are being considered. The first approach is to reintroduce cells into the patient that might correct the patient's disease (**Figure 23.63**). In this case, iPS cells would be generated from the patient's fibroblasts and converted to specific cell types, depending on the disease. These fully functional differentiated cells would then be reintroduced into the patient. In some cases, a disease-causing mutation might first be corrected in the iPS cells before generating the differentiated cell types. Two disease states being considered for iPS cell–directed personalized medicine are treatment of type 1 diabetes and Parkinson disease. In the case of type 1 diabetes, it may be possible to generate pancreatic β cells from the patient's modified iPS cells that now produce insulin, which would be reintroduced into the diabetic patient. Similarly, patients with

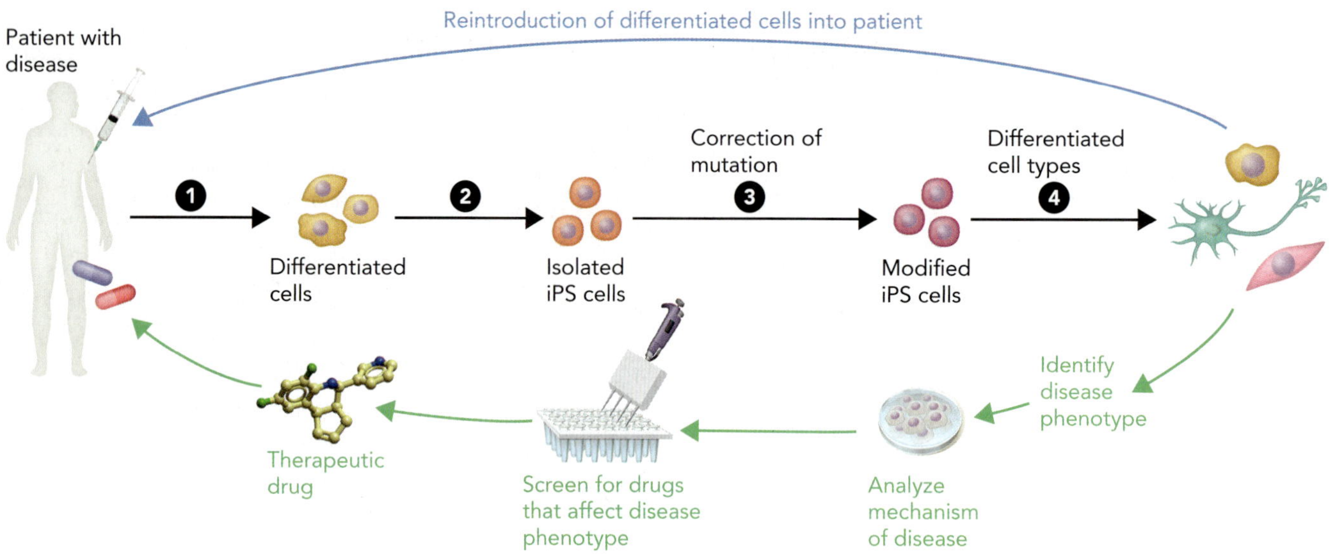

Figure 23.63 Two examples are shown of how iPS cells could be used to treat human diseases. One possibility is to reintroduce cells into patients as a therapeutic treatment for disease, and a second is drug discovery and pharmaceutical intervention.

Parkinson disease, which is caused by the loss of neuronal brain cells that produce dopamine, could be treated by generating dopamine-producing cells that would be put back into the patient.

The second approach is to use iPS cells or differentiated derivatives to help identify disease mechanisms and to screen for drugs that correct disease phenotypes. In this set of applications, iPS cells from a patient are differentiated to the desired cell type and then characterized either to investigate the biochemical basis for the disease phenotype or to use for drug discovery assays to correct the disease phenotype. These approaches work best when the disease operates on a cell-by-cell basis, as opposed to cases in which the defect lies in communication between cells, especially cells of different types.

These medical advances are still on the horizon, but it is important to note that they have only been made possible by decades of research into basic mechanisms of gene regulation and development. This work led to identification of the transcription factors now being used to create iPS cells and to cause them to develop into specialized cell types. This example illustrates once again the value of basic research, and indicates that we never know which basic findings are going to have practical applications.

concept integration 23.3

How is transcription factor recruitment used in eukaryotic gene regulatory circuits?

Recruitment refers to the property of trans-acting factors, bound to cis-acting sites near the genes to be regulated, to interact with other proteins or protein complexes and bring them to the vicinity of the gene to be regulated. Three different classes of proteins or complexes are recruited. The first is proteins that act to modify the side chains of histones in nucleosomes. The second is chromatin remodeling machines, which move nucleosomes around, usually making additional cis-acting sites available. The third is components of the general transcription machinery, such as the mediator complex and TFIID, which are brought to the vicinity of the gene so that a pre-initiation complex can assemble at the transcription start site.

chapter summary

23.1 Principles of Gene Regulation

- Gene regulation can occur at many stages in the pathway of expression. The most common point of control is the decision to initiate transcription, but gene expression is also regulated at the level of RNA processing, RNA degradation, protein synthesis, protein modifications, and protein degradation.
- Gene regulation is mediated by proteins called trans-acting factors (or, more commonly, transcription factors) that bind to specific sites on DNA called cis-acting sites. Transcription factor protein binding to cis-acting sequences occurs primarily via interaction between amino acid side chains on the protein and the nucleotide bases.
- Binding of transcription factors to sites near each other is often cooperative, where the binding of one protein to a region of DNA assists in the binding of another.
- The activity of transcription factors can be affected in many ways. For instance, their synthesis can be regulated; their activity can be modulated allosterically by binding of small-molecule ligands; they can undergo posttranslational modification; and they can be degraded in a controlled manner.
- The two major types of gene regulation are positive control and negative control. In positive control, transcriptional activator proteins stimulate the transcription of target genes, whereas in negative control, transcriptional repressor proteins inhibit transcription.
- If the protein represses its own expression, it shows negative feedback, which maintains a constant level of protein. If the protein activates its own expression, the resulting positive feedback drives the system into either a fully off or fully on state.
- Positive self-regulation can lead to very stable gene regulatory states called epigenetic states. These maintain themselves through multiple generations. Often, they have an establishment phase, in which the regulatory state is set up, and a maintenance phase, in which regulatory events continue to stabilize the state of the circuit.
- Regulated overexpression of recombinant proteins in heterologous systems (human genes in bacterial cells) and analysis of gene expression using reporter genes are two examples of biochemical applications that exploit our knowledge of gene regulation.

23.2 Mechanisms of Prokaryotic Gene Regulation

- In prokaryotes, gene regulation almost always occurs by the action of transcription factors working locally at the promoter of the gene being regulated. These proteins operate directly on the RNA polymerase to activate or repress its action.
- In prokaryotes, genes involved in common pathways are often organized into single transcription units called operons. Expression of the entire set of genes can then be regulated by action at a single promoter. A regulon is a set of genes that are coordinately regulated but not physically linked to each other as in an operon.
- Genes are commonly regulated in ways that appear to be economical uses of cellular resources. Proteins involved in catabolic pathways are only expressed when those nutrients are available, and proteins involved in biosynthetic pathways are only made when the end product is not abundant in the environment.
- The *lac* operon is subject to two types of controls. It is expressed only when lactose (or a similar compound) is present in the environment and when glucose is absent.
- In the presence of lactose, the lac repressor dissociates from a site near the promoter, allowing RNA polymerase access to the promoter. In the absence of glucose, the level of cAMP is high; cAMP binds to the CRP activator, and the cAMP–CRP complex binds near the *lac* promoter to stimulate its expression.
- The lac repressor is a tetramer, and it binds to two sites near the promoter, leading to much more complete repression than if only one site were present. DNA-bound cAMP–CRP protein makes contacts with RNA polymerase to facilitate its binding to the promoter.
- The SOS regulatory circuit of *E. coli* controls the response of the cell to conditions that damage DNA or inhibit DNA replication. During normal growth, the LexA repressor protein decreases expression of the SOS regulon. DNA damage activates the RecA protein to a form that stimulates LexA cleavage, inactivating LexA and turning on the SOS genes.

- Bacteriophage λ can establish an association with the bacterial host, setting up a highly stable epigenetic state called the lysogenic state. This state is stabilized by the action of a transcriptional regulatory protein called CI, which activates its own expression and represses the expression of a second regulatory protein called Cro. Increased Cro expression inhibits CI expression and induces the lytic pathway.
- The *trp* operon of *E. coli* has two regulatory controls. The first is negative control by the Trp repressor, which binds tryptophan and turns off transcription of the operon. The second mechanism is called attenuation, and it acts after initiation of transcription.
- The mechanism of *trp* operon attenuation involves the formation of alternate mRNA structures as a consequence of coupled transcription and translation. One mRNA structure is a terminator, which causes RNA polymerase to terminate; the other is an antiterminator, which prevents the terminator from forming. The antiterminator forms when charged $tRNA^{Trp}$ is not available, leading to read-through into the rest of the operon.

23.3 Mechanisms of Eukaryotic Gene Regulation

- Eukaryotic DNA is packaged into DNA–protein structures called nucleosomes, in which 147 bp of DNA wrap around a protein core called a histone octamer. Nucleosomes can be further compacted into higher-order structures, some of which inhibit gene expression.
- The default state of eukaryotic genes is that they are unavailable for transcription because they are packaged in chromatin. Expression of eukaryotic genes is an active process that modifies chromatin in various ways to allow access by the transcription machinery.
- Eukaryotic transcriptional activator proteins do not interact directly with RNA polymerase II. Instead, they act to recruit other proteins to the genes being regulated. RNA polymerase II becomes part of the pre-initiation complex at a late stage.
- Transcriptional activator proteins have two major roles in transcription. First, they antagonize the repressive effects of chromatin by recruiting modifying proteins. Second, they recruit protein components that function in the general transcription machinery.
- One major class of nucleosome modification is chemical modification of the N-terminal histone tails. Common modifications are acetylation of lysines, phosphorylation of serines, and methylation of lysines. Enzymes that carry out these modifications are recruited by transcriptional activator and repressor proteins.
- Lysine acetylation is carried out by histone acetyltransferases (HATs). This modification is associated with actively transcribed genes. Deacetylation by histone deacetylases (HDACs) accompanies transcriptional repression of genes. Lysine methylation has different effects on transcription, depending on which residue is methylated. For instance, methylation at one particular site leads to a highly compacted structure for the chromatin.
- In highly compacted chromatin, the DNA itself is often methylated on cytosine residues. The resulting 5-methylcytosine is usually found in the palindromic sequence CG, and both strands are methylated. This pattern of methylation can readily be passed on during the process of DNA replication and is a highly stable epigenetic state.
- Nucleosomes can be laterally moved along the DNA or removed from it by chromatin remodeling machines, which are recruited to genes by activator proteins. These machines are large multiprotein complexes and require ATP. Their action usually results in exposure of cis-acting sites for other transcription factors.
- Transcriptional activator proteins often bind at cis-acting sites called enhancers, which can function at variable distances from the promoter and in either orientation. They interact with the promoter by looping the DNA to assemble the pre-initiation complex.
- In multicellular eukaryotes, most genes are turned off in cells in which their products are not needed, usually by mechanisms that permanently silence them by placing them in highly condensed chromatin. Not all mammalian genes are controlled in this way: Some genes are responsive to transient extracellular signals, such as steroid hormones.
- In single-celled eukaryotes, expression of genes is typically controlled by environmental signals. One example of an environmental signal in yeast is the presence or absence of the sugar galactose. When galactose is present, it acts to remove an inhibitory protein from the activator protein GAL4. This allows GAL4 to activate several target genes.
- The GAL4 activator has a modular organization with a series of functional and structural domains. GAL4 contains a DNA binding domain and a separate activation domain.
- In *Drosophila* development, a large number of transcription factors is expressed in a defined sequence of events; in each case, the combination of factors dictates which transcription factors are turned on and off at later times.
- The expression of the *Drosophila even-skipped* (*eve*) gene occurs in a series of seven stripes along the axis of the embryo, where each stripe has a separate set of regulatory controls. The *eve* stripe 2 enhancer region contains cis-acting sites for four different transcription factors (Bicoid, Hunchback, Giant, and Krüppel) that function together to restrict expression of the *eve* structural gene to a small number of cells.
- Induced pluripotent stem (iPS) cells are generated by introducing DNA for four transcription factor genes (*Oct4*, *Sox2*, *Klf4*, and *c-Myc*), which function together to convert differentiated cells into pluripotent cells. Special culture conditions can be used to differentiate iPS cells into a number of distinct cell types.

biochemical terms

(in order of appearance in text)
trans-acting factor (p. 1146)
cis-acting site (p. 1146)
helix-turn-helix motif (p. 1149)
leucine zipper motif (p. 1150)
helix-loop-helix motif (p. 1150)
cooperativity (p. 1153)
epigenetic states (p. 1157)
lac repressor (p. 1158)
reporter gene (p. 1160)
lac operon (p. 1161)
allolactose (p. 1162)
SOS regulatory system (p. 1166)
LexA repressor (p. 1166)
regulon (p. 1166)
RecA (p. 1168)
lysogeny (p. 1169)
lysis (p. 1169)
trp operon (p. 1170)
attenuation (p. 1170)
Trp repressor (p. 1171)
enhancer (p. 1175)
histone tails (p. 1175)
histone acetyltransferase (HAT) (p. 1176)
histone deacetylase (HDAC) (p. 1176)
bromodomain (p. 1177)
chromodomain (p. 1177)
Swi/Snf complex (p. 1179)
pioneer factor (p. 1180)
mediator complex (p. 1182)
insulator sequence (p. 1183)
GAL4 activator protein (p. 1183)
eve stripe 1185 enhancer (p. 1185)
pluripotent state (p. 1187)
somatic stem cell (p. 1187)
embryonic stem (ES) cell (p. 1187)
induced pluripotent stem cell (iPS cell) (p. 1187)

review questions

1. Regulation can occur at many steps in the pathway of gene expression. Give at least one example of a regulator at the level of (**a**) transcription initiation, (**b**) protein stability, and (**c**) protein posttranslational modification.
2. What are trans-acting factors and cis-acting sites?
3. How are DNA binding proteins able to recognize a specific double-stranded DNA sequence?
4. The activity of regulatory proteins can be modulated in many different ways. Give at least one example of regulation by each mechanism.
   a. Protein is blocked by an inhibitor.
   b. Protein requires a binding partner.
   c. Active protein is unmodified.
   d. Active protein is a ligand-bound form.
   e. Active protein is a form not bound to ligand.
5. What are positive regulation and negative regulation? Give at least two examples of each for both prokaryotes and eukaryotes.
6. List at least six different ways in which eukaryotes and prokaryotes differ, with emphasis on issues bearing on transcription and gene regulation.
7. Compare and contrast the action of transcriptional activators in eukaryotes and prokaryotes.
8. Compare and contrast the action of transcriptional repressors in eukaryotes and prokaryotes.
9. What are the two primary roles for activators in eukaryotes? Give several different examples of each.
10. What is an epigenetic state?
11. What is cooperative DNA binding? Why might it be useful in stabilizing epigenetic states?
12. Regulatory systems that control anabolic or catabolic pathways operate in ways that conserve energy for the cell. Give an example of each type of pathway, and indicate how the regulation serves this purpose.
13. What are eukaryotic pioneer factors, and why is their action important? How are they able to operate? Give an example.
14. What are iPS cells? How are they produced?

challenge problems

1. Consider the regulation of the *trp* operon in *E. coli*. If this system were present in yeast instead of *E. coli*, would you expect it to function properly? Explain.
2. The modified base 5-methylcytosine is used in higher eukaryotes as an epigenetic mark of particular chromatin states. But 5-methylcytosine is also known to be a "hot spot" for mutation; that is, it has a higher rate of mutation from C to T than most C residues. Explain (on the basis of your understanding of DNA repair) why this is the case.
3. In the *trp* attenuation system of *E. coli*, predict the effects of the following mutations on the operation of

the regulatory system, and give your reasons. To simplify matters, assume that the Trp repressor is not active. Three broad possibilities exist for whether the operon is transcribed in the absence or presence of charged $tRNA^{Trp}$: behavior is normal, transcription is always low ("noninducible"), or transcription is always high ("constitutive").

a. Mutation in region 4 so it cannot pair with region 3.

b. Mutation in region 3 so it can pair with region 4 but not with region 2.

c. Change of the two Trp codons to glycine codons.

d. Change of the AUG codon of the leader peptide to AUA.

e. Combination of the first mutation with each of the other three.

4. You are studying the regulation of a *Drosophila* gene, and you have available the genome sequences of several related *Drosophila* species. By comparing these sequences in the vicinity of your gene, you identify several regions that appear to be conserved. Moreover, you identify potential binding sites for known regulatory proteins. You speculate that this region is an enhancer. Suggest how you might test this idea. Assume that you are competent at recombinant DNA work and that you have a method to introduce desired constructs into *Drosophila*. (Hint: One of the defining properties of an enhancer is that it works from variable distances and in both orientations.)

5. Bacteriophages are often studied by the following approach. First, cells are mixed with the bacteriophages; after a period of time to allow the virus to bind the cells, the cells are mixed with a small amount of liquid agar, and the mixture is poured onto the surface of a plate containing solid agar. The top agar quickly hardens. Eventually, the cells grow and form a dense layer of cells in the top agar layer. Because the viruses multiply much more rapidly than the cells, however, they eat a hole, or "plaque," in the bacterial layer. Each plaque develops from a single infected cell, so this is a way to count the number of viruses. It is also a way to do genetics, as was done in the early days of studying bacteriophage λ.

Plaques of bacteriophage λ differ from those of most viruses. Most viruses kill all the cells in the plaque, so there are no surviving cells and the plaques are "clear"; you can see through them. In contrast, plaques of bacteriophage λ are cloudy or "turbid" because lysogens (bacterial cells in which the virus is in the lysogenic stage) arise in the plaque and can grow.

a. In the plaque, lysogens are often infected by other λ viruses present in the plaque. Why don't these newly infecting viruses grow lytically and kill the cells?

b. Mutants of λ can arise that form clear plaques. Which virus genes are likely to be mutated?

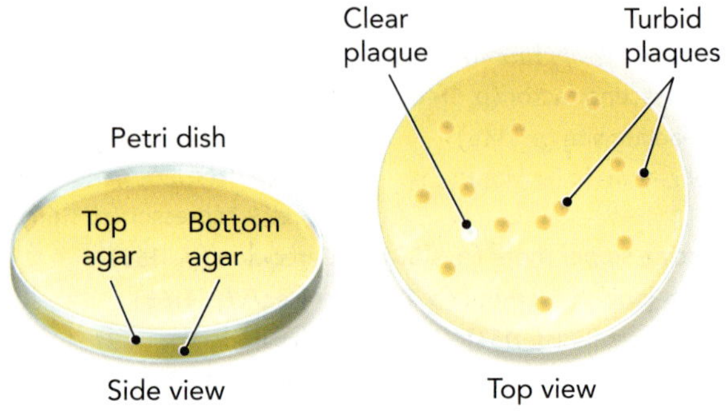

6. Your colleague is studying the regulation of an *E. coli* gene, and she purifies a protein that stimulates expression of the gene in an *in vitro* transcription system. In this system, the gene is expressed at a very low level in the absence of this protein and at a high level in its presence. She interprets these data to mean that it is an activator protein. Can you suggest another possibility, along with one or more experiments that would distinguish between the two models?

7. A nucleosome is positioned as shown in the diagram here, with two cis-acting sites lying on the DNA close to the exit point of the DNA. At a low frequency, the two sites are transiently exposed by dissociation of a segment of DNA from the histones. If the second site (in green) is exposed, the green factor can bind to it, as shown. What effect will this have on the ability of the red trans-acting factor to bind to its site? Can this same effect on binding occur by other mechanisms?

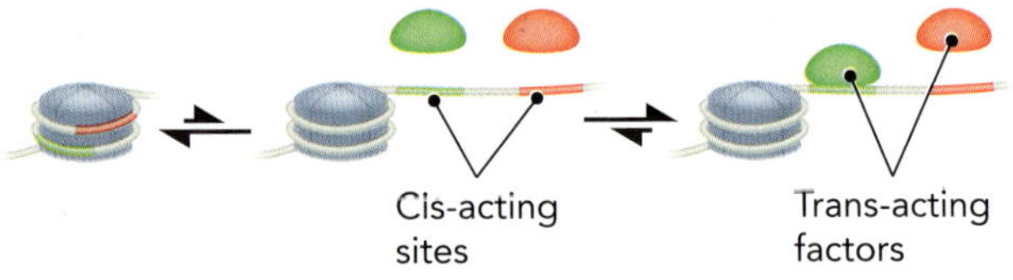

8. Assume that you have identified all the cis-acting sites in a prokaryotic genome for a particular gene regulatory protein. Can you predict the consequences of the regulatory protein binding to these sites? Give your reasons, with examples.

9. Histone acetylation is generally associated with promoter regions for genes that are actively being transcribed. Suggest several mechanistic reasons why this might be the case.

10. In the galactose regulatory system of yeast, what would be the behavior of mutants that lack one of the following components: GAL3, GAL4, or GAL80? Consider the expression of the target genes with and without the presence of galactose. What would be the behavior of a GAL3–GAL80 double mutant?

suggested reading

Books and Reviews

Hahn, S., and Young, E. T. (2011). Transcriptional regulation in *Saccharomyces cerevisiae*: transcription factor regulation and function, mechanisms of initiation, and roles of activators and coactivators. *Genetics, 189*, 705–736.

Kouzarides, T. (2007). Chromatin modifications and their function. *Cell, 128*, 693–705.

Levine, M. (2010). Transcriptional enhancers in animal development and evolution. *Current Biology, 20*, R754–R763.

Levine, M. (2013). Computing the away the magic? *eLife, 2013(2)*, e01135.

Levo, M., and Segal, E. (2014). In pursuit of design principles of regulatory sequences. *Nature Reviews Genetics, 15*, 453–468.

Ptashne, M. (2004). *A genetic switch: phage lambda revisited* (vol. 3). Cold Spring Harbor, NY: Cold Spring Harbor Laboratory Press.

Ptashne, M. (2009). Binding reactions: epigenetic switches, signal transduction and cancer. *Current Biology, 19*, R234–R241.

Young, R. A. (2011). Control of the embryonic stem cell state. *Cell, 144*, 940–954.

Zaret, K. S., and Carroll, J. S. (2011). Pioneer transcription factors: establishing competence for gene expression. *Genes & Development, 25*, 2227–2241.

Primary Literature

Bell, C. E., and Lewis, M. (2000). A closer view of the conformation of the Lac repressor bound to operator. *Nature Structural Biology, 7*, 209–214.

Benoff, B., Yang, H. W., Lawson, C. L., Parkinson, G., Liu, J. S., Blatter, E., Ebright, Y. W., Berman, H. M., and Ebright, R. H. (2002). Structural basis of transcription activation: the CAP-αCTD-DNA complex. *Science, 297*, 1562–1566.

Bothma, J. P., Garcia, H. G., Esposito, E., Schlissel, G., Gregor, T., and Levine, M. (2014). Dynamic regulation of *eve* stripe 2 expression reveals transcriptional bursts in living *Drosophila* embryos. *Proceedings of the National Academy of Sciences USA, 111*, 10598–10603.

Brent, R., and Ptashne, M. (1985). A eukaryotic transcriptional activator bearing the DNA specificity of a prokaryotic repressor. *Cell, 43*, 729–736.

Hanna, J., Wernig, M., Markoulaki, S., Sun, C.-W., Meissner, A., Cassady, J. P., Beard, C., Brambrink, T., Wu, L.-C., Townes, T. M., and Jaenisch, R. (2007). Treatment of sickle cell anemia mouse model with iPS cells generated from autologous skin. *Science, 318*, 1920–1923.

Luo, Y., Pfuetzner, R. A., Mosimann, S., Paetzel, M., Frey, E. A., Cherney, M., Kim, B., Little, J. W., and Strynadka, N. C. J. (2001). Crystal structure of LexA: a conformational switch for regulation of self-cleavage. *Cell, 106*, 585–594.

McConnell, B. B., and Yang, V. W. (2010). Mammalian Kruppel-like factors in health and disease. *Physiological Reviews, 90*, 1337–1381.

Small, S., Blair, A., Levine, M. (1992). Regulation of *even-skipped* stripe 2 in the *Drosophila* embryo. *EMBO Journal, 11*, 4047–4057.

Von Hippel, P. H., and Berg, O. G. (1986). On the specificity of DNA-protein interactions. *Proceedings of the National Academy of Sciences USA, 83*, 1608–1612.

Suggested reading

Answers

Chapter 1

REVIEW QUESTIONS

1. Buchner used an *in vitro* reaction to show that brewer's yeast can produce carbon dioxide and ethyl alcohol from sugar. This discovery introduced the idea of enzymes, or catalysts, which can increase the rate of chemical reactions in living systems.
2. Most enzymes are proteins, though sometimes they are RNA molecules. Enzymes catalyze reactions in respiration, fermentation, nitrogen metabolism, energy conversion, and programmed cell death.
3. Biochemistry is the study of biological processes at the molecular and cellular level. Its goal is to understand the structure and function of biological molecules using hypothesis-driven experiments designed to answer specific questions.
4. Amino acids, nucleotides, simple sugars, and fatty acids.
5. Elements, functional groups, biomolecules, biopolymers, metabolism, cells, organisms, and ecosystems.
6. It must have at least one carbon atom. Carbon can make up to four covalent bonds, and provide stable structural backbones for complex organic molecules.
7. Energy conversion, coenzymes, information storage, and signaling.
8. A nitrogenous base, a five-carbon ribose/deoxyribose sugar, and one to three phosphate groups.
9. Base stacking involves noncovalent aromatic ring interactions between adjacent nitrogenous bases of the DNA helix. Hydrogen bonding between complementary base pairs also stabilizes the DNA helix.
10. Messenger RNA is the template for protein synthesis. Small nuclear RNA has a role in RNA processing. Micro RNA regulates translation. Ribosomal RNA is a component of ribosomes. tRNA provides amino acids to the ribosome for translation.
11. Mutations in a gene can change codons in the mRNA transcript, which lead to changes in the amino acids incorporated into protein. Changes in amino acid sequence can lead to protein denaturation (unfolding) or loss of function, which could affect function of biochemical pathways and even lead to disease or death.
12. Gene duplication might lead to evolutionary change, for example if doubling the amount of gene product (protein) increased fitness for an organism, then the duplicated copy would be maintained. If there was no fitness advantage from doubling the amount of product, the duplicated gene could be mutated and possibly confer new fitness advantages or be lost by natural selection if it conferred deleterious qualities.

CHALLENGE PROBLEMS

1. (a) Pyruvate $\rightarrow$ acetaldehyde + CO_2 (pyruvate decarboxylase). (b) Acetaldehyde + NADH + H^+ $\rightarrow$ ethanol + NAD^+ (alcohol dehydrogenase). (c) Acetaldehyde is a product in (a) and a reactant in (b).
2. Buchner (1) used a different strain of yeast than Pasteur used, (2) prepared the yeast extract using quartz mixed with diatomaceous earth rather than glass, and (3) his extract buffer contained glucose, the carbon source for fermentation.
3. The total number of dodecanucleotides is $4^{12} = 16{,}777{,}216$. The total number of tetrapeptides is $20^4 = 160{,}000$.
4. Amylose is a glucose polymer with $\alpha(1\rightarrow4)$ glycosidic bonds that can be cleaved by the human enzyme amylase. Cellulose is a glucose polymer with $\beta(1\rightarrow4)$ glycosidic bonds, cleaved by cellulase, an enzyme not found in humans. Horse's intestines have bacteria that secrete cellulase, which aids the horse to digest plant material.
5. Liver cells and skeletal muscle cells both contain insulin receptors, but only liver cells contain glucagon receptors; skeletal muscle cells do not.
6. mRNA: 5′ – AAAAAAUUUAAAUUU – 3′
   Protein: NH_3^+-Lys-Lys-Phe-Lys-Phe-COO^-
7. Germ-line mutations occurring outside of protein-coding sequences can persist in the germ line if they do not affect genome stability or regulation; these are neutral mutations. However, germ-line mutations occurring in protein-coding sequences will be selected against by natural selection if they alter reproductive rates.
8. Paralogous genes result from gene duplication and can diverge within the same species in both structure and function over time, whereas orthologous genes encode proteins with the same structure and function in different species.
9. Not much can be predicted about amino acid sequences with <20% identity; they could have a similar protein structure, or a similar protein function, or neither. In contrast, two proteins that are >80% identical are very likely to have the same structure and function.

Chapter 2

REVIEW QUESTIONS

1. Chemical work: biosynthesis of new organic molecules; osmotic work: differential solute concentrations across biological membranes; mechanical work: muscle contraction or flagellar rotation.
2. Redox reactions are a form of chemical work used, for example, to synthesize biomolecules with energy derived from electron transfer, and for maintenance of homeostasis.
3. This process (photophosphorylation) is paramount to the survival of all organisms: it uses redox energy from photooxidation to transform light energy to usable chemical energy, ultimately forming ATP that can be used to convert carbon dioxide into glucose, consumed by plants or the animals that eat those plants.
4. $\Delta G = \Delta H - T\Delta S$ (ΔG is the free energy change between reactants and products of a reaction, ΔH is the change in enthalpy, T is the temperature, and ΔS is the change in entropy). A favorable reaction has a negative value for ΔG. The equation shows that both ΔH and ΔS contribute to the change in free energy and that the temperature (T) amplifies or decreases the ΔS of the reaction.
5. The reaction might still proceed in a cell, because the initial concentrations of reactants and products contribute to the actual free energy change. In cells, if product concentrations are low and reactant concentrations are high ([products]/[reactants] < 1), the second term of Equation 2.14 will be negative. If it is sufficiently negative to overcome a positive $\Delta G°'$, then the reaction will proceed.

6. (1) Water is less dense as a solid (ice) than as a liquid (water). If ice did not float, then it would cause an upwelling of cold water in the ocean, which would also freeze and sink until all water was frozen solid. (2) Water is liquid over a wide range of temperatures, which helps maintain oxygen levels for life—production of oxygen on Earth is largely dependent on photosynthetic algae growing in (liquid) water. (3) Water is a fantastic solvent because of its polar nature and hydrogen-bonding capabilities.
7. (1) *Hydrogen bonds* form between a donor group: a hydrogen atom attached to an electronegative atom, and an acceptor group (an electronegative atom). Common acceptors are oxygen, nitrogen, and sulfur. (2) Electrostatic *ionic interactions* occur between two atoms with opposite charges; the strength of interaction depends upon the distance between ions and the environment around them. (3) *van der Waals interactions* occur between nonpolar molecules and arise from temporary dipoles caused by fluctuation in electron clouds. If dipoles of opposite signs align at the appropriate distance, an interaction can occur. These interactions are quite weak, but collectively among several atoms, they have a strong cumulative effect. (4) The *hydrophobic effect* is driven by the tendency for hydrophobic molecules to pack together in solution. By packing closely, hydrophobic molecules decrease overall surface area, decreasing the number of water molecules forming ordered (entropically unfavorable) structures around the hydrophobic molecules.
8. Colligative properties are those that depend on the number of solute particles, including freezing point depression, boiling point elevation, vapor pressure lowering, and osmotic pressure.
9. The pH scale is a logarithmic scale. A value of 0–6.5 is considered acidic; 7.5–14 is considered basic. Anything in between those two ranges (6.5–7.5) is considered pH neutral.
10. Amphipathic molecules contain hydrophobic and hydrophilic regions. Amphipathic lipids form biological membranes, whose hydrophobic core is relatively impermeable to polar molecules. This maintains separation of the inside of cells from the environment and allows compartmentalization within cells where specific reactions can occur.
11. (1) *Phospholipid monolayers* have polar head groups pointed toward the water and tails pointed toward the air. (2) *Phospholipid bilayers*, such as cell membranes, form a hydrophobic barrier between two aqueous compartments. (3) *Micelles* have hydrophobic tails packed in the center of a sphere and polar head groups facing outward toward water. (4) *Liposomes* are spherical structures bound by a lipid bilayer with an aqueous center. All these structures are organized to increase entropy, consistent with the hydrophobic effect.
12. (1) The *plasma membrane* is the lipid bilayer surrounding a cell. (2) The *endomembrane* consists of structurally related intracellular membrane networks and vesicles. (3) *Chloroplast and mitochondrial membranes* enclose the enzymes that perform energy conversions and separate the environment inside of these organelles from the cytoplasm.

CHALLENGE PROBLEMS

1. Plants (at night) and animals (all of the time) metabolize nutrients (carbohydrates) by the process of aerobic respiration.
2. All natural processes in the universe tend toward dispersal of energy (entropy) in the absence of energy input. Many biochemical reactions are maintained at a steady state far from equilibrium; to do this, they are using chemical energy to limit entropy and avoid reaching equilibrium with the environment (death).
3. (a) $\Delta G°'$, expressed in kilojoules (kJ), refers to the biochemical standard free energy change of a reaction. This is the change from initial reactant and product concentrations of 1 M each, to steady-state concentrations at equilibrium under physiological conditions (298 K, 1 atm, pH 7). ΔG refers to the actual free energy change in a reaction, calculated from its $\Delta G°'$ value and the initial concentrations of reactants and products in a real system. (b) When a reaction is at equilibrium, $\Delta G = 0$, and $\Delta G°'$ is described by

$$\Delta G = \Delta G°' + RT\ln\frac{[\text{B}]_{\text{actual}}}{[\text{A}]_{\text{actual}}} = 0$$

$$\Delta G°' = -RT\ln\frac{[\text{B}]_{\text{eq}}}{[\text{A}]_{\text{eq}}}$$

(c) Enzymes are catalysts and only affect the rates of chemical reactions. ΔG, $\Delta G°'$, and K_{eq} are all unchanged in the presence of an enzyme.
4. (a) $K_{eq} = [\text{B}]/[\text{A}] = 1 \times 10^5 = 100{,}000$. Rearrange to get $[\text{B}] = 100{,}000[\text{A}]$. The total number of moles of A plus B must stay constant throughout the reaction. Therefore, $[\text{A}] + [\text{B}] = 1.000\text{ M} + 0.001\text{ M} = 1.001\text{ M}$ and now $[\text{A}] + 100{,}000[\text{A}] = 1.001\text{ M}$. $[\text{A}] = 1.000989 \times 10^{-5}\text{ M}$ and thus $[\text{B}] = 1.000989\text{ M}$. (b) At equilibrium, $\Delta G = 0$ and therefore

$$\Delta G = \Delta G°' + RT\ln\frac{[\text{B}]_{\text{eq}}}{[\text{A}]_{\text{eq}}} = 0\text{, and}$$

$$\Delta G°' = -(8.3 \times 10^{-3}\text{ kJ K}^{-1}\text{mol}^{-1})(298\text{ K})\ln 1 \times 10^5$$

$$\Delta G°' = -RT\ln K_{eq}$$

$$\Delta G°' = -28.5\text{ kJ mol}^{-1}$$

(c) $$\Delta G = \Delta G°' + RT\ln\frac{[\text{B}]_{\text{actual}}}{[\text{A}]_{\text{actual}}}$$

$$\Delta G = -28.5\text{ kJ mol}^{-1} + (8.3 \times 10^{-3}\text{ kJ K}^{-1}\text{mol}^{-1}) \times (298\text{ K})\ln\frac{1.5 \times 10^{-2}\text{ M}}{5 \times 10^{-5}\text{ M}}$$

$$\Delta G = -28.5\text{ kJ mol}^{-1} + 14.1\text{ kJ mol}^{-1}$$

$$\Delta G = -14.4\text{ kJ mol}^{-1}$$

5. (a) Pathway 1: $\text{A} \rightleftharpoons \text{B} \rightleftharpoons \text{E} \rightleftharpoons \text{F}$; Pathway 2: $\text{A} \rightleftharpoons \text{B} \rightleftharpoons \text{C} + \text{D} \rightleftharpoons \text{F}$. (b) Pathway 1 has a $\Delta G°' = -10$ kJ/mol; pathway 2 has a $\Delta G°' = +4$ kJ/mol. Pathway 1 is more likely to proceed (favorable $\Delta G°'$).
6. (a) Energy charge (EC) $= \dfrac{[\text{ATP}] + 0.5[\text{ADP}]}{[\text{ATP}] + [\text{ADP}] + [\text{AMP}]}$

$$\text{EC} = \frac{1.25\text{ mM} + 0.5(0.35\text{ mM})}{1.25\text{ mM} + 0.35\text{ mM} + 0.12\text{ mM}} = 0.83$$

(b) $$K_{eq} = \frac{[\text{ADP}][\text{ADP}]}{[\text{ATP}][\text{AMP}]}$$

$$K_{eq} = \frac{(0.35\text{ mM})^2}{(1.25\text{ mM})(0.12\text{ mM})} = 0.81$$

7. The energy charge of the cell refers to the relative concentrations of ATP, ADP, and AMP. At high energy charge, ATP concentrations are high relative to AMP and ADP. Anabolic pathways require ATP, so a high energy charge leads to increased flux through anabolic pathways. In contrast, at low energy charge, flux is increased through catabolic pathways to replenish ATP levels.
8.

$$\begin{array}{ll} \text{Glucose} + P_i \rightleftharpoons \text{Glucose-6-phosphate} & \Delta G^{\circ\prime} = +13.8 \text{ kJ/mol} \\ \text{ATP} \rightleftharpoons \text{ADP} + P_i & \Delta G^{\circ\prime} = -30.5 \text{ kJ/mol} \\ \hline \text{Glucose} + \text{ATP} \rightleftharpoons \text{Glucose-6-phosphate} + \text{ADP} & \Delta G^{\circ\prime} = -16.7 \text{ kJ/mol} \end{array}$$

9. The water molecules hydrogen bond into a cage-like structure around the hydrophobic limonene, increasing the order of the water molecules. This is entropically unfavorable, and therefore ΔS is negative.
10. Water molecules in ice are oriented such that maximum hydrogen bonding between water molecules occurs. Protons move through ice (and water) via proton hopping.
11. (1) Hydrogen bonds, (2) ionic interactions, (3) van der Waals interactions, and (4) hydrophobic effects. Hydrogen bonds form directly between H_2O and biomolecules and between H_2O molecules themselves, whereas hydrophobic effects are indirectly caused by H_2O due to "water-avoiding" interactions between nonpolar molecules.
12. (a) Using the Henderson–Hasselbalch equation:

$$pH = pK_a + \log \frac{[A^-]}{[HA]}$$

$$7.4 = 6.1 + \log \frac{[HCO_3^-]}{[CO_2(aq)]}$$

$$7.4 - 6.1 = 1.3 = \log \frac{[HCO_3^-]}{[CO_2(aq)]}$$

$$10^{1.3} = 10^{\frac{[HCO_3^-]}{[CO_2(aq)]}}$$

$$\frac{[HCO_3^-]}{[CO_2(aq)]} = \frac{20}{1}$$

(b) $0.025 \text{ M} = 2.5 \times 10^{-2} \text{ M} = [HCO_3^-] + [CO_2(aq)]$

$$\frac{2.5 \times 10^{-2} \text{ M}}{20 + 1} = [CO_2(aq)] = 1.19 \times 10^{-3} \text{ M}$$

$[HCO_3^-] = 2.5 \times 10^{-2} \text{ M} - 1.19 \times 10^{-3} \text{ M} = 2.38 \times 10^{-2} \text{ M}$

13. The proteins laterally diffuse in the plane of the membrane so that eventually the bleached molecules diffuse out of the laser-treated area and become diluted among fluorescent molecules. The rate of lateral protein diffusion can be determined using this approach.
14. To form a micelle, the fatty acid must have a charged polar head group. This will only occur when the carboxyl group has ionized ($COOH \rightleftharpoons COO^- + H^+$), which only occurs at pH values higher than the pK_a.

Chapter 3

REVIEW QUESTIONS

1. A nucleoside contains a ribose sugar and base; a nucleotide contains a ribose sugar, base, and phosphoryl groups.
2. Chargaff's rule that A = T and G = C in DNA molecules led Watson and Crick to focus on base pairing. The base pair rule allowed Watson and Crick to reduce the number of possible structures that fit data from Rosalind Franklin's X-ray crystallography work, and led them to identify the double helix model.
3. A-, B-, and Z-form DNA, all of which are antiparallel double helices. The covalent linkages between nucleotides are the same, but conformations differ, most distinctly in width of the helices: A-form is the widest helix, and Z-form is the narrowest. The number of base pairs per helical turn is greater in A-DNA than in B-DNA. Z-DNA contains 12 bp per helical turn because even though it is narrow, it is stretched relative to A- and B-forms. A- and B-DNA are right-handed helices; Z-DNA is a left-handed helix.
4. Linking number (*Lk*), twist (*Tw*), and writhe (*Wr*). $Lk = Tw + Wr$. *Tw* and *Wr* can change as long as *Lk* remains constant.
5. Primary structure is the nucleotide sequence of each DNA strand. Secondary structure is the formation of a double helix.
6. Type I topoisomerases cleave only one strand of the DNA; type II enzymes cleave both strands.
7. Both involve the winding and/or unwinding of DNA, leading to positive supercoiling and topological strain. Topoisomerases relieve the positive supercoiling so that the topological strain does not inhibit further unwinding.
8. The 2′-hydroxyl group of ribose in RNA can lead to autocleavage of the phosphodiester backbone. DNA does not have this 2′-hydroxyl group and is therefore not prone to autocleavage.
9. The increase in light absorbance at 260 nm when DNA strands separate, allowing the extent of DNA strand separation to be monitored under different conditions.
10. Examples: N^2,N^2-dimethylguanosine, 1-methyladenosine, pseudouridine, and inosine (see Figure 3.25). These bases can facilitate tertiary structures or influence function. Inosine is often used in tRNA anticodons to allow pairing to uracil, adenine, or cytosine of mRNA during protein synthesis (see Chapter 22).
11. Formation of the nucleosome: DNA is wrapped around eight histones and, sealed by a linker histone. Nucleosomes pack together into 30-nm chromatin fibers.
12. Euchromatin is more loosely packed than heterochromatin. Euchromatin allows easier access by DNA binding proteins and contains coding sequences. Heterochromatin is mostly in noncoding regions.
13. Centromeres are heterochromatin in the center of the chromosome, where the kinetochore proteins assemble during cell division. These proteins assist in the separation of sister chromatids during mitosis. Telomeres are short, repetitive sequences on the ends of chromosomes that stabilize the ends of the chromosomal DNA where DNA polymerase cannot replicate. Telomeric repeats ensure that the polymerase is able to finish replicating the entire chromosomal DNA without losing a bit of the chromosomal end at each round of cell division.
14. Monocistronic or polycistronic transcription units. Monocistronic genes contain a promoter followed by a single coding region; polycistronic genes contain a promoter followed by multiple coding regions, usually related in their cellular function.
15. Introns and exons are two types of sequences found in eukaryotic genes. Exons contain the coding sequences, and introns contain noncoding sequences.
16. Plasmids are circular, self-replicating DNA molecules. They confer a survival advantage for bacteria by expressing antibiotic-resistance genes.

17. Denaturation, to separate the DNA strands; annealing, to match up the primers with complementary regions on the template; and primer extension, in which DNA polymerase synthesizes a complementary strand starting from the primer.
18. Blunt ends result from cleavage in the middle of the recognition sequence. Overhang ends (5′ or 3′) are generated by cleavage after the first or before the last base of the recognition sequence, respectively.
19. When dideoxynucleoside triphosphates are incorporated into growing DNA strands, strand synthesis is terminated because they lack a ribose 3′-hydroxyl group. For sequencing, strand synthesis is sequentially stopped at each nucleotide, leading to products that differ in length by one base. The sequence is determined by separating the molecules by size, then reading the order of dideoxynucleoside triphosphates ending each product.

CHALLENGE PROBLEMS

1. Hydrogen bonding occurs between bases within a DNA double helix, but when DNA strands are denatured, the hydrogen bonds are replaced by hydrogen bonds between bases and water that are similar in strength.
2. Topoisomerase II inhibitors such as etoposide prevent the repair of double-strand breaks. Cancer cells are rapidly dividing and are therefore replicating and generating DNA breaks. Failure to repair breaks can lead to cell death, so it disproportionately kills cancer cells. However, the inhibitors also affect normal cells, and can result in the loss of portions of chromosomes or in joining of different chromosomes. When such deletions or translocations occur in certain genes, they can result in certain types of leukemia.
3. Whether a protein binds to DNA or RNA in a sequence-specific manner or nonspecifically depends on the function of the protein. Histone proteins bind to any DNA in order to compact it into chromosomes; they recognize the negatively charged backbone that is present in any DNA sequence. By contrast, proteins such as transcription factors need to bind to specific nucleotide sequences to initiate transcription of particular genes.
4. Every time a cell divides, the telomere length is reduced. Because somatic cells have little to no telomerase activity, these cells cannot restore telomere length, and older cells generally have shorter telomeres. Scientists feared that an embryo created from a somatic cell would have the telomere length, and therefore "age," of the somatic cell.
5. The promoter is a region of DNA that contains recognition sequences for transcription factors that recruit RNA polymerase to initiate transcription on DNA. Separation of DNA strands in regions rich in A-T requires less energy, so are often found in promoters.
6. Introns must be spliced out of mRNA before translation can occur. Because the majority of prokaryotic genes do not contain introns, the mRNA does not need to be processed and can be translated as it is transcribed, leading to faster cell division.
7. Human genomes contain many random mutations that have little to no effect on protein expression or function. Some are spontaneous, whereas SNPs and STRs are inherited. Simply comparing the DNA of two individuals would result in the identification of hundreds to thousands of differences. If the affected gene is known, then the sequence of that gene from an affected individual can be compared to unaffected individuals, including parents or siblings, in order to help eliminate inherited variations as potential disease-causing mutations.
8. Forensic analysis of DNA involves the comparison of common variable regions among genomes. Short tandem repeats (STRs) are inherited from maternal and paternal DNA. Siblings, parents, and children of the suspect should have a significant number of STRs in common with the suspect, and so would be a good comparison in lieu of the suspect's DNA. More distant relatives should have more STRs in common with the suspect than with the general population, but not as many as a closer relative, and would not make as convincing a case.
9. When a plasmid is transformed into a culture of bacterial cells, not every cell will take up the plasmid. The use of antibiotic selection prevents bacteria without the plasmid from growing, so a bacterial colony that is antibiotic resistant can be assumed to have taken up the plasmid successfully.
10. The G-C content of a primer contributes more to the T_m than the A-T content because G-C base pairs require more heat to disrupt the duplex. Therefore, the higher the G-C content, the higher the T_m and the annealing temperature.
11. The primer sequence has 6 A, 6 T, 10 C, and 4 G. The T_m calculated using this method is $2\ °C(6 + 6) + 4\ °C(10 + 4) = 80\ °C$. There are a variety of online calculators; all use similar methods. The T_m using one of these calculators will be in the range 61–65 °C. Factors that influence this value are: number of base pairs compared with the %GC, the monovalent cation concentration, and the base stacking (nearest neighbor) interactions based on sequence.
12. Eukaryotic genes contain many introns, so genes cloned from genomic DNA would include intron sequences. Because prokaryotic genes do not contain introns and lack splicing proteins, the introns are not removed in bacteria and may lead to abortive protein synthesis. The student needs to synthesize cDNA from mRNA, which will then only contain the gene-coding sequence because the introns were removed during the processing of mRNA.

Chapter 4

REVIEW QUESTIONS

1. Isoelectric point of a protein: The pH at which it carries no net charge; term: zwitterions.
2. Chemical properties of nearby functional groups can alter the pK_a of specific amino acids if it is energetically favorable. For example, because positive or negative charges in the hydrophobic interior are energetically unfavorable, the pK_a value of a side chain may be altered to favor the neutral state over the charged state at physiological pH.
3. Charged: aspartate, glutamate, lysine, arginine, histidine; hydrophobic: glycine, alanine, proline, valine, leucine, isoleucine, methionine; hydrophilic: serine, threonine, cysteine, asparagine, glutamine; aromatic: phenylalanine, tyrosine, tryptophan.
4. Peptide bond: joins the carboxylic acid group of one amino acid and the amine group of another amino acid; formed by condensation reactions catalyzed by the ribosome. The reaction is unfavorable and requires ATP hydrolysis.
5. Phi (ϕ): The torsional angle between the α carbon and the amide nitrogen of an amino acid; psi (ψ): The angle between the α carbon and the carbonyl carbon. The values of ϕ and ψ affect the conformation of the peptide backbone.
6. Primary structure: The specific amino acid sequence of a protein; secondary structure: The regular repetitive arrangements of local regions of the polypeptide (α helices, β strands, and β turns); tertiary structure: The spatial location of all atoms in a polypeptide chain; quaternary structure: Multi-subunit protein complexes, which can involve multiple copies of the same polypeptide or of different polypeptide chains.
7. Alpha helices are amphipathic when residues that are hydrophobic (or hydrophilic) are placed three to four amino acids

away from each other. Because there are 3.6 amino acids per turn, amino acids that are three to four amino acids apart will lie on the same side of the helix.

8. (1) They are formed from combinations of ϕ and ψ angles that minimize steric hindrance; (2) they allow maximum hydrogen bonding interactions in the polypeptide backbone.

9. α helix: glutamate, alanine, and leucine; β strands: valine, isoleucine, and tyrosine; β turns: glycine, asparagine, and proline.

10. Complexes can provide structural properties not present in individual subunits and can be a mechanism for regulation of protein function through conformational changes affecting subunit interfaces. Also, bringing functional components into proximity can increase efficiency of biochemical processes.

11. Preferred folding pathways. Random folding would be too time consuming; for example, a protein of 100 amino acids sampling three possible ϕ and ψ angles for each amino acid would take billions of years to fold.

12. Protein folding introduces favorable enthalpic changes because there are more interactions in the folded state than in the unfolded state. These changes result from noncovalent interactions (or disulfide bond formation) because covalent bonds other than disulfide bonds do not change during folding. It is entropically unfavorable because it restricts the number of conformations a polypeptide can take, but the folded protein causes an increase in the disorder of water molecules, which is entropically favorable.

13. (1) Hydrophobic collapse model: Hydrophobic residues form the interior of the protein due to the hydrophobic effect, causing a loosely defined tertiary structure called a molten globule. Then, proximal residues in the molten globule interact to form well-ordered secondary and tertiary structures through van der Waals interactions and hydrogen bonding. (2) Framework model: Initially, local secondary structures form independently. Then, local secondary structures interact to form tertiary structures. (3) Nucleation model: Random interactions lead to a localized region of correct three-dimensional structure, which facilitates the formation of the surrounding tertiary and secondary structures.

14. (1) The degradation of a misfolded protein, known as a loss-of-function effect because the activity of the particular protein is missing. (2) Protein aggregation, known as a gain-of-function effect because these proteins add a process to the cell. Gain-of-function protein-folding phenotypes can result from missense mutations or from accumulation of misfolded wild-type proteins.

CHALLENGE PROBLEMS

1.

Functional group	Charge at pH 7	Charge at pH 11
N-terminal amino group of chain	+1	0
C-terminal carboxyl group of chain	−1	−1
Side chain of Lys	+1	0
Side chain of Tyr	0	−1
Side chain of Glu	−1	−1
Side chain of Asn	0	0
Side chain of His	0	0
Net charge on peptide	0	−3

2. (a) It is energetically unfavorable for the α amino group to be charged in the hydrophobic interior, thus the pK_a would be lower. The proton can be released, as H_3O^+, before the amino group enters the interior during the folding process because the strong electrostatic attraction between the lone pair electrons in the N atom and the proton is outweighed by the unfavorable condition of a charge in a hydrophobic environment. (b) The pK_a for this α amino group would increase relative to the example in (a), and would be close to the pK_a of ~8. This ionic bond neutralizes both the NH_3^+ charge and the COO^- charge, reducing the effect of having a charged group in the hydrophobic environment. Ionic bonds are stronger in a hydrophobic environment because it is unfavorable to have unbalanced charges.

3. Changing an Ala to a Val would introduce a bulkier side chain, taking up more volume in the protein interior; the resulting structural adjustments in the tertiary structure must be serious enough to cause the enzyme to lose activity. The replacement of an Ile residue with a Gly allows a tertiary structure close enough to the original structure for partial enzyme activity.

4. When the amide and carbonyl groups are involved in hydrogen bonding, such as in α-helix and β-sheet structures, this serves to minimize the effects of these polar groups.

5. Lys-Ser-Phe

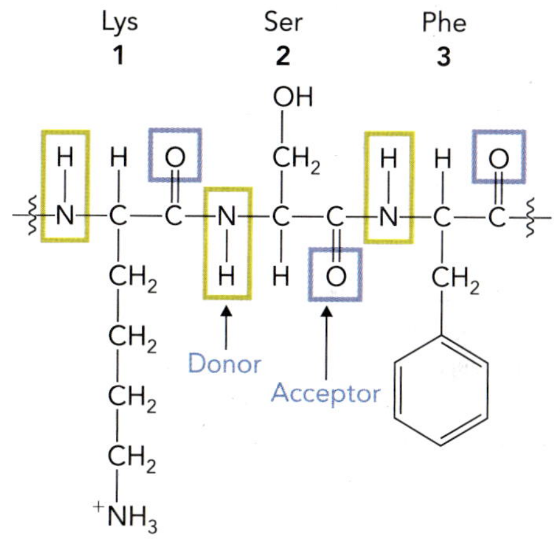

6. Ionic interactions, hydrogen bonds, van der Waals interactions, and hydrophobic interactions.

7. (a)–(c)

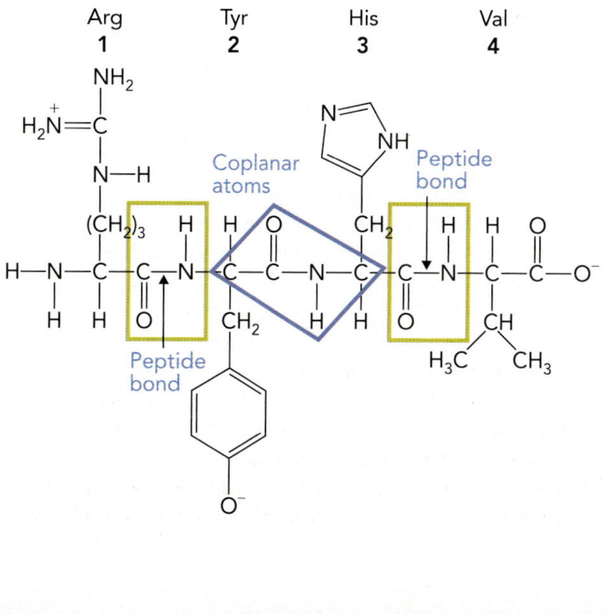

(d) Based on the pK_a values for the three ionizable residues (Arg, ~12.5; Tyr, ~10.9; His, ~6.0) and the N-terminal (~8.0) and C-terminal (~3.1) ionizable groups, a pH of 11–12 would result in the charges shown.

8. Globular proteins will have some polar backbone groups (amide NH and carbonyl O groups) buried inside the protein. When these groups are in the protein's interior, they have lost their favorable hydrogen bond interactions with H_2O. The enthalpy change (ΔH) for this process would be unfavorable (breaking bonds) if new hydrogen bonds were not made within the protein by secondary structures (hydrogen bonding within α helices and between β strands to form β sheets).

9. The polypeptide strands are antiparallel and together form an antiparallel β sheet stabilized by two hydrogen bonds:

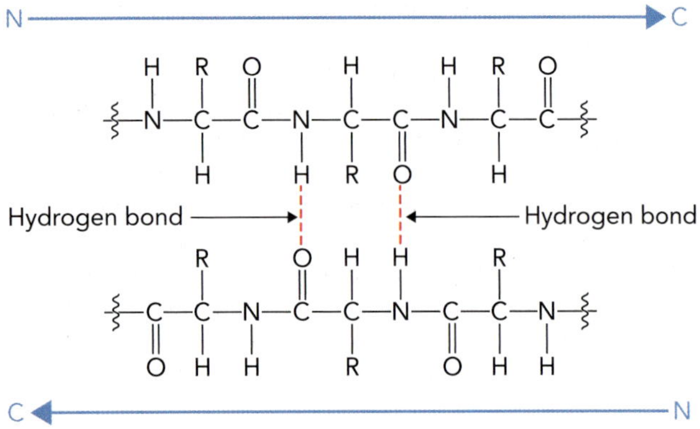

10. The zwitterion of Glu is represented by **c**. The letter(s) corresponding to the ionic forms of Glu that occur at each pH value are written above the arrows:

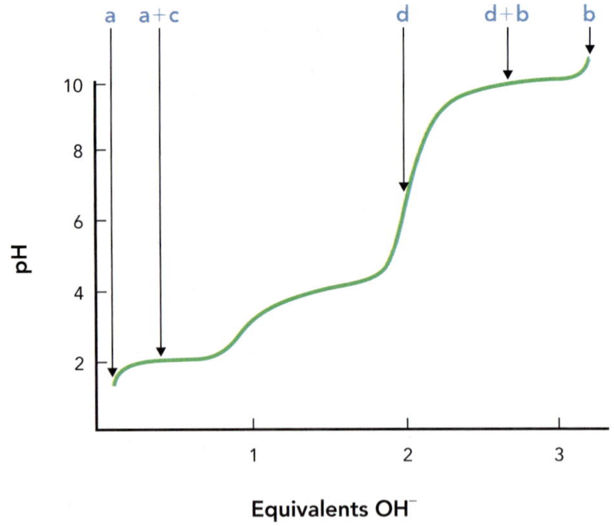

11. Cys1–Cys2 and Cys3–Cys4; Cys1–Cys4 and Cys2–Cys3. Only the Cys1–Cys2 and Cys3–Cys4 combination can explain the observation that two oligopeptides (with disulfide bonds preserved) are generated by trypsin cleavage; the other two combinations would result in one polypeptide.

12. Urea is a denaturing agent: It unfolds proteins by disrupting noncovalent interactions; β-mercaptoethanol is a reducing agent: It breaks disulfide bonds. With sufficient amounts of both urea and β-mercaptoethanol, ribonuclease protein was completely unfolded. The important conclusion was that the primary structure has all the information necessary to specify the tertiary structure.

13. Unfolded at pH 1 (the charge repulsion from the numerous positively charged α amino groups); mostly unfolded at pH 7 (even though the pK_a of the α amino groups in the α-helix would be lower than the normal ~10.8, it would not be as low as pH 7, and therefore still mostly protonated); α helical at pH 11 (above the pK_a of ~10.8, so slightly more than half of the α amino groups would be deprotonated and uncharged).

14. (a) 3 (it is the only one with a Gly-Pro sequence for a β turn); (b) 2 (it is the only one that contains alternating hydrophobic and hydrophilic side chains: Ile, Met, Leu, Val, Leu, Phe alternate with Asp, Glu, Lys, Asn, Asp, Arg); (c) 1 (every third or fourth residue is Leu [heptad repeat motif]); (d) 4 (Phe, Val, Leu, Ile, Phe, Met, Ala on the hydrophobic surface; Asn, Ser, Gln, Asp, Glu, Gln, Ser, Cys on the hydrophilic face).

Chapter 5

REVIEW QUESTIONS

1. 25,000; differences in RNA processing and protein modifications lead to many more actual proteins.

2. (1) Physical properties, such as the total number of amino acids and the sequence of side chains; (2) relative abundance in a specific cell type.

3. (1) Sonication disrupts the cell membrane through vibrational forces of ultrasonic waves. (2) Shearing uses a tight-fitting Teflon plunger in a glass vessel, a syringe, or a mechanical device called a French press to force cells through a small opening. (3) Mild detergents incubated with the cell sample can disrupt cell membranes.

4. The total amount of activity of a target protein divided by the total amount of protein in the fraction.

5. (1) Gel filtration chromatography uses porous carbohydrate beads to separate proteins on the basis of their size. The beads contain pores that allow small proteins to enter the beads while larger proteins flow around the beads, such that larger proteins flow through the column faster than smaller proteins. (2) Ion-exchange chromatography exploits the charge differences between proteins. Ion-exchange column matrices bind proteins based on charge: Negatively charged proteins bind to anion-exchange resins, and positively charged proteins bind cation-exchange resins. Once the protein sample is loaded onto the column, oppositely charged proteins and resin interact to suspend the target protein in the column. Extensive washing with loading buffer removes unbound protein, and then a buffer with a high concentration of an appropriate competing ion can be used to displace the bound target protein. (3) Affinity chromatography exploits specific binding properties of the target protein. A high-affinity ligand for the target protein is covalently linked to a matrix bead and the protein sample is passed through the column under optimal protein–ligand binding conditions. The target protein specifically associates with the matrix and other protein are washed away. The target protein is eluted using a buffer with a competing ligand.

6. Polyacrylamide gel electrophoresis (PAGE) separates proteins on the basis of their mass and charge. Sodium dodecyl sulfate (SDS) surrounds proteins, leading to a uniform negative charge, whose magnitude depends on the length, and thus the mass, of the protein. Samples are added to a vertical gel apparatus, in which gel matrix is sandwiched between two glass plates, with buffer in the top and bottom chambers. Current flows between the two chambers, passing through the gel matrix, causing the negatively charged proteins to migrate toward the anode (bottom chamber). Smaller proteins migrate faster than larger ones.

7. After each round of cleavage, a new sample of protein had to be added because the acid hydrolysis destroyed the sample.
8. The N-terminal amino acid is tagged using phenylisothiocyanate (PITC), then trifluoroacetic acid is used to cleave the peptide bond between the first and second amino acid, releasing a thiazolinone derivative and the cleaved polypeptide. The thiazolinone derivative is converted to a phenylthiohydantoin (PTH) derivative that can be identified by chromatography using known standards.
9. Mass spectrometry measures the mass-to-charge ratio of molecules. The mass of a charged molecule in an applied magnetic field exerts a force that affects acceleration of the molecule, described by Newton's second law, $F = ma$. The mass of the molecule is calculated by measuring its acceleration along a curved path with known applied magnetic force.
10. The C-terminal amino acid is covalently attached to a resin molecule; this is the first amino acid in the chain, with an N-terminal blocking group, abbreviated as Fmoc. Fmoc is removed with base, and a new Fmoc-blocked amino acid—activated at the carboxyl group by dicyclohexylcarbodiimide (DCC)—forms a new peptide bond. The amino acid side chains are chemically protected to prevent reactions, so that the amino terminus of the resin-bound amino acid is the only group available to react with the incoming DCC-activated amino acid.
11. X-ray crystallography is not as limited by constraints to protein size as is NMR spectroscopy.
12. An X-ray beam interacts with homogeneous protein crystal and the X-rays are diffracted off electron-dense regions and scatter in different directions. The intensities and directions of diffracted X-rays are recorded by X-ray detectors, computational tools calculate a map of the electron-dense regions, and a model of the protein is built.
13. Growing diffraction-quality crystals and determining the phase of the diffracted X-rays.
14. NMR spectroscopy utilizes intrinsic magnetic properties of certain atoms (e.g., ^1H, ^{15}N, and ^{13}C) to derive the relative locations of atoms in a protein in a highly concentrated solution. NMR instruments use large magnetic fields to align the nuclear spins from NMR-active nuclei and then perturb these spins using short radio-frequency pulses. Information about the perturbations and the return to the ground state is used to interpret the environment of each nucleus.
15. Large, multi-subunit proteins that bind antigens (often proteins) with high specificity and affinity.
16. Polyclonal and monoclonal antibodies. Polyclonal antibodies are a heterogeneous mixture of antibodies that recognize one or more epitopes on an antigen. Monoclonal antibodies are homogeneous and recognize only a single epitope on the antigen.
17. Highly antigenic amino acid sequences, created using recombinant DNA technology, for analysis of gene-encoded proteins.
18. A primary (or "capture") antibody, specific to a single epitope on the antigen, is covalently attached to the bottom of a 96-well plate. Sample is added to each well and incubated to promote antigen–antibody binding. Unbound proteins are washed away, then the second ("detection") antibody that recognizes a distinct epitope on the same antigen, is added to each well, forming a molecular sandwich. A secondary antibody, linked to an enzyme such as horseradish peroxidase (HRP), recognizes the detection antibody as an antigen. This and a chromogenic or fluorogenic HRP substrate are added and samples with antigenic protein produce HRP product, which is detected spectrophotometrically.

CHALLENGE PROBLEMS

1. (a) First step: 2×10^4 units/mg protein; fourth step: 2×10^8 units/mg protein. (b) Second step: 1×10^5 mg; third step: 1×10^3 mg. (c) 4×10^9 units/1×10^{10} units = 40% yield. (d) ~25,000 kDa; the native protein is a homotetramer: four subunits of 25,000 kDa each.
2. (a) Protein C (smallest-mass native protein). (b) Protein E (largest-mass denatured protein). (c) Protein D (most negatively charged at pH 6.5; highest affinity to DEAE). (d) Protein D is a homotrimer: three subunits of 25,000 Da each.
3. The enzyme precipitated by heat denaturation will have little, if any, activity upon resolubilization, whereas the protein precipitated by ammonium sulfate should retain most of its activity upon resolubilization. The heat-denatured protein is nearly impossible to purify further because the activity assay will not work.
4. Size-exclusion column: The 80-kDa protein migrates faster (less interaction with the resin). SDS-PAGE: The 20-kDa protein migrates faster (smaller frictional coefficient).
5. On the basis of the pI values, at pH 7 cytochrome c is positively charged and cytochrome c peroxidase is negatively charged. DEAE is an anion-exchange resin (positively charged), so cytochrome c cannot be directly bound to resin; it must be bound to cytochrome c peroxidase, which is in turn bound to resin. Increasing the salt concentration weakens the interaction of cytochrome c with cytochrome c peroxidase, causing it to elute first.
6. Run the protein through a gel filtration column using a low-ionic-strength buffer (10 mM NaCl) at pH 6, which is a pH greater than the pI of protein A and protein B, but below the pI of protein C. Protein B will elute more rapidly from the column than A or C, which will likely co-elute. Because the pH is midway between the pI of A and C, protein A will have a negative charge and C will have a positive charge. Running these proteins through a DEAE column (positively charged) would cause protein A to bind tightly to the column, while C should elute. Protein A could then be eluted by increasing either ionic strength or pH of the buffer.
7. Trypsin cleaves on the carboxyl side of lysine or arginine residues. Because the 1st or 14th amino acid must be Q (it cannot be the 7th or 8th amino acid based on the substrate specificity of trypsin), the 6th or 7th amino acid and the 13th or 14th amino acid must be K or R. Chymotrypsin cleaves on the carboxyl side of tyrosine, tryptophan, and phenylalanine residues and generated G-I-F; cyanogen bromide cleaves on the carboxyl side of methionine and generated G-I-F-M, thus the N-terminal sequence must be G1-F2-I3-M4, and the C-terminal amino acid must be Q14. The V-8 protease cleaves on the carboxyl side of glutamate and aspartate, so the C-terminal end must be P12-R13-Q14, and the sixth amino acid must be K6, as only G is represented twice. The V-8 tetrapeptide must be adjacent to this C-terminal end and be G8-Y9-N10-D11. The V-8 heptapeptide is the first seven amino acids: G1-F2-I3-M4-X5-K6-E7, in which X5 must be cysteine (C5), the only amino acid remaining.

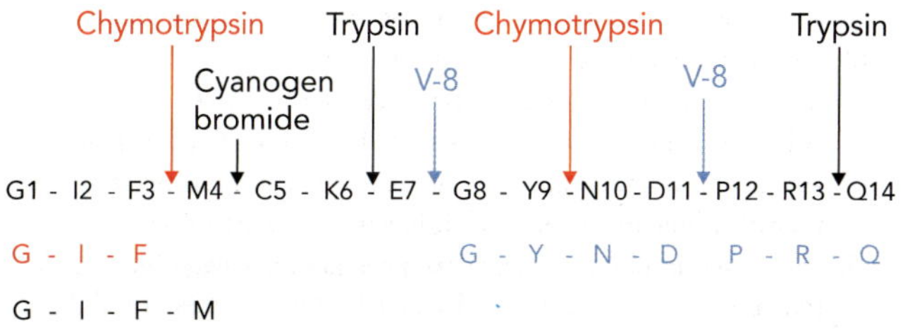

8. (a) Structural heterogeneity of the protein can inhibit crystal growth. (b) The main advantage of orthologous proteins is that sequence differences between closely related proteins can result in subtle chemical and physical changes that allow diffractable crystals to form. A disadvantage is that the two proteins may have significant structure–function differences that limit the data interpretation (see Figure 1.30). (c) Recombinant DNA techniques used to cause minimal changes in amino acid sequence, not likely to alter protein function, may change chemical and physical properties to permit crystallization. For example, a few amino acid changes in hydrophobic regions or in ionic interactions can sometimes be enough to facilitate crystallization. (d) An alternative method is nuclear magnetic resonance (NMR) spectroscopy, which identifies the relative locations of atoms in a concentrated solution of purified protein, providing spatial information used to decipher three-dimensional structures. The main advantage of NMR is that it uses soluble protein in its native form; the disadvantage is that it is limited to proteins <30 kDa.

9. Polyclonal antibodies are a collection of immunoglobulin proteins that recognize different epitopes on the same antigen with various specificities and affinities. A monoclonal antibody is a single immunoglobulin that recognizes only one epitope. Monoclonal antibodies are suitable for human diagnostic and clinical applications for two main reasons: (1) the antibody–antigen interaction is highly specific and amenable to quality control, and (2) monoclonal antibodies are secreted by immortalized cell lines and can be produced indefinitely.

10. SDS-PAGE analysis separates denatured proteins and is more likely to expose the antigen to the antibody during Western blotting. In contrast, immunofluorescent staining traps proteins in the cell under conditions that could mask antigen sites as a result of intramolecular or intermolecular interactions.

Chapter 6

REVIEW QUESTIONS

1. Metabolic enzymes, structural proteins, transport proteins, signaling proteins, and genomic caretaker proteins.

2. Enzymes increase the rate of a reaction by lowering the activation energy of that reaction. Enzymes do not change the equilibrium constant of the reaction.

3. Actin and tubulin: Actin polymers are called thin filaments; tubulin polymers are called microtubules.

4. Passive transporters: Energy independent and function as membrane pores dependent on chemical gradients.
Active transporters: Require energy input to induce protein conformational changes that open and close a gated channel.

5. Genomic caretaker proteins maintain the integrity of genomic DNA over the life of the cell and ensure that expression of specific genes is tightly controlled as per the biochemical needs of the cell.

6. Myoglobin and hemoglobin: Hemoglobin transports heme-bound O_2 from lungs to tissues; myoglobin functions as a storage depot for O_2 in muscle.

7. The affinity of carbon monoxide for the Fe^{2+} in hemoglobin is 200 times higher than that of oxygen.

8. The binding curve for myoglobin is hyperbolic, while the binding curve for hemoglobin is sigmoidal. Sigmoidal binding is indicative of cooperative binding; that is, binding of the first molecular oxygen to hemoglobin facilitates the binding of more oxygen.

9. (1) Ligand binding is a reversible process involving weak noncovalent interactions. (2) Ligand binding induces or stabilizes structural conformations in target proteins. (3) The equilibrium between ligand-bound protein and ligand-free protein can be altered by effector molecules, which induce conformational changes in the protein that increase or decrease ligand affinity.

10. In the tetramer complex of fetal hemoglobin, the normal adult β subunit is replaced by γ. In this subunit, His143 is replaced by a serine, lowering affinity for 2,3-BPG by eliminating two of the six positive charges in the 2,3-BPG binding site. Decreased 2,3-BPG affinity facilitates transfer of oxygen from the mother's hemoglobin to the fetal hemoglobin.

11. Glutamic acid is changed to valine at position 6 of the β-globin polypeptide (β-E6V), which causes the recessive genetic disease sickle cell anemia. Hemoglobin with this mutated β globin is pathological in low O_2, primarily in tissues such as the kidney and spleen where erythrocytes circulate more slowly and are often destroyed. It is thought that *Plasmodium* infection of red blood cells leads to a reduction in pH, which induces aggregation of HbS tetramers and cell sickling, and removal of sickled cells by the spleen. Because *Plasmodium*-infected cells are preferentially depleted from circulation, heterozygous HbS individuals are protected from severe malaria.

12. (1) Membrane receptor proteins transduce extracellular signals across the plasma membrane through ligand-induced conformational changes. (2) Membrane-bound metabolic enzymes, including the inner mitochondrial membrane enzymes involved in the oxidation–reduction reactions and ATP synthesis. (3) Membrane transporter proteins facilitate the transfer of polar molecules across the hydrophobic membrane. Active transporters require energy to transport molecules against their concentration gradient; passive transporters do not require energy and transport molecules along their concentration gradient.

13. (1) Primary active transporters use ATP hydrolysis to drive conformational changes that serve to pump molecules across the membrane. (2) Secondary active transporters use energy from movement of one molecule down its concentration gradient to transport a different molecule against its concentration gradient. This co-transport can be symport or antiport and is usually coupled to a primary active transport mechanism.

14. Zoloft targets the serotonin transporter, a secondary active symporter that couples sodium ion transport with the reuptake of serotonin from the synaptic cleft into presynaptic neurons. Inhibition of the serotonin transporter results in elevated levels of serotonin and thus elevated levels of serotonin receptor activation, associated with feelings of well-being and happiness.

15. In the presence of Ca^{2+}, the troponin complex undergoes a conformational change that moves the TnI subunit and tropomyosin complex away from the myosin binding site on the actin polymer duplex, allowing the myosin head to bind tightly to actin, leading to contraction.

16. (1) Ca^{2+} induces conformational changes in actin filaments, facilitating binding of ADP + P_i–bound myosin heads; (2) release of P_i from the myosin head causes a large myosin movement (power stroke), pulling the actin along the myosin; (3) ADP release from myosin frees up the nucleotide binding site; (4) ATP binds the myosin head causing myosin to disengage from actin; (5) ATP hydrolysis induces the recovery conformation by myosin, now in ADP + P_i form and ready for another round of the cycle.

CHALLENGE PROBLEMS

1. 2,3-bisphosphoglycerate (2,3-BPG), H^+ (decreased pH), and CO_2—all negative effectors of O_2 binding. O_2 is a positive homotropic effector because binding of one O_2 molecule induces a conformational shift from the T to R state, which has a much higher affinity for O_2.

2. The 2,3-BPG-depleted red cells contain Hb that binds O_2 too strongly and would not deliver O_2 to the tissues.

3. (a) Inositol hexaphosphate

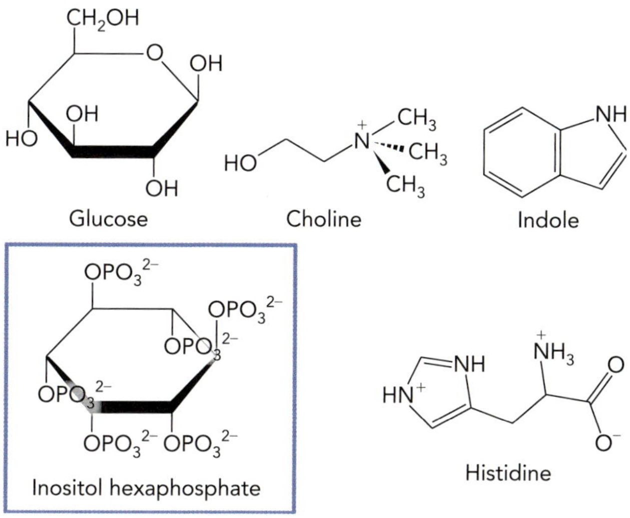

(b) It should bind in the central cavity between the four globin subunits, by ionic interactions with positively charged residues. (c) The 2,3-BPG analog should decrease O_2 affinity by shifting the equilibrium to the T-state conformation.

4. The four graphs here show the O_2 binding curves of the various Hb proteins.

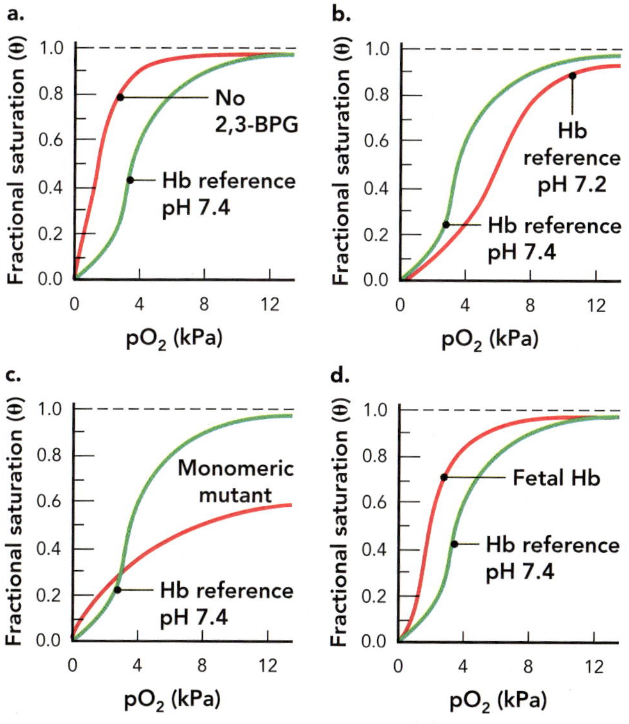

(a) In absence of 2,3-BPG, R–T equilibrium shifts towards R and the curve is slightly more hyperbolic (less sigmoidal) because of reduced cooperativity of O_2 binding; the P_{50} will be lower (curve is shifted to the left). (b) Higher $[H^+]$ (lower pH) shifts the R–T equilibrium toward T; the curve is more sigmoidal with a higher P_{50} (curve shifted to the right). (c) Monomeric Hb has no cooperativity; the plot is purely hyperbolic with higher P_{50} (curve shifted to the right). (d) Fetal hemoglobin has one less positive charge on each γ subunit in the central cavity, so 2,3-BPG binds less tightly and more hemoglobin is in the R state; the curve is mostly sigmoidal and shifted to the left (lower P_{50}).

5. The α and β subunits of hemoglobin are more hydrophobic because subunit interactions use hydrophobic amino acids.

6. This mutant would have a lower O_2 binding affinity because the R–T equilibrium will be mostly shifted toward T.

7. HbF contains two α and two γ chains ($\alpha_2\gamma_2$); HbA has two α chains and two β chains ($\alpha_2\beta_2$). The γ chains in HbF contain a Ser143 residue in place of the His143 residue found in the β subunit. With two fewer positively charged groups on HbF (the γ chains), 2,3-BPG binds more weakly to HbF than to HbA, so that HbF has a higher O_2 affinity than does maternal HbA. At comparable 2,3-BPG concentrations in maternal and fetal blood, HbF will bind O_2 more tightly than HbA, and O_2 from maternal Hb can transfer to fetal Hb in the placenta, thus being "delivered" to the HbF through the placenta.

8. (d), which has alternating hydrophobic and hydrophilic residues. The β strands must alternate in this way in order for the membrane side of the porin β barrel to be hydrophobic and the channel side of the β barrel to be hydrophilic.

9. 33.4 ms:

$$\frac{2 \times 10^5 \text{ monomers}}{4 \text{ monomers/tetramer}} = 5 \times 10^4 \text{ tetramers}$$

$$(5 \times 10^4 \text{ tetramers})(5 \times 10^8 \text{ molecules/tetramer} \cdot \text{sec})$$
$$= 2.5 \times 10^{13} \text{ molecules/s}$$

$$\text{Half volume} = 2.5 \times 10^{-11} \text{ mL} = 2.5 \times 10^{-14} \text{ L}$$

If water = half volume, then

$$(55.5 \text{ mol/L})(2.5 \times 10^{-14} \text{ L})(6.02 \times 10^{23} \text{ molecules/s})$$
$$= 8.35 \times 10^{11} \text{ molecules}$$

$$\frac{8.35 \times 10^{11} \text{ molecules}}{2.5 \times 10^{13} \text{ molecules/s}} = 3.34 \times 10^{-2} \text{ s} = 33.4 \text{ ms}$$

10. $\Delta G = -10.1$ kJ/mol:

$$\Delta G = RT\ln\left(\frac{C_2}{C_1}\right) + ZF\Delta V$$

Glucose is uncharged, so

$$\Delta G = RT\ln\left(\frac{C_2}{C_1}\right) + 0$$

$$= RT\ln\left(\frac{[\text{lactose}]_{\text{inside}}}{[\text{lactose}]_{\text{outside}}}\right)$$

$$= (8.314 \times 10^{-3} \text{ kJ/mol K})(310 \text{ K})\ln\left(\frac{5 \times 10^{-4}\text{ M}}{1 \times 10^{-3}\text{ M}}\right)$$

$$= -10.1 \text{ kJ/mol}$$

11. (a) $\Delta G = +3.66$ kJ: The Na^+–K^+ ATPase imports K^+ ions *into* the cell, therefore

$$\Delta G = RT\ln\left(\frac{C_2}{C_1}\right) + ZF\Delta V$$

$$= (8.314 \times 10^{-3} \text{ kJ/mol K})(310 \text{ K}) \ln\left(\frac{14 \times 10^{-2}\text{ M}}{5 \times 10^{-3}\text{ M}}\right)$$
$$+ [(+1) \times 95 \text{ kJ/V mol} \times -0.070 \text{ V}]$$

(Note: The membrane potential is negative inside relative to the outside. The membrane potential here is −0.070 V, thus $ZF\Delta V$ is negative. It is favorable electrically to transport a positive ion into a negative environment.)

$$\Delta G = 8.59 \text{ kJ/mol} - 6.76 \text{ kJ/mol} = 1.83 \text{ kJ/mol}$$

To transport 2 mol of K^+, $\Delta G = 1.83 \text{ kJ/mol} \times 2 \text{ mol} = +3.66$ kJ.
(b) $\Delta G = +41.22$ kJ: The Na^+–K^+ ATPase protein exports Na^+ ions *out of* the cell, so

$$\Delta G = RT\ln\left(\frac{C_2}{C_1}\right) + ZF\Delta V$$

$$= (8.314 \times 10^{-3} \text{ kJ/mol K})(310 \text{ K}) \ln\left(\frac{15 \times 10^{-2} \text{ M}}{1 \times 10^{-2} \text{ M}}\right) + [(+1) \times 95 \text{ kJ/V mol} \times 0.070 \text{ V}]$$

(Note: The membrane potential is negative inside relative to the outside. The membrane potential here is 0.070 V, making $ZF\Delta V$ positive. It is unfavorable electrically to transport a positive ion into a positive environment.)

$$\Delta G = 6.98 \text{ kJ/mol} + 6.76 \text{ kJ/mol} = 13.74 \text{ kJ/mol}$$

For 3 mol of Na^+, $\Delta G = 13.74 \text{ kJ/mol} \times 3 \text{ mol} = +41.22$ kJ.
(c) The Na^+–K^+ ATPase *exchanges* 2 mol of K^+ imported for every 3 mol of Na^+ exported, and so hydrolysis of a single ATP is sufficient (i.e., 50 kJ is greater than the combined 3.66 kJ for K^+ transport and 41.22 kJ for Na^+ transport).

12. 47.9:1, as shown here:

$$\Delta G = RT\ln\left(\frac{C_2}{C_1}\right) + ZF\Delta V$$

Lactose is uncharged, then

$$\Delta G = RT\ln\left(\frac{C_2}{C_1}\right) + 0$$

$$= -10 \text{ kJ/mol} = RT\ln\left(\frac{C_2}{C_1}\right)$$

$$= RT\ln\left(\frac{[\text{lactose}]_{\text{inside}}}{[\text{lactose}]_{\text{outside}}}\right)$$

$$= -10 \text{ kJ/mol} = (8.314 \times 10^{-3} \text{ kJ/mol K})(310 \text{ K}) \times \ln\left(\frac{[\text{lactose}]_{\text{inside}}}{[\text{lactose}]_{\text{outside}}}\right)$$

$$= -10 \text{ kJ/mol} = 2.58 \text{ kJ/mol} \times \ln\left(\frac{[\text{lactose}]_{\text{inside}}}{[\text{lactose}]_{\text{outside}}}\right)$$

$$\frac{10 \text{ kJ/mol}}{2.58 \text{ kJ/mol}} = 3.87$$

$$\left(\frac{[\text{lactose}]_{\text{inside}}}{[\text{lactose}]_{\text{outside}}}\right) = e^{3.87} = 47.9{:}1 \text{ maximum concentration ratio}$$

13. Actin and ATP must bind to different conformational states of the myosin head, so actin binding stabilizes a conformational state with low affinity for ATP, whereas ATP binding stabilizes a state with low affinity for actin.

14. Ca^{2+} binds to troponin C (TnC), changing the conformation of troponin complex subunits such that TnC, TnI, and TnT all rotate tropomyosin away from the myosin binding site on actin, so that myosin can bind actin and initiate the contraction cycle.

Chapter 7

REVIEW QUESTIONS

1. (1) Substrates bind with high affinity and specificity to the enzyme active sites; (2) substrate–active site binding often involves structural changes to maximize weak interactions between substrate and enzyme; (3) enzyme activity is highly regulated to maximize energy balance between catabolic and anabolic pathways and to respond to environmental stimuli.

2. By increases or decreases in enzyme levels (bioavailability) and by changes in catalytic efficiency. Bioavailability can be at the level of gene expression or protein synthesis, whereas catalytic efficiency is regulated by posttranslational modifications or binding of allosteric regulators.

3. By lowering the activation energy. They provide favorable conditions to lower the transition energy barrier, which increases the rate of the reaction without altering overall ΔG.

4. Cofactors are small molecules that aid in catalytic reaction mechanisms by providing additional chemical groups to supplement the chemistry of amino acid functional groups. These include inorganic ions such as Fe^{2+}, Mg^{2+}, Mn^{2+}, Cu^{2+}, and Zn^{2+}, or coenzymes, which are organic compounds. Coenzymes can be loosely or tightly associated with enzymes, or even covalently attached. Most vitamins from nutritional sources are coenzymes. Coenzymes that are permanently associated with enzymes, such the heme of catalase, are called prosthetic groups.

5. (1) By providing an alternative path to product formation, e.g., forming stable reaction intermediates that are covalently attached to the enzyme; (2) by lowering the activation barrier through stabilization of the transition state; (3) by increasing the substrate's ground state energy by limiting bond rotation within the active site, lowering the energy difference between the ground state and the transition state.

6. A set of three amino acids that form a hydrogen-bonded network required for catalysis. In chymotrypsin, these three amino acids are Ser195, His57, and Asp102.

7. First, a covalent acyl-enzyme intermediate is formed between Ser195 and the polypeptide substrate, to promote cleavage of the scissile peptide bond and release the amino terminal polypeptide fragment. Second, the enzyme is regenerated in a series of steps resulting in deacylation and release of the carboxyl-terminal polypeptide fragment.

8. Eight amino acids have ionizable side chains (can gain or lose a proton, depending on the local environment and effective pK_a; see Figure 7.17). Changes in pH alter active-site chemistry by protonation and deprotonation of the side chains, disrupting critical tertiary structures within the protein, altering enzyme activity.

9. (1) A competitive inhibitor competes with substrate for binding to the active site (see Figure 7.45a). This shifts the apparent K_m value for the enzyme to the right as a function of increasing inhibitor concentration, [I], but does not change the v_{max} of the reaction. (2) Uncompetitive inhibitors bind to enzyme–substrate complexes (ES) and alter the active site conformation, reducing catalytic activity (see Figure 7.45b), and resulting in a decrease in the K_m and v_{max} without altering the K_m/v_{max} ratio. (3) Mixed inhibitors also bind to sites distinct from the active site, but they bind to *both* the enzyme and the enzyme–substrate complex (see Figure 7.45c). A mixed inhibitor decreases v_{max} and may increase or decrease K_m, depending on the relative K_I and K_I' values.

10. Feedback inhibition occurs when the end product of a pathway functions as an inhibitor of the first enzyme in the pathway.

11. Phosphorylation of Ser, Thr, and Tyr residues by kinase enzymes, which adds negative charge to enzymes through the addition of phosphoryl groups (PO_3^{2-}).

CHALLENGE PROBLEMS

1. The bulk of the protein is necessary to produce the correct tertiary structure for the substrate binding site and for proper orientation of the catalytic residues.

2. You make the transition state bind more tightly in order to lower the activation energy of the reaction. Binding the substrate more tightly would make a higher activation energy.
3. An induced-fit mechanism requires some energy to drive conformational change in the protein upon substrate binding, which reduces the energy that could be used to bind the transition state and lower the activation energy.
4. (a) Rate enhancement $= \dfrac{k_F \text{ catalyzed}}{k_F \text{ uncatalyzed}}$

$$= \frac{5 \times 10^6 \text{ s}^{-1}}{1 \times 10^{-2} \text{ s}^{-1}}$$

$= 5 \times 10^8$-fold enhancement

(b) Catalysts do not change the K_{eq}, so the K_{eq} remains 1×10^3.
5. The inhibitor loses activity as a result of deprotonation at pH 8.5 relative to pH 6.5, on a group found in the inhibitor molecule.
6. (a) No effect; (b) decrease; (c) increase; (d) no effect; (e) decrease.
7. This would decrease rate of product formation because stabilization of ES increases the activation energy (even though the transition state energy is unchanged).
8. (a) $K_{eq} = \dfrac{k_{F(uncat)}}{k_{R(uncat)}} = \dfrac{10^{-5} \text{ s}^{-1}}{10^{-2} \text{ s}^{-1}} = 10^{-3}$ (unitless)

(b) $k_{R(cat)} = \dfrac{k_{F(cat)}}{K_{eq}} = \dfrac{10^7 \text{ s}^{-1}}{10^{-3}} = 10^{10} \text{ s}^{-1}$

(c) $k_{cat} = \dfrac{v_{max}}{[E_t]} = \dfrac{5 \times 10^3 \text{ M s}^{-1}}{2 \times 10^{-6} \text{ M}} = 2.5 \times 10^9 \text{ s}^{-1}$

9. (a) The imidazole of the active site His residue. (b) The pH at which [base] = [acid], so protein would be 50% active at pH 6.8.

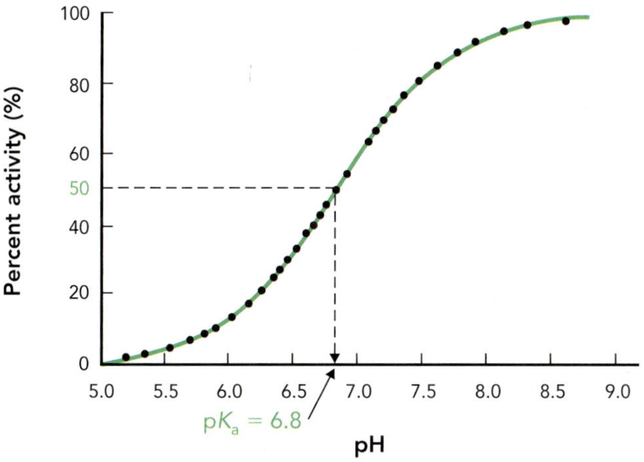

(c) In the first catalytic step of the mechanism, the active site His is unprotonated because it acts as a general base to accept a proton from the active site Ser.

10.

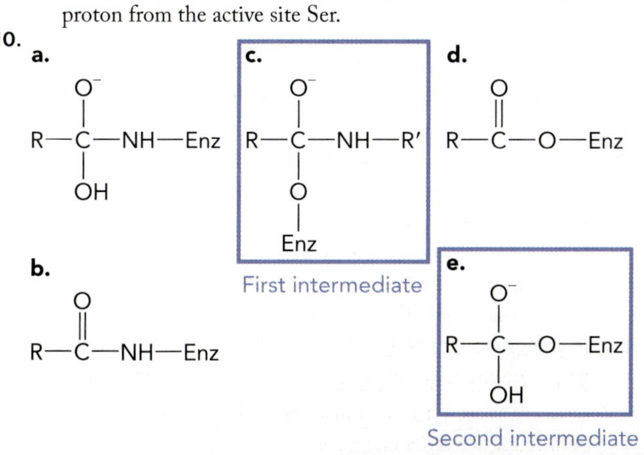

11. (a) $\dfrac{-1}{K_m} = -8.0 \text{ mM}^{-1}$, therefore $K_m = 0.125$ mM

(b) $\dfrac{1}{v_{max}} = 4(\text{mM min}^{-1})^{-1}$, therefore, $v_{max} = 0.25 \text{ mM min}^{-1}$

(c) $[E_t]$ = concentration of total enzyme active sites

$= 1.2 \times 10^{-7}$ M $= 1.2 \times 10^{-4}$ mM

$$k_{cat} = \frac{v_{max}}{[E_t]} = \frac{0.25 \text{ mM min}^{-1}}{1.2 \times 10^{-4} \text{ mM}}$$

$$k_{cat} = 2.08 \times 10^3 \text{ min}^{-1} \times \frac{1 \text{ min}}{60 \text{ s}} = 34.7 \text{ s}^{-1}$$

(d) $\theta = \dfrac{[ES]}{[E_t]} = \dfrac{v_0}{v_{max}} = \dfrac{[S]}{(K_m + [S])}$

$$= \frac{2.5 \times 10^{-4} \text{ M}}{(1.25 \times 10^{-4} \text{ M} + 2.5 \times 10^{-4} \text{ M})}$$

$$\theta = \frac{[ES]}{[E_t]} = \frac{2.5}{3.75} = 0.67$$

12. (a) Yes, Y is a competitive inhibitor of X; the apparent K_m for X will increase. (b) No, the v_{max} for X will be unaffected. (c) Yes, the pH dependence of v_{max} reflects the ionization of catalytic site residues, whereas pH dependence of v_{max}/K_m will reflect the ionization state of substrate binding residues.
13. (a) v_{max}[−I] = 0.1 μmol/min, using $1/v_{max} = 10$ (μmol/min)$^{-1}$; (b) Iv_{max}[+I] = 0.04 μmol/min, using $1/v_{max} = 25$ (μmol/min)$^{-1}$; (c) K_m[−I] = 0.25 mM, using $-1/K_m = -4$ (mM)$^{-1}$ (± inhibitor); (d) K_m[+I] = 0.25 mM, using $-1/K_m = -4$ (mM)$^{-1}$ (± inhibitor); (e) mixed (noncompetitive) inhibition because v_{max} decreased in the presence of inhibitor, but K_m was unchanged in presence of the inhibitor.
14.

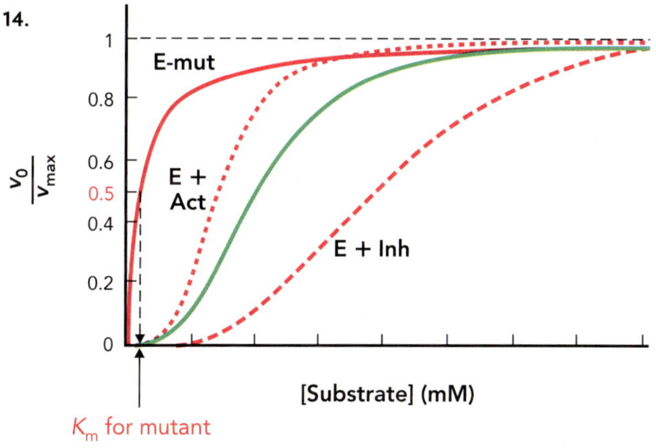

15. Methanol concentration in the body: (100 mL × 0.79 g/mL)/(32 g/mol × 40 L) = 0.062 M. Ethanol will behave as a competitive inhibitor: v (uninhibited) = v_{max}[methanol]/K_m+ [methanol]; v (inhibited) = v_{max}[methanol]/αK_m + [methanol], where α = (1 + [ethanol]/K_I). Need v(inhibited)/v(uninhibited) = 0.05 = (K_m + [methanol])/(αK_m + [methanol]). 0.05 = (0.01 + 0.062)/(0.01α + 0.062); α = 138 = [ethanol]/K_I; [ethanol] = 0.137 M. 0.137 M in 40 L = 5.48 mol; 46 g/mol × 5.48 mol = 252 g; (252 g)/(0.79 g/mL) = 319 mL. Whiskey is 50% ethanol, therefore the person needs to consume 638 mL of whiskey.

Chapter 8

REVIEW QUESTIONS

1. (1) First messengers (e.g., hormones such as insulin or epinephrine); (2) receptor proteins (e.g., insulin or adrenergic

receptors); (3) upstream signaling proteins, such as heterotrimeric G proteins and adenylate cyclase; (4) second messengers, such as cyclic AMP, diacylglycerol, and calcium; (5) downstream signaling proteins, such as cAMP-dependent protein kinase A, protein kinase C, or calmodulin; and (6) target proteins (e.g., metabolic enzymes such as phosphorylase kinase, transcription factors, cytoskeletal proteins, and transport proteins).

2. Endocrine signaling involves hormones produced by secretory glands and transported through the circulatory system to reach target cells expressing appropriate receptors. Paracrine signaling involves localized production of first messengers, such as nitric oxide or acetylcholine, to activate receptors on nearby cells. Autocrine signaling activates receptors on the same cells that produce the first messenger, a mechanism common in immune system cells.
3. Activation of a single extracellular receptor by a first messenger, such as epinephrine, activates an upstream signaling protein, such as adenylate cyclase or phospholipase C, producing 100 second messengers. Each second messenger activates a downstream signaling protein with catalytic activity, such as a protein kinase capable of phosphorylating 100 protein targets; signal amplification at this point is $10^2 \times 10^2 = 10^4$. Because most protein targets are themselves catalytic (or transcription factors), the signal is amplified at least 100-fold more ($10^4 \times 10^2 = 10^6$).
4. GPCRs are transmembrane receptors that undergo a conformational change in response to ligand binding. This dissociates the heterotrimeric G protein complex ($G_{\alpha\beta\gamma}$), which activates enzymes that generate second messengers. RTKs are enzymes with an extracellular ligand binding domain and an intracellular catalytic domain that phosphorylates tyrosine residues in the receptor C-terminal tail and in receptor-associated target proteins. TNF receptors transmit extracellular signals by forming trimers that promote cytosolic adaptor protein complexes. Nuclear receptors are ligand-regulated transcription factors that modulate gene expression. Ligand-gated ion channels control the flow of ions across membranes in response to ligand binding, controlling membrane depolarization.
5. The G_α subunit is a GTPase containing a lipid anchor embedded in the plasma membrane (cytosolic face). GPCR activation leads to dissociation of the heterotrimeric G complex and to GDP–GTP exchange in G_α, generating the active GTP-bound form. Stimulation of GTPase activity in G_α reverts it to the inactive GDP-bound form. The G_β subunit is not membrane associated but is tightly bound to the G_γ subunit, which is anchored to the plasma membrane by a lipid anchor. The $G_{\beta\gamma}$ dimer functions to sequester G_α in the GDP-bound inactive form. However, upon receptor stimulation, $G_{\beta\gamma}$ functions as a membrane-associated signaling complex and activates downstream signals.
6. The human genome contains 17 G_α, 5 G_β, and 12 G_γ genes, providing for distinct heterotrimeric G protein complexes depending on which subunits are present. These 34 genes can form ~1,000 different $G_{\alpha\beta\gamma}$ complexes ($17 \times 5 \times 12 = 970$).
7. The key differences in each of these pathways are ligand selectivity of the respective GPCR and the identity of the associated G_α protein. Light absorption by the rhodopsin retinal activates $G_{t\alpha}$, stimulating a cGMP phosphodiesterase, causing ion-channel closure and neuronal signaling. Odorant binding to olfactory receptors activates $G_{s\alpha}$, stimulating adenylate cyclase to make cAMP, causing ion-channel opening and neuronal signaling. Ligand binding to taste receptors on tongue cells activates $G_{q\alpha}$, stimulating phospholipase C to produce DAG and IP_3, causing ion-channel closure and neuronal signaling.
8. The efficiency is controlled by G_α regulatory proteins and by proteins that control GPCR turnover (GRK and β-arrestin). G_α proteins are regulated by GEFs, which promote GDP–GTP exchange. GEF activity can be mediated by the ligand-activated GPCR itself or by other associated proteins. In contrast, GAPs stimulate the intrinsic GTPase activity and inactivate G_α. GPCR GAPs are called regulators of G protein signaling; the target protein can be a regulator of G protein signaling (RGS). Turnover is controlled by GRKs, which phosphorylate the GPCR cytoplasmic domains on serine and threonine residues, marking them for recycling. GRK-mediated phosphorylation of GPCRs allows docking of β-arrestin, which prevents the GPCR from reassociating with the $G_{\alpha\beta\gamma}$ complex.
9. GRB2 is an EGFR adaptor with two types of protein–protein interaction domains: an SH2 domain that binds phosphotyrosine residues on the cytosolic tail of activated EGFRs, and an SH3 domain (there are two SH3 domains in GRB2) that binds to a downstream GEF protein (son of sevenless; SOS). The function of GRB2-activated SOS is to activate GDP–GTP exchange in the Ras G protein, which then activates a downstream Raf kinase. Once this happens, subsequent phosphorylation reactions mediated by the MAP kinase family of signaling proteins are stimulated.
10. The EGF and insulin pathways both stimulate the MAP kinase pathway through upstream GRB2–SOS–Ras signaling. One difference is that the GRB2 adaptor protein Shc is linked to the ligand-activated insulin receptor through a PTB domain. Another difference is that the insulin receptor also stimulates a completely separate PTB-mediated pathway of insulin receptor substrate (IRS-1), which is an adaptor protein that activates the downstream signaling protein PI-3K.
11. Trimerization of the TNF receptor initiates the assembly of an intracellular adaptor protein complex. In the complex are upstream signaling proteins with death domains (DDs), which associate with two types of downstream adaptor proteins. In the cell death apoptotic pathway, the TNF receptor DD sequences recruit an adaptor protein called TRADD, which in turn recruits FADD, a protein with a death effector domain (DED) that interacts with procaspase 8. The DED-mediated adaptor complex shifts the conformation of procaspase 8 resulting in autocleavage generating proteolytically active CASP8. Activated CASP8 cleaves and activates CASP3, leading to cell death. In the cell survival pathway, TRADD recruits TRAF2 and the downstream signaling kinase RIP. TRAF2 recruits NIK, which along with RIP phosphorylates and activates IKK. IKK phosphorylates the inhibitor protein IκBα, leading to activation of the NFκB transcription complex, which translocates to the nucleus and induces the expression of anti-apoptotic genes. The cell death versus cell survival outcome in a TNF receptor–stimulated cell depends on the relative expression level of downstream signaling proteins in the pro-apoptotic and anti-apoptotic pathways.
12. Nuclear receptors are ligand-activated transcription factors. The two major types can be distinguished by their mode of DNA binding to cognate response element DNA sequences located near the promoters of target genes. Steroid receptors bind specifically to inverted repeat DNA sequences as head-to-head homodimers. These include nuclear receptors that bind to cholesterol-derived hormones such as cortisol, estradiol, progesterone, and testosterone. Metabolite receptors bind as

head-to-tail heterodimers to direct repeat DNA sequences. PPARs are metabolite receptors that selectively bind dietary unsaturated fatty acids and form heterodimers with RXRs. The three determinants of nuclear signaling are (1) ligand bioavailability, (2) cell-specific expression of nuclear receptors and coregulatory proteins, and (3) genomic accessibility of target gene DNA sequences in chromatin.

CHALLENGE PROBLEMS

1. Second messengers are generated rapidly by signal-activated enzymes. Two examples are adenylate cyclase, which converts ATP to cAMP, and phospholipase C, which converts PIP_2 into DAG and IP_3.
2. If each process amplifies the signal 250-fold, the net amplification would be $250^3 = 15{,}625{,}000$. If process A were defective and only amplified the signal by 100-fold, then the net level of amplification would be $100 \times 250^2 = 6{,}250{,}000$, which is a reduction of 60%.
3. Sildenafil (Viagra) is a cGMP analog that inhibits the enzyme cGMP phosphodiesterase, resulting in elevated levels of cGMP and stimulation of cGMP-dependent protein kinase G, which is required for vasodilation. Sildenafil only works if nitric oxide signaling is activated by neuronal input to the brain, which is necessary to activate guanylyl cyclase and produce cGMP.
4. Glucagon binding a GPCR results in $G_{s\alpha}$ stimulation of adenylate cyclase, resulting in elevated levels of cAMP and activation of PKA signaling. Epinephrine binding the β2-adrenergic receptor in liver cells activates the same pathway through $G_{s\alpha}$, a shared signaling pathway. In contrast, epinephrine binding to α1-adrenergic receptors is a GPCR mechanism in which $G_{q\alpha}$ stimulates phospholipase C, resulting in elevated levels of DAG, IP_3, and Ca^{2+}; this is an example of parallel signaling pathways.
5. When you are hungry, glucagon stimulates cAMP production through $G_{s\alpha}$ signaling, but it is unlikely that epinephrine will be released. In contrast, when you almost step on a snake, acute stress stimulates epinephrine signaling through both β2 and α1 adrenergic receptors, leading to increased intracellular levels of the second messengers cAMP, DAG, IP_3, and Ca^{2+}.
6. EGF signaling normally stimulates exchange of GDP for GTP to produce active Ras (Ras–GTP); cleavage of GTP by Ras GTPase converts it back to the inactive Ras–GDP form. However, mutations that decrease the activity of Ras GTPase—either mutations in the catalytic site itself or mutations that alter GAP binding to Ras–GTP—result in formation of a constitutive "activated" Ras–GTP even in the absence of upstream EGF signaling.
7. The G protein cycle for Ras illustrates that SOS functions as a GEF and RasGAP simulates intrinsic GTPase activity in Ras.

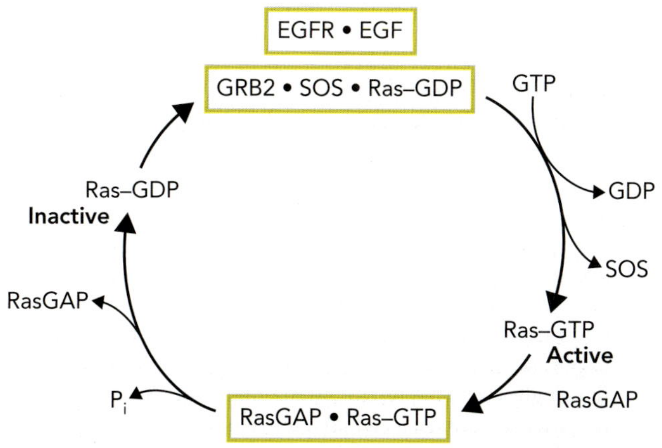

8. First, GAPs, which in the case of GPCR signaling pathways are called regulators of G protein signaling, stimulate the intrinsic GTPase activity of G_α to terminate signaling by converting the active form (G_α–GTP) into the inactive form (G_α–GDP). Second, regulatory proteins called GRKs phosphorylate serine and threonine residues in the GPCR cytoplasmic domain, which provides a docking site for an inhibitory protein called β-arrestin. These β-arrestin–GPCR complexes prevent reassociation of $G_{\alpha\beta\gamma}$ heterotrimers with the receptor, and moreover, β-arrestin–GPCR complexes are removed from the plasma membrane and translocated to endocytic vesicles. Once this happens, the GPCRs are either degraded by lysosomes or returned to the plasma membrane.
9. Conformational change is critical for many steps in signal transduction. (1) EGF binding to the extracellular domain of EGFR induces conformational change in the receptor that facilitates dimerization. (2) Receptor dimerization facilitates protein–protein interactions, promoting activation of the receptor tyrosine kinase activity and phosphorylation of partner receptors. Receptor phosphorylation is a sequential process in which first one receptor intramolecularly phosphorylates the other (EGFR1 phosphorylates EGFR2; see Figure 8.36), then another protein conformational change permits reciprocal intermolecular phosphorylation of the first receptor.
10. Order, starting at the top: 7, 2, 10, 4, 6, 1, 3, 8, 5, 9.
11. (1) The TNF receptor cytoplasmic region contains a DD that initiates adaptor complex assembly; (2) SODD, an inhibitory protein that binds to the DDs of the TNF receptor in the absence of ligand to prevent adaptor complex formation; (3) TRADD, the primary adaptor protein, has a DD and a TRAF interaction domain; (4) FADD, the DD-containing protein that links TRADD and the TNF receptor to the cell death pathway; and (5) RIP, a DD-containing kinase that activates a phosphorylation cascade leading to cell survival.
12. There are several good answers to the question; here are a few. (1) Proteases destroy proteins, the workhorses of the cell—without proteins, the cell cannot survive. (2) Proteases are enzymes that function catalytically and therefore can do a lot of damage in a very short amount of time. (3) Proteases must be activated by proteolytic cleavage to convert the inactive zymogen to the active form—this provides a means to control protease activity by regulating the initiating cleavage reaction. (4) Proteases, especially caspases, have preferred substrate recognition sites for cleavage, and therefore can preferentially degrade other proteins without undergoing high rates of autocleavage (caspases lack accessible substrate recognition sequences and are protected from degradation).
13. Steroid receptors bind as homodimers to inverted repeat DNA sequences separated by N_3 nucleotides, whereas metabolite receptors bind as heterodimers to direct repeat DNA sequences separated by N_{1-5} nucleotides. Because metabolite receptors are heterodimers and bind two different ligands, one of which is usually 9-*cis*-retinoic acid (binds to RXR), there is potential for tighter control over ligand-dependent activation compared to that of steroid receptors, which, as homodimers, bind only one type of ligand. In addition, the metabolite receptors bind to direct repeat DNA sequences separated by 1–5 nucleotides (depending on the type of heterodimer), which provides higher specificity in the target gene binding site.
14. Cell-specific glucocorticoid signaling is due to one or all of these parameters: (1) cell-specific expression of coregulatory proteins, (2) cell-specific bioavailability of glucocorticoids, and

(3) cell-specific chromatin accessibility of target gene DNA sequences to glucocorticoid receptor binding.

Chapter 9

REVIEW QUESTIONS

1. Catabolic: capture energy as ATP and reducing potential (NADH and $FADH_2$) by degrading macromolecules obtained in food. Anabolic: build macromolecules from small metabolites using ATP and reducing ability (NADPH) as source of energy.
2. Metabolic flux describes the overall reaction rates in metabolic pathways and is primarily controlled by available enzyme activity, a function of enzyme abundance and catalytic activity. Flux is also affected by availability of substrates and utilization of products.
3. Before: glucagon signaling in the liver elevates glucose efflux by increasing flux through liver glycogen degradation and gluconeogenesis. After: insulin signaling stimulates glucose influx into the liver by increasing flux through the glycogen synthesis and glycolysis.
4. Key macromolecules: proteins, nucleic acids, carbohydrates, and lipids. Primary metabolites: amino acids, nucleotides, fatty acids, glucose, pyruvate, and acetyl-CoA. Key small molecules: NH_4^+, CO_2, NADH, $FADH_2$, O_2, ATP, and H_2O.
5. An unfavorable reaction (one with a positive ΔG) may occur if the sum of the standard free energy change ($\Delta G°'$) and $RT \times \ln Q$ is sufficiently less than zero (Q is the mass action ratio), resulting in an overall negative ΔG.
6. (1) Glycolytic enzymes are evolutionarily conserved in essentially all living organisms; (2) glycolysis is the only anaerobic pathway for ATP generation; (3) glycolytic intermediates are shared metabolites in numerous pathways.
7. Glucose (aldose sugar)

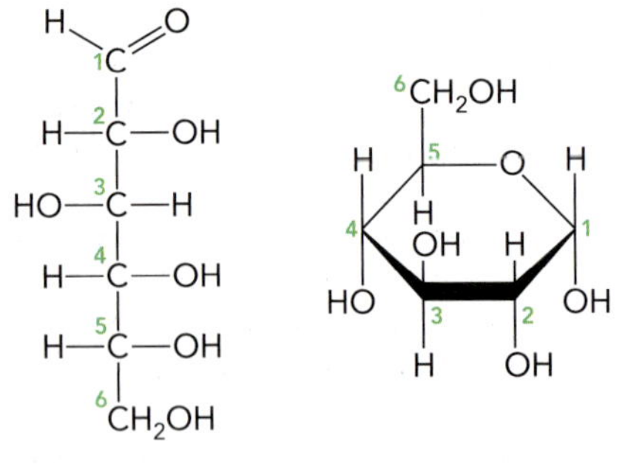

Fructose (ketose sugar)

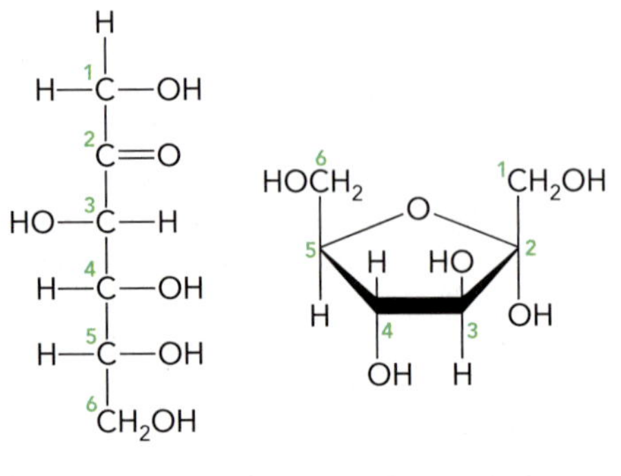

8. Glucose + 2 NAD^+ + 2 ADP + 2 P_i →

$$\text{2 Pyruvate} + \text{2 NADH} + 2\,H^+ + \text{2 ATP} + 2\,H_2O$$

Stage 1: ATP investment phase includes five reactions, requiring the input of 2 ATP to generate fructose-1,6-bisphosphate, which is cleaved, followed by an isomerization reaction to yield two molecules of glyceraldehyde-3-phosphate for each glucose. Stage 2: ATP earnings phase consists of five reactions and generates a total of four molecules of ATP for each molecule of glucose. Because 2 ATP were invested in stage 1, the net ATP yield of glycolysis is 2 ATP per glucose.
9. Substrate-level phosphorylation is a reaction in which ATP is generated from ADP by transfer of a phosphoryl group from a phosphorylated metabolite (does not require ATP synthase complex). The substrate-level phosphorylation reactions in glycolysis are reaction 7, catalyzed by phosphoglycerate kinase, and reaction 10, catalyzed by pyruvate kinase.
10. Both phosphorylate glucose on the C-6 position in the first reaction of the glycolytic pathway. Hexokinase, however, has a very high affinity for glucose (100 times higher than that of glucokinase); can phosphorylate fructose; is allosterically inhibited by glucose-6-phosphate; and is expressed in all tissues. Glucokinase is highly specific for glucose, but binds with low affinity; is not allosterically inhibited by glucose-6-phosphate; is expressed primarily in liver and pancreas; and functions as a glucose sensor in pancreatic β cells.
11. PFK-1 is a homotetramer with two conformational structures: the T state (inactive) and the R state (active). Allosteric regulation shifts the equilibrium between states. PFK-1 activity is stimulated by a low energy charge (low ATP relative to AMP and ADP), and inhibited by high energy charge. PFK-1 is allosterically activated by AMP, ADP, and fructose-2,6-bisphosphate, and allosterically inhibited by ATP and citrate.
12. Maltose is a disaccharide of glucose, cleaved by maltase to yield two molecules of glucose used by glycolysis. Lactose is cleaved by lactase to yield glucose and galactose; galactose is phosphorylated by galactokinase, converted to glucose-1-phosphate, and then isomerized to glucose-6-phosphate used by glycolysis. Sucrose is cleaved by sucrase into glucose and fructose. In muscle, fructose is directly phosphorylated by hexokinase to fructose-6-phosphate for glycolysis, whereas in liver cells, fructose is phosphorylated by fructokinase to yield fructose-1-phosphate, which is converted to eventually yield two molecules of glyceraldehyde-3-phosphate.
13. Under aerobic conditions, pyruvate is decarboxylated and converted to acetyl-CoA in organisms that use the citrate cycle. This yields the maximum number of ATP from glucose oxidation. Under anaerobic conditions, pyruvate is converted to lactate by lactate dehydrogenase in order to regenerate NAD^+ for the glyceraldehyde-3-phosphate dehydrogenase reaction in glycolysis. Alternatively, fermenting organisms, such as yeast, convert pyruvate to CO_2 and ethanol under anaerobic conditions.

CHALLENGE PROBLEMS

1. Before breakfast, after fasting at night, blood glucose levels are low and trigger the release of glucagon from the pancreas. Glucagon receptors activated in liver cells lead to increase in glycogen degradation and in glucose synthesis from pyruvate (gluconeogenesis). Together, these two metabolic pathways increase glucose levels to those appropriate for brain function (~5 mM). After breakfast, high blood glucose levels lead to insulin release from the pancreas and stimulation of insulin receptor signaling in liver cells, causing the metabolic flux of glucose to be shifted toward influx into the liver, where it is stored as glycogen and used for energy conversion by glycolysis.

2. Fructose, a ketose, is converted to glucose by the base (OH^-) by Benedict's reagent. Then glucose, an aldose, is readily oxidized by Cu^{2+}, resulting in a false-positive result in Benedict's test.
3. No. The two reactions in glycolysis can occur under anaerobic conditions when pyruvate is converted to lactate.
4. Reaction: 2-phosphoglycerate ↔ phosphoenolpyruvate
$\Delta G°' = +1.7$ kJ/mol

$$\Delta G = (1.7 \text{ kJ/mol}) + (8.314 \times 10^{-3} \text{ kJ/mol K})(298 \text{ K})\left(\ln \frac{1}{10}\right)$$

$$\Delta G = (1.7 \text{ kJ/mol}) + (8.314 \times 10^{-3} \text{ kJ/mol K})(298 \text{ K})(-2.3)$$

$$\Delta G = (1.7 \text{ kJ/mol}) + (-5.69 \text{ kJ/mol}) = \mathbf{-4 \text{ kJ/mol}}$$

The ΔG value remains the same; enzymes only change the reaction rate, not the ΔG.
5. $\Delta G°' = -RT \ln K_{eq}$
$\Delta G°' = -(8.314 \times 10^{-3} \text{ kJ/mol K})(298 \text{ K})(\ln 6.8 \times 10^{-5})$
$\Delta G°' = \mathbf{+23.8 \text{ kJ/mol}}$
There is no effect on the K_{eq} value when twice as much aldolase is added to the reaction; enzymes are catalysts and don't change the equilibrium constant of a reaction.
6. (a)

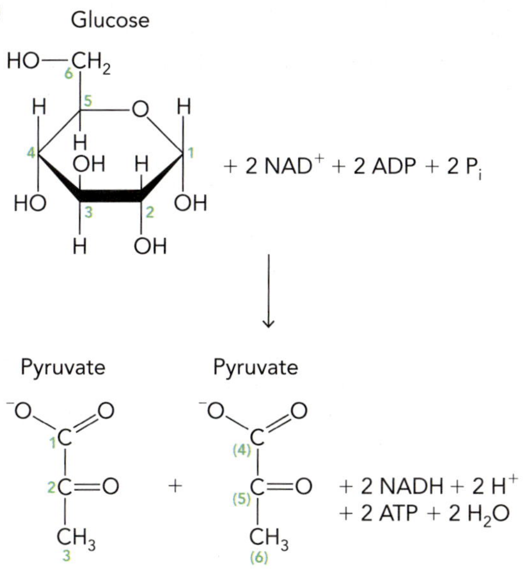

(b) 2 Glyceraldehyde-3-P + 2 P_i + 2 NAD^+ ⇌ 2 1,3-Bisphosphoglycerate + **2 NADH** + 2 H^+

2 2-Phosphoglycerate ⇌ 2 Phosphoenolpyruvate + **2 H_2O**

7. Net reaction: 4 ATP(equivalents) + 4 H_2O → 4 ADP(equivalents) + 4 P_i
Control of the two opposing pathways is critical so that ATP is not continually hydrolyzed. Coordinated control of glycolytic and gluconeogenic enzymes shifts the metabolic flux appropriately in response to glucose availability to prevent futile cycling.
8. Hexokinase is in all cells, whereas glucokinase is liver specific; glucokinase has a much lower affinity (higher K_m) for glucose than that of hexokinase; and hexokinase is feedback inhibited by glucose 6-P, whereas glucokinase is not inhibited.
9. When blood glucose levels are high, glycolysis produces high levels of fructose-1,6-bisphosphate, which activates pyruvate kinase—the last step of the glycolytic pathway—to stimulate metabolic flux in the direction of pyruvate formation.
10. PFK-1 is critical to maintaining flux through the glycolytic pathway. Thus, complete loss of this enzyme would be lethal at early stages of embryonic development, resulting in a miscarriage early in the pregnancy.
11. (a) The lack of phosphoglycerate mutase blocks production of pyruvate from glucose. During intense exercise, pyruvate is converted to lactate to replenish the necessary NAD^+ for energy production by glycolysis. (b) No. Because the individual is not producing enough pyruvate, there will not be very much, if any, lactate.
12. Inherited fructose intolerance is due to defects in aldolase B (fructose-1-P aldolase), inhibiting conversion of fructose-1-P to dihydroxyacetone-P and glyceraldehyde. This traps available phosphate in fructose-1-P, so no ADP + P_i → ATP reactions occur. The resulting low energy charge in liver cells shuts down membrane pumps that prevent the buildup of toxins.
High-fructose foods and beverages to be avoided: processed foods, soda or fruit drinks containing "fructose sweetener" (from corn syrup), and food/beverages containing sucrose or honey (both contain fructose).
13. Lactase converts lactose to glucose and galactose. One lactose is metabolized to four pyruvate molecules, two each from glucose and galactose. Anaerobic metabolism would generate four ATP from lactose; aerobic metabolism would generate up to 64 ATP.
14. (a) Aerobic conditions during the first phase of beer production facilitate rapid growth from the yeast starter culture to grow sufficient amounts of yeast (aerobic growth is much faster than growth under anaerobic conditions because of differences in efficiency of ATP energy generation). Then the beer mixture is shifted to anaerobic conditions to initiate the fermentation process, using the sugars from the plant materials. (b) To bottle the beer, the fermentation vessel must be opened, and the CO_2 is lost. To make more CO_2, because the original yeast culture is no longer growing rapidly, a fresh yeast culture needs to be added, along with glucose, to start the fermentation process again in the capped bottles.

Chapter 10

REVIEW QUESTIONS

1. Commercially, citric acid is purified from either lemon extracts or fermentation cultures of *Aspergillus niger*.
2. (1) It decarboxylates citrate to remove two carbons from acetyl-CoA; (2) it transfers 8 e^- to 3 NAD^+ and 1 FAD; (3) it generates 1 GTP (ATP) by substrate-level phosphorylation; and (4) it regenerates oxaloacetate.
3. The name citrate cycle reflects that all three carboxylate groups on citrate are deprotonated at physiologic pH. Citrate is the conjugate base of citric acid. The name Krebs cycle pays tribute to Hans Krebs, but it is not very descriptive, especially as Krebs also discovered the urea cycle.
4. The energy available from redox reactions is due to differences in the electron affinity of two reactants: compounds that accept electrons (oxidants) are reduced in the reaction, and compounds that donate electrons (reductants) are oxidized. The standard reduction potential, $E°'$, of a conjugate redox pair is measured experimentally against hydrogen, which has an $E°'$ value of 0 V. Oxidants with a higher affinity for electrons than H^+ have positive $E°'$ values, and oxidants with lower affinity have negative values. The amount of energy available from a coupled redox reaction, $\Delta E°'$, is determined by subtracting the $E°'$ of the reductant (e^- donor) from the $E°'$ of the oxidant (e^- acceptor). Using the $\Delta E°'$ value for a coupled redox reaction, we can calculate the change in free energy using the equation $\Delta G°' = -nF\Delta E°'$.
5. The net PDH reaction is
Pyruvate + CoA + NAD^+ → Acetyl-CoA + CO_2 + NADH
Five coenzymes: (1) NAD^+, derived from the vitamin niacin (B_3), found in poultry, fish, and vegetables; (2) FAD, derived from riboflavin (B_2), found in dairy products, almonds, and asparagus;

(3) CoA, derived from the water-soluble vitamin pantothenic acid (B_5), found in brown rice, lentils, chicken, yogurt, and avocados; (4) TPP, derived from thiamine (B_1), found in cooked brown rice and thiamine-fortified processed foods; and (5) α-lipoic acid (lipoamide), which is synthesized in plants and animals and therefore is not an essential nutrient (but is found in broccoli, liver, spinach, and tomatoes).

6. (1) E1 subunit (pyruvate dehydrogenase) decarboxylates pyruvate resulting in the formation of hydroxyethyl-TPP and the subsequent release of CO_2. (2) E1 reacts with the disulfide of the lipoamide group on the N-terminal domain of the E2 subunit (dihydrolipoyl acetyltransferase) forming thioester-linked acetyl-dihydrolipoamide. (3) E2 lipoamide domain carries the acetyl group from the E1 catalytic site to the E2 catalytic site, where it reacts with CoA to yield acetyl-CoA. (4) E2 lipoamide domain swings to the E3 subunit (dihydrolipoyl dehydrogenase), is reoxidized and transfers 2 e^- and 2 H^+ to the E3-linked FAD moiety. (5) E3-$FADH_2$ coenzyme intermediate is reoxidized and transfers 2 e^- to NAD^+ to form NADH + H^+.

7. Pyruvate dehydrogenase (PDH) converts pyruvate to acetyl-CoA, used in the citrate cycle or in fatty acid synthesis. PDH is allosterically controlled in response to energy charge; it is inhibited by high ratios of NADH/NAD^+ and acetyl-CoA/CoA, as well as by serine phosphorylation. Pyruvate dehydrogenase kinase (PDK phosphorylates and inactivates PDH in response to elevated NADH, acetyl-CoA, and ATP, whereas pyruvate dehydrogenase phosphatase-1 dephosphorylates PDH, activating it in response to elevated Ca^{2+} levels. PDK activity is, in turn, inhibited by NAD^+, CoA, ADP, and Ca^{2+}, which maintains PDH in an active dephosphorylated state.

8. The net reaction of the citrate cycle is

$$\text{Acetyl-CoA} + 3\,NAD^+ + FAD + GDP + P_i + 2\,H_2O \rightarrow \text{CoA} + 2\,CO_2 + 3\,NADH + 3\,H^+ + FADH_2 + GTP$$

Eight enzymatic reactions: (1) citrate synthase catalyzes a condensation forming citrate from oxaloacetate and acetyl-CoA; (2) aconitase isomerizes citrate to isocitrate; (3) isocitrate dehydrogenase catalyzes an oxidative decarboxylation converting isocitrate to α-ketoglutarate, generating CO_2 and NADH; (4) α-ketoglutarate dehydrogenase catalyzes a second oxidative decarboxylation, requiring CoA and converting α-ketoglutarate to succinyl-CoA, yielding CO_2 and NADH; (5) succinyl-CoA synthetase catalyzes substrate-level phosphorylation generating GTP and succinate and CoA; (6) succinate dehydrogenase oxidizes succinate to form fumarate, reducing FAD to $FADH_2$; (7) fumarase hydrates fumarate to generate malate; and (8) malate dehydrogenase oxidizes malate to form oxaloacetate and NADH.

9. Three enzymes, citrate synthase, isocitrate dehydrogenase, and α-ketoglutarate dehydrogenase, are inhibited by high NADH/NAD^+ ratios, a signal that the electron transport system is inhibited because of elevated ATP/(ADP + AMP) ratios. Citrate synthase, isocitrate dehydrogenase, and α-ketoglutarate dehydrogenase are inhibited by NADH and ATP, whereas citrate synthase and α-ketoglutarate dehydrogenase are both inhibited by succinyl-CoA. Citrate synthase is also potently inhibited by citrate, which is exported from the mitochondria when it builds up.

10. It provides biosynthetic precursors to both catabolic and anabolic pathways. Pyruvate carboxylation is anaplerotic because it restores a citrate cycle intermediate: oxaloacetate. Phosphoenolpyruvate carboxylase and malic enzyme reactions are also anaplerotic sources of oxaloacetate.

CHALLENGE PROBLEMS

1. (a) Malate dehydrogenase reaction:

$$\text{Malate} + NAD^+ \rightarrow \text{Oxaloacetate} + NADH + H^+$$

$$\begin{aligned}\Delta E^{\circ\prime} &= (\Delta E^{\circ\prime}{}_{e^-\text{ acceptor}}) - (\Delta E^{\circ\prime}{}_{e^-\text{ donor}})\\ &= (-0.32\text{ V}) - (-0.17\text{ V}) = -0.15\text{ V}\\ \Delta G^{\circ\prime} &= -nF\Delta E^{\circ\prime}\\ &= -(2)(96.48\text{ kJ/mol V})(-0.15\text{ V})\\ &= +28.9\text{ kJ/mol}\end{aligned}$$

Yes, this value is within ~1 kJ/mol of the value given in Table 10.2. $\Delta G < 0$ kJ/mol for this reaction because oxaloacetate is converted to citrate in the citrate synthase reaction; i.e., the mass action ratio (Q) is less than 1. (b) Succinate dehydrogenase reaction:

$$\text{Succinate} + FAD \rightarrow \text{Fumarate} + FADH_2$$

$$\begin{aligned}\Delta E^{\circ\prime} &= (+0.03\text{ V}) - (0.0\text{ V}) = +0.03\text{ V}\\ \Delta G^{\circ\prime} &= -nF\Delta E^{\circ\prime}\\ &= -(2)(96.48\text{ kJ/mol V})(+0.03\text{ V})\\ &= -5.8\text{ kJ/mol}\end{aligned}$$

No, this value is not within ~1 kJ/mol of the value given in Table 10.2. This is because the $E^{\circ\prime}$ value for FAD bound to succinate dehydrogenase is more positive than the value of ~0 V listed in Table 10.1. In order for the $\Delta G^{\circ\prime}$ of the isocitrate dehydrogenase reaction to be +0.4 kJ/mol, the $E^{\circ\prime}$ for bound FAD would have to be close to 0.03 V. ΔG is ~0 kJ/mol for this reaction.

2. Pyruvate is decarboxylated by the pyruvate dehydrogenase reaction to yield $^{14}CO_2$ and acetyl-CoA. The reverse reaction is highly unfavorable and does not occur in the cell.

3. TPP: decarboxylation of pyruvate using the thiazolium ring to create a Schiff base. Lipoic acid/lipoamide: oxidation and reduction. Reduced as it accepts the acetyl group from TPP and oxidized as it passes electrons to FAD one electron at a time. It holds the acetyl group in place until it is replaced by CoA. FAD/$FADH_2$: electron carrier. It accepts electrons one at a time from reduced lipoamide and passes electrons to NAD^+.

4. (a) Individuals with beriberi cannot readily metabolize pyruvate to acetyl-CoA because they lack the pyruvate dehydrogenase coenzyme TPP. This leads to a buildup of pyruvate in the blood when glycolysis is stimulated. (b) Thiaminase degrades thiamine pyrophosphate, but cooking denatures the thiaminase and inactivates it. (c) Without the pyruvate dehydrogenase phosphatase, the dehydrogenase would remain in its phosphorylated, or inhibited, state. This, the symptoms are not caused by thiamine deficiency.

5. (a) The first CO_2 contains C-6. (b) Isocitrate dehydrogenase is the enzyme. (c) The carbons present in oxaloacetate after the first turn are C-1, C-2, C-3, and C-4.

6. (a) The radioactive carbon became incorporated into all of the cycle intermediates at atom positions that could only be explained by products also serving as substrates. Metabolism of oxaloacetate produces a different radioactively labeled product depending on the number of cycles; this would not be true for a linear pathway. (b) Inorganic phosphate is required by one of the reaction steps (succinyl-CoA synthetase) that couples GTP synthesis to thioester bond cleavage. Without P_i, this reaction is inhibited, so radioactive carbon would only be found in intermediates that precede this step.

7. The rate of oxygen consumption is limited by the rate of acetyl-CoA metabolism in the citrate cycle. Addition of fumarate increases the concentration of citrate cycle intermediates, this increasing the capacity of the cycle to oxidize acetyl-CoA, hence a large increase in oxygen consumption.

8. Addition of citrate increased the capacity of the citrate cycle to metabolize acetyl-CoA by increasing the concentration of all cycle intermediates—particularly oxaloacetate. The addition of acetyl-CoA to this system had little effect because oxaloacetate levels were limiting. The cycle was working at full capacity, so having more substrate available had little effect on the rate of glucose metabolism.

9. Pyruvate dehydrogenase reaction: 1 NADH
Citrate cycle: 3 NADH + 1 $FADH_2$ + 1 GTP (ATP)
4 NADH total = ~10 ATP
1 $FADH_2$ = ~1.5 ATP
1 GTP = 1 ATP
Total = ~12.5 ATP (~13 ATP)
10. If pyruvate carboxylase is not activated by acetyl-CoA, then the oxaloacetate becomes rate limiting, and the excess acetyl-CoA is not fully used.
11. (a) When blood glucose levels are low, oxaloacetate is converted to phosphoenolpyruvate and used in gluconeogenesis to produce more glucose. (b) When mitochondrial acetyl-CoA levels are high, oxaloacetate is efficiently converted to citrate in the citrate cycle to raise energy charge.
12. The coordinate regulation would be disrupted, because normally citrate is exported from the mitochondrion when citrate levels are high. Citrate in the cytosol inhibits phosphofructokinase, shutting down glycolysis when citrate cycle activity is high and there is less need for energy production.
13. The defective enzyme is fumarase: This causes a buildup of fumarate and succinate (intermediates before the block) and a lack of malate (after the block). High lactate and low citrate are a consequence of low citrate cycle activity.

Chapter 11

REVIEW QUESTIONS

1. Peter Mitchell proposed the chemiosmotic theory: Energy from redox reactions or light absorption is used to form an electrochemical proton gradient across a membrane. This gradient, which has both a chemical (ΔpH) and a membrane potential ($\Delta\psi$) component, generates the proton-motive force whereby protons flow down the electrochemical proton gradient. The energy from this flow is harnessed by the ATP synthase complex as protein conformational changes that catalyze the reaction ADP + P_i $\rightarrow$ ATP.
2. In 1973, Racker and Stoeckenius published the results of their experiments using reconstituted membrane vesicles containing a bacterial light-activated bacteriorhodopsin proton pump and an animal mitochondrial ATP synthase complex. In the presence of light, bacteriorhodopsin pumped protons into the vesicle, generating an electrochemical proton gradient across the vesicle membrane. This gradient made it favorable for protons to cross the membrane through ATP synthase, resulting in ATP production, as predicted by Mitchell.
3. Two mechanisms have been proposed. The redox loop mechanism involves a redox reaction in which the resulting H^+ and e^- are separated within the electron transport system protein and deposited on opposite sides of the inner mitochondrial membrane. The e^- reduces an oxidant in the mitochondrial matrix, and the H^+ is released into the intermembrane space. This is thought to be the mechanism for the Q cycle. The proton pump mechanism is based on redox-driven conformational changes in the electron transport system protein such that matrix-derived protons are translocated across the inner mitochondrial membrane by altering pK_a values of functional groups located on the inner and outer faces of the membrane. This is thought to be the mechanism used by complexes I and IV.
4. Three of the electron carriers are membrane-spanning protein complexes that translocate protons across the inner mitochondrial membrane (complexes I, III, IV); one is a membrane protein that transfers electrons directly from the citrate cycle (complex II); and two are mobile carriers (cytochrome c and coenzyme Q). Complex I (NADH–ubiquinone oxidoreductase) oxidizes NADH and translocates 4 H^+ from the mitochondrial matrix to the intermembrane space. Complex III (ubiquinone–cytochrome c oxidoreductase) transfers electrons from ubiquinol (QH_2) to cytochrome c and translocates 4 H^+ across the inner mitochondrial membrane. Complex IV (cytochrome c oxidase) transfers electrons from cytochrome c to oxygen to form water and translocates 2 H^+ across the inner mitochondrial membrane. Complex II (succinate dehydrogenase) is a citrate cycle enzyme that transfers electrons from $FADH_2$ to coenzyme Q (Q; ubiquinone) to form QH_2 without translocating protons. Cytochrome c is localized to the intermembrane space and uses a heme prosthetic group to transfer 1 e^- from complex III to complex IV. Coenzyme Q is a hydrophobic membrane-localized electron carrier that transfers electrons from complexes I and II to complex III.
5. The three components are (using the naming of the yeast system): (1) a protein rotor consisting of γ, δ, and ε subunits and a c ring that rotates as protons enter and exit; (2) a protein catalytic headpiece made from the $\alpha_3\beta_3$ hexamer, where each β subunit is a catalytic site for ATP synthesis; and (3) a protein stator made from the a, b, d, h, and OSCP subunits that stabilizes the entire complex and provides the proton half-channels needed to drive rotation.
6. The binding change mechanism of ATP synthesis, proposed by Boyer, states that the rotation of the γ subunit changes the conformation of the β subunits and sequentially alters nucleotide binding affinities with the catalytic active sites, thus affecting ATP production. Protein contacts between the γ subunit and each β subunit are unique and mediate three distinct protein conformations, referred to as loose (L), tight (T), and open (O), where the enzyme active site is either ADP + P_i bound (L), ATP bound (T state), or unbound (O). Proton flow through the ATP synthase complex c ring induces rotation of the γ subunit, such that with each 120° rotation, ~3 H^+ cross the inner mitochondrial membrane, and β subunits undergo a conformational change sequence of L $\rightarrow$ T $\rightarrow$ O $\rightarrow$ L. A 360° rotation of the γ subunit produces ~3 ATP and requires the translocation of ~9 H^+ from the intermembrane space to the mitochondrial matrix.
7. One is the ATP/ADP translocase, also called the adenine nucleotide translocase, which exports 1 ATP for every ADP imported. The second translocase protein is the phosphate translocase, which imports 1 P_i and 1 H^+ into the matrix by an electrically neutral import mechanism. This translocase can have either symporter or antiporter functions; for example, when $H_2PO_4^-$ accompanies an H^+ across the inner mitochondrial membrane down the proton gradient, it is acting as a symporter because both molecules are translocated in the same direction. This is an electrically neutral translocation because the two charges ($H_2PO_4^-$ and H^+) cancel each other out.
8. Both shuttles serve to donate electrons from mitochondrial NADH into the cytosol to regenerate NAD^+ for the glycolytic pathway. The malate–aspartate shuttle in liver cells uses NADH to reduce cytosolic oxaloacetate and generate NAD^+ and malate, which is shuttled into the matrix and oxidized to produce NADH and oxaloacetate. The glycerol-3-P shuttle in

muscle and brain cells catalyzes a redox reaction that oxidizes cytosolic NADH and reduces FAD to yield cytosolic NAD^+ and mitochondrial $FADH_2$. Oxidation of the $FADH_2$ within the inner mitochondrial membrane transfers 2 e^- directly to coenzyme Q. The malate–aspartate shuttle in liver cells converts 1 $NADH_{cytosol}$ to 1 $NADH_{matrix}$, the maximum yield of redox energy. In contrast, the glycerol-3-P shuttle in muscle and brain cells provides less ATP (oxidation of $FADH_2$ in the electron transport system results in the translocation of fewer H^+ across the inner mitochondrial membrane).

9. The ATP currency exchange ratio is based on the number of H^+ translocated across the inner mitochondrial membrane and the number of H^+ flowing through the ATP synthase complex to generate each ATP. In the case of complex I oxidation of mitochondrial NADH, and subsequent electron transfer through complexes III and IV, a total of 10 H^+ are translocated (4 H^+ + 4 H^+ + 2 H^+ = 10 H^+). However, because $FADH_2$ oxidation donates a pair of electrons directly to coenzyme Q and bypasses complex I, a total of only 6 H^+ are translocated by complexes III and IV combined (4 H^+ + 2 H^+ = 6 H^+). Because 1 ATP is synthesized for every 4 H^+ translocated back into the mitochondrial matrix from the intermembrane space (3 H^+ for each 120° rotation and 1 H^+ for the P_i translocase = 4 H^+/ATP), then ~2.5 ATP/NADH are generated (10 H^+/4 H^+ = 2.5), and ~1.5 ATP/$FADH_2$ are generated (6 H^+/4 H^+ = 1.5).

10. (1) Those that inhibit electron flow through the electron transport system, such as rotenone (inhibits e^- flow through complex I), antimycin (inhibits e^- flow through complex III), and cyanide (inhibits e^- flow through complex IV). (2) Proton gradient uncouplers, which allow protons to cross the mitochondrial membrane without going through the ATP synthase complex; for example, 2,4-dinitrophenol. (3) Those that interfere with ATP synthesis, either with the activity of the ATP synthase complex (oligomycin) or with the translocation of ATP and ADP by the ATP/ADP translocase (bongkrekic acid).

CHALLENGE PROBLEMS

1. Cytochrome c, required for electron transport and localized to the inner mitochondrial membrane, should not be in the cytosol. High levels of cytochrome c in the cytosol are a signal that mitochondria are not functioning properly and induce apoptosis.

2. The phosphate translocase requires that 1 H^+ accompanies each P_i brought into the matrix.

3.

$$\Delta G = RT\ln(C_2/C_1) + ZF\Delta V$$

$$\Delta G - RT\ln(C_2/C_1) = ZF\Delta V$$

$$\frac{\Delta G - RT\ln(C_2/C_1)}{ZF} = \Delta V$$

$$\frac{(21.8\ \text{kJ/mol}) - [(8.314 \times 10^{-3}\ \text{kJ/mol K})(298\ \text{K})\ln(25)]}{(+1)(96.48\ \text{kJ/mol V})} = \Delta V$$

$$\Delta V = 0.14\ \text{V} = 140\ \text{mV}$$

4. When the H^+ binds to a single c subunit, it causes a rotation of the c ring of 30° (1/12 of a circle). The three-sided γ subunit rotates along with the c ring and interacts directly with the three β subunits, changing from L, T, or O conformation for every 120° rotation of the c ring. Because an ATP is synthesized for every turn of 120°, 4 H^+ must cross through the ATP synthase complex to synthesize 1 ATP.

5. (a) There is no proton-motive force in this system, and the ATP synthase complex functions as an ATPase when unconstrained, so the direction of the headpiece rotation would be opposite (clockwise) that seen in intact cells when viewed from the same orientation. (b) ATP levels would decrease as a result of ATP hydrolysis mediated by the ATP synthase complex.

6. (a) Cyanide blocks electron transfer in the electron transport system (ETS), resulting in the loss of the proton-motive force; therefore, ATP synthesis rates decrease. Adding 2,4-dinitrophenol to this cyanide-inhibited mitochondrial suspension has no effect on ATP synthesis rates because the ETS is blocked and there is no proton-motive force. (b) Oligomycin blocks proton flow through the ATP synthase complex, resulting in a buildup of the proton gradient, eventually inhibiting proton translocation and decreasing O_2 consumption. Adding 2,4-dinitrophenol to this oligomycin-inhibited system leads to increased rates of O_2 consumption because the ETS is able to resume as protons travel down the gradient via 2,4-dinitrophenol.

7.

Inhibitor(s) added	Electron transport system activity?	ATP synthesis?	Explain
Myxothiazol	None	Decrease	ETS is blocked; H^+ gradient is depleted
FCCP	Increase	Decrease	FCCP is an uncoupler, and H^+ flow through the ATP synthase complex decreases
Venturicidin	Decrease	None	ATP synthase complex is inhibited. ETS activity is low due to excessive H^+ gradient
Venturicidin + FCCP	Increase	None	FCCP is an uncoupler independent of ATP synthase complex activity

8. Thermogenin uncouples the electron transport system from oxidative phosphorylation. Heat created by an uncoupled electron transport system in these special cells keeps the animal alive even in the middle of winter. Hibernating animals depend on stored fat for energy to keep cells alive and on thermogenin-mediated futile cycling to create sufficient thermoregulation.

9. Cytosolic NADH in muscle donates 2 e^- to $FADH_2$ by the glycerol phosphate shuttle, and mitochondrial $FADH_2$ oxidation results in the production of 1.5 ATP/NADH, or 3 ATP/2 NADH. In contrast, liver cells use the malate–aspartate shuttle, where redox reactions transfer 2 e^- from cytosolic NADH to mitochondrial NAD^+, producing mitochondrial NADH, which is used to generate 2.5 ATP/NADH, or 5 ATP/2 NADH.

10. The ^{3}H is transported inside the mitochondria by shuttle systems (either glycerol-3-P shuttle or malate shuttle), whereas NAD cannot cross the inner mitochondrial membrane so the ^{14}C label does not get into the matrix.

11. The mitochondrial matrix produces 25 ATP from oxidation of 2 molecules of pyruvate, which does not include any energy conversion reactions from glycolysis.

8	NADH →	20	ATP
2	$FADH_2$ →	3	ATP
		2	GTP (ATP)
		25	*Total ATP*

12. (a-4) The canister shell represents the proton-impermeable inner mitochondrial membrane. (b-3) The spray nozzle represents the ATP synthase complex, which relies on the pressure in the canister. (c-2) The pump handle represents reductants that provide redox energy to establish the gradient, analogous to internal pressure in the canister. (d-1) The release valve represents an uncoupler that permits protons to cross the membrane in response to the electrochemical gradient without passing through the ATP synthase complex.

Chapter 12

REVIEW QUESTIONS

1. Photosynthesis generates ATP and NADPH using the energy obtained from light absorption. Photosynthetic organisms are autotrophs, able to use stored chemical energy when light energy is not available by converting the chemical energy in ATP and NADPH into carbohydrates (sucrose and starch). This process is called carbon fixation and takes place inside chloroplasts of plants. In the absence of light, photosynthetic autotrophs metabolize stored carbohydrates using aerobic respiration in plant mitochondria. Heterotrophs cannot convert light energy into chemical energy; they consume stored carbohydrates from autotrophs as nutrients that they metabolize by aerobic respiration to generate ATP. Many heterotrophs also consume autotrophs in order to obtain essential amino acids, fatty acids, and coenzymes (vitamins). Taken together, autotrophs are completely self-sufficient and use both photosynthesis and aerobic respiration, whereas heterotrophs are completely dependent on consuming autotrophs and use aerobic respiration as their only means of energy conversion.
2. Chloroplasts are photosynthetic organelles of eukaryotes. Like mitochondria, they likely evolved from an endosymbiotic relationship between an ancestral eukaryotic cell and a bacterium. The two key metabolic pathways contained in chloroplasts are the photosynthetic ETS, including the chloroplast ATP synthase, and the Calvin cycle, which converts CO_2 into glyceraldehyde-3-phosphate using ATP and NADPH. Chloroplasts contain three membranes: (1) outer chloroplast membrane, permeable to most metabolites; (2) inner chloroplast membrane, containing metabolite transporter proteins; (3) thylakoid membrane, a proton-impermeable membrane and the site of photosynthetic electron transport and ATP synthesis. The thylakoid membrane is analogous to the inner mitochondrial membrane with regard to redox energy conversion and ATP synthesis. The stacks of thylakoid membranes are called grana. The chloroplast compartment outside the grana is the stroma and contains all of the enzymes required for the Calvin cycle. Similar to mitochondria, chloroplasts form a barrier between two compartments within the chloroplast: the thylakoid lumen (a continuous aqueous chamber inside the thylakoid membrane) and the stroma. Similar to the inner mitochondrial membrane, chloroplasts contain their own DNA and carry out protein synthesis within the organelle.
3. Light absorption by a pigment such as chlorophyll excites an electron, lifting it to a higher orbital. If the pigment is in the PSII or PSI reaction centers, then an electron is transferred to a nearby electron acceptor (known as photooxidation because the chlorophyll is oxidized after excitation by light and the acceptor molecule is reduced). Oxidation of H_2O by the oxygen-evolving complex replaces the lost electron in chlorophyll, returning it to the ground state. Absorption of light by chlorophylls in the light-harvesting complexes results in energy transfer to a nearby chlorophyll molecule by a process called resonance energy transfer, but not oxidation. There are ~300 energy transfer reactions by light-harvesting complexes for every photooxidation reaction at PSII and PSI reaction centers.
4. $$2\,H_2O + 8\text{ photons} + 2\,NADP^+ + {\sim}3\,ADP + {\sim}3\,P_i \rightarrow O_2 + 2\,NADPH + {\sim}3\,ATP$$

   The photosynthetic ETS consists of three transmembrane proteins (PSII, cytochrome b_6f, and PSI) and three soluble proteins (plastocyanin, ferredoxin-$NADP^+$ reductase, and ferredoxin). PSII contains 20 protein subunits, ~40 pigment molecules, two plastoquinones (PQ_A and PQ_B), and the oxygen-evolving complex with a single Mn_4 cluster. PSII undergoes photooxidation in which it absorbs 4 photons at 680-nm wavelength and transfers electrons to pheophytin. Cytochrome b_6f complex is a proton pump and transfers 8 H^+ from the stromal side of the thylakoid membrane to the thylakoid lumen. Plastocyanin carries the electrons from cytochrome b_6f to PSI. PSI contains 18 protein subunits, ~190 pigment molecules, two phylloquinones, and three 4 Fe–4 S clusters. PSI undergoes photooxidation in which it absorbs 4 photons at 700 nm and transfers 4 electrons to nearby chlorophylls, where they reduce the 2 Fe–2 S cluster in ferredoxin, which transfers the electrons one at a time to ferredoxin-$NADP^+$ reductase, leading to the reduction of 2 $NADP^+$ to generate 2 NADPH.
5. In cyclic photophosphorylation, the electron normally transferred from PSI to ferredoxin is transferred instead to plastoquinone (PQ). This provides for the maintenance of an electrochemical proton gradient by cytochrome b_6f and ATP synthesis, but generates less NADPH so the ATP-to-NADPH ratio increases. A balance of electron transfer through cyclic photophosphorylation and standard photophosphorylation enables the cell to meet its needs for ATP and NADPH for use in biosynthetic pathways.
6. The PSI and ATP synthase are concentrated in unstacked lamellar regions, whereas PSII complexes are mostly localized to the stacked membranes (grana). When PSI is not being oxidized by light absorption, plastocyanin and PQ_BH_2 levels build up. In order to stimulate electron flow from PSI, phosphorylation of light-harvesting complex II (LHC II) causes redistribution into the lamellae, where they can harvest photons for PSI. Increased levels of photooxidation by PSI complexes not only stimulates electron flow through PSII by oxidizing plastocyanin but also leads to higher levels of cyclic photophosphorylation and ATP synthesis when $NADP^+$ levels are limiting.
7. Flux through the Calvin cycle pathway is severely restricted in the absence of light, because photosynthetic electron transport is required to supply the Calvin cycle with sufficient ATP and NADPH to maintain flux through this energy-dependent carbon fixation pathway. Moreover, light activates many of the enzymes in the Calvin cycle.
8. (1) A CO_2 fixation reaction by ribulose-1,5-bisphosphate carboxylase/oxygenase (rubisco), in which ribulose-1,5,-bisphosphate is combined with CO_2. (2) The 3-phosphoglycerate is reduced to glyceraldehyde-3-phosphate. (3) The ribulose-1,5-bisphosphate is regenerated though a series of carbon shuffle reactions that recombine C_3 molecules to produce C_5 molecules with no net loss of carbon.
9. Enzyme activity is controlled by (1) inhibitor proteins present at higher levels in the dark; (2) elevated pH and increased Mg^{2+} levels in the stroma during photosynthetic electron transport, which stimulate enzyme activity; and (3) thioredoxin-mediated activation through reduction of disulfide bridges.
10. It produces 2-phosphoglycolate, which must be converted to 3-phosphoglycerate through the glycolate pathway, requiring numerous enzyme reactions in chloroplasts, peroxisomes, and mitochondria, and the investment of 1 ATP.
11. *C_4 plant* refers to plants that store CO_2 in the form of malate, a C_4 metabolite, under conditions when rubisco oxygenase rates are elevated because of high temperatures and a high O_2-to-CO_2 ratio. It is an adaption for growth in high-temperature climates

by trapping CO_2 into malate, limiting the exposure of rubisco to elevated O_2. When conditions are favorable for carbon fixation, malate is decarboxylated and released CO_2 is used by rubisco to generate 3-phosphoglycerate. Two variations of C_4 metabolism in plants have been described: (1) the Hatch–Slack pathway in tropical plants with two separate cell types, one for CO_2 uptake and the other for rubisco-mediated carboxylation; and (2) the CAM pathway found in desert succulents, which captures CO_2 into malate at night when transpiration rates are low, then releases the CO_2 during the day when the Calvin cycle is favored. In CAM plants the stomata are open at night to capture the CO_2 but are closed during the day to prevent O_2 from entering the cell and stimulating oxygenase activity.

12. The glyoxylate cycle converts acetyl-CoA into succinate, a carbon source for glucose biosynthesis, not found in animals. It is a mechanism for fats stored in plant seeds to be converted to sucrose for export to developing plant tissues prior to the onset of photosynthesis. Two unique glyoxylate enzymes, isocitrate lyase and malate synthase, are localized to glyoxysomes in plant cells (and bacteria). Succinate is the central metabolic intermediate in this pathway, transporting the four carbons obtained from two acetyl-CoAs to the mitochondria. There succinate is converted to malate, which is then exported to the cytosol and oxidized to oxaloacetate, a substrate for gluconeogenesis.

CHALLENGE PROBLEMS

1. (a) Both PSI and PSII and the mitochondrial ETS convert redox energy into proton-motive force. (b) ATP and NADPH. (c) CO_2. (d) Plant cells have mitochondria in order to generate high levels of ATP at night to sustain metabolic activity. Per glucose, mitochondrial ETS produces many more ATP than glycolysis.
2.

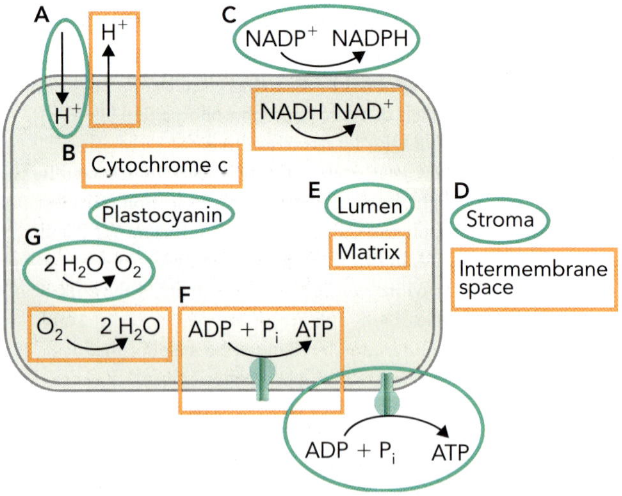

3. The oxidation of 2 H_2O to form $O_2 + 4\,H^+ + e^-$.
4. Plastocyanin is oxidized by PSI in response to light. If electron flow through cytochrome b_6f is inhibited because no oxidized plastocyanin is available, then proton translocation is also inhibited.
5. The oxidation of 2 H_2O generates 4 e^-; however, 4 photons are required at each of the PSII and PSI reaction centers to move the e^- through the photosynthetic electron transport system.
6. The e^- carrier molecule that *accepts* the e^- from chlorophyll must have a higher reduction potential (the standard reduction potential is more positive than chlorophyll).
7. The oxidation of H_2O generates the e^- that reduces chlorophyll.
8. Paraquat is a strong oxidant (electron acceptor) that prevents electrons from reaching the final electron acceptor in the PSI ETS (ferredoxin-$NADP^+$ reductase).
9. C_5: ribulose-1,5-bisphosphate; C_3: 3-phosphoglycerate. The abundance of rubisco reflects the importance of carbon fixation and that it catalyses this reaction slowly. It is found in plants as well as algae and cyanobacteria.
10. One mechanism to activate some Calvin cycle enzymes is increased stromal pH and $[Mg^{2+}]$, caused by increased flux through the photosynthetic ETS. A second mechanism is thioredoxin-mediated reduction of disulfide bonds.
11. Photorespiration is a "wasteful" reaction because 2-phosphoglycerate must be salvaged by the glycolate pathway at the expense of ATP hydrolysis. Therefore, the C_4 plant has an advantage in summer because it minimizes loss of carbon to photorespiration when CO_2-to-O_2 ratios are low (elevated soluble O_2). However, the C_4 pathway requires an investment of ATP to regenerate metabolites in a reaction that is not required by C_3 plants. Therefore, at low temperatures, when the CO_2-to-O_2 ratio is still relatively high, this extra ATP investment by C_4 plants comes at an energy cost, thereby giving C_3 plants a growth advantage.
12. ATP and NADPH will be quickly depleted, and the Calvin cycle enzymes, most importantly rubisco, are inactivated in the dark.
13. Disulfide bridges are spontaneously oxidized in the absence of reduced thioredoxin.

Chapter 13

REVIEW QUESTIONS

1. (1) Simple sugars consisting of monosaccharides, disaccharides, and oligosaccharides primarily function as readily available chemical energy or as the sugar moiety in nucleotides, including ATP; (2) polysaccharides consisting of glucose homopolymers or disaccharide heteropolymers containing a hexosamine primarily function as structural elements in multicellular organisms (cellulose and chitin) or as forms of stored energy (starch and glycogen); and (3) glycoconjugates (proteins or lipids with covalently attached glycans) function as proteoglycan cell walls of bacteria and extracellular matrix materials in animals (aggrecan, syndecan).
2. Milk oligosaccharides contain between 3 and 22 sugar residues from a core lactose disaccharide; they are degraded by bacteria in the infant's intestinal tract. Their probable function is as probiotics supporting the growth of beneficial bacteria or as soluble glycan decoys that are structurally related to glycoconjugates of intestinal epithelial cell surfaces and prevent pathogenic bacteria from binding to these cells. The raffinose series of plant oligosaccharides—raffinose, stachyose, and verbascose—are derived from sucrose. Most mammals cannot digest raffinose oligosaccharides because they lack the α-galactosidase enzyme; however, these can be metabolized by intestinal bacteria, which ferment the glycans to produce methane, carbon dioxide, and hydrogen gases, all of which can lead to gastrointestinal discomfort.
3. Plant cell walls consist of cellulose, a glucose polysaccharide linked via β-1,4 glycosidic bonds. Extensive hydrogen bonding between individual cellulose strands results in a polysaccharide fibril that is itself hydrogen bonded to the hemicellulose and pectin. Hemicellulose is a branched polysaccharide consisting of up to six different sugar residues, whereas pectin is a homopolymer of galacturonic acid.
4. Starch and glycogen contain linear branched glucose polymers consisting of α-1,4 and α-1,6 glycosidic bonds. Plants make amylose, a linear homopolymer containing ~100 glucose units linked by α-1,4 glycosidic bonds, and amylopectin, an α-1,6–branched glucose polymer with the same structure as glycogen.

Amylose forms a left-handed helix of six glucose residues per turn, stabilized by intrastrand hydrogen bonds. The α-1,6 bonds in amylopectin and glycogen create branch points resulting in many free ends in the homopolymeric molecule. Amylopectin contains an α-1,6 glycosidic bond about once every 30 glucose units, whereas glycogen contains the same α-1,6 glycosidic bond about once every 10 residues. Amylopectin contains one free glucose at the reducing end, while glycogen contains a covalently linked protein called glycogenin.

5. Lectins are proteins that bind to glycan groups on glycoconjugates and mediate two types of glycoconjugate binding interactions in human immune cells: (1) intrinsic glycoconjugate binding between glycans and lectins on human cells, and (2) extrinsic glycoconjugate binding between glycans and lectins on human cells and pathogen cells. The intrinsic binding mediates cell recognition functions on immune cells and facilitates cell attachment and migration. Extrinsic binding between immune cells and pathogens, such as bacteria, viruses, or fungi, protects host cells from infection by inducing immune responses such as antibodies and hydrolytic enzymes that neutralize or kill the pathogen.
6. Glycan linkages occur through either the amide nitrogen atom of asparagine (the N-linked oligosaccharides) or the oxygen atom of serine or threonine residues (the O-linked oligosaccharides). One N-glycosidic bond in glycoproteins is between certain asparagine residues and GlcNAc. O-glycosidic bonds do not use a preferred recognition sequence and are usually between the amino acid and GalNAc. Both N-linked and O-linked glycoproteins contain a core glycan structure, which serves as a scaffold for the addition of other glycans.
7. The basis is determined by genetic variants of the glycosyltransferases (GTs), which attach either a GalNAc or Gal to the O antigen, present on glycoproteins and glycolipids on red blood cells. The α-1,3-*N*-acetylgalactosaminyltransferase (GTA) variant attaches GalNAc, an *N*-acetyllactosamine disaccharide, yielding A-type blood, whereas the α-1,3-galactosyltransferase (GTB) variant attaches Gal in the same position to specify B-type blood. If there is no GT enzyme, then the O-antigen glycan group is not modified resulting in O-type blood. An individual with the GTA variant from one parent and the GTB variant from the other parent has AB-type blood.
8. Penicillin blocks bacterial cell wall synthesis by inhibiting transpeptidase, which forms the oligopeptide linkages between hexosamine polysaccharide chains in the peptidoglycan. Penicillin forms a suicide inhibitor complex between a serine residue in transpeptidase and a carbonyl carbon in the β-lactam ring of penicillin. Penicillin-treated bacteria lack sufficient transpeptidase activity during cell division, and cells die. Bacteria resistant to penicillin produce β-lactamase, which hydrolyzes the β-lactam ring so that it is no longer a suicide inhibitor substrate for transpeptidase. Methicillin is a penicillin derivative that is not hydrolyzed by β-lactamase and therefore is effective even in bacteria that are resistant to penicillin; however, methicillin-resistant cells do arise that express a variant transpeptidase that does not bind methicillin.
9. (1) Identification of glycan group structures on purified glycoproteins by liquid chromatography and mass spectrometry, and (2) applications of high-throughput, array-based screening to identify biologically relevant glycan binding interactions.
10. (1) Glycans have multiple bonding arrangements between as many as 11 different sugars; (2) several sugar stereoisomers have identical masses; and (3) glycan structures can differ in subtle ways. Liquid chromatography in combination with specific glycosidase enzymes is one approach, and the other is mass spectrometry, which is able to detect mass differences accurately.
11. Lectin arrays contain hundreds of different lectins covalently attached to a solid support and arranged in a grid. These are routinely used to screen (1) fluorescently labeled glycoproteins isolated from cell extracts; (2) fluorescently labeled glycoconjugates on the surface of eukaryotic viruses; and (3) bacteria expressing a fluorescent protein (GFP). After incubation with experimental samples, the array is scanned by laser to generate a pattern of positive and negative signals that are interpreted on the basis of the known location of specific lectins. Antibody arrays can be used to identify specific glycan groups or the protein moiety of glycoproteins. Glycan binding antibodies are used to detect fluorescently labeled glycoproteins in an experimental sample in an array that functions in much the same way as a lectin array. Glycoprotein binding antibodies can be used to investigate glycan modifications on glycoproteins when combined with fluorescently labeled lectins of known specificity. Glycan arrays are used to analyze all cellular glycan groups in a sample and are constructed using HPLC fractions of cell extracts covalently linked to a grid through protein-specific linkages. Fluorescently labeled lectins or glycan-specific antibodies identify the glycan groups. Alternatively, arrays of chemically synthesized glycan groups of known structure can be constructed and used for investigating the glycan specificity of purified lectins and monoclonal antibodies on the basis of their affinities for different glycan groups on the array.

CHALLENGE PROBLEMS

1. Water disrupts the intrafiber hydrogen bonds that hold the cellulose fibers together, whereas oil has no effect on hydrogen bonds.
2. The trisaccharide has many more possible structures. For example, 3 amino acid residues in a tripeptide can have a maximum of 6 possible sequences ($3 \times 2 \times 1$), all in a linear chain connected by peptide bonds. In contrast, a trisaccharide can have a minimum of 384 different structures. Consider that the trisaccharide has 6 possible linear sequences, but with multiple arrangements of glycosidic bonds possible at the two linkage sites. Specifically, each of the two glycosidic bonds can be an α or β linkage at C-1, which is connected to the hydroxyl group at C-2, C-3, C-4, or C-6. In total, 6 linear sugar arrangements multiplied by 2 bond arrangements (α or β) of the first glycosidic bond, which can be made with connections to any one of 4 carbons of the second sugar, and multiplied by 2 bond arrangements (α or β) of the second glycosidic bond, which can be made with connections to any one of 4 carbons of the third sugar, gives us $6 \times 2 \times 4 \times 2 \times 4 = 384$.
3. (a) Pigs and other mammals lack the enzyme α-galactosidase and therefore cannot break the glycosidic bonds in raffinose oligosaccharides to metabolize this carbohydrate fully. By predigesting the raffinose oligosaccharides with α-galactosidase, the feed has more nutritional value (broken down to simple sugars) and the pigs gain more weight in less time. (b) The flatulence comes from intestinal bacteria that digest the raffinose oligosaccharides contained in broccoli, cabbage, and soybeans using their own α-galactosidase enzyme. Some of the products of anaerobic bacterial metabolism are hydrogen and methane gas. In contrast, potatoes, squash, and corn contain low amounts of raffinose but high amounts of starch, which is easily metabolized by animals using the enzyme α-amylase.

4. (a)

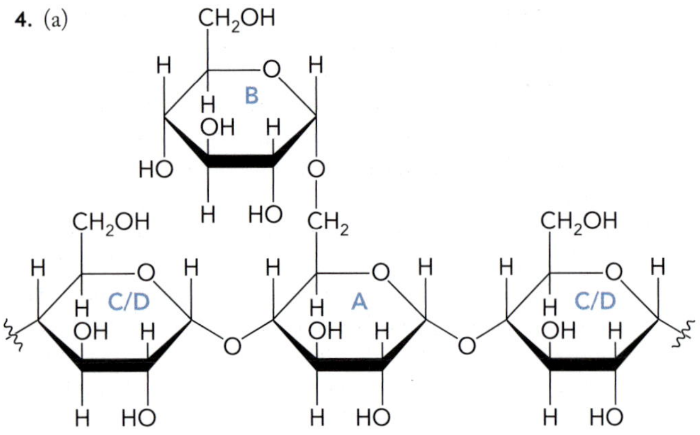

(b) $\frac{0.5 \text{ g}}{162 \text{ g/mol}} = 3.0 \times 10^{-3}$ mol of glucose in sample

$\frac{0.025 \text{ g}}{208 \text{ g/mol}} = 1.2 \times 10^{-4}$ mol of 2,3-dimethylglucose in sample

$\frac{1.2 \times 10^{-4} \text{ mol}}{3.0 \times 10^{-3} \text{ mol}} \times 100\% = 4.0\%$ of the sample is 2, 3-dimethylglucose

$\frac{100\%}{4\%} = 25{:}1$ α-1,4 glycosidic bonds compared to α-1,6 glycosidic bonds

(c) 3,000 glucose residues with a branch point every 25 residues = 120 nonreducing ends.

5. Catalytic activity is not the only consideration for determining the *in vivo* function of an enzyme. First, the glycan groups may be required for targeting the enzyme to the proper secretion or folding pathway inside the cell, which would not be required in an *in vitro* protein synthesis system. Second, the glycan groups may be required for functional associations with the extracellular matrix, which were not tested in the *in vitro* enzyme assay. Third, the glycan groups may determine serum survival time for the enzyme, and again, this was not tested.

6. (a) The A blood group contains a terminal GalNAc residue on the core glycan group called the O antigen. The B blood group contains a terminal Gal residue in place of GalNAc, whereas half of the AB blood group glycans contain a terminal GalNAc residue and the other half contain a terminal Gal residue. Lastly, the O blood group glycan lacks a terminal residue on the O-antigen glycan group. (b) Humans contain three functional variants of the glycosyltransferase enzyme that differ in key amino acid residues in the enzyme active site. The GTA variant of the enzyme binds UDP-GalNAc in the enzyme active site and catalyzes a reaction linking GalNAc to the O-antigen glycan group; the GTB variant of the enzyme binds UDP-Gal in the enzyme active site and catalyzes a reaction linking Gal to the O-antigen glycan group; and the third variant of the enzyme is defective, which gives rise to the O blood group consisting of the O-antigen glycan group. Individuals containing paternal and maternal alleles corresponding to the GTA and GTB variants have the AB blood group.

(c)

Blood group	Allele 1 (paternal or maternal)	Allele 2 (paternal or maternal)
A blood group	GTA	Defective enzyme *or* GTA
B blood group	GTB	Defective enzyme *or* GTB
AB blood group	GTA	GTB
O blood group	Defective enzyme	Defective enzyme

(d) Type O blood group is the most useful at blood banks because type O red blood cells lack both the GTA and GTB glycan groups, so they do not stimulate an immune reaction when used in otherwise compatible transfusions. In contrast, type AB blood group is the best to have when in need of a transfusion because serum in type AB individuals lacks anti-A and anti-B antibodies, so blood can be donated from any group and it won't stimulate an immune reaction.

7. (a) Heparin binds with high affinity to antithrombin, which activates the inhibitory activity of antithrombin. This leads to inhibition of the blood-clotting cascade through antithrombin inactivation of thrombin and Factor Xa. (b) Protamine sulfate is positively charged and binds to the negatively charged heparin, which prevents heparin from binding to and activating antithrombin. This leads to inactivation of antithrombin and normal regulation of the blood-clotting cascade.

8. (a) Fleming noticed that bacteria failed to grow in an area surrounding a large mold colony on the plate, and he hypothesized that the mold secreted an antibacterial compound. Fleming identified the compound and named it penicillin after the antibiotic-producing mold *Penicillium notatum*. (b) Penicillin targets the enzyme transpeptidase, which is required for bacterial cell wall synthesis. If bacteria cannot synthesize a functional cell wall at each cell division, the cell cannot divide and proliferate. (c) The most common mechanism of penicillin resistance in bacteria is expression and secretion of the enzyme β-lactamase, which cleaves the lactam ring of penicillin and destroys its transpeptidase-inhibiting activity. (d) MRSA is resistant to a wide range of antibiotics, including methicillin, which until recently was one of the most potent antibiotics available. Methicillin resistance in MRSA is due to expression of a functional transpeptidase enzyme that does not bind methicillin or penicillin. This methicillin-resistant transpeptidase was acquired by lateral gene transfer from another bacterial species.

9. After releasing the N-linked glycan group from the antibody-bound glycoprotein with PNGaseF, the purified glycan would be labeled with the fluorescent dye 2-AB for use in HPLC analysis. In this method, the glycan is treated sequentially with different glycosidases having known specificities, and the reaction products after each enzyme digestion are then separated by HPLC. The elution profiles of the glycan products are compared to the elution profiles of glucose polymers of known sizes. By combining the information gained from glycosidase specificities with the size of digestion products as determined by elution profiles, a probable glycan structure is proposed.
A second method to determine the glycan structure is to use mass spectrometry to determine the precise mass of the glycan before and after various treatments, which could be enzyme digestion or, more likely, fragmentation by electrospray ionization (ESI) in combination with collision-induced dissociation (CID). On the basis of predicted molecular weights of various glycan derivatives compared to observed molecular weights, a probable glycan structure is proposed and compared to that predicted from HPLC.

10. (a) A synthetic glycan array could be constructed containing variations of common glycan groups most likely to be present on uroplakin. By comparing the binding capacity of fluorescently labeled pathogenic (expresses FimH) and nonpathogenic *E. coli* strains to the glycan array, it should be possible to identify one or more glycan groups on the array with selective binding to FimH-expressing pathogenic *E. coli* strains. If glycan structural analysis of uroplakin were available using HPLC or mass

spectrometry, the predicted glycan structure should be consistent with results from the unbiased glycan array. (b) A glycan array could be constructed containing the FimH target glycan, as well as other glycans that serve as positive and negative controls. Positive controls would be glycans known to be binding sites for *E. coli* lectins not associated with pathogenicity. This would control for overall binding of bacteria to the array in case the drug candidate nonselectively blocked all binding. In contrast, negative controls would be structurally related glycans that are known to have only very low affinity for FimH, which should not be affected by the presence of drug candidates in the binding reaction.

Chapter 14

REVIEW QUESTIONS

1. Although NADH and NADPH are 2 e^- carriers in metabolic reactions, the phosphate on the ribose C-2 of NADPH distinguishes it from NADH and is recognized by enzymes in redox reactions of biosynthetic and detoxification pathways. NADH is more often used in energy-converting reactions such as catabolic pathways.
2. (1) Low NADPH, (2) low nucleotides, and (3) low ATP.
3. Glucose-6-P + $NADP^+ \rightleftharpoons$ 6-Phosphogluconolactone + NADPH + H^+. Enzyme: glucose-6-phosphate dehydrogenase; $\Delta G°' = -17.6$ kJ/mol.
4. In the oxidative phase, 12 molecules of NADPH are generated from 6 molecules of glucose-6-P, giving rise to 5 molecules of glucose-6-P in the nonoxidative phase. The six carbons are lost as CO_2 during the oxidative phase.
5. The enzyme glucose-6-phosphate dehydrogenase is activated by $NADP^+$ when production of NADPH is needed. It is inhibited by NADPH to decrease flux through this rate-limiting reaction.
6. Glutathione reductase requires NADPH, primarily generated in red blood cells by the glucose-6-phosphate dehydrogenase reaction in the pentose phosphate pathway. Reduced glutathione is a strong reductant in redox reactions and protects red blood cells from oxidizing agents. A deficiency in this enzyme compromises this protection, causing elevated reactive oxygen species levels that lead to sensitivity to primaquine, but also inhibiting the growth of malarial parasites.
7. Each mole of glucose generated by gluconeogenesis requires 2 moles of pyruvate. Hydrolysis of GTP is counted as ATP hydrolysis, yielding a total of 4 ATP required to convert 2 pyruvate to 2 phosphoenolpyruvate.
8. The citrate cycle. The pyruvate carboxylase reaction is responsible for maintaining high oxaloacetate levels under conditions of high acetyl-CoA metabolism and low energy charge.
9. Cytosolic phosphofructokinase-1 is in glycolysis and is allosterically inhibited by high levels of cytosolic citrate, present when energy charge in the cell is high and citrate is shuttled out of mitochondria into the cytosol. At the same time, high cytosolic citrate activates flux through gluconeogenesis by activating frutose-1,6-bisphosphatase-1. This reciprocal regulation by citrate prevents futile cycling and net ATP consumption.
10. Fructose-2,6-BP is a potent allosteric regulator of flux through both gluconeogenesis and glycolysis, in reciprocal fashion. It is not a metabolite in either pathway; fructose-2,6-BP activates the glycolytic phosphofructokinase-1 and inhibits the gluconeogenic fructose-1,6-bisphosphatase-1.
11. The dual-function enzyme phosphofructokinase-2/fructose-2,6-bisphosphatase is a single polypeptide encoding two catalytic activities. Kinase activity encoded by the phosphofructokinase-2 domain is inhibited by phosphorylation on Ser32, whereas phosphatase activity encoded by the fructose-2,6-bisphosphatase domain is activated by Ser32 phosphorylation.
12. Glycolysis in muscle cells yields 2 ATP per glucose metabolized to 2 lactate under anaerobic conditions. Conversion of 2 lactate into 1 glucose in liver cells requires the investment of 6 ATP, thus the net ATP cost of running the Cori cycle between muscle and liver cells is 4 ATP.
13. The frequency of α-1,6 glycosidic bonds is about once every 10 glucose residues, therefore the best answer is ~2,500 ends.
14. High levels of inorganic phosphate (HPO_4^{2-}) in cells leads to a mass action ratio (Q) that is much less than 1 and an ln Q value less than zero. Therefore, the positive $\Delta G°'$ value of +3.1 kJ/mol gives rise to a negative ΔG value of –6 kJ/mol using the Gibbs free energy equation ($\Delta G = \Delta G°' + RT \ln Q$).
15. Glucose-6-phosphatase dephosphorylates glucose-6-P generated from gluconeogenesis in liver so that glucose can be exported to other tissues. Muscle cells lack this enzyme and cannot export glucose.
16. Generation of glucose-1-P by glycogen phosphorylase does not require ATP because the phosphate group is derived from inorganic phosphate (P_i). Moreover, phosphoglucomutase conversion of glucose-1-P to glucose-6-P is also energy independent. Thus, metabolism of glucose-6-P derived from glycogen yields 3 net ATP from glycolysis, compared to 2 ATP derived from metabolism of dietary glucose. The latter needs to be phosphorylated by hexokinase into glucose-6-P, which requires ATP.
17. Glucagon signals low blood glucose levels and activates glycogen degradation by stimulating protein kinase A and subsequent phosphorylation of phosphorylase kinase, which activates glycogen phosphorylase. Protein kinase A also phosphorylates and inactivates glycogen synthase to prevent futile cycling and cause a net efflux of glucose out of liver cells into blood. Insulin signals high blood glucose levels and activates glycogen biosynthesis by stimulating protein phosphatase 1 activity, which dephosphorylates and activates glycogen synthase, and dephosphorylates and inactivates glycogen phosphorylase, causing glucose influx into liver and muscle, lowering blood glucose levels.
18. The inability to add branches to glycogen chains drastically reduces the number of nonreducing ends, making it difficult to store and retrieve glucose in liver cells to maintain safe blood glucose levels. In contrast, lacking debranching enzyme does not affect the ability to store glucose and has only a minor effect on glucose retrieval from glycogen, as 90% of the glucose is linked by α-1,4-glycosidic bonds and cleaved by glycogen phosphorylase.

CHALLENGE PROBLEMS

1. (a) Glucose-6-phosphate dehydrogenase is required in the pentose phosphate pathway to generate reducing equivalents as NADPH, required to sustain a high level of reduced glutathione in red blood cells. Fava beans contain a toxic compound (vicine), reduced by glutathione. Individuals who cannot maintain NADPH levels are highly susceptible to favism, a diet-induced form of anemia. (b) Geographic overlap in Africa of glucose-6-phosphate dehydrogenase deficiencies and malarial resistance suggest a correlation. Also, growth of the malarial parasite *Plasmodium* is inhibited by an oxidizing environment in host cells. Glucose-6-phosphate dehydrogenase deficiencies result in an oxidative environment because of decreased levels of reduced glutathione, so individuals with this deficiency likely have red blood cells that are unsuitable for *Plasmodium* growth.
2. Glucose-1-P is converted to glucose-6-P, which then bypasses the first investment phase of glycolysis (hexokinase). Because only 1 ATP is invested, the net ATP is 3 (4 total ATP are generated in glycolysis).

3. (1) Glucose $\rightleftharpoons$ glucose-6-phosphate. Enzymes: hexokinase and glucose-6-phosphatase; controlled by substrate availability and feedback inhibition of hexokinase (most tissues lack the phosphatase). (2) Fructose-6-phosphate $\rightleftharpoons$ fructose-1,6-bisphosphate. Enzymes: phosphofructokinase and fructose-1,6-bisphosphatase. Phosphofructokinase-1 is inhibited by ATP and citrate and is activated by AMP and fructose-2,6-BP; fructose-1,6-bisphosphatase is inhibited by fructose-2,6-BP and AMP. (3) PEP to pyruvate (glycolysis) and PEP to OAA then pyruvate (gluconeogenesis). Acetyl-CoA, ATP, alanine, and cAMP-dependent phosphorylation inhibit pyruvate kinase during gluconeogenesis. Acetyl-CoA stimulates pyruvate carboxylase during gluconeogenesis.
4. The finding that glycerol addition to the biopsied liver sample leads to glucose production suggests that fructose-1,6-bisphosphatase and glucose-6-phosphatase are normal. The defect could be in phosphoenolpyruvate carboxykinase or pyruvate carboxylase. Because malate, converted to oxaloacetate by malate dehydrogenase, does not lead to glucose production suggests that the defective gluconeogenic enzyme is phosphoenolpyruvate carboxykinase. A defect in pyruvate carboxylase would still allow glucose production after malate addition.
5. This high level of NADH produced by alcohol dehydrogenase drives the lactate dehydrogenase reaction in the direction of lactate production, leading to depletion of pyruvate. Moreover, NAD^+ becomes limiting for glyceraldehyde-3-phosphate dehydrogenase, inhibiting glycolysis, further depleting pyruvate and thus inhibiting gluconeogenesis. The excess NADH meanwhile stimulates malate dehydrogenase to produce malate, depleting the gluconeogenic substrate oxaloacetate.

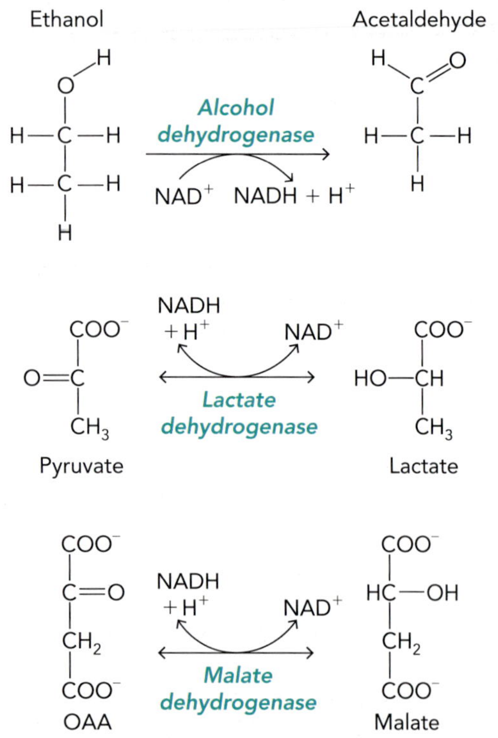

6. (a) The net yield is 3 ATP, not 2, because isomerization of glucose-1-P to glucose-6-P generates the glycolytic metabolite without the investment of 1 ATP by hexokinase. (b) This reaction requires 2 ATP equivalents because glucose must first be phosphorylated by hexokinase to generate glucose-6-P, then isomerized to glucose-1-P and activated by UTP to form UDP-glucose. To regenerate UTP, a second ATP is required by nucleotide diphosphate kinase. (c) The net ATP investment is 5 ATP per glucose from glycogen in muscle cells:

+3 ATP yield per glucose derived from glycogen to lactate in muscle cells
−6 ATP to convert 2 lactate to 1 glucose in liver cells
−2 ATP to add back each glucose into glycogen in muscle cells
−5 ATP net cost

7. Liver cells use glycogen as a source of exporting glucose. Following isomerization of glucose-1-P to glucose-6-P, the phosphate must be removed for export. Muscle cells use the glucose-6-P derived from glycogen as a source of metabolic energy for muscle contraction and need it to remain phosphorylated so that it is not exported.
8. In individuals with von Gierke disease, glucose-6-P accumulates in liver because it cannot be exported. In addition to glucose-6-P derived from gluconeogenesis (also stimulated by low blood sugar), high glucose-6-P levels stimulate glycogen synthase in liver. Glucagon stimulates glycogen breakdown, and the product of debranching enzyme is free glucose, which is released into the blood (~10% of available glucose in glycogen is contained in α-1,6 branch points).
9. Muscle glycogen phosphorylase releases glucose from stored muscle glycogen, needed for energy conversion processes (ATP generation) to sustain muscle contraction.
10. Muscle phosphorylase is activated by AMP, a signal for a low energy state. Glucose inhibits liver phosphorylase by feedback inhibition, signaling that glucose is not being exported rapidly and reducing glycogen degradation. Muscle phosphorylase is not inhibited by glucose because muscle cells use glucose for energy production. Degradation of glycogen in the liver is independent of the energy needs of the liver cell.
11. (a) Protein kinase A activation results in the phosphorylation and inactivation of glycogen synthase, and in activation of glycogen phosphorylase through the phosphorylation and activation of phosphorylase kinase. (b) Protein kinase A activation results in decreases in stored glycogen because glycogen synthase is inactivated and glycogen phosphorylase is activated. (c) Glucagon signals a need for glucose in between meals, epinephrine signals a need for glucose when there is a need for quick energy. In contrast, insulin signals the fed state and a need to replenish glycogen stores.
12. (a) Phosphorylase kinase and glycogen phosphorylase are both activated by the covalent addition of a phosphate. (b) This can be calculated from: 1 glucagon receptor activates 10 $G_{s\alpha}$−GTP. 10 $G_{s\alpha}$−GTP subunits activate 10 adenylate cyclase enzymes. 10 adenylate cyclase enzymes generate 10,000 cAMP per second, but 1,000 cAMP per second are degraded by cAMP phosphodiesterase. 9,000 cAMP activate 4,500 protein kinase A catalytic subunits per second. 4,500 protein kinase A catalytic subunits activate 4,500,000 phosphorylase kinase subunits per second. 4,500,000 phosphorylase kinase subunits per second activate 4,500,000,000 glycogen phosphorylase subunits per second. 4,500,000,000 glycogen phosphorylase subunits per second generate 4,500,000,000,000 glucose-1-phosphate per second.

$= 4.5 \times 10^{12}$ glucose-1-phosphate formed per second

Chapter 15

REVIEW QUESTIONS

1. Energy storage (fatty acids and triacylglycerols), cell membranes (glycerophospholipids and sphingolipids), and receptor-mediated endocrine signaling (steroids and eicosanoids).

2. The answers are contained in the following table.

Carbon bonds	Molecular formula	Common name at (pH 3.0)	Common name at (pH 7.2)
cis 18:1(Δ^9)	$CH_3(CH_2)_7CH{=}CH(CH_2)_7COOH$	Oleic acid	Oleate
18:0	$CH_3(CH_2)_{16}COOH$	Stearic acid	Stearate
cis 16:1(Δ^9)	$CH_3(CH_2)_5CH{=}CH(CH_2)_7COOH$	Palmitoleic acid	Palmitoleate
16:0	$CH_3(CH_2)_{14}COOH$	Palmitic acid	Palmitate
cis 18:2($\Delta^{9,12}$)	$CH_3(CH_2)_4CH{=}CHCH_2CH{=}CH(CH_2)_7COOH$	Linoleic acid	Linoleate
trans 18:1(Δ^9)	$CH_3(CH_2)_7CH{=}CH(CH_2)_7COOH$	Elaidic acid	Elaidate

3. Triglycerides have a higher stored energy potential per gram because fats are essentially unhydrated compared to glycogen.
4. They are palmitic acid ester of myricyl alcohol, hexaeicosanoic acid ester of myricyl alcohol, and palmitic acid ester of cetyl alcohol. The fully saturated long hydrocarbon chains pack tightly together, leading to high melting temperatures. In jojoba oil, large amounts of erucic acid ester of erucyl alcohol pack less tightly because of their monounsaturated chains, resulting in lower melting points.
5. Glycerophospholipids: phosphatidylcholine, phosphatidylserine, phosphatidylethanolamine, and phosphatidylinositol. Sphingolipids: sphingomyelin, cerebrosides, and gangliosides. Cholesterol content is 25% to 40% of the membrane lipid in most plasma membranes. Cholesterol is small and disrupts the packing of membrane lipids with long hydrophobic tails so that low concentrations of cholesterol increase membrane fluidity, but high cholesterol concentrations decrease membrane fluidity because of the packing of cholesterol with itself.
6. Glycerophospholipids are a source for fatty acid–derived signaling molecules, which are released by phospholipase cleavage; for example, phospholipase C cleaves the ester bond at C-3 in phosphatidylinositol-4,5-bisphosphate (PIP_2) to generate inositol-1,4,5-trisphosphate (IP_3) and diacylglycerol (DAG). Phospholipase A_2 releases arachidonate, an important mediator of inflammatory signaling, from glycerophospholipids.
7. Ceramide is the sphingolipid precursor and consists of sphingosine and an attached fatty acid. Linkage of phosphocholine to ceramide generates sphingomyelin, a phospholipid chemically similar to phosphatidylcholine.
8. The glycan structures of ABO gangliosides and glycoproteins are attached to membrane lipids and proteins, respectively, by the same two variant GTA and GTB glycosyltransferase enzymes. Thus, the blood group antigen profile in any one individual has the same four combinations (A, B, AB, O) on erythrocyte lipids and proteins.
9. Progesterone: prepares uterus for ovum. Androgens: regulate male secondary sex characteristic development. Estrogens: regulate female secondary sex characteristic development. Glucocorticoids: regulate gluconeogenesis and fat mobilization. Mineralocorticoids: regulate salt balance and kidney function.
10. Steroid agonists bind to steroid hormone receptors and activate the same biological response as the natural steroid hormone. Steroid antagonists inhibit the biological response either by competing with the natural steroid for receptor binding or by inactivating the receptor. Nandrolone is an androgen agonist that mimics dihydrotestosterone and stimulates muscle growth. Bicalutamide is an androgen antagonist that blocks dihydrotestosterone signaling to treat prostate cancer.
11. Ultraviolet light activates a pathway converting 7-dehydrocholesterol present in skin cells to vitamin D_3, also called cholecalciferol. Humans also obtain vitamin D_3 directly in their diet from vitamin D–fortified foods, as well as from fish and dairy products. Regardless of the source of vitamin D_3, it is first converted to 25-hydroxyvitamin D_3 by the enzyme P450C25 hydroxylase in liver cells, and then the active form of vitamin D, called 1,25-hydroxyvitamin D_3, is synthesized in kidney cells by the enzyme P450C1 hydroxylase.
12. They are prostaglandins, prostacyclins, thromboxanes, and leukotrienes. Prostaglandins: regulate blood flow by modulating smooth muscle contraction and relaxation; stimulate inflammatory responses; and modulate the secretion of proteoglycans that protect the stomach lining from low pH. Prostacyclins: control platelet aggregation and blood clot formation and stimulate vasodilation. Thromboxanes: regulate blood vessel constriction. Leukotrienes: pro-inflammatory molecules that also regulate smooth muscle contraction.

CHALLENGE PROBLEMS

1. (a) Amount of stored fat = 90 kg × 0.15 = 13.5 kg

$$(13.5\text{ kg}) \times (1{,}000\text{ g/kg}) \times (37\text{ kJ/g}) = 5 \times 10^5\text{ kJ}$$

(b) If this amount of energy were stored as hydrated glycogen, the man would weigh $(5 \times 10^5\text{ kJ}) \times (\text{g}/6\text{ kJ}) = 8 \times 10^4\text{ g} = 80\text{ kg}$ of glycogen − 13.5 kg of fat = 66 kg of extra weight, which makes him weigh 90 kg + 66 kg = 156 kg = 344 lb.

2. (a) The molecular mass of KOH is 56 g/mol, so the number of moles of KOH required is

$$\frac{193.2\text{ g}}{56\text{ g/mol}} = 3.5\text{ mol}$$

Because 3 mol of KOH are required to saponify each mole of triacylglycerol, the number of moles of triacylglycerols in the sample of beef fat must be

$$\frac{3.5\text{ mol}}{3} = 1.2\text{ mol of triacylglycerols}$$

The average molecular mass (MM) of the triacylglycerols in 1 kg of beef fat is

$$\text{MM} = \frac{\text{Mass in grams}}{\text{Number of moles}} = \frac{1{,}000\text{ g}}{1.2\text{ mol}} = 833\text{ g/mol}$$

(b) Molecular masses of the fatty acids and glycerol component:
MM of palmitate = 255.4 g/mol
MM of stearate = 283.5 g/mol
MM of oleate = 281.5 g/mol
MM of glycerol component of triacylglycerol (C_3H_5) = 41 g/mol
Because a triacylglycerol (TAG) contains three fatty acids, you need to calculate the total molecular mass of 4 triacylglycerol

molecules containing the following distribution of fatty acids: 6 oleate : 3 palmitate : 3 stearate, which reduces to 2 oleate : 1 palmitate : 1 stearate.

TAG1 = Oleate + Stearate + Palmitate + C_3H_5 = 281.5 + 283.5 + 255.4 + 41 = 861.4 g/mol

TAG2 = Oleate + Palmitate + Oleate + C_3H_5 = 281.5 + 255.4 + 281.5 + 41 = 859.4 g/mol

TAG3 = Oleate + Palmitate + Stearate + C_3H_5 = 281.5 + 255.4 + 283.5 + 41 = 861.4 g/mol

TAG4 = Oleate + Stearate + Oleate + C_3H_5 = 281.5 + 283.5 + 281.5 + 41 = 887.5 g/mol

Add these molecular masses and divide by 4 to get the average molecular mass:

$$\text{TAG1} + \text{TAG2} + \text{TAG3} + \text{TAG4} = \frac{3{,}469.7}{4} = 867.4 \text{ g/mol}$$

(c) Yes, these calculations are consistent with the measured values in Table 15.1, as oleate is present in beef tallow at 43%, with 24% palmitate and 19% stearate. The average molecular mass of 870 g/mol based on saponification (part 2b) is similar to an average molecular mass based on an approximate fatty acid ratio of 2 oleate : 1 palmitate : 1 stearate.

3. Hepatocytes synthesize fatty acids from dietary carbohydrate and protein and assemble them into triacylglycerols using three fatty acids and glycerol. These lipids are packaged into very-low-density lipoproteins that are exported into the circulatory system. In contrast, adipocytes store triacylglycerols in lipid droplets and release free fatty acids into the circulatory system in response to hormonal signaling. The free fatty acids are transported to peripheral tissues by serum albumin protein.

4. The bacteria grown at higher temperature will have the higher saturated/unsaturated fatty acid ratio. They need more saturated fatty acids to maintain optimal fluidity of their membrane lipids because of the higher ambient temperature.

5. Moles of glycerophospholipids = 3.7×10^{-4} mol

Moles of sphingoglycolipids = 1.2×10^{-4} mol

Moles of sphingomyelin = 7.3×10^{-5} mol

Moles of cholesterol = 4.0×10^{-4} mol

Sample calculation: Relative amounts of lipid from Figure 15.25 is equal to 17 + 3 + 8 + 9 = 37. Amount of glycerophospholipids = 17/37 = 0.46 × 650 mg = 299 mg. Average molecular mass (MM) of glycerophospholipid is 800 g/mol (calculated average from MM of all four). Therefore, the number of moles of glycerophospholipid in the 650-mg membrane fraction sample is (0.299 g)/(800 g/mol) = 3.7×10^{-4} mol of glycerophospholipids. The same calculations can be used to determine the number of moles for other lipids.

Sphingoglycolipid: 8/37 = 0.22 × 650 mg = 140.5 mg of lipid
(0.141 g)/(1,192 g/mol) using average MM of cerebroside and GM1 ganglioside
= 1.2×10^{-4} mol of sphingoglycolipid

Sphingomyelin: 3/37 = 0.22 × 650 mg = 52.7 mg of lipid
(0.053 g)/(731 g/mol) using average MM of cerebroside and GM1 ganglioside
= 7.3×10^{-5} mol of sphingomyelin

Cholesterol: 9/37 = 0.22 × 650 mg = 158.1 mg of lipid
(0.158 g)/(387 g/mol) using average MM of cerebroside and GM1 ganglioside
= 4.0×10^{-4} mol of sphingomyelin

6. (a) The hydrocarbon chain would span the bilayer with free hydroxyl groups on either side. (b) The compound is stable at high temperature because of extensive van der Waals interactions and the stability from spanning the bilayer; the ether linkages are chemically stable to hydrolysis.

7. (a) The cleavage products are IP_3 and DAG, which induce glucose export from liver cells by stimulating glycogen degradation and inhibiting glycogen synthesis (see Figure 8.29).

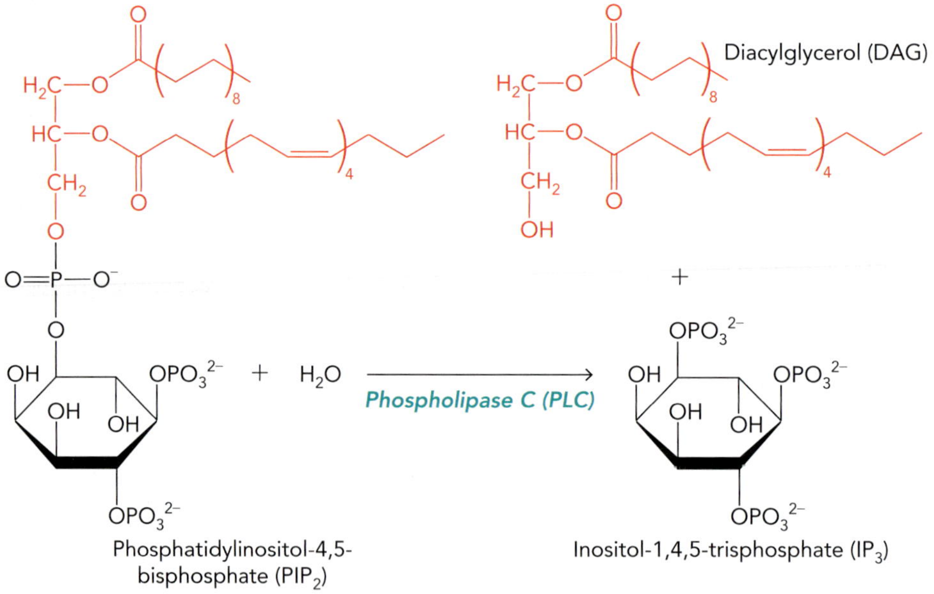

(b) Cleavage of phosphatidylinositol-4,5-bisphosphate shown in Figure 15.30 by phospholipase A_2 releases archidonate, the precursor molecule for synthesis of inflammatory eicosanoids (see Figure 15.41).

8. (a) Tay-Sachs disease is a neurologic disorder caused by a deficiency in hexosaminidase A, which is required for sphingolipid metabolism, resulting in buildup of the ganglioside GM2 substrate, which is not degraded. (b) If both parents are heterozygous, the probability of having a child with Tay-Sachs disease is 25%. If one parent is heterozygous and the other homozygous normal, the probability of having a child that is heterozygous for the Tay-Sachs mutation is 50%. (c) Tay-Sachs disease is rarely seen today because prenatal diagnosis can detect fetuses with the disease prior to birth, providing an opportunity to terminate the pregnancy. Moreover, individuals identified by genetic screening as Tay-Sachs carriers often choose not to have children.

9. (a) Taurocholate is an emulsifying agent. It works by associating with hydrophobic molecules in the diet, such as triacylglycerols and other oils. (b) The structures of taurocholate and glycocholate are shown below. Both are amphipathic molecules with nonpolar hydrophobic cholesterol ring structures and polar head groups. They function as the body's soap by forming micelle-like structures around hydrophobic compounds in food.

Glycocholate

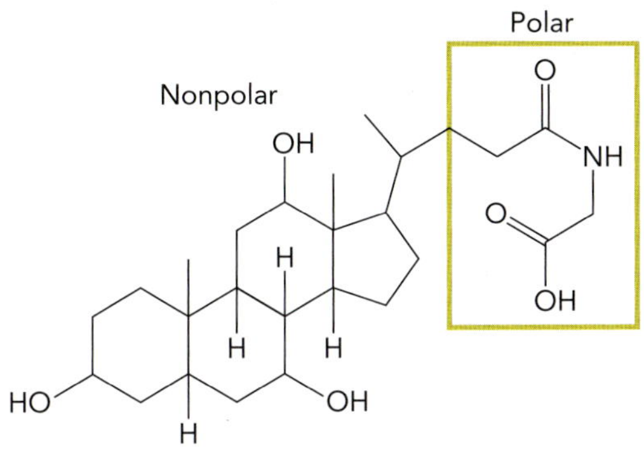

Taurocholate

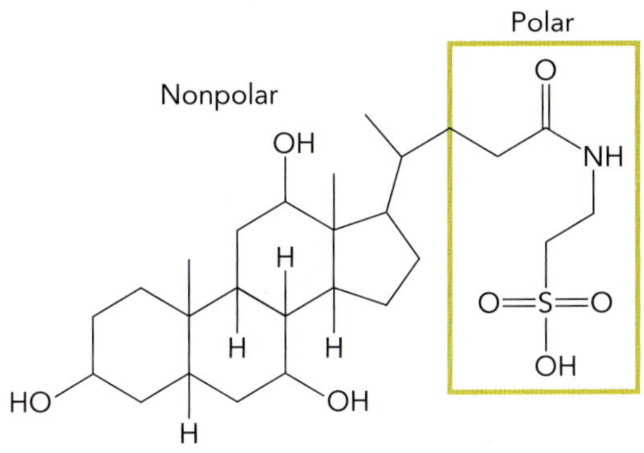

(c) Bile soap would be more effective at removing Italian dressing stains because salad dressing contains olive oil, made up of lipids that can be emulsified by bile acids and soap. Stains caused by red wine are removed with oxidizing agents such as hydrogen peroxide.

10. (a) A deficiency in 21-hydroxylase prevents conversion of progesterone to cortisol and corticosterone, increasing the production of androstendione, and in turn adrenal androgens, through both substrate availability and elevated adrenocorticotropic hormone (in response to decreased cortisol), which stimulates adrenal steroidogenesis. This causes females to overproduce androgens and develop masculine features. This deficiency also decreases aldosterone (mineralocorticoid) production, causing kidney malfunction and electrolyte imbalance (in both males and females). (b) 5α-Reductase is required to convert testosterone to dihydrotestosterone, which is required during prenatal reproductive development in males and is a more potent androgenic hormone than testosterone. At puberty, sufficient amounts of testosterone are produced to cause an androgen surge that promotes masculinization. Females are asymptomatic because dihydrotestosterone is not required for female reproductive development.

11. Prostate cell growth is stimulated by androgens. Blocking 5α-reductase activity with finasteride inhibits production of the physiologic androgen dihydrotestosterone and the prostate shrinks, providing relief from benign prostatic hypertrophy. Androgens also inhibit the growth of hair follicles, and so blocking dihydrotestosterone production decreases androgen signaling, stimulating hair to grow.

12. (a) NSAIDs such as aspirin are generally small molecules that bind to the active sites of both COX-1 and COX-2. Celecoxib is large but only binds to the COX-2 active site. Because COX-1 inhibition is associated with stomach bleeding, this NSAID side effect is not seen with celecoxib. (b) Rofecoxib is large and can partially bind to the prostacyclin synthase active site to inhibit enzyme activity. Aspirin does not bind to or inhibit prostacyclin synthase (the aspirin molecule is too small).

Chapter 16

REVIEW QUESTIONS

1. Fatty acids are amphipathic and are transported across the inner mitochondrial membrane by the carnitine cycle. ATP-dependent activation of fatty acids by fatty acyl-CoA synthetase neutralizes the carboxyl group and provides a high-energy thioester linkage used in attachment to carnitine.

2. (1) Oxidation catalyzed by acyl-CoA dehydrogenase; (2) hydration catalyzed by enoyl-CoA hydratase; (3) oxidation catalyzed by 3-L-hydroxyacyl-CoA dehydrogenase; and (4) thiolysis catalyzed by β-ketoacyl-CoA thiolase.

3. Each stearyl-CoA yields 9 acetyl-CoA, 8 $FADH_2$, and 8 NADH. Stearyl-CoA yields 90 ATP from oxidation of acetyl-CoA in the citrate cycle and electron transport system (ETS), 12 ATP from oxidation of $FADH_2$ in the ETS, and 20 ATP from oxidation of NADH in the ETS, which equals 122 ATP. Subtracting the 2 ATP needed to activate cytosolic stearate with coenzyme A results in a net ATP yield of 120.

4. X-ALD is caused by a defect in a transporter that translocates very-long-chain fatty acids (VLCFAs) across the peroxisomal membrane to be oxidized by the peroxisomal β-oxidation pathway. Accumulation of VLCFAs in the cytosol results in export to the circulatory system, where they damage the myelin sheath surrounding neuronal cells. Lorenzo's oil is a 4:1 mixture of glycerol derivatives of the monounsaturated fatty acids oleate and erucidate, prepared from olive and rapeseed oils. It is thought to decrease VLCFA production, limiting myelin sheath damage in the early stages of the disease.

5. Ketoacidosis is caused by increased flux through the ketogenic pathway when carbohydrate availability is low and demand for energy from fatty acid degradation is high. Abnormally high acetoacetate and D-β-hydroxybutyrate production causes low blood pH, hypoglycemia, delirium, nausea, vomiting, and a fruity odor on the breath.

6. Both pathways require a four-step reaction cycle that removes (fatty acid degradation) or adds (fatty acid synthesis) C_2 units attached to coenzyme A. Fatty acid degradation differs from fatty acid synthesis by its cellular location (mitochondrial matrix), use of coenzymes (FAD and NAD^+ are oxidants), large number of enzymes required, use of CoA as the acetyl group anchor, and is controlled by rate of fatty acyl-CoA transport into mitochondria. Fatty acid synthesis occurs in the cytosol, uses NADPH as a reductant, is catalyzed by a multifunctional enzyme, uses acyl carrier protein (ACP) anchor, and is rate-limited by production of malonyl-CoA by acetyl-CoA carboxylase.

7. (1) Condensation reaction by β-ketoacyl-ACP synthase (KS); (2) reduction by β-ketoacyl-ACP-reductase (KR); (3) dehydration by β-hydroxyacyl-ACP dehydratase (DH); and (4) reduction by enoyl-ACP reductase (ER).

8. The priming reaction of palmitate synthesis uses acetyl-CoA to generate the KS enzyme–bound acetyl group, which then condenses with malonyl-ACP in decarboxylation. Production of C_{16} palmitate requires six more malonyl-CoA, bringing the total malonyl-CoA required to seven.

9. Palmitate is the substrate for elongation and desaturation. Elongation reactions use coenzyme A as the carrier molecule, and palmitoyl-CoA is extended using C_2 groups from malonyl-CoA. Desaturation via mixed function oxidases use molecular oxygen (O_2) as the oxidant. For example, palmitate, 16:0, can generate stearate, 18:0, which is desaturated to form oleate, 18:1(Δ^9).
10. Sphingomyelin, a sphingophospholipid, is both a phospholipid and a sphingolipid. The phosphate group comes from attachment of phosphocholine to ceramide.
11. High levels of intracellular glucose lead to high levels of acetyl-CoA from glycolysis and pyruvate dehydrogenase. Acetyl-CoA is combined with oxaloacetate to generate citrate by the citrate synthase reaction, which is exported to the cytosol by the citrate shuttle, where the reaction is reversed by citrate lyase. Cytosolic acetyl-CoA is converted to malonyl-CoA by acetyl-CoA carboxylase, and then to palmitate by fatty acid synthase. Triacylglycerol synthesized in the cytosol by glycerol metabolizing enzymes and acyltransferases is packaged into VLDL particles and exported. Flux through these pathways is maintained by conversion of oxaloacetate to malate, which is transported back into mitochondria, and used to replenish oxaloacetate. Cytosolic malate can also be converted to pyruvate by malic enzyme in a $NADP^+$-dependent reaction, and pyruvate is then transported into mitochondria and converted to oxaloacetate by pyruvate carboxykinase.
12. (1) Synthesis of mevalonate; (2) conversion of mevalonate to isopentenyl diphosphate and dimethylallyl diphosphate; (3) formation of squalene from isopentenyl diphosphate and dimethylallyl diphosphate; and (4) squalene cyclization to form cholesterol.
13. (1) It is stored as cholesterol esters in lipid droplets; (2) it is packaged into lipoproteins and exported to the circulatory system; and (3) it is converted into bile acids, secreted into the small intestine through the bile duct.
14. (1) Cholesterol input (diet and *de novo* biosynthesis); (2) cholesterol recycling (returning tissue cholesterol to the liver); and (3) cholesterol output (excretion of bile acids).

CHALLENGE PROBLEMS

1. The child might have a defect in unloading of triacylglycerols from chylomicrons in the circulatory system, most likely defects in lipoprotein lipase, needed to unload triacylglycerols from chylomicrons, or in chylomicron surface protein apolipoprotein CII, needed to activate lipoprotein lipase.
2. Without carnitine acyltransferase, no fatty acid oxidation takes place and no acetyl-CoA is produced for ketogenesis or ATP to drive gluconeogenesis, resulting in hypoglycemia.
3. Fat mobilized from adipose tissue is transported as free fatty acids bound to serum albumin, whereas dietary triacylglycerols are transported by chylomicrons.
4. Palmitate has 31 hydrogens, of which 21 are lost during β oxidation, leaving three hydrogens on the methyl carbon, and one each on the even-numbered carbons (which become the methyl carbons of acetyl-CoA). The maximum expected specific radioactivity would be:
$$(10/31) \times (3.0 \times 10^8 \text{ cpm}/\mu\text{mol}) = 0.97 \times 10^8 \text{ cpm}/\mu\text{mol}$$
5. If carnitine acyltransferase I is inhibited, palmitoyl-CoA is not converted into palmitoylcarnitine for transport into mitochondria by the carnitine transporter, but added palmitoylcarnitine can be transported.
6. The molecular mass of palmitic acid is 256 g/mol, and therefore
$$1 \text{ g} \times 1 \text{ mol}/256 \text{ g} = 3.9 \times 10^{-3} \text{ mol palmitate}$$
$$3.9 \times 10^{-3} \text{ mol palmitate} \times 130 \text{ mol } H_2O/\text{mol palmitate} = 0.5 \text{ mol } H_2O$$
$$0.5 \text{ mol} \times 18 \text{ g/mol} = 9.0 \text{ g } H_2O$$
The density of water is 1 g/mL at 4 °C, therefore 9 mL of water will be generated per gram of palmitate.
7. The product of odd-chain fatty acid degradation, propionyl-CoA, is converted to succinyl-CoA, and then to gluconeogenic precursor oxaloacetate via citrate cycle.
8. Each mole of triglyceride contains 1 mol of glycerol, converted to dihydroxyacetone phosphate for gluconeogenesis. Humans do not contain the enzymes to convert fatty acids to gluconeogenic intermediates.
9. Palmitoyl-CoA is not transferred directly into the mitochondria; it is first converted to a carnitine ester for transport. In mitochondria the carnitine ester is replaced by acetyl-CoA. The radioactive CoA in the cytosol never enters the mitochondria.
10. Step 1: Pentanoyl-CoA → propionyl-CoA + acetyl-CoA (4 ATP; one cycle of β oxidation → 1 $FADH_2$ and 1 NADH)
Step 2: Propionyl-CoA → succinyl-CoA (cost is 1 ATP)
Step 3: Succinyl-CoA → 4 CO_2 (20 ATP; two turns of TCA cycle)
Step 4: Acetyl-CoA → 2 CO_2 (10 ATP; one turn of TCA cycle)
Total is 33 net ATP (34 ATP yield − 1 ATP invested in step 2)
11. (a) The liver contains a limited amount of CoA; when β oxidation produces acetyl-CoA faster than the citrate cycle consumes it, hepatocytes quickly run out of CoA and rates of β oxidation decrease. Ketogenesis maintains flux through the β-oxidation pathway to ensure that cells obtain energy from acetyl-CoA via ketone bodies. (b) Ketogenesis results from high rates of fatty acid oxidation and low levels of citrate cycle intermediates to oxidize acetyl-CoA in the mitochondria. Ketogenesis is due to an overflow of acetyl-CoA in the mitochondria of the liver. (c) High-fat/low-carbohydrate diets lead to acetyl-CoA accumulation from fatty acid degradation because carbohydrates are limiting, resulting in low oxaloacetate levels via pyruvate carboxylase and increased flux through the ketogenic pathway.
12. (a) Hydrolysis of ATP to ADP and P_i. (b) Release of CO_2 coupled with hydrolysis of a high-energy thioester bond. (c) Reduction of β-ketoacyl-ACP by NADPH and reduction of enoyl-ACP by NADPH.
13. (a) When excess acetyl-CoA is available, mitochondrial citrate is sent to the cytoplasm and converted to acetyl-CoA for fatty acid biosynthesis. High levels of cytoplasmic citrate stimulate acetyl-CoA carboxylase activity through conversion from monomer to active polymerized form, resulting in malonyl-CoA production. Citrate also inhibits phosphofructokinase-1 of glycolysis, which shifts carbohydrate metabolism toward the pentose phosphate pathway to generate the additional NADPH needed for fatty acid synthesis. (b) Glucagon signaling activates AMP-dependent kinase, which phosphorylates acetyl-CoA carboxylase, inactivating it by depolymerization.
14. When energy levels are high, excess citrate is transported into the cytosol and cleaved by citrate lyase into oxaloacetate and acetyl-CoA, the latter a substrate for fatty acid synthesis. Citrate is an allosteric activator of acetyl-CoA carboxylase, producing malonyl-CoA from acetyl-CoA. The citrate shuttle provides net NADPH for fatty acid synthesis by the cytosolic reactions converting oxaloacetate → malate → pyruvate.
15. (1) It links energy charge to this anabolic pathway, because it is activated from ATP. (2) The carboxylation to form malonyl-CoA provides an intermediate that is separate from β oxidation and carries energy in the reactive carboxyl group of malonyl-CoA. (3) Malonyl-CoA is an inhibitor of the carnitine transacetylase reaction (cytosolic side), such that β oxidation is slowed when flux through the fatty acid synthesis pathway is high.

16. (1) Inability of the LDL receptors to remove LDL (the major source of serum cholesterol) from the serum. (2) Cholesterol biosynthesis (also VLDL production) in the liver is higher than normal because HMG-CoA reductase is not feedback-inhibited by intracellular cholesterol levels.
17. (a) There will be no apoB-containing lipoproteins, chylomicrons, VLDL or LDL. (b) Cholesterol biosynthesis is upregulated because, with no LDL to deliver cholesterol to the cells, cells must make all their cholesterol. (c) Carbohydrate is converted to fatty acids and then triacylglycerol in the liver, but they cannot package the lipid into VLDL particles, so it accumulates.
18. (a) The test depends on fatty acid oxidation to CO_2, but in order for that to occur, the triacylglycerol must be digested and the fatty acids absorbed. (b) No. It can only tell whether the fatty acid is malabsorbed. To make that distinction would require assessing the chemical nature of the fat in the stool.

Chapter 17

REVIEW QUESTIONS

1. Nitrogen fixation (1) catalyzed by nitrogenase, found in some soil and aquatic microorganisms; (2) catalyzed by the Haber process, in which N_2 and H_2 gases are reacted under extreme temperature and pressure to produce NH_3; and (3) an atmospheric process in which energy from lightning combines N_2 with O_2 to form nitrogen oxides that are dissolved in rain and fall to Earth.
2. Each time the Fe protein in nitrogenase is reduced by ferredoxin, 2 ATP are required. Because a total of 6 e^- are required to convert $N_2 \rightarrow 2\ NH_3$, and each round of nitrogen reduction by the MoFe protein transfers 1 e^-, then 12 ATP should be needed. However, the nitrogenase reaction is inefficient, and 2 e^- are lost to a side reaction that converts $2\ H^+ \rightarrow H_2$, thus an additional 2 e^- and 4 ATP are required, for a total of 16 ATP.
3. (1) *Klebsiella pneumoniae* only produces the nitrogenase complex when it is in an anaerobic environment; (2) *Azotobacter vinelandii* rapidly decreases local O_2 concentrations by increasing flux through the electron transport system; and (3) *Sinorhizobium meliloti* invades the roots of leguminous plants, which make leghemoglobin that sequesters O_2 away from the nitrogenase complex.
4. The nitrogen cycle redistributes soil NH_4^+, NO_2^-, and NO_3^- between plants and the atmosphere through decomposition, nitrification, and denitrification. Beginning with decomposition of dead cellular materials by soil bacteria and fungi, organic nitrogen is converted to NH_4^+, used by *Nitrosomonas* bacteria to generate NO_2^- (nitrite). The *Nitrobacter* species convert NO_2^- to NO_3^- (nitrate), which together with NH_4^+ are reincorporated into plants through the soil. Some of the nitrogen returns to the atmosphere as N_2 through denitrification, which is carried out by *Pseudomonas* ($NO_3^- \rightarrow N_2$) bacteria.
5. The combined action of glutamine synthetase and glutamate synthase generates glutamate from α-ketoglutarate and NH_4^+.

   Glutamine synthetase: Glutamate + NH_4^+ + ATP → Glutamine + ADP + P_i

   Glutamate synthase: Glutamine + α-Ketoglutarate + NAD(P)H + H^+ → Glutamate + Glutamate

   Net: α-Ketoglutarate + NAD(P)H + H^+ + NH_4^+ + ATP → Glutamate

   Glutamate dehydrogenase catalyzes the same reaction, but only under very high NH_4^+ levels, as the reaction is unfavorable ($\Delta G^{\circ\prime} = +30$ kJ/mol). Under physiologic conditions, this reaction deaminates glutamate.
6. E1 enzymes attach ubiquitin to E2 enzymes, E2 enzymes conjugate ubiquitin to target proteins, and E3 enzymes recognize target proteins and facilitate ubiquitination by interacting directly with E2–ubiquitin and the target protein. E3 proteins must recognize a large number of target proteins, and thus there are hundreds of different E3 genes. In contrast, E1 needs to recognize only ubiquitin and E2 and therefore fewer genes are needed.
7. The Krebs bicycle is a set of reactions that link the urea cycle and citrate cycle through fumarate. This shunt pathway recycles the carbon backbone of aspartate, which donates one of the two nitrogens to urea. Fumarate produced by the urea cycle is used by the citrate cycle to generate oxaloacetate for the aspartate aminotransferase reaction. In this way, the amino group of glutamate is transferred to urea via aspartate and fumarate.
8. Glucogenic amino acids are metabolized to form pyruvate or citrate cycle intermediates, precursors for gluconeogenesis. Ketogenic amino acids are converted into acetyl-CoA or acetoacetyl-CoA, which give rise to ketone bodies. Isoleucine, phenylalanine, threonine, tryptophan, and tyrosine are both glucogenic and ketogenic.
9. "Inborn errors of metabolism" was first used to describe alkaptonuria, a genetically inherited defect in homogentisate-1,2-dioxygenase. Homogentisate is a metabolite of tyrosine degradation that accumulates in alkaptonuria and is responsible for the name black urine disease. Two other genetic diseases related to phenylalanine degradation are phenylketonuria, a defect in phenylalanine hydroxylase, and albinism, a defect in tyrosinase.
10. Essential amino acids must be obtained in our diet because we lack the necessary enzymes for their synthesis. Nonessential amino acids are those we can synthesize and do not necessarily need in our diet. As plants and bacteria must be able to synthesize all 20 amino acids for survival, but humans eat diverse foods that provide many amino acid requirements, an evolutionary explanation is humans have lost the amino acid biosynthetic pathways that require multiple steps and are the most complex and energy demanding; that is, pathways for the 10 essential amino acids.
11. (a) Glyphosate inhibits 5-enolpyruvylshikimate-3-phosphate (EPSP) synthase, which converts shikimate-3-phosphate to EPSP in chorismate biosynthesis. Animal cells do not synthesize chorismate, and therefore enzyme inhibitors of this pathway are considered animal-safe. (b) It takes 7 days to see the effects of glyphosate because the plant has to use up all available phenylalanine to become dependent on phenylalanine biosynthesis. (c) Roundup Ready soybeans contain a bacterial gene coding for the enzyme CP4 EPSP synthase, which does not bind glyphosate and thereby provides glyphosate resistance to the plant.
12. (a) Recessive genetic mutations in protein-coding genes result in loss of protein function, either loss of expression or a loss of functional activity. Dominant mutations in protein-coding genes lead to gain of protein function; for example, unregulated activity (not subject to negative control) or relaxed substrate specificity. Recessive phenotypes usually require homozygosity (both copies of the gene are defective), whereas dominant phenotypes require only one mutated gene because the variant protein can function even in the presence of wild-type protein. (b) A homozygous with phenylketonuria (PKU) father and homozygous normal mother will produce 100% PKU carriers as offspring because the father will donate a defective gene to all offspring. (c) A heterozygous mother with acute intermittent porphyria will have a 50% chance of having a child with the disease because she can donate either a normal copy of the gene or a mutant copy encoding a gain-of-function mutation.

CHALLENGE PROBLEMS

1. Nitrogen fixation converts atmospheric N_2 into NH_3 (ammonia or NH_4^+). Nitrogen assimilation incorporates ammonia or NH_4^+ into amino acids.
2. (a) NH_4^+ is preferred because it is already reduced, so no additional energy is required to convert NO_3^- to NH_4^+, using nitrite reductase and nitrate reductase, before it can be assimilated into amino acids. (b) After nitrogen starvation, the plant needs nitrogen and obtains it from NO_3^- at the expense of the additional redox energy needed to convert NO_3^- to NH_4^+. (c) Any NH_4^+-based fertilizer would be the best choice because plants use it as a nitrogen source more efficiently than NO_3^--based fertilizers.
3. Root nodules are the sites of nitrogen fixation, so large amounts of ammonia accumulate that need to be assimilated into glutamine by this enzyme.
4. One mechanism is cumulative allosteric inhibition by end products of pathways in which glutamine is a precursor; for example, AMP, tryptophan, carbamoyl phosphate, CTP, histidine, and glucosamine-6-phosphate. Alanine and glycine are also allosteric inhibitors because of the central role of glutamine in amino acid metabolism. A second mechanism of regulation is inhibition by adenylylation and deadenylylation. Adenylylation is indirectly stimulated by glutamine and P_i and inhibited by α-ketoglutarate and ATP.
5. (a) Aspartate is a common amino group donor for transamination, such as:

   Aspartate + α-Keto acid → Oxaloacetate + α amino acid

   Because the α-keto acid in the equation can come from any amino acid, the ^{15}N from aspartate is rapidly transferred to other amino acids. (b) The cofactor is pyridoxal phosphate, which forms a Schiff base between the cofactor carbonyl carbon and an amino group from either a lysine of the enzyme or a free amino acid.
6. (a) Fasting lowered blood glucose, so that the fed amino acids are deaminated to form substrates for gluconeogenesis. (b) Arginine is required for the urea cycle, thus the experimental animals were unable to remove ammonia via the urea cycle.
7. Tadpoles can excrete ammonia directly into the water, but frogs that live on land need to conserve water and excrete ammonia as urea. Arginase is essential for the production of urea from arginine.
8. An enzyme defect in carbamoyl phosphate synthetase would block the urea cycle at the first step of citrulline production from ornithine and carbamoyl phosphate. This would cause higher than normal alanine and glutamine concentrations because they cannot be metabolized and lower than normal arginine concentration because the urea cycle is not functioning (no arginine produced by argininosuccinase). However, a defective arginase would cause arginine to accumulate, and alanine and glutamine concentrations would be closer to normal, assuming excess nitrogen is excreted as argininosuccinate and arginine.
9. (a) Ammonia is removed from muscle tissue by the glucose-alanine cycle, which converts pyruvate to alanine using glutamate as the nitrogen donor. (b) Ammonia is removed from liver tissue by the production of urea in the urea cycle using glutamate and glutamine as nitrogen donors. (c) Ammonia is removed from all other tissues by export of glutamine produced by glutamine synthetase.
10. An alternative way to excrete nitrogen-rich products is glycine depletion by formation of hippuric acid. This process drives ammonia into amino acid synthesis reactions to replace the lost glycine, thereby lowering the level of free ammonia.
11. Using amino acids as the only source of carbon causes increased urea production. Urea is used to remove ammonia resulting from amino acid deamination to generate pyruvate, oxaloacetate, and α-ketoglutarate for energy generation. Because a large volume of water is required to excrete urea, individuals become dehydrated and weigh less in the short term if they do not drink large amounts of water to compensate.
12. Glutamine carries nitrogen from most tissues to the liver for disposal after transamination reactions, whereas alanine carries nitrogen from muscle to the liver. The alanine-glucose cycle is responsible for removing excess muscle nitrogen that is due to protein degradation.
13. (a) Kwashiorkor is caused by a lack of protein in the diet. The decreased pigmentation is caused by a deficiency in tyrosine, which is the precursor to melanin pigments. Moreover, because phenylalanine is converted to tyrosine by phenylalanine hydroxylase, phenylalanine deficiency also contributes to the skin and hair condition. (b) In such a case, the individual may have a defect in an enzyme required to convert tyrosine to melanin.
14. (a) The level of phenylalanine in the body can be controlled by diet. There is no safe way to continually add pigments to cells through diet or drugs. (b) Individuals with phenylketonuria have tyrosinase and obtain tyrosine in their diet.
15. Porphyrias are genetically inherited diseases caused by defects in enzymes for heme biosynthesis. Acute intermittent porphyria is a genetic defect in porphobilinogen deaminase causing a buildup of bilirubin, which leads to yellow skin, or jaundice. Because some of the heme biosynthesis intermediates are chemically related to heme degradation products, defects in heme biosynthesis can lead to a buildup of substrates converted to bilirubin to cause jaundice. Not all porphyrias lead to an accumulation of bilirubin, therefore not all porphyrias are associated with jaundice.

Chapter 18

REVIEW QUESTIONS

1. Energy conversion reactions, signal transduction, coenzyme-dependent reactions, and genetic information storage and transfer. Guanine is used in energy conversion as part of GTP, produced by succinyl-CoA synthetase in the citrate cycle. Guanine in signal transduction includes GTP or GDP bound to G proteins and cGMP as second messenger. All of the common coenzymes (NAD^+, FAD, CoA) are based on adenine; but, CTP is required by phosphopantothenate–cysteine ligase for biosynthesis of coenzyme A (see Figure 10.9). All five bases (adenine, cytosine, guanine, uracil, and thymine) are used for genetic information and storage.
2. (a) uridine, (b) deoxyadenosine-5′-triphosphate, (c) cytidine-5′-diphosphate, (d) thymine, and (e) guanylate (or guanosine-5′-monophosphate).
3. The degradation of cellular RNA, a transient nucleic acid required for gene expression. Because RNA does not contain thymine, salvaged dTMP (thymidylate) from DNA is the only source of thymine and thymidine.
4. Purine bases are synthesized directly on the ribose, whereas pyrimidine bases are first synthesized as a closed ring before attaching to ribose.
5. The four nitrogens in purine bases come from aspartate (N-1), glycine (N-7), and the amide nitrogen of glutamine (N-3 and N-9). The five carbons come from glycine (C-4 and C-5), HCO_3^- (C-6), and formyl groups from N^{10}-formyl-tetrahydrofolate (C-2 and C-8). Inosine-5′-monophosphate is the precursor to adenylate and guanylate.

6. Purinosomes are multifunctional complexes in eukaryotes containing all catalytic activities required for a 10-step reaction leading to the synthesis of IMP. Purinosomes were proposed on the basis of experiments in which green fluorescent protein (GFP) fusions to enzymes in the *de novo* purine biosynthetic pathway were observed to form fluorescent cytosolic protein aggregates in purine-depleted cell cultures. Biochemical approaches to identify purinosomes under physiological conditions include chemical cross-linking, coupled with protein purification and mass spectrometry, to characterize "nearest neighbor" proteins, which should be other enzymes in the same pathway.
7. Flux through the purine biosynthetic pathway is controlled by inhibition of the PRPP synthetase and glutamine-PRPP amidotransferase. AMP and GMP synthesis is controlled by both feedback inhibition of individual branches in the pathway, and by ATP and GTP cross-talk regulation.
8. They all degrade excess purines into uric acid. Primates (humans), insects (cockroaches), reptiles (rattlesnakes), and birds (flamingos) excrete the uric acid. In contrast, nonprimate mammals (Siberian tigers) metabolize uric acid to allantoin for excretion. Bony fish (tuna) convert allantoin to allantoic acid for excretion, whereas cartilaginous fishes (sharks) and amphibians (salamanders) convert allantoic acid to urea for excretion. Lastly, marine invertebrates (lobsters) use urease to break down urea into NH_4^+.
9. ADA–SCID is caused by defects in adenosine deaminase (ADA), leading to adenosine accumulation and elevated levels of dATP. Because dATP inhibits ribonucleotide reductase, rapidly dividing B and T cells are starved of deoxyribonucleotides and fail to proliferate, resulting in severe immunodeficiency. Treatment for ADA–SCID patients involves frequent intravenous infusions of immunoglobulins or recombinant ADA. ADA–SCID can be cured only by a bone marrow transplant, requiring the correct match to avoid tissue rejection. Human gene therapy using a virus containing the human ADA gene is a promising experimental procedure.
10. Pyrimidine biosynthesis is regulated by feedback inhibition and allosteric activation in *E. coli* and humans. Aspartate transcarbamoylase is the key regulated enzyme in *E. coli* cells, activated by ATP and inhibited by CTP. In contrast, in human cells it is allosteric regulation of the trifunctional CAD enzyme and regulation of UMP synthase and CTP synthetase.
11. Dihydropyrimidine dehydrogenase deficiencies are fairly common (~5% of the human population) and can cause 5-fluorouracil toxicity in cancer patients undergoing chemotherapy because 5-fluorouracil is not degraded as efficiently in the liver. Normally, ~80% of 5-fluorouracil is degraded, so that the effective concentration of 5-fluorouracil is only ~20% of the supplied dose. However, in dihydropyrimidine dehydrogenase–deficient individuals, much less is degraded so the effective dose is much higher and can harm healthy cells. Cancer patients being considered for 5-fluorouracil treatment are now routinely prescreened for dihydropyrimidine dehydrogenase deficiencies in order to adjust their drug dose appropriately.
12. This hypothesis proposes that RNA was the first genetic molecule in early life on Earth, and RNA-based catalysts called ribozymes mediated reactions, rather than protein-based reaction catalysts, which arose later. Following the emergence of ribonucleotide reductase, deoxyribonucleotides were polymerized into DNA, which is better suited to function as genetic material. The best evidence to date for the RNA world hypothesis is that RNA is responsible for the catalytic activity in peptide bond formation during protein synthesis, not protein.
13. Ribonucleotide reductase reduces C-2′ on nucleoside diphosphates to convert NDPs to dNDPs, requiring a pair of electrons from NADPH, and producing H_2O. The reduction of the sulfhydryls in ribonucleotide reductase is not directly by NADPH, but rather by a redox circuit requiring intermediary proteins (thioredoxin or glutaredoxin). Ribonucleotide reductase has two subunits, R1 and R2, in a tetrameric complex ($R1_2R2_2$). Substrate specificity is regulated by allosteric binding of dTTP, dGTP, dATP, or ATP to a specificity site on R1. The overall enzyme activity is regulated by allosteric binding of ATP (activator) or dTTP (inhibitor) to the regulatory site.
14. (1) Dephosphorylation of dUTP by dUTP diphosphohydrolase; (2) deamination of dCTP by dCTP deaminase to generate dUTP, then converted to dUMP by dUTP diphosphohydrolase; or (3) phosphorylation of deoxyuridine by thymidine kinase to yield dUMP.
15. (1) Point mutations in dihydrofolate reductase coding sequence to lower methotrexate binding affinity; (2) dihydrofolate reductase gene amplification to increase expression of the enzyme; and (3) overexpression of the multidrug resistance protein, which rapidly exports methotrexate.

CHALLENGE PROBLEMS

1. (1) Nucleotide salvage pathway (base + PRPP forms nucleoside monophosphates); (2) *de novo* purine biosynthesis (IMP is synthesized beginning with PRPP); and (3) *de novo* pyrimidine biosynthesis (orotate + PRPP forms OMP).
2. In defect A, large amounts of ^{15}N are incorporated in uric acid because glycine contributes a nitrogen to the purine ring in *de novo* synthesis. In defect B, uric acid should be relatively free of ^{15}N because much of the excess uric acid is due to purines in the diet, not as a result of *de novo* purine biosynthesis.
3. An HGPRT deficiency prevents salvage biosynthesis of purines; therefore, *de novo* synthesis is stimulated in order to maintain sufficient purine levels.
4. (a) The defective enzyme is HGPRT and the ~50% level of radioactive DNA suggests this individual is heterozygous for HGPRT; that is, he or she is an asymptomatic carrier of the HGPRT defect. (b) Female. Lesch–Nyhan syndrome is an X-linked recessive genetic disorder, so females can be heterozygous asymptomatic, whereas males are either normal (for a male) or symptomatic for Lesch–Nyhan syndrome. (c) A single copy of HGPRT, such as males have (with one X chromosome), is sufficient to remove excess metabolic intermediates and avoid symptoms of Lesch–Nyhan syndrome. (d) For an X-linked recessive disorder such as Lesch–Nyhan syndrome, a daughter will not develop Lesch–Nyhan syndrome, but has a 50% chance of being a carrier; a son has a 50% chance of developing Lesch–Nyhan syndrome.
5. The synthesis of AMP from IMP (Figure 18.11) generates fumarate from skeleton of aspartate, providing increased citrate cycle intermediates to stimulate flux through this pathway in muscle cells. Production of fumarate can be sustained by AMP deaminase, which deaminates AMP to regenerate IMP (Figure 18.14).
6. (a) The disease ADA–SCID (ADA: adenosine deaminase; SCID: severe combined immunodeficiency disease) is characterized by a defective immune system, resulting from extremely low levels of infection-fighting B and T cells. (b) This defect leads to accumulation of dATP, which inhibits ribonucleotide reductase, and thus deoxynucleotide biosynthesis. This is only a problem

in rapidly dividing cell types, such as immune cells that need to proliferate quickly in response to antigen-mediated stimulation (B and T cells).

7. (a) It is in the *de novo* pyrimidine biosynthesis pathway and the reaction it catalyzes is

$$\text{Orotate} + \text{PRPP} \rightarrow \text{OMP} + \text{PP}_i$$

(b) OMP is the precursor to UTP and CTP, which are not produced when orotate phosphoribosyl transferase is defective. However, the salvage pathway can use uridine and cytidine to make UTP and CTP to provide nucleotides for the synthesis of blood cell precursors to alleviate anemia. A decrease in orotic acid excretion may occur because the fed nucleotides inhibit *de novo* pyrimidine biosynthesis by feedback inhibition, preventing further accumulation of orotate.

8. The substrate for thymidylate synthase is dUMP, which is derived from dUDP.

9. Cells that lack thymidine kinase are not able to phosphorylate 5-bromodeoxyuridine, which prevents 5-bromodeoxyuridine incorporation that mutates A-T base pairs into G-C base pairs.

10. (a) Xanthine oxidase inactivates 6-mercaptopurine by converting it to 6-thiouric acid, which lowers its effective concentration as a competitive inhibitor of adenylosuccinate synthetase and IMP dehydrogenase in leukemic cells. Therefore, by including allopurinol in the treatment, inactivation of 6-mercaptopurine by xanthine oxidase is repressed, leading to a higher effective concentration. (b) A leukemia patient with xanthine oxidase deficiency should be treated with a reduced dose of 6-mercaptopurine to avoid possible toxic side effects. Addition of allopurinol to this treatment would likely have no effect and should not be done. This is a similar problem to treating a cancer patient who has a deficiency in dihydropyrimidine dehydrogenase with 5-fluorouracil, as illustrated in Figure 18.26.

11. Addition of high amounts of thymidine to cultured cells leads to the accumulation of dTTP through the salvage pathway. dTTP binding to the specificity site of ribonucleotide reductase inhibits reduction of CDP to dCDP, thus inhibiting DNA replication due to insufficient substrate.

12. Addition of methotrexate inhibits dihydrofolate reductase, blocking the production of new tetrahydrofolate, a critical metabolite for numerous pathways. Because thymidylate synthase is active, the remaining tetrahydrofolate pool is quickly depleted. Adding thymidine to the culture does not rescue the cells because it is the low levels of tetrahydrofolate that cause cell death, not lack of thymidine. In contrast, thymidylate synthase–deficient cells do not use up the tetrahydrofolate pool when methotrexate is added, so that other tetrahydrofolate-dependent pathways can function for a time. Moreover, added thymidine rescues the thymidylate synthase deficiency, so these cells are able to survive longer than normal cells.

Chapter 19

REVIEW QUESTIONS

1. Safe blood glucose levels of 4.5 mM are maintained by release of glucose between meals to compensate for the drop in glucose levels due to glucose-dependent brain cells and erythrocytes. Glucose from the liver is made by liver glycogen breakdown and gluconeogenesis. After a meal, the liver helps reduce spikes in blood glucose from carbohydrate-rich foods by promoting glucose influx through the GLUT2 glucose transporter and glucose phosphorylation by hexokinase and glucokinase, which trap glucose-6-P inside liver cells. Glucose levels below 3.5 mM cause light-headedness; below 2.5 mM the person becomes hypoglycemic (fainting, perspiring); and levels below 2.2 mM can result in coma and death.

2. The key metabolite is glucose-6-P, which has four metabolic fates: (1) glucose released into the blood; (2) glucose-1-P for glycogen synthesis; (3) 6-phosphogluconolactone to generate NADPH by the pentose phosphate pathway; and (4) fructose-6-P, used in glycolysis to produce pyruvate. Metabolic flux of glucose-6-P among these pathways is controlled by the activity of rate-limiting enzymes and metabolite concentrations.

3. The largest amount of muscle in the human body is skeletal muscle, which uses free fatty acids, glucose, and ketone bodies for metabolic fuel. The other major muscle type is cardiac muscle, which primarily uses fatty acids and ketone bodies.

4. They are visceral fat and subcutaneous fat. The primary function of both types of fat is energy storage in the form of triacylglycerol; visceral fat also functions as an endocrine organ to release adipokine hormones in response to metabolic cues. Visceral fat lies within the abdominal cavity (apple body type); subcutaneous fat lies just below the skin surface in the thighs, buttocks, arms, and face (pear body type).

5. (1) Body mass index (BMI), calculated by dividing a person's weight in kilograms by the square of his or her height in meters; (2) waist-to-hip ratio, based on circumferences of the waist and hips; and (3) percent body fat, determined using whole-body bioelectrical impedance. A retrospective clinical study applies statistical analyses on data that were already collected to identify disease associations. Because BMI takes into account only height and weight, it cannot distinguish between different body types. The waist-to-hip ratio can distinguish body types if the ratios are significantly higher or lower than 1.0, but it cannot distinguish between a muscular individual with small hips and an obese individual with large hips and waist. The best approach for a study in which patients are not available for additional measurements is to combine percent body fat, BMI, and waist-to-hip ratio to get an accurate profile of the few "outlier" individuals.

6. This cycle exchanges triacylglycerols synthesized in adipose tissue with those synthesized in liver cells using the same pool of fatty acids. The systemic component of the cycle exchanges free fatty acids and triacylglycerols between adipose tissue and liver cells, whereas the intracellular component of the cycle interconverts pools of free fatty acids and triacylglycerols in adipocytes. Most fatty acids and triacylglycerols are recycled back to the adipose tissue, unless needed by peripheral tissues for energy conversion or membrane biosynthesis. When dietary glucose levels are low, glyceroneogenesis in adipocytes and hepatocytes synthesizes glycerol-3-P from dihydroxyacetone phosphate, which is generated from amino acids and lactate in order to maintain triacylglycerol synthesis using free fatty acids.

7. Insulin and glucagon are secreted from the pancreas and have opposing actions for regulating blood glucose levels by activating cognate receptors on multiple target tissues. Insulin levels rise in response to high blood glucose levels after a carbohydrate-rich meal, whereas glucagon levels rise in response to low blood glucose levels, usually between meals. Insulin signaling in liver, skeletal muscle, and adipose tissue stimulates glucose uptake and storage of glycogen and lipid, whereas insulin signaling in the brain stimulates anorexigenic neurons to decrease appetite and increase energy expenditure. Glucagon in liver and adipose tissue stimulates glycogen and triacylglycerol degradation. Skeletal muscle and brain cells lack glucagon receptors.

8. PPARs regulate gene expression in response to low-affinity, fatty acid–derived nutrients such as polyunsaturated fatty acids and eicosanoids. This makes PPARs ideal metabolic sensors of lipid homeostasis for long-term control of pathway flux by directly altering the steady-state levels of key proteins. PPAR-regulated genes control fatty acid oxidation, fatty acid synthesis, adipose differentiation, energy uncoupling, and insulin sensitivity.
9. The two metabolic objectives are to supply (1) the brain with glucose to maintain ATP-dependent ion pumps and (2) the heart with fatty acids and ketone bodies to maintain rhythmic contraction and blood circulation. The four major changes in metabolic flux under starvation conditions are (1) increased triacylglycerol hydrolysis in adipose tissue, (2) increased gluconeogenesis in liver and kidney cells, (3) increased ketogenesis in liver cells, and (4) protein degradation in skeletal muscle tissue.
10. The hypothesis states that humans have metabolic gene variants that provide protection against famine by maximizing fat storage during times of feasting. These same gene variants contribute to obesity and diabetes in developed countries where high-Calorie food is readily available and a sedentary lifestyle is common. Observed differences in rates of obesity and type 2 diabetes in two populations of genetically related Pima Indians support this idea. The Pima Indians of southern Arizona have a sedentary lifestyle and high-Calorie diet, which leads to higher rates of obesity and diabetes than those found for the Pima Indians of northern Mexico, who cultivate their fields using manual labor.
11. Leptin is produced at a level directly related to the amount of stored lipid. Leptin binds leptin receptors on first-order POMC and NPY/AGRP neurons, which produce neuropeptides (α-MSH, AGRP, and NPY) that bind to their cognate receptors on anorexigenic and orexigenic second-order neurons. The net effect of leptin signaling is eating less and metabolizing more. Insulin is secreted from the pancreas in response to elevated blood glucose and has the same effect on neuronal signaling as leptin. Ghrelin is synthesized at high levels in the stomach between meals and binds to ghrelin receptors on NPY/AGRP first-order neurons that promote signaling in second-order neurons to eat more, metabolize less. $PYY_{3\text{-}36}$ is synthesized in the colon and counters the effect of ghrelin by sending signals that food is in the digestive tract and it is time to eat less, metabolize more.
12. Type 1 diabetes is due to insufficient insulin production and is treatable with insulin injections. Type 2 diabetes is due to insulin resistance, characterized by high levels of circulating insulin and desensitization of insulin receptor signaling in muscle, liver, and adipose tissue. The symptoms of diabetes include frequent urination, thirst, and blurry vision, which are caused by the osmotic effects of high glucose concentrations. Glucose tolerance tests confirm diabetes diagnoses by measuring glucose clearance from the blood after ingestion of a glucose solution. Type 1 and type 2 diabetics are unable to lower blood glucose levels within 2 hours. Type 1 and type 2 diabetes can be distinguished by an insulin sensitivity test, which shows a decrease in blood glucose levels for type 1 diabetics, but not for type 2 diabetics.
13. (1) α-Glucosidase inhibitors (miglitol), (2) sulfonylurea drugs to inhibit the pancreatic ATP-dependent K^+ channel (glipizide), (3) drugs that stimulate the activity of AMPK (metformin), and (4) ligand agonists of the nuclear receptor PPARγ (pioglitazone).
14. (1) Ephedrine increases basal metabolic rates; side effects include nausea, nervousness, and insomnia. (2) Lorcaserin targets neuronal signaling in the brain to control appetite (serotonin receptors); side effects are mild and include headache, dizziness, and nausea. (3) Orlistat inhibits the activity of pancreatic lipase in the small intestine; side effects include stomach pain, fecal urgency, and nausea. (4) Olestra is a zero-Calorie food substitute containing fatty acid side chains covalently linked to sucrose and is not a substrate for digestive lipases; side effects include intense diarrhea and anal leakage. A drug-free alternative to weight loss is to eat less and exercise more.
15. AMPK-mediated serine/threonine phosphorylation of proteins in muscle cells leads to increased ATP concentration by three mechanisms: (1) stimulation of flux through glycolysis; (2) stimulation of flux through fatty acid oxidation; and (3) increased oxidative phosphorylation.

CHALLENGE PROBLEMS

1. (a) Assuming all 56 g of starch is converted to 56 g of glucose:
$$56 \text{ g} \times 3.7 \text{ kcal/g} = 207.2 \text{ kcal}$$
$$207.2 \text{ kcal} \times 4.18 \text{ kJ/kcal} = 866 \text{ kJ}$$
$$866 \text{ kJ} \times \text{mol ATP}/30.5 \text{ kJ} = 28.3 \text{ mol ATP} \times 0.65$$
$$= 18.5 \text{ mol of ATP } (\sim 19 \text{ mol})$$
(b) (Recall that 1 Calorie equals 1 kcal.) Time a female would need to garden:
$$207.2 \text{ Calories} \times 1 \text{ min gardening}/4 \text{ Calories}$$
$$= 51.8 \text{ minutes of gardening}$$
Time a male would need to vacuum:
$$207.2 \text{ Calories} \times 1 \text{ min vacuuming}/4.3 \text{ Calories}$$
$$= 48.1 \text{ minutes of vacuuming}$$
2. (a) Lower levels; (b) lower levels; (c) higher levels; (d) higher levels; (e) lower levels
3. Eating six donuts will cause a sharp rise in insulin levels and stimulate fatty acid synthesis in adipose tissue but inhibit fatty acid release. The short-lived glucose rise at the beginning of the race is detrimental later, when muscle and liver glycogen levels are depleted and fat utilization from adipose tissue is needed to make energy for muscle cells.
4. The pattern results from the maintenance of blood glucose levels during the first days of the fast. Glucose comes initially from glycogen degradation, but once glycogen is depleted, gluconeogenesis in the liver and kidneys occurs using amino acids from protein hydrolysis as the source of carbon. After the brain adapts to using ketones as a component of its energy needs, flux through gluconeogenesis decreases, and protein is spared until survival is the highest priority.
5. (a) Yes, this would be a candidate because malonyl-CoA would be synthesized to higher levels than normal (no feedback inhibition) and thereby increase the total amount of stored triacylglycerol in adipose tissue. (b) Yes, this would be a candidate because even at low insulin levels, fatty acid synthesis and lipogenesis would increase fat stores. (c) No, this would not be a candidate because uncoupling protein overexpression would lead to more thermogenesis and high rates of fatty acid oxidation, raising basal metabolic rates. (d) Yes, this would be a candidate because it would scavenge as much lipid as possible from circulating lipoprotein particles to increase fat storage in adipose tissue. (e) No, this would not be a candidate because it would continually hydrolyze triacylglycerol in adipose tissue and make it difficult to accumulate stored fat.
6. The uncoupler protein dissipates the proton gradient in the mitochondria, allowing higher fatty oxidation rates than would be possible in coupled mitochondria. Such high rates of oxidation use up some of the excess fatty acids, reducing the amount stored in nonadipose tissue.

7. OB mutant mice lack leptin, and leptin injections enable them to better maintain energy balance and metabolic homeostasis. Most obese humans have chronically high leptin levels rather than a mutation in the leptin gene like the OB mice. In fact, most obese individuals are leptin insensitive. This is analogous to type 2 diabetics at initial diagnosis, who often have high levels of insulin but are insulin insensitive.
8. Eating slowly gives ghrelin and $PYY_{3\text{-}36}$ a chance to tell your brain that food is in the stomach and you feel full. If you eat too quickly, you eat more than you need.
9. (a) The K_m of fructokinase for fructose is 20 times lower than the K_m of glucokinase for glucose, so fructose is efficiently converted to fructose-1-P in liver cells and converted to acetyl-CoA, the substrate for fatty acid synthesis in the liver. In contrast, glucokinase requires more glucose to achieve the same flux to acetyl-CoA, and some glucose is converted to glucose-6-P, which can be used for glycogen synthesis (after conversion to glucose-1-P). An equimolar amount of sucrose contains 50% fructose and 50% glucose, which is 5% less fructose than that in high-fructose corn syrup.
   (b) 110 g carbohydrate/soda + 56 g carbohydrate/chips = 166 g carbohydrate × 4 Calories/g carbohydrate = 664 Calories
   664 Calories/2,000 Calories = 33%
   of recommended Calories in this late-night snack.
10. Sulfonylurea is used on the assumption that the insulin receptor in a diabetic individual is less sensitive to insulin; therefore, increasing the insulin concentration should increase binding to the receptor and upregulate blood glucose uptake. Metformin is used on the assumption that decreased flux through gluconeogenesis lowers blood glucose levels.
11. Diabetics cannot respond to high glucose levels in the blood because of the inability to produce or respond to insulin. Under these conditions, the body responds to decreased glucose levels by increasing fatty acid degradation with excess acetyl-CoA converted to ketone bodies in the liver.
12. Diet no. 1 is the most complete because proteins are required (nine essential amino acids), and fats can be synthesized from carbohydrates, but carbohydrates cannot be synthesized from fats.
13. On a low-carbohydrate diet, oxaloacetate, α-ketoglutarate, and pyruvate come from amino acid degradation. The acetone breath indicates that these reactions are not sufficient to keep the citrate cycle functioning at full capacity and acetyl-CoA builds up, leading to the production of ketone bodies and acetone.
14. Dinitrophenol is an uncoupler of mitochondrial ATP synthesis, leading to high rates of fatty acid oxidation to compensate for the resulting low energy charge. Dinitrophenol can lead to weight loss because it mimics the "starved" state and fat is catabolized at a high rate, but it has been reported that long-term dinitrophenol use causes fatal liver damage due to hepatocyte apoptosis.
15. High-glycemic-index foods (jelly beans) cause a rapid rise in insulin levels, stimulating fatty acid synthesis. Low-glycemic-index foods (kidney beans) release carbohydrates slowly, so insulin levels are gradually elevated, resulting in a more balanced metabolic response.
16. Epidemiologic studies found an association between improved cardiovascular health and diets low in saturated fats and high in unsaturated fats, hence the term "good fat." People who favor bad fats over good fats show the opposite trends in cardiovascular disease and death.
17. Muscle cells use oxygen to replenish ATP by oxidative phosphorylation to drive synthesis of phosphocreatine. Liver cells use oxygen to replenish ATP to convert lactate to glucose via gluconeogenesis.

Chapter 20

REVIEW QUESTIONS

1. Semiconservative replication creates two daughter DNA molecules, each containing one template strand and one newly synthesized strand. Conservative replication would yield one DNA duplex made up of both templates and one made of two newly synthesized strands. Dispersive replication would result in both DNA molecules made of mixed portions of template and newly synthesized DNA, template paired with template, and new DNA paired with new DNA.
2. DNA segments produced by discontinuous replication of the lagging strand. One is started each time a new β clamp loads onto the lagging strand template and is completed when DNA polymerase encounters the preceding Okazaki fragment.
3. The 3′-hydroxyl of the previously added nucleotide is the nucleophile for DNA polymerase. Initiation requires a primer, a short stretch of RNA produced by primase, to provide the nucleophile.
4. In prokaryotes: DNA polymerase III; in eukaryotes: DNA polymerase δ (lagging strand synthesis) and DNA polymerase ε (leading strand synthesis).
5. HIVRT uses the single-stranded RNA of the HIV genome as a template to produce a DNA–RNA duplex, using the 3′ end of the host cell Lys-tRNA as a primer. HIVRT has RNase activity and degrades the RNA strand once DNA is produced. RNA fragments serve to prime synthesis of the complementary DNA strand to generate double-stranded DNA.
6. Ahead of the replication fork, topoisomerase adds negative supercoils to relieve the positive supercoiling caused by separation of strands. Helicase and primase are coupled at the front of the replication fork to separate the strands and synthesize RNA primers, respectively. Single-stranded binding proteins maintain single-stranded DNA and protect it from damage. DNA polymerase III, tethered to helicase, consists of a core enzyme and a β clamp, which increases processivity of the polymerase. DNA polymerase I and ligase follow the replication fork to remove RNA primers and seal nicks in the newly formed DNA.
7. Binding of DnaA begins DNA melting and guides helicase to single-stranded DNA (with the help of DnaC). Gyrase allows further opening of the double helix by relieving torsion created by the single-strand region, and single-stranded DNA binding proteins maintain the single-stranded state. DNA Pol III binds to release DnaA and start the replication process.
8. Human telomerase contains a 451-nucleotide RNA with a tandem RNA at the 3′ end that is complementary to the chromosome end. After association of the RNA with the end of the chromosome, telomerase extends the chromosome using the RNA as template. When the end of the template is reached, telomerase releases from the chromosome.
9. The Ames test uses a mutant strain of *Salmonella* that cannot produce histidine to test the mutagenicity of various molecules. The test molecule is incubated with *Salmonella* and liver extract to simulate mammalian metabolism. If colonies are produced on minimal-histidine media, then a back mutation has occurred in the histidine synthesis genes so the pathway is functional. Substances that are mutagenic to bacteria can be identified, which are potentially carcinogenic in humans.
10. Cytosine can be spontaneously deaminated, producing uracil. Uracil is removed by uracil DNA glycosylase to produce an abasic

site that is repaired by base excision repair, where a nick is created in the DNA by AP endonuclease. The nucleotide is replaced by short-patch repair or a short section of nucleotides is replaced by long-patch repair.

11. Both of these can distort the DNA helix and present a roadblock for access to individual bases. The replication fork stalls at such lesions or skips over them and attempts to place an incorrect nucleotide. RNA polymerase either stalls or skips over lesions, putting a potentially incorrect nucleotide at the site.
12. (1) Silent mutations do not change the coded amino acid due to degeneracy of the genetic code; (2) nonsense mutations result in premature stop codons and generally yield nonfunctional proteins; (3) missense mutations incorporate incorrect amino acids into the protein. Both nonsense and missense mutations can be detrimental to a cell, depending upon the protein altered and the degree of alteration.
13. The methylation product is shown in Figure 20.42. O^6-methylguanine-DNA methyltransferase (MGMT) is required to remove O^6-methylguanine. It is not a true enzyme, as it is irreversibly modified during the process. A cysteine in the active site performs a nucleophilic attack upon the methyl group, removing it from the guanine and converting the cysteine to homocysteine. The protein is then degraded to metabolize homocysteine.
14. The *oriC* region has 11 sites methylated by Dam methylase to distinguish the template strand from the newly synthesized strand. This highly methylated region is a SeqA binding site. The SeqA–*oriC* complex is sequestered at the membrane to prevent further initiation until the current replication process is at least a third completed. In DNA repair, methylation sites allow recognition of the template strand for mismatch repair by MutS–MutL–MutH. These proteins locate mismatches and use the methylation site to determine the template strand.
15. The meiotic homologous recombination (MHR) process exerts control over where strand breaks occur, whereas DNA damage-induced, double-strand breaks might occur anywhere within the chromosome. During MHR, the break is catalyzed by Spo11, with ends that are fairly easily repaired using Spo11, Rad51, and Dmc1. DNA-damaging breaks yield DNA ends that must be extensively processed, often involving resection of DNA to produce repairable ends, repaired by Rad51, BRCA1, and BRCA2.
16. Initially the phage binds the cell and injects its DNA genome into the cell. The DNA is circularized and integrated into the host chromosome using integrase and integration host factor. Once integrated, bacterial replication passively replicates the phage DNA as well. If the cell is exposed to a high dose of UV, a lytic cycle is initiated due to the DNA damage: The prophage excises from the host DNA using integrase, integration host factor, and Xis (an excision-specific phage protein), then replicates many times and produces phage proteins. Phage virions assemble and cell lysis releases the virions.
17. V(D)J recombination produces specific antibody-heavy chains. With 51 V segments, 27 D segments, and 6 J segments, more than 8,000 different heavy chains are possible by combining one of each segment. Pairing with a light chain produces even greater potential for variety. The process involves first combining D and J segments, with all intervening segments removed. Next, the V segment is combined with the DJ product, once again removing intervening segments. Removal of the remaining V and J segments is not done on a DNA level, but rather by splicing the mRNA.

CHALLENGE PROBLEMS

1. The average Okazaki fragment is 2,000 bases; the *E. coli* chromosome is 4,638 kb. Therefore, ~2,319 Okazaki fragments form per replication.
2. The difference would basically occur following the actual polymerization process. Following the core polymerase, a pair of helicases would be required to separate the template and daughter strands, followed by a protein with the ability to rejoin the template strands and move the daughter strands together for double-strand formation. Mismatch repair would need to occur before this separation of the template–daughter strands. Thus, conservative replication would require increased complexity over the semiconservative process.
3. Any decrease in the error rate of HIVRT would result in an overall decrease in the rate of viral mutations. This would probably not be an advantage for the virus, as the relatively high error rate often helps to thwart the use of antiviral drugs.
4. DNA polymerase has two Mg^{2+} ions complexed to active site aspartate residues. When a dNTP binds, the Mg^{2+} withdraws electron density from phosphoryl groups, making the α phosphate more susceptible to nucleophilic attack by the 3′-hydroxyl. Electron density is also withdrawn from the β and γ phosphoryl groups, making the pyrophosphate a better leaving group. Absence of Mg^{2+} would significantly decrease DNA polymerase activity.
5. Because helicase requires 1 ATP per two bases, it requires 2.3×10^6 molecules of ATP to replicate 4.6×10^6 bases.
6. The average Okazaki fragment is 2,000 bases and synthesized in 2 seconds (1,000 bases/second). A typical primer is about 10–12 bases; thus, primase would need a rate of 5–6 nucleotides/second to avoid being the rate-limiting step of replication.
7. Yeast cells contain an average of 80 copies of repeats about 10 nucleotides in length, so the average maximum telomere is 800 nucleotides. If each Okazaki fragment is about 200 nucleotides in length, then 200 nucleotides are lost from the telomere per replication; telomeres would disappear after four replications.
8. The results suggest the suspected mutagen is mutagenic without liver metabolism, and even more mutagenic after metabolism by liver enzymes.
9. The guanine would become 8-hydroxyguanine and tautomer, 8-oxoguanine. These base pair with adenine, resulting in a G-A mismatch that is not likely to be repaired by DNA repair mechanisms. After a second round of replication, the result is a G→C substitution in one of the daughter strands.

First round of replication:	5′-AAC*G*C-3′	5′-AACGC-3′
(8-hydroxyguanine = *G*):	3′-TTG**A**G-5′	3′-TTGCG-5′
Second round of replication:	5′-AAC**C**C-3′	5′-AACGC-3′
	3′-TTGAG-5′	3′-TTGCG-5′
	5′-AACGC-3′	5′-AACGC-3′
	3′-TTGAG-5′	3′-TTGCG-5′

10. 5′-ACG**TC**AG**TT**ACGTA**CT**GACGT. The TC photoproduct is likely to be a (6-4) photoproduct, which can stall replication forks if not repaired. The TT and CT photoproducts would most likely be pyrimidine dimers and, if unrepaired, would most likely result in adenine on the daughter strand. The TT photoproduct would not result in a mutation, but CT would result in a mismatch, with high potential for mutation, depending upon when repair occurs. If the daughter strand were used to correct the lesion, then the C would be converted to a T.
11. Because Xis is required for λ phage DNA excision, a mutation rendering Xis nonfunctional would prevent excision and prevent initiation of a lytic cycle.

12. This would prevent degradation of the RNA–DNA hybrid formed by reverse transcriptase, which would prevent the production of the double-stranded DNA that is required for insertion into the host genome and thus prevent an infection by HIV. However, an RNase inhibitor might also prevent degradation of human RNA, reducing the recycling of RNA nucleotides that provides a cell with building blocks for RNA. This would place tremendous energy demands upon the cell.

Chapter 21

REVIEW QUESTIONS

1. They are mRNA, tRNA, and rRNA, essential in all organisms for converting the genetic code into proteins.
2. Ribosomes are essential for protein synthesis, and rRNA is a structural and catalytic component. Each ribosome translates only one mRNA at a time; so the more ribosomes present, the more protein synthesis. More complex organisms (that is, yeast, humans) have a greater number of proteins expressed and so require more active ribosomes than organisms such as bacteria.
3. The antisense strand is used as the template for transcription to make a complementary strand. The mRNA sequence would be

   CGCGGAUCCUUGAAUUCUAAAUAAACC
   AUUUACCACCAUGACC

4. Co-transcriptional folding is necessary because of the size of the transcripts; folding of the 5′ end as soon as it exits the polymerase decreases the number of structural variations possible making it less likely that the RNA would be trapped in an unfavorable conformation. The secondary structure also generates binding sites for ribosomal proteins, and such interactions during transcription help "lock" the structures into place and protect the RNA.
5. The –35 and –10 boxes are underlined. Transcription starts at the ATG after the –10 box (in gray).

   CAGGCGATGATCAAATCTCCGTTGTACTT
   TGTTTCGCGCGTTGGTATAATCGCT
   GGGGTCAAG**ATG**AGT

6. Bacterial σ factors are transcription factors that direct RNA polymerase to specific gene promoters. The holoenzyme can interact with any DNA strand, but σ increases holoenzyme affinity for –10 and –35 boxes. Because bacteria use different genes to respond to changing nutrient and environmental conditions, they have σ factors that preferentially increase transcription at specific sets of these genes as needed.
7. There are two Mg^{2+} ions in the active site of RNA polymerase that interact with the negatively charged oxygens on the α and β phosphoryl groups of the incoming NTP and that lower the pK_a of the 3′-OH of the last ribose in the existing transcript. These interactions facilitate nucleophilic attack by positioning the NTP and assisting deprotonation of the 3′-OH.
8. Rho-dependent termination involves destabilization of the transcription bubble by Rho binding to a C-rich sequence in mRNA, resulting in transcriptional pausing and termination. Rho-independent termination also destabilizes the interaction between RNA polymerase and the transcript, however structural elements within the mRNA interfere with the transcription bubble: Conserved sequences at the 3′ end of the transcript fold into a hairpin loop leading to release of the polymerase.
9. TATA-less promoters often contain the downstream promoter element that may interact with TBP.
10. Phosphorylation of the conserved heptapeptide repeat in the CTD is important because it mediates interaction of capping and splicing factors and polymerase activity. Certain serine residues are phosphorylated during initiation and elongation, but upon termination the CTD is dephosphorylated, facilitating release of the RNA polymerase II complex from the nascent transcript.
11. The 5′ m^7G cap and poly(A) tail protect mRNA from digestion. During mRNA decay, removal of either (or both) allows exonucleases to completely digest the mRNA. Other RNA decay mechanisms exist in which the transcript is first cleaved within the sequence, but exonuclease digestion is more common.
12. Most biological reactions require a catalyst to occur in a physiologic time frame suitable for the organism. The RNA world hypothesis relies on the catalytic ability of RNA. Hammerhead ribozyme and group I introns are examples of catalytic RNA that can function without the assistance of any protein components, suggesting that life is possible without proteins.
13. In both types of introns, splicing is initiated by a nucleophilic attack on the 3′ splice site. In group I introns this attack occurs from an essential guanosine cofactor, whereas in group II introns it is initiated from a conserved adenosine in the branch site between the 5′ and 3′ splice sites. The released group II intron has a lariat structure, whereas the released group I intron is linear. The group II intron mechanism closely resembles a spliceosome-catalyzed reaction, and there are also sequence similarities between the splice and branch sites.
14. Base modifications in tRNA make the correct tertiary structure possible and enable specific binding between tRNAs and their aminoacyl-tRNA synthetases, as primary sequence and secondary structure are conserved in all tRNAs and cannot be used for specific interactions. By contrast, rRNA sequence and structure vary enough to be used for specific interactions.
15. The U1 and U2 snRNAs in the U1 and U2 snRNPs recognize and bind the 5′ splice site and branch point through base pairing.
16. Point mutations can cause alternative splicing through elimination of a splice site (mutation in one of the conserved U1 or U2 recognition sequences) or introduction of a cryptic 5′ or 3′ splice site, leading to splicing within an exon.
17. RNAi and miRNA differ primarily in the structure of the initiating double-stranded RNA and in the mechanism of translational repression. The siRNAs in RNAi are perfectly matched double-stranded RNA fragments, whereas miRNAs may contain loops and bulges. Thus, siRNAs are generally a perfect match to their target, leading to mRNA cleavage, but miRNAs are often an imperfect match with mRNA, resulting in translational repression.

CHALLENGE PROBLEMS

1. Eukaryotic and prokaryotic promoters have different sequences for RNA polymerase, and so the bacterial σ factor would most likely not bind the eukaryotic promoter sequence, thus RNA polymerase would not associate with the DNA, preventing transcription and thus protein expression. Cloning the gene behind a bacterial promoter sequence could solve this problem.
2. If ribosomes were the source of genetic material, the bacteriophage would synthesize its own ribosomes soon after infection as a way of propagating itself. This would occur in the normal medium, and so these ribosomes would not contain heavy isotopes. However, if ribosomes were only sites of protein synthesis, then the viral mRNA would be translated by bacterial ribosomes, which were synthesized while the heavy isotopes were present. The presence of isotopes in ribosomes enabled Jacob and Meselson to confirm that ribosomes were not the viral genetic material.

3. Continued association with σ factor would make RNA polymerase unable to continue down the template. Dissociation of σ factor increases the affinity of the holoenzyme for general DNA, enabling the enzyme to release from the –35 and –10 boxes and move along the template.
4. RNaseP recognizes the secondary structure of tRNA and rRNA rather than the primary structure. Such structures form co-transcriptionally, so the primary transcript is cleaved on the basis of the presence of these structures and not any individual sequences.
5. The structure of amylopectin is similar to that of glycogen, as both contain an α(1→4) chain of glucose molecules with α(1→6) branches. The spacing between branches varies, so it is reasonable to conclude that starch-branching enzymes and glycogen-branching enzymes that catalyze formation of the same bond are similar. If siRNA against a starch-branching enzyme was transmitted to mammals, it could result in repression of a glycogen-branching enzyme, and the cell would store only straight-chain glycogen, reducing the rate of glycogen breakdown and the capacity to store glycogen. These effects would negatively affect ATP production and could lead to symptoms similar to those observed for glycogen storage disorders.
6. Bacterial RNA polymerase is highly conserved; therefore, an inhibitor of its activity would likely affect all bacteria, meaning that the antibiotic can be used without knowing the species of bacteria causing the infection, an advantage because most bacterial infections require treatment as soon as possible. The structure of RNA polymerase is conserved from bacteria to eukaryotes, but these enzymes do not share a high degree of sequence conservation. The interaction between the antibacterial molecule and RNA polymerase depends on specific base contacts, and differences at the primary sequence level means that the eukaryotic RNA polymerases do not form the same contacts, and therefore the inhibitor has no effect.
7. Both are small noncoding RNAs with roles in RNA processing found in protein complexes in which they are critical for activity and for binding to the target RNA. They differ in their cellular location and RNA target. Small nuclear RNAs are part of the snRNP complexes of the spliceosome, located in the nucleus. Each spliceosome subunit (that is, U1, U2, and so forth) is named after the snRNA bound within the complex. The U1 and U2 snRNAs are responsible for binding the intron sequences. The snoRNAs are located in the nucleolus: The two main classes of snoRNAs facilitate base modification of rRNAs and bind to the ribonucleoprotein complexes that assemble on rRNA as it is transcribed.
8. The crystal structure of mouse guanylyltransferase (Mce1) bound to an 18-AA CTD phosphopeptide shows that the phosphoryl group bound to Ser5 interacts with Arg386, Arg330, and Lys331. The presence of these amino acid contacts increases the affinity of Mce1 for the RNA polymerase II CTD. Elimination of these contacts would not completely prevent interaction because Mce1 contains additional residues (Cys383, Val372, and Cys375) whose interaction does not depend on the phosphorylation state of Ser5. In the crystal structure, the Ser2 phosphoryl group is pointed away from the binding pocket and cannot contribute to binding. But if pointed toward the binding site, it still would not increase the affinity because interactions between the Ser2 phosphoryl group and Cys375 would replace existing van der Waals interactions between Cys375 and C_α of Ser2 and C_δ of Pro3, not add new ones.
9. U2 and U6 snRNA base pair, leaving a bulge in the center of the structure. Residues in this bulge are then free to base pair with the mRNA intron. U2 snRNA binds the conserved RY sequence in the branch point, leaving the catalytic adenine unpaired and extruding from the structure. The 5′ splice site is stabilized through base pairing with the conserved U6 snRNA sequence ACAGAGA, bringing the branch-point adenine in close proximity to the 5′ splice site for the first transesterification.
10. Both DNA and RNA polymerase read 3′ to 5′, resulting in a 5′ to 3′ polymerization reaction. When both replication and transcription complexes proceed down the same region of DNA, the DNA polymerase can overtake RNA polymerase because transcription is much slower than replication; this results in a codirectional collision.
11. Poly(A) polymerase interacts with the 2′-OH in the three adenines at the end of the existing poly(A) tail (positions –1, –2, and –3). Lack of 2′-OH in a single-stranded DNA template would decrease the affinity for the substrate.
12. RNA synthesis requires NTPs; the first nucleoside added has three phosphoryl groups. The reaction to add each subsequent nucleotide is initiated by a nucleophilic attack on the α phosphate, resulting in cleavage and release of PP_i.
13. Strong bacterial promoters differ only slightly from the conserved sequences at the –35 and –10 boxes, whereas weak promoters can contain several base pair changes. These are the sites of σ-factor binding, so the closer the –35 and –10 boxes are to the conserved sequences, the stronger the association between RNA polymerase and the promoter. Therefore, strong promoters are more likely to be bound by σ factor and transcribed than weak promoters.

Chapter 22

REVIEW QUESTIONS

1 Crick did not support these models because in eukaryotes, DNA and protein synthesis occur in different cellular compartments. Also, on a physicochemical basis, nucleotide structure seemed insufficiently complex to act directly as a template for 20 amino acids.
2. They are Pro (CCA), Gln (CAA), or Lys (AAA).
3. The wobble hypothesis states that noncanonical base pairs can exist between the third (3′) position in the codon and the first (5′) position in the anticodon. Wobble base pairing explains how some mRNA codons that differ by the nucleotide at the third position can bind to the same tRNA, facilitated by the presence of modified nucleotides such as inosine at the wobble position. Inosine can form hydrogen bonds with A, C, or U.
4. tRNA forms a cloverleaf secondary structure with three arms: D, TΨC, and anticodon. The top stem of tRNA is the acceptor site for amino acid attachment. The D and TΨC loops further interact to form the L-shaped tertiary structure.
5. Aminoacylation is a two-stage reaction: First the amino acid is activated through the formation of an aminoacyl-AMP moiety; second, the amino acid is transferred to the acceptor stem of tRNA. Class I aminoacyl-tRNA synthetases transfer the amino acid to the 2′-hydroxyl on the last conserved adenosine residue in the acceptor stem. Class II enzymes transfer the amino acid to the 3′-hydroxyl of this nucleotide residue.
6. The aminoacyl (A) site is the entry point during elongation for aminoacylated tRNAs in the ribosome. The peptidyl (P) site is where peptide bond formation occurs between the tRNA-bound amino acid and the growing polypeptide chain attached to the P-site tRNA. The uncharged tRNA leaves the ribosome from the exit (E) site.

7. When a termination codon reaches the A site, a release factor binds the A site, preventing further tRNA binding and leading to dissociation of the ribosomal subunits.
8. Puromycin is not a useful antibiotic because it affects both bacterial and eukaryotic protein synthesis. It would lead to cell death in humans as it would in bacteria.
9. Proteins that contain a specific N-terminal signal sequence are directed to the endoplasmic reticulum during translation, when the N-terminal signal sequence is recognized and bound by the signal recognition particle. Translation halts until the ribosomal complex is associated with the endoplasmic reticulum. Signal recognition particle binding to a receptor on the endoplasmic reticulum membrane allows the nascent polypeptide to enter and translation to resume.
10. Methylation and acetylation are not found on the same regions of histones because they have opposite effects on histones. Methylation leads to tighter compaction of histones, often found in transcriptionally inactive regions. Acetylation is often found in histones that are not as tightly condensed and are therefore more accessible by the transcriptional machinery.
11. The three types are COPI, COPII, and clathrin. COPI vesicles mediate transport within the Golgi apparatus; COPII vesicles mediate transport between the endoplasmic reticulum and the Golgi apparatus; and clathrin vesicles mediate transport between the Golgi apparatus and plasma membrane.

CHALLENGE PROBLEMS

1. Gamow concluded that three nucleotides were the minimum number required for each codon because a two–nucleotide codon would result in only 4^2, or 16, possible combinations. Because there are 20 standard amino acids, a two–nucleotide codon would have required more than one amino acid to be associated with the same codon. However, a codon of three nucleotides gives 4^3, or 64, possible combinations.
2. Khorana's method of nucleotide synthesis was the first procedure that allowed directed synthesis, not simply random combinations of nucleotides. This was essential to cracking the genetic code, as it allowed Khorana and others to test specific triplet combinations in filter-binding and *in vitro* translation experiments.
3. Normally, AUGCACAGU is translated to produce the tripeptide Met-Thr-Gly. The sixth or ninth nucleotides can be changed without changing the amino acid sequence, as there are four codons for Thr and Gly that only differ at the third position. It would not be possible to change the third nucleotide because there is only one Met codon.
4. This is probably due to its ancestral origin. It is widely believed that mitochondria evolved from bacteria taken up by early nucleated cells. Because mitochondrial DNA has different evolutionary origins from chromosomal DNA, it contains a separate protein translation system, including tRNA genes.
5. All tRNA molecules must have similar structure in the region where they bind to elongation factors and the A, P, and E sites on the ribosome. These interactions are not dictated by the particular amino acid charged to the tRNA, so all tRNAs have similar structures.
6. The initiator tRNA charged with methionine begins the protein sequence. Because Met is the first amino acid, it must bind directly to the P site so that a second tRNA can bind to the A site and the first peptide bond can be formed.
7. A bacterial mRNA may encode more than one polypeptide; each translational start site has a corresponding Shine–Dalgarno sequence.
8. Translation initiation is different in prokaryotes and eukaryotes because the structure and organization of mRNA is different. Transcription and translation occur simultaneously in prokaryotes, and the mRNA generally does not contain significant secondary structure. Eukaryotic mRNA travels to the cytosol from the nucleus, so it is modified to ensure its stability [5′ cap and 3′ poly(A) tail] and may contain secondary structures. The size of the 5′ untranslated region can vary because there are additional factors in eukaryotic systems that bind the 5′UTR in the pre-initiation complex.
9. There are no nuclear transport complexes analogous to the ER signal recognition particle–receptor complex that work co-translationally.
10. It is a ribozyme because it contains a catalytic RNA (23S rRNA in *E. coli*) that catalyzes the formation of a peptide bond between amino acids.
11. Lipid modifications can be used for strong association of a protein with a membrane. As soluble proteins fold, the hydrophobic regions are mostly sequestered in the interior of the protein, while hydrophilic regions remain exposed to the polar, aqueous environment of the cell. Membranes have hydrophobic fatty acid tails arranged in the center of the membrane. The addition of lipids to the exterior of a protein adds hydrophobic regions that can insert within the membrane and therefore firmly associate with the membrane.

Chapter 23

REVIEW QUESTIONS

1. (a) *lac* operon, yeast galactose genes, *Drosophila eve* gene; (b) LexA protein; (c) metabolic regulation through phosphorylation, methylation, or acetylation.
2. Trans-acting factors can operate on DNA sites anywhere in the genome; not limited to the chromosome that encodes them. Cis-acting sites are binding sites for trans-acting factors, generally located at or near the genes on which the trans-acting factors act. The actions of trans-acting factors are highly specific because of these sites.
3. Specific DNA binding usually involves interaction of the protein with the major groove of the DNA at the site. The edges of the base pairs present different patterns of hydrogen-bond donors and acceptors between the major and minor grooves. In the major groove, all four base pairs have distinct patterns; in the minor groove, AT and TA patterns cannot be distinguished, nor can GC and CG. Hence the major groove is more suitable for sequence-specific interactions. Additionally, protein α helices fit into the major groove, but are too large for the minor groove. This means that some side chains projecting from α helices can make specific contacts with the edges of base pairs in the major groove. Some protein–DNA interactions involve less direct contact, such as a well-positioned water molecule forming hydrogen-bond networks between the protein and the DNA, or even van der Waals interactions, but the latter are usually less specific.
4. (a) Yeast GAL4; (b) mammalian Oct4 and Sox2; (c) yeast PHO4; (d) CRP, Trp repressor; (e) lac repressor.
5. Positive regulation: gene expression is stimulated by the action of regulatory proteins; prokaryotes: cAMP–CRP in *lac* operon, CI activation of λ *cI*; eukaryotes: GAL4 activation of GAL genes, Bicoid and Hunchback activation of *eve*. Negative regulation: gene expression is inhibited by the action of regulatory proteins; prokaryotes: lac repressor in *lac* operon, Trp repressor in *trp* operon, LexA in SOS regulon; eukaryotes: TR inhibition in

absence of ligand by recruited SMRT complex, HPI inhibition of DNA, Giant and Krüppel inhibition of *eve* stripe 2.

6.

Prokaryotes	Eukaryotes
No nucleus	Eukaryotes have a nucleus
Transcription and translation can be coupled	Transcription and translation in separate compartments
One RNA polymerase	Three types of RNA polymerase
RNA polymerase binds promoter directly	Transcription factors bind promoters (usually) and recruit RNA polymerase II
Very little RNA processing	Extensive RNA processing, and splicing
No chromatin	Chromatin
Activators and repressors work directly on RNA polymerase	Activators and repressors work indirectly by effects on chromatin and recruiting other factors
Enhancers rare	Enhancers very common, especially in higher eukaryotes
Default state: genes are accessible	Default state: are inaccessible due to chromatin

7. In prokaryotes, activators generally interact directly with RNA polymerase, recruiting it to a promoter or increasing the rate of open complex formation. In eukaryotes, activators act on other transcription machinery components, nearly always recruiting proteins to particular genes where they open up the chromatin, or recruit components of the general transcription machinery.
8. In prokaryotes, repressors affect RNA polymerase directly; usually they block access to the promoter or affect later steps in transcription. Repressors in eukaryotes work indirectly to set up repressive chromatin states. They recruit HDACs to deacetylate histones, or they recruit histone methyltransferases, which in turn recruit HP1.
9. The first role is to antagonize repressive chromatin states; for example, by recruiting HATs to acetylate histones and open up chromatin, enabling acetylated histones to recruit more components. Activators can also recruit chromatin remodeling complexes to move or remove nucleosomes, exposing other cis-acting sites for additional factors. The second role is to recruit components of transcription machinery, including the large protein complexes mediator and TFIID. Function of these complexes requires the promoter region to be relatively free of nucleosomes.
10. It is a long-lasting pattern of gene expression, almost always defined as passing from one generation to the next. It does not involve a change in DNA sequence, but may involve modification of cytosine to form 5-methylcytosine.
11. When the binding of one regulatory protein at its site increases the occupancy of another molecule at a different site, usually close to the first one. This usually involves direct interaction between the two bound molecules but it can involve dimerization in solution, followed by dimer binding, or more indirectly by one protein making the binding site for the second one available (e.g., fluctuations "breathing" in nucleosome structure). In cooperative DNA binding, occupancy of the second site goes from near zero to nearly 100% over a very narrow range of concentrations of the DNA binding molecules. This confers switch-like behavior on the regulatory circuit, and can help stabilize epigenetic states.
12. Regulatory systems save energy by ensuring that gene products are made only when needed. If the product of the biosynthetic pathway is present, there is no need to make more; if a metabolite is not available, there is no need to make enzymes to catabolize it. Examples: Anabolic pathway, tryptophan or histidine biosynthesis. Catabolic pathway: lactose metabolism in *E. coli*, galactose metabolism in yeast.
13. Pioneer factors are the first trans-acting factors to bind to an enhancer or promoter, even in the presence of nucleosomes in these regions. Binding can lead to availability of other cis-acting sites thereby setting in motion a long chain of events that assemble a pre-initiation complex and subsequent transcription. They may bind because their binding site is not covered by a nucleosome or because the nucleosomes have variable positions and the site is sometimes free. Alternatively, they bind to DNA while it is wrapped around the histone core, or when their site becomes transiently available due to fluctuations in nucleosome structure. Pioneer factors include yeast PHO4, and Oct4 and Sox2, which can convert a range of cell types to iPS cells.
14. iPS cells are induced pluripotent stem cells that have the potential to develop into almost any cell type, given the right conditions and stimuli. They can be produced from many cell types by adding the genes for four transcription factors, Sox2, Oct4, Klf4, and c-Myc, either as free DNA or as viruses. A small fraction of cells treated in this way are converted to iPS cells.

CHALLENGE PROBLEMS

1. It would not work in eukaryotes because this mechanism depends on transcription and translation occurring simultaneously; in eukaryotes these processes occur in different cellular compartments.
2. Cytosine residues are deaminated at a very low rate to form uracil, which will lead to a uracil-guanine (U-G) base pair. Because U is not a normal component of DNA, this is quickly repaired to a C-G base pair by uracil *N*-glycosylase and subsequent events. Deamination of 5-MeC would lead to formation of a T-G base pair, and the cell would not rapidly repair this, as there is no way to be sure which base is the correct base. If DNA replication occurs before the mismatch is repaired, it fixes a C-to-T transition mutation in one of the progeny duplexes.
3. (a) Mutation in region 4 so it cannot pair with region 3: would be constitutive, because no terminator is formed. (b) Mutation in region 3 so it can pair with region 4 but not with region 2: would be noninducible; antiterminator cannot form in the absence of Trp-tRNA^{Trp}, so terminator would always form. (c) Change of the two Trp codons to glycine codons: noninducible, because the ribosome would not pause at the Gly codons in the presence of low Trp-tRNA^{Trp}. It might pause if glycine levels are low, but glycine is rarely low. (d) Change of the AUG codon of the leader peptide to AUA: noninducible, because the ribosome would not be able to stabilize the antiterminator. (e) Combination of the first mutation (a) with each of the other three (b, c, or d). Constitutive, because if the terminator cannot form, it does not matter what happens upstream. For the mutation (a) and (d) combined: pausing at the 1–2 stem–loop is too brief to reduce the rate of transcription significantly. This latter is an "epistasis" test, often used to order the events in a regulatory pathway.
4. To validate your hypothesis, you would use recombinant DNA to place the proposed enhancer upstream of a reporter gene in

Drosophila embryos. Several constructs should be prepared, with the enhancer in either orientation and at a variable distance from the promoter for the reporter. If the DNA region includes an enhancer, you should see expression of the reporter. See Levine and colleagues' study to analyze the *eve* stripe 2 enhancer (see Figure 23.54).

5. (a) A lysogen in a λ plaque can be superinfected by other λ virions, but the lysogen makes CI, which will bind to the lytic promoters of the incoming viral DNA, repressing them and preventing them from growing lytically. This is termed "immunity." (b) Clear plaques have mutations in the cI gene; they cannot make CI and cannot establish or maintain the lysogenic state. Clear plaques led to the discovery of this gene and two others; hence the name of the gene *cI* as clear plaque, class I.
6. Another possible explanation is that the purified protein is an alternate σ factor. Sigma determines the promoter binding specificity, and there are multiple σ factors. To distinguish these models, the *in vitro* transcription system could use an RNA polymerase without the usual σ^{70} factor, called the "core" RNA polymerase. The new-sigma model predicts that the target gene would be transcribed; the activator model predicts no transcription.
7. Binding of the green factor would enhance the ability of the red factor to bind, as its binding site is now accessible, not bound up in the nucleosome. This is a form of cooperativity, which can also involve protein–protein interactions between the two proteins bound to DNA or in solution.
8. No, you cannot be certain what the effect of protein binding will be. Numerous examples indicate that the effect of a bound trans-acting factor depends on its context. Bound bacteriophage λ CI protein and CRP–cAMP both can act as either activators or repressors. The mammalian Oct4–Sox2 heterodimer activates several ES-specific genes and works to repress developmental genes.
9. Three models have been suggested for this correlation, which are not mutually exclusive. First, acetylation neutralizes the charge on the lysine side chain, which may weaken interaction of the histone tails with the DNA in a nucleosome. Second, charge neutralization may weaken interaction between nucleosomes, disrupting higher-order structure. Third, acetylated histones serve as docking sites for other proteins and complexes. Although these interactions do not necessarily favor activation, this is found to be true in many cases.
10. GAL3 and GAL4 mutants will not express the target genes in the presence of galactose. A GAL80 mutant and a GAL3–GAL80 double mutant will express the target genes in the absence of galactose.

Glossary

2,4-dinitrophenol A chemical that uncouples electron transport from ATP synthesis by disrupting the proton gradient.

3′ poly(A) tail The addition of adenine-containing nucleotides to the 3′ end of mRNA, which contributes to the stability and translational efficiency of mRNA.

3′UTR A region of a eukaryotic gene that is not translated but instead contains sequences necessary for the termination of transcription by RNA polymerase.

5′ cap See **7-methylguanylate cap (m^7G cap)**.

5′UTR A region of a eukaryotic gene that is not translated but instead contains sequences that when transcribed into RNA will facilitate interaction with the protein translational machinery.

5-fluorodeoxyuridine An analog of 5-fluorouracil and anticancer drug that irreversibly inhibits thymidylate synthase.

5-fluorouracil A pyrimidine analog and anticancer drug that irreversibly inhibits thymidylate synthase.

5-methylcytosine A methylated form of cytosine whose presence in the regulatory region of genes is associated with decreased rates of transcriptional initiation.

7-methylguanylate cap (m^7G cap) A modification to the 5′ end of mature eukaryotic RNA that facilitates interactions between the mRNA and the ribosome.

A

α amino acid A small molecule containing both a primary amino group and a carboxyl group; the building blocks of proteins.

α/β barrel See **TIM barrel fold**.

α helix A right-handed helical conformation of a polypeptide chain; one of the most common elements of protein secondary structure.

α-ketoglutarate dehydrogenase A multi-subunit complex that catalyzes the conversion of α-ketoglutarate to succinyl-CoA. It is a key regulator of the citrate cycle.

α-lipoic acid (lipoamide) A coenzyme carrier of acetyl groups; it contains a reactive thiol group.

abasic site A ribose-5′-phosphate within a DNA sequence.

ABO blood groups A system for typing blood based on the presence of GalNAc and/or galactose bound to the O antigen of glycoproteins and glycolipids found on the surface of red blood cells.

acetyl-CoA carboxylase A cytosolic enzyme that carboxylates acetyl-CoA to form malonyl-CoA; this is the first step in fatty acid synthesis.

acid dissociation constant (K_a) The equilibrium constant for the dissociation of an acid:

$$K_a = \frac{[H^+][A^+]}{[HA]}$$

acidosis A metabolic condition of low pH (usually of the blood), reducing the body's ability to buffer H^+.

acinar cell A cell within the islet of Langerhans that secretes digestive proteases destined for the duodenum.

actin An abundant structural protein in animal cells that functions in molecular cables, mediating muscle movement through calcium-stimulated muscle contraction.

activation energy The difference in energy between the ground state of a reactant and the transition state in a reaction.

active transporter protein A membrane transport protein that requires energy, usually from ATP hydrolysis or an ionic gradient, to induce protein conformational changes that open or close a gated channel.

acyl carrier protein (ACP) A protein with a serine-linked phosphopantetheine group that serves as the attachment site for the intermediates during fatty acid chain synthesis.

adenine (A) A purine base found in DNA and RNA that hydrogen bonds with thymine and uracil, respectively. It is also the base in ATP, the key energy currency molecule in cells, and is a component of NADH, $FADH_2$, and CoA.

adenosine deaminase deficiency An autosomal recessive mutation in the gene for adenosine deaminase, a purine-degradation enzyme that converts adenosine to inosine.

adenylate cyclase An enzyme that produces the second messenger cAMP from ATP.

adenylate system A group of several phosphoryl transfer reactions that interconvert ATP, ADP, and AMP.

adenylylation A covalent modification of an enzyme by addition of AMP that serves to regulate catalytic activity.

adipocyte A fat cell; site of triacylglycerol storage.

adipokine Peptide hormones secreted by adipose tissue that regulate metabolic homeostasis.

adiponectin An adipokine hormone that modulates glucose and lipid metabolism in skeletal muscle and liver by stimulating AMP-activated protein kinase signaling.

aerobic respiration A set of metabolic processes and reactions that uses oxygen to generate ATP.

affinity chromatography A column chromatography method that exploits specific binding properties of the target protein to separate it from other cellular proteins that lack this binding function.

alanine A small and abundant hydrophobic amino acid.

alanine–glucose cycle A mechanism for transporting amino groups from muscle to liver in the nontoxic form, alanine.

albinism A group of inherited conditions characterized by the complete or partial absence of skin, hair, and eye pigments.

albumin Highly abundant serum protein that, among other functions, carries hydrophobic molecules such as free fatty acids in the blood.

aldose A monosaccharide that contains an aldehyde functional group.

alkalosis A metabolic condition of high pH (usually of the blood), reducing the body's ability to buffer H^+.

alkaptonuria Also called black urine disease. A genetic disease of phenylalanine catabolism, in which homogentisate-1,2-dioxygenase is defective causing homogentisate to accumulate.

allantoic acid A degradation product of uric acid produced in nonprimate mammals.

allantoin A degradation product of uric acid produced in bony fish.

allolactose A β-1,6–linked disaccharide that is made by β-galactosidase. It serves as an inducer of lac operon gene expression by binding to the lac repressor.

allopurinol A purine that is a competitive inhibitor of xanthine oxidase. It is prescribed for the treatment of gout.

allostery Refers to interactions that occur within a protein between spatially distant sites.

Ames test A test for evaluating the mutagenic potential of a chemical by looking for bacterial histidine revertants that are able to grow on a minimal defined medium lacking histidine.

amino acid Nitrogen-containing molecules (substituted carboxylic acids) that function as the building blocks for proteins and precursors to numerous nitrogen-containing biomolecules.

amino acid side chain A functional group attached to an amino acid; the 20 common functional groups or side chains differ in chemical structure, polarity, and charge.

aminoacyl site (A site) The binding site on a ribosome for all aminoacyl-tRNA molecules except the initiator tRNA.

aminoacyl-tRNA synthetase A group of enzymes that catalyzes the covalent attachment of amino acids to their respective tRNA molecules.

aminopeptidase A group of enzymes that remove amino acids from the amino terminus of proteins and peptides.

ammonia assimilation The process of incorporating NH_4^+ into the amino acids glutamine and glutamate.

AMP-activated protein kinase (AMPK) An AMP-sensitive regulatory protein that functions as a key regulator of numerous metabolic pathways, including fatty acid synthesis.

amphibolic pathway A metabolic pathway that functions in both catabolism and anabolism; for example, the citrate cycle.

amphipathic Polar and nonpolar chemical properties contained within the same molecule.

amyloid plaques The buildup of amyloid protein around neurons, for example in regions of the brain where misfolded proteins accumulate in association with Alzheimer's disease.

amylopectin An α-1,6 branched form of starch.

amylose An α-1,4 linear form of starch.

anabolic pathway A metabolic pathway for the biosynthesis of biomolecules from smaller precursors.

anaplerotic reaction Reaction that replenishes key intermediates that have been shunted to other metabolic pathways.

Andersen disease A glycogen storage disease resulting from a defect in the liver glycogen branching enzyme that results in the formation of large linear glycogen molecules.

anemia Clinical term for reduced numbers of red blood cells per unit of blood, which results in decreased oxygen transport from the lungs to the tissues.

angstrom (Å) A unit of length equal to 10^{-10} meter typically used to describe the length of a chemical bond.

anomer One of two stereoisomers that differ in configuration at the anomeric carbon; for example, the cyclic molecule D-glucose exists as two anomers: α-D-glucose and β-D-glucose.

anomeric carbon The carbon molecule in a cyclic monosaccharide derived from the carbonyl carbon of the linear form of the monosaccharide.

anorexigenic Neurons that control processes resulting in a decrease in appetite and an increase in expenditure of energy.

antibiotic A molecule that can kill or inhibit the growth of certain microorganisms.

antibody A defense protein synthesized by a vertebrate's immune system that can bind to an antigen.

antigen A foreign (non-self) molecule within an organism.

antiparallel β sheet A β-sheet structure in which adjacent β strands are oriented in opposite directions with regard to amino to carboxyl termini.

antiporter An active membrane transport protein that can move molecules across a membrane in opposite directions.

apoenzyme An enzyme without its cofactor.

apolipoprotein A protein associated with lipoproteins that promotes lipoprotein particle formation in the endomembrane system.

apoptosis Programmed cell death in which a cell causes its own death and lysis.

aquaporin A passive membrane transport protein responsible for transporting water molecules across hydrophobic cell membranes.

arginine An amino acid that is often positively charged at pH 7.

asparagine A hydrophilic amino acid that can both donate and accept hydrogen atoms to form hydrogen bonds.

aspartame The artificial sweetener aspartyl-phenylalanine-methyl ester.

aspartate An amino acid that is often negatively charged at pH 7.

aspirin A name for acetylsalicylate. It is an anti-inflammatory drug that is used to reduce pain, swelling, and fever.

association constant (K_a) An equilibrium constant for the binding of two molecules, such as protein and ligand forming a protein–ligand complex.

astrocyte Glial cells that surround neurons and are in close contact with the vasculature of the brain; forms the blood–brain barrier.

atherosclerosis Arterial blockage due to the accumulation of fibrous tissue—rich in cholesterol, fat, and foam cells—in arterial walls.

ATP binding cassette (ABC) transporter An active membrane transport protein that uses energy from ATP hydrolysis to drive large conformational changes and pump molecules across the membrane.

ATP synthase complex The enzyme complex responsible for the synthesis of ATP from ADP and inorganic phosphate using the energy of proton flow down a concentration gradient.

ATP/ADP translocase An antiporter membrane transport protein that exports ATP from the mitochondrial matrix while it imports ADP.

attenuation A bacterial mechanism of regulating transcription in which the conformation of a translated leader peptide regulates the transcription of downstream structural genes.

autocrine A hormone that functions over short distances to activate receptors on the same type of cell.

autofluorescent Biological structures that naturally emit light.

B

β-adrenergic receptor kinase (βARK) A kinase enzyme that phosphorylates the β2-adrenergic receptor to terminate the G protein–coupled receptor signal mediated by this receptor.

β-arrestin A transport protein that binds to phosphorylated G protein–coupled receptors to initiate internalization of the receptor by endocytic vesicles.

β clamp A DNA replication fork protein that facilitates DNA polymerase III attachment to DNA.

β-lactamase An enzyme produced by penicillin-resistant bacteria that inactivates penicillin by hydrolyzing the β-lactam ring in penicillin.

β-oxidation pathway A fatty acid oxidation pathway that removes two-carbon units from a fatty acid chain, producing $FADH_2$, NADH, and acetyl-CoA.

β pleated sheet See **β sheet.**

β sheet Also called *β pleated sheet.* A common secondary structure of proteins formed from β strands hydrogen-bonded together.

β strand A secondary structure of proteins consisting of an extended polypeptide chain with side chains positioned above and below it.

β turn A turn or loop in a polypeptide chain that connects two β strands in an antiparallel β sheet.

backbone The chain of carbon and nitrogen atoms in a polypeptide, exclusive of side chains.

bacteriophage A virus that infects bacteria; useful as a tool in nucleic acid biochemistry and molecular biology.

bacteroid A differentiated rhizobial cell within a plant host cell. The bacteroid fixes nitrogen into NH_4^+, which it shares with the host plant.

base excision repair A DNA repair mechanism that removes damaged bases creating an apurinic site, which is followed by repair synthesis.

base pair Two nucleotides in nucleic acid chains that are paired by hydrogen bonding, such as C-G, A-T, or A-U.

base stacking Stabilizing interactions between the aromatic rings of the nucleotide bases within the interior of the DNA helix.

Benedict's test A method for detecting the presence of reducing sugars in solution based on reduction of cupric ion (Cu^{2+}) to generate a red cuprous (Cu^+) oxide precipitate.

beriberi A disease caused by thiamine deficiency, resulting in anorexia, cardiovascular maladies, and neurologic problems.

bile acid Polar molecules derived from cholesterol that are secreted into the intestines where they emulsify dietary lipids, which aids in lipid absorption.

binding change mechanism A model of how the ATP synthase complex converts conformational changes (mechanical energy) into the production of ATP from ADP and inorganic phosphate.

bioavailability The amount of a nutrient or enzyme present in a cell that is capable of participating in a biochemical process.

biochemical standard conditions Reaction conditions of 1 atm pressure, 298 K temperature, 1 M concentrations of reactants and products, pH of 7, and concentration of water at 55.5 M.

biochemical standard reduction potential ($E°'$) A measure of the tendency of a compound to accept or donate electrons under the conditions of 1 M concentration, 25 °C, pH 7, and 1 atm pressure.

bioenergetics Energy conversion processes in biological systems, including transformation of solar energy into chemical energy and interconversion of chemical energy through oxidation and reduction of organic molecules.

bioinformatics The use of computational tools to probe and analyze large data sets of biological information, typically whole genomes or proteomes.

biologic Shorthand for biological medical product, it is any pharmaceutical that is prepared from a biological source rather than being synthesized in a laboratory.

biological membrane A physical barrier in a living system, most often consisting of nonpolar molecules with hydrophobic properties to partition aqueous compartments.

biomolecule An organic compound essential to living organisms; major types include amino acids, nucleotides, simple sugars, and fatty acids.

biosphere All the living organisms on Earth, considered as a whole.

biotin A small molecule that functions as a carrier of carboxyl groups (or CO_2) in enzymatic reactions.

blood–brain barrier A physical barrier formed by astrocytes between the blood vessels and tissues of the brain. Controls the movement of substances from the blood to the neural tissues of the brain.

body mass index (BMI) An estimation of total body fat in a person calculated as weight divided by the square of height:

$$\text{Body mass index (BMI)} = \frac{\text{Weight (kg)}}{[\text{Height (m)}]^2}$$

Bohr effect The dependence of oxygen binding in hemoglobin upon pH and carbon dioxide concentration.

bomb calorimeter A device in which a compound is combusted by a spark at constant volume in the presence of pure oxygen (completely oxidized); the amount of heat exchanged between the reaction chamber and a surrounding water jacket is a measure of enthalpy change.

branching and debranching enzymes Enzymes required for glycogen metabolism to mediate the formation (branching) and cleavage (debranching) of α-1,6 glycosidic bonds between glucose molecules.

bromodomain A common protein domain constructed of a bundle of α helices that binds to acetylated histone proteins.

buffer An aqueous solution that resists changes in pH because of the protonation or deprotonation of an acid–base conjugate pair.

bundle sheath cell Chloroplast-containing cells arranged in tight formation around the veins of a leaf.

C

C_4 pathway A carbon fixation process that averts the problems of photorespiration by isolating rubisco in oxygen-free bundle sheath cells and delivering CO_2 to rubisco in the form of malate and aspartate made in mesophyll cells.

Ca^{2+}-ATPase A P-type transporter responsible for transporting Ca^{2+} ions from the cytoplasm into the lumen of the sarcoplasmic reticulum to promote muscle relaxation.

calmodulin A signaling protein that binds with four Ca^{2+} ions, undergoing a large conformational change and then binding to and activating a wide variety of target proteins.

calorie (cal) A unit of energy; the amount of heat energy required to raise 1 gram of water by 1 °C using a calorimeter; a Calorie is 10^3 calories or 1 kilocalorie (kcal), which is equal to 4.184 kilojoules (kJ).

Calvin cycle The metabolic pathway of carbon fixation in which the CO_2-condensing step is catalyzed by the enzyme rubisco.

CAM pathway A carbon fixation process that collects CO_2 at night, storing it as malate. During the day, the stomata remain closed and the malate is converted to CO_2, which enters the Calvin cycle.

cAMP phosphodiesterase An enzyme that hydrolyzes cAMP to produce AMP.

cAMP receptor protein (CRP) A bacterial transcription regulatory protein that upon binding cAMP binds to specific DNA sequences and stimulates transcription.

capsule An outer layer, usually of polysaccharides, that surrounds some prokaryotic cells, particularly bacteria.

carbamoyl phosphate A small molecule that transfers NH_4^+ into the urea cycle by transferring a carbamoyl group to ornithine to form citrulline.

carbamoyl phosphate synthetase I An ATP-dependent urea cycle enzyme that forms carbamoyl phosphate from NH_4^+ and bicarbonate (HCO_3^-).

carbohydrate An organic compound of carbon, hydrogen, and oxygen that can be a simple sugar or a polymer of sugars; the typical formula is $(CH_2O)_n$, where n is at least 3.

carbon fixation The conversion of carbon dioxide into other organic compounds, particularly glucose; often considered as part of photosynthesis.

cardiac muscle Striated muscle that forms the contractile tissue of the heart.

carnitine acyltransferase I A mitochondrial outer membrane enzyme that converts fatty acyl-CoA molecules to fatty acyl-carnitine for transport into the mitochondria.

carnitine transport cycle The three-step process that translocates fatty acids across the inner mitochondrial membrane.

catabolic pathway A metabolic pathway that converts energy-rich compounds into energy-depleted compounds, releasing energy for the cell.

catalytic efficiency A quantitative measure of enzyme activity.

catalytic triad In serine proteases, a set of three amino acids that form a hydrogen-bonded network required for catalysis.

catecholamines A class of first messenger hormones derived from tyrosine.

cell extract An isotonic suspension of lysed cells, used for protein isolation and identification.

cell signaling pathway A linked set of biochemical reactions that are initiated by ligand-induced activation of a receptor protein and terminated by a measurable cellular response.

cellobiose A disaccharide made of two glucose residues joined by a β-1,4 glycosidic bond.

cellotetraose A tetrasaccharide of four glucose residues; the product of cellulase digestion of cellulose.

cellulase The enzyme that hydrolyzes cellulose into cellotetraose, a tetrasaccharide of four glucose residues.

cellulose A polysaccharide made of repeating units of the disaccharide cellobiose. Cellulose is the major component of plant cell walls and is thus the most abundant carbohydrate on Earth.

central dogma The description of information transfer in molecular biology, which flows from DNA to RNA to proteins.

centrifugation A process that separates particles of different sizes or densities by spinning solution samples in a rotor at high speeds.

centromere The region of connection between sister chromatids, composed of heterochromatin; the site of attachment for the mitotic or meiotic spindle.

ceramide The precursor of cerebrosides, gangliosides, and sphingophospholipids that consists of sphingosine covalently linked to a fatty acid.

cerebroside A type of sphingoglycolipid containing either glucose or galactose bound to the terminal hydroxyl group.

cGMP phosphodiesterase An enzyme that hydrolyzes cGMP.

chaperone A protein that facilitates the formation of stable three-dimensional structures through the process of protein folding.

chaperonin protein A cytoplasmic protein that assists in protein folding.

Chargaff's rule In DNA from any cell of any organism, the percentage of adenine equals the percentage of thymine (%A = %T) and the percentage of guanine equals the percentage of cytosine (%G = %C).

chemiosmosis The generation of ATP as a result of the flow of protons across a membrane in response to a concentration gradient.

chemiosmotic theory Peter Mitchell's theory that energy captured by coupled redox reactions in an electron transport system translocates protons across a membrane, which creates an electrochemical gradient that is used to drive ATP synthesis via proton flow through ATP synthase.

chiral center The atom within a molecule that makes that molecule chiral. A chiral molecule exists as two optical isomers, known as enantiomers.

chitin A linear polysaccharide consisting of *N*-acetylglucosamine units linked by β-1,4 glycosidic bonds. Chitin is a major component of insect and crustacean exoskeletons and fungal cell walls.

chitinase An enzyme that hydrolyzes chitin by cleaving the β-1,4 glycosidic bond.

chlorophyll The primary chromophore in chloroplasts and cyanobacteria.

chloroplast In plant cells, an organelle that converts light energy into chemical energy.

cholate The major bile acid formed in the liver by the conversion of cholesterol.

cholesterol A steroid with a hydrocarbon tail and a hydroxyl polar end. Cholesterol is a key modulator of membrane fluidity and the precursor for all of the steroid hormones.

chondroitin sulfate A glycosaminoglycan made of alternating units of sulfated *N*-acetylgalactosamine and glucuronic acid.

chorismate The precursor to tyrosine, phenylalanine, and tryptophan; it is formed in the shikimate pathway.

chromatin In eukaryotes, a complex of DNA and proteins that constitutes chromosomes.

chromodomain A region found in many proteins involved in chromatin remodeling; protein chromodomains bind methylated histone proteins.

chromophore A light-absorbing molecule found as part of photosensitive biological proteins or complexes.

chromosome A DNA molecule that functions to store and transmit genetic information.

chylomicron Large lipoprotein particles that transport triacylglycerols from the intestines to tissues throughout the body.

chyme A slurry of food and gastric juice generated in the stomach and delivered to the small intestines to facilitate nutrient absorption.

chymotrypsin A protease that cleaves a polypeptide chain on the carboxyl side of tyrosine, tryptophan, and phenylalanine residues; used to provide shorter polypeptide chains for Edman degradation.

circular dichroism (CD) A method of studying protein folding that measures differences in the absorption of right-handed versus left-handed circularly polarized light.

cis-acting site DNA sequences that interact with trans-acting factors (protein or RNA) to regulate gene expression.

citrate cycle Also called *Krebs cycle* or *tricarboxylic acid (TCA) cycle*. A central pathway in oxidative respiration that converts acetyl units to CO_2 with the concomitant production of $FADH_2$, NADH, and ATP.

citrate lyase A cytosolic enzyme that cleaves an acetyl group from citrate to form oxaloacetate and acetyl-CoA; linked to the citrate shuttle.

citrate shuttle A mechanism for transporting acetyl-CoA groups via citrate from mitochondria to the cytosol, where they are used for fatty acid synthesis.

citrate synthase The first enzyme in the citrate cycle; it converts four-carbon oxaloacetate into five-carbon citrate and regenerates CoA.

clathrin A protein that coats the surface of protein transport vesicles that bud from either the trans-Golgi network or the plasma membrane.

clofibrate A drug prescribed for the reduction of high levels of cholesterol and triacylglycerol in the blood that functions by activating the enzyme lipoprotein lipase.

closed complex Refers to the inactive form of bacterial RNA polymerase holoenzyme bound to DNA at a gene promoter sequence.

co-suppression Inhibition of translation from both an endogenous gene and a transgene after introduction of the transgene into a cell or organism.

coatomer protein Proteins that function in the eukaryotic COPI, COPII, and clathrin transport processes by coating the cytosol-facing surface of membrane transport vesicles.

coding strand A DNA sequence that has the same 5′ to 3′ polarity as the corresponding mRNA transcript; it is complementary to the DNA template strand.

codominant Refers to a genetic trait in which a heterozygote exhibits equally the phenotype of each allele; for example, inheritance of the ABO blood group.

coenzyme A (CoA) A metabolite that carries acetyl groups; participates in numerous essential metabolic pathways.

coenzyme An organic enzyme cofactor.

coenzyme Q (Q) Also called *ubiquinone*. Mobile electron carrier that moves laterally within the inner membrane between protein complexes of the electron transport system.

cofactor A small inorganic molecule, often a metal ion, that aids in the catalytic reaction mechanism within the enzyme active site.

coiled coil Two helical polypeptides that are wrapped around each other.

colipase A regulatory protein associated with pancreatic lipase. It enhances lipase function by binding bile acids, which inhibit lipase function.

colligative properties Physical properties of a solution that depend on the number of solute particles, such as freezing-point depression or osmotic pressure.

column chromatography A technique that separates proteins on the basis of differential physical or chemical interactions with a solid gel matrix.

competitive inhibitor A molecule that inhibits substrate binding at the active site.

complementary DNA (cDNA) A DNA molecule, usually synthesized by the enzyme reverse transcriptase, that is complementary to a given mRNA; used in DNA cloning.

concerted model Also called *symmetry model*. A model of cooperative binding behavior in a protein complex in which the probability of being in one of two states of the complex is affected by ligand binding.

condensation A reaction that combines two chemically similar substrates to form a larger molecule, with the loss of a smaller molecule, usually water.

conjugate base The deprotonated species of an acid; the anion of a weak acid.

conjugate redox pair The oxidized and reduced states of a molecule or ion; for example, NAD^+ and NADH.

conjugation (plasmid) Part of the bacterial mating process in which a donor bacterium transfers a copy of the plasmid to a recipient cell.

Consortium for Functional Glycomics (CFG) An international initiative that serves as a clearinghouse, repository, and resource center for scientists working with glycans, glycoconjugates, and associated molecules.

constant domain A protein domain found within immunoglobulin subunits.

contractile vacuole An organelle in some unicellular organisms that helps regulate osmosis by collecting and expelling water from the cell.

cooperative binding See **cooperativity**.

cooperativity Also called *cooperative binding*. The phenomenon in which the binding of a molecule (e.g., O_2 or a trans-acting factor) to a macromolecule lowers the energy of binding of subsequent molecules to that macromolecule.

coregulatory protein Also called a coactivator or corepressor protein, often it interacts with ligand-activated nuclear receptors and modulates rates of transcription by modifying chromatin-associated proteins.

Cori cycle A mechanism in which lactate produced by anaerobic glycolysis in muscle is converted in liver cells to glucose via the gluconeogenesis pathway.

Cori disease A glycogen storage disease resulting from a defect in the liver and muscle glycogen debranching enzyme that limits the ability to fully degrade glycogen molecules.

coronary artery bypass graft A surgical procedure in which veins are grafted to the myocardium to restore blood flow to regions of the heart suffering from decreased oxygen delivery due to atherosclerotic coronary arteries.

crista Folded areas of the mitochondrial inner membrane; greatly increases surface area for chemiosmotic production of ATP.

cyanobacteria Oxygen-producing phototrophic bacteria that contain chlorophylls a and b.

cyclic AMP (cAMP) 3′,5′-Cyclic adenosine monophosphate; a second messenger that activates numerous signaling proteins and target proteins.

cyclic GMP (cGMP) 3′,5′-Cyclic guanosine monophosphate; a second messenger for numerous signaling pathways, including synaptic transmission in light-stimulated vision and the control of arterial blood flow through vasodilation.

cyclic pathway A metabolic pathway containing several metabolites that regenerate during each turn of the cycle, serving as both reactants and products.

cyclic photophosphorylation An alternative electron pathway in chloroplasts that cycles electrons in photosystem I back to cytochrome b6f, resulting in increased ATP production and reduced NADPH production.

cyclin A group of regulatory proteins responsible for controlling progression through distinct stages of the cell cycle.

cyclin-dependent protein kinase (CDK) Signaling enzymes that phosphorylate downstream proteins in response to cyclin binding.

cysteine A hydrophilic amino acid that contains sulfur and can form disulfide bonds.

cysteine–aspartate protease (caspase) A class of proteins that initiates a proteolytic cascade leading to protein degradation and cell death.

cystic fibrosis A disease that results from thickened mucus in the lungs, caused by a functional defect in an ABC transporter protein that helps control movement of water across cell membranes.

cytochrome b_6f A protein complex that links photosystems II and I. Electron flow through cytochrome b_6f results in proton translocation from the stroma to the thylakoid lumen.

cytochrome c Mobile electron carrier protein that moves within the mitochondrial intermembrane space between protein complexes III and IV of the electron transport system.

cytochrome c oxidase Complex IV in the electron transport system; accepts electrons from cytochrome c and ultimately donates them to O_2 to form water.

cytoplasm A cell's contents within the plasma membrane (but not including the nucleus in eukaryotic cells), including organelles; the site of most cellular activities.

cytoplasmic coat I (COPI)-coated vesicle A protein transport vesicle that shuttles proteins from the Golgi apparatus to the endoplasmic reticulum and within the Golgi apparatus.

cytoplasmic coat II (COPII)-coated vesicle A protein transport vesicle that carries proteins from the endoplasmic reticulum to the Golgi apparatus.

cytosine (C) A pyrimidine base found in DNA and RNA that hydrogen bonds with guanine.

cytoskeletal proteins Structural proteins that are responsible for cell shape, cell migration, and cell signaling.

cytoskeleton In eukaryotic cells, a network of intracellular filaments, consisting of oligomeric proteins, that maintains cell structure.

D

D isomer A right-handed enantiomer.

Dam methylase The enzyme DNA adenine methylase, which is responsible for methylating the parental strand after daughter strand synthesis, to aid in DNA mismatch repair.

deadenylation The removal of adenine residues from RNA or protein.

death domain (DD) A region in the cytoplasmic tail of TNF receptors that functions as a protein–protein interaction module.

death effector domain (DED) A protein–protein interaction module that enables the TRADD–FADD complex to bind with procaspase 8 as part of the cell death pathway.

dehydration A reaction that cleaves a substrate into two products by the removal of water.

dehydrogenase A type of oxidoreductase that also results in the release of hydrogen ions.

denaturation Partial or complete unfolding of the conformation of a protein or nucleic acid chain.

denitrification The reduction of nitrate (NO_3^-) and nitrite (NO_2^-) to N_2 by using nitrate as the terminal electron acceptor in anaerobic respiration.

deoxyribonucleic acid (DNA) A polymeric macromolecule consisting of nucleotides covalently linked through phosphodiester bonds; the biomolecule encoding inheritable genetic information.

diacylglycerol (DAG) A second messenger signaling lipid that binds to and activates protein kinase C (PKC).

dialysis A diffusion-based process that uses a semipermeable membrane to allow small molecules to cross the membrane but not large molecules (usually proteins).

dideoxynucleoside triphosphate (ddNTP) A key reagent in the Sanger DNA sequencing method that terminates DNA synthesis reactions.

dihydrofolate The oxidation product formed when N^5,N^{10}-methylenetetrahydrofolate donates a methyl group to form dTMP from dUMP.

diisopropylfluorophosphate A highly reactive compound that serves as an irreversible enzyme inhibitor by forming a covalent link with serine residues, blocking protease and phospholipase enzymes.

dipeptidase A group of intestinal enzymes that hydrolyze dipeptides into individual amino acids.

direct repair Refers to DNA repair mechanisms that correct the damaged nucleotides without requiring repair replication.

discontinuous synthesis The mechanism used by cells to synthesize the lagging strand during DNA replication.

dissociation constant (K_d) The inverse of the association constant for a given reaction.

distal histidine A histidine residue in globin proteins that forms a hydrogen bond with oxygen, stabilizing its interaction with the heme group.

disulfide bond Also called *disulfide bridge*. A covalent cross-link between two cysteine residues that can stabilize the three-dimensional structure of a protein or connect polypeptide subunits in a protein.

disulfide bridge See **disulfide bond.**

DNA bioinformatics The use of DNA sequence information to predict the function or evolutionary relatedness of a given DNA segment.

DNA ligase An enzyme that cells use to repair a nick in the phosphodiester backbone; it catalyzes the formation of a phosphodiester bond.

DNA methylase An enzyme that methylates DNA at specific sequences.

DNA polymerase A group of enzymes that catalyze the synthesis of new DNA molecules.

DNA replication The enzyme-mediated process of doubling the DNA content of a cell during division.

DNA transcription The process of generating RNA from a DNA template.

DNase I footprinting An *in vitro* laboratory method for identifying DNA sequences that can function as binding sites for sequence-specific DNA binding proteins.

docosahexaenoic acid (DHA) An ω-3 fatty acid available in ω-3–rich foods such as fish and nuts. It can also be synthesized from α-linolenic acid.

domain An independent folding module within a polypeptide chain.

dominant mutation When referring to somatic mutations, also called gain-of-function mutation.

double-strand DNA break A region of DNA that lacks phosphodiester bonds on both strands of the DNA double helix.

E

ecosystem A geographic community where different organisms have adapted to share resources and waste management in a linked system of energy flows and nutrient cycles.

Edman degradation A protein sequencing method based on labeling and cleaving the N-terminal residue without disrupting the rest of the polypeptide chain.

eicosanoid A group of immune-system signaling molecules derived from long-chain polyunsaturated fatty acids.

eicosapentaenoic acid (EPA) An ω-3 fatty acid available in ω-3–rich foods such as fish and nuts. It can also be synthesized from α-linolenic acid.

electrochemical cell An apparatus used to measure the reduction potential of a test redox pair (such as NAD^+/NADH); typically measured under standard conditions against a reference standard of H^+/H_2.

electrochemical gradient A concentration gradient of an ion (such as protons) combined with a charge difference (due to the uneven distribution of the ion) across the membrane.

electron transport system A series of proteins embedded in a membrane (e.g., the inner mitochondrial membrane) that converts the energy of redox reactions to a proton potential (a proton gradient across the membrane).

electrospray ionization A method for preparing proteins for mass spectrometry, providing highly charged peptide ions in the gas phase.

elongation factors (EFs) Proteins in bacteria that facilitate aminoacyl tRNA binding to the ribosome A site and translocating the ribosome by one codon in the 3′ direction.

elution buffer In ion-exchange chromatography, a buffer solution with a high concentration of an appropriate competing ion to displace the bound protein.

embryonic stem (ES) cell A class of pluripotent cells found in animal embryos that are capable of differentiating into every cell type over the course of development.

enantiomers Stereoisomers that are mirror images of each other and whose structures cannot be superimposed.

endergonic A chemical reaction that requires energy and is unfavorable in the direction written with a $\Delta G > 0$.

endocrine A hormone secreted into the circulatory system by a gland that can bind to a receptor protein on a target cell.

endomembrane system An intracellular network of lipid bilayers that is used to exchange material through vesicle transport.

endoplasmic reticulum In eukaryotic cells, highly invaginated membrane structures that sequester ribosomes for protein synthesis.

endoprotease An enzyme that cleaves a peptide bond within a protein.

endosymbiotic theory The theory proposing that eukaryotic cells evolved about 1.5 billion years ago as a result of large predatory cells engulfing aerobic bacteria or cyanobacteria, giving rise to mitochondria and chloroplasts, respectively.

endothermic A reaction that absorbs heat and $\Delta H > 0$.

endotoxin A lipopolysaccharide in the outer membrane of Gram-negative bacteria that becomes toxic to the host after the bacteria have lysed.

energy balance Refers to the ratio of energy input in an organism to the amount of energy it expends.

energy charge (EC) A measure of the energy state of a cell in terms of ATP, ADP, and AMP ratios:

$$EC = \frac{[ATP] + 0.5[ADP]}{[ATP] + [ADP] + [AMP]}$$

enhancer DNA cis-acting sites that operate at a distance from eukaryotic gene promoters by functioning as binding sites for trans-acting transcriptional regulatory proteins.

enteropeptidase An intestinal protease released into the duodenum that activates several pancreatic proteases.

enthalpy (*H*) The heat content of a system.

entropy A measure of the spreading of energy; also a measure of the disorder (randomness) in a system.

enzyme active site The region of an enzyme where the catalytic reaction takes place.

enzyme kinetics The quantitative study of the chemical reactions performed by enzymes.

enzyme Protein or RNA biomolecules that function as reaction catalysts to increase the rates of biochemical reactions.

enzyme-linked immunosorbent assay (ELISA) A high-throughput biochemical assay that uses monoclonal antibodies to detect small amounts of antigen in aqueous samples.

ephedrine A weight-loss drug that acts by enhancing the release of noradrenaline and stimulating adrenergic receptors.

epidermal growth factor (EGF) A serum growth factor that binds to the EGF receptor (EGFR) and stimulates receptor dimerization on the cell surface.

epigenetic states Inheritable patterns of gene expression due to chemical changes in DNA or DNA binding proteins.

epimer One of a pair of monosaccharides that differ only in the position of a hydroxyl group around a single carbon atom.

epinephrine A hormone also called adrenaline (the "fight or flight" hormone) that reduces the recovery time between muscle contractions and also controls metabolic flux in response to acute stress.

epitope tag A highly antigenic amino acid sequence added to the protein-coding sequence of cloned genes that is recognized by commercially available antibodies.

equilibrium constant (K_{eq}) A measure of the directionality of a reaction under standard conditions, where all products and reactants start at 1 M and proceed to their equilibrium concentrations.

equilibrium The state of a system in which no net change occurs.

equivalent The amount of base (OH^-) needed to deprotonate an acid; the amount of substance that will react with 1 mol of electrons in a redox reaction.

erythrocyte A red blood cell; contains hemoglobin for oxygen transport in an organism.

essential fatty acid Fatty acids that humans must obtain in the diet because the human body lacks the enzymes to synthesize them.

euchromatin A region of chromatin that is loosely packed with nucleosomes and associated with actively transcribed genes.

eukaryote A cell that contains a nucleus and other organelles bounded by membranes, creating microenvironments for biochemical reactions.

***eve* stripe 2 enhancer** A regulatory region of DNA in the *Drosophila* embryo that binds multiple interacting transcription factor proteins. It is essential for the localized expression of the transcription factor gene called *even-skipped.*

exergonic A chemical reaction that releases energy and is favorable in the direction written with a $\Delta G < 0$.

exon A coding region of a eukaryotic gene; exons are separated from one another by introns.

exon shuffling The mixing and matching of protein-coding sequences during evolution to generate genes with novel functions.

exosome An exonuclease complex that cleaves RNA molecules targeted for destruction.

exothermic A reaction that releases heat and $\Delta H < 0$.

exportin A transport protein that ferries proteins out of the nucleus. It forms a ternary complex with its cargo and a small GTPase protein called Ran.

extracellular signal–regulated kinase (ERK) A protein in the MAP kinase signaling pathway that is phosphorylated by the kinase enzyme MEK and then translocates to the nucleus where it phosphorylates target proteins.

F

Fabry disease A genetic disease of sphingolipid metabolism due to defect in the enzyme α-galactosidase A, which causes ceramide trihexosides to accumulate.

familial hypercholesterolemia (FH) A recessive genetic disorder in which individuals have reduced numbers of LDL receptors, resulting in high levels of serum LDL and associated cardiovascular disease.

Fas-associated death domain (FADD) A protein complex that binds with TRADD as part of the cell death signaling pathway.

fatty acid An organic molecule consisting of a polar carboxyl group covalently linked to a nonpolar extended hydrocarbon chain.

fatty acid synthase A multifunctional protein (in eukaryotes) or a multienzyme complex (in prokaryotes) that is responsible for the synthesis of fatty acids such as palmitate.

fatty acyl-CoA synthetase A set of three enzymes responsible for the formation of fatty acyl-CoA molecules.

favism An acute hemolytic anemia resulting from the ingestion of fava beans by individuals who have a glucose-6-phosphate dehydrogenase deficiency.

Fe protein A protein subunit in the bacterial nitrogenase enzyme complex responsible for converting N_2 to NH_3 in the process of nitrogen fixation.

feedback inhibition An enzyme regulatory mechanism where the end product of a pathway functions as an inhibitor of the first enzyme in the pathway.

FERM domain fold A large protein fold (300 amino acids) containing three distinct structural components that are found in some proteins that bind to membrane-associated biomolecules.

fermentation The conversion of rotting fruit or grain into alcohol and CO_2 through the action of yeast enzymes under anaerobic conditions.

ferredoxin (Fd) The final electron carrier in photosystem I. It is a soluble protein located in the thylakoid stroma, which ultimately reduces $NADP^+$.

fibroin heavy chain A 350-kDa protein containing large numbers of β sheets, part of silkworm fibroin.

fibroin light chain A protein subunit of silkworm fibroin, covalently linked to the heavy chain by disulfide bonds.

first law of thermodynamics In any physical or chemical change, the total amount of energy in the universe remains the same, even though the form of energy may change.

first messenger An extracellular ligand that binds to a receptor protein, activating a signaling pathway.

first-order reaction A reaction in which the rate varies as the first power of the reactant concentration.

flagellum An extracellular structure used for cell movement by bacteria and sperm cells.

flavin adenine dinucleotide ($FAD/FADH_2$) A coenzyme that mediates redox reactions at carbon–carbon bonds.

flavin mononucleotide ($FMN/FMNH_2$) A coenzyme that mediates redox reactions at carbon–carbon bonds.

flavoprotein An enzyme that requires the coenzyme FAD or FMN for its activity; for example, pyruvate dehydrogenase.

flippase A membrane protein that uses energy available from ATP hydrolysis to catalyze phospholipid flipping.

fluid mosaic model Model of cell membrane organization proposing that a membrane is a two-dimensional solution in which many membrane proteins are both oriented across the membrane and free to move laterally through the membrane.

fluorocitrate A toxin that is formed from the conversion by citrate synthase of plant-derived fluoroacetate. The toxin blocks aconitase activity, thus halting the citrate cycle and rapidly killing the cell.

forked pathway A metabolic pathway that generates two products, each of which undergoes a different metabolic fate.

four-helix bundle A common protein fold, consisting of four helices linked together.

frameshift mutation A single nucleotide mutation in DNA that changes the codon register, resulting in a new amino acid sequence.

French press A mechanical device that forces cells through a small opening in a metal attachment to break cell walls and generate a cell extract.

fructose intolerance A serious genetic malady caused by the inability to make aldolase B, an enzyme that catalyzes the reversible reaction fructose-1-phosphate to glyceraldehyde and dihydroxyacetone phosphate.

fructose-1,6-bisphosphatase-1 Enzyme in the gluconeogenic pathway that converts fructose-1,6-bisphosphate to fructose-6-phosphate and opposes the reaction catalyzed by phosphofructokinase-1 in the glycolytic pathway.

fructose-2,6-bisphosphate (fructose-2,6-BP) A molecule derived from fructose-6-phosphate that reciprocally regulates glycolysis and gluconeogenesis; fructose-2,6-BP stimulates glycolysis and inhibits gluconeogenesis.

furanose A cyclic sugar formed from either a ketohexose or an aldopentose; so called because it resembles the five-member furan ring.

G

G protein cycle The sequential stimulation of G protein signaling by guanine exchange factors (GEFs) activity, with subsequent activation of its intrinsic GTPase activity by GTPase activating proteins (GAPs).

G protein family Related proteins that function in signal transduction pathways and encode an intrinsic GTPase activity that converts the G protein to the inactive state. See also *GTPase*.

G protein–coupled receptor (GPCR) A receptor protein that upon activation causes dissociation of the heterotrimeric G protein complex, leading to downstream signaling that can include the production of second messengers.

G protein–coupled receptor kinase (GRK) A regulatory protein that phosphorylates the GPCR cytoplasmic domain on serine and threonine residues, which marks it for recycling.

GAL4 activator protein A yeast transcription factor protein that binds to cis-acting DNA sites and induces transcription of genes required for galactose metabolism.

galactosemia A genetic defect that blocks the conversion of galactose to glucose due to the absence of galactose-1-phosphate uridyltransferase.

ganglioside A cerebroside with an oligosaccharide bound to the terminal glucose or galactose molecule.

gastrin A peptide hormone secreted by mucosal cells in the stomach lining that stimulates the secretion of hydrochloric acid and digestive enzymes shortly after eating.

gel electrophoresis A biochemical technique that separates proteins on the basis of charge and size.

gel filtration chromatography Also called *size-exclusion chromatography*. A column chromatography method that uses porous hydrocarbon beads to separate proteins on the basis of size.

gene A segment of DNA that codes for a transcribed RNA molecule; the functional unit of heredity.

gene amplification An evolutionary mechanism that produces multiple copies of a gene or section of a genome.

gene duplication A mechanism by which duplication of a region of DNA containing a gene can lead to evolution of new genetic material.

gene expression microarray A solid surface, often a microchip, that contains covalently linked deoxyoligonucleotide segments that are used to identify complementary mRNA (or cDNA) sequences in a sample.

gene silencing A group of processes that interfere with RNA expression through either inhibition of mRNA translation or degradation of mRNA transcripts.

genome All of the genetic information in a cell or virus contained in DNA or RNA.

genomic caretaker protein A protein that functions to ensure the integrity of genomic DNA throughout the life of the cell, which includes DNA repair enzymes, DNA synthesis and recombination enzymes, and proteins required for RNA synthesis.

germ-line cell Cells that give rise to an egg or sperm cell in eukaryotes.

Gibbs free energy (G) A measure of the spontaneity of a reaction, defined as the difference between the enthalpy and entropy of a system at a given temperature ($G = H - TS$).

globin fold A protein fold containing eight α helices; found in hemoglobin and myoglobin.

glucagon A peptide hormone that signals low glucose levels in the blood and sometimes called the "I am hungry" hormone; glucagon signaling stimulates glycogen degradation and gluconeogenesis pathways in liver cells.

glucocorticoid response element (GRE) The DNA cis-acting sequence located near glucocorticoid-regulated genes that functions as the binding site for ligand-activated glucocorticoid receptors.

glucogenic Refers to amino acids whose carbon chains can be used to form glucose and glycogen via gluconeogenesis.

gluconeogenesis Metabolic pathway used for the production of glucose from noncarbohydrate sources.

glucose tolerance test A medical test to measure glucose clearance kinetics in the blood of a patient.

glucose-6-phosphatase Enzyme in the gluconeogenic pathway that converts glucose-6-phosphate to glucose and opposes the reaction catalyzed by hexokinase or glucokinase in the glycolytic pathway.

glucose-6-phosphate dehydrogenase (G6PD) The enzyme that catalyzes the initial step in the pentose phosphate pathway, converting glucose-6-phosphate to 6-phosphogluconolactone.

glutamate An amino acid that is often negatively charged at pH 7.

glutamate dehydrogenase Enzyme that converts glutamate into α-ketoglutarate and NH_4^+ in the presence of NAD^+ or $NADP^+$; a key enzyme in nitrogen metabolizing pathways.

glutamate synthase Enzyme that converts glutamine and α-ketoglutarate into glutamate in the presence of NADH or NADPH; a key enzyme in nitrogen metabolizing pathways.

glutamine A hydrophilic amino acid that can both donate and accept hydrogen atoms to form hydrogen bonds.

glutamine synthetase An ATP-dependent enzyme that converts glutamate and NH_4^+ into glutamine in a two-step reaction; a key enzyme in nitrogen metabolizing pathways.

glutaredoxin The bacterial protein that provides the reducing power for ribonucleotide reductase to form deoxyribonucleotides from ribonucleotides.

glutathione The reduced form of γ-glutamylcysteinylglycine serves as an electron donor in numerous redox reactions.

glycan A synonym for carbohydrate, a class of molecules with the formula $(CH_2O)_n$ where n is greater than 2.

glycemic index A value applied to various carbohydrate-containing foods that correlates with the rate of glucose release into the blood after ingestion of that food.

glycerol-3-phosphate shuttle A mechanism used predominantly in muscle cells that enables the NADH formed in glycolysis to feed its electrons into the electron transport system by reducing FAD to form $FADH_2$ in the mitochondrial intermembrane space.

glyceroneogenesis The synthesis of glycerol-3-phosphate by an abbreviated version of gluconeogenesis, which is used in the triacylglycerol cycle.

glycerophospholipid A membrane lipid composed of glycerol linked to two fatty acid chains and a phosphate group, which is bound to a polar head moiety.

glycine The smallest and least chemically active amino acid.

glycobiology The study of the biological functions of glycans.

glycocholate The major bile acid in the human body.

glycoconjugate A protein or lipid with covalently linked glycans.

glycogen particle An aggregation of glycogen core complexes found in liver and muscle cells; visible in tissue sections using periodic acid–Schiff staining.

glycogen phosphorylase Enzyme that catalyzes a phosphorolysis reaction utilizing inorganic phosphate to remove glucose residues from the nonreducing ends of glycogen.

glycogen synthase Enzyme that adds glucose residues to the nonreducing ends of glycogen using UDP-glucose as a substrate.

glycogenin A protein linked to the reducing end of glucose; it functions as a glycosyltransferase enzyme and also a primer to initiate the synthesis of glycogen core particles.

glycolate pathway A salvage pathway in plants that recovers carbon from 2-phosphoglycolate for use in the Calvin cycle; also called the C_2 pathway.

glycolipid A major class of glycoconjugates characterized by glycans covalently linked to lipid molecules.

glycolysis The catabolic pathway of glucose oxidation to pyruvate. One glucose molecule that enters the pathway yields two molecules of pyruvate, two ATP, and two NADH.

glycome The complete set of glycan groups in a cell or organism.

glycoprotein A major class of glycoconjugates characterized by glycans covalently linked to protein molecules.

glycosaminoglycan A linear polysaccharide made of repeating hexosamine disaccharides, often found in the glycoconjugates known as proteoglycans.

glycosidase An enzyme that catalyzes the removal of glycan units from glycoconjugates.

glycosyltransferase An enzyme that catalyzes the addition of glycan units to proteins and lipids to form glycoconjugates.

glyoxylate cycle A metabolic pathway that converts acetyl-CoA molecules into succinate, which provides source molecules for carbohydrate biosynthesis.

glyoxysome A plant organelle that is the site of the glyoxylate cycle.

glyphosate The active ingredient in the herbicide Roundup; it is a competitive inhibitor of EPSP synthase, thus blocking the *de novo* synthesis of aromatic amino acids.

Golgi apparatus In eukaryotic cells, a membranous structure required for protein translocation within the cell and in facilitating protein secretion at the plasma membrane.

gout A disease of the joints and kidneys caused by an excess of uric acid in the blood leading to formation of uric acid crystals.

Gram-negative bacteria Bacteria possessing a thin peptidoglycan layer that does not retain the Gram stain leaving cells a pink-red color after counterstaining.

Gram-positive bacteria Bacteria possessing a thick peptidoglycan layer that retains the Gram stain leaving cells a dark purple color after staining.

gramicidin A A passive membrane transporter in bacteria that serves as an ion channel for Na^+ and K^+ ions to leak from bacterial cells.

granum *pl.* **grana** A stack of thylakoids.

Greek key fold A common protein fold consisting of four or more β strands linked together to form β-sheet structures.

green fluorescent protein A protein from jellyfish that emits green light after absorption of blue light from either a jellyfish protein called aequorin or artificial light.

greenhouse effect The process by which specific gases in Earth's atmosphere capture and recycle thermal energy radiated by Earth, thus further warming the surface of Earth.

group I introns A class of catalytic RNAs that use a guanosine cofactor to self-splice the introns and join the exons to form mature RNA.

group II introns A class of catalytic RNAs that undergo self-splicing without the need for a guanosine cofactor, and instead use a cis-acting adenine residue within the intron.

growth factor receptor–bound 2 (GRB2) An SH2/SH3 adaptor protein that binds to the EGF receptor.

GTPase activating protein (GAP) A protein that stimulates the intrinsic GTP hydrolyzing activity of G proteins to inhibit signal transduction; countered by guanine nucleotide exchange factor (GEF) proteins.

GTPase An enzyme that cleaves GTP to form GDP plus inorganic phosphate; the G_α subunit of heterotrimeric G protein is a GTPase as is the signaling protein Ras.

guanine (G) A purine base found in DNA and RNA that hydrogen bonds with cytosine.

guanine nucleotide exchange factor (GEF) A protein that promotes GDP–GTP exchange and activates signaling; countered by GTPase activating proteins (GAPs).

gyrase A topoisomerase that relieves the torsional strain generated by helicase as it unwinds the DNA double helix.

H

H^+–K^+ ATPase A P-type transporter responsible for transporting H^+ into the stomach and lowering the pH of gastric juices to aid in digestion.

Haber process The industrial nitrogen fixation process in which N_2 is hydrogenated by methane under extreme heat and pressure to form NH_3.

half-life ($t_{1/2}$) The time it takes for half of a reactant to decompose.

hammerhead ribozyme A catalytic RNA capable of both cis and trans cleavage.

heat shock family A group of proteins that provide a way for the cell to recover from heat denaturing by helping proteins refold when the temperature returns to normal.

heat shock protein 70 (Hsp70) A well-characterized clamp-type chaperone protein.

heat shock protein 90 (Hsp90) A chaperonin protein involved in many pathways including nuclear receptor signaling.

helicase (DnaB) An enzyme that unwinds DNA ahead of DNA polymerase at the replication fork.

helix–loop–helix motif A DNA binding motif found in many transcription regulatory proteins that form dimers. It is characterized by a dimerization α helix and a DNA binding α helix separated by a loop of amino acids.

helix–turn–helix motif A common DNA binding motif in transcription factors, consisting of two α helices connected by a short turn.

heme An Fe^{2+} porphyrin complex that functions as a prosthetic group, binding oxygen.

hemiacetal A molecule formed by the reaction of an alcohol and an aldehyde-containing compound (such as an aldose sugar); the product is both an ether and an alcohol.

hemicellulose A class of branched heteropolysaccharide that forms a matrix with cellulose and pectin to form plant cell walls.

hemiketal A molecule formed by the reaction of an alcohol and a ketone-containing compound (such as a ketose sugar); the product is both an ether and an alcohol.

hemodialysis A medical procedure in which the blood of a patient in renal failure is circulated through a machine containing a semipermeable membrane and a dialysis solution.

hemoglobin A tetrameric globular transport protein that transports oxygen from the lungs to the tissues through the circulatory system.

Henderson–Hasselbalch equation A useful relationship between pH and pK_a:

$$pH = \frac{pK_a + \log[A^-]}{[HA]}$$

heparan sulfate A glycosaminoglycan made of alternating units of *N*-acetylglucosamine and a uronic acid. There is a variable amount of sulfation among heparan sulfate molecules.

heparin A heavily and variably sulfated glycosaminoglycan made of alternating units of a uronic acid and a hexosamine.

Hers disease A glycogen storage disease caused by a defect in liver glycogen phosphorylase.

heterochromatin A dense form of chromatin composed of mostly noncoding DNA.

heterodimer A protein complex with two subunits derived from distinct polypeptides (different gene products).

heterotrimeric G protein A membrane-bound protein complex associated with G protein–coupled receptors that dissociates upon receptor ligand binding to initiate downstream signaling pathways.

heterotroph An organism that cannot convert solar energy to chemical energy directly, but must depend on nutrients obtained from autotrophs and other heterotrophs as a source of energy.

hexokinase A key enzyme in glycolysis that catalyzes the conversion of glucose to glucose-6-phosphate.

hexosamine A six-carbon monosaccharide that contains an amino group.

high-density lipoprotein (HDL) A class of cholesterol-transporting molecules that transports cholesterol from the peripheral tissues to the liver.

high-performance liquid chromatography (HPLC) A high-resolution version of gravity-based gel filtration chromatography, using high pressure to force buffer and protein through the column.

histidine An amino acid that can be positively charged or neutral at pH 7.

histone acetyltransferase (HAT) A group of enzymes that catalyzes the addition of acetyl groups to lysine residues on histone proteins.

histone deacetylase (HDAC) A class of enzymes that removes acetyl groups from lysine residues on histone proteins.

histone protein A group of small basic eukaryotic proteins around which DNA wraps to form nucleosomes.

histone tail A region of the amino terminus of a histone that protrudes from the nucleosome and is the site of histone acetylation.

HMG-CoA reductase The enzyme responsible for the formation of mevalonate from acetyl-CoA in the cholesterol biosynthetic pathway.

Holliday junction An intermediate formed during homologous recombination consisting of four double-stranded DNA regions joined together.

holoenzyme The assembled and catalytically active form of a multi-subunit enzyme or an enzyme that requires a cofactor.

homeostasis The use of energy to maintain a dynamic steady state of an organism that can adjust to changing environmental conditions.

homodimer A protein complex containing two identical protein subunits encoded by the same gene.

homologous recombination The physical exchange of DNA sequences between homologous chromosomes.

hormone A biologically active compound that is released into the circulatory system and binds to hormone receptors contained in target cells.

horseradish peroxidase An enzyme attached to a secondary antibody that catalyzes a reaction generating a chromogenic or fluorogenic product to identify antigenic proteins bound by a primary antibody.

housekeeping gene One of a class of genes that encode proteins and RNA involved in essential biochemical processes common to most all cells.

Hutchinson–Gilford progeria syndrome (HGPS) The most common in a group of fatal disorders that cause rapid aging in children as a result of DNA damage.

hyaluronic acid A glycosaminoglycan made of alternating residues of glucuronic acid and *N*-acetylglucosamine; it is a component of proteoglycans found in the extracellular matrix.

hybridoma cell A cell fused from antibody-expressing B cells and immortalized tumor cells, which can be cultured *in vitro* and screened for antibody production.

hydration A chemical reaction in which H_2O (OH^- and H^+) is added to convert a carbon–carbon double bond to an alcohol.

hydrogen bond A weak noncovalent bond in which hydrogen is shared between two electronegative atoms.

hydrogen ion (H^+) A proton; the cation of water and a key component of biochemical reactions.

hydrogenation A process of heating unsaturated fatty acids in the presence of hydrogen in order to reduce carbon–carbon double bonds, thus increasing the degree of saturation.

hydrolysis A reaction that cleaves a substrate into two products by the addition of water.

hydronium ion (H_3O^+) A hydrated hydrogen ion.

hydrophilic Polar molecules with an attraction for hydrogen bonds to water.

hydrophobic Nonpolar molecules that tend to pack close together away from water.

hydroxyl ion (OH^+) The anion of water; a key component of biochemical reactions.

hyperammonemia Elevated levels of ammonia in the blood.

hyperbolic curve A graph line in the shape of a hyperbola, typical of a simple protein–ligand interaction.

hyperchromic effect The increase in light absorbance at 260 nm as double-stranded DNA unwinds and separates.

hyperglycemia A physiologic condition in which glucose levels are elevated in the blood after a 12-hour fast.

hyperlipidemia A physiologic condition in which fatty acid levels are elevated in the blood after a 12-hour fast.

hypertonic A solution with a higher solute concentration compared to that of a solution with lower solute concentration such as the inside of a cell.

hypoglycemia A medical condition in which blood glucose levels are much lower than normal, typically dropping from 80 mg/dL (4.4 mM) to less than 50 mg/dL (2.8 mM); associated with mental confusion and fainting.

hypoglycin A An amino acid derivative found in high concentrations in unripe ackee fruit; hypoglycin A is a potent inhibitor of liver mitochondrial acyl-CoA dehydrogenases and responsible for the symptoms of Jamaican vomiting sickness.

hypothalamus A region of the brain important for the autonomic nervous system, the endocrine system, and for the maintenance of homeostasis.

hypotonic A solution with a lower solute concentration compared to that of a solution with higher solute concentration such as the inside of a cell.

hypoxanthine The purine base for the nucleoside monophosphate inosine, which is the precursor of the nucleoside monophosphates AMP and GMP.

I

immunofluorescence A technique to detect proteins within cells that uses fluorescent microscopy to locate cells that react with a primary antibody.

immunoglobulin A quaternary protein complex linked by disulfide bonds; also known as an antibody.

immunoglobulin fold A protein fold contained in immunoglobulin molecules consisting of two β sheets separated by a Greek key protein fold.

immunoprecipitation A physical separation technique in which a monoclonal antibody is covalently linked to a carbohydrate bead and used to isolate protein complexes containing an antigen.

importin A transport protein that imports proteins from the cytosol to the nucleus. It forms a ternary complex with its cargo and a small GTPase protein called Ran.

induced pluripotent stem cell (iPS cell) A pluripotent cell created *in vitro* by the addition of genes encoding specific transcription factors that reverse cell fate resulting in dedifferentiation.

induced-fit model A model of enzyme catalysis in which the enzyme conformation changes in response to substrate binding; induced fit is in contrast to the lock and key model of enzyme catalysis.

initial velocity (v_0) The reaction rate at the beginning of a reaction, before the substrate concentration has changed significantly.

initiation factor (IF) Proteins in bacteria that facilitate the assembly of mRNA, the initiator tRNA, and the ribosomal subunits to complete the initiation phase of protein synthesis.

initiator tRNA A specialized methionine-carrying tRNA (written as $Met\text{–}tRNA_i^{Met}$) that initiates translation by being able to bind to an AUG codon in an otherwise empty ribosome.

inosine-5′-monophosphate (IMP) The precursor of the nucleoside monophosphates AMP and GMP; the base in IMP is hypoxanthine.

inositol-1,4,5-trisphosphate (IP_3) A second messenger that activates Ca^{2+} channels in the endoplasmic reticulum.

insulator sequence A DNA sequence that represses the action of an enhancer on a gene promoter by serving as the cis-acting binding site for trans-acting repressor proteins.

insulin A peptide hormone secreted by pancreatic β cells that controls blood glucose levels by binding to the insulin receptor and activating downstream pathways that remove glucose from the blood.

insulin receptor substrate (IRS) A class of signaling proteins that bind to phosphorylated insulin receptors through phosphotyrosine binding domains.

intermediate filament A type of cytoskeletal protein complex that is critical to cell structure and function.

intracellular signaling protein A protein that functions as a molecular switch, undergoing conformational changes in response to incoming signals, such as receptor activation.

intron A noncoding region of DNA in a eukaryotic gene that is flanked by DNA sequences present in exons.

ion channel A type of passive transporter protein that allows ions such as H^+, K^+, Na^+, Cl^-, and Ca^{2+} across a cell membrane.

ion-exchange chromatography A protein purification method that exploits charge differences between proteins.

ionic interaction A weak interaction between oppositely charged atoms or groups.

iron–sulfur (Fe–S) centers Electron carrier prosthetic groups found in complexes I, II, and III of the electron transport system.

irreversible inhibition An enzyme regulatory mechanism in which an inhibitory molecule forms a covalent bond with catalytic groups in the enzyme active site.

islet of Langerhans Regions of the pancreas where insulin, glucagon, somatostatin, and digestive proteases are synthesized and secreted.

isoacceptor tRNA A group of related tRNA molecules that contain the same linked amino acid but bind different codons; can also be a single tRNA that binds to different codons because of the 5′ wobble position in the tRNA anticodon (usually inosine).

isocitrate dehydrogenase Catalyzes the oxidative decarboxylation of isocitrate to form α-ketoglutarate, NADH, and CO_2.

isoelectric focusing A type of polyacrylamide gel electrophoresis that separates proteins on the basis of charge as a function of pH.

isoelectric point The pH at which a given amino acid has no net charge.

isoforms Functionally distinct proteins transcribed from the same gene.

isoleucine A hydrophobic amino acid, usually found in the interior of a protein.

isomerization A reaction that does not change the molecular formula of the product compared to that of the substrate.

isomorphous replacement A procedure that determines the phases of diffracted X-rays by comparing them to X-ray diffraction from a crystal containing an electron-rich element.

isopentenyl diphosphate The activated intermediate molecule that is central to cholesterol biosynthesis.

isotonic A solution with the same solute concentration as another solution such as the inside of a cell.

isozymes Functionally related enzymes encoded by different genes.

J

jaundice A medical condition characterized by buildup in the skin of bilirubin, a yellow heme degradation product.

joule (J) The SI unit of energy; the amount of work done (or energy transferred) when a force of 1 Newton displaces an object by 1 meter in the direction of the force.

K

KEGG (Kyoto Encyclopedia of Genes and Genomes) A set of databases that map the integrated systems of molecular interactions that occur within a cell or organism.

keratan sulfate A glycosaminoglycan made of alternating units of galactose and sulfated *N*-acetylglucosamine.

ketoacidosis A side effect of excess ketone body formation, resulting in low blood pH.

ketogenesis A mechanism in liver cell mitochondria that converts excess acetyl-CoA into acetoacetate and D-β-hydroxybutyrate, also known as ketone bodies.

ketogenic Refers to amino acids whose carbon chains can be used to form ketone bodies.

ketone bodies Acetoacetate and D-β-hydroxybutyrate formed by ketogenesis in the liver and used elsewhere in the body to make acetyl-CoA for energy conversion reactions.

ketose A monosaccharide that contains a ketone functional group.

kilobase One thousand nucleotide bases.

kinase An enzyme that catalyzes the transfer of a phosphoryl group, usually from ATP to a substrate.

kinetochore A protein complex, assembled at the centromere, that is necessary for proper separation of the chromosomes during cell division.

Klenow fragment A 68-kDa proteolytic fragment of *E. coli* DNA polymerase I that contains polymerase and 3′→5′ exonuclease activities. It lacks the endogenous 5′→3′ exonuclease activity found in the DNA polymerase I holoenzyme.

Kozak sequence In eukaryotes, a consensus sequence in some mRNA transcripts that contains the initiation AUG codon.

Krebs bicycle Another name for the aspartate–arginosuccinate shunt connecting the urea cycle to the citrate cycle.

Krebs cycle See **citrate cycle**.

L

L isomer A left-handed enantiomer.

***lac* operon** Bacterial genes that are physically linked and encode proteins the are required for lactose metabolism.

lac repressor A bacterial DNA binding protein that binds to the *lac* operator (a cis-acting site) in the presence of allolactose or IPTG and inhibits expression of the *lac* operon.

lactate dehydrogenase deficiency A genetic disorder caused by defects in the lactate dehydrogenase gene that limits the regeneration of NAD^+ under anaerobic conditions and thereby decreases metabolic flux through the glycolytic pathway.

lactose intolerance A condition caused by decreased expression of the enzyme lactase, which is required for the metabolism of lactose in the gastrointestinal system; reduced lactase leads to a buildup of lactose in the intestines, leading to diarrhea and flatulence.

lactose permease A bacterial transport protein that imports lactose from the periplasmic space to the cytosol.

lagging strand The DNA strand that is synthesized in a discontinuous process and generates Okazaki fragments, which are short DNA fragments made in the 5′ to 3′ direction.

lamella *pl.* lamellae A region of thylakoid membrane that is not stacked.

lateral gene transfer The movement of genes between species of bacteria through the transfer of plasmids, via bacteriophages, or by other mechanisms.

Le Châtelier's principle The equilibrium of a reaction shifts in the direction that reduces change resulting from altered reactant concentration, temperature, pressure, or volume.

leading strand The DNA strand that is synthesized continuously in the 5′ to 3′ direction.

lectin A glycan (carbohydrate) binding protein.

leghemoglobin A heme-containing plant protein that sequesters O_2 to maintain an anoxic environment within cells where nitrogenase is present; O_2 inhibits the efficiency of nitrogen fixation by nitrogenase.

leptin An adipocyte peptide hormone (an adipokine) that regulates neuronal signaling in the brain leading to decreased appetite and increased energy expenditure.

Lesch–Nyhan syndrome A rare, X-linked neurologic disorder caused by mutations in the gene for hypoxanthine-guanine phosphoribosyltransferase.

leucine A hydrophobic amino acid, usually found in the interior of a protein.

leucine zipper motif A DNA binding motif that contains regularly spaced leucine residues in each of two α helices in the protein.

LexA repressor A transcriptional repressor protein that inhibits expression of the *E. coli* SOS operon in the absence of DNA damage.

ligand A small molecule that is often a metabolite, hormone, or peptide and which binds to target proteins and alters their structure and function to control biochemical processes.

ligand-gated ion channel A receptor that controls the flow of ions, such as K^+, Na^+, and Ca^{2+}, across cell membranes in response to ligand binding.

light-harvesting complex (LHC) A protein–chromophore complex that harvests photons and transfers them via resonance energy transfer to a photosynthetic reaction center.

linear metabolic pathway A metabolic pathway in which each reaction generates only a single product, which is a reactant for the next reaction in the pathway.

Lineweaver–Burk equation An algebraic transformation of the Michaelis–Menten equation that allows a double reciprocal plot of the enzyme data to be drawn.

Lineweaver–Burk plot Linear graphs of the Lineweaver–Burke equation from which Michaelis–Menten parameters can be determined.

linking number The number of times a strand of DNA winds in the right-handed direction around the helix axis when the axis lies in an imaginary plane.

lipase A water-soluble enzyme that hydrolyzes triacylglycerols into free fatty acids and glycerol.

lipid droplet Aggregate of triacylglycerol molecules surrounded by a phospholipid monolayer in the cytoplasm of adipocytes.

lipid raft An aggregation of membrane proteins that form densely packed complexes thought to serve as sites for cell signaling, membrane trafficking, and neurotransmission.

lipidomics Large-scale investigation of the structures and functions of the complete set of lipids in a biological system.

lipoamide The oxidized form of the coenzyme lipoic acid.

lipopolysaccharide (LPS) A class of glycolipids found in the outer leaflet of the outer membrane of Gram-negative bacteria.

lipoprotein A molecular complex composed of a core of hydrophobic lipids surrounded by a shell of polar lipids and apolipoproteins.

liposome A spherical vesicle bounded by a lipid bilayer and containing an aqueous center.

lipoteichoic acid A negatively charged polymer of ribitol phosphate or glycerol phosphate, found in the cell wall of Gram-positive bacteria.

lock and key model A model of enzyme catalysis in which rigid physical and chemical complementarity between the reactant and the enzyme is required for the reaction to proceed.

long noncoding RNA (lncRNA) A class of noncoding RNAs that are greater than 200 nucleotides in length.

long terminal repeat (LTR) A repeated nucleic acid sequence found at both the 5′ and 3′ ends of a viral genome that integrates into host DNA.

lorcaserin A weight-loss drug that suppresses appetite by its action on 5-HT_{2C} serotonin receptors.

low-density lipoprotein (LDL) A class of cholesterol-transporting molecules that moves cholesterol from the liver through the blood to peripheral tissues.

luciferase An enzyme found in bioluminescent organisms that catalyzes a reaction of luciferin to generate visible light.

luciferin The oxidation of luciferin to oxyluciferin generates light and was one of the earliest biochemical assays.

lye An alkaline liquid rich in potassium carbonate leached from wood ashes; often used in soap making.

lysine An amino acid that is often positively charged at pH 7.

lysis Rupture of a cell by breaking through the cell membrane.

lysogeny A viral life cycle in which the viral genome integrates into the host genome where it replicates, until signals induce the genome to switch to a lytic life cycle.

lysosome In eukaryotic cells, a membrane-bound organelle involved in degradation and detoxification of macromolecules.

M

macromolecule An organic polymer such as protein, nucleic acid, or polysaccharide with high molecular weight (from a few thousand to several million daltons).

macronutrient A nutrient required in relatively large amounts, such as fats, proteins, and carbohydrates.

macular corneal dystrophy A rare genetic disorder that is due to a mutation in the carbohydrate sulfotransferase 6 gene, whose protein product adds sulfate groups to keratan sulfate in corneal cells.

major groove In the DNA double helix, a groove running along the outside of the helix where the distance between the phosphate backbones is larger than in the minor groove.

malaria A mosquito-transmitted disease caused by the parasite microorganism *Plasmodium* that first invades liver cells and then red blood cells in the animal host.

malate–aspartate shuttle A mechanism used predominantly in liver cells that enables the NADH formed in glycolysis to transfer electrons into the electron transport system.

MAP/ERK kinase (MEK) The second protein in the MAP kinase signaling pathway, it is phosphorylated by Raf and in turn phosphorylates ERK.

mass action ratio (Q) The ratio of the concentration of product over the concentration of substrate, under actual conditions in a cell; the mass action ratio is used to calculate ΔG for a reaction.

mass spectrometry A method of measuring the mass-to-charge ratio of molecules, which is then used to deduce the molecular mass.

matrix-assisted laser desorption/ionization (MALDI) A method for generating peptide ions for mass spectrometry that uses a laser to ionize the peptide fragments.

maximum velocity (v_{max}) A point in a reaction where the reaction velocity cannot increase any further, even with the addition of more substrate.

McArdle disease A glycogen storage disease due to a defect in muscle glycogen phosphorylase.

mediator complex A large multiprotein eukaryotic transcription factor complex that interacts with activator proteins and is necessary for expression of genes transcribed by RNA polymerase II.

melanin A group of pigments found in the skin, hair, eyes, scales, and feathers of animals; formed from tyrosine.

melanocyte Melanin-producing skin cells.

melting point (MP) The temperature at which a substance changes from a solid to a liquid state.

melting temperature The temperature (T_m) at which half of the DNA molecules in a sample become denatured.

membrane receptor A transmembrane protein that changes conformation upon binding of a cognate ligand molecule.

mesophyll cell A class of cell that fills most of the interior tissue of a leaf.

messenger RNA (mRNA) A molecule of RNA that serves as a template for protein synthesis.

metabolic enzyme A protein that catalyzes biochemical reactions involved in energy conversion pathways.

metabolic flux The rate at which reactants and products are interconverted in a metabolic pathways.

metabolic homeostasis The process of maintaining optimal metabolite concentrations and managing chemical energy reserves within an organism.

metabolic pathway Sequences of biochemical reactions coordinated and controlled by a cell in response to available energy.

metabolic syndrome A set of medical conditions that puts a person at risk for diabetes and cardiovascular disease. They are: excess visceral fat, hypertension, hyperlipidemia, and hyperglycemia.

metabolism The collection of biochemical reactions in an organism that convert chemical energy into work.

metabolite Any of a group of small biomolecules that serve as reactants and products in biochemical reactions within cells.

metabolite receptor A type of nuclear receptor, activated by ligands often derived from dietary nutrients, that binds to direct repeat DNA sequences.

metalloenzyme An enzyme containing a tightly bound metal ion cofactor.

metformin A drug prescribed for controlling type 2 diabetes, which works by elevating AMP levels that then activate AMP-activated protein kinase.

methicillin A β-lactam antibiotic that is not a substrate for the bacterial enzyme β-lactamase, which inactivates the antibiotic penicillin.

methionine A hydrophobic amino acid containing an unreactive sulfur atom.

methotrexate A dihydrofolate reductase inhibitor and anticancer drug.

mevalonate An intermediate in cholesterol biosynthesis, formed by the condensation of three acetyl-CoA molecules.

micelle A structure in which hydrophobic tails are in the center of a globular sphere and polar head groups face outward toward aqueous solvent.

Michaelis constant (K_m) A numerical value that relates the rate constants of breakdown and formation of the enzyme–substrate complex for a given enzyme reaction:

$$K_m = \frac{k_{-1} + k_2}{k_1}$$

Michaelis–Menten equation An equation describing the hyperbolic relationship between the initial reaction velocity v_0 and the substrate concentration [S]:

$$v_0 = \frac{v_{max}[S]}{K_m + [S]}$$

Michaelis–Menten kinetic model A model of reaction kinetics in which the focus is on the formation of an enzyme–substrate complex and the subsequent catalytic step, at an early time in the reaction when no appreciable product has been generated.

micro RNA (miRNA) Genome encoded small RNA molecules that regulate gene expression and mRNA translation.

microtubule Cable-like component of the cytoskeleton that enables an animal cell to move by extending the plasma membrane in one direction while retracting it at the opposite end of the cell.

minor groove In the DNA double helix, a groove running along the outside of the helix where the distance between the phosphate backbones is smaller than in the major groove.

mismatch repair The enzyme-mediated process by which mismatched base pairs occurring in DNA as a result of faulty DNA replication are removed and replaced by the correct base pair.

missense mutation A single nucleotide mutation in DNA that changes one amino acid in a protein.

mitochondrial matrix The aqueous interior of a mitochondrion; the matrix is surrounded by the inner mitochondrial membrane.

mitochondrion *pl.* **mitochondria** The eukaryotic organelle responsible for many of the metabolic reactions involved in energy conversion and production of ATP.

mitogen-activated protein kinases (MAP kinases) A trio of related kinases that activate a phosphorylation cascade leading to increased rates of eukaryotic cell division.

mixed inhibitor A molecule that binds to sites distinct from the enzyme active site, but can bind to both the enzyme and the enzyme–substrate complex.

mixed-function oxidase A group of desaturating enzymes that convert fatty acyl-CoA molecules, derived from palmitate, into longer fatty acids, such as stearate.

MoFe protein A heterotetramer protein associated with bacterial nitrogenase complex that contains a large molybdenum–FeS cofactor required for catalysis.

molecular replacement A procedure that compares the phases of diffracted X-rays with those from X-ray diffraction of a similar protein of known structure.

molten globule An intermediate stage in a model of globular protein folding in which hydrophobic residues first form the interior of the protein.

monocistronic A gene containing a promoter sequence followed by a single protein-coding region.

monoclonal antibody A homogeneous immunoglobulin species that recognizes only a single epitope on an antigenic protein.

monounsaturated fatty acid Fatty acids containing one carbon–carbon double bond in the hydrocarbon chain.

motif A small but distinct structural unit of a protein fold.

mRNA translation The process by which a ribosome decodes a molecule of mRNA and synthesizes a corresponding protein.

mucin O-linked glycoproteins that are secreted and bound to the cell surface, which aid in the maintenance of mucosal barriers around epithelial cells.

multidrug resistance The ability of pathogenic cells or organisms to resist the detrimental effects of drugs directed against them.

multiple cloning site (MCS) A segment of DNA that can be cleaved by several sequence-specific endonucleases called restriction enzymes to facilitate gene cloning.

muscarinic acetylcholine receptor A group of G protein–coupled receptors, found in a number of tissues, that respond to the ligand acetylcholine.

myoblast A large fused muscle cell containing many nuclei and sharing a common plasma membrane.

myocardial infarction A heart attack.

myofibril A bundle of muscle fibers within a myoblast.

myoglobin A globular transport protein that is concentrated in muscle tissue and functions in oxygen storage.

myosin An abundant structural protein in animal cells that functions in molecular cables, mediating muscle movement through calcium-stimulated muscle contraction.

myristoylation Modification of a protein by attaching a 14-carbon fatty acid chain to an amino-terminal glycine.

N

N-end rule The observation that the half-life of a protein in the cytosol is greatly determined by its amino-terminal amino acid residue.

N-linked oligosaccharide Describes the linkage of short chains of glycans to the amide nitrogen atom in the side chain of asparagine in a glycoprotein.

Na^+–I^- symporter A secondary active transport protein in humans that imports I^- ions into thyroid cells for thyroid hormone synthesis.

Na^+–K^+ ATPase A P-type transporter responsible for maintaining an Na^+ gradient across a membrane.

NADH–ubiquinone oxidoreductase Complex I in the electron transport system; catalyzes the oxidation of NADH and the reduction of FMN, which leads to the translocation of 4 H^+ across the inner mitochondrial membrane.

$NADP^+$/NADPH (nicotinamide adenine dinucleotide phosphate) A coenzyme that mediates redox reactions that take place at carbon–oxygen bonds.

native PAGE An alternative to SDS-PAGE that uses nondenaturing conditions in the absence of SDS.

natural selection The change in the frequency of genes in a population under conditions that favor some genes over others.

Nernst equation An equation for calculating the reduction potential under nonstandard conditions:

$$E = E^{\circ\prime} + \frac{RT}{nF} \ln \frac{[e^- \text{acceptor}]}{[e^- \text{donor}]}$$

niacin The water-soluble vitamin B_3, also called nicotinic acid; a precursor of NAD^+.

nicotinamide adenine dinucleotide (NAD^+/NADH) A coenzyme derived from the vitamin niacin; a required component in many redox reactions involving dehydrogenase enzymes.

nicotinic acetylcholine receptor A ligand-gated ion channel that mediates ion transport in response to the neurotransmitter acetylcholine.

Niemann–Pick disease A genetic disease of sphingolipid metabolism due to a mutation in the gene for sphingomyelinase. Lack of the enzyme blocks the conversion of sphingomyelin to ceramide.

nitrate reductase Enzyme that reduces nitrate to nitrite.

nitric oxide (NO) A first messenger molecule that can rapidly diffuse into smooth muscle cells, resulting in muscle relaxation, vasodilation, and increased blood flow.

nitrification Oxidation of reduced nitrogen compounds to nitrate or nitrite.

nitrite reductase Enzyme that reduces nitrite to nitric oxide, nitrous oxide, or ammonium ion.

nitrogen assimilation The processes used by plants and prokaryotes to incorporate nitrogen (usually as NH_4^+) into organic compounds.

nitrogen balance Refers to an organism whose daily intake of nitrogen equals the amount of nitrogen it excretes.

nitrogen cycle Biochemical processes that together lead to nitrogen balance in the biosphere.

nitrogen fixation An ATP-dependent process carried out by select bacteria and archaea in which N_2 gas is reduced to ammonium ion (NH_4^+).

nitrogenase The enzyme that catalyzes nitrogen fixation.

NMR spectroscopy A method of structure determination that detects nuclear spin properties of certain atoms (^1H, ^{13}C, ^{15}N) to deduce their relative locations.

noncoding RNA (ncRNA) Refers to a diverse set of RNA molecules that do not encode proteins and are distinct from the three types of RNA involved in protein synthesis (mRNA, tRNA, rRNA).

noncompetitive inhibition A special case of mixed inhibition in which a molecule has equal affinity for both the free enzyme and the enzyme–substrate complex.

nonreducing sugar A carbohydrate that is unable to reduce an oxidizing agent, such as cupric ion (Cu^{2+}); sucrose is a nonreducing disaccharide sugar.

nonsense mutation A single nucleotide mutation in DNA that results in a termination codon in the corresponding mRNA (UGA, UAA, UAG).

nonsteroidal anti-inflammatory drug (NSAID) An inhibitor of prostaglandin H_2 that is used to reduce inflammation; examples include aspirin, naproxen, and ibuprofen.

nuclear export signal (NES) A sequence of amino acids in a protein that serves as a binding site for exportin protein, thereby stimulating the process of nuclear export.

nuclear localization signal (NLS) A sequence of amino acids in a protein that serves as a binding site for importin protein, thereby stimulating the process of nuclear import.

nuclear pore complex Large protein complexes within the nuclear membrane through which all macromolecular traffic flows in and out of the nucleus.

nuclear receptor A transcription factor that regulates gene expression in response to ligand binding; examples include the glucocorticoid and vitamin D receptors.

nucleoid In bacteria, a region containing the chromosome (without any surrounding membrane).

nucleolus In eukaryotic cells, a part of the nucleus where ribosomes are assembled from ribosomal RNA and protein.

nucleoside An organic molecule consisting of a purine or pyrimidine base covalently linked to a five-membered sugar (ribose or deoxyribose).

nucleosome A DNA strand wound around a histone protein core that serves for packaging chromatin.

nucleotide An organic molecule consisting of a purine or pyrimidine base, a five-membered sugar (ribose or deoxyribose), and one to three phosphate groups.

nucleotide excision repair (NER) A DNA repair mechanism that removes damaged bases and then repairs the region with a DNA repair replication process.

nucleus In eukaryotic cells, a membrane-bound organelle that contains chromosomes.

O

ω-3 fatty acid An unsaturated fatty acid in which the double bond is located three carbons from the terminal methyl carbon.

ω-6 fatty acid An unsaturated fatty acid in which the double bond is located six carbons from the terminal methyl carbon.

O-glycosidic bond The covalent bond that links two monosaccharides together through oxygen; can be in the α (down) or β (up) configuration.

O-linked oligosaccharide Describes the linkage of short chains of glycans to the oxygen atom in the side chain of serine or threonine in a glycoprotein.

O_2-evolving complex (OEC) A region of photosystem II (PSII) that contains a cluster of four manganese ions, a chloride ion, and a reactive tyrosine residue; the OEC extracts electrons from H_2O to generate O_2 and provide electrons to the photooxidized PSII reaction center.

Okazaki fragment Short fragments of DNA that form the lagging strand during DNA synthesis.

olestra A fatty acyl-sucrose derivative that tastes like triglycerides, but is not digestible and therefore has no Caloric value.

oligomycin An antibiotic that blocks the synthesis of ATP by inhibiting the activity of the F_o proton channel of the ATP synthase complex.

oligopeptide An amino acid polymer chain of less than 40 amino acids.

oncogene A mutated version of a normal gene that can cause cancer; can be a gain-of-function (dominant) or loss-of-function (recessive) cancer causing mutation.

open complex Refers to the active form of bacterial RNA polymerase holoenzyme bound to DNA at a gene promoter sequence that has been unwound to allow transcription to proceed.

operon A polycistronic gene (and its regulatory sequences) that contains coding sequences for proteins involved in a single biochemical process or pathway. It is transcribed as a single mRNA.

orexigenic Neurons that stimulate an increase in appetite and a decrease in expenditure of energy.

origin of replication A DNA sequence at which DNA replication initiates.

orlistat A drug that inhibits pancreatic lipase, thus it reduces the uptake of dietary fats.

orotate A pyrimidine base that is the precursor to the nucleoside monophosphate UMP and ultimately of the nucleoside triphosphates UTP, CTP, and TTP.

orotidine-5′-monophosphate (OMP) An intermediate in pyrimidine nucleotide biosynthesis, OMP is the precursor to the nucleoside monophosphate UMP.

orthologous gene One of a set of highly conserved gene sequences that arose from a common ancestral gene and encode proteins with the same function in different species.

osmosis The diffusion of solvent molecules from a region of lower solute concentration to one of higher solute concentration.

osmotic pressure A difference in pressure across a semipermeable membrane caused by osmosis across the membrane.

oxidant The molecule in a redox reaction that gains electrons by oxidizing a different molecule.

oxidation reaction The half of a redox reaction in which a molecule or functional group loses electrons.

oxidative decarboxylation A redox reaction that generates CO_2.

oxidative phosphorylation A metabolic pathway that oxidizes nutrients, particularly glucose, to generate ATP from ADP.

oxidoreductase An enzyme that catalyzes a biochemical redox reaction.

oxyanion hole A region of an enzyme active site that can transiently accept a negatively charged oxygen ion, which can then form hydrogen bonds with nearby groups; the chymotrypsin reaction mechanism involves formation of an oxyanion hole.

P

ϕ (phi) and ψ (psi) torsional angles The torsional angle ϕ between the amide nitrogen and the α carbon, and the torsional angle ψ between the α carbon and the carbonyl carbon, determine the structure of the peptide backbone as a function of bond rotation around the α carbon.

P-type transporter An active membrane transport protein that uses energy from ATP hydrolysis to drive large conformational changes and pump molecules across the membrane.

P450 monooxygenase A class of heme-containing oxidoreductase enzymes that alter side groups on substrate molecules in reactions involving oxygen.

palindrome A string of letters or numbers that reads the same in both directions; many restriction endonuclease cleavage sites are palindromes when considering both strands of DNA read in the same 5′ to 3′ direction such as the *EcoRI* restriction site 5′-GAATTC-3′.

palmitoylation Modification of a protein by attaching a 16-carbon fatty acid chain to a cysteine residue.

pancreatic β cell A cell in the pancreas responsible for the secretion of insulin into the blood.

pantothenic acid The water-soluble vitamin B_5; a precursor of coenzyme A.

paracrine A hormone that functions over short distances to activate receptors on nearby cells.

parallel β sheet A β-sheet structure in which β strands lie in the same amino to carboxyl orientation.

paralogous genes Highly conserved genes within the same species; most often derived from the process of gene duplication.

Parkinson disease A severe neurologic disease resulting from decreased numbers of dopamine-producing cells in the substantia nigra region of the brain.

passive transporter protein A transport protein that allows specific molecules to move across a membrane down their chemical gradients.

pectin A homopolysaccharide of galacturonic acid monomers that forms a matrix with cellulose and hemicellulose to form plant cell walls.

pellagra A disease caused by a severe deficiency of niacin (vitamin B_3), resulting in a skin rash, diarrhea, and neurologic problems; niacin is required for synthesis of the coenzymes NAD and FAD.

penicillin An antibiotic, derived from the *Penicillium* mold, that inhibits bacterial cell wall synthesis by interfering with transpeptidase enzyme.

pentose phosphate pathway Also called *phosphogluconate pathway*. A metabolic pathway in which glucose-6-phosphate is oxidized and then decarboxylated, generating ribulose-5-phosphate and NADPH.

peptide bond A covalent bond between the α amino group of one amino acid and the α carboxyl group of another amino acid.

peptidoglycan A bacterial proteoglycan made of peptide-linked chains of repeating hexosamines; the major component of bacterial cell walls.

peptidyl site (P site) The binding site on the ribosome where the peptidyl transferase active site is located.

percent body fat A measurement of the total fat mass divided by the total body mass of an organism.

perilipin A monolayer-embedded protein forming the shell of adipocyte lipid droplets. It prevents degradation of triacylglycerol by the endogenous adipocyte lipases.

periplasmic space An aqueous compartment located between the outer and inner membranes of Gram-negative bacteria.

peroxisome In eukaryotic cells, an organelle containing enzymes for forming or destroying peroxides.

peroxisome proliferator–activated receptor (PPAR) A member of the nuclear receptor superfamily; can be characterized as a metabolite receptor to distinguish it from a steroid receptor.

pH The negative logarithm of the hydrogen ion concentration of an aqueous solution.

phenylalanine A hydrophobic aromatic amino acid that only weakly absorbs UV light.

phenylketonuria (PKU) A genetic disease of phenylalanine catabolism, in which phenylalanine hydroxylase is defective causing phenylalanine and its metabolites to build up to high levels and cause neurologic damage.

pheophytin (Pheo) The first molecule in the photosynthetic electron transport system that accepts an electron from activated chlorophyll in the photosystem II photooxidation reaction.

phosphatase An enzyme that removes a phosphoryl group from a biomolecule.

phosphatase and tensin homolog (PTEN) A phosphatase enzyme that removes a phosphoryl group from PIP_3 to regenerate PIP_2, thereby disrupting downstream insulin signaling.

phosphate translocase A membrane transport protein that imports inorganic phosphate and protons into the mitochondrial matrix.

phosphatidate The precursor for many common glycerophospholipids consisting of glycerol linked to two fatty acid chains and a phosphate group.

phosphatidylcholine A glycerophospholipid with choline as the polar head group.

phosphatidylethanolamine A glycerophospholipid with ethanolamine as the polar head group.

phosphatidylinositol A glycerophospholipid with inositol as the polar head group.

phosphatidylinositol-3,4,5-trisphosphate (PIP_3) A molecule cells from PIP_2 by the lipid kinase PI-3K; it recruits proteins with a pleckstrin homology domain to the plasma membrane.

phosphatidylinositol-4,5-bisphosphate (PIP_2) A membrane phospholipid that can be hydrolyzed to produce the second messengers DAG and IP_3.

phosphatidylserine A glycerophospholipid with serine as the polar head group.

phosphocreatine A molecule in muscle cells that is used to carry out substrate-level phosphorylation to generate ATP.

phosphodiester bond A chemical grouping resulting from two hydroxyl groups in phosphoric acid forming bonds with hydroxyl groups on two other molecules to form ester bonds; the primary linkage between nucleotides in DNA and RNA.

phosphoenolpyruvate carboxykinase Enzyme that catalyzes a decarboxylation reaction in the gluconeogenic pathway that converts oxaloacetate to phosphoenolpyruvate and carbon dioxide.

phosphofructokinase-1 A key enzyme in glycolysis that converts fructose-6-phosphate to fructose-1,6-bisphosphate. A key regulated enzyme in the glycolytic pathway.

phosphofructokinase-2/fructose-2,6-bisphosphatase (PFK-2/FBPase-2) A dual-function enzyme that catalyzes the interconversion of fructose-6-phosphate and fructose-2,6-bisphosphate.

phosphogluconate pathway See **pentose phosphate pathway**.

phosphoinositide-3 kinase (PI-3K) An SH_2-containing lipid kinase that phosphorylates PIP_2 to generate PIP_3, thereby initiating a downstream signaling pathway.

phospholipase C A membrane-associated enzyme that hydrolyzes PIP_2 to form DAG and IP_3.

phospholipid bilayer A component of biological membranes in which hydrophilic polar head groups orient toward the aqueous environment and hydrophobic nonpolar hydrocarbon tails form a water-impermeable barrier in the interior of the membrane.

phospholipid Lipids containing a simple organic molecule attached to a negatively charged phosphoryl group and two fatty acids.

phosphoribosyl pyrophosphate (PRPP) An intermediate in the purine and pyrimidine biosynthetic pathways; also an important component in nucleotide salvage pathways.

phosphorylase A class of enzymes that catalyze a phosphorolytic cleavage reaction.

phosphotyrosine binding (PTB) domain A protein domain that binds to phosphotyrosine residues on target proteins.

photolyase A DNA repair enzyme that repairs ultraviolet-induced DNA damage.

photooxidation Oxidation caused by light; used particularly for oxidation of chlorophyll, resulting in the transfer of an electron from chlorophyll to an acceptor molecule such as pheophytin.

photophosphorylation The conversion of ADP to ATP coupled to the transfer of electrons in photosynthesis.

photoproduct Damaged regions of DNA resulting from exposure to ultraviolet light.

photorespiration An energy-consuming pathway in which the oxygenase activity of rubisco uses O_2 as a substrate to generate 3-phosphoglycerate and 2-phosphoglycolate, of which the 2-phosphoglycolate must be converted to 3-phosphoglycerate by the glycolate salvage pathway at the expense of ATP.

photosynthesis The use of solar energy to oxidize water, capture chemical energy, and generate oxygen.

photosynthetic autotroph An organism that can use photosynthesis to oxidize water and produce oxygen, generating chemical energy in the form of glucose.

photosynthetic electron transport system An electron transport system in which a light-absorbing pigment is photoexcited and transfers an electron to another molecule as a result of photooxidation.

photosystem I A component of the photosynthetic electron transport system containing the reaction center P700 whose terminal electron acceptor is $NADP^+$.

photosystem II A component of the photosynthetic electron transport system containing the reaction center P680 that is associated with the O_2-evolving complex.

phylloquinone An electron carrier in photosystem I; also known as vitamin K.

pilus A hair-like appendage on the surface of bacteria, used in cell movement and reproduction.

pioneer factor A transcription factor protein that binds to regulatory DNA sequences associated with eukaryotic gene promoters that control cell fate; pioneer factors activate silent chromatin.

pK_a Acid dissociation constant expressed as a negative logarithm, similar to pH.

plasmid A circular DNA molecule that replicates independently of the chromosome; bacterial plasmids encode genes for cell mating, antibiotic resistance, and pathogenesis.

plastocyanin (PC) The soluble electron carrier that transfers electrons from cytochrome b_6f to the reaction center of photosystem I.

plastoquinone (PQ) A pair of quinone-type electron carriers found in the electron transport system of photosystem II.

pleckstrin homology (PH) domain A binding domain in signaling proteins that binds to PIP_3 in the plasma membrane.

pluripotent state Refers to a cell that has the potential to differentiate into a specific cell type as a result of developmental process or laboratory manipulations.

poly(A) binding protein A protein that binds short tracts of adenines in the 3′ end of mRNA and facilitates the process of mRNA polyadenylation.

poly(A) polymerase A special RNA polymerase that synthesizes the poly(A) tail on protein-coding eukaryotic mRNA molecules.

polyacrylamide gel electrophoresis (PAGE) A gel electrophoresis technique that uses a solid-support molecular sieve made of polyacrylamide to separate molecules in an electric field on the basis of charge and size.

polycistronic A gene containing a promoter sequence followed by multiple coding regions.

polyclonal antibody A heterogeneous mixture of immunoglobulin proteins that recognize one or more epitopes on an antigenic protein.

polyethylene glycol (PEG) A polymer of ethylene oxide ($C_{2n}H_{4n+2}O_{n+1}$) with many industrial uses; in biochemistry it serves as water-soluble polymer that can be used to identify hydrophobic channels in enzyme active sites and to facilitate precipitation of bacteriophage particles collected from cell media.

polyglutamine track expansion A gene mutation characterized by multiple tandem copies of the repeating trinucleotide CAG encoding the amino acid glutamine; polyglutamine track expansion is associated with protein aggregation.

polymerase chain reaction (PCR) A method for *in vitro* DNA replication to generate multiple copies of a specific target DNA segment based on the positioning of opposing single-strand DNA primers.

polypeptide A short linear chain of amino acids linked by peptide bonds.

polypeptide chain An extended polypeptide with many amino acids linked by peptide bonds such as found in a protein.

polyprotic acid A weak acid with more than one dissociable H^+; amino acids such as glutamate and aspartate are polyprotic acids.

polysaccharide A linear or branched polymer (carbohydrate) of monosaccharide units.

polyunsaturated fatty acid Fatty acids that contain multiple carbon–carbon double bonds in the hydrocarbon chain.

Pompe disease A glycogen storage disease caused by a deficiency in lysosomal α-1,4-glucosidase, which results in the accumulation of glycogen in lysosomes and a shortage of glucose for ATP production.

porin A passive membrane transport protein that has a β-barrel structural motif and is abundant in the outer membrane of bacteria, mitochondria, and chloroplasts.

porphyrias A group of diseases characterized by the accumulation of heme precursors in the blood due to deficiencies in the enzymes responsible for heme biosynthesis.

portal vein Carries blood from the gastrointestinal tract and the spleen to the liver.

positron emission tomography A neuroimaging technique that measures metabolic activity or blood flow changes in the brain by monitoring the distribution of radioactive glucose molecules.

posttranslational modification The attachment of a functional group or molecule to specific amino acids within a protein; examples include phosphorylation, methylation, and addition of lipids and glycan groups.

PQ cycle The translocation of reduced plastoquinol to cytochrome b_6f, where it donates its electrons, and the movement of oxidized plastoquinone PQ_B back to photosystem II.

pre-steady-state condition An initial time in an enzyme reaction when enzyme–substrate complex formation is linear over time and changes in [S] are negligible.

precision medicine The use of biochemical markers obtained from a patient such as DNA sequence information or metabolite concentrations in blood and urine for the purpose of designing a patient-specific treatment regimen.

primaquine An antimalarial drug that induces oxidative stress in red blood cells and thereby inhibits reproduction of the malarial parasite in humans; however, in people with a glucose-6-phosphate dehydrogenase deficiency, primaquine treatment leads to acute hemolytic anemia.

primary antibody A protein-specific antibody that recognizes one or more epitopes on an antigenic protein; primary antibodies are used in Western blotting and immunofluorescence.

primary structure The unique arrangement of monomeric subunits in a single polypeptide chain or nucleic acid.

primase An enzyme required in DNA replication that synthesizes RNA primers needed for discontinuous DNA synthesis.

primer An oligonucleotide that binds to the 3′ end of the DNA template strand, thus providing the free 3′-OH needed for DNA polymerase to initiate synthesis.

primosome The combined action of helicase and primase required in DNA synthesis.

prion hypothesis The idea that normal prion protein can be converted to an infectious, misfolded version of the same protein without infection with a DNA or RNA virus.

prion protein A protein that can be misfolded and lead to protein aggregation in the brain, causing diseases such as kuru and mad cow disease in humans.

processivity Refers to the ability of DNA polymerase to perform many rounds of catalysis and nucleotide addition without dissociating from the template DNA strand.

prokaryote A single-celled organism that lacks a nucleus and other membrane-bound organelles; this category includes all bacteria.

proliferating cell nuclear antigen (PCNA) Part of the replisome in eukaryotes; it provides a 3′→5′ exonuclease proofreading function and enhances Pol δ processivity.

proline A hydrophobic amino acid containing nitrogen.

promoter A specific DNA sequence that usually occurs on the 5′ side of genes that are transcribed by the enzyme RNA polymerase; transcription factor proteins bind to promoter DNA sequences and recruit RNA polymerase to initiate RNA synthesis.

prostacyclin An eicosanoid derived from prostaglandin H_2 that regulates platelet aggregation and blood clotting.

prostaglandin A class of eicosanoids that have numerous local effects, including regulating blood flow, stimulating inflammation, controlling ion transport, and initiating contractions during birth.

prosthetic group A coenzyme that is permanently associated with an enzyme.

protease An enzyme that catalyzes hydrolysis of a peptide bond.

proteasome A large protein complex in cells consisting of an inner chamber lined with proteases that degrade ubiquitinated proteins targeted to the proteasome.

protein A polymer of amino acids represented by an extended polypeptide chain.

Protein Data Bank (PDB) A Web-based archive of atomic coordinates of protein structures that are maintained by a consortium of scientists.

protein fold Collections of secondary structures that describe the spatial arrangement of a polypeptide chain.

protein kinase A (PKA) A signaling protein that activates numerous target proteins and enzymes in the cAMP signaling pathway.

protein phosphatase 1 An enzyme that dephosphorylates specific protein targets; protein phosphatase 1 activity is stimulated by insulin signaling.

proteoglycan A glycoconjugate composed of short polypeptide segments linked to large numbers of glycan units.

proteome The complete collection of proteins in a cell or organism.

proton circuit The cyclical movement of protons across a membrane as a function of the electron transport system or photosynthesis, which then flow back down their concentration gradient across the same membrane through a protein channel.

proton hopping A series of hydrogen-bond exchanges between adjacent H_2O molecules leading to the transient formation of hydronium ions (H_3O^+).

proton-motive force The potential energy of the proton concentration gradient (ΔpH) plus the charge difference across the membrane ($\Delta\psi$).

proximal histidine A histidine residue in globin proteins that coordinates with the Fe^{2+} of the porphyrin ring.

purine A nine-member heterocyclic compound. Its substituted forms include adenine and guanine.

purinosome A multienzyme complex in animal cells thought to be responsible for the coordinated synthesis of purine nucleotides.

pyranose Member of the class of cyclic hexose sugars; so called because it resembles the six-member pyran ring.

pyridoxal phosphate (PLP) A prosthetic group of numerous amino acid transformation enzymes; it is derived from vitamin B_6.

pyrimidine A six-member heterocyclic compound. Its substituted forms include cytosine, thymine, and uracil.

pyruvate carboxylase An ATP-dependent enzyme that carboxylates pyruvate to generate oxaloacetate and thereby maintains flux through the citrate cycle in the presence of excess acetyl-CoA.

pyruvate dehydrogenase A large multi-subunit enzyme that converts pyruvate to acetyl-CoA and CO_2 with the reduction of NAD^+ to form NADH; pyruvate dehydrogenase links the glycolytic pathway to the citrate cycle.

pyruvate kinase A key enzyme in glycolysis that catalyzes the conversion of phosphoenolpyruvate and ADP to pyruvate and ATP.

Q

quaternary structure The structure of a protein complex containing more than one polypeptide chain.

R

Raf A kinase enzyme that phosphorylates target proteins on serine and threonine residues; the first protein in the MAP kinase signaling pathway.

Ramachandran plot A graph showing the allowable angles for amino acid residues on the basis of steric hindrance; certain regions of the Ramachandran plot identify allowable angles for amino acids associated with α helices and β strands.

Ras A member of the G protein family of signaling proteins; Ras is one of the best-characterized oncogenes because Ras mutations have a gain-of-function phenotype.

Ras-related nuclear protein (Ran) A small G protein that interacts with exportin and importin proteins to transport proteins and RNA between the nucleus and the cytosol.

RasGAP A GTPase activating protein (GAP) that binds to Ras and stimulates GTP hydrolysis to inactivate its signaling functions.

rate constant of a reaction (*k*) A numerical constant that reflects how quickly a substrate molecule is converted to product as a function of time under a defined set of conditions.

rate limiting An (essentially) irreversible reaction step in a metabolic pathway; so called because the level of activity of the enzyme that catalyzes the step can be tightly controlled.

reaction center A specialized pair of chlorophyll molecules in a photosystem that are able to convert photon energy to electrochemical energy via photooxidation.

reaction coordinate diagram A graph illustrating changes in energy as a chemical reaction proceeds.

RecA A component of the *E. coli* SOS regulatory system that induces autocleavage of LexA repressor; also involved in induction of the bacteriophage λ lysogenic cycle.

receptor agonist A compound that binds and activates a receptor protein similar to the natural ligand for that receptor.

receptor antagonist A compound that binds to a receptor protein and blocks receptor activation by preventing binding of the natural receptor ligand.

receptor interacting protein (RIP) A protein recruited to the TNF receptor complex to bind with the TRADD–TRAF2 complex as part of the cell survival pathway.

receptor protein A protein that stimulates a cellular response after protein structural changes initiated by binding of a ligand.

receptor tyrosine kinases (RTKs) Receptor proteins containing an extracellular domain that binds ligands and an intracellular domain that phosphorylates tyrosine residues in target proteins to initiate a downstream signaling pathway.

recessive mutation Requires that both copies of the gene be mutated to display a phenotype; also called loss-of-function mutation.

recombinant DNA DNA molecules from different sources that have been linked together using laboratory methods.

redox reaction An oxidation–reduction reaction in which electrons are transferred from a compound of lower reduction potential (more negative) to one of higher reduction potential (more positive) as in the electron transport system.

reducing sugar A carbohydrate that reduces an oxidizing agent, such as cupric ion (Cu^{2+}); for example, glucose or fructose.

reductant The molecule in a redox reaction that loses electrons by reducing a different molecule.

reduction reaction The half of a redox reaction in which a molecule or functional group gains electrons.

regulated intramembrane proteolysis A general mechanism of protein activation via proteolytic cleavage of an inactive precursor.

regulator of G protein signaling (RGS) A GTPase activating protein (GAP) that functions with G proteins associated with GPCRs.

regulon A set of genes located in different regions of a genome that are coordinately regulated.

release factor (RF) Proteins that bind to one of three stop codons when it appears in the ribosome A site, which promotes disassembly of the ribosome to terminate protein synthesis.

renaturation Refolding of a denatured protein or nucleic acid chain back to its native structure and function.

replication fork The region of the chromosome that is being unwound during DNA synthesis.

replisome The molecular machine made up of DNA polymerase and replication accessory proteins needed for DNA replication.

reporter gene A gene that encodes a biological marker that is easy to detect; used in experiments to quantify or locate an activity or molecule of interest.

residue An amino acid within a polypeptide chain or a simple sugar in a carbohydrate polysaccharide.

resonance energy transfer A mechanism in which an excited electron returns to its ground state by transferring the energy of excitation to a nearby molecule, causing the acceptor molecule to shift to an excited state.

restriction endonuclease An enzyme that cleaves DNA at specific sequences; isolated from bacteria that use restriction endonucleases to protect against invading bacteriophage.

retinitis pigmentosa A disease of the retina resulting in progressive retinal degeneration, which in some cases is caused by mutations in genes that encode spliceosome components.

retinoid X receptor (RXR) The heterodimeric binding partner of several types of nuclear receptors, it facilitates head-to-tail binding to direct repeat sequences of DNA.

retrotransposon A transposon that uses reverse transcriptase to carry out transposition through an RNA intermediate.

reverse transcriptase An RNA-dependent DNA polymerase required for retrovirus replication; used in the laboratory to generate complementary DNA (cDNA).

reverse transcription The process that converts RNA into DNA, most often related to replication of retroviruses.

reversible inhibition An enzyme regulatory mechanism that involves the noncovalent binding of small biomolecules or proteins to an enzyme subunit.

Rho-dependent termination A bacterial transcription termination mechanism that involves binding of the protein Rho to the newly transcribed mRNA.

Rho-independent termination A bacterial transcription termination mechanism that requires a GC-rich region near the terminus of the newly transcribed mRNA.

rhodopsin A G protein–coupled receptor (GPCR) consisting of seven α helices and a bound retinal molecule that absorbs light.

ribbon diagram A graphical representation of the polypeptide backbone associated with a protein's secondary structure.

riboflavin The water-soluble vitamin B_2; a precursor of FAD and FMN.

ribonuclease A An enzyme that cleaves RNA, also the first protein shown to be capable of denaturing and refolding *in vitro*.

ribonucleic acid (RNA) A nucleic acid polymer consisting of nucleoside monophosphates with unmodified adenine, guanine, cytosine, or uridine nucleotide bases or modified nucleotide bases.

ribonucleotide reductase Enzyme that catalyzes the reduction of ribonucleotides to generate deoxyribonucleotides.

ribosomal RNA (rRNA) A type of RNA that forms the major component of ribosomes.

ribosomes Large RNA–protein complexes that mediate protein synthesis in prokaryotic and eukaryotic cells.

riboswitch A hairpin structure found in some RNA molecules that interacts with metabolites and undergoes a conformational change to alter its function.

ribozyme An RNA molecule with catalytic activity; an RNA enzyme.

rickets A disease due to a vitamin D deficiency that results in inadequate calcium deposition in cartilage and bone.

RNA world A proposed point in the evolution of life on Earth in which RNA, rather than DNA, was the hereditary molecule specifying biochemical functions.

RNA-induced silencing complex (RISC) A multiprotein complex that associates with siRNA molecules and translocates to target mRNAs that are complementary to the siRNA.

RNA-seq A gene expression assay that uses PCR amplification to generate cDNA fragments, which are then sequenced by high-throughput DNA sequencing.

RNase A group of enzymes that catalyze the hydrolysis of RNA.

Rossmann fold Two regions of alternating α helices and β strands that fold together into a compact structural domain; found in several proteins that bind nucleotides.

RS system A method of describing the absolute spatial configuration of a molecule.

rubisco Ribulose-1,5-bisphosphate carboxylase/oxygenase is the enzyme responsible for fixing CO_2 in the first step of the Calvin cycle.

S

σ factor A protein that binds bacterial RNA polymerase and is required for the initiation of transcription.

S-adenosyl-L-methionine A molecule that functions as the methyl donor in a variety of biochemical reactions.

salting out A protein separation method that involves adding increasing amounts of a saturated ammonium sulfate solution to the protein solution.

salvage pathway A mechanism used by cells in which nucleic acids are degraded and the resulting free bases are recycled into new nucleoside monophosphates.

Sanger DNA sequencing A method for determining the nucleotide sequence of DNA based on the incorporation of dideoxynucleoside triphosphate molecules in an *in vitro* DNA synthesis reaction.

saponification The process of making soap from fat by alkaline hydrolysis to produce glycerol and fatty acid salts.

sarcomere An overlapping arrangement of thick and thin filaments that forms a repeating structure in skeletal muscle and decreases in length during muscle contraction.

sarcoplasmic reticulum An organelle that stores Ca^{2+} ions that are released when muscle fiber is stimulated.

saturated fatty acid Fatty acids in which the hydrocarbon chain contains only fully reduced methylene groups.

scissile peptide bond A peptide bond that can be broken by an enzyme; a site of cleavage.

scurvy A condition caused by vitamin C deficiency in which the connective tissue in skin and blood vessels breaks down due to structural defects in collagen.

SDS-PAGE Polyacrylamide gel electrophoresis (PAGE) that includes the detergent sodium dodecyl sulfate (SDS).

second law of thermodynamics In the absence of an energy input, all spontaneous processes in the universe tend toward dispersal of energy (disorder), and moreover, the measure of this disorder, called entropy, is always increasing in the universe.

second messenger A molecule that works with signaling proteins to transmit, amplify, and terminate a biochemical signal.

second-order reaction A bimolecular reaction in which the rate is proportional to the product of the substrate concentrations.

secondary antibody In Western blotting, a detection antibody that recognizes the primary antibody as an antigen.

secondary structure The local conformation of a polymer backbone; common types of protein secondary structures are α helix, β strand, and β turns.

selectivity channel A narrow opening within the interior of a channel protein complex that acts as a filter to allow only selected ions to pass across the membrane in different conditions.

semiconservative replication The model of DNA replication in which each new double helix contains one parental strand and one newly synthesized daughter strand.

sequential model A two-state model of cooperative binding behavior in a tetrameric protein complex in which binding of ligand to one subunit causes that subunit to switch to a new state.

serine A hydrophilic amino acid that can form hydrogen bonds and can be phosphorylated by kinases.

serine protease An enzyme that cleaves the peptide backbone of proteins, in which serine is the nucleophilic amino acid in the enzyme active site.

serine/threonine kinase An enzyme that can phosphorylate serine or threonine residues within proteins.

serotonin-selective reuptake inhibitor An antidepressant drug that targets serotonin transporter proteins to inhibit serotonin import and thereby lead to increased serotonin levels in the neuron synapse.

set point In reference to fat storage in the body, it is the average amount of adipose tissue that the body maintains at physiologic homeostasis.

severe combined immunodeficiency disease (SCID) A group of immune disorders in which the patient lacks functional T and B cells.

shikimate pathway The aromatic amino acid anabolic pathway in plants, fungi, and prokaryotes.

Shine–Dalgarno sequence In bacteria, a short nucleotide sequence upstream of the AUG start codon in the mRNA that binds to the 16S rRNA sequence, thus correctly positioning the mRNA for translation.

short tandem repeat (STR) A class of repetitive DNA elements consisting of a variable number of tandem repeat sequences of ~2–10 nucleotides in each repeat.

sickle cell anemia A recessive genetic disease involving substitution of valine for glutamate in a hemoglobin polypeptide, leading to anemia.

sigmoidal curve A graph line in the shape of the letter S, typical of oxygen binding to hemoglobin; indicator of cooperativity.

signal peptide sequence An amino-terminal stretch of amino acids on a nascent polypeptide in eukaryotes that is bound by a signal recognition particle and directed for import into the endoplasmic reticulum.

signal recognition particle (SRP) A protein–RNA complex in eukaryotes that binds to a signal peptide sequence on a newly formed polypeptide, and then, by binding to an SRP receptor on the ER membrane, directs the polypeptide into the lumen of the ER.

signal transduction Mechanisms that facilitate communication between cells, often initiated through the binding of small molecules (ligands) to proteins called receptors.

signaling protein A protein that transmits a biochemical signal from a receptor protein to a second messenger (upstream signaling protein) or from a second messenger to a target protein (downstream signaling protein).

silent mutation A single nucleotide mutation in DNA that does not change the sequence of amino acids in a protein.

simple sugar A monomeric carbohydrate used in the cell for functions such as energy storage, structural organization, or cell recognition.

single nucleotide polymorphism (SNP) A single nucleotide change in the genome that can occur when an error is made during DNA replication and is not repaired.

single-strand DNA break A region of DNA that lacks phosphodiester bonds on one strand of the DNA double helix.

single-stranded DNA binding protein (SSB) A protein that binds to single-strand DNA at the replication fork to prevent reformation of double-strand DNA until DNA synthesis is complete.

sister chromatids Two identical copies of replicated DNA contained in a mitotic chromosome.

size-exclusion chromatography See **gel filtration chromatography**.

skeletal muscle Striated muscle that attaches to the skeletal system and is responsible for voluntary movement.

sliding filament model A model of muscle contraction in which actin and myosin proteins slide over one another using chemical energy provided by ATP hydrolysis.

small nuclear ribonucleoprotein (snRNP) A complex of small nuclear RNAs and associated proteins that compose a spliceosome.

small nuclear RNA (snRNA) Small RNA molecules involved in RNA processing.

small nucleolar RNA (snoRNA) Small RNA molecules that form part of the ribonucleoprotein complexes in the nucleolus that aid in processing rRNA molecules.

sodium dodecyl sulfate (SDS) An amphipathic molecule used in gel electrophoresis to give proteins a net negative charge.

solar energy Energy produced through thermonuclear reactions in the Sun.

solid-phase peptide synthesis A technique to synthesize peptide antigens for antibody production and to manufacture peptide-based therapeutic drugs to treat a variety of diseases.

soluble NSF attachment protein receptor (SNARE) Proteins on the surface of membrane transport vesicles and target membranes that ensure vesicles deliver their cargo to the proper location.

somatic cell Any cell in an organism other than a germ-line cell.

somatic mutation Changes to the DNA sequence within the genome of a cell other than a germ cell.

somatic stem cell Also called an adult stem cell; responsible for replenishing dying cells in differentiated tissues, such as skin, intestine, and bone marrow.

somatostatin A paracrine hormone produced and secreted by δ cells in the islet of Langerhans, which controls secretion of insulin, glucagon, and the acinar cell proteases.

sonication A cell extract homogenization technique that disrupts cell membranes through the vibrational effects of ultrasonic waves.

SOS regulatory system Two bacterial regulatory proteins, LexA repressor and RecA, that control the expression of genes involved in the response to DNA damage.

specific activity The total amount or activity of a target protein, divided by the total amount of protein in the sample.

specificity constant (k_{cat}/K_m) A ratio that measures the catalytic efficiency of two enzyme reactions or of the same enzyme with two different substrates.

sphingolipid A molecule containing sphingosine bound to a single fatty acid chain.

sphingophospholipid A sphingolipid linked to a phosphate group that is connected to a polar head group such as choline.

sphingosine A long-chain amino alcohol derived from palmitate and serine.

spliceosome A ribonucleoprotein complex that excises introns from mRNA and splices together the exons to form a mature mRNA.

Src kinase homology-2 (SH2) A protein domain of about 100 amino acids that contains a binding site for a specific amino acid sequence containing a phosphotyrosine residue.

Src kinase homology-3 (SH3) A protein domain of about 70 amino acids that binds to specific proline-rich sequences.

standard free energy change (Δ*G*°) A reference point for comparing chemical reactions under a defined set of conditions (1 atm pressure, 298 K, 1 M concentration of reactants and products):

$$\Delta G^\circ = \Delta H^\circ - T\Delta S^\circ$$

statin drug A class of drugs that bind to and inhibit the liver enzyme HMG-CoA reductase thereby indirectly resulting in decreased serum cholesterol levels and a reduced risk of cardiovascular disease.

steady-state condition A state of a reaction in which the concentration of product ES is relatively constant after an initial reaction time.

stereoisomers Molecules with the same molecular formula and atomic connectivity, but different three-dimensional orientations of their atoms.

steroid receptor A class of nuclear receptor proteins that is activated by physiologic hormones derived from cholesterol such as cortisol, estrogen, and progesterone.

steroid receptor coactivator (SRC) A family of nuclear receptor coregulatory proteins.

steroidogenesis The collective set of pathways that convert cholesterol into steroid hormones.

sterol regulatory element (SRE) DNA transcription regulatory sequences (cis-acting elements) within the LDL receptor gene (and related genes) that bind sterol regulatory element binding proteins (SREBPs).

sterol regulatory element binding protein (SREBP) A dimeric DNA-binding protein (trans-acting factor) that binds to sterol regulatory elements and regulates gene expression.

stigmasterol A sterol found in plants that is analogous to cholesterol.

stomate *pl.* stomata A gas-exchange pore in plants composed of two guard cells that change shape to control the pore opening.

stop-flow kinetics A very rapid enzyme assay method that permits measurements in the millisecond range.

stroma The aqueous region within a chloroplast that is between the inner membrane and the thylakoid membrane.

Structural Organization of Proteins (SCOP) A formal hierarchical organization of protein structures, developed to categorize protein folds.

structural proteins The most abundant proteins in living organisms, they function as the architectural framework for individual cells, as well as for tissues and organs.

subcutaneous fat Adipose tissue located just below the skin in the thighs, buttocks, arms, and face.

substrate-level phosphorylation A mechanism for the production of ATP by direct transfer of a phosphoryl group from a high-energy donor (such as phosphoenolpyruvate) to ADP.

subunit A polypeptide chain that is part of a protein containing several chains.

supercoil A coiled molecule, such as DNA, folded upon itself; a coiled coil.

Svedberg value A measure of particle size based on the particle's sedimentation rate in a tube subjected to centrifugal force.

Swi/Snf complex A protein complex that displaces nucleosomes along the DNA, which aids in transcription activation by exposing DNA binding sites.

symbiont An organism that lives in close association with another organism.

symmetry model See **concerted model**.

symporter An active membrane transport protein that can co-transport molecules across a membrane in the same direction.

systems biology The study of complex chemical reaction networks in cells.

T

T7 RNA polymerase A highly efficient bacteriophage RNA polymerase used in a variety of biochemical methods.

***Taq* DNA polymerase** A heat-stable DNA polymerase isolated from a thermophilic microorganism and used routinely in the polymerase chain reaction (PCR).

target protein An intracellular protein that is modified as a result of an upstream signaling pathway.

TATA binding protein (TBP) A eukaryotic transcription initiation factor protein that binds to the sequence 5′-TATAAAA-3′ located approximately 30 nucleotides upstream of the RNA polymerase II transcription start site.

Tay–Sachs disease A genetic disease of sphingolipid metabolism due to defective hexosaminidase A, which causes GM2 gangliosides to accumulate.

telomerase A eukaryotic reverse transcriptase that synthesizes telomeric DNA (short repetitive sequences) from an RNA template in order to protect the termini of chromosomes.

telomere A specialized region of heterochromatin located at the ends of chromosomes.

template strand A DNA sequence that has the opposite 5′ to 3′ polarity as the corresponding mRNA transcript; it is complementary to the DNA coding strand.

tertiary structure The spatial location of all the atoms in a polypeptide chain.

tetrahedral intermediate A transient stage in an enzyme reaction mechanism in which a covalent fragment of the polypeptide chain resembles a transition state.

tetrahydrobiopterin (BH_4) A cofactor that transfers electrons to reactants; for example, in the reaction that converts phenylalanine to tyrosine.

tetrahydrofolate An intermediate in the pathway that regenerates N^5,N^{10}-methylenetetrahydrofolate.

thermogenin See **UCP1 uncoupling protein**.

thiamine pyrophosphate A coenzyme derived from vitamin B_1; required in decarboxylation reactions, such as that catalyzed by the pyruvate dehydrogenase complex.

thiamine The water-soluble vitamin B_1; a precursor of thiamine pyrophosphate.

thiazolidinedione A drug prescribed for the treatment of type 2 diabetes; ligand for the nuclear receptor PPAR (peroxisome proliferator–activated receptor).

thin filament An actin polymer that forms part of a network to control cell shape and cell migration.

thioredoxin A ubiquitous small, cysteine-containing protein that is an essential regulator of numerous biological processes.

threonine A hydrophilic amino acid that can form hydrogen bonds and can be phosphorylated by kinases.

thrifty gene hypothesis An explanation that genes may have been advantageous in a culture when times of famine were routine but have become detrimental in modern society where food is in abundance.

thylakoid lumen The aqueous space enclosed by the thylakoid membrane. Site of higher proton concentration (compare with *stroma*).

thylakoid membrane Highly organized membrane that forms stacks of disks (grana) connected by lamellae. Site of the photosynthetic electron transport system and the ATP synthase complex.

thymidine A deoxyribonucleoside containing the pyrimidine base thymine.

thymine (T) A pyrimidine base found in DNA that hydrogen bonds with adenine.

TIM barrel fold Also called *α/β barrel*. An alternating α helix/β strand fold first identified in the glycolytic enzyme triose phosphate isomerase.

titin A protein in the I-band of muscles that functions as a flexible spring connecting myosin proteins to Z-disk proteins.

titration curve A plot of experimental data showing the pH of a solution as a function of the amount of base added.

TNF receptor–associated death domain (TRADD) A downstream signaling protein that binds to TNF receptors in the absence of SODD proteins, regulating apoptosis.

TNF receptor–associated factor 2 (TRAF2) A protein complex that binds with TRADD as part of the cell survival pathway alternative to apoptosis.

topoisomerase An enzyme that catalyzes the cleavage of one or both DNA strands and relaxes positive supercoiled regions, allowing DNA to return to its relaxed state.

topoisomers Different forms of circular DNA that differ only in linking number.

trans fat An unsaturated fatty acid in which the carbon–carbon double bond is in the trans configuration rather than the more common cis configuration.

trans-acting factor Transcription regulatory protein that interacts with cis-acting sites on DNA to regulate gene expression.

transaldolase An enzyme that catalyzes the transfer of three-carbon units among sugars in the pentose phosphate pathway.

transcription factor A protein that binds a specific cis-acting DNA sequence to regulate gene expression.

transcriptome The collection of DNA transcripts (RNA products) generated by DNA transcription.

transduction (viral) A process in which a virus mediates the transfer of genetic material between host cells.

transfer RNA (tRNA) The adaptor molecule in protein synthesis that delivers an amino acid to the ribosome.

transformation (plasmid) A process in which DNA is taken up by bacteria and incorporated into the genome or contained on a DNA plasmid.

transition curve midpoint (T_m) A denaturing condition in which 50% of the proteins are fully folded and 50% are fully unfolded.

transition state analog Stable molecules that mimic the proposed transition state to an enzyme active site.

transition state theory The idea that the conversion of substrate to product involves a high-energy transition state in which a molecule can either become a product or remain a substrate.

transketolase An enzyme that catalyzes the transfer of two-carbon units among sugars in the pentose phosphate pathway.

transpeptidase An enzyme that catalyzes the formation of oligopeptide cross-bridges between hexosamine polysaccharide chains in the peptidoglycan cell wall of bacteria.

transport protein A protein that spans the width of a cell membrane and permits polar or charged molecules to enter or exit the cell.

transposon A segment of DNA capable of translocating within a chromosome or between chromosomes.

triacylglycerol A neutral (uncharged) lipid that contains three fatty acid esters covalently linked to glycerol.

triacylglycerol cycle A two-component system for keeping fatty acids in circulation. Fatty acids and triacylglycerol circulate between adipose tissue and the liver, while within adipocytes, fatty acids and triacylglycerol undergo interconversion.

tricarboxylic acid (TCA) cycle See **citrate cycle**.

tropomyosin A thin filament protein that is a coiled-coil α helix arranged head-to-tail along the entire length of the actin polymers in muscle.

troponin A heterotrimeric protein complex that mediates Ca^{2+} regulation of muscle contraction.

***trp* operon** A bacterial operon that contains the genes for tryptophan biosynthesis.

Trp repressor A bacterial DNA binding protein that binds to the *trp* operon and blocks transcription of the *trp* operon.

trypsin A protease that cleaves a polypeptide chain on the carboxyl side of lysine or arginine residues.

tryptophan The largest amino acid, it is hydrophobic, aromatic, and strongly absorbs light.

tumor necrosis factor (TNF) receptor A membrane receptor protein that activates signaling pathways controlling inflammation and apoptosis.

tumor suppressor A gene that normally functions to inhibit uncontrolled cell proliferation.

turnover number (k_{cat}) The maximum catalytic activity under saturating levels of substrate.

two-dimensional differential in-gel electrophoresis (2-D DIGE) An application of 2-D PAGE that enables detection of proteins that differ in abundance, charge, or molecular mass.

two-dimensional polyacrylamide gel electrophoresis (2-D PAGE) A protein separation technique that combines isoelectric focusing and SDS-PAGE to separate proteins on the basis of both pI and molecular mass.

type 1 diabetes A disease characterized by reduced uptake of glucose from the blood into tissues due to low production of insulin in the pancreas.

type 2 diabetes A disease characterized by reduced uptake of glucose from the blood into tissues due to insulin receptors that bind insulin but fail to activate downstream signaling pathways.

tyrosine A hydrophobic, aromatic, amphipathic amino acid that absorbs light; can be phosphorylated by kinases.

U

ubiquinone See **coenzyme Q.**

ubiquinone–cytochrome c oxidoreductase Complex III in the electron transport system that catalyzes the oxidation of ubiquinol.

ubiquitin A small protein that when attached to other proteins tags the ubiquitinated protein for destruction in a proteasome.

ubiquitin ligase A large group of enzymes that catalyzes the transfer of ubiquitin from E2 to the targeted protein.

UCP1 uncoupling protein Also called *thermogenin.* A component of brown adipose tissue that uncouples electron transport from ATP synthesis, resulting in the generation of heat.

uncompetitive inhibitor A molecule that binds to enzyme–substrate complexes and alters the active site conformation, thus rendering the enzyme less catalytically active.

unsaturated fatty acid Fatty acids in which one or more carbon–carbon double bonds are present in the hydrocarbon chain.

uracil (U) A pyrimidine base found in RNA that hydrogen bonds with adenine.

uric acid A purine that is the product of hypoxanthine and xanthine oxidation during degradation of nucleotides.

uridylylation A process that regulates enzyme activity by controlling adenylylating and deadenylylating activity.

V

V(D)J recombination The mechanism used in developing T cells and B cells that recombines different gene segments to generate protein chains that combine to form functional antibodies and T-cell receptors.

vacuole A membrane-bound organelle in many types of eukaryotic cells, particularly plant cells, that stores metabolites and also isolates molecules that might be harmful to the cell.

valine A small hydrophobic amino acid.

van der Waals interaction A weak interaction between the dipoles of nearby electrically neutral molecules.

van der Waals radius The optimal distance between neutral atoms for van der Waals interactions, resulting in the lowest potential energy.

variable domain A protein domain within immunoglobulin subunits that binds to antigen molecules.

variable number tandem repeat (VNTR) Noncoding regions of DNA containing repetitive sequences; often used for identifying individuals or members of a family.

velocity of the reaction (v) The product of the rate constant of a reaction k and [S], the concentration of substrate.

verapamil A calcium channel blocker used to treat a variety of cardiovascular conditions; has been used to inhibit multidrug resistance in cancer cells.

vicine The toxic glycoside in fava beans that induces oxidative stress in red blood cells, which in people with a glucose-6-phosphate dehydrogenase deficiency leads to favism.

visceral fat Layers of adipose tissue in the abdominal cavity; secretes adipokines.

vitamin D_3 A precursor to the biologically active form of vitamin D; vitamin D_3 is formed by the photoconversion of 7-dehydrocholesterol.

void volume The volume of a chromatographic column exclusive of the column matrix; generally about 35% of the column is void volume.

von Gierke disease A glycogen storage disease caused by a deficiency in liver glucose-6-phosphatase. Glycogen accumulates in the liver thereby preventing glucose from being released into the blood.

W

waist-to-hip ratio Used as an indicator of health based on the ratio of the circumference of the waist to that of the hips.

water ionization constant (K_w) The dissociation constant for water:

$$K_w = [H^+][OH^-] = 1.0 \times 10^{-14}\ M^2$$

waxes A lipid composed of long-chain fatty alcohols esterified to long-chain fatty acids.

weak acid An acid that is only partially ionized in aqueous solution.

Western blotting A method to identify proteins that have been separated by SDS-PAGE (one-dimensional or two-dimensional) and are recognized by an antigen-specific antibody.

wild-type A fully functional protein-coding sequence without any mutations.

wobble hypothesis Francis Crick's proposal that certain bases in the first position of a tRNA anticodon (5′ nucleotide of the anticodon) could undergo noncanonical hydrogen bonding with several different mRNA codons.

X

X-inactive specific transcript (XIST) A long noncoding RNA that regulates X-chromosome inactivation in females.

X-linked adrenoleukodystrophy (X-ALD) A genetic disease caused by inheritance of a mutant X-linked *ABCD1* gene, which encodes the peroxisomal adrenoleukodystrophy transport protein.

Z

Z scheme The visually descriptive name for the photosynthetic electron transport system composed of reaction centers P680 and P700 linked by the electron transport system of photosystem II, cytochrome b_6f, and plastocyanin.

Zellweger syndrome A set of rare and fatal genetic diseases due to defects in the PEX genes, which are required for normal formation of peroxisomes during fetal development.

zinc finger A zinc-mediated tertiary structure in DNA binding proteins that contains four coordination sites to the protein through two cysteine and two histidine residues.

zwitterion An electrically neutral but dipolar molecule that contains both positive and negative charges.

zymogen An inactive enzyme precursor that is activated by a proteolytic cleavage reaction.

Index

Note: Page numbers followed by f or t denote figures or tables, respectively.

Selected Conversion Factors

Length	1 cm = 10 mm = 10^4 μm = 10^7 nm = 10^8 Å
Mass	1 g = 10^{-3} kg = 10^3 mg = 10^6 μg = 3.53×10^{-2} oz 1 oz = 28.3 g
Temperature	°C = 5/9(°F – 32) K = °C + 273
Energy	1 J = 10^7 erg = 0.239 cal 1 cal = 4.184 J
Pressure	1 torr = 1 mm Hg = 1.32×10^{-3} atm = 1.333×10^2 Pa 1 atm = 758 torr = 1.01×10^5 Pa

Selected Physical Constants, with Symbols and Values

Atomic mass unit (dalton)	amu	1.661×10^{-24} g
Avogadro's number	N	6.022×10^{23}/mol
Boltzmann constant	k_B	1.381×10^{-23} J/K; 3.298×10^{-24} cal/K
Electron volt	eV	1.602×10^{-19} J; 3.828×10^{-20} cal
Faraday constant	F	96,480 J/V · mol
Gas constant	R	8.314 J/mol · K; 1.987 cal/mol · K
Planck's constant	h	6.626×10^{-34} J · s; 1.584×10^{-34} cal · s

Amino Acids

Name	Chemical structure	Molecular mass (D)	Ionizable side chain pK_a
Alanine Ala A	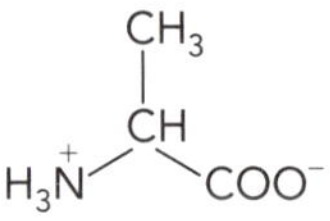	89	—
Arginine Arg R	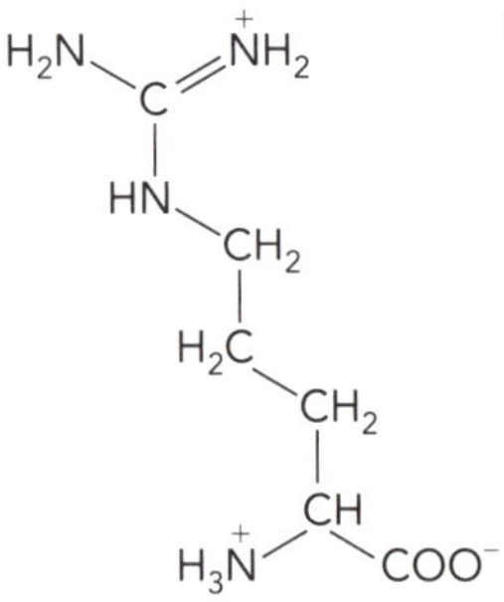	174	12.5
Asparagine Asn N	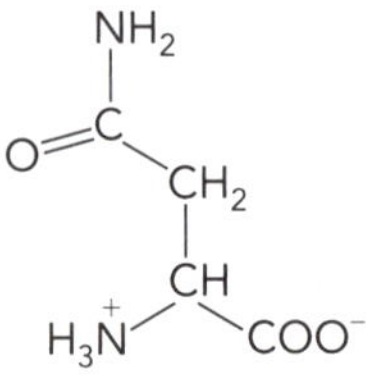	132	—
Aspartate Asp D	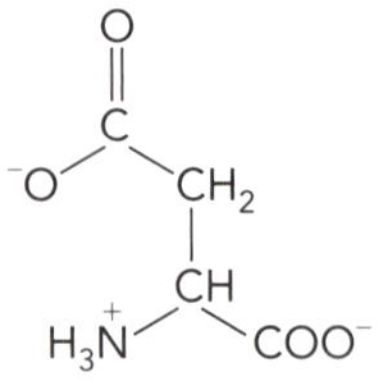	133	3.8
Cysteine Cys C	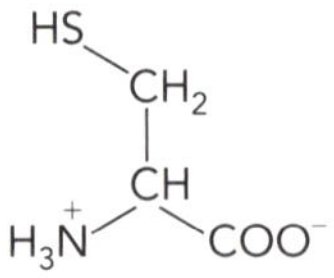	121	8.3
Glutamate Glu E	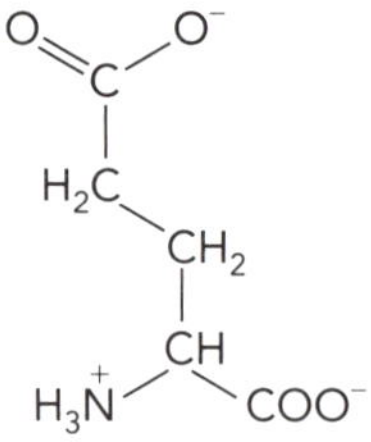	147	4.3
Glutamine Gln Q	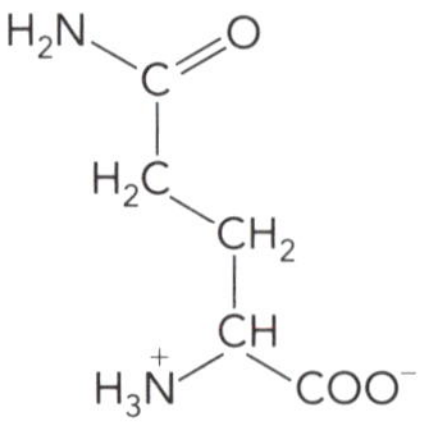	146	—
Glycine Gly G	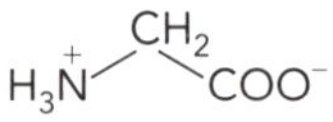	75	—
Histidine His H	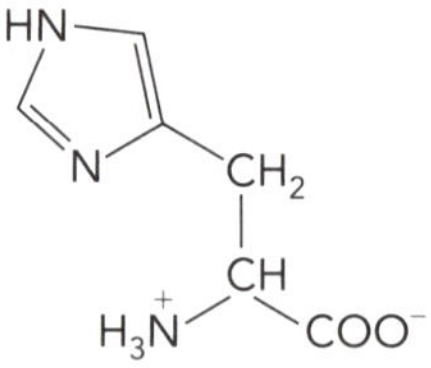	155	6.0
Isoleucine Ile I	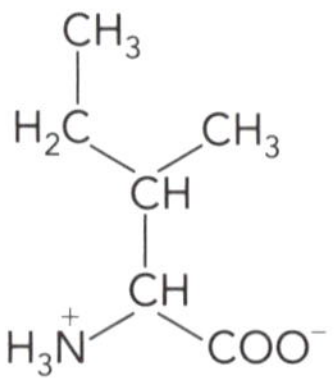	131	—

Name	Chemical structure	Molecular mass (D)	Ionizable side chain pK_a
Leucine Leu L	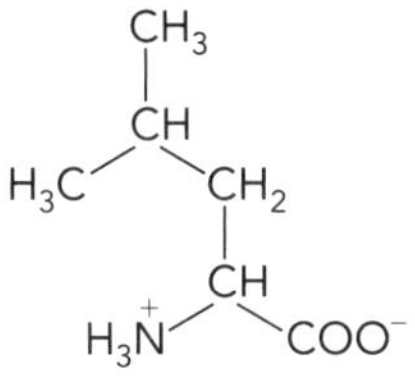	131	—
Lysine Lys K	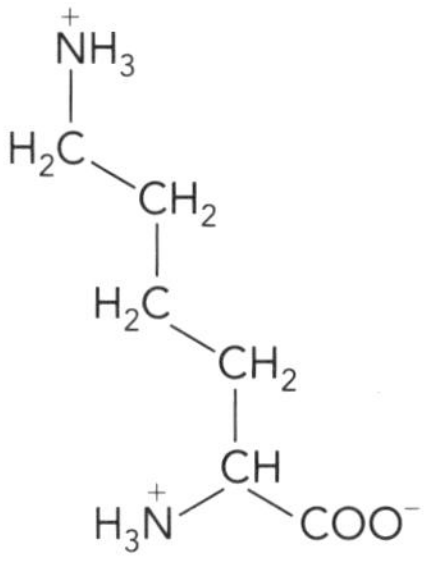	146	10.5
Methionine Met M	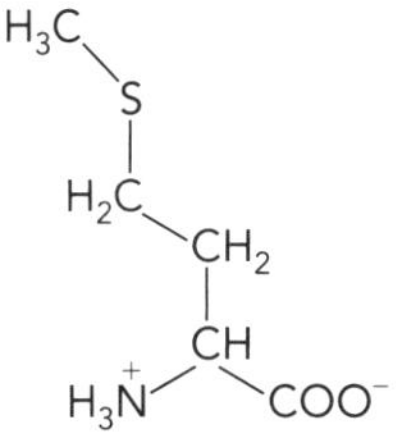	149	—
Phenylalanine Phe F	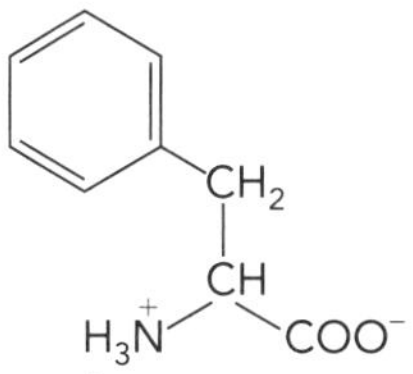	165	—
Proline Pro P	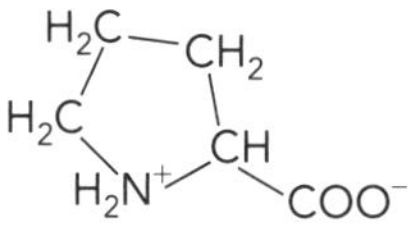	115	—

Name	Chemical structure	Molecular mass (D)	Ionizable side chain pK_a
Serine Ser S	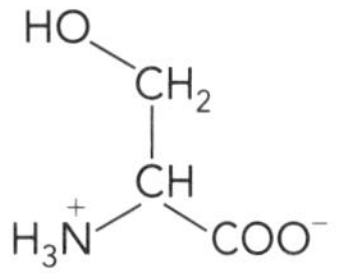	105	—
Threonine Thr T	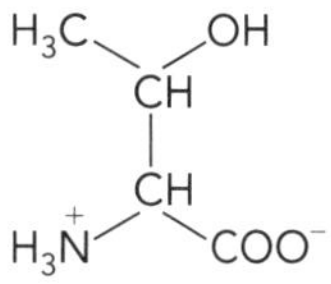	119	—
Tryptophan Trp W	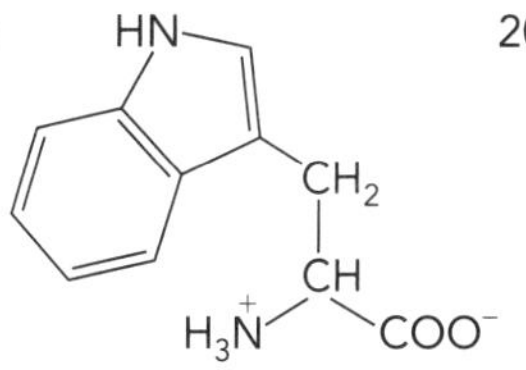	204	—
Tyrosine Tyr Y	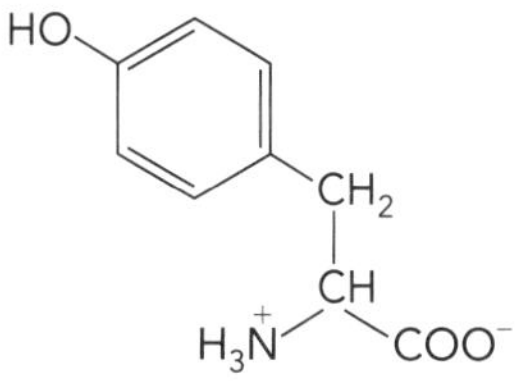	181	10.1
Valine Val V	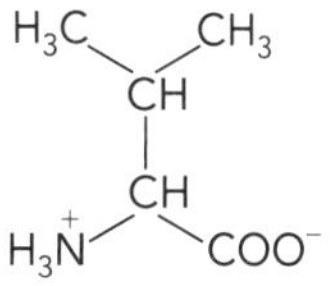	117	—